REGISTERED PROFESSIONAL ENGINEER
MICHAEL J. O'CONNOR
No. C 33404
CIVIL
STATE OF CALIFORNIA

D1560055

CIVIL ENGINEERING HANDBOOK

McGRAW-HILL HANDBOOKS

Abbott and Smith · National Electrical Code Handbook, 9th ed.
Aljian · Purchasing Handbook
American Institute of Physics · American Institute of Physics Handbook
American Society of Mechanical Engineers
ASME Handbook: Engineering Tables
ASME Handbook: Metals Engineering—Design
ASME Handbook: Metals Engineering—Processes
ASME Handbook: Metals Properties
American Society of Tool Engineers · Die Design Handbook
American Society of Tool Engineers · Tool Engineers Handbook
Beeman · Industrial Power Systems Handbook
Berry, Bollay, and Beers · Handbook of Meteorology
Brady · Materials Handbook, 8th ed.
Cockrell · Industrial Electronics Handbook
Compressed Air and Gas Institute · Compressed Air Handbook, 2d ed.
Condon and Odishaw · Handbook of Physics
Considine · Process Instruments and Controls Handbook
Crocker · Piping Handbook, 4th ed.
Croft · American Electricians' Handbook, 7th ed.
Davis · Handbook of Applied Hydraulics, 2d ed.
Etherington · Nuclear Engineering Handbook
Factory Mutual Engineering Division · Handbook of Industrial Loss Prevention
Fink · Television Engineering Handbook
Harris · Handbook of Noise Control
Henney · Radio Engineering Handbook, 5th ed.
Hunter · Handbook of Semiconductor Electronics
Johnson and Auth · Fuels and Combustion Handbook
Juran · Quality-control Handbook
Ketchum · Structural Engineers' Handbook, 3d ed.
King · Handbook of Hydraulics, 4th ed.
Knowlton · Standard Handbook for Electrical Engineers, 9th ed.
Kurtz · The Lineman's Handbook, 3d ed.
Labberton and Marks · Marine Engineers' Handbook
Landee, Davis, and Albrecht · Electronic Designers' Handbook
Laughner and Hargan · Handbook of Fastening and Joining of Metal Parts
Le Grand · The New American Machinist's Handbook
Liddell · Handbook of Nonferrous Metallurgy, 2 vols., 2d ed.
Magill, Holden, and Ackley · Air Pollution Handbook
Manas · National Plumbing Code Handbook
Mantell · Engineering Materials Handbook
Marks and Baumeister · Mechanical Engineers' Handbook, 6th ed.
Markus and Zeluff · Handbook of Industrial Electronic Circuits
Markus and Zeluff · Handbook of Industrial Electronic Control Circuits
Maynard · Industrial Engineering Handbook
Merritt · Building Construction Handbook
Morrow · Maintenance Engineering Handbook
O'Rourke · General Engineering Handbook, 2d ed.
Pacific Coast Gas Association · Gas Engineers' Handbook
Perry · Chemical Business Handbook
Perry · Chemical Engineers' Handbook, 3d ed.
Shand · Glass Engineering Handbook, 2d ed.
Staniar · Plant Engineering Handbook, 2d ed.
Terman · Radio Engineers' Handbook
Truxal · Control Engineers' Handbook
Urquhart · Civil Engineering Handbook, 4th ed.
Yoder, Heneman, Turnbull, and Stone · Handbook of Personnel Management and Labor Relations

CIVIL ENGINEERING HANDBOOK

LEONARD CHURCH URQUHART, C.E., Editor-in-Chief

Consulting Engineer, Porter, Urquhart, McCreary & O'Brien
Newark, N.J.; Los Angeles, San Francisco, and Sacramento, Calif.

FOURTH EDITION

McGRAW-HILL BOOK COMPANY, INC.

New York Toronto London

1959

CIVIL ENGINEERING HANDBOOK

Library of Congress Catalog Card Number: 58-11195

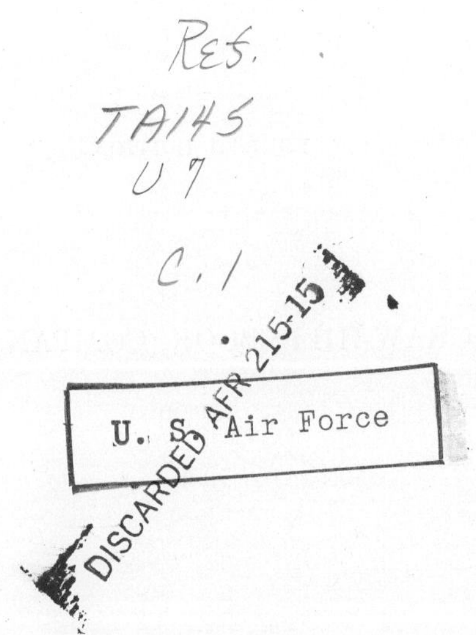

CONTRIBUTORS

Harold E. Babbitt, M.S., Professor of Sanitary Engineering Emeritus, University of Illinois. *Water Supply and Purification.*

John B. Babcock, 3d, S.B., Professor of Railway Engineering, and
Alexander J. Bone, S.M., Associate Professor of Highway and Airport Engineering, Massachusetts Institute of Technology. *Railway, Highway and Airport Engineering.*

Carlton T. Bishop, C.E., Late Professor of Civil Engineering, Yale University. *Steel Design.*

Ernest F. Brater, Professor of Hydraulic Engineering, University of Michigan. *Hydraulics.*

Albert E. Cummings, B.S.C.E., Late Director of Research, Raymond Concrete Pile Company. *Soil Mechanics and Foundations.*

Raymond E. Davis, M.S., C.E., Dr. Eng., Professor of Civil Engineering Emeritus, University of California. *Surveying.*

Herbert J. Gilkey, C.E., Professor of Theoretical and Applied Mechanics, Iowa State College. *Cement and Concrete.*

Linton Hart, C.E., Raymond Concrete Pile Company, Detroit, Michigan. *Soil Mechanics and Foundations.*

Jesse B. Kommers, M.E., Professor of Mechanics Emeritus, University of Wisconsin. *Mechanics of Materials.*

Frederick S. Merritt, B.S.C.E., Senior Editor, *Engineering News-Record. Stresses in Framed Structures.*

William McGuire, B.S.C.E., M.C.E., Associate Professor of Structural Engineering, Cornell University. *Reinforced Concrete.*

Richard G. Tyler, C.E., Professor of Sanitary Engineering Emeritus, University of Washington. *Sewerage and Sewage Disposal.*

PREFACE

This edition, like the previous ones, aims to present in one volume the fundamentals of the various subdivisions of civil engineering in such a manner that graduates and practicing engineers will find it extremely useful when confronted with problems outside of their specialized field. In this edition each section has been critically reviewed and revised wherever possible to increase its effectiveness.

In the section on surveying, the subject of photogrammetric surveying has been considerably expanded, and revisions have been made in the subjects of precision, errors, leveling, triangulation, land surveying, map drafting, and construction surveying.

In the section on railway, highway, and airport engineering, new material has been added on aerial photogrammetric route surveys, railway signaling, and railway motive power and tonnage rating. The text on highway engineering has been extensively rewritten, particularly the portions pertaining to interstate highway system requirements, traffic and planning surveys, intersection design, grade separations, flexible pavement design bituminous pavement mixes, and concrete pavement design and construction. Statistics and cost data have been brought up to date. The section on airports includes new material on runway configurations, soils classifications, pavement design, airport terminals, and heliports.

The section on hydraulics has been modernized in respect to the portions dealing with the flow in pipes, the flow through orifices, flow over submerged weirs, and the flow through culverts, and additional numerical examples have been added.

The section on stresses has been completely rewritten with less attention given to railroad structures and more to continuous beams and frames, with the material on the various methods of stress analysis amplified and extended. The section on steel construction has had additional material added on welded structures.

The section on concrete introduces new material on the reproportioning of mixtures for air entrainment and discusses additions and admixtures at greater length and the effect of various substances on concrete, while that on reinforced concrete has been entirely rewritten and expanded to provide more design aids for the reader's use.

The section on foundations has been brought up to date with the introduction of some matters which did not appear or were very briefly treated in the previous edition.

The section on sewerage has been modernized with regard to estimating population growths, the improved processes and efficiencies in aeration and sewage treatment, the progress in the activated sludge process, and the increasing use of chlorine around the disposal plant.

The section on water supply has been entirely rewritten and brought up to date with many new illustrations added.

Leonard Church Urquhart

CONTENTS

CIVIL ENGINEERING HANDBOOK

Section 1

SURVEYING

By Raymond E. Davis

GENERAL

The Earth. The earth is an oblate spheroid of revolution, the length of its polar axis being about 1/297 less than that of its equatorial axis, as follows:

	Polar axis, ft	Equatorial axis, ft
Clarke (1866), used in public-land surveys...	41,710,242	41,852,124
Hayford (1909)...........................	41,711,920	41,852,860
Adopted by International Geodetic and Geophysical Union (1924)..................	41,711,940	41,852,860

Units of Measurement. In all English-speaking countries the common units of linear measurement are the *yard, foot,* and *inch.* On most surveys in these countries, distances are measured in feet, tenths, and hundredths; and surveyor's tapes are usually graduated in these units. The *Gunter's chain,* the unit of length for public-land surveying, is 66 ft long and is divided into 100 links each 7.92 in. long.

Many other civilized countries employ the *meter* as the unit of length: 1 m = 39.37 in. = 3.2808 ft = 1.0936 yd. The meter is the unit of length employed by the U.S. Coast and Geodetic Survey. The *vara* is a Spanish unit of measure used in Mexico and several other countries falling under early Spanish influence: 1 vara = 32.993 in. (Mexico), 33 in. (California), or 33⅓ in. (Texas).

In the United States the units of area commonly used are the *square foot* and the *acre;* formerly the *square rod* and the *square Gunter's chain* were also used: 1 acre = 43,650 sq ft = 160 sq rods = 10 sq Gunter's chains.

The units of volumetric measurement are the *cubic foot* and the *cubic yard.*

Care of Instruments. In tightening the various clamp screws, adjusting screws, and leveling screws, bring them only to a firm bearing. *The general tendency is to tighten these screws far more than necessary.* Such a practice may strip the threads, twist off the screw, bend the connecting parts, or place undue stresses in the instrument so that the setting may not be stable.

Remove grit from exposed movable parts such as the threads of tangent screws by wiping them with an oiled rag. If the threads or slides work hard, clean them in gasoline and oil lightly. Use no abrasives.

Use only the best quality of clock or watch oil. Oil sparingly, and wipe off the excess oil. In cold weather, it may be necessary to use graphite for lubrication.

Never rub the objective or the eyepiece of a telescope with the fingers or with a rough cloth. Use a camel's-hair brush to remove dust, or use clean chamois or lint-free soft cloth if the dust is caked or damp. Occasionally the lenses may be cleaned with a mixture of equal parts of alcohol and water. Keep oil off the lenses.

If the level vial becomes loose, it can be reset with plaster of paris, or wedged lightly with strips of paper or with toothpicks.

Broken cross hairs can be replaced in the field. Threads from ordinary spider webs are too rough, coarse, and dirty for use as cross hairs. The best spider thread is freshly spun from a small spider, but commonly the thread from a cocoon—preferably a brown cocoon—is used. If the thread is stretched too tightly it will break easily; if too loose, it will sag in wet weather. The thread is handled by means of a pair of dividers or a forked stick. It is wetted, stretched moderately, and held securely in position on the marks of the cross-hair ring while a drop of shellac is placed on each end and left to dry.

The cross-hair ring is removed from the tube as follows: Two opposite capstan-headed screws are removed, and the ring is rotated 90° about the remaining two screws by means of a pointed stick inserted through the end of the telescope. The stick is then inserted in a screw hole, the remaining screws are taken out, and the ring is withdrawn without damage. The operations of replacing the cross-hair ring are in the reverse order of those employed in removing it.

The adjusting pin should fit the hole in the capstan head. If the pin is too small, the head of the screw is soon ruined.

Preferably make adjustments with the instrument in the shade.

In order to do up a tape which is not wound on a reel, the chainman takes the zero end in the left hand, and—allowing the tape to slide loosely through the right hand—extends his arms. As the 5-ft mark is reached, he grasps it with the right hand. He brings the hands together and lays the 5-ft mark in the fingers of the left hand without permitting the tape to turn over, thus forming a figure 8. He grasps this loop with the left hand, again extends the arms for another 5-ft length, lays the 10-ft mark on top of the 5-ft mark; and so on. When the last mark of the tape is reached, he ties the loop tightly where the first and last marks come together, by means of the rawhide thongs. He then grasps one side of the loop at the crossing of the figure 8 with one hand, and the other side of the loop at the crossing with the other hand. Finally he twists the loop in such a manner that it will be thrown into circular form, with diameter (about 10 in.) half that of the loop.

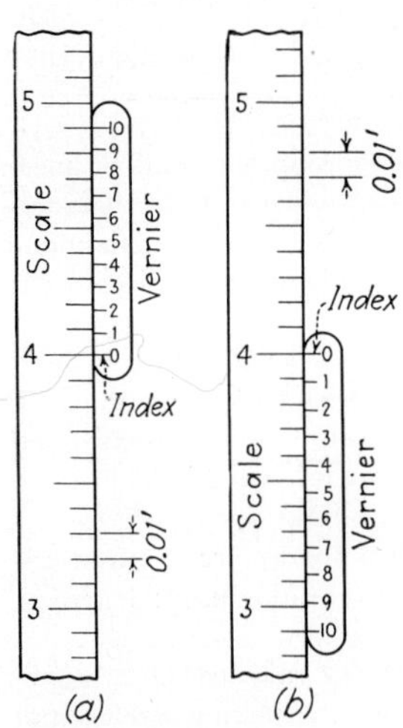

FIG. 1. (*a*) Direct vernier. (*b*) Retrograde vernier.

Verniers. A vernier (Fig. 1) is a short auxiliary scale placed parallel and in contact with the main scale, by means of which fractional parts of the least division of the main scale can be measured precisely. The scale may be either straight (as on a leveling rod) or curved (as on the horizontal and vertical circles of a transit). The zero of the vernier scale is the index for the main scale.

The *least count*, or fineness of reading, of a vernier is equal to the difference in length between one scale space and one vernier space. Further, the number n of spaces on the vernier is equal to the number of equal parts into which one space s on the main scale can be subdivided by reading the vernier. It follows that the least count is s/n.

Verniers are of two types: the $\left\{\begin{matrix}\text{direct}\\ \text{retrograde}\end{matrix}\right\}$ vernier which has spaces slightly $\left\{\begin{matrix}\text{shorter}\\ \text{longer}\end{matrix}\right\}$

than those of the main scale; n spaces on the vernier are equal in length to $\begin{Bmatrix} n-1 \\ n+1 \end{Bmatrix}$ spaces on the main scale. The use of the two types is identical, and they are equally sensitive and equally easy to read.

Precision of Measurements and Computations. As defined by the American Society of Civil Engineers, accuracy is "nearness to the truth" whereas precision is "degree of fineness of reading in a measurement, or, the number of places to which a computation is carried." As defined by the U.S. Coast and Geodetic Survey, accuracy is "degree of conformity with a standard" whereas precision is "degree of refinement in the performance of an operation or in the statement of a result."

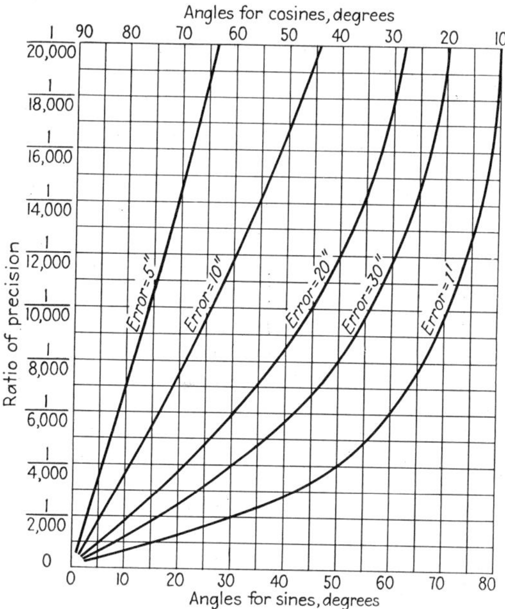

FIG. 2. Ratios of precision for sines and cosines.

It is the duty of the surveyor to maintain a degree of precision as high as is justified by the purpose of the particular survey, but not higher. It is common practice to fix the permissible error of linear measurement. So far as practicable, the precision of angles should be consistent with the precision of related distances. A consistent relation between angles and distances will be maintained if the tangent or sine of the allowable error in angles equals the allowable error, expressed as a ratio, in the distances. The ratio between linear and angular error can be determined by simple trigonometry; e.g., at 1,000 ft an error of 01′ in angle corresponds to a linear error of 0.3 ft.

The curves of Figs. 2 and 3 show the ratios of precision corresponding to various angular errors from 05″ to 01′ for sines, cosines, tangents, and cotangents. For the function under consideration, these curves may be used to determine the ratio of precision corresponding to a given angular error and angle, or to determine the maximum or minimum angle that for a given angular error will furnish the required ratio of precision, or to determine the precision with which angles of a given size must be measured to maintain a required ratio of precision in computations.

Measured quantities are not exact, and the number of digits that have meaning is limited strictly by the precision with which the measurement has been made. In calculations involving multiplication or division, the precision is governed by the number of *significant figures;* in addition or subtraction, by the number of *decimal places.* A product cannot properly have a greater number of significant figures than those in its least precise factor; and a sum cannot properly have a greater number of decimal places than those in the least precise number being added. In computations it is advisable to carry out the intermediate results to one figure or decimal place more than that desired in the final result, but it is useless and time-consuming to employ more than one.

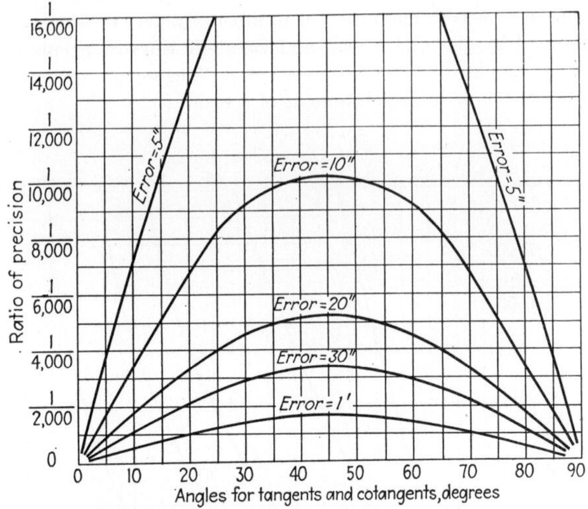

FIG. 3. Ratios of precision for tangents and cotangents.

ERRORS

Sources of Error. In the measurements of surveying, *instrumental errors* arise from imperfections or faulty adjustment of the devices with which measurements are taken; *personal errors* occur through the observer's inability to read the instruments exactly; and *natural errors* occur from variations in the phenomena of nature such as temperature, humidity, wind, gravity, refraction, and magnetic declination.

Kinds of Error. A *mistake* is an unintentional fault of conduct arising from poor judgment or from confusion in the mind of the observer. Mistakes have no place in a discussion of the theory of errors. They are detected and eliminated by checking all work.

The *resultant error* is the difference between a measurement and the true value. It is made up of individual errors from a variety of sources.

A *discrepancy* is the difference between two measurements of the same quantity.

A *systematic error* is one that, so long as conditions remain unchanged, always has the same magnitude and the same algebraic sign (which may be either positive or negative). If conditions do not change during a series of measurements, the error is termed a *constant* systematic error; e.g., a line may be measured with a tape that is too short. If conditions change, resulting in corresponding changes in the magnitude of the error, it is termed a *variable* systematic error; e.g., a line may be chained during a period in which the temperature varies. A systematic error always follows some definite mathematical

or physical law, and a correction can be determined and applied. The error may be instrumental, personal, or natural. The total systematic error in any given number of measurements is the algebraic sum of the individual errors of the individual measurements.

An *accidental error* is an error due to a combination of causes beyond the ability of the observer to control and for which it is impossible to make correction; for each observation the magnitude and algebraic sign of the accidental error are matters of chance and hence cannot be computed as can the magnitude and algebraic sign of a systematic error. However, accidental errors taken collectively obey the law of probability. Since each accidental error is as likely to be positive as negative, a certain compensative effect exists; and accidental errors are sometimes called "compensating" errors. Accidental errors are also termed "irregular" or "erratic" errors. As an example of the occurrence of an accidental error, in chaining it is impossible to set the chaining pin exactly at the proper graduation on the tape. Accidental errors remain after mistakes have been eliminated by checking and systematic errors have been eliminated by correction. According to the mathematical theory of probability, accidental errors tend to increase in proportion to the square root of the number of opportunities for error.

In general, the more refined the methods used, the smaller the systematic errors as compared with the accidental errors.

If the discrepancy between a check measurement and the original measurement is small, it is an indication that no mistakes have been made and that the accidental errors are small, but it is not an indication that the systematic errors are small.

Probable Value. If systematic errors are so far eliminated as to be a negligible factor, the *most probable value* is the adjusted value of a quantity that, according to the principles of least squares, has more chances of being correct than has any other.

For a series of measurements of the same quantity made under identical conditions, the most probable value is the mean of the measurements.

For related measurements (taken under identical conditions) the sum of which should equal a single measurement taken under the same conditions, the most probable values of the several related quantities and of the measured sum are obtained by dividing the discrepancy equally among all the measurements including the sum. If the correction is added to each of the related measurements, it is subtracted from the measurement representing their sum, and vice versa.

Probable Error of a Single Measured Quantity. If all systematic errors are assumed to have been eliminated from a given measurement, there is left an accidental error which can be neither eliminated nor exactly determined.

Probable error is a plus or minus quantity within which limits the actual accidental error is as likely as not to fall. It is an indication of precision and does not signify either the actual error or the error most likely to occur.

When a series of observations of a single quantity is made under uniform conditions and in such manner as to eliminate systematic errors, the mean of the series is the most probable value. For the purpose of determining the probable error this mean value is mathematically regarded as being the most likely value (based on this series of observations), and the difference between each of the individual measurements and the mean value is determined. These differences are termed *residuals* or *deviations*.

The probable error of a *single observation* is calculated by the expression

$$E = 0.6745 \sqrt{\frac{\Sigma v^2}{n - 1}} \tag{1}$$

in which Σv^2 is the sum of the squares of the residuals, and n is the number of observations.

The probable error of a single observation may be calculated approximately by the expression

$$E = \frac{0.845\Sigma v}{\sqrt{n(n-1)}} \text{ (approx)} \tag{2}$$

in which Σv is the sum of the residuals without regard to signs.

The probable error of a single observation may be calculated with about the same degree of approximation by the expression

$$E = 0.845\bar{v} \text{ (approx)} \tag{3}$$

in which $\bar{v}$ is the mean value of the residuals without regard to signs. The term $\bar{v}$ is also called the *average deviation.*

The probable error of the *mean* of a number of observations is calculated by the equation

$$E_m = \frac{E}{\sqrt{n}} \tag{4}$$

Weighted Observations. In the foregoing discussion it has been assumed that all observations are equally reliable. Frequently in surveying, however, it is required to combine the results of measurements which are not made under similar conditions and which therefore have different degrees of reliability. In such cases it is necessary to determine the degree of reliability, or *weight* (as nearly as it can be determined), that applies to each of the separate measurements. Weights are relative, not absolute.

Weights may be determined (1) by assuming that they are proportional to the number of like observations, (2) arbitrarily according to the judgement of the observer, or (3) from the probable errors by means of the relation that weights are inversely proportional to the square of the corresponding probable errors. If W is the weight and E the probable error, then

$$W_1E_1^2 = W_2E_2^2 = \cdots = W_nE_n^2 \tag{5}$$

The most probable value of a quantity for which measurements of different reliability have been made is the *weighted mean.* The weighted mean is computed by multiplying each value by its weight, adding the products, and dividing by the sum of the weights.

When the sum of measured values having different weights must equal a known value, either measured or exact, the most probable values of the measurements (including the sum) are the observed values, each corrected by an appropriate portion of the discrepancy or of the total error. The corrections to be applied are inversely proportional to the corresponding weights. If the correction needs to be added to each of the related measurements, it is subtracted from the measurement representing their sum, and vice versa. No correction, of course, is made to an exact quantity.

Error in Computed Quantities. The probable error of the *sum* of independent measurements $Q_1, Q_2 \ldots Q_n$, for which the probable errors are $E_1, E_2 \ldots E_n$, is

$$E_s = \sqrt{E_1^2 + E_2^2 + \cdots + E_n^2} \tag{6}$$

The probable error of the *difference* between two independent measurements Q_1 and Q_2, for which the probable errors are E_1 and E_2, is

$$E_d = \sqrt{E_1^2 + E_2^2} \tag{7}$$

The probable error of the *product* of a constant or known quantity K and a measured quantity Q, for which the probable error is E, is

$$E_p = KE \tag{8}$$

If E_1 and E_2 represent respectively the probable errors of lengths l_1 and l_2, the probable error of the area representing the product of these two lengths is

$$E_a = \sqrt{l_1^2E_2^2 + l_2^2E_1^2} \tag{9}$$

Summary of Principal Relations. In each of the four cases which arise in practice, the most probable value is as shown in the following tabulation:

Measurements	Most probable value	
	Same quantity	Related quantities
Of equal reliability....	Mean	Each observed value corrected equally
Of different reliability..	Weighted mean	Each observed value corrected by an amount inversely proportional to its weight

The corresponding probable errors are as follows:

Measurements	Probable error of most probable value	
	Same quantity	Related quantities
Of equal reliability....	$E_m = 0.6745\sqrt{\frac{\Sigma v^2}{n(n-1)}}$ $= \frac{E}{\sqrt{n}}$	$E_s = E\sqrt{n}$
Of different reliability	$E_{wm} = 0.6745\sqrt{\frac{\Sigma(Wv^2)}{(\Sigma W)(n-1)}}$	$E_s = \sqrt{E_1^2 + E_2^2 + \cdots + E_n^2}$

MEASUREMENT OF DISTANCE

Methods. Table 1 classifies the principal methods of measuring distance according to the usual degree of precision obtained. Electronic methods are also in use and developing rapidly.

Horizontal Measurements over Smooth Level Ground. The following represents the usual practice when the measurements are of ordinary precision (say 1/5,000): The tape is supported throughout its length, and the only requirement is that the distance between two fixed points (as the corners of a parcel of land) be determined.

The rear chainman, *with one pin*, stations himself at the point of beginning. The head chainman, with the zero (graduated) end of the tape and 10 pins, advances toward the distant point. When the head chainman has gone nearly 100 ft, the rear chainman calls "chain." The rear chainman holds the 100-ft mark at the point of beginning and lines in a chaining pin (held by the head chainman) with the range pole or other signal marking the distant point. The head chainman sets the pin vertically on line and a short distance to the rear of the zero mark. With his other hand, he then pulls the tape taut and, making sure that it is straight, brings it in contact with the pin. The rear chainman, when he observes that the 100-ft mark is at the point of beginning, calls "stick" or "all right." The head chainman pulls the pin and sticks it at the zero mark of the tape, with the pin sloping away from the line. As a check, he again pulls the tape taut and notes that the zero point coincides with the pin at its intersection with the ground. He then calls

TABLE 1. GENERAL METHODS OF MEASURING DISTANCE

Method	Usual precision	Use	Instrument for measuring angles with corresponding precision
Pacing	1/100–1/200	Reconnaissance; small-scale mapping; checking tape measurements	Hand compass; peep-sight alidade
Stadia	1/300–1/1,000	Location of details for mapping; rough traverses; checking more precise measurements	Transit; telescopic alidade of plane table; surveyor's compass
Ordinary chaining	1/1,000–1/5,000	Traverses for land surveys and for control of route and topographic surveys; ordinary construction work	Transit (angles doubled)
Precision chaining	1/10,000–1/30,000	Traverses for city surveys; base lines for triangulation of intermediate precision; precise construction work	Transit (angles by repetition)
Base-line measurement	1/50,000–1/1,000,000	Triangulation of high precision for large areas, city surveys, or long bridges and tunnels	Repeating theodolite; direction instrument

"stuck" or "all right," the rear chainman releases the tape, the head chainman moves forward as before, and so the process is repeated. As the rear chainman leaves each intermediate point, he pulls the pin. Thus there is always one pin in the ground, and the number of pins held by the rear chainman at any time indicates the number of hundreds of feet, or *stations*, from the point of beginning to the pin in the ground.

Horizontal Measurements over Uneven or Sloping Ground. The measurements are carried forward by holding the tape horizontal, and the plumb line is used by either chainman, or at times by both chainmen, for projecting from tape to pin, or vice versa.

Where the slope is less than 5 or 6 ft in 100, the head chainman advances a full tape length at a time, and pins are set by him and collected by the rear chainman as described in the preceding article.

Where the course is steeper and is downhill, the head chainman advances a full tape length and then returns to an intermediate point from which he can hold the tape horizontal. He suspends the plumb line at a foot mark, is lined in by the rear chainman, and sets a pin at the indicated point. The rear chainman comes forward, *gives the head chainman a pin*, and holds the tape at the foot mark from which the plumb line was previously suspended. The head chainman proceeds to another point from which he can hold the tape horizontal, and so the process is repeated until the zero mark on the tape is reached. At each *intermediate* point of a tape length the rear chainman gives the head chainman a pin, but not at the point marking the full tape length.

Slope Measurements. Where the ground is fairly smooth, measurements on the slope may sometimes be made more accurately and quickly than horizontal measurements; and slope measurements are generally preferred. Either the clinometer for measuring slope or the hand level for measuring difference in elevation between successive stations or breaks in slope may be used.

Corrections for Slope. For measurements of ordinary precision when the slope is not greater than about 20 in 100, the correction C_h to distances measured on the slope to

give horizontal distances may be calculated approximately by the expression

$$C_h = s - d = \frac{h^2}{2s} \text{ (approx)} \tag{10}$$

in which h is the difference in elevation, d is the horizontal distance, and s is the slope distance.

Where the *angle* of slope is determined,

$$C_h = 0.015S(\theta°)^2 \text{ (approx)} \tag{11}$$

in which $\theta°$ is the slope angle, deg, and S is the slope distance in 100-*ft stations.*

The exact formula for slope correction is

$$C_h = s - d = s \text{ vers } \theta \text{ (exact)} \tag{12}$$

Correction for Sag. The correction C_s for sag is given with sufficient precision for most surveys by the expression

$$C_s = \frac{w^2 l^3}{24P^2} = \frac{W^2 l}{24P^2} \tag{13}$$

in which w is the weight of tape, lb per ft, W is the total weight of the tape *between supports,* lb, l is the distance between supports, ft, P is the applied tension, lb, and C_s is the correction between points of support, ft.

Correction for Tension. If the tension is greater or less than that for which the tape is standardized, the tape is correspondingly too long or too short. The correction C_p for variation in tension in a steel tape is

$$C_p = \frac{(P - P_o)l}{AE} \tag{14}$$

in which P is the applied tension, lb, P_o is the tension for which the tape is standardized, lb, l is the length, ft, A is the cross-sectional area, sq. in., C_p is the correction per distance l, ft, and E is the modulus of elasticity of the steel, usually taken as 30,000,000 psi.

Elimination of Effect of Sag. Normal tension P_n, the pull that will counteract the shortening due to sag, is

$$P_n = \frac{0.204W\sqrt{AE}}{\sqrt{P_n - P_o}} \tag{15}$$

This equation is solved by trial.

Correction for Temperature. The coefficient of thermal expansion of steel is about 0.0000065 per 1°F. If the tape is standard at a temperature of $T_o°$ and measurements are taken at a temperature of $T°$, the correction C_x for change in length is

$$C_x = 0.0000065l(T - T_o) \tag{16}$$

in which l is the measured length.

Summary of Errors. Table 2 summarizes the various errors and corrections in chaining.

Mistakes in Chaining. Some of the mistakes commonly made by inexperienced chainmen are (1) adding or dropping a full tape length, (2) adding a foot in measuring the fractional part of a foot at the end, (3) taking wrong points as 0 or 100-ft marks on tape, (4) reading numbers incorrectly, as "6" for "9," or (5) calling numbers incorrectly, as "fifty-three" for 50.3.

TABLE 2. ERRORS AND CORRECTIONS IN CHAINING

NOTE: In *measuring* a distance with a tape that is *too long*, *add* the correction

Error	Source		Amount	Error of 0.01 ft per 100-ft tape length caused by	Makes tape too	Importance in ordinary chaining	Procedure to eliminate or reduce
Systematic	Erroneous length		....	....	Long or short	Usually small, but should be checked	Standardize tape and apply computed correction
	Temperature		$C_x = 0.0000065l(T - T_o)$	15°F	Long or short	Of consequence only in hot or cold weather	Measure temperature and apply computed correction. In precise work, chain at favorable times and/or use invar tape
	Pull or tension (change in)		$C_p = \dfrac{(P - P_o)l}{AE}$	15 lb (for 1½-lb tape)	Long or short	Negligible	Apply computed correction. In precise work, use spring balance. May use normal tension, $P_n = \dfrac{0.204W\sqrt{AE}}{\sqrt{P_n - P_o}}$
	Sag		$C_s = \dfrac{w^2l^3}{24P^2} = \dfrac{W^2l}{24P^2}$	0.6 ft	Short	Large, especially with heavy tape	Apply computed correction. May be avoided by chaining on slope. May use normal tension, $P_n = \dfrac{0.204W\sqrt{AE}}{\sqrt{P_n - P_o}}$
	Slope		$C_h = h^2/2s$ (approx) $= 0.015S(\theta^\circ)^2$ (approx) $= s \text{ vers } \theta$ (exact)	1.4 ft	Short	....	At breaks in slope, determine difference in elevation or slope angle; apply computed correction
	Imperfect alignment	Horizontal	Same as slope	1.4 ft	Short	Not serious	Use reasonable care in sighting. Keep tape taut and reasonably straight
		Grass	Twice slope	0.7 ft	Short	Not serious	
		Wind	Same as sag	0.6 ft			
		Vertical	Same as slope	1.4 ft	Short	Often large	Level tape
Accidental	Manipulation	Plumbing	In rough country, breaking tape, 0.05 to 0.10 ft per tape length		±	Large, but accidental	May avoid by slope chaining. Use reasonable care to reduce
		Marking ends of tape; reading graduations	0.01 ft per tape length		±	Not serious	.... Use reasonable care to reduce
		Error in pull	Same as pull	15 lb (1½-lb tape)	±	Not serious	.... Use reasonable care to reduce
		Error in determination of slope	Amount varies with slope		±	Not serious	.... Use reasonable care to reduce

MEASUREMENT OF DIFFERENCE IN ELEVATION

Barometric Leveling. Since the pressure of the earth's atmosphere varies inversely with the elevation, the barometer may be employed for making observations of difference in elevation. However, the pressure is likely to vary over a considerable range in the course of a day or even an hour. Under ordinary conditions elevations determined by barometric leveling are likely to be several feet in error; with sensitive barometers, however, elevations can be determined within a foot or so.

The mercurial barometer is cumbersome, and for field work an aneroid barometer is used. A single barometer is sometimes used by topographers on small-scale surveys where the contour interval is large. Stops are made at frequent intervals during the day, and the rate of change in atmospheric conditions is observed in order to determine suitable corrections. It is desirable also to return to the starting point.

Usually barometric observations are taken at a fixed station during the same period that observations are made on a second barometer which is carried from point to point in the field. This procedure makes it possible to correct the readings of the portable barometer for atmospheric disturbances.

One type of extremely sensitive barometer is graduated in feet instead of units of pressure. It is used in the following method which employs two of the instruments at fixed bases and one or more instruments carried from point to point over the area being surveyed. One fixed instrument is located at a point of known elevation near the highest elevation of the area, and one near the lowest elevation. A third instrument is carried to the point whose elevation is desired, and a reading is taken. Readings on the fixed instruments are taken either simultaneously or at fixed intervals of time. The elevation of the portable instrument is then determined by interpolation. The horizontal location of each point at which a reading is taken is determined by conventional methods.

Curvature of the Earth; Refraction. The combined effect of the earth's curvature and atmospheric refraction is given by the expression

$$h' = 0.57K^2 = 0.021M^2 \text{ (approx)} \qquad (17)$$

in which K is the distance from the point of tangency (station of the observer), miles, M is the same distance, thousands of feet, and h' is the combined effect of the earth's curvature and of atmospheric refraction, ft. The effect of the earth's curvature alone is about $0.66K^2$. The effect of refraction alone is about $0.09K^2$ in the opposite direction.

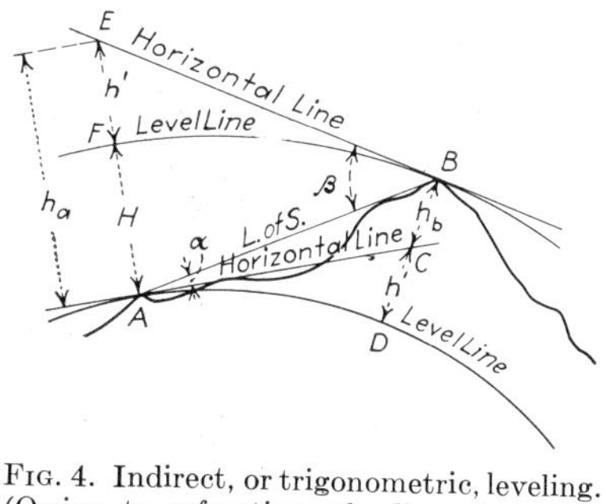

FIG. 4. Indirect, or trigonometric, leveling. (Owing to refraction, the line of sight is slightly curved.)

Indirect Leveling. In Fig. 4, A represents a point of known elevation and B a point the elevation of which is desired. In indirect or trigonometric leveling, the vertical angle α at A is measured, and the distance AD is determined by some method of measurement. The difference in elevation is therefore

$$H = 0.021\left(\frac{AC}{1{,}000}\right)^2 + AC \tan \alpha \qquad (18)$$

If the vertical angle β is also taken from B to A,

$$H = \frac{h_a + h_b}{2} = \frac{AC}{2}(\tan \alpha + \tan \beta) \qquad (19)$$

From Eq. (19) may be deduced the general rule that, *when vertical angles are measured from (and to) each of two points whose difference in elevation is desired, the difference in elevation is one-half the horizontal distance between them multiplied by the sum of the tangents of the angles*, and the effect of the earth's curvature and atmospheric refraction is thereby eliminated. This procedure is employed in precise indirect leveling.

Direct Leveling. In Fig. 5, A represents a point of known elevation, and B represents a point the elevation of which is desired. In the method of direct or spirit leveling, the level is set up at some intermediate point as L, and the vertical distances AC and BD are observed by holding a leveling rod first at A and then at B, the line of sight of the

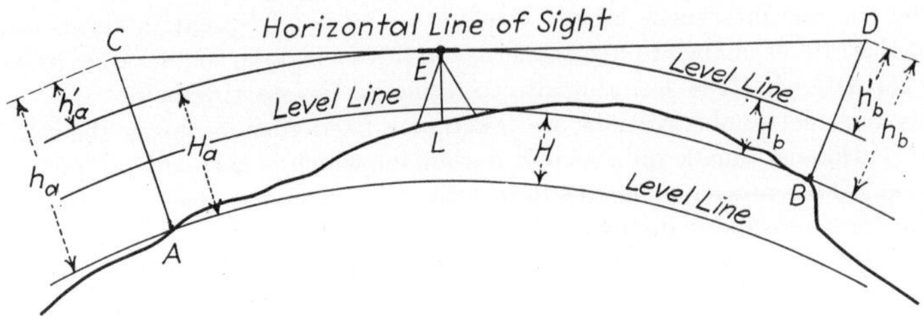

FIG. 5. Direct leveling. (Owing to refraction, the line of sight is slightly curved.)

instrument being horizontal. (Owing to refraction, the line of sight is slightly curved.)

The difference in elevation H between A and B is

$$H = h_a - h_b - h_a' + h_b' \tag{20}$$

If the backsight distance LA is equal to the foresight distance LB, then $h_a' = h_b'$, and

$$H = h_a - h_b \tag{21}$$

Thus, if backsight and foresight distances are balanced, the difference in elevation between two points is equal to the difference between the rod readings taken to the two points, and no correction for earth's curvature and atmospheric refraction is necessary.

Use of the Level and Rod. The engineer's level is set up with tripod legs well spread and firmly pressed into the ground and with the tripod head nearly level. The telescope is brought over one pair of opposite leveling screws, and the bubble is centered approximately; the process is then repeated with the telescope over the other pair. By repetition of this procedure the leveling screws are manipulated until the bubble remains centered, or nearly so, for any direction in which the telescope is pointed. If the instrument is in adjustment, the line of sight is then horizontal.

The rod is held vertical. For accurate observations, the rod is held on some well-defined point. The leveler sights at the rod, focuses the objective, and carefully centers the bubble. If the self-reading rod is used, the leveler records the reading of the cross hair on the rod; and, as a check, he again observes the bubble and the rod. If the target rod is used, the target is set by the rodman as directed by the leveler. To eliminate error caused by the rod being tipped forward or backward, the rodman *waves the rod*, or tilts it slowly forward and backward, and the leveler takes the least reading.

For less precise leveling, as when rod readings for ground points are determined to the nearest 0.1 ft, usually less care is exercised in keeping the bubble centered and the rod vertical, and the observations are not checked.

Some levels have a micrometer screw which permits tilting the telescope with its attached level tube and thus permits centering the bubble while sighting at the rod.

Adjustment of the Dumpy Level. 1. *To Make the Axis of the Level Tube Perpendicular*

to the Vertical Axis. Approximately center the bubble over each pair of opposite leveling screws; then carefully center the bubble over one pair. Rotate the level end for end about its vertical axis. If the bubble is displaced, bring it back halfway to the center by means of the capstan nuts at one end of the level tube. Relevel the instrument with the leveling screws, and repeat the process until the adjustment is perfected.

2. *To Make the Horizontal Cross Hair Lie in a Plane Perpendicular to the Vertical Axis (and Thus Horizontal When the Instrument Is Level).* Sight the horizontal cross hair on some clearly defined point, and rotate the instrument slowly about its vertical axis. If the point appears to depart from the cross hair, loosen two adjacent capstan screws and rotate the cross-hair ring until by further trial the point appears to travel along the cross hair. Tighten the same two screws. The instrument need not be level when the test is made.

3. *To Make the Line of Sight Parallel to the Axis of the Level Tube (Two-peg Test).* Method A. Set two pegs 200 to 300 ft apart on approximately level ground. Set up and level the instrument in a location such that the eyepiece is ½ in. or less in front of the rod held on one of the pegs as at A (Fig. 6). With the rod held at A, take a rod reading

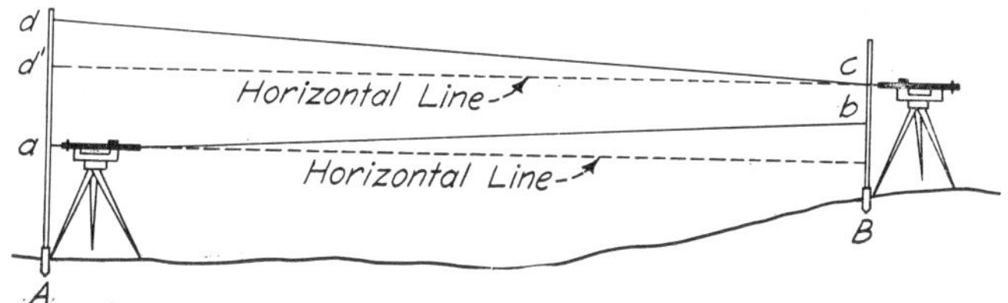

FIG. 6. Two-peg test, Method A.

a by sighting through the objective end of the telescope (with the eyepiece next to the rod). The center of the field of view may be determined within one or two thousandths of a foot by holding the point of a pencil on the rod.

Move the rod to the other peg B, and take a rod reading b with level at A, in the usual manner.

Move the instrument to B, set up as before, and take rod readings c and d.

If $(a - b) = (d - c)$, the line of sight is in adjustment. If not, the correct rod reading at A for instrument with position unchanged at B is

$$d' = c + \text{true diff. el.} = c + \frac{(a - b) + (d - c)}{2} \tag{22}$$

Equation (22) must be solved with due regard to signs. Strictly, the effect of curvature and refraction should be included; but for ordinary leveling the quantities are negligible.

The adjustment is made by moving the cross-hair ring vertically until the line of sight cuts the rod at d'. The preceding steps are then repeated as a check.

Instead of viewing the near rod through the objective end of the telescope, the level may be set up a short distance (say, 6 or 8 ft) beyond each near rod and the near-rod reading observed in the customary manner. The adjustment should be considered as a first approximation and should be repeated for precise results.

Method B. Set two pegs 200 to 300 ft apart on approximately level ground, and designate as A the peg near which the second setup will be made (Fig. 7); call the other peg B. Set up and level the instrument at any point M equally distant from A and B, that is, in a vertical plane bisecting the line AB. Take rod readings a on A and b on B; then $(a - b)$ will be the true difference in elevation, since any error would be the same

for the two equal sight distances L_m. Due account must be taken of signs throughout the test.

Move the instrument to a point P near A, preferably but not necessarily on line with the pegs; set up as before, and measure the distances L_a to A and L_b to B. Take rod readings c on A and d on B. Then $(c - d)$, taken in the same order as before, is the indicated difference in elevation; if $(c - d) = (a - b)$, the instrument is in adjustment. If not, $(c - d)$ is called the "false" difference in elevation, and the inclination (error) of

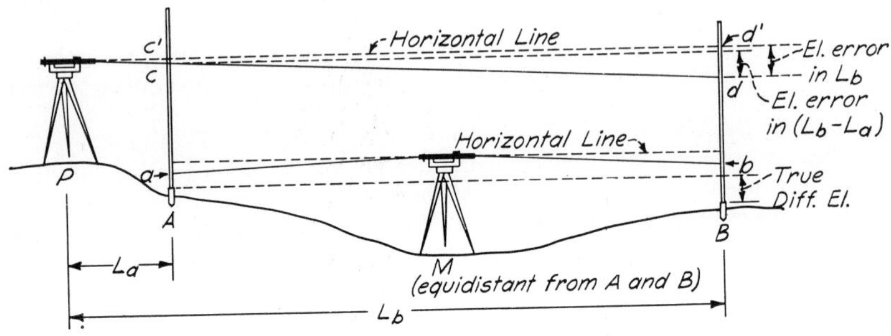

FIG. 7. Two-peg test, Method B.

the line of sight in the net distance $(L_b - L_a)$ is equal to $(a - b) - (c - d)$. By proportion, the error in the reading on the far rod is $\dfrac{L_b}{L_b - L_a}[(a - b) - (c - d)]$. Subtract algebraically the amount of this error from the reading d on the far rod to obtain the correct reading d' at B for a horizontal line of sight with the position of the instrument unchanged at P. Set the target at d' and bring the line of sight on the target by moving the cross-hair ring vertically.

As a partial check on the computations, the correct rod reading c' at A may be computed by proportion; the difference in elevation computed from the two corrected rod readings c' and d' should be equal to the true difference in elevation observed originally at M.

Theoretically a correction for earth's curvature and atmospheric refraction should be added numerically to the final rod reading d', although in practice it is usually considered negligible.

Some surveyors prefer to set up at P within 6 to 8 ft of A and to consider $[(a - b) - (c - d)]$ as being the *total* error in elevation, to be subtracted directly from d. This serves as a first approximation; the procedure for the setup at P is then repeated.

Adjustment of the Wye Level. 1. *To Make the Axis of the Level Tube Lie in the Same Plane with the Axis of the Wyes.* Raise the wye clips, level the instrument, and rotate the telescope a few degrees in the wyes. If the bubble is displaced, bring it back to the center by means of the lateral adjusting screws at one end of the level tube.

2. *To Make the Axis of the Level Tube Parallel to the Axis of the Wyes.* Raise the wye clips; level the instrument carefully; lift the telescope from the wyes, and turn it end for end. If the bubble is displaced, bring it back halfway to the center by means of the vertical adjusting nuts at one end of the level tube. Relevel the instrument by means of the leveling screws, and repeat the process until the adjustment is perfected.

3. *To Make the Horizontal Cross Hair Lie in a Plane Perpendicular to the Vertical Axis (and Thus Horizontal When the Instrument Is Level).* This adjustment is the same as for the dumpy level, except that for some instruments the adjustment may be made by rotating the telescope in the wyes instead of rotating the cross-hair ring in the telescope.

4. *To Make the Line of Sight Coincide with the Axis of the Wyes (and Thus Parallel to the Axis of the Level Tube).* Raise the wye clips, sight the intersection of the cross hairs on some well-defined point, and clamp the vertical axis. Revolve the telescope 180° (about its own axis) in the wyes. If the line of sight is displaced from the point, adjust the cross-hair ring until the line of sight is midway between its two former positions. Repeat the process until the proper relation is obtained.

5. *To Make the Axis of the Level Tube (and Thus the Axis of the Wyes) Perpendicular to the Vertical Axis.* Inasmuch as the preceding adjustments have established parallelism or coincidence between the axis of the level tube, the line of sight, and the axis of the wyes, this adjustment does not add to the precision of observations. The adjustment makes it possible, however, to level the instrument so that the bubble will remain centered for any direction in which the telescope may be pointed. Level the instrument, and rotate the telescope end for end about the vertical axis. If the bubble is displaced, bring it back halfway to the center by means of the capstan nuts controlling the vertical position of one of the wyes.

DIFFERENTIAL LEVELING

Definitions. A *bench mark* (BM) is a definite point of known elevation and location and of more or less permanent character.

A *turning point* (TP) is an intervening point between two bench marks, on which point both foresight and backsight rod readings are taken. No record is made of its location.

A *backsight* (BS) is a rod reading taken on a point of known elevation.

A *foresight* (FS) is a rod reading taken on a point the elevation of which is to be determined.

The *height of instrument* (HI) is the elevation of the line of sight of the telescope when the instrument is leveled.

Procedure in Differential Leveling. In Fig. 8, BM_1 represents a point of known elevation (bench mark) and BM_2 represents a bench mark to be established some distance away. It is desired to determine the elevation of BM_2. The rod is held at BM_1 and

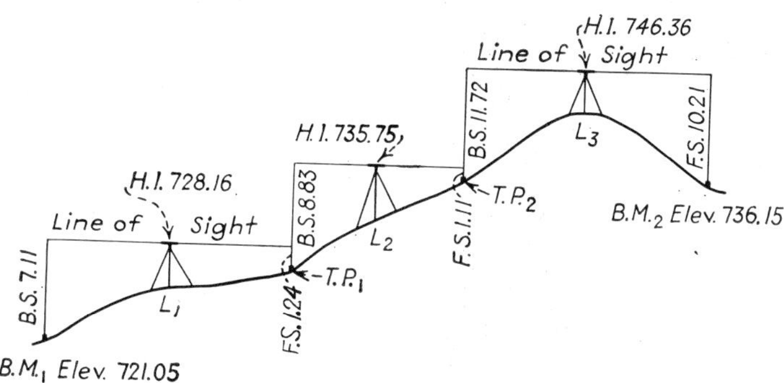

FIG. 8. Differential leveling.

the level is set up in some convenient location, as L_1, along the general route but not necessarily on the direct line joining BM_1 and BM_2 (unless it is desired to obtain rough measurements of distance between bench marks by stadia). A backsight is taken on BM_1. The rodman then goes forward and, as directed by the leveler, chooses a turning point TP_1 at some convenient spot within the range of the telescope along the general route BM_1 to BM_2. It is desirable, but not necessary, that each foresight distance be made approximately equal to the preceding backsight distance. The rod is held on the turning point, and a foresight is taken. The leveler then sets up the instrument at some

favorable point, as L_2, and takes a backsight to the rod held on the turning point; the rodman goes forward to establish a second turning point TP_2; and so the process is repeated until finally a foresight is taken on the terminal point BM_2.

In Fig. 8, it is seen that a *backsight added to the elevation* of a point on which the backsight is taken gives the height of instrument, and that a *foresight subtracted from the height of instrument* determines the elevation of the point on which a foresight is taken. As a check, the difference between the sum of all backsights and the sum of all foresights gives the difference in elevation between the bench marks.

In mines, tunnels, or construction work, sometimes the rod is held, inverted, on points higher than the telescope. In these cases the backsights are *subtracted* and the foresights are *added.*

No special attempt is made to balance *each* foresight distance against the corresponding backsight distance. The effect of the earth's curvature and atmospheric refraction is slight unless there is an abnormal difference between the backsight and foresight distances. Any error due to lack of parallelism between the line of sight and the axis of the level tube is eliminated if, between bench marks, the *sum* of the foresight distances is made equal to the *sum* of the backsight distances.

Differential-level Notes. For ordinary differential leveling in which no special effort is made to equalize backsight and foresight distances between bench marks, usually the record of field work is kept in the form indicated by Fig. 9.

LEVELS FOR BENCH MARKS ALONG RIDGE ROAD

Dec. 31, 1951 Fair — J.G. Sutter, ⊼ — W.R. Knowles, Rod

Sta.	B.S.	H.I.	F.S.	Elev.		Remarks
B.M.$_1$	7.11	728.16		721.05		Top of Hydrant Cor. Oak St.
T.P.$_1$	8.83	735.75	1.24	726.92		Curb
T.P.$_2$	11.72	746.36	1.11	734.64		
B.M.$_2$	4.32	740.47	10.21	736.15		Spike in Pole North of Williams House Marked B.M. 736.15
T.P.$_3$	3.06	733.57	9.96	730.51		
T.P.$_4$	2.74	727.40	8.91	724.66		Stone
T.P.$_5$	0.81	716.59	11.62	715.78		
B.M.$_3$			12.42	704.17		Concrete Monument No. of Road; County Line
ΣB.S.=	38.59	ΣF.S.=	55.47	721.05		
			38.59			
		Diff=	16.88 =	16.88;	ck.	

Fig. 9. Differential-level notes.

Mistakes in Leveling. Some common mistakes in leveling are (1) confusion of numbers in reading the rod; (2) recording backsights in foresight column, and vice versa; (3) faulty additions and subtractions; (4) rod not held on same point for both foresight and backsight; (5) not having the long rod fully extended; (6) wrong reading of target vernier; and (7) when the long target rod is used, not having the target vernier set exactly the same as the vernier on the back of the rod when the rod is short.

Errors in Leveling. Errors in leveling are summarized in Table 3. If the proper leveling procedure is observed, the important errors are accidental, and the resultant error may be expected to vary as the square root of the distance.

TABLE 3. ERRORS IN LEVELING

Source	Type	Cause	Remarks	Procedure to eliminate or reduce
Instrumental	Systematic	Line of sight not parallel to axis of level tube	Error of each sight proportional to distance[1]	Adjust instrument, or balance sum of backsight and foresight distances
		Rod not standard length (throughout length)[2]	May be due to manufacture, moisture, or temperature. Error usually small	Standardize rod and apply corrections, same as for tape
Personal	Accidental	Parallax	....	Focus carefully
		Bubble not centered at instant of sighting	Error varies as length of sight	Check bubble before making each sight
		Rod not held plumb	Readings are too large. Error of each sight proportional to square of inclination[1]	Wave the rod, or use rod level
		Faulty reading of rod or setting of target	....	Check each reading before recording. For self-reading rod, use fairly short sights
		Faulty turning points	....	Choose definite and stable points
Natural	Accidental	Temperature	May disturb adjustment of level	Shield level from sun
		Earth's curvature	Error of each sight proportional to square of distance[1]	Balance *each* backsight and foresight distance; or apply computed correction
		Atmospheric refraction	Error of each sight proportional to square of distance[1]	Same as for earth's curvature; also take short sights, well above ground, and take backsight and foresight readings in quick succession
	Systematic	Settlement of tripod or turning points	Observed elevations are too high	Choose stable locations; take backsight and foresight readings in quick succession

[1] The error of *each sight* is systematic, but the resultant error is the difference between the systematic error for foresights and that for backsights; hence the resultant error tends to be accidental.
[2] Uniform wear of the bottom of the rod causes no error.

Precision of Differential Leveling. 1. *Rough leveling,* as practiced on rapid reconnaissance or preliminary surveys. Sights up to 1,000 ft. Rod readings to tenths of feet. No particular attention paid to balancing backsight and foresight distances. Maximum error in feet, $\pm 0.4\sqrt{\text{distance in miles}}$.

2. *Ordinary leveling,* as necessary in connection with most engineering works. Sights up to 500 ft. Rod readings to hundredths of feet. Backsight and foresight distances roughly balanced only when running for long distances uphill or downhill. Turning points on solid objects. Maximum error in feet, $\pm 0.1\sqrt{\text{distance in miles}}$.

3. *Excellent leveling* for the principal bench marks on extensive surveys. Sights up to 300 ft. Rod readings to thousandths of feet with either the target rod or the self-reading rod. Backsight and foresight distances measured by pacing and approximately balanced between bench marks. Rod waved for large rod readings. Bubble carefully centered before each sight. Turning points on well-defined points of solid objects. Tripod set on firm ground. Maximum error in feet, $\pm 0.05\sqrt{\text{distance in miles}}$.

4. *Precise leveling* for establishing bench marks with great accuracy at widely distributed points. High-grade level equipped with stadia hairs and with sensitive bubble. Adjustments carefully tested daily. Rod standardized frequently. Sights up to 300 ft. Rod readings of three horizontal hairs to thousandths of feet. Level protected from the sun. Turning points on metal pin or plate. Two rodmen. Backsights and following foresights taken in quick succession. Bubble very carefully centered and under observation at instant of taking sight. Rod plumbed with rod level. Backsight and foresight distances balanced between bench marks by stadia readings. Level set up securely on firm ground. Levels not run when air is boiling badly or during high winds. Maximum error in feet, $\pm 0.02\sqrt{\text{distance in miles}}$.

Adjustment of Elevations. When a line of levels makes a complete circuit, the final elevation of the initial bench mark as computed from the level notes will not agree with the initial elevation of this point. The difference is called the *error of closure.* The elevations of intermediate bench marks will also be in error, and it is necessary to adjust their elevations in proportion to their distances from the point of beginning. Thus if E_c is the error of closure of a level circuit of length L, and if $C_a, C_b \ldots C_n$ are the respective corrections to be applied to the observed elevations of bench marks $A, B \ldots N$ whose respective distances from the point of beginning are $a, b \ldots n$, then

$$C_a = -\frac{a}{L}E_c \qquad C_b = -\frac{b}{L}E_c \qquad \ldots \qquad C_n = -\frac{n}{L}E_c \tag{23}$$

Levels over Different Routes. When the elevation of a bench mark is established by lines of levels run over several routes, there will be as many observed elevations as there are lines terminating at the point. The weight to be applied to a given observed elevation will vary inversely as the length of the corresponding line. The most probable value of the elevation will be the weighted mean of the observed values.

Level Net. Where elevations of bench marks in an interconnecting network of level circuits are to be adjusted, a simple and accurate method is that of successive approximations. It consists in adjusting each separate figure in the net in turn, with the adjusted values for each circuit used in the adjustment of adjacent circuits; the process is repeated for as many cycles as necessary to balance the values for the whole net. Within each circuit the error of closure is normally distributed to the various sides in proportion to their lengths, as previously explained. The following example shows a method of solution suggested by Prof. Bruce Jameyson.

Example: Figure 10 represents a level net made up of the circuits $BCDEB$, $AEDA$, and $EABE$. Along each side of the circuit is shown the length in miles and the observed

TABLE 4. ADJUSTMENT OF LEVEL NET BY SUCCESSIVE APPROXIMATIONS

Ci cuit	Side	Distance		Cycle I			Cycle II			Cycle III			Cycle IV		
		Miles	Per-cent	DE *	Corr.†	Corr. DE ‡	DE *	Corr.†	Corr. DE ‡	DE *	Corr.†	Corr. DE ‡	DE *	Corr.†	Corr. DE ‡
BCDEB	*BC*	12	17	+10.94	+0.07	+11.01	+11.01	−0.02	+10.99	+10.99	−0.01	+10.98	+10.98	0	+10.98
	CD	28	40	+21.04	+0.16	+21.20	+21.20	−0.05	+21.15	+21.15	−0.01	+21.14	+21.14	−0.01	+21.13
	DE	13	19	−27.15	+0.07	−27.08	−27.02	−0.03	−27.05	−27.03	−0.01	−27.04	−27.03	0	−27.03
	EB	17	24	− 5.23	+0.10	− 5.13	− 5.06	−0.03	− 5.09	− 5.07	−0.01	− 5.08	− 5.08	0	− 5.08
	Total	70	100	− 0.40	+0.40	0	+ 0.13	−0.13	0	+ 0.04	−0.04	0	+ 0.01	−0.01	0
AEDA	*AE*	11	22	−17.91	−0.06	−17.97	−17.93	−0.01	−17.94	−17.93	0	−17.93	−17.93		
	ED	13	26	+27.08	−0.06	+27.02	+27.05	−0.02	+27.03	+27.04	−0.01	+27.03	+27.03		
	DA	26	52	− 8.92	−0.13	− 9.05	− 9.05	−0.04	− 9.09	− 9.09	−0.01	− 9.10	− 9.10		
	Total	50	100	+ 0.25	−0.25	0	+ 0.07	−0.07	0	+ 0.02	−0.02	0	0		
EABE	*EA*	11	26	+17.97	−0.04	+17.93	+17.94	−0.01	+17.93	+17.93	0	+17.93	+17.93		
	AB	15	35	−22.93	−0.06	−22.99	−22.99	−0.01	−23.00	−23.00	−0.01	−23.01	−23.01		
	BE	17	39	+ 5.13	−0.07	+ 5.06	+ 5.09	−0.02	+ 5.07	+ 5.08	0	+ 5.08	+ 5.08		
	Total	43	100	+ 0.17	−0.17	0	+ 0.04	−0.04	0	+ 0.01	−0.01	0	0		

* Difference in elevation.
† Correction.
‡ Corrected difference in elevation.

difference in elevation in feet between terminal bench marks; the sign of the difference in elevation corresponds with the direction indicated by the arrows. Within each circuit is shown its length and the error of closure computed by summing up the differences in elevation in a clockwise direction.

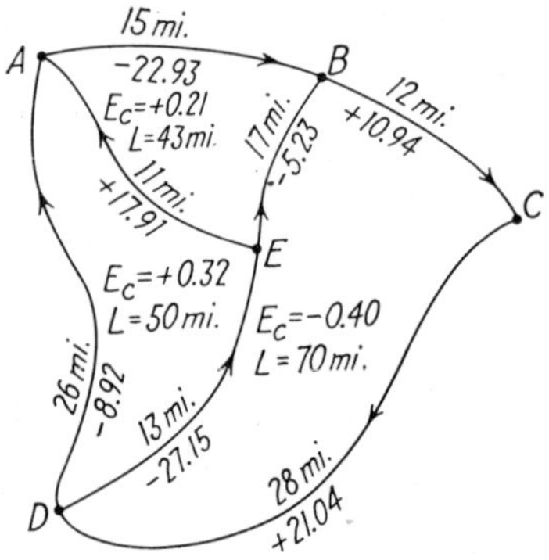

Fig. 10. Adjustment of level net.

Table 4 shows the computations required to balance the net. For each circuit are listed the sides, the distances (expressed in miles and in percentages of the total), and the differences in elevation. For circuit *BCDEB* the error of closure is −0.40 ft. This is distributed among the lines in proportion to their lengths; thus for the line *BC* the correction is $\frac{12}{70} \times 0.40$ or $0.17 \times 0.40 = 0.07$ ft, with sign opposite to that of the error of closure. The corrections are applied to the differences in elevation to obtain the values of "corrected difference in elevation" shown in the seventh column. The line *DE* in circuit *BCDEB* is the same as the line *ED* in circuit *AEDA*. Hence, in listing the differences in elevation for circuit *AEDA*, the difference in elevation for *ED* is taken, not as the observed value (27.15), but as the adjusted value (27.08) from circuit *BCDEB*, with opposite sign. The error of closure for circuit *AEDA* is then +0.25 ft. The error is distributed as before. Similarly, in circuit *EABE* the differences in elevation listed for *EA* and *BE* are the adjusted values from the previous circuits. In Cycle II the process for Cycle I is repeated, and the latest value from previously adjusted circuits is always listed before the new error of closure is computed. And so the cycles are continued until the corrections become zero.

The order in which the various circuits and lines are taken is immaterial, although the optimum order may reduce the number of cycles required. If desired, the computations may be based on the *corrections* rather than on the differences in elevation as shown. The sides of a given circuit, or a given circuit as a whole, may be weighted as desired. The elevations of intermediate bench marks are adjusted as previously described.

PROFILE LEVELING; CROSS SECTIONS

Profile-level Notes. The notes for profile leveling may be recorded as shown in Fig. 11, where foresights to turning points and bench marks are in a separate column from intermediate foresights to ground points. Some surveyors keep all foresights in one column and tabulate the elevations of ground points in the fifth column and elevations of BM's and TP's in the sixth column. Others record all foresights in the fourth column and all elevations in the fifth.

Route Cross Sections. Preliminary cross-section notes may be kept in the form shown in Fig. 12. The center line of the right-hand page represents the traverse line, and to the right and left of this line are recorded the observed distances and rod readings and the computed elevations.

Figure 13 illustrates a suitable form of final cross-section notes.

The method usually employed to determine cuts and fills is as follows: Figure 14 shows at *A* the level in position above grade, for taking rod readings at a section in fill. The height of instrument (HI) has been determined; the elevation of grade at the particular station is known. The leveler computes the difference between the HI and grade elevation, a difference known as the *grade rod;* i.e., HI − el. of grade = grade rod. The rod is held at any point for which the fill is desired, and a reading, called the *ground rod* is, taken. The difference between the grade rod and the ground rod is equal to the fill or,

PROFILE LEVELS FOR I. N. RY. LOCATION

Cox Brook to Big Forks

Buff Dumpy Level — J.C. Brown, ⊼ — F. Graham, Rod — Sept. 16, 1951 — Fair

Sta.	B.S.	H.I.	I.F.S.	F.S.	Elev.		
B.M.28	1.56	565.87			564.31	On spruce root 50 ft. lt. Sta. 605.	
605			0.7		565.2		
606			2.9		563.0		
607			3.5		562.4		
608			6.7		559.2		
609			11.9		554.0		
T.P.	0.41	554.65		11.63	554.24	On stone.	
609+50			3.2		551.5		
610			8.6		546.1		
+40			9.0		545.7	Bank Cox Brook.	
+50			12.2		542.5	Ctr. " "	water 1.5 ft. deep.
+65			9.3		545.4	Bank " "	
611			8.4		546.3		
612			7.1		547.6		
613			8.4		546.3		
614			7.0		547.7		
615			4.1		550.6		
T.P.	8.02	559.94		2.73	551.92	On plug.	
616			9.7		550.2		
+40			6.3		553.6	Ctr. highway to St. Leonards.	
	9.99		564.31	14.36			
			559.94	9.99			
			4.37 =	4.37	ck.		

Fig. 11. Profile-level notes.

PRELIMINARY CROSS-SECTIONS C. & R. EXTENSION

O.H. Ellis, ⊼ — C.O. Lord, Rod — J.A. Crum, Tape — Jan. 20, 1951 — Cold, Snow

Sta.	B.S.	H.I.	F.S.	Elev.		Left	℄	Right
405	12.4	633.0		620.6	(Dist.) (Elev.) (Rod)	300 210 123 80 632.1 630.8 627.0 626.7 0.9 2.2 6.0 6.3	620.6	
405	0.6	621.2		620.6				50 160 250 350 617.0 612.3 610.0 609.7 4.2 8.9 11.2 11.5
406	12.1	628.9		616.8		90 50 628.2 624.3 0.7 4.6	616.8	
406	11.5	639.7	0.7	628.2		280 200 120 638.3 635.6 632.2 1.4 4.1 7.5		
406	1.9	618.7		616.8				60 100 200 300 615.5 611.2 610.9 609.3 3.2 4.5 7.8 9.4
407	4.7	615.9		611.2		300 180 100 615.6 614.7 612.6 0.3 1.2 3.3	611.2	75 155 270 606.2 604.5 602.9 9.7 11.4 13.0
408	10.6	615.9		605.3		280 200 100 611.6 609.2 607.7 4.3 6.7 8.2	605.3	100 200 300 604.1 603.2 603.0 11.8 12.7 12.9

Fig. 12. Preliminary route cross-section notes.

CROSS-SECTIONS FOR I.N.RY. FINAL LOCATION

Cox Brook to Big Forks | Dec. 4, 1951 | F.F. Smith, ⊼

Roadbed 20 ft. in Cut, 16 ft. in Fill | Cloudy | J. Richie, Rod

Slope 1½:1 | O. Byram, Tape

Sta.	B.S.	H.I.	F.S.	Elev.	Grade	Left	Ctr.	Right	Remarks
B.M.28	2.67	566.98		564.31					50 ft. Lt. Sta. 605
605			1.9	565.1	556.00	C8.6/22.9	C9.1	C11.2/26.8	Gravel in this hill.
606			4.0	563.0	555.60	C5.0/17.5	C7.4	C8.4/22.6	
607			4.5	562.5	555.20	C4.6/16.9	C7.3	C8.0/22.0	
608			8.0	559.0	554.80	C1.8/12.7	C4.2	C5.1/17.7	
+25			9.2	557.8	554.70	0.0/10.0	C3.1	C2.6/13.9	
T.P.	1.94	557.19	11.73	555.25					On plug.
+90			2.8	554.4	554.44	F1.8/10.7	0.0	C2.4/13.6	
609			3.4	553.8	554.40	F3.2/12.8	F0.6	0.0/4.0 C1.0/11.5	
+20			5.6	551.6	554.32	F3.6/13.4	F2.7	0.0/8.0	
610			11.1	546.1	554.00	F8.2/20.3	F7.9	F6.3/17.5	
+40			11.2	546.0	553.84	F8.0/20.0	F7.8	F7.8/19.7	Top of bank Cox Brook.
+45			14.6	542.6	553.82	F11.2/24.8	F11.2	F11.2/24.8	In brook.
+60			14.5	542.7	553.76	F11.1/24.7	F11.1	F11.1/24.7	" "
+65			11.6	545.6	553.74	F8.2/20.3	F8.1	F8.0/20.0	Top of bank.
611			10.9	546.3	553.60	F7.8/19.7	F7.3	F7.2/18.8	
612			9.7	547.5	553.20	F7.6/19.4	F5.7	F3.1/12.7	
613			10.8	546.4	552.80	F7.8/19.7	F6.4	F2.0/11.0	
T.P.	11.96	559.69	9.46	547.73					On stump.
	16.57	564.31	21.19						
		559.69	16.57						
		4.62=4.62 ck.							

FIG. 13. Cross-section notes for final location of roadway.

for a section in cut, is equal to the cut. If the HI is *below* grade, as at B in Fig. 14, it is clear that the fill is the *sum* of the grade rod and the ground rod.

The plotting of cross sections is described in Section 2.

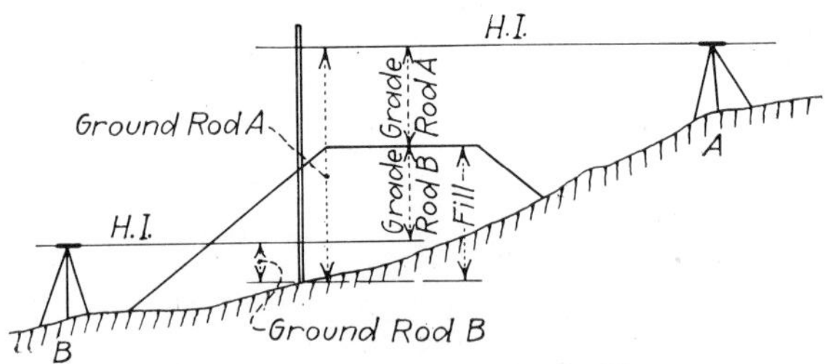

FIG. 14. Road cross section in fill.

Setting Slope Stakes. If w is the width of roadbed, d the measured distance from center to slope stake, s the side-slope ratio (ratio of horizontal distance to drop or rise), and h the cut (or fill) at the slope stake, then when the slope stake is in the correct position,

$$d = \frac{w}{2} + hs \tag{24}$$

Slope stakes are set side to the line, sloping outward in fill and inward in cut. On the back of the stake is marked the station number. On the front (side nearest the center line) are marked the cut or fill at the stake and sometimes the distance from center to

slope stake. The numbers read down the stake. In cut, the slope stakes may be set at a fixed distance, say, 2 ft, back from the edge of the slope. The cut marked on the stake applies to the elevation of the ground at the stakes thus offset.

Setting Grades. The operation of setting grades is similar to profile leveling. The grade rod to be employed in setting a given grade stake to grade is computed by subtracting the established grade elevation (taken from the profile) from the HI. The rodman starts the stake and holds the rod on its top. The leveler reads the rod and calls out the approximate distance that the stake must be driven to reach grade. The rodman drives the stake nearly the desired amount, and a second rod reading is taken; and so the process is continued until the rod reading is made equal to the grade rod. Sometimes the rod is held alongside of the stake, and the position of grade is indicated by a crayon mark or a nail. Usually grade elevations are determined to hundredths of feet. The notes are kept as in profile leveling except that the right-hand column of the left-hand page is for grade elevations.

The procedure of setting grades for typical works of construction is described under "Construction Surveying," later in this section.

DIRECTION WITH THE MAGNETIC COMPASS

Magnetic Declination. The angle between the true meridian and the magnetic meridian is called the "magnetic declination." If the north end of the compass needle points to the east of the true meridian, the declination is said to be east; if it points to the west of the true meridian, the declination is said to be west.

If a true north-south line is established, the mean declination of the needle for a given locality can be determined by compass observations extending over a period of time. The declination may be estimated with sufficient accuracy for most purposes from the isogonic chart of the United States shown in Fig. 15.

The true meridian is established by astronomical observations. On compass surveys, the true meridian may be established with sufficient precision by ranging two plumb lines with Polaris.

Isogonic Chart. This chart (Fig. 15) shows lines of equal declination for the date Jan. 1, 1950, as based on observations made by the U.S. Coast and Geodetic Survey at stations widely scattered throughout the country. The *agonic line*, or line of zero declination, is the heavy full irregular line extending in a southeasterly direction from the Great Lakes. The *isogonic lines*, or light full lines, when east of the agonic line mark the paths where the declinations were on the above date 1° west, 2° west, etc.; similarly, those west of the agonic line show the routes along which the declinations were 1° east, 2° east, etc. The dash lines indicate the rate of change in magnetic declination from year to year.

Direction with Surveyor's Compass. In order that *true* bearings may be read directly, some compasses are so designed that the compass circle may be rotated with respect to the box in which it is mounted. When the circle is in its normal position, the line of sight as defined by the vertical slits in the sight vanes is in line with the N and S points of the compass circle, and the observed bearings are magnetic. If the circle is turned through an angle equal to the magnetic declination, the observed bearings will be true.

When the direction of a line is to be determined, the compass is set up on line and is leveled. The needle is released, and the compass is rotated about its vertical axis until a range pole or other object on line is viewed through the slits in the two sight vanes. When the needle comes to rest, the bearing is read. Ordinarily the sight vane at the end of the compass box marked S is held next to the eye, in which case the bearing is given by the north end of the needle. When a compass traverse is run, only alternate stations need be occupied.

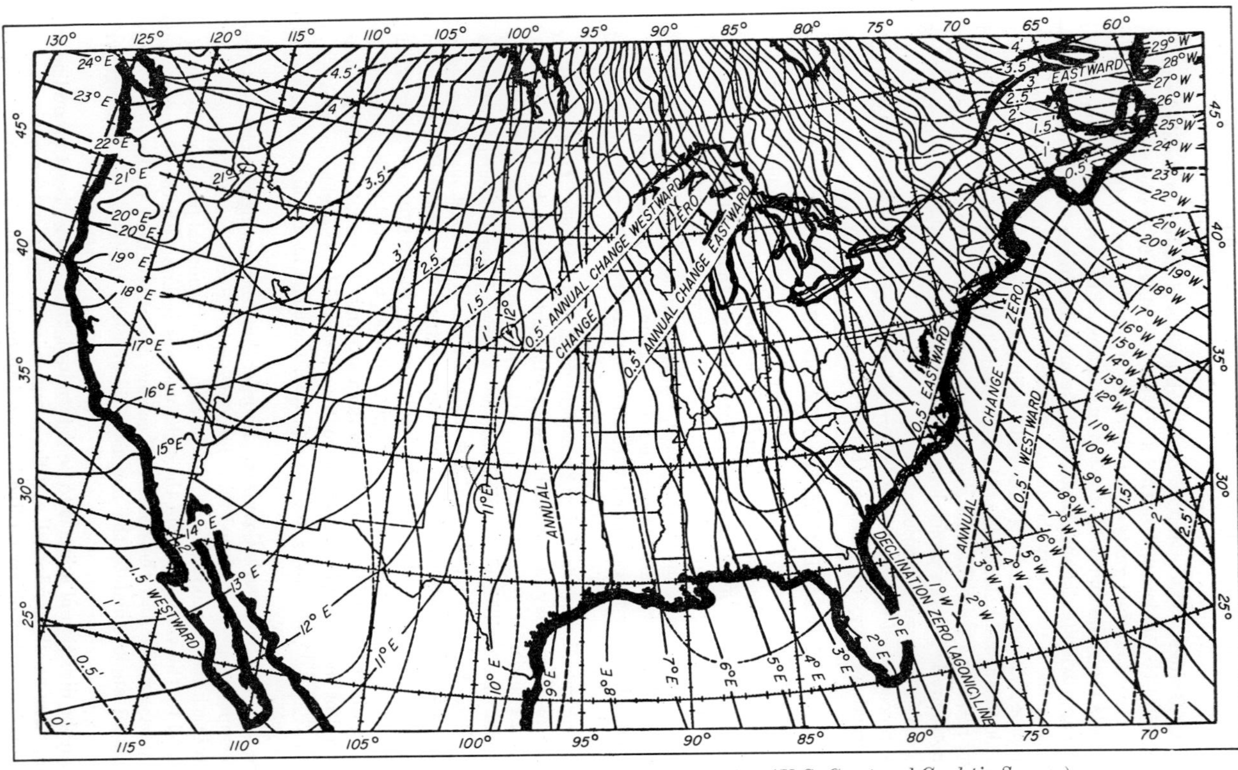

FIG. 15. Isogonic chart of the United States for Jan. 1, 1950. (*U.S. Coast and Geodetic Survey.*)

Correction for Local Attraction. If local attraction from a fixed source exists at any station in a traverse, both the back and the forward bearings taken from that station will be affected by the same amount. Disregarding for the time being the accidental errors due to observing, it is probable that the terminal points of any line, as AB, are free from local attraction if the back bearing from B is the reverse of the forward bearing from A. Keeping in mind that the computed *angle* between the forward and back lines from any station can be determined correctly from the observed bearings taken from that station regardless of fixed local attraction, the direction of a line free from local attraction may be chosen as a basis, the traverse angles may be computed from observed bearings, and, starting from the unaffected line, the correct bearings of successive lines may be computed.

Errors in Compass Work. Common sources of error and corrections are as follows:

1. Needle bent. Readings for both ends may be averaged, or needle may be straightened with pliers.

2. Pivot bent. Readings for both ends may be averaged, or pivot may be bent until end readings of needle are 180° apart for any direction of pointing.

3. Plane of sight not vertical, or graduated circle not horizontal. Vanes may be tested by sighting at a plumb line. Level tubes may be tested by reversal.

4. Sluggish needle. The glass on top of the compass box should be tapped lightly before each reading. If the needle is weak, it may be remagnetized by drawing its ends over a bar magnet. If the pivot point is blunt, it may be sharpened by means of a fine-grained oilstone.

5. Reading the needle. Ends of needle should be in plane parallel to horizontal circle, and eye of observer should be above the coinciding graduation and in line with the needle.

6. Undetected magnetic variations. Fixed local attraction can be corrected as previously described. Iron and steel objects should be kept away from the compass while it is in use, and the observer should remain on the same side of the instrument. Static charges of electricity on the glass cover may be removed by touching the glass with a moist finger.

SURVEYS WITH TRANSIT AND TAPE

Measuring a Horizontal Angle. If a horizontal angle, as AOB, is to be measured, the transit is set up over O. The upper motion is clamped with one of the horizontal verniers near zero, and by means of the upper tangent screw the vernier is set at 0°. The telescope is sighted approximately to A, the lower motion is clamped, and by turning the lower tangent screw the line of sight is set exactly on a range pole or other object marking the point. The upper clamp is loosened, and the telescope is turned until the line of sight cuts B. The upper clamp is tightened, and the line of sight is set exactly on B by turning the upper tangent screw. The reading of the vernier which was initially set at 0° gives the value of the angle.

Measuring a Horizontal Angle by Repetition. By means of the transit, a horizontal angle may be mechanically multiplied, and the product can be read with the same precision as the single value. The precision increases directly with the number of repetitions up to six or eight; beyond this number the precision is not appreciably increased by further repetition, because of accidental errors.

To repeat an angle, as AOB, the transit is set up at O, and the single value of the angle is observed as previously described. The vernier setting is left unaltered; the instrument is turned on its lower motion; and a second sight is taken to the first point, as A. The upper clamp is loosened, and the telescope is again sighted to B. In this way the process is continued until the angle has been multiplied the required number of times. The vernier is read, and the value of the angle is determined by dividing the difference between initial and final readings by the number of times the angle was turned.

Usually the angle is multiplied four to six times, half the observations being made with the telescope normal and half with it inverted. Both verniers are read, and the mean values are used in the computations.

Sample notes for measuring the angles about a point by repetition are shown in Fig. 16. For each angle, five "repetitions" are taken with telescope normal and five with telescope inverted, always measuring clockwise. The vernier is set at zero at the beginning but not thereafter; the error of closure (called the "horizon closure") is thus obtained directly as a check on the computations, and errors in setting the vernier are avoided. Rough computations on the right-hand page serve as a check on the number

ANGLES ABOUT △A

Between Sta.	Tel. Rep.	Vernier A	Vernier B	Vernier Mean	Mean Angle
B-Y	N 0	0°00′00″	180°00′00″	0°00′00″	
	1	20 32 00			
	卌 5	102 38 30	282 38 30		
	I 卌 10	205 16 00	25 16 30	205 16 15	20°31′37.″5
Y-R	N 0	205°16′00″	25°16′30″	205°16′15″	
	1	286 03 00			
	卌 5	249 11 00	69 11 30		
	I 卌 10	293 01 00	113 01 30	293 01 15	80°46′30″
			Diff.	87 45 00	
			+	720	
			Sum	807 45 00	
R-B	N 0	293°01′00″	113°01′30″	293°01′15″	
	1	191 43 00			
	卌 5	146 32 30	326 33 00		
	I 卌 10	359 58 30	179 58 30	359 58 30	258°41′43.″5
			Diff.	66 57 15	
			+	2520	
			Sum	2586 57 15	
Horizon closure = $\frac{1'30''}{10}$ = 09″					
(Correction: +03″ per ∠)				Sum	359°59′51″

BY REPETITION

Adj. Angle		
		H.O. Ward ⊼
		E. B. Erhart, Notes
		Aug. 10, 1951
		Cool; air steady
	Start 9:05 A.M.	Berger Transit No. 8
		30″ Vernier
20°31′41″		
80°46′33″		
258°41′46″	Finish 9:40 A.M.	
360°00′00″		

258°41′46″
20°31′41″
80°46′33′
A
B
Y
R

B-Y 20°32′
× 5
✓ 102 40 (5 Rep.)
102 40
✓ 205 20 (10 Rep.)

Y-R 286°03′
205 16
80 47
× 5
403 55
205 16
609 11
360
✓ 249 11 (5 Rep.)
403 55
653 06
360
✓ 293 06 (10 Rep.)

R-B 191°43′
360
551 43
293 01
258 42
× 5
1293 30
293 01
1586 31
1440
✓ 146 31 (5 Rep.)
1293 30
1440 01
1440
✓ 0 01 (10 Rep.)

FIG. 16. Notes for measuring angles by repetition.

of repetitions and detect appreciable mistakes in turning the wrong tangent screw. The recorded value for five repetitions is used only as a check; and the *B* vernier is used only as a check, except with regard to the number of seconds. The final adjusted values of the angles (to the nearest second) are recorded on the sketch, for ready reference in further computations.

A horizontal angle may be laid off by repetition in a similar manner. First the angle is laid off by a single sighting, and this angle is measured by repetition. The correction is then applied at the point to be established by measuring off a linear offset equal to the distance from transit to point times the tangent (or sine) of the correction angle.

Measuring a Vertical Angle. The vertical angle to a point is its angle of elevation (+) or depression (−) from the horizontal. The transit is set up and leveled as when measuring horizontal angles.

For a transit having a fixed vertical vernier, the plate bubbles should be centered carefully. The telescope is sighted approximately at the point, and the horizontal axis is clamped. The horizontal cross hair is set exactly on the point by turning the telescope tangent screw, and the angle is read by means of the vertical vernier.

For a transit having a movable vertical vernier with control level, the telescope is sighted on the point as described above, the vernier control bubble is centered, and the angle is read.

In ordinary trigonometric leveling, vertical angles are taken by sighting usually at a leveling rod, the line of sight being directed at a rod reading equal to the height of the horizontal axis of the transit above the station over which the transit is set up. Sights may be taken to points defined by signals erected at the distant stations.

For leveling with the transit, for astronomical observations, or for measurement of horizontal angles requiring steeply inclined sights, usually it is desired to level the transit with greater precision than that which is possible through the use of the plate levels. In such cases the vertical axis is made truly vertical by means of the telescope level, the plate levels being disregarded.

Double-sighting. For a transit having a full vertical circle, sights to determine vertical angles can be taken with the telescope either normal or inverted. The method of *double-sighting* consists in reading once with the telescope normal and once with it inverted, and taking the mean of the two values thus obtained. It eliminates the effect of certain instrumental errors and reduces the personal error of observation. In traversing, a similar result is obtained by measuring the vertical angles of each traverse line from each end, with the telescope the *same side up* for the two observations, and taking the mean of the two values.

Index Error. *Index error* is the error in an observed angle due to (1) lack of parallelism between the line of sight and the axis of the telescope level, (2) displacement (lack of adjustment) of the vertical vernier, and/or (3) for a transit having a fixed vertical vernier, inclination of the vertical axis.

The effect of index errors due to lack of adjustment of the instrument can be eliminated either by double-sighting for each observation or by applying to each observation a correction determined (by double-sighting) for the instrument in its given condition of adjustment. For the common type of transit having a fixed vertical vernier, the effect of imperfect leveling cannot be eliminated by double-sighting, but—provided the line of sight is in adjustment—for each direction of pointing a correction can be determined (as described later) and applied. The index correction is equal in amount but opposite in sign to the index error. Methods of determining the index error (and, therefore, the correction) are given in the following paragraphs.

1. *Lack of Parallelism between Line of Sight and Axis of Telescope Level.* If the axis of the telescope level is not parallel to the line of sight and if the vertical vernier reads zero when the bubble is centered (Fig. 17*a*), an error in vertical angle results. This error can be rendered negligible for ordinary work by careful adjustment of the instrument. The combined error due to this cause and to displacement of the vertical vernier (see following paragraph) can be eliminated by double-sighting. The index error due to the two causes can be determined by comparing a single reading on any given point with the mean of the two readings obtained by double-sighting to the same point.

2. *Displacement of Vertical Vernier.* Displacement of the vertical vernier (Fig. 17*b*) introduces a constant index error. The error can be rendered negligible by careful adjustment. The combined index error due to this cause and to lack of parallelism between

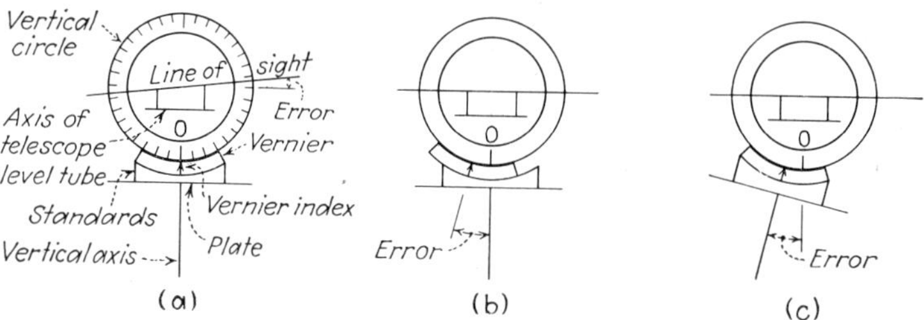

FIG. 17. Sources of error in measurement of vertical angles.

the line of sight and the axis of the telescope level can be eliminated by double-sighting; or the combined error can be determined as described in the preceding paragraph. For a transit having a fixed vertical vernier, the error due to displacement of the vertical vernier alone can be determined—provided the line of sight is in adjustment—by leveling the transit carefully, leveling the telescope, and reading the vertical vernier. For a transit having a movable vertical vernier with control level, the error due to displacement of the vertical vernier alone can be determined by leveling both the telescope level and the vernier level, and reading the vertical vernier.

3. *Inclination of Vertical Axis.* For a transit having a fixed vertical vernier, any inclination of the vertical axis (Fig. 17*c*) due to erroneous leveling of the instrument introduces an index error which varies with the direction in which the telescope is pointed and which is equal in amount to the angle through which the fixed vertical vernier is displaced about the horizontal axis while the instrument is directed toward the point. This index error can be rendered negligible by careful leveling of the transit before each observation, making sure that the plate-level bubbles remain in position for any direction of pointing. It is not eliminated by double-sighting, since the condition causing the error is not changed by reversal (and plunging) of the instrument (see Fig. 17*c*). If the line of sight and the vertical vernier are in adjustment, the index error due to inclination of the vertical axis alone can be determined for each direction of pointing by leveling the telescope and reading the vertical vernier.

For a transit having a movable vertical vernier with control level, any moderate inclination of the vertical axis does not introduce an appreciable error in vertical angles, provided the instrument is in adjustment and provided the vernier control-level bubble is centered each time an observation is made. On topographic surveys or similar work where many horizontal and vertical angles are to be observed, the use of the movable vertical vernier with control level results in a considerable saving of time.

Setting a Monument. To set a subsurface monument, a string is stretched along each of two reference lines intersecting at the surface location of the point. The location of the mark on the monument is then projected below the surface by plumbing.

Adjustment of the Transit. 1. *To Make the Vertical Cross Hair Lie in a Plane Perpendicular to the Horizontal Axis.* Sight the vertical cross hair on a well-defined point not less than 200 ft away, and swing the telescope through a small vertical angle. If the point appears to depart from the vertical cross hair, loosen two adjacent capstan screws and rotate the cross-hair ring in the telescope tube until the point traverses the entire length of the hair. Tighten the same two screws.

2. *To Make the Axis of Each Plate Level Lie in a Plane Perpendicular to the Vertical Axis.* Rotate the instrument about the vertical axis until each level tube is parallel to a pair of opposite leveling screws. Center the bubbles by means of the leveling screws. Rotate the instrument end for end about the vertical axis. If the bubbles become displaced, bring them *halfway* back by means of the adjusting screws.

3. *To Make the Line of Sight Perpendicular to the Horizontal Axis.* Level the instrument. Sight on a point *A* about 500 ft away, with telescope normal. With both horizontal motions of the instrument clamped, plunge the telescope and set another point *B* on the line of sight and about the same distance away on the opposite side of the transit. Unclamp the upper motion, rotate the instrument end for end about the vertical axis, and again sight at *A* (with telescope inverted). Clamp the upper motion. Plunge the telescope as before; if the line of sight does not fall on *B*, set a point *C* on the line of sight beside *B*. Mark a point *D*, *one-quarter* of the distance from *C* to *B*, and adjust the cross-hair ring (by means of the two opposite horizontal screws) until the line of sight passes through *D*.

4. *To Make the Horizontal Axis Perpendicular to the Vertical Axis.* Set up the transit near a building or other object on which is some well-defined point *A* at a considerable

vertical angle. Level the instrument very carefully, thus making the vertical axis truly vertical. Sight at the high point A, and with the horizontal motions clamped depress the telescope and set a point B on or near the ground. If the horizontal axis is perpendicular to the vertical axis, A and B will be in the same vertical plane. Plunge the telescope, rotate the instrument end for end about the vertical axis, and again sight on A. Depress the telescope as before; if the line of sight does not fall on B, set a point C on the line of sight beside B. A point D, halfway between B and C, will lie in the same vertical plane with A. Sight on D; elevate the telescope until the line of sight is beside A; loosen the screws of the bearing cap; and raise or lower the adjustable end of the horizontal axis until the line of sight is in the same vertical plane with A.

5. *To Make the Axis of the Telescope Level Parallel to the Line of Sight.* Proceed the same as for the two-peg adjustment of the dumpy level (see p. 1–3), except as follows: With the line of sight set on the rod reading established for a horizontal line, the correction is made by raising or lowering one end of the telescope level tube until the bubble is centered.

6. (For Transit Having a Fixed Vertical Vernier) *To Make the Vertical Circle Read Zero When the Telescope Bubble Is Centered.* With the plate bubbles centered, center the telescope bubble. If the vertical vernier does not read zero, loosen it, and move it until it reads zero.

6*a*. (For Transit Having a Movable Vertical Vernier with Control Level) *To Make the Axis of the Auxiliary Level Parallel to the Axis of the Telescope Level When the Vertical Vernier Reads Zero.* Center the telescope bubble, and by means of the vernier tangent screw move the vertical vernier until it reads zero. By means of the capstan screws, adjust the level tube attached to the vertical vernier until the bubble is centered.

Errors in Transit Work. Except in field astronomy, a measured angle is always closely related to a measured distance. On surveys of ordinary precision, it usually requires much more care to keep *linear* errors within prescribed limits than to maintain a corresponding degree of *angular* precision. Often undue attention is paid to securing accuracy in angular measurements.

Instrumental Errors. The adjustments, even though carefully made, are never exact. Likewise the graduations are not perfect, and the centers are not absolutely true.

Errors in horizontal angles due to nonadjustment of plate levels or of horizontal axis become large as the angle of inclination of the sights increases.

Nonadjustment of the line of sight becomes of consequence only when the telescope is plunged.

Errors due to instrumental imperfections or nonadjustment are all systematic. By proper methods of procedure, usually by double-sighting, they may be eliminated or reduced to a negligible quantity. The systematic part of the error due to inclination of the vertical axis is eliminated by double-sighting and recentering the plate bubbles between sights.

Personal Errors. Personal errors arise from the limitations of the human eye in setting up and leveling the transit and in making observations. The transit may not be set up exactly over the point; the plate bubbles may not be centered exactly; the verniers may not be set or read accurately; parallax may exist in focusing; and the line of sight may not be directed exactly at the point.

All the personal errors are accidental and hence cannot be eliminated. They form a large part of the resultant error in transit work. They can be kept within reasonable limits by care in observing.

Natural Errors. Sources of natural errors are (1) settlement of the tripod; (2) unequal atmospheric refraction; (3) unequal expansion of parts of the telescope due to temperature changes; and (4) wind, producing vibration of the transit or making it difficult to plumb correctly.

In general, the errors resulting from natural causes are not large enough to affect appreciably the measurements of ordinary precision. However, when the transit is set on boggy or thawing ground, large errors may arise from settlement accompanied by displacement about the vertical axis.

Precision of Angular Measurements. Many factors influence the precision of transit work, and no rigid rules can be formulated to ensure a required precision. The following values represent, in a general way, the *maximum* error likely to occur in measuring a horizontal angle under average conditions of practice, instruments being in fair condition and in fair adjustment except as otherwise stated. The *average* angular error will be materially less. Also, as the errors are largely accidental, the resultant error in the sum of a series of measured angles may be expected to vary as the square root of the number of angles involved.

Case 1. Short sights, point indicated by range pole obscured near ground. Range pole plumbed by eye. Single observation of angle. Maximum error 02′ to 04′.

Case 2. Long sights, but otherwise as stated for case 1. Maximum error 01′ to 02′.

Case 3. Unobscured sights on well-defined points; sights not steeply inclined. Single observation of angle, vernier reading to minutes. Maximum error 30″ to 01′.

Case 4. Unobscured sights on well-defined points. Sights not steeply inclined. Verniers reading to 30″. Single observation of angle represented by means of readings of both verniers. Transit in excellent condition and in good adjustment. Maximum error 15″ to 30″.

Case 5. As for case 4, but verniers reading to 10″. Also, instrument set up with great care. Maximum error 10″ to 15″.

Case 6. Unobscured sights on well-defined points. Instrument set up with great care. Sights not steeply inclined. Transit in excellent condition and adjustment. Vernier reading to 30″. Angle repeated six times with telescope normal and six times with telescope inverted. Maximum error 02″ to 04″.

Case 7. As for case 6, but transit reading to 10″. Observations taken at favorable times. Maximum error 01″ to 02″.

Transit Surveys. The field work of surveying with the transit may ordinarily be divided as follows:

1. Establishing transit stations and lines by angular and linear measurements. The transit lines may be said to form the skeleton of the survey and are called the *control* or *horizontal control*.

2. Locating objects and points with respect to the transit lines, thus furnishing the *detail* with which the transit lines are clothed.

Transit surveys may be made by *radiation, intersection, traversing, triangulation,* or some combination of these.

Radiation. This method by itself is applicable only to surveys covering small areas. The transit is set up at any convenient station from which can be seen all points that it is desired to locate. The distance from the transit station to each of the points is measured, and the horizontal angle is observed. The angles between successive points may be measured; or the true, magnetic, or assumed bearing or azimuth of each of the lines joining the points with the transit station may be observed.

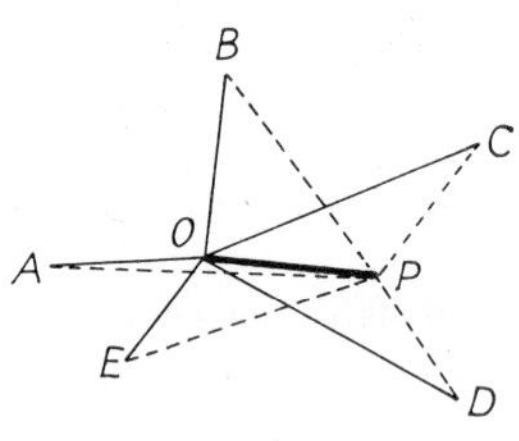

Fig. 18. Intersection.

Intersection. In Fig. 18, let the points A, B, C, etc., represent objects which it is desired to locate, and let OP be a convenient line from both ends of which the unknown points are visible. The length of the base line OP is measured with the tape. The transit is set up at O, and angles to the unknown points are observed; these may be

expressed as azimuths, as bearings, or as angles between successive points. A similar series of observations is made with the transit at P. In this manner each of the unknown points becomes the vertex of a triangle of which the base line OP is the side of measured length and in which the angles adjacent thereto are observed values; the locations of the unknown points are thus defined.

Sights may be taken simultaneously by means of two transits set up at opposite ends of a line the length and direction of which are known. In hydrographic surveying, this method is useful for locating soundings. In the construction of bridges and dams, it is used to establish points for piers and other structural parts that are difficult of access.

A variation of the method of intersection is the method of *resection*, by means of which the transit may be set up at a station of unknown location and its location determined by sighting on points of known location (see three-point problem, p. 1–49). The principle of resection employing the transit is as described for the plane table later in this section.

Traversing. Following is a general description of the work of running a *closed* traverse, the transit stations being established in advantageous locations as the survey progresses, and the distances being measured between successive transit stations: In Fig. 19, let A and B be selected locations for transit stations marking the first line of a traverse. Hubs defining the points are driven and properly identified by guard stakes. The transit is set up at B, the horizontal vernier is set to a given angular value, a backsight is taken on a range pole at A, and the lower motion is clamped. The line AB is then chained, the head chainman being lined in by the transitman. The distance AB is recorded. The location of C is selected, and the transit point is established. The transit is turned on the upper motion until a foresight is secured on C. The upper motion is clamped, and the angular value is read and recorded. The distance BC is chained and recorded. The transit is moved forward to C, a backsight is taken to B, the point D is chosen, a foresight is taken to D, and the angle is read. The line CD is chained. The process is repeated for point E, etc., until the traverse is finally brought to closure on the initial point A.

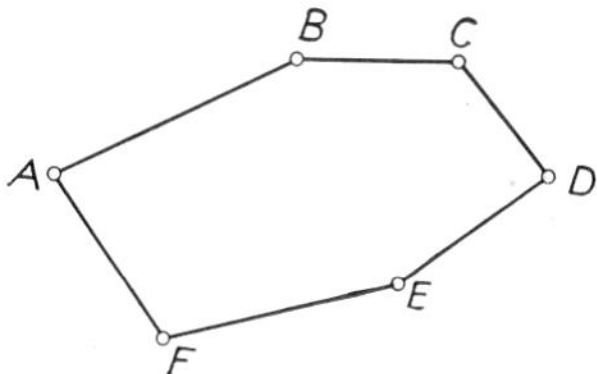

FIG. 19. Closed traverse.

A continuous, or *open*, traverse may be run in exactly the same manner, except that of course there is no closure. The cumulative angular error may be determined by beginning and closing on previously established lines, or on long traverses by astronomical observations at each end, with account taken of convergency of meridians.

Deflection-angle Traverse. Successive transit stations are occupied, and at each station a backsight is taken with the A vernier set at zero and the telescope inverted. The telescope is then plunged, the foresight is taken by turning the instrument about the vertical axis on its upper motion, and the deflection angle is observed. The angle is recorded as right R or left L, according to whether the upper motion is turned clockwise or counterclockwise.

Figure 20 illustrates the field notes for a portion of an open deflection-angle traverse where stakes are set every 100 ft.

Azimuth Traverse. The azimuth of the initial line of the traverse may be referred to either a true or an assumed meridian. Successive stations are occupied, beginning with the line of known azimuth. At each station, the transit is *oriented* by setting the A vernier to read the back azimuth (forward azimuth $\pm 180°$) of the preceding line and then backsighting to the preceding transit station. The instrument is then turned on the upper motion, and a foresight on the following transit station is secured. The reading indicated by the A vernier is the azimuth of the forward line.

Figure 21 illustrates the notes for a short closed azimuth traverse.

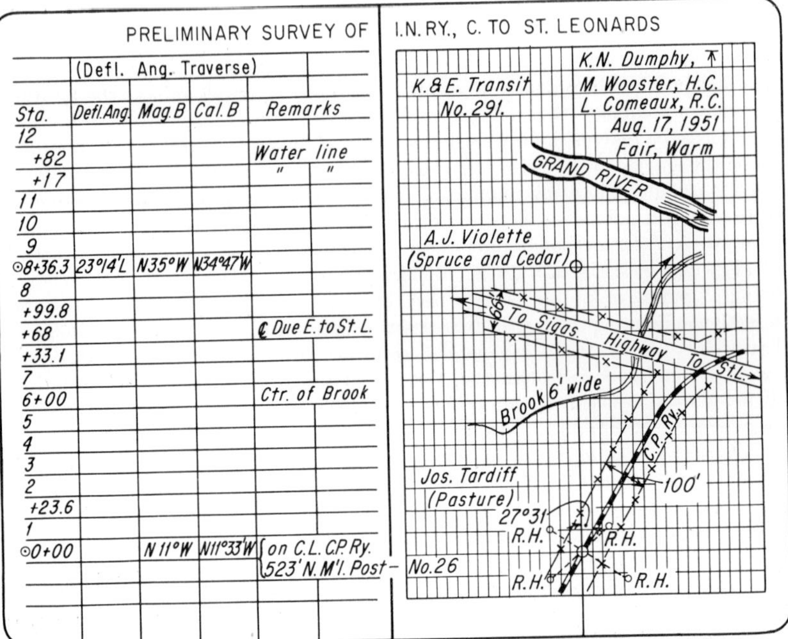

Fig. 20. Notes for open deflection-angle traverse.

AZIMUTH TRAVERSE AT HIGH-WATER LINE

Proposed Mill Pond, El. 741.36
Silver Creek, Penn.
(For Land Damage Est.)

Sta.	Obj.	Dist.	Azimuth	Mag. B.	Cal. Bear.
1	5		270°28'	N 80½° W	N 89°32' W
	2	689.32	350°30'	N	N 9°30' W
2	1		170°30'	S	S 9°30' E
	3	509.66	303°05'	N 48° W	N 56°55' W
3	2		123°05'	S 48¼° E	S 56°55' E
	4	678.68	236°13'	S 65½° W	S 56°13' W
4	3		56°13'	N 65¼° E	N 56°13' E
	5	572.50	177°58'	S 7° W	S 2°02' E
5	4		357°58'	N 7¼° E	N 2°02' W
	1	1082.71	90°29'	S 80° E	S 89°31' E
			Error =	01'	

Gurley transit No. 191

J. Stanbois ⊼
F. Lowe
June 15, 1951
Cloudy, Warm

3 Old Dam 2
N
4 Byram Bros.
Proposed Dam
SILVER CREEK
Byram Bros.
5 1

True azimuth of line 1-5 found by solar observation.
Mag. declination = 9° 10' W

Fig. 21. Notes for short closed azimuth traverse.

Specifications for Traversing with Transit and Tape. The precision of transit traverses is affected by both linear and angular errors of measurement. Many factors affect the precision, and it can be expressed only in very general terms. The following specifications give approximately the *maximum* linear and angular errors to be expected when the methods stated are followed. If the surveys are executed by well-trained men, with instruments in good adjustment, and under average field conditions, in general the error of closure should not exceed *half* the specified amount. The specifications apply to traverses of considerable length. It is assumed that a standardized tape is used.

Class 1. Precision sufficient for many preliminary surveys, for horizontal control of surveys plotted to intermediate scale, and for land surveys where the value of the land is low.

Transit angles read to the nearest minute. Sights taken on a range pole plumbed by eye. Distances measured with a 100-ft steel tape. Pins or stakes set within 0.1 ft of end of tape. Slopes under 3 percent disregarded. On slopes over 3 percent, distances either measured on the slope and corrections roughly applied, or measured with the tape held level and with an estimated standard pull.

Angular error of closure not to exceed $1'30''\sqrt{n}$, in which n is the number of observations. Total linear error of closure not to exceed 1/1,000.

Class 2. Precision sufficient for most land surveys and for location of highways, railroads, etc. By far the greater number of transit traverses fall in this class.

Transit angles read carefully to the nearest minute. Sights taken on a range pole carefully plumbed. Pins or stakes set within 0.05 ft of end of tape. Temperature corrections applied to the linear measurements if the temperature of air differs more than 15°F from standard. Slopes under 2 percent disregarded. On slopes over 2 percent, distances either measured on the slope and corrections roughly applied, or measured with the tape held level and with a carefully estimated standard pull.

Angular error of closure not to exceed $1'\sqrt{n}$. Total linear error of closure not to exceed 1/3,000.

Class 3. Precision sufficient for much of the work of city surveying, for surveys of important boundaries, and for the control of extensive topographic surveys.

Transit angles read twice with the instrument plunged between observations. Sights taken on a plumb line or on a range pole carefully plumbed. Pins set within 0.05 ft of end of tape. Temperature of air determined within 10°F, and corrections applied to the linear measurements. Slopes determined within 2 percent, and corrections applied. If tape is held level, the pull kept within 5 lb of standard, and corrections for sag applied.

Angular error of closure not to exceed $30''\sqrt{n}$. Total linear error of closure not to exceed 1/5,000.

Class 4. Precision sufficient for precise surveying in cities and for other especially important surveys.

Transit angles read twice with the instrument plunged between readings, each reading being taken as the mean of both A and B vernier readings. Verniers reading to 30″. Instrument in excellent adjustment. Sights taken with special care. Pins set within 0.02 ft of end of tape. Temperature of tape determined within 5°F, and corrections applied. Slopes determined within 1 percent, and corrections applied. If tape is held level, the pull kept within 3 lb of standard, and corrections for sag applied.

Angular error of closure not to exceed $15''\sqrt{n}$. Total linear error of closure not to exceed 1/10,000.

Details from Transit Lines. The precision with which details are located depends upon the purpose of the survey. In retracing property lines the actual lines may be obstructed by hedges or buildings, so that the actual corners must be located by measurements from other transit lines; such measurements should be taken with a precision as great as that for the transit line. If details are located solely for map-making purposes,

generally the required precision of measurements to details is less than that for the transit lines.

All well-defined objects should be correctly shown within the scale of the map, it being borne in mind that points cannot be plotted within less than perhaps 1/100 in. Thus, if the map scale were 1 in. = 1,000 ft, there would be no particular advantage in taking measurements closer than the nearest 10 ft to details; but if the scale were 1 in. = 10 ft, measurements should be taken to 0.1 ft.

Angular measurements to details are usually made with the transit, with angles read to minutes. Often the angles used in mapping details are estimated to the nearest 05′ without the aid of the vernier. Where details are located with respect to stations intermediate between transit stations, often some hand instrument is employed to measure the angles or directions.

Linear measurements to details are made with the 100-ft steel tape, with the 50-ft metallic tape, by stadia, or sometimes by pacing, according to the precision required and the convenience of measuring.

On most surveys the details are located as the traverse is run. The observations with the transit, called *side shots*, are made immediately after the foresight to the following station has been taken.

STADIA SURVEYING

Equipment. Equipment for measuring distances by stadia as described herein consists of a telescope with two horizontal *stadia hairs* used for sighting on a graduated *stadia rod* which is held vertical. (In European practice, a horizontal rod called a *subtense bar* is often used, and the horizontal angle between the ends of the bar is read on the horizontal circle of the transit.)

Horizontal Sights. The horizontal distance from principal focus to rod is $d = \frac{f}{i}s = Ks$, in which $K = \frac{f}{i}$ is a coefficient called the *stadia interval factor* which for a particular instrument is a constant so long as conditions remain unchanged, and s is the *stadia interval* apparently intercepted by the stadia hairs on the rod. Thus for a horizontal sight the distance from principal focus to rod is obtained by multiplying the stadia interval factor by the stadia interval.

The principal focus is at a fixed distance in front of the objective. Then the horizontal distance from center of instrument to rod is

$$D = Ks + (f + c) \qquad (25)$$

in which the focal distance f is a constant for a given instrument, and c (the distance from objective lens to center of instrument), though a variable depending upon the position of the objective, may for all practical purposes be considered a constant. This formula is employed in transforming stadia readings to horizontal distances when sights are horizontal and the rod is held vertical.

Stadia Constants. Usually the value of $(f + c)$ is determined by the manufacturer and is stated on the inside of the instrument box. The focal distance f can be determined with all necessary accuracy by focusing the objective on a distant point and then measuring the distance from the cross-hair ring to the objective. Similarly a mean value of c can be determined by measuring the distance from the vertical axis to the objective when the objective is focused for an average length of sight. Usually $(f + c)$ is nearly 1 ft and under ordinary circumstances may be considered as 1 ft without error of consequence. For internal-focusing telescopes, $(f + c)$ is zero or nearly so.

Interval Factor. The nominal value of the stadia interval factor $K = f/i$ is usually 100. With an instrument having fixed stadia hairs, the interval factor can be determined

by observation. The usual procedure is to set up the instrument in a location where a horizontal sight can be obtained and with a tape to lay off, from a point distant $(f + c)$ in front of the center of the instrument, distances of 100, 200 ft, etc., up to perhaps 1,000 ft, stakes being set at the points thus established. The stadia rod is then held on each of the stakes, and the stadia interval is read. The mean of the computed stadia interval factors is taken as the most probable value.

Inclined Sights. Figure 22 illustrates an inclined line of sight, AB being the stadia interval on the vertical rod.

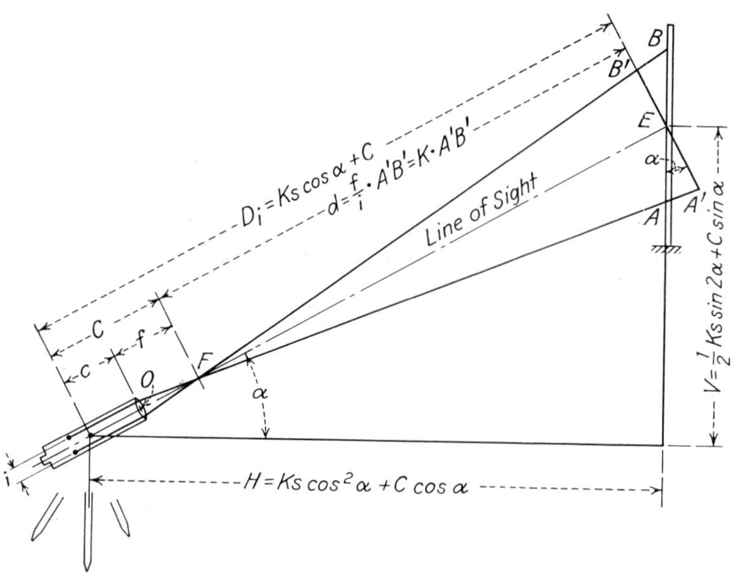

FIG. 22. Inclined stadia sight.

The horizontal component H of the inclined distance D is

$$H = Ks \cos^2 \alpha + (f + c) \cos \alpha \tag{26}$$

which is the general equation for determining the horizontal distance from center of instrument to rod, when the line of sight is inclined.

The vertical component V of the inclined distance is

$$V = \tfrac{1}{2}Ks \sin 2\alpha + (f + c) \sin \alpha \tag{27}$$

which is the general equation for the difference in elevation between the center of the instrument and the point where the line of sight cuts the rod. Equations (26) and (27) are known as the "stadia formulas for inclined sights."

Permissible Approximations. More approximate forms of the stadia formulas are sufficiently precise for most stadia work. Generally distances are computed only to feet, and elevations to tenths of feet. Under these conditions, for side shots where vertical angles are less than 3°, Eq. (26) may properly be reduced to the form

$$H = Ks + (f + c) \tag{28}$$

But for traverses of considerable length, owing to the systematic error introduced, this approximation should not be made for vertical angles greater than perhaps 2°.

Owing to unequal refraction and to accidental inclination of the rod, observed stadia intervals are in general slightly too large. To offset the systematic errors from these sources, frequently on surveys of ordinary precision the constant $(f + c)$ is neglected. Hence in any ordinary case, Eq. (26) may with sufficient precision be expressed in the form

$$H = Ks \cos^2 \alpha \text{ (approx)} \tag{29}$$

Also for small angles, Eq. (27) may generally be expressed with sufficient precision for ordinary work in the form

$$V = \tfrac{1}{2}Ks \sin 2\alpha \text{ (approx)} \tag{30}$$

Differences in elevation obtained by using this approximate formula will be correct to the nearest 0.1 ft for angles of less than 25°.

The determination of horizontal distances and differences in elevation by algebraic, graphical, or mechanical methods is considerably simplified by making Eqs. (29) and (30) the basis of calculations; hence these forms of the stadia formulas are the ones generally employed. More nearly accurate results can be obtained either by adding 0.01 ft to the observed stadia interval s or by adding 1 ft to the observed stadia distance Ks, if the telescope is not of the internal-focusing type. When K is 100, the common practice is to multiply mentally the stadia interval by 100 at the time of observation and to record this stadia distance Ks in the field notebook. Thus, if the stadia interval were 7.37 ft, the stadia distance recorded would be 737 ft.

Some instruments are equipped with a *stadia arc* on the vertical circle, by means of which horizontal distances and differences in elevation are determined without reading vertical angles and without tables or diagrams. The stadia arc has no vernier. For some instruments, horizontal distance is computed by *multiplying* the observed stadia distance by the reading of the "Hor." stadia arc, expressed as a percentage. For other instruments, horizontal distance is computed by *subtracting* from the observed stadia distance an observed correction, in percent, times the stadia distance. Vertical distance is computed by multiplying the observed stadia distance by the net reading of the "Vert." stadia arc, expressed as a percentage.

Transit-stadia Surveying. The field procedure of locating points consists in observing directions usually by azimuths and distances by stadia. In addition, differences in elevation are determined either by direct leveling when it is practicable to do so or, more usually, from observed vertical angles and stadia distances. The field party consists of a transitman, one or more rodmen, and usually a recorder.

In topographic surveying, this method may be employed merely for the collection of details, the horizontal and vertical control being established by other means; or it may be employed for establishing control as well as for details.

If details only are to be located, the instrument is set up at a traverse or triangulation station the elevation and location of which are known. The height of the instrument (HI) above the station over which it is set is measured with a rod or tape. The transit is oriented by sighting along a line the azimuth of which is known, this azimuth having been set off on the horizontal circle. The upper motion is unclamped, and sights to desired points are taken. Figure 23 shows notes of observations taken from station C of a traverse, the elevation of the station having been previously determined as 423.9.

In measuring vertical angles it is customary, wherever practicable, to sight at a rod reading equal to the height of instrument above the station over which the transit is set.

Where the detail to be observed is at nearly the same elevation as the point over which the transit is set, there is a marked advantage in determining difference in elevation by direct leveling. The notes of Fig. 23 show that object 12 was observed in this manner.

Where the required precision is not high, the stadia traverse with elevations of traverse

stations determined by vertical angle and stadia distance is a rapid means of establishing both horizontal and vertical control. The procedure is the same as that already described; in addition, vertical angles are observed for both the backsight and the foresight from each station, the telescope being sighted at a rod reading equal to the height of instrument above the station over which the transit is set up. The mean values for a given backsight and the preceding foresight are used in the computations.

TOPOGRAPHIC DETAILS, BLACK ESTATE 32

Inst. at C; El. 423.9; H.I.=4.4

Wisconsin Transit; (f+c)=1.25; K=100.2 — G. Burke, ⊼; M.D. Rand, Notes; F.J. & K.D., Rods; Apr. 4, 1951; Cloudy, Cold

Obj.	Az.	Rod Int.	Vert. Ang.	Hor. Dist.	Diff. El.	Elev.	
B	176°14′						
1	10°21′	7.23	-3°11′	723	-40.2	383.7	Water's Edge-Corner
2	3°14′	7.02	-3°17′	702	-40.2	383.7	" " -On Line
3	352°45′	5.64	-4°11′	563	-40.9	383.0	" "
4	7°18′	5.76	-4°04′	575	-40.9	383.0	" "
5	349°10′	-(7.14×4) on 7.7		717	-31.9	392.0	Line (Intervals)
6	16°55′	5.50	-2°50′	551	-27.3	396.6	"
7	315°20′	-(7.86×5) on 1.9		789	-36.8	387.1	" (")
8	349°15′	4.13	-5°46′	410	-41.4	382.5	Water's Edge
9	339°30′	5.40	-4°22′	539	-41.1	382.8	Bank Brook 6′ wide
10	0°05′	3.71	-4°12′	371	-27.2	396.7	
11	344°40′	4.85	-4°54′	484	-41.4	382.5	" " 15′ "
12	25°00′	2.86	0° on 3.2	288	+1.2	425.1	Direct Levels
13	307°45′	4.88	-4°56′	487	-42.0	381.9	Water's Edge
14	319°10′	4.02	-5°56′	400	-41.6	382.3	" "
15	309°45′	5.80	-3°00′	581	-30.7	393.2	
16	318°25′	3.27	-4°36′	327	-26.3	397.6	
B	176°15′	ck.					
17	340°00′	6.34	-3°08′	635	-34.7	389.2	
18	278°35′	2.51	-5°43′	250	-25.0	398.9	
19	276°20′	3.07	-7°56′	303	-42.3	381.6	Water's Edge
20	277°40′	4.24	-5°40′	422	-41.9	382.0	" "

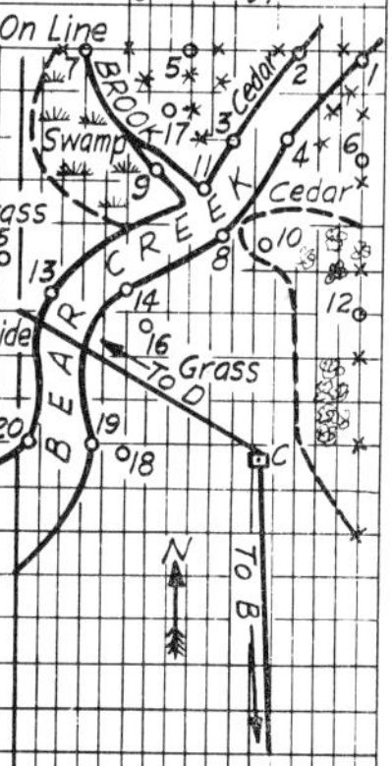

FIG. 23. Stadia notes for location of details, with elevations.

Errors in Stadia Surveying. Many of the errors in stadia surveying are those common to all similar operations such as direct leveling, indirect leveling, and measurement of horizontal and vertical angles. Sources of error in horizontal and vertical distances computed from observed stadia intervals are (1) stadia interval factor not that assumed, (2) rod not standard length, (3) incorrect observed stadia interval, (4) rod not plumb, and (5) unequal refraction.

Precision of Stadia Surveying. Many factors influence the precision of stadia surveying, and no definite statement of the precision for a given procedure can be made. Following are estimates believed to be fairly representative of several classes of stadia work, these estimates being based on the results secured on surveys run under a variety of conditions.

1. For side shots where a single observation is taken with sights steeply inclined and with no particular care taken to ensure the rod's being plumb, horizontal distances may have a precision lower than 1/100, and individual differences in elevation may be in error 2 ft or more per 1,000 ft of horizontal distance.

2. Under the same conditions as in 1 but with small vertical angles and reasonable care used in approximately plumbing the rod and with lengths of sight between 200 and 1,500 ft, the precision of horizontal distances should be not lower than 1/200. The error

in difference in elevation per 1,000 ft of horizontal distance need not be more than 0.3 ft if vertical angles are observed to 01′, or more than 1 ft if vertical angles are observed to 05′.

3. For a rapid stadia traverse of considerable length run through rough country with numerous long sights, angles being measured to minutes but without special precaution to eliminate systematic errors, the error of closure may be as low as 25 ft per mile in plan and 3 ft per mile in elevation.

4. For conditions as in 3 but for country fairly level so that all vertical angles are small, the error of closure ought not to exceed 15 ft per mile in plan and 0.5 ft $\sqrt{\text{distance in miles}}$ in elevation.

5. For rough country with vertical angles up to 15°, angles to minutes, rod standardized, rod plumbed with level, sights limited to 1,500 ft and taken forward and back from each transit station, and interval factor carefully determined, the error of closure may be less than 15 ft $\sqrt{\text{distance in miles}}$ in plan and 1 ft $\sqrt{\text{distance in miles}}$ in elevation.

6. For conditions as in 5 but for level country so that all vertical angles are small, the error of closure may be as small as 6 ft $\sqrt{\text{distance in miles}}$ in plan and 0.3 ft $\sqrt{\text{distance in miles}}$ in elevation.

7. For conditions as in 5 but stadia intervals determined by use of a target rod with two targets and observations made during cloudy days, the error of closure in plan should not exceed 4 ft $\sqrt{\text{distance in miles}}$.

TRIANGULATION OF ORDINARY PRECISION

Classification. Triangulation systems are classified according to (1) the average angular error of closure in the triangles of the system and (2) the discrepancy between the measured length of a base line and its length as computed through the system from an adjacent base line. Triangulation for the extensive surveys of the United States government is classified as follows:

	First order	Second order	Third order	Fourth order
Average error of triangle closure, seconds	1	3	6	>6
Check on base	1/25,000	1/10,000	1/5,000	>1/5,000

First- and second-order triangulation require methods of high precision not often necessary except on very extensive surveys. Third- and fourth-order triangulation require methods of ordinary or low precision, sufficient for control of the usual intermediate and large-scale surveys of limited area. The discussion herein generally pertains to triangulation methods of ordinary precision for surveys of limited extent.

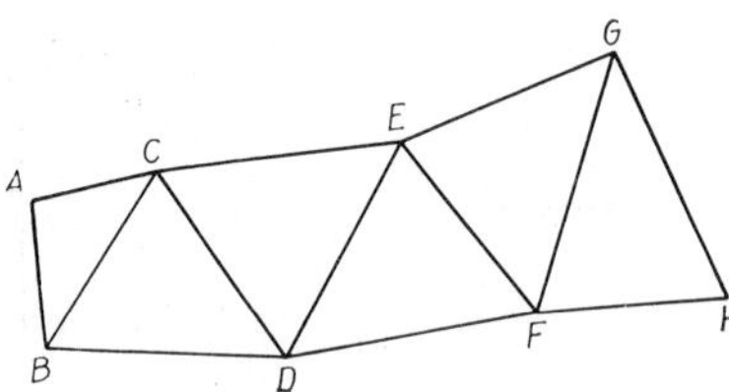

Fig. 24. Chain of single triangles.

Triangulation Figures. In a narrow triangulation system a chain of figures is employed, consisting of *single triangles*, *polygons*, *quadrilaterals*, or combinations of these figures. A wide system consists of similar combinations, irregularly overlapping and intermingling so that several routes exist for computing the length of triangle sides and the length of each base line from an adjacent base line.

Chain of Single Triangles. In the chain of single triangles (Fig. 24) there is but one route by which distances can be computed through the chain. As the sum of the measured angles in each triangle normally will not equal exactly 180°, the angles are adjusted to satisfy this requirement before the distances are computed.

Chain of Polygons. In triangulation, a polygon or "central-point figure" is composed of a group of triangles, the group being bounded by three or more sides and having within the figure a triangulation station at a vertex common to all the triangles. Figure 25 illustrates a chain composed of a six-sided polygon *EGJKIF* and two five-sided polygons. The sum of the measured angles in each triangle of the polygon should equal 180°; also, the sum of the angles about the central point should equal 360°. Further, the length of any side may be computed by two routes, and these two computed lengths should agree. The observed angles in each polygon are so adjusted that these three conditions exist.

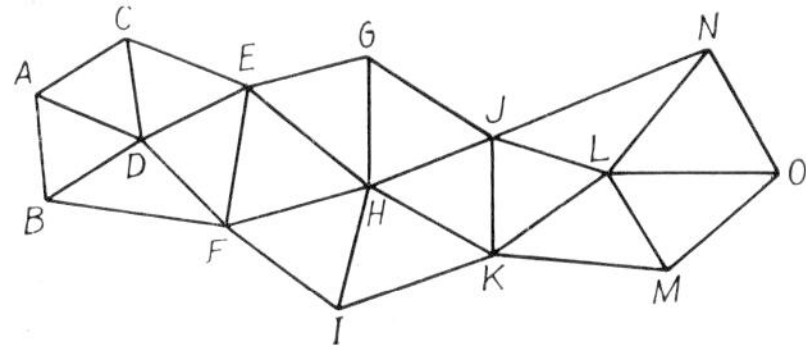

FIG. 25. Chain of polygons.

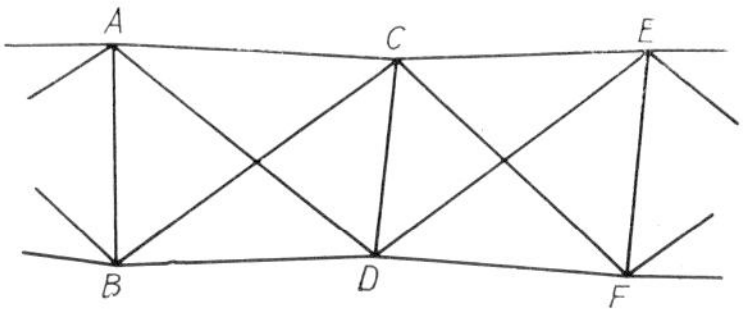

FIG. 26. Chain of quadrilaterals.

Chain of Quadrilaterals. A quadrilateral differs from a four-sided polygon in that there is no triangulation station within the figure. In Fig. 26, consider one of the quadrilaterals, as *ACDB*. The measured angles give values for four triangles *ACB*, *ACD*, *ADB*, and *CDB*, in each of which the sum of the angles should equal 180°. In addition, the length of any line should be the same when computed by one route as when computed by another.

Choice of Figure. The chain of triangles is the simplest but does not afford so many checks as the other forms; hence, for a given precision the base lines would need to be placed closer together. The chain of triangles is satisfactory for work of low precision. For more precise work, generally quadrilaterals or polygons are used; the quadrilaterals are best adapted to long narrow systems, and the polygons to wide systems.

Strength of Figure. In the trigonometric computations of triangulation, nearly always the sine function is used. As for a given error in angle the error in sine is relatively large for angles near 0° or 180°, angles near 0° and 180° are undesirable. It has been found in practice that satisfactory results can be secured for most purposes if the angles *used in the computations* fall between 30° and 150°. However, many angles measured in the field are not used in computing the length of the sides of the system.

As an aid in deciding which of several alternative figures (or chains of figures) is to be used in triangulation, the relative strengths of the figures can be determined by computations based on the size of the angles, the number of directions to be observed, and the number of geometric conditions that must be satisfied. Considerations of economy sometimes render one figure more desirable than another even though it may be the weaker of the two.

The relative strength of figure can be evaluated quantitatively in terms of a factor R based on the theory of probability; the lower the value of R, the stronger the figure. Strength of figure is a factor to be considered in establishing a triangulation system for which the computations can be maintained within a desired degree of precision. For example, for third-order triangulation it is desirable that R for a single figure not exceed 25 and that R between two base lines not exceed 125. In some cases it may not be

necessary to occupy all the stations of the system or to observe all the lines in both directions. Further, by means of computed strengths of figure, alternative routes of computation (chains of elemental triangles) can be compared and the best route chosen. The following brief treatment gives the essential relations for computing R.

Let

C = number of conditions to be satisfied in figure.

n = total number of lines in figure, including known line.

n' = number of lines observed in both directions, including known line if observed.

s = total number of stations.

s' = number of occupied stations.

D = number of directions observed (forward and/or back), excluding those along known line.

δ_A, δ_B = respective logarithmic differences of the sines, expressed in units of the sixth decimal place, corresponding to a change of 1 second in the "distance angles" A and B of a triangle. The distance angles are those opposite the known side and the side required.

$\Sigma(\delta_A^2 + \delta_A\delta_B + \delta_B^2)$ = summation of values for the particular chain of triangles through which the computation is carried from the known line to the line required. Values of $(\delta_A^2 + \delta_A\delta_B + \delta_B^2)$ for a triangle are given in published tables.

Then

$$C = (n' - s' + 1) + (n - 2s + 3) \tag{31}$$

$$R = \frac{D - C}{D}\Sigma(\delta_A^2 + \delta_A\delta_B + \delta_B^2) \tag{32}$$

Base Nets. In practice, for economic reasons usually the base lines are much shorter than the average length of the sides of the triangles in the main triangulation system. In order to obtain the required precision in the computed length of the sides of the main

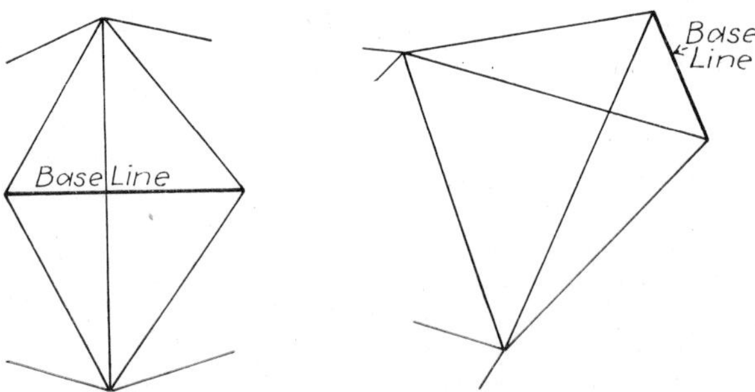

FIG. 27. Base nets.

triangles, it is necessary to expand the base line through a group of smaller triangles called the *base net*. Figure 27 shows examples of base nets affording quick and accurate expansion of the base line to the longer sides of the system. In triangulation of ordinary precision, base lines are placed perhaps 20 to 60 triangles apart, the distance depending on the strength of the figures.

Methods. The work of triangulation consists of the following steps:

1. Reconnaissance, to select the location of stations.

2. Erection of signals and, in some cases, tripods or towers for elevating the signals and/or the instrument.

3. Measurement of angles between the sides of triangles.

4. In many cases, astronomical observations at one or more triangulation points, in order to determine the true meridian to which azimuths are referred.

5. Measurement of the base lines.

6. Computations, including the adjustment of the observations, the computation of the length of each triangle side, and the computation of the coordinates of the stations.

Herein the description of methods will be concerned principally with triangulation of ordinary precision, corresponding roughly to the upper range of fourth-order triangulation. Triangulation of low precision differs from that of ordinary precision in the following respects: There is practically no reconnaissance, and often the stations are selected as the work progresses. The stations are marked with a stake, pole, or portable tripod. The base lines are measured by the ordinary methods of chaining or sometimes even by stadia. The angles of the triangles are not necessarily adjusted to meet the known geometric and trigonometric conditions. No correction is made when the instrument is not set up exactly over the station. No astronomical observations are made. Often the method of graphical triangulation with the plane table is employed (p. 1–52).

Reconnaissance. The chief of the party examines the terrain, choosing the most favorable sites for stations. In this connection are considered the number and shape of the triangles, the intervisibility and accessibility of stations and their usefulness in later work, the cost of necessary signals, and the convenience of base-line measurements. Angles and distances to adjacent stations are roughly determined—angles, either directly by use of the prismatic compass or similar hand instrument or graphically by use of the plane table; distances, either directly by pacing or odometer or graphically by use of the plane table. Existing maps are a great aid in reconnaissance.

Signals and Instrument Supports. At an instrument station, it is desirable to have the signal of a type that will permit placing the instrument directly over the station when angles are to be measured. In a small triangulation system with triangle sides only a few hundred feet in length, portable signals such as poles or light tripods with a plumb line may be used. In larger systems, generally the signal is a permanent tripod.

Where the instrument must be elevated to secure visibility, a combined observing tower and signal may be built. A central tripod supports the instrument; around this is a three- or four-sided structure supporting the platform upon which the observer stands. Thus the instrument tower is free from the vibrations caused by movements of the observing party.

In addition to the *major stations* from which observations are taken, often *minor stations* for local control are established by observations from the major stations. Minor stations may be signals erected at desirable locations, or they may be such objects as lone trees, spires, flagstaffs, and chimneys.

The best time for observing is in the late afternoon or at night. For night observations an electric lamp is used as a signal.

Angle Measurements. For triangulation of intermediate precision, usually the angles in the system are measured by means of a repeating theodolite, which if of American manufacture is similar in general design to the ordinary engineer's transit but is of larger size and of a higher grade of workmanship. The horizontal circle is 7 or 8 in. in diameter, and commonly the verniers read to 10″. For triangulation of ordinary and low precision, the ordinary transit may be used.

In triangulation of ordinary precision, angles are measured by the method of repetition, the number of repetitions and procedure employed depending on the required precision. The instrument, so far as possible, should be protected from sun and wind; the air should be clear; and great care should be taken in setting up the instrument and

in observing. In triangulation of low precision, the angles are usually doubled, with the telescope plunged between measurements.

If precise results are to be obtained, the instrument must be manipulated with care. The plate bubbles should be kept centered, but the leveling screws should not be disturbed except between repetitions. When turning on the lower motion, the hands should be in contact with the lower plate; when turning on the upper motion, they should be in contact with the upper plate and not the telescope. The last motion of the tangent screws should be clockwise or against the opposing spring. To eliminate the effect of twist in the tripod, after each repetition the instrument should be rotated on its lower motion in the same direction in which it was turned on its upper motion. That is, the direction of movement is always either clockwise or counterclockwise. Owing to the possibilities of unequal settlement of the tripod and to unequal expansion of the parts of the telescope, it is desirable that the observations should be made as rapidly as is consistent with careful work.

Azimuth Determinations. In computing the coordinates of triangulation stations a meridian of reference (either true or assumed) is used, and azimuths of all lines in the system are computed from corrected angles. For an extensive system the true meridian is employed, and account must be taken of the convergency of meridians. The direction of the true meridian or the true azimuth of any line may be determined by astronomical observations or from reference lines established by governmental agencies.

Base-line Measurement. For base-line measurements of ordinary precision either the steel tape or the invar tape may be employed, but for measurements of higher precision the invar tape is always used. For long base lines, such as are used in an extensive triangulation system, often a "long" tape (length 50 m to 500 ft) is used. It is desirable that the tape be standardized under the conditions of tension and support that will be employed in the field.

Where the base line is along a paved highway or a railroad, usually measurements are made with the tape supported over its entire length and at a time when the temperature of the supporting surface (highway or rail) is not appreciably different from that of the surrounding air.

Where the base line is over uneven ground, end supports for the tape are provided usually by substantial posts, perhaps 2 in. by 4 in., driven firmly in the ground. These are placed on a transit line at intervals of one tape length, as nearly as can be determined by careful preliminary measurements. A strip of copper or zinc is tacked to the top of the post to receive the markings. Profile levels are run over the top of the posts to determine the gradient from post to post. The tape is usually supported at one, two, or three points between the end supports. These intermediate points are placed accurately on the grade line by driving nails at grade in stakes placed on line at the proper intervals.

The equipment for base-line measurement includes at least one standardized tape; two stretcher devices for applying tension, one of which is equipped with a spring balance or a weight and pulley; two or three thermometers; a finely divided pocket scale; dividers; and a marking awl or needle.

The party consists of four to six men whose duties are indicated by the following description of the procedure. The proper tension is applied to the tape by means of the stretchers, with the spring balance at the forward end of the tape. When the rear end of the tape is observed to coincide with the previously established mark and when the proper tension is applied, the position of the forward end of the tape is marked by a fine line engraved by means of the awl or needle on the metal strip on top of the post. Thermometers fastened to the tape, one near each end and sometimes one near the mid-point, are read at the instant that the tape length is marked on the forward post.

The tape is then carried forward without being allowed to drag on the ground, and the process is repeated. After a few measurements, the end of the tape will probably fall

beyond or short of the limits of the metal strip of the next forward post; accordingly, it will be necessary to use *set backs* or *set forwards*, which are measurements of small distances made by means of a finely divided pocket scale and a pair of dividers. The conditions of measurement are recorded in detail, and notes are kept in the following form:

From post no.	To post no.	Temperature, °F			Set forward	Set back
		Forward	Middle	Rear		

Errors; Corrections. The various errors and corrections in ordinary measurements with the tape are discussed previously herein. In the measurement of base lines the effect of temperature is the most serious source of error; hence, in more precise work it is customary to use an invar tape and to measure the base line on cloudy days or at night, when the air and the ground are at nearly the same temperature. Corrections are made for length of tape, temperature, and slope. They are also made for sag and tension when conditions of use make such corrections necessary. Errors due to imperfect marking of the tape lengths on the metal strips are reduced to a minimum by careful manipulation.

Specifications. For a base line not less than ½ mile in length measured over uneven ground, satisfactory results for third-order triangulation will be obtained under the following specifications: The tape shall be standardized by the Bureau of Standards; the mean temperature of the tape shall be determined within a maximum error of 3°F; the elevations of adjacent marking posts shall be determined within a maximum error of 0.3 ft; the tape shall be supported, if possible, under the same conditions as those existing when it was standardized; the tension of the tape shall not vary more than 2 oz from the standard tension adopted for the work; and the errors in marking shall not exceed 0.002 ft.

If the line is measured along a pavement or a railroad track, the foregoing conditions will apply except that the permissible variation in tension may be increased to 1 lb.

For base lines of lower precision, corresponding modifications may be made.

Reduction to Sea Level. It is sometimes necessary to reduce the length of the base line to the equivalent length at mean sea level. The correction C_l to be subtracted from the actual length is given by the equation

$$C_l = \frac{LA}{R} \tag{33}$$

in which L is the length of the base line, A is the mean altitude of the base line above sea level, and R is the radius of the earth (mean R = 20,889,000 ft, log R = 7.31992).

Reduction to Center. At certain triangulation stations it is difficult, if not impossible, to place the instrument vertically beneath the object which has been observed from adjacent stations. At such a place, the instrument is set over any convenient point near the principal station, and angles to the adjacent stations are measured with the same precision as other angles in the system. These angles will not be the same as those which would be observed if the instrument were occupying the exact location of the main station; to obtain the corresponding main-station values, corrections are computed and applied to the measured angles. This procedure of correcting the observed angles is called *reduction to center*.

In addition to the measurement of the angles to adjacent stations, measurements are made of (1) the distance from the main station to the occupied station, and (2) the (clockwise) angle at the occupied station between the main station and an adjacent

station in the system. The situation is illustrated by Fig. 28, where O represents a main station in the system $OABCD$, and T represents the point occupied by the instrument; the distance d and the angle ATO are measured. The lengths of all sides in the main system, as, for example, t_1 = 8,659 ft, are known approximately from the angles that have been measured at the stations A, B, C, D and from the known sides AB, BC, etc.

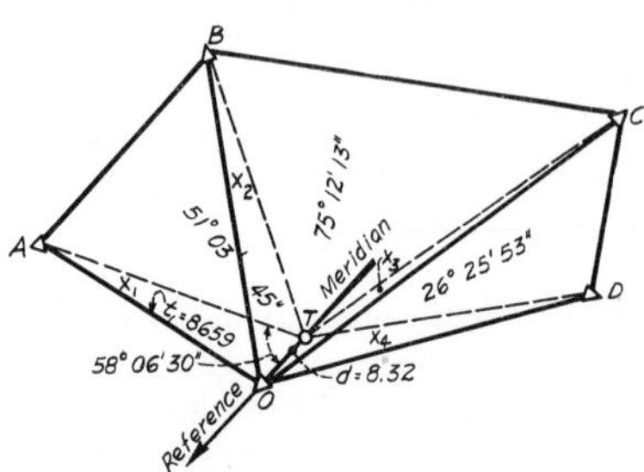

Fig. 28. Reduction to center.

In the triangle AOT, the angle T and the two sides t_1 and d are known; hence, the angle x_1 can be computed. This angle is seen to be the difference in direction at station A between the lines AT and AO. Therefore, if the direction (azimuth) of AT is known with respect to any reference meridian, the direction of AO with respect to the same meridian can be computed. Likewise, the directions of the lines BO, CO, and DO can be found; and since these directions are referred to the same meridian, the correct angles at O between these stations can be determined.

The value of x_1 in triangle AOT is given by the equation

$$\sin x_1 = \frac{d \sin T}{t_1} \tag{34}$$

and since the angle x_1 is a small angle for which the sine is nearly equal to the arc, the value of x_1 will be given in seconds of arc if both members of the equation are divided by the sine of 1″, or

$$x_1'' = \frac{d \sin T}{\sin 1'' t_1} \text{(approx)} \tag{35}$$

It will be noted that $d/\sin 1''$ is a constant for a given station, so that once its value has been determined the successive correction angles x_1, x_2, etc., can be computed by a single multiplication. Since the correction angles are usually small, the slide rule will ordinarily render values correct to seconds.

The method of correcting angles for which one of the sights has been taken to an eccentric signal is the same as that just described for an eccentric instrument station.

Correction for Spherical Excess. Since the measured angles are spherical angles, each triangle will contain more than 180°. The amount greater than 180° is termed the *spherical excess* and is about 1″ for each 75 sq miles of area of triangle. More accurately,

$$E = \frac{a}{C}(1 - e^2 \sin^2 \phi)^2 \tag{36}$$

in which E = spherical excess, seconds
a = area, sq miles
ϕ = latitude at center of triangle

$$\log e^2 = 7.8305026 - 10$$
$$\log C = 1.8787228$$

It is clear that no correction for spherical excess will be necessary unless the triangles are very large, and then only in the most precise work. One-third of the correction is subtracted from each of the angles.

Adjustment of a Chain of Triangles. A single chain of triangles is adjusted in two steps: (1) the *station adjustment*, to make the sum of the angles about each point equal 360°; and (2) the *figure adjustment*, to make the sum of the three angles in each triangle equal 180°. In precise triangulation these two adjustments are made in one operation by the method of least squares, but the following approximate solution yields results sufficiently precise for most cases of triangulation of ordinary precision.

To make the sum of the angles about each point equal 360°, the observed angles are added together, and the sum is subtracted from 360°. The resulting difference is divided by the number of angles, and the quantity so found is added algebraically to each angle. To make the sum of the angles in each triangle equal 180°, a similar plan is followed, using the angles obtained by the station adjustment; i.e., the three angles of each triangle are added together, and their sum is subtracted from 180°. One-third of the difference is added algebraically to each of the three angles.

This method of adjustment assumes that all the angles were observed in the same way and with the same precision and is applicable only when such is the case. If certain angles are measured with a higher precision than certain others, the method may be modified readily by weighting the observations of the several angles within the system.

Adjustment of a Quadrilateral. In the figure adjustment of the angles of a quadrilateral, two conditions are considered: (1) the *geometric condition* that the sum of the interior angles of a plane rectilinear figure is equal to $(n - 2)180°$, in which n is the number of sides of the figure; and (2) the *trigonometric condition* that in any triangle the sines of the angles are proportional to the lengths of the sides opposite. First the station adjustment is made; then the geometric condition is satisfied by adjustment of the angles of the four overlapping triangles forming the quadrilateral. Then the trigonometric condition is satisfied by using the adjusted angles to compute the length of an unknown side of the quadrilateral from that of a known side by the four possible routes, the angles being further adjusted if necessary to make the length of the computed side independent of the route used.

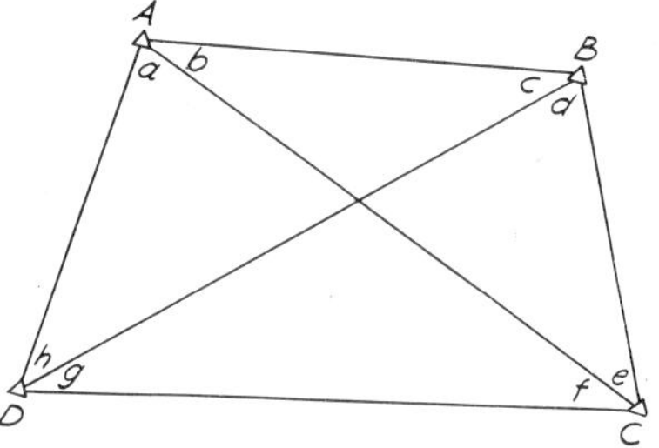

Fig. 29. Adjustment of a quadrilateral.

Geometric Condition. When all angles in a quadrilateral are measured, there are four overlapping triangles. These are shown as triangles ABC, ACD, ABD, and BCD in Fig. 29. In each of these triangles, the sum of the three angles must be 180°.

Hence, from the figure,

$$b + c + d + e = 180° \tag{37}$$
$$a + f + g + h = 180° \tag{38}$$
$$a + b + c + h = 180° \tag{39}$$
$$d + e + f + g = 180° \tag{40}$$

Also, the sum of the eight lettered angles in the figure must equal 360°, since they form the interior angles of a closed figure of four sides. This may be derived also by the addition of Eqs. (37) and (38) or (39) and (40).

$$a + b + c + d + e + f + g + h = 360° \tag{41}$$

Further, since the opposite angles at the intersection of the diagonals must be equal, it follows that

$$b + c = f + g \tag{42}$$
$$d + e = h + a \tag{43}$$

If any three of these seven equations are satisfied, the other four must of necessity be satisfied also. Equations (41), (42), and (43) are the ones most convenient to use. The following procedure is suggested:

1. Make the station adjustment as follows: Adjust the angles around each point to make their sum equal to 360°.

2. Using the values resulting from the station adjustment 1, add the eight angles a, b, c, d, e, f, g, and h, and subtract their sum from 360°. Divide the difference by 8, and algebraically add the result to each of the eight angles, thus satisfying the conditions of Eq. (41).

3. Using the adjusted values from 2, find the difference between the sums $(b + c)$ and $(f + g)$, and divide that difference by 4. Apply the result as a correction to each of the four angles, increasing each of the two whose sum is the smaller and decreasing each of the two whose sum is the larger, thus making these angles satisfy Eq. (42) without disturbing the adjustment for Eq. (41). Proceed in the same way with each of the four angles involved in Eq. (43).

Trigonometric Condition. If the length of one line, as AB, is known, and the length of the opposite side CD is to be computed, the computer may select one or another series of triangles for use in accomplishing this result. For example, a solution of triangle ABC gives the length of AC, and then from triangle ACD the required length of CD is found; or in the triangle ABC the length BC is found, and then in BCD the length CD is computed. There are four possible choices of route through the figure, and it now remains to be seen whether the angles, as so far adjusted, are so related as to make the value of the length of a computed side independent of the route used. Assume that the length of AB is known and that the length of CD is to be found.

$$AD = AB\frac{\sin c}{\sin h} \tag{44}$$

$$CD = AD\frac{\sin a}{\sin f} = AB\frac{\sin a \sin c}{\sin f \sin h} \tag{45}$$

Similarly,

$$CD = AB\frac{\sin b \sin d}{\sin e \sin g} \tag{46}$$

Equating these two values of CD,

$$\frac{\sin a \sin c}{\sin f \sin h} = \frac{\sin b \sin d}{\sin e \sin g} \tag{47}$$

or

$$\frac{\sin a \sin c \sin e \sin g}{\sin b \sin d \sin f \sin h} = 1 \tag{48}$$

Expressed in logarithmic form this is

$$(\log \sin a + \log \sin c + \log \sin e + \log \sin g) - (\log \sin b + \log \sin d + \log \sin f + \log \sin h) = 0 \tag{49}$$

The angles are tested for satisfaction of this equation by adding the logarithmic sines in the two groups as indicated and by finding the difference between the two sums.

Various adjustments by which this difference may be reduced to zero are possible. The "least squares" adjustment gives the most probable values to the adjusted angles, but it is somewhat more elaborate than is necessary for most surveys. A simple approximate method, which gives an equal correction to each angle and which does not disturb the geometric condition, is as follows: (1) record the log sines; (2) for each angle record the tabular logarithmic sine difference for 1″ opposite each logarithm; (3) find the average

required change α in log sine by dividing the difference between the sums by 8; (4) find the average difference β for 1″; (5) the ratio α/β gives the number of seconds of arc to be applied as a correction to each angle; (6) add this correction to each of the four angles the sum of whose log sines is the smaller, and subtract it from each of the angles the sum of whose log sines is the larger, and thus the corrected values are obtained.

If one or more of the angles is greater than 90°, the trigonometric adjustment is made as just described, without disturbing the geometric relations. However, since the sine of an obtuse angle decreases as the angle increases, the corresponding log sines will be changed in the direction opposite to that desired. Usually the error introduced by this condition will be negligible for this approximate adjustment; if not, the trigonometric adjustment should be repeated.

Adjustment of a Chain of Figures between Two Base Lines. If two base lines are measured an additional condition is introduced, namely, that the length of each side in the connecting chain of triangles or quadrilaterals must be the same when computed from one base line as when computed from the other. An exact solution is possible only by the method of least squares, but the following approximate methods may be used in triangulation of ordinary precision in the case of a single chain of figures.

The figures (triangles or quadrilaterals) are adjusted individually as previously described. The lengths of the sides are then computed from each line to a common line about midway between the base lines. This common line may then be corrected to reconcile the two computed values of its length, with equal or different weights being assigned to the two computed values as desired, based on the known conditions. The effect of this correction may then be carried back through each half of the chain, as follows:

If the precision of the *angular* measurements is relatively high as compared with that of the linear measurements, the lines of each half of the chain are corrected in proportion to their lengths as compared with the length of the common line, leaving the angles unchanged. In effect, this procedure shrinks one entire half of the chain (including its base line) by a fixed proportion, and swells the other half (including its base line) by a fixed proportion.

If, however, the precision of the *linear* measurements is relatively high, the lengths of the base lines may be assumed to be correct. In this case, the correction is tapered off from the full amount at the common line to zero at each base line, the correction to each line being not only proportional to the length of the line but also roughly proportional to the relative distance of the given line from the base line. This procedure changes the values of the angles, and the new values of the angles are used in further computations.

Between these two extremes, the procedure depends on the relative precision of the angular and the linear measurements, and weights may be assigned accordingly.

Computation of Triangles and Coordinates. In computing the lengths of the sides and the coordinates of the stations in a triangulation system, it is desirable to follow an orderly procedure to expedite the work and to avoid mistakes. Convenient arrangements for these computations for plane triangulation are given in Figs. 30 and 31.

Triangles. A sketch of the figure is drawn (Fig. 30), and the vertices are lettered as A, B, and C in a clockwise direction, beginning with the side the length of which is known. The sides opposite the vertices are indicated by the corresponding lower-case letters as a, b, and c. The sine relation states that

$$b = c\frac{\sin B}{\sin C} \qquad \text{or} \qquad \log b = \log c - \log \sin C + \log \sin B \tag{50}$$

Accordingly, if the logarithms are recorded in the column of logarithms in the order $\log c$, colog $\sin C$, $\log \sin A$, and $\log \sin B$, then $\log a$ is found by covering $\log \sin B$ with a

Station or line	Angle or distance	Logarithm	Figure
c	1,432.58 ft.	3.156119	Given: 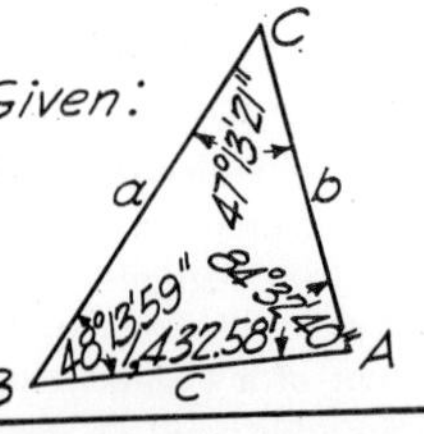
C	47°13′21″	0.134306 (colog)	
A	84°32′40″	9.998028	
B	48°13′59″	9.872657	
a	1,942.94 ft.	3.288453	
b	1,455.74 ft.	3.163082	

Fig. 30. Computation of triangle.

narrow strip of paper and adding the other three values. Also, log b is found by covering log sin A and adding the other three values. Finally the distances a and b are found as the numbers corresponding to their respective logarithms.

From Station B		From Station A	
Given:		Given:	
Z =	34°32′54″	Z =	12°40′27″
BC =	1,942.94	AC =	1,455.74
Total lat. B =	+661.36	Total lat. A =	+841.37
Total dep. B =	−1,590.94	Total dep. A =	−169.71

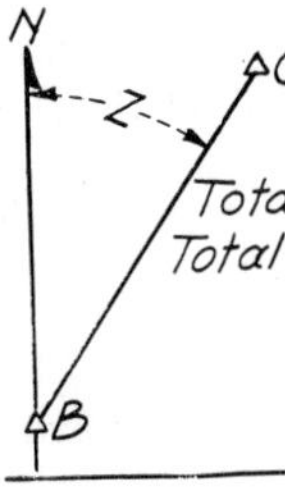

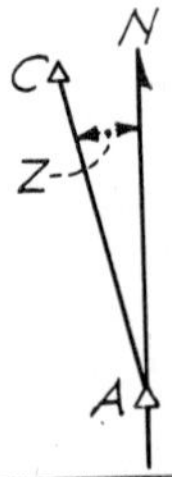

Mean Total Latitude C = +2,261.64
(Check)

Total lat. C	+2,261.64	Total lat. C	+2,261.63
Total lat. B	+661.36	Total lat. A	+841.37
L cos Z	+1,600.28	L cos Z	+1,420.26
log L cos Z	3.204195	log L cos Z	3.152369
log cos Z	9.915742	log cos Z	9.989287
log L	3.288453	log L	3.163082
log sin Z	9.753661	log sin Z	9.341249
log L sin Z	3.042114	log L sin Z	2.504331
L sin Z	+1,101.83	L sin Z	−319.40
Total dep. B	−1,590.94	Total dep. A	−169.71
Total dep. C	−489.11	Total dep. C	−489.11

(Check)
Mean Total Departure C = −489.11

Fig. 31. Computation of coordinates.

An example of the computations is shown in the tabulation of Fig. 30, for which the known data are given on the sketch of the triangle.

Coordinates. Figure 31 shows a form for computing the coordinates of a station C from each of the stations B and A. The known plane coordinates (total latitude and total

departure) of B and A and the known bearings and lengths of BC and AC are shown at the top of the figure. The computation is carried out as indicated, with due regard to signs. Beginning with log L in the tabulation, computations for total latitude are made reading upward and computations for total departure are made reading downward.

Three-point Problem. When the main triangulation has been completed, frequently it is desired to determine the location of additional points which are to be used as instrument stations of a topographic survey or for other purposes. In triangulation work the position of an instrument station as O (Fig. 32) is determined by measuring each of the two angles subtended by lines to three visible stations, as A, B, and C, and by solving the triangles involved; this is known as solving the *three-point problem.* Thus, in Fig. 32 all parts of the triangle formed by the stations A, B, and C are known. The angles α and β are measured at the station O. The problem is solved when the values of the angles x and y have been determined, for the remaining parts in each of the triangles ABO and ACO can then be computed. A check is afforded if the same value for the side AO results from each of these triangles.

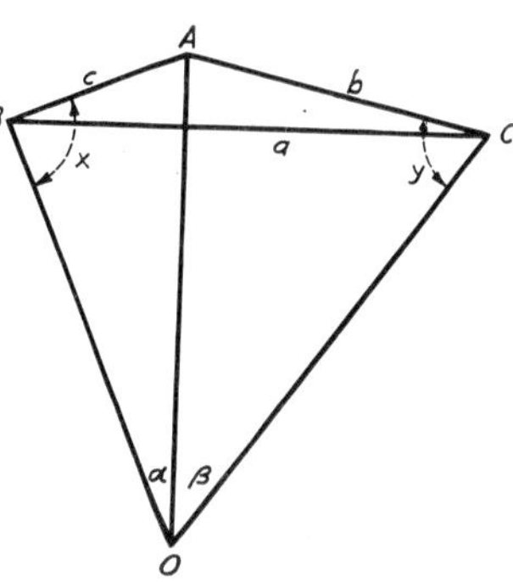

Fig. 32. Three-point problem.

The problem is indeterminate if the station O lies on or near the great circle passing through the stations A, B, and C. This condition will be evidenced by the condition that $\alpha + \beta + A = 180°$.

Solution: Given the sides b and c and the angle A, also the observed angles α and β (Fig. 32). Let

$$S = 180° - \tfrac{1}{2}(A + \alpha + \beta) = \tfrac{1}{2}(x + y)$$

If the stations A and O lie on the same side of the side a and if the station O is outside the triangle ABC, then

$$S = \tfrac{1}{2}(A - \alpha - \beta) = \tfrac{1}{2}(x + y) \qquad (51)$$

for which case the solution by this method is impossible when $\alpha + \beta = A$. Let

$$\tan \phi = \frac{c \sin \beta}{b \sin \alpha}$$

and let

$$\Delta = \tfrac{1}{2}(x - y)$$

then

$$\tan \Delta = \cot (\phi + 45°) \tan S \qquad (52)$$

If $\tan \Delta$ is positive, then $x = S + \Delta$ and $y = S - \Delta$.

If $\tan \Delta$ is negative, then $x = S - \Delta$ and $y = S + \Delta$.

State Systems of Plane Coordinates. A great opportunity for local use of the national triangulation system has come about through the adoption of state systems of plane coordinates, whereby one set of plane rectangular coordinates is made to serve the whole area of a small state or a portion (usually half) of the area of a large state. Many triangulation stations have been established and monumented by the U.S. Coast and Geodetic Survey. A map projection has been chosen for the state, or portion thereof, such that the errors of projection will rarely exceed 1/10,000 and, therefore, will be negligible for most local surveys. For states of greater extent east and west a Lambert conformal conic projection is used, while for states of greater extent north and south a transverse Mercator projection is employed. Reference axes for each zone are such

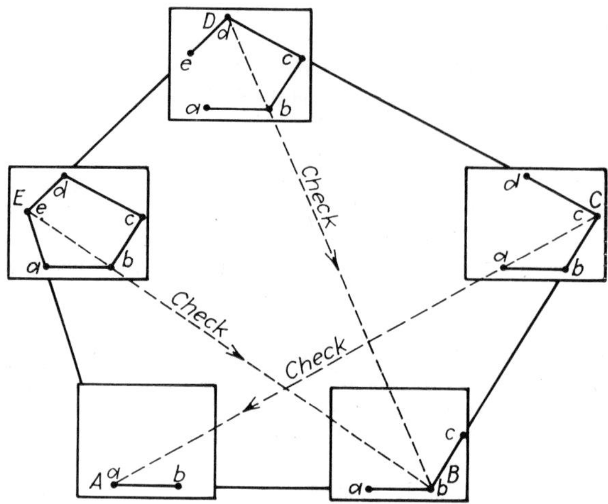

Fig. 34. Traverse with plane table.

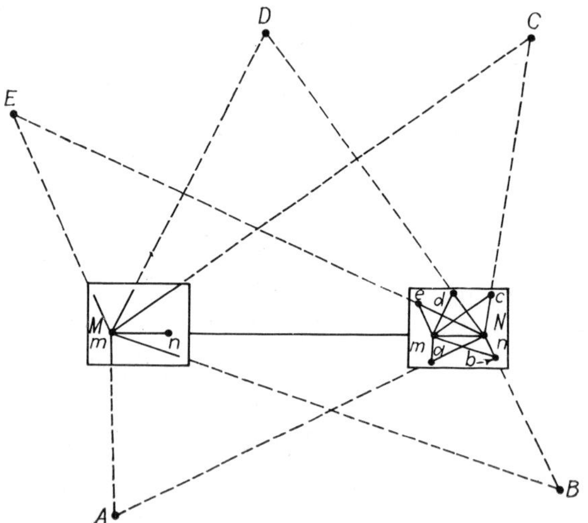

Fig. 35. Intersection with plane table.

c, etc. Distances to the objects are not measured but may be scaled from the map. If the angle between the intersecting rays is small, the location will be indefinite.

Graphical Triangulation. Graphical triangulation achieves the same results as triangulation with the transit, but the procedure differs in that the plotted locations of the distant signals are determined graphically on the plane-table sheet. The method involves both intersection and, where necessary, resection.

In employing this method, at least two and preferably three stations as *A*, *B*, and *C* (Fig. 36), the locations of which are known, must be capable of being occupied by the instrument and must be marked by signals. Their locations are plotted on the plane-table sheet at *a*, *b*, and *c* prior to going to the field. The field procedure is as follows: The plane table is set up as at *A*, and is oriented by sighting at *B* and *C*; rays are drawn

THE

toward other stations such as U, V
up and oriented at stations B and C in
The correct plotted location of each sta
rays, as by those drawn from stations

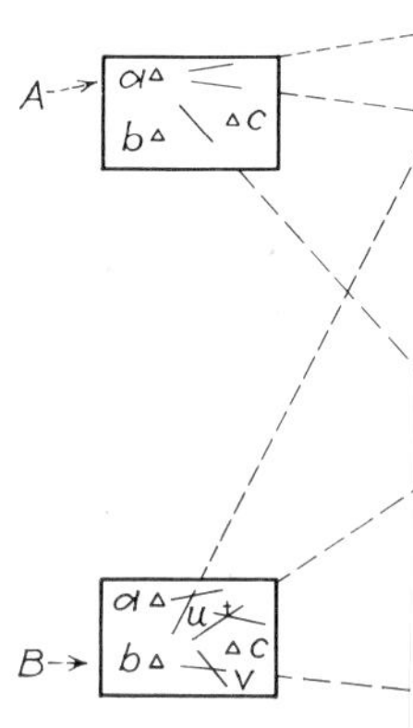

FIG. 36.

rays drawn toward a given object ar
third rays drawn from station C.

Resection. Resection is the proce
occupied by the instrument, by mean
of which have been plotted. Resecti
plane-table stations which have not
With the plane table oriented ove
map), two or more objects of know
line of indefinite length is drawn thr
The intersection of these lines mark
The table may be oriented by any
that, for the methods of orientation
chart, resection can be accomplished
is by the three-point problem or
point problem, orientation and rese
accomplished in the same operation.

Resection after Orientation by
If the plane table has been ori
means of the compass, the method
tion is as follows: Let P be the stati
known location occupied by the pla
and let A and B be two visible
which have been plotted on the sh
and b. Then the plotted location
determined by drawing a line, or
through a in the direction of A ar
ing through b in the direction of
point p where the two (or more) lin
instrument station P.

departure) of B and A and the known bearings and lengths of BC and AC are shown at the top of the figure. The computation is carried out as indicated, with due regard to signs. Beginning with log L in the tabulation, computations for total latitude are made reading upward and computations for total departure are made reading downward.

Three-point Problem. When the main triangulation has been completed, frequently it is desired to determine the location of additional points which are to be used as instrument stations of a topographic survey or for other purposes. In triangulation work the position of an instrument station as O (Fig. 32) is determined by measuring each of the two angles subtended by lines to three visible stations, as A, B, and C, and by solving the triangles involved; this is known as solving the *three-point problem.* Thus, in Fig. 32 all parts of the triangle formed by the stations A, B, and C are known. The angles α and β are measured at the station O. The problem is solved when the values of the angles x and y have been determined, for the remaining parts in each of the triangles ABO and ACO can then be computed. A check is afforded if the same value for the side AO results from each of these triangles.

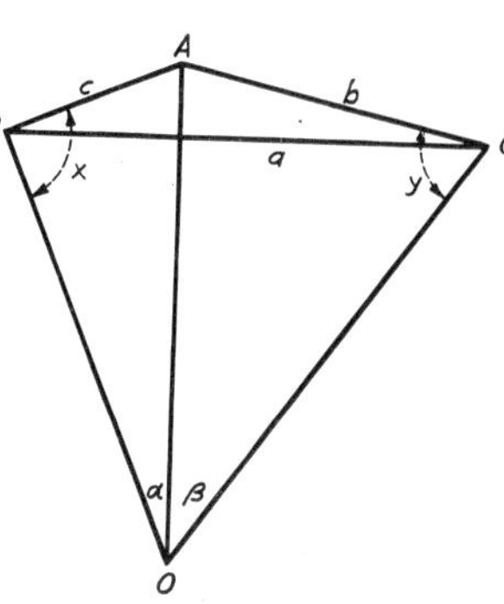

FIG. 32. Three-point problem.

The problem is indeterminate if the station O lies on or near the great circle passing through the stations A, B, and C. This condition will be evidenced by the condition that $\alpha + \beta + A = 180°$.

Solution: Given the sides b and c and the angle A, also the observed angles α and β (Fig. 32). Let

$$S = 180° - \tfrac{1}{2}(A + \alpha + \beta) = \tfrac{1}{2}(x + y)$$

If the stations A and O lie on the same side of the side a and if the station O is outside the triangle ABC, then

$$S = \tfrac{1}{2}(A - \alpha - \beta) = \tfrac{1}{2}(x + y) \qquad (51)$$

for which case the solution by this method is impossible when $\alpha + \beta = A$. Let

$$\tan \phi = \frac{c \sin \beta}{b \sin \alpha}$$

and let

$$\Delta = \tfrac{1}{2}(x - y)$$

then

$$\tan \Delta = \cot (\phi + 45°) \tan S \qquad (52)$$

If $\tan \Delta$ is positive, then $x = S + \Delta$ and $y = S - \Delta$.
If $\tan \Delta$ is negative, then $x = S - \Delta$ and $y = S + \Delta$.

State Systems of Plane Coordinates. A great opportunity for local use of the national triangulation system has come about through the adoption of state systems of plane coordinates, whereby one set of plane rectangular coordinates is made to serve the whole area of a small state or a portion (usually half) of the area of a large state. Many triangulation stations have been established and monumented by the U.S. Coast and Geodetic Survey. A map projection has been chosen for the state, or portion thereof, such that the errors of projection will rarely exceed 1/10,000 and, therefore, will be negligible for most local surveys. For states of greater extent east and west a Lambert conformal conic projection is used, while for states of greater extent north and south a transverse Mercator projection is employed. Reference axes for each zone are such

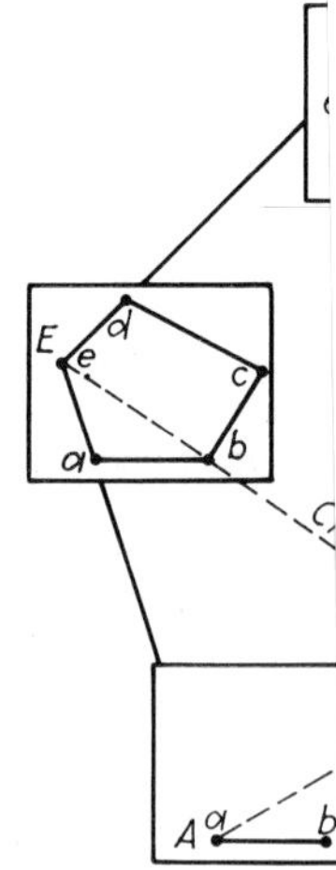

Fig. 3

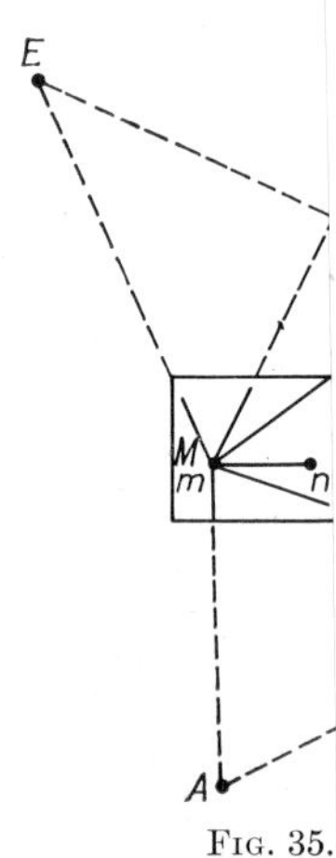

Fig. 35.

c, etc. Distances to the objects a
the angle between the intersectin

Graphical Triangulation. Gra
angulation with the transit, but t
distant signals are determined g
volves both intersection and, whe

In employing this method, at l
(Fig. 36), the locations of which
instrument and must be marked
table sheet at a, b, and c prior to
The plane table is set up as at A,

Resection after Orientation by Backsighting. If the table is oriented by backsighting along a line the direction of which has been plotted but the length of which may be unknown, the method of orientation and resection is as follows: Suppose that the topographer wishes to occupy station C (Fig. 37) the location of which has not been plotted but from which can be seen two points as A and B the locations of which have been plotted at a and b. He orients the board at one of the known stations as B, takes a foresight to C, and draws through b a ray of indefinite length. He then sets up the plane table at station C, orients it by backsighting to B, and resects from A through a. The intersection c of the ray from b and the resection line from a is the plotted location of the plane-table station C, since the triangles abc and ABC are similar. If the angle between the ray and the resection line is small, the location will be indefinite; for strong location the acute angle between these lines should be greater than 30°.

Resection and Orientation: Three-point Problem. Frequently the topographer wishes to occupy an advantageous station which has not been located on the map and toward which no ray from located stations has been drawn, and at the same time orientation by use of the compass is not sufficiently accurate. If three located stations are visible the three-point problem offers a convenient method of orienting and resecting in the same operation. There are several solutions of the three-point problem. In the United States, experienced topographers commonly employ a method of direct trial, guided by rules (see following paragraphs). The mechanical or tracing-cloth solution (p. 1–56) is simpler to understand but is not so satisfactory or so expeditious under the usual field conditions. For solution of the three-point problem by computation, employed in transit work, see p. 1–49.

Trial Method. The plane table is set up over the station of unknown location and is oriented approximately either by compass or by estimation. Resection lines from the

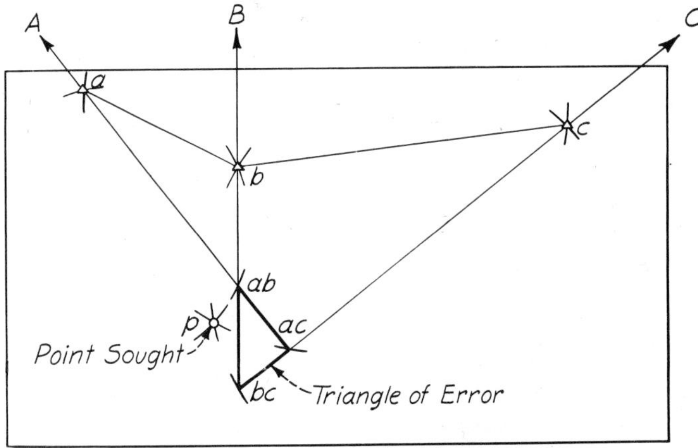

Fig. 38. Three-point problem with plane table.

three stations of known location are drawn through the corresponding plotted points. These lines will not intersect at a common point unless the trial orientation happens to be correct. (An exception to this statement is discussed later in this article.) Usually a small triangle called the *triangle of error* is formed by the three lines (Fig. 38). The correct plotted location of the plane-table station, called the *point sought*, is then estimated more closely. One method is to draw arcs of circles through the points shown (through a, b, and point ab; b, c, and bc; and a, c, and ac); the circles will intersect at p, the point sought. Usually, however, the correct location of the point sought is estimated more

1. *Length and Bearing of One Side Unknown.* If the algebraic sum of the latitudes and the algebraic sum of the departures of the known sides are respectively designated by ΣL and ΣD, then the length S of the unknown side is

$$S = \sqrt{(\Sigma L)^2 + (\Sigma D)^2} \tag{56}$$

and the tangent of the bearing angle is

$$\tan \alpha = \frac{-\Sigma D}{-\Sigma L} \tag{57}$$

with due regard to sign.

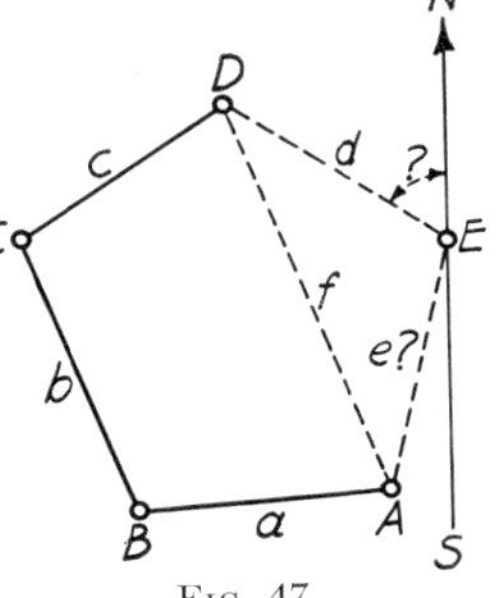

Fig. 47

2. *Length of One Side and Bearing of Another Side Unknown.* Figure 47 represents a closed traverse for which the direction of the line $DE = d$ and the length of the line $EA = e$ are not determined by field measurements. Let an imaginary line extend from D to A, cutting off the unknown sides from the remainder of the traverse. Then $ABCDA$ forms a closed traverse for which the side $DA = f$ is unknown in both direction and length. By the method of the preceding paragraph,

$$\text{tan bearing angle of } f = \frac{\text{dep. } f}{\text{lat. } f} = \frac{-(\text{dep. } a + \text{dep. } b + \text{dep. } c)}{-(\text{lat. } a + \text{lat. } b + \text{lat. } c)} \tag{58}$$

and

$$\text{Length of } f = \frac{\text{lat. } f}{\text{cos bearing angle of } f} = \frac{\text{dep. } f}{\text{sin bearing angle of } f} \tag{59}$$

In computing the length of f by Eq. (59), it is desirable to use the larger of the two quantities, latitude or departure.

The angle between the lines e and f in the triangle ADE is

$$\angle DAE = \text{azimuth of } AE - \text{azimuth of } AD \tag{60}$$

In the triangle ADE, the length of the two sides d and f and one angle DAE are known. By the relation that sines of angles are proportional to sides opposite,

$$\sin DEA = \sin DAE \frac{f}{d} \tag{61}$$

With angle DEA known, angle ADE can be computed, and the remaining unknown length is given by the equation

$$e = f\frac{\sin ADE}{\sin DEA} = d\frac{\sin ADE}{\sin DAE} \tag{62}$$

Also, the azimuth of DE = azimuth of $DA - \angle ADE$.

This method of solution is generally applicable even though two partly unknown courses are not adjoining. Obviously the latitude and the departure of any line of fixed direction and length are the same for one location of the line as for any other. Also, regardless of the order in which the lines of a closed figure are placed, the algebraic sum of the latitudes and the algebraic sum of the departures must be zero. Hence, when two partly unknown sides of a closed traverse are not adjoining, one of the sides is considered as moved from its location to a second location parallel with the first, such that the two partly unknown sides adjoin; the solution then becomes identical with that just described. To simplify the problem the data are usually plotted roughly to small scale.

When the length of one side and the bearing of another are unknown, the solution

described in this article will generally render two values of each of the unknowns. Often it is impossible to tell which are the correct values unless the general direction of the side of unknown bearing is observed.

As the angle between the partly unknown lines approaches 90° the solution becomes weak; and as the angle between these lines becomes small the solution becomes strong.

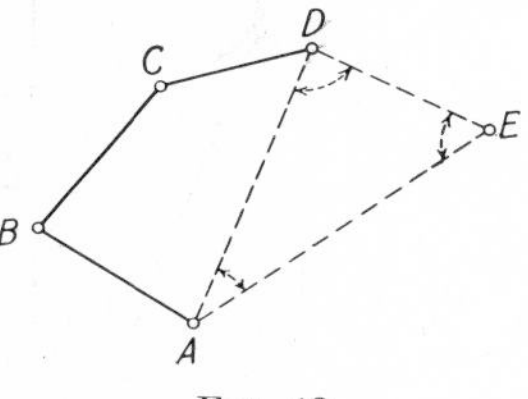

Fig. 48

3. *Length of Two Sides Unknown.* In Fig. 48, $ABCD$ represents the portion of a closed traverse for which the courses are known in both direction and length, and the lines DE and EA are courses for which the direction is known but the length is unknown. From the latitudes and departures of the known sides, the length and bearing of the closing line DA are computed; and in the triangle ADE the angles A, D, and E are computed from the known directions of the sides. The lengths DE and EA are determined through the relation

$$\frac{DE}{\sin A} = \frac{EA}{\sin D} = \frac{DA}{\sin E} \tag{63}$$

If the two lines are not adjoining, the problem may be solved as though they were, as explained in 2. As the angle between the partly unknown lines approaches 90° the solution becomes strong; and as the angle approaches 0° or 180° the solution becomes weak, the problem being indeterminate when the lines are parallel.

4. *Direction of Two Sides Unknown.* The solution is similar to that described in the preceding paragraph. In Fig. 48, if DA is the closing side of the known portion of the traverse, its direction and length are computed; then the lengths of the three sides of the triangle ADE are known, and the angles A, D, and E can be computed.

The general direction of at least one of the partly unknown lines must be observed, since the values of the trigonometric functions merely determine the shape of the triangle but do not fix its position.

CALCULATION OF AREAS OF LAND

Area by Coordinates. In Fig. 49, $ABCDF$ represents a tract the area of which is to be determined, SN a reference meridian, and WE a reference parallel. The coordinates of $A, B \ldots F$ are known; for any point these coordinates are the perpendicular distance from the reference meridian, defined as the total departure or the *meridian distance*, and the perpendicular distance from the reference parallel, defined as the total latitude or the *parallel distance*. Thus, for A the meridian distance is $aA = m_1$ and the parallel distance is $a'A = p_1$. Meridian distances are regarded as positive or negative according to whether they lie east or west of the reference meridian; parallel distances are regarded as positive or negative according to whether they lie north or south of the reference parallel. In the figure all meridian and parallel distances are positive.

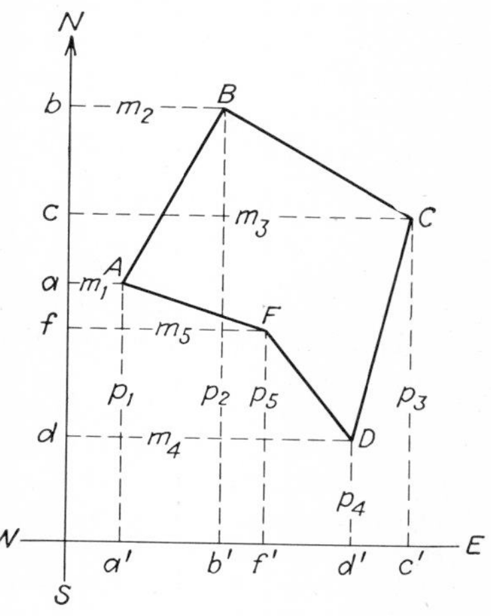

Fig. 49. Area by coordinates.

The area of the tract can be computed by summing algebraically the areas of the trapezoids formed by projecting the lines upon

the reference meridian; thus

$$\text{Area } ABCDF = \tfrac{1}{2}(m_2 + m_3)(p_2 - p_3) + \tfrac{1}{2}(m_3 + m_4)(p_3 - p_4) - \tfrac{1}{2}(m_4 + m_5)(p_5 - p_4) - \tfrac{1}{2}(m_5 + m_1)(p_1 - p_5) - \tfrac{1}{2}(m_1 + m_2)(p_2 - p_1) \quad (64)$$

By multiplication and a rearrangement of terms in the above equation, there is obtained

$$2 \text{ area} = -[p_1(m_5 - m_2) + p_2(m_1 - m_3) + p_3(m_2 - m_4) + p_4(m_3 - m_5) + p_5(m_4 - m_1)] \quad (65)$$

Rule: To determine the area of a tract of land when the coordinates of its corners are known, multiply the parallel distance, or ordinate, of each corner by the difference between the meridian distances, or abscissas, of the following and the preceding corners, always algebraically subtracting the following from the preceding. One-half of the algebraic sum of the resulting products is the required area.

A result identical except for sign would be obtained by always subtracting the preceding from the following. The sign of the area is not significant.

Equation (64) can also be expressed in the form

$$2 \text{ area} = m_2p_1 + m_3p_2 + m_4p_3 + m_5p_4 + m_1p_5 - m_1p_2 - m_2p_3 - m_3p_4 - m_4p_5 - m_5p_1 \quad (66)$$

When this form is employed, computations can be made conveniently by tabulating each parallel distance below the corresponding meridian distance as follows:

$$\frac{m_1}{p_1} \quad \frac{m_2}{p_2} \quad \frac{m_3}{p_3} \quad \frac{m_4}{p_4} \quad \frac{m_5}{p_5} \quad \frac{m_1}{p_1} \quad (67)$$

Then in Eq. (67) the difference between the sum of the products of the coordinates joined by full lines and the sum of the products of the coordinates joined by dotted lines is equal to twice the area of the tract.

The method may be modified by interchanging the words "parallel" and "meridian." When the corners are not all in one quadrant, particular attention must be given to the algebraic signs of meridian and parallel distances.

Principles of the Double-meridian-distance Method. In computing area by the double-meridian-distance method, the latitudes and departures of all the courses are determined, and the survey is balanced. A reference meridian is then assumed to pass through some corner of the tract; the double meridian distance of each line is computed; and double the areas of the trapezoids or triangles formed by orthographically projecting the several traverse lines upon the meridian are computed. The algebraic sum of these double areas is double the area within the traverse.

The perpendicular distance from the meridian to any point in the traverse is called the total departure or the *meridian distance* of the point. Meridian distances are considered as being positive if they lie east of the reference meridian and negative if they lie west of the reference meridian.

The sum of the meridian distances of the two extremities of a straight line is called the *double meridian distance* (DMD) of the line. If the meridian passes through the most westerly corner of the traverse, the double meridian distance of all lines will be positive, which is a convenience (although not a necessity) in computing.

The length of the orthographic projection of a line upon the meridian is the latitude of the line, latitudes being considered as positive if the direction of the line has a northerly component and negative if the direction of the line has a southerly component.

Each trapezoid or triangle for which a course in the traverse is one side is bounded on the north and south by meridian distances and on the west by the latitude of that course.

Therefore the double area of any triangle or trapezoid formed by projecting a given course upon the meridian is the product of the double meridian distance of the course and its latitude, or

$$\text{Double area} = \text{DMD} \times \text{latitude} \tag{68}$$

In computing double areas, account is taken of signs. If the meridian extends through the most westerly point, all double meridian distances are positive; hence the sign of a double area is the same as that of the corresponding latitude.

Following are three convenient rules for determining DMD's:

1. The DMD of the first course (reckoned from the point through which the reference meridian passes) is equal to the departure of that course.
2. The DMD of any other course is equal to the DMD of the preceding course, plus the departure of the preceding course, plus the departure of the course itself.
3. The DMD of the last course is numerically equal to the departure of the course but with opposite sign.

The first two rules are employed in computing values. The third rule is useful as a check on the correctness of the computations. Due regard must be given to signs.

Area within Closed Traverse by DMD Method. Following is a summary of the steps employed in calculating by the DMD method the area within a closed traverse when the lengths and bearings of the sides are known.

1. Compute the latitudes and departures of all courses as described on p. 1–61.
2. Find the error of closure in latitude and in departure as described on p. 1–62.
3. Balance the latitudes and departures in accordance with one of the rules given on p. 1–62.

AREA OF BALSAM PARK, ISLAND POND, VERMONT
by D.M.D. Method 7

Field Notes
Book No. 3
Page 47

Computations
Aug. 17, 1951
Computed and
Checked by S. D. M.

Line	Calc. Bearing	Dist., 66-ft. Ch.	Latitudes N	Latitudes S	Departures E	Departures W	Corrected Lats.	Corrected Deps.	D.M.D.'s	Double Areas +	Double Areas −
AB	S 80°29½′ W	34.464		5.694		33.991	−5.693	−33.990	61.812		351.89
BC	S 33 04 W	25.493		21.364		13.911	−21.361	−13.911	13.911		297.15
CD	S 33 46¾ E	33.934		28.205	18.867		−28.201	+18.867	18.867		532.06
DE	N 87 58¼ E	28.625	1.013		28.607		+1.013	+28.608	66.342	67.21	
EA	N 0 27 E	54.235	54.234		0.426		+54.242	+0.426	95.376	5173.51	
		176.751	55.247	55.263	47.900	47.902	ΣL=0	ΣD=0		5240.72	1181.10
				55.247		47.900				1181.10	
				0.016		0.002				2 ⌊ 4059.62	
										2029.81 Sq. Ch.	
										or 202.98 Ac.	

$E = \sqrt{0.016^2 + 0.002^2} = 0.016$ Chains

$\text{E. of C.} = \dfrac{0.016}{176.751} = \dfrac{1}{11{,}000}$

Line	AB	BC	CD	DE	EA
Lat.	5.694	21.364	28.205	1.013	54.234
Log Lat.	0.75542	1.32968	1.45032	0.00584	1.73427
Log cos	9.21805	9.92326	9.91969	8.54899	9.99999
Log Dist.	1.53737	1.40642	1.53063	1.45674	1.73428
Log sin	9.99399	9.73689	9.74509	9.99973	7.89535
Log Dep.	1.53136	1.14331	2.27572	1.45647	9.62964
Dep.	33.991	13.911	18.867	28.607	0.426
Log Cor. Lat.	0.75534	1.32962	1.45026	0.00584	1.73434
Log D.M.D.	1.79107	1.14336	1.27570	1.82179	1.97944
Log D.A.	2.54641	2.47298	2.72596	1.82763	3.71378
Double Area	351.89	297.15	532.06	67.21	5173.51

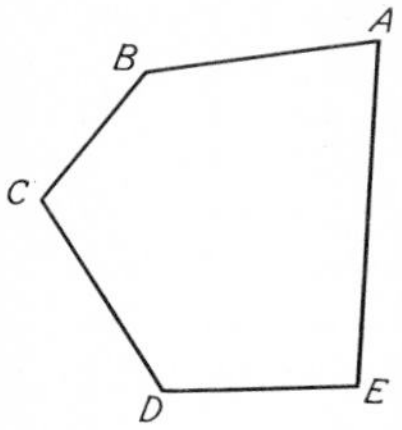

Note:
Survey Balanced by Transit Rule.

A B C D E

FIG. 50. Computations for area by DMD method.

4. Assume that the reference meridian passes through the most westerly point in the survey, and calculate the DMD's by the rules of the preceding article, using the corrected departures.

5. Compute the double areas by multiplying each DMD by the corresponding corrected latitude.

6. Find the algebraic sum of the double areas, and determine the area by dividing this sum by 2.

The foregoing steps are illustrated in the computations shown by Fig. 50 in which the area of a tract within a transit traverse is determined.

Area of Tract with Irregular or Curved Boundaries. Often an irregular boundary is not a sharply defined line; and if offsets from a traverse line are taken sufficiently close

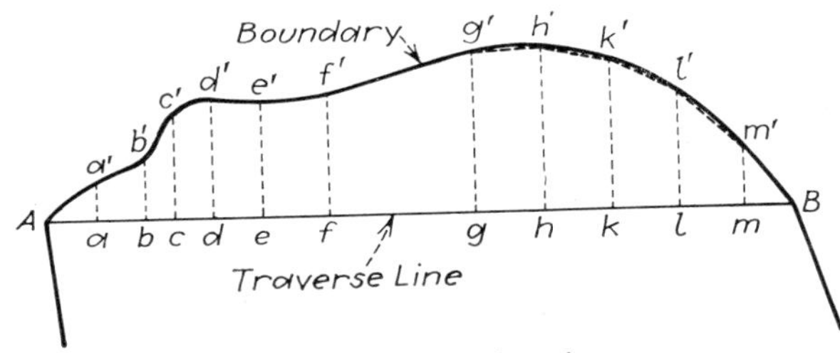

FIG. 51. Irregular boundary.

together, the error involved in considering the boundary as straight between offsets is small as compared with the inaccuracies of the measured offsets. When this assumption is made, as is usually the case, the assumed boundary takes some such form as that illustrated by the dotted lines $g'h'$, $h'k'$, etc., in Fig. 51, and the areas between offsets are of trapezoidal shape. Under such an assumption, irregular areas are said to be calculated by the *trapezoidal method* or the *trapezoidal rule.*

Where the curved boundaries are of such definite character as to make it justifiable, the area may be calculated somewhat more accurately by assuming that the boundary is made up of segments of parabolas and computing the area by *Simpson's one-third rule.*

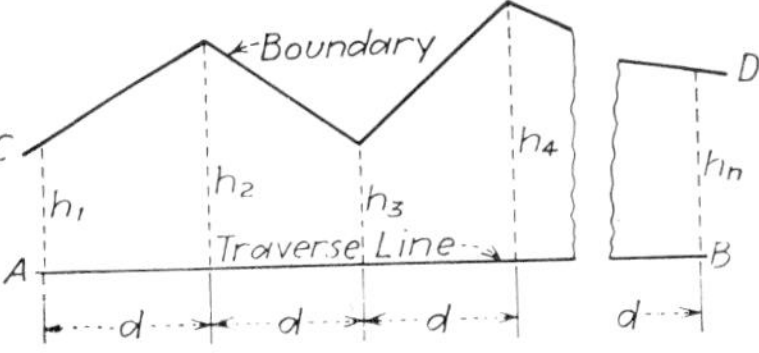

FIG. 52. Area by trapezoidal rule.

Offsets at Regular Intervals: Trapezoidal Rule. Let Fig. 52 represent a portion of a tract lying between a traverse line AB and an irregular boundary CD, offsets h_1, h_2 . . . h_n having been taken at the regular intervals d. The summation of the areas of the trapezoids comprising the total area is

$$\text{Area} = \frac{h_1 + h_2}{2} d + \frac{h_2 + h_3}{2} d + \cdots \frac{h_{(n-1)} + h_n}{2} d \tag{69}$$

$$\text{Area} = d\left(\frac{h_1 + h_n}{2} + h_2 + h_3 + \cdots h_{(n-1)}\right) \tag{70}$$

Equation (70) may be expressed conveniently by the following rule:

Trapezoidal Rule. Add the average of the end offsets to the sum of the intermediate offsets. The product of the quantity thus determined and the common interval between offsets is the required area.

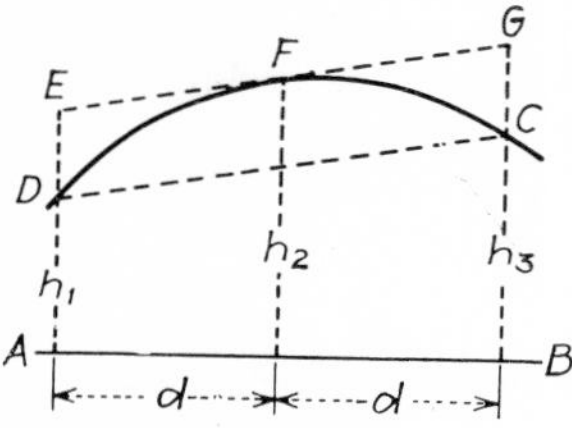

FIG. 53. Area by Simpson's rule.

Offsets at Regular Intervals: Simpson's Rule. In Fig. 53, let AB be a portion of a traverse line, DFC a portion of the curved boundary assumed to be the arc of a parabola, and h_1, h_2, and h_3 any three consecutive rectangular offsets from traverse line to boundary taken at the regular interval d.

The area between traverse line and curve may be considered as composed of the trapezoid $ABCD$ plus the area of the segment between the parabolic arc DFC and the corresponding chord DC. One property of a parabola is that the area of a segment (as DFC) is equal to two-thirds the area of the enclosing parallelogram (as $CDEFG$). Then the area between the traverse line and curved boundary within the length of $2d$ is

$$A_{1,2} = \frac{h_1 + h_3}{2} 2d + \left(h_2 - \frac{h_1 + h_3}{2}\right) 2d \frac{2}{3} \tag{71}$$

$$= \frac{d}{3}(h_1 + 4h_2 + h_3) \tag{72}$$

Similarly for the next two intervals,

$$A_{3,4} = \frac{d}{3}(h_3 + 4h_4 + h_5) \tag{73}$$

The summation of these partial areas for $(n - 1)$ intervals, n being an odd number and representing the number of offsets, is

$$\text{Area} = \frac{d}{3}[h_1 + h_n + 2(h_3 + h_5 + \cdots h_{(n-2)}) + 4(h_2 + h_4 + \cdots h_{(n-1)})] \tag{74}$$

which is conveniently expressed by the following rule applicable to any case where the number of offsets is odd and the interval between the offsets is uniform:

Simpson's One-third Rule. Find the sum of the end offsets, plus twice the sum of the odd intermediate offsets, plus four times the sum of the even intermediate offsets. Multiply the quantity thus determined by one-third of the common interval between offsets, and the result is the required area.

If the total number of offsets is *even*, the partial area at either end of the series of offsets is computed separately.

Offsets at Irregular Intervals. The method of coordinates described on p. 1–66 may be applied to this problem by assuming the origin as being on the traverse line and at the point where the first offset is taken. The coordinate axes are then the traverse line and a line at right angles thereto. The rule given on p. 1–67 may then be modified to the following: *Multiply the distance (along the traverse) of each intermediate offset from the first by the difference between the two adjacent offsets, always subtracting the following from the preceding. Also, multiply the distance of the last offset from the first by the sum of the last two offsets. The algebraic sum of these products, divided by two, is the required area.*

Area Cut Off by a Line between Two Points. In Fig. 54, let $ABCDEFG$ represent a tract of land to be divided into two parts by a line extending from A to D. It is desired to determine the length and direction of the cutoff line AD without additional field measurements, and to calculate the areas of each of the two parts into which the tract is divided.

Either of the two parts may be considered as a closed traverse with the length and bearing of one side DA unknown. Considering the part $ABCDA$, the latitudes and

departures of AB, BC, and CD are given; hence the latitude, departure, length, and bearing of DA may be determined as described on p. 1–65. The area of either part can then be found by the coordinate method or the DMD method.

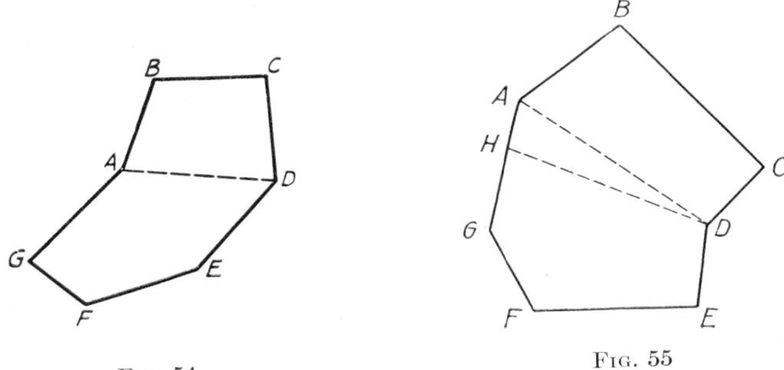

FIG. 54

FIG. 55

Area Cut Off by a Line Running in a Given Direction. In Fig. 55, $ABCDEFG$ represents a tract of known dimensions, for which the corrected latitudes and departures are given; and DH represents a line, running in a given direction, which passes through the point D and divides the tract into two parts. It is desired to calculate from the given data the lengths DH and HA and the area of each of the two parts into which the tract is divided.

Either of the two parts may be considered as a closed traverse for which the lengths of two sides are unknown; these lengths can be computed as described on p. 1–66. Considering the part $ABCDHA$, the latitudes and departures of AB, BC, and CD are known; from these the length and bearing of DA are computed. In the triangle ADH, the lengths of the sides DH and HA are found, and their latitudes and departures are computed. The area of $ABCDHA$ is then calculated by the coordinate method or the DMD method.

To Cut Off a Required Area by a Line through a Given Point. In Fig. 56, $ABCDEF$ represents a tract of known dimensions, for which the corrected latitudes and departures are given; and G represents a point in the boundary, through which a line is to pass

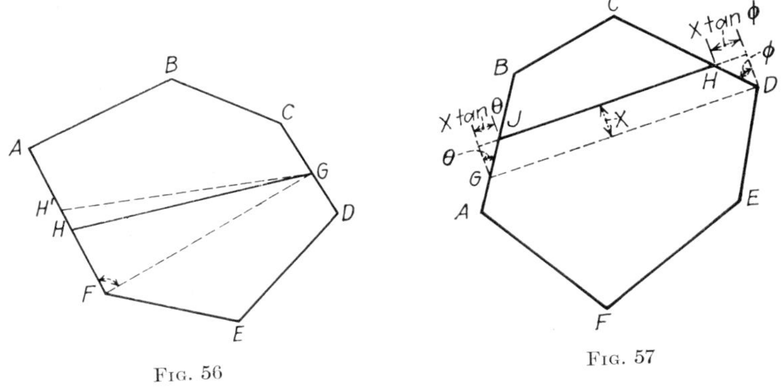

FIG. 56

FIG. 57

cutting off a required area from the tract. It is assumed that the area within the tract has been computed by the coordinate method or the DMD method, and that a sketch of the tract has been prepared.

To find the length and direction of the dividing line, the procedure is as follows:

A line *GF* is drawn to that corner of the traverse which, from inspection of the sketch, will come nearest being on the required line of division. The latitude and departure of *CG* are computed. Then in the traverse *ABCGFA* all sides are known except *GF*. By the methods described on p. 1–65, the latitude, departure, length, and bearing of *GF* are determined. The area of *FABCG*, the amount cut off by the line *FG*, is computed. The difference between this area and that required is found.

In the figure, it is assumed that *FABCG* has an area greater than the desired amount, *GH* being the correct position of the dividing line. Then the triangle *GFH* represents this excess area; and, as the angle *F* may be computed from known bearings, there are given in this triangle one side *FG*, one angle *F*, and the area. The length *HF* is computed from the relation, area = ½ *ab* sin *C*:

$$HF = \frac{2 \times \text{area } GFH}{FG \sin F} \tag{75}$$

The triangle is then solved for angle *G* and length *GH*. From the known direction of *GF* and the angle *G*, the bearing of *GH* is computed. The latitudes and departures of the lines *FH*, *GH*, and *HA* are then computed.

To Cut Off a Required Area by a Line Running in a Given Direction. In Fig. 57, *ABCDEF* represents a tract of known dimensions and area, which is to be divided into two parts, each of a required area, by a line running in a given direction. The corrected latitudes and departures are known.

Through the corner that seems likely to be nearest the line cutting off the required area, a trial line *DG* is drawn in the given direction. Then in the closed traverse *GBCDG* the latitudes and departures of *BC* and *CD* and the bearings of *DG* and *GB* are known, and the lengths *DG* and *GB* are unknown. By the methods given on p. 1–66 these unknown quantities are found, and the latitudes and departures of the courses are determined. The area cut off by the trial line is calculated. The difference between this area and that required is represented in the figure by the trapezoid *DGJH* in which the side *DG* is known. The angles at *D* and *G* can be computed from the known bearings of adjacent sides, and in this way θ and ϕ are determined. Then

$$\text{Area of trapezoid} = DG \cdot x + \frac{x^2}{2}(\tan\theta + \tan\phi) \tag{76}$$

in which $\tan\theta$ or $\tan\phi$ is positive or negative according as θ or ϕ lies within or without the trapezoid, and x is the altitude of the trapezoid. (In the figure, both angles lie without the trapezoid, and hence both tangents are negative.) The value of x is found by solving this equation. Then $GJ = x \sec\theta$; $DH = x \sec\phi$; and $JH = DG + x(\tan\theta + \tan\phi)$, in which the signs of $\tan\theta$ and $\tan\phi$ are as given above.

LATITUDE AND AZIMUTH

Latitude by Observation on Sun at Noon. The latitude of a given station can be determined with a fair degree of precision, ordinarily within 01′, by observing with the engineer's transit the altitude of the sun at local apparent noon, when the sun crosses the meridian. If the longitude of the place is roughly known, it is unnecessary to observe the time; but if the longitude is unknown, the standard time of the observation must be taken. The problem consists in determining the true altitude h of the center of the sun above the celestial horizon and computing the apparent declination δ of the sun at the instant of sighting. Then the latitude ϕ is

$$\phi = 90° - h + \delta \tag{77}$$

The horizontal cross hair is sighted continuously on either the lower or the upper limb of the sun until the sun reaches its maximum altitude and begins its apparent descent. The true altitude of the sun's center is determined by applying to the observed altitude the corrections for index error, semidiameter (see ephemeris), and refraction and parallax. The latitude is then determined by the equation given above. It should be noted that the sign of the correction for refraction and parallax is always negative. The sign of the

LATITUDE OF TOWN HALL 14

(Observation on Sun at Apparent Noon)

Field Work

Circle	Obj.	Time	V.Circle	Index E.	
L	A		+2°58'30"		h_1
R	A		+2°55'30"		h_2
				+0°01'30"	
L	☉	$11^h34^m01^s$	47°51'00"		h'

Computations

Watch time	$11^h34^m01^s$
" slow	29^s
G.C.T. (E.S.T.+5^h)	$16^h34^m30^s$
Eq. of time (from Ephemeris)	+ 4^m37^s
G.A.T.	$16^h39^m07^s$
δ_0 (declin. at G.A. Noon)	+ 3°14'27"
$\Delta\delta$ (57.6"×4.66^h)	− 4'28"
δ	+ 3°09'59"
Obs. h' on ☉	47°51.0'
Index correction	− 01.5'
Ref. & parallax (Table I)	− 00.7'
Sun's semidiameter	+ 15.9'
h (corrected altitude)	48°04.7'
δ	3°10.0'
$\phi = 90° - h + \delta$	45°05.3'

Wm. Bolton, H. L. Brown — Observers
Sept. 15, 1951
Remarks — Fair, Warm, Calm
"A" is mark on barn 400 ft. south
See note — B.& B. Transit No. 142
By watch — Waltham Watch, 29^s slow E.S.Time
16.65^h — Note:- Index error found by reversal on point "A".
Index E = +(2°58'30") − (2°55'30") / 2
" " = +0°01'30" with circle left
Ephemeris gave values for G.A. Noon
Latitude of Town Hall

Fig. 58. Latitude by observation on sun at noon.

declination is negative from September to March and is positive for the remainder of the year.

The field notes and computations are made in a form similar to that shown in Fig. 58. For these observations the longitude of the place was unknown, and the standard time was recorded. Only the sun's lower limb was sighted. The index error was determined by double-sighting at a mark; the letters L and R in the column headed "Circle" indicate whether the vertical circle was left or right and therefore whether the telescope was normal or inverted. As the available ephemeris gave values of declination at Greenwich apparent noon, the watch time was converted into Greenwich apparent time. In the line beginning "$\Delta\delta$," the change in declination during the 4.66 hr that elapsed since Greenwich apparent noon was computed by multiplying the elapsed time by the variation per hour (57.6″) taken from the ephemeris.

Azimuth by Direct Solar Observation. The azimuth of a line can be determined by a single observation of the sun at any time when it is visible, provided the latitude of the place is known. With the engineer's transit, under ordinary conditions the azimuth of the line can be determined within about 01′.

At a known instant of time the sun is observed, and the altitude of the sun and the horizontal angle from the sun to a given reference line are measured. The declination

of the sun at the given instant is found from a solar ephemeris. With the declination δ, latitude ϕ, and altitude h known, the azimuth A of the sun from south is given by the equation

$$\cos A = \tan h \tan \phi - \frac{\sin \delta}{\cos h \cos \phi} \tag{78}$$

The azimuth of the sun and the horizontal angle from its position to the line being known, the azimuth of the line is readily computed.

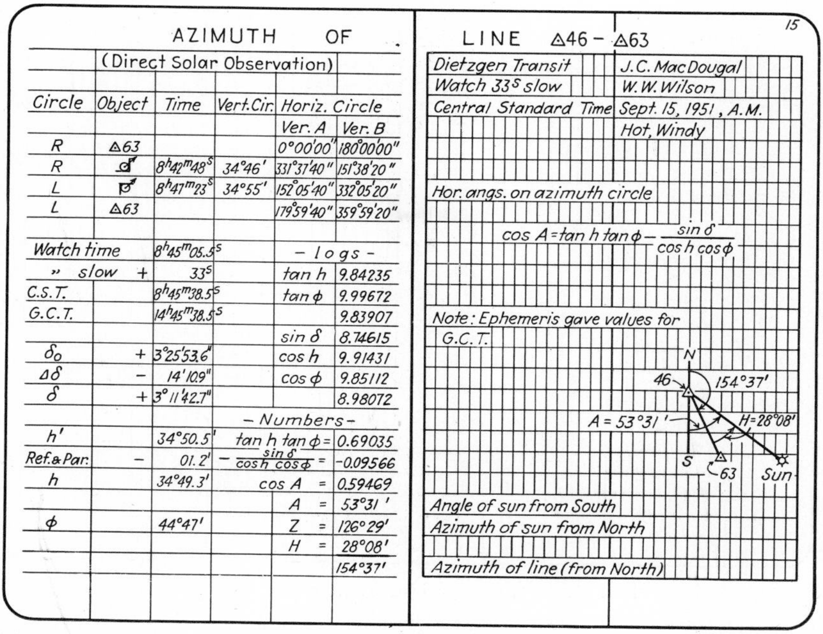

AZIMUTH OF LINE △46 – △63

(Direct Solar Observation)

Circle	Object	Time	Vert. Cir.	Horiz. Circle Ver. A	Ver. B
R	△63			0°00′00″	180°00′00″
R	☉	$8^h42^m48^s$	34°46′	331°37′40″	151°38′20″
L	☉	$8^h47^m23^s$	34°55′	152°05′40″	332°05′20″
L	△63			179°59′40″	359°59′20″
Watch time		$8^h45^m05.5^s$		– logs –	
″ slow	+	33^s		tan h	9.84235
C.S.T.		$8^h45^m38.5^s$		tan φ	9.99672
G.C.T.		$14^h45^m38.5^s$			9.83907
				sin δ	8.74615
δ_0	+	3°25′53.6″		cos h	9.91431
$\Delta\delta$	–	14′10.9″		cos φ	9.85112
δ	+	3°11′42.7″			8.98072
			– Numbers –		
h′		34°50.5′		tan h tan φ =	0.69035
Ref. & Par.	–	01.2′		$-\frac{\sin\delta}{\cos h \cos\phi}$ =	–0.09566
h		34°49.3′		cos A =	0.59469
				A =	53°31′
φ		44°47′		Z =	126°29′
				H =	28°08′
					154°37′

15

Dietzgen Transit — J.C. MacDougal
Watch 33^s slow — W. W. Wilson
Central Standard Time — Sept. 15, 1951, A.M.
Hot, Windy

Hor. angs. on azimuth circle

$\cos A = \tan h \tan \phi - \dfrac{\sin \delta}{\cos h \cos \phi}$

Note: Ephemeris gave values for G.C.T.

N
46
154°37′
A = 53°31′
H = 28°08′
S
63
Sun

Angle of sun from South
Azimuth of sun from North
Azimuth of line (from North)

FIG. 59. Azimuth by observation on sun.

The usual procedure is as follows: The transit is set up and very carefully leveled over one end of the line. The A vernier is set at zero on the horizontal circle, and the reading of the B vernier is noted. A sight is taken along the given line with the telescope in, say, the normal position, and the lower motion is clamped. The upper motion is loosened, and a sight is taken at the sun, the vertical and horizontal cross hairs being brought tangent in the upper left-hand quadrant of the field of view if the observation is in the morning or in the upper right-hand quadrant if the observation is in the afternoon. At the instant of tangency, the time is observed. As quickly as convenient, vertical and horizontal circle readings are taken on all verniers. The upper motion is then loosened, the telescope is plunged, and a second pointing is made, this time with the sun in the diagonally opposite quadrant from that of the first observation. The time of tangency is observed, and the vertical and horizontal angles are measured as before. The upper motion is loosened, and the field work is completed by again sighting along the line and reading the horizontal circle with the telescope still inverted. The mean of the two observed altitudes is taken as the apparent altitude of the sun's center at the mean of

the two observed times. Similarly, the mean of the two horizontal angles is taken as the horizontal angle to the sun's center at the mean of the observed times. To the mean of the observed altitudes is applied the correction for refraction and parallax. The sun's apparent declination is found from a solar ephemeris.

Figure 59 illustrates a suitable form for notes, the observations being taken with a transit having a vertical circle reading to 01′ and a horizontal circle reading to 20″. It is seen that horizontal angles are read from the azimuth circle. An ephemeris giving declinations for 0^h Greenwich civil time was used.

Polaris. Figure 60 shows the position of Polaris with respect to the pole. The seven most brilliant stars in the constellation of Ursa Major are known as the "Great Dipper," and the two stars forming the part of the bowl farthest from the handle are called the *pointers* because a line through these stars points very nearly to the celestial north pole. The constellation of Cassiopeia is on the same side of the pole as Polaris, so that, when Cassiopeia is above the pole, Polaris is near upper culmination, etc.

The "Ephemeris of the Sun, Polaris, and Other Selected Stars," published annually by the U.S. Bureau of Land Management, gives values of the declination of Polaris and the Greenwich mean time of culmination at the meridian of Greenwich.

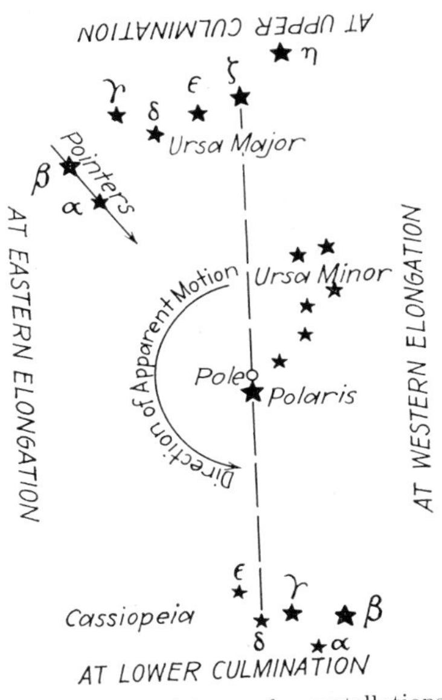

FIG. 60. Positions of constellations near the North Pole when Polaris is at culmination and elongation.

Latitude by Observation on Polaris. If h is the true altitude of any circumpolar star as it crosses the meridian, then the latitude ϕ is

$$\phi = h \pm p \tag{79}$$

in which the sign preceding p, the polar distance, is positive or negative according as the star is at lower or upper culmination. By this method, the latitude of a station is determined by measuring the altitude of Polaris when at either upper or lower culmination and by applying to this altitude, corrected for refraction, the star's polar distance as computed from the declination given in Polaris tables.

The usual procedure of determining latitude by observation on Polaris at culmination is as follows: The standard time of culmination at the given station is determined within perhaps 5 min by use of an ephemeris. A few minutes before the estimated time of culmination the transit is set up and is leveled very carefully. The telescope is focused for a star, and the estimated latitude of the place is laid off on the vertical circle to facilitate finding Polaris. When Polaris is sighted, the cross hairs are illuminated if necessary, and the star is continuously bisected with the horizontal cross hair. When during a period of 3 or 4 min Polaris no longer appears to move away from the hair but moves horizontally along it, the star is practically at culmination. The vertical angle is read with dispatch; the transit is carefully releveled; the telescope is plunged; and a second observation on the star is taken with the telescope inverted. Usually the instrument is releveled, and a second pair of observations is made. The mean of the altitudes, corrected for refraction and index error, is taken as the true altitude of the star. Finally the latitude is computed by applying to the true altitude the polar distance with proper sign. Under ordinary conditions the latitude can be determined within about 01′, or

less if the mean of several observations is taken. It is not essential that the altitude be observed at the exact instant the star crosses the meridian.

Observed altitudes of Polaris for positions other than culmination are reduced to the altitude at culmination by applying a correction, which, for latitudes within the United States, is given approximately by Table 7.

TABLE 7. CORRECTIONS TO BE APPLIED TO ALTITUDE OF POLARIS NEAR CULMINATION TO GIVE ALTITUDE AT CULMINATION

Interval from culmination, minutes of time	Change in altitude from culmination, seconds of arc	Interval from culmination, minutes of time	Change in altitude from culmination, seconds of arc
3	00	18	12
6	01	21	17
9	03	24	22
12	06	30	34
15	09		

Azimuth by Observation on Polaris at Elongation. The azimuth of a circumpolar star at elongation is determined from the equation

$$\sin Z = \frac{\sin p}{\cos \phi} \tag{80}$$

in which Z is the azimuth reckoned east or west of north according as the star is at eastern or western elongation, p is the star's polar distance, and ϕ is the latitude of the place. It is necessary that the latitude of the place and the polar distance of the star be known. In the field, the direction of Polaris from a given station is established by projecting a vertical plane from the star to the earth and then measuring the horizontal angle between a terrestrial line in this plane and the given station. The star's polar distance is found from an ephemeris. The latitude is determined by observation, as explained in a preceding paragraph. Inasmuch as the star is apparently traveling in a vertical line when at either elongation, it is not essential to the precision of azimuth determination that the time of elongation be found or that the sight to Polaris be taken at the exact instant of elongation; but the work of observing is facilitated if the approximate time is known.

The usual procedure is as follows: The standard time of elongation at the given station is determined within perhaps 5 min by use of an ephemeris. A few minutes before the estimated time of elongation the transit is set up over a given station and is very carefully leveled. The telescope is focused for a star, and the latitude of the place is laid off on the vertical circle to facilitate finding Polaris. When Polaris is sighted, the horizontal and vertical motions are clamped, the cross hairs are illuminated if necessary, and the star is continuously bisected with the vertical cross hair. When during a period of 2 or 3 min Polaris no longer appears to move away from the hair but moves vertically along it, the star is practically at elongation. The telescope is depressed, and a point on the line of sight is marked on a stake or other reference monument 300 ft or more away. The telescope is then plunged, and another sight is taken on Polaris. The line of sight is again depressed, and second point is set on the stake beside the first. Usually the transit is releveled, and a second pair of observations is made.

Later the mean of the points is found and marked on the stake, thus defining the direction from the observer's station to Polaris at elongation. The azimuth of Polaris at elongation either can be computed by Eq. (80) or can be found from an ephemeris. The azimuth of any other line through the station can be determined by measuring the angle

between the two lines, by the method of repetition if necessary to secure the required precision. A true meridian can be established by a perpendicular offset from the established point on the stake.

By the method just described, the azimuth of any line can be determined within perhaps 0.5′, or less if the mean of several observations is taken. If an error of 01′ is allowable, points need not be set beneath the star, and the angle may be measured directly.

LAND SURVEYING

Methods. Nearly all land surveys are run with the transit and tape, by methods already described. Ordinarily distances are measured to feet and decimals, and angles are measured to the nearest minute. On the United States public-land surveys, all distances are in Gunter's (66-ft) chains as prescribed by law. Usually the directions of lines are expressed as true bearings.

Formerly the surveyor's compass and 66-ft link chain were used extensively, and the descriptions in many old deeds are given in terms of magnetic bearings and Gunter's chains. In retracing old surveys of this character, allowance must be made for change in magnetic bearing since the time of the original survey. Also, it must be kept in mind that the compass and link chain used on old surveys were relatively inaccurate instruments and that great precision was not regarded as necessary, since generally the land values were low.

Wherever possible, the lengths of boundary lines and the angles between boundaries are determined by direct measurement. Therefore the land survey is in general a traverse, the transit stations being at corners of the property and the traverse lines coinciding with property lines. Where obstacles render direct measurement of boundaries impossible, a traverse is run as near the property lines as practicable, and measurements are made from the traverse to property corners; the lengths and directions of the property lines are then calculated. Higher precision is required in urban than in rural areas.

Corners; Monuments. It is customary to mark the corners of landed property by visible monuments. On early original surveys, many of the corners were marked by natural objects such as trees and large stones. In general, however, the corner monuments are established by the surveyor either to mark the intersections of boundaries already in existence or to define new boundaries. Examples of markers of a permanent character are an iron pipe or bar driven in the ground; a concrete or stone monument with drill hole, cross, or metal plug marking the exact corner; a stone with identifying mark, placed below the ground surface; charcoal placed below the ground surface; a mound of stones; and a metal marker set in concrete below the surface, reached through a covered shaft. The U.S. Bureau of Land Management has recently adopted as the standard for the monumenting of the public-land surveys a post made of iron pipe filled with concrete, the lower end of the pipe being split and spread to form a base, and the upper end being fitted with a brass cap with identifying marks.

Where a corner falls in such location as to make it impossible or impracticable to establish a monument in its true location, it is customary to establish a *witness corner* on one or more of the boundary lines leading to the corner, as near to the true corner as practicable.

In surveying along the shore of a lake or the bank of a stream a traverse called a *meander line* is run, roughly following the shore or bank line. Meander lines are not property lines except in the rare case where they are specifically stated as property lines in the deed. If more accurate location of the shore line is desired, offsets are taken from the meander line.

Boundary Records. Descriptions of the boundaries of real property may be found from deeds, official plats or maps, or notes of original surveys.

Records of the transfer of land from one owner to another are usually kept in the county registry of deeds, exact copies of all deeds of transfer being filed in deed books. These files are open to the public and are a frequent source of information for the land surveyor. An alphabetic index is kept, usually by years, giving in one part the names of *grantors*, or persons selling property, and in the other part the names of *grantees*, or persons buying property. Usually the preceding transfer of the same property is noted on the margin of the deed.

As the United States public lands are subdivided, official plats are prepared showing the dimensions of subdivisions and the character of monuments marking the corners. When all public lands within a state have been surveyed, all records are given to the state. An exception is Oklahoma, for which the United States survey records are filed with the Director of the Bureau of Land Management at Washington, D.C. States in possession of records have them on file at the state capitol. Information concerning these records can be secured from the secretary of state of the particular state. Photographic copies of the official plats are obtainable at nominal cost.

In most cases the deeds of transfer of city lots give only the lot and block number and the name of the addition or the subdivision. The official plat or map showing the dimensions of all lots and the character and location of permanent monuments is on file either in the office of the city clerk or in the county registry of deeds; copies are also on file in the offices of city and county assessors.

Some organizations, generally called *title companies*, for a fee will search the records for boundary descriptions.

Description of Rural Land. The following is illustrative of a description by metes and bounds typical of rural lands in the Eastern states and of isolated grants in the Western states where the subdivision of lands has been outside the rectangular system of the public-land surveys:

"Beginning at an iron pipe sunk in the ground thirty-three and no tenths (33.0) feet west of the center line of the concrete highway extending from Fryeberg to North Conway and twelve hundred and seventy-three and four-tenths (1273.4) feet south of the north face of the south abutment of the bridge over Jordon Brook in the town of Redstone, New Hampshire, at the southeast corner of land now or formerly belonging to William Bancroft, and running by fence and land of said Bancroft south 89°26′ west a distance of fourteen hundred and seventy-nine and four-tenths (1479.4) feet to a cross cut in the surface of granite ledge; . . . thence turning and running along the west boundary of said road north 2°15′ west a distance of nineteen hundred and eighty-nine and two-tenths (1989.2) feet to the point of beginning; all bearings being referred to the true meridian, and the parcel of land containing a calculated area of fifty-nine and sixty-seven hundredths (59.67) acres more or less."

Following is an example of the legal description of a 40-acre tract comprising a full quarter-quarter section:

"The north-east quarter of the south-west quarter of section ten (10), Township four (4) south, Range six (6) east, of the Initial Point of the Mount Diablo Meridian, containing forty (40) acres, more or less, according to the United States Survey."

In some states the locations of land corners are legally described by their coordinates with respect to the state-wide plane-coordinate system (p. 1–49). The Tennessee Valley Authority describes land by metes and bounds and by coordinates, with further reference to the United States public-land survey; a typical map of a tract is shown in Fig. 61.

Original Survey of Rural Land. With the desired boundaries of the land given, the surveyor establishes monuments at the corners and runs a closed traverse around the property, measuring the lengths of lines and the angles between intersecting lines. Where boundaries are not straight, offsets from transit line to curved boundary are measured at known intervals; and where obstructions make direct measurement along

boundaries impossible, the traverse is run as close to the boundary as convenience will allow, and measurements are taken from transit stations to corners of the tract, so that the length and bearing of the boundary line may be calculated. Angular measurements may be taken by any of the methods described on pp. 1–31 to 1–33, but most often the interior angles are observed. Preferably the corners should be referenced to permanent

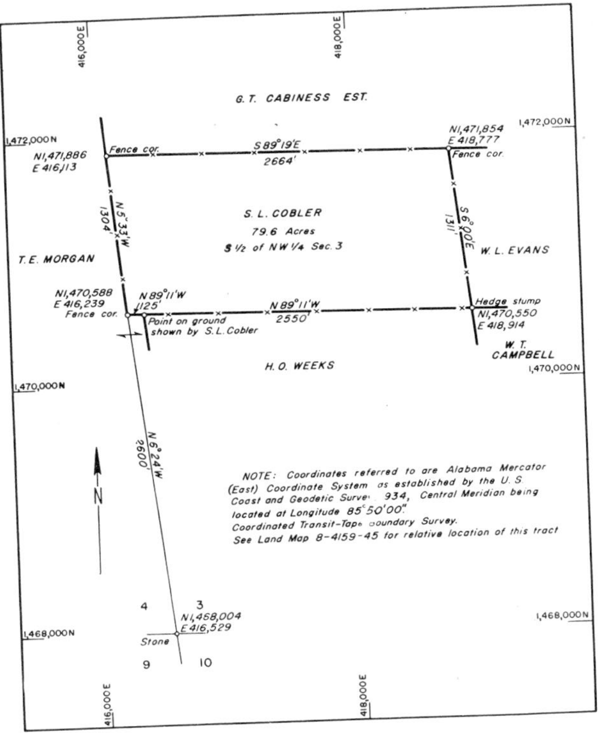

Fig. 61. Land map.

objects. The direction of the true meridian is determined, usually by a solar observation (see p. 1–73).

Resurvey of Rural Land. The purpose of a resurvey is to reestablish boundaries in their original positions. To guide him, the surveyor has available a description contained in the deed or obtained from old records, and descriptions of adjoining property.

As a first step, the surveyor critically examines the descriptions for gross errors; he then computes the latitudes and departures of the several courses as given in the description, determines the error of closure, and plots the boundaries of the tract to scale. If original bearings are magnetic, the magnetic declination at the time of the original survey is found, and true bearings are computed. If true bearings cannot be found in this manner and if one or more boundaries can be positively identified, observations are made to determine the true bearings of these known lines; and by a comparison of true and original magnetic bearings the declination at the time of the original survey is estimated, and the true bearings of the other lines are computed.

One or More Boundaries Evident. If one or more boundary lines can be identified from the monuments or from reliable reference marks, a comparison is obtained between the

length of the chain or tape used on the original survey and that to be used on the resurvey; the proportionate lengths of other sides of the tract are then computed. With the calculated directions and proportionate lengths the surveyor starts from a known corner and reruns the courses; at each estimated location of a corner, he seeks physical evidence of the location of the original corner. Thus, if a stake had originally been set at the corner, careful slicing of the topsoil with a shovel might reveal rotted wood, a hole in the ground, or even discolored earth, which might be considered rather positive evidence of the old location. If such evidence is found, and if the old monument is not in good condition, he sets a new monument. If the location of the corner does not agree with that in the description, he makes new measurements to the established monument.

At any point where physical evidence as to the original location of a corner is entirely lacking, the corner is located temporarily from the description. The survey is then continued until positive evidence of the location of a succeeding corner is found or until the traverse is brought to a closure at the initial point. If a succeeding corner is found, the traverse between the previously located line and that corner is adjusted to meet the known conditions.

If no physical evidence except one boundary is found, the survey is run to the point of beginning, and the linear error of closure is measured. The survey is then balanced, and the computed corrections are applied by moving the preceding temporary monuments and establishing them as permanent. Finally the lengths and bearings of the adjusted courses are measured in the field.

One Corner Evident. Where only a single corner can be found, the true bearings are determined either directly from the description or by computation from magnetic bearings given in the description. If the date of the survey from which the description by magnetic bearings is derived is unknown, it is estimated as closely as possible, and the corresponding magnetic declination is found for computation. By use of the true bearings and the lengths of lines given in the description, the latitudes and departures of the boundaries are computed, and the linear error of closure of the original survey is determined. If this error is reasonably small (say not greater than $\frac{1}{300}$ if the old survey was run with a compass), it is indicated that there are no mistakes in the lengths and bearings given in the description, although this check does not detect systematic errors (which may be large) in chainage. About the only course open to the surveyor is to establish the true meridian and to rerun the survey in accordance with the old description, distributing the error of closure proportionately among the several courses.

No Corner Evident. If a description is available but all evidence of the location of original corners is lost, the surveyor will find it expedient to search the records for descriptions of adjoining property and by means of these descriptions to reestablish by measurement as many corners of the tract in question as seems feasible. It is possible that these locations may be considerably in error. A corner may be reestablished by measurements from several sources, each resulting in a different location; in such cases the surveyor is called upon to exercise his judgment as to the most probable location.

Sometimes it is possible to determine the location of an obliterated corner through evidences of previously existing lines such as fences and roads. Occasionally the surveyor may find it desirable to consult old settlers who were familiar with the original boundaries; but although such persons are usually very positive in their opinions, the information is seldom of much value and is frequently misleading.

Having thus tentatively fixed the location of one or more corners, the surveyor attempts to reconcile these locations with the description of the given tract. Readjustments are made to conform to the judgment of the surveyor in the light of the information that he obtains as the survey progresses.

Report. When a resurvey has been completed, it is the duty of the surveyor to render a report to his client stating exactly what he found and what course of procedure he

toward other stations such as U, V, and the church spire. The table is then set up and oriented at stations B and C in succession, and the same objects are sighted again. The correct plotted location of each station may be determined by the intersection of two rays, as by those drawn from stations A and B; but the location is proved if the three

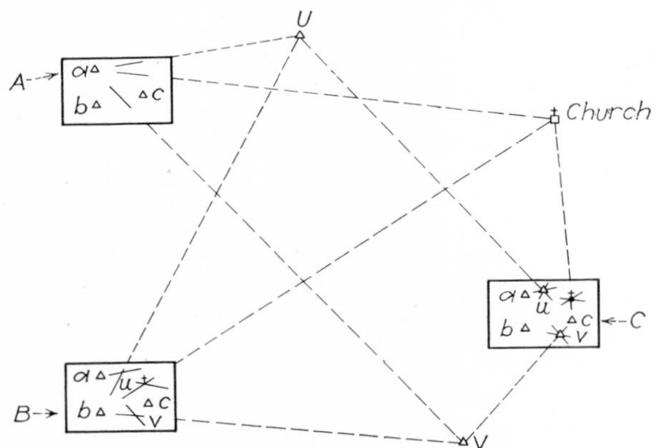

FIG. 36. Graphical triangulation.

rays drawn toward a given object are found to pass through a point, as shown by the third rays drawn from station C.

Resection. Resection is the process of determining the plotted location of a station occupied by the instrument, by means of sights taken toward known points the locations of which have been plotted. Resection enables the topographer to select advantageous plane-table stations which have not been plotted previously.

With the plane table oriented over the desired station of unknown location (on the map), two or more objects of known location are sighted; as each object is sighted, a line of indefinite length is drawn through the plotted location of that object on the map. The intersection of these lines marks the plotted location of the plane-table station.

The table may be oriented by any of the methods stated on p. 1–50. It is emphasized that, for the methods of orientation by magnetic compass, by backsighting, or by solar chart, resection can be accomplished only *after the board has been oriented.* If resection is by the three-point problem or the two-point problem, orientation and resection are accomplished in the same operation.

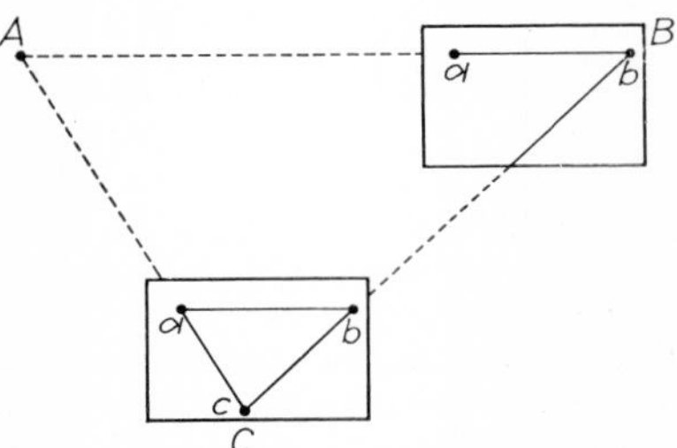

FIG. 37. Resection after orientation by backsighting.

Resection after Orientation by Compass. If the plane table has been oriented by means of the compass, the method of resection is as follows: Let P be the station of unknown location occupied by the plane table, and let A and B be two visible stations which have been plotted on the sheet at a and b. Then the plotted location of P is determined by drawing a line, or resecting, through a in the direction of A and resecting through b in the direction of B. The point p where the two (or more) lines cross, or resect, marks the plotted location of the instrument station P.

Resection after Orientation by Backsighting. If the table is oriented by backsighting along a line the direction of which has been plotted but the length of which may be unknown, the method of orientation and resection is as follows: Suppose that the topographer wishes to occupy station C (Fig. 37) the location of which has not been plotted but from which can be seen two points as A and B the locations of which have been plotted at a and b. He orients the board at one of the known stations as B, takes a foresight to C, and draws through b a ray of indefinite length. He then sets up the plane table at station C, orients it by backsighting to B, and resects from A through a. The intersection c of the ray from b and the resection line from a is the plotted location of the plane-table station C, since the triangles abc and ABC are similar. If the angle between the ray and the resection line is small, the location will be indefinite; for strong location the acute angle between these lines should be greater than 30°.

Resection and Orientation: Three-point Problem. Frequently the topographer wishes to occupy an advantageous station which has not been located on the map and toward which no ray from located stations has been drawn, and at the same time orientation by use of the compass is not sufficiently accurate. If three located stations are visible the three-point problem offers a convenient method of orienting and resecting in the same operation. There are several solutions of the three-point problem. In the United States, experienced topographers commonly employ a method of direct trial, guided by rules (see following paragraphs). The mechanical or tracing-cloth solution (p. 1–56) is simpler to understand but is not so satisfactory or so expeditious under the usual field conditions. For solution of the three-point problem by computation, employed in transit work, see p. 1–49.

Trial Method. The plane table is set up over the station of unknown location and is oriented approximately either by compass or by estimation. Resection lines from the

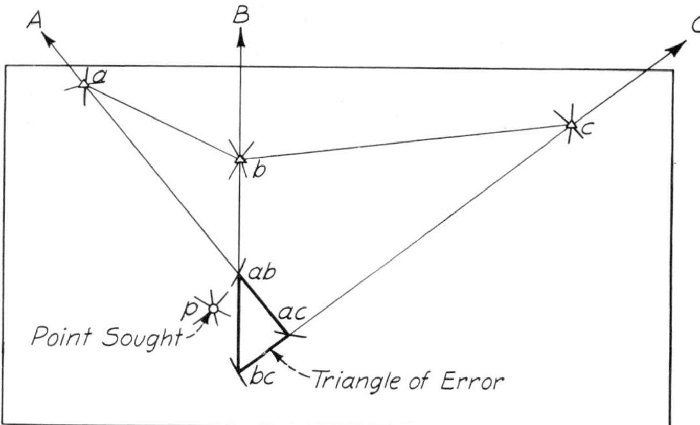

Fig. 38. Three-point problem with plane table.

three stations of known location are drawn through the corresponding plotted points. These lines will not intersect at a common point unless the trial orientation happens to be correct. (An exception to this statement is discussed later in this article.) Usually a small triangle called the *triangle of error* is formed by the three lines (Fig. 38). The correct plotted location of the plane-table station, called the *point sought*, is then estimated more closely. One method is to draw arcs of circles through the points shown (through a, b, and point ab; b, c, and bc; and a, c, and ac); the circles will intersect at p, the point sought. Usually, however, the correct location of the point sought is estimated more

conveniently by trial (Fig. 39), aided by Rules 1 and 2, given below. The board is then reoriented by backsighting through the estimated location of p toward one of the known stations (preferably the most distant); and the orientation is checked by resecting from the other two known stations. If the three lines still do not meet at a point, the process is repeated until they do; the orientation is then correct, and the common intersection of the three lines is the correct plotted location of the plane-table station.

Rule 1. *The point sought is on the same side of all resection lines.* That is, it lies either to the right of each line (as the observer faces the corresponding station) or to the left of each line.

Rule 2. *The distance from each resection line to the point sought is proportional to the length of that line.* By "length" is meant either the actual distance from plane-table station to known station or the corresponding plotted distance.

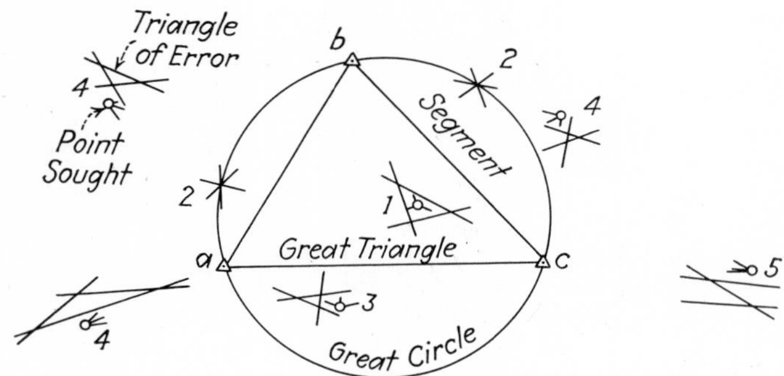

FIG. 39. Three-point problem, solution by trial.

Rules 1 and 2 are general and apply to any location of the plane-table station except on the *great circle* passing through the three known stations (Fig. 39). In this case, regardless of the orientation of the table the lines will meet in a common point (see points 2 in Fig. 39) which will not necessarily be the point sought. If it is suspected but not known that the plane-table station is on the great circle, either the great circle should be plotted on the plane-table sheet or the orientation of the board should be changed slightly and a second trial made. If it is found that the station is on the great circle, one (or more) of the three known stations must be replaced by a known station (or stations) suitably located.

Rules 1 and 2 are supplemented by the following auxiliary rules which apply to particular locations of the plane-table station:

a. If the new station is inside the great triangle, the point sought is within the triangle of error and is in the same position relative to the triangle of error that the triangle of error occupies in the great triangle (Fig. 39, point 1).

b. If the new station is in one of the three segments of the great circle formed by the sides of the great triangle, the point sought is on the *opposite* side of the resection line through the middle known point from the intersection of the other two lines (Fig. 39, point 3).

c. If the new station is outside the great circle, the point sought is always on the *same* side of the resection line from the most distant point as the point of intersection of the other two lines (Fig. 39, points 4).

d. If the new station is outside the great triangle, of the six sectors formed by the resection lines there are only two in which the point sought can be on the same side (right or left) of all lines.

e. If the new station is so located that the triangle of error is not formed within the limits of the plane-table sheet, that is, if two of the resection lines are almost parallel (Fig. 39, point 5), the foregoing rules still apply.

f. If the new station is on line between two of the known stations, the resection lines drawn from those two stations will be parallel. The foregoing rules still apply.

The strength of the determination varies with the location of the plane-table station, as described below. The strength of determination should be considered not only in selecting the most favorable of the available known stations but also in deciding whether the three-point problem can satisfactorily be used with the plane table at a given location.

When the new station is inside the great circle, the nearer the new station to the center of gravity of the great triangle the stronger the determination.

When the new station is on the great circle, its location is indeterminate.

When the new station is near the great circle, the determination is weak.

When the new station is outside the great circle, for given angles the nearer the new station to the middle known station the stronger the determination.

Either when one angle is small or when the new station is on line with two known stations, the larger the angle for the third known station (up to 90°) the stronger the determination. The two known stations near or on line should not be near each other.

Tracing-cloth Method. A piece of tracing cloth or tracing paper is fastened on the plane table over the map. Any convenient point on the tracing cloth is chosen to represent the unknown station over which the plane table is set, and from it rays are drawn toward the three known stations or objects. Then the cloth is loosened and is shifted over the map until the three rays pass through the corresponding plotted points. The intersection of the rays marks the plotted location of the plane-table station. It is pricked through onto the map, and the table is oriented by backsighting on one of the known stations (preferably the most distant).

Resection and Orientation: Two-point Problem. The purpose of the two-point problem is to orient the table and to locate the station occupied when only two stations are visible and when it is impossible or undesirable to occupy either of them. To accomplish this, it is necessary that two setups be made, the first at a convenient distance from the station to be occupied, and the second at that station. Owing to the length of the procedure, this method is not practical except where other methods cannot be used.

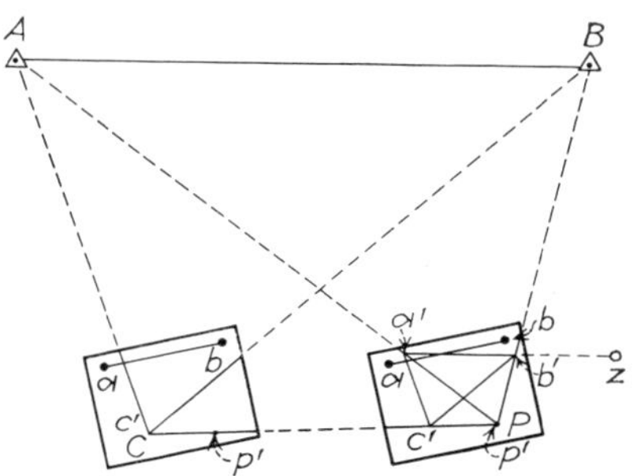

Fig. 40. Two-point problem, graphical solution.

One graphical solution of the two-point problem is as follows: The locations of the two stations A and B are plotted on the plane-table sheet at a and b (Fig. 40). The plane table is set up over some ground point C from which can be seen A, B, and the point P whose plotted location is desired. The board is oriented as nearly as possible either by compass or by estimation. A point c' corresponding to C on the ground is plotted by estimation on the plane-table sheet (Fig. 40, left). Foresights are taken from c' on A, B, and P. The distance from C to P is estimated, and the corresponding estimated location of P is plotted at p'. The table is taken to station P and is oriented (tentatively) by a backsight on C. Foresights are taken from p' on A and B; the intersections of these rays with the corresponding rays from c' are a' and b' (Fig. 40, right). The line $a'b'$ is parallel to a line between the

corresponding actual points A and B. With the straightedge of the alidade along the line $a'b'$, a point Z at some distance from the table is set (or selected) on the line of sight. The alidade is moved to the line ab, and the board is turned until the same point Z is sighted. The plane table is now oriented correctly, and by resection through a and b the correct location of the plane-table station is plotted at p.

Details with Plane Table. The operation of plotting details by radiation with the plane table is commonly known as taking *side shots*. Table 5 gives the sequence of

TABLE 5. SEQUENCE OF OPERATIONS FOR SIDE SHOT WITH PLANE TABLE
Alidade equipped with stadia arc and fixed vertical vernier

Rodman	Topographer	Computer	Recorder
Sets rod on ground point	Sights on rod, with lower cross hair on a foot mark		
	Reads and calls stadia interval	Sets index of slide rule on stadia distance	Records stadia interval (or distance)
	Draws ray		
	Levels telescope, and sets stadia arc to zero (or 50)[1]		
	Sights on rod, with stadia index set on some mark of V arc		
	Calls V reading, H reading, and reading of middle cross hair of rod	....	Records V, H, and rod reading
	Waves rodman ahead	Computes and calls horizontal distance	Records horizontal distance
Starts to new ground point	Plots point	Computes and calls vertical distance	Records vertical distance
		Computes and calls elevation	Records elevation
	Records elevation on sheet		
	Interpolates and sketches contours and features		

[1] If the alidade is equipped with a vertical vernier control level, the topographer centers the control-level bubble instead.

operations for taking a side shot when the alidade is equipped with a stadia arc but not with a vertical control level. It is assumed that the party consists of a plane-table man (called a *topographer*), one or more rodmen, and a computer. In actual practice usually no record is kept of the side shots. In this article, however, it is assumed that a recorder is employed for purposes of training, checking, or record; and his operations are also listed. Each line of the table reads from left to right, and the operations in each line follow those of the preceding line. It is evident that the progress of the party depends

largely upon systematic operation and upon the dispatch with which the topographer performs his duties.

If the alidade is equipped with a vertical vernier control level, it is not necessary for the topographer to level the telescope; instead he centers the control-level bubble. If the alidade is not equipped with a stadia arc or vertical control level, he levels the telescope, reads the index error, and then reads the vertical angle; the details of procedure are modified accordingly.

Adjustment of the Plane-table Alidade. 1. *To Make the Axis of Each Plate Level Parallel to the Plate.* Center the bubble (or bubbles) of the plate level (or levels) by manipulating the board. On the plane-table sheet mark a guide line along one edge of the straightedge. Turn the alidade end for end, and again place the straightedge along the guide line. If the bubble is off center, bring it back *halfway* by means of the adjusting screws. Again center the bubble by manipulating the board, and repeat the test.

2. *To Make the Vertical Cross Hair Lie in a Plane Perpendicular to the Horizontal Axis.* Sight the vertical cross hair on a well-defined point not less than 200 ft away, and swing the telescope through a small vertical angle. If the point appears to depart from the vertical cross hair, loosen two adjacent screws of the cross-hair ring, and rotate the ring in the telescope tube until by further trial the point sighted traverses the entire length of the hair. Tighten the same two screws.

3. (For Alidade of Tube-in-sleeve Type) *To Make the Line of Sight Coincide with the Axis of the Telescope Sleeve.* Sight the intersection of the cross hairs on some well-defined point. Rotate the telescope in the sleeve through 180° (usually the limits of rotation are fixed by a shoulder and a lug). If the cross hairs have apparently moved away from the point, bring each hair halfway back to its original position by means of the capstan screws holding the cross-hair ring. The adjustment is made by manipulating opposite screws, bringing first one cross hair and then the other to its estimated correct position. Again sight on the point, and repeat the test.

4. (For Alidade of Tube-in-sleeve Type) *To Make the Axis of the Striding Level Parallel to the Axis of the Telescope Sleeve (and Hence Parallel to the Line of Sight).* Place the striding level on the telescope, and center the bubble. Remove the level; turn it end for end; and replace it on the telescope tube. If the bubble is off center, bring it back halfway by means of the adjusting screw at one end of the level tube. Again center the bubble (by means of the tangent screw), and repeat the test.

4*a*. (For Alidade of Fixed-tube Type) *To Make the Axis of the Telescope Level Parallel to the Line of Sight.* This adjustment is the same as the two-peg adjustment of the dumpy level, except as follows: With the line of sight set on the rod reading established for a horizontal line, the correction is made by raising or lowering one end of the telescope level tube (by means of the capstan screws) until the bubble is centered.

5. (For Alidade Having a Fixed Vertical Vernier) *To Make the Vertical Vernier Read Zero When the Line of Sight Is Horizontal.* With the board level, center the bubble of the telescope level. If the vertical vernier does not read zero, loosen it and move it until it reads zero.

5*a*. (For Alidade Having a Movable Vertical Vernier with Control Level) *To Make the Axis of the Vernier Control Level Parallel to the Axis of the Telescope Level When the Vertical Vernier Reads Zero.* Center the bubble of the telescope level, and move the vernier by means of its tangent screw until it reads zero. If the control-level bubble is off center, bring it to center by means of the capstan screws at the end of the control-level tube.

5*b*. (For Alidade Having Tangent Movement to Vertical Vernier Arm) This type of vernier needs no adjustment, because for each direction of sighting the vernier is set at the index (by means of the tangent screw) when the telescope has been leveled.

MAP DRAFTING

Drawings. The drawings of surveying consist of maps, profiles, and cross sections; their usefulness depends largely upon the accuracy with which points and lines are projected on paper. Few dimensions are shown. Distances are measured with the engineer's scale, and angles with a protractor. Points are plotted within $\frac{1}{50}$ to $\frac{1}{100}$ in. by means of a well-sharpened hard pencil; with a fine needle and reading glass and using great care, points can be plotted within $\frac{1}{200}$ in. The most common forms of lettering are Reinhardt slope letters (Fig. 41) and Reinhardt vertical letters (Fig. 42).

ABCDEFGHIJ
KLMNOPQ
RSTUVWXYZ
abcdefghijklmn
opqrstuvwxyz
1234567890
NORMAL
Hickory Tree 10 ft.
COMPRESSED
WASHINGTON Sta. 71+43.8
EXTENDED
NEVADA

FIG. 41. Reinhardt letters, slope form.

ABCDEFGHI
JKLMNOPQR
STUVWXYZ
abcdefghijklmn
opqrstuvwxyz
1234567890
NORMAL
Excavation 23 cu. yd.
COMPRESSED
RICHARDSON ESTATE 300 Ac.
EXTENDED
RED RIVER

FIG. 42. Reinhardt letters, vertical form.

Maps. In general the information that should appear on a map that is to become a part of public records of land division should include the following:

1. The length of each line.
2. The bearing of each line or the angle between intersecting lines.
3. The location of the tract with reference to established coordinate axes.
4. The number of each formal subdivision, such as a section, block, or lot.
5. The location and kind of each monument set, with distances to reference marks.
6. The location and name of each road, stream, landmark, etc.
7. The names of all property owners, including owners of property adjacent to the tract mapped.
8. The direction of the meridian (true or magnetic or both).
9. A legend or key to symbols shown on the map.
10. A graphical scale with a corresponding note stating the scale at which the map was drawn.
11. A full and continuous description of the boundaries of the tract by bearing and length of sides; and the area of the tract.
12. The witnessed signatures of those possessing title to the tract mapped; and, if the tract is to be an addition to a town or city, a dedication of all streets and alleys to the use of the public.
13. A certification by the surveyor that the plat is correct to the best of his knowledge.
14. A neat and explicit title showing the name of the tract, or its owner's name, its location, the scale of the drawing (unless this is shown elsewhere), the surveyor's name, the draftsman's name, and the date.

On maps that represent natural and artificial features, and on topographic maps, the following should always appear:

1. The direction of the meridian.

2. A legend or key to symbols used, if they are other than the common conventional signs.

3. A graphical scale of the map with a corresponding note stating the scale at which the map was drawn.

4. A neat and appropriate title generally stating the kind or purpose of the map, the name of the tract mapped or the name of the project for which the map is to be used, the location of the tract, the scale of the drawing (unless this is shown elsewhere), the contour interval, the name of the engineer or draftsman or both, and the date.

5. On topographic maps, a statement of the contour interval.

Conventional Signs. Objects are represented on a map by signs or symbols, many of which are conventional. For many purposes, it would lessen the usefulness of the map if all the objects were shown for which conventional signs are available. On the other hand, maps made for special purposes often show features not included among the common conventional signs. A chart of more than 300 standard symbols adopted by the U.S. Board of Surveys and Maps is published and is for sale by the U.S. Geological Survey, Washington, D.C. The size of the symbols should be proportioned somewhat to the scale of the map.

Scales. A map scale may be stated either numerically or graphically, as follows:

1. One inch on the drawing represents some whole number of feet on the ground, as 1 in. = 200 ft.

2. Some whole number of inches on the drawing represents 1 mile on the ground, as 6 in. = 1 mile.

3. One unit of length on the drawing represents a stated number of the *same* units of length on the ground, as 1/62,500. This ratio is called the *representative fraction.* The scale is independent of the units of measurement.

4. A line on the map is subdivided into map distances corresponding to convenient units of length on the ground, forming a *graphical scale.* If the map shrinks or swells, or is reproduced in another size, the graphical scale remains in correct proportion. The units of measurement should always be stated; but if the map is to be reproduced in another size it should be made clear that the numerical scale is that at which the map was drawn; for example, "Original scale 1 in. = 200 ft."

MAP PLOTTING

Plotting Control by Protractor. Where a traverse is to be plotted by protractor, the position of the first line is fixed by estimation, and its length (as AB) is laid off by measurement. The protractor is oriented at the forward point (as B), the deflection angle to the succeeding line is laid off, and a light line of indefinite length is drawn. Along this line is laid off the given distance (as BC) to the succeeding traverse point (as C); and so on.

When the directions of lines of a traverse are given either as azimuths or as bearings, a meridian line is drawn through each station, and the direction of the succeeding line is laid off with respect to the meridian.

Plotting Control by Tangent Offsets. In principle, this method is similar to the protractor method, with the difference that an angle is laid off by linear measurement which is a constant times the natural tangent of the angle. In Fig. 43, AB represents the plotted position of the initial line of a traverse, and α is the deflection angle at B which it is desired to lay off in order to determine the direction of BC. By the tangent method, the line last established (in this case AB) is prolonged some convenient distance, usually 10 in., to form a base line Bb, at the end of which a perpendicular bb' is erected; and

the distance bb' is laid off equal to the length of the base line Bb multiplied by the natural tangent of α. A line drawn from B through b' defines the direction of BC.

If the deflection angle is greater than 45°, usually the base line is established as a perpendicular to the preceding line at the point last plotted, and the cotangent of α is used instead of the tangent.

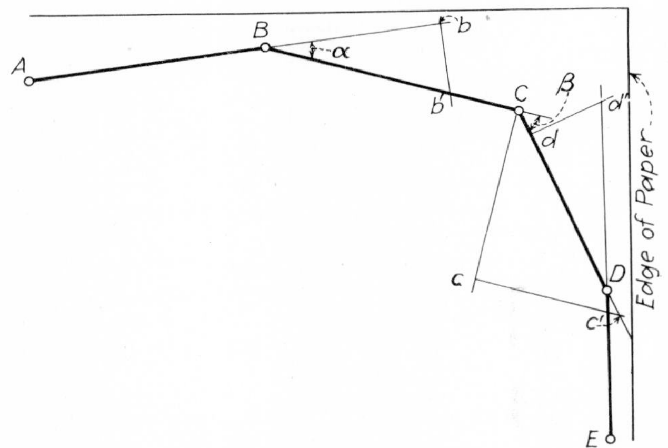

FIG. 43. Plotting by tangent offsets.

Plotting Control by Rectangular Coordinates. This method, also known as the "method of total latitudes and departures," is the only practical one for plotting extensive systems of horizontal control. It is considerably more accurate than the methods just described for plotting traverses. Rectangular coordinates are also employed for calculating areas (see p. 1–66).

The coordinate axes are a *reference meridian* (true, magnetic, or assumed) and a line at right angles thereto called a *reference parallel.* The intersection of these lines, marking the origin, may be any point in the survey or may be a point entirely outside the survey. With the direction of each survey line determined and its length known, computations are made to determine its *latitude* and *departure* or the lengths of its orthographic projection upon the meridian and upon the parallel respectively. The origin having been chosen, the coordinates for the several control points are computed by using the latitudes and departures. The coordinate of a point measured normal to the parallel is called the *total latitude* of the point; the coordinate measured normal to the meridian is called the *total departure.* With the coordinate axes established on paper, a point is plotted by laying off its total latitude and its total departure to the required scale. The following paragraphs are devoted to a more detailed discussion of the processes involved.

Latitudes and Departures. In Fig. 44, AB represents any line the latitude and departure of which it is desired to determine, and the lines NS and EW represent any meridian and any parallel. The line AB makes the angle α with the meridian.

As the latitude of a line is the orthographic projection of the line upon a meridian, the latitude of AB is $A'B' = Ab = AB \cos \alpha$. And as the departure of a line is its orthographic projection upon a parallel, the departure of AB is $A''B'' = Aa = AB \sin \alpha$. Stated in the form of a general rule applicable to any line,

$$\text{Latitude} = \text{length} \times \text{cosine bearing angle} \tag{53}$$

$$\text{Departure} = \text{length} \times \text{sine bearing angle} \tag{54}$$

Latitudes are designated as *North* or *positive* for all lines having a northerly bearing and *South* or *negative* for all lines having a southerly bearing. Departures are designated as *East* or *positive* for lines having an easterly bearing and *West* or *negative* for lines having a westerly bearing. Thus, in Fig. 45, for the line AB the latitude is North or +, and the departure is East or +. For BC the latitude is South or −, and the departure is East or +. For CD the latitude is South or −, and the departure is West or −. For DE the latitude is North or +, and the departure is West or −.

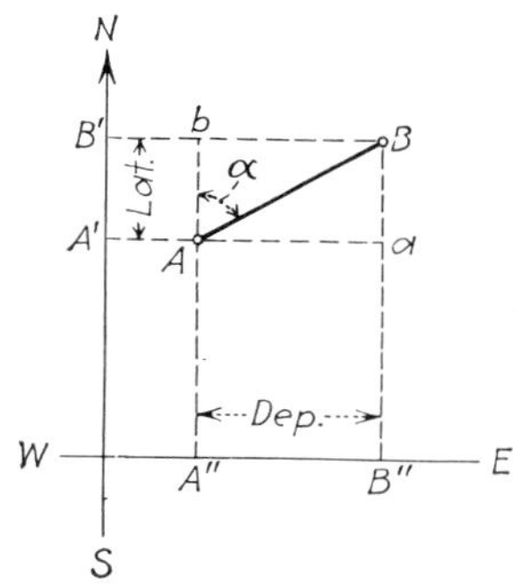

Fig. 44. Latitude and departure of line AB.

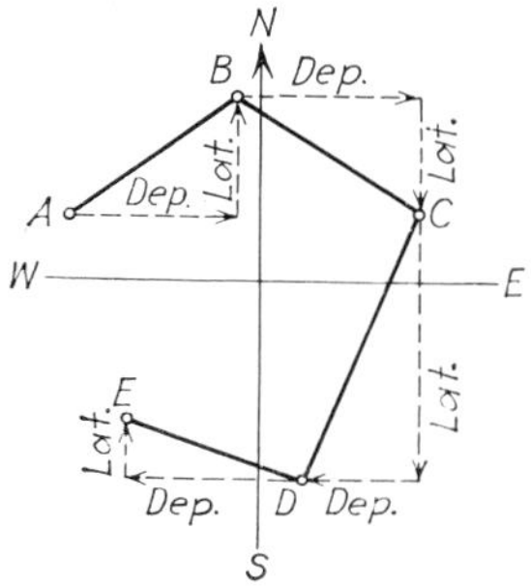

Fig. 45. Directions of latitudes and departures.

Adjustment of Angular Error. Before the latitudes and departures of the lines of a closed traverse are computed, the angular error of closure is determined by the known geometrical conditions. The angles or bearings are then so adjusted or corrected that the sum of the interior angles is $(n - 2)180°$, in which n is the number of sides. Often this adjustment is arbitrary, based on a knowledge of the field conditions; but if all angles have been measured under like conditions, the error is distributed equally to each angle in the traverse. The angular adjustment does not yield true values, but it yields the most probable values that can be assigned.

Error of Closure. For any closed traverse, the algebraic sum of the latitudes ΣL should be equal to zero, and the algebraic sum of the departures ΣD should be equal to zero. However, owing to errors in field measurements of both angles and distances, in general an unadjusted traverse will not close on paper even though the plotting is without error. The conditions stated above make it possible to determine the linear error of closure e by means of the computed latitudes and departures, as follows:

$$e = \sqrt{(\Sigma L)^2 + (\Sigma D)^2} \tag{55}$$

The direction of the side of error is given by the relation $\tan e = (-\Sigma D)/(-\Sigma L)$, with due regard to sign and the quadrant in which the bearing lies.

Balancing the Survey. When the linear error of closure has been determined as described in the preceding article, usually corrections are made so that the traverse will form a mathematically closed figure, and the corrections are applied to the latitudes and departures in such manner as to make their algebraic sum equal zero. This operation is called *balancing the survey* or *balancing the traverse.* In many cases a careful arbitrary distribution of corrections based on knowledge of field conditions is the best that can be made.

Rules for Balancing the Survey. The *compass rule* states that the correction to be applied to the $\left\{\begin{matrix}\text{latitude}\\\text{departure}\end{matrix}\right\}$ of any course is to the total correction in $\left\{\begin{matrix}\text{latitude}\\\text{departure}\end{matrix}\right\}$ as

employed in attempting to reestablish missing corners. The report should be accompanied by a plat similar to that of an original survey. In addition, it should indicate which are original monuments and which are monuments established at the time of the resurvey. Mistakes in the original description should be pointed out, but the surveyor should clearly understand that it is his function to reestablish boundaries of a given tract in as nearly as possible their original location.

Description of Urban Land. The manner of legally describing the boundaries of a tract of land within the corporate limits of a city depends on conditions attached to the survey by which the boundaries of the tract were first established, as indicated by the following classification:

1. *By Lot and Block.* If the boundaries of the tract coincide exactly with a lot which is a part of a subdivision or addition for which there is recorded an official map, the tract may be legally described by giving the lot and block numbers and the name and date of filing of the official map, as illustrated by the following:

"Lot 15 in Block 5 as said lots and block are delineated and so designated upon that certain map entitled *Map of Thousand Oaks, Alameda County, California,* filed August 23, 1909, in Liber 25 of Maps, page 2, in the office of the County Recorder of the said County of Alameda."

2. *By Metes, Bounds, and Lots.* If the boundaries of a given tract within a subdivision for which there is a recorded map do not conform exactly to boundaries shown on the official map, the tract is described by metes and bounds, with the point of beginning referred to a corner shown on the official map. Also, the numbers of the lots of which the tract is composed are given. Following is an example of a description of this kind:

"Beginning at the intersection of the Northern line of Escondido Avenue, with the Eastern boundary line of Lot 16, hereinafter referred to; running thence Northerly along said Eastern boundary line of Lot 16, and the Eastern boundary line of Lot 17, eighty-nine (89) feet; thence at right angles Westerly, fifty-one (51) feet; thence South 12°06′ East, seventy-five (75) feet to the Northern line of Escondido Avenue; thence Easterly along said line of Escondido Avenue, fifty-three and 13/100 (53.13) feet, more or less, to the point of beginning.

"Being a portion of Lots 16 and 17, in Block 5, as said lots and block are delineated and so designated upon that certain map entitled *Map of Thousand Oaks, Alameda County, California,* filed August 23, 1909, in Liber 25 of Maps, page 2, in the office of the County Recorder of the said County of Alameda."

3. *By Metes and Bounds to City Monuments.* A number of the cities of the United States have, by precise surveys, established an elaborate system of reference monuments and have determined their coordinates with respect to an arbitrarily assumed initial point. State-wide systems of plane coordinates have been established in various states. When the tract cannot be defined by descriptions such as the preceding, the point of beginning may be definitely fixed by stating its direction and distance from an official reference monument and by describing the monument that marks the corner. The boundaries of the tract may then be described by metes and bounds.

The location of corners may also be defined by rectangular coordinates referred to the origin or initial point of the system.

If the tract is within a city not so monumented, the point of beginning of the boundary description may be referred by direction and distance to the intersection of the center lines of streets. It is not good practice to refer to the intersection of sidewalk or curb lines, for these are apt to be changed from time to time. In sections of the country within the rectangular system of United States surveys, the point of beginning of a boundary description may properly be referred to section lines and corners.

Subdivision Survey of Urban Land. For large and important developments the work of originating the general plan is often carried out by persons specializing in city

planning and landscape architecture, under whose direction the surveyor works. For the ordinary real-estate development the owner usually calls for the services of an engineer or surveyor who has had experience in such work. The surveyor confers with the owner, and they discuss a general plan. The surveyor makes a resurvey of the entire property; and if the character of the topography is irregular, he usually makes certain preliminary surveys for the purposes of finding the location and elevation of the governing features of the terrain. In some cases a complete topographic survey may be made. With the general plan fixed, and having studied the results of the field investigation, the surveyor works out a detailed plan on paper, showing on the drawing the names of all streets and the numbers of all blocks and lots, the dimensions of all lots, the width of streets, the length and bearing of all street tangents, and the radius and length of all street curves. He also prepares a report which, in addition to a discussion of the plan of subdivision, may consider the cost of subdividing, including not only the establishing of boundaries but also the work of grading, paving, constructing sewers, and landscaping. This detailed plan, when approved by the owner, is submitted to the governing body in the municipality. If it meets with the requirements of this body, it is approved.

Upon the authority of the owner, the surveyor then proceeds to execute the necessary subdivision surveys, including the laying out of roads, walks, blocks, and lots. When the surveys are completed, the map of the subdivision is revised to show minor changes made during the survey, together with the location and character of permanent monuments. A tracing is submitted to the municipality, and this, when duly signed by those in authority, becomes the official map of the subdivision. It then becomes a part of the public records and is usually filed in the registry of deeds of the county in which the municipality lies. Upon this approval, if the subdivision is outside the corporate limits of the municipality, they are extended to include it.

Legal Interpretation of Deed Description. A deed description may contain errors or mistakes of measurement, or mistakes of calculation or record, thus introducing inconsistencies which cannot be reconciled completely when retracement becomes necessary. In such cases, where uncertainty has arisen as to the location of property lines, it is a universal principle of law that the endeavor is to make the deed effectual rather than void and to execute the intentions of the contracting parties. The following general rules are pursuant to this principle.

1. *Monuments.* It is presumed that visible objects, which marked the corners when a conveyance of ownership was made, indicated best the intentions of the parties concerned; hence it is agreed that a corner is established by an existing material object or or by conclusive evidence as to the previous location of the object. A corner thus established will prevail against all other conflicting evidence.

2. *Distance and Direction vs. Area.* In the case of discord between the described courses and the calculated area of a tract, the deed-description requirements, or "calls," for distances or directions of courses will prevail against the call for area—again on the assumption that the boundary lines are more visible and actual evidence of the intentions of the parties than is the calculated area of the tract.

3. *Mistakes.* It is a well-established principle that a deed description which taken as a whole plainly indicates the intentions of the parties concerned will not be invalidated by evident mistakes or omissions. For example, such obvious mistakes as the omission of a full tape length in a dimension or the transposition of the words "northeast" for "northwest" will have no effect on the validity of a description, provided it is otherwise complete and consistent or provided its intention is manifest.

4. *Purchaser Favored.* In the case of a description that is capable of two or more interpretations, that one will prevail which favors the purchaser.

5. *Ownership of Highways.* Land described as being bounded by a highway or street conveys ownership to the center of the highway or street. Any variation from this interpretation must be explicitly stated in the description.

6. *Original Government Surveys Presumed Correct.* Errors found in original surveys do not affect the location of the boundaries established under those surveys, and the boundaries remain fixed as originally established.

Legal Authority and Liability of the Surveyor. A resurvey may be run to settle a controversy between owners of adjoining property. The surveyor should understand that, although he may act as an arbiter in such cases, it is not within his power legally to fix boundaries without the mutual consent and authority of all interested parties. A competent surveyor by wise counsel can usually prevent litigation; but if he cannot bring his clients to an agreement, the boundaries in dispute become valid and defined only by a decision of the court. In boundary disputes the surveyor is an expert witness, not a judge.

The right to enter upon property for the purpose of making public surveys is generally provided by law, but there is no similar provision regarding private surveys. The surveyor (or his employer, whether public or private) is liable for damage caused by cutting trees, destroying crops or fences, etc.

It has been held in court decisions that county surveyors and surveyors in private practice are members of a learned profession and may be held liable for incompetent services rendered. If the surveyor knows the purpose for which the survey is made, he is liable for damages resulting from an erroneous survey. The general principle is that the surveyor is bound to exhibit that degree of prudence, judgment, and skill which may reasonably be expected of a member of his profession. Ruling Case Law says, " . . . a person undertaking to make a survey does not insure the correctness of his work, nor is absolute correctness the test of the amount of skill the law requires. Reasonable care, honesty, and a reasonable amount of skill are all he is bound to bring to the discharge of his duties."

Laws Relating to United States Public-land Surveys. Following are the provisions of the public-land laws in which the surveyor is principally interested:

1. All responsibility for the surveying and sale of the public lands of the United States is placed in the hands of the Director of the Bureau of Land Management, who is authorized to carry into execution every part of the public-land laws not otherwise specially provided for.

2. When the surveys and records of a state are completed, all the field notes, maps, and records pertaining to land titles are delivered to the secretary of state of that state.

3. Any agent of the United States, acting on the authority of the Director of the Bureau of Land Management, has free access to public records delivered to any state, but no transfer of such records is made to any state until the state has enacted legislation providing for the safekeeping of such records and for the allowance of free access thereto by authorities of the United States.

4. It is required that all surveys and resurveys of public lands under the supervision of the Director of the Bureau of Land Management are to be made by surveyors selected by the Bureau of Land Management. (Prior to 1910, surveys were made by contract.) The field work is now performed by a permanent corps of engineers under civil-service regulations.

5. It is provided that resurveys may be made by the government under certain conditions.

6. Boundaries of public lands, when established by duly authorized surveyors and when approved by the director, are unchangeable.

7. The original corners established by the surveyors stand as the true corners they were intended to represent, whether in the place shown by the field notes or not.

8. The unit of length is the 66-ft or Gunter's chain divided into 100 links.

9. The surveyors are required to place appropriate monuments at all established corners, to cause to be marked on a tree near each corner within each section where possible to do so the number of township and section, to record in their field books the lengths and directions of all lines and the character and markings of all monuments and objects to which the monuments are referred, and to note the situations of water courses, springs, mineral deposits, etc., which come to their attention, as well as the quality of the lands traversed.

10. Quarter-quarter-section corners not established by the original surveys are to be on the line joining the section and quarter-section corners and midway between them, except in the northern and western half miles of the township.

11. The center lines of sections are to be straight between opposite quarter-section corners.

12. In a fractional section where no opposite quarter-section corner has been or can be established, the center line of such section is to be run from the quarter-section corner as nearly in a cardinal direction as due parallelism with section lines will permit to the meander line, reservation, or other boundary of such fractional section.

13. Lost or obliterated corners of the approved surveys are to be restored to their original location, if possible.

Scheme of United States Public-land Subdivision. Since the time of the earliest surveys, townships and sections have been located with respect to principal axes passing

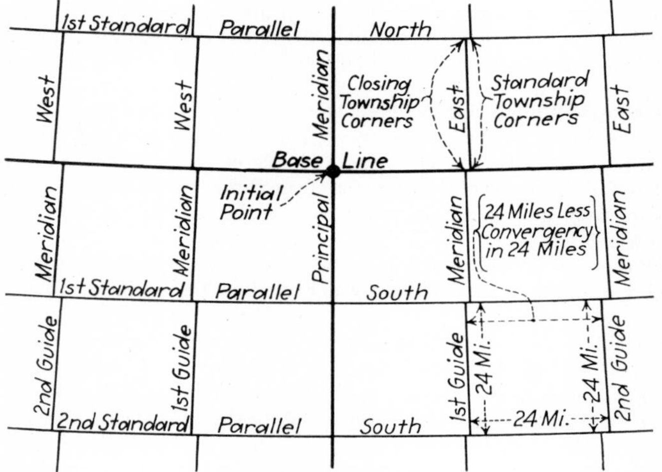

FIG. 62. Standard lines.

through an origin called an *initial point;* the north-south axis is a true meridian called the *principal meridian*, and the east-west axis is a true parallel of latitude called the *base line*. The principal meridian is given a name to which all subdivisions are referred. The extent of the surveys which are referred to a given initial point is shown on a map published by the Bureau of Land Management.

Secondary axes are established at intervals of 24 miles east or west of the principal meridian and at intervals of 24 miles north or south of the base line, thus dividing the

tract being surveyed into quadrangles bounded by true meridians 24 miles long and by true parallels, the south boundary of each quadrangle being 24 miles long, and the north boundary being 24 miles long less the convergency of the meridians in that distance. The secondary parallels are called *standard parallels* or *correction lines*, and each is continuous throughout its length. The secondary meridians are called *guide meridians*, and each is broken at the base line and at each standard parallel.

A typical system of principal and secondary axes is shown in Fig. 62. The base line and standard parallels, being everywhere perpendicular to the direction of the meridian, are laid out on the ground as curved lines, the rate of curvature depending on the latitude. The principal meridian and guide meridians, being true north-south lines, are laid out as straight lines but converge toward the north, the rate of convergency depending on the latitude.

Standard parallels are counted north or south of the base line; thus the *second standard parallel south* indicates a parallel 48 miles south of the base line. Guide meridians are counted east or west of the principal meridian; thus the *third guide meridian west* is 72 miles west of the principal meridian.

The principal meridian, base line, standard parallels, and guide meridians are called *standard lines*. Along the base line and each standard parallel is established at full intervals a series of corners called *standard corners;* and another series of corners called *correction corners* is established later during the process of subdivision into townships and sections, being at smaller intervals owing to the convergency of the meridians. Hence the term *correction line* is used to designate either a base line or a standard parallel.

Townships. The division of the 24-mile quadrangles created by the standard lines into townships is accomplished by laying off true meridional lines called *range lines* at

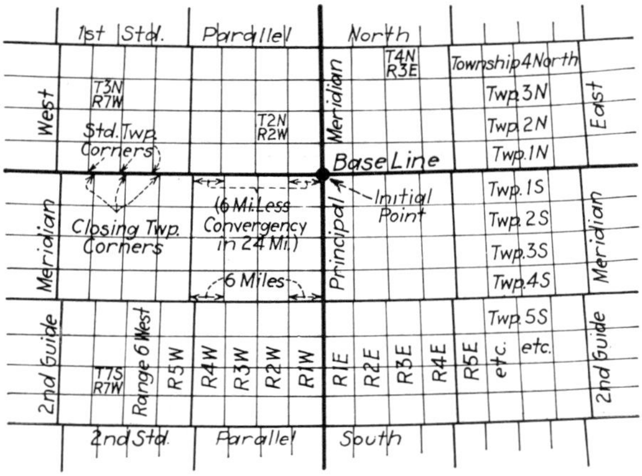

FIG. 63. Township and range lines.

intervals of 6 miles along each standard parallel, the range line extending north 24 miles to the next standard parallel, and by joining the township corners established at intervals of 6 miles on the range lines, guide meridians, and principal meridian with latitudinal lines called *township lines*. The east and west boundaries are 6 miles long but the north and south boundaries, except south boundaries of townships lying north of a correction line, are less than 6 miles long owing to the convergency of the meridians.

The plan of subdivision is illustrated by Fig. 63. A row of townships extending north and south is called a *range*, and a row extending east and west is called a *tier*. Ranges

are counted east or west of the principal meridian, and tiers are counted north or south of the base line. Usually for purposes of description the word "township" is substituted for "tier." A township is legally described by giving the number of its tier and range and the name of the principal meridian, e.g., T7S, R7W (read "Township seven south, range seven west") of the Third Principal Meridian.

Sections. The division of townships into sections is performed by establishing, at intervals of 1 mile, lines parallel to the east boundary of the township and by joining the section corners established at intervals of 1 mile with straight latitudinal lines. These lines, called *section lines,* divide each township into 36 sections, as illustrated by Fig. 64.

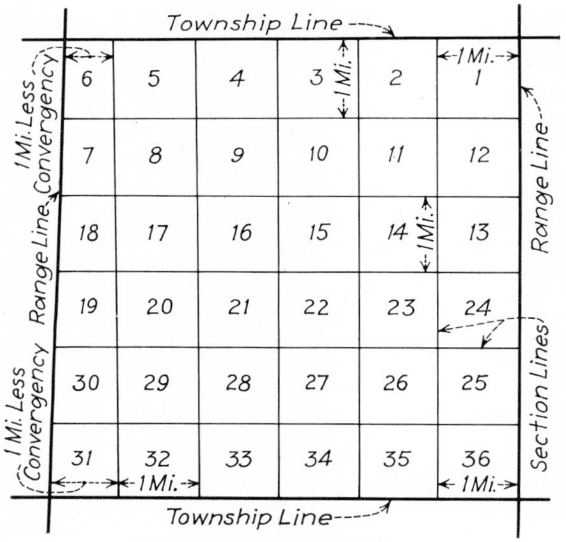

Fig. 64. Numbering of sections.

The sections are numbered consecutively from east to west and from west to east, beginning with No. 1 in the northeast corner of the township and ending with No. 36 in the southeast corner. A section is legally described by giving its number, the tier and range of the township, and the name of the principal meridian, e.g., Section 16, T7S, R7W, of the Third Principal Meridian.

On account of the convergency of the range lines (true meridians) forming the east and west boundaries of townships, the latitudinal lines forming the north and south boundaries are less than 6 miles in length, except for the south boundary of townships that lie just north of a correction line. As the north-south section lines are run *parallel to the east boundary* of the township, it follows that, if the surveys are without error, all sections except those adjacent to the west boundary will be 1 mile square, but that those adjacent to the west boundary will have a latitudinal dimension less than 1 mile by an amount equal to the convergency of the range lines. Any excess or deficiency in the measured distance between the $\left\{\begin{matrix}\text{south and north}\\ \text{east and west}\end{matrix}\right\}$ boundaries of the township is placed in the most $\left\{\begin{matrix}\text{northerly}\\ \text{westerly}\end{matrix}\right\}$ half mile. If the section is further subdivided, all the excess or deficiency is placed in the most $\left\{\begin{matrix}\text{northerly}\\ \text{westerly}\end{matrix}\right\}$ quarter mile.

Subdivision of Sections. The *regular* subdivisions of a section are the quarter section (½ mile square), the half-quarter section (¼ mile by ½ mile), and the quarter-quarter section (¼ mile square); the last contains 40 acres and is the legal minimum for purposes of disposal under the general land laws.

Of the 36 sections in each normal township (Fig. 64), 25 are returned as containing 640 acres each; 10 sections adjacent to the north and west boundaries (comprising sections 1 to 5, 7, 18, 19, 30, and 31) each contain regular subdivisions totaling 480 acres and in addition 4 *fractional lots* each containing 40 acres plus or minus definite differences to be determined in the survey; and 1 section (section 6) in the northwest corner contains regular subdivisions totaling 360 acres and in addition 7 fractional lots. Sections or lots may also be made fractional on account of meanderable bodies of water, mining claims, and other segregated areas within their limits.

On all section lines, monuments are placed at intervals of 40 chains, thus marking all section corners and all exterior quarter-section corners. Further subdivision of sections must be done by surveyors in private practice, but such subdivision must conform to the established dimensions of the official plat, including certain lines of subdivision which are shown on the plat but are not laid off on the ground.

A section cannot be legally subdivided until the section and exterior quarter-section corners have been either found or restored, and until the resulting courses and distances have been determined in the field. When the opposite quarter-section corners have been located, the interior quarter-section corner may be established. If the boundaries of quarter-quarter sections or of fractional lots are to be established on the ground, it is necessary to measure the boundaries of the quarter section and to fix thereon the quarter-quarter-section corners at distances in proportion to those given on the official plat; then the interior quarter-quarter-section corner may be placed. The subdivisional lines of fractional quarter sections are run from properly established quarter-quarter-section corners, with courses governed by the conditions represented on the official plat.

Convergency of Meridians. Let l be the length in miles of the meridian between two parallels whose mean latitude is ϕ, and let the mean distance between meridians, in miles measured along a parallel, be d. Then the linear convergency c_f in feet measured along a parallel, is approximately

$$c_f = \tfrac{4}{3} dl \tan \phi \tag{81}$$

Values of convergency for various latitudes are given in published tables.

Secant Method of Laying Off a Parallel of Latitude. As the base line, standard parallels, and latitudinal township lines are true parallels of latitude, they are curved lines when established on the surface of the earth. Ordinarily a latitudinal line is laid off on the ground by the method of offsets from a straight line either tangent to or intersecting the parallel. The amounts of the offsets either may be calculated from the convergency of the meridians or may be obtained from published tables.

The secant method of laying off a parallel of latitude is recommended by the Bureau of Land Management for its simplicity of execution and for the proximity of the straight line (secant) to the true latitude curve. In Fig. 65, the secant is a straight line 6 miles in length, which intersects the true parallel at the end of the first and fifth miles from the point of beginning. For the latitude of the given parallel, the offsets (in links) from secant to parallel are given in the figure, at intervals of ½ mile.

The procedure employed in establishing a true parallel 6 miles long by this method is as follows: The initial point on the secant is located by measuring south of the beginning corner a distance equal to the secant offset for 0 mile (5 links in the figure). The transit is set up at this point, and the direction of the secant line is established by laying off from true north the azimuth, either calculated or taken from published tables; for the condi-

tions illustrated by Fig. 65 the bearing of the secant which extends east from the point of beginning is N89°57.3′E. (Owing to the convergency of meridians, the azimuth of the secant—a straight line—varies along its length.) The secant is then projected in a straight line for 6 miles; and, as each 40 chains (½ mile) is laid off along the secant, the

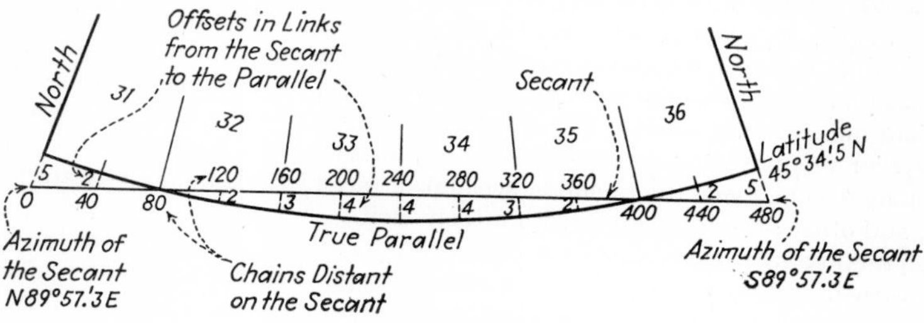

FIG. 65. Parallel of latitude by secant method.

proper offset is taken to establish the corresponding section or quarter-section corner on the true parallel.

Restoration of Lost Corners. Many corner marks become obliterated with the progress of time. It is one of the important duties of the local or county surveyor, in the relocation of property lines or in the further subdivision of lands, to examine all available evidence and to identify the official corners if they exist. Should a search of this kind result in failure, then it is the duty of the surveyor to employ a process of field measurement that will result in the obliterated corner's being restored to its most probable original location. If the original location of a corner cannot be determined beyond reasonable doubt, the corner is said to be *lost;* and it is then restored to its original location, as nearly as possible, by processes of surveying that involve the retracement of lines leading to the corner. Where linear measurements are necessary to the restoration of a lost corner, the principle of *proportionate measurement* must be employed. Single proportionate measurement consists in first comparing the resurvey measurement with the original measurement between two existing corners on opposite sides of the lost corner and then laying off a proportionate distance from one of the existing corners to the lost corner. Double proportionate measurement consists in single proportionate measurement on each of two such lines perpendicular and intersecting at the lost corner. Detailed instructions for the relocation of lost corners are given in the manual of the Bureau of Land Management.

CITY SURVEYING

Methods. The term *city surveying* is frequently applied to the surveying operations within a municipality with regard to mapping its area, laying out new streets and lots, and constructing streets, sewers and other public utilities, and buildings. Although the principles of city surveying are not different from those of ordinary surveying, there are some differences in the details of the methods employed. Some features pertinent to city surveying are as follows:

1. Measurements are made with a greater degree of refinement than for land of less value.

2. Some cities maintain a standard of length with which tapes may be compared.

3. Usually the horizontal control of the survey for the map of a city is by triangulation rather than by traversing, which would be employed for an equal area outside the city.

4. A system of reference points and bench marks is established at points a few blocks apart (usually at street intersections), and preferably this system is tied in with the United States precise surveys. Points are located in the street, at the curb, or on the

sidewalk, one such point being sufficient for each chosen intersection. On subsequent surveys, it is good practice to tie in to more than one of these established points, as monuments may have been moved.

5. The established points are well referenced to more or less permanent objects such as building corners, curb or walk lines, centers of street intersections, and manhole covers. In undeveloped districts, these points are referenced to stakes.

6. Maps showing the location of proposed sewers, street extensions, and other improvements usually show, to scale and in figures, the exact dimensions of adjacent lots and of all other lots that will be benefited by, or assessed for, the proposed improvement.

7. Sometimes separate maps are made of surface and underground utilities such as car lines, sewers, water lines, gas lines, electric power and telephone lines, conduits, and tunnels, both for convenient reference and in order to avoid interference in the location of new projects.

City Survey. Recently the term *city survey* has come to mean an extensive coordinated survey of the area in and near a city for the purposes of fixing reference monuments, locating property lines and improvements, and determining the configuration and physical features of the land. Such a survey is of value for a wide variety of purposes, particularly for planning city improvements.

Briefly, the work consists in

1. Establishing horizontal and vertical control, as described for topographic surveying. The primary horizontal control is usually by triangulation, supplemented as desired by precise traversing. Secondary horizontal control is by traversing of appropriate precision. Primary vertical control is by precise leveling.

2. Making a topographic survey and topographic map. Usually the scale of the topographic map is 1 in. = 200 ft. The map is divided into sheets which cover usually 60″ of longitude and 35″ or 40″ of latitude. Points are plotted by rectangular plane coordinates.

3. Monumenting a system of selected points at suitable locations such as street corners, for reference in subsequent surveys. These monuments are referred to the plane-coordinate system and to the city datum.

4. Making a property map. The survey for the map consists in (*a*) collecting recorded information regarding property; (*b*) determining the location on the ground of street intersections, angle points, and curve points; (*c*) monumenting the points so located; and (*d*) traversing to determine the coordinates of the monuments. Usually the scale of the property map is 1 in. = 50 ft. The map is divided into sheets which cover usually 15″ of longitude and 10″ of latitude, thus bearing a convenient relation to the sheets of the topographic map. The property map shows the length and bearing of all street lines and boundaries of public property, coordinates of governing points, control, monuments, important structures, natural features of the terrain, etc., all with appropriate legends and notes.

5. Making a wall map which shows essentially the same information as the topographic map but which is drawn to a smaller scale; preferably the scale should be not less than 1 in. = 2,000 ft. The wall map is reproduced in the usual colors—culture in black, drainage in blue, wooded areas in green, and contours in brown.

6. Making a map, or maps, to show underground utilities. Usually the scale of the underground map and the size of the map sheets are the same as those for the property map. The underground map shows street and easement lines, monuments, surface structures and natural features affecting underground construction, and underground structures (with dimensions), all with appropriate legends and notes.

Cadastral Surveying. Cadastral surveying is a general term referring to extensive surveys relating to land boundaries and subdivisions, whether they are city surveys as

described in the preceding article or surveys of rural land including the public lands. A cadastral map shows individual tracts of land with corners, length and bearing of boundaries, acreage, ownership, and sometimes the cultural and drainage features. The surveying methods are the same as those described for topographic surveying for maps of intermediate and large scale. The term *cadastral* is indefinite and is not in common use; specific terms such as *land*, *public-land*, or *property* survey are preferable.

TOPOGRAPHIC MAPS

Sources of Maps. Existing maps and/or aerial photographs are an aid to any survey, even though they may not be of the particular nature or scale desired. The central source of information regarding all Federal maps and aerial photographs is the Map Information Office, U.S. Geological Survey, Washington, D.C. Many maps and/or photographs are available from state, county, and city agencies.

Characteristics of Contour Lines. The principal characteristics of contour lines are as follows:

1. The horizontal distance between contour lines is inversely proportional to the slope; hence, on steep slopes, the contour lines are spaced closely.
2. On uniform slopes, the contour lines are spaced uniformly.
3. Along plane surfaces, contour lines are straight and parallel to one another.
4. As contour lines represent level lines, they are perpendicular to the lines of steepest slope. They are perpendicular to ridge and valley lines where they cross such lines.
5. As all land areas may be regarded as summits or islands above sea level, all contour lines must close upon themselves either within or without the borders of the map. It follows that a closed contour line on a map always indicates either a summit or a depression. If water lines or the elevations of adjacent contour lines do not indicate which condition is represented, a depression is indicated by a hachured contour line, called a *depression* contour.
6. As contour lines represent contours of different elevation on the ground, they cannot merge or cross one another on the map, except in the rare cases of vertical or overhanging ground surfaces.
7. A single contour line cannot lie between two contour lines of higher or lower elevation.

Contour Interval. The vertical distance between contours is called the *contour interval*. The choice of interval depends on the purpose and scale of the map and on the character of terrain represented; three principal considerations are the desired accuracy of elevations, the character and texture of the terrain, and legibility. For large-scale maps of flat country, the interval may be taken as small as ½ ft; for small-scale maps of rough country, the interval may be 50 ft, 100 ft, or more. For maps of intermediate scale, such as are used for many engineering studies, the interval is usually 2 or 5 ft (see also Table 8).

Contour-map Construction. Normally the construction of a topographic map consists of three operations: (1) the plotting of the horizontal control, or skeleton upon which the details of the map are hung; (2) the plotting of details, including the map location of points of known ground elevation, called *ground points*, by means of which the relief is to be indicated; and (3) the construction of contour lines at a given contour interval, the ground points being employed as guides in the proper location of the contour lines. A ground point on a contour is called a *contour point*.

Any contour line must be drawn, to some degree, by estimation. Skill and judgment are required to the end that the contour lines may best represent the actual configuration of the ground surface.

Contour lines are shown for elevations which are multiples of the contour interval; usually each fifth contour line is made heavier than the rest and is numbered.

Contours ordinarily change direction most sharply where they cross ridge and valley lines; also, the gradients of ridges and valleys are generally uniform. Hence ridge and valley lines are important aids to the correct drawing of contour lines.

Spot Elevations. Elevations of significant points such as summits, depressions, sharp breaks in slope, bridges, water surfaces, and road intersections should be shown on the map as numerals.

Interpolation of Contours. *By Estimation.* On intermediate- and small-scale maps, often the desired precision can be obtained if the interpolation is made by careful estimation supplemented by approximate mental computations.

By Computation. Where considerable precision is desired in the map, the computations may be made with the aid of a slide rule.

By Graphical Means. Various means of graphical interpolation are in use. One of these is illustrated by Fig. 66. A number of parallel lines are drawn at equal intervals

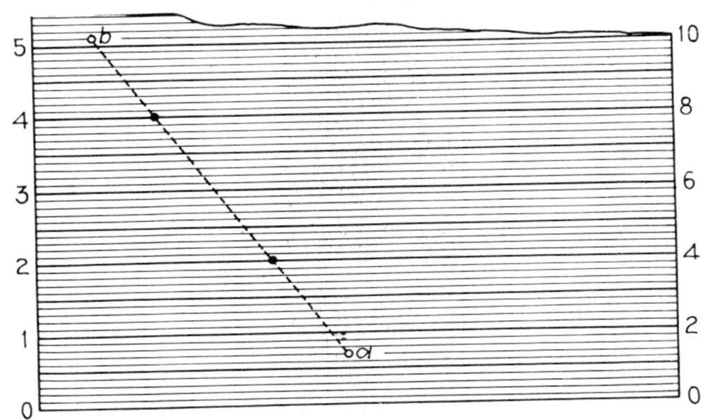

FIG. 66. Graphical interpolation of contour lines.

on tracing cloth, each fifth or tenth line being made heavier than, or of a different color from, the rest and being numbered as shown. If it is desired to interpolate the position of, say, the 52- and 54-ft contours between *a* with elevation of 50.7 and *b* with elevation of 55.1, the line on the tracing cloth corresponding to 0.7 ft (scale at left end) is placed over *a*, and the tracing is turned about *a* as a center until the line corresponding to 5.1 ft (scale at left end) covers *b*. The interpolated points are at the intersections of lines 2.0 and 4.0 (representing elevations 52 and 54) and the line *ab* and may be pricked through the tracing cloth.

Another convenient graphical means of interpolation is through the use of a rubber band graduated at equal intervals with lines forming a scale. The band is stretched between two plotted points so that these points fall at scale divisions corresponding to their elevations.

Systems of Ground Points. The typical systems of ground points commonly used in the preparation of topographic maps are the *controlling-point*, *cross-profile*, *checkerboard*, and *trace-contour* systems.

1. *Controlling Points.* Ground points are chosen as necessary to define the summits, depressions, valley and ridge lines, and all important changes in slope. This irregular system is plotted as in Fig. 67, and the contour lines are located by interpolation.

2. *Cross Profiles.* The cross-profile method is most frequently used in connection with route surveys. The field surveys determine the location either of all contour points or of all points of change in slope, along selected lines normal to the route traverse line. The

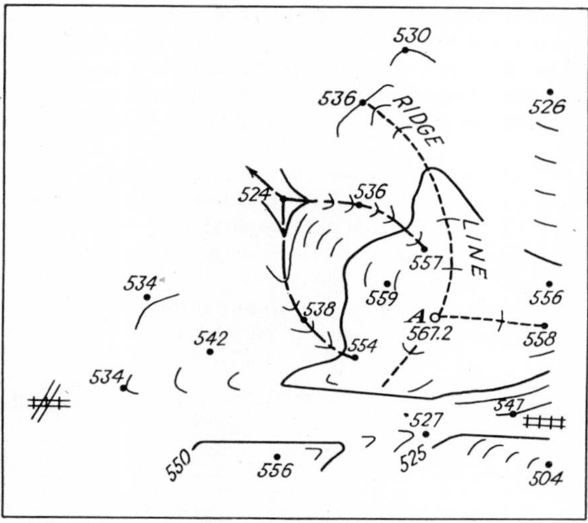

Fig. 67. Controlling points.

traverse, cross-profile lines, and ground points are plotted, and the contour lines are drawn, as illustrated in Fig. 68.

3. *Checkerboard.* A system of squares or rectangles is plotted, as in Fig. 69, and near each corner is written its elevation. Also, the locations of valley and ridge lines are shown. The contour crossings are interpolated on the valley and ridge lines and on the sides of the squares, and the contour lines are drawn.

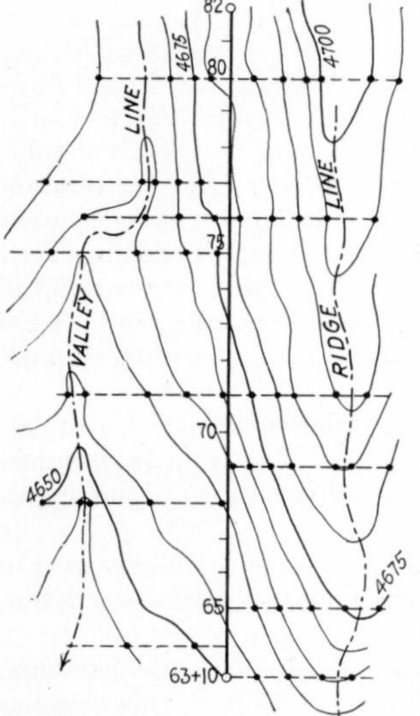

Fig. 68. Cross profiles.

4. *Tracing Contours.* A number of points on a given contour are located on the ground, and their corresponding locations are plotted on the map. The contour line is then drawn through these plotted points.

Tests for Accuracy. A topographic map can be tested for accuracy, both in plan and in elevation. In this discussion it is assumed that the errors in field measurement may be disregarded and that a graphical scale is provided on the map to render negligible any effect of shrinkage of the paper.

The test for horizontal dimensions consists in comparing distances scaled from the map and distances measured on the ground between the corresponding points. The precision with which distances may be scaled from a map depends on the scale of the map and on the size of the plotting errors. Thus, if for a map scale of 1 in. = 100 ft it is known that the error in location of any one point with respect to any other on the map is $\frac{1}{40}$ in., then the error represents 2.5 ft on the ground.

One test for elevations consists in comparing, for selected points, the elevations determined by field levels and the corresponding elevations taken from the map. Usually

the points are taken at 100-ft stations along traverse lines crossing typical features of the terrain. A more searching test is to plot selected profiles of the ground surface as determined by the field levels and the corresponding profiles taken from the map. The presence of systematic errors will be evidenced if the map profile is above or below the ground profile for an undue proportion of its length.

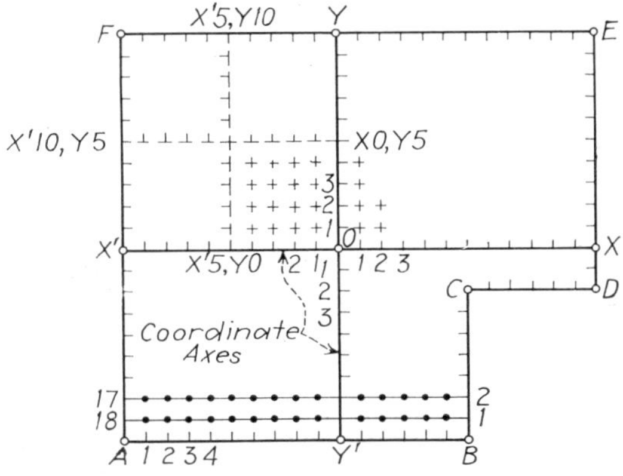

Fig. 69. Checkerboard system.

Choice of Map Scale. From a consideration of the tests described in the preceding article it is possible to choose a map scale consistent with the purpose of the survey if the approximate size of the plotting error is known. For example, if it is known that (with reasonable care in plotting) the average error in distance between any two definite points on the map is $\frac{1}{40}$ in., and if it is known that the purpose of the survey will be met if the average error in scaled distances is 10 ft, these conditions are satisfied by a map scale of 1 in. = 400 ft. By a similar course of reasoning, a map scale may be chosen that will represent a given *area* within desired limits of plotting error.

Choice of Contour Interval. The contour interval may be thought of as the scale by which the vertical distances or elevations are measured on a map. The choice of a proper contour interval for a topographic survey and map is based upon three principal considerations, as follows:

1. The greater the desired accuracy of elevations read from the map the smaller should be the contour interval. In general, the map error in elevation of any point should not exceed one-half of one contour interval.

2. The contour interval should generally be smaller for a map of flat country than for a map of hilly country. If a given kind of terrain is irregular rather than smooth, the contour interval should be relatively small in order to show the complexity of configuration.

3. To ensure legibility of the map, contour lines should not be spaced more closely than about 30 to the inch. Under this limitation the smaller the scale of the map the larger should be the contour interval.

Table 8 represents good practice in the selection of contour interval under usual conditions.

Specifications for Topographic Maps. The average errors in scaled horizontal dimensions between definite points chosen at random may be limited to a stated value, or the percentage of error in areas scaled from the map may be limited. The accuracy of contour lines may be specified by assigning maximum values to the average error in

TABLE 8. RELATION BETWEEN SCALE OF MAP, SLOPE OF GROUND, AND CONTOUR INTERVAL

Scale of map	Range of scale	Slope of ground	Interval, ft
Large	1 in. = 100 ft or less	Flat Rolling Hilly	0.5 or 1 1 or 2 2 or 5
Intermediate	1 in. = 100 to 1,000 ft	Flat Rolling Hilly	1, 2, or 5 2 or 5 5 or 10
Small	1 in. = 1,000 ft or more	Flat Rolling Hilly Mountainous	2, 5, or 10 10 or 20 20 or 50 50, 100, or 200

elevations taken from the map, the maximum error indicated by random test profiles, and the ratio of the length of the map profile that lies above the ground profile to the length that lies below the ground profile.

For example, the accuracy of a given topographic map might be specified as follows: The average error in distances between definite points as scaled from the map shall not exceed 8 ft; the average error in elevations read from the map shall not exceed 1 ft; the maximum error indicated by random test profiles shall not exceed 4 ft; and the ratio of the length of the map profile that lies above the ground profile to the length that lies below the ground profile shall be between ⅓ and 3.

National standard map accuracy requirements adopted by the principal Federal organizations engaged in making maps include the following:

1. *Horizontal Accuracy.* At the publication scale, for maps of scale larger than 1 in. = 1,670 ft, not more than 10 percent of the well-defined points tested shall be in error by more than $\frac{1}{30}$ in.; for maps of scale 1 in. = 1,670 ft or more, $\frac{1}{50}$ in.

2. *Vertical Accuracy.* Not more than 10 percent of the elevations tested shall be in error by more than one-half the contour interval.

TOPOGRAPHIC SURVEYING

Classes of Surveys. Topographic surveys fall roughly into three classes, according to the scale of the map, as follows:

Large scale: 1 in. = 100 ft or less.

Intermediate scale: 1 in. = 100 ft to 1 in. = 1,000 ft.

Small scale: 1 in. = 1,000 ft or more.

General Field Methods. The topographic survey of a tract consists in (1) establishing over the area a system of key stations connected by measurements of relatively high precision, a system called the *control;* and (2) locating the *details,* including the selected ground points, by measurements of lower precision from the control stations.

In the choice of field methods for topographic surveying, the items to be considered are the *intended use of the map* (more refined methods should be used for surveys for detailed maps than for surveys for maps of a general character); the *area of the tract* (control measurements for a large area should be more precise than those for a small area); the *map scale* (the probable errors in the field measurements should be considerably less than the probable errors in plotting at the given scale); and the *contour interval* (the smaller the contour interval, the more refined should be the field methods).

The principal instruments used are the engineer's transit, the plane table, the engineer's level, the hand level, and the clinometer. The use of the transit has advantages

over the use of the plane table where there are many definite points to be located or where the ground cover limits the visibility and requires many setups. Conditions favorable to the use of the plane table are open country and many irregular lines to be mapped. Sometimes the transit and the plane table, or the transit and the engineer's level, may be used together to advantage. Through dense woods, elevations of details are determined most advantageously by means of the hand level or the clinometer, and distances are usually determined by chaining.

The horizontal control is established by triangulation or traversing, and the vertical control is established by leveling.

The locations of the selected ground points in plan are determined in the same manner as for definite details, usually by radiation. The ground points may or may not be contour points. The elevations of ground points are determined usually by trigonometric leveling or, where the terrain is flat, by direct leveling. The stadia is used extensively except on large-scale surveys, for which the errors in stadia distances are large compared with the errors of plotting; on large-scale surveys usually the distances to definite details are measured with the tape. The details may be located either at the time of establishing the control or later. For the four systems of ground points commonly employed in locating details, the general field methods are as follows:

1. Where the *controlling-point* system is used, the ground points form an irregular system along ridge and valley lines and at other critical features of the terrain (Fig. 67). The ground points are located in plan by radiation or intersection with transit or plane table, and their elevations are determined commonly by trigonometric leveling or sometimes by direct leveling.

2. Where the *cross-profile* system is used, as on route surveys, the ground points are on relatively short lines transverse to the main traverse (Fig. 68). The distances from traverse to ground points are measured with the tape, and the elevations of ground points are determined by direct leveling, often with the hand level.

3. Where the *checkerboard* system is used, as where the scale is large and the tract is wooded or the topography is smooth, the tract is divided into squares or rectangles with stakes set at the corners (Fig. 69). The elevation of the ground is determined at these corners and at critical intermediate points where changes in slope occur, usually by direct leveling.

4. Where the *trace-contour* system is used, the contours are traced out on the ground. The various contour points occupied by the rod are located by radiation with transit or plane table. Frequently the engineer's level is employed as an auxiliary instrument.

The following statements summarize the use of the various systems of ground points employed in locating details:

Intermediate-scale Surveys. Generally the controlling-point system is used on hilly or rolling ground, and the cross-profile system is used on flat ground or for route surveys.

Large-scale Surveys. Generally the trace-contour system is used if the required accuracy is high and the ground is somewhat irregular in form, and the checkerboard system is used if the ground is smooth and the contour lines may be generalized to some extent.

Small-scale Surveys. The controlling-point system is used almost universally. A relatively small number of ground points are located, often by triangulation with the plane table; their elevations are determined by trigonometric leveling, the horizontal distances used in computing the difference in elevation often being scaled from the map.

Control. Control consists of two parts: (1) *horizontal control*, for which by triangulation and/or traversing the control stations are located in plan; and (2) *vertical control*, for which by leveling the bench marks are established and the control stations are located in elevation.

On surveys of wide extent a few stations distributed over the tract are connected by more precise measurements, forming the *primary control;* within this control system other control stations are located by less precise measurements, forming the *secondary control.* For small areas only one control system is necessary, corresponding in precision to the secondary control for larger areas. The terms "primary" and "secondary" are purely relative. Table 9 gives approximate values of the limits of permissible error for control measurements suitable to the different map scales.

Another classification of control—either triangulation, traversing, or leveling—with regard to precision is by *orders.* The various orders are absolute, not relative. The extensive surveys executed by the Federal agencies include *first-order, second-order, third-order,* and *fourth-order* control; roughly these correspond, respectively, to primary, secondary, tertiary, and quaternary control for small-scale maps.

Horizontal Control. The horizontal control may consist of a traverse system, a triangulation system, or a combination of the two. The required precision of horizontal control depends on the size and shape of the tract and on the scale of the map; in Table 9 are given approximate values of permissible errors on ordinary surveys. The values given in the table for secondary control of large areas may be applied to the control of small areas.

Traversing. Convenient routes are chosen that will result in the advantageous location of stations. Because of the cumulative effects of the errors in transit-tape traversing, in primary traversing it is desirable as a check to arrange closed circuits of length not to exceed perhaps 10 miles, dividing the tract into roughly equal areas. If closed circuits cannot be secured conveniently, checks for distance should be applied as the work proceeds, and checks for azimuth by astronomical observations should be applied at intervals of perhaps 10 miles. In making the check for azimuth, account must be taken of the convergency of meridians.

Wherever secondary traverses are required to establish the instrument stations from which the details are located, an area within a closed primary traverse may be divided by means of the secondary traverses into a series of roughly parallel strips. Secondary traverses are usually run with the transit but are sometimes run with the plane table and occasionally with the surveyor's compass. The lengths of the traverse lines are determined commonly by stadia or, if greater precision is required, by means of the tape.

Triangulation. The field conditions favorable to the use of triangulation to establish the horizontal control are a fairly extended area in an open, hilly region; a city where traversing is difficult because of street traffic; or a rugged, mountainous region where traversing would be slow and laborious.

A general layout of the scheme of primary triangulation is planned on an existing small-scale map; the field stations are established on summits where visibility is good; and signals are erected. One or more base lines are established and measured, and their true azimuths are determined by astronomical observations. Observations of angles are made on (1) major stations, which are marked by signals and which are to be occupied by the instrument, and (2) minor stations marked by such objects as trees, spires, and chimneys. When the field measurements have been completed, the necessary computations and adjustments are made; then the coordinates of each station are determined for use in plotting.

Although secondary control is commonly established by traversing, triangulation is employed where instrument stations can be advantageously located by this method, particularly in open, rough country where chaining would be difficult. For small areas, usually it is possible to obtain the required precision by the method of graphical triangulation employing the plane table (p. 1–52). This method has the advantages that no computations are required and that resection can be readily accomplished.

Table 9. Topographic Survey Control Data (Approximate Values)

Scale of map	Kind of control, for given scale	Triangulation					Traverse			Levels	
		Length of sight, miles	Average error of closure in triangles	Distance between bases, miles	Probable error in base measure	Maximum discrepancy between bases	Length of traverse, miles	Maximum error of angles	Maximum linear error of closure	Length of circuit, miles	Maximum error of closure
Small	Primary	10 to 200	1″	100 to 500	$\frac{1}{1,000,000}$	$\frac{1}{25,000}$	50 to 500	2″	$\frac{1}{25,000}$	50 to 500	0.017 ft $\sqrt{\text{miles}}$
	Secondary	5 to 20	3″	50 to 200	$\frac{1}{500,000}$	$\frac{1}{10,000}$	25 to 200	5″	$\frac{1}{10,000}$	....	0.035 ft $\sqrt{\text{miles}}$
	Tertiary	1 to 10	6″	10 to 100	$\frac{1}{250,000}$	$\frac{1}{5,000}$	10 to 100	30″	$\frac{1}{5,000}$	10 to 100	0.05 ft $\sqrt{\text{miles}}$
	Quaternary	½ to 2	1′ or graphical	2 to 10	....	....	1 to 2	2′ or compass	$\frac{1}{1,000}$	1 to 10	0.1 to 0.5 ft $\sqrt{\text{miles}}$
Intermediate	Primary	1 to 5	10″ to 20″	5 to 50	$\frac{1}{10,000}$ to $\frac{1}{40,000}$	$\frac{1}{1,000}$ to $\frac{1}{4,000}$	1 to 20	10″ to 1′	$\frac{1}{1,000}$ to $\frac{1}{5,000}$	1 to 25	0.05 to 0.3 ft $\sqrt{\text{miles}}$
	Secondary	½ to 2	Graphical	1 to 5	....	....	1 to 5	30″ to 3′	$\frac{1}{500}$ to $\frac{1}{2,500}$	1 to 5	0.1 to 0.5 ft $\sqrt{\text{miles}}$
Large	Primary	1 to 5	2″ to 10″	2 to 20	$\frac{1}{20,000}$ to $\frac{1}{80,000}$	$\frac{1}{2,000}$ to $\frac{1}{5,000}$	1 to 5	30″ to 1′	$\frac{1}{5,000}$ to $\frac{1}{20,000}$	1 to 10	0.05 ft to 0.1 ft $\sqrt{\text{miles}}$
	Secondary	¼ to 1	5″ to 20″	1 to 5	....	....	½ to 3	30″ to 2′	$\frac{1}{1,000}$ to $\frac{1}{5,000}$	1 to 3	0.05 ft to 0.1 ft $\sqrt{\text{miles}}$

Vertical Control. Bench marks are established at convenient intervals over the area, to serve as points of departure and closure for the leveling operations of the topographic parties when locating details.

Primary and secondary level routes are required in about the same amount, and bear about the same relation to each other, as do the primary and secondary traverses of triangulation systems. Often the level routes follow the traverse lines, the traverse stations being used as bench marks. In Table 9 are given approximate values of permissible error on ordinary surveys.

Ordinarily, the vertical control is accomplished by direct leveling, but for small areas or in rough country frequently the vertical control is established by trigonometric leveling.

Location of Details. The precision required in locating such definite objects as buildings, bridges, and boundary lines should be consistent with the precision of plotting, which may be assumed to be a map distance of about $\frac{1}{50}$ in. Such less definite objects as shore lines, streams, and edges of woods are located with a precision corresponding to a map distance of perhaps $\frac{1}{30}$ or $\frac{1}{20}$ in. For use in maps of the same relative precision, more located points are required for a given area on large-scale surveys than on intermediate-scale surveys; hence the location of details is relatively more important on large-scale surveys.

The veracity with which contour lines represent the terrain depends on the accuracy and precision of the observations, the number of observations, and the distribution of the points located. Ground points are definite; but, as the contour lines must necessarily be generalized to some extent, it would be inappropriate to locate the points with refined measurements. The error of field measurement in plan should be consistent with the error in elevation, which in general should not exceed one-fifth of a contour interval; thus generally the error in plan should not exceed one-fifth of the horizontal distance between contours. The purpose of the survey will be better served by locating a greater number of points with less precision, within reasonable limits, than by locating fewer points with greater precision.

Controlling-point Method. Details may be located by the controlling-point method employing the transit and stadia, the plane table, or the transit and plane table together. The distances are usually measured by stadia; but, on large-scale surveys, distances to definite details may be measured with the tape.

Transit and Stadia. In locating ground points, usually the vertical angles are observed more precisely than the horizontal angles.

The transit is set up and oriented at a control station. Details in the vicinity of the station are located by angle and distance measurements. Notes may be kept in the form shown in Fig. 23. Sketches are used if necessary to make the location of all points clear to the map draftsman.

The rodmen choose ground points along valley and ridge lines and at summits, depressions, important changes in slope, and definite details. The selection of points is important, and the rodmen should be instructed and trained for their work. They should follow a systematic arrangement of routes such that the entire area is covered. They should observe the terrain carefully and report important features that cannot be seen from the transit station.

Plane Table. Before the plane table is taken into the field, the horizontal control is adjusted and plotted on the plane-table sheet. The elevations of all bench marks either are recorded on the sheet or are in the hands of the computer.

The instrumentman sets up and orients the plane table at a control station. He then directs the rodmen to the controlling points of the terrain, as just described for the transit. When a rodman holds the rod on a ground point, the instrumentman sights on the rod to determine the direction, stadia interval, and vertical angle; the computer

calculates the horizontal distance from station to ground point and the elevation of the ground point; and the instrumentman plots the point and records its elevation near its plotted location (see also Table 5). As rapidly as sufficient data are secured, the instrumentman sketches the contour lines. Other objects of the terrain are located and are drawn either in finished form or with sufficient detail so that they may be completed in the office.

Transit and Plane Table. For large-scale maps and where many details are to be sighted, sometimes it is advantageous to use both the transit and the plane table. This method saves time in the field, but it may not reduce the total cost, as a larger party is required than for the plane table alone.

The transit is set up and oriented at the control station, the location of which is plotted on the plane-table sheet. The plane table is set up and oriented nearby, and its location is plotted on the map in its correct relation to the transit station. When a rodman has selected a ground point, the transitman observes the stadia distance and vertical angle to it; the plane-table man sights in the direction of the point, draws a ray toward it from the plotted location of the plane-table station, plots the point at the correct distance scaled from the plotted location of the transit station, and records on the map the elevation (computed by the transitman or the computer) of the plotted point.

Cross-profile Method. For the cross-profile method of locating details, the party consists of a topographer and usually two men (herein called "chainmen") who act either as chainmen or as rodmen. Sometimes only one chainman is employed, and the topographer assists in chaining. The equipment consists of a leveling rod, a steel or metallic tape, a hand level or a clinometer, and usually a cross-ruled, wide-page sketchbook or sketch sheets mounted on a board. Sometimes a Jacob's staff or other rod about 5 ft long is used as a support for the hand level or clinometer while sights are being taken.

The control points are the 100-ft stations of the transit traverse; these points have been marked on the ground by stakes, and their elevations have been determined by profile leveling and have been furnished to the topography party.

The ground points are points on relatively short crosslines transverse to the traverse line, as illustrated in Fig. 68. They are either contour points or more commonly points of change in slope; in the latter case the intermediate contour points are located by interpolation.

The party proceeds from station to station along the traverse. At each station, the topographer notifies the chainmen of the elevation of the station. The head chainman carrying the rod moves out on a line estimated to be at right angles with the traverse line until the rod is on the next contour (either higher or lower) from the station, as determined by the rear chainman employing the hand level; the distance out to the contour is then measured with the tape. The rear chainman then goes out to the point occupied by the rod, and the head chainman again moves out until the next contour is reached; and so the process is repeated until all contour points are located out to the edge of the strip being surveyed. A similar procedure is followed on the other side of the traverse line. Usually the trends or directions of the contours are sketched at each crossline and along ridge and valley lines, but on the field sheets the contour lines are not sketched for their full length. Definite details are located with relation to the transit line by tape measurements. If the topography is regular, sometimes the sketches are omitted, and the distances from traverse to contour points are recorded numerically.

For relatively small-scale maps, sometimes the clinometer is employed to determine the elevations of controlling points of change in slope on the crosslines by rough indirect leveling, and the distances are measured by pacing. The method is considerably faster than that just described for the hand level and tape, but is less precise.

For large-scale surveys or in flat country where the contour interval is small, usually elevations are determined either by direct leveling with the engineer's level or by indirect stadia leveling with the transit or the plane table.

If the ground cover is dense, either the hand level or the clinometer has a definite advantage over the transit or the level, because hand-level or clinometer sights can be taken through small openings in the underbrush; the area is surveyed in a series of overlapping strips. The maximum lengths of hand-level or clinometer sights should be limited to about 100 ft and, if the contour interval is small (2 ft or less), to about 50 ft.

Checkerboard Method. The checkerboard method of locating details is well adapted for large-scale surveys, as the points are located in plan by tape measurements.

The tract is staked off into squares or rectangles—usually 50- or 100-ft squares. The ground points and other details are then located with reference to the stakes and connecting lines. The usual procedure is first to run a rectangular transit-tape traverse near the perimeter of the tract, with stakes set at each 100-ft station. The error of closure becomes apparent in the field; if this is greater than the permissible error the stakes are reset. The interior of the figure bounded by this rectangular traverse is then filled in with stakes set at the corners of the 100-ft squares; each stake is marked usually with a letter and a number indicating its position with respect to a pair of coordinate axes, as illustrated in Fig. 69. By direct leveling, the ground elevation at each stake is determined. Sketch sheets are prepared, on which are shown the elevations of the corners of the squares. The location of ground irregularities or other details inside the squares is determined by measurements either from adjacent coordinate points or from the sides of the squares. The elevation of such details is determined by use of the hand level. The map is constructed in the office.

If many irregular features are to be mapped, the plane table may be used advantageously. Before the plane table is taken into the field, the corners of the squares are established on the ground with transit and tape, their elevations are determined by direct leveling, and the plane-table sheet is prepared showing the elevations of the corners of the squares, all as described earlier in this article. The plane table is then set up over the corner of a square and is oriented by backsighting along one of the control lines marked by stakes. Directions to details inside the squares are determined usually with a peep-sight alidade, and distances to these details are determined usually by chaining either from the instrument station or from a convenient corner or line. Only as many stations are occupied by the plane table as are necessary to cover the area.

Trace-contour Method. The trace-contour method of locating contour points on the ground is commonly used on large-scale surveys, and sometimes on intermediate-scale surveys where the ground is irregular. Under these conditions, if visibility is good, the trace-contour method is more rapid and more accurate than the checkerboard method.

Although the transit may be used in this work, either alone or with the engineer's level, the plane table is commonly used. Often the plane table and the engineer's level are used together. In this case, the levelman sets up the level at a convenient location and directs the rodman up or down the slope until a point on a given contour is located. This point is immediately sighted by the plane-table man and is plotted on the plane-table sheet. The rodman then moves to another contour point, usually on the same contour. The distances from plane-table station to contour points are measured by stadia; if the scale is large, definite objects may be located by taped distances.

CONSTRUCTION SURVEYING

Methods. Surveys for construction generally involve (1) a topographic survey of the site, to be used in the preparation of plans for the structure; (2) the establishment on the ground of a system of stakes or other markers, both in plan and in elevation, from which measurements of earthwork and structures can be taken conveniently by the construc-

tion force; (3) the giving of line and grade as needed either to replace stakes disturbed by construction or to reach additional points on the structure itself; and (4) the making of measurements necessary to verify the location of completed parts of the structure and to determine the volume of work actually performed up to a given date (usually each month), as a basis of payment to the contractor. In connection with construction, often it is necessary to make property-line surveys as a basis for the acquisition of lands or rights of way.

The detailed methods employed on construction surveys vary greatly with the type, location, and size of structure and with the preference of the engineering and construction organizations. Much depends on the ingenuity of the surveyor to the end that the correct information is given without confusion or needless effort. In the following paragraphs are given some general methods applicable to all types of structures.

Alignment. At the corners and along the sides of the proposed structure, outside the limits of excavation or probable disturbance but close enough to be convenient, transit stations are set and well referenced. Permanent targets or marks called *foresights* may be erected as convenient means of orienting the transit on the principal lines of the structure and for sighting along these lines by eye. Stakes or markers are set on all important lines in order to mark clearly the limits of the work.

In many cases, line and grade are given more conveniently by means of batter boards than by means of stakes. A batter board is a board (usually 1 in. by 6 in.) nailed to two substantial posts (usually 2 in. by 4 in.) with the board horizontal and its top edge preferably either at grade or at some whole number of feet above or below grade. The alignment is fixed by a nail driven in the top edge of the board. Between two such batter boards a stout cord or wire is stretched to define the line and grade.

Often it is impracticable to establish permanent markers on the line of the structure. In such cases the survey line is established parallel to the structure line, as close as practicable and with the offset distance some whole number of feet.

Grade. A system of bench marks is established near the structure in convenient locations that will probably not be subject to disturbance. From time to time these bench marks should be checked against one another to detect any disturbance.

The various grades and elevations are defined on the ground by means of pegs and/or batter boards, as a guide to the workmen. The grade pegs may or may not be the same as the stakes used in giving line. When stakes are used, the vertical measurements may be taken from the top of the stake, from a keel mark or a nail on the side of the stake, or (for excavation) from the ground surface at the stake; in order to avoid mistakes, only one of these bases for measurement should be used for a given kind of work. When batter boards are used, the vertical measurements are taken from the top edge of the board.

Precision. For purposes of excavation only, usually elevations are given to the nearest 0.1 ft. For points on the structure, usually elevations to 0.01 ft are sufficiently precise. Alignment to the nearest 0.01 ft will serve the purposes of most construction, but greater precision may be required for prefabricated steel structures or members.

It is desirable to give dimensions to the workmen in feet, inches, and fractions of an inch. Ordinarily measurements to the nearest ¼ or ⅛ in. are sufficiently precise, but certain of the measurements for the construction of buildings and bridges should be given to the nearest 1/16 in.

Establishing Points by Intersection. Where conditions render the use of the tape difficult or impossible, often points are established at the intersection of two transit lines by simultaneous sighting with two transits in known locations. By this method points may be located in elevation as well as in plan.

Highways. Generally just prior to the beginning of construction of a section of highway, the located line is rerun, missing stakes are replaced, and hubs are referenced.

Borrow pits (if necessary) are staked out and cross-sectioned. Lines and grades are staked out for bridges, culverts, and other structures. If slope stakes have not already been set during the location survey, they are set except where clearing is necessary; in that case they are set when the right of way has been cleared. For purposes of clearing, only rough measurements from the center-line stakes are necessary. Additional slope stakes may be offset a uniform distance away from the work, with appropriate marking to indicate the offset. If intercepting ditches are to be placed along the cuts, these are staked out also.

Where the depth of cuts and fills does not average more than about 3 ft, the slope stakes may be omitted; in this case the line and grade for earthwork may be indicated by a line of pegs (with guard stakes) along one side of the road and offset a uniform distance such that they will not be disturbed by the grading operations. Pegs are usually placed on both sides of the road at curves, and may be so placed on tangents; when this is done, measurements for grading purposes may be taken conveniently by sighting across the two pegs or by stretching a line or tape between them.

When rough grading has been completed, a line of finishing stakes is set on both sides of the roadway at the edge of the shoulder, as a guide in trimming the slopes. For fills, it should be understood whether these include any allowance for settlement, or whether they represent the final grade.

To give line and grade for the pavement, a line of stakes is set along each side, offset a uniform distance (usually 2 ft) from the edge of the pavement. The grade of the top of the pavement, at the edge, is indicated either by the top of the stake or by a nail or line on the side of the stake. The alignment is indicated on one side of the roadway only, by means of a tack in the top of each stake. For concrete highways, pegs may be set so that the side forms may be placed directly upon them, and a line of stakes set near one edge to give line for the forms. The distance between stakes in a given line is usually 100 or 50 ft on tangents at uniform grade and half the normal distance on horizontal or vertical curves. The dimensions of the finished subgrade and of the finished pavement are checked by the construction inspector, usually by means of a template.

Streets. For street construction the procedure of surveying is similar to that just described for highways. Ordinarily the curb is built first. The line and grade for the top of each curb are indicated by pegs driven just outside the curb line, usually at 50-ft intervals. The grade for the edge of the pavement is then marked on the face of the completed curb; or for a combined curb and gutter it is indicated by the completed gutter. Ground pegs are set on the center line of the pavement, either at the grade of the finished subgrade or with the cut or fill indicated on the peg or on an adjacent stake. Where the street is wide, an intermediate row of pegs may be set between center line and each curb. It is usually necessary to reset the pegs after the street is graded. Where driving stakes is impractical because of hard or paved ground, nails or spikes may be driven or marks may be cut or painted on the surface.

Railroads. Prior to construction the located line is rerun, missing stakes are replaced, hubs are referenced, borrow pits are staked out, slope stakes are set, and lines and grades for structures are established on the ground. When rough grading is completed, finishing stakes are set to grade at the outer edges of the roadbed.

When the roadbed has been graded, alignment is established precisely by setting tacked stakes along the center line at full stations on tangents and usually at fractional stations on horizontal and vertical curves. Spiral curves are staked out at this time. An additional line of pegs is set on one side of the track and perhaps 3 ft from the proposed line of the rail, with the top of the peg usually at the elevation of the top of the rail. Track is usually laid on the subgrade and is lifted into position after the ballast has been dumped.

Sewers and Pipelines. The center line for a proposed sewer is located on the ground with stakes or other marks set usually at 50-ft intervals where the grade is uniform and as close as 10 ft on vertical curves. At one side of this line, just far enough from it to prevent being disturbed by the excavation, a parallel line of ground pegs is set. A guard stake is driven beside each peg, with the side to the line; on the side of the guard stake farthest from the line is marked the station number and offset, and on the side nearest the line is marked the cut (to the nearest ⅛ in.).

When the trench has been excavated, batter boards are set across the trench at the intervals used for stationing. The top of the board is set at a fixed whole number of feet above the sewer invert (inside surface of bottom of sewer); and a measuring stick of the same length is prepared. A nail is driven in the top edge of each batter board to define the line. As the sewer is being laid, a cord is stretched tightly between these nails, and the free end of each tile is set at the proper distance below the cord as determined by measuring with the stick. If the trench is to be excavated by hand, the side pegs may be omitted and the batter boards set at the beginning of excavation.

For pipelines, the procedure is similar to that for sewers, but the interval between grade pegs or batter boards may be greater, and less care need be taken to lay the pipe at the exact grade.

Tunnels. For a short tunnel, a traverse and a line of levels are run between the terminal points; and the length, direction, and grade of the connecting line are computed. A coordinate system is particularly appropriate for tunnel work. Outside the tunnel, on the center line at both ends, permanent monuments are established. As construction proceeds, the line at either end is given by setting up at the permanent monument outside the portal, taking a sight at a fixed point on line, and then setting points along the tunnel, usually in the roof. Grade is given by direct levels taken to points in either the roof or the floor, and distances are measured from the permanent monuments to stations along the tunnel. If the survey line is on the floor of the tunnel, it is usually offset from the center line to a location relatively free from traffic and disturbance; from this line a rough temporary line is given as needed by the construction force. The dimensions of the tunnel are usually checked by some form of template transverse to the line of the tunnel but may be checked by direct measurement with the tape.

Bridge Sites. Normally the location survey will provide sufficient information for use in the design of culverts and small bridges; but for long bridges and for grade-separation structures usually a special topographic survey of the site is necessary. The site map should show all the data of the location survey, including the line and grade of the roadway and the marking and referencing of all survey stations. The usual map scale is 1 in. = 100 ft, and the usual contour interval is 5 ft on steep slopes and 2 ft over flat areas.

The preliminary report submitted with the site map should give all available information necessary for economic design, such as the character of the watershed and the steam bed, the elevation of the highest water, and the character of the foundation. For a grade-separation structure, the required information is similar except that it relates to the intersecting roadway and its traffic instead of the intersecting stream and its flow. Photographs are useful adjuncts to the report.

Bridges. For a short bridge with no offshore piers, first the center line of the roadway is established, the stationing of some governing line such as the abutment face is established on the located line, and the angle of intersection of the face with the located line is turned off. This governing crossline may be established by two well-referenced transit stations at each end of the crossline beyond the limits of excavation or, if the face of the abutment is in the stream, by a similar transit line offset on the shore. Similarly, governing lines for each of the wing walls are established on shore beyond the limits

of excavation, with two transit stations on the line prolonged at one, or preferably both, ends of the wing-wall line. If the faces are battered, usually one line is established for the bottom of the batter and another for the top. Stakes are set as a guide to the excavation and are replaced as necessary. When the foundation concrete is cast, line is given on the footings for the setting of forms and then by sighting with the transit for the top of the forms. As the structure is built up, grades are carried up by leveling, with marks on the forms or on the hardened portions of the concrete. Also, the alignment is established on completed portions of the structure. The data are recorded in field books kept especially for the structure, principally by means of sketches.

For long sights or for work of high precision, as in the case of offshore piers, various transit stations are established on shore by a system of triangulation, such that favorable intersection angles and checks will be obtained for all parts of the work. To establish the offshore piers, simultaneous sights are taken from the ends of a line of known length.

Where cofferdams are used, reference points are established on the cofferdams for measurements to the pier.

When the structure has been completed, permanent survey points are established and referenced for use in future surveys to determine the direction and extent of any movement.

Culverts. At the intersection of the center line of the culvert with the located line, the angle of intersection is turned off, and a survey line defining the direction of the culvert is projected for a short distance beyond its ends and is well referenced. At (or offset from) each end of the culvert, a line defining the face is turned off and referenced. If excavation is necessary for the channel to and from the culvert, it is staked out in a manner similar to that for a roadway cut. Bench marks are established nearby, and pegs are set for convenient leveling to the culvert. Line and grade are given as required for the particular type of structure.

Building Sites. In the preparation of his plans for a building, the architect requires a large-scale map of the site to show the information necessary to the proper location of the building both in plan and in elevation. Such maps are usually drawn to the scale of 1 in. = 10 ft or 1 in. = 20 ft.

The party consists of an instrumentman and one or two chainmen, and the equipment includes that of a transit party. Because of the large number of elevations to be determined, frequently an engineer's level is also used. The notes may be kept in a transit or topography notebook; or a sketch board may be used to good advantage.

First the lot corners are located, and permanent markers are set at these points. Then, property lines being used as reference lines, all objects are located, usually by tape measurements only, the instrumentman recording the data and drawing such sketches as are necessary. On extensive sites, or where it is not convenient for the chainmen to locate objects by coordinate measurements, transit angles and taped distances may be used.

The details should be shown and described as follows: lot corners, state kind; property lines, give dimensions and the distances from the walks; street lines, show widths; sidewalks and drives, give kind and widths; pavements, give kind and widths; gas and water mains, state size and show exact location; manholes and storm and sanitary sewers, give size and kind of pipe; trees, state kind and size; poles of all kinds; fire hydrants; and existing structures on or near the site, state materials of construction. Also, give the elevations of inverts of sewer outlets from manholes and the gradients of the sewers; the reference bench mark, with description; points along the sidwalks, curbs, and lot lines at intervals at 50 ft; and ground points at the corners of 50-ft squares. The elevations of the sewer inverts, sidewalks, and curbs should be taken to hundredths, and all ground points to tenths, of a foot; also contour lines are shown if the ground is irregular.

The map should give the legal description of the tract and the other information ordinarily shown on the plot of an urban land survey.

Buildings. At the beginning of excavation. the corners of the building are marked by stakes, which will of course be lost as excavation proceeds. Sighting lines are established and referenced on each outside building line and line of columns, preferably on the center line of wall or column. A batter board is set at each end of each outside building line, about 3 ft outside the excavation. If the ground permits, the tops of all boards are set at the same elevation; in any event the boards at opposite ends of a given line (or portion thereof) are set at the same elevation so that the cord stretched between them will be level. The elevations are chosen at some whole number of feet above the bottom of the excavation. When the board has been nailed on the posts, a nail is driven in the top edge of the board on the building line, which is given by the transit. Carpenter's lines stretched between opposite batter boards define both the line and the grade. If the space around the building is obstructed so that batter boards cannot be set, other means of marking the line and grade are substituted to meet the requirements of the situation.

When excavation is completed, grades for column and wall footings are given by ground pegs driven to the elevation of either the top of the footing or the top of the floor. Lines for footings are given by batter boards set in the bottom of the excavation. Column bases and wall plates are set to grade directly by the leveler. The position of each column or wall is marked in advance on the footing; and when a concrete form, a steel member, or a first course of masonry has been placed on the footing, its alignment and grade are checked directly.

In setting the form for a concrete wall, the bottom is aligned and fixed in place before the top is aligned.

Similarly, at each floor level the governing lines and grade are set and checked, except that for prefabricated steel framing the structure as a whole is plumbed by means of the transit at every second or third story level.

Throughout the construction of large buildings, selected key points are checked by means of stretched wires, plumb lines, transit, or level.

Dams. Prior to the design of a dam, a topographic survey is made to determine the feasibility of the project, the approximate size of the reservoir, and the optimum location and height of the dam. To provide information for the design, a topographic survey of the site is made. Extensive soundings and borings are made, and topography is taken in detail sufficient to define not only the dam itself but also the appurtenant structures, necessary construction plant, roads, and perhaps a branch railroad. A property-line survey is made of the area to be covered by, or directly affected by, the proposed reservoir.

Prior to construction, a number of transit stations, sighting points, and bench marks are permanently established and referenced upstream and downstream from the dam, at advantageous locations and elevations for sighting on the various parts of the structure as work proceeds. These reference points are usually established by triangulation from a measured base line on one side of the valley, and all points are referred to a system of rectangular coordinates, both in plan and in elevation. To establish the horizontal location of a point on the dam, as for the purpose of setting concrete forms or of checking the alignment of the dam, simultaneous sights are taken from two transits set up at reference stations, each transit being sighted in a direction previously computed from the coordinates of the reference station and of the point to be established. The elevation of the point is usually established by direct leveling. However, it may be established by setting off on one (or, as a check, both) of the transits the computed vertical angle, the height of instrument being known.

A traverse is run around the reservoir, above the proposed shore line, and monuments

are set for use in connection with property-line surveys and for future reference. Similarly, bench marks are established at points above the shore line.

Aircraft Jigs. In the manufacture of aircraft it is necessary to employ very large jigs which are held to extremely close tolerances of measurement. It is difficult and time-consuming to align such large structural assemblies by means of the plumb lines and stretched wires which are used for smaller jigs; therefore the alignment is often accomplished by sighting with surveying instruments. Horizontal alignment (elevation) is established usually by direct leveling with the engineer's level or with the telescope level of the transit; and vertical alignment by use of the transit telescope, with the horizontal axis carefully leveled. The special techniques required are largely in the field of tool engineering. A special form of the transit, called a "jig collimator," has been developed for use in this work.

MINE AND MINERAL-LAND SURVEYING

Underground Surveying. Distinctive features of underground surveying are that stations are usually in the roof instead of the floor; the object to be sighted and the cross hairs of the telescope must be illuminated; distances are usually measured on the slope; either the transit tripod has adjustable legs or a trivet is used; and often an auxiliary telescope is attached to the transit, either at one end of the horizontal axis or above the main telescope, with the line of sight of the auxiliary telescope parallel to that of the main telescope.

Horizontal and vertical distances are computed from slope distances and vertical angles. The transit is set up at one station, being centered by plumbing, and the vertical distance from station to horizontal axis of transit is measured. A plumb bob is hung at the next station, with a point on the plumb line marked by some form of clamping target. The vertical angle to the point so marked is measured, and the distance from horizontal axis to target is taped. The transit should be in excellent adjustment and carefully leveled.

The offset of a side auxiliary telescope requires a correction to observed horizontal angles, and the offset of a top auxiliary telescope requires a correction to observed vertical angles. The process of computing the correct angle from the observed angle is called *reduction to center* (see also p. 1–43).

Connections between surface and underground surveys require no special methods unless the tunnel is steeply inclined. Direction can be projected down a vertical shaft by means of two plumb lines suspended in the shaft, as far apart as practicable. In the case of a shallow and wide shaft, it is sometimes possible to set two stations at the bottom of the shaft from a transit set up at the surface. If possible, the survey should be carried into at least two openings and closed within the mine. If there are two vertical shafts, a single plumb line may be suspended in each shaft, and surface and underground traverses run between the plumb lines.

In a vertical shaft, differences in elevation are best measured by means of the steel tape.

Mineral-land Surveying. Ordinarily ownership of land implies ownership within vertical planes through the boundaries. However, in order to encourage the development of mining, the United States government has passed laws modifying in certain cases the usual rule of vertical planes and specifying the manner in which the person discovering a mineral vein or lode on government land may acquire title to the vein and thereby profit by his discovery. It is provided that a mining claim of specified maximum dimensions may be located on the surface, and that after certain requirements designed to prove the serious intent of the claimant have been satisfied, the United States government will give the claimant a patent carrying a clear title to the land claimed, this title carrying ownership of the vein beneath.

The Federal laws dealing with lode claims are based upon the concept of a relatively thin vein or lode, limited between surfaces that are essentially plane. According to the law, the claim is to be located along the outcrop of the vein (its intersection with the ground surface) and is limited to a maximum length of 1,500 ft and to a maximum width of 600 ft. Any state may by law reduce, but not increase, these dimensions. The outcrop must cross the end lines but not the side lines.

The ownership of a properly located lode claim carries with it ownership of the vein anywhere between vertical planes through the end lines. This holds even though the inclination of the vein from the vertical carries the vein underground beyond the side lines of the claim. The effect of this is to modify the usual rule of bounding vertical planes; the owner of the claim on the outcrop owns the vein even if it passes beyond a vertical plane through either of the side lines of his claim.

The end lines of the claim must be parallel straight lines, and the length of the claim must be measured along the center line of the claim. Except as limited above, the claim may be of any shape.

In locating a mining claim to secure the maximum dimensions and area allowed by law, the following procedure is suggested:

When the maximum allowable length of the center line of a claim is fixed, this line is so located as to follow the outcrop closely with as few breaks as practicable. Deflection angles are read at breaks in the center line, after hubs are established at these points and at the two ends of the line. The sum of the various segments of the center line should be equal to the maximum allowable length of the claim (as 1,500 ft). The transit is set up at one end of the center line, the proper angle is turned from the center line, and the corners are set at the two ends of the end line. A similar procedure is followed at the other end of the claim and at breaks on the center line. The direction and length of each line of the traverse bounding the claim are computed. Then as a check on the location, a traverse is run through the points forming the boundary. Such care in the location survey is not a legal requirement, but it is desirable from the point of view of the locator.

The location survey may be made by the claimant or by someone employed by him. The final survey for patent must be made by a United States mineral surveyor, commissioned by the United States to do that work.

HYDROGRAPHIC SURVEYING

Purpose. Hydrographic surveys are those which are made in relation to any considerable body of water, such as a bay, harbor, lake, or river. These surveys are made for the purposes of (1) determination of channel depths for navigation; (2) determination of quantities of subaqueous excavation; (3) location of rocks, sand bars, lights, and buoys for navigation purposes; and (4) measurement of areas subject to scour or silting. In the case of rivers, surveys are made for flood control, power development, navigation, water supply, and water storage.

Control. Since a certain amount of shore location is included in most hydrographic surveys, a single control survey is located on shore to serve both for soundings and for shore details.

As in topographic surveying, the horizontal control is a series of connected lines whose azimuths and lengths have been determined. The control may be established either by traversing or by triangulation of appropriate precision. Long narrow inlets or rivers are usually surveyed from a single traverse on shore. Where the shore line is obscured by woods, a system of triangulation is used. The control for large lakes and ocean shore lines consists of a network of triangles on shore.

A chain of bench marks is established to serve as a vertical control. These bench

marks are near the shore line and are located at frequent intervals so that gages may be set conveniently.

Most hydrographic surveys require the location of all irregularities in shore line, all prominent features of topography and culture, and all lighthouses, buoys, etc., in order that these points may be used for reference in sounding work.

Soundings. The determination of the relief of the bottom of a body of water is made by soundings. The depth of the sounding is referred to water level at the time it is made, and later is corrected to the datum water level through the use of gage readings. Before the corrected soundings can be plotted on the map their location with reference to the shore traverse is determined by one of the following methods:

1. By taking soundings on a known range line and reading one angle either from a boat or from a fixed point on shore.
2. By rowing at a uniform rate along a known range line and taking soundings at equal intervals of time.
3. By taking soundings from a boat at the intersections of known range lines.
4. By reading two angles simultaneously from two fixed points on shore, i.e., by intersection.
5. By radiation with transit and stadia.
6. By taking soundings at known distances along a wire stretched between stations.
7. By reading two angles from a boat to three fixed points on shore, by means of a sextant.
8. By reading a direction and a vertical angle simultaneously from an elevated point on shore.

Where current velocities are low, soundings are made with rods up to depths of perhaps 16 ft. For greater depths, nonstretching lines and weights of 3 to 25 lb are used. So far as possible, soundings are taken without stopping the boat. Soundings may be taken through holes bored in the ice. Electronic methods of sounding and of locating soundings are used on extensive surveys.

Before soundings are plotted, they are reduced to datum by subtracting (algebraically) from the sounding the corresponding gage correction. If corrections are necessary for wind, current, or erroneous length of sounding line, they are made at this time.

Map. A hydrographic map should show the datum, high and low water lines, soundings, lines of equal depth interpolated from soundings, and land and water features indicated by conventional signs. The methods used in plotting the soundings are the inverse of the field methods used in locating the soundings. Ingenious methods of plotting include the use of two polar protractors, two tangent protractors, a three-armed protractor, a tracing-cloth method similar to that used for resection (p. 1–56), and prepared plotting charts covering the working area.

Volumes. If dredged material is to be measured in place, soundings on a fixed section are taken before and after dredging, and the change in cross section is determined either by computation or through the use of the planimeter; volumes are then computed as described in Section 2. Excavated material in scows is measured by change in displacement during the loading, by capacity of scow pockets, or by measurement of the dimensions of the pile deposited on the deck.

The capacity of existing lakes or reservoirs is measured by means of either contours or cross sections. Usually the volume is expressed in acre-feet, 1 acre-ft being 43,560 cu ft.

Snow Surveys. In areas which for their water supply depend to a considerable degree on melted snow from mountainous regions, snow surveys are made annually as an aid to forecasting the runoff of streams. Snow courses are established at key locations representative of the entire area, preferably in the early winter when some snow has fallen.

The locations are marked by poles or by boards nailed to trees, and are recorded. The depth and density of snow at any time are determined by means of a metal sampling tube. The tube is inserted and withdrawn, and the weight of its contents of snow and ice is determined and is expressed in terms of inches of water.

PHOTOGRAMMETRIC SURVEYING [1]

Photogrammetry. *Photogrammetry* is the science of measurement by means of photographs. The scale and position of objects in photographs vary according to the distance and position of the corresponding actual objects relative to the camera. Photogrammetric surveying is accomplished by measurement of these differences in scale and displacements in position in single or overlapping photographs, in order to determine the location and elevation of ground points and other details.

Although oblique photographs taken from ground stations or from airplanes are used to some extent, the principal development has been in the field of photographs taken from airplanes with the axis of the camera vertical (as nearly as may be). In connection with limited ground surveys made for the purpose of accurately establishing visible control points, vertical aerial photogrammetry has become the most rapid and accurate method of topographic surveying except perhaps where the ground is relatively flat, where elevations must be determined within less than 5 ft, or where the area is small. Its advantages are the speed with which the field work is accomplished, the wealth of detail secured, and the use in locations otherwise difficult or impossible of access.

Aerial photogrammetry is used for general topographic surveys, route surveys, and surveys of forest and agricultural areas (see also p. 1–117). Usually the actual photography and sometimes the mapping are done under contract by organizations specializing in that work. Considerable areas of the United States have already been photographed, and in many cases the photographs are available to surveyors and others for a nominal fee. The central source of information regarding all Federal maps and aerial photographs is the Map Information Office, U.S. Geological Survey, Washington, D.C.

Terrestrial Photogrammetry. Photographs for terrestrial (ground) photogrammetry are taken with a *phototheodolite*, which is a camera mounted on a transit tripod and having a transit telescope with its optical axis parallel to that of the camera. The photographs are later inserted in an automatic plotting machine. Terrestrial photographs are taken in pairs from the ends of a measured base line. The pairs are generally made with the two positions of the camera axis parallel to each other. Usually the camera stations are at the higher points in the area, in order that the direction of pointing may be as nearly normal to the slope as possible.

General Procedure of Aerial Surveying. Aerial surveying consists in *advance planning*, *photography*, *ground control*, and *compilation* of control and details. The general procedure is briefly as follows.

Advance planning includes selection of the scale and coverage of the photographs, and computation of the corresponding flight altitude, interval between lines of flight, and interval between exposures. A "flight map" of the area is prepared, showing the lines along which flights are to be made.

Aerial photographs of the area to be surveyed are taken with specially designed automatic cameras, with the axis of the camera vertical (as nearly as may be). The airplane is flown at a constant altitude (as nearly as may be) above sea level, as indicated by corrected barometer readings. The area is covered in a series of parallel strips, with photographs of the same strip overlapping about 60 percent and photographs of adjacent strips overlapping 30 to 50 percent. From the photographs an index map is prepared, showing the relative position and the coverage of each photograph.

Ground control consists of a relatively small number of points the location and eleva-

[1] By B. B. Talley.

tion of which are known. Ground-control points either may be visible stations established by other surveys such as the Federal triangulation and land surveys or may be established especially for the photogrammetric survey. With adjacent ground-control points identified on a photograph or series of photographs and with the flight altitude and the focal length of the camera known, the scale of the photographs can be computed. A series of photographs can be brought to the same scale by enlargement or reduction of the individual photographs.

Secondary control consists of selected image points on photographs, which points are used to bring a series of photographs between adjacent ground-control stations into the desired relations for plotting to scale. The scale of the secondary control determines that for the photographs.

With the secondary control plotted on a transparent sheet and with the photographs brought to the same scale and placed successively under the sheet, the map details surrounding each control point are traced onto the sheet. Conventional signs are employed as desired. Alternatively, the details may be brought to the required scale by graphical means.

For purposes of representing relief, or *contouring*, use is made of the difference in parallax displacement between selected ground points of unknown elevation and control points of known elevation. To compute the displacements of all the image points of controlling points of the terrain would be unduly laborious, and in practice it is customary to employ automatic plotting machines based upon the principle of stereoscopic observation. If desired, contours may be drawn directly on the photographs. Contouring may be accomplished in the field with the plane table, a copy of the planimetric map being used as the plane-table sheet.

Aerial Photographs. If the photographs were truly vertical and if the terrain were flat, the photographs would be true maps to some scale. In general, however, the features are displaced and their size is varied because of longitudinal and transverse inclination (tip and tilt) of the camera axis and because of the relief of the ground. The effect of tip and tilt can be corrected either by photographic methods or (more laboriously) by computation; usually the combined average tip and tilt is less than 3°, and often it is neglected. The displacement, or parallax, of features due to ground relief makes necessary special methods of plotting to secure a true scale map; but it is useful, in fact necessary, as a means of determining the elevations of points by computation or by stereoscopic measurement.

The scale of a photograph can be computed from the relationship of a measured distance on the photograph to the corresponding distance on the ground. In many instances, a number of identifiable points of known position will appear on the photographs, and the known distances between these points offer means of determining scale. Scale may also be computed approximately as the ratio of the focal length of the aerial camera to the flight altitude, or height of the airplane above the average elevation of the terrain. Conversely, the flight altitude to produce photographs of a desired scale can be computed from the same relationship.

Coverage of the area to be surveyed is obtained by photographing the area in parallel strips, which are spaced at predetermined distances to ensure the desired overlap (usually 30 to 50 percent) of photographs for adjacent strips. Photographs are taken at the proper intervals along each strip to give the desired overlap (usually about 60 percent) of photographs for the given strip. The required time interval between exposures is computed, being directly proportional to the intervals of distance and inversely proportional to the ground speed of the airplane. For preliminary estimates, the required number of photographs is computed by dividing the total area to be photographed by the *net* area covered by a single photograph.

An *index map* of the photographs is prepared, as follows: for large jobs, the position

and coverage of each photograph are plotted on the flight map; for small jobs, the photographs are laid in the form of a mosaic and rephotographed.

Photographic Airplanes. Any airplane which is stable, which may be flown hands-off, which has the required service ceiling, and which has sufficient space for pilot, photographer, and aerial camera can be used for aerial photography. In addition, the airplane should allow excellent visibility for the pilot, permit operation from small and relatively unimproved fields, and have economical speed and performance characteristics. For flying at high altitudes, the airplane should afford protection from the cold and should have provision for oxygen for the crew.

Aerial Cameras. Aerial cameras may be of the *single-lens* or the *multiple-lens* type. Usually the airplane is flown at sufficient altitudes to permit shutter speeds of $\frac{1}{150}$ sec or slower, and between-the-lens shutters are employed in preference to focal-plane shutters. However, one type synchronizes the movement of the ground image across the plate of the camera with the film speed, thereby producing a motion-stopping effect; the exposure method resembles that of a focal-plane-shutter camera with the exception that the slit is stationary and the film moves. The cameras are fully automatic, being operated by electricity; the operator has only to concern himself that the camera is level and fully aligned with the line of flight.

Determination of Elevations by Measurement of Parallax. The object of measuring the parallax difference between two ground points is to determine the difference in their

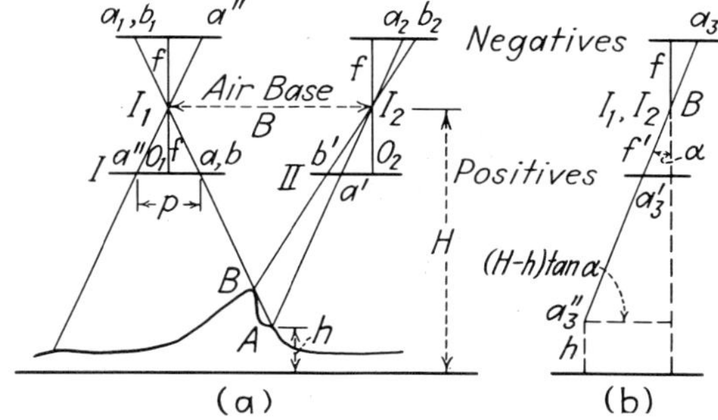

FIG. 70. Determination of elevations by measurement of parallax.

elevations and thus to determine their respective heights above sea level. In Fig. 70, it is desired to measure the ground elevations above sea level of points A and B. Point A appears on photograph positive I at a and on positive II at a'. Line I_1a'' is constructed parallel to line I_2a', making a triangle $a''I_1a$ similar to triangle I_1AI_2. Then

$$p = \frac{fB}{H - h} \tag{82}$$

where H is the altitude of flight, h is the elevation of point A above sea level, B is the "air base," or distance the plane flew between exposures, and the distance p is the *absolute parallax* of point A.

Similarly the absolute parallax of point B can be determined. The difference between the absolute parallax of A and that of B is a measure of the difference between their elevations.

For purposes of contouring, use is made of Eq. (83) which is derived from Eq. (82) and which expresses the value of the change in parallax Δp (in millimeters) for a cor-

responding change in elevation Δh (in feet) between two points or between two contours on the photograph.

$$\Delta p = \frac{B_m \, \Delta h}{H - h} \tag{83}$$

where H and h are in feet and B_m (in millimeters) is the measured stereoscopic base of the photographs $[B_m = Bf/(H - h)]$.

In an aerial photograph, the displacement due to relief is a natural phenomenon and should not be considered an error. Were it not for the parallax displacement due to relief, stereoscopic measurement would be impossible.

Map Control. In addition to the necessary ground control, photogrammetric maps are based upon *secondary control* which is usually obtained graphically in a drafting room. Secondary control is said to tie the pictures to the ground. It may be established either by the radial-line method or by aerial triangulation (p. 1–113).

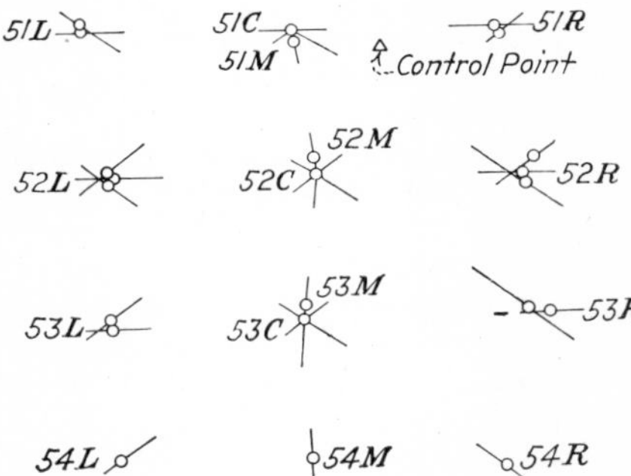

FIG. 72. Plotted map-control points.

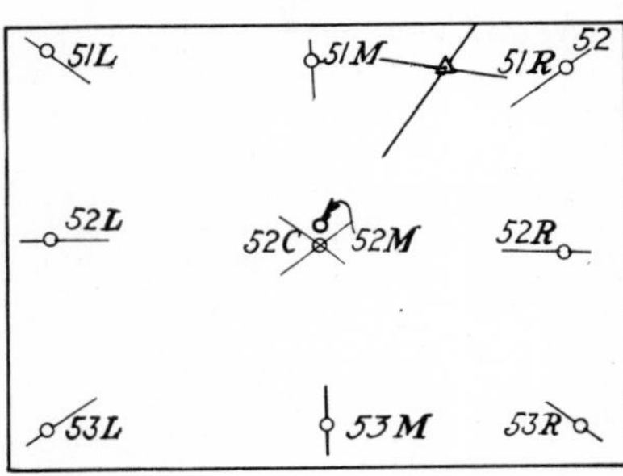

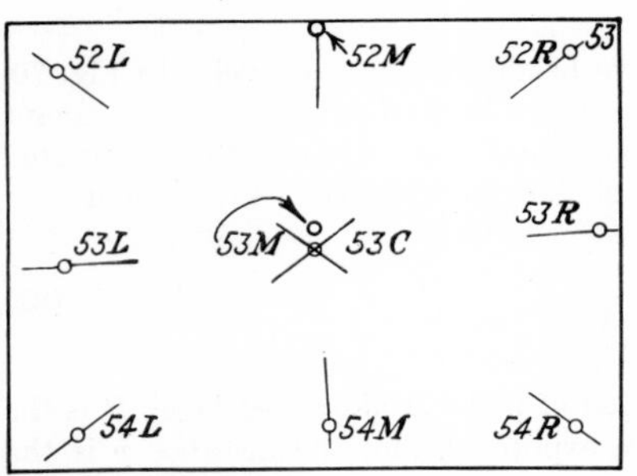

FIG. 71. Photographs marked for radial-line control.

Radial-line Method of Control. The radial-line method of providing secondary map control from vertical aerial photographs is based on the principle that a visible point included in two properly overlapping photographs is located at the intersection of rays drawn through that point from the principal point of each photograph. Briefly, the procedure is as follows.

A group of several successive photographs is selected which includes at least two ground-control points—one near each end of the group. On each photograph is marked the principal point, as, for example, 51C (Fig. 71). Near the principal point of the first photograph is chosen a visible point, as 51M, which appears also on the adjoining overlapping photograph (52); and this "substitute center" is marked on both photographs.

Points 51*R* and 51*L* are chosen at objects near the middle of the right and left edges of the first photograph, and are marked on both photographs; and so on for points 52*R* and 52*L* which are near the lower corners of the first photograph. The five visible points are also marked on photograph 52. Similarly, points are marked on photographs 52 and 53, and so on until another photograph is reached which includes a control point. Radial lines are drawn on each photograph from the principal point to all marked points including the control points.

A large sheet of transparent film base (cellulose acetate) is laid over the first photograph, and points 51*C*, 51*M*, and 52*M* are traced, as well as the radial lines to all other points. Photograph 51 is removed, the film base is placed over photograph 52 so that the traced position of 52*M* falls on its position as it appears on photograph 52, and the film base is then oriented. Radial lines to the other points are traced. The correct location of points 51*R*, 51*L*, 52*R*, and 52*L* are thus determined by intersection, as shown in Fig. 72, regardless of parallax displacements and of differences in scale as between the individual photographs. And so the process is continued for succeeding photographs of the group. The plotted locations of the various points may or may not coincide with their pictured location on either photograph.

The scale of the data assembled on the film base is then determined by measuring the distance between the two control points of known location. Proper reduction or enlargement can be accomplished if desired.

Slotted-template Method. A variation of the radial-line method of secondary map control is known as the *slotted-template* method. Instead of the rays being drawn on the photographs as heretofore explained, the points are selected, marked on the photographs, and transferred to cardboard templates. Slots representing rays radiating from the principal point to the selected photo points are cut into the templates by means of a mechanical slot cutter; the slotted templates are approximately oriented on the sheet; and movable metal studs are inserted through the slots which intersect at each selected photo point. Those studs which correspond to known control points are then fixed in position, and the system of slotted templates is shifted slightly about these fixed points until the arrangement having the least apparent residual strain is found. The movable studs are then in the most probable position for the corresponding photo points.

An alternative slotted-template method involves the use of slotted strips of spring steel radiating about a center bolt in the form of "spiders," with each strip in the direction of a ray to a selected photo point. The procedure of orienting, connecting, and adjusting the system is similar to that just described.

Section-line Method. Surveys either of small areas or of large areas of purely local interest are often tied into the United States public-land surveys by means of visible indications (such as highways and fence lines) of section lines and corners. However, the relatively low precision of the public-land surveys must be kept in mind. The use of section lines and corners as control is usually limited to the joining (or the holding) of two photographs together, to fixing the positions of photographs for surveys of relatively low precision, or to the laying of mosaics.

Aerial Triangulation. Aerial triangulation is a method of secondary map control employed with stereoscopic plotting instruments. For surveys of medium and small scale, the ground control need appear only in every third, fifth, or in some cases fifteenth photograph; the intervening distances are bridged by aerial triangulation. The method involves a precise adjustment of the first stereoscopic model (in the plotting machine) to the existing ground control; additional photographs are then adjusted and oriented relative to those already adjusted.

Compilation of Detail from Photographs. In an aerial photograph, unimportant details are shown with the same degree of intensity as the important. The purpose of map compilation is to separate all these features and to represent them by conventional

signs according to their importance to the task at hand. No feature should be out of its relative location by a greater distance than the dimensions of the conventional sign used to represent it.

First the secondary control is transferred to the compilation sheet, and the number of each photograph and of each control point is shown. The sheet is then oriented over each photograph in turn, and the detail is traced onto the sheet by the radial-line method, as follows:

1. Beginning at the center of the photograph, symbols are traced in for all the features in the central area, halfway to the nearest radial-line points.

2. The sheet is shifted slightly (if necessary) so that the location of one of the nearest radial-line points is exactly over the corresponding point on the photograph, the sheet is oriented, and the detail is traced in for the area surrounding the point, halfway to the adjacent radial-line points.

3. The detail between the two areas is connected, adjustments being made if necessary so that no feature is out of position by more than about half the dimensions of the conventional sign.

4. The foregoing process is continued for all the other points in the area of the photograph, and for successive photographs in the series.

Contouring. Contouring can be accomplished in several ways. A *field sheet* can be made by printing the compilation sheet onto a plane-table sheet; the field sheet is then taken into the field, and contouring is accomplished with the plane table. A field check with the plane table is desirable even when other methods of contouring are employed.

Contouring can be done on the stereocomparagraph (p. 1–116) on a separate template for each photograph; later the contours are compiled on the planimetric sheet in the same manner as the detail of the photograph.

In the compilation of contours on the multiplex projector (p. 1–115) and certain other automatic plotting machines, the radial-line method is not used; all features of the map are compiled in a single operation.

Plotting Machines. A plotting machine for aerial photogrammetry consists essentially of a viewing system, a measuring system, and a drawing system. The elevation of a given point shown on the photographs is determined by manipulating an index mark of the machine until it apparently touches that point in the three-dimensional stereoscopic image.

A wide variety of stereoscopic plotting machines are used for the compilation of topographic maps. Some of these are precise, complicated, and so expensive that only a governmental agency or a large mapping concern can afford to own and operate one. With this type the ground points and map details are located in plan and elevation simultaneously, and no secondary control is required. Examples are the aerocartograph, the autograph, the multiplex projector, and the stereoplanigraph.

To fill the need for less expensive equipment, a variety of simple and relatively inexpensive plotting machines have recently been developed in the United States; these are less precise than the more elaborate machines but are satisfactory for contouring after the secondary control has been plotted by other means such as an elaborate machine or the radial-line method previously described. Examples are the contour finder and the stereocomparagraph.

In the following articles are briefly described one elaborate and one simple plotting machine.

Multiplex Projector. The multiplex projector is a stereoscopic plotting machine for producing topographic maps by means of the simultaneous projection in complementary colors of overlapping aerial photographs. As shown in Fig. 73, it consists of a series of projectors which are small-scale reproductions of the taking cameras. A small diapositive, or positive transparency, is used in the projectors. A colored filter may be

inserted in each projector in order to project with either red or blue-green light, as may be necessary.

In operation, the diapositives are inserted in the projector and are then mutually adjusted so that their projected images will intersect in space above the drafting table and form a spatial model which is a true small-scale reproduction of the landscape photographed. The spatial model is obtained by projecting one photograph of an overlapping pair in red light and the other in blue-green light, and by observing the combination of colors through spectacles containing one red and one blue-green lens. Only the red image is seen with one eye, and only the blue-green image is seen with the other eye;

FIG. 73. Multiplex projector with tracing stand and voltage regulator.

thus the condition of stereoscopic vision is fulfilled. When the instrument is properly adjusted, the intersections of the bundles of rays from the projectors form a true image, in space, of the original landscape.

As the fusion of the images occurs in space above the drafting table, the image can be cut and measured at any desired height. An index mark, or "floating mark," is carried in the center of a circular disk on the tracing table as shown in Fig. 73. This disk is raised and lowered by means of a screw on the center post at the back of the tracing stand. On the left post of the tracing stand is a millimeter scale on which is read the height of the disk above the drafting table, which may be considered as the datum plane. Carried in the tracing stand directly below the floating mark is the drawing pencil which traces on the plotting sheet the horizontal movements of the floating mark. The height of any point is measured by bringing the floating mark into contact with the spatial model at the point, and reading the elevation in millimeters (at the plotting scale) directly on the millimeter scale on the tracing stand.

In contouring, the floating mark is set at the correct height of the contour (the value of the contour, in millimeters, multiplied by the respresentative fraction of the plotting

scale), and the pencil is lowered onto the plotting sheet. The tracing stand is then moved about over the sheet while the floating mark is held in contact with the spatial model. Although this process may seem difficult, it is easily accomplished on the instrument.

The projectors can be adjusted for tip and tilt and in all directions so that they are located and oriented into the same relative positions as those of the aerial camera at the instant of exposure of the several photographs.

Stereocomparagraph. The stereocomparagraph (Fig. 74) is a simple automatic plotting instrument for contouring directly from vertical aerial photographs. It was

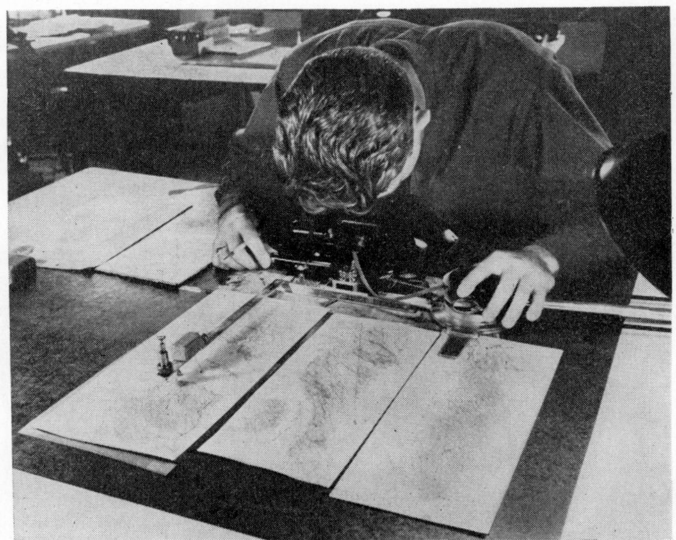

FIG. 74. Fairchild stereocomparagraph.

invented by B. B. Talley in 1934. The viewing system is a magnifying mirror stereoscope; the measuring system consists of two index marks engraved in the centers of two meniscus lenses actuated by a micrometer screw; and the drawing system consists of a pencil mounted at the end of an arm fixed rigidly to the base of the instrument. In operation, two aerial vertical photographs are oriented beneath the instrument so that stereoscopic vision is obtained throughout the area of their overlap.

The lenses containing the index marks are in contact with the photographs. The left index mark is fixed to the base; the right index mark may be moved either to the right or to the left by means of a micrometer screw which also serves to indicate the extent of the movement. The two index marks fuse into a single image, or floating mark, which appears to rise or fall in space as the distance between the two lenses is decreased or increased respectively. As indicated by Eq. (83), the movement of the micrometer (Δp), in millimeters, corresponding to an apparent change in height of the floating mark is equal to the mean stereoscopic base (B_m), in millimeters, of the photographs multiplied by the difference in elevation (Δh), in feet, between the two pictured points on which the floating mark is held, divided by the altitude of the airplane ($H - h$), in feet, above these points.

In practice, the difference in elevation of points on aerial photographs is determined on the stereocomparagraph by measuring directly the difference in parallax of the points

and by converting this difference into feet by reference to a parallax table which is constructed by solving Eq. (83) for an assumed stereoscopic base, for given increments of difference in elevation, and for variations in flight altitude throughout the range likely to be encountered.

For contouring, the micrometer is set at a value corresponding to the elevation of the contour to be drawn, and the floating mark is maintained in contact with the spatial model while the instrument is moved about over the photographs. Each photograph is contoured separately, and the contours are compiled onto the map sheet as previously described for the cultural detail of the photographs. Thus the scale is made uniform and the horizontal displacements of the contours due to relief are adjusted in the same operation. Control is required in every photograph.

Applications of Photogrammetry. Aerial photographs and/or photogrammetry are widely used in many fields including the following: military, topographic mapping, tax equalization (property maps), petroleum geology, geophysical prospecting, mining, forestry (including classification of timber and determination of tree heights), rural rehabilitation, soil conservation, agricultural land use, city and regional planning and development, route location, and site studies.

For route location, mosaics are used in reconnaissance, paired stereoscopic prints are used for detailed studies, and contour maps are employed for quantitative analyses. Application of aerial photography to highway location is accomplished in four stages: (1) reconnaissance of a wide area, with the mosaic used as a plotting sheet; (2) comparison of routes with the aid of larger-scale photographs, and selection of best route; (3) preliminary survey and design, which involve the plotting and use of a contour map; and (4) location survey and construction survey, which consist in staking the alignment, grade, cross sections, and structures on the ground. Mosaics and paired stereoscopic prints are also useful for displays, negotiations for routes or for rights of way, studies of drainage, or exploration for suitable construction materials.

Bibliography

ADAMS, K. T.: "Hydrographic Manual," *U.S. Coast and Geodetic Survey, Special Pub.* 143, Government Printing Office, 1942.

AMERICAN SOCIETY OF CIVIL ENGINEERS: *Manuals of Engineering Practice:* 10, "Technical Procedure for City Surveys," 1934; 14, "Location of Underground Utilities," 1937; 16, "Land Subdivision," 1939; 20, "Horizontal Control Surveys to Supplement the Fundamental Net," 1940; and 34, "Definitions of Surveying, Mapping, and Related Terms," 1954.

AMERICAN SOCIETY OF PHOTOGRAMMETRY: "Manual of Photogrammetry," 2d ed., 1952.

BIRDSEYE, C. H.: "Topographic Instructions of the United States Geological Survey," *U.S. Geological Survey, Bull.* 788, Government Printing Office, 1928.

CLARK, F. E.: "Law of Surveying and Boundaries," 2d ed., The Bobbs-Merrill Company, Inc., 1939.

DAVIS and FOOTE: "Surveying, Theory and Practice," 4th ed., McGraw-Hill Book Company, Inc., 1953.

DURHAM, E. B.: "Mine Surveying," McGraw-Hill Book Company, Inc., 1927.

HODGSON, C. V.: "Manual of Second and Third Order Triangulation and Traverse," *U.S. Coast and Geodetic Survey, Special Pub.* 145, Government Printing Office, 1935.

HOSMER and ROBBINS: "Practical Astronomy," 4th ed., John Wiley & Sons, Inc., 1948.

KIELY, R. E.: "Surveying Instruments," Columbia University Press, 1947.

LOW, JULIAN W.: "Plane Table Mapping," Harper & Brothers, 1952.

MITCHELL and SIMMONS: "The State Coordinate Systems: A Manual for Surveyors," *U.S. Coast and Geodetic Survey, Special Pub.* 235, Government Printing Office, 1945.

SLOANE and MONTZ: "Elements of Topographic Drawing," 2d ed., McGraw-Hill Book Company, Inc., 1943.

SWAINSON, O. W.: "Topographic Manual," *U.S. Coast and Geodetic Survey, Special Pub.* 144, Government Printing Office, 1928.

Swanson, L. W.: "Topographic Manual: Part II, Photogrammetry," *U.S. Coast and Geodetic Survey, Special Pub.* 249, Government Printing Office, 1949.

Talley and Robbins: "Photographic Surveying," Pitman Publishing Corporation, 1945.

U.S. Bureau of Land Management: "Ephemeris of the Sun, Polaris, and Other Selected Stars," Government Printing Office, annual.

U.S. Bureau of Land Management: "Manual of Instructions for the Survey of the Public Lands of the United States," Government Printing Office, 1947.

U.S. Coast and Geodetic Survey: Various publications.

U.S. Geological Survey: Various publications.

U.S. Naval Observatory: "American Ephemeris and Nautical Almanac" and "American Nautical Almanac," Government Printing Office, annual.

Wright and Hayford: "The Adjustment of Observations," D. Van Nostrand Company, Inc., 1906.

Section 2

RAILWAY, HIGHWAY, AND AIRPORT ENGINEERING

By John B. Babcock, 3d
and Alexander J. Bone

ROUTE SURVEYS

Location. The establishment of the location of a railway, highway, or other transportation facility warrants the most thorough study. The importance of adequate surveys and plans becomes evident when it is realized that the *final location* is the basis for construction of a *permanent artery of traffic.* Subsequent improvements may be made when justified by traffic requirements, provided the basic location is satisfactory. The principles of *route location* are essentially the same for all types of land transportation, viz., the *transportation service* which the facility should provide, its relation to other lines and to the system as a whole, and the standards of design and construction which should be adopted. Primary factors for railways and highways are length, gradients, curvature, and right of way. *Engineering standards* for these basic elements vary considerably with type of transport and character and amount of traffic, and they are treated separately for railways and highways. "Route Surveys" includes primarily those methods which are applicable to the location of *railways* and *highways. Airport* location is discussed under "Airport Engineering."

Route Surveys. The selection of the *final location* (alignment and grades) for a new section of highway or railway involves a series of coordinated steps similar in principle but differing somewhat in method of execution. Prior to the development of *aerial surveying* and *photogrammetry, route surveys* were conducted by *ground-surveying* methods exclusively. In recent years, *aerial surveys* and *photogrammetric methods* have taken the place of a large amount of ground surveys; these methods, however, still require ground surveying for horizontal and vertical control, special surveys, and staking of final line. On minor projects, where photogrammetric methods may not be justified, ground-survey methods will still be used. The principal applications of aerial surveying and photogrammetry to route surveys have been made by state highway departments and U.S. Bureau of Public Roads. Photogrammetric methods are now being supplemented by use of *electronic devices* for rapid computation of traverses and calculation of earthwork quantities. Important results in *rapidity and economy of route surveys* may be expected with further development of these processes. The following description of route-surveying methods applies to both ground and photogrammetric types of surveys. Particular attention is given to the more modern methods since information as to complete ground surveys is available in various textbooks on route surveying.

Basically, the steps in a route survey are: (1) *reconnaissance,* (2) *preliminary survey,* and (3) *final-location survey. Reconnaissance* includes two stages, (*a*) broad study of an *area* to determine *all feasible routes* and (*b*) comparison of these *alternate routes* as

to terrain, valley and ridge lines, water courses, approximate grades, service and convenience, and costs of right of way and construction in sufficient detail to determine which route appears to offer the most advantages. A *preliminary survey* of the general route selected constitutes the next step. This survey is made to obtain detailed information as to topography, drainage, soil conditions, river crossings, land use, etc., necessary for establishing the final location and estimating quantities and costs. *Final-location survey* consists of staking out on the ground the center-line location together with such other ground surveys as are needed in advance of construction. "Line and grade" surveying is considered under "Construction Surveying," p. 1–100.

In practice, these steps may not be carried out exactly as indicated. For some projects, reconnaissance and preliminary survey may be combined; where the approximate location of final line is fairly evident (or definitely limited in choice) separate steps are not required. Some route surveys represent a combination of ground- and aerial-surveying methods, wherein the reconnaissance is primarily based on a study of aerial photographs but ground surveys are used for obtaining detailed information for locating the final line.

Ground-survey Method. *Reconnaissance* may be made by a study of maps available from different sources, supplemented by pocket compass, hand level, aneroid barometer, or rough stadia surveys. This was typical of early railway locations in rugged and generally wooded terrain, where satisfactory maps were seldom available. Topographic (quadrangle) sheets, prepared by U.S. Geological Survey, are available for a considerable portion of the United States. A reconnaissance would seldom be made without making some use of air photos which are available for the entire country.[1]

Preliminary Survey. There is no standard form of preliminary survey by *ground* methods. In *railway* practice it comprised a transit and tape "open" traverse (along the general route indicated by the reconnaissance) for locating property lines, buildings, roads, etc.; bench-mark and profile levels; and location of contours (usually by hand level) for about 300′ each side of traverse. Alternate methods included stadia survey by transit or plane table for obtaining contours. Trial location lines were laid out on a topographic map prepared from preliminary survey. Ground profiles were plotted, tentative grades established, and earthwork quantities computed (usually based on center-line heights, sometimes with allowance for transverse ground slopes). The final "paper" location was then laid out on the ground, making such changes as were found necessary.[2]

Final-location survey is much the same whether complete ground surveys or photogrammetric methods are used in preliminary stages. *Final-location line on the ground* comprises a transit line, stationed throughout, with "line" stakes at each 100′ station (often 50′ apart on curves) and transit points (hubs) at PC's, PT's, and at intermediate points where needed. If preliminary traverse points are available in the field (ground-survey type), a number of points on final location can be staked from ties to preliminary survey and a complete final line run connecting them.

[1] For information as to maps and air photos, consult USGS Map Information Office, Washington, D.C., which serves as a coordinating office for the availability of maps of all United States mapping agencies.

USGS quadrangle sheets have been made for about 70 percent of the United States. Although some sheets are not up to date, probably at least 40 percent of the country is covered by modern maps. Many of these have a scale of 1:62,500 (approx 1″ = 1 mile) with 20′ contours; for mountainous terrain, scale may be 1:125,000 with 50′ contours. Those made in recent years usually have a scale of 1:24,000 (or 1:31,680) with 10′ or 20′ (sometimes 5′) contours. Quadrangle sheets are obtainable from USGS, Washington, D.C. Index sheets for maps in each state are furnished without charge.

Vertical aerial photographs (stereocoverage) are available for the entire country (scale usually 1:20,000). These are obtainable from Commodity Stabilization Service, U.S. Department of Agriculture. Index sheets by counties are available. Air photos have also been flown for parts of the United States by USGS, and along coastal areas by USCGS.

[2] In highway practice, it was more common to stake a tentative location line directly on the ground. Cross sections were taken along this line and a grade established on plotted cross sections. In open country, where not too rugged, the trial line was often satisfactory for the final location. If modifications in alignment were necessary, revised lines were laid out, cross-sectioned and quantities computed, until a satisfactory final line was secured.

When *aerial photogrammetric* methods are used for preliminary survey, *horizontal-control* monuments are available on the ground and their location is shown on topographic map. The control survey is usually tied in to the state coordinate system and the coordinates of control points determined. (Sometimes the horizontal control is tied in to a "grid" system for the project instead of to a state-wide system.) To "fix" the final line on map, it is only necessary to establish coordinates of PI's (vertices of curves) by *scaling* on map. *Stationing* of PC's, PT's, and any intermediate points on final line can then be determined, and their coordinates calculated. Convenient points on final location line can be staked on the ground by *ties* from horizontal-control monuments.[1] The complete location is then run between them and curves laid out connecting the tangents. Important transit points on final line should be "referenced" so that they can be relocated during construction. Bench marks are set and profile levels run. Property lines are usually tied in with final line for use in securing right of way. Cross sections may be taken along the final line although methods have now been developed for computing them by *photogrammetric* and *electronic* means. Grades are established on center-line stakes (or on parallel "offset" lines). "Slope stakes" may be set on projects involving heavy grading. Additional ground surveys may be needed for bridge sites, drainage, etc.

AERIAL PHOTOGRAMMETRIC ROUTE SURVEYS[2]

Photogrammetric Methods. The application of *aerial photogrammetry* to route surveying involves an understanding of some basic terminology and principles. *Oblique aerial* photographs are taken looking toward horizon; everything is seen in perspective. Although not used in the photogrammetric process, they are helpful in visualizing land use, right-of-way takings, culture, design of interchanges, etc. *Vertical aerial* photographs (usually 9″ by 9″) are taken with optical axis of lens pointing vertically downward; *they are fundamental to photogrammetry.* These photos are taken with 55 or 60 percent *overlap* in direction of line of flight; 30 percent *side overlap* is common when parallel flight lines are flown. Adjacent pairs of photographs, when viewed through a stereoscopic device, give the observer a three-dimensional view as though looking at a *relief model.* This principle of *stereovision* is the foundation of photogrammetry. Study of *stereopaired* photographs by simple stereoscope in reconnaissance or preliminary route surveys is an invaluable aid in selecting available routes for more detailed investigation. A *stereoplotter* is an instrument for creating and measuring a spatial model and recording the results.[3] This device, of which there are various types ranging from $4,000 to $75,000 each, makes possible the determination of elevations with reference to a system of vertical control and is the basic equipment used for plotting contours, determining cross sections, etc. A *mosaic* (sometimes called *photo-map*) is an assembly of adjoining vertical photographs, matched to give a complete photograph of an area. *Uncontrolled mosaic* is one made from photographs which have not been brought to uniform scale

[1] If it is impossible to set final line points by *direct* ties from horizontal-control monuments because of terrain or wooded ground, traverses can be run from the monuments to set them. If adjacent points, established on final line, are not mutually intervisible, it may be necessary to run trial traverses between them with subsequent minor adjustments to set stations on final line.

[2] For information on photogrammetric principles and practices, see AMERICAN SOCIETY OF PHOTOGRAMMETRY, "Manual of Photogrammetry," 2d ed., 1952. *Photogrammetry Engineering*, the journal of the society, provides excellent material on current methods and uses. Textbooks on photogrammetry (and on surveying) are also available.

References on aerial photogrammetric route surveys: PRYOR, W. T., Experience of U.S. Bureau of Public Roads in Highway Surveys, *Proc. ASCE*, vol. 82, No. SU3, December, 1956. BAKER, W. O., The Use of Photogrammetry to Civil Engineers, *Proc. ASCE*, vol. 82, No. SU3, December, 1956. FUNK, L. L., Improved Methods in Highway Location and Design, *Jour. Div. of Highways*, California Department of Public Works, November–December, 1956. To Make Location Maps: Go Up 20,000 Feet (location of 75 miles of railway in northern Minnesota), *Railway Age*, Nov. 5, 1956. OSOFSKY, S., Electric Computers Save Time on Traverses and Earthwork, *Civil Eng.*, November, 1956. PRYOR, W. T., Highway Engineering Applications of Photogrammetry, *Photogrammetric Eng.*, June, 1954. WILLIAMS, F. J., Photogrammetry Locates 208 Miles of Penn. Turnpike Extensions, *Civil Eng.*, December, 1950. (See footnote on p. 2–6 for references on air-photo soil interpretation.)

[3] MILLER, C. L., The Spatial Model Concept of Photogrammetry, *Photogrammetric Eng.*, March, 1957.

or corrected for tilt of camera. Although not accurate, this type is usually satisfactory for reconnaissance of alternate routes in an area. In a *controlled mosaic*, photographs have been brought to uniform scale and corrected for tilt; this type is generally used where accuracy is desired. Modern stereoplotters include devices which make it possible to adjust in the machine for uniform height and correct for tilt of each photograph. (*Diapositive* plates are generally used in stereoplotter.)

Control for Aerial Surveys. *Horizontal* and *vertical control* are essential for aerial photogrammetry. Although *horizontal control* is essential in the first instance for photogrammetry, the current trend is to use it also in establishing final location line on ground. To accomplish this, it is often desirable to establish more horizontal-control monuments, and to require a greater degree of accuracy, than would be needed for mapping control alone. (See p. 2–3 for use of horizontal-control monuments in staking location line.) *Control surveys* are made after photographs are available; it is then possible to select control points which are readily identifiable on photographs.

Accuracy of Topographic Map. Although not completely standardized, a common requirement for *horizontal accuracy* of topographic maps made by photogrammetric methods is as follows: Distance between 90 percent of any two well-defined cultural features, scaled from map, should check distance measured on ground within 1/40″ (map scale), and no inaccuracies should exceed 1/20″ (map scale). (For 1″ = 200′ map, these tolerances would be 5′ and 10′ respectively.) For *vertical accuracy*, common requirement is that 90 percent of all contours plotted should be correct within *one-half* of contour interval and no inaccuracy should exceed *one* contour interval, allowance being made for shifting of contour 1/40″ (map scale) if necessary to meet the requirements. If a contour map is desired where terrain is very heavily wooded, a vertical tolerance is sometimes allowed up to one-half average height of vegetation or two contour intervals, whichever is greater. The cost of checking maps by ground measurements, currently borne by the *user*, may be a considerable expense if it is done thoroughly.[1]

Aerial Reconnaissance. The first step in reconnaissance of an area is assembly of all available maps, including USGS quadrangle sheets where available, geologic and soil maps, and maps from other sources. Vertical air photographs are usually available (see footnote, p. 2–2). Stereoscopic examination of these photographs will furnish information as to topography and other features. The available maps and photographs should make it possible to determine limits of area to be studied. At this stage, it is common to have the area flown at a scale between 2,000′ and 1,000′ per in. for widths of 10 to 50 miles, and 3 to 10 miles respectively. An *uncontrolled mosaic* is prepared from photographs. Stereoscopic study of these photographs and use of mosaic should enable the locating engineer to determine *all feasible routes.* These alternate routes are then compared as to their relative advantages and disadvantages on basis of grades and curvature, directness of route, approximate quantities and costs, character of soil, drainage, bridge structures, land use, etc. Usually the route can thus be selected for *preliminary survey.* Occasionally a small-scale topographic map is made by photogrammetric methods to a scale of 1″ = 800′ with 20′ contours to determine the best route for preliminary survey.

Aerial preliminary survey is now made of the route selected. Flight strips are established covering sufficient width to include probable location of final line. Widths may vary from a few hundred yards to a mile or more; normally one flight strip will cover the desired width although occasionally parallel strips may be needed. From vertical photographs, a topographic map is made by stereoplotting equipment. Scale should be large enough to furnish topographic and land-use detail; usually map will

[1] An analysis of the survey cost of six freeway projects in California indicated that the average cost of ground checking was $220 per mile where the original cost of aerial survey and photogrammetric contour maps was $1,165 per mile.

be 100 or 200 ft per in. with 5′ contours, or perhaps in urban areas with 2′ contours. Figure 1 shows a section of topographic map together with one of stereopaired photographs used in preparing it. From thorough investigation of map, *supplemented by careful ground inspection,* preliminary location lines may be plotted and compared. Profiles and cross sections may be determined and preliminary quantities of earthwork calculated. (Attention has been called to modern electronic equipment which will enable the computations to be made rapidly and economically.) When final location

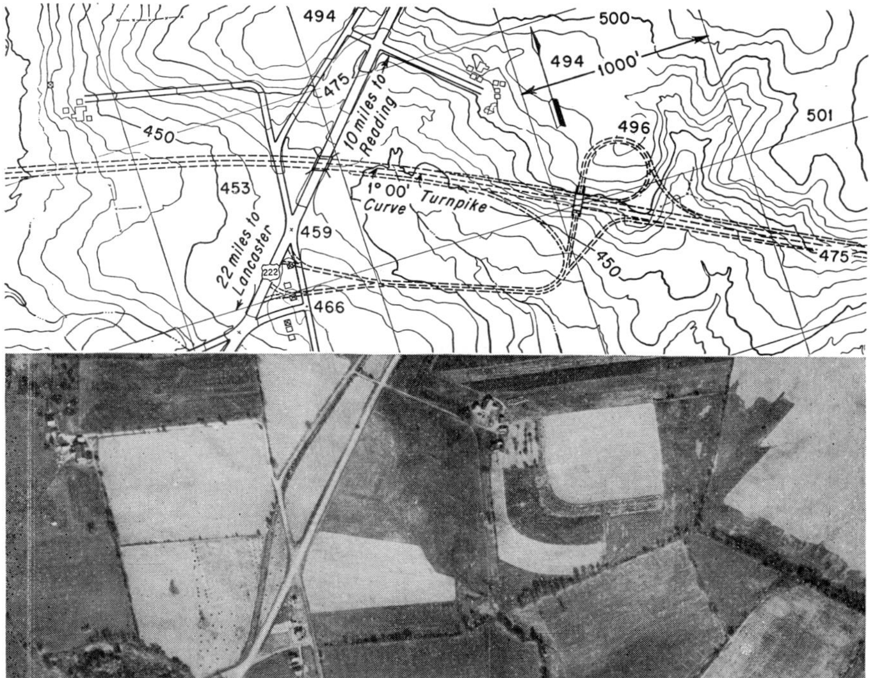

Fig. 1. Topographic map by photogrammetric method. *Upper half:* contour map of section of eastern extension of Pennsylvania Turnpike; dashed lines indicate preliminary location of trumpet-type interchange. *Lower half:* one of photographs used in making map by stereoplotter. (*Photo from Aero Service Corp., Philadelphia, Pa.*)

line has been plotted on map it is staked on the ground. For detailed design of the project it is often desirable to have larger-scale maps along the final line with scales of 1″ = 100′ with 2′ contours, or 1″ = 40′ or 50′ with 1′ or 2′ contours. For bridge-site studies, scale might be 1″ = 20′ with 1′ contours. For flat land, *spot* elevations may be taken to 0.2′ instead of showing contours.[1]

Final-location Survey. The staking of final location line is done by ground surveying. Attention has been called to modern practice of locating line by coordinates on map and then staking points on ground from horizontal-control monuments (see p. 2–3). The other field-work details of final location are substantially the same for both *aerial-photogrammetric* and *complete-ground-survey* types of route surveys.

[1] Scale of vertical photographs depends on map scale to be used. For map of 1″ = 100′ or 200′, photographic scale would be 1″ = 500′ to 1,000′; for larger-scale maps, photographic scale is from 1″ = 250′ to 500′; for reconnaissance study (without topographic map) photographic scale may be 1″ = 2,000′ or more.

Quantities of earthwork for bidding purposes have usually been based on cross sections taken on the ground along final location line. But it is becoming more common to determine these quantities from map or from vertical photographs; it is now practicable to compute cross-section areas and volumes and mass-diagram ordinates by photogrammetric and electronic methods without plotting cross sections. However, *typical cross sections* would be shown on construction plans for information of bidders. Latest developments indicate the possibility of determining *final pay quantities* for earthwork from these basic data; where final measurements are required for determination of quantities, additional photographs of completed work could be taken to provide necessary information.[1]

Route surveys for other transportation facilities include some of the foregoing methods. (See p. 2–26 for airport *location*.) Route surveys for *pipelines* and *transmission lines* are well adapted to photogrammetric methods. Surveys for these facilities include a reconnaissance survey of available routes. In general, *directness of route* is of prime importance. A preliminary survey by photogrammetric methods may not be needed since topography is not so important; a center-line profile may be sufficient. For pipelines, a study of underground conditions (avoidance of ledge rock) is essential. Planimetric surveys (without contours) and aerial photographs (often oblique) are helpful in connection with obtaining right of way and in securing approval of location by regulatory bodies. Photographic methods are also useful for location of *canals* and *improvement* of waterways.

Aerial photographic interpretation of *soil* and *construction conditions*, although not limited to route surveys, is important in that field. It includes the invariable use of stereoscope, examination of all elements of air-photo pattern and an evaluation of them by an interpreter who has a technical and practical geologic background. Examination of all air-photo elements includes evaluation of landform, surface-drainage network, erosion, vegetation, etc. Photo interpretation of these gives *qualitative* information as to location, origin, texture of soils and unconsolidated deposits; type and structural characteristics of ledge rock; delineation of areas having good internal drainage and imperfect drainage; and location of probable construction problems or adverse conditions. From these air-photo studies it is possible to determine locations where detailed soil investigations should be made at the site. An important feature of air-photo interpretation is its use in locating sources of sand and gravel and potential rock quarries for construction materials. The incorporation of air-photo soil and construction material interpretation with standard route-surveying methods represents an important development in recent years.[2]

Comparison of Methods. The *aerial photogrammetric* method has many advantages over complete *ground surveys* for highway and railway location. These include speed, economy, a more *over-all coverage* of *reconnaissance areas*, and *wider strip coverage* of topography on *preliminary surveys*. Other advantages include the elimination of staking tentative routes on the ground before final location is determined. Unnecessary publicity at preliminary stages may be avoided. Photographs are invaluable for securing approval by public authorities, and for obtaining right of way. Aerial methods afford opportunity for study of present and future land use. Possible disadvantages may be inaccuracies in map if ground is densely wooded and difficulties of photography

[1] Extensive research on the application of photogrammetry to highway location, design, and determination of preliminary and final quantities of earthwork is being carried on by the Photogrammetry Laboratory of Department of Civil and Sanitary Engineering, Massachusetts Institute of Technology under sponsorship of Massachusetts Department of Public Works and U.S. Bureau of Public Roads. Publication 111, August, 1957, "Digital Terrain Model Approach to Highway Earthwork Analysis," and Publication 115, "Earthwork Data Procurement by Photogrammetric Methods," are typical reports of the research under way.

[2] For references see Weder, D. R., Aerial Photographic Interpretation, *The Highway Magazine*, August, 1956; and Gray, H., Applications of Air-photo Interpretation to Soil Surveys, *Proc. Conference on Modern Highways*, MIT, June, 1953. See also *Photogrammetric Eng.*

when foliage is out; the latter can usually be overcome by planning work so that photography can be done in late fall (after foliage is gone and before excessive snow cover) and in early spring (before foliage is out and after most of snow has melted).

Relative *costs per acre* will *always* be cheaper by photogrammetric method, but *cost per route mile* is sometimes about the same. Width of topography is, however, much greater by aerial method. (Where possible route is closely limited, increased width may not be needed.) Speed of photogrammetric survey is very much greater; flights can be made in short period of time and topographic maps furnished relatively soon after photography and ground control have been completed. The following costs of route surveys by photogrammetric method with estimated cost of complete ground survey are available for six freeways in California.[1]

	Average cost per mile by photogrammetric method			
	Low	High	Average	Percent
Contour maps	$945	$1,520	$1,165	56
Checking	165	355	220	11
Additional field data	40	415	205	10
Staking center line	295	1,035	480	23
Total cost			$2,070	100

The estimated cost by complete ground surveys was $3,530 per mile, indicating a saving by photogrammetric method of 41 percent. Average cost of ground surveying (for aerial method) was $905 per mile (including checking, additional field work, and staking); comparison of this with estimated cost of $3,530 per mile for complete ground surveys indicated a saving in manpower (to the state) of 74 percent.[2] Costs per acre for photogrammetric survey on extensions of Pennsylvania Turnpike (1950) were from $1.00 to $1.29 per acre (200′ scale map with 5′ contours, 1 mile width of coverage). Current costs would probably be at least $2.00 per acre.

CIRCULAR CURVES

Simple Curves. The alignment of a railway or highway consists of straight portions (tangents) connected by circular curves; on main-line railway tracks and on some high-speed highways, spiral transition curves are inserted between tangents and circular curves. Circular curves are classed as simple, compound, and reversed. A *simple curve* is a single circular arc connecting two tangents. The point where the curve starts is the PC, and the end of the curve is the PT. (PC is at the end toward station 0 + 00.) The intersection of the tangents extended is the PI (or V). The intersection angle at PI, which equals the central angle, is I. Important elements of a simple curve are intersection angle I, radius R, degree of curve D, length of curve L, tangent distance T from PC (or PT) to PI, long chord C from PC to PT, external distance E from PI to mid-point of curve, and middle ordinate M from mid-point of curve to mid-point of long chord C. The following formulas apply (Fig. 2):

$$T = R \tan \tfrac{1}{2}I \qquad E = R \operatorname{exsec} \tfrac{1}{2}I = R(\sec \tfrac{1}{2}I - 1)$$
$$C = 2R \sin \tfrac{1}{2}I \qquad M = R \operatorname{vers} \tfrac{1}{2}I = R(1 - \cos \tfrac{1}{2}I)$$

[1] Funk, L. L., "Photogrammetry in Highway Location and Design," Division of Highways, California Department of Public Works, 1956.

[2] On a recent photogrammetric survey for railway relocation, cost for vertical photographs, mosaic, and topographic map was $4,600 (not including checking, staking line, etc.) compared with an estimated cost by complete ground surveys of about $9,000. The time required for this method was much less than for ground surveys.

Degree of Curve. In railway and highway practice (except for city streets), the degree of curve has been generally adopted as a unit of sharpness. The degree is defined either (1) as the central angle subtended by a *chord* of 100 ft, or (2) as the central angle subtended by an *arc* of 100 ft. The *chord* definition has been universally adopted for railway curves. Although both are in use on highways, the *arc* definition is gradually replacing the earlier *chord* definition.

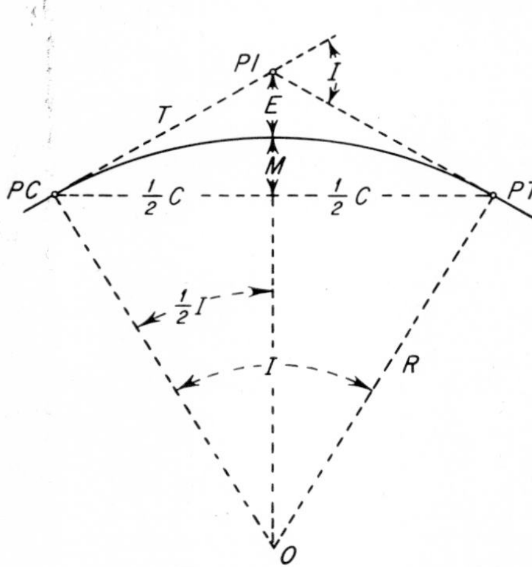

FIG. 2. Simple curve.

The important relations among D, R, and L are as follows:

Chord Definition. $R_c = 50/\sin \frac{1}{2}D$. R_c (for 1° curve) = 5,729.65′. When D is small, $R_c = 5{,}730/D$ (approx). $L = I/D$ (stations) $= 100I/D$ (ft). The 100′ *chord* station is unit of length; L is the sum of successive chords from PC to PT including subchords at ends.

Arc Definition.[1] $D/360 = 100/(2\pi R_a)$; $R_a = 18{,}000/\pi D = 5{,}729.58/D$. R_a (for 1° curve) = 5,729.58′. $L = I/D$ (stations) $= 100I/D$ (ft). The 100′ *arc* station is unit of length; L is actual *arc* length from PC to PT. Difference between arc and chord $(a - c) = a^3/24R^2$; this formula is approximate but may be used when a does not exceed 100′.

Table 1 gives R for different values of D for both *chord* and *arc* definitions; also difference between arcs and chords for arcs of 100′ and 50′.

TABLE 1. RADII FOR DEGREE BY CHORD AND ARC DEFINITIONS AND DIFFERENCE BETWEEN ARCS AND CHORDS FOR 100′ AND 50′ ARCS

Deg D	Radius R_c (chord definition), ft	Radius R_a (arc definition), ft	Arc − chord		Deg D	Radius R_c (chord definition), ft	Radius R_a (arc definition), ft	Arc − chord	
			100′ arc, ft	50′ arc, ft				100′ arc, ft	50′ arc, ft
1°	5,729.65	5,729.58			9°	637.27	636.62	0.10	0.01
1°30′	3,819.83	3,819.72			10°	573.69	572.96	0.13	0.02
2°	2,864.93	2,864.79	0.01		11°	521.67	520.87	0.15	0.02
2°30′	2,292.01	2,291.83	0.01		12°	478.34	477.47	0.18	0.02
3°	1,910.08	1,909.86	0.01		13°	441.68	440.74	0.21	0.03
3°30′	1,637.28	1,637.02	0.02		14°	410.28	409.26	0.25	0.03
4°	1,432.69	1,432.40	0.02		15°	383.06	381.97	0.29	0.04
4°30′	1,273.57	1,273.24	0.03		16°	359.26	358.10	0.33	0.04
5°	1,146.28	1,145.92	0.03		17°	338.27	337.03	0.37	0.05
6°	955.37	954.93	0.05	0.01	18°	319.62	318.31	0.41	0.05
7°	819.02	818.51	0.06	0.01	19°	302.94	301.56	0.46	0.06
8°	716.78	716.20	0.08	0.01	20°	287.94	286.48	0.51	0.06

Functions of 1° Curve. For *arc* definition: $T_a = T_1/D$ and $E_a = E_1/D$, where T_1 and E_1 are the values for a 1° curve. For *chord* definition: $T_c = T_1/D$ (approx) =

[1] A modification sometimes used is $R = 5{,}730/D$ as an *exact* relation, results are nearly the same as by *arc* definition.

T_1/D plus "correction," and $E_c = E_1/D$ (approx). Most textbooks on route surveying contain tables giving functions of 1° curves for different values of I; their use simplifies the computations.

For city streets, D is seldom used; the arc length is obtained directly from R and I. The radius is usually chosen as a multiple of 10′, 20′, 50′, or 100′. Table 2 provides a simple method of finding the length of arc.

TABLE 2. LENGTHS OF CIRCULAR ARCS FOR UNIT RADIUS

Degrees	Length	Minutes	Length	Seconds	Length
1	0.0174533	1	0.0002909	1	0.0000048
2	0.0349066	2	0.0005818	2	0.0000097
3	0.0523599	3	0.0008727	3	0.0000145
4	0.0698132	4	0.0011636	4	0.0000194
5	0.0872665	5	0.0014544	5	0.0000242
6	0.1047198	6	0.0017453	6	0.0000291
7	0.1221730	7	0.0020362	7	0.0000339
8	0.1396263	8	0.0023271	8	0.0000388
9	0.1570796	9	0.0026180	9	0.0000436
10	0.1745329	10	0.0029089	10	0.0000485
20	0.3490659	20	0.0058178	20	0.0000970
30	0.5235988	30	0.0087266	30	0.0001454
40	0.6981317	40	0.0116355	40	0.0001939
50	0.8726646	50	0.0145444	50	0.0002424
60	1.0471976				
70	1.2217305				
80	1.3962634				
90	1.5707963				
100	1.7453293				
110	1.9198622				
120	2.0943951				

Example: $I = 82°17'53''$, $R = 750'$. Find arc length.

80°	1.3962634
2°	0.0349066
10′	0.0029089
7′	0.0020362
50″	0.0002424
3″	0.0000145

750 × 1.4363720 = 1077.279′

Deflection-angle Method (Fig. 3). Curves are usually laid out by deflection angles, whereby points are set by angles laid off from the PC (or intermediate hub) and chord lengths taped from preceding points. Unless obstructions interfere, curves are run in from PC to PT with transit at PC. Deflection angles for 100′ lengths are multiples of ½D. Deflection angle *in minutes* for a length less than 100′ equals $0.3cD$ (*chord* method) or $0.3aD$ (*arc* method); the formula is exact for *arc* definition and involves a

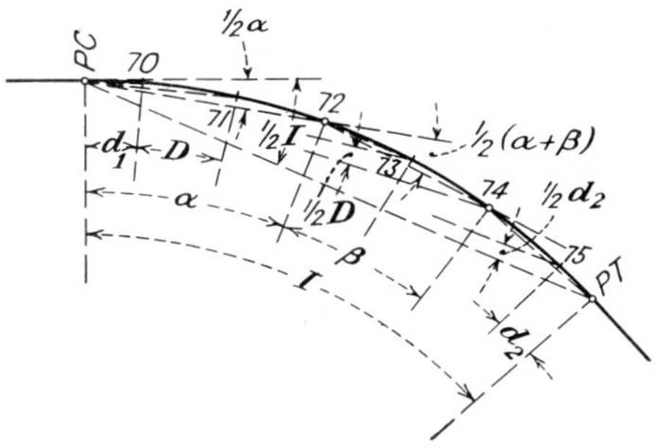

FIG. 3. Deflection-angle method of laying out curve.

negligible approximation for *chord* definition. Total deflection angle to PT equals $\frac{1}{2}I$ and provides a check on the computations. When *arc* definition is used, chord lengths may be found by formula: $a - c = a^3/24R^2$. Points are often needed less than 100, apart, especially on sharp curves. When *chord* definition is used, intermediate points (not at full stations) may be taped from preceding intermediate points (or from full station), but full stations should be taped from preceding full stations.

Intermediate Setups. When not practicable to lay out entire curve from PC, a hub (transit point) is set at a convenient station, and curve continued from there. Two methods are available: (1) With transit at intermediate hub, take backsight on preceding hub, with plate set at the deflection angle from PC to the hub *on which backsight* is taken. (If transit is at *first* intermediate hub, backsight 0° on PC.) Foresight angles for remaining stations may then be taken from the *original* notes (i.e., as though transit were at PC). (2) With transit at intermediate hub, take backsight on preceding hub with plate set *back of* (*or beyond*) 0° by one-half the central angle between transit point and the hub used for backsight. (This orients transit at 0° on the *local tangent* at transit point.) Foresight angle to next point is one-half the central angle between transit point and the point to be set. When intermediate hub is at a full station, foresight angles to following full stations are multiples of $\frac{1}{2}D$. Method of computing and laying out a curve by deflection angles is shown by the following example.

Example: Laying Out Curves by Deflection Angles. Given $I = 67°14'$. $D = 11°$ (*chord* definition). Station PI = 72 + 83.55. $R_c = 50/\sin \frac{1}{2}D = 521.67'$ (or from Table 1). $T_c = R_c \tan \frac{1}{2}I = 346.82'$.[1] $L = 100I/D = 611.21'$. PC = PI − T_c = 69 + 36.73. PT = PC + L = 75 + 47.94. $\frac{1}{2}d_1 = 0.3 \times 63.27 \times 11 = 209'$ = 3°29'; $\frac{1}{2}d_2 = 0.3 \times 47.94 \times 11 = 158' = 2°38'$; $\frac{1}{2}D = 5°30'$.

Without Intermediate Setup

Transit at PC BS 0° on PI

Station	Chord	Deflection angle
70	63.27	3°29'
71	100	8°59'
72	100	14°29'
73	100	19°59'
74	100	25°29'
75	100	30°59'
PT	47.94	33°37'($\frac{1}{2}I$)

With Intermediate Setups [1]

Stations 70, 71, and 72 set from PC

Transit set up at 72. *BS* 0° *on PC*
Deflection angle to 73 19°59'
Deflection angle to 74 25°29'

Transit set up at 74. *BS* 14°29' *on* 72.
Deflection angle to 75 30°59'
Deflection angle to PT 33°37'

[1] *Alternate method:* Transit at 72, BS—14°29' (*back of* 0°) on PC. Deflection angles to 73 and 74 are 5°30' and 11°00'. Transit at 74, BS—11°00' (*back of* 0°) on 72. Deflection angles to 75 and PT are 5°30' and 8°08'.

NOTE: If $D = 11°$ (*arc* definition). $R_a = 5{,}729.58/11 = 520.87'$ (or from Table 1). $T_a = 346.29'$. $L = 611.21'$. PC = 69 + 37.26. PT = 75 + 48.47. Deflection angles = 3°27', 8°57', 14°27', 19°57', 25°27', 30°57', and 33°37'. Chord at PC = 62.70'; chord at PT = 48.45'; "station" chord = 99.85'.

Offset from Tangent. When PC or PT is inaccessible, it may be desirable to set a hub on the curve by an offset from the main tangent and then lay out the rest of the curve by deflection angles. In Fig. 4, assume PC is inaccessible and that a hub is to be set at F. Set G by taping from PI; G-PI = $T - R \sin \Delta$. With transit at G, set hub at F by right-angle offset $GF = R$ vers Δ. Set transit at F and backsight on G with plates set at $(90 + \Delta)$ *back of* 0°. Transit is now oriented at 0° on tangent to curve at F; the remainder of curve may be run by deflection angles, checking on PT which was previously set from PI. (If GF is too short a backsight for accurate work,

[1] If table for T_1 is available: $T_c = T_1/D$ + correction = 3,809.2/11 + 0.53 = 346.82'.

transit may be oriented by backsighting on a "swing offset" (equal to GF) at PI, with plates set at Δ *back of* 0°.) The method of *tangent offsets* is often used to compute coordinates to points on a curve, referred to main tangent at PC or PT.

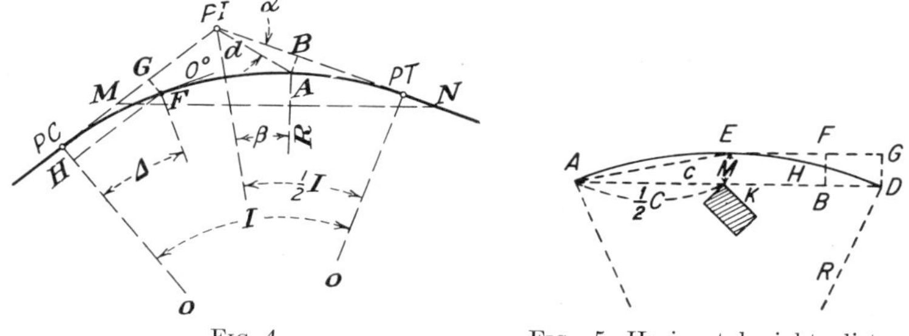

FIG. 4

FIG. 5. Horizontal sight distance.

Curve through a Given Point (Fig. 4). It is desired to find R for a curve to pass through A, which has been located by distance d and angle α from PI (or by distance PI-B and right-angle offset AB). Angle O-PI-A $= 90° - \frac{1}{2}I - \alpha$; O-PI $= R/\cos \frac{1}{2}I$; $\sin(O\text{-}A\text{-PI}) = \dfrac{\cos(\frac{1}{2}I + \alpha)}{\cos \frac{1}{2}I}$; $\beta = 180° - (O\text{-PI-}A) - (O\text{-}A\text{-PI})$; $R = \dfrac{d \sin(O\text{-PI-}A)}{\sin \beta}$ $= \dfrac{AB}{\text{vers}(A\text{-}O\text{-PT})} = \dfrac{d \sin \alpha}{\text{vers}(\frac{1}{2}I - \beta)}$.

Curve When PI Is Inaccessible (Fig. 4). Measure MN (between two points on the tangents) and angles PI-M-N and PI-N-M; the sum of these angles equals I. Compute PI-M and PI-N in triangle M-PI-N. Determine D by any desired method. Set PC by $[T - (\text{PI-}M)]$ from M and set PT by $[(\text{PI-}N) - T]$ from N.

Horizontal Sight Distance (Fig. 5). If K is corner of a building or other obstruction, sight distance from A on curve is AD. Measure EK at an elevation of 4.5′ above the pavement which is height of eye for a car driver. $EK = M$ (middle ordinate of curve AD). $M = c^2/2R$ (exact) $= C^2/8R$ (approx). Sight distance $AD = \sqrt{8MR}$ (approx).

For other ordinates from a chord, the following formulas are useful (Fig. 5): $HB = M - FH = M - (M)(EF^2)/(\frac{1}{2}C)^2$ (approx). Also $HB = (AB \times BD)/2R$ (approx).

Change of Location (Fig. 6). Three tangents are originally connected by curves 1 and 2. Required to move the middle tangent into a parallel position p distant from

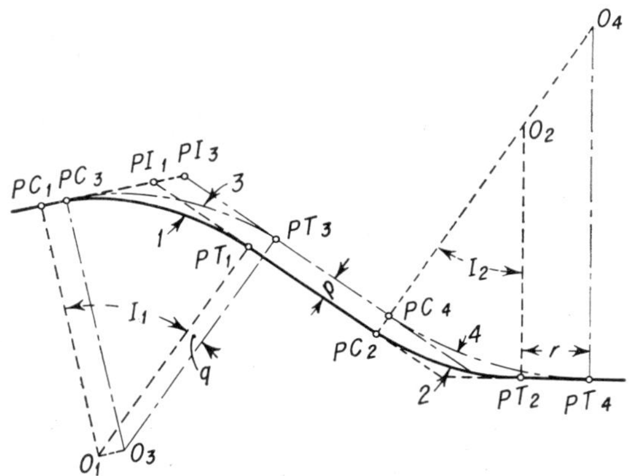

FIG. 6. Change of location.

original location. Assume that curve 3 is to have the same degree as curve 1 and that degree of curve 4 is to be such that PC_4 will be opposite PC_2. PC_1-PC_3 = PI_1-PI_3 = $p/\sin I_1$. $q = p \cot I_1$. $T_4 = T_2 + p \cot I_2$. $R_4 = T_4/\tan \frac{1}{2}I_2$. D_4 may be obtained from R_4. Compute station PT_4 based on the *new alignment* via curves 3 and 4. $r = T_4 + p/\sin I_2 - T_2 = p/\sin I_2 + p \cot I_2$. Station $PT_2 + r$ will give location of PT_4 based on *original* stationing. A "chainage equation" should be shown at PT_4 giving its stationing on both *original* and *revised* alignments. (NOTE: Degrees of curves 3 and 4 could have been chosen in other ways, but the general method of solution would be the same.)

Compound Curves. A compound curve is formed by two circular curves having a common tangent and a common point of tangency PCC, the curves lying on the same side of the common tangent. Conditions may be such that no single (simple) curve will

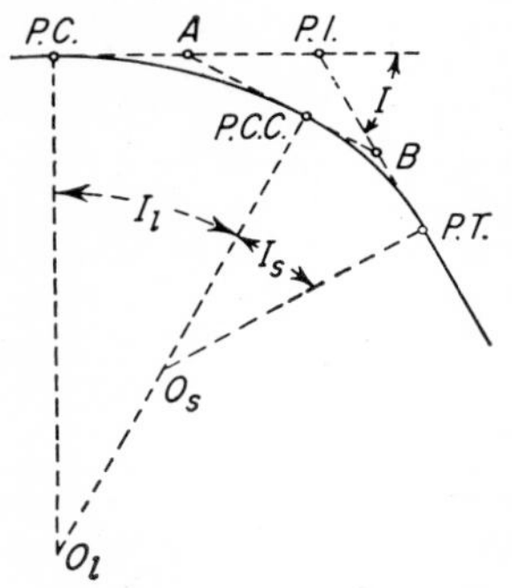

FIG. 7. Compound curve.

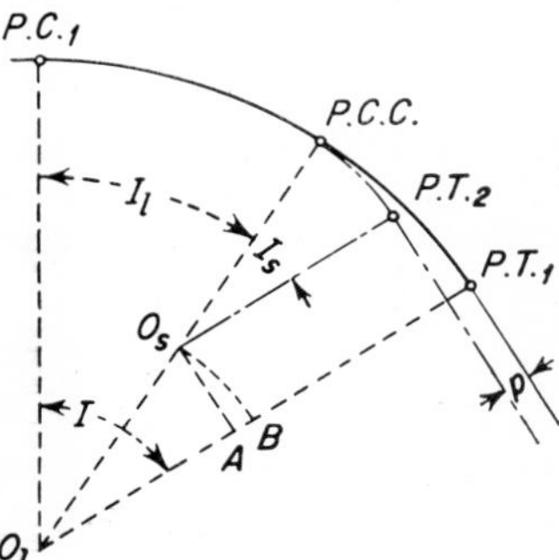

FIG. 8. Change of location.

be satisfactory and it is necessary to use a compound curve. In Fig. 7, R_l, D_l, etc., refer to flatter curve, and R_s, D_s, etc., to sharper one. T_l and T_s are *total* tangent distances from PC (or PT) to PI. $I = I_l + I_s$. T_l is always greater than T_s. *Given* R_l, R_s, I_l, I_s. To *find* T_l, T_s, and station PC, PCC, and PT. PC-A = A-PCC = $R_l \tan \frac{1}{2}I_l$. PT-B = B-PCC = $R_s \tan \frac{1}{2}I_s$. Solve for A-PI and B-PI. T_l = PC-A + A-PI. T_s = PT-B + B-PI. Station PC = PI − T_l. Station PCC = PC + $100I_l/D_l$. Station PT = PCC + $100I_s/D_s$.

A compound curve may be used to revise the location at one end of a simple curve. In Fig. 8, PC_1-PT_1 is original curve. The tangent through PT_1 is to be moved *inward* to a parallel position p distant from its original location, and the simple curve is to be compounded to fit the revised tangent. Station PCC may be selected and new radius R_s computed, or vice versa. In either case, $AB = p = (R_l - R_s)$ vers I_s. $I_l = D_l$ (PCC-PC_1)/100. $I_s = I - I_l$. Station PT_2 = PCC + $100I_s/D_s$. $O_sA = (R_l - R_s) \sin I_s$. If tangent is moved *outward*, new curve is flatter than original simple curve; the same type of solution may be used.

Reversed Curves. A reversed curve is formed by two circular curves having a common tangent and a common point of tangency PRC, the curves lying on opposite sides of the common tangent. Reversed curves are often used to connect parallel lines as shown in Fig. 9. In this case $I_l = I_s = I$, and PRC lies on straight line connecting PC and PT. $p = a + b = (R_l + R_s)$ vers I. PC-A = $p \cot \frac{1}{2}I = (R_l + R_s) \sin I$. Ordinarily, equal radii R_c are used: $R_c = p/(2 \text{ vers } I)$; PC-$A$ = $2R_c \sin I$. If PC-A is the limiting condition, $\tan \frac{1}{2}I = p/(\text{PC-}A)$, and $R_c = p/(2 \text{ vers } I)$.

When necessary to connect nonparallel tangents, the problem may be simplified by reducing to a case of parallel tangents and solving as above. In Fig. 10, tangents make

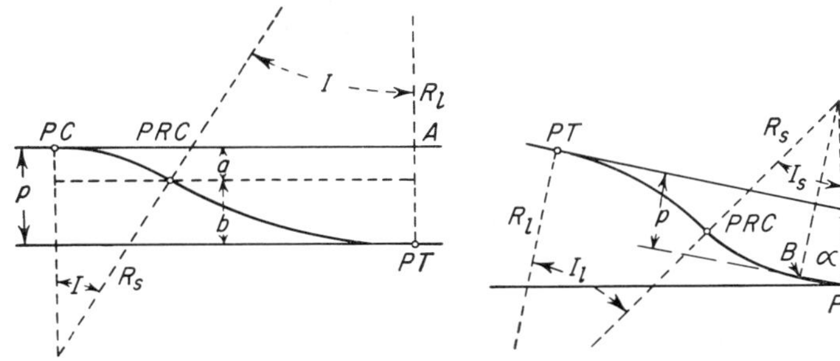

FIG. 9. Reversed curve (parallel tangents).

FIG. 10. Reversed curve (nonparallel tangents).

an angle of α with each other. Given R_l, R_s, α, and location of PC, $p = R_s \text{ vers } \alpha + (\text{PC-}O) \sin \alpha$.

$$\text{vers } I_l = p/(R_l + R_s) \qquad I_s = I_l + \alpha$$

PT-O = (PC-O) $\cos \alpha + R_s \sin \alpha + (R_l + R_s) \sin I_l$.

Compound and reversed curves are also used to connect the sides of intersecting streets when one or both streets have curved alignment.

Highway Geometric Design. Many highway-design problems are involved in grade separations and interchanges. Methods of computation involve the principles of circular curves as presented here but the geometric design is more complicated. The use of coordinates often facilitates the computations and the subsequent staking of points in the field. Figure 37 shows some typical layouts for which detailed calculations are often needed.

VERTICAL CURVES

Parabolic vertical curves are used to connect grade lines of railway, highway, and airport runway profiles. Other civil engineering applications include their use for transverse crown of a pavement, spillway section of a dam, parabolic arch of a bridge, etc. *Horizontal* parabolic curves are used occasionally for roads in landscape work when *unequal* tangent lengths are desired; such curves can be laid out by *oblique* offsets from the tangents.

Method of Computation. Vertical curve elevations are needed at ends of curve (*bvc* and *evc*), at all full stations, and often at additional points. These elevations are usually obtained by applying vertical offsets to the tangent (grade-line) elevations. A *sag curve* occurs when a descending grade is followed by an ascending grade or by a flatter descending one; a *summit curve* occurs when an ascending grade is followed by a descending grade or by a flatter ascending one. Grades (g_1 and g_2) are expressed as percentages of rise to horizontal distance. Length L is horizontal distance (in 100-ft stations) from *bvc* to *evc*; vertical curves usually extend $\frac{1}{2}L$ (stations) each side of vertex V.[1] The algebraic difference in gradients $A = g_1 - g_2$. Since the slope of a parabola changes uniformly throughout its length, the rate of change of grade r per station equals A/L. The vertical curve bisects the line from V to mid-point of chord connecting *bvc* and *evc*. Referring to Fig. 11, $M = EV = \frac{1}{2}DV$. (As a check $M = AL/8$.) Other offsets vary as the squares of the distances from *bvc* (or *evc*), $y = Mx^2/(\frac{1}{2}L)^2$. Elevations on vertical curve, as computed by tangent offsets, may be checked

[1] If vertical curve extends an *unequal* number of stations each side of vertex, separate parabolas may be used from *bvc* and *evc* respectively. These will meet at a point vertically above (or below) vertex on a *common tangent* parallel to chord connecting *bvc* and *evc*. Elevation of this point is the average of that of vertex and of point on chord *bvc-evc* directly above (or below) vertex.

by the principle of *constant* "second" differences since the slope of the parabola changes uniformly from initial to final tangent. (For 100-ft stations, the constant "second" difference is A/L or r.)

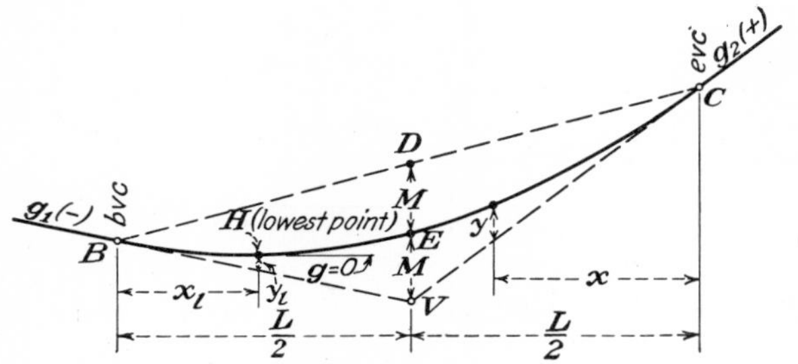

FIG. 11. Vertical curve connecting grade lines.

Lowest or Highest Point on Vertical Curve. It may be necessary to find the station and elevation of the lowest point on a *sag curve* or the highest point on a *summit curve.* When grades are equal, these points are directly above or below V; but when grades are unequal (*and of opposite sign*) the points lie between V and the end of curve which is on the flatter gradient. The slope of the tangent is zero at these points. Referring to Fig. 11, $-g_1 + rx_l = 0$ or $x_l = g_1/r = g_1L/A$. $y_l = Mx_l^2/(\frac{1}{2}L)^2$.

Example: (See Fig. 11, and accompanying table.) *Sag curve.* $g_1 = -3.28$; $g_2 = +4.40$. $L = 8$ stations (800′). V = elevation 91.78. $A = -3.28 - 4.40 = (-)7.68$. $r = A/L = 0.96$. Elevation B = 104.90; elevation C = 109.38; elevation $D = (104.90 + 109.38)/2 = 107.14$. $M = EV = \frac{1}{2}DV = 7.68$. ($M$ may be checked by $AL/8 = 7.68$.) y_1 (stations 17 and 23) $= M/(4)^2 = 0.48$; y_2 (stations 18 and 22) $= M/(2)^2 = 1.92$; etc. For lowest point: $x_l = g_1L/A = 3.42$ stations (342′). $y_l = M(3.42)^2/(4)^2 = 5.61$. Lowest point is elevation 99.29 at station 19 + 42.

The selection of the length of vertical curve is discussed for railways on p. 2–26 and for highways on p. 2–71.

Station	Elevation on tangent	Offset (+) for sag curve	Elevation on vertical curve	First difference	Second difference
16	104.90	....	104.90		
17	101.62	0.48	102.10	+2.80	0.96
18	98.34	1.92	100.26	+1.84	0.96
19	95.06	4.32	99.38	+0.88	0.96
20	91.78	7.68	99.46	−0.08	0.96
21	96.18	4.32	100.50	−1.04	0.96
22	100.58	1.92	102.50	−2.00	0.96
23	104.98	0.48	105.46	−2.96	0.96
24	109.38	....	109.38	−3.92	
19 + 42	93.68	5.61	99.29 (lowest point)		

SPIRAL TRANSITION CURVES

Spiral transition curves are used between tangents and circular curves and between branches of compound curves on main-line railway tracks. Although not standard in highway practice, spirals are used to some extent on modern high-speed routes.

Spiral Curve. Although several types of spirals are used, most of them are based on a cubic curve having the fundamental property that the *degree of the spiral increases directly with the length.* Other common spiral relations are (1) central or spiral angle varies as the square of length along the spiral; (2) deflection angle from TS (or ST) to any point on the spiral is one-third the spiral angle at that point;[1] (3) offset from main tangent (extended) to spiral varies as the cube of length along the spiral; and (4) length of circular curve extended back from SC (toward TS) through total spiral angle is one-half the length of spiral. Figure 12 shows application of spirals to a circular curve.

Notation.[2] D = degree of circular curve; d = degree at any point on spiral. L = total spiral length; l = length from TS (or ST) to any point on spiral. S = length L in 100-ft stations; s = length l in stations. Δ = central (or spiral) angle of whole spiral; δ = spiral angle from TS to any point on spiral. A = deflection angle of whole spiral; a = deflection angle from TS to any point on spiral. k = increase in degree of curvature per 100-ft station along the spiral. B = orientation angle from tangent at SC to TS. *Functions are in feet or degrees unless otherwise noted.*

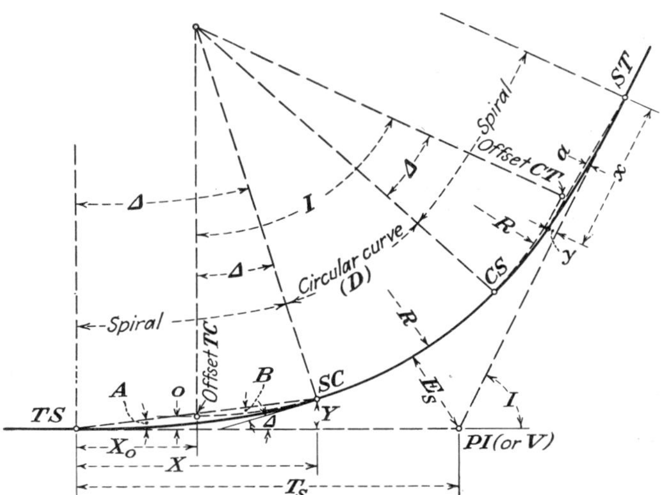

Fig. 12. Spirals applied to circular curve.

Formulas for AREA Spiral. $D = kS = kL/100$; $d = ks = kl/100$. $\Delta = DL/200 = \frac{1}{2}kS^2$ ($\Delta = L/2R$ in radians); $\delta = dl/200 = \frac{1}{2}ks^2$. $A = \frac{1}{3}\Delta = \frac{1}{6}kS^2$. $a = \frac{1}{3}\delta = \frac{1}{6}ks^2 = Al^2/L^2$. $B = \frac{2}{3}\Delta = 2A$. $b = \frac{2}{3}\delta = 2a$. $o = 0.1454\Delta S$. $X = L - 0.003048\Delta^2 S$. $X_o = X - R \sin \Delta$.[3] $Y = 0.582\Delta S - 0.00001264\Delta^3 S$. $E_s = (R + o)$ exsec $\frac{1}{2}I + o$. $T_s = X_o + T_c + o \tan \frac{1}{2}I$ (where $T_c = R \tan \frac{1}{2}I$).

Example (Fig. 12): Given $D = 5°$ (*chord* definition); $I = 52°18'$; $L = 200'$; PI = 73 + 14.62; $k = 2.5$; $\Delta = 5°$; $o = 1.45$; $X = 199.85$; $X_o = 199.85 - 99.90 = 99.95$; $T_s = 99.95 + 562.79 + 0.71 = 663.45$; station TS = (73 + 14.62) − 663.45 = 66 + 51.17; SC = 68 + 51.17; $L_c = 100 \times 42.30/5 = 846.00$; CS = 76 + 97.17;

[1] This involves a slight approximation, negligible when spiral angle does not exceed 15°.
[2] No standard nomenclature has been adopted for spirals. Nomenclature in this text is that used in "AREA Manual."
[3] When *arc* definition of D is used, X_o also equals $\frac{1}{2}L - 0.000508\Delta^2 S$.

ST = 78 + 97.17; A = 1°40′; B = 3°20′. For 10-*chord* spiral: a_1 = 0°01′; a_2 = 0°04′; a_3 = 0°09′, etc.[1]

Textbooks on route surveying usually contain tables giving values of Δ, o, X_o, X, Y, and deflection angles for the 10-chord spiral for various combinations of L and D.[2] The *chord* definition of D is used in railway work, whereas the *arc* definition is increasingly used on highways.

Laying Out Spirals by Deflection Angles. After T_s is computed, TS and ST are located by taping from PI (or from other hubs on the tangents). The first spiral is run by deflection angles from TS to SC. Transit is then set at SC and backsight taken on TS with plate reading $-2A$ (i.e., *back of* 0°). Full stations on circular curve and CS are then set by deflection angles. Transit is then set at ST and spiral run to CS using the same angles as for first spiral; the field-work check is obtained at CS.

In staking by deflection angles, it is convenient to divide spiral into a number of equal chords (10 is common). The initial deflection angle a_1 is calculated for the first chord point; deflection angles for following chord points are a_1 times the chord numbers *squared.* If it is necessary to continue spiral from an intermediate transit point on spiral, the following method may be used. Backsight on TS with an angle set off (*back of* 0°) equal to twice the deflection angle from TS to transit point. (This orients transit on *local tangent* at that point.) For a following spiral point, the deflection angle for a circular curve having the *same degree as the spiral at the transit point* and a length equal to distance from transit to spiral point is then calculated; to this is added the deflection angle for the same length of spiral but calculated as it would be from TS.

Staking Spirals by Offsets. The spiral may be staked to the mid-point by right-angle offsets from the main tangent (extended) and from there to the SC by normal (radial) offsets from circular curve (extended back from SC to the "offset TC"). The offset at the mid-point equals ½o, and the other offsets vary as the cubes of the distances from TS or SC. (Offsets from tangent are equal to those from circular curve for equal distances from TS and SC.)

Applying Spiral to Compound Curve. The length of spiral is based on the difference in degrees of curvature of the branches of the compound curve. Elements of spiral may be found by the formulas given, using the L selected and for D the difference in degrees of curvature. Spiral extends ½L each side of the offset point of compound curvature. Spiral deflects from inside of flatter curve and from outside of sharper curve at the same rate as it would from a tangent. Spiral may be staked by deflection angles from either end. If transit is located at spiral point on flatter curve, reading 0° when sighting along the tangent to the circular curve, the deflection angles to set points on the spiral are equal to the deflection angles for corresponding points on the circular curve (extended) plus the deflection angles for the spiral. As an alternate method, spiral can be staked by offsets from the two circular curves in a manner similar to that previously described.

The selection of the length of spiral is discussed for railways on p. 2–27 and for highways on p. 2–69.

SUPERELEVATION OR BANKING

A vehicle traveling around a curve of constant radius at a uniform velocity exerts an outward or centrifugal force F (pounds) equal to Wv^2/gR, in which W is weight of vehicle (pounds); v, velocity (feet per second); g, acceleration due to gravity (feet per second per second); and R, radius of curve (feet). The outer rail of a curved track is *superelevated* (elevated above inner rail), and pavement of a curved highway is *sloped* or

[1] If *arc* definition of D is used: X_o = 99.97; T_c = 562.62; T_s = 99.97 + 562.62 + 0.71 = 663.30; station TS = 66 + 51.32; SC = 68 + 51.32; CS = 76 + 97.32; ST = 78 + 97.32.

[2] Barnett's "Transition Curves for Highways" (Government Printing Office, Washington, D.C.) contains tables for a *unit* length spiral, based on *arc* definition of D.

banked to counteract in whole or in part the centrifugal force of the moving vehicle. For *equilibrium* superelevation, the centrifugal force must be wholly resisted by the component of the weight of vehicle parallel to the plane of superelevation. Assuming that this resisting force acts horizontally (a very slight approximation), $F/W = e$, in which e is the slope of the plane of superelevation or banking (foot of rise per foot of width). Substituting 32.16 for g and changing from v (feet per second) to V (mph), the following formula for *equilibrium* superelevation is obtained: $e = 0.067V^2/R$ (sometimes expressed as $e = V^2/15R$).

For standard gage railway track, the formula for *equilibrium* superelevation becomes

$$E = (0.067DV^2 \times 4.958 \times 12)/5{,}730 = 0.0007DV^2$$

in which E is superelevation of outer rail, in., and D the degree of curve.[1] In *highway* practice slope or banking is expressed in feet (or inches) per foot of pavement width.

Whenever the actual superelevation is made less than that required for *equilibrium* superelevation for vehicles traveling at maximum authorized speed on a railway track (or at probable maximum speed at which automotive vehicles are likely to be operated on the curve), superelevation or banking is referred to as "unbalanced." The use of some form of *unbalanced* superelevation is common in both railway and highway practice (see p. 2–27 for railways and p. 2–68 for highways).

EARTHWORK [2]

Roadbed Sections. The width of subgrade depends upon the character and dimensions of the superstructure—ballast and track for *railways;* pavement, base courses, and shoulders for *highways* and *airport runways.* Figure 13 shows simple cross sections for

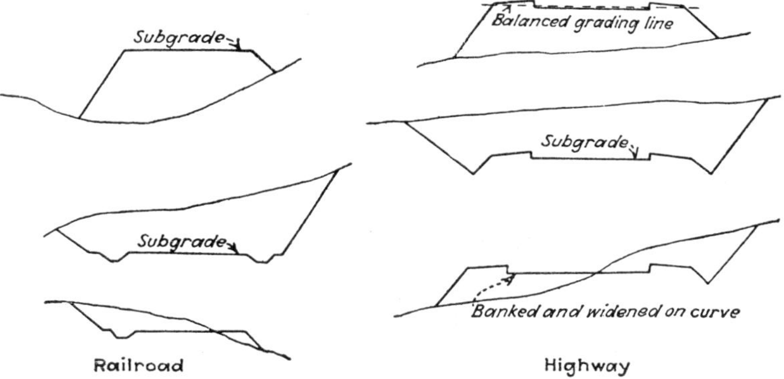

FIG. 13. Roadbed sections.

railways and highways.[3] Side slopes for railways are commonly 1½ (horizontal) to 1 (vertical) for earth cuts and fills, ½ to 1 for cuts in loose rock, and ¼ to 1 in solid rock. Rock cuts are blasted to 6″ to 12″ below subgrade. For highways, slopes of 2 to 1 are preferred for high cuts and fills, and 3 to 1 or 4 to 1 for low sections. Airport runway sections sometimes have side slopes of 5:1 or 7:1; or even flatter where plane clearances require. On railway work subgrade is either flat or crowned slightly from the center; superelevation on curves is added in the ballast. In highway practice, pavement and

[1] Center to center of rails is 4′11½″ (4.958′).
[2] For discussion of grading methods and classification, see p. 2–29 (railways), and p. 2–89 (highways).
[3] For other sections, see Fig. 18 (railways), Figs. 35, 39, and 46 (highways), and Fig. 60 (airports).

base materials are usually placed in a trenched section; subgrade is banked on curves and is widened if the pavement is widened on sharp curves. Pavement base courses are sometimes carried across the full width of the subgrade eliminating the trench-type section, and the slope and ditch intersections are sometimes rounded off with vertical curves (Fig. 39).

Cross Sections and Slope Stakes.[1] Cross sections are taken of original ground at full stations (or +50's); at intermediate points where ground profile changes; at "grade" points where subgrade passes from cut to fill; at PC's and PT's of curves; and at culverts, stream crossings, road intersections, etc. Sufficient points are taken on each cross section to represent the ground properly, dimensions and elevations being recorded to 0.1 ft. These ground cross sections together with the proposed grading sections are the basis for preliminary earthwork calculations. A final set of cross sections is usually taken after grading is completed for determining final "pay" quantities. On railway work and major highway grading projects, slope stakes are set where side slopes intersect the original ground; "0.0" stakes are set at "grade" points, i.e., where grade passes from cut to fill (Fig. 14). Three intermediate sections are taken: (1) where cut runs out on downhill side, (2) where center line is at grade, and (3) where fill runs out on uphill side. (Cuts are recorded as +, and fills as −.) Notes applicable to railway practice and corresponding to plan shown in Fig. 14 are given below:

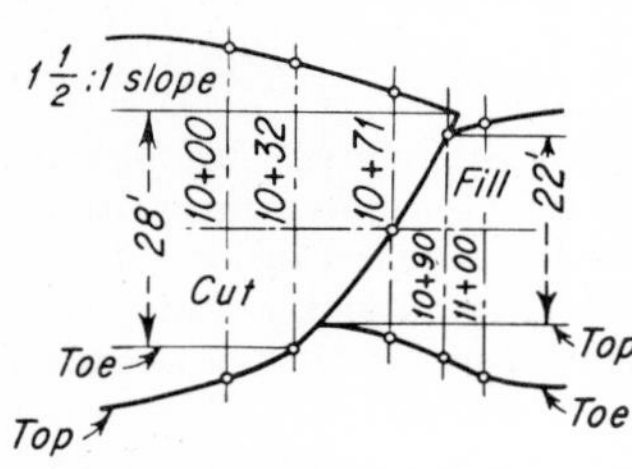

FIG. 14. Plan of earthwork (passing from cut to fill).

BASES: FILL 22′ AND CUT 28′. SLOPES 1½:1

Station	Surface elevation	Grade elevation	Cross section
11	98.6	101.70	$\frac{13.9}{-1.9}$ -3.1 $\frac{18.8}{-5.2}$
+90	100.3	101.61	$\frac{11}{0}$ -1.3 $\frac{16.6}{-3.7}$
+71	101.4	101.44	$\frac{18.2}{+2.8}$ 0 $\frac{13.4}{-1.6}$
+32	104.3	101.09	$\frac{23.6}{+6.4}$ $+3.2$ $\frac{14}{0}$
10	105.9	100.80	$\frac{25.1}{+7.4}$ $+5.1$ $\frac{17.3}{+2.2}$

Computation of Volumes.[2] *End-area Method.* Earthwork volumes are usually computed by the average-end-area method: $V = \frac{1}{2}(A_1 + A_2)l/27$; V in cu yd, A_1 and A_2 in sq ft, l in ft. Areas may be computed from cross-section notes or obtained by planimeter from plotted cross sections. Areas and volumes of cuts and fills must be computed separately since payment is based on "cut" quantities. For highways, cross sections are plotted to a scale of 1″ = 10′ (or 5′), and the entire area, including ditches, determined by double run of a planimeter (allowable error ½ to 1 percent for large areas, 2 percent

[1] For method of taking cross sections and setting slope stakes, see Section 1, "Surveying," pp. 1–20 and 1–22.

[2] See p. 2–6 for photogrammetric methods of determining cross sections from photographs and electronic methods of calculating areas and volumes.

for small areas). To facilitate plotting, templates may be cut to the shape of the grading section; one for cut and one for fill. Sometimes a simple "balanced grading section" (Fig. 13) is used with grade and width so adjusted that grading quantities will be equal to that for the actual subgrade. For rough calculations, "level sections" (subgrade and ground line both horizontal) are sometimes assumed with heights equal to those shown on center-line profile.

A mathematical method for computing areas from cross-section notes which is applicable to any shape of section is as follows [1] (d = distance from center line; h = height above or below base elevation): "(*a*) The notes must show values of d to each edge of base. (*b*) Use arbitrarily the minus sign for values of d to left of center, plus sign to right of center. (*c*) Use minus sign for any value of h below the base grade in cuts (as for side ditches). (*d*) Notes for points on original surface of ground should appear in brackets. The rule is: (1) Start at any point; use every value of h in order, proceeding clockwise around the figure. (2) Multiply each value of h by $(d_a - d_b)$, using algebraic values. (Here d_a represents the value of d for point next in advance, and d_b for point next back.) (3) Find the algebraic sum of these, and *divide* by 2."

$$\left(\frac{21.0}{+7.0}\right) \qquad \left(\frac{6.0}{+8.0}\right) \qquad \left(\frac{30.0}{+10.0}\right)$$

$$\frac{17.0}{+4.0} \quad \frac{10.0}{-1.0} \quad \frac{8.0}{0.0} \quad \frac{0.0}{0.0} \quad \frac{12.0}{0.0} \quad \frac{14.0}{-1.0} \quad \frac{15.0}{-1.0}$$

		−	+
$-\ 1.0\,[-17.0 - (-8.0)]$	=		9.0
$+\ 4.0\,[-21.0 - (-10.0)]$	=	44.0	
$+\ 7.0\,[6.0 - (-17.0)]$	=		161.0
$+\ 8.0\,[30.0 - (-21.0)]$	=		408.0
$+10.0\,[15.0 - 6.0]$	=		90.0
$-\ 1.0\,[14.0 - 30.0]$	=		16.0
$-\ 1.0\,[12.0 - 15.0]$	=		3.0

$$-44.0 + 687.0 = 643.0$$
$$\tfrac{1}{2}(643.0) = 321.5 \text{ sq ft}$$

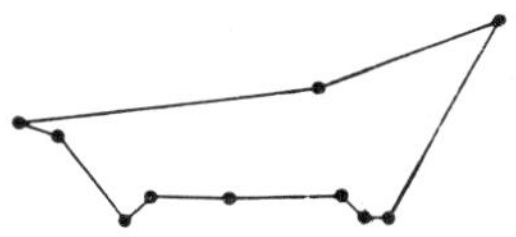

Prismoidal formula may be used for volumes when the average-end-area method does not give the desired precision; such as concrete volumes which carry a high unit price, and long tapering earth volumes (like bridge approaches) when computed in one operation. The formula is $V = l(A_1 + 4A_m + A_2)/6$, where A_1 and A_2 are areas of end sections, A_m is area of *mid-section*,[2] and l is length.

Computing earthwork volumes between original and final contours provides an alternate method which is particularly applicable to grading for highway interchanges, etc., where the use of vertical (parallel) cross sections is often rather complicated.

Borrow Pits. Two methods are available for computing volume of material excavated from a *borrow pit:* (1) taking original and final ground cross sections at right angles to a convenient base line, computing areas between these sections and obtaining volumes by average-end-area method in a manner similar to that for road sections; (2) dividing pit into a system of squares, usually 25 or 50 ft on a side, elevations being taken at the corners before and after excavation. (Borrow-pit method is applicable to determination of volumes of fill in similar manner.)

[1] From Allen, "Railroad Curves and Earthwork," 7th ed., Arts. 230, 231.

[2] Area of mid-section is based on dimensions of a section midway between end sections and is not the average of end areas except for volumes which could be computed exactly by the average-end-area method.

Borrow pit (Fig. 15) may be divided into *ABCDEFGHI* (nine squares *a* on a side), triangle *HGL*, and trapezoid *IHLJ*:

Vol. *ABCDEFGHI* (cu yd) $= a^2(\Sigma d_1 + 2\Sigma d_2 + 3\Sigma d_3 + 4\Sigma d_4)/(4 \times 27)$ where Σd_1 is the sum of heights (depth of excavation) of corners occurring *once*, Σd_2 is the sum of heights occurring *twice*, etc.; e.g., *A* is a corner used once, *H* twice, *F* three times, *M* four times. Vol. $HGL = \frac{1}{2}ab(H + G + L)/(3 \times 27)$. Trapezoid *IHLJ* may be computed by drawing diagonal *JH* or *IL* and computed by triangular prisms, as *HGL*. A better method is to interpolate for height at *K*; then vol. $IHLK = ab(I + H + L + K)/(4 \times 27)$. Vol. $KLJ = \frac{1}{2}a(a - b)(K + L + J)/(3 \times 27)$.

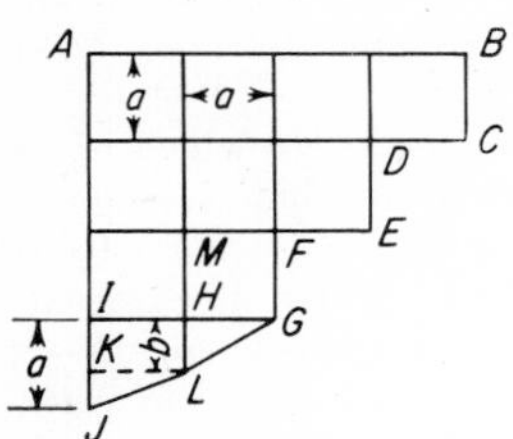

Fig. 15. Borrow pit.

Haul and Mass Diagram.[1] The total amount of haul is the product of volume *V* moved and distance $d_{c.g.}$ (in 100-ft stations) between centers of gravity in cut and in fill. *Overhaul* (provided for in some contracts) $= V \times (d_{c.g.} - fhl)$ in which fhl = free haul limit. Since it is not practicable to locate the center of gravity (c.g.) of an entire cut, haul usually is computed for sections of earthwork 100 ft long. The *mass diagram* furnishes a graphical means for determining in advance the most economical method of handling the work and/or finding the actual amount of overhaul that has been done by the contractor. In highway practice, the grade line is adjusted so that quantities of cut and fill (with proper allowance for shrinkage or swellage) will balance, as far as practicable, within relatively short distances.

The mass diagram is a cumulative plot of earthwork quantities (after adjustment for shrinkage, etc.), cuts being designated "plus" and fills "minus." The ordinate at any point is the cumulative volume of cut and fill to that point; i.e., *BJ* in Fig. 16 represents the cut between *A* and *B* as indicated on the profile. The difference in ordinates, such as to *B* and to *C* on mass diagram is the cut from *B* to *C* on profile, difference between ordinates to *C* and *D* is measure of fill between *C* and *D*, etc. Cut and fill are balanced between *B* and *D*, between *D* and *F*, and between overall balance points *A* and *G*. Areas of loops *BCD* and *DEF* and rectangle *BHIJ* represent free haul. Overhaul is sum of shaded areas *ABJ* and *HFGI*, in units of yard-stations.[2] Areas on mass diagram are obtained by planimeter or by dividing into rectangles, trapezoids, and triangles; tops and bottoms of loops may be taken as parabolas (area = ⅔ base × altitude). Arrows on profile indicate direction of haul of material.

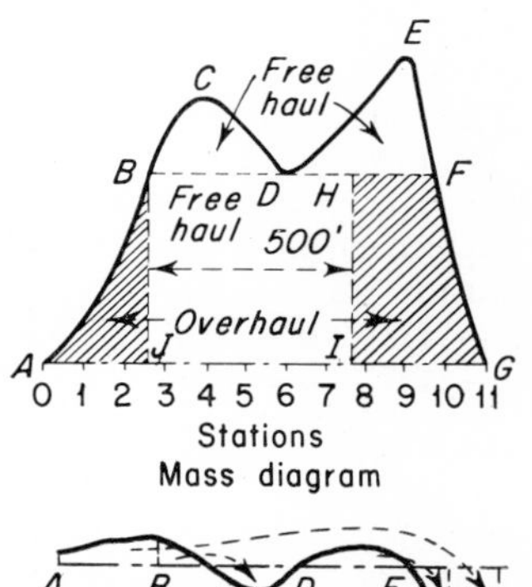

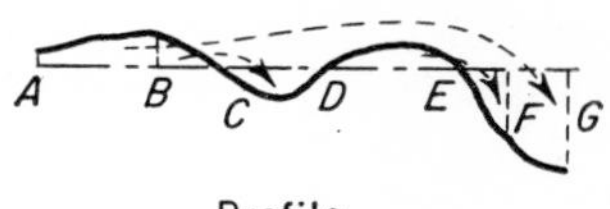

Fig. 16. Mass diagram.

Figure 16 represents a case where the actual cut was found to be just sufficient (after adjustment) to make the fill (i.e., stations 0 and 11 were "balance points"). Computed volume of cut from cross sections was 1,958 cu yd, computed volume of fill was 1,697 cu yd. Before plotting, fills were increased by 15.4 percent for shrinkage as shown in table.

[1] ALLEN's "Railroad Curves and Earthwork" gives more extended treatment.
[2] Overhaul is sometimes paid for on the basis of "yard-miles" or "yard-half-miles."

Station	Theoretical volume		Fills and allowance for shrinkage[1]	Mass diagram
	Cut	Fill		
0				0
1	+ 238			+ 238
2	+ 416			+ 654
3	+ 592			+1,246
4	+ 103			+1,349
5		− 109	−126	+1,223
6		− 211	−244	+ 979
7	+ 147			+1,126
8	+ 252			+1,378
9	+ 210			+1,588
10		− 785	−905	+ 683
11		− 592	−683	0
	+1,958	−1,697	(1,958)	

1 261/1,697 = 15.4 percent to be added to fills for shrinkage allowance.

RAILWAY ENGINEERING

Railway Transportation. The railways of United States are owned and operated by private companies. They provide a coordinated service for transportation of freight and passengers to every city and seaport and to thousands of smaller communities throughout the country. For about ninety years (1830–1920) railways furnished the only common carrier service, except for a limited amount of transport on navigable inland and coastal waters. During this period the railways developed a network of a quarter of a million *road* (route) miles (totaling about 400,000 miles of track). Although several different gages were used in the early period, the gage has been uniform (4′8½″) for the past seventy years, making possible interchange of locomotives and cars throughout the entire country.

As a result of technological progress in recent decades, additional modes of transportation have become available, competing with railways in freight and passenger services. These include automotive transportation (private car, bus, and truck), commercial aviation, and petroleum pipelines (crude oil and refined products). Since 1920, inland waterways of United States have been improved by deepening of channels and by construction of dams and locks, and water transportation of freight has increased considerably. Of these agencies, only railways and pipelines *provide* and *maintain* their own transportation routes.

TABLE 3. ESTIMATED DISTRIBUTION OF INTERCITY FREIGHT AND PASSENGER TRAFFIC[1]

	Freight ton-miles, percent				Passenger-miles, percent				
Year	Rail	Truck	Water[2]	Pipelines	Rail	Bus	Air	Water[2]	Electric railways
1930	75.2	3.9	15.5	5.4	68.5	18.1	0.2	7.1	6.1
1940	62.1	10.2	18.0	9.7	64.5	26.5	2.8	3.6	2.6
1945[3]	68.0	6.6	12.9	12.5	72.9	21.4	2.7	1.6	1.4
1950	57.0	16.2	14.5	12.3	45.3	37.7	14.3	1.7	1.0
1956[4]	49.2	17.3	16.8	16.7	34.9	30.9	31.9	2.1	0.2

1 Based on table in 1957 ed., "Yearbook of Railroad Information," Eastern Railroad Presidents Conference.
2 Includes rivers, canals, and Great Lakes.
3 "War year" traffic distribution is not representative of the 25-year trend.
4 Estimated.

The railways' share of intercity freight ton-miles and passenger-miles has declined considerably since 1930, as indicated in Table 3 on p. 2–21.

In recent years railways have concentrated on increasing the capacity and efficiency of their plants rather than in extending their lines. Roadbed and track structure have been strengthened to carry heavier loads at higher speeds; mechanized methods have been adopted for more economical maintenance; critical grades have been lowered and curvature reduced; improved signaling and communication systems have been applied; yards and terminals have been rebuilt to provide speedier operation; diesel-electric locomotives have been substituted for steam; new types of passenger equipment have been installed; and faster freight and passenger service has been provided. Research has been carried out in practically all phases of railway activity. The results of these improvements, particularly those involving maintenance of way, signals, and motive power, are shown subsequently.

Railway Plant and Equipment. United States railways comprise a network of 220,000 *road* (route) miles. Total mileage of track, including additional main tracks, passing tracks, yards, and sidings, is 390,000. This includes over 6,000 miles of track electrification. Motive power of Class I railways on Dec. 31, 1956, consisted of 30,433 locomotive units (steam, 3,714; diesel-electric, 26,081; electric and other locomotive units, 638). The railways owned 31,000 passenger-train cars (exclusive of those belonging to Pullman Co.); about half of these are baggage, mail, and express cars. Railway freight car ownership in 1956 was 1,707,000 (exclusive of those owned by private car lines), by types as follows: box cars, 728,000; flat cars, 56,000; gondola and hopper cars, 827,000; stock cars, 37,000; tank cars, 6,500; refrigerator cars, 20,000; other freight-carrying cars, 32,500. Average freight-car capacity is 54 tons. Average number of railway employees in 1956 was 1,043,000. Investment in road and equipment (less depreciation) for all United States railways in 1956 was $27.2 billion; which amounted to $124,000 per *road* mile or $70,000 per *track* mile (including yards and sidings). Net railway capitalization of all United States railways, except switching and terminal companies, was $15.2 billion in 1955; this represents total railway securities (stock and unmatured funded debt) in hands of the public. Stock amounted to $6.8 billion and funded debt $8.4 billion.

Railway Revenues and Expenses.[1] In 1956, total operating revenues amounted to $10.6 billion (84.8 percent freight; 7.2 percent passenger; 8.0 percent all other). Operating expenses were $8.1 billion (17.3 percent maintenance of way; 23.3 percent maintenance of equipment; 50.0 percent transportation; 9.4 percent all other). Operating ratio (expenses to revenues) was 76.9 percent. Net railway operating income was $1,068 million; this is operating revenue less operating expenses, taxes, and joint facility and equipment rentals. Net income was $876 million after fixed charges (principally interest on debt, rental of leased lines, etc.), available for payment of dividends on stock, investment in road and equipment, or addition to surplus.

Railway Traffic and Operations.[2] *Freight* traffic statistics for 1956 were: revenue tons originated, 1,447 million; cars of revenue freight loaded, 37.8 million; revenue ton-miles, 647 billion; revenue per ton-mile, 1.37 cents; tons of freight carried per train, 1,420; average haul per ton, 431 miles; freight cars per train, 66 (42 loaded and 24 empty); tons per loaded car, 33.0 (all traffic) and 43.1 (carload traffic alone); car-miles per serviceable freight car-day, 48.2; ton-miles per serviceable freight car-day, 1,023; ton-miles of freight carried per freight train-hour, 26,119; miles per hour (between terminals), 18.6; miles per day for serviceable freight locomotives, 147. *Passenger* traffic statistics for 1956 were: revenue passengers carried, 429 million; revenue passenger-miles, 28.2 billion (including commutation) and 23.3 billion (excluding commutation); average trip

[1] United States Class I railways. (See ICC classification of railways in footnote, p. 2–25.)
[2] United States Class I railways.

per passenger, 66 miles; revenue per passenger-mile, 2.68 cents (all passengers), 2.21 cents (commutation passenger), 2.56 cents (other coach passengers), 3.40 cents (Pullman passengers, exclusive of charge for Pullman accommodations); miles per hour (excluding motorcar trains), 41.9; miles per day for passenger locomotives, 326.

Freight traffic is classified as carload (CL) and less-than-carload (LCL). CL traffic in 1956 constituted 99.5 percent of total originated tonnage and accounted for 97 percent of total freight revenue. Percentages of total freight tonnage originated and of freight revenue were: products of agriculture, 9.5 percent of tonnage and 12.6 percent of freight revenue; animal products, 0.9 and 3.2; products of mines, 55.0 and 25.2; products of forests, 6.1 and 7.4; manufactures and miscellaneous, 28.0 and 48.7; LCL freight, 0.5 and 2.9. A CL shipment is loaded by shipper at a *private siding* or a *public delivery* (team) track of originating railway; car is then sealed and hauled by railway (and connecting lines) to its destination; it is unloaded by consignee at a *private siding* or *public delivery* track of terminating railway. An LCL shipment is delivered to a railway freight house where railway employees place it in a car with other shipments; this car may go directly to freight house at destination, or shipment may be transferred to other LCL cars en route; shipment is unloaded at freight house by terminating railway. Trucking to and from freight house is done by shipper (and consignee) or railway may furnish *pickup and delivery* service (store-door delivery) under which railway (or its agent) does the trucking to and from freight house.

Many railways now offer a new form of freight service, commonly referred to as *piggyback* service (officially TOFC—*trailer on flat car*). Under this system loaded truck trailers are placed on flat cars (often specially designed for this service) at a railway yard or station and transported in freight trains to terminating railway yard or station where trailers are taken off the cars and hauled to their final destinations. *Piggyback* service is being used in several different forms: (1) trailers owned by trucking companies and hauled by them to railway for loading on cars and, at terminating railway point, hauled away by trucking companies for final delivery; (2) railways providing a complete service by picking up freight in their own trailers (or those of a railway agent), and transporting trailers on flat cars, and then hauling trailers to consignee. Freight is moved on either *class* or *commodity* rates, major portion of tonnage being carried on the latter. LCL rates are considerably higher than CL rates, because of smaller weights of individual shipments, necessity in many cases of transferring shipments en route, and the fact that LCL cars are loaded and unloaded by railway employees.

Passenger traffic is classed as *commutation, other coach,* and *first class. Commuters* are those traveling regularly to and from business on monthly or multiple-ride tickets; the average length of trip (one-way) is about 20 miles. *Other coach* passengers are those riding in coaches on one-way (or round-trip) tickets. *First-class* passengers are those traveling in Pullman-type sleeping or parlor cars. In 1956, passenger mileage was distributed as follows: commuter, 5 billion; other coach, 17 billion; first class, 6 billion.

Railway Organization. The basic corporate organization of a railway is similar to that of other corporations but details of organization vary with magnitude, character, and geographical extent of the operation. Stockholders (owners of the company) elect a board of directors to represent them in its management. Corporate officers include a president, secretary, treasurer, and several vice-presidents. Each vice-president is responsible for a specific field of railway activity. Departments common to most railway organizations are: operating, traffic, accounting, legal, and purchases and stores. Operation of a railway also involves such matters as real estate, personnel, public relations, taxes, industrial development, and research. Individual bureaus usually handle these activities within the framework of major departments although in some cases their operation may be supervised by separate vice-presidents. *Operating* department, under

direction of a vice-president, who may also serve as general manager, is largest organization on the railway. It is directly responsible for transportation of freight and passengers and for upkeep of plant and equipment. It has three functional branches: transportation, maintenance of way (engineering), and maintenance of equipment (mechanical). *Transportation* department (often referred to as "operating" department) is usually headed by a general superintendent of transportation. This department is responsible for receipt, transportation, and delivery of freight and transportation of passengers. Its activities include preparation of schedules (in cooperation with traffic department), assignment of equipment, car service, and direct supervision of all those engaged in transportation operations, such as train dispatchers, station and yard employees, engine and train crews, and signalmen. *Engineering* department, under a chief engineer with a staff of assistants (engineer maintenance of way, bridge engineer, signal engineer, etc.), is responsible for design, construction, and maintenance of all *fixed* facilities. *Mechanical* department, under a chief mechanical officer, is responsible for design, inspection, and maintenance of locomotives and cars.

The complete organization of the operating department from system (staff) level to local level is rather complex because of geographical extent of a railway, which often includes several thousand miles of line. The basic operating unit is usually a *division*, which may include from 100 to 500 miles. A division superintendent is responsible for its operation. Transportation functions are under immediate control of trainmasters reporting to the superintendent. Division engineer is responsible for maintenance of fixed property; he has as subordinate officers supervisors of track, bridges and buildings, and signals and communications. Division master mechanic is in charge of inspection and maintenance of locomotives and cars.[1]

A general superintendent, with a staff of engineering, mechanical, and transportation officers, is often responsible for operation of a *district* which comprises several operating divisions. Large railways are often divided into two or more *regions* headed by vice-presidents or general managers.[2]

Traffic department is the sales organization. It is responsible for development and solicitation of traffic and establishment of rates (subject to ICC approval). Usually there are separate freight and passenger traffic managers reporting to a traffic vice-president. Within traffic department there are subsidiary officers in charge of industrial development, mail and express, coal traffic, etc.

Accounting department, headed by a vice-president or a comptroller, is responsible for accounting and auditing of receipts and disbursements. Compilation of statistics needed by officers of railway and those required by ICC and other regulatory bodies is carried out in this department, often through a bureau of statistics. *Legal* department, under a vice-president or general counsel, is responsible for contracts, settlement of claims, and for handling matters before courts, public utility commissions, and ICC, and for all matters of a legal nature affecting railway. *Purchases and stores* department, often headed by a vice-president, does the buying for railway through a purchasing agent. A general storekeeper is responsible for storage of materials until requisitioned by the *using* departments.

Although only the divisional organizations of operating department have been mentioned specifically, the other departments usually have representatives located at division points for carrying out their work effectively.

[1] Local organization may be along *divisional* or *departmental* lines. In the former, division engineer and master mechanic report directly to superintendent; they are also responsible to engineering and mechanical staff officers for standards, budgets, etc. Under departmental system, divisional engineering and mechanical officers work in cooperation with division superintendent but report directly to their respective staff officers. The present tendency is toward the use of divisional system although on many railways the organization is a combination of both.

[2] In a reorganization (1955) of Pennsylvania Railroad, the system of operating divisions was eliminated and railway was divided into nine relatively large regions. Regional officers are in *direct* charge of all operations within the region.

Association of American Railroads (AAR). Coordination of railway activities is effected through Association of American Railroads which was established in 1934 by merger of a number of individual railway organizations. Its operations are administered through a board of directors elected annually by member railways, which include principal railways of United States, Canada, and Mexico. AAR functions through departments of operations and maintenance; law; research; traffic; and finance, accounting, taxation, and valuation. Within the framework of these departments there are numerous divisions, sections, and committees for carrying out their varied activities.[1]

Interstate Commerce Commission (ICC). Railways of United States are regulated with respect to interstate and foreign commerce by Interstate Commerce Commission, established in 1887. This regulation includes such matters as service, rates, safety, issuance of securities, and construction and abandonment of lines. ICC activities are carried out by its 11 commissioners through bureaus dealing with different phases of its work. Publications of ICC include an annual report, annual statistics of railways and other carriers, reports of individual bureaus, and reports of special studies and investigations.[2]

ICC also regulates to a lesser extent *interstate* activities of other transportation agencies, including motor carriers, water carriers, freight forwarders, and pipelines for petroleum products. ICC regulation of railways is more complete than it is for other carriers since activities of railways are principally interstate and of a common-carrier nature. Contract and private carriers, which constitute a large portion of the activities of motor and water carriers, are subject to much less ICC regulation. Various state departments of public utility, railway commissions, etc., are also involved in regulation of railways with respect to their *intrastate* activities in individual states.

RAILWAY ENGINEERING STANDARDS

American Railway Engineering Association (AREA). Engineering standards for location, design, construction, and maintenance of way and structures of United States railways have been developed over a period of 60 years by American Railway Engineering Association, an organization comprising 3,300 railway engineers and executives in United States and foreign countries. AREA serves concurrently as construction and maintenance section in engineering division of AAR.[3] Reports of its committees, issued in bulletins during the year, are presented for discussion at annual meeting in March. "AREA Manual for Railway Engineering" includes material adopted by the association as *recommended practice*.[4] The "Manual" and the annual proceedings furnish authoritative information on current railway engineering practice. Frequent reference to AREA material is made throughout this section of the handbook, and grateful acknowledgement is made of the permission given for its use.[5] A number of individual railways have also developed maintenance-of-way standards and specifications. Important research on roadway, ballast, ties, rail, and structural matters is currently underway. This work, sponsored by AREA committees, is carried out by research staff of engineering division, AAR, making use of facilities of AAR Research

[1] Information as to organization and operations of Association of American Railroads may be obtained from its headquarters: Transportation Building, Washington 6, D.C.

[2] By ICC direction, United States railways are grouped by classes depending on magnitude of their operation. Class I line-haul railways (*through* 1955) included those with annual operating revenues exceeding $1,000,000; Class II those with annual operating revenues from $100,000 to $1,000,000; and Class III those below $100,000. (Class I railways operated 95 percent of the railway mileage and earned 99 percent of the revenue.) Effective Jan. 1, 1956, Class I railways include those with annual operating revenues of $3,000,000 or more; and Class II all other railways.

[3] Engineering division, AAR, comprises construction and maintenance section, electrical section, and signal section.

[4] The "Manual" is issued in loose-leaf form, revised annually. An AREA "Portfolio of Trackwork Plans and Specifications," published separately for convenient use, is part of the manual; both publications (or separate parts) may be purchased from AREA, 59 East Van Buren St., Chicago 5, Ill.

[5] Page references to "AREA Manual" are to 1956 edition.

Center at Chicago and of several college and research organizations. Much of the research is based on field investigations made under practical operating conditions.[1]

Grades.[2] Rate of grade (gradient) is inclination of profile grade line from the horizontal, expressed as a percentage; in early practice it was stated in feet of rise per mile. *Ruling* grade is the one which determines tonnage that can be handled by a single locomotive over a particular engine district. Maximum tonnage which can be hauled on a ruling grade is different for each class of power. Ruling grades in opposite directions on an engine district are seldom the same because of differences in the terrain. *Compensated* grades are those which have been reduced to keep *total* resistance the same on curves and tangents. Compensation may be taken as 0.04 percent per degree of curve for average conditions. If a 5° curve occurs on a maximum grade on a district with a ruling grade of 0.72 percent, compensated (actual track) grade on curve should not exceed 0.52 percent ($0.72 - 5 \times 0.04$). Ruling grade is not always the maximum gradient on an engine district because of possible use of helper and momentum grades at some locations. *Helper* (pusher) grades are those on which *helper* engines are needed to assist road locomotives; for economical operation, a helper grade should be steep enough to utilize the combined capacity of road and helper engines, which may be taken as 90 percent of the sum of the individual locomotive capacities. *Momentum* grades are those *so situated* that the kinetic energy of a train (due to its speed at the foot of the grade) will enable locomotive to haul train to the top without a reduction of speed below 10 or 12 mph. Momentum grades are undesirable where stops or restricted speeds are apt to occur; nor should grade be so steep that a locomotive could not haul a tonnage train up in *two parts if it became stalled on the grade.*[3]

No standards are available for establishing the ruling grade for a particular district. It depends on terrain, amount and character of traffic, and desirability of keeping ruling grades about the same on adjacent engine districts (thus avoiding the breaking up of trains for reduction or increase in tonnage). On *main lines not in rugged country,* ruling grade is usually not much over 1 percent. Where very heavy traffic (such as coal or ore) is a predominant feature, it may be desirable to have a compensated ruling grade as low as 0.3 or 0.4 percent in direction of loaded car movement. In mountainous country, ruling grades are steeper, and stretches of helper grades are often required to get over summits. Several lines crossing the Rocky Mountains have grades up to 2.2 percent or occasionally more. On *minor* lines, grades up to 2 percent are not uncommon. The *minimum* gradient through a cut should be at least 0.3 percent to provide satisfactory drainage of side ditches. Grades on industrial sidings may be 3 or 4 percent if necessary.

Vertical Curves.[4] Grade lines are connected by vertical parabolic curves when difference in gradients is appreciable. Length of vertical curve depends on algebraic difference in grades and on whether the curve is in a sag or on a summit. Conditions are more severe through sags than over summits, because of greater speed going downhill and because the "slack" of draft gear is "running out" as front of train reaches ascending grade. For *high-speed main* tracks, AREA recommends a length of vertical curve such that *rate of change of grade* will not be more than 0.05 ft per station (of 100 ft)

[1] Annual reports of AAR Research Center describe important features of research currently underway by engineering division, mechanical division, container and loading research and development laboratory and sanitation research and development unit.

"Application of Modern Scientific Research on Railroads of the United States" by Loyd J. Kiernan gives a comprehensive account of all phases of research in the railway field. Copies may be obtained from AAR, Washington, D.C.

[2] Relation of grades to locomotive capacity, train resistance, and tonnage rating is discussed on p. 2–56.

[3] Ruling grade is very important since it limits the tonnage of a train and thereby affects the number of trains required to move a given total tonnage over an engine district. When steam locomotives provided the source of motive power, railways often made large capital expenditures to reduce ruling grades on important routes. However, use of diesel-electric locomotives, which may be made up of several units, provides more flexibility in selection of motive power. Now it is usually found more economical to vary the amount of motive power than to make costly grade reductions. Furthermore diesel-electric (and electric) locomotives can operate at considerable overload capacity for short periods of time, which is important where an individual ruling grade is not too long.

[4] See p. 2–13 for method of calculating elevations on vertical curves.

in sags and not more than 0.10 ft per station on summits; for *secondary main* tracks, the respective rates of change of grade may be twice as much. For tracks *of lesser importance* vertical curve may be relatively short. Grade-line changes usually occur at full stations and vertical curves are made an *even* number of stations in length, thereby simplifying calculations.

Curves.[1] Maximum degree of curvature on *main* lines, except in rugged terrain, is usually not over 6°; in mountainous territory this may be exceeded. On *minor* lines, degrees up to 8° or 10° are common. On *high-speed* lines, sharp curvature will limit speed of passenger trains since it is not practicable to elevate outer rail for speeds as high as would be maintained on the tangents. Objectionable features of curvature are (1) increased wear and tear on track and equipment, (2) increased train resistance, (3) higher cost of track maintenance, and (4) reduction in allowable speed when full superelevation of the outer rail cannot be made. Railways have spent large amounts to reduce maximum degree to 2° (or less) where high-speed passenger service was desired. In *yards and sidings*, degree should not exceed 14° to 16° if road locomotives are used; on tracks where switching engines are used exclusively, curves may be as sharp as 30°. If necessary to adopt curves with a radius of 150′ or less in an industrial yard, a small diesel-electric (or gasoline) engine with a short wheel base is used.

Spirals.[2] On main-line tracks, spiral transition curves are used between tangents and circular curves and between branches of compound curves. Length of spiral should be such that the rate of change of elevation of outer rail will be 1¼″ per sec for *maximum* speed of train operation on curve, in accordance with AREA formula: $L = 1.17EV$, where L is minimum desirable length of spiral, ft; E, superelevation of outer rail, in.; and V, maximum speed, mph. For length of spiral between branches of a *compound* curve, the same formula may be applied if E is taken as the *difference* in superelevations of adjacent branches of compound curve.

Elevation of Curves.[3] Elevation of outer rail on curves, often called *superelevation*, depends on radius of curve, bearing distance between rails, and speed of train operation. For *equilibrium* elevation, in which the resultant of the weight and the centrifugal force is perpendicular to plane of the track, following AREA formula may be applied: $E = 0.0007V^2D$, in which E = elevation, in. of outer rail; V = speed, mph; and D = degree of curve.

If it were practicable to operate all classes of traffic at same speed on a curve, the ideal condition for smooth riding and minimum rail wear would be obtained by elevating for *equilibrium*. However, curved track must handle traffic operating at various speeds, which results in slow trains causing excessive wear on inside rail and high-speed trains causing accelerated wear on outside rail. For practical reasons *maximum* superelevation on standard-gage track is limited to 6″ or 6½″. If allowable passenger-train speeds were based on *equilibrium* elevation, resulting speeds would be considerably below those desired for high-speed service. Safety and comfort limit speed with which a passenger train may negotiate a curve. Any speed which gives comfortable riding on a curve is well within limits of safety. Experience has shown that conventional passenger-train equipment will ride comfortably around a curve at a speed which would require an elevation about 3″ higher for *equilibrium* elevation. For that reason it is customary to use formula $E = 0.0007V^2D - 3''$ for determining maximum allowable passenger-train speed.[4]

[1] *Degree of curve D* in railway practice is central angle subtended by a 100′ chord. The number of inches in middle ordinate of a 62′ chord is approximately equal to D; this provides a simple way of finding degree of an existing curved track. (See p. 2–7 for theory of circular curves and method of laying them out.)

[2] See p. 2–15 for theory of spiral transition curves and method of laying them out.

[3] See p. 2–17 for theory involved.

[4] Passenger-train equipment having large center bearings, roll stabilizers, and outboard swing hangers can negotiate curves comfortably at greater than 3″ unbalanced elevation because there is less car-body roll. It is suggested that, where *complete* passenger trains are equipped with cars utilizing these refinements, "lean" tests be made to determine the body roll ("AREA Manual," 1956 ed., p. 5-3-11). If roll angle is less than 1°30′, a 4½″ unbalanced elevation may be used.

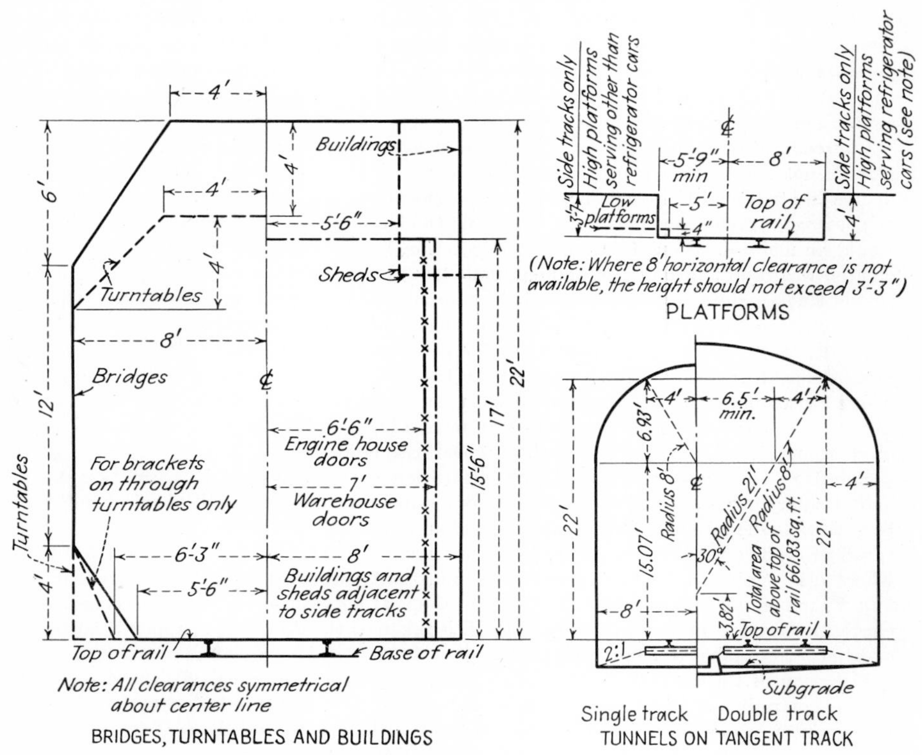

FIG. 17. AREA clearance diagrams.

Profile grade is maintained on inner rail throughout curve. Where *spirals are used,* superelevation starts at TS, and increases uniformly along the spiral, the full amount being attained at SC. If *spirals are not used,* it is necessary to start the superelevation at some point on tangent (back of PC) and attain full superelevation at a point on curve a short distance from PC.

Gage. *Standard* gage of railway track is 4′8½″ between inside heads of rails (measured ⅝″ below top of rail).[1] This gage is standard in North America, the British Isles (except Ireland), and much of Europe, as well as in many other parts of the world. *Wider* gages in use are: 5′0″ in Russia and several Baltic countries; 5′3″ in Ireland, Australia, and a few other countries; and 5′6″ in India, Spain, Portugal, and the Argentine. Of the *narrower* gages, most common are 3′6″ in Japan, Newfoundland, and various parts of Africa; "meter" gage in North Africa, Burma, Western China, and parts of Europe; and 3′0″ in Central American countries. Several minor lines in foreign countries use 2′0″ and 2′6″ gages.

Track Spacing. For standard-gage tracks, distance between center lines of main tracks should not be less than 13′ (increased on curves by 2″ per degree). It is quite common to specify 14′ centers for new construction. A number of states require 13′6″ or 14′0″ between main tracks.[2]

[1] Some railways widen gage up to ½″ on curves sharper than 10°.
[2] A table of Legal Requirements for Clearances (as of Dec. 1, 1957), which gives the individual states' requirements for track centers, is shown in *Proc. AREA*, vol. 59, p. 660.

Clearances. AREA has adopted clearance diagrams for bridges, turntables, tunnels, buildings, and other fixed structures adjacent to tracks, as shown on Fig. 17.[1] Although an overhead clearance of 22′ is desirable, especially on electrified lines with overhead trolley wires, many existing structures have considerably less clearance (often as low as 16′).[2] Several railways have "clearance" cars with devices for measuring and recording the clearance of fixed structures adjacent to tracks.

Equipment Dimensions.[3] *Diesel-electric locomotives: Weight per driving axle:* 50,000 to 62,500 lb. *Height* (above top of rail): 15′ (max). *Width* (out to out): 10′8″ (max). *Length:* 50′ per unit for 1,750-hp units and up to 70′ for larger units. *Electric locomotives:* weight per driving axle: up to 70,000 lb, new types 58,000 to 62,500 lb. *Height:* 16′ (max). *Length:* usually from 60′ to 75′ per unit. *Steam locomotives: Weight per driving axle:* 70,000 to 72,000 lb (max); 62,000 to 67,000 lb more common. *Height:* 16′ occasionally, 15′6″ common maximum, less for older locomotives. *Width:* 11′2″ maximum; 10′6″ common. *Length* (engine and tender): 130′ for large articulated types, 100′ to 115′ for nonarticulated.

Freight cars: Weight per axle: usually not over 40,000 to 45,000 lb. *Height:* 13′6″ to 14′. *Width:* 10′6″ max. *Length:* 40′ to 50′; special types up to 70′.

Passenger cars: Weight per axle: under 35,000 lb. *Height:* 13′6″ usually; for vista-dome or two-level cars, up to 15′10″. *Width:* 9′6″ to 10′. *Length:* 70′ for older coaches, modern cars 75′ to 85′. (Talgo-train cars are 109′ long (for 3 units), 10′10″ high.)

RAILWAY ROADBED, TRACK, AND STRUCTURES [4]

Roadbed. Width of roadbed at subgrade is dependent on ballast section. Twenty feet is common for embankments for single-track lines; for sidings and minor lines, 14′ to 18′ may be used. For multiple tracks, width is increased by track spacing. A roadbed shoulder of at least 18″ should be maintained outside toe of ballast slope. Subgrade is usually flat under ties and sloped down (about ½ in. per ft) beyond ties to facilitate drainage; occasionally subgrade is sloped downward both ways from center line. Shape of subgrade is the same for straight and for curved track, provision for superelevation being made in ballast section. The section in excavation is based on standard embankment section with additional width to provide for side ditches. Roadway side slopes in excavation depend on material encountered—usually 1½ (horizontal) to 1 (vertical) for earth, ½ to 1 for loose rock, and ¼ to 1 for solid (ledge) rock; side slopes on fill are usually 1½ to 1 (they may be steeper if fill is made of rock). Modern methods of constructing railway fills include placing material in 6″ to 12″ layers and rolling to secure proper compaction.[5]

Drainage.[6] Adequate drainage is of fundamental importance. Culverts and bridges should provide sufficient openings to carry normal flood flows of natural waterways

[1] Recommendations in "AREA Manual," 1956 ed.: (1) Clearances shown are for tangent track and for new construction. Clearances for reconstruction work or for alterations are dependent on existing physical conditions and, where reasonably possible, should be improved to meet requirements for new construction; (2) on curved track clearances shall be increased to allow for overhang and tilting of a car 85′ long, 60′ between centers of trucks, and 14′ high; (3) superelevation of outer rail shall be in accordance with recommended practice of AREA; (4) distance from top of rail to top of ties shall be taken as 8″; (5) legal requirements to govern when in conflict with dimensions shown.

[2] It should be noted in connection with elimination of grade crossings that the clearance of a highway bridge above a railway track should be at least 18′ (22′ desirable), whereas a 13′ or 14′ clearance is sufficient when railway passes over highway.

[3] Detailed information may be found in "Locomotive Cyclopedia" and "Car Builders Cyclopedia" (Simmons-Boardman Publishing Corporation), new editions of which are issued every 3 years.

[4] AREA recommended practice for railway engineering is given in "AREA Manual," 1956 ed. (see footnote, p. 2–25). "Railway Track and Structures Cyclopedia" compiled by editorial staff of Simmons-Boardman Publishing Corporation in cooperation with AREA and signal section, AAR, provides an excellent source of reference for tracks, bridges, buildings, signals, and water, oil, and sanitation facilities. A new edition of the cyclopedia is published every 3 years.

[5] Recommendations and Specifications for Formation of the Roadway are shown in "AREA Manual," 1956 ed. (pp. 1-1-1 to 1-1-15). See p. 2–90 for methods of embankment compaction. Construction methods and equipment used on recent railway grading projects are described in various reports in AREA proceedings.

[6] See "AREA Manual," 1956 ed., pp. 1-1-45 to 1-1-51 for roadway drainage; pp. 1-3-1 to 1-3-20 for natural waterways; and pp. 1-4-1 to 1-4-35 for culverts.

without "ponding" against the roadbed. Where tracks are built in river valleys, line should be located high enough to avoid being washed out by floods, although it may not always be practicable to protect it against the most extreme floods. *Roadway drainage* includes artificial surface and subsurface drainage of cuts and fills which make up the roadbed. In locating lines it is desirable to avoid as far as practicable (1) cuts in wet springy ground, (2) long cuts on low gradients, and (3) fills across swampy ground which cannot be readily drained. *Surface drainage* is provided by *side ditches* in cuts which drain roadbed and intercept water from side slopes, and by *intercepting ditches. Side ditches* should be at least 1 ft wide at bottom and 1 ft deep below subgrade, with side slopes depending upon the material. They should have a minimum gradient of 0.3 percent; in long cuts grade of side ditches will be approximately that of the track. Wide ditches are desirable where trouble from slides, sand, or falling rock occurs, both to afford storage space and to permit use of "off-track" equipment for cleaning. *Intercepting ditches* may be needed along upper side of cuts if ground slopes toward the roadbed, and at least 15′ from toes of fills which rest on soils that may become unstable when saturated. *Except in impervious soils*, effective *subdrainage* of wet cuts and of saturated soil upon which embankments rest may be attained through use of pipe drains. In cuts pipes should be laid parallel to and 9′ from center line of adjacent track on either one or both sides. Along embankments they should be located about 10′ from toe of slope. Pipe gradient should be at least 0.2 percent, preferably 0.4 percent or more where practicable. Trenches should be backfilled with permeable material. In wet cuts it may also be necessary to install lateral pipes under the track, spaced from 10′ to 40′ apart with a minimum gradient of 4 percent (preferably 8 percent); laterals drain into the longitudinal pipes. Vitrified-clay sewer pipe with bell ends or perforated corrugated galvanized-iron pipes are commonly used for subsurface drainage. *French* or rock drains, consisting of stones loosely placed in trenches, have been used for draining embankments which were made of water-retaining materials. These are 4′ or 5′ wide and built across fill at intervals, usually extending from track to toe of slope. If needed, perforated corrugated-iron pipes can be placed at bottom of these trenches. No standard remedy is available for drainage of *impervious* soils (containing more than 20 percent clay). Lowering of water table in such soils will improve the bearing power of subgrade. Drain pipes are of no value in the denser clays. Water should be kept away from soils of this kind wherever possible.

Roadbed Stabilization.[1] "Soft" track requiring constant maintenance to keep it in proper "surface" is caused primarily by *soft spots, water* (*ballast*) *pockets*, or other *unstable conditions* in roadbed materials or in the ground on which the roadbed is built. This condition may arise below the subgrade in cuts but it is more common in fills. *Soft spots* are due to instability of the material when saturated with water, which may be held in soil either by capillary action, or by lack of drainage, or both. A *water pocket* is usually a depression in an impervious subgrade down into which ballast has been forced. Constant churning, under traffic, of ballast in the water which is confined there increases size of pocket and much of the softened material is forced outward and upward. This results in unstable track requiring expensive maintenance. Fills made of unsuitable material or built on sloping ground may develop tendencies to sheer off or slide. Stabilization of roadbed may be accomplished by (1) improved drainage, (2) pressure grouting, (3) driving of poles or ties vertically along the track near ends of ties, and (4) application of sand piles (drains) to remove excess water. Studies have also been made of use of chemicals to stabilize the soil and of a patented electro-osmosis process (in which by passing an electric current through clay, water can be made to flow through the clay, consolidating and strengthening it).

[1] See "AREA Manual," 1956 ed., pp. 1-1-17 to 1-1-35 for roadbed stabilization and roadway protection.

Pressure grouting consists of forcing a "slurry" of cement and sand into subgrade or embankment either by use of compressed air or by hydraulic pressure. The pneumatic process is more economical for short sections of track requiring numerous "moves," and hydraulic method is better on *out-of-face* work (in a continuous stretch) and for stabilizing fills.[1] A number of railways report very satisfactory results in the use of pressure grouting at locations where troublesome conditions had previously existed.[2] On new lines, particular care has been taken to investigate soil conditions and apply modern soil-mechanics principles.

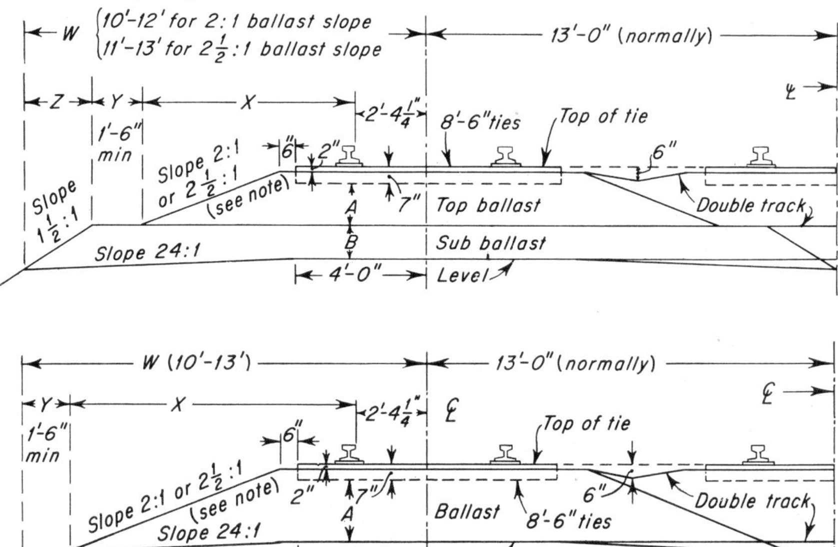

Note: Ballast slopes
2:1 for prepared materials – stone, slag and gravel having over 20% crushed particles
2½:1 for prepared gravel having 0 to 20 % crushed particles and pit run gravel

Depth of section to use will depend on conditions peculiar to each railroad or location
Values for A, B, W, X, Y and Z shown in AREA manual

FIG. 18. AREA ballast sections for tangent track.

Ballast. Ballast holds track in line and surface, provides drainage, and distributes the load uniformly to subgrade. Stone, gravel, slag, cinders, and a few local materials are used for ballast.[3] Stone is the most satisfactory material because of its excellent drainage qualities, freedom from dust, and the fact that it can be cleaned periodically; it is used

[1] For *pneumatic grouting* a mixer and an injector pressure tank are used; for *hydraulic pressure grouting*, a mud jack, or other machine, which will mix the slurry mechanically and supply grout-line pressure through direct piston action, is required. *Injection points* are driven into subgrade or embankment, their length and spacing having been determined through a preliminary investigation in which location and depth of ballast pockets or soft spots were found. Best results are obtained with the use of a rather fine sand such as beach sand. The proportions of cement and sand in the slurry vary widely dependent on conditions to be alleviated. For *pneumatic grouting* range has been from 1 part cement to from 1 to 7 parts of sand. For *hydraulic grouting*, mixtures have sometimes been as lean as 1 part cement to 32 parts of sand.

[2] Costs and maintenance data on grouting projects are shown in *Proc. AREA*, vol. 54, p. 1136.

[3] "AREA Manual," 1956 ed., includes specifications for prepared stone, slag, and gravel ballast and for pit-run gravel ballast, pp. 1-2-2 to 1-2-5.

for heavy-traffic lines whenever its cost can be justified. Figure 18 shows AREA ballast sections for tangent track.[1] Upper figure shows the use of an inferior subballast upon which is placed a layer of selected top ballast; lower figure shows a more common section using one grade of ballast. Ballast slopes (near ends of tie) are often rounded using a 4′ radius. Superelevation on curves is provided for in the ballast section, normal depth being maintained under tie at lower rail with additional depth of ballast on high side. Depth of ballast depends on character and amount of traffic, kind of ballast, spacing and length of ties, and roadbed conditions.[2]

Concrete-supported Track. Concrete slabs are occasionally placed on subgrade to provide better support for ballast and ties under turnouts and crossings. Special track construction is often used in stations and tunnels in which rails rest on treated wooden blocks set in recesses cast in a reinforced concrete base slab.

Ties. *Wooden crossties* (sawn or hewn) are standard in this country. They hold track in line and surface, maintain gage, and transmit load to ballast and subgrade. For heavy-traffic lines, 7″ by 8″ or 7″ by 9″ ties are generally used (laid flat); on minor tracks, 6″ by 7″ or 6″ by 8″ are common. Ties are usually 8′ or 8′6″ long, although a few railways use 9′ ties, where subgrade conditions are poor. (AREA recommends use of 9′ ties in heavy traffic lines.) Longer ties distribute the load better and would probably reduce cost of track maintenance enough to justify the additional cost. Average tie spacing (c to c) of the nearly one billion ties in United States railway tracks is 21¼″ (3,000 ties per mile). For main tracks, 21 to 24 ties are used per 39′ rail; for sidings, 16 to 18 per 39′ rail or 14 to 16 per 33′ rail.

Over 23 million new ties were used in replacement on Class I railways in 1956; 98.8 percent of these were *treated* ties. Average cost of treated ties was $3.45 (before placing in track); for untreated ties, $2.01. Annual tie renewals averaged 79 per mile of maintained track for 1952 to 1956. During 1955, 26,688,000 crossties were preservatively treated in this country; 54 percent of these were oak, 9 percent Douglas fir, 7 percent gum, 20 percent mixed hardwoods, and 10 percent of other woods. Fifty-seven percent were treated with a creosote and a creosote–coal tar mixture, and 43 percent with a creosote-petroleum mixture. The "empty-cell" method of treatment is used for crossties.[3] *Untreated* ties last from 5 to 7 years depending on quality of tie, method of protection from rail cutting, and intensity of traffic. *Treated* ties have a service life of at least 30 years if thoroughly protected from mechanical wear and track well maintained. Tie renewals are necessitated by decay, mechanical disintegration of wood fibers under rail, injuries due to spiking, or checking, splitting, or derailments. Decay can be substantially eliminated if ties are properly seasoned, adzed and bored before treatment, and if sufficient preservative (6 to 8 lb retained per cu ft) is applied by a satisfactory process. Mechanical wear can be reduced to a minimum by adzing for plates, preboring for spikes, and using canted tie plates of sufficient size to distribute the load. A number of railways fasten tie plates independently to the ties. In recent years considerable attention has been given to use of tie pads under tie plates as well as to application of coatings and sealing compounds to surface of tie. *Tie pads* under investigation include a number of materials, variously processed. *Substitute* ties of steel or concrete have been installed occasionally on an experimental basis in this country but have not been adopted; they are used to some extent in foreign countries where satisfactory woods are scarce.

Switch timbers are furnished in sets to support turnouts and crossovers. They are usually 7″ by 9″ treated timbers, in lengths from 9′ to 16′6″ for turnouts and 9′ to 21′6″

[1] "AREA Manual," 1956 ed., pp. 1-2-7 to 1-2-12, shows ballast sections for straight and curved track; quantities of ballast per mile of track for varying depths are included.

[2] Replies to an AREA questionnaire indicated following *total* depths of ballast (under tie) as desirable: for 131-lb rail, 22″ to 30″; for 112-lb rail, 21″ to 27″; for 90-lb rail, 20″ to 24″. Ballast depths on light and medium traffic lines are considerably less. On sidings, only enough ballast is needed to hold track in place and provide adequate drainage.

[3] Ties are treated either by "Lowry" or "Rueping" process. Specifications for treatment of ties are given in "AREA Manual," 1956 ed., pp. 17-4-1 to 17-4-18.

for crossovers. AREA and individual railways have prepared standard layouts with bills of material for turnouts and crossovers for different frog numbers.

Rails. *Sections.* The tee rail (Fig. 19) is standard for open track in United States; grooved girder rails are used to some extent in paved streets. Rail weights are stated in pounds per yard. Prior to 1910, ASCE section was in common use. For this section height equaled width of base, and proportions in head, web, and base were constant for all weights: 42, 21, and 37 percent respectively. With increased wheel loadings, denser traffic, and higher speeds, heavier sections with more girder strength were needed. The trend has been to increase height, strengthen web, and improve design of head and head-fillet radii to reduce stress concentration under wheel loads. Table 4 gives dimensions of typical rail sections. Principal weights now being rolled are "medium heavy" (112 to 119 lb) and "heavy" (127 to 140 lb).[1]

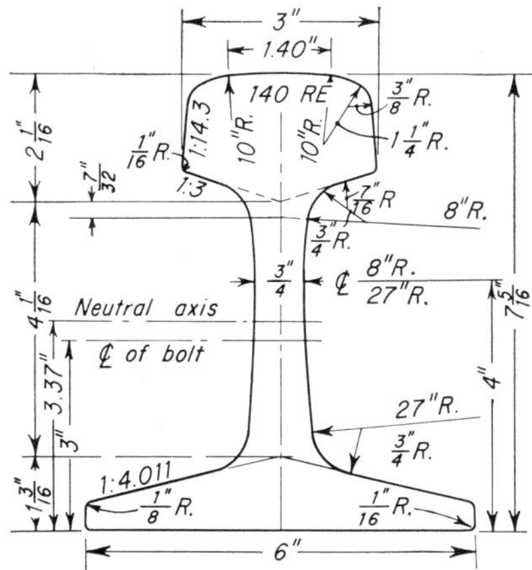

Fig. 19. 140-lb AREA rail section. (See Table 4 for other properties of this section.)

The economic value of various sizes of rail has been the subject of much study. Optimum weight to use on a particular section of line is dependent on traffic conditions

Table 4. Dimensions and Properties of Typical Tee-rail Sections

Section	Nominal weight per yd, lb	Dimensions, in.						Distribution of metal, percent			Moment of inertia	Section modulus (base)
		Height	Base width	Fishing	Max head width	Depth of head	Min web thickness	Head	Web	Base		
PS[1]	155	8	6 3/4	4 21/32	3	2 1/16	3/4	33.1	27.9	39.0	130.9	37.4
RE[2]	140	7 5/16	6	4 1/16	3	2 1/16	3/4	36.7	28.0	35.3	96.8	28.7
CF & I[3]	136	7 5/16	6	4 3/16	2 15/16	1 15/16	11/16	36.4	27.1	36.5	94.9	28.3
RE[2]	133	7 1/16	6	3 15/16	3	1 15/16	11/16	36.3	26.5	37.2	86.0	27.0
RE[2]	132	7 1/8	6	4 3/16	3	1 3/4	21/32	34.1	28.3	37.6	88.2	27.6
CB & Q[4]	129	7 5/16	6	4 9/32	2 5/8	1 27/32	21/32	32.9	28.7	38.4	90.4	28.1
NYC[5]	127	7	6 1/4	4 5/32	3	1 11/16	21/32	34.2	26.4	39.4	81.6	26.4
CF & I[3]	119	6 13/16	5 1/2	3 13/16	2 21/32	1 7/8	5/8	37.1	26.1	36.8	71.4	22.9
RE[2]	115	6 5/8	5 1/2	3 13/16	2 23/32	1 11/16	5/8	34.8	27.1	38.1	65.6	22.0
CB & Q[4]	112	6 3/4	5 1/2	3 7/8	2 1/2	1 3/4	5/8	34.3	26.7	39.0	67.0	22.3
CF & I[3]	106	6 3/16	5 1/2	3 3/8	2 21/32	1 3/4	19/32	38.3	23.9	37.8	53.6	18.8
NYC	105	6	5 1/2	3 13/32	3	1 5/8	5/8	40.9	24.0	35.1	49.9	17.3
RE[2]	100	6	5 3/8	3 9/32	2 11/16	1 21/32	9/16	38.2	22.6	39.2	49.0	17.8
ASCE	100	5 3/4	5 3/4	3 5/64	2 3/4	1 45/64	9/16	42.0	21.0	37.0	44.0	16.1
RA-A[2]	90	5 5/8	5 1/8	3 5/32	2 9/16	1 15/32	9/16	36.3	24.0	39.7	38.7	15.2
ASCE	85	5 3/16	5 3/16	2 3/4	2 9/16	1 35/64	9/16	42.0	21.0	37.0	30.1	12.2
ASCE	75	4 13/16	4 13/64	2 35/64	2 15/32	1 27/64	17/32	42.0	21.0	37.0	22.9	9.9

[1] Pennsylvania Railroad.
[2] AREA recommended rail section.
[3] Colorado Fuel & Iron Corp.
[4] Torsion-resisting rail section.
[5] Modified Dudley rail section.

(density, wheel loads, and speed), subgrade, ballast, and ties. Although adoption of heavier rail increases first cost of rail and track fastenings, there are substantial savings

[1] Average weight of rail on main tracks of United States railways is 105 lb per yd. Rail weighing 110 lb or over is used on 47 percent of mileage. Heaviest *tee* rail rolled is 155-lb PS section used by Pennsylvania RR on its heavy-traffic lines. RE sections, recommended by AREA, have been adopted as standard by a large number of railways.

due to longer life of rail, reduction in tie and ballast renewal, and in labor costs of track maintenance.[1]

Length. Standard length of rail is 39 ft, earlier standards being 33 and 30 ft. Adoption of longer rails would reduce cost of track maintenance considerably since the upkeep of joint bars and joint ties constitues a major portion of the expense. Several railways have installed 78-ft rails (made up by welding two standard rails together).[2] Several railways have made installations of long stretches of *continuous welded* rail; this type of construction is described on p. 2–35.[3]

Manufacture. Rails are rolled from open-hearth steel. Specifications have been adopted by AREA and by American Society for Testing Materials (ASTM). AREA requirements for percentages of constituents are as follows: *carbon* 0.67 to 0.80 for 91- to 120-lb rail, and 0.69 to 0.82 for 121-lb and over; *manganese* 0.70 to 1.00 for rail of 91 lb and over; *phosphorus* (maximum) 0.04; *silicon* 0.10 to 0.23. Tensile strength increases with *carbon* content, and ductility decreases. *Manganese* increases strength and toughens resistance to abrasion. *Phosphorus* is an impurity which reduces resistance to impact and makes the rail brittle. *Silicon* improves quality of steel, increasing its density. *Control cooling* of rail, now standard practice in rail manufacture, has resulted in production of rails of more uniform quality, which are free from "shatter" cracks. *End hardening* of rails provides increased resistance to rail batter (wearing down of railhead near joint).

Rail failures are classified as transverse fissure, compound fissure, detail fracture (including shelly spots and head checks), engine burn fracture, horizontal split head, vertical split head, crushed head, piped rail, split web, broken base, ordinary break (square or angular), and damaged rail (broken or injured by wrecks; broken, flat, or unbalanced wheels; slipping or similar causes). The transverse fissure has been a cause of serious trouble since it develops under traffic from imperfections (shatter cracks) that cannot be detected at the mill. However, current practice of control cooling of rail has practically eliminated the development of transverse fissures in rails rolled in recent years. Most railways check their tracks periodically with a "detector" car. This makes use of an electrical process for detection of transverse fissures and other flaws, making it possible to remove such rails before actual failures occur. Apparatus has also been developed to locate any flaws in rail within joint bars.[4]

Alloy steel is "steel containing more than 1.65 percent manganese or more than 0.60 silicon or other elements added for the purpose of modifying or improving the mechanical or physical properties normally possessed by plain carbon steel" (AREA). *Manganese steel* rail is used in construction of frogs and crossings and may be used on sharp curves to reduce rail wear. High-silicon and chrome-vanadium steel alloys have been used on test sections of track to reduce shelly spots and head checks.

Continuous Welded Rail. Since maintenance of the rail joint amounts to nearly half the cost of maintaining track, the elimination of joints through the use of *continuous welded* rail has received increasing consideration. Since 1930, 721 track miles have been laid on open track (exclusive of about 90 miles by Pennsylvania RR in a large yard). Advantages of its use include: elimination of joint bars and accessories; increased life of rail through elimination of rail batter; reduction in labor costs for track surfacing; re-

[1] A comparison of 112-lb and 131-lb rail on a test section on Illinois Central RR showed an average saving in "annual" cost (including investment charges) of 12.9 percent over an 11-year period. Tests on other railways have also indicated "annual" savings due to installation of heavier rails.

[2] AREA rail committee reported that in 1955 nine railways had 351.5 miles of 78-ft rail on their lines; exclusive of many short stretches of 78-ft (or 127-ft) rail used through highway crossings, along station platforms, etc.

[3] As a possible alternative to installation of continuous welded rail, tests are underway in *laying rail tight with frozen joints.* Under this procedure, rails are laid without any joint gap and with sufficient bolt tension to eliminate any subsequent rail contraction.

[4] Annual reports of AREA rail committee include statistics on "detected" and "service" failures. Current research is underway on cause and method of preventing *shelly spots* and *head checks,* which have been a source of much trouble on curves located on heavy-traffic lines.

duction in mechanical wear of joint ties with resulting increased life of these ties; more efficient track signal circuits without use of bonds between adjacent rails; and reduced maintenance of locomotives and cars. Possible disadvantages include: first cost of weld being greater than that of joint-bar assembly; increased cost of "changing out" rail in case of a break; and probable limitation of track maintenance to temperatures below that at which welded rail was laid. A study by six railways, which had installed continuous welded rail (131-lb or heavier section), indicated average first cost of welded track to be nearly $4,000 more per mile than for jointed track, but the *net annual* saving was nearly $700 per mile. Life of welded rail was estimated to be 26 years, compared with 19 years for jointed track. Cost of weld was $10 to $12 each.[1]

Early welding of rail in United States was done by Thermit process in which Thermit metal (mainly iron oxide and aluminum) was fused with the rail metal. The processes now principally used are oxyacetylene pressure welding and electric flash butt welding.[2] Standard-length control-cooled rails are welded together in lengths of 1200′ to 1500′ at a plant set up at a yard along the railway. These "strings" of welded rail are usually hauled to track location on flat cars, although other methods of transporting them have been used. Rails are unloaded by anchoring them at one end and pulling cars out from under them. Rails can be placed by a rail crane (or other method) and "barred" into place. These long rails are then welded in the track to adjacent ones to form continuous welded rails of length desired. Length is governed by location of turnouts and by distance between signals, since insulated joints must be provided at ends of signal circuits. Length is usually from 5,000 to 6,000 ft although welded rails nearly 20,000 ft long have been installed. Standard bolted joints are used where continuous welded rails are joined to each other (or to adjacent sections of jointed track).

Anchorage. Continuous welded rail requires adequate provision for anchorage of rails to ties in order to develop sufficient restraint to eliminate expansion or contraction due to temperature changes. If rail is rigidly anchored to ties, the only change in length of rail that can occur is that corresponding to movement of ties in the ballast. Six rail lengths (234′) are usually needed at each end of welded rail in order to build up full tie restraint. There is some change in length of rail throughout this distance which is required to "fix" the rail. For remainder of rail, there is no expansion or contraction, but a uniform stress (of compression or tension) is set up in rail because of full restraint.[3] Rail anchorage may be obtained by use of rail anchors (Fig. 23) or by adoption of special *clip* fastenings of rail to tie. If these are used, it is common to alternate *clip* fastenings with standard *cut* spikes. For middle portion of welded rail (where no expansion or contraction occurs), it is desirable to have sufficient anchorage to eliminate creeping of rail and to allow for possible rail breaks, which require temporary use of bolted joints. Little trouble has been found due to expansion and contraction if rail has been laid in fairly warm weather (70° to 90°F).

[1] Report of Special Committee on Continuous Welded Rail, *Proc. AREA*, vol. 54, p. 1170. Other reports of this committee since 1951 cover fabrication, laying, fastenings, maintenance, and economics of continuous welded rail.

[2] Descriptions of these methods are contained in reports of special committee on continuous welded rail. For oxyacetylene process, see *Proc. AREA*, vol. 53, p. 684; and for electric flash butt welding, see vol. 57, p. 776.

[3] Following method may be used to find s, stress, psi, in rail, for a temperature change of t, °F, above or below temperature at which rail was laid; F, total force, lb, which must be resisted to attain full restraint; N, number of ties needed to develop full restraint; L, distance, ft, needed to "fix" rail; and l, change in length, in., in L. A is cross-section area of rail, sq in.; $E = 30 \times 10^6$ psi, modulus of elasticity; $n = 6.5 \times 10^{-6}$, coefficient of expansion; c, spacing of ties (c to c), in.; T, resistance per tie, lb; and J, estimated joint restraint, lb.

Assuming no joint restraint: $s = Ent = 195t$; $F = 195tA$; $N = F/T$; $L = cN/12$; $l = cF^2/2TAE$. If joint is assumed to provide additional restraint, $(F - J)$ should be substituted for F in formulas for N, L, and l. s and F are independent of length of welded rail provided it is equal to or greater than $2L$. If welded rail is less than $2L$, above formulas do not apply; in such a case there is some change in length up to middle point of rail and the full stress s is not developed.

Example: Assume a 5,000′ welded rail, 132-lb RE section ($A = 12.95$), laid at 70°F, with $T = 800$, $c = 21.3$ (22 ties per 39′ rail). For a temperature change of 50° above or below 70°F, and *without joint restraint*, results are as follows: $s = 9{,}750$; $F = 126{,}300$; $N = 158$; $L = 280'$; $l = 0.55''$. If a *joint restraint* of 50,000 lb is assumed: $N = 95$; $L = 169'$; $l = 0.20''$.

Experimental work is being carried on in *laying rail tight with frozen joints* (see p. 2–37) If this is found to be successful in reduction of joint maintenance, it may provide an alternative to installation of continuous welded rail.

Track Fastenings.[1] *Fastenings* include joint bars and assemblies (bolts, nuts, spring washers), spikes, tie plates, and rail anchors. "A rail joint should fulfill the following general requirements: (1) It should so connect the rails that they will act as a continuous girder with uniform surface and alignment; (2) its resistance to deflection should approach, as nearly as practicable, that of the rail to which it is to be applied; (3) it should prevent vertical or lateral movement of the ends of the rails relative to each other and permit longitudinal movement necessary for expansion and contraction; (4) it should be as simple and of as few parts as possible to be effective." (AREA).

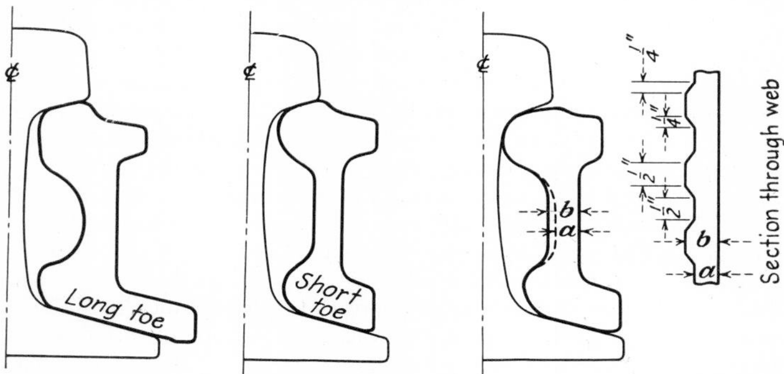

FIG. 20. Head contact joint bars (plain web).

FIG. 21. Head-free joint bar (corrugated web).

Joint bars are made of high-carbon steel or quenched carbon steel. Joint bar fits "fishing" space (between underside of rail head and top of rail base). Top of bar has a fairly heavy section to provide rigidity under head of rail; bottom may have a "short toe" (*toeless type*) or a "long toe" (*flanged type*). (Toeless type is recommended by AREA for rails weighing over 100 lb per yd.) Web of bar is either plain (uniform thickness) or corrugated. Joint bars may be *head contact* (Fig. 20) or *head free* (Fig. 21) depending on type of contact made with underside of head of rail. Either 4-hole bars (24″ long) or 6-bolt bars (at least 36″ long) may be used. For rails of 115 lb or more, 6-bolt bars are commonly used. Bars are drilled with alternate oval and circular holes. Neck of bolt fits the oval hole; alternation of oval and circular holes is provided so that not all the bolts will be sheared off in event of a derailed wheel. Net weight of a *pair* of AREA joint bars is as follows: *For* 100-*lb rail,* 67 lb for 24″ bars and 100 lb for 36″ bars; *for* 115-*lb rail,* 63 lb for 24″ bars and 94 lb for 36″ bars; *for* 132- *and* 133-*lb rail,* 110 to 115 lb for 36″ bars; and *for* 140-*lb rail,* 120 lb for 36″ bars. An *insulated joint* is needed at the end of a track circuit; this is accomplished by inserting a layer of fiber between joint bar and rail and bolts. *Compromise* (step) joints are used to connect rails of different sections or weights; such bars have a transition section at middle of bar.

Oval neck or *elliptic neck track bolts* may be made of medium-carbon steel, heat-treated high-carbon steel, or alloy steel. One-inch-diameter bolts are generally used for 100-lb or heavier rail, although some railways use 1⅛″ bolts for heavy rails. Length of bolt (under the head) varies from 5¼″ to 5¾″ for 90- to 140-lb rail. Figure 22 shows AREA track bolts. *Square nuts,* made of low- or medium-carbon steel are recommended by

[1] AREA specifications for joint bars, bolts and nuts, spikes, tie plates, and spring washers are given in "Manual."

AREA. Hexagon nuts are occasionally used. Recessed nuts, with a depression of ⅛″ to ¼″ to provide one or two unused threads for tightening if the joint becomes loose, are sometimes used. Special self-locking nuts may be used to prevent the nuts backing off; but these are more commonly applied to frogs and crossings than to joint bars.

Adequate track bolt tension is essential for proper functioning of rail joint. Under service conditions there is loss of bolt tension so that periodic tightening is necessary, usually about once a year; except that newly applied joint bars should be retightened

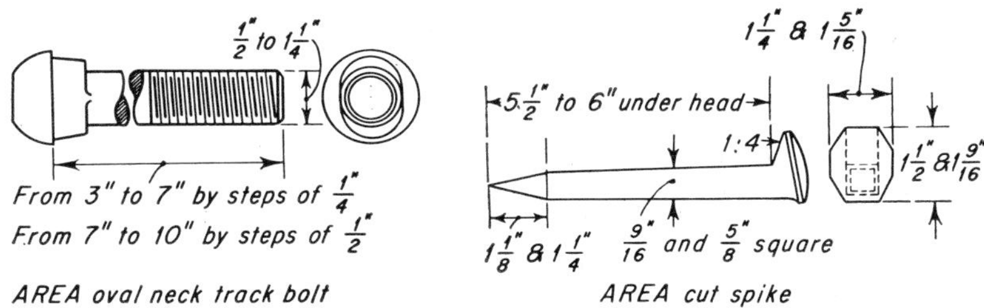

FIG. 22. AREA track bolts and spikes.

within 1 to 3 months after installation. AREA recommendations are: 20,000 to 30,000 lb *per bolt* at initial application; and from 15,000 to 25,000 lb for subsequent tightenings to provide reserve tension over the period between tightenings. (In test installations where rail was *laid tight with frozen joints* a high bolt tension of 35,000 to 40,000 lb per bolt was used to prevent any slippage of rail within joint bars.) Loss of bolt tension in service is due to gradual removal of mill scale from joint bars, wear of "fishing" surfaces, and occasionally stretching of bolt. Most railways include some form of *spring washer* under the nut of each track bolt to help maintain bolt tension.[1] The device commonly used is a *coiled* (deformed) *spring washer* made of high-carbon heat-treated steel or alloy steel. An alternate device is a *rail joint spring* comprising a short steel bar of arched contour to provide spring tension.

Rail is usually fastened to tie with *hook-head cut spikes* (Fig. 22). Spike is square in cross section with a chisel point (parallel to rail). For rails of 100 lb or over, ⅝″ spikes are used, 6″ long (under the head). On straight track, 2 spikes per tie per rail are used; on curves additional spikes are applied. Spikes are driven into circular holes, prebored in tie.[2] *Screw spikes* are sometimes used to attach plate independently to tie. When standard hook-head spike is used, it is necessary to provide rail anchors to prevent creepage of rail (relative to tie). Several alternate methods have been used for attaching rail to tie in order to provide a constant firm bearing on top of rail base, resisting rail slippage and loosening of tie plate. One of these is a *compression clip* which consists of a spring clip held firmly against top of base of rail by a nut on an inverted bolt whose head is hooked under tie plate. This type has been used to some extent on continuous welded track as well as on standard jointed track, compression clips usually being alternated with cut spikes. Other devices include spring-type spikes and studs of various designs.

Tie plates are used under the rail on treated ties, on curves, on all heavy traffic lines, and in turnouts and crossovers. They may be made of low-carbon steel or of hot-worked high-carbon steel. They should be of sufficient length and width to distribute load and thick enough to prevent bending of plate. The top is canted *inward* at about 1:40 under base of rail; bottom may be flat or have shallow ribs. Double shoulders, or a

[1] AREA Specifications for Spring Washers are given in the "Manual," 1956 ed., p. 4-2-19, although AREA makes no recommendations for or against use of spring washers.

[2] AREA specifications are given in the "Manual" for soft-steel cut track spikes (tensile strength, 55,000 psi) and for high-carbon-steel track spikes (tensile strength, 70,000 psi).

single shoulder on outside, are provided against which edge of rail base rests. Plates have four ¾″ square holes for rail-holding spikes and two to four additional holes so that plate can be fastened independently to tie if desired.[1]

Rail anchors, made of high-carbon steel or alloy steel, are used to prevent rail creeping. Rail anchor (Fig. 23) grips base of rail, and the device bears against the side of tie so that rail cannot move (creep) without movement of tie and ballast. Rail creepage is principally in the direction of traffic; the amount will vary with kind of ballast and with local conditions.[2] If *continuous welded* rail is fastened to ties by ordinary track spikes, rail anchors must be used to prevent rail movement relative to ties; see p. 2–35.

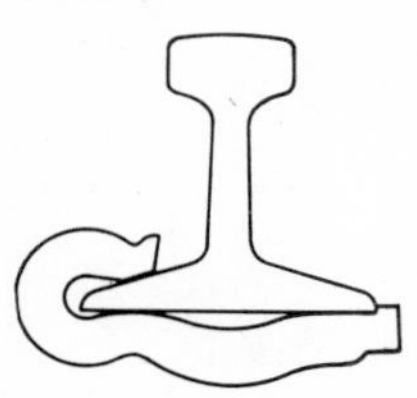

Fig. 23. Rail anchor.

Guardrails are needed in turnouts (unless flanged frogs are used) to prevent wheel flanges from striking frog point and on bridges to prevent derailed wheels from running off ties. They are also used occasionally on sharp curves. Adjustable *rail braces* are used in turnouts and sometimes on sharp curves to provide against overturning or outward movement of rail.

Derails are placed on sidings (adjacent to main tracks) to prevent a train or cars from passing the "clearance" point, which is usually located where track centers are 11 ft apart. *Block-type* derails, permanently fastened to and operated simultaneously with switch leading to siding, are generally used. Derails are sometimes used at interlocking plants or at movable bridges, although present practice does not favor their use on main tracks; when installed at such locations a *switch-point* type is used. *Bumpers* are permanent structures of wood and steel members or of massive concrete. In wood and steel bumpers, the track rails are usually extended and raised, converging so that they may be attached to bumper face at height of car couplers.

Fencing. The right of way is fenced to indicate property lines, to prevent ingress of trespassers and animals, and often to meet legal requirements. Standard right-of-way fence is 4′6″ high and consists of galvanized-steel wires attached to posts; wooden posts are most common although concrete or metal posts are also used. Tight board fences may be used in thickly settled districts. *Snow* fences are used to form artificial eddies on windward side of cuts and are placed at sufficient distance to cause snow to deposit between snow fence and top of cut. They are usually of portable type, the commonest being of wood-slat construction. *Snowsheds* are expensive to build and maintain and constitute a serious fire hazard. Their use should be confined to locations where protection from mountain snowslides is needed.[3]

Signs may be grouped broadly as highway-crossing signs and signals, and roadway signs. The former are for information and warning of the public and the latter for proper operation and maintenance of the railway.[4]

[1] "AREA Manual," 1956 ed. (pp. 5-1-7 to 5-1-22) includes designs for tie plates for use with all AREA rail sections for light, medium, heavy, and extra-heavy traffic. For 115-lb rail and over, tie plates are 7¾″ wide and from 11″ to 16″ long with double shoulders (15″ and 16″ plates are designed for use on curved track only). Approximate weight *per tie plate* is as follows (assuming plates with 6 punched holes): for 12″ tie plate, 15 to 16 lb; for 13″, 18 to 20 lb; for 14″, 22 to 23 lb; for 15″, 27 lb; for 16″, 30 lb. Specifications for low-carbon-steel tie plates and for hot-worked high-carbon-steel tie plates are shown in "AREA Manual."

[2] AREA recommendations for number and position of rail anchors on jointed track are as follows: *For track with essentially one direction of traffic.* Use a minimum of 8 anchors per 39′ rail in direction of traffic, spaced about uniformly along the rail with anchors against the same ties on both rails; and to provide for occasional reversal of traffic and to prevent excessive opening in case of a broken rail, use at least 2 back-up anchors per rail length, boxed in around the tie with 2 of the forward anchors near the quarter points of rail. *For track with traffic in both directions.* Use 16 rail anchors per 39′ rail, 8 to resist movement in each direction for balanced traffic; if traffic is much lighter in one direction, number of anchors in that direction may be reduced. Rail anchors should be spaced about uniformly along the rail with anchors against the same tie on both rails. Anchors to resist movement in the two directions should be placed in pairs and boxed around the same tie.

[3] For design and specifications of right-of-way fences and snow fences, see "AREA Manual," 1956 ed., pp. 1-6-1 to 1-6-24.

[4] For highway-crossing signals and crossing gates, see "AREA Manual," 1956 ed., pp. 9-2-1 to 9-5-2; for roadway signs, pp. 1-7-1 to 1-7-12.

Highway-Railway Grade Crossings. The following types of crossings are in common use: bituminous, wood planks, prefabricated treated timber, precast concrete slab and monolithic concrete. "AREA Manual" (1956 ed., pp. 9-1-1 to 9-1-14) gives specifications for these crossings, together with recommendations for their use on high-speed and on low-speed tracks for varying traffic conditions on highway.

Track Construction.[1] No track shall be laid or track material placed on roadbed until subgrade is completed. Ties shall be laid with heart side down, spaced uniformly, and laid at right angles to rail. Ties shall be adzed, giving full bearing with minimum cutting of ties. Tie plates shall be laid at time of rail laying and so placed that the shoulder is in contact with base of rail or joint bar for full length of shoulder. Metal, fiber, or wood expansion shims shall be used to provide proper openings between rails.[2] All shims shall be removed to within 12-rail lengths of rail laying. (Rails are usually canted inward, inclined tie plates being used.) All joints shall be full bolted and rail drilled where necessary. An initial bolt tension of 20,000 to 30,000 lb per bolt shall be secured. Rail joints shall be staggered, short rail lengths being used to maintain stagger throughout curves. Joints shall be kept out of highway crossings as far as possible. (AREA recommends oiling of track bolts, joint bars, and turnout fixtures.) Rails shall be fully spiked (usually with two spikes per tie per rail on tangents with an additional spike in each tie on inside of each rail on curves). Rail anchors shall be given a firm bearing against the ties. Rail anchors [3] shall be "boxed" around ties when used to resist movement in opposite directions. Outer rail shall be elevated properly on curves, particular care being given to elevations along spiral transitions. Switches must be left in proper adjustment, special care being given to bending of stock rail. Ballast shall be unloaded by dumping or plowing, and spread uniformly to proper section. Preliminary surfacing shall be carried out in such a way that a final lift of between 1″ and 2″ will bring track to finished grade. Ties shall be well packed or tamped from a point 12″ under each rail for 8′ ties, 15″ inside each rail for 8′6″ ties, and 18″ inside each rail for 9′ ties, on both sides of ties to end of ties. When track has been raised to within 1″ or 2″ of final grade and properly compacted by traffic, the finishing lift shall be made. Track shall be placed in good alignment before final lift is made, but a final lining shall be made. Ballast shall be trimmed to standard section and roadbed shoulder properly dressed; surplus ballast after trimming shall be removed.[4]

Track Maintenance. The section foreman, under direction of roadmaster or track supervisor, is responsible for routine maintenance-of-way operations such as daily track inspection, maintaining line and surface of track, renewal of ties, maintenance of rail joints, cleaning of ditches and culverts, cutting grass and brush in right of way, and maintenance of railway-highway grade crossings. Extra (floating) gangs are usually employed for large-scale operations such as "out-of-face" (continuous) track lifts, renewal of rails, renewal or cleaning of foul ballast, roadbed consolidation by grouting methods, oiling of track and roadbed, and widening of cuts and deepening of ditches. *Mechanized* equipment, both "on-track" and "off-track" types, is available for these operations. Repairs (other than minor ones) to bridges and other structures are made by bridge and building forces. Periodic rail inspection is usually made with a "detector" car. Oiling of wheel flanges near curves reduces rail and wheel-flange wear; mechanical rail lubricators are often installed near ends of curves for this purpose.

Ballast and Tie Renewals. When gravel becomes foul (dirty), it should be renewed; stone ballast may be cleaned by mechanical equipment of various types. *Tie renewals*

[1] Condensed from Specifications for the Laying of New Track, "AREA Manual," 1956 ed., pp. 5-4-1 to 5-4-4; and from other AREA recommendations.

[2] For 39′ rails: 0° to 25°F, ¼″; 26° to 50°F, 3⁄16″; 51° to 75°F, ⅛″; 76° to 100°F, 1⁄16″; over 100°F, none.

[3] See p. 2–38 for recommendations as to number and location of rail anchors.

[4] Power equipment is available for such operations as adzing of tie plates, bolt tightening, spike driving, jacking up of track, and tie tamping. Information on such equipment will be found in annual reports of committee on maintenance of way work equipment, published in *Proc. AREA*.

are made in spring and early summer. Individual ties to be replaced are determined by section foreman (perhaps checked by track supervisor).

Rail Renewal. When rails are worn enough to require renewal or when it is desired to relay with a heavier section, the work is done "out of face," i.e., in a continuous stretch.[1] The rail removed is used as *relayer* rail on branch lines or sidings. Reconditioning of rail ends by welding, grinding, or cropping is recommended by AREA as good practice.

Maintenance of Joints. Proper maintenance of rail joint constitutes a large part of the cost of track maintenance. Cracked or broken joint bars should be replaced promptly. Bolts should be kept tight enough to transmit loads properly but not tight enough to prevent expansion or contraction of rails. Metal that has rolled out from end of rail should be chipped off. Low joints should be raised to proper surface. Oiling of joint bar assemblies is often adopted. *Rail batter*, wearing down of railhead near joint, causes poor track. When rail is badly worn, the ends can be built up by welding. End hardening of rail at the mill or in the track reduces rail batter considerably.

String lining is commonly used in maintaining alignment on curved track instead of relining by use of surveying instruments. This method consists of (1) laying off on gage side of outer rail a series of equal chords (usually 62′, although 78′ is sometimes used) throughout curve, (2) measuring and tabulating the ordinates from *each* point to middle of chord that connects the point in the rear and the one in advance, (3) adjusting these middle ordinates until they are substantially equal (with proper allowance for *uniform variation* of ordinates along the spirals). After some practice, adjustment of ordinates may be made without difficulty, and setting of points for revision of track alignment is simple.[2]

Cost of Railway Track. Main track laid with new 115- to 140-lb rail and fastenings, treated ties, tie plates, rail anchors, and stone ballast costs from $15 to $20 per foot. Cost of material varies with weight of rail, joint bars, and tie plates; grade of ties; number of anchors per rail; and depth and quality of ballast. Labor cost per foot of track, which is about one-third of total cost, varies considerably depending on class of track, amount to be built, and labor rates. *For sidings* with 100- to 130-lb "relayer" (used) rail and fastenings, cost is from $10 to $13 per foot depending on depth of ballast, size of ties, and whether tie plates are used. These costs do *not* include grading, culverts, or other track appurtenances.[3]

Railway Structures.[4] Bridges, buildings, and other structures on United States railways represent an investment of several billion dollars and require an annual expenditure of at least $200 million for maintenance. It is estimated that there are about 192,000 railway bridges in United States and Canada, with a total length exceeding

[1] See Specifications for Laying Rail, "AREA Manual," 1956 ed., pp. 5-5-1 to 5-5-3.

[2] Method of string-lining curves is given in "AREA Manual," 1956 ed., pp. 5-3-5 to 5-3-9; and in textbooks on route surveying. Special devices are available for string-lining computations.

[3] *Prices of track material* (December, 1956) as follows: rail, $101.50 per net (2,000-lb) ton. Other track hardware per cwt, *joint bars*, $6.35; *track bolts*, $13.10; *cut spikes*, $8.775; *tie plates*, $6.025. *Rail anchors*, 35 to 40 cents each; *spring washers*, 6 to 8 cents each. *Treated* ties (United States average), $3.45 each; *untreated ties*, $2.01. *Stone ballast*, $2.00 to $2.50 per cubic yard; *gravel ballast*, $1.00 to $1.50. *Labor cost of tracklaying and surfacing:* $3.00 to $5.00 per foot of track (exclusive of cost of distributing track materials). *Turnouts installed:* new No. 10, 140-lb rail, stone ballast, $3,800 to $4,800 each (the lower cost on existing track, the higher cost on new track); new No. 15, 140-lb rail, stone ballast, $5,700 to $7,000; "relayer" No. 8, 100-lb rail, cinder ballast, $2,300 to $2,800; "relayer" No. 10, 130-lb rail, stone ballast, $3,000 to $3,600.

Quantities of track material: Ballast, ¾ to 1¼ cu yd per ft of main track; ½ cu yd for sidings. *Ties*, 3,200 per mile of main track; 2,600 to 2,800 for sidings. *Rail*, net (2,000-lb) tons per mile of track: 100-*lb rail*, 176.0; 115-*lb*, 202.4; 133-*lb*, 234.1; 140-*lb*, 246.4. *Joint bars:* 278 joints per mile of track, for 89 percent 39′ and 11 percent 32′ rail (for weights of bars, see p. 2–36). *Bolts* (*and nuts*) (1″ by 5¾″): 10.5 kegs per mile of track (200 lb per keg) for 39′ rail and four-hole bars; 16 kegs for six-hole bars. *Spikes:* 54 kegs per mile of track (200 lb per keg) for ⅝″ by 6″ spikes based on 3,200 ties per mile and four spikes per tie. *Tie plates* (2 per tie) (for weights, see p. 2–38). *Rail anchors*, 8 to 16 per 39′ rail (see p. 2–38).

[4] "AREA Manual," 1956 ed.: *Culverts*, pp. 1-4-1 to 1-4-35. *Buildings:* general and specifications, pp. 6-1-1 to 6-19-9; passenger stations and freight houses, pp. 6-20-1 to 6-20-10; engine houses, pp. 6-21-1 to 6-21-8; fueling stations, pp. 6-22-1 to 6-22-8; locomotive sanding facilities, pp. 6-23-1 to 6-23-9; locomotive cinder pits, pp. 6-24-1 to 6-24-2; shops, pp. 6-25-1 to 6-25-19; miscellaneous railway buildings, pp. 6-26-1 to 6-26-9. *Wood bridges and trestles*, pp. 7-1-1 to 7-M-8. *Masonry*, pp. 8-1-1 to 8-16-2. *Water, oil, and sanitation services*, pp. 13-1-1 to 13-M-8. *Iron and steel structures*, pp. 15-1-1 to 15-M-34. *Waterproofing*, pp. 29-1-1 to 29-4-5.

4,000 miles; about 60 percent, or 2,500 miles, are trestles. The rapid development of diesel-electric locomotives in substitution for steam locomotives has required the replacement or substantial rebuilding of a large number of repair shops and engine terminals. Railway buildings and other structures include passenger stations and office buildings, freight houses, engine houses, locomotive- and car-repair shops, fuel and water stations, coal and ore wharves, powerhouses, signal towers, and miscellaneous minor structures such as storehouses, pumphouses, and section houses.

Terminal Facilities. *Passenger*, *freight*, and *locomotive* terminal facilities vary widely depending on physical layout and traffic requirements. "AREA Manual" includes considerable information on basic requirements of design of these facilities. Reports of committees published in *annual proceedings* contain additional material. Descriptions of individual terminals appear currently in *Railway Age*.[1]

RAILWAY TURNOUTS AND CONNECTING TRACKS

Turnouts. A turnout is "an arrangement of a switch and frog with closure rails, by means of which rolling stock may be diverted from one track to another" (AREA). There are three types: "split switch," standard on American railways; "tongue switch," sometimes used in paved streets; and "stub switch," an early form of turnout now used only on industrial tramways.

Split-switch Turnout (Figs. 24 and 25). The essential parts of a split-switch turnout are a pair of movable switch (or point) rails with necessary fixtures, closure rails, frog, guardrails, and a set of switch timbers. Switch rails vary from 11 to 39 ft in length; they

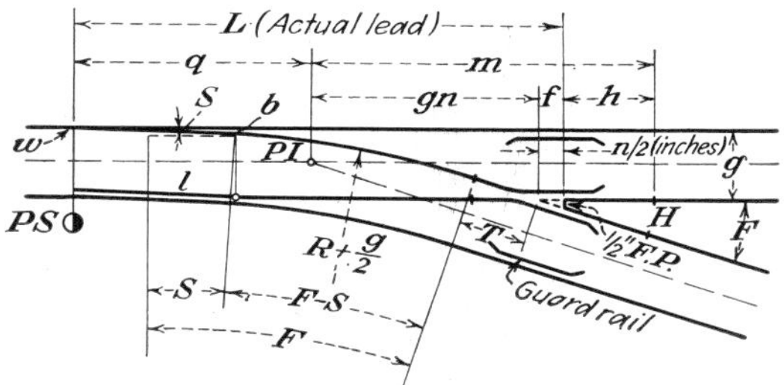

Fig. 24. Split-switch turnout.

are usually straight, although curved rails are sometimes used. The throw of a switch is 4¾ in.; and distance between gage lines at heel of switch rail is 6¼ in. Switch fixtures include switch rods for connecting the switch rails, braces, slide plates, heel plates, and turnout plates. Switch stands are used except when switches are operated from remote points as in interlocking and centralized traffic control. Switch stands are bolted or spiked to two head blocks (extra-long switch timbers).

Closure rails (straight and curved) connect heels of switch rails with toe of frog. A frog is "a track structure used at the intersection of two running rails to provide support for wheels and passageways for their flanges, thus permitting wheels on either rail to cross the other" (AREA). The frog is made up of point rails, wing rails, base plate, flangeway filler blocks, foot guards, together with necessary assembly of bolts. The intersection of point rails is cut back to a thickness of ½ in., forming the actual (½″) frog point. Frogs are either rigid or spring rail. Rigid frogs do not have any moving parts; they are always used in yards, and many railroads use them exclusively. Spring-

[1] "AREA Manual," 1956 ed. *Terminals* (general), pp. 14-1-1, 14-1-2. *Passenger terminals*, pp. 14-2-1 to 14-2-19. *Freight terminals*, pp. 14-3-1 to 14-3-16. *Locomotive terminals*, pp. 14-4-1 to 14-4-13. *Scales*, pp. 14-5-1 to 14-5-109. See also footnote on p. 2–40 for references to structures required in connection with terminal facilities.

rail frogs have a movable wing rail which is normally held against the point rail by springs, thus making an unbroken running surface for wheels using the principal track, whereas flanges of wheels on the other track force the movable wing rail away from the point rail to provide a passageway; these are often used in turnouts leading from a main track. A rigid self-guarded (or flange) frog, which does not require guardrails, is often used in yards. Rigid frogs include bolted frogs built essentially of rolled rails, with fillers between the rails, and held together with bolts; rail-bound manganese-steel frogs consisting of a manganese-steel body casting fitted into and between rolled rails and held together with bolts; and solid manganese cast-steel frogs.

Guardrails are needed opposite the frog on both tracks (unless self-guarded frogs are used) to form a flangeway with rail and hold wheels in proper alignment while passing through the frog. The central portion of guardrail is parallel to stock rail forming a 1⅞-in. flangeway; the ends are flared out by bending or planing.

Spring switches are sometimes used where it is desired to permit flanges of trailing wheels to pass along the other track from that for which the points are set for facing movements. To accomplish this a spring device is incorporated in the operating mechanism so that the points automatically return to their normal position after the trailing wheels have passed through the switch.

Tongue switches are often used in paved streets. The tongue switch consists of a movable tongue with a suitable enclosing and supporting body structure designed for use on one side of track while on the other side either another tongue switch or a fixed mate is used. For heavy traffic, double-tongue switches are often used and for light traffic and street-railway service a single-tongue switch and mate is more common. Girder rail is sometimes used when tracks are in paved streets.

Stub switches, used to some extent on narrow-gage industrial tramways, consist of a pair of short switch rails, held only at or near one end and free to move at other end to meet rails of straight or diverging track. It is assumed that switch rail (when set for diverging route) and closure rail form a circular curve with PC at fixed end of switch and PT at point of frog.

Turnout Design.[1] Turnouts are designated by the number of the frog. The frog number n is the ratio of the length (measured along the bisector of the frog angle F) to the spread between gage lines on opposite sides of frog. $n = \frac{1}{2} \cot \frac{1}{2}F$. Standard frogs are built for full numbers from 5 to 24; occasionally a No. 30 frog is used for high-speed operation. AREA recommendations are as follows: for yards and sidings, No. 8; for main-line slow speed, 12 or 10; and for high speed, 16 or 20. The allowable speed (mph) through the diverging route of a turnout is at least twice the frog number.

The lead (or turnout) curve extends from heel of switch rail to toe of frog; it is tangent to switch rail when it is set for the diverging route. The lead (actual) is the length between actual point of switch and ½″ point of frog measured along the parent track. Occasionally the turnout is modified by extending the tangent either at heel of switch rail or at toe of frog in order to change the "lead" and simplify the cutting of closure rails. (Note columns 15 and 16 in Table 5 and Fig. 25.)

[1] The design of a split-switch turnout with curved closure rail tangent at heel of switch rail and at toe of frog (Fig. 24) may be condensed as follows: g, gage of track (4′8½″ is standard); n, frog number; F, frog angle; T, distance from toe to theoretical frog point; l, length of switch rail; w, thickness of switch point (⅛″ or ¼″); b, spread at heel of switch rail; S, switch angle between gage lines of switch and stock rails; R, radius of center line of lead curve; L, actual lead to ½″ frog point; f, distance between theoretical and ½″ frog points. $\sin S = (b - w)/l$. $f = \frac{1}{2}n$ (inches).

$$R + \frac{1}{2}g = \frac{g - b - T \sin F}{2 \sin \frac{1}{2}(F + S) \sin \frac{1}{2}(F - S)}$$

$$L = l + \frac{g - b - T \sin F}{\tan \frac{1}{2}(F + S)} + T \cos F + f$$

$$\text{Curved closure} = (R + \tfrac{1}{2}g)\,[\text{angle } (F - S)]$$

$$\text{Straight closure} = \frac{g - b - T \sin F}{\tan \frac{1}{2}(F + S)} - T \operatorname{vers} F$$

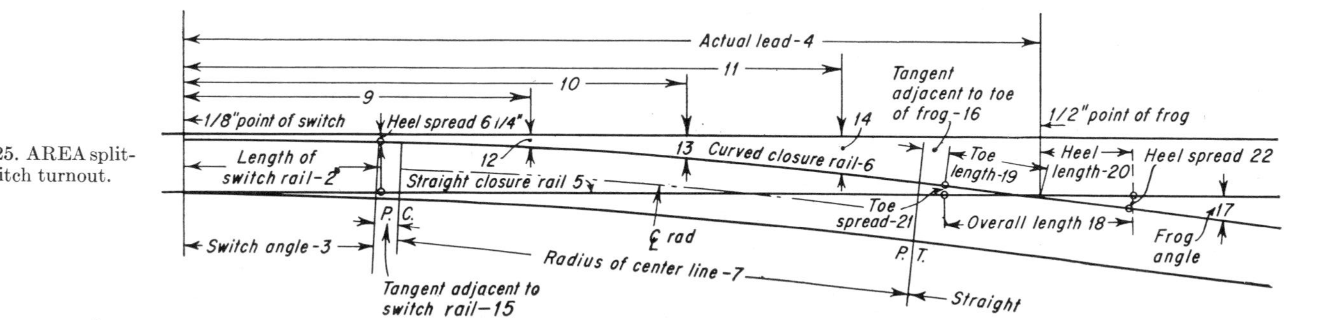

FIG. 25. AREA split-switch turnout.

TABLE 5. AREA TURNOUT DATA FOR STRAIGHT SPLIT SWITCHES

	Prop. of switches			Closure distance		Lead curve		Gage line offsets								Properties of frogs					
1	2	3	4	5	6	7	8	9	10	11	12	13	14	15	16	17	18	19	20	21	22
Frog No.	Length of switch rail	Switch angle	Actual lead	Straight closure rail	Curved closure rail	Radius of center line	Degree of curve									Frog angle	Overall length	Toe length	Heel length	Toe spread	Heel spread
	Ft In.	Deg Min Sec	Ft In.	Ft In.	Ft In.	Feet	Deg Min Sec	Ft In.	Ft In.	Ft In.	Inches	Inches	Ft In.	Feet	Feet	Deg Min Sec	Ft In.	Ft In.	Ft In.	Inches	Inches
5	11-0	2-39-34	42- 6½	28- 0	28-4	177.80	32-39-56	18- 0	25- 0	32- 0	11 13/16	20⅝	2- 8⅞	0.00	0.78	11-25-16	9- 0	3-6½	5- 5½	7 15/16	13 9/16
6	11-0	2-39-34	47- 6	32- 9	33-0	258.57	22-17-58	19- 2¼	27- 4½	35- 6¾	12⅜	21⅝	2-10	0.00	1.75	9-31-38	10- 0	3-9	6 3	7	13
7	16-6	1-46-22	62- 1	40-10½	41-1¼	365.59	15-43-16	26- 2¼	35-10½	45- 6¾	11⅜	19 9/16	2- 6⅞	0.01	0.00	8-10-16	12- 0	4-8½	7- 3½	7 9/16	13
8	16-6	1-46-22	68- 0	46- 5	46-7½	487.28	11-46-44	27- 7¼	38- 8½	49- 9¾	11⅞	20 9/16	2- 8 5/16	0.64	0.00	7-09-10	13- 0	5-1	7-11	7⅛	12⅜
9	16-6	1-46-22	72- 3½	49- 5	49-7¼	615.12	9-19-30	28-10¼	41- 2½	53- 6¾	12 5/16	21⅜	2- 9 7/16	0.00	0.17	6-21-35	16- 0	6-4½	9- 7½	8	13 5/16
10	16-6	1-46-22	78- 9	55-10	56-0	779.39	7-21-24	29-11¾	43- 5½	56-11¼	12¼	21	2- 8⅝	2.0	0.00	5-43-29	16- 6	6-5	10- 1	7 3/16	12⅝
1	22-0	1-19-46	91-10¼	62-10¼	63-0	927.27	6-10-56	37- 8½	53- 5	69- 1½	12¼	21⅜	2- 9¾	0.00	0.13	5-12-18	18- 8½	7-0	11- 8½	7⅛	13¼
12	22-0	1-19-46	96- 8	66-10½	67-0	1104.63	5-11-20	38- 8½	55- 5	72- 1½	12 7/16	21⅝	2- 9⅞	0.00	0.50	4-46-19	20- 4	7-9½	12- 6½	7 5/16	13 1/16
14	22-0	1-19-46	107- 0¾	76- 5¼	76-6¾	1581.20	3-37-28	41- 1¼	60- 2½	79- 3¾	12⅞	22 5/16	2-10½	0.24	0.00	4-05-27	23- 7	8-7½	14-11½	6⅞	13 5/16
15	30-0	0-58-30	126- 4½	86-11½	87-0¾	1720.77	3-19-48	51- 9	73- 6	95- 3	12⅛	21¼	2- 9¾	1.56	0.00	3-49-06	24- 4½	9-5	14-11½	7	12 7/16
16	30-0	0-58-30	131- 4	91-11	92-0	2007.12	2-51-18	53- 0	76- 0	99-0	12 7/16	21 13/16	2-10 5/16	0.66	0.00	3-34-47	26- 0	9-5	16- 7	6 9/16	12 15/16
18	30-0	0-58-30	140-11½	99-11	100-0	2578.79	2-13-20	55- 0	80- 0	105- 0	12¾	22⅛	2-10 7/16	0.57	0.00	3-10-56	29- 3	11-0½	18-2½	6⅞	12⅝
20	30-0	0-58-30	151-11½	110-11	111-0	3289.29	1-44-32	57- 9	85- 6	113- 3	13 1/16	22 11/16	2-11 3/16	2.47	0.00	2-51-51	30-10½	11-0½	19-10	6⅛	12⅜

Turnout standards, prepared by individual railways and by AREA, give the track foreman complete data for installing the turnout. AREA Turnout Data for Straight Split Switches are shown in Fig. 25 and Table 5.[1]

Laying Out Turnout. The engineer usually locates the frog point or the heel of frog and the foreman constructs the turnout in accordance with standard turnout plans. When a turnout is to be installed in an existing track, the heel of frog is usually placed at an existing joint on inside rail of parent track. The inside rail is removed between this joint and the switch point, the outside rail of the original track being undisturbed. The frog, switch points, guardrails, and fixtures are then installed. Curved closure rail is located by offsets from gage line of outside rail of parent track. It is also necessary to remove the track ties and install a standard set of switch timbers.

Connecting Tracks. The design and field work of laying out tracks beyond the heel of frog may be simplified by using the *PI of turnout* as a reference point; the PI is the intersection of center lines parallel to both sides of frog (Fig. 24). The following formulas are useful: distance from PI to point on center line opposite the theoretical frog point (on either track) $= gn$; $m = gn + f + h$, from PI to point on center line opposite heel of frog; and $q = L - f - gn$, from PI to switch point. ($f = \frac{1}{2}n$ in inches; $h =$ frog heel to ½″ FP; $L =$ lead to ½″ FP.) With transit at PI, track centers may be set for diverging track if it is straight beyond frog, or the PC may be set if diverging track is curved.

Turnout to Parallel Track (Fig. 26*a*). $PI_1\text{-}V_1 = p/\sin F = pn + p/4n$. $d = pn + p/4n - m - R \tan \frac{1}{2}F$. $AB = q + pn - p/4n + R \tan \frac{1}{2}F$.

For slow-speed turnouts, the connecting curve may be selected as a *full* degree not sharper than the lead curve, and *d* computed. For high-speed turnouts, PC may be

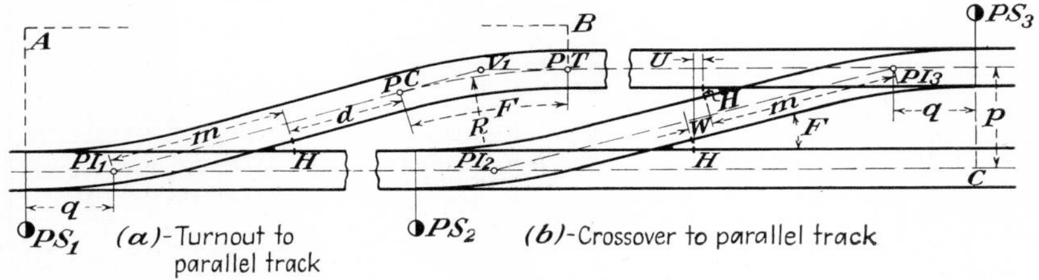

FIG. 26

tentatively assumed opposite heel of frog ($d = 0$), and *D* computed. This will give an odd value for *D* and usually the next larger *full* degree is adopted and *d* computed.[2]

Crossover between Parallel Tracks (Fig. 26*b*). $PI_2\text{-}PI_3 = p/\sin F = pn + p/4n$. $PS_2\text{-}C = 2q + pn - p/4n$. $U = pn - p/4n - 2m$. $W = pn + p/4n - 2m$.

Above formulas are based on the use of equal frogs and a straight crossover. If *p* is large, reversed curves may be used between heels of frogs to reduce the length of crossover. If unequal frogs are used, it is necessary to insert a curve beyond heel of frog having the smaller angle.

Turnout to Diverging Line (Fig. 27). $PI\text{-}M = m + w$. PI-*V* and *MV* may be computed. $d = (PI\text{-}V) - m - R \tan \frac{1}{2}(I\text{-}F)$. $M\text{-}PT = MV + R \tan \frac{1}{2}(I\text{-}F)$.

M (intersection of center lines of main and diverging lines) is located and *I* measured. A tentative location of frog heel *H* may be selected at an existing joint in main track and *w* measured. Usually a *full* degree is tentatively selected for *D*, and *d* is computed.

[1] AREA Trackwork Plans and Specifications include a similar table for turnouts with curved switch rails. Complete details and bill of material are also shown for switches, frogs, and appurtenances. "Railway Track and Structures Cyclopedia" and manufacturers' catalogues contain additional information.

[2] Degree of curve in railway practice is based on *chord* definition (p. 2–8).

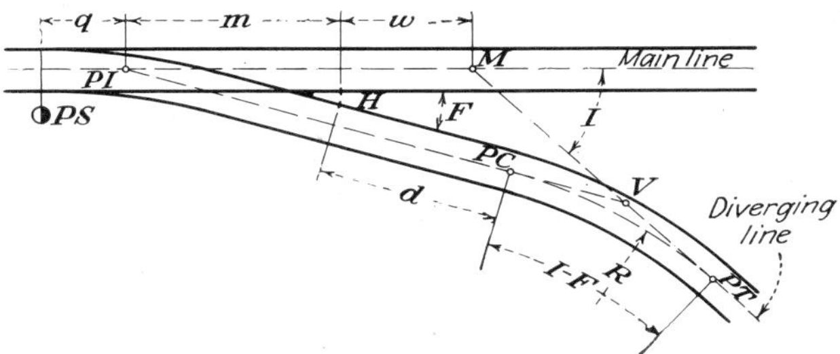

FIG. 27. Turnout to diverging line.

If d is found to be negative, it is necessary either to use a sharper curve or to move H farther from M, i.e., to increase w. (If H is moved, PI-V and MV must be recomputed.)

Turnout from Curved Track. When a split switch is installed in a curved track it is assumed that lead, closure rails, etc., are *unchanged in length* but that their curvature is changed *by the degree* of the parent track. Straight frogs are generally used; switch rails may be straight or curved. This "bending" process is not precise but it may be used in many cases; for high-speed turnouts and crossovers a more precise design may be desirable. If turnout is installed on inside of a curved track, a frog flatter than the standard one is often used to reduce the sharpness of outside curved closure rail.

Connecting Track beyond Curved Turnout. For yards and sidings the "bending" process may be applied as follows: Compute degree and length of a "hypothetical" connecting curve *as though parent track were straight.* To this degree is added (or subtracted) the degree of parent track; length of curve is unchanged from that determined for the "straight-track" layout.[1]

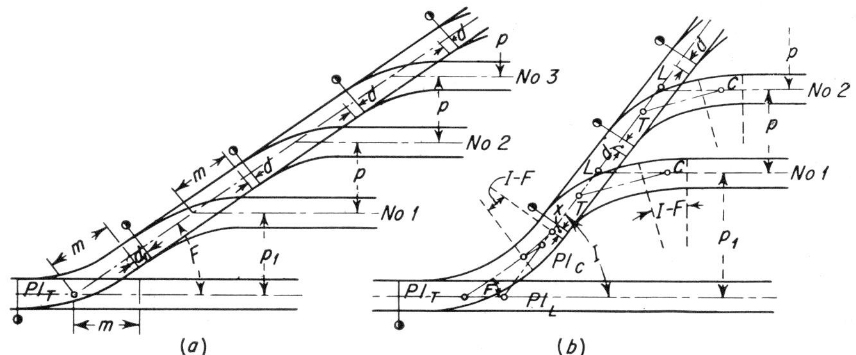

FIG. 28. Ladder tracks.

Ladder Tracks. A ladder track is used to connect a group of parallel "body" tracks. Normal spacing of body tracks is 13 ft; at least 16 ft is desirable between the main (or running) track and the first body track. The simplest layout is obtained when the ladder makes an angle F (frog angle) with body tracks as shown in Fig. 28a. PS-PI_T =

[1] A more precise method is to assume a trial location for frog point on gage line of curved parent track. Set transit at this point, orient on local tangent, and lay off frog angle. This establishes the inside gage line of frog tangent for connecting track. Then set transit on center line of connecting track opposite frog point, orient parallel to frog tangent, and run this line to an intersection with center line of the track which is to be connected with the parent track. It is then possible to determine what curve can be used. If the trial location of frog does not provide a satisfactory solution, frog point may be moved forward or back along the parent track and the process repeated.

$q = L - f - gn$. $m = gn + f + h$. d_1 (for first body track) $= p_1 n + p_1/4n - L - h$. d (for other body tracks) $= pn + p/4n - L - h$.

When ladder angle equals F, space is wasted on the ladder between the heel of each frog and the next switch point and the effective length of the body tracks is reduced. Additional length of body tracks may be obtained by increasing ladder angle a small amount to I, and using a short curve beyond the main-line frog as shown in Fig. 28*b*. Assuming $d = 4$ ft, the limiting value of $I = \sin^{-1} p/(L + h + 4)$.

The center line of ladder (extended) intersects main line at PI_L. $PI_T\text{-}PI_C = m + R \tan \frac{1}{2}(I\text{-}F)$. $PI_T\text{-}PI_L$ and $PI_L\text{-}PI_C$ may be computed. x (PT of first curve to PS for first body track) $= p_1/\sin I - (PI_L\text{-}PI_C) - R \tan \frac{1}{2}(I\text{-}F) - q - (PI_T\text{-}PI_L)$. If x is negative but small, the PS (for first body track) may be placed on the curve. But if x is negative and large, it may be desirable to connect the first body track with the main (or running) track back of heel of the ladder frog instead of directly with the ladder.

To lengthen body tracks still more, I may be made slightly greater than $2F$, limited by $I = \sin^{-1} 2p/(L + h + 4)$, and only the *alternate* body tracks connected directly with ladder. The other body tracks are taken off the adjacent tracks beyond heels of frogs. Body track No. 1 is usually taken off main (or running) track; No. 3 is taken off No. 2, No. 5 is taken off No. 4, etc.

Lap-switch turnouts may be used to secure a compact layout in classification yards, the switch point for one turnout being placed between heel of switch and toe of frog of the other turnout. The frog where the outside curved closure rails of turnouts cross may be a standard frog or of special design.

Crossings. A crossing track is "a structure used where one track crosses another at grade, and consisting of four connected frogs" (AREA). The two frogs at opposite ends of the short diagonal are "center" frogs; the other two are "end" frogs. Bolted rolled rail, manganese-steel insert, or solid manganese-steel frogs may be used; they may be made of single-rail, two-rail, or three-rail design depending on angle of crossing, traffic, etc. When the crossing angle is small, a continuous running surface may be obtained by using movable points for the center frogs.

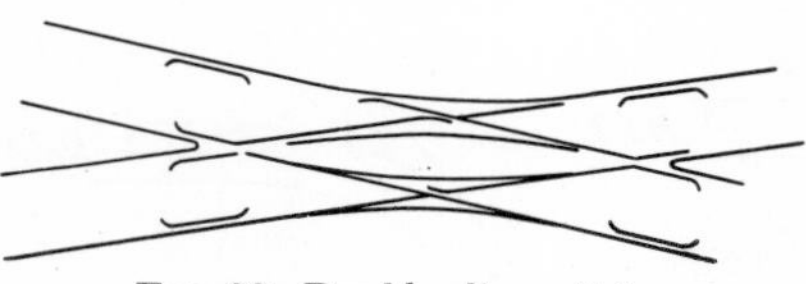

FIG. 29. Double-slip switch.

Slip switches (Fig. 29) are often used on interior ladders or crossing tracks so that a train may be switched to each body track; such switches are called "single" if operating in one direction only, and "double" if operating in both directions. Slip switches consist essentially of two end frogs, protected by guardrails; two center frogs with movable points; and two pairs of switch points for "single" (or four pairs for "double") slip switches. Switch points and movable point frogs are usually operated from an interlocking plant.

RAILWAY SIGNALING AND COMMUNICATIONS [1]

Railway signals are used to *direct* and *control* the operation of trains on main tracks and at crossings, junctions, and terminals. *Signaling* includes the operation of associated switches, crossovers, car retarders, etc., which are coordinated with the signals. Originally adopted solely to provide safety in railway operation, modern signaling increases traffic capacity, speeds up train movements, and furnishes direct economic benefits as well as providing maximum safety. *Signaling* comprises interlocking, block signaling,

[1] For additional information, see publications of Signal Section, AAR: *Ann. Proc.*, "Manual of Recommended (Signaling) Practice," and separate booklets on "American Railway Signaling Principles and Practices"; also "Railway Track and Structures Cyclopedia," and *Railway Signaling and Communications* (monthly), Simmons-Boardman Publishing Corporation; annual "Tabulation of Statistics Pertaining to Signals, etc.," ICC Bureau of Safety and Service; and bulletins of General Railway Signal Co. and Union Switch & Signal Co.

centralized traffic control, automatic train control and cab signals, classification yard control (car retarders), and train communication systems. Interstate Commerce Commission (ICC) has authority over signal practices on United States railways, and prescribes rules for installation and maintenance of signaling systems and appliances.

Signal Aspects and Indications. *Aspect* of a signal is *what is seen* by the engineer when he views the blades or lights in their relative positions or colors. *Indication* is what the engineer is *instructed to do* in controlling his train in compliance with aspect displayed. *Semaphore* signals comprise one or more blades, operated in three positions in upper-right quadrant for *daylight* aspects. Horizontal position indicates *stop;* at 45° it indicates *caution* (or *approach*); in vertical position it indicates *clear*. *Night* aspects of semaphores are shown by colored lights: red, yellow, and green. *Light* signals, now used for new installations, give day and night aspects by *color* or *position* of lights (or occasionally by both color and position). For *color* lights, red indicates stop; yellow, caution; and green, clear. *Position* lights (usually slightly amber) are lighted in "group" positions corresponding to positions of blades on semaphores. *High* signals govern normal train movements; they are mounted on masts to right of track governed, or on overhead signal bridges. *Dwarf* signals on very short masts control low-speed movements in yards and "back-up" movements on main tracks. An *absolute* (*stop* and *stay*) signal is one which (in restrictive position) indicates that a full stop must be made and train must not proceed until a change of aspect is shown (or hand signal given). A *permissive* (*stop then proceed*) signal indicates that train must come to a full stop but may then proceed slowly (after a specified interval—usually 1 min) prepared to stop within "range or limit of vision." *Approach* (or *distant*) signals are needed to give an engineer sufficient warning if signal ahead is displaying a restrictive indication. Poor weather conditions may reduce visibility to such an extent that a signal cannot be seen until the train has practically reached it. The distance between an *approach* and a *home* signal should be sufficient for "braking" the train from high speed to stop (or to a restricted speed). Distance needed to stop a high-speed passenger train or a heavy freight train may be 6,000 to 8,000 ft. *Protective* signals may be used at special locations; these include *dragging equipment detectors, slide detection fences, high water* (*flood*), and *fire detectors*. When these are actuated because of an unsafe condition, automatic block or interlocking signals in that area display their most restrictive aspects.

Track Circuits. Signaling systems are based on closed circuits, the track being divided into electrically isolated sections by insulated rail joints. When no train is in a section, current flows continuously through the circuit and the relay contact keeps signal battery circuit closed. A train on the circuit causes current to flow from one rail to the other through car wheels and axles, causing track-relay armature to open a contact breaking flow of current to signal, and causing it to display a restrictive aspect. Similarly an open switch or a broken rail (if completely broken away) puts signal in a restrictive position. The basic principle in all signaling is that, in case of failure of a circuit or other apparatus, *the signal goes to danger*. D-C track circuits are generally used on nonelectrified tracks, although alternating current may be used; on electrified tracks alternating current is used. *Coded circuits* have received increasing application. Their use reduces the number of *line* wires needed. In automatic block signaling, coded circuits provide a track current which is *continuously interrupted* so that track relay "makes and breaks" its contacts in response to the code cycle. Different codes are used to control the signals; for example, 75 "on-off" interruptions per minute for *caution* (approach); 180 for proceed, and *no code* (steady current) or absence of current puts signal to stop.

Interlocking. Signals and switches are interlocked to prevent conflicting movements at grade crossings between railways and at junctions. This is effected at the control point by a combination of mechanical and electrical "locks" or by a more recent *all-relay* system in which the locking is done by interconnection of circuits through contacts on

relays. In *mechanical* interlocking plants, switches and signals are operated *manually* from signal tower by pipes to controlled units. This was the earliest form of interlocking and is still used at many points. In *power* interlocking, the *control* is electric; switches and signals are *operated* by electric motors (*electric* system) or by compressed-air cylinders (*electropneumatic* system). An *electromechanical* interlocker is one using both electrical and mechanical levers; usually nearby switches are operated manually and signals by electric motors. *Automatic* interlocking is often used at grade crossings of railways and at simple junctions, the signals being controlled directly by the presence of a train in interlocked zone.

All-relay interlocking has been an outstanding development in railway signaling. The interlocking is accomplished through a network of circuits and relays. Compact panels at control point include small levers (or knobs) for operating switches and signals, and a miniature geographical layout of the track in the control area. Position of switches is shown on track panel by electrically operated "indicators." Lights at signal levers show when a signal has been cleared; lights on track diagram indicate location of trains in interlocked area. The operator has a complete picture of entire zone at all times. There are two types of all-relay interlocking systems. One is an "individual lever" system with a separate lever for each signal and switch (or crossover), in which the operator "sets up" each switch and signal for a desired route. The other is an "entrance-exit" system in which it is only necessary to operate two levers (or knobs) to set up an entire route.

Remote control is defined as a method of controlling outlying signal apparatus from a designated point. It is based on *all-relay* interlocking and was first used to control switches and signals at ends of passing tracks, etc. It has made practicable the consolidation of several interlocking plants in a terminal area into a single all-relay installation operated by one man.[1]

Block Signaling. This system provides a *space interval* between "following" trains on single and double track and prevents "opposing" trains from entering the same section (block) on single track. Early block-signal systems were of *manual* type, controlled by an operator at end of each block; a modification of this is *controlled manual* system which includes electric locks between adjacent block stations. Under *manual* system, block is several miles long, operators usually being located at railway stations. Its use is generally restricted to districts where traffic is light.

Automatic block signals are now installed on most double-track lines and on a large mileage of single track. Under automatic block system, trains actuate signals through *track circuits* (see p. 2–47). Signals are usually of three-aspect type (stop, approach, clear) providing control for two blocks in advance of train. Automatic block signals are usually spaced 1 to 1½ miles apart, depending on local conditions (grade, curvature, etc.). On a few high-speed lines (and often on "rapid-transit" lines) a four-aspect signal is used (comprising two semaphore blades or two light signals), controlling operation over three blocks. On *double-track lines*, automatic block system normally provides for "following" movements only, although in some cases *either-direction* signaling is provided on one or both tracks to facilitate train movements where traffic is heavy or predominantly in one direction at certain periods. *Single-track* automatic block signaling controls (1) entrance of trains into single-track sections and (2) "following" movements of trains in same direction. The *absolute-permissive* system is the most efficient type. Under this system *absolute* signals are placed at "entering" ends of single-track sections and *permissive* signals are used for intermediate signals, operating in the same way as for "following" movements on double track.

[1] Number and type of interlocking plants on United States railways (Jan. 1, 1957): automatic, 651; electric, 1,043; electromechanical, 266; electropneumatic, 365; mechanical, 977; remotely controlled, 945; total number of interlockers, 4,247.

ICC requires *automatic* block signals on all sections of track over which passenger trains are operated at 60 mph or more, or over which freight trains are operated at 50 mph or more.[1]

Centralized Traffic Control (CTC).[2] Practically all train movements on single track were originally governed by timetable and train orders. Although substantial installations of centralized traffic control, a system of "*train operation by signal indication*," have been made in recent years, the major portion of single-track mileage is still operated under the old system. Regular trains are given timetable rights by "class" and "direction," and meeting points are shown on timetable. For all *extra* trains (freight trains are usually so classed) and for regular trains much behind schedule, train orders are issued prescribing meeting points. These are transmitted from dispatcher to local operators, who deliver them to crews of all trains affected. This method involves considerable time and labor and may cause delays at meeting points; it also includes the possibility of "human" errors, resulting in head-on collisions.

The development of *remote control* of switches and signals by *all*-relay interlocking, *coded track circuits*, etc., has made it practicable to direct train operation entirely by *signal indications* given by a train director (or CTC operator) at a central control point. CTC may be applied to a small area where traffic is heavy and bottlenecks occur or to an entire operating district. A panel of the type described under *all-relay* interlocking is installed at control point which may be anywhere in the territory. All switches and signals (except intermediate automatic block signals for "following" movements) are *controlled* and *operated* by CTC operator. Interlockings at crossings and junction points, previously operated individually, can be brought into CTC control. The operator is automatically provided with information as to position of trains as they pass specific points on the track. *Dual-type* switches are usually installed at ends of intermediate sidings where local switching is done; these switches can be operated by train crews with permission of CTC operator. Since the operator knows the position of trains at all times, he can arrange "meets" so that little delay is encountered; in many cases the "meets" are *nonstop*. *Centralized traffic control* has resulted in speeding up traffic, reducing "bottlenecks," and in providing many operating economies. It is now quite common for railways to revert from double-track to single-track CTC operation on parts of their lines. Some railways have installed CTC on portions of double-track lines, providing *either-direction* signaling on one or both tracks. This provides increased flexibility in train operation.[3]

Automatic Train Control. Three systems are in operation on United States railways. They are *automatic train stop* (ATS), *automatic train control* (ATC), and *automatic cab signals* (ACS). Both ATS and ATC include automatic setting of brakes if engineer fails to control his train in accordance with signal indications. ACS provides continuous indications in locomotive cab but does *not* include automatic setting of brakes. ATS system is based on *intermittent inductive control, operative only* at *wayside signals.* If engineer fails to obey a signal indicating *stop*, brakes are automatically set and cannot be released until the train is stopped.[4] ATS system was installed in accordance with an ICC requirement in 1922. A number of railways have since been permitted to abandon ATS and substitute cab signals (ACS) without brake-setting features. A few railways

[1] Block-signal statistics on United States railways (Jan. 1, 1957): *automatic*, 54,944 miles on single track; 56,193 track miles on "two or more" tracks; total, 111,137. *Nonautomatic*, 27,324 miles on single track; 1,759 track miles on "two or more" tracks; total, 29,083. *Total block signals* (automatic and nonautomatic), 140,220 track miles.

[2] ICC has designated this as *Traffic Control System* (TCS). Since the term CTC is in common usage in railway practice it is used here.

[3] On United States railways (Jan. 1, 1957), 25,246 road (29,588 track) miles were operated by *signal indication without train orders.* Most of this is CTC, although a small portion of it represents a modified form in which a train dispatcher *directs* the movement of trains but the *signal indications* are given by local operators on telephonic instructions from dispatcher.

[4] However, automatic brake setting is not operative if engineer actuates a *forestalling* lever indicating that he has recognized the restrictive indication displayed.

tions. An *inductive* type, in which a carrier current is superimposed on line wires, and in part, on the rails, is used to some extent on road installations. Although train communication systems provide additional safety, they are not substituted for railway signaling but serve to supplement it.[1]

LOCOMOTIVES, TRAIN RESISTANCE, AND TONNAGE RATING[2]

Locomotives. Motive power on United States railways includes principally diesel-electric, steam, electric, and gas-turbine-electric (oil-fired) locomotives; multiple-unit electric cars; and rail diesel cars (RDC). Research is being conducted on gas-turbine-electric locomotives using pulverized coal. The possibility of developing an atomic-powered locomotive is being investigated. Although steam locomotives constituted the principal motive power until about 1945, diesel-electric locomotives have practically supplanted them.[3]

Locomotives may be considered in three groups (1) reciprocating steam engines mechanically connected to driving axles; (2) diesel-electric and gas (or steam) turbine electric; and (3) electric. The first two groups are similar in that locomotive carries its own power plant whereas electric locomotives have a relatively unlimited source of power available in the distribution system. Groups 2 and 3 are similar in that the *final drive* is electric. Advantages of electric drive are *large tractive effort* at starting and low speeds, smooth starting due to *uniform torque*, good *tracking* qualities, relatively *low axle loads*, and opportunities for *electric braking.*[4]

Locomotive Capacity. The final output of a locomotive is a combination of *speed* (V, mph) and *tractive effort* (TE, pull, lb). Power developed or exerted by a locomotive may be expressed in *horsepower* (hp). For uniformity in comparing different locomotives, hp and TE at the *driving axles* should be used.[5]

The capacity of a locomotive is limited in two ways: first, by *amount of power it can use*, and second, by *amount of power it can develop.* At starting and low speeds, the first limit is set by *adhesion* between drivers and rail. This depends upon rail and weather conditions and on uniformity of torque. Available adhesion may vary from 5 percent of weight on drivers under extremely poor rail conditions to 40 percent under exceptionally good conditions. Normally adhesive limit is taken as 25 percent of weight on

[1] Train communication on United States railways (Jan. 1, 1957). *Road installations.* Road miles: radio, 82,188; inductive, 7,739; miscellaneous (including wire intercommunication, public radio telephone service, etc.), 5,313. Wayside stations, 1,125; locomotives, 6,204; caboose and other mobile units, 3,550; portable pack sets, 2,679. *Yard and terminal installations.* Number of installations, 569; wayside stations, 690; locomotives, 2,959; cabooses and other mobile units, 377; portable pack sets, 942.

[2] "AREA Manual," 1956 ed., includes material by AREA committee on economics of railway location and operation as follows: *Part* 1, *Location*, pp. 16-1-1 to 16-1-17; *Part* 2, *Operation*, pp. 16-2-1 to 16-2-24; *Part* 3, *Power*, pp. 16-3-1 to 16-3-35; *Part* 4, *Complete Roadway and Track Structure*, pp. 16-4-1, 16-4-2. *Part* 1, economics, grade and line revisions. *Part* 2, velocity profiles, methods of operation, economies of flat and hump yard switching, freight-train performance charts. *Part* 3, power characteristics of steam, electric, oil-electric (diesel) locomotives and rail cars, performance and fuel, speed-time-distance calculations. *Part* 4, traffic classification of main tracks.

[3] Locomotives are classified under either Whyte or AAR symbols. In *Whyte system* (steam locomotives) the number of trucks and arrangement of wheels are indicated. Thus, 4-8-4 refers to a locomotive supported on 3 trucks: pilot (guiding) truck has 4 wheels, driving (rigid) truck has 8 driving wheels, and trailing truck 4 wheels. A switching locomotive would be 0-6-0 or 0-8-0. In *AAR system*, commonly adopted for locomotives having electric drive, letters indicate number of *driving* axles (A for 1, B for 2, etc.), and numerals represent *idle* axles. For example, 4-8-4 locomotive (under Whyte system) would be 2-D-2. Most diesel-electric units are B-B or C-C. If middle axles on 3-axle trucks are idle (nondriver), unit would be classed A1A-A1A.

[4] There are two forms of electric braking which convert energy stored in train into electric energy, thus reducing heating and wear of brake shoes and wheels. In *dynamic* braking (applicable to diesel-electric or turbine-electric types) electric energy is wasted in heating resistances carried on locomotive. In *regenerative* braking (applicable *only* to electric locomotives) electric energy is returned to distribution system.

[5] For *steam* locomotives, *driving-axle horsepower* equals cylinder (or indicated) horsepower less deductions for mechanical transmission losses (cylinders to driving axles); see p. 2–56. For *diesel-electric* (or *turbine-electric*) locomotives, *rated horsepower* is usually defined as power *available for propulsion.* The prime mover (diesel engine or turbine) has sufficient capacity (above rated horsepower) to operate auxiliaries needed on locomotive. The combined loss through generator, traction motors, and gears *averages* about 18 percent so that *driving-axle horsepower* is usually taken as 82 percent of *rated horsepower.* For *electric* locomotives, horsepower rating is an arbitrary one determined by heating of motors, and usually based on rules prescribed by American Institute of Electrical Engineers.

drivers for steam locomotives; for locomotives with electric drive 30 percent is often used.[1]

The *power capacity* of *steam* locomotives depends on various factors of locomotive design. For single-expansion reciprocating steam locomotives, maximum (rated) cylinder TE = $\frac{1}{2}n(0.85pd^2s/W)$ where n is number of cylinders; p, boiler pressure, psi; d, cylinder diameter, in.; s, piston stroke, in.; W, diameter of driving wheels, in. Rated TE can be developed only at starting and very low speeds; for running speeds it must be

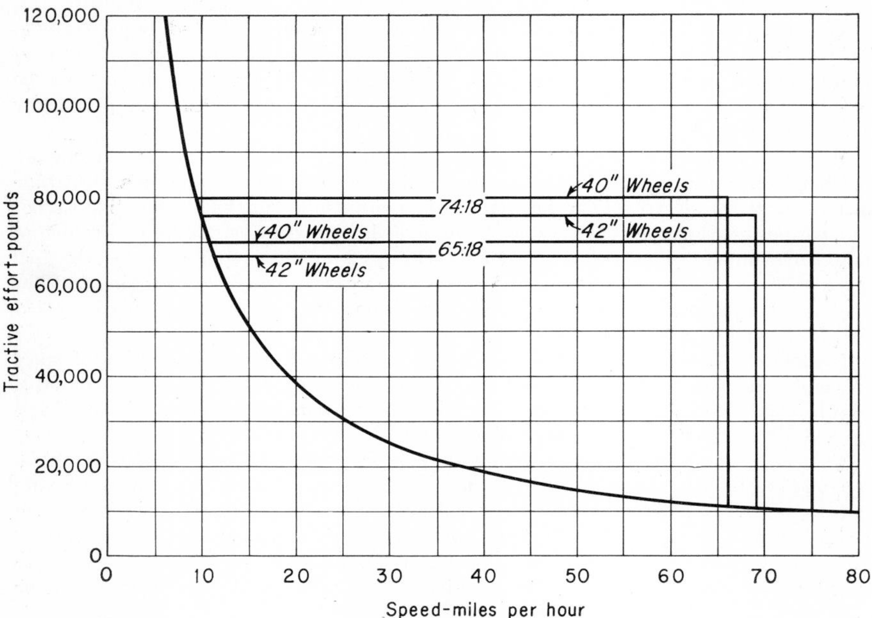

FIG. 30. Tractive effort curve for diesel-electric locomotive. Curve is for 2,400-hp "Train Master" locomotive (Fairbanks-Morse). Maximum continuous TE, lb, shown by horizontal lines; vertical lines indicate maximum allowable speeds. $\left(\text{TE based on } \frac{\text{hp} \times 308}{\text{mph}}.\right)$

reduced by a speed factor.[2] *Cylinder* TE thus found must be reduced to allow for mechanical transmission losses to obtain *driving axle* TE.

Horsepower equals TE, lb, multiplied by V, mph, divided by 375; hp = TE × V/375 or TE = 375 hp/V. For *diesel-electric* locomotives, *driving-axle* TE = 308 hp/V, in which hp is *rated horsepower for propulsion* and 82 percent is used for transmission efficiency. Figure 30 shows *driving-axle* TE for a typical diesel-electric locomotive. For locomotives in which final drive is electric, the heating of armatures in traction motors establishes a limit to *maximum continuous* TE at a corresponding *minimum speed.* Operation at a lower speed (with increased TE) will heat the motors beyond the desired limit. However, this is usually allowable for *short* periods of time. For electric locomo-

[1] Recent tests have indicated that application of a new chemical to the surface of rails will improve locomotive traction and reduce slippage. Studies indicate that wheel slippage is increased by an almost invisible oil film spread on the rails by moisture; the new chemical apparently breaks this film, thereby improving adhesion under all weather conditions.

[2] "Locomotive Data," Baldwin-Lima-Hamilton Corp., Philadelphia, Pa., includes curves showing percentage speed factors.

tives (and often for diesel-electric units), TE ratings are given for periods of 1 hr or 30 min, etc., as well as for *continuous* operation. This additional TE is often helpful in operating heavy trains on ruling grades.[1] Maximum speed at which locomotive should be operated depends on the safe rotational speed of traction motor armature.

Drawbar pull (DBP) is the net TE available for pulling the cars. DBP of a locomotive for specific speed, grade, and curve conditions is found by deducting *all locomotive resistances* from *driving-axle* TE. These always include level-tangent resistance and may also include grade and curve resistances. DBP = TE − LR, where LR is the sum of all locomotive resistances between driving axles and rail. (See p. 2–56 for method of determining LR.)

Diesel-electric locomotives have the following advantages compared with steam locomotives: better fuel economy, lighter axle loads, more uniform torque for starting trains, higher TE at low speeds, quicker acceleration and deceleration (including dynamic braking), higher availability due to reduction in servicing needed at terminals, more flexibility due to use of different numbers and sizes of locomotive units, standardization of units, and simplified maintenance and repairs due to use of removable subassemblies. Each horsepower of *diesel-electric* capacity installed has supplanted at least 2 hp of steam capacity. First cost of diesel-electric locomotive units of road capacity is a little over $100 per rated hp; for smaller (yard switching) units cost per hp is somewhat higher. *Diesel-electric* locomotives are usually built in 600- to 1,200-hp units for passenger, freight, road switching, and transfer service.

Gas-turbine-electric (*oil-fired*) locomotives are used to some extent for heavy freight service by Union Pacific; the first ones were 4,500 hp capacity and more recent ones 8,500 hp. Not enough experience has been obtained yet for a thorough comparison of these with diesel-electric locomotives, although it is likely that the use of gas-turbine electrics will increase considerably.

Electric Traction. Electrification of a limited track mileage (6,200) on United States railways has been made to meet special conditions. These include terminal and suburban passenger service, tunnel operation, and heavy main-line traffic. Electrification was adopted at large passenger terminals with tracks underground (such as New York City) to avoid operating steam locomotives into them. Their extension into suburban areas, where commuter traffic was heavy, provided improved service; suburban operation is usually by multiple-unit trains. Other electrified sections were installed in tunnels where use of steam locomotives was undesirable. (Electric traction has since been abandoned at some of these locations and diesel-electric locomotives used.) A few main-line installations have been made where traffic was heavy, notably on the Pennsylvania and New Haven; a long main-line electrification was made on the Milwaukee to take advantage of low-cost hydroelectric power. The investment cost of power plants and distribution systems is large and can be justified only where operating economies can be shown or where electric operation is much superior to other methods (as in passenger-terminal areas). The development of diesel-electric and gas-turbine-electric locomotives, which are flexible units and readily transferred to other sections to meet changing traffic requirements, makes it doubtful whether much additional electric traction will be installed.

Electric locomotives may be operated with either d-c or a-c motors, but d-c motors have operating advantages. Transmission lines may be either by trolley lines of high-tension alternating current (11,000 volts), or low-tension direct current (3,000 volts); or by d-c (660-volt) third rail. High-tension a-c lines provide the most economical transmission. Recent developments have made possible the use of a-c transmission and d-c motors,

[1] Driving axles of diesel-electric locomotives are geared to motor pinions. For switching or heavy freight service, where large TE is needed (at low speeds), gear ratio (motor pinion to driving axle) is usually about 1:4; for general-purpose road locomotive 1:3 or 1:3.5; and for high-speed passenger locomotives 1:2.5 or 1:3. The gear ratio may be changed to modify the operating characteristics for a different service.

rectification being made in the locomotive by means of *ignitron* rectifiers. A motor-generator locomotive has also been used to take advantage of a-c transmission and d-c motors. The ability to operate electric locomotives for a limited period with TE considerably higher than the *continuous* rating is important.

Rail Diesel Cars. A noteworthy development in passenger service (beginning in 1950) has been the *rail diesel car* (RDC), designed and built by Budd Co. These are stainless-steel cars each powered by two diesel engines placed under car floor; power transmission is through a torque converter. Diesel engines originally used were 275 hp each; subse-

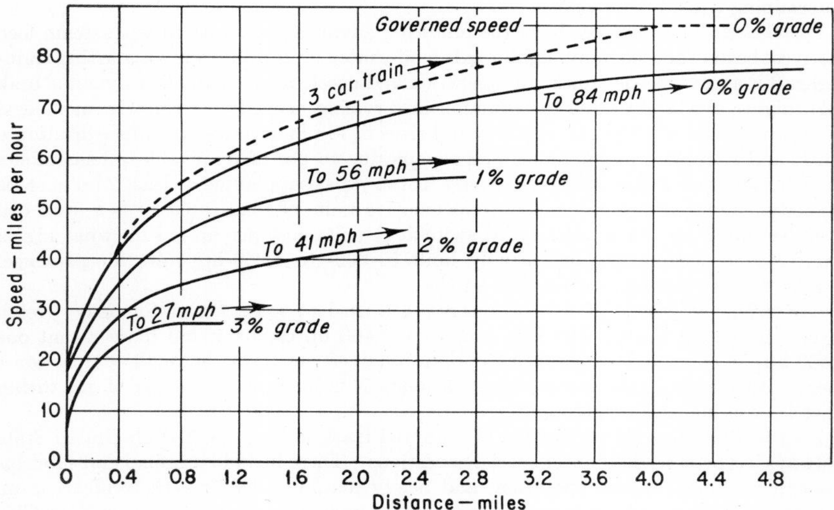

FIG. 31. Speed-distance diagram for RDC cars. Solid lines are acceleration curves for 600-hp Budd RDC single car on different grades; dashed line is for three-car train on level track.

quently engine ratings were increased to 300 hp. Controls are located at each end of car, and cars may be operated singly or multiple-unit.[1] RDC cars can run at speeds up to 85 mph; acceleration rate is about 1.4 mph per sec; rapid deceleration is obtained through use of disk brakes. (See Fig. 31 for acceleration characteristics.) Standard designs are available for several models (*a*) passengers only (90 seats); (*b*) passengers (71 seats) and baggage; (*c*) passengers (49 seats), baggage, mail; and (*d*) mail, baggage, express. The first three types are 85′ long; type (*d*) 73′10″ long. RDC cars weigh from 115,000 to 120,000 lb, depending on interior arrangement. Cost (1957) is about $130,000 for standard type. RDC cars are now used by a large number of United States railways in various types of service.[2] Operating results have shown economies in replacement of standard passenger trains on light-traffic lines by 2-unit RDC trains, costs having been cut by 50 percent in some cases.

Passenger-train Equipment. Several new types of passenger trains have been placed in operation in an effort to provide faster service at lower operating and investment costs. Although the cars differ considerably in design they all are based on light weight

[1] A trailer-type car has also been developed with a single 300-hp engine to operate with standard RDC cars in commuter service.

[2] On Boston & Maine, they have been substituted for standard passenger trains for most of the suburban service. Number of cars per train varies from 2 to 6; each RDC train averages 11 trips per day. On Western Pacific, they have operated on 928-mile runs.

and low center of gravity to reduce operating costs and permit higher speeds, particularly on curves. It is expected that their investment cost will be considerably less than for standard types. A TALGO-type train, first developed for use in Spain, has been modified for use in this country. The first train was placed in operation on the Rock Island in 1956; others were scheduled for early use.[1] A different type will be installed on the New York Central and the New Haven.[2] Other types are in process of development and in trial operation. In some cases cars are designed for use only in "complete" trains; others have standard couplers and can be used in any train.

Train resistance of locomotives and cars comprises an "inherent" *level-tangent* resistance and "incidental" resistances due to grade, curvature, winds, etc. *Level-tangent* resistance (often termed *tractive, frictional* or *rolling* resistance) varies with speed, axle load, cross-section area, streamlining, type of journals, condition of track, temperature, and winds. The following formula for *level-tangent* resistance, presented by W. J. Davis, Jr., in 1926, is commonly used.[3] The basic formula is

$$R = 1.3 + 29/w + kV + KAV^2/wn$$

in which R is *level-tangent* resistance, lb per ton; V, speed, mph; w, average load per axle, tons; n, number of axles; wn, average weight of locomotive or car, tons; A, cross-section area, sq ft; k and K are coefficients applicable to different types of equipment.

Type of equipment	k	K [1]	A
Locomotives	0.03	0.0024	120
Freight cars (and loco tenders)	0.045	0.0005	85–90
Passenger cars (vestibuled)	0.03	0.00034	120
Multiple-unit trains (vestibuled), leading car	0.045	0.0024	100–110
Multiple-unit trains (vestibuled), trailing cars	0.045	0.00034	100–110
Motorcars	0.09	0.0024	80–100

[1] For streamlined equipment, 0.0016 may be substituted conservatively for 0.0024, and 0.00027 for 0.00034.

Important tests to determine *level-tangent* resistance for freight and passenger cars of varying weights at different speeds were made at University of Illinois from 1910 to 1916 (for freight cars with speeds up to 40 mph, for passenger cars with speeds up to 75 mph). These tests furnished much of the data used in developing the *Davis* formula. Subsequent tests (1948) for freight cars at higher speeds indicate that values found by *Davis* formula are low for speeds between 50 and 70 mph. However, general use of this formula appears justified for freight cars since their resistances are important principally at moderate speeds in determining tonnage ratings. Other investigations have been

[1] TALGO train consists of four cars, each of which is 109′ long (on four axles) and made up of three articulated units. Its weight per seat is about half that of standard passenger cars. Floor height is 2′4″ above top of rail (compared with 4′3″ on standard cars).

[2] In this type, a single-unit car (on two axles) is placed in the middle of train. The cars in front of it, and behind it, are two-unit articulated cars (one axle per unit). Seating capacity is 392 for a train with central car and four two-unit cars. On the New Haven, locomotives will be used on both ends for convenience in operating in either direction between terminals.

[3] The *Davis* formula (*Gen. Elec. Rev.*, October, 1926, pp. 685–707) is based on an analytical study of important tests and investigations made up to that time. It is applicable at average temperatures (above freezing) to locomotives or cars moving at uniform speed on *level-tangent* track. The first two terms represent journal friction almost entirely. The third term comprises flange friction, concussion, swaying, and miscellaneous resistances directly proportional to speed. The last term gives air resistance (through *still* air), no allowance being made for head or strong side winds. The investigations on which the formula is based applied to *standard* equipment (see footnote for values of K applicable to streamlined equipment). For roller-bearing journals, at speeds from 5 to 35 mph, substitute $0.9(1.3 + 29/w)$ for first two terms in formula. For axle loads of less than 5 tons, substitute $(9.4/\sqrt{w} + 12.5/w)$ for $(1.3 + 29/w)$. The *locomotive* formula gives *level-tangent* resistance to be deducted from tractive effort at driving axles. For steam locomotives, 20 lb per ton of weight on drivers should be allowed for mechanical resistance between cylinders and driving axles.

made (including model tests in wind tunnels) to determine level-tangent resistances for high-speed streamlined passenger trains.[1]

Level-tangent resistance at starting may be as high as 25 or 30 lb per ton when ordinary *friction-bearing* journals are used, but it drops off rapidly to its minimum value at about 5 mph. *Roller bearings* reduce *starting* resistance very much. At speeds from 5 to 35 mph, roller bearings reduce journal resistance about 10 percent; at speeds above 35 mph no appreciable reduction is obtained. Low temperatures increase level-tangent resistance when friction-bearing journals are used, particularly if train has been *standing* a long time. Allowance must be made for this in tonnage ratings of freight trains on *low-gradient* lines since ratio of *level-tangent* resistance to *total train* resistance is large on low gradients. The effect of low temperature is of *minor* importance on steep grades. (See footnote, p. 2–57.) *Head* or *strong side winds* increase *level-tangent* resistance but the amount is uncertain; in practice, tonnage rating is reduced somewhat arbitrarily where such conditions are expected.

AREA recommends use of following formula for *level-tangent* resistance in comparing operation of freight trains at speeds between 7 and 35 mph on different gradients and locations: *Total level-tangent resistance of train*, lb $= 2.2T + 121.6C$, where T is total weight of cars and contents, tons, and C, number of cars in train.

Grade resistance is 20 lb per ton for each 1 percent of gradient. It is of considerable importance on steep-gradient lines, where it constitutes the major part of total train resistance. (On *down* grades, it becomes an accelerating force.) *Curve* resistance amounts to about 0.8 lb per ton per degree of curvature. When ruling grade is *compensated*, grade resistance *includes* curve resistance. If grades are *uncompensated, curve* resistance may be expressed as *equivalent grade resistance* (see p. 2–26).

The *total resistance* of locomotive (and tender) includes *level-tangent* resistance together with *grade* resistance and *curve* resistance (if grade is uncompensated).[2]

Example: Find DBP for 4,800-hp diesel-electric locomotive at 50 mph on 0.60 percent "uncompensated" grade on which a 5° curve occurs. Locomotive consists of three 1,600-hp units, each weighing 123 tons and having 4 axles. $A = 120$ sq ft.

$$\text{TE (driving axle)} = 308 \times 4{,}800/50 = 29{,}570 \text{ lb}$$
$$\text{Grade and curve resistance} = [0.60 + (5 \times 0.04)]20 = 16 \text{ lb per ton}$$
LR (leading unit):
$$= 123[16 + 1.3 + 29/30.75 + 0.03 \times 50 + 0.0024 \times 120 \times 2{,}500/123]$$
$$= 123(16 + 9.59) = 3{,}150 \text{ lb}$$
LR (2 following units):
$$= 2 \times 123[16 + 1.3 + 29/30.75 + 0.03 \times 50 + 0.00034 \times 120 \times 2{,}500/123]$$
$$= 246 \times (16 + 4.57) = 5{,}060 \text{ lb}$$
$$\text{DBP} = 29{,}570 - (3{,}150 + 5{,}060) = 21{,}360 \text{ lb}$$

Tonnage Rating. The following formula shows the relation between drawbar pull, car resistance, and accelerating (or decelerating) force.

$$\text{TE} - \text{LR} = \text{DBP} = \text{CR} \pm \text{AF}$$

in which TE is tractive effort at driving axles; LR, *total* resistance of locomotive; CR, *total* resistance of cars in train; and AF, accelerating force, lb. (When AF is negative, it indicates a decelerating force.) For a train moving at *uniform* speed, DBP = CR.

[1] SCHMIDT, E. C., "Freight Train Resistance," *Univ. Illinois Eng. Expt. Sta. Bull.* 43, 1910; SCHMIDT, E. C., and F. W. MARQUIS, "Effect of Cold Weather on Train Resistance and Tonnage Rating," *Bull.* 59, 1912; SCHMIDT, E. C., and H. H. DUNN, "Passenger Train Resistance," *Bull.* 110, 1916; TUTHILL, J. K., "High Speed Freight Train Resistance," *Bull.* 376, 1948. Other references: DEBELL, G. W., and A. I. LIPITZ, Air Resistance of Passenger Trains, *Ry. Mech. Eng.*, December, 1935; and TOTTEN, A. I., Resistance of Light Weight Passenger Trains, *Ry. Age*, vol. 103, July 17, 1937.

[2] Method of determining *locomotive level-tangent* resistance is as follows: For *diesel-electric and electric* locomotives, apply *Davis* "locomotive" formula to *leading* unit and "passenger-car" formula to *following* units. For *steam* locomotives (after deducting mechanical resistance of 20 lb per ton of weight on drivers), apply "locomotive" formula to engine and "freight-car" formula to tender.

The *tonnage rating* for a specific class of locomotive is the total weight of cars and contents which it can haul at a desired speed on *ruling* grade of engine district. In *passenger* service, the *number* of cars which can be hauled equals DBP divided by *total* resistance of one car, provided all cars are of substantially uniform type, weight, and axle loading. If cars vary in character and weight, *tonnage* of cars may be found by dividing DBP by *total resistance per ton* for cars.[1]

For *freight* trains, the determination of *tonnage rating* involves additional factors. Cars in an ordinary freight train vary in weight from *empty* cars (25 to 35 tons) to *heavily* loaded cars (80 tons or more including contents). At 20 mph, *level-tangent* resistance varies from 9.5 lb per ton for a 25-ton car to 4.2 lb for 80-ton car. On *low-gradient* lines, *level-tangent* resistance constitutes a large part of the *total car resistance* (including *grade* and *curve* resistances) and it is desirable to adopt a fairly precise method of *tonnage rating*. Theoretically, *total resistance* of cars should be equated with DBP of locomotive at desired speed on *ruling* grade. However, this entails considerable labor and a simpler *adjusted* (*equated*) *tonnage rating* method is often used.[2] When the ruling grade is steep, a *single* tonnage rating may be used, obtained by dividing DBP by *total resistance per ton* for an *average* freight car (using 5 or 6 lb per ton for *level-tangent* resistance).

Level-tangent resistance increases in cold weather and *tonnage ratings* should be reduced. The foregoing discussion is based on A (warm weather) *tonnage ratings* (35°F or above). B ratings are often used between 20° and 35°F, C ratings from 0° to 20°F, and D ratings below 0°F. On *steep-gradient* lines (2 percent, for example), B, C, and D ratings are respectively 98, 96, and 93 percent of A rating. But in *low-gradient* lines (0.4 percent, for example), the corresponding percentages of A rating are 93, 85, and 76.[3]

Acceleration and Retardation. Tonnage ratings are based on *uniform* speeds on specific grades. But in many problems in railway operation it is desired to know *distance* and/or *time* required to *accelerate* (or *decelerate*) from one speed to another. Formula DBP = TF − LR = CR ± AF expresses the basic relations. AF is force (total lb) available for *acceleration* at a specific speed on the grade; it is a *retarding* force if negative. Since DBP *decreases* and CR *increases* with an *increase* in speed (and vice versa for decrease in speed), theoretically AF applies only momentarily (i.e., before the speed changes). However, if small increments of speed changes are used (5 mph or less), accompanying values of AF may be considered constant throughout the respective increments. The following formulas may be used:[4]

$$S = 70.2(V_2^2 - V_1^2)/a \qquad \text{and} \qquad t = 95.7(V_2 - V_1)/a$$

where S is distance, ft; t = time, sec; V_2 and V_1, speeds, mph; and a, accelerating force, lb per ton of total train weight *including* locomotive.

[1] In this method, an *average* unit *level-tangent* resistance (at desired speed) is used; although not exact, it usually gives satisfactory results. A precise method would be to equate DBP with *total* resistance of all cars; this would require the determination of *level-tangent* resistance of *each* car.

[2] This method provides satisfactorily for the difference between *unit level-tangent* resistances of light and heavy cars. An *adjustment* or *car factor* CF (in tons), is added to *actual* weight of each car. It is determined as follows: CF = $(W_L - W_E)/(N_E - N_L)$ and used to nearest ton. W_L is tonnage which could be hauled if all cars in train were heavily loaded and N_L is corresponding number of cars; W_E and N_E are similar figures for a *tonnage* train made up of empty cars. (W_L equals DBP divided by *total unit resistance* per ton of car.) *Adjusted tonnage rating* (ATR) = $W_L + N_L \times$ CF = $W_E + N_E \times$ CF. To determine number of cars of *miscellaneous* weights which can be hauled, $W_A + N_A \times$ CF is made equal to ATR. W_A is total weight of cars being made up into a *tonnage* train and N_A, number of cars. CF depends almost entirely on rate of *ruling* grade and is therefore the same for *all* locomotives on engine district. ATR is a constant for a specific class of locomotive operating on engine district. Unit *level-tangent* resistances of light and heavy cars (for determining W_L and W_E) may be computed by *Davis* "freight-car" formula or taken from a table in *Univ. Illinois Eng. Expt. Sta. Bull.* 43.

[3] "AREA Manual," 1956 ed., p. 16-1-8, gives a table showing CF and B, C, and D percentages of A rating for *adjusted tonnage rating* method for grades from 0.1 to 3.0 percent (CF varies from 29 tons on 0.1 percent grade to 2 tons on 3.0 percent grade). In this method, CF is kept constant for all temperature ratings and reduction made in ATR. An *alternate* method is used by some railways in which ATR is kept constant (for each class of locomotives) for all temperatures and CF is increased progressively for B, C, and D ratings.

[4] These formulas allow 5 percent for rotative energy of axles and wheels.

Example: Assume that train with total weight of 1,000 tons has accelerating force AF of 20,000 lb at 30 mph on level grade, and 18,000 lb at 35 mph. *Find* distance and time to increase speed from 30 to 35 mph on *level* grade. Average AF = 19,000 lb or $a = 19$ lb per ton. $S = 70.2(1{,}225 - 900)/19 = 1{,}200$ ft; $t = 95.7(35 - 30)/19 = 25$ sec.[1]

If train had a speed of 30 mph on a 0.4 percent *ascending* grade, a longer distance would be required to increase speed to 35 mph. In this case, a would be *reduced* by 8 lb per ton (0.4×20). $S = 70.2(1{,}225 - 900)/(19 - 8) = 2{,}074$ ft; $t = 95.7(35 - 30)/(19 - 8) = 44$ sec. If grade were 0.4 percent *descending* (assuming operation at full power without any brake application), distance required would be less than on level grade. In this case a would be *increased* by 8 lb per ton. $S = 70.2(1{,}225 - 900)/(19 + 8) = 845$ ft; $t = 95.6(35 - 30)/(19 + 8) = 18$ sec. If train is *drifting* on a descending grade *without power or brake application*, the value of a would be equal to 20 times the track gradient (percent) *minus* average train resistance (lb per ton). If brakes were being applied it would be necessary to know the characteristics of the *braking* curve for the train.

Velocity Profile. An alternate method of finding S is by the use of a *velocity* (*virtual*) *profile*. *Velocity head VH* at a given speed is the height through which a train may be raised vertically because of its kinetic energy. VH (ft) $= 0.0351V^2$, in which 5 percent is allowed for rotative energy of axles and wheels. *Velocity heads*, corresponding to known speeds at various locations along the track, may be plotted above the *track profile*. The elevations of these points are *velocity heights* and the line connecting them is the *velocity profile*. This profile is theoretically a curved line, since V is changing momentarily. However, if speed increments do not exceed 5 mph, a *velocity profile* may usually be considered as straight between adjacent plotted points. Speeds at intermediate points may be found by scaling *velocity heads* (ordinates between *track* and *velocity profiles*) at desired locations; corresponding velocities are found from VH $= 0.0351V^2$, or from a table of *speeds* and corresponding *velocity heads*. It is often desired to find the point at which some definite speed is reached; in this case a template of *velocity heads* is moved along the profile until the desired ordinate is found.[2]

Although the *velocity profile* method is usually applied graphically, it is possible to illustrate its principles analytically for comparison with the *formula* method previously shown.[3] Tables may be prepared which simplify calculations for *accelerating forces*, *velocity grades*, and *velocity heads*, etc.[4]

Speed-distance-time Diagrams. Studies in railway operation may include such matters as (1) effect of grade and/or line revisions on train operation, (2) comparison of lightweight and standard passenger trains, (3) comparison of different types or capacities of motive power, and (4) possible changes in tonnage ratings. These and similar problems usually involve an analysis of speeds and accelerating distances and elapsed times. The plotting of *velocity profiles* over a district often provides the best method of determining these factors. However, it may be desirable to prepare *speed-distance* or *speed-time diagrams* for particular trains operating on different grades. A typical *speed-distance diagram* is plotted with speeds (mph) as ordinates and distances in 100-ft stations

[1] More precise values would be obtained if S and t were summations of distances and times for 1 mph increments, but calculations would be more laborious.

[2] See "AREA Manual," 1956 ed., pp. 16-2-1 to 16-2-9 for method of plotting and using a *velocity profile*.

[3] Assume 1,000-ton train (of previous example) on level grade. Average accelerating force $a = 19$ lb per ton is equivalent to *velocity* (*accelerating*) grade of $+0.95(19 \div 20)$. Difference of *velocity heads* for 35 and 30 mph speeds is 11.41 ft $(43.00 - 31.59)$. Distance needed to increase speed from 30 to 35 mph (on level track) is $11.41/0.55 = 12$ stations (1,200 ft). Time required (using average speed of 32.5 mph) is $1{,}200/(32.5 \times 1.467) = 25$ sec (1.467 is ft per sec for 1 mph). These values check S and t obtained in previous formulas. If train is on a $+0.40$ percent grade, difference between *velocity* and *track grades* is 0.55 percent $(0.95 - 0.40)$. $S = 11.41/0.55 = 20.74$ stations (2,074 ft). If grade is -0.40 percent (and full power is used without brake application), difference between *velocity* and *track grades* is 1.35 percent $(0.95 + 0.40)$. $S = 11.41/1.35 = 8.45$ stations (845 ft).

[4] A table of this character is shown in "AREA Manual," 1956 ed., p. 16-2-2. Typical columns would be as follows: (1) TF, (2) LR, (3) DBP, (4) CR, (5) AF (total lb), (6) a (lb per ton of total train weight) for each speed, (7) *average a* applicable through speed increment, (8) *average velocity* (*accelerating*) *grade* (equal to *average a* divided by 20), (9) VH for each speed, (10) difference between adjacent VH's. Additional columns may be included for *increment* values of S and t for level grade and *summated* values of S and t from $V = 0$ up to each speed. Similar columns can be used for *increment* and *summated* values of S and t for other grades if desired.

(or miles) as abscissa. Separate curves are required for each type of train on each grade. Figure 31 shows a *speed-distance diagram* for 600-hp rail diesel cars (RDC) on different grades. Solid lines are for a single-unit RDC unit, dashed line is for three-car multiple-unit train. A *speed-time diagram* would be of the same character; times (in seconds or minutes) would be used for abscissa. Tabulations (as previously described) would provide data for plotting the curves. Diagrams of this kind are particularly useful in studying possible changes in grades, motive power, and cars where precise results are not needed. The *velocity-profile* method is usually more satisfactory if it is desired to study the *time schedule* for a particular train over an engine district.[1]

HIGHWAY ENGINEERING

Sources of Highway Information.[2] Current practice and policies in highway engineering and administration are available in publications of the following agencies:

Highway Research Board, a division of the National Research Council, publishes annual proceedings, bulletins, and special reports, which include papers on economics, finance, administration, design, materials, construction, maintenance, traffic, and soil mechanics. Research on airport-runway design and construction is also reported.

U.S. Bureau of Public Roads (BPR) of Department of Commerce publishes a magazine *Public Roads,* containing results of research of the BPR in the fields of planning, design, and construction of highways. Statistical tables are issued annually giving road mileages, motor-vehicle tax receipts, and highway expenditures.

American Association of State Highway Officials (AASHO), composed of officials of state highway departments and the BPR, has issued specifications for highway materials and testing [3] and for highway bridges and pavement types.[4] It has also published policies on design and maintenance of highways.[5] Current topics are published in its quarterly magazine *American Highways.* Proceedings of convention meetings are published annually.

Publications are available from American Association of State Highway Officials, National Press Building, Washington 4, D.C.

American Road Builders' Association, an association of engineers, equipment manufacturers, and contractors engaged in highway work, publishes bulletins on a wide range of highway and airport subjects.

University Bulletins. Highway research conducted at engineering colleges is published in bulletins or research reports. Some universities publish annual proceedings of road schools or highway conferences.

Technical magazines featuring highway subjects are *Public Roads, Roads and Streets, Roads and Engineering Construction, Better Roads, Engineering News-Record, Civil Engineering, American City, Public Works,* and *Traffic Engineering.*

Trade bulletins and manuals on construction methods and specifications are published by the Portland Cement Association, Asphalt Institute, Barrett Division of Allied Chemical and Dye Corp. (tars), Calcium Chloride Association, Armco Culvert Manufacturers Association, and others.

[1] The Pennsylvania has developed methods of determining train performance by use of an "analog" computer and as a later improvement by means of a "digital" computer. These methods provide needed information without burdensome calculations.

[2] See also Bibliography at end of section.

[3] Standard Specifications for Highway Materials and Methods of Sampling and Testing; Part I, Specifications; Part II, Tests.

[4] Standard Specifications for Highway Bridges, for Movable Bridges, for Concrete Pavement Construction, and for Bituminous Surface Treatment. Also Manual of Highway Construction Practices and Methods, and Specifications for General Provisions.

[5] Highway "Policies" as follows: Geometric Design of Rural Highways; Arterial Highways in Urban Areas; Maintenance of Roadway Surfaces; Maintenance of Shoulders, Road Approaches, Roadsides, and Sidewalks; and Maintenance of Safety and Traffic Control Devices and Related Traffic Services.

have adopted ATC system which is based on *continuous inductive control.* ATC differs from ATS in that it includes *speed control* as well as automatic stop. Under this system, if engineer fails to operate his train in accordance with speeds indicated by the signals, speed is automatically reduced (by necessary amount of braking) to permitted speed.

Automatic cab signals (ACS) require *continuous inductive control.* Cab signals furnish engineer with *continuous* information as to wayside signal indications. Whenever a more restrictive aspect is shown, engineer receives an audible warning as well as visual aspect of cab signal. They increase safety because signal will change instantaneously to a more restrictive indication if a switch is opened or a complete rail break occurs while train is in the block. If engineer (and fireman) fail to see a wayside signal because of bad weather conditions, both visual and audible warnings will be given if it has a restrictive aspect. ACS also speeds up train operations because cab signals indicate when a less restrictive condition occurs because a train ahead is passing into next block. In 1947, ICC required railways to install either ATS, ATC, or ACS on lines over which any trains are operated at 80 mph or more.[1]

Classification Yard Control. Switching of freight cars on industrial tracks and in many freight yards is done by *drilling* in which locomotive pushes and pulls individual cars (or groups) into proper tracks. A *gravity* (or *hump*) type was developed for larger yards in which a train to be classified was pushed slowly over a hump where cars (or "cuts") were uncoupled. These "cuts" ran down a steep grade and were switched to appropriate tracks, each car or cut being accompanied by a "rider" who *braked* the car to control its speed. Cost of switching in such yards was large because of number of riders employed. A subsequent development was the use of *car retarders,* near foot of the steep grade (below hump) and on tracks leading to classification tracks, to provide requisite retardation. Car retarders, consisting of track brakes which apply friction to sides of wheels, control the speed at which cars enter classification tracks. Car retarders and track switches are actuated by electric motors or compressed air and are operated from control towers. The operators can actuate car retarders with varying degrees of retardation. The "hump" foreman and control-tower operators are furnished lists giving car initials and numbers, track destination, and indication as to car weight (whether empty, loaded, etc.).

Car-retarder yards built recently include numerous improvements which provide for considerable *automation* in classification yard operation with resulting efficiency and economy. *Dragging-equipment detectors* and *inspection booths* (under and alongside of cars) provide for checking of cars *before* they reach crest of hump; *automatic switching* by which a single control button positions all the switches leading to destination track (with ability to "store" route information for five "cuts" of cars); *electronic devices* which measure weight and speed of car and automatically control degree of retardation, and operate track switches in proper sequence; and *track-fullness lights* which indicate extent to which a classification track is occupied. These improvements make possible the concentration of control at one point near the "hump." New yards include many of these devices, although the extent of "automation" varies.

Train Communication Systems. These provide means by which members of a train crew can talk with each other, with other train crews, and with train dispatchers, yardmasters, etc. They include *road* installations and *yard* and *terminal* installations. The former provides increased operating efficiency and safety in main-line operations; the latter has been developed to facilitate switching and terminal operations. A *radio* type system, operating on frequencies allocated by Federal Communications Commission, is used in practically all yard and terminal installations and on 85 percent of road installa-

[1] On Jan. 1, 1957, United States railways had installations as follows: ATS, 15,084 track miles with 5,289 locomotives or motorcars equipped; ATC, 1,951 track miles with 828 units equipped; ACS, 8,560 track miles with 3,577 units equipped.

HIGHWAY STATISTICS, ADMINISTRATION, AND FINANCE

Classes of Highways. There are about 3,418,000 miles of roads and streets in the United States. State-controlled highways (1955) comprise 619,000 miles, including 387,000 miles of primary state highways, 222,000 miles of county roads and secondary state highways, 10,000 miles of other rural roads (including toll roads and roads in forest and park reservations), and 42,000 miles of urban extensions of state highways. Of the total state-controlled highways, 7.4 percent are nonsurfaced, 20.0 percent have low-type surfaces, 33.8 percent have intermediate-type, and 38.8 percent have high-type. Included within the state primary system and its urban extensions is the national system of interstate highways, comprised of 41,000 miles of the main traffic arteries of the nation.

Local roads include 2,333,000 miles (1955) of which 1,738,000 miles are under county control, 562,000 miles are town and township roads, and 33,000 miles are other local roads. There are also 93,000 miles of roads under Federal control in national parks and reservations. Of all local roads 42.3 percent are nonsurfaced, 45.3 percent have granular or soil surfaces, 9.8 percent are of low-bituminous type, and 2.6 percent are of high type.

City streets comprise about 331,000 miles exclusive of urban extensions of state highways, of which about 87 percent are surfaced. City streets may be divided functionally into residential, business, and arterial, the latter including through city streets, particularly those of expressway or freeway design.

In 1956 there were 1,848 miles of toll road in operation, 1,256 miles under construction, 5,525 miles authorized but not started, and 4,900 miles under study.

Administration. *State highways* are administered by a state highway commission, state highway department, public works department or, in some states, by a single director or secretary of highways.[1] The engineering organization is headed by a chief engineer under whom are such major divisions as accounting, surveys and plans, design, right of way, estimating, construction, bridge design, maintenance, traffic, materials, and testing. Each state is divided into highway districts for handling the work locally. On Federal aid work, the U.S. Bureau of Public Roads administers the distribution of Federal funds and approves plans and specifications submitted by the states. The BPR also conducts research and supervises the construction of certain roads in public lands and national parks.

Local roads are under county and town highway departments. In some states, the counties have strong highway organizations; in others the work is handled mostly by the towns. In a few states, a considerable mileage of local roads is under state control. In other states, financial aid is available for the construction of important local roads outside the state highway system. Such work is usually supervised by the state highway department.

City streets are administered by street or public works departments, except for those urban extensions of the state highway system which are under the state highway department. Where state or Federal aid is received for construction of arterial streets, the work is carried out by the state highway department. In some metropolitan areas, special authorities are set up to administer parkway systems, express roads, bridges, or tunnels.

Toll roads are usually administered by a toll or turnpike authority authorized by state legislature to finance, build, and maintain a particular highway or system of highways. In some instances toll roads are maintained and operated by the state highway department.

Finance. The sources of highway and street funds in 1955 [2] were as follows: Federal

[1] State Highway Administrative Bodies, 1952, and State Highway Organization Charts, *Special Report* 20, 1954, published by Highway Research Board.

[2] Highway Finance 1946–55 issued by U.S. Bureau of Public Roads, Dept. of Commerce.

government $784,000,000, highway user imposts (gasoline taxes, license fees, etc.) $3,587,000,000, tolls $263,000,000, property taxes $1,231,000,000, miscellaneous $158,000,000, and proceeds from bond issues $1,587,000,000. Of the $7,837,000,000 expended on highways and streets in 1955, $5,107,000,000 was spent on state highways, $1,434,000,000 on county and local roads, $1,218,000,000 on urban streets, and $78,000,000 by Federal government on highways not classified by system. Capital outlay amounted to $4,818,000,000 and maintenance and operation to $2,587,000,000.

Motor-vehicle fuel taxes and fees levied by the states are the largest source of highway revenue. In 1955 motor-fuel taxes yielded $2,503,000,000 and license fees $1,401,000,000. In addition, motor-vehicle owners paid over $2,500,000,000 in taxes not specifically designated for highway pruposes, such as automobile property taxes, and Federal excise taxes on sale of vehicles and accessories, gasoline, oil, and tires.

Federal aid has been available for primary state highway construction in rural areas since 1916, for secondary roads since 1933, and for urban extensions of state highways since 1944. In 1952 a separate authorization was made for the interstate system. The Federal funds are matched by the states, usually on a 50-50 basis. The allotments for each class of highway are apportioned among the states by formulas which take into account the relative area, population, and road mileage of each state. The Federal Aid Act of 1956 greatly expanded Federal participation in highway construction. The sum of $24,800,000,000 was authorized for the completion of the national system of interstate and defense highways within 13 years, representing 90 percent of the cost; the remaining 10 percent is to be matched by the states, making a total program of $27,600,000,000. Annual authorizations for other Federal aid systems were also increased. For example, for the fiscal year ending June 30, 1959, the authorizations are: primary system $393,750,000 (45 percent), secondary system $262,500,000 (30 percent), and urban system $218,750,000 (25 percent). These funds are to be matched by the states on a 50-50 basis. In addition $103,000,000 is provided for roads in the public domain (parks, reservations, etc.). For the first 3 years of the program the act provides that funds for the interstate system be apportioned to the states on the basis of two-thirds weight to the state's population, one-sixth to its area, and one-sixth to its rural road mileage. In subsequent years the apportionment is to be on the basis of remaining interstate system needs as determined from joint surveys by the Bureau of Public Roads and the state highway departments. The formulas for apportioning regular Federal aid remain as in the past: for primary system, one-third in ratio of area, one-third in ratio of population, and one-third in ratio of mileage of rural mail-delivery routes; for secondary roads, one-third in ratio of area, one-third in ratio of rural population (places under 5,000 population), and one-third in ratio of mileage of rural mail-delivery routes; for urban projects, in ratio of population in municipalities of 5,000 population or more. Federal aid funds are to be used for construction and right of way but not for maintenance. An amount not to exceed 3.75 percent of the sums authorized is deducted for expense of administration and for carrying on research. The states may match up to 1.5 percent for highway-planning studies and research. To finance the highway program proposed in the 1956 act, the Federal government levied an increase of 1 cent in the Federal gasoline tax, increases in taxes on tires, and a new use tax of $1.50 per 1,000 lb of gross weight on heavy vehicles weighing over 26,000 lb. The proceeds from these special taxes are to be paid into a highway trust fund.

Bond issues are commonly used to finance highway and bridge construction. Although state highway construction is financed largely from motor-vehicle revenues, some states have borrowed extensively to accelerate their highway program. Others borrow only for special projects, such as long-span bridges, tunnels, or high-cost expressways. Cities and other local governments, which do not have direct access to motor-vehicle revenue, often borrow for all major street improvements. Toll turnpikes are financed exclusively

from bond issues, the interest and principal payments being paid from toll revenue. Highway bonds are usually of the serial type, whereby the principal is paid off in regular installments. The term of the bond issue should be less than the life of the improvement; 15 to 25 years is common for highways and up to 30 years for big bridges and turnpikes. State highway bonds are nearly always serviced from motor-vehicle revenue. Municipal bonds are generally serviced from local general taxes.

Tolls are sometimes levied by state or local governments to finance the construction of costly bridges or tunnels. Since the Pennsylvania Turnpike was first opened in 1940, tolls have been increasingly used to finance superhighways in locations where existing facilities are so inadequate that the public will pay tolls for the superior service offered by the toll facility. Most toll-road authorizations provide that the road become toll-free when the bonds are paid off. There has been a tendency for toll-road authorities to use surpluses to expand the toll mileage rather than to accelerate the rate of paying off debt. The BPR advocates a toll-free system of interstate highways developed progressively to high standards of design from current revenue sources.[1] The Federal Highway Act of 1956 provides funds for carrying out such a policy.

General taxes are a principal source of revenue for financing city streets and an important source for local roads. In recent years allocations of motor-vehicle revenue for local roads have supplemented general taxes. The allocations to cities have been relatively small. However, the states have assumed much of the cities' major highway problem by financing expressways in urban areas.

Assessments are usually levied against abutting property for the initial improvement of residential streets in cities and towns. Maintenance and reconstruction costs are usually borne by the community. Assessments are commonly based on the linear frontage of the owners. Sometimes a zoning system is used which takes into consideration area and depth of lots as well as frontages.

Traffic and Planning Surveys.[2] Traffic data are necessary for the efficient planning of a highway system, for location and design of specific projects, and for traffic control. The states and larger municipalities have traffic divisions which conduct traffic surveys for such purposes. *Traffic counts* are taken by a sampling process. Sufficient counts are made at key locations to determine hourly, daily, and seasonal trends in traffic flow. Short counts (6, 8, 12, and 24 hr) are taken at other locations of similar traffic characteristics and are expanded to longer periods using factors developed at key locations. Counts may be taken manually and recorded on forms prepared for that purpose, or by means of automatic counters operated from tubes or other devices stretched across the roadway. Counts are tabulated by hours or half hours. When turning movements are needed, such as for traffic-control studies, or when vehicle classifications are desired (by passenger car, bus, and trucks of different weights), the counts are taken manually. The common unit of traffic designation is the average daily traffic in a specified year (ADT). For design and highway capacity analyses hourly volumes are required. The hourly volume for design (DHV) is usually taken as the thirtieth highest hourly volume (30 HV) in a future year. The 30 HV is determined from an analysis of long-time counts, or by applying an appropriate percentage to the ADT. The percentage varies from 10 to 20 depending upon conditions. It is lower in urban areas and higher in rural areas, particularly for roads having seasonal peaks of recreational traffic. The projection to a future year is made on the basis of trends in motor-vehicle registrations, population, land use, and other pertinent factors. *Traffic-flow maps* are usually drawn. These show traffic flow by varying widths of bands superimposed

[1] Progress and Feasibility of Toll Roads and Their Relation to the Federal-Aid Program, House Document 139, 84th Congress, 1st Session, 1955.

[2] "Manual of Traffic Engineering Studies" published by the Accident Prevention Department of the Association of Casualty and Surety Companies, 60 John St., New York, N.Y. "Traffic Engineering Handbook" published by Institute of Traffic Engineers. Matson, T. M., W. S. Smith, and F. W. Hurd, "Traffic Engineering," McGraw-Hill Book Company, Inc., 1955.

on a map of the highways to which they apply. Flow maps at intersections show all through and turning movements plotted to a scale of traffic volume.

Origin and destination surveys are essential for planning a highway program in a particular area. The principal methods used are roadside interviews, matching license numbers at different locations, distribution of post-card questionnaires to be returned postpaid, and home interviews in which a sample of residents is visited and a record obtained of the travel of the occupants on a particular day. The data are reduced to an average day and tabulated by various combinations of origin and destination. The daily trips between zones of origin and destination are commonly shown as *desire lines*. These are bands plotted to a traffic volume scale and drawn as straight lines between pairs of origins and destinations. In this way the basic desires of traffic are revealed more accurately than by counts taken on existing streets or highways. Where existing facilities are inadequate the traffic often seeks other less direct routes.

State-wide *highway-needs studies* have been made in a number of states. These include a consideration of all classes of highways and streets. Future transportation needs of the state are analyzed with respect to expected economic and population growth. An inventory is made of existing highway facilities and an estimate prepared of the cost of bringing the highway and street systems up to acceptable standards. A 10-, 15-, or 20-year program of improvements is set up and recommendations made for legislative action to implement the plan. In establishing priorities of improvement for specific routes, *sufficiency ratings* are commonly used by means of which projects are rated on the basis of physical condition, design deficiencies, traffic adequacy, accident experience, etc.

A number of municipalities have made transportation surveys for studying means for relieving highway traffic congestion and improving transportation facilities. Such studies include a consideration of population trends, land uses, adequacy of public transportation, terminal facilities for railroads, trucks, and busses, a survey of street traffic including origin and destination, and an investigation of the parking situation. Detailed studies are also made of traffic flow and accident occurrence at important intersections for the purpose of recommending improvements in street design or in traffic-control methods.

Sizes and Weights of Motor Vehicles. The maximum dimensions and weights of motor vehicles are prescribed by the individual states. In an effort to obtain uniformity, the AASHO has recommended the following limits:[1] width 96 in.; height 12.5 ft; length, single vehicle 35 ft, single three-axle bus 40 ft, tractor semitrailer 50 ft, other combinations (not more than two units) 60 ft; single-axle weight 18,000 lb, and on a pair of axles less than 8 ft apart 32,000 lb; gross weight not over $1{,}025(L + 24) - 3L^2$, in which L is distance, ft, between the first and last axles of a vehicle or combination of vehicles. The gross-weight formula is intended to protect bridges from excessive loads. The axle-load limits are important in the design of pavements. A number of states allow higher loading than recommended by the AASHO. The maximum single-axle load allowed in any state (1956) is 20,400 lb, and the maximum tandem-axle load 40,000 lb. For design purposes the AASHO has set up four design vehicles with typical dimensions and turning characteristics, as follows: passenger cars (P), single-axle truck (SU), semitrailer combinations—intermediate (C43), and semitrailer combinations—large (C50).

Motor-vehicle Registrations and Use.[2] In 1955 a total of 62,794,000 motor vehicles were registered in the United States as follows: 52,174,000 passenger automobiles, 10,365,000 trucks, and 255,000 busses. In addition, 3,262,000 trailers of all types

[1] Policy Concerning Maximum Dimensions, Weights and Speeds of Motor Vehicles to Be Operated over the Highways of the United States, adopted Apr. 1, 1946, by AASHO.

[2] Statistics of motor-vehicle registration, weights, and mileages are issued annually by the U.S. Bureau of Public Roads.

were registered, of which about 140,000 were commercial trailers or semitrailers. Gasoline consumption for highway purposes in 1955 was 48.4 billion gal. Travel by motor vehicles of all types amounted to about 603 billion vehicle-miles of which 336 billion was rural and 267 billion was urban travel. The average annual mileage of passenger cars was 9,400. Motor trucks of all types averaged 10,700 and busses 17,600 miles. Commercial trucks and busses in intercity service will travel from 40,000 to 80,000 miles per year.

Highway Traffic Capacity. A knowledge of highway capacity is essential to plan and design facilities adequate for traffic needs. A thorough treatise on this subject will be found in the "Highway Capacity Manual."[1] In that manual highway capacity is defined in three ways. *Basic capacity* is the maximum number of passenger cars that can pass a given point on a lane or roadway during 1 hr under the most nearly ideal roadway and traffic conditions which can possibly be attained; *possible capacity* is the maximum number of vehicles that can pass a given point on a lane or roadway during 1 hr, under the prevailing roadway and traffic conditions; *practical capacity* is the maximum number of vehicles that can pass a given point on a roadway or in a designated lane during 1 hr without the traffic density being so great as to cause unreasonable delay, hazard, or restriction to the driver's freedom to maneuver under prevailing roadway and traffic conditions.

Under ideal conditions the basic capacity of a two-lane road is 2,000 passenger cars per hour (both lanes), for a three-lane road 4,000 (all lanes), and for multilane highways 2,000 per lane per hour in direction of heavy flow. For ideal conditions possible capacities are the same as basic capacities. They represent an upper limit of capacity near the verge of congestion and breakdown. For purposes of design the practical capacity is used. The difference between practical and possible capacities provides a factor of safety for infrequent and unusual demands. Practical capacities derived in the "Capacity Manual" for ideal conditions are as follows: on two-lane roads, 900 per hour in rural areas and 1,500 in urban areas (both lanes); on three-lane roads, 1,500 in rural areas and 2,000 in urban areas (all lanes); on multilane roads, 1,000 per lane in rural areas and 1,500 in urban areas in direction of heavier flow. Ideal conditions imply uninterrupted traffic, no commercial vehicles, traffic lanes 12 ft wide, shoulders or side clearances at least 6 ft wide, and no restricted-sight distances of less than 1,500 ft. When conditions are not ideal, deduction factors must be applied. On multilane highways one commercial vehicle, having dual tires on one or more rear axles, has approximately the same effect as two passenger cars on level terrain, and of four passenger cars on rolling terrain. On two-lane highways the effect of commercial vehicles is about 25 percent greater.

Other deductions are given in Tables 6*A*, *B*, and *C*. To determine the practical capacity of a 20-ft-wide two-lane rural road on rolling terrain with 4-ft shoulders, 20 percent of length with sight distances less than 1,500 ft and 6 percent of dual-tired or larger vehicles, the following reduction factors would be applied: for 20-ft width apply 77 percent taken from Table 6*A*; for 4-ft shoulders apply 19.2/20 from Table 6*B* or 96 percent; for restricted sight distances, apply 860/900 from Table 6*C* or 95.6 percent; for 6 percent commercial vehicles on rolling terrain assume each commercial vehicle displaces three passenger cars, and apply 100/118 or 85 percent. The resulting practical capacity is $900 \times 0.77 \times 0.96 \times 0.956 \times 0.85 = 541$ vehicles per hour. In urban areas the capacity of streets is usually limited by the capacity of the intersections. It is also affected by the use of outer lanes for parking, streetcars or bus traffic, and pedestrian crossings. The capacities of signalized intersections of various types are

[1] "Highway Capacity Manual" by Committee on Highway Capacity, Department of Traffic and Operations of Highway Research Board, published by U.S. Bureau of Public Roads. May be purchased from Superintendent of Documents, Washington 25, D.C. See also Design Capacity Charts for Specialized Street and Highway Intersections, *Public Roads*, for February, 1951.

TABLE 6. EFFECT OF DESIGN FEATURES ON PRACTICAL CAPACITIES[1]

A. Effect of Lane Width

Lane width, ft	Percent of 12-ft lane capacity	
	2-lane rural roads	Multilane urban expressways
12	100	100
11	86	97
10	77	91
9	70	81

B. Effect of Restricted Lateral Clearance

Clearance from pavement edge to obstruction, ft	Effective width of 2 lanes, ft					
	Obstruction on one side			Obstruction on both sides		
	12-ft lanes	11-ft lanes	10-ft lanes	12-ft lanes	11-ft lanes	10-ft lanes
6	24.0	22.0	20.0	24.0	22.0	20.0
4	23.5	21.5	19.6	23.0	21.0	19.2
2	22.5	20.6	18.8	21.0	19.3	17.5
0	21.0	19.3	17.5	18.0	16.5	15.0

C. Effect of Passing Sight-distance Restriction on 2-lane Highways When Adequate Stopping Sight Distances Are Always Present

Sight distance restricted to less than 1,500 ft: Percentage of total length of highway	Practical capacity, passenger cars per hr	
	For operating speed[2] of 45–50 mph	For operating speed[2] of 50–55 mph
0	900	600
20	860	560
40	800	500
60	720	420
80	620	300
100	500	160

[1] From "Highway Capacity Manual," U.S. Bureau of Public Roads.
[2] Average speed for drivers trying to travel at maximum safe speed.

treated in detail in the "Highway Capacity Manual." Also covered are capacities of interchange ramps and weaving sections.

HIGHWAY-DESIGN STANDARDS

The American Association of State Highway Officials has adopted standards of highway design which reflect current practice of the state highway departments, particularly on Federal aid projects. The following standards are for highways of major traffic importance.[1]

[1] Taken from Geometric Design Standards for National System of Interstate and Defense Highways, adopted by AASHO, July, 1956.

Geometric Design Standards for National System of Interstate and Defense Highways.[1] *Traffic Basis.* Interstate highways shall be designed to serve safely and efficiently the volume of passenger vehicles, busses, and trucks, including tractor-trailer and semitrailer combinations and corresponding military equipment, estimated to be that which will exist in 1975, including attracted, generated, and development traffic on the basis that the entire system is completed. The peak-hour traffic used as a basis for design shall be as high as the thirtieth highest hourly volume of the year 1975, hereafter referred to as the design hourly volume, "DHV (1975)." Unless otherwise specified, DHV is the total, two-direction volume of mixed traffic.

Control of Access. On all sections of the interstate system, access shall be controlled by acquiring access rights outright prior to construction or by the construction of frontage roads, or both. Control of access is required for all sections of the interstate system. Under certain conditions, intersections at grade may be permitted in sparsely settled rural areas which are a sufficient distance from municipalities or other traffic-generating areas to be outside their influence, and where no appreciable hazard is created thereby. Where a grade separation is called for under these standards and extraordinary conditions exist under which a grade separation would not be in the public interest, an intersection at grade may be permitted through agreement between the state highway department and the Secretary of Commerce.

Railroad Crossings. Railroad grade crossings shall be eliminated for all through-traffic lanes.

Intersections. All at-grade intersections of public highways and private driveways shall be eliminated, or the connecting road terminated, rerouted, or intercepted by frontage roads, except as otherwise provided under Control of Access.

Design Speed. The design speed of all highways on the system shall be at least 70, 60, and 50 mph for flat, rolling, and mountainous topography, respectively, and depending upon the nature of terrain and development. The design speed in urban areas should be at least 50 mph.

Curvature, Superelevation, and Sight Distance. These elements and allied features, such as transition curves, should be correlated with the design speed in accordance with the Policy on Geometric Design of Rural Highways of the American Association of State Highway Officials (see p. 2–59). On two-lane highways, sections with sufficient sight distance for safe passing should be frequent enough and the total length of such sections be a sufficient percentage of the highway length to accommodate the DHV. Where it is not feasible to provide enough passing opportunities, a divided highway should be provided instead.

Gradients. For design speeds of 70, 60, and 50 mph, gradients generally shall be not steeper than 3, 4, and 5 percent, respectively. Gradients 2 percent steeper may be provided in rugged terrain.

Width and Number of Lanes. Traffic lanes shall not be less than 12 ft wide. Where the DHV (1975) exceeds 700 or exceeds a lower two-lane design capacity applicable for the conditions on a particular section, the highway shall be a divided highway. For lower volumes, the highway shall be a two-lane highway so designed and located on the right of way that an additional two-lane pavement can be added in the future to form a divided highway. Efficiency and capacity of two-lane highways may be increased by providing added climbing lanes on up grades where critical lengths of grade are exceeded or by providing more frequent and longer sections safe for passing.

Medians. Medians in rural areas in flat and rolling topography shall be at least 36 ft wide. Medians in urban and mountainous areas shall be at least 16 ft wide. Narrower medians may be provided in urban areas of high right-of-way cost, on long and costly

[1] The AASHO Policy on Geometric Design of Rural Highways, the Policy on Arterial Highways in Urban Areas, and the Standard Specifications for Highway Bridges shall be used as design guides where they do not conflict with these standards.

bridges, and in rugged mountain terrain, but no median shall be less than 4 ft wide. Curbs or other devices may be used where necessary to prevent traffic from crossing the median. Where continuous barrier curbs are used on narrow medians, such curbs shall be offset at least 1 ft from the edge of the through-traffic lane. Where vertical elements more than 12 in. high, other than abutments, piers, or walls, are located in a median, there shall be a lateral clearance of at least 3½ ft from the edge of through-traffic lane to the face of such element.

Shoulders. Shoulders usable by all classes of vehicles in all weather shall be provided on the right of traffic. The usable width of shoulder shall be not less than 10 ft. In mountainous terrain involving high cost for additional width, the usable width of shoulder may be less but at least 6 ft. Usable width of shoulder is measured from edge of through-traffic lane to intersection of shoulder and fill or ditch slope ecxept where such slope is steeper than 4:1 where it is measured to beginning of rounding.

Slopes. Side slopes should be 4:1 or flatter where feasible and not steeper than 2:1 except in rock excavation or other special conditions.

Right of Way. Fixed minimum widths of right of way are not given because wide widths are desirable, conditions may make narrow widths necessary, and right of way need not be of constant width. The following minimum widths are given as guides. In rural areas right-of-way widths should be not less than the following, plus additional widths needed for heavy cuts and fills:

Type of highway	Minimum width, ft	
	Without frontage roads	With frontage roads
2-lane	150	250
4-lane divided	150	250
6-lane divided	175	275
8-lane divided	200	300

In urban areas right-of-way width shall be not less than that required for the necessary cross-section elements, including median, pavements, shoulders, outer separations, ramps, frontage roads, slopes, walls, border areas, and other requisite appurtenances.

Culverts. All culverts shall be of sufficient length to accommodate the pavements, median, and shoulders.

Bridges and Other Structures. The following standards apply to interstate highway bridges, overpasses, and underpasses. Standards for crossroad overpasses and underpasses are to be those for the crossroad.

Bridges and overpasses, preferably of deck construction, should be located to fit the overall alignment and profile of the highway. The clear height of structures shall be not less than 14 ft over the entire roadway width, including the usable width of shoulders. Allowance should be made for any contemplated resurfacing. The width of all bridges, including grade-separation structures, of a length of 150 ft or less between abutments or end-supporting piers shall equal the full roadway width on the approaches, including the usable width of shoulders. Barrier curbs on bridges longer than 150 ft between abutments or end-supporting piers and curbs on approach highways if used shall be offset at least 2 ft. Offsets to face of parapet or rail shall be at least 3½ ft measured from edge of through-traffic lane and apply on right and left. The lateral clearance from the edge of through-traffic lanes to the face of walls or abutments and piers at underpasses shall be the usable shoulder width but not less than 8 ft on the right and 4½ ft on the left. A safety walk shall be provided in tunnels and on long-span

structures on which the full approach roadway width, including shoulders, is not continued.

The above standards are minimums and do not preclude higher standards. Many freeways and toll roads are built to higher standards. For the New York Thruway the minimum design speed is 70 mph in rural areas and 50 mph in urban or restricted areas; minimum radius of curve is 2,800 ft for 70 mph and 1,500 ft for 50 mph; minimum sight distance is 1,000 ft in rural and 800 ft in restricted areas; maximum grade is 3 percent; lane widths are 12 and 13 ft; shoulder widths are 9 and 10 ft; and median widths are 6 ft in urban and 20 ft in rural areas. For the New Jersey Turnpike minimum design speeds are 60 and 70 mph in northern sections and 75 mph in southern sections; minimum radius of curve is 3,000 ft, maximum grade 3 percent, lane widths 12 ft, shoulder widths 16 ft on outside and 5 ft on inside (next to median), median widths are 20 and 26 ft, and right-of-way widths are 250 and 300 ft.

Design Standards for Secondary and Feeder Roads. Table 7 gives these standards as adopted by the AASHO.[1]

TABLE 7. MINIMUM DESIGN STANDARDS FOR SECONDARY AND FEEDER ROADS

Design control	Annual average daily traffic volume [1]					
	Under 100		100–400		400–1,000	
	Minimum	Desirable	Minimum	Desirable	Minimum	Desirable
Design speed, mph:						
Flat topography	40	....	45	55	50	60
Rolling topography	30	....	35	45	40	50
Mountainous topography	20	....	25	35	30	40
Sharpest curve, deg:						
Flat topography	14	....	11	7	9	6
Rolling topography	25	....	18	11	14	9
Mountainous topography	56	....	36	18	25	14
Maximum gradient, percent:						
Flat topography	8	5	8	5	7	5
Rolling topography	12	7	10	7	8	6
Mountainous topography	15	10	12	9	10	7
Nonpassing sight distance, [2] ft:						
Flat topography	....	....	315	415	350	475
Rolling topography	....	....	240	315	275	350
Mountainous topography	....	....	165	240	200	275
Width of surfacing or pavement, ft	12, if any	....	16	20	18	20
Width of roadbed, ft	20	....	24	28	26	30
New bridges:						
Clear width, ft	14	20	22	24	24 *	
Design load, AASHO	H-10	H-15	H-15	....	H-15	
Bridges to remain:						
Clear width, ft	....	....	15	....	18	
Safe load, posting basis, tons	....	....	6	....	10	
Width of right of way, ft	40 †	....	40 †	80	50	80

[1] Design peak hour capacity is assumed to be approximately 10 percent of the annual average daily traffic.
[2] See stopping sight distance (Table 8).
* Minimum of 24 ft or 4 ft more than approach pavement width.
† Minimum of 40 ft or as required for construction.

Elements of Design.[2] *Curvature.* The relation between radius of curve and design speed is established by *superelevation* and *friction factors* as expressed in the formula

[1] Design Standards: Secondary and Feeder Roads, published by AASHO, March, 1949.
[2] See Chap. III, A Policy on Geometric Design of Rural Highways, AASHO.

$e + f = 0.067V^2/R$, in which e is rate of superelevation and f is the side-friction factor between tires and pavement. Values of e for design of rural highways vary from 0.12 where no snow or ice is encountered to 0.08 where icing occurs. In urban and slow-speed areas 0.06 is often used. Side-friction factors are chosen which are not only safe but will cause no discomfort to occupants of vehicle. Values adopted for f vary from 0.16 at 30 mph to 0.12 at 70 mph. In designing intersections, lower values of e (0 at 15 mph to 0.08 at 35 mph) and higher values of f (0.32 at 15 mph to 0.16 at 40 mph) are suggested by AASHO.

Method of Banking. For curves that are neither widened nor spiraled, the superelevation is begun at some arbitrary distance before the PC, and full banking attained at the PC or a short distance beyond on the curve. For widened curves, the banking usually is accomplished concurrently with the widening. On spiraled curves, it is attained over the length of the spiral, and full banking carried from SC to CS. A gradual transition is made from a crowned section to a sloping plane. Some standards specify that the center-line profile be held and the inside edge of pavement lowered and the outside raised. This method creates a depression along the inside edge which may cause drainage difficulties in cuts. Other standards call for holding the normal grade of the inside edge of the pavement and sloping the pavement from this line. The profile of runoff of superelevation from zero to full banking may consist of a pair of reversed vertical (parabolic) curves, or a short straight runoff with vertical curves at each end. The individual highway departments have prepared tables giving data for laying out the transition from normal to banked section.

Highway Spiral.[1] On major highways, spiral transition curves are sometimes used, the length being determined by the formula $L = 3.15V^3/RC$, in which L is length of spiral, ft, V is velocity, mph, R is radius, ft, and C is rate of increase of centripetal acceleration, ft per sec^3. The value of C is determined from a consideration of safety and comfort in making transition from tangent to curve. A value of 2 is commonly used although lesser values have been suggested (1.0, 1.25, and 1.35), especially when superelevation is taken into account. Spirals are usually omitted on very flat curves, such as those less than 1.5°, because the spiral effect is so small that vehicles can describe their own spiral within normal width of traffic lane. In some practice a compound curve is used as a transition between tangent and curve, the radius of the compound curve being about twice that of the central curve. The theory of the spiral is given on p. 2–15.

Widening. Pavements on two-lane roads narrower than 24 ft are commonly widened on sharp curves and at higher design speeds. Widening is also applied on curved ramps at traffic interchanges. The amount of extra width required depends upon the original width of pavement, sharpness of curve, design speed, wheel base and turning characteristics of vehicle, safe clearance required between vehicles passing on curve, and an allowance for the difficulty of driving on a curve. Widening for a 20-ft pavement will vary from 1 ft at 70 mph on a 1.5° curve to 4 ft at 30 mph on a 24° curve. On a 22-ft pavement the range is from 0.5 ft at 70 mph on a 2° curve to 3 ft at 30 mph on a 24° curve.[2]

The added width is usually placed on the inside of the pavement, although it is sometimes divided between the inside and outside when spiral transitions are used. Full (concentric) widening is maintained substantially around the curve with transition curves to connect with the straight edge of the pavement. The transitions may be accomplished by a compound curve, by an arbitrary curve of parabolic type, or by the use of spirals.

[1] See BARNETT, JOSEPH, "Transition Curves for Highways," Bureau of Public Roads, Government Printing Office; MEYER, CARL F., "Route Surveying," 2d ed., Chap. 8, International Textbook Company.
[2] See A Policy on Geometric Design of Rural Highways, AASHO, Table III-15.

Sight Distance. There are two definitions: *stopping sight distance* (sometimes called "nonpassing" sight distance) in which a driver can bring his vehicle to a stop before reaching an obstacle in the roadway; and *passing sight distance* which is distance in which the driver of an overtaking vehicle should be able to see ahead before safely starting a passing maneuver. The stopping sight distance is determined by formula $SSD = 1.47Vt + V^2/30(f \pm g)$, in which V = assumed initial speed, mph, t = perception plus reaction time, sec, f = coefficient of friction between tires and roadway, and g = percent of grade ÷ 100. Minimum stopping distances adopted by AASHO are shown in Table 8. These are based upon a level highway with values of V somewhat

TABLE 8. MINIMUM SIGHT DISTANCES[1]

Design speed, mph	Stopping sight distance		Passing sight distance		
	Initial speed, mph	Distance, ft	Assumed passing speed	Distance 2-lane highway, ft	Distance 3-lane highway, ft
30	28	200	30	800	
40	36	275	40	1,300	
50	44	350	48	1,700	1,200
60	52	475	55	2,000	1,400
70	59	600	60	2,300	1,600

[1] From A Policy on Geometric Design of Rural Highways, AASHO, Tables III—1, 4, 5 (rounded values).

less than design speeds, t = 2.5 sec, and f (for wet pavement) varying from 0.36 at 30 mph to 0.29 at 70 mph.

The passing sight distance for two-lane roads is made up of four distances, assuming that the overtaking vehicle is trailing overtaken vehicle before passing: (1) distance traveled during perception plus reaction time and during initial acceleration to point of encroachment on left lane, (2) distance traveled while passing vehicle occupies the left lane, (3) distance between the passing vehicle at end of its maneuver and an opposing vehicle in left lane, and (4) distance traveled by the opposing vehicle for two-thirds of the time the passing vehicle occupies the left lane, or two-thirds of (2). The passing distance for three-lane roads is taken as the sum of (1) + (2) + (3). The opposing vehicle distance is omitted since it is assumed that the center lane will be available for passing.[1]

In Table 8 passing speeds somewhat less than design speeds are used since experience indicates that most drivers pass at less than design speed. The passed vehicle is assumed to be traveling 10 mph slower than the passing vehicle, and the opposing vehicle at the same speed as the passing vehicle.

Minimum stopping sight distances should be provided on all highways. Passing sight distances are desirable over as great a length as possible on two- and three-lane roads.

Horizontal curves should be flat enough or obstructions should be cut back to allow a clear line of sight between ends of sight distance measured along center line of inside lane. The middle ordinate from an obstruction to center line of inside lane is $m = R \text{ vers } 28.65S/R$, in which R = radius of center line of inside lane, ft, and S = sight distance, ft. In profile, the lengths of vertical curve are designed to give at least minimum stopping sight distance. The latter is measured between height of eye of driver

[1] For evaluations of elements of passing sight distance see A Policy of Geometric Design of Rural Highways, Chap. III.

at 4.5 ft and height of object of 4 in. Passing sight distances are measured between two points each 4.5 ft above pavement.

Highway Gradients. Grades should be kept as low as possible on highways carrying a considerable portion of motor-truck traffic. Heavy vehicles lack the power to maintain speed on grades over 2 or 3 percent, and they impede other traffic, particularly on two- and three-lane roads. Short grades may be somewhat steeper than long ones, since momentum will take the heavier vehicles over these grades with only moderate speed reduction. The AASHO suggests that lengths of up grade be limited to those causing a speed reduction of not less than 15 mph below average speed approaching the grade. On two- and three-lane roads where these lengths cannot be obtained and truck volume is high, consideration should be given to constructing separate climbing lanes for trucks.[1] To facilitate drainage in cut sections a minimum grade of 0.5 percent (0.35 percent for paved gutters) is desirable. The cost of grading is important in determining the practical gradient; therefore, higher maximum grades are used in mountainous and rolling country than in flat terrain. The extra cost necessary to flatten grades should be weighed against the resulting savings in motor-vehicle operating costs. Obviously a greater expenditure is justified on heavily traveled roads carrying a high percentage of motor trucks than on lightly traveled roads (see "Highway Economics," p. 2–16).

Vertical Curves. Straight grade lines are connected by vertical parabolic curves (see p. 2–14 for theory). The length of curve is frequently determined by sight-distance

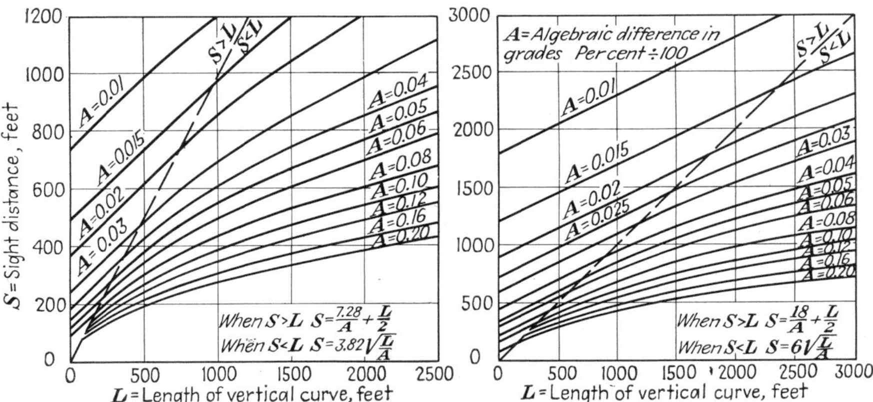

Fig. 32. Stopping sight distance. (Height of eye 4.5 ft; of object, 4 in.)

Fig. 33. Passing sight distance. (Height of eye and of object, 4.5 ft.)

requirements. The relations between sight distance at summits and length of vertical curve derived from formulas adopted by AASHO[2] are shown for different combinations of grades and for stopping and passing sight distances, respectively, in Figs. 32 and 33. Vertical curves in sags should be long enough to allow the rays of headlights to pick up an object on the road at a distance equal to stopping sight distance. Formulas for L in terms of S for condition shown in Fig. 34 are:[3] $L = AS^2/(500 + 3.5S)$ when S is less than L, and $L = 2S - (500 + 3.5S)/A$ when S is greater than L, in which L and S are in feet and A is in percent. Where sight-distance requirements do not

[1] "Vehicle Climbing Lanes," *Bull.* 104, Highway Research Board, 1955.
[2] See A Policy on Geometric Design of Rural Highways, Chap. III, pp. 173–176, AASHO.
[3] Noble, Charles M., Design Features—Pennsylvania Turnpike, *Civil Eng.*, July, 1940. Also Thompson, Donald, Sight Distance on Sag Vertical Curves, *Civil Eng.*, January, 1944.

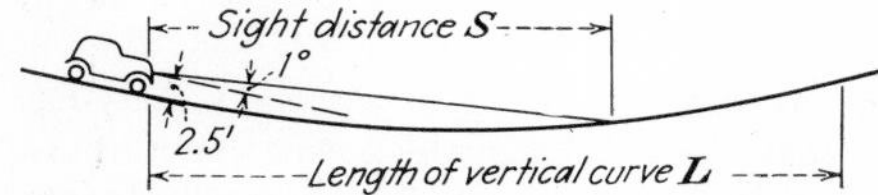

FIG. 34. Headlight sight distance on vertical curves in sags.

control, vertical curves are chosen to fit the topography in such a way as to keep cost of construction to a minimum. They should not be so short as to cause an abrupt break in grade. A minimum length (in feet) of three times design speed is recommended. A smooth-flowing profile with long vertical curves is desired.

Cross Sections. Pavements are sloped or crowned to provide proper drainage. The smoother, high-type surfaces are crowned at rates of ⅛ to ¼ in. per ft, intermediate surfaces at 3/16 to ⅜ in., and low type at ¼ to ½ in. A usable shoulder width of 10 ft is standard on primary highways. It may be wider on heavily traveled routes, and narrower in rugged terrain where costs are high or on secondary roads carrying light traffic. Where shoulders are narrower than 10 ft turnout areas are advisable at frequent intervals. Cross slopes of shoulders vary from ⅜ to ½ in. per ft if bituminous treated, ½ to ¾ in. per ft if gravel or stone, and 1 in. per ft if turf. The surface and shoulders

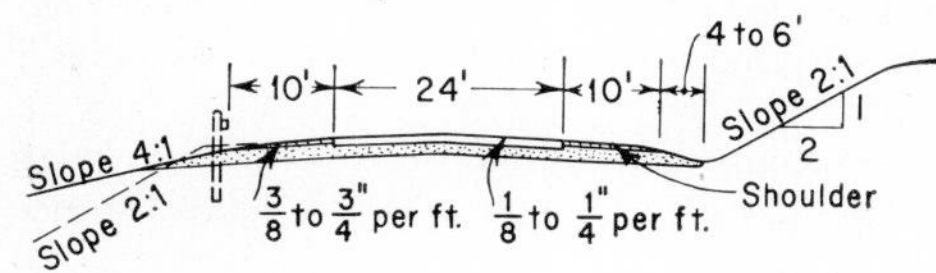

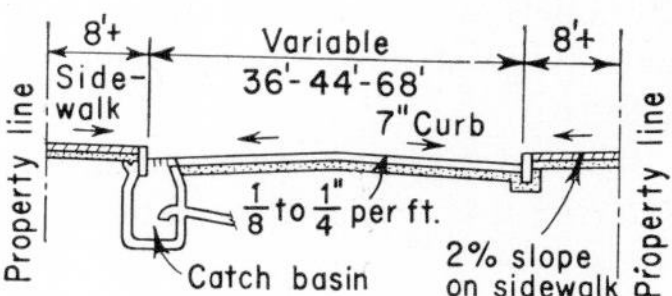

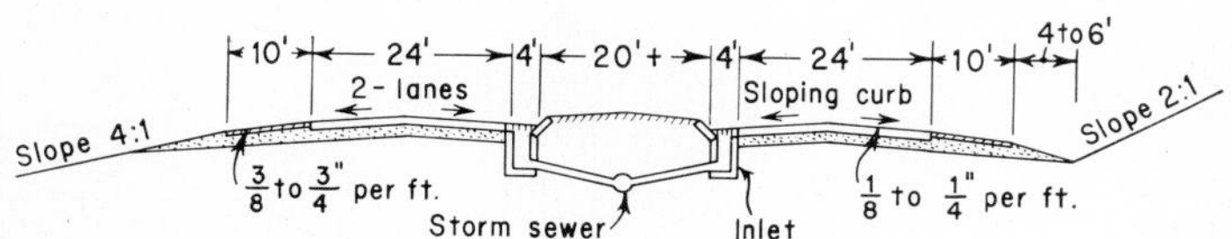

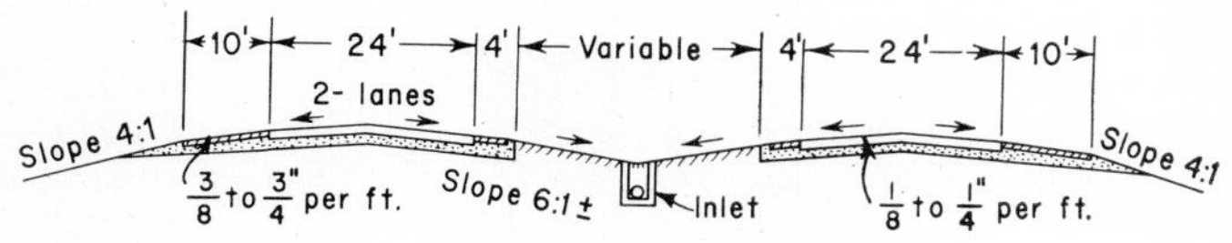

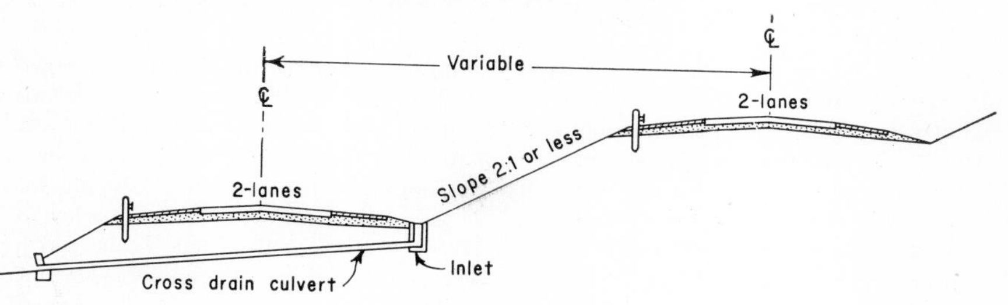

FIG. 35. Highway and street cross sections.

should contrast in color and texture, such as light-colored concrete pavement and dark bituminous shoulder, or smooth bituminous pavement and rough-textured macadam or surface-treated shoulder. Where pedestrian traffic exceeds 100 to 150 per day and traffic volume exceeds 100 vehicles per hour a sidewalk should be considered on one side. If pedestrian volume exceeds 300 to 500 per day sidewalks should be considered on both sides. Figure 35 shows geometric characteristics of typical cross sections. Figure 35*a* is a typical two-lane primary highway partly in cut and partly in fill. Figure 35*b* is a typical urban street. Note that sidewalks are sloped away from property line to drain into gutters where flow is collected in a storm-sewer system (Fig. 45). Urban street widths depend upon parking requirements and traffic lanes required. A residential street with two 10-ft lanes for moving traffic and two parking lanes at 8 ft will be 36 ft wide. An important business or feeder street with two 12-ft lanes for moving traffic and two parking lanes at 10 ft will be 44 ft wide; for four traffic lanes and two parking lanes the width becomes 68 ft. Figure 35*c* is a divided highway with raised median designed to give positive restraint against vehicles crossing over it. In climates where snow and ice are encountered the best design is to provide surface drainage at both the median and at the roadsides. Figure 35*d* shows a divided highway with depressed median. This type is economical in drainage and construction and suitable for wide medians. The interior ditch section is made deep enough to discourage crossing but not so deep as to be a hazard. Fences are sometimes erected in narrow depressed medians to prevent vehicles from making U turns. Figure 35*e* is a divided highway with roadways on separate locations. This type of design with the two roadways varying in both profile and alignment is often economical in construction as well as attractive to the driver and safer to drive on than a uniform cross section of monotonous alignment. Side slopes are made as flat as practical. For fills under 10 ft high, 4:1 is desirable, for fills 10 to 15 ft, 3:1, and for fills over 15 ft, 2:1. In rugged country, slopes may be steepened to 1¾:1 or to 1½:1. Guardrails are installed where side slopes are greater than 4:1 (see p. 2–113).

Highway-location Surveys and Plans. The location of a highway offers few problems where the improved road follows the line of an old road. Some line and grade changes may be necessary to meet modern design standards, and the right of way may have to be widened for additional lanes. In municipalities, the streets are usually well established, the lines and grades being a matter of record in the city engineering department. Radical changes in line and grade are seldom made because of the resulting damages to abutting property.

When the cost of grading and property damage involved in modernizing an old route is great, consideration is given to locating a new route roughly parallel to the old one. The old road may then be abandoned or retained for local use. Traffic surveys and cost estimates are required to justify the choice of route. If a new route is decided upon, the problem becomes one of studying the terrain and locating a line which will keep construction and land costs to a minimum and at the same time satisfy design requirements for present and anticipated future traffic requirements. In cities, location problems arise in connection with the design of express highways for the relief of traffic congestion. Interchange ramps should be located where they will give access to areas of greatest traffic demand, while at the same time avoiding high property damages. Express highways are often located in blighted areas. Property takings are sometimes linked with urban redevelopment. The principles of location and the surveys employed to locate the final line are described under "Route Surveys," pp. 2–1 to 2–7. Aerial surveys are particularly advantageous in highway location (pp. 2–3 to 2–7). In urban areas photographs and mosaics are of great value for studying alternate routes and estimating property costs.

The final line is laid out with transit and tape, and profile levels taken. Adjustments

may be made if the field survey shows that improvements can be made which will reduce the cost. For rural highway surveys the center line is usually laid out, and the transit points of the survey referenced to fixed objects outside the limits of the proposed work. In city surveys, the lines are often run on a parallel or offset line in the sidewalk area. Cross sections are taken for computing earthwork quantities (see p. 2–17). On new locations soil surveys are desirable to determine foundation conditions which influence the design of pavements and drainage facilities.

The construction plans for a highway project usually consist of the following parts: title sheet with index, conventional signs, signatures, and a small-scale map showing location of project; tabulation of engineer's estimate of quantities upon which the contractor's bid is based; standard cross section of proposed road showing pavement details, shoulders, ditches, slopes, etc.; plans of drainage structures, walls, guard fences, etc.; combined plan and profile sheets showing in plan (top half of sheet) alignment, pavement edges, drainage, property lines, right-of-way lines, and all physical features affecting the construction, and showing in profile (bottom half of sheet) the ground line and finished grade line with elevations, rates of grade, and vertical curve data; earthwork cross sections showing ground line and proposed sections at full stations and other critical points throughout the length of the project. These plans are not only used for construction purposes, but are also used by the right-of-way division in obtaining titles to lands taken and for easements taken for drainage, slopes, or other facilities on property not purchased.

INTERSECTION DESIGN

Channelization.[1] Traffic flow at intersections can be greatly facilitated by the use of traffic islands to "channelize" traffic into desirable paths. Islands are of three types: divisional, directional, and refuge. A median strip is an example of a divisional type. Intersection islands are directional, and islands providing protection for pedestrians are refuge, although all types of islands may provide refuge. Medians present special problems where traffic must make left or U turns through them. Such medians should be wide enough to shelter a turning vehicle from traffic on through lanes, and also wide enough to provide sufficient radius to allow vehicles, including large trucks, to make turns through them. Figure 36*a* illustrates one type of treatment where left-turn lanes have been provided in the median to accommodate turning vehicles. The lanes should be made long enough to accommodate a number of waiting vehicles, and reduce speed for turning. Each intersection usually presents a different problem requiring islands to be tailored to the traffic flow. At new intersections the traffic flows must be estimated. At old intersections being redesigned, a thorough preliminary study should be made of physical conditions, traffic flow, and accident experience. Often an examination of the traffic stains on the pavement will reveal the natural channels of traffic. The absence of stains delineates little-used areas which can be made into islands. The traffic pattern is often well defined on aerial photographs or by paths made in light snow or the gathering of sand after ice treatment. In laying out islands certain design criteria should be kept in mind. Width and radii of all channels must be sufficient to accommodate all vehicles including large trucks. On curbed layouts without shoulders the width of roadways should be sufficient for the design vehicle to pass a stalled vehicle. Radii required for turning vary with the central angle. For angles about 90° the minimum radius for large trucks at slow speed is about 50 ft; for passenger cars about 30 ft.[2] The shape of islands should be such as to produce the desirable paths for traffic. In general, turning movements are separated from through traffic where practical. Conflicting

[1] Channelization—The Design of Highway Intersections at Grade, *Special Report* 5, Highway Research Board, 1952. A Policy on Arterial Arteries in Urban Areas and A Policy on Geometric Design of Rural Highways, Chap. VIII, At-grade Intersections by AASHO.

[2] A Policy on Geometric Design of Rural Highways, Chap. VII.

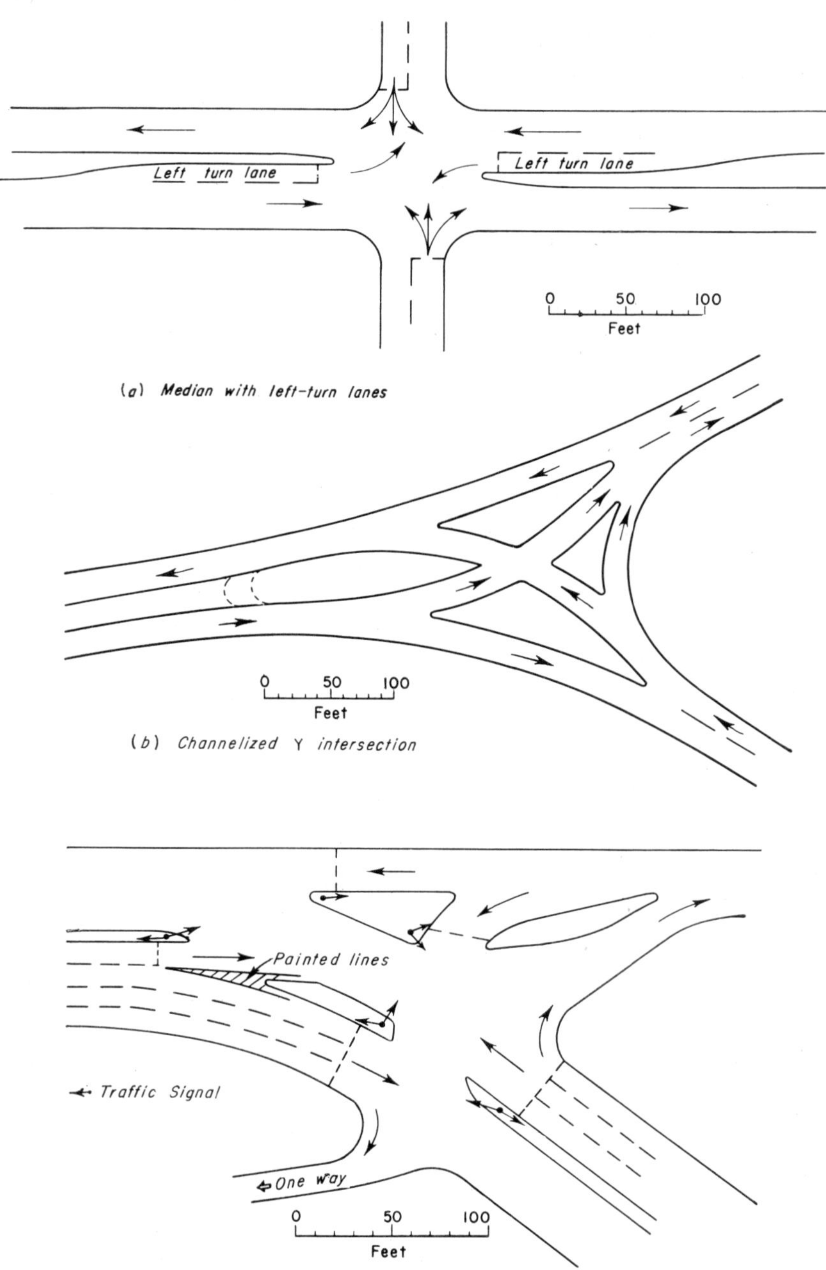

FIG. 36. Channelized intersections.

movements are brought to a focus where they can be controlled by stop signs or traffic signals. Sheltered roadway areas are provided between islands where vehicles can stand safely while waiting for an opportunity to cross other traffic. The islands should be as large as practicable. Small islands should be avoided since they are hard to see and constitute a traffic hazard. In general, islands should be at least 50 sq ft and preferably over 75 sq ft in area. The stopping sight distance should be available throughout the layout. Islands are commonly delineated by white, slanting (mountable) curbs. These should be set back 1 or 2 ft from normal traffic-lane width. Channelization designs for Y-type and signalized intersections are shown in Fig. 36*b* and *c*.

A *traffic rotary* [1] is a special type of channelization in which all entering traffic circulates one-way on the rotary. Cross movements are accomplished by weaving. The design is well adapted to multiple-leg intersections where weaving capacity of the rotary is not exceeded. This capacity depends upon length of weaving section, width of rotary roadway, speed of operation, percentage of commercial vehicles, and relative volumes of weaving and through traffic in any section. On rotaries of 200 to 300 ft diameter, the practical weaving capacity is usually reached at about 1,500 crossing vehicles per hour. When the volumes exceed practical weaving capacity the rotary breaks down and the traffic can better be served by a grade separation or signalization. Rotaries are not well suited to signal control unless redesigned to provide direct crossings (usually through the circle) for major flows of traffic. Rotaries are not necessarily circular, but can be adjusted in shape to give the longest weaving lengths where they are most needed. Principles of the rotary, and capacity relationships for weaving sections are developed in "Highway Capacity Manual." They apply to weaving sections at interchanges as well as to rotaries.

Grade Separations and Interchanges.[2] On limited-access highways and on other heavy-volume highways cross movements of traffic are eliminated by separating grades. At very minor crossings, where traffic cannot be rerouted to other roads, a simple bridge without ramps is constructed. Where a minor road crosses a major highway, and connections are warranted, a partial cloverleaf is constructed requiring only right turns to and from the major road, but permitting left turns on minor road (Fig. 37*a*, *b*). Where both highways carry heavy traffic, a full cloverleaf is used, left turns being accomplished by loops (Fig. 37*e*). Weaving is required between on and off ramps. At T-type intersections, interchanges of the "Y" (Fig. 37*c*) or "trumpet" type (Fig. 37*d*) are used, the latter being well adapted to toll-road connections where toll can be collected from both ingoing and outbound traffic at the stem of the interchange. For very heavy traffic expressways, direct connections are preferred whereby vehicles may make turns directly in the direction they wish to go (or nearly so) and without weaving. Usually only the major turning movements are provided with direct connections, loops being used for minor turns (Fig. 37*f*). Interchanges with all direct connections are very expensive, requiring three- or four-level bridges or lengthy viaducts. Where several local roads converge on an expressway a combination rotary and grade separation is sometimes used, the rotary acting as a collector and distributor for expressway traffic (Fig. 37*g*).

A design traffic is adopted for each element of the interchange, and the appropriate standards of sight distance, curvature, road width, grade, etc., applied to both main roadways and ramps. The layout is usually influenced by topographic features which affect the cost of construction. Any compromise with design standards for economy reasons is usually made on minor roads rather than on major elements. In making cost comparisons of alternate designs, weight should be given to user benefits resulting

[1] A Policy on Geometric Design of Rural Highways, Chap. VIII, Rotaries, pp. 354–364.
[2] See A Policy on Geometric Design of Rural Highways, Chap. IX, pp. 365–420. Also Appendix A on intersection-design procedures.

from reductions in delay, saving in operating cost, and accident reduction. The minimum vertical clearance at structures is 14 ft 3 in. over traffic lanes. The 3 in. is to allow for future resurfacing. Where wide medians are employed two-span structures are

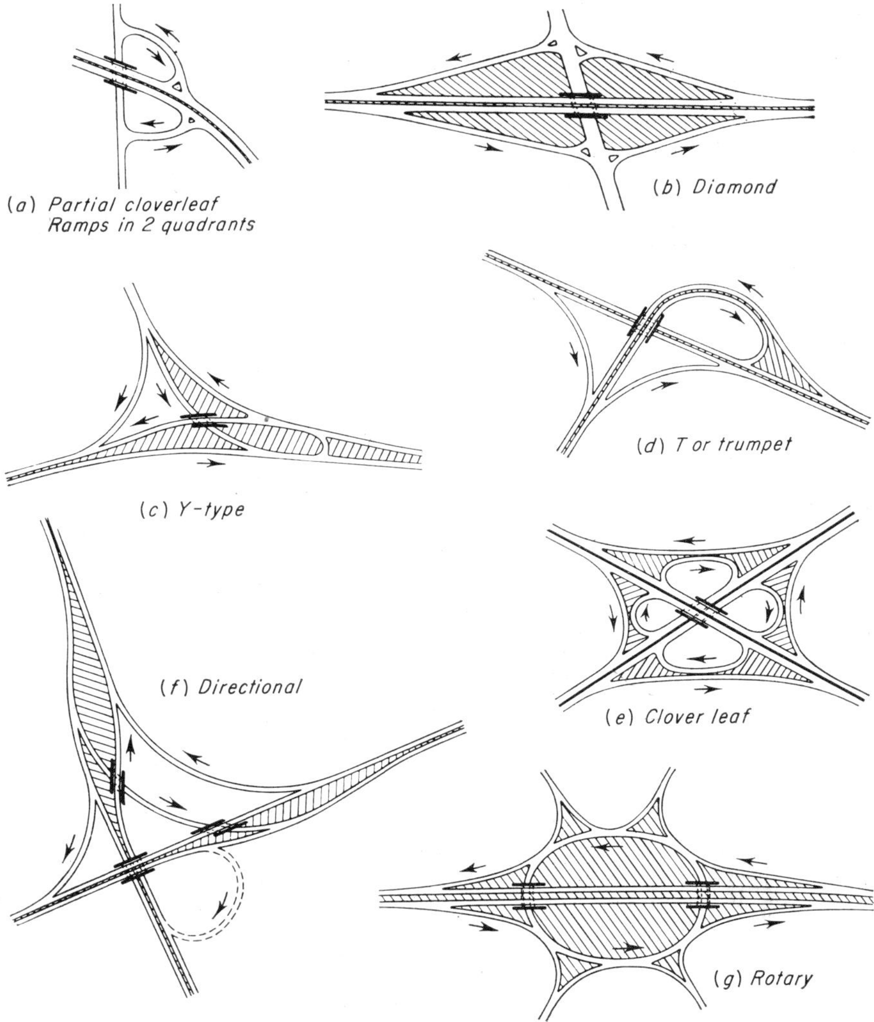

Fig. 37. Typical types of interchanges. (*From "A Policy on Geometric Design of Rural Highways," AASHO.*)

usually built with center pier in median. Clearance on right of roadway should be at least normal shoulder width, but not less than 6 ft. Clearance to face of pier or abutment on left side should be same as on right side, but not less than 4.5 ft. Structures and approaches should be laid out so as to provide required sight distances at all points.

Speed-change lanes[1] are required at ramp ends to make a transition from high-speed through traffic to slower ramp-design speed and vice versa (Fig. 38). Generally it is assumed that a turning vehicle will leave or enter a through lane at a somewhat lower speed than the design speed of the through highway. Lengths of deceleration lanes depend upon speed difference, turn-off distance, and reasonable braking ability of an

TABLE 9. DESIGN LENGTHS OF SPEED-CHANGE LANES[1]

Flat grades—2 percent or less

Design speed of turning roadway curve, mph . . .		Stop condition	15	20	25	30	35	40	45	50
Minimum curve radius, ft.			50	90	150	230	310	430	550	690
Design speed of highway, mph	Length of taper, ft	Total length of deceleration lane, including taper, ft, all main highways								
40	175	300	250	250	200	175	*			
50	200	400	350	350	300	250	250	200	*	
60	225	450	400	400	350	350	300	250	225	*
70	250	500	450	450	400	400	350	350	300	250
Design speed of highway, mph	Length of taper, ft	Total length of acceleration lane, including taper, ft								
Case I. High-volume highways:										
40	175	. . .	450	400	350	250	*			
50	200	. . .	700	650	600	500	400	250	*	
60	225	. . .	1,000	950	900	800	700	550	400	250
70	250	. . .	1,300	1,200	1,100	1,000	900	800	700	500
Case II. Other main highways:										
40	175	. . .	300	250	200	*				
50	200	. . .	500	450	350	300	*			
60	225	. . .	750	700	600	500	400	250	*	
70	250	. . .	950	900	800	750	600	500	350	*

[1] From A Policy on Geometric Design of Rural Highways, AASHO.
* Less than length of taper; use compound curve or partial taper.

average vehicle. They are longer on down grades than on up grades. Acceleration lanes depend upon speed difference and accelerating ability of design vehicle. They are longer on up grades and shorter on down grades. Deceleration lanes may start with a taper or they may begin abruptly. The acceleration lane starts at end of ramp curve, follows parallel (or nearly so) to the main highway (sometimes separated by a short median in first one-third or one-half of length) and then merges with through-traffic lane in a long taper. The AASHO, after a study of several theoretical approaches, recommends the speed-change lengths shown in Table 9.

[1] See A Policy on Geometric Design of Rural Highways, Chap. VII, pp. 276–288.

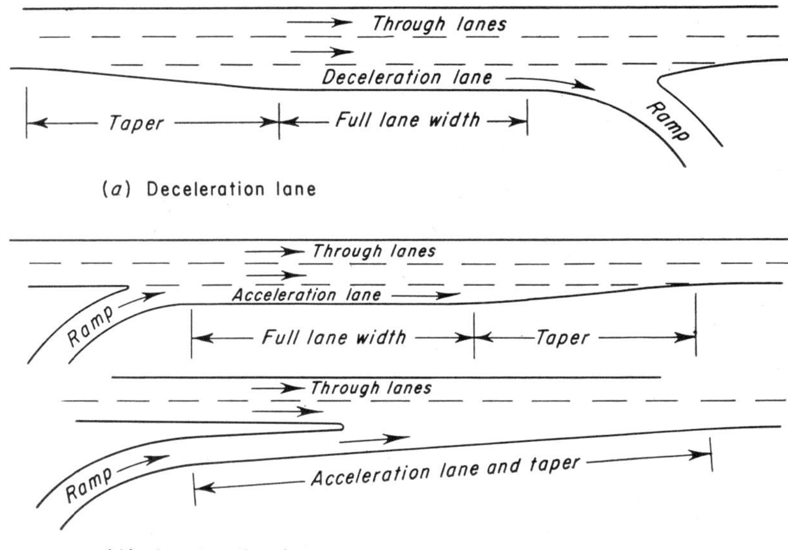

FIG. 38. Speed-change lanes.

HIGHWAY MATERIALS AND TESTS

Nonbituminous materials used in highway construction are natural soil, sand gravel, slag, crushed stone, brick, portland cement, pipe, timber, structural steel, and many miscellaneous materials. The suitability of these materials for different purposes is determined by tests standardized by the American Association of State Highway Officials (AASHO) and the American Society for Testing Materials (ASTM). Many of the standard tests of these two agencies are identical.

Soils are subject to sampling and testing to determine their behavior as subgrades and in embankments. Common tests are the *mechanical analysis* [1] to determine grain size distribution, *limit tests* [2] to determine the moisture contents in percent of dry weight at which the characteristics of the soil change from liquid to plastic, to semi-solid, and to solid, respectively, and *density tests* [3] to determine the compaction characteristics of the soil. Other tests applied particularly to foundation problems are for consolidation, shearing strength, permeability, and capillarity.[4] (Also see Table 18 and Section 8, Foundations.)

Aggregate for highway purposes is usually tested for gradation, quality, and durability. The gradation is specified by the percentages passing designated screen sizes, the maximum size and distribution of sizes depending upon the purpose for which the stone is to be used (see Table 10). The quality of the aggregate is determined by either the Los Angeles abrasion test [5] or the Deval abrasion test. The *percent of wear* by

[1] Mechanical analysis ASTM D422 and AASHO T88.

[2] Liquid limit, ASTM D423 and AASHO T89; plastic limit, ASTM D424 and AASHO T90; plasticity index (liquid limit − plastic limit), AASHO T91, shrinkage factors, ASTM D427 and AASHO T92; field moisture equivalent, ASTM D426 and AASHO T93; centrifuge moisture equivalent, ASTM D425 and AASHO T94.

[3] Compaction and density of soils, ASTM D698 and AASHO T99.

[4] Procedures of Testing Soil, ASTM, July, 1950. Also Compendium on Soil Testing Apparatus, *Highway Research Board Proc. 18th Annual Meeting*, Part II, 1938.

[5] In the Los Angeles abrasion test (ASTM C131 and AASHO T96) a charge of 5 kg of aggregate together with a specified number of cast-iron spheres (depending upon grading) is loaded in a steel drum with an interior projecting shelf and rotated for 500 revolutions at 30 to 33 rpm.

TABLE 10. STANDARD SIZES OF COARSE AGGREGATE

Size No.	Nominal size square openings [2]	Amounts finer than each laboratory sieve (square openings), percentage by weight														
		4	3½	3	2½	2	1½	1	¾	½	⅜	No. 4	No. 8	No. 16	No. 50	No. 100
1	3½–1½	100	90–100	...	25– 60	......	0– 15	......	0– 5							
2	2½–1½	...	......	100	90–100	35– 70	0– 15	......	0– 5							
24	2½–¾	...	......	100	90–100	......	25– 60	......	0– 10	0– 5						
3	2–1	...	......	...	100	95–100	35– 70	0– 15	......	0– 5						
357	2–No. 4	...	......	...	100	95–100	......	35– 70	......	10– 30	......	0– 5				
4	1½–¾	...	......	...	......	100	90–100	20– 55	0– 15	......	0– 5					
467	1½–No. 4	...	......	...	......	100	95–100	......	35– 70	......	10– 30	0– 5				
5	1–½	...	......	...	......	......	100	90–100	20– 55	0– 10	0– 5					
56	1–⅜	...	......	...	......	......	100	90–100	40– 75	15– 35	0– 15	0– 5				
57	1–No. 4	...	......	...	......	......	100	95–100	......	25– 60	......	0– 10	0– 5			
6	¾–⅜	...	......	...	......	......	......	100	90–100	20– 55	0– 15	0– 5				
67	¾–No. 4	...	......	...	......	......	......	100	90–100	......	20– 55	0– 10	0– 5			
68	¾–No. 8	...	......	...	......	......	......	100	90–100	......	30– 65	5– 25	0–10	0– 5		
7	½–No. 4	...	......	...	......	......	......	......	100	90–100	40– 70	0– 15	0– 5			
78	½–No. 8	...	......	...	......	......	......	......	100	90–100	40– 75	5– 25	0–10	0– 5		
8	⅜–No. 8	...	......	...	......	......	......	......	......	100	85–100	10– 30	0–10	0– 5		
89	⅜–No. 16	...	......	...	......	......	......	......	......	100	90–100	20– 55	5–30	0–10	0–5	
9	No. 4–No. 16	...	......	...	......	......	......	......	......	......	100	85–100	10–40	0–10	0–5	
10	No. 4–0 [3]	...	......	...	......	......	......	......	......	......	100	85–100	......	......	......	10–30

[1] AASHO Specification M43–49.
[2] In inches, except where otherwise indicated. Numbered sieves are those of the U.S. Standard sieve series.
[3] Screenings.

either test is the weight of the material passing the No. 12 sieve after test multiplied by 100 and divided by the original weight. Using the Los Angeles test, a percent of wear of not over 50 is commonly specified for base course aggregates, and not over 40 for surface courses. In bituminous macadam, using hard tough traprock, 35 percent is specified for base courses and 25 percent for surfaces. Using the Deval test,[1] a percent of wear of not over 5 is specified for better grades of base-course stone; 3 to 5 percent for base courses in bituminous macadam and not over 3 percent for surface courses.

The durability of stone (or gravel), i.e., its resistance to weathering, may be determined by the sodium sulfate soundness test [2] in which a sample is submitted to alternate cycles of saturation in sodium sulfate (or magnesium sulfate) solution and drying in an oven at 105° to 110°C. In general terms, the "loss" is the weighted percentage by weight of the particles passing from one sieve size to the next smaller size as the result of the test. Durability is sometimes measured by alternate cycles of freezing and thawing.[3] A common requirement for aggregates is that the percentage loss shall not exceed 15 percent for five cycles of the sodium sulfate test, or after 50 cycles of freezing and thawing. For proportioning bituminous concrete and portland cement concrete mixtures the specific gravity of aggregates must be determined.[4]

Bituminous materials [5] used in highway construction are asphalt and tar. The principal source of asphalt is from the distillation of asphaltic-base petroleum. Asphalt

[1] In the Deval abrasion test (ASTM D2 and AASHO T3 for rock; ASTM D289 and AASHO T4 for gravel) the 5 kg charge is placed in a sealed bucket inclined at 30° with a horizontal shaft and rotated for 10,000 revolutions at 30 to 33 rpm. When applied to gravel, 6 cast-iron spheres 1⅞ in. in diameter are added to 5 kg of gravel which must conform to one of five specified gradings. A percent of wear of 15 is satisfactory for most purposes.

[2] Sodium sulfate soundness test AASHO T104 and ASTM C88.

[3] Soundness of aggregates by freezing and thawing, AASHO T103.

[4] Specific gravity of fine aggregate, AASHO T84 and ASTM C128; coarse aggregate, AASHO T85 and ASTM C127.

[5] The following definitions have been adopted by the ASTM (D8): *Bitumens.* Mixtures of hydrocarbons of natural or pyrogenous origin, or combinations of both, frequently accompanied by their nonmetallic derivatives, which may be gaseous, liquid, semisolid, or solid, and which are completely soluble in carbon disulfide. *Bituminous material.* A substance which is characterized by the presence of bitumen, or one from which bitumen can be derived. *Bituminous emulsion.* (*a*) A suspension of minute globules of bituminous material in water or in an aqueous solution. (*b*) A suspension of minute globules of water or of an aqueous solution in a liquid bituminous material. *Cutback products.* Petroleum or tar residua which have been blended with distillates. *Flux.* A bituminous material, generally liquid, used for softening other bituminous materials. *Liquid bituminous materials.* Those having a penetration at 25°C (77°F), under a load of 50 g applied for 1 sec of more than 350. *Semisolid bituminous materials.* Those having a penetration at 25°C (77°F), under a load of 100 g applied for 5 sec of more than 10, and a penetration at 25°C (77°F), under a load of 50 g applied for 1 sec. of not more than 350. *Solid bituminous materials.* Those having a penetration at 25°C (77°F), under a load of 100 g applied for 5 sec of not more than 10. *Asphalt.* A dark brown to black cementitious material, solid or semisolid in consistency, in which the predominating constituents are bitumens which occur in nature as such or are obtained as residual in refining petroleum. *Asphalt cement.* A fluxed or unfluxed asphalt specially prepared as to quality and consistency for direct use in the manufacture of bituminous pavements, and having a penetration at 25°C (77°F) of between 5 and 300, under a load of 100 g applied for 5 sec. *Native asphalt.* Asphalt occurring as such in nature. *Rock asphalt.* A naturally occurring rock formation, usually limestone or sandstone, impregnated throughout its mass with a minor amount of bitumen. *Asphaltenes.* The components of the bitumen in petroleums, petroleum products, malthas, asphalt cements, and solid native bitumens, which are soluble in carbon disulfide but insoluble in paraffin naphthas. *Carbenes.* The components of the bitumen in petroleums, petroleum products, malthas, asphalt cements, and solid native bitumens, which are soluble in carbon disulfide but insoluble in carbon tetrachloride. *Tar.* Brown or black bituminous material, liquid or semisolid in consistency, in which the predominating constituents are bitumens obtained as condensates in the destructive distillation of coal, petroleum, oil-shale, wood, or other organic materials, and which yields substantial quantities of pitch when distilled. *Coal tar.* Tar produced by the destructive distillation of bituminous coal. *Gashouse coal tar.* Coal tar produced in gashouse retorts in the manufacture of illuminating gas from bituminous coal. *Coke-oven tar.* Coal tar produced in by-product coke ovens in the manufacture of coke from bituminous coal. *Oil-gas tars.* Tars produced by cracking oil vapors at high temperatures in the manufacture of oil gas. *Refined tar.* Tar freed from water by evaporation or distillation which is continued until the residue is of desired consistency; or a product produced by fluxing tar residuum with tar distillate. *Pitches.* Black or dark-brown solid cementitious materials which gradually liquefy when heated and which are obtained as residua in the partial evaporation or fractional distillation of tar. *Straight-run pitch.* A pitch run to the consistency desired in the initial process of distillation and without subsequent fluxing. *Free carbon in tars.* Organic matter which is insoluble in carbon disulfide. *Normal temperature.* As applied to laboratory observations of the physical characteristics of bituminous materials, is 25°C (77°F). *Fixed carbon.* The organic matter of the residual coke obtained upon burning hydrocarbon products in a covered vessel in the absence of free oxygen. *Penetration.* The consistency of a bituminous material expressed as the distance that a standard needle vertically penetrates a sample of the material under known conditions of loading, time, and temperature. Where the conditions of test are not specifically mentioned, the load, time, and temperature are understood to be 100 g, 5 sec, and 25°C (77°F), respectively, and the units of penetration to indicate hundredths of a centimeter. *Viscosity.* The measure of the resistance to flow of a bituminous material, usually stated as the time of flow of a given amount of the material through a given orifice.

also occurs in nature in surface deposits (lake asphalt), and impregnated in porous rock (rock asphalt). Tars are produced as by-products in the manufacture of illuminating gas, coke, and carbureted water gas.

Important properties of bituminous road materials are specific gravity for weight-volume conversions, consistency at different temperatures, distillates present in liquid products, temperature of ignition (flash point), and bitumen content. These properties are identified by standard tests of the ASTM and AASHO.[1] Semisolid and solid asphalts (asphalt cements) and heavier tars require heating to become liquid. The asphalt cements are graded according to the penetration test, and the heavier tars by the float test. Other bituminous products include sufficient lighter distillates to render them liquid at ordinary temperatures or with moderate heating. These materials are graded primarily by viscosity. Permanence and fluidity of these materials are controlled by the amount and nature of the distillates present. Liquid asphaltic road materials are divided into classes depending upon the degree of hardening or cementitiousness which is ultimately desired in the road and the rapidity with which it is developed. The three classes are designated as slow-curing SC, medium-curing MC, and rapid-curing RC. The SC materials contain nonvolatile oils. They harden very slowly and develop low cementing values. MC and RC products are cutbacks, the solvent being kerosene for MC and gasoline for RC. These cure or harden through

TABLE 11. SPECIFICATIONS FOR ASPHALT CEMENTS—THE ASPHALT INSTITUTE

Characteristics	Paving cements		
General requirements	The asphalt shall be prepared by the refining of petroleum. It shall be uniform in character and shall not foam when heated to 350°F		
Flash point (Cleveland open cup), °F	450+	425+	350+
	Penetration grades		
Penetration, 77°F, 100 g, 5 sec	40–50 50–60 60–70 70–85 85–100	100–120 120–150 150–200	200–300
Loss on heating 325°F, 5 hr, percent	1−	2−	2−
Penetration after loss on heating:			
77°F, 100 g, 5 sec, percent of original	70+	70+	60+
Ductility 77°F, cm	100+	60+	
Solubility in carbon tetrachloride, percent	99.5+	99.5+	99.5+
Mixing temperature, °F	275°–325°	275°–325°	200°–275°

Working temperatures for spraying 275° to 350°F.

the loss of the volatile solvents, leaving in place a residue of semisolid asphalt. Each class is further subdivided into viscosity grades 0 to 5, the zero grade being quite fluid and the 5 grade very viscous.

[1] Standard tests for bituminous materials: specific gravity ASTM D71 and D70, AASHO T43; consistency by penetration test ASTM D5 and AASHO T49, Saybolt-Furol viscosity for asphalts ASTM D88 and AASHO T72, by Engler viscosity for tars AASHO T54, by float test ASTM D139 and AASHO T50; softening point ASTM D36 and AASHO T53; ductility ASTM D113 and AASHO T51; flash point ASTM D92 and D93, AASHO T48, T79, and T73; volatilization ASTM D6 and AASHO T47; distillation (tars) ASTM D20 and AASHO T52 (asphalt cutbacks) ASTM D402 and AASHO T78; test for bitumen content, solubility in carbon disulfide ASTM D4 and AASHO T44; solubility in carbon tetrachloride ASTM D165 and AASHO T45; water and sediment ASTM D95 and AASHO T55.

Asphalt may be rendered temporarily liquid by emulsification with water in the presence of an emulsifying agent of a soapy nature. After the emulsion is applied to the road, it "breaks," i.e., the water separates from the asphalt and evaporates or runs off, leaving the asphalt cement on the aggregate. Three types of emulsions are available: rapid-setting RS, medium-setting MS, and slow-setting SS. The RS emulsion is used for surface treatments when no manipulation of the aggregate is required. The MS and SS products are used where there is manipulation of coarse aggregate and of well-graded aggregates, respectively.[1] Specifications for asphaltic products are given in Tables 11 to 15, and for tars in Table 16. The tars are graded from RT-1 to RT-12 according to increasing viscosity. The RTCB products are cutback tars used primarily for patching mixtures.

TABLE 12. SPECIFICATIONS FOR SLOW-CURING ASPHALT MATERIALS—THE ASPHALT INSTITUTE

Specification designation	SC-0	SC-1	SC-2	SC-3	SC-4	SC-5
General requirements.......			The material shall be free from water			
Flash point (Cleveland open cup), °F..........	150+	150+	175+	200+	225+	250+
Furol viscosity at 77°F, sec	75–100					
Furol viscosity at 122°F, sec	......	75–100				
Furol viscosity at 140°F, sec	......	......	100–200	250–500		
Furol viscosity at 180°F, sec	......	......	....	....	125–250	300–600
Water, percent............	0.5−	0.5−				
Distillation:						
Total distillate to 680°F..	15–40	10–30	5–25	2–15	10−	5−
Float test on residue at 122°F, sec............	15–100	20–100	25–100	50–125	60–150	75–200
Asphalt residue of 100 penetration, percent..........	40+	50+	60+	70+	75+	80+
Ductility asphalt residue at 77°F..................	100+	100+	100+	100+	100+	100+
Solubility in carbon tetrachloride, percent.........	99.5+	99.5+	99.5+	99.5+	99.5+	99.5+
Application temperature, °F.	50–120	80–200	150–200	175–250	175–250	200–275
Typical uses (not part of specifications)	Dust layer		Road mix Dense graded aggregate		Cold-laid plant mix Dense graded aggregate	

[1] Standard tests and specifications for asphalt emulsions: ASTM D244 and D977, and AASHO T59 and M140.

TABLE 13. SPECIFICATIONS FOR MEDIUM-CURING ASPHALTIC MATERIALS—THE ASPHALT INSTITUTE

Specification designation	MC-0	MC-1	MC-2	MC-3	MC-4	MC-5
General requirements......	The material shall be free from water					
Flash point (open Tag.), °F.	100+	100+	150+	150+	150+	150+
Furol viscosity at 77°F, sec.	75–100					
Furol viscosity at 122°F, sec.	....	75–150				
Furol viscosity at 140°F, sec.	....	....	100–200	250–500		
Furol viscosity at 180°F, sec.	....	....	....	....	125–250	300–600
Distillation:						
Distillate (percent of total distillate to 680°F):						
To 437°F.............	25−	20−	10−	5−	0	0
To 500°F.............	40–70	25–65	15–55	5–40	30−	20−
To 600°F.............	75–93	70–90	60–87	55–85	40–80	20–75
Residue from distillation to 680°F:						
Volume percent, by difference.............	50+	60+	67+	73+	78+	82+
Tests on residue from distillation:						
Penetration 77°F, 100 g, 5 sec...............	120–300	120–300	120–300	120–300	120–300	120–300
Ductility, 77°F [1]........	100+	100+	100+	100+	100+	100+
Solubility in carbon tetrachloride, percent....	99.5+	99.5+	99.5+	99.5+	99.5+	99.5+
Working temperature, °F:						
Spraying..............	50–120	80–150	100–200	175–250	200–275	225–275
Mixing................	50–120	80–150	100–200	150–200	175–225	200–250
Typical uses (not part of specifications)	Primer			Cold-laid plant mix		
			Seal coat and surface treatment			
			Road mix			
			Cold patch			

[1] If penetration of residue is more than 200 and its ductility at 77°F is less than 100, the material will be acceptable if its ductility at 60°F is 100+.

TABLE 14. SPECIFICATIONS FOR RAPID-CURING ASPHALTIC MATERIALS—THE ASPHALT INSTITUTE

Specification designation	RC-0	RC-1	RC-2	RC-3	RC-4	RC-5
General requirments	The material shall be free from water					
Flash point (open Tag.), °F	......	......	80+	80+	80+	80+
Furol viscosity at 77°F, sec	75–150					
Furol viscosity at 122°F, sec	......	75–150				
Furol viscosity at 140°F, sec	......	......	100–200	250–500		
Furol viscosity at 180°F, sec	......	......	....	....	125–250	300–600
Distillation:						
Distillate (percent of total distillate to 680°F):						
To 374°F	15+	10+				
To 437°F	55+	50+	40+	25+	8+	
To 500°F	75+	70+	65+	55+	40+	25+
To 600°F	90+	88+	87+	83+	80+	70+
Residue from distillation to 680°F:						
Volume percent by difference	50+	60+	67+	73+	78+	82+
Tests on residue fron distillation:						
Penetration 77°F, 100 g, 5 sec	80–120	80–120	80–120	80–120	80–120	80–120
Ductility 77°F	100+	100+	100+	100+	100+	100+
Solubility in carbon tetrachloride, percent	99.5+	99.5+	99.5+	99.5+	99.5+	99.5+
Working temperature, °F:						
Spraying	50–120	80–150	100–175	150–200	175–250	200–275
Mixing	50–120	80–125	80–150	125–175	150–200	175–225
Typical uses (not part of specifications)	Seal coat and surface treatment					
			Road mix	Cold-laid plant mix		Penetration macadam

TABLE 15. SPECIFICATIONS FOR EMULSIFIED ASPHALT—THE ASPHALT INSTITUTE

Specification designation	Rapid setting		Medium setting	Slow setting
	RS-1	RS-2	MS-2	SS-1
Tests on emulsion:				
Viscosity, Saybolt Furol:				
At 77°F, sec	20–100	....	100+	20–100
At 122°F, sec	....	75–400		
Residue by distillation, percent	57–62	62–69	62–69	57–62
Settlement, 5 days, percent	3−	3−	3−	3−
Demulsibility:				
35 ml of 0.02N $CaCl_2$, percent	60+	50+		
50 ml of 0.10N $CaCl_2$, percent	....	....	30−	
Sieve test (retained on No. 20), percent	0.10−	0.10−	0.10−	0.10−
Cement-mixing test, percent	....	....	....	2.0−
Tests on residue:				
Penetration at 77°F, 100 g, 5 sec	100–200	100–200	100–200	100–200
Soluble in CS_2, percent	97.5+	97.5+	97.5+	97.5+
Ductility at 77°F, cm	40+	40+	40+	40+
Working temperature, °F	Spraying 50–140; mixing 50–140			
Typical uses (not part of specifications)	Surface treatment Tack coat Penetration macadam		Plant mix—open Road mix Cold patch	Plant mix—dense Soil mix

TABLE 16. STANDARD SPECIFICATION FOR TAR FOR ROAD CONSTRUCTION [1]

Test requirements (AASHO)	Grades (+ indicates minimum allowable; − indicates maximum allowable)								
	RT-1	RT-2	RT-3	RT-4	RT-5	RT-6	RT-7	RT-8	RT-9
Consistency:									
Specific viscosity, Engler, at 40°C	5–8	8–13	13–22	22–35					
Specific viscosity, Engler, at 50°C	......	......	......	......	17–26	26–40			
Float test at 32°C	......	......	......	......	......	......	50–80	80–120	120–200
Float test at 50°C									
Specific gravity at 25°C/25°C	1.08+	1.08+	1.09+	1.09+	1.10+	1.10+	1.12+	1.14+	1.14+
Total bitumen, percent by weight	88+	88+	88+	88+	83+	83+	78+	78+	78+
Water, percent by volume	2.0−	2.0−	2.0−	2.0−	1.5−	1.5−	1.0−	0.0	0.0
Total distillates, percent by weight:									
To 170°C	7.0−	7.0−	7.0−	5.0−	5.0−	5.0−	3.0−	1.0−	1.0−
To 200°C									
To 235°C									
To 270°C	35.0−	35.0−	30.0−	30.0−	25.0−	25.0−	20.0−	15.0−	15.0−
To 300°C	45.0−	45.0−	40.0−	40.0−	35.0−	35.0−	30.0−	25.0−	25.0−
Softening point of residue, °C	30–60	30–60	35–65	35–65	35–70	35–70	35–70	35–70	35–70
Sulfonation index (when specified), total distillation, percent by weight:									
To 300°C	8.0−	7.0−	6.0−	6.0−	5.0−	5.0−			
300° to 355°C	1.5−	1.5−	1.5−	1.5−	1.5−	1.5−			
Typical uses and suggested temperatures for application (not part of specification)	Prime coat; 60° to 125°F		Prime coat and surface treatment; 80° to 150°F		Surface treatment and road mix; 80° to 150°F		Surface treatment, road mix premix, and seal coat; 150° to 225°F		

	RT-10	RT-11	RT-12	RTCB-5	RTCB-6
Consistency:					
Specific viscosity, Engler, at 40°C	......	....	....	17–26	26–40
Specific viscosity, Engler, at 50°C					
Float test at 32°C					
Float test at 50°C	75–100	100–150	150–220		
Specific gravity at 25°C/25°C	1.15+	1.16+	1.16+	1.09+	1.09+
Total bitumen, percent by weight	75+	75+	75+	80+	80+
Water, percent by volume	0.0	0.0	0.0	1.0−	1.0−
Total distillates, percent by weight:					
To 170°C	1.0−	1.0−	1.0−	2.0–8.0	2.0–8.0
To 200°C	......	....	....	5.0+	5.0+
To 235°C	......	....	....	8.0–18.0	8.0–18.0
To 270°C	10.0−	10.0−	10.0−		
To 300°C	20.0−	20.0−	20.0−	35.0−	35.0−
Softening point of residue, °C	40–70	40–70	40–70	40–70	40–70
Typical uses and suggested temperatures for application (not part of specification)	Surface treatment premix, seal coat, penetration and crack filler; 175° to 250°F			Surface treatment road mix and premix for low-temperature application and quick setting; 60° to 120°F	

[1] Specification M-52-42 of the AASHO.

ROADS AND PAVEMENTS

Soil Investigations. The subgrade or foundation is an important element in highway construction. Subgrade defects are sure to result in poor riding surfaces and high maintenance costs. A thorough survey of soil conditions should be made in advance of construction, since it is easier and cheaper to correct these defects before expensive pavements are built. Test samples should be taken along the line, particularly where poor subgrade conditions appear probable, and field notes made of all conditions that may influence soil behavior and drainage requirements. For purposes of identification the Bureau of Public Roads has developed a soil classification,[1] which groups soils according to their probable behavior in highway subgrade. There are eight major classes, A-1 to

[1] See Classification of Soils and Procedures Used in Construction of Embankments, *Public Roads*, February, 1942. Also Report of Committee on Classification of Materials for Subgrades and Granular Type Roads, *Highway Research Board, 25th Proc.*, 1945, pp. 375–392. Also, CASAGRANDE, ARTHUR, Classification and Identification of Soils, *Trans. ASCE*, vol. 113, pp. 901–987, 1948. For airfield soil classifications see pp. 2–138 and 2–140. Soil classifications based on gradation and texture are described in Section 8, pp. 8–3.

Table 17. Classification of Soils and Soil-aggregate Mixtures (with Suggested Subgroups)[1]

General classification	Granular materials (35 percent or less passing No. 200)							Silt-clay materials (more than 35 percent passing No. 200)			
Group classification	A-1		A-3	A-2				A-4	A-5	A-6	A-7
	A-1-a	A-1-b		A-2-4	A-2-5	A-2-6	A-2-7				A-7-5: A-7-6
Sieve analysis, percent passing:											
No. 10	50 max										
No. 40	30 max	50 max	51 min								
No. 200	15 max	25 max	10 max	35 max	35 max	35 max	35 max	36 min	36 min	36 min	36 min
Characteristics of fraction passing No. 40:											
Liquid limit[3]	....		....	40 max	41 min	40 max	41 min	40 max	41 min	40 max	41 min
Plasticity index	6 max		N.P.	10 max	10 max	11 min	11 min	10 max	10 max	11 min	11 min[2]
Group index	0		0	0		4 max		8 max	12 max	16 max	20 max
Usual types of significant constituent materials	Stone fragements gravel and sand		Fine sand	Silty or clayey gravel and sand				Silty soils		Clayey soils	
General rating as subgrade	Excellent to good							Fair to poor			

[1] From Standard Specifications, M145–49, AASHO.
[2] Plasticity index of A-7-5 subgroup is equal to or less than LL minus 30. Plasticity index of A-7-6 subgroup is greater than LL minus 30.
[3] See Section 8, p. 8–12, for limit tests.

A-8, ranging from materials that are predominantly granular to the fine-grained silts and clays. The A-8 class is organic peat, which is wholly unsuitable for supporting a highway. A modified form of BPR classification is shown in Table 17. The classes are defined by physical properties, sieve analysis, and limit tests performed on that portion passing No. 40 sieve. A group index is included for the finer, more plastic soils. It is derived from empirical formula $GI = 0.2a + 0.005ac + 0.01bd$, in which a = portion of percentage passing No. 200 sieve greater than 35 percent and not exceeding 75 percent, expressed as a positive whole number (1 to 40); b = that portion of percentage passing No. 200 sieve greater than 15 percent and not exceeding 55 percent, expressed as a positive whole number (1 to 40), c = that portion of numerical liquid limit greater than 40 and not exceeding 60, expressed as a positive whole number (1 to 20); d = that portion of numerical plasticity index greater than 10 and not exceeding 30, expressed as positive whole number (1 to 20).

The results of a soil survey are frequently plotted on the profile of the proposed road.[1] The several strata of soil encountered for a depth of 5 or 6 ft below the proposed grade line are shown by symbols together with the maximum height of ground-water level and depth to ledge. From a study of this profile and other field data, recommendations are made with respect to the depth of base course and the locations where subdrainage is needed.

In northern climates an important object of the soil survey is to detect areas where the soil is subject to frost heave.[2] In certain silty-clay soils (Group A-4, for example), lenses of pure ice form in the soil at different depths during the freezing season. These layers are formed from water brought up by capillarity from the water table (ground-water level). As the ice lenses grow they push the pavement upward, causing the heave. When the ice melts in the spring, the subgrade becomes supersaturated, causing cracking of concrete pavements and the disintegration of bituminous and gravel surfaces. To avoid frost heave, adequate drainage should be installed to lower the ground water, and, if at all possible, the frost-heave type of material should be removed to the depth of normal frost penetration and replaced with granular material not subject to ice formation. A mechanical analysis is the principal method for detecting frost-heave material. Nonuniform soil with more than 3 percent finer than 0.02-mm grain size, and uniform soil with more than 10 percent finer than 0.02 mm are usually subject to frost heave. Soils with less than 1 percent finer than 0.02 mm are free from frost heave.

Grading. Highway grading includes clearing and grubbing, forming cuts and fills for the finished grade, together with excavation and backfill for ditches, trenches, culverts, walls, bridges, and drainage pipe. Clearing includes cutting, removal, and disposal of trees and brush. Grubbing consists of removing stumps and roots. One price per acre is usually paid for both items. Under low embankments (3.5 ft or less) all stumps and roots are removed; for higher fills trees are cut off close to the ground. Roadway excavation is usually classified as *rock excavation*, which includes solid rock that requires blasting for its removal and detached boulders over a specified size (usually ½ cu yd); or *earth excavation*, which includes all types of material not otherwise specified. The unit of measurement is the cubic yard, and the price paid includes the excavation of material and its disposal in fills or in other designated areas. In some contracts the excavation is unclassified, i.e., covered by a single unit price. Separate classifications may be used for special types of excavation, such as trench, bridge, channel change, and peat. When more embankment material is needed than is available from cuts, the additional material is classified as earth or ordinary *borrow* and paid for by cubic yard. Other types of borrow are gravel for base courses, stone or gravel for backfilling drains or

[1] Surveying and Sampling Soils for Highway Subgrades, ASTM Standards D420 and AASHO Standard Specifications T86.

[2] Frost Heave in Highways and Its Prevention, *Public Roads*, March, 1934; also Frost Action in Roads and Airfields: A Review of the Literature, 1765–1951, *Special Report* 1, Highway Research Board.

walls, and loam for landscaping. Highway contracts frequently do not specify separate payment for haul, the cost being included in the unit price of excavation. In some contracts overhaul is paid for beyond a specified limit (such as 1,000 ft). The unit is usually the yard-station, or 1 cu yd hauled 100 ft; sometimes the yard-mile is used. An item of *fine grading* per square yard may be included to cover the final shaping of subgrade and shoulders. Typical bid quantities on highway projects are given on p. 2–121.

Control of Settlement of Embankments. An important consideration in placing embankments is to compact them sufficiently to avoid subsequent settlement with resulting pavement failures and costly maintenance. Granular materials, sand and gravel, usually compact readily under rolling, although vibration is sometimes desirable. Fine materials, particularly those with over 55 percent passing a No. 200 mesh sieve, require special treatment. For these soils there is an optimum moisture content at which the greatest density can be obtained for a given compactive effort. A test procedure, known as the "Proctor method," is used to determine the optimum moisture content for maximum density of a soil when compacted in a $\frac{1}{30}$ cu ft cylindrical mold under prescribed conditions of compaction.[1] Specifications require that soil in fills be compacted to 90, 95, or 100 percent of maximum dry density found by test, depending upon height of fills and nature of soil (see Table 18).

The required compaction can be obtained by spreading the fill in thin layers of 6 to 8 in., keeping the moisture content near the optimum by sprinkling or drying as needed, and rolling with special equipment, such as rubber-tired rollers or sheep's-foot rollers. The latter have blunt prongs which penetrate the loose fill and compact it from the bottom up, and are well adapted for compacting clayey-type soils. The compaction characteristics of BPR soil classes are given in Table 18.

TABLE 18. COMPACTION CHARACTERISTICS OF BPR SOIL CLASSES [1]

Group	A-1	A-2		A-3	A-4	A-5	A-6	A-7	A-8
		Friable	Plastic						
Compaction characteristics:									
Max dry weight, lb per cu ft	130 min	120–130	120–130	120–130	110–120	80–110	80–110	80–110	90 (max)
Optimum moisture, percentage of dry weight (approx)	9	9–12	9–12	9–12	12–17	22–30	17–28	17–28	
Max field compaction required, percentage of max dry weight, lb per cu ft	90	90	90	90	95	100	100	100	Waste
Rating for fills 50 ft or less in height	Excellent	Good	Good	Good	Good to poor	Poor to very poor	Fair to poor	Fair to poor	Unsatisfactory
Rating for fills more than 50 ft in height	Good	Good to fair	Good to fair	Good to fair	Fair to poor	Very poor	Very poor	Very poor	Unsatisfactory
Required total thickness for subbase, base, and surfacing, in.	0–6	0–6	2–8	0–6	9–18	9–24	12–24	12–24	

[1] From *Public Roads*, February, 1942. See Table 17 for description of classes.

Shrinkage, Swell, and Subsidence. When earth is excavated, it at first becomes less dense (or swells), but later when compacted in the fill it occupies less space than in the original state. The difference in volume is called *shrinkage* and is often expressed as a percentage of the original volume. Shrinkage varies for different materials and for

[1] In the Proctor or "standard" AASHO test (ASTM D698 and AASHO T99) the soil is placed in the mold in three equal layers, each being compacted by 25 blows of a rammer with a 2-in.-diameter face weighing 5.5 lb and dropped from a height of 1 ft. In the "modified" AASHO test, used extensively in airfield construction, the soil is placed in five equal layers, each tamped with 25 blows using a rammer weighing 10 lb and dropped from a height of 18 in. For procedures see: Classification of Soils and Control Procedures Used in Construction of Embankments, *Public Roads*, February, 1942. Research on Construction of Embankments, *Public Roads*, July-August-September, 1944. "Compaction of Embankments, Subgrades and Bases," *Bull.* 58, Highway Research Board, 1952.

different methods of compaction. The usual limits are between 10 and 20 percent. It can be determined precisely by making density determinations of cut materials in their natural state and comparing them with densities in the compacted fill. When rock is excavated it *swells* from 25 to 40 percent. In making estimates for balancing cuts and fills, allowance must be made for both shrinkage and swell.

When fills are placed on soft ground such as muck or peat, *subsidence* occurs; i.e., the fill settles below the original ground level as its weight consolidates or displaces a portion of the soft material below. Minor roads are sometimes "floated" on peat by placing them on a mat of logs or saplings. Widened fills with flat side slopes are sometimes constructed to restrain the lateral displacement of the peat. On major highways, peat deposits should, where possible, be avoided in location. If this is impracticable, the peat should be removed and replaced with granular material. Shallow depoists are commonly excavated with cranes using clamshell- or dragline-type buckets. Deeper deposits are sometimes displaced by blasting.[1] The fill material is placed over the peat and charges set off in the peat underneath. The disturbed peat is forced out partly by the force of the blast and partly by the weight of the fill on top. The process is repeated until the bottom of the new fill rests on solid foundation. *Sand drains* [2] are sometimes used to facilitate consolidation of soft material. A blanket of sand is placed over the peat or muck. Hollow pile casings are driven through this blanket into the soft material, cleaned out, filled with coarse sand, and then removed. The fill is built up several feet above final grade causing a surcharge load which forces pore water from the soft material into sand columns and out into the sand blanket. When soft material has consolidated to a density at which it can support the surcharge and fill without appreciable settlement, the surcharge is removed and the surface placed.

Grading Methods. The site is cleared of the larger trees which are cut into convenient lengths for disposal. Bulldozers are used for light clearing, for pushing roots and shrubbery into piles for burning, and for removing undesirable topsoil. Plows or rooters loosen the soil and dig out roots. Tractor-drawn and powered scrapers with cutting edges and bowls for carrying earth are well adapted for making cuts and fills where the hauls are not great and the soil is free of large rocks and easy to dig. These scrapers cut, haul, dump, and spread in one cycle of operation. In deep cuts and harder digging, power shovels are used, the usual bucket size ranging from ¾ to 2 cu yd capacity. Shovels load into trucks or into larger tractor-drawn wagons, which transport the material to its place in fills where it is spread by bulldozers. Compaction is obtained under the treads of the earth-handling equipment and by the use of sheep's-foot and rubber-tired rollers. The final surface grading is done with blade graders. Hoe-type shovels and ditch diggers are used for trench excavation, and clamshell or orange-peel bucket cranes for bridge and peat excavation.

Drainage.[3] Adequate surface and subsurface drainage is essential for a stable subgrade. Most soils offer good support when relatively dry, but many soften and deform in the presence of excess moisture. *Surface drainage* is accomplished by crowning the roadway and sloping the shoulders to side ditches or gutters. Surface runoff from adjacent lands is also gathered in these ditches. In high cuts, ditches are provided at the top of slope to intercept surface flow and to prevent excessive erosion of the cut face. These ditches are led into natural water courses or into paved channels brought down

[1] Accelerated Settlement of Embankments by Blasting, *Public Roads*, December, 1939. Also Methods and Costs of Peat Displacement in Highway Construction, *Highway Research Board, Proc. 14th Ann. Meeting*, 1934, p. 315. Treatment of Soft Foundations for Highway Embankments, *Highway Research Board, Proc. 31st Ann. Meeting*, 1952, pp. 601–621. Bibliography on Survey and Treatment of Marsh Deposits, Highway Research Board, 1954.

[2] "Vertical Sand Drains for Stabilization of Embankments," *Bull.* 115, Highway Research Board, 1956.

[3] See papers on drainage, *Highway Research Board, Proc. 24th Ann. Meeting*, and "Surface Drainage of Highways," Research Reports 6-B, 11-B, and 15-B, Highway Research Board, 1948, 1950, and 1953; "Subsurface Drainage," *Bull. 45*, Highway Research Board, 1951; and Highway Subdrainage, *Highway Research Board, Proc.*, vol. 31, pp. 543–606, 1952. Also "Handbook of Culvert and Drainage Practice," Armco Culvert Manufacturers Assoc., Middletown, Ohio.

the face of the cut. On high fills, a berm is commonly built along the edge of the fill to keep surface flow on the top of the embankment except at points where paved channels are provided to carry the flow down the embankment (Fig. 39). Side ditches drain into natural stream crossings or into inlets with pipe outlets to the lower slopes. Culverts are provided for cross drainage at low points in the ground profile. On important roads, shallow ditch sections are used with frequent outlets. Ditch sections and tops of slopes are usually rounded with vertical curves. This reduces erosion, facilitates planting of

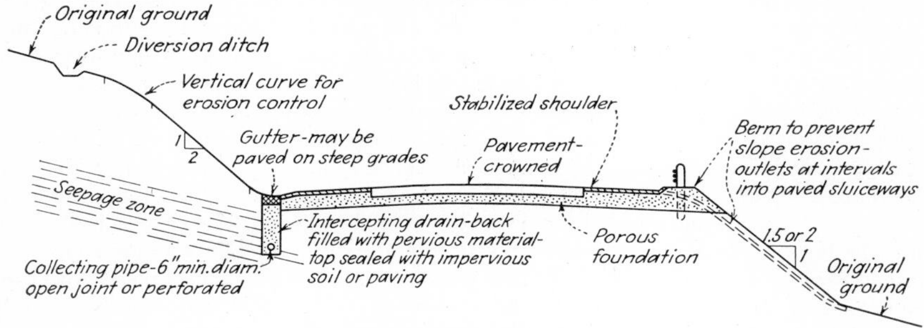

FIG. 39. Drainage facilities on side-hill location.

slopes, and improves the appearance of the roadside. In built-up areas, paved gutters are provided at the outer edges of the pavement; these drain into catch basins with pipe outlets into storm sewer system. (See Section 9 for design of storm sewers.) Sidewalks are sloped about 2 percent from the property line to the curb (Fig. 45).

Subsurface drainage is needed where the ground-water level is likely to be near the surface at any time. At such locations, intercepting drains are constructed under the shoulders to lower the ground-water level and keep moisture from reaching the subgrade.

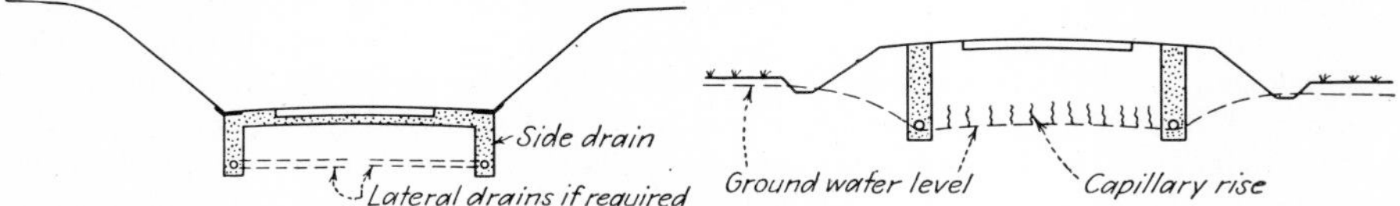

FIG. 40. Drainage in through cut.

FIG. 41. Drainage to lower ground water in low areas.

In soils subject to capillarity the water level should be 4 to 6 ft below the surface to prevent the rise of moisture into the subgrade by capillary action. Side drains are trenches backfilled with porous material, such as gravel or graded broken stone, with an open-joint or perforated pipe at the bottom to collect the flow. Clay pipe with open joints or perforations, or perforated corrugated-metal pipe is commonly used. The top of the trench is sealed with impervious soil, such as a compacted loam clay, to keep surface water out of the drain. The backfill should be a well-graded gravel or broken stone. It should be free of fines, but not so coarse as to allow the passage of silt through the voids in the backfill where it will enter the pipe and cause clogging.[1] Where the drain is to carry storm water as well as seepage, or where the surrounding soil is silty, the perforations are placed "up," and the pipe embedded in impervious material. Where no surface

[1] Clogging Hazards in Underdrains, *Eng. News-Record*, Mar. 12, 1942. Also Underdrain Practice of the Connecticut Highway Department, *Highway Research Board, Proc. 24th Ann. Meeting.*, pp. 377–389.

flow is admitted and the soil is not silty, the holes are placed "down," and the pipe embedded in pervious material (Fig. 42). Typical facilities for surface and subsurface drainage on side-hill location are shown in Fig. 39. In cuts where seepage planes run longitudinally with the roadway, side drains are constructed on both sides of the road (Fig. 40), and lateral branch drains are placed under the central portion of the roadbed, where needed. In flat country where the ground-water table is high, side drains constructed on both sides will lower the ground water (Fig. 41.) In low swampy areas where outlets cannot readily be found for side drains, and in soils that hold water by capillarity (will not drain by gravity flow), a layer of porous material (sand, cinders, or gravel)

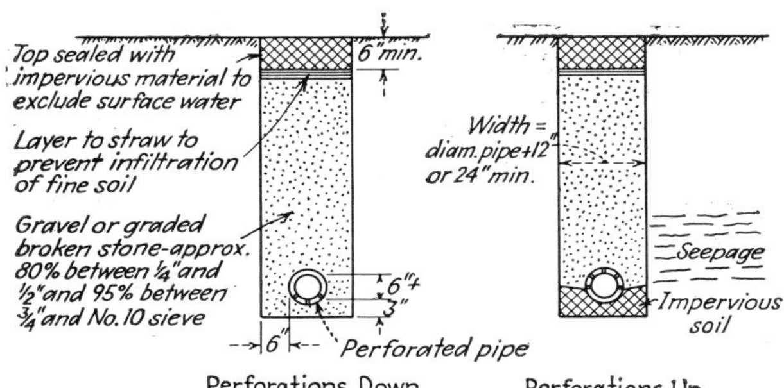

FIG. 42. Details of side drains.

should be placed under the pavement to intercept capillary water. This porous layer should extend under the shoulders to a free outlet at the side drains or in the face of the fill. It is good practice to provide such a foundation under all pavements, as shown in Fig. 39. Where porous materials are scarce, the width of base may be narrowed to a little more than the pavement width, and porous outlets, sometimes called "bleeder drains," provided at low points in the profile to prevent ponding of water in the base. Subdrainage is usually not needed in cities because the water level is kept down by adjacent basement-floor levels. In residential districts, however, subdrainage of the type mentioned above may be desirable.

Drainage Structures. The area of waterway opening [1] depends upon discharge to be handled at maximum runoff. This may be computed by the rational formula explained in Section 9. An approximation may be had from Talbot's formula $a = C\sqrt[4]{A^3}$, in which a is the area of opening required in square feet, A is drainage area in acres, and C is a coefficient depending upon character of drainage area. Values for C for different topography are 1.00 for mountains, 0.80 to 0.60 for hilly, 0.50 to 0.40 for rolling, and 0.30 to 0.20 for flat land. An examination of existing openings, if any, at nearby sites is also valuable in determining the size of opening required. A study of the stream crossing should be made to determine the best location and angle of crossing, and the advisability of modifying the stream channel to obtain the best alignment of the bridge or culvert.

Pipe culverts [2] are commonly used for small openings (up to 20 sq ft or 60 in. diameter); multiple pipes with common headwalls may be used for larger areas. *Reinforced-concrete pipe* is made in diameters from 12 to 108 in., usually in laying lengths of 4 ft with bell-and-spigot or flat hub and spigot ends. *Cast-iron culvert pipe* may be obtained in diame-

[1] See Section 9 for method of computing flow from drainage area.

[2] Specifications for pipe culverts: cast iron, ASTM A142 and AASHO M64; reinforced concrete, ASTM C76 and AASHO M41; corrugated metal, AASHO M36. Also "AREA Manual," 1956 ed., pp. 1-4-5 to 1-4-32, for all types of pipe.

ters up to 72 in.; the common range is 12 to 48 in. *Corrugated-metal pipe* culverts have corrugations spaced about 2½ in. apart and ½ in. deep; gage of metal is No. 16 for 8

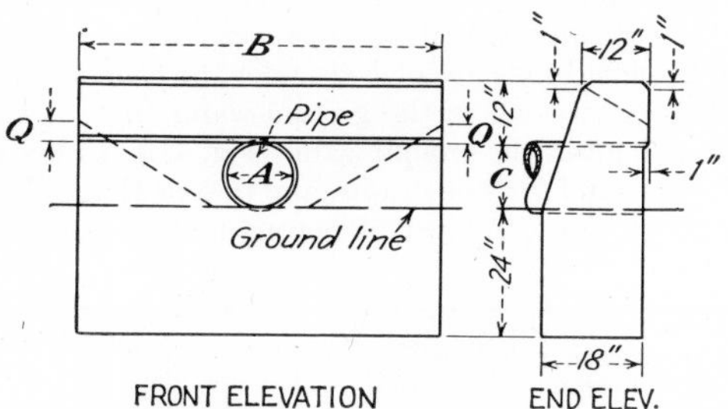

Fig. 43. Headwall for pipe culvert. (*Massachusetts Department of Public Works.*)

1½:1 and 2:1 slope		1½:1 slope		2:1 slope	
A	*C*	*B*	Cu yd	*B*	Cu yd
8″	8″	4′2″	0.77	5′10″	1.08
10″	10″	4′10″	0.92	6′8″	1.28
12″	12″	5′6″	1.08	7′6″	1.49
15″	15″	6′6″	1.34	8′9″	1.82
16″	16″	6′10″	1.42	9′2″	1.94
18″	18″	7′6″	1.61	10′0″	2.17
20″	20″	8′0″	1.76	10′10″	2.42
24″	24″	9′3″	2.15	12′6″	2.95
30″	30″	10′6″	2.61	15′0″	3.83
Q	4″ for 1½:1 slope				
	6″ for 2:1 slope				

to 21 in., No. 14 for 24 and 30 in., No. 12 for 36 to 54 in., No. 10 for 60 and 72 in. and No. 8 for 84 in. diameter. To prolong life, the invert section, or sometimes the entire pipe, is asphalt-coated. Perforated designs are available for use as subdrains.

1½:1 and 2:1 slopes											1½:1 slope		2:1 slope	
D	*E*	*G*	*H*	*I*	*J*	*K*	*L*	*M*	*N*	*P*	*F*	Cu yd	*F*	Cu yd
30″	4′0″	4′0″	5′6″	12″	3′6″	24″	1′6″	18″	2′0″	5′3″	3′0″	2.60	4′3″	3.16
36″	4′6″	4′3″	6′0″	12″	4′0″	24″	1′8″	18″	2′3″	5′11″	3′6″	3.35	5′0″	4.15
42″	5′0″	4′6″	6′6″	12″	4′6″	24″	1′10″	18″	2′6″	6′6″	4′0″	4.20	5′9″	5.25
48″	5′6″	4′9″	7′0″	12″	5′0″	24″	2′0″	18″	2′9″	7′2″	4′6″	5.19	6′6″	6.50
54″	6′0″	5′0″	7′6″	12″	5′6″	24″	2′2″	18″	3′0″	7′10″	5′0″	6.26	7′3″	7.88
60″	6′6″	5′3″	8′0″	12″	6′0″	24″	2′4″	18″	3′3″	8′5″	5′6″	7.43	8′0″	9.37
72″	7′6″	5′9″	9′0″	12″	7′0″	24″	2′8″	18″	3′9″	9′9″	6′6″	10.25	9′6″	12.99
84″	8′6″	6′3″	10′0″	12″	8′0″	24″	3′0″	18″	4′3″	11′0″	7′6″	13.49	11′0″	17.32

Quantities in cubic yards are for one end.

Headwalls for Pipe Culverts. Figure 43 shows a type in which the headwall is parallel to center line of embankment and earth fill forms a cone about the ends. Figure 44 shows headwall with wings flared at 45° with axis of pipe. This design is used with large-diameter pipes.

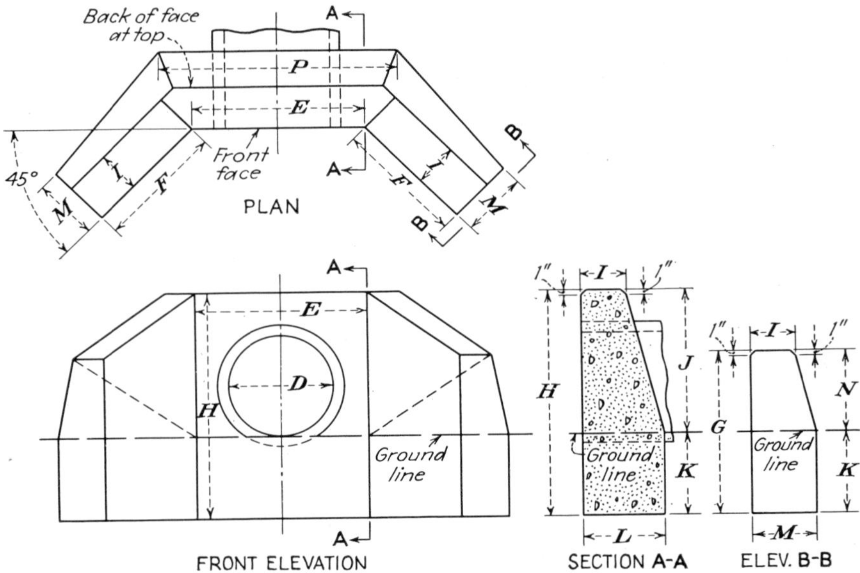

FIG. 44. Headwall and wings for pipe culvert. (*Massachusetts Department of Public Works.*)

Box Culverts. Reinforced concrete box culverts are commonly used for waterway areas exceeding 12 to 16 sq ft. The type may be used up to 12 by 12 ft, although twin boxes are frequently built instead of one box for the large sizes. *Arch culverts* are sometimes used for spans up to 20 ft and waterway openings up to 300 sq ft. *Slab-top* rein-

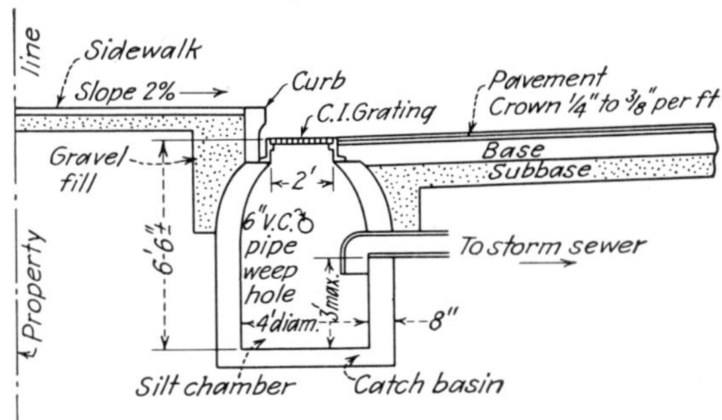

FIG. 45. Urban cross section at catch basin.

forced concrete culverts are often used for spans from 10 to 20 ft. For greater spans, T-beam and concrete deck-girder spans may be used. Structures with spans over 20 ft are usually classed as "bridges."[1] State highway departments have standard plans for culverts and small bridges which are used wherever practicable. *Catch basins* are installed in gutter sections to collect surface water from the street and sidewalk (Fig. 45).

[1] See "Standard Plans for Highway Bridge Superstructures," Bureau of Public Roads, available from Superintendent of Documents, Government Printing Office.

The outlet pipe is located above the bottom of the structure creating a basin in which silt and debris are caught before entering the pipe. Periodic clean-outs are required.

Unsurfaced Roads. Many miles of rural roads are of natural earth with little or no improvement. These are maintained in usable condition by occasional blading and by building up soft spots with earth or gravel (Fig. 46*a*). The condition of such roads depends on the nature of the soil and the effectiveness of drainage. Where the natural soil is gravelly or sand with some clay, the surface will be fairly stable; in silty or clay soils, the surface will be muddy during rains and dry out into ruts at other times.

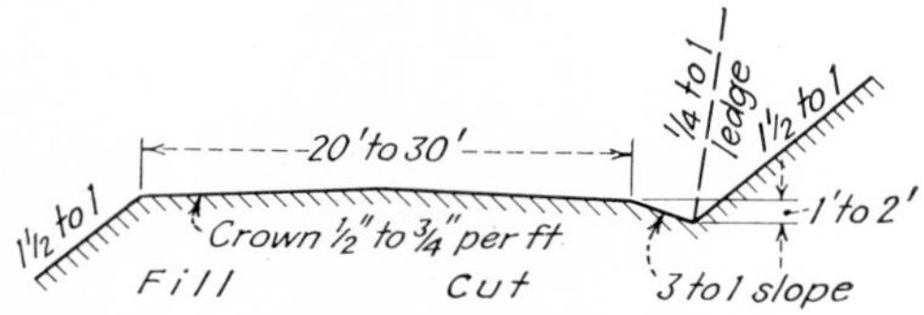

(a) Earth Road-Graded and Drained

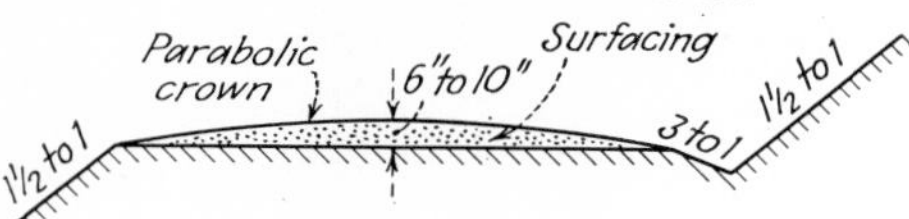

(b) Feather-Edge Section-Sand-Clay, Gravel or Stabilized

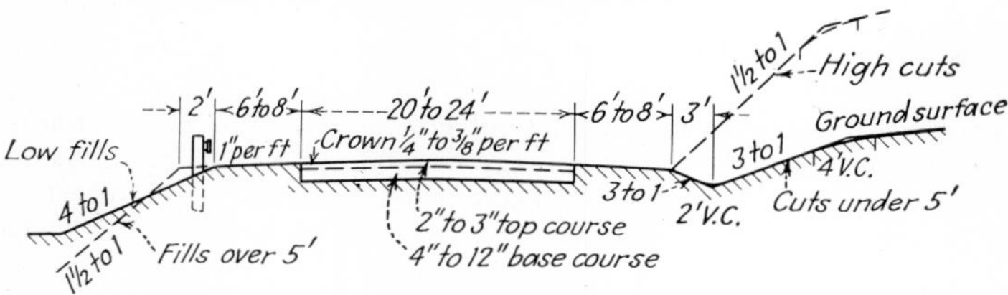

(c) Trench Section-Gravel, Stabilized, Road-Mix or Macadam

FIG. 46. Cross sections of two-lane low- and medium-traffic roads.

Untreated Surfaces. These are constructed of natural aggregates, crushed stone or slag, without bituminous or other surfacing. The *sand-clay* surfaces consist of an intimate mixture of sand and clay in about the proportions of 12 to 18 percent clay, 5 to 15 percent silt, and 65 to 80 percent sand. If the natural topsoil occurs in about these proportions, it can be used for the surface; or clay may be added to a sandy soil, or sand to a clayey soil. The sand and clay are mixed on the roadbed by harrowing and blading, consolidated by traffic or preferably by rolling, and shaped with blade graders to a crowned section (Fig. 46*b*). *Gravel surfaces* are commonly used for minor roads where natural gravel is plentiful. The gravel is spread on the subgrade, compacted by rolling or under traffic, and bladed to a crowned surface (Fig. 46*b*). The gravel may be placed in one layer, but better results are obtained if pit-run gravel is used in the base course and a selected gravel for the surface (Fig. 46*c*). Sizes up to 3 in. may be used in the base. The top course should be screened gravel under 1¼ in., and contain 10 to 15 percent binding material (passing 200-mesh sieve), such as iron oxide, limestone dust, clay, or loam. If a satisfactory binder is not present in the natural gravel, it may be obtained from another source and mixed into the top course during construction. The surface course is 2 to 3 in. thick, and the base course 4 to 18 in. thick. Thick courses are required on soft subgrades and in places where frost heave is likely to occur. The surface is maintained by

dragging and blading. When it becomes rough, it may be scarified and reshaped. New gravel is needed from time to time to replace that washed away or blown away as dust. Gravel surfaces are dusty and often develop corrugations and potholes. These conditions can be partly prevented by the choice of a cohesive binder as described above, by treating the surface with a light oil to lay the dust, or by sprinkling it with calcium chloride to keep the surface damp. Two or more applications of ½ lb per sq yd are required per season. Gravel roads form excellent bases for bituminous surface treatments. If surface treatments are to be applied, a highly cohesive soil binder is undesirable since it may cause softening of road under the treatment due to an accumulation of capillary moisture.

Stabilized roads and bases [1] are obtained by "stabilizing" such natural materials as sand-clay mixtures and gravels by controlling their gradation, amount of binder soil, moisture content, and thoroughness of compaction. Admixtures are usually employed to increase and to retain the stability obtained by the above controls. Natural gravels are often lacking or overabundant in certain sizes. The grading is corrected in the stabilizing process by combining materials from different sources and mixing them thoroughly on the roadbed by harrowing and blading, or in mechanical mixers which travel along the roadway. Grading limits for stabilized mixtures have been standardized by the AASHO.[2] These are shown in Table 19. The liquid limit and the plasticity index

TABLE 19. GRADING REQUIREMENTS FOR SOIL-AGGREGATE MATERIALS [1]

Sieve designation	Percentage by weight passing square mesh sieves					
	Grading A	Grading B	Grading C	Grading D	Grading E	Grading F
2-in.	100	100				
1-in.	.....	75–95	100	100	100	100
⅜-in.	30–65	40–75	50–85	60–100		
No. 4	25–55	30–60	35–65	50–85	55–100	70–100
No. 10	15–40	20–45	25–50	40–70	40–100	55–100
No. 40	8–20	15–30	15–30	25–45	20–50	30–70
No. 200	2–8	5–20	5–15	10–25	6–20	8–25

[1] From AASHO Standard Specifications for Materials for Soil-aggregate Subbase, Base and Surface Courses, Designation M147-55.

NOTES: Fraction passing No. 200 sieve shall not be greater than two-thirds that passing No. 40 sieve. Fraction passing No. 40 sieve shall have liquid limit not greater than 25 and plasticity index not greater than 6. Any of gradings A to F may be specified for subbase or base course material. Lower percentages passing No. 200 sieve may be required where experience shows them necessary to prevent damage from frost action. Any of gradings C to F may be specified for surface-course material. Where a soil-aggregate surface is to be maintained for several years without bituminous surface treatment, a minimum of 8 percent passing No. 200 sieve should be specified for gradings C, D, or E, and a maximum liquid limit of 35 and a plasticity index range of 4 to 9 specified for gradings C, D, E, or F.

(see p. 2–88) are also a part of the specifications. The plasticity index is a measure of the cohesive qualities of the binder soil. In general, dense graded mixtures, such as sand-clay, require a higher plasticity index than the coarser mixtures. Likewise more cohesion is desirable in an untreated surface course than in a base course with a bituminous surface. The upper value of the liquid limit is that at which softening may occur. Common admixtures applied to stabilized surfaces are calcium chloride, liquid asphalts, tars, and portland cement. The calcium chloride, being deliquescent, absorbs moisture

[1] See Granular Stabilized Roads, *Wartime Road Problems*, 5, February, 1943, and Soil-bituminous Roads, *Current Road Problems*, 12, September, 1946, "Soil Stabilization," *Bull.* 69, 1953, Stabilization of Soils, *Bull.* 98, 1955, and "Soil and Soil-aggregate Stabilization," *Bull.* 108, 1955, all by Highway Research Board; also brochures on stabilized roads by Calcium Chloride Association, Asphalt Institute, Barrett Division, Allied Chemical and Dye Corp. (tars), and Portland Cement Association.

[2] AASHO Standard Specifications—Base Course M56, Surface Course M61. Also ASTM Tentative Specification D1241.

from the air, thus keeping the soil moist and preserving the natural cohesion. Bituminous materials waterproof the mixtures and add cohesion. Portland cement sets the mix in a weak concrete. Calcium chloride may be applied to the surface or mixed with the aggregate; other products are thoroughly mixed into the road during construction.

Traffic and water-bound macadam surfaces are composed of crushed stone or slag and a natural binder, such as cementitious stone dust or screenings. In the traffic-bound process, layers of crushed stone 1 to 2 in. thick are impregnated with subgrade material or screenings as they are compacted under traffic. In the water-bound process, common in the preautomobile era, a slurry of stone screenings and water was pressed into a layer of broken stone under the action of the roller. Upon drying, the slurry set up into a weak mortar, binding the stones together. This type is unsuited for any appreciable motor traffic unless surface treated with bituminous materials.

Bituminous pavements are of four general types: *surface treatments*, consisting of a thin bituminous layer usually less than ¾ in. over a prepared road surface; *road mixes* in which the bituminous material is mixed with the aggregate by manipulation on the roadbed; *bituminous macadam* in which the top course of broken stone is penetrated with bituminous binder; and *bituminous concrete* and *sheet asphalt* in which the aggregates and bituminous material are mixed under controlled conditions at a central plant.

Bituminous surface treatments [1] (Fig. 47*a*) are used to some extent as dust layers, but their primary use is for binding and waterproofing macadam, gravel, and soil surfaces. In maintenance they are used on all types of bituminous surfaces, and on badly worn cement concrete, stone, and brick types. Light asphaltic oil of the slow-curing type (SC-1 or SC-2; see Table 12) is suitable as a dust layer on earth or gravel roads. An application of from 0.3 to 0.7 gal per sq yd (gpsy) is required depending upon the amount which the surface will absorb. Tars, cutback asphalts, asphalt emulsions, and soft asphalt cements are used for surface treatments (see Tables 11 to 16). Some of these products may be applied cold; others must be heated. Practice in applying surface treatments varies considerably in different localities and for different materials. The essential steps for an initial treatment are as follows: (1) Patch, shape, and compact the untreated surface to obtain a smooth surface; it is difficult to correct irregularities after treatment. (2) Sweep off any loose dust without disturbing the natural bond in the surface. (3) Apply prime coat, MC-0, or MC-1 or RS-1 for asphalts and RT-1, 2, or 3 for tars. The application varies from 0.2 to 0.5 gpsy depending upon the amount that will be readily absorbed, less being needed for bonded than for loose surfaces. The prime coat is allowed to penetrate for at least 24 hr. (4) Apply binder coat of 0.25 to 0.50 gpsy of RC-1 or RC-2 or RS-1 for asphalts, and RT-5 or 6 for tars. (5) Cover binder coat with 25 to 50 lb per sq yd of fine gravel or stone, usually finer than ½ in., drag with broom-drag and roll (a rough rule is that 10 lb of cover stone is required for each 0.1 gpsy). (6) Apply seal coat of 0.125 to 0.25 gpsy of RC-1 (or RT-6), cover with 10 to 15 lb per sq yd of cover stone under ⅜ in., broom-drag, and roll. This step may be delayed until the surface shows signs of raveling. All bituminous applications are made with a pressure distributor. Maintenance seal coats and additional cover stone will be required at intervals of from 3 to 5 years. A sharp granular cover is preferable to a fine sand cover because the former gives a more nonskid texture. Usually all treatments are either of asphalt products or of tar, although tar prime coats are sometimes used with asphaltic binders. The tar primer has high penetrating qualities and is also toxic to vegetation that may be seeded in the base course. For absorbent aggregates, MC products may be substituted for RC in the binder coat. A hot application of soft asphalt cement (150 to 200 penetration) may be substituted for a cutback in binder and seal-coat applications.

[1] "Asphalt Handbook" and "Construction Specifications," both by The Asphalt Institute. Also Specifications for Bituminous Surface Treatment of Highways, AASHO: "Tarvia Manual" (tars), Barrett Division, Allied Chemical and Dye Corp.

Road-mix surfaces [1] are of two general types: "graded aggregate" in which bituminous material is mixed on the roadbed with a well-graded gravel, and "macadam aggregate" in which the aggregate is broken stone. The first type derives stability from compaction to high density, whereas the second type depends upon the strength developed by the interlocking of stone fragments under rolling. The road-mix process is also known as "retread," "oil-processed," or "mixed in place." It produces a wearing surface from 1 to 4 in. thick, 2 to 2.5 in. being most common (Fig. 46*c*). In brief, the steps are (1) shape base (rolling if necessary) and apply a prime coat of bituminous material, windrowing surface material to the roadsides if necessary to expose the base; (2) spread surface aggregate in sufficient quantity to produce the compacted thickness desired in the final road; (3) treat aggregate with bituminous binder using pressure distributor and mix with harrows, road graders, and drags; (4) repeat step 3 once or twice until the total amount of binder required has been incorporated; (5) blade mixture to required crown and compact by rolling; (6) apply seal coat of bituminous material, cover with small aggregate, and roll.

Aggregate for the *graded aggregate type* may be obtained by scarifying the existing road or by adding new material. The grading of such material should fall within the following limits: passing 1-in. screen, 100 percent; No. 4 sieve, 50 to 70 percent; No. 10 sieve, 35 to 60 percent; and No. 200 sieve, 5 to 14 percent. If necessary, materials from two sources may be blended by mixing with harrows and blades on the roadbed. When asphalt products are used, the primer of MC-0 is applied at the rate of 0.3 to 0.5 gpsy, depending upon the amount which the base will readily absorb. (For open- or porous-textured bases, MC-1 or MC-2 may be substituted for MC-0.) The binder of MC-3 is applied at rate of 0.5 gpsy and mixed into the aggregate. Additional applications of 0.5 gpsy are mixed in until the total required amount of binder has been obtained. For a 2-in. wearing course 1 to 1.5 gpsy are required, depending upon absorption characteristics of aggregate. After the mixing has been completed and the surface shaped and rolled, a seal coat of 0.2 to 0.3 gpsy of MC-3 is applied, then lightly covered with sand or fine stone aggregate and rolled. Using tar, the primer is RT-1, 2, or 3 applied at rate of 0.3 to 0.5 gpsy and the binder is RT-5 or 6 applied in applications of 0.3 gpsy until the total amount required is mixed in. (A rough rule is that 0.3 gpsy is needed per inch of depth.) The seal coat of RT-6 or 7 is applied at rate of 0.17 gpsy and covered with sand, pea gravel, or stone chips.

Aggregate for the *macadam aggregate type* is crushed stone or crushed gravel. For the coarse type, the size range is from 1½ in. to No. 8 sieve; for the fine type it is from ¾ in. to No. 8 sieve. When asphalt products are used, the prime coat is 0.3 to 0.5 gpsy of MC-0. The binder is RC-3 applied in applications of 0.3 to 0.4 gpsy each, followed by mixing. For a 2-in. wearing course, two applications and mixings are required. After the final mixing the surface is shaped to the desired cross section and rolled. Within a week, 15 to 18 lb per sq yd of fine aggregate is spread into the surface voids and followed by a seal coat of 0.2 to 0.3 gpsy of RC-3, broom-dragged and rolled. An asphalt emulsion, such as MS-2, may be substituted for the RC binder. Emulsions can be applied to damp surfaces, whereas cutbacks require a dry surface for best results. For tar types, or "retread," the primer is 0.25 gpsy of RT-2. The binder is 0.8 gpsy of RT-6, 7, or 8 for a 2.5-in. layer of loose aggregate. After mixing, the surface is bladed to the desired cross section and rolled. Choke stone (No. 4 to No. 16) is swept into the surface voids, and a seal coat of 0.3 gpsy of RT-6, 7, or 8 is applied and covered with (No. 4 to No. 16) stone and rolled. A final seal coat of 0.25 gpsy is applied, covered with fine aggregate (No. 4 to No. 16), and rolled.

There are many variations of the road-mix process. The precise amount and kind of

[1] Construction Specifications RM-1, 2, 3 of Asphalt Institute. Also see Tables 12 to 16, and "Tarvia Manual," Barrett Division, Allied Chemical and Dye Corp.

bituminous material to be used must be determined by the conditions of the job. As a rule, the more fluid slow-curing and medium-curing products are used with dense graded aggregates, and the more viscous and rapid-curing types with open mixes. More bituminous material is needed per square yard for a given thickness of graded mix with a high percentage of fines than for open types. Preliminary tests must be made to determine the needs of a particular aggregate and mixing process.[1] Traveling plants [2] have been developed which take in aggregates from the roadbed, mix them with bituminous material, and discharge the mix onto the base. By this means, a more uniform mix is obtained, and the lengthy process of mixing with blade graders and harrows is eliminated. A maintenance seal coat of 0.125 to 0.25 gpsy of the same product used in original construction is required on road-mix surfaces every 3 to 5 years.

Rock asphalt surfaces are used in localities where suitable natural rock asphalts are available. Two common types are Uvalde (Texas), which is an impregnated shell limestone, and Kentucky, which is an impregnated sandstone. The rock asphalt is crushed and blended to give the desired asphalt content (6 to 8 percent by weight). If the rock is deficient in asphalt, a heavy asphaltic oil is added. If the natural asphalt is too hard, as is the Uvalde, a flux oil is added to soften it. The crushed rock asphalt is spread uniformly on a firm base previously primed with RC-0 or RC-1, shaped and rolled.

Bituminous macadam is a broken-stone wearing surface penetrated with bituminous material applied by a pressure distributor. The term is used rather loosely to cover a wide range of types from a light surfacing for a residential street or secondary road to

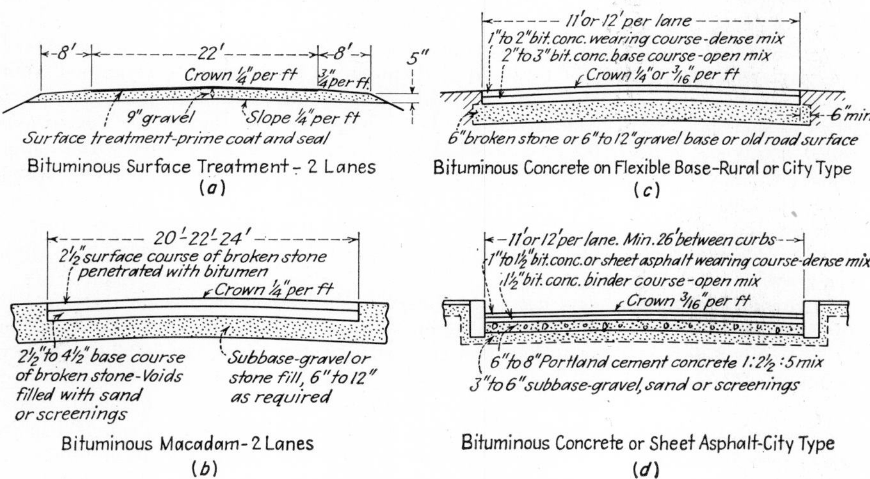

FIG. 47. Bituminous pavement cross sections.

a layered roadway nearly 2 ft thick. In certain states where strong tough stone is available, bituminous macadam is used successfully on heavy-traffic roads. The heavy-duty design shown in Fig. 47*b* is built up as follows: [3]

1. Subgrade is compacted and shaped parallel to the finished surface.
2. A minimum foundation course of 12 in. of gravel is spread and compacted in two layers. If field stone is prevalent, the foundation may be of 8 to 10 in. of stone fragments

[1] A Symposium on Research Features of Flexible-type Bituminous Roads, *Highway Research Board Proc. 14th Ann. Meeting,* Part II, 1934.
[2] "Bituminous Construction Handbook," Barber-Greene Co.
[3] Briefed from Massachusetts specifications for bituminous macadam.

roughly fitted together and with the voids filled with screenings or sand. An old roadbed may also serve as foundation where such exists.

3. A base course of crushed stone 2½ to 1¼ in. size is spread and compacted to 4½ in. thickness. The base stone should have a percent of wear of not over 35 by Los Angeles abrasion test. Stone is spread uniformly by means of a spreader box drawn behind the stone truck or by a machine spreader. The spreader box has an adjustable slot which may be set to spread the desired loose thickness. The power spreader strikes off the stone with a screed to the desired profile. The compacted depth is usually 20 to 25 percent less than loose depth. The shoulders are built up with gravel along the edges to form a trench to retain the stone. The rolling is commenced with one-half the rear wheel of the roller overlapping the shoulder, and then proceeds longitudinally from edges to center, thus preserving the desired crown. The rolling should continue until the wheel marks are rolled out but not to the point of crushing aggregate. Three-wheel rollers weighing not less than 12 tons are used. Stone screenings or sand is swept into voids of base course after rolling.

4. A wearing course of crushed stone 2 to ¾ in. size is spread to give a compacted thickness of 2½ in. Surface stone should have a percent of wear of not over 25 in the Los Angeles abrasion test. It is rolled lightly at this stage to lock the stone but not sufficiently to crumble any pieces. The voids should be open and free from all dust, dirt, leaves, or twigs.

5. Surface is penetrated with 2 gpsy of asphalt cement heated to 300° to 350°F. Penetration grade 85 to 100 is used for warm weather conditions, and 100 to 120 for cool weather. The surface must be dry at time of application of the asphalt cement.

6. Keystone of ½ to No. 8 size is spread over surface to fill surface voids, and the course thoroughly rolled while still warm.

7. After the road has set, usually within a day or two, all excess keystone is swept off and a prime coat of 0.375 gpsy of the same asphalt cement is applied. Cover stone of ½ to No. 8 size is spread and rolled.

If asphalt emulsion is used instead of hot asphalt cement, smaller size aggregate should be used in the wearing course, since the emulsion is more fluid and the voids should be smaller. Also the aggregate should be damp rather than dry. For a 1½-in. wearing course using an RS-1 emulsion,[1] a 1½- to ¾-in. aggregate is penetrated with 0.5 to 0.6 gpsy. This is covered with 20 to 30 lb per sq yd of ½ to No. 8 size stone, and followed by a second penetration of 0.6 to 0.7 gpsy covered with 10 to 15 lb per sq yd of No. 4 to No. 16 stone. A seal coat of 0.25 to 0.35 gpsy is applied and covered with 8 to 12 lb of No. 4 to No. 80 cover stone. Bituminous macadam may also be bound with cutback, an RC-5 being substituted for asphalt cement when the aggregate runs up to 2-in. size. For smaller aggregate, RC-4 or 3 may be used.

If tar is used in the wearing course the process is briefly as follows:[2]

1. Spread aggregate (2- to 1-in. size) and roll until stones are locked and bound together.

2. Apply 1.7 gpsy of RT-12 (temperature 175° to 250°F).

3. Fill surface voids with ½-in. to No. 8 choke stone; sweep and roll until surface voids are filled.

4. Apply first seal coat of 0.75 gpsy of RT-12.

5. Cover with 30 lb per sq yd of ½-in. to No. 8 stone, and sweep and roll cover into surface.

6. Open to traffic for 30 to 60 days and then sweep and apply second seal coat of 0.33 gpsy of RT-6 or RT-8 and cover with 25 lb of ⅜ to No. 8 aggregate per square yard.

[1] Modified Penetration Emulsified Asphalt Surface Course (MP-1), "Construction Specifications," The Asphalt Institute.

[2] Briefed from "Tarvia Manual," p. 19.

With tar type, a tight impervious seal is important to preserve the binding qualities of the tar. Best results are obtained with any type of bituminous macadam when laid in the early summer, thus allowing the bituminous material to work into the road during the hot weather. Maintenance seal coats are required at periods of from 3 to 5 years.

Bituminous Concrete.[1] This type includes plant-mix bituminous surfaces consisting of graded aggregates, mineral filler (fine material passing No. 200 sieve), and a bituminous cementing agent. When the mix is bound with asphalt, it is properly an asphaltic concrete. Tar concretes are less common. They are used for special purposes, such as on airfields, since tar is more resistant to concentrated spillage of aircraft fuel. Bituminous mixes are classified by maximum size of aggregate and by type of grading. The three principal groups are coarse aggregate bituminous concrete (under 2½ in.), fine aggregate bituminous concrete (under ½ in.), and sheet asphalt (under No. 4 sieve). Dense graded mixes have well-graded aggregates and low void content. Open mixes usually predominate in coarser sizes of aggregate and have high void content. Hot mixes are those in which the asphalt and aggregates are heated to obtain required consistency for mixing and placing. Those in which fluidity is obtained by the use of cutbacks or emulsions are called *cold mixes*. Dense mixes are usually hot mixes; open mixes may be hot or cold. Hot dense-graded mixes are most commonly used for primary highways and city street paving. Cold mixes are more common for secondary roads and patching mixtures. Open mixes are employed in binder and leveling courses (Fig. 47*c* and *d*). When open mixes are used in the surface course a seal coat is desirable to prevent water penetration. Certain aggregates have a greater affinity for water than asphalt. When water is present, the asphalt will strip from such aggregates. This tendency can be overcome by designing a dense, waterproof mix, by introducing an additive with the asphalt, or for surface treatments by precoating stone with a light oil.

Hot-mix Bituminous Concrete.[2] Typical specifications for hot-mix bituminous concrete recommended by The Asphalt Institute are shown in Table 20. Nos. I and II are suitable for lower (base and leveling) courses and for intermediate (binder) courses. No. III is suitable for either surface or binder course and Nos. IV and V are surface course mixes. The several state highway departments and local highway agencies have specifications which may differ in some details from those of The Asphalt Institute. These should be consulted for practice in local areas. The specifications provide only upper and lower limits of grading and asphalt content. Within these limits a specific mix (job mix) is designed using the aggregates and asphalt to be supplied on the work. This mix must possess sufficient stability to resist rutting or shoving under traffic in warm weather, yet be pliable enough to adjust to temperature changes and to conform to slight movements of the subgrade without cracking. Long life and a nonskid surface texture are also desired.

Stability depends upon the characteristics and gradation of the aggregates and the degree of compaction (density) of mix. It is also influenced by the amount and consistency of asphalt. A high asphalt content or the use of a soft asphalt will increase plasticity and therefore decrease stability. Durability depends upon the nature of asphalt and the thickness of asphalt films binding the aggregates. In general, the highest asphalt content that can be used without seriously impairing stability is desirable for long life. The desired surface texture can be obtained by control of grading. Since the asphalt is the slippery component of the mix, an excess should be avoided. Sufficient voids should be left in the mix to allow for expansion of asphalt and for some consolidation to take place under traffic without flushing asphalt to the surface. The job mix is determined from trial mixes. Aggregate grading and asphalt content are varied until

[1] "Manual on Hot-mix Asphaltic Concrete Pavement," and "Construction Specifications," both by The Asphalt Institute.

[2] "Mix Design Methods for Hot-mix Asphalt Paving," The Asphalt Institute. Symposium: Investigation of the Design and Control of Asphalt Paving Mixtures and Their Role in Structural Design of Flexible Pavements, *Research Report* 7-*B*, Highway Research Board, 1949. "Bituminous Paving Mixtures: Fundamentals for Design," *Bull.* 105, Highway Research Board, 1955.

TABLE 20. TYPICAL SPECIFICATIONS FOR ASPHALT CONCRETE MIXES—THE ASPHALT INSTITUTE[1]

Mix No.	I	II	III	IV	V
Courses for which recommended	Lower or intermediate		Intermediate or surface	Surface	
Min thickness individual course	3″	2½″	2″	1½″	1″
Max thickness individual course	3½″	3½″	3″	2½″	2″

Mineral Aggregate Combination

Constituent	Passing sieve	Retained on sieve	Percent by weight	Percent by weight	Percent by weight	Percent by weight	Percent by weight
Coarse aggregate	2″	1″	15–45				
	1½″	¾″	.....	14–48			
	1″	½″	3–45	.....	17–52		
	¾″	⅜″	.....	3–45	.....	18–50	
	½″	⅜″	.....	.....	.....	.....	8–39
	½″	No. 4	.....	.....	6–42		
	⅜″	No. 4	.....	.....	.....	3–36	8–45
	No. 4	No. 10	5–15	5–15	5–15	9–22	9–27
Subtotal	....	No. 10	60–80	55–75	55–70	50–65	50–65
Fine aggregate filler	No. 10	No. 40	3–19	3–21	4–20	5–22	5–22
	No. 40	No. 80	5–22	6–25	8–25	9–27	9–27
	No. 80	No. 200	3–15	3–16	4–16	5–18	5–18
	No. 200	....	3–5	4–6	4–7	5–8	5–8
Subtotal	No. 10	....	20–40	25–45	30–45	35–50	35–5
Total mineral matter			100	100	100	100	100

Total mix

	I	II	III	IV	V
Total mineral aggregate	93.5–95.5	93.0–95.0	93.0–95.0	92.0–94.0	92.0–94.0
Asphalt cement (bitumen)[2]	4.5– 6.5	5.0– 7.0	5.0– 7.0	6.0– 8.0	6.0– 8.0
Total mix	100	100	100	100	100

[1] From "Manual on Hot-mix Asphaltic Concrete Pavement," The Asphalt Institute.
[2] For highly absorptive aggregates the upper limit may be raised.

desired conditions of stability, void content, and density are obtained. The more common stability measuring procedures are Hubbard Field, Marshall, Hveem, and triaxial. They are conducted at 140°F to simulate hot pavements in summer sun. The job mix is converted to a plant-mix formula which specifies for each batch the weight of aggregate to be drawn from each bin and the weight of asphalt to be added. The following tolerances are allowed from the job mix: coarser than No. 4 sieve ±5 percent; No. 4 to No. 10 sieve ±3 percent; No. 10 to No. 200 sieve ±2 percent; under No. 200 sieve ±1.0 percent; asphalt ±0.3 percent.

Construction Procedure.[1] The desired proportions are obtained at the plant by combining sizes of coarse aggregate, sand, and filler. The aggregates are assembled and blended, run through a dryer and fed over screens into elevated bins from which the pro-

[1] "Bituminous Construction Handbook," Barber-Greene Co. "Asphalt Handbook," The Asphalt Institute.

portions for each batch are weighed into the mixer (pug mill). The hot asphalt is then added. The mixing time is usually 45 to 60 sec. Long mixing times and excessive temperatures are to be avoided since they tend to harden the asphalt and shorten the life of pavement. The batches are dumped into trucks, transported to the job, and placed while still hot. The mixing temperature should not exceed about 325°F and the temperature of application should not be below 225°F. The trucks dump the mixture into a hopper in the rear of a mechanical spreader where it is reworked by revolving paddles or a screw device, and fed uniformly to the strike-off screed. The latter is set to spread the desired uncompacted thickness. Some spreaders tamp as well as spread the mixture. The width of spread is usually adjustable from 10 to 14 ft. To obtain specified compaction the pavement is laid in layers, usually between 1 and 2 in. thick, and rolled with tandem two-wheel or three-wheel rollers weighing from 10 to 15 tons. Rolling starts at the edges and works toward the center, overlapping at least one-half wheel track on each pass. It continues until all wheel marks are ironed out and the design density has been attained. When high densities are desired, as for heavy aircraft, rubber-tired rollers are used for compaction and tandem rollers for smoothing.

Bituminous concrete can also be produced in continuous-mix plants, where the proportions of aggregate are controlled by volume as they pass through adjustable gates. The asphalt is incorporated in the mix as it moves continuously over the mixing paddles. A continuous flow of mix is fed into waiting trucks. These mixes are usually placed in base and binder courses where the control of the mix is somewhat less critical than for surface courses. For secondary-road construction, a roughly controlled mix is produced by a traveling plant which picks up graded aggregate from the roadbed.

Sheet asphalt (Fig. 47*d*) is laid in two courses. The bottom, or binder course, has the following composition limits: aggregate, between 1-in. screen and No. 4 sieve, 52 to 72 percent; No. 4 to No. 10 sieve, 8 to 20 percent; passing No. 10 sieve, 15 to 35 percent; bitumen, 4 to 6 percent. The surface course limits are: aggregates, between No. 10 and No. 40 sieve, 10 to 40 percent; No. 40 to No. 80 sieve, 20 to 45 percent; No. 80 to No. 200 sieve, 12 to 32 percent; passing No. 200 sieve, 10 to 20 percent; bitumen, 9.5 to 12 percent; voids in compacted mix, less than 5 percent. Sheet asphalt is a dense hot mix prepared under carefully controlled proportioning methods in a central mixing plant. It is spread, rolled, and compacted in the same manner as other bituminous concretes.

Design of Thickness of Flexible Pavements. The thickness of a flexible pavement depends upon the magnitude and frequency of load applications, tire pressure and wheel arrangements, the physical characteristics of pavement and base materials, and the strength of the subgrade. Many design procedures [1] have been proposed. Those most commonly used are based upon tests applied to subgrades or to subgrade materials. The relation between these test results and the thickness required is obtained from experience with pavements in service, and from test road sections submitted to controlled traffic.[2]

A consolidated design chart [3] based upon several design criteria and correlated with subgrade soil classifications is given in Fig. 48. The Highway Research Board Classifications and Group Index are explained on pp. 2–88 and 2–89. The "Unified Soil Classification" is an airfield classification given on p. 2–138. The Civil Aeronautics Administration classification is given on p. 2–140. The resistance value is derived from Hveem Stabilometer test.[4] The California Bearing Ratio test is described on p. 2–140. The bearing value [5]

[1] Thickness of Flexible Pavements for Highway Loads, *Current Road Problem* 8-*R*, and Design of Flexible Pavements, *Research Report* 16-*B*, both by Highway Research Board. Also see airport section, p. 2–140.

[2] Such as the WASHO road test reported in *Highway Research Reports* 18 *and* 22, Highway Research Board.

[3] "Thickness Design—Flexible Pavements for Streets and Highways," The Asphalt Institute, 1956.

[4] See above references, and also CARMANY, R. M., and F. N. HVEEM, Factors Underlying the Rational Design of Pavements, *Highway Research Board, Proc.* 28*th Ann. Meeting*, 1948.

[5] See Method of Test Designation D1195, ASTM Standards, 1955; and BENKLEMAN, A. C., and R. F. OLMSTEAD, A Cooperative Study of Structural Design of Non-rigid Pavements, *Public Roads*, vol. 25, No. 2, December, 1947.

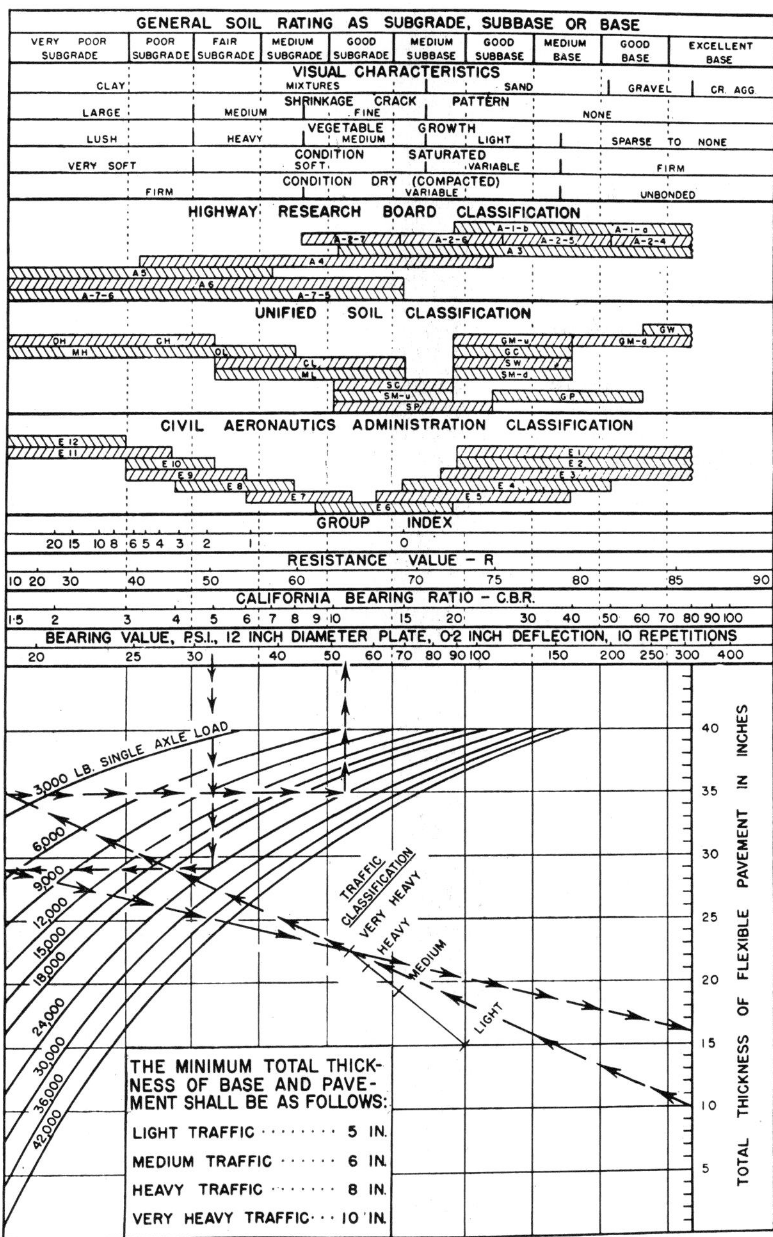

FIG. 48. Flexible pavement thickness design chart. (*The Asphalt Institute.*)

of a subgrade or subbase is obtained by applying increments of load to a circular plate and measuring resulting deflections of the plate. A certain maximum allowable deflection is adopted for design depending upon the manner in which the test is conducted. It is usually taken as 0.1 or 0.2 in. deflection.

To find total thickness of flexible pavement from Fig. 48, the subgrade is evaluated by one of methods given in upper part of the chart. A line is then dropped vertically to intersect the design axle load curve, then carried horizontally to the left margin of the chart. From that intersection it is drawn through the appropriate traffic classification to the right-hand scale where the total thickness is read. The traffic classification is by volumes per lane per day. Light traffic is defined as fewer than 25 passenger cars and light trucks of which fewer than 5 are commercial trucks and busses. For medium traffic these limits are 500 and 25. For heavy traffic the number of passenger cars and light trucks is unlimited but the number of commercial trucks and busses must be less than 250. For very heavy traffic the number of both classes of vehicles is unlimited.

The total thickness is divided into surface, base, and subbase courses, each of which may consist of one or more layers of materials (see Fig. 47). The minimum total thickness of surface and base for each class of traffic is noted on the chart. The type of surface will vary from surface treatment or road mix for very light traffic to bituminous concrete for heavier traffic classes. When asphaltic concrete is used, The Asphalt Institute recommends 3-in. thickness for medium and heavy traffic, and 4 in. for very heavy traffic. The base will vary from stabilized gravel for light traffic to crushed stone or macadam bases for heavy traffic. Subbases may be of different types of material as economy warrants, so long as the depth to the top of each layer is at least equal to thickness required by that layer to support the design load. The materials must also meet requirements for free drainage and absence of frost susceptibility.

Example: Design traffic 5,000 passenger cars and 300 commercial trucks and busses per lane per day; design axle load 24,000 lb; subgrade CBR = 5; available subbase materials CBR = 15 and CBR = 30; base course material CBR = 80.

For subgrade of CBR = 5, axle load 24,000 lb, and very heavy traffic, the total thickness from Fig. 48 is 16 in. The chart specifies a minimum of 10 in. for pavement and base. The surface will be of 4-in. asphaltic concrete leaving 6 in. for base course (CBR = 80). The remaining 6 in. (16 − 10) is subbase. Using the chart in the reverse order, we find that the minimum CBR for very heavy traffic at a 10-in. depth is 11. Therefore either the CBR = 15 or the CBR = 30 material is satisfactory for full 6-in. depth of subbase, provided it meets drainage and frost-resistant requirements. The CBR = 30 is of better quality and would be used unless the CBR = 15 material was more economical. When a bituminous macadam base is used, an asphaltic concrete surface course of 3 in. laid in two layers is usually satisfactory (see Fig. 47*c*).

The subject of flexible pavement design is complex and still in the development stage. The reader is directed to references quoted in footnotes and to current literature for a more complete treatment of the subject.

Portland Cement Concrete. This is a rigid pavement composed of slabs of concrete which possess elastic properties and can be subjected to structural analysis. Loads are distributed through slab action over large areas of subgrade. The design theory is complicated, however, by the flexible nature of the subgrade and ever-changing climatic conditions.

Design.[1] The principal causes of stress in concrete pavements are wheel loads, restrained expansion, contraction and warping, uneven subgrade support, restrained moisture-volume changes and horizontal forces due to braking and acceleration. Moisture-volume changes and horizontal forces are usually not critical and are commonly

[1] See also section of design of concrete pavements for airports, p. 2–144.

neglected. Uneven subgrade can be avoided by construction methods which eliminate excessive settlement. Well-compacted granular bases or stabilized subgrades are needed to ensure uniform support. Fills must be thoroughly compacted and all soft muck or peat removed or stabilized to avoid settlement. Some variation in subgrade support will result from warping due to differential temperatures between top and bottom of slab. On hot days, the surface becomes warmer than the bottom causing the slab edges to curl downward. On cool nights after warm days the slab edges will tend to curl upward. The weight of the concrete restrains this warping, thus introducing stresses in the concrete. The friction of the subgrade restrains horizontal movement, inducing compression stress when the slab is expanding and tensile stress when it is contracting. Since concrete is weaker in tension than compression, the contracting condition is more critical. Temperature stresses increase with the length of slab and are most effectively controlled by choice of slab dimensions. Slabs are formed by transverse and longitudinal joints built into pavement. Distributed reinforcement is also used to control cracks between joints. This reinforcement does not add materially to strength of concrete but serves to keep cracks tightly closed.

Wheel-load stresses are dependent upon magnitude and position of the load. The maximum tension stress occurs when the wheel load is applied along a free edge or at a free corner (not connected to an adjacent slab). The smaller stresses occur in the interior of the slab at some distance from any joint or crack. This difference is the basis for thickened-edge designs in which wheel-load stresses are approximately equalized by varying the thickness of pavement cross section. Warping stresses increase with depth of slab, and often add to wheel-load stresses, thus counteracting some of the advantage of thickening the edge. Although many two-lane roads have been built with thickened edges, the trend is to use uniform thickness designs 8, 9, or 10 in. thick, especially for multilane highways.

A rational theory for the design of concrete pavements has been developed by H. M. Westergaard.[1] In the original theory three conditions of loading were investigated; at a free corner, at a free edge, and in the interior of slab. The load is spread over a circle of area equivalent to that of the tire imprint. The subgrade reactions (in psi) on the slab under load are assumed to be everywhere proportional to the deflection of the slab (in inches). The proportionality factor k is called modulus of subgrade reaction and is expressed in psi per inch. It may be evaluated in the field by a plate bearing test, in which the pressure on the plate is divided by the resulting deflection. Computations for k are usually made from data obtained at 0.05 or 0.10 in. deflection. Equivalent CBR ratings for k values of 100, 200, 300, and 400 are roughly 3, 10, 27, and 38, respectively.

The Westergaard formulas are written for stress in terms of wheel load, thickness of slab, radius of circular loaded area, modulus of subgrade reaction, modulus of elasticity, and Poisson's ratio. Over the years modifications have been made in the formulas to make them conform more closely to research findings,[2] or to take into account variations in subgrade support due to warping. The design curves in Fig. 49 are based upon a modified form of Westergaard formulas adopted by the Portland Cement Association.[3] These curves give thickness required for loads placed at free corners of slabs which are not connected; i.e., "unprotected." Unless otherwise specified, the thickness d is the uniform thickness of slab. If an equivalent thickened edge section is desired, the PCA recommends an interior thickness $d_i = 0.85d$. The edge thickness depends upon width

[1] Stresses in Concrete Pavements Computed by Theoretical Analysis, *Public Roads*, April, 1926, and Analytical Tools for Judging Results of Structural Tests of Concrete Pavements, *Public Roads*, December, 1933. Also see airport pavement design, p. 2–144.

[2] The Structural Design of Concrete Pavements, *Public Roads*, October, November, and December, 1935; September and October, 1936; and April–May–June, 1943. Also Application of the Results of Research to the Structural Design of Concrete Pavements, *Public Roads*, July and August, 1939.

[3] "Concrete Pavement Design," Portland Cement Association.

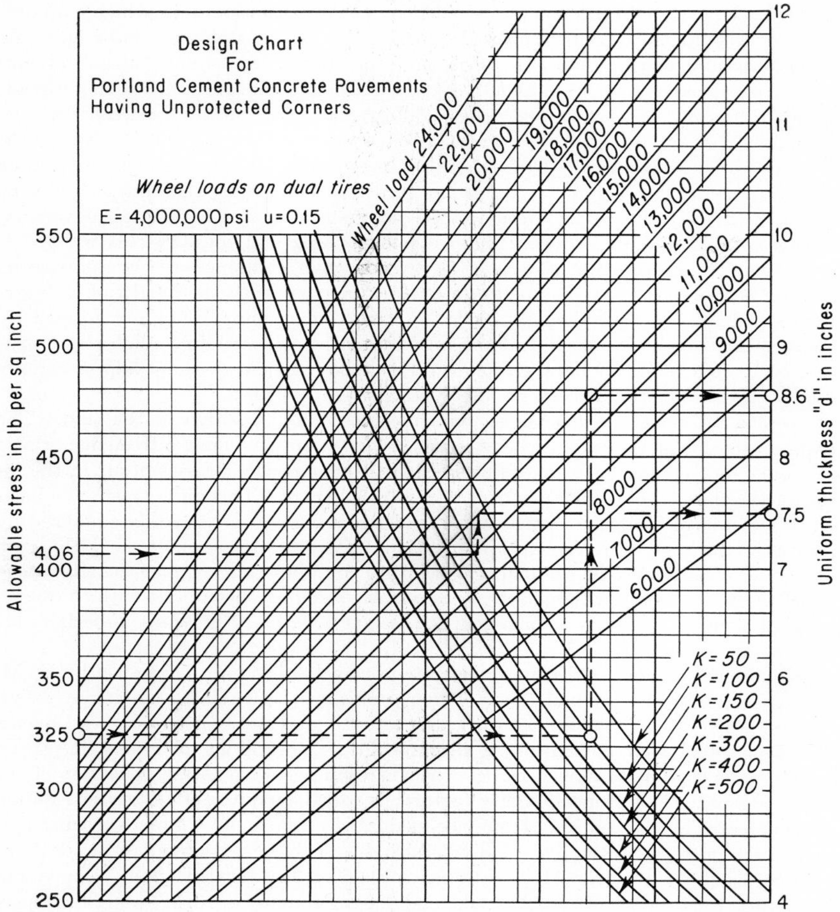

FIG. 49. Design chart for thickness of portland cement concrete pavements. (*The Portland Cement Association.*)

of taper used (Fig. 50). For a 2.5-ft taper width, the edge thickness $d_e = 1.275d_i$. For wider tapers it is somewhat less.

When adjacent slab corners are connected by dowels or other load-transfer devices forming "protected corners," the PCA finds that the stress under a given wheel load is reduced by 20 percent. When using Fig. 49 to find thickness for *protected* corners, the allowable stress should be increased in ratio of 100:80.

Fatigue tests have shown that concrete specimens will withstand an indefinite number of load repetitions if individual stress repetitions do not exceed one-half the modulus of rupture. When they do exceed this amount a decreasing number of applications will cause failure. For capacity operation the allowable stress is commonly taken as one-half the ultimate. In areas where applications of design load are infrequent, the PCA recommends the use of stresses greater than one-half the ultimate depending on number of repetitions actually expected.[1]

[1] "Concrete Pavement Design," Portland Cement Association.

Impact reactions are usually small on smooth surfaces and for vehicles with low-pressure pneumatic tires. A factor of 1.2 is suggested to allow for impact effect of roughness which may normally develop over the life of a pavement.

Example: Design wheel load = 11,000 lb; k = 100 lb per in.^{2}; allowable stress = 325 psi. Unprotected corners. In Fig. 49 enter with 325 psi, cross horizontally to k = 100, then vertically to 11,000-lb wheel load, and then horizontally to thickness scale where uniform thickness of 8.6 in. is read. If corners are protected by dowels, the chart is entered with 325(100/80) = 406 psi, and a thickness of 7.5 in. is found.

An equivalent thickened edge section with protected corners would have interior thickness $d_i = 0.85d = 0.85 \times 7.5 = 6.4$ in.; and edge thickness, $d_e = 1.275d = 1.275 \times 7.5 = 9.6$ in. Since standard height of forms is in full inches, the nearest higher full inch thickness would be used for edges (10 in.) and uniform slabs (8 in.). The interior thickness, which is shaped by templates, is rounded to nearest higher ½ in. (6.5 in.).

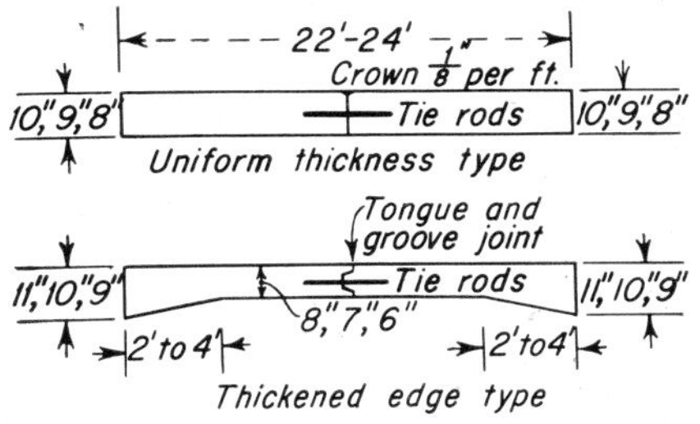

FIG. 50. Typical cross sections of concrete pavement.

Reinforcing is used to keep slabs in alignment and to control cracking. Dowels or other load-transfer devices are placed across transverse joints (Fig. 52) and tie rods of reinforcing steel across longitudinal joints (Fig. 51). In some designs a layer of welded wire fabric, or well-distributed reinforcing of closely spaced small rods, is placed over entire slab area between transverse contraction or expansion joints, particularly where joint spacing is over 30 ft. The weight of fabric is about 60 lb per 100 sq ft with the greater weight of wire (closer wire spacing) placed longitudinally in the slab.

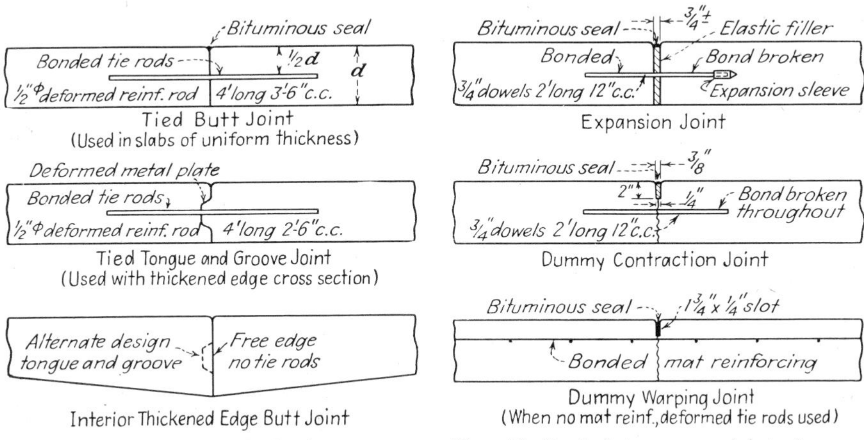

FIG. 51. Typical longitudinal joints for concrete pavement.

FIG. 52. Typical transverse joints for concrete pavement.

Joints.[1] A *longitudinal* joint is formed at the center line of two-lane pavements and between all adjacent lanes in pavements of more than two-lane width (Fig. 51). These

[1] Structural Design Considerations for Pavement Joints and Design Considerations for Concrete Reinforcing for Crack Control, *Jour. Am. Concrete Inst.*, July and October, 1956.

joints prevent irregular and unsightly longitudinal cracks; they also serve as traffic lines and facilitate lane-at-a-time construction. Longitudinal joints may be formed by butting two slabs of uniform thickness together, by impressing or sawing a groove in concrete, or by placing a deformed metal plate at the joint, thus producing a tongue-and-groove connection between slabs. The latter is usual practice in thickened-edge design. Adjoining slabs are held together by bonded tie rods, usually ½-in. rods 3 ft long extending equal lengths into adjacent slabs, and spaced about 2′6″ apart. *Transverse joints* are provided to control cross cracking that would otherwise occur because of restrained contraction and expansion and warping of the pavement. There are three types (Fig. 52): *expansion joints* which are about ¾ in. wide and contain an elastic filler, *contraction joints* which are formed by a groove in the pavement about one-quarter of the depth, and *warping joints* which are grooved joints with bonded reinforcing running across them. The groove creates a weakened section in the pavement where cracks are expected to occur when the pavement contracts. Expansion joints are installed adjacent to structures or other fixed objects. They are omitted in the pavement in many designs or placed at long intervals of 400 to 600 ft. Since most concrete is placed in warm weather and shrinks some in setting, there is little need for expansion space. When laid in cold weather, expansion joints are advisable to avoid buckling of slabs (blowups). Contraction or warping joints are spaced at 15 to 25 ft (Fig. 53*a*, *c*) depending upon the expansion characteristics of the concrete. When wire fabric reinforcing is placed between contraction joints the spacing may be increased, Fig. 53*b*. Expansion joints are filled with pliable materials such as bituminous impregnated fabric, cork, rubber, and rubber latex. Redwood fillers have also been used. These offer resistance to the closing of the joint, thus inducing some compression in the slab which counteracts wheel-load and warping tension stress and also serves to keep contraction joints and cracks closed. Smooth round dowels or special load-transfer devices are placed across expansion joints. Dowels are ¾ to 1¼ in. diameter, 15 to 24 in. long, and spaced 12 to 15 in. along the joint. One-half the dowel is oiled to break the bond, and a cap or sleeve placed over the oiled end to allow movement in or out of the concrete. Contraction joints are commonly filled with hot-poured rubber-asphalt or with cold-applied mastic sealer. Dowels are usually placed across these joints with the bond broken by oiling but with no cap on the end. In some designs, the dowels are omitted and the interlock of the rough fractured faces of concrete is relied upon to transfer some load and keep the adjacent slabs in alignment. Warping joints are a hinged type installed to relieve the tension stresses set up when the pavement attempts to warp or curl.

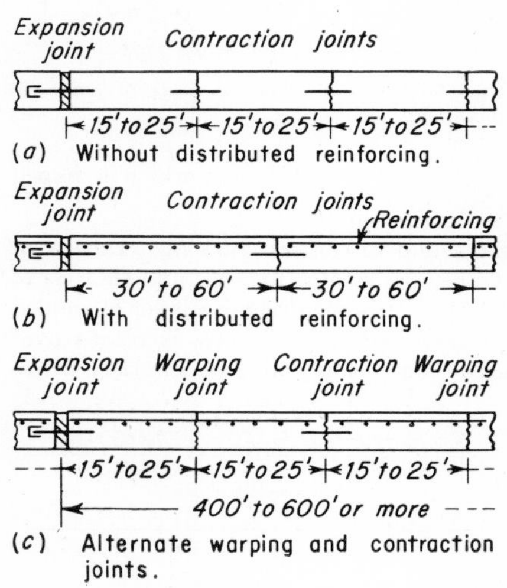

Fig. 53. Longitudinal sections showing joints and reinforcing for concrete pavement.

Concrete. The nominal proportions of concrete are about 1:2:3 or 1:2:3½ by volume for surfaces, and 1:2½:5 or 1:3:6 for base courses. The exact proportions are made by weight and depend upon the nature of the aggregates and the strength and workability desired. (See Section 7 for cement specifications and the design of concrete mixes.) An "air-entraining" cement or an air-entraining agent is used to improve the workability

and durability of the concrete.[1] The entrained air forms minute voids throughout the concrete which resist water penetration and minimize scaling and spalling under repeated cycles of freezing and thawing. The water-cement ratio is specified within the range of 5.0 to 6.0 gal per bag of cement, and the minimum cement content (cement factor) from 1.25 to 1.65 bbl (5.0 to 6.5 bags) per cu yd of concrete produced. The exact amounts specified depend upon the design of the mix. The moisture content of the aggregates is determined and deducted from the water added at the mixer. The aggregates and the cement are usually assembled and proportioned at a material yard near the source of supply, a railroad siding, gravel pit, or quarry. The sand and stone are loaded by a crane into elevated bins and weighed into trucks in the proper batch proportions. To obtain a well-graded aggregate, the coarse aggregate is stock-piled and proportioned in two sizes, separated on the ¾-in. screen. Cement is added to the dry batch in bags; or, when bulk cement is used, it is weighed into the batch. Dry batches are transported to the mixer in compartments in trucks. With bulk cement, batch box containers are used, or the batch compartment is tightly covered. If transit-mix trucks are used, the dry ingredients are charged into the mixer at the material yard and the water added just before the job is reached. In city work, ready-mix concrete is often purchased from a concrete plant and delivered to the job in agitator trucks.

Construction Methods.[2] Steel side forms are set to true line and grade and firmly bedded on the subgrade or foundation course so that they will not settle or move during construction. Mechanical tampers are used for this purpose. The area between forms is "fine-graded," compacted with a light roller, and tested for true cross section with a template riding on the forms. The subgrade is sprinkled immediately in front of the paver. Standard concrete pavers are 27E (1 cu yd) and 34E (34 cu ft) capacities. The 34E model is usually made with dual drums, allowing two batches to be mixing simultaneously. The pavers are end-dump. An overload of 10 percent of capacity is allowed. The dry batches are dumped from trucks into a steel apron or "skip" at the rear of the mixer. The skip is elevated, sliding the materials into the revolving mixing drum. The specified amount of water (after allowing for moisture in the aggregates) is added, and the concrete mixed for 1 to 1¼ min. It is then discharged into a bucket suspended from a boom which moves out over the subgrade and deposits the concrete. In crowded locations the mixer may travel between the forms; preferably it should travel at one side where it will not interfere with laying reinforcing or otherwise disturb the finished grade. Sample beams and cylinders are cast of concrete as it comes from the mixer, cured in the same manner as paving concrete, and later tested for strength. The concrete is spread between the forms by a mechanical spreader which distributes the concrete laterally between the forms and roughly strikes it off a little above grade. When mat reinforcement is to be installed, the concrete is roughly struck off about 2 in. below top of forms and the mats set in place. The concrete is then brought up to the top of the forms. The concrete is worked around reinforcing and joint assemblies by spading or by hand vibrators. A mechanical finishing machine, running on the forms, strikes off the concrete and shapes it transversely to the proper cross section. When vibrators are used, frequently as an attachment to the finishing machine, a dry concrete can be used, and the required strength obtained with less cement. The surface is finished longitudinally with a 12-ft float worked mechanically from a machine spanning the forms, or by hand from temporary bridges across the fresh pavement. Any excess moisture or laitance appearing on the surface is scraped off and deposited outside of the forms. The surface is tested against a 10-ft straightedge. Smoothness requirements are no irregularity over ⅛ in. in 10 ft. The surface is then roughened by brooming or burlap drag to

[1] Use of Air-entraining Concrete in Pavements and Bridges, *Highway Research Board, Current Road Problems* 13.

[2] "Concrete Pavement Manual," Portland Cement Association. Also "Specifications for Concrete Pavement Construction," AASHO.

give a nonskid texture. The slab and joint edges are then rounded with an edging tool.

Curing. As soon as the surface is finished, concrete is covered with a protective material such as burlap and kept wet. Within 24 hr, the burlap is replaced with damp earth or straw which is kept moist for about 10 days. Other curing agents that retain water in the concrete by preventing evaporation are heavy fiber mats and white pigmented membranes applied to the fresh concrete. Traffic is usually not allowed on the concrete for about 2 weeks, or until the test beams taken at the time of pouring show a modulus of rupture of 550 to 600 psi. High-early-strength cement may be used in special cases where it is necessary to open the pavement to traffic as soon as possible.

Traffic Signs and Signals.[1] *Regulatory signs* are used to require a full stop on minor streets entering a through way or at isolated, dangerous intersections and to indicate permissible speeds at different locations, restricted turning movements, restricted passing, one-way traffic, parking limitations, etc. *Warning signs* are placed in advance of unusual conditions requiring a reduction in speed or special caution, such as sharp curves, steep hills, signal lights, traffic circles, thickly settled districts, schools, pedestrian or animal crossings, and railroad crossings. The more dangerous the hazard, the more prominent should be the sign. *Guide signs* indicate route numbers, directions and mileages to places, names of streams and landmarks, etc. Route signs should be placed at and in advance of all intersections and at frequent intervals along the route. Changes of direction at a junction should be indicated in advance. Where more than one route follows the same road, all route numbers should be shown, although not necessarily all on the same post. The design and color of traffic signs and markings have been standardized in "Manual on Uniform Traffic Control Devices for Streets and Highways." These standards are generally followed by the states and municipalities. *Regulatory signs* are rectangular in shape with longest dimension in vertical direction, except for the stop sign, which is octagonal. *Warning signs* are diamond-shaped, except railroad-grade-crossing warning signs, which are circular. *Yield right of way signs* are triangular with base at top. *Guide or directional signs* are rectangular with the longest dimension in horizontal direction. Red color means stop and is used as background for stop signs. Yellow, denoting warning, is used on warning signs. Guide signs usually have black letters on white background, although considerable experimentation is being made with different types of lettering and different colored backgrounds, especially on high-speed highways where visiblity at long distances is essential. Signs are made luminous at night by reflectorizing message, background, or both. Neon signs or flood lighting is sometimes employed at dangerous locations. *Pavement markings* are used to indicate the center line of roadway, limits of traffic lanes, edges of pavement, streetcar clearances, boundaries of pedestrian cross walks, safety zones, stop lines at signalized intersections, approaches to obstacles, etc. *Flashing beacons* are used to give warning at dangerous intersections; a flashing yellow light indicates caution; a flashing red light means stop before entering.

Traffic signals are of several types: *fixed-time* which operate continuously on a preset timing, *coordinated* which refers to groups of interconnected signals which may operate simultaneously or on a preset sequence (progression); *programmed* in which more than one fixed timing can be displayed, such as one setting for morning peak, one for off-peak and another for evening peak traffic; *semitraffic actuated* in which green remains on main street until "called" by vehicle passing over a detector in cross street; *fully actuated* in which signal timing is controlled by relative volumes of traffic arriving on the various

[1] For a complete treatise on this subject, see "Manual on Uniform Traffic Control Devices for Streets and Highways," prepared by Joint Committee on Uniform Traffic Control Devices of AASHO, National Safety Council, and Institute of Traffic Engineers, published by Public Roads Administration, Revised 1954, and "Policy on Maintenance of Safety and Traffic Control Devices and Related Traffic Services," AASHO.

Also see "Traffic Engineering Handbook," joint publication of Institute of Traffic Engineers and Association of Casualty and Surety Cos., New York, and MATSON, SMITH, and HURD, "Traffic Engineering," McGraw-Hill Book Company, Inc., 1955.

approaches; and *pedestrian activated* in which a pedestrian interval is given when pedestrian pushes signal call button. Detectors may be of the pressure plate, magnetic field, or radar-beam intercept type. In timing signals, the object is to select a cycle length and to distribute green time so that all vehicles reaching the intersection in one cycle will be able to pass through on the first green phase. Automatic signals can best meet this requirement. Fixed-time types, although most commonly used, cannot adjust to varying traffic demands. They are usually timed for peak-hour traffic and thus give cycles which are too long in other hours. The type of control adopted should be the least restrictive that will handle the traffic with safety. Unwarranted signals cause unnecessary delays, encourage violations, and often cause accidents. Minimum conditions warranting traffic signals are recommended in "Manual on Uniform Traffic Control Devices for Streets and Highways." The minimum traffic volume warranting fixed-time signals at the intersection of two-lane, two-way streets on highways in urban areas is prescribed as more than 750 vehicles per hour for each of 8 hr of an average day entering the intersection from all approaches including at least 250 vehicles per hour for the same 8 hr entering from the minor street or streets. In rural areas, the traffic figures are reduced to 500 and 125, respectively. When traffic volumes fall below 50 percent of the above minima for two consecutive hours, signals should be changed to flashing operation, such as flashing yellow for caution on main street and flashing red on side street for stop and enter control. Lesser traffic volumes warrant signal control if there is considerable left-turn movement, heavy pedestrian traffic, unusual accident hazard, or when the signal is one of a coordinated group. The minimum accident warrant is five or more accidents of types susceptible of correction by signals occurring within a year and involving personal injury or property damage to the extent of $50 or more. Signals should be installed only after a thorough study of traffic movements, approach speeds, accident experience, and physical conditions at the intersection.

In large cities extensive coordinated signal systems are in operation. The cycle length is controlled at a central station and the time distribution at each separate signal installation is adjusted to the traffic demand at that intersection. On main arteries an effort is made to coordinate signals to give progression in both directions; i.e., vehicles entering the system will be able to proceed at the set speed of progression and meet a green light at each intersection. Frequently the timing of progressive systems can be varied to fit the different traffic demands in morning, evening, and off-peak hours. Coordination is usually accomplished by interconnecting cables. Radio interconnection is also possible. Since signal mechanisms are operated by synchronous motors, they can be coordinated by setting desired offsets in cycle starting time manually using a stop watch. The system will then remain coordinated so long as there is no interruption in power service.

Highway Appurtenances. *Guardrail*[1] is needed along embankments over about 10 ft high where side slope is steeper than 4:1, along bodies of water, and at all places where the driving public requires protection. They are mounted on wooden or concrete posts about 6 ft long, spaced 10 to 12 ft apart. Different types of rail are used, such as wood planks, wire cable, woven-steel wire fabric, and smooth or corrugated steel plate. The function of the guard is to deflect the vehicle back onto the road, rather than to stop it suddenly with severe damage to vehicle and injury to passengers. Hence, the rails are mounted low enough to make contact with bumpers, wheels, and the stronger parts of the vehicle. The guard should be painted a light color with reflectorized paint. When cable is used, the posts should be painted a light color or equipped with light reflecting devices. *Heavy wooden planks* attached to posts at hub height have taken the place of the earlier 30-in.-high wooden rails. *Wire cable*, consisting of two or three ¾-in. galvanized-steel wire cables mounted on wooden or concrete posts, is a common type. It is strong

[1] Specifications for Highway Guards, AASHO, 1952.

yet yields slightly and distributes the impact of a collision to several posts. *Woven-wire* (wire-tape) guard consists of drawn annealed-steel wire of high tensile strength, woven into a continuous fabric and mounted on posts. This type is more visible than the standard ¾-in. cable type. *Metal-plate* guardrail consists of a continuous galvanized-steel plate of No. 16 gage, 12 in. wide, attached to posts by means of steel springs. Metal and cable guards are offset from posts to increase impact absorption and to cause vehicle to strike and slide along guard rather than to contact posts. The ends of the guard are anchored so that it will hold even when individual posts are knocked out.

Curbs, Gutters, and Catch Basins. Curbs used in urban areas are of granite or portland cement concrete. Stone curbs are usually 6 to 8 ft long, 5 to 8 in. thick, and 19 to 21 in. in depth. Concrete curbs are cast in lengths up to 10 ft and widths of 6 or 7 in. The base of these curbs is usually wider than the top with the front face battered. Concrete curbs may be cast integrally with the pavement or gutter. The height of curb above pavement varies from 5 to 7 in. It should be low enough not to contact running gear of vehicle, yet high enough to contain flow in gutter. Sloping curbs of light-colored concrete or inclined stone slabs are used along median strips and around traffic islands. These curbs should not offer too substantial an obstruction to vehicles forced over them in an emergency. They should have high visibility at night. At hazardous locations curbs are often painted with reflecting material. Lip curbs of bituminous concrete are sometimes used on rural highways or on residential streets. No special gutter section is needed on hard-surfaced pavements with curbs. Paved gutters of bituminous concrete, cement concrete or grouted rubble are built along rural highways where needed to prevent excessive erosion. Gutter flow is discharged into drop inlets or catch basins. The former are box type about 3 ft deep with grating on top and pipe leading out from the bottom. Catch basins are set with grating adjacent to curb. They are 6 to 8 ft deep with outlet about 3 ft above the bottom leaving the lower part of basin to collect sediment and debris (Fig. 45). In urban areas gutters should have a minimum slope of 0.3 to 0.5 percent to ensure prompt flow to catch basins. The latter feed into a storm-sewer system, which is designed as explained in Section 9.

Sidewalks [1] are built of gravel, bituminous pavement, portland cement concrete, and brick. The gravel type consists of about 6 in. of compacted gravel with a surface covering of stone screenings. The portland cement concrete type is laid 4 or 5 in. thick on a gravel or cinder base. A mix of 1:2:3 is recommended for a one-course walk, and a 1:2.5:4 base and a 1:2 mortar top for two-course construction. Widths are 4 or 5 ft, and the walk is divided into 5-ft lengths by contraction joints. Expansion joints filled with bituminous material are advisable at 50-ft intervals and where a long length of walk abuts a cross curb, as at street intersections. Bituminous sidewalks are common in rural and residential districts. The foundation course is 4 in. of stone, slag, cinders, or gravel. Where there is no adjacent curb, the base should be 4 to 6 in. wider than the surface course, which consists of 1½ to 2 in. of plant-mixed bituminous concrete, either hot or cold mix. The sidewalks are sloped toward curb and gutter for drainage. The usual cross slope is 1½ to 2 percent but is sometimes steepened to 6 percent in making adjustments around corners of streets intersecting on steep grades.

Roadsides.[2] Much attention has been given in recent years to developing roadside beauty and controlling erosion of slopes by planting grasses and shrubs. Existing growth is cut to provide vistas of scenic beauty or to provide the required sight distance on curves. Trees and other landmarks within the right of way are preserved, where they do not present a traffic hazard. Some states have passed legislation controlling the use of the roadsides for commercial purposes and for advertising signs.

[1] "Concrete Pavement Manual" and "Concrete Walks, Driveways and Steps," Portland Cement Association. Also "Asphalt Handbook," The Asphalt Institute.

[2] Roadside Development Reports, Highway Research Board.

Maintenance.[1] Prompt and efficient maintenance of highways and streets is essential to retain the value of the original investment and to keep the road in serviceable condition at all times. There are three classes of maintenance: that pertaining to the surface, or roadway; that on the shoulders of roadsides and roadside structures; and that for the benefit of traffic, such as maintenance of signs, markings, and signals, and the removal of snow and sanding of slippery surfaces. In cities, cleaning of streets, gutters, and catch basins is also required. The maintenance of untreated surfaces consists of blading or dragging, and the addition of new material. Bituminous surfaces are maintained by patching and surface treatments. Bituminous patching mixtures are commonly used to repair surface defects, although, in the case of high-type pavement, it is better practice to place permanent patches of the same material as the original surface. Patching mixtures consist of fine aggregate mixed with either cutback asphalt (RC-2 for warm and MC-2 for cold weather), asphalt emulsion (MS-3), or cutback tar (CBRT-5 or 6). The aggregate size is usually ¾ in. to No. 8. Patching mixtures remain plastic for some time and may be stock-piled and used as required. Small areas in surface-treated or bituminous macadam roads that show signs of raveling may be skin-patched by applying a surface treatment to that area. Holes in bituminous macadam should be cleaned out and squared off, the edges painted with cutback, the hole filled with aggregate of the same dimensions as in the original top course, and penetrated and sealed in the same manner as in the original pavement. Good results have also been obtained by patching bituminous macadam with plant-mix bituminous concrete. Defects in bituminous concrete types should be repaired by removing a rectangular-shaped area of the top course, including the defect, and replacing this area with a patch of the same composition as the original surface course. Shallow depressions and failures in cement concrete pavements can be repaired with dense, fine-graded bituminous concrete. This is necessary because thin patches of cement concrete will not adhere to old concrete. If the disintegration extends through the depth of the slab, a rectangular portion (not less than about 2 ft on a side) may be chiseled out and a patch of dry-mix cement concrete vibrated or tamped into place and finished to grade. The joints and cracks in the pavement are filled periodically with asphaltic, tar, or rubber-bituminous materials applied with a pressure nozzle or pouring can. Concrete pavements that have settled badly can be raised to grade by "mud-jacking." Small holes are drilled through the settled slabs and a mixture of earth, water, and cement is forced through these holes onto the subgrade. Asphalt is also used for this purpose in a process called "subsealing."[2] It not only raises the slabs but also seals the subgrade against water penetration. Air-oxidized asphalt having a softening point of 160°F and penetration of 30 to 45 is heated to about 400°F and forced through holes drilled in the pavement. Service cuts in city pavements should be backfilled with granular material, if possible, or the earth should be compacted by tamping in layers or by vibrating. Thorough compaction is essential to avoid subsequent settlement. The base course of the pavement should be cut back 6 to 12 in. from the edge of the trench and replaced with new base of the same character. Similarly, the surface course, or courses, should be stepped back from the course below. Roadside maintenance includes care of shoulders, ditches, fences, slopes, culverts, walls, bridges, and landscaping. An efficient organization is essential for prompt snow removal and sanding. The sand is frequently mixed with about 100 lb of calcium chloride per cubic yard, which facilitates the handling of the sand in cold weather and also helps to embed the sand on a slippery surface. In municipalities, rock salt without sand is used to keep streets free of ice. This eliminates accumulation of sand on streets and in catch basins.

[1] "Policy on Maintenance of Roadway Surfaces," AASHO. Also "Asphalt Handbook," Chapter on Maintenance and Resurfacing, and "Tarvia Manual," Barrett Division, Allied Chemical and Dye Corp.

[2] Maintenance Methods for Preventing and Correcting Pumping Action of Concrete Slabs, *Current Road Problems* 4-*R*, Highway Research Board. Field Experimentation with Bituminous Undersealing Materials, *Highway Research Board, Proc. 22d Ann. Meeting*, 1953, pp. 343–354.

HIGHWAY ECONOMICS

The broad objective of a highway program is to provide adequate highway transportation at the least overall cost considering both the highway and the vehicle. Motor-vehicle performance is an important factor in determining design standards. It also determines grade-climbing ability, particularly the speed which heavy vehicles can maintain on grades. For highway problems the net brake horsepower (bhp) or the torque T at the flywheel is used. This power is transmitted to the rim of the tires as tractive effort (T.E.), the propelling force. In terms of horsepower, T.E. $= 375e$ bhp$/V$, in which e is an efficiency factor allowing for loss of power between the flywheel and rim of tires. It depends upon type of transmission, differential gears, axle, wheel and tire types. In direct gear it may vary from 0.85 to 0.90 for passenger cars and 0.75 to 0.85 for trucks. In terms of torque, T.E. $= eTG_TG_D/r$, in which T is torque, ft-lb, e is efficiency factor as above, r is rolling radius of tire, ft, G_T is the transmission-gear ratio, and G_D is the differential-gear ratio. The torque is usually given in the form of a curve of torque vs. rpm. Tractive resistance is made up of rolling resistance, air resistance, and (plus or minus) grade resistance. Rolling resistance is usually expressed in pounds per ton of total weight of vehicle. On hard-surfaced roads in good condition, it varies from 20 to 30 lb per ton,[1] the lower value being for light vehicles at low speeds, and the higher value for heavy vehicles at high speeds. On rough gravel roads, the resistance may be as high as 50 lb per ton, and on earth roads in muddy condition, it may be over 100 lb per ton. Total air resistance is expressed by the formula KAV^2, in which K is a coefficient depending upon the shape of the vehicle (degree of streamlining), A is projected area of vehicle, sq ft normal to direction of motion, and V is velocity, mph, relative to the air i.e., the velocity of the vehicle corrected for head or tail wind. The value of K averages about 0.0012 for passenger cars and 0.0020 for trucks. Grade resistance in pounds per ton equals $20g$, where g is rate of grade in percent. The relation between tractive effort and resistance is as follows:

$$\text{T.E.} = WR_R + KAV^2 \pm 20gW \pm \text{accelerating force}$$

in which W is total weight of vehicle including load in tons and R_R is rolling resistance, lb per ton. The grade resistance is plus on up grades and minus on down grades. The acceleration force is positive when there is an excess of tractive effort over resistances and negative (deceleration) when tractive effort is less than resistances. The principal application of motor-vehicle power and resistances in the highway field is to determine the grade-climbing ability of trucks; thus it is a factor in the selection of grades [2] (see "Gradients," p. 2–71). There is usually ample friction between tires and pavement surfaces to develop the desired tractive effort except on icy surfaces. For clean dry road surfaces, the coefficient of sliding friction between tires and road surface is from 0.6 to 0.9; for wet pavements, it is from 0.4 to 0.8; on ice or packed snow, it may be as low as 0.1. At high speeds, the coefficients are 20 to 25 percent less than these values for most types of pavement.[3]

Vehicle operating costs are an important factor in the economic analysis of highways since they are influenced by design features such as road surface roughness, grades, curves, and length of route, and by such operating factors as speed, accident hazard, and traffic congestion. Vehicle-operating costs vary widely with weight of vehicle, weight-power ratio, standard of maintenance, driver habits, and annual mileage. Operating costs may be divided into fixed and variable. The fixed costs include those not affected by use, such as for registrations, property taxes, insurance, garaging, interest

[1] Hill Climbing Ability of Motor Trucks, *Public Roads*, May, 1942.
[2] "Vehicle Climbing Lanes," *Bull.* 104, Highway Research Board, 1955.
[3] "Roughness and Skid Resistance," *Bull.* 37, and Skid Measurements on Virginia Pavements, *Research Report* 5-*B*, both by Highway Research Board.

on investment, and depreciation when charged on a time basis. Variable costs (or mileage costs) are those which vary directly with use, such as for fuel, oil, tires, repairs, and depreciation when charged on mileage basis. In general, depreciation would be charged on a time basis for passenger cars which depreciate by obsolescence, and on a mileage basis for large trucks and busses which have large annual mileages. Sometimes depreciation is charged partly on time basis and partly on mileage basis. For passenger cars operating 10,000 miles per year under average conditions the total operating cost is about 8 cents per mile of which about 4 cents is for mileage items. For trucks the total costs will range from 10 to 50 cents per mile of which 5 to 15 cents is for mileage cost. In truck accounting, fixed costs, wages, and overhead costs are often reduced to time costs in dollars per hour (see Table 23). Motorbus costs vary with weight and number of seats from about 20 cents per mile for light busses to 40 cents per mile for heavy intercity-type busses. In economic studies a time cost is usually added to passenger-car costs to evaluate savings to road users from highway improvements.

A detailed analysis of the effect of road features and operating conditions on vehicle mileage costs is given in "Road User Benefit Analyses for Highway Improvements" by AASHO. Surface type has little effect on operating costs so long as the surface is firm and smooth. Tests [1] have shown that the difference in operating cost on nonskid texture bituminous surface treatments is only 0.06 cent greater per mile than on cement concrete. Between untreated gravel and concrete the difference is 0.3 to 0.5 cent per mile, and between earth and good gravel, about 0.4 cent. Shortening a route by relocation results in a saving on all mileage costs for the distance saved. Decreased travel distance will also result in a time saving. Little economy is gained by reducing grades below 6 percent for passenger-car traffic. For motor trucks the saving is substantial in both fuel and time. Table 21 gives composite gasoline costs for different gross weights of

TABLE 21. COMPOSITE GASOLINE COST IN CENTS PER MILE FOR DIFFERENT RATES OF RISE AND FALL [1]

Based on gasoline at 30 cents per gallon

Rate of rise and fall	Passenger cars at 40 mph	Motor trucks at average speeds for gross weights					
		10,000 lb	20,000 lb	30,000 lb	40,000 lb	50,000 lb	60,000 lb
0	1.62	2.7	3.6	4.2	4.8	5.4	5.7
1	1.65	2.7	3.9	4.8	5.7	6.6	7.2
2	1.68	2.7	4.2	5.4	6.6	7.8	9.0
3	1.74	2.7	4.8	6.3	7.8	9.3	10.8
4	1.80	3.0	5.1	7.2	9.3	11.1	13.2
5	1.86	3.3	6.0	8.7	11.4	13.5	16.2
6	1.92	3.9	7.2	10.5	13.8	16.8	19.8

[1] Passenger-car fuel-consumption data from "Highway-user Taxation," *Bull.* 92, Highway Research Board, p. 29. Motor-truck data from *Research Report* 9-A, Highway Research Board.

trucks on rates of rise and fall between 0 and 6 percent. Composite gasoline cost is based upon composite gasoline consumption, which is the total number of gallons of gasoline required to travel in both directions over a section of highway divided by twice the length of that section in miles. Rate of rise and fall is the total feet of rise plus the total feet of fall in a section divided by the length of the section in hundreds of feet.

[1] Vehicle Costs, Road Roughness and Slipperiness of Various Bituminous and Portland Cement Concrete Surfaces, *Highway Research Board, Proc. 22d Ann. Meeting*, 1942. Motor Vehicle Operating Costs and Related Characteristics on Untreated Gravel and Portland Cement Concrete Road Surfaces, *Highway Research Board, Proc. 19th Ann. Meeting*, 1939.

The time required for heavy vehicles to operate over different rates of rise and fall is given in Table 22. It depends upon the weight-power ratio, which is the gross weight

TABLE 22. COMPOSITE TRAVEL TIME IN MINUTES FOR VARIOUS WEIGHT-POWER RATIOS FOR DIFFERENT RATES OF RISE AND FALL[1]

Rate of rise and fall	Weight-power ratios					
	100	200	300	400	500	600
0	1.15	1.15	1.15	1.15	1.15	1.15
1	1.17	1.25	1.33	1.42	1.51	1.60
2	1.23	1.43	1.63	1.83	2.04	2.25
3	1.35	1.70	2.05	2.40	2.76	3.12
4	1.48	2.02	2.56	3.10	3.65	4.20
5	1.62	2.40	3.18	3.96	4.74	5.52
6	1.80	2.87	3.94	5.01	6.08	7.65

[1] Compiled from *Research Report* 9-*A*, Highway Research Board.

of vehicle divided by the net horsepower (rated horsepower less power consumed by engine accessories). Both gasoline costs and travel times are an average of operation up and down grades in either direction. If the traffic is of the same weight and volume in each direction, Tables 21 and 22 give net cost of fuel and travel time, respectively, per unit of traffic per mile. For longer or shorter lengths, the values may be increased or decreased proportionally. When the traffic is not balanced in each direction, "directional" gasoline consumption and travel-time data are available in *Bulletin* 9-*A* for different percentages that rise is of the total rise and fall in one direction of travel.

Fuel consumption and tire wear increase at higher speeds. The decrease in fuel mileage will vary for different types of vehicles. For a light passenger car traveling at sustained speeds the trend is about as follows: 21.5 mpg at 30 mph, 19.5 mpg at 40 mph, 17.5 mpg at 50 mph, and 15.2 mpg at 60 mph. Tire wear at 50 mph is about 50 percent greater than at 30 mph, and 100 percent greater at 60 mph.

Traffic congestion[1] requires frequent starts, stops, and long idling periods, which increase fuel consumption, tire wear, and maintenance. On the most congested streets in downtown areas gasoline mileage for passenger cars may be as low as 9 mpg when average speed is only 8 mph. On less congested streets permitting an average speed of 20 mph, gasoline mileage is about 14 mpg. Accident costs are also higher on congested streets and highways, because of higher accident rates. Considerable time is lost by delays on congested highways. This time loss has real value to commercial operators. The value of time to passenger-car drivers may be debatable, but it has been demonstrated that motorists will pay for time savings through tolls or in other ways, if a definite advantage is gained thereby. Table 23 gives time costs for different types of vehicles under different operating conditions.[2] For application to any specific problem the reader should consult the paper upon which Table 23 is based and adjust data to cost levels prevailing in the area under study.

Comparison of Pavement Types. The cost of construction, annual maintenance, service life, salvage value at replacement, and vehicle-operating cost should all be considered in determining the relative economic value of different surface types for a specific project. The all-inclusive unit of comparison is the cost of highway transportation, which

[1] "Vehicle Operation as Affected by Traffic Control and Highway Type," *Bull.* 107, Highway Research Board.

[2] LAWTON, LAWRENCE, Evaluating Highway Improvements on a Mileage-and-Time-Cost Basis, *Traffic Quarterly*, Eno Foundation for Highway Traffic Control, January, 1950. Time and Gasoline Consumption in Motor Truck Operation as Affected by the Weight and Power of Vehicles and the Rise and Fall in Highways, *Research Report* 9-*A*, Highway Research Board, 1950.

TABLE 23. MILEAGE COSTS FOR DIFFERENT TYPES OF VEHICLES ON DIFFERENT TYPES OF STREETS AND HIGHWAYS[1]

Type of highway	Passenger cars, cents	Light trucks, cents	Heavy trucks, cents	Tractor trailer, cents
Congested streets:				
Gasoline	3.4	6.3	10.3	16.6
Mileage	2.3	3.3	5.1	8.5
Accidents	1.0	1.0	1.0	1.0
	6.7	10.6	16.4	26.1
Through city streets:				
Gasoline	2.0	3.6	6.0	10.0
Mileage	2.3	3.3	5.1	8.5
Accidents	1.0	1.0	1.0	1.0
	5.3	7.9	12.1	19.5
Arterial streets (coordinated signals):				
Gasoline	1.7	3.1	4.8	8.1
Mileage	2.3	3.3	5.1	8.5
Accidents	0.4	0.4	0.4	0.4
	4.4	7.0	10.3	17.0
Expressways:				
Gasoline	1.4	2.5	4.0	6.6
Mileage	2.3	3.3	5.1	8.5
Accidents	0.2	0.2	0.2	0.2
	3.9	6.0	9.3	15.3
Time Cost per Hour				
All types	$1.10	$1.89	$3.24	$5.43

[1] Based upon method in LAWTON, LAWRENCE, Evaluating Highway Improvements on a Mileage-and-Time-Cost Basis, *Traffic Quarterly*, Eno Foundation for Highway Traffic Control, January, 1950. Gasoline costs adjusted to 30 cents per gallon. Mileage costs adjusted to 1956 by price index.

is the sum of highway cost and vehicle (or road-user) cost, expressed as either an annual cost or cost per mile. The highway cost is usually expressed as an annual cost for renewal in perpetuity, as follows:

$$C = Ar + M + \frac{(A - S)r}{(1 + r)^n - 1}$$

in which A is first cost of road, r is rate of interest applicable to highway financing, M is annual maintenance, and S is salvage value at end of service life n in years.[1] The last term is an annuity that will earn an amount $(A - S)$ for replacement every n years in perpetuity. A similar term may be set up for any periodic maintenance that may be required (such as a seal coat every 5 years). The formula is also useful for comparing alternate types of structures designed to serve the same purpose.

In selecting a surface type other factors to be considered are attractiveness, tire noise, nonskid characteristics, visibility at night, use of local materials, availability of contractors skilled in particular types of construction, aid to local industries, and public preference.

[1] Life Characteristics of Surfaces on Primary Rural Roads, *Highway Research Board, Proc. 20th Ann. Meeting*, 1940. Also *Public Roads*, March, 1941; August, 1949; and June, 1956.

Selection of Alternate Locations. The cost of highway transportation, as defined above, is a useful tool for making economic comparisons of alternate routes, since it brings in all elements of cost both for the highway and to the road users. Thus the effect of road surface, distance, grades, traffic congestion, cost of construction, and property damages will enter the calculations. A physical inventory of each route is also helpful, comparing length of line, number and sharpness of curves, length and steepness of grades, number of traffic-control points, number of intersecting roads, length of minimum sight distances and frequency of occurrence, number of stream crossings, etc. Other considerations are the relative amount of traffic that will be attracted to each route, number of dwellings and communities served, probable effect on highway accident loss, and effect on property values along the route and in the region as a whole.

The AASHO has developed a procedure for comparing alternate locations or for justifying proposed new highways on the basis of road-user benefit ratios.[1] This ratio is the difference in road-user costs on alternate sections of highway divided by the difference in highway costs for those sections. The road-user costs include mileage costs for fuel, oil, tires, repairs, and depreciation plus time costs. The AASHO also recommends the inclusion of accident costs where reliable records are available and suggests a cost for absence of "comfort and convenience" on substandard roads. The highway costs are annual costs including items of interest, maintenance, and replacement as outlined above.

COST OF ROADS AND PAVEMENTS

The cost of roads and pavements varies from year to year, from place to place, and from job to job, depending upon availability of materials, wage scales, etc. Approximate cost ranges for the principal types of surfaces are given in Table 24. These are for base

TABLE 24. APPROXIMATE COST OF CONSTRUCTION AND MAINTENANCE FOR DIFFERENT TYPES OF PAVEMENT[1]

Surface and base only—excluding grading, drainage, and structures

Type	Construction cost		Annual maintenance cost		Approx traffic limit, vehicles per day	Probable life of surface, years
	Per mile of 24-ft width	Per sq yd	Per mile of 24-ft width	Per sq yd		
Gravel	$ 6,000– 9,000	$0.43–0.64	$400–900	$0.028–0.064	300	5+
Bituminous surface treated gravel	10,000–14,000	0.71–0.99	350–600	0.025–0.043	600	8+
Road mix	16,000–20,000	1.13–1.42	300–500	0.021–0.035	1,000	12
Bituminous macadam	20,000–30,000	1.42–2.13	250–500	0.018–0.035	3,000	17
Bituminous concrete on flexible base	35,000–45,000	2.48–3.19	200–400	0.014–0.028	Full capacity	20
Bituminous concrete on concrete base	65,000–75,000	4.61–5.32	200–350	0.014–0.025	Full capacity	20
Portland cement concrete 9" thick	60,000–80,000	4.26–5.67	150–300	0.011–0.021	Full capacity	25

[1] Cost ranges are for year 1956, when *Engineering News-Record* construction cost index was 690, and U.S. Bureau of Public Roads index for composite mile was 167.

and surface courses above the subgrade. In addition, there will be cost of grading, drainage, bridges, and highway accessories. The cost of these items will vary widely depending upon the nature of the terrain and whether the job is the reconstruction of an old road or is on new location. For a two-lane width the other than surface costs will run from $20,000 to $75,000 per mile and may reach $150,000 or more in mountainous terrain. The costs for three- and four-lane roads will be proportionately greater. For divided highways of limited-access type the total cost, including right of way, grading, drainage, surface, and grade-separation structures, will run about $1,000,000 per mile

[1] "Road User Benefit Analyses for Highway Improvement," AASHO.

in rural areas, and range from $5,000,000 to $25,000,000 per mile in thickly settled urban areas.

Maintenance costs are also subject to considerable variations, even for the same type of surface, being influenced by volume of traffic, climatic conditions, thoroughness of construction, age of surface, and economic conditions. In addition to surface costs shown in Table 24, there are costs for maintaining shoulders, ditches, slopes, drainage structures, fences, and for cutting brush along the right of way. These costs range from $100 per mile on local roads to $500 per mile on major two-lane roads. For divided highways the cost may run $2,000 or over per mile, depending upon the amount of landscaping undertaken. The cost of removing snow and sanding slippery surfaces ranges from $100 to $500 per mile per lane, depending upon the amount of snowfall and the frequency of icing conditions.

Unit prices are commonly used for estimating purposes. Highway departments keep a careful record of these costs and the trends of material, labor, and equipment costs, upon which they depend. Unit prices are subject to variation from locality to locality and are directly affected by changing economic conditions. Table 25 gives average unit prices bid on work in one locality in 1954.

TABLE 25. HIGHWAY CONSTRUCTION PRICES IN MASSACHUSETTS, 1954[1]

Selected items based on weighted average of contract unit prices on state highway Federal aid work

Item	Unit	Price	Item	Unit	Price
Clearing and grubbing	Acre	$328.00	Structural steel	Lb	$ 0.161
Excavation:			Catch basins	Each	156.00
Roadway earth	Cu yd	0.62	Frame and grate	Each	164.00
Rock	Cu yd	2.39	Manholes, excluding		
Bridge	Cu yd	3.47	castings	Each	126.00
Trench, earth	Cu yd	2.01	Frames and cover	Each	54.00
Trench, rock	Cu yd	7.03	Pipe, 12″, heavyweight		
Channel	Cu yd	1.83	cast iron	Lin ft	6.51
Peat	Cu yd	0.53	Pipe, 12″, asphalt-		
Borrow:			coated corrugated		
Ordinary	Cu yd	0.67	metal	Lin ft	3.50
Gravel	Cu yd	1.13	Pipe, 24″, reinforced		
Sand	Cu yd	2.73	concrete	Lin ft	6.71
Fine grading	Sq yd	0.09	Pipe, 10″, vitreous-clay		
Gravel for surfacing	Cu yd	1.55	sewer	Lin ft	3.06
Crushed stone:			Highway guard:		
Base course	Ton	3.18	Anchors, end	Each	47.00
Surface course	Ton	3.60	Cable, 3 strand	Lin ft	1.78
Bituminous material:			Fence, stock	Lin ft	0.41
For base course	Gal	0.137	Concrete edging,		
For surface treat-			straight	Lin ft	4.12
ment	Gal	0.158	Granite curb:		
Bituminous concrete,			Straight	Lin ft	3.28
Type I—hot mix	Ton	7.50	Curved	Lin ft	4.68
Concrete pavement	Cu yd	20.00	Riprap	Cu yd	7.50
Steel reinforcement			Waterways, paved		
for	Sq yd	1.07	with Type I bitumi-		
Sawed contraction			nous concrete	Sq yd	3.32
joints	Lin ft	0.98	Seeding	Sq yd	0.09
Concrete masonry,			Sidewalks:		
Class B	Cu yd	34.50	Bituminous concrete	Ton	13.20
Steel reinforcement for			Concrete, 2-course	Sq yd	4.05
structures	Lb	0.118			

[1] From "Summary of Unit Bid Prices for Highway and Bridge Contracts," 1954, Massachusetts Department of Public Works.

Cost indexes have been developed which are useful for making comparisons between costs incurred in different years and for bringing costs to date. One of the best known is the *Engineering News-Record* Construction Cost Index, tabulated monthly in the *Engineering News-Record* magazine. The Bureau of Public Roads has developed an index particularly for highway construction costs based upon bid prices on Federal aid work. Using an average of years 1925–1929 as a base 100, index numbers are computed quarterly for a "composite mile" of highway consisting of representative quantities of excavation, surfacing, structural steel, and structural concrete. Index numbers at 5-year intervals are as follows: 1940, 71.6; 1945, 109.0; 1950, 137.7; 1955, 152.8. Several state highway departments compile indexes based on their own construction costs.[1]

AIRPORT ENGINEERING

Sources of Airport Information.[2] Current practice and policies in airport design and construction are available from the following:

Civil Aeronautics Administration (CAA) of the U.S. Department of Commerce has published bulletins covering airport classifications, planning, design specifications, layout, choice of site, drainage, paving, buildings, and lighting. Typical drawings and construction specifications have also been issued. The services of the CAA are available for advice on airport problems. Frequent references will be made to CAA standards throughout this text.

Department of the Air Force has issued Air Force Regulations Nos. 86-1 to 6 covering airfield design criteria, air-approach clearances, and planning and development procedures.

Corps of Engineers, Department of the Army. The Engineering Division of the Corps of Engineers has conducted extensive research on airport-pavement design and construction, much of which is available in reports. Current practice is in "Engineering Manual for Military Construction," Office of Chief of Engineers, Department of the Army.

Highway agencies, such as the Highway Research Board, U.S. Bureau of Public Roads, and American Road Builders Association, have published papers and bulletins on airport subjects, particularly soils investigations, grading methods, and pavement design.

Technical magazines are *Aviation Age, Aviation Week, Aero Digest, American Aviation,* and *Aeronautical Engineering Review.* Also see highway magazines, p. 2–59, and architectural periodicals.

Definitions. Terms frequently used in discussing airports are: *landing strip,* a graded strip 200 to 500 ft wide; *runway,* a paved strip in the central portion of the landing strip provided specifically for landings and take-offs; *taxiway,* a strip (usually paved) connecting runways with each other and with apron; *apron,* a paved portion of the airport immediately adjacent to terminal area or hangars, which is used for loading, unloading, or parking planes; *warm-up apron,* paved area near end of a runway where a plane may stand while engines are being tested or while waiting for take-off; *overrun strip,* a graded area in extension of runway to allow for emergency overruns or short landings; *approach zone,* the air space at each end of a landing strip which should be free of obstructions, the lower boundary being a plane sloping upward from the end of the runway overrun strip at a specified slope known as the "glide angle" and defining the "glide path" or "approach path"; *turning zone,* air space between approach zones that should be free of obstructions; *contact landing,* one made under conditions of good visibility when the pilot is operating under "visual flight rules" (VFR); *instrument landing,* one made in overcast weather when the pilot is operating under "instrument flight rules" (IFR)

[1] See Annual Construction Cost numbers of *Engineering News-Record* for summary of highway, general construction, and building-cost indexes.

[2] See also "Sources of Highway Information," p. 2–59, and Bibliography at end of section.

and must rely on instruments in the aircraft and radio directions from airport control tower, or on "instrument landing systems" (ILS) or "ground-control approach" (GCA) to guide his plane to the runway for a landing; *wind rose*, a diagram showing the relative frequency and the velocity range of winds blowing from different directions in a given region; *wind coverage*, percentage of the time that runway directions will permit landings and take-offs without exceeding a specified allowable cross-wind component; *parallel runway*, one of a pair of parallel runways laid out in the same direction; *preference runway*, the runway best suited for landings and take-offs and used at all times except when cross winds are excessive; *cross-wind runway*, used only when cross-wind components are exceeded on other runway or runways; *instrument runway*, the one equipped with instrument-landing systems and used when landings are made under instrument-flight rules.

AIRPORT CLASSIFICATION AND DESIGN STANDARDS

Airport Classifications.[1] The CAA groups civil airports into eight classes based upon the type of air service to be accommodated and runway lengths required (Table 26). *Personal* airports are those capable of handling light (up to 3,000 lb) aircraft for small communities or urban areas. *Secondary* airports are for larger (2,000- to 15,000-lb) aircraft used in nonscheduled flying activities. *Feeder* or *local-service* air-line airports are to serve smaller cities on air-line trunk routes. *Express* airports are at important cities or junction points on trunk routes. *Continental* airports are to serve aircraft making long nonstop domestic flights. *Intercontinental* airports are at termini of long international flights. *Intercontinental express* airports are those serving the highest type of transoceanic flights.

Military airports [2] are classified by the Department of Air Force according to type of aircraft to be accommodated, such as bombardment, fighter, cargo and troop carrier, and training. Runway lengths are designed for the particular type of aircraft to be operated from each base. In general, minimum runway lengths at sea level under standard conditions range from 5,000 ft for basic trainers to 10,000 ft and more for heavy bombers.

Airport Design Standards. The CAA civil airport design standards for different classes of air service are summarized in Table 26. These specifications are the minimum which the CAA considers acceptable for safe operation. Since aircraft and airport requirements are constantly changing, the CAA specifications are subject to revision. The designer should consult the CAA for current requirements. The runway lengths were set as the maximum for which the CAA would match local construction funds with Federal aid. Greater lengths may be and are frequently built by airport owners.

Runway lengths are specified at sea level. For higher altitudes, the length must be increased, because the velocity required to develop the lift for take-off increases as the density of the air decreases. Longer runways are also required as the air density decreases with rising temperature, and when aircraft must climb ascending grades on the take-off run. The CAA prescribes the following corrections to be applied to runway length for other than sea level, standard temperature, and zero grade conditions: increase the runway length by 7 percent for each 1,000 ft of elevation above sea level; increase the corrected length by 0.5 percent for each degree which the mean temperature of the hottest month of the year exceeds the corrected standard temperature; further increase the length corrected for elevation and temperature by 20 percent for each 1 percent of effective runway gradient (see footnote, Table 26). The standard temperature is 59°F at sea level and should be decreased by 3.566°F for each 1,000 ft above sea level.

[1] "Airport Design," Civil Aeronautics Administration, Government Printing Office, Washington, D.C.
[2] Air Force Regulation 86-5, Department of Air Force.

TABLE 26. CIVIL AERONAUTICS ADMINISTRATION STANDARDS FOR PHYSICAL CHARACTERISTICS OF DIFFERENT CLASSES OF AIRPORTS [a]

Type of service	Personal	Secondary	Feeder	Trunk line	Express	Continental	Intercontinental	Intercontinental express
Length of runways, ft [b]	1,500–2,300	2,301–3,000	3,001–3,500	3,501–4,200	4,201–5,000	5,001–5,900	5,901–7,000	7,001–8,400
Width of landing strips, ft	200	250	300	400	500	500	500	500
Width of runways, ft	50 [c]	75 [c]	100	150	150	150	200	200
Width of taxiways, ft	20 [c]	30 [c]	40	50	60	75	75	100
Minimum distance between:								
Center lines of parallel runways, contact operations, ft	150	300	500	500	500	500	700	700
Runway center line and taxiway center line, ft	75	150	200	250	300	350	400	450
Center lines of parallel taxiways, ft	60	125	150	200	250	275	300	325
Center line of taxiways to aircraft parking area, ft	50	110	130	175	220	240	260	280
Center line of taxiway to obstruction, ft	50	75	100	100	125	150	175	200
Center line of runway to building line:								
Instrument operations, ft	......	......	750	750	750	750	750	750
Noninstrument operations, ft	150	225	300	350	425	500	575	650
Maximum runway grades, percent:								
Effective [d]	2	1½	1	1	1	1	1	1
Longitudinal [e]	3	2	1½	1½	1½	1½	1½	1½
Transverse [f]	3	2	1½	1½	1½	1½	1½	1½
Approach zones—obstruction clearance line ratios:								
Noninstrument runways	20:1	20:1	40:1	40:1	40:1	40:1	40:1	40:1
Instrument runways	......	......	50:1	50:1	50:1	50:1	50:1	50:1
Pavement loading per wheel (in 1,000 lb):								
Single wheel	g	g	15	30	45	60	75	100
Dual wheel	g	g	20	40	60	80	100	125

[a] "Airport Design," Civil Aeronautics Administration, January, 1949. Consult CAA for revisions.
[b] For sea-level elevation, standard sea-level temperature of 59°F and 0 percent effective gradient.
[c] No pavement required.
[d] Maximum effective gradient obtained by dividing the maximum difference in runway center line elevation by the total length of runway.
[e] When necessary longitudinal taxiway grades may be as high as 3 percent.
[f] Percentages shown are for pavement. To improve runoff, the slopes on unpaved areas may be increased to 2 percent; and to 5 percent for a distance of 10 ft from the edge of pavement.
[g] When pavement is used it shall be designed in accordance with Fig. 67.

The safe runway length for a given airplane depends on its weight and performance characteristics. The Civil Air Regulations [1] specify three requirements for civil air transports, each of which must be met: (1) the runway length should be sufficient for the plane to accelerate to the point of take-off and then, in case of failure of one engine, be braked and brought to a stop within the limits of the runway (or usable landing strip); (2) if engine failure occurs at point of take-off, the plane should be capable of take-off on the operating engine (or engines) and be able to clear the end of the runway at an elevation of 50 ft; (3) in landing, the plane should clear the end of the runway by 50 ft and be landed and brought to a stop within 60 percent of the available runway length.

Runway-numbering system adopted by CAA designates the ends of runways by the reverse of their azimuth, measured clockwise 0° to 360° from magnetic north. For simplicity the numbers are expressed in 10° units of azimuth. For example, if a runway has a magnetic azimuth of 32°, the end lying in that direction is numbered 21 (32° + 180°

[1] Civil Air Regulations, Part 04, Airplane Airworthiness, Sec. 04.75, Performance Requirements for Transport Category Airplanes, CAA.

= 212°). The other end is numbered 3 (for 32°), and the runway is referred to as 3-21. The object of the system is to have the number facing a landing plane correspond (in 10° units) to magnetic course of plane.

Grades of runways and taxiways are restricted to low maxima (Table 26), since up grades increase power required for take-off and down grades increase braking distance. Abrupt and frequent changes in grade are undesirable, because they make precise landing difficult, may cause bouncing of fast-moving planes, and may obscure the view along the runway. The CAA standards require that longitudinal grade changes (algebraic differences in grade) should not exceed 3 percent for personal and secondary airports and 1½ percent for feeder and larger airports. Longitudinal intersecting grades should be connected by vertical curves when the grade change is greater than 0.4 percent. The length of vertical curves should be 300 ft for each 1 percent change in grade for personal or secondary airports, and 500 ft for each 1 percent change in grade for feeder and larger airports. The distance between grade-intersection points should be at least 25,000 times the sum of the algebraic differences in grade (expressed in decimals) at the adjacent intersection points for personal and secondary airports; for feeder and larger airports the multiplier should be 50,000 instead of 25,000.

Approach-zone and turning-zone limits are prescribed for civil airports by the CAA.[1] These are shown in diagrammatic form in Fig. 54. Two sets of conditions are illustrated for approach clearances, one for instrument-landing runway and the other for noninstrument runway. For the former, the approach zone starts 200 ft from the end of runway with a width of 1,000 ft and rises at 50:1 for 10,000 ft where it is 4,000 ft wide. It then rises at 40:1 for an additional 40,000 ft where it is 16,000 ft wide. From the sides of glide planes, and from the runway lateral clearance lines, a side-clearance plane rises on a slope of 7:1. The clearance requirements for noninstrument approaches vary with the class of airport as noted in tabulation of dimensions a, b, c, etc. The turning zone consists of a horizontal circular plane of radius c (measured from airport reference point) at an elevation of 150 ft above established airport elevation (usually the highest point on landing area). It is bounded by a conical surface of width d rising on a 20:1 slope. The airport reference point is a centrally located point chosen for the purpose of defining the geographical location of the airport. The turning-zone clearances apply to all areas outside the approaches. The CAA has recommended that where practical the clearance regulations be applied to imaginary surfaces at heights above the terrain varying from 120 ft at 3 miles to 250 ft at 10 miles from the runway ends (see TSO-N18). Approaches should be protected from future encroachment by taking air easements or by zoning [2] ordinances based upon clearance requirements.

Design standards for military airports [3] differ somewhat from those of the CAA, the minima usually being higher. Runway clearance zone is 1,500 ft wide, and an overrun strip 1,000 ft long is required at ends of all runways, its width being equal to that of the runway plus two 200-ft shoulders. Maximum lingitudinal runway grade is 1 percent. Maximum rate of change of grade is 0.167 per 100 ft. Minimum length of vertical curve is 600 ft for each 1 percent of algebraic difference in grade. At least 1,000 ft is required between grade-change points on runway profile. Along runway edges and shoulders a maximum grade change of 0.4 percent is permitted to adjust grades at runway and taxiway intersections.

Approach zones start with a width of 1,500 ft at end of overrun strip and rise on 50:1 slope for 10,000 ft to a width of 4,000 ft at an elevation of 200 ft above the airport and then extend level and 4,000 ft wide for an additional distance of 15,000 ft, at which point the side-clearance slope of 7:1 is extended around the end of the approach zone. The turning zone consists of a horizontal circular plane of 10,000 ft radius at an elevation of

[1] Technical Standard Order TSO-N18, CAA, Apr. 26, 1950, amended July 30, 1952.
[2] The CAA and the Air Force have prepared model zoning regulations for approach protection.
[3] Department of Air Force Regulations, Series 86.

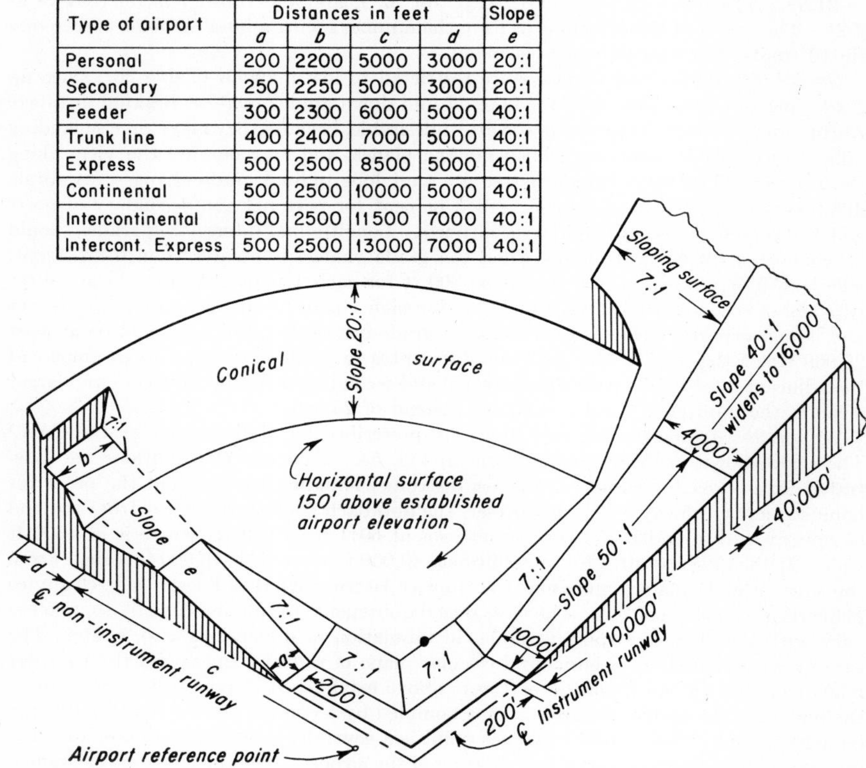

Type of airport	Distances in feet				Slope
	a	b	c	d	e
Personal	200	2200	5000	3000	20:1
Secondary	250	2250	5000	3000	20:1
Feeder	300	2300	6000	5000	40:1
Trunk line	400	2400	7000	5000	40:1
Express	500	2500	8500	5000	40:1
Continental	500	2500	10000	5000	40:1
Intercontinental	500	2500	11500	7000	40:1
Intercont. Express	500	2500	13000	7000	40:1

FIG. 54. CAA runway and turning zone clearance limits.

150 ft above the airfield elevation. This is bounded by a conical surface 7,000 ft in width rising on a 20:1 slope to an elevation of 500 ft above the airfield, whence it extends horizontally to the limits of a radius of 50,000 ft from airfield reference point. For some types of operation glide slopes flatter than 50:1 are required.

SELECTION OF AIRPORT SITE

The selection of an airport site is influenced by a number of factors, such as the area required, possibility for expansion, accessibility to the community, absence of obstructions in approaches, freedom from fog and smoke, nature of terrain, nearness of other airports, and the cost of development.

Area needed is determined by runway length and layout and by terminal-area requirements. A small airport of the personal class may be located on 50 to 150 acres. Airports of the intercontinental express class may cover as much as 4,000 to 5,000 acres.

Possibility for expansion should be ensured by the selection of a site that is not hemmed in by built-up property, railroad yards, mountains, rivers, harbors, or other features, which prohibit enlargement except at excessive cost. Ample undeveloped land should be available adjacent to the site and protected by zoning against uncontrolled growth of industrial or residential property that will block runway extensions or terminal-area expansion.

Terrain should be relatively flat to avoid high grading costs. Elevated sites are preferable to those in lowlands, since the former are usually free from obstructions in approach zones, less subject to fog and erratic winds, and easy to drain.

Approaches to the proposed runway layout should be free of obstacles, such as mountains, hills, tall buildings, transmission lines, factory chimneys, and radio towers.

Control of site and its surroundings by zoning should be investigated to ensure protection of approach and possibility of expansion. If the airport is located outside the community to be served, arrangements must be made with local government to guarantee proper control.

Nature of soil influences cost of construction. If possible, the site should be cleared ground that is easily drained and has sandy- or gravelly-type soil that offers a satisfactory foundation for runway pavements without excessively thick subbases and costly subdrainage systems.

Accessibility to the community is essential to preserve the advantage of the speed of air transportation. In general, accessibility is measured in time rather than distance. Sites near modern express highways are to be sought, and those bounded by traffic-congested streets avoided. On the other hand, the site should not be so remote from the community as to require excessive transportation time.

Effect on land values and tax assessments may be adverse or beneficial depending upon the nature of the site. If the airport is located near residential property, the value may go down because of noise nuisance and commercial atmosphere of the airport. If located on undeveloped land, the airport will increase value of land for industrial sites and for other uses related to the airport.

Availability of utilities, such as electric power, telephone, water, sewers, and public transportation, should be investigated.

Spacing of airports is a consideration, since airports should not be spaced so closely that their air-approach traffic patterns overlap. The CAA recommends that the minimum distance between airports for contact operations should be equal to the sum of the radii of their traffic patterns. The radii of airport traffic patterns are as follows: personal and secondary airports, 1 mile; feeder, 2 miles; trunk line and express, 3 miles; continental and larger, 4 miles. A spacing of 14 miles is recommended for the larger airports where scheduled air lines make frequent instrument landings.

Costs of development and operation are major considerations. Alternate sites that meet operational requirements should be compared on the basis of cost; not only the first cost, but the long-range economic cost, which includes interest on investment, amortization, and annual maintenance and operating expenses (see "Economics," p. 2–119). In making this comparison the cost of the ultimate design as worked out on a master plan should be considered as well as the cost of the initial stages of the work.

Reconnaissance for airport sites involves the use of U.S. Geological Survey topographic maps supplemented by a study of aerial photographs, using stereopairs for studying relief. Preliminary runway patterns are drawn on the topographic map, and the approaches checked for obstructions. If soils maps [1] are available, they may be used for foundation and drainage studies. Promising sites are investigated in the field and careful notes made of all factors influencing the desirability of the site. Detailed studies are then made of the most desirable site.

LAYOUT OF AIRPORT

Runway Layout. The choice of runway pattern is influenced by the necessity of obtaining clear approaches, the desirability of covering winds, and the necessity for fitting the layout to the topography in such a way as to secure low grading and drainage

[1] The Origin, Distribution and Airphoto Identification of United States Soils, *CAA Tech. Development Rept.* 52, May, 1946 (with bibliography).

costs. The shape and location of the terminal area also influence the layout. Short and direct taxiing distances are desired between runways and the airport terminal. The number of runways will depend upon wind-coverage requirements and traffic volume to be handled. To increase capacity, the layout must permit simultaneous use of two or more runways. The orientation of the runways depends upon obstacle-clearance requirements and wind coverage. The instrument runway should, if possible, be aligned with the airway pattern so as to permit direct approaches from direction of heaviest traffic. Also, the approaches should be chosen to fall, if possible, over sparsely settled areas where the public will be the least inconvenienced by noise and least exposed to hazard of aircraft accidents.[1]

Wind Coverage. The CAA specifies that runways should be oriented so that aircraft may be landed at least 95 percent of the time with cross-wind components not to exceed 15 mph. This is considered the average maximum cross wind which can be safely accepted by light- and medium-weight aircraft. Larger transport aircraft can be landed safely with cross-wind components of 20 to 30 mph. Since most airports are used by light planes as well as transports, compliance with a 15 mph component is recommended wherever practical. However, with increased landing speeds, the general adoption of tricycle landing gears on large aircraft, and the development of castered landing gears for smaller aircraft, the factor of wind coverage has become less critical. The trend is toward one or possibly two directional layouts. In some localities, where the prevailing winds are consistently in one direction or the reverse, a single runway will meet CAA requirements. One-runway layouts are sometimes adopted when wind-coverage requirements are not fully met but the approaches are excellent and other factors are satisfied.

Wind coverage is determined from a *wind rose.* A simple type consists of bars radiating in the several compass directions, each representing to scale the percent of time that the wind blows from the direction in which that bar points. For a mathematical computation of wind coverage on the basis of cross-wind component, a wind rose similar to that shown in Fig. 55 is required. This rose gives the percent of time by velocity ranges as well as by directions. The small figures on the diagram represent the percentages of time during which the wind blows from the several compass directions between designated velocities. For the wind rose in Fig. 55, the percentages of winds were known for velocity ranges of 0 to 4 mph (calms), 4 to 15 mph, 16 to 31 mph, 32 to 47 mph, and over 47 mph. Winds over 47 mph accounted for less than 0.1 percent and were neglected. This wind rose may be used to determine the maximum wind coverage for a one-, two-, or three-directional runway layout, or it may be used to check the wind coverage for a layout adopted after a study of obstacles in approaches and other factors. For finding the maximum wind coverage possible for a given runway, a transparent plot is made of the runway center line, and parallel lines are drawn representing the limits of 15-mph cross-wind components on each side of the center line. This plot is then superimposed on the wind rose with the center line passing through the center of the rose, and rotated until a direction is found in which the greatest percentage of wind is included within the 30-mph-wide band. If the layout has more than one runway, bands are plotted for each runway and shifted about the center of the wind rose until the direction for each runway is found such that the total percentage of wind coverage by all runways is a maximum. In Fig. 55, a two-runway layout is to be checked for wind coverage; first for Runway A alone, and then for both Runways A and B. The runway center lines are plotted on the wind rose in their proper compass directions, and lines drawn parallel to each center line representing to the scale of the wind rose the limits of all cross-wind components of 15 mph. For simplicity, the percentage of

[1] See "The Airport and Its Neighbors," Report of the President's Airport Commission, May 16, 1952, published by Government Printing Office.

winds not covered is computed and deducted from 100. The percentages and fractions of percentages outside the limits of coverage (dashed lines in Fig. 55) for Runway A are as follows: in directions NW to E, 0.4 × 0.1 + 0.0 + 0.6 × 0.7 + 0.1 + 0.9 × 0.8 + 0.0 + 1.1 + 0.2 + 2.3 + 0.0 + 0.8 × 0.1 + 0.6 × 0.1 + 0.1 × 0.2; from SE to W, 0.4 × 0.1 + 0.0 + 0.5 × 0.1 + 0.0 + 0.9 × 0.4 + 0.1 + 1.2 + 0.1 + 0.9 × 0.5 + 0.0 + 0.6 × 1.0 + 0.6 × 0.1 + 0.1 × 1.6 = 8.16, or 91.84 percent coverage. The addi-

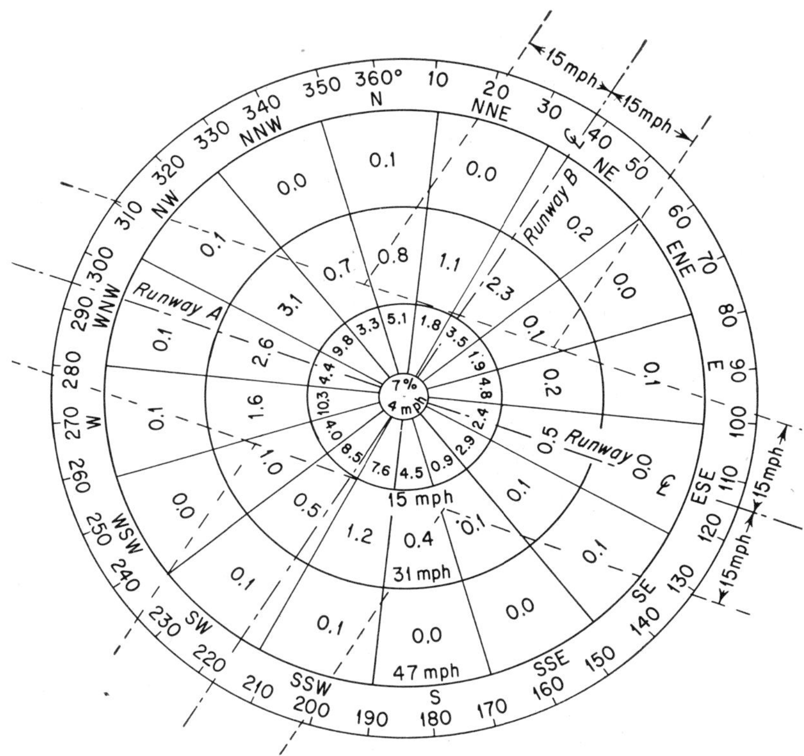

FIG. 55. Wind coverage for cross-wind component of 15 mph.

tion of Runway B will add the following coverage: from N to ENE, 0.5 × 0.8 + 0.0 + 1.1 + 0.2 + 2.3 + 0.0 + 0.6 × 0.1 and from S to WSW 0.5 × 0.4 + 0.8 × 0.1 + 1.2 + 0.1 + 0.9 × 0.5 + 0.0 + 0.4 × 1.0 = 6.49, giving total coverage for two runways of 98.33 percent. The analysis may be refined by using more wind-velocity groups if they are available, or it may be applied for other cross-wind components. The Air Force Regulations call for coverage of a 13-mph cross-wind component for 95 percent of the time for all-purpose airports, 90 percent for medium-bomber, and 75 percent for heavy-bomber airports.

The wind rose usually employed for study purposes is plotted on an annual basis. In locations where the wind distribution varies during the year, roses should be plotted for the different seasons and the fluctuations taken into account in design, particularly if the airport is used mostly in certain seasons. For selecting the instrument-runway orientation, a wind rose for low-visibility conditions is useful.

Checking Obstacle Clearance. To test the approach zones for obstacle clearances, an accurate topographic map of the airport site and its environs is required for a radius of at least 5 miles from the airport boundary. A convenient method for testing approaches is to prepare a plot on transparent paper or cloth showing the extension of a runway center line, the limits of the approach zone and contour lines representing the elevations of the prescribed glide plane and side-clearance planes. Figure 56 illustrates a template

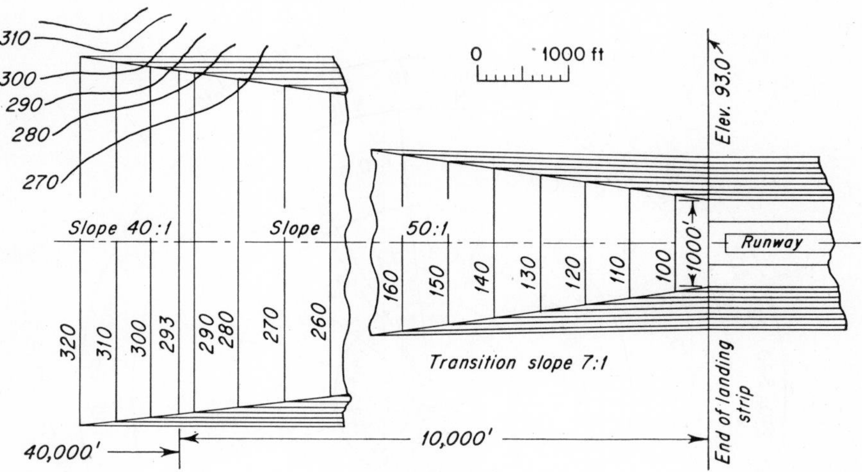

FIG. 56. Template for checking approach-zone clearance.

for testing clearances for an instrument-runway approach. The approach-zone template is fitted to the end of each runway, and the ground-surface contours compared with those of the glide plane. Any high places or man-made features on the ground that will protrude into the glide plane are noted. The runway layout is adjusted, if necessary, to avoid obstacles with a minimum of sacrifice of wind coverage. The turning-zone clearances are examined in a similar manner. All obstructions in the turning zone are

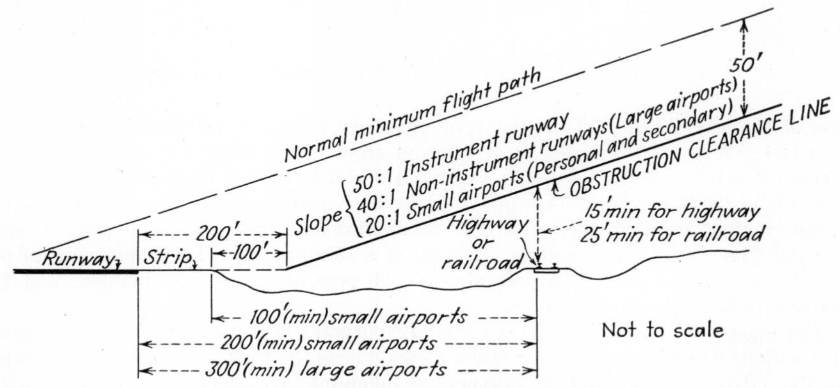

FIG. 57. CAA standards for clearance over highways and railroads.

spotted and measures taken to remove as many as possible, and to mark and light those which cannot be removed. Detailed plans are made of critical areas in approach zones showing heights of trees, poles, buildings, etc., that come near the glide plane. Steps are then taken to obtain control of these areas by easement or purchase, so that the obstructions may be removed. Clearances for railroads and highways are shown in Fig. 57.

Runway Configurations. The simplest layout is a single runway with parallel taxiway and centrally located terminal area as shown by full lines in Fig. 58*a*. Two directions of

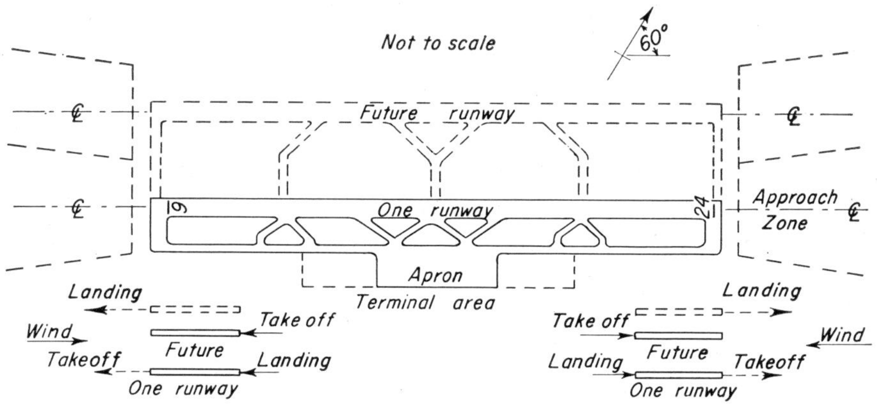

(*a*) Single runway airport layout with future parallel runway.

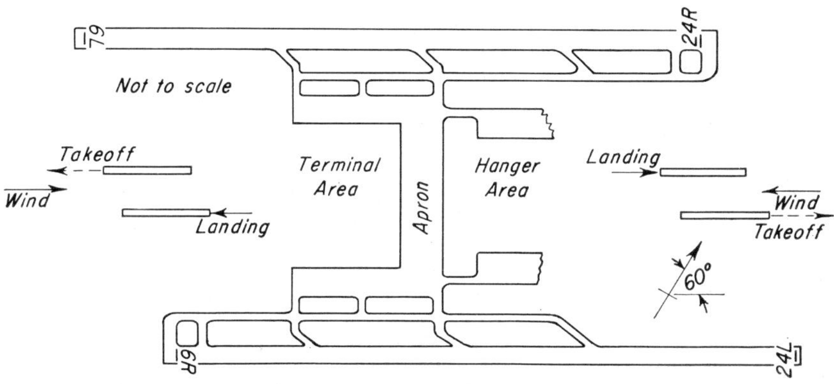

(*b*) Separated parallel runway layout.

Fig. 58. Single- and parallel-runway layouts.

operation are possible, 6-24 or 24-6. Only one landing or take-off can be made at a time. Under these conditions the capacity of the runway is about 40 movements per hour (including both landings and take-offs). When more capacity is needed, a second parallel runway may be built as shown by dashed lines in Fig. 58*a*. In this design, the original runway can be used for take-offs while the "future runway" is used for landings. The capacity under visual flight rules will be raised to about 80 movements per hour.

Landing traffic will have to cross the take-off runway under control from tower. Figure 58*b* shows parallel runways 3,000 to 4,000 ft apart with the terminal area between runways. This arrangement has definite operational advantages over the layout in Fig. 58*a*. Taxiways do not cross runways, the terminal area is centrally located with ample room for expansion, and the wide separation of runway approaches will increase capacity under conditions of low visibility. A larger area will be required than in Fig. 58*a*. The two parallel runways need not be opposite each other. Increasing the offset from the terminal area will decrease taxiing distance but may increase land and construction

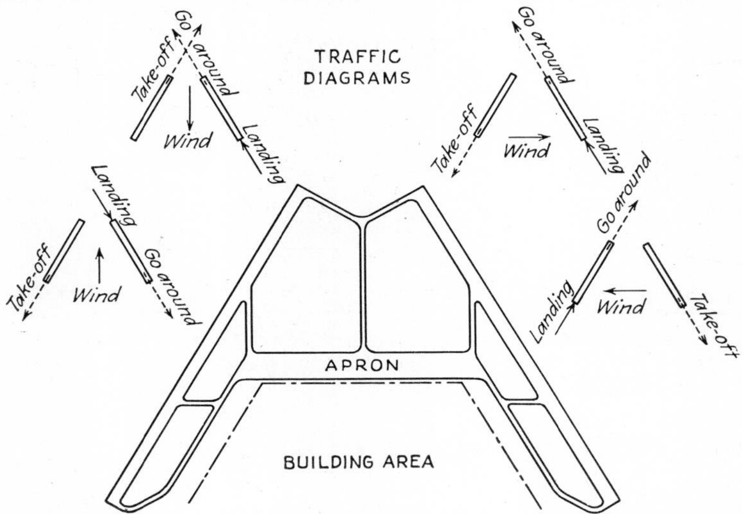

Fig. 59. Two-directional layout with nonintersecting runways. (*Civil Aeronautics Administration.*)

costs. Taxiways may be extended to the runway ends to provide exits for incompleted take-offs, or, in case of emergency, to facilitate landings and take-offs on the same runway.

In Fig. 59 an open V-type layout is shown. This layout gives four directions of wind coverage and also allows simultaneous operation of runways in most directions when wind velocities are not unusually high. The traffic diagrams indicate a separation of landings and take-offs in three of four wind directions. In the one situation where the landing go-around path intersects take-off path the landings and take-offs will have to be made separately. The V shape permits a centrally located terminal area with room for expansion. In some designs the angle of the V is made about 90°. When additional capacity is required, the designs in Figs. 58*b* and 59 may be expanded by building a parallel runway outside of each of the original runways at a spacing of 700 to 1,000 ft. Two runways would then be available for landings and two for take-offs at all times in Fig. 58*b*, and for most of the time in Fig. 59. Most existing airports have crossing runways. At some locations it is impractical to build nonintersecting runways. When winds are not critical, the capacity of these designs can be improved over single-runway operations by using one runway for take-offs and another for landings. The movements are alternated under directions from the traffic-control tower. Under instrument-landing procedures, capacity is reduced and delays to landings occur. Improvements in air-traffic control are ultimately expected to increase landing rates in overcast weather to nearly equal those in good weather.

Taxiway Systems. Taxiways are laid out to connect the terminal area with ends of runways for take-offs and to tap the runways at several points to provide exits for landing aircraft. Landings usually do not require the full length of the runway. To clear runway of landing planes as rapidly as possible, easy turns are introduced at taxiway exits (Fig. 58). At the ends of runways the taxiways join the runway at about 90° to give the pilot a view of the runway and its extension in both directions. Additional pavement is added to make room for waiting planes and to allow one plane to pass another in the take-off sequence. Taxiway widths and clearances are given in Table 26. Widths of 75 or 100 ft are usually provided at the large commercial airports. At some of the smaller airports, runways or portions of runways are used as taxiways. On personal-type airports, taxiing is often done over turf.

Aprons. The apron or "ramp" adjacent to the terminal is used for loading and unloading airplanes, for fueling, and for minor servicing and checkup. The dimensions of the apron depend upon the number of loading positions required and the size and turning characteristics of aircraft. The number of spaces depends upon the time of occupancy per aircraft, the time being longer at terminal airports than at en route stops. A study of apron operations by the CAA [1] developed an average ratio of 0.45 loading or "gate" positions for each hourly plane movement. For 40 movements per hour 18 loading positions are indicated. The study points out that apron operations could be performed in much less time, but practical considerations, such as standing over between schedules and allotment of gate positions to individual companies, prevent the attainment of maximum efficiencies. The airplane positions are designated by circles depending upon size of airplane (see p. 2–148). A diameter of 150 ft will take care of aircraft with wing spans up to about 120 ft. Larger aircraft will require 175-ft-diameter circles. Aprons may be marked to provide desirable combinations of 150-ft, 175-ft, or other size circles. The minimum length of apron for 40 movements per hour, with 18 places at 150 ft diameter is 2,700 ft; at 175 ft diameter it would be 3,150 ft. To provide such lengths of frontage without requiring excessive walking distances from the terminal, special shapes are designed as described under terminals, p. 2–147. The width of apron depends upon size of loading positions and taxiing requirements. At small airports the widths will be 200 to 300 ft. At larger airports 500 to 600 ft is required.

Master Plan.[2] Each airport should have a master plan showing ultimate development, even though the actual construction is to be in stages, as in Fig. 58*a*. The ultimate plan will provide a basis for acquiring ample land and for determining zoning required to protect future approaches. The plan should be flexible enough to permit modifications between stages of construction to meet the changing demands of air transportation.

Military air bases [3] require a master plan not only for the airfield but also for many related activities. The base is usually located at some distance from settled areas, and so requires all the facilities of a small city: roads, housing, water supply, sewerage system, power, etc., as well as large storage facilities for supplies and ammunition. The master plan requires an allocation of space for the ultimate needs of all operations with provisions for stage development.

Airport Construction Plans.[4] Airfield construction plans include a location and site plan, master plan, layout plan, clearing plan, borings and soils-exploration plot, grading and drainage plan, runway and taxiway profiles, access-road plans and profiles, drainage-line profiles, pavement cross sections, drainage structures, lighting and conduit plan, turfing plan, and summary of construction quantities. Plans are also required for development of terminal area, parking lots, and for construction of terminal buildings.

[1] Stafford, P. H., and W. Carsel, Aircraft Ramp Time, *Aviation Age*, vol. 17, Nos. 4, 5, and 6, April, May, and June, 1952.
[2] "Airport Planning," CAA, July, 1952.
[3] Installation Planning and Development, Air Force Regulation 86-4.
[4] "Model Set of Airport Plans," available from CAA.

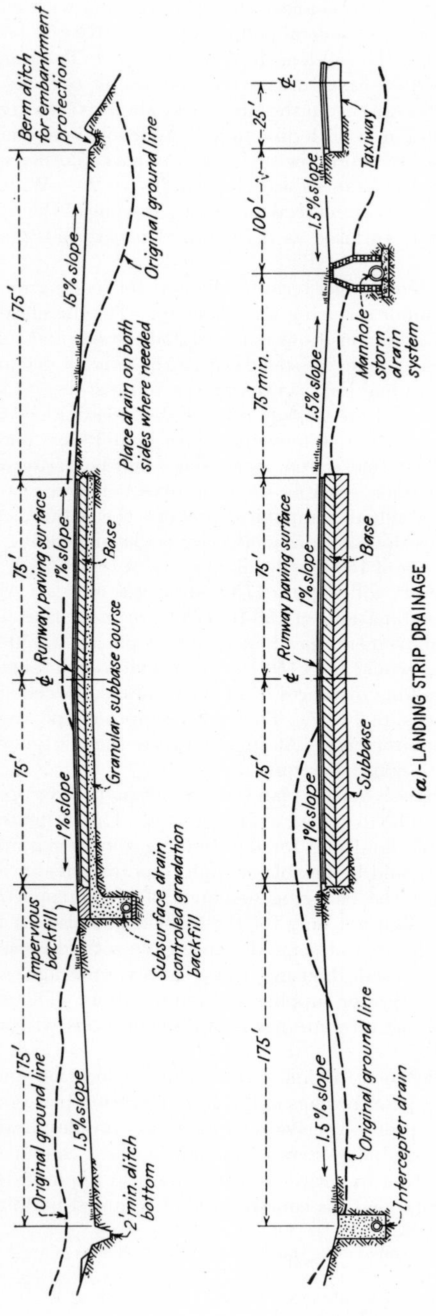

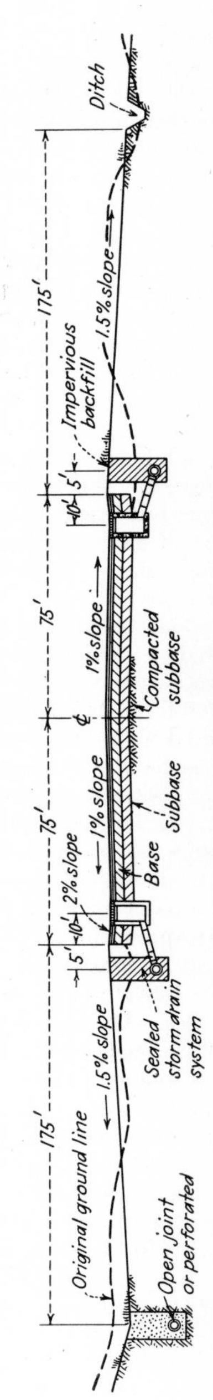

Fig. 60. Typical runway cross sections. (*Civil Aeronautics Administration.*)

AIRFIELD DESIGN AND CONSTRUCTION

Runway Cross Section. Runways and taxiways are sloped transversely each way from the center line to provide for surface drainage. Paved surfaces are sloped 1.0 to 1.5 percent, and graded areas of the landing strip are sloped to 1.5 to 2.0 percent. Paved runways are of bituminous types or portland cement concrete, the thickness depending upon the weights of aircraft to be accommodated (see "Pavement Design," pp. 2–140 and 2–144). Side slopes of cuts and fills should be as flat as possible. In cut they should not encroach on lateral clearance ratio, measured normal to edge of landing strip. For smaller airports the ratio is 5:1, for larger airports 7:1, and for military fields 10:1.

For *military* air base runways a minimum cross slope of 1.5 percent is required on paved runways and taxiways, and a minimum of 2 percent and maximum of 3 percent on turfed shoulders. In the lateral safety zone outside the shoulders no grades over 10 percent are permitted. The shoulders are 200 ft wide instead of 175 ft as in Fig. 60. The safety zone extends 750 ft on each side of runway center line.

Drainage. *Surface drainage* is collected along the edges of the landing strip in shallow ditches leading to inlets piped to storm sewers (Fig. 60*a*). At some of the larger airports, with wide paved runways, the surface water is also collected along the edge of the runways (Fig. 60*b*), particularly in northern climates where snowbanks along the edges of the runway block drainage across the landing strip. Surface-drainage inlets may be placed just outside the edges of runways, or they may be set in a shallow depression built in the outer edge of the pavement (Fig. 61). Inlets are usually spaced from 200 to 300 ft apart along the runway (or taxiway).

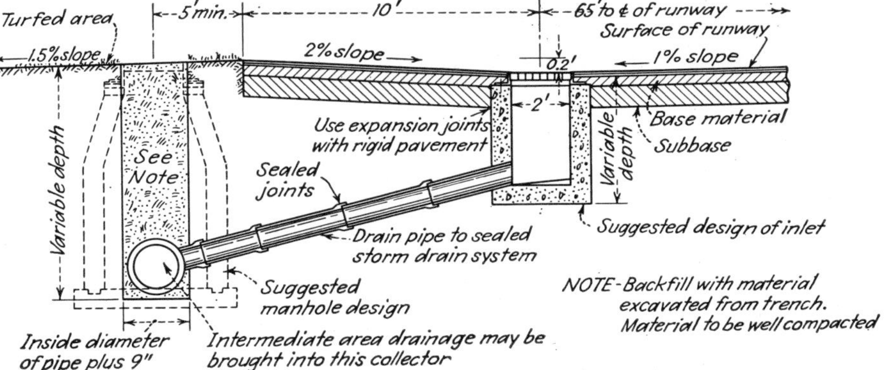

Fig. 61. Section showing details of pavement gutter drain. (*Civil Aeronautics Administration.*)

Subsurface drainage is obtained by the use of interceptor drains and pervious base-course layers in much the same way that highways are drained (see "Highway Drainage," p. 2–92). Some of the smaller turfed fields are drained by a network of subdrains covering the entire area. At airports with

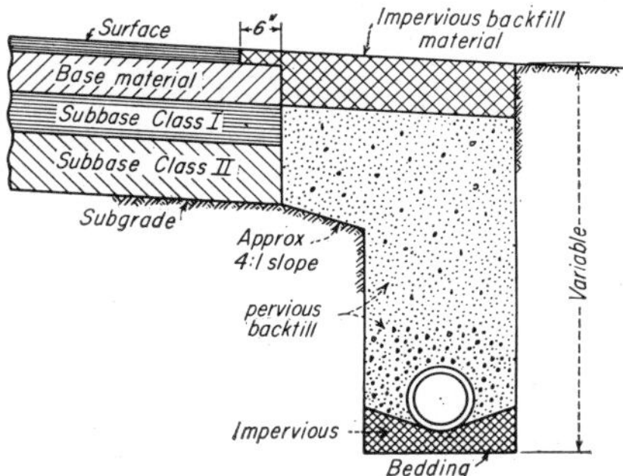

Fig. 62. Typical section of subbase drain. (*Civil Aeronautics Administration.*)

paved runways, subdrains are usually placed along the edges of the runways where soil conditions indicate that drainage is needed to lower the ground-water level. A combined interceptor and base drain is often used (Fig. 62).

Design of Drainage System.[1] A system of underground pipes is required to carry surface water from inlets and subdrains to outlets into natural waterways. In low areas,

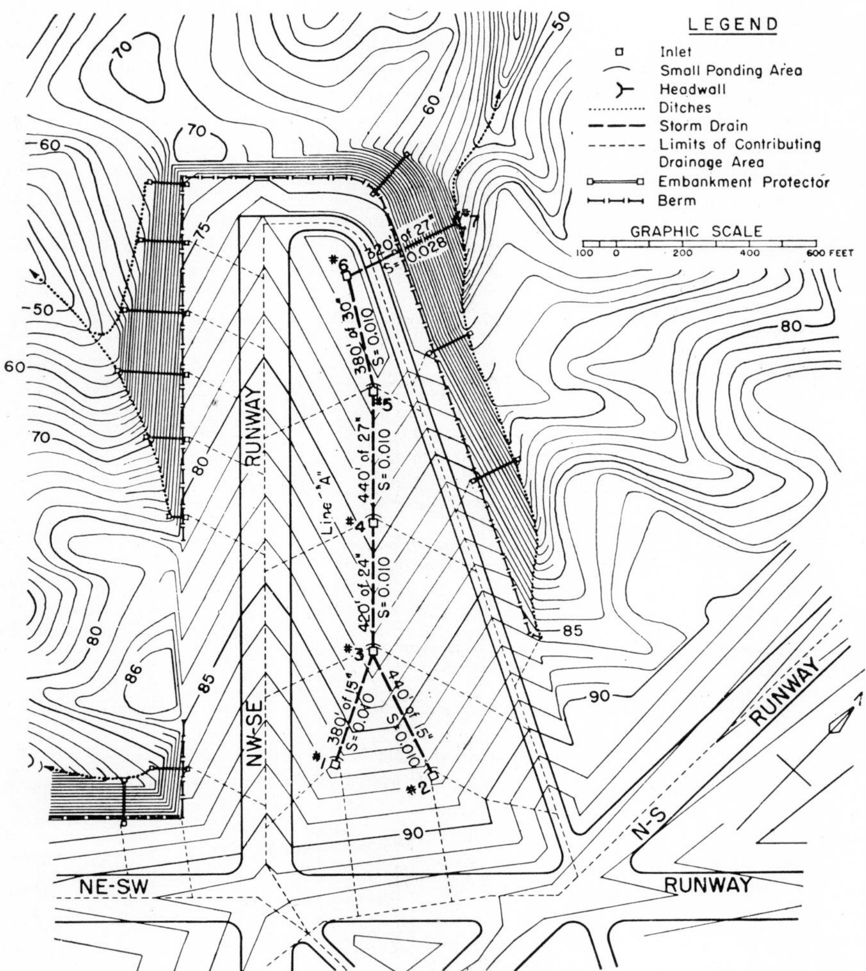

Fig. 63. Portion of airport showing drainage design details. (*Civil Aeronautics Administration.*)

surface waters are sometimes drained into ditches or canals running around the perimeter of the airport. For the design of the drainage system, a topographic map to the scale of 200 ft or less to the inch is required, upon which is plotted the proposed layout of runways, taxiways, aprons, and the terminal plan. The proposed surface grades of these

[1] See "Airport Drainage," CAA, Government Printing Office.

features are shown by contours of small interval; 0.1 or 0.2 ft for paved areas, and 0.5 or 1.0 ft for turfed areas. Inlet locations and subdrains are plotted, and storm drain lines laid out to collect the discharge from them. The system should be as direct as possible to avoid excessive lengths of pipe; frequent changes in pipe size should also be avoided. Crossings of pipes under runways should be held to a minimum.

Figure 63 shows a portion of an airport drainage system. The pipe sizes are computed to accommodate the discharge from the design storm, which may be taken as that expected once in every 2 to 10 years, depending upon how serious an effect an occasional flooding may have on airplane operations. In some designs, a certain amount of ponding is permitted in areas outside the runways. The design of storm-sewer systems by the "rational" method is covered in Section 9. For military air bases the "overland-flow" method of computation is recommended.[1]

Intersection studies are required to determine pavement grades where runways or taxiways intersect. Center-line grades are held constant and the grades of the outer portions of the runway adjusted so that there will be no abrupt changes in grade in the path of planes.[2] The surface should also have sufficient slope to drain properly. A separate plot is made to the scale of 1 in. = 50 ft and the final grades represented by contours using an interval of 0.1 ft.

Soils Investigations. These are required for designing runway pavements and base courses, and for planning drainage and subdrainage systems. Soils testing is also required for the control of the compaction of fills and base courses so that there will be no detrimental settlement under heavy plane loads. The procedures for soil sampling and testing are much the same as for highways (see p. 2–87). The results are plotted on soils profiles or on a boring plan which shows locations of borings with respect to proposed runway layout, and also individual profiles of the soil layers encountered at each location with a description of each soil type. Soil classifications for airports have been adopted by the Corps of Engineers and by the CAA. These are useful for correlating soil types and conditions with runway-pavement and base-thickness requirements (see Tables 27 and 28). The classifications are supplemented by charts (Fig. 64) to aid in classifying fine-grained soils. The A line in the Corps of Engineers Unified Soil Classification is an empirical boundary between inorganic clays (CL-CH groups) and organic colloids (OL-OH groups). The vertical line at liquid limit of 50 separates soils of low and high compressibility. In the CAA classification the soils are arranged in order of decreasing desirability as subgrades. The relation between soil classes and soil description is given in Fig. 48.

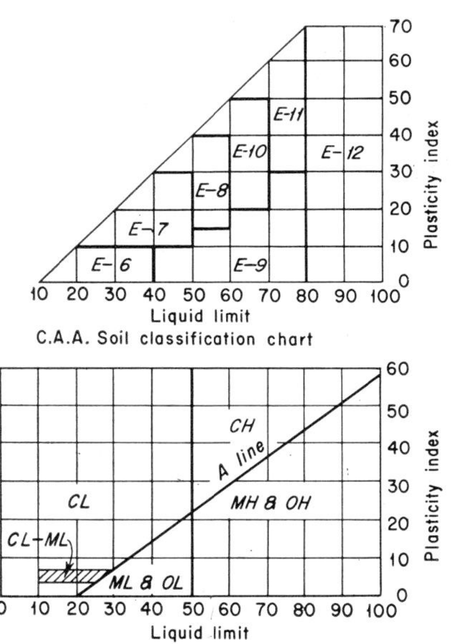

FIG. 64. Classification charts for fine-grained soils.

[1] Surface Drainage Facilities for Airfields, "Engineering Manual for Military Construction," Part XIII, Chap. 1, Office of Chief of Engineers, Department of Army. Also Military Airfields, A Symposium, *Trans. ASCE*, vol. 72, No. 2, Part 3, pp. 697–733, 1945.

[2] "Engineering Manual for Military Construction" (Office of Chief of Engineers, Department of Army) requires no rate of change of grade over 0.4 percent along runway edges at intersections.

TABLE 27. UNIFIED SOIL CLASSIFICATION—CORPS OF ENGINEERS [1]

Group symbol [2]	Description [3]	Unit dry weight, lb per cu ft [4]	Field CBR	Subgrade modulus k, lb per cu in.
	Coarse-grained Gravelly Soils			
GW	Well-graded gravels or gravel-sand mixtures, little or no fines. Excellent foundation, none to very slight frost action potential, excellent drainage characteristics	125–140	60–80	300 or more
GP	Poorly graded gravels or gravel-sand mixtures, little or no fines. Good to excellent foundations, none to very slight frost action, excellent drainage	110–130	25–60	300 or more
GM-d	Silty gravels, gravel-sand-silt mixtures. Good to excellent foundation, slight to medium frost action, fair to poor drainage	135–145	40–80	300 or more
GM-u	Silty gravels, gravel-sand-silt mixtures. Good foundation, slight to medium frost action, poor to practically impervious drainage	120–140	20–40	200–300
GC	Clayey gravels, gravel-sand-clay mixtures. Good foundation, slight to medium frost action, poor to practically impervious drainage	120–140	20–40	200–300
	Coarse-grained Sand and Sandy Soils			
SW	Well-graded sands or gravelly sands, little or no fines. Good foundation, none to very slight frost action, excellent drainage	110–130	20–40	200–300
SP	Poorly graded sands or gravelly sands, little or no fines. Fair to good foundation, none to very slight frost action, excellent drainage	100–120	10–25	200–300
SM-d	Silty sands, sand-silt mixtures. Good foundation, slight to high frost action, fair to poor drainage	120–135	20–40	200–300
SM-u	Silty sands, sand-silt mixtures. Fair to poor foundation, slight to high frost action, poor to practically impervious drainage	105–130	10–20	200–300
SC	Clayey sands, sand-clay mixtures. Fair to good foundation, slight to high frost action, poor to practically impervious drainage	105–130	10–20	200–300

TABLE 27. UNIFIED SOIL CLASSIFICATION—CORPS OF ENGINEERS [1] (*Continued*)

Group symbol [2]	Description [3]	Unit dry weight, lb per cu ft [4]	Field CBR	Subgrade modulus k, lb per cu in.
	Fine-grained Soils—Silt and Clay [5]			
ML	Inorganic silts and very fine sands, rock flour, silty or clayey fine sands or clayey silts with slight plasticity. Fair to poor foundation, medium to high frost action, fair to poor drainage	100–125	5–15	100–200
CL	Inorganic clays of low to medium plasticity, gravelly clays, sandy clays, silty clays, lean clays. Fair to poor foundation, medium to high frost action, practically impervious	100–125	5–15	100–200
OL	Organic clays of medium to high plasticity, organic silts. Poor foundation, medium to high frost action, poor drainage	90–105	4–8	100–200
	Fine-grained Soils—Silts and Clays [6]			
MH	Inorganic silts, micaceous or diatomaceous fine sandy or silty soils, elastic silts. Poor foundation, medium to high frost action, fair to poor drainage	80–100	4–8	100–200
CH	Inorganic clays of high plasticity, organic silts. Poor to very poor foundation, medium frost action, practically impervious	90–110	3–5	50–100
OH	Organic clays of medium to high plasticity, organic silts. Poor to very poor foundation, medium frost action, practically impervious	80–105	3–5	50–100
Pt	Peat and other highly organic soils. Not suitable for foundation, slight frost action, fair to poor drainage, very high compressibility			

[1] Condensed from Technical Manual 3-357, "Unified Soil Classification System," Waterways Experimental Station, Corps of Engineers, Department of Army. For original airfield classification see CASAGRANDE, ARTHUR, Classification and Identification of Soils, *Trans. ASCE*, vol. 113, pp. 901–907, 1948.

[2] The suffix *d* applies when the liquid limit is 28 or less and plasticity index is 6 or less. The suffix *u* applies when liquid limit is greater than 28.

[3] Coarse-grained soils are those with more than one-half the material larger than No. 200 sieve size. Quality of soil in foundations as first stated is when not subjected to frost action, and not directly under bituminous pavement.

[4] Unit dry weights are for compacted soil at optimum moisture content for modified AASHO compactive effort.

[5] Liquid limit less than 50.

[6] Liquid limit greater than 50.

TABLE 28. CIVIL AERONAUTICS ADMINISTRATION CLASSIFICATION OF SOILS FOR AIRPORT CONSTRUCTION [1]

Soil group [2]	Mechanical analysis [3]				Liquid limit	Plasticity index	Subgrade class			
	Material retained on No. 10 sieve, percent	Material finer than No. 10 sieve, percent					Good drainage		Poor drainage	
		Coarse sand passing No. 10 retained on No. 60	Fine sand passing No. 60 retained on No. 270	Combined silt and clay passing No. 270			No frost	Severe frost	No frost	Severe frost
E-1	0–45	40+	60−	15−	25−	6−	Fa or Ra	Fa or Ra	Fa or Ra	Fa or Ra
E-2	0–45	15+	85−	25−	25−	6−	Fa or Ra	Fa or Ra	F1 or Ra	F2 or Ra
E-3	0–45			25−	25−	6−	F1 or Ra	F1 or Ra	F2 or Ra	F2 or Ra
E-4	0–45			35−	35−	10−	F1 or Ra	F1 or Ra	F2 or Rb	F3 or Rb
E-5	0–45			45−	40−	15−	F1 or Ra	F2 or Rb	F3 or Rb	F4 or Rb
E-6	0–55			45+	40−	10−	F2 or Rb	F3 or Rb	F4 or Rb	F5 or Rc
E-7	0–55			45+	50−	10–30	F3 or Rb	F4 or Rb	F5 or Rb	F6 or Rc
E-8	0–55			45+	60−	15–40	F4 or Rb	F5 or Rc	F6 or Rc	F7 or Rd
E-9	0–55			45+	40+	30−	F5 or Rc	F6 or Rc	F7 or Rc	F8 or Rd
E-10	0–55			45+	70−	20–50	F5 or Rc	F6 or Rc	F7 or Rc	F8 or Rd
E-11	0–55			45+	80−	30+	F6 or Rd	F7 or Rd	F8 or Rd	F9 or Re
E-12	0–55			45+	80+	....	F7 or Rd	F8 or Re	F9 or Re	F10 or Re
E-13	Muck and peat—field examination						Not suitable for subgrade			

[1] From "Airport Paving," CAA, October, 1956.
[2] Soils E-1 to E-5 are classed as "granular" and E-6 to E-12 as "fine-grained."
[3] Determination of sand, silt, and clay fractions is made on that portion of sample passing No. 10 sieve. If percentage of material retained on No. 10 sieve exceeds that shown, the classification may be raised provided such material is sound and fairly well graded.

Pavement Design—Flexible Types. The thickness of flexible pavements (bituminous surfaces and bases) is determined from information obtained from subgrade soil tests, field-loading tests, and experience with pavements in service (see highway pavements, p. 2–104). One approach is to determine from loading tests on circular plates (usually 30 in. diameter) the thickness required to limit the deflection to a value that will not cause detrimental settlement in the pavement under the expected number of repetitions of the design wheel load. Plate-bearing data may also be used in connection with theoretical analyses for determining thicknesses of layers of base and paving material to support design load.[1]

The Corps of Engineers has adopted the California Bearing Ratio (CBR) method of design in which the thickness is obtained from empirical curves based on a penetration test on a compacted and saturated sample of subgrade (or base course) material.[2] The CBR is the ratio (expressed in percent) of the bearing value of the material in psi at a deflection of 0.1 in. to that of a standard crushed-stone sample at the same deflection. The design curves [3] (Figs. 65 and 66) give the relationships between total thickness and wheel loads for different landing-gear arrangements. These curves are based upon extensive field tests and research with static and moving loads. In design, the maximum gross weight of the aircraft is usually divided equally between the two main landing-gear struts. For the larger airplanes with tricycle landing gear, 90 percent of the total weight may be assigned to the main gear. An impact allowance is not added to wheel loads. The static or slowly moving load is considered critical for design of aprons, taxiways, and runway ends, where there is a concentration of traffic. On the center portions of runways (excluding taxiway or runway crossings) a reduction of 10 percent is allowed from thick-

[1] Airfield Pavement, *Tech. Pub.*, Bureau of Yards and Docks, Department of Navy.
[2] Airfield Pavement Design—Flexible Pavements, "Engineering Manual for Military Construction," Part XII, Chap. 2, Office of Chief of Engineers, Department of Army. Also see MCFADDEN, GAYLE, Airfield Pavement Design of Corps of Engineers, *Proc. ASCE*, vol. 80, Sep. No. 458, July, 1954; and A Comparison of Design Methods for Airfield Pavements, *Proc. ASCE*, vol. 78, Sep. No. 163, December, 1952.
[3] Design curves have been consolidated and condensed from those in Part XII, Chap. 2, of "Engineering Manual for Military Construction."

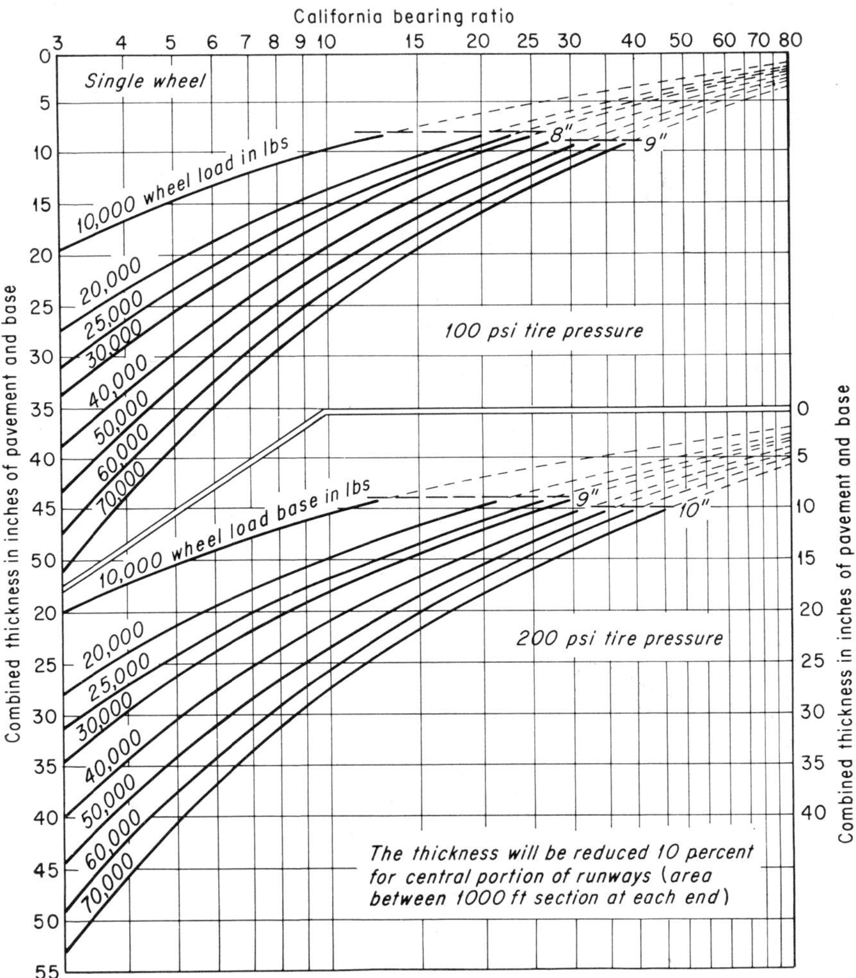

Fig. 65. Design curves for flexible pavements for taxiways, aprons, and runway ends—single wheels. (*Corps of Engineers.*)

ness shown in Figs. 65 and 66. This is permissible because the applied loads are dispersed and moving rapidly. The curves give total thicknesses which are to be further divided into surface, base, and subbase courses. The minimum surface and base-course thicknesses are indicated by horizontal dashed lines near top of curves. The surface course is usually of bituminous concrete which varies in thickness with magnitude of load and tire pressure. A minimum of 2 in. is recommended for single wheel loads of less than 30,000 lb with tire pressure of 100 psi and for dual assemblies under 45,000 lb. Three inches is required for single wheels above 30,000 lb with 100 psi tire pressure, for single wheels below 30,000 lb with 200 psi, for dual assemblies between 45,000 and 60,000 lb, and for twin tandem assemblies of 125,000 lb and less. Four inches is required for single wheel loads above 30,000 lb with 200 psi tire pressure, for duals above 60,000 lb, and

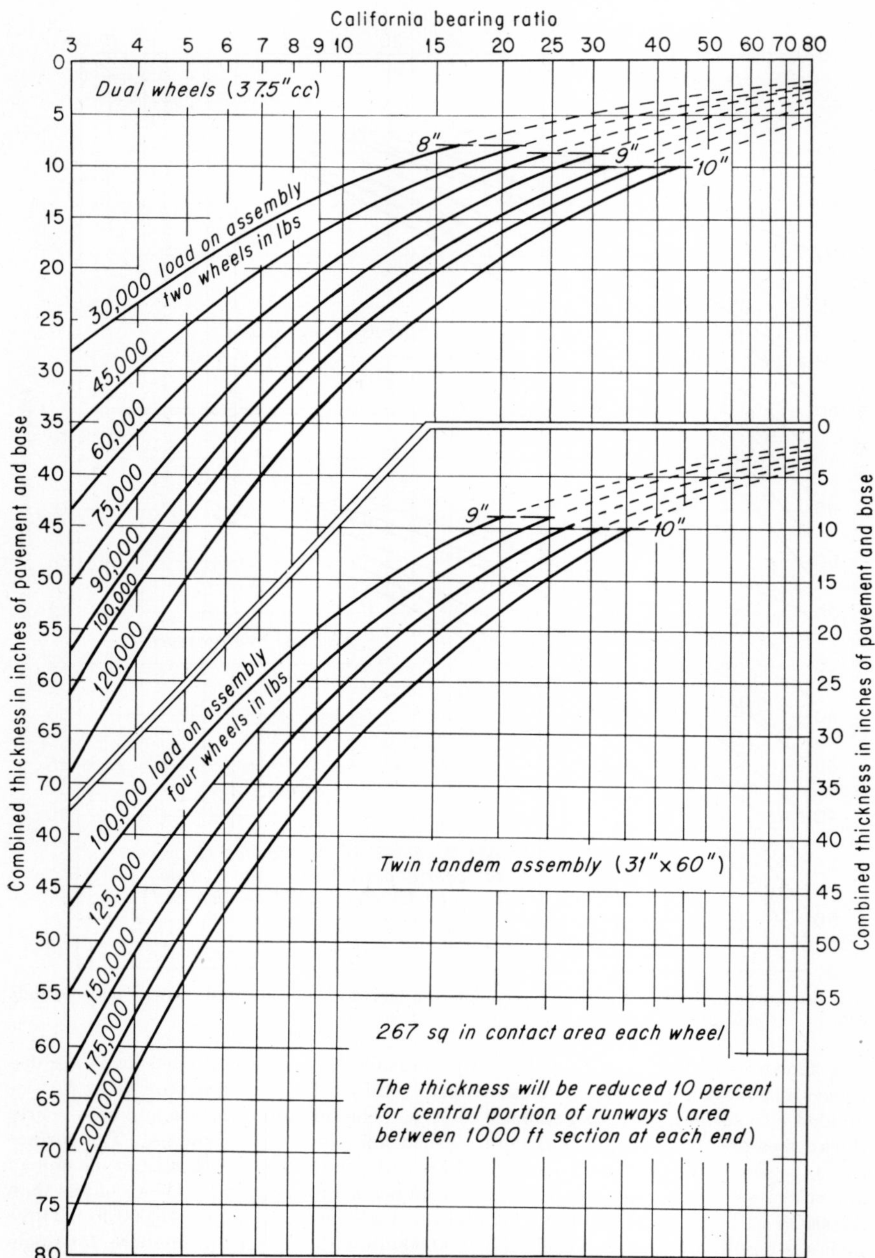

Fig. 66. Design curves for flexible pavements for taxiways, aprons, and runway ends—dual and twin tandem wheels. (*Corps of Engineers.*)

for twin tandem above 125,000 lb. Base courses directly below pavement should consist of high-quality crushed stone, slag, or stabilized gravel of CBR 80 or over. The subbase may be built up of several layers of different materials of lesser CBR, if economy may be gained thereby, provided that the total thickness of all courses above any one layer is at least equal to that required by the CBR of that layer (see p. 2–106). The Corps of

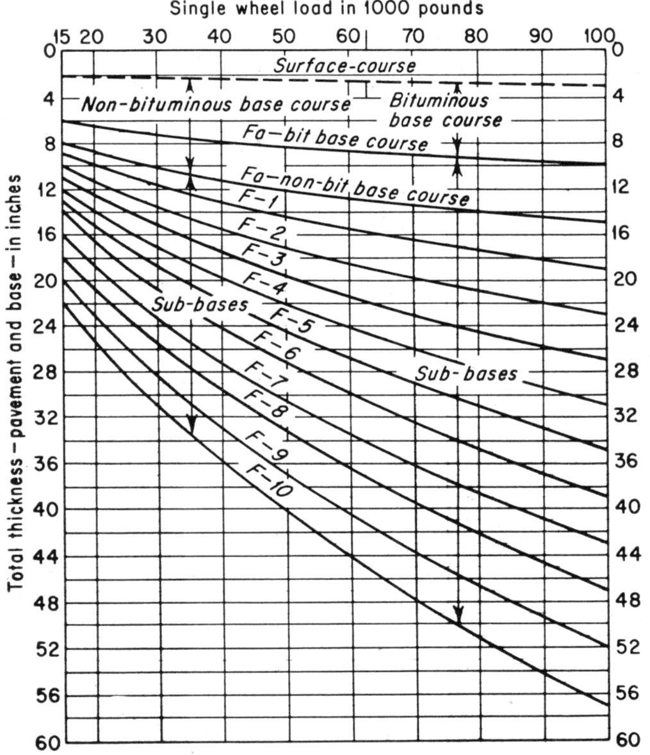

Fig. 67. Design curves for flexible pavements—taxiways, aprons, and runway ends—CAA method.

Engineers has also developed criteria for thickness required for protection against frost action, in soils of varying degrees of frost susceptibility.[1]

The Civil Aeronautics Administration has developed an empirical relation between their soil classes and thickness of surface course, base course, and subbase course required for different wheel loads and different conditions of drainage and frost action (Table 28). Design curves for flexible pavements are plotted in Fig. 67, based on more detailed curves in the CAA publication "Airport Paving." The "F" numbers refer to those given for soil classes in Table 28. Good drainage implies that surface water will be removed rapidly and that the ground-water level is low and there will be no accumulation of water in the soil by either percolation or capillarity. Poor drainage indicates conditions where the subgrade may become unstable because of saturation. Severe

[1] Airfield Pavement Design—Frost Conditions, "Engineering Manual for Military Construction," Part XII, Chap. 4, Office of Chief of Engineers, Department of Army.

frost classification applies when the depth of frost penetration for a particular site is greater than the anticipated total thickness as determined for no frost and appropriate drainage condition. For center portions of runways the total thicknesses may be reduced about 15 percent for 15,000-lb load to 20 percent for 100,000-lb load. Spreading the load over dual or twin tandem wheel assemblies will reduce the thickness required by varying amounts depending upon wheel spacing and type of subgrade.[1] For strong subgrades requiring thin pavements, the thickness required is only slightly more than for one wheel of the assembly. For poor subgrades and thick pavements, the thickness approaches that required for a single load of same magnitude as total assembly load. Single and dual wheel loads for different classes of civil airports are given in Table 26.

Rigid-pavement Design. The thickness of portland cement concrete pavements is derived from Westergaard formulas [2] or from empirical modifications of these formulas. The subgrade support is expressed by the modulus of subgrade reaction k which is a proportionality factor between pressure on soil under pavement in psi and deflection of slab under load in inches, and is expressed in pounds per cubic inch. It is determined from plate-bearing tests on the subgrade, and is usually computed at 0.05-in. or 0.10-in. deflection. The Corps of Engineers has developed design curves for concrete aprons, taxiways, and runway ends for different concrete strengths, k values, tire pressures, and wheel-load arrangements.[3]

The Portland Cement Association has also prepared design curves (Fig. 68) based upon a modification of Westergaard theory and the use of influence charts by means of which stresses may be found for a wide range of tire pressures, contact areas, and wheel arrangements.[4] The curves are for interior thicknesses where the load is some distance from a free edge. All interior joints are assumed to have road transfer devices. Where free edges occur in traffic areas such as at approaches to buildings or at runway intersections and expansion joints, the outer slab should be thickened gradually to an additional depth of 20 percent at the free edge. The allowable stress is obtained by dividing the ultimate stress in bending by a factor of safety. The PCA recommends using the ultimate strength at 3 months, which is roughly 110 percent of 28-day strength. The choice of factor of safety depends upon frequency of load applications (see p. 2–108). For unlimited repetitions the factor is 2.0. Load applications are more concentrated on taxiways, aprons, and runway ends (last 1,000 ft) than on the center portion of runways, where they are dispersed and partly air-borne. The PCA therefore recommends the following range of factors of safety: for aprons, taxiways, hardstands, runway ends, and hangar floors, 1.7 to 2.0; for center portion of runways, 1.25 to 1.50. The charts in Fig. 68 are computed for a modulus of elasticity $E = 4{,}000{,}000$ psi and a Poisson's ratio $\mu = 0.15$, typical of paving concrete. Normal variations from these values have little effect on thickness.

Example: Find taxiway thickness for a flexural strength of 700 psi, $k = 100$ lb in.^{3}, single wheel load 60,000 lb, and tire pressure 100 psi. Using a factor of safety of 2.0, enter Fig. 68 (chart a) at left with $700 \div 2 = 350$ psi, cross horizontally to curve $k = 100$, then vertically to wheel load = 60,000 lb with tire pressure of 100 psi, then horizontally to right edge of graph and read thickness of 13.9 in. A similar solution for 60,000 lb on dual wheels and 100 psi tire pressure using chart b gives a thickness of 11.8 in. The other charts may be used in a similar manner.

Base courses are required under concrete pavements placed on fine-grained soils to prevent pumping action of soil through joints and cracks, to improve drainage, and to

[1] CAA "Airfield Pavements" gives method for determining equivalent single wheel loads.

[2] Westergaard, H. M., New Formulas for Stresses in Concrete Pavements of Airfields, *Trans. ASCE*, vol. 113, pp. 425–443, 1948. Also see footnote on p. 2–107.

[3] Airfield Pavement Design—Rigid Pavements, "Engineering Manual for Military Construction," Part XII, Chap. 3, Office of Chief of Engineers, Department of Army.

[4] Design of Concrete Airport Pavement, The Portland Cement Association, and Pickett, G. O., and G. K. Ray, Influence Charts for Concrete Pavements, *Trans. ASCE*, vol. 116, 1951.

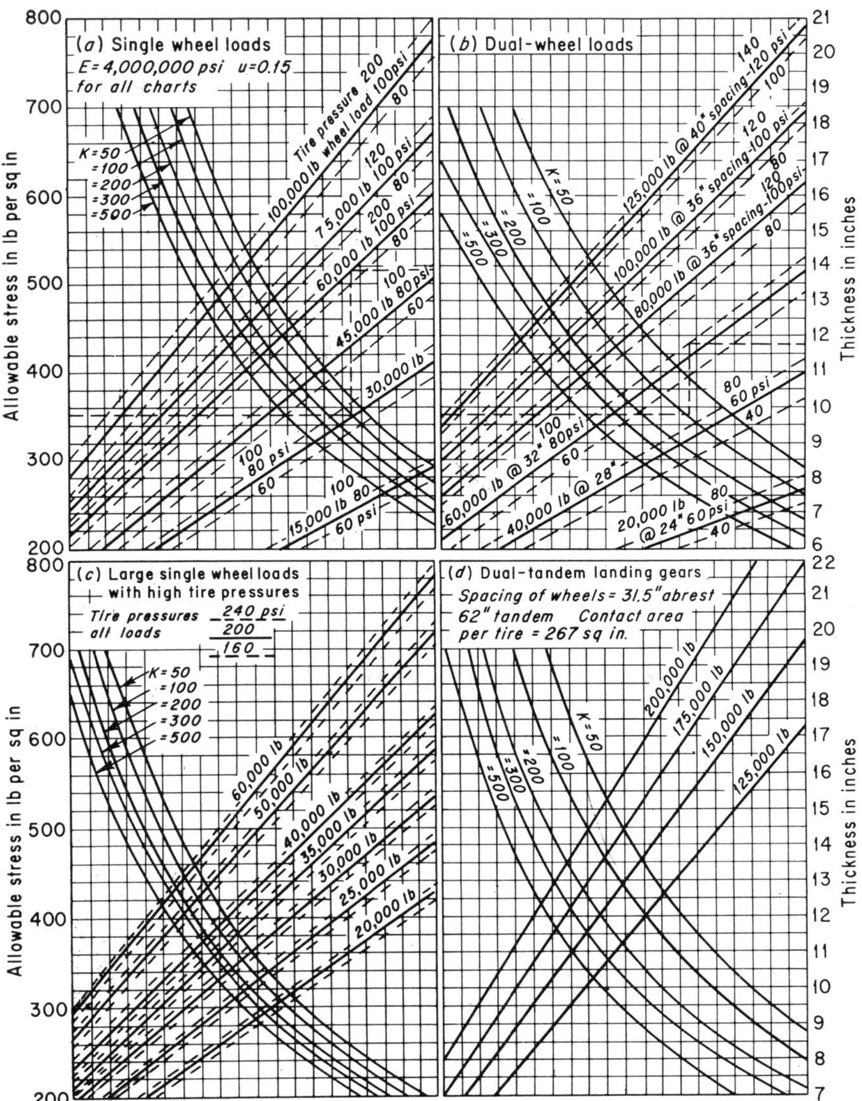

FIG. 68. Design curves for concrete runways, taxiways, and aprons—Portland Cement Association method.

prevent frost heave. The base thickness is usually based upon the above requirements rather than on wheel loads alone.

The CAA has developed design curves for rigid pavements and bases correlated with the CAA soil classification in Table 26. The curves plotted in Fig. 69 are for taxiways, aprons, etc. They are based upon more detailed charts given in CAA publication "Airport Paving." For center portions of runways, 80 percent of these thicknesses are rec-

ommended. Lesser thicknesses are required for dual wheels or dual tandem wheel assemblies, depending upon wheel spacing, tire imprint, and physical characteristics of concrete. The CAA suggests that for average conditions an equivalent single wheel load may be found by dividing total dual gear load by 1.35 and the total dual tandem load by 3.0.

Types of joints and reinforcing used in airport concrete are similar to those used in highway pavements (Figs. 51 to 53), except that wider slabs and larger dowels are used for thick pavement. Longitudinal joints are of the dummy groove or tongue-and-groove

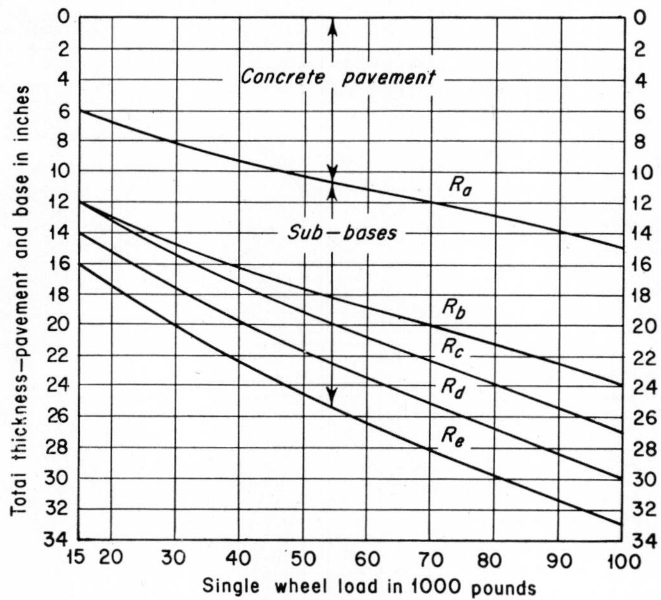

Fig. 69. Design curves for rigid pavements—taxiways, aprons, and runway ends—CAA method.

types, spaced 12.5 ft for slabs less than 10 in. thick, and up to 25 ft apart for thicker slabs. Longitudinal expansion joints are advisable at runway and taxiway intersections and next to structures. Transverse contraction or warping joints are spaced 15 to 25 ft apart in unreinforced pavement and 50 to 75 ft apart when distributed reinforcing is used. Transverse expansion joints are not generally used except at intersections. The CAA recommends a maximum spacing of 1,500 ft. Dowels are used across expansion and also across contraction joints in some designs. Tie bars or bonded reinforcing are carried across "hinged" or warping joints. Construction joints between runs of paving are tongue-and-groove or butt types with dowels. Diameters of dowels vary from ¾ in. for 6- to 7-in. slabs to 1½ in. for 16- to 20-in. slabs. Lengths range from 16 to 24 in. and spacings from 12 to 15 in.

Airport Paving.[1] Foundations and subgrades for runways and taxiways and aprons should be thoroughly compacted and stabilized, particularly those designed for the heavier airplanes. The Corps of Engineers has adopted the modified Proctor compaction test for military airports [2] which requires a higher standard of compaction than con-

[1] See Standard Specifications for Construction of Airports, CAA, Government Printing Office.

[2] Airfield Pavement Design—Flexible Pavements, "Engineering Manual for Military Construction," Part XII, Chap. 2, Office of Chief of Engineers, Department of Army.

sidered necessary for most highway construction (p. 2–90). High bearing capacity is obtained by thoroughly compacting base and subbase courses of stabilized gravel or crushed stone in preference to building excessively thick bituminous surface courses.

Turf surfaces [1] are used for landings and take-offs at small airports, and on the unpaved areas of landing strips at larger airports. A tough thickly matted grass is required in these areas, the type of grass depending upon the soil characteristics and the climate at the site. If tests show the soil to be deficient in nutrient elements, these may be supplied by appropriate fertilizers. When a fertile topsoil must be removed during grading operations, it should be stock-piled and later spread on the areas to be turfed. A vegetable cover is also desirable on embankment and cut slopes and other interior areas of the airport to prevent dusting and erosion. Turfed surfaces should be kept mowed to a height of 3 to 5 in. and fertilizer applied as needed.

Bituminous pavements for airports are similar to those used for highways (see p. 2–98). The smaller airports may be surfaced with biuminous surface treatments or road-mix surfaces. For the larger classes, plant-mix bituminous concrete is generally used and placed on a crushed-stone or bituminous macadam base. For the heaviest aircraft and for high-pressure tires greater densities and stabilities are required than for highways. For example, stabilities of 500 to 1,000 lb by Marshall test (see p. 2–102) are usually adequate for highway traffic and light aircraft. For larger jet aircraft with high-pressure tires stabilities of 1,500 to 2,000 are required where traffic is concentrated, such as on taxiways, aprons, and runway ends. On aprons and other areas subject to jet-fuel spillage, tar or tar-rubber concretes are sometimes used or the surface is treated with a solvent-resisting coating.

Concrete pavements are constructed in same manner as on highways (see p. 2–111). They are commonly used for aprons since they are unaffected by spillage of fuel, oil drippings, or shearing stresses caused by turning aircraft. They are also more resistant to jet blast than are bituminous types.

Overlays. Bituminous or concrete pavements may be resurfaced with bituminous concrete overlays. The purpose is to restore a smooth riding surface and to strengthen the pavement. The thickness is seldom less than 3 in. and will vary with the unevenness of the old surface and the added strength required. Design criteria have been developed by the Corps of Engineers and CAA.

AIRPORT TERMINALS

Terminal facilities at small airports usually consist of an administrative building with pilots' lounge and office space for operators basing planes at the airport. Storage hangars are required for small planes and maintenance hangars and shops for servicing these planes. At airports serving scheduled and nonscheduled airlines, a terminal building capable of handling passengers and cargo is required, together with hangars and shops for servicing aircraft of the type using and expected to use the airport. The size and layout of facilities should be based on a thorough study of the potential traffic of all kinds: scheduled, nonscheduled, charter, training, and personal.[2] Ample provision should be made in the design for future expansion. Some airports will be wholly commercial, some will serve all classes of traffic, and others will serve private planes only.

The *terminal building* [3] should be laid out to provide a smooth flow of passengers and cargo from land vehicles to aircraft. This is often accomplished by a separation of passengers and cargo, and a further separation of ingoing and outgoing traffic. A typical airline terminal building for a moderately large city may be described as follows: baggage, mail, express, and air freight are handled on the ground floor at apron level. Trucks

[1] "Airport Turfing" and "Airport Landscape Planting," CAA, Government Printing Office.
[2] Airport Planning, CAA, Government Printing Office.
[3] Airport Terminal Buildings, CAA, Government Printing Office.

make deliveries at platforms at the rear, from which cargo is processed and transferred to tractor-drawn carts for delivery to the apron for loading into airplanes. Airline operations offices and building utilities are located on this floor. Passengers enter lobby on first floor where the waiting room, ticket counters, baggage scales, chute to baggage room below, telephones, public conveniences, concessions (stores, newsstand, snack bar or coffee shop, etc.), post office and entrance to observation deck are located. Passenger corridors, ramps, or stairways lead to gates at plane-loading positions. On the second floor are restaurant, other concessions, general offices, airline reservations, pilot room, and airport management offices. On third floor are found such activities as CAA, traffic control, and weather bureau. The airport traffic-control tower usually tops the building. Designs for airports in large cities envision the terminal as a center of commercial and recreational activity producing revenue from concessions catering to air passengers and visitors, as well as from rentals obtained from airlines and other aviation activities. Space requirements for various functions are based on entering and leaving passengers in peak hour of design year. The type of building described above is known as a *centralized* design in which the accommodations for all airlines are contained in one central building. Another type of operation is the *unit*-type terminal in which each airline has its own independent station. The units are usually contiguous, but may be in separate buildings. The unit type provides a convenience for patrons of the airline it serves, but introduces delays in transferring from one airline to another.

Loading-apron size depends upon the number and size of aircraft to be accommodated as determined from a forecast of peak-hour aircraft movements. The aircraft loading positions are designated by circles of 100, 150, or 175 ft diameter, depending on wing span, length, and turning radius of aircraft. The length and width of apron depend upon the size and number of positions. See p. 2–133. In order to keep walking distances to a minimum aprons are sometimes laid out in semicircular form or nose-shaped with terminal in partially enclosed area. In other designs, covered passageways or *fingers* are built out from the front or sides of the terminal building, and gate positions arranged around these fingers. At some of the larger airports, underground systems are installed for fuel, compressed air, electric power, telephones, etc. These services are available from pits near the loading positions.

Hangar size depends on the dimensions and numbers of aircraft to be serviced. The gross weight, wing span, length, and height of selected aircraft are as follows:

Name and model	Gross weight, lb	Wing span, ft	Length, ft	Height, ft
Piper PA12	1,750	35.5	23.0	6.8
Beech D-18-S	8,750	47.6	34.0	9.3
Convair 440	49,100	105.3	79.2	28.2
Douglas DC-3	25,200	95.0	64.5	16.9
DC-6B	107,000	117.5	105.6	28.7
DC-7	122,200	117.5	108.9	28.6
DC-8	265,000	139.7	148.8	42.3
Lockheed 1049C	130,000	123.0	113.5	24.7
Boeing 377	142,500	141.2	110.3	38.3
707	295,000	130.8	146.7	38.6

Arch-type structures or cantilevered roof construction are employed to obtain necessary door widths and clear working space.[1] Floor layouts are arranged to accommodate

[1] KYLE, J. M., Design Concepts Change at New York International Airport, *Civil Eng.*, vol. 25, No. 1, November, 1955. Also Kelly Air Force Base Maintenance Hangar; Planning, Engineering Design Features and Construction Features, *Proc. ASCE*, vol. 81, papers Nos. 852, 853, and 854, December, 1955.

a maximum combination of types of aircraft to be serviced. Offices, shops, and supply rooms are provided at ends of arches or in the towers supporting roof cantilevers. For large aircraft in temperate climates, nose-in-type hangars are designed to enclose only the propellers, engines, and front portion of the fuselage. For small private airplanes, individual T-shape hangars are advocated.[1] Other buildings are required to house maintenance equipment, heating or power plants, transformers, and crash and fire-fighting equipment. With the growing importance of air freight, separate terminal buildings are being built to handle freight business.

Adequate *highway approaches* are needed to the terminal building and to the several truck-loading platforms and to parking lots. A separation of passenger and commercial traffic is desirable, as well as a separation of patron, spectator, and employee traffic. Ample parking facilities are required for patrons, employees, and spectators. The latter traffic often predominates on Sundays and when special events occur at the airport. In general, 250 to 300 sq ft of parking lot (including drives) is needed per car.

Heliports.[2] The characteristics of helicopters, including those in the development stage, fall roughly within the following limits: passenger capacity 3 to 50, number of engines (or rotors) 1 or 2, rotor diameter 44 to 90 ft, gross weight 5,300 to 40,000 lb, maximum wheel load 1,875 to 16,000 lb, tire pressure 45 to 100 psi, cargo capacity 1,400 to 12,000 lb, and speed 85 to 160 mph. Design specifications for heliports are tentative pending more experience with helicopters in commercial service. The landing area must be large enough to handle loading and unloading of helicopters as well as a limited number of positions for parking and servicing. The surface must be clean, well drained, and strong enough to support impact landing loads. Although it is possible for helicopters to land and take off vertically, for economical operation with maximum pay load they must take off on a sloping ascent preferably into the wind. Consequently approach-zone clearances must be established similar to those for fixed-wing aircraft but with steeper slopes. The Port of New York Authority has recommended the heliport specifications given in Table 29.

Helicopter sites may be on rooftops, on the ground, or on floats over water surfaces. Location considerations include accessibility to traffic-generating areas, economy of construction, maximum safety of operation, and minimum noise nuisance to adjacent residence and office personnel. Rooftop locations usually have advantages of accessibility, and economy when combined with other commercial uses of the building. Clear approaches are usually available. Disadvantages are the limited space on roof, absence of any overrun area for emergency landings after take-off, lower visibility than on the ground, possible cost of strengthening the structure, and the need of transporting passengers, cargo, supplies, and fuel vertically. Ground sites are usually more economical to construct, and provide more flexibility in planning facilities. Elevators, etc., are unnecessary. The problem is usually to find adequate ground space close to centers of urban areas. Where cities are located near bodies of water, a water-front location can often be found which has advantages of accessibility as well as of clear approaches over water surfaces and the other advantages of a ground site.

Lighting.[3] A system of field lighting is required for night operations at all major airports. The airport is marked by a rotating beacon with beams of clear and green light 180° apart. It is mounted on a tower or top of a building, and sometimes supplemented by a flashing code beam, where two or more airports are close together. Wind direction is indicated by a wind cone illuminated from above, by a wind tee with green lights, or

[1] "Small Airports," CAA, Government Printing Office.

[2] Transportation by Helicopter—1955–1975, The Port of New York Authority, Aviation Department. What Governs Heliport Design? *Eng. News-Record*, Sept. 22, 1955. Horonjeff, R., and H. S. Lapin, Planning for Urban Heliports, *Research Report* 19, The Institute of Transportation and Traffic Engineering, University of California, also Urban Heliports Are on the Way, *Civil Eng.*, February, 1955.

[3] See manuals and technical orders of CAA pertaining to airport lighting. Also "Aerodromes Annex 14" of International Civil Aviation Organization, Montreal, Canada.

TABLE 29. HELIPORT DESIGN REQUIREMENTS—PORT OF NEW YORK AUTHORITY[1]

Dates	1955	1960	1965	Ultimate
Heliport-area requirements:				
Size of landing and take-off area, ft				
Major heliport	100 × 100 [3]	200 × 400	200 × 400	200 × 400
Secondary heliport	100 × 100 [3]	150 × 300	150 × 400	150 × 400
Number of parking positions, major heliport [2]	4	4	4	5
Size, ft:				
Major heliport:				
a. If helicopters are positioned mechanically	....	85 × 135	85 × 135	85 × 135
b. If helicopters are taxied	80 × 125	100 × 160	100 × 160	100 × 160
Secondary heliport, helicopters taxied	60 × 60	60 × 110	60 × 110	85 × 135
Approach-zone characteristics:				
Approach-zone width at landing area, ft:				
Major heliport	200	300	300	300
Secondary heliport	200	250	250	250
Approach-zone width outward from area	Width increases at 15° angle symmetrically on both edges to width 1,000 ft and continues at 1,000 ft			
Approach-zone slope:				
a. With emergency landing areas	1:8	1:8	1:6	1 4
b. When no emergency landing areas are provided	1:20	1:8	1:6	1:4
Curved approach zone—minimum radius of turn, ft	650	650	650	650
Approach-zone transition areas	1:2	1:2	1:2	1:2
Helicopter weights:				
Intercity and suburban helicopter maximum weight, lb	15,000	45,000	50,000	50,000
Aerocab helicopter, lb	15,000	20,000	25,000	25,000
Passengers per aircraft—intercity and suburban maximum	15	50	60	60
Passengers per aircraft—aerocab	15	22	27	30
Cargo capacity—intercity, lb	3,000	10,000	12,000	12,000
Loading on landings:				
Impact loading, lb, per main gear:				
Major heliport	10,500	31,500	35,000	35,000
Secondary heliport	10,500	14,000	17,500	17,500
Maximum tire pressures, psi	70	100	100	100
Area, sq in., on which impact load to be applied:				
Major heliport	144	405	450	450
Secondary heliport	144	180	225	225

[1] From *Eng. News-Record*, vol. 155, No. 12, pp. 22–23, Sept. 22, 1955.
[2] In addition, space should be provided for one disabled aircraft, or means should be provided for removing it from the landing area.
[3] If heliport elevation is such that loss of ground cushion occurs on take-off immediately after leaving heliport, size must be increased sufficiently to assure that operations can be safely conducted in compliance with the height-velocity diagrams for the helicopters using the heliport.

by a tetrahedron outlined by green and red lights. The tee is set by the wind, or, at times of low winds, it may be set from the control tower to be consistent with the runway in use. At small all-purpose airports boundary lights of white color mounted 2.5 to 5.0 ft above the ground are used to outline the landing area. They are placed at angle points in the boundary and about 300 ft apart in between. At major airports the runways are delineated by runway lights placed 200 ft apart and usually placed about 15 ft outside the edge of pavement on each side. On wide runways they may be placed just inside edge of pavement. These lights are of high intensity, usually adjustable in stages, which can be controlled from the tower. The mountings are low and sufficiently flexible to yield under impact of aircraft, but also easy to replace. Part of the illumination is cast upward and sidewise, but the major beam is directed toward a pilot making a landing. The color is white in the central portion of the runway length, and half white and half yellow at each end. The colors are displayed so that an approaching pilot sees white lights for all but the last 1,500 ft of runway where the yellow lights give him warning that he is nearing the end of the runway. The runway ends are marked by a row of green threshold lights placed just beyond the end of pavement. Blue taxiway lights are placed at about 200-ft intervals (closer on curves) along both sides of taxiways. Green and red traffic-control lights, operated from the tower, are sometimes used at runway and taxiway intersections. On instrument-approach runways, approach lights are installed for a distance of about 3,000 ft in extension of the runway center line. One system consists of horizontal bars of white lights placed 100 ft apart supplemented by a 100-ft-long horizon-orientation cross bar of white lights placed at 1,000 ft from the runway, by red terminating bars 200 ft from runway, red wing bars at 100 ft from runway, and green threshold bars on each side at the beginning of runway pavement. Another feature is a string of condenser-discharge lights installed along the center of the approach lane which flash a brilliant, instantaneous light in sequence toward the runway, giving the impression of a flashing streak of light pointing to the runway. A panel in the control tower shows the runway layout in miniature with small bulbs indicating corresponding lights on the field. Working areas on the apron and around the terminal and hangars are illuminated by flood lights. Red obstruction lights are placed on objects protruding into approach and turning zones and on other isolated objects which represent a real or apparent hazard in the path of flight.

Markings are required to delineate runway and taxiway center lines and turning lines at intersections. The runway number is painted at each end together with symbols denoting runway length. Transverse bars are sometimes placed at 500-ft intervals near ends of runways and coded to indicate distances from runway ends. Signs are erected at taxiway turn-offs and other intersections to guide pilots.

Bibliography

Route Surveys

American Society of Photogrammetry: "Manual of Photogrammetry"; *Photogrammetry Engineering.*

Hickerson, T. F.: "Route Location and Surveying," 3d ed., McGraw-Hill Book Company, Inc., 1953.

Meyer, C. F.: "Route Surveying," 2d ed., International Textbook Company, 1956.

Rubey, H.: "Route Surveys and Construction," 3d ed., The Macmillan Company, 1956.

Railways

American Railway Engineering Association: "Manual"; *Ann. Proc.*

Signal Section, American Association of Railroads: "Manual"; *Ann. Proc.*

"Railway Track and Structures Cyclopedia," 8th ed., Simmons-Boardman Publishing Corporation, 1955.

Railway Age; Railway Track and Structures; Railway Signaling and Communications, Simmons-Boardman Publishing Corporation.

HAY, W. W.: "Railroad Engineering," vol. 1, John Wiley & Sons, Inc., 1953.

Highways [1]

BRUCE, A. G., and J. CLARKESON: "Highway Design and Construction," 3d ed., International Textbook Company, 1950.

HEWES, L. I., and C. H. OGLESBY: "Highway Engineering," John Wiley & Sons, Inc., 1954.

MATSON, T. M., W. S. SMITH, and F. W. HURD: "Traffic Engineering," McGraw-Hill Book Company, Inc., 1955.

RITTER, L. J., JR., and R. J. POQUETTE: "Highway Engineering," The Ronald Press Company, 1951.

Airports [1]

Civil Aeronautics Administration: manuals on Airport Design, Airport Planning, Airport Drainage, Airport Paving, Airport Turfing, Airport Landscape Planting, Airport Buildings, Standard Specifications for Construction of Airports.

American Society of Civil Engineers: Journal of the Air Transport Division, *Proc. ASCE*, vol. 83, AT2, December, 1957, "Jet Age Airport Conference."

FROESCH, C., and W. PROKOSCH: "Airport Planning," John Wiley & Sons, Inc., 1946.

General

HENNES, R. G., and M. I. EKSE: "Fundamentals of Transportation Engineering," McGraw-Hill Book Company, Inc., 1955.

[1] See also sources of information: highways, p. 2–59; airports, p. 2–122.

Section 3

MECHANICS OF MATERIALS

By Jesse B. Kommers

TENSION, COMPRESSION, AND SHEAR

Stress. Figure 1 shows a clamp carrying a load as indicated. At a section such as AA, the portion D of the clamp must exert a force or forces on the portion C in order to produce equilibrium. By the stress at the section AA is meant the total force exerted by the portion D upon C or the equal and opposite force exerted by C upon D.

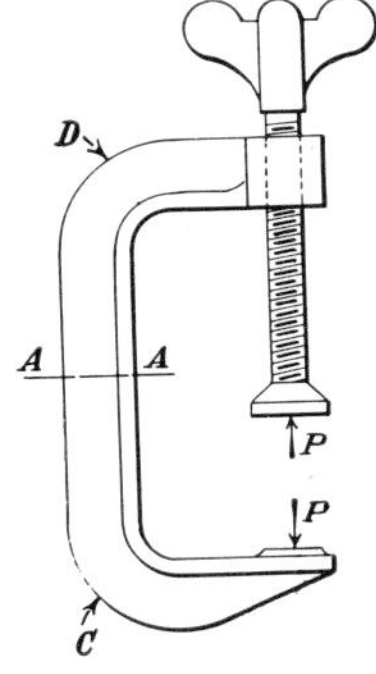

Fig. 1. Clamp under stress.

Unit Stress. The total stress acting on a cross section of a structural element is not nearly so significant to the engineer as the *intensity* of stress acting there, for the total stress will usually not cause failure if the cross section is made large enough. By intensity of stress at a cross section is meant the stress per unit area; and when the engineer knows the intensity of stress and the kind of material, he can immediately form a judgment as to whether the part in question is safely or dangerously stressed. Instead of "intensity of stress," the term "unit stress" is commonly employed, and it has been widely adopted even though careful analysis shows that it is not particularly appropriate.

In simple cases of loading in which the intensity of stress, or unit stress, is uniformly distributed over an area, the unit stress may be calculated by dividing the load by the area. In more complicated loadings under which the unit stress varies over the area, the unit stress at any point is defined as the limit of the load on the area divided by the area as the latter approaches zero.

In the United States, loads, forces, and stresses are usually measured in pounds; in England, in long tons; and on the Continent, in kilograms. Correspondingly, the unit stresses are measured in pounds per square inch, tons per square inch, and kilograms per square centimeter.

Strain. When a body or a structure is subjected to loads, it usually suffers changes of shape or size, and such change in shape or size is called "strain." The most common use of the word strain is in connection with the change that occurs in a linear dimension. Thus, when a bar subjected to a pull stretches 0.01 in 8 in., the total stretch is called the strain or *deformation.* By *unit strain* is meant the total strain divided by the original dimension in which the strain occurred. If the unit strain varies at different points on a cross section, it is defined as the limit of the total strain divided by the original dimension

as the latter approaches zero. The term strain may also be used in connection with changes of area and of volume.

Axial Stresses. When a straight bar is subjected to forces at the ends so that the resultant forces coincide with the axis of the bar, the stress is called an "axial stress." When the forces at the ends of the bar, as shown in Fig. 2, are such as to cause the bar to stretch, the stress is called a *tension*. When the forces, as shown in Fig. 3, are such as to cause the bar to shorten, the stress is called a *compression*. In both cases, the total

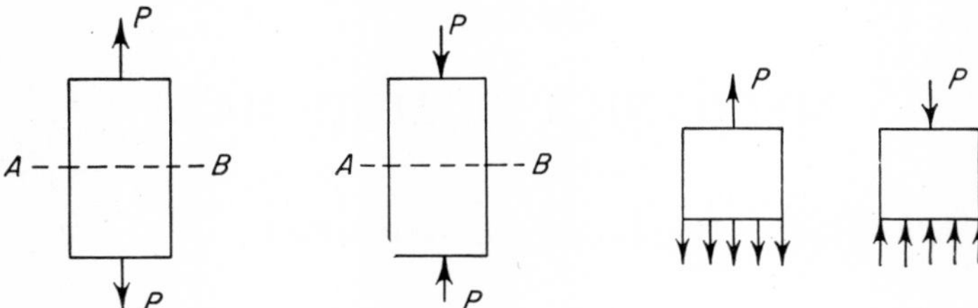

FIG. 2. Axial tension. FIG. 3. Axial compression. FIG. 4. Axial stresses.

stress acting on any cross section, such as AB, must be equal to the load P in order to produce equilibrium. In both cases, also, the unit stress at the cross section AB is uniformly distributed as shown in Fig. 4.

In the case of compressive stress, there is a distinction that is commonly made between compressive and bearing stress. When the stress acts at a section such as AB in Fig. 3, in which the area separates two portions of the same bar, the term *compressive stress* is used. When the stress acts at an area of contact between two different bodies the term *bearing stress* is used.

In Fig. 4, since the unit stress is uniformly distributed over the cross-sectional area normal to the load, the unit stress may be calculated by the simple formula

$$s = \frac{P}{A} \tag{1}$$

in which s denotes the unit stress on the area, psi, P denotes the load on the bar, lb, and A denotes the area of the bar on a plane normal to the load, sq in.

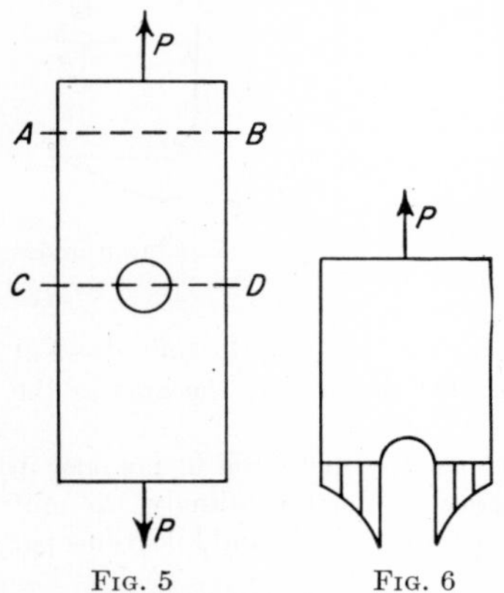

FIG. 5 FIG. 6
FIG. 5. Plate with a drilled hole.
FIG. 6. Stress distribution at the hole in a drilled plate.

Effect of a Hole. Numerous experiments made in various ways, and also mathematical investigations, have shown that, when a bar or plate is subjected, say, to an axial tensile load, the unit stress on the cross section will not be uniformly distributed when a bar has a hole drilled through it as shown in Fig. 5.

A section AB at a considerable distance from the hole may have the unit stress distributed quite uniformly across the width of the plate; but, as the section is taken nearer the hole, the distribution becomes more and more nonuniform. Figure 6 shows the kind of stress distribution that may be expected on the section of the plate through the center of the hole. From the experiments that he made in England, Coker developed an approximate expres-

sion for the unit stress at the edge of the hole, as follows:

$$s = \frac{3K}{K + 1} s_1 \tag{2}$$

in which s denotes the unit stress at the edge of the hole, psi, K denotes the ratio of the width of the plate to the diameter of the hole, and s_1 denotes the average unit stress on the cross section at the hole, psi. This formula is approximately correct when s is within the elastic limit and K is greater than 2. According to this formula, when K is 3, s is 2.25 times the average unit stress.

It has been shown by numerous experiments that high localized unit stresses such as are produced at the edges of a hole in a tension plate may also be produced by notches or fillets at the sides of the plate or, indeed, by any abrupt change of cross section. Such high localized stresses may be present owing to tensile, compressive, bending, or torsion stresses.

When such high localized stresses occur in a ductile material like steel under a steady load, they are not considered to be particularly dangerous. When the maximum local unit stress exceeds the yield point in a ductile material, it is possible for the material to yield locally and thus cause a more uniform distribution of stress. This property is in fact one of the important advantages of a ductile material. When the material is brittle, on the other hand, there is no possibility of much local yielding, and failure may result. It will be shown later that, for both ductile and brittle materials subjected to repeated or reversed stresses, the repetitions of stress may develop a crack due to the fatigue action. For these reasons, abrupt changes of cross section should be avoided as much as possible.

FIG. 7. Shear stress in a rivet.

Shear Stress. One of the most familiar examples of shear stress is that produced in a riveted joint. In Fig. 7, two plates are held together by a rivet, and the effect of the forces P is to cut the rivet in two. In such cases, the shearing unit stress is assumed to be uniformly distributed, and the formula for unit stress is

$$s_s = \frac{P}{A} \tag{3}$$

in which s_s denotes the shearing unit stress, psi, P denotes the load producing shear, lb, and A denotes the area parallel to the force P, sq in.

The above formula is similar to the one given previously for computing the unit stress due to axial tension or compression. It should be noted that under axial stress the area involved in the formula is normal to the force P, while in the case of shear the area is parallel to the force P.

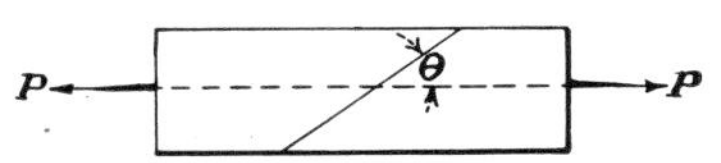

FIG. 8. Shear stress on an oblique plane.

Shear Stress on Oblique Planes. When a bar is subjected to axial tension or compression, it is simultaneously subjected to shearing stresses on all planes oblique to the load. In Fig. 8, the bar is subjected to axial tension. The unit shear stress on any oblique plane making an angle θ with the axis of the bar can be shown [1] to be as follows:

$$s_s = \frac{P}{2A} \sin 2\theta \tag{4}$$

[1] BOYD and FOLK, "Strength of Materials," 5th ed., p. 30. TIMOSHENKO, "Strength of Materials," 3d ed., Part I, p. 38.

in which s_s denotes the unit shear stress on the oblique plane, psi, P denotes the axial load applied, lb, A denotes the area of the bar normal to the load P, sq in., and θ denotes the angle between the oblique plane and the axis of the bar.

It is evident from the formula that the shear stress is a maximum when θ equals 45° and that the maximum value is equal to one-half of the axial unit stress. Brittle materials under compression, like cast iron or concrete, will usually fail by shear on an oblique plane because the ultimate shear strength is low.

In Fig. 8, the tensile unit stress which acts normal to the oblique plane is given by the formula

$$s_t = \frac{P}{A} \sin^2 \theta \tag{5}$$

in which the symbols have the same meaning as in Eq. (4). This tensile stress or compressive stress, normal to the oblique planes, is usually not important, because it is always less in magnitude than the axial tensile or compressive unit stress when θ is 90°.

Normal Stresses Due to Shear. Consider a small parallelepiped, such as is shown in Fig. 9. The top and bottom faces are subjected to shearing forces P, and the end faces to shearing forces Q. Such shearing forces P and Q may be produced by torsion in a shaft, and it will be shown later that they may also exist in an ordinary loaded beam. Since the moment effect of the Q couple must be balanced by that of the P couple, it can be shown that the unit stress s_h due to the force P must equal the unit stress s_v due to the force Q.

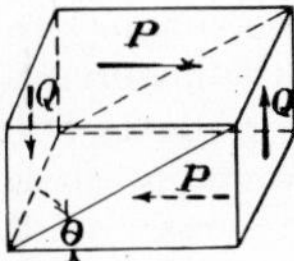

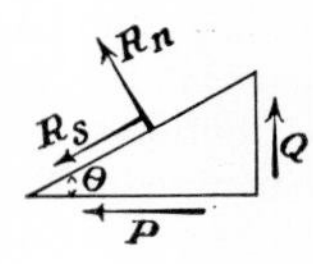

Fig. 9. Normal stress due to shear.

On every oblique plane making an angle θ with the horizontal, there exist normal unit stresses, whose value is given [1] by the following formula:

$$s_n = s_s \sin 2\theta \tag{6}$$

in which s_n denotes the unit stress normal to the oblique plane, psi, s_s denotes the unit shear stress due to the forces P and Q, psi, and θ denotes the angle that the oblique plane makes with the horizontal.

It is evident from the formula that s_n will attain a maximum value when θ equals 45° and that this maximum unit normal stress is equal to the original shearing stress s_s. In Fig. 9, when θ is 45°, the normal unit stress would be tensile; when θ is 135°, the normal unit stress would be compressive. Figure 10 shows how in the one case P and Q may compound to produce a tension and in the other to produce a compression.

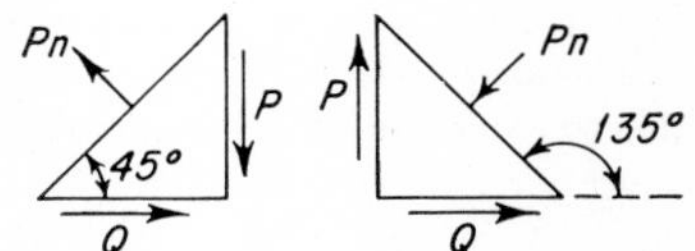

Fig. 10. Normal stresses due to shear.

The above discussion shows, therefore, that when a specimen is subjected, say, to torsion by couples applied to the ends, there are brought into action stresses in tension, compression, and shear, which are all equal. When the torsion couple is increased to produce failure, the kind of fracture that results depends on which of the three ultimate strengths of the material is the least. This action is strikingly illustrated by twisting a piece of chalk. For this brittle material, the tensile ultimate strength is the least; when the chalk fails owing to tension, it will crack on a line normal to the maximum stress and therefore at 45° with the axis of the specimen. If the experiment is performed, with an end of the chalk held in each hand, failure will occur as suggested.

[1] Boyd and Folk, "Strength of Materials," 5th ed., p. 33.

In Fig. 9, the shearing unit stress acting on the oblique plane is given by the formula

$$s_s' = s_s \cos 2\theta \tag{7}$$

in which the notation is the same as in Eq. (6). This shearing unit stress s_s' is always less than the original shear stress s_s, except, of course, for values of θ equal to zero or 90°.

Biaxial Stresses. Figure 11 shows a parallelepiped subjected to two pairs of axial, coplanar, and mutually perpendicular forces. These forces may be both tension, both compression, or one tension and one compression. They produce normal stresses and shear stresses on oblique planes at an angle θ with the horizontal. It can be shown [1] that the equation for the normal unit stress is as follows:

$$s_n = s_1 \sin^2 \theta + s_2 \cos^2 \theta \tag{8}$$

in which s_n denotes the unit stress normal to the oblique plane, psi, s_1 denotes the unit stress due to P, psi, s_2 denotes the unit stress due to Q, psi, and θ denotes the angle that the oblique plane makes with the horizontal.

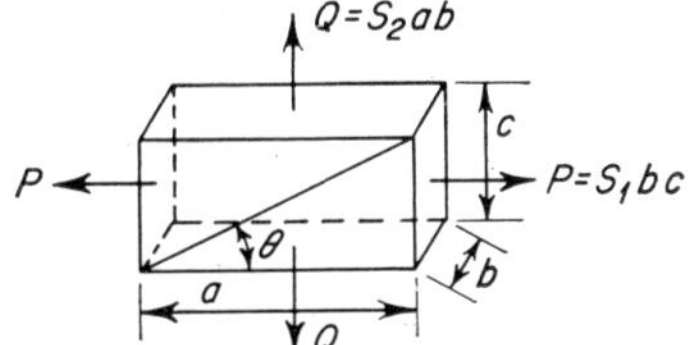

FIG. 11. Biaxial stresses.

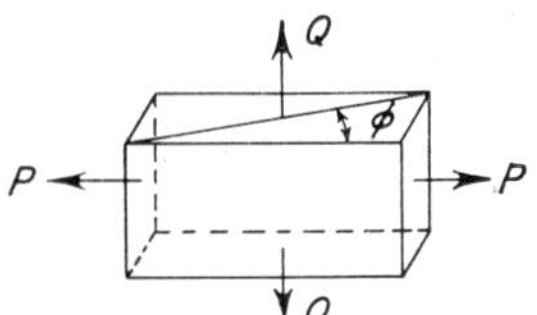

FIG. 12. Shear stress caused by biaxial stresses.

If s_1 is greater than s_2, the normal unit stress is a maximum when θ is 90° and is equal to s_1. If s_2 is greater than s_1, the normal unit stress is a maximum when θ is zero and is equal to s_2.

The equation for the shearing unit stress on any oblique plane is as follows:

$$s_s = \frac{s_1 - s_2}{2} \sin 2\theta \tag{9}$$

in which the notation is the same as for Eq. (8).

In Eqs. (8) and (9), s_1 and s_2 are called plus when they are tensile stresses. Either s_1 or s_2 is called minus if it is compressive. When s_1 and s_2 are of opposite sign, Eq. (9) gives the maximum shearing unit stress when θ equals 45°. However, when s_1 and s_2 are of the same sign, there is another plane upon which the maximum unit shear stress will act and which is not given by Eq. (9). If, for instance, s_1 is greater than s_2, the maximum unit shear stress will act on the plane indicated in Fig. 12, when ϕ is equal to 45°, and this maximum unit stress will be equal to one-half of s_1. That this is true is evident from the fact that Q is parallel to the plane shown and therefore produces no shear on this plane. The only shearing unit stress, therefore, is that due to P.

DEFORMATION AND STRESS

Elasticity and Plasticity. When a material is stressed, it suffers strain; and when the stress is removed, the material will recover its original form more or less completely. If the material returns to its original dimensions after the stress is removed, it is said to

[1] BOYD and FOLK, "Strength of Materials," 5th ed., p. 36. TIMOSHENKO, "Strength of Materials," 3d ed., Part I, p. 45.

be elastic; if it retains a permanent set after small stresses are applied and removed, it is *plastic.* Many materials are elastic up to a certain unit stress and more or less plastic beyond that stress.

The limiting unit stress up to which a material is elastic is of the greatest practical importance, for in designing structures or machines it is usually required that there shall be very little if any permanent deformation.

Hooke's Law. In 1678, Robert Hooke discovered that the ratio of the load applied to a material to the corresponding deformation is constant under low unit stresses. His statement of the law was: "As the extension, so the force." This is known as "Hooke's law."

Elastic Limit. The unit stress up to which a material may be stressed without suffering permanent deformation when the stress is removed is called the "elastic limit." The elastic-limit determination is not made very commonly, because the stresses must be applied and removed by increasing increments, the test being thus an expensive and tedious one. It should also be noted in passing that the elastic limit of a material is not so clear-cut as the definition implies. Experiments with very sensitive apparatus indicate that most materials show a very slight deformation even at very low stresses. In determining the elastic limit, therefore, it is necessary to agree in advance upon the arbitrary amount of permanent set that shall be used as a criterion of lack of elasticity. In past years there has been a great amount of discussion in the technical press on the question of practical means for determining the limit of elasticity.

Proportional Limit. The proportional limit of a material is defined as that unit stress up to which the unit stress is proportional to the corresponding unit deformation. It is much easier to make a test to determine the proportional limit than to determine the elastic limit, because in the former test the stresses do not have to be removed after each increment of stress is applied. There is, however, a difficulty even in the determination of the proportional limit. The more sensitive the apparatus which is used to determine deformations and stresses, the lower will be the unit stress at which there is a deviation from the linear relation of stress and strain. Here again it is desirable to adopt some standardized method for making the determination.

For practical purposes, it is probably sufficiently correct to regard the elastic limit and the proportional limit of a material as being equal in value.

Modulus of Elasticity. Within the elastic limit, the ratio of the unit stress to the corresponding unit deformation is constant, and this ratio is called the modulus of elasticity. The value of the modulus of elasticity in tension or compression (Young's modulus) may be computed from any of the following expressions:

$$E = \frac{s}{\epsilon} = \frac{PL}{Ae} = \frac{sL}{e} = \frac{P}{A\epsilon} \tag{10}$$

in which E denotes the modulus of elasticity, psi, s denotes the unit stress, psi, ϵ denotes the unit deformation, in. per in., e denotes the total deformation in the gage length L, in., L denotes the gage length in which the deformations are measured, in., A denotes the area of cross section, sq in., and P denotes the total axial load on the specimen, lb.

Table 1 gives some average values of modulus of elasticity for various engineering materials.

The modulus of elasticity is often spoken of as a measure of the stiffness of a material, stiffness being defined as the ability to resist deformation. Since ϵ equals s/E, it is evident that for a given unit stress the material with the largest modulus of elasticity will suffer the smallest unit deformation and will therefore be the stiffest.

Lateral Strain. Experiments show that, when a material is subjected to axial stress within the elastic limit, it deforms not only longitudinally but also laterally. Under

TABLE 1. AVERAGE VALUES OF MODULUS OF ELASTICITY

Material	Modulus of elasticity, psi	
	Tension or compression E	Shear E_s
Aluminum	10,000,000	3,700,000
Brass	15,500,000	6,200,000
Bronze	15,000,000	5,600,000
Manganese	16,000,000	6,000,000
Copper, cast	15,000,000	5,600,000
drawn	17,000,000	6,400,000
Iron, cast	15,000,000	6,000,000
malleable	22,000,000	8,800,000
wrought	27,000,000	10,800,000
Steel	30,000,000	12,000,000
Zinc	12,000,000	
Wood	1,500,000	
Concrete	3,000,000	

tension the lateral dimensions diminish, and under compression they increase. The ratio of the unit lateral deformation to the unit longitudinal deformation is called *Poisson's ratio.* The equation for Poisson's ratio is as follows:

$$\lambda = \frac{\epsilon'}{\epsilon} \tag{11}$$

in which λ denotes Poisson's ratio, ϵ' denotes unit lateral deformation, in. per in., and ϵ denotes unit longitudinal deformation, in. per in.

Table 2 gives some average values of Poisson's ratio for various engineering materials.

TABLE 2. AVERAGE VALUES OF POISSON'S RATIO

Material	Poisson's Ratio
Steel	0.25
Wrought iron	0.25
Cast iron	0.25
Brass	0.33
Copper	0.33
Concrete	0.08–0.16
Glass	0.24

Elastic Deformations. 1. *Axial Stress.* In Fig. 13 is shown a unit cube subjected to axial stresses in the y direction. From what has been said previously regarding Poisson's ratio, it is evident that the unit deformations in the x, y, and z directions would be as follows:

$$\epsilon_x = \frac{s_1\lambda}{E} \tag{12}$$

$$\epsilon_y = \frac{s_1}{E} \tag{13}$$

$$\epsilon_z = \frac{s_1\lambda}{E} \tag{14}$$

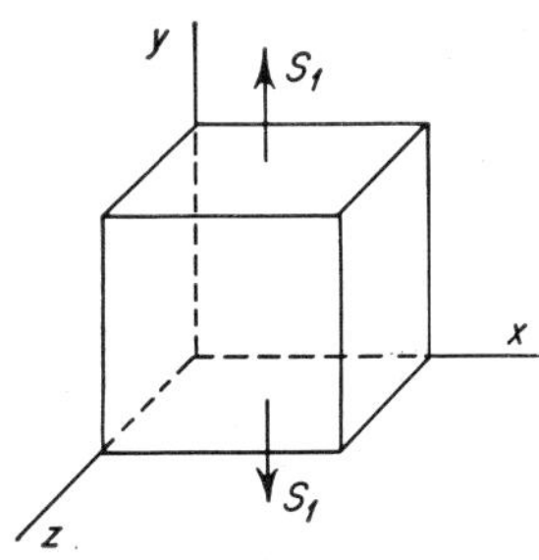

FIG. 13. Axial stress.

The original volume of the cube was unity, and the new volume may be computed by noting that the new dimensions are as follows:

$$\text{New length} = 1 + \frac{s_1}{E} = 1 + \epsilon_y \tag{15}$$

$$\text{New width} = 1 - \frac{s_1\lambda}{E} = 1 - \epsilon_y\lambda \tag{16}$$

$$\text{New thickness} = 1 - \frac{s_1\lambda}{E} = 1 - \epsilon_y\lambda \tag{17}$$

In these equations, an increase of dimension is given the plus sign and a decrease the minus sign. The new width times the new thickness gives an area equal to $1 - 2\epsilon_y\lambda$, when the terms involving ϵ_y^2 are neglected, as they may be. This area times the new length gives a volume equal to $1 - 2\epsilon_y\lambda + \epsilon_y$. This volume will be larger than the original volume for most engineering materials, because λ is less than one-half. The new volume minus the original volume gives the change in volume per cubic inch, which is

$$\text{Change in volume per cubic inch} = \epsilon_y(1 - 2\lambda) \tag{18}$$

This change in volume will be an increase for axial tension and a decrease for axial compression.

2. *Biaxial Stress.* In Fig. 14 is shown a unit cube subjected to biaxial stresses along the x and y directions. By considering the effect of the stresses one at a time, it is evident that the unit deformations in the x, y, and z directions are as follows:

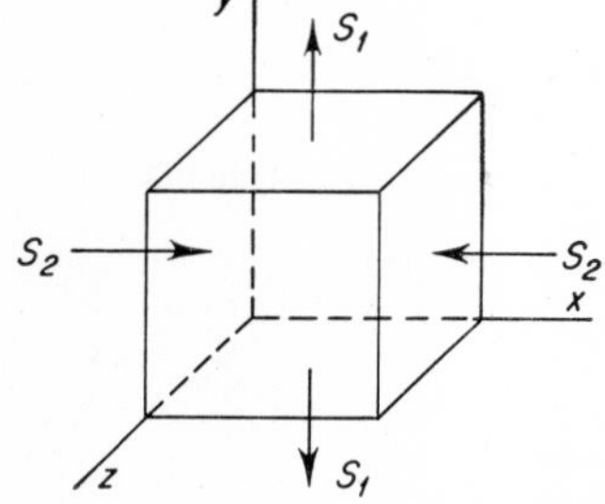

FIG. 14. Biaxial stresses.

$$\epsilon_x = -\frac{s_2}{E} - \frac{s_1\lambda}{E} \tag{19}$$

$$\epsilon_y = \frac{s_1}{E} + \frac{s_2\lambda}{E} \tag{20}$$

$$\epsilon_z = -\frac{s_1\lambda}{E} + \frac{s_2\lambda}{E} \tag{21}$$

The terms in the above equations are given the plus or minus sign according as the deformation produced by the stress, considered by itself, would produce an increase or a decrease of dimension parallel to the axis under discussion.

3. *Triaxial Stress.* In Fig. 15 is shown a unit cube subjected to triaxial stresses along the x, y, and z directions. The unit deformations along the three axes are as follows:

$$\epsilon_x = -\frac{s_2}{E} - \frac{s_1\lambda}{E} - \frac{s_3\lambda}{E} \tag{22}$$

$$\epsilon_y = \frac{s_1}{E} + \frac{s_2\lambda}{E} - \frac{s_3\lambda}{E} \tag{23}$$

$$\epsilon_z = \frac{s_3}{E} + \frac{s_2\lambda}{E} - \frac{s_1\lambda}{E} \tag{24}$$

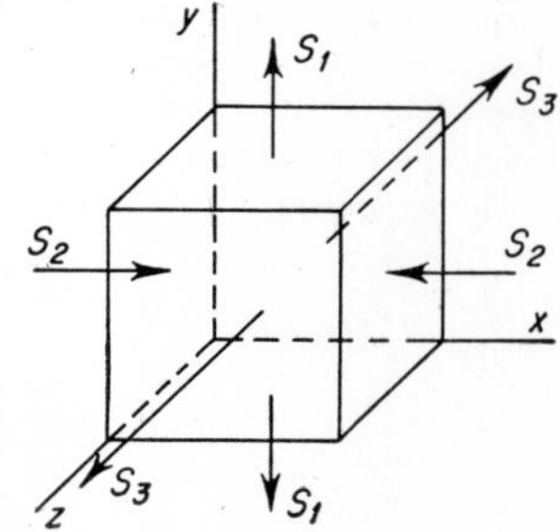

FIG. 15. Triaxial stresses.

There is a certain difference, which should be carefully noted, between the case of axial stress and the

cases of biaxial and triaxial stress. In Fig. 13, the stress was in the y direction; and, if the net deformation in the y direction is multiplied by E, the result is the unit stress in that direction. For the cases of biaxial and triaxial stress, none of the net deformations multiplied by E will give the correct unit stress in the direction involved. In numerical problems involving biaxial and triaxial stress, it is usually most convenient to write down the equations for the net deformations along the axes and then solve for the unit stresses that may be unknown.

It can be shown that the change in volume per unit volume, whether the case is single stress, biaxial stress, or triaxial stress, is given by the following equation:

$$\text{Change in volume per unit volume} = \epsilon_x + \epsilon_y + \epsilon_z$$

Plus signs are used for the unit deformations when they are increases and minus signs when they are decreases.

Modulus of Rigidity. The shearing modulus of elasticity, or modulus of rigidity, is defined by the following equation:

$$E_s = \frac{s_s}{\epsilon_s} \tag{25}$$

in which E_s denotes the modulus of rigidity, psi, s_s denotes the shearing unit stress, psi, and ϵ_s denotes the shearing unit deformation, or unit detrusion, in. per in.

In Fig. 16, let $ABCD$ be a prism of unit thickness with sides equal to L. The side AD is considered fixed, and the prism deforms under the shear couples as shown. The shear deformation, or detrusion, is BB' in the length L. The unit deformation is

$$\epsilon_s = \frac{BB'}{L} = \phi \tag{26}$$

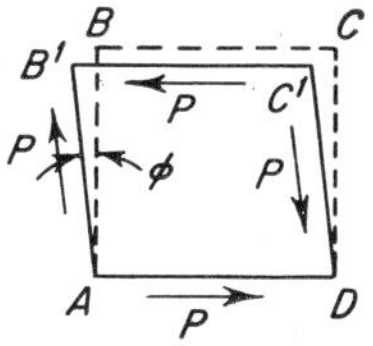

FIG. 16. Prism under shearing forces.

It is evident, therefore, that the unit detrusion is equal to the angle ϕ in radians.

As has been shown in a previous discussion on p. 3–4, the shearing stresses set up tensile stresses along the diagonal DB' and compressive stresses along the diagonal AC', and these unit stresses are equal to the original applied shearing unit stress. On this basis, it can be shown[1] that the relation between the modulus of elasticity in tension and compression and the modulus of rigidity is as follows:

$$E_s = \frac{E}{2(1 + \lambda)} \tag{27}$$

This equation shows that, if Poisson's ratio is 1/4, the modulus of rigidity would be equal to $0.4E$. The modulus of rigidity is usually determined from a torsion test, and many such tests have shown that the modulus of rigidity for steel is about 12,000,000 psi.

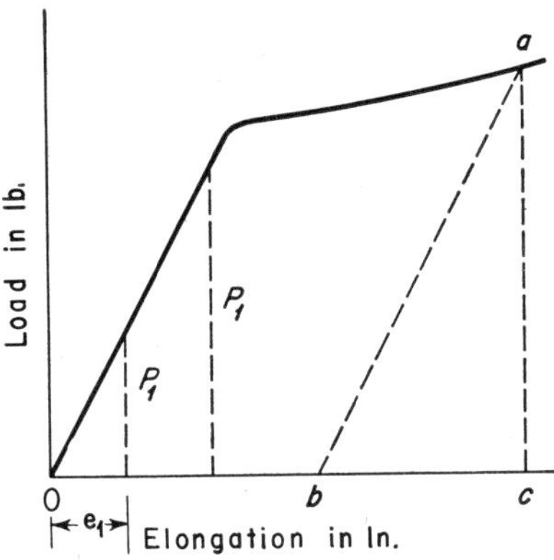

FIG. 17. Load-deformation diagram.

Resilience. For stresses within the elastic limit, it is possible to calculate the work done when the material is deformed by the application of loads. Figure 17 shows the load-

[1] BOYD and FOLK, "Strength of Materials," 5th ed., p. 39. SINGER, "Strength of Materials," 1st ed., p. 17.

elongation diagram for an axial-tension test. When the load applied is P_1 and the corresponding elongation is e_1, the average force applied during the elongation is $P_1/2$, and the work done is

$$\text{Total work done} = \frac{P_1 e_1}{2} \tag{28}$$

expressed in inch-pounds. Since P_1 equals $s_1 A$, and e_1 equals $\epsilon_1 L$, the work done may also be expressed as

$$\text{Total work done} = \frac{s_1 \epsilon_1}{2} \cdot AL \tag{29}$$

Since AL equals the volume, evidently $s_1\epsilon_1/2$ is the work done per cubic inch measured in inch-pounds per cubic inch. Also, since ϵ_1 equals s_1/E, the work per cubic inch may be expressed as follows:

$$\text{Work per cubic inch} = \frac{s_1 \epsilon_1}{2} = \frac{s_1^2}{2E} \tag{30}$$

It is evident from Eq. (28) and from Fig. 17 that the total work done up to the load P_1 is equal to the area under the load-elongation curve. Also, the area under the $s - \epsilon$ curve is equal to the work done per cubic inch.

If, in Fig. 17, it is desired to obtain the work done when the load changes from P_1 to P_2, it may be expressed as follows:

$$\text{Work done} = \frac{P_1 + P_2}{2}(e_2 - e_1) = \frac{s_1 + s_2}{2}(\epsilon_2 - \epsilon_1)V = \frac{(s_2^2 - s_1^2)V}{2E} \tag{31}$$

in which V denotes the volume of the specimen.

For stresses within the elastic limit, the energy that is absorbed by the specimen can be again released, and this stored energy is often called the *resilience*. The maximum resilience is evidently determined by the value of the elastic limit, and the work per cubic inch absorbed when the unit stress changes from zero to the elastic limit is called *modulus of resilience.* This energy is a measure of the capacity of the material to absorb the work done by a suddenly applied load or a blow, without danger of suffering a permanent deformation.

The amount of energy which can be stored in engineering materials is not large enough so that it can be used to drive a machine, unless the amounts of energy required are relatively small. Practical devices in which elastic energy is thus used are clock springs, watch springs, and the ordinary spring motor.

When the load in Fig. 17 is increased beyond the elastic limit, say, to the point a, part of the work done is used in permanently deforming the material. Upon release of the load to zero, the curve ab is approximately parallel to the original straight portion of the load-elongation curve. The total work done in stressing the material up to the point a is represented by the area $OacO$, and the area $bacb$ represents the work that is again released when the load is reduced to zero.

MECHANICAL PROPERTIES OF MATERIAL

Definitions. In describing the behavior of materials under the application of loads, certain descriptive terms are commonly used, which may be more or less overlapping rather than mutually exclusive. The terms *elastic, plastic,* and *stiff* have already been defined.

Ductile materials are those which can be permanently drawn out without rupture. This ductility refers particularly to the drawing out that is possible while the volume remains practically constant. In a tensile test, the ductility commonly reported is that

which is determined after actual rupture has taken place. Soft steel, brass, and aluminum are examples of ductile metals.

Brittle materials may be looked upon as the opposite of ductile. They can be deformed very little without rupture, and their rupture is characterized by a sudden, shattering failure. Cast iron, concrete, and brick are examples of brittle materials.

Malleable materials are capable of being beaten into thin sheets. Evidently materials that are ductile are also likely to be malleable. Gold, tin, and lead are examples of malleable metals.

Flexible materials are capable of being bent without breaking. Malleable iron is more flexible than gray iron, soft steel is more flexible than hard steel, and hickory is more flexible than basswood.

Hardness is a quality that has been defined in a number of arbitrary ways. The most common scale of hardness used for engineering materials is the indentation hardness, like the Brinell hardness, which is measured by the indentation of a standard ball under a standard load. Scratch hardness is measured by the Mohs scale for minerals. Rebound hardness, measured by a standard hammer dropped from a standard height, is determined by the Shore scleroscope. Cutting hardness and abrasion hardness are other measures of hardness that are of practical importance.

Resilient materials are those which are capable of absorbing large amounts of energy without suffering permanent deformation. Evidently the combination of low modulus of elasticity with high elastic limit would produce highly resilient materials.

Tough materials are capable of absorbing large amounts of energy before rupture. Toughness is evidently dependent upon a combination of high strength and high ductility or high strength and high flexibility. Toughness measures the ability to absorb a shock or sudden blow without rupture.

Stress-Strain Diagrams. When an extensometer tension test is made, an instrument called an extensometer is attached to the specimen so that the elongation of the specimen may be measured. The extensometer must be capable of measuring small deformations, and the least reading on the instrument is usually 0.0001 in. or smaller. The distance between the points of attachment of the instrument, on which the deformation is measured, is called the *gage length*. Readings of load and deformation are taken for a series of load increments. The load divided by the cross-sectional area of the specimen gives the unit stress, and the total deformation divided by the gage length gives the unit deformation. The curve drawn with unit stress as the ordinate and unit deformation as the abscissa is called the stress-strain or *stress-deformation diagram*. Similar diagrams may of course be drawn also for other types of test, such as compression and torsion tests.

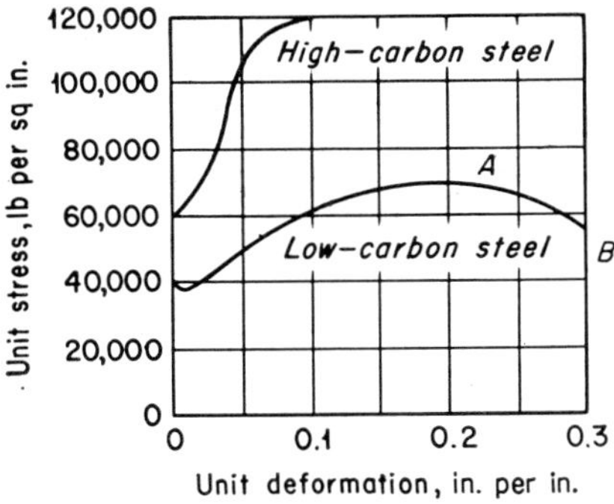

FIG. 18. Stress-deformation curves for steels.

Figure 18 shows stress-strain curves for a low-carbon and a high-carbon steel. In order to show the complete deformation up to rupture, the scale for abscissa is such as to make negligible the deformation within the elastic limit. If the unit stress at the elastic limit were 30,000 psi the corresponding unit deformation would be only 0.001 in. per in., while at rupture it might be as much as 0.25 in. per in. Figure 19 shows to an enlarged scale of abscissas the portion of a diagram that goes only slightly beyond the elastic limit.

For some engineering materials, like the steels, the first portion of the diagram is an inclined straight line, and the proportional limit is fairly well defined. For other materials, like cast iron and concrete, the stress-strain diagrams take a form similar to that shown in Fig. 20. Here the curve deviates from a straight line almost from the beginning, and the proportional limit is poorly defined.

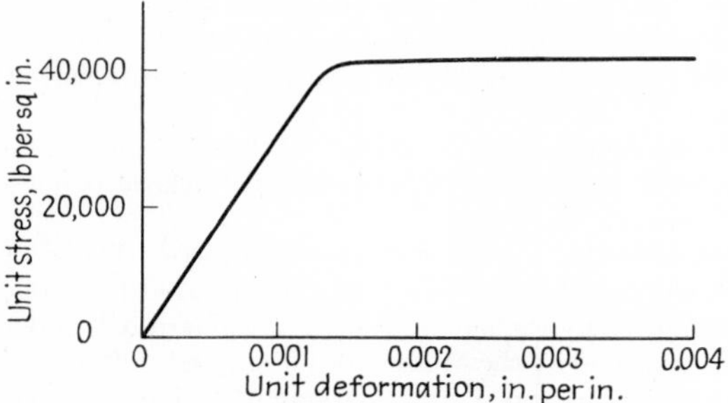

FIG. 19. Stress-deformation curve for steel.

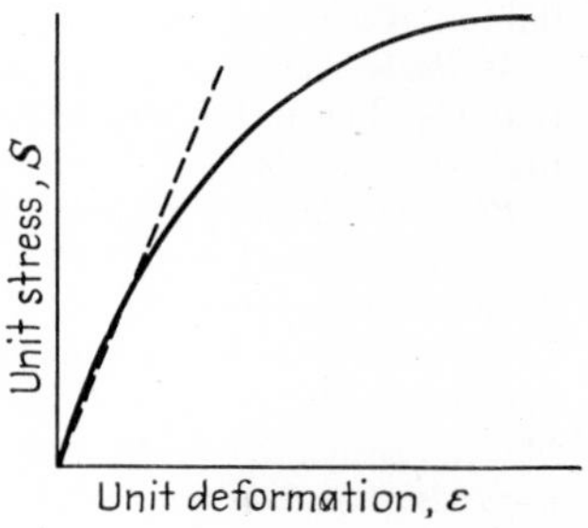

FIG. 20. Stress-deformation curve for cast iron.

Yield Point. Materials like wrought iron and the softer steels deform relatively rapidly at a unit stress slightly beyond the proportional limit. The yield point is defined as that unit stress, slightly beyond the proportional limit, at which the material deforms with little or no increase in unit stress. Sometimes, as shown in Fig. 18 in the lower curve, there is an actual dip in the curve beyond the yield point. This may occur in the softer steels and indicates that, when the yielding of the material has once begun, it will continue to yield somewhat even at a lower unit stress.

Yield Strength. Nonferrous metals and alloys, like aluminum and brass, do not exhibit the yield point that is characteristic of wrought iron and the softer steels. For these materials the stress-strain curve usually consists of an initial straight line which gradually bends to the right at the higher stresses. The ASTM specifications suggest the use of the term "yield strength" for such materials, this value being determined as shown in Fig. 21. The line *mn* is drawn parallel to the straight portion of the stress-strain curve with an offset *Om*, which may be 0.1 percent, 0.2 percent, or whatever value is specified for a given material. The intersection of the line *mn* with the stress-strain curve determines the yield strength S_y as indicated, this value being the unit stress corresponding to a given percentage of permanent set.

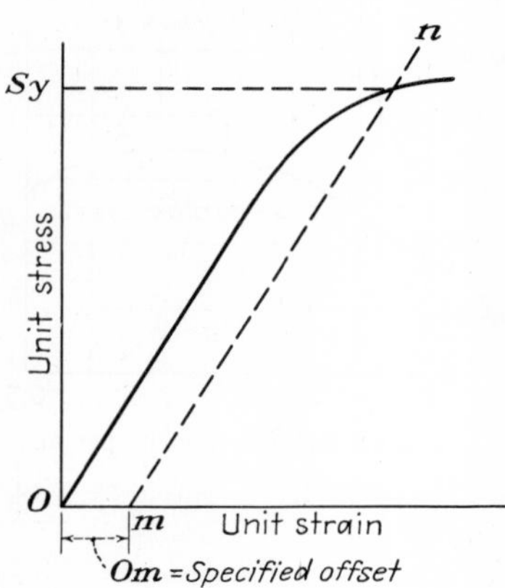

FIG. 21. Determination of yield strength.

If unit stress is plotted against unit strain, a 0.1 percent offset would correspond to a unit strain of 0.001. If total load is plotted against total strain, a 0.1 percent offset would mean 0.1 percent of the gage length used in making the test.

Ultimate Strength. The maximum unit stress attained before rupture is called the ultimate strength of the material. This unit stress is indicated by the point *A* in Fig.

18. Up to the ultimate strength, the diminution in cross section in a tension test is uniform along the length of the specimen, and the volume may be considered as practically constant. For the softer metals, the tension specimen begins to *neck down* after the ultimate strength is reached; and, when rupture finally occurs, at B in Fig. 18, the unit stress may be considerably less than the ultimate strength. The unit stress at B is called the *breaking strength*.

For materials like cast iron, which are not ductile, there is no necking down, and the ultimate strength and the breaking strength coincide, as shown in Fig. 20.

Although the cross-sectional area in a tension test is smaller at the ultimate than it was originally, it is customary to calculate the ultimate strength on the basis of the original area and not on the reduced area.

Percentage of Elongation. The elongation of a material up to rupture is a valuable indication of the ability of the material to deform greatly under load without producing a shattering failure. The percentage of elongation, measured after rupture, is defined as follows:

$$\text{Percentage of elongation} = \frac{\text{final length} - \text{original length}}{\text{original length}} \times 100 \tag{32}$$

The percentage of elongation is regarded as a measure of ductility.

When *necking down* occurs, the percentage of elongation is of course greatest for that portion of the specimen near the fracture, and it is for this reason that the percentage of elongation will be different if different gage lengths are employed. It is customary to adopt a standard gage length in reporting percentage of elongation; in American practice, the 2- and the 8-in. gage lengths are most common.

Percentage of Reduction in Area. The percentage of reduction in area, measured after rupture, is also regarded as a measure of ductility and is defined as follows:

$$\text{Percentage of reduction in area} = \frac{\text{original area} - \text{final area}}{\text{original area}} \times 100 \tag{33}$$

The final area is calculated from the minimum diameter after rupture has taken place.

Tension Test. From an ordinary commercial tension test, made without the use of an extensometer, the results that are commonly reported are:

1. Ultimate strength.
2. Yield point.
3. Percentage of elongation.
4. Percentage of reduction in area.

For brittle materials like cast iron, there will be no yield point, and items 3 and 4 are so small as to be negligible.

The *energy of rupture*, or the work per cubic inch absorbed up to rupture, may be calculated roughly by assuming that the stress-strain diagram is a trapezoid with the base equal to the unit elongation at rupture and the sides equal respectively to the yield point and ultimate strength. Then it follows that

$$\text{Energy of rupture} = \frac{s_y + s_u}{2} \times \epsilon_u \tag{34}$$

in which s_y denotes the yield point, psi, s_u denotes ultimate strength, psi, and ϵ_u denotes unit deformation at rupture, in. per in.

When an extensometer is used in a tension test, several other results may be determined besides those mentioned above. These are the modulus of elasticity, the pro-

portional limit, and the modulus of resilience. If the stress-strain curve is complete up to rupture, the energy of rupture may be computed from the area under the curve.

Compression Test. The compression test is more commonly made on brittle than on ductile materials. For ductile materials, the only significant result obtained from a compression test is the yield point, beyond which the specimen merely shortens without actually fracturing. For these materials, therefore, there is no ultimate compressive strength, unless the yield point is so considered.

When a compression test is made on a brittle material without using a compressometer, only the ultimate strength is determined. If a compressometer is used, it is possible to determine also the proportional limit, the modulus of elasticity, and the energy of rupture. For a stress-strain curve as shown in Fig. 20, it is customary to determine the modulus of elasticity from the slope of a straight line drawn through the origin of coordinates and a value of working, or safe, unit stress.

Shear Test. When the shear test is made on metals, a shear tool is employed in which the specimen is clamped so as to prevent bending action as much as possible. The shear test determines only the ultimate shearing strength.

The shear test is commonly made on wood to determine the shear strength parallel to the grain. One type of specimen which has been found satisfactory by the Forest Products Laboratory is shown in Fig. 22. The shear tool used for this test provides an ⅛-in. offset between the inner edge of the supporting surface and the plane along which the failure occurs. The grain of the wood runs vertically, and the ultimate strength is determined by shearing off the end of the specimen.

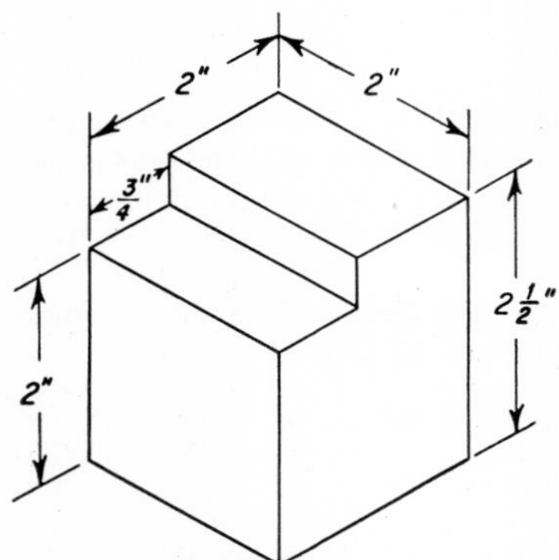

Fig. 22. Wood shear specimen.

Allowable Stress. The allowable stress for a material is the maximum unit stress that may be safely applied. The allowable stress and the *working stress* should be of equal magnitude in a proper design.

The *factor of safety* is the ultimate load divided by the safe load, or the ultimate strength divided by the allowable stress.

Allowable stresses and factors of safety are determined by competent engineering judgment applied to the results of experience and experiment. The building codes of cities usually specify allowable stresses for construction covered by the code. In large designing offices, much of the work is done on the basis of definite specifications. When the organization in which an engineer is employed is small, or when some more or less original design is contemplated, he may be called upon to use his own judgment with respect to allowable stresses and factors of safety.

The choice of allowable stresses and factors of safety depends on the materials themselves and on the circumstances under which the structure or machine must operate. The more homogeneous and reliable materials may be employed with smaller factors of safety than materials not so reliable. When failure of a structure or machine would endanger life and limb, the factor of safety should be ample. Parts that are subject to sudden blows, shocks, or uncertain hazardous service are designed with correspondingly larger factors of safety.

Table 3 gives average values of some of the mechanical properties of engineering materials.

TABLE 3. MECHANICAL PROPERTIES OF SOME ENGINEERING MATERIALS

Material	Equivalent	Ultimate strength, psi			Yield point, tension, psi	Modulus of elasticity, tension or compression, psi	Modulus of elasticity, shear, psi	Weight per cu in., lb
		Tension	Compression *	Shear				
Steel, forged-rolled:								
C, 0.10–0.20	SAE 1015	60,000	39,000	48,000	39,000	30,000,000	12,000,000	0.28
C, 0.20–0.30	SAE 1025	67,000	43,000	53,000	43,000	30,000,000	12,000,000	0.28
C, 0.30–0.40	SAE 1035	70,000	46,000	56,000	46,000	30,000,000	12,000,000	0.28
C, 0.60–0.80	...	125,000	65,000	75,000	65,000	30,000,000	12,000,000	0.28
Nickel	SAE 2330	115,000	...	92,000	...	30,000,000	12,000,000	0.28
Cast iron:								
Gray	ASTM 20	20,000	80,000	27,000	...	15,000,000	6,000,000	0.26
Gray	ASTM 35	35,000	125,000	44,000	...	...	...	0.26
Gray	ASTM 60	60,000	145,000	70,000	...	20,000,000	8,000,000	0.26
Malleable	SAE 32510	50,000	120,000	48,000	...	23,000,000	9,200,000	0.26
Wrought iron	...	48,000	25,000	38,000	25,000	27,000,000	...	0.28
Steel, cast:								
Low C	...	60,000	...	...	...	...	...	0.28
Medium C	...	70,000	...	...	...	...	...	0.28
High C	...	80,000	45,000	...	45,000	...	...	0.28
Aluminum alloy:								
Structural, No. 350	...	16,000	5,000	11,000	5,000	10,000,000	3,750,000	0.10
Structural, No. 17ST	...	58,000	35,000	35,000	35,000	10,000,000	3,750,000	0.10
Brass:								
Cast	...	40,000	...	...	...	...	...	0.30
Annealed	...	54,000	18,000	...	18,000	...	...	0.30
Cold-drawn	...	96,700	49,000	...	49,000	15,500,000	6,200,000	0.30
Bronze:								
Cast	...	22,000	...	...	...	...	...	0.31
Cold-drawn	...	85,000	...	...	...	15,000,000	6,000,000	0.31
Brick, clay:								
Grade SW	ASTM	...	3,000 (min) †	...	...	...	...	0.072
Grade MW	ASTM	...	2,500 (min)					
Grade NW	ASTM	...	1,500 (min)					
Concrete, 1:2:4 (28 days)	...	...	2,000	...	...	3,000,000	...	0.087
Stone	...	...	8,000	...	...	...	...	0.092
White oak:								
Parallel to grain	...	...	7,440	2,000	4,760 ‡	1,780,000	...	0.028
Across grain	...	800	...	...	1,320 ‡			
White pine:								
Parallel to grain	...	...	4,840	860	3,680 ‡	1,280,000	...	0.015
Across grain	...	300	...	...	550 ‡			
Southern longleaf pine:								
Parallel to grain	...	...	8,440	1,500	6,150 ‡	1,999,000	...	0.024
Across grain	...	470	...	...	1,199 ‡			

* The ultimate strength in compression for ductile materials is usually taken as the yield point. The bearing value for pins and rivets may be much higher, and for structural steel is taken as 90,000 psi.

† Average of five bricks.

‡ Proportional limit in compression.

REPEATED STRESSES

Fatigue. The subject of repeated stresses has become more and more important because of the higher speeds that have become common in machines like the automobile, the airplane, and the turbine. Experiment has shown that a material will fail under repeated stresses not only at unit stresses less than the ultimate strength but even at stresses below the elastic limit. Numerous experiments have shown that, when a material is subjected to repeated stresses of sufficient magnitude, the action of the stresses is such as to start a microscopic crack, which, under continued application of stress, spreads until failure occurs. In the literature on the subject, the name "fatigue" has been given to this behavior, although a better description of the action which is going on would be to say that it is progressive failure due to repeated stresses.

The case of repeated stresses may be illustrated by a pair of wheels and axle under an ordinary railroad freight car. Consider the portion of the axle projecting beyond the wheels. The upper fibers of the axle are under tensile stress, and the lower fibers are under the same amount of compressive stress. As the wheels roll, the fiber that was

at the top of the axle is at the bottom at the next instant. In such a case, the unit stresses are reversed from tension to compression for each revolution of the wheels; this is called a *cycle* of stress. In other cases of repeated loads, the unit stresses may be only partly reversed; and, in still other cases, the unit stresses may be of the same kind but changing from a minimum to a maximum value.

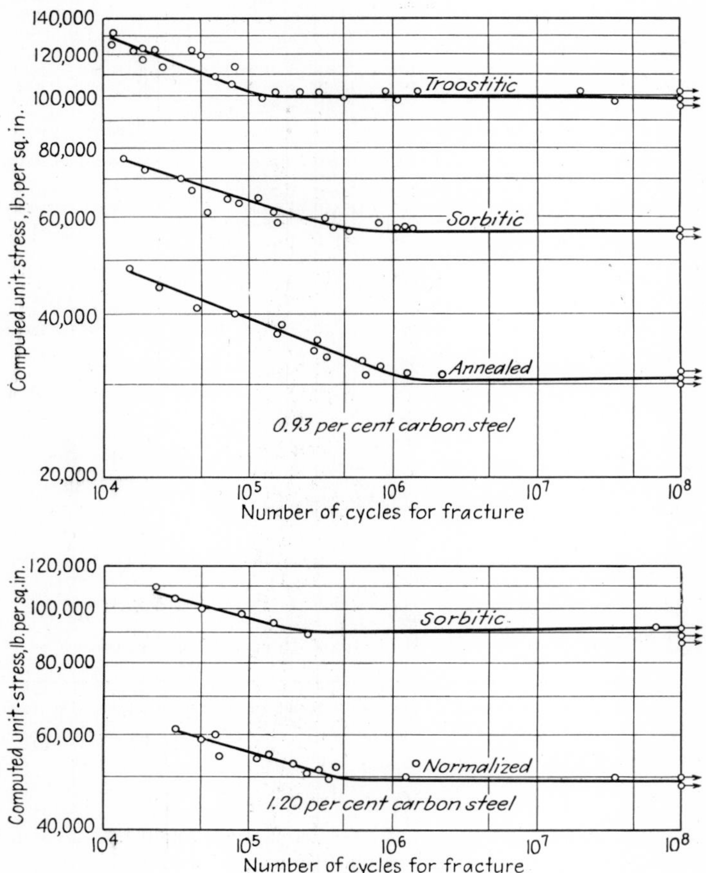

FIG. 23. *S-N* diagrams obtained from fatigue tests. (*Moore and Kommers.*)

Figure 23 shows typical endurance curves, or *S-N diagrams*, in which the unit stress to which the specimen was subjected is plotted as ordinate, and the number of cycles of stress necessary to produce rupture is plotted as abscissa. The unit stress that the material can withstand apparently indefinitely, which is shown on the diagram by the horizontal part of the curve, is called the *endurance limit*. For materials subjected to repeated stresses, it is evident that the endurance limit is a significant property, and machines and structures must be so designed that the working unit stress is less than this limit.

Numerous experiments on steel, cast iron, malleable iron, nonferrous metals, concrete, and wood indicate that engineering materials in general have *S-N* diagrams similar to

those shown in Fig. 23. Sometimes the portion of the diagram for the higher stresses is a straight line, and sometimes it is curved, but in all cases the curve tends to become asymptotic to the cycles axis. For many materials, a rather abrupt break in the curve occurs at 1,000,000 cycles or less; but for some of the nonferrous metals, it is necessary to carry out the tests to 50,000,000, 100,000,000, or even more before the curve becomes asymptotic.

Goodman Diagram. Figure 24 shows a diagram for wrought ferrous metals that is known as the Goodman diagram and that was developed independently by both Goodman and J. B. Johnson. The ordinate to the line *EB* represents the static ultimate tensile strength of the metal. The minimum stresses to which a specimen is subjected during a cycle are plotted along a 45° line *DOB*. The horizontal line through *O* is the line of zero stress, tensile stress being plotted above the line and compressive stress below it. According to the "dynamic theory" of suddenly applied loads,[1] the minimum or dead-load stress plus twice the live-load stress equals the static ultimate strength, and the maximum applied stress during a cycle should fall on a line *CAB*, such that the point *A* is five-tenths of the ultimate static strength. Goodman plotted endurance limits obtained by various investigators after the material had been subjected to over 4,000,000 cycles of stress, and he found that these experimental points fell fairly well on the straight line *CAB*.

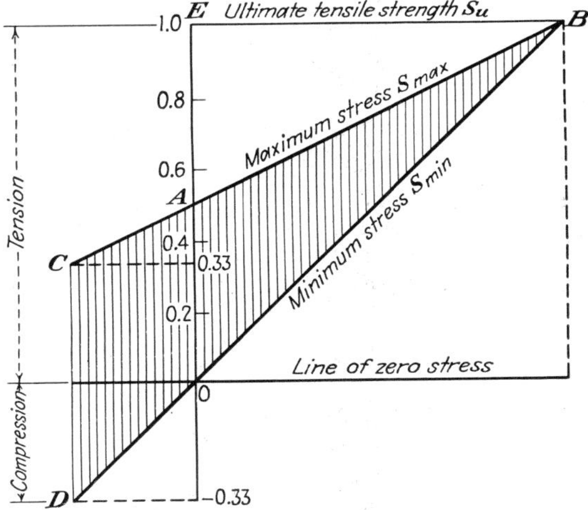

FIG. 24. The Goodman diagram.

As the diagram shows, when the minimum stress is zero, the maximum stress for indefinite endurance should be five-tenths of the ultimate static strength. When the stress is completely reversed at *CD*, the plus and minus stresses should be one-third of the ultimate strength. Presumably a diagram similar to the one shown in Fig. 24 would hold when the stress above the zero line is compressive and that below is tensile. Experimental data, however, are very meager for these combinations of compressive stress.

According to the Goodman diagram, the *range* of stress (algebraic difference between maximum and minimum) is greatest for reversed stresses and decreases as the maximum stress is increased above one-third of the ultimate strength, being actually zero when the maximum stress coincides with the ultimate. In other words, as the maximum

[1] GOODMAN, "Mechanics Applied to Engineering," 9th ed., p. 634. FIDLER, "Practical Treatise on Bridge Construction."

stress is increased, the minimum stress must be decreased algebraically in order that the material may be stressed indefinitely without failure.

J. B. Johnson developed a formula, which was later simplified by Barr, based on a diagram similar to Goodman's. The formula is

$$s_{\max} = \frac{0.5s_u}{1 - 0.5r} \tag{35}$$

in which $s_{\max}$ denotes the maximum unit stress during the cycle, psi, s_u denotes the static ultimate tensile strength, psi, and r denotes the range ratio for the cycle of stress, i.e., the ratio of the minimum to the maximum stress. It should be remembered that r, the range ratio, in Eq. (35), is positive if the stress limits of a cycle are both tensile or both compressive but is negative if one limit is tensile and the other compressive.

Experiments on ferrous metals have shown that s_{-1}, the endurance limit for completely reversed stresses, is about five-tenths of the ultimate tensile strength, instead of one-third as indicated by the Goodman diagram.

The Goodman diagram shows that the ratio of $s_0:s_{-1}$ has a value of 1.5, s_0 being the endurance limit when the minimum stress is zero, and s_{-1} having the same meaning as above. Experiments show that it is reasonably safe to assume that such a ratio exists.

Moore and Kommers[1] have suggested a modification of the J. B. Johnson formula that is not based on any assumed ratio of s_{-1} to s_u but rather on an experimentally determined value of s_{-1} for each metal. The value of 1.5 for the ratio of $s_0:s_{-1}$ is retained, and the formula is

$$s_{\max} = \frac{1.5s_{-1}}{1 - 0.5r} \quad \text{or} \quad \frac{s_{\max}}{s_{-1}} = \frac{3}{2 - r} \tag{36}$$

in which the notation is the same as before. If r equals zero, $s_{\max}/s_{-1}$ equals 1.5, the Goodman ratio.

It should be remembered that elastic failure is as important as fatigue failure, and therefore $s_{\max}$ determined by Goodman's diagram or any formula for endurance stresses should never be permitted to exceed the elastic limit of the material.

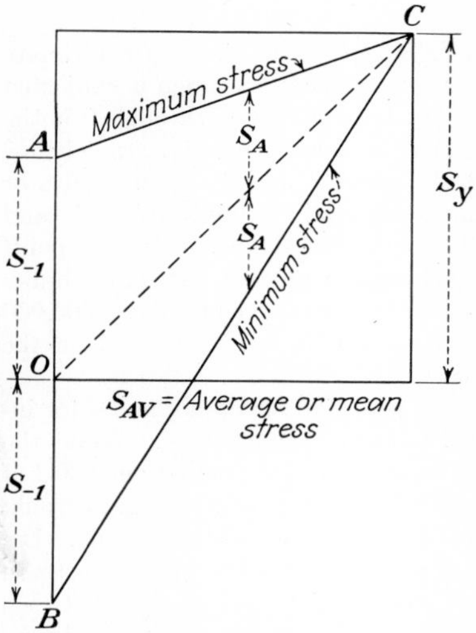

FIG. 25. Modified fatigue stress diagram.

A Modified Diagram. There are a number of objections to the Goodman diagram, shown in Fig. 24, and the formulas derived from it. Comparatively few experiments have been made to verify the correctness of the diagram for the higher maximum stresses, and for ferrous metals and alloys the average ratio of endurance limit to tensile strength is more nearly 50 percent rather than 33 percent. A further objection is the fact that it would probably never be desirable to use a maximum stress in a cycle which exceeds the yield point of the material, because such a stress would cause undesirable permanent deformation.

Experiments made in Europe have suggested a modified diagram,[2] as shown in Fig. 25. The point C in the figure represents the value of the yield point, and OA and OB the endurance limit for completely reversed stress. The

[1] MOORE and KOMMERS, "Fatigue of Metals," p. 185.
[2] See KOMMERS, J. B., Design-stress Diagrams, *Product Eng.*, October, 1938, p. 395.

abscissas represent the average, mean, or steady unit stress in the cycle. When the same scale is used for ordinates and abscissas the line OC will be a 45° line. Two straight lines are drawn from the yield point to the plus and minus values of the endurance limit for completely reversed stress, and these lines represent the values of maximum and minimum stress. The diagram has the further advantage that it does not depend upon an arbitrary ratio of endurance limit for completely reversed stress to tensile strength.

The formula that expresses the relationships in such a diagram is as follows:

$$s_{\max} = \frac{2s_{-1}}{(1 - r) + (s_{-1}/s_y)(1 + r)} \tag{37}$$

in which the notation is the same as before, and s_y is the yield point or yield strength.

In connection with Fig. 25, the following relationships are of interest. $\pm s_a$ = alternating unit stress in a cycle, s_{av} = average, mean, or steady unit stress in a cycle:

$$s_{\max} = s_{av} + s_a \qquad s_{\min} = s_{av} - s_a \qquad \text{and} \qquad s_{av} = \frac{s_{\max} + s_{\min}}{2}$$

When the endurance limit for completely reversed stress and the yield point or yield strength are known, a diagram like Fig. 25 may be easily constructed and Eq. (37) may be used. It should be understood that the use of this diagram and formula is advocated only when the designer is working with materials for which complete information regarding the various fatigue limits for different ratios of r is not available.

Haigh Diagram. Figure 26 shows the diagram suggested by B. P. Haigh for plotting fatigue results. In this diagram, the alternating stress is used as the ordinate, and the steady stress as the abscissa. By "alternating stress" is meant the amount of fluctuation of stress on either side of the steady or average stress in a cycle, the average stress being the algebraic sum of the maximum and minimum stress divided by 2. For instance, if the maximum unit stress is 30,000 psi tension, and the minimum stress is 6,000 psi tension, the steady stress is 18,000 tension, and the alternating stress is 12,000 psi tension.

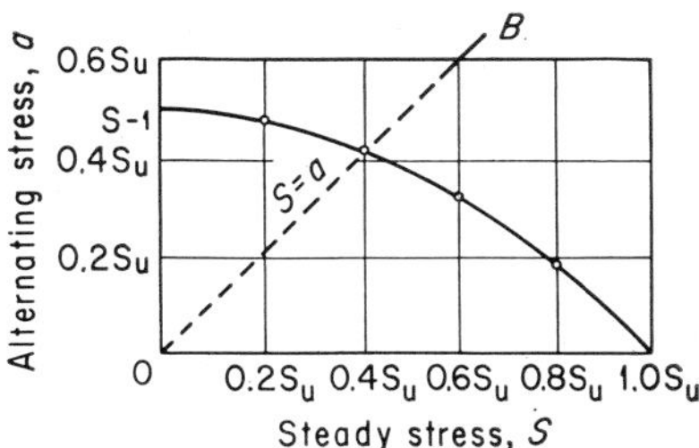

FIG. 26. The Haigh diagram.

The equation that Haigh used in connection with the diagram in Fig. 26 is as follows:

$$a = s_{-1}\left(1 - \frac{s^n}{s_u}\right) \tag{38}$$

in which a denotes the alternating unit stress, psi, s denotes the steady unit stress, psi, s_{-1} denotes the endurance limit for completely reversed stresses, psi, n denotes an experimental constant, and s_u denotes the ultimate tensile strength, psi.

Haigh states that the constant n lies between the limits 1 and 2, varying with different metals. In the above equation, when n equals 2, it can be shown that the equation is the same as Gerber's parabola,[1] which was proposed in 1872 to show the effect of range of stress on endurance limit. When n equals 1, it can be shown that the results would plot in a diagram similar to Goodman's. In Fig. 26, the diagram has been plotted for n equal to 2 and s_{-1} equal to $0.5s_u$. For points on the diagram lying above the 45° line OB, the

[1] Relation between the Superior and Inferior Stresses of a Cycle of Limiting Stress, *Zeit. Bayerischen Arch. Ing.-Vereins*, 1874.

cycle of stress will be partly reversed, because for such a point a will be greater than s. A straight line connecting s_{-1} and $1.0s_u$ would represent the results as given by a diagram similar to Goodman's.

As a picture explaining fatigue results on range of stress, either the Goodman diagram or the modified diagram of Fig. 25 is easily interpreted and can be readily understood even by one not particularly familiar with fatigue data. In the opinion of the author, the Haigh diagram is, next to the diagram in Fig. 25, probably most generally useful and easily interpreted. Also, the equation given in Eq. (38) for plotting Haigh's diagram has the advantage of being expressed in terms of s_{-1}, so that s_{-1} need not be a particular or constant fraction of the ultimate tensile strength. In the exponent n, the formula has a constant that should prove useful in adapting the equation to various kinds of metals and other materials.

Fatigue Fracture. Fracture under repeated stresses is characterized by the initiation of a crack which, under the action of the repeated stresses, spreads until failure results. Even ductile materials under fatigue stresses behave as though they were brittle, and the characteristic fatigue fracture occurs suddenly without warning.

Ewing and Rosenhain [1] observed that, when a metal is subjected to a static stress sufficiently great, the crystals of which the metal is composed yield by slipping on certain gliding planes within the crystal. This slipping has the effect of breaking up the polished surface of a grain into microscopic elevations and depressions in the nature of steps. Under vertical illumination, the steps show as dark lines, which Ewing and Rosenhain called "slip bands."

Ewing and Humfrey [2] carried on a similar study on slip bands when specimens were subjected to reversed bending stresses. When the stress was sufficiently high, slip lines appeared on the crystals. With increase in the number of cycles of stress, additional slip lines appeared, and the original ones showed a tendency to broaden. As the number of cycles increased, the broadening process continued, until parts of the surface became covered by groups of dark markings. At this stage, it was found that an actual crack had opened up along some of the broadened slip lines. Once a crack was formed, it developed rapidly because of concentration of stress at the end of the crack.

More recent experiments have shown that it is possible for slip to occur under fatigue stresses without producing failure. It has also been shown that repeated stresses at or just below the endurance limit may actually produce a considerable increase in strength. This result seems to be due to the repetitive cold working on minute amounts of material. There is at present no commonly accepted theory as to the process that initiates failure by fatigue. Experiments have shown, however, that failure often starts at flaws or discontinuities either on the surface or in the interior of a metal.

Correlation of Fatigue and Static Tests. There has been considerable discussion as to the correlation between the results of static tests and of fatigue tests. The determination of the endurance limit of a material is expensive because of the time consumed and because a series of specimens is necessary to make the determination. If the endurance limit could be predicted from a static test, a considerable saving would be effected.

For ferrous metals subjected to completely reversed stresses, the best correlation seems to be between tensile strength and endurance limit. The endurance limit, on the average, is about 50 percent of the ultimate tensile strength. Since, however, the results may vary between 40 and 60 percent, this relation cannot be relied upon to give more than a rough approximation. The Brinell and Rockwell hardness numbers also show a fair correlation with endurance limit. Proportional limit, yield point, ductility, and impact values of wrought ferrous metals show very poor correlation with endurance limit.

[1] *Roy. Soc. London Phil. Trans.*, vol. 193A, p. 352, 1899.
[2] *Roy. Soc. London Phil. Trans.*, vol. 200A, p. 241, 1902.

For the nonferrous metals, there is no very good correlation between ultimate strength and endurance limit, since the ratio of the latter to the former may vary from 18 to 50 percent.

The results of endurance tests on wood are meager. The endurance limit seems to be approximately 30 percent of the modulus of rupture.

Considerable work has been done on the fatigue strength of concrete. Beam tests show that, for cycles of stress ranging from zero to a maximum, the endurance limit is about 50 percent of the modulus of rupture. Tests on compression cylinders show that, for cycles of stress ranging from zero to a maximum, the endurance limit is about 50 to 55 percent of the static compressive strength.

Abrupt Change of Cross Section. One of the commonest causes of fatigue failure is the presence of an abrupt change of cross section, which may be caused by a square shoulder, a V notch, a screw thread, a keyway, a toolmark, a scratch, or an internal or external flaw in the material. Experiments by Moore and Kommers [1] on steel showed that a square corner may reduce the endurance limit by as much as 50 percent and a V notch as much as 60 percent.

The cause of this action is the presence of high localized stresses at the bottom of the sharp corners or notches. Both theoretical and experimental investigations have proved the existence of these high localized stresses. With a steady stress and a ductile material, such high local stress is not considered particularly dangerous; but, when the stresses are repeated, a crack may be initiated, even though the high stress is present on only a very small area.

In Table 4 are given some experimental results obtained on both ferrous and nonferrous metals, showing the percentage by which the notch reduced the endurance limit.

D. J. McAdam, Jr., and R. W. Clyne have discussed [2] the effect of chemically and mechanically formed notches on the fatigue of metals. In their discussion, they use the decrease in endurance limit of a notched specimen, as compared with a smooth specimen, as a measure of the damage done by the notch. They show that for ordinary steels the sensitiveness to notches increases with increase in tensile strength, but at a gradually decreasing rate.

McAdam and Clyne, however, point out that there are some exceptions to the general rule. They say that local work-hardening during the process of forming a notch may strengthen the region of highest stress concentration and thus diminish the apparent damage due to the notch. There is evidence also that increase in ultimate strength due to cold-working causes greater increase in notch sensitivity than does the same increase in strength due to composition and heat-treatment.

Whenever it is necessary to have changes of cross section, the shoulders or notches should be provided with generous fillets in order to prevent as much as possible the production of high local stress. For the same reason, it has been found that a smooth finish is more desirable than a rough one. Even the indentation caused by marking a member has been known to start a fatigue failure.

Corrosion Fatigue. B. P. Haigh [3] showed in 1917 that the simultaneous action of repeated stress and a corroding agent may very greatly reduce the fatigue strength of a metal. D. J. McAdam in this country, has studied this subject very exhaustively, and a large amount of work has been done on corrosion fatigue, both here and abroad.

With fresh water as the corroding agent, the endurance limit, or corrosion-fatigue limit, for carbon steels and ordinary alloy steels ranges from about 66 to less than 15 percent of the endurance limit in air. With salt water as the corroding agent, the corrosion-fatigue limit is still lower. The corrosion-fatigue limit seems to be improved

[1] MOORE and KOMMERS, *Univ. Illinois Expt. Sta. Bull.*, 124, 1921.
[2] MCADAM and CLYNE, *Nat. Bur. Standards Research Paper* 725, 1934
[3] *Jour. Brit. Inst. Metals*, Sept. 19, 1917, p. 55.

TABLE 4. EFFECT OF NOTCHES ON ENDURANCE LIMIT, ROTATING BEAM TESTS

Kind of material	Kind of notch	Ultimate tensile strength, psi	Endurance limit		Decrease in column 5 compared with column 4, %	Investigator
			Standard specimen	Notched specimen		
1	2	3	4	5	6	7
Gray cast iron	90° square shoulders	20,000	9,300	9,300	0	*K*
Gray cast iron	90° square shoulders	23,200	11,800	11,200	5	*K*
Gray cast iron	90° square shoulders	30,300	15,000	13,700	9	*K*
Gray cast iron	90° square shoulders	37,100	19,500	16,300	16	*K*
Gray cast iron	90° square shoulders	42,500	24,100	19,100	21	*K*
Malleable cast iron	90° square shoulders	49,800	25,600	19,400	24	*K*
0.27% C steel	90° square shoulders	78,500	34,000	21,600	37	*K*
0.48% C steel	90° square shoulders	123,900	51,200	24,600	52	*K*
0.62% C steel	90° square shoulders	134,500	48,800	30,400	38	*K*
0.49% C steel (sorbitic)	90° square shoulders	96,900	48,000	23,500	51	*M*
0.49% C steel (sorbitic)	60° V notch	96,900	48,000	19,200	60	*M*
Armco iron	90° square shoulders	42,400	26,000	13,500	48	*M*
Armco iron, hot-rolled	90° square shoulders	44,300	26,200	18,500	29	*K*
Armco iron, cold-drawn	90° square shoulders	73,100	33,300	25,400	24	*K*
Mg + 4% Al	V notch [1]	35,200	12,000	8,000	33	*N*
Mg + 4% Al + 0.25% Mn	V notch [1]	39,000	15,000	10,000	33	*N*
Mg + 6.5% Al	V notch [1]	41,300	13,000	10,000	23	*N*
Mg + 10% Cu	V notch [1]	39,000	11,000	6,000	45	*N*
Extruded aluminum bronze (heat-treated)	V notch [1]	77,500	34,000	20,000	41	*N*
Cast aluminum bronze (heat-treated)	V notch [1]	77,800	26,000	22,000	15	*N*
1050 steel	V notch [1]	106,900	57,000	24,000	28	*N*

[1] The V notch used by R. R. Moore was a 60° V notch, 0.038 in. deep, with a fillet at the bottom of the notch having a radius of 0.01 in.

K, J. B. Kommers.
M, H. F. Moore and J. B. Kommers.
N, R. R. Moore.

very little either by heat-treatment or by chemical composition, unless such heat-treatment or composition affects corrosion resistance. For "stainless" (high-chromium) and other corrosion-resistant steels, the corrosion-fatigue limit is higher than for carbon steels and other alloy steels.

The study of protection against corrosion fatigue is proceeding along three lines:

1. A study of alloying elements that increase the resistance of metals to corrosion.
2. A study of the effectiveness of protective coatings for metals and alloys.
3. The introduction into the water or other corroding medium of chemical inhibitors or retardants of corrosion.

IMPACT STRESSES

Tensile Impact. Figure 27 shows an ordinary tension bar, supported at the upper end and subjected to the stresses produced when a weight W is allowed to drop from a height h. It can be shown [1] that, under these circumstances, the stress produced in the bar due to the drop of the load is as follows:

$$s_1 = s + s\sqrt{1 + \frac{2h}{e}} \tag{39}$$

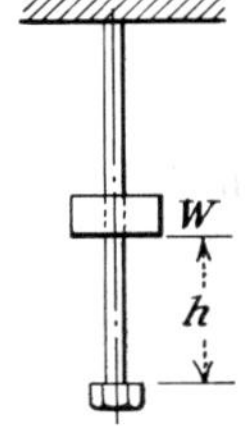

FIG. 27. Tensile impact.

in which s_1 denotes the unit stress produced by the impact, psi; s denotes the unit stress produced in the bar, if W were a quiescent load, psi; e denotes the total deformation in the bar, due to W acting as a quiescent load, in.; and h denotes the distance, through which W falls, in.

The formula shows that when h is zero, i.e., when the load W is "suddenly applied" without falling through any distance, the unit stress is twice as great as that due to W applied as a quiescent load. It is obvious also that, with increase of h, the unit stress s_1 increases very rapidly, and it follows that disastrously high stresses may be produced by impact loads.

If the bar in Fig. 27 is held in a horizontal position and the weight W is projected against the end with a velocity V, it can be shown that the unit stress produced in the bar is as follows:

$$s_1 = s\sqrt{\frac{2h}{e}} \tag{40}$$

in which s_1 denotes the unit stress produced in the bar by the impact, psi; s denotes the unit stress in the bar, if W were a quiescent load, psi; $h = V^2/2g$, V being the velocity of the weight W, in. per sec, and g being the acceleration of gravity, in. per sec per sec; and e denotes the total deformation in the bar, due to W acting as a quiescent load, in.

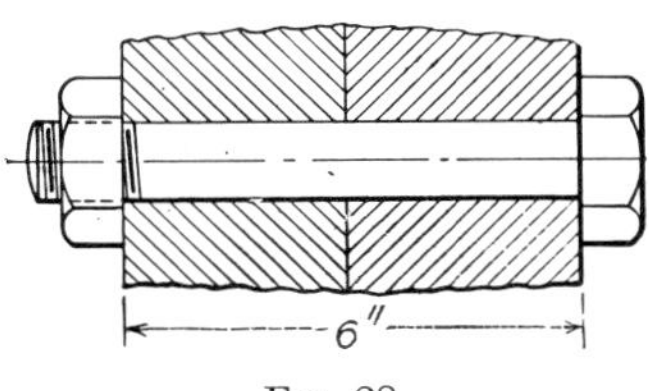

FIG. 28

Effect of Shape on Resilience. The capacity to absorb a blow, such as was discussed in the previous paragraphs, is measured by the modulus of resilience, which is the energy per cubic inch that a material can absorb when stressed to its elastic limit.

The shape of a member may affect the amount of energy that can be absorbed safely, for the reason that resilience is the product of stress and deformation. Suppose that in Fig. 28 the steel bolt must withstand tension, and let the maximum unit stress permitted at the root area of the threads be 16,000 psi. A 1-in. bolt has a diameter at the root of

[1] SINGER, "Strength of Materials," p. 385. TIMOSHENKO, "Strength of Materials," 3d ed., Part I, p. 305.

the thread of 0.837 in.; therefore the unit stress in the shank of the bolt would be only 11,200 psi. The shank of the bolt could therefore absorb an amount of energy equal to

$$\text{Energy} = \frac{11{,}200^2}{2 \times 30{,}000{,}000} \times \frac{\pi \times 1^2}{4} \times 6 = 9.84 \text{ in.-lb}$$

If the design of the shank of the bolt were changed as shown in Fig. 29, so that the smaller diameter of the shank were equal to that at the root of the threads, then the energy that could be absorbed by the bolt would be

$$\text{Energy} = \frac{16{,}000^2}{2 \times 30{,}000{,}000} \times \frac{\pi \times 0.837^2}{4} \times 4.5 + \frac{11{,}200^2}{2 \times 30{,}000{,}000} \times \frac{\pi \times 1^2}{4} \times 1.5$$

$$= 10.55 + 2.46 = 13.01 \text{ in.-lb}$$

This shows that the bolt in Fig. 29 can absorb 32 percent more energy than the one in Fig. 28. This is due to the fact that with the higher unit stress in the smaller diameter of the shank goes a greater elongation and that these together produce a greater capacity for absorbing energy.

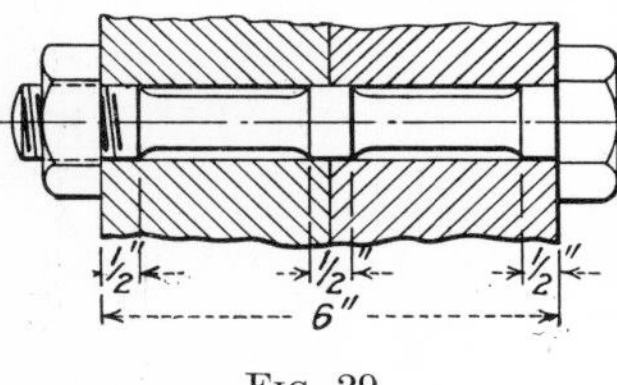

FIG. 29

Repeated Impact. Structural and machine parts are sometimes subjected to repeated impacts, and it is of interest to inquire in what way the effect of repeated impacts is related to the result obtained from an ordinary endurance test and also how it is related to the case of a specimen ruptured by a single blow.

McAdam [1] has made tests on various carbon and alloy steels, determining the ordinary rotating-beam endurance limit, an impact-endurance limit, the energy of rupture for a single blow in a modified Charpy machine, and the energy of rupture in slow bending.

Two types of specimen were used in the impact-endurance machine: Type A, having a diameter of 0.5 in., diameter at bottom of notch 0.4 in., notch sides parallel, radius at bottom 1⁄16 in.; Type B, diameter 0.75 in., diameter at bottom of notch 0.6 in., angle between sides of notch 60°, radius at bottom ¼ mm. The specimens were supported at the ends in the impact-endurance machine, as a simple beam, and subjected to reversed impacts by the drop of a hammer.

The purpose of the impact-endurance test was to determine the relation between the energy of blow and the number of blows necessary to cause fracture. By having various hammers and using various heights of drop, the number of blows necessary to cause fracture could be varied from about 500 to many millions.

The results from the impact-endurance tests were plotted, energy of blow being used as ordinate and number of blows to produce fracture as abscissa. The curves were similar to those obtained in a rotating-beam endurance test, the horizontal asymptote being reached in the neighborhood of 10,000,000 impacts. The ratio of the rotating-beam endurance limit to the impact-endurance limit was found to be fairly constant.

McAdam says: "Evidently, therefore, the ordinate at the extreme left of the impact-endurance graph depends on the impact properties, and the ordinates at the extreme right depend on the endurance properties of the metal. Between these two extremes, the influence of the impact properties decreases, and that of the endurance properties increases with increase in the number of blows necessary to cause fracture."

Presumably, therefore, if the unit stress due to the impact were accurately known and were plotted against number of blows to produce rupture, the resulting diagram would probably coincide with the *S-N* diagram obtained from a rotating-beam endurance test.

[1] *Proc. ASTM*, vol. 23, p. 56, 1923.

COMBINED STRESSES

Combined Axial Stresses. The cases of combined axial stresses acting normally to each other have been previously discussed under biaxial and triaxial stress.

Axial Stress and Flexural Stress. The case of axial stress combined with flexural stress, which occurs in prisms under eccentric loads and in beams that carry an axial load or an oblique load, is discussed in later sections.

Axial Stress and Shear Stress. Figure 30 shows an element subjected to axial tension combined with shear, the shear acting on the two end surfaces and on the top and bottom. There are many cases in practice in which structural or machine members are subjected to such combinations of stress. In a reinforced concrete beam, there is tension due to flexure and shear due to beam shear. In many shafts, there are tension and compression due to flexure or direct stress, and shear due to torsion.

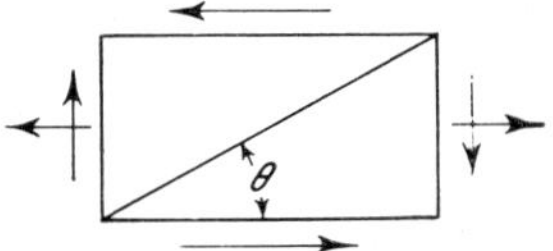

FIG. 30. Axial tension combined with shear.

For the element in Fig. 30, it can be shown [1] that the unit normal stress s_n, acting normal to the inclined surface, and the shear stress s_p, acting parallel to the surface at any angle θ, are given by the following formulas:

$$s_n = s\frac{1 - \cos 2\theta}{2} + s_s \sin 2\theta \tag{41}$$

$$s_p = \frac{s}{2}\sin 2\theta + s_s \cos 2\theta \tag{42}$$

in which s_n denotes the unit stress acting normal to any plane making an angle θ with the horizontal, psi, s denotes the unit axial tensile stress, psi, s_s denotes the unit shear stress, psi, θ denotes the angle that any oblique plane makes with the horizontal, and s_p denotes the unit stress acting parallel to any plane making an angle θ with the horizontal, psi.

Maximum Normal Stress. The maximum and minimum normal unit stresses s_n' occur on planes given by the following criterion:

$$\tan 2\theta = -\frac{2s_s}{s} \tag{43}$$

The two planes for the angle 2θ are 180° apart, and the two planes for the angle θ are 90° apart. The maximum and minimum unit stresses acting on these planes have the following values:

$$s_n' = \frac{s}{2} \pm \sqrt{\left(\frac{s}{2}\right)^2 + s_s^2} \tag{44}$$

the notation being the same as before.

In Eq. (44), the plus sign gives the maximum unit stress, which would be tensile for the case shown in Fig. 30, and the minus sign gives the minimum unit stress, which would be compressive. When tensile stress is combined with shear, the maximum normal unit stress is tensile; when compressive stress is combined with shear, the maximum unit stress is compressive. The minimum normal unit stress is always of the opposite kind to the maximum unit stress.

If, in Fig. 30, the axial stress is compressive instead of tensile or the shear is reversed in direction, Eqs. (43) and (44) may still be used. If the axial stress is compressive,

[1] BOYD and FOLK, "Strength of Materials," 5th ed., p. 296. SINGER, "Strength of Materials," p. 296.

change the sign of s to minus in Eqs. (43) and (44); if s_s is reversed, change the sign of s_s to minus.

The angles for the planes of maximum and minimum normal unit stress may be determined from Eq. (43). After these planes have been determined, there may still be some question as to which one corresponds to the maximum unit stress and which to the minimum. This question may be answered by using Eq. (43) to determine the sine and cosine of 2θ and then substituting in Eq. (41). However, the question may be answered more quickly by means of a sketch such as is shown in Fig. 31. The sketch shows that the shears acting alone would produce a tensile stress normal to the plane AA. The stress s acting alone would produce a tensile stress on the plane BB. Therefore, the combination of shear and tensile forces will produce a maximum unit tensile stress on a plane such as CC lying between planes AA and BB. Thus the plane for maximum normal unit stress lies in the first and third quadrants, and the plane for minimum normal unit stress lies in the second and fourth quadrants, the two planes being at right angles to each other.

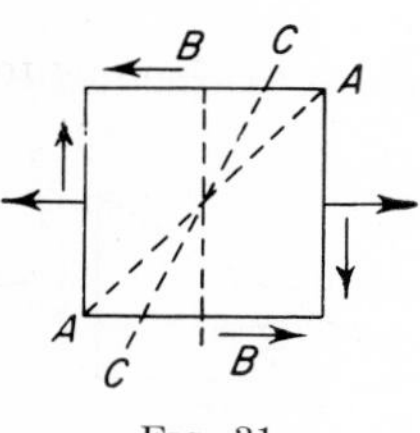

FIG. 31

Principal Stresses. The maximum and minimum normal unit stresses s_n' are called *principal stresses* and the planes on which they act *principal planes.* It can be shown from Eqs. (42) and (43) that there is no shear acting on the principal planes.

Maximum Shear Stress. The maximum and minimum shear stresses occur on planes given by the following criterion:

$$\tan 2\theta = \frac{s}{2s_s} \tag{45}$$

There are two planes for the angle 2θ, and they are 180° apart, and the two planes for the angle θ are 90° apart. The maximum and minimum unit shear stresses acting on these planes have the following value:

$$s_p' = \pm \sqrt{\left(\frac{s}{2}\right)^2 + s_s^2} \tag{46}$$

in which the notation is the same as before.

From Eqs. (43) and (45), it is evident that the planes for maximum and minimum shearing unit stress lie 45° away from the planes for maximum and minimum normal unit stress. Therefore, when the latter planes have been found by means of Eq. (43), it is unnecessary to apply Eq. (45) to determine the position of the shear planes.

Maximum-stress Theory of Failure. From time to time, various theories of failure of materials under combined stresses have been suggested, failure being assumed to have occurred when the stress in the material reaches the yield point. The maximum-stress theory, or *Rankine's theory* as it is sometimes called, assumes that, whatever the ratio of stresses in two directions at right angles to each other and whether they are of like or of opposite sign, failure will occur when one of the stresses reaches the value corresponding to the yield point in simple tension or compression, as the case may be. This theory takes no account of the effect of Poisson's ratio on strength and assumes that a material is neither weakened nor strengthened by the addition of a second stress at right angles to the first. If this theory is correct, the material should reach its yield point when the greater stress reaches the yield point, as determined under unidirectional loading.

Maximum-strain Theory of Failure. In the mathematical theory of elasticity, three equations of the following type are derived:

$$E\epsilon_1 = s_1 - \lambda(s_2 - s_3) \tag{47}$$

in which E denotes the modulus of elasticity, assumed to be constant in all directions, psi, s_1, s_2, and s_3 denote the three unit stresses at right angles to each other, psi, and λ denotes Poisson's ratio. The expression $E\epsilon_1$ is called *reduced* stress, *true* stress, or *ideal* stress. Expressions similar to this may be written for $E\epsilon_2$ and $E\epsilon_3$, and thus three equations for the true stresses are obtained. The maximum-strain theory, or *St. Venant's theory*, assumes that, whatever the combination of stresses, the material will fail when the maximum strain reaches a value equal to that at the yield-point stress in simple tension or compression.

The maximum-strain theory assumes that, when a material is subjected to two or three stresses at right angles to each other, the strength is increased when the stresses are of like sign and diminished when they are of unlike sign. Thus, for two unequal tensions or compressions, the deformation due to the smaller stress, because of Poisson's ratio, would tend to decrease the deformation along the direction of the larger stress, and the strength would be increased. Equation (47) shows that, if a material is subjected to two equal stresses s_1 and s_2 in tension, the strength will be increased 43 percent if Poisson's ratio is 0.3. If one stress is tension and the other compression, the strength will be decreased 23 percent.

Maximum-shear Theory. The maximum-stress theory and the maximum-strain theory both assume that failure takes place by yielding in tension or compression. The maximum-shear theory, or *Guest's law*, assumes that failure occurs by yielding due to shear when the shearing unit stress reaches the shearing yield point. If two specimens are tested so that one is subjected to simple tension and the other to combined stresses, but the shearing unit stress is kept the same in both specimens, failure will be identical in the two specimens due to yielding in shear.

The maximum-shear theory holds that, when two of the three principal stresses are zero, failure is still due to shear. Thus a steel bar subjected to axial tension would fail by shear. In this case, the maximum shearing unit stress occurs on a 45° plane, and its value is equal to one-half of the tensile unit stress. If yielding is to occur owing to shear, then the shearing yield point must be less than one-half of the tensile yield point.

In a case of tension combined with compression at right angles to it, the maximum unit shear stress is one-half the sum of the tensile and compressive unit stresses, as shown on p. 3–5. If the tensile and compressive unit stresses are equal, the intensity of the shearing unit stress is equal to the tensile or compressive unit stress, and failure will occur due to shear, unless the shearing yield point is greater than that in either tension or compression.

Experimental Results. In a bulletin of the Engineering Experiment Station of the University of Illinois (1916), A. J. Becker discusses the three more common theories of failure given above and also several others. Becker made tests on various cases of biaxial loading and drew the following conclusions:

1. With increasing values of the biaxial stresses, the yield-point strength follows the maximum-strain theory until the value of the shearing stress reaches the shearing yield point; then the shearing stress controls according to a maximum-shear theory. There are thus two independent laws each dominant within proper limits instead of a single law as has heretofore been assumed.

2. Because these two laws govern the strength of ductile materials under biaxial loading, the ratio for simple stresses of the shearing yield-point stress to the tensile yield-point stress is important.

3. The results of the tests reported by previous investigators conform better to the two laws of strength than to any single law.

In a bulletin of the Engineering Experiment Station of the University of Illinois (1919), F. B. Seely and W. J. Putnam reported tests on the relations between the elastic strengths

of steel in tension, compression, and shear. Some of their conclusions that are pertinent to the above discussion are as follows:

The correct value of the elastic shearing strength of steel as measured by the proportional limit or the useful limit point is from fifty-five to sixty-five hundredths (0.55 to 0.65) of the elastic tensile strength and may be taken with reasonable accuracy as six-tenths (0.6) of the elastic tensile strength. The maximum-shear theory of the failure of elastic action of ductile steel (sometimes called Guest's law) is, therefore, not an accurate statement of the law of elastic breakdown, since Guest's law assumes that the elastic shearing strength is one-half (0.5) of the elastic tensile strength. The maximum-shear theory as expressed by Guest's law, however, is of much use in obtaining approximate results.

More recent experiments indicate that the shear energy criterion of yielding, suggested by von Mises and also by Hencky and Huber, fits results of experiments better than any other. The equation for yielding is as follows:

$$(s_1 - s_2)^2 + (s_2 - s_3)^2 + (s_3 - s_1)^2 = 2s_y^2 \tag{48}$$

in which s_1, s_2, and s_3 are the principal stresses, and s_y is the yield point for simple tension.

However, since the criterion of the maximum-shear theory gives safer stresses than the shear-energy theory, it is probable that the former theory will continue to recommend itself for design work.

EXAMPLES OF STRESSES IN BODIES

Thin Pipe under Internal Pressure. In a thin pipe under internal pressure, the assumption is made that the circumferential unit stress, which would tend to make the pipe fail along a longitudinal element, is uniformly distributed across the thickness of the pipe. This approximation is sufficiently correct if the ratio of the diameter of the pipe to the thickness of the metal is at least from 10 to 15, but it would not be a good approximation for thick cylinders such as gun tubes.

It can be shown [1] that the unit circumferential stress in a thin pipe is

$$s_t = \frac{pD}{2t} \tag{49}$$

in which s_t denotes the unit stress in the metal, psi, p denotes the internal pressure, psi, D denotes the inside diameter of the pipe, in., and t denotes the thickness of the pipe, in.

The longitudinal unit stress in a thin pipe, which would tend to cause failure along a circumferential element, is

$$s_t = \frac{pD}{4t} \tag{50}$$

in which s_t denotes the longitudinal unit stress, psi, and the other notation is the same as for Eq. (49). It should be noted that the unit stress in Eq. (50) is only one-half of that in Eq. (49).

Thin Sphere under Internal Pressure. In a thin sphere under internal pressure, when the ratio of the diameter to the thickness of the shell is at least 10 or 15, the following formula gives the unit stress in the metal:

$$s_t = \frac{pD}{4t} \tag{51}$$

in which the notation is the same as in Eq. (49).

[1] SINGER, "Strength of Materials," p. 40. TIMOSHENKO, "Strength of Materials," 3d ed., Part I, p. 45.

Stresses in Riveted Joints. Stresses in riveted joints, such as are found in boiler shells, are usually calculated on the assumption that the stresses involved are simple tension, bearing, and shear. The further assumption is made that each rivet carries its proportionate share of the load in shear and in bearing, although this is not quite true for safe loads when there are more than two rows of rivets parallel to the seam. In American practice, the frictional resistance between the plates is neglected, as is also the bending which occurs in a lap joint.

The diameter of the rivet is $\frac{1}{16}$ in. less than that of the hole in which it is to fit, to ensure easy entrance of a hot rivet. In boiler calculations, the rivet is assumed to fill the hole after driving, and the bearing, shearing, and net tensile areas are all based on the diameter of the hole.

In calculating the bearing unit stress, it is assumed that the bearing pressure between the plate and rivet is uniformly distributed in a radial direction, as in the case of a fluid. The bearing unit stress will then be the load divided by the projected area, which is the thickness of the plate times the diameter of the rivet hole.

When holes are punched, there is a loss of strength in the plate unless the holes are reamed, although annealing the plate will restore its strength to some degree. This loss due to punching is estimated at 10 percent for $\frac{1}{4}$-in. plates, 20 percent for $\frac{1}{2}$-in. plates, and 30 percent for $\frac{3}{4}$-in. plates.

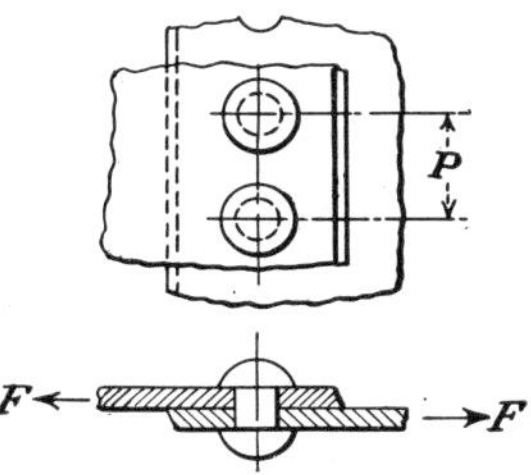

FIG. 32. Lap joint in tension.

In structural work, other than pressure vessels, punched holes are permitted, punching being cheaper, and the insertion of a few extra rivets in structural joints perhaps more than offsets any weakness due to the use of punched holes. For work in which the plates are punched but not reamed, the practice is to punch the hole $\frac{1}{16}$ in. larger than the diameter of the rivet. For purposes of calculation, the bearing and shearing areas are based upon the diameter of the rivet, and the net tensile or compressive area is based upon the diameter of the rivet plus $\frac{1}{8}$ in.

Figure 32 shows a single-riveted lap joint in which the pitch of the rivets is called p. In calculating the unit stresses that would be developed by the load F, the following formulas result, the load F being carried on the pitch distance p, which is called a *unit strip:*

$$s_t = \frac{F}{t(p - d)} \tag{52}$$

in which s_t denotes the unit tensile stress, psi, F denotes the force acting on the unit strip, lb, t denotes the thickness of the plate, in., p denotes the pitch of the rivets, in., and d denotes the diameter of the hole, in., in the case of boilers, or the diameter of the rivet plus $\frac{1}{8}$ in., in the case of structural punched plates.

The shearing unit stress is

$$s_s = \frac{F4}{\pi d^2} \tag{53}$$

The bearing unit stress is

$$s_b = \frac{F}{td} \tag{54}$$

in which d denotes the diameter of the rivet hole in the case of boilers and the diameter of the rivet in the case of structural work. The other notation is the same as for Eq. (52).

Figure 33 shows a triple-riveted lap joint. It is evident that, in the pitch distance p, three rivets carry the load. At row A in the lower plate or row C in the upper plate, the full load must be taken care of, and the unit tensile stress is

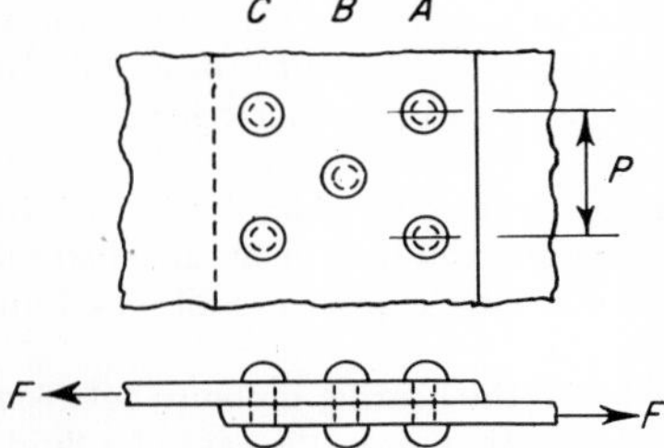

Fig. 33. Triple-riveted lap joint.

$$s_t = \frac{F}{t(p - d)} \tag{55}$$

At row B in the lower or upper plate, only two-thirds of the full load acts in tension, for one-third of the load was transferred from the lower to the upper plate at row A. The unit stress is therefore less at row B than at row A, for the tensile areas are the same at both rows. For a similar reason, the tensile unit stress at row C in the lower plate or row A in the upper plate will be less and therefore need not be calculated.

The load F is carried by three rivets in single shear, and the unit stress is

$$s_s = \frac{F4}{3\pi d^2} \tag{56}$$

The load F is carried by three rivets in bearing, and the unit stress is

$$s_b = \frac{F}{3td} \tag{57}$$

Figure 34 shows a two-strap butt joint. The formulas for the unit stresses in tension, shear, and bearing would be similar to the ones for lap joints, except that each rivet has two areas in shear instead of one.

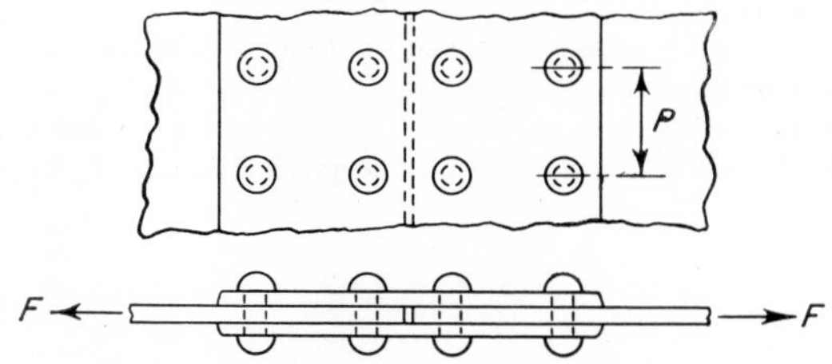

Fig. 34. Double-riveted, two-strap butt joint.

Relative Strength of Joints. The relative strength of a riveted joint, often called its *efficiency*, is the ratio of the strength of the joint to the strength of the unperforated

Table 5. Relative Strength of Riveted Joints, Percent

Joints	Minimum	Maximum	Average
Single-riveted lap joint	52	64	55
Double-riveted lap joint	64	78	70
Triple-riveted lap joint	64	84	78
Single-riveted butt joint	53	64	60
Double-riveted butt joint	64	78	75
Triple-riveted butt joint	78	88	80
Quadruple-riveted butt joint	94	96	95

solid plate. Table 5 shows the approximate relative strengths that may be expected from various types of riveted joints.

Factor of Safety. The factor of safety that is recommended by the Boiler Code of the ASME is 5. The allowable unit stresses for design then become

$$s_t = \frac{55,000}{5} = 11\ 000 \text{ psi}$$

$$s_s = \frac{44,000}{5} = 8,800 \text{ psi}$$

$$s_b = \frac{95,000}{5} = 19,000 \text{ psi}$$

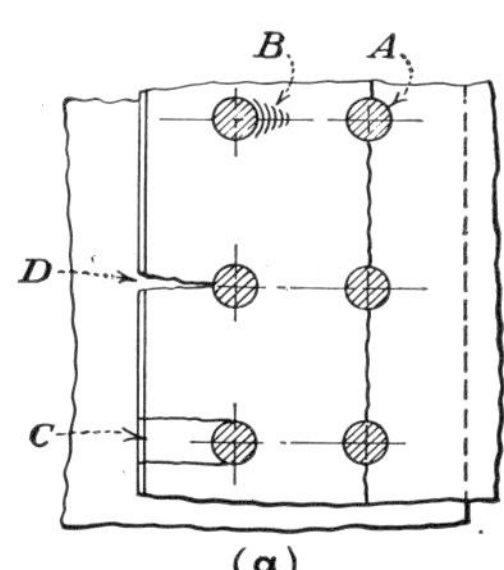

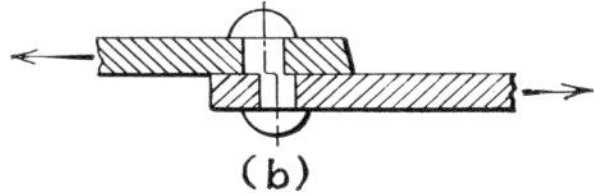

FIG. 35. Types of failure.

Types of Failure. A single-riveted lap boiler joint may fail due to any one of the following causes:

1. Shearing of the rivet as shown by Fig. 35*b*.
2. Tearing of the plate through the rivet holes as at *A* (Fig. 35*a*).
3. Crushing of the plate in front of a rivet as at *B*.
4. Crushing of the rivet.
5. Tearing out of the plate as shown at *C*.
6. Splitting the plate as at *D*.

If the joint is a complex one, failure may of course be due to a combination of the above causes.

Temperature Stresses. If a steel rail is lying on the ground and is subjected to changes of temperature, there is no stress developed in the steel, for the rail is quite free to expand or contract. However, if the rail were fixed at the ends so that it could not contract or expand, then changes of temperature would produce stress. If the coefficient of expansion of a material is η, the unit deformation due to a change of temperature of $t°$ is

$$\epsilon = \eta t \tag{58}$$

The corresponding unit stress is

$$S = E\epsilon = E\eta t \tag{59}$$

Shrinkage of Hoops. Sometimes it is desirable to heat a hoop and slip it onto a cylinder that is slightly larger in diameter than the hoop. Upon cooling, a tensile stress is produced in the hoop. As in the case of a thin pipe under internal pressure, in this case also the assumption is made that the diameter of the hoop is large compared with the thickness of the metal in the hoop. If the diameter of the cylinder is d and the normal diameter of the hoop is d_1, the unit change in the length of the hoop after it is shrunk on the cylinder is

$$\epsilon = \frac{\pi d - \pi d_1}{\pi d_1} = \frac{d - d_1}{d_1} \tag{60}$$

The corresponding unit stress in the hoop is

$$s_t = E \times \frac{d - d_1}{d_1} \tag{61}$$

Compound Prisms. Consider a prism made up of two or more different materials, such as a concrete compression prism reinforced with longitudinal steel rods. If the length of the rods is equal to the length of the prism, the unit deformation of both materials will be the same when the prism is loaded. If ϵ_c and ϵ_s are the unit deforma-

tions of the concrete and the steel respectively, then it follows that

$$\epsilon_c = \epsilon_s \tag{62}$$

Also

$$P_c + P_s = P \tag{63}$$

in which P_c denotes the load carried by the concrete, lb, P_s denotes the load carried by the steel, lb, and P denotes the total load on the prism, lb. Also, since

$$s_c = E_c\epsilon_c \qquad s_s = E_s\epsilon_s \qquad P_c = s_cA_c \qquad \text{and} \qquad P_s = s_sA_s$$

it is evident that the unit stress in each material and the load carried by each material may be determined.

STRESSES IN BEAMS

Use of Beams. A beam is a structural member that is subjected to loads and reactions acting transversely to its longitudinal dimension. Common examples of beams are the wooden joists and steel I beams used to support floors in buildings, plate girders used in plate-girder bridges, and the axles on railroad freight trucks, which transfer the loads to the wheels and so to the tracks. In some cases of beam action, the loads will not be perpendicular to the long dimension of the beam but will act more or less obliquely, as in the case of rafters used for supporting a roof. In this case, the rafters are inclined to the horizontal, and the resultant of the dead load and the wind load is not usually normal to the long axis of the rafter.

The so-called "common case" of beam action will be discussed first, because the loads and reactions are fairly simple. The more complicated cases will be discussed in a later section. In the common case of beam action, the longitudinal axis of the unloaded beam is straight, and the beam has at least one plane of symmetry. The loads and reactions lie in this plane and are perpendicular to the longitudinal axis of the beam.

Kinds of Beams. Beams are commonly classified on the basis of the way in which they are supported. Figure 36 shows a *simple* beam, one that is supported near its ends.

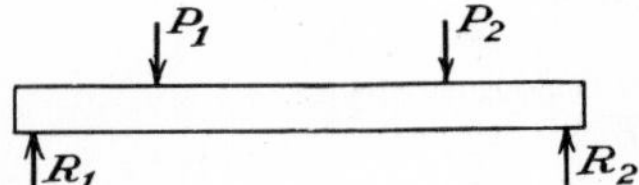

FIG. 36. Simple beam.

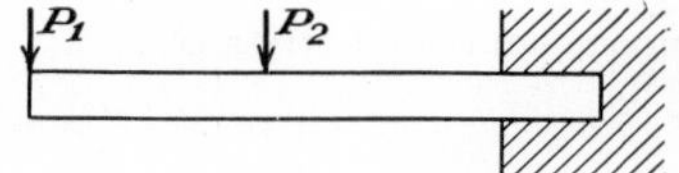

FIG. 37. Cantilever beam.

Figure 37 shows a *cantilever* beam, which is free at one end and *fixed* or *built-in* at the other. A beam supported at only one place between its ends might be a double cantilever. Figure 38 shows a *continuous* beam, which has more than two supports. Figure 39 shows a *propped* cantilever, which is supported at one end and fixed or built-in at the other. Figure 40 shows a beam which is *fixed* at both ends.

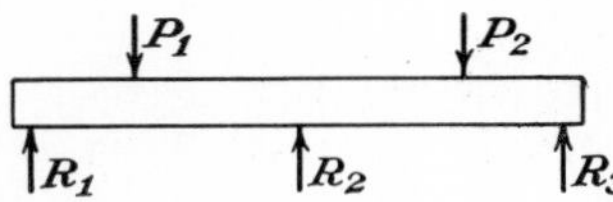

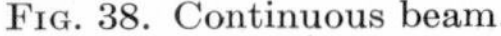

FIG. 38. Continuous beam.

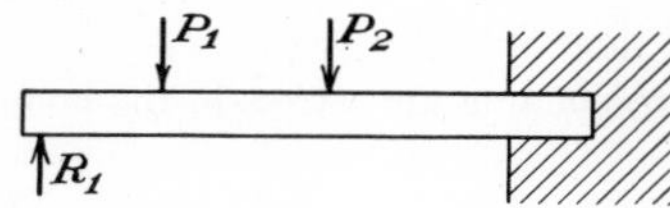

FIG. 39. Propped-cantilever beam.

In the cases cited above, the supporting forces may be calculated by the principles of statics for the simple and the cantilever beam, and these are said to be statically determinate. For the continuous, the propped-cantilever, and the fixed-ended beam, the

supporting forces cannot be calculated by the principles of statics alone, and these are said to be statically indeterminate.

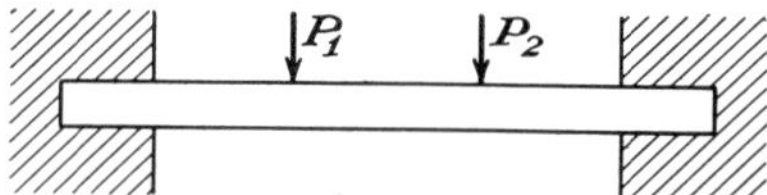

Fig. 40. Beam fixed at both ends.

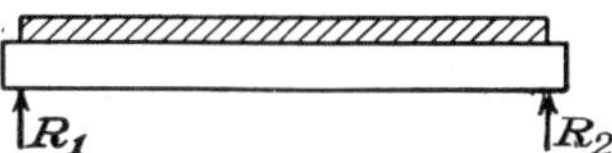

Fig. 41. Uniformly distributed load.

Kinds of Loading. Most of the loads and reactions are applied to a beam in such manner that there is involved a certain amount of contact area. In many cases, however, this area of contact is so small that the loads may be considered as applied at a line or at a point. These are called *concentrated* loads and reactions; Fig. 36 shows such a case.

One of the assumptions of loading on the basis of which many beams are designed is *uniformly distributed* loading, which is usually stated in pounds per running foot of beam or in pounds per square foot of floor space. Figure 41 shows such a loading on a simple beam. In other cases, there may be various kinds of nonuniformly distributed loads, but these need not be considered at this time.

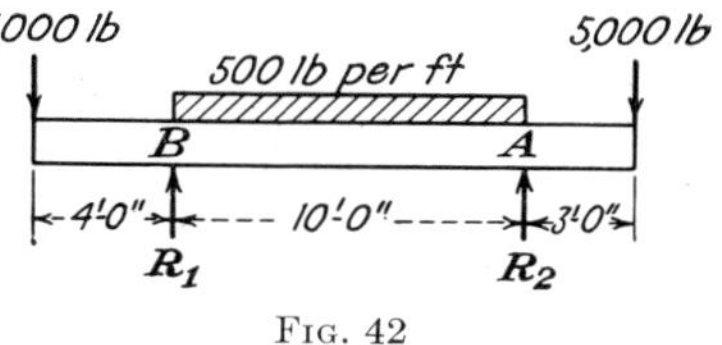

Fig. 42

Calculation of Reactions. Some beams are like simple beams except that they carry loads on portions that overhang the supports, as shown in Fig. 42. These beams and also simple and cantilever beams are statically determinate, and the reactions may be computed by the principles of static equilibrium. The conditions of equilibrium when the loads and reactions are coplanar and parallel are: (1) The algebraic sum of the forces must equal zero. (2) The algebraic sum of the moments of the forces about any point in their plane must equal zero.

In Fig. 42, the reactions R_1 and R_2 may be determined as follows:

$$\Sigma V = R_1 + R_2 - 1{,}000 - (500 \times 10) - 5{,}000 = 0$$
$$\Sigma M_A = (R_1 \times 10) - (1{,}000 \times 14) - (500 \times 10 \times 5) + (5{,}000 \times 3) = 0$$

From these equations, $R_1 = 2{,}400$ lb and $R_2 = 8{,}600$ lb.

R_1 and R_2 could also be obtained by writing two moment equations with origins at A and at B. Then, as a check, the sum of the loads must equal the sum of the reactions.

In determining the reactions for statically indeterminate beams, it is necessary to make use of the equations of the elastic curve in addition to the static conditions of equilibrium. These matters are discussed in a later section.

Stresses and Moments at a Section. If, in Fig. 42, the overhanging portion of the beam to the left of the point B is considered as a body in equilibrium, it is evident that neither a single horizontal force nor a single vertical force, acting at the cross section 4 ft from the end of the beam, could hold that portion of the beam in equilibrium. When the overhanging portion of the beam bends, one knows from experience that the bottom fibers of the beam will be compressed and the top fibers will be stretched. Experiments have shown that, in such a case, the unit strains vary linearly from the top to the bottom of the beam. Somewhere between the top and bottom surfaces of the beam, the fibers neither shorten nor elongate, and this surface is called the *neutral surface.*

In the analysis of the stresses at a cross section of a loaded beam, it is usual to divide the total stress into normal and tangential components. The normal stresses will be those of tension and compression, and the tangential stress will be the shear stress.

Figure 43 shows an arrangement of the tensile, compressive, and shear forces on the cross section that would make it possible for this portion of the beam to satisfy the conditions of equilibrium for such a system of coplanar, nonconcurrent, nonparallel forces. These conditions are:

1. The algebraic sum of the forces along the horizontal must equal zero.
2. The algebraic sum of the forces along the vertical must equal zero.
3. The algebraic sum of the moments of the forces about any point in their plane must equal zero.

It is evident from Fig. 43 that the tensile and compressive forces, called T and C, would constitute a couple, because there are no other horizontal forces. The tensile and compressive unit stresses, or *fiber stresses* as they are called, must produce a TC couple which is equal and opposite to the moment of the 1,000-lb force with respect to a horizontal axis in the cross section. The shear stresses must produce an upward force V' which is equal and opposite to the 1,000-lb load.

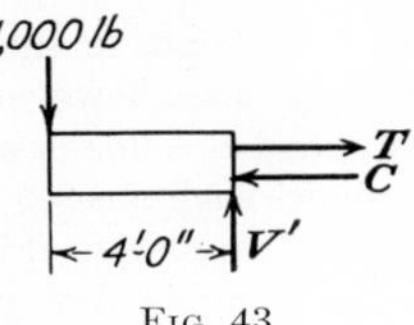

Fig. 43

The algebraic sum of the moments of all loads and reactions to the left or to the right of a given cross section is called the *bending moment* at that section. The bending moment must be balanced by the TC couple produced by the internal stresses at the section, and this is called the *resisting moment*. It is convenient to represent the bending moment by the symbol M and the resisting moment by M'.

The algebraic sum of all the loads and reactions to the left or to the right of a given cross section is called the *shearing force* at that section. The shearing force must be balanced by a force produced by the internal stresses at the section, and this is called the *resisting shear*. It is convenient to represent the shearing force by the symbol V, and the resisting shear by V'.

Shear Diagrams. It will be shown later that there is an important relation existing between the variation of shear and the variation of bending moments at points along the length of a beam. Because of this relation and also because shear sometimes determines the strength of beams, it is desirable to know how the shearing force varies from point to point along the beam. The diagram that shows this variation is called the *shear diagram*, and the ordinates on the diagram represent the values of the shearing force. Figure 44 shows a loaded beam and the corresponding shear diagram.

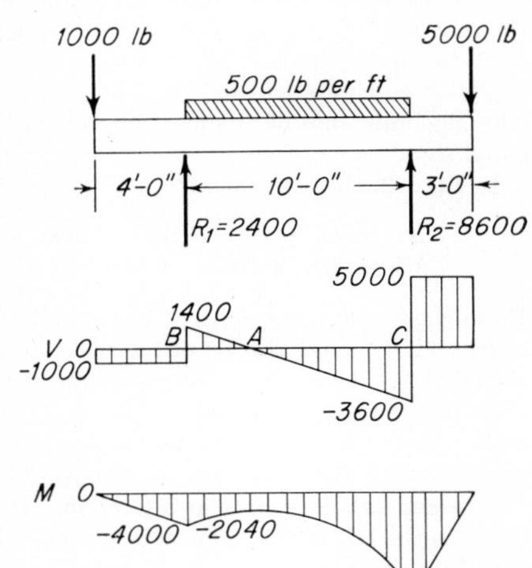

Fig. 44. Shear and moment diagrams.

It is convenient to adopt a convention for sign in calculating the shear at a section of a beam. The convention here adopted will be to take the sum of all loads and reactions to the *left* of any section and call downward forces negative and upward forces positive.

It is often desirable to have the equations for the shearing force for various portions of a beam. Unless there is no change in loading along the beam, it is impossible to write a shear equation that will hold for the full length of the beam. The result is that shear equations usually are valid for only a portion of the beam. For the beam in Fig. 44, it is

necessary to write three separate equations, as given below, the distance x being measured from the left end of the beam:

From 0 to 4 ft,

$$V_x = -1{,}000$$

From 4 to 14 ft,

$$V_x = -1{,}000 + 2{,}400 - 500(x - 4)$$
$$= +3{,}400 - 500x$$

From 14 to 17 ft,

$$V_x = -1{,}000 + 2{,}400 + 8{,}600 - 5{,}000$$
$$= +5{,}000$$

Such equations may be used to determine accurately the point where the shear diagram passes through zero. A later paragraph will explain the significance of points of zero shear. In Fig. 44, the point A on the shear diagram may be determined by setting the shear equation for that portion of the beam equal to zero.

$$V_x = 3{,}600 - 500x = 0$$
$$x = 6.8 \text{ ft}$$

Another simple way of locating the point A is to note that the ordinate on the shear diagram is 1,400 lb at B and that beyond B the load drops off at the rate of 500 lb per ft. It is necessary to go a distance from B equal to 1,400 divided by 500, or 2.8 ft, to bring the ordinate to zero.

It will be found that, for those portions of the beam on which there are only concentrated loads and reactions, the shear diagram is made up of horizontal straight lines. For those portions of the beam on which there are uniformly distributed loads and reactions, the shear diagram is made up of inclined straight lines. The shear equations verify these statements.

Moment Diagrams. The ordinates on the moment diagram represent the value of the bending moment at various points along the beam. The lower portion of Fig. 44 shows such a diagram.

It is convenient to adopt a convention for the sign of the bending moment. The convention here used is to calculate the algebraic sum of the moments of all loads and reactions to the *left* of any section, the plus sign being given to the moments of positive loads and reactions and the minus sign to the moments of negative loads and reactions. Plus moments, therefore, produce clockwise rotation, and negative moments counterclockwise rotation.

Moment equations may be written for various portions of the beam. For the beam in Fig. 44 three sets of moment equations are necessary, as follows:

From 0 to 4 ft,

$$M_x = -1{,}000x$$

From 4 to 14 ft,

$$M_x = -1{,}000x + 2{,}400(x - 4) - \frac{500(x - 4)^2}{2}$$
$$= 3{,}400x - 13{,}600 - 250x^2$$

From 14 to 17 ft,

$$M_x = -1{,}000x + 2{,}400(x - 4) + 8{,}600(x - 14) - 500 \times 10(x - 9)$$
$$= 5{,}000x - 85{,}000$$

It will be found that for those portions of the beam upon which are acting only con centrated loads and reactions the moment diagram is made up of inclined straight

lines, while for those portions having uniformly distributed loads and reactions the moment diagram is made up of curved lines. The equations for moment will verify these statements. Uniformly distributed loads acting down will produce curved lines for moment that are convex upward, while those acting up will produce curved lines which are convex downward.

Relations between Shear and Moment. It can be shown [1] that the rate of change of moment with respect to x at any point in a beam is equal to the shear at that point, or, stated mathematically,

$$\frac{dM}{dx} = V \tag{64}$$

Since dM/dx is the slope at any point on the moment diagram, it is evident that this slope is equal to the shear at that point.

If the above equation is integrated between two limits of x, the following equation results:

$$M_2 - M_1 = \int_{x_1}^{x_2} V\,dx \tag{65}$$

It is evident from this equation that the area of the shear diagram between the values of x_1 and x_2 is equal to the increment of moment between the two sections of the beam represented by x_1 and x_2. If M_1 is zero at x_1, then the area of the shear diagram up to x_2 is equal to the moment at x_2. These relations provide a convenient method for making check calculations for the moment at any section.

Maximum Shear and Maximum Moment. Inspection of the shear diagram will show where the greatest value of the shear occurs, and this maximum value will be found to occur near a support.

To determine the maximum and minimum values of moment, it is necessary to set the first derivative of moment with respect to x equal to zero. But Eq. (64) has shown that this derivative of moment is equal to the shear. Hence it follows that the maximum and minimum moments will occur where the shear is zero. The points of zero shear may be found by an inspection of the shear diagram, and it is then necessary to calculate the values of moment at the places of zero shear. A comparison of these moment values will determine which is numerically the greatest.

In Fig. 44, the shear is zero at B, A, and C. The moment calculations show that the greatest moment occurs at C and is equal to 15,000 ft-lb.

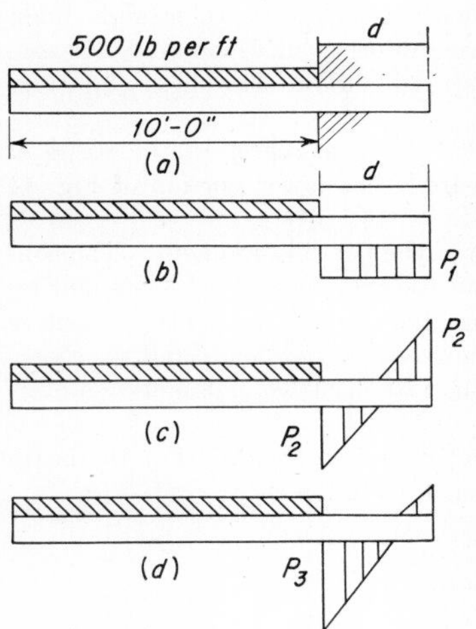

FIG. 45. Cantilever beam.

Cantilever Beam. The cantilever beam, which may be built into masonry or concrete at the fixed end, presents some unique features which should be discussed. Figure 45 shows a cantilever beam which projects 10 ft from the wall and carries a uniformly distributed load of 500 lb per ft. For the portion of the beam that projects from the wall, the moment equation is $M = wx^2/2$, and this is a maximum at the wall.

The portion of the beam within the wall must have such forces acting on it that the whole beam is held in equilibrium. If the sum of the forces along the vertical is to be

[1] BOYD and FOLK, "Strength of Materials," 5th ed., p. 124. SINGER, "Strength of Materials," p. 105.

zero, there must be acting upward a force equal to the 5,000 lb that acts downward. This is shown in Fig. 45*b* as a distributed force on the portion of the beam within the wall, and the unit pressure is called p_1. To balance the moment of the load with respect to the mid-point of the distance *d*, there must be acting another equal and opposite moment. It is usually assumed that this moment is supplied by forces that have a linear distribution as shown in Fig. 45*c*. The resultant distribution of forces on the portion of the beam within the wall would then be as shown in Fig. 45*d*. Here the unit pressure p_3 would equal $p_1 + p_2$.

With this distribution of forces, it can be shown that the maximum moment would really occur at a slight distance within the wall. The maximum moment, however, is assumed to come at the edge of the wall, and the discrepancy involved is small enough so that it may be neglected.

Suppose then that it is desired to determine the distance *d* of the beam within the wall and that the wall is made of concrete and the beam is a steel I beam. If the maximum moment due to the loads and the allowable unit stress for steel are known, the size of the I beam could be determined. The width of the flange of the I beam being known and an allowable unit compressive stress p_3 for the concrete being assumed, it would be possible to determine the distance *d* by means of the conditions of static equilibrium given above.

Assumptions in Beam Theory. The assumptions that are made in the development of the ordinary beam theory and that are justified by experiment and experience, are as follows:

1. Cross sections of a beam that are plane before bending are plane after bending. From this it follows that the tensile and compressive fiber strains are proportional to their distances from the neutral surface.
2. Layers of fibers in the beam are free to contract and expand without interference from other layers.
3. The material of the beam is homogeneous and isotropic and obeys Hooke's law in tension and compression, at least up to allowable unit stresses.

Figure 46 shows a cross section of a beam and the distribution of the unit fiber stresses on that cross section. The line *AB* is a trace of the neutral surface and is called the *neutral axis*. If the fiber strains are proportional to their distances from the neutral surface, then the unit fiber stresses are also proportional to these distances so long as the elastic limit is not exceeded.

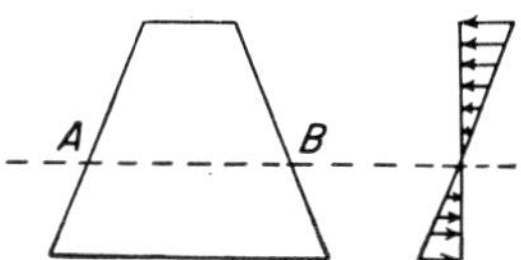

Fig. 46. Distribution of unit stresses.

It can be shown [1] that, if the distribution of unit fiber stresses is linear, as shown in Fig. 46, then the neutral axis passes through the centroid of the cross section.

Fiber-stress Formula. It can be shown [2] that the formula for the resisting moment M' at any cross section of a beam is equal to sI/c. For equilibrium, this resisting moment must balance the bending moment M at the same section. The following equation therefore connects bending moment and unit fiber stress:

$$M = \frac{sI}{c} \qquad (66$$

in which M denotes the bending moment at the section of a beam, in.-lb, s denotes the unit fiber stress (tensile or compressive) at the outside fiber of the section, psi, I denotes the moment of inertia of the cross section with respect to the neutral axis, in.^{4}, and c denotes the distance from the neutral axis to the outside fiber, in.

[1] Boyd and Folk, "Strength of Materials," 5th ed., p. 135. Singer, "Strength of Materials," p. 127.
[2] Boyd and Folk, "Strength of Materials," 5th ed., p. 134. Singer, "Strength of Materials," p. 127.

Although the above formula has been expressed in terms of the unit stress at the extreme fiber in the cross section, it may be used to calculate any unit stress s' at a distance c' from the neutral axis. This follows from the fact that s'/c' is equal to s/c as long as the distribution of fiber stress is linear.

The term I/c in Eq. (66) is entirely dependent on the shape and size of the cross section and is usually called *section modulus* or *section factor*.

Equation (66) shows that the fiber stress s is directly proportional to the bending moment M at the section of the beam in question and inversely proportional to the section modulus I/c. For the case of a beam of constant cross section, the section modulus would also be constant, and the maximum fiber stress would occur at that section where the bending moment is the greatest.

Whether the unit stress at a given fiber in the cross section of a beam is tensile or compressive may be determined by the sign of the bending moment at that cross section. Consider a simple beam supported at the ends. For such a beam, the bending moment is positive throughout the length of the beam, and, when the beam bends, the top fibers are in compression and the bottom ones in tension. Consider next a cantilever beam projecting from a wall with the free end at the left. Such a beam will have negative bending moment throughout its length, and the top fibers will be in tension and the bottom ones in compression. It is evident, therefore, that in a horizontal beam, when the bending moment is positive, the top fibers are in compression and the bottom ones in tension. When the bending moment is negative, the top fibers are in tension and the bottom ones in compression.

Moment of Inertia. Equation (66) shows that it is necessary to know the moment of inertia of the cross section of a beam with respect to the neutral axis. The moment of inertia of an area is given by the following mathematical expression:

$$I_x = \int dA\, y^2 \tag{67}$$

In this integral, each elementary area must be multiplied by the square of its distance from the neutral axis, and the products summed up for the upper and lower limits of the area. The formulas for the moment of inertia of some of the simpler areas are given in Table 6.

TABLE 6. MOMENT OF INERTIA OF SOME AREAS.

Section	Moment of Inertia	Section	Moment of Inertia
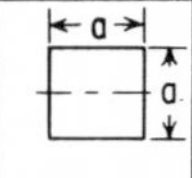 a; a	$\frac{a^4}{12}$	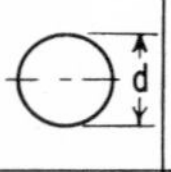 d	$\frac{\pi d^4}{64}$
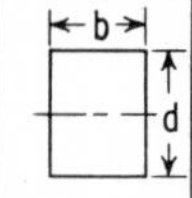 b; d	$\frac{bd^3}{12}$	$C_1=0.288d$; $C_2=0.212d$; 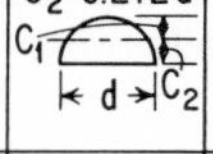 C_1; d; C_2	$0.00686d^4$
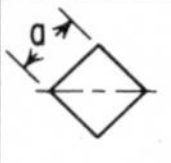 a	$\frac{a^4}{12}$	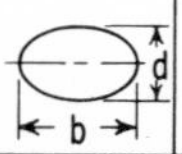 d; b	$\frac{\pi bd^3}{64}$
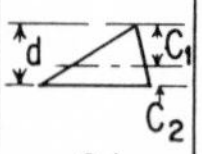 d; C_1; C_2; $C_1=\frac{2d}{3}$; $C_2=\frac{d}{3}$	$\frac{bd^3}{36}$	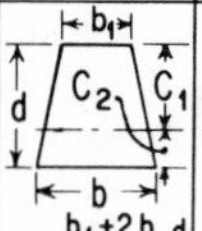 b_1; d; C_2; C_1; b; $C_1=\frac{b_1+2b}{b+b_1}\frac{d}{3}$; $C_2=\frac{b+2b_1}{b+b_1}\frac{d}{3}$	$\frac{b^2+4bb_1+b_1^2}{36(b+b_1)}d^3$

When the formulas for the moments of inertia of the simpler areas are known, it is possible to find the moment of inertia of a composite area by dividing it into smaller, simpler areas the moments of inertia of which may be computed. Figure 47 shows the cross section of an American standard steel I beam and the way it may be broken up into rectangles and triangles so that the moment of inertia of the entire section may be determined.

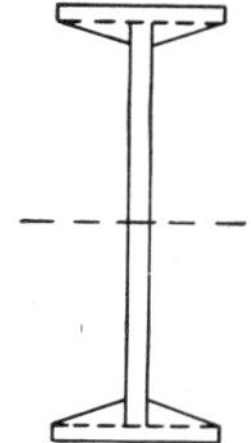

Fig. 47. Cross section of I-beam.

Figure 47 also illustrates the need of the so-called *transfer formula* for computing the moment of inertia with respect to an axis that is parallel to the centroidal axis. The transfer formula takes the following form:

$$I_x = \bar{I} + Ad^2 \tag{68}$$

in which I_x denotes the moment of inertia of the area with respect to a parallel axis at a distance d from the centroidal axis, in.^{4}, $\bar{I}$ denotes the moment of inertia of the area with respect to the centroidal axis, in.^{4}, A denotes the area, in.^{2}, and d denotes the distance from the centroidal axis to the other parallel axis, in.

Stress on Partial Area. It is sometimes desirable to determine the total stress on a portion of a beam cross section. In Fig. 48, a portion of the section is shown shaded. The total stress on the shaded area A' is given by the following formula:[1]

$$\text{Total stress on } A' = \frac{s}{c} A'y' \tag{69}$$

in which A' denotes the partial area, in.^{2}, s denotes the unit stress on the outer fiber, psi, c denotes the distance from the neutral axis to the outer fiber, in., and y' denotes the distance from the neutral axis to the centroid of the partial area, in.

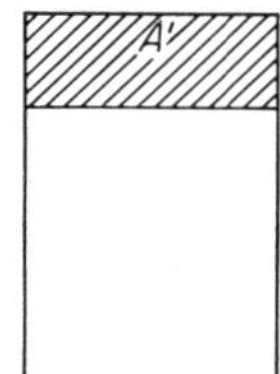

Fig. 48. Stress on partial area.

The resisting moment contributed by the stress on a partial area A' is given by the following formula:

$$\text{Moment of stress on } A' = \frac{s}{c} I' \tag{70}$$

in which A', s, and c have the same meaning as in Eq. (69), and I' denotes the moment of inertia of the partial area A' with respect to the neutral axis, in.^{4}

Calculation will show that the upper and lower quarters of the cross section of a rectangular beam contribute seven times as much resisting moment as the two inner quarters. The two inner quarters are therefore relatively inefficient, for their moments of inertia are small and the unit stresses on these quarters are also small. These facts immediately suggest why the I-beam shape was developed for steel beams. For a given depth of beam, the moment of inertia may be increased by putting the material in two outstanding flanges and removing the inefficient material near the neutral axis. In this way, the flanges of an I beam may contribute as much as 85 percent of the total resisting moment of the beam.

When, as in the case of cast iron, the allowable unit stress in compression is much greater than the allowable unit stress in tension, it is not desirable to have the beam symmetrical with respect to the neutral axis. In such a case, it is desirable to make the distance from the neutral axis to the extreme compressive fiber several times greater than the distance to the extreme tensile fiber. This may be accomplished by using a T

[1] Withey and Washa, "Materials of Construction," 1954, p. I-28.

section, having the stem of the T on the compressive side and the flange of the T on the tensile side.

Horizontal Shear in Beams. The discussion of shear diagrams has made it plain that transverse shear exists in a beam. It is not so evident that longitudinal shear is also present in beams, although a previous discussion has shown that, when a parallelepiped is subjected to a shearing couple in the two end faces, there must also be a shearing couple in the top and bottom faces to hold the element in equilibrium.

Figure 49 shows a simple beam carrying a uniformly distributed load. The beam is divided into four parts, and one of these parts is shown separately. It is evident that at the center of the span there would be compressive fiber stresses which would exert a

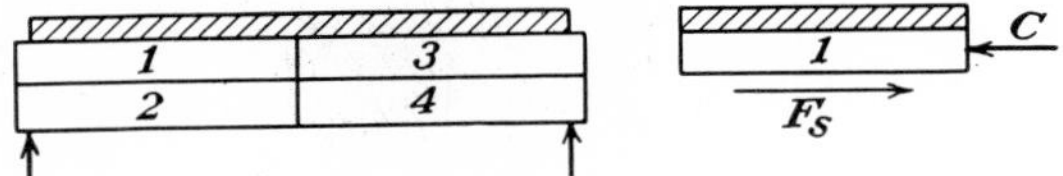

FIG. 49. Horizontal shear in beams.

horizontal force C. In order that this portion of the beam may be in equilibrium under the horizontal forces, there must be a shearing force F_s acting to the right as shown.

As an illustration of the effect of horizontal shearing forces on the strength of a beam, two cases will be considered, one in which the load is carried by three separate boards placed one on top of the other, and the other a solid beam of the same width and depth as the three boards. Calculations will show that the three separate boards can carry only one-third as much load as the solid beam. In the case of the separate boards, there is no longitudinal shear acting at the contact surfaces of the boards, while, in the solid beam, longitudinal shear would be present.

Let Fig. 50 represent the cross section of a loaded beam. It can be shown that the shearing unit stress at any point in the section is given by the following formula:[1]

$$s_s = \frac{V}{Ib} A'y' \tag{71}$$

in which s_s denotes the shearing unit stress at a point y distant from the neutral axis, psi; V denotes the total shearing force at the cross section in question, lb; I denotes the moment of inertia of the entire cross section with respect to the neutral axis, in.^{4}; b denotes the net width of the cross section at the point where s_s is wanted, in.; A' denotes the area above or below the point where s_s is wanted, in.^{2}; and y' denotes the distance from the neutral axis to the centroid of A', in.

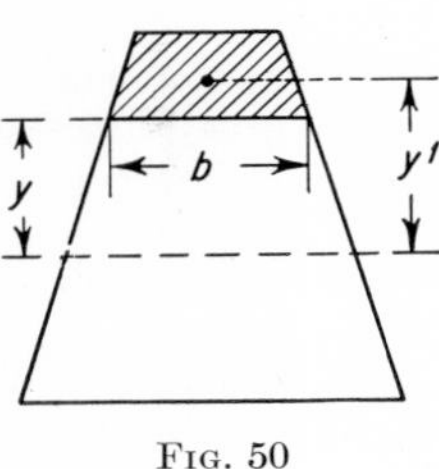

FIG. 50

Equation (71) gives the value of the unit longitudinal and also the unit transverse shearing stress at any point in a beam. That this is true is evident from a previous discussion of a parallelepiped subjected to shearing forces on four faces.

Variation of Shear Stress. It is evident from Eq. (71) that s_s varies directly with V, so that s_s will be a maximum at that cross section of the beam at which V is a maximum. The place of maximum V is most readily determined from a shear diagram. The product $A'y'$, or "statical moment," varies for different points in the cross section. At the top and bottom fiber of the beam, this product is zero, so that s_s will also be zero there. The width b in the equation may be a variable for different points in the cross section, and Eq. (71) shows that a small value of b tends to make s_s large.

[1] BOYD and FOLK, "Strength of Materials," 5th ed., p. 144. SINGER, "Strength of Materials," p. 152.

For a beam of rectangular cross section, I and b are both constant, so that the maximum shearing unit stress will occur at that point in the section for which the product $A'y'$ is a maximum. This is evidently at the neutral axis. Substituting in Eq. (71):

$$s_s = \frac{12V}{bd^3 \times b} \cdot \frac{bd^2}{8} = \frac{3}{2}\frac{V}{bd} \tag{72}$$

This result shows that for the rectangular beam the maximum unit stress at the neutral axis is 50 percent greater than the average on the entire cross section.

In the same way, for the circular cross section, substituting in the formula:

$$s_s = \frac{4V}{\pi r^2 \times 2r} \cdot \frac{\pi r^2}{2} \times \frac{4r}{3\pi} = \frac{4}{3}\frac{V}{\pi r^2} \tag{73}$$

This result indicates that for the circular beam the maximum unit shear stress at the neutral axis is 33 percent greater than the average on the entire cross section.

For many of the more common cross sections, the maximum unit shear stress occurs at the neutral axis, but this is not always the case. For the square in the position shown in Fig. 51, the maximum stress occurs at a distance of $C/4$ from the neutral axis and varies as indicated in the figure.

Shear in I Beams. Figure 52 shows the variation of the unit shear stress in a loaded I beam. The figure indicates that the flanges of an I beam contribute very little to the total shearing resistance of the cross section. Since the unit shear stress is fairly uniform on the web, the usual practice is to assume that the web of the I beam has a shear area equal to the full depth of the beam times the thickness of the web and that the shear stress is uniformly distributed on this area. This is the assumption commonly made in the tables found in steel handbooks.

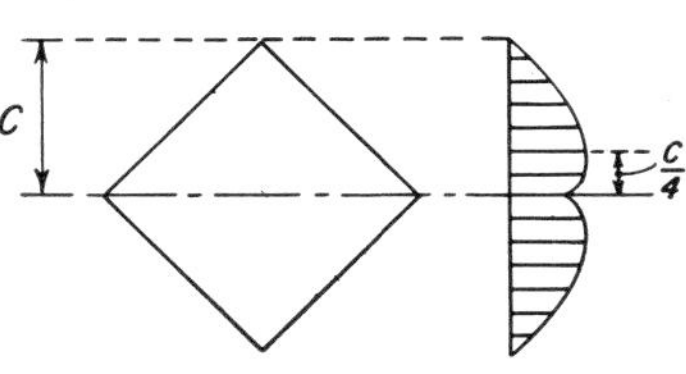

Fig. 51

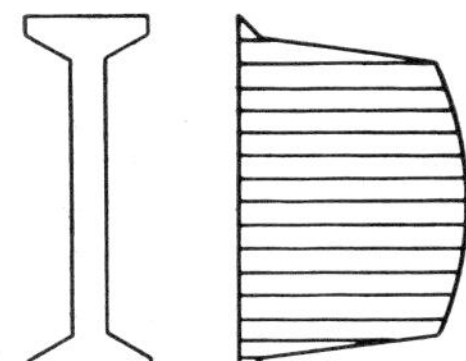

Fig. 52. Distribution of shear stress.

Strength of Beams. The strength of a beam is measured by the load it can carry, and in determining the safe loads for beams the latter may be classified as long and short. Long beams are those in which the fiber stress determines the load, and short beams are those in which the shearing stress determines the load.

The following table lists four common cases of loading, giving the maximum moment, the maximum shear, and the load as based on fiber stress:

Table 7. Some Common Cases of Loading

Loading	Maximum moment	Maximum shear	Maximum load based on fiber stress
Simple beam, center load	$Pl/4$	$P/2$	$4sI/cl$
uniform load	$Wl/8$	$W/2$	$8sI/cl$
Cantilever beam, end load	Pl	P	sI/cl
uniform load	$Wl/2$	W	$2sI/cl$

From the last column in Table 7, it is evident that the strength of a beam when determined by fiber stress depends directly on the magnitude of s and I, inversely upon c and l, and on a constant determined by the expression for the maximum moment.

When the safe load is governed by the shearing unit stress, it is not possible to write a general expression for the load, except perhaps for some of the simpler cases. For instance, in the case of a simple beam, uniformly loaded, and of rectangular cross section,

$$s_s = \frac{W}{2Ib} A'y'$$

and

$$W = \frac{2s_s b^2 d^3 8}{12bd^2} = \frac{4}{3} s_s bd$$

Types of Beam Problems. For beams as well as for other structural members, there are three types of problems that the engineer is called upon to solve: (1) investigation, i.e., determining the maximum unit stresses in the beam and the factors of safety; (2) safe loads, i.e., determining the loads that a beam may safely carry; and (3) design, i.e., determining the size of beam necessary for carrying the loads.

Investigation. Problem 1: A 10-in. 35-lb steel I beam is 12 ft long and supported at the ends. It carries a uniformly distributed load of 1,000 lb per ft, including the weight of the beam, and a center load of 9,000 lb. Determine the factor of safety.

Since the maximum moment for each load separately will come at the center of the span, the maximum moment for the combination also occurs at the center and is

$$M = \frac{PL}{4} + \frac{WL}{8} = \frac{9{,}000 \times 12}{4} + \frac{12{,}000 \times 12}{8} = 45{,}000 \text{ ft-lb}$$

From the handbook of the American Institute of Steel Construction, I is found to be 145.8 in.^{4} Then:

$$s = \frac{Mc}{I} = \frac{45{,}000 \times 12 \times 5}{145.8} = 18{,}500 \text{ psi}$$

If the ultimate strength of the steel is taken as 60,000 psi, the factor of safety is equal to 3.2.

The beam will now be examined for maximum shearing unit stress. It will be assumed that the web of the I beam carries the entire shearing force on the area bd, b from the handbook being 0.594 in. The maximum V occurs at the supports and is equal to 10,500 lb. Then:

$$s_s = \frac{V}{bd} = \frac{10{,}500}{10 \times 0.594} = 1{,}770 \text{ psi}$$

The ultimate shearing strength of steel being taken at 50,000 psi, the factor of safety is 28. This result indicates that the beam is quite safe against shear and that the factor of safety for the beam is that determined by fiber stress.

Problem 2: A 5-in. by 3⅛-in. by ½-in. steel T beam is 13 ft long and supported at the right end and 3 ft from the left end, the flange of the beam being on top. The beam carries a uniformly distributed load of 200 lb per ft, including its own weight, and a concentrated load at the left end of 200 lb. Determine the maximum unit fiber stress in tension and in compression.

The handbook shows that I is 2.7 in.^{4}, and the distances from the neutral axis to the extreme fibers are as shown in Fig. 53. The reactions are first calculated, and the shear diagram drawn as shown in the figure.

Calculating the moments at 3 and at 8.75 ft, the points of zero shear, we have

At 3 ft,

$$M = -200 \times 3 - 200 \times 3 \times 1.5 = -1{,}500 \text{ ft-lb}$$

At 8.75 ft,

$$M = -200 \times 8.75 - 200 \times 8.75 \times 4.38 + 1{,}950 \times 5.75 = +1{,}780 \text{ ft-lb}$$

The unit stresses are:

At 3 ft,

$$\text{Top fiber } s = \frac{1{,}500 \times 12 \times 0.76}{2.7} = 5{,}070 \text{ psi } (T)$$

$$\text{Bottom fiber } s = \frac{1{,}500 \times 12 \times 2.36}{2.7} = 15{,}720 \text{ psi } (C)$$

At 8.75 ft,

$$\text{Top fiber } s = \frac{1\ 780 \times 12 \times 0.76}{2.7} = 6{,}020 \text{ psi } (C)$$

$$\text{Bottom fiber } s = \frac{1{,}780 \times 12 \times 2.36}{2.7} = 18{,}620 \text{ psi } (T)$$

This problem illustrates the point that, in a beam whose cross section is not symmetrical with respect to the neutral axis, the maximum unit compressive and maxi-

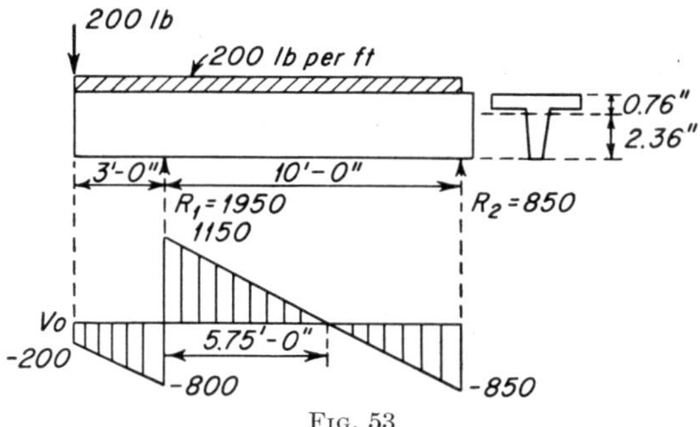

FIG. 53

mum unit tensile stresses do not necessarily occur at the same cross section. In this case, the maximum compressive stress occurs at the bottom fiber 3 ft from the left end, and the maximum tensile stress occurs at the bottom fiber 8.75 ft from the left end. The maximum compressive stress occurs at 3 ft from the left end, even though the moment is numerically smaller there, because the distance from the neutral axis to the compressive fiber more than overbalances the difference in bending moment.

Safe Loads. Problem 1: An 8-in. by 10-in. Douglas-fir beam rests on end supports 10 ft apart and carries a uniformly distributed load over its entire length. If the allowable unit fiber stress is 1,400 psi and the allowable unit shear stress is 100 psi, determine the safe load for the beam.

The nominal dimensions of wooden beams are usually multiples of 2 in., although beams 3 in. in width are used.

The actual dressed dimensions differ in different localities, but much has been done to standardize them, so that dimensions less than 6 in. are ⅜ in. less than nominal, and larger dimensions are ½ in. less. Therefore, section moduli should be based on actual dimensions.

Since the maximum moment for this loading is $Wl/8$, the safe load based on fiber stress is as follows:

$$\frac{Wl}{8} = \frac{sI}{c}$$

Since I/c for a rectangle is $bd^2/6$, then,

$$\frac{W \times 10 \times 12}{8} = \frac{1{,}400 \times 7.5 \times 9.5^2}{6}$$

$$W = 10{,}520 \text{ lb}$$

It being recalled that for a rectangular beam the maximum unit shear stress is three halves of the average, the safe load based upon shear is as follows:

$$s_s = \frac{3V}{2bd}$$

Since $V = W/2$,

$$100 = \frac{3 \times W}{2 \times 2 \times 7.5 \times 9.5}$$

$$W = 9{,}500 \text{ lb}$$

These results show that the safe load for the beam is governed by the low value of unit shear stress and is equal to 9,500 lb.

Problem 2: A 10-in. 35-lb steel I beam 14 ft long is supported at the right end and 4 ft from the left end. It carries a concentrated load of 1,000 lb at the left end. If

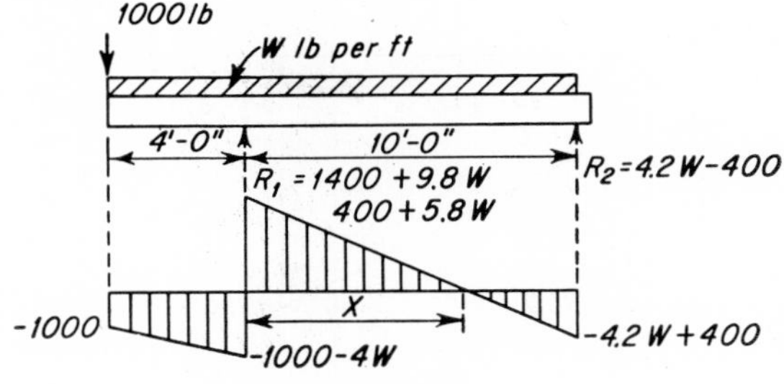

Fig. 54

the allowable fiber stress is 18,000 psi and the allowable shear stress is 10,000 psi, determine the additional distributed load per foot over the entire length that the beam could safely carry.

In this problem, the reactions must be calculated in terms of w, the uniform load per foot, and the shear diagram may then be drawn. Figure 54 shows the reactions and the shear diagram.

Solving for the position of maximum positive moment:

$$x = \frac{400 + 5.8w}{w} = \frac{400}{w} + 5.8$$

The maximum moments will come at 4 ft and at $\left(\frac{400}{w} + 9.8\right)$ ft.

At 4 ft,

$$M = -1{,}000 \times 4 - 4w \times 2 = -4{,}000 - 8w$$

At $\left(\dfrac{400}{w} + 9.8\right)$ ft,

$$M = -1{,}000\left(\frac{400}{w} + 9.8\right) - \frac{w}{2}\left(\frac{400}{w} + 9.8\right)$$

$$+ (1{,}400 + 9.8w)\left(\frac{400}{w} + 5.8\right) = \frac{80{,}000}{w} - 1{,}675 + 8.8w$$

On the basis of fiber stress, using the maximum negative moment and noting that I is equal to 145.8 in.^{4}:

$$18{,}000 = \frac{(4{,}000 + 8w)12 \times 5}{145.8}$$

$$w = 4{,}975 \text{ lb per ft}$$

Using the maximum positive moment:

$$18{,}000 = \left(\frac{80{,}000}{w} - 1{,}675 + 8.8w\right)\frac{12 \times 5}{145.8}$$

$$w = 5{,}160 \text{ lb per ft}$$

With w equal to 4,975 lb per ft, the beam may now be investigated for shear. If w is known, the numerical value of the ordinates on the shear diagram may be found. The maximum ordinate is the positive ordinate at the left support and is equal to 29,200 lb.

Since $s_s = V/bd$ and $b = 0.594$ in.,

$$s_s = \frac{29{,}200}{10 \times 0.594} = 4{,}930 \text{ psi}$$

This unit stress is less than the allowable unit shear stress, and the conclusion may be drawn that the safe load for the beam is not governed by shear. The safe load is that determined on the basis of fiber stress at the maximum negative moment and is equal to 4,975 lb per ft.

Design. Problem 1: A Douglas-fir beam is to be of rectangular section, 12 ft long, supported at the ends, and is to carry a uniformly distributed load of 600 lb per ft. Determine the size of the beam if the allowable fiber stress is 1,400 psi and the allowable shear stress is 100 psi.

The maximum moment occurs at the center of the span and is $Wl/8$. Then,

$$s = \frac{Mc}{I}$$

and

$$1{,}400 = \frac{7{,}200 \times 12 \times 12 \times 6}{8bd^2}$$

$$bd^2 = 557$$

Since both b and d are unknown, the problem may be solved by trial-and-error methods. Furthermore, since a beam of rectangular section is much stronger with the large dimension of the beam used as the depth, it is desirable to use the beam in this position from the standpoint of economy.

If a beam 6 in. by 8 in. is tried, $bd^2 = 384$, which is too small. For an 8-in. by 10-in. beam, $bd^2 = 800$, which is ample. The actual dressed dimensions of an 8-in. by 10-in. beam are 7.5 by 9.5. This makes $bd^2 = 677$, which is still ample. An 8-in. by 8-in. square beam is found to be too small.

The 8-in. by 10-in. beam will now be investigated for shear.

$$s_s = \frac{3V}{2bd} = \frac{3 \times 3{,}600}{2 \times 7.5 \times 9.5} = 76 \text{ psi}$$

This unit stress is less than the allowable shear stress, which shows that shear does not govern the design.

Problem 2: Determine the size of an I beam that would safely carry the loads shown in Fig. 53, the allowable stresses being 18,000 psi fiber stress and 10,000 psi shear stress.

The maximum moment for this beam was found to be 1,780 ft-lb. Then:

$$s = \frac{Mc}{I}$$

$$18{,}000 = \frac{1{,}780 \times 12 \times c}{I}$$

and

$$\frac{I}{c} = 1.19 \text{ in.}^3$$

The tables of properties for steel I beams in the handbook show that a 3-in. 5.7-lb I beam has a section modulus of 1.7 in.^{3}, which is ample.

This beam will now be investigated for shear. Figure 53 shows that the maximum shear occurs at the left support and is equal to 1,150 lb. The above I beam has a web thickness of 0.17 in. Then:

$$s_s = \frac{1{,}150}{3 \times 0.17} = 2{,}260 \text{ psi}$$

Since this result is less than the allowable shear stress, it is evident that shear does not govern the design (see Section 6 for further treatment of steel beams).

FAILURE OF BEAMS

There is a distinction to be made between the relation of loads and unit stresses in the case of beams as compared with the case of bars loaded in axial tension or compression. In both cases, when the unit stresses are within the elastic limit, an increase in load produces the same percentage of increase in unit stress. When the elastic limit is exceeded in the case of axially loaded bars, an increase of load still produces the same percentage increase in unit stress, for the original area is used in calculating the unit stress. In beams, the calculation of the unit fiber stress in the beam formula assumes that a linear variation of unit stress exists from the neutral axis to the outer fiber. This is true for stresses within the elastic limit but is not true when the elastic limit is exceeded. Hence it follows that after the elastic limit is exceeded, an increase of load does not produce the same percentage of increase in unit stress. If the beam formula is used to calculate the unit stress after the elastic limit is exceeded, the calculated unit stress will be higher than the true stress that exists in the outer fiber of the beam.

Modulus of Rupture. The name modulus of rupture has been given to the unit stress calculated from the beam formula $s = Mc/I$, when the bending moment used in the formula is the maximum bending moment before rupture. Since the variation of unit stress in this case is no longer linear, the modulus of rupture is really a fictitious value, which is higher than the true stress actually present in the beam. In spite of this fact, the modulus of rupture is calculated in beam tests and serves some useful purposes.

Factor of Safety. In axially loaded bars, the factor of safety will be the same whether calculated by dividing the ultimate load by the working load or the ultimate strength by the working unit stress. In beams, the factor of safety will be the same whether the ultimate load is divided by the working load or the modulus of rupture is divided by the working unit stress. However, if the true unit stress corresponding to the maximum moment were divided by the working unit stress, the factor of safety would not be the same as that based on the ratio of loads.

Uses of Modulus of Rupture. The Forest Products Laboratory of the U.S. Department of Agriculture has found the modulus of rupture useful in computing the ultimate beam strength of various species and grades of wood.

If a laboratory beam test is to be made and it is desired to determine a load-deflection curve plotted from 15 or 20 points, the modulus of rupture is useful in calculating the probable ultimate load for the beam. The increments of load to be used during the progress of the test would then be known.

The modulus of rupture is useful also in determining safe unit stresses for the design of beams. The modulus of rupture divided by a proper factor of safety would give the allowable unit stress. It will be shown, however, that judgment must be exercised in this case, for the same material may give various values of modulus of rupture depending on the size and shape of the beam.

Failure of Beams. When beams made of a ductile material like structural steel are tested to failure, the stress distribution will be about the same in tension and compression. When the yield point is reached on the outer fiber, the beam may be said to have failed, for under increasing loads the beam supports the loads for a short time only and then gradually deflects under the load. The beam does not actually break in two but becomes inadequate because of excessive deflection.

This type of failure of beams made of ductile material may be considered the normal way in which they fail. When beams are made of a material like structural steel in the form of solid I beams or built-up I beams, it will be shown later that there are several other ways in which they may fail.

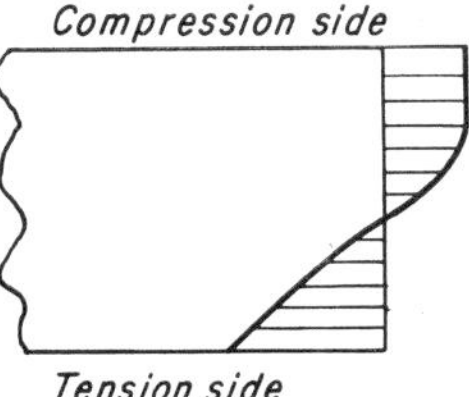

Fig. 55. Stresses in a wood beam at failure.

Failure of a Wood Beam. Since the ultimate compressive unit stress of wood is less than the ultimate tensile stress, a wood beam will start to fail by the fibers buckling on the compression side. As the load is increased, the neutral axis gradually shifts toward the tensile side, in an effort to keep the total tension on the cross section equal to the total compression. Finally the tensile fibers reach their ultimate strength, and the beam fails by rupturing more or less violently. Figure 55 shows the probable distribution of fiber unit stress in a wood beam when failure occurs. Shear failure of beams will be discussed later.

Failure of a Brittle Beam. For brittle materials like cast iron, concrete, or brick, the maximum unit tensile strength is much smaller than the compressive strength. When the maximum unit tensile stress in beams of these materials reaches the ultimate, the beams fail by cracking on the tensile side, and the failure is a more or less shattering failure. When the elastic limit in tension is exceeded, the neutral axis begins to shift toward the compression side. Figure 56 shows the probable distribution of unit stress in a cast-iron beam when failure occurs.

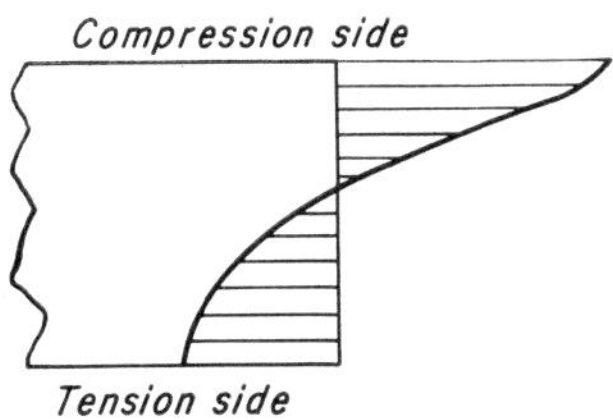

Fig. 56. Stresses in a cast-iron beam at failure.

Effect of Shape and Size on Modulus of Rupture.

Many tests on various materials have shown that both shape and size may influence the result obtained for modulus of rupture in a beam test.

Tests on wooden beams by the Forest Products Laboratory [1] indicate that a beam of increasing depth has a modulus of rupture which is smaller than that for a beam 2 in. in depth. A beam 10 in. deep had a modulus of rupture which was 91 percent of that for the smaller beam.

The same laboratory showed that when the shape of the cross section of a wooden beam is in the form of a T, I, box, or round, the modulus of rupture is different from that for a square beam 2 in. by 2 in. in cross section. The I and box sections had moduli of rupture which were only about 70 percent of that of the standard 2-in. by 2-in. beam.

Tests on cast-iron beams have shown [2] that the modulus of rupture is somewhat less when the span is increased from 12 to 24 in. Also, tests on cast-iron beams of various shapes but of about equal cross-sectional area have shown [3] values of modulus of rupture varying from 18,900 to 33,300 psi.

The practical conclusion to be drawn from the above tests is that the modulus of rupture determined from the results of tests must be applied with judgment. The modulus of rupture determined for a certain material may be appropriately used for similar material in beams of the same shape and size but may not be at all appropriate for beams of different shape or size.

Shear Failure of Beams. Unless beams are relatively short, they do not usually fail by shear. However, wood is a material which is particularly weak in shear along the grain, and it follows that wood beams may fail owing to the fact that the horizontal unit shear in the beam reaches its ultimate strength.

The formula for the horizontal unit shear in a beam was derived on the assumption that the variation in the fiber-stress distribution on the cross section was linear. However, in many cases, the place of maximum unit shear stress in the beam, occurring, as it does, near the supports, will be at places where the moment may be relatively small, and the unit fiber stresses may still be within the elastic limit. In such cases, therefore, the formula for the unit shear stress may still be appropriately used in calculating the ultimate strength of the beam as based upon shear.

Failure in Compression Flange. Several of the more abnormal types of failure of I beams will now be discussed. When a beam of this kind has a comparatively long span, so that the ratio of the length of the compression flange to its width is large, failure may be due to column action which causes the compression flange to buckle laterally.

The American Institute of Steel Construction handbook uses the following formula as a criterion to investigate this behavior:

$$s_c = \frac{12{,}000{,}000}{\dfrac{ld}{bt}} \tag{74}$$

in which s_c denotes the safe compressive unit stress due to column action, psi, l denotes the length of the unbraced compression flange, in., d denotes the depth of the beam, in., b denotes the width of the flange, in., and t denotes the thickness of the flange, in.

This formula [4] permits the use of a unit fiber stress of 20,000 psi, the ordinary stress used in beam design, for values of ld/bt of 600 or less. For values greater than 600 Eq. (74) gives the allowable unit stress that should be used to prevent lateral buckling of the compression flange.

[1] NEWLIN and TRAYER, *National Advisory Comm. Aeronaut. Rept.* 181.
[2] MATHEWS, *Proc. ASTM*, 1910, vol. 10, p. 303.
[3] BENJAMIN, *Machinery*, May, 1906, p. 437.
[4] For derivation of this formula consult *Proc. ASCE*, September, 1946.

Failure by Web Buckling. It has been shown on p. 3–4 that an element subjected to shearing unit stresses develops a compressive unit stress on a 45° line, which is equal to the shearing unit stress. Near the supports of an I beam, therefore, these compressive unit stresses in the web may cause the web to fail between the flanges owing to column action.

The American Institute of Steel Construction handbook uses the following formula as a criterion to investigate this behavior:

$$s_s = \left(\frac{8{,}000}{h/t}\right)^2 \tag{75}$$

in which s_s denotes the safe unit shear stress, psi, h denotes the clear height between flanges, in., and t denotes the thickness of the web, in.

When h/t is 70, this formula[1] allows a value of s_s equal to 13,000 psi, the ordinary value used in beam design. For values of h/t greater than 70, the stress is reduced according to Eq. (75). For plate girders, when h/t is greater than 70, intermediate stiffeners are required.

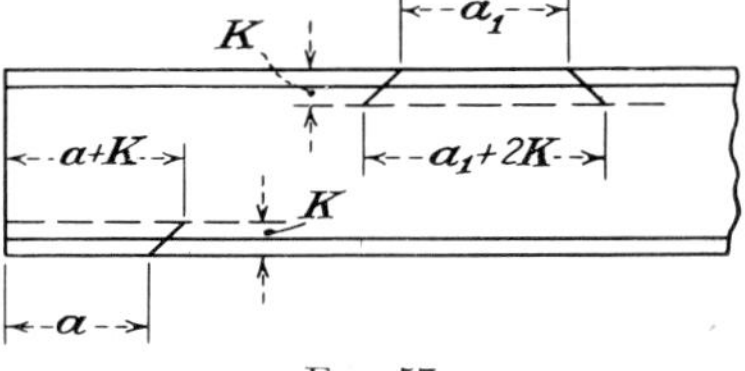

Fig. 57

Failure by Web Crippling. The third way in which an I beam may fail by column action is at the supports or at points of concentrated loading. Such loads may cause the web to fail between the flanges because of the large local compressive unit stress.

The American Institute of Steel Construction handbook uses the following formulas to control this action (see Fig. 57):

$$\text{Maximum end reaction} = 24{,}000t(a + k) \tag{76}$$

$$\text{Maximum interior load} = 24{,}000t(a_1 + 2k) \tag{77}$$

in which t denotes the thickness of the web, in., k denotes the distance from the outer face of the flange to web toe of fillet, in., a denotes length of bearing, in., and a_1 denotes length of concentrated load, in. The unit compressive stress of 24,000 psi is not to be exceeded, and when necessary the webs of the beams should be reinforced, or the length of the bearing increased.

BEAMS OF UNIFORM STRENGTH

General Principles. A very common design for machine and structural parts that are subjected to flexure is to make them of approximately uniform strength. In the beams discussed thus far, the cross section has been constant throughout the length of the beam, while the maximum fiber unit stress in a cantilever, for instance, would occur only at the fixed end. It is evident, therefore, that the cross sections at other points along the length of the beam are larger than necessary. In a beam of uniform strength, the maximum unit fiber stress at any section is kept constant, and the cross section is varied in size according to the requirements of the fiber-stress formula for beams

$$\frac{I}{c} = \frac{M}{s}$$

Some Common Cases. It will be sufficient to illustrate the principle used in beams of uniform strength, by discussing cantilever beams of rectangular cross section, carrying

[1] For derivation of this formula consult *Trans. ASCE*, vol. 101, p. 857, 1936.

the more common loadings. Figure 58 represents a cantilever beam with a concentrated load P at the free end, and the design will be based on a cross section of constant width and varying depth.

The bending moment on any cross section at a distance x from P is Px, and, according to the fiber-stress formula for beams,

$$Px = \frac{sI}{c} = \frac{sBd^2}{6}$$

$$d = \sqrt{\frac{6Px}{sB}} \tag{78}$$

in which s denotes the allowable fiber stress, psi, d denotes the varying depth of the beam, in., B denotes the constant width of the beam, in., P denotes the concentrated load at the end of the beam, lb, and x denotes the distance from the free end of the beam to any cross section, in. By assigning various values of x in Eq. (78) and computing the depth d, the profile of the beam may be determined.

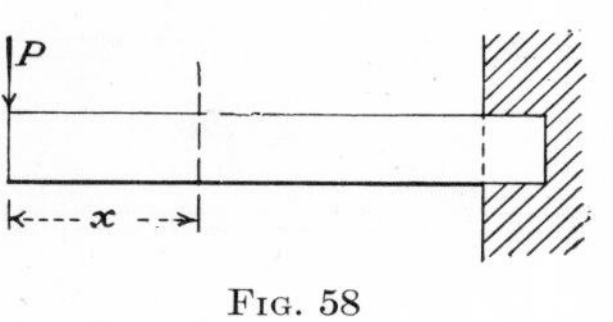

Fig. 58

For the beam just discussed, the depth at the free end might be zero as far as bending moment is concerned, but this is not the case if shear is to be provided for. For a rectangular cross section, the maximum shearing unit stress occurs at the neutral axis and is equal to three-halves of the average, or

$$s_s = \frac{3V}{2A} = \frac{3P}{2Bd}$$

and

$$d = \frac{3P}{2Bs_s} \tag{79}$$

The profile of the beam would therefore look like that shown in Fig. 59. The notch at the free end could be retained, or the profile could be made a smooth curve.

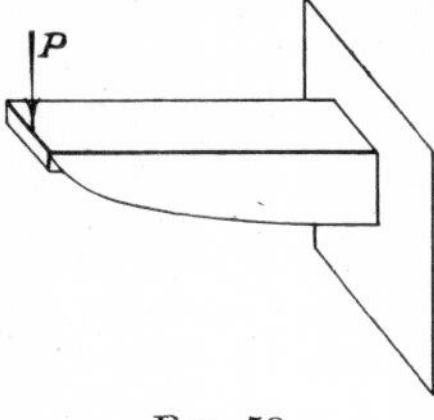

Fig. 59

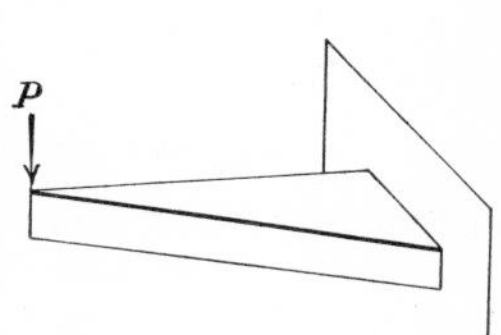

Fig. 60. Concentrated load.

Figures 60, 61, and 62 show other profiles of cantilever beams of uniform strength, loaded in various ways.

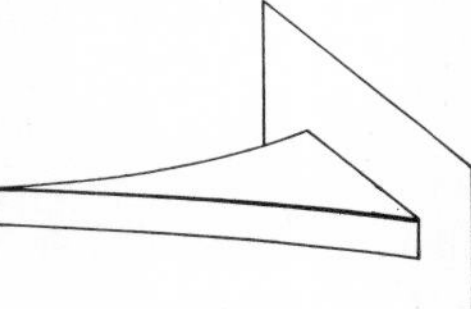

Fig. 61. Uniform load.

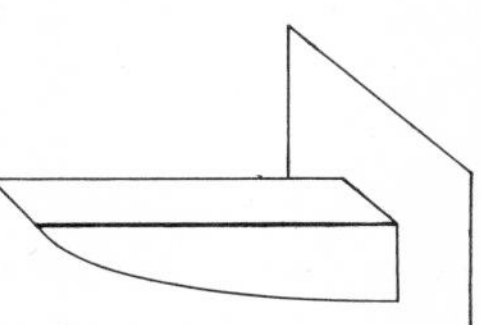

Fig. 62. Uniform load.

Even for cross sections much more complicated than the rectangular, it is possible to determine a beam of approximately uniform strength. The formula for fiber stress may be applied at every half foot or every foot along the length of the beam, enough points being thus located to determine the profile of the beam.

One of the commonest cases of beams of approximately uniform strength in structural work is the plate girder. In a plate girder supported at the ends, the design at the supports would be governed by the shear to be provided for there; at other points along the span, the fiber stress would govern the design. The usual design consists of a built-up I-beam type, a deep plate being used for the web, while the flanges consist of angle sections. As the moment increases toward the center of the span, additional plates are provided at the top and bottom of the beam, in order that the value of the fiber stress may not exceed a predetermined value. The result is a design of approximately uniform strength.

Deflections of Uniform-strength Beams. The general subject of deflections of beams will be discussed in succeeding paragraphs; it is the intention now to mention briefly some matters that are of interest in connection with uniform-strength beams.

A cantilever beam, of constant width and varying depth, carrying a concentrated load at the free end and designed for uniform strength, will have the same dimensions at the fixed end as would a beam of constant cross section. At all other sections, however, the dimensions would be smaller for the uniform-strength beam. The result is that such a beam, having less material in it, will be less stiff than the beam of constant cross section. Table 8 gives formulas for the maximum deflections of simple and cantilever beams designed for uniform strength and carrying some of the more common loadings. In this table y denotes the maximum deflection in inches, P denotes the load on the beam in pounds, E denotes the modulus of elasticity in tension and compression in pounds per square inch, b denotes the maximum breadth in inches, d denotes the maximum depth in inches, and l denotes the length of the beam in inches.

TABLE 8. DEFLECTION OF UNIFORM-STRENGTH BEAMS

Kind of Loading	Maximum Deflection, in.
1. Simple beam, supported at ends Concentrated load at center Constant breadth, varying depth	$y = \frac{P}{2bE}\left(\frac{l}{d}\right)^3$
2. Simple beam, supported at ends Concentrated load at center Constant depth, varying breadth	$y = \frac{3P}{8bE}\left(\frac{l}{d}\right)^3$
3. Cantilever beam Concentrated load at free end Constant breadth, varying depth	$y = \frac{8P}{bE}\left(\frac{l}{d}\right)^3$
4. Cantilever beam Concentrated load at free end Constant depth, varying breadth	$y = \frac{6P}{bE}\left(\frac{l}{d}\right)^3$

As an example, take case 3 in Table 8. The maximum deflection of the uniform-strength beam is

$$y = \frac{8P}{bE}\left(\frac{l}{d}\right)^3$$

The maximum deflection of the same beam when of constant cross section throughout its length would be

$$y = \frac{Pl^3}{3EI} = \frac{12Pl^3}{3Ebd^3} = \frac{4P}{bE}\left(\frac{l}{d}\right)^3$$

These two beams would have the same dimensions at the fixed end, and the result shows that the beam of uniform strength would deflect twice as much as the beam of constant cross section.

In deriving equations for the maximum deflection of a beam of uniform strength, such as those given in Table 8, account must be taken of the fact that the moment of inertia of these beams is a variable and must be expressed in terms of x.

DEFLECTION OF BEAMS

For many cases of loaded beams, the maximum deflection of the beam rather than the maximum unit stress governs the design; and, in many beam problems, it is desirable to be able to calculate the deflection. Methods have therefore been developed for determining the equation of the *elastic curve*, which is the curve that the axis of the beam assumes in the deflected position of the beam. In developing the relations existing between the bending moment and the elastic curve, it can be shown [1] that the following formula holds:

$$R = \frac{EI}{M} \tag{80}$$

in which R denotes the radius of curvature of the elastic curve at any point, in., M denotes the bending moment at the same point, in.-lb, E denotes the modulus of elasticity in tension and compression, psi, and I denotes the moment of inertia of the cross section of the beam with respect to the neutral axis, in.^{4}

Equation (80) is useful in determining the fiber unit stress in such problems as the bending of a band saw over a wheel. For this case, since

$$M = \frac{SI}{c}$$

then

$$R = \frac{Ec}{s} \tag{81}$$

Equation (81) shows that the fiber stress in the saw may be computed when the radius of the wheel and the thickness of the saw are known. In this case, R would be the distance from the center of the wheel to the neutral surface of the saw, and c would be the half thickness of the saw.

The general equation of the elastic curve of any loaded beam may be obtained from the following formula: [2]

$$EI\frac{d^2y}{dx^2} = M \tag{82}$$

In applying Eq. (82) to a specific case, the bending moment M will usually not be a constant but a function of x. Since M has been expressed in terms of x, the equation is integrated twice to obtain the equation between x and y of the elastic curve. The double integration will involve two constants of integration. These constants may be determined from known boundary conditions of slope and deflection at various points in the beam. From the general equation between x and y, the maximum deflection may be determined.

[1] Boyd and Folk, "Strength of Materials," 5th ed., p. 157. Singer, "Strength of Materials," p. 127.
[2] Boyd and Folk, "Strength of Materials," 5th ed., p. 159. Singer, "Strength of Materials," p. 170.

In using Eq. (82), M is given sign according to the convention adopted in the discussion of moment diagrams on p. 3–35. Investigation will show that this convention for sign gives M the correct mathematical sign, so that, when the second derivative is minus, M has the minus sign, and, when the second derivative is plus, M has the plus sign.

Moment-area Method. It can be shown [1] that the change of slope between two points on the elastic curve of a beam is given by the following equation:

$$\theta = \frac{1}{EI} \int_0^{x_1} M\,dx \tag{83}$$

That is, the change of slope between two points is given by the area of the bending-moment diagram between them, divided by EI, the cross-sectional area of the beam

[1] The line AB of Fig. Xa represents a portion of the elastic curve of a member in flexure. An elementary length ds of the member is shown in the figure. The angle between the radii at the ends of ds is denoted by $d\theta$. The linear deformation of a fiber at a distance c from the neutral surface is $c\,d\theta$, and the unit de-

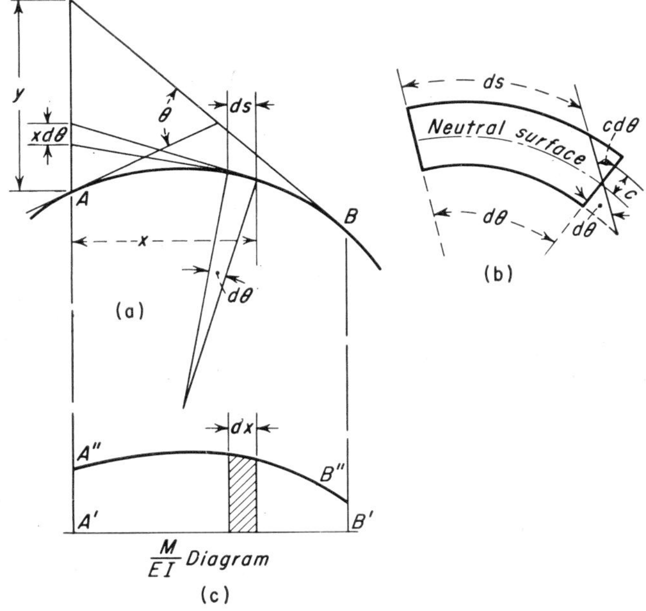

formation of the same fiber is $c\,d\theta/ds$. The unit stress in the fiber is $f = Mc/I$, in which M is the resisting moment and I the moment of inertia of the section.

Since the modulus of elasticity is the ratio of unit stress to unit deformation,

$$E = \frac{Mc}{I} \div \frac{c\,d\theta}{ds}$$

from which

$$d\theta = \frac{M}{EI}\,ds$$

In a well-designed beam, the curvature and slope are small, so that dx may be substituted for ds without appreciable error.

Then

$$d\theta = \frac{M}{EI}\,dx$$

In Fig. Xc, an ordinate measured between the curve $A''B''$ and the straight line $A'B'$ at any point between A' and B' represents, to a scale, the moment in the member AB at that point, divided by EI; or $A''B''B'A'$

being constant in this case. If the moment of inertia is not constant, the difference in slope is equal to the area of the M/I diagram, divided by E. From a point of zero slope to any other point, the area of the bending-moment diagram, divided by EI, is equal to the actual slope at the second point.

It can also be shown that the difference in deflection between two points on a beam is given by the following equation:

$$y = \frac{1}{EI} \int_{x_1}^{x_2} Mx\,dx \tag{84}$$

Equation (84) represents the deflection at x_2 from the tangent at x_1, when the origin of coordinates is taken at x_1. If the slope of the elastic curve is zero at x_1, the formula gives the actual deflection of the beam at x_2. The integral in Eq. (84) represents the moment of the bending-moment diagram from x_1 to x_2 with respect to x_2.

As a simple example of the moment-area method, consider the case of a concentrated load at the end of a cantilever beam, as shown in Fig. 63. Since the slope of the elastic curve is zero at the wall and since the area of the moment diagram is $Pl^2/2$, the slope at the free end of the beam, according to Eq. (83) is

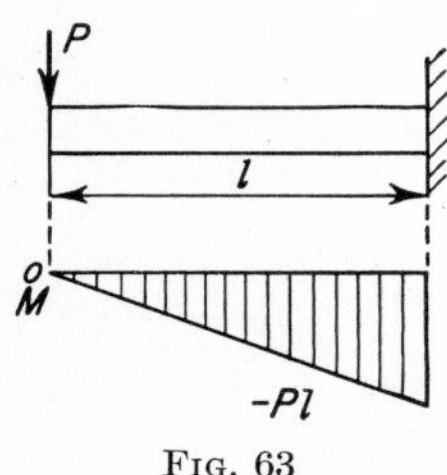

FIG. 63

$$\text{Slope} = \frac{Pl^2}{2EI}$$

Also, the centroid of the moment diagram is at two-thirds of l from the free end, and the moment of the area of the bending-moment diagram with respect to the free end of the beam is $Pl^2/2$ times $2l/3$, or $Pl^3/3$. Then, according to Eq. (84), the deflection at the free end of the beam, since the slope is zero at the wall, is

$$y = \frac{Pl^3}{3EI}$$

---

is the M/EI diagram for the member AB. The area of the diagram for the length dx is $\frac{M}{EI}\,dx$, and the area of the diagram $A''B''B'A'$ is $\int_A^B \frac{M}{EI}\,dx$. But $d\theta = \frac{M}{EI}\,dx$, and the angle between the tangents to the elastic curve at A and B is $\theta = \int_A^B d\theta = \int_A^B \frac{M}{EI}\,dx$. Hence: *The change in the slope of the elastic curve between any two points is equal to the area of the M/EI diagram for the portion of the member between those two points.*

In Fig. *Xa*, the tangents at the extremities of the elementary length ds are extended until they intersect the vertical line through A. Since the angles are small, the intercept between these tangents is practically equal to $x\,d\theta$. The total vertical distance y is the algebraic sum of all the intercepts between the tangents to the curve between A and B. That is;

$$y = \int_A^B x\,d\theta$$

Substituting, for $d\theta$, its value as previously determined:

$$y = \int_A^B \frac{M}{EI}\,x\,dx$$

In the $\frac{M}{EI}$ diagram of Fig. *Xc*, $\frac{M}{EI}\,dx$ is the area of the diagram for the length dx, and $\frac{M}{EI}\,dx$ times x is the moment of this area about the point A. The moment of the entire area of the M/EI diagram between the points A and B may be expressed as

$$\int_A^B \frac{M}{EI}\,x\,dx$$

which is equal to the expression developed above for y. Hence: *The distance of any point on the elastic curve from a tangent to the curve at any other point measured in a direction normal to the initial position of the member is equal to the moment of the area of the M/EI diagram, included between the two points, about the first point.*

Propped Cantilever. Figure 64 shows a cantilever, uniformly loaded, which is supported at the free end. This is sometimes called a "propped" cantilever. Figure 65 shows the forces that support the beam. Since the unknown forces are four in number and since only three independent equations of equilibrium can be written for this system of forces, the beam would be called a statically indeterminate case.

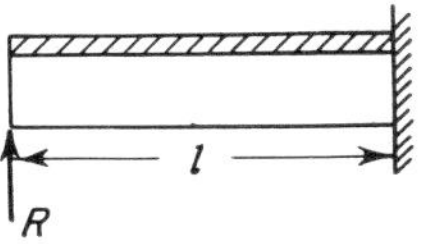

FIG. 64. Propped-cantilever beam.

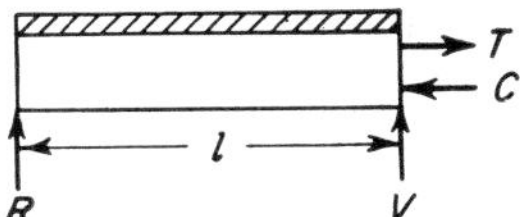

FIG. 65. Forces on a propped-cantilever beam.

If only the strength of this beam is to be investigated, the reaction R may be determined by making use of our information about deflection of beams. The argument which may be used is that the deflection downward at the free end of the beam, due to the uniform load acting alone, must be balanced by the deflection upward at the end of the beam due to R. The deflection due to the uniform load is

$$y = \frac{Wl^3}{8EI}$$

The deflection due to R is

$$y = \frac{Rl^3}{3EI}$$

Equating these deflections and solving for R:

$$R = \frac{3W}{8}$$

If the elastic curve and the maximum deflection of the beam are to be determined also, this may be accomplished by making use of Eq. (82) in the regular way.

Beams Fixed at the Ends. Beams that are fixed or built-in at both ends, as shown in Fig. 66, are also statically indeterminate. Figure 67 shows the forces that support the beam. Here M_0 and V_0 are the moment and shear respectively at the wall. At any section at a distance x from the left wall, the moment is

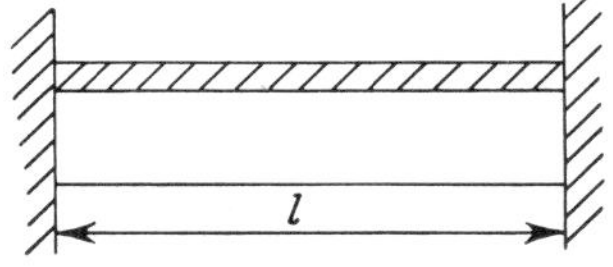

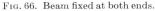

FIG. 66. Beam fixed at both ends.

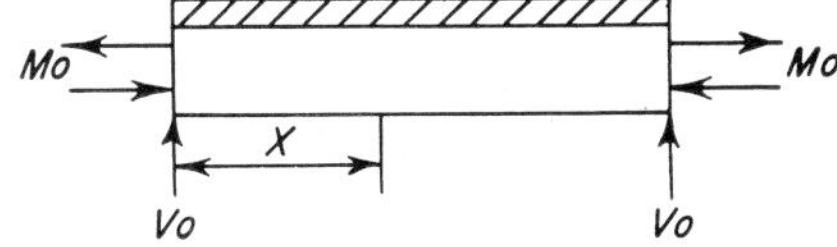

FIG. 67. Forces on fixed-ended beam.

$$M_x = M_0 + V_0x - \frac{wx^2}{2}$$

This expression for moment may be substituted in Eq. (82), and the problem may then be solved in the usual way. There will be sufficient boundary conditions involving slope and deflection so that the values of M_0, V_0, and the two constants of integration may be determined. As a matter of fact, it is evident from inspection that V_0 must equal one-half of the total load on the beam.

Table 9 lists a number of the more common cases of loading, giving the reactions, shear and moment diagrams, maximum shear, maximum moment, and maximum deflection.

TABLE 9. DATA ON COMMON BEAM LOADINGS

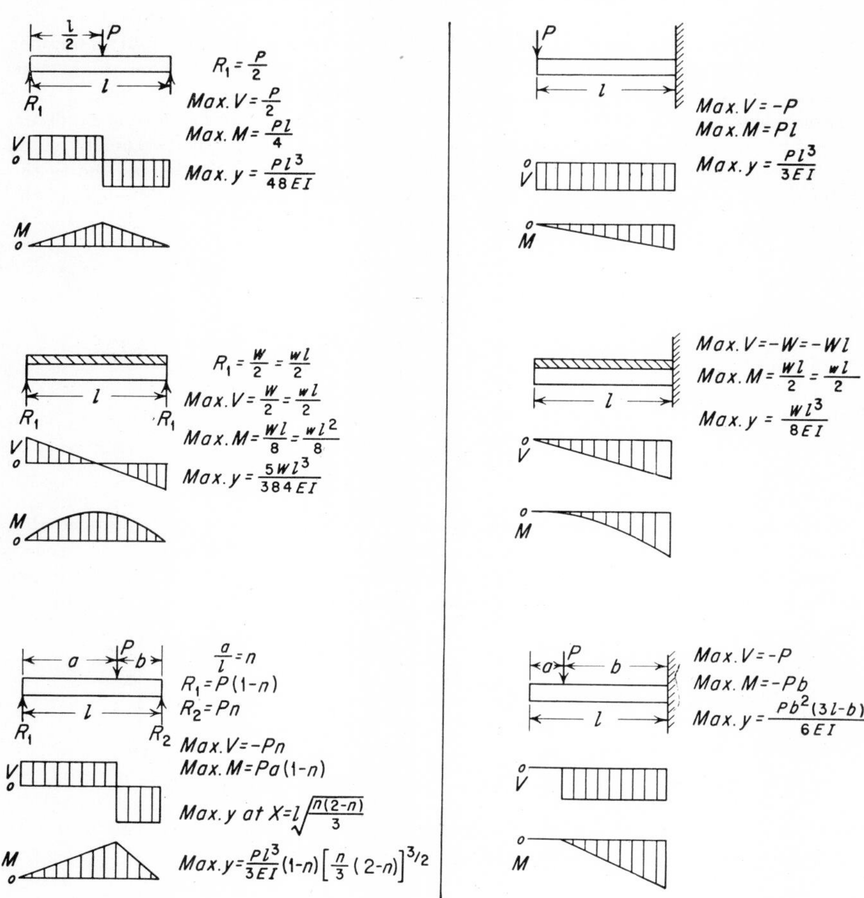

TABLE 9. DATA ON COMMON BEAM LOADINGS (*Continued*)

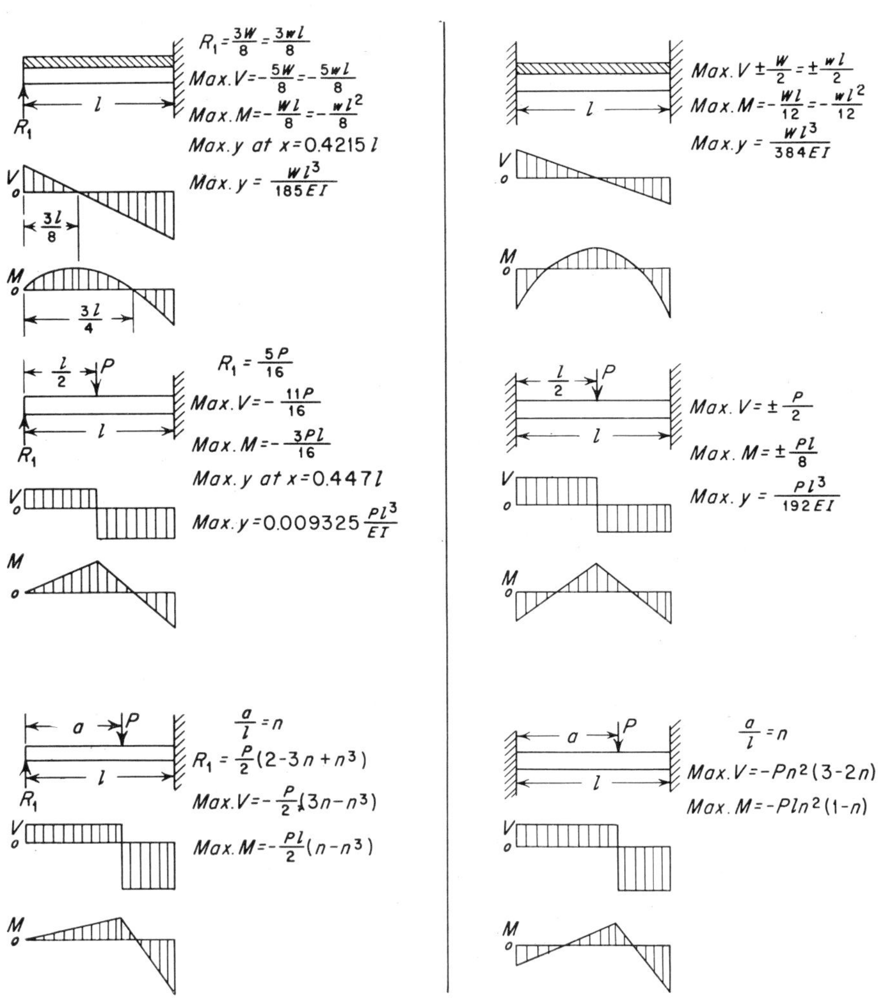

CONTINUOUS BEAMS

A *continuous* beam is one that has more than two supports. When any beam is subjected to vertical loads and reactions, it is possible to write only two independent equations of equilibrium, and a continuous beam is therefore statically indeterminate. If the general equation for discussing elastic curves [Eq. (82)] is to be used for discussing continuous beams, the number of equations to be written and the number of constants to be determined make the work very laborious if there are more than two spans.

Theorem of Three Moments. If, in continuous beams, the elastic curves may be neglected and it is desired to determine only the reactions, shears, and moments, the

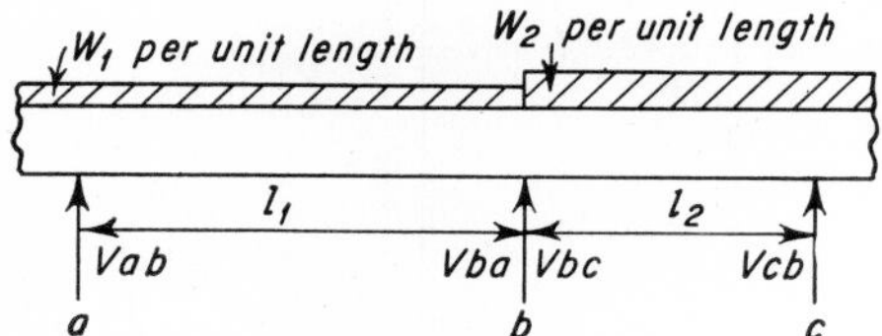

FIG. 68. Continuous beam, uniform loads.

theorem of three moments may be used to advantage. In Fig. 68 are shown three supports for a typical continuous beam carrying two different uniformly distributed loads. For this case, it can be shown [1] that the following equation holds:

$$M_a l_1 + 2M_b(l_1 + l_2) + M_c l_2 = -\tfrac{1}{4}(w_1 l_1^3 + w_2 l_2^3) \tag{85}$$

[1] From the upper portion of Fig. Y, it can be seen that $\dfrac{H_2 - (H_3 + D_2)}{l_2} = \dfrac{(H_1 + D_1) - H_2}{l_1}$ or $l_1 D_2 + l_2 D_1 = (H_2 - H_3)l_1 + (H_2 - H_1)l_2$. From the footnote on p. 3–54, D_1 is equal to the moment of the area of the M/EI diagram for the span l_1 about g (lower portion of figure), and D_2 is equal to the moment of the M/EI diagram for the span l_2 about i. $M_1/EI_1 = ag$, $M_2/EI_1 = hc$, $M_2/EI_2 = hd$, and $M_3/EI_2 = if$.

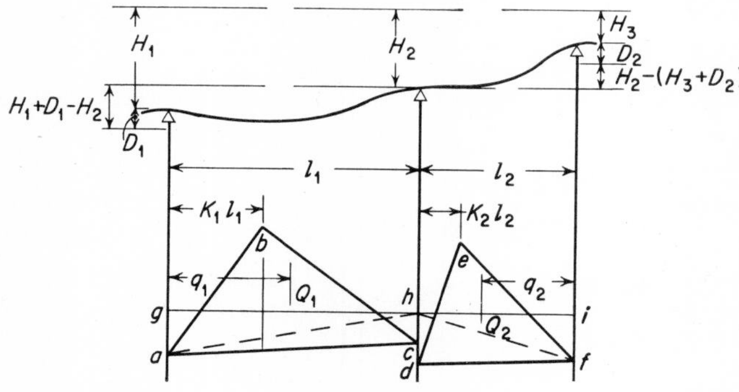

The area gbc, the normal moment (M/EI) diagram is designated Q_1 and the area dei, Q_2. Then, for a concentrated load of P_1 in the first span a distance $k_1 l_1$ from the left support and for a load P_2 in the second span a distance $k_2 l_2$ from the second support,

$$D_1 = Q_1 q_1 + \left(\frac{M_1 l_1}{2EI_1} \times \frac{l_1}{3}\right) + \left(\frac{M_2 l_1}{2EI_1} \times \frac{2l_1}{3}\right)$$

$$D_2 = Q_2 q_2 + \left(\frac{M_2 l_2}{2EI_2} \times \frac{2l_3}{3}\right) + \left(\frac{M_3 l_2}{2EI_2} \times \frac{l_2}{3}\right)$$

and

$$\frac{M_1 l_1}{I_1} + 2M_2\left(\frac{l_1}{I_1} + \frac{l_2}{I_2}\right) + \frac{M_3 l_2}{I_2} = -6E\left(\frac{Q_1 q_1}{l_1} + \frac{Q_2 q_2}{l_2}\right) + 6E\left(\frac{H_2 - H_3}{l_2} + \frac{H_2 - H_1}{l_1}\right)$$

In Fig. 68, the three supports are labeled a, b, and c; and M_a, M_b, and M_c are the moments at the three supports. In general, there is a shear on either side of each support; and, in this case, V_{ab} is the shear on the right of a, V_{ba} and V_{bc} are the shears to the left and right of b, and V_{cb} is the shear to the left of c. The theorem of three moments shows the relation between the moments at three successive supports, provided that the beam is of uniform cross section and the supports are on the same level. For a beam with three supports, the theorem of three moments supplies one independent equation, and it is necessary to know two of the moments or to have two other independent equations. For a beam with four supports, the theorem of three moments supplies two independent equations, and an additional two independent relations would be necessary to solve the problem.

When the spans are equal and the distributed load is the same per unit length on each span, Eq. (85) becomes

$$M_a + 4M_b + M_c = -\frac{wl_1^2}{2} \tag{86}$$

Calculation of Reactions and Moments. Figure 69 shows a uniformly loaded beam having three equal spans, for which the reactions and moments at the four supports are to be determined.

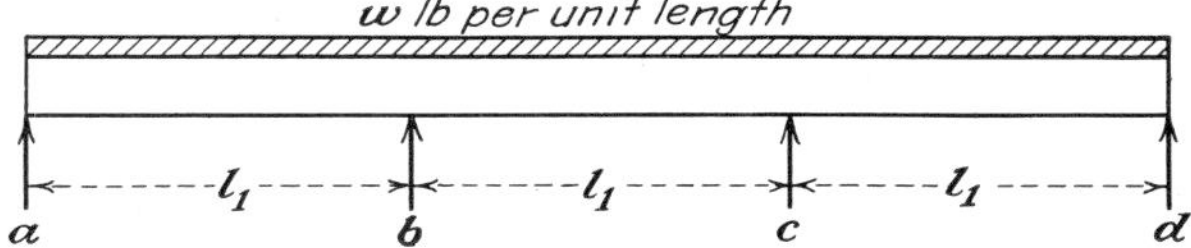

FIG. 69. Continuous beam, uniform load.

In this case, M_a and M_d are each equal to zero, and from Eq. (86) for the first two spans:

$$4M_b + M_c = -\frac{wl_1^2}{2}$$

For the second and third span:

$$M_b + 4M_c = -\frac{wl_1^2}{2}$$

The ordinate $bj = \frac{P_1(l_1 - k_1l_1)k_1l_1}{EI_1l_1} = \frac{P_1l_1k_1(1-k_1)}{EI_1}$ and

$$Q_1q_1 = \frac{P_1l_1k_1(1-k_1)}{2EI_1}\left[k_1l_1 \times \frac{2k_1l_1}{3} + (1-k_1)l_1\left(k_1 + \frac{1-k_1}{3}\right)l_1\right] = \frac{P_1l_1^3}{6EI_1}(k_1 - k_1^3)$$

In a similar manner, Q_2q_2 may be shown to be equal to $\frac{P_2l_2^3}{6EI_2}(2k_2 - 3k_2^2 + k_2^3)$.

Therefore, the general three-moment equation for concentrated load is

$$\frac{M_1l_1}{I_1} + 2M_2\left(\frac{l_1}{I_1} + \frac{l_2}{I_2}\right) + \frac{M_3l_2}{I_2} = -\sum\frac{P_1l_1^2}{I_1}(k_1 - k_1^3) - \sum\frac{P_2l_2^2}{I_2}(2k_2 - 3k_2^2 + k_2^3) + 6E\left(\frac{H_2 - H_3}{l_2} + \frac{H_2 - H_1}{l_1}\right)$$

If a uniform load of w_1 is added on the span l_1 and a load of w_2 on l_2, additional quantities Q_1q_1/l_1 and Q_2q_2/l_2 must be added to the right-hand side of the above equation.

$$Q_1q_1 = \frac{w_1l_1^2}{8EI} \times \frac{2l_1}{3} \times \frac{l_1}{2} = \frac{w_1l_1^4}{24EI}$$

and similarly, $Q_2q_2 = \frac{w_2l_2^4}{24EI}$

and the right-hand side of the equation has the additional terms of $-\frac{w_1l_1^3}{4I_1}$ and $-\frac{w_2l_2^3}{4I_2}$.

With all supports on the same level (the usual case), the term containing H_1, H_2, and H_3, becomes zero; for equal moments of inertia, the equation is further simplified; and where the spans are also equal and the loads symmetrical, a still further simplification follows.

Solving these simultaneous equations:

$$M_b = M_c = -\frac{wl_1^2}{10}$$

The moment at b, calculated in the usual way, is

$$M_b = R_a l_1 - \frac{wl_1^2}{2} = -\frac{wl_1^2}{10}$$

Solving for R_a:

$$R_a = \frac{4wl_1}{10}$$

The moment at c is

$$M_c = R_a 2l_1 + R_b l_1 - 2wl_1^2 = -\frac{wl_1^2}{10}$$

Solving for R_b:

$$R_b = \frac{11wl_1}{10}$$

The moment at d is

$$M_d = R_a 3l_1 + R_b 2l_1 + R_c l_1 - 3wl_1 \cdot 1.5l_1 = 0$$

Solving for R_c:

$$R_c = \frac{11wl_1}{10}$$

$$\Sigma V = -3wl_1 + R_a + R_b + R_c + R_d = 0$$

Solving for R_d:

$$R_d = \frac{4wl_1}{10}$$

From symmetry it is evident that $R_a = R_d$ and $R_b = R_c$, and as a check the sum of the reactions must equal the total load on the beam.

Concentrated Loads on Continuous Beams. Figure 70 shows two spans of a continuous beam loaded with a different concentrated load on each span. It can be shown [1] that the

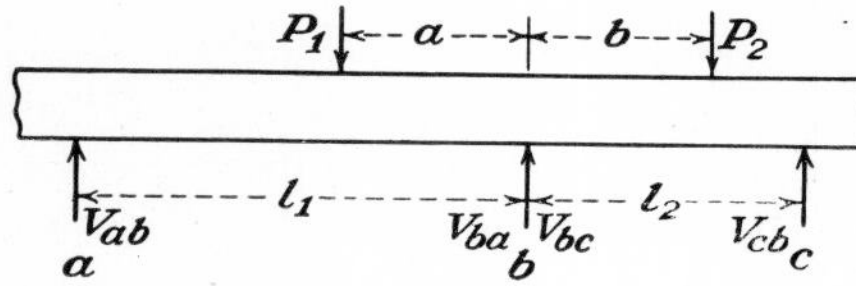

FIG. 70. Continuous beam, concentrated loads.

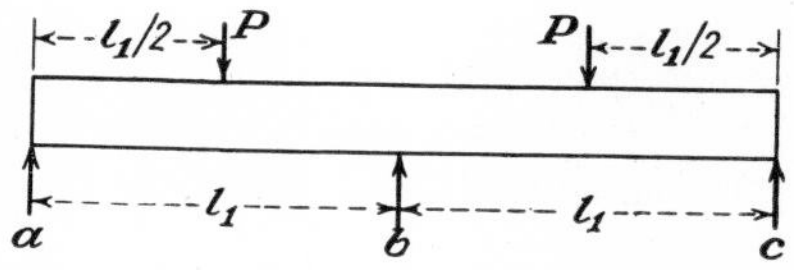

FIG. 71. Continuous beam, concentrated loads.

equation showing the relation between the three moments is as follows:

$$M_a l_1 + 2M_b(l_1 + l_2) + M_c l_2 = -P_1 a\left(2l_1 - 3a + \frac{a^2}{l_1}\right) - P_2 b\left(2l_2 - 3b + \frac{b^2}{l_2}\right) \quad (87)$$

In Fig. 70, if l_1 is equal to l_2, P_1 is equal to P_2, and each load is at the center of its span, Eq. (87) becomes

$$M_a + 4M_b + M_c = -\frac{3Pl_1}{4} \quad (88)$$

Calculation of Reactions and Moments. In Fig. 71 is shown a continuous beam of two equal spans, each span carrying a load P at the center. In this case, M_a and M_c

[1] BOYD and FOLK, "Strength of Materials," 5th ed., p. 221.

are equal to zero, and, according to Eq. (88):

$$M_b = -\tfrac{3}{16}Pl_1$$

The moment at b is

$$M_b = R_a l_1 - \frac{Pl_1}{2} = -\frac{3Pl_1}{16}$$

$$R_a = \frac{5}{16}P$$

From the summation of forces along the vertical:

$$R_b = \frac{11}{8}P$$

Alternative Solution: The reaction R_b may be determined from the argument that the deflection downward at b due to the two loads must be balanced by the deflection upward due to the reaction R_b. The deflection at the center of a simple beam of length l, due to a single load P, is as follows:

$$y = \frac{Pb(3l^2 - 4b^2)}{48EI}$$

In this case, b is equal to $l_1/2$, and l is equal to $2l_1$. Since there are two similarly placed loads, the total deflection is twice that given above. Equating this downward deflection to the upward deflection due to R_b, we have

$$\frac{2Pl_1}{2 \times 48EI}\left(3 \times 4l_1^2 - \frac{4l_1^2}{4}\right) = \frac{R_b(2l_1)^3}{48EI}$$

and

$$R_b = \frac{11}{8}P$$

CURVED BEAMS

The discussion of the stresses in beams in previous sections may be correctly applied when the neutral surface of the unloaded beam is a plane or in the common case when the top and bottom of the beam are straight and parallel to each other. It is not correct to compute the flexural stress by this method for cases like C frames or hooks in which the

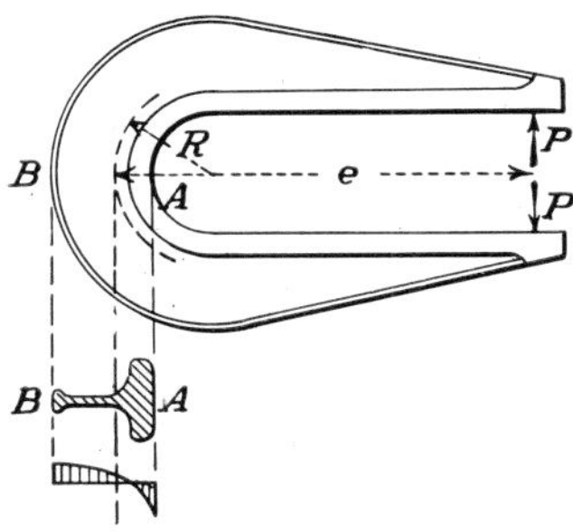

FIG. 72. Curved beam.

sides of the beams are curved, as shown in Fig. 72. The unit flexural stress at a point such as A depends on the radius of curvature R in such manner that, the smaller the radius R, the larger the unit flexural stress.

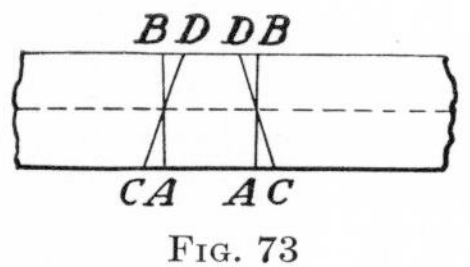

FIG. 73

The reason for this may be illustrated by Figs. 73 and 74. In Fig. 73, two parallel planes such as AB before loading might take the position CD after loading, the deformations being proportional to the distance from the neutral axis. The deformation BD and the deformation CA would both occur in an original length, which is the same in each case, being the distance between the original planes AB. The unit deformations would therefore be equal for two fibers equally distant from the neutral axis.

In Fig. 74, the planes AB before loading might take the position CD after loading. The deformation DB, however, would have occurred in an original length BB, while the deformation EF would have occurred in an original length EE, which is smaller. Therefore, even for fibers equally distant from the neutral axis, the unit deformation would be greater for fibers to the right of the neutral axis than for fibers to the left. It is evident, therefore, that, for curved beams, the unit deformations and also the unit stresses will not vary linearly but will vary somewhat as shown in the lower part of Fig. 72.

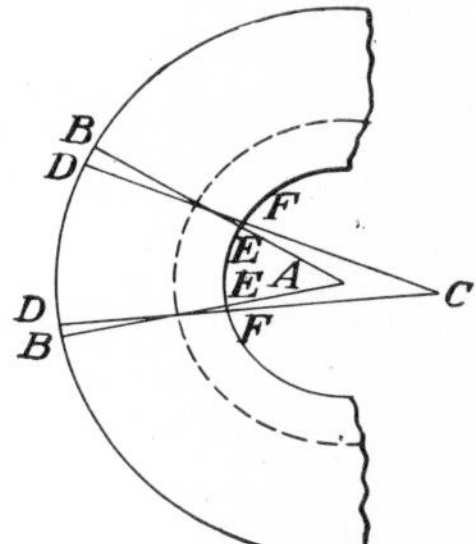

FIG. 74

For curved beams, therefore, the formula $Mc/I = s$, based upon beams with straight sides, gives maximum unit stresses that are too small. It can be shown [1] that the formula for the maximum unit stress in curved beams is as follows:

$$s = \frac{Pe}{R(A' - A)}\left(\frac{R}{R - c} - \frac{A'}{A}\right) + \frac{P}{A} \qquad (89)$$

$$s = \frac{Pe}{R(A' - A)}\left(\frac{A'}{A} - \frac{R}{R + c}\right) - \frac{P}{A} \qquad (90)$$

in which Pe denotes bending moment, in.-lb; A denotes area of cross section, in.^{2}; R denotes radius of curvature for the centroidal axis, in.; c denotes distance from centroidal axis to place where unit stress is wanted, in.; s denotes the maximum unit stress, psi; and A' denotes a factor depending upon the shape and size of the cross section.

For Fig. 72, Eq. (89) would be used for computing the tensile stress to the right of the centroidal axis, and Eq. (90) for computing the compressive stress to the left of the centroidal axis.

It can be shown that the factor A' has the following value:

$$A' = R\int\frac{dA}{R - c'}$$

the notation being the same as above, except that c' is variable.

The value of A' for regular geometrical areas like the rectangle, triangle, or circle may be computed by the calculus. For irregular areas, recourse is had to graphical and semigraphical methods.[2]

OBLIQUE LOADING ON BEAMS

In the common cases of flexure that have been discussed hitherto, the beam had a longitudinal plane of symmetry, and all the loads and reactions lay in this plane. In the case of *oblique* loading, the beam may not have a longitudinal plane of symmetry,

[1] BOYD and FOLK, "Strength of Materials," 5th ed., p. 363. SINGER, "Strength of Materials," p. 406.
[2] MAURER and WITHEY, "Strength of Materials," 2d ed., p. 218. MORLEY, "Strength of Materials," 3d ed., p. 339.

or, if it has, the loads may not lie in this plane. Figure 75 shows an example of the former, and Fig. 76 of the latter case. The beam in Fig. 75 is an unequal-legged angle section and has no plane of symmetry. The beam in Fig. 76 has a plane of symmetry, but the load on the beam does not lie in this plane.

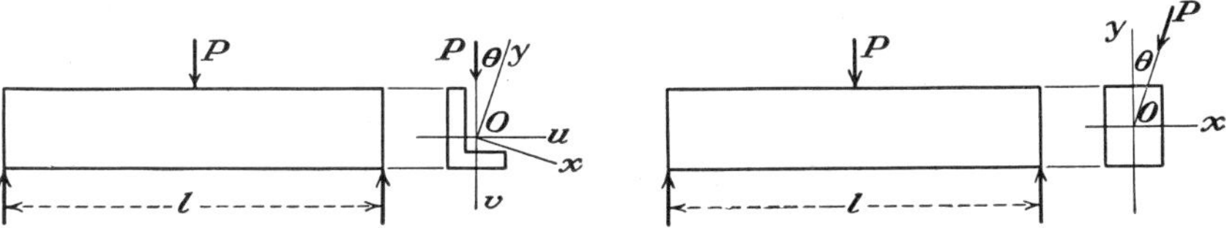

FIG. 75. Oblique loading.

FIG. 76. Oblique loading.

Position of Neutral Axis. For the common case of flexure in which the beam has a longitudinal plane of symmetry and the loads lie in this plane, the neutral axis is perpendicular to the load plane. For the cases shown in Figs. 75 and 76, the neutral axis would not be perpendicular to the load plane.

It can be shown [1] that, if the plane of the bending moment contains the principal axis of a cross section of a beam, the neutral axis of that cross section is perpendicular to the plane of the bending moment. The principal axes for any point in an area are the axes through that point for which the moments of inertia are respectively greatest and least, and they are always at right angles to each other. The methods of determining the position of the principal axes and the values of the principal moments of inertia may be found in standard textbooks on strength of materials.[2]

The angle α (not shown in Fig. 75), which gives the inclination of the neutral axis to the principal axis ox, is given by the following equation:

$$\tan \alpha = \frac{M_y I_x}{M_x I_y} \tag{91}$$

in which M_y denotes the component of bending moment with respect to the y axis, in.-lb, M_x denotes the component of bending moment with respect to the x axis, in.-lb, I_x denotes the moment of inertia with respect to the principal axis ox, in.^{4}, and I_y denotes the moment of inertia with respect to the principal axis oy, in.^{4}

Calculation of Fiber Stress. In order to calculate the fiber stresses for the cases shown in Figs. 75 and 76, it is first necessary to determine the position of the principal axes ox and oy. The components of bending moment with respect to the principal axes are then

$$M_x = M \cos \theta \tag{92}$$
$$M_y = M \sin \theta \tag{93}$$

If the distance of any fiber from the x axis is y and its distance from the y axis is x, the resultant unit fiber stress is

$$s = \pm \frac{M_x y}{I_x} \pm \frac{M_y x}{I_y} \tag{94}$$

in which s denotes the resultant unit fiber stress, psi, I_x denotes the principal moment of inertia for the axis x, in.^{4}, I_y denotes the principal moment of inertia for the axis y, in.^{4}, and the other notation is the same as before.

The method employed in Eq. (94) consists in calculating the unit fiber stresses produced by the separate components of moment and adding them algebraically. The two

[1] SINGER, "Strength of Materials," p. 400.
[2] SINGER, "Strength of Materials," pp. 400, 440.

terms in the formula would be given the plus or minus sign according as the unit stress produced is tensile or compressive.

Combined Axial and Flexural Stress. On p. 3–25, the case of axial combined with flexural stress was referred to. A cantilever beam, for instance, might be subjected to an end-compressive load P_1, parallel to the axis of the beam, and a vertical load P producing flexure in the usual way. An approximate solution of the problem, which neglects the deflection that may occur in the beam, consists in calculating separately the unit compressive stress due to P_1 and the flexural stress due to P and adding them algebraically. In this case, the unit stress due to P_1 is P_1/A, and the unit stress due to P is Mc/I. At the bottom fiber at the wall, the two unit stresses would be compressive and would be added. At the top fiber, the two unit stresses would be of opposite sign and would be subtracted.

If a single end load P on a cantilever beam is not vertical but is inclined at an angle θ with the axis of the beam, then one component of P will produce axial stress and the other component flexural stress. If the force is resolved into a vertical and a horizontal component at the axis of the beam, the component producing axial stress is $P \cos \theta$, and the component producing flexural stress is $P \sin \theta$.

The above method gives an approximate solution of the maximum stresses, which is sufficiently correct in many cases. More accurate solutions may be found in textbooks on strength of materials.

The handbook of the American Institute of Steel Construction specifies the following formula to be used for the case of combined stresses. Members subjected to both axial and bending stresses shall be so proportioned that the quantity

$$\frac{f_a}{F_a} + \frac{f_b}{F_b} \text{ shall not exceed unity}$$

in which F_a = axial unit stress that would be permitted if only axial stress existed
F_b = bending unit stress that would be permitted if only bending stress existed
f_a = axial unit stress (actual) = axial stress divided by area of member
f_b = bending unit stress (actual) = bending moment divided by section modulus of member

Stress Concentration. The effect of abrupt changes of cross section has been mentioned in discussing repeated stresses. The stress-concentration factor K is defined as the ratio of the maximum to the average stress in the section.

$$K = \frac{S_m}{S} = \text{stress-concentration factor}$$

where S_m = maximum stress in the section
S = average stress in the section

Stress-concentration factors have been determined by photoelastic methods, reported by Max M. Frocht.[1] Curves are shown for the effect on the stress-concentration factor of grooves and fillets of varying depth, and of holes, in axial tension and compression. Shown also are the effects of grooves and fillets of varying depth in pure bending.

COLUMNS

Compression members may be divided into three classes: (1) those which are so short that practically no bending takes place and the stress is uniformly distributed over the cross section; (2) those which are so long that the bending effect is the most significant action; and (3) those which are intermediate in length, so that the resultant maximum

[1] Factors of Stress Concentration Photoelastically Determined, *Trans. ASME*, vol. 57, p. A67, 1935.

stress is due partly to bending and partly to direct stress. The word *column* is usually restricted to long compression members that would fall into Classes 2 and 3, above.

Slenderness Ratio. It will be found that the load which a column can carry is largely influenced by its slenderness, which is measured by the *slenderness ratio*, i.e., the ratio of the length of the column in inches to the least radius of gyration of the column cross section in inches.

End Conditions. The end conditions of a column affect the load that can be carried, by the restraint that the ends offer in preventing the bending of the column. Figure 77 shows various end conditions of columns. Round ends, as shown in Fig. 77*a*, do not restrain the column at the ends. Figure 77*b* shows a pin-ended column bending in a plane perpendicular to the axis of the pins. In this case, the restraint would be due to the friction at the pins. For bending in a plane parallel to the axis of the pins (Fig. 77*c*), the column would be very considerably restrained and would approach the square-ended or fixed-ended condition. Square ends or flat ends (Fig. 77*d*) offer considerable restraint at the ends, and fixed ends (Fig. 77*e*) offer even greater restraint than square ends.

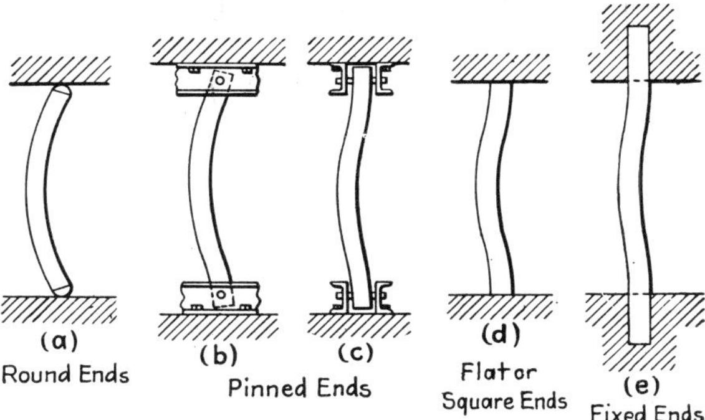

FIG. 77. End conditions of columns.

Euler's Formula. It can be shown[1] that Euler's formula for long columns is the appropriate one to apply to columns of Class 2, as follows:

$$P = \frac{mEI}{l^2}$$

or

$$\frac{P}{A} = \frac{mE}{(l/r)^2} \tag{95}$$

in which P denotes the maximum load that the column will carry, lb, E denotes the modulus of elasticity in tension and compression, psi, I denotes the least moment of inertia of the cross section, in.^{4}, r denotes the least radius of gyration of the cross section, in., l denotes the length of the column, in., and m denotes a constant depending upon end conditions.

Experiments have shown that the constant m has an average value of about 10 for round-ended columns, 15 for pin or hinged ends, and 30 for square or flat ends.

It will be noted that Eq. (95) does not determine the unit stress that exists in the column; and, indeed, the derivation of the formula shows that for an axially loaded column neither the unit stress nor the deflection of the column can be computed mathematically. When the loads and deflections are measured in an experiment on a column,

[1] BOYD and FOLK, "Strength of Materials," 5th ed., p. 251. SINGER, "Strength of Materials," p. 350

it is possible to compute the unit stress. Since Euler's formula gives the ultimate load on the column, it must be divided by a proper factor of safety for design work.

Rankine Formula. Most of the columns used in bridges and buildings fall into Class 3, and the formulas that are used in discussing such columns have in them experimental constants determined from tests on columns. The Rankine-type formula is one of the oldest that has been used in discussing column strength. The column formula used by the American Institute of Steel Construction is of the Rankine type and takes the following form:

$$\frac{P}{A} = \frac{18{,}000}{1 + \dfrac{1}{18{,}000}\left(\dfrac{l}{r}\right)^2} \tag{96}$$

This formula is a safe-load formula and is limited to values of slenderness ratio of 120 to 200 for secondary members.

Straight-line Formula. One of the simplest column formulas, which was very commonly used in the United States in the past, is the so-called "straight-line" formula, first suggested in 1886 by T. H. Johnson. Johnson found that a straight line made tangent to Euler's curve for long columns fitted experimental points quite well. The following formulas for duralumin are of the straight-line type. They are recommended by the Aluminum Company of America for alloy 17ST or 25ST.

For fixed ends,

$$\frac{P}{A} = 15{,}000 - 61.5\,\frac{l}{r} \qquad \frac{l}{r} < 162 \tag{97}$$

$$\text{Euler}\ \frac{P}{A} = \frac{132{,}000{,}000}{(l/r)^2} \qquad \frac{l}{r} > 162 \tag{98}$$

For round ends,

$$\frac{P}{A} = 15{,}000 - 123\,\frac{l}{r} \qquad \frac{l}{r} < 81 \tag{99}$$

$$\text{Euler}\ \frac{P}{A} = \frac{33{,}000{,}000}{(l/r)^2} \qquad \frac{l}{r} > 81 \tag{100}$$

Parabolic Formula. When a parabola, instead of a straight line, is made tangent to Euler's curve, a column formula is obtained which fits experimental results very satisfactorily. The formulas recommended by the ASCE Special Committee on Column Research are of this type:

For structural steel, flat ends,

$$\frac{P}{A} = 36{,}000 - 0.36\left(\frac{l}{r}\right)^2 \qquad \frac{l}{r} < 224 \tag{101}$$

$$\frac{P}{A} = \frac{900{,}000{,}000}{(l/r)^2} \qquad \frac{l}{r} > 224 \tag{102}$$

For structural steel, pin ends,

$$\frac{P}{A} = 36{,}000 - 0.72\left(\frac{l}{r}\right)^2 \qquad \frac{l}{r} < 158 \tag{103}$$

$$\frac{P}{A} = \frac{450{,}000{,}000}{(l/r)^2} \qquad \frac{l}{r} > 158 \tag{104}$$

The above formulas are for ultimate loads.

The American Institute of Steel Construction column formulas are a combination of the parabolic and Rankine types.

For main and secondary members,

$$\frac{P}{A} = 17{,}000 - 0.485\left(\frac{l}{r}\right)^2 \qquad \frac{l}{r} > 120 \tag{105}$$

For bracing and secondary members,

$$\frac{P}{A} = \frac{18{,}000}{1 + \dfrac{1}{18{,}000}\left(\dfrac{l}{r}\right)^2} \qquad \frac{l}{r} < 120 \tag{106}$$

For main members,

$$\frac{P}{A} = \frac{18{,}000}{1 + \dfrac{1}{18{,}000}\left(\dfrac{l}{r}\right)^2}\left(1.6 - \frac{\dfrac{l}{r}}{200}\right) \qquad \frac{l}{r} > 120$$

These formulas give the safe loads.

Cast-iron Columns. J. B. Johnson's formula for the ultimate strength of cast-iron columns is as follows:

$$\frac{P}{A} = 34{,}000 - 88\frac{l}{r} \tag{107}$$

The New York formula is a satisfactory one for calculating the safe loads for cast-iron columns:

$$\frac{P}{A} = 9{,}000 - 40\frac{l}{r} \qquad \frac{l}{r} < 70 \tag{108}$$

Criterion for Columns. It is usually necessary to determine whether a column falls into Class 2 or Class 3. The criterion for slenderness ratio that is applied for the straight-line–Euler combination is

$$\frac{l}{r} = \sqrt{\frac{3mE}{s}} \tag{109}$$

and for the parabolic-Euler combination is

$$\frac{l}{r} = \sqrt{\frac{2mE}{s}} \tag{110}$$

in which the notation is the same as before, and s denotes an experimental constant based on the yield point or ultimate compressive strength of the material.

The slenderness ratio given by Eq. (109) is the value at which Johnson's straight-line formula is tangent to Euler's curve for long columns, and Eq. (110) is the value at which the parabolic formula is tangent to Euler's curve. If the slenderness ratio of a given column is greater than that given by Eq. (109) or (110), it falls into Class 2 and should be discussed by means of the Euler formula. If the slenderness ratio is less than that given by Eq. (109) or (110), the column falls into Class 3 and should be discussed by means of a straight-line or a parabolic formula.

Because of the difficulty of estimating the restraint caused by various end conditions, the possibility of initial eccentricity in the column, and the further possibility of eccentricity in the position of the load, it is customary for structural designers to make

use of one set of end conditions. The usual practice is to assume that steel columns are pinned at the ends and that cast-iron and wood columns have flat ends. However, the Forest Products Laboratory formulas for wood columns, to be given later, are based on round-end conditions.

Wood Columns. The Forest Products Laboratory of the U.S. Department of Agriculture is the best source of information on the strength properties of wood columns. The recommendations of the laboratory may be found in "Wood Handbook," Handbook No. 72, 1955. The FPL column formula is a fourth-power parabolic equation made tangent to Euler's curve, as follows:

$$\frac{P}{A} = C\left[1 - \frac{1}{3}\left(\frac{l}{Kd}\right)^4\right] \tag{111}$$

in which C is the allowable compressive strength for short columns, psi, d is the lesser dimension of a rectangular cross section, in., and K denotes the largest ratio of l/d for which the column formula is appropriate. The factor of safety in these formulas is approximately 4 for short prisms and between 3 and 4 for longer members.

These formulas are used for values of l/d falling between the following limits:

$$\frac{l}{d} > 10 < K \tag{112}$$

The limit K is determined from the following formula:

$$K = \frac{\pi}{2}\left(\frac{E}{6C}\right)^{1/2} \tag{113}$$

The formulas have been so arranged that the coefficient of d is equal to the limit K. Thus a glance at the formula shows immediately the maximum value of l/d that may be employed.

For values of l/d greater than that given by Eq. (113), the following Euler formula, employing a factor of safety of 3, is used;

$$\frac{P}{A} = \frac{\pi^2 E}{36(l/d)^2} \tag{114}$$

Equations (111) and (114) are not to be used for l/d greater than 50.

The values of C, K, and E to be used in Eqs. (111) and (113) are given in Table 10. These values are for continuously dry locations.

Taking as an example the values for dense Douglas fir in Table 10, Eq. (111) becomes

$$\frac{P}{A} = 1{,}300\left[1 - \frac{1}{3}\left(\frac{l}{22.5d}\right)^4\right] \tag{115}$$

The above formulas for wood columns are to be used for rectangular cross sections in which d is the smaller dimension. For other cross sections, the formulas may be modified by substituting $0.289l/r$ for l/d, r being the least radius of gyration.

Typical Column Problems. *Problem* 1: *Safe Loads.* Determine the safe load that can be carried by a 10-in. 35-lb American standard I beam, 9 ft long.

The handbook shows that the least radius of gyration is 0.91, giving a slenderness ratio of 119. The area of the cross section is 10.22 sq in. Using Eq. (96):

$$P = \frac{10.22 \times 18{,}000}{1 + \dfrac{119^2}{18{,}000}} = 103{,}000 \text{ lb}$$

TABLE 10. VALUES OF C, K, AND E

Grades and species	C	K	E
Dense Douglas fir (coast and inland)	1,300	22.5	1,600,000
Dense longleaf or dense shortleaf southern pine	1,300	22.5	1,600,000
Tidewater red cypress	1,200	20.3	1,200,000
Close-grained Douglas fir (coast)	1,200	23.4	1,600,000
Dense longleaf or dense shortleaf southern pine	1,200	23.4	1,600,000
Close-grained redwood	1,200	20.3	1,200,000
Douglas fir (coast)	1,100	24.4	1,600,000
Oak	1,100	23.7	1,500,000
Close-grained redwood	1,100	21.2	1,200,000
Port Orford cedar	1,000	22.2	1,200,000
Tidewater red cypress	1,000	22.2	1,200,000
Oak	1,000	24.8	1,500,000
Dense longleaf southern pine	1,000	25.6	1,600,000
Close-grained redwood	1,000	22.2	1,200,000
Port Orford cedar	900	23.4	1,200,000
Dense shortleaf southern pine	900	27.0	1,600,000
Western red cedar	800	22.7	1,000,000

Problem 2: *Design.* Determine the size of the I beam, 10 ft long, that will carry safely a load of 150,000 lb.

Using the AISC formula:

$$\frac{P}{A} = \frac{18{,}000}{1 + \dfrac{1}{18{,}000}\left(\dfrac{l}{r}\right)^2}$$

It is evident that there are two unknowns A and r in the formula and that for the I-beam shape there is no simple relation between these quantities. The result is that the problem must be solved by trial-and-error methods. The formula shows that short prisms could carry a unit load of 18,000 psi. The load of 150,000 lb divided by 18,000 results in a required area of 8.3 sq in. This would be the area required if the member were a short prism and may be looked upon as the minimum limit of area. Since the I-beam shape is particularly weak for bending with respect to the axis parallel to the web, the first trial will be made with an area considerably greater than 8.3 sq in. A 12-in. 40.8-lb I beam has an area of 11.84 sq in. and a least radius of gyration of 1.08 in., the slenderness ratio being thus 111. This will be tried in the formula

$$P = \frac{11.84 \times 18{,}000}{1 + \dfrac{111^2}{18{,}000}} = 127{,}000 \text{ lb}$$

This I beam is somewhat too small. If the radius of gyration is about the same for the next I beam that is tried, the area should be increased by the ratio of 150,000: 127,000. This would indicate a required area of about 14 sq in. A 12-in. 50-lb I beam, with an area of 14.57 sq in. and a radius of gyration of 1.05, will be tried. The corresponding slenderness ratio is 114.

$$P = \frac{14.57 \times 18{,}000}{1 + \dfrac{114^2}{18{,}000}} = 153{,}000 \text{ lb}$$

This size will be satisfactory.

Problem 3: *Design.* Determine the size of a rectangular wood column of dense Douglas fir, 15 ft long, to carry safely a load of 90,000 lb.

Equation (115) will be used. The load of 90,000 being divided by 1,300, it is found that a short prism would require an area of 69 sq in. Since an 8-in. by 8-in. cross section will be too small, a 10-in. by 10-in. size will be used for the first trial. The actual dressed dimensions of 9.5 in. by 9.5 in. will be used.

$$P = 9.5 \times 9.5 \times 1{,}300 \left[1 - \frac{(15 \times 12)^4}{3(22.5 \times 9.5)^4} \right] = 97{,}500 \text{ lb}$$

The value of l/d for this column is 18.9, which indicates that the value of l/d is within the limit of $K = 22.5$ given in the formula. The formula is appropriate, and the size is satisfactory. If a nominal size 8 in. by 10 in. is tried, it will be found to be too small.

In all column problems it is necessary to calculate the slenderness ratio in order to determine whether it is within the limits for which the formula is appropriate.

Eccentric Loads on Columns. In the discussion of axially loaded columns, it was stated that the formula did not make possible the calculation of either the unit stress in the column or the deflection of the column. However, when the load is applied eccentrically to the column, it is possible to calculate both the unit stress and the deflection.

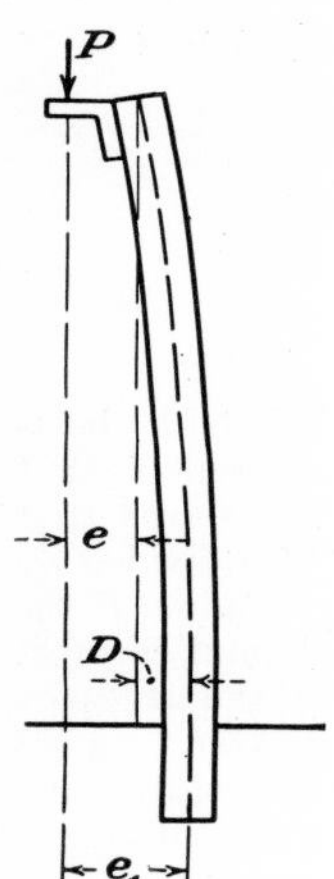

FIG. 78. Eccentrically loaded column.

Figure 78 shows a column that has the lower end fixed and the upper end free. The eccentricity of the load at the top of the column is e, the maximum eccentricity is e_1, and the maximum deflection is D. It can be shown [1] that the expressions for the eccentricity e_1 and the maximum unit stress s are given by the following formulas:

$$e_1 = e \sec \left(\frac{Pl^2}{EI}\right)^{1/2} \tag{116}$$

$$s = \frac{P}{A} + \frac{Mc}{I} = \frac{P}{A}\left(1 + \frac{e_1 c}{r^2}\right) \tag{117}$$

$$s = \frac{P}{A}\left[1 + \frac{ec}{r^2} \sec \left(\frac{Pl^2}{EI}\right)^{1/2}\right] \tag{118}$$

$$s = \frac{P}{A}\left\{1 + \frac{ec}{r^2} \sec \left[\frac{P}{AE}\left(\frac{l}{r}\right)^2\right]^{1/2}\right\} \tag{119}$$

in which s denotes the maximum unit stress on the concave side of the column, psi; P denotes the load on the column, lb; A denotes the area of the column cross section, in.^{2}; e denotes the eccentricity of the load at the top of the column, in.; c denotes the distance from the center of the column to the extreme fiber on the concave side, in.; r denotes the radius of gyration of the column cross section, with respect to the axis about which bending occurs, in.; E denotes the modulus of elasticity in tension and compression, psi; and l denotes the length of the column, in. The foregoing formulas involve the secant of an angle, which is given in radians.

It is evident from Fig. 78 that the maximum deflection D of the column is

$$D = e_1 - e \tag{120}$$

For both ends round, as shown in Fig. 79, the formulas for e_1 and s are as follows:

[1] BOYD and FOLK, "Strength of Materials," 5th ed., p. 249. SINGER, "Strength of Materials," p. 372.

$$e_1 = e \sec \left(\frac{Pl^2}{4EI}\right)^{1/2} \tag{121}$$

$$s = \frac{P}{A}\left[1 + \frac{ec}{r^2} \sec \left(\frac{Pl^2}{4EI}\right)^{1/2}\right] \tag{122}$$

$$s = \frac{P}{A}\left\{1 + \frac{ec}{r^2} \sec \left[\frac{P}{4EA}\left(\frac{l}{r}\right)^2\right]^{1/2}\right\} \tag{123}$$

Eccentric Loads on Prisms. A short prism, such as is shown in Fig. 80, may be subjected to a tensile or compressive load P, acting at a distance e from the axis of the prism. The action of this load is equivalent[1] to a direct axial load P and a bending moment Pe.

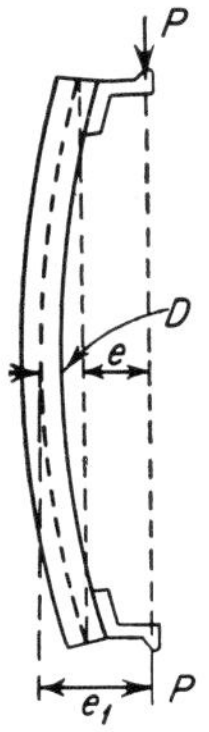

FIG. 79. Eccentrically loaded column.

The direct axial load produces a uniformly distributed stress on the cross section of the prism, which equals

$$s = \frac{P}{A}$$

The bending moment Pe produces a flexural stress which, according to the fiber stress formula, equals

$$s = \frac{Mc}{I} = \frac{Pec}{I}$$

The resultant unit stress is

$$s' = \frac{P}{A} \pm \frac{Pec}{I} \tag{124}$$

$$s' = \frac{P}{A}\left(1 \pm \frac{ec}{r^2}\right) \tag{125}$$

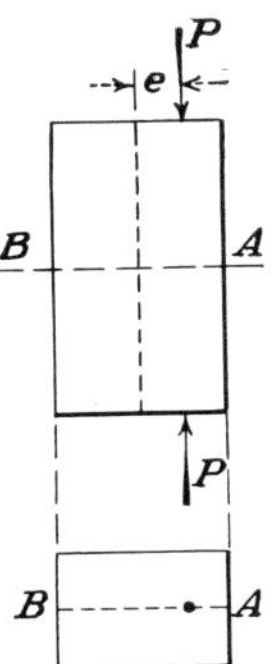

FIG. 80. Eccentrically loaded prism.

in which s' denotes the resultant unit stress, psi; P denotes the load on the prism, lb; A denotes the area of cross section, sq in.; e denotes the eccentricity of the load with respect to the axis about which bending occurs, in.; c denotes the distance from the axis to the extreme fiber, in.; I denotes the moment of inertia of the cross section with respect to the axis about which bending occurs, in.^{4}; and r denotes the radius of gyration corresponding to I, in. For the case shown in Fig. 80, the flexural unit stress would be compressive at A and tensile at B. The resultant stress s' at A would therefore be obtained by using the plus sign in Eq. (125) and the unit stress at B by using the minus sign.

Double Eccentricity. The case of an eccentrically loaded prism which has been discussed above assumes that the load line intersects one of the principal axes of the cross section. In such cases, the other principal axis is the neutral axis for calculating the fiber stresses. Figure 81 shows a case in which the load is eccentric with respect to both principal axes.

In this case, the moment producing bending about the x axis is Py_1, called M_x, and the moment producing bending about the y axis is Px_1, called M_y. The unit stress at any point, such as A,

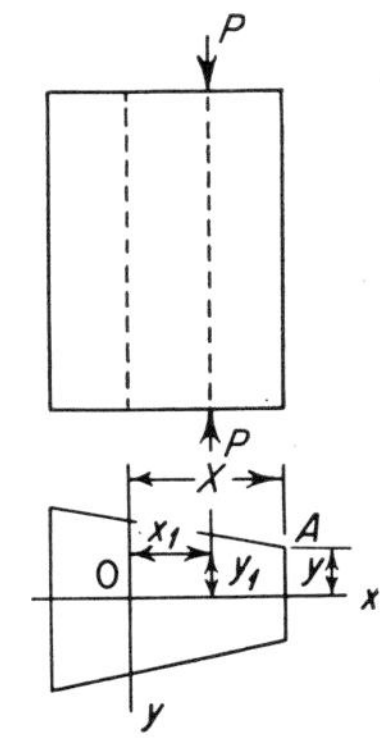

FIG. 81. Eccentrically loaded prism.

[1] BOYD and FOLK, "Strength of Materials," 5th ed., p. 233. SINGER, "Strength of Materials," p. 291.

at distances x and y from the axes will be

$$s = \frac{P}{A} + \frac{M_y x}{I_y} + \frac{M_x y}{I_x} \tag{126}$$

in which the notation is the same as before, and I_y and I_x are the moments of inertia with respect to the principal axes y and x respectively.

TORSION

Torsion Formula. When a machine or structural element is subjected to a twisting moment as shown in Fig. 82, the member is said to be subjected to *torsion*, and the stresses developed by torsion are shearing stresses.

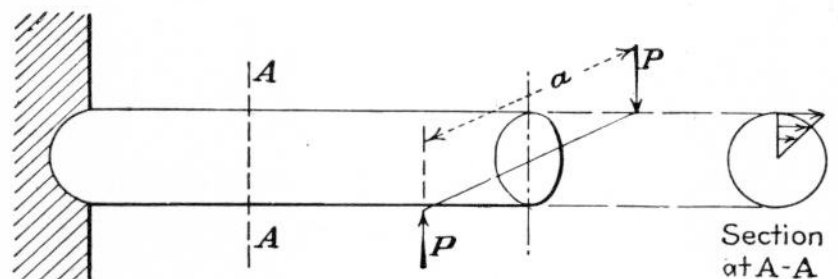

FIG. 82. Torsion in a shaft.

Experiment has shown that a longitudinal element on the surface of a shaft will be twisted into a helix, and at any cross section the unit shear stresses, which are produced by the external *twisting moment* or *torque*, vary as the distance from the axis of the shaft.

It can be shown [1] that the relation existing between torque and unit stress is as follows:

$$T_m = \frac{S_s J}{r} \tag{127}$$

in which T_m denotes the torque or twisting moment, in.-lb, J denotes the polar moment of inertia of the cross section, in.^{4}, r denotes the distance from the axis of the shaft to the outer fiber, in., and s_s denotes the unit stress on the outer fiber, psi.

It should be noted that this formula is correct only for solid or hollow circular shafts, which are the most commonly used sections for transmitting torque in machines. Textbooks on strength of materials give formulas for the ellipse and the rectangle; but for complicated sections, like rolled I beams, there are no theoretical formulas, and recourse must be had to experiment or to empirical formulas based on experiment.

Just as was the case for the fiber-stress formula for beams, Eq. (127) may be used to calculate the unit shear stress at any point of the cross section of a shaft. If r' is the distance from the axis of the shaft to the fiber where the unit stress is s_s', then

$$T_m = \frac{s_s' J}{r'} \tag{128}$$

Angle of Twist. Another important formula for torsion problems is that which gives the angle of twist of the shaft between two cross sections a distance l apart. The formula [2] is

$$\theta = \frac{57.3 T_m l}{E_s J} \tag{129}$$

in which θ denotes the angle of twist, deg; T_m denotes the torque acting on the shaft, in.-lb; l denotes the length over which the twist occurs, in.; E_s denotes the modulus of elasticity in shear, psi; and J denotes the polar moment of inertia of the cross section, in.^{4}

[1] BOYD and FOLK, "Strength of Materials," 5th ed., p. 100. SINGER, "Strength of Materials," p. 68.
[2] BOYD and FOLK, "Strength of Materials," 5th ed., p. 98. SINGER, "Strength of Materials," p. 68.

Torque and Horsepower. The most common case of torsion is that of a shaft transmitting horsepower. It can be shown [1] that the following formula gives the relation between torque and horsepower:

$$T_m = \frac{63{,}024\ hp}{N} \tag{130}$$

in which T_m denotes the torque, in.-lb, hp denotes horsepower, and N denotes revolutions per minute.

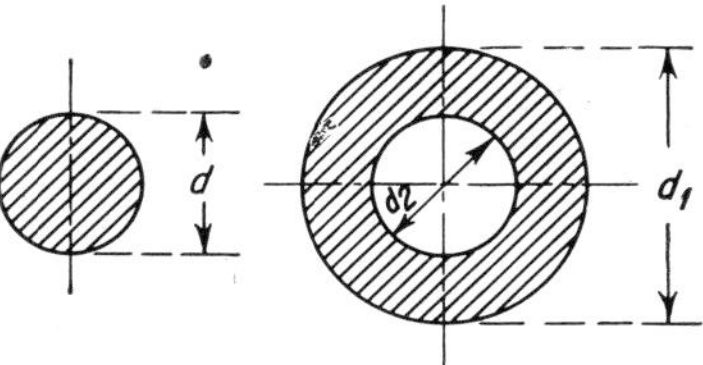

Fig. 83. Solid and hollow shafts.

Solid and Hollow Shafts. Large shafts, especially those for marine work, are often made hollow. Hollow shafts are stronger per pound of material than solid shafts, because the material near the axis of a solid shaft has low stresses acting on it and is therefore relatively ineffective in producing resisting moment. Hollow shafts may also be forged on a mandrel, the material being thus rendered more homogeneous than would be possible in a solid shaft.

Referring to Fig. 83, the expressions for J and J/r are as follows:

For a solid shaft,

$$J = \frac{\pi d^4}{32} \tag{131}$$

For a solid shaft,

$$\frac{J}{r} = \frac{\pi d^3}{16} \tag{132}$$

For a hollow shaft,

$$J = \frac{\pi(d_1^4 - d_2^4)}{32} \tag{133}$$

For a hollow shaft,

$$\frac{J}{r} = \frac{\pi(d_1^4 - d_2^4)}{16d_1} \tag{134}$$

If a hollow shaft is to be equal in strength to a solid shaft, the resisting moment of the one must be equal to that of the other; hence:

$$\frac{\pi(d_1^4 - d_2^4)}{16d_1} \times s_s = \frac{\pi d^3}{16} \times s_s' \tag{135}$$

If the two shafts are of the same material so that s_s is equal to s_s', then:

$$d_1^3 - \frac{d_2^4}{d_1} = d^3 \tag{136}$$

Shaft Couplings. Shaft couplings are fastenings used to join lengths of shafting together, so that motion from one section may be transmitted to another. A *flange coupling* is adapted for severe service; when pressed and keyed onto the shaft, it is practically a part of the shaft, and its use ensures ease of alignment and permanence. This form of coupling is used almost exclusively on shaft sizes 4 in. in diameter and larger. The coupling consists of two cast-iron flanges which are bolted together, each part being keyed to a shaft end. Alignment is secured by extending one shaft end a short distance

[1] Singer, "Strength of Materials," p. 69.

beyond one part and into the second part of the coupling, as shown in Fig. 84*a*; or the faces may be provided with a concentric ring and recess as shown in Fig. 84*b*.

In designing the coupling, it is usually assumed that the unit stress on the bolts is uniformly distributed over their cross-sectional areas. The resisting moment of the coupling is then

$$T_m = \frac{\pi d_1^2}{4} s_s' R n \tag{137}$$

in which T_m denotes the resisting moment, in.-lb, s_s' denotes the allowable unit shear stress in the bolts, psi, d_1 denotes the diameter of the bolts, in., R denotes the distance

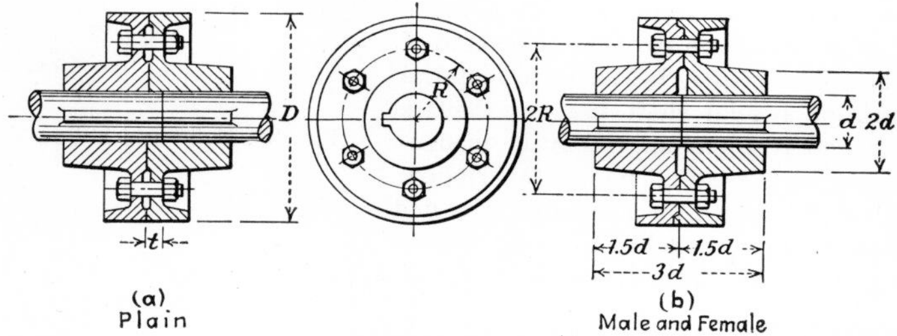

FIG. 84. Shaft couplings.

from the axis of the shaft to the bolt centers, in., and n denotes the number of bolts.

If the strength of the coupling is to be equal to the strength of the shaft,

$$\frac{\pi d_1^2}{4} s_s' R n = \frac{s_s \pi d^3}{16}$$

and

$$n = \frac{s_s d^3}{4 d_1^2 s_s' R} \tag{138}$$

in which the notation for the coupling is the same as before, s_s denotes the allowable unit stress in the solid shaft, psi, and d denotes the diameter of the solid shaft, in.

Rupture by Torsion. If the formula for the unit stress in a shaft $s_s = Tr/J$ is applied when the shaft is subjected to its maximum torque before rupture, the value of the computed s_s is fictitious, because the stress distribution is no longer linear. This value of s_s is called the *modulus of rupture* in torsion. It has the same uses as described for the modulus of rupture for beams. Experiments show that the shape of the torsion specimen affects the value of the modulus of rupture obtained, so that, when a solid and a hollow specimen of the same material are tested to destruction, the modulus of rupture for the solid specimen may be quite different from that for the hollow one. This indicates that computations involving the use of modulus of rupture should be made with judgment.

Torsion Combined with Bending. One of the common cases of design for shafting is that of a shaft which is simultaneously subjected to a twisting moment and a bending moment. This gives rise to a combination of shear and flexure stresses, the shear stresses being due to the torque and the flexure stresses being due to the bending moment. This combination of stresses was discussed on p. 3–25, and the resultant normal and shear stresses were shown to be as follows:

$$s_n' = \frac{s}{2} + \sqrt{\left(\frac{s}{2}\right)^2 + s_s^2} \tag{139}$$

$$s_s' = \sqrt{\left(\frac{s}{2}\right)^2 + s_s^2} \tag{140}$$

These equations may be rewritten as follows, since

$$s_s = \frac{Tr}{J} = \frac{Tr}{2I}$$

and

$$s = \frac{Mc}{I} = \frac{Mr}{I}$$

$$s_n' \frac{I}{r} = \frac{1}{2}[M + (M^2 + T^2)^{1/2}] \qquad (141)$$

$$s_s' \frac{J}{r} = (M^2 + T^2)^{1/2} \qquad (142)$$

In Eqs. (141) and (142), the term $\frac{1}{2}[M + (M^2 + T^2)^{1/2}]$ may be called the equivalent bending moment, and the term $(M^2 + T^2)^{1/2}$ may be called the equivalent torque.

It should be carefully noted that, when the allowable unit normal stress for a shaft is, say, 16,000 psi, it is s_n' which is to be taken as 16,000 and not s. Also, when the allowable unit shear stress is 12,000 psi, it is s_s' which is 12,000 and not s_s.

If the allowable unit stresses in a design problem are 16,000 for s_n' and 12,000 for s_s', these values would be substituted in Eqs. (141) and (142) respectively and the diameters determined. Of the two diameters determined by the two formulas, the larger would govern the design. It can be shown that, when the ratio of the allowable unit stress s_n' to the allowable unit stress s_s' equals 2, the diameter as determined by the shearing unit stress will govern the design. In the Code for Design of Transmission Shafting, recommended by the ASME and approved by the American Engineering Standards Committee in 1927, the ratio of the allowable s_n' to the allowable s_s' is equal to 2, so that shear governs the design.

Shock and Fatigue Factors. A machine part subjected to shock or fatigue is much more likely to fail than a part subjected to steady loads only. The following factors, K_t for torque and K_m for bending moment, are recommended in the above-mentioned code for various types of loading, and they are to be applied in every case to the twisting and bending moments as shown in Table 11.

TABLE 11. SHOCK AND FATIGUE FACTORS

Nature of loading	Values for K_m	Values for K_t
Stationary shafts:		
Gradually applied load	1.0	1.0
Suddenly applied load	1.5 to 2.0	1.5 to 2.0
Rotating shafts:		
Gradually applied or steady loads	1.5	1.0
Suddenly applied loads, minor shocks only	1.5 to 2.0	1.0 to 1.5
Heavy shocks	2.0 to 3.0	1.5 to 3.0

Torsion Combined with Axial Stress. Propeller shafts may be subjected to compression and torsion, and vertical turbine shafts may be subjected to compression and torsion or tension and torsion. The resultant normal and shear stresses produced by such combinations may be found by means of Eqs. (139) and (140). For a solid shaft, the value of s, the axial stress, is $4P/\pi d^2$; and the value of s_s, the shear stress, is $16T/\pi d^3$.

Substituting these values in Eqs. (139) and (140), we have

$$s_n' = \frac{2}{\pi d^2}\left\{P + \left[P^2 + \left(\frac{8T}{d}\right)^2\right]^{1/2}\right\} \tag{143}$$

$$s_s' = \frac{2}{\pi d^2}\left[P^2 + \left(\frac{8T}{d}\right)^2\right]^{1/2} \tag{144}$$

SPRINGS

There are various applications of springs to structures and machines, for it is possible to control the distortion of the spring under a known load. Springs may be used to absorb energy, as in railway cars and in automobiles; to store energy, as in clocks and other spring motors; to measure forces, as in spring balances; to hold machine parts in place, as in cams and their followers; and to produce normal pressure in friction devices, as in automobile clutches. The two types of spring most commonly used are *helical* and *laminated* springs.

Helical Springs. The helical spring is commonly made by wrapping a rod of circular cross section around a cylindrical surface, the axis of the rod being thus formed into a helix, as shown in Fig. 85.

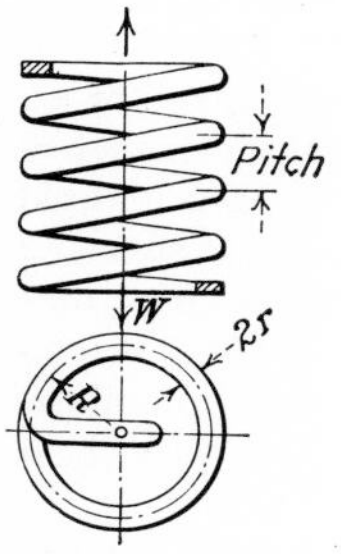

If the spring is supporting an axial load W and the mean radius of the coil is R, then it is evident that every cross section of the coil is subjected to a twisting moment of WR. The stress produced on the cross sections is a torsion stress, and its value may be found from the torsion formula

$$WR = \frac{s_s J}{r} \tag{145}$$

For a rod or wire of circular cross section,

$$s_s = \frac{2WR}{\pi r^3} \tag{146}$$

Fig. 85. Helical spring.

in which W denotes the axial load on the spring, lb, R denotes the mean radius of the coil, in., r denotes the radius of the wire or rod, in., J denotes the polar moment of inertia of the wire or rod cross section, in.^{4}, and s_s denotes the unit shear stress in the wire or rod, psi.

The above analysis may be used for closely coiled springs in which the obliquity of the helix may be neglected. When the coils are open so that the obliquity is appreciable, there will be produced a bending moment as well as a twisting moment.[1]

To determine the deflection of a close-coiled spring, consider the action as the twist of a wire of a length equal to the length of the spring, which permits the load W to descend. The angle of twist under torsion, from Eq. (129) (p. 3–72), is

$$\theta = \frac{T_m L}{E_s J} \qquad \text{in radians}$$

The amount d, which the load W will descend, is $R\theta$, and the length l is the length of the spring, or the number of coils n multiplied by $2\pi R$. Hence,

$$d = \frac{R \times WR \times n \times 2\pi R \times 2}{E_s \times \pi r^4} = \frac{4WR^3 n}{E_s r^4} \tag{147}$$

The formula shows that, for a given load and number of coils, the deflection varies as the

[1] Morley, "Strength of Materials," 3d ed., p. 291.

cube of R. A large value of R therefore means a weak, flexible spring, and a small value of R means a strong, stiff spring.

Design of Helical Springs.[1] It is well to use a rod or wire of the smallest diameter, consistent with the other requirements, since the wires of smaller diameter have a higher elastic limit. The torsional modulus of elasticity, however, is constant and may be taken as 12,000,000 psi for steel and 6,000,000 psi for brass. From the standpoint of strength, it can be shown that a square wire with a side equal to the diameter of a round wire is 6 percent stronger but has in it 22 percent more material. From the standpoint of economy, therefore, it is usually desirable to use a round cross section.

From tests made by E. T. Adams at Sibley College on governor springs made from steel wire, varying in diameter from ⅜ to ¾ in., it was found that the maximum safe stress for such springs was as follows:

$$s_s = 40{,}000 + \frac{7{,}500}{r} \tag{148}$$

Problem 1: A helical spring is 4 in. outside diameter and has eight effective coils of ½-in. round wire. The stress should not exceed 63,000 psi. What deflection may be expected with a load of 1,000 lb?

From Eq. (147),

$$d = \frac{4 \times W \times R^3 \times n}{E_s \times r^4} = \frac{4 \times 1{,}000 \times 1.75^3 \times 8}{12{,}000{,}000 \times 0.25^4} = 3.66 \text{ in.}$$

From Eq. (146), the unit stress will be

$$s_s = \frac{2 \times 1{,}000 \times 1.75}{\pi \times 0.25^3} = 71{,}300 \text{ psi}$$

This unit stress seems to be excessive. Using Eq. (148):

$$s_s = 40{,}000 + \frac{7{,}500}{0.25} = 70{,}000 \text{ psi}$$

This value indicates that the spring should probably not be subjected to a load so great as 1,000 lb.

Problem 2: Using wire of circular cross section, design a helical spring that will deflect 4 in. under a load of 1,000 lb. The wire is made of steel, and the unit stress shall not exceed 60,000 psi.

Substituting in Eqs. (146) and (147):

$$60{,}000 = \frac{2 \times 1{,}000 \times R}{\pi r^3}$$

$$4 = \frac{4 \times 1{,}000 \times R^3 n}{12{,}000{,}000 \times r^4}$$

It is desirable to have a whole number for n and convenient standard dimensions for r and R. Assuming $r = ¼$ in.:

$$R = \frac{60{,}000 \times \pi \times 0.25^3}{2 \times 1{,}000} = 1.47 \text{ in.}$$

[1] A discussion by A. M. Wahl on the use of more exact formulas for the design of heavy helical springs may be found in *Trans. ASME*, 1929, 1930, 1935.

Use 1.5 in. and solve for n:

$$n = \frac{4 \times 12{,}000{,}000 \times 0.25^4}{4 \times 1{,}000 \times 1.5^3} = 13.9 \text{ coils}$$

Use 14 coils. The outside diameter of the coils will be 3.5 in., and the spring may be formed on an arbor 2.5 in. in diameter.

Data on Helical Springs. It is evident from the above discussion that, when only the load W and the deflection d are specified, there are a large number of springs which might meet the requirements of a design.

High-class compression springs have the ends ground off square with the axis of the spring, as shown by Figs. 86*c* and *d*, and the height should not be more than three times

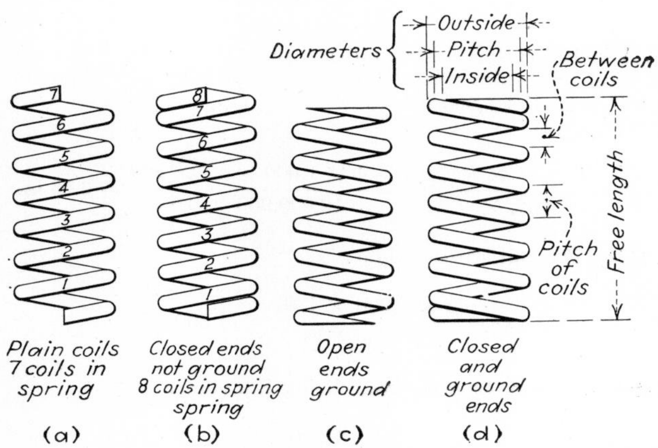

FIG. 86. Helical compression springs.

the outside diameter of the coil. Tension springs with end hooks and coils are shown in Figs. 87*a*, *b*, and *c*. Steel springs are often japanned and baked to protect the material from corrosion. Springs are heat-treated after forming.

Some of the special alloy steels used for springs, such as vanadium steels, have elastic limits ranging from 180,000 to 225,000 psi.

Brass is used for springs that must resist corrosion from moisture. These springs are more expensive than those made of steel, not only because brass weighs and costs more but also because the permissible unit stress for brass is smaller, and the springs must therefore be larger for the same capacity.

Phosphor-bronze wire used in helical springs may have a maximum allowable unit stress ranging from 30,000 to 40,000 psi.

Springs that act occasionally may carry a stress close to the elastic limit of the material; but when the push or pull is frequently repeated, a factor of safety must be used. The maximum unit torsional stress in a steel spring, closing the valve in a gas engine, should not exceed 40,000 psi, which means that it should have a factor of safety of at least 3.

Single-leaf Springs. Flat springs consisting of a single leaf may be used as cantilever beams or as beams with more than one support. The unit stresses that result will be tensile on one outer fiber and compressive on the other. Such springs may be designed

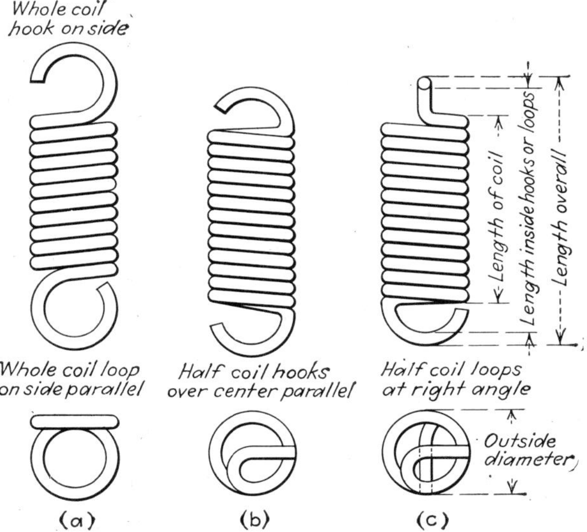

Fig. 87. Helical tension springs.

by means of the beam theory given in previous sections, and they may be of uniform cross section throughout or may be designed as beams of approximately uniform strength.

The deflection formulas for beams of uniform strength and rectangular cross section are given in Table 12 for some of the common cases.

Table 12. Deflections of Uniform-strength Beams

Kind of Beam and Loading	Maximum Deflection, in.
1. Simple beam, supported at the ends Concentrated load at center Constant breadth, varying depth	$d = \frac{P}{2bE}\left(\frac{l}{h}\right)^3$
2. Simple beam, supported at the ends Concentrated load at center Constant depth, varying breadth	$d = \frac{3P}{8bE}\left(\frac{l}{h}\right)^3$
3. Cantilever beam Concentrated load at free end Constant breadth, varying depth	$d = \frac{8P}{bE}\left(\frac{l}{h}\right)^3$
4. Cantilever beam Concentrated load at free end Constant depth, varying breadth	$d = \frac{6P}{bE}\left(\frac{l}{h}\right)^3$

In the foregoing table, d denotes the maximum deflection in inches, P denotes the load on the beam in pounds, E denotes the modulus of elasticity in tension or compression in pounds per square inch, b denotes the maximum breadth in inches, h denotes the maximum depth in inches, and l denotes the length of the beam in inches.

Problem: A flat steel spring is to carry a load of 500 lb and is to deflect 1 in. The length of the spring is 24 in., the maximum unit stress shall not exceed 50,000 psi,

the spring is supported at the ends and loaded with a concentrated load at the middle, and is to be of constant depth and varying breadth.

The material will be steel having a modulus of elasticity of 30,000,000 psi.

From the stress formula for beams,

$$s = \frac{Mc}{I}$$

and

$$M = \frac{Pl}{4}$$

$$50{,}000 = \frac{500 \times 24 \times 6}{4bh^2}$$

From Table 12,

$$d = \frac{3P}{8bE}\left(\frac{l}{h}\right)^3$$

$$1 = \frac{3 \times 500 \times 24^3}{8 \times b \times 30{,}000{,}000 \times h^3}$$

Solving for the breadth b in each case and putting these values equal to each other, we have,

$$\frac{500 \times 24 \times 6}{4 \times 50{,}000 \times h^2} = \frac{3 \times 500 \times 24^3}{8 \times 30{,}000{,}000 \times h^3}$$

$$h = 0.24 \text{ in.}$$
$$b = 6.25 \text{ in.}$$

The spring would have a maximum breadth of 6.25 in. at the center, tapering to a point at the ends, and the constant depth would be approximately ¼ in.

Laminated Springs. It is possible to design springs using several plates and thus obtain a given deflection for a certain load without having excessive maximum breadth of spring.

Figure 88 represents a laminated spring having five plates. The plates are assumed to have been cut from a single diamond-shaped plate, shown in Fig. 89, which is a beam

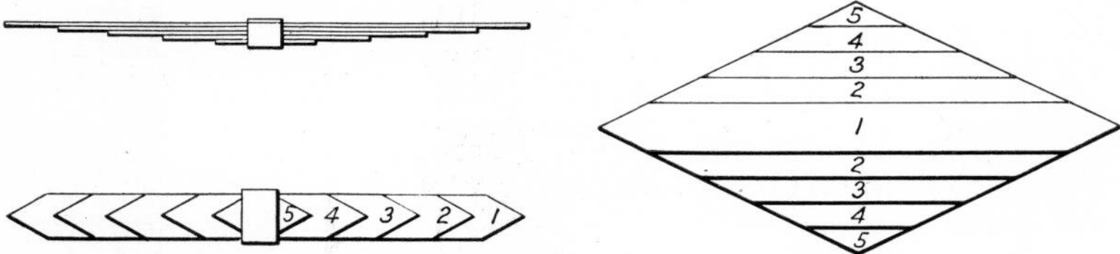

FIG. 88. Laminated spring.

FIG. 89. Laminated spring.

of uniform strength with constant depth and varying breadth. The spring in Fig. 88 will therefore be a beam of approximately uniform strength, the maximum deflection and unit stress of which will be the same as for the plate in Fig. 89.

The formulas for unit stress and deflection in terms of the new breadth and number of plates are

$$s = \frac{3Pl}{2nb'h^2} \tag{149}$$

$$d = \frac{3Pl^3}{8Enb'h^3} \tag{150}$$

in which n denotes the number of plates or leaves in the spring and b' denotes the breadth of the laminated spring in inches.

If we take the spring designed in the preceding section and replace it with five plates, the new width would be 6.25 in. divided by 5, or 1.25 in., which would make a much more practical design.

In laminated springs as actually used, the full-length leaf must usually have a square end by means of which it can be fastened to its supports. Sometimes the shorter leaves are also cut square at the ends, as shown in Fig. 90. This change makes no error so far

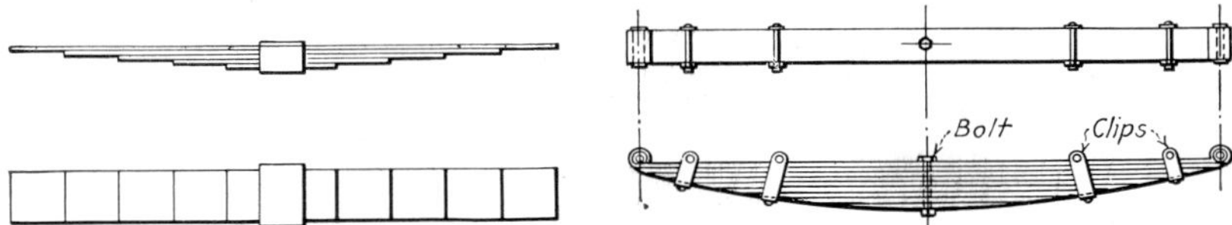

FIG. 90. Spring with square ends.

FIG. 91. Automobile spring.

as Eq. (149) is concerned, but the deflection as given by Eq. (150) might not be quite correct, for the ends of the leaves are not pointed. The ends of the shorter leaves are sometimes rounded and made thinner so as to approximate the pointed condition. Figure 91 shows an automobile spring with the top leaf bent at the ends to form shackles.

J. B. Peddle [1] has shown that the formula for the deflection of full-elliptic leaf springs is

$$d = \frac{4l^2S}{tE}K \tag{151}$$

and for semielliptic springs is

$$d = \frac{2l^2S}{tE}K \tag{152}$$

The notation is the same as before, and K is a factor that depends on the ratio r which is the ratio of the number of full-length leaves to the total number of leaves; i.e.,

$$r = \frac{\text{number of full-length leaves}}{\text{total number of leaves}} \tag{153}$$

$$K = \frac{1}{(1-r)^3}\left[\frac{1-r^2}{2} - 2r(1-r) - r^2\log_e r\right] \tag{154}$$

In Fig. 92a, the width w, where the leaves are held together by a band, is neglected, and the length of the spring l is taken as indicated. Peddle points out that it is necessary to have the points of the shortened leaves tapered in width or in thickness, or both, as shown in Fig. 92b, so as to make the transition from one leaf to the next gradual.

The unit stress s to be used in Eqs. (151) and (152) may be calculated from the following formula, the load P being the one indicated in Fig. 92a:

$$s = \frac{3lP}{nbh^2} \tag{155}$$ [2]

It is evident from Eqs. (151) and (152) that the deflection of the full-elliptic spring for the same load is double that of the semielliptic spring.

[1] *Am. Machinist*, Apr. 17, 1913, p. 645.
[2] Use l, as shown in Fig. 92.

The leaves of a laminated spring are often given an initial curvature, so that they will tend to straighten under the load and be straight when fully loaded. When this is done, the deflection due to full load would determine the curvature of the plates.

It will be noted that the full-length plate is a beam of constant cross section, whereas the shorter ones are not, with the result that the unit stress in the long plate will be greater than in the short plates. To overcome this defect, E. R. Morrison [1] has suggested that the shorter plates be given an additional initial curvature, equal to the difference in deflection which might be expected between the stiffer long plate and the shorter plates. After assembling the spring, the short plates would then have an initial

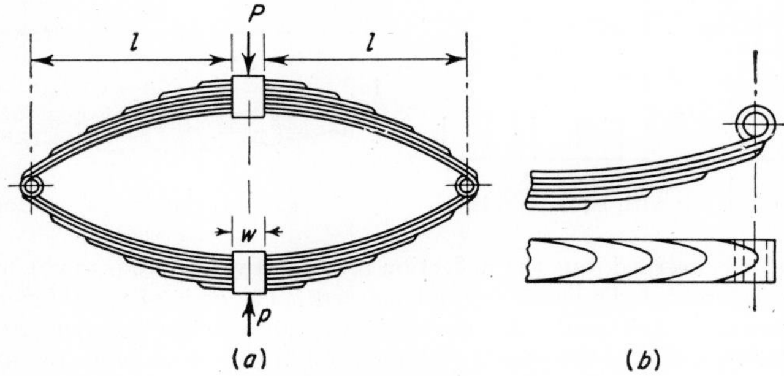

FIG. 92. Elliptical spring.

stress, which, in addition to the stress induced in them by the load, would cause a uniformity of stress in the spring unit.

Data on Leaf Springs. Figure 92*a* shows a full-elliptic spring, so called on account of its form. A semielliptic spring would mean the upper or lower half of the spring shown in the figure. Springs with rounded ends for the shorter leaves are used in automobiles, and springs with square ends are used for locomotives.

Springs subjected to suddenly applied loads are often provided with rubber blocks to prevent metal-to-metal shock, the rebound being absorbed by snubbers of various designs.

A high-grade alloy steel is used in the manufacture of elliptic springs, and the heat-treatment of the plates is an important part of the process of manufacture.

The maximum unit stress allowed in springs depends on the thickness of the plates. It is usually taken as 100,000 psi for $3/16$-in. plates and 80,000 psi for $3/8$-in. plates.

The rational design of all classes of springs should be checked by performance tests. This is especially true if the design contemplates the use of springs that are to function accurately with respect to load and deflection.

THICK CYLINDERS

When the ratio of the diameter to the thickness of a cylinder is 15 or greater, the maximum unit stress due to internal pressure is not very much greater than the average unit stress, and the assumption of uniform stress across the thickness of the cylinder wall may be safely made. This assumption cannot be made for *thick cylinders* under internal pressure, because there is a considerable difference between the maximum tensile unit stress on the inside fiber and the minimum stress on the outside fiber. Besides the circumferential tensile stress, there exists also a radial compressive stress. When a thick cylinder is subjected to external pressure, the circumferential stress is of course compressive.

[1] *Machinery*, January, 1910.

There are several formulas that are used to determine the stresses in thick cylinders, those of Barlow and Lamé having been used considerably by design engineers. Lamé's formula is used by many engineers for thick-walled cast-iron and steel cylinders, and Barlow's formula for medium thick-walled cylinders.

Barlow's Formula. Barlow's formula assumes that the volume of metal in a cylinder does not change during the expansion of the cylinder, so that the unit stress s varies inversely as the square of the radius. Therefore,

$$sr^2 = s_1 r_1^2$$

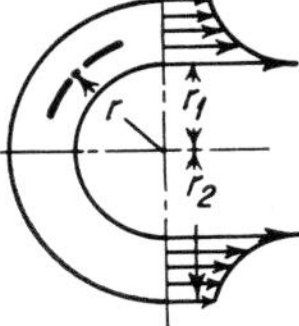

FIG. 93

in which s_1 denotes the unit stress at the inside fiber, psi, r_1 denotes the inside radius, in., s denotes the unit stress at any radius r, psi, and r denotes any radius, in. For equilibrium in Fig. 93,

$$2r_1 p_1 = 2\int_{r_1}^{r_1+t} s\,dr$$

and

$$p_1 = \frac{s_1 t}{r_1 + t} \tag{156}$$

in which p_1 denotes the internal pressure, psi, and t denotes the thickness of the cylinder wall, in.

Equation (156) may be expressed as

$$p_1 = \frac{2s_1 t}{d_2} \tag{157}$$

in which d_2 is the outside diameter, in. It will be noted that Eq. (157) is the same as that for thin cylinders, except that the outside diameter d_2 is used instead of the inside diameter.

Lamé's Formula. For very thick cylinders, many design engineers make use of Lamé's formula. In developing the formula, Lamé assumed that each particle of metal is subjected to a radial compressive stress and to a longitudinal and circumferential tensile stress. The formula is as follows:[1]

$$s = p\,\frac{r_2^2 + r_1^2}{r_2^2 - r_1^2} \tag{158}$$

in which s denotes the maximum unit circumferential stress on the inside of the cylinder, psi, p denotes the internal pressure, psi, r_2 denotes the outside radius, in., and r_1 denotes the inside radius, in.

Solving for r_2:

$$r_2 = r_1\sqrt{\frac{s+p}{s-p}}$$

Substituting $r_1 + t$ for r_2 and solving for t:

$$t = r_1\left(\sqrt{\frac{s+p}{s-p}} - 1\right) \tag{159}$$

When cast iron is used for hydraulic cylinders, the iron should be close-grained, and for such cast iron an allowable tensile fiber stress as high as 5,000 psi may be used; for steel cylinders, an allowable tensile fiber stress of 15,000 psi is recommended. Hydraulic cylinders are usually lined with brass or bronze bushings to prevent the corrosive action of water on iron.

[1] MORLEY, "Strength of Materials," 3d ed., p. 303.

Problem: A cast-iron hydraulic cylinder has a bore of 10 in. and is subjected to an internal pressure of 1,200 psi. Determine the wall thickness if the tensile stress is not to exceed 5,000 psi.

According to Barlow's formula [Eq. (157)]:

$$t = \frac{1{,}200d_2}{2 \times 5{,}000} = 0.12d_2$$

since

$$t = \frac{d_2 - d_1}{2}$$

$$d_2 - 10 = 0.24d_2$$
$$d_2 = 13.2 \text{ in.}$$
$$t = 1.58 \text{ in.}$$

According to Lamé's formula [Eq. (159)]:

$$t = 5\left(\sqrt{\frac{5{,}000 + 1{,}200}{5{,}000 - 1{,}200}} - 1\right)$$

$$t = 1.40 \text{ in.}$$

Steam Cylinders. Steam-engine and pump cylinders are not classified as thick or thin cylinders, although it is probably safer to apply the thick-cylinder formulas in design. The following empirical formulas have been used:

J. H. Barr [1] gives

$$t = 0.05d + 0.3 \text{ in.} \tag{160}$$

Kent's "Mechanical Engineers' Handbook" gives

$$t = 0.0004dp + 0.3 \text{ in.} \tag{161}$$

Mark's "Mechanical Engineers' Handbook" gives

$$t = 0.0005dp + 0.3 \text{ in.}^{2} \tag{162}$$

In the above formulas, t denotes the wall thickness, in., d denotes the bore of the cylinder, in., and p denotes the steam pressure, psi.

FLAT PLATES

Circular Flat Plates. Cylinder heads, pistons, and other circular flat plates are designed according to rational formulas which have been developed from the investigations of Grashof, Bach, and others. These formulas give results that agree within reasonable limits.

For circular flat plates, *supported at the edges* and uniformly loaded: [3]

$$s = \frac{39}{32} p \frac{r^2}{t^2} \tag{163}$$

in which s denotes the maximum radial and tangential unit stress, psi, which occurs at the outer fibers at the center of the plate; p denotes the pressure on the plate, psi; r denotes the radius of the plate, in.; and t denotes the thickness of the plate, in. Equation (163) is based on a value of 0.25 for Poisson's ratio.

[1] *Trans. ASME*, vol. 18, p. 741.

[2] This value of t allows for reboring the cylinder owing to wear.

[3] Morley, "Strength of Materials," 3d ed., p. 382. Seely and Smith, "Advanced Mechanics of Materials," 2d ed., p. 225.

For circular flat plates *fixed at the edges* and uniformly loaded, the formula is

$$s = \frac{3}{4} p \frac{r^2}{t^2} \tag{164}$$

Equation (164) gives the maximum unit stress in the plate, which is the radial stress at the circumference.

For circular flat plates *supported at the edges* and with a load uniformly distributed over a small circle of radius r_0 at the center of the plate, the formula is

$$s = \frac{P}{\pi t^2}\left(\frac{3}{2} + \frac{15}{8}\log_e \frac{r}{r_0} - \frac{9}{32}\frac{{r_0}^2}{r^2}\right) \tag{165}$$

in which P denotes the center load.

Equation (165) gives the maximum unit stress in the plate, which is the radial and the circumferential stress at the center.

Equation (165) may be written as

$$s = \frac{P}{\pi t^2} K \tag{166}$$

in which K equals the parentheses in Eq. (165).

Table 13 shows the value of K as the ratio of r to r_0 changes.

TABLE 13

Ratio r/r_0	Value of K
10	5.79
20	7.12
30	7.86
40	8.42
50	8.83

The above formulas on flat plates do not take into account the practice of adding reinforcing ribs to large cylinder heads. These added stiffening members reduce the maximum stress and are on the side of safety.

Rectangular Flat Plates. Rectangular flat plates are usually ribbed for added strength and stiffness and are so proportioned that the casting will be sound and true. Ribbed flat plates, such as steam-chest covers, are designed according to empirical rules.

Rectangular flat plates without ribs may be designed according to the following rational formulas:

For rectangular plates *supported at the edges* and uniformly loaded:

$$s = \frac{pl^2b^2}{2t^2(l^2 + b^2)} \tag{167}$$

in which s denotes the maximum unit stress, psi, p denotes the pressure on the plate, psi, l denotes the length of the plate, in., b denotes the width of the plate, in., and t denotes the thickness of the plate, in.

For rectangular flat plates *fixed at the edges* and uniformly loaded:

$$s = \frac{pl^2b^2}{3t^2(l^2 + b^2)} \tag{168}$$

HOOKS

In the discussion of curved beams, it was mentioned that a hook is an example of a curved beam. The formulas derived for curved beams may therefore be used in the design of hooks.

In Fig. 94, the load P produces a direct tension and a flexure stress on the section AB. The proportions that are commonly used in the design of hooks may be found in the various engineering handbooks.[1]

Problem: For the hook shown in Fig. 94, the load is 30,000 lb, and the hook has been given dimensions that are commonly used. The maximum unit stresses in tension and compression are to be determined.

The dimensions of the cross section at BA in Fig. 94 are shown in Fig. 95. For convenience of calculation, the cross section of the hook is transformed into the

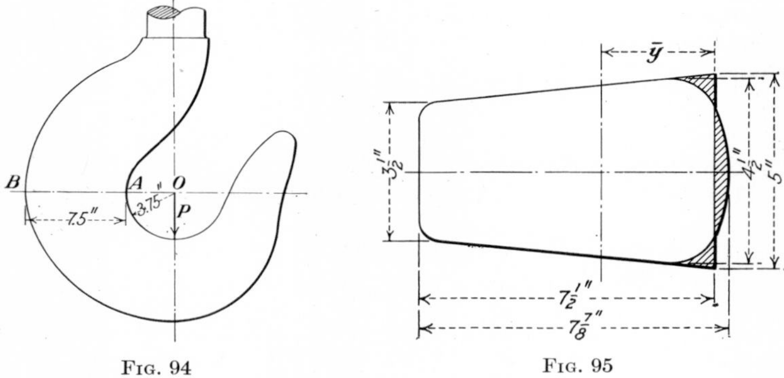

Fig. 94

Fig. 95

trapezoidal section as indicated, the area that is subtracted being made approximately equal to the area that is added.

$$\text{Area} = \frac{(5 + 3.5) \times 7.5}{2} = 31.8 \text{ sq in.}$$

$$\bar{y} = \frac{(5 + 2 \times 3.5) \times 7.5}{3(5 + 3.5)} = 3.53 \text{ in.}$$

$$A' = R\int \frac{dA}{R + y} = 7.28 \int \frac{b'\,dy}{7.28 + y}$$

The variable width $b' = 4.3 - 0.2y$.

$$A' = 7.28 \int_{-3.53}^{+3.97} \frac{(4.3 - 0.2y)\,dy}{7.28 + y} = 7.28 \int_{-3.53}^{+3.97} \left(-0.2 + \frac{5.76}{y + 7.28}\right) dy$$

$$= 7.28 \left(-1.5 + 5.76 \log_e \frac{11.25}{3.75}\right) = 35 \text{ sq in.}$$

[1] Halsey, "Handbook for Machine Designers and Draftsmen," 2d ed., p. 486.

Substituting in Eqs. (89) and (90):

$$s_t = \frac{30{,}000 \times 7.28}{7.28(35 - 31.8)}\left(\frac{7.28}{7.28 - 3.53} - \frac{35}{31.8}\right) + \frac{30{,}000}{31.8}$$

$$= 7{,}860 + 940 = 8{,}800 \text{ psi}$$

$$s_c = \frac{30{,}000 \times 7.28}{7.28(35 - 31.8)}\left(\frac{35}{31.8} - \frac{7.28}{7.28 + 3.97}\right) - \frac{30{,}000}{31.8}$$

$$= 4{,}230 - 940 = 3{,}290 \text{ psi}$$

Halsey states that for a high grade of wrought iron, properly heat-treated, a maximum unit tensile stress of 17,000 psi is permissible. For wrought iron with an ultimate tensile strength of 48,000 psi, this cannot be considered safe, especially when it is remembered that a hook may be subjected to a shock load and that there is considerable danger to life and limb in case of failure. With a factor of safety of 7, the allowable unit stress is about 7,000 psi for wrought iron and about 8,500 psi for steel. The foregoing computed maximum unit stress may therefore be considered safe for a hook that is made of a good grade of steel.

If the curvature is neglected in the above problem, the maximum unit tensile stress is as follows:

$$I_y = \frac{5^2 + (4 \times 5 \times 3.5) + 3.5^2}{36(5 + 3.5)} \times 7.5^3 = 148 \text{ in.}^4$$

$$s_t = \frac{Mc}{I} + \frac{P}{A} = \frac{30{,}000 \times 7.28 \times 3.53}{148} + \frac{30{,}000}{31.8}$$

$$= 5{,}220 + 940 = 6{,}160 \text{ psi}$$

This calculation shows that the flexural unit stress is much too low when determined by this approximate method.

Sometimes the cross section of a hook is of such shape that it cannot be conveniently broken up into areas for which the factor A' may be computed. In such cases, the value of A' may be determined by employing graphical or semigraphical methods.[1]

CASTIGLIANO'S ENERGY METHOD

The theorems of Castigliano may be stated as follows:

1. The partial derivative of the strain energy with respect to the deformation at the point of application of one of the external forces gives the value of this external force.

2. The partial derivative of the strain energy with respect to one of the external forces gives the deformation of the point of application of this external force.

3. The values of the unknown internal and external forces correspond to the minimum of strain energy, and one can determine the values of these forces when the partial derivatives of the strain energy with respect to these unknown forces are set equal to zero.

For axial stress the energy U may be expressed as

$$U = \int_0^l \frac{s^2}{2E} A \, dx$$

in which s is the unit stress, psi, E is the modulus of elasticity in tension or compression, psi, A is the area of cross section, sq in., and l is the length of the bar, in.

[1] Morley, "Strength of Materials," 3d ed., p. 339.

For energy caused by bending moment the following expression may be used:

$$U = \int_0^l \frac{M^2}{2EI} dx \tag{169}$$

For energy caused by torsion the following expression may be used:

$$U = \int_0^l \frac{T^2}{2E_s J} dx \tag{170}$$

For energy caused by beam shear the following equation may be used:

$$U = \iint \frac{s_s^{\,2}}{2E_s} dA\, dx \tag{171}$$

It can be shown [1] that, when the partial derivative of the energy is taken with respect to the moment M of a couple applied at a point on a beam, the result will give the angular displacement or slope at the point where the couple is applied. In the same manner, in torsion, when the partial derivative of the energy is taken with respect to the torque T the result will give the angular displacement or angle of twist.

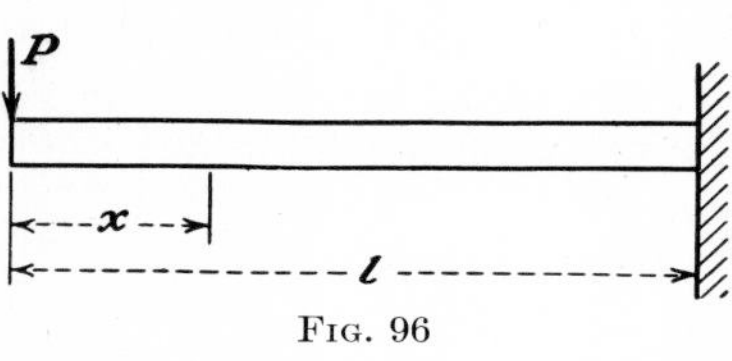

FIG. 96

The energy method will be illustrated by working out a number of examples.

Example 1: A cantilever beam carries a single concentrated load at the free end. Determine the deflection of the beam at the free end. As shown in Fig. 96, the value of the bending moment M at any section of the beam is equal to Px. Then using Eq. (169),

$$U = \int_0^l \frac{P^2x^2}{2EI} dx$$

The partial derivative of the energy with respect to P will give the deflection of the beam where P is located.

$$\text{Deflection} = \frac{\partial U}{\partial P} = \frac{P}{EI}\int_0^l x^2\, dx = \frac{P}{EI}\left(\frac{x^3}{3}\right)_0^l = \frac{Pl^3}{3EI}$$

An alternative method is the following:

$$U = \int_0^l \frac{M^2\, dx}{2EI}$$

$$\text{Deflection} = \frac{\partial U}{\partial P} = \int_0^l \frac{M(\partial M/\partial P)\, dx}{EI}$$

Now

$$M = Px \qquad \text{and} \qquad \frac{\partial M}{\partial P} = x$$

Therefore,

$$\text{Deflection} = \int_0^l \frac{Px\cdot x\cdot dx}{EI} = \frac{Pl^3}{3EI}$$

Example 2: A cantilever beam is loaded with w lb per unit length, and the maximum deflection of the beam is to be determined.

[1] TIMOSHENKO, "Strength of Materials," 3d ed., Part I, p. 331.

In this case there is no load at the free end where the deflection is desired, and it is necessary to introduce a dummy load P', which is later allowed to equal zero. From Fig. 97,

$$M = P'x + \frac{wx^2}{2} \qquad \text{and} \qquad \frac{\partial M}{\partial P'} = x$$

Then

$$\text{Deflection} = \frac{\partial U}{\partial P'} = \int_0^l \frac{M(\partial M/\partial P')\,dx}{EI}$$

Now let P' equal zero and substitute in the equation

$$\text{Deflection} = \frac{\partial U}{\partial P'} = \int_0^l \frac{wx^2}{2} \cdot \frac{x\,dx}{EI} = \frac{w}{2EI}\left(\frac{x^4}{4}\right)_0^l = \frac{wl^4}{8EI} = \frac{Wl^3}{8EI}$$

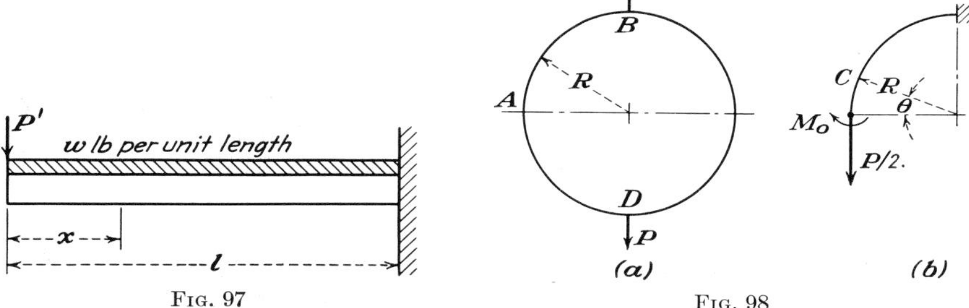

FIG. 97

FIG. 98

Example 3: In Fig. 98*a* a circular ring is shown whose cross-sectional dimensions are small compared with the radius R. The moments at A and at B are desired.

In this case it is convenient to consider the forces on one-quarter of the ring, as shown in Fig. 98*b*. It is evident from Fig. 98*a* that the slopes at A and at B do not change. Therefore, the partial derivative of the energy with respect to M_0 may be placed equal to zero. The moment at any section C is

$$M = M_0 - \frac{P}{2}(R - R\cos\theta)$$

also

$$\frac{\partial M}{\partial M_0} = 1 \qquad \text{and} \qquad ds = R\,d\theta$$

$$\frac{\partial U}{\partial M_0} = \int \frac{M(\partial M/\partial M_0)\,ds}{EI} = \int_0^{\frac{\pi}{2}} \frac{[M_0 - (PR/2)(1 - \cos\theta)]R\,d\theta}{EI} = 0$$

$$= \frac{R}{EI}[M_0\theta - PR(\theta - \sin\theta)]_0^{\frac{\pi}{2}} = \frac{M_0R\pi}{2EI} - \frac{PR^2}{2EI}\left(\frac{\pi}{2} - 1\right) = 0$$

From which

$$M_0 = \frac{PR}{2}\left(1 - \frac{2}{\pi}\right) = 0.182PR$$

The moment at B is then as follows:

$$M_B = M_0 - \frac{PR}{2} = 0.182PR - 0.5PR = -0.318PR$$

Example 4: In Fig. 98 determine the increase in the vertical diameter from B to D.

In this case we determine the energy for the whole ring, and since there are two equal loads P, the partial derivative of energy with respect to P will give the movement at both B and D where the loads P are applied.

$$\delta = \frac{\partial U}{\partial P} = 4\int_0^{\frac{\pi}{2}} \frac{M(\partial M/\partial P)R\,d\theta}{EI}$$

Since

$$M = M_0 - \frac{PR}{2}(1 - \cos\theta) = \frac{PR}{2}\left(1 - \frac{2}{\pi}\right) - \frac{PR}{2}(1 - \cos\theta)$$

$$= \frac{PR}{2}\left(\cos\theta - \frac{2}{\pi}\right)$$

and

$$\frac{\partial M}{\partial P} = \frac{R}{2}\left(\cos\theta - \frac{2}{\pi}\right)$$

$$\delta = 4\frac{PR^3}{4EI}\int_0^{\frac{\pi}{2}}\left(\cos\theta - \frac{2}{\pi}\right)^2 d\theta = \frac{PR^3}{EI}\int_0^{\frac{\pi}{2}}\left(\cos^2\theta - \frac{4}{\pi}\cos\theta + \frac{4}{\pi^2}\right) d\theta$$

$$= \left(\frac{\pi}{4} - \frac{2}{\pi}\right)\frac{PR^3}{EI} = 0.149\frac{PR^3}{EI}$$

Example 5: Assume that the arch rib of a Quonset hut is a semicircle, that it is pinned at the bottom, and that the load is uniformly distributed along the length of the arch. The cross-sectional dimensions of the rib are small compared with the radius R. Neglecting the energy of shear and direct stress, determine the horizontal thrust at the bottom in terms of the total load carried. Figure 99 shows the forces.

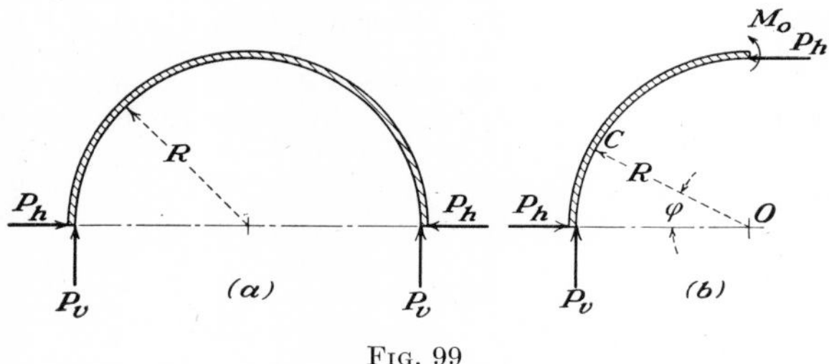

Fig. 99

For the load on the arc subtending the angle ϕ, the centroid is on the bisector of ϕ at a distance from 0 equal to

$$\bar{x} = \frac{2R\sin\frac{\phi}{2}}{\phi}$$

From the sum of forces along the vertical,

$$P_v = \frac{w\pi R}{2}$$

The bending moment at any section C is

$$M = P_vR(1 - \cos\phi) - P_hR\sin\phi - wR\phi\left(\bar{x}\cos\frac{\phi}{2} - R\cos\phi\right)$$

Because there is no movement at P_h, the partial derivative of the energy with respect to P_h may be placed equal to zero. Hence

$$\int_0^{\frac{\pi}{2}} \frac{M(\partial M/\partial P_h)R\,d\phi}{EI} = 0$$

Since

$$\frac{\partial M}{\partial P_h} = -R \sin \phi$$

$$0 = \frac{1}{EI}\int_0^{\frac{\pi}{2}} \Big(-P_vR^3 \sin \phi + P_vR^3 \cos \phi \sin \phi + P_hR^3 \sin^2 \phi + wR^3\phi \cdot \frac{2R}{\phi} \sin \frac{\phi}{2} \cos \frac{\phi}{2} \sin \phi - wR^4\phi \cos \phi \sin \phi\Big)\,d\phi$$

$$\text{First integral} = -\frac{P_vR^3}{EI}$$

$$\text{Second integral} = \frac{P_vR^3}{2EI}$$

$$\text{Third integral} = \frac{P_hR^3\pi}{4EI}$$

$$\text{Fourth integral} = \frac{wR^4\pi}{4EI}$$

$$\text{Fifth integral} = \frac{wR^4\pi}{8EI}$$

Setting the sum of these integrals equal to zero and solving for P_h:

$$P_h = \frac{2P_v}{\pi} - \frac{wR}{2} = \frac{wR}{2} = \frac{W}{2\pi} = 0.159W$$

This result shows that the horizontal thrust is about 16 percent of the total load on the arch rib.

Bibliography

ASTM Standards, 1955.

Boyd and Folk: "Strength of Materials," 5th ed., McGraw-Hill Book Company, Inc., 1950.

Miner and Seastone: "Handbook of Engineering Materials," John Wiley & Sons, Inc., 1955.

Morley, A.: "Strength of Materials," 3d ed., Longmans, Green & Co., Inc., 1914.

Roark, R. J.: "Formulas for Stress and Strain," 3d ed., McGraw-Hill Book Company, Inc., 1954.

Seely and Smith: "Resistance of Materials," 4th ed., John Wiley & Sons, Inc., 1956.

Seely and Smith: "Advanced Mechanics of Materials," 2d ed., John Wiley & Sons, Inc., 1952.

Singer, F. L.: "Strength of Materials," Harper & Brothers, 1951.

Timoshenko, S.: "Strength of Materials," Parts I and II, 3d ed., D. Van Nostrand Company, Inc., 1956.

Withey and Washa: "Materials of Construction," John Wiley & Sons, Inc., 1954.

Section 4

HYDRAULICS

By Ernest F. Brater

This section will treat of methods of solving problems involving fluids at rest and in motion. Fluids differ from solids in their inability to resist shear stresses except by continuous deformation. Liquids are fluids which are relatively incompressible. When the volume of a liquid is less than the volume of its container, a free surface is formed. Gases are fluids which are easily compressible and which fill the entire volume of their container. Many phases of this subject will be treated in a manner sufficiently general to permit the solution of problems involving any fluid. Because the civil engineer is primarily concerned with problems involving water, most of the illustrative problems will deal with this fluid.

Dimensions and Units. Any quantity used in hydraulics may be expressed in terms of the three primary dimensions: length L, mass M, and time T. In the foot-pound-second system, the basic units of these three dimensions are feet, slugs, and seconds, respectively. Acceleration a may, for example, be described dimensionally as feet per second per second (ft per sec^2) or as L/T^2. As a further example, consider the dimensions of force. The unit of force in the foot-pound-second system is the pound, which is defined as the force necessary to accelerate a mass of 1 slug at the rate of 1 ft per sec^2. The basic definition of force is expressed mathematically in Eq. (1).

$$F = Ma \tag{1}$$

If the dimensions of acceleration given above are inserted in this equation, the following dimensional equation is obtained:

$$F = \frac{ML}{T^2} \tag{1a}$$

Expressed in words, the dimensions of force are seen to be slugs-feet per second^2. The dimensions of the quantities commonly used in hydraulics are given in Table 1, column 3. These dimensions apply to any system of units, whereas the units given in column 4 are those of the foot-pound-second system. The dimensions of column 3 are used in *dimensional analysis*, which is a method of arranging the quantities upon which a particular phenomena depends into a *dimensionally homogeneous* form. Dimensional homogeneity requires that an equality have the same dimensions on either side of the equal sign. For example, the expression relating pressure intensity p to depth below a liquid surface h is

$$p = wh$$

where w is the specific weight of the liquid. If the dimensions of these quantities, from

Table 1, are inserted into the above expression, the following dimensional equation is obtained:

$$\frac{M}{LT^2} = \frac{M}{L^2T^2} \times L$$

TABLE 1

Quantity (1)	Symbol (2)	Dimensions (3)	Basic units (ft-lb-sec system) (4)
Length	l, d, D, h, H	L	Ft
Area	a	L^2	Sq ft
Volume	V	L^3	Cu ft
Time	T, t	T	Sec
Velocity	V, v, u	L/T	Ft per sec
Angular velocity	ω (omega)	T^{-1}	Radians per sec
Acceleration	a	L/T^2	Ft per sec per sec
Kinematic viscosity	ν (nu)	L^2/T	Sq ft per sec
Discharge	Q	L^3/T	Cu ft per sec (cfs) (sec-ft)
Mass	M	M	Slugs
Density	ρ (rho)	M/L^3	Slugs per cu ft
Force, weight	F, W	ML/T^2	Pounds (lb)
Specific weight (unit weight)	w	M/L^2T^2	Lb per cu ft
Viscosity	μ (mu)	M/LT	Slugs per ft-sec
Surface tension	σ (sigma)	M/T^2	Lb per ft
Modulus of elasticity (bulk)	E	M/LT^2	Lb per sq ft
Pressure intensity	p	M/LT^2	Lb per sq ft
Shear stress	τ (tau)	M/LT^2	Lb per sq ft
Momentum or impulse	....	ML/T	Slugs-ft per sec, or pound sec
Energy of work	....	ML^2/T^2	Ft-lb
Power	....	ML^2/T^3	Ft-lb per sec
Specific gravity	sp gr		

It will be seen that the two sides of the equation are dimensionally equivalent, thus showing that the expression for pressure is dimensionally homogeneous. A number of other commonly used formulas, as, for example, the Chezy formula (p. 4–67) are not dimensionally homogeneous. One advantage of a dimensionally homogeneous expression is that it can be used in any system of units without changing numerical constants. Sometimes, lack of dimensional homogeneity indicates that an expression is not basically correct and that it should therefore not be used outside of the experimental range.

It will be noted that such commonly used units as inches and gallons do not appear in the list given in Table 1, column 4. Formulas may be altered to permit the use of such nonbasic units, but it is usually more desirable to solve problems with the units of column 4 and to make any desired conversions on the answer. Three nonbasic units are used so frequently that they deserve special mention. A convenient unit in dealing with large storage volumes is the acre-foot, which is a volume 1 ft deep on area of 1 acre. One acre-foot is therefore equivalent to 43,560 cu ft. A discharge of 1 cu ft per sec for a period of 1 day will produce a volume of approximately 2 acre-ft. Two commonly used units of discharge are gallons per minute and million gallons per day. Discharge at the rate of 1 cu ft per sec is equivalent to 448.8 gal per min and to 0.646 million gal per day.

Properties of Fluids. Many of the quantities shown in Table 1 require no further explanation. However, those quantities which describe properties of the fluid will be discussed briefly.

The unit of *mass* M is defined as follows: A force of 1 lb will accelerate a mass of 1 slug at the rate of 1 ft per sec per sec. In the case of a 1-lb weight falling freely, a force of 1 lb produces an acceleration of 32.2 ft per sec per sec. Therefore, because the magnitude of the force must be equal to the product of the mass and acceleration, it follows that the 1-lb weight possesses a mass of 1/32.2 slug. Consequently, to convert *weight* W, in pounds, to mass it is necessary to divide by the gravitational acceleration g, as follows:

$$M = \frac{W}{g} \tag{2}$$

Specific weight w is defined as the weight per unit of volume of a substance. The mass per unit of volume is called *density* ρ. It follows from the above discussion of *mass* that, in the foot-pound-second system, specific weight in pounds per cubic foot may be converted to density in slugs per cubic foot by dividing by the gravitational acceleration as shown by Eq. (2*a*).

$$\rho = \frac{w}{g} \tag{2a}$$

If either of the above equations were solved for g it would be seen that g represents the weight per unit of mass. In other words, a mass of 1 slug weighs 32.2 lb. Although g varies slightly with both latitude and altitude, the value 32.2 ft per sec per sec is sufficiently accurate for all hydraulic computations. The densities and specific weights of water and some other fluids are given in Table 2.

TABLE 2[1]

Fluid	Temperature, °F	Specific gravity (sp gr)	Specific weight w, lb per cu ft	Viscosity μ, slugs per ft-sec	Kinematic viscosity ν, sq ft per sec	Pressure of saturated vapor p_v, lb per sq ft
Water........	32	0.9999	62.42	0.00003746	0.00001931	12.7
	39.2	1.0000	62.427	0.00003274	0.00001687	16.9
	50	0.9997	62.41	0.00002735	0.00001410	25.6
	60	0.9990	62.37	0.00002359	0.00001217	36.8
	70	0.9980	62.30	0.00002050	0.00001059	52.3
	80	0.9966	62.22	0.00001799	0.00000930	73.0
	100	0.9931	62.00	0.00001424	0.00000739	136
	200	0.9630	60.12	0.00000637	0.00000341	1,650
Sea water.....	.....	....	64			
Mercury......	32	13.60	......	0.0000355		
	68	13.55	......	0.0000328		
	80.6	13.53				
Castor oil.....	50	....	......	0.0505		
	59	0.969	......	0.0315		
	68	....	......	0.0206		
Dry air at 14.7 psi.........	32	0.00129	......	0.355×10^{-6}		
	50	0.00125	......	0.372×10^{-6}		
	59	0.00122	......	0.376×10^{-6}		
	68	0.00121	......	0.380×10^{-6}		

[1] Values given in this table were taken from "Handbook of Chemistry and Physics," "International Critical Tables," and "ASCE Manual for Hydraulic Laboratory Studies." Many of the values are averages of those obtained from various sources. Other sources of similar data are Smithsonian Physical Tables and Standard Density and Volumetric Tables of U.S. Bureau of Standards.

The *specific gravity* (sp gr) of a substance is obtained by dividing the weight of a certain volume of a substance by the weight of an equal volume of water at 39.2°F. Values of specific gravity of several fluids are given in Table 2.

The *viscosity* μ of a fluid is a measure of the relative ease or difficulty with which a particle of the fluid may be deformed. For example, heavy oil has a greater viscosity than water, and water is more viscous than air. The viscosity of a liquid decreases as temperature increases, but the opposite is true of a gas. Viscosity may be defined as follows: Assume that the space between the two horizontal plates shown in Fig. 1 is filled with a fluid and that the upper plate is in motion with the velocity V while the lower plate is stationary. Those particles of fluid which are in contact with a boundary surface will have the same velocity as the boundary. It follows that the velocity of the fluid will vary from V at the top to zero at the bottom, the velocity of any layer of fluid being different from the adjacent layers. As a result, all particles of fluid will be continuously deformed by the shear stresses between adjacent layers as illustrated in Fig. 1 by two positions of a particle at a distance y from the lower plate. It has been shown by laboratory tests that for a particular fluid at a given temperature the ratio of the shear stress τ to the rate of deformation du/dy is constant. This constant is called the coefficient of viscosity, the dynamic viscosity, or simply the viscosity. The definition of viscosity is expressed mathematically by Eq. (3).

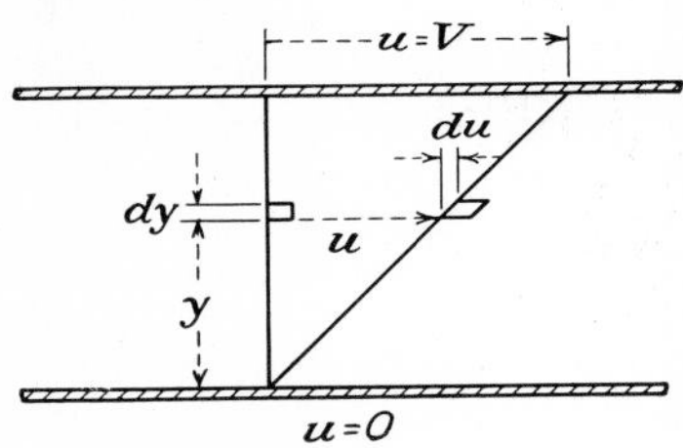

FIG. 1. Viscous deformation.

$$\mu = \frac{\tau}{du/dy} \tag{3}$$

By inserting the dimensions of the terms on the right side of Eq. (3) as follows,

$$\mu = \left(\frac{M}{LT^2} \cdot \frac{T}{L} \cdot L\right) = \left(\frac{M}{LT}\right) \tag{4}$$

the units of viscosity are seen to be slugs per foot-second. If in the above operation the dimensions of τ are taken as pounds per L^2 rather than as M/LT^2 it may be readily seen that pound-seconds per square foot are equivalent units of viscosity. The corresponding units in the metric system, grams (mass) per centimeter-second, are called poises. One slug per foot-second is equivalent to 478.8 poises. The viscosities of a number of fluids are given in Table 2.

The *kinematic viscosity* ν is obtained by dividing the viscosity by the density as shown by Eq. (5).

$$\nu = \frac{\mu}{\rho} \tag{5}$$

The dimensions are obtained in Eq. (6),

$$\nu = \left(\frac{M}{LT} \cdot \frac{L^3}{M}\right) = \left(\frac{L^2}{T}\right) \tag{6}$$

from which it may be seen that the units of kinematic viscosity are square feet per second. The corresponding units in the metric system (square centimeters per second) are called stokes. One square foot per second is equivalent to 929.0 stokes.

Surface tension is the property of liquids which causes capillary action in small tubes. If the liquid "wets" the surface of the tube, the liquid surface will be concave upward

and the liquid will rise as shown by Fig. 2*a*. If the liquid does not "wet" the tube, the surface will be concave downward, causing the surface to be depressed as shown in Fig. 2*b*. Where pressures are measured by manometers, care must be exercised to avoid errors resulting from capillarity. A number of experimental values of capillary rise or fall for glass tubes of various sizes as determined by Folsom[1] are plotted in Fig. 3. The straight line shown on the graph indicates the rise that might be expected with very pure water in contact with air. Surface-tension effects, other than those mentioned above, are usually of minor importance in hydraulic problems.[2]

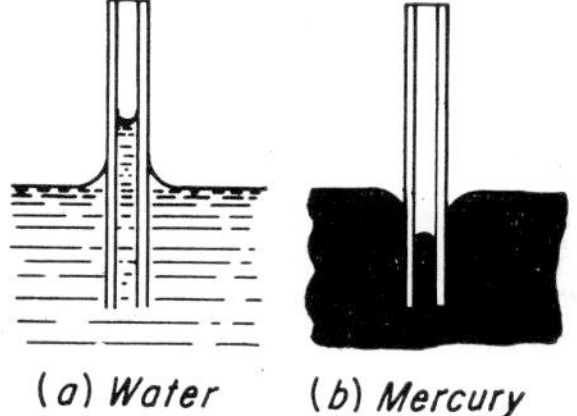

FIG. 2. Capillary action.

The *modulus of elasticity* E of liquids is defined as the change in pressure intensity divided by the corresponding change in volume per unit volume as shown by Eq. (7).

$$E = -\frac{\Delta p}{\Delta V/V} \tag{7}$$

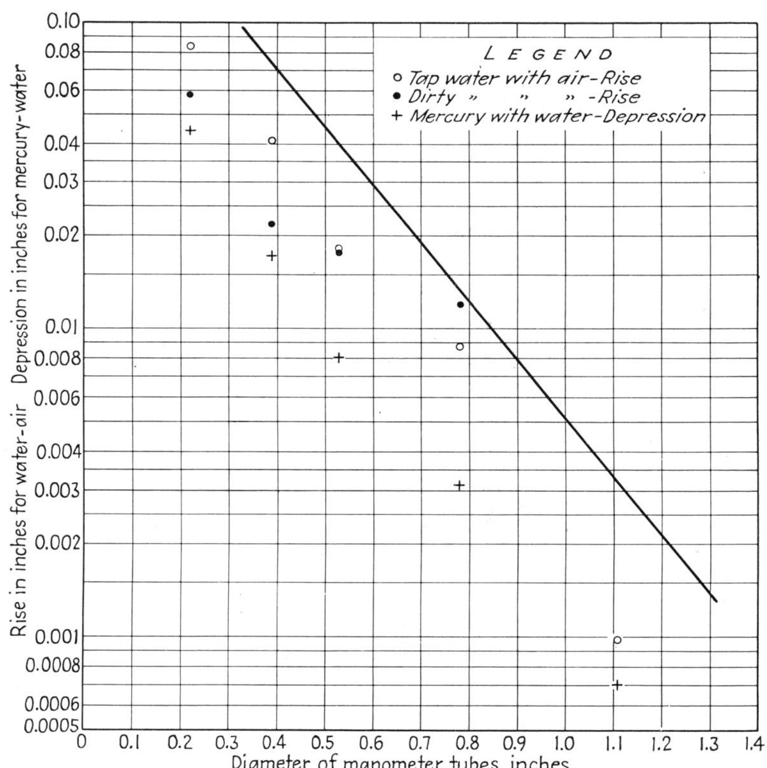

FIG. 3. Capillary rise and depression in glass tubes.

[1] FOLSOM, RICHARD G., Manometer Errors Due to Capillarity, *Instruments*, February, 1936, pp. 36–37.

[2] For a more complete discussion of surface tension see GIBSON, "Hydraulics and Its Applications," p. 6; PRANDTL and TIETJENS, "Fundamentals of Hydro- and Aeromechanics," pp. 60–65; WISLER and BRATER, "Hydrology," p. 201; ADAM, N. K., "The Physics and Chemistry of Surfaces," 3d ed., pp. 1–16.

The value of E for water is approximately 300,000 psi, varying slightly with temperature.[1] Values of E are sufficiently high to permit the assumption that liquids are incompressible in most hydraulic problems.

HYDROSTATICS

The force per unit area acting on a real or imaginary surface within a fluid is called *intensity of pressure* p or simply *pressure.* It may be demonstrated that the pressure at any point in a fluid acts equally in all directions.[2] The force on a boundary surface resulting from fluid pressure must be normal to the surface at all points because of the inability of fluids at rest to transmit shear. The pressure variation with depth within a liquid may be evaluated by considering the forces acting on the vertical prism

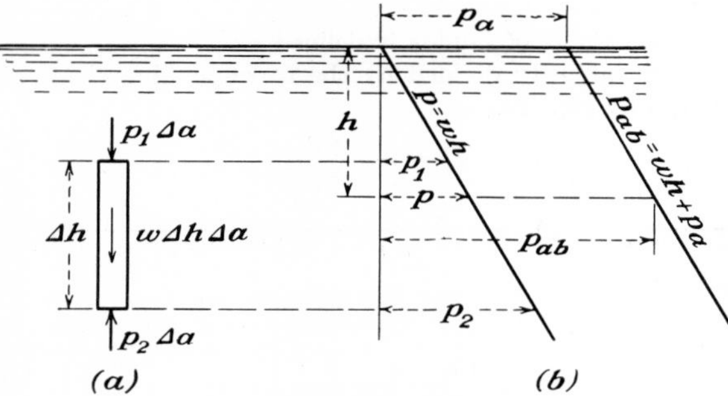

Fig. 4. Pressure in a liquid.

of height Δh and cross-sectional area Δa shown in Fig. 4a. The summation of all the forces acting on this prism in the vertical direction, as well as in all other directions, must be equal to zero. The vertical forces consist of the weight and the force due to the pressure p_1 at the top and that due to p_2 at the bottom. The summation of these forces gives

$$p_2 \,\Delta a = w \,\Delta h \,\Delta a + p_1 \,\Delta a$$

and

$$p_2 = w \,\Delta h + p_1 \tag{8}$$

the downward direction of h being taken as positive. Equation (8) may be solved for Δh to give

$$\Delta h = \frac{p_2}{w} - \frac{p_1}{w} \tag{9}$$

If p_1 is taken as the pressure at the liquid surface, Δh becomes h, the vertical distance below the liquid surface to the point where the pressure is p_2. Furthermore, when the pressure at the liquid surface is atmospheric pressure (p_a), p_2 is the absolute pressure (p_{ab}) at that point. The following expression for absolute pressure may now be obtained from Eq. (8).

$$p_{ab} = wh + p_a \tag{10}$$

More commonly used in engineering work is the gage pressure p. The gage-pressure scale is obtained by designating atmospheric pressure as zero. The following expressions for gage pressure may be obtained from Eq. (10) by letting p_a become zero:

$$p = wh \tag{11}$$

and

$$h = \frac{p}{w} \tag{12}$$

[1] "Handbook of Chemistry and Physics."

[2] King, Wisler, and Woodburn, "Hydraulics," 5th ed., p. 5.

Equations (10) and (11) are illustrated graphically in Fig. 4*b*. The expression p/w is called the "pressure head." It expresses the depth in feet of a liquid of specific weight w required to produce a pressure p. Values of absolute pressure are always positive, whereas gage pressure may be positive or negative depending upon whether the pressure

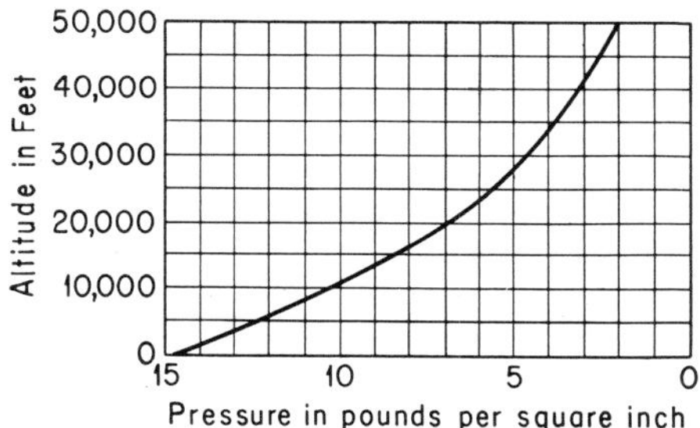

FIG. 5. Mean atmospheric pressure at different altitudes.

is greater or less than atmospheric. Negative gage pressures indicate that a partial vacuum exists.

Atmospheric Pressure. The variation of pressure with elevation in the atmosphere is illustrated by Fig. 5. The straight-line variation derived for liquids does not apply to the atmosphere (or any other gas) because the density varies with the altitude.[1] The average value of atmospheric pressure at sea level is usually taken as 14.7 psi, or 2,116 lb per sq ft.

The pressure gages used to measure atmospheric pressure are called *barometers*. A barometer of the type shown in Fig. 6 may be made by filling the tube with liquid, temporarily sealing the open end, inverting the tube, and releasing the seal beneath the surface of the liquid in the pan. If the tube is sufficiently long, there will be a space at the top that is empty except for the vapor of the liquid. Evaporation will take place at the liquid surface in the tube until a saturated vapor and corresponding vapor pressure p_v exist above the liquid surface. Example 1 illustrates that water is an impractical liquid for barometers. The liquid usually used is mercury. Mercury not only permits the use of a shorter tube, but its vapor pressure is so low [2] that it may be neglected.

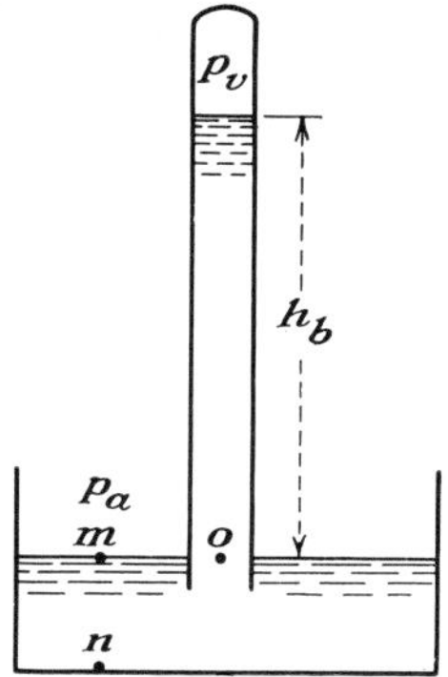

FIG. 6. Barometer.

Example 1: Determine the value of h_b in Fig. 6 if the liquid is water at a temperature of 70°F, and p_a is 14.7 psi. If Eq. (8) is written first from m to n and then from n to o, it will be seen that $p_o = p_m$ and therefore that the pressure at o is atmospheric. Equation (9) may now be written as follows:

$$\Delta h = h_b = \frac{p_a}{w} - \frac{p_v}{w}$$

The value of p_a/w is $\dfrac{14.7 \times 144}{62.3}$, or 33.98 ft, the value of w being obtained from Table 2. The value of p_v is given in Table 2 as 52.3 lb per sq ft. Therefore, p_v/w is

[1] For a detailed discussion of pressure variations in gases see PRANDTL and TIETJENS, "Fundamentals of Hydro- and Aeromechanics," pp. 29–35.

[2] "Handbook of Physics and Chemistry."

52.3/62.3, or 0.84 ft; hence,

$$h_b = 33.98 - 0.84 = 33.14 \text{ ft}$$

Example 2: Determine the value of h_b in Fig. 6 if the liquid is mercury at a temperature of 70°F and p_a is 14.7 psi.

The vapor pressure of mercury is negligible so that Eq. (9) becomes

$$\Delta h = h_b = \frac{p_a}{w} + 0$$

$$h_b = \frac{14.7 \times 144}{13.54 \times 62.42} = 2.50 \text{ ft}$$

which is equivalent to 30.1 in., or 764 mm.

Manometers. Manometers are tubes attached to reservoirs, pipes, or channels to measure the pressure. The equations of hydrostatics are used to determine pressures

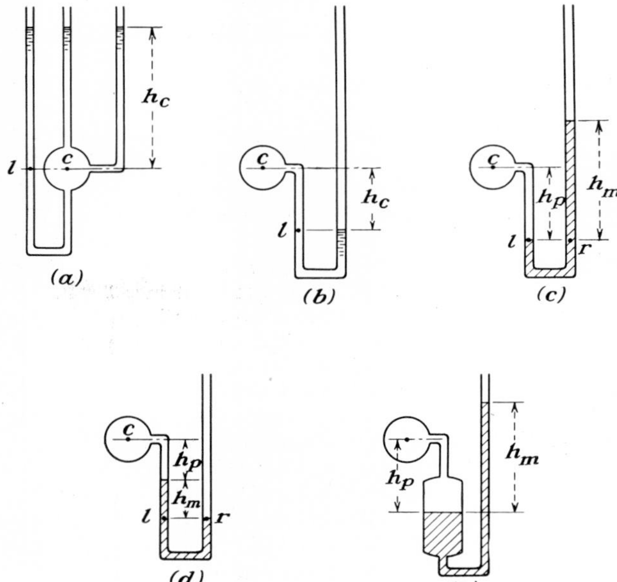

Fig. 7. Manometers.

from manometer readings even though manometers are most frequently used to measure pressures in moving fluids. To ensure against including forces due to acceleration in the manometer readings it is necessary to install the tube in a wall which is parallel to the flow lines in such a manner that the flow pattern is not disturbed by the opening. When the manometers contain only the fluid in the conduit as illustrated by Figs. 7*a* and 7*b*, they are frequently called *piezometers*. In Fig. 7*a*, the pressure at l, p_l, is equal to the pressure at the center of the pipe p_c; consequently the value of p_c may be obtained from either of the three tubes using Eq. (11) arranged as follows:

$$p_c = wh_c \tag{13}$$

The left tube of Fig. 7*a* and the one shown in 7*b* may be used to measure negative gage

pressures as well as positive ones. Equation (13) may be applied to Fig. 7*b*, knowing the pressure at l to be zero and that values of h_c measured upward from the zero datum are negative.

Example 3: Determine the value of h_c in Fig. 7*a* when the gage pressure at c is 50 psi if the pipe contains water.

Solving Eq. (13),

$$h_c = \frac{p_c}{w} = \frac{50 \times 144}{62.4} = 115 \text{ ft}$$ [1]

Example 4: Determine the pressure at c in Fig. 7*b* if h_c is 6.3 ft and the pipe contains oil having a specific gravity of 0.95.

$$p_c = wh_c = -0.95 \times 62.4 \times 6.3 = -373 \text{ lb per sq ft or } -2.59 \text{ psi}$$

Piezometers are very sensitive pressure gages but are impractical for the measurement of high pressures because of the excessive length of tube required, as illustrated by Examples 3 and 4. More satisfactory manometers for the measurement of high pressures are shown in Figs. 7*c*, 7*d*, and 7*e*. The manometer liquid most commonly used is mercury. Any liquid of greater density than the pipe fluid may be used provided that the two fluids are not miscible. Manometers of this type may be used whether the pipe fluid is a gas or a liquid and for both positive and negative pressures.

Values of h_p and h_m in Fig. 7*c* are usually read from a scale placed between the two legs of the U tube. The location of the zero of this scale fluctuates greatly with variations in the pipe pressure, thus causing some inconvenience in making the readings. This objectionable feature may be partly eliminated by the arrangement shown in Fig. 7*e*. The amount of movement of the mercury surface in the left and right tubes will be inversely proportional to the square of the diameters of the tubes. If the difference in diameters is made large, the mercury surface in the left tube will be nearly stationary and the value of h_p will then be nearly constant. For many practical purposes the scale may be marked off to give the pressure at c directly from the values of h_m.

Example 5: Determine the value of h_m in Fig. 7*c* if p_c is 50 psi gage, h_p is 6 ft, the pipe fluid is water, and the manometer fluid is mercury.

Let w_p be the specific weight of the pipe fluid and w_m the specific weight of the manometer fluid. From Eq. (8),

$$p_l = w_p h_p + p_c = 62.4 \times 6 + 50 \times 144$$

and from Eq. (11)

$$p_r = w_m h_m = 13.6 \times 62.4 \cdot h_m$$

Knowing p_l and p_r to be equal,

$$13.6 \times 62.4 \cdot h_m = 62.4 \times 6 + 50 \times 144$$

from which h_m is found to be 8.9 ft.[2]

Example 6: Determine the value of p_c in Fig. 7*d* if h_p is 8 in. and h_m is 20 in. The pipe fluid is water and the manometer fluid is mercury.

Equation (8) gives

$$p_l = w_m h_m + w_p h_p + p_c = 0$$

or

$$(13.6 \times 62.4 \times {}^{20}\!/_{12}) + (62.4 \times {}^{8}\!/_{12}) + p_c = 0$$

from which

$$p_c = -1{,}460 \text{ lb per sq ft gage or } -10.1 \text{ psi gage}$$

[1] Unless some particular temperature is stated, the specific weight of water will be taken as 62.4 lb per cu ft in numerical examples.

[2] Unless some particular temperature is stated, the specific gravity of mercury will be taken as 13.6 in numerical problems.

Differential manometers of the type shown in Fig. 8 are employed to measure the difference in pressure between two points. In both Fig. 8*a* and 8*b* it is desired to measure $(p_1 - p_2)$. In Fig. 8*a* the specific weight of the manometer fluid w_m is greater than that of the pipe fluid w_p, while in Fig. 8*b* the reverse is true. In the arrangement shown in Fig. 8*a* the pipe fluid may be liquid or gas whereas in Fig. 8*b*, it must be a liquid. The

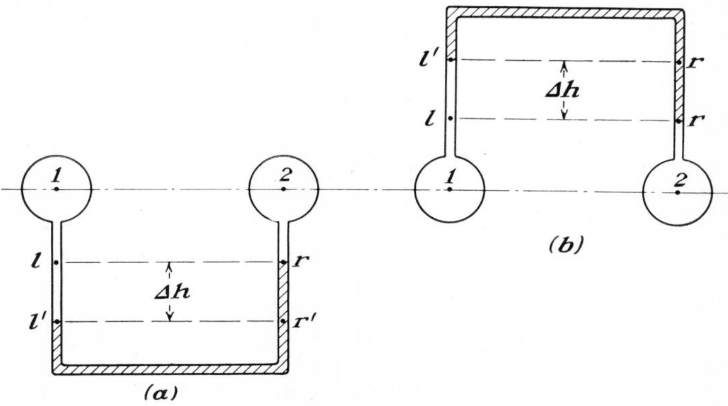

FIG. 8. Differential manometers.

manometer fluid in Fig. 8*b* may, however, be either liquid or gas. An application of the equations of hydrostatics will show that, for either U tube of Fig. 8,

$$p_{l'} = p_{r'} \tag{14}$$

and

$$(p_l - p_r) = (p_1 - p_2) \tag{15}$$

for Fig. 8*a*, Eq. (8) gives

$$p_{l'} = w_p \, \Delta h + p_l \qquad \text{and} \qquad p_{r'} = w_m \, \Delta h + p_r$$

From Eq. (14),

$$w_p \, \Delta h + p_l = w_m \, \Delta h + p_r$$

or

$$p_l - p_r = \Delta h(w_m - w_p)$$

and from Eq. (15),

$$p_1 - p_2 = \Delta h(w_m - w_p) \tag{16}$$

The corresponding equation for a U tube of the type shown in Fig. 8*b* will be

$$p_1 - p_2 = \Delta h(w_p - w_m) \tag{17}$$

Equations (16) and (17) show that the gage becomes more sensitive as values of w_p and w_m become more nearly equal. Numerical examples of differential gage problems will be found on pp. 4–25 and 4–27.

Pressure Forces on Plane Surfaces. In Fig. 9, *LM* represents any immersed plane surface which, if extended, would cut the liquid surface at *O* with an angle θ. Considering *O* the origin and *OM* the *y* axis, the pressure force *dP* on an elementary area *dA* is $wh\, dA = wy \sin\theta \, dA$. The total pressure force on the area *LM* is then

$$P = w \sin\theta \int y \, dA \tag{18}$$

If $\bar{y}$ is the distance of the center of gravity of the surface from O then

$$\int y\,dA = A\bar{y} \qquad \text{and} \qquad P = w\bar{y}\sin\theta A$$

or letting $\bar{h} = \bar{y}\sin\theta$,

$$P = w\bar{h}A \tag{19}$$

The *center of pressure* C is the point on the immersed surface through which the resultant pressure acts. Because the intensity of pressure increases with the depth, the center

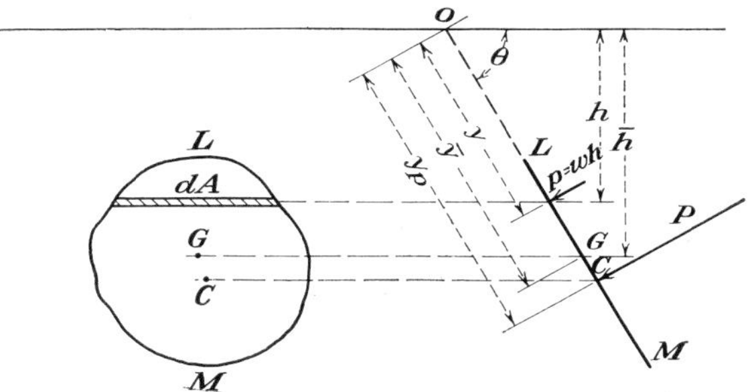

FIG. 9. Pressure forces on plane surfaces.

of pressure is below the center of gravity. The distance y_p from an axis through O, formed by the intersection of the plane surface LM with the water surface, to the center of pressure may be found from a summation of the moments of the elementary forces about this axis as follows:

$$y_p = \frac{\int y\,dP}{P} \tag{20}$$

which from the above may be written

$$y_p = \frac{w\sin\theta \int y^2\,dA}{w\sin\theta \int y\,dA} = \frac{\int y^2\,dA}{\int y\,dA} \tag{21}$$

The numerator of the fraction in the above equation is the moment of inertia of the surface LM about the axis through O. This moment of inertia is equal to $Ak^2 + A\bar{y}^2$, where k is the radius of gyration of the surface. The denominator is equal to $A\bar{y}$. Substituting these values into Eq. (21) and reducing,

$$y_p = \bar{y} + \frac{k^2}{\bar{y}} \tag{22}$$

For surfaces such as are shown in Fig. 10, where the locus of the mid-points of horizontal lines is a straight line, the center of pressure lies on this straight line. For other

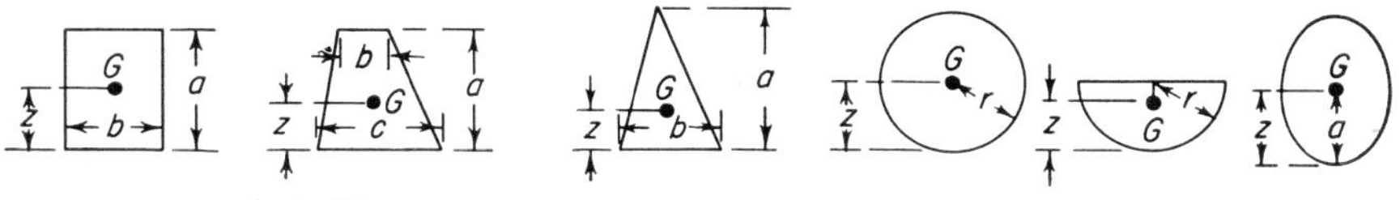

$z = \frac{a}{2}$, $k^2 = \frac{a^2}{12}$ $\qquad$ $z = \frac{a(c+2b)}{3(c+b)}$, $k^2 = \frac{a^2}{18}\left[1 + \frac{2bc}{(b+c)^2}\right]$ $\qquad$ $z = \frac{a}{3}$, $k^2 = \frac{a^2}{18}$ $\qquad$ $z = r$, $k^2 = \frac{r^2}{4}$ $\qquad$ $z = 0.5756r$, $k^2 = 0.06987r^2$ $\qquad$ $z = a$, $k^2 = \frac{a^2}{4}$

FIG. 10. Positions of centers of gravity and values of squares of radii of gyration.

surfaces, the horizontal position of the center of pressure can be obtained by taking moments as above about an axis perpendicular to the one through O. The use of the

above equations is illustrated in the following examples. Problems involving fluid pressures on plane surfaces may also be solved more directly by considering the surface to be a slab with a nonuniform load equal in magnitude to the volume of the pressure-intensity diagram. Examples 7 and 8 give numerical applications of this method.

Example 7: Determine the magnitude and location of the resultant water-pressure force on a 1-ft section of the gate *ab* shown in Fig. 11 (*a*), by the use of Eqs. (19) and (22) and (*b*), by loading the gate with the pressure volumes.

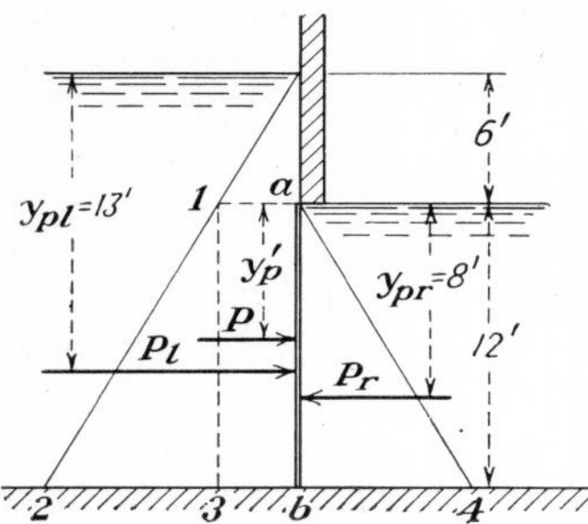

FIG. 11. Pressure forces on a vertical gate.

a.

$$P_l = w\bar{h}A = 62.4 \times 12 \times 12 = 9{,}000 \text{ lb}$$

$$y_{pl} = \bar{y} + \frac{k^2}{\bar{y}} = 12 + \frac{(12)^2}{12 \times 12} = 13 \text{ ft}$$

$$P_r = w\bar{h}A = 62.4 \times 6 \times 12 = 4{,}500 \text{ lb}$$

$$y_{pr} = \bar{y} + \frac{k^2}{\bar{y}} = 6 + \frac{(12)^2}{12 \times 6} = 8 \text{ ft}$$

Note that the latter value might have been determined by inspection because the force on the right side of *ab* results from a triangular pressure distribution, the center of gravity of which would therefore be two-thirds of its altitude from the water surface, i.e.,

$$y_{pr} = \tfrac{2}{3} \times 12 = 8 \text{ ft}$$

The resultant force P is then obtained by adding P_l and P_r as follows:

$$P = 9{,}000 - 4{,}500 = 4{,}500 \text{ lb} \qquad \text{(to the right)}$$

The location of P may be found by taking moments about *a*.

$$y_p' \times 4{,}500 = 7 \times 9{,}000 - 8 \times 4{,}500$$

and

$$y_p' = 6 \text{ ft}$$

b. Inspection of Fig. 11 will show that the force resulting from the triangular pressure distribution 123 on the left is exactly balanced by the force on the right resulting from the pressure diagram *ab*4. Consequently, the only unbalanced pressure is a uniform intensity of 6*w* lb per sq ft on the left as shown by the pressure diagram 13*ba*. Therefore,

$$P = 6 \times 62.4 \times 12 = 4{,}500 \text{ lb}$$

Because P is the resultant of a uniform pressure distribution, it must act on the center of gravity of the gate, i.e., 6 ft from *a*.

Example 8: Determine the magnitude and location of the pressure force on the rectangular sloping gate shown in Fig. 12 by three methods.

a. Using Eqs. (19) and (22),

$$P = w\bar{h}A \qquad (19)$$
$$= 62.4 \times 8 \times 60 = 30{,}000 \text{ lb}$$

$$y_p = \bar{y} + \frac{k^2}{\bar{y}} \qquad (22)$$

$$= 10 + \frac{100}{12 \times 10} = 10.83 \text{ lb}$$

b. Using Eqs. (18) and (21),

$$P = w \sin \theta \int y\, dA \tag{18}$$

$$= 62.4 \times 0.8 \int_5^{15} 6y\, dy$$

$$= 62.4 \times 0.8 \times 6 \left(\frac{y^2}{2}\right)_5^{15} = 30{,}000 \text{ lb}$$

$$y_p = \frac{\int y^2\, dA}{\int y\, dA} \tag{21}$$

$$= \frac{6\int_5^{15} y^2\, dy}{6\int_5^{15} y\, dy} = \frac{(y^3/3)_5{}^{15}}{(y^2/2)_5{}^{15}} = 10.83 \text{ ft}$$

c. The problem may also be solved by determining the magnitude and location of two components of the pressure force directly from the pressure loading diagrams of

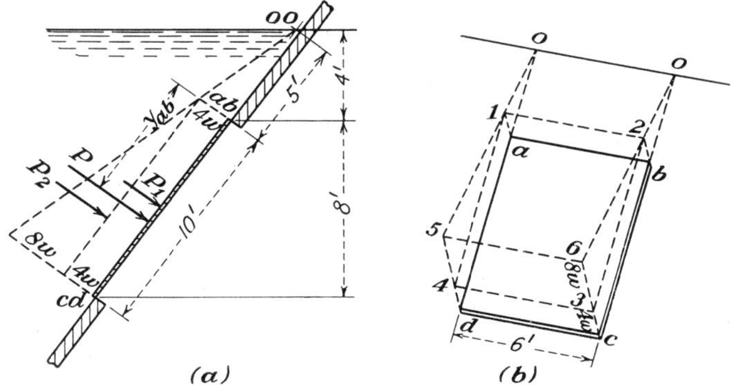

FIG. 12. Pressure force on a sloping gate.

Fig. 12. Let P_1 be the force resulting from the uniform pressure of $4w$; then

$$P_1 = 4wA = 4 \times 62.4 \times 60 = 15{,}000 \text{ lb}$$

Note that this force is equal to the volume 1234*dabc* shown in Fig. 12*b*.

Because P_1 is produced by a pressure of uniform intensity, it must act at the center of gravity of the gate, i.e., 5 ft from *ab*.

Let P_2 be the force resulting from the triangular pressure volume 123456. The intensity of pressure varies from 0 to $8w$, the average being $4w$. Therefore

$$P_2 = 4wA = 4 \times 62.4 \times 60 = 15{,}000 \text{ lb}$$

Because this force is produced by triangular load, its location will be $\frac{2}{3} \times 10 =$ 6.67 ft from *ab*.

Having P_1 and P_2, $P = P_1 + P_2 = 30{,}000$ lb, and the location of P may be determined by taking moments about *ab* as follows:

$$y_{ab}P = 5P_1 + 6.67P_2$$

from which $y_{ab} = 5.83$ ft.

Pressure Forces on Curved Surfaces. The resultant pressure force on a curved surface is made up of the sum of the small elements of force $p\ dA$ each acting perpendicular to the surface. The magnitude and location of the resultant of these elementary forces is not easily determined by the methods used for plane surfaces. However, the horizontal and vertical components of the resultant may be readily determined and then combined vectorially.

Consider the forces acting on the prism of liquid shown in Fig. 13 bounded by the liquid surface *ao*, by the vertical plane surface *ob*, and by the curved surface *ab*. Acting downward on this volume is its weight W and acting from right to left on *ob* is the horizontal force $P_H = w\bar{h}A$ where A is the area of the imaginary vertical plane surface one edge of which is *ob*. These forces must be put into equilibrium by equal and opposite forces from the curved surface *ab*. It follows then that the horizontal component of the pressure force on a curved surface is equal to and acts at the same point as the force on a vertical plane surface formed by projecting the curved surface horizontally. Furthermore, the vertical component of the force on a curved surface is equal to the weight of the liquid above the surface and acts at its center of gravity. Similar reasoning will show that when the liquid is beneath the curved surface the vertical component is equal to the weight of the imaginary volume of liquid above the surface and acts upward through its center of gravity. For example, the vertical component of the pressure force on the taintor gate of Fig. 14 is equal to the volume represented by *LNM* and acts upward through *G* as shown.

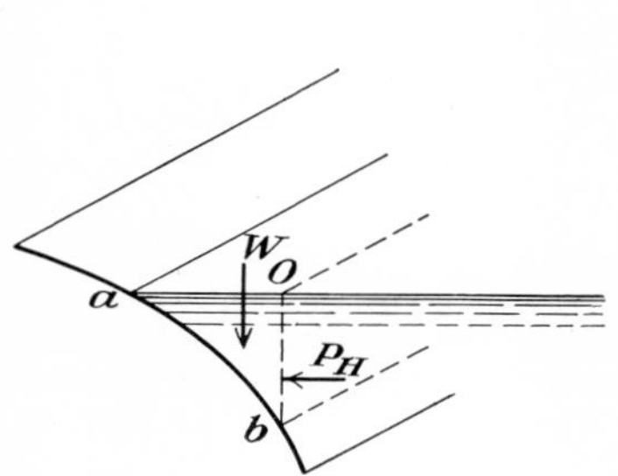

Fig. 13. Pressure forces on curved surfaces.

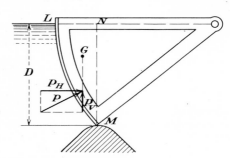

Fig. 14. Taintor gate.

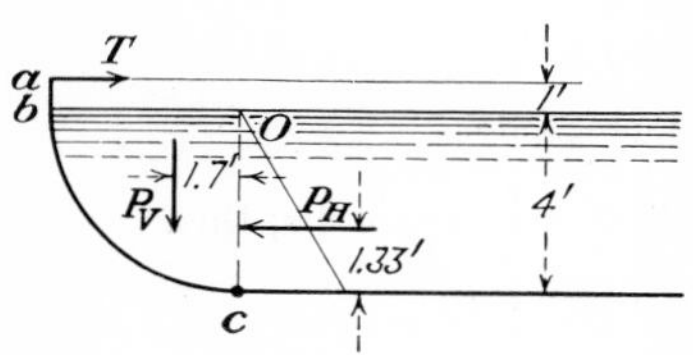

Fig. 15. Section through a water tank.

Example 9: Figure 15 shows a section through a water tank 20 ft long. The wall of the tank *abc* is hinged at *c* and supported by a horizontal tie rod at *a*. The segment of the wall *bc* is a quarter circle having a 4 ft radius.

a. Determine the force T in the tie rod.

$$P_H = w\bar{h}A = 62.4 \times 2 \times 80 = 10{,}000 \text{ lb}$$

The location of P is $\frac{1}{3} \times 4 = 1.33$ ft above c.

$$P_V = W = \frac{20 \times \pi r^2 \times w}{4} = \frac{20 \times \pi \times 4^2 \times 62.4}{4} = 15{,}700 \text{ lb}$$

P_V acts at the center of gravity of the quarter circle which is to left of *oc* by the amount

$4r/3\pi = 1.7$ ft. T may then be obtained by taking moments about the hinge as follows:

$$5T = \tfrac{4}{3} \times 10{,}000 + 1.7 \times 15{,}700 \qquad \text{and} \qquad T = 8{,}000 \text{ lb}$$

b. Determine the resultant water-pressure force on the gate.

$$P = \sqrt{(10{,}000)^2 + (15{,}700)^2} = 18{,}600 \text{ lb}$$

The direction and location of P may be found by combining P_V and P_H vectorially at their intersection. Since all the elementary components of P are perpendicular to the gate and therefore pass through point o, it follows that P must also pass through o.

c. Determine the resultant force on the hinge, neglecting the weight of the wall. From $\Sigma H = 0$, the horizontal component is found to be 2,000 lb, and from $\Sigma V = 0$, the vertical component is found to be 15,700 lb. The resultant force on the hinge is therefore

$$\sqrt{(15{,}700)^2 + (2{,}000)^2} = 15{,}800 \text{ lb}$$

The pressure distribution on the upstream face of overflow spillways is illustrated by the dotted line 1–5 of Fig. 16. The hydrostatic pressure distribution on the face of the dam would be as shown by 1234. The actual pressure at the top is less than this because

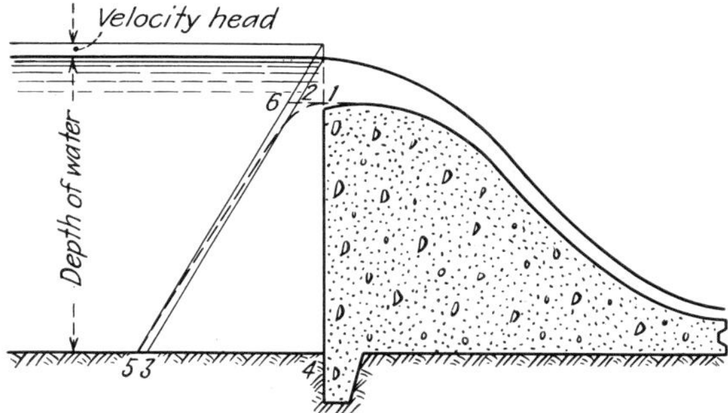

FIG. 16. Pressure distribution on overflow spillways.

the intensity of pressure at the crest of the spillway is approximately zero. Near the bottom, the actual pressure is greater than 2–3 if the velocity of approach is of appreciable magnitude, the effect being the same as at the stagnation point of a pitot tube (see p. 4–23). The pressure distribution 1654 is frequently used in making computations involving the stability of dams.[1]

TRANSLATED AND ROTATED LIQUIDS

Acceleration in Straight Line. If a vessel containing a liquid is moved in a straight horizontal path at a uniform velocity, the forces acting on any particle are balanced and the surface is horizontal, all internal pressures remaining the same as for a liquid at rest. The condition of a vessel containing liquid having uniform acceleration a in a horizontal straight line to the right is illustrated in Fig. 17. Consider any small mass of liquid M, at the surface. P_x and P_y are respectively the horizontal and vertical components of a resultant pressure force P exerted on the mass by the surrounding water. Since this particle is in relative equilibrium, the resultant pressure must be normal to the free sur-

[1] A more detailed discussion of this subject may be found in HINDS, CREAGER, and JUSTIN, "Engineering for Dams," vol. 2, pp. 255–260.

face. P_y is balanced by the weight Mg of the mass. P_x is the unbalanced force producing acceleration, and therefore $P_x = Ma$.

Hence

$$\frac{P_x}{P_y} = \frac{a}{g} \tag{23}$$

Constructing the parallelogram of forces, making P_x and P_y proportional respectively to a and g, gives the direction of the resultant force P. Since the resultant acts normal to the free surface and M is any mass on the surface, the liquid surface must be a plane. From the figure, θ, the angle between P and P_y, equals the angle of inclination of the liquid surface, and

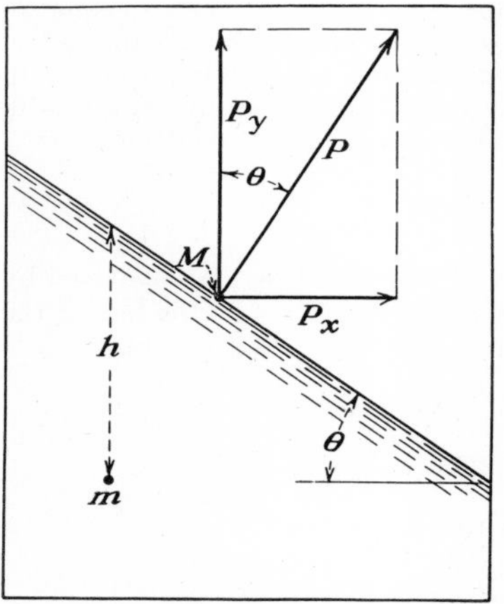

Fig. 17. Acceleration of liquid in horizontal straight line.

$$\tan\theta = \frac{P_x}{P_y} = \frac{a}{g} \tag{24}$$

From this equation, the slope of the surface of a liquid for any uniform horizontal acceleration can be obtained.

Since in Fig. 17 the liquid is not accelerated vertically, the intensity of pressure at any point m, where w is the unit weight of the liquid, is $p = wh$, the same as for liquids at rest.

In order that a vessel containing liquid may be accelerated vertically upward, a force in excess of the weight of the liquid and container must be applied. Any force in excess of this weight produces an acceleration a. In this case, the liquid surface will be level, and the intensity of pressure at any point m (Fig. 18) will be

$$p = wh + wh\frac{a}{g} = wh\left(1 + \frac{a}{g}\right) \tag{25}$$

If the force producing acceleration equals the weight, $a = g$, and the intensity of pressure is twice that due to gravity alone. If the acceleration is downward, the last term in Eq. (25) becomes negative; and for a liquid falling freely in a vacuum, $a = -g$, and the pressure throughout the liquid is zero.

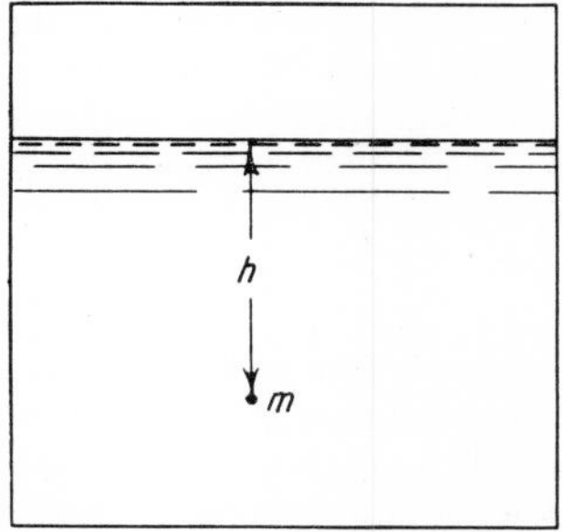

Fig. 18. Acceleration of liquid in vertical line.

The condition of a liquid being accelerated along a straight line inclined upward at an angle α with the horizontal is illustrated in Fig. 19. The resultant pressure force P, acting on the small mass M, has the components P_α and P_y along inclined axes. The accelerating force P_α is the vector difference between the total force applied and the force required to overcome gravity. The direction of P can, as for horizontal acceleration, be obtained by plotting P_α and P_y in their proper directions proportional to a and g and completing the parallelogram of forces. As before, the surface of the liquid is a plane perpendicular to P.

The angle θ which the plane makes with the horizontal can be determined by scaling from the diagram or by solving the triangle of forces.

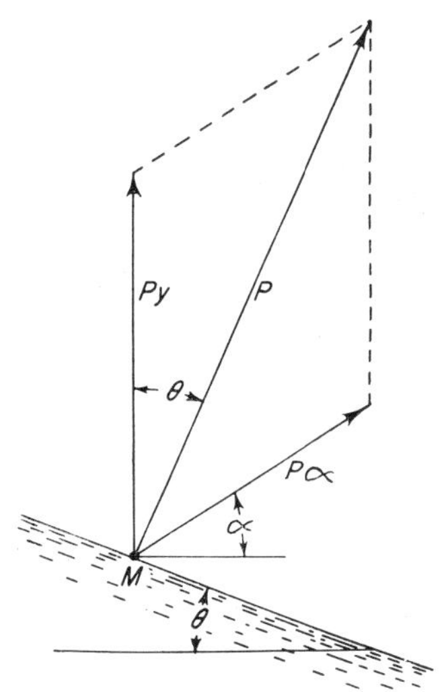

FIG. 19. Acceleration of liquid in inclined straight line.

Rotating Liquids. A cylindrical vessel containing liquid rotating at a constant velocity about a vertical axis OY is shown in Fig. 20.

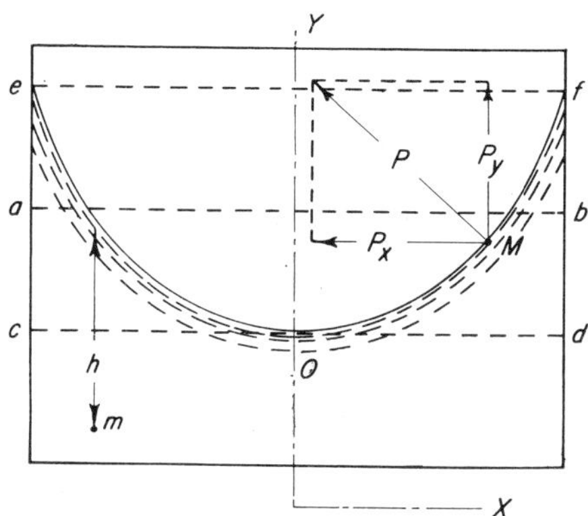

FIG. 20. Rotating liquid with free surface.

The original liquid surface ab is changed by rotation to the curved surface eOf. The velocity of rotation is ω radians per second. The horizontal and vertical components of the resultant force P acting on the small mass M on the surface of the liquid at a distance x from the axis are respectively P_x and P_y. The latter is balanced by the weight Mg of the mass. The unbalanced force P_x, producing acceleration toward the center of rotation, is the centripetal force, and therefore $P_x = M\omega^2 x$. If θ is the angle which the tangent to the curve makes with the horizontal, and x and y are taken as coordinates of M, referred to the origin O,

$$\tan\theta = \frac{dy}{dx} = \frac{P_x}{P_y} = \frac{M\omega^2 x}{Mg} = \frac{\omega^2 x}{g} \tag{26}$$

After integration, Eq. (26) becomes

$$y = \frac{\omega^2 x^2}{2g} \tag{27}$$

This is the equation of a parabola, and the surface is a paraboloid of revolution about the axis Y. Since the volume of a paraboloid is equal to one-half the volume of the circumscribed cylinder and the volume of the liquid remains constant, $ca = ae$ (Fig. 20); i.e., the maximum depression of the liquid surface is equal to the maximum rise. The linear velocity at M is $v = \omega x$. Substituting v for ωx in Eq. (27) gives $y = v^2/2g$; i.e., any point on the surface of the liquid will rise above the point of greatest depression an amount equal to the velocity head (see p. 4–22) at that point.

Since there is no vertical acceleration, the intensity of pressure at any point m (Fig. 20) within the liquid is $p = wh$, the same as for a liquid at rest. It follows that the ordinates to curve eOf represent graphically the pressures on the bottom of the containing

vessel and that the pressures on the sides of the vessel are the same as for a vessel at rest filled to the level *ef*.

If a cylindrical vessel with closed top (Fig. 21) is completely filled with liquid and revolved about a vertical axis, the liquid, being confined, cannot assume a parabolic surface. By methods analogous to those described above, it can be shown, however, that intensities of pressure against the top of the container are given by the formula $p/w = \omega^2 x^2/2g$ or by Eq. (27). This means that the intensity of pressure is zero at the center and varies as the ordinates of a parabola.

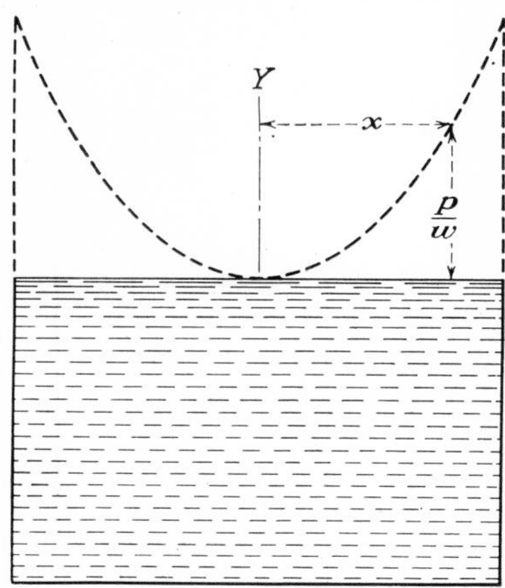

FIG. 21. Rotating liquid in closed cylinder.

IMMERSED AND FLOATING BODIES

Buoyancy. Any body when immersed in a liquid suffers a loss in weight equal to the weight of liquid displaced. In the immersed body shown in Fig. 22, the horizontal forces acting on the two ends of the small horizontal prism *ab* are equal and opposite in direction. Since the entire body can be divided into such prisms, the horizontal components of pressures acting on the body must balance. There is a resultant pressure acting upward against the bottom of the small vertical prism *cd* equal to $w(h_2 - h_1)$ multiplied by the cross-sectional area of the prism, w being the unit weight of the liquid. This is equal and opposite in direction to the weight of a similar prism of the liquid. Since the entire body can be divided into small vertical prisms, the whole upward pressure against it, or the *buoyant force*, must be equal to the weight of a volume of the liquid of the same volume as the body. If the weight of a body is greater than its buoyant force, it will sink, but its weight will be decreased by an amount equal to the weight of displaced liquid. If the weight of a body is less than the buoyant force, it will float, displacing a volume of liquid having the same weight as the body.

The buoyant force acts through the center of gravity of the displaced liquid at a point termed the *center of buoyancy*. If a body is homogeneous and wholly immersed, its center of gravity will coincide with the center of buoyancy. Under other conditions, the two points do not usually coincide.

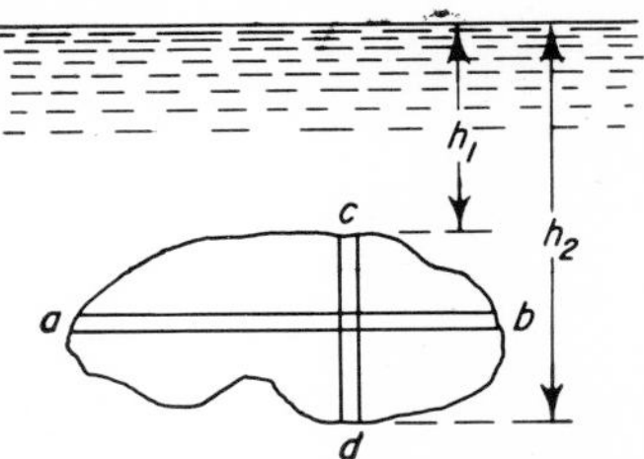

FIG. 22. Immersed body.

Equilibrium of Floating Bodies. The weight and buoyancy of floating bodies are equal, but the former acts at the center of gravity of the body and the latter acts at the center of buoyancy. In order that the body may be in equilibrium, these two points must lie in the same vertical line. Figure 23*a* illustrates the cross section of a ship floating in equilibrium. The center of gravity is at G, and the center of buoyancy at B, both on the axis of symmetry. In Fig. 23*b*, the ship is heeled through an angle θ. The center of gravity remains at G if the cargo does not shift, but the center of buoyancy is moved to B'. In this position, the two equal and opposing forces form a righting couple tending to restore equilibrium. The point M, where a vertical line through the center of buoyancy intersects the axis of symmetry, is

termed the *metacenter*, and its distance MG from the center of gravity of the ship is termed the *metacentric height*. If M is above G, there is a righting moment; and if below G, an overturning moment. The metacentric height is a measure of the stability

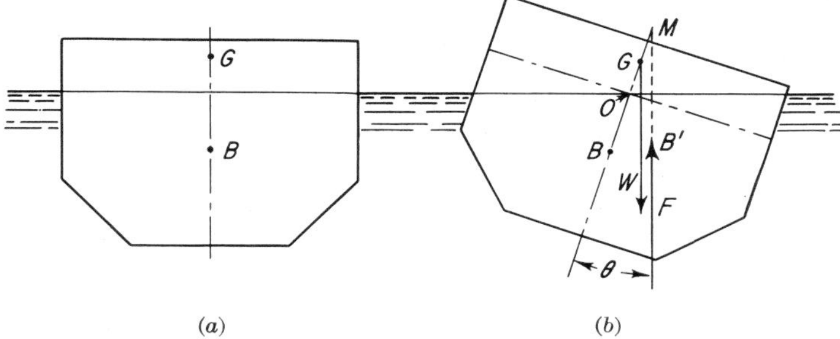

FIG. 23. Equilibrium of floating bodies.

of a ship. It changes for different values of θ; but for values not greater than 10° or 15°, it is nearly constant.

The metacentric height can be computed by the formula[1]

$$GM = \frac{I}{V} \pm GB \tag{28}$$

in which I is the moment of inertia of the water-line section about the longitudinal axis through O and V is the displacement volume. The sign of the last term of the equation is positive if G falls below B and negative if it falls above.

THE FLOW OF FLUIDS

The science that treats of fluids in motion is often divided into *kinetics*, which describes the motion itself, and *dynamics*, which deals with the forces producing motion. The quantities which describe the motion of a fluid (kinetic quantities) are those shown in Table 1 which have dimensions involving length and time, whereas those which include mass in their dimensions are dynamic quantities.

A particle of fluid will remain at rest as long as the vector sum of all forces acting on it is zero. When the summation of forces results in a vector other than zero, motion will occur in the direction of the unbalanced force. This force may be the result of a pressure difference such as occurs near an orifice in the wall of a tank containing a liquid or of an unbalanced component of the gravity force such as occurs when liquid is deposited on an inclined surface. If there were no resistance to the motion, the particle of fluid would be accelerated at a rate proportional to the applied force and liquid on an inclined plane could continue to accelerate indefinitely as in the case of a body falling freely through a vacuum. However, as has been mentioned previously, motion in fluids is associated with shear stresses which tend to retard the acceleration and eventually produce a uniform velocity, much as the gravity force on a toboggan sliding downhill increases its velocity only to the point where the resistance of the snow is equal to the applied force. The above analogy is not perfect because in fluid motion there is a continuous deformation causing viscous stresses which resist motion within the fluid itself rather than only at its boundaries as in the case of solids. Having concluded that forces are necessary to maintain fluids in constant motion and that the motion is in the direction

[1] For a derivation of this formula and further discussion of this subject, see KING, WISLER, and WOODBURN, "Hydraulics," also STREETER, V. L., "Fluid Mechanics."

of the force, it follows that work is being done or energy is being expended in the process. This energy may be drawn from the initial potential energy of the fluid, as in the case of a liquid flowing from a higher to a lower elevation, or from the work being done by a mechanical device such as a pump. In either case energy is expended in the form of heat, raising the temperature of the fluid, its container, and the surrounding atmosphere. In a number of cases of flow, energy losses are relatively small as, for example, flow from a sharp-edged orifice or the tangential impingement of a jet of water upon a vane; but even in these cases the results obtained by neglected energy losses in computations are usually too much in error for engineering purposes. As a result, mathematical derivations of flow laws based on the assumption of a frictionless (ideal) fluid are not of direct use to the engineer. Such computations, however, are often of great indirect value in

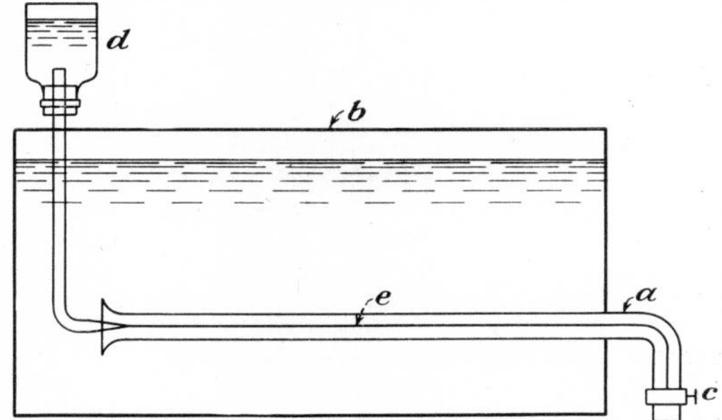

FIG. 24. Reynolds experiment on laminar flow.

providing a mathematical framework upon which to place the correction factors determined by tests. The energy loss due to viscous shear may be treated mathematically, as will be shown for flow through pipes and open channels. The derived equations agree with the results of tests as long as flow is laminar.

Laminar flow is the type described in the definition of viscosity, i.e., each layer of fluid travels parallel to the adjacent layers and the paths of individual particles do not cross. This may be illustrated by the apparatus shown in Fig. 24, in which water is drawn through the glass tube *a* from the large tank *b*, the flow being regulated by the valve at *c*. A fine stream of dye may then be injected from *d*. When the velocity in the glass tube is low, the dye will be carried through the tube as a single thread *e*. As the velocity is increased, a point is reached when the dye will mingle with the water throughout the cross section of the tube, indicating that cross-current velocities have developed. When cross-current velocities are present the flow is called *turbulent flow*. The rate of energy dissipation for a given discharge is much greater in turbulent flow than in laminar flow. Turbulent flow is characterized by the presence of eddies which in their formation and disintegration are largely responsible for the additional energy loss.

The existence of the two types of flow was described by Hagen in his publications (1839 to 1869), but the law governing the transition between laminar and turbulent flow was stated by Osborne Reynolds (1883).[1] He showed that the point at which pipe flow changes from laminar to turbulent depends upon the velocity, density, and viscosity of the fluid and upon the diameter of the pipe, arranged in the following dimensionless form, later called the "Reynolds number."

$$R = \frac{dv\rho}{\mu} \tag{29}$$

[1] The history of the early developments in the science of flowing fluids is given in PRANDTL and TIETJENS, "Applied Hydro- and Aeromechanics," Chap. III.

The significance of the Reynolds number lies in the fact that it is the ratio of the inertial force to the viscous force acting on a unit of fluid mass. When R is very large the inertial or disturbing effects are large compared with the stabilizing viscous effects and flow is fully turbulent. When R is very small the viscous forces are in full control and flow is laminar. For intermediate values of R there exists a transition range in which flow may be either laminar or turbulent. The usefulness of the Reynolds number is not restricted to pipe flow, for it is the all-important unifying factor in plotting test results for many other flow conditions.

Flow may be classified as steady or unsteady. Flow is *steady* if (1) the velocity at any point does not vary with time, and (2) the continuity equation is satisfied. *Unsteady flow* does not satisfy one or both of these conditions. The *continuity equation* requires that the mass discharge ρQ be constant at all points in a conduit. In other words, it means that there is no opening in the conduit for fluid to be gained or lost and that the flow has stabilized in such a manner that any change of state occurring within the conduit is taking place at a constant rate. If v is the average velocity at any cross section through a conduit and a the cross-sectional area, then $Q = av$ and the continuity equation may be written

$$\rho Q = \rho_1 a_1 v_1 = \rho_2 a_2 v_2 = \cdots \rho_n a_n v_n \tag{30}$$

If the fluid is incompressible ρ is constant and the equation may be written

$$Q = a_1 v_1 = a_2 v_2 = \cdots a_n v_n \tag{31}$$

The continuity equation is sometimes written for a small portion of the area of the conduit da which is bounded by streamlines. *Streamlines* are paths of individual fluid particles which are assumed not to cross each other as in the case of laminar flow. Although streamlines do not exist in turbulent flow the analogy is helpful in studying the general nature of the flow pattern. A series of such streamlines arranged systematically across the entire cross section of a conduit is called a *flow net.*[1]

Flow may be further classified as uniform or nonuniform. In *uniform flow*, the area and shape of a cross section through the flowing stream are the same at successive points such as in the case of flow in a pipe of constant diameter. In *nonuniform* flow the fluid is being accelerated not only in the direction of flow but in the normal direction as in the case of flow through a venturi meter (p. 4–25).

Energy and the Bernoulli Equation. The energy of a flowing stream may be evaluated by considering a small particle of weight W as shown in Fig. 25. This particle is located at a point in a closed conduit where the pressure is p' and the velocity is u. The kinetic energy of the particle is then $Wu^2/2g$ ft-lb. The potential energy due to the elevation of the particle above the datum z' is Wz' ft-lb. If the motion in the conduit is steady, the particle has available the additional potential energy Wp'/w ft-lb, which could be converted to elevation energy or kinetic energy. The total energy of this particle is then

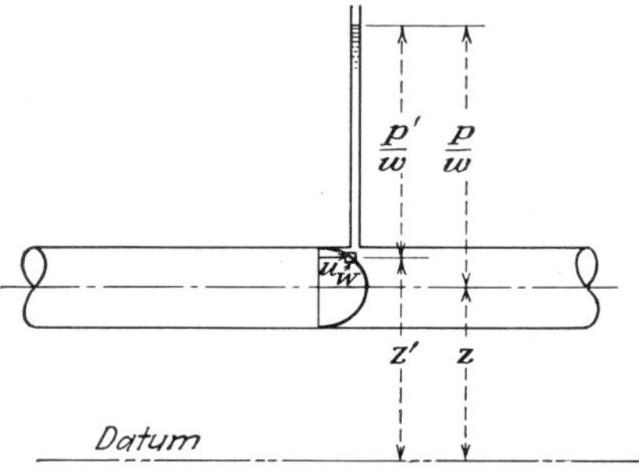

FIG. 25. Energy of a flowing fluid.

$$E = \frac{Wu^2}{2g} + \frac{Wp'}{w} + Wz' \text{ ft-lb} \tag{32}$$

[1] For a detailed description of flow nets see ROUSE and HOWE, "Basic Mechanics of Fluids" and STREETER, V. L., "Fluid Mechanics."

and at that particular point in the conduit the energy per unit weight is

$$\frac{E}{W} = \frac{u^2}{2g} + \frac{p'}{w} + z' \text{ ft-lb per lb} \tag{33}$$

For the purpose of solving flow problems it is necessary to obtain an expression for the average energy per pound passing a particular cross section of a conduit. An examination of Fig. 25 will show that the sum of the latter two terms in Eq. (33) is constant for all points in the cross section. Thus these two terms are already perfectly general and may be replaced by p/w and z, which apply to the center of the pipe. However, the term in Eq. (33) for kinetic energy $u^2/2g$ applies only to a particular portion of the conduit and the determination of an average value requires further consideration. The expression for the average kinetic energy is usually written as shown by Eq. (34), where α is a number greater than one and v is the average velocity in the cross section.

$$\text{KE per pound} = \alpha \frac{v^2}{2g} \tag{34}$$

The average kinetic energy per pound for the entire conduit section may be obtained by adding the small quantities of energy passing through each elementary area da per second and then dividing by the total flow in pounds per second as follows:

$$\text{KE per pound} = \frac{\int wu\, da \frac{u^2}{2g}}{wva} \tag{35}$$

If a velocity-distribution curve is available, the integration may be carried out either mathematically or graphically [1] and the value of α determined by equating Eqs. (34) and (35) (see p. 4–61). The value of α is approximately 1.1 for turbulent flow in a straight conduit but may be as high as 2 near obstructions.[2] For laminar flow in pipes it is 2 (p. 4–62) and in open channels approximately 1.5. For ordinary hydraulic computations dealing with turbulent flow, α is usually taken as 1 either for lack of information concerning its value in particular problems or because the error created thereby is negligible. By substituting $\alpha v^2/2g$, p and z for $u^2/2g$, p' and z' respectively in Eq. (33), and taking $\alpha = 1$, the following general expression for the average energy per pound of fluid is obtained:

$$H = \frac{v^2}{2g} + \frac{p}{w} + z \text{ ft-lb per lb} \tag{36}$$

Because each term in Eq. (36) represents energy per unit weight, i.e., foot-pounds per pound, the dimensions of each term are equivalent to length. This may be verified by examining the terms dimensionally by the method illustrated on p. 4–4. The letter H is used in Eq. (36) because, being equivalent to length, the terms are frequently called "head." An expression for the energy per unit of volume may be obtained by multiplying the terms of Eq. (36) by the specific weight w as shown in Eq. (37).

$$wH = \frac{wv^2}{2g} + p + wz \text{ ft-lb per cu ft} \tag{37}$$

In the latter expression w may be replaced by ρg as follows:

$$\rho g H = \frac{\rho v^2}{2} + p + \rho g z \text{ ft-lb per cu ft} \tag{38}$$

[1] O'BRIEN and JOHNSON, Velocity-head Correction for Hydraulic Flow, *Eng. News-Record*, Aug. 16, 1934.
[2] KING, H. W., "Handbook of Hydraulics," 4th ed., p. 7-12.

Any of the above expressions is called the *Bernoulli constant* after Daniel Bernoulli, who was probably the first to discuss the forms of flow energy (1738). For a frictionless (ideal) fluid the sum of three terms in Bernoulli's constant is the same at all points in the stream. Mathematically this may be stated by the *Bernoulli equation*

$$\frac{v^2}{2g} + \frac{p_1}{w} + z_1 = \frac{v_2^2}{2g} + \frac{p_2}{w} + z_2 \tag{39}$$

Because fluids are not ideal but viscous, and all fluid movement is associated with energy dissipation, the energy at the downstream point 2 will be less than at the upstream point 1. Consequently the above equation must be corrected for the loss as follows:

$$\frac{v_1^2}{2g} + \frac{p_1}{w} + z_1 = \frac{v_2^2}{2g} + \frac{p_2}{w} + z_2 + h_l \tag{40}$$

Bernoulli's equation will first be applied to a tank of liquid at rest. The velocity term is then zero and Eq. (36) becomes

$$H = \frac{p}{w} + z \tag{41}$$

These terms are shown graphically in Fig. 26, from which it may be seen that H is the same for all the liquid in the tank. If the vertical distance below the water surface $(H - z)$ is replaced by h in Eq. (41), it becomes

$$h = \frac{p}{w} \tag{12}$$

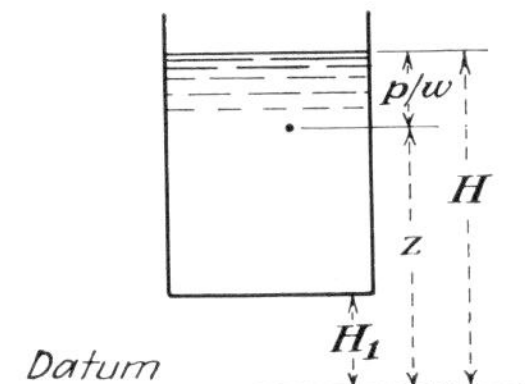

FIG. 26. Tank of liquid at rest.

which was derived in the study of hydrostatics. If the liquid in the tank were utilized to do work, as by directing a jet at a 100 percent efficient turbine placed at the datum plane, the work done by each pound of liquid would be H provided that there were a source of supply which would keep the liquid surface at a constant elevation. If on the other hand there were no supply and the tank were emptied in the process, only the first particle of liquid would have the energy H and all others would have less, the final particle having only H_1 ft-lb per lb. Obviously in the second case Eq. (41) cannot be used to determine the total energy of the liquid in the tank, thus illustrating why, in the discussion of energy, the pressure term was restricted to steady motion.[1]

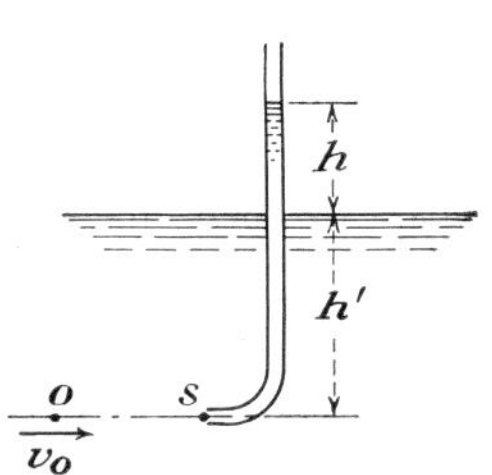

FIG. 27. Pitot tube.

Pitot Tube. In its simplest form, the pitot tube is a pipe with open ends having a right-angle bend near one as shown in Fig. 27. When it is immersed in a flowing stream with its open end directed against the current, water will rise in the tube to a height h above the water surface. Writing Bernoulli's equation from o to s in Fig. 27, neglecting energy losses,

$$\frac{v_o^2}{2g} + \frac{p_o}{w} + z_o = \frac{v_s^2}{2g} + \frac{p_s}{w} + z_s$$

if the datum plane is taken through points o and s, z_o and z_s will be zero. Also, p_o/w is equal to h' and p_s/w is equal to $(h' + h)$. Finally, v_s is zero because s is the stagnation point where particles of fluid are brought to a complete stop. Then

[1] The derivation of Bernoulli's constant by other methods may serve to clarify this matter further. See, for example, KING, WISLER, and WOODBURN, "Hydraulics," 5th ed., p. 94; STREETER, V. L., "Fluid Mechanics," p. 81.

$$\frac{v_o^2}{2g} + h' + o = o + (h' + h) + o$$

and

$$v_o = \sqrt{2gh} \tag{42}$$

This simple method of measuring velocity has little practical value, since the height to which water will rise in the tube cannot be accurately measured. If the tube shown in the figure is turned through 180° so that the opening points downstream, a suction is created. The distance that the water in the tube will then be below the water surface in the stream has been found experimentally to be about $0.43v^2/2g$. If a straight tube with open ends is inserted vertically into a stream, a similar suction is created. For this reason, piezometer tubes (see p. 4-8) used to measure static pressure in streams should not project into the water but should be made flush with the conduit walls.

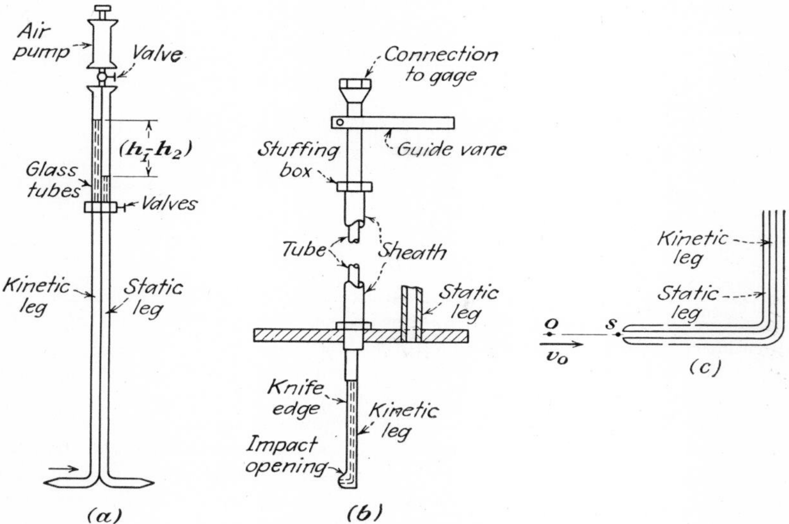

FIG. 28. Three arrangements of pitot tubes.

Several arrangements for utilizing the pitot tube principle for measuring velocities have been devised. All these devices have a kinetic leg directed against the current and also a static leg, the difference in pressure in the two legs being a measure of the velocity. Figure 28*a* illustrates a design by Darcy for measuring velocities in open channels. Glass tubes are connected to the upper ends of the legs and also to an air suction pump through a common air chamber containing a valve. There are two valves just below the glass tubes. Since this device is symmetrical, either leg can be pointed against the current and thus serve as the kinetic leg. By operating the pump, air is rarefied sufficiently to draw both water columns into the glass tubes. The valves are then closed and the instrument is lifted from the water, and the difference in heights of columns $h_1 - h_2$ is read. Then, $v = C\sqrt{2g(h_1 - h_2)}$. The coefficient C, a constant for each instrument, should be obtained from a rating in still water.

The pitot tube shown in Fig. 28*b* is adapted to measuring velocities in pipes.[1] With

[1] This design was recommended in the report of the Special Research Committee of the ASME on "Fluid Meters, Their Theory and Application," 1924.

this arrangement, the static leg is entirely separate from the kinetic leg. The static leg should be flush with the inner surface of the pipe.

Another type of pitot-static tube is shown in Fig. 28*c*. If Bernoulli's equation is written from *o* to *s* (Fig. 28*c*) it takes the following form:

$$\frac{v_o^2}{2g} + \frac{p_o}{w} + o = o + \frac{p_s}{w} + o$$

and

$$v_o = \sqrt{2g\left(\frac{p_s}{w} - \frac{p_o}{w}\right)} \tag{43}$$

This equation applies to the arrangements shown in Fig. 28*b* and 28*c*, either of which may be used to measure velocities in gases as well as in liquids. The value of $\left(\frac{p_s}{w} - \frac{p_o}{w}\right)$ may be obtained by the use of a U tube containing a liquid heavier than the flowing fluid. Pitot-static tubes of the type shown in Fig. 28*c* are frequently used as air-speed indicators for airplanes. It should be noted that v_o is the velocity of the fluid with respect to the tube, either of which may be in motion. In fact, pitot tubes are often calibrated by moving them at known velocities through fluids at rest. All pitot tubes must be calibrated to determine a correction coefficient to be applied to Eq. (43) as follows:

$$v = C\sqrt{2g\left(\frac{p_s}{w} - \frac{p_o}{w}\right)} \tag{44}$$

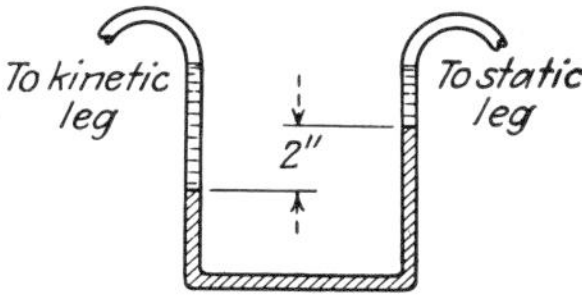

FIG. 29. Mercury U tube.

Example 10: Assume that the pitot-static tube of Fig. 28*b* is installed in a water pipe and that its legs are connected to the mercury U tube shown in Fig. 29. Determine the theoretical velocity of the water striking the pitot tube.

From Eq. (16),

$$\begin{aligned} p_s - p_o &= \Delta h(w_m - w_p) \\ &= \tfrac{2}{12}(13.6 \times 62.4 - 62.4) = 131 \text{ lb per sq ft} \end{aligned}$$

Therefore,

$$\left(\frac{p_s}{w} - \frac{p_o}{w}\right) = \frac{131}{62.4} = 2.1 \text{ ft}$$

Then from Eq. (44),

$$v_o = C\sqrt{2g\left(\frac{p_s}{w} - \frac{p_o}{w}\right)}$$

$$= 1 \times 8.02\sqrt{2.1} = 11.6 \text{ ft per sec}$$

Venturi Meter. A section of a venturi meter reproduced from a drawing contained in the report of the Special Committee on Fluid Meters of the ASME is shown in Fig. 30. It consists of a constricted portion of a pipe which accelerates the water and lowers its static pressure. Starting at the upstream end, there is first a short cylindrical portion which contains a piezometer-tube connection. There then follows the entrance cone of about 21° total angle, which leads to the throat, a short cylinder containing a connection for a second piezometer tube. The diameter of the throat is usually between one-half and one-fourth of the entrance or pipe diameter. The end of the throat leads into the

exit cone or diffuser, having a total angle of 5° to 7°, which terminates in the following pipe line. A pressure connection is sometimes provided at the downstream end to determine the overall loss of head in the venturi. The function of the long diverging cone is to decelerate the water smoothly and restore the pressure as nearly as practicable

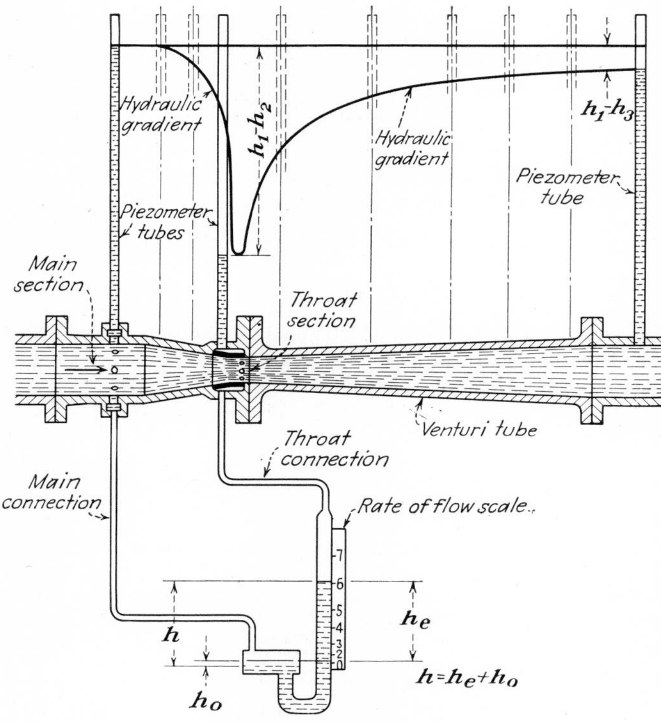

FIG. 30. Venturi meter.

to the entrance pressure. The overall loss of head is 10 to 20 percent of the differential from entrance to throat.

If in Fig. 30 point 1 is at the center of the main section and point 2 is at the center of the throat section, the Bernoulli equation applied from 1 to 2 is

$$\frac{v_1^2}{2g} + \frac{p_1}{w} + z_1 = \frac{v_2^2}{2g} + \frac{p_2}{w} + z_2 + h_l \tag{40}$$

If the datum plane is taken through points 1 and 2, the z terms are zero and from Eq. (31),

$$v_1 = \left(\frac{d_2}{d_1}\right)^2 v_2 \tag{45}$$

Equation (40) may now be written

$$\left(\frac{d_2}{d_1}\right)^4 \frac{v_2^2}{2g} + \frac{p_1}{w} + o = \frac{v_2^2}{2g} + \frac{p_2}{w} + o + h_l$$

and

$$v_2 = \sqrt{\frac{2g\left[\left(\frac{p_1}{w} - \frac{p_2}{w}\right) - h_l\right]}{1 - \left(\frac{d_2}{d_1}\right)^4}} \tag{46}$$

It is convenient to take care of the energy loss by means of a coefficient C having a value less than 1 rather than by subtracting its actual value as in Eq. (46). Inserting C and multiplying by a_2,

$$Q = a_2v_2 = Ca_2\sqrt{\frac{2g\left(\frac{p_1}{w} - \frac{p_2}{w}\right)}{1 - \left(\frac{d_2}{d_1}\right)^4}} \tag{47}$$

For any particular ratio of throat diameter to pipe diameter, values of C vary with the velocity, the throat diameter, the viscosity, and the density. It is to be expected that the combined effect of these four quantities on C can be best determined by arranging

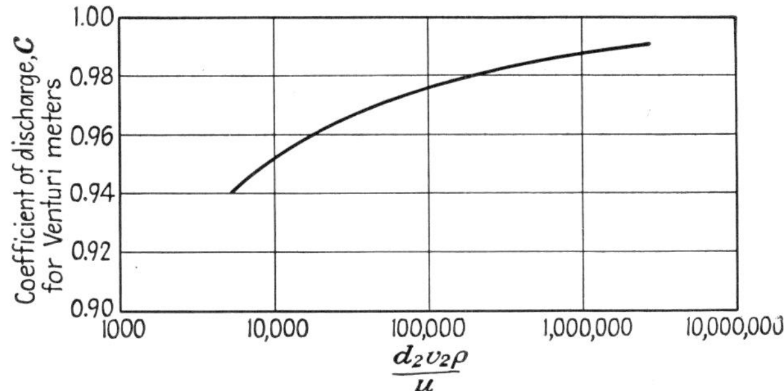

FIG. 31. Venturi meter coefficients.

them in the dimensionless form of Reynolds number. Values of C plotted against Reynolds number for meters having a ratio of pipe to throat diameter varying from 2:1 to 3:1 result in the curve shown in Fig. 31.[1] These values are applicable not only to liquids but to gases if the pressure drop between points 1 and 2 (Fig. 30) is not more than 1 percent of the pressure at 1. For larger pressure drops, corrections for density changes must be made.[1]

Example 11: Determine the discharge of water at 80°F through the venturi shown in Fig. 30 if the throat diameter is 6 in., the pipe diameter is 12 in., h is 6 in., and the manometer fluid is mercury.

From Eq. (16),

$$\begin{aligned} p_1 - p_2 &= \Delta h(w_m - w_p) \\ &= \tfrac{6}{12}(13.5 \times 62.2 - 62.2) = 389 \text{ lb per sq ft} \end{aligned}$$

and

$$\left(\frac{p_1}{w} - \frac{p_2}{w}\right) = \frac{389}{62.2} = 6.25 \text{ ft}$$

[1] "Fluid Meters, Their Theory and Application," 3d ed., ASME, 1931.

The theoretical value of v_2 is, from Eq. (46),

$$v_2 = \sqrt{\frac{2g\left(\frac{p_1}{w} - \frac{p_2}{w}\right)}{1 - \left(\frac{d_2}{d_1}\right)^4}}$$

$$= 8.02\sqrt{\frac{6.25}{1 - (\frac{1}{2})^4}} = 20.7 \text{ ft per sec}$$

Therefore

$$R = \frac{d_2 v_2}{\nu} = \frac{\frac{6}{12} \times 20.7}{0.0000093} = 1{,}110{,}000$$

From Fig. 31,

$$C = 0.99$$

and

$$Q = Ca_2 v_2 = 0.99 \times 0.196 \times 20.7 = 4.01 \text{ cu ft per sec}$$

ORIFICES

Flow of water through an orifice is illustrated in Fig. 32. Water approaches the orifice through the channel *abcd*, the quantity of water flowing in the channel being the same as the quantity flowing through the orifice, so that the flow is steady. The orifice as shown is *sharp-edged;* i.e., it has a sharp upstream corner so that the water in passing touches only a line. The stream of water issuing from the orifice is termed the *jet.* If the orifice discharges into the air, it is said to have *free discharge;* and if it discharges under water, it is said to be *submerged.* The *channel of approach* is the channel leading up to the orifice, and the mean velocity v_1 in this channel is the *velocity of approach.* Orifices may be circular, square, rectangular, or of any other regular form.

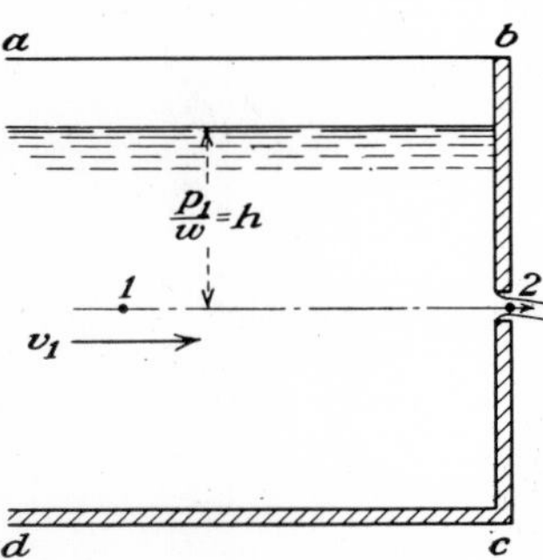

Fig. 32. Flow through orifice.

If the orifice is not too close to the sides or bottom of the channel or to the surface of the water, the water particles approach it in uniformly converging paths from all directions; and since it is not possible for the water particles abruptly to change their directions immediately upon leaving the orifice, they continue to travel for some distance in curvilinear paths, causing the jet to contract. The section where contraction ceases (2, Fig. 32) is called the *vena contracta.* The vena contracta for a circular orifice is about one-half the diameter of the opening downstream from the plane of the orifice.

The Bernoulli equation written from any point in the liquid such as 1 in Fig. 32 to the vena contracta 2 taking the datum plane through the center of the orifice is

$$\frac{v_1^2}{2g} + \frac{p_1}{w} = \frac{v_2^2}{2g} + \frac{p_2}{w} + h_l$$

and

$$v_2 = \sqrt{2g\left(\frac{p_1}{w} - \frac{p_2}{w} + \frac{v_1^2}{2g} - h_l\right)} \tag{48}$$

Point 2 being located where the jet has ceased to contract, its pressure is that of the surrounding fluid. For discharge into the atmosphere p_2 is therefore zero on the gage scale.

For large tanks, v_1 is so small that it may be neglected. Replacing p_1/w with h and dropping the subscript of v_2, Eq. (48) may now be written

$$v = \sqrt{2g(h - h_l)} \tag{49}$$

Neglecting energy losses, the equation for the theoretical velocity becomes

$$v_t = \sqrt{2gh} \tag{50}$$

The energy loss may be taken care of by applying a coefficient of velocity C_v to the theoretical velocity as follows:

$$v = C_v\sqrt{2gh} \tag{51}$$

Smith and Walker [1] found values of C_v to vary from 0.954 to 0.991 for orifices varying in diameter from 0.75 in. to 2.5 in. respectively. They also found a small variation with head, the above values being averages for heads varying from 1 to 60 ft.

The discharge through an orifice is obtained from the product of the velocity and the area at the vena contracta. The area at the vena contracta a_2 is less than the area of the orifice a, the ratio between the two being called the coefficient of contraction C_c. Therefore,

$$a_2 = C_c a \tag{52}$$

and

$$Q = a_2 v_2 = C_c a C_v \sqrt{2gh} \tag{53}$$

Values of C_c for circular sharp-edged orifices have been found to vary from approximately 0.67 for ¾-in. orifices to 0.614 for 2.5-in. orifices when the head is 2 ft or more.[1] Values are slightly larger for lower heads.

The product of C_c and C_v is called the coefficient of discharge C. Equation (53) may therefore be written

$$Q = Ca\sqrt{2gh} \tag{54}$$

One of the earliest experimenters on sharp-edged orifices was Hamilton Smith, Jr.[2] His values of the coefficient of discharge for round and square orifices are given in Table 3. There have been many subsequent investigations of circular orifices, not all of which

TABLE 3. SMITH'S COEFFICIENTS OF DISCHARGE FOR CIRCULAR AND SQUARE ORIFICES WITH FULL CONTRACTION

Diameter of circular orifices, ft							Head, ft	Side of square orifices, ft						
0.02	0.04	0.07	0.1	0.2	0.6	1.0		0.02	0.04	0.07	0.1	0.2	0.6	1.0
	0.637	0.624	0.618	.....	.....	.....	0.4	.....	0.643	0.628	0.621			
0.655	0.630	0.618	0.613	0.601	0.593	.....	0.6	0.660	0.636	0.623	0.617	0.605	0.598	
0.648	0.626	0.615	0.610	0.601	0.594	0.590	0.8	0.652	0.631	0.620	0.615	0.605	0.600	0.597
0.644	0.623	0.612	0.608	0.600	0.595	0.591	1	0.648	0.628	0.618	0.613	0.605	0.601	0.599
0.637	0.618	0.608	0.605	0.600	0.596	0.593	1.5	0.641	0.622	0.614	0.610	0.605	0.602	0.601
0.632	0.614	0.606	0.604	0.599	0.597	0.595	2	0.637	0.619	0.612	0.608	0.605	0.604	0.602
0.629	0.612	0.605	0.603	0.599	0.598	0.596	2.5	0.634	0.617	0.610	0.607	0.605	0.604	0.602
0.627	0.611	0.604	0.603	0.599	0.598	0.597	3	0.632	0.616	0.609	0.607	0.605	0.604	0.603
0.623	0.609	0.603	0.602	0.599	0.597	0.596	4	0.628	0.614	0.608	0.606	0.605	0.603	0.602
0.618	0.607	0.602	0.600	0.598	0.597	0.596	6	0.623	0.612	0.607	0.605	0.604	0.603	0.602
0.614	0.605	0.601	0.600	0.598	0.596	0.596	8	0.619	0.610	0.606	0.605	0.604	0.603	0.602
0.611	0.603	0.599	0.598	0.597	0.596	0.595	10	0.616	0.608	0.605	0.604	0.603	0.602	0.601
0.601	0.599	0.597	0.596	0.596	0.596	0.594	20	0.606	0.604	0.602	0.602	0.602	0.601	0.600
0.596	0.595	0.594	0.594	0.594	0.594	0.593	50	0.602	0.601	0.601	0.600	0.600	0.599	0.599
0.593	0.592	0.592	0.592	0.592	0.592	0.592	100	0.599	0.598	0.598	0.598	0.598	0.598	0.598

[1] SMITH and WALKER, Orifice Flow, *Proc. Inst. Mech. Engrs.* (*London*), 1923, pp. 23–36.
[2] "Hydraulics," 1886.

are in agreement. Recent investigations by Medaugh and Johnson [1] check Smith's coefficients for orifices larger than ¼ in. diameter within ⅓ of 1 percent. Values of the coefficient of discharge for a 1-in. orifice as determined by various investigators and plotted by Medaugh and Johnson are shown in Fig. 33. The differences between the values are undoubtedly not entirely due to experimental errors. Many other factors may contribute, as for instance the ratio of the orifice diameter to the dimensions of the

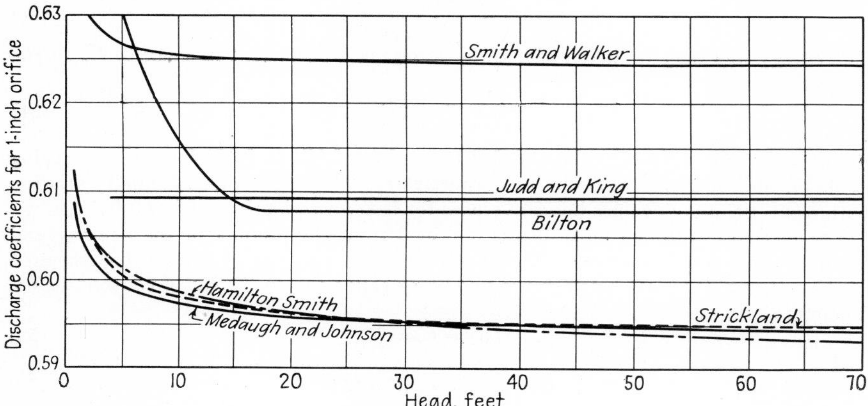

FIG. 33. Comparison of orifice coefficient determinations by various experimenters.

tank wall, the sharpness of the edge of the orifice, the roughness of the inner surface, the orifice plate, and the temperature of the water. The effect of having the tank wall approach the orifice is to suppress contraction and therefore to make C approach the value of C_v. Proximity of a side of an orifice to a surface parallel to the direction of flow (Fig. 34*b*) prevents the free approach of water and restricts contraction on this side. If an orifice has any portion of its edge flush with the bottom or a side of the supply

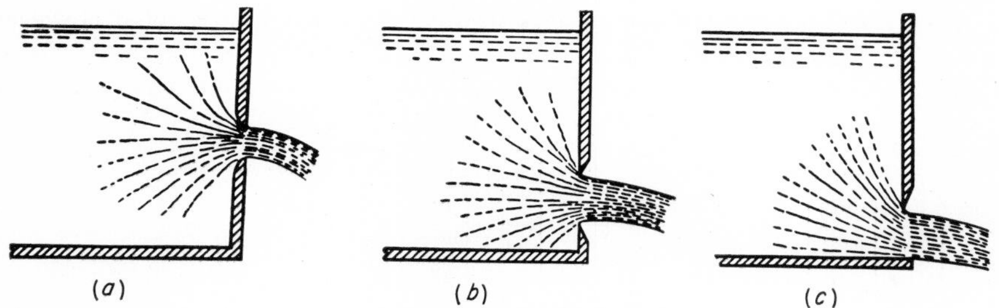

FIG. 34. Suppression of contraction.

reservoir or channel of approach (Fig. 34*c*), contraction for that part of the orifice is entirely suppressed.

The coefficient of contraction of an orifice does not vary directly with the length of perimeter subject to contraction. If there is suppression of contraction on one side and opportunity for complete contraction on the other sides, more water will approach with velocity components parallel to the face of the orifice on these sides and cause increased

[1] MEDAUGH and JOHNSON, Investigation of the Discharge and Coefficient of Small Circular Orifices, *Civil Eng.*, July, 1940, pp. 422–424.

contraction. This to a large extent will compensate for loss of contraction on the other side. Williams[1] found, for rectangular orifices 30 in. wide and 2 to 4 in. high with full contraction at the top and completely suppressed contraction on the two sides and bottom, the average coefficient of discharge was 0.607. This value corresponds closely to the coefficient for orifices with complete contraction. For orifices having full contraction at the top, one side a sharp edge 6 in. from the side of the channel, and contraction suppressed at one side and the bottom, Williams secured an average coefficient of 0.611. With the above orifices, except that the top was beveled to an angle of 45°, he obtained values of C of 0.776 and 0.755 respectively. Table 4, from results compiled by Smith,[2] indicates the effect of suppression of contraction for small orifices. In this table, "suppressed contraction" means that the edge of the orifice coincides with the side of the channel, and "partly suppressed" means that the distance of the edge of the orifice from the side of the channel was 0.066 ft. A special case of suppressed contraction is the pipe orifice which will be discussed later.

TABLE 4. COEFFICIENTS OF DISCHARGE FOR RECTANGULAR ORIFICES 0.656 FT WIDE WITH PARTIALLY SUPPRESSED CONTRACTION

Description of contraction	Height, ft	Head, ft 1	Head, ft 3	Head, ft 5	Description of contraction	Height, ft	Head, ft 1	Head, ft 3	Head, ft 5
Complete contraction.	0.656	0.598	0.604	0.603	Suppressed at bottom and partly on one side.	0.656	0.633	0.636	0.637
	0.328	0.616	0.615	0.611		0.328	0.658	0.656	0.654
	0.164	0.631	0.627	0.620		0.164	0.676	0.673	0.672
	0.098	0.632	0.628	0.623		0.098	0.682	0.683	0.681
	0.033	0.652	0.634	0.620		0.033	0.708	0.705	0.695
Suppressed at bottom only.	0.656	0.620	0.624	0.625	Suppressed at bottom and partly on two sides. . . .	0.656	0.678	0.664	0.663
	0.328	0.649	0.647	0.643		0.328	0.680	0.675	0.672
	0.164	0.671	0.668	0.666		0.164	0.687	0.680	0.673
	0.098	0.680	0.677	0.677		0.098	0.693	0.688	0.683
	0.033	0.710	0.705	0.696		0.033	0.708	0.705	0.698
Suppressed on both sides only.	0.656	0.632	0.628	0.628					
	0.328	0.637	0.630	0.630	Suppressed at bottom and two sides.	0.656	0.690	0.677	0.672
	0.164	0.641	0.634	0.635					
	0.098	0.653	0.643	0.639					
	0.033	0.682	0.667	0.655	Complete suppression	0.656	. . . . .	0.950	

TABLE 5. COEFFICIENTS OF DISCHARGE OF VARIOUS-SHAPED ORIFICES WITH COMPLETE CONTRACTION

Fanning's coefficients for vertical rectangular orifices 1 ft wide							Bovey's coefficients for various-shaped orifices, each 0.196 sq in. area							
Height of orifice, ft						Head, ft	Circle	Square		Rectangle, ratio of sides 4:1		Rectangle, ratio of sides 10:1		Triangle
0.125	0.25	0.5	1	2	4			Sides vertical	Diagonal vertical	Long sides vertical	Long sides horizontal	Long sides vertical	Long sides horizontal	
0.622	0.616	0.611	0.605	. . . . .	. . . . .	1	0.620	0.627	0.628	0.642	0.643	0.663	0.664	0.636
0.619	0.614	0.609	0.604	0.609	. . . . .	2	0.613	0.620	0.628	0.634	0.636	0.650	0.651	0.628
0.614	0.610	0.607	0.603	0.606	0.608	4	0.608	0.616	0.618	0.628	0.629	0.641	0.642	0.623
0.610	0.608	0.604	0.601	0.604	0.605	6	0.607	0.614	0.616	0.626	0.627	0.637	0.637	0.620
0.608	0.606	0.603	0.601	0.603	0.604	8	0.606	0.613	0.614	0.623	0.625	0.634	0.635	0.619
0.606	0.604	0.602	0.601	0.602	0.603	10	0.605	0.612	0.613	0.622	0.624	0.632	0.633	0.618
0.607	0.603	0.601	0.601	0.602	0.603	15	0.604	0.610	0.611	0.620	0.622	0.630	0.630	0.617
0.607	0.604	0.602	0.601	0.602	0.603	20	0.603	0.609	0.611	0.620	0.621	0.629	0.628	0.616
0.609	0.604	0.603	0.601	0.603	0.605	30								
0.614	0.607	0.605	0.602	0.606	0.609	50								

[1] Unpublished experiments performed at the University of Michigan in 1928.
[2] "Hydraulics," pp. 65–67, 1886.

When the inner edge of the orifice is completely rounded as in Fig. 34*a*, all contraction is suppressed, and C_c becomes 1 and therefore C is equal to C_v. Values of C_v for such orifices are approximately the same as for sharp-edged orifices.

Roughening the inner surface of the orifice plate retards the velocity components along the wall which cause contraction and thus tends to increase the values of both C_c and C. Similar effects would result from having the orifice plate differ from a perfectly plane surface. If it were slightly warped outward, C_c would be increased, whereas an inward curvature of the plate would have the opposite effect (see reentrant tubes p. 4–37).

The effect of water temperature is to change the viscosity and density. It has already been shown in Table 3 that C varies with the velocity and the orifice diameter. It therefore appears that here again the Reynolds number is likely to be the coordinating

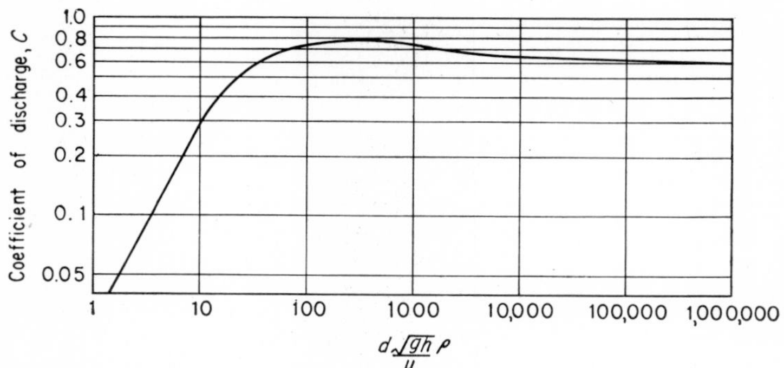

Fig. 35. Coefficient of discharge for circular sharp-edged orifices in tanks.

factor which will show the effect on C of all four of the above-mentioned factors. This has been demonstrated by Lea,[1] who plotted more than one hundred experimental values of C against $d\sqrt{gh}\rho/\mu$. This quantity is closely related to the Reynolds number because the value $\sqrt{gh}$ is a function of and has the dimensions of velocity. It may be seen from Eq. (50) that the $\sqrt{gh}$ multiplied by $\sqrt{2}$ is the theoretical velocity at the vena contracta. The author's curve derived from Lea's plotted points is shown in Fig. 35. The fluids used in the tests were water, various mixtures of water and glycerin, and a number of oils. Flow is laminar for $d\sqrt{gh}\rho/\mu < 12$ and fully turbulent for values greater than 10,000, intervening values corresponding to a transition range. The points plotted by Lea show a spread of about 3 percent except in the transition range where the spread is 15 percent. The tests of Medaugh and Johnson [2] agree very well with Lea's curve within the range of values of $d\sqrt{gh}\rho/\mu$ covered by both studies (10,000 to 1,000,000).

Example 12: Determine the discharge from a large tank through a 2-in.-diameter sharp-edged orifice if the liquid surface is 4 ft above the center of the orifice and the orifice discharges into the atmosphere. The fluid is water at a temperature of 100°F.

From Table 2, p. 4–3,

$$w = 62.0 \text{ lb per cu ft}$$

and

$$\mu = 0.00001424 \text{ slugs per ft sec}$$

[1] Lea, F. C., "Hydraulics," 6th ed., Longmans, Green & Co., Inc., 1938.
[2] Medaugh and Johnson, Investigation of the Discharge and Coefficient of Small Circular Orifices *Civil Eng.*, July, 1940, pp. 422–424.

Then

$$\frac{d\sqrt{gh\rho}}{\mu} = \frac{\frac{2}{12}\sqrt{32.2 \times 4} \times 62.0/32.2}{0.00001424} = 25{,}600$$

From Fig. 35, $C = 0.63$, and from Eq. (54)

$$Q = 0.63 \times 0.0218\sqrt{2 \times 32.2 \times 4} = 0.22 \text{ cfs}$$

Investigations by Bilton [1] showed the coefficient of discharge of a circular orifice under any given head to be the same whether the jet be horizontal, vertical, or at any intermediate angle.

The accompanying extracts from tables by Fanning [2] and Bovey [3] (Table 5) give coefficients of discharge for various shapes and arrangements of orifices with complete contraction. Fanning's table, compiled from experiments from several sources, contains coefficients for orifices 1 ft wide, of various heights, and under a wide range of heads. Bovey's table, prepared from his own experiments on orifices of different shapes, each with the area of a circle ½ in. in diameter, indicates the effect of shape of opening on the coefficient.

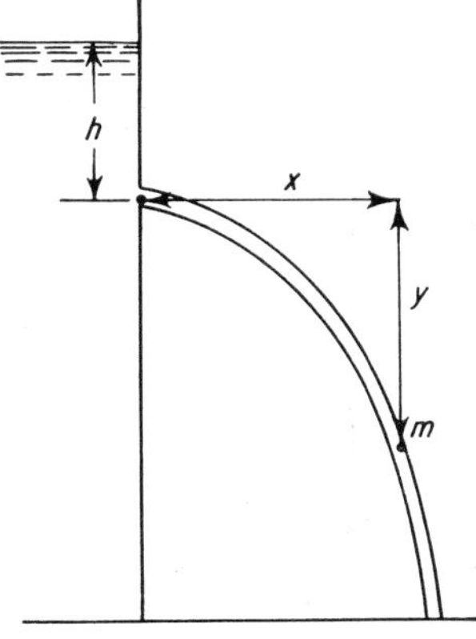

FIG. 36. Path of jet.

Path of Jet. In Fig. 36, x and y are the coordinates of any point m in a jet discharging from a vertical orifice under a head h. If v_o is the velocity in the vena contracta, at the end of time t, $x = v_o t$; and from the law of falling bodies, $y = gt^2/2$. Eliminating t from these equations,

$$x^2 = \frac{2v_o^2}{g} y \tag{55}$$

which is the equation of a parabola having its vertex at the center of the orifice. Substituting for v_o the value given in Eq. (51),

$$x^2 = 4C_v^2 hy \tag{56}$$

Head Lost in Discharge. From Eq. (51), it appears that the actual velocity v_o in the vena contracta is less than the theoretical velocity as given by Eq. (50). This is because the water sustains a loss of head due to friction in passing through the orifice. Transposing Eq. (51), the total head producing discharge becomes

$$h = \frac{1}{C_v^2} \cdot \frac{v_o^2}{2g} \tag{57}$$

The lost head h_l is equal to the total head less the velocity head corresponding to v_o, or

$$h_l = \left(\frac{1}{C_v^2} - 1\right)\frac{v_o^2}{2g} \tag{58}$$

Since $v_o^2/2g = hC_v^2$, substituting in (58), $h_l = (1 - C_v^2)h$; and for sharp-edged orifices, if $C_v = 0.98$, $h_l = 0.041v_o^2/2g = 0.040h$.

[1] Coefficients of Discharge through Circular Orifices, from paper read before Victorian Institute of Engineers, April, 1908, *Eng. News*, July 9, 1908.
[2] "Water-supply Engineering," pp. 205–206, D. Van Nostrand Company, Inc., New York, 1906.
[3] "Hydraulics," 2d ed., p. 39, 1909.

Discharge under Falling Head. Water is discharging from a vessel through an orifice (Fig. 37) having an area a and a discharge coefficient C. There is no compensating inflow, so the depth of water will gradually decrease. At any depth y, the area of water surface is A, and the increment of time required to lower the water surface the infinitesimal distance dy is

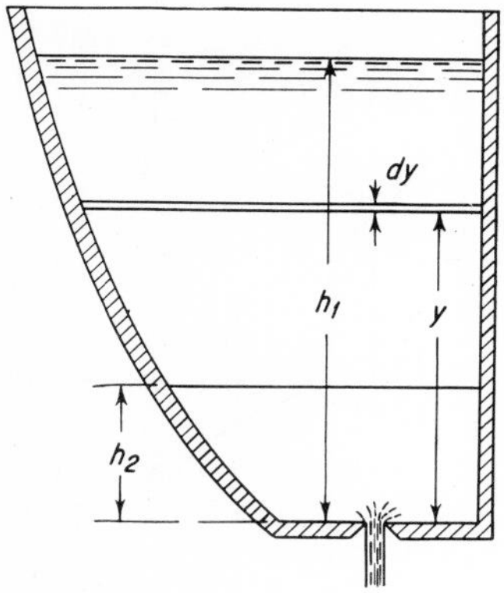

$$dt = \frac{A\,dy}{Ca\sqrt{2gy}} \tag{59}$$

This equation can be solved only when A can be expressed in terms of y. When this can be done, integrating between the limits h_1 and h_2 gives the time required to lower the water surface the distance $h_1 - h_2$. Placing $h_2 = 0$ gives the time of emptying the vessel. The formula applies to vertical or inclined orifices as long as they flow full. For a cylinder or prism, A is constant, and

$$t = \frac{2A}{Ca\sqrt{2g}}(\sqrt{h_1} - \sqrt{h_2}) \tag{60}$$

FIG. 37. Orifice under falling head.

Discharge under Low Heads. Equation (54) assumes that discharge occurs under the average head. For low heads, there is an appreciable difference between the true theoretical discharge and the discharge given by Eq. (54). Because it affords the simplest treatment, the rectangular orifice (Fig. 38) will be investigated. The effect of velocity of approach will be neglected. The orifice has a width L and a height M, h_1 and h_2 being the respective heads on the upper and lower edges of the orifice. The theoretical discharge through any elementary strip of area $L\,dy$, discharging under a head y, is $dQ_t = CL\sqrt{2gy}\,dy$, which, when integrated between the limits h_2 and h_1, gives as the discharge through the orifice

$$Q = C\tfrac{2}{3}L\sqrt{2g}\,(h_2^{3/2} - h_1^{3/2}) \tag{61}$$

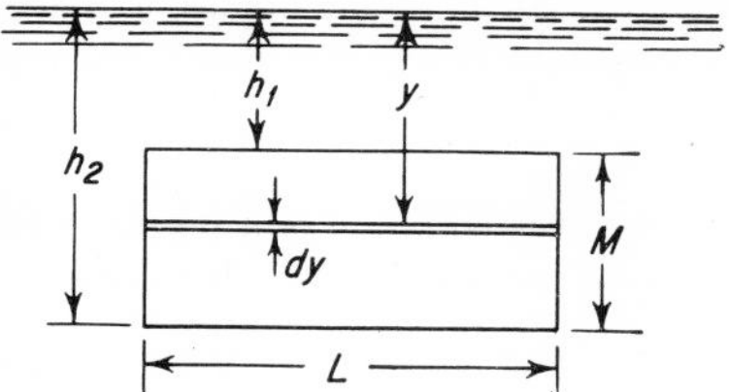

FIG. 38. Rectangular orifice under low head.

When $h_1 = 0$, $Q = C\tfrac{2}{3}L\sqrt{2g}h_2^{3/2}$, which is the formula, without velocity of approach correction, for discharge over a weir (see p. 4–46).

The formula giving the true theoretical discharge for vertical circular orifices is quite complicated and has no practical value. For $h_1 = M$, Eq. (61) gives results about 1 percent greater than Eq. (54); and for $h_1 = 2M$, about 0.3 percent greater. There are similar differences between the corresponding formulas for circular orifices. Equation (54) is generally employed for all orifices, including those discharging under low heads, deviation from the more precise form of formula being corrected for in the coefficient.

Submerged Orifices. Since the water particles at all points in the plane of a submerged orifice are subjected to the same unbalanced pressure resulting from the difference in level of water surfaces (h, Fig. 39), the theoretical velocity through the orifice is $v_t = \sqrt{2gh}$. Including a coefficient of discharge C, the discharge of a submerged orifice area a is

$$Q = Ca\sqrt{2gh} \tag{54}$$

There are but a few experiments available for determining C for submerged orifices. For sharp-edged circular and square submerged orifices 0.01 sq ft and less in area, dis-

charging under heads of 0.3 to 4 ft, Smith obtained an average C of 0.602. Ellis obtained practically this same value for a circle 1 ft in diameter and also for an orifice 1 ft square, under heads of 2 to 18 ft. For a submerged orifice 1 ft square with rounded corners, under heads of 3 to 18 ft, Ellis obtained an average C of 0.945. All experiments indicate that the value of the coefficient is not greatly affected by submergence.

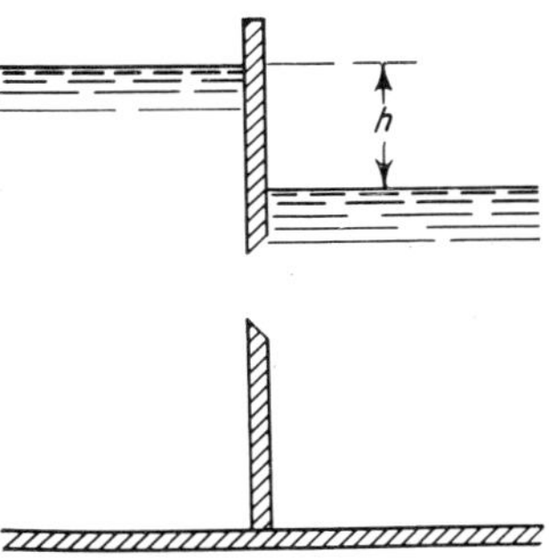

FIG. 39. Submerged orifice.

Gates. A gate is an opening in a dam or other structure to permit the passage of water, usually including means of regulating the outflow. The discharge may be either free or submerged. Gates have the hydraulic properties of orifices, the formula for discharge being

$$Q = Ca\sqrt{2gh} \tag{54}$$

in which h is the head on the center of the orifice for free discharge and the difference in elevation of water surfaces for submerged discharge. Since there are no standards of design, gates have varying degrees of contraction with corresponding variations in C. The results of many experiments on gates have been published, but each set of experiments is applicable only to one particular type of structure and to the special conditions under which the experiments were performed. Even experiments on the same structure, discharging under apparently similar conditions, frequently show an inexplicably wide range in results. In selecting a discharge coefficient for purposes of design, the engineer should be conservative and choose a minimum rather than an average value. The size of opening can be reduced by operation of the gate, but it cannot be increased.

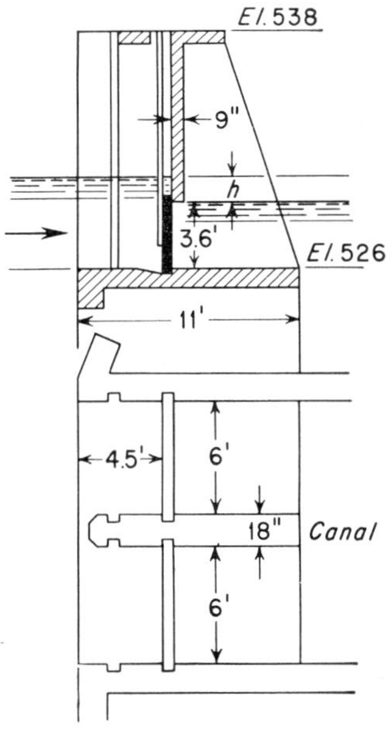

FIG. 40. Canal head gate.

The more common features embodied in ordinary canal head gates are illustrated in Fig. 40. At the bottom and sides of the opening, contraction is usually completely suppressed, while at the top the contraction will vary with the amount that the upstream bottom edge of the gate or curtain wall is rounded. Experimental values of C, corresponding to different heads in feet, by Newell [1] for the Maxwell canal head gate (Fig. 40) are as follows: $h = 0.11$, $C = 0.83$; $h = 0.16$, $C = 0.83$; $h = 1.08$, $C = 0.685$; $h = 1.98$, $C = 0.715$. Other experiments on gates of this type show a similar but sometimes wider range in values and even more glaring inconsistencies.

Pipe Orifices. A circular sharp-edged pipe orifice is shown in Fig. 41. The velocity at the vena contracta 2 is given by Eq. (48). Replacing the energy-loss term h_l with C_v as was done for orifices in tanks, Eq. (48) becomes

$$v_2 = C_v \sqrt{2g\left(\frac{p_1}{w} - \frac{p_2}{w} + \frac{v_1^2}{2g}\right)} \tag{62}$$

[1] *U.S. Bur. Reclamation Experimental Data* 9, 1916.

In pipe orifices the velocity of approach factor ($V_1^2/2g$) cannot be omitted. Again letting $a_2 = C_c a$ and $C = C_c C_v$, the expression for the discharge becomes

$$Q = Ca\sqrt{2g\left(\frac{p_1}{w} - \frac{p_2}{w} + \frac{v_1^2}{2g}\right)} \tag{63}$$

In this form the equation requires a cut-and-try solution. However by replacing v_1 with

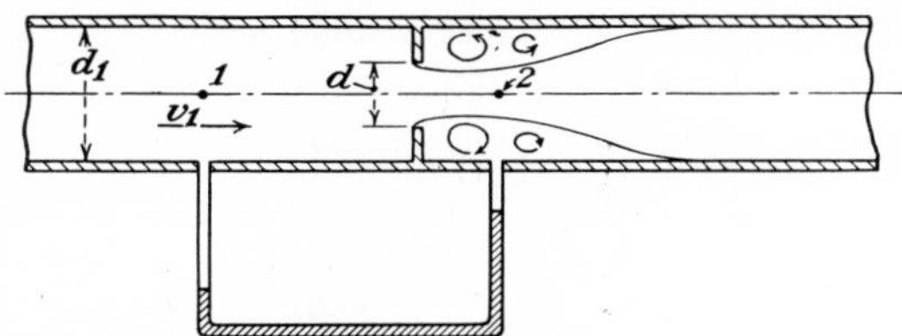

FIG. 41. Circular sharp-edged orifices in pipes.

its value obtained from the equation of continuity (Q/a_1), a_1 being the area of the pipe, the following expression for Q is derived.

$$Q = Ca\sqrt{\frac{2g\left(\frac{p_1}{w} - \frac{p_2}{w}\right)}{1 - C^2(a/a_1)^2}} \tag{64}$$

The value of C varies not only with the factors which affected orifices in tanks but also with the ratio of the orifice diameter d to the pipe diameter d_1 and with the location of the pressure taps. Tests indicate that the location of the vena contracta varies from

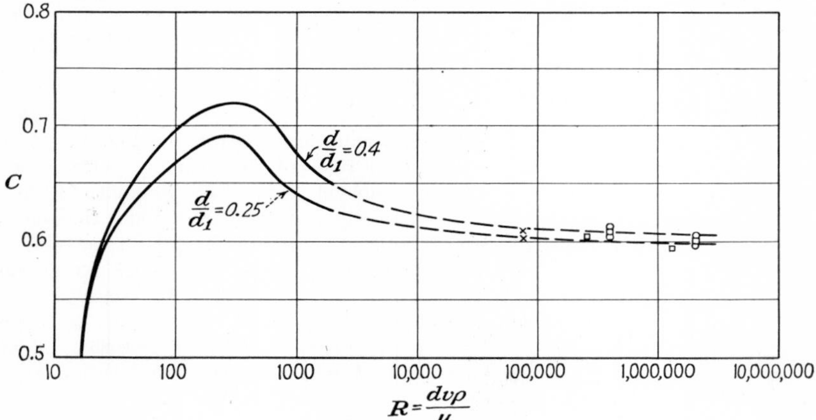

FIG. 42. Orifice coefficients.

Symbol	Fluid	Experimenter	Source	Remarks
—	Castor oil	Kowalke, Bain and Moss	King, Wisler, and Woodburn, "Hydraulics"	
×	Air	Bean, Buckingham, and Murphy	King, Wisler, and Woodburn, "Hydraulics"	Values of R were estimated
○	Water	Blackburn	King, Wisler, and Woodburn, "Hydraulics"	Values of R were estimated
□	Water	Bailey	"Fluid Meters: Their Theory and Application," ASME Research Pub.	Values of R were estimated

0.32 pipe diameters to 0.87 pipe diameters below the orifice plate as the ratio of the orifice diameter to the pipe diameter (d/d_1) varies from 0.8 to 0.2.[1] When the orifice is at the end of a pipe, p_2 in Eq. (64) is zero and only a single pressure tap is required. Curves showing the relation between C and the Reynolds number are plotted in Fig. 42. The velocity v used in computing the Reynolds number is the average velocity in the plane of the orifice. Portions of the curves are shown as dashed lines because values of R are uncertain for the meager data available in that region. Additional information on the location of the vena contracta and the value of the orifice coefficient is given by King [2] and in the ASME Power Test Codes.[3]

TUBES, CULVERTS, AND SIPHONS

Cylindrical Tubes. A short cylindrical tube 2½ to 3 diameters in length with sharp upstream corners (Fig. 43) is called a *standard short tube.* The sharp corner causes the jet to contract, but it soon expands and fills the tube. The space s surrounding the contracted jet is filled with air and eddying water. Since the velocity at m in the contracted

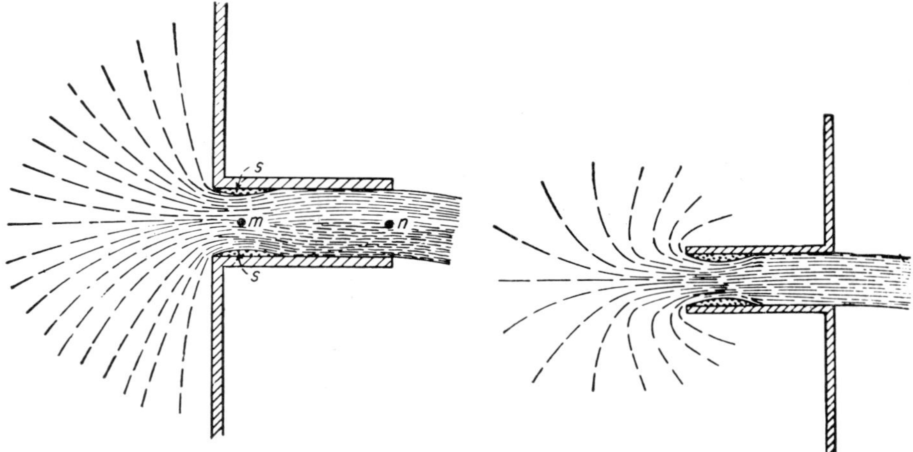

FIG. 43. Standard short tube. FIG. 44. Reentrant tube.

jet is greater than at the outlet n where pressure is atmospheric, it can be shown by Bernoulli's equation that the pressure at m is less than atmospheric pressure. The discharge of this tube is therefore greater than the discharge of a sharp-edged orifice of the same diameter under the same head. Since the tube flows full, the coefficient of contraction is unity, and the coefficient of discharge [Eq. (54)] equals the coefficient of velocity. From experiments, the average coefficient of discharge C is 0.82. Substituting this value of C, or C_v, in Eq. (58) gives as the head lost in discharge

$$h_l = 0.50 \frac{v^2}{2g} \tag{65}$$

where v is the mean velocity in the tube. The conditions of flow in a short tube correspond to the conditions at the entrance to a pipe.

A tube projecting into a body of water (Fig. 44) is called a *reentrant tube.* The contraction produced by this tube is greater than for the standard short tube. The behavior

[1] ASME "Fluid Meters, Their Theory and Application," Part 1, 4th ed., 1937.
[2] "Handbook of Hydraulics," 4th ed., The McGraw-Hill Book Company, Inc., 1954.
[3] ASME "Flow Measurement by Means of Standard Nozzles and Orifice Plates," Power Test Codes, Supplement on Instruments and Apparatus, Part 5, Chap. 4, 1949.

of the water in the two kinds of tubes is otherwise similar. From experiments, the coefficient of discharge $C = 0.75$ and the lost head $h_l = 0.78v^2/2g$.

Conical Tubes. A tube having the form of a truncated cone with water flowing toward the smaller base is called a *converging tube.* Such tubes may have a sharp-cornered entrance (Fig. 45) or a rounded entrance (Fig. 46). The jet contracts slightly at a just beyond the outlet. The coefficient of contraction C_c decreases as the angle of convergence θ increases until $\theta = 180°$ and $C_c = 0.62$, the coefficient for a sharp-edged orifice.

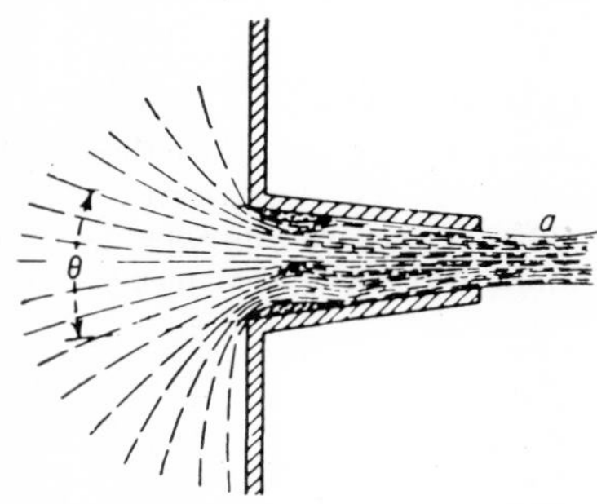

FIG. 45. Converging tube with sharp cornered entrance.

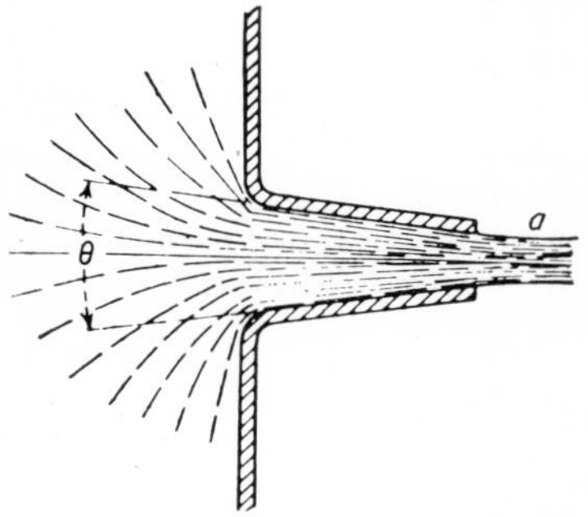

FIG. 46. Converging tube with rounded entrance.

The coefficient of velocity C_v increases with θ. Discharge is given by Eq. (54), in which $C = C_cC_v$. The accompanying table of coefficients (Table 6) for converging tubes with sharp-cornered entrance shows the variation in the coefficient. These values are based upon experiments on small tubes and are not generally applicable. The coefficient of velocity and therefore the coefficient of discharge will be increased by rounding the entrance.

TABLE 6. COEFFICIENTS FOR SHARP-CORNERED CONVERGING TUBES (FIG. 45)

θ	0°	5°	10°	15°	20°	25°	30°	40°	50°
C_v	0.829	0.911	0.947	0.965	0.971	0.973	0.976	0.981	0.984
C_c	1.000	0.999	0.992	0.972	0.952	0.935	0.918	0.888	0.859
C	0.829	0.910	0.939	0.938	0.924	0.911	0.896	0.871	0.845

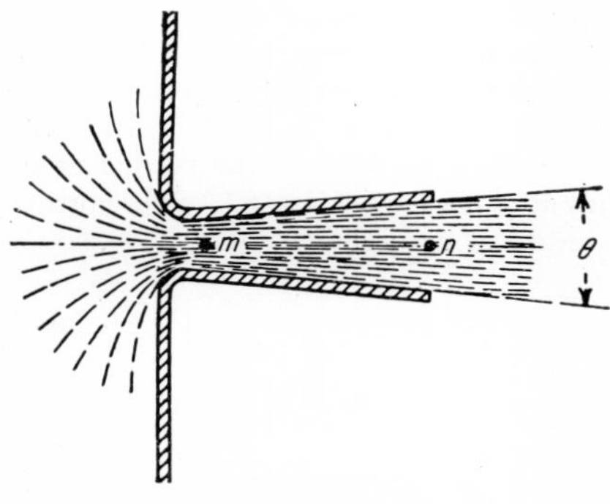

FIG. 47. Diverging tube.

The conical tube shown in Fig. 47 with water flowing toward the larger base is called a *diverging tube* or *venturi tube.* The entrance is rounded so that changes in velocity occur gradually. If the angle of flare θ is not too great, the tube will flow full. Experiments by Venturi show the most efficient value of θ to be about 5°. A discharge two to three times that of a sharp-edged orifice having the same diameter as the throat may be obtained with a tube of this kind. That the pressure in the throat is less than atmospheric pressure can be shown by writing Bernoulli's equation between points m and n (Fig. 47).

Nozzles. A nozzle is a converging tube attached to the outlet of a hose or pipe to

increase the velocity. Two shapes of nozzles having a high efficiency are shown in Fig. 48. Each of these nozzles terminates with a cylindrical ring which prevents contraction. In a, the converging part is conical; in b, it is curved inward. The following are coefficients of discharge obtained by Freeman [1] for nozzles of different diameters similar to a: ¾ in., $C = 0.983$; ⅞ in., $C = 0.982$; 1 in., $C = 0.980$; 1⅛ in., $C = 0.976$; 1¼ in., $C = 0.971$; 1⅜ in., $C = 0.959$.

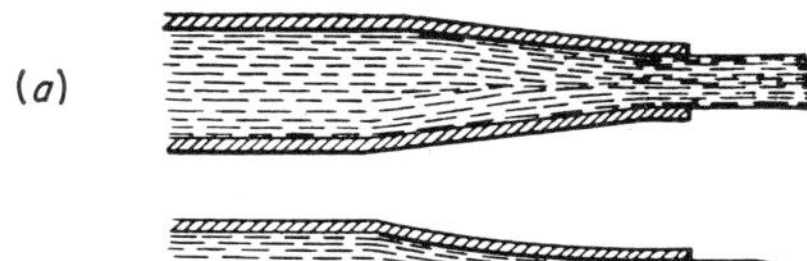

Fig. 48. Nozzles.

Submerged Tubes. Although experiments to show comparative discharges are not available, it is probable that, as is the case with orifices, the coefficient of discharge for tubes [C in Eq. (54)] is not materially changed by submergence. Large submerged tubes frequently form important parts of engineering structures. Such tubes may have lengths ten or more times their perimeters and are in reality short pipes, but in many instances where their length is not too great, it is found convenient to consider them as tubes and to include the loss of head due to pipe friction in the coefficient of discharge. The loss of head in a submerged tube is affected by entrance conditions, being greatest for a reentrant end (Fig. 44) and least for a bellmouthed entrance. Even moderate rounding of the entrance to a tube materially reduces loss of head. To obtain maximum discharge, the intake must be submerged to a depth not less than the sum of the velocity head and loss of head at entrance (see p. 4–71). For a square-cornered entrance, the minimum submergence of entrance is commonly taken as $1.5v^2/2g$, v being the mean velocity in the tube. Examples of large submerged tubes are culverts, inverted siphons, and siphon spillways. Values of the coefficient of discharge C compiled from experiments by Stewart [2] and Rogers and Smith [3] on submerged tubes 0.5 to 4 ft square are given in Table 7. These experiments show C to vary with the length and size of tube but not with the head. In the table, L is the length of tube and p is the perimeter of a cross section of the tube.

Culverts. The most common purpose of culverts is to pass natural drainage through an embankment. The important consideration is that the capacity will be sufficient to pass all ordinary floods without damage. It is therefore necessary to investigate conditions of maximum discharge. The ultimate capacity of the culvert is reached when water on the upstream side of the embankment stands at the highest safe level. Under this condition, the outlet will ordinarily be submerged. A field examination will usually be necessary to estimate the two water levels. The head h (Fig. 49) is the difference in

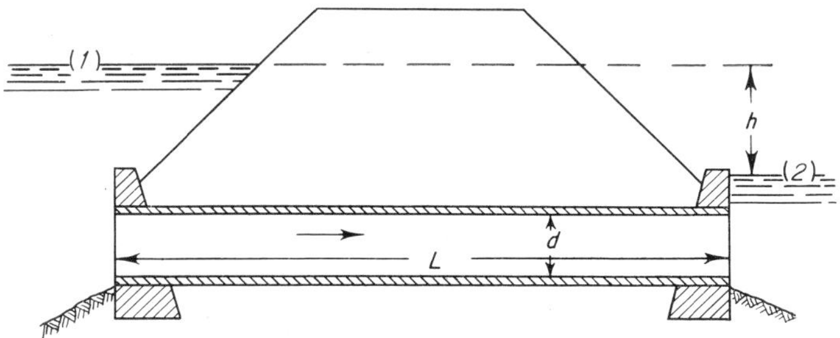

Fig. 49. Pipe culvert.

[1] Experiments Relating to Hydraulics of Fire Streams, *Trans. ASCE*, vol. 21, pp. 303–482, 1888.
[2] Investigation of Flow through Large Submerged Orifices and Tubes, *Univ. Wis. Bull.* 216, 1908.
[3] Experiments with Submerged Orifices and Tubes, *Eng. News*, Nov. 2, 1916.

TABLE 7. COEFFICIENTS OF DISCHARGE FOR LARGE SUBMERGED TUBES

Description of contractions	L/p = length ÷ perimeter				
	0.02	0.1	0.2	0.3	1.0
All corners square	0.61	0.66	0.74	0.79	0.80
Suppressed on bottom only	0.63	0.67	0.73	0.77	0.81
Suppressed on bottom and one side	0.68	0.69	0.74	0.79	0.82
Suppressed on bottom and two sides	0.77	0.73	0.73	0.83	0.85
Suppressed on bottom, two sides, and top	0.95	0.93	0.92	0.91	0.90

TABLE 8. COEFFICIENTS OF DISCHARGE FOR CULVERTS

Description of culvert	L \ d	1	1.5	2	3	4	5	6
Concrete pipe, beveled-lip entrance	10	0.86	0.89	0.91	0.92	0.93	0.94	0.94
	20	0.79	0.84	0.87	0.90	0.91	0.92	0.93
	30	0.73	0.80	0.83	0.87	0.89	0.90	0.91
	40	0.68	0.76	0.80	0.85	0.88	0.89	0.90
	50	0.65	0.73	0.77	0.83	0.86	0.88	0.89
square-cornered entrance	10	0.80	0.81	0.80	0.79	0.77	0.76	0.75
	20	0.74	0.77	0.78	0.77	0.76	0.75	0.74
	30	0.69	0.73	0.75	0.76	0.75	0.74	0.74
	40	0.65	0.70	0.73	0.74	0.74	0.74	0.73
	50	0.62	0.68	0.71	0.73	0.73	0.73	0.72
Vitrified-clay pipe, bell end upstream	10	0.90	0.91	0.91	0.89	0.85	0.80	0.75
	20	0.83	0.86	0.87	0.86	0.83	0.79	0.75
	30	0.77	0.82	0.84	0.84	0.82	0.78	0.74
	40	0.72	0.78	0.81	0.82	0.80	0.77	0.73
	50	0.69	0.75	0.78	0.80	0.79	0.76	0.72
Corrugated-metal pipe	10	0.67	0.74	0.77	0.79	0.80	0.80	0.79
	20	0.55	0.63	0.68	0.73	0.75	0.76	0.76
	30	0.48	0.56	0.62	0.68	0.71	0.73	0.74
	40	0.43	0.51	0.57	0.64	0.68	0.70	0.71
	50	0.39	0.47	0.53	0.61	0.65	0.68	0.69
	L \ r	0.2	0.3	0.4	0.6	0.8	1.0	1.5
Concrete box, rounded-lip entrance	10	0.85	0.89	0.92	0.94	0.95	0.96	0.96
	20	0.76	0.83	0.87	0.91	0.92	0.94	0.95
	30	0.70	0.78	0.82	0.88	0.90	0.92	0.94
	40	0.65	0.73	0.79	0.85	0.88	0.90	0.93
	50	0.60	0.70	0.75	0.82	0.86	0.89	0.92
square-cornered entrance	10	0.79	0.82	0.83	0.84	0.84	0.83	0.82
	20	0.72	0.77	0.79	0.81	0.82	0.82	0.81
	30	0.67	0.73	0.76	0.79	0.80	0.81	0.81
	40	0.62	0.69	0.73	0.77	0.79	0.80	0.80
	50	0.58	0.66	0.71	0.75	0.77	0.78	0.79

these levels, and discharge is obtained from Eq. (54). Results of more than 3,000 experiments to determine the coefficient of discharge C for concrete, vitrified-clay, corrugated-metal pipe culverts and concrete box culverts have been published by Yarnell, Nagler, and Woodward.[1] The experiments were conducted on pipes 12 to 30 in. in diameter and on box culverts, both square and rectangular, of 2 to 16 sq ft cross-sectional area. Lengths of culverts experimented upon were 24, 30, and 36 ft. From the experiments, formulas were derived for determining C. Table 8 gives values of C computed from these formulas. In the table, d is the diameter of pipe culverts, r (see p. 4–67) the hydraulic radius of box culverts, and L the length of culvert, all in feet. For culverts more than 50 ft long, the head lost in the total length, less 50 ft, should be computed by a pipe or open-channel formula. The value of h to be used in Eq. (54) is then the difference in elevation of water surfaces minus the computed loss of head, and the value of C to be used is that given in the table for a culvert 50 ft long.

A study of entrance and outlet conditions of culverts made in connection with the investigation of Yarnell, Nagler, and Woodward indicates that 45° wing walls used in connection with pipe culverts increase the discharge 1 to 10 percent over that obtained with a straight end wall and that wing walls are more efficient when set flush with the edge of the pipe than when set 6 in. back from the edge. V-type wing walls used in connection with vitrified-clay pipe produced a slightly smaller discharge than that obtained with a straight end wall. The beveled lip at the entrance to concrete culverts greatly reduced entrance loss. It was found that a 24-in. clay pipe 38 ft long with a straight end wall and the bell end upstream carried about 10 percent more water than the same culvert with a square-cornered entrance. Rounding the entrance of a 24-in. vitrified-clay pipe culvert increased the capacity 13 percent. It was found also that the discharge is decreased slightly by projecting the pipe through the head wall. There is little difference whether a pipe with square-cornered entrance projects 3 in. or 2 ft or more. An 18-in. corrugated-metal pipe with a 3-in. projection carries slightly more water than the same pipe with a greater projection. Doubling the end area of the submerged outlet of an 18-in. vitrified-clay pipe culvert by attaching a 10° diverging bell was found to increase the discharge by 40 percent over that obtained through the same culvert without the diverging end.

Culvert discharge may also be computed by considering the culvert as a short pipe. Bernoulli's equation written from point (1) to point (2), Fig. 49, taking the datum plane at the elevation of the water surface at (2), gives

$$h = h_e + h_f + h_o \tag{66}$$

In the above equation h_e is the entrance loss, h_f is the friction loss, and h_o is the loss at the outlet. The entrance and outlet losses are usually expressed in terms of the velocity head as follows:

$$h_e = K_e \frac{V^2}{2g} \tag{67}$$

and

$$h_o = K_o \frac{V^2}{2g} \tag{68}$$

where V is the average velocity in the pipe. The friction loss may be determined from the Manning formula or the Darcy-Weisbach equation as described on p. 4–62.

Straub and Morris [2] have reported values of K_e and K_o as well as values of the Manning coefficient n and the Darcy-Weisbach coefficient f, based on a series of laboratory tests

[1] Flow of Water through Culverts, *Univ. Iowa Studies in Engineering*, 1926.

[2] Straub, L. G., and H. M. Morris, "Hydraulic Tests on Concrete Culvert Pipes," *Tech. Paper No.* 4, *Series B*, St. Anthony Falls Hydraulic Laboratory, Minneapolis, 1950, and "Hydraulic Tests on Corrugated Metal Culvert Pipes," *Tech. Paper No.* 5, *Series B*, St. Anthony Falls Hydraulic Laboratory, Minneapolis, 1950.

on concrete and corrugated-metal-pipe culverts. Full-scale tests were made on culverts 193 ft long for pipes flowing full and for the case of uniform flow with a free surface. Round pipes of 18 in., 24 in., and 36 in. diameters were used as well as arched corrugated-metal pipes having the same perimeters as the round pipes. The dimensions of the corrugated pipe refer to the inside diameter, and the height of the corrugations was ½ in. The entrance to the concrete culverts consisted of the groove end of a length of tongue-and-groove pipe. Tests were made with the inlet and flush with the wall of the head-water pool and also with a projecting inlet.

For the concrete pipe, design values of the entrance-loss coefficient K_e were found to be 0.10 for the flush inlet and 0.15 for the projecting inlet. These values apply to full flow and to part-full flow conditions. For corrugated-metal pipes flowing full, practical design values of K_e were found to be 0.5 for a flush entrance and 0.9 for the projecting entrance. For the case of flow with a free surface in corrugated pipes, the corresponding values of K_e were 0.4 and 0.7, respectively.

The outlet-loss coefficient K_o was found to be nearly 1.0 in all cases.

Values of the Manning n varied somewhat with the pipe size, the shape of the pipe, and the Reynolds number. Practical design values taken near the upper end of the range are 0.011 for concrete pipe and 0.025 for corrugated-metal pipe. Similar values of the Darcy-Weisbach f for concrete pipe are 0.018 for the 18-in. pipe and 0.016 for the 24- and 36-in. pipe. Corresponding values of f for corrugated-metal pipe are 0.105 for the 18-in. size, 0.090 for the 24-in. size, and 0.075 for the 36-in. size. The results obtained from Eq. (66), using the coefficients given above, are in good agreement with values computed from Eq. (54), using appropriate values of C from Table 8.

Example 13: Determine the head required to produce a discharge of 30 cfs through a concrete culvert having a beveled lip, a length of 50 ft, and a diameter of 2 ft. Assume that the culvert is flowing full.

a. Using Eq. (54) and taking the value of C from Table 8,

$$Q = Ca2gh$$

or

$$h = \frac{Q^2}{C^2a^22g} = \frac{(30)^2}{(0.77)^2(3.142)^264.4} = 2.39 \text{ ft}$$

b. Using Eq. (66),

$$V = \frac{Q}{a} = \frac{30}{3.142} = 9.55 \text{ ft per sec}$$

and

$$V^2/2g = 1.42 \text{ ft}$$

Then, from Eq. (67),

$$h_e = K_e\frac{V^2}{2g} = 0.10 \times 1.42 = 0.14 \text{ ft}$$

and, from Eq. (68)

$$h_o = K_o\frac{V^2}{2g} = 1.0 \times 1.42 = 1.42 \text{ ft}$$

The value of h_f may be obtained from Eq. (123), p. 4–62, using $f = 0.016$

$$h_f = f\frac{1}{d}\frac{V^2}{2g} = 0.016 \times \frac{50}{2} \times 1.42 = 0.57 \text{ ft}$$

Then,

$$h = 0.14 + 0.57 + 1.42 = 2.13 \text{ ft}$$

Short Siphons. When the central portion of a culvert is depressed below the entrance and outlet levels, the structure is termed an *inverted siphon* or, more frequently, a *siphon*.

Short siphons are commonly used to convey canal waters under highways, railways, and other canals (Fig. 50). The discharge is given by Eq. (54), and the head h is the difference in level of water surfaces. If all changes in grade are made with easy curves so that there are no abrupt changes in velocity or in direction of flow, it is probable that the values of the coefficient of discharge C given in Table 8 for culverts will apply quite closely. In conveying canal waters, the discharge is usually fixed by the capacity of the canal, and the head necessary to force this quantity of water through a siphon of a given

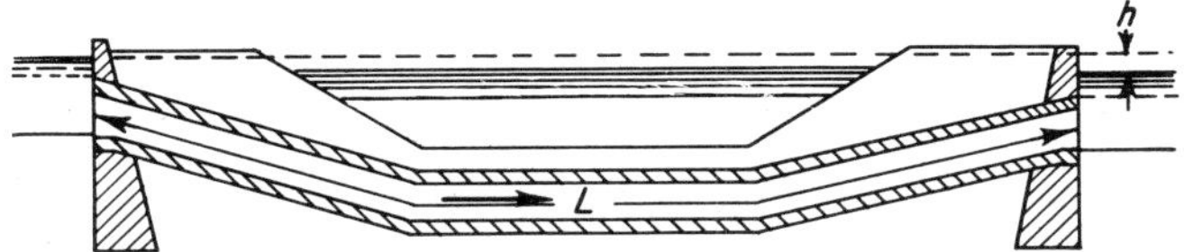

FIG. 50. Inverted siphon under canal.

cross-sectional area is required. Transposing Eq. (54), the required head or difference in elevation of water surfaces is

$$h = \frac{Q^2}{2gC^2a^2} = \frac{1}{C^2} \cdot \frac{v^2}{2g} \tag{69}$$

which corresponds with Eq. (57).

An investigation by Stewart[1] of short road siphons of the U.S. Bureau of Reclamation, having a square cross section and four right-angled turns (one turn to direct the water vertically downward, one to direct it horizontally, one to direct it vertically upward, and finally one to discharge the water horizontally at the outlet), showed a loss of head of three to five times the velocity head. These results show that abrupt changes in direction of flow produce excessive losses of head or, if the head is fixed, that they cause a corresponding reduction in discharge.

Siphon Spillways. The principle governing the operation of siphon spillways is illustrated in Fig. 51. Under normal conditions, both intake and outlet are submerged, the space within the tube being sealed. When the upper water level gets high enough for flow through the siphon to begin, air within the enclosed space is carried out by the

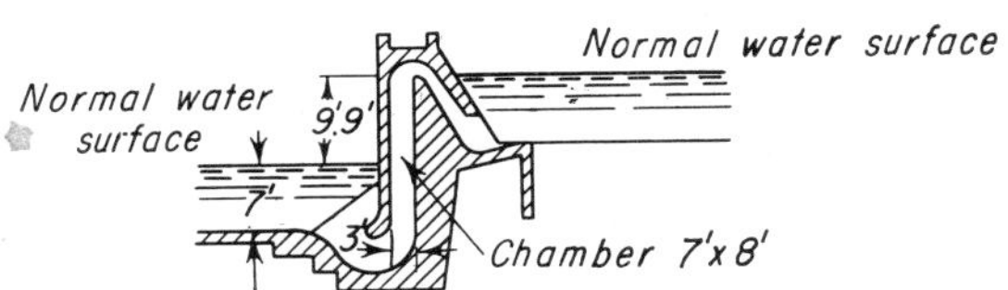

FIG. 51. Siphon spillway.

moving water, and in a short time the passageway is flowing full. Flow continues until the water surface is lowered enough to expose the top of the intake and break the seal. Such siphons, if of airtight construction, have the hydraulic properties of submerged tubes. Discharge is given by Eq. (54), the head h being the difference in elevation of water surfaces (9.9 ft in figure). From experiments on the five siphons of a spillway on the Yuma project of the U.S. Bureau of Reclamation (Fig. 51 is a cross section of one of these siphons), Baron[2] obtained values of the coefficient of discharge C ranging from 0.62 to 0.74.

[1] *U.S. Bur. Reclamation Experimental Data* 41, 42, 43, 44, 1923.
[2] *U.S. Bur. Reclamation Experimental Data* 6. rev. 1925.

WEIRS AND DAMS

Flow of water through a notch or weir is illustrated in Fig. 52. Water is led up to the weir through a channel called the *channel of approach*, the mean velocity in this channel being the *velocity of approach*. The edge or surface over which the water flows is called the *crest* of the weir. The overflowing water after leaving the weir is termed the *nappe*. The weir shown in the figure is *sharp-crested;* i.e., it has a sharp upstream corner such that the water in passing touches only a line. In passing over all other weirs, the water touches a surface instead of a line, and such weirs for want of a better term are classed as *weirs not sharp-crested*. A weir is said to have *free discharge* when, as in Fig. 52, it discharges freely into the air. If the discharge is partially under water, as shown in Fig. 56, the weir is said to be *submerged* or *drowned*. Classified in accordance with shape of notch, weirs may be *rectangular*, *triangular* or *V-notch*, *trapezoidal*, or of any other regular form. Sharp-crested weirs provide a useful means of measuring flowing water, but they serve no other purpose. Weirs not sharp-crested are commonly incorporated in hydraulic structures where they may be used to measure water, but this is usually a secondary function.

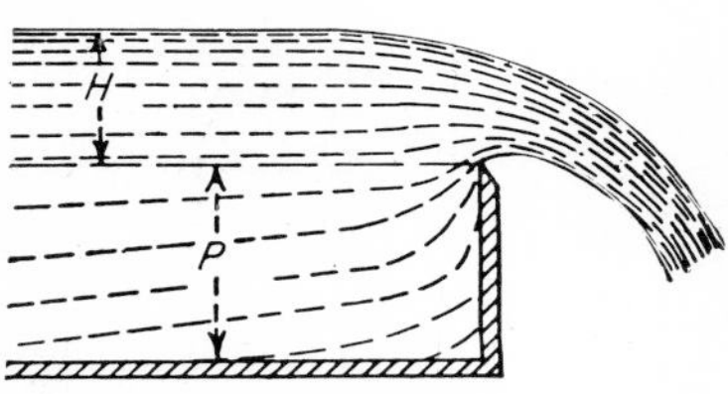

FIG. 52. Sharp crested weir.

The difference in elevation between the upstream water surface and the weir crest H (Fig. 52) is called the *head*. The significance of the term "head" as applied to weirs differs from the same term as it is commonly employed in connection with orifices where the head corresponds to the velocity head due to the theoretical mean velocity through the orifice. For weirs, the head corresponds to the velocity head for the depth H only. The theoretical discharge over a rectangular weir of length L, not including the effects of velocity of approach, is shown on p. 4–46 to be

$$Q_t = \tfrac{2}{3}\sqrt{2g}\, LH^{3/2} \tag{70}$$

If LH is taken as the cross-sectional area of the stream in the plane of the weir, the formula for the theoretical mean velocity is $v_t = \tfrac{2}{3}\sqrt{2gH}$, or two-thirds the velocity through an orifice under an average head H.

Velocity of Approach. The effective head producing discharge over a weir is the potential or measured head H (Fig. 52) plus the kinetic or velocity head resulting from velocity of approach V. As explained on p. 4–22, this velocity head would be $V^2/2g$ if the velocities in a cross section of the channel of approach were equal. Such velocities are, however, never equal, and the kinetic head is always greater than the velocity head due to mean velocity. The total effective head is $H + \alpha \dfrac{V^2}{2g}$, α being an empirical coefficient the value of which varies with distribution of velocities in the channel of approach but which is always greater than unity. Since the variation in velocities is most erratic, there is always uncertainty regarding the value of α. This is one reason for lack of precision in measurements of discharge by weirs, particularly if velocities of approach are high.

Contractions. The nappe after leaving a sharp-crested weir suffers a contraction similar to the contraction of a jet issuing from a sharp-edged orifice (see p. 4–29). If the distance P (Fig. 52) of the crest of the weir above the bottom of the channel is not at least $2.5H$, the contraction along the weir crest, or the *crest contraction*, will be partly suppressed. A drop in water surface, called *surface contraction*, begins about $2H$ up-

stream and continues to the plane of the weir. The *vertical contraction* of the nappe includes both surface and crest contraction.

Rectangular Sharp-crested Weirs. *End Contractions.* If the sides of a rectangular weir have sharp upstream edges, the nappe is contracted in width, and the weir has end contractions. If the crest length is the same as the width of the channel, the sides of the weir coincide with the sides of the channel, and the weir has *suppressed end contractions.* For complete end contractions, the ends of the weir should be at least $2.5H$ from the sides of the channel. End contractions reduce the effective length of a weir. From experiments performed in 1852, Francis [1] found

$$L = L' - 0.1NH \tag{71}$$

in which L is the effective length of weir, L' the measured length, and N the number of contractions. For complete end contractions, $N = 2$; for contractions suppressed at one end, $N = 1$. Fteley and Stearns and other later investigators have found that the effects of end contractions do not vary directly with the head. Francis' correction is obviously in error for very short weirs. No better general formula than (71) has, however, been suggested. Because of uncertainty regarding the effect of end contractions, weirs with end contractions suppressed should be used wherever practicable.

Derivation of Discharge Formula. Consider that water having a velocity of approach V discharges over the weir crest A (Fig. 53), the measured head being H. The effective head is equal to the measured head plus $\alpha \frac{V^2}{2g}$, α being an empirical coefficient of which the value depends on the distribution of velocities in the channel of approach, as

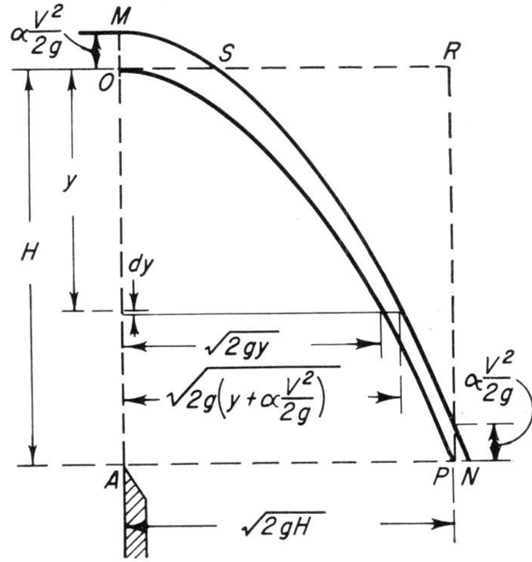

Fig. 53. Discharge over weir.

explained above. If the velocity head of each filament of water in the channel of approach is assumed to equal the average velocity head of all filaments, the velocity v at any depth y is

$$v = \sqrt{2g\left(y + \alpha \frac{V^2}{2g}\right)} \tag{72a}$$

[1] Francis, J. B., "Lowell Hydraulic Experiments," 4th ed., D. Van Nostrand Company, Inc., New York, 1883.

If the velocity of approach is so small as to be negligible, $V = 0$, and

$$v = \sqrt{2gy} \tag{72b}$$

The parabolas MN and OP are graphs of Eqs. (72*a*) and (72*b*) and show respectively velocities at different depths with and without velocity of approach.

Taking the origin at O (Fig. 53), a distance H vertically above the weir crest A, the discharge per unit length of weir through the elementary strip dy is $dQ_1 = v\,dy$, and, substituting the value of v from Eq. (72*a*),

$$dQ_1 = \sqrt{2g\left(y + \alpha\frac{V^2}{2g}\right)}\,dy \tag{73}$$

Integrating this expression between the limits H and zero gives as the theoretical discharge over a weir 1 ft long

$$Q_1 = \frac{2}{3}\sqrt{2g}\left[\left(H + \alpha\frac{V^2}{2g}\right)^{3/2} - \left(\alpha\frac{V^2}{2g}\right)^{3/2}\right] \tag{74}$$

This formula also expresses the area of the surface $OSNA$. The formula for discharge per unit length of weir with velocity of approach taken as zero is $Q_1 = \frac{2}{3}\sqrt{2g}H^{3/2}$, which is also the area of the surface OPA. Figure 53 is drawn to scale for the values $\alpha = 1.5$, $H = 1$, and $V = 2$. The difference in the two areas therefore shows the effect of velocity of approach for these assumed conditions. The relative amount that the subtractive term $(\alpha V^2/2g)^{3/2}$ affects the discharge is indicated by the area $OSNP$.

Writing the symbol h for $V^2/2g$ in Eq. (74), the theoretical discharge for a weir of length L is

$$Q_t = \tfrac{2}{3}\sqrt{2g}\,L[(H + \alpha h)^{3/2} - (\alpha h)^{3/2}] \tag{75}$$

This formula can be transposed to

$$Q_t = \frac{2}{3}\sqrt{2g}\,LH^{3/2}\left[\left(1 + \alpha\frac{h}{H}\right)^{3/2} - \left(\alpha\frac{h}{H}\right)^{3/2}\right] \tag{76}$$

In the right-hand member of the above equation, the expression outside the brackets gives the theoretical discharge without velocity of approach, and the term within the brackets is the factor correcting for velocity of approach.

As is the case with orifices, derived weir formulas must be corrected to include the modifying influences of friction and contraction of the nappe to bring them approximately into accordance with experimental results. Combining the necessary corrective factor and $\frac{2}{3}\sqrt{2g}$ into a single coefficient C, called the *weir coefficient*, and applying this coefficient to Eq. (76), the formula for discharge becomes

$$Q = CLH^{3/2}\left[\left(1 + \frac{\alpha h}{H}\right)^{3/2} - \left(\frac{\alpha h}{H}\right)^{3/2}\right] \tag{77}$$

Experimental investigations have shown that C is not a constant. It varies with H and possibly with other factors. In order to avoid the necessity of a variable coefficient and also to provide a simpler solution of weir problems, it is desirable to modify the form of Eq. (77).

Modified Weir Formula. The following formula:

$$Q = CLH^{3/2}\left[1 + \frac{3}{2}\frac{\alpha h}{H} + \frac{3}{8}\left(\frac{\alpha h}{H}\right)^2 \cdots - \left(\frac{\alpha h}{H}\right)^{3/2}\right] \tag{78}$$

is Eq. (77) with the first term in parentheses expanded by the binomial theorem. Dropping all terms within the brackets excepting the first two, Eq. (78) becomes

$$Q = CLH^{3/2}\left(1 + \frac{3\alpha}{2}\frac{h}{H}\right) \tag{79}$$

That Eq. (79) is very nearly equivalent to (77) can be seen from the fact that all the terms of the expanded series that are dropped are of opposite sign from and their sum is less than the term $(\alpha h/H)^{3/2}$. From an examination of Fig. 53, the relative unimportance of this latter term can be seen (see also preceding paragraph).

By definition of velocity of approach, $V = Q/A = CLH^{3/2}/A$ (approximately), where A is the cross-sectional water area in the channel of approach. The value of Q as substituted is an approximation, since it assumes no velocity of approach. It is, however, used in correcting a term which is itself a relatively small corrective factor, making the error introduced by this approximation of negligible importance. Using this value of V, $h = V^2/2g = C^2L^2H^3/2gA^2$, and substituting for h in Eq. (79) and reducing,

$$Q = CLH^{3/2}\left[1 + \frac{3\alpha C^2}{4g}\left(\frac{LH}{A}\right)^2\right] \tag{80}$$

or, writing a single coefficient C_1 for $3\alpha C^2/4g$,

$$Q = CLH^{3/2}\left[1 + C_1\left(\frac{LH}{A}\right)^2\right] \tag{81}$$

Equation (81) can also be written

$$Q = CLH^{3/2}\left(1 + C_1\frac{H^2}{d^2}\right) \tag{82}$$

in which $d = A/L$ for all weirs, and for weirs with suppressed end contractions it is the depth of water in the channel of approach.

Although two approximations were introduced in obtaining the above formulas, it can be shown by substituting the same numerical values of C and α in Eqs. (77) and (80) that, within the range of conditions that occur in practice, they are practically identical expressions. Equations (81) and (82) are of convenient form. Many of the weir formulas are written in this form, and most of the others can be reduced to it. The main advantage of formulas in this form is that they provide a direct solution for Q, while formulas of other forms require a trial-and-error method of correcting for velocity of approach. Values of C and C_1 or of C and α must be determined from experiments.

Experiments. There are many published results of experiments [1] on sharp-crested rectangular weirs performed in the United States, France, and Germany. The earliest experiments were conducted in France more than one hundred years ago. Although these early experiments were all performed with comparatively small quantities of water, for the range of conditions covered, they agree very closely with later experiments. In 1852, Francis investigated weirs 8 and 10 ft long under heads of 0.6 to 1.6 ft. He

[1] The following publications contain full details of the available experimental data on sharp-crested rectangular weirs: Horton, Weir Experiments, Coefficients, and Formulas, *U.S. Geol. Survey Water Supply and Irrigation Paper* 200, 1907. This paper gives a very complete description of most of the weir experiments performed prior to 1907, including the experiments of Francis, Fteley, and Stearns, and Bazin. Francis, "Lowell Hydraulic Experiments," 4th ed., 1883; also *Trans. ASCE*, vol. 13, 1884. Fteley and Stearns, Flow of Water over Weirs, *Trans. ASCE*, vol. 12, 1883. Bazin, *Ann. ponts et chaussées*, October, 1888, trans. by Marichal and Trautwine, *Proc. Eng. Club*, Philadelphia, January, 1890; also, *Ann. ponts et chaussées*, *Mémoires* 7, 1894. Frese, Versuche über den Abfluss des Wassers bei vollkommenen Uberfallen, *Zeit. Ver. deut. Ing.*, 1890. Rehbock, Discussion of "Precise Weir Measurements," by Schoder and Turner, *Trans. ASCE*, vol. 93, 1929; also, "Handbuch der Ingenieurwissenschaften," Pt. 3, No. 2, vol. 1, Abteilung, p. 58, Leipzig, 1912. Schoder and Turner, Precise Weir Measurements, *Trans. ASCE*, vol. 93, 1929. The author has not been able to find reference to any published results of an investigation of sharp-crested rectangular weirs by the Swiss Society of Engineers and Architects.

had facilities for using larger quantities of water than the earlier investigators. Following Francis, experiments were performed by Fteley and Stearns, 1877–1879; Bazin, 1886; Frese, 1890; Rehbock, 1903–1929; and Schoder and Turner, 1904–1920. Francis and Fteley and Stearns experimented on weirs with end contractions and also on weirs with end contractions suppressed. The later experiments were performed exclusively on weirs with suppressed end contractions. Altogether these experiments cover a wide range of conditions as regards height and length of weir. Schoder and Turner investigated heads as great as 2.5 ft, but with this exception all heads were less than 1.7 ft. These various experiments present many annoying inconsistencies, which are reflected by discrepancies in the several weir formulas derived from them.

Weir Formulas. Formulas for discharge over sharp-crested rectangular weirs have, in general, been obtained by determining from experiments the values of the coefficients C and α [Eq. (77)] or C and C_1 [Eq. (81) or (82)]. The experiments show that C decreases as H increases. The coefficients in weir formulas correct for the effects of friction, contraction of nappe, unequal velocities in channel of approach, partial suppression of crest contraction for low weirs, and probably other factors. Below are given the most important weir formulas that have been developed in the United States and Europe. The formulas are all written in the form for suppressed end contractions. The effective length for weirs having end contractions is given by Eq. (71). The European formulas are based entirely upon experiments for weirs with end contractions suppressed and are not intended to apply to weirs with end contractions. It is probable, however, that with slight modification they can be made to apply to weirs with end contractions as well as the American formulas. The European formulas, originally written in metric units, are here given in English units. Referring to Fig. 53, the following nomenclature is used:

Q = discharge.
H = measured head.
L = effective length of weir (see p. 4–45).
P = height of weir.
A = cross-sectional water area in channel of approach.
$V = Q/A$ = velocity of approach.
$h = V^2/2g$ = velocity head due to mean velocity.
V_a = mean velocity in channel of approach above crest level.
V_b = mean velocity in channel of approach below crest level.
$d = A/L$ = depth of water $(H + P)$ for weirs with end contractions suppressed.
g = acceleration due to gravity = 32.16 (approximately).
K = weir factor.

In the following list of formulas for sharp-crested rectangular weirs, the names given are the names of the authors of the formulas. "Swiss Society" is an abbreviation for Swiss Society of Engineers and Architects.

Francis:

$$Q = 3.33L[(H + h)^{3/2} - h^{3/2}] \tag{83}$$

Fteley and Stearns:

$$Q = 3.31L(H + 1.5h)^{3/2} + 0.007L \tag{84}$$

Bazin:

$$Q = \frac{2}{3}\sqrt{2g}\,LH^{3/2}\left(0.6075 + \frac{0.01476}{H}\right)\left(1 + 0.55\frac{H^2}{d^2}\right) \tag{85}$$

Frese:

$$Q = \frac{2}{3}\sqrt{2g}\,LH^{3/2}\left(0.615 + \frac{0.00689}{H}\right)\left(1 + 0.55\frac{H^2}{d^2}\right) \tag{86}$$

Rehbock:

$$Q = \frac{2}{3}\sqrt{2g}\,LH^{3/2}\left(0.605 + \frac{1}{320H - 3} + \frac{0.08H}{P}\right) \tag{87}$$

King:

$$Q = 3.34LH^{1.47}\left(1 + 0.56\,\frac{H^2}{d^2}\right) \tag{88}$$

Swiss Society:

$$Q = \frac{2}{3}\sqrt{2g}\,LH^{3/2}\left(0.615 + \frac{0.615}{305H + 1.6}\right)\left(1 + 0.5\,\frac{H^2}{d^2}\right) \tag{89}$$

Schoder:

$$Q = 3.33L\left[\left(H + \frac{V_a^2}{2g}\right)^{3/2} + \frac{V_b^2}{2g}\,H\right] \tag{90}$$

Each of the above formulas excepting Eq. (90) can be put in the form

$$Q = KLH^{3/2} \tag{91}$$

the following being expressions for K:

Francis:

$$K = 3.33\left[\left(1 + \frac{h}{H}\right)^{3/2} - \left(\frac{h}{H}\right)^{3/2}\right] \tag{92}$$

Fteley and Stearns:

$$K = 3.31\left(1 + 1.5\,\frac{h}{H}\right)^{3/2} + \frac{0.007}{H^{3/2}} \tag{93}$$

Bazin:

$$K = \left(3.248 + \frac{0.079}{H}\right)\left(1 + 0.55\,\frac{H^2}{d^2}\right) \tag{94}$$

Frese:

$$K = \left(3.288 + \frac{0.0368}{H}\right)\left(1 + 0.55\,\frac{H^2}{d^2}\right) \tag{95}$$

Rehbock:

$$K = 3.235 + \frac{1}{60H - 0.56} + 0.428\,\frac{H}{P} \tag{96}$$

King:

$$K = \frac{3.34}{H^{0.03}}\left(1 + 0.56\,\frac{H^2}{d^2}\right) \tag{97}$$

Swiss Society:

$$K = \left(3.288 + \frac{1}{92.8H + 0.49}\right)\left(1 + 0.5\,\frac{H^2}{d^2}\right) \tag{98}$$

Table 9 gives values of K, for different heads and heights of weir, computed with the above formulas. For intermediate values of H and P, K can be interpolated well within the limits of uncertainty that exist in results given by any weir formula. To determine Q, therefore, by any of the formulas, it is necessary only to take from the table the K corresponding to the proper head and height of weir and to apply this K in Eq. (91).

Comparison of Formulas. Table 9 shows comparative discharges given by the various weir formulas. To aid in making comparisons, the formulas in closest agreement are placed adjacent to each other. The formulas listed in the table are all based upon experiments on weirs not more than 5 ft high and with heads less than 1.7 ft. Within these limits, the Fteley and Stearns, Rehbock, and Swiss Society formulas give values of K which, in general, do not differ from each other by more than 1 percent, while the Bazin,

TABLE 9. VALUES OF K IN THE FORMULA $Q = KLH^{3/2}$ FROM VARIOUS WEIR FORMULAS

Name of formula	P \ H	0.2	0.5	1.0	1.5	2.0	3.0	4.0	5.0
Francis	0.5	3.40	3.55	3.72	3.84	3.92	4.03	4.10	4.15
Fteley and Stearns	....	3.50	3.74	4.32	5.25				
Rehbock	....	3.49	3.70	4.11	4.53	4.95	5.81	6.66	7.51
Swiss Society	....	3.48	3.72	4.03	4.22	4.35	4.50	4.59	4.65
Frese	....	3.63	3.82	4.14	4.34	4.47	4.63	4.73	4.79
King	....	3.67	3.89	4.17	4.34	4.44	4.56	4.62	4.66
Bazin	....	3.81	3.87	4.14	4.32	4.44	4.60	4.69	4.75
Francis	1.0	3.35	3.43	3.55	3.64	3.72	3.84	3.92	3.98
Fteley and Stearns	....	3.43	3.47	3.72	4.00	4.32	5.31		
Rehbock	....	3.41	3.48	3.68	3.89	4.10	4.52	4.95	5.38
Swiss Society	....	3.39	3.49	3.71	3.89	4.03	4.22	4.34	4.43
Frese	....	3.53	3.57	3.78	3.97	4.11	4.32	4.46	4.55
King	....	3.56	3.62	3.81	3.96	4.09	4.25	4.35	4.42
Bazin	....	3.70	3.61	3.78	3.95	4.09	4.29	4.42	4.51
Francis	2.0	3.34	3.36	3.43	3.49	3.55	3.64	3.72	3.79
Fteley and Stearns	....	3.40	3.38	3.47	3.59	3.72	4.00	4.31	4.69
Rehbock	....	3.37	3.38	3.47	3.57	3.67	3.88	4.09	4.31
Swiss Society	....	3.35	3.38	3.48	3.60	3.70	3.88	4.02	4.13
Frese	....	3.49	3.44	3.53	3.65	3.76	3.95	4.10	4.22
King	....	3.52	3.49	3.55	3.64	3.73	3.88	4.00	4.09
Bazin	....	3.66	3.48	3.53	3.63	3.74	3.92	4.07	4.18
Francis	3.0	3.33	3.35	3.38	3.43	3.47	3.55	3.61	3.67
Fteley and Stearns	....	3.39	3.36	3.40	3.47	3.54	3.72	3.90	4.08
Rehbock	....	3.35	3.34	3.39	3.46	3.53	3.67	3.81	3.95
Swiss Society	....	3.34	3.34	3.40	3.48	3.56	3.70	3.83	3.93
Frese	....	3.48	3.40	3.44	3.52	3.60	3.75	3.89	4.00
King	....	3.51	3.45	3.46	3.50	3.56	3.68	3.79	3.88
Bazin	....	3.65	3.44	3.44	3.50	3.58	3.72	3.85	3.97
Francis	5.0	3.33	3.34	3.35	3.38	3.40	3.45	3.50	3.55
Fteley and Stearns	....	3.39	3.34	3.34	3.38	3.42	3.51	3.61	3.72
Rehbock	....	3.34	3.31	3.34	3.37	3.41	3.50	3.58	3.66
Swiss Society	....	3.34	3.32	3.34	3.38	3.43	3.52	3.62	3.70
Frese	....	3.48	3.38	3.38	3.41	3.46	3.56	3.66	3.75
King	....	3.51	3.43	3.39	3.40	3.42	3.49	3.56	3.63
Bazin	....	3.65	3.42	3.38	3.40	3.44	3.53	3.62	3.71
Francis	10.0	3.33	3.33	3.34	3.34	3.35	3.38	3.40	3.43
Fteley and Stearns	....	3.39	3.33	3.33	3.33	3.35	3.38	3.42	3.46
Rehbock	....	3.33	3.29	3.29	3.31	3.33	3.37	3.41	3.45
Swiss Society	....	3.34	3.31	3.31	3.32	3.34	3.38	3.43	3.47
Frese	....	3.47	3.37	3.34	3.34	3.36	3.40	3.45	3.50
King	....	3.51	3.41	3.36	3.33	3.32	3.33	3.35	3.38
Bazin	....	3.64	3.41	3.34	3.33	3.34	3.37	3.40	3.46
Francis	∞	3.33	3.33	3.33	3.33	3.33	3.33	3.33	3.33
Fteley and Stearns	....	3.39	3.33	3.32	3.31	3.31	3.31	3.31	3.31
Rehbock	....	3.32	3.27	3.25	3.25	3.24	3.24	3.24	3.24
Swiss Society	....	3.34	3.31	3.30	3.30	3.29	3.29	3.29	3.29
Frese	....	3.47	3.36	3.33	3.31	3.31	3.30	3.30	3.30
King	....	3.51	3.41	3.34	3.30	3.27	3.23	3.20	3.18
Bazin	....	3.64	3.41	3.33	3.30	3.29	3.27	3.27	3.26

King, and Frese formulas give values 1 to 5 percent greater than the other formulas. These discrepancies are explained by the fact that, in deriving the last three formulas, much weight was attached to the Bazin experiments which, for the same dimensions, give greater discharges than the experiments of Fteley and Stearns, Rehbock, and the Swiss Society of Engineers and Architects. The experiments by Francis were all conducted under conditions giving low velocities of approach. This explains why his formula, which was necessarily based entirely upon his own experiments, gives smaller discharges for low weirs and for high heads than any of the other formulas.

Discrepancies that have appeared in carefully conducted experiments on sharp-crested weirs by different investigators have not been satisfactorily explained. It is known, however, that the discharge is affected by (1) the sharpness of the crest, (2) the smoothness of the upstream face of the weir, (3) the distribution of velocities in the channel of approach, and (4) the ventilation beneath the nappe. The experiments of Schoder and Turner indicate that if the weir has a sharp edge and polished upstream face and if there is no abnormal distribution of velocities in the channel of approach and the nappe is well ventilated, the discharge is more accurately given by the formulas of Fteley and Stearns, Rehbock, and the Swiss Society than by those of Frese, King, and Bazin. A departure from any one of these conditions, however, will cause an increased discharge. Schoder found that velocities in the channel of approach above and below crest level affect the discharge differently and designed his formula to correct for this condition. Rehbock claims that slight irregularity in velocities has but little effect on discharge but that large variations in magnitude of velocity and direction of flow may have an appreciable effect.

The experiments of Schoder and others show many of the conditions that modify the flow over weirs, but standard conditions under which the sharp-crested weir can be used for the accurate measurement of flowing water have not been well determined. The standard weir has a sharp crest and smooth upstream face, but there are different degrees of sharpness and smoothness; and even assuming that a weir can be set up to conform to prescribed standards, it is difficult to maintain it in this condition. It appears that in the present state of the art, the sharp-crested weir cannot be called a precise instrument for measuring water. In a laboratory where great refinement in setup and methods of measurement is possible, a higher degree of precision can be expected than under conditions ordinarily encountered in the field. Since there are few if any really precise methods of measuring flowing water, in spite of its apparent disadvantages, the sharp-crested weir remains one of the best methods available. If care is taken in erecting and using a weir, it should give an accuracy in measurements of possibly 2 percent; but in the author's opinion, claims of an accuracy of 1 percent or less for a particular formula are not justified.

Use of Weirs. In order to obtain maximum accuracy in the use of sharp-crested rectangular weirs, they should be installed with vertical face and level crest. Preferably, end contractions should be suppressed. The upstream face of the weir should be smooth, the crest sharp, and the nappe well ventilated. Upstream from the weir there should be as long a reach of straight smooth channel as practicable, so as to avoid excessively irregular velocities in the channel of approach. In order that the weir may be used within the approximate limits of experimental data, the length of the weir should be at least three times the head, and preferably the head should not be greater than 2.0 or less than 0.2 ft. The head should be measured at least $2.5H$ upstream from the weir in order to be beyond the effects of surface contraction. Francis and Fteley and Stearns measured heads 6, and Bazin 16.4 ft upstream from the weir. Other experimenters used intermediate distances. The difference in head measurements at these two points is very small under all ordinary conditions and inappreciable for low velocities of approach. The point of measurement should, however, conform as closely as practicable to experi-

mental conditions and therefore will vary with the formula used for determining discharges. Heads are usually measured in a stilling well by means of a hook gage. The pipe connecting the well and channel of approach should not project beyond the channel surface. The gage reading for zero head should be determined with great care and checked frequently. The head used to compute discharge should be the mean of at least 10 and preferably 20 separate measurements made at equal intervals of about 30 sec.

The selection of a weir formula should be made after careful consideration of all conditions. From the evidence of experiments, it appears that for a very smooth face of weir and sharp crest, with adequate provision for ventilating the nappe and with reasonably uniform velocities in the channel of approach, the Fteley and Stearns, Rehbock, Swiss Society, or (provided it is feasible to measure velocities in the channel of approach) the Schoder formula will apply more accurately. If all these conditions do not obtain, the Frese, King, or Bazin formula will probably be more suitable. Schoder is of the opinion that a combination of a slight rounding of the crest and rusting of the upstream face of a weir could produce just such a percentage increase over the results of his experiments as Bazin's measurements show. The evidence does not, however, indicate that Bazin's weir lacked either a sharp crest or a smooth face.

Sharp-crested Weirs with Sloping Sides. *V-notch Weirs.* A triangular or V-notch weir is shown in Fig. 54. The head, measured above the bottom of the notch, is H, and the distance between the edges of the weir at the water-surface level is l. The sides make equal angles with the vertical. The area of an elementary strip, dy in thickness and of length l', is $l'\,dy$. Neglecting velocity of approach, the theoretical velocity through this strip under any head y is $\sqrt{2gy}$, and the theoretical discharge is $dQ_t = l'\sqrt{2gy}\,dy$. From similar triangles, $l' = l(H - y)/H$. Substituting this value of l' and integrating between the limits H and zero and reducing, $Q_t = \frac{4}{15}\sqrt{2g}\,lH^{3/2}$. The slope, horizontal to vertical, is $z = \frac{1}{2}l/H$ or $l = 2zH$. Substituting for l,

$$Q_t = \frac{8}{15}\sqrt{2g}\,zH^{5/2} \tag{99}$$

Applying a weir coefficient which includes the constant terms, the formula for discharge over V-notch weirs becomes

$$Q = CzH^{5/2} \tag{100}$$

in which the coefficient C must be determined experimentally. Most of the available experiments have been performed on right-angled notches. For this angle, $z = 1$. Experiments show that C decreases with increasing heads.

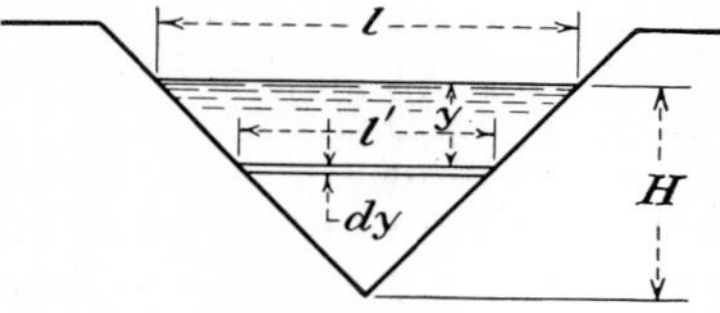

FIG. 54. V-notch weir.

A more general equation for expressing discharge over right-angled V-notch weirs is

$$Q = CH^n \tag{101}$$

When, from a series of experiments on a given weir, logarithms of Q and H respectively are plotted against each other, the plotted points have been found to lie on a straight line. This provides a simple graphical method of obtaining C and n. From experiments at the University of Michigan [1] with heads of 0.2 to 1.8 ft, the following formula was obtained:

$$Q = 2.52H^{2.47} \tag{102}$$

From experiments by Barr: [2]

$$Q = 2.48H^{2.48} \tag{103}$$

[1] *Univ. Mich. Tech.*, October, 1916, p. 190.
[2] Experiments on the Flow of Water over Triangular Notches, *Engineering*, Apr. 8, 15, 1910.

The experiments at the University of Michigan were performed on a sharp-edged right angled notch cut in a large sheet of commercial steel plate. The face of the weir was not therefore a polished surface. Barr used a polished brass plate. He found that by applying emery dust over fresh varnish to the face of his weir he could increase the discharge about 2 percent. From the conditions under which the two sets of experiments were taken, it would appear that Eq. (102) will apply more accurately if the upstream face of the weir is unpolished; and Eq. (103), if it is polished. Ordinarily V-notch weirs are not materially affected by velocity of approach, because of the small relative area of the weir opening in comparison with the cross-sectional area of the channel of approach.

The right-angled V-notch weir is to be preferred to the rectangular weir for measuring flows of 1 cu ft per sec or less; and for discharges of 1 to 10 cu ft per sec, it is at least as accurate as any other type of weir. One advantage of the V-notch weir is that, being narrower, small discharges occur under comparatively high heads, and the tendency of the nappe to adhere to the crest, which occurs when rectangular weirs discharge under low heads, is eliminated. Another advantage is that the V-notch weir is more sensitive to changes in flow than the rectangular weir and that therefore a smaller error in discharge results from a given error in measuring head. For example, in measuring 1 cu ft per sec of water, with a rectangular weir 2 ft long an error of 2.7 percent in discharge results from an error of 0.005 ft in measuring head; while for a right-angled V-notch weir, for the same error in measuring head, the corresponding error in discharge is 1.8 percent.

Trapezoidal Weirs. The trapezoidal weir shown in Fig. 55 has a horizontal crest length L and side slopes, horizontal to vertical, of $e/H = z$. As for V-notch weirs, $dQ_t = l'\sqrt{2gy}\,dy$. Expressing l' in terms of y and integrating and reducing,

$$Q_t = \tfrac{2}{3}\sqrt{2g}\,LH^{3/2} + \tfrac{8}{15}\sqrt{2g}\,zH^{5/2} \tag{104}$$

This formula is also obtained by adding the theoretical discharges for rectangular and V-notch weirs. With coefficients included, the formula for discharge is

$$Q = C_1LH^{3/2} + C_2zH^{5/2} \tag{105}$$

There are no experiments giving C_1 and C_2.

The Cipolletti Weir. A trapezoidal weir (Fig. 55) having a value of $z = e/H$ of $\tfrac{1}{4}$ is called a Cipolletti weir. This value of z is approximately that required to secure a discharge through the triangular portion of the opening that equals the decrease in discharge resulting from end contractions. The advantage claimed for this weir is that it does not require correction for end contractions. The theoretical discharge through the triangular portion of the weir is $Q_t = \tfrac{8}{15}\sqrt{2g}\,zH^{5/2}$, and the theoretical decrease in discharge resulting from end contractions, from Eq. (71), is $Q_t = \tfrac{2}{3} \times 0.2\sqrt{2g}\,H^{5/2}$. Assuming the same coefficient of discharge for these equations and equating their right-hand members, there results $z = \tfrac{1}{4}$. The formula for discharge given by Cipolletti for a weir of length L discharging under a head H is

FIG. 55. Trapezoidal weir.

$$Q = 3.367LH^{3/2} \tag{106}$$

Cipolletti weirs should have a crest length of $4H$ or more. Experiments by Flinn and Dyer and others give a decreasing coefficient for an increasing head, indicating that z should be greater than $\tfrac{1}{4}$. The Cipolletti weir has been used extensively to measure irrigation water. It should not be used when any degree of precision is required. No

method of correcting for velocity of approach has been suggested. It is probable that any one of the formulas [(84) to (89), p. 4–48] for weirs with suppressed end contractions will apply as accurately to Cipolletti weirs as Eq. (106) and more accurately if the velocity of approach is high.

Submerged Sharp-crested Weirs. A profile through a submerged sharp-crested weir is illustrated in Fig. 56. The discharge Q over such a weir is related not only to the head on the upstream side of the weir (H_1) but also to the head on the downstream side (H_2) and, to a lesser extent, to the height of the weir crest above the floor of the channel (P).

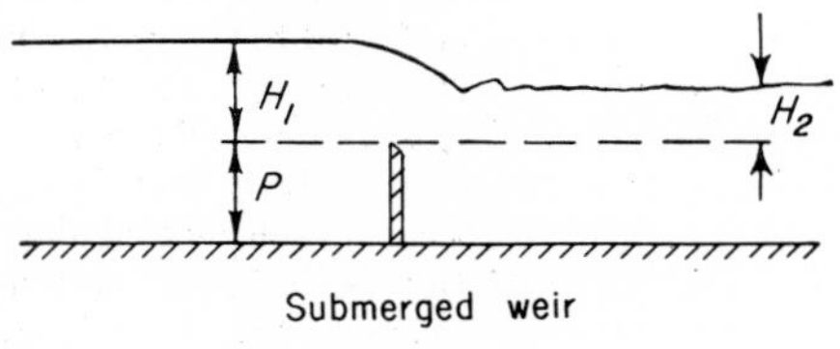

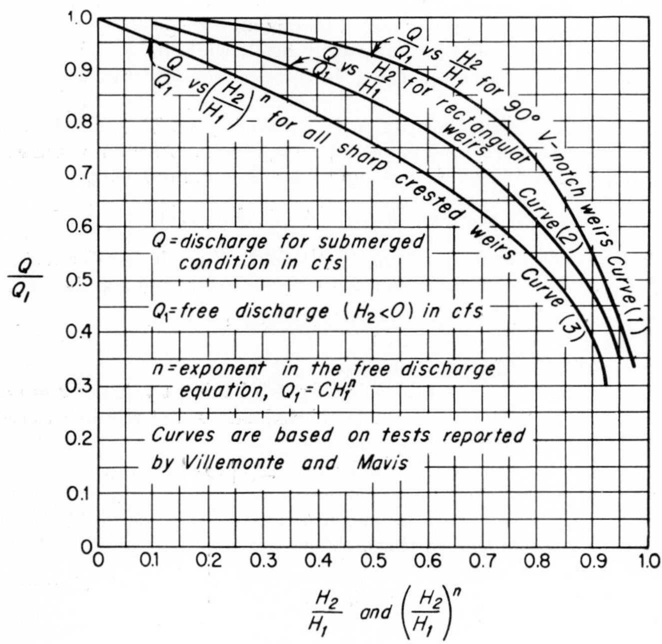

FIG. 56. Flow over submerged weirs.

Early experiments by Francis (1848, 1883), Fteley and Stearns (1882), Bazin (1894), Cone (1916), and Cox (1928) have been summarized by Vennard and Weston.[1] They showed that the various test data could be presented in an orderly manner by selecting as variables Q/Q_1 and H_2/H_1, where Q_1 is the discharge at the head H_1, computed from the equation for free discharge (unsubmerged), which is expressed in general terms as follows:

[1] VENNARD, JOHN K., and RAY F. WESTON, Submergence Effect on Sharp-crested Weirs, *Eng. News-Record*, June 3, 1943, p. 818.

$$Q_1 = CH_1^n \tag{107}$$

By plotting Q/Q_1 against H_2/H_1 they found that the various data tended to fall on a single curve, except for small values of P/H_1.

In 1947, Villemonte [1] presented the results of a series of tests on submerged sharp-crested weirs. He conducted tests on rectangular, triangular, parabolic, cusped, and porportional weirs. He showed that the results for all types could be represented by the single equation

$$\frac{Q}{Q_1} = \left[1 - \left(\frac{H_2}{H_1}\right)^n\right]^{0.385} \tag{108}$$

where n is the exponent in the free-discharge equation [Eq. (107)] and the other terms are as previously defined. This equation was found to satisfy all test results with a maximum deviation of 5 percent for some of the individual test results.

In 1949, Mavis [2] presented results of tests on rectangular, triangular, parabolic, circular, sutro, and cusped weirs. He found that a single equation could be used to express the results for all the tests. His equation with subscripts changed to conform with usage in this book is

$$\frac{Q}{Q_1} = 1 - \left[0.45S + \frac{0.40}{2^{(10-10S)}}\right] \tag{109}$$

where S is defined as follows:

$$S = \frac{a_2\sqrt{H_2}}{a_1\sqrt{H_1}} \tag{110}$$

In Eq. (110), a_2 is the weir area corresponding to H_2 and a_1 is the weir area corresponding to H_1. Mavis also presented some interesting data which resulted from tests made in 1717 by Poleni. It was found that Poleni's results agreed within 2 to 4 percent with the Mavis data.

The author has plotted the curves shown in Fig. 56 based on the results of the work of Villemonte and Mavis. Information regarding the experimental arrangements for the 90° V-notch weirs and rectangular weirs is given in the following table. It may be noted

	Mavis	Villemonte		
Channel width..........	4 ft 0 in.		3.02 ft	
P for 90° V-notch weirs...	1 ft 6 in.		2.0 ft	
P for rectangular weirs...	1 ft 10 in.	2.0 ft	1.0 ft	1.25 ft
Widths of notches of rectangular weirs........	1 ft 3 in.	3.02 ft	0.5 ft	1.00 ft

that the channel widths differed for the two sets of tests, that P was different for all cases, and that rectangular weirs of four different widths were tested. Curves 1 and 2 are composite curves based on the results of the two investigators for the 90° V-notch weirs and the rectangular weirs, respectively. Curves 1 and 2 differ by no more than 1 percent from the test results.

Because Eqs. (108) and (109) both indicate that Q/Q_1 is a function of $(H_2/H_1)^n$, the author has prepared curve 3 which is an average of results obtained from Eqs. (108) and

[1] Villemonte, James R., Submerged-weir Discharge Studies, *Eng. News-Record*, Dec. 25, 1947, p. 866.
[2] Mavis, F. T., How to Calculate Flow over Submerged Thin-plate Weirs, *Eng. News-Record*, July 7, 1949, p. 65.

(109). Results obtained from either equation differ by less than 1 percent from curve 3. Curve 3 may be used to compute the discharge of a submerged sharp-crested weir of any shape. This curve is also in reasonable agreement with the results of the investigations summarized by Vennard and Weston, as well as with data presented by Stevens.[1] It should be noted, however, that for some of the weirs tested the results could be represented more closely by an equation differing slightly from Eqs. (108) and (109) and by a curve differing slightly from curves 1, 2, and 3. Therefore, if great accuracy is essential, it is recommended that the particular weir or a similar one be tested in a laboratory under conditions comparable to field conditions. In using the curves shown in Fig. 56 it is recommended that H_1 be measured at least $2.5H_1$ upstream from the weir and that H_2 be measured beyond the turbulence caused by the nappe.

Example 14: Determine the discharge of a 90° V-notch weir if H_1 is 0.9 ft, H_2 is 0.3 ft, and $Q_1 = 2.5H_1^{2.5}$.

a. Use curve 1 of Fig. 56.

$$Q_1 = 2.5 \times 0.9^{2.5} = 1.92 \text{ sec-ft}$$
$$H_2/H_1 = 0.3/0.9 = 0.333$$
$$Q/Q_1 = 0.972 \text{ (from curve 1)}$$
$$Q = 0.972 \times 1.92 = 1.86 \text{ sec-ft}$$

b. Use curve 3 of Fig. 56.

$$(H_2/H_1)^n = (0.333)^{2.5} = 0.064$$
$$Q/Q_1 = 0.972 \text{ (from curve 3)}$$
$$Q = 0.972 \times 1.92 = 1.86 \text{ sec-ft}$$

Example 15: Determine the discharge of a parabolic weir if H_1 is 0.8 ft, H_2 is 0.4 ft, and $Q_1 = 2.0H_1^{2.0}$.

$$Q_1 = 2.0 \times (0.8)^{2.0} = 1.28 \text{ sec-ft}$$
$$(H_2/H_1)^n = (0.4/0.8)^{2.0} = 0.25$$
$$Q/Q_1 = 0.89 \text{ (from curve 3)}$$
$$Q = 0.89 \times 1.28 = 1.14 \text{ sec-ft}$$

Weirs Not Sharp-crested. Weirs without sharp crests have a relation to sharp-crested weirs analogous to the relation of tubes to sharp-edged orifices. The crest of a dam or a spillway from a canal or reservoir is a weir of this type. Weir crests are built of various dimensions and shapes, there being no standards of practice in this regard. The relation of discharge to head being different for different shapes, to determine this relation for any particular form of weir crest requires an independent experimental investigation. The number of sections that have been investigated experimentally is necessarily limited. There are available, however, the results of several series of experiments on different sections (see Fig. 58), which furnish much valuable information relative to discharges over weirs of the same or similar shapes.

Conditions of Discharge. The effect of the shape of crest on the nappe form is illustrated in Fig. 57. The rectangular section as shown in *a* has a sharp upstream corner, and because of the narrow crest width *b*, the nappe springs clear and discharge occurs as for a sharp-crested weir. Section *d* is the same as *a* except that the width is greater. In this case, free fall of the nappe is interfered with, the conditions of discharge corresponding to those for a standard short tube (see p. 4-37). In general, if *b* is about $\frac{1}{2}H$ or less, the nappe will spring clear; otherwise it will touch the top of the crest. Discharge is increased by rounding the upstream edge of the crest, as illustrated in *b*. The nappe continues to contract for a short distance after leaving the triangular section shown in *c*. The ogee section *e* is so designed that the nappe adheres to the downstream face.

[1] Stevens, J. C., Experiments on Small Weirs and Modules, *Eng. News*, Aug. 18, 1910.

It appears from experiments that, for any of the sections investigated, there is no simple relation between head and discharge. The condition is further complicated by the fact that for a given crest the nappe may assume a variety of forms and that for each form there is a different relation of head to discharge. The nappe may (1) discharge

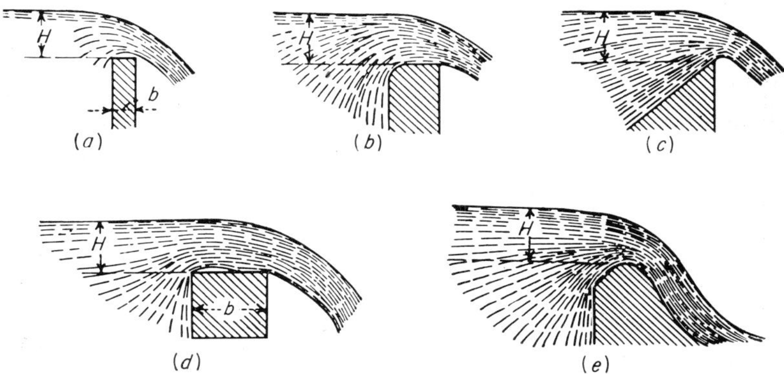

FIG. 57. Cross sections of weir crests.

freely, touching only the upstream crest edge; (2) adhere to top of crest; (3) adhere to downstream face; (4) adhere to both top of crest and downstream face; (5) remain detached and become "wetted underneath"; (6) adhere to top, be detached from face, and become wetted underneath; (7) be depressed with air imprisoned underneath at less than atmospheric pressure. The nappe may undergo several of these modifications in succession as the head is varied. There are certain types of weir, such as are shown in Fig. 57*e*, for which, from the nature of their section, the nappe can assume only one form.

Discharge Formula. The following nomenclature is used:

Q = discharge.
H = measured head (see Fig. 57).
L = length of weir.
A = cross-sectional area of channel of approach.
V = velocity of approach.
C = weir coefficient.

The basic formula

$$Q = CLH^{3/2} \tag{111}$$

is assumed for all weirs, experimental values of C corresponding to different values of H being required for each form of crest. The more important crest forms for which experimental data are available are shown in Fig. 58. This figure and the accompanying table of coefficients are from results published by Horton.[1] Before computing his values of C from experiments, Horton corrected for velocity of approach by adding $V^2/2g$ to the measured head. This same method of correcting for velocity of approach should, therefore, be used with his values of C. With velocity-of-approach correction, Eq. (111)

[1] *U.S. Geol. Survey Water Supply and Irrigation Paper* 200, 1907. In this paper are contained the results of *an extensive investigation of all weir experiments then available.*

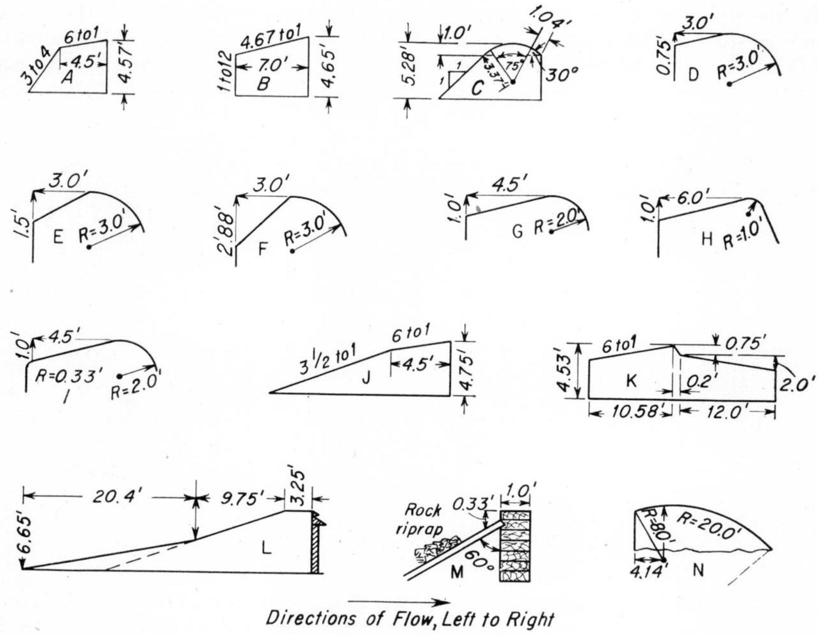

Fig. 58. Weir-crest models.

Weir Coefficient C for Different Heads H of Crest Models in Fig. 59

0.5	1	1.5	2	3	4	H Model	H Model	0.5	1	1.5	2	3	4
	3.46	3.45	3.42	3.32	3.37	A	H	3.28	3.50	3.54	3.52	3.31	3.30
				3.39	3.38	B	I	3.18	3.27	3.43	3.52	3.64	3.70
	3.26	3.28	3.32	3.47	3.59	C	J			3.44	3.35	3.32	3.38
3.29	3.29	3.32	3.36	3.43	3.53	D	K		3.12	3.20	3.22	3.22	3.22
3.27	3.38	3.46	3.51	3.58	3.67	E	L	3.12	3.14	3.10	3.14	3.26	3.36
3.15	3.45	3.63	3.75	3.87	3.88	F	M	....	3.80				
3.18	3.30	3.38	3.42	3.49	3.53	G	N	3.10	3.10	3.33			

becomes

$$Q = CL\left(H + \frac{V^2}{2g}\right)^{3/2} \tag{112}$$

Or, following the method of reduction given on p. 4–47,

$$Q = CLH^{3/2}\left[1 + 0.024C^2\left(\frac{LH}{A}\right)^2\right] \tag{113}$$

Broad-crested Weirs. The weir section illustrated in Fig. 59 has a broad and approximately level crest, a sharp upstream corner, and vertical faces. The sharp corner produces a contraction of the nappe. It appears that, if the crest width b is sufficient to

prevent the nappe from springing clear, the coefficient for weirs of this type has a very nearly constant value of 2.63 or 2.64 for heads of 1 ft or more and for all crest widths. Woodburn[1] found (see Table 10) that this constant value is not reached below heads of 1.4 ft.

Rounding the upstream corner of a broad-crested weir reduces contraction and increases the weir coefficient. Woodburn found that for heads of 1.5 and less there was no appreciable benefit from rounding the corner on a radius greater than 2 in. (see Table 10). Curvature on radii of 2, 3, 6, and 8 in. respectively gave practically

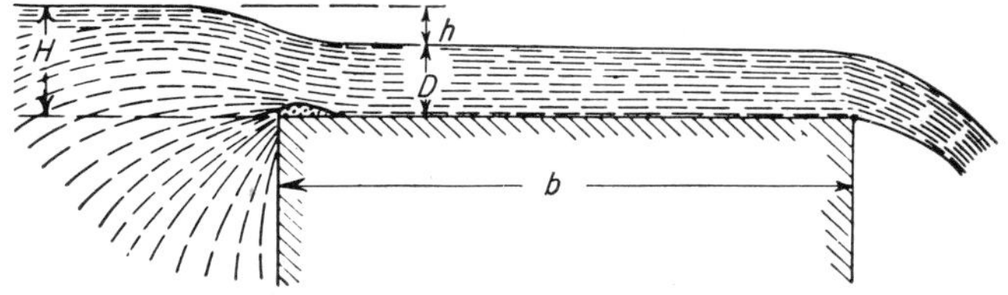

Fig. 59. Broad crested weir.

the same coefficient. Also, elliptical entrances gave practically the same coefficients as circular entrances. If the upstream corner is rounded sufficiently to prevent contraction and the crest is given sufficient slope to offset loss of head due to friction, flow occurs at critical depth (see p. 4–85), and the rational formula for discharge is

$$Q = 3.087LH^{3/2} = 5.67LD^{3/2} \tag{114}$$

where D (Fig. 59) is the depth of water on the weir crest below the downdrop curve. Woodburn found from experiments in a smooth wooden flume 2 ft wide with models made of planed lumber that the greatest discharges he could obtain with a crest slope of 0.026 were 1 to 3 percent less than are given by Eq. (114). Table 10 giving values of the co-

Table 10. Coefficient C for Broad-crested Weirs for Different Heads H

Upstream corner	Crest \ H	0.5	0.6	0.7	0.8	0.9	1.0	1.2	1.4	1.5
Sharp	Level	2.60	2.60	2.60	2.60	2.60	2.60	2.61	2.63	
2-in. radius	Level	2.78	2.79	2.80	2.81	2.83	2.84	2.85	2.85	2.85
3-in. radius	Level	2.77	2.79	2.80	2.81	2.83	2.84	2.85	2.85	2.85
6-in. radius	Level	2.78	2.79	2.80	2.81	2.82	2.83	2.85	2.85	2.85
8-in. radius	Level	2.78	2.79	2.80	2.81	2.82	2.83	2.84	2.85	2.85
3- by 6-in. ellipse	Level	2.78	2.79	2.80	2.82	2.83	2.84	2.85	2.85	2.85
8- by 4-in. ellipse	Level	2.77	2.78	2.80	2.81	2.82	2.83	2.85	2.86	2.86
6-in. radius	Slope, 0.004	2.95	2.94	2.93	2.92	2.91	2.90	2.88	2.87	2.87
6-in. radius	Slope, 0.026	3.07	3.06	3.05	3.04	3.03	3.02	3.00	2.99	

efficient C [Eq. (111)] for broad-crested weirs 10 ft wide is taken from Woodburn's experiments. The coefficient for all the models with rounded entrance corner appears to approach a common value of about 2.85 which remains constant for the higher heads.

The drop in water surface (Fig. 59), neglecting velocity of approach and entrance losses, is $h = v^2/2g$, where v is the velocity on the weir crest just below the downdrop

[1] Tests of Broad-crested Weirs, with appendix by A. R. Webb, *Proc. ASCE*, September, 1930. This paper describes an extensive experimental study of conditions of flow over broad-crested weirs. A wide variety of models was investigated.

curve. In this case, $v = \sqrt{2gh}$, and, including a discharge coefficient, $v = C\sqrt{2gh}$. If there is appreciable velocity of approach, neglecting entrance losses, h is the difference in velocity heads due to velocity below the downdrop curve and velocity of approach. The broad-crested weir is, in reality, a short channel, and the conditions of flow correspond to the conditions at the entrance to a canal receiving water freely from a reservoir or other comparatively large body of water.

Submerged Weirs and Dams. The general laws governing the flow of water over submerged sharp-crested weirs, described on p. 4–54, undoubtedly apply to submerged dams and weirs not sharp-crested, but, since discharge is affected by the shape of the weir section, each shape presents a separate problem, and each requires a separate experimental investigation. From the few experiments available, it is impossible to develop a general formula. The following rules will be helpful in some instances: (1) If the depth of submergence is not greater than 0.2 of the head, ignore the submergence and treat the weir as though it had free discharge. (2) For narrow weirs having a sharp upstream corner, use a submerged-weir formula for sharp-crested weirs. (3) Broad-crested weirs are not affected by submergence up to approximately 0.66 of the head. (4) For weirs with narrow rounded crests, increase discharges obtained by a formula for submerged sharp-crested weirs by 10 percent or more. Of the above rules, 1, 2, and 3 probably apply quite accurately, while 4 is simply a rough approximation.

Falls and Drops. If a channel having a rectangular section terminates abruptly, allowing the water to discharge freely into the air, a condition analogous to that illustrated in the right-hand portion of Fig. 59 is produced, and discharge is given by the right-hand member of Eq. (114). The depth D should be measured a short distance upstream from the crest of the fall, well above the effects of surface contraction. Canal drops are required to drop water to a lower elevation and at the same time maintain a given depth in the canal. In such structures, flow occurs at critical depth, and discharges can be determined by the principles and formulas given on pp. 4–84 to 4–89.

FLOW IN PIPES

Many investigators have sought to determine the laws governing the flow of fluids in pipes. From the time of Chezy (1775) test results gave information which permitted the solution of practical problems involving the flow of water. The early formulas such as the Chezy formula (p. 4–67) were based on the assumption that the pressure drop in a pipeline was related only to the velocity, the dimensions of the conduit, and the roughness. The work of Hagen (1839), Poiseuille (1840), and Reynolds (1883) showed clearly

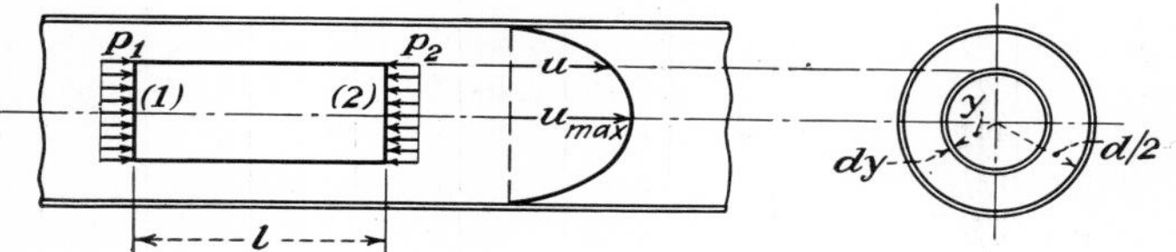

Fig. 60. Laminar flow in round pipes.

that the density and viscosity of the fluid also influence the pressure drop. Finally it has become generally recognized that roughness is a relative quantity, i.e., two pipes have the same roughness only if the ratio between some average dimension of the irregularities of the wall surface and the diameter of the pipe is the same.

Laminar Flow in Pipes. The laws of laminar flow in pipes were determined experimentally, independently by Hagen and Poiseuille. The Hagen-Poiseuille law can be developed from fundamental principles as follows: Consider the forces acting on a cylinder of fluid of length l and radius y as shown in Fig. 60. If steady motion exists,

the force caused by the difference in pressure on the ends of the cylinder must be exactly balanced by the force resulting from the shear stress at the boundaries as expressed by Eq. (115).

$$(p_1 - p_2)\pi y^2 = \tau 2\pi y l \tag{115}$$

Simplifying and introducing the value of τ from Eq. (3), the above equation becomes

$$(p - p_2)y = -2\mu \frac{du\, l}{dy}$$

the minus sign being included because increments of u and y are opposite in sign as shown in Fig. 60. The above equation may be solved for du and integrated as follows:

$$u = \int du = \frac{-(p_1 - p_2)y^{-}}{4\mu l} + C \tag{116}$$

The value of C may be obtained from the boundary condition that $y = d/2$ when $u = 0$. Then

$$C = \frac{(p_1 - p_2)d^2}{16\mu l}$$

and

$$u = \frac{(p_1 - p_2)}{4\mu l}\left(-y^2 + \frac{d^2}{4}\right) \tag{117}$$

This equation shows the velocity distribution for laminar flow in circular pipes to be a paraboloid of revolution. The value of u_{max} may be determined by letting $y = 0$.

$$u_{max} = \frac{(p_1 - p_2)d^2}{16\mu l} \tag{118}$$

The discharge is obtained by integration as follows:

$$Q = \int dQ = \int u\, dA = \int u 2\pi y\, dy$$

Substituting the value of u from Eq. (117),

$$Q = \frac{\pi(p_1 - p\ )}{2\mu l}\int_0^{\frac{d}{2}}\left(-y^3\, dy + \frac{d^2 y\, dy}{4}\right)$$

and

$$Q = \frac{\pi(p_1 - p_2)d^4}{128\mu l} \tag{119}$$

The average velocity v may be obtained from Eq. (119) by dividing by the area of the pipe.

$$v = \frac{(p_1 - p_2)d^2}{32\mu l} \tag{120}$$

It may be seen from Eqs. (118) and (120) that for laminar flow the maximum velocity is twice the average velocity. The expression for the change in piezometric head in a length l may be obtained by solving for $(p_1 - p_2)$ and dividing by the specific weight.

$$\frac{p_1 - p_2}{w} = h = \frac{32\mu l v}{d^2 \rho g} \tag{121}$$

Introducing Reynolds number, the above expression becomes

$$h = \frac{64}{R}\frac{l}{d}\frac{v^2}{2g} \qquad (122)$$

If in Eq. (122) the roughness coefficient f is used in place of $64/R$, the expression becomes

$$h = f\frac{l}{d}\frac{v^2}{2g} \qquad (123)$$

which is the well known Darcy-Weisbach formula for flow in pipes, h being the energy loss in feet of pipe in foot-pounds per pound. It is clear from the above that for laminar flow f is completely independent of roughness but varies only with Reynolds number, i.e., with the relative strength of the viscous and inertial forces. The straight line shown at the left side in Fig. 61 is a graphical representation of the relation

$$f = \frac{64}{R} \qquad (124)$$

The experimental work of Hagen and Poiseuille and tests by many later investigators have established the accuracy of this relationship beyond question. So well does the relation hold that the application of Eq. (121) to circular tubes is one of the most basic methods of determining viscosity.

The Hagen-Poiseuille law, Eq. (119), or the corresponding relation between f and R, Eq. (124), apply when R is less than 2,000. In the range of Reynolds numbers from 2,000 to 4,000, flow changes from laminar to turbulent. Values of f are uncertain in this range. If a pipeline were to be designed for flow in this range, the only safe procedure would be to assume that flow is turbulent and select f by extending the curves shown in Fig. 61.

The *kinetic energy* of laminar flow may be evaluated from Eqs. (34) and (35) letting $a = \pi d^2/4$, $da = 2\pi y\,dy$ and substituting values of u and v from Eqs. (117) and (120) respectively, as follows

$$\alpha\frac{v2}{2g} = \frac{1}{2gav}\int_0^{\frac{d}{2}} u^3\,da$$

Integration of the above expression gives $\alpha = 2$.

Turbulent Flow in Pipes. When flow occurs at Reynolds numbers greater than 4,000, values of f in the Darcy-Weisbach formula [Eq. (123)] vary with roughness as well as with viscosity and density. Turbulent flow may be divided into three categories, namely, flow in smooth pipes, flow in the relatively rough pipes at high velocities, and flow in the transition zone between the first two categories.

For the case of flow in very smooth pipes, values of f vary with R, as shown by the lower curve of Fig. 61. It may be seen that this curve never becomes horizontal, which shows that the fluid properties influence the flow throughout the entire range of Reynolds numbers. Glass pipes or very smooth drawn tubing would fall into this category.

Flow in rough pipes at large values of R is illustrated in Fig. 61 by the zone above and to the right of the broken line. This is called the zone of fully developed turbulence. In this zone, the f curves become horizontal, thus showing that flow is entirely independent of the fluid properties. Nikuradse [1] showed that values of f in this zone depend only on the relative roughness (ϵ/d) where ϵ is the dimension of the roughness and d is the pipe diameter in the same units. He determined this by testing pipes which had been artificially roughened with sand of a uniform size.

[1] NIKURADSE, J., "Stromungsgesetze in rauhen Rohren," Forschungsheft 361, V.D.I., 1933.

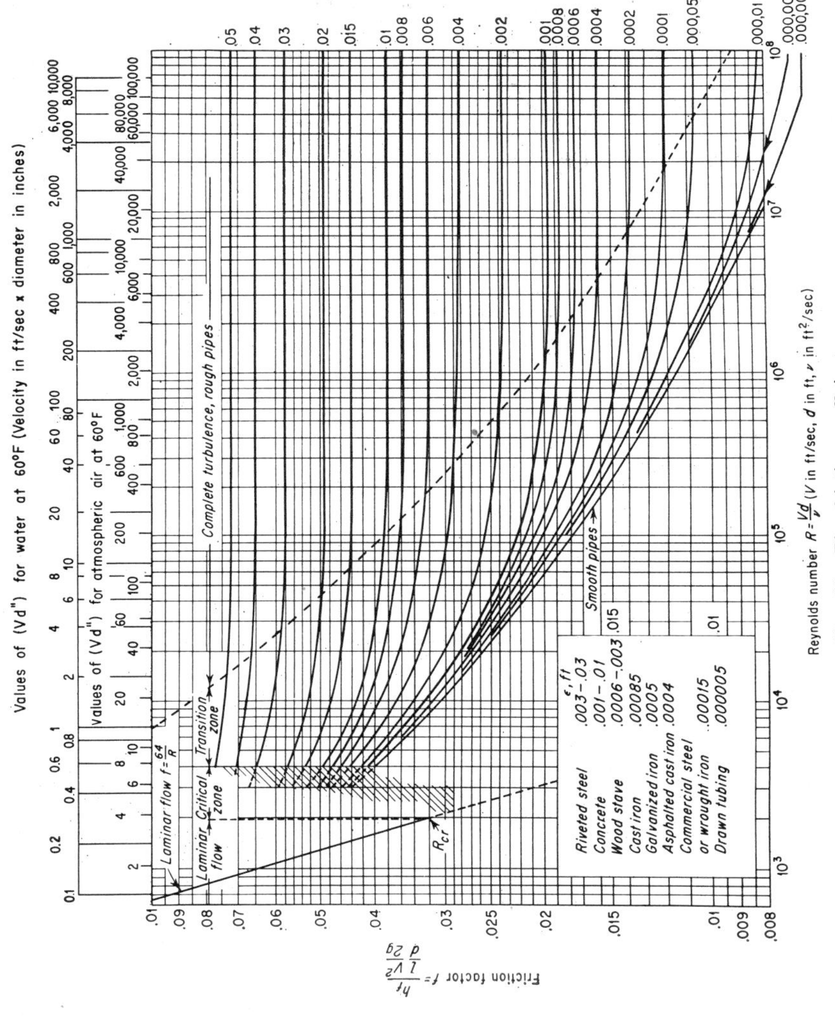

FIG. 61. Pipe friction coefficients.

The third category of turbulent flow in pipes occurs when values of f fall in the zone between the curve for smooth pipes and the broken line in Fig. 61. Flow in commercial pipe usually occurs in this category. In this zone, the f curves for various values of relative roughness depart from the smooth pipe curves at successive locations and become horizontal as they enter the zone of fully developed turbulence. It was difficult at first to apply the principle of relative roughness to commercial pipe because such a small number of experimental points extended into the zone of fully developed turbulence. This difficulty was overcome by Colebrook and White,[1] who developed a relationship between f and R, Eq. (125), which applies reasonably well to all the transition curves. With the aid of this equation, Colebrook [2] was able to assign roughness values to various commer-

$$\frac{1}{\sqrt{f}} = -2 \log \left(\frac{\epsilon}{3.7d} + \frac{2.51}{R\sqrt{f}}\right) \qquad (125)$$

cial pipe materials by comparing values of f for these pipes with Nikuradse's values for artificially roughened pipe. Rouse [3] carried the work further and presented the table of values of roughness shown as an insert in Fig. 61. The broken line separating the transition zone from the zone of fully developed turbulence in Fig. 61 was also suggested by Rouse.[3] The various curves shown in Fig. 61 were presented in this form by Moody.[4] Auxiliary scales at the top of the figure permit one to enter the diagram with only the value vd if it is assumed that the water or air temperature is 60°F. Values of kinematic viscosity ν to be used in computing R may be obtained from Table 2, p. 4–3. Values of f for water at ordinary temperatures flowing in new cast-iron or steel pipe may also be obtained from Table 11.[5] Values of f for fire hose computed from test data of the Underwriters Laboratories, Inc., are shown in Table 12.[6]

Roughness increases with age in pipes which are subject to corrosion. Ippen [7] re-

TABLE 11. VALUES OF f IN DARCY-WEISBACH FORMULA

d_i \ v	2	5	10	20	d_i \ v	2	5	10	20
1	0.032	0.027	0.025	0.023	24	0.019	0.018	0.018	0.017
3	0.029	0.025	0.024	0.022	36	0.017	0.016	0.016	0.015
6	0.026	0.023	0.022	0.021	48	0.015	0.014	0.014	0.014
12	0.023	0.021	0.020	0 019	60	0.013	0.013	0.013	0.012
18	0.021	0.020	0.019	0.018	72	0.011	0.011	0.011	0.011

TABLE 12. VALUES OF f FOR WATER FLOWING IN COTTON RUBBER-LINED FIRE HOSE

Nominal diameter, in.	Velocity, ft per sec					
	4	6	10	15	20	25
1½	0.024	0.023	0.023	0.022	0.021	
2½	.....	0.020	0.019	0.018	0.018	0.018

[1] COLEBROOK, C. F., Turbulent Flow in Pipes, with Particular Reference to the Transition Region between Smooth and Rough Pipe Laws, *J. Inst. Civil Engrs. (London)* February, 1939.

[2] COLEBROOK, *op. cit.*

[3] ROUSE, HUNTER, "Evaluation of Boundary Roughness," Proceedings of the Second Hydraulics Conference, *Univ. Iowa, Bull.* 27, 1943.

[4] MOODY, L. F., Friction Factors for Pipe Flow, *Trans. ASME*, November, 1944.

[5] FANNING, J. T., "Water-supply Engineering," D. Van Nostrand Company, Inc., New York, 1906.

[6] KING, H. W., C. O. WISLER, and J. G. WOODBURN, "Hydraulics," 5th ed., John Wiley & Sons, Inc., New York, 1948.

[7] MOODY, L. F., Friction Factors for Pipe Flow, Discussion by A. T. Ippen, *Trans. ASME*, November, 1944.

ported one case in which observations were made on galvanized steel pipe in which it was found that the value of ϵ was doubled within 3 years as the result of moderate conditions of use. Freeman[1] tested a number of new wrought-iron pipes and some "old, rusty" wrought-iron pipes of the same size. The roughness of his pipes was evaluated by the author, by comparison with the Nikuradse data, with the following results:

Type of Pipe	ϵ, ft
2-in. new wrought-iron	0.00014
2-in. old, rusty wrought-iron	0.003
3-in. new wrought-iron	0.00015
3-in. slightly rusty wrought-iron	0.00044
3-in. very rusty, old wrought-iron	0.0031
4-in. new wrought-iron	0.00016
4-in. old, rusty wrought-iron	0.0095

It may be noted that the roughness size is from twenty to sixty times larger for very old, rusty pipe than for new pipe. If the relative roughness is obtained, however, and inserted in Fig. 61, it will be seen that the values of f for the rusty pipe are only two or three times as great as those for new pipe.

Example 16: The discharge through 3,000 ft of 12-in. new cast-iron pipe is 4 cu ft per sec. The water is delivered to the pipe by a pump at point 1 and the pipe discharges into the atmosphere at point 2. Point 2 is 50 ft higher than point 1. Determine p_1 the pressure at which the pump must deliver water to the pipe.

The Bernoulli equation for points 1 and 2 is

$$\frac{v_1^2}{2g} + \frac{p_1}{w} + z_1 = \frac{v_2^2}{2g} + \frac{p_2}{w} + z_2 + f\frac{l}{d}\frac{v^2}{2g}$$

Taking the datum through point 1 and knowing $v_1 = v_2 = v$, and $p_2 = 0$, the equation becomes

$$\frac{p_1}{w} = 50 + f\frac{l}{d}\frac{v^2}{2g}$$

The value of v is determined from Q/a,

$$v = \frac{4}{0.785} = 5.1 \text{ ft per sec}$$

and the value of f, from Table 11, is 0.021. Then

$$\frac{p_1}{w} = 50 + 0.021 \times \frac{3{,}000}{1} \times \frac{(5.1)^2}{64.4} = 75.5 \text{ ft}$$

and

$$p_1 = 75.5w = 4{,}700 \text{ lb per sq ft}$$

Example 17: Castor oil at 68°F enters 300 ft of 1-in. pipe at point 1 under a pressure of 10 psi gage and discharges into the atmosphere at point 2. Determine the discharge if point 2 is 5 ft higher than point 1.

Taking the datum through point 1 and canceling out equal velocity-head terms, the Bernoulli equation from 1 to 2 becomes

$$\frac{p_1}{w} = 5 + h_l$$

[1] FREEMAN, JOHN R., "Flow of Water in Pipes and Pipe Fittings," ASME, New York, 1944.

From Table 2 the viscosity is 0.0206 slugs per ft-sec and assuming the specific gravity to be 0.969 the Reynolds number is

$$R = \frac{dv\,\rho}{\mu} = \frac{\left(\frac{1}{12}\right)\left(\frac{0.969 \times 62.4}{32.2}\right)v}{0.0206} = 7.6v$$

Although v is not yet known, it is obvious from the above that flow is laminar. Therefore, Eq. (122) may be used for determining the energy loss h_l. Numerical values may now be substituted in the Bernoulli equation as follows:

$$\frac{10 \times 144}{0.969 \times 62.4} = 5 + \frac{64}{7.6v} \times \frac{300}{1/12} \times \frac{v^2}{64.4}$$

and

$$v = \frac{23.8 - 5}{470} = 0.040 \text{ ft per sec}$$

then

$$Q = av = 0.00545 \times 0.040 = 0.000218 \text{ cu ft per sec, or } 0.098 \text{ gal per min}$$

Example 18: Determine the discharge of water at a temperature of 50°F through 12,000 ft of new 8-in. cast-iron pipe. The water enters the pipe at a pressure of 20 psi gage at point 1 and leaves at atmospheric pressure at point 2, which is 15 ft lower than point 1.

The Bernoulli equation from 1 to 2 taking the datum through 2 and canceling equal velocity-head terms gives

$$\frac{p_1}{w} + 15 = f\frac{l}{d}\frac{v^2}{2g}$$

or

$$f\frac{l}{d}\frac{v^2}{2g} = \frac{20 \times 144}{62.41} + 15 = 61.2 \text{ ft}$$

The value of R in terms of v is

$$R = \frac{dv}{\nu} = \frac{8/12\,v}{0.0000141} = 47{,}300v$$

and the value of ϵ/D is 0.00085/0.67 = 0.00127. Assuming v to be 5 ft per sec the first trial value of R is 235,000 and the value of f, from Fig. 61, is 0.0215. Then

$$0.0215 \times \frac{12{,}000}{8/12} \times \frac{v^2}{64.4} = 61.2$$

and

$$v = 3.2 \text{ ft per sec}$$

The corrected Reynolds number is 47,300 × 3.2 = 151,000 and the corresponding value of f is 0.022. The new value of v is 3.15 ft per sec. No further trials are necessary, and the discharge may be computed as follows:

$$Q = av = 0.349 \times 3.15 = 1.1 \text{ cu ft per sec}$$

Other Pipe Formulas. Of the many published pipe formulas, only a few will be discussed. The symbols commonly used in pipe and open-channel formulas are given in the following list:

l = length of pipe, ft.
h = energy loss in 1 ft of pipe, ft-lb per lb.
$s = h/l$ = energy loss per foot length of pipe. (For channels and horizontal pipes, s is also the slope of the energy gradient.)
d = diameter of pipe, ft.
d_i = diameter of pipe, in.
a = area of cross section of stream, sq ft.
v = mean velocity in pipe, ft per sec.
$Q = av$ = discharge, cu ft per sec.
p = wetted perimeter = πd for round pipes flowing full.
$r = a/p$ = hydraulic radius = $d/4$ for round pipes flowing full.

The Chezy Formula. This formula [Eq. (126)] was the earliest attempt to express

$$v = c\sqrt{rs} \tag{126}$$

energy loss in conduits algebraically. If in this formula r is replaced by $d/4$ and s by h/l, the value of h is given by the expression

$$h = \frac{4}{c^2}\frac{l}{d}v^2 \tag{127}$$

from which it can be seen that the Darcy-Weisbach formula [Eq. (123)] is a rearrangement of the Chezy formula, the roughness coefficients being related as follows:

$$f = \frac{8g}{c^2} \tag{128}$$

From the previous discussion concerning Fig. 61 it may be concluded that the Chezy formula would give excellent results for flow in rough conduits at large Reynolds numbers, where the exponent of v is approximately 2. When later investigators found that this formula did not adhere to test results for smooth pipe with low velocities other empirical formulas were devised to satisfy each particular group of tests. Only in recent years has it beome generally recognized that all such tests can be unified by means of Reynolds number.

A dimensional investigation of the Chezy formula will show that the left term has the dimension L/T whereas the right side is simply $L^{1/2}$. The expression is therefore not dimensionally homogeneous and can be used in the above form only in the foot-pound-second system. A similar investigation of the Darcy-Weisbach formula will show it to be dimensionally correct.

Hazen-Williams Formula. This formula, designed for both pipes and open channels, has been used extensively for pipes. The selection of exponents was made with a view to obtaining a minimum variation in the coefficient C_1 for all conduits of the same degree of roughness. The formula as published by its authors [1] is

$$v = C_1 r^{0.63} s^{0.54} 0.001^{-0.04} \tag{129}$$

The expression $0.001^{-0.04}$, which equals 1.318, was introduced to equalize the value of C_1 with the value of c in the Chezy formula. The authors of the formula state: "If exponents could be selected agreeing perfectly with the facts, the value of C_1 would depend upon the roughness only, and for any given degree of roughness C_1 would be a constant. It is not possible to reach this actually, because the values of the exponents vary with different surfaces, and also their values may not be exactly the same for large diameters and for small ones and for steep slopes and for flat ones. Exponents can be

[1] WILLIAMS and HAZEN, "Hydraulic Tables," 3d ed., John Wiley & Sons, Inc., New York, 1920.

selected, however, representing approximately average conditions, so that the value of C_1 for a given condition of surface will vary so little as to be practically constant. Several such 'exponential' formulas have been suggested. These formulas are among the most satisfactory yet devised, but their use has been limited by the difficulty in making computations by them. This difficulty was eliminated by the use of a slide rule constructed for the purpose."

The following are values of C_1: For extremely smooth, straight pipes, $C_1 = 140$; for very smooth pipes, $C_1 = 130$. For smooth wooden or wood-stave pipes, $C_1 = 120$; for new riveted-steel pipes, $C_1 = 110$; for vitrified pipes, $C_1 = 110$. In estimating discharges where the carrying capacity after a series of years is the controlling factor, for cast-iron pipe $C_1 = 100$; and for riveted steel, $C_1 = 95$. For old iron pipes in bad condition, $C_1 =$ 80 to 60; and for small pipes badly tuberculated, C_1 may be as low as 40.

Manning[1] *Formula.* This is a general formula which can be applied to all types of conduits. In the form for open channels, it is usually written

$$v = \frac{1.486}{n} r^{2/3} s^{1/2} \tag{130a}$$

It is applied more conveniently to pipes in one of the following forms:

$$v = \frac{0.590}{n} d^{2/3} s^{1/2} \tag{130b}$$

$$Q = \frac{0.463}{n} d^{8/3} s^{1/2} \tag{130c}$$

$$h = 2.87n^2 \frac{lv^2}{d^{4/3}} \tag{130d}$$

$$h = 4.66n^2 \frac{lQ^2}{d^{16/3}} \tag{130e}$$

$$d = \left(\frac{2.159Qn}{s^{1/2}}\right)^{3/8} \tag{130f}$$

$$d_i = \left(\frac{1{,}630Qn}{s^{1/2}}\right)^{3/8} \tag{130g}$$

The coefficient n increases with the degree of roughness. The reason the formula is expressed in the form (130a) instead of the obviously equivalent expression $v = Kr^{2/3}s^{1/2}$ is that in the form used it gives practically the same results as the Kutter formula when the same value of n is used in each formula. Values of n to be used with the Manning formula are given in Table 13. Most of the experiments published give results lying between the extremes of these values. Smooth pipes of small diameter take a somewhat smaller coefficient than those of large diameter. For rough pipes, the coefficient does not appear to vary with the diameter. Since iron and steel pipes deteriorate with age, for estimating discharges of such pipes where the carrying capacity after a series of years is the controlling factor, the larger values of n should be used.

Selection of a Pipe Formula. The Darcy-Weisbach formula has the distinct advantage of being applicable to any system of units and to all fluids flowing in round conduits whether flow is laminar or turbulent. There seems to be little question that because of its versatility it will eventually take precedence over all other formulas. For use with

[1] MANNING, Flow of Water in Open Channels and Pipes, *Trans. Inst. Civil Engrs. (Ireland)*, vol. 20, 1890.

TABLE 13. VALUES OF n TO BE USED WITH THE MANNING FORMULA

Kind of pipe	Variation		Use in designing	
	From	To	From	To
Clean uncoated cast-iron pipe	0.011	0.015	0.013	0.015
Clean coated cast-iron pipe	0.010	0.014	0.012	0.014
Dirty or tuberculated cast-iron pipe	0.015	0.035		
Riveted-steel pipe	0.013	0.017	0.015	0.017
Lock-bar and welded pipe	0.010	0.013	0.012	0.013
Galvanized-iron pipe	0.012	0.017	0.015	0.017
Brass and glass pipe	0.009	0.013		
Wood-stave pipe	0.010	0.014		
small diameter	.....	.....	0.011	0.012
large diameter	.....	.....	0.012	0.013
Concrete pipe	0.010	0.017		
with rough joints	.....	.....	0.016	0.017
"dry mix," rough forms	.....	.....	0.015	0.016
"wet mix," steel forms	.....	.....	0.012	0.014
very smooth	.....	.....	0.011	0.012
Vitrified sewer pipe	0.010	0.017	0.013	0.015
Common-clay drainage tile	0.011	0.017	0.012	0.014

water at ordinary temperatures no one of the above formulas possesses a marked advantage from the standpoint of accuracy, if it is used with care and understanding. Each requires the selection of a coefficient within about the same range of variation. The main consideration is that the engineer know the formula and be thoroughly familiar with the coefficients that apply in each case. The Manning formula has the advantage of being applicable to both pipes and open channels, and its exponents lend themselves more easily to algebraic manipulation than those of the Hazen-Williams formula.

Minor Losses. In addition to the friction loss taking place throughout the length of a conduit, there are local losses at each point where an acceleration occurs. Such losses are called "minor losses" because in very long pipes they are negligible compared with the friction losses. However, in pipes less than 1,000 ft long, the minor losses may be of the same order of magnitude as the friction loss. The most common minor losses are due to sudden enlargements, sudden contractions, bends, and valves. In all cases it is common practice to state the loss as a function of the velocity head as follows:

$$h_m = \frac{K_m v^2}{2g} \qquad (131)$$

If there is a permanent change in pipe diameter, the velocity in the above expression is the one in the smaller pipe.

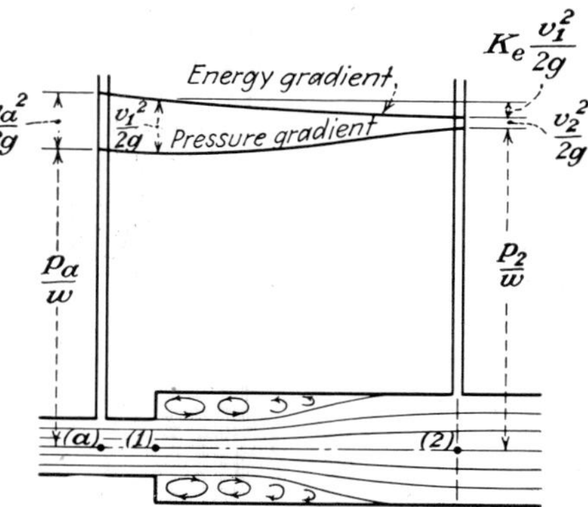

FIG. 62. Sudden enlargement in a pipe.

The loss due to *sudden enlargement* may be evaluated from basic principles. In Fig. 62 is shown a section through a round pipe in which the diameter is suddenly increased. The fluid between points 1 and 2 is undergoing a change in momentum which must be

equal to the applied force. Neglecting wall shear, this may be expressed mathematically as follows:

$$F = \frac{M(v_1 - v_2)}{t}$$

which, for the case shown in Fig. 62, becomes

$$(p_2 - p_1)a_2 = \frac{Qw}{g}(v_1 - v_2) \tag{132}$$

The Bernoulli equation between points 1 and 2 is

$$\frac{v_1^2}{2g} + \frac{p_1}{w} = \frac{v_2^2}{2g} + \frac{p_2}{w} + h_e \tag{133}$$

Equations (132) and (133) may be solved simultaneously for h_e with the result:

$$h_e = \frac{(v_1 - v_2)^2}{2g} \tag{134}$$

This equation agrees very well with the Archer[1] formula [Eq. (135)] which was derived experimentally.

$$h_e = 1.10\frac{(v_1 - v_2)^{1.92}}{2g} \tag{135}$$

A special case of sudden enlargement occurs at the outlet of a pipe beneath the water surface of a reservoir. For this case the value of v_2 is approximately zero and therefore the value of h_e becomes nearly equal to the velocity head in the pipe. Values obtained from Eq. (135) reduced to the form of Eq. (131) result in the values of K_e given in Table 14. Values of K_e for gradual enlargement computed from test of Andres and Gibson[2] are given in Table 15.

TABLE 14. COEFFICIENT K_e FOR SUDDEN ENLARGEMENT IN PIPES

Velocity in smaller pipe, v	Ratio of smaller to larger diameter									
	0.0	0.1	0.2	0.3	0.4	0.5	0.6	0.7	0.8	0.9
2	1.00	1.00	0.96	0.86	0.74	0.60	0.44	0.29	0.15	0.04
5	0.96	0.95	0.89	0.80	0.69	0.56	0.41	0.27	0.14	0.04
10	0.93	0.91	0.86	0.77	0.67	0.54	0.40	0.26	0.13	0.04
20	0.86	0.84	0.80	0.72	0.62	0.50	0.37	0.24	0.12	0.04
40	0.81	0.80	0.75	0.68	0.58	0.47	0.35	0.22	0.11	0.03

TABLE 15. COEFFICIENT K_e FOR GRADUAL ENLARGEMENT IN PIPES

Angle between axis and surface of pipe	Ratio of smaller to larger diameter								
	0.1	0.2	0.3	0.4	0.5	0.6	0.7	0.8	0.9
5°	0.04	0.04	0.04	0.04	0.04	0.04	0.03	0.02	0.01
15°	0.16	0.16	0.16	0.16	0.16	0.15	0.13	0.10	0.06
30°	0.49	0.49	0.48	0.48	0.46	0.43	0.37	0.27	0.16
45°	0.64	0.63	0.63	0.62	0.60	0.55	0.49	0.38	0.20
60°	0.72	0.72	0.71	0.70	0.67	0.62	0.54	0.43	0.24

[1] ARCHER, W. H., Loss of Head Due to Enlargements in Pipes, *Trans. ASCE*, vol. 76, pp. 999–1026, 1913.
[2] KING, H. W., "Handbook of Hydraulics," 4th ed., p. 6-16, 1954.

Table 16 gives values of K_c for *sudden contraction* based on the work of Merriman and Brightmore.[1] Tests by Gibson [2] give values that are of the same order of magnitude. Values based on tests by Weisbach are approximately the same as those of Table 16 for a velocity of 40 ft per sec.[3] A special case of sudden contraction occurs where water enters a pipe from a reservoir. Values of K_c for this condition are given in Table 17.

TABLE 16. COEFFICIENT K_c FOR SUDDEN CONTRACTION IN PIPES

Velocity in smaller pipe, v	Ratio of smaller to larger diameter									
	0.0	0.1	0.2	0.3	0.4	0.5	0.6	0.7	0.8	0.9
2	0.49	0.49	0.48	0.45	0.42	0.38	0.28	0.18	0.07	0.03
5	0.48	0.48	0.47	0.44	0.41	0.37	0.28	0.18	0.09	0.04
10	0.47	0.46	0.45	0.43	0.40	0.36	0.28	0.18	0.10	0.04
20	0.44	0.43	0.42	0.40	0.37	0.33	0.27	0.19	0.11	0.05
40	0.38	0.36	0.35	0.33	0.31	0.29	0.25	0.20	0.13	0.06

TABLE 17. COEFFICIENT K_c FOR ENTRANCE LOSS IN PIPES

Description of Entrance	Coefficient
Inward projecting	0.78
Sharp-cornered	0.50
Slightly rounded	0.23
Bellmouth	0.04

Average values of K_g for losses at *gate valves* as determined by Corp and Ruble [4] are given in Table 18.

TABLE 18. LOSS OF HEAD DUE TO GATE VALVES

$$\text{Values of } K_g \text{ in } h_g = K_g \frac{V^2}{2g}$$

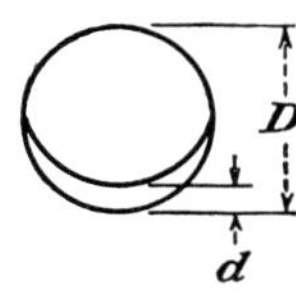

Nominal diameter of valve, in.	Ratio of height d of valve opening to diameter D of full valve opening					
	⅛	¼	⅜	½	¾	1
½	450	60	22	11	2.2	1.0
¾	310	40	12	5.5	1.1	0.28
1	230	32	9.0	4.2	0.90	0.23
1½	170	23	7.2	3.3	0.75	0.18
2	140	20	6.5	3.0	0.68	0.16
4	91	16	5.6	2.6	0.55	0.14
6	74	14	5.3	2.4	0.49	0.12
8	66	13	5.2	2.3	0.47	0.10
12	56	12	5.1	2.2	0.47	0.07

[1] *Ibid.*, p. 6–17.
[2] GIBSON, A. H., "Hydraulics and Its Applications," 3d ed., p. 99, 1925.
[3] O'BRIEN and HICKOX, "Applied Fluid Mechanics," p. 211, 1937.
[4] CORP and RUBLE, "Experiments on Loss of Head in Valves and Pipes of One-half to Twelve Inches Diameter," *Bull. Univ. Wis., Eng. Series*, vol. IX, No. 1, p. 59, 1922.

The minor loss at a *bend* results from a distortion of the velocity distribution, thereby causing additional shear stresses within the fluid. The normal velocity distribution may not be restored for a distance as great as 25 diameters below the bend. The bend loss is considered to be the loss in excess of the loss for an equal length of straight pipe. Tests by Beij [1] indicate that the loss is a function of the relative radius (R/d = bend radius ÷ pipe diameter) and the roughness of the pipe. Throughout the range of his tests, for Reynolds numbers varying from 20,000 to 300,000, he found that the bend loss was independent of the Reynolds number. Figure 63 is the curve plotted by Beij through his test results. These tests were run on 4-in. pipe having a roughness such that the value of f in the Darcy-Weisbach formula varied from 0.020 to 0.025. The meager data available indicate that very smooth pipe will have values of K_b which are approximately 70 percent of those given in Fig. 63, while rougher pipe will have correspondingly larger values.

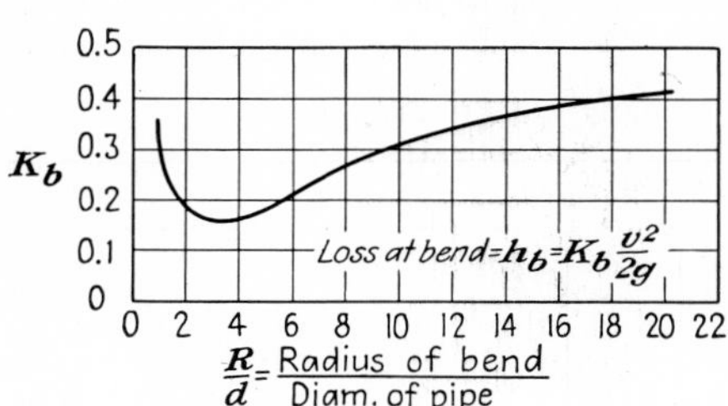

FIG. 63. Energy loss in pipe bend.

Example 19: Determine the pressure at point 2 of Fig. 62 if the pressure at 1 is 20 psi. The average velocity at 1 is 10 ft per sec., d_1 is 4 in. and d_2 is 10 in.

The Bernoulli equation between points 1 and 2, taking the datum through the center of the pipe is

$$\frac{v_1^2}{2g} + \frac{p_1}{w} = \frac{v_2^2}{2g} + \frac{p_2}{w} + K_e \frac{v_1^2}{2g}$$

From the equation of continuity,

$$v_2 = \frac{a_1 v_1}{a_2} = \frac{16 \times 10}{100} = 1.6 \text{ ft per sec}$$

From Table 14, $K_e = 0.67$.
Inserting numerical values in the Bernoulli equation it becomes

$$\frac{(10)^2}{64.4} + \frac{20 \times 144}{62.4} = \frac{(1.6)^2}{64.4} + \frac{p_2}{w} + 0.67 \frac{(10)^2}{64.4}$$

Then

$$\frac{p_2}{w} = 45.7 \text{ ft}$$

and

$$p_2 = 2{,}910 \text{ lb per sq ft}$$

Hydraulic and Energy Gradients. Losses of head are illustrated graphically in Fig. 64, which shows the system of pipes *ABCD* of different diameters connecting two reservoirs and discharging under the total head *H*. The piezometer tubes *ttt* may have any positions in the pipe line. The locus of the elevations to which water will rise in such piezometer tubes is termed the *hydraulic gradient.* The hydraulic gradient thus shows the pressure head at any section of the pipe. The line showing the pressure head plus the velocity head (i.e., the total head) at any section of the pipe is termed the *energy gradient.* The two lines are parallel for pipes of uniform diameter but not for those of varying diameter.

[1] BEIJ, K. H., Pressure Losses for Fluid Flow in 90° Pipe Bends, *Nat. Bur. Standards, Research Paper* RP1110, July, 1938.

The loss of head due to friction H_1 in a given length of pipe is always the drop (at a practically constant rate) in the energy gradient, and for pipes of uniform diameter it is also the drop in the hydraulic gradient.

At the entrance A, the drop in the hydraulic gradient represents the loss of head at entrance plus the velocity head. At B, where the pipe changes to a smaller diameter, the drop equals the loss of head due to contraction plus the increase in velocity head. At C, the pipe changes to a larger diameter, and the hydraulic gradient rises an amount

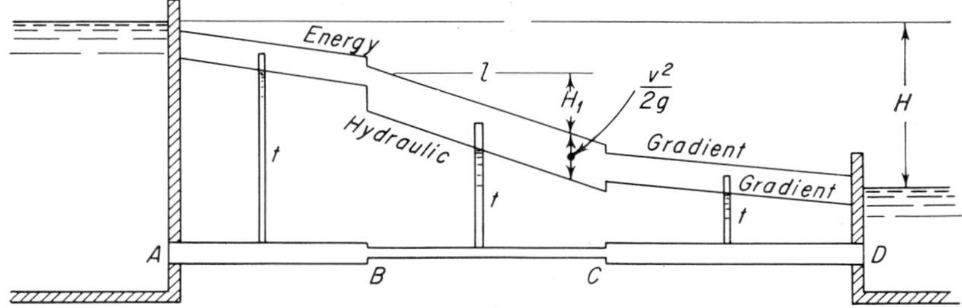

FIG. 64. Hydraulic and energy gradients.

equal to the decrease in velocity head minus the loss of head due to enlargement. At the outlet D, there is a loss of head corresponding to the loss due to enlargement, and the hydraulic gradient is slightly below the water surface in the reservoir. A short distance below the outlet, the hydraulic gradient coincides with the water surface, as the velocity head is soon dissipated in turbulence. The energy gradient is affected only by loss of head, not by changes in velocity head. Since there can be no flow without loss of head, the energy gradient drops progressively. Where there is only loss of head due to friction, the drop is gradual, but there is an additional abrupt drop where minor losses occur.

Pipe Systems. Problems involving flow in *parallel* pipes may be solved by means of the following principles. Because the pipes begin and end at a common point, it follows that the pressure drop must be the same in each pipe. It has previously been shown that for pipes of uniform diameter the energy loss is equal to the drop in pressure head. Therefore, neglecting minor losses, it can be stated with reference to Fig. 65 that

$$f_2 \frac{l_2}{d_2} \frac{v_2^2}{2g} = f_3 \frac{l_3}{d_3} \frac{v_3^2}{2g} \tag{136}$$

Furthermore, the sum of the discharges in the parallel pipes must be equal to the dis-

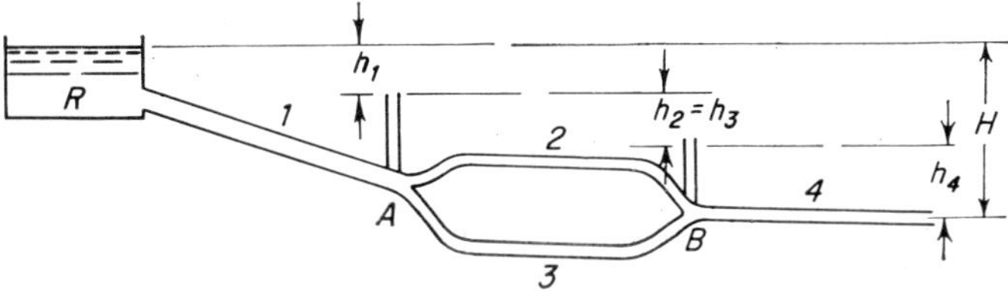

FIG. 65. Divided flow.

charge in the approach or outlet pipe, so that, for the system shown in Fig. 65,

$$Q_1 = Q_4 = Q_2 + Q_3 \tag{137}$$

and

$$a_1v_1 = a_4v_4 = a_2v_2 + a_3v_3 \tag{138}$$

Finally, neglecting minor losses, the total pressure drop H may be evaluated as follows:

$$H = f_1 \frac{l_1}{d_1} \frac{v_1^2}{2g} + f_2 \frac{l_2}{d_2} \frac{v_2^2}{2g} + f_4 \frac{l_4}{d_4} \frac{v_4^2}{2g} \tag{139}$$

If, for example, the value of H in Fig. 65 is known and it is desired to compute the discharge in each pipe, Eqs. (136) and (139) together with the two equations shown as Eq. (138) provide the four equations required for the determination of the four values of average velocity. If great refinement is desired, several trials may be necessary in order to determine the exact values of f. If on the other hand the discharge through the system is known and it is desired to find H, Eqs. (136), (137), and (138) would first be solved for Q_2, Q_3, and either h_2 or h_3. The value of H would then be determined from Eq. (139). Any other suitable pipe formula might have been used in Eqs. (136) and (139).

Problems involving a *system of three pipes* such as is illustrated by Fig. 66 may be solved by first assuming that the pressure at the junction of the three pipes is such that there will be no flow into or out of the intermediate reservoir. With reference to Fig. 66,

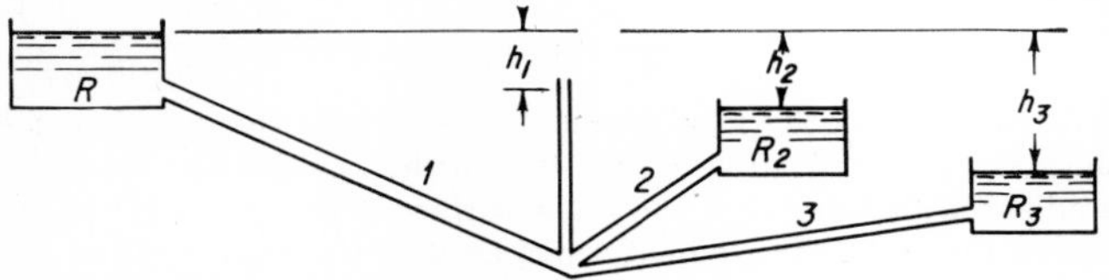

FIG. 66. Branches discharging at different elevations in compound pipes.

the above assumption would require zero discharge in pipe 2. On the basis of this assumption the discharge in pipes 1 and 3 may be determined from Eq. (123) or any other pipe formula. If Q_1 is greater than Q_3, flow must be into reservoir 2 and

$$Q_1 = Q_2 + Q_3 \tag{140}$$

If Q_1 is less than Q_3, flow is out of 2 and

$$Q_1 + Q_2 = Q_3 \tag{141}$$

The solution may then be completed by successive approximations, assuming in each case either a value of the discharge in one of the pipes or the pressure at the junction of the three pipes, until Eq. (140) or (141) is satisfied.

FLOW WITH A FREE SURFACE

Much of the earliest work on flow in conduits was applicable to open channels as well as pipes. The Chezy formula [Eq. (126)], for example, has been used as much for the design of sewers and canals as for the design of pipes. This formula, with modifications which will be discussed later, is still used for the solution of open-channel problems. There is less need for a knowledge of the effect of the fluid properties of open-channel flow than on pipe flow, primarily because the open-channel problems encountered in engineering work deal mainly with water at ordinary temperatures flowing under fully turbulent conditions. In certain important cases, however, a more comprehensive understanding of flow with a free surface is necessary. One such case is the flow encountered in small-scale hydraulic models. Another is the flow in thin sheets produced by rainfall on flat surfaces. These and other instances are of sufficient importance to warrant a brief discussion of laminar flow in channels.

Laminar Flow with a Free Surface. The law of laminar flow with a free surface for the case of wide rectangular channels may be developed in the same manner as for pipes. Consider the free body of fluid shown in Fig. 67, having a width of 1 ft, a length l, and a

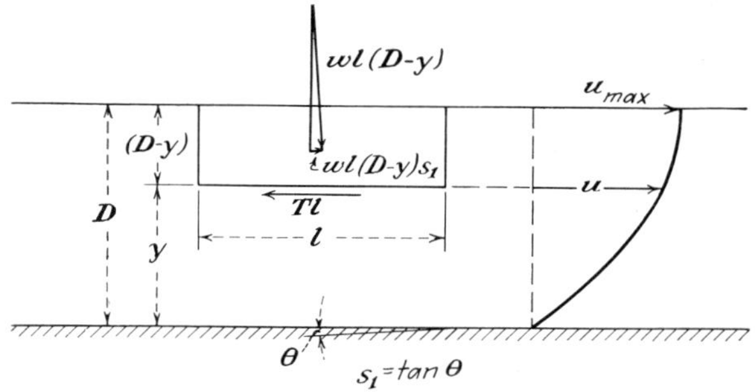

FIG. 67. Laminar flow in open channels.

height $(D - y)$. If s_1 is the slope of the bottom, the summation of forces in the direction of flow gives

$$\tau l = w(D - y)ls_1 \tag{142}$$

Replacing τ with its value from Eq. (3), rearranging the terms, and simplifying, the above equation becomes

$$du = \frac{ws_1}{\mu}(D\,dy - y\,dy) \tag{143}$$

and integration gives

$$u = \frac{ws_1}{\mu}\left(Dy - \frac{y^2}{2}\right) + \overset{O}{\overset{\|}{C}} \tag{144}$$

The value of C in Eq. (144) must be zero to satisfy the condition that $u = 0$ when $y = 0$. The maximum velocity may be obtained by letting $y = D$, then

$$u_{\max} = \frac{ws_1D^2}{2\mu} \tag{145}$$

The discharge is obtained from a summation of small elements of discharge, utilizing the value of u from Eq. (144) as follows:

$$Q = \int dQ = \int u\,dy = \frac{ws_1}{\mu}\int_0^D \left(Dy\,dy - \frac{y^2\,dy}{2}\right)$$

from which

$$Q = \frac{ws_1D^3}{3\mu} \tag{146}$$

and the average velocity is

$$v = \frac{Q}{a} = \frac{ws_1D^2}{3\mu} \tag{147}$$

and

$$s_1 = \frac{3\mu v}{wD^2} \tag{148}$$

Finally the Reynolds number $(Dv\rho/\mu)$ may be introduced to give

$$s_1 = \frac{3v^2}{RgD} \tag{149}$$

It will be recalled that in the discussion of pipe flow it was convenient to change the Hagen-Poiseuille law for laminar flow into the form of a commonly used turbulent-flow formula (Darcy-Weisbach). In the same manner Eq. (149) will be related to the Manning formula (p. 4–80) for flow in open channels. When flow is uniform, $s = s_1$, and the Manning formula gives

$$s_1 = \frac{n^2v^2}{2.21r^{4/3}}$$

For rectangular channels which are wide with respect to their depth, the hydraulic radius is nearly equal to the depth. Replacing r in the above expression with D and setting it equal to the value of s_1 given in Eq. (149), the following relation between the roughness coefficient n, the depth, and the Reynolds number is obtained:

$$\frac{n^2}{D^{1/3}} = \frac{0.206}{R} \tag{150}$$

This relationship corresponds to Eq. (124) for flow in pipes. It is plotted in Fig. 68

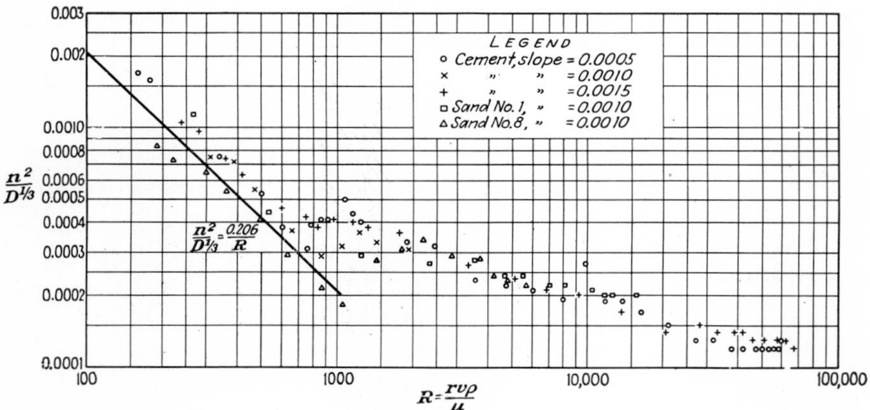

FIG. 68. Variation of roughness coefficient with Reynolds number for open channels.

together with a number of points determined experimentally.[1] It will be seen that the points plotted in Fig. 68 follow a trend similar to those for smooth pipes shown in Fig. 61.

The transition between laminar and turbulent flow takes place in a range of Reynolds numbers varying from 600 to 1,200. These values are approximately one-fourth of the corresponding values for pipes (p. 4–63), which might have been expected because values of Reynolds number for pipes based on the hydraulic radius $(d/4)$ rather than on the diameter would be one-fourth as large as the values plotted in Fig. 61.

Turbulent Flow with a Free Surface. In the turbulent range the Manning formula

[1] "Studies of River Bed Materials and Their Movement, With Special Reference to the Lower Mississippi River," *U.S. Waterways Experiment Station, Paper* 17, January, 1935.

is used most frequently.[1] Manning proposed the following value of c in the Chezy formula

$$c = \frac{1.486 r^{1/6}}{n} \tag{151}$$

where n is the same coefficient of roughness used in the earlier Ganguillet-Kutter formula.[2] Values of n prepared by Horton [3] are given in Table 19.

TABLE 19. HORTON'S VALUES OF n TO BE USED WITH THE MANNING FORMULA

Surface	Best	Good	Fair	Bad
Vitrified sewer pipe	{0.010 / 0.011}	0.013 *	0.015	0.017
Common-clay drainage tile	0.011	0.012 *	0.014 *	0.017
Glazed brickwork	0.011	0.012	0.013 *	0.015
Brick in cement mortar; brick sewers	0.012	0.013	0.015 *	0.017
Neat cement surfaces	0.010	0.011	0.012	0.013
Cement mortar surfaces	0.011	0.012	0.013 *	0.015
Concrete pipe	0.012	0.013	0.015 *	0.016
Wood-stave pipe	0.010	0.011	0.012	0.013
Plank flumes:				
Planed	0.010	0.012 *	0.013	0.014
Unplaned	0.011	0.013 *	0.014	0.015
With battens	0.012	0.015 *	0.016	
Concrete-lined channels	0.012	0.014 *	0.016 *	0.018
Cement-rubble surface	0.017	0.020	0.025	0.030
Dry-rubble surface	0.025	0.030	0.033	0.035
Dressed-ashlar surface	0.013	0.014	0.015	0.017
Semicircular metal flumes, smooth	0.011	0.012	0.013	0.015
corrugated	0.0225	0.025	0.0275	0.030
Canals and ditches:				
Earth, straight and uniform	0.017	0.020	0.0225 *	0.025
Rock cuts, smooth and uniform	0.025	0.030	0.033 *	0.035
jagged and irregular	0.035	0.040	0.045	
Winding sluggish canals	0.0225	0.025 *	0.0275	0.030
Dredged earth channels	0.025	0.0275 *	0.030	0.033
Canals with rough stony beds, weeds on earth banks	0.025	0.030	0.035 *	0.040
Earth bottom, rubble sides	0.028	0.030 *	0.033 *	0.035
Natural stream channels:				
1. Clean, straight bank, full stage, no rifts or deep pools	0.025	0.0275	0.030	0.033
2. Same as 1, but some weeds and stones	0.030	0.033	0.035	0.040
3. Winding, some pools and shoals, clean	0.033	0.035	0.040	0.045
4. Same as 3, lower stages, more ineffective slope and sections	0.040	0.045	0.050	0.055
5. Same as 3, some weeds and stones	0.035	0.040	0.045	0.050
6. Same as 4, stony sections	0.045	0.050	0.055	0.060
7. Sluggish river reaches, rather weedy or with very deep pools	0.050	0.060	0.070	0.080
8. Very weedy reaches	0.075	0.100	0.125	0.150

* Values commonly used in designing.

This formula as it is usually written is

$$v = \frac{1.486}{n} r^{2/3} s^{1/2} \tag{152a}$$

[1] Symbols used in open-channel formulas are defined on p. 4–67.
[2] KING, H. W., "Handbook of Hydraulics," 4th ed., p. 7–15, 1954.
[3] HORTON, R. E., Some Better Kutter's Formula Coefficients, *Eng. News*, Feb. 24, May 4, 1939.

Table 20. Values of K for Trapezoidal Channels in the Formula $Q = \frac{K}{n} D^{8/3} s^{1/2}$

D = depth of water and b = bottom width of channel

D/b	Side slopes of channel, ratio of horizontal to vertical									
	Vertical	¼–1	½–1	¾–1	1–1	1½–1	2–1	2½–1	3–1	4–1
0.01	146.8	147.2	147.6	148.0	148.3	148.8	149.2	149.5	149.9	150.5
0.02	72.4	72.9	73.4	73.7	74.0	74.5	74.9	75.3	75.6	76.3
0.03	47.6	48.2	48.6	49.0	49.3	49.8	50.2	50.6	50.9	51.6
0.04	35.3	35.8	36.3	36.6	36.9	37.4	37.8	38.2	38.6	39.3
0.05	27.9	28.4	28.9	29.2	29.5	30.0	30.5	30.9	31.2	32.0
0.06	23.0	23.5	23.9	24.3	24.6	25.1	25.5	25.9	26.3	27.1
0.07	19.5	20.0	20.4	20.8	21.1	21.6	22.0	22.4	22.8	23.6
0.08	16.8	17.3	17.8	18.1	18.4	18.9	19.4	19.8	20.2	21.0
0.09	14.8	15.3	15.7	16.1	16.4	16.9	17.4	17.8	18.2	19.0
0.10	13.2	13.7	14.1	14.4	14.8	15.3	15.7	16.2	16.6	17.4
0.11	11.83	12.33	12.76	13.11	13.42	13.9	14.4	14.9	15.3	16.1
0.12	10.73	11.23	11.65	12.00	12.31	12.8	13.3	13.8	14.2	15.0
0.13	9.80	10.29	10.71	11.06	11.37	11.9	12.4	12.8	13.3	14.1
0.14	9.00	9.49	9.91	10.26	10.57	11.1	11.6	12.0	12.5	13.4
0.15	8.32	8.80	9.22	9.57	9.88	10.4	10.9	11.4	11.8	12.7
0.16	7.72	8.20	8.61	8.96	9.27	9.81	10.29	10.75	11.20	12.1
0.17	7.19	7.67	8.07	8.43	8.74	9.28	9.77	10.22	10.68	11.6
0.18	6.73	7.19	7.61	7.96	8.27	8.81	9.30	9.76	10.21	11.1
0.19	6.30	6.78	7.19	7.54	7.85	8.39	8.88	9.34	9.80	10.7
0.20	5.94	6.40	6.81	7.16	7.47	8.01	8.50	8.97	9.43	10.3
0.21	5.60	6.06	6.47	6.82	7.13	7.67	8.16	8.63	9.09	10.00
0.22	5.30	5.75	6.16	6.51	6.82	7.36	7.85	8.32	8.79	9.70
0.23	5.02	5.47	5.88	6.22	6.54	7.08	7.57	8.04	8.51	9.43
0.24	4.77	5.22	5.62	5.96	6.28	6.82	7.32	7.79	8.26	9.18
0.25	4.54	4.99	5.38	5.73	6.04	6.58	7.08	7.56	8.03	8.95
0.26	4.32	4.77	5.16	5.51	5.82	6.37	6.87	7.34	7.81	8.74
0.27	4.12	4.57	4.96	5.31	5.62	6.17	6.67	7.14	7.62	8.54
0.28	3.95	4.38	4.77	5.12	5.43	5.98	6.48	6.96	7.43	8.36
0.29	3.78	4.20	4.60	4.95	5.25	5.81	6.31	6.79	7.26	8.19
0.30	3.62	4.05	4.44	4.78	5.09	5.64	6.15	6.63	7.11	8.04
0.31	3.48	3.91	4.29	4.63	4.94	5.49	6.00	6.48	6.96	7.89
0.32	3.34	3.76	4.15	4.49	4.80	5.35	5.86	6.34	6.82	7.75
0.33	3.21	3.64	4.02	4.36	4.67	5.22	5.73	6.21	6.69	7.62
0.34	3.09	3.51	3.89	4.23	4.54	5.09	5.60	6.09	6.57	7.50
0.35	2.98	3.40	3.78	4.12	4.43	4.98	5.49	5.97	6.45	7.39
0.36	2.88	3.29	3.67	4.01	4.32	4.87	5.38	5.86	6.34	7.28
0.37	2.78	3.19	3.57	3.90	4.21	4.76	5.27	5.76	6.24	7.18
0.38	2.68	3.09	3.47	3.81	4.11	4.66	5.17	5.66	6.14	7.09
0.39	2.59	3.00	3.38	3.71	4.02	4.57	5.08	5.57	6.05	6.99
0.40	2.51	2.92	3.29	3.62	3.93	4.48	4.99	5.48	5.96	6.91
0.41	2.43	2.83	3.21	3.54	3.85	4.40	4.91	5.40	5.88	6.83
0.42	2.36	2.76	3.13	3.46	3.77	4.32	4.83	5.32	5.80	6.75
0.43	2.29	2.68	3.05	3.38	3.69	4.24	4.75	5.25	5.73	6.67
0.44	2.22	2.61	2.98	3.31	3.62	4.17	4.68	5.17	5.65	6.60
0.45	2.15	2.55	2.91	3.24	3.55	4.10	4.61	5.11	5.59	6.54

TABLE 20. VALUES OF K FOR TRAPEZOIDAL CHANNELS IN THE FORMULA $Q = \frac{K}{n} D^{8/3} s^{1/2}$
(*Continued*)

D/b	Side slopes of channel, ratio of horizontal to vertical									
	Vertical	¼–1	½–1	¾–1	1–1	1½–1	2–1	2½–1	3–1	4–1
0.46	2.09	2.48	2.85	3.18	3.48	4.04	4.55	5.04	5.52	6.47
0.47	2.03	2.42	2.79	3.12	3.42	3.97	4.49	4.98	5.46	6.41
0.48	1.98	2.36	2.73	3.06	3.36	3.91	4.42	4.92	5.40	6.35
0.49	1.92	2.31	2.67	3.00	3.30	3.85	4.37	4.86	5.34	6.29
0.50	1.87	2.26	2.61	2.94	3.25	3.80	4.31	4.81	5.29	6.24
0.51	1.82	2.20	2.56	2.89	3.19	3.75	4.26	4.75	5.24	6.19
0.52	1.78	2.16	2.51	2.84	3.14	3.70	4.21	4.70	5.19	6.14
0.53	1.73	2.11	2.46	2.79	3.09	3.65	4.16	4.65	5.14	6.09
0.54	1.69	2.06	2.42	2.74	3.05	3.60	4.11	4.61	5.09	6.04
0.55	1.65	2.02	2.37	2.70	3.00	3.55	4.07	4.56	5.05	6.00
0.56	1.61	1.98	2.33	2.66	2.96	3.51	4.02	4.52	5.00	5.96
0.57	1.57	1.94	2.29	2.61	2.91	3.47	3.98	4.48	4.96	5.92
0.58	1.53	1.90	2.25	2.57	2.87	3.43	3.94	4.44	4.92	5.88
0.59	1.50	1.86	2.21	2.53	2.83	3.39	3.90	4.40	4.88	5.84
0.60	1.46	1.83	2.17	2.50	2.80	3.35	3.86	4.36	4.85	5.80
0.61	1.43	1.79	2.14	2.46	2.76	3.31	3.83	4.32	4.81	5.76
0.62	1.40	1.76	2.11	2.43	2.73	3.28	3.79	4.29	4.77	5.73
0.63	1.37	1.73	2.07	2.39	2.69	3.24	3.76	4.25	4.74	5.69
0.64	1.34	1.70	2.04	2.36	2.66	3.21	3.72	4.22	4.71	5.66
0.65	1.31	1.67	2.01	2.33	2.63	3.18	3.69	4.19	4.68	5.63
0.66	1.28	1.64	1.98	2.30	2.60	3.15	3.66	4.16	4.64	5.60
0.67	1.26	1.61	1.95	2.27	2.57	3.12	3.63	4.13	4.61	5.57
0.68	1.23	1.58	1.92	2.24	2.54	3.09	3.60	4.10	4.59	5.54
0.69	1.21	1.56	1.89	2.21	2.51	3.06	3.58	4.07	4.56	5.52
0.70	1.18	1.53	1.87	2.19	2.48	3.03	3.55	4.04	4.53	5.49
0.71	1.16	1.51	1.84	2.16	2.46	3.01	3.52	4.02	4.50	5.46
0.72	1.14	1.48	1.82	2.13	2.43	2.98	3.49	3.99	4.48	5.44
0.73	1.12	1.46	1.79	2.11	2.41	2.96	3.47	3.97	4.45	5.41
0.74	1.10	1.44	1.77	2.09	2.38	2.93	3.45	3.94	4.43	5.39
0.75	1.08	1.41	1.75	2.06	2.36	2.91	3.42	3.92	4.41	5.36
0.76	1.056	1.39	1.73	2.04	2.33	2.88	3.40	3.90	4.38	5.34
0.77	1.037	1.37	1.70	2.02	2.31	2.86	3.38	3.87	4.36	5.32
0.78	1.018	1.35	1.68	2.00	2.29	2.84	3.35	3.85	4.34	5.30
0.79	1.000	1.33	1.66	1.97	2.27	2.82	3.33	3.83	4.32	5.28
0.80	0.982	1.31	1.64	1.95	2.25	2.80	3.31	3.81	4.30	5.26
0.82	0.949	1.28	1.60	1.92	2.21	2.76	3.27	3.77	4.26	5.22
0.84	0.917	1.24	1.57	1.88	2.17	2.72	3.23	3.73	4.22	5.18
0.86	0.887	1.21	1.53	1.84	2.14	2.68	3.20	3.70	4.18	5.14
0.88	0.858	1.18	1.50	1.81	2.10	2.65	3.16	3.66	4.15	5.11
0.90	0.831	1.15	1.47	1.78	2.07	2.62	3.13	3.63	4.12	5.07
0.92	0.805	1.12	1.44	1.75	2.04	2.58	3.10	3.60	4.08	5.04
0.94	0.781	1.10	1.41	1.72	2.01	2.55	3.07	3.57	4.05	5.01
0.96	0.758	1.07	1.39	1.69	1.98	2.53	3.04	3.54	4.03	4.98
0.98	0.736	1.05	1.36	1.66	1.95	2.50	3.01	3.51	4.00	4.96
1.00	0.714	1.02	1.34	1.64	1.93	2.47	2.99	3.48	3.97	4.93
∞	....	0.091	0.274	0.500	0.743	1.24	1.74	2.23	2.71	3.67

It can also be transposed to the following forms:

$$Q = \frac{1.486}{n} ar^{2/3}s^{1/2} \tag{152b}$$

$$s = \frac{(nv)^2}{2.2082r^{4/3}} \tag{152c}$$

$$s = \frac{(nQ)^2}{2.2082r^{4/3}a^2} \tag{152d}$$

In Eq. (152*b*), for similar sections, the value of the expression $ar^{2/3} = a^{5/3}/p^{2/3}$ varies as the eight-thirds power of corresponding linear dimensions. It can therefore be written in the equivalent form

$$Q = \frac{K}{n} D^{8/3} \cdot s^{1/2} \tag{153a}$$

where K, which is termed the *discharge factor*, depends for its value on the ratio of the depth of water to some other linear dimension of the cross section.

For trapezoidal channels (including rectangular and triangular sections)

$$K = \frac{1.486\left(\frac{1}{x} + z\right)^{5/3}}{\left(\frac{1}{x} + 2\sqrt{1+z^2}\right)^{2/3}} \tag{153b}$$

where $x = D/b$ = ratio of depth of water to bottom width of channel (Fig. 69) and $z = e/d$ = side slopes of channel, ratio of horizontal to vertical. Table 20 contains values of K corresponding to D/b for various side slopes.[1]

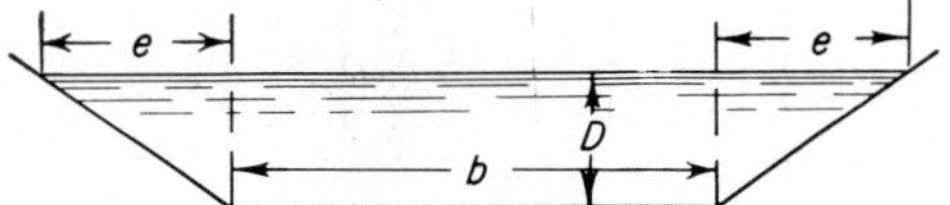

FIG. 69. Trapezoidal canal section.

Values of K for circular conduits flowing partly full, corresponding to different ratios of depth of water to diameter of conduit, are given in Table 21.

Example 20: Determine the width of trapezoidal concrete-lined channel required to carry 200 cu ft per sec at a depth of 4 ft. The channel has a grade of 1 ft per 1,000 ft and side slopes of 1 to 1.

The value of n from Table 19 is 0.014. Solving for K in Eq. (153*a*),

$$K = \frac{Qn}{D^{8/3}s^{1/2}} = \frac{200 \times 0.014}{40.3 \times 0.0316} = 2.20$$

From Table 20,

$$\frac{D}{b} = 0.825$$

and

$$b = \frac{D}{0.825} = 4.85 \text{ ft}$$

[1] More complete tables for the solution of open channel problems are given by KING in "Handbook of Hydraulics," 4th ed., 1954.

Table 21. Values of K for Circular Channels in the Formula $Q = \frac{K}{n} D^{8/3} s^{1/2}$

D = depth of water and d = diameter of channel

D/d	0.00	0.01	0.02	0.03	0.04	0.05	0.06	0.07	0.08	0.09
0.0	.....	15.02	10.56	8.57	7.38	6.55	5.95	5.47	5.08	4.76
0.1	4.49	4.25	4.04	3.86	3.69	3.54	3.41	3.28	3.17	3.06
0.2	2.96	2.87	2.79	2.71	2.63	2.56	2.49	2.42	2.36	2.30
0.3	2.25	2.20	2.14	2.09	2.05	2.00	1.96	1.92	1.87	1.84
0.4	1.80	1.76	1.72	1.69	1.66	1.62	1.59	1.56	1.53	1.50
0.5	1.470	1.442	1.415	1.388	1.362	1.336	1.311	1.286	1.262	1.238
0.6	1.215	1.192	1.170	1.148	1.126	1.105	1.084	1.064	1.043	1.023
0.7	1.004	0.984	0.965	0.947	0.928	0.910	0.891	0.874	0.856	0.838
0.8	0.821	0.804	0.787	0.770	0.753	0.736	0.720	0.703	0.687	0.670
0.9	0.654	0.637	0.621	0.604	0.588	0.571	0.553	0.535	0.516	0.496
1.0	0.463									

Section of Greatest Efficiency. The cross section of greatest hydraulic efficiency is the one that, for a given slope and area, has the maximum capacity and therefore the shortest wetted perimeter. This can be seen from an examination of Eq. (152*b*) (p. 4–80). Though there are usually practical objections to using the theoretically most efficient section, particularly for earth canals, the dimensions of such sections should be known and adhered to as closely as conditions appear to justify.

Of all possible cross sections for open channels, for a given area, the semicircle has the shortest perimeter and therefore the highest hydraulic efficiency. The half hexagon has the highest efficiency of all trapezoidal sections. The rectangular section of highest efficiency has a depth of water equal to one-half the width. A general expression for wetted perimeter of a trapezoidal section (see Fig. 69 for nomenclature) can be expressed in terms of b/D and $z = e/D$. The first derivative of this expression with respect to b/D equated to zero reduces to the following formula for determining the dimensions of the most efficient trapezoidal section:

$$b = 2D(\sqrt{1 + z^2} - z) \qquad (154)$$

Transitions in Open Channels. In open channels, there are losses corresponding to minor losses in pipes; but, where water flows with a free surface, any condition producing a loss of head causes a drop in water surface that must be provided for in designing the channel. Also, wherever a change in velocity occurs, the design must provide for the change in water-surface elevation resulting from difference in velocity heads. The short reaches of channel that are used to connect conduits of different form are termed *transition structures*. Examples are the connections between flumes and canals, siphons and canals, tunnels and canals, and canals of different cross-sectional area. It is usually important to keep losses of head at transitions as small as practicable.

If a conduit carrying water with a velocity v_1 changes to a smaller cross section in which the velocity is v, the total drop in water surface is

$$h = \frac{v^2}{2g} - \frac{v_1^2}{2g} + K_c\left(\frac{v^2}{2g} - \frac{v_1^2}{2g}\right) \qquad (155)$$

The maximum value of the coefficient K_c may be as great as 0.5, the value for sharp-cornered entrance to pipes (see p. 4–71), but this value can be greatly reduced by proper design. It should be noted, however, that even with K_c zero there is still a drop in water

surface equal to the difference in velocity heads. It is of the utmost importance that this drop in water surface be provided for in designing transitions. If the discharge is from a reservoir or other body of comparatively still water, v_1 may be considered zero.

If a conduit carrying water with a velocity v changes to a larger cross section in which the velocity is v_1, the total rise in water surface is

$$h = \frac{v^2}{2g} - \frac{v_1^2}{2g} - K_e\left(\frac{v^2}{2g} - \frac{v_1^2}{2g}\right) \tag{156}$$

If the transition is made abruptly, K_e approaches unity, making h approximately zero (see loss of head due to enlargement in pipes, p. 4–69). In a properly designed transition, however, a large part of the difference in velocity heads can be converted into static head.

Properly and improperly designed transitions are illustrated in Fig. 70. This drawing is reproduced from a paper by Hinds [1] describing in detail practices in the U.S. Bureau

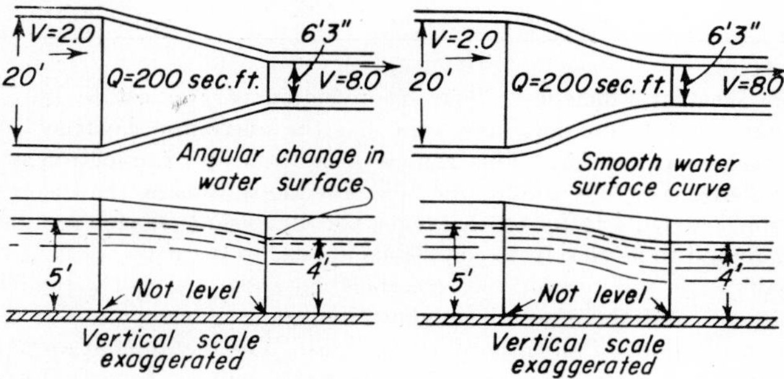

FIG. 70. Transition structures in open channels.

of Reclamation. If the change in section is accomplished with straight sides, as shown to the left in the figure, there is an angular change in water surface where the transition ends which produces turbulence and results in loss of head. In a properly designed transition, shown to the right in the figure, the water surface is a smooth continuous curve approximately tangent to the water surfaces in the two channels. There are no fixed rules for designing transitions. A satisfactory water surface can be assumed, and then structure dimensions can be computed to conform. Inlet and outlet structures (for contraction or expansion of stream) may be given the same form of transition. The value of K_c [Eq. (155)] in a well-designed inlet may be less than 0.05, and a value of 0.1 is safe for use in designing. K_e, Eq. (156), for a well-designed outlet is likely to be less than 0.1, and a value of 0.2 is safe for use in designing.

Information which is useful for the design of transitions for high-velocity flow has been reported by a number of investigators.[2]

Obstructions and Bends in Open Channels. Obstructions to flow in open channels may be caused by submerged weirs and gates, which are treated in earlier articles (see pp. 4–60 and 4–35). Other common obstructions are piers and abutments, which produce losses of head resembling transition losses, described in the preceding article. Such losses may be divided into three parts: (1) the transition loss where the channel changes to the reduced section; (2) the increase in friction loss, due to higher velocity between

[1] The Hydraulic Design of Flume and Siphon Transitions, *Trans. ASCE*, vol. 92, 1928. This paper also gives many examples of transition structures suited to different conditions.

[2] IPPEN, KNAPP, DAWSON, ROUSE, BHOOTA, and HSU, High Velocity Flow in Open Channels, A Symposium, *Trans. ASCE*, 1951.

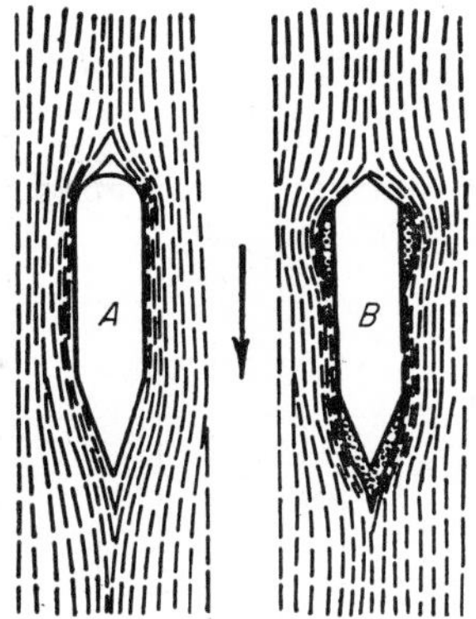

FIG. 71. Models of bridge piers.

piers; and (3) a second transition loss where the channel changes back to normal section. Equations (155) and (156) apply in determining losses at the two ends of piers and abutments. It is probable that K_c and K_e can be kept to 0.1 and 0.2 respectively by careful design. For square corners, they may approach 0.5 and unity. Experiments by Nagler[1] showed model A (Fig. 71) to produce a minimum of turbulence, while model B produced excessive turbulence.

The loss of head due to *curves* or *bends* in open channels is small if velocities are low, and ordinarily no correction is made for this loss unless curves are frequent and sharp. If velocities are relatively high, a greater slope should be provided for curved reaches than for tangents. It is common to allow for loss of head at curves in selecting the coefficient n for the Kutter or Manning formulas. The selection of the value of n should be made only after careful consideration of all conditions and a study of available experimental data.

Stages of Equal Energy. Assume water to be flowing in an open channel, Fig. 72, at a depth D and mean velocity v. The energy head H_e at any given cross section or the energy per pound of water, referred to the bottom of the channel, is

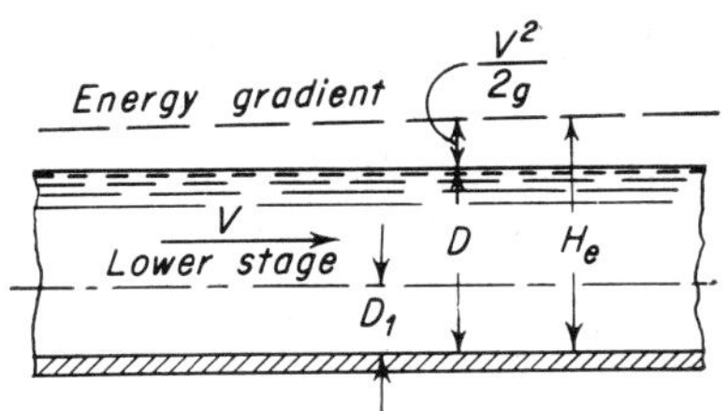

FIG. 72. Depths of equal energy.

$$H_e = D + \frac{v^2}{2g} \tag{157}$$

or, for a discharge Q in a channel of cross-sectional area a,

$$H_e = D + \frac{Q^2}{2ga^2} \tag{158}$$

The curve shown in Fig. 73 is a graph of Eq. (157) or (158) for a rectangular channel of unit width with H_e assigned a constant value of 10. In this case, $a = D$. It appears from the curve that there are two depths illustrated respectively by D and D_1, Fig. 72, at which any given discharge will flow with the selected energy head H_e, or, what is the same thing, with the same content of energy. For example, if the discharge per foot of width $Q_1 = 50$, the depths of equal discharge as determined from a solution of Eq. (158) are respectively 9.58 and 2.24. This quantity of water cannot flow in this channel with this energy head at any other depth. It appears that as Q_1 increases the two depths at which the water has the same energy content approach each other. They reach a common value of $6\frac{2}{3}$ when $Q_1 = 97.7$. For any greater Q_1, the values of D given by Eq. (158) become imaginary. This is therefore the greatest discharge possible in this channel with $H_e = 10$. The depth at which maximum discharge occurs is termed *critical depth*.

A graph of Eq. (157) or (158) with the discharge per foot of width given a constant value of 50, is shown in Fig. 74. From this curve, it appears that, within the limits indicated, for any given discharge, there are two depths at which the energy will be the same. As H_e decreases, the two depths of equal energy approach each other, and they reach a common value of 4.27 when $H_e = 6.40$. This is the minimum energy head with

[1] Obstruction of Bridge Piers to Flow of Water, *Trans. ASCE*, vol. 82, 1918. This paper contains experiments on 34 different models of piers.

which 50 cu ft per sec can flow in this channel. This depth of 4.27 ft, the critical depth, is thus also the depth of minimum energy. Diagrams of similar form and exhibiting all the general characteristics of Figs. 73 and 74 can be constructed for any shape of channel.

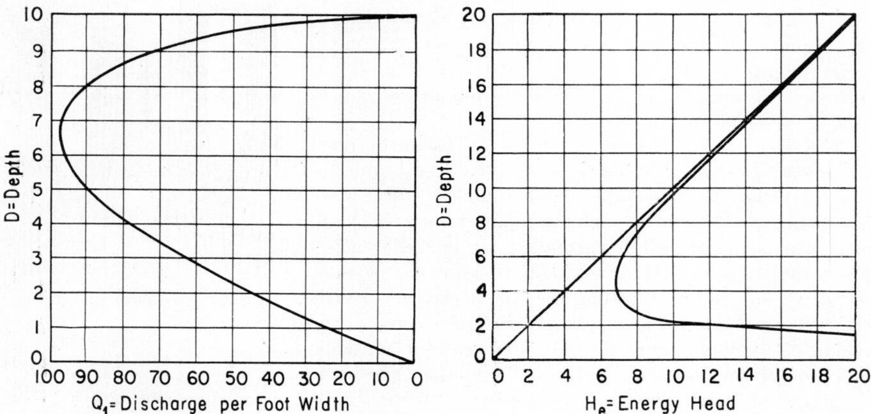

FIG. 73. Graph of Eq. (158) with $H_e = 10$ ft. FIG. 74. Graph of Eq. (158) with $Q_1 = 50$ cfs.

Critical Depth. As illustrated respectively by Figs. 73 and 74, critical depth can be defined as (1) the depth at which, for a given energy content of the water in a channel, maximum discharge occurs; or (2) the depth at which in a given channel a given quantity of water flows with the minimum content of energy. It will be observed from Fig. 74 that at or near critical depth a relatively large change in depth corresponds to a relatively small change in energy. Flow in this region is therefore quite unstable, as is usually indicated by characteristic water-surface undulations.

There follows a derivation of a criterion for determining when, in any given channel, flow occurs at critical depth. Equation (158) can be written

$$H_e = D + \frac{1}{a^2} \cdot \frac{Q^2}{2g} \tag{158}$$

In this equation, H_e is the energy head, D is the maximum depth of water, Q is the discharge, and a is the cross-sectional area of the stream. Since a is a function of D, Eq. (158) can be written

$$H_e = D + \frac{1}{f^2(D)} \cdot \frac{Q^2}{2g} \tag{159}$$

For any given discharge, critical depth is the depth at which H_e is a minimum. An expression for critical depth is thus obtained by differentiating the above equation with respect to D and equating to zero. Then,

$$dH_e/dD = -\frac{f'(D)}{f^3(D)} \cdot \frac{Q^2}{g} \tag{160}$$

From Fig. 75, if T is the top width of the cross section of the stream, $\Delta a = T\Delta D$, and in the limit

$$da/T\,dD$$

Therefore, Eq. (160) can be written

$$1 - \frac{T}{a^3} \cdot \frac{Q^2}{g} = 0 \tag{161}$$

or

$$\frac{a^3}{T} = \frac{Q^2}{g} \tag{162}$$

If we call a/T the mean depth D_m, Eq. (162) can be expressed

$$D_m = \frac{Q^2}{ga^2} = \frac{v^2}{g} \tag{163}$$

Equations (163) and (162) are general and applicable to all channels. They indicate the relation between cross section and discharge that must exist for flow to occur at critical depth. Special formulas for flow at critical depth in a given form of channel are obtained by expressing a and T or a and D_m in terms of dimensions applying to the particular cross section.

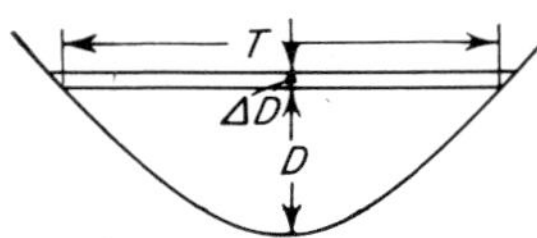

FIG. 75. Open channel section.

Rectangular Sections. A rectangular channel has a depth of water D equal to the mean depth D_m and a bottom width b equal to the top width T. For this cross section, from Eqs. (162) and (163),

$$D_c = \frac{v^2}{g} \tag{164}$$

and

$$v = \sqrt{gD_c} \tag{165}$$

If Q_1 is the discharge per foot width of channel,

$$D_c = \sqrt[3]{\frac{Q_1{}^2}{g}} \tag{166}$$

and

$$Q_1 = \sqrt{g}\, D_c{}^{3/2} \tag{167}$$

For any rectangular channel of width b ($g = 32.16$), the discharge is

$$Q = 5.67bD_c{}^{3/2} \tag{168}$$

Eliminating D_c from Eqs. (164) and (166), if flow is at critical depth,

$$v = \sqrt[3]{gQ_1} \tag{169}$$

From Eqs. (157) and (164),

$$H_e = \tfrac{3}{2} D_c \tag{170}$$

and

$$D_c = \tfrac{2}{3} H_e \tag{171}$$

Substituting this value of D_c in (167),

$$Q_1 = \sqrt{g}\, (\tfrac{2}{3})^{3/2} H_e{}^{3/2} \tag{172}$$

or for any rectangular channel of width b ($g = 32.16$), if flow is at critical depth, the discharge is

$$Q = 3.087bH_e{}^{3/2} \tag{173}$$

Trapezoidal Sections. The trapezoidal section shown in Fig. 69 has a depth D and bottom with b. The ratio, horizontal to vertical, of the slope of the sides of the channel is z. Expressing the mean depth D_m in Eq. (163) in terms of channel dimensions, we obtain the following relations between critical depth D_c and mean velocity v:

$$D_c = \frac{v^2}{g} - \frac{b}{2z} + \sqrt{\frac{v^4}{g^2} + \frac{b}{4z^2}} \tag{174}$$

and

$$v = \sqrt{\frac{b + zD_c}{b + 2zD_c} \cdot gD_c} \tag{175}$$

Also, in terms of critical depth and discharge,

$$D_c^3 = \frac{b + 2zD_c}{(b + zD_c)^3} \cdot \frac{Q^2}{g} \tag{176}$$

or

$$Q = \sqrt{\frac{(b + zD_c)^3}{b + 2zD_c} \cdot g}\; D_c^{3/2} \tag{177}$$

From Eqs. (157) and (163), expressing D_m in terms of channel dimensions,

$$H_e = \frac{3b + 5zD_c}{2b + 4zD_c} \cdot D_c \tag{178}$$

or, in the transposed form,

$$D_c = \frac{4zH_e - 3b + \sqrt{16z^2H_e^2 + 16zH_eb + 9b^2}}{10z} \tag{179}$$

If H_e and channel dimensions are known, the above value of D_c substituted in (177) will give Q.

Since rectangles and triangles are special cases of trapezoids, the former having $z = 0$ and the latter $b = 0$, formulas applying to sections of these forms can be obtained from Eqs. (174) to (179) by substituting the proper value in each equation and reducing. The solution of the critical-depth equations for trapezoidal channels, as well as similar relationships for circular and parabolic channels, may be simplified by means of appropriate tables.[1]

Critical Slope. Uniform flow at critical depth will occur when the grade or slope of the channel is just equal to the loss of head per foot resulting from flow at this depth. In any channel for any given discharge there is one grade that will maintain uniform flow at critical depth. This is termed the *critical slope.* For any grade flatter or steeper than critical slope, the depth of flow will be respectively greater or less than critical depth.

In cases where nonuniform (accelerated) flow passes through the critical stage, critical depth will occur at the section at which the energy gradient has critical slope. The approximate position of the section where critical depth will occur can thus usually be predicted. Examples of flow passing through critical stage are contained in the following pages.

Substituting the value of Q in terms of D_m given by Eq. (163) in (152*b*) and reducing, we have

$$s_c = \frac{14.56n^2D_m}{r^{4/3}} \tag{180}$$

[1] King, H. W., "Handbook of Hydraulics," 4th ed., 1954.

In this equation, s_c is the critical slope, r is the hydraulic radius, D_m is the mean depth of channel, and n is the coefficient of roughness in the Manning formula. Since s_c varies as the square of this coefficient, slopes computed by Eq. (180) are likely to show considerable variance from actual conditions.

Increase of Slope or Section. The channel illustrated in Fig. 76*a* has a gradually increasing grade. At A, where critical slope occurs, the actual energy gradient coincides with the minimum energy gradient, and flow is at critical depth.

The channel shown in Fig. 76*b* has an abrupt change at B to a grade greater than critical slope. Upstream from B where the grade is less than critical slope, the water surface drops faster than the canal bottom to critical depth D_c. It cannot drop below

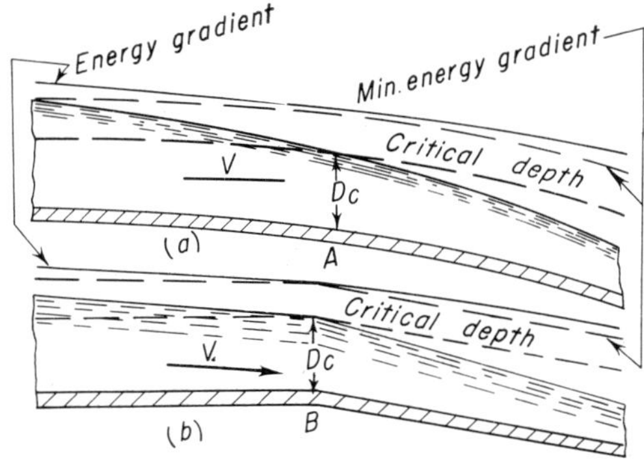

FIG. 76. Examples of flow at critical depth

critical depth since this is the stage of minimum energy and any further decrease in depth would require the addition of energy from an outside source. If the grade downstream from B is critical slope, flow will be uniform and at critical depth; otherwise it will be accelerated and at less than critical depth.

If a channel having a constant grade that is less than critical slope suddenly expands to a sufficiently great width, conditions controlling the water surface upstream from the section where the expansion begins are much the same as those upstream from the abrupt change in slope. In each case, removal of restriction causes the water to accelerate, and critical depth will be at the place where the change in channel conditions occur. Below this section, as the stream expands in width there is a corresponding decrease in depth which causes the water to discharge under an increased head and produces an accelerating velocity.

Channel Entrance. The source of supply may be either a reservoir or another channel whose stage is greater than critical depth. The entrance shown in Fig. 77 is rounded so as to reduce losses of head. The energy head is H_e, and the head under which water enters the channel is $h = H_e - D$, D being the depth of water and h the difference in level between the energy gradient and the water surface just downstream from the *entrance drop*. The discharge into the channel is

$$Q = a\sqrt{2g(H_e - D)} \tag{181}$$

a being the area of the cross section at which D is measured. For a rectangular entrance

the discharge per foot width is

$$Q_1 = D\sqrt{2g(H_e - D)} \tag{182}$$

If the channel has a grade equal to or greater than critical slope, discharge occurs at critical depth and D becomes D_c. Actual discharges will always be somewhat less than the theoretical discharges given by the above formulas.

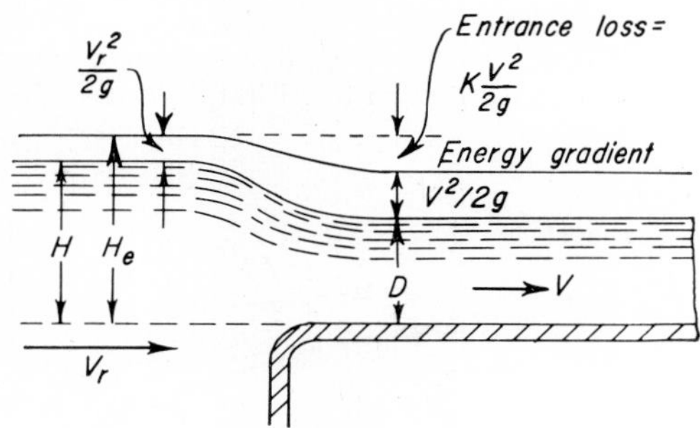

FIG. 77. Channel entrance.

Constrictions. Curves of the type shown in Fig. 74 serve to show the effect of a *rise in a channel bottom* on the water-surface profile. In Fig. 78 is shown the result of raising the bottom 0.2 ft in a rectangular channel in which the depth was originally greater than critical depth. Neglecting energy losses, which are relatively small for a smoothly rounded rise, raising the bottom reduces the specific energy and thus reduces the depth from 0.9 ft to 0.62 ft with the resulting water-surface profile shown. The amount of additional rise necessary to cause flow at critical depth is indicated by the dashed line.

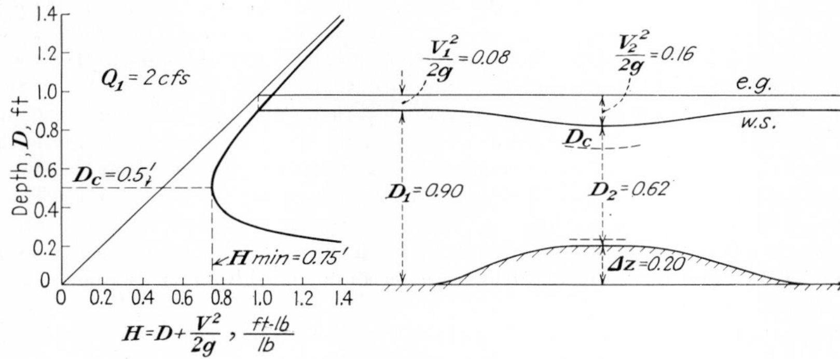

FIG. 78. Flow over rise in channel bottom.

A consideration of Fig. 78 will show that, if the original depth were less than D_c, a similar rise in the bottom would cause an increase in depth, thus producing a standing wave. Waves of this type may be observed wherever water flows swiftly over a rocky stream bed.

If the rise is carried beyond the point shown by the dashed line (Fig. 78), backwater must occur because the minimum value of specific energy has been reached. Under these conditions critical depth will still occur and a hydraulic jump will be formed below

the rise as shown in Fig. 79. Arrangements of this type are called *critical depth meters* and are frequently used for measuring the discharge in open channels. A gage is placed upstream from the rise to measure the depth H, shown in Fig. 79. This value is inserted into Eq. (173) to obtain a first trial value of the discharge. Based on this value of Q, an approximate value of v_1 and $v_1^2/2g$ can be computed. The second value of Q is com-

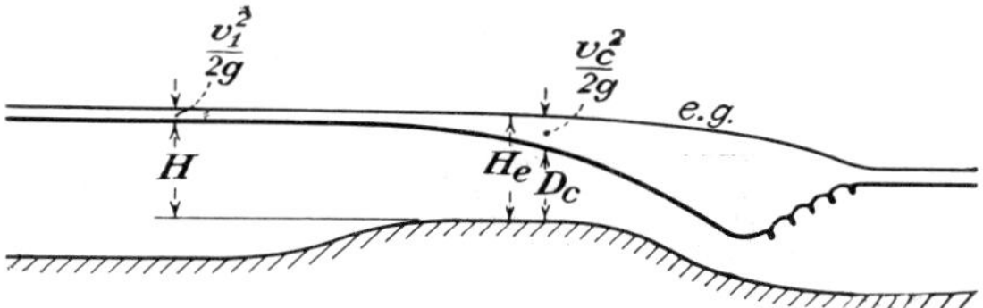

FIG. 79. Critical depth meter.

puted on the basis of $H_e = H + v_1^2/2g$. Several sets of computations are usually required in order to determine an exact answer. It is impossible to measure D_c directly because its exact location is not known.

In Fig. 80 a curve of the type shown in Fig. 73 is used to illustrate the effect of a *constriction* in the channel width on the water-surface profile when the original depth was greater than D_c. The constriction increases the discharge per foot width, thus causing

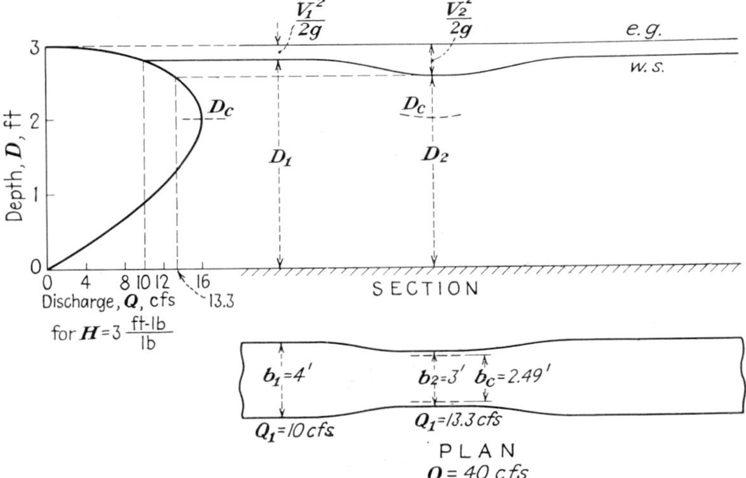

FIG. 80. Flow past a constriction in channel width.

the depth to decrease. If the original depth had been below critical, it may be seen from Fig. 80 that a constriction would increase the depth.

Gradually Varied Flow. A number of cases of rapidly varying nonuniform flow have been discussed under transitions and constrictions. In this section the type of accelerated or retarded flow which often extends over long reaches of channel will be discussed. The Bernoulli constant as given by Eq. (36) may be written as follows for open channels:

$$H = \frac{v^2}{2g} + D + z \tag{183}$$

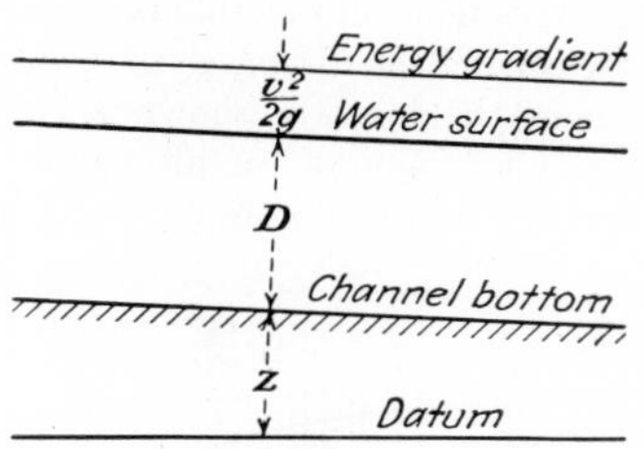

FIG. 81. Profile through open channel.

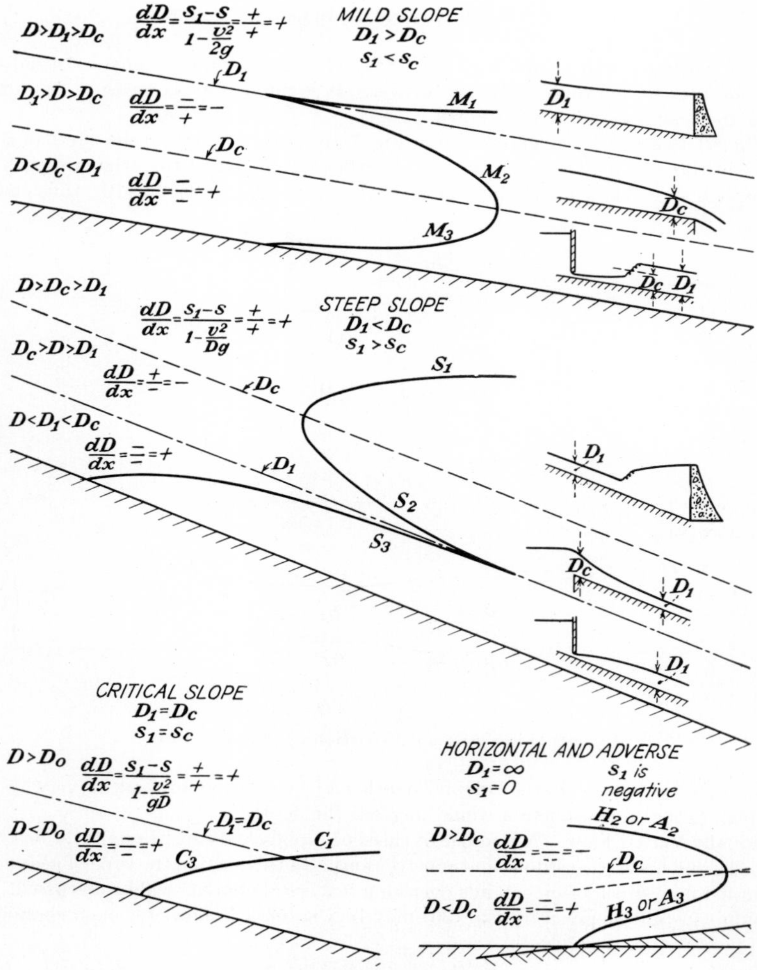

FIG. 82. Twelve possible regimes of open channel flow.

The values in Eq. (183) are shown graphically in Fig. 81. Differentiation with respect to distance along the channel x gives

$$-\frac{dH}{dx} = \frac{d(v^2/2g)}{dx} + \frac{dD}{dx} - \frac{dz}{dx} \tag{184}$$

the negative signs being included so that downward slopes in the direction of flow may be considered to be positive. For the particular case of a wide rectangular channel, $v = Q_1/D$ and

$$\frac{d(v^2/2g)}{dx} = \frac{d(Q_1^2/2gD^2)}{dx} = \frac{Q_1^2 dD}{gD^3 dx} = \frac{v^2}{gD} \cdot \frac{dD}{dx}$$

Also, dH/dx is the slope of the energy gradient s and dz/dx is the slope of the bottom s_1. The general differential equation for nonuniform flow in wide rectangular channels may now be obtained by inserting the above values in Eq. (184) and rearranging terms. Then

$$\frac{dD}{dx} = \frac{s_1 - s}{1 - v^2/gD} \tag{185}$$

When flow is uniform, the channel bottom and the energy gradient are parallel ($s_1 = s$) and dD/dx becomes zero. When flow is at critical depth, $v^2/gD = 1$ [Eq. (164)] and dD/dx is infinity. While this latter condition is obviously not attained, flow at D_c is characterized by a wavy surface. A careful study of Eq. (185) will show 12 possible regimes of flow as illustrated by Fig. 82. The scheme for identifying the various regimes is patterned after that used by Bakhmeteff.[1] In Fig. 82, D_1 is the depth of normal or uniform flow, i.e., the depth at which the energy gradient, the water surface, and the channel bottom are parallel.

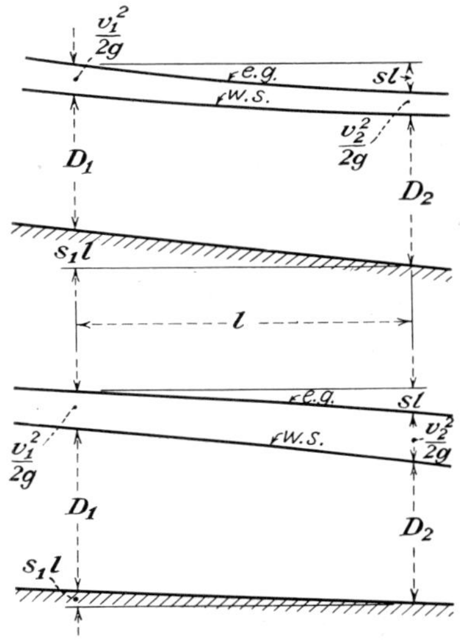

FIG. 83. Profile through open channel.

Determination of the Water-surface Profile. Equation (185) is not in a convenient form for the computation of backwater or accelerated-flow profiles. If for either of the profiles shown in Fig. 83 the Bernoulli equation is written from point 1 to point 2, the following expression is obtained.

$$\frac{v_1^2}{2g} + D_1 + s_1 l = \frac{v_2^2}{2g} + D_2 + sl \tag{186}$$

Solving for l,

$$l = \frac{\left(D_2 + \frac{v_2^2}{2g}\right) - \left(D_1 + \frac{v_1^2}{2g}\right)}{s_1 - s} = \frac{H_2 - H_1}{s_1 - s} \tag{187}$$

An examination of Eq. (187) in connection with Fig. 83 will show that when H_2 is greater than H_1, s_1 is greater than s and when H_2 is less than H_1, s_1 is less than s, with the result that the value of l is always positive.

[1] BAKHMETEFF, B. A., "Hydraulics of Open Channels," 1932.

The value of s in Eq. (187) is determined from the Manning formula [Eq. (153a)] using the average depth in the reach. Although the Manning formula was derived for uniform flow, it has been found that it gives satisfactory results for gradually varied flow, provided that reaches are kept short. It is difficult to give a rule concerning the maximum length of reach to be used. In general it may be stated that velocity changes should not exceed 20 percent nor should the change in depth be more than 1 ft. The most direct method of using Eq. (187) is to compute the distance from some known depth to some other assumed depth. Having determined s from the average conditions, l may be determined directly. If the length of the reach and the depth at one end are known, the depth at the other end must be determined by successive approximations. The latter method is used for natural streams where it is necessary to compute the depth at predetermined sections.

Canal with Steep Slope. If water enters a canal from a reservoir or from another canal in which the velocity of approach is less than the velocity at critical depth and the slope s_1 of the receiving channel is greater than critical slope, water will enter the receiv-

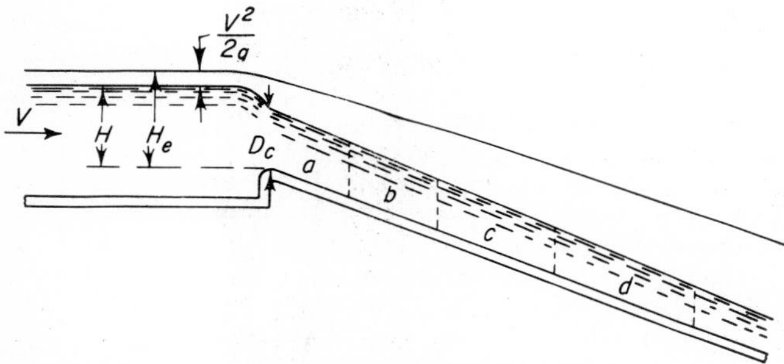

FIG. 84. Profile through channel having steep slope.

ing channel under conditions of maximum discharge, i.e., at critical depth. If the entrance has a rectangular cross section, discharge is given by Eq. (173) (p. 4–85). For a trapezoidal channel, the discharge can be obtained by substituting the value of D_c given by Eq. (179) in Eq. (177).

If it is required to design a canal with steep slope to carry a given discharge, the entrance dimensions for flow at critical depth are first determined, and computations for successive reaches (a, b, c, etc., Fig. 84) can then be made. Since the velocity is accelerating, successively smaller cross sections of canal will be required. From Eq. (187), the distance to any assumed cross section can be determined. After the dimensions at the lower end of reach a have been decided on, this section can be taken as the upper end of reach b. Computations for reaches b, c, d, etc., proceed the same as for reach a. If extended far enough, the channel will approach a minimum cross section the dimensions of which can be determined from the Manning formula (see p. 4–77).

Short Canal with Flat Slope. A short canal with unrestricted outlet, such as might be used as a spillway for a reservoir, is shown in Fig. 85. If the slope of the bottom of the canal is greater than critical slope, discharge is determined by conditions at the intake and is independent of the slope. This case is described in the preceding paragraph. If the canal does not have slope enough to carry water at critical depth, flow at critical depth D_c occurs a short distance upstream from the outlet. The discharge under these conditions must be determined by trial solutions as follows:

A discharge Q is first assumed. Discharge at intake is a function of $h = H_e - D_a$ (see p. 4–87). From the given entrance dimensions, h is computed for the assumed Q,

and D_c at the outlet is computed from Eq. (166) or (177). Trial elevations of water surface at the two ends of the channel can now be determined. By dividing the channel into as many reaches as may be found necessary and using Eq. (187), the slope of water surface can be computed. Trial solutions are continued until the assumed discharge is such that the computed values of D_c and h give the slope of water surface required for this discharge.

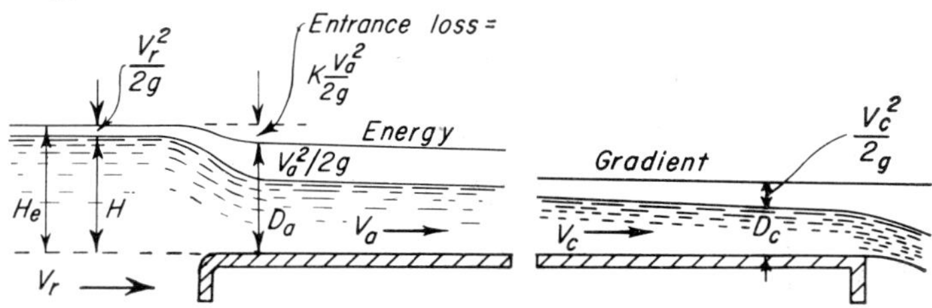

FIG. 85. Canal with flat slope.

Hydraulic Jump. It has been shown on p. 4–83 that, when water is flowing in an open channel at any depth other than critical depth, there is another depth or stage at which this quantity of water will have the same content of energy and that at all intermediate stages the energy content is less than for these two depths. Water cannot be made to change abruptly either from one stage of equal energy to the other or from the higher stage to a depth less than critical depth. Water flowing at a depth less than critical depth will, under certain conditions, change suddenly to a depth greater than critical depth but less than the other depth of equal energy. This change of stage produces a *stationary wave*, commonly called the hydraulic jump.[1] A hydraulic jump will occur only where water flowing at a depth less than critical depth enters a stream having a slope less than critical slope. It is commonly observed where water, after passing at high velocity over the spillway of a dam (Fig. 88), encounters the tail water.

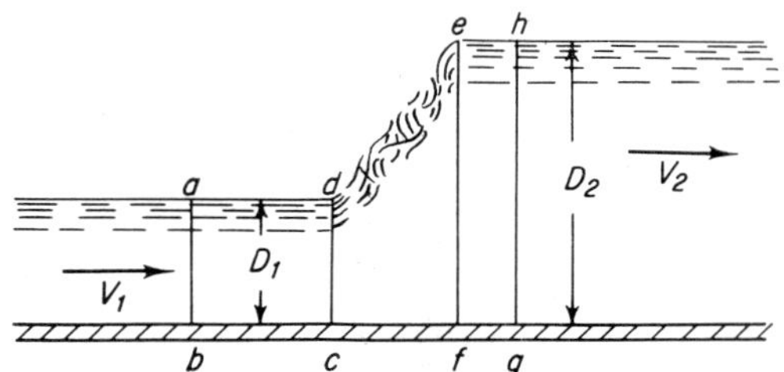

FIG. 86. Hydraulic jump.

As shown in Fig. 86, water flowing with a velocity V_1 and depth D_1 jumps to a depth D_2 with a corresponding velocity V_2. In all cases, D_1 is less than critical depth, and D_2 is greater than critical depth but less than the depth at which the energy content of the water is the same as for depth D_1. There is thus a loss of energy (or head) in the jump. In a short interval of time, the mass of water *abcd* moves to the position *efgh* and in

[1] The following are valuable discussions of the hydraulic jump: KENNISON, The Hydraulic Jump in Open-channel Flow at High Velocity, *Trans. ASCE*, vol. 80, 1916. WOODWARD, Theory of Hydraulic Jump and Backwater Curves; also, RIEGEL and BEEBE, The Hydraulic Jump as a Means of Dissipating Energy, *Miami Conservancy District Tech. Repts.*, Part III, 1917. HINDS, The Hydraulic Jump and Critical Depth in the Design of Hydraulic Structures, *Eng. News-Record*, Nov. 25, 1920. KING, "Handbook of Hydraulics," Chap. VIII, 4th ed., 1954. BAKHMETEFF and MATZKE, The Hydraulic Jump in Terms of Dynamic Similarity, *Trans. ASCE*, 1936. SCOBEY, Notes on the Hydraulic Jump, *Civil Eng.*, August, 1939. KINDSVATER, C. E., The Hydraulic Jump in Sloping Channels, *Trans. ASCE*, 1944.

changing positions loses much momentum. According to Newton's second law of motion, the rate of loss of momentum must be equal to the unbalanced force acting on the moving mass to retard its motion. Otherwise stated, this unbalanced force must equal the rate in change of momentum. Considering the unbalanced force F acting on the mass M for the time t, the rate of discharge being Q and the unit weight of water w,

$$F = \frac{MV_1 - MV_2}{t} = \frac{Qw}{g}(V_1 - V_2) \tag{188}$$

The unbalanced force is the difference between the hydrostatic forces corresponding to the depths D_2 and D_1 or, for rectangular channels (see p. 4–11),

$$F = \frac{D_2^2 w}{2} - \frac{D_1^2 w}{2} \tag{189}$$

Equating the values of F in (188) and (189) and substituting V_1D_1 for Q and V_1D_1/D_2 for V_2 in (188) and reducing,

$$V_1^2 = \frac{gD_2}{2D_1}(D_2 + D_1) \tag{190}$$

The solution of Eq. (190) for D_2 gives

$$D_2 = -\frac{D_1}{2} + \sqrt{\frac{2V_1^2 D_1}{g} + \frac{D_1^2}{4}} \tag{191}$$

Similarly, from Eq. (190) after substituting V_2D_2/D_1 for V_1,

$$D_1 = -\frac{D_2}{2} + \sqrt{\frac{2V_2^2 D_2}{g} + \frac{D_2^2}{4}} \tag{192}$$

Equation (191) gives the depth of water after a jump in rectangular channels where V_1 and D_1 are known. Equation (192) is useful in determining the position of the jump where V_2 and D_2 are known. More general relationships which are applicable to channels of any cross-sectional shape can be derived in a similar manner.[1]

Position of Jump. Hydraulic jump can occur only when water flowing below critical stage enters a channel in which flow is normally above critical stage and where the requirements expressed by Eqs. (191) or (192) can be fulfilled. If at entrance the normal depth exceeds the upper conjugate depth, there can be no jump. If this normal depth is less than the upper conjugate depth, the stream will continue in the lower stage until, through loss of head and consequent increase in depth, D_1 and D_2 are brought into conjugate positions and the jump must occur. In passing through the jump, the flow may change abruptly from nonuniform to uniform or from uniform to nonuniform, or it may be nonuniform both before and after the jump. The method of determining the position of the jump for each of these three conditions is discussed below.

A case where nonuniform flow becomes uniform after the jump is illustrated by Fig. 87. A given discharge Q enters a canal of uniform cross section through the gate G at less than critical stage, the grade of the canal being less than critical slope and the uniform flow depth D_2 being less than the upper conjugate depth corresponding to that at which water enters the canal. Since the grade of the canal is not sufficient to compensate for loss of head due to friction, the water increases in depth as it passes downstream from G, and the water surface slopes upward. A jump will occur at the section where the depth D_1 becomes the lower conjugate depth corresponding to D_2. Since D_2 and Q are known, D_1 can be computed from formula (192) for rectangular channels. With the

[1] King, H. W., "Handbook of Hydraulics," 4th ed., 1954.

velocity and depth at G known, the distance downstream from the gate to the section of depth D_1, where the jump occurs, can be obtained from Eq. (187). If the velocity change is not too great, computations may be made for the single reach; otherwise the distance should be divided into two or more reaches. For the case where the uniform flow depth D_2 is greater than the upper conjugate depth, the jump will drown out the jet and move

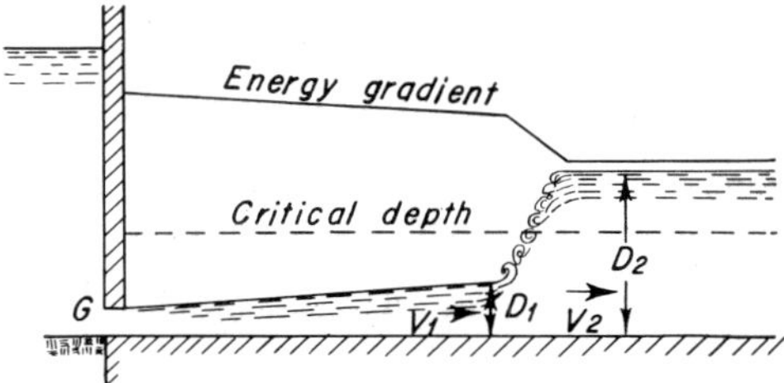

FIG. 87. Hydraulic jump downstream from gate.

up to the gate. The solution of the problem under such conditions is described elsewhere.[1]

Uniform flow changes abruptly in a hydraulic jump to nonuniform flow when a channel having a grade steeper than critical slope (see Fig. 82) encounters backwater. In this case, by a method corresponding to that described above for computing D_1, if D_1 and Q are known, D_2 can be computed from formula (191) for rectangular channels; the distance upstream from the dam to the section of depth D_2, where the jump occurs, can be obtained from Eq. (187).

An example of nonuniform flow occurring both before and after a jump is afforded by the overflow dam illustrated in Fig. 88. The slope of the apron is insufficient to provide

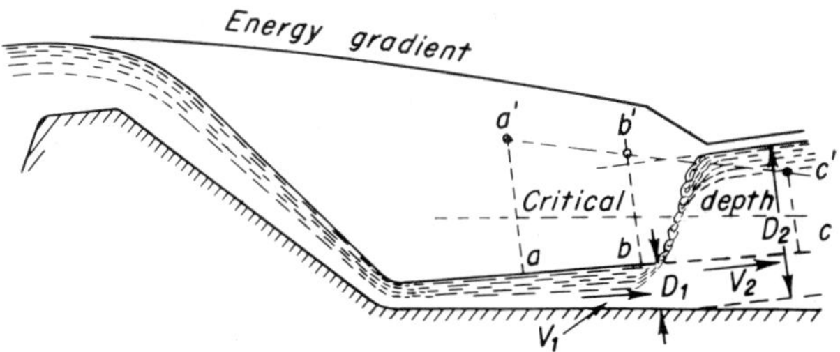

FIG. 88. Hydraulic jump downstream from spillway.

for loss of head due to friction before the jump, but it is more than sufficient after the jump. In both cases, therefore, the depths increase with the distance downstream and flow is nonuniform. To determine the position of the jump, the discharge Q must be known, and data must be available for determining the profile of water surface in the higher stage projected some distance upstream from the place where the jump will occur. The water surface in the lower stage must be computed downstream from a known depth, as from the toe of the dam, to a place beyond the position of the jump. Trial values of D_2 should be computed for three sections a, b, and c which will locate three points a', b', and c' to which the water must jump at the respective sections. One of the points should be on the opposite side of the jump from the other two, and, preferably,

[1] KING, H. W., "Handbook of Hydraulics," 4th ed., p. 3-14, 1954.

the middle point should be near the jump. The intersection of the line $a'b'c'$ with the higher water-surface profile gives the position of the jump.

Refinement in computations for determining the position of the jump is not justified, since loss of head varies as the square of the coefficient of roughness n, formula (152c), and errors in estimating its value are magnified when the computed loss of head is substituted in formula (187). It also should be understood that the jump does not occur at a cross section, as has been assumed, but that it has usually, according to experiments by Bakhmeteff and Matzke,[1] a length of four to five times the depth after jump. As a matter of precaution, in designing aprons or other structures to resist erosion, the computed position should be considered the beginning of the jump with the end about $5D_2$ farther downstream.

Abrupt Translatory Waves. The gate G (Fig. 89) controls the entrance of water into a canal of rectangular cross section which has been carrying Q_1 cu ft per sec per ft width of channel at a depth D_1 and velocity V_1. The gate opening is assumed to be suddenly

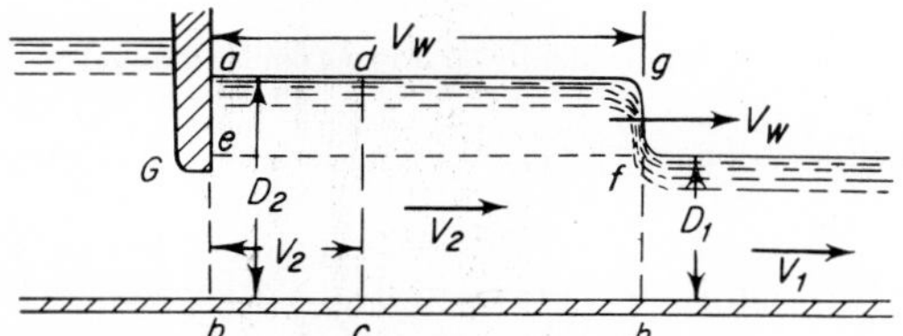

FIG. 89. Abrupt wave traveling downstream.

increased sufficiently to increase the discharge per foot width to Q_2. This causes an abrupt wave of acceleration to travel down the canal. Upstream from the wave, the modifying effects of friction and channel slope being neglected, each cross section has the same depth D_2 and the same velocity V_2. Downstream from the wave, the depth D_1 and the velocity V_1 remain the same as before the additional water was admitted.

Conditions at the end of 1 sec are illustrated in the figure. The accelerating wave, which travels with a velocity v_w, is at a distance v_w below the gate. During the first second of increased discharge, the total volume of water that entered the canal was $Q_2 = D_2V_2$. This volume is represented in the figure by the area $abcd$. The increase in volume over that which would have passed through the gate in its original position, or $Q_2 - Q_1$, is represented by the area $aefg$. Expressed algebraically,

$$Q_2 - Q_1 = v_w(D_2 - D_1) \tag{193}$$

or after substituting V_2D_2 and V_1D_1 respectively for Q_2 and Q_1 and transposing,

$$V_2 = (D_1V_1 + D_2v_w - D_1v_w)\frac{1}{D_2} \tag{194}$$

The mass of water $dchg$ (Fig. 89) has had its velocity increased from V_1 to V_2, and its momentum has thereby been increased. The mass being called M, the unbalanced force required to change the momentum in 1 sec is

$$F = M(V_2 - V_1) = \frac{(V_2 - V_1)(v_w - V_2)D_2w}{g} \tag{195}$$

[1] BAKHMETEFF, B. A., and A. E. MATZKE, The Hydraulic Jump in Terms of Dynamic Similarity, *Trans. ASCE*, 1936.

The unbalanced force is equal to the difference in hydrostatic pressure corresponding to the depths D_2 and D_1, or

$$F = \frac{D_2^2 w}{2} - \frac{D_1^2 w}{2} \tag{196}$$

Equating the values of F in (195) and (196) and reducing,

$$\frac{g}{2D_2}(D_2^2 - D_1^2) = (V_2 - V_1)(v_w - V_2) \tag{197}$$

Substituting V_2 from (194) and making algebraic transformations,

$$(v_w - V_1)^2 = \frac{gD_2}{2D_1}(D_2 + D_1) \tag{198}$$

The similarity of Eqs. (190) and (198) shows that the hydraulic jump and abrupt translatory wave are different manifestations of the same phenomenon.

Equation (198) can also be written

$$v_w = \pm\sqrt{\frac{g}{2} D_2 \left(1 + \frac{D_2}{D_1}\right)} + V_1 \tag{199}$$

For the case illustrated in Fig. 89, the sign before the radical is positive. From Eq. (199), with V_1, D_1, and D_2 known, the velocity of the wave can be determined. If the difference between D_2 and D_1 is small in comparison with D_2, Eq. (199) becomes approximately

$$v_w = \pm\sqrt{gD_2} + V_1 \tag{200}$$

This approximate formula has long been known and is commonly used. It is interesting to note that, with $V_1 = 0$, Eq. (200) becomes identical with (165) (p. 4–85), the formula for velocity corresponding to a given critical depth.

If D_1 in Eq. (199) is zero, v_w becomes infinite. This means that there can be no abrupt wave when water is suddenly admitted into an empty channel. It has been commonly observed that, following the failure of a dam, the flood approaches in the form of a "wall" of water. This observation can be correct only to the extent that an abrupt wave forms to accelerate the water already in the stream. The height of wave depends on the depth of water in the channel before the additional water arrives.

The complete solution of a problem of the type illustrated in Fig. 89 requires that, for given values of Q_1 and Q_2, the depth D_2 as well as the velocity of the wave be determined. By eliminating v_w from Eqs. (193) and (198) and substituting Q_1/D_1 for V_1, the following relation is obtained:

$$\left(\frac{Q_2}{D_2} - \frac{Q_1}{D_1}\right)^2 = \frac{g}{2D_1D_2}(D_2^2 - D_1^2)(D_2 - D_1) \tag{201}$$

This equation can be solved by trial for D_2, and then v_w can be obtained from Eq. (199). It will usually be easier, however, to solve (193) and (198) simultaneously by trial solutions for v_w and D_2.

If instead of the gate being opened instantaneously, as indicated in Fig. 89, it is opened gradually, the additional water can be assumed to pass down the channel in a series of waves a, b, c, d, etc. (Fig. 90), the depths of water at these waves being respectively D_a, D_b, D_c, etc., and velocities in the channel at the waves, V_a, V_b, V_c, etc. The velocity of wave a, from Eq. (200), is $\sqrt{gD_a} + V_a$ and of wave b is $\sqrt{gD_b} + V_b$, etc. Since the depth at any wave is greater than the depth at the wave next below it, the velocity of any wave is greater than the velocity of the wave next below it. Each wave

therefore tends to overtake the waves farther downstream and to become progressively shorter. If flow continues far enough, all the small waves will combine into one larger wave. The tendency, therefore, is for flow ultimately to occur with one abrupt wave, as illustrated in Fig. 89, even when the gate is opened gradually.

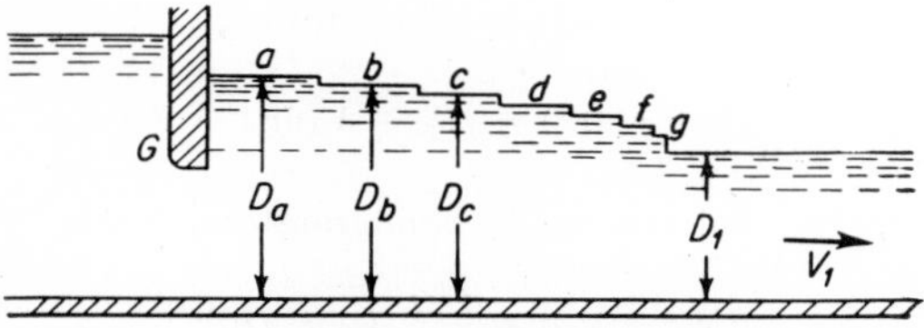

FIG. 90. Wave caused by gradual opening of gate.

The general principles applying to sudden increase in discharge at the intake of a channel apply also to sudden decrease in discharge at the outlet. The latter case is illustrated in Fig. 91. The gate G controls the discharge from the canal which has been carrying Q_1 cu ft per sec per ft width of channel at a depth D_1 and velocity V_1. The gate opening is assumed to be reduced instantaneously, and the discharge per foot width of channel to be reduced thereby to Q_2. An abrupt wave of deceleration travels up the

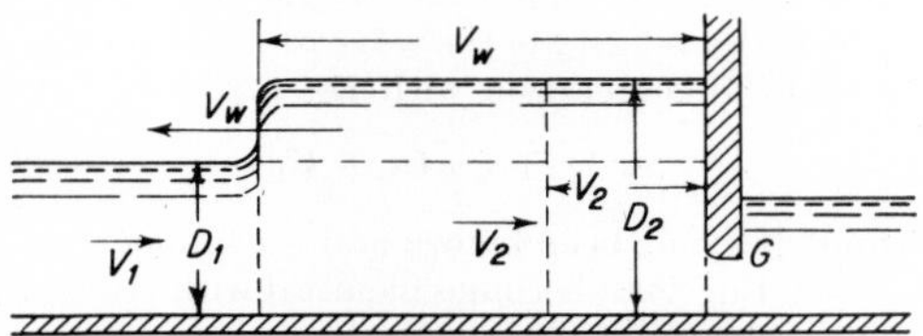

FIG. 91. Abrupt wave traveling upstream.

canal with a velocity v_w. Downstream from the wave, friction and channel slope being neglected, each cross section has the same depth D_2 and the same velocity V_2. Upstream from the wave, the depth D_1 and the velocity V_1 remain the same as before the flow was reduced. The figure illustrates conditions at the end of 1 sec. From principles analogous to those described for a wave traveling downstream, the following formulas, corresponding respectively to (193), (195), and (198), apply to an abrupt wave traveling upstream:

$$Q_1 - Q_2 = v_w(D_2 - D_1) \tag{202}$$

$$F = M(V_1 - V_2) = \frac{(V_1 - V_2)(v_w + V_2)D_2 w}{g} \tag{203}$$

$$(v_w + V_1)^2 = \frac{gD_2}{2D_1}(D_2 + D_1) \tag{204}$$

In these formulas, the numerical value only of v_w is considered. If velocities in the direction of flow are taken as positive, v_w when traveling upstream is negative. Considered in this way, Eqs. (193) to (201) are general and apply to all abrupt waves. For waves traveling downstream, the sign before the radical in Eqs. (199) and (200) is positive; for waves traveling upstream, the sign is negative.

Regarding waves of translation, in general, it may be stated that an accelerating (or decelerating) wave forms in a channel whenever the discharge is suddenly changed. Behind this wave, the water immediately acquires a new depth and new velocity cor-

responding to the new discharge. If the change in discharge acts to increase the depth, the wave has an abrupt face and is in reality a traveling hydraulic jump. If the change in discharge acts to decrease the depth, the wave has a sloping face. The latter case is discussed in the following paragraph.

Sloping Translatory Waves. The gate G (Fig. 92) controls the entrance of water into a canal of rectangular cross section which has been carrying Q_2 cu ft per sec per ft width of channel at a depth D_2 and velocity V_2. The gate opening is assumed to be decreased

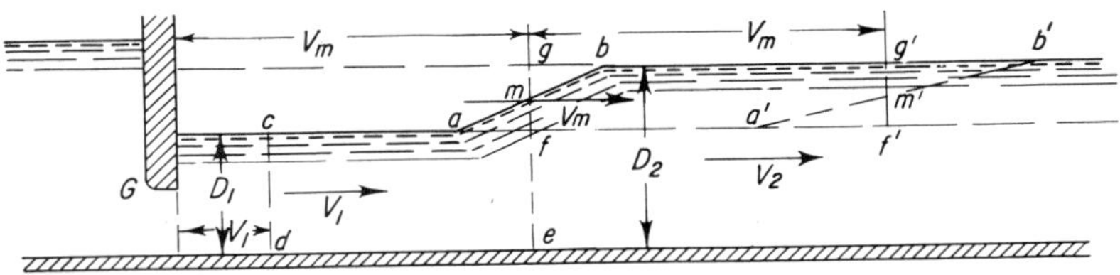

FIG. 92. Sloping wave traveling downstream.

instantaneously sufficient to decrease the discharge per foot width of channel to Q_1. This causes a sloping wave of deceleration ab (Fig. 92) to travel down the canal. Upstream from a, the effects of friction and channel slope being neglected, each cross section has the same depth D_1 and the same velocity V_1. Downstream from b, the depth D_2 and the velocity V_2 remain the same as before the flow was reduced. Between a and b, deceleration is still in progress.

Conditions at the end of 2 sec are illustrated in the figure. The line fg, which represents the mean position of the wave, is so drawn that the area $amf = bmg$. The mean velocity of the wave is v_m. During the first second of decreased discharge, the volume of water that entered the canal was $Q_1 = V_1 D_1$. The decrease in volume from that which would have passed through the gate in its original position is

$$Q_2 - Q_1 = v_m(D_2 - D_1) \tag{205}$$

Assuming that the mass of water $cdef$ (Fig. 92) has had its velocity reduced from V_2 to V_1 and following the method described in the preceding article for abrupt waves, we have

$$v_m = \pm\sqrt{\frac{gD_1}{2D_2}(D_2 + D_1)} + V_2 \tag{206}$$

The assumptions made in deriving Eq. (206) depart widely from actual conditions for large differences in depth, and for this reason the formula is not generally applicable. In the limiting condition, for waves of infinitesimal height, the difference between actual and assumed conditions disappears, and the velocity of the wave is

$$v_w = \pm\sqrt{gD} + V \tag{207}$$

in which D and V are respectively the depth and velocity of water.

The wave in Fig. 92 can be assumed to be made up of an infinite number of small waves (opposite but similar to Fig. 90) of which the bottom wave has a velocity $v_a = \pm\sqrt{gD_1} + V_1$ and the top wave a velocity $v_b = \pm\sqrt{gD_2} + V_2$. If a straight-line variation between v_a and v_b is assumed, the mean velocity of the wave is

$$v_m = \tfrac{1}{2}(v_a + v_b) = \tfrac{1}{2}(\pm\sqrt{gD_1} + V_1 \pm\sqrt{gD_2} + V_2) \tag{208}$$

Since D_2 is greater than D_1, v_b is greater than v_a. Intermediate velocities also decrease

with the depth. The slope of the face *ab* therefore decreases with the distance that the wave travels. At the end of 4 sec, the wave has the position $a'b'$. After a short time, the wave can be discerned only by a gradual lowering of the water surface. For the case illustrated in Fig. 92, if velocities in the direction of flow are called positive and velocities in the opposite direction negative, the sign before the radical is positive. For a sloping wave traveling upstream, such as would be produced if the gate opening in Fig. 91 were suddenly increased instead of being suddenly reduced, the sign before the radical is negative.

Since the wave has a curved face, Eq. (208) is not exact. It has, however, the advantage of being simple of application and will be found exact enough for ordinary problems. The simultaneous solution of Eqs. (205) and (208) for given values of V_2, D_2 and Q_1 give D_1 and v_m; also, v_a and v_b can be obtained.

Limitations of Wave Formulas. In developing the above expressions for waves, the effect of shear stresses at the boundaries has been neglected. For small waves the shear effects are negligible and waves will travel great distances at the velocities given by the above formulas. For larger waves the energy dissipation is much greater and the effect of friction in modifying the original velocity and height of the wave is more pronounced.

It is important to keep in mind that the sketches upon which the wave derivations were based picture conditions as they might be 1 or 2 sec after the formation of the wave. It is quite likely that after several seconds have elapsed the picture would be entirely different. The velocity and depth after the formation of the wave as given by the above formulas are obviously of only temporary duration because flow must eventually become stable in accordance with the inlet and outlet conditions and the Manning formula. It may take minutes or even hours for a completely stable condition to develop because of the recurrence of reflected waves.

DYNAMIC ACTION OF FLOWING WATER

Force Exerted by Jet. A constant force F acting in the direction of the original velocity during a period t upon a mass M changes the momentum of the mass from Mv_0 to Mv, where v_0 and v are respectively velocities at the beginning and end of the period. The force is equal to the rate of change of momentum, as expressed by the following equation.

$$F = \frac{Mv - Mv_0}{t}$$

This equation may be arranged in the form shown below for convenience in application to flowing fluids,

$$F = \frac{M}{t}(v - v_0)$$

in which M/t is the mass per unit of time undergoing a change in velocity from v to v_0. If the discharge of the jet is Q, then $M/t = Qw/g$ and the expression for the force becomes

$$F = \frac{Qw}{g}(v - v_0) \tag{209}$$

in which w is the unit weight of the fluid and g the gravitational acceleration. Equation (209) gives the force of the vane on the jet. The force of a jet on a vane would be equal and opposite in direction.

Both force and velocity are vector quantities. Therefore, the force acting in a particular direction can be obtained from Eq. (209) by using the original and final components of the velocity in the selected direction. A convenient method of separating the force

into two components is illustrated in Fig. 93 where the X axis is taken parallel to the original direction of the jet. Forces and velocities parallel to the X axis are designated as positive if they are directed to the right, whereas, for the Y axis, the upward direction is positive.

Jets Impinging against Deflecting Surfaces. In Fig. 93a, a jet in the plane of the coordinate axes X and Y is shown impinging tangentially against the stationary vane AB.

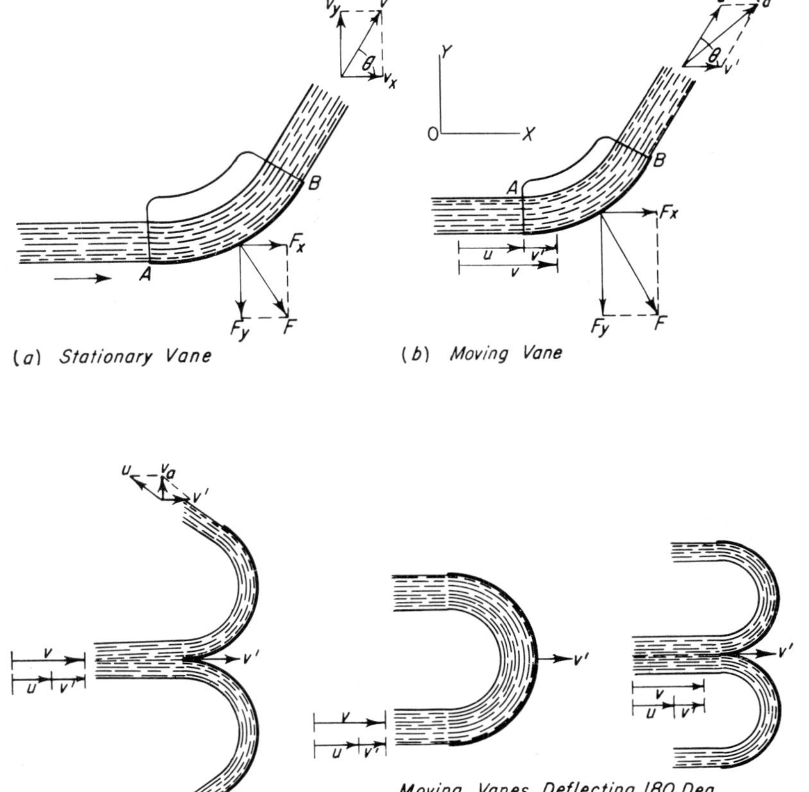

Fig. 93. Jets impinging against deflecting surfaces.

The surface of the vane is assumed to be frictionless; so that the jet leaves the vane with its original velocity v but with its direction deflected through the angle θ. The X component F_x of the force exerted on the vane, acting in the original direction of the velocity, equals the mass impinging per second times the change in velocity along the X axis. Similarly, F_y equals the mass impinging per second times the change in velocity along the Y axis. The resultant force acting on the vane is

$$F = \sqrt{F_x^2 + F_y^2} \tag{210}$$

and the tangent of the angle that this resultant makes with the X axis is F_y/F_x. If the velocity of the water in a given direction is decreased, the force on the vane is exerted in this direction; if the velocity in a given direction is increased, the force on the vane is exerted in the opposite direction. Thus in Fig. 93a, F_x acts to the right and F_y acts downward.

The component of the velocity in its original direction after leaving the vane is v_x, and the change in velocity in this direction is $v - v_x$. Since $v_x = v \cos \theta$, the force exerted upon the vane in its original direction, from Eq. (209), is

$$F_x = \frac{Qw}{g}(v - v \cos \theta) = \frac{Qwv}{g}(1 - \cos \theta) \tag{211}$$

Similarly, since the change in velocity in a vertical direction is from zero to $v \sin \theta$,

$$F_y = \frac{Qwv}{g} \sin \theta \tag{212}$$

and, from Eq. (210),

$$F = \frac{Qwv}{g} \sqrt{2(1 - \cos \theta)} \tag{213}$$

If the jet strikes normally against a stationary flat plate (Fig. 94), $\cos \theta = 0$, and, from (211),

$$F_x = F = \frac{Qwv}{g} \tag{214}$$

Since the jet is turned by the plate through 90° equally in all directions, there is no resulting Y component. The plate simply destroys all the momentum in the original direction of the velocity. If a jet is deflected through 180° by either a single- or a double-cusped stationary vane, $\cos \theta = -1$, and

$$F_x = F = 2\frac{Qwv}{g} \tag{215}$$

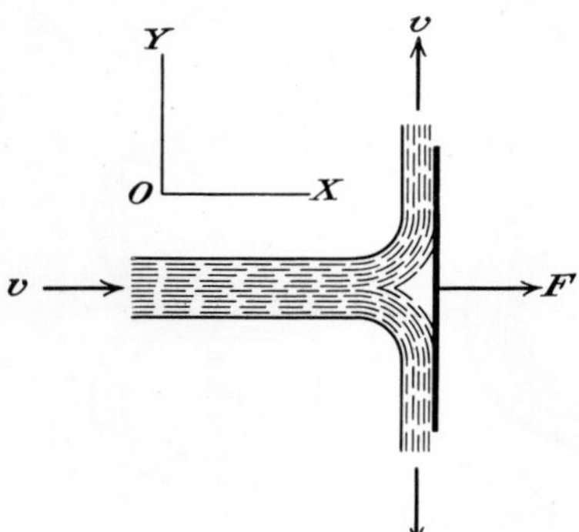

FIG. 94. Jet impinging against flat plate.

Since there is no change in velocity in a vertical direction, there is no F_y component.

As shown in Fig. 93b, the jet having an absolute velocity v impinges against a vane moving with a velocity v'. The relative velocity of the jet with respect to the vane is $u = v - v'$. The mass impinging per second, which in the above formulas is Qw/g, is reduced through movement of the vane, in the ratio of relative to absolute velocity, or as $u:v$. The velocity v in these formulas becomes the relative velocity u. For a moving vane, therefore, from (211),

$$F_x = \frac{Qwu^2}{vg}(1 - \cos \theta) \tag{216}$$

Corresponding changes in Eqs. (212) to (215) make them applicable to moving vanes. The absolute velocity and direction of the jet as it leaves the vane are shown by the vector v_a, which is the resultant of the relative velocity u and the velocity v' of the vane.

If a series of vanes are so arranged on the periphery of a wheel (as, for example, a Pelton water wheel) that the entire jet directed tangentially to the circumference strikes

successive vanes, the mass again becomes Qw/g, and Eq. (216) becomes

$$F_x = \frac{Qwu}{g}(1 - \cos\theta) \tag{217}$$

Only the tangential component F_x tends to produce rotation.

Work Done on Moving Vanes. Work is equal to force times distance. Since any one of the moving vanes b, c, d, or e (Fig. 93) advances the distance v' in 1 sec, work done per second in the direction of the jet is $G = F_x v'$. Hence, from Eq. (216), substituting $v - v'$ for u

$$G = \frac{Qw(v - v')^2}{vg}(1 - \cos\theta)v' \tag{218}$$

This equation makes $G = 0$ when $v' = v$ and also when $v' = 0$. The value of v' which makes G a maximum is obtained by equating the first derivative of G with respect to v' to zero and solving for v'. This solution gives as the relation of velocity of vane to velocity of jet for maximum work $v' = v/3$.

Similarly for a jet impinging tangentially against successive buckets on the periphery of a wheel, from Eq. (217),

$$G = \frac{Qw(v - v')}{g}(1 - \cos\theta)v' \tag{219}$$

As with (218), this equation makes $G = 0$ when $v' = v$ and also when $v' = 0$. Equating the first derivative to zero, we find that G is a maximum when $v' = v/2$. Equation (219) is of fundamental importance in determining the best operating conditions for water wheels. If the speed of maximum work $v/2$ is substituted for v', with $\theta = 180°$ ($\cos\theta = -1$), Eq. (219) reduces to

$$G = \frac{Qwv^2}{2g} = \frac{Mv^2}{2} \tag{220}$$

the total kinetic energy of the jet being converted into work. This also appears from considering that the relative velocity of the jet as it leaves the vane $v/2$ is also the velocity of the vane and that these velocities, being equal and opposite in direction, have a resultant of zero. The water thus leaves the vane with zero velocity, signifying that all its original energy has been utilized in performing work. The angle θ of water-wheel vanes must be somewhat less than 180°, so that the discharging water will not interfere with succeeding vanes. If $\theta = 172°$, the force exerted on the wheel is only 0.5 percent less than for $\theta = 180°$.

Water Hammer. If a valve at the outlet of a pipe is closed suddenly, the velocity is arrested and the kinetic energy is transformed into dynamic pressure. This dynamic pressure is called *water hammer*. On the assumption that both the pipe and the water are inelastic, if dv/dt represents the rate of change of velocity of the mass M of water in the pipe, the pressure due to water hammer is

$$P = M\frac{dv}{dt} \tag{221}$$

Elasticity modifies the results given by this equation. It causes a series of pressure waves and reversals of flow to traverse the pipe, there being four periods of different types of wave in a complete cycle.

The four periods of the wave cycle are as follows: (1) When the outlet valve is closed, a wave of increase pressure is transmitted back through the pipe with constant velocity and

intensity. Behind the wave, the water is compressed, and the pipe walls are stretched. The instant the wave reaches the intake, the entire pipe is expanded and all the water is compressed, but there is no longer any moving mass of water to maintain this high pressure. (2) Beginning at the intake, the water now expands and the pipe contracts, and a wave of normal pressure travels down the pipe. Flow toward the intake accompanies the decrease in pressure. The momentum of this moving mass causes an amount of water to flow back greater than the excess stored at the end of the first period. When, therefore, the wave of normal pressure reaches the outlet valve, pressure at the valve drops not only to normal static pressure but also below it. (3) A wave of pressure less than normal static pressure now passes up the pipe. When it reaches the intake, the entire pipe is under less than static pressure; but since the water is again all at rest, pressure at the intake immediately returns to normal static pressure. (4) During the final period of the cycle, a second wave of normal static pressure passes down the pipe. Since pressure is being increased, flow toward the outlet is being established. The instant that the wave reaches the valve, pressure is normal, and conditions throughout the pipe resemble those when the valve was first closed.

Instantly, after the first cycle is completed, a second cycle begins; but, because of friction and imperfect elasticity of the pipe and the water, the velocity of the water and the resulting water hammer are reduced. Successive cycles, similar to the first, follow each other until the waves finally die out.

Equation (221) shows that, on the assumption that the pipe and water are inelastic, dynamic pressure is directly proportional to the rate at which velocity is arrested, being infinite for instantaneous closure or for any finite reduction in area of valve opening in zero time. Instantaneous movement of any valve or gate is obviously impossible. The mass M [Eq. (221)] which has its velocity changed in the time dt is the mass of the length of water column traversed by the pressure wave in the time dt. Intensity of water hammer is not affected by amount of closure; e.g., the same dynamic pressure is produced by reducing velocity from 7 to 4 ft per sec as by entirely stopping a velocity of 3 ft per sec. With respect to elasticity of the pipe and the water, if the valve is closed before the first pressure wave returns to the valve or if the time of closing is less than $2L/v_w$, L being the length of pipe and v_w the velocity of the wave, the pressure will continue to increase up to the time of complete closure, and the resulting pressure will be the same as if the valve had been closed instantaneously. If the time of closure is greater than $2L/v_w$, the earlier waves of low pressure return and tend to reduce the rise in pressure resulting from the later stages of valve closure.

The kinetic energy contained in the moving mass of water in the pipe is used in the work of compressing the water and expanding the pipe. If the valve is closed before the first pressure wave returns to the valve, the entire energy of the mass of water is utilized during this time interval, and maximum pressure is produced. By expressing algebraically the relation: Total kinetic energy of water in pipe before closure = energy used in work of compressing the water + energy used in work of expanding the pipe, and making transformations and reductions, the following formulas can be obtained:

$$v_w = \frac{4{,}660}{\sqrt{1 + \dfrac{E'd}{Eb}}} \tag{222}$$

$$h = \frac{v_w v}{g} \tag{223}$$

In the foregoing formulas, v_w is the velocity of pressure wave, h the maximum possible head due to water hammer (excess over static head), v the velocity of water in pipe before

closing valve, d the inside diameter of pipe, b the thickness of pipe walls, E the modulus of elasticity of pipe walls in pounds per square inch, E' the modulus of elasticity of water in pounds per square inch, and g the acceleration due to gravity. Since sound is transmitted by means of pressure waves, the velocity v_w in any pipe is the velocity of sound through water in that pipe.[1] Joukovsky found from experiments that the maximum pressure due to water hammer in pounds per square inch obtainable in a straight pipe without branches in time less than $2L/v_w$ is about fifty-seven times the velocity in the pipe.

Several formulas and methods for determining water hammer where the time T of closing the valve is greater than $2L/v_w$ have been derived. The formulas are largely empirical and take a variety of forms. A method proposed by Gibson consists of tracing the action of the pressure wave instant by instant and of computing the pressure head at the end of successive time intervals. It is claimed that this method has received quite satisfactory experimental verification.

In order to guard against the destructive effects of water hammer, pipes should, wherever practicable, be equipped with slow-closing gates. The time of closure should decrease proportionally with the length of pipe. Air chambers, surge tanks, and other devices near the outlet of pipes are in special cases employed to counteract partly the effects of water hammer.

HYDRAULIC MODELS

Many problems encountered by the hydraulic engineer are too complex to be solved analytically. Hydraulic-model tests afford a means of gaining at least qualitative and sometimes quantitative answers to such problems. Hydraulic models must be designed and the test results interpreted in accordance with the principles of similitude.

The *principles of similitude* require that there be geometric, kinematic, and dynamic similiarity between model and prototype. *Geometric similarity* is provided by constructing the model so that all linear dimensions of the model have some predetermined ratio L_r to the corresponding dimensions of the prototype. *Kinematic similarity* requires that corresponding particles of fluid traverse homologous paths in the proper time ratios in model and prototype. Kinematic similarity can be attained if the model is geometrically similar to the prototype and if the ratio of resisting to impelling forces is the same at all points in the fluid. The latter condition is called *dynamic similarity*. The impelling force is the inertial force resulting from the velocity and mass of the fluid. The resisting forces which tend to retard or change the direction of the fluid velocity may be viscous forces, gravity forces, surface-tension forces, or elastic forces. Only the first two are ordinarily of importance in hydraulic-model studies. The Reynolds number

$$R = \frac{Lv\rho}{\mu} \tag{29}$$

which is the ratio between the inertial and viscous forces has already been discussed (p. 4–20). A similar ratio between the inertial forces and the gravity forces [Eq. (224)] is called the Froude number F.

$$F = \frac{v}{\sqrt{gL}} \tag{224}$$

Just as the length term in the Reynolds number was varied to fit the flow conditions

[1] Equations (222) and (223) were first derived and experimentally verified by Prof. N. Joukovsky in 1898. Miss O. Simin published an account of the researches of Professor Joukovsky in the *Proc. Am. Waterworks Assoc.*, 1904. Other valuable contributions to the literature on water hammer are: GIBSON, Pressure in Penstocks Caused by the Gradual Closing of Turbine Gates, *Trans. ASCE*, vol. 83, 1919; also, WARREN, Penstock and Surge-tank Problems, *Trans. ASCE*, vol. 79, 1915. The solution of a problem by the Gibson method is given in KING, "Handbook of Hydraulics," 4th ed., 1954; also in KING and WISLER, "Hydraulics," 5th ed., 1948. MCNOWN, "Engineering Hydraulics," Chap. VII, 1950.

being described, so L in the Froude number is taken as the most significant dimension in any particular case, as, for example, the depth in open-channel flow.

It follows from the above discussion that to obtain dynamic similarity it is necessary that flow in model and prototype occur with equal values of R if viscous forces control or equal values of F if gravity forces control. The most important example of flow controlled entirely by viscous forces is that in pipes. In fact, the principles of similitude may be illustrated by referring to material already covered under flow in pipes (p. 4–60). The importance of geometric similarity was illustrated by the tests of Nikuradse which showed that two pipes of different sizes but having the same ratio of roughness magnitude to diameter produced identical sets of values of f which varied only with R. Other tests have shown that the velocity distribution pattern is also a function of R, thus showing that kinematic similarity is also attained.

Cases where the gravity force is important and the viscous forces are negligible occur where there is fully developed turbulent flow with a free surface, as, for example, flow over weirs and through orifices and flow in channel constrictions and transitions, in the hydraulic jump, and in wave motion. The importance of the Froude number in open-channel flow has been illustrated in the general differential equation for gradually varied flow [Eq. (185)], where it will be seen that the denominator is $(1 - F^2)$. Furthermore it should be noted that each depth in the graphs of Figs. 78 and 80 corresponds to a particular value of F and that the corresponding changes in depth resulting from constrictions depend upon the magnitude of the original value of F.

The conversion ratios required for designing models and for converting model results to the prototype may be obtained by equating values of R or F for model and prototype. If the Froude law applies, the procedure is as follows: Let $F_m = F_p$, then

$$\frac{V_m}{\sqrt{g_m L_m}} = \frac{V_p}{\sqrt{g_p L_p}} \tag{225}$$

and

$$V_r = \frac{V_m}{V_p} = \frac{\sqrt{g_m L_m}}{\sqrt{g_p L_p}} = g_r^{1/2} L_r^{1/2} \tag{226}$$

The value of g_r is so nearly 1 (p. 4–3) that it can be omitted, with the result that

$$V_r = L_r^{1/2} \tag{227}$$

The discharge ratio can now be obtained as follows:

$$Q_r = \frac{Q_m}{Q_p} = \frac{a_m V_m}{a_p V_p} = \frac{L_m{}^2 V_m}{L_p{}^2 V_p} = L_r{}^2 V$$

and inserting the value of V_r from Eq. (227),

$$Q_r = L_r^{5/2} \tag{228}$$

The ratios of all other quantities may be obtained in a similar manner. All such ratios for the Reynolds as well as the Froude law are tabulated in various publications.[1]

In many cases of flow, both the Froude and Reynolds laws apply. Even in the case of orifice flow, it was shown (Figs. 35 and 42) that the coefficient of discharge varies with the smaller values of R. The most difficult case to handle is open-channel flow, which in its longer reaches depends on R, as shown in Fig. 68, whereas local phenomena such as the hydraulic jump and flow at constrictions depend on F. The question arises as to whether it is possible to satisfy both laws at the same time. To do this it would be

[1] FREEMAN, J. R., "Hydraulic Laboratory Practice," ASME, 1929. "Hydraulic Models," Manual 25, ASCE, 1942.

necessary for the conversion factors for any quantity to be the same whether determined from the Froude law or from the Reynolds law. Consequently, the conditions that must be satisfied can be obtained by setting any two corresponding ratios equal to each other with the result shown in Eq. (229). A study of Eq. (229)

$$L_r = \left(\frac{\mu_r}{\rho_r}\right)^{2/3} \tag{229}$$

will show that, even if warm mercury were used as the model fluid, the scale ratio would be approximately 1:4, which would result in a model that would be far too large to be feasible. It is therefore desirable to build models in which one or the other law is predominant. Scale ratios based on the Manning formula have been found to give satisfactory results in many river models.[1]

In order to operate at higher Reynolds numbers and at the same time avoid the effects of surface tension, which would involve still a third model law, models of rivers are often built with a distorted scale, the vertical scale being larger than the horizontal scale. At the same time, in order to reproduce the prototype water-surface profile, the slope and the roughness may be distorted. In such cases the discharge ratio may be determined by test rather than from a derived law.

Movable-bed models have been used with considerable success to predict the effect of various structures on shoaling or scouring in rivers and harbors. Although the results are qualitative only, they are usually sufficiently conclusive to point out the type of structure which most effectively meets the requirements. Movable-bed models are usually "verified" by beginning with a relatively flat bed and the existing flow conditions and determining whether normal operation will duplicate the present bed configuration. If so, it is believed that a change in the flow conditions brought about by the proposed construction will be duplicated qualitatively in the prototype.

The design of models is governed by the cost, the space available, the capacity of the laboratory water supply, and the accuracy desired. Generally speaking, the larger the model, the more accurate will be the results and the greater will be the cost. For models requiring flowing water, the model discharge is limited by the available pumping capacity and hence the largest possible scale ratio is determined by Eq. (228) if the Froude law applies. Undistorted models are commonly built at scale ratios varying from 1:25 to 1:150. Distorted models of rivers and harbors usually have vertical scale ratios varying from 1:25 to 1:150 while the horizontal scales may vary from 1:50 to 1:2,000.

Bibliography

BAKHMETEFF, B. A.: "Hydraulics of Open Channels," McGraw-Hill Book Company, Inc., 1932.

HINDS, CREAGER, and JUSTIN: "Engineering for Dams," vol. II, John Wiley & Sons, Inc., 1944.

KING, H. W.: "Handbook of Hydraulics," 4th ed., McGraw-Hill Book Company, Inc., 1954.

PRANDTL and TIETJENS: "Fundamentals of Hydro- and Aeromechanics," and "Applied Hydro- and Aeromechanics," McGraw-Hill Book Company, Inc., 1934.

ROUSE and HOWE: "Basic Mechanics of Fluids," John Wiley & Sons, Inc., 1953.

STREETER, V. L.: "Fluid Mechanics," McGraw-Hill Book Company, Inc., 1958.

[1] VOGEL, H. D., Practical River Laboratory Hydraulics , *Trans. ASCE* ,1935

Section 5

STRESSES IN FRAMED STRUCTURES

BY FREDERICK S. MERRITT

COMMON TYPES OF FRAMING

Beam and Girder Framing. Bridge decks and floors and roofs of buildings frequently are supported on a rectangular grid of flexural members. Different names often are given to the components of the grid, depending on the type of structure and the part of the structure supported on the grid. In general, though, the members spanning between main supports are called **girders** and those they support are called **beams** (Fig. 1). Hence, this type of framing is known as beam and girder.

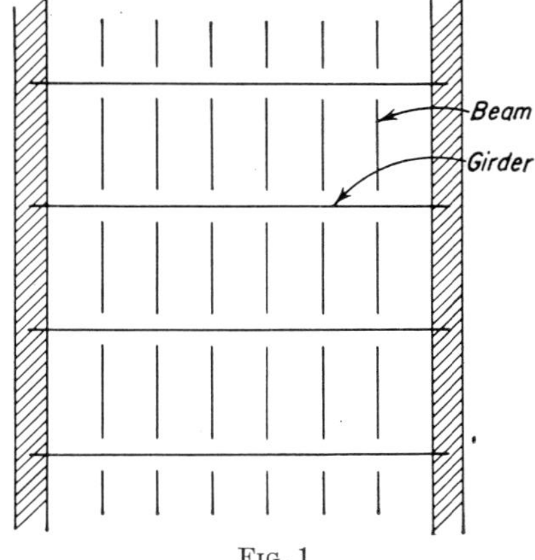

FIG. 1

In bridges, the smaller structural members parallel to the direction in which traffic moves may be called **stringers** and the transverse members **floor beams.** In building roofs, the grid components may be referred to as **purlins** and **rafters.** And in floors, they may be called **joists** and **girders.**

Beam and girder framing usually is used for relatively short spans and where shallow members are desired to provide ample headroom underneath.

Trusses. Where headroom is not critical and where long spans are desired, trusses may be used. A **truss** is a coplanar system of structural members joined together at their ends to form a stable framework. Neglecting small changes in the lengths of the members due to loads, the relative positions of the joints cannot change.

Three bars pinned together to form a triangle represent the simplest type of truss. Some of the more common types of roof and bridge trusses are illustrated in Fig. 2.

In these trusses, the top members are called the **upper chord;** the bottom members the **lower chord;** and the verticals and diagonals **web members.**

Trusses act like long, deep girders with cutout webs. Roof trusses not only have to carry their own weight and the weight of roof framing and roofing but also wind loads, snow loads, suspended ceilings and equipment, and a live load to take care of construction, maintenance, and repair loading. Bridge trusses have to support their own weight and that of deck framing and deck, live loads imposed by traffic (automobiles, trucks,

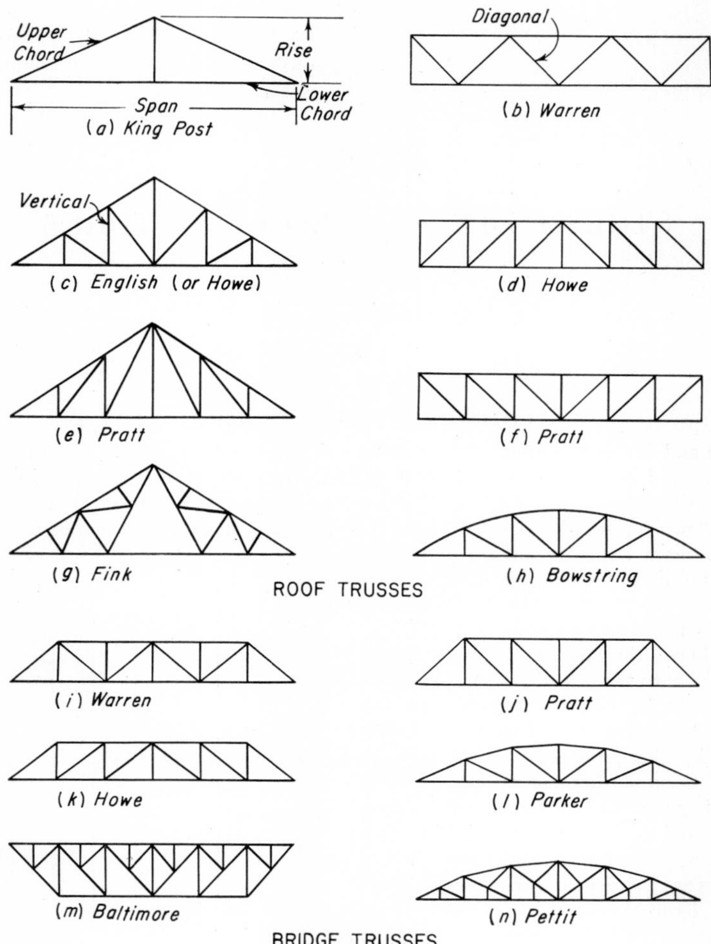

Fig. 2

railroad trains, pedestrians, etc.), and impact caused by the live load. **Deck trusses** carry the live load on the top chord, and **through trusses** on the bottom chord.

Loads generally are applied at the intersection of the members, or panel points, so that the members will be subjected principally to direct stresses. To simplify stress analysis, the weight of the truss members is apportioned to upper- and lower-chord panel points. The members are assumed to be pinned at their ends, even though this may actually not be the case. If, however, the joints are of such nature as to restrict relative rotation substantially, then the "secondary" stresses set up as a result should be computed and superimposed on the stresses obtained with the assumption of pin ends.[1]

[1] Timoshenko, S., and D. H. Young, "Theory of Structures," McGraw-Hill Book Company, Inc., New York, 1945.

Framing with roof trusses is similar to beam and girder framing, with the trusses substituted for the girders. Since truss spans usually are long and roof decks are made of light material, the roof must be braced against both lateral and longitudinal forces with additional horizontal and vertical trusses. Frequently, in addition, trusses are stiffened in their own vertical planes by inserting knee braces at both ends between the bottom chord and supporting columns, and purlins carrying the roof deck are fastened to the top chords, which are in compression, thus providing lateral bracing to them.

Trussed roof bracing may be placed in the plane of the top or bottom chords. Putting it in the plane of the top chords offers the advantages of simpler details, shorter unsupported length of diagonals, and less sagging of bracing, because it can be connected to the purlins at all intersections. It is seldom necessary to brace both top and bottom chords with separate trussed systems. However, the bottom chord should be braced at frequent intervals, even though it is a tension member, to reduce its unsupported length.

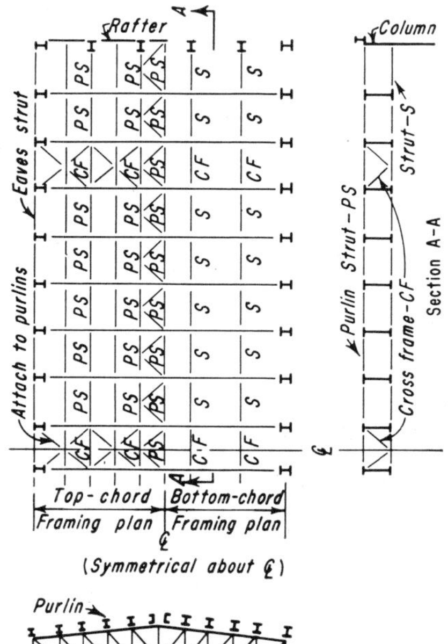

FIG. 3

Typical bracing for a mill-building roof is illustrated in Fig. 3. Diagonal bracing is placed in the plane of the top chord in three bays, assuming that the purlins will be sufficiently well connected to the trusses to transmit longitudinal forces from the unbraced trusses to the braced bay. Not more than five unbraced bays should be permitted between braced trusses.

Struts are shown between lower chords at every other panel point. At corresponding top-chord panel points, the purlins should be designed to be able to carry compressive forces in addition to vertical loading. These struts between the upper and lower chords should transmit longitudinal forces to the laterally braced bays, where cross frames are placed between the trusses in the plane of the struts, as indicated in Fig. 3, to prevent the trusses from tipping over.[1]

In bridges, two parallel trusses generally are used (Fig. 4). They must be adequately braced because a bridge structure is subjected to lateral forces due to wind pressure and vibration caused by the impact of moving loads. These stresses are resisted by lateral trusses placed between the chords of the main vertical trusses, and portal and sway bracings placed between each pair of end posts and verticals, respectively. When the chords of the main trusses are horizontal, the lateral trusses are horizontal trusses the chords of which are the chords of the main trusses. The sway bracing lies in a vertical plane and the portal bracing in the plane of the end posts (Fig. 4). The type of truss used for the lateral trusses depends on the size and type of the main structure. In short spans where the diagonal length is not great, a bracing of the Warren type is often used.

[1] DUNHAM, C. W., "Planning Industrial Structures," McGraw-Hill Book Company, Inc., New York, 1948.

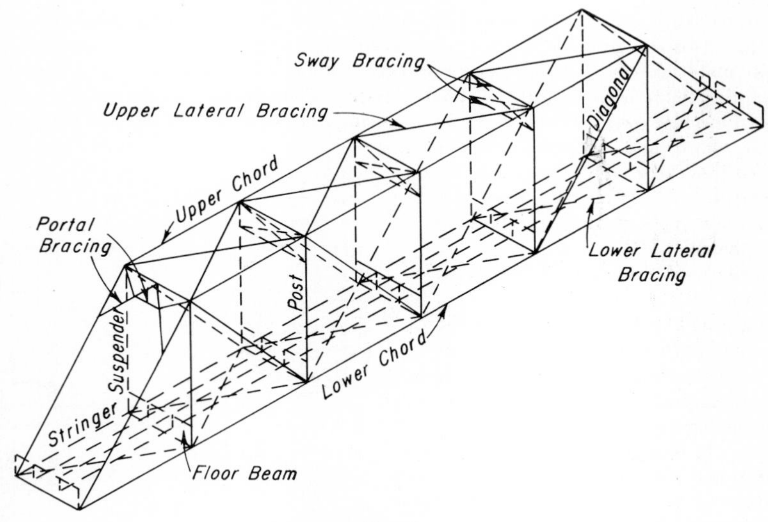

Fig. 4

This same type, too, is usually found in deck-plate girder bridges, where often only one set of laterals (the upper) is used, and the lower flange is stiffened by the transverse bracing or cross frames. The usual type of lateral truss for through bridges is illustrated in Fig. 5. The upper lateral system is shown in *a*, and the lower lateral system in *c*. The lateral forces acting upon the upper lateral truss are carried by that truss to the points *BB′* and *FF′* and taken thence through the portal bracing and the end posts shown in *d*

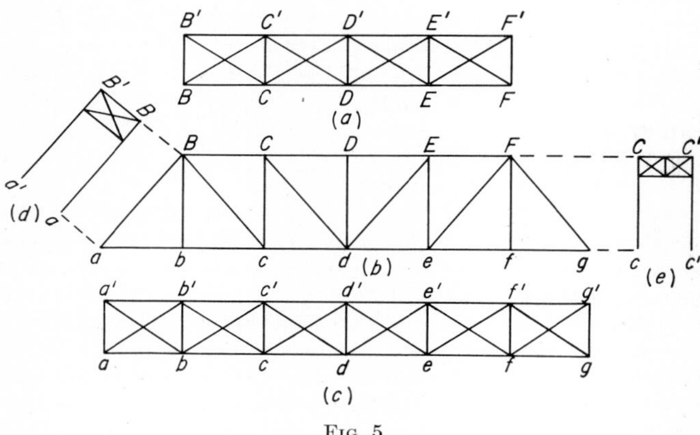

Fig. 5

to the abutments of the bridge at *aa′* and *gg′*. In a through truss, this compels a portion of the end posts to act as beams, inasmuch as the portal bracing cannot extend to their lower extremities. Loads on the lower lateral systems are carried directly to the abutments at *aa′* and *gg′*.

The diagonals are usually made rigid, i.e., capable of resisting both compressive and tensile stresses; and the lateral shear in each panel is assumed to be distributed equally

to the two diagonals. If the unsupported diagonal length is long, it is often better to consider the diagonals as resisting tensile stresses only.

In the upper lateral system of a through bridge, the upper member of the sway bracing (CC' of e) acts as the lateral strut; in the lower lateral system, the floor beams perform this function.

In the smaller and lighter trusses, the sway bracing is often omitted, the lateral strut being the only transverse member between the intermediate panel points.

Space Trusses. When the basic truss triangles are combined in such a manner that they do not all lie in the same plane, the assemblage is called a space truss. Tall towers, both guyed and free-standing, are frequently designed as space trusses, and sometimes floor and roof trusses of that type are used. Several dome-shaped structures also have been constructed as space trusses.

In towers, the legs form the chords and the bracing between them, the web members. In floor and roof systems, three-chord trusses have been used to advantage. Such space structures are economical because the web members serve as both load-carrying members and bracing and the arrangement of the chords gives the structure stiffness in all directions.

Rigid Frames. With beam and girder framing, trussed bracing is not necessary if equilibrium is maintained by joint restraint at the connections of girders to columns. When girders are so connected to columns that both members must rotate through the same angle, the structure is called a rigid frame.

In tall buildings of skeleton-frame construction, wind, earthquake, and other lateral forces generally are resisted by rigid-frame action. Though the beams and girders may

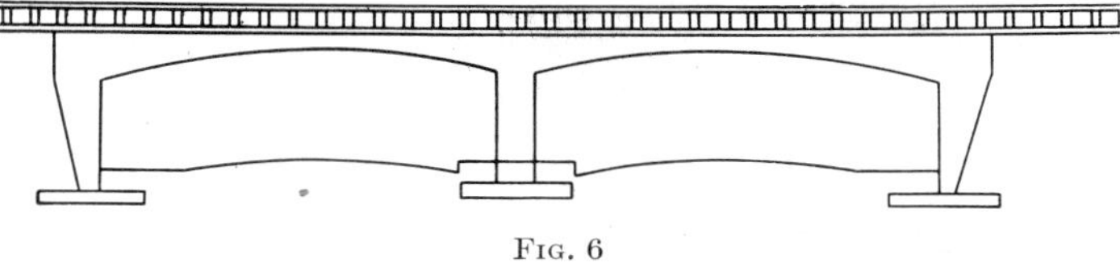

FIG. 6

be designed as simply supported for dead and live loads, "wind" connections, capable of transmitting bending moments, are used at the columns and the members connected are designed for the lateral-force bending moments, as well as for dead- and live-load moments.

Rigid frames are often used for highway bridges because of their attractive appearance and because they can span 50 ft or more with less depth than required for simply supported girders or trusses. Usually, the bottoms of rigid-frame girders are given a pleasing curve, providing ample haunches to resist the bending moments at the columns, and the verticals are tapered from a wide section at the top to a minimum at the base (Fig. 6).

Single-span rigid frames behave structurally very much like arches. Precautions must be taken at the base for the horizontal components of the reactions as well as the vertical, as is done with arches, and all the members are subjected to both bending and axial stresses.

Continuous rigid frames frequently are used to support the roofs of one-story industrial buildings. The girders may be straight, arched, or composed of straight segments for monitors or sawtooth construction.

Arches. An arch is a curved girder or truss. It develops inclined reactions even under vertical loads. The supports must be capable of resisting the horizontal components of the reactions as well as the vertical. If the foundations cannot take the thrust, tie rods must be inserted between opposite sides of the arch.

Unlike straight girders and trusses, arches can be constructed with three hinges and still be stable (Figs. 7 and 8). Fixed-end or hingeless arches can be used only where the foundations are capable of providing suitable support—usually only on good rock. When foundation conditions are less satisfactory, two-hinge or three-hinge designs are preferred.

Roof arches are usually of the general form shown in Fig. 7. To avoid the massive abutments that would normally be required to resist the horizontal thrust, tie rods are placed under the floor between the two base hinges. The dotted members at the crown in Fig. 7 are inserted to provide a smooth roof surface; they permit necessary movement of the arch and do not transmit stress.

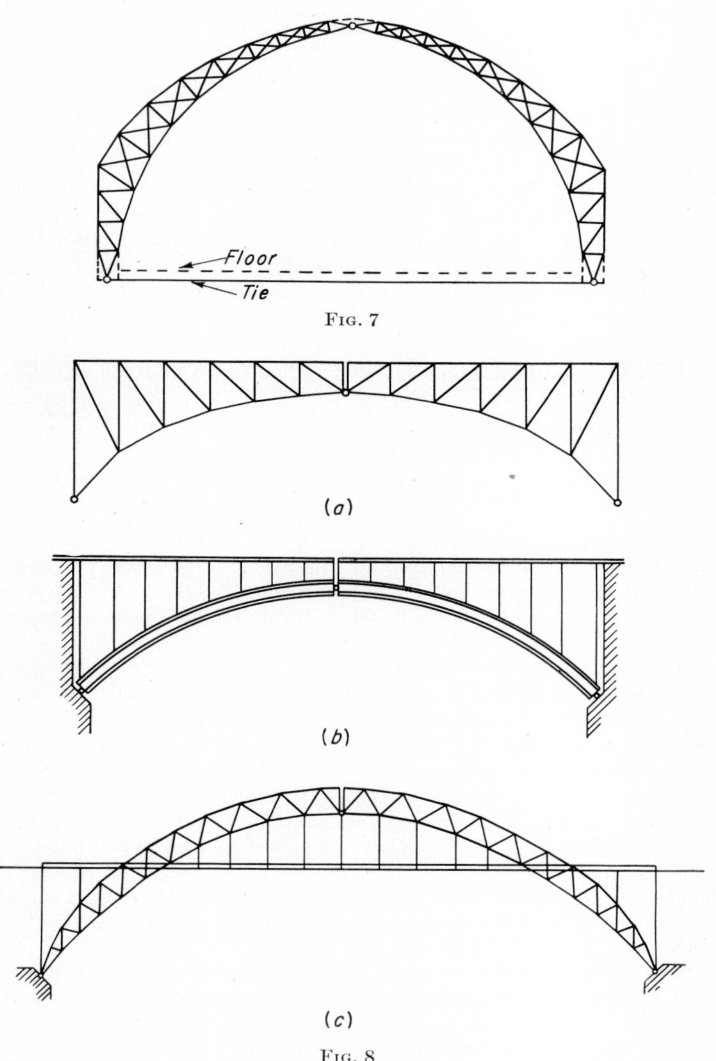

Fig. 7

Fig. 8

Three common forms of bridge arches are shown in Fig. 8. In (*a*), the bracing extends from the curved lower chord to the straight upper chord, which supports the bridge deck. Arches of this type are called spandrel-braced arches. The two verticals at the center hinge permit rotation there, a sliding plate in the deck allowing for movements of the upper chord. In (*b*) and (*c*) the deck is supported on vertical posts or hangers, which are carried by the arch.

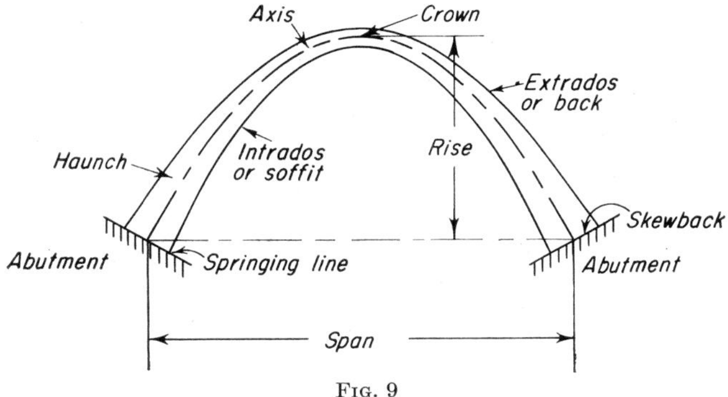

FIG. 9

In Fig. 9 are indicated the names usually given to the various parts of an arch.

Thin-shell Structures. Shell action is distinguished from arch action by the fact that load is resisted not only by the thrust, moment, and shear on a radial section, but also by tangential shears on a transverse section. In thin shells, bending stresses may be negligible over a relatively large portion; i.e., the shell may be considered as a membrane, subject only to direct stresses.

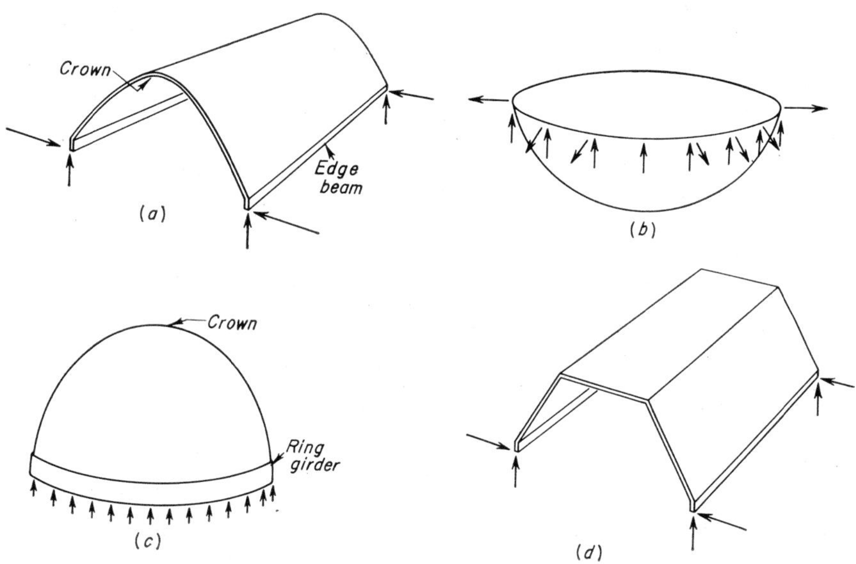

FIG. 10

Shells may be singly curved, such as cylindrical or barrel arches, or doubly curved, such as domes, catenaries, or hyperbolic paraboloids, or composed of interconnected planes, or folded plates. Doubly curved shells are much more efficient in supporting loads than arches or singly curved shells.

A cylindrical arch is illustrated in Fig. 10*a*. Its edges are stiffened by beams. In addition, crossties are required to take the thrust, and frequently, the shell is stiffened transversely by incorporation of arched ribs at regular intervals. This type of shell acts very much like a beam with a curved cross section.

A catenary (Fig. 10*b*) is a membrane that is suspended from its supports. It is in tension throughout. The supports must be capable of resisting the inward pull, and provision must be made for its stability under unbalanced vertical loads and wind, as well as to prevent flutter.

A dome, in effect, is an inverted catenary and is in compression throughout. As shown in Fig. 10*c*, a ring girder usually is placed along the bottom edge to resist the thrust; it is in tension.

A folded-plate roof is shown in Fig. 10*d*. Its structural behavior is very much like that of a cylindrical arch when it consists of several plates, and the same measures for stiffening and resisting thrust must be taken. When it consists of only two plates, its components may be considered as two deep, tilted girders.

LOADS ON STRUCTURES

Types of Loads. External loads on structures may be classified in several different ways. In one classification, they may be considered as static, repeated, and impact loads.

Static loads are forces that are applied slowly and then remain nearly constant. One example is the weight, or dead load, of a floor or roof system.

Repeated loads are forces that are applied a number of times, causing a variation in the magnitude, and sometimes also in the sense, of the internal forces in a structure. A good example is the load on a highway bridge from automobile and truck traffic or on a railway bridge from trains.

Impact loads are forces that require a structure or its components to absorb energy in a short interval of time. An example is a truck bouncing over rough pavement on a bridge.

External forces may also be classified as distributed and concentrated.

Uniformly distributed loads are forces that are, or for practical purposes may be considered, constant over the length or surface of the supporting member. Dead weight of a rolled wide-flange steel beam is a good example.

Concentrated loads are forces that have such a small contact area as to be negligible compared with the surface area of the supporting member. A beam supported on a girder, for example, may be considered, for all practical purposes, a concentrated load on the girder.

Another common classification for external forces labels them axial, eccentric, and torsional.

An **axial load** is a force whose resultant passes through the centroid of a section under consideration and is perpendicular to the plane of the section.

An **eccentric load** is a force whose resultant is perpendicular to the plane of the section under consideration but does not pass through the centroid of the section, thus bending the supporting member.

Torsional loads are forces that are offset from the shear center of the section under consideration and are inclined to or lie in the plane of the section, thus twisting the supporting member.

TABLE 1. MINIMUM DESIGN DEAD LOADS [1]

	Lb per sq ft
Walls:	
4-in. clay brick, high absorption	34
4-in. clay brick, medium absorption	39
4-in. clay brick, low absorption	46
4-in. sand-lime brick	38
4-in. concrete brick, heavy aggregate	46
4-in. concrete brick, light aggregate	33
4-in. brick, 4-in. load-bearing structural clay tile backing	60
8-in. load-bearing structural clay tile	42
12-in. load-bearing structural clay tile	58
8-in. concrete block, heavy aggregate	55
12-in. concrete block, heavy aggregate	85
8-in. concrete block, light aggregate	35
12-in. concrete block, light aggregate	55
2-in. furring tile, one side of masonry wall, add to above figures	12
Partitions:	
3-in. clay tile	17
4-in. clay tile	18
6-in. clay tile	28
8-in. clay tile	34
10-in. clay tile	40
2-in. facing tile	15
4-in. facing tile	25
6-in. facing tile	38
2-in. gypsum block	9½
3-in. gypsum block	10½
4-in. gypsum block	12½
5-in. gypsum block	14
6-in. gypsum block	18½
2-in. solid plaster	20
4-in. solid plaster	32
4-in. hollow plaster	22
4-in. concrete block, heavy aggregate	30
Wood studs 2 by 4, unplastered	4
Wood studs 2 by 4, plastered one side	12
Wood studs 2 by 4, plastered two sides	20

	Lb per sq ft
Glass-block masonry:	
4-in. glass-block walls and partitions	18
Split furring tile:	
1½-in.	8
2-in.	8½
Concrete slabs:	
Concrete, reinforced-stone, per inch of thickness	12½
Concrete, reinforced-cinder, per inch of thickness	9¼
Concrete, reinforced, lightweight, per inch of thickness	9

	Finish floor to top slab, in.	Lb per sq ft
Floor finish and fill:		
3-in. wood block on mastic, no fill	3	10
⅞-in. wood block on stone-concrete fill	4	40
1-in. cement finish on stone-concrete fill	4	40
1-in. terrazzo on stone-concrete fill	4	48
Clay tile on stone-concrete fill	4	48
Marble and mortar on stone-concrete fill	4	50
Linoleum on stone-concrete fill	4 5	46 58
Linoleum on light-concrete fill	4 5	27 34

	Thickness, in.	Lb per sq ft
Floor finish:		
1½-in. asphalt mastic flooring	1½	18
3-in. wood block on ½-in. mortar base	3½	16

[1] Weights of masonry include mortar but not plaster. For plaster, add 5 lb per sq ft for each face plastered. Values given represent averages. In some cases there is a considerable range of weight for the same construction. ASA A58.1-1955.

TABLE 1. MINIMUM DESIGN DEAD LOADS (*Continued*)

	Thickness, in.	Lb per sq ft
Floor finish:		
Solid flat tile on 1-in. mortar base	2	23
2-in. asphalt block, ½-in. mortar....	2½	30
1-in. terrazzo, 2-in. stone concrete...	3	38
Waterproofing:		
5-ply membrane...	½	5
5-ply membrane mortar, stone concrete........	5	55
2-in. split tile, 3-in. stone concrete...	5	45

	Lb per sq ft	
	12-in. spacing	16-in. spacing
Wood-joist floors (no plaster) double wood floor:		
Joist sizes		
2 by 6..........	6	5
2 by 8..........	6	6
2 by 10........	7	6
2 by 12........	8	7
3 by 6.........	7	6
3 by 8.........	8	7
3 by 10........	9	8
3 by 12.........	11	9
3 by 14........	12	10

	Lb per sq ft
Ceilings:	
Plaster on tile or concrete.....	5
Suspended metal lath and gypsum plaster...............	10
Suspended metal lath and cement plaster.............	15
Plaster on wooden lath.......	8
Roof and wall coverings:	
Asbestos-cement shingles......	4
Asphalt shingles.............	6
Cement tile.................	16
Clay tile (for mortar add 10 lb):	
2-in. book tile.............	12
3-in. book tile.............	20
Roman....................	12
Spanish...................	19
Ludowici..................	10
Composition:	
Three-ply ready roofing.....	1
Four-ply felt and gravel....	5½
Five-ply felt and gravel.....	6
Copper or tin................	1
Corrugated asbestos-cement roofing....................	4
Corrugated iron.............	2
Fiberboard, ½-in.............	¾
Gypsum sheathing, ½-in......	2
Skylight, metal frame, ⅜-in. wire glass................	8
Slate, 3/16-in.................	7
Slate, ¼-in..................	10
Wood sheathing, per inch thickness................	3
Wood shingles...............	3

Building Loads. Components of a building usually are required to support the following types of loads: dead load due to the weight of materials and equipment, including plumbing stacks and risers, electrical feeders and ventilating and air-conditioning systems; live loads; impact loads; snow, wind, and earthquake loads; and soil and hydrostatic pressures.

Under the sponsorship of the National Bureau of Standards, the American Standards Association has adopted certain minimum design loads as a standard (Minimum Design Loads in Buildings and Other Structures, A58.1-1955). Table 1 contains the recommendations for dead loads given in an appendix of that standard. Table 2 presents minimum live loads; Table 3 design wind pressures and Table 4 coefficients for computing earthquake forces. The coefficient C given in the last table is for use in the equation $F = CW$, where F is the horizontal lateral load, lb, and W the total dead load, lb, tributary to the point under consideration. For warehouses and tanks, W is the total dead load plus live load.

Standard A58.1 recommends that live loads be reduced for members supporting large floor areas when it is improbable that the whole area will be fully loaded with the total design load. For live loads of 100 lb per sq ft or less, the design live load on any member

Table 2. Minimum Uniformly Distributed Live Loads

Occupancy or use	Live load, lb per sq ft	Occupancy or use	Live load, lb per sq ft
Armories and drill rooms	150	Dwellings:	
Assembly halls and other places of assembly:		Second floor and habitable attics	30
Fixed seats	60	Uninhabitable attics	20
Movable seats	100	Hotels:	
Bowling alleys, poolrooms, and similar recreational areas	75	Guest rooms	40
Corridors:		Public rooms	100
First floor	100	Corridors serving public rooms	100
Other floors, same as occupancy served except as indicated		Public corridors	60
Dance halls	100	Private corridors	40
Dining rooms and restaurants	100	Schools:	
Garages (passenger cars)	100	Classrooms	40
Floors shall be designed to carry 150 percent of the maximum wheel load anywhere on the floor.		Corridors	100
Trucks, with load, 3 to 10 tons	150 *	Sidewalks, vehicular driveways, and yards, subject to trucking	250
Trucks, with load, above 10 tons	200 *	Stairs, fire escapes, and exitways	100
Gymnasiums, main floors and balconies	100	Storage warehouse, light	125
Hangars	150 *	Storage warehouse, heavy	250
Hospitals:		Storage, hay or grain	300 †
Operating rooms	60	Stores:	
Private rooms	40	Retail:	
Wards	40	First floor, rooms	100
Libraries:		Upper floors	75
Reading rooms	60	Wholesale	125
Stack rooms	150	Telephone exchange	150 ‡
Corridors	100 †	Theaters:	
Manufacturing	125	Aisles, corridors, and lobbies	100
Offices	80	Orchestra floors	60
Lobbies	100	Balconies	60
Penal institutions:		Stage floors	150
Cell blocks	40	Dressing rooms	40
Corridors	100	Grid-iron floor or fly gallery:	
Residential:		Grating	60
Multifamily houses:		Well beams, 250 lb per lin ft per pair	
Private apartments	40	Header beams, 1,000 lb per lin ft	
Public rooms	100	Pin rail, 250 lb per lin ft	
Corridors	60	Projection room	100
Dwellings:		Toilet rooms	60
First floor	40	Transformer rooms	200 ‡
		Vaults, in offices	250 †
		Yards and terraces, pedestrians	100

* Also subject to not less than 125 percent maximum axle load.
† Increase when occupancy exceeds this amount.
‡ Use weight of actual equipment when greater.

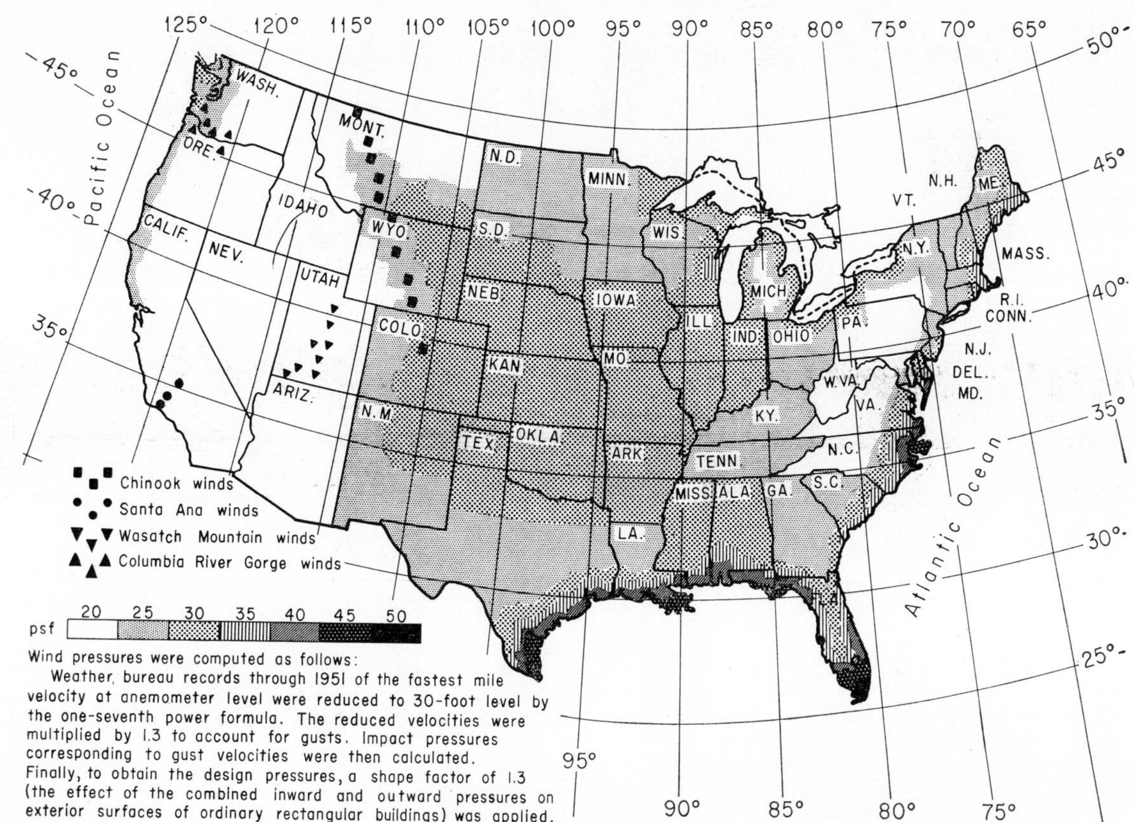

FIG. 11. Minimum allowable resultant wind pressures (ASA A58.1-1955).

TABLE 3. WIND PRESSURES FOR VARIOUS HEIGHT ZONES ABOVE GROUND [1]

Height zone, ft	Wind-pressure-map areas, lb per sq ft						
	20	25	30	35	40	45	50
Less than 30	15	20	25	25	30	35	40
30 to 49	20	25	30	35	40	45	50
50 to 99	25	30	40	45	50	55	60
100 to 499	30	40	45	55	60	70	75
500 to 1,199	35	45	55	60	70	80	90
1,200 and over	40	50	60	70	80	90	100

[1] Reference should be made to Fig. 11 and that wind-pressure column in the table should be selected which is headed by a value corresponding to the minimum permissible resultant wind pressure indicated for the particular locality in Fig. 11.

The figures given are recommended as minimum. These requirements do not provide for tornadoes. Chimneys, tanks, and solid towers shall be designed and constructed to withstand the above pressures multiplied by the following factors (ASA A58.1-1955):

Horizontal Cross Section	Factor
Square or rectangular	1.00
Hexagonal or octagonal	0.80
Round or elliptical	0.60

TABLE 4. HORIZONTAL FORCE FACTORS

Part or portion	Value of C *	Direction of force
Floors, roofs, columns, and bracing in any story of a building or the structure as a whole †	$\frac{0.15}{N + 4\frac{1}{2}}$ ‡	Any direction horizontally
Bearing walls, nonbearing walls, partitions, free-standing masonry walls, over 6 ft in height ¶	0.05	Normal to surface of wall
Cantilever parapet and other cantilever walls, except retaining walls	0.25	Normal to surface of wall
Exterior and interior ornamentations and appendages	0.25	Any direction horizontally
When connected to or a part of a building: towers, tanks, towers and tanks plus contents, chimneys, smokestacks, and penthouses	0.05	Any direction horizontally
Elevated water tanks and other tower-supported structures not supported by a building	0.025	Any direction horizontally

ASA A58.1-1955.

* The values given C are minimum and are intended for locations subject to infrequent seismic disturbances of relatively weak intensity. For areas subject to more severe hazard, the values should be doubled or multiplied by 4, according to the risk involved. Information on evaluation of this risk may be obtained from the Coast and Geodetic Survey, U.S. Department of Commerce, Washington 25, D.C.

† Where specified wind load would produce higher stresses, this load shall be used in lieu of the factor shown.

‡ N is the number of stories above the story under consideration, provided that, for floors or horizontal bracing, N shall be only the number of stories contributing loads.

¶ The design dead load of walls and partitions shall be taken as being not less than 100 lb per sq ft.

supporting 150 sq ft or more may be reduced at the rate of 0.08 percent per sq ft of area supported by the member. (No reductions are permitted for roofs or for areas to be occupied as places of public assembly.) However, the reduction cannot exceed either 60 percent or

$$R = 100 \times \frac{D + L}{4.33\,L}$$

where R is the maximum reduction, percent, D the dead load per square foot of area supported by the member and L the design live load per square foot of area supported by the member. For live loads exceeding 100 lb per sq ft, no reduction is permitted, except for columns, for which a 20 percent live-load reduction is allowed.

In addition to the uniformly distributed live loads in Table 2, A58.1 recommends the following concentrated loads:

Elevator-machine-room grating (on area of 4 sq in.), lb	300
Finish light floor plate construction (on area of 1 sq in.), lb	200
Office floors,[1] lb	2,000
Scuttles, skylight ribs, and accessible ceilings, lb	200
Sidewalks,[1] lb	8,000
Stair treads (on tread center),[1] lb	300
Roof truss lower chord (any panel point), lb	2,000

Also, primary structural members supporting roofs over garage, manufacturing, and storage floors should be capable of carrying safely a suspended concentrated load of at least 2,000 lb. Craneways should be designed to resist a horizontal transverse force equal to 25 percent of the crane capacity plus the weight of the trolley applied one-half at the top of each runway rail for impact. They should also be able to carry a horizontal longitudinal force equal to 12.5 percent of the total of the maximum wheel loads applied at the top of each rail.

All moving elevator loads should be increased 100 percent for impact, and the structural supports should be designed within the limits of deflection prescribed by the American Standard Safety Code for Elevators, Dumbwaiters and Escalators (A17.1) and American Standard Inspection of Elevators (A17.2). The weight of heavy machinery should be increased at least 25 percent for impact.

To provide for loads incidental to construction and repair, for sleet loads and minor snow loads, ordinary roofs should be designed for a live load of at least 20 lb per sq ft of horizontal projection. Where snow loads may exceed 20 lb per sq ft, the roof should be designed for the maximum anticipated. If the roof is pitched, snow loads over 20 lb per sq ft may be reduced for each degree of pitch over 20° by $S/40 - 0.5$, where S is the total snow load, lb per sq ft.

Roofs used for incidental promenade purpose should be designed for a minimum live load of 60 lb per sq ft. When used for roof gardens or assembly purposes, they should be designed for 100 lb per sq ft.

Exterior walls should be designed for the wind pressures indicated in Table 3. Roofs, however, should be able to withstand pressures acting outward normal to the surface equal to 1.25 times those specified in the table for their height zone (above the ground). The outward pressures should be assumed to act on the entire roof area. Roofs with slopes greater than 30° should be designed to withstand wind pressures, acting inward normal to the surface, equal to those specified in the table for their height zone and applied to the windward slope only. Overhanging eaves and cornices should be designed for outward pressures equal to twice those given in Table 3.

[1] Assume these loads to act on an area of 2.5 sq ft and located to produce maximum stress.

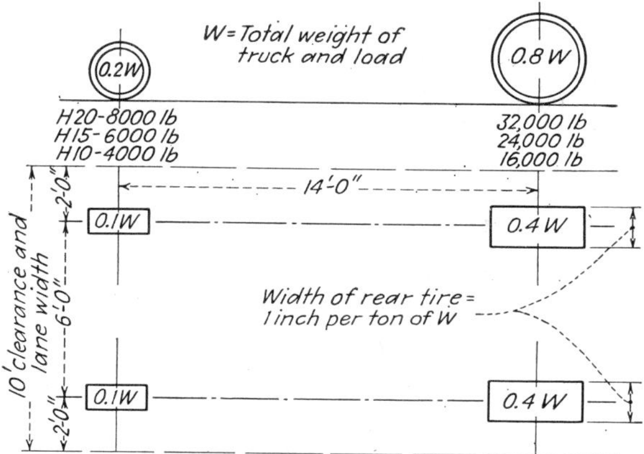

Fig. 12. H-truck loadings.

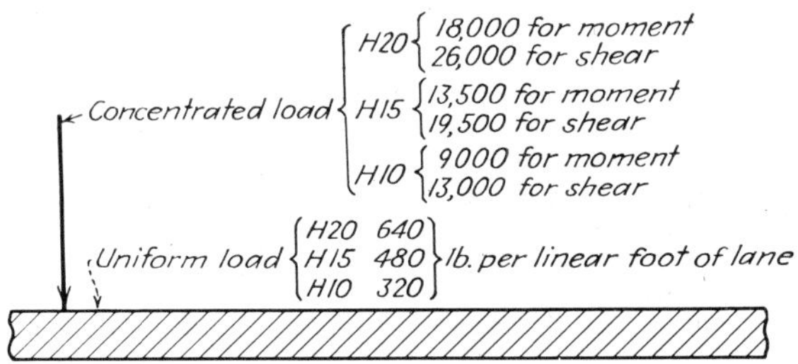

Fig. 13. H-lane loadings (lane width 10 ft).

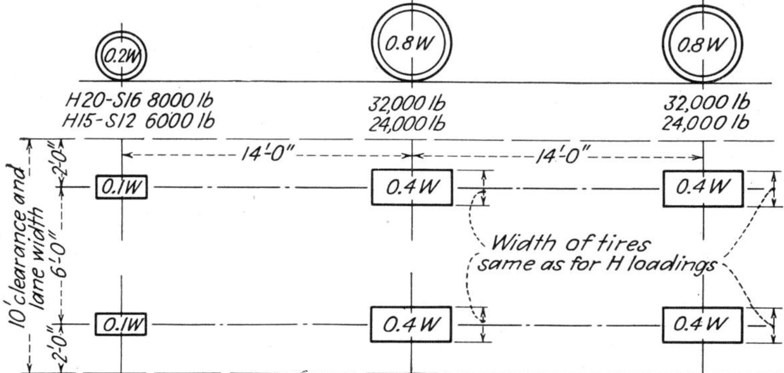

Fig. 14. H-S truck loadings.

Highway-bridge Loads. Highway bridges should be designed for dead loads, live load, impact, dimensional and temperature changes, wind, traction and braking forces, and on curves, for centrifugal forces. Dead load on the main members consists of the weight of the deck, framing supporting the deck and transmitting the load to the main members, bracing and weight of the main members themselves.

The most widely used specifications for highway-bridge design are those of the American Association of State Highway Officials. In these specifications five classes of loadings are given: H20, H15, H10, H20-S16, and H15-S12. The H loadings are illustrated in Figs. 12 and 13, and the H-S loadings in Figs. 14 and 15.

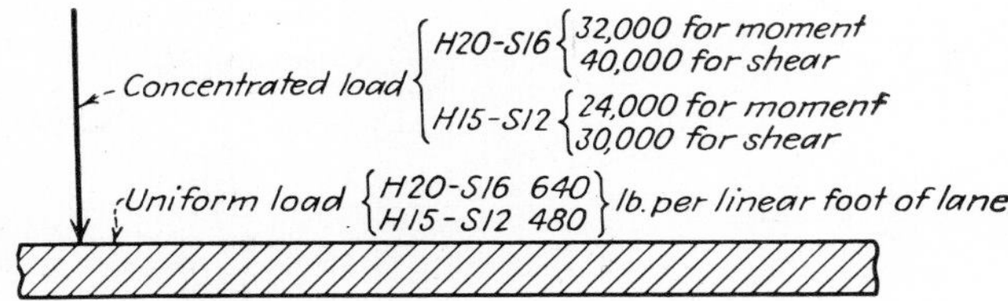

Fig. 15. H-S lane loadings (lane width 10 ft).

The H loadings consist of a four-wheel truck or the corresponding lane loading. The H-S loadings consist of a tractor truck with semitrailer or the corresponding lane loading. The H-S truck loading is used for loaded lengths of 40 ft or less and the H-S lane loading for greater lengths. Either the H truck loading or the H lane loading is used for all spans depending upon which gives the larger stress. For bridges carrying more than two lanes of traffic, in view of the improbable simultaneous maximum loading of all lanes, the following percentages of the resultant live-load stresses are used:

Three lanes................	90 percent
Four lanes or more..........	75 percent

In 1956, the Bureau of Public Roads recommended an alternate loading for spans under 40 ft "to overcome known deficiencies in floor systems of bridges" designed for H20-S16 loading. The recommended loading consists of two 12-ton axles 4 ft apart.

When highway bridges carry electric-railway traffic, the railway loading is determined on the basis of the class of traffic which the bridge may be expected to carry. The possibility that the bridge may be required to carry freight cars should be given consideration. The electric-railway loading on each track is a train of two electric cars followed by, or preceded by, or both followed and preceded by, a uniform load per foot of track corresponding to the H or H-S highway loading specified. Electric-railway loadings are shown in Fig. 16 and freight-car loadings in Fig. 17.

Sidewalk floors, stringers, and their immediate supports are designed for a live load of 85 lb per sq ft of sidewalk area. Girders, trusses, arches, and other members for bridge spans less than 25 ft in length are designed for this same load, while for bridge spans of greater length the sidewalk live load may be decreased to 60 lb per sq ft or less in designing these members.

For highway bridges, the AASHO specifications require a proportion of the live-load stress equal to $I = 50/(L + 125)$, where L is the loaded length, ft, and I the impact coefficient by which the live-load stress is multiplied to obtain the impact (maximum $I = 0.30$).

For wind loads, the AASHO specifies that:

1. The wind force on trusses and arches shall be assumed as a moving horizontal load equal to 75 lb per sq ft and on girders and beams 50 lb per sq ft on the area of the

structure as seen in elevation, including the floor system and railings (based on a wind velocity of 100 mph).

2. The lateral force due to the moving live load and the wind pressure against it shall be considered as acting 6 ft above the roadway and shall be as follows:

Highway bridges, 100 lb per lin ft.

Highway bridges carrying electric railway traffic, 300 lb per lin ft.

3. The total assumed wind force shall be not less than 300 lb per lin ft in the plane of the loaded chord and 150 lb per lin ft in the plane of the unloaded chord on truss spans and not less than 300 lb per lin ft on girder spans.

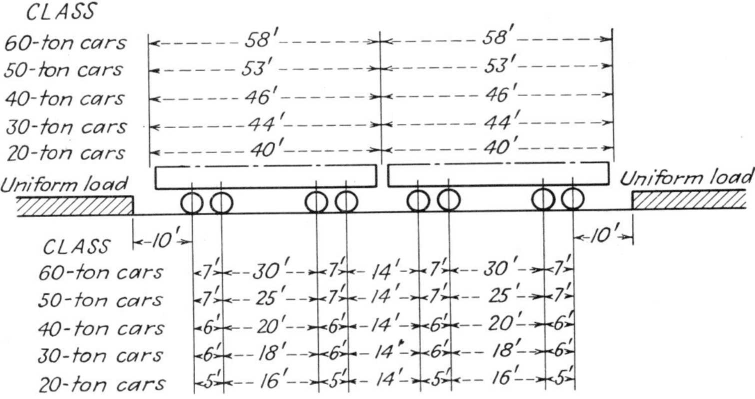

Fig. 16. Electric-railway loadings.

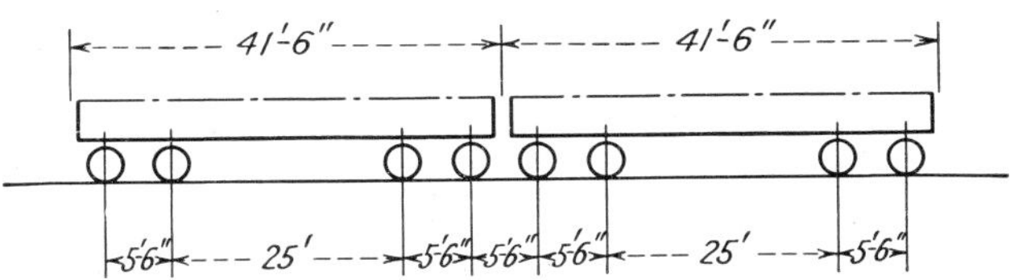

Fig. 17. Freight-car loadings. Total loaded weight per car with 10 percent over load: 40-ton capacity, 132,000 lb; 70-ton capacity, 212,000 lb.

Railroad-bridge Loads. For railroad bridges, the live load consists of the weight of locomotive and train, which acts as a system of concentrated rolling loads. The maximum load usually specified consists of two locomotives coupled together, followed by a uniform load representing the train load. A great variety of engine loadings has been used by the various railroad systems, each generally selecting a loading based on the weights of the heaviest locomotive in service or to be anticipated during the life of the structures under consideration. With such great divergence in specifications, the calculation of stresses by exact methods produced variable and questionable results.

Of the innumerable ones proposed, the only compromise loading that has had any extensive use is that proposed by Theodore Cooper in 1894. In this loading, the load on each driving axle determines the class of loading; i.e., in Fig. 18, which shows diagrammatically Cooper's E 60 loading, each driving axle carries 60,000 lb, or 60 kips. Any other class of Cooper's loading may be obtained directly from this one, as the axle loads are proportional and the wheel spacings the same for each class. The current

specifications of the American Railway Engineering Association recommend Cooper's E 72 loading for main-line bridges.

To some extent, "equivalent uniform loads" are used in stress calculations. They are somewhat simpler in their application than concentrated loads and, if properly chosen, give results agreeing closely with those obtained by the use of concentrated loads. For the simpler type of bridge, the calculations of stresses, using any one of the conventional wheel-load systems, is not difficult and is to be preferred; but, for trusses with more in-

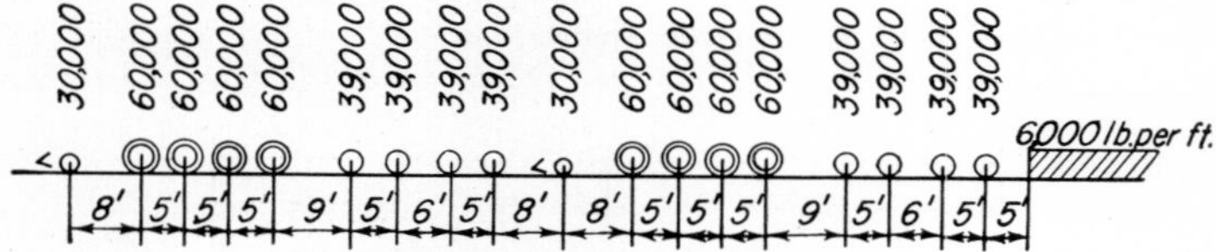

FIG. 18. Cooper's E 60 loading.

tricate systems of webbing, arches, cantilevers, and suspension bridges, the equivalent load offers a readier solution.

For impact on railroad bridges the AREA specifies (1) 10 percent of the live load stresses, acting downward on one rail and upward on the other, due to the rolling effect, plus (2) for steam locomotives a percentage of the live-load stress equal to $100 - 0.60L$ for spans less than 100 ft in length and $\frac{1{,}800}{L - 40} + 10$ for spans greater than 100 ft, where L is the length, ft, center to center of supports of stringers, longitudinal girders, and trusses (for chords and main members), or (for floor beams, floor-beam hangers, and subdiagonals of trusses) the length in feet of floor beams. For electric locomotives the plus percentage is $\frac{360}{L} + 12.5$.

For lateral forces on railroad bridges, the AREA specifies that:

1. The lateral force on the structure shall be a moving load of 30 lb per sq ft on 1.5 times its vertical projection on a plane parallel with its axis but not less than 200 lb per lin ft at the loaded chord or flange and 150 lb per lin ft at the unloaded chord or flange.
2. The lateral force on the train shall be a moving load of 300 lb per lin ft on one track, applied 8 ft above the top of rail.
3. The lateral force to provide for the effect of the nosing of locomotives, in addition to the lateral forces specified in paragraphs 1 and 2, shall be a moving load of 20,000 lb applied at the top of rail, in either horizontal direction at any point of the span.

When a train sets its brakes while crossing a bridge, a horizontal force is exerted on the track through the friction of the braked wheels. This force is usually taken as 20 percent of the vertical live load and is assumed to be applied 6 ft above the top of the rail. However, when because of continuity of the rails part of the force is transmitted to the roadbed beyond the bridge, a 50 percent reduction in the force is permitted.

When a train crosses a bridge on a curve, it exerts centrifugal forces on the structure. The amount of centrifugal force F developed by a load P moving in a curve with a curvature of D deg is

$$F = 0.0000117v^2DP$$

where v is the velocity of the train.

Moving Loads and Influence Lines. One of the most helpful devices for solving problems involving moving loads is an influence line. Whereas shear and moment diagrams evaluate the effect of loads at all sections of a structure, an influence line indicates the effect at a given section of a unit load placed at any point on the structure.

For example, to plot the influence line for bending moment at some point A on a beam, a unit load is applied at some point B. The bending moment at A due to the unit load at B is plotted as an ordinate to a convenient scale at B. The same procedure is followed with the load at every point along the beam and a curve is drawn through the points thus obtained.

Actually, the unit load need not be placed at every point. The equation of the influence line can be determined by placing the load at an arbitrary point and computing the bending moment in general terms.

Suppose we wish to plot the influence line for reaction at A for a simple beam AB (Fig. 19a). We place a unit load at an arbitrary distance of xL from B. The reaction at A due to this load is $xL/L = x$. Then $R_A = x$ is the equation of the influence line. It represents a straight line sloping upward from zero at B to unity at A (Fig. 19a). In other words, as the unit load moves across the beam, the reaction at A increases from zero to unity in proportion to the distance of the load from B.

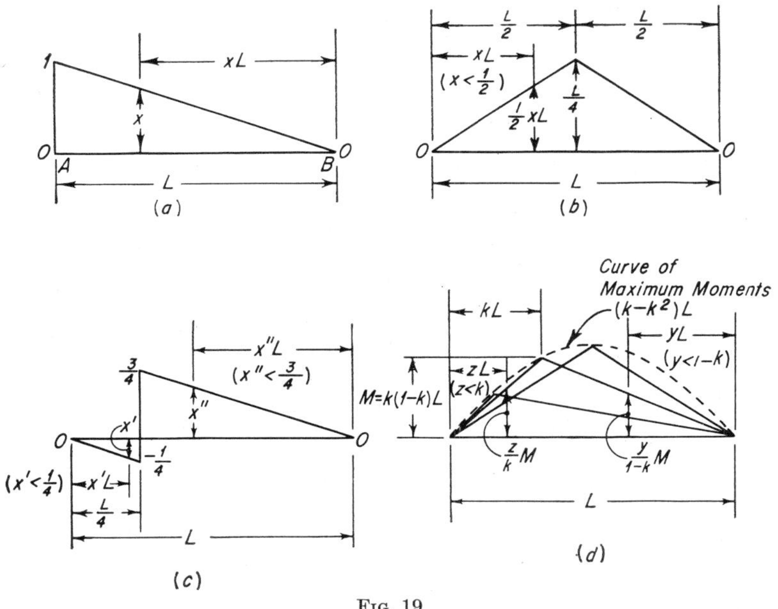

FIG. 19

Figure 19b shows the influence line for bending moment at the center of a beam. It resembles in appearance the bending-moment diagram for a load at the center of the beam, but its significance is entirely different. Each ordinate gives the moment at midspan for a load at the corresponding location. It indicates that, if a unit load is placed at a distance xL from one end, it produces a bending moment of $\frac{1}{2}xL$ at the center of the span.

Figure 19c shows the influence line for shear at the quarter point of a beam. When the load is to the right of the quarter point, the shear is positive and equal to the left reaction. When the load is to the left, the shear is negative and equal to the right reaction.

The diagram indicates that, to produce maximum shear at the quarter point, loads should be placed only to the right of the quarter point, with the largest load at the

quarter point, if possible. For a uniform load, maximum shear results when the load extends from the right end of the beam to the quarter point.

Suppose, for example, that the beam is a crane girder with a span of 60 ft. The wheel loads are 20 kips and 10 kips, respectively, and are spaced 5 ft apart. For maximum shear at the quarter point, the wheels should be placed with the 20-kip wheel at that point and the 10-kip wheel to the right of it. The corresponding ordinates of the influence line (Fig. 19*c*) are $\frac{3}{4}$ and $\frac{40}{45} \times \frac{3}{4}$. Hence, the maximum shear at the quarter point is $20 \times \frac{3}{4} + 10 \times \frac{40}{45} \times \frac{3}{4} = 51.7$ kips.

Figure 19*d* shows influence lines for bending moment at several points on a beam. It is noteworthy that the apexes of the diagrams fall on a parabola, as shown by the dashed line. This indicates that the maximum moment produced at any given section by a single concentrated load moving across a beam occurs when the load is at that section, and the magnitude of the maximum moment increases when the section is moved toward mid-span, in accordance with the equation shown in Fig. 19*d* for the parabola.

Maximum Bending Moment. When there is more than one load on a span, the influence line is useful in developing a criterion for determining the position of the loads for which the bending moment is a maximum at a given section:

Maximum bending moment will occur at a section C of a simple beam as loads move across it when one of the loads is at C. The proper load to place at C is the one for which the expression $W_a/a - W_b/b$ (Fig. 20) changes sign as that load passes from one side of C to the other.

When several loads move across a simple beam, the maximum bending moment produced in the beam may be near but not necessarily at mid-span. To find the maximum

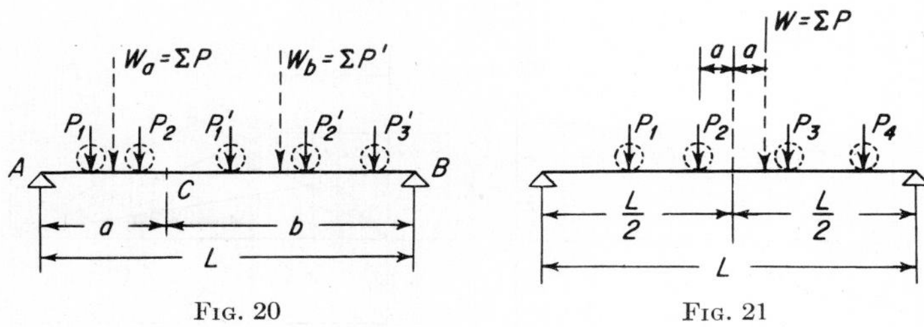

Fig. 20 Fig. 21

moment, first determine the position of the loads for maximum moment at mid-span. Then, shift the loads until the load P_2 (Fig. 21) that was at the center of the beam is as far from mid-span as the resultant of all the loads on the span is on the other side of mid-span. Maximum moment will occur under P_2.

When other loads move on or off the span during the shift of P_2 away from mid-span, it may be necessary to investigate the moment under one of the other loads when it and the resultant are equidistant from mid-span.

GRAPHIC STATICS

Representation of a Force. Since a force is completely determined when it is known in amount, direction, and point of application, any force may be represented by the length, direction, and position of a straight line. The length of the line is fixed by the relation between the amount of the given force and a prearranged unit of force. If the unit of force is taken as 1 lb to the inch, a line 3 in. long represents a force of 3 lb.

The graphical representation of a force may be designated by marking the line representing the force with the letter P, followed, in some instances, by a subscript (Fig.

22*a*), or each extremity of the line may be indicated by a letter, and the force referred to by means of these letters (Fig. 22*b*). The direction of the force may be given, where necessary, by an arrow on the line representing the force. In the second method of designation, given above, the order of the letters given in referring to a force indicates its direction; i.e., in Fig. 22*b*, referring to the upper force as *ab* indicates that it acts from *a* toward *b*, while reference to the lower force as *ca* indicates that it acts from *c* toward *a*. A diagram such as Fig. 22 (*a* or *b*) which shows the analytical relation of the forces acting —their positions, directions, and amounts—is known as a "force diagram."

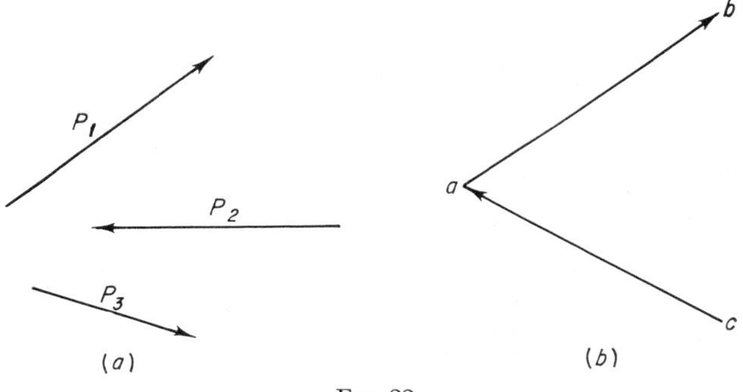

FIG. 22

Composition and Resolution of Forces. In Fig. 23, P_1 and P_2 represent two forces in the same plane acting at the point *a*. The amount of the resultant of the two forces is given by the length of the diagonal of the parallelogram *acbd* constructed upon the two given forces as sides. A single force applied at the point *a*, acting in the direction of point *b*, of an amount represented by the length of the line *ab*, will replace P_1 and P_2 in so far as accomplishing work is concerned. A force of the same amount but acting in the opposite direction will hold the forces P_1 and P_2 in equilibrium, i.e., at rest. If the force represented by the line *ab* acts from *a* toward *b*, it is the resultant; if it acts from *b* toward *a*, it is the equilibrant of the given forces. If another force P_3 were

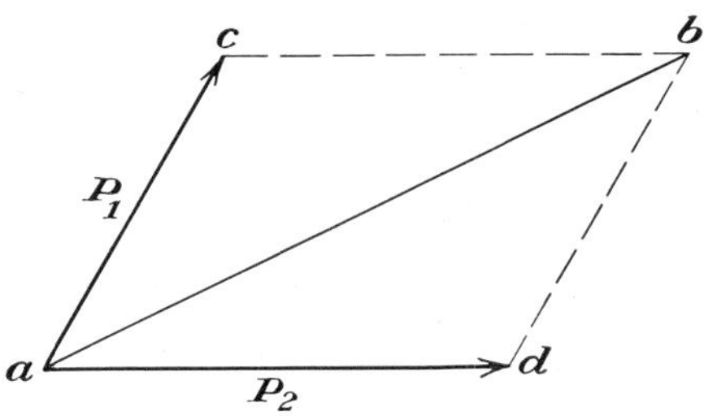

FIG. 23

acting at point *a*, it could be combined in a similar manner with the resultant of P_1 and P_2 to determine the resultant of P_1, P_2, and P_3. The process of replacing forces P_1 and P_2 (and more if present) with the single force *ab* is called the *composition of forces*—two (or more) forces are combined into one. On the other hand, any single force may be resolved into two components acting in any given directions. The solution of this problem, called the *resolution of forces*, is the reverse of that employed in the composition of forces.

The Force Triangle. Reference to Fig. 23 shows that, in order to determine the amount and direction of the resultant of P_1 and P_2, it is not necessary to construct the complete parallelogram *acbd*. From *c* of the force P_1, a line *cb* is drawn parallel and equal to P_2. The triangle *acb* is called the *force triangle* and represents graphically the relation between the forces P_1, P_2, and *ab*.

Thus it is seen that, if in addition to forces P_1 and P_2 a third force *ba* is applied at point *a* in the direction from *b* to *a*, these three forces will be in equilibrium. The point *a* will then remain at rest. It may now be concluded that, *if three forces meeting in a point are in equilibrium, they will form a closed triangle;* and not only will the three lines parallel and equal to the three forces respectively form a closed, three-sided figure but also the arrows representing the direction of the forces will all point in the same direction around the triangle.

By applying this principle, any one of the three forces may be determined in amount and direction if the other two are fully known, or any two of the forces may be obtained in amount if their directions are given and the third force is fully known.

The Force Polygon. If instead of two forces, three or more were acting at point *a*, the resultant of these forces could be determined by the successive application of the principle of the force triangle; two forces are first composed into one, and this one combined with the third force; this process is continued until the last force is combined with the resultant of all the other forces to determine the resultant of the entire group.

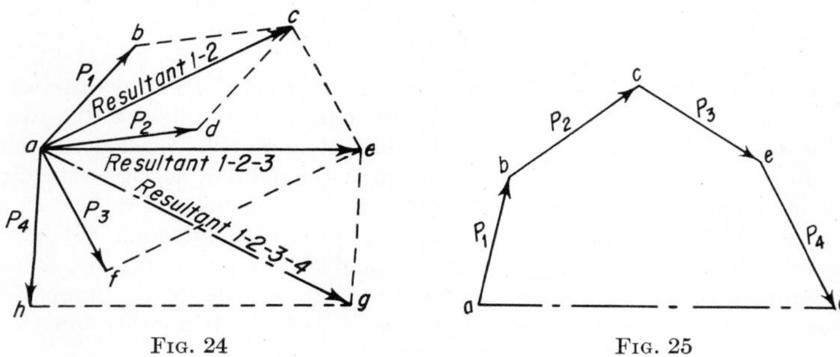

FIG. 24 FIG. 25

Figure 24 represents four forces P_1, P_2, P_3, and P_4 acting in a common point *a*. Assuming that they are not in equilibrium, let it be required to find the force necessary to produce a condition of equilibrium. Forces P_1 and P_2 are combined as described in the preceding article by completing the parallelogram *abcd*. The resultant of these two forces is the force *ac* acting in the direction from *a* to *c*. It is now assumed that P_1 and P_2 have been replaced with the one force *ac* and there are but three forces acting at *a*, i.e., P_3, P_4, and *ac*. The force *ac* is combined with P_3 by constructing the parallelogram *acef*, the diagonal *ae* being the force necessary to replace *ac* and P_3. The forces *ae* and P_4 are then combined to determine *ag*, the resultant of the four original forces.

A study of Fig. 24 shows that it is not necessary to complete all the parallelograms in order to obtain *ag*. The polygon *abceg*, of which four sides are respectively parallel and equal to forces P_1, P_2, P_3, and P_4, gives the required information. This polygon is repeated for clearness in Fig. 25. The line drawn from the beginning *a* to the end *g* of the last known line *eg* of the polygon, contrary to the order of the forces, gives the intensity and direction of the resultant of the four given forces. A force of equal intensity applied in the opposite direction, from *g* to *a* (in the direction of the forces), would hold the four given forces in equilibrium. Point *a* would remain at rest under the action of five forces P_1, P_2, P_3, P_4, and *ga*.

The polygon *abceg*, formed by the successive laying off of lines parallel and equal to the given forces, is called the *force polygon*—the force triangle expanded. By its use, any one of a number of forces acting at a point in a static structure may be determined in amount and direction if the others are completely known, or any two may be obtained in amount if they are known in direction and the other forces in amount and direction. The force polygon serves a similar purpose for a series of forces not meeting in a point; however, it does not in this case determine the point of application of the unknown force.

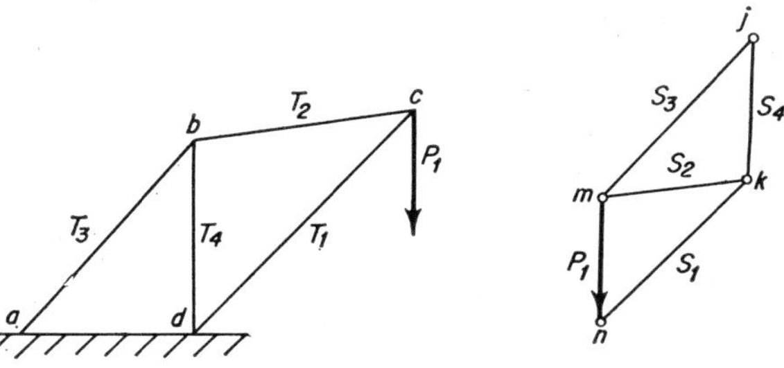

FIG. 26

Application of the Force Polygon. The crane truss of Fig. 26 consists of a vertical post *bd*, supporting a boom *dc*, from which the external load P_1 is suspended. The boom is anchored to the post by the tie rod *bc*, and the post is in turn anchored at its upper end by the backstay *ab* fastened to the ground. In constructing the force polygon, P_1 $(=mn)$ is laid off equal to the external load. By drawing *mk* and *nk* parallel to T_2 and T_1, respectively, the values of the stresses in these two members are obtained. Similarly, *mj* and *kj* determine the stresses in T_3 and T_4. In the triangle *mnk*, the direction of P_1 is known. Following around the triangle in the direction indicated by P_1, s_1 acts from *n* to *k*; and, transferring this direction to the truss diagram, the stress in T_1 acts toward the joint *c* and is therefore compression. Proceeding in the direction from *k* to *m*, the stress in T_2 is tension. In the triangle *kmj*, since T_2 is in tension and acts away from the joint *b*, the direction to be taken around the triangle is from *m* to *k*. This indicates tension in T_3 and compression in T_4.

The Equilibrium Polygon. The preceding paragraphs have dealt with concurrent forces only. With nonconcurrent forces, the methods so far used become cumbersome and with a system of parallel forces do not furnish a solution. A general method, applicable to all cases involving coplanar forces, is illustrated in Fig. 27. The forces P_1, P_2, P_3, and P_4 acting upon the given body are not in equilibrium. The amount and di-

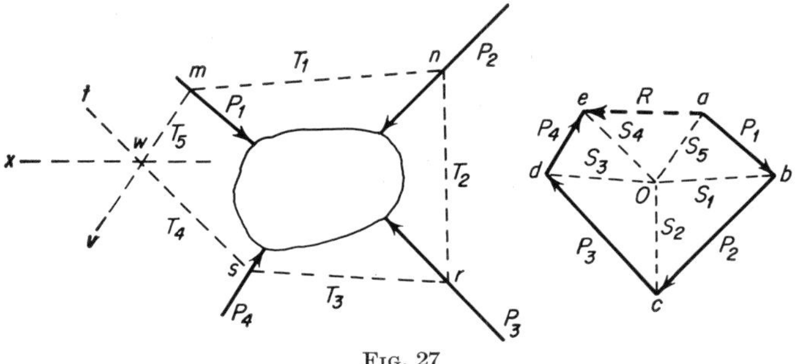

FIG. 27

rection of their resultant are obtained from the force polygon *abcde*. The line of action of this resultant may be determined as follows.

From any point O in the force polygon, a line may be drawn to each of the vertices of the polygon. Since the lines Oa and Ob form a closed triangle with the force P_1, they represent two forces that will hold P_1 in equilibrium—two forces that may replace P_1 in the force diagram. In Fig. 27, at any point m on the line of action of P_1, lines mn and mv drawn parallel to S_1 and S_5 respectively represent the lines of action of these two forces. Similarly, S_1 and S_2 represent two forces that may replace P_2. The line of action of S_1 has already been fixed by the line mn. From the point of intersection n of the line mn with the force P_2, nr is drawn parallel to S_2. From r, rs is drawn parallel to S_3; and, from s, st is drawn parallel to S_4. Lines mv and st, which are parallel to S_5 and S_4 respectively, represent the lines of action of S_5 and S_4. These latter forces together with the resultant ae form a closed force triangle. S_5, S_4, and ae therefore represent a series of three forces in equilibrium. To fulfill this condition where three forces only are involved, the three forces must meet at a point. The line of action of the resultant ae must therefore pass through the point of intersection w of the lines mv and st. The resultant of the four given forces is thus fully determined. A force of equal magnitude but acting in the opposite direction, i.e., from e to a (in the direction of the forces in the force polygon), will hold P_1, P_2, P_3, and P_4 in equilibrium and keep the body on which they act at rest.

The polygon *mnrsw* represents a jointed frame which, by means of the stresses of tension and compression in its various members, could hold the given forces in equilibrium. It is called, therefore, an *equilibrium polygon*. The amount and character of stress in each of the members of the jointed frame are given by the lines $S_1 \ldots S_5$ in the force polygon. The point O is called the "pole," and the lines $S_1 \ldots S_5$ are called the "rays" of the force polygon. Obviously, since an infinite number of poles may be selected and an infinite number of starting points may be used, an infinite number of equilibrium polygons may be drawn for a given group of forces; the final result will, however, be the same for all; i.e., the line of action of the resultant of any given force group, determined by one equilibrium polygon, will coincide with that found by any of the other possible polygons.

Resultant of Parallel Forces. In the case of a system of parallel forces, the force polygon becomes a straight line. Figure 28 represents a system of four such forces, not

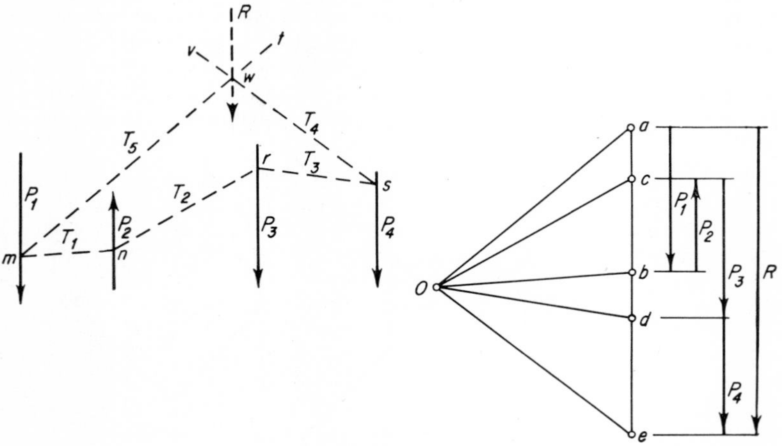

Fig. 28

in equilibrium. The force polygon *abcde* is a straight line; and *ae*, its closing line, gives the magnitude and direction of the resultant. For convenience in the following construction, the forces represented by each of the lines of the force polygon in Fig. 28 are indicated at the right of the polygon. To hold the given forces in equilibrium, a force equal in amount to *ea*, acting upward from *e* to *a*, is necessary. The line of action of this force is determined from the equilibrium polygon.

The pole O is selected and the rays drawn to the vertices of the force polygon. From any point on the line of action of P_1, lines *mn* and *mt* are drawn parallel to *Ob* and *Oa* (the rays that form with P_1 a closed force triangle) respectively. From *n*, the point of intersection of *mn* and the line of action of P_2, a line *nr* is drawn parallel to *Oc*; from *r*, a line *rs* is drawn parallel to *Od*; and from *s*, a line *sv* is drawn parallel to *Oe*. The point of intersection *w* of the lines *mt* and *sv* is a point in the line of action of the resultant force. An upward force equal and parallel to *ea*, applied so that its line of action passes through point *w*, will hold the forces $P_1 \ldots P_4$ in equilibrium. A downward force of the same amount will replace the given forces.

STRESSES IN TRUSSES

Stresses in roof trusses due to dead and snow loads may be computed analytically, or graphical methods may be used. For all trusses except those having a simple regular form with few web members, the latter method usually furnishes the readier solution.

Graphical Analysis of Trusses. The stresses in the truss of Fig. 29 due to dead and snow loads are to be determined graphically. The combined dead and snow panel load is 4,880 lb. The diagram of the truss is repeated in Fig. 30. In this diagram, the trusses are lettered in the spaces between the members and the loads, and each member or load is designated by means of the letters in the adjoining spaces; i.e., the first vertical is *EF*, the left lower-chord member is *ER*, and the left reaction *AR*.

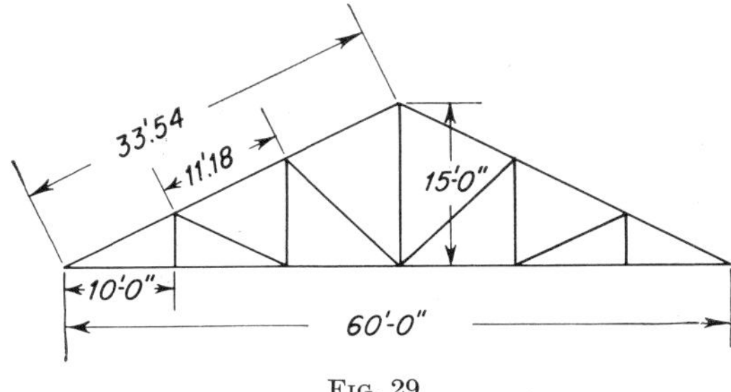

Fig. 29

For equilibrium, the external forces on the truss, consisting of the downward panel loads and the upward reactions, must form a closed force polygon. On account of the symmetry of the loads, both in amount and in position, each reaction equals one-half of the total downward load on the truss. Since the external forces are parallel, the sides of the force polygon lie in a straight line $a \ldots a'ra$ (Fig. 30), constructed by laying off the panel loads in regular order from left to right and the reactions equal to $a'r$ and ra at the right and left ends respectively, point r being midway between a and a'.

At the left support, the external forces *AB* and *AR* are held in equilibrium by the internal stresses in the members *BE* and *ER*. These four forces therefore form a closed force polygon. The vertices of the force polygon are lettered with the lower-case letters corresponding to the capital letters of the truss diagram. Through the point *b*, a line *be* is drawn parallel to *BE*; through *r*, a line *re* is drawn parallel to *RE*. The lengths of the

lines *be* and *re*, measured to the scale of the load line, give the magnitudes of the stresses in *BE* and *RE* respectively. To determine the character of the stresses, it is necessary only to trace the forces around the polygon in regular order. For equilibrium, the forces must act in the same direction around the polygon. The reaction *ra* acts upward, and the load *ab* downward; hence, the stress *be* must act from *b* to *e*, and the stress *er* from *e* to *r* to close the polygon. Transferring these two directions to the truss diagram, the

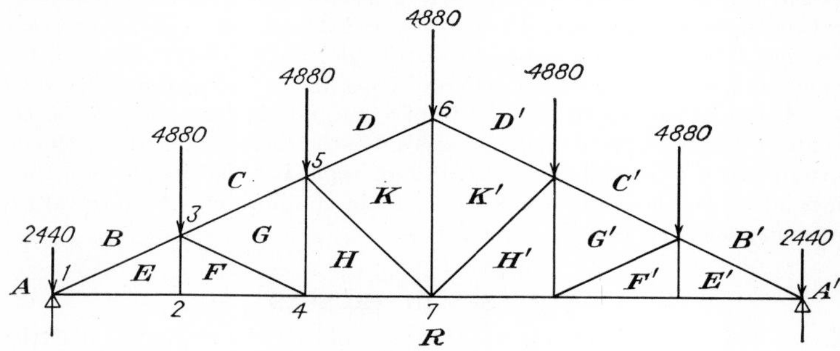

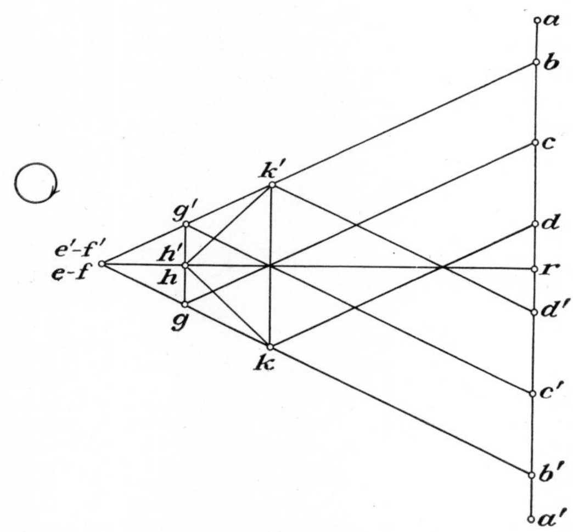

FIG. 30

stress in *BE* acts downward toward the joint under consideration and the stress in *ER* acts away from this joint. *BE* is therefore in compression and *ER* in tension.

The stress diagram is completed for the given truss by constructing a force polygon for each of the remaining joints in regular order. Joint 3 must be considered before joint 4 in order to determine the stress in *FG*; otherwise, there would be three unknowns at joint 4, and the solution of this joint would be impossible. The complete stress diagram is shown in Fig. 30. When measured to the scale of the load line *aa'*, each of the lines in this diagram represents the magnitude of the combined dead- and snow-load stress in the corresponding member of the truss.

It is possible to determine the character of the stress in each member of the truss without reference to any other member or load and thus save considerable time in the stress determination. In Fig. 30, the loads and reactions were laid off on the load line in the order of *AB*, *BC* ... *A′R*, *RA*—clockwise around the truss. This is indicated by the circular arrow on the stress diagram. In passing around joint 3 in this same direction, the members and external load occur in the order *BC*, *CG*, *GF*, *FE*, *EB*. The force polygon for this joint is *bcgfeb* and must be followed in this same order, since the load *bc* acts downward and is laid off from *b* to *c*. Hence, the stress in *GF* acts from *g* to

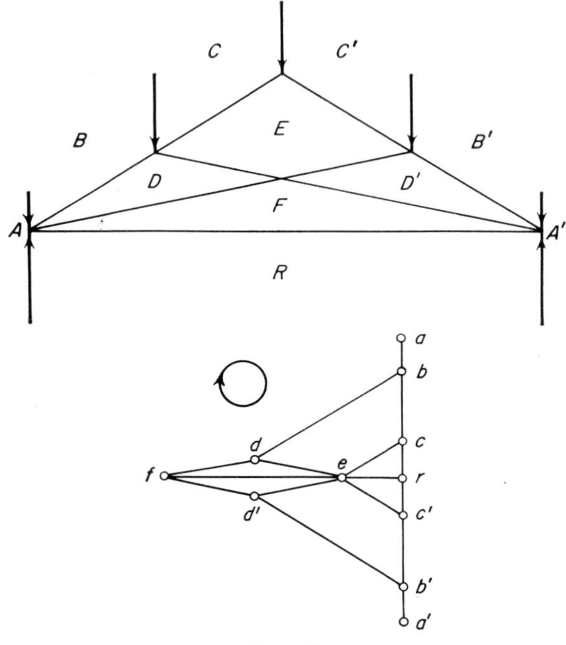

FIG. 31

f; when transferred to the truss diagram, this direction is toward the joint, and *GF* is in compression. In like manner, the character of the stress in any member may be obtained without reference to any other. This method of determining the character of the stresses as obtained from the stress diagram is known as the *principle of the circular arrow*.

It is not always possible to begin the stress diagram at one of the end joints. In the truss shown in Fig. 31, the stress diagram is begun by cutting a circular section at the peak joint. In this section, only two unknown quantities exist, and the determination of the stresses in *CE* and *C′E* is possible. If the construction were attempted by first cutting a section around the end joint, three unknown stresses would make the analysis of the joint impossible. By starting at the peak and working toward the end, this condition is obviated. The complete stress diagram for this truss is shown in Fig. 31. All the members are in compression except the tie *FR*, which is in tension.

Analytical Analysis of Roof Trusses. The net reaction at *A* in Fig. 30 is $2\frac{1}{2} \times 4{,}880 = 12{,}200$ lb. Applying the first two conditions of static equilibrium, i.e., $\Sigma X = 0$ and

$\Sigma Y = 0$, and taking successive joints as free bodies, we may determine the stresses as follows:

At joint 1.. $\begin{cases} BE \times \dfrac{33.54}{15} + 12{,}200 = 0 & BE = -27{,}300 \\ ER - \dfrac{33.54}{30} \times 27{,}300 = 0 & ER = +24{,}400 \end{cases}$

At joint 2. . . . . $EF = 0$

At joint 3.. $\begin{cases} +27{,}300 \times \dfrac{15}{33.54} + CG \times \dfrac{15}{33.54} - FG \times \dfrac{15}{33.54} - 4{,}880 = 0 \\ -27{,}300 \times \dfrac{30}{33.54} - CG \times \dfrac{30}{33.54} - FG \times \dfrac{30}{33.54} = 0 \end{cases}$

$FG = -5{,}500$
$CG = -21{,}800$

At joint 4.. $GH - 5{,}500 \times \dfrac{15}{33.54} = 0$ $\qquad GH = +2{,}500$

At joint 5.. $\begin{cases} +21{,}800 \times \dfrac{15}{33.54} + DK \times \dfrac{15}{33.54} - HK \times \dfrac{10}{14.14} - 4{,}880 - 2{,}500 = 0 \\ -21{,}800 \times \dfrac{30}{33.54} - DK \times \dfrac{30}{33.54} - HK \times \dfrac{10}{14.14} = 0 \end{cases}$

$HK = -6{,}900$
$DK = -16{,}300$

At joint 6.. $+2 \times 16{,}300 \times \dfrac{15}{33.54} - 4{,}880 - KK' = 0$ $\qquad KK' = +9{,}700$

Reactions in Roof Trusses Due to Wind Loads. Since the wind pressure on sloping roofs acts normal to the roof surface, the load line is not vertical, and the reactions are not equal. If both ends of the truss are anchored to the walls in such a manner that each wall furnishes resistance to a horizontal thrust, both reactions are inclined. Since there are four unknowns (amount and direction of each reaction) and only three equations available for the solution ($\Sigma X = 0$, $\Sigma Y = 0$, and $\Sigma M = 0$), an assumption must be made as to the distribution of the horizontal thrust between the two reactions. One assumption often made is that the horizontal components of the two reactions are equal. Another assumption is that both reactions are parallel to the resultant wind loads on the truss. The latter assumption is as reasonably accurate as the former and permits of a somewhat easier graphical solution.

In steel trusses, especially for the longer spans, provision is usually made for expansion and contraction by allowing one end to rest on rollers in such a manner that the truss is free to move horizontally at this end.[1] In such cases, the reaction at the free end must be vertical, since no provision is made for resisting a horizontal pressure at this end. The entire horizontal component of all the wind panel loads must be resisted by the other support, which is fixed rigidly to the wall in order to furnish the required horizontal resistance. The reaction at the fixed end will therefore be inclined at an angle, unknown until the complete analysis of the external forces is made.

In steel trusses of shorter spans, the provision for expansion and contraction is made by merely allowing one end of the truss to rest freely on a steel plate securely fastened

[1] This provision is often more theoretical than practical. Once the truss is erected, very little attention is given to the maintenance of the roller bearings in a frictionless condition. Where a ceiling is suspended from the trusses, there is often no provision made for the effect of a movement of the free end of the truss upon the ceiling. If it were not for the fact that the walls supporting the trusses are considerably more elastic than they are theoretically considered, a good many more cracked ceilings would result.

to the wall. Friction being neglected, the reaction at the free end in this type of construction will be vertical, as in the preceding type.

Both Ends Fixed. Ordinarily the analytical solution for determining truss reactions is somewhat easier than the graphical solution and also more exact. The values of the reactions are determined by taking moments first about one reaction point and then about the other. The graphical solution, however, may prove more desirable if the lever arms of the given loads about the centers of moments are not easily obtained.

The graphical solution for determining wind-load reactions for trusses with both ends fixed, each reaction being assumed parallel to the resultant wind load, is illustrated in Fig. 32. The truss is that of Fig. 29.

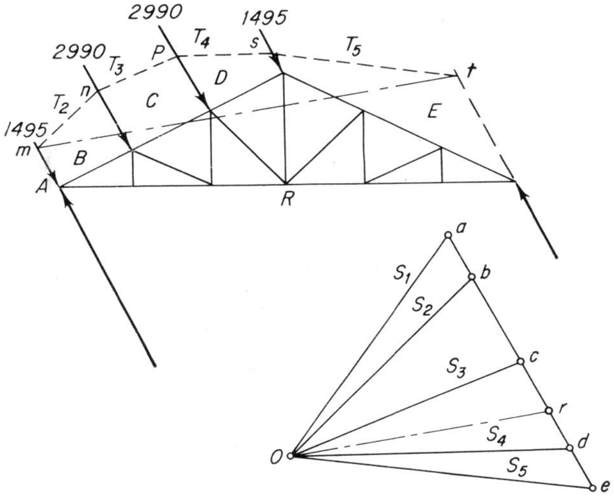

Fig. 32

Considering the truss as a free body acted upon by the loads and reactions as shown, the load line *ae* is constructed by laying off the loads *AB* . . . *DE* in order.

A convenient pole *O* is selected, and the rays *Oa* . . . *Oe* are drawn to the load line. The equilibrium polygon *mnpst* is constructed for the pole *O* by drawing, on the truss diagram, lines parallel to the rays S_1, S_2 . . . S_5 in order. Each of these lines terminates on the lines of action of the wind panel loads corresponding to the loads on either side of the parallel ray in the force polygon—for example, line T_2, which is parallel to the ray S_2, is drawn from the line of action of *AB* to that of *BC*. This polygon may be commenced anywhere on the line of action of the force *AB*. Since the lines of action of *AB* and *AR* coincide, the line of the equilibrium polygon which is parallel to the ray S_1 becomes a point. This explains the apparent omission of the parallel to S_1 from the construction. The line *Or*, parallel to the closing line *mt* of the equilibrium polygon, divides the load line into the required reactions, *er* representing the right reaction and *ra* the left reaction. The scale of the original drawing was 100 lb to the inch. Line *er* measured 2.80 and line *ra* 6.17 in. The right reaction is therefore 2,800, and the left reaction 6,170 lb.

Figure 33 shows the construction where all the wind loads are not parallel. Since the lines of action of *AB* and *AR* are not parallel, the construction of the equilibrium polygon must be begun at their intersection if the readiest solution is desired.

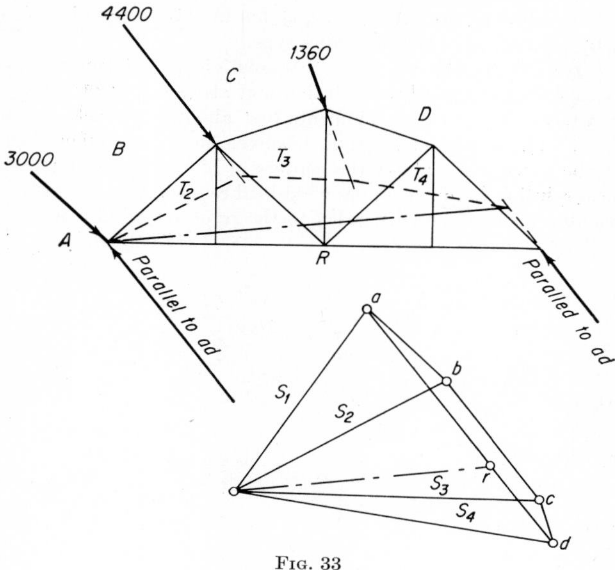

FIG. 33

One End Free. With one end of the truss free to move horizontally, the reaction at the free end is vertical, and the entire horizontal component of the wind pressure is resisted at the fixed end. Since the wind pressure may be exerted in either direction and the stresses in the truss members must be determined for each condition, the reactions

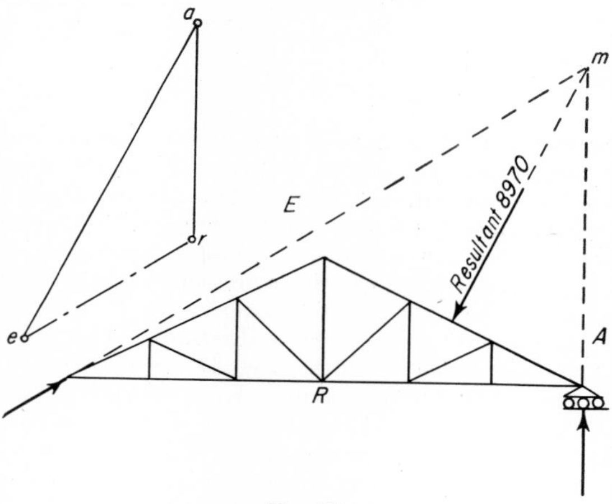

FIG. 34

for wind on the fixed side and on the free side must alike be determined. In a truss with unbroken chords and equal panels, the vertical reaction may be obtained readily by the method of moments, the center of moments being the fixed end of the truss. With this

reaction known, as well as the value of the resultant wind load, the construction of a force triangle determines the amount and direction of the fixed reaction.

Graphically the reactions may be obtained by treating the truss as a whole as a three-force piece, as illustrated in Fig. 34. In this construction, the sum of all the wind loads is concentrated at their center of gravity. The direction of the reaction at A being known, as well as the amount and direction of the resultant, the force triangle *aer* gives the amount and direction of the reaction ER and the amount of the reaction AR.

Another graphical method is illustrated in Fig. 35. Here, since the reaction at the right end is known to be vertical, the equilibrium polygon is constructed between its vertical line of action and the inclined lines of action of the wind loads. Since the line of action of the reaction at A is unknown, *the equilibrium polygon must be commenced at the point* A, which is the only known point on the line of action of that reaction. The ray *or* in the force polygon, drawn parallel to the closing line *mn* of the equilibrium polygon, intersects a vertical through *e* and determines the amount and direction of the reaction AR and the amount of the reaction ER.

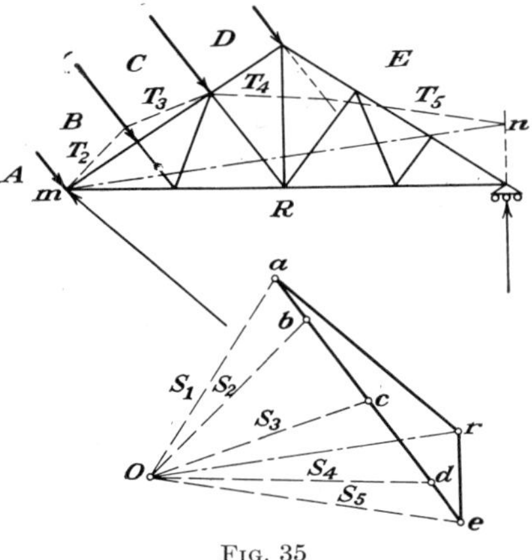

Fig. 35

Wind-load Stresses in Roof Trusses. The reactions being known, the wind-load stresses are determined by the application of the principle of the force polygon to each joint in succession, in a manner similar to that used for stresses due to vertical loads. A broken upper chord or an inclined lower chord or both do not materially complicate the internal stress determination. Figure 36 shows the complete construction necessary for the determination of the wind-load stresses in a truss of this type. In this truss, all the members are stressed by the wind; while in trusses similar to those of Fig. 35, the web members on the leeward side have no wind-load stresses. A check on the accuracy of the construction of a wind-load diagram is that the last line drawn in the stress diagram must be parallel to the corresponding member of the truss diagram. In Fig. 36, *jr*, connecting the points *j* and *r*, previously located in the construction, must be parallel to the member JR in the truss diagram to show absolute accuracy of construction.

The Fink Truss. One type of Fink truss is shown in Fig. 37. In determining its stresses graphically, it becomes necessary temporarily to replace two of its members with one member in order to complete the solution. In Fig. 37, the stress diagram is begun in the usual manner, joints 1, 2, and 3 being isolated in order and the portion *abgrjhc* of the diagram drawn. At either the next upper- or lower-chord joint, three unknown stresses exist, and the construction cannot be continued in the usual manner.

Removing the members KL and LM and replacing them with one member XM, as shown in the left-hand part of Fig. 37, allows the construction of the complete stress diagram for such a truss. The diagram *abgrjhcdxmnre* is that diagram. Comparing the two trusses it is seen that, around the peak joint, no changes have been made in the amount of external load or in the position of the internal members. Therefore the stresses in EM and MN in either truss are respectively equal to the stresses in the corresponding members of the other truss. The stress in member EM of the original truss is, therefore, represented by the line *em* in the stress diagram for the revised truss. The

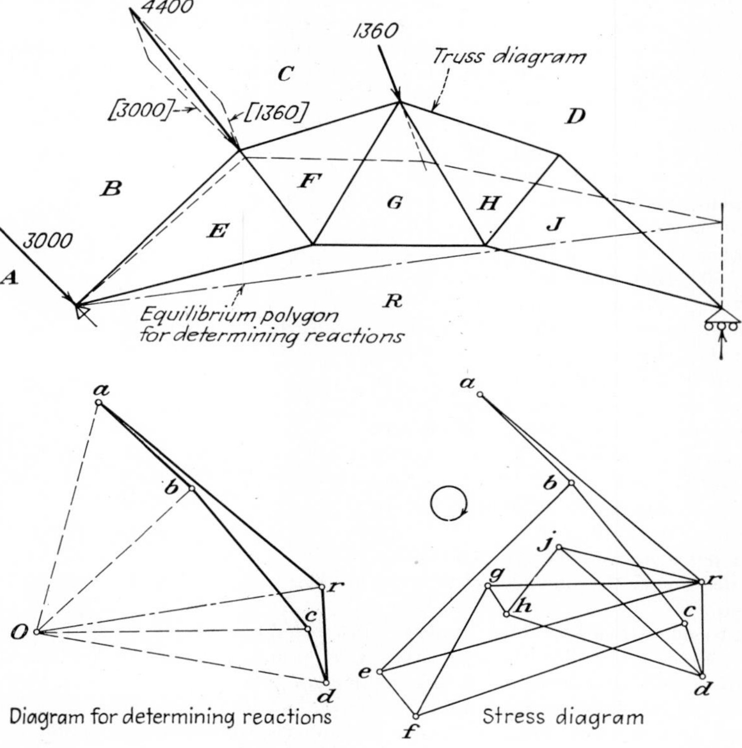

Fig. 36

stresses in *LM* and *DL* may now be determined by drawing the force polygon *deml* for the forces acting at joint 8. Similarly the force polygon *klmn* at joint 9 determines *KL* and *KN*, and the force polygon *nrjk* at joint 10 determines *JK*. The complete stress diagram for the original truss is shown in Fig. 37, the point *x* and the dotted line being a part of the auxiliary construction.

Trusses with Ceiling Loads. When a ceiling is suspended from the lower chord the load brought by it to the truss must be considered in determining the total stresses in the truss members. Since the ceiling is a fixed part of the structure, its weight may be combined with the other vertical loads in the dead-load stress diagram, thus avoiding the necessity of drawing a separate ceiling-load diagram. In order to make the solution of this combined diagram possible, it is necessary that the loads and reactions be laid off in regular order around the truss. Thus, in Fig. 38, the force polygon *afgka* is drawn by laying off the loads *AB*, *BC*, *CD*, *DE*, and *EF*, downward in that order; then the reaction *FG* upward; next the downward loads *GH*, *HJ*, and *JK*; followed by the upward reaction *KA* to complete the force polygon. This sequence is shown at the right of the load line in the figure.

After the loads and reactions have been laid off in the proper order, the construction of the stress diagram is completed in the usual manner; i.e., each joint is isolated in succession from the rest of the truss, and a series of connected force polygons drawn,

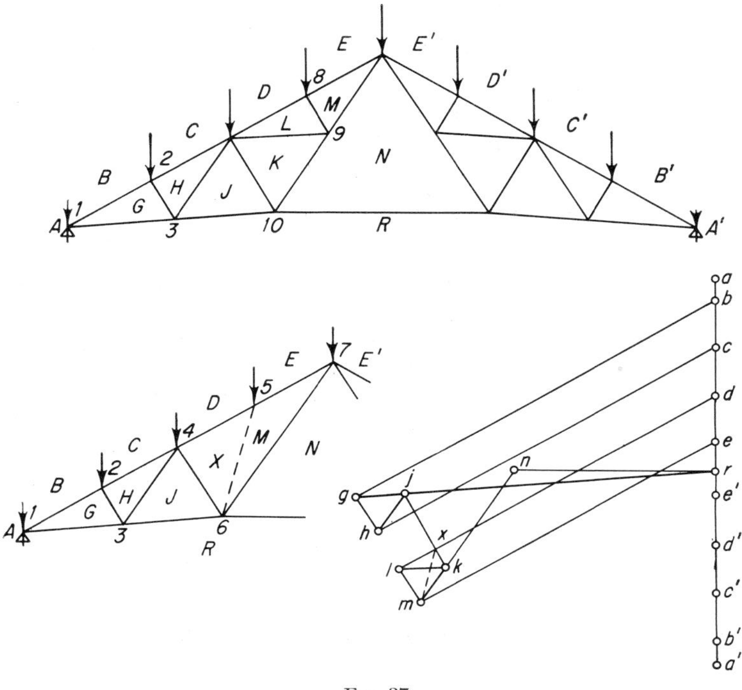

FIG. 37

one for each joint. The complete stress diagram for a truss with ceiling loads in addition to the usual vertical loads is shown in Fig. 38.

Maximum Stresses in Roof Trusses. The greatest stress that occurs in any member of a roof truss is the algebraic sum of the individual stresses caused by the simultaneous action of the dead load, snow load, wind load, ceiling load, or special loads. Since the wind may come from either side, the character of the stress in any one member may be variable, so that a complete analysis of any roof truss requires that the stresses due to the wind be determined for both conditions.

It is generally assumed that full snow-load stresses are not likely to occur simultaneously with full wind-load stresses. The exact pitch of the roof has an effect on the proportion of snow-load stresses to be used with the full wind-load stresses. One-half of the snow-load stresses is commonly used in such a combination.

The value of the probable snow load, varying as it does in different latitudes, indicates that there are three possible combinations of loads that may produce the maximum stresses. These are dead load plus snow load, dead load plus wind load, and dead load plus one-half snow load plus wind load. In the middle latitudes, the last combination usually produces the maximum stresses.

Dead-load Stresses in Bridge-truss Web Members. *Parallel Chords and Single-web Systems.* In this type of truss, there are two classes of web members. The first and simplest class is illustrated by members *Bb* of Fig. 39. Such a member is known as a *suspender* or *subvertical.* From an inspection of the section *MM*, it is seen that the stress is dependent only upon, and is equal to, the load P_1.

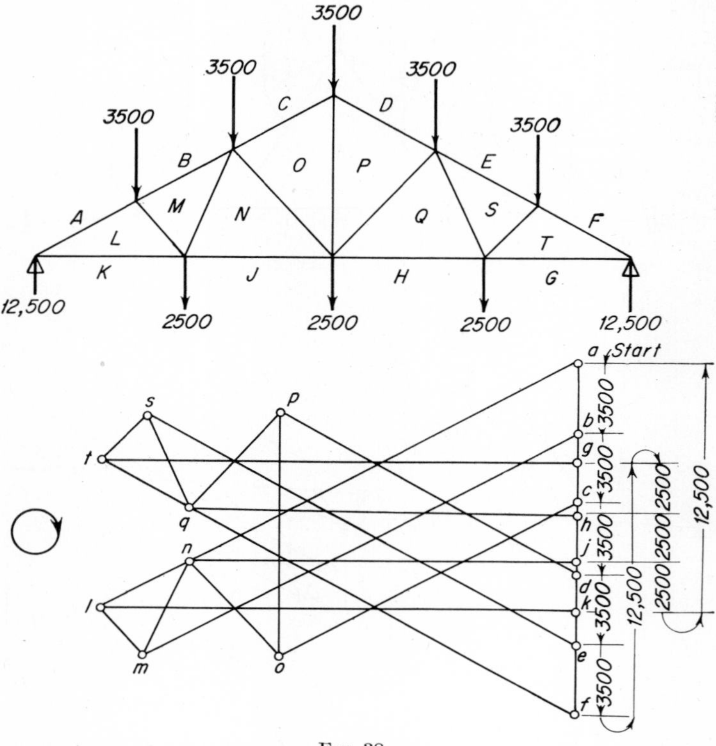

Fig. 38

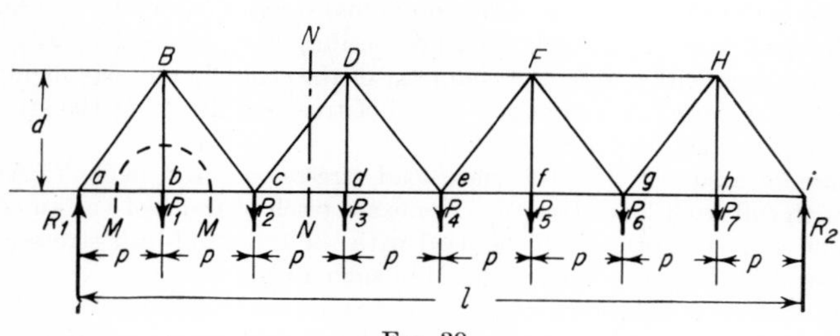

Fig. 39

The other class of web member is illustrated by the member cD. Since the internal stresses in any section of a truss hold in equilibrium the external forces on either side of the section, a summation of the forces acting on that portion of the truss to the left of the section NN determines the vertical component of the stress in the member cD. Taking that portion of the truss as a free body, as shown in Fig. 40, with θ as the acute angle between cD and a vertical:

$$R_1 - P_1 - P_2 + S \cos \theta = 0$$

But $R_1 - P_1 - P_2$ is the algebraic sum of all the external vertical forces on the left of the section and is the *vertical shear* in the section. It may be designated as V. Therefore,[1]

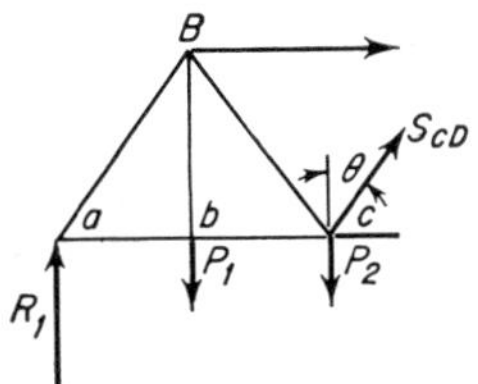

FIG. 40

$$V + S\cos\theta = 0 \qquad \text{or} \qquad S = -V\sec\theta$$

From this it follows that, for trusses with horizontal chords and single-web systems, the stress in any web member, other than subverticals, is equal to the vertical shear in the member multiplied by the secant of the angle that the member makes with the vertical.

Inclined Chords and Single-web Systems. In trusses of this type, since the inclined chord has a vertical component, the method developed above cannot be used without modification. The method of moments offers a simpler solution.

Figure 41 represents a portion of a Parker truss. To determine the stress in Bc, the section is cut as shown. The chord BC produced intersects the chord bc produced at the point O, which is therefore the center of moments for the member Bc. The lever arm is Om, the perpendicular distance from O to Bc produced. From the similar triangles Omc and Bbc,

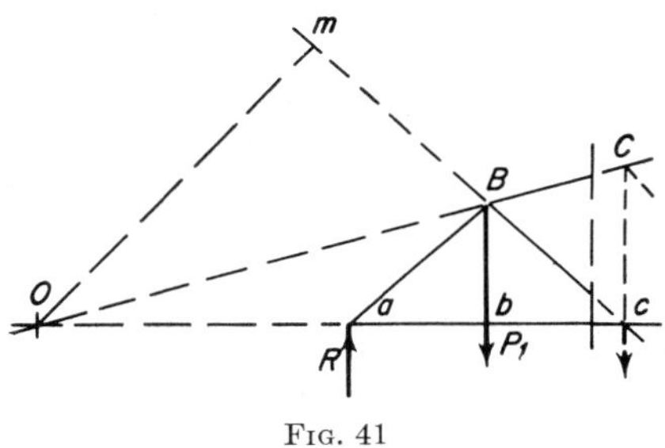

FIG. 41

$$Om = \frac{Oc \times Bb}{Bc}$$

With the net reaction at a, and the panel loads at B and b known, taking moments about O

$$(S_{Bc} \times Om) - (R \times Oa) + (P_1 \times Ob) = 0$$

from which the stress in Bc may be determined.

Although there is no difficulty involved in this method, it can be somewhat simplified by determining first the vertical component of the diagonal and from it the stress. The lever arm is then the same as that for the adjacent vertical and often a multiple

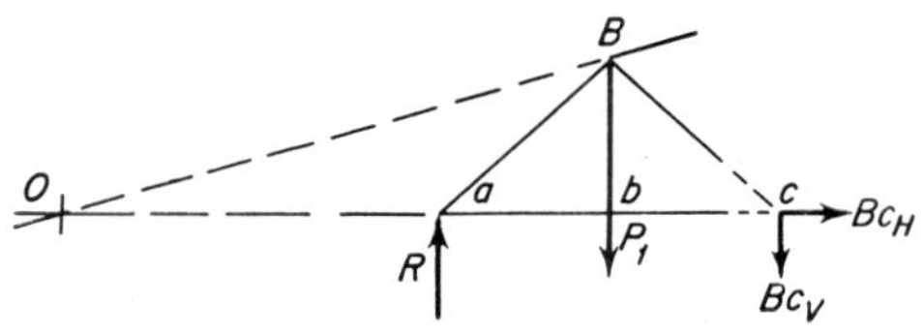

FIG. 42

of the panel length. In Fig. 42, consider the unknown stress in Bc resolved into two components applied at c, one horizontal Bc_H, and the other vertical Bc_V. The line of action of the horizontal component passes through the center of moments O. Taking moments about O, as before, $(Bc_V \times Oc) - (R \times Oa) + (P_1 \times Ob) = 0$ from which Bc_V may be determined. The actual stress in Bc is Bc_V multiplied by the secant of the angle that Bc makes with the vertical. The stresses in verticals, such as Cc, are determined in a manner similar to that used for the diagonals, except that the stress is determined directly, since there is no horizontal component to be considered.

[1] If the unknown stress is always assumed to act away from the section under consideration, a positive result indicates a tensile stress and a negative result a compressive stress.

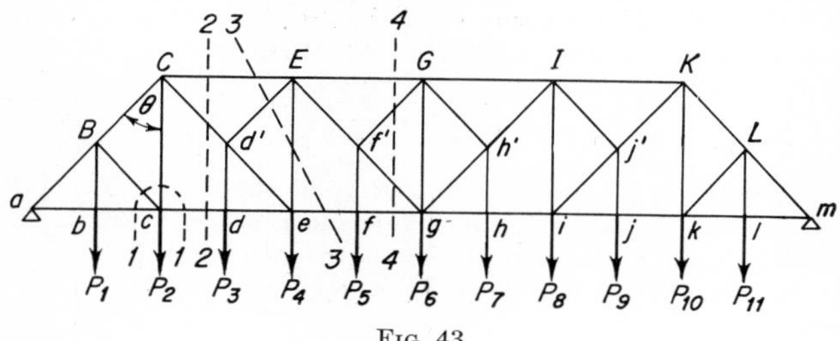

FIG. 43

Parallel Chords and Subdivided Panels. In this type of truss, the subdiagonals may be in either tension or compression. In the truss of Fig. 43, the subdiagonal Bc is in compression, and $d'E$ is in tension. The vertical component of the stress in any subdiagonal is equal to one-half of the load applied at the panel points in the vertical plane of the intersection of the subdiagonal with the main diagonal. This may be proved as follows: If a circular section is taken around the joint d' of Fig. 43, as shown in Fig. 44, the stresses in both the upper and lower portions of the main diagonal Ce are tensile stresses. The stress in the vertical $d'd$ is also tension and is equal to the load applied at d. The direction of the stresses in Cd', $d'e$, and $d'd$ with reference to the joint d' is shown in the figure. The unknown stress $d'E$ may be resolved into components at any point along its line of action, such as E. Taking moments about C, the lever arms of Cd', $d'e$ and $d'E_H$ are zero, and

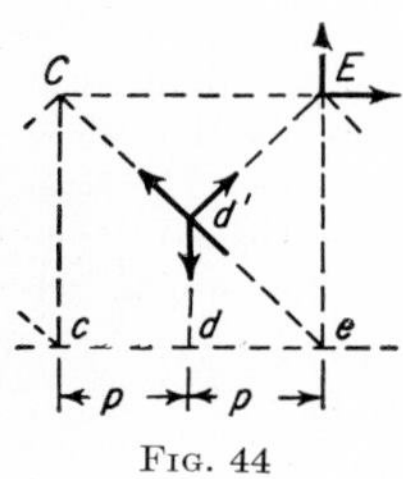

FIG. 44

$$(d'd \times p) - (d'E_V \times 2p) = 0 \qquad \text{or} \qquad d'E_V = d'd \div 2$$

In determining the stresses in the verticals and the main diagonals the method of resolution of forces may be used, but the effect of stresses in the subdiagonals must be considered. In section 1-1, the equation for the stress in Cc is $Cc - \dfrac{P_1}{2} - P_2 = 0$. In section 2-2, the equation for the stress in Cd' is the same as for a diagonal of a truss with a single system of web members, i.e., $Cd = V_2 \sec \theta$. In section 3-3, the equation for the stress in the vertical Ee is $Ee + \dfrac{P_3}{2} + V_3 = 0$. In section 4-4, the equation for the stress in $f'g$ is $-f'g \cos \theta + \dfrac{P_5}{2} + V_4 = 0$.

Inclined Chords and Subdivided Panels. Since the slopes of the main diagonals and the subdiagonals are not the same in this type of truss, the vertical components of the

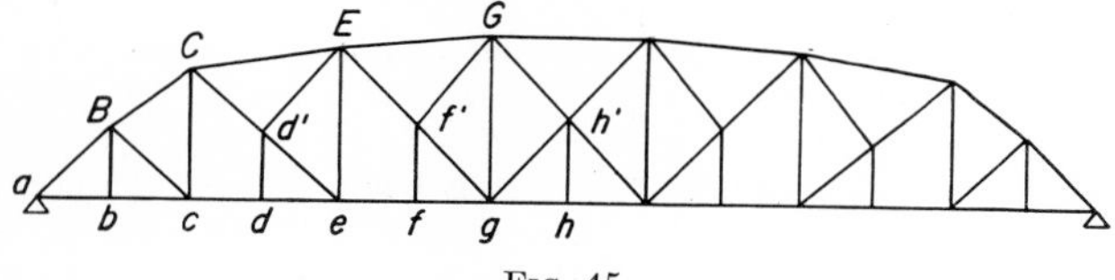

FIG. 45

stresses in the latter are not equal to one-half of the panel loads as in the previous analysis. In the truss of Fig. 45, considering a circular section around the joint d', as shown in

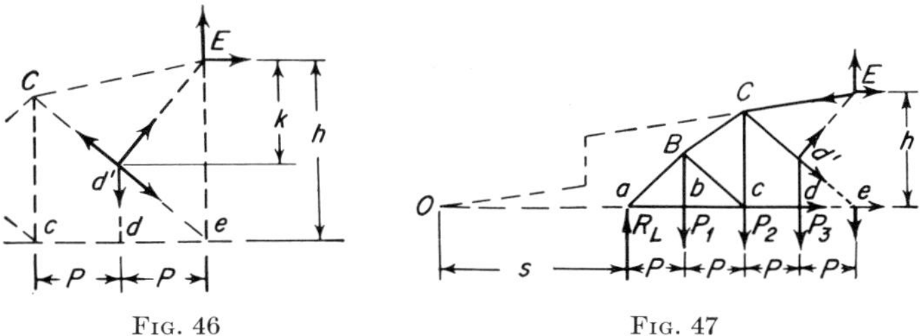

FIG. 46 FIG. 47

Fig. 46, we find that an analysis similar to that for Fig. 44, with the center of moments at e, gives

$$(d'E_H \times h) - (d'd \times p) = 0 \qquad \text{or} \qquad d'E_H = d'd \times \frac{p}{h} \qquad \text{or} \qquad d'E_V = d'd \times \frac{k}{h}$$

from which

$$d'E = d'd \times \frac{p}{h} \times \frac{l}{p} = d'd \times \frac{l}{h}$$

where l is the length of $d'E$.

The stresses in the verticals and diagonals not affected by the stresses in the subdiagonals are obtained as for a truss with inclined chords and a single system of web members. For a member such as $d'e$, the stress is determined as follows: The stress in $d'E$ and its horizontal and vertical components are obtained as above; then in Fig. 47, with the center of moments at O and the stress in $d'E$ resolved into its components at E whereas that in $d'e$ is resolved into its components at e,

$$d'e_V = \frac{(R_L \times s) - (d'E_H \times h) + d'E_V(s + 4p) - P_1(s + p) - P_2(s + 2p) - P_3(s + 3p)}{s + 4p}$$

The stresses in the verticals are obtained in a similar manner.

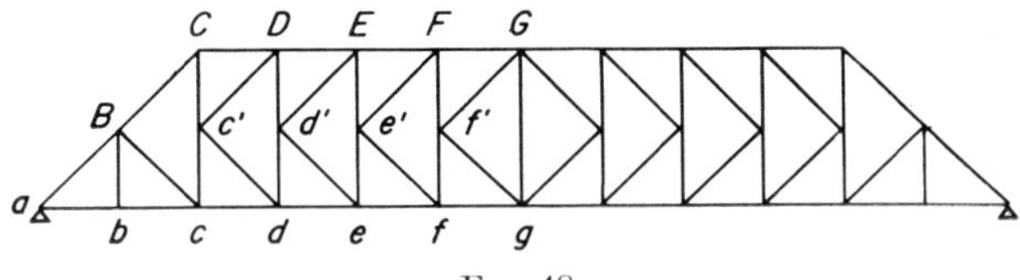

FIG. 48

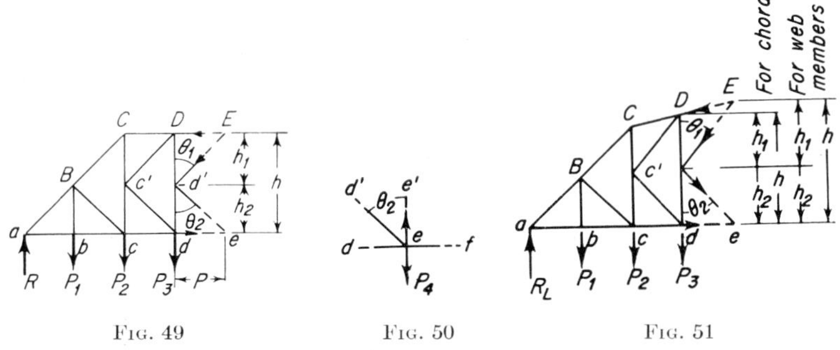

FIG. 49 FIG. 50 FIG. 51

The K Truss. This type of truss is used either with parallel or with inclined chords. For a truss with parallel chords, such as is shown in Fig. 48, the stresses in the diagonals may be obtained by considering sections like that of Fig. 49. From a summation of the vertical forces in this section, $R_L - P_1 - P_2 - P_3 = V = d'E \cos\theta_1 + d'e \cos\theta_2$. From a summation of the horizontal forces at d', $d'E \sin\theta_1 = d'e \sin\theta_2$. With the length of $d'E$ taken as l_1 and that of $d'e$ as l_2, $\sin\theta_1 = p/l_1$, $\sin\theta_2 = p/l_2$, $\cos\theta_1 = h_1/l_1$, and $\cos\theta_2 = \dfrac{h_2}{l_2}$. Then $d'E = d'e \times \dfrac{p}{l_2} \times \dfrac{l_1}{p} = d'e\dfrac{l_1}{l_2}$, and $d'e = d'E\dfrac{l_2}{l_1}$; and, in the equation of the summation of vertical forces, $V = d'E\dfrac{h_1}{l_1} + \left(d'E\dfrac{h_2}{l_2} \times \dfrac{l_2}{l_1}\right) = d'E\dfrac{h_1}{l_1} + d'E\dfrac{h_2}{l_1} = d'E\dfrac{h}{l_1}$, or $d'E = \dfrac{l}{h}V$, and similarly $d'e = \dfrac{l}{h}V$, where l is the length of the diagonal and V is the shear in the panel de.

Since the lower diagonals are in tension, the lower portion of the verticals must be in compression. At this point e, a summation of the vertical forces (Fig. 50) gives $e'e - P + d'e \cos\theta_2 = 0$, or, the value of $d'e$ being substituted from above, $e'e = P - \dfrac{h_2}{h}V$. The stress in the upper portion of the vertical is tension and may be obtained by taking a summation of the vertical forces about the joint e'.

Where the upper chord is inclined, the summation of the vertical forces in the section includes the vertical component of the stress in the inclined chord, i.e., in Fig. 51, $V - DE_V - d'E \cos\theta_1 - d'e \cos\theta_2 = 0$. The stress in the chord, therefore, must be known before the stress in the diagonals can be obtained (see next paragraph). The stresses in the verticals are determined in the same manner as for parallel chords.

Stresses in Chord Members of Bridge Trusses. Chord stresses in simple trusses may always be obtained by the application of the principle of moments. For example, in Fig. 39, in section NN, the stress in the chord BD may be determined by taking the center of moments at c, *the intersection of the other members cut by the section NN*, and solving

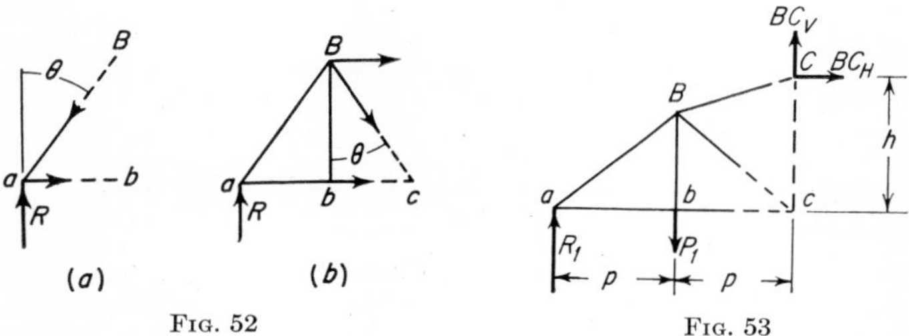

FIG. 52 FIG. 53

the equation of moments for all the forces on one side of the section, i.e., $(R_1 \times 2p) - (P_1 \times p) - (BD \times d) = 0$.

For trusses with parallel chords, the stresses may be determined more readily by the resolution of forces. In Fig. 52*a*, which represents the left end of the truss of Fig. 39, the stresses in aB and ab are in equilibrium with the reaction R (or in general, the shear V). A summation of horizontal components gives $-aB \sin\theta + ab = 0$; and since $aB = -V \sec\theta$, $ab = +V \tan\theta$. By inspection, $ab = bc$; and in Fig. 52*b*, the summation of horizontal forces gives $BD + Bc \sin\theta + bc = 0$ or $BD = -(ab + Bc \sin\theta) = -(ab + V \tan\theta)$. In a similar manner, by passing successive sections, the stress in each chord member may be determined by *adding* the horizontal component ($V \tan\theta$) of the diagonal cut by the section to the chord stress previously obtained.

In a truss with subdivided panels, care must be taken to ascertain the direction of each known stress when making a summation of the horizontal forces in a section. For example, in Fig. 43, in section 3-3, CE, the chord of which the stress is known, is in compression, while the subdiagonal $d'E$ is in tension. Hence, with the unknown stress in eg assumed to act away from the section toward the right, the summation of horizontal forces in the section gives $eg + d'E \sin\theta - CE = 0$ or $eg = CE - d'E \sin\theta = CE - d'E_V \tan\theta$. In this case, the horizontal component of the diagonal is *subtracted* from the known chord stress.

In trusses with inclined chords, the stress in the inclined chord may be determined by resolving it into its horizontal and vertical components where its line of action cuts a vertical plane through the center of moments. For example, in Fig. 53, with the center of moments at c, $BC_H = \dfrac{(-R \times 2p) + (P_1 \times p)}{h}$. The stresses in the horizontal chords are determined directly by similar analysis.

In the K truss with parallel chords, the procedure is as follows: In Fig. 49, with the center of moments at d', $M = (ED \times h_1) + (de \times h_2)$. The summation of horizontal forces in the section gives $ED + d'E_H - d'e_H - de = 0$; but since $d'E_H = d'e_H$, $ED = de =$ moment of external forces $\div (h_1 + h_2) = M \div h$. In the K truss with inclined chords, the same method may be followed, the horizontal component of the stress in the inclined chord being equal to the stress in the horizontal chord.

Uniform Live Load on Bridge Trusses. The position of the uniform load for maximum chord stresses is obvious; for since a load at any point on the span produces a positive moment, the whole span should be loaded for the maximum moment or maximum stresses.

For web members, the conventional method of calculation of the stresses due to uniform load is to assume all panel points on one side of the panel cut by the section through the member fully loaded, while no load is considered on the other side.

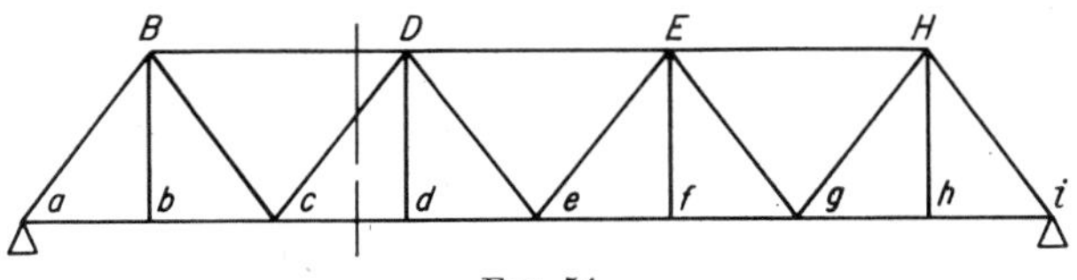

FIG. 54

The maximum and minimum shears are obtained under these two possible loadings. For instance, in Fig. 54, if all panel points to the right of the section are loaded with the load P, moments being taken about the right support, the reaction at a is $^{15}\!/_8P$, and, since there is no load between a and the section, the shear in the section is $^{15}\!/_8P$. If a load were added at c, the reaction would become $^{21}\!/_8P$, but the shear in the section would be $^{21}\!/_8P - {}^8\!/_8P = {}^{13}\!/_8P$, which is less than the value obtained with only the panel points to the right of the section loaded. It follows, therefore, that the largest positive shear occurs when the live load extends from the section to the right support, and the largest negative shear when the live load extends from the section to the left support.

This assumed condition of loading is actually an impossible one, for in order to have a full panel load at the panel point d the panels cd and de must be fully loaded. Since the load on a panel is transferred through the stringers and floor beams to the panel points of the truss, the load on the panel cd would cause a partial panel load at the point c. This would, in case of loading for maximum positive shear, bring a one-half panel load to the point c on the left of the section, provided that a full load were desired at d.

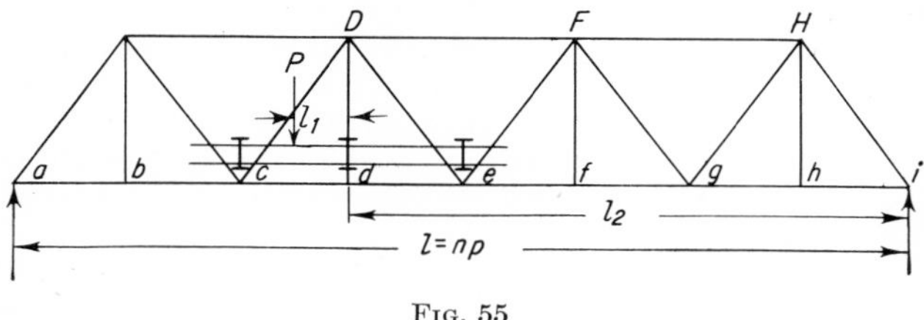

Fig. 55

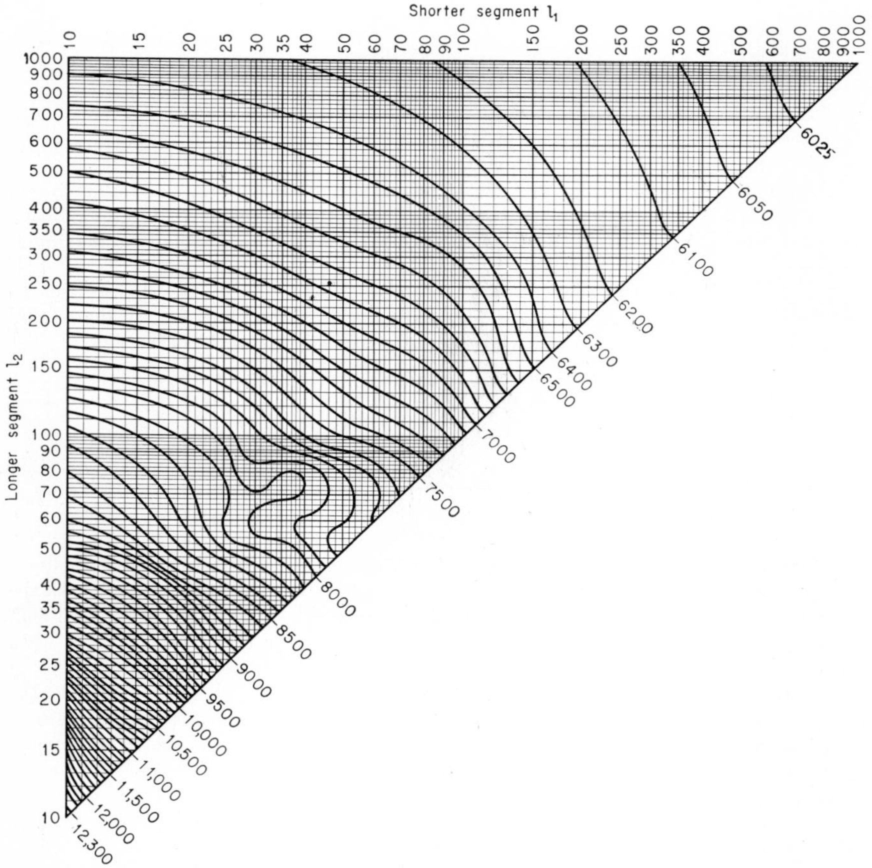

Fig. 56. Curves of equivalent uniform loads for moments and shears. Plotted for Cooper's E-60 loading, values are given in pounds per lin ft per track. (From *Locomotive Loadings for Railway Bridges*, by D. B. Steinman in *Trans.* ASCE, vol. 86, p. 606.)

The true maximum shear in the panel *cd* will evidently occur when the uniform load extends from the right beyond *d* some distance into the panel *cd*.

With a single load P applied on the stringer in the panel *cd* (Fig. 55) at a distance l_1 from the right end of the panel,

$$V_{cd} = R_a - r_c = \frac{M_i}{l} - \frac{M_d}{p} = \frac{P(l_2 + l_1)}{l} - \frac{Pl_1}{p} = P\left(\frac{l_2 + l_1}{l} - \frac{l_1}{p}\right)$$

If the position is such that $\dfrac{l_2 + l_1}{l} = \dfrac{l_1}{p}$ or $l_1 = \dfrac{pl_2}{l - p}$, the shear in the panel is 0, so that the uniform load should extend into the panel a distance of l_1, which for $l = np$ is $l_1 = \dfrac{l_2}{n - 1}$. With the same uniform load per foot of bridge, the exact stresses obtained with the load in the above position are smaller than those obtained by the conventional method.

Figure 56 is a chart from which the proper uniform live load as a substitute for Cooper's E 60 loading may be obtained. For chord members, l_1 is the distance from the center of moments to the right support, and l_2 the corresponding distance to the left support. For web members of trusses with parallel chords, l_2 is the distance shown in Fig. 55 and l_1 is determined as above. If this method is used for trusses with inclined chords, the error is not great.

It is evident from the above that no single uniform load may be substituted for a wheel-load system in calculating the stresses in the various members of a truss but that a special load must be selected for each member.

Counters. In long-span bridges, it is often more economical to design the diagonals of the trusses for tension only, especially in a pin-connected truss. Under such conditions, it becomes necessary to place in some of the panels another diagonal, which crosses the main diagonal, for the purpose of taking care of the resultant negative shear due to the combination of live and dead loads. Such diagonals are known as counters. Let Fig. 57 represent any panel of a truss of which the diagonals can sustain tensile stress only. With the resultant of all the external forces on the left negative, i.e., acting downward, the quadrilateral $CDdc$ tends to assume the shape $C'D'd'c'$, since Cd cannot resist compression. Another tensile diagonal cD prevents deformation occurring.

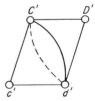

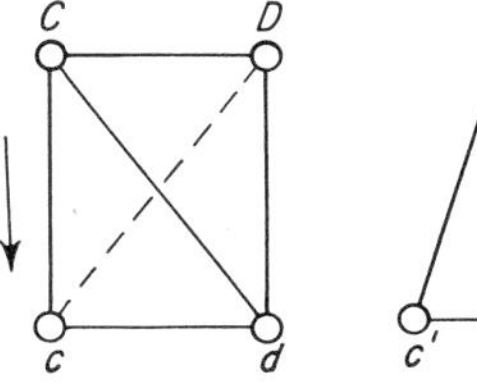

Fig. 57

The two diagonals never act together, the stress in one being zero when the other one is stressed. Although the maximum stresses in the main members of a truss are the same whether or not counters are used, the minimum stresses in the verticals are affected by the presence of counters. In most trusses, however, the minimum stresses in the verticals where counters are used are of the same sign as the maximum stresses and hence have no significance.

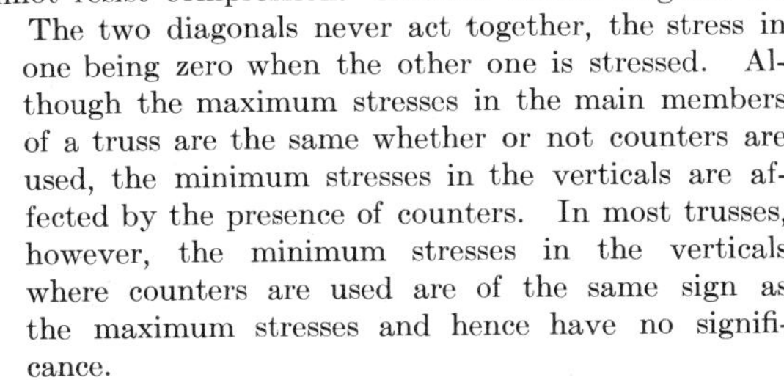

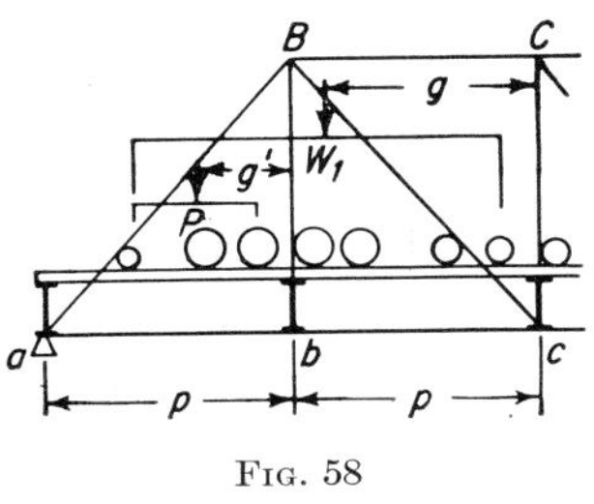

Fig. 58

Concentrated Loads on Bridge Trusses. Where concentrated loads are used in determining the live-load stresses in the members of a truss, certain criteria are helpful in locating the position of the loads producing the maximum stresses.

Position of Loads for Maximum Floor-beam Reaction. In Fig. 58, let P be the resultant of all the loads between a and b, and g' the distance from P to b. Similarly, let W_1 be the resultant of all the loads from a to c, and g the distance from W_1 to c. Moments about b give $r_ap - Pg' = 0$, or $r_ap = Pg'$. Similarly, moments about c give $r_a2p + r_bp - W_1g = 0$. Substituting Pg' for r_aP, we have $r_b = \frac{W_1g - 2Pg'}{p}$. If the loads are moved to the left a distance x, the change in r_b is $\frac{W_1x - 2Px}{p}$, and the rate of change per unit of length is $\frac{W_1 - 2P}{p}$. Since p is a constant, W_1 and P are the only quantities that can change the value of this expression. With the loads so placed that W_1 is greater than $2P$, a movement to the left increases the value of r_b; but when the point is reached where $2P$ becomes equal to W_1, a further movement in the same direction decreases the value of r_b. Therefore, for the maximum floor-beam reaction (= maximum live-load stress in Bb), the loads should be placed so that $W_1 = 2P$, and the reaction itself is, from the foregoing equation for r_b, equal to *the moment of the loads in the two panels about the right end of the two panels minus twice the moment of the loads in the left-hand panel about the right end of the left-hand panel divided by the panel length.*

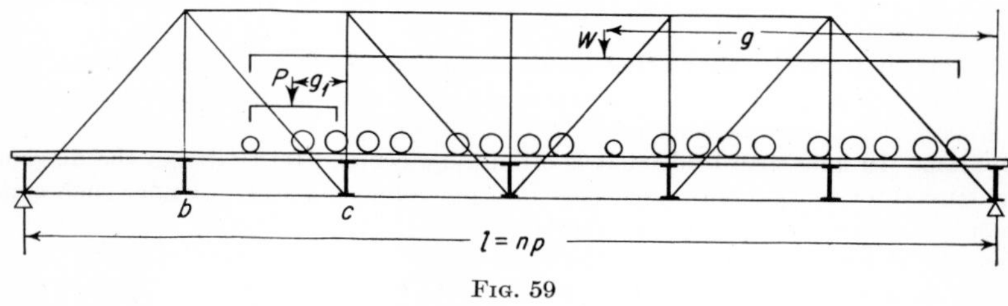

FIG. 59

Position of Loads for Maximum Shear. Trusses with Parallel Chords and Single-web Systems. In Fig. 59, let P be the resultant of all the loads to the left of c and g_1 the distance of P from c. Similarly, let W be the resultant of all the loads on the span and g the distance of W from the right support. The shear in the panel bc is $R_L - r_b$, where R_L is the left reaction of the truss and r_b is the reaction of the floor beam bc due to the loads between b and c. But $R_L = Wg/l$, and $r_b = Pg_1/p$; therefore, $V_{bc} = \frac{Wg}{l} - \frac{Pg_1}{p}$. If the loads are moved to the left a distance x, the change in shear is[1] $\frac{Wx}{l} - \frac{Px}{p}$, and the rate of change per unit of length is $\frac{W}{l} - \frac{P}{p}$. In the usual case, the panels are equal and $l = np$; therefore, for maximum shear, the loads should be placed so that $W = nP$.

This criterion may also be used without modification for the diagonals of a K truss with parallel chords. For the verticals of such a K truss, reference to the equation for $e'e$ on p. 5–38 shows that the criterion must be changed to $W = 2nP$, in order to determine the position of loads for maximum compression in the lower portion of the verticals.

Position of Loads for Maximum Shear. Trusses with Inclined Chords and Single-web Systems. In Fig. 60, let s be the distance from the center of moments to the left support and v the distance from the left support to the left end of the panel through which the section cutting the web member is passed. Let W be the total load on the span and g

[1] If loads were placed in the panel ab, considering that the distance x is small, the expression for change in shear would be the same as above.

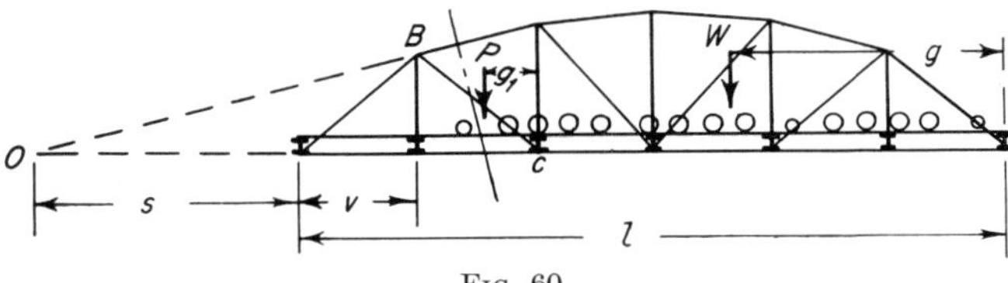

FIG. 60

the distance of its center of gravity from the right support. Let P be the load in the panel through which the section is passed and g_1 the distance of its center of gravity from the right end of the panel. Let p be the panel length and l the span; then $M_o = -\frac{Wg}{l}s + \frac{Pg_1}{p}(s + v)$.

If the loads are moved to the left a short distance x, the change in M_o is $-\frac{Wx}{l}s + \frac{Px}{p}(s + v)$, and the rate of change per unit of length $-\frac{W}{l}s + \frac{P}{p}(s + v)$, so that, for maximum moment at O, $W = \frac{Pl}{p}\frac{(s + v)}{s}$.

In the usual case, with the panels equal and $l = np$, the criterion for the maximum stress in a web member due to any system of concentrated loads is $W = nP\frac{(s + v)}{s}$.

Minimum Stresses in Web Members. Generally, it is easier to determine minimum stresses in web members on the left half of the truss by considering the symmetrical member on the right half with the load coming on from the right. The criteria developed above may be used without modification, except (1) for the lower portion of the verticals of a K truss, where the criterion becomes $W = 2nP_1 - P_2$, where P_1 is the load in the panel to the left of the vertical and P_2 the load in the panel to the right of the vertical, with W and n having the same significance as in previous criteria; and (2) in the Parker truss, where the criterion is $W = nP\frac{(s - v)}{s}$, with s being the distance from the center of moments (now on the right-hand side of the truss) to the left reaction, as before.

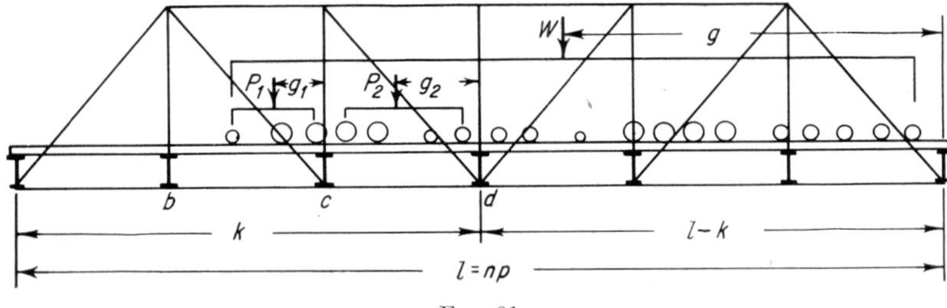

FIG. 61

Position of Loads for Maximum Moment. In all trusses with verticals, the position of the loads may be determined by using the criterion $\frac{k}{l}W = P$, which for equal panels becomes $\frac{m}{n}W = P$. With reference to Fig. 61, moments about d give $M_d = \frac{Wgk}{l} -$

$(r_b \times 2p) - (r_c \times p)$. But $r_b = \dfrac{P_1 g_1}{p}$, and $r_c = \dfrac{P_1(p - g_1)}{p} + \dfrac{P_2 g_2}{p}$. Substituting the values of r_b and r_c, $M_d = \dfrac{Wgk}{l} - P_1(p + g_1) - P_2 g_2$.

If the loads are moved to the left a short distance x, the change in moment is $\dfrac{Wkx}{l} - P_1 x - P_2 x$. If P is the summation of all the loads to the left of the center of moments, this expression becomes $\dfrac{Wkx}{l} - Px$, and the rate of change per unit of length is $\dfrac{Wk}{l} - P$; or for maximum stress the loads must be so placed that $\dfrac{k}{l} W = P$ or, for equal panels, $\dfrac{m}{n} W = P$.

In a truss without verticals, the position of the loads for the maximum stress in the loaded chord cannot be determined by the above criterion.

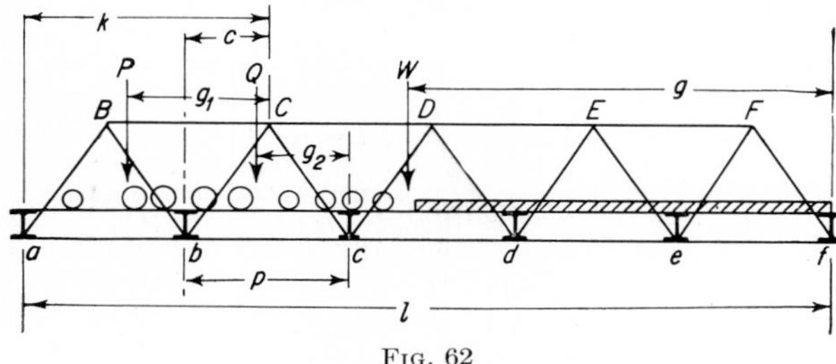

FIG. 62

With reference to Fig. 62, the moment center for the chord bc is the panel point C. Let P be the load to the left of the panel bc, Q the load in the panel bc, and W the total load on the span. Let g_1 be the distance from the center of gravity of the loads P to the center of moments, g_2 the distance from the center of gravity of the loads Q to the right end of the panel, p the panel length, c the distance of the center of moments from the left end of the panel, k the distance of the center of moments from the left reaction, and g the distance of the center of gravity of the loads W from the right support. Then the moment at C is $M_C = \dfrac{Wg}{l} k - Pg_1 - \dfrac{Qg_2}{p} c$. If the load advances a distance x, the increase in moment is $\dfrac{Wk}{l} x - Px - Q \dfrac{c}{p} x$, whence, from previous similar deductions, the maximum moment is obtained when $\dfrac{k}{l} W = P + \dfrac{c}{p} Q$.

The load Q changes in value whenever a wheel load passes either the point b or the point c, so that the position of the loads which causes the maximum moment at C and hence the maximum chord stress in bc occurs with a wheel load at either b or c. When a load passes c, Q may be the only quantity to change in value, while a load passing b causes both of the values P and Q to change.

For the usual case, $c = \frac{1}{2}p$, so that the criterion for maximum moment becomes $\dfrac{k}{l} W = P + \dfrac{1}{2} Q$, and the expression for the moment, $\dfrac{Wg}{l} k - Pg_1 - \dfrac{1}{2} Qg_2$. The position of the loads for maximum stresses in the opposite chord is obtained by the use of

simpler criterion, as all the loads to the left of the center of moments have the same direct effect on the value of the moment.

Stresses in Lateral Trusses. Wind loads may be assumed as all applied on the windward side or as applied equally on the two sides of the lateral truss between the chords of main vertical trusses. In the former case, the stresses in the lateral struts are one-half panel load greater than if the latter assumption were made; but this is of no practical consequence.

Where the diagonals are considered as tension members only, counter stresses need not be computed, since reversal of wind direction gives greater stresses in the members concerned than any partial loading from the opposite direction. When a rigid system of diagonals is used, the two diagonals of a panel may be assumed to be equally stressed. Stresses in the chords of the lateral truss should be combined with those in the chords of the main trusses due to dead and live loads.

Since the stresses in the lateral system of the loaded chord are caused by the effect of both the wind on the truss and the wind on the live load, and the latter cannot occur except when the live load covers the structure, the actual determination of stresses is more easily obtained by using the *method of coefficients.*

In this method, all panel loads are assumed equal to unity and so placed on the truss as to produce the maximum stresses in the respective members. The coefficients of the web members are numerically equal to their respective shears. The actual stresses in the lateral struts are obtained by multiplying the coefficient by the computed panel load, and those in the diagonals by multiplying the panel load times sec θ where θ is the angle between the diagonal and the lateral strut. The coefficient of each chord member equals the sum of the coefficients[1] of all the diagonals on its left and the coefficient of the diagonal at its left panel point. The chord stress is obtained by multiplying the coefficient by the panel load times tan θ.

The coefficients for a rigid system of lateral bracing for a six-panel through Pratt truss with 25-ft panels are given in Fig. 63. A reversal of the direction of the wind causes a change of sign of stress in every member of the system. In all cases, one-half of the load is considered applied on the windward and one-half on the leeward truss.

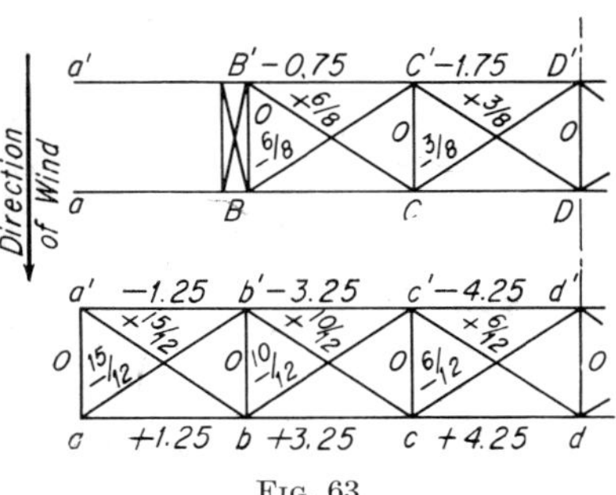

Fig. 63

In bridges of which the trusses have inclined chords, the lateral system of the chords that are inclined lies in several planes, and the exact determination of all the wind stresses is rather difficult. The stresses in the lateral members, however, may be determined by considering the truss flattened out into one plane. The panel lengths will vary, but the panel loads will be equal and may be determined from the horizontal panel length. The resulting chord stresses are not the exact stresses, but the error is not great enough to be of any importance.

For the lateral truss of Fig. 63, assume the panel load on the upper lateral system is $25 \times 150 = 3{,}750$ lb $= 3.75$ kips. For the lower lateral system, assume the panel load due to the wind on the structure is $25 \times 200 = 5{,}000$ lb $= 5.0$ kips; and that due to the wind on a train is $25 \times 300 = 7.5$ kips. Sec $\theta = 1.74$ and tan $\theta = 1.43$. The coefficients of Fig. 63 being used, the stresses are computed and are shown on the upper three diagrams of Fig. 64. The stresses shown on the lower diagram are those due to the applica-

[1] These coefficients are not the maximum coefficients but those obtained with live loads over the whole structure.

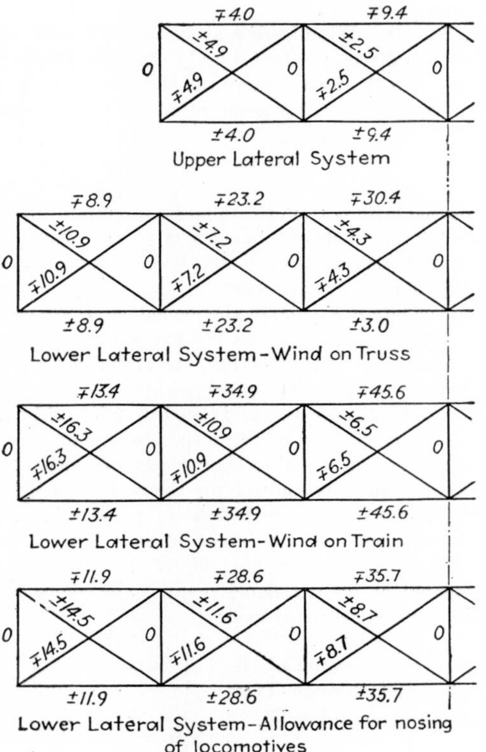

Fig. 64

Fig. 65

Fig. 66

tion of the 20-kip load, specified by the AREA for the locomotive nosing, applied at any point of the span.

Stresses in Main Trusses Due to Lateral Forces. Since some of the lateral forces are applied considerably above the horizontal plane of the end supports of the bridge, these forces tend to overturn the structure.

The lateral forces of the upper lateral system are carried to the portal struts, and the horizontal loads at these points produce an overturning moment about the horizontal plane of the supports. In Fig. 65, P represents the horizontal load brought to each portal strut by the upper lateral bracing, h the depth of the truss, and c the distance between trusses. The overturning moment produced at each end of the structure is Ph, which is balanced by a reaction couple Rc. The value of the reaction R is then Ph/c, and the same effect is produced on the main trusses as is caused by the lateral force on the truss if loads equal to Ph/c are applied to the main trusses at the panel points B and F, as shown in Fig. 66. These loads produce stresses in the end posts and in the lower chord members, but the web members are not stressed.

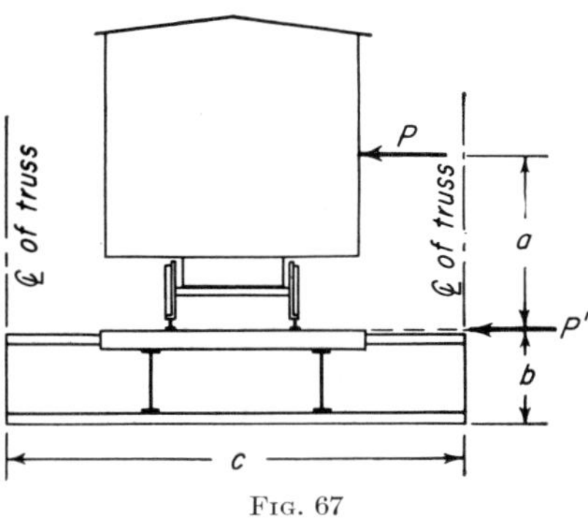

Fig. 67

The lateral force on a train is considered to be applied some distance above the top of the rail, usually 7 or 8 ft, and the top of the rail in a through truss is usually several feet above the plane of the end supports. Hence, another overturning moment is produced by this force, which may be treated in a similar manner to that caused by the lateral force on the upper lateral system. There is a difference, however, as far as the web members of the main trusses are concerned. Since the lateral force on the train produces an effect corresponding to the position of the train load on the bridge, it is necessary to determine equivalent vertical panel loads rather than equivalent reactions. From Fig. 67, it can be readily seen that the equivalent vertical panel load is $P(a + b)/c$, where P is the horizontal panel load due to the lateral force on the train.

The effect of the lateral force specified to compensate for the nosing of locomotives is determined in a similar manner, the equivalent vertical panel load being $P'b/c$, where P' is the horizontal panel load.

Applying the foregoing principles to the lateral truss of Fig. 63, the equivalent vertical loads and reactions producing the same effect as the lateral forces on the upper lateral system are $\dfrac{3{,}750 \times 5 \times 30}{2 \times 17.5} = 16{,}100$ lb $= 16.1$ kips,[1] and the stresses produced in the truss farthest in line of action of the lateral forces or the leeward truss are:

$$\begin{aligned} &\text{End posts:} \quad 16.1 \times 1.30 = -20.9 \text{ kips} \\ &\text{Lower chords:} \quad = +13.4 \text{ kips} \end{aligned}$$

The stresses in the corresponding members of the windward truss have opposite signs. The same specifications being used as for the lateral forces on the trusses, the horizontal force on the train per panel is $300 \times 25 = 7{,}500$ lb. The equivalent vertical panel loads producing the same effect as these horizontal forces due to the lateral force on the train and the lateral force specified to allow for the nosing of locomotives are (considering

[1] Considering a full horizontal panel load at BB' and FF' to allow for the lateral forces on the end posts and portal bracing.

the top of rail 5 ft above the plane of the end supports):

$$\text{For lateral force on train: } \frac{7{,}500 \times (8 + 5)}{17.5} = 5{,}600 \text{ lb} = 5.6 \text{ kips}$$

$$\text{For nosing of locomotives: } \frac{20{,}000 \times 5}{17.5} = 5{,}700 \text{ lb} = 5.7 \text{ kips}$$

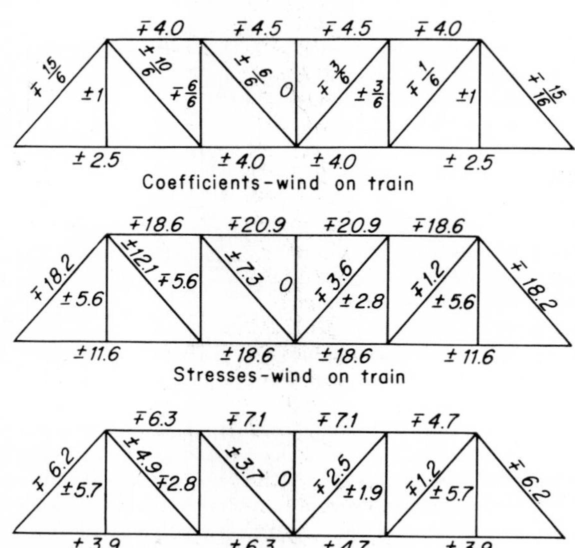

Fig. 68

The first panel load being treated as a moving load, the coefficients of the maximum stresses produced are marked on the upper diagram of Fig. 68 and the stresses are given on the middle diagram. The maximum stresses produced by the single equivalent vertical panel load for nosing of locomotives are marked on the lower diagram. The upper sign preceding the values of the coefficients and stresses is for the leeward and the lower for the windward truss.

Portal Bracing. The chief function of the portal bracing is to transfer the reactions of the upper lateral system to the supports of the bridge. In addition, it stiffens the end posts against vibration. In order to reduce the flexure in these posts to a minimum, the portal framing should extend as low as the headroom will permit. The bases of the end posts are partly fixed by virtue of the direct stress due to the vertical load. The exact location of the point of inflection in the end posts cannot be determined until the design has been made and the distance center to center of pin bearings of each end post is known.[1]

Maximum and Minimum Stresses. The stresses for which the several members of a truss are designed are the summation of all the simultaneous stresses caused by the dead load, live load, impact, and lateral forces. In those members where no reversal of stress is possible under any possible combination of forces, the minimum stress has no

[1] For a more detailed discussion of portal bracing and lateral forces on bridge trusses, see Urquhart and O'Rourke, "Stresses in Simple Structures," McGraw-Hill Book Company, 1932.

significance. It is usual in specifications to allow certain increases in the allowable unit stresses when the full effect of the lateral forces is considered. It therefore becomes necessary to have two tabulations of maximum and minimum stresses.

GENERAL TOOLS FOR STRUCTURAL ANALYSIS

For some types of structures, the equilibrium equations are not sufficient to determine the reactions or the internal stresses. These structures are called statically indeterminate.

For the analysis of such structures, additional equations must be written based on a knowledge of the elastic deformations. Hence, methods of analysis that enable deformations to be evaluated in terms of unknown forces or stresses are important for the solution of problems involving statically indeterminate structures. Some of these methods, like the method of virtual work, are also useful in solving complicated problems involving statically determinate systems.

Virtual Work. A virtual displacement is an imaginary, small displacement of a particle consistent with the constraints upon it. Thus, at one support of a simply supported beam, the virtual displacement could be an infinitesimal rotation $d\theta$ of that end, but not a vertical movement. However, if the support is replaced by a force, then a vertical virtual displacement may be applied to the beam at that end.

Virtual work is the product of the distance a particle moves during a virtual displacement by the component in the direction of the displacement of a force acting on the particle. If the displacement and the force are in opposite directions, the virtual work is negative. When the displacement is normal to the force, no work is done.

Suppose a rigid body is acted upon by a system of forces with a resultant R. Given a virtual displacement ds at an angle α with R, the body will have virtual work done on it equal to $R \cos \alpha \, ds$. (No work is done by internal forces. They act in pairs of equal magnitude but opposite direction, and the virtual work done by one force of a pair is equal and opposite in sign to the work done by the other force.) If the body is in equilibrium under the action of the forces, then $R = 0$, and the virtual work also is zero.

Thus, the **principle of virtual work** may be stated: *If a rigid body in equilibrium is given a virtual displacement, the sum of the virtual work of the forces acting on it must be zero.*

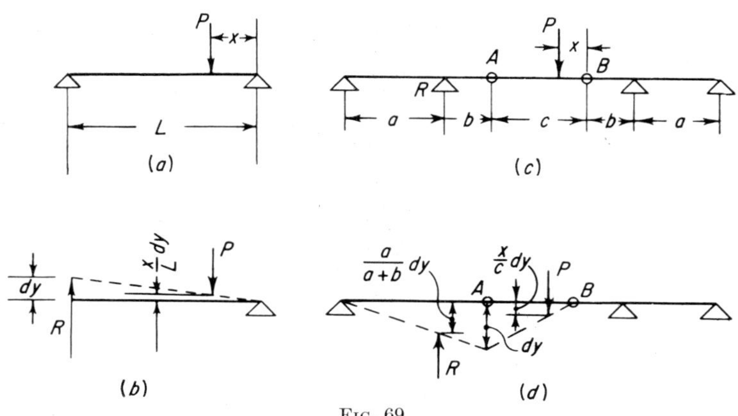

FIG. 69

As an example of how the principle may be used, let us apply it to the determination of the reaction R of the simple beam in Fig. 69a. First, replace the support by an unknown force R. Next, move that end of the beam upward a small amount dy as in Fig. 69b.

The displacement under the load P will be $x\,dy/L$, upward. Then, by the principle of virtual work, $R\,dy - Px\,dy/L = 0$, from which $R = Px/L$.

The principle may also be used to find the reaction R of the more complex beam in Fig. 69*c*. The first step again is to replace the support by an unknown force R. Next, apply a virtual downward displacement dy at hinge A (Fig. 69*d*). The displacement under the load P will be $x\,dy/c$, and at the reaction R will be $a\,dy/(a + b)$. According to the principle of virtual work, $-Ra\,dy/(a + b) + Px\,dy/c = 0$; thus $R = Px(a + b)/ac$. In this type of problem, the method has the advantage that only one reaction need be considered at a time and internal forces are not involved.

When an elastic body is deformed the virtual work done by the internal forces is equal to the corresponding increment of the strain energy taken with negative sign. Utilizing this concept, the principle of virtual work can be adapted for the solution of statically indeterminate structures.

Assume a constrained elastic body acted upon by forces $P_1, P_2 \ldots$, for which the corresponding deformations are $e_1, e_2, \ldots$. Application of the principle of virtual work yields $\Sigma P_n de_n + (-dU) = 0$, in which the first term on the left side of the equation gives the work done by the external forces on a virtual displacement and the second term is the work done by the internal forces. The increment of the strain energy due to the increments of the deformations is given by

$$dU = \frac{\partial U}{\partial e_1} de_1 + \frac{\partial U}{\partial e_2} de_2 + \cdots$$

In solving a specific problem, a virtual displacement that is most convenient in simplifying the solution should be chosen. Suppose, for example, a virtual displacement is selected that affects only the deformation e_n corresponding to the load P_n, other deformations being unchanged. Then, the principle of virtual work requires that

$$P_n\,de_n - \frac{\partial U}{\partial e_n} de_n = 0$$

This is equivalent to

$$\frac{\partial U}{\partial e_n} = P_n$$

which states that the partial derivative of the strain energy with respect to a specific deformation gives the corresponding force.

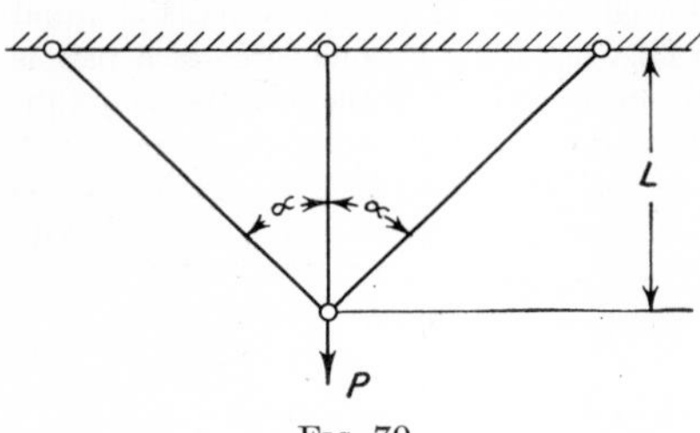

FIG. 70

Suppose, for example, the stress in the vertical bar in Fig. 70 is to be determined. All bars are made of the same material and have the same cross section A. If the vertical bar stretches an amount e under the load P the inclined bars will each stretch an amount $e \cos \alpha$. The strain energy in the system is

$$U = \frac{AE}{2L}(e^2 + 2e^2 \cos^3 \alpha)$$

and the partial derivative of this with respect to e must be equal to P; that is,

$$P = \frac{AE}{2L}(2e + 4e \cos^3 \alpha) = \frac{AEe}{L}(1 + 2 \cos^3 \alpha)$$

Noting that the force in the vertical bar equals AEe/L, we find from the above equation that the required stress equals $P/(1 + 2 \cos^3 \alpha)$.

Castigliano's Theorem. It can also be shown that, if the strain energy is expressed as a function of statically independent forces, the partial derivative of the strain energy with respect to one of the forces gives the deformation corresponding to that force.[1]

$$\frac{\partial U}{\partial P_n} = e_n$$

This is known as Castigliano's first theorem. (His second theorem is the principle of least work.)

Method of Least Work. If deformation of a structure is prevented, as at a support, the partial derivative of the strain energy with respect to that supporting force must be zero, according to Castigliano's first theorem. This establishes his second theorem:

The strain energy in a statically indeterminate structure is the minimum consistent with equilibrium.

As an example of the use of the method of least work, we shall solve again for the stress in the vertical bar in Fig. 70. Calling this stress X, we note that the stress in each of the inclined bars must be $(P - X)/2 \cos \alpha$. The strain energy in the system can be expressed in terms of X as follows:

$$U = \frac{X^2L}{2AE} + \frac{(P - X)^2L}{4AE \cos^3 \alpha}$$

Hence, the internal work in the system will be a minimum when

$$\frac{\partial U}{\partial X} = \frac{XL}{AE} - \frac{(P - X)L}{2AE \cos^3 \alpha} = 0$$

Solving for X gives the stress in the vertical bar as $P/(1 + 2 \cos^3 \alpha)$, as before.

Dummy Unit-load Method. The strain energy for pure bending is $U = M^2L/2EI$. To find the strain energy due to bending stress in a beam, we can apply this equation to a differential length dx of the beam and integrate over the entire span. Thus,

$$U = \int_0^L \frac{M^2\,dx}{2EI}$$

If M represents the bending moment due to a generalized force P, the partial derivative of the strain energy with respect to P is the deformation d corresponding to P. Differentiating the strain-energy equation gives:

$$d = \int_0^L \frac{M}{EI}\frac{\partial M}{\partial P}\,dx$$

The partial derivative in this equation is the rate of change of bending moment with the load P. It is equal to the bending moment m produced by a unit generalized load applied at the point where the deformation is to be measured and in the direction of the deformation. Hence, the last equation can also be written as

$$d = \int_0^L \frac{Mm}{EI}\,dx$$

To find the vertical deflection of a beam, we apply a dummy unit load vertically at the point where the deflection is to be measured and substitute the bending moments due to this load and the actual loading in this equation. Similarly, to compute a rotation, we apply a dummy unit moment.

As a simple example, let us apply the dummy unit-load method to the determination of the deflection at the center of a simply supported, uniformly loaded beam of constant

[1] Timoshenko, S., and D. H. Young, "Theory of Structures," McGraw-Hill Book Company, Inc., New York, 1945.

moment of inertia (Fig. 71*a*). As indicated in Fig. 71*b*, the bending moment at a distance x from one end is $\frac{wL}{2}x - \frac{w}{2}x^2$. If we apply a dummy unit load vertically at the center of the beam (Fig. 71*c*), where the vertical deflection is to be determined, the moment at x is $x/2$, as indicated in Fig. 71*d*. Substituting in the deflection equation gives

$$d = 2\int_0^{L/2} \left(\frac{wL}{2}x - \frac{w}{2}x^2\right)\frac{x}{2}\frac{dx}{EI} = \frac{5wL^4}{384EI}$$

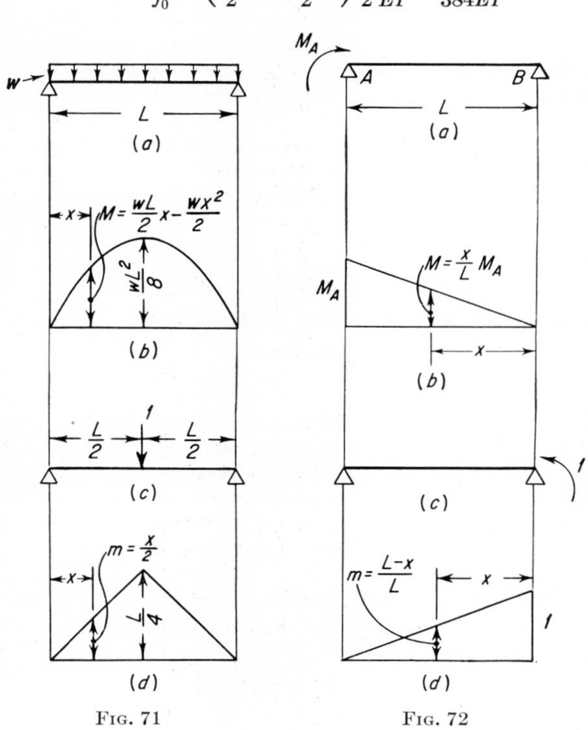

FIG. 71

FIG. 72

As another example, let us apply the method to finding the end rotation at one end of a simply supported, prismatic beam produced by a moment applied at the other end. In other words, the problem is to find the end rotation at B, θ_B, in Fig. 72*a*, due to M_A. As indicated in Fig. 72*b*, the bending moment at a distance x from B due to M_A is $M_A x/L$. If we apply a dummy unit moment at B (Fig. 72*c*), it will produce a moment at x of $(L - x)/L$ (Fig. 72*d*). Substituting in the deflection equation gives

$$\theta_B = \int_0^L M_A \frac{x}{L}\frac{L-x}{L}\frac{dx}{EI} = \frac{M_A L}{6EI}$$

To determine the deflection of a beam due to shear, Castigliano's theorem can be applied to the strain energy in shear:

$$U = \iint \frac{v^2}{2G}\,dA\,dx$$

where v is the shearing unit stress, G the modulus of rigidity, and A the cross-sectional area.

Truss Deflections by Dummy Unit-load Method. The dummy unit-load method may also be adapted for determination of the deformation of trusses. The strain energy in a truss is given by

$$U = \sum \frac{S^2L}{2AE}$$

which represents the sum of the strain energy for all the members of the truss. S is the stress in each member due to the loads, L the length of each, A the cross-sectional area, and E the modulus of elasticity. Applying Castigliano's first theorem and differentiating inside the summation sign yields the deformation:

$$d = \sum \frac{SL}{AE}\frac{\partial S}{\partial P}$$

The partial derivative in this equation is the rate of change of axial stress with the load P. It is equal to the axial stress in each bar of the truss u, produced by a unit load applied at the point where the deformation is to be measured and in the direction of the deformation. Consequently, the last equation can also be written

$$d = \sum \frac{SuL}{AE}$$

To find the deflection at any point of a truss, apply a dummy unit vertical load at the panel point where the deflection is to be measured and substitute in the deflection equation the stresses in each member of the truss due to this load and the actual loading. Similarly, to find the rotation of any joint, apply a dummy unit moment at the joint, compute the stresses in each member of the truss, and substitute in the deflection equation. When it is necessary to determine the relative movement of two panel points in the direction of a member connecting them, apply dummy unit loads in opposite directions at those points.

It should be noted that members that are not stressed by the actual loads or the dummy loads do not enter into the calculation of a deformation.

Illustrative Problem. As an example of the application of the deflection equation, let us compute the mid-span deflection of the truss in Fig. 73. The stresses due to the 20-kip load at every panel point are shown in Fig. 73*a*, and the ratio of length of members in inches to their cross-sectional area in square inches is given in Table 5. We apply a dummy unit vertical load at L_2, where the deflection is required. Stresses u due to this load are shown in Fig. 73*b* and Table 5.

TABLE 5. MID-SPAN DEFLECTION OF A TRUSS

Member	L/A	S	u	SuL/A
L_0L_2	160	+40	$+\frac{2}{3}$	4,267
L_0U_1	75	−50	$-\frac{5}{6}$	3,125
U_1U_2	60	−53.3	$-\frac{4}{3}$	4,267
U_1L_2	150	+16.7	$+\frac{5}{6}$	2,083
				13,742

$$d = \sum \frac{SuL}{AE} = \frac{2 \times 13{,}742{,}000}{30{,}000{,}000} = 0.916 \text{ in.}$$

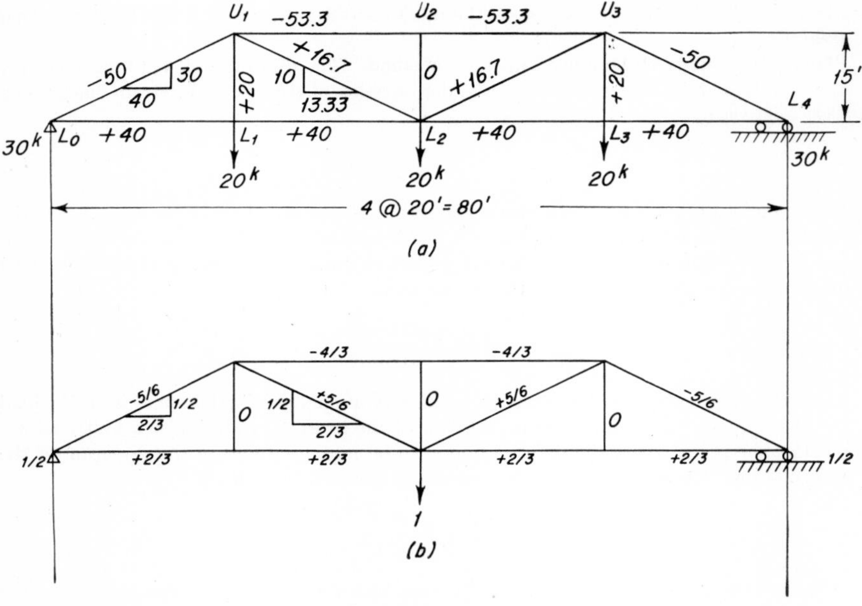

Fig. 73

The computations for the deflection also are given in Table 5. Members not stressed by the 20-kip loads or the dummy unit loads are not included. Taking advantage of the symmetry of the truss, we tabulate the values for only half the truss and double the sum. Also, to reduce the amount of calculation, we do not include the modulus of elasticity E, which is equal to 30,000,000, until the very last step, since it is the same for all members.

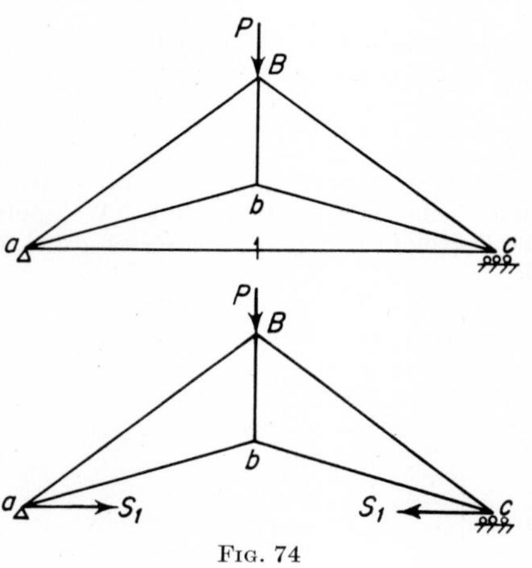

Fig. 74

Statically Indeterminate Trusses. A truss is statically indeterminate when the number of unknown quantities exceeds the number of independent equations of static equilibrium that may be written for the structure or portions of the structure. Let m equal the number of members, n the number of joints, and ΣR the number of reaction conditions.

The number of possible equations is $2n$, and the number of unknown quantities $m + \Sigma R$. Where $2n = m + \Sigma R$, the structure is statically determinate; where $2n < m + \Sigma R$, the structure is statically indeterminate.

The truss of Fig. 74 is statically indeterminate for $m = 6$, $n = 4$, $\Sigma R = 3$, since $2 \times 4 < 6 + 3$. Without the member ac, the truss is statically determinate; for then $m = 5$, $n = 4$, $\Sigma R = 3$, and $2n = m + \Sigma R$.

With member 1 cut, a force of S_1 is applied at both a and c. Let the stress in any member due to P and S_1, be $S' + uS_1$, where S' is the stress due to P, u, the stress in any member due to unit stress in member 1, and uS_1, the stress due to S_1. The displacement of c (on the assumption that a is fixed) due to both P and S_1, is

$$\delta = \sum_2^n \frac{(S' + uS_1)ul}{AE}$$

and this displacement must equal the elongation of member 1, which[1] is $-\dfrac{S_1 l_1}{A_1 E}$. Therefore,

$$\sum_2^n \frac{S'ul}{AE} + \sum_2^n \frac{S_1 u^2 l}{AE} = -\frac{S_1 l_1}{A_1 E}$$

In member 1, $u = 1$ and the right-hand term of the foregoing equation may be written $S_1 1^2 l_1/A_1E$ or $S_1 u^2 l_1/A_1 E$. Hence,

$$\sum_2^n \frac{S'ul}{AE} + S_1 \sum_1^n \frac{u^2 l}{AE} = 0$$

and

$$S_1 = -\frac{\displaystyle\sum_2^n \frac{S'ul}{AE}}{\displaystyle\sum_1^n \frac{u^2 l}{AE}}$$

or with E a constant

$$S_1 = -\frac{\displaystyle\sum_2^n \frac{S'ul}{A}}{\displaystyle\sum_1^n \frac{u^2 l}{A}}$$

With S_1 known, the stress in each member of the truss may be determined, it being equal to $S' + uS_1$.

Illustrative Problem. In Fig. 75, the member 1 is assumed as the redundant member. The pair of unit forces is applied near the upper end of the member in such a direction

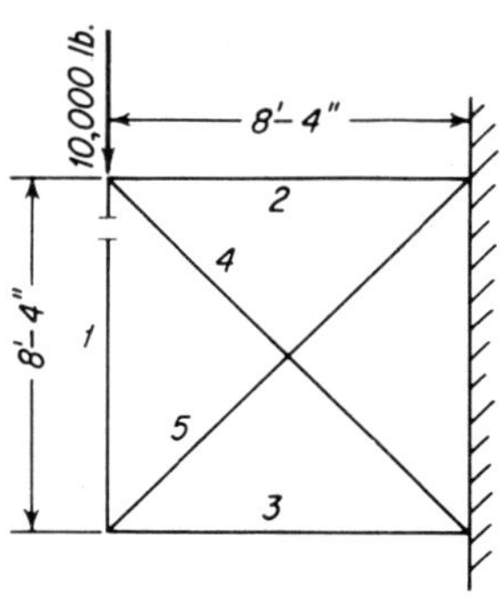

FIG. 75

as to cause tension in the member. From the following table, with E a constant, $S_1 = -717{,}000/168.4 = -4{,}300$.

[1] Minus since the movement is opposite to that indicated in the figure.

The true stress in each member of the complete frame is given in the last column of the table.

Member	A, sq in.	l, in.	S', lb	$S'l/A$	u, lb	$S'ul/A$	u^2l/A	S_1u	$S = S' + S_1u$, lb
1	4	100	0	....	+1.00	....	+25.0	−4300	−4300
2	4	100	+10,000	+250,000	+1.00	+250,000	+25.0	−4300	+5700
3	4	100	0	....	+1.00	....	+25.0	−4300	−4300
4	6	141	−14,100	−331,000	−1.41	+467,000	+46.7	+6000	−8100
5	6	141	0	....	−1.41	....	+46.7	+6000	+6000
					Σ	+717,000	+168.4		

Trussed Beams. In a truss-and-beam combination, both the bending and the direct stress must be considered. The general equation for any redundant quantity, such as an intermediate reaction in a continuous beam, is

$$S_r = \frac{\int \frac{M'm\,dx}{EI}}{\int \frac{m^2\,dx}{EI}}$$

in which M' is the bending moment in the determinate structure due to the external loads and m the bending moment due to a unit force applied at the point of the intermediate reaction.

But since from the previous article, where no bending occurs,

$$S_r = \frac{\Sigma \frac{S'ul}{AE}}{\Sigma \frac{u^2l}{AE}}$$

for combined bending and direct stress,

$$S_r = \frac{\Sigma \frac{S'ul}{AE} + \int \frac{M'm\,dx}{EI}}{\Sigma \frac{u^2l}{AE} + \int \frac{m^2\,dx}{EI}}$$

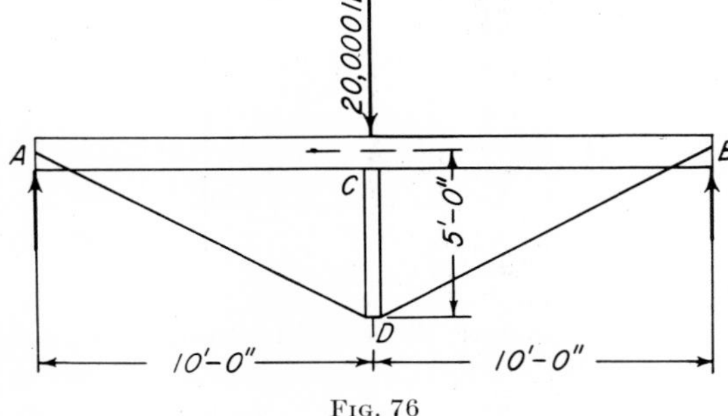

Fig. 76

Illustrative Problem. In Fig. 76, which represents a trussed-steel beam, the member CD is assumed as the redundant member cut near the upper end. In the resultant determinate structure, the load P is carried to the supports entirely by the bending in AB. Under these conditions, there is no truss action, and the term $S'ul/AE$ vanishes from the foregoing general equation. Also, provided that the unit forces applied on the cut faces of CD are so directed as to cause a unit tensile stress in that member, $M' = +Pm$, and with E a constant,

$$S_r = S_{CD} = -\frac{O + P\int_A^B \frac{m^2\,dx}{I}}{\sum \frac{u^2 l}{A} + \int_A^B \frac{m^2\,dx}{I}}$$

Member	Section	A, sq in.	l, in.	u, lb	u^2	u^2l/A
AB	12 in. 31.8 lb I	9.26 (I = 215.8 in.)	240.0	+1.0	1.00	25.9
CD	5 in. 10.0 lb I	2.87	60.0	+1.0	1.00	20.9
AD }	1 − 1 in. ϕ	0.79	134.3	−1.12	1.25	212.5
DB }	rod (upset)	0.79	134.3	−1.12	1.25	212.5
					$\sum \frac{u^2l}{A}$	471.8

$$\int_0^{240} \frac{m^2\,dx}{I} = \frac{2}{215.8}\int_0^{120}\left(\frac{1}{2}x\right)^2 dx = \frac{2}{12 \times 215.8}\,[x^3]_0^{120} = 1{,}320$$

$$S_{CD} = -\frac{20{,}000 \times 1{,}320}{471.8 + 1{,}320} = -\ 14{,}750 \text{ lb}$$

It follows that

$$S_{AD} = S_{BD} = S' + S_r u = 0 + [-14{,}750 \times (-1.12)] = +16{,}500 \text{ lb}$$
$$S_{AB} = 0 + [-14{,}750 \times 1.0] = -14{,}750 \text{ lb}$$

The maximum bending moment in AB, its own weight being neglected, is

$$M_{AB} = \frac{20{,}000 - 14{,}750}{2} \times 10 = 26{,}250 \text{ ft-lb}$$

Displacement Diagrams. *The Displacement of a Joint.* If any member of a framed structure is subjected to a stress or a change in temperature, its length is changed. When the change is caused by stress, its amount is $\lambda = Sl/AE$, in which λ is the change in length, S the total stress in the member, l its length, A its cross section, and E the modulus of elasticity of the material. When the change in length is caused by a change in temperature, its amount is $\lambda_t = \omega tl$, in which ω is the coefficient of linear expansion per degree of temperature change and t the number of degrees of change.

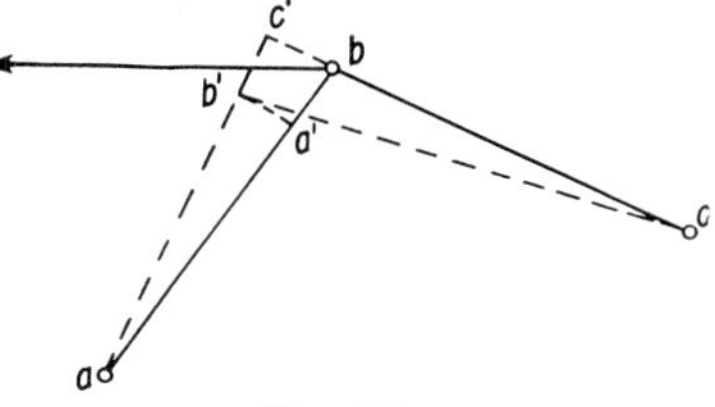

FIG. 77

In Fig. 77, consider the points a and c fixed as to horizontal or vertical movement, and the frame abc hinged at its three joints. The effect of a load applied at b, as shown, is to shorten the member ab and lengthen the member bc owing to the respective compressive and tensile stresses produced in them. Let their lengths under this load be aa' and cc' respectively. With a as a center and

aa' as a radius, let an arc be described. The point b must lie somewhere on this arc. Similarly, an arc is described with c as a center and cc' as a radius. The intersection of the two arcs b' is the position of the point b; and the frame abc, under the action of the load shown, assumes the shape $ab'c$.

The deformations, as shown in the figure, are very much exaggerated; for if they were laid off to the same scale as the lengths, they would not be visible. Since they are so small in comparison with the lengths of the members, the tangents to the arcs may be substituted for the arcs themselves without appreciable error. Therefore, the determination of the position of the point b' may be made by erecting perpendiculars to ab and bc at a' and c' respectively, their intersection being the position of the point b'.

The Displacement Diagram. In the previous discussion, two points of the frame abc were considered fixed. If, however, this frame were considered merely as a portion of a larger frame the stresses in which cause a shifting in the positions of the points a and c, the final position of b would also be affected. In Fig. 78, the magnitude and direction of the displacements of the points a and c respectively are represented by the lines aa' and cc'. The position of b as affected by the change in position of a being first considered, it is seen that b takes the position b_1; but the shifting in the position of c to c' considered alone would cause b to assume the position b_2. The deformation in the members ab and bc causes a further shifting of b. To proceed as in the previous article, the change in the length of ab is laid off from b_1 *toward* a', since the change in length is a shortening; its magnitude is represented by the heavy line. Similarly, from b_2 the elongation of cb is laid off *away from* c'. At the extremities of these two deformations, perpendiculars are erected, their intersection locating the point b'. The final displacements of the three points a, b, and c are bb_1, bb', and bb_2 respectively.

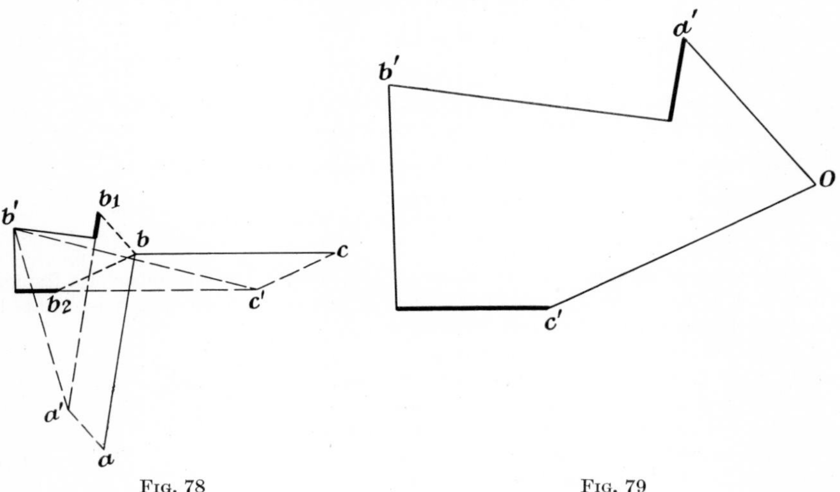

Fig. 78

Fig. 79

The displacements of the points a and c, the deformations of ab and bc, together with the perpendiculars erected at the extremities of the deformations, form a closed polygon. Therefore, that portion of the diagram may be constructed separately in order to determine the final displacements. This is desirable on account of the exceedingly small values of the deformations compared with the lengths of the members. Figure 79 is the necessary diagram drawn to a larger scale. Such a diagram is called a "displacement diagram."

The Displacement Diagram for a Truss. In Fig. 80 the changes in length of the several members of the truss of (a) due to a single load applied at d are marked on the (a) diagram. The support a is fixed, while the other support c is free to move in the inclined direction indicated.

The displacement diagram may be constructed by considering any joint fixed in position, and one of the members making the joint may be considered fixed in direction. In the following construction, a and the direction of ad are regarded as fixed:

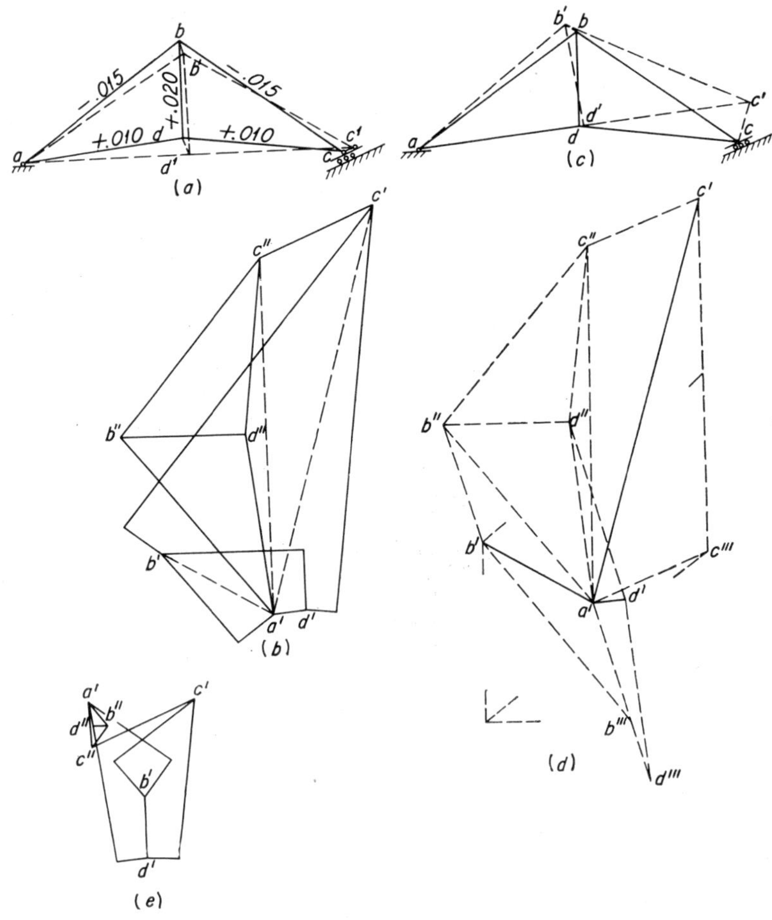

Fig. 80

To begin at a' in (b), the elongation of the member ad is laid off in the direction of a toward d, since in the final position of the deformed truss d is pulled away from a. With d' thus determined, the displacement of b is found by regarding a and d in the triangle abd as fixed. The elongation of bd is laid off from d' in the direction of d toward b, and the shortening of ab from a' in the direction of b toward a. The intersection of the perpendiculars, erected at their extremities, is the point b'. In the same manner, c'

is located by laying off the shortening of bc from b' and the elongation of dc from d' and erecting the perpendiculars. The lines $a'b'$ and $a'c'$ are the displacements of the points b and c respectively.

The deformation of the truss, under the assumed conditions of (a) and the direction of ad being fixed, is shown in (c), where the displacements obtained in (b) are laid off to a smaller scale and the corresponding panel points joined by broken lines. The deformation is greatly exaggerated in order to show the general effect.

The original conditions, however, require that c shall move only on the inclined plane of the support, and therefore the whole truss must be revolved about a as a center until c' falls into a line drawn through c parallel to the constrained line of motion. As the arc thus described by c' is very small compared with the radius ac' and as the direction of ac' is practically the same as that of ac, a perpendicular from c' to ac may be substituted for the arc.

In (d), in which the displacements $a'b'$, $a'c'$, and $a'd'$ are shown without the construction lines of (b), the corresponding path of rotation of c is represented by $c'c'''$, which is drawn perpendicular to ac of (a) to an intersection with a line drawn through a' parallel to the constrained line of motion of c. The actual displacement of c is $a'c'''$. The displacement of b caused by the rotation of the truss about a is $b'b'''$, which is drawn perpendicular to ab of (a) and the length of which bears the same relation to $c'c'''$ as their respective radii of rotation bear to one another. That is, $b'b''':c'c''' = ab:ac$. The length of $b'b'''$ may be determined by similar triangles as follows: Through the mid-point of $c'c'''$ draw a line through b', and from c''' another parallel line to its intersection with one drawn from b' perpendicular to ac of (a). From this intersection, a line perpendicular to bd of (a) locates b''' at its intersection with the line drawn from b' perpendicular to ab of (a). The construction is indicated on the diagram. The displacement of d caused by the rotation of the truss is determined in a similar manner. The resultant displacements are then represented in amount and direction by $a'b'''$, $a'c'''$, and $a'd'''$. Since a is fixed, there is no displacement of that joint, and a' and a''' coincide.

In (a), the final position of the deformed truss is shown in broken lines, the resultant displacements $a'b'''$, $a'c'''$, and $a'd'''$ being laid off to a smaller scale from the points b, c, and d respectively. As in (c) the actual displacements are greatly exaggerated.

For the purpose of simplifying the construction, the parallelograms of (d) are completed; i.e., $b'b''$ is drawn parallel to $a'b'''$, $a'b''$ parallel to $b'b'''$, $c'c''$ parallel to $a'c'''$, $a'c''$ parallel to $c'c'''$, $d'd''$ parallel to $a'd'''$, and $a'd''$ parallel to $d'd'''$. Since $b''b' = a'b'''$, $c''c' = a'c'''$, and $d''d' = a'd'''$, $b''b'$, $c''c'$, and $d''d'$ represent the respective final displacements of the points b, c, and d. Also, since $a'b''$, $a'c''$, and $a'd''$ are respectively perpendicular to ab, ac, and ad of (a), if the points a', b'', c'', and d'' are joined, they form a truss similar to the original truss of (a). It follows, therefore, that the actual displacements may be obtained from the diagram of (b) as follows: From c', draw $c'c''$ parallel to the constrained line of motion of the point c to an intersection with a perpendicular from a' to the radius of rotation of c, i.e., to ac of a. On $a'c''$ as a base, construct a truss diagram similar to the original diagram. The required displacements are then given by the directions and distances of b', c', and d' from b'', c'', and d'' respectively. The necessary construction is shown in (b) in full lines.

Since the displacement diagram may be drawn, on the assumption that any joint is fixed and the direction of any one of the members making that joint is also fixed, it is advisable to choose the joint and the direction that will make the diagram as compact as possible. This allows the use of a larger scale and reduces the probability of error. Such a result may be secured by selecting a member that suffers the minimum change in direction under the applied load. In a simple truss, one of the chords in the center panel or the middle vertical should be chosen. In the truss whose displacement diagram

has been constructed as described above, if the direction of the middle vertical and either one of its extremities is considered fixed, the resultant diagram is as shown in (*e*). If the constrained line of motion were horizontal, the truss diagram $a'b''c''d''$ would be reduced to a point, since a' and c'' would coincide.

The Deflection of a Bridge Truss. When live load covers the entire span of a simple truss bridge, the elongation and shortening of the various members of the trusses are due to the stresses caused by the live and dead load, and the several panel points are displaced an appreciable amount from their theoretical positions with no stresses in the members. If this displacement were not considered in the design of the trusses there would be, in any but the shortest spans, a noticeable sag at the mid-point of the span. In order to prevent any point of the lower chord falling below a line joining the two supports, the tension members are shortened by an amount equal to or greater than their elongation as computed with full live load on the bridge. Similarly the compression members are lengthened a corresponding amount. Since full live load does not cause the maximum stresses in most of the web members of a simple truss, it is not necessarily the maximum stresses that are used in determining the various elongations and shortenings.

With the selections and lengths of the members known, together with their stresses under live load, the change in length in each member may be computed, and the displacement diagram constructed. Generally, the vertical components of the displacements of the lower chord panel points are all that are required. These components are the deflections.

Reciprocal Theorem and Influence Lines. Consider a structure loaded by a group of independent forces A, and suppose that a second group of forces B are added. The work done by the forces A acting over the displacements due to B will be W_{AB}.

Now, suppose the forces B had been on the structure first, and then load A had been applied. The work done by the forces B acting over the displacements due to A will be W_{AB}.

The reciprocal theorem states that $W_{AB} = W_{BA}$.

Some very useful conclusions can be drawn from this equation. For example, there is the **reciprocal deflection relationship:** *The deflection at a point A due to a load at B is equal to the deflection at B due to the same load applied at A. Also, the rotation at A due to a load (or moment) at B is equal to the rotation at B due to the same load (or moment) applied at A.*

Another consequence is that deflection curves may also be influence lines, to some scale, for reactions, shears, moments, or deflections **(Muller-Breslau principle)**. For example, suppose the influence line for a reaction is to be found; that is, we wish to plot the reaction R as a unit load moves over the structure, which may be statically indeterminate. For the loading condition A, we analyze the structure with a unit load on it at a distance x from some reference point. For loading condition B, we apply a dummy unit vertical load upward at the place where the reaction is to be determined, deflecting the structure off the support. At a distance x from the reference point, the displacement is d_{xR} and over the support the displacement is d_{RR}. Hence, $W_{AB} = -1 \cdot d_{xR} + Rd_{RR}$. On the other hand, W_{BA} is zero, since loading condition A provides no displacement for the dummy unit load at the support in condition B. Consequently, from the reciprocal theorem, $W_{AB} = W_{BA} = 0$:

$$R = d_{xR}/d_{RR}$$

Since d_{RR} is a constant, R is proportional to d_{xR}. Hence, the influence line for a reaction can be obtained from the deflection curve resulting from a displacement of the support (Fig. 81). The magnitude of the reaction is obtained by dividing each ordinate of the deflection curve by the displacement of the support.

Similarly, the influence line for shear can be obtained from the deflection curve produced by cutting the structure and shifting the cut ends vertically at the point for which the influence line is desired (Fig. 82).

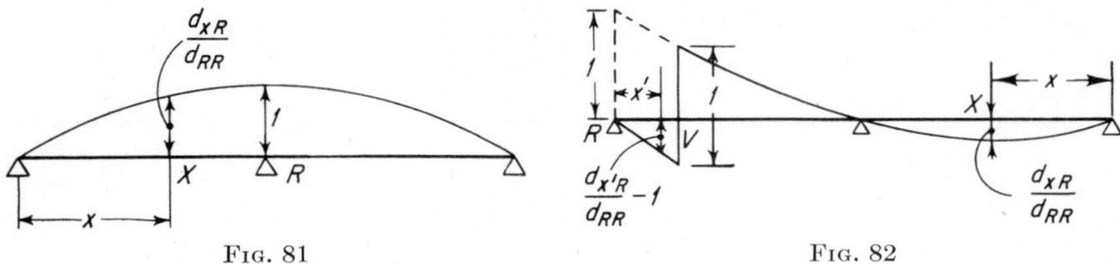

Fig. 81 Fig. 82

The influence line for bending moment can be obtained from the deflection curve produced by cutting the structure and rotating the cut ends at the point for which the influence line is desired (Fig. 83).

And finally, it may be noted that the deflection curve for a load of unity at some point of a structure is also the influence line for deflection at that point (Fig. 84).

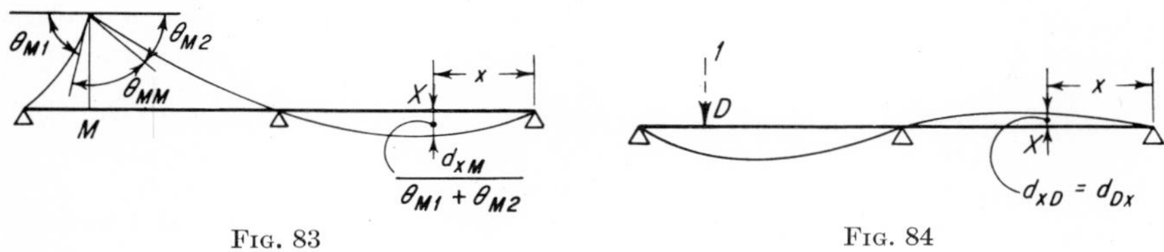

Fig. 83 Fig. 84

CONTINUOUS BEAMS AND FRAMES

General Method of Analysis. Continuous beams and frames consist of members that can be treated as simple beams, the ends of which are prevented by moments from rotating freely. Member LR in the continuous beam in Fig. 85*a*, for example, can be isolated, as shown in Fig. 85*b*, and the elastic restraints at the ends replaced by couples M_L and M_R. In this way, LR is converted into a simply supported beam acted upon by transverse loads and end moments.

The bending-moment diagram for LR is shown at the left in Fig. 85*c*. Treating LR as a simple beam, we can break this diagram down into three simple components, as shown at the right of the equal sign in Fig. 85*c*: Thus, the bending moment at any section equals the simple-beam moment due to the transverse loads, plus the simple-beam moment due to the end moment at L, plus the simple-beam moment due to the end moment at R.

Obviously, once M_L and M_R have been determined, the shears may be computed by taking moments about any section. Similarly, if the reactions or shears are known, the bending moments can be calculated.

A general method for determining the elastic forces and moments exerted by redundant supports and moments is as follows: Remove as many redundant supports or members as necessary to make the structure statically determinate. Compute for the actual loads the deflections or rotations of the statically determinate structure in the direction of the forces and couples exerted by the removed supports and members. Then, in terms of these forces and couples, compute the corresponding deflection or rotations the forces and couples produce in the statically determinate structure. Finally, for each redundant support or member write equations that give the known rota-

tions or deflections of the original structure in terms of the deformations of the statically determinate structure.

For example, one method of finding the reactions of the continuous beam $ALRBC$ in Fig. 85a is to temporarily remove supports L, R, and B. The beam is now simply supported between A and C, and the reactions and moments can be computed from the laws of equilibrium. Beam AC deflects at points L, R, and B, whereas we know that continuous beam $ALRBC$ is prevented from deflecting at these points by the supports

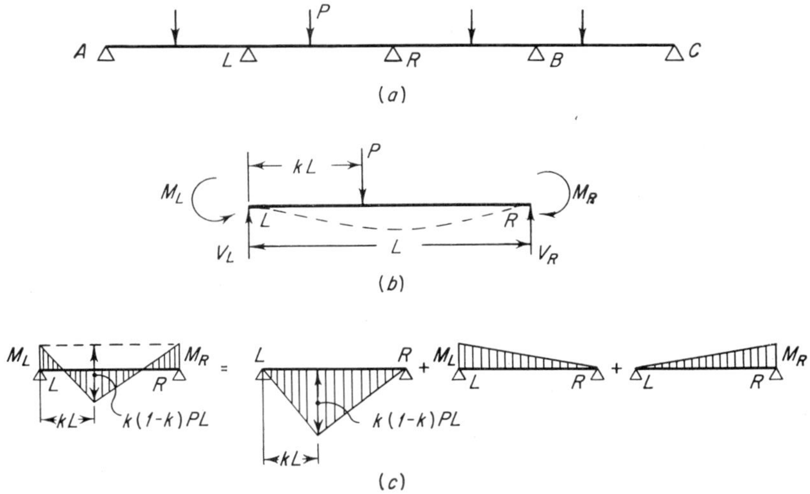

Fig. 85

there. This information enables us to write three equations in terms of the three unknown reactions that were eliminated to make the beam statically determinate.

So we first compute the deflections of simple beam AC at L, R, and B due to the loads. Then, in terms of the unknown reactions, we compute the deflection at L when AC is loaded with only the three reactions. The first equation is obtained by noting that the sum of this deflection and the deflection at L due to the loads must be zero. Similarly, equations can be written making the total deflection at R and B, successively, equal to zero. In this way, three equations with three unknowns are obtained, which can be solved simultaneously to yield the required reactions.

For continuous beams and frames with a large number of redundants, this method becomes unwieldy because of the number of simultaneous equations. Special methods, like moment distribution, are preferable in such cases.

Sign Convention. For moment distribution, the following sign convention is most convenient: A moment acting at an end of a member or at a joint is positive if it tends to rotate the joint clockwise, negative if it tends to rotate the joint counterclockwise. Hence, in Fig. 85, M_R is positive and M_L is negative.

Similarly, the angular rotation at the end of a member is positive if in a clockwise direction, negative if counterclockwise. Thus, a positive end moment produces a positive end rotation in a simple beam.

For ease in visualizing the shape of the elastic curve under the action of loads and end moments, bending-moment diagrams will be plotted on the tension side of each member. Hence, if an end moment is represented by a curved arrow, the arrow will point in the direction in which the moment is to be plotted.

Fixed-end Moments. A beam so restrained at its ends that no rotation is produced there by the loads is called a fixed-end beam, and the end moments are called fixed-end moments. Actually, it would be very difficult to construct a beam with ends that are truly fixed. However, the concept of fixed ends is useful in determining the moments in continuous beams and frames.

Fixed-end moments may be expressed as the product of a coefficient and WL, where W is the total load on the span L. The coefficient is independent of the properties of other members of the structure. Thus, any member can be isolated from the rest of the structure and its fixed-end moments computed. Then, the actual moments in the beam can be found by applying a correction to each fixed-end moment.

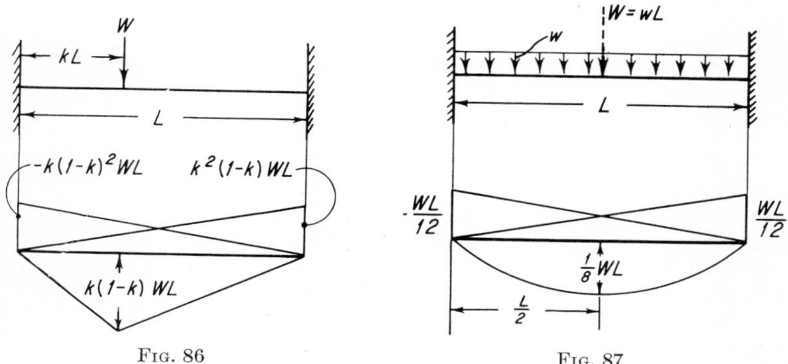

Fig. 86

Fig. 87

Fixed-end moments may be determined conveniently by the moment-area method (see p. 3–53). For frames, the column analogy [1] is frequently very useful. This method takes advantage of the mathematical identity between the moment produced by continuity and the fiber stresses in a short column eccentrically loaded.

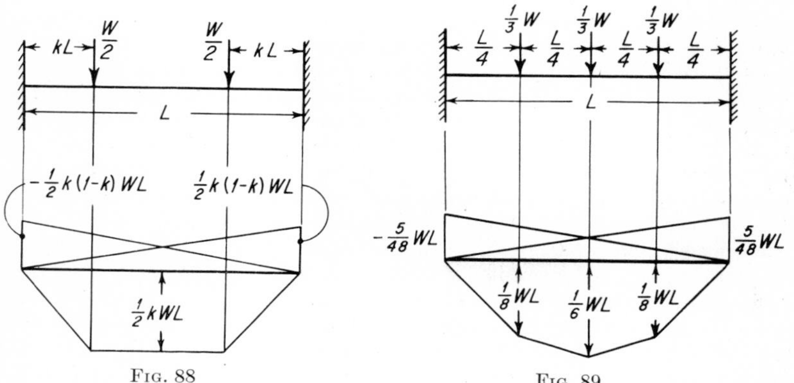

Fig. 88

Fig. 89

To find statically indeterminate moments by the column analogy choose any convenient bending-moment diagram statically consistent with the loads, apply this diagram as the loading on the analogous column, and compute the fiber stresses ($f = P/A \pm Mc/I$). The total moment at any point, thus computed, is the difference be-

[1] Cross and Morgan, "Continuous Frames of Reinforced Concrete," John Wiley & Sons, Inc., New York.

tween the moments statically determined and the indeterminate moments, or fiber stresses. The analogous column has the same dimensions as the frame except that at each point of its cross section the width is equal to $1/EI$, where E is the modulus of elasticity and I the moment of inertia of the frame cross section at that point.

Fixed-end moments for several common types of loading on beams of constant moment of inertia (prismatic beams) are given in Figs. 86 to 89. Also, the curves in Fig. 91 enable fixed-end moments to be computed easily for any type of loading on a prismatic beam. Before the curves can be entered, however, certain characteristics of the loading must be calculated. These include $\bar{x}L$, the location of the center of gravity of the loading with respect to one of the loads; $G^2 = \dfrac{\Sigma b_n^2 P_n}{W}$, where b_nL is the distance from each load P_n to the center of gravity of the loading (taken positive to the right); and $S^3 = \dfrac{\Sigma b_n^3 P_n}{W}$. (See case 8, Fig. 90.) These values are given in Fig. 90 for some common types of loading.

The curves in Fig. 91 are entered at the bottom with the location a of the center of gravity with respect to the left end of the span. At the intersection with the proper G curve, proceed horizontally to the left to the intersection with the proper S line, then vertically to the horizontal scale indicating the coefficient m by which to multiply WL to obtain the fixed-end moment. The curves solve the equations:

$$m_L = \frac{M_L^F}{WL} = G^2[1 - 3(1 - a)] + a(1 - a)^2 + S^3$$

$$m_R = \frac{M_R^F}{WL} = G^2(1 - 3a) + a^2(1 - a) - S^3$$

where M_L^F is the fixed-end moment at the left support and M_R^F at the right support.

As an example of the use of the curves, find the fixed-end moments in a prismatic beam of 20-ft span carrying a triangular loading of 100 kips, similar to the loading shown in case 4, Fig. 90, distributed over the entire span, with the maximum intensity at the right support.

Case 4 gives the characteristics of the loading: $y = 1$; the center of gravity is $L/3$ from the right support, so $a = 0.67$; $G^2 = 1/18 = 0.056$; and $S^3 = -1/135 = -0.007$. To find M_R^F, we enter Fig. 91 at the bottom with $a = 0.67$ on the upper scale and proceed vertically to the estimated location of the intersection of the coordinate with the $G^2 = 0.06$ curve. Then, we move horizontally to the intersection with the line for $S^3 = -0.007$, as indicated by the dash line in Fig. 91. Referring to the scale at the top of the diagram, we find the coefficient m_R to be 0.10. Similarly, with $a = 0.67$ on the lowest scale, we find the coefficient m_L to be 0.07. Hence, the fixed-end moment at the right support is $0.10 \times 100 \times 20 = 200$ ft-kips, and at the left support $-0.07 \times 100 \times 20 = -140$ ft-kips.

Fixed-end Stiffness. To correct a fixed-end moment to obtain the end moment for the actual conditions of end restraint in a continuous structure, the end of the member must be permitted to rotate. The amount it will rotate depends on its stiffness, or resistance to rotation.

The **fixed-end stiffness** of a beam is defined as the moment required to produce a rotation of unity at the end where it is applied, while the other end is fixed against rotation. It is represented by K_R^F in Fig. 92.

For prismatic beams, the fixed-end stiffnesses for both ends are equal to $4EI/L$, where E is the modulus of elasticity, I the moment of inertia of the cross section about the centroidal axis, and L the span (generally taken center to center of supports). When

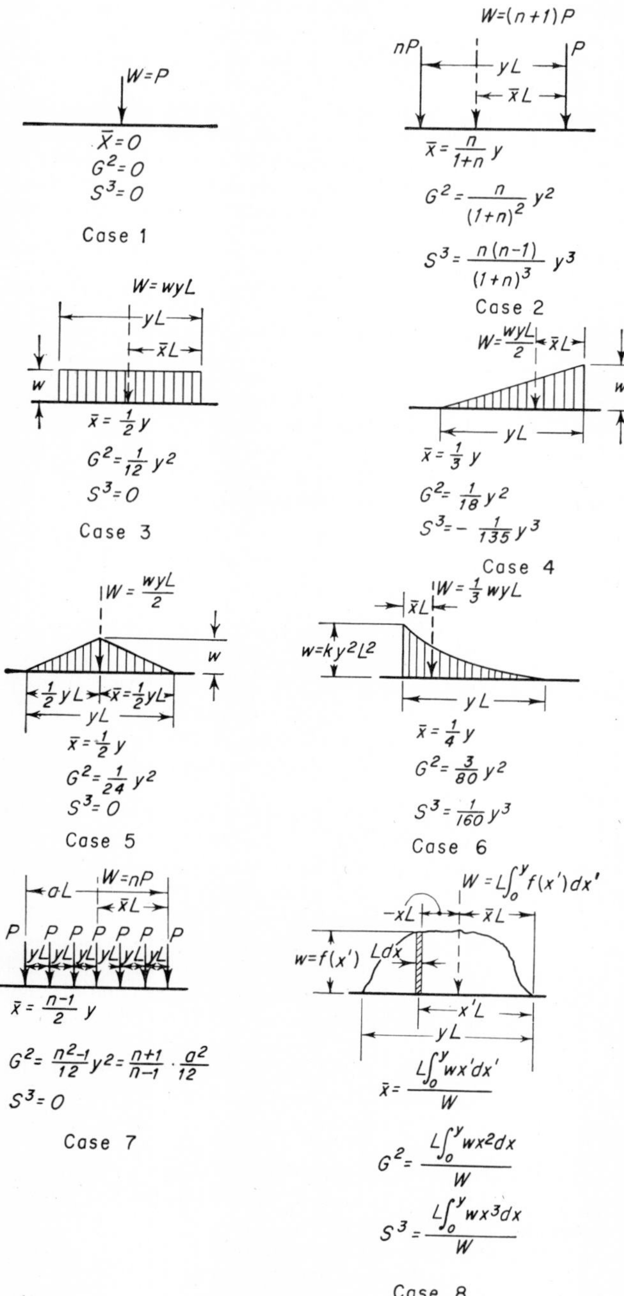
W=P
X̄=0
G²=0
S³=0
Case 1
W=(n+1)P
nP
P
yL
x̄L
x̄ = n/(1+n) y
G² = n/(1+n)² y²
S³ = n(n−1)/(1+n)³ y³
Case 2
W=wyL
yL
x̄L
w
x̄ = 1/2 y
G² = 1/12 y²
S³=0
Case 3
W=wyL/2
x̄L
w
yL
x̄ = 1/3 y
G² = 1/18 y²
S³ = − 1/135 y³
Case 4
W = wyL/2
w
1/2 yL
x̄ = 1/2 yL
yL
x̄ = 1/2 y
G² = 1/24 y²
S³=0
Case 5
W = 1/3 wyL
x̄L
w=ky²L²
yL
x̄ = 1/4 y
G² = 3/80 y²
S³ = 1/160 y³
Case 6
aL
W=nP
x̄L
P P P P P P P
yL yL yL yL yL yL
x̄ = (n−1)/2 y
G² = (n²−1)/12 y² = (n+1)/(n−1) · a²/12
S³=0
Case 7
W = L∫₀^y f(x′)dx′
−xL
x̄L
w=f(x′)
Ldx
x′L
yL
x̄ = L∫₀^y wx′dx′ / W
G² = L∫₀^y wx²dx / W
S³ = L∫₀^y wx³dx / W
Case 8

FIG. 90

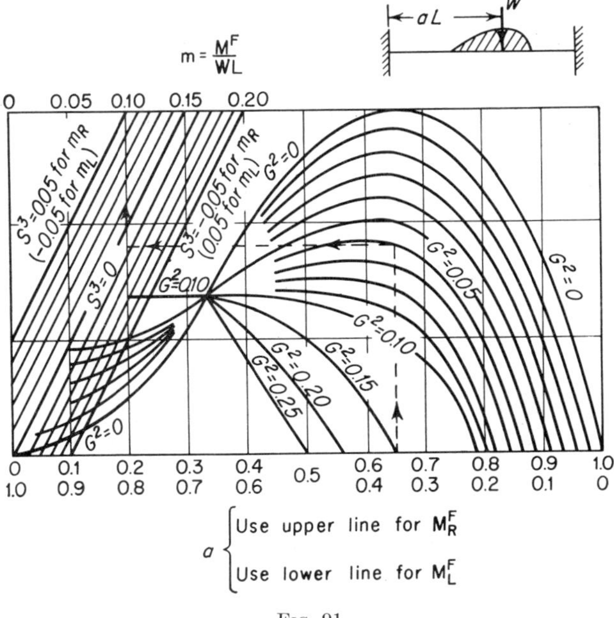

FIG. 91

deformations are not required to be calculated, only the relative values of K^F for each member need be known; hence, only the ratio of I to L has to be computed. (For prismatic beams with a hinge at one end, the actual stiffness is $3EI/L$, or three-fourths the fixed-end stiffness.)

For beams of variable moment of inertia, the fixed-end stiffness may be calculated by methods presented later in this section or obtained from tables, such as those in the "Handbook of Frame Constants," published by the Portland Cement Association, Chicago, Ill.

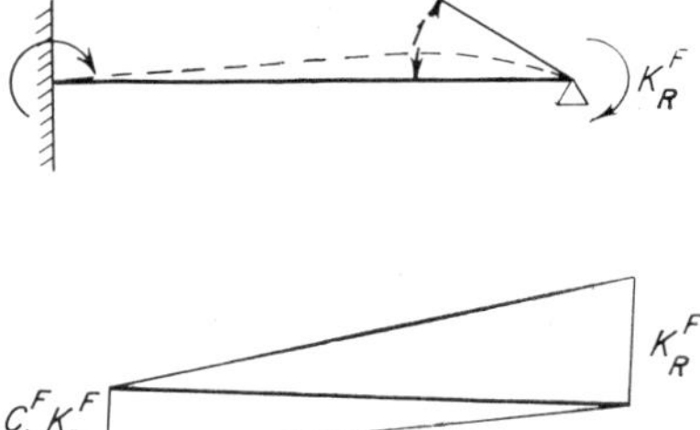

FIG. 92

Fixed-end Carry-over Factor. When a moment is applied at one end of a beam, a resisting moment is induced at the far end if that end is restrained against rotation (Fig. 92). The ratio of the resisting moment at a fixed end to the applied moment is called the fixed-end carry-over factor C^F.

For prismatic beams, the fixed-end carry-over factor toward either end is 0.5. It should be noted that the applied moment and the resisting moment have the same sign (Fig. 92); i.e., if the applied moment acts in a clockwise direction, the carry-over moment also acts clockwise.

For beams of variable moment of inertia, the fixed-end carry-over factor may be calculated by methods presented later in this section or obtained from tables, such as those in the "Handbook of Frame Constants," published by the Portland Cement Association, Chicago, Ill.

Moment Distribution by Converging Approximations. The frame in Fig. 93 consists of four prismatic members rigidly connected together at O and fixed at ends A, B, C, and D. If an external moment U is applied at O, the sum of the end moments in each member at O must be equal to U. Furthermore, all members must rotate at O through the same angle θ, since they are assumed to be rigidly connected there. Hence, by the definition of fixed-end stiffness, the proportion of U induced in the end of each member

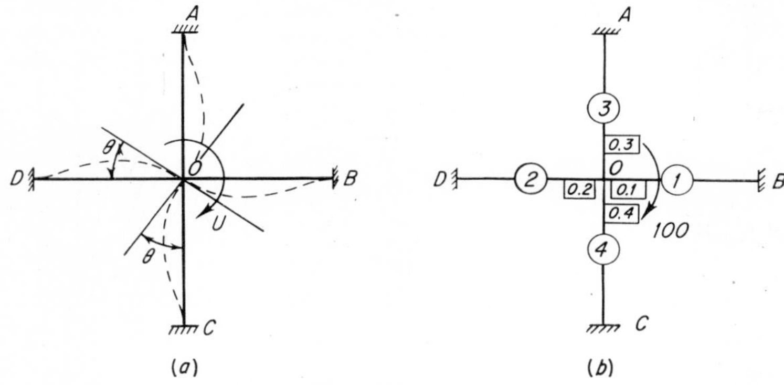

FIG. 93

at O is equal to the ratio of the stiffness of that member to the sum of the stiffnesses of all the members at the joint.

Suppose a moment of 100 ft-kips is applied at O, as indicated in Fig. 93*b*. The relative stiffness (or I/L) is assumed as shown in the circle on each member. The distribution factors for the moment at O are computed from the stiffnesses and shown in the boxes. For example, the distribution factor for OA equals its stiffness divided by the sum of the stiffnesses of all the members at the joint: $3/(3 + 1 + 4 + 2) = 0.3$. Hence, the moment induced in OA at O is $0.3 \times 100 = 30$ ft-kips. Similarly, OB gets 10 ft-kips, OC 40 ft-kips, and OD 20 ft-kips.

Because the far ends of these members are fixed, one-half of these moments are carried over to them. Thus, $M_{AO} = 0.5 \times 30 = 15$; $M_{BO} = 0.5 \times 10 = 5$; $M_{CO} = 0.5 \times 40 = 20$; and $M_{DO} = 0.5 \times 20 = 10$.

Most structures consist of frames similar to the one in Fig. 93, or even simpler, joined together. Though the ends of the members are not fixed, the technique employed for the frame in Fig. 93 can be applied to find end moments in such continuous structures.

Before the general method is presented, one short cut is worth noting. Advantage can be taken when a member has a hinged end to reduce the work in distributing moments. This is done by using the true stiffness of the member instead of the fixed-end stiffness. (For a prismatic beam, the stiffness of a member with one end hinged is three-fourths the fixed-end stiffness; for a beam with variable moment of inertia, it is equal to the fixed-end stiffness times $1 - C_L^F C_R^F$.) Naturally, the carry-over factor toward the hinge is zero.

When a joint is neither fixed nor pinned, but restrained by elastic members connected there, moments can be distributed by a series of converging approximations. All joints are locked against rotation at first. As a result, the loads will create fixed-end moments at the ends of every loaded member. At each joint, a moment equal to the algebraic sum of the fixed-end moments there is required to hold it fixed. Then, one joint is unlocked at a time by applying a moment equal but opposite in sign to the moment that was needed to prevent rotation. The unlocking moment must be distributed to the

members at the joint in proportion to their fixed-end stiffnesses and the distributed moments carried over to the far ends.

After all joints have been released at least once, it generally will be necessary to repeat the process—sometimes several times—before the corrections to the fixed-end moments become negligible. To reduce the number of cycles, the unlocking of joints should start with those having the greatest unbalanced moments.

Suppose the end moments are to be found for the continuous beam $ABCD$ in Fig. 94. The I/L values for all spans are equal; therefore, the relative fixed-end stiffness for all members is unity. However, since A is a hinged end, the computation can be shortened by using the actual relative stiffness, which is ¾. Relative stiffnesses for all members

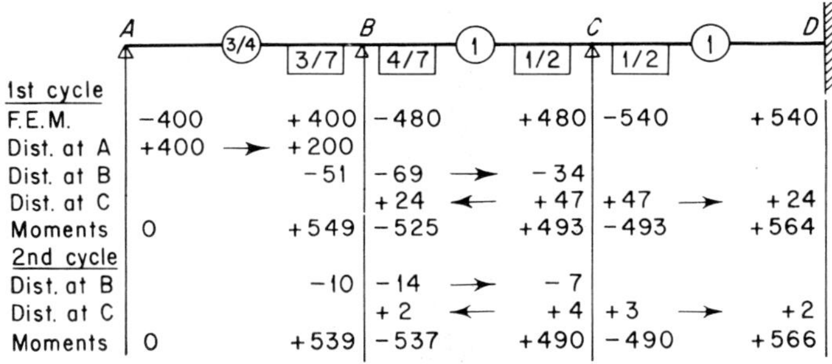

A B C D
3/4 1 1
3/7 4/7 1/2 1/2
1st cycle
F.E.M. −400 +400 −480 +480 −540 +540
Dist. at A +400 +200
Dist. at B −51 −69 −34
Dist. at C +24 +47 +47 +24
Moments 0 +549 −525 +493 −493 +564
2nd cycle
Dist. at B −10 −14 −7
Dist. at C +2 +4 +3 +2
Moments 0 +539 −537 +490 −490 +566

FIG. 94

are shown in the circle on each member. The distribution factors are shown in the boxes at each joint.

The computation starts with determination of fixed-end moments for each member. These are assumed to have been found and are given on the first line in Fig. 94. The greatest unbalanced moment is found from inspection to be at hinged end A; so this joint is unlocked first. Since there are no other members at the joint, the full unlocking moment of +400 is distributed to AB at A and one-half of this is carried over to B. The unbalance at B now is +400 − 480 plus the carry-over of +200 from A, or a total of + 120. Hence, a moment of −120 must be applied and distributed to the members at B by multiplying by the distribution factors in the corresponding boxes.

The net moment at B could be found now by adding the entries for each member at the joint. However, it generally is more convenient to delay the summation until the last cycle of distribution has been completed.

The moment distributed to BA need not be carried over to A, because the carry-over factor toward the hinged end is zero. However, half the moment distributed to BC is carried over to C.

Similarly, joint C is unlocked and half the distributed moments carried over to B and D, respectively. Joint D should not be unlocked, since it actually is a fixed end. Thus, the first cycle of moment distribution has been completed.

The second cycle is carried out in the same manner. Joint B is released, and the distributed moment in BC is carried over to C. Finally, C is unlocked, to complete the cycle. Adding the entries for the end of each member yields the final moments.

Continuous Frames. In practice, the problem is to find the maximum end moments and interior moment produced by the worst combination of loading. For maximum moment at the end of a beam, live load should be placed on that beam and on the beam adjoining the end for which the moment is to be computed. Spans adjoining these two should be assumed to be carrying only dead load.

For maximum mid-span moments, the beam under consideration should be fully loaded, but adjoining spans should be assumed to be carrying only dead load.

The work involved in distributing moments due to dead and live loads in continuous frames in buildings can be greatly simplified by isolating each floor. The tops of the upper columns and the bottoms of the lower columns can be assumed fixed. Furthermore, the computations can be condensed considerably by following the procedure recommended in "Continuity in Concrete Building Frames" (Portland Cement Association, Chicago, Ill.) and indicated in Fig. 95.

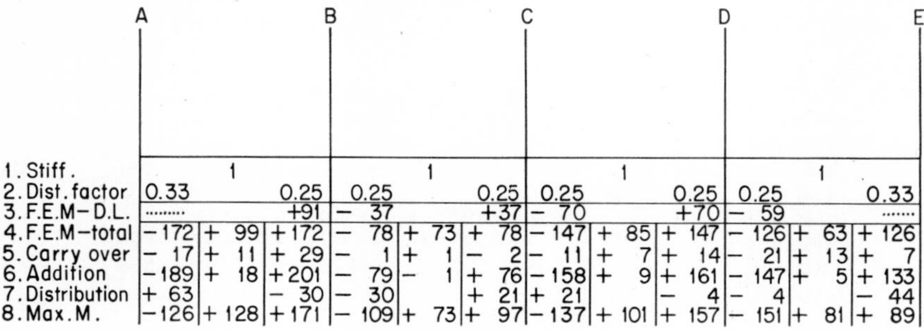

	A			B			C			D		E
1. Stiff.		1			1			1			1	
2. Dist. factor	0.33		0.25	0.25		0.25	0.25		0.25	0.25		0.33
3. F.E.M–D.L.	....		+91	− 37		+37	− 70		+70	− 59		....
4. F.E.M–total	−172	+ 99	+172	− 78	+ 73	+ 78	−147	+ 85	+ 147	− 126	+ 63	+ 126
5. Carry over	− 17	+ 11	+ 29	− 1	+ 1	− 2	− 11	+ 7	+ 14	− 21	+ 13	+ 7
6. Addition	−189	+ 18	+201	− 79	− 1	+ 76	−158	+ 9	+ 161	− 147	+ 5	+ 133
7. Distribution	+ 63		− 30	− 30		+ 21	+ 21		− 4	− 4		− 44
8. Max. M.	−126	+ 128	+171	− 109	+ 73	+ 97	−137	+ 101	+ 157	− 151	+ 81	+ 89

Fig. 95

Figure 95 presents the complete calculation for maximum end and mid-span moments in four floor beams, AB, BC, CD, and DE. Building columns are assumed to be fixed at the story above and below. None of the beam or column sections is known to begin with; so as a start, all members will be assumed to have a fixed-end stiffness of unity, as indicated on the first line of the calculation.

On the second line, the distribution factors for each end of the beams are shown; column moments will not be computed until moment distribution to the beams has been completed. Then, the sum of the column moments at each joint may be easily computed, since they are the moments needed to make the sum of the end moments at the joint equal to zero. The sum of the column moments at each joint can then be distributed to each column there in proportion to its stiffness. In this example, each column will get one-half the sum of the column moments.

Fixed-end moments at each beam end for dead load are shown on the third line, just above the horizontal line, and fixed-end moments for live plus dead load on the fourth line. Corresponding mid-span moments for the fixed-end condition also are shown on the fourth line, and like the end moments will be corrected to yield actual mid-span moments.

For maximum end moment at A, beam AB must be fully loaded, but BC should carry dead load only. Holding A fixed, we first unlock joint B, which has a total-load fixed-end moment of $+172$ in BA and a dead-load fixed-end moment of -37 in BC. The releasing moment required, therefore, is $-(172 - 37)$, or -135. When B is released, a moment of -135×0.25 is distributed to BA. One-half of this is carried over to A, or $-135 \times 0.25 \times 0.5 = -17$. This value is entered as the carry-over at A on the fifth line in Fig. 95. Joint B then is relocked.

At A, for which we are computing the maximum moment, we have a total-load fixed-end moment of -172 and a carry-over of -17, making the total -189, shown on the sixth line. To release A, a moment of $+189$ must be applied to the joint. Of this, 189×0.33, or 63, is distributed to AB, as indicated on the seventh line of the calculation. Finally, the maximum moment at A is found by adding lines 6 and 7: $-189 + 63 = -126$.

For maximum moment at B, both AB and BC must be fully loaded, but CD should carry only dead load. We begin the determination of the moment at B by first releasing joints A and C, for which the corresponding carry-over moments at BA and BC are $+29$ and $-(+78 - 70) \times 0.25 \times 0.5 = -1$, shown on the fifth line in Fig. 95. These bring the total fixed-end moments in BA and BC to $+201$ and -79, respectively. The releasing moment required is $-(201 - 79) = -122$. Multiplying this by the distribution factors for BA and BC when joint B is released, we find the distributed moments, -30, entered on line 7. The maximum end moments finally are obtained by adding lines 6 and 7: $+171$ at BA and -109 at BC. Maximum moments at C, D, and E are computed and entered in Fig. 95 in a similar manner. This procedure is equivalent to two cycles of moment distribution.

The computation of maximum mid-span moments in Fig. 95 is based on the assumption that in each beam the mid-span moment is the sum of the simple-beam mid-span moment and one-half the algebraic difference of the final end moments (the span carries full load but adjacent spans only dead load). Instead of starting with the simple-beam moment, however, we begin with the mid-span moment for the fixed-end condition and apply two corrections. In each span, these corrections are equal to the carry-over moments entered on line 5 for the two ends of the beam multiplied by a factor.

For beams with variable moment of inertia, the factor is $\pm \frac{1}{2}\left(\frac{1}{C^F} + D - 1\right)$, where C^F is the fixed-end carry-over factor toward the end for which the correction factor is being computed and D the distribution factor for that end. The plus sign is used for correcting the carry-over at the right end of a beam, and the minus sign for the carry-over at the left end. For prismatic beams, the correction factor becomes $\pm\frac{1}{2}(1 + D)$.

For example, to find the corrections to the mid-span moment in AB, we first multiply the carry-over at A on line 5, -17, by $-\frac{1}{2}(1 + 0.33)$. The correction, $+11$, is also entered on the fifth line. Then, we multiply the carry-over at B, $+29$, by $+\frac{1}{2}(1 + 0.25)$ and enter the correction, $+18$, on line 6. The final mid-span moment is the sum of lines 4, 5, and 6: $+99 + 11 + 18 = +128$. Other mid-span moments in Fig. 95 are obtained in a similar manner.

Moment-influence Factors. In certain types of problems, particularly those in which different types of loading conditions must be investigated, it may be more convenient to find maximum end moments from a table of moment-influence factors. This table is made up by listing for the end of each member in a structure the moment induced in that end when a moment (for convenience, $+1{,}000$) is applied to each joint successively. Once this table has been prepared, no additional moment distribution is necessary for computing the end moments due to any loading condition.

For a specific loading pattern, the moment at any beam end M_{AB} may be obtained from the moment-influence table by multiplying the entries under AB for the various joints by the actual unbalanced moments at those joints divided by 1,000, and summing.

As an example of the use of the technique, let us determine the moments for the continuous beam and loading in Fig. 94. As indicated in Fig. 96a, moments of $+1{,}000$ are applied successively at joints A, B, and C, the resulting moments being the moment-influence factors.

To release joint A, which has a fixed-end moment of -400, a moment of $+400$ must be applied, as shown in Fig. 96b. That moment is then distributed by using the moment-influence factors in Fig. 96a for $+1{,}000$ at A: at AB, $+400 \times 1.000 = +400$; at BA, $+271 \times 400/1{,}000 = +108$; at CB, $-76 \times 400/1{,}000 = -30$; at DC, $+38 \times 400/1{,}000 = +15$.

Joint B has an unbalance of -80 due to fixed-end moments of $+400$ and -480. To release this joint, a moment of $+80$ must be applied at B; it is distributed by multiplying the moment-influence factors in Fig. 96a for $+1{,}000$ at B by $80/1{,}000$. The results are

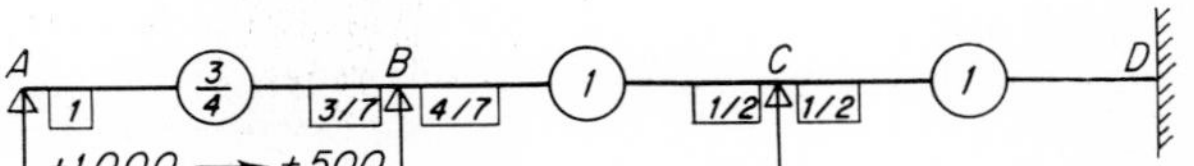

	A [1]	(3/4)	B [3/7]	B [4/7]	(1)	C [1/2]	C [1/2]	(1)	D
+1,000 at A	+1,000	→	+500						
			-214	-286	→	-143			
				+ 36	←	+ 72	+71	→	+ 36
			- 15	- 21	→	- 10			
						+ 5	+ 5	→	+ 2
Influence factors	+1,000		+271	-271		- 76	+76		+ 38
+1,000 at B			+429	+571	→	+286			
				- 71	←	-143	-143	→	- 71
			+ 30	+ 41	→	+ 20			
				- 5	←	- 10	- 10	→	- 5
			+ 2	+ 3					
Influence factors			+461	+539		+153	-153		- 76
+1,000 at C				+250	←	+500	+500	→	+250
			-107	-143	→	- 72			
				+ 18	←	+ 36	+ 36	→	- 18
			- 8	- 10	→	- 5			
						+ 3	+ 2	→	+ 1
Influence factors			-115	+115		+462	+538		+269

(a) MOMENT INFLUENCE FACTORS

	A	B	B	C	C	D
$M^F_{AB} = -400$	-400					
	+400	+ 108	- 108	- 30	+ 30	+ 15
$M^F_{BA} = +400$		+400	-480			
$M^F_{BC} = -480$		+ 37	+ 43	+ 12	- 12	- 6
$M^F_{CB} = +480$				+480	-540	
$M^F_{CD} = -540$		- 7	+ 7	+ 28	+ 32	+ 16
$M^F_{DC} = +540$						+540
	0	+538	-538	+490	-490	+565

(b) FINAL MOMENTS

FIG. 96

given in Fig. 96*b*, as well as for similar distribution of the moments at *C* and *D*. The final moments are obtained by summing up the entries for the ends of each member.

If now, another system of loads were applied to this beam, the final moments could be obtained by a calculation similar to that in Fig. 96*b*, using the moment-influence factors in Fig. 96*a*.

Deflection of Supports. For some problems, it is convenient to know the effect of a deflection of a support normal to the original position of a continuous beam. But the moment-distribution method is based on the assumption that such movement of a support does not occur. However, the method can be modified to evaluate the end moments resulting from a support movement.

The procedure is to distribute moments, as usual, assuming no deflection at the supports. This implies that additional external forces are exerted at the supports to prevent movement. These forces can be computed. Then, equal and opposite forces are applied to the structure to produce the final configuration, and the moments that they produce are distributed as usual. These moments added to those obtained with undeflected supports yield the final moments.

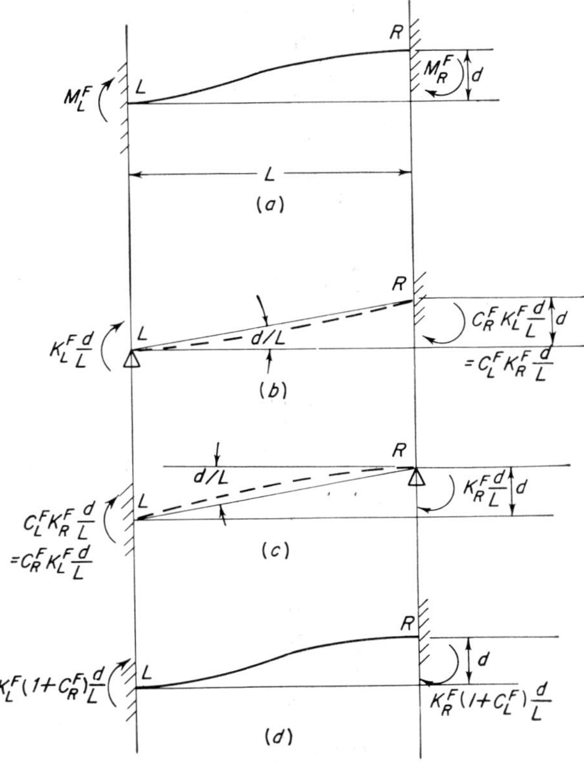

Fig. 97

To apply this procedure, it is first necessary to know the fixed-end moments for a beam with supports at different levels. In Fig. 97*a*, the right end of a beam with span L is at a height d above the left end. To find the fixed-end moments, we first deflect the beam with both ends hinged, then fix the right end, leaving the left end hinged, as in Fig. 97*b*. Noting that a line connecting the two supports makes an angle approximately equal to d/L (its tangent) with the original position of the beam, we apply a moment at the hinged end to produce an end rotation there equal to d/L. By the definition of stiffness, this moment equals $K_L^F d/L$. The carry-over to the right end is C_R^F times this.

By the law of reciprocal deflections, the fixed-end moment at the right end of a beam due to a rotation of the other end is equal to the fixed-end moment at the left end of the beam due to the same rotation at the right end. Therefore, the carry-over moment for the right end is also equal to $C_L^F K_R^F d/L$ (see Fig. 97*c*). By adding the end moments for the loading conditions in Fig. 97*b* and *c*, we obtain the end moments in Fig. 97*d*, which is equivalent to the deflected beam in Fig. 97*a*:

$$M_L^F = K_L^F(1 + C_R^F)\frac{d}{L}$$

$$M_R^F = K_R^F(1 + C_L^F)\frac{d}{L}$$

In a similar manner, the fixed-end moment can be found for a beam with one end hinged and the supports at different levels:

$$M^F = K\frac{d}{L}$$

where K is the actual stiffness for the end of the beam that is fixed; for beams of variable moment of inertia K is equal to the fixed-end stiffness times $(1 - C_L^F C_R^F)$.

Procedure for Sidesway. The problem of computing sidesway moments in rigid frames is conveniently solved by the following method:

1. Apply forces to the structure to prevent sidesway while the fixed-end moments due to loads are distributed.
2. Compute the moments due to these forces.
3. Combine the moments obtained in Steps 1 and 2 to eliminate the effect of the forces that prevented sidesway.

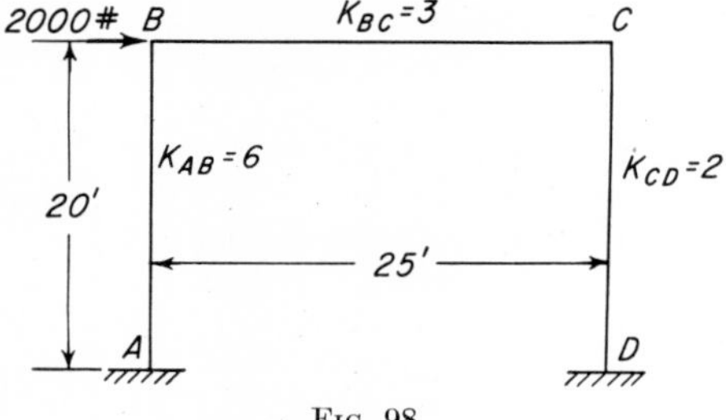

FIG. 98

Suppose the rigid frame in Fig. 98 is subjected to a 2,000-lb horizontal load acting to the right at the level of beam BC. The first step is to compute the moment-influence factors by applying moments of +1,000 at joints B and C, assuming sidesway prevented.

MOMENT-INFLUENCE FACTORS FOR FIG. 98

Member	+1,000 at B	+1,000 at C
AB	351	−105
BA	702	−210
BC	298	210
CB	70	579
CD	−70	421
DC	−35	210

MOMENT-COLLECTION TABLE FOR FIG. 98

Remarks	AB	BA	BC	CB	CD	DC
1. Sidesway—FEM.	−3,000M	−3,000M	....	....	−1,000M	−1,000M
2. Distribution for B.	+1,053M	+2,106M	+894M	+210M	−210M	−105M
3. Distribution for C.	−105M	−210M	+210M	+579M	+421M	+210M
4. Final sidesway M.	−2,052M	−1,104M	+1,104M	+789M	−789M	−895M
5. For 2,000-lb horizontal.	−17,000	−9,100	+9,100	+6,500	−6,500	−7,400
6. 4,000-lb vertical FEM.	....	....	−12,800	+3,200		
7. Distribution for B.	+4,490	+8,980	+3,820	+897	−897	−448
8. Distribution for C.	+336	+672	−672	−1,853	−1,347	−673
9. Moments with no sidesway.	+4,826	+9,652	−9,652	+2,244	−2,244	−1,121
10. Sidesway M.	−4,710	−2,540	+2,540	+1,810	−1,810	−2,060
11. For 4,000-lb vertical.	+116	+7,112	−7,112	+4,054	−4,054	−3,181

Since there are no intermediate loads on the beam and columns, the only fixed-end moments that need be considered are those in the columns due to lateral deflection of the frame caused by the horizontal load.

This deflection, however, is not known initially. So we assume an arbitrary deflection, which produces a fixed-end moment of $-1{,}000M$ at the top of column CD. M is an unknown constant to be determined from the fact that the sum of the shears in the deflected columns must be equal to the 2,000-lb load. The same deflection also produces a moment of $-1{,}000M$ at the bottom of CD.

From the geometry of the structure, we furthermore note that the deflection of B relative to A is equal to the deflection of C relative to D. Then, according to the equations developed in Fig. 97, the fixed-end moments in the columns of this frame are proportional to the stiffnesses of the columns, and hence, are equal in AB to $-1{,}000M \times 6/2 = -3{,}000M$. The column fixed-end moments are entered in the first line of the Moment Collection Table for Fig. 98.

In the deflected position of the frame, joints B and C are unlocked. First, we apply a releasing moment of $+3{,}000M$ at B and distribute it by multiplying by 3 the entries in the column marked "+1,000 at B" in the table, Moment-influence Factors for Fig. 98. Similarly, a releasing moment of $+1{,}000M$ is applied at C and distributed with the aid of the moment-influence factors. The distributed moments are entered in the second and third lines of the moment-collection table. The final moments are the sum of the fixed-end moments and the distributed moments and are given in the fourth line, in terms of M.

Isolating each column and taking moments about one end, we find that the overturning moment due to the shear is equal to the sum of the end moments. We have one such equation for each column. Adding these equations, noting that the sum of the shears equals 2,000 lb, we obtain

$$-M(2{,}052 + 1{,}104 + 789 + 895) = -2{,}000 \times 20$$

from which we find $M = 8.26$. This value is substituted in the sidesway totals in the moment-collection table to yield the end moments for the 2,000-lb horizontal load.

Suppose now a vertical load of 4,000 lb is applied to BC of the rigid frame in Fig. 98, 5 ft from B. The same moment-influence factors and moment-collection table can again be used to determine the end moments with a minimum of labor:

The fixed-end moment at B, with sidesway prevented, is $-12{,}800$, and at C, $+3{,}200$. With the joints still locked, the frame is permitted to move laterally an arbitrary amount, so that in addition to the fixed-end moments due to the 4,000-lb load, column fixed-end moments of $-3{,}000M$ at B and $-1{,}000M$ at C are induced. The moment-collection table already indicates the effect of relieving these column moments by unlocking joints B and C. We now have to superimpose the effect of releasing joints B and C to relieve the fixed-end moments for the vertical load. This we can do with the aid of the moment-influence factors. The distribution is shown in lines 7 and 8 of the moment-collection table. The sums of the fixed-end moments and distributed moments for the 4,000-lb load are shown on line 9.

The unknown M can be evaluated from the fact that the sum of the horizontal forces acting on the columns must be zero. This is equivalent to requiring that the sum of the column end moments equals zero:

$$-M(2{,}052 + 1{,}104 + 789 + 895) + 4{,}826 + 9{,}652 - 2{,}244 - 1{,}120 = 0$$

from which $M = 2.30$. This value is substituted in line 4 of the moment-collection table to yield the sidesway moments for the 4,000-lb load. The addition of these moments to the totals for no sidesway (line 9) gives the final moments.

This procedure enables one-story bents with straight beams to be analyzed with the necessity of solving only one equation with one unknown regardless of the number of bays. If the frame is several stories high, the procedure can be applied to each story. Since an arbitrary horizontal deflection is introduced at each floor or roof level, there are as many unknowns and equations as there are stories.

The procedure is more difficult to apply to bents with curved or polygonal members between the columns. The effect of the change in the horizontal projection of the curved or polygonal portion of the bent must be included in the calculations.[1] In many cases it may be easier to analyze the bent as a curved beam (arch) or by the column analogy.[1]

Single-cycle Moment Distribution. In the method of moment distribution by converging approximations, all joints but the one being unlocked are considered fixed. In distributing moments, the stiffnesses and carry-over factors used are based on this assumption. However, if actual stiffnesses and carry-over factors are employed, moments can be distributed throughout continuous frames in a single cycle.

Formulas for actual stiffnesses and carry-over factors can be written in several simple forms. The equations given in the following text were chosen to permit the use of existing tables for beams of variable moment of inertia that are based on fixed-end stiffnesses and fixed-end carry-over factors.[2]

Considerable simplification of the formulas results if they are based on the simple-beam stiffness of members of continuous frames. This value can always be obtained from tables of fixed-end properties by multiplying the fixed-end stiffness by $(1 - C_L^F C_R^F)$, in which C_L^F is the fixed-end carry-over factor to the left and C_R^F is the fixed-end carry-over factor to the right.

To derive the basic constants needed, we apply a unit moment to one end of a member, considering it simply supported (Fig. 99*a*). The end rotation at the support where the moment is applied is α, and at the far end, the rotation is β. By the dummy-load method, if x is measured from the β end:

$$\alpha = \int_0^L \frac{x^2}{EI_x}\,dx$$

$$\beta = \int_0^L \frac{x(L - x)}{EI_x}\,dx$$

in which I_x is the moment of inertia at a section a distance of x from the β end, and E is the modulus of elasticity.

The simple-beam stiffness K of the member is the moment required to produce a rotation of unity at the end where it is applied (Fig. 99*b*). Hence, at each end of a member, $K = 1/\alpha$.

For prismatic beams, K has the same value for both ends and is equal to $3EI/L$. For haunched beams, K for each end can be obtained from tables for fixed-end stiffnesses, as mentioned previously, or by numerical integration of the equation for α.

While the value of α, and consequently of K, is different at opposite ends of an unsymmetrical beam, the value of β is the same for both ends, in accordance with the law of reciprocal deflections. This also is evident from the integral for β, where $L - x$ can be substituted for x without changing the value of the integral.

Now, if we apply a moment J at one end of a simple beam to produce a rotation of unity at the other end (Fig. 99*c*), this moment will be equal to $1/\beta$, and will have the same value regardless at which end it is applied. K/J is equal to the fixed-end carry-over factor.

[1] CROSS and MORGAN, "Continuous Frames of Reinforced Concrete," John Wiley & Sons, Inc., New York; "Gabled Concrete Roof Frames Analyzed by Moment Distribution," Portland Cement Association, Chicago, Ill.

[2] "Handbook of Frame Constants," Portland Cement Association, Chicago, Ill.

J is equal to $6EI/L$ for prismatic beams. For haunched beams, it can be computed by numerical integration of the equation for β.

The actual stiffness S of the end of an unloaded span is the moment producing a rotation of unity at the end where it is applied when the other end of the beam is restrained against rotation by other members of the structure (Fig. 99d).

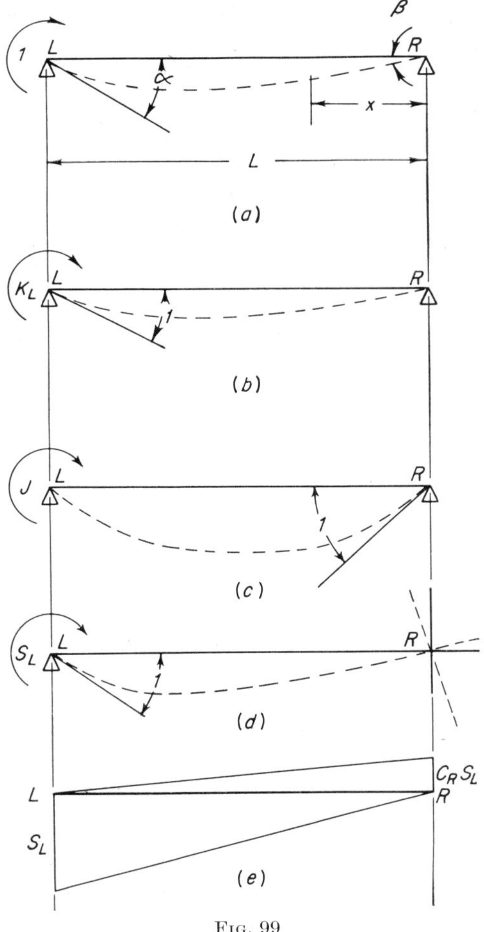

FIG. 99

The bending-moment diagram for a moment S_L applied at the left end of a member of a continuous frame is shown in Fig. 99e. As indicated, the moment carried over to the far end is C_RS_L, where C_R is the carry-over factor to the right. At L, the rotation produced by S_L alone is S_L/K_L, and by C_RS_L alone is $-C_RS_L/J$. By definition of stiffness, the sum of these angles must equal unity:

$$\frac{S_L}{K_L} - \frac{C_RS_L}{J} = 1$$

Solving for S_L and noting that $K_L/J = C_L^F$, the fixed-end carry-over factor to the left, we find the formula for the stiffness of the left end of a member:

$$S_L = \frac{K_L}{1 - C_L^F C_R}$$

Similarly, the stiffness of the right end of a member is

$$S_R = \frac{K_R}{1 - C_R^F C_L}$$

For prismatic beams, the siffness formulas reduce to

$$S_L = \frac{K}{1 - C_R/2} \qquad \text{and} \qquad S_R = \frac{K}{1 - C_L/2}$$

where $K = 3EI/L$.

When the far end of a prismatic beam is fully fixed against rotation, the carry-over factor equals ½. Hence, the fixed-end stiffness equals $4K/3$. This indicates that the effect of partial restraint on prismatic beams is to vary the stiffness between K for no restraint and $1.33K$ for full restraint. Because of this small variation, in many cases, an estimate of the actual stiffness of a beam may be sufficiently accurate.

The restraint R at the end of an unloaded beam in a continuous frame is the moment applied at that end to produce a unit rotation in all the members of the joint. Since the sum of the moments at the joint must be zero, R must be equal to the sum of the stiffnesses of the adjacent ends of the members connected to the given beam at that joint.

Furthermore, the moment induced in any of these other members bears the same ratio to the applied moment as the stiffness of the member does to the restraint. Consequently, *end moments are distributed at a joint in proportion to the stiffnesses of the members.*

Actual carry-over factors can be computed by modifying the fixed-end carry-over factors. In Fig. 99*d* and *e*, by definition of restraint, the rotation at joint R is $-C_R S_L/R_R$, which must be equal to the rotation of the beam at R due to the moments at L and R. The rotation due to $C_R S_L$ alone is equal to $C_R S_L/K_R$, and the rotation due to S_L alone $-S_L/J$. Hence $-C_R S_L/R_R = C_R S_L/K - S_L/J$. Solving for C_R and noting that $K_R/J = C_R^F$, the fixed-end carry-over factor to the right, we find the actual carry-over factor to the right:

$$C_R = \frac{C_R^F}{1 + K_R/R_R}$$

Similarly, the actual carry-over factor to the left is

$$C_L = \frac{C_L^F}{1 + K_L/R_L}$$

In analyzing a continuous beam, we generally know the carry-over factors toward the ends of the first and last spans. Starting with these values, we can calculate the rest of the carry-over factors and the stiffnesses of the members. However, in many frames, there are no end conditions known in advance. To analyze these structures, we must assume several carry-over factors.

This will not complicate the analysis, because in many cases it will be found unnecessary to correct the values of C based on assumed carry-over factors of preceding spans. The reason is that C is not very sensitive to the restraint at far ends of adjacent members.

When carry-over factors are estimated, the greatest accuracy will be attained if the choice of assumed values is restricted to members subject to the greatest restraint.

A very good approximation to the carry-over factor for prismatic beams may be obtained from the following formula, which is based on the assumption that far ends of adjacent members are subject to equal restraint:

$$C = \frac{\Sigma K - K}{2(\Sigma K - \delta K)}$$

where ΣK is the sum of the K values of all the members at the joint toward which the carry-over factor is acting; K is the simple-beam stiffness of the member for which the carry-over factor is being computed; and δ is a factor that varies from zero for no restraint to $\frac{1}{4}$ for full restraint at the far ends of the connecting members. Since δ varies within such narrow limits, it affects C very little.

To illustrate the estimation and calculation of carry-over factors, the carry-over factors and stiffnesses in the clockwise direction will be computed for the frame in Fig. 100*a*. Relative I/L, or K values, are given in the circles.

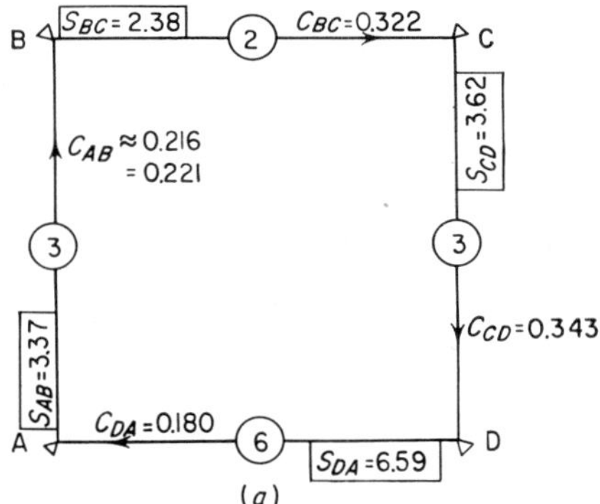

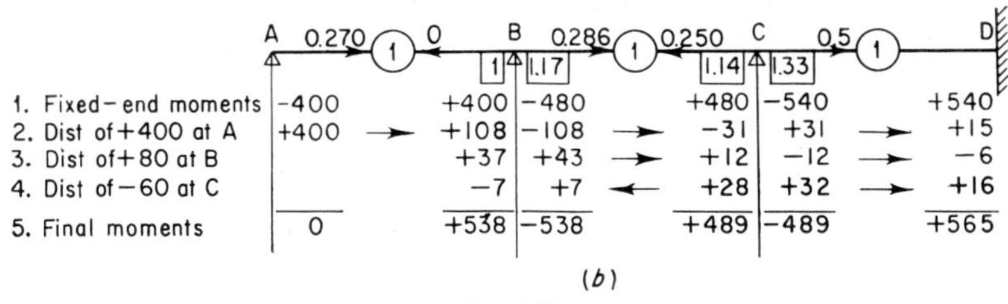

FIG. 100

A start will be made by estimating C_{AB}. Taking $\delta = \frac{1}{8}$, we apply the approximate formula for C with $K = 3$ and $\Sigma K = 3 + 2 = 5$ and find $C_{AB} = 0.216$, as shown in Fig. 100*a*. The stiffness S_{AB} then is equal to $3/(1 - 0.216/2) = 3.37$. Noting that $R_{AD} = S_{AB}$, we can now use the exact formula to obtain the carry-over factor from D to A: $C_{DA} = 0.5/(1 + 6/3.37) = 0.180$. Continuing around the frame in this manner, we return to C_{AB} and recalculate it with the exact formula, obtaining 0.221. This differs only slightly from the estimated value. The change in C_{DA} due to the new value of C_{AB} is negligible.

If a bending moment of 1,000 ft-lb were introduced at A in AB, it would induce a moment of $1{,}000C_{AB} = 221$ ft-lb at B; $221 \times 0.322 = 71$ ft-lb at C; $71 \times 0.343 = 24$ ft-lb at D; $24 \times 0.180 = 4$ ft-lb at A, etc.

To demonstrate how the moments in a continuous beam would be computed, the end moments will be determined for the beam in Fig. 100*b*, which is identical with the one in Fig. 94, for which moments were obtained by converging approximations. Relative I/L, or K values, are shown in the circles on each span.

Since A is a hinged end, $C_{BA} = 0$. $S_{BA} = 1/(1 - 0) = 1$. Since there is only one member joined to BC at B, $R_{BC} = S_{BA}$, and $C_{CB} = 0.5/(1 + 1/1) = 0.250$. With this value, we compute $S_{CB} = 1.14$. To obtain the carry-over factors for the opposite direction, we start with $C_{CD} = 0.5$, since we know D is a fixed end. This enables us to compute $S_{CD} = 1.33$ and the remainder of the beam constants.

The fixed-end moments are given on the first line of calculations in Fig. 100*b*. We start the distribution by unlocking A by applying a releasing moment of $+400$. Since A is unrestrained, the full 400 is given to A and $400 \times 0.270 = 108$ is carried over to B. If several members had been connected to AB at B, this moment, with sign changed, would be distributed to them in proportion to their stiffnesses. But since only one member is connected at B, the moment at BC is -108. Next, $-108 \times 0.286 = -31$ is carried over to C. Finally, a moment of $+15$ is carried to D.

Then, joint B is unlocked. The unbalanced moment of $400 - 480 = -80$ is counteracted with a moment of $+80$, which is distributed to BA and BC in proportion to their stiffnesses (shown in the boxes at the joint). BC, for example, gets $80 \times 1.17/(1.17 + 1) = 43$. The carry-over to C is $43 \times 0.286 = 12$, and to D, -6.

Similarly, the unbalanced moment at C is counteracted and distributed to CB and CD in proportion to the stiffnesses shown in the boxes at C, then carried over to B and D. The final moments are the sum of the fixed-end and distributed moments.

On occasion, advantage can be taken of certain properties of loads and structures to save work in distribution by using carry-over factors as the ratio of end moments in loaded members. For example, suppose it is obvious, from symmetry of loading and structure, that there will be no end rotation at an interior support. The part of the structure on one side of this support can be isolated and the moments distributed only in this part, with the carry-over factor toward the support taken as C^F.

Again, suppose it is evident that the final end moments at opposite ends of a span must be equal in magnitude and sign. Isolate the structure on each side of this beam and distribute moments only in each part, with the carry-over factor for this span taken as 1.

Method for Checking Moment Distribution. End moments computed for a continuous structure must be in accordance with both the laws of equilibrium and the requirements of continuity. At each joint, therefore, the sum of the moments must be equal to zero (or to an external moment applied there), and the end of every member connected there must rotate through the same angle. It is a simple matter to determine whether the sum of the moments is zero, but further calculation is needed to prove that the moments yield the same rotation for the end of each member at a joint. The following method not only will indicate that the requirements of continuity are satisfied but also will tend to correct automatically any mistakes that may have been made in computing the end moments.

Consider a joint O made up of several members OA, OB, OC, etc. The members are assumed to be loaded and the calculation of end moments to have started with fixed-end moments. For any one of the members, say OA, the end rotation at O for the fixed-end condition was

$$0 = \frac{M_{OA}{}^F}{K_{OA}} - \frac{M_{AO}{}^F}{J_{OA}} - \phi_{OA}$$

where $M_{OA}{}^F$ is the fixed-end moment at O, $M_{AO}{}^F$ is the fixed-end moment at A, K_{OA} is the simple-beam stiffness at O, J_{OA} is the moment required at A to produce a unit rotation at O when the span is considered simply supported, and ϕ_{OA} is the simple-beam end rotation at O due to loads on OA.

For the final end moments, the rotation at O is

$$\theta = \frac{M_{OA}}{K_{OA}} - \frac{M_{AO}}{J_{OA}} - \phi_{OA}$$

Subtracting the first equation for rotation at O from the second and multiplying by K_{OA} yields

$$K_{OA}\theta = M_{OA} - M_{OA}{}^F - C_{AO}{}^F M_{OA}'$$

in which the fixed-end carry-over factor toward O, $C_{AO}{}^F$, has been substituted for K_{OA}/J_{OA}, and M_{OA}' for $M_{AO} - M_{AO}{}^F$. An analogous expression can be written for each of the other members at O. Summing these equations, we obtain

$$\theta \Sigma K_O = \Sigma M_O - \Sigma M_O{}^F - \Sigma C_O{}^F M_O'$$

With this value of θ we can solve for each of the final end moments at O and thus determine the equations that will check the joint for continuity. For example,

$$M_{OA} = M_{OA}{}^F + C_{AO}{}^F M_{OA}' - m_{OA}$$

$$m_{OA} = \frac{K_{OA}}{\Sigma K_O}(-\Sigma M_O + \Sigma M_O{}^F + \Sigma C_O{}^F M_O')$$

Similar equations can be written for the other members at O by substituting the proper letter for A in the subscripts.

If the calculations based on these equations are carried out in table form, the equations prove to be surprisingly simple (see tables in the following example).

For prismatic beams, the terms $C^F M'$ become ½ the change in the fixed-end moment at the far end of each member at a joint.

Example of Continuity Check: Suppose we want to check the end moments in the beam in Fig. 100*b*. Each joint and the ends of the members connected there are listed in Table 6, and a column is provided for the summation of the various terms for each joint. K values are given on line 1, the end moments to be checked on line 2, and the fixed-end moments on line 3. On line 4 is entered one-half the difference obtained when the fixed-end moment is subtracted from the final moment at the far end of each member. $-m$ is placed on line 5 and the corrected end moment on line 6. The $-m$ values are obtained from the summation columns by adding line 2 to the negative of the sum of lines 3 and 4 and distributing the result to the members of the joint in proportion to the K values. The corrected moment M is the sum of lines 3, 4, and 5.

Assume that a mistake was made in computing the end moments for Fig. 100*b* giving the results shown on line 2 of the second part of Table 6 for the fixed-end moments on line 3. The correct moments can be obtained as follows:

At joint B, the sum of the incorrect moments is zero, as shown in the summation column on line 2. The sum of the fixed-end moments at B is -80, as indicated on line 3. For BA, $\frac{1}{2}M'$ is obtained from lines 2 and 3 of the column for AB: $\frac{1}{2} \times (0 + 400) = +200$, which is entered on line 4. The line 4 entry for BC is obtained from CB: $\frac{1}{2} \times (450 - 480) = -15$. The sum of the line 4 values at B is, therefore, $200 - 15 = +185$. Entered in the summation column, this is then added to the summation value on line 3, the sign is changed and the number on line 2 (in this case zero) added to the sum, giving -105, which is noted in the summation column on line 5. The

TABLE 6. CONTINUITY CHECK FOR FIG. 100*b*

	A		*B*			*C*			*D*	
	AB	Σ	*BA*	*BC*	Σ	*CB*	*CD*	Σ	*DC*	Σ
1. *K*	1	1	1	1	2	1	1	2	1	∞
2. *M*	0	0	+538	−538	0	+489	−489	0	+565	
3. M^F	−400	−400	+400	−480	−80	+480	−540	−60	+540	
4. $C^F M'$			+200	+5	+205	−29	+13	−16	+26	
5. $-m$			−62	−63	−125	+38	+38	+76	0	
6. Check	0	0	+538	−538	0	+489	−489	0	+566	
	Check when moments are incorrect									
2. Wrong *M*	0	0	+560	−560	0	+450	−450	0	+530	
3. M^F	−400	−400	+400	−480	−80	+480	−540	−60	+540	
4. $C^F M'$			+200	−15	+185	−40	−5	−45	+45	
5. $-m$			−53	−52	−105	+53	+52	+105	0	
6. New *M*	0	0	+547	−547	0	+493	−493	0	+585	
	Second cycle									
7. Trial *M*	0	0	+547	−547	0	+493	−493	0	+585	
8. M^F	−400	−400	+400	−480	−80	+480	−540	−60	+540	
									+24	
9. $C^F M'$			+200	+3	+203	−34	+12	−22		
10. $-m$			−62	−61	−123	+41	+41	+82	0	
11. New *M*	0	0	+538	−538	0	+487	−487	0	+564	

values for *BA* and *BC* on line 5 are obtained by multiplying −105 by the ratio of the *K* value of each member to the sum of the *K* values at the joint; that is, for *BA*, $-m = -105 \times \frac{1}{2} = -53$. The corrected moment for *BA*, the sum of lines 3, 4, and 5, is $+400 + 200 - 53 = +547$. Since this differs from the value on line 2, it indicates that one or more of the moments on that line were incorrect. The other corrected moments are found in the same way and are shown on line 6.

A comparison with Fig. 100*b* shows that the new moments, though incorrect, are closer to the right answer than those with which we started. Even closer results can be obtained by repeating the calculations. However, convergence can be obtained much more quickly by starting first with fixed ends and joints that appear to have been most in error. Then, use the corrected values obtained for these in correcting adjacent joints.

For example, in the second cycle shown in Table 6, calculations were started with joint *D*, which is a fixed end. Using the values obtained at the end of the first cycle to compute M', we find the corrected value for *DC* to be +564, which is very close to the exact final moment. Then, we move to joint *C*. For *M*, we use the value obtained at the end of the first cycle. But M' for *CD* is based on the end moment just computed: $+564 - 540 = +24$, and half of this is placed on line 9 under *CD*. Continuing in this manner, we obtain moments that check closely the final moments in the first part of the table.

The procedure is useful also for estimating the effect of changing the stiffness of one or more members.

The checking equations can be generalized to include the effect of the movement *d*

of a support in a direction normal to the initial position of a span of length L:

$$M_{OA} = M_{OA}{}^F + C_{AO}{}^F M_{OA}' - K_{OA}\frac{d}{L} - m_{OA}$$

$$m_{OA} = \frac{K_{OA}}{\Sigma K_O}\left(-\Sigma M_O + \Sigma M_O{}^F + \Sigma C_O{}^F M_O' - \Sigma K_O \frac{d}{L}\right)$$

For each span with a support movement, the term Kd/L can be obtained from the fixed-end moment due to this deflection alone; for OA, for example, by multiplying moment $M_{OA}{}^F$ by $(1 - C_{AO}{}^F C_{OA}{}^F)/(1 + C_{OA}{}^F)$. For prismatic beams, this factor reduces to ½.

In Table 7, the solution for the bent in Fig. 98 is checked for the condition in which a 4,000-lb vertical load was placed 5 ft from B on span BC. The computations are similar to those in Table 6, except that the terms $-Kd/L$ are included for the columns to account for sidesway. These values are obtained from the sidesway fixed-end moments in the Moment-collection Table for Fig. 98 (p. 5–74). For BA, for example, $Kd/L = ½ \times 3{,}000M$, with $M = 2.30$, as found in the solution. The check indicates that the original solution was sufficiently accurate for a slide-rule computation. If line 7 had contained a different set of moments, the shears would have had to be investigated again. A second cycle could be carried out by distributing the unbalance to the columns to obtain new Kd/L values.

TABLE 7. CONTINUITY CHECK FOR FIG. 98

	A		*B*			*C*			*D*	
	AB	Σ	*BA*	*BC*	Σ	*CB*	*CD*	Σ	*DC*	Σ
1. K	6	∞	6	3	9	3	2	5	2	∞
2. M	+116	....	+7,110	−7,110	0	+4,050	−4,050	0	−3,180	
3. M^F	0	....	0	−12,800	−12,800	+3,200	0	+3,200	0	
4. $C^F M'$	+3,555	....	+60	+430	+490	+2,840	−1,590	+1,250	−2,030	
5. $-Kd/L$	−3,450	....	−3,450	0	−3,450	0	−1,150	−1,150	−1,150	
6. $-m$	0	....	+10,510	+5,250	+15,760	−1,980	−1,320	−3,300	0	
7. M	+105	....	+7,120	−7,120	0	+4,060	−4,060	0	−3,180	

STRESSES IN THREE-HINGED ARCHES

Analytical Determination of Reactions. In Fig. 101, the load P causes inclined reactions R_A and R_B at A and B, respectively. In the analysis of arch structures, it is more convenient to resolve the inclined reactions into their horizontal and vertical components. To consider the entire arch as a free body;

$$\Sigma H = 0: \qquad H_A = H_B = H$$
$$\Sigma M_B = 0: \qquad V_A = P(1 - k)$$
$$\Sigma M_A = 0: \qquad V_B = Pk$$

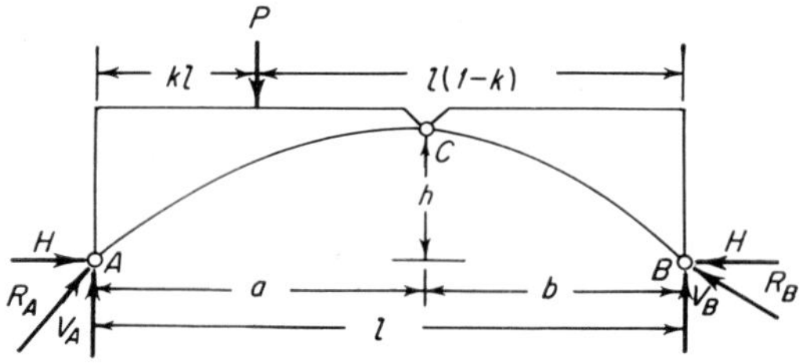

FIG. 101

Because of the hinge construction at C, $M_C = 0$; hence, the right half of the arch of Fig. 101 being considered as a free body, $+Hh - V_B b = 0$, and $H = V_B b/h = Pkb/h$.

If the arch is symmetrical, $a = b = L/2$, and $H = PkL/2h$; and if the load P is placed at C, $kL = L/2$, and $H = PL/4h$. If the load P is on the right portion of the arch at a distance $k'L$ from the right support, an analysis similar to the above results in the equation $H = Pk'a/h$ for the unsymmetrical arch, and $H = Pk'L/2h$ for the symmetrical arch, with the maximum value of $H = PL/4h$ when $k'L = b = L/2$. The influence line[1] for H is obviously a straight line, varying from zero for loads at A and B to a maximum of Pab/Lh for a load at C.

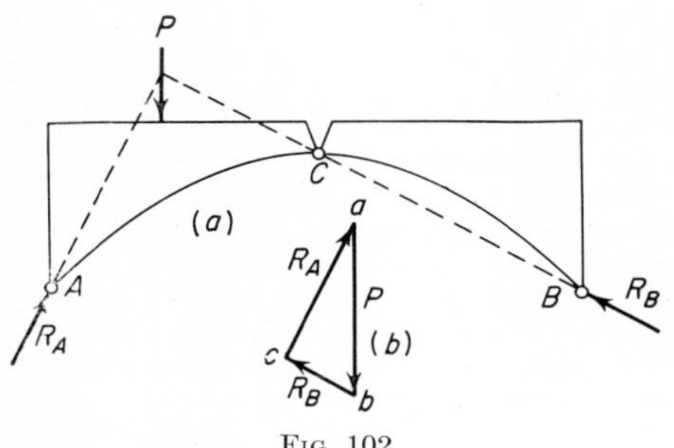

FIG. 102

The reaction components due to any number of loads may be computed by adding the values of V_A, V_B and H for each load considered separately. The amount and direction of the inclined reactions may be obtained by the principles of the composition of forces. If the load P is not vertical but is inclined in the plane of the arch, the reaction components H and V are found in a similar manner, due consideration being given to the lever arms of the load, these being measured to the left and right supports and at right angles to the line of action of the load.

Graphical Determination of Reactions. In Fig. 102, the line of action R_B in (a) must pass through the hinge C; otherwise, there would be a moment at C. (If the load P were on the right portion of the arch, the line of action of R_A would pass through A and C.) Whenever three forces act on a body that is in equilibrium, the lines of action of these three forces must meet in a point. Hence, the line of action of R_A must pass through the intersection of R_B and P. In (b), lay off ab equal and parallel to the given load P. Through either end of ab, draw ac parallel to the line of action of R_A as determined in (a); through the other end of ab, draw bc parallel to the line of action of R_B. The point of intersection c of the lines ac and bc determines the amounts of the two reactions, ac representing R_A and bc representing R_B.

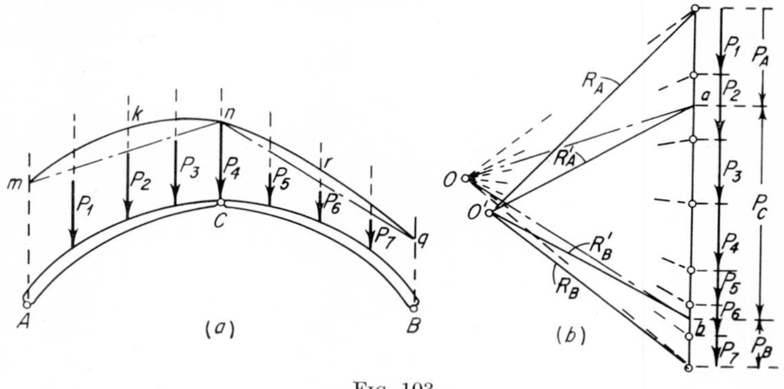

FIG. 103

Where a series of loads (vertical or inclined) acts on the arch, more direct solution is obtained by the use of the equilibrium polygon. In Fig. 103, the loads $P_1 \ldots P_7$ in (a) are laid off in regular order in (b) to form a load line. With any convenient pole O, two

[1] See p. 5-89.

equilibrium polygons *mkn* and *nrq* are drawn in (*a*), one for each segment of the arch. *Oa* is drawn parallel to the closing line *mn* of the equilibrium polygon *mkn*; similarly, *Ob* is drawn parallel to *nq*. Rays *Oa* and *Ob* divide the load line into three parts P_A, P_C, and P_B, which may be considered to act at the hinges *A*, *C*, and *B* respectively in determining the reactions. From *a*, R_A' is drawn parallel to a line through hinges *A* and *C*; and, from *b*, R_B' is drawn parallel to a line through hinges *B* and *C* to intersect R_A' at *O'*. R_A' and R_B' represent the reactions at *A* and *B* caused by the load P_C acting at the hinge *C*. Loads P_A and P_B are then combined with R_A' and R_B' respectively, as shown in (*b*), to obtain the true reactions R_A and R_B for the entire series of loads. This method of solution is also valid if the loads cover only one segment; in this case, however, one equilibrium polygon is sufficient. If the loads cover both segments and some or all of the loads are inclined, it may be desirable to use separate poles for the two equilibrium polygons in order to avoid polygons of undesirable proportions.

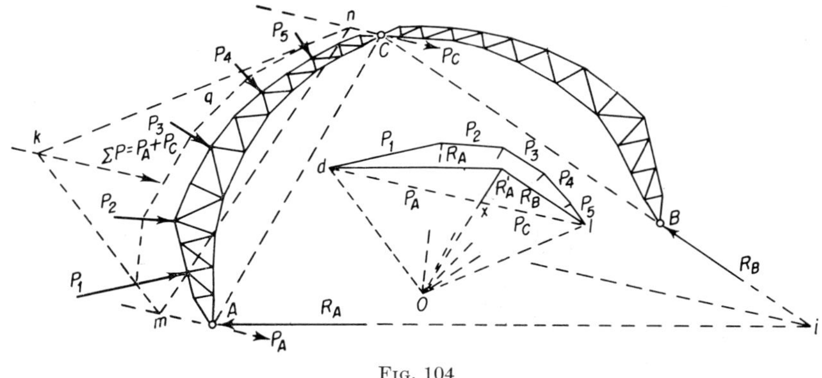

FIG. 104

Stresses in Roof Arches. The dead, wind, and snow panel loads are computed in a manner similar to that used in computing similar loads on ordinary roof trusses. The dead load is ordinarily assumed to act at the upper-chord panel points. With the reactions known, the stresses in the various members may be determined analytically by the method of moments. For preliminary designs, the lever arms of the external loads and internal stresses may be scaled from an accurately made large-scale drawing. For more precise work, these lever arms should be computed analytically.

The graphical solution of stresses in roof arches is generally the most expeditious, the stress diagram being completed in a manner similar to that described for roof trusses. Figure 104 shows the graphical method of determining the reactions.

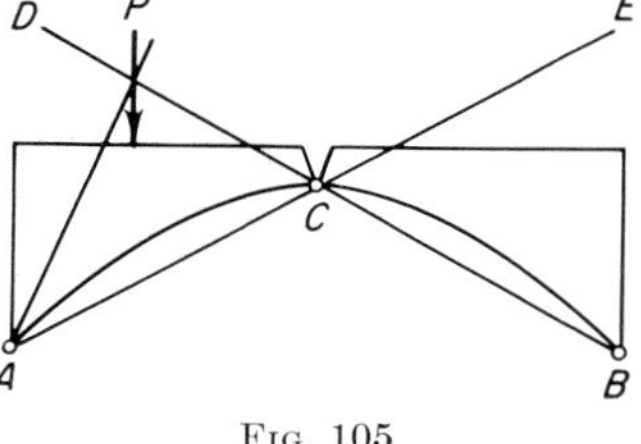

FIG. 105

Loading for Maximum Stresses in Bridge Arches. To determine the position of the live load producing maximum tension or compression in any member of an arch truss, reaction locus lines or influence lines may be used. In Fig. 105, with a load *P* placed at any point on the left segment of the arch, the line of action of the right reaction R_B must pass through the center hinge *C*. The line of action of the left reaction R_A must pass through the intersection of *BCD* and the line of action of the load *P*. The portion *CD* of the line *BCD* is the *reaction locus* for loads on the left segment. Similarly, *CE* is the reaction locus for loads on the right segment.

In Fig. 106, the position of the loads for the maximum tension and compression in L_2L_3 is determined as follows: With section ① taken as shown, U_2 is the center of moments for L_2L_3. A line through U_2 from A intersects the reaction locus CD at d_1. Any vertical load applied so that its line of action passes through d_1 produces *no stress* in L_2L_3. Any load to the right of d_1 causes compression, since, the portion of the truss on the left of the section being considered as a free body, the line of action of R_A passes to the right of the center of moments, indicating counterclockwise rotation, which necessitates, for equilibrium, an equal clockwise rotation for the internal stress L_2L_3. Such a direction indicates a compressive stress in that member. A similar analysis, using the portion of the truss to the right of the section, indicates tension for loads to the left of d_1.

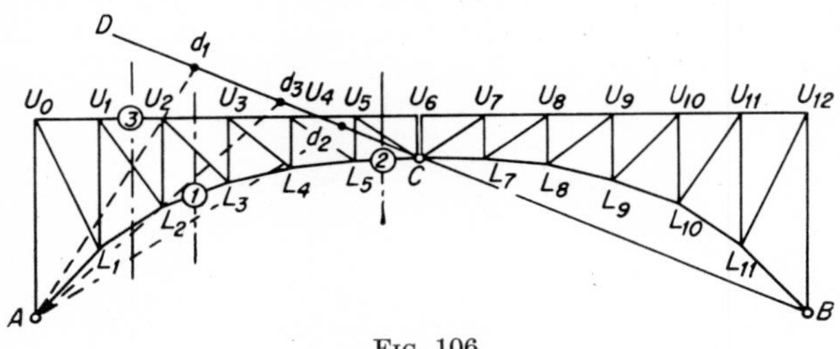

Fig. 106

With section ② taken as shown, the center of moments U_5 for L_5C is on the right of the reaction locus CD, and an analysis similar to the foregoing indicates that d_2 is not a point of inflection for L_5C and that any vertical load on the arch causes compression in that member.

A similar analysis is used for the upper chords. Section ③ through U_1U_2 shows L_2 to be the center of moments and that loads to the right of d_3 produce tension in U_1U_2 and loads to the left produce compression.

For the diagonal U_1L_2 (Fig. 107), section ① shows the center of moment to be c_1. A line Ac_1 intersects the reaction locus CD at d_1. Loads on the right of d_1 produce compression in the diagonal; and those on the left, tension.

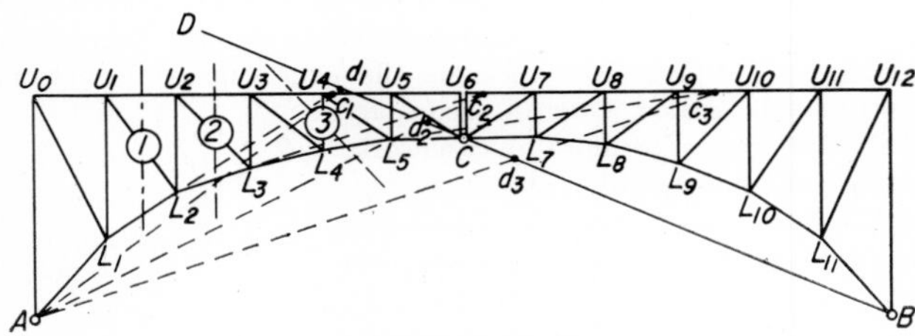

Fig. 107

For the diagonal U_2L_3, the center of moments is at c_2. Since c_2 is on the right of the reaction locus CD, a second point of division occurs in the panel cut by section ②. Considering the segment AC of the arch as a simple truss, with reactions at A and C, the point of intersection of the chords cut by section ② lies outside the lines of action of the reaction, and, as in any simple truss, loads on the panel points immediately adjacent to the section produce stresses of opposite signs in the diagonal. For U_2L_3, therefore, the maximum tension occurs with loads at U_3, U_4, and U_5 and the maximum compression, with loads at U_1, U_2, $U_6 \ldots U_{11}$.

For the member U_4L_4, the point d_3 falls below the center line and therefore does not indicate a point of division. Since, however, c_3 is on the right of BCD there is a point of division in the panel cut by section ③, and the maximum tension is provided with loads at U_1, U_2, and U_3, loads on the right of the section producing the maximum compression.

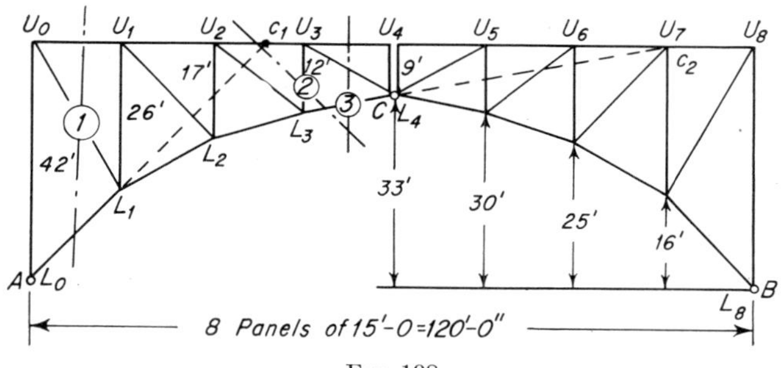

Fig. 108

Illustrative Problem. For the arch of Fig. 108, the dead panel load is 60 kips, and the live panel load 20 kips. The vertical and horizontal components of the left reaction due to loads at the various panel points are computed and tabulated below.

Load at panel point	Dead panel load = 60 kips		Live panel load = 20 kips	
	V_A	H	V_A	H
U_1	52.5	13.6	17.5	4.5
U_2	45.0	27.3	15.0	9.1
U_3	37.5	40.9	12.5	13.6
U_4	30.0	54.5	10.0	18.2
U_5	22.5	40.9	7.5	13.6
U_6	15.0	27.3	5.0	9.1
U_7	7.5	13.6	2.5	4.5
Full load.....	210.0	218.1	70.0	72.6

The panel points that shall be loaded for maximum compression and maximum tension in each member are tabulated below. These were determined by the analysis given on this and the preceding page.

Web members			Chords		
Member	Maximum live-load compression	Maximum live-load tension	Member	Maximum live-load compression	Maximum live-load tension
U_0L_0	0, 1, 2	3, 4, 5, 6, 7	L_0L_1	1, 2, 3, 4, 5, 6, 7	None
U_1L_1	1, 2, 3	4, 5, 6, 7	L_1L_2	2, 3, 4, 5, 6, 7	1
U_2L_2	2, 3, 4, 5, 6, 7	1	L_2L_3	3, 4, 5, 6, 7	1, 2
U_3L_3	3, 4, 5, 6, 7	1, 2	L_3L_4	1, 2, 3, 4, 5, 6, 7	None
U_4L_4	4	None	U_0U_1	1, 2	3, 4, 5, 6, 7
U_0L_1	3, 4, 5, 6, 7	1, 2	U_1U_2	1, 2, 3	4, 5, 6, 7
U_1L_2	1, 4, 5, 6, 7	2, 3	U_2U_3	1, 2, 3	4, 5, 6, 7
U_2L_3	1, 2	3, 4, 5, 6, 7	U_3U_4	None	None
U_3L_4	1, 2, 3	4, 5, 6, 7			

Diagonal U_0L_1. Vertical section through U_0L_1. Portion on left of section as free body. Summation moments about c_1, the intersection of U_0U_1 and L_0L_1. Lever arm of internal stress about $c_1 = 34.1$ ft.

Dead Load: $(+210.0 \times 39.4) - (218.1 \times 42) - (S \times 34.1) = 0$

$$S = -26.0 \text{ kips}$$

Maximum Live-load Compression: Loads U_3, U_4, U_5, U_6, U_7.

$$V_A = 12.5 + 10.0 + 7.5 + 5.0 + 2.5 = 37.5$$

$$H = 13.6 + 18.2 + 13.6 + 9.1 + 4.5 = 59.0$$

$$(+37.5 \times 39.4) - (59.0 \times 42) - (S \times 34.1) = 0$$

$$S = -29.5 \text{ kips}$$

Maximum Live-load Tension: Loads U_1, U_2.

$$V_A = 17.5 + 15.0 = 32.5$$

$$H = 4.5 + 9.1 = 13.6$$

$$(+32.5 \times 39.4) - (13.6 \times 42) - (S \times 34.1) = 0$$

$$S = +20.8 \text{ kips}$$

$$\text{Maximum stress} = -26.0 - 29.5 = -55.5 \text{ kips}$$

$$\text{Minimum stress} = -26.0 + 20.8 = -5.2 \text{ kips}$$

Vertical U_3L_3. Section through U_3L_3 parallel to diagonals U_2L_3 and U_3L_4. Portion on left of section as free body. Summation moments about c_2, the intersection of U_2U_3 and L_3L_4. Lever arm of internal stress about $c_2 = 60.0$ ft.

Dead Load: $(+210.0 \times 105) - (218.1 \times 42) - 60(75 + 90) + (S \times 60) = 0$

$$S = -49.8 \text{ kips}$$

Maximum Live-load Compression: Loads U_3, U_4, U_5, U_6, U_7.

$$V_A = 12.5 + 10.0 + 7.5 + 5.0 + 2.5 = 37.5$$

$$H = 13.6 + 18.2 + 13.6 + 9.1 + 4.5 = 59.0$$

$$(+37.5 \times 105) - (59.0 \times 42) + (S \times 60) = 0$$

$$S = -24.3 \text{ kips}$$

Maximum Live-load Tension: Loads U_1, U_2.

$$V_A = 17.5 + 15.0 = 32.5$$

$$H = 4.5 + 9.1 = 13.6$$

$$(+32.5 \times 105) - (13.6 \times 42) - 20(75 + 90) + (S \times 60) = 0$$

$$S = +7.7 \text{ kips}$$

$$\text{Maximum stress} = -49.8 - 24.3 = -74.1 \text{ kips}$$

$$\text{Minimum stress} = -49.8 + 7.7 = -42.1 \text{ kips}$$

Vertical U_0L_0. This stress may be obtained from that in U_0L_1. Section around U_0. Summation vertical components. The one-half panel load (dead and live) at U_0 must be taken into consideration.

Dead Load: Vertical component of $U_0L_1 = -26.0 \times \frac{26}{30.1} = -22.4$

$$+22.4 - {}^{60}\!/_{2} - S = 0; \qquad S = -7.6 \text{ kips}$$

Maximum Live-load Compression: Loads U_0, U_1, U_2.

$$\text{Vertical component of } U_0L_1 = \left(+20.8 \times \frac{26}{30.1}\right) = +18.0$$

$$-18.0 - {}^{20}\!/_{2} - S = 0; \qquad S = -28.0 \text{ kips}$$

Maximum Live-load Tension: Loads U_3, U_4, U_5, U_6, U_7.

$$\text{Vertical component of } U_0L_1 = -29.5 \times \frac{26}{30.1} = -25.5$$

$$+25.5 - S = 0; \qquad S = +25.5 \text{ kips}$$

$$\text{Maximum stress} = -7.6 - 28.0 = -35.6 \text{ kips}$$

$$\text{Minimum stress} = -7.6 + 25.5 = +17.9 \text{ kips}$$

Lower Chord L_3L_4. Vertical section through L_3L_4. Portion on left of section as free body. Summation moments about U_3. Lever arm of internal stress about $U_3 = 11.8$ ft.

Dead Load: $(+210.0 \times 45) - (218.1 \times 42) - 60(15 + 30) - (S \times 11.8) = 0$

$$S = -204.2 \text{ kips}$$

Maximum Live-load Compression: Since U_3 is on the right of the reaction locus CD, there can be no live-load tension in L_3L_4, and the maximum live-load compression occurs with full load.

$$S = {}^{20}\!/_{60}(-204.2) = -68.1 \text{ kips}$$

$$\text{Maximum stress} = -204.2 - 68.1 = -272.3 \text{ kips}$$

$$\text{Minimum stress} = -204.2 \text{ kips}$$

Influence Lines for Three-hinged Arches. Influence lines may be used to advantage in obtaining the position of loads for maximum and minimum stresses and in determining the amounts of these stresses in three-hinged arches. For example, for the member U_3L_3 of Fig. 108, the stress could be computed for unit load at each panel point successively and the resulting values plotted as ordinates, due consideration being given to sign. This tedious process is not necessary, however, for usually one or two ordinates coupled with a knowledge of the geometric structure of the line furnish sufficient information for the construction of the complete influence line.

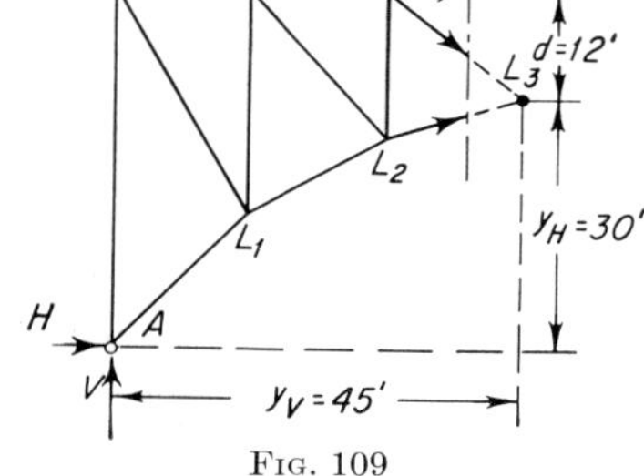

FIG. 109

In a three-hinged arch, the total stress in any member is the algebraic sum of the stresses caused by the vertical forces and the horizontal thrust at the supports. The influence line for the stress in any member, therefore,

may be constructed by superimposing the influence line for the stress caused by the vertical load and reactions upon that for the stress caused by the horizontal thrust at the supports. This method is illustrated in Fig. 110, the arch of that diagram having the same dimensions as the arch of Fig. 108.

Upper Chord U_2U_3. With a unit load placed at U_4, $H = 120 \div (4 \times 33) = 0.909$ (see p. 5–84). In Fig. 109, moments about L_3 give $S_H = H \times y_H \div d = 0.909 \times {}^{30}\!/_{12} = 2.273$. Since H varies uniformly from 0.909 to zero as the load moves from the center

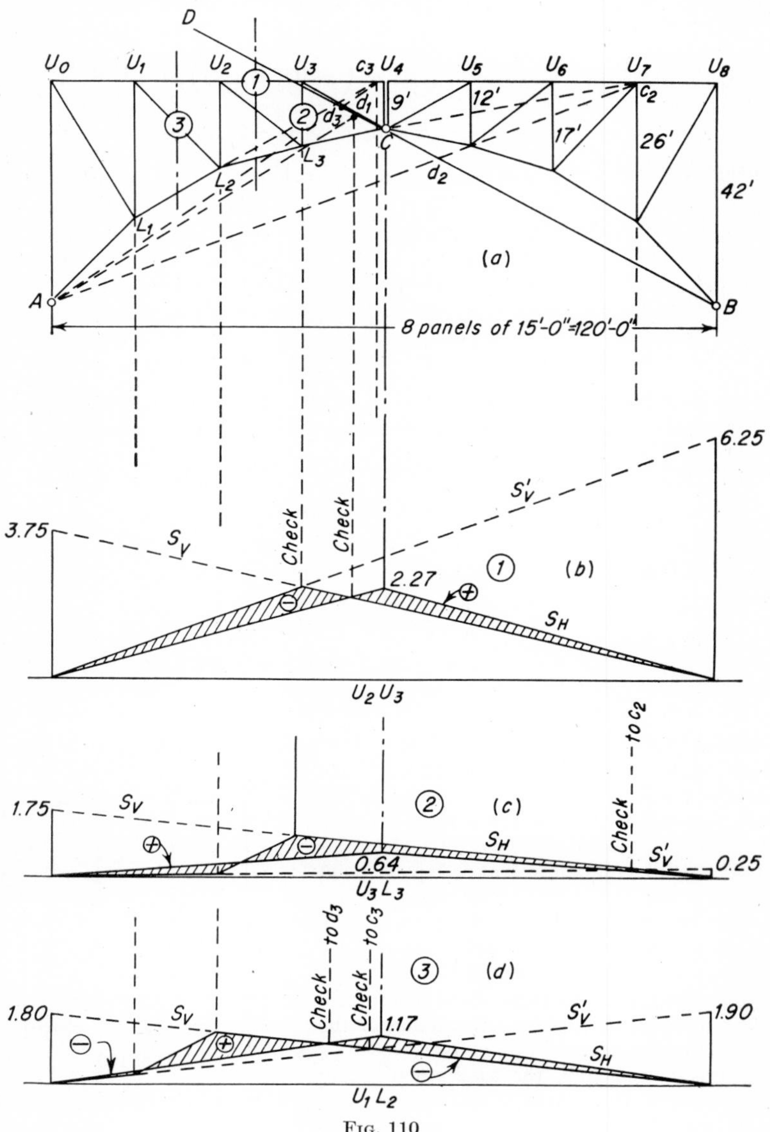

Fig. 110

toward either support, the influence line for S_H varies from +2.273 to zero, as shown in Fig. 110*b*.

A unit vertical reaction being assumed at A, with no load between the center of moments and A, $S_V = -1 \times y_V \div d$ (Fig. 109) $= -1 \times 45/12 = -3.75$. In order that this vertical reaction of unity may exist at A, a unit load must be placed at U_0. As this load moves toward the right, the reaction at A decreases uniformly to zero, and S_V varies accordingly. The line S_V of Fig. 110*b* is the influence line for the stress due to vertical forces from U_3 to the right support. To the left of U_3, it has no significance, since, in the equation for S_V, above, the load producing the reaction has not been considered. By a similar analysis of the right portion of the arch, $S_V' = -1 \times 75/12 = -6.25$, and the complete influence line for the vertical forces is constructed as shown in Fig. 110*b*, its maximum ordinate being directly under the center of moments.

The stress due to H is tension, and that due to the vertical forces, compression; hence, by plotting both influence lines above the horizontal reference line, the algebraic sum of the stresses due to a load at any point may be obtained by measurement between the H and V influence lines. Wherever the H line is above the V line, the resultant stress is tension, and vice versa. The maximum live-load tension occurs with panel loads at U_4, U_5, U_6, and U_7, and the amount of this tension may be obtained by adding the ordinates of the influence diagram (the shaded area) at these points and multiplying by the panel load. The maximum live-load compression is obtained in a similar manner.

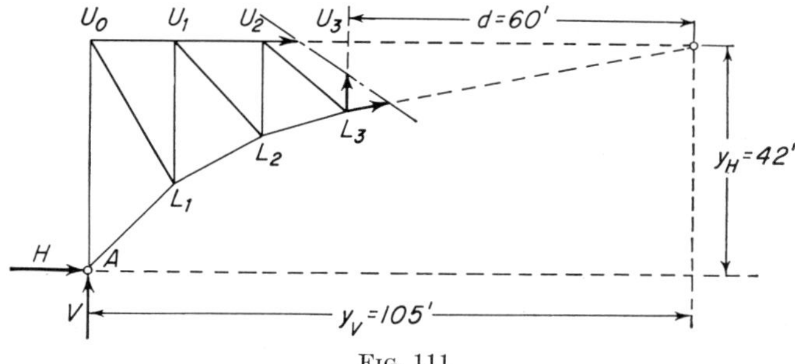

FIG. 111

Vertical U_3L_3. With unit load at U_4, $H = 0.909$, and $S_H = 0.909 \times 42/60 = +0.64$ [Fig. 110*c*], and the influence line for S_H is a triangle with its maximum ordinate under C, as shown in Fig. 110*c*. With a unit vertical reaction at A (Fig. 111), $S_v = -1 \times 105/60 = -1.75$. Similarly, the portion of the arch to the right of the section, $S_V' = -1 \times 15/60 = -0.25$. The influence lines for S_V and S_V' are plotted in Fig. 111*c*. In the panel U_2U_3 (the panel cut by the section), the influence line for stress due to vertical loads is a straight line joining the lines S_V and S_v', since a load at any point in the panel is distributed to the arch only at the ends of the panel.

Diagonal U_1L_2. The center of moments is at c_3 (Fig. 110*a*) 58.33 ft from the left end of the arch, and the lever arm of the stresses in the member is 32.50 ft. $S_H = -0.909 \times \dfrac{42}{32.50} = -1.17$, $S_V = +1 \div \dfrac{58.33}{32.50} = +1.80$, and $S_V' = +1 \times \dfrac{61.67}{32.50} = +1.90$. The influence diagram is shown in Fig. 110*d*.

Where the center of moments for a diagonal or vertical falls on the right of B, the stress S_V' will have the same sign as S_H, and the line S_V' should be plotted below rather than above the horizontal reference line.

STRESSES IN TWO-HINGED ARCHES

Exact Analysis. The two-hinged arch is a statically indeterminate structure for it develops one more unknown reaction condition than can be determined by the three equations of equilibrium. If one of the horizontal reactions of such an arch is considered removed, as, for example, the reaction H_B of the arch ring of Fig. 112, it is apparent that point B will be displaced horizontally to the right a distance that may be designated δ_B. Considering a horizontal force of unity applied at B in the direction tend-

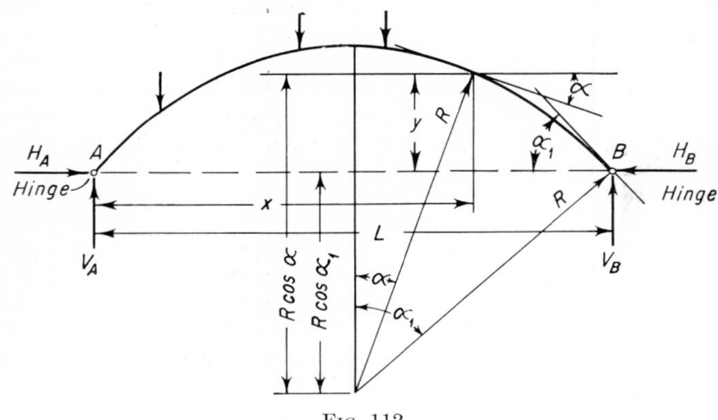

FIG. 112

ing to restore the original form of the arch, let the horizontal displacement so produced be δ_{1B}. Then obviously $-\delta_B = H_B \times \delta_{1B}$, and

$$H_B = -\frac{\delta_B}{\delta_{1B}}$$

Consequently, if the values of δ_B and δ_{1B} can be calculated, the redundant reaction is at once obtained, and the stresses at any point on the arch may then be determined by statical conditions.

From the theory of curved beams and the foregoing equation, the expression for the horizontal displacement (which is, of course, actually zero) of one of the reaction points may be written thus:

$$0 = \delta_B + H\delta_{1B} = \int_A^B \frac{Mm\,ds}{EI} - \int_A^B \frac{Nn\,ds}{AE} + \int_A^B \frac{Nm\,ds}{AE\rho}$$

where M = moment at any section (x, y) along the arch due to the applied loads
N = normal thrust on any section (x, y) due to the applied loads
m = moment at any section (x, y) due to a pair of horizontal unit forces acting inward at A and B with no other loads acting
n = normal thrust on any section (x, y) due to the unit horizontal forces at A and B with no other loads acting
I = moment of inertia of any section along the arch
A = area of any section along the arch
ρ = radius of curvature of the arch axis at any section
E = modulus of elasticity in tension and compression of the material of the arch

From Fig. 112, it is apparent that $m = y$, $n = \cos\alpha$, and $ds = dx \sec\alpha$. Further, it may be assumed that $N = H \sec\alpha$, since in a well-designed arch the line of thrust

closely parallels the arch axis. Substituting these values,

$$0 = \int_A^B \frac{My\,ds}{EI} - H\left[\int_A^B \frac{dx \sec \alpha}{AE} - \int_A^B \frac{y\,ds \sec \alpha}{AE\rho}\right]$$

If the radius of curvature is approximately uniform, as is usually the case, we may write $y = \rho \cos \alpha - \rho \cos \alpha_1$ (see Fig. 112), from which the bracketed quantity of the above equation may be simplified as follows:

$$\int_A^B \frac{dx \sec \alpha}{AE} - \int_A^B \frac{y\,ds \sec \alpha}{AE\rho} = \int_A^B \frac{dx \sec \alpha}{AE} - \int_A^B \frac{ds}{AE} + \int_A^B \frac{ds \sec \alpha \cos \alpha_1}{AE} = \int_A^B \frac{ds \sec \alpha \cos \alpha_1}{AE}$$

Let M' be the moment at any section (x, y) due to the applied loads with the redundant reaction removed. Then $M = M' - Hy$, and the equation for H is

$$H = \frac{\displaystyle\int_A^B \frac{M'y\,ds}{EI}}{\displaystyle\int_A^B \frac{y^2\,ds}{EI} + \int_A^B \frac{ds \sec \alpha \cos \alpha_1}{AE}}$$

In this expression, the second term in the denominator represents the effect of the rib shortening due to the axial thrust throughout the arch, and except in very flat arches it may be neglected. If the common assumption is made that the sectional area varies as secant α, this equation becomes

$$H = \frac{\displaystyle\int_A^B \frac{M'y\,ds}{EI}}{\displaystyle\int_A^B \frac{y^2\,ds}{EI} + \frac{S \cos \alpha}{A_c E}}$$

where S is the length of the arch axis and A_c the sectional area at the crown.

If the arch ring is divided into a number of small finite divisions of length ΔS, an expression for H closely approximating the above equation is

$$H = \frac{\displaystyle\sum \frac{M'y\,\Delta S}{EI}}{\displaystyle\sum \frac{y^2\,\Delta S}{EI} + \frac{S \cos \alpha_1}{A_c E}}$$

and since the denominator of this expression is a constant regardless of the loading, finally,

$$H = \frac{\displaystyle\sum \frac{M'y\,\Delta S}{EI}}{K}$$

The analysis of two-hinged arches is most easily effected by the use of influence diagrams for the various unknown forces. The first step is therefore the construction of the influence diagram for the horizontal reaction H. If a vertical load of unity is applied at point q, it follows from Maxwell's principle of reciprocal deflections that the horizontal deflection of A or B due to this loading is numerically equal to the vertical

deflection of q due to a pair of unit horizontal loads at A and B. Consequently, if a curve of vertical deflections for all points on the arch due to a pair of unit horizontal forces at A and B be constructed, it will also be, to a scale, the H influence line, since it represents the variation of the quantity $\sum \frac{M'y\,\Delta S}{EI}$ due to a unit vertical load moving across the structure. To obtain the true H influence diagram, it is necessary only to correct the scale by dividing the ordinates to this diagram by the constant K. The resulting diagram is then really a graph of the equation for H, when M' is the moment at the desired section due to a load of unity, and with the redundant reaction removed so that the arch acts as a simple beam.

Now from the principle of elastic weights,[1] it follows that the deflection curve for a simple beam may be obtained by constructing the bending-moment diagram for the beam, if it is considered as loaded with the M/EI diagram for the applied loads. Since $\Delta S = \Delta x \sec \alpha$, the curvature for unit horizontal loads at A and B equals:

$$\frac{y\,\Delta S}{EI} = \frac{y\,\Delta x}{EI \cos \alpha}$$

Therefore, if a fictitious loading that is equal at any point to $\frac{y}{EI \cos \alpha}$ be considered applied to the arch, the moment diagram for this condition will be the desired H influence diagram when the ordinates have been divided by the constant K. The procedure is as follows:

1. The dimensions of the arch being assumed, the rib is divided into a number of small segments of length ΔS. Usually, 12 or 15 segments will be sufficient.

2. The quantity $\frac{y\,\Delta x}{EI \cos \alpha}$ is computed for each segment and considered applied as a vertical load through the center of each segment.

3. The bending-moment diagram is constructed for this condition of loading. This may be done either graphically or analytically.

4. The ordinates of this diagram are divided by $K \left(= \sum \frac{y^2\,\Delta S}{EI} + \frac{S \cos \alpha_1}{A_c E} \right)$, and with the resulting values for ordinates another curve is constructed which will be the true H influence diagram. It will be noted that the second term in the expression for K, which represents the rib-shortening effect, is usually so small that it may safely be neglected.

[1] In a simple beam, the curvature

$$\frac{d^2y}{dx^2} = \frac{M}{EI} \tag{1}$$

from which, by integration, the deflection at any point

$$y = \iint \frac{M}{EI} \cdot dx \cdot dx \tag{2}$$

It is also demonstrated in treatises on strength of materials that the load

$$w = \frac{d^2M}{dx^2} \tag{3}$$

which integrated gives the bending moment

$$M = \iint w \cdot dx \cdot dx \tag{4}$$

The analogy between Eqs. (1) and (2) and Eqs. (3) and (4) is at once evident, and it will be seen that the load curve must bear the same relation to the bending-moment diagram that the curvature diagram does to the deflection curve. Then if the curvature ($= M/EI$) diagram is conceived as applied to the beam as a load, the bending-moment diagram for this condition will be the true deflection curve for the beam due to the actual loads. In the foregoing equations, proper limits of integration must of course be introduced.

5. With the H influence diagram determined, the arch becomes, in effect, statically determinate, and the various stresses may be determined by the conditions of static equilibrium.

If the structure is subjected to large variations in temperature, it will be necessary to consider the effect of this on the stresses. Recalling that $H = -\frac{\delta_B}{\delta_{1B}}$, it is clear that the horizontal thrust so produced will be $H_t = \pm ktl/K$, where k is the coefficient of expansion of the material of the arch and t the estimated plus or minus variation of the temperature from that at which the arch is erected.

Similarly a horizontal yielding Δ_H of an abutment will produce a horizontal reaction $H_y = \Delta_H/K$.

Approximate Solution for Ribbed Arches. The parabolic curve is by far the most common form for two-hinged arch ribs, since this form usually follows the line of thrust for dead load very closely. Other curves which may be employed generally depart from the parabolic so slightly that they may safely be analyzed as such. If this and the further assumptions noted below are made, a simplified solution may be developed which will be sufficiently accurate for the majority of two-hinged arch ribs.

It is customary to increase the rib section gradually from the crown to the hinges, and in this analysis it is assumed that the moment of inertia I varies as sec α; i.e., $I = I_c \sec \alpha$, where I_c is the moment of inertia of the crown section. Comparison shows that, even if this is only roughly approximate, the effect on the final values will be small. Further, it may be shown that the effect of rib shortening is usually very small, so that for practical purposes it may be assumed negligible. Then for a parabolic arch rib, these assumptions being made, $H = \frac{5Pl}{8h}(k^4 - 2k^3 + k)$ where P is a concentrated load distance kl from A. Plotting this equation for P equals unity and for values of k between zero and 1 gives the influence diagram for H.

It remains to construct the influence diagrams for the various internal stresses and moments. Since the true moment at any section $M = M' - Hy$, it is clear that the moment influence diagram for any section may be constructed by combining that for M' (the simple beam moment) with an H influence diagram of which the ordinates have been multiplied by y for the section considered. It is, however, inconvenient to compute the quantity Hy for each section, and the process may be simplified by writing the expression $\frac{M}{y} = \frac{M'}{y} - H$. To obtain the moment influence diagram for any section CC', in Fig. 113, the H influence diagram, which is shown in (a), is combined with that for M_q/y_q, as in (c), and the ordinates to the shaded area multiplied by y_q are then the true influence-line values.

With reference to Fig. 114, the unit stress at the extrados $S_e = \frac{N}{A} + \frac{Mc_e}{I}$, where N is the component of the thrust perpendicular to the section considered and A is the area of the section. If r is the radius of gyration of the section about the neutral axis for bending, $S_e = \frac{Nr^2}{I} + \frac{Ntc_e}{I} = \frac{N}{I}(r^2 + tc_e) = N\left(\frac{r^2}{c_e} + t\right)\frac{c_e}{I}$; i.e., the stress equals $\frac{c_e}{I}$ times the moment $N\left(\frac{r^2}{c_e} + t\right)$. This last term represents the moment of the thrust about the extradosal "kernel point" of the section, which, as is seen from Fig. 114, lies a distance r^2/c_e below the neutral axis for bending. Similarly the intradosal kernel point is $\frac{r^2}{c_i}$ above the neutral axis. If in the equation $\frac{M}{y} = \frac{M'}{y} - H$, the y term is re-

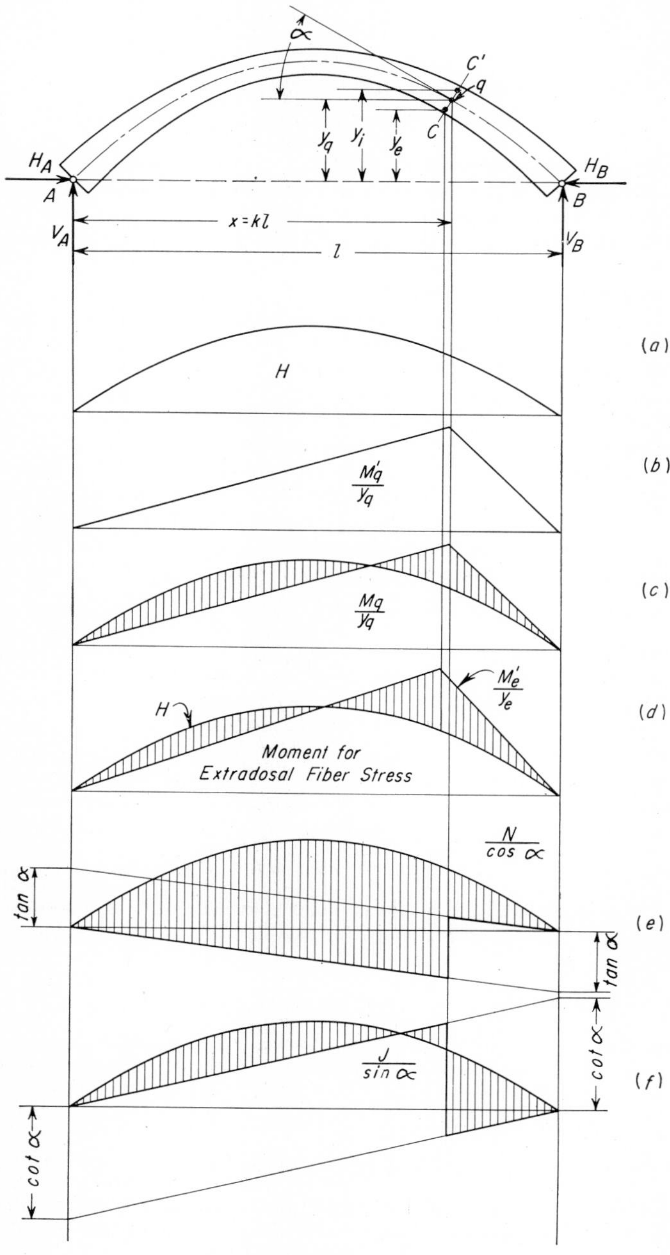
α
C'
q
C
y_q
y_i
y_e
H_A
H_B
A
B
x = kl
V_A
V_B
l
(a)
H
(b)
$\frac{M'_q}{y_q}$
(c)
$\frac{M_q}{y_q}$
(d)
H
$\frac{M'_e}{y_e}$
Moment for
Extradosal Fiber Stress
$\frac{N}{\cos\alpha}$
tan α
(e)
tan α
cot α
(f)
$\frac{J}{\sin\alpha}$
cot α

FIG. 113

placed by the ordinate of the proper kernel point, an influence diagram for the moment about that point may be constructed exactly as above. An ordinate to this diagram multiplied by c_e/I will give the extradosal fiber stress at the section considered due to a unit vertical load at the ordinate measured. The construction is illustrated in Fig. 113*d*.

The influence diagram for axial thrust at any section may easily be constructed. If the load is placed at the right of a section, $N = H_A \cos \alpha - V_A \sin \alpha$; and if just to the left, $N = H_B \cos \alpha + V_B \sin \alpha$. But $V_A = 1(1 - k)$, $V_B = 1 \cdot k$, and $H_A = H_B$. Substituting these values in the above equations, $\dfrac{N}{\cos \alpha} = H - (1 - k) \tan \alpha$ (unit load to right of section) and $\dfrac{N}{\cos \alpha} = H + k \tan \alpha$ (unit load to left of section). From these relations, the influence diagram is constructed as in (*e*) where the shaded area represents the desired diagram.

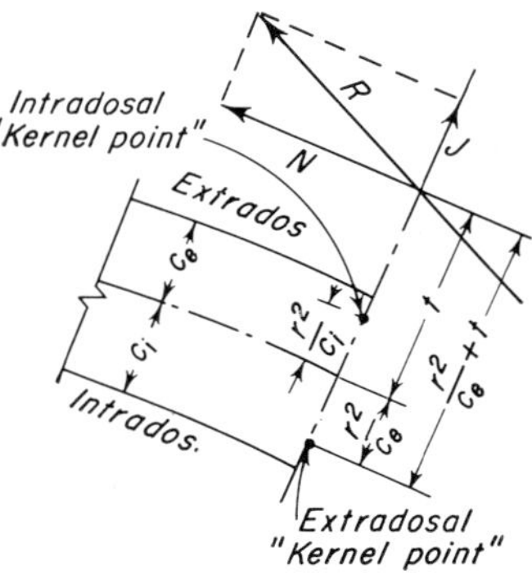

FIG. 114

The shear influence diagrams, though not of so great importance, may be similarly constructed. If the unit load is placed just to the right of the section considered, the shear $J = V_A \cos \alpha + H \sin \alpha$; if just to the left, $J = V_B \cos \alpha - H \sin \alpha$. Substituting the unit load values of V_A and V_B in the above equations, as before, $\dfrac{J}{\sin \alpha} = (1 - k) \cot \alpha + H$ (unit load to right of section) and $\dfrac{J}{\sin \alpha} = k \cot \alpha - H$ (unit load to left of section). The construction is shown in (*f*).

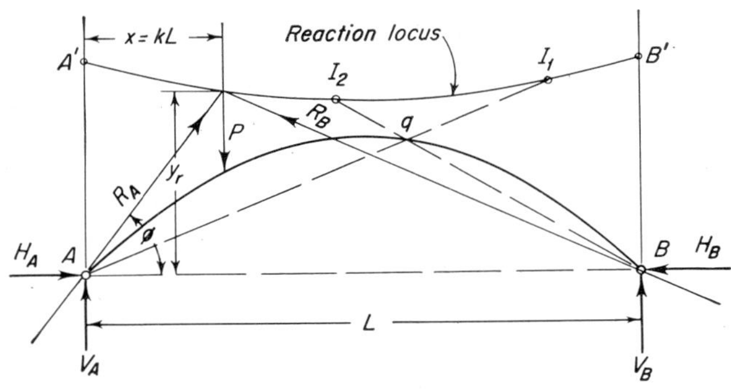

FIG. 115

From the fundamental condition of statics that the summation of moments about any point must equal zero, it is obvious that the line of action of the reactions R_A and R_B (Fig. 115) must intersect on the line of action of the resultant of the applied loads. The curve described by the intersection of these lines with the line of action of a vertical load of unity as the latter moves across the structure is known as the *reaction locus*. It affords a convenient method for studying the action of a single concentrated moving load. From Fig. 115, it is apparent that $H_A = R_A \cos \phi$ and $V_A = R_A \sin \phi$. Then $V_A/H_A = \tan \phi = y_r/kl$, where y_r is the ordinate of the intersection point. Transposing

and substituting the value of H, $y_r = V_Akl/H = \dfrac{P(1-k)kl}{\dfrac{5Pl}{8H}(k-2k^3+k^4)} = \dfrac{8h}{5(1+k-k^2)}$,

which is the equation for the reaction locus for a parabolic rib.

The properties of the reaction locus may further be employed as follows: Let it be required to determine the portions of the arch to be uniformly loaded to produce maximum positive or negative moment at any point q. If the lines AqI_1 and BqI_2 are drawn, any load to the left of I_2 or to the right of I_1 produces negative moment at q, and any load between I_1 and I_2, positive moment. A uniform load distributed between the points noted produces maximum positive or negative moment respectively. The value of H for a load distributed between k_1l and k_2l may be obtained from the equation

$$H = \frac{5wl}{8h}\int_{k_1l}^{k_2l}(k-2k^3+k^4)\,d(kl) = \frac{5wl^2}{8h}\int_{k_1}^{k_2}(k-2k^3+k^4)\,dk = \frac{wl^2}{16h}\Big[5(k^2-k^4)+2k^5\Big]_{k_1}^{k_2}$$

where w is the uniform load per unit length.

Approximate Solution for Spandrel-braced Arches. For a spandrel-braced arch,

$$H = \frac{\sum \dfrac{S'ul}{AE}}{\sum \dfrac{u^2l}{AE}}$$

where S' is the stress in any member due to the applied load with the redundant reaction removed, u the stress in any member due to a pair of unit horizontal forces acting inward at a and a' (Fig. 116), and A the sectional area of the member considered.

In order to construct an influence diagram for H, it would be necessary to make summations of the numerator of the foregoing equation for each symmetrical panel point. In an arch of many panels, such summations become very tedious. The graphical method described below will expedite the work.

1. Assume that the reaction a is free to move horizontally and that the reaction a' is fixed. Apply a horizontal force of unity at a, acting toward a', and determine by means of a stress diagram the stresses in all the members of the arch truss due to this force. Assume that the sectional areas of all the members are equal. Then the relative elongations and shortenings due to unit load applied at a are ul.

2. With the values of ul as determined above, a displacement diagram for the arch truss may be drawn (the one shown is drawn assuming the center vertical to be fixed in direction).

3. Divide the vertical deflection at each load point by the horizontal displacement of the reaction points a and a'. The resulting values are the ordinates to the H influence diagram.[1]

[1] The proof of this is as follows: With a horizontal force of unity applied at a, δ_c due to this force is obtained by applying a unit load at C and is equal to $\sum \dfrac{S_a'u_cl}{AE}$. Similarly, the horizontal displacement of a or $\delta_a = \sum \dfrac{S_a'u_al}{AE}$. Since the horizontal force applied at a is of unit value

$$\delta_c = \sum \frac{u_au_cl}{AE} \qquad \text{and} \qquad \delta_a = \sum \frac{u_a{}^2l}{AE}$$

Now $\delta_c = \sum \dfrac{u_au_cl}{AE}$ is equal either to the displacement at C due to unit load applied at a, or to the displacement at a due to unit load applied at C. If the latter condition is considered, $S' = u_c$ and H due to unit load applied at C is

$$H = \frac{\sum \dfrac{u_au_cl}{AE}}{\sum \dfrac{u_a{}^2l}{AE}} = \frac{\delta_c}{\delta_a}$$

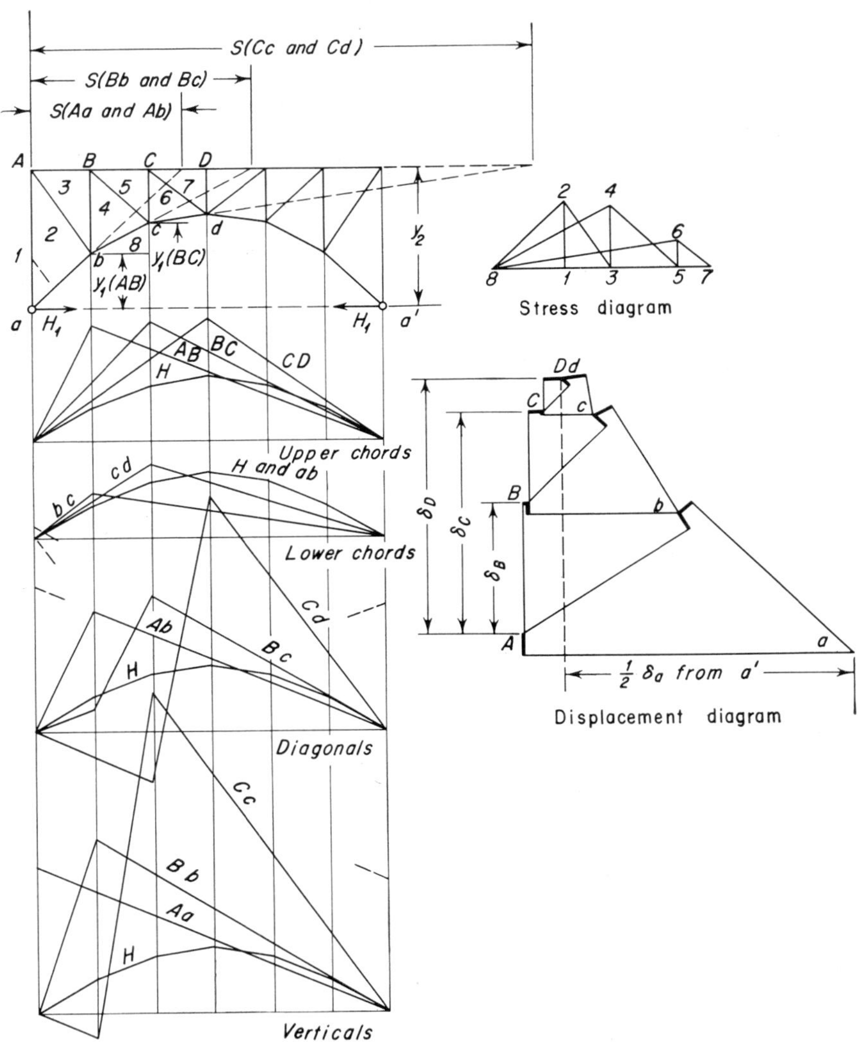

FIG. 116

4. With the influence diagram for H constructed as described, H is known for all positions of loading and the various stresses may be computed by ordinary methods.

5. With approximate sectional areas determined, relative elongations and shortenings based on $ul \div A$ may be computed and a second displacement and influence diagram for H constructed. Usually a third approximation will give sufficiently accurate results.

Complete influence diagrams for all members of an arch truss are shown in Fig. 116. The method of constructing them is as follows:

Upper-chord Members. For the chord BC, the center of moments is at c and $M_c = M' - Hy_1$, where M' is the moment due to the vertical loads. Since the influence dia-

gram for H has already been determined, it simplifies the construction to write this equation as $M_c = \left(\frac{M'}{y_1} - H\right) y_1$. For unit load above the center of moments, $\frac{M'}{y_1} = \frac{(l - a)a}{ly_1}$, where l is the span and a the distance from the center of moments to the left reaction. This value is easily obtained for each chord member and the influence diagram for the vertical loads constructed. The stresses are obtained by taking the algebraic summation of the ordinates of the complete diagram (that portion between the M'/y_1 lines and the H lines). This summation is multiplied by the load and y_1 and divided by the lever arm.

Lower-chord Members. The same procedure is followed except that y_2 is used instead of y_1.

Diagonals and Verticals. The center of moments is taken at the intersection of the two chords cut by the section. The distance from this center of moments to the left reaction is s. Then $M_s = Vs - Hy_2$ or $\left(\frac{Vs}{y_2} - H\right) y_2$, where V is the vertical component of the left reaction. If V is of unit value, a unit load is directly over the left reaction and Vs/y_2 becomes s/y_2, which is easily obtained for each member. The left reaction decreases uniformly to zero as the unit load moves across the span from the left to the right. Therefore if s/y_2 is plotted at the left reaction and zero at the right reaction, a line joining these two is the influence diagram for the vertical loads to the right of the section. Similarly, the influence diagram for the vertical load to the left of the section may be constructed by plotting at the right reaction $\frac{l - s}{y_2}$ and connecting this point to the left reaction by a straight line. In the section the influence diagram is a straight line. The stresses are obtained from the complete diagram in the same manner as for chord members.

STRESSES IN FIXED-END ARCHES

Method of Analysis. An arch is considered fixed when rotation is prevented at the supports. Such an arch is statically indeterminate; there are six reaction components and only three equations are available from conditions of equilibrium. Three more equations must be obtained from a knowledge of the deformations.

One way to determine the reactions is to consider the arch cut at the crown, forming two cantilevers. First, the horizontal and vertical deflections and rotation produced at the end of each half arch by the loads are computed (Fig. 117*a*). Next, the deflection components and rotation at those ends are determined for unit vertical force, unit horizontal force, and unit moment applied separately at the ends (Fig. 117*b*).

These deformations, multiplied, respectively, by V, the unknown shear; H, the unknown horizontal thrust at the crown; and M, the unknown moment there, yield the deformations caused by the unknown forces at the crown. Adding these deformations algebraically to the corresponding ones produced by the loads gives the net movement of the free end of each half arch. Since these ends must deflect and rotate the same amount to maintain continuity, three equations can be obtained in this manner for the determination of V, H, and M.

The various deflections can be computed by the dummy unit-load method, as has been demonstrated for two-hinged arches.[1]

When the unknowns have been evaluated, the normal and tangential forces and the bending moment at any section can be found by applying the equations of equilibrium.

[1] For typical solutions by this method, see S. Timoshenko and D. H. Young, "Theory of Structures," McGraw-Hill Book Company, Inc., New York, 1945. For solution by the column analogy, see H. Cross and N. D. Morgan, "Continuous Frames of Reinforced Concrete," John Wiley & Sons, Inc., New York.

Computation of stresses is similar to that for two-hinged arches. Stresses in single-bay rigid frames also may be calculated in a similar manner.

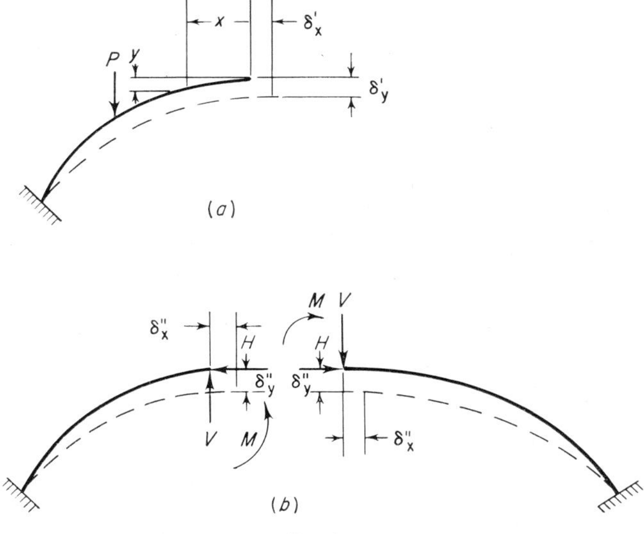

FIG. 117

CYLINDRICAL SHELL STRUCTURES

Shell Analysis. Stress analysis of shells may involve the solution of differential equations and much tedious calculation. However, the work can be simplified and shortened by the methods and tables presented in "Design of Cylindrical Shell Roofs," Manual of Engineering Practice No. 31, American Society of Civil Engineers.[1]

The procedure employed is to start with the assumption that the shell is statically determinate and the loads are carried by direct stresses only (membrane theory). This condition can be met, however, only when transverse and shearing forces at the boundaries are countered by equal and opposite reactions. When the required reactions are not supplied, radial shearing forces and moments are required to maintain equilibrium. The stresses created by actual edge and end conditions must be added to those obtained by the membrane theory.

The relative importance of the boundary conditions depends on the ratio of the radius r of the shell to the distance between longitudinal supports L. When r/L is less than 0.25, the longitudinal stresses are given fairly accurately by simple-beam theory, treating the barrel as a beam with a curved cross section. In shells of these proportions, stresses due to deformations of the transverse supports, such as stiffeners or end beams, fade out rapidly, whereas stresses due to deflection of longitudinal edges are felt throughout the shell and influence the stress distribution in the shell substantially. For larger values of r/L, the longitudinal stresses are much larger than those given by simple-beam theory.

As an example of the application of simple-beam theory to a cylindrical arch, assume a 5-in.-thick roof curved to a radius of 8 ft 4 in. to be designed to support a uniform vertical load of 100 lb per sq ft, with columns spaced 16 ft on centers transversely and

[1] See also "Design of Barrel Shell Roofs," Portland Cement Association, Chicago, Ill.

40 ft longitudinally. Basically, the shell will act as a beam in the long direction and as an arch in the short direction. This arch, however, is not supported at the ends but by shearing forces along its edges. If a 1-ft-wide strip is cut transversely out of the shell, as in Fig. 118*a*, it will be supported by the algebraic sum of the shearing forces acting on both faces:

$$(S + \Delta S) - S = \Delta S$$

For convenience in computing the stresses in the curved element, it is divided into 12 units, each 21.46 in. long. From the geometric properties of these units, the section

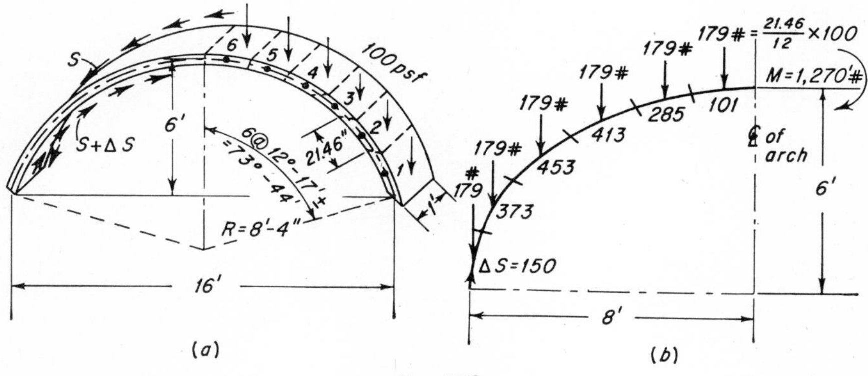

Fig. 118

moduli of the curved element acting as a beam are found to be

$$I/c \text{ (top)} = 23{,}860 \text{ in.}^3$$
$$I/c \text{ (bottom)} = 12{,}934 \text{ in.}^3$$

The total load on the shell is: $W = 12 \times 100 \times 21.46/12 = 2{,}146$ lb, from which the maximum longitudinal bending moment is computed:

$$M = 2{,}146 \times 40^2/8 = 429{,}200 \text{ ft-lb}$$

and the longitudinal stresses in the top and bottom fibers are

$$\text{Top: } f = 429{,}200 \times 12/23{,}860 = 216 \text{ psi}$$
$$\text{Bottom: } f = 429{,}200 \times 12/12{,}934 = 398 \text{ psi}$$

For computing the transverse bending moment, the shearing forces ΔS acting on each unit may be calculated from the standard formula for shearing stresses, $S = VQ/It$, where V is the total shear on the section, Q is the moment about the center of gravity, of the areas on either side of the fiber for which the shear is to be computed, I is the moment of inertia of the section, and t is the thickness of the section. For the arch element:

$$\Delta S = (\Delta V)Q/It$$

Values of ΔS computed for each element are shown in Fig. 118*b*. Each of these forces acts tangent to the curve. The net moment of the load and these forces about the crown figures out to be 1,270 ft-lb.[1]

Surface loads supported by a shell are transmitted by tangential and radial shears to the stiffeners. Since the shell and each stiffener are continuous, the shell participates

[1] *Eng. News-Record*, Sept. 2, 1954, p. 40.

in the bending action of the stiffener. As a result, tangential and radial shears determined from the shell analysis are changed. A two-step procedure is required to compute the final stresses:

1. Initially, the stiffeners are assumed to be infinitely rigid. Stresses and reactions of the shell are computed for this condition. Equivalent surface loads are next computed having the same effect as the radial and tangential shears.

2. These loads are applied to the stiffener, which is considered a T beam. The flanges represent the effect of the shell on the bending of the curved stiffener.

As in the case of arches, careful consideration should be given to secondary stresses. Included in these are stresses due to uniform volume change of the rib and shell, rib shortening, unequal settlement of footings, and differential volume change between rib and shell.

FOLDED-PLATE ROOFS

When a thin-shell roof is formed by joining two or more flat plates along their edges, the result is a folded-plate roof (Fig. 119*b*).

Approximate Method of Analysis. Just as a curved thin-shell roof can be analyzed as a beam with a curved cross section, so can a folded-plate roof. Transverse tie rods

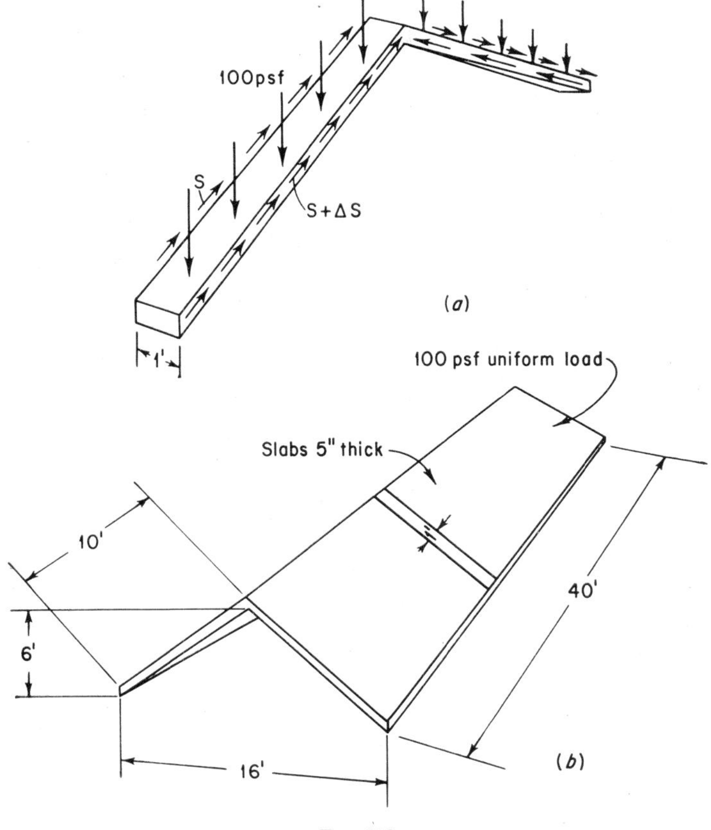

Fig. 119

or stiffeners are required at the ends of the span or at intermediate points to take the thrust.

In Fig. 119*a* is shown an element 1 ft wide cut out of the folded-plate roof in Fig. 119*b*. Spans and rise are the same as for the curved roof in Fig. 118. The element carries a uniform vertical load of 100 lb per sq ft of sloping surface. It is supported by the algebraic sum of the shearing forces acting on both faces ΔS.

Since the vertical component of the sum of the ΔS forces must equal the sum of the applied vertical forces:

$$\Sigma\,\Delta S = 100 \times 10 \times 10/6 = 1{,}667 \text{ lb}$$

for each plate.

The transverse bending moment at any section of the element is the moment about the section of the applied load on one side of it, less the moment of the ΔS forces on that side. The moment at the crown, noting that the ΔS forces act at the center of the section and go through the crown, thus producing no moment, is

$$M = WL/2 = 100 \times 10 \times 8/2 = 4{,}000 \text{ ft-lb}$$

With the folded plate acting as a simple beam on a 40-ft span, the longitudinal bending moment in each plate is that due to the shear $\Sigma\,\Delta S$ acting on a simple beam:

$$M = 1{,}667 \times 40^2/8 = 333{,}333 \text{ ft-lb}$$

The section modulus of the 120-in.-deep tilted girder is

$$I/c = bd^2/6 = 5 \times 120^2/6 = 12{,}000 \text{ in.}^3$$

Hence, the unit stress in the top or bottom fiber is

$$f = 333{,}333 \times 12/120 = 333 \text{ psi}$$ *

STRESSES IN DOMES AND CATENARIES

Circular Domes. A spherical surface is often used for domes, with an integral edge member provided at the base to resist hoop tension there. Such domes are called circular, because the surface is formed by revolving an arc of a circle about a vertical axis.

Theoretically, the shells are assumed to carry only uniformly distributed, symmetrical loading.

In Fig. 120*b*, the surface area of the strip of the dome between points A and B is given by

$$A = 2\pi r^2(\cos \phi_A - \cos \phi_B)$$

For a constant load w per square foot of area, the total load is

$$W_u = wA = 2\pi r^2 w(\cos \phi_A - \cos \phi_B)$$

If the load increases from zero at point Z at a rate of w' per radian (57.3°), the total load on the strip between A and B then is

$$W_v = 2\pi r^2 w'[\sin \phi_B - \sin \phi_A - \cos \phi_B(\phi_B - \phi_A)]$$

Another type of load considered to act on a dome is a collar load uniformly distributed around a circle of latitude. Let W_c be the total load from this source. Then, the total load acting on the strip between A and B is

$$W = W_u + W_v + W_c$$

* *Eng. News-Record*, Sept. 2, 1954, p. 40. For a more complete analysis, see also GEORGE WINTER and MINGLUN PEI, Hipped Plate Construction, *Jour. Am. Concrete Inst.*, January, 1947, p. 505.

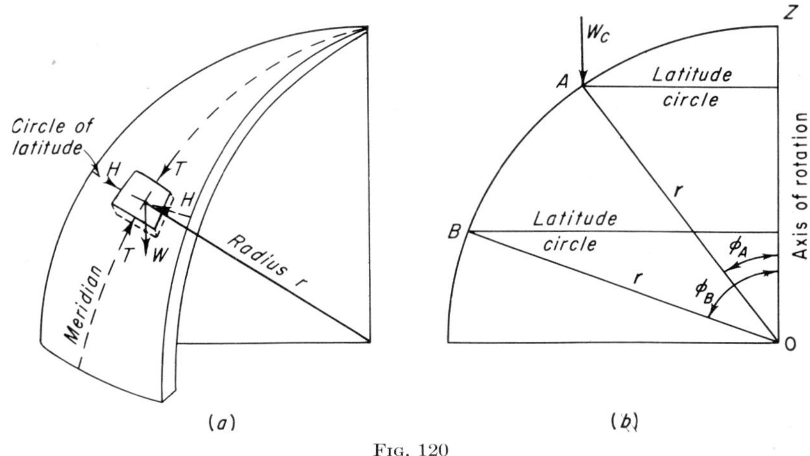

FIG. 120

As indicated in Fig. 120*a*, the vertical load at any point on the dome is resisted by a meridional thrust T and a hoop force H. Along the circle of latitude through B, the total meridional thrust is

$$T = \frac{W}{2\pi r \sin^2 \phi_B}$$

Accompanying it is a hoop force,

$$H = -T + [w + w'(\phi_B - \phi_A)]r \cos \phi_B$$

If the dome is discontinued along a circle of latitude through point B, the edge member will be subjected to a ring tension equal to

$$S = \frac{W \cos \phi_B}{2\pi \sin \phi_B}$$

If the dome is omitted above a circle of latitude through point A, the collar load will produce a ring compression along the edge of the opening:

$$S' = \frac{W_c \cos \phi_A}{2\pi \sin \phi_A}$$

At the top of a solid dome:

$$T = H = \tfrac{1}{2}wr$$

Similar approximate formulas for the design of conoidal and elliptical domes are given in a pamphlet published by the Portland Cement Association, Chicago, Ill.

Stresses in Cables. When a cable is suspended from two points of support, the end reactions have a horizontal as well as a vertical component. If the cable carries only vertical loads, the horizontal component of the tension at any point in the cable is equal to the horizontal component of the end reaction. The vertical components of the reactions can be computed by taking moments about the supports in the same way as for beams, provided the cable supports are at the same level. When the supports are not at the same level, another equation, derived from a knowledge of the shape of the cable, is required.

Generally, it is safe to assume that the thickness of the cable is negligible compared with the span. Hence, bending stresses may be neglected, and the bending moment at any point is equal to zero. As a consequence, the shape assumed by the cable is similar to that of the bending-moment diagram for a simply supported beam carrying the same loads. Stresses along the cable are directed along the axis.

At the lowest point of a cable, the vertical shear either is zero or changes sign on either side of the low point.

The horizontal component of the reaction can be computed from the fact that the bending moment is zero at a point on the cable at which the sag is known (usually the low point).

For example, suppose a cable spanning 30 ft between supports at the same level is permitted to sag 2 ft in supporting a 12-kip load at a third point. Taking moments about the supports, we find the vertical components of the reactions equal to 8 kips and 4 kips, respectively. Setting the bending moment at the low point equal to zero, we can write the equation $8 \times 10 - 2H = 0$; from which the horizontal component H of the reaction is found to be 40 kips. The maximum tension in the cable then is

$$T = \sqrt{8^2 + 40^2} = 40.8 \text{ kips}$$

WIND AND SEISMIC STRESSES IN TALL BUILDINGS

Resistance to Lateral Forces. Buildings must be designed to resist horizontal forces as well as vertical loads. In tall buildings, the lateral forces must be given particular attention, because if they are not properly provided for, they can cause collapse of the structure.

Horizontal forces are generally taken care of with X bracing, shear walls, or wind connections. X bracing, usually adopted for low industrial-type buildings, transmits the forces to the ground by truss action. Shear walls are vertical slabs capable of transmitting the forces to the ground without shearing or buckling; they generally are so deep in the direction of the lateral loads that bending stresses are easily handled. Wind connections are a means of establishing continuity between girders and columns. The structure is restrained against lateral deformation by the resistance to rotation of the members at the connections.

The continuous rigid frames thus formed can be analyzed by such methods as moment distribution and dummy unit load. However, such solutions are so laborious that they are seldom attempted for tall buildings. Generally, approximate methods are used.

It is noteworthy that for most buildings even the "exact" methods are not exact. In the first place, the forces acting are not static loads, but generally dynamic; they are uncertain in intensity, direction, and duration. Earthquake forces, usually assumed as a percentage of the weight of the building above each level, act at the base of the structure not at each floor level as is assumed in design. Also, at the beginning of a design, the sizes of members are not known; so the exact resistance to lateral deformation cannot be calculated. Furthermore, floors, walls, and partitions help resist the lateral forces in a very uncertain way.

Portal Method of Analysis. Since an exact analysis is impossible, most designers prefer a wind-analysis method based on reasonable assumptions and requiring a minimum of calculations. One such method is the so-called "portal method."

It is based on the assumptions that points of inflection (zero bending moment) occur at the mid-points of all members and that exterior columns take half as much shear as do interior columns. These assumptions enable all moments and shears throughout the building frame to be computed by the laws of statics.

Consider, for example, the roof level (Fig. 121*a*) of a tall building. A wind load of 600 lb is assumed to act along the top line of girders. To apply the portal method, we

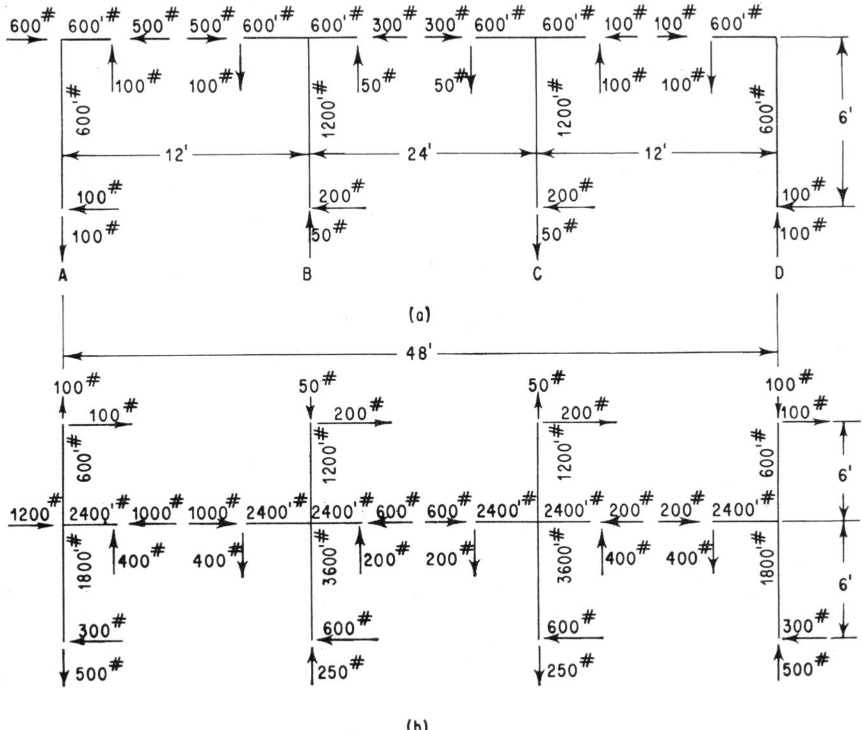

FIG. 121

cut the building along a section through the inflection points of the top-story columns, which are assumed to be at the column mid-points, 6 ft down from the top of the building. We need now consider only the portion of the structure above this section.

Since the exterior columns take only half as much shear as do the interior columns, they each receive 100 lb, and the two interior columns, 200 lb. The moments at the tops of the columns equal these shears times the distance to the inflection point. The wall end of the end girder carries a moment equal to the moment in the column. (At the floor level below, as indicated in Fig. 121*b*, that end of the end girder carries a moment equal to the sum of the column moments.) Since the inflection point is at the mid-point of the girder, the moment at the inner end of the girder must be the same as at the outer end. The moment in the adjoining girder can be found by subtracting this moment from the column moment, because the sum of the moments at the joint must be zero. (At the floor level below, as shown in Fig. 121*b*, the moment in the interior girder is found by subtracting the moment in the exterior girder from the sum of the column moments.)

Girder shears then can be computed by dividing girder moments by the half span. When these shears have been found, column loads can be easily computed from the fact that the sum of the vertical loads must be zero, by taking a section around each joint through column and girder inflection points. As a check, it should be noted that the column loads produce a moment that must be equal to the moments of the wind loads

above the section for which the column loads were computed. For the roof level (Fig. 121a), for example, $-50 \times 24 + 100 \times 48 = 600 \times 6$.

Cantilever Method of Analysis. Another wind-analysis procedure that is sometimes employed is the cantilever method. Basic assumptions here are that inflection points are at the mid-points of all members and that direct stresses in the columns vary as the distances of the columns from the center of gravity of the bent. The assumptions are sufficient to enable shears and moments in the frame to be determined from the laws of statics.

The results obtained from this method generally will be different from those obtained by the portal method. In general, neither solution is correct, but the answers provide a reasonable estimate of the resistance to be provided against lateral deformation.[1]

Bibliography

Crane, T.: "Architectural Construction," John Wiley & Sons, Inc., New York.

Cross and Morgan: "Continuous Frames of Reinforced Concrete," John Wiley & Sons, Inc., New York.

Dunham, C. W.: "Planning Industrial Structures," McGraw-Hill Book Company, Inc., New York, 1948,

Hoff, N. J.: "The Analysis of Structures," John Wiley & Sons, Inc., New York.

Merritt, F. S.: "Building Construction Handbook," McGraw-Hill Book Company, Inc., 1958.

Parcel and Maney: "Statically Indeterminate Structures," John Wiley & Sons, Inc., New York.

Sutherland and Bowman: "Introduction to Structural Theory and Design," John Wiley & Sons, Inc., New York.

Timoshenko and Young: "Theory of Structures," McGraw-Hill Book Company, Inc., New York, 1945.

Urquhart and O'Rourke: "Stresses in Simple Structures," McGraw-Hill Book Company, Inc., New York, 1932.

Wilbur and Norris: "Elementary Structural Analysis," McGraw-Hill Book Company, Inc., New York, 1948.

[1] See also *Trans. ASCE*, vol. 105, pp. 1713–1739, 1940.

Section 6

STEEL DESIGN

BY CARLTON T. BISHOP

SHEAR AND BENDING MOMENT

Introductory. *Shear* and *bending moment* are terms used in the design of beams, girders, and similar members which resist bending. They have to do with *external* forces, such as the applied loads or the reactions at the supports. Since the majority of external forces for which beams are designed are either horizontal or vertical, it is convenient to resolve other forces into horizontal and vertical components; other coordinate axes may be chosen for special cases. For simplicity, only horizontal beams with vertical loads will be considered here, but the same principles can be adapted to other cases quite easily. The elementary conceptions of forces, shear, and bending moment are discussed in Section 3 on "Mechanics of Materials." A few of the more important facts are repeated here as a basis for the more advanced discussion of moving loads.

Forces. The external forces acting on a beam are in equilibrium, and any unknown external force, such as a reaction at a support, may be obtained by applying the equations of equilibrium to all the external forces acting on the entire beam. If a beam were cut by a transverse section, it would fall. The tendency to fall is resisted by internal forces or stresses acting in the fibers cut by the section. The equations of equilibrium may be applied to a portion of a beam, provided both external and internal forces acting on that portion are considered. The internal forces depend on the form of cross section and the strength of the material. It is convenient to treat the external forces on a portion of a beam or girder separately and to find the shear or the bending moment, it being realized, however, that these forces are not by themselves in equilibrium but are held in equilibrium by the internal forces.

Reactions. After the magnitudes and positions of the loads on a beam are determined, the next step is to find the reaction, or balancing force, at one of the supports. If a cantilever beam is supported at one end only, the reactions are often indeterminate; but fortunately shears and bending moments may be found without using a reaction. The reactions of a simple beam, i.e., one supported at both ends with all loads between the supports, are considered to act at the centers of the supports, and the effective length of the beam is the distance from center to center of the supports. The reactions of an overhanging cantilever beam, which has loads beyond one or both supports, are also considered to act at the centers of the supports. If the loads are symmetrical in magnitude and position, each reaction is equal to one-half the sum of the loads. If the loads are unsymmetrical, by taking moments about a point in the line of action of one of the reactions, its lever arm and therefore its moment becomes zero, and the other re-

action may be found from a single equation containing the moments of all external forces on *both sides* of the point of moments.

Shear. *The shear on a segment of a beam is the algebraic sum of all the external forces that act upon that segment.* It is not a force but a *sum* of forces which act not at any one point but on *one side* of a given section plane; it is better, therefore, to speak of the shear *for* a section taken at a given point rather than the shear *at* a given section. For any section, the shear on one segment of a beam is numerically equal to the shear on the other segment but with opposite sign. In designing a beam, it seldom matters whether a shear is positive or negative; it is important to be precise in using signs in order to obtain the correct algebraic sum, but the sign may then be dropped before equating the shear to the expression for shear resistance of the internal forces. A section should not be taken in the line of action of an external force, but it may be considered infinitely close on either side, so that the force is wholly on one segment. For simplicity, only the left-hand segment will be considered here. When a load is uniformly distributed, only that portion of the load at the left of the section should be considered.

Bending Moment. *The bending moment at any point of a beam is the algebraic sum of the moments of all the external forces on one side of the given point, the center of moments being taken at that point.* The bending moment found from the forces on the left-hand segment is numerically equal to that found from the forces on the right-hand segment but with opposite sign, and only the left-hand segment will be considered here, even though the right-hand segment may have fewer terms. When a load is uniformly distributed, that portion of the load at the left of the point of moments may be replaced by an equivalent concentrated load applied at the center of the portion replaced. It is usually preferable and often essential to find the bending moments for uniformly distributed loads separately from the bending moments for any concentrated loads which may be on the same beam, the sum of the two giving the total bending moment. If a uniform load extends the full length of a simple beam, it is well to find the bending moment by the short-cut method explained below instead of by the general method. Bending moments are used constantly in both foot-pounds and inch-pounds or foot-kips and inch-kips; so it is imperative to indicate the units. Since distances are usually expressed in feet, it is simpler to find all bending moments in foot-pounds (ft-lb) or foot-kips (ft-kips); thus, when there are several bending moments to be combined, they can be added before they are multiplied by 12, and the factor need be used only once when the bending moment is equated to the resisting moment in design.

Weights of Beams. Every beam must be self-supporting, and a properly designed beam must have sufficient reserve strength to provide for its own weight as well as the superimposed loads. Since the weight of a beam is not known until after the beam is designed, either a preliminary design is made to determine the probable size of beam and a second design made including the weight, or else an assumed weight is included in the loads for which the beam is first designed. With a little experience, one can assume the weight of a beam so accurately that a redesign is seldom necessary; one can usually tell by inspection whether such redesign would result in a change in the size of the beam. As a matter of fact, the weight of a steel beam is relatively small compared with other loads; hence, a little variation does not affect the total bending moment seriously unless the beam is embedded in concrete or other heavy fireproofing material; the weight of such fireproofing depends on the size of the beam, and so the combined weight should be made to correspond to the size of the beam to be used.

Short-cut Rule, Uniform Loads. Since every beam is self-supporting, it must support its uniformly distributed weight in addition to the superimposed loads which may be concentrated or uniformly distributed, or both. A simple beam, i.e., one supported at both ends with all loads between the supports, is used so commonly that it is convenient to use a short-cut rule for finding the bending moment under a full-length

uniform load. This rule is as follows: *The bending moment at any point of a simple beam with a full-length uniformly distributed load is equal to one-half the unit load multiplied by the product of the segments.* The unit load is usually expressed in pounds per foot (lb per ft), and the segments in feet; so the resulting bending moment is in foot-pounds. At the center, the bending moment is $\frac{w}{2}\left(\frac{l}{2}\right)^2$. This is equivalent to the more usual expression $wl^2/8$ but is often more convenient because of smaller factors; furthermore, it is just a special application of the more general rule "one-half the unit load times the product of the segments," which applies at any point of the beam, whereas the $wl^2/8$ gives the bending moment at the center only. The rule may be proved as follows: In Fig. 1 are shown the original beam, the left-hand segment uniformly loaded, and the same with the uniform load replaced by an equivalent load concentrated at the center of the segment. The bending moment is the algebraic sum of the moments of the reaction and this equivalent load.

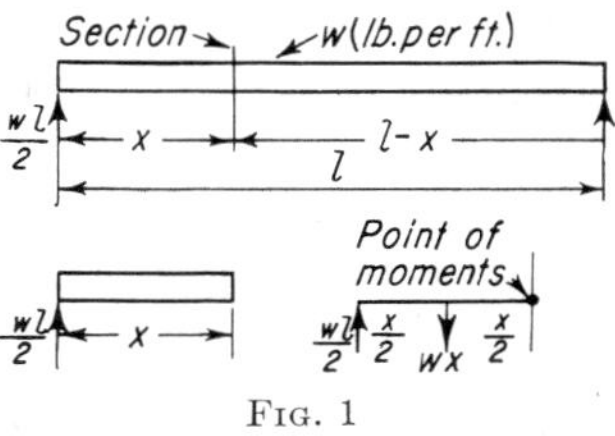

Fig. 1

$$M = \frac{wl}{2} \times x - wx \times \frac{x}{2} \qquad \text{or} \qquad M = \frac{w}{2} x(l - x)$$

Maximum Shear and Maximum Bending Moment. Since most beams are made of uniform cross section throughout their lengths, they must be designed to resist the greatest shear or bending moment to which they are to be subjected. In a *simple beam,* the maximum shear and the maximum bending moment do not occur at the same section; on the contrary, the maximum shear occurs for the section where the bending moment is zero, and the maximum bending moment occurs for the section where the shear is zero. In general, the maximum shear for a simple beam occurs for a section infinitely close to the larger reaction, and the shear is equal to that reaction minus any fixed concentrated load which may be applied directly above the reaction. For unsymmetrical loads, it is usually obvious which reaction is larger. For a system of concentrated loads, the maximum bending moment in a simple beam occurs in the line of action of one of the concentrated loads, the one at the section for which the shear is positive (or zero) when the load is considered on one segment and negative (or zero) when it is considered on the other segment; it is usually apparent that the maximum occurs under one of two loads, and it is often simpler to find bending moments under both loads than to compute the shears. For a uniform load, the maximum bending moment in a simple beam occurs at a section for which the shear is zero; for symmetrical loads, this is at the center. For combined concentrated and uniform loads, the maximum bending moment in a simple beam occurs either at a section for which the shear is zero or else in the line of action of one of the concentrated loads at the section for which the shear is positive when that load is considered on one segment and negative when it is considered on the other segment.

In a *cantilever beam* supported at one end only, both the maximum shear and the maximum bending moment occur for a section at the face of the support. In a cantilever beam which overhangs one or both supports, the maximum negative shear is for a section just outside one of the reactions, and the maximum positive shear is for a section just inside one of the reactions. In an overhanging beam, the maximum negative bending moment occurs at one of the supports, and the maximum positive bending moment occurs as in a simple beam, the loads beyond the support being taken into consideration as well as those between.

Live Loads. Loads which may be placed on a structure temporarily or which may be changed in position are termed *live loads* or *moving loads*, to distinguish them from fixed *dead loads* or *static loads*. The live loads may consist of trains, trucks, people, water, etc.; the dead loads include the weight of the structure or member and any other constant load such as tracks, floors, and tanks. When the live loads are placed in a certain position and stopped, they become static loads, and shears and bending moments may be found in the same manner. Inasmuch as the beams or girders must support the live loads in any position as they move from end to end, they must be designed for the positions which cause the greatest stresses. The placement of different types of load is discussed in the following paragraphs.

Live-load Shear. *Simple Beams.* For any *given section* of a simple beam, the maximum shear occurs when the longer segment is fully loaded and usually when the shorter segment is unloaded; the *maximum* shear in the beam is for a section taken infinitely close to the reaction with the entire beam loaded to make that reaction as large as possible. Thus a single concentrated load causes the greatest shear for a given section when it is placed on the longer segment infinitely close to the section; the maximum shear in the beam equals the reaction when the load is placed infinitely close to that reaction—so close that the reaction virtually equals the load itself. Similarly, two concentrated loads cause a maximum shear when placed on the longer segment as close together as possible, with the larger load infinitely close to the section. Three or more concentrated loads are placed in a similar manner, unless one or more of the end loads is considerably smaller than the following loads, as, for example, the pilot wheel of a locomotive; in this case, the maximum shear may occur when the first heavy wheel is placed next to the section, even though the smaller wheel falls on the shorter segment and has to be subtracted from the reaction to give the shear. A uniform live load is placed throughout the longer segment, the shorter segment being left unloaded. The shear for any segment is thus equal to the reaction, but the reaction differs for each section; the maximum shear is for a section just inside the reaction when the beam is fully loaded.

Cantilever Beams. The maximum shear on a beam supported at one end only is for a section taken at the face of the support with the full load on the beam, regardless of position. Similarly, the maximum negative shear for a cantilever beam that overhangs one or both supports is for a section taken just outside the reaction when the full load is on the overhanging portion, whether or not the portion of the beam between supports is loaded. The maximum positive shear for this overhanging beam is for a section just inside the reaction when the portion extending beyond this reaction is fully loaded, with the loads placed as far from the reaction as possible; when the portion between supports is fully loaded, with the loads placed as near this reaction as possible; and when no load is on the portion of the beam which may extend beyond the other reaction.

Live-load Bending Moment. *Simple Beams. Concentrated Loads.* The maximum bending moment for a simple beam supporting moving concentrated loads is under one of the loads. For a single load, the maximum bending moment for a given section occurs when the load is at that section, and the maximum moment for the beam is at the center when the load is at the center. For two moving concentrated loads, the maximum bending moment for a given section occurs when the larger load is at that section and the smaller one is on the longer segment as close to the first load as possible. *The absolute maximum bending moment for the beam occurs under the larger load when the center of the beam is midway between that load and the center of gravity of both loads, the loads being placed as close together as possible.*[1]

For more than two moving concentrated loads, the maximum bending moment oc-

[1] If the span is less than $(1 + \sqrt{r})$ times the distance between loads, where r is the ratio of the larger load to the sum of the loads, the maximum occurs at the center when the larger load is at the center and the other load is off the beam. When the distance between loads is more than one-half the span, it is usually simpler to find the bending moments for both positions and compare them than to use this ratio.

curs under one of the loads, but which one is not always evident. There are criteria for determining the critical load, but often more than one load satisfies a criterion. Rather than to apply such a criterion, it is better to place each load which is likely to cause the maximum and then calculate the bending moments and compare them. To find the maximum bending moment at a given point the critical load should be placed at that point. The absolute maximum bending moment in the beam occurs under one of the concentrated loads when the center of the beam is midway between that load and the center of gravity of all the loads on the beam. The critical load is always one of the two loads adjacent to the center of gravity, and usually the nearer load. The position of the center of gravity may change as the loads are moved, for additional loads may come on one end and others move off the other end. Enough trials must be made to make sure that the bending moment under one load is greater than under either adjacent load when each is placed for maximum. Railroad bridges are usually designed to support trains headed by two conventional locomotives. The train loads are considered as uniformly distributed, but the locomotives are usually taken as a series of concentrated loads. For special suggestions regarding conventional locomotive loadings, see Section 5.[1]

Uniform Loads. The maximum bending moment at any given point of a simple beam occurs when the uniform load extends the full length of the beam and the absolute maximum occurs at the center under the full load. In case the uniform load can extend over only a portion of the beam, the maximum bending moment occurs at the center when the load is centrally located; the maximum bending moment at any point a distance b from the nearest support occurs when the nearest end of the uniform load is $b - \frac{bc}{l}$ from that support, c being the length of the uniform load and l the length of the beam.

Cantilever Beams. The maximum bending moment for a beam supported at one end only occurs at the edge of the support when all loads are placed as far from the support as possible. For beams which overhang one or both supports, the maximum negative bending moment occurs at one of the supports when the overhanging end is fully loaded and the loads are as far from the supports as possible; the maximum positive bending moment occurs with the full load between the supports and with no loads on the overhanging ends, the bending moment being the same as for a simple beam.

Illustrative Problem. *Two Moving Loads.* To find the maximum bending moment for a 20-ft beam due to two concentrated live loads of 8,000 and 6,000 lb respectively, spaced 7 ft apart.

Distance from larger load to center of gravity (Fig. 2)
$6,000 \times 7 \div (6,000 + 8,000) =$ 3.0
Distance from left end to larger load .... $(20 - 3) \div 2 =$ 8.5
R_L .... $(8,000 \times 11.5 + 6,000 \times 4.5) \div 20 =$ 5,950
M under larger load .... $5,950 \times 8.5 =$ 50,600 ft-lb

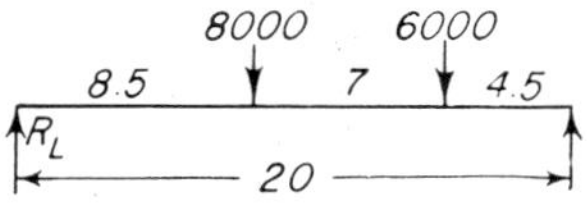

Fig. 2

Combined Bending Moments. *Simple Beams.* Concentrated loads are found only in conjunction with uniform loads, for at least the weight of a beam is uniformly dis-

[1] Also Bishop, "Structural Design," p. 166, 1938, or Tracy, "Stresses Statically Determined," p. 369, 1929.

tributed. The maximum bending moment for a series of concentrated loads occurs at one of the loads. For symmetrical loads, the bending moment at the center is the same as that under the nearest load, and so the maximum combined bending moment is at the center; for unsymmetrical loads, the maximum combined bending moment may occur at one of the concentrated loads or between this point and the center, wherever the combined shear is zero or changes from plus to minus. For moving concentrated loads, the maximum bending moment is so near the center and usually so large compared with the bending moment for the uniform loads that it is customary to combine the maximum bending moment for the concentrated loads and the maximum bending moment for the uniform loads even though they do not occur at the same point and hence must be computed separately; this is on the side of safety and is usually quite accurate, for the curve of bending moments for uniform loads is so nearly horizontal between the center and the critical concentrated load.[1]

Floor-beam Reactions. Beams and girders which support other beams or girders are naturally designed to carry the maximum loads which can come on them. In case the loads are fixed or uniformly distributed, there is no question as to load distribution. When moving concentrated loads are used, they must be placed so that the maximum concentrations reach the supporting members. Thus the transverse floor beams of a railroad bridge support the longitudinal stringers, which in turn support the track and hence the live load. The maximum concentrations that reach a floor beam through the stringers of the adjacent panels occur when the center of gravity of as many heavy wheel loads as can be placed on the two panels is nearly over the floor beam. This position is different from that which causes either maximum shear or maximum bending moment on the stringers.[2]

Impact. The effect of moving loads on a structure is considerably greater than if they were standing still. This is usually taken into account by increasing the live load by a certain percentage called "impact." For crane-runway beams and girders, 25 percent is often specified; for railway bridges and most highway bridges, the percentage is determined from some formula that is a function of the loaded length; for some highway bridges, the conventional truck loads or uniform loads specified include an impact allowance so that no further amount need be added unless trolley tracks are supported.

Statically Indeterminate Beams. For the calculation of shears and bending moments for continuous beams and for fixed-ended beams, see Section 3. For rigid frames, see Section 5.

BEAMS

The size of beam required to satisfy given conditions is usually determined by bending, with due regard to lateral stiffness. Steel grillage beams, or beams which support heavy concentrated loads near the supports should be tested for buckling. In beams that support shafting, deflection is an important consideration, but ordinarily, excessive deflection is provided against by using the minimum depths specified in building codes and specifications. In Section 3, the theory of design of beams is given for bending, for shear, for buckling, for web crippling, and for deflection, together with sufficient illustrative problems. Only a few additional suggestions for the practical designer are given below.

Beam Proportions. Beams of different shapes and sizes may be selected which will meet given conditions, and judgment is needed in making the best selection. The deeper beams resist bending and deflection better; the wider beams are not so likely to deflect laterally and are less likely to buckle or be overturned unless supported laterally. The wider steel beams require more fireproofing, however, and the additional weight may be objectionable. The heavier beams of a certain depth often weigh more

[1] See Sections 3 and 5 for bending moments in restrained and continuous beams.
[2] For further information, see Bishop, *op. cit.*, p. 172, or Tracy, *op. cit.*, p. 401.

than the lighter beams of the next deeper group. The depth of a beam is often determined by its relation to other parts of a structure.

Design Tables. The "Manual" of the American Institute of Steel Construction gives tables which are essential for the design of steel beams. The properties are arranged according to sizes, but in addition, a table of section moduli of the different beams aids in the selection of the most economical beam. The tables of safe loads for different sizes of beams and for different spans have four limitations which should be understood. They apply only to (1) beams which are simply supported, (2) beams which are uniformly loaded for the full length, (3) beams which are laterally supported, and (4) beams designed for a single unit stress. The safe loads given are total loads for the entire lengths of the beams, and are expressed in thousands of pounds (kips).

Lateral Supports. The specified unit stresses for bending in steel beams are allowed only in case the compression flange of the beam is supported laterally to prevent it from buckling. Such support may be furnished by wooden flooring or sheathing or by any form of solid-floor construction, if properly attached, or by tie rods, diaphragms, etc. Unless some such lateral support is provided, the allowed unit stress should be reduced by the use of a reduction formula given in the specifications. Such formulas are expressed in terms of l/b or ld/bt, in which l is the length between lateral supports, d the depth, b and t the breadth and thickness of the compression flange. Beams which are subjected to a lateral thrust due to a floor arch on only one side should receive special consideration.

Flange Holes. Holes or notches should not be cut where they will impair the strength of a beam unless the beam is designed accordingly. A simple beam of uniform cross section can be punched or notched at the ends for ordinary connections without weakening it; but if the holes or notches are near the point where the bending moment is maximum, the net moment of inertia must be used in designing the beam.

Illustrative Problems. To design a 20-ft laterally supported beam to resist a bending moment of 60,000 ft-lb in addition to that due to its own weight, using a unit stress of 18,000 psi.

M due to 40-lb weight assumed .... $40/2 \times (20/2)^2 = 2{,}000$ ft-lb

$$12 \times 62{,}000 = 18{,}000S$$

$$S = 41.3$$

The following beams have satisfactory section moduli as indicated:

15 ⊏ 33.9 .... 41.7
12 I 40.8 .... 44.8
15 I 42.9 .... 58.9
12 WF 36 .... 45.9

The channel is the lightest section, but it may not be appropriate; a 12 WF 36 would probably be selected.

PLATE GIRDERS

Types. Plate girders are used extensively in every form of steel construction because of their adaptability. They resist transverse bending like beams, but they are used for heavier loads, for longer spans, or for conditions for which single rolled beams or open-webbed joists are not well adapted. Each is made with a web plate to which flanges are riveted or welded at the top and bottom edges; the most common forms of cross section are shown in Fig. 3, type *a* being welded, the others being either riveted or welded.

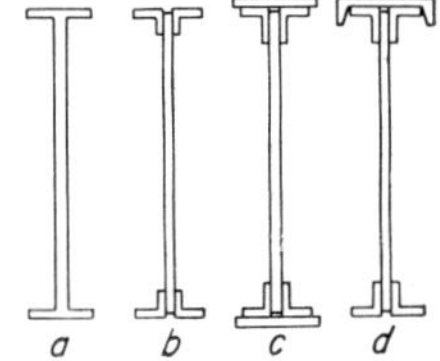

Fig. 3

Depth. The depth of a plate girder is often predetermined

by specific requirements. Within practical limits, flanges may be designed for any depth of girder; the deeper the web, the lighter the flanges. The most economical depth is from one-seventh of the length for short spans to one-twelfth of the length for long spans; in the absence of other data an average of one-tenth of the length may be chosen. The maximum depth is limited to about 10 ft 6 in. or 11 ft by shipping clearances unless shipped in sections with horizontal splices. The depth of the web plate is usually a multiple of 2 in. and preferably a multiple of 6 in. Flange angles are usually placed so that they project ¼ in. beyond the edge of the web to save chipping any projections in the plate resulting from rapid shearing at the mill. In case this projection might leave a rain pocket exposed to the elements, as in outdoor girders without cover plates, the angles are placed flush with the edge of the plate on the top side. Thus, in a riveted girder the depth from back to back of angles is either ½ or ¼ in. more than the nominal depth of the web plate. In a welded girder, however, better results are obtained in welding if the flange plates are placed in contact with the web plate; to ensure this, the web plate is ordered somewhat wider and planed to the required width.

Web Thickness. The usual thickness of a web plate is from 5⁄16 to ⅝ in., with a minimum thickness of 1/170 of the clear distance between flange angles. The web plate must have sufficient area to resist the maximum shear. The maximum shear intensity at the neutral axis is substantially equivalent to the value found by dividing the maximum shear by the gross area of cross section of the web plate, the flange angles and cover plates being neglected; this shear intensity should not exceed 13,000 psi or other value specified. In riveted girders the web plate should be thick enough to furnish sufficient bearing for the flange rivets and sufficient strength between rivets, as explained later. In welded girders the web plate is usually made thick enough to develop the fillet welds on both sides.

Flange Angles. The angles most used for plate girders are 6 × 4, 6 × 6, 8 × 6, and 8 × 8; smaller angles are not used so much as they were before the larger wide-flanged beams were available. The unequal-legged angles have the shorter legs against the web and the longer legs outstanding, because the girder is stronger that way, both vertically and horizontally. When angles are used in the compression flange without cover plates, the thickness of each should not be less than one-twelfth the length of the outstanding leg, or one-tenth if ties are to rest directly upon the angles.

Cover Plates. Cover plates are used in riveted girders to give additional strength in case 6 × 4 × ¾ or 6 × 6 × ¾ angles are insufficient; 8 × 8 angles are seldom used without cover plates. They are usually 14 in. wide on 6-in. angles and 18 in. wide on 8-in. angles. Cover plates should not have an area larger than the flange angles unless 6 × 6 × ¾ or larger angles are used. For convenience, the plates of each flange are treated as a single plate in design, but the thicker plates in riveted work must be subdivided so that no plate exceeds the maximum punching thickness or is less than ⅜ in.; usually the smallest number of plates conforming to these limits is used, in order to save the extra punching and handling. The plates should be of the same thickness unless it is desired to save metal by making part of the plates 1⁄16 in. less than the others, in which case the thicker plates should be placed next to the angles. In welded girders, a single flange plate without angles is usually placed in each flange, regardless of thickness, but if more than one plate is used, the plates should differ in width by 1 in. in order to facilitate welding. In long girders thinner flange plates may be used near the ends to save metal. They are butt-welded to the thicker central plates.

Methods of Design. Plate girders may be designed by the moment-of-inertia method and some specifications permit the use of the flange-area method. Unless exhaustive tables of moments of inertia of plate girders with angles are available, this method of design is likely to become tedious; it is often quicker to use the flange-area method as a preliminary design and then verify or modify the results by the moment-of-inertia

method. Since a girder must support itself in addition to the superimposed loads it is well to assume a weight in the preliminary design and then correct it in the final design. The weight of a girder is the weight of the main component parts plus about 10 percent in riveted work to allow for stiffening angles, fillers, rivetheads, and splices, or about 7 percent in welded work to allow for stiffening plates and splice plates. The web plate is usually designed first so its weight is determined, but the angles and cover plates may be estimated. In the final design the weight of the girder should be found from the sizes resulting from the preliminary design.

Moment-of-inertia Method. *Riveted Girder.* This is an application of the general formula for flexure, $M = fI/c$. According to the older specifications, the tension flange of a riveted girder was considered to be weakened by the holes for rivets, but now some specifications allow the use of the full gross section unless the holes exceed a certain percentage. In any case, holes to be left open for bolts or for other purposes should be deducted. When rivet holes are to be considered, the tension flange is designed first and then the compression flange is made the same unless a larger section should be needed because of a reduced unit stress in the compression flange which is based upon the ratio between the width of the flange to the length between lateral supports. In a riveted girder the rivet holes weaken the tension half only, but for convenience a symmetrical girder is considered to have holes deducted from both halves so as to avoid the useless refinement of using the precise center of gravity and the corresponding c distance for the resulting unsymmetrical section. Similarly, the web plate is taken as symmetrical even if the top edge is ¼ in. nearer the backs of the angles than the bottom. For a truly unsymmetrical girder such as the type d in Fig. 3, it is necessary to find the center of gravity of the gross section in order to locate the neutral axis and find the corresponding moment of inertia.

The moment of inertia of the entire girder about the neutral axis is found by combining the moments of inertia of the different component parts about the same axis. In general, the moment of inertia of the net section is found by deducting the moments of inertia of the holes from the moment of inertia of the gross section. The moment of inertia of the web plate may be found directly from the tables (or $bd^3/12$). The moments of inertia of the other component parts may be found by transfer from a centroidal axis to a parallel axis, although many tables are available which will simplify this work. For a given area, the moment of inertia about an axis through its center of gravity is less than about any parallel axis; the moment of inertia I_a about any axis is found from the moment of inertia I_c about a parallel axis through the center of gravity by adding the product of this area and the square of the distance between the axes, thus: $I_a = I_c + Ax^2$. The moment of inertia of one angle about its own center of gravity may be found from the usual tables of properties. The moment of inertia of the cover plates may be taken as if all the plates of one flange were combined; the moment of inertia of these plates about their own center of gravity may be neglected for it is relatively small.

The holes for the rivets which connect the angles to the web of a riveted girder are placed on the standard gage lines, two lines being used for 6-in. legs or over. It is assumed that these rivets will be staggered, and near the point of maximum bending moment the rivets will usually be far enough apart so that only the one nearer the back of the angle need be considered. The area of this hole is the combined thickness of the web and the two angles multiplied by the diameter of the hole, which is taken as ⅛ in. larger than the nominal size of the rivet. The moment of inertia is the product of this area and the square of the distance from its center to the neutral axis, the moment of inertia about its own center being negligible. Similarly, when cover plates are riveted to the flange angles, two additional holes are deducted. The combined thickness of the angle and the cover plate is used in determining the area, and the distance from the

neutral axis to the center of the hole may be found by subtracting one-half of this thickness from the c distance. If stiffening angles or splice plates are to be riveted to the web near the section of maximum bending moment, holes should be taken out of the web plate at an average distance of 4 in.; this is not very important and is considered too great a refinement by many designers. Stiffening angles are required if the clear distance between flange angles is more than 60 times the web thickness, and at points of concentrated loads.

Illustrative Problem. *Moment-of-inertia Method. Riveted Girder.* To design a 50-ft girder composed of a 72-in. web plate, 6 × 6 angles, 14-in. cover plates, and ⅞-in. rivets to support three concentrated loads at the center and the quarter points of 145,000 lb each, as shown in Fig. 4. The specified unit stresses are as used below. In order to reduce the number of trial designs, let us assume that the girder was designed by the flange-area method illustrated on p. 6–11 and that the resulting preliminary section was a 72 × ⅜ web, 4 *Ls* 6 × 6 × ¾, and 4 cover plates 14 × ⅝, two on each flange. This section will now be investigated by the moment-of-inertia method.

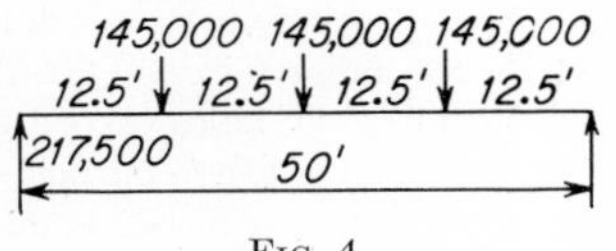

Fig. 4

R_L for concentrated loads (Fig. 4)........$3 \times 145{,}000 \div 2$	=	217,500 lb
M for concentrated loads...$217{,}500 \times 25 - 145{,}000 \times 12.5$	=	3,625,000 ft-lb
Weight of assumed section plus 10 percent for details $(92 + 4 \times 28.7 + 2 \times 59.5)1.1$	=	360 lb per ft
Maximum shear......................$217{,}500 + 360 \times 25$	=	226,500 lb
Web thickness required for shear...$217{,}500 \div 13{,}000 \times 72$	=	¼ in.
Minimum web thickness............$(72.5 - 2 \times 6) \div 170$	=	⅜ in.
M due to weight.........................$360/2 \times 25^2$	=	112,500 ft-lb
c................................$72.5/2 + 2 \times 5/8$	=	37.5 in.
fI/c................................$20{,}000\ I \div 37.5$	=	$12 \times 3{,}737{,}500$ in.-lb
Net I required, considering tension........................	=	84,090 in.^{4}
I of web..	=	11,660
I of 4 *Ls* (Fig. 5)..................$4(28 + 8.44 \times 34.47^2)$	=	40,220
I of cover plates, both flanges $2 \times 14 \times 1.25(36.25 + 0.62)^2$	=	47,580
Total I of gross section................................	=	99,460
I of holes in angles and cover plates $4 \times 1 \times 2(37.5 - 1)^2$	=	10,660
I of holes in angles and web............$2 \times 1 \times 1.88 \times 34^2$	=	4,350
I of other holes in web $2 \times 1 \times 0.38(28^2 + 24^2 + 20^2 + 16^2 + 12^2 + 8^2 + 4^2)$	=	1,700
Total I of holes......................................	=	16,710
Total I of net section.................................	=	82,750

This is somewhat less than the 84,090 required. By increasing each pair of cover plates from 1¼ to 1$\frac{5}{16}$, the moment of inertia of the net section would be increased by $2(14 - 2 \times 1)0.06 \times 37.53^2 = 2{,}030$ making 84,780. The c distance would increase to 37.56 and the weight would also increase a negligible amount, but the resulting net I required would increase only to 84,230 which is under the 84,780 furnished.

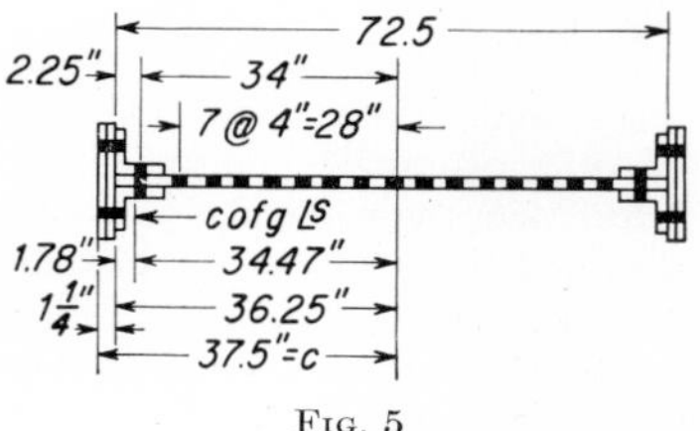

Fig. 5

The compression flange should be like the tension flange unless the gross I found from a new unit stress is larger than that of the section used above, $99{,}460 + 2 \times 14 \times 0.06 \times 37.53^2$. The new unit stress must be found from the

formula specified for the compression flange in terms of the width of the flange and the distance between lateral supports. In this case, assuming that the girder is supported laterally at each concentrated load, the $ld/bt = \dfrac{12.5 \times 12 \times 75.12}{14 \times 2.06}$ does not exceed the 600 allowed (or the distance between lateral supports 12.5×12 does not exceed 40 times the flange width 14) so no increase is needed.

Moment-of-inertia Method. *Welded Girder.* The design of a welded plate girder is relatively simple when composed of a web plate and two flange plates. After the web plate is selected thick enough to develop the flange welding on both sides, the thickness of the flange plates may be assumed for weight and for the c distance. The required moment of inertia may be found from the flexure formula and the actual moment of inertia from the assumed size; then both may be revised until in adjustment. Only permanent holes need be considered.

The strength of a fillet weld per linear inch for the usual unit stress is $13{,}600 \times 0.707$ times the size of the weld, as explained on p. 6–25. To develop welds on both sides the web must be twice as strong. At the corresponding unit stress in shear on the web, this means that the thickness of the web must be $\dfrac{2 \times 13{,}600 \times 0.707}{13{,}000}$ or approximately $1\frac{1}{2}$ times the size of the weld to be used. Usually a $\frac{5}{16}$-in. weld is considered the largest that can be made with a single pass of the electrode.

Illustrative Problem. *Moment-of-inertia Method. Welded Girder.* To design a 50-ft welded girder composed of a 72-in. web plate and two 14-in. flange plates to satisfy the same conditions as the riveted girder on p. 6–10. If we select $\frac{5}{16}$-in. continuous welds on both sides, the web plate should be $1\frac{1}{2} \times \frac{5}{16} = \frac{1}{2}$ in. thick; this provides amply for the shear requirements. Let us assume $1\frac{5}{8}$-in. flange plates. A preliminary design by the flange-area method seems hardly worth while because a few extra trials can be made so easily.

M for concentrated loads (from preceding problem)		=	3,625,000 ft-lb
Weight of assumed section plus 7 percent for details	$(122 + 2 \times 77.4)1.07$	=	300 lb per ft
M due to weight	$300/2 \times 25^2$	=	93,800 ft-lb
c	$72/2 + 1.62$	=	37.62 in.
fI/c	$20{,}000\ I \div 37.62$	=	$12 \times 3{,}718{,}800$
I required		=	83,940 in.^{4}
I of web		=	15,550
I of both flange plates	$2 \times 14 \times 1.62(36 + 0.81)^2$	=	61,460
Total I		=	77,010
This is less than the 83,940 required; so we add $\frac{1}{8}$ to each plate			
I of flange plates revised	$2 \times 14 \times 1.75(36 + 0.88)^2$	=	66,650

The revised total 82,200 is still under the 83,940, which will be increased as the c distance is increased, but it seems obvious that another $\frac{1}{16}$ will suffice. The final section is thus a $72 \times \frac{1}{2}$ web and two $14 \times 1\frac{13}{16}$ flange plates which weighs 315 or about 50 lb less than the riveted girder. In addition there are fewer parts to handle, no fillers, and no holes to punch; so there would be a substantial saving. A further saving in weight may be made by using a thinner flange plate at the ends where the bending moment is less.

Flange-area Method. *Riveted Girder.* When the use of this method of design is permitted by the specifications it is somewhat simpler than the moment-of-inertia method; it may be used also in conjunction with the latter to approximate the section to be checked by the moment-of-inertia method. As in the latter, some specifications require the deduction of rivet holes in the tension flange, while some do so only when

they exceed 15 percent of the area. For completeness, the full rivet deduction is presented here but the method can be adapted readily to the requirements of the different specifications.

The stress in each flange is assumed to act at the center of gravity of the flange, and a web equivalent is counted as flange area; these two stresses must be equal and opposite to satisfy the H equation of equilibrium (unless there are external horizontal forces), and they form a couple the moment of which is the resisting moment of the girder. The lever arm of this couple is the distance from center to center of gravity of the gross section of the flanges and is called the effective depth d_g. The gross area is used here for convenience, for it is simpler and gives substantially the same results. In a girder without cover plates, the effective depth can be determined with sufficient accuracy from an assumed angle of intermediate thickness, say $\frac{1}{2}$ of $\frac{9}{16}$. The effective depth is the distance from back to back of angles minus twice the distance from the center of gravity of the angles to the backs of the outstanding legs. The distance from back to back of angles is $\frac{1}{2}$ in. more than the depth of the web plate (or $\frac{1}{4}$ in., see p. 6–8). When cover plates are used, it is necessary to find the center of gravity of the angles and cover plates of each flange in order to find the effective depth. For a preliminary design it is close enough to use the depth of the web as d_g but if equal-legged angles are used it would probably be closer to use it as 1 in. less. A depth greater than the distance back to back of angles is not allowed.

The total force acting at the center of gravity of *one* flange is found by dividing the maximum bending moment by the effective depth, and this divided by the allowed unit stress will give the flange area required. This is made up of the area of the flange angles and cover plates of *one* flange plus the equivalent portion of the web counted as flange area. For simplicity this portion is taken as one-sixth the gross area of the whole web plate if no holes are to be deducted, or one-eighth the gross area of the web if the net section is used in the angles and cover plates.[1] Unless $6 \times 6 \times \frac{3}{4}$ angles or larger are used, the area of the cover plates should not exceed that of the angles. Cover plates which exceed the maximum punching thickness should be subdivided as in the other method.

Illustrative Problem. *Flange-area Method. Riveted Girder.* To design a 50-ft girder composed of a 72-in. web plate, 6×6 angles, 14-in. cover plates, and $\frac{7}{8}$-in. rivets to satisfy the same conditions as in the problem on p. 6–11 illustrating the moment-of-inertia method.

R_L for concentrated loads (p. 6–11) . . . . = 217,500 lb

M for concentrated loads (p. 6–11) . . . . = 3,625,000 ft-lb

Weight of assumed $\frac{3}{8}$ web, $\frac{3}{4}$ angles, cover plates of equal weight, plus 10 percent for details . . . . $(92 + 4 \times 28.7 \times 2)1.1$ = 350 lb per ft

Maximum shear . . . . $217{,}500 + 350 \times 25$ = 226,300 lb

Web thickness required for shear $226{,}300 \div 13{,}000 \times 72$ . . . . = $\frac{1}{4}$ in.

Minimum web thickness . . . . $(72.5 - 2 \times 6) \div 170$ = $\frac{3}{8}$ in.

M due to weight . . . . $350/2 \times 25^2$ = 109,400 ft-lb

Total net flange area required in tension flange (d_g being assumed 1 in. less than the 72-in. web) $12 \times 3{,}734{,}400 \div 20{,}000 \times 71$ = 31.6 sq in.

$\frac{1}{8}$ gross area of the web . . . . $\frac{1}{8} \times 72 \times \frac{3}{8}$ = 3.4 sq in.

Net area 2 *Ls* $6 \times 6 \times \frac{3}{4}$ . . . . $2(8.44 - 2 \times \frac{3}{4} \times 1)$ = 13.9 sq in.

Net area required in cover plates in tension flange $31.6 - 3.4 - 13.9$ = 14.3 sq in.

Total thickness of cover plates in tension flange $14.3 \div (14 - 2 \times 1)$ = $1\frac{1}{4}$ in.

[1] This is based upon the resisting moment of the rectangular web, $\frac{1}{6} ftd^2$, divided by the unit stress and the depth. Since it is a relatively small amount the difference between the effective depth and the depth of the web is unimportant; the ratio of the net area to gross area through a line of rivet holes for stiffeners is roughly $\frac{3}{4}$; so holes may be taken into consideration by multiplying the $\frac{1}{6}$ by $\frac{3}{4}$ to make $\frac{1}{8}$.

Revised weight. . . . .$(92 + 4 \times 28.7 + 2 \times 59.2)1.1$ = 360 lb per ft
Revised bending moment due to weight. . . . .$360/2 \times 25^2$ = 112,500 lb-ft
Distance from back of angle to center of gravity (Fig. 6)

$$\frac{2 \times 8.44 \times 1.78 - 17.5 \times 0.62}{2 \times 8.44 + 17.5} = 0.56 \text{ in.}$$

Revised d_g. . . . .$72.5 - 2 \times 0.56$ = 71.4 in.
Revised net area required. . . . .$12 \times 3{,}737{,}500 \div 20{,}000 \times 71.4$ = 31.4 sq in.
This is $31.6 - 31.4 = 0.2$ less than for the first trial and may be taken from the area of the cover plate, leaving the web and angles unchanged
Revised thickness of cover plates. . . . .$14.1 \div (14 - 2 \times 1)$ = 1 3/16 in.

The compression flange should be like the tension flange unless the gross area found by dividing the bending moment by the effective depth and by the unit stress required in the compression flange exceeds the gross area of the two angles and the cover plates of one flange plus one-sixth the gross area of the web. In this case, assuming that the girder is supported laterally at each concentrated load, the $ld/bt = \frac{12.5 \times 12 \times 74.88}{14 \times 1.94}$ does not exceed the 600 allowed (or the distance between lateral supports 12.5×12 does not exceed 40 times the flange width 14); so no increase is needed.

On account of punching the holes for the rivets, the plate in each flange should be subdivided into one plate 14 × 5/8 next to the angles and one plate 14 × 9/16.

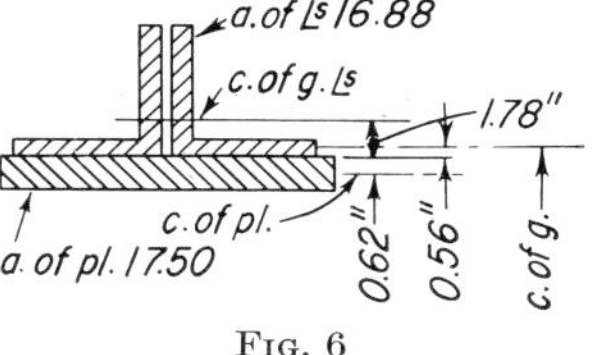

Fig. 6

Flange-area Method. *Welded Girder.* The design of a welded girder is relatively simple when composed of a web plate and two flange plates. After the web plate is selected thick enough to develop the flange welding on both sides, the thickness of each flange plate may be assumed for weight and effective depth. The gross area required in each flange may be found by dividing the bending moment by the effective depth and by the allowed unit stress, one-sixth the gross area of the web being counted as flange area. The resulting size may be revised until in adjustment.

Illustrative Problem. *Flange-area Method. Welded Girder.* To design a 50-ft welded girder composed of a 72-in. web plate and two 14-in. flange plates to satisfy the same conditions as the riveted girder on p. 6–12. If we select 5/16-in. continuous fillet welds on both sides, the web plate should be 1/2 in. thick as on p. 6–25. Let us assume 1 5/8-in. flange plates.

M for concentrated loads (from p. 6–11). . . . . = 3,625,000 ft-lb
Weight of assumed section plus 7 percent for details
$(122 + 2 \times 77.4)1.07$ = 300 lb per ft
M due to weight. . . . .$300/2 \times 25^2$ = 93,800 ft-lb
Total flange area required. . . $12 \times 3{,}718{,}800 \div 20{,}000(72 + 1.62)$ = 30.3 sq in.
1/6 gross area of web. . . . .$1/6 \times 72 \times 0.5$ = 6.0
Thickness of plate in each flange. . . . .$(30.3 - 6.0) \div 14$ = 1 3/4 in.

This is close enough to the assumed size so that no revision is necessary.

TENSION AND COMPRESSION MEMBERS

Tension Members. A tension member is designed to transmit tensile stresses in a direction parallel to its principal axis; these stresses tend to elongate the member. The member must be so designed that it is strong enough at its weakest point to carry

the total stress. The effective net area required is found by dividing the total stress by the allowed unit stress.

Rods. Rods are used as light tension members in bracing systems or to tie different parts of a structure together. They are usually threaded for nuts, clevises, or turnbuckles. The threads are cut at the ends of a rod, so that the effective area is reduced; the effective area at the root of the thread is given in tables of threads, and the diameter of the rod is so selected that the corresponding "root area" equals or exceeds the required area. Sometimes the threaded end of the rod is upset to a larger diameter so that the root area will exceed the gross area of the main rod, and only the latter need be considered in the design. Similarly, a loop rod is made by bending the end of a rod back upon the rod and welding it to form a loop which fits around a cylindrical pin. The loops are stronger than the main part of the rod, so that the designer simply has to determine the size of the square or round rod, the gross area of which equals or exceeds the required area. Square rods furnish better bearing on a pin, but round loop rods are also used. Rods are available in multiples of ⅛ in.

Eyebars. Flat-rolled steel bars are sometimes heated and upset at the ends in special presses to form eyebars. The enlarged heads are kept of the same thickness as the original bars and large holes are punched in the ends, while hot, to form eyes; later these holes are bored accurately at the proper distances apart to fit cylindrical pins used for connections. Large eyebars were formerly used for the tension members of pin-connected bridge trusses, but they have been largely superseded by riveted or welded members. Only bars up to 6 in. in width and 2 in. thick are now available in this form, but larger bars may be built by welding reinforcing plates at the ends of flat bars and cutting and punching them to simulate forged eyebars. Eyebars are used in the anchorages of suspension bridges and for certain types of hangers. The heads of either the upset or welded eyebars are made large enough to more than develop the effective area of the main parts of the bars.

A tension member may consist of one or more eyebars, depending upon conditions. A member of a truss should be composed of an even number of bars, in order to keep the truss symmetrical about its central plane. The number and the size of a bar must be such that the total area of cross section of the main portion of the bars equals or exceeds the required area found by dividing the total stress by the unit stress allowed in tension. The available widths and thicknesses are given in the handbooks, the thicknesses varying by sixteenths of an inch between the minimum and maximum values given in the table. The design of the main part of a welded bar is similar, except that much wider plates are used.

Riveted Tension Members. One or two angles are used for the lighter tension members of roof trusses, latticed girders, or bracing systems; heavier members are made of beam sections, channels, four angles, plates and angles, or plates and channels. The component parts of a single member are fastened together at intervals; stitch rivets are used to hold two angles together, washers being used if necessary to maintain a uniform distance apart; tie plates, lattice bars, or continuous plates are used for larger members. The effective cross section of a tension member is reduced by holes, even though these holes are later filled with rivets; a hole reduces the cross-sectional area by the area of a rectangle, one dimension being the thickness of metal and the other the diameter of the hole. The diameter of a hole for a rivet or bolt is taken ⅛ in. larger than the nominal size of the rivet or bolt; this allows for punching the holes ¹⁄₁₆ in. larger and also for reaming holes if required because of inaccurate punching. Unless rivet holes in adjacent lines are less than a certain distance apart, as explained in the next paragraph, the weakest section of an ordinary tension member is through the largest number of holes, usually near the ends where the member is connected to other members of the structure. The details must be anticipated so that the correct number

of holes in the critical section may be deducted; usually, holes for rivets in tie plates or lattice bars may be so placed that they will not come at the critical section. The net area may be found by combining the net areas of the component parts or by subtracting the areas of the holes from the gross area of the whole member. The gross area of the different component parts may be found from tables of properties, and the areas of holes may be found from tables or by computation. The net area of a plate is equivalent to the product of the thickness by the *net width,* the latter being the gross width minus the diameters of all the holes in the cross section. Usually the design is an indirect process, the section being approximated, its net area being found and compared with the required area, and the section being revised if necessary until the minimum size is found which meets the requirements. According to some specifications, only part of the unconnected leg of a single angle is considered effective. A welded member may or may not have holes deducted.

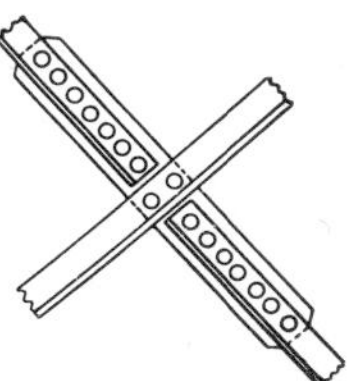

FIG. 7

Illustrative Problem. One angle 5 × 3½ × 5⁄16 of diagonal cross bracing is cut and spliced as shown in Fig. 7 so as not to interfere with the other angle. How thick must a 6-in. splice plate be to develop the full tensile strength of the angle, the rivets being ¾ in.?

Net area of angle through one rivet hole . . . . 2.56 − (⅞ × 5⁄16) = 2.29
Net width of plate through two rivet holes . . . . 6 − (2 × ⅞) = 4.25
Thickness of plate required . . . . 2.29 ÷ 4.25 = 9⁄16

Effect of Staggered Rivets. The least net section is not necessarily a right section, for rivets in adjacent lines may be close enough to make a zigzag section weakest. Even though the zigzag section has the same net area as a right section, it would fail first because the diagonal portions are not so effective in resisting axial stress, although this fact is sometimes ignored. In all the different methods of determining the effective section when rivets are staggered, the minimum value is found when the staggered rivets are considered directly opposite with the full holes deducted, and the maximum value is found when the staggered rivets are considered far enough away from the critical section to have no effect. Between these two limits, a fraction x of the area of the staggered holes is deducted as $x = 1 - \frac{s^2}{4gh}$ or $x = 1.3 - \frac{s}{g}$, in which s is the stagger or longitudinal distance from center to center of rivets, g is the gage or transverse distance between rivet lines, and h is the diameter of the hole. It is important that the proper effective section be used where the stress is maximum. At the ends, after a portion of the stress has been transmitted by rivets to the connection plates, the section may be reduced and still carry the remaining stress; recent tests indicate, however, that the strength depends on the weakest section, and little benefit is derived by the omission of rivets near the center.

Combined Tension and Bending. When tension members are subjected to transverse bending, the combined unit stresses must not exceed the allowed value, for tension.

Compression Members. A compression member is designed to transmit compressive stresses in a direction parallel to its principal axis; these stresses tend to shorten the member and also to buckle it transversely, since the resultant of all forces cannot be assumed to act absolutely along the gravity axis. The strength of a compression member depends, therefore, on its form of cross section and its length, as well as on its area of cross section; for a given amount of material, a pipe is stronger than a solid bar, and a short member is stronger than a long one. The allowed unit stress is not a fixed amount

but depends upon the slenderness ratio l/r, in which l is the length of the member between lateral supports and r the least radius of gyration. The specifications to be used for a given structure show the maximum value of l/r to be used for main members that are stressed by principal loads, as, for example, 100 for railway bridges and 120 for other structures; larger values are given for bracing and other secondary members. For a discussion of different compression formulas, see Section 3.

Forms of Members. Single and double angles are used for light compression members or "struts." The least radius of gyration of a single angle is about a diagonal axis, but the least radius of any other ordinary structural member is about one of two perpendicular axes which are parallel to its principal faces and which pass through the center of gravity. Usually the longer legs of two angles are placed together, either in contact or separated by the thickness of a connection plate because the least radius of gyration is more favorable than when the shorter legs are connected. When members are supported laterally in one direction at shorter distances than in the other direction, two different ratios of slenderness should be considered in determining the allowed unit stress; usually the member should be turned so that the larger r corresponds to the larger l, and often the component parts may be rearranged to advantage, as, for example, by connecting the shorter instead of the longer legs of two angles. Properties of one and two angles, of wide-flanged beams, and of other rolled shapes are given in the tables.

The radius of gyration $r = \sqrt{I/A}$ of a built-up member may be found from the combined moment of inertia of the component parts, remembering that for any area the moment of inertia about any axis is the moment of inertia of that area about a parallel axis through its own center of gravity plus the area times the square of the distance between the two axes; thus $I_a = I_c + Ax^2$. The area and moment of inertia of one angle, channel, or other rolled shape about its own center of gravity may be found from the tables. The moment of inertia of a plate about its own center of gravity is $bd^3/12$, in which d is the dimension perpendicular to the axis; this term is negligible when d is a fraction of an inch. If a member is unsymmetrical, the position of its center of gravity may be found by moments, areas being used as forces. It is usually most convenient to find the eccentricity from the center of gravity to the center of the web, thus: multiply the area of each component part by the distance from its center of gravity to the center of the web, add all these area moments on one side of the point of moments, subtract the area moments on the other side, and divide by the total area of the member. All parts which are symmetrical about the center of the web will cancel out, but their areas must be included in the total area. The moment of inertia of the entire member may be found first about an axis through the center of the web and then this may be *reduced* by subtracting the product of the total area and the square of the eccentricity to give the least moment of inertia about the center of gravity. Often it will be apparent which axis will have the smaller moment of inertia; otherwise the calculation of both must be carried far enough to determine the least.

Design. Unless appropriate tables of safe loads are available, the design of a compression member is an indirect process, necessitating the assumption of an approximate section and then the verification or modification until the total safe load meets the requirements. The total safe load is the product of the allowed unit stress and the gross area of the member, but the area of any holes which are not to be filled by rivets should be deducted. The preliminary assumption is largely a matter of experience and judgment, although the form of member is often determined by its relation to other members in the structures. Some of the practical points are considered later in this section, but the theory of design is illustrated by the following problems. The tables in the manual give the safe loads for certain specifications for many of the more common types of compression members, and they give also the areas and radii of gyration which facilitate the design for other specifications.

Illustrative Problems. 1. *Two Angles.* Design an 8-ft strut of two angles to support a load of 100,000 lb at $15{,}000 - \frac{1}{4}l^2/r^2$. With no intermediate support, the most favorable arrangement is with the longer legs of unequal-legged angles connected. The angles are ⅜ in. apart to provide for the insertion of connection plates at the ends. Assuming the unit stress for a small member as 13,000, the area required would be 7.7. The area of 2*Ls* 6 × 4 × 7/16 is 8.36, and the smaller radius of gyration is 1.63. The allowed unit stress would be $15{,}000 - \frac{1}{4}(8 \times 12)^2/1.63^2 = 14{,}130$, and the safe load 118,000. 2*Ls* 6 × 4 × ⅜ would have the same unit stress and an area of 7.22, which would give a safe load of 102,000 and would be satisfactory.

2. *Rolled Section.* Design a 20-ft column to carry a load of 300,000 lb at $\dfrac{18{,}000}{1 + (l^2/18{,}000r^2)}$. The tables give the safe load for a different formula as 319,000 for a 14-78 wide-flanged beam section. This has an area of 22.94 and a least radius of 3.00. Substituting in our formula, we find a unit stress of 13,270 and a safe load of 304,000.

3. *Unsymmetrical Section.* Find the safe load at $17{,}000 - 0.485\dfrac{l^2}{r^2}$ of the 20-ft top-chord section composed of 2 webs 18 × ½, 1 cover plate 21 × ½, 2∠ 3 × 3 × ⅜, and 2∠ 4 × 3 × ½ arranged as shown in Fig. 8. From the table of properties, the distances from the backs of the angles to their centers of gravity are found and recorded directly upon the sketch, also the corresponding distances from the center of the plate to these centers of gravity. With these distances, the algebraic sum of the moments of the component areas is found and divided by the total area in order to find the eccentricity e; by taking the point of moments at the center line of the webs, the moments of the upper and lower halves of the web balance, so that they need not be found, but the web area is included in the total area.

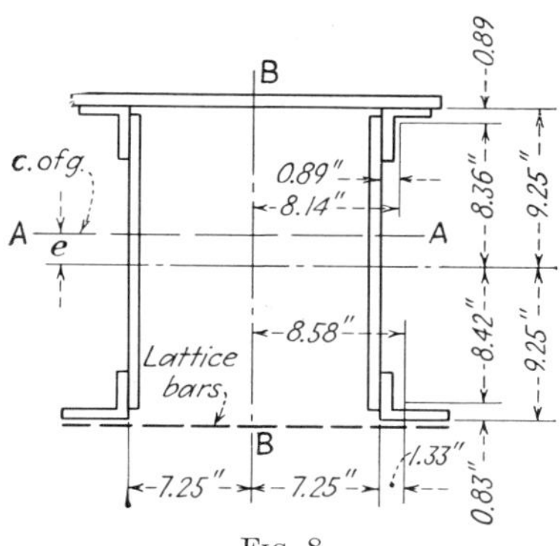

FIG. 8

$$e = \frac{(10.5 \times 9.5) + (2 \times 2.11 \times 8.36) - (2 \times 3.25 \times 8.42)}{10.5 + 2(2.11 + 3.25 + 9.0)} = \frac{80.4}{39.2} = 2.05$$

I of webs about horizontal center line of webs	2×243.0	=	486
I of cover plate	10.50×9.50^2	=	948
I of top angles	$2[1.8 + (2.11 \times 8.36^2)]$	=	299
I of bottom angles	$2[2.4 + (3.25 \times 8.42^2)]$	=	466
Total I about axis through centers of webs		=	2,199
Transfer Ae^2	39.2×2.05^2	=	165
Total I_A about horizontal axis AA through center of gravity		=	2,034
I of cover plate about vertical axis BB		=	386
I of webs	$2 \times 9.0 \times 7.0^2$	=	882
I of top angles	$2[1.8 + (2.11 \times 8.14^2)]$	=	283
I of bottom angles	$2[5.1 + (3.25 \times 8.58^2)]$	=	488
Total I_B about vertical axis BB through center of gravity		=	2,039
Least radius of gyration, squared	$2{,}034 \div 39.2$	=	51.9
Allowable unit stress	$17{,}000 - \dfrac{0.485(20 \times 12)^2}{51.9}$	=	16,460
Total safe load	$16{,}460 \times 39.2$	=	645,200

Combined Compression and Bending. When compression members are subjected to transverse bending, the combined unit stresses must not exceed the allowed value. The unit stress due to compression is found by dividing the maximum compressive stress by the gross area of cross section; the unit stress due to bending is found by equating the resisting moment fI/c to the bending moment and solving for f, c being the distance from the neutral axis to the extreme *compressive* fiber and I being the moment of inertia of the gross cross section. The allowed unit stress is determined by the least radius of gyration; however, if the bending stress is the principal component and the direct compressive stress is relatively small, it is permissible to use the radius of gyration about an axis perpendicular to the plane of bending even though it is somewhat larger; under no circumstances should the direct stress alone exceed the allowed value as determined by the least radius. A better method is to divide the unit stress developed in compression by the unit stress allowed in compression, and to divide the unit stress developed in bending by the unit stress allowed in bending; the sum of these two quotients must not exceed unity.

When the bending stresses are caused by wind, the allowed unit stress is often increased by a specified percentage in view of the fact that the maximum gravity loads and wind loads are not likely to occur simultaneously. Any holes that are not to be filled with rivets must be taken into consideration in both area and moment of inertia.

Lattice Bars and Tie Plates. The component parts of a compression member must be fastened together to prevent the parts from buckling separately. Stitch rivets with or without washers, or short welds, are used for two-angle members, and lattice bars or continuous plates for larger members; stitch rivets are spaced from 1 ft 6 in. to 2 ft apart. Lattice-bar systems should have tie plates at each end and at points where the latticing is interrupted; the plates at the ends should extend longitudinally at least as far as the transverse distance between the lines of rivets which attach them to the member, but intermediate plates need be only half as large; the thickness of tie plates should not be less than 1/50 the distance between the rivet lines. If the distance between rivet lines is not more than 1 ft 3 in., single lattice bars are used with an inclination of at least 60° with the axis for primary members or 45° for secondary bracing members; for wider members double lattice bars are required, riveted at their intersections and inclined at least 45° with the axis. The l/r of no component part between lateral supports should exceed ¾ (or ⅔) of the l/r of the whole member. Continuous plates ¼ in. thick are often preferred to lattice bars because at least a portion of the plate can be counted in the effective area; if the distance between rivet lines exceeds 30 times the thickness of the plate, the effective area of the plate is taken as if this limiting distance were used. When component parts of a compression member are riveted together, the maximum pitch should be 16 times the thickness of the outside plate or 20 times the thickness of the thinnest enclosed plate, and for a distance equal to 1½ times the width of the member the spacing should not exceed 4 times the diameter of the rivets.

The minimum widths of lattice bars are commonly specified as follows: 2½ in. for ⅞-in. rivets, 2¼ in. for ¾-in. rivets, and 2 in. for ⅝-in. rivets; ⅞-in. rivets are used for latticing flanges 3¼ in. or over. One rivet is generally used at each end unless the flanges are 5 in. or more in width, when two are used. The minimum thickness of single lattice bars is 1/40 of the length from center to center of end holes, and the minimum thickness for double lattice bars is 1/60 of such length. For secondary bracing members these ratios may be taken as 1/50 and 1/75 respectively.

Lattice bars or continuous plates are supposed to resist the total shear in a compression member, dividing the shear equally among all parallel planes in which there are resisting elements, whether bars or plates; the bars act much like the web members of a Warren truss in resisting shear. Different plans have been devised to determine the amount of shear to be resisted. The preferred method seems to be to consider the shear as 2

percent of the total axial stress in the member, plus any shear due to the weight of the member or to eccentric loading. Higher percentages have been advocated, but tests [1] would indicate that 2 percent is ample. The specifications of the American Railway Engineering Association and of the American Association of State Highway Officials give a formula for percentage in terms of l/r, but only when $l/r < 48$, as in the larger members, do the results deviate from 2 percent by more than 0.2.

The stress in a single bar is determined from its slope and from the shear explained above; when single bars are used making 60° with the axis of the member, the stress in each bar is 1.15 times the shear; when double latticing is used making an angle of 45° with the axis, the stress in each bar is 0.707 times the shear. Since the bars resist both tension and compression, they must be strong enough for either; usually the compression determines the size. The same compression formula is used as in the design of the whole member; the unsupported length of a single bar is taken from center to center of end rivets, but for double bars only 70 percent of this distance because they are riveted at their intersection. The radius of gyration of a rectangular bar in terms of its thickness is $r^2 = t^2/12$ or $r = 0.289t$.

For example, in the chord section of the illustrative problem on p. 6–17 the distance between rivet lines in the bottom angles is 19.5 = 2(7.25 + 2.5) which exceeds 15 so double latticing is used. The shear of 2 percent of the axial stress 645,200 is divided equally between the cover plate and the lattice bars, and the stress in each bar is 4,560 = 6,450 × 0.707. Assuming ⅞-in. rivets, the minimum width is 2½ and the minimum thickness is $\frac{1}{2} = \frac{19.5 \times 1.41}{60}$. For a length of 19.2 = 19.5 × 1.41 × 0.70 the strength in compression is

$$10{,}500 = \left[17{,}000 - \frac{0.485 \times 19.2^2}{\frac{1}{12}(\frac{1}{2})^2}\right] 2.5 \times \frac{1}{2}$$

and in tension is 15,000 = 20,000(2.5 − 1.0)½, both being sufficient. The l/r of one bottom angle and the bottom half of the web plate is found less than ⅔ of the $33 = \frac{20 \times 12}{\sqrt{51.9}}$ of the whole member. This is found as follows: the distance from the back of the angle to the center of gravity is $0.41 = \frac{3.25 \times 1.33 - 9 \times 0.5 \times 0.25}{3.25 + 4.5}$. The radius of gyration is

$$1.12 = \sqrt{\frac{5.0 + 3.25(1.33 - 0.41)^2 + 4.5(0.25 + 0.41)^2}{7.75}}$$

and the $l/r = 17.1 = 19.2/1.12$.

Perforated Cover Plates. The function of lattice bars and tie plates may be supplied by continuous cover plates perforated by a succession of access holes provided that the length of hole along the axis of the member shall not be over twice the width of the hole, and the clear distance between holes shall not be less than the transverse distance between the nearest lines of connecting rivets or welds.

RIVETING AND WELDING

Shop and Field Connections. The component parts of each member are fastened together in the structural shop by either shop rivets or welds, and the members themselves are connected together in the finished structure by field rivets, welds, or high-

[1] Final Report, Committee on Steel Column Research, *Trans. ASCE*, 1933.

strength bolts. Ordinary bolts are often permitted for roof purlins, intermediate beams, and secondary members which do not affect the rigidity of the main structure. Rivet and bolt holes are cold-punched in the separate parts, and the parts to be riveted together are held by bolts until the heated rivets are "driven." Each rivet has a head at one end and a shank long enough to extend through the parts connected with enough metal protruding to form the second head and completely fill the enlarged hole when upset under pressure. Members are held together in the field by erection bolts until enough permanent rivets, bolts, or welds are provided; then the temporary bolts are replaced. Some structures are fully riveted, some are fully welded, and some are shop riveted and erected with field welds or high-strength bolts.

Structural Riveting. The strength of a rivet in *shear* in a structural joint depends on the area of cross section of the shank of the rivet taken as a circle with a diameter equal to the nominal diameter of the rivet before it is upset in driving. If all the parts which tend to move in one direction are on the *same side* of the parts which act in the opposite direction, the rivets tend to shear in one plane, as shown in Fig. 9*a*, and they are said to be in "single shear." If the parts which act in one direction are *between* the parts which act in the opposite direction, as in Fig. 9*b*, the rivets tend to shear in two planes and they are said to be in "double shear." No more than double shear need be considered, for it is assumed that a rivet will not shear in more than two planes simultaneously. The unit stress in shear is given in the specifications, and the strength of each rivet in single shear is the product of the unit stress and the area of cross section; the strength of a rivet in double shear is twice what it would be in single shear, so far as shearing is concerned.

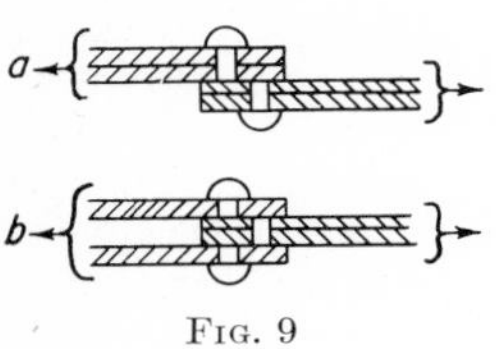

Fig. 9

The *bearing value* of a rivet in the parts connected may be less than the shear value and hence may determine the strength of the rivet. Even though the surface of contact is cylindrical, it can offer no more resistance to forces parallel to the line of stress than a flat surface, and the effective area is considered to be a rectangle, the dimensions of which are the nominal diameter of the rivet and the thickness of the metal in which it bears. If two plates of different thickness are connected, the rivets would naturally tend to tear through the thinner plate and its thickness would determine the strength in bearing. If more than two plates are connected by the same rivets, the bearing value of each rivet would be determined by the thinner combined thickness of all the plates which act in one direction and not necessarily by the thinnest plate; this is true whether the rivets are in single or double shear. The unit stress allowed in bearing is often larger for rivets in double shear than for those in single shear because of the resistance of the greater friction between the parts connected. The bearing value of each rivet is equal to the product of the unit stress, the nominal diameter of the rivet, and the limiting thickness explained above. Web thicknesses for beams, channels, and other rolled shapes may be found in the tables.

Number of Rivets. The number of rivets required in an ordinary connection is found by dividing the total stress by the limiting value of one rivet. First it should be determined whether the rivets act in single or double shear, and from the table of rivet values the corresponding shear value should be found for the given diameter of rivet and for the specified unit stress. Next the combined thickness of the parts which act in one direction should be compared with the total thickness of the parts which act in the opposite direction, and the smaller chosen; from the table of rivet values, the bearing value of one rivet in this thickness of metal should be found for the given diameter of rivet and for the specified unit stress. The limiting value of one rivet is the smaller of these values found for shear and for bearing. Bearing values which ex-

ceed double shear are not shown in the tables. At least two rivets are used in each connection, except at the end of a lattice bar.

Illustrative Problems. 1. *Truss.* To find the numbers of ¾-in. rivets required to connect different members of a roof truss to the connecting gusset plate shown in Fig. 10, unit stresses being used in pounds per square inch of 15,000 shear, 32,000 bearing in single shear, and 40,000 bearing in double shear. The stresses in the members in thousands of pounds are indicated in the figure. Each web member has two ¼-in. angles connected to a 5⁄16-in. plate between them, and so the rivets are in double shear; the combined thickness of the angles exceeds the plate thickness, so that the latter determines the bearing value. From the tables, the double-shear value of one ¾-in. rivet is 13,250, and the bearing value is 9,380; the latter is the limiting value, since it is smaller. The numbers of rivets required in the diagonals are 18,000 ÷ 9,380 = 2 and 20,000 ÷ 9,380 = 3, as shown. In a bottom-chord joint, where there is no external load, the vertical components of the web members balance each other, but their horizontal components may accumulate, their algebraic sum equaling the difference in stress in the two horizontal chord members. If the bottom chord were made continuous, the 40,000 stress would pass from one panel to the other through the angles, and rivets would be required merely for the balance of 60,000 − 40,000 = 20,000. When the angles are spliced as shown, part of the 40,000 may be carried to the other chord angles by means of a plate which splices the outstanding legs, the rest being taken by rivets which connect the angles to the gusset plate, at least enough being used in the left-hand angles to take the stress which comes through the plate from the web members, viz., 20,000. The splice plate may be a separate plate but more often serves also as a connection plate for bottom-chord bracing. The rivets in the splice plate are in single shear, and the shear value of one rivet is 6,630, which governs, since it is less than the bearing value in the 5⁄16 plate; if four rivets were used in each pair of angles, this would account for 26,500 of the stress in each member. Even if more rivets were used in one-half, they could not be counted, for the stress carried directly from one member to the other can be no greater than the weaker connection; this does not apply to the rivets in the gusset plate because other forces are involved. The number required to connect the ⅜-in. angles to the plate is (60,000 − 26,500) ÷ 9,380 = 4, and to connect the 5⁄16-in. angles (40,000 − 26,500) ÷ 9,380 = 2, both being limited by the bearing value in the plate.

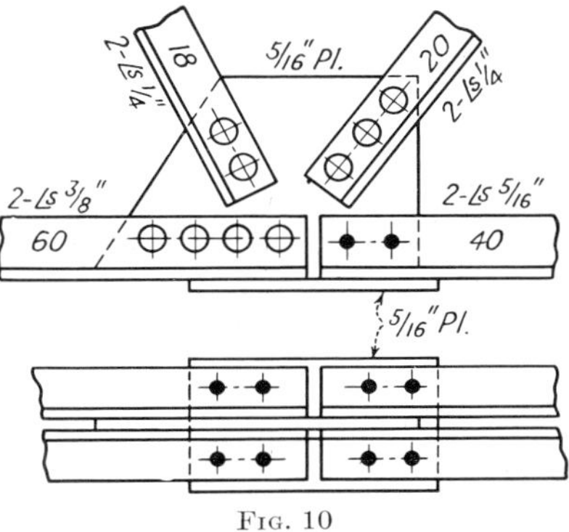

Fig. 10

2. *Beam.* To find the number of ⅞-in. rivets required to connect a 10-21 to the web of a 12-85, as shown in Fig. 11, if the maximum end reaction of the 10-in. beam is 20,000 lb, unit stresses in pounds per square inch being used as 15,000 shear, 32,000 bearing in single shear, and 40,000 bearing in double shear. Connections such as this are standardized for usual conditions, and the safe loads of standard angles are given in the tables; often special connections must be designed, and this problem illustrates the method. Two ⅜ angles would be used like the standard. The shop rivets connecting the angles to the web of the 10-in. beam are in double shear; the shear value of one rivet is 18,040, and the bearing value in the 0.24-in. web is 8,400; the number required is 20,000 ÷ 8,400 = 3. The rivets connecting the angles to the web of the 12-in. beam are in single shear; the shear value is 9,020 which is less than the bearing value in the ⅜ angles; the number of

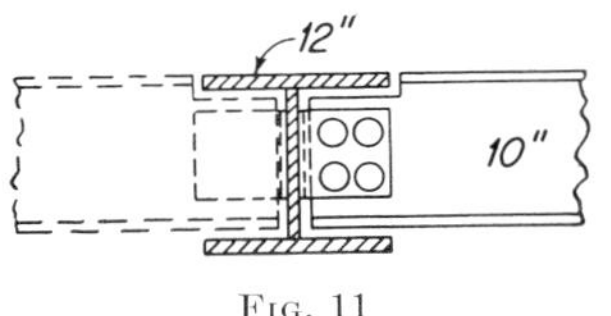

Fig. 11

rivets required is 20,000 ÷ 9,020 = 3. In case another similar 10-in. beam is connected on the opposite side of the 12-in. beam by the same rivets as indicated by the dotted lines, the load is doubled, and the rivets are in double shear; the bearing value of one rivet in the 0.50 web is 17,500, which is less than the double-shear value; the number of rivets is 2 × 20,000 ÷ 17,500 = 3, 4 being used for symmetry.

Eccentric Connections. A connection which has considerable tendency to rotate is eccentric, and the rivet stresses due to moment are often larger than those due to the direct load. Connection angles are treated as eccentric connections only when the rivets are farther from the backs of the angles than the usual gage distances. The rivets are not equally effective in resisting rotation, but those farthest from the center of rotation are stressed most. It is usual to assume the number and the spacing of the rivets and then determine whether the conditions are most satisfactorily met or whether the number or spacing of the rivets should be modified. No rule can be given to aid the designer in making his first assumption, but each trial serves as a guide for subsequent trials.

Theory of Eccentric Connections. The theory is based on making the algebraic sum of the moments of the stresses in the rivets equal to the moment of the force or forces to be resisted, the moments being taken about the center of rotation. For convenience, the method is modified so that moments may be taken about the center of gravity of the group of rivets, for the lever arms are more easily determined. In Fig. 12, let e be the eccentricity of the force P from G the center of gravity of the group of rivets and R the center of rotation located an unknown distance k from G measured perpendicularly to the force P and on the opposite side of G. For any rivet, let x and y represent the horizontal and vertical coordinates from G and z the corresponding distance from R. If E is the resultant stress in a rivet at unit distance from R, the resultant stress in the rivet at z distance in Ez acting normal to the line from R to the rivet. By similar right triangles, the *horizontal* component of the stress Ez is Ey, and the vertical component is $E(k \pm x)$. If the rivets resisted rotation only, the center of rotation would coincide with the center of gravity, and by similar triangles the vertical component of the stress would be Ex; the rest of the vertical component Ek has to do with supporting the load as in a concentric connection; it is constant for all the rivets of a connection and equal to P/N, where N is the total number of rivets. The moment of the resultant stress Ez about R is Ez^2, and the sum of all such moments must equal the moment of the force P; thus,

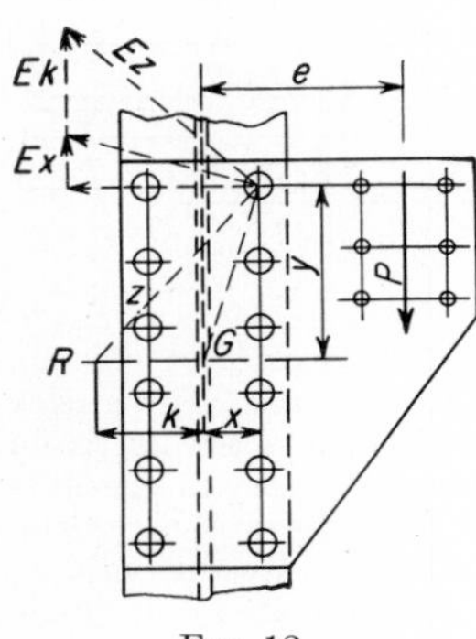

Fig. 12

$$\Sigma Ez^2 = P(k + e)$$

From the figure,

$$z^2 = (k \pm x)^2 + y^2$$

and since E is constant, we have

$$\Sigma E[(k \pm x)^2 + y^2] = P(k + e)$$
$$\Sigma Ek^2 \pm \Sigma 2Ekx + \Sigma Ex^2 + \Sigma Ey^2 = Pk + Pe$$

But $\Sigma Ek = P$; hence, ΣEk^2 cancels Pk; $\Sigma x = 0$, since G is at the center of gravity; hence, $\Sigma 2Ekx = 0$, and

$$\Sigma Ex^2 + \Sigma Ey^2 = Pe \qquad \text{or} \qquad E = \frac{Pe}{\Sigma x^2 + \Sigma y^2}$$

The critical rivet is the one farthest from the center of rotation, and the maximum stress $r' = Ez_m$ found from the two components should not exceed the limiting value r of one rivet as determined in the usual manner. The critical rivet is the one at a maximum distance z_m from R, and its coordinates x_m and y_m from G are usually but not necessarily the maximum values of x and y. The horizontal component of the stress on the maximum rivet is Ey_m, and the vertical component is the sum of Ex_m and P/N. If the rivets are in a single vertical line, x becomes zero and the problem is somewhat simplified, but the P/N part of the vertical component must be considered.

Illustrative Problem. Does the arrangement of ¾ rivets shown in Fig. 13 satisfy the conditions, if the allowed unit stress in shear is 13,500 psi? The center of gravity is midway between the two central rivets. The value of x is 2¾ for each rivet, and the value of y is zero for the two central rivets, 8 for the four outer rivets, and 4 for the remaining four. The expression for E is

$$E = \frac{20{,}000 \times 10}{10 \times (2\tfrac{3}{4})^2 + 4(4^2 + 8^2)} = 505$$

Horizontal component of stress in critical rivet =

$$Ey_m = 505 \times 8 = 4{,}040$$

$$Ex_m = 505 \times 2\tfrac{3}{4} = 1{,}390$$

Vertical component of stress in critical rivet =

$$1{,}390 + \frac{20{,}000}{10} = 3{,}390$$

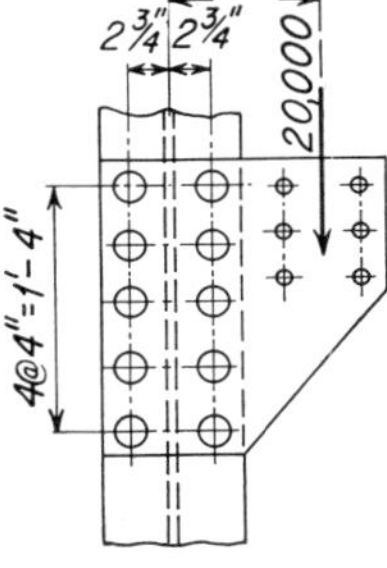

FIG. 13

The resultant stress is 5,270, which is considerably less than the 5,960 limiting value in single shear. The spacing of the rivets might be reduced if desired. If two rivets were removed, all expressions would be changed; thus, $E = \dfrac{20{,}000 \times 10}{8 \times (2\tfrac{3}{4})^2 + 4(2^2 + 6^2)}$; the horizontal component is $E \times 6 = 5{,}450$; the vertical component, $E \times 2\tfrac{3}{4} + \dfrac{20{,}000}{8} = 5{,}000$; and the resultant obviously would exceed the 5,960 allowed.

Rivets in Tension. It is often expedient to subject rivets to tensile stresses, even against the judgment of some engineers. They have felt that the initial stresses caused by shrinkage during cooling add enough to the calculated tension to cause the heads to pull off. Tests show, however, that little trouble is experienced from full-sized heads, and flattened rivets need not be used in tension. Tests show that the initial tension does not increase the total stress because no additional stress due to applied load is registered until after the load equals the initial tension. The proper unit stress to allow for tension depends upon the type of connection and upon the judgment of the designer, but for most work it appears that the tensile strength of a rivet can be considered equal to the single-shear value.

Flattened and Countersunk Rivets. It is often necessary to flatten rivetheads to ⅜ or ¼ in. in height in order to furnish clearance for erecting other members. Usually a flattened rivet may be considered as strong as an ordinary button-headed rivet unless it is subjected to tension. Countersunk rivets are used where projecting heads would interfere with smooth bearing or with erection. The holes are reamed out for the conical heads of the rivets. The heads may not fit the reamed holes exactly and they may have to be chipped off if smooth bearing is required. Countersunk rivets are sometimes counted as full value in shear and also in bearing if the plate is thick enough to allow the shank to bear the necessary amount.

Eccentric Connections Which Involve Tension. Although it is better to avoid the use of rivets in tension whenever possible, there are cases where their use is justified. In the bracket shown in Fig. 14, the rivets which connect to the supporting column are subjected to tension as well as shear. The center of rotation may be taken at the bottom rivets, although this is sometimes taken one-seventh of the depth of the bracket above the lower edge. The uppermost rivets are subjected to the greatest tension t, and intermediate rivets receive proportionate amounts ty/Y. The sum of the moments of the tensile stresses in the rivets must equal the eccentric moment Pe. For two rows of rivets as shown, $2\sum \frac{ty}{Y} y = Pe$, or $t = PeY/2\Sigma y^2$. The shearing stress in each rivet is $s = P/N$, where N is the total number of rivets in both rows. The Y of the critical rivets in terms of N and the rivet pitch p is $Y = \frac{1}{2}(N - 2)p$. The general expression for the sum of squares is $(1^2 + 2^2 + 3^3 \cdots n^2) = \frac{1}{6}n(n + 1)(2n + 1)$. For two rows of rivets $N = 2(n + 1)$, whence $\Sigma y^2 = \frac{1}{24}N(N - 1)(N - 2)p^2$. The usual formula for total tension due to combined tension and shear is $T = \frac{1}{2}t + \sqrt{s^2 + \frac{t^2}{4}}$. By substitution, we find the maximum total tensile stress in each critical rivet to be

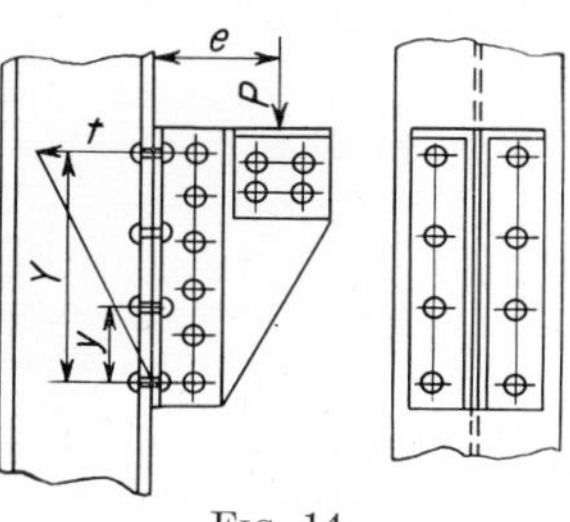

Fig. 14

$$T = \frac{P}{N}\left[\frac{3e}{(N - 1)p} + \sqrt{1 + \left(\frac{3e}{(N - 1)p}\right)^2}\,\right]$$ [1]

This maximum stress should not exceed the single-shear value of one rivet.

Welding. Fusion welding in steel structures is allowed by most codes as a substitute for riveting. Welding is done by oxyacetylene torch or by electric arc, the latter being more common. Direct or alternating current is used, the arc being formed between the base metal to be welded acting as one electrode, and the fusion steel wire, in a suitable holder, acting as the other; the heat from the arc fuses the wire to form the weld and also fuses the base metal enough to ensure adhesion. The edges or ends of two plates can be joined flush by a *butt weld*, as shown in Fig. 15*a*; if the plates are more than ¼ in. thick, one or both must be beveled to make a single V or double V in order to permit complete fusion. A lap joint or a connection of two pieces at right angles is made with a *fillet weld*, as shown in Fig. 15*b*. A *plug weld* is made by welding through a hole punched in one of the two lapped plates connected, and similarly a *slot weld* is made by welding through a slot cut in one plate. A *stitch weld* is a small weld made to hold component parts of a member together. In structural work, the fillet weld is the most common. The *throat* of a fillet weld is the altitude of the isosceles triangle forming the

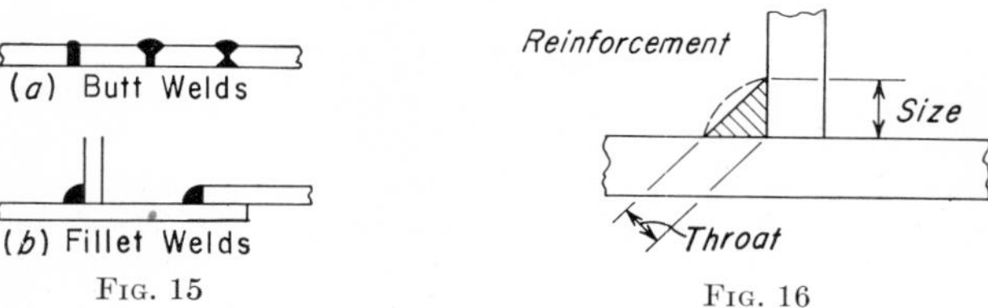

Fig. 15

Fig. 16

[1] Adapted from a similar expression developed by W. R. Osgood in *Eng. and Contr.*, December, 1928.

weld, and the *size* is the width of contact with each part connected, as shown in Fig. 16. The throat of a butt weld is equal to the thickness of the thinner part connected, and the size is equal to the throat. The *length* of weld is measured along the intersection of the surfaces connected. A weld is designated by its size and its length and is designed accordingly. Additional metal is often deposited in a weld to ensure the development of the required section; this extra metal is called *reinforcement.*

Strength of Welded Joints. The effective area of a welded joint is the product of the throat and the length. The unit stresses recommended by the American Welding Society in a butt weld are 20,000 in tension or compression and 13,000 in shear, and in a fillet weld 13,600 shear. A fillet weld is usually placed along the sides of the part connected so that the weld acts in shear, or at the end of the part connected so that one-half acts in either tension or compression and the other half in shear; the unit stress in shear is thus the determining factor. The strength of the weld per linear inch is the unit stress in shear times the throat; in the usual equal-legged fillet weld the throat is 0.707 times the nominal size of the weld. The strength of a 5⁄16-fillet weld is 13,600 × 0.707 × 5⁄16 = 3,000 lb per lin in. The strengths of other sizes are proportional, a difference of 1,200 for each eighth or 600 for each sixteenth. The size of the fillet weld along the edge of a plate or toe of an angle should usually be 1⁄16 in. less than the nominal thickness of plate or angle end or along the toe of an angle three-quarters of the thickness, although in some cases the weld can be built up to the full thickness. A 5⁄16-in. weld is usually considered the largest that can be made with a single pass of the electrode.

Design of Welded Connections. The total number of inches of weld required to connect one member or part to another is found by dividing the total stress to be carried by the strength of the weld per linear inch. The distribution of the weld depends upon the type of connection and the available space, but so far as feasible it should be arranged to avoid secondary stress or a tendency to twist. A seat angle should be connected at both ends symmetrically, with or without welding along the lower edge. A truss member or bracing member composed of one or two angles should be connected so that the center of gravity of welding coincides with the center of gravity of the member, most of the weld being placed along the backs of the angles. The length is usually expressed in quarters of an inch.

Illustrative Problems. 1. *Beam Seat.* To connect a 6 × 4 × 3⁄8 seat angle to the face of a column to support a beam with an end reaction of 30,000 lb. If a 5⁄16-in. fillet weld at 3,000 lb per in. is used, the total length of weld is 30,000 ÷ 3,000 = 10 in. Either 5 in. can be placed at each end, or 3 in. at each end and 4 in. at the bottom. (See Fig. 17.)

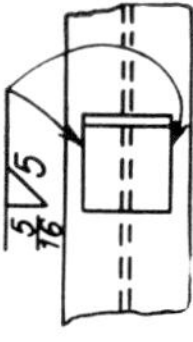

Fig. 17

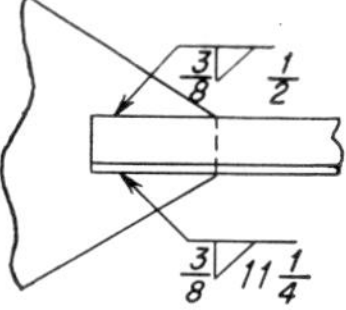

Fig. 18

2. *Truss Angle.* To connect a 6 × 4 × 1⁄2 angle to a gusset plate for a stress of 60,000 lb. Using a 3⁄8 weld which is 3⁄4 of the 1⁄2 thickness, the total length is 60,000 ÷ 3,600 = 16¾. The longer leg is usually connected, and the center of gravity is 1.99 from the back of the shorter leg. The length is proportioned so that 16¾ × 1.99/6 = 5½ is placed along the toe and the balance 11¼ along the back of the angle. (See Fig. 18.)

Eccentric Welds. A weld which has a tendency to twist as in Fig. 19, may be checked from an assumed length L. The vertical component per linear inch of weld is P/L, and the horizontal component is $6Pe/L^2$, found by dividing the eccentric moment by the section modulus of a rectangle. The resultant of these two components must not exceed the allowed strength of the size weld to be used. A diagram for the direct determination of the length in terms of constants found by dividing P by the strength of weld can be constructed by anyone who has frequent need.

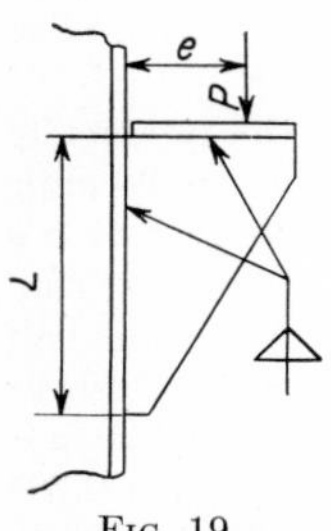

Fig. 19

RIVETED GIRDERS

End-connection Angles. The connection of a girder to the face of a supporting column or girder by means of connection angles is like a beam connection, except that the connection angles are not in contact with the web but are separated from it by the vertical legs of the flange angles and by fillers of the same thickness. The rivets which pass through both connection angles and flange angles are usually fully developed in transmitting flange stress, as explained later; so it is well not to count upon them to carry part of the vertical reaction. The rivets which pass through loose fillers are not so effective as those which connect parts in direct contact, for the parts are more liable to slide and to bend the rivets. Specifications commonly require that, if rivets carrying stress pass through fillers, the fillers shall be extended beyond the connected member, and the extension secured by enough additional rivets to develop the value of the filler. The number of rivets in the angles, not counting those through the flange angles, is determined by the double-shear value, and the total number in the angles and fillers is at least the number determined by the bearing value in the web. This provides for no increase in number beyond those which would be used if the fillers did not extend beyond the angles, but the arrangement is better because the fillers are fastened to the web and the rivets are less liable to bend. The extra rivets are placed opposite some of the rivets in the connection angles, usually at the top and bottom and arranged symmetrically.

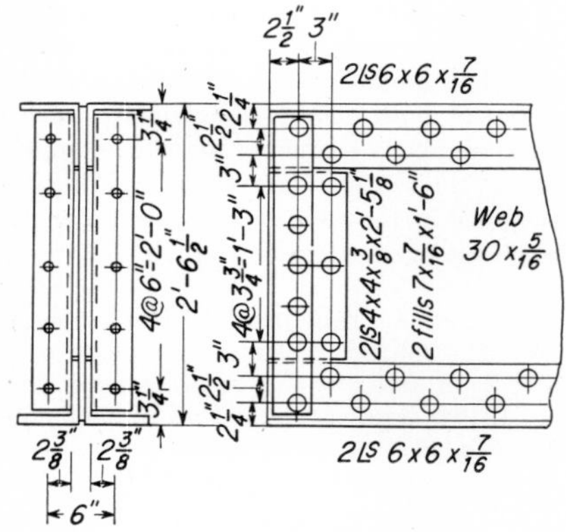

Fig. 20

Illustrative Problem. To find the number of ¾-in. rivets required to connect the ⅜-in. connection angles to the girder shown in Fig. 20, if the maximum reaction is 65,000 lb and the unit stresses are 15,000 shear and 40,000 bearing.

Number required in angles, determined by double shear. . . . . .65,000 ÷ 13,250 = 5
Number required in angles and fillers, determined by bearing in 5/16 web
65,000 ÷ 9,380 = 7
Rivets in fillers outside of angles. . . . . .7 − 5 = 2
Three are used, as shown, to provide excess and to simplify spacing
Number required in outstanding legs, determined by single shear 65,000 ÷ 6,630 = 10

End-stiffening Angles. When a girder rests on a supporting column, a pedestal, or masonry, the shearing stresses in the web plate are transmitted to the support by means of end-stiffening angles or stiffeners. These angles act as columns restrained in one di-

rection by the web, but this is not an important factor, since the stresses from the web are cumulative through the rivets and reach the full load only at the bottom rivet. The entire reaction is transmitted through the bearing of the stiffening angles on the outstanding legs of the flange angles. The required bearing area is found by dividing the reaction by the allowed unit stresses in bearing; this is the same as the bearing value allowed for rivets in single shear unless a different value is specified. Since the ends of the stiffeners must be cut to clear the curved fillets of the flange angles, this portion cannot be counted in bearing, because contact cannot be assured; and even if it could, the bearing on a curved surface would not be effective. The radius of the flange-angle fillets is ½ in. (⅝ in. for 8 × 8∠s); so unless the stiffeners are thicker than this amount, no part of the web legs can be counted. The web legs are 3, 3½, or 4 in., unless a double row of rivets is required; then 6-in. legs are used. The outstanding legs are usually the largest commercial size which will not extend beyond the flange angles, but sometimes the same size is used even though the legs project beyond the flange angles or have to be planed off. No part of the stiffeners which may project beyond the flange angles can be counted in bearing. The thickness of the stiffeners must be sufficient to give the required effective area; the minimum thickness is ⅜ in., or one-twelfth the length of the outstanding leg. Stiffeners are used in pairs, more than one pair being used if necessary.

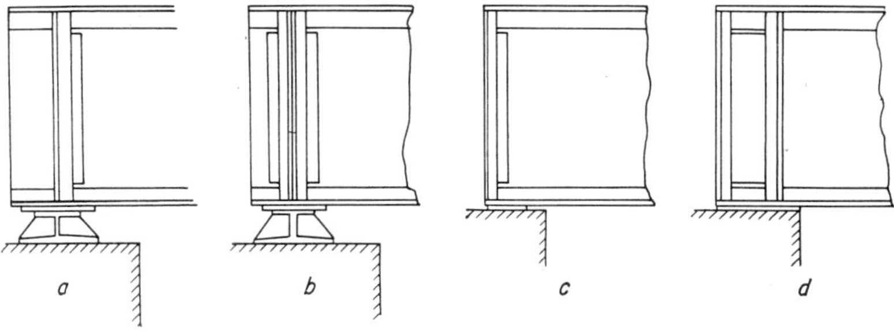

Fig. 21

The outstanding legs are placed at the center of bearing preferably, as in Fig. 21*a* or *b*. Sometimes they are placed at the extreme end of the girder, as in *c*, but full bearing cannot be assured with ordinary shop methods. When two pairs of stiffeners are placed at opposite ends of the bearing plate, as in *d*, the distribution of stress is problematical; owing to the deflection of the girder, the inner pair is often assumed to take twice as much of the reaction as the pair at the extreme end.

The number of rivets is determined as for end-connections angles. The fillers are made to extend outside the stiffeners to provide for additional rows of rivets; they are usually cut to allow ¼-in. clearance at each end. It is common practice to maintain the same vertical spacing throughout the length of the girder; if the web is to be spliced, it is desirable to have closer spacing near the flange angles than in the central portion, because the rivets in the splice plates are more effective when farther from the neutral axis.

Illustrative Problem. To design the end stiffeners for the girder shown in Fig. 22, if the maximum reaction is 360,000 lb and the unit stresses are 13,500 shear, 24,000 bearing (single shear), and 30,000 bearing (double shear). Use ⅞-in. rivets. For 6-in. flange angles, we first try 5-in. stiffeners, using 5 × 3½. The effective length of bearing for one pair of angles is 2(5 − ½) = 9, and the thickness required is 360,000 ÷ (9 × 24,000) = 1¹¹⁄₁₆. Even with two pairs, the thickness exceeds the ¾-in.

maximum for punching, but part of the web leg can be counted; the effective area of each angle is $3\frac{1}{2} \times \frac{1}{2}$ less than the gross area A; thus, $4(A - 3\frac{1}{2} \times \frac{1}{2})24{,}000 = 360{,}000$, whence $A = 5.50$. Two pairs of $5 \times 3\frac{1}{2} \times \frac{3}{4}$ angles suffice; this implies that the stiffeners are cut in such a manner that the remainder of the web legs will actually bear. The number of rivets required in the angles, not counting those through the flange angles, is $360{,}000 \div 16{,}240 = 23$, determined by double shear. The number required in the angles and fillers is $360{,}000 \div 11{,}480 = 32$, determined by the bearing in the $\frac{7}{16}$-in. web. With the same number of rivets in each pair of angles, at least

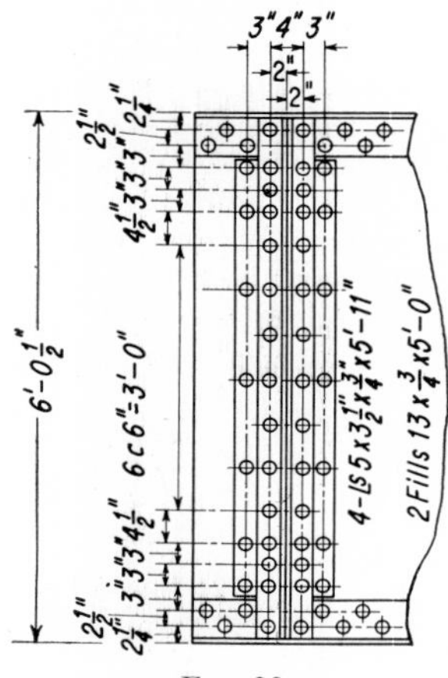

FIG. 22

24, or 12 in each pair, must be used; as a matter of detail, it is better to use an odd number, and so 13 are used, which allows the omission of alternate rivets in the extended fillers and intermediate stiffeners while symmetrical spacing is still maintained, so that the stiffeners are interchangeable instead of being rights and lefts. If the web plate is to be spliced, it is desirable to place the rivets closer together near the flange angles than in the central portion, so that the same vertical spacing can be used as in the splices where the rivets farthest from the neutral axis are most effective, as explained later. The final spacing is shown in the figure.

Intermediate Stiffening Angles. Stiffeners with fillers are used at points of concentration to transmit the loads to the web. Stiffeners with fillers or crimped stiffeners without fillers are used at other points of deeper girders to stiffen the web plate against buckling when the clear vertical distance between flange angles exceeds 60 (or 70) times the web thickness. The spacing of intermediate stiffeners in the clear should not exceed 7 ft or the distance in inches determined by a formula in the specifications, as, for example, $\dfrac{11{,}000t}{\sqrt{v}}$, in which t is the web thickness, in., and v the shear intensity, psi, found by dividing the shear for the section under consideration by the gross area of the web plate. Standard angles with unequal legs are used, with the outstanding longer legs one or two sizes less than the outstanding legs of the flange angles; the minimum allowed thickness is usually sufficient. The rivets line up with those in the end stiffeners, but alternate rivets may be omitted if the resulting spaces do not exceed eight times the diameter of the rivets. Stiffeners which support concentrated loads must be cut to fit the top flange angles and designed for bearing if the load rests upon the angles. Similarly, stiffeners at the intermediate reactions of cantilever girders

must be cut to bear at the bottom. When the concentrated loads are connected to the outstanding legs of the stiffeners bearing is not required, but an extra angle may be used on one or both sides of the girder if the number of rivets requires it. Unloaded intermediate stiffeners are often cut with 1/4-in. clearance at each end.

Web Splices. The web plates of all but the shorter plate girders must be spliced, for the wider plates are not rolled long enough to extend the entire length, and for convenient handling no single plate should weigh over 3,000 lb. The flange angles usually extend full length. When more than one splice is used, they are spaced symmetrically about the center line; the splice for maximum stress is designed first, and the other splices are usually made like it. A splice is preferably located under a pair of stiffeners; the angles stiffen the splice, there is one less line of rivets to be driven, and metal is saved by the use of thinner fillers. A single pair of splice plates may be designed to provide both shear and moment requirements, but in some of the heavier girders a pair of plates is placed near each flange to transmit stresses due to bending moment, and another pair is placed between them to transmit the stresses due to shear. In order to develop fully the web plate in shear, the gross area of the two splice plates must equal or exceed the gross area of the web plate; this condition will be met when, in order fully to develop the web plate in moment, the section modulus of the two splice plates equals or exceeds the section modulus of the web plate; thus, the square of the depth of the splice plates multiplied by their combined thickness must not be less than the square of the depth of the web plate multiplied by its thickness. The thickness of each splice plate must not be less than the minimum thickness of metal allowed by the specifications, usually 3/8 or 5/16 in.

Two to four rows of rivets are used on each side of the splice; the resultant shearing stress acts upward at the center of gravity of the rivets in one half, and an equal resultant shearing stress acts downward at the center of gravity of the rivets in the other half; these two forces form a couple, one-half the moment of which, as well as the direct stress, must be resisted by the rivets in each half. The rivets in each half must also resist that part of the bending moment carried by the web plate, and this depends on the method used in design. Theoretically, if the moment-of-inertia method is used, the bending moment to be resisted bears the same relation to the total bending moment at the point of splice as the net moment of inertia of the web plate bears to the net moment of inertia of the whole cross section at that point; and if the flange-area method is used, the bending method to be resisted bears the same relation to the total bending moment at the point of splice as one-eighth the gross area of the web bears to one-eighth the gross area of the web plus the net area of the angles and cover plates of one flange at that point. It is more convenient and quite accurate to develop the web plate. Thus, in the moment-of-inertia method, the moment to be resisted is the net moment of inertia of the web plate times the unit stress in bending divided by the distance from the neutral axis to the extreme fiber in the flange; and, in the flange-area method, the moment to be resisted is one-eighth the gross area of the web times the unit stress in bending times the effective depth from center to center of gravity of the flanges. The horizontal spacing between rivets is usually 3 in. except that the inner rows are separated by twice the gage distance of the stiffeners; this allows ample edge distance on the web segments and a space between them. Three rows of rivets in each half are first assumed, with the spacing the same as in the end stiffeners, and then these rivets are tested by the method of eccentric connections explained on p. 6–22; if the number is found excessive, alternate rivets may be omitted from one row, or an entire row may be omitted, and the splice tested for each revision until the stress in the critical rivet is just below the limiting value of one rivet bearing in the web. Since the rivets near the flanges are more effective in resisting moment than those near the neutral axis, it is better to use a few 3-in. spaces near each flange, unless that results in an excessive number in the stiffeners.

Illustrative Problem. To design the web splice at a section for which the shear is 100,000 lb in the girder for which the end stiffeners were designed on p. 6–27. Assume three rows of rivets in each half, as shown in Fig. 23, the vertical spacing like that in the end stiffeners. Assume that the girder was designed by the moment-of-inertia method and that two ⅝-in. cover plates are required in each flange at the section taken.

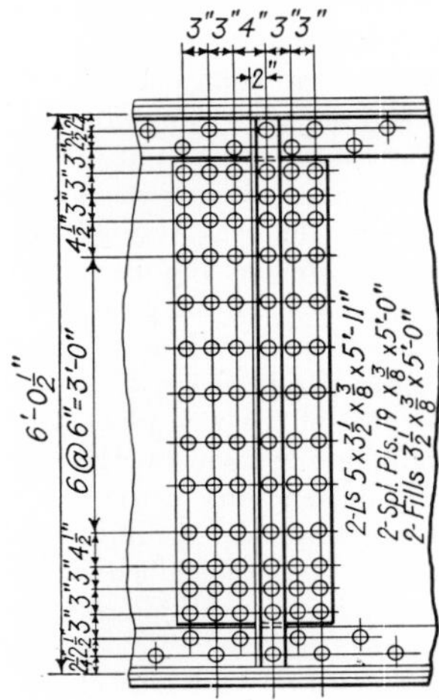

Fig. 23

Depth of splice plates . . . . $72.5 - 2(6 + \frac{1}{4}) =$ 60 in.
Thickness of each plate required . . . . $(72^2 \times \frac{7}{16}) \div (2 \times 60^2) = \frac{5}{16}$
Minimum thickness specified . . . . $= \frac{3}{8}$
Eccentric moment due to shear . . . . $\frac{1}{2} \times 100{,}000(3 + 4 + 3) =$ 500,000 in.-lb
Gross I of web . . . . $=$ 13,610
I of holes
$1 \times \frac{7}{16} \times 2(6^2 + 12^2 + 18^2 + 22.5^2 + 25.5^2 + 28.5^2 + 34^2) =$ 3,180
Net I of web . . . . $=$ 10,430
c . . . . $(\frac{1}{2} \times 72.5) + (2 \times \frac{5}{8}) = 37.5$
Moment resisted by web plate . . . $fI/c = (18{,}000 \times 10{,}430) \div 37.5 =$ 5,020,000 in.-lb
Combined moment to be resisted . . . . $500{,}000 + 5{,}020{,}000 =$ 5,520,000 in.-lb
$\Sigma x^2 + \Sigma y^2 = (2 \times 13 \times 3^2) + 6(6^2 + 12^2 + 18^2 + 22.5^2 + 25.5^2 + 28.5^2) =$ 15,070
E . . . . $5{,}520{,}000 \div 15{,}070 =$ 366
H component of stress on critical rivet . . . . $366 \times 28.5 =$ 10,430
V component of stress on critical rivet
$(366 \times 3) + 100{,}000 \div (3 \times 13) =$ 3,660
Resultant stress . . . . $=$ 11,050

This is less than the 11,480 allowed but not enough less to reduce the number of rivets.

Separate Splice Plates. When the single pair of splice plates described on p. 6–29 proves impractical, three pairs may be used as shown in Fig. 24. The plates near the flange angles are designed to carry the horizontal stresses due to bending moment, and the intermediate plates to carry the shear. Usually a clearance of ¼ in. is left between the moment and shear plates, and between the moment plates and flange angles. The width and length of the plates depend upon the number and the spacing of the rivets. The vertical spacing in the moment plates is usually 3 in., and three rows of rivets may

be assumed at first. The resisting moment of the two pairs of moment plates must equal or exceed the resisting moment of the web plate (see p. 6–30). The distance from center to center of splice plates d_s is so much less than the extreme depth $2c$ used in the moment-of-inertia method of design or the effective depth d_g used in the flange-area method, that the developed unit stress is only $d_s/2c$ or d_s/d_g of that used in the design of the flanges. The resisting moment of the two pairs of moment plates is equal to the net area of each pair of plates multiplied by the developed unit stress and by the lever arm of the couple d_s. By equating this to the resisting moment of the web, the required net area of one pair of plates may be found, and the combined thickness may be found by dividing this by the net width. No plate should be thicker than the flange angles; the thickness may be reduced by increasing the width. The number of rivets in each half of a pair of splice plates should be sufficient to develop fully the tensile strength of the plates selected at the same reduced unit stress used in their design. These rivets may be placed opposite one another vertically (making the plates shorter) or staggered (making the plates thinner), as best suits the conditions. After the moment plates are determined, the central shear plates may be designed in the manner of a single pair of splice plates except that the resisting moment of the web need not be added to the eccentric moment due to shear.

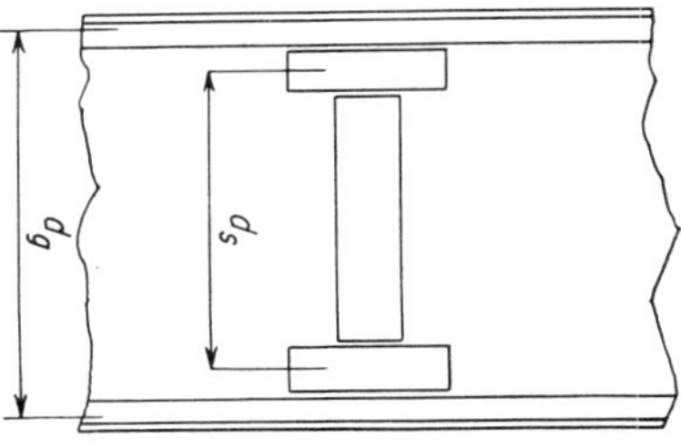

FIG. 24

Illustrative Problem. To design three pairs of splice plates for the conditions given in the problem on p. 6–30. Assume 9-in. moment plates with three rows of rivets, as shown in Fig. 25.

Distance center to center of splices. . . $d_s = 72\frac{1}{2} - 2(6 + \frac{1}{4}) - 9 = 51.0$ in.
Extreme depth. . . . . $2c = 72\frac{1}{2} + 4 \times \frac{5}{8} = 75.0$ in.
Moment resisted by web plate, as found above. . . . . $= 5{,}020{,}000$ in.-lb
To this we equate an expression for the resisting moment of the splice plates:

$$18{,}000 \times \frac{51.0}{75.0} \times 51.0 \times 2(9 - 3 \times 1)t = 5{,}020{,}000$$

Thickness of each plate. . . . . $t = \frac{11}{16}$

Developed strength of 2 plates. . $18{,}000 \times \dfrac{51.0}{75.0} \times 2(9 - 3 \times 1)\frac{11}{16} = 101{,}000$

Number of $\frac{7}{8}$-in. rivets in each half, bearing in $\frac{7}{16}$ web. . . . $\dfrac{101{,}000}{11{,}480} = 9$

This number of rivets makes a rather short plate, but plates with two rows would have to be thicker than the flange angles; so the change is not made. The depth of the central shear plates would be $41.5 = 51.0 - 9 - 2 \times \frac{1}{4}$, and the thickness of each plate to give a cross section equal to that of the web would be $\dfrac{3}{8} = \dfrac{72 \times \frac{7}{16}}{2 \times 41.5}$. In the central portion, $\frac{3}{8}$-in. fillers would be added to make up the total thickness of the flange angles; the moment plates are not enough thinner to justify a separate filler; so they would be increased from $\frac{11}{16}$ to $\frac{3}{4}$, no increase in the number of rivets being necessary. The vertical spacing of the rivets is assumed as shown in Fig. 25 with due regard to symmetry and to the number required in the stiffeners (Fig. 22), and the strength of

the assumed number is investigated. The center of gravity of the rivets in each half of the shear plates is $1.1 = 3 \times 4/11$ from the inner row.

One-half the eccentric moment due to shear............. $100{,}000(1.1 + 2) = 310{,}000$

$\Sigma x^2 + \Sigma y^2$.................. $4 \times 1.9^2 + 7 \times 1.1^2 + 4(6^2 + 18^2) + 2 \times 12^2 = 1{,}751$

Horizontal component in critical rivet..................... $\dfrac{310{,}000}{1{,}751} \times 18 = 3{,}190$

Vertical component in critical rivet.............. $\dfrac{310{,}000}{1{,}751} \times 1.1 + \dfrac{100{,}000}{11} = 9{,}290$

(Note that the x of the critical rivet is not the maximum value of x.)

The resultant of these components, 9,820, is less than 11,480, the value of a rivet bearing in the 7/16-in. web plate, but it would not be feasible to try a smaller number of rivets because the spacing would exceed 6 in.

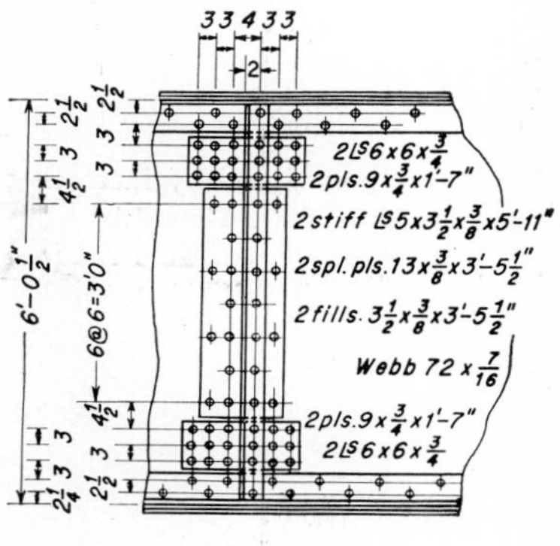

Fig. 25

Flange Rivets. The rivets or welds which fasten the flange angles to the web of a plate girder transmit the flange stresses from the web to the flange angles and cover plates. This flange stress is cumulative from the point of zero bending moment to the point of maximum bending moment, but the rate of increase is greater near the ends than near the center, because the increase in the bending moment is greater. Since it is impractical to use different spaces throughout, it is customary to determine the longitudinal spacing, or "pitch," required at the outer end of each panel and to use that pitch throughout the panel, even though the number of rivets is thus increased. The pitch is the longitudinal distance from center to center of rivets, whether on a single line or staggered on two lines. It is always changed at the point of application of a concentrated load (except at the center of a symmetrical girder); if there is no uniform load except the weight of the girder, it is hardly worth while to change the pitch between points of concentration. If a girder supports additional uniform loads or moving loads, the pitch is calculated at each pair of stiffeners or at distances apart approximately equal to the depth of the girder, preferably the depth between rivet lines. It is not so important to choose the panels at definite points as it is to space the rivets in accordance with the calculated pitches. Though it is on the side of safety to extend any pitch farther from the end than necessary, it should be realized that because of symmetry four extra rivets must be driven for every extra space, and 12 extra holes must be punched. The controlling factors which determine the pitches of flange rivets are: (1) the form of loading, (2) the application of the loads to the web or to the flange, (3) the method of design, (4) the depth between rivet lines, (5) the limiting value of each rivet, (6) the strength of the web plate between rivets, and (7) the general rules which govern rivet spacing. A pitch may have to be increased next to a stiffener in order to provide sufficient driving clearance, but otherwise no pitch should exceed the calculated value.

Determination of Rivet Pitch. *Web Loads.* The equal and opposite flange stresses in the top and bottom flanges of a plate girder form a couple, the moment of which equals the bending moment. Since the stresses are transmitted from the web to the flanges through the rivets, the lever arm is equal to the depth between rivet lines d_r, the mean depth being used when there are two lines in each flange, i.e., for 6×6 angles or larger; more rivets are thus required than if they could be placed at the centers of

gravity of the flanges, where they would be more effective. The increase in bending moment between two sections is equal to the vertical shear V multiplied by the distance between the sections; the increase between one rivet and the next is therefore Vp, where p is the pitch or longitudinal distance between rivets, and the corresponding increase in flange stress is Vp/d_r. Since part of this flange stress is resisted by the web, the rivets transmit only the remainder to the angles and cover plates; the stress to be carried by one rivet, the limiting value of which is r, is found by multiplying the increase in flange stress per rivet by the ratio which the resistance of the angles and cover plates bears to the resistance of the angles, cover plates, and web plate, or

$$r = \frac{Vp}{d_r} \times \text{ratio}$$

whence

$$p = \frac{rd_r}{V \times \text{ratio}}$$

If d_r is taken in inches, the resulting pitch p is in inches. If the girder is designed by the moment-of-inertia method, the ratio is the ratio which the moment of inertia of the angles and cover plates of *both* flanges bears to the moment of inertia of the entire cross section including the web plate. If the girder is designed by the flange-area method, the ratio is the ratio which the area of the angles and cover plates of *one* flange bears to the area of the angles and cover plates of one flange plus one-eighth (or one-sixth) of the gross area of the web plate. These moments of inertia or areas may be taken for either the net section or the gross section, according to which was used in the design of the girder. The difference in the pitches found by the net section or by the gross section is not great if consistent values are used. Where cover plates are cut off to reduce the section, the ratio changes, but this refinement is usually disregarded, provided that the largest ratio found from the largest section is used. Moving loads should be placed to give the maximum shear for each section. Pitches are usually expressed to the nearest ¼ in., with a maximum of 6 in.

Illustrative Problem. To find the pitches of ⅞-in. rivets at 40,000 psi bearing in the 50-ft girder composed of a 72 × ⅜ web, 4 *Ls* 6 × 6 × ¾, 2 cover plates 14 × 11/16, and 2 cover plates 14 × ⅝, designed on p. 6–10. It weighs 360 lb per ft and supports three concentrated loads of 145,000 lb.

Limiting value of one rivet bearing in ⅜-in. web . . . . r = 13,100 lb
d_r . . . . 72.5 − (2 × 2¼) − 2½ = 65.5
Shear at end of first panel . . . . 217,500 + (360 × 25) = 226,500 lb
Shear at end of second panel . . . . 226,500 − 145,000 − (360 × 12.5) = 77,000 lb
Total net I . . . . = 84,780 in.^{4}
Net I of web plate . . . . $11{,}660 - \left(4{,}350 \times \frac{3/8}{1\frac{7}{8}}\right) - 1{,}700 = 9{,}090$
Ratio, by moment-of-inertia method . . . . (84,780 − 9,090) ÷ 84,780 = 0.89
Ratio, by flange-area method
[13.9 + (12 × 1$\frac{3}{16}$)] ÷ [13.9 + (12 × 1$\frac{3}{16}$) + 3.4] = 0.89
Pitch in first panel . . . . (13,100 × 65.5) ÷ (226,500 × 0.89) = 4¼ in.
Pitch in second panel . . . . (13,100 × 65.5) ÷ (77,000 × 0.89) = 6 in. max

Determination of Rivet Pitch. *Flange Loads.* Loads which rest directly or indirectly on the top flange are transmitted to the web through the rivets, subjecting them to vertical components as well as the same horizontal components which they would have if the loads were transmitted to the web by other means; the resultant stress must not exceed the limiting value r. Both components must be in equivalent units before the resultant is taken, the stress per linear inch being most convenient. The rivet pitch in any panel is found by dividing the value of one rivet by the resultant per linear inch.

The horizontal component is V times ratio/d_r, which varies with the shear. The vertical component is the stress per inch which tends to shear the rivets vertically and is constant; it includes the weight of the top-flange angles and cover plates, any rail, floor, or other dead load, and the live load. Weights of continuous material per foot may be combined and divided by 12; weights of rails per yard may be divided by 36; crane-wheel loads may be considered to be distributed by the rail and top flange over 30 in.; concentrated engine loads may be considered to be distributed over three ties (from 32 to 36 in.). A tabular form may be used to advantage in the solution of problems in flange rivets, especially when both live and dead loads are involved or when there are many panels. Live-load shears for uniform loads may be found from the end shear by the proportion of the squares of the distances from the opposite end of the girder, the shear curve being a parabola; for equal panels, panels may be used instead of distances. In order that the vertical component may be properly distributed among the rivets, the pitch should not exceed 4½ in. (4 in. for crane girders).

Illustrative Problem. To find the pitches of ¾-in. rivets at 24,000 psi bearing in a 40-lb girder composed of a 48 × 7⁄16 web, 4∠6 × 4 × ¾, and 4 cover plates 14 × ½. It carries on the top flange a uniform dead load of 1,000 lb-ft and a uniform live load of 7,000 lb per ft.

Limiting value of one rivet bearing in 7⁄16-in. web r . . . . = 7,880
d_r . . . . 48.5 − (2 × 2½) = 43.5
Weight per foot . . . . [71.4 + (4 × 23.6) + (3 × 23.8)]1.1 = 290
Difference in dead load per panel length d_r . . . . (290 + 1,000)43.5/12 = 4,700
Number of panels . . . . 40 × 12 ÷ 43.5 = 11
Vertical component . . . . [7,000 + 1,000 + (2 × 23.6) + (2 × 23.8)] ÷ 12 = 680
Ratio, using net areas . . . . (11.3 + 12.2) ÷ (11.3 + 12.2 + 2.6) = 0.90

By inspection of the pitches of the first four panels, it is apparent that the pitch for the fifth panel will be limited by the maximum value, so that no computation is required.

Panel	Shear = V			$\frac{V \times 0.90}{43.5}$	Resultant	Pitch = $\frac{7,880}{\text{Resultant}}$
	Dead load	Live load	Total			
End	1,290 × 20 = 25,800	7,000 × 20 = 140,000	165,800	3,430	3,500	2¼
Second	25,800 − 4,700 = 21,100	140,000 × 10²/11² = 115,700	136,800	2,830	2,910	2¾
Third	21,100 − 4,700 = 16,400	140,000 × 9²/11² = 93,700	110,100	2,280	2,380	3¼
Fourth	16,400 − 4,700 = 11,700	140,000 × 8²/11² = 74,000	85,700	1,770	1,900	4¼
Fifth	. . . . . .	. . . . . .	. . . .	. . . .	. . . .	4½ max

Minimum Pitch of Flange Rivets. If the calculated pitch is less than the minimum space allowed, the design should be modified by increasing the web thickness or the rivet diameter, or otherwise. The usual minimum of "three diameters" applies to rivets on a single line, but for staggered rivets the minimum depends on the shearing strength of the web plate between adjacent rivets on the inner line. The unit stress in shear on the web s is usually specified in terms of the gross section, and it is assumed that the unit stress on the net section will be greater in inverse proportion, or $\frac{2ps}{2p - (d + \frac{1}{8})}$, in which p is the rivet pitch and d the rivet diameter. The net section $[2p - (d + \frac{1}{8})]t$ times the unit stress should equal the limiting values of two rivets, usually the bearing value in the web, or $2dtb$, in which b is the unit stress in bearing, whence $p = db/s$.

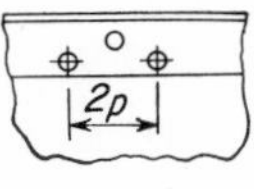

FIG. 26

Lengths of Cover Plates. Beams and girders are designed to resist the maximum bending moments, and unless cover plates are used they are generally made of uniform cross section throughout their lengths, regardless of the fact that less metal is needed

near the ends where the bending moments are much less. It is not practical to reduce the sections of rolled beams or plate girders without cover plates, for the cost of splicing would offset the saving in material, and the resulting girders would not be so satisfactory. When cover plates are used to furnish part of the flange area, they may be placed only where needed and be discontinued beyond the points where the remaining section is sufficient to meet the requirements. If a girder is exposed to the weather, one cover plate of the top flange should extend the full length in order to keep water from getting between the angles and the web. Similarly a crane-runway girder should have the top cover plate or channel extend full length to give uniform bearing for the rail. The rivets in cover plates except in the heavier girders are determined by maximum spacing values and need not be calculated. Close spacing of four diameters or 3 in. is used at the end of each plate for a distance equal to about 1½ times its width.

Uniform Loads. The curve of bending moments for a simple girder with a uniformly distributed load is a parabola, as shown in Fig. 27, the properties of which are such that the vertical ordinates from the vertex of the curve at the center vary as the squares of the corresponding abscissas, or $X^2/X_1^2 = Y/Y_1$. Since moments of inertia and also flange areas are proportional to bending moments, use may be made of these relations in finding the length of cover plates, depending on the method of design used. If X_1 is the half length of the girder and Y_1 the total moment of inertia or flange area required at the center, X is the half length of the longest cover plate which extends to the point where the moment of inertia or area of the web and angles $Y_1 - Y$ just equals the required amount. By doubling the lengths, the proportion remains the same, and we have: Theoretical length of cover plate $= L\sqrt{I''/I}$ or $L\sqrt{A''/A}$ depending on the method of design used. L is the total length of girder, I the total moment of inertia at the center, I'' the moment of inertia of the plates in both flanges of which the length is required plus the moment of inertia of all plates that are outside of these plates; A is the total area of one flange, and A'' the area of the plate of which the length is required plus any plates that are outside of this plate. The moments of inertia or the flange areas of either the gross section or the net section may be used, whichever is more convenient, so long as values in both numerator and denominator are consistent. In order not to weaken a girder by cutting off the cover plates and to offset any approximations in the method of calculation, they are made about 3 ft longer than the theoretical lengths, usually taken to the nearest half foot. This method of finding the lengths of cover plates is applied to uniformly distributed loads whether fixed or moving, but it does not apply to concentrated loads. For a system of moving concentrated engine loads, such as Cooper's or Steinman's, it is usually impractical to construct an accurate curve of bending moments just to determine the lengths of cover plates; the curve of bending moments resembles a parabola so closely that this method may be used, provided that an extra foot or two is added to the length of each plate to offset the discrepancy between the curves.

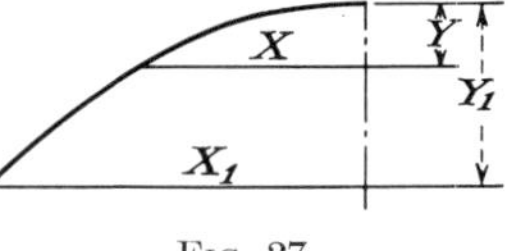

Fig. 27

Illustrative Problem. 1. *Moment-of-inertia Method.* A 40-ft girder uniformly loaded is composed of a 48 × 7/16 web, 4*Ls* × 4 × ¾, 4 cover plates 14 × 7/16, and ¾ rivets. The total net moment of inertia is 28,870 of which the web and angles furnish 15,800, the net moment of inertia of the two inner plates 6,420, and of the two outer plates 6,650. To find the lengths of the cover plates.

Theoretical length of inner plates . . . . $40\sqrt{(6{,}420 + 6{,}650) \div 28{,}870} = 26.9$
Length to be used . . . . $26.9 + 3.1 = 30.0$
Theoretical length of outer plates . . . . $40\sqrt{6{,}650 \div 28{,}870} = 19.2$
Length to be used . . . . $19.2 + 2.8 = 22.0$

2. *Flange-area Method.* The total net area in the flange of a 70-ft girder uniformly loaded is 27.0 sq in., ⅛ the web is 6.3, the net area of 2*Ls* is 10.9, of one plate 14 × 7/16 is 5.3, and of one plate 14 × ⅜ is 4.5. To find the lengths of the plates.

Theoretical length of inner plate	$70\sqrt{(5.3 + 4.5) \div 27.0} = 42.2$
Length to be used	$42.2 + 2.8 = 45.0$
Theoretical length of outer plate	$70\sqrt{4.5 \div 27.0} = 28.6$
Length to be used	$28.6 + 2.9 = 31.5$

Lengths of Cover Plates. *Concentrated Loads.* The curve of bending moments for a series of concentrated loads is a series of straight lines with a break at each point of

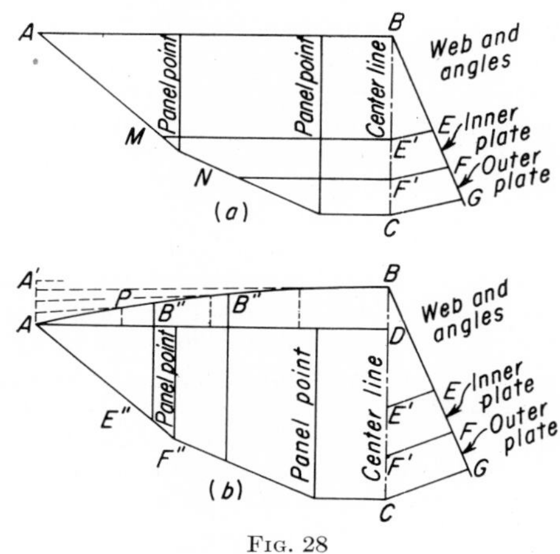

FIG. 28

concentration. No general formula is applicable because of the number of variables, and so it is usually more convenient to determine the lengths of cover plates graphically. If the loads are symmetrical about the center, only one-half need be plotted. If the uniform load is relatively small, as when there is no other uniform load than the weight of the girder itself, the combined bending moment may be found at each point of concentration, and the curve plotted accordingly as a series of straight lines, as in Fig. 28*a*. When enough uniform load is added to make appreciable curves in the diagram, separate curves should be plotted for the bending moments due to the uniform and the concentrated loads,[1] as in Fig. 28*b*, for it is not sufficiently accurate and not on the safe side to consider the combined curve a series of straight lines. After the bending moments have been plotted, the maximum ordinate *BC* for which the girder was designed can be subdivided in proportion to the resisting moments of the component parts. This can be done graphically, as follows: Lay off on a random line *BG* equal to the total moment of inertia or the total flange area depending on the method of design used; lay off *BE* equal to the moment of inertia of the web and all angles, or one-eighth the gross area of the web plus the area of the angles of one flange; lay off *EF*, *FG*, etc., equal to the moment of inertia of the successive cover plates in pairs, or the areas of the successive cover plates

[1] The two curves are plotted on opposite sides of the base line to simplify scaling total bending moments from curve to curve. The curve for uniform loads is a parabola, the maximum bending moment at the center being calculated and plotted as *DB*. The parabola may be drawn through *B* and *A* as follows: Draw $AA' = DB$, and subdivide it into any number of equal parts. Connect each of these points with *B*. Divide *AD* into the *same number* of equal parts, and draw a vertical line at each subdivision point. The intersection of the diagonal and vertical nearest *A* gives point *P* on the curve; as many other points are found in a similar manner as needed to determine the curve with the desired accuracy. The points may be joined by means of a curved ruler. Draw only enough of the construction lines to show the intersections.

of one flange; either gross section or net section may be used, as preferred. Connect G to C, and draw EE' and FF' parallel to GC; the intercepts on the line BC show the portions of the total bending moment that are resisted by the different component parts. For small uniform loads combined with the concentrated loads, as in Fig. 28*a*, a horizontal line drawn through E' cuts the curve M, showing the point where the resisting moment of the web and angles equals the bending moment and theoretically where the first plate must begin; between the end and this point, the web and angles suffice; similarly, the point N where the horizontal line through F' cuts the curve shows where the second plate must begin. Where uniform loads are large enough to be plotted separately, as in Fig. 28*b*, horizontal lines cannot be used, but the corresponding points are found where the vertical intercepts $B''E''$ and $B''F''$ from curve to curve just equal BE' and BF'. These intercepts may be plotted on the edge of a separate piece of paper and slipped along the curves. For symmetrical girders, the length of each plate is about 3 ft longer than twice the distance from its theoretical end to the center.

Illustrative Problem. To find the lengths of the cover plates of the girder designed on p. 6–11 by the moment-of-inertia method. There is no uniform load other than that due to the weight of the girder; so a curve may be drawn for the combined bending moments, as shown in Fig. 29.

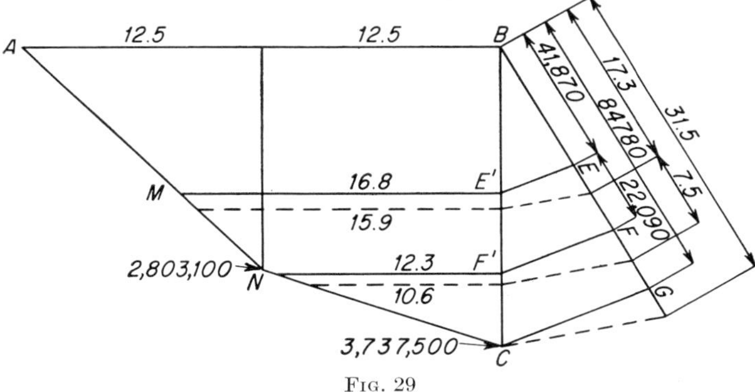

FIG. 29

Total maximum bending moment at center........................ = 3,737,500 ft-lb

Total bending moment at quarter point

$(217{,}500 \times 12.5) + (360/2 \times 12.5 \times 37.5) = 2{,}803{,}100$

Net I of four cover plates.......$2[14 - (2 \times 1)]1\frac{5}{16}(36.25 + 0.66)^2 =$ 42,910 in.^{4}

Net I of two inner plates........$2[14 - (2 \times 1)]\frac{11}{16}(36.25 + 0.34)^2 =$ 22,090

Net I of web and angles..........................84,780 − 42,910 = 41,870

By plotting $BG = 84{,}780$, $BE = 41{,}870$, and $EF = 22{,}090$ on a random line, joining GC and drawing EE' and FF' parallel to GC, the horizontal line $E'M = 16.8$ is one-half the theoretical length of the inner (thicker) plate, and the total length of the $14 \times \frac{11}{16}$ plate is $(2 \times 16.8) + 2.9 = 36.5$. Similarly, the total length of the $14 \times \frac{5}{8}$ plate is $(2 \times 12.3) + 2.9 = 27.5$.

To find the lengths of the cover plates of the same girder designed on p. 6–12 by the flange-area method, we can use the same curve of bending moments by plotting along the random line $BG = 31.5$, $BE = 3.4 + 13.9 = 17.3$, and $EF = (14 - 2 \times 1)\frac{5}{8} = 7.5$; following the same method of procedure, we find that the length of the $14 \times \frac{5}{8}$ plate is $(2 \times 15.9) + 3.2 = 35.0$, and of the $14 \times \frac{9}{16}$ plate $(2 \times 10.6) + 2.8 = 24.0$.

Unsymmetrical Loads. When the concentrated loads on a plate girder are unsymmetrical in either position or magnitude, the entire length of girder should be plotted

in order to locate both ends of each plate. When the concentrated loads are fixed in position but variable in magnitude because of the passage of the live load, as in a railroad through bridge, separate curves should be plotted for different positions of the live load. Thus for each point of concentration of the half girder, the loads should be placed to give the maximum bending moment at that point, and the corresponding bending moments at the other points should be found; the curve plotted from each set of these bending moments will be outside the other curves for at least one point. The bending moment at each point should include the constant dead-load bending moment at that point. The length of each plate should be determined by the intersection of the horizontal line with the outermost curve.

WELDED GIRDERS

Stiffeners. In a welded girder, the stiffeners are made of plates instead of angles because the web legs required for riveting may be dispensed with. The end stiffeners of a girder which rests upon a seat are designed to transmit the maximum reaction in bearing, the ends being cut accurately to bear upon the bottom flange plates. The width of plate is usually a multiple of 1 in., and not enough to project beyond the flange. The thickness must be such that the bearing area of the pair of plates is as large as the required area found by dividing the maximum reaction by the unit stress in bearing. The bottom ends of the stiffeners are welded to the flange plates by short welds conveniently placed, but the sides must be welded to the web plate by intermittent or continuous welds sufficient to transmit the whole reaction. The welds should not be larger than two-thirds of the web thickness unless staggered.

Intermediate stiffeners made of narrower and thinner plates with shorter welds are used when necessary. Since the webs of welded girders are usually thicker than those of riveted girders, fewer girders need stiffening. The ends of intermediate stiffeners may be cut to bear on the flange plates and be welded to them, especially at the compression flange, or they may be cut short with clearance.

Illustrative Problem. To design the end stiffeners of the plate girder designed on p. 6–10 at a unit stress of 30,000 psi.

Maximum reaction. . . . . $217{,}500 + 315 \times 25 = 225{,}400$ lb

Thickness of 6-in. stiffeners. . . . . $\dfrac{225{,}400}{30{,}000 \times 2 \times 6} = \frac{5}{8}$ in.

Number of inches of $\frac{5}{16}$-in. welds on each side of web. . . . . $\dfrac{225{,}400}{2 \times 3{,}000} = 37\frac{1}{2}$ in.

This may be subdivided into ten 4-in. welds $7\frac{1}{2}$ in. center to center, or any equivalent which gives convenient spacing with the proper total length.

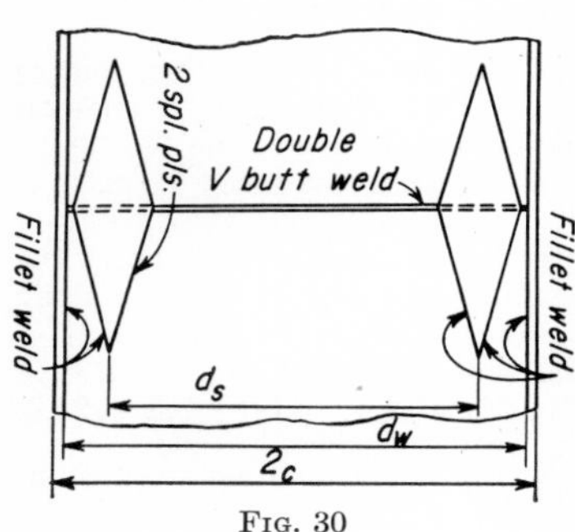

Fig. 30

Web Splices. A long web plate must be spliced to resist both the shearing stresses and the stresses due to bending moment. The shearing stresses may be provided for by butt-welding the web sections together, usually with double-V welds. The moment stresses may be provided for by a pair of plates welded near each flange, or by lug angles welded to the different web sections and bolted together. The plates may be rectangular or diamond-shaped, but the latter are proportional to the stresses and furnish a better distribution of welds (see Fig. 30). The resisting moment of the web plate is $\frac{1}{6}ftd_w^2$ and

to this is equated the moment of the couple formed by the two pairs of splice plates, the lever arm of which is d_s from center to center of splices. The unit stress developed at the splice plates is only $d_s/2c$ part of the unit stress at the extreme fiber; so $d_s/2cfAd_s = \frac{1}{6}ftd_w^2$, in which A is the area required in one pair of splice plates. Since the butt weld undoubtedly carries some of the moment, it is close enough to let $2c = d_w$, whence $A = td_w^3/6d_s^2$. For an assumed width of plate, d_s is determined and the area of the two plates found; if the resulting thickness does not seem satisfactory, the width may be modified. The stress to be taken by the welds in each half of each plate is equal to the area of one plate multiplied by fd_s/d_w; the corresponding length of weld is found by dividing this stress by the strength of the weld per linear inch. Unless a short vertical edge is left at each end, the length of the weld should be made slightly greater than the computed amount lest the plate section near the point prove insufficient to develop the welds.

Illustrative Problem. To design the splice for the 72 × ½ web plate of the girder designed on p. 6–11 at a point in the second panel.

Shear....................217,500 + 315 × 25 − 145,000 − 315 × 12.5 = 76,450 lb

Strength of a ½-in. butt weld......................72 × ½ × 13,000 = 468,000 lb

which exceeds the shear.

d_s for 8-in. plates assumed, allowing 1 in. for welding 72-8-2............ = 62 in.

Area of two 8-in. plates.................................. $\frac{\frac{1}{2} \times 72^3}{6 \times 62^2} = 8.1 \text{ in.}^2$

Thickness of each plate.................................. $\frac{8.1}{2 \times 8} = \frac{1}{2} \text{ in.}$

This is so much more than one-half the web thickness that a 12-in. plate is assumed.

Area of two 12-in. plates.................................. $\frac{\frac{1}{2} \times 72^3}{6 \times 58^2} = 9.2 \text{ in.}^2$

Thickness of each plate.................................. $\frac{9.2}{2 \times 12} = \frac{3}{8} \text{ in.}$

Stress in each plate........................ $12 \times \frac{3}{8} \times 20{,}000 \times \frac{58}{75.6} = 69{,}000$ lb

Length of ⁵⁄₁₆-in. fillet weld.................................. $\frac{69{,}000}{3{,}000} = 23 \text{ in.}$

one-half of which would be the *horizontal* length along each edge of each plate.

Flange Welding. The flange plates of a girder must be welded to the web plate by welds sufficient to transmit the flange stress from the web to the flange plates. Either continuous welding or intermittent welds may be used. Some designers prefer continuous welding on girders exposed to the weather in order to prevent water from seeping in between the web and the flange plates, particularly in the bottom flange. Other designers maintain that close contact is assured by intermittent welds along the planed edge of the plate, so that paint will seal the joint. More welding is needed near the ends of the girder than in the central portion because the increase in flange stress due to bending moment is usually more rapid. At the ends, welds may be made longer, closer together, or even continuous. Where continuous welds are used throughout it may be necessary to have larger welds at the ends. The size of continuous weld should not exceed one-half the web thickness, and there is not much occasion to use larger welds intermittently, although if the welds are staggered it would be allowed.

By analogy, formulas may be developed for welding much as they were for flange rivets (p. 6–34), taking the form $p = \frac{ud_wI}{VI'}$ or $p = \frac{ud_wA}{VA'}$, in which p is the distance center to center of intermittent welds, u is the strength of the intermittent welds on both sides of the web, d_w is the depth of the web plate, V is the maximum shear for a section where the spacing is computed, I is the moment of inertia of the entire cross section including the web plate and the flange plates of *both* flanges, I' is the moment of inertia of the flange plates of *both* flanges, A is the area of the flange plates of *one* flange plus one-sixth the area of the web plate, and A' is the area of the flange plate of *one* flange only. These moments of inertia or areas may be taken for either the *net* section or the *gross* section, according to which was used in the design of the girder. The length of weld may be assumed 3, 4, 5, or 6 in. If the distance p is only slightly more than the length of weld, it may be preferable to use continuous welding; if it should be less, either the size of weld would have to be increased, if possible, or the girder designed with a thicker or deeper web. The spacing should be figured for each sudden change in shear as at concentrated loads, and at as many intermediate points as justified by changes in length or spacing of welds. The clear distance between welds should not exceed 12 in. Deck loads are carried to the web plate through bearing on the planed edge; so no vertical component on the welds need be considered.

Illustrative Problem. To design the flange welding in the end panel of the girder designed on p. 6–11. The strength of two $\frac{5}{16}$-in. welds 3 in. long is $2 \times 3 \times 3{,}000 = 18{,}000$ lb. $I = 84{,}590$, $I' = 69{,}040$, and the shear in the end panel is $217{,}500 + 315 \times 25 = 225{,}400$ lb. Spacing center to center of 3-in. welds in the end panel is

$$p = \frac{18{,}000 \times 72 \times 84{,}590}{225{,}400 \times 69{,}040} = 7 \text{ in.}$$

Cover Plates. When two or more plates are used in one flange, they should differ in width by about 1 in. to provide for means of welding them together. If the end portions of the flange plate are made thinner than the central portion, the points of change can be determined much as in the case of riveted girders, as explained on p. 6–36.

PINS AND REINFORCING PLATES

Pins. The members of the large bridge trusses and of some of the smaller ones may be pin-connected. Only one pin is used at a point, and greater flexibility is thus obtained than in a truss with fully riveted or welded joints which cause secondary stresses. Similar small pins are used in conjunction with loop rods, U bolts, and clevises; these small pins may be rough or turned bolts, or cotter pins. The larger bridge pins are turned cylinders with threaded ends of reduced diameter upon which pilot nuts and driving nuts are screwed during erection; these nuts are then replaced by permanent nuts or caps. The pins are driven cold without upsetting, the holes in the members being only $\frac{1}{64}$ or $\frac{1}{32}$ in. larger than the diameters of the pins.

Tension members of pin-connected trusses are composed chiefly of groups of bars, the bars of one member alternating on a pin with those of an adjacent member. Large eyebars were formerly used, but these are no longer available; so wider plates with the ends reinforced by riveting or welding additional plates to them are made to simulate eyebars. Compression chords are made of plates and channels or plates and angles, and the pins pass through the webs; in order to furnish sufficient bearing area on the pins, the webs are reinforced near the pins by reinforcing plates or pin plates, which are riveted or welded to the webs. Compression web members may be made of wide-flanged beam sections or built-up sections. The pins may go through the reinforced webs or through pin plates fastened to the flanges and projecting beyond the main members.

The thickness of these plates depends on the diameter of the pin, and, conversely the size of the pin depends on the thickness of the bearing; one must be assumed and the other determined to correspond, and then the first must be verified or modified.

Not every pin in a truss need be designed, because it is impractical to use many different sizes; it is preferable to make several pins alike, in order to reduce the number of different members to be made and the number of different sizes of pinholes to be bored. In a small Pratt bridge truss, it is customary to calculate the sizes of the pins at the shoe, at the hip, at the top-chord joint next to the hip, and at the bottom-chord joint nearest the center. These are critical pins, and the remaining top-chord pins may be made like the one next to the hip, and the remaining bottom-chord pins like the central one; in fact, the upper and lower pins may often be made alike without much waste.

Design of Pins. Pins are designed as cylindrical beams to resist both bending and shear, the former usually determining the size. For convenience in finding the bending moment, each force due to the stress in a member or part of a member is assumed to act upon the pin as if concentrated at the center of bearing, i.e., at the center of the surface of contact. Since the forces which act upon a pin do not lie in the same plane, each force is resolved into horizontal and vertical components. The position of the point of maximum bending moment is usually not apparent and so it is necessary to find the bending moment due to the horizontal components and that due to the vertical components at each point of concentration. The resultant bending moment at each point is the square root of the sum of the squares of the horizontal and vertical bending moments at that point. The maximum bending moment on the pin often occurs at a point where there is no vertical bending moment.

The arrangement, or "packing," of the different component parts of the members on a pin affects the size required; sometimes the results of different arrangements must be compared in order that the best one may be selected. The packing depends on adjacent pins, because each bar must be kept parallel to the plane of the truss or slope not more than $\frac{1}{16}$ in. per ft. The packing is made symmetrical, and members composed of bars should have even numbers of bars, and riveted chord members or posts should have two or more webs. A single counter or stirrup is sometimes placed at the center. No two bars of the same member should be placed in contact, as it would be impossible to paint between them; they should be separated by a 1-in. collar, unless a bar of an opposing member is placed between them. In determining the distances between the centers of bearing, a clearance of at least $\frac{1}{16}$ in. should be used between bars, and $\frac{1}{4}$ in. between a bar and a riveted member, due allowance being made for rivetheads; sometimes these rivets are flattened or countersunk, and the flanges of channels or angles are notched to clear the bars, provided the webs are properly reinforced. The bars of diagonal members are placed next to vertical posts in order to reduce the vertical bending moment on the pin.

A stress diagram shows the maximum stress for each member, but these maxima do not occur simultaneously. The maximum chord stresses occur when the truss is fully loaded, but the maximum web-member stresses are found when the truss is only partly loaded. The pin at the shoe and at the hip and usually the bottom-chord pins will be stressed most under a full load that causes the maximum chord stresses and the corresponding stresses in the web members. The intermediate top-chord pins are designed for maximum web-member stresses and the corresponding chord stresses. Counters are not stressed for either condition of loading, but they must be considered, because they may affect lever arms and therefore bending moments. The stresses which act upon a pin at any one time must be in equilibrium. Since the forces on a pin are symmetrically placed, it is necessary only to determine the bending moments on one-half of the pin.

The bending moment at one point of concentration is found from the bending moment at an adjacent point by adding algebraically the product of the shear for a section between the two points by the distance between the points. When finding the resultant bending moment between the vertical bending moment and the corresponding horizontal bending moment, it should be noted that the latter is constant between the horizontal force nearest the center and the corresponding force on the opposite side of the center, the shear being zero. Since the actual position of the bars on the pin may vary slightly from the position used in the design, it is consistently accurate to use lever arms to the nearest ⅛ in. and shears and bending moments to the nearest thousand. It is convenient to arrange the computation in tabular form.

After the maximum bending moment is determined, the corresponding diameter of pin for the specified unit stress may be taken from a table of bending moments if available or from the section modulus of $0.098d^3$ for a beam of circular cross section. The allowed unit stress in bending on pins is usually about 50 percent greater than for other beams. The pin selected for bending moment should be tested to see if it satisfies the shear requirements. The maximum shear intensity, found from the maximum shear V on the pin, is $16V/3\pi d^2$, and this must not exceed the allowed value, which is usually the same as for power-driven rivets.

Illustrative Problem. To design the pin at the joint $L3$ of the truss shown in Fig. 31, on the assumption that the size will be determined when the chord stresses are maximum.

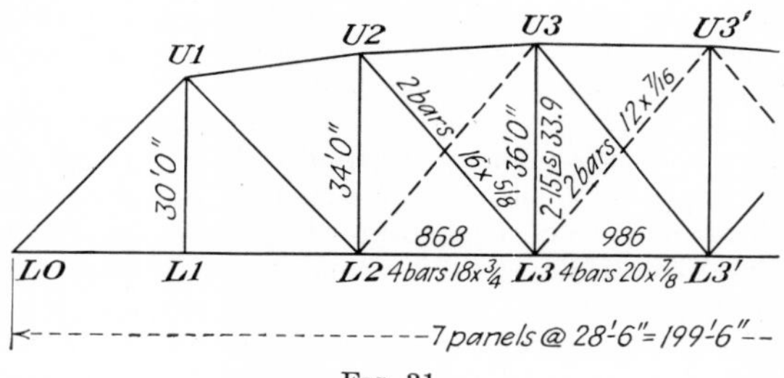

Fig. 31

More than one arrangement should be tried and the results compared. The inclined bars (shown with section lines in Fig. 32) are placed next to the post, and the smallest

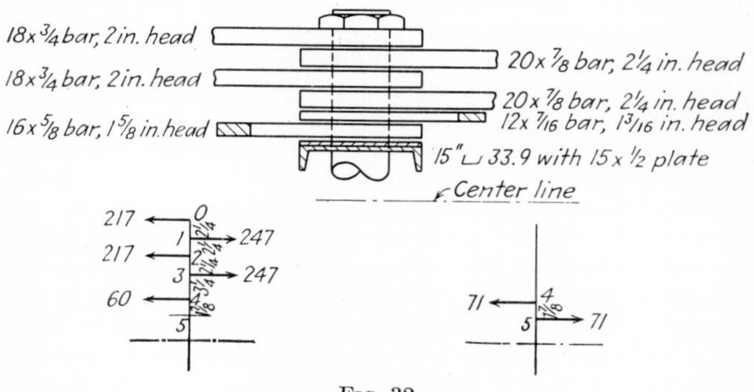

Fig. 32

horizontal bar is placed at the end of the pin. Separate small sketches are drawn to show the horizontal and vertical forces acting upon the half pin. The stress in thousands

of pounds in each of the four 18 × ¾ bars is one-fourth of the stress 868 given on the diagram, or 217; and in each of the 20 × ⅞ bars is $^{986}/_4$ = 247. In order to satisfy the conditions of equilibrium, the horizontal component in each bar of the main diagonal is (2 × 247) − (2 × 217) = 60; the corresponding vertical component is 60 × 34.0/28.5 = 71, and the force in one-half of the post must be the same. The lever arms are found as follows:

$$\tfrac{1}{2}(2 + 2\tfrac{1}{4}) + \tfrac{1}{16} = 2\tfrac{1}{4}$$
$$\tfrac{1}{2}(2\tfrac{1}{4} + 1\tfrac{5}{8}) + 1\tfrac{3}{16} + (2 \times \tfrac{1}{16}) = 3\tfrac{1}{4}$$
$$\tfrac{1}{2}(1\tfrac{5}{8} + \tfrac{1}{2} + 0.4) + \tfrac{3}{8} + \tfrac{1}{4} = 1\tfrac{7}{8}$$
(assuming ½-in. reinforcing plates and rivets flattened to ⅜ in.)

Horizontal components				Point of moments	Vertical components			
Shear, kips	Lever arm, in.	Product, in.-kips	Bending moment, in.-kips		Shear, kips	Lever arm, in.	Product, in.-kips	Bending moment, in.-kips
−217	2¼	−488	−488	1				
+ 30	2¼	+ 68	−420	2				
−187	2¼	−421	−841	3				
+ 60	3¼	+195	−646	4				
0	. . .	. . . . .	−646	5	−71	1⅞	−133	−133

The maximum bending moment for this arrangement is 841 at point 3, since this is obviously greater than the resultant $\sqrt{646^2 + 133^2}$ at point 5. Using a unit stress of 30,000 psi, we find the diameter required for bending is $\sqrt[3]{\dfrac{841}{30 \times 0.098}} = 6\tfrac{3}{4}$ in., which is more than required for shear at 15,000 psi or $\sqrt{\dfrac{16 \times 217}{3\pi \times 15.0}}$. The presence of the counter does not affect the size in this case, because the maximum moment does not fall at point 4 or 5 where the increased lever arm is involved.

Reinforcing Plates. Reinforcing plates are used to strengthen the weaker parts of members in order to develop fully the strength of the remaining parts. The method of designing the reinforcing plates of pin-connected trusses, often called "pin plates," typifies the design of all reinforcing plates. Eyebars need no reinforcement, provided that the pins are large enough to give sufficient bearing area to develop the bars. A built-up member, whether riveted or welded, may have one or more web or plate which bears upon the pin, the pin being located at the center of the web, at the center of gravity, or between the two. Reinforcing plates are used on these webs or plates to furnish additional bearing and to help distribute the stresses from the pin to other parts of the member. In a tension member, the reinforcing plates also increase the net section to prevent failure at the pinholes. Similarly, plates are used to reinforce a member that is weakened by having part of the flanges cut away for clearance. At least one plate should be as wide as the available space permits, and each plate should be as long as it is wide in order to develop properly the rivets along the edges. Some specifications require that one plate should extend to the farther edge of a stay plate and the others 6 in. beyond the nearer edge. When the ends of two compression members bear on opposite sides of a pin, the outer plates on one member and the inner plates on the other member may pass around the pin to hold the members in position during erection and to make them less liable to be dislocated by accident.

Design for Bearing. One-half the total stress in a member with two webs is imparted to the pin through each web and its reinforcing plates. When there are more than two

webs, the proportion is approximately equal to the relative cross-sectional areas between lines drawn midway between the webs. The bearing area of each web and its reinforcing plates is found by dividing the corresponding stress (one-half or other portion of the total stress) by the unit stress allowed for bearing on pins. The combined thickness of the web and its reinforcing plates is found by dividing this bearing area by the diameter of the pin, which must be predetermined or assumed. The web thickness is subtracted from this combined thickness, leaving the thickness of reinforcing plates required for bearing; this may be subdivided to suit specific conditions, usually with a minimum of 3/8 and a maximum of 3/4. If it is subdivided, and if it is to be riveted, it is desirable to place part on each side of the web, so that the rivets will be in double shear and hence more effective.

Design for Tension. Not only must the reinforcing plates for a tension member be designed for bearing as for a compression member, but they must also reinforce the member for tension, for the pin bears on the outer surface of the hole and tends to pull the member apart at the pin. Not only must the net area of cross section through the pinholes fully develop the net area for which the main member is designed, but most specifications require an excess of 25 to 40 percent. The net area for which the member is designed is generally through the row of rivets in the reinforcing plate nearest the center of the member. The diameter of the pinhole is so slightly larger than the pin that it may be taken the same in computing the net section. Due allowance must be made for any reduction in area due to flanges of channels or angles being cut away for clearance. The member must extend far enough so that the net area beyond the pin measured longitudinally is from 90 to 100 percent of the net area of cross section of the main member. Some tension members are cut short of the pin, and the entire stress is carried by the pin plates.

Rivets Required. The rivets should develop the full bearing value of the reinforcing plates, and in a tension member they should develop the tensile strength of the plates as well. The developed stress in bearing is the thickness of the plates required in bearing multiplied by the diameter of the pin and by the unit stress in bearing. The developed stress in tension is the net area required in the plates multiplied by the unit stress in tension. If a thicker plate is needed for tension than for bearing, the developed stress in bearing is that part of the total stress which the thickness of plate used bears to the thickness of plate and web combined. If the plates are all on one side of the web, the total number of rivets is determined by the single-shear value, unless the bearing value in a thin channel web is less. The total number of rivets must pass through the thickest plate, placed next to the web, but the outer plates may be made shorter, each including only the rivets required to develop that plate and all plates outside it. As far as feasible, the rivets should be located so as to distribute the stress proportionately among the component parts of the member. If about half the plates are placed on the opposite side of the web, the rivets are in double shear. If these plates are of different thickness, they may be of different lengths, each containing enough rivets to develop the plate; the portion of the limiting value of the double-shear rivets which passes through plates on both sides is proportional to the plate thicknesses, but the full value of the single-shear rivets is used when they pass through plates on one side only.

At the end of a compression member, all the rivets may be placed on the side of the pin toward the center of the member; in fact, it is often necessary to do so when an opposing member bears on the opposite side of the pin. The distribution of rivets on opposite sides of an intermediate pin in a compression member is determined much as in a tension member. In a tension member, the number of rivets placed on the side of the pin toward the center of the member must conform to certain minimum and maximum limits, and the remainder of the total number is placed beyond the pin. The minimum number to be placed between the pin and the center of the member must develop the

tensile strength of the plates *required* in tension, and the maximum number must develop the tensile strength of the plates actually used. Often, some rivets must be flattened or countersunk to clear eyebars or the nuts on the pin.

Welding. If the reinforcing plates are to be welded in place, the design is much the same as if they were to be riveted. Plates may be thicker. With two or more plates it is desirable to make them of different widths and lengths in order to facilitate welding. Even though they are on opposite sides of the web, they should be so placed that the welds do not come directly opposite; otherwise, the size of each weld is limited to two-thirds the web thickness.

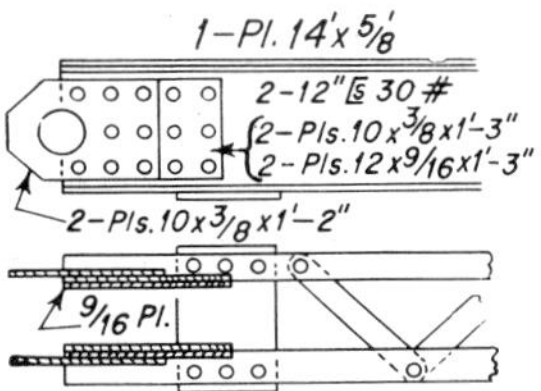

FIG. 33

Illustrative Problems. 1. *Compression Member.* To design the reinforcing plates at the end of a top-chord member composed of 2-12 [S] 30 and 1 cover plate $14 \times \frac{5}{8}$, with a maximum stress of 460,000 lb, as shown in Fig. 33. The pin is 4 in. and the rivets $\frac{3}{4}$ in. Unit stresses are 15,000 psi in shear, 40,000 psi in bearing on the rivets, and 32,000 psi in bearing on the pin.

Stress in each web and its reinforcement . . . . $460{,}000 \div 2 = 230{,}000$ lb
Combined thickness of web and its plates . . . . $230{,}000 \div (4 \times 32{,}000) = 1.80$ in.
Thickness of reinforcing plates . . . . $1.80 - 0.51 = 1\frac{5}{16}$ in.

Part of the total thickness is placed on the back of the web and made full depth, the remainder on the outer face being limited to 10 in. by the flanges. Two $\frac{3}{8}$-in. plates are used on the outside so that one may be extended to surround the pin, leaving $\frac{9}{16}$ in. on the inside.

The total number of rivets required is $1\frac{5}{16} \times 4 \times 32{,}000 \div 13{,}250 = 13$, limited by double shear. The outer $\frac{3}{8}$-in. plate need include only one-half of the rivets, since $\frac{3}{8}$ is one-half of the thickness of plates on that side of the web, but the spacing is better if an additional rivet is included as shown.

2. *Compression Member.* To design the reinforcing plates at the hip end of a top-chord member composed of two webs $24 \times \frac{5}{8}$, two bottom angles $6 \times 4 \times \frac{5}{8}$, two top angles $4 \times 4 \times \frac{7}{16}$, and one cover plate $29 \times \frac{5}{8}$, with a total stress of 900,000 lb, as shown

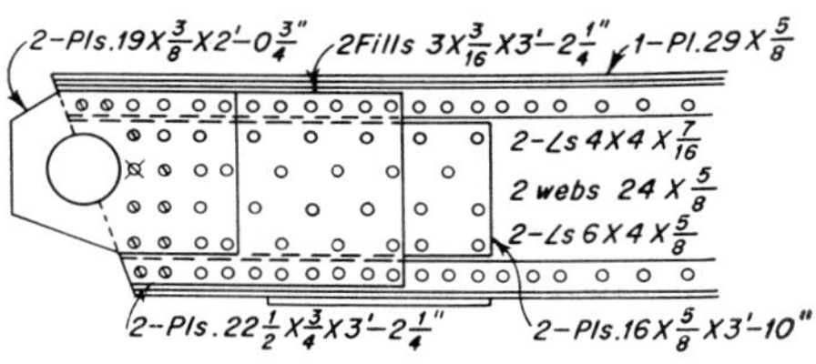

FIG. 34

in Fig. 34. Use an 8-in. pin with a unit stress in bearing of 24,000 psi and $\frac{7}{8}$-in. rivets with a unit stress in shear of 13,500 psi and in bearing 27,000 psi.

Stress in each web and its reinforcement . . . . $900{,}000 \div 2 = 450{,}000$
Combined thickness of web and plates . . . . $450{,}000 \div (8 \times 24{,}000) = 2\frac{3}{8}$
Thickness of reinforcing plates . . . . $2\frac{3}{8} - \frac{5}{8} = 1\frac{3}{4}$

In arranging the plates so that the center of bearing comes nearly opposite the center of bearing of the end post and with the outer $19 \times \frac{3}{8}$ plate of the top chord and the inner $\frac{3}{8}$ plate of the end post surrounding the pin without interference, all the plates are placed outside the web. One $16 \times \frac{5}{8}$ plate is placed between the angles and made

as thick as the bottom angles to serve as a filler. An intermediate plate $22\frac{1}{2} \times \frac{3}{4}$ extending over the vertical legs of the angles makes up the remainder of the required thickness. A $3 \times \frac{3}{16}$ filler is placed along the top angles to make up the difference in thickness in the angles. The area of the half member is $(24 \times \frac{5}{8}) + (\frac{1}{2} \times 29 \times \frac{5}{8}) + 3.3 + 5.9 = 33.3$; the stress in the bottom angle is $450{,}000 \times 5.9/33.3 = 80{,}000$, and the corresponding number of rivets at single shear is $80{,}000/8{,}120 = 10$. This determines the length of the intermediate plate and provides more than enough rivets in the top angles. The number required in the outer $\frac{3}{8}$ plate is $\frac{3}{8} \times 8 \times 24{,}000 \div 8{,}120 = 9$, and the total in all plates is $1\frac{3}{4} \times 8 \times 24{,}000 \div 8{,}120 = 42$. The 16-in. plate is extended to the outer stay plate on the bottom, and the practical spacing of the rivets results in a somewhat larger number than the total required. Some of the rivets near the pin are countersunk but not chipped, to clear the inner plate of the end post; one is countersunk outside to clear the nut on the pin.

3. *Welded Compression Member.* If the plates for the member of the preceding problem were to be welded, the arrangement and the design would be substantially the same. The fillers can be made $16\frac{1}{2} \times \frac{5}{8}$ and welded along the edges to the web and to the angles also. The 3-in. filler should be welded along each edge. The $\frac{3}{4}$-in. plate can be made $22\frac{3}{4}$ in. wide and welded to the curved fillets or made narrower and welded to the vertical legs of the angles. The 19-in. plate can be welded on two or three sides; it would require $\frac{3}{8} \times 8 \times 24{,}000 \div 3{,}600 = 20$ in. of $\frac{3}{8}$-in. fillet weld. The welds along the top and bottom edges of the $\frac{3}{4}$-in. plate to develop this plate and the outer one would be $1\frac{1}{8} \times 8 \times 24{,}000 \div 3{,}600 = 60$ in. Some of this can be placed at the end if desired, provided that the filler is attached to the web with sufficient welding; either the filler or the $\frac{3}{4}$-in. plate must be welded to the bottom angle to develop its 80,000 stress.

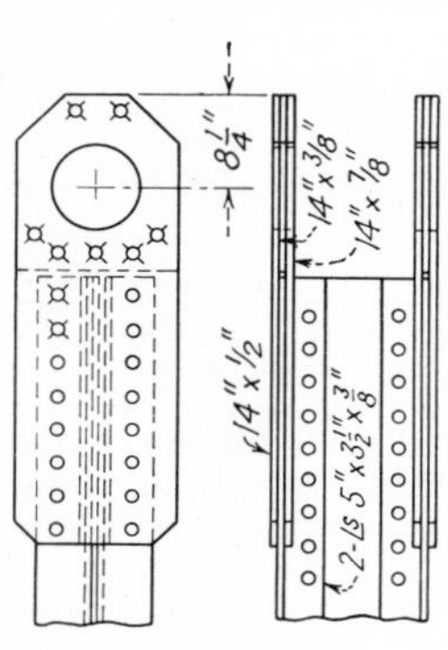

Fig. 35

4. *Tension Member.* To design the plates to connect a hip vertical composed of one plate $12 \times \frac{3}{8}$ and $4Ls\ 5 \times 3\frac{1}{2} \times \frac{3}{8}$ to an 8-in. pin. The stress is 250,000 lb, the unit stress in tension 18,000 psi, in bearing on the pin 24,000 psi, and in bearing on $\frac{3}{4}$-in. rivets 30,000 psi. The member is cut off below the pin, and the plates carry the full load, as shown in Fig. 35.

Stress in each half member . . . . $250{,}000 \div 2 = 125{,}000$
Thickness of plates required for bearing . . . . . . $125{,}000 \div (8 \times 24{,}000) = \frac{11}{16}$
Net area of half section used, through rivet holes
$(\frac{1}{2} \times 4.5) + (2 \times 3.05) - (\frac{7}{8} \times 1\frac{1}{8}) = 7.37$
40 percent greater net area required through pinhole . . . . . $7.37 \times 1.40 = 10.32$
Width of plate required if $\frac{11}{16}$ thick . . . . $(10.32 \div \frac{11}{16}) + 8 = 23$

This would be too wide, and so a plate is placed on each side of the angles with a $\frac{3}{8}$-in. filler between, the inner plate being notched over the web legs. 14-in. plates projecting about 2 in. beyond the angles would have to be $10.32 \div (14 - 8) = 1\frac{3}{4}$ at the pinhole, including the fillers. Assuming that only $\frac{1}{2}$ in. can be used outside the angles because of clearance, the thickness of the inner plate must be $1\frac{3}{4} - \frac{1}{2} - \frac{3}{8} = \frac{7}{8}$ at the pinhole and $\dfrac{7.37 - [14 - (2 \times \frac{7}{8})] \times \frac{1}{2}}{14 - 1\frac{3}{4} - (2 \times \frac{7}{8})} = \frac{1}{8}$ at the connection to the angles.

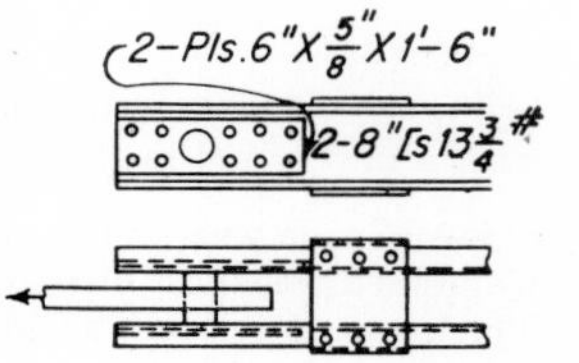

Fig. 36

The number of rivets required to connect the plates to the angles, bearing in $\frac{3}{8}$, is $7.37 \times 18{,}000 \div 8{,}440 = 16$. The number of rivets to develop the fillers, placed below the center of the pin, is $(14 - 8)\frac{3}{8} \times 18{,}000 \div 8{,}440 = 5$, two being placed above the pin to hold the

plates together. These are countersunk to clear the chord members. The net length above the pin is $7.37 \div 1\frac{3}{4} = 4\frac{1}{4}$.

5. *Tension Member.* To design the 6-in. reinforcing plates at the end of a hanger composed of two channels 8 in. 13.75 lb, from which a load of 140,000 lb is suspended by means of a 3-in. pin, as shown in Fig. 36. The unit stresses are: bearing 30,000 psi, shear on $\frac{3}{4}$-in. rivets 15,000 psi, and tension 20,000 psi.

Stress in each web and its reinforcement. . . . . $140{,}000 \div 2 = 70{,}000$
Combined thickness of web and reinforcing plate. . . . . $70{,}000 \div (3 \times 30{,}000) = 0.78$
Thickness of plate required for bearing. . . . . $0.78 - 0.30 = \frac{1}{2}$
Net area required at pin, determined by rivet holes in the flanges
$$[4.02 - (2 \times \tfrac{7}{8} \times \tfrac{3}{8})]1.4 = 4.70$$
Thickness of plate required for tension
$$\{4.70 - [4.02 - (3 \times 0.30)]\} \div (6 - 3) = \tfrac{9}{16}$$
Stress developed in bearing. . . . . $70{,}000 \times \dfrac{\frac{9}{16}}{\frac{9}{16} + 0.30} = 45{,}700$
Stress developed in tension. . . . . $(6 - 3) \times \frac{9}{16} \times 20{,}000 = 33{,}800$
Total number of $\frac{3}{4}$-in. rivets required in single shear. . . . . $45{,}700 \div 6{,}630 = 7$
Minimum number required above the pin. . . . . $33{,}800 \div 6{,}630 = 6$

When the thickness of the plate is determined by tension as here, the maximum number of rivets that may be placed above the pin is the same as the minimum number, except that the number actually developed should be expressed by a decimal; this should be subtracted from the total number in decimal form to determine the munber below the pin. This gives $6.9 - 5.1 = 2$, or one more than $7 - 6$. The net length of member below the pin should be at least $[4.02 - (2 \times \frac{7}{8} \times 0.30)] \div (0.62 + 0.30) = 3\frac{3}{4}$.

BEARING PLATES AND GRILLAGE BEAMS

Bearing Plates. Rectangular steel plates are used under the ends of wall-bearing beams to distribute the loads over sufficient areas so that the allowed bearing values of the masonry will not be exceeded. Most companies have standard sizes of plates for usual conditions, but special plates are designed for beams with relatively large reactions, for roof trusses, and for light plate girders or latticed girders. The required area of the plate is equal to the maximum reaction divided by the allowed bearing value of the masonry, which is usually from 200 psi for brick to 600 for concrete and 800 for granite. Usually the beam extends entirely across the plate; but if too long it may extend through the wall, and if too short the load may crack the wall. Common values are 8 in. for beams from 7 to 10 in. deep, 12 in. for beams from 12 to 15 in., and 16 in. for deeper beams. The dimension at right angles to the axis of the beam is taken to the nearest inch to give the necessary bearing. The bearing plates of light trusses and girders are usually bolted to the masonry either through the angles or through the plates beyond the angles, room being left to turn the nuts; this may require the use of longer plates.

The thickness of the plate depends on the projection p beyond the edge of the superimposed metal and should be sufficient to develop the bearing value of the masonry at every point. The maximum bending moment per inch along the edge of the superimposed metal due to the uniform pressure b on the underside is $pb \times p/2$, which is resisted by a beam of rectangular cross section 1 in. wide and t in. thick, the resisting moment of which is $\frac{1}{6} ft^2$. If these are equated, the usual expression for the thickness is found to be $t = p\sqrt{3b/f}$. When the load is applied to the plate by means of angles, as in the case of a roof truss or light girder, the combined thickness of the plate and the angle must be sufficient to prevent bending at the edges of the vertical legs of the angles; this combined thickness is found from the same expression by taking the projection p from the face of the vertical leg, and the plate thickness is found by subtracting the thickness of the angle, the plate being made thick enough to meet both conditions. The

thickness is usually taken to the nearest ⅛ in. At one end of a light girder or truss, two plates may be used instead of one, the surfaces of contact being planed to allow for expansion, slotted holes being used in the upper plate; each plate need be only $\frac{7}{10}$ as thick as a single plate.

Illustrative Problem. To design a bearing plate for a roof truss which rests upon a brick wall, as shown in Fig. 37. The bottom-chord angles are $5 \times 3\frac{1}{2} \times \frac{3}{8}$, and they are separated by a ⅜ gusset plate. The maximum reaction is 22,000 lb and the allowed unit stresses are 200 psi bearing and 18,000 psi bending. The bearing area required is $22{,}000/200 = 110$. A 10×11 plate would suffice; but if ¾-in. anchor bolts are used, the holes would be punched about 2 in. from the edges of the angles to allow for the turning of the nuts, and the edge distance would be 1½ in., making a minimum length of $\frac{3}{8} + 2(3\frac{1}{2} + 2 + 1\frac{1}{2}) = 14\frac{1}{2}$. It would be better to increase the area rather than reduce the 10-in. bearing of the truss, and a corresponding reduction in the developed bearing b may be made to $22{,}000 \div (10 \times 14.5) = 152$. The projection p beyond the angles is $\frac{1}{2}(14\frac{1}{2} - \frac{3}{8}) - 3\frac{1}{2} = 3.56$, and beyond the vertical legs is $\frac{1}{2}(14\frac{1}{2} - \frac{3}{8}) - \frac{3}{8} = 6.68$. The thickness of the plate at the edge of the angle is $3.56\sqrt{3 \times 152/18{,}000} = \frac{5}{8}$, and at the vertical leg is $6.68\sqrt{3 \times 152/18{,}000} - \frac{3}{8} = \frac{3}{4}$; the size to be used is $10 \times \frac{3}{4} \times 1'2\frac{1}{2}''$.

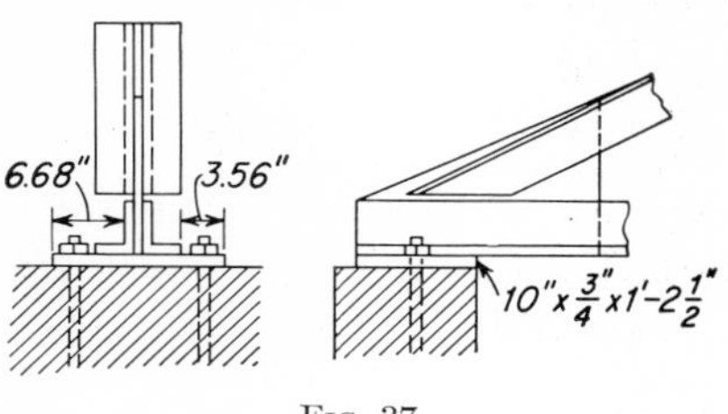

FIG. 37

Pedestals. Bridge girders are commonly supported by cast-iron or cast-steel pedestals, as shown in Fig. 21. These pedestals are fastened to the masonry by anchor bolts, and the girders are bolted to them. Slotted holes are provided in one end of girders up to about 60-ft span to allow for expansion, the tops of the pedestals and the bottoms of the bearing plates on the girders being planed. Segmental rollers are placed under the pedestals at one end for greater spans. Special hinged shoes or rockers are used at the ends of some of the longer girders to prevent unequal distribution of the load upon the masonry as the girder deflects.

Shoes. Bridge trusses are usually supported at the ends by pins, even though the rest of the truss is riveted or welded, in order to keep the reactions vertical regardless of the amount of deflection. The pins are supported by cast-steel shoes or by shoes built of plates and angles, those at one end resting upon roller nests to provide for expansion. The webs of the shoes must be thick enough to furnish proper bearing on the pins, and they must be stiffened by angles to prevent buckling. Angles are used to transmit the stresses from the webs to the bearing plates of the shoes, rivets or welding being proportioned accordingly. The end floor beams are usually connected to the shoes, particularly at the expansion ends, so that the bridge will expand and contract as a unit, which is not so likely when the end stringers have separate bearings on the piers or abutments.

Column Bases. Column bases are of two principal types. In mill buildings, the column loads are comparatively light, and the bases must be anchored to the masonry piers to prevent displacement by accident and to resist the overturning effect of the wind. Cross bracing can be used only at the ends of mill buildings, and the wind pressure on the sides must be resisted largely by columns acting as beams fixed at the lower ends. The effect of this, combined with the effect of eccentric loading on the columns, is to cause unequal distribution of the load on the masonry and may even cause uplift on one side so that anchor bolts are required. In order to make the anchors effective, the base plates must be securely riveted to the columns, so that the rivets carry much

or all of the loads. Columns which support crane runways are also milled to bear on the base plates. This type of base is discussed more fully under "Industrial Buildings" (p. 6-72). In office buildings, the stresses are relatively large; and, except for tall narrow buildings or towers, the dead loads exceed the uplift due to wind, so that no anchors are required. From the nature of the building, there is slight chance of displacement of the column bases by accident, and the column load transmitted by direct bearing is usually sufficient to hold the column in place without anchoring. Rolled-steel slabs are most used for this type of base, although cast-steel bases are still used by some engineers. The method of design for a slab is similar to that for a bearing plate, only the slab projects on all four sides and the thickness is determined by the maximum projection. The bottom face of the column is milled, and the corresponding surface of the slab is planed to ensure uniform bearing. The load from the column is assumed to have a uniform intensity of bearing over a rectangle the dimensions of which are 0.95 of the column depth and 0.8 of the column flange width. It is usually desirable to proportion the slab so that the projections are approximately equal. Slab widths, chiefly in multiples of 4 in., and slab thicknesses in inches or half inches are standardized as shown in the tables. The lengths may be flame cut to dimension.

Illustrative Problem. To design a steel slab for a 14-95 column supporting a load of 450,000 lb, with an allowed bearing value of 560 psi and fiber stress of 20,000 psi.

$$\text{Bearing area} = \frac{450{,}000}{560} = 800$$

$$\text{Size of square base} = \sqrt{800} = 28.3 \text{ in.}$$

The nearest standard width is 28 in., and the corresponding length is $^{800}/_{28} = 29$. The depth of column is $14\frac{1}{8}$ and the flange width $14\frac{1}{2}$, the longest projection being $[29 - (0.8 \times 14.5)] \div 2 = 8.7$. The thickness is $8.7\sqrt{3 \times 560/20{,}000} = 2.5$, but the nearest standard thickness for a 28-in. slab is 3 in.; so a $28 \times 3 \times 2'5''$ is used.

Grillage Beams. Steel grillage beams may be used to distribute a column load over the proper area. The purpose is primarily to distribute heavy loads to rock, but they may be used to distribute smaller loads to soil, although reinforced concrete footings are usually cheaper. A layer of concrete is laid below the beams to provide a smooth level bearing at the proper elevation. When this concrete mat is on soil, the dimensions are determined by the allowed bearing value of the soil; it is made 1 ft thick and this is sufficient to distribute some of the pressure so that the lower tier of beams need extend only within 6 in. of the edge of the mat. In a rock-bearing grillage, however, the dimensions of the lower tier of beams are determined by the allowed bearing on the concrete leveling mat, which is made to extend from 4 to 6 in. beyond the beams.

A footing is approximately square, unless a rectangle is required because of space limitations such as property lines. Grillage beams are usually in one or two tiers, those in the second tier being at right angles to those in the first. The beams are kept properly spaced by rods with gas-pipe separators. The spaces between them are filled with concrete; then the beams and the slab are encased in concrete at least 4 in. thick to hold the parts in position and to protect the steel from fire and corrosion. The minimum clear distance between the flanges of the upper tier is 1 in. and in the lower tier 2 in., the maximum in the lower tier being $1\frac{1}{4}$ or $\frac{3}{4}$ times the flange width. When a slab is proportioned from the bearing value of concrete, it may be grouted in position, even though grillage beams are used. When grillage beams are used to reduce the size of the slab or base plate, the beams must be accurately set, and the slab or plate must be planed if necessary to ensure contact of the steel surfaces. One dimension of the slab must be

at least 30 percent of the length of the upper-tier beams, and the other dimensions must extend at least ¾ in. beyond the center of each outer beam. Similarly, the upper-tier beams must extend at least ¾ in. past the center of each outer beam of the lower tier; it may be assumed that the upper beams will extend within 1 in. of each outer edge of the lower tier, but the lengths may have to be revised after the lower beams are designed. I beams are used rather than wide-flanged beams because of the narrower flanges and thicker webs.

Design of Grillage Beams. The upper-tier beams are first designed for bending, and then their resistance to buckling and shear is investigated; the lower-tier beams are usually determined by bending requirements. The beams of each tier support a total column load P which is distributed uniformly throughout a distance L', equal to one dimension of the slab or to the extreme width of a superimposed tier of beams; the beams are supported by an equal force uniformly distributed throughout the entire length L. The maximum bending moment is at the center. The downward forces acting at the left of the center may be replaced by a single resultant force $P/2$ acting a distance $L'/4$ from the center, and similarly the resultant $P/2$ of the upward forces acts at a distance $L/4$ from the center. An expression for the bending moment of all the beams in the tier is found from the moments of these resultant forces; thus, $\left(\frac{P}{2} \times \frac{L}{4}\right) - \left(\frac{P}{2} \times \frac{L'}{4}\right)$ or $P(L - L') \div 8$. The combined section modulus of all the beams is found from this total bending moment by dividing by the unit stress allowed in bending, and this should be equaled or exceeded by the product of the number of beams used and the section modulus of a single beam. Larger beams are required in the upper tier than in the lower, because there cannot be so many placed in the available space.

The webs have a tendency to buckle at the junction of the web and the curved fillet of the top flange. The effective area is taken as the product of the number of beams, the web thickness, and the effective length, which is the slab dimension L' plus twice the vertical distance from the flange to the junction of the web and the fillet. The total column load divided by this area must not exceed the value specified for web crippling.

The resistance of each beam to shear is found by multiplying the unit stress specified by the area of the web, i.e., the clear depth between the curved fillets of the flanges times the thickness. The resistance must equal or exceed the shear found by multiplying the distance in inches that the beam projects beyond the superimposed load by the upward force per linear inch of beam. This upward force is found by dividing the total column load by the number of the beams and by the length of each beam in the tier.

Illustrative Problem. To design grillage beams to distribute the column load of 450,000 lb on the 28-in. by 29-in. slab designed above, over a soil supporting 6,500 lb per sq ft, as shown in Fig. 38. Use unit stresses as indicated. The approximate size of square concrete mat is $\sqrt{450{,}000/6{,}500} = 8.3$ ft. Assume that the weight of the beams, slab, and concrete encasing them is approximately equivalent to a 2½-ft block of concrete 8½ ft square. This is less than the total depth of mat, two tiers of beams, slab, and a rough approximation is as close as necessary for an error in weight has relatively small effect.

Assumed weight. . . . .$8.5^2 \times 2.5 \times 150$ = 27,000 lb
Revised size of mat. . . . . $\sqrt{477{,}000/6{,}500} = 8.5$ ft = 102 in.
Length and width of the lower tier of beams. . . . . 102 − (2 × 6) = 90 in.
Section modulus of all beams in upper tier
450,000(90 − 2 − 29) ÷ (8 × 20,000) = 166
Section modulus of 5-12 *Is* 31.8. . . . . = 180

Clear space between beam flanges........[$28 - (2 \times \frac{3}{4}) - (4 \times 5)] \div 4 =$ 1.6

Webs resist buckling because......$450{,}000 \div 5 \times 0.35[29 + (2 \times 1\frac{1}{8})] =$ 8,230
which is less than 24,000 allowed

Shear on each beam........................$\dfrac{450{,}000}{5(90 - 2)} \times \dfrac{90 - 2 - 29}{2} = 30{,}200$ lb

which is less than..........................$13{,}000 \times 9.75 \times 0.35 = 44{,}400$

L' of the lower tier = extreme width of upper tier....$28 - (2 \times \frac{3}{4}) + 5 = 31.5$ in.

Section modulus of all the beams in the lower tier
$450{,}000(90 - 31.5) \div (8 \times 20{,}000) =$ 165

Section modulus of 16-7 *I*s 15.3.......................................= 166

Clear space between beams..................[$90 - (16 \times 3.66)] \div 15 =$ 2.1

The maximum distance from the end of the upper beam to the edge of the lower beam is $\dfrac{3.66}{2} - 0.75 = 1.08$; this is so close to the 1 in. assumed that no revision is necessary.

The buckling is $450{,}000 \div 16 \times 0.25[31.5 + (2 \times 1\frac{3}{16})] = 3{,}400$ and the shear on each beam is $\dfrac{450{,}000}{16 \times 90} \times \dfrac{90 - 31.5}{2} = 9{,}140$, which is less than $13{,}000 \times 5.37 \times 0.25 = 17{,}450$. Usually if the webs meet the buckling and shear requirements in the upper tier they need not be investigated in the lower tier.

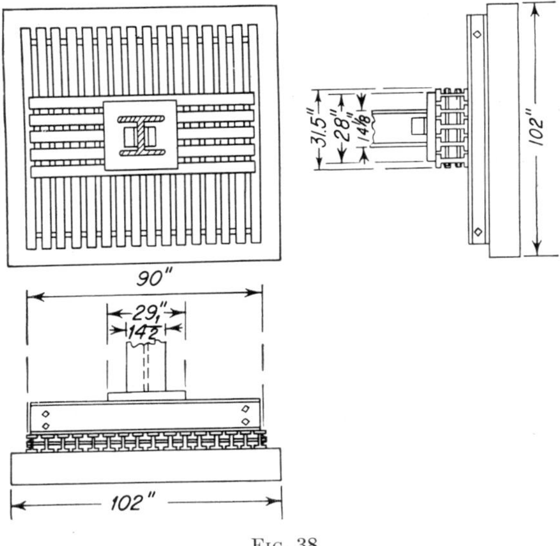

FIG. 38

BRIDGES

The various component parts of a bridge, the types of truss more commonly used, the loads sustained, and the methods of computing stresses are explained in Section 5. In this section, the design of the various parts of a bridge will be illustrated with as little stress computation included as possible. In order to show the interrelation of the component parts, an illustrative problem will be carried progressively along with the text.

Stringers. The stringers of a through railroad bridge are designed to support the

locomotive loading with its impact allowance, the track, the stringers themselves, and often ballast with its supporting trough. The loads per track are divided equally among the stringers, unless the track is curved. Two rolled beams or two stringers composed of web plates and angles are commonly used to support the ties of open-floor construction. The depth is approximately one-seventh of the span, because, if not relatively deep, cover plates may be required and it may be difficult to provide sufficient rivets in the flanges and in the connections to the floor beams. The stringers of a highway bridge are designed to support the conventional concentrated loads specified, together with the pavement, the slabs, or other supporting medium, the stringers themselves, and any spiking pieces or any protective covering which may be used. The distribution of the wheel loads to the stringers depends on the stringer spacing and the type of flooring, as specified. Highway stringers may be of reinforced concrete or wood, but most often in a steel bridge they are rolled beams. Wooden stringers are also called "joists." The minimum depth of highway stringers should be as specified, from one twenty-fifth to one-fifteenth of the span.

Floor Beams. The effective length of a floor beam is the distance from center to center of girders or trusses. At each stringer connection is a concentrated load which comes from the stringers in adjacent panels, including the dead load from the stringer and the track or pavement which it supports and the corresponding live load and impact. The maximum bending moment due to these concentrated loads occurs at the stringer point nearest the center. The bending moment due to dead loads concentrated at the stringer points of a highway bridge may often be found more conveniently by means of an equivalent uniform load, provided that the bending moment is taken at the stringer point nearest the center, rather than at the center. Similarly, bending moments may be found for uniformly distributed live loads. The bending moment due to the weight of the floor beam itself is maximum at the center. The live loads on adjacent stringers are placed to cause the maximum concentrations. The maximum concentration or "floor-beam reactions" in a railroad bridge are found from a special position of the locomotive as explained in Section 5. Rolled beams or built girders are used for the floor beams of highway bridges, with the stringers connected to the webs or to the top flanges; built girders are used for railroad bridges.

Girders. The tracks of deck railroad bridges rest directly upon the girders, and the position of the live loads that cause maximum bending moment is the same as if the loads rolled across the tops of the girders. In a highway girder bridge or a through railroad girder bridge, the concentrated loads are placed to give the maximum bending moment at the floor beam nearest the center, and for this position the different corresponding concentrations at the other floor beams are computed. An odd number of panels is preferred so that the maximum bending moment will be less. The maximum for concentrated loads, both live and dead, will occur at the floor beam nearest the center; the maximum total bending moment, including that due to the weight of the girder is usually at this same point, but it may be nearer the center at the point of zero shear. (For the common types and proportions of girders, see p. 6–63). Heavier girders have vertical flange plates between the webs and angles, and sometimes four angles are used in each flange.

Truss Bridge. The design of typical members of the highway truss bridge shown in Fig. 39 will serve to illustrate typical bridge design. The Specifications for Steel Highway Bridges recommended by the AASHO, 1953, will be used with H20 loading.

Design of Stringers. It is assumed that a 2-in. asphalt pavement is carried by an 8-in. reinforced concrete slab crowned to 10 in. at the center. The maximum bending moment in the stringer occurs at the center when the rear wheel of a 20-ton truck is placed there. The wheel load is $40{,}000 \times 0.4 = 16{,}000$, but the load per stringer is increased by dividing by 5 ft and multiplying by the distance between stringers. The

impact percentage of the live load is $\frac{50}{L + 125} = 35$, where L is the stringer length 16 ft but the maximum value of 30 percent is used. The bending moment due to concentrated load and impact $= \frac{16{,}000}{2} \times \frac{5.5}{5} \times 8 \times 1.3 = 91{,}500$ ft-lb. Each stringer carries also a dead load of 5.5 ft of asphalt and concrete, equivalent to 1 ft of concrete,

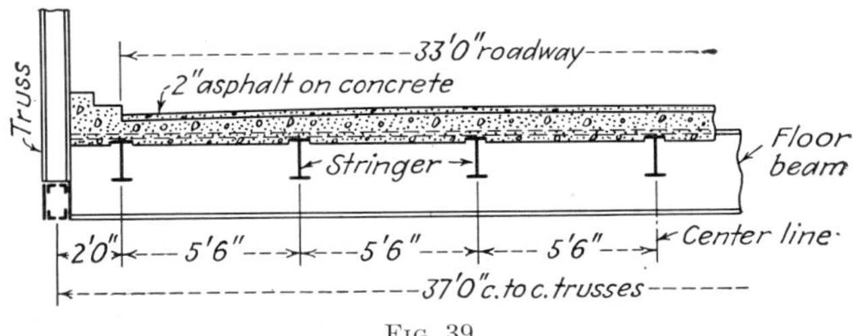

Fig. 39

plus the weight of the stringer, assumed to be 50 lb per ft. The bending moment due to dead load $= \frac{(150 \times 5.5) + 50}{2} \times 8^2 = 28{,}000$ ft-lb. The section modulus required $=$ $\frac{12(91{,}500 + 28{,}000)}{18{,}000} = 80$, which requires a 16-50 wide-flanged beam.

Design of Floor Beams. The maximum bending moment due to live load is at the stringer nearest the center due to a uniform load of 640 lb per ft of 10-ft lane plus a concentrated load of 18,000 lb per 10-ft lane. With a stringer at the center and three lanes symmetrically loaded as shown in Fig. 40, the total live load is $[(640 \times 16) + 18{,}000]3 = 84{,}700$ and the bending moment $= 42{,}350(18.5 - 7.5) = 465{,}900$ ft-lb. The impact allowance is found for an L of 37 which gives more than the 30 percent maximum. Only 90 percent of the live load is used according to the specifications because it is improbable that all three lanes will be loaded at once. The bending moment due to dead load can be taken as uniformly loaded from center to center of trusses because the weight of the curb and connection angles makes up for the fact that the slab and floor beam extend only to the face of the truss and any difference is so near the reaction that it has little effect; the bending moment is equivalent only at points of concentration at the stringer points. The weights of the stringers may be taken as equivalent uniform loads by dividing by the distance between stringers.

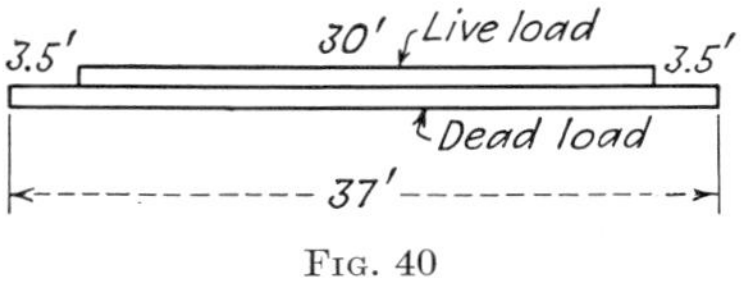

Fig. 40

The dead load per foot $= \left(150 + \frac{50}{5.5}\right) 16 + 200 = 2{,}750$, and the bending moment $\frac{2{,}750}{2} \times 18.5^2 = 470{,}600$.

The total bending moment is $(465{,}900 \times 0.9 \times 1.3) + 470{,}600 = 1{,}016{,}000$ ft-lb. The section modulus $= 12 \times 1{,}016{,}000 \div 18{,}000 = 677$, which requires a 33-220. This weighs more than the 200 assumed but is strong enough to carry the extra load. The compression flange is laterally supported by the concrete slab.

Loads on the Trusses. The maximum live loads on one truss are obtained when the three lane loads are adjacent to the curb on that side, as shown in Fig. 41. The live load per panel due to the uniform load is $640 \times 16 \times 3 \times {}^{20}\!/_{37} \times 0.9 = 15{,}000$. The panel

load due to the concentrated live load is $18{,}000 \times 3 \times {}^{20}\!/_{37} \times 0.9 = 26{,}200$ for moment and $26{,}000 \times 3 \times {}^{20}\!/_{37} \times 0.9 = 38{,}000$ for shear. The impact for the hip vertical $L_1 - U_1$ is 30 percent as for the floor beam, but for all other members is 21 percent found from the formula for $L = 112$.

The dead load from each floor beam is $2{,}770 \times {}^{37}\!/_{2} = 51{,}200$. The weight of each truss, including bracing, per panel, is approximately equal to the total length in feet times the panel load plus one-ninth the total load from each floor beam, or $112 \times 16 + \{51{,}200 + [15{,}000 + (26{,}200 \div 7)]1.21\} \div 9 = 10{,}000$. This makes the total dead load per panel 61,200. The stresses are found by any of the methods described in Section 5; the total stresses in kips, including dead load, live load, and impact, are shown in Fig. 38. The stresses due to lateral forces need not be considered in the design of the chord members, because they are less than 25 percent of the chord stresses due to vertical loads and hence would not increase the sections because of the 25 percent increase in unit stress allowed for combined stresses.

FIG. 41

Interrelation of Members. The gusset plates of a riveted or welded truss are usually shipped with the chord members or end post, and the transverse dimensions of the web members must be such that they can be slipped between or outside the plates with suitable clearance. This necessitates an agreement between the top- and bottom-chord members. The floor beams connect to the gusset plates, and vertical members must be made to connect to the floor beams or provide independent connections. The hip-vertical tension hanger should provide for the same floor-beam connection as the other verticals, although this does not necessitate the same type of member. In order to keep the floor-beam connections alike, some verticals may be made larger than required for the stresses. Pin-connected trusses must have top chords large enough to house properly the upper pins and eyebars, and adjacent joints must be arranged so that the bars are parallel. Either the floor beams of pin-connected bridges must connect to the posts above the eyebars, or else the bottoms of the floor beams must be cut away to clear the bars.

Web Members. It is well to design the verticals and perhaps the other web members first, because the width of truss is thus determined. Common sections are shown in Fig. 42. Types *a*, *b*, and *c* are used for tension, for compression, or for members in which there is a reversal of stress. Type *d* may be used as a vertical post with the floor beam connecting to the flanges or to the web; if connected to the web, a diaphragm must be inserted to transmit some of the stress to the other channel. Type *e* may serve as a

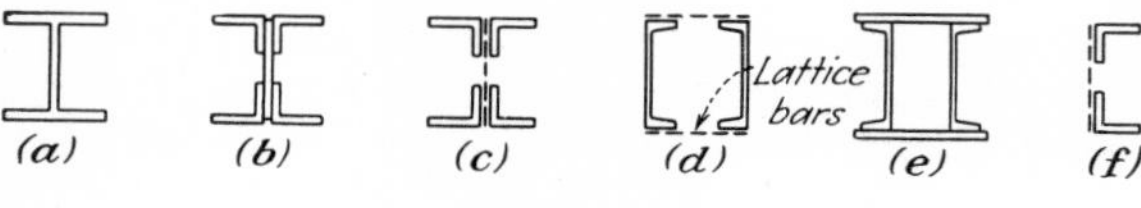

FIG. 42

diagonal compression member or as a vertical post; the floor beam can connect to the flanges or to the cover plate between the channels. Type *f* is used for light diagonals with the angles either turned as shown to go between gusset plates or turned the other way to go outside the plates; counters are often turned one way and the main diagonals the other way so that they can pass at their intersection without being spliced. Lattice bars are often used on compression members, and either lattice bars or tie plates on tension members. There is not much economy in using lattice bars in the verticals, because solid plates must be used for the full depth of the floor-beam connection and the

depth of the bracing bracket at the top; very little metal can be saved in between. The ratio l/r should not exceed 120 for compression or 200 for tension. In pin-connected trusses, eyebars are used for the tension diagonals; adjustable eyebars or loop rods are used for counters.

The verticals have compressive stresses of 113,000 and 35,000 and a tensile stress of 125,000 respectively. For one post, some metal might be saved by using four angles or two channels, but a wide-flanged beam section requires no rivets or welds except at the ends, and a little excess is justified in one member to save shop work on many members. An 8-35 is the smallest with a 5⁄16 minimum web, and it has a 5½-in. gage which simplifies the floor-beam connection. The area is 10.30, and the least radius of gyration 2.03. The safe load at $15{,}000 - \frac{l^2}{4r^2}$ is $\left(15{,}000 - \frac{(21 \times 12)^2}{4 \times 2.03^2}\right) 10.3 = 118{,}400$, which is sufficient for both L_2U_2 and L_3U_3. It is also sufficient for the hanger L_1U_1 and the diagonal L_3U_2, for the safe load in tension, 4 holes being deducted for ¾-in. rivets, is $[10.30 - (4 \times 7/8 \times 0.49)]18{,}000 = 155{,}000$. An 8-58 must be used for the 244,000 stress of diagonal L_2U_1. Two angles $2\frac{1}{2} \times 2\frac{1}{2} \times \frac{3}{8}$ may be used for the central diagonals, with a net area of at least $45{,}000/18{,}000 = 2.5$. The angles of one diagonal may be placed inside the plates, and those of the other outside to avoid cutting.

Compression Chords. The usual types of top-chord and end-post sections are shown in Fig. 43g and h; type i is used for some light highway half-through bridges which are not deep enough to permit the use of overhead bracing. Lattice bars are used across the

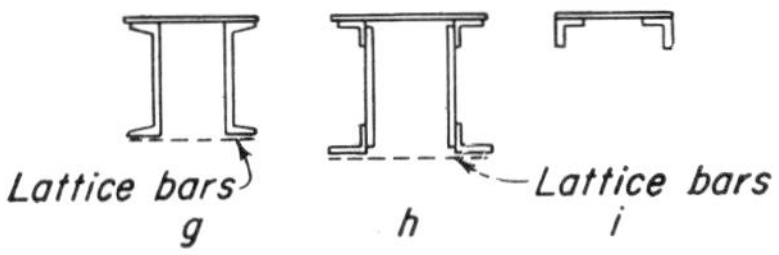

Lattice bars
g
h
Lattice bars
i

FIG. 43

j
k
Tie
l
Plates
m

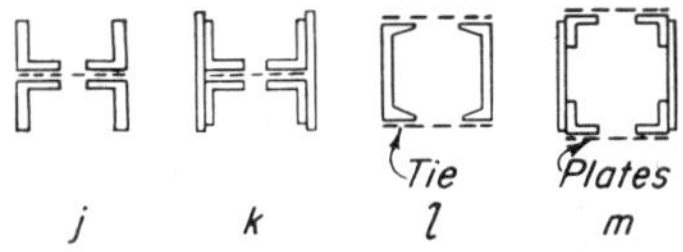

FIG. 44

bottoms, because solid plates could not extend full length without interfering with the gusset plates or diagonals. It is desirable to keep the radii of gyration about the two axes nearly equal, but this is not always feasible. The clear distance between webs must provide for the gusset plates and web members with about 1⁄16-in. clearance each side. The depth of a pin-connected member must provide room for the heads of the diagonal eyebars without interference with the cover plates; the pin should be placed at the center of gravity of the section, but in a small member it may have to be placed at or near the center of the web to make room for the eyebars, even though secondary stresses are caused by such eccentricity. The amount of eccentricity may be minimized by making the cover plate relatively thin or by using light equal-legged angles at the top and heavier unequal-legged angles at the bottom. The minimum thickness of metal, including channel webs, is usually ⅜ for railroad bridges and either 5⁄16 or ⅜ for highway bridges; a cover plate should not be thinner than 1⁄40 the distance between rivet lines unless the extra width is neglected in computing the effective area of cross section. A uniform depth and width should be maintained in all top-chord and end-post sections, the difference in stress being provided for by changes in thickness.

Using a member of type g, the channels must be 9⅝ in. back to back to allow for the largest web member 8-58, which is 8¾ deep, ⅜-in. gusset plates, and for 1⁄16-in. clearance on each side. Even with 12-in. channels the least radius of gyration is about a horizontal axis, so it is economical to use them in preference to 10-in. channels; 15-in. channels may be used if the 12 is not large enough. The lightest 12 with a 5⁄16 minimum web is 12-25, and the lightest cover plate which will cover the flanges and have a

thickness of $\frac{1}{40}$ the distance between rivet lines is $16 \times \frac{3}{8}$. The safe load of this section is found as follows:

Area	$2 \times 7.32 + 6.00 =$	20.64
Eccentricity	$6.00 \times 6.19 \div 20.64 =$	1.80
I about horizontal axis through center of web	$(2 \times 143.5) + (6.00 \times 6.19^2) =$	517
Transfer	$20.64 \times 1.80^2 =$	67
I about axis through center of gravity	$=$	450
r^2	$450/20.64 =$	21.8
Allowed unit stress	$15{,}000 - \dfrac{(16 \times 12)^2}{4 \times 21.8} =$	14,580
Safe load of section assumed	$14{,}580 \times 20.64 =$	301,000

This is less than 336,000 required in $U_{1\text{-}2}$ and the area must be increased by $\dfrac{336{,}000 - 301{,}000}{14{,}580} = 2.4$, changing the channels to 12-30. The lattice bars should be designed as explained on p. 6–18 for one-half of 2 percent of the stress 336,000 or 3,360 lb. The transverse distance between rivet lines is $9\frac{5}{8} + 2 \times 1\frac{3}{4} = 13\frac{1}{8}$ in. and the length if placed at 60° with the axis is 15.2 in. The width for $\frac{3}{4}$-in. rivets is $2\frac{1}{4}$ in. and the minimum thickness $15.2/40 = \frac{3}{8}$ in. The strength of a $\frac{3}{8}$-in. lattice bar in tension is $(2\frac{1}{4} - \frac{7}{8})\frac{3}{8} \times 18{,}000 = 9{,}280$ lb and in compression is $15{,}000 - \dfrac{15.2^2}{4 \times \dfrac{\frac{3}{8}^2}{12}} 2\frac{1}{4} \times \frac{3}{8}$ $= 8{,}500$ lb, both of which exceed the 3,360. Similarly, the other top-chord members require two 12-30 channels, and a $16 \times \frac{5}{8}$ plate. The end post with a length of 26.4 has a unit stress of 13,850 and a stress of 349,000, which requires two 12-30 and $16 \times \frac{1}{2}$. In some bridges, the bending due to portal action must be considered, even though an increase is allowed in the unit stress.

Tension Chords. Common types of riveted or welded bottom-chord section are shown in Fig. 44. The members are placed inside the gusset plates to allow for better floor-beam connections. Tie plates or lattice bars are used in horizontal planes instead of plates, in order to provide better drainage. In pin-connected trusses, eyebars are used for the bottom-chord members, except in the two panels at each end of a single-track railroad bridge, where stiff members may be needed to provide greater resistance to overturning.

A $3\frac{1}{2}$-in. leg is the largest that can be used horizontally in type j, k, or m, since 8-in. web members are used. Four angles are not large enough for the central member; so type k or m is used. Rivets in tie plates need not be placed opposite those in the vertical legs, so only one hole for a $\frac{3}{4}$ rivet need be deducted from each angle. Chord members are usually spliced independently of the gusset plates as close to them as feasible and on the side of the smaller stress. The member $L_{3\text{-}3'}$ in the central panel must be designed for a stress of 404,000, but it would extend past the gusset plates into the adjacent panels where it need be spliced for only 336,000. For a net area of $404{,}000 \div 18{,}000 = 22.4$, four angles $4 \times 3\frac{1}{2} \times \frac{5}{8}$ and two $10 \times \frac{1}{2}$ plates may be used. Similarly, four angles $4 \times 3\frac{1}{2} \times \frac{1}{2}$ and two plates $10 \times \frac{7}{16}$ may be used for $L_{2\text{-}3}$, and four angles $4 \times 3\frac{1}{2} \times \frac{1}{2}$ without plates for $L_{0\text{-}2}$.

Lateral Bracing. The bracing between trusses stiffens the bridge as a whole and resists forces which tend to distort the bridge laterally such as wind and the swaying of the locomotives. Modern practice calls for heavier bracing than is usually required by the stresses, and the bracing is more or less standardized by each company so that it

is unnecessary to calculate the stresses in the ordinary bridge.[1] Cross bracing in the plane of the bottom chords is connected by plates to the floor beams and to the girders or trusses; connections are made also to the bottoms of railroad-bridge stringers to relieve the floor beams of traction stresses which might buckle them. Single angles are used for the bottom laterals in girder bridges and the lighter truss bridges, but double angles in the heavier bridgework, $5 \times 3\frac{1}{2} \times \frac{3}{8}$ being a common size. The diagonals of top-chord cross bracing are made of latticed members of suitable depth to be connected to the tops and bottoms of the chord sections, two angles about $3\frac{1}{2} \times 2\frac{1}{2} \times \frac{5}{16}$ being used for the average highway bridge and four angles $3\frac{1}{2} \times 3\frac{1}{2} \times \frac{3}{8}$ for the average railroad bridge. Vertical transverse sway bracing is used at each panel point of a through bridge between the two trusses; this is at least twice as deep as the chord members and as much deeper as the clearance allows. There are usually two or four angles at the top and two angles at the bottom of each sway strut, with diagonal angles between them. Portal struts are used at the ends of through bridges to transmit top lateral stresses to the abutments through the end posts acting as girders; each strut has two angles at the top and two at the bottom, with either a solid web or double-angle diagonals between them, and knee braces carrying the stresses as far down the end posts as feasible. The portal should be made as large as the clearance permits, and the depth of the truss should be sufficient to accommodate a generous portal strut. In deck bridges, vertical cross bracing between trusses is used at each panel, reducing the sizes of the bottom laterals and making sway braces and portal braces unnecessary.

Camber. Bridge trusses are built with a slight camber or vertical curve so that the track or floor will assume a horizontal position under the full load. This is usually effected by making each section of the top chord slightly longer than the calculated length, approximately $\frac{1}{8}$ in. for each 10 ft. A highway bridge with inclined approaches is built with considerably more camber, in order to provide a smooth vertical curve in the roadway and also the necessary underclearance at the center with minimum heights of abutments and fills.

MULTI-STORY BUILDINGS

Method of Design. The steel skeleton frame of an office-building type of structure is designed for the floor loads, the wind forces and the dead loads of the entire building including the panel or curtain walls. The wind forces are unimportant in buildings of moderate height with relatively large bases, but they become increasingly important as the ratio of height to width increases. Ordinary office buildings, and loft buildings for light manufacturing, are designed by one of the approximate "portal" methods or "cantilever" methods; modern tower buildings are designed by more precise methods, which are beyond the scope of this book. The approximate portal method outlined here is probably used as much as any other. In this method, it is assumed that the shear in any story is resisted equally by the interior columns of a bent and one-half as much by each exterior column. This is much simpler than to proportion the shear in accordance with the moments of inertia of the column sections, and the method has been used with satisfaction in the design of many buildings for conditions which are not extreme.

Loads. The floor and roof loads for which a building is designed must be taken for the proper class of structure from the building code of the city where the building is to be located.[2]

Except in buildings for storage purposes, the following reductions in assumed total floor live loads are permissible in designing all columns, piers or walls, foundations, trusses, and girders.

[1] For the method of calculation, see Section 5.
[2] For a complete tabulation of dead and live loads see pp. 5–9 to 5–11 of Section 5.

Reduction of Total Live Loads

Carrying	Percent
One floor	0
Two floors	10
Three floors	20
Four floors	30
Five floors	40
Six floors	45
Seven or more floors	50

For determining the area of footings, the full dead loads plus the live loads, with reductions figured as permitted above, shall be taken; except that in buildings for human occupancy, a further reduction of one-half the live load as permitted above may be used.

The floor slabs are usually of reinforced concrete or a combination of reinforced concrete and hollow cores of a lighter material such as gypsum or terra cotta. The fireproofing around steel beams and columns may be of hollow tile, gypsum, or concrete, a minimum thickness of 2 in. of concrete or 4 in. of tile being used. The exterior walls are carried by the beams at every floor or at alternate floors. Light partitions are placed after the floor slabs and are often moved to suit new tenants; the slabs and beams may be designed to support a single partition in any position, or an equivalent load of about 20 lb per sq ft of floor may be added; light fixed partitions of gypsum and plaster weigh about 18 lb per sq ft of vertical projection and tile about 25 lb. Special loads for corridors, stairs, elevators, etc., must be provided for. The wind pressure is taken from the building code or specifications; it is usually either a fixed amount of 15, 20, or 30 lb per sq ft or else a graduated amount depending upon the height above ground. The steel frame should be braced to resist the entire wind, neglecting the stiffness of the walls and partitions. It is not expected that the maximum wind pressure and the maximum gravity loads will occur simultaneously, and so it is customary to increase the allowed unit stresses by 33⅓ percent or other percentage when designing for wind alone or for wind and gravity stresses combined. Similarly the live loads on office-building columns are discounted, for it is not likely that all floors will have the maximum loads at the same time.

Wind Stresses. Each transverse bent of columns and beams may resist its share of the wind on the side of the building, or the entire wind pressure may be resisted by relatively few braced bents. The end bents are braced bents, because the brackets can be hidden in the walls; similarly, braced bents are placed where permanent partitions can screen the brackets. Intermediate bents, with ordinary top and bottom angles or similar connections of beams to columns, may not be counted on to carry wind stresses unless the proportions of the building are such that such connections are sufficient without special wind bracing. Architectural considerations, such as offset or omitted columns, also have their effect upon the location of bracing systems. When the braced bents have been located, a large sectional view should be drawn showing the beams and columns of each different bent, upon which wind stresses can be recorded as determined; it is usually simpler to compute and tabulate all wind stresses first and refer back to them as needed in the design. At every floor and roof line except those below the ground, indicate a concentrated wind load based upon the pressure tributary to that panel point; the vertical distance is from the mid-point of the story below to that above, except that no wind blows below the ground level, and above the roof there is a 4-ft parapet or fire wall; the horizontal distance is that portion of the total length which is to be resisted by the bent in question. Points of contraflexure are assumed to be midway between floors and midway between columns. The bent can be cut at any points of

contraflexure, provided that the effect of the remaining parts at each cut is compensated for by horizontal and vertical components external to the portion of the bent cut out, and the equations of equilibrium are applied to such components and to any wind forces acting on the portion considered.

Data for Typical Design. To design the steel skeleton frame of the office building shown in Fig. 45, on the assumption that architectural limitations require that brackets for wind bracing can be used only along the end bents and two intermediate bents, as indicated in the plan by the heavier beam lines, the intermediate bents taking twice as much as the end bents.

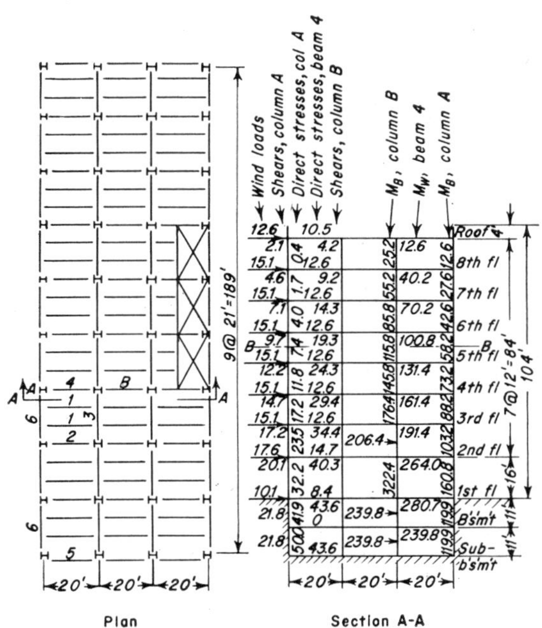

FIG. 45

LOADS

	Lb per sq ft
Live load on roof	25
Live load on intermediate floors	50
Live load on first and basement floors	100
Wind load	20
Weight of tar and gravel roofing	10
Weight of suspended ceiling under roof	10
Weight of ceiling plaster under each floor	5
Weight of wooden floors (intermediate)	5
Weight of tile floor (first floor only)	12
Weight of concrete slabs and fireproofing	150
Weight of cinder concrete fill (per cubic foot)	100
Weight of brick curtain walls (per cubic foot)	120
Weight of tile fireproofing around columns (per cubic foot)	70
Weight of fixed partitions (vertical projection)	25
Equivalent floor load to provide for movable partitions	20
Weight of windows and metal sash	8

Use specifications of the American Institute of Steel Construction, 1946.

For the braced bent shown in section AA, the external forces are found as follows and recorded on the sketch in thousands of pounds (kips):

		Lb
Roof	$63(4 + {}^{12}\!/_{2})20$	= 12,600
Each floor above second	$63 \times 12 \times 20$	= 15,100
Second floor	$63 \times \dfrac{12 + 16}{2} \times 20$	= 17,600
First floor	$63 \times {}^{16}\!/_{2} \times 20$	= 10,100

Shearing Stresses. First the entire bent is cut by a series of horizontal planes through the columns of each story at their points of contraflexure. For any story, the total shear is the sum of the external forces above the section, and this is resisted by the columns, the shear in each interior column of a symmetrical bent being twice that in each exterior column. Thus, in section BB, the total shear is $12.6 + (3 \times 15.1) = 57.9$, one-sixth of which, or 9.7 is taken by each wall column and two-sixths, or 19.3, by each interior column. Similarly the shears are found for each story. The shears must reach all columns except the windward column through the beams. Thus the shear in the top story is 12.6, one-sixth of which, or 2.1, is taken by the windward column, the remaining five-sixths, or 10.5, passing through beam 4 as a direct compressive stress; two-sixths of this is left in the next column, and three-sixths passes through the next beam, etc., but this diminution has little effect, and so the beam designed for the first panel is used throughout the width of the average building. In each successive story, the direct stress in the beam is five-sixths of the additional force at the last floor, since five-sixths of the rest of the shear has already been taken over by the upper beams; thus, the direct stresses are alike when the external forces are alike.

Bending Moments. When a column is cut at the point of contraflexure, the forces acting upon the cut end may be resolved into horizontal and vertical components, the former being the shear as explained above; the vertical component is the direct stress in the column, but this is not required in determining the bending moment, because its line of action passes through the point of moments. The bending moment at any point in a column is equal to the shear at the point of contraflexure multiplied by the distance from the given point to the point of contraflexure; this increases by direct proportion until the maximum is reached at the intersection of the center lines of the column and the beam either at the top or at the bottom of the story in question. The bending moments in the intermediate columns are twice those in the wall columns, because the corresponding shears are twice as much. The bending moments in thousands of foot-pounds are tabulated only on the right-hand half of the sketch, since the stresses are symmetrical. To save duplication, each bending moment is recorded once in each story. But it should be remembered that this is the bending moment at the center line of the beam above or the beam below; that the bending moment at the mid-story is zero; that values between may be found by proportion; similarly, that there are two values given for the bending moment at each beam, one for the column above, and the other for the column below.

The bending moments in the beams vary from zero at the mid-points to maximum at the column centers, the maximum value M_W in each beam being equal to the sum of the bending moments in the upper and lower exterior columns at the points of intersection. This may be proved by cutting small portions of the bent at adjacent points of contraflexure and applying the equations of equilibrium. In Fig. 46a, showing the connection of the roof beam to the wall column, the external force, the direct stress in the beam, and the direct stress in the column all pass through the point of moments

and hence do not affect the moment equation. Only the horizontal shear in the column and the vertical shear in the beam are left, and their moments must be equal; the bending moment in the beam equals the bending moment in the column (a special case of the sum of two column moments where one is zero). The shear in the beam is 2.1 × 6 ÷ 10 = 1.26, and the corresponding direct stress in the column is equal to it, but these need not be determined in this way or tabulated; they are shown here in order to complete the stress diagrams. In *b*, showing the connection of the roof beams to the intermediate

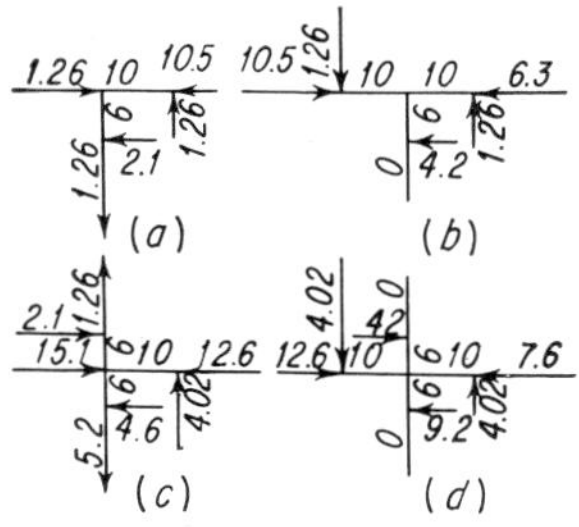

Fig. 46

column, the vertical shear in the left-hand beam is equal and opposite to that in the other half of the beam in *a*, the moment thus being made negative. To satisfy the moment equation, the bending moment in the right-hand beam must be (4.2 × 6) − (1.26 × 10) = 12.6, the same as in the left-hand beam. The shear is also the same, and so there is no vertical component in the intermediate column. In *c*, showing the connection of the eighth-floor beam to the wall column, the bending moment in the beam must balance the two column moments (2.1 × 6) + (4.6 × 6), for they act in the same direction, the shear 2.1 used for the lower half column being opposite that used when the upper half is considered. The shear in the beam is the bending moment 40.2 divided by the half length of beam, and the corresponding direct stress in the column is found from the V equation; this may not be the direct stress used in design, as explained in the next paragraph. Similarly *d* shows the connection of the eighth-floor beams to the intermediate column.

Direct Stresses in Columns. According to the assumption of shear distribution, there are no direct stresses in the intermediate columns, the wall columns resisting the entire tendency for the whole structure to overturn. Instead of using the direct stresses in the wall columns as found in the preceding paragraph, it is reasonable to assume that all the wall columns resist overturning even though not every bent is designed as a braced bent to prevent distortion; by this assumption the stiffness of the walls, floors, and partitions is counted on to distribute certain stresses to the braced bents, which is not always justifiable. Regardless of whether the overturning resistance is based upon the pressure on a single panel or more, it is much simpler as well as more accurate to find the direct stress in each story by treating the bent as a cantilever truss projecting above the point of contraflexure with a uniformly distributed load extending from the point of contraflexure to the top of the parapet wall. The corresponding bending moment is resisted by the moment of a couple, each force of which is the direct stress in the wall column, tension on the windward side, and compression on the leeward side; the lever arm is the distance between the centers of the wall columns. An expression may be derived for each building which gives the direct stress in terms of the height H from the point of contraflexure to the top of the wall, and the stresses for all stories except those below the ground may be found from a single setting

of the slide rule. Thus, for a horizontal section through any point of contraflexure above the ground, the resultant force is the pressure per linear foot times H, and the bending moment is this force times $H/2$; this is equated to 60 times the direct stress required. The pressure per linear foot is 20×63 if only the braced bent resists overturning or 20×21 if each bent is considered, as we shall assume. The direct stress is therefore $20 \times 21H^2 \div (2 \times 60)$ or $3.5H^2 = 0.0035H^2$ in kips; for section BB, $H = 4 + (3 \times 12) + 6 = 46$, and the direct stress is $0.0035 \times 46^2 = 7.4$. This expression does not apply to the basement columns, because the wind pressure is not applied below the ground level. For all basement columns, the resultant force is found from the exposed distance from the ground to the top of the wall, but the lever arm is one-half this distance plus the distance from the ground to the point of contraflexure. Thus, the direct stress in the basement story is $\frac{20 \times 21 \times 104}{60 \times 1{,}000} \times \frac{104 + 11}{2} = 41.9$; in the subbasement story, $\frac{20 \times 21 \times 104}{60 \times 1{,}000} \times \left(\frac{104}{2} + 11 + \frac{11}{2}\right) = 49.9$.

Design of Beam 1. Beam 1 is simply supported, and beam 2 is like it except for end connections; beam 4 has the same gravity loads but resists wind also, as explained later. In this design, reinforced concrete slabs are used with concrete fireproofing around the beams in order to show complete analysis; modern practice trends toward some form of core floor which is much lighter but with which reinforced concrete is used instead of intermediate steel beams, the steel design being thus simplified. Unless the steel beams extend into the slabs, they are considered laterally unsupported and must be designed accordingly; in this problem, it is assumed that the tops of the slabs are 2 in. above the tops of the beams, the remainder of the slab giving lateral support to the beams. Deflection is usually not serious if roof beams are at least one-thirtieth of the span in depth, and floor beams one-twentieth of the span. The effective length of a beam in an inner panel is the distance center to center of columns; in the outer panel, the length may be greater if beam 1 connects to a single wall girder 6 which is placed off center to support better the curtain wall; this may require a heavier beam than needed for an inner panel unless girder 6 is made of two beams, one at the column center to support beam 1, the other at the wall center to support the wall. Even though the live load does not extend through to the support, the bending moment is substantially the same as if it did. In the present instance, a single wall girder will be used, assumed to be 6 in. beyond the column center, the effective length being thus 20.5.

Live load per square foot of roof	=	25
Roofing	=	10
Suspended ceiling	=	10
4-in. minimum slab	$\frac{4}{12} \times 150$ =	50
Total load per square foot	=	95
Superimposed load per linear foot	95×7 =	665
Weight of 10-in. beam assumed	=	29
Concrete around beam	$\frac{(10.2 + 2 + 2 - 4)[5.8 + (2 \times 2)]}{144} \times 150$ =	105
Total load per foot	=	800
Bending moment	$\frac{800}{2} \times 10.25^2$ =	42,000 ft-lb
	$12 \times 42{,}000$ =	$20{,}000S$
	S =	25

This requires a 10-25 wide-flanged beam, which is deeper than the minimum $20.5 \times 12 \div 30 = 9$. For the 20-ft central panel the same size beam is required.

The minimum depth for floor beams is $20.5 \times 12 \div 20 = 12.3$ or 14. The total load per linear foot of beam 1 in an intermediate floor, including 20 lb per sq ft allowed for movable partitions and 20 lb per sq ft for 2-in. cinder concrete fill around conduits, is $(50 + 5 + 5 + 20 + 50 + 30)7 + 30 + \dfrac{(13.9 + 2 + 2 - 4)[6.7 + (2 \times 2)]}{144} \times 150 = 1{,}240$. This requires a 14-30 beam. The total load per linear foot of beam 1 in the first floor where the slab is 6 in. is $[100 + 12 + 5 + 20 + (6/12 \times 150) + 20]7 + 40 + \dfrac{(16 + 2 + 2 - 6)[7 + (2 \times 2)]}{144} \times 150 = 1{,}820$. This requires a 16-40 beam.

Design of Intermediate Girder 3. Each concentrated load is equal to the total load per foot on beam 1 multiplied by the sum of the half spans on opposite sides; the half span in the outer panel is one-half the effective length, even though the opposite end is not fully loaded, for the difference is negligible. The effective length of girder is from center to center of columns. The maximum bending moment is at the center, but for two symmetrical concentrated loads the bending moment is constant between the loads and may be found more readily at one of the loads; to this is added the bending moment at the center due to the weight of the girder and its fireproofing. An intermediate girder should be enough deeper than the beams to permit the use of seat-angle connections, preferably without coping. In the case here considered, each concentrated load and the corresponding reaction is $800(20.5 + 20) \div 2 = 16{,}200$. The bending moment due to concentrated loads is $16{,}200 \times 7 = 113{,}400$ ft-lb. The weight of a 16-in. girder assumed with its fireproofing is $50 + \dfrac{(16.3 + 1 + 2 - 4)[7.1 + (2 \times 2)]}{144} \times 150 = 230$ and the bending moment $230/8 \times 10.5^2 = 12{,}700$. This gives a section modulus of 76, which requires a 16-50 beam. In an intermediate floor, each concentrated load is $\dfrac{1{,}240(20.5 + 20)}{2} = 25{,}100$, which requires a 21-62 beam. In the first floor, the concentrated load is $\dfrac{1{,}820(20.5 + 20)}{2} = 36{,}900$, which requires a 21-82 beam.

Design of Beam 4. Beam 4 supports the same gravity loads as beam 1 or 2 and, in addition, resists a bending moment and direct stress due to wind. The bending moment due to wind is maximum at the column centers, zero at the point of contra-

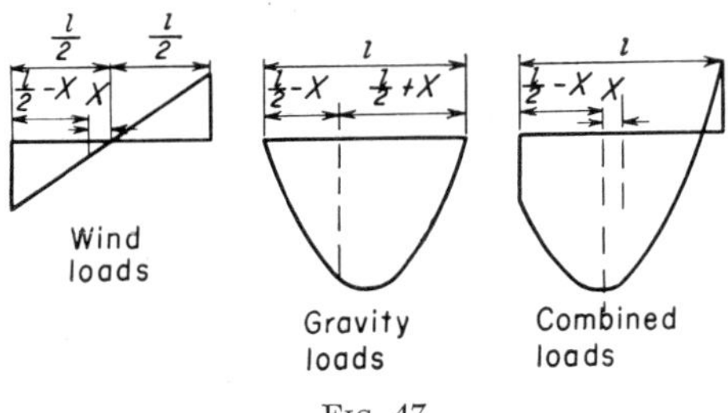

FIG. 47

flexure at the center of the beam, and proportionate amounts at intermediate points; the bending moment due to gravity loads is maximum at the center of the beam and zero at the ends, varying as the ordinates of a parabola. The curve of combined bending moments is of the form shown in Fig. 47, and the point of maximum bending moment

may be found by placing the first derivative of the expression for combined bending moment equal to zero and solving for the distance X from the center of the beam; this is equivalent to finding the point for which the shear is zero. The bending moment due to wind is that proportion of the bending moment M_W already tabulated that X bears to $l/2$ or $2XM_W/l$; the bending moment due to gravity loads is one-half the load per foot w times the product of the segments or $\frac{w}{2}\left(\frac{l}{2} - X\right)\left(\frac{l}{2} + X\right)$. The value of X that makes the combined moment maximum is $X = 2M_W/wl$ and is substituted in the foregoing expressions for bending moment to get the maximum. For combined wind and gravity loads, the allowed unit stress is 33⅓ percent (or other percentage specified) greater than the normal value. The beam selected must have sufficient reserve strength so that the developed unit stress in bending, plus the direct wind stress already tabulated divided by the area of cross section, does not exceed the allowed value; or, by the specifications used for this problem, the unit stress developed in bending divided by the unit stress allowed in bending, plus the direct stress in compression divided by the unit stress allowed in compression, should not exceed unity. In the roof, the wind stresses are negligible, and the beam would be made like beam 1; in the floors, the size of the beam may be determined either by the combined loads with the increased unit stress or by the gravity loads alone with the normal unit stress, the maximum bending moment at the center being used as for beam 1. The weights of fixed partitions should replace an equivalent area of live load on floor beams.

Gravity load per foot on beam 4 of the seventh floor, including fixed partition, is $1{,}240 + 25(12 - 1) - 50 \times 0.5 = 1{,}490$.

Bending moment for gravity loads. $1{,}490/2 \times 10^2 = 74{,}500$ ft-lb

$$12 \times 74{,}500 = 20{,}000S$$
$$S = 45$$

This requires a 14-34 beam for gravity loads alone. Considering wind, X for beam 4 of seventh floor $= 2M_W/wl = \dfrac{2 \times 70{,}200}{1{,}490 \times 20} = 4.7$ and the maximum combined bending moment is $\dfrac{2 \times 4.7 \times 70{,}200}{20} + \dfrac{1{,}490}{2}(10.0 - 4.7)(10.0 + 4.7) = 91{,}000$ ft-lb.

Unit stress developed in bending. . . . . $12 \times 91{,}000 \div 48.5 = 22{,}500$

Unit stress developed in compression. . . . . $12{,}600 \div 10 = 1{,}300$

Unit stress allowed in bending. . . . . $20{,}000 \times \frac{4}{3} = 26{,}700$

Unit stress allowed in compression (laterally supported)

$$\left[17{,}000 - 0.485\left(\frac{20 \times 12}{5.83}\right)^2\right]\frac{4}{3} = 21{,}600$$

The sum of fractions is less than unity. . . . . $\dfrac{22{,}500}{26{,}700} + \dfrac{1{,}300}{21{,}600} = 0.90$

This shows that the beam is satisfactory as far as the maximum stress in the extreme compression fibers is concerned; and if welded brackets are used, this is the final result. If brackets are riveted to the flanges, however, as in Fig. 52, the holes so reduce the area of the tension flange that a larger section may be required through the holes nearest the center of the beam, even though the bending moment is considerably less at this point. The location of these holes cannot be determined until both the bracket and the column have been designed, but the position may be assumed, the beam designed, and later the bracket and column designed; then the beam and perhaps the bracket may be revised if necessary. For purposes of illustration, it may be assumed that 1-in. holes for ⅞-in. rivets are 1′6″ from the center of the column. According to the AISC specifi-

cations, no deduction need be made for holes unless their area exceeds 15 percent of the gross flange area, when the excess is deducted from both flanges. If the developed unit stress in tension due to bending exceeds 26,700, the 1,300 unit direct stress added above may be *deducted,* because it tends to reduce the tension, but the net unit stress must not exceed the allowed amount.

Bending moment $1\frac{1}{2}$ ft from column center

$$\frac{2 \times (10 - 1.5) \times 70{,}200}{20} + \frac{1{,}490}{2} \times 1.5 \times 18.5 = 80{,}400$$

Ratio of holes to flange

$2 \times 1/6.75 = 30$ percent, an excess of 15 percent over the 15 percent allowed

Effective I of 14-34. . . . . $339 - 2 \times 6.75 \times 0.45 \times 0.15(7 - 0.22)^2 =$ 297

Unit tensile stress developed in bending. . . . . $12 \times 80{,}400 \times 7 \div 297 = 22{,}700$

This is less than the 26,700 allowed without subtracting the 1,300. When the net section governs the design of beam 4 in any floor, it will govern the design in all floors below; so it is unnecessary to find the maximum bending moments in these lower beams; even though the maximum is on the column side of the critical net section, it does not affect the design, because the brackets reinforce the beam sufficiently.

Beam 5 is designed in a similar manner to beam 4, except that the wall replaces the fixed partition, the fireproofing, and one-half the floor load. There is usually enough wall below the windows to distribute the pier loads between windows, so that no great error is made by treating the whole load of wall and windows as uniformly distributed, especially if both veneer and backing are of brick.

Design of Wall Girder 6. In the roof, girder 6 supports the concentrated loads from beam 1 and a uniform load due to the weight of the girder itself and the 4-ft parapet wall. It is also subject to bending moment and direct stress due to wind on the end of the building; but usually these need not be considered, for they are not large enough to govern the size of the beam in view of the increased unit stress allowed for combined stresses. In a floor, girder 6 supports the concentrated loads from beam 1, a uniform load due to the weight of the girder, and a full story (sometimes two) of wall and windows. A girder may be made of two beams instead of one if the loads or spacing justify this; one is then placed at the column center to support the concentrated loads from beam 1, and the other is placed best to support the wall, or both may support the wall. A beam with a top cover plate may be used to support the wall; but this is not so necessary as in lintel work, for the wall extends below the beam, and cracks due to overhang are less likely to occur. In the upper floors, the gravity loads with normal unit stress may determine the size of girder; but as the bending moment and direct stress due to wind on the end of the building increase, the size is determined by combined stress with increased unit stress. The point of maximum bending moment may be at the concentrated load or between the load and the center, depending upon the relative magnitudes of the loads. The point of maximum bending moment may be found most simply for the section where the shear is equal to zero or else where it changes sign; this is equivalent to equating to zero the first derivative of the sum of the bending moments due to concentrated loads, uniform load, and wind. After a beam is found for the maximum bending moment, it should be ascertained whether or not it has sufficient reserve to carry the direct stress also. It is quite possible that all intermediate floors require the same size girder. It is well to design the top-floor girder for gravity loads and the second-floor girder for combined loads; if the latter is larger, intermediate floors must be investigated to determine where the change is to be made. In the present case, the inner face of the panel wall is approximately at the center of the column, and so it is close enough to compute the concentrated load from beam 1 for a 10-ft half span.

Concentrated load from beam 1 of roof . . . . 800 × 10 = 8,000 lb
Bending moment due to concentrated loads . . . . 8,000 × 7 = 56,000 ft-lb
Weight of girder and 12-in. parapet above and around girder
40 + (5.5 × 120) = 700 lb per ft
Bending moment due to uniform load . . . . $\frac{700}{2} \times 10.5^2$ = 38,600 ft-lb
12 × 94,600 = 20,000S
S = 57

This requires a 16-40 for the roof girder 6.

The concentration from beam 1 in an intermediate floor is (1,240 × 10) = 12,400. Allowing for three 4-ft by 6-ft windows per story per panel, we find that the uniform load per foot for wall, windows, and girders is

$$\frac{[(21 \times 12) - (3 \times 4 \times 6)]120 + 3 \times 4 \times 6 \times 8}{21} + 55 = 1{,}110$$

Bending moment at center due to concentrated loads . . . . 12,400 × 7 = 86,800
Bending moment at center due to uniform loads . . . . 1,110/2 × 10.5^2 = 61,200
12 × 148,000 = 20,000S
S = 89

This requires an 18-50 beam.

With wind on the end of the building resisted by the two wall bents, the bending moment in girder 6 of the second floor due to wind is found to be 43,500 ft-lb. The corresponding shear is 43,500 ÷ 10.5 = 4,100, and the direct stress is 8,200. The shear just to the left of the concentrated load is −4,100 + 12,400 + 1,110(10.5 − 7) = +12,200, and just to the right is 12,200 − 12,400 = −200, making the maximum bending moment at the point of concentration of the 12,400 load where the shear changes sign.

Bending moment due to wind . . . . 43,500 × 3.5/10.5 = 14,500 ft-lb
Bending moment due to concentrated loads . . . . 12,400 × 7 = 86,800 ft-lb
Bending moment due to uniform loads . . . . 1,110/2 × 7 × 14 = 53,900 ft-lb
Unit stress developed in bending . . . . 12 × 155,200 ÷ 89 = 20,900
Unit stress developed in compression . . . . 8,200 ÷ 14.7 = 600
Unit stress allowed in compression . . . $\left[17{,}000 - 0.485\left(\frac{21 \times 12}{7.38}\right)^2\right]\frac{4}{3} = 21{,}900$
The sum of the fractions is less than unity . . . . $\frac{20{,}900}{26{,}700} + \frac{600}{21{,}900} = 0.81$

An 18-50 can be used for all intermediate floors. For convenient erection, it might be desirable to use seat angles for beam 1 on girder 6 like those on girder 3; if this were done, a 21-62 would be used for all floors above the first.

Design of Intermediate Column B. Columns are usually erected in two-story lengths, with splices 1′6″ to 2′0″ above the finished floor line; for an odd number of stories, the

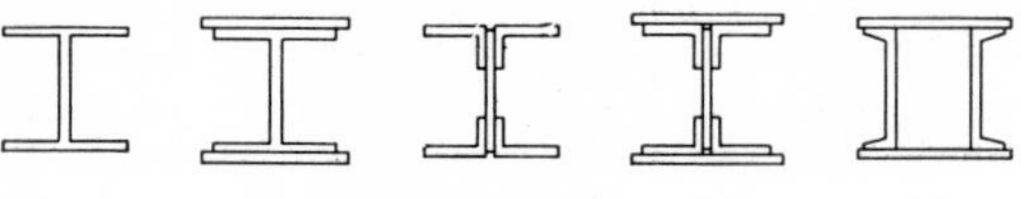

Fig. 48

top section is made either one or three stories in length. Common types of column for buildings of moderate height are shown in Fig. 48. The total column loads may be found by combining the reactions of the beams and girders supported and the weight of the column and its fireproofing. All building codes allow live-load discounts in buildings

not used for storage, for it is unlikely that more than one floor will be fully loaded at the same time. The fireproofing may be of concrete, with a minimum thickness of 2 in., or it may be hollow tile and concrete, with a minimum thickness of 4 in. and an average weight of 70 lb per cu ft. The fireproofed column may be either circular or rectangular in cross section. The columns are placed so that the larger section moduli resist the bending moments when this is feasible, but it may be better to turn the columns the other way if in so doing the heaviest beams are made to connect to the flanges instead of to the webs. Although the maximum wind moment is at the centers of the beams, the floor loads are transmitted to the column through seat angles or brackets. For gravity loads, the critical section is at the splice; but for combined gravity and wind loads the critical section is at the floor above the splice, even though the weight of the column and its fireproofing is somewhat less. The maximum wind moment is at the center of the transverse beam or girder, but the full gravity load does not reach the column at the same point. It is on the side of safety to combine the maximum stresses, but this may be excessive when large beam or girder loads are supported on seat connections or connected with top and bottom brackets. The refinement taken in the selection of the critical section depends upon the size of the beam and the relative magnitudes of the loads as well as upon the type of connection.

The total gravity load with live-load reduction, divided by the gross area of cross section and by the normal unit stress found from the column formula, plus any fiber stress in bending due to eccentric loads, divided by the normal unit stress in bending, must not exceed unity; the unsupported length is usually one story height and the least radius of gyration is used, with due regard to the maximum allowed unit stress. Also the axial fiber stresses due to gravity and wind, divided by the increased unit stress allowed in compression, plus the fiber stresses due to wind and eccentricity, divided by the increased unit stress allowed in bending, must not exceed unity. The handbooks show total areas, moments of inertia, radii of gyration, and section moduli about both axes for some of the common types shown above. In this case, the two top sections of intermediate column B will be designed, using wide-flanged column sections.

Roof load from girder 3	$(16{,}200 \times 2) + (230 \times 21) =$	37,200
Roof load from beam 4	$800 \times 20 =$	16,000
Eighth-floor load from girder 3	$(25{,}100 \times 2) + (310 \times 21) =$	56,700
Eighth-floor load from beam 4	$1{,}490 \times 20 =$	29,800
Total gravity load except weight of column		139,700

From the table of safe loads, select a 12-45, which has a safe load of 190,000 for a 12-ft length, somewhat larger than 139,700 to provide for bending stresses. The area is 13.24, the larger section modulus 58.2, and the least radius of gyration 1.94. The allowed unit stress is $17{,}000 - 0.485\left(\frac{12 \times 12}{1.94}\right)^2 = 14{,}300$, which, increased $33\frac{1}{3}$ percent for combined stresses, is 19,100.

Weight of column and fireproofing, less duplicated partition weight is $\left[45 + \frac{(12+8)(8+8)70}{144} - \frac{25 \times 20}{12}\right][12 - 1] = 1{,}700$; this item is subject to variation depending on the method of calculation, but it has relatively little effect on the design. Because floor and beam loads are all taken to the column centers, there is probably sufficient overlapping to take care of the weights of beam connections, splices, and column; so about 1 ft is deducted from the story height and an equivalent amount is deducted for the fixed partition replaced by the column fireproofing. The bending moment due to wind at the center of the eighth-floor beam was found to be 55,200

as tabulated; the bending moment at the critical section is found by proportion, assuming the critical section to be 1 ft from the center of the 14-in. beam 4.

Unit stress due to gravity loads . . . . 141,400 ÷ 13.24 = 10,700
(which is less than the 14,300 allowed)

Unit stress due to bending . . . . $\frac{6-1}{6} \times 55{,}200 \times 12 \div 58.2 =$ 9,500

The sum of the fractions is less than unity . . . . $\frac{10{,}700}{19{,}100} + \frac{9{,}500}{26{,}700} = 0.92$

For the next section, supporting the seventh and sixth floors,

Load from upper section, including full live load . . . . = 141,400
Loads from girder 3, seventh and sixth floors . . . . 56,700 × 2 = 113,400
Load from beam 4, seventh floor . . . . 1,490 × 20 = 29,800
Load from beam 4, sixth floor . . . . 1,500 × 20 = 30,000

Assuming a 12-79 with an area of 23.22, section modulus of 107.1, and least radius of gyration 3.05, the allowed unit stress is 15,900 and the increased value for combined stresses 21,200.

Weight of two more stories of column and fireproofing

$$\left[79 + \frac{(12.4+8)(12+8)70}{144} - \frac{25 \times 20.4}{12}\right] 2 \times 11 = \quad 5{,}200$$

Total load, including full live load . . . . = 319,800

20 percent live-load reduction for column supporting three floors (the 21-ft panel dimension is reduced 6 in. because of the fixed partition)
$3 \times 50 \times 20 \times 20.5 \times 0.20 =$ 12,300

Design load . . . . = 307,500
Unit stress due to gravity loads . . . . 307,500 ÷ 23.22 = 13,200

Unit stress due to bending . . . . $\frac{6-1}{6} \times 115{,}800 \times 12 \div 107.1 =$ 10,800

The sum of the fractions is more than unity . . . . $\frac{13{,}200}{21{,}200} + \frac{10{,}800}{26{,}700} = 1.02$

This would probably be accepted, since the excess is so slight and the unit stress due to gravity loads 13,200 is so much under the 15,900 allowed.

Design of Wall Column A. A wall column supports approximately one-half the floor load that an intermediate column carries and a wall load on the opposite face of the column; these two do not usually balance, and so there is a moment due to eccentricity as well as the moment due to wind. The effect of the eccentricity from one floor is assumed to be dissipated before the next floor is reached, and at each floor the eccentric moment is considered to be resisted by the column above and the column below in proportion to the story heights; it is often neglected. The wall column is an intermediate column of the braced bent which resists wind on the end of the building, and the strength about each axis must be considered. The lateral support furnished by the wall is usually neglected unless the whole column is solidly embedded. The position of the column and of the girder in the wall depends upon fireproofing requirements and upon architectural features. If no outside pilaster is used, the face of the column must be at least 4 in. inside the face of the wall. If the column centers are continuous for the

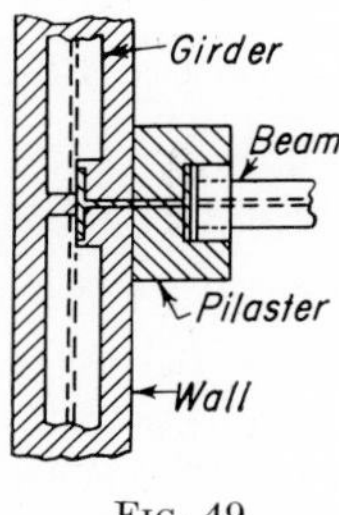

FIG. 49

entire height of the building, allowance must be made for the largest column; but if the columns are offset so that the outer faces are in line, this is not necessary. The girder flanges must be at least 2 in. inside the face of the wall and preferably more; it is well to place the girder approximately in the center of the wall, unless a plate or angle projects on one side to support the wall. In the present case, the girder is placed in the center of the 12-in. wall, and the face of the column 6¼ in. from the face of the wall to permit the direct connection of the girder web to the column flange, as shown in Fig. 49. In practice, such details must be made in advance of the design, because the walls are usually fixed and the column centers must conform.

Roof load from girder 6..................(8,000 × 2) + (700 × 21) = 30,700
Roof load from beam 4..................................800 × 10 = 8,000
Eighth-floor load from girder 6..........(12,400 × 2) + (1,110 × 21) = 48,100
Eighth-floor load from beam 4..........................1,490 × 10 = 14,900
Assuming a 12-40 with an area of 11.77, section moduli 51.9 and 11.0, and least radius of gyration 1.94, the allowed unit stress is 14,300 and the increased value for combined stresses 19,000.
Weight of column and fireproofing, less duplicated partition weight

$$\left[40 + \frac{(6+4)[8+(2\times 4)]70}{144} - \frac{25\times 10}{12}\right](12-1) = 1{,}100$$

Total gravity load.. = 102,800
Direct stress due to wind, as tabulated.......................... = 1,700
Moment due to eccentricity...............(48,100 − 14,900)6 ÷ 2 = 99,600 in.-lb

Moment due to wind on side................ $\frac{6-1}{6} \times 27{,}600 \times 12 = 276{,}000$ in.-lb

Moment due to wind on end................ $\frac{6-1}{6} \times 9{,}100 \times 12 = 91{,}000$ in.-lb

Unit stress due to gravity loads...................102,800 ÷ 11.77 = 8,700
Unit stress due to eccentricity.....................99,600 ÷ 51.9 = 1,900
Unit stress due to direct wind stress, wind on side......1,700 ÷ 11.77 = 100
Unit stress due to moment, wind on side............276,000 ÷ 51.9 = 5,300
Unit stress due to moment, wind on end..............91,000 ÷ 11.0 = 8,300
The sum of the fractions for gravity loads is less than unity

$$\frac{8{,}700}{14{,}300} + \frac{1{,}900}{20{,}000} = 0.71$$

The sum of the fractions for wind on side is less than unity

$$\frac{8{,}700 + 100}{19{,}000} + \frac{1{,}900 + 5{,}300}{26{,}700} = 0.73$$

The sum of the fractions for wind on end is less than unity

$$\frac{8{,}700}{19{,}000} + \frac{1{,}900 + 8{,}300}{26{,}700} = 0.84$$

This section is larger than necessary, but the next lighter section is too small.

Design of Beam Connections. The usual type of connection of a beam to a girder is a seat and side angle, as shown in Fig. 50. This facilitates erection because the beam is not connected to the 4 × 3½ × ⅝ seat angle which is connected to the girder by three rivets as a minimum. The 4 × 4 × ⅜ side angle is bolted to each beam by two bolts regardless of the size of the beam. The number of rivets in the 21-62

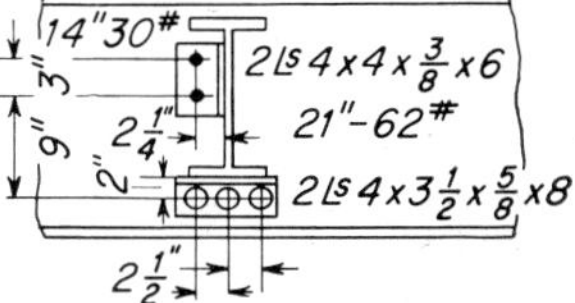

Fig. 50

girder 3 required to support the two 14-30 beams 1 is only two determined by the bearing of ⅞-in. rivets in the web $1{,}240 \times \dfrac{20 + 20.5}{2}$ 14,000.

A typical seat connection of girder 3 to a 12-40 column is shown in Fig. 51. The combined loads from two girders is 55,000 lb and at 10,500 bearing in the web requires six rivets. Either one or two stiffeners are used on each side. More often the beams and girders are arranged so that the heavier girders are connected to the flanges of the column and the lighter beams to the web. The 4 × 3 × ⅜ top angles are placed ⅛ in. above the top of the girder and bolted for shipment so that it can be removed for erection.

Wind brackets are made in many different forms, and they may be riveted or welded. A riveted bracket of the type shown in Fig. 52 is chosen to illustrate the connection of the 14-34 beam 4 of the sixth floor to the 12-79 column, because it shows the full number of steps; most other types of bracket are designed by the same method, but many in-

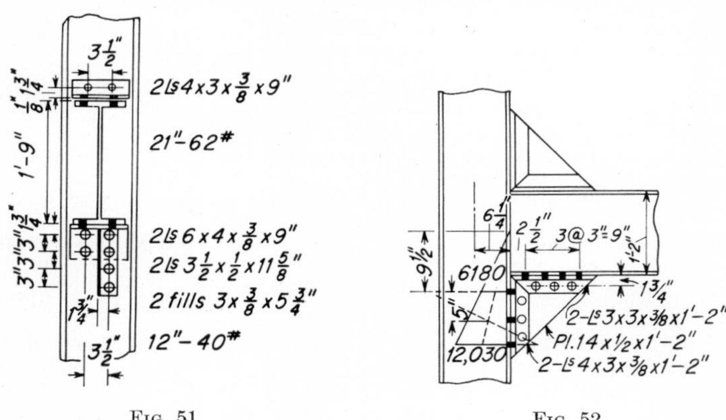

Fig. 51

Fig. 52

volve only one or two steps. The wind bending moment at the center of the column was found to be 100,800, but at the face of the 12-in. column it is only 100,800 × 9.5/10.0 = 95,700. As the beam tends to bend relative to the column, one flange of the beam tends to pull away from the column and the other tends to move toward it; these shearing forces are equal and opposite, forming a couple, the moment of which equals the bending moment. The value of one ⅞-in. rivet in single shear at 15,000 psi is 9,020, which can be increased 33⅓ percent for wind stresses to 12,030. The lever arm is equal to the depth of the beam, and the number of rivets required in each flange is (12 × 95,700) ÷ (14 × 12,030) = 7, one-half or four in each angle. Three-inch spacing is used, the distance from the column to the first rivet being determined by layout so that the end rivet in the vertical leg has sufficient edge distance when it is placed midway between the rivets in the outstanding leg to allow driving clearance. Since the beam flange is 6¾ in., the outstanding leg is made 3 in., and the vertical leg 3 in.

The rivets in the vertical leg are in double shear; they must resist the same moment, but the lever arm is greater, being 14 + (2 × 1.75) = 17.5. Instead of assuming a plate thickness and finding the number of rivets, it is often better to assume the number of rivets which best corresponds to the number in the outstanding leg and then find the thickness of plate required. Thus, if four rivets are used in the outstanding leg, three would be used in the plate. The horizontal stress in each rivet is (12 × 95,700) ÷ (3 × 17.5) = 21,900; the vertical shear due to gravity is 15,000 and that due to wind

100,800/10 = 10,000; these are distributed among the six rivets of the two brackets, or 25,200/6 = 4,200. The resultant 22,300 equals 40,000 × ¾ × ⅞t, whence t, the thickness of plate, = ½. This is thicker than usual for such a small bracket, and some might prefer to use a ⅜-in. plate with four rivets even if this means extra rivets in the beam. A corresponding number of rivets is used in the vertical angles, although a smaller number may suffice because of the reduced stress due to the larger lever arm.

The rivets which connect the angles to the column are not equally stressed, for one bracket is in tension while the other is in compression, there being no stress at the center of the beam; the value of the rivets farthest from the center of the beam is limited by the tensile strength, usually taken equal to the single-shear value, and the others are stressed in proportion to their distances from the center of the beam. The limiting value of the lower rivets is 12,030, as found above; they are 18½ in. from the center of the beam, being the same distance from the corner of the bracket as the outer rivets in the beam connection. The upper rivets are 9½ in. from the center of the beam, and they may be stressed 12,030 × 9.5/18.5 = 6,180; the average stress in rivets is the mean value 9,100. The resultant stress in the group of rivets acts at the center of gravity of the stress trapezoid; this may be found graphically at the intersection of the median line connecting the bases and the line joining the points found by prolonging the line of each base a distance equal to the other base as shown. The lever arm is (5.0 + 9.5) × 2 = 29.0, and the number of rivets required is (12 × 95,700) ÷ (29.0 × 9,100) = 5. No attention need be given to vertical shear due to gravity or wind or to horizontal compression due to wind, for there is sufficient reserve strength in the rivets that are not fully developed in tension and in the rivets of the compression bracket that are otherwise unstressed. Three rivets on each side are sufficient if the gage is large enough to give clearance for driving the middle rivets; otherwise, they must be staggered as in the beam connection. The angles must be thick enough to transmit the stress to the rivets without bending. A satisfactory angle can be selected usually without designing; but if the gage is large because of column conditions, the thickness may be determined by the method used for bearing plates. Stiffening angles should be riveted to the diagonal edge of the plate if the clear distance between angles exceeds 30 times the thickness of the plate.

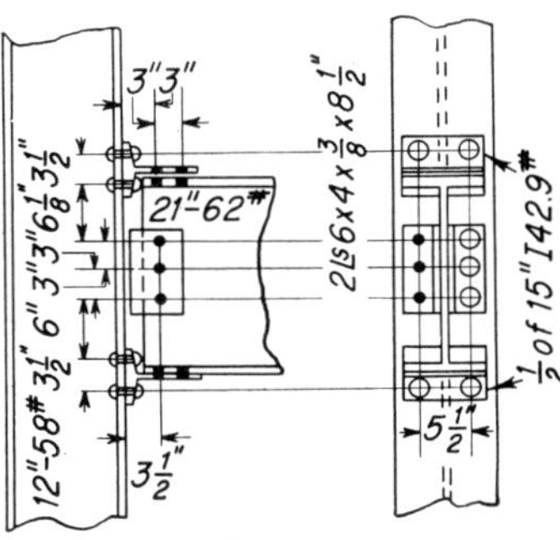

Fig. 53

An alternate type of bracket is shown in Fig. 53. The shear is carried by a web connection and the moment by split I beams which are better suited than wide-flanged beams because the flanges are narrower and stiffer. As the beam tends to bend relative to the column, one flange of the beam tends to pull away from the column and the other to move toward it; these two shearing forces are equal and opposite and form a couple. The moment of this couple must equal the total bending moment at the face of the column. The rivets in each flange of the beam are in single shear and the lever arm of the couple is equal to the depth of the beam. The same number of rivets connect each bracket to the column, rivets in tension being taken as having the same value as in single shear.

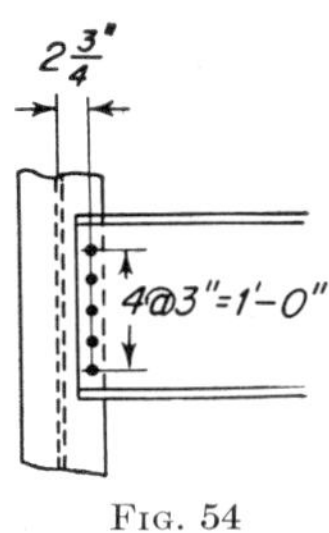

Fig. 54

When the web of a beam is connected directly to the column, the riveting is designed by the method of eccentric connections, because the rivets resist both moment and shear. A single row of rivets is used for the upper beams, and a double row lower down as the

moment increases, provided that the column flange will permit; a connection plate is used when sufficient rivets cannot be provided otherwise; in the latter case, there is a moment due to eccentricity as well as to wind. In the seventh floor of the case under consideration, the 18-50 is connected to the 12-40 column by a single row of rivets on the standard-gage line, as shown in Fig. 54.

Bending moment at rivet line, due to wind....... $11{,}600 \times 10.3 \div 10.5 = 11{,}400$ ft-lb
Shear due to gravity loads.................. $12{,}400 + (1{,}110 \times 10.5) = 24{,}100$
Shear due to wind.................................... $11{,}600 \div 10.5 = 1{,}100$
Number of rivets required for gravity loads............. $24{,}100 \div 9{,}020 = 3$
Number of rivets assumed, to allow for moment also.................. $= 5$

Horizontal component on critical rivet, including direct stress per
rivet............................... $\left(\frac{12 \times 11{,}400}{2(3^2 + 6^2)} \times 6\right) + \frac{7{,}000}{5} = 10{,}500$
Vertical component due to shear................ $(24{,}100 + 1{,}100) \div 5 = 5{,}040$
Resultant, which is less than 12,030 allowed........ $\sqrt{10{,}500^2 + 5{,}040^2} = 11{,}600$

INDUSTRIAL BUILDINGS

Characteristics. A modern industrial building may be in one of many forms, depending upon its purpose. A multi-story factory building is similar to an office building in design but has relatively few partitions. Small industrial buildings, or larger buildings with saw-tooth roofs for lighting, are made of rigid-frame construction; wide-flanged beams are used for rafters and for columns, and they are reinforced at their junction by cutting and bending one flange and welding in a plate to stiffen the joint. Garages, steel mills, structural shops, and similar buildings are of the mill-building type with long-span trusses and few if any intermediate columns. Flat roofs for tin or composition roofing have the top chords slope just enough to provide drainage, and steep-pitched roofs for other types of roofing have a center rise of one-quarter of the span ("quarter pitch") or something similar. The walls of an industrial building may be of brick, stone, concrete, or concrete blocks, with or without columns embedded in them. Corrugated iron or steel may be used on the roofs and sides of buildings in which heating is not an important factor, as in furnace buildings or foundries.

Purlins. Purlins are the longitudinal members which transmit the roof loads to the trusses. They support vertical gravity loads due to snow, the roofing, and the purlins themselves; they may support the sheathing or slabs necessary to carry certain types of roofing or wooden spiking pieces to which sheathing is nailed. They also resist wind pressure, which is commonly taken normal to the roof surface.[1] The combinations of loads used depends on their magnitudes and the pitch of the roof. Wooden sheathing and often corrugated iron are considered so stiff that the slope bending perpendicular to the web can be neglected. If the slope bending is to be considered, the vertical force is resolved into components normal and parallel to the roof, the size of purlin is determined that will resist the normal component, and this purlin is investigated to see if there is sufficient reserve so that the extreme fiber is not overstressed when the slope component parallel to the roof is considered. Purlins that support corrugated steel are supported at one or two points between trusses by sag rods. Most specifications allow an increase in the allowed unit stress when wind stresses are included. A channel is commonly used as a purlin, the flanges facing up the slope when no spiking piece is used but down the slope when a spiking piece is bolted to the web. The purlin depth should be at least one-thirtieth of the span. Purlins are preferably placed at the panel points of the truss; otherwise, the top chord must be designed for bending as well as

[1] See Section 5.

compression. The purlin spacing depends also on the gage of the corrugated steel or the thickness of the sheathing.

Illustrative Problem. To design a purlin to support a sleet load of 10 lb per sq ft, slate roofing [1] at 6.5 lb per sq ft, spiking pieces at 4 lb per lin ft and a normal wind pressure on a one-fourth pitch roof of 15 lb per sq ft corresponding to 20 lb horizontal pressure. The purlins are 4 ft apart and 20 ft long. The normal unit stress is 20,000 psi, but this is increased 33⅓ percent for wind and gravity loads combined. The minimum depth is $20 \times 12 \div 30 = 8$.

Vertical load per linear foot, assuming 8 ⊏ 11.5		
	$(10 + 6.5 + 8)4 + 4 + 11.5 =$	114
Normal component, including wind	$(114 \times 0.89) + (15 \times 4) =$	161
Slope component	$114 \times 0.45 =$	51
Bending moment, normal component	$\frac{161}{2} \times 10^2 =$	8,050 ft-lb
Bending moment, slope component, considered supported at mid-point by sag rods	$\frac{51}{2} \times 5^2 =$	640 ft-lb
	$12 \times 8{,}050 =$	$26{,}700S$
	$S =$	3.6
which requires an 8 ⊏ 11.5, as assumed.		
Developed stress, normal component	$12 \times 8{,}050 \div 8.1 =$	11,900
Developed stress, slope component	$12 \times 640 \div 0.8 =$	9,600

The sum, 21,500, is less than the 26,700 allowed.

Roof Trusses. Stresses are determined, as explained in Section 5, for panel loads due to wind, weight of snow, roofing, sheathing or slabs, spiking pieces, purlins, and proportionate amount of the weight of the truss assumed or determined from some empirical formula. Single angles are used for the shorter web members. A 2½ leg is the smallest which can be riveted with the usual ¾-in. rivet, and the minimum angle is 2½ × 2 if only one leg is riveted; the same size may be considered a minimum for welded trusses even though no holes have to be deducted. The minimum thickness may be 3⁄16, ¼, or 5⁄16, depending on the class of work and the designer. The strength of the minimum size in tension and also in compression for different unsupported lengths, at the increased unit stress, may be compared with the stresses found from the stress diagram in order to determine which members can be made of this section. Two angles of the same size may be used where the stress requires them or where the ratio of slenderness would otherwise exceed the specified amount. Two larger angles may be needed for the main tension and compression web members; the main tension members may have reversal of stress due to wind on the opposite side of the truss, especially if knee braces are used under the truss; such members may be unduly large, unless it is assumed that connecting web members give lateral support about both axes. Bottom-chord members may extend full length unless the truss is shipped in sections; two angles are used, with the longer legs vertical. A single-angle hanger is used at the center of the larger trusses to give intermediate support to the bottom chord, which is in compression during erection. The top chord is made of two angles with the longer legs outstanding unless subjected to bending, in which case the angles are turned the other way or else a web plate is inserted or a wide-flange beam is split to make a T-shaped member; the unsupported length in a vertical plane is one panel length but about the other axis is usually taken as two purlin spaces. In general, minimum sections are used for truss members as far as possible; but for larger stresses it is better to increase the size of leg rather than the thickness, because the slenderness ratio is made more favorable and the truss stiffer.

[1] It is unusual to have a slate roof on a building with corrugated steel sides, but the design is made more complete in this manner. The design of a purlin for corrugated steel roofing would be simpler than of one with slate and sheathing, and the girts, the eave struts, and sway bracing would be omitted if solid masonry wall were used for bearing.

Crane Girders. The crane-runway girders carry the rail and the two wheels which support one end of the traveling crane, the percentage of the live load specified for impact being included. The wheel loads and spacing must be obtained from the manufacturers of the crane to be used; these wheel loads allow for the weight of the crane and for the capacity load when it is as near one end of the crane as possible. The maximum bending moment for the concentrated loads will occur under one of the loads, when it is either at the center of the span or at a distance from the center equal to one-fourth the distance between wheels, whichever gives the larger value. The bending moment due to the weight of the girder and the rail is taken at the center and added to the other value, even though they may not occur at the same point. A wide-flanged rolled beam is now generally used for moderate spans instead of the standard I beam or built girder; it may have sufficient lateral stiffness to resist the thrust due to the motion of the crane and the swaying of the suspended load, or it may need a cover channel on the top flange. Boltholes must be provided for special adjustable clamps for holding the rail in position. The clamps are adapted to 7½ in. gage, but this may be modified slightly if desired. The depth of a crane girder is usually about $\frac{1}{10}$ the span. The lateral bending moment is sometimes taken as $\frac{1}{20}$ of the vertical bending moment due to live load and impact, but the specifications of the American Institute of Steel Construction state that the lateral force shall be 20 percent of the sum of the weights of the lifted load and of the crane trolley (exclusive of other parts of the crane), one-half on each side of the runway. The combined unit stress must not exceed that specified for the compression flange.

Illustrative Problem. To design a 20-ft girder to support a 75 lb per yd rail and two crane wheels 10.8 ft apart, each with a maximum load of 44,500 lb with 25 percent impact allowance. These wheel loads are determined as the reactions due to the 25-ton load and trolley as near one end as possible, plus the weight of the crane. For lateral force we shall assume that one-half of 20 percent of the 50,000 load plus trolley is 5,200, one-half at each wheel.

Distance from center to point of maximum bending moment. . 10.8 ÷ 4 = 2.7
Distances from right end to wheels. . . . . = 1.9 and 12.7
Maximum bending moment for live load and impact

$$44{,}500 \times 1.25 \times \frac{1.9 + 12.7}{20} \times 7.3 = 296{,}400 \text{ ft-lb}$$

This is greater than for a single load at the center.

Assuming a 24-130 beam, bending moment for dead load

$$\frac{130 + 25}{2} \times 10^2 = 7{,}800 \text{ ft-lb}$$

Total vertical bending moment. . . . . = 304,200

Horizontal bending moment. . . . . $2{,}600 \times \frac{1.9 + 12.7}{20} \times 7.3 = 13{,}900$

S of net section, horizontal axis

$$[4{,}010 - 4 \times \tfrac{7}{8} \times 0.9(12.12 - 0.45)^2] \div 12.12 = 302$$

S of net section, compression flange, vertical axis. . . . . . . $0.9 \times 14^2 \div 6 = 29.4$
Developed unit stress, vertical bending moment. . . $12 \times 304{,}200 \div 302 =$ 12,100
Developed unit stress, horizontal bending moment. . $12 \times 13{,}900 \div 29.4 =$ 5,700
The total 17,800 must not exceed the allowed value of 20,000 for values of ld/bt not over 600. In this case $20 \times 12 \times 24.25 \div (14 \times 0.9) = 462$. The next smaller size is found to be too small because of the narrower flange.

Girts. Angle girts are used to support corrugated steel siding and the corresponding window frames. They resist primarily horizontal wind pressure, but the combined stresses in any fiber due to wind and gravity must not exceed the increased unit stress

allowed. Some designers take liberties in designing girts, because they are secondary members which might actually fail without wrecking the main structure; they sometimes design for a smaller wind pressure than they use for the main bents and sometimes for horizontal bending moments only. Angles most commonly used are 4 × 3, 5 × 3, and 5 × 3½, with the longer legs horizontal; bottom girts are 5/16, but intermediate girts may be thicker. Sag rods or bars are used in the longer spans to give intermediate support vertically.

Illustrative Problem. To design an intermediate girt to support 22-gage corrugated steel and resist a wind pressure of 20 lb per sq ft. The girts are 4 ft apart and 20 ft long. The normal unit stress is 20,000, but this is increased to 26,700 for wind and gravity loads combined. The girts are supported by sag rods at their mid-points.

Horizontal bending moment........................ $\frac{20 \times 4}{2} \times 10^2 = 4{,}000$ ft-lb

$$12 \times 4{,}000 = 26{,}700S$$
$$S = 1.8$$

A 5 × 3½ × 5/16 has a section modulus of 1.9 and weighs 8.7.

Vertical bending moment...................... $\frac{(1.7 \times 4) + 8.7}{2} \times 5^2 = 193$

Developed unit stress, horizontal bending........... $12 \times 4{,}000 \div 1.9 = 25{,}300$

Developed unit stress, vertical bending............... $12 \times 193 \div 1.0 = 2{,}300$

The combined stress, 27,600, exceeds 26,700; so a 5 × 3½ × ⅜ is used.

End Columns. The end roof trusses of a mill building are preferably made like the intermediate ones, unless masonry walls are used, so that the building can be extended without removing the whole end panel of roof and side covering. In this case, the end columns simply support the girts below the truss and resist the wind pressure; for they are connected to the trusses with slotted connections, and so they receive no roof

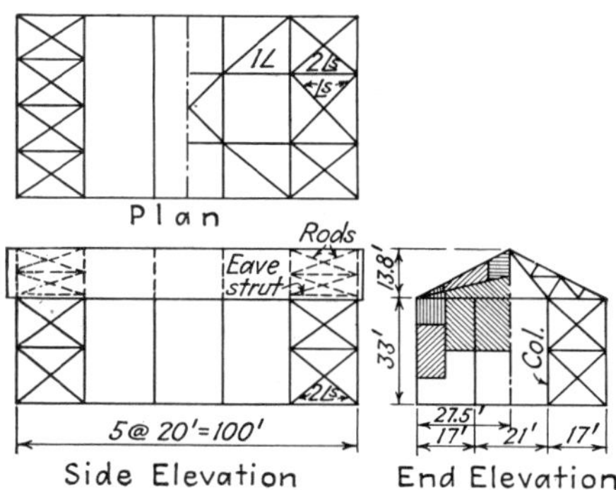

Fig. 55

loads as the trusses deflect. The columns are spaced so that they connect to the trusses approximately at the principal panel points which are supported laterally by the bracing in the plane of the bottom chords. The size of each column may be approximated by designing it as a simple beam for wind on the end of the building, with a small reserve to provide for the vertical loads from the girts and corrugated iron which amount to only a few hundred pounds per square inch.

Illustrative Problem. To design the 33-ft end column of the mill building shown in Fig. 55 for a wind pressure of 20 lb per sq ft. The end columns are approximately 21 ft apart and 17 ft from the centers of the outside columns.

Bending moment due to wind. . . . . $\frac{20 \times 19}{2} \times 16.5^2 = 51{,}700$ ft-lb

$$12 \times 51{,}700 = 26{,}700S$$
$$S = 23$$

This requires a 12-27, or a 10-25.

Bracing Systems. The end bays and every third or fourth intermediate bay of a mill building are fully braced in the planes of both chords of the roof trusses and in a vertical plane between columns; monitors are also braced in the same panels, both vertically and in the planes of the top chords. In the top-chord bracing, ¾-, ⅞-, or 1-in. rods are used as indicated in the left-hand half of the plan (Fig. 55), either single crosses or double crosses being used, whichever makes the angles more nearly 45°. Purlins with extra connections form the compression members of the system. In the plane of the bottom chords, a continuous line of struts extends the full length of the building in line with the end columns; these struts are made of two angles 5 × 3½ × 5⁄16 or 6 × 4 × ⅜ in the braced bays and single angles, like the diagonals, in the intermediate bays; these struts in the braced bays divide the cross bracing into three sets of diagonals, as indicated in the right-hand half of the plan. In the intermediate bays, enough additional angles are placed to form a large system of diagonals extending the entire width of the building to prevent distortion of the building as a whole. When the number of panels is such that there are three bays between braced bays, a full-width cross can be used as shown; otherwise, the two systems overlap, some of the members in the braced bays acting also as part of the larger system. The angle sway bracing between columns is made of a single cross or a double cross, depending on the height of the building. A double-angle strut is used between double crosses, but a more elaborate eave strut is used at the tops of the columns to serve as a compression member for all three systems of bracing. Angle sway bracing is also used at the ends of the building between the main and the end columns. The diagonal angles are usually 3½ × 3 × 5⁄16, 4 × 3 × 5⁄16, or 5 × 3½ × 5⁄16.

The shaded areas show the wind pressure resisted by the different systems of the end bay. The nuts on the rods are tightened, and the angles made a little short and drawn into place by drift pins in order to give them enough initial tension to ensure immediate action. It is not necessary to calculate the stresses in the bracing systems, for larger sections have been more or less standardized in modern practice to give rigidity. The angles are placed with their longer legs vertical to reduce sag, and bottom-chord bracing angles are turned upward to give crane clearance. Top-chord bracing is often omitted when sheathing is used, but care must be taken to prevent collapse during erection. Sway bracing is omitted when masonry walls are used.

Eave Struts. An eave strut is a horizontal member connecting the tops of the columns, serving as a girt, a purlin, and a compression member of three systems of bracing; it may also serve as a support for the sag rods for the girts. An eave strut is commonly made of a channel with a stiffening angle along the top of the web, two angles connected with lattice bars, four angles latticed, or two channels latticed. The section is usually determined by stresses due to wind on the side of the building rather than wind on the end. There is a direct compressive stress caused by the initial tension in each system of diagonals. The intermediate eave struts are made like those in the braced bays.

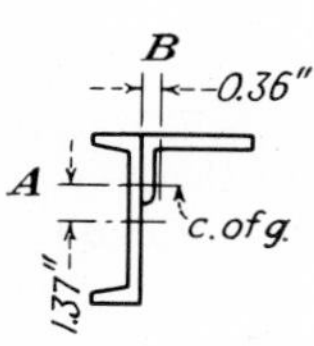

Fig. 56

Illustrative Problem. To design an eave strut for the mill building

shown in Fig. 56, using an 8 ⊏ 11.5, the same size as the purlins but with a 5 × 3½ angle, as shown in Fig. 56.

Weight of sleet, slate, and sheathing per linear foot, using a 1-ft overhang and one-half of a 4-ft space to the next purlin........$(10 + 6.5 + 8)3 = 73.5$

Weight of channel, 5/16 angle assumed, spiking piece, and 2 ft of corrugated steel............................$11.5 + 8.7 + 4 + 1.7 \times 2 = 27.6$

Vertical component of normal wind pressure on roof

$$15 \times 3 \times 27.5 \div 30.9 = 40.0$$

Total vertical component per foot.................................. $= 141$

Weight of windows, corrugated steel, and girts carried by sag rods at mid-point..$\{(2 \times 8 \times 8 \times 12) + 1.7[(20 \times 31) - (2 \times 8 \times 12)] + (6 \times 10.4 \times 20)\} \div 2 = 1{,}760$

Vertical bending moment...............$(^{141}\!/_{2} \times 10^2) + (1{,}760/2 \times 10) = 15{,}900$ ft-lb

Horizontal component due to wind $15 \times 3 \times 14 \div 30.9 + (20 \times 2) = 60$

Horizontal bending moment.............................$^{60}\!/_{2} \times 10^2 = 3{,}000$ ft-lb

Direct stress due to initial tension in three sets of diagonals

$$5{,}000\left(\frac{20}{26.2} + \frac{20}{25.3} + \frac{20}{26}\right) = 11{,}600 \text{ lb}$$

Vertical distance from center of web to center of gravity $\dfrac{2.56(4 - 0.84)}{2.56 + 3.36} = 1.37$

Horizontal distance from back of web to center of gravity

$$\frac{(2.56 \times 1.59) - (3.36 \times 0.58)}{2.56 + 3.36} = 0.36$$

I about horizontal axis A

$$32.3 + 3.36 \times 1.37^2 + 2.7 + 2.56(4 - 0.84 - 1.37)^2 = 49.5$$

I about vertical axis B

$$1.3 + 3.36(0.58 + 0.36)^2 + 6.6 + 2.56(1.59 - 0.36)^2 = 14.8$$

Unit stress due to vertical bending....$15{,}900 \times 12(4 - 1.37) \div 49.5 = 10{,}100$

Unit stress due to horizontal bending...$3{,}000 \times 12(2.26 + 0.36) \div 14.8 = 6{,}400$

Unit stress due to initial tension.......................$11{,}600 \div 5.92 = 1{,}960$

r^2...$14.8 \div 5.92 = 2.5$

Unit stress allowed in compression $1/r > 120$........ $\dfrac{18{,}000 \times \frac{4}{3}}{1 + \dfrac{(20 \times 12)^2}{18{,}000 \times 2.5}} = 10{,}500$

The sum of the fractions.................. $\dfrac{10{,}100 + 6{,}400}{26{,}700} + \dfrac{1{,}960}{10{,}500} = 0.81$

This is satisfactory because a 4 × 3½ × 5/16 angle would be too small.

Main Columns. The lower part of a column which supports crane-runway girders should project beyond the upper portion to allow sufficient clearance for the crane to pass the upper portion which supports the roof. The upper part must be spliced to the lower part by enough rivets or welds to carry the direct stress and the bending; it may not be necessary to use both web and flange splices. Some type of seat is usually provided for the crane girders with a distributing plate and supporting angles and stiffeners. Knee braces are used also. A diaphragm is used to connect the top of the girder to the column to prevent the girder from overturning. It is not satisfactory to connect the girder web directly to the column flange or to use a deep diaphragm connecting to the girder web, because the rivets or welds are likely to be loosened by the continuous-beam action. Separate columns for the crane runway are also used and tied to the columns which support the roof. The columns are designed to resist the overturning tendency

of the wind on the side of the building. The sway bracing at the ends of the building is sufficient to prevent a short structure from overturning, but even with bottom-chord bracing it is not very satisfactory in preventing the collapse of the intermediate bents of a long structure. The connection of the truss to the column even with a large gusset plate is not especially rigid unless a knee brace is used between the column and one of the panel points of the truss, or unless the truss is deep enough at the end to have two connections to the column.

With knee braces, each bent is designed to resist the wind pressure on one panel. The restraint caused by the knee braces and by the anchored bases makes the columns bend in two directions, as shown in Fig. 57. The determination of the point of con-

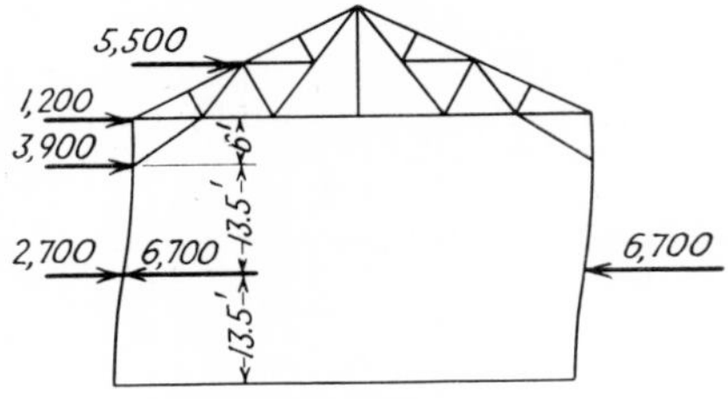

FIG. 57

traflexure is not exact because of too many uncertain elements, such as the amount of restraint and the fact that the column is not of uniform width; it is close enough to assume it midway between the knee brace and the base. When the bent as a whole is designed for horizontal pressure on the column and normal pressure on one side of the roof, temporary members are used, intersecting opposite the knee-brace connections, to replace the column by a simple truss so that stresses in the knee braces and truss members may be found graphically; later, modified stresses in the column after the temporary members are removed may be found. Since so many of the truss members are determined by minimum section or by slenderness ratio, it is usually better to design the truss for equivalent vertical loads which provide for both gravity and wind loads, applied to both sides of the truss. It is then difficult to determine just what wind forces to consider in the design of the column. Since the vertical stresses have relatively little effect upon design, it is close enough to take the truss reaction as determined from the symmetrical roof loads, without adding anything for the overturning due to the horizontal wind forces. The bending moments may be obtained from

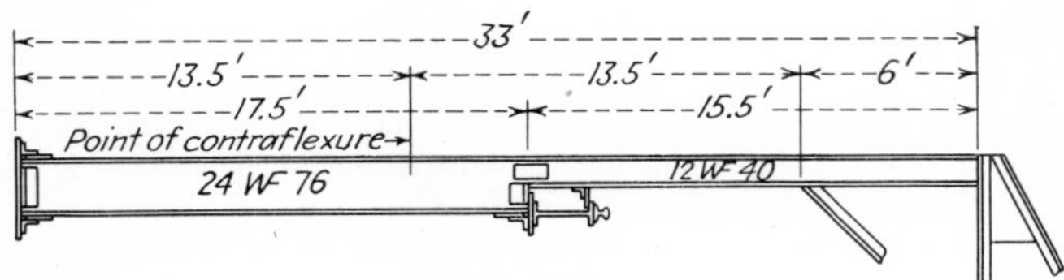

FIG. 58

the horizontal forces indicated in the figure, including one at the mid-height of the truss based upon the vertical projection of the roof, 20 × 20 × 13.8 = 5,500. The pressures at the heel, the knee brace, and the point of contraflexure are 20 × 20 times the vertical distances ½ × 6, ½(6 + 13.5), and ½ × 13.5. The horizontal components of the reactions are assumed to be equal, each being ½(5,500 + 1,200 + 3,900 + 2,700)

= 6,700. The maximum bending moment in the upper portion of the column is at the foot of the knee brace on the leeward side, or 6,700 × 13.5 = 90,400. In the windward column it would be only (6,700 − 2,700)13.5. The horizontal component of the compression in the leeward knee brace may be found from the reaction in the leeward column by taking moments about the heel, or 6,700 × 19.5 ÷ 6.0. The unsupported length of the upper section is the length of the narrow portion, 15.5 (Fig. 58). Assuming a 12-40 section,

Unit stress developed in bending.......................... $\dfrac{12 \times 90{,}400}{51.9} = 20{,}900$

Vertical load at knee brace due to roof, truss, siding, girts, and column

$$(44 \times 20 \times 31) + [(6 \times 1.5) + 10.4]20 + 40 \times 6 = 28{,}000$$

Unit stress developed in compression...................... $28{,}000 \div 11.77 = 2{,}400$

Unit stress allowed in compression..... $\left[17{,}000 - 0.485\left(\dfrac{15.5 \times 12}{1.94}\right)^2\right] \tfrac{4}{3} = 16{,}200$

The sum of the fractions.............................. $\dfrac{20{,}900}{26{,}700} + \dfrac{2{,}400}{16{,}200} = 0.93$

This is satisfactory, being less than unity, but the next size smaller is found to be insufficient.

The lower section of the column must be about 11 in. wider than the upper section in order to provide proper clearance for the passage of the crane; this means that a 24-in. column must be used. The flange of the column directly under the crane girder should carry the full crane load at the normal unit stress found from the length from the base to the crane seat and a radius of gyration for that portion about an axis parallel to the web; for this portion designed for maximum crane load, the knee braces would cause rather than resist bending. The radius of gyration for the rectangular flange is 0.289 times the flange width.

Maximum crane load, one load at column

$$\left(44{,}500 \times 1.25 \times \frac{20 + 9.2}{20}\right) + (155 \times 20) = 84{,}300$$

Allowed unit stress, assuming 9-in. flange......$17{,}000 - 0.485\left(\dfrac{17.5 \times 12}{9 \times 0.289}\right)^2 = 13{,}900$

Thickness of flange required to give proper area................ $\dfrac{84{,}300}{13{,}900 \times 9} = 0.68$

Try a 24-76.

Considering the column as a whole, the stresses due to vertical loads, wind moments, and eccentric moment must be combined at the critical section, which may be taken about 1 ft above the base if the reinforcement furnished by the base angles and wing plates seems to justify it. The windward column is designed so that the compression due to wind will act on the same side as the crane. The effect of the wind above the point of contraflexure is concentrated at that point, and the effect of the wind below the point of contraflexure may be considered as uniformly distributed. In determining the eccentricity, it is close enough to consider the girt loads (taken as twice the sag-rod load), ¾ of the truss load and the weight of the upper-column section, and ½ of the weight of the lower section to act at one face of the column, while the crane load, ¼ of the truss load and weight of upper section, and ½ of the weight of the lower section act at the face of the opposite flange. The length is the full length of the lower section and the radius of gyration is the least radius.

Load at outer face

$(1{,}760 \times 2) + [28{,}000 + (40 \times 9.5)]0.75 + (76 \times 17.5 \times 0.5) = 25{,}500$

Load at inner face. . $84{,}300 + [28{,}000 + (40 \times 9.5)]0.25 + (76 \times 17.5 \times 0.5) = 92{,}100$

Moment due to eccentricity (12-in. lever arm). . . . . $(92{,}100 - 25{,}500)1 = 66{,}600$

Moment due to wind. . . . . $(6{,}700 - 2{,}700)12.5 + (20 \times 20 \times 12.5^2/2) = 81{,}200$

Unit stress due to bending. . . . . $\dfrac{12(66{,}600 + 81{,}200)}{175.4} = 10{,}100$

Unit stress due to vertical loads. . . . . $\dfrac{25{,}500 + 92{,}100}{22.37} = 5{,}300$

Unit stress allowed in compression. . . . . . $\left[17{,}000 - 0.485\left(\dfrac{17.5 \times 12}{1.85}\right)^2\right]\dfrac{4}{3} = 14{,}300$

Sum of the fractions. . . . . $\dfrac{10{,}100}{26{,}700} + \dfrac{5{,}300}{14{,}300} = 0.75$

which is less than unity. This is the lightest 24-in. section; so it is selected.

Columns Bases. All crane columns are milled to ensure contact with the base plate, but much of the load must be transmitted to the base angles and plate by rivets or welds, as shown in Fig. 59, so that the anchor bolts can hold the column in place. Trucks or swinging crane loads are liable to strike the column, and the vertical load is usually not sufficient to prevent displacement. The anchor bolts also prevent uplift due to the overturning tendency of the wind, and the angles must be fastened to the column in order to function in this manner. The method of design is based upon the assumption that the footing is rigidly embedded in suitable material so that it is stable enough to develop a fixed end for the anchored column under the assumed loading. This condition may be far different from the actual conditions realized, and the designer often proportions the base with judgment based on experience without computation. If he wishes to supplement his judgment with figures, the method given may be of interest. A base angle is connected to each flange of the column by four rivets or equivalent welding; additional rivets may be used if a wing plate is placed between the angle and the flange, the wing plate and the angle extending beyond the flange far enough to provide driving clearance and edge distance for the rivets. Corresponding rivets must connect the wing plate to the flange above the angle, and additional rivets may be used because of the bearing of the wing plate on the base plate. Sometimes an inverted angle is placed above the base angle, and the anchor bolts extend through both outstanding legs with stiffening angles between. The width b of the base plate is assumed the same as the wing plates and angles, and the length d great enough to extend slightly beyond the angles, both dimensions being a multiple of 1 in. The total gravity load P causes a uniform unit pressure equal to P/bd. The total overturning moment m, in inch-pounds, due to wind and eccentricity, causes a variable downward pressure on one-half the base and an equal uplift on the other half, the maximum unit pressure or uplift at the end of the plate being $\dfrac{6m}{bd^2}$. The combined pressure at one end of the plate is $\dfrac{P}{bd} + \dfrac{6m}{bd^2}$, which must not exceed the allowed pressure on the masonry increased for combined loads. The resultant uplift T, if

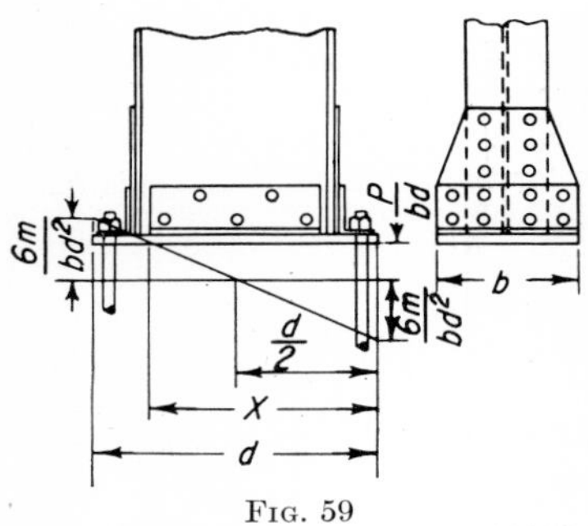

Fig. 59

any, must be resisted by anchor bolts. By similar triangles, the distance to the point of zero pressure is $x = \frac{d}{2}\left(\frac{\frac{P}{bd} + \frac{6m}{bd^2}}{6m/bd^2}\right) = \frac{d}{2}\left(\frac{Pd}{6m} + 1\right)$, and the total pressure is $\frac{bx}{2}\left(\frac{P}{bd} + \frac{6m}{bd^2}\right)$. The total uplift or tension in the rods at one end is the difference between this total pressure and the total vertical load P. By substituting the value of x and simplifying, this expression becomes $T = \frac{P^2d}{24m} + \frac{3m}{2d} - \frac{P}{2}$. The thickness of the base plate, or the combined thickness of plate and angles, may be found for the corresponding projection p from the usual expression $t = p\sqrt{3b'/f}$, in which b' is the unit pressure at the critical point found by proportion from the maximum pressure and f is the increased unit stress in bending.

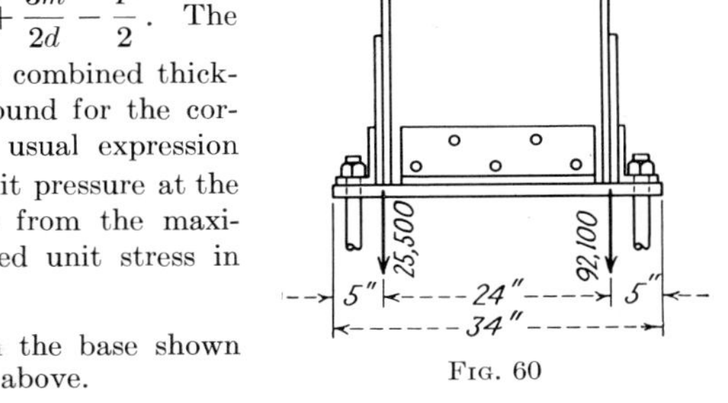

Fig. 60

Illustrative Problem. To design the base shown in Fig. 60 for the column designed above.

Minimum width of base plate, b .... $9 + (2 \times 3) =$ 15 in.
Minimum length, d .... $24 + 2(4 + 3/8) + 1\frac{1}{4} =$ 34 in.
Total load, P .... $25{,}500 + 92{,}100 =$ 117,600
Bending moment due to wind
$(6{,}700 - 2{,}700)13.5 + (20 \times 20 \times 13.5^2/2) =$ 90,400 ft-lb
Eccentric moment .... $(92{,}100 - 25{,}500)1 =$ 66,600 ft-lb
Total overturning moment .... $(90{,}400 + 66{,}600)12 =$ 1,884,000 in.-lb
Maximum unit pressure .... $\frac{117{,}600}{15 \times 34} + \frac{6 \times 1{,}884{,}000}{15 \times 34^2} =$ 882
This exceeds $600 \times 4/3 = 800$; so the width is increased to 17, and the resulting pressure is thus reduced to 779, which is satisfactory.
Maximum tension, T ... $\frac{117{,}600^2 \times 34}{24 \times 1{,}884{,}000} + \frac{3 \times 1{,}884{,}000}{2 \times 34} - \frac{117{,}600}{2} =$ 34,800
Area required at root of thread .... $34{,}800 \div 26{,}700 =$ 1.30
Diameter, if single bolt is used .... $= 1\frac{1}{2}$
Projection, p, beyond vertical leg of ¾-in. flange angle .. $5\frac{3}{8} - \frac{3}{4} =$ 3.9
Thickness of plate and angle .... $3.9\sqrt{3 \times 779/26{,}700} = 1\frac{1}{4}$
Thickness of plate under flange angle .... $1\frac{1}{4} - \frac{3}{4} = \frac{1}{2}$
Projection, p', beyond vertical leg of ¾-in. web angle
$\frac{17 - 0.44}{2} - \frac{3}{4} =$ 7.5
Distance to point of zero pressure .... $x = \frac{34}{2}\left(\frac{117{,}600 \times 34}{6 \times 1{,}884{,}000} + 1\right) =$ 23.0
Pressure 8 in. from right end .... $\frac{23 - 8}{23} \times 779 =$ 507
Thickness of plate and angle .... $7.5\sqrt{3 \times 507/26{,}700} =$ 1.8
Thickness of plate under angle .... $1.8 - \frac{3}{4} = 1\frac{1}{8}$

The selection of the critical section is a matter of judgment. The pressure under the web angle is greatest at the right end and is zero 6 in. to the left of the center. The base plate is prevented from bending by the flange, the wing plate, and the flange

angle; so it is necessary to find the thickness at the end of the web angle, or in the space between this and the flange. The full length of the web angle may help to resist bending, but the right end may bend first and cause a progressive failure. It seems reasonable to find the thickness for a section approximately 3 in. from the wing plate, as was done above. The 1⅛ is thicker than the ½ required at the end and is therefore selected.

Bibliography

American Institute of Steel Construction: "Structural Steel Drafting," 3 vols., American Institute of Steel Construction, 1950–1955.

American Welding Society: "Design for Welding," American Welding Society, 1950.

Bishop, C. T.: "Structural Design," John Wiley & Sons, Inc., 1938.

Bishop, C. T.: "Structural Drafting," John Wiley & Sons, Inc., 1941.

Cissel, J. H.: "Stress Analysis and Design of Elementary Structures," John Wiley & Sons, Inc., 1948.

Dencer, F. W.: "Detailing and Fabricating Structural Steel," McGraw-Hill Book Company, Inc., 1930.

Dunham, C. W.: "Planning Industrial Structures," McGraw-Hill Book Company, Inc., 1948.

Fuller and Kerekes: "Analysis and Design of Steel Structures," D. Van Nostrand Company, Inc., 1937.

Green, R. S.: "Design for Welding," Lincoln Electric Company, 1948.

Grinter, L. E.: "Design of Modern Steel Structures," The Macmillan Company, 1941.

Hauf and Pfisterer: "Design of Steel Buildings," John Wiley & Sons, Inc., 1949.

Huntington, W. C.: "Building Construction," John Wiley & Sons, Inc., 1941.

Ketchum, M. S.: "Handbook of Standard Structural Details for Buildings," Prentice-Hall, Inc., 1956.

Kidder-Parker: "Architects' and Builders' Handbook," John Wiley & Sons, Inc., 1945.

Kirkham, J. E.: "Structural Engineering," McGraw-Hill Book Company, Inc., 1933.

Lincoln Electric Company: "Procedure Handbook for Welding," 1945.

Shedd, T. C.: "Structural Design in Steel," John Wiley & Sons, Inc., 1934.

Sutherland and Bowman: "Structural Design," John Wiley & Sons, Inc., 1938.

Williams and Harris: "Structural Design in Metals," The Ronald Press Company, 1949.

Section 7

Part 1: CEMENT AND CONCRETE

By Herbert J. Gilkey

INTRODUCTION

Definitions and Scope. In its broader sense *concrete* may be defined as an artificially made conglomerate of inert particles held by a matrix of cementing or binding material. Without qualification, concrete has come to mean *portland cement concrete*, but there are other cements and other concretes. For the purpose of this treatment the *cement* is restricted to a mineral, usually in powder form, which can be mixed with water to form a plastic mass and which hardens chemically by combination and by gel and crystal formation rather than by cooling, drying, or vitrification. These limitations exclude clay, probably the oldest of the cementing materials of construction. They also exclude the asphaltic concretes and the organic *adhesives*—pastes and glues—many of which are called cements. In addition to the *portland cements* the scope does include *lime* products, *gypsum* products, *Roman* or *pozzolan* cement, *natural* or *Rosendale* cement, *alumina* or *Lumnite* cement, masonry cements, *plastic magnesia* (*magnesium oxychloride*) or *Sorel cement* and, more recently, *slag cement*, all of which are accorded introductory mention.

Sources of Information; Miscellaneous Items. The field is broad and constantly undergoing change. Since about 1940, for example, some of the major developments with regard to cement and concrete have greatly modified several well-established concepts and practices. The treatment here aspires only to outline the subject as of this date but in doing so to indicate authoritative sources for more complete coverage and to point out publication channels where the records of future progress may be expected to appear. At the end of this section are listed a few general sources of information with which anyone interested in cement and/or concrete needs to be conversant. Specific references are cited in legends for tables or figures or parenthetically in the text.

In Fig. 1 are shown representative relative compositions for all the cements mentioned except the magnesium oxychloride or Sorel cement (the only one in which calcium is not a primary constituent) and the calcium sulfate or gypsum derivatives (plaster of paris, Keene's cement, and other hard plasters). Incidentally, the pumicite of Fig. 1 is not a cement; it is a cementing material that acquires cementing properties in combination with the lime, usually present in excess in portland and other cements. Table 1 supplies similar information for hydraulic cements and pozzolanic materials. In 1953 the United States produced 264 million bbl of portland cement in 156 plants, this being more than one-fourth the world production. Types I and II (see Tables 2 and 3) accounted for about 83 percent and air-entraining cement (much of it Type I or II) constituted another 12.6 percent. Type III (high-early-strength), portland pozzolan, oil-well, and white cement amounted to 3.0, 0.9, 0.7, and 0.4 percent respectively.

Estimated 1955 United States production of portland cement was 293 million bbl. The United States production of nonportland hydraulic cements, including natural, masonry (natural), slag lime pozzolan, and alumina, totaled in 1953 only about 1.3 percent as much as the portlands.

NONHYDRAULIC CEMENTING MATERIALS

Lime Mortars and Plasters. Lime (ASTM C51) mortars and plasters are of ancient origin and continue to be important construction materials. Lime is used both as a primary cementing material and admixed with gypsum plasters and portland cement. The *quicklime* CaO obtained from the dehydration (calcination) of limestone [calcium carbonate ($CaCO_3$)] is either prehydrated to form the dry *hydrated lime powder* [$Ca(OH)_2$] (much more stable than the CaO) or is water slaked on the job to form the wet *hydrate* or *lime* putty ready for mixing with sand and more water as needed. The hardening is a return to the carbonate ($CaCO_3$) necessitating access to the air for the acquisition of carbon dioxide (CO_2). Lime is hydraulic (will harden under water) only if impurities such as clays or silicates happen to be present (see Fig. 1) in which case it is called *hydraulic lime.* Lime plasters shrink considerably upon hardening and drying, and gypsum plasters, some of which swell in hardening, are often admixed to reduce shrinkage as well as to speed up hardening and/or to increase the strength. Portland cement may be admixed to provide greater strength, although strength is rarely of primary importance for a lime mortar. The compressive stress at the base of the 555-ft Washington monument is less than 300 psi. Lime contributes "fatness" or smooth workability under the trowel and is often added as a plasticizer to both gypsum and portland cement mixtures. Sand-lime brick are sometimes competitive with vitrified clay brick and to some extent with precast concrete blocks.

Gypsum Plasters and Concretes. Gypsum plaster work over 3,500 years old is found in the pyramids and 2,400 years old in the Temple of Apollo at Bassae. Gypsum products (ASTM C11) include plaster of paris (variously known as wall plaster, finishing plaster, molding plaster, pottery plaster, and dental plaster) and the hard or hard-finish plasters (also called flooring plaster and tiling plaster) among the better known of which are *Keene's cement* (ASTM C61), *Parian,* and *Martin's cement.*

The lack of shrinkage and the ease with which plaster products can be cast to form rapid-hardening (from ½ hr to 2 or 3 days) strong lightweight products establish the importance of gypsum plasters over a wide range of use. Gypsum plasters are hydraulic but are slowly soluble and cannot be used satisfactorily in moist locations. Slater (*Proc. ASTM,* vol. 19, Pt. 1, 1919; *Jour. Western Soc. Engrs.,* vol. 24, 1919) evolved design procedures and formulas for *reinforced gypsum* design, and reinforced gypsum "concrete" (ASTM C317) is used in subbases for floors and roofs. Much of the gypsum used in construction is in the form of precast blocks, tile, and slabs. Cast-in-place and precast gypsum products are widely used as "fireproofing" (thermal insulation) for structural-steel construction.

Both lime and gypsum as construction materials have always belonged to the building trades, proportioning and use being in the hands of artisans. Satisfactory results depend more on personal skill and workmanship than on design and formal control. In some degree this aspect is undergoing change.

Magnesium Oxychloride, Plastic Magnesia, or Sorel Cement (ASTM C376). This cement may be used with a variety of aggregates, from mineral to such filler materials as sawdust for strong quick-hardening concrete of limited application. In from 1 to 7 days strengths comparable to the 28-day strengths of portland cement mixtures can be developed. The cement consists of magnesium chloride ($MgCl_2$) and calcined magnesia. The mortar is used as a base for interior floorings, such as terrazzo and asphalt tile, especially in railroad cars. It is suitable for use only in dry locations, and difficulty

may be encountered in attaining uniformity. Mixed with asbestos fiber (ASTM C193) it forms a weak cement much used for insulating around furnaces and steam pipes. It withstands temperatures up to about 600°F. Although it is not a new cement, only in 1947 was ASTM Committee C2 organized to formulate specifications, testing methods, and definitions governing magnesium oxychloride cements and their products.

HYDRAULIC CEMENTS OTHER THAN PORTLAND

The true hydraulic cements are the various *pozzolanic* and *natural cements, alumina cement, slag cement,* and *portland cement.* Other cements and cementing materials display hydraulic properties in that they will harden under water, but their slow solubility makes them unsuited for use under water or in damp locations. Table 1 supplies typical compositions for hydraulic cements including portland cement, as does also Fig. 1.

TABLE 1. TYPICAL PERCENTAGE COMPOSITIONS OF VARIOUS HYDRAULIC CEMENTS AND POZZOLANIC MATERIALS [1]

Kind of cement	SiO_2	Al_2O_3	Fe_2O_3 *	CaO	MgO	SO_3	K_2O †	Na_2O †	Insoluble †	Loss †
Portland Type I	21.3	6.0	2.7	63.2	2.9	1.8	...	...	0.2	1.3
Portland Type II	22.3	4.7	4.3	63.1	2.5	1.7	...	...	0.1	0.8
Portland Type III	20.4	5.9	3.1	64.3	2.0	2.3	...	...	0.2	1.2
Portland Type IV	24.3	4.3	4.1	62.2	1.8	1.9	...	...	0.2	0.9
Portland Type V	25.0	3.4	2.8	64.1	1.9	1.6	...	...	0.2	0.9
Portland white	25.5	5.9	0.6	65.0	1.1	0.1	...	...	n.d.	n.d.
Portland-pozzolan ASTM C340	21.0	5.0	3.0	63.0	3.0	1.5	...	...	...	2.0
Slag ASTM C358	36.0	14.0	1.0	45.0	4.0	1.0				
Aluminous	5.3	39.8	14.6	33.5	1.3	0.1	...	...	4.8	0
Natural	27.8	5.5	4.3	35.6	21.2	0.5	...	...	n.d.	4.1
Pozzolan	26.0	6.9	3.6	52.3	4.2	1.8	...	...	9.4	4.8
Trass 30/70 (German) ‡	22.3	8.3	2.5	46.8	1.9	1.8	...	...	9.5	3.4
Trass 50/50 (German) §	25.1	8.9	3.3	35.3	1.2	1.1	...	...	15.9	5.8
Eisenportland (German) ‖	23.5	8.5	1.7	57.2	3.3	2.0	...	...	n.d.	n.d.
Pozzolanic materials:										
Pumicite	72.3	13.3	1.4	0.7	0.4	Trace	5.4	1.6	...	4.2
Burned clay	58.2	18.4	9.3	3.3	3.9	1.1	3.1	0.8	...	1.6
Burned shale	51.7	22.4	11.2	4.3	1.1	2.1	2.5	1.2	...	3.2
Blast-furnace slag	33.9	13.1	1.7	45.3	2.0	Trace	n.d.	n.d.	...	n.d.
Santorin earth	63.2	13.2	4.9	4.0	2.1	0.7	2.6	3.9	...	4.9
Rhenish trass	55.2	16.4	4.6	2.6	1.3	0.1	5.0	4.3	...	10.1
Burned gaize	83.9	8.3	3.2	2.4	1.0	0.7	n.d.	n.d.	...	0.4

[1] Excerpted mainly from Ref. 1, Tables 2.3 and 2.6, pp. 32–33.
* Includes any FeO present, calculated as Fe_2O_3.
† n.d. = not determined.
‡ Trass cement having a trass—clinker ratio of 30:70.
§ Trass cement having a trass—clinker ratio of 50:50.
‖ Eisenportland is a German portland slag cement, 65 to 70 percent clinker, 35 to 30 percent slag.

Pozzolanic and Natural Cements. Of the true hydraulic cements (*natural, portland,* and *alumina*) the oldest is the ancient Roman *natural* cement produced by mixing lime with a volcanic ash called *pozzolana* (also pozzuolana, puzzolana, or puzzolano) after the town of Pozzuoli near Naples where it was first found. It has since been found in many other places. The *trass* of Germany and Holland and the *Santorin* cement of Greece (from the island of Santorin) are of similar origin and composition. Artificial *pozzolana* or *slag cements* (ASTM C358) (not to be confused with *portland cement* manufactured from slag, ASTM C205) are made from pulverized blast-furnace slags and from highly siliceous minerals such as diatomite and pumicite.

Under the Romans the concrete art became highly developed, and much excellent concrete was placed as judged even by today's standards. It must be recognized, however, that neither the climatic exposure nor the rigors of use were at all comparable with those encountered by a modern pavement in the northern United States. The

writings of Vitruvius indicate a recognition of many of the problems of concrete that we think of today as being modern developments.

Following the Romans, the cement arts had a long period of dormancy or retrogression and it was not until the classic studies of John Smeaton in preparation for rebuilding the Edystone (Smeaton's spelling) lighthouse (1756–1759) that progress was resumed. Smeaton evolved a process of manufacture for *natural* cement whereas the Romans had merely used materials already fully prepared except for the addition of lime. He recognized that impure limestone when calcined and ground had hydraulic properties. The mortar he used was *hydraulic lime* or virtually a *natural cement.*

Smeaton's primary advances over Roman practice were his recognition that impurities in the natural limestone accounted for the hydraulic properties and his use of calcination and the subsequent grinding of the calcined product. The Roman cements had been largely manufactured by nature through volcanic calcination and were already preground in the form of volcanic dust or ash. In his careful selection of the impure limestone through much experimentation, Smeaton secured a manufactured natural cement of excellent quality but missed the discovery of portland cement because of:

1. His reliance upon natural rather than upon controlled artificial proportioning.
2. The low calcining temperature employed; high enough for complete decarbonization but not high enough for vitrification or incipient fusion.
3. His failure to recognize that the hard-burned particles when ground formed the best cement; instead these were discarded.

Natural cement (ASTM C310) is less uniform, less strong, and slower setting than is portland cement.

About 1818, as the construction of the Erie Canal was begun, 6 years prior to Aspdin's British patent on the manufacture of portland cement, a large deposit of uniform natural cement rock was discovered in New York. The use in the canal structures of natural cement manufactured from these deposits contributed to the continued widespread use of natural cement in this country for many additional years. The relative importance of natural cement decreased rapidly, however, after the turn of the century, and since about 1910 the use of Rosendale (natural) cement has been mainly local and largely as a constituent of masonry mortars. In the late 1930's interest in natural cement was revived as a blend with portland for highway pavement construction because of the apparent added resistance it gave against the damage incident to freezing and the use of active chlorine salts in snow removal. Subsequent tests indicate that the beneficial effect of the natural cement blends is probably due solely to a limited contribution of air entrainment (see "Air Entrainment" and "Admixtures") from a slight oil contamination and the use of tallow as a grinding aid. Better results can be secured by using "air-entraining portland cement" (ASTM C175) or by using air-entraining admixtures in the concrete.

Alumina Cement. Alumina cement, currently the most important of the nonportland cements, was developed in France about 1908 as *ciment fondu* (so called because of the ease with which it fuses). It also carries such names as *high-alumina cement, calcium aluminate cement, aluminous cement,* and *Lumnite* (the proprietary trade name under which it is manufactured and sold in the United States by the Universal Atlas Portland Cement Co.). During World War I, *alumina* cement came into European prominence because of its marked resistance to active waters and soils. Manufacture started in England, Germany, and the United States about 1924. In this country its initial appeal was for use in high-early-strength or 24-hr concrete having about the same strength as 28-day portland cement concrete, making it valuable for emergency repairs, for the setting of machinery, and for strategic construction locations to expedite erection schedules. More recently it has assumed an important place in some phases of oil-well cementing (grouting) and for corrosion-resistant concrete exposed not only to naturally

active soils and waters but to industrial wastes, including use in linings for tanks and in wrought iron and steel pipes. High-alumina cement concrete is not resistant to strong caustics or to acetic, hydrochloric, and nitric acids. Alumina cement is valuable for cold-weather concreting both because of the short period over which it is vulnerable to freezing and because its high heat of hydration provides added protection during the limited period of vulnerability. It is becoming increasingly important for heat-resistant and refractory concrete, crushed (old) firebrick being often used for aggregate. Whereas portland cement is virtually dehydrated and disintegrated at 500°C (900°F) (see Fig. 26), alumina cement retains enough of its combined water for stability and moderate strength up to about 1600°F, at which temperature it begins to melt slightly, attaining a fired or fused bond which provides renewed or added strength. If finely ground topaz, silica flour, or potter's flint is used as an admixture, the strength increases greatly between 1600° and 2200°F.

For many purposes the basis of selection between portland and alumina cement concrete is economic, the choice lying between alumina and use of a richer concrete, possibly Type III portland cement with or without steam curing. (Steam curing is unsuitable for use with alumina cement.) Its high rapid heat of hydration makes alumina cement entirely unsuited to massive construction, and it must be kept thoroughly wet throughout its short period of hydration to meet cooling as well as hydrating needs. In combination with portland cement, alumina cement is subject to flash set, and care should be exercised to avoid use of the same mixing equipment without thorough cleaning between uses.

The high percentage of alumina (see Table 1 and Fig. 1) differentiates the alumina cement from the natural, pozzolanic, and portland cements. Alumina cement cannot be manufactured from limestone and clay but requires bauxite, an impure hydrated alumina for mixing with the limestone. It is several times as costly as ordinary portland cement. There is currently (1957) no ASTM standard for alumina cement.

PORTLAND CEMENT

History and Development. After a great deal of supplementary and additional work by others subsequent to the studies of Smeaton (much of it repetitive), it presumably remained for Joseph Aspdin, a bricklayer of Leeds, England, to recognize and capitalize upon the harder burning and artificial proportioning of the argillaceous and calcareous constituents and in 1824 to patent a process for the manufacture of "portland" cement, reportedly so named because of the resemblance of the hardened mortar to the oölitic limestone on Portland Island. Another version ascribes the name to a quotation from Smeaton's writings: ". . . to make a cement that would equal the best merchantable portland stone in solidity and durability." (See Refs. 1 and 2 for a more detailed recital of historical pros and cons.)

The high cost of importing portland cement and the well-established use of natural cement retarded familiarity with, and demand for, portland cement in the United States, and it was not until 1871 that the first mill of importance was put in operation by David O. Saylor at Coplay, Pa. Cement from this mill was specified in 1878 for use by James B. Eads in the construction of the jetties at the mouth of the Mississippi River (the first important public work in this country for which portland cement was used). The first portland cement concrete street was built by George W. Bartholomew in 1891 at Bellefontaine, Ohio, and it remains in excellent condition. Although much used for sidewalks, concrete did not begin to be widely used for highways until some 20 years later.

From the beginning portland cement has conformed to one general definition but has nevertheless undergone continual evolutionary change in actual proportions and in manufacturing techniques. Aspdin's "portland" cement probably bore little resem-

TABLE 2. PORTLAND CEMENTS COMPARED ON BASIS OF ASTM CHEMICAL AND PHYSICAL REQUIREMENTS AND ON AVERAGE VALUES FOR COMPOUND COMPOSITIONS AND COMBINATIONS [1]

	Description and use	Type									
		I				II		III		IV	V
		General use				Moderate low heat and sulfate resisting		High early strength		Low heat	High sulfate resisting
		Regular portland cement		Portland blast-furnace slag cement		Regular portland cement					
	ASTM Specification that covers	C150-55	C175-55T	C205-53T		C150	C175	C150	C175	C150	C150
	Subtype (A = air entrained; S = slag)	I	IA	IS	IS-A	II	IIA	III	IIIA	IV	V
1	ASTM Specification requirements (partial)										
2	Chemical per ASTM C114:										
3	SiO_2, min percent	....	....	....	....	21.0	21.0				
4	Al_2O_3, max percent	....	....	....	....	6.0	6.0	....	....	....	4.0
5	Fe_2O_3, max percent	....	....	....	....	6.0	6.0	....	....	6.5	5.0
6	MgO, max percent	5.0	5.0	5.0	5.0	5.0	5.0	5.0	5.0	5.0	4.0
7	SO_3, max percent	2.5	2.5	2.5	2.5	2.5	2.5	3.0	3.0	2.3	2.3
8	Mn_2O_3, max percent	....	....	1.5	1.5						
9	S, max percent	....	....	2.0	2.0						
10	Loss on ignition, max percent	3.0	3.0	3.0	3.0	3.0	3.0	3.0	3.0	2.3	3.0
11	Insoluble residue, max percent	0.75	0.75	1.0	1.0	0.75	0.75	0.75	0.75	0.75	0.75
12	C_3S (tricalcium silicate), max percent	....	....	....	....	50	50	....	....	35	50
13	C_2S (dicalcium silicate), min percent	....	....	....	....	....	....	....	....	40	
14	C_3A (tricalcium aluminate), max percent	....	....	....	....	8	8	15	15	7	5
15	Physical requirements per ASTM:										
16	Fineness 325 (44 microns) sieve (C115)	....	....	12	12						
17	Fineness specific surface permeability, min (C204)	....	....	3,400	3,400						
18	Fineness specific surface turbidimeter, min (C115)	1,600	1,600	....	....	1,700	1,700	....	....	1,800	1,800
19	Soundness, autoclave, max percent (C151)	0.50	0.50	0.20	0.20	0.50	0.50	0.50	0.50	0.50	0.50
20	Time of set per ASTM										
21	Initial (Gillmore), min (mins) (C266)	60	60	....	....	60	60	60	60	60	60
22	Final (Gillmore), max (hr)	10	10	....	....	10	10	10	10	10	10
23	Initial (Vicat), min (mins) (C191)	45	45	45	45	45	45	45	45	45	45
24	Final (Vicat), max (hr)	10	10	7	7	10	10	10	10	10	10
25	Air content mortar, percent by volume (C185)	....	19 ± 3	....	16 ± 4	....	19 ± 3	....	18 ± 3		
26	Compressive strength, psi, 1 day (C109)	....	....	....	....	....	....	1,700	1,100		

Description and use	Type									
	I				II		III		IV	V
	General use				Moderate low heat and sulfate resisting		High early strength		Low heat	High sulfate resisting
	Regular portland cement		Portland blast-furnace slag cement		Regular portland cement					
ASTM Specification that covers	C150-55	C175-55T	C205-53T		C150	C175	C150	C175	C150	C150
Subtype (A = air entrained; S = slag)	I	IA	IS	IS-A	II	IIA	III	IIIA	IV	V
27 Compressive strength, psi, 3 days (C109)	1,200	900	900	750	1,000	750	3,000	2,200		
28 Compressive strength, psi, 7 days (C109)	2,100	1,500	1,800	1,500	1,800	1,400	......	......	800	1,500
29 Compressive strength, psi, 28 days (C109)	3,500	2,800	3,000	3,000	3,500	2,800	......	......	2,000	3,000
30 Tensile strength, psi, 1 day (C190)	......	......	......	......	......	......	275			
31 Tensile strength, psi, 3 days (C190)	150	......	150	......	125	......	375			
32 Tensile strength, psi, 7 days (C190)	275	......	275	......	250	......	......	......	175	250
33 Tensile strength, 28 days (C190)	350	......	350	......	325	......	......	......	300	325
34 Compounds average percent[1]										
35 $C_3S = 3CaO \cdot SiO_2$	45	......	......	......	44	......	53	......	28	38
36 $C_2S = 2CaO \cdot SiO_2$	27	......	......	......	31	......	19	......	49	43
37 $C_3A = 3CaO \cdot Al_2O_3$	11	......	......	......	5	......	11	......	4	4
38 $C_4AF = 4CaO \cdot Al_2O_3 \cdot Fe_2O_3$	8	......	......	......	13	......	9	......	12	9
39 $CaSO_4$	3.1	......	......	......	2.8	......	4.0	......	3.2	2.7
40 MgO	2.9	......	......	......	2.5	......	2.0	......	1.8	1.9
41 Free CaO	0.5	......	......	......	0.4	......	0.7	......	0.2	0.5
42 Totals	(97.5)	......	......	......	(98.7)	......	(98.7)	......	(98.2)	(99.1)
43 $C_3S + C_2S$	72	......	......	......	75	......	72	......	77	81
44 $C_3S + C_2S + C_3A$	83	......	......	......	80	......	83	......	81	85
45 $C_3S + C_2S + C_3A + C_4AF$	91	......	......	......	93	......	92	......	93	94
46 $C_3S + C_4AF$	19	......	......	......	18	......	20	......	16	13

Notes: All air-entrained cements contain an interground air-entraining addition as per footnotes to ASTM C175 and C205. All ASTM Standards are subject to continual revision, and for use the latest version of a designated standard or method of test should be consulted. This tabulation omits certain qualifying notes. Where more than one test is indicated for the same property (as for *fineness, time of setting,* and *strength*—compressive or tensile) an option is open to the purchaser. See Fig. 23 for rates of heat generation (calories per gram of cement) for each of the five types. See Fig. 24 for comparative rates of strength gain under *standard* and *mass* curing conditions. Requirements for Portland-Pozzolan Cement, Types IP and IP-A (ASTM C340) differ little from those for Type I (C150) and Type IA (C175).

[1] Ref. 1, Table 2.4.

blance to the cement of today, and as recently as 1920 conventional 28-day standard-cured concrete strengths ranged from 1,000 to 3,000 psi. The strength corresponding to 7.5 gal water per sack of cement (approximating the old 1:2:4 concrete) was 2,000 psi, whereas today's comparable mixture develops about 3,000 psi and strengths range from a minimum of 2,000 psi for the leanest mixtures upward to 6,000 or even 8,000 psi, with 4,000 to 5,000 psi concretes being common for a wide range of adaptations. Strengths above 10,000 psi are easily attainable.

The long-standing basic definition of portland cement has remained essentially constant as: ". . . the product obtained by finely pulverizing clinker produced by calcining to incipient fusion an intimate and properly proportioned mixture of argillaceous and calcareous materials, with no additions subsequent to calcination excepting water and calcined or uncalcined gypsum."

The clause prohibiting additions, other than gypsum, subsequent to calcination was probably wise in preventing irresponsible tampering with what was intended to be a standardized product. The inference was that any such addition would be injurious or at least nonbeneficial. During the late 1930's cement manufacturers came to recognize that a very small amount of tallow or resinous material introduced with the clinker facilitated final grinding without apparent harm to the cement. ASTM Committee C1 on Cement introduced a footnote clause to the portland cement specification providing for the certification that certain proved additions were "not harmful" (and therefore admissible) when used within stipulated maximum amounts. A grinding-aid material known as "T.D.A." was so approved about 1938, and a second grinding-aid addition known as "109B" was approved in 1947, each as a "nonharmful addition" under the ASTM Specification C150 for Portland Cement.

Shortly thereafter it became apparent that slightly larger amounts of certain resinous or greasy materials interground with the clinker gave air-entrainment properties to the cement. Although tests showed that concretes from air-entraining cements were likely to have somewhat lower strengths than the non-air-entrained concrete, they showed greater plasticity (at even a lowered water content), freedom from segregation, and remarkable increases in durability against freezing and thawing and the use of chlorine salts in snow removal (see "Air Entrainment").

In spite of the obvious benefits, an addition that lowered the strength could scarcely be certified as nonharmful, and a separate ASTM specification, C175, was issued in 1942 for air-entraining portland cement with a footnote permitting the use of certain "air-entraining additions" after having, from tests, been pronounced by ASTM Committee C1 as "acceptable." Under this clause "Vinsol resin" was accepted about 1942 and "Darex" about 1944. Others followed.

Chemical formulas and methods for quantitative determination are given in the appropriate ASTM designations for all "additions" that have been pronounced either "nonharmful" under ASTM C150 or "acceptable" under ASTM C175. Air-entraining agents are now introduced and sold both as "additions" integral with air-entraining cement and as "admixtures" to be introduced with the other concrete ingredients at the time of mixing (see "Admixtures").

Within the broad general definition of portland cement there has always existed considerable leeway for variability of product, which may be influenced greatly not only by the chemical proportions of respective constituents but also by the hardness of the burning and the fineness of the grinding.

From the initial ASTM standard for portland cement (adopted in 1904 as C9-04 and revised in 1908, 1909, 1916, 1920, 1926, 1930, 1937, and 1938) until about 1930, the cement industry and the ASTM recognized but one *portland* cement. In part, perhaps, because of the introduction in this country of *Lumnite* cement (1924), the first high-early-strength American-made portland cement was marketed in 1927 and

a separate ASTM Standard for High Early Strength Portland Cement became current in 1930 as C74-30T. Under the Hoover Dam stimulus for lowered heat of hydration in massive concrete structures, a low-heat portland cement was developed almost simultaneously and produced under a U.S. Bureau of Reclamation government specification. Serious deterioration of concrete under exposure to alkali soils and active waters created demands for still another special type of portland cement, a sulfate-resisting cement.

Current Portland Cements. These varied demands led to the development about 1940 of ASTM C150 recognizing five distinct types of portland cement (Tables 1 to 3 and Figs. 1 and 2). This standard replaced the two ASTM Standards C9 and C74.

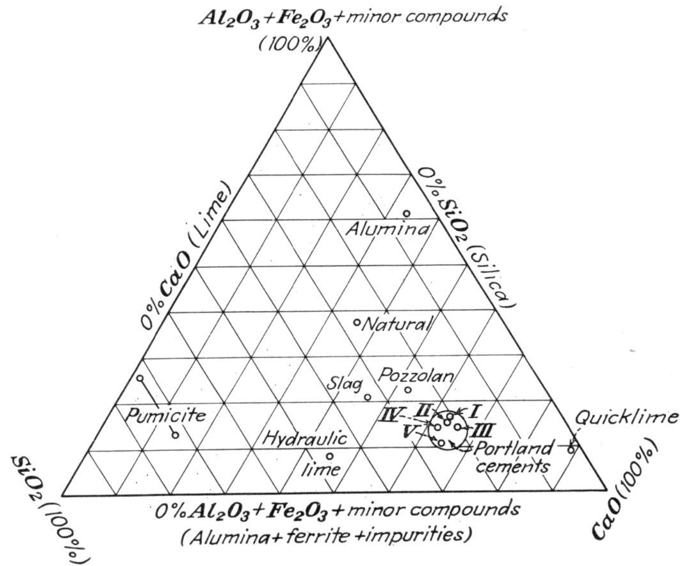

Fig. 1. Relative compositions of representative calcium cementing materials exclusive of gypsum products. Plotted mainly from data of Ref. 1 (Table 2.3, p. 32).

Since 1940 portland cement developments have been rapid, and at present (1958) there are four separate ASTM specifications for portland cement. The specification requirements and compound compositions corresponding to three of these standards (C150, C175, and C205) are compared in Table 2. The fourth (ASTM C340-55) was first issued in 1954 and is discussed under "Blends and Substitutions."

Portland Cement (ASTM C150):

Type I	For use in general concrete construction when the special properties specified for Types II, III, IV, and V are not required.
Type II	For use in general concrete construction exposed to moderate sulfate action or where a somewhat lower heat of hydration is required.
Type III	For use when high early strength is required.
Type IV [1]	For use when a low heat of hydration is required.
Type V [1]	For use when high sulfate resistance is required.

[1] Neither Type IV nor Type V is regularly carried in stock and, prior to specifying either of these cements, availability should be investigated.

TABLE 3. APPROXIMATE AVERAGE VALUES FOR FINENESS OF THE CEMENT AND FOR RELATIVE COMPRESSIVE STRENGTHS OF THE CONCRETE AS AFFECTED BY TYPE OF PORTLAND CEMENT [1]

Type of portland cement (ASTM C150)	I Normal	II Modified	III High early	IV Low heat	V Sulfate resisting
Fineness (Wagner turbidimeter, sq cm/g)	1,710	1,990	2,730	1,880	1,960
Fineness (percent passing 325-mesh)	90.7	94.7	99.5	93.1	93.2
Relative strength of concrete (percent) 3-day	100	80	190	50	65
Relative strength of concrete (percent) 28-day	100	85	130	65	65
Relative strength of concrete (percent) 3-month	100	100	115	90	85

[1] Excerpted from Tables 2 and 3 *Portland Cement Assoc. Bull.* T-12, 10th ed. For specification limits of fineness and the required minimum strength of mortars (ASTM C109 and C190) see Table 2, lines 16–18 and 27–33. For average compound compositions of the cements see Table 2, lines 34–46.

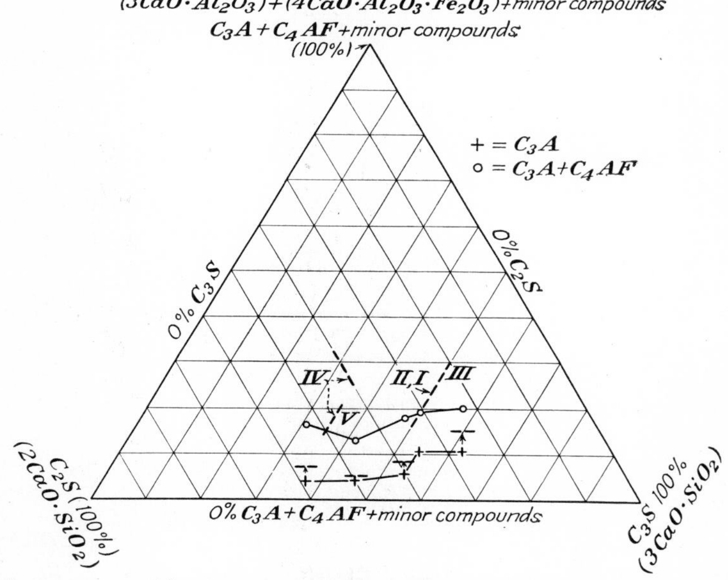

FIG. 2. Average composition for each of the five current types of portland cement (——— = specified limiting amounts as per ASTM C150-47). Typical contents for C_3A and C_4AF per Ref. 1, p. 32, Table 2.4. (A more complete listing of typical amounts of the several constituents appears in lines 34 to 46 of Table 2.)

Type	I	II	III	IV	V
C_3A, percent	11	5	11	4	4
C_4AF	8	13	9	12	9
$C_3A + C_4AF$, percent	19	18	20	16	13
Maximum permissible per ASTM C150-55:					
C_3A, percent	No limit	8	15	7	5
C_4AF			No limits specified		

Air-entraining Portland Cement (ASTM C175):

Types IA, IIA, and IIIA correspond to Types I, II, and III of C150 except for the presence of an "acceptable" interground air-entraining addition (see "Air Entrainment").

Portland Blast-furnace Slag Cement IS, and IS-A (air-entrained) (ASTM C205).

Portland-Pozzolan Cement IP, and IP-A (air-entrained) (ASTM C340).

Slag Cement S and SA (air-entrained) (ASTM C358):

Generally used as a blend with portland cement in making concrete or as a blend with hydrated lime in making masonry mortar.

Blends and Substitutions. European experience long had indicated possible economies with little sacrifice of strength from partial replacements of portland cement by finely interground or otherwise intermixed suitable pozzolanic materials. These might occur naturally, or as by-products. Apparently the free lime in the cement combines with the pozzolan to contribute a compensatory cementing action with good workability and little evolution of heat.

On the basis of favorable indications from partial replacements of calcined opaline shale in piers of the San Francisco Bay bridge (1932–1935) and supplementary experimentation, Type II portland cement with a 25 percent mill-produced pozzolanic replacement was evolved for the Bonneville Dam (1935–1937). Subsequently it was further discovered that, additional to low heat, good placement quality, and possible economies, certain pozzolanic replacements would greatly reduce the possibility of adverse cement-aggregate reactivity. These discoveries led to the prompt adoption of replacement portland-pozzolanic cement for such other major structures as Friant Dam, 1945 (using a 25 percent replacement of a natural pumicite with Type IV cement), and Hungry Horse Dam, about 1950 (using a 35 percent replacement of Type II cement with Chicago fly ash. Fly ash is fine flue dust, a combustion waste product of the powdered coal, used in the generation of steam power).

ASTM C340-55T calls for a clinker corresponding to Type I or Type II (Type I if unspecified by the purchaser). The specified replacement ranges between 15 and 50 percent by weight of the portland-pozzolan cement. The cement may be either non-air-entrained IP or IIP or air-entrained IP-A or IIP-A.

Natural pozzolans include certain types of clays, opaline shales, diatomaceous earths, cherts, volcanic tuffs, and pumicites. Artificial pozzolans include fly ash, silica fume, powdered brick, burnt oil shales, and certain slags. To become active some pozzolans require calcination at 800° to 2000°F. Most of them require grinding. Whether plant or job processed, an intimate uniform blend is essential. Pozzolans vary greatly in their suitability, not all the pertinent differences being understood.

Optimum portland-pozzolanic combinations supply: improved workability, low heat of hydration with its attendant reduction in thermal distortions and stresses, increased resistance to sulfate soils, reduced adverse alkali reactivity, and in most cases, reduced costs. With Type II clinker C340 appears destined to supplant Type IV cement. With sufficiently high cement factors, C340 with Type II appears also to be a good substitute for Type V.

The portland-pozzolan cements of ASTM C340 should not be confused with the portland blast-furnace slag cements of ASTM C205 or with the slag cements of ASTM 358.

Several other cements for which there is no present ASTM Designation fall or may fall within the portland cement bracket:

White Portland Cement. White portland cement is of essentially the same composition as other portland cement (but more finely ground), being made from pure calcite limestones and white clays which exist in eastern Pennsylvania and in France. White coarse aggregate and white quartz sands similar to the "standard sand" of Ottawa, Ill., may be used with white cement for obtaining concretes and mortars of maximum whiteness. The colors of cements and their concretes are due mainly to iron oxide impurities and are usually negligible in their effect on the physical properties of the concrete. In the manufacture of white portland cement even the fuel must be free of discoloring oxides. White mortars and concretes are used mainly in decorative work or for such other applications as center lines and reflecting curbs. The cost is more than twice that of the regular portland cements.

Low-alkali Portland Cement. Portland cements low in alkali are now specified whenever a question exists regarding possible alkali reactance of aggregate (see "Defective and Reactive Aggregates"). The U.S. Bureau of Reclamation stipulates that the Na_2O equivalent of the Na_2O plus K_2O be below a value now set at about 0.6 percent. Some commercial cements regularly meet this requirement; others would not qualify without special manufacturing precautions.

Job-processed Portland Cements. Job-processed portland cements are sometimes used. During the period from 1912 to 1916 the U.S. Reclamation Service (now the U.S. Bureau of Reclamation) interground about equal volumes of portland cement with silica sand, primarily as a measure of economy. Such *sand-cement* was employed for the then large Arrowrock and Elephant Butte Dams. The relatively low cement factor had an advantage in reducing the heat of hydration generated (not then recognized as one of the major problems of mass concrete, however). The same device was subsequently considered for consciously reducing the heat of hydration for Hoover (Boulder) Dam but was not adopted mainly because of the surface nondurability exhibited by the earlier sand-cement structures.

Tufa cement formed by intergrinding about equal parts of volcanic tufa rock with portland cement was used in the construction of the Los Angeles Aqueduct about 1905–1910. The tufa contained 65 to 70 percent silica and 10 to 15 percent alumina, with some of the silica soluble. In northern Europe trass (Table 1) has been used in a manner similar to tufa, producing a cement claimed to offer high resistance to sea-water attack by virtue of neutralization of the lime in the cement by the silica-rich trass.

Waterproof Portland Cements. Waterproof portland cements are marketed under various brand names (see "Admixtures"). In general, no form of integral waterproofing beyond regular portland cement is necessary to produce a waterproof concrete if the aggregates are properly graded and proportioned and cracking is not present.

Plastering and Stucco Portland Cements. Plastering and stucco portland cements are specified under American Standards Association designations ASA 42.2 and 42.3. These standards are issued as separate reprints. There are no current (1957) ASTM Designations covering these portland cements.

Masonry Cement; Thermal Insulating Cements. Masonry cement (ASTM C91) and thermal insulating cement (ASTM C194, C195, C196, and C197) may or may not use portland cement as the binding material. Current masonry cements usually include interground limestone or hydrated lime as a plasticizer, gypsum for setting time regulation, and air entrainment for both plasticity and water retentiveness. The lime should be low in magnesia to avoid the pop-out type of difficulty from delayed response to mixing.

Oil-well Cements. Such operations as the drilling and sealing of oil and gas wells to depths attaining 8,000 ft or more create many difficult-to-meet situations. Pressures up to 18,000 psi and temperatures as high as 400°F are encountered, and cement mixtures may need to remain fluid for as much as 4 hr, then harden rapidly. Normal portland

cements are modified according to the nature of the cementing action required. Slow set is promoted by drastic reduction in the tricalcium aluminate (almost to zero) and retarders used may be starches or cellulose products, sugars, and acids or salts of acids. Materials may also be introduced to reduce water losses in the slurry and to increase the density. Oil-well cements are specialized and of considerable importance. See also "Admixtures—Grouting Agents."

Expanding Cements. Since about 1940 expanding cements have been used to some extent to offset or more than compensate for normal shrinkage. An expanding cement 1:3 mortar approximately 0.1 in. in thickness has been claimed to impart leakproofness to gasoline and water storage tanks. These cements are mainly a foreign development and because of prediction and control difficulties some of their alleged usefulness is subject to question.

Hydrophobic Cements. British researchers (see *Concrete and Constructional Engineering,* August, 1953, p. 255) indicate the possibility of safeguarding cement against the hazards of dampness in storage by intergrinding with the clinker less than 1 percent of oleic acid. The chemical not only functions as a grinding aid and an air-entrainment addition in some degree but coats the cement grains giving them water repellency that will enable them to float on water for weeks. It protected from deterioration indefinitely a ½-in. layer of cement exposed to 90 percent humidity. The concrete-mixing operation apparently scrubs off the surface coating and the set and strength of the concrete are not affected. The British have experimented with other chemicals and the Russians have reported progress along similar lines. Hydrophobic cements are still in the developmental stage.

Antibacteria Cements. Use of an interground bactericide to safeguard stored materials (see D. Leviowitz, *Food Engineering,* vol. 24, p. 6, June, 1952) appears to be a questionable expedient. Pending convincing evidence to the contrary, both the accessibility of enough of the bactericide and its stability or retention of effectiveness would seem improbable.

Compounds in Hardened Portland Cement; Comparisons. The four principal compounds present in hardened portland cement are:

Compound	Abbreviation or formula			
	1 *	2	3	4
Tricalcium silicate	C_3S	3CS	$3CaO \cdot SiO_2$	Ca_3SiO_5
Dicalcium silicate	C_2S	2CS	$2CaO \cdot SiO_2$	Ca_2SiO_4
Tricalcium aluminate	C_3A	3CA	$3CaO \cdot Al_2O_3$	$Ca_3Al_2O_6$
Tetracalcium aluminoferrite	C_4AF	4CAF	$4CaO \cdot Al_2O_3 \cdot Fe_2O_3$	$Ca_4Al_2Fe_2O_{10}$

* The widely used abbreviations of column 1 can be confused with chemical formulas, and Witt (Ref. 3, p. 193) proposes those of column 2 as preferable.

From Table 2 (lines 43 to 45) it is observed that C_3S and C_2S constitute together about 70 to 80 percent of the cement. They contribute most of the desirable strength-giving characteristics. The four principal compounds account for more than 90 percent of the cement, with overall values among the five basic types differing but slightly. Although the C_3A does add to the early strength of portland cement it is the least desirable of the major compounds; it contributes most to heat of hydration and volume change. From the manufacturing standpoint its elimination would be costly, but (from line 37, Table 2) it is apparent that the C_3A is kept low in Types II, IV, and V. The C_2S (line 36) gains strength slowly but is low in its contribution to heat and volume change (see also Figs. 22, 23, and 24).

From the foregoing and Fig. 2 and Table 2 it is clear that Type III (high early strength) and Type IV (low heat) represent the two extremes from the standpoints of rate of hardening, heat of hydration, and volume change.

The high early strength of Type III cement stems from three sources: the higher CaO (or lime) content, being high in both C_3S and C_3A (it is called a high-lime cement), lines 35 and 37, Table 2; the higher C_3A permitted, line 14 (a limit of 15 percent against 8, 7, and 5 percent for Types II, IV, and V); and the finer grinding.[1] Whereas minimum finenesses are specified for Types I, II, IV, and V, neither a maximum nor a minimum fineness is stipulated for Type III. To meet the added early strength requirements necessitates finer grinding than for any of the other four cements (see Table 3). Because of grinding costs and the tendency of an overly fine cement to harden in storage, the manufacturer is not likely to overgrind beyond a safe minimum for securing the strength required.

The higher sulfate resistance for Types II and V is accomplished mainly by holding down the C_3A; the low heat generation for Type IV by limiting both the C_3A and the C_3S, the reduced rate of strength increase not being serious for the massive concrete structures for which Type IV cement was evolved.

The air-entraining cements of ASTM C175 currently include only Types I, II, and III of C150, since the benefits of air entrainment (added durability against frost action and active salts used in snow removal, secondarily better workability and greater freedom from segregation) are of minor concern in the types of structure and exposures for which Type IV and often Type V were evolved. Portland blast-furnace slag cements are currently less important than the other portland cements.

Acceptance Tests. Acceptance tests for cements have in recent years undergone considerable change. For thermal insulating and masonry cements the tests are wholly physical. A cement of virtually any composition (portland or other) that will meet the stipulated requirements of use, or simulated use, is satisfactory, and a wide variety of types or combinations may qualify. For the portland cements the acceptance requirements are chemical and physical rather than functional. Most portland cement is used in concrete rather than in neat pastes or mortars; the tests cannot easily be linked directly to the service conditions they are designed to control. *Fineness, soundness, time of setting* (initial and final), and various *strength tests* are the primary ones for the cements of C150, C205, and C340, whereas the *percentage of air entrained* is an added requirement for the air-entraining equivalents. Reference to original ASTM or equivalent sources is necessary not only for the conduct of a test but to check on changes and/or detailed requirements.

Materials and Manufacture. Portland cement is manufactured by either the *wet* or the *dry* process, the dry process being the one adapted to the materials most generally available and commonly used in this country. The calcareous material in the dry process is usually limestone and in the wet process marl, being chiefly calcium carbonate ($CaCO_3$) in either case. The *argillaceous* material contributing silica (SiO_2) and alumina (Al_2O_3) for either process may be shale, clay, cement rock (argillaceous limestone), or blast-furnace slag (basic).

The dry process embodies the following operations: (1) preliminary grinding of dry raw materials separately, (2) proportioning, (3) pulverizing the properly proportioned mixture, (4) burning to incipient fusion forming the clinker, (5) cooling and seasoning the clinker, (6) addition of gypsum (calcined or uncalcined) for control of rate of setting, (7) grinding of the clinker to a fine powder that meets the fineness requirements for the cement, and (8) storage in bins for seasoning prior to package

[1] Fineness of grinding (Fig. 25) adds to cost but increases the rate of strength gain (and also rate of heat generation). It increases the later strengths and total heat generated only slightly. Hard-burned clinker produces stronger better cement, but the hard burning requires greater outlay for both fuel and grinding.

or bulk shipment. Where the calcareous material occurs as marl, the wet process is commonly employed: (1) the marl is stored in vats as a thin mud or slurry; (2) the clay or other argillaceous material is reduced to a fine powder; (3) the ingredients are proportioned; and (4) the ingredients are mixed through a pug mill, after which the burning, cooling, and seasoning, addition of gypsum to control set, final grinding, and storage are carried out as in the dry process.

The vertical or stack kilns used in parts of Europe have been replaced in the United States by slightly sloping rotary kilns 100 to 500 ft long and 8 to 15 ft in diameter. The dry powder or thin slurry is fed uniformly into the upper end of the kiln into which is also fed powdered coal, fuel oil, or gas. Air blasts bring the temperature at the lower end of the kiln to 1400° to 1500°C (2550° to 2730°F). As the charge moves slowly down the kiln, there is first drying, then loss of organic matter, followed by the various chemical reactions so essential to the proper compound composition. About one-third of the initial weight of the charge is volatilized and 20 to 30 percent of the mass is converted to liquid in the course of clinker formation. Underburned clinker will probably produce unsound cement of below normal strength because of uncombined lime. Overburning adds to cost of fuel and grinding without justifiable gain.

Storage and Marketing. *Mill Storage.* The newly ground cement is stored at the mill in bins or silos. For large work the purchaser's representative samples and tests the cement at the mill, and sealed bins are allocated to the particular job. Dry cement behaves much as a fluid and increasingly is "pumped" from bin to car or truck for bulk shipment. Smaller quantities are sacked in either cloth or paper bags of 1 cu ft nominal volume. The actual filling is by weight, at 94 lb per sack. Barrels of 4 cu ft (4 sacks) are no longer used domestically except as a term equivalent to 4 sacks of cement. The specific gravity of portland cement lies between 3.10 and 3.20, giving a solid weight of about 196.6 lb per cu ft, indicating that a compact bulk cubic foot of cement as sacked is about half air voids. Loose as pumped or scooped lightly into a container, the bulk weight per cubic foot is considerably below that of sacked cements. The fluffier it is, the more fluid is its action.

Warehouse or Job Storage. Dry storage is of utmost importance (see "Hydrophobic Cement"). Temporary storage structures should be weathertight; if possible, moisture-tight. Cement should never be stored against an exterior wall or within capillary contact with the ground. Covering with tarpaulins should be without direct contact if the tarpaulins are likely to become wet; they should be weighted to prevent disturbance by wind. The finer the cement is ground, the more responsive it is to moisture. High-early-strength or other fine-ground cement is likely to be specially susceptible to lumping and other deterioration under storage conditions. Hardened lumps should not be used, either as lumps or after recrushing.

PORTLAND-CEMENT CONCRETE

History and Introduction. The rapid continuous expansion of portland-cement concrete in volume and variety of adaptations since 1900 has been augmented by such important overlapping developments as: (1) recognition of the inherent advantages of reinforced concrete and the development of appropriate design and construction techniques for its use, (2) the coming of the automobile with its accompanying demands for paved highways, and (3) the era of large dams and appurtenant structures.

The use of reinforcement in concrete dates from about 1850, 25 years after Aspdin's initial patent of portland cement. In 1855 Lambot, a Frenchman, took out an English patent on a small reinforced concrete boat about the same time as Coignet (also French) took out English patents on applications of reinforced concrete to structural units. Thaddeus Hyatt, an American, described as "a lawyer by education but an inventor by nature" published in London in 1877 a treatise on bond between concrete and steel.

As recently as 1900, except for some progress in the design and use of reinforced concrete arches, reinforced concrete was little used abroad and almost not all at in the United States. It was with the extensive experimental work of Talbot closely followed by many contemporary investigators and disciples that the spectacular sustained advance in the use of concrete got under way. Techniques continue to improve and applications to multiply.

The recognized need for a code of practice for both plain and reinforced concrete led to the organization in 1904 of the First Joint Committee on Specifications for Concrete and Reinforced Concrete. Its final report, published in 1916 (Ref. 13) set the pattern for concrete practice throughout the United States and much of the world. In due course both a second and a third joint committee were organized, their final reports being issued in 1924 (Ref. 14) and 1940 (Ref. 15) respectively. Between 1920 and 1940, European countries, notably England and Germany, published manuals of practice comparable to our joint committee reports. The American Concrete Institute (ACI) was organized in 1905 (as the National Association of Cement Users) and has assumed leadership in the field, although it functions in close cooperation with other interested organizations. The American Society for Testing Materials (ASTM) in particular, through its active Committees C1 on Cement and C9 on Concrete and Concrete Aggregates is continually making important contributions by its published researches and its standardizing activities.[1]

As regards portland cement products there is no definite line of cleavage between the publication scope of ASTM and of ACI, but in their standardizing activities ASTM covers materials and tests and ACI covers products. Thus the standards and recommended practices of ACI relate to such subjects as building regulations, design of mixtures, concrete pavements and bases, cast stone, precast floor units, farm silos, detailing, winter concreting methods, bond, and curing.

The American Standards Association (ASA) approves and publishes standards, many of which have been originated by such an organization as ASTM. The American Railway Engineering Association (AREA) and American Association of State Highway Officials (AASHO) evolve and publish standards applicable within their respective fields of specialization.

Supplementing the work of the ACI and ASTM and the others mentioned are the research activities of such government laboratories as those of the National Bureau of Standards, Bureau of Public Roads, Bureau of Reclamation, and the Army Engineers. The laboratories of state highway departments and leading educational institutions have contributed much. The Engineering Foundation and the Highway Research Board of the National Research Council have sponsored important research projects and published extensively. The Portland Cement Association, one of the earliest and most fruitful of the great trade associations, has since 1918 made many research contributions through its central laboratory at Chicago and its constructive sponsorship of cooperative research with government and professional agencies. Contributions from other countries have largely been by scientists operating under government auspices. See list of selected references covering these and other recognized sources for readily available concrete information.

Uses. From the standpoint of suitability and cost, concrete is without important competition from steel, timber, and other structural materials for such construction as high dams, heavy-duty pavements, air strips, curbs, sidewalks, reservoirs and outdoor swimming pools, underground storage tanks, abutments, piers and retaining walls, tunnel and canal linings, dry docks, harbor works, and many military installations.

[1] In like manner, ASTM Committees C2 on Magnesium Oxychloride Cement, C7 on Lime, C11 on Gypsum, C12 on Mortars for Unit Masonry, C13 on Concrete Pipe, C15 on Manufactured Masonry Units, C17 on Asbestos-cement Products, and D4 on Road and Paving Materials are contributing to other aspects of the general field.

For much other construction portland cement products are both supplementary and competitive, as for medium height and low dams (in competition with earth and rock fill); buildings of low and medium height (up to about twenty stories); airplane hangars; moderate-duty highways and city pavements; foundations and footings; stadiums; band shells; terminals and warehouses; highway and railway bridges; arches up to almost 900-ft span; large water pipe, culverts, and conduits; barges (and even ships under conditions of speed-up and war shortages); precast products such as burial vaults, posts, piling, pipe and drain tile; mats for river-bank protection; silos, feeding platforms, and other farm structures; paved tennis courts; sound and heat insulation; and smoke-stacks. The adaptations to thin-shell construction, prestress, and to atomic shielding add immeasurably to the range of possible applications.

Other special-purpose portland-cement concretes include heavy-aggregate counterweight and shielding concretes, lightweight concretes, often for sound or heat insulation; nailing concretes (soft-aggregate concrete into which nails can be driven); and porous or porous-base concretes for tile or underdrainage. Such variations may be secured through the use of special aggregates (ranging from iron and iron ore to cinders or sawdust, or by adding a small amount of some frothing agent such as the aluminum powder in aerocrete,[1] or by "poor" grading or use of one-size aggregate to produce honeycombing and a pervious structure.

Besides being the active constituent of concrete, portland cement is used in masonry mortars and stuccos, oil-well and foundation grouts, soil-cement stabilization mixtures, and asbestos-cement mixtures covering a variety of products including shingles, pipe, corrugated board, and sheet material. A thin layer of portland cement grout has proved useful as a lining for cast-iron water pipe, reducing corrosion and hydraulic friction.

Concrete has even assumed an important role as a decorative and sculptural medium (Ref. 12, pp. 196–211); ornamental as well as utilitarian is the exposed-aggregate construction of a terrazzo floor. The sculptural phase was introduced by Lorado Taft in his Black Hawk Monument at Oregon, Ill. (1912), and his Fountain of Time on the Chicago Midway (1921). The latter is but one of the many notable structures illustrating the artistic concrete craftsmanship of J. J. Earley, others being the replica of the Parthenon at Nashville, Tenn. (1921 to 1931), and Baha'i Temple at Wilmette, Ill. In decorative monolithic exterior wall construction sculptural figures are cast against plaster molds which are made an integral part of the form for the exterior face of the wall as in the Life Sciences Building of the University of California at Berkeley. Southern California abounds with all manner of decorative concrete. Shape, texture, and color can be controlled through the forming, choice of aggregates, use of pigments, and surface treatments of several sorts. Figure 3 indicates the great range of surface type available from forming alone. (See *Proc. ACI*, vol. 19, pp. 178–185, 1923; vol. 30, pp. 251–278, 407–421, 1934.)

Scarcely an important structure exists that does not rest upon concrete foundations. If the primary framework is not of concrete the structure contains auxiliary concrete construction in such forms as footings, basement walls, stairways, terrazzo floors, and encasement for columns. Thus the world's highest structure, the Empire State Building, a monument in steel, contains 100,000 cu yd of concrete; the Rockefeller Center group of buildings contains three times this amount. In our great steel bridges the mass of the concrete in the piers, anchorages, abutments, and floors usually exceeds that of the steel in the superstructure.

As a structural material, concrete owes its important place to the fact that it is eco-

[1] This lightweight insulating concrete with foamy spongelike texture is not the air-entrained concrete of similar but much finer texture used for added durability against the action of frost and salts. The air-entrained concrete normally contains only 3 to 5 percent of air voids and is discussed subsequently. Aerocrete may contain 50 percent or more of void space.

FIG. 3. Ornamental concrete surfaces cast against forms.

nomically utilitarian, highly resistant to fire, wind, water, earthquake, and war, and lends itself readily to almost any type of decorative or architectural expression ranging from the conservative to the fantastic.

Physical Properties and Factors That Influence Them. The following discussions qualify the values given as ranges for the physical properties listed in Table 4.

Color. Concrete can be made of any desired color or surface texture by the use of mineral pigments or by embedding selected aggregates near the surfaces and exposing them by grinding or treatment with dilute hydrochloric acid. With proper precautions, concrete surfaces can be painted satisfactorily with both cement-water and oil paints (see "Waterproofing and Painting").

TABLE 4. USUAL CHARACTERISTICS AND PHYSICAL PROPERTIES OF PLAIN CONCRETE

Color	Some shade of slate or gray
Unit weight (structural concrete), lb per cu ft	145–155
Unit weight (light structural, filler, and insulating), lb per cu ft	50–140
Strength,[1] compressive (ultimate) (proportional limit about 40 percent ultimate), psi	2,000–8,000
Strength, flexural (modulus of rupture) (proportional limit about 70 percent ultimate), psi	400–1,200
Strength, tensile ($\frac{1}{8}$–$\frac{1}{12}$ compressive) (proportional limit about 70 percent ultimate), psi	250–700
Strength, torsional (modulus of rupture) (proportional limit about 70 percent ultimate), psi	300–1,000
Strength, fatigue (endurance limit flexure)	About 50 percent modulus of rupture
Strength, bond (sliding against embedded steel, plain bars)	5–40 percent ultimate compressive strength
Durability (see discussion)	
Deformability (see Table 5)	
Stiffness, modulus of elasticity (compression, tension, flexure), psi	2,000,000–6,000,000
Stiffness, modulus of rigidity (torsion), psi	1,000,000–2,500,000
Poisson's ratio (0.20 average)	0.15–0.24
Coefficient of thermal expansion per °F (differs little from that for steel, 0.0000065 average)	$4\text{–}7(10^{-6})$
Specific heat (mass concrete), Btu per lb per °F	0.20–0.24
Thermal conductivity (mass concrete), Btu per sq ft per in. per hr per °F	14.4–24.0
Thermal conductivity (building concrete), Btu per sq ft per in. per hr per °F	10.8–12.6
Thermal conductivity (cinder aggregate), Btu per sq ft per in. per hr per °F	5.0
Thermal conductivity (expanded burned-clay aggregate), Btu per sq ft per in. per hr per °F	2.3
Cement content, sacks per cu yd	4.0–8.0
Corresponding yield, cu ft concrete per sack of cement	6.8–3.4
Mixing water per cu yd plastic concrete, gal	28–40
Mixing water per sack of cement, gal	4–9

[1] Rate of testing has an influence on strength obtained. Nominal rate of load application is 20 to 50 psi per sec (ASTM C39). At a very rapid rate (almost instant application of load) the strength is about 15 percent greater. When applying load very slowly (up to 4 hr per specimen), the strength is reduced 12 to 15 percent. The long-time sustained compressive load may be expected to be not more than 80 or 85 percent of that secured from a regulation short-time compressive test. Long-time test data are meager.

Unit Weight. Unit weights vary with proportions and specific gravities of the aggregates, a fair average specific gravity for the usual quartz and limestone aggregates being 2.65. Natural rocks used for aggregates have specific gravities varying from 2.4 to 3.0. Special lightweight aggregates such as cinders, burned clay or haydite, pumice, coral, perlite, vermiculite, and granulated slag have specific gravities from 0.10 upward. Frothy concretes having a spongelike structure can be produced at less than half the unit weight of ordinary concrete. The solidity ratio, often called density (ratio of volume of solids to space occupied, i.e., unity minus the voids ratio) of the concrete affects the unit weight, the solidity ratio in turn being a function of both the maximum size and the grading of the aggregates (see also Table 19). Counterweight concrete up to 270 lb per cu ft has been made using steel punchings and iron ore for aggregates; even heavier concrete is used for atomic shielding.

Strength. Unqualified, the strength of concrete refers in the United States to the 28-day compressive strength of a standard cylinder having height twice its diameter, which has been kept moist (by immersion, sprays, or coverage with wet burlap or moist sand) at approximately 70°F from the time it was cast.

The compressive strengths of representative specimens fabricated, cured, and tested in accordance with ASTM C192 or C31 constitute the long-accepted primary criterion of excellence for concrete. If the testing is of a research nature or preliminary to the design of a mixture C192 (laboratory specimens) will control. If the testing relates to work in progress the procedures of C31 (field specimens) will be followed. If the specimens are to be drilled cores from hardened concrete then ASTM C42 governs.

There are two classes of "field" (ASTM C31) specimens: (1) *quality control* to indicate the potential characteristics of the concrete based on its quality just prior to deposition in the form and (2) *job control* intended to indicate the characteristics of the hardened concrete on the job at the time the specimen is tested. After fabrication the "quality-control" specimen is stored and tested under essentially the standard laboratory conditions. When tested the results should fairly indicate the potential quality of the concrete as it entered the form. The job-control specimen, on the other hand, is, as nearly as may be, accorded the same treatment (exposure, temperature, curing, and other conditions), as is the concrete of the structure. Actually these conditions are difficult to simulate with assurance, and the "job-control" specimen should not be expected to indicate accurately the strength of the concrete it is supposed to represent.

Tests on drilled cores from the actual structure constitute the best basis for estimating job strengths but such tests are time-consuming and costly. The core must be taken carefully and processed in accordance with ASTM C42 which entails moisture conditioning (being saturated prior to test). It must also be capped on both ends and the tested strength will usually require a correction to compensate for the departure of height-diameter ratio from that of the standard compressive cylinder (which is always 2); see Fig. 12 and ASTM C42.

On the other hand, the drilled core, aside from constituting an excellent specimen for a representative job-strength test, supplies the basis for a definite check on the actual thickness of floors, pavements, and walls. Moreover, the core can likewise be used as a valid sample for the determination of the constituents of the hardened concrete, as is sometimes necessary or desirable. (See *ASTM Special Technical Publication* 169, p. 221, 1955).

Ingenuity and analogy have given rise to several attempts to evolve simple empirical predictive tests for the early strength development as criteria for form removal, termination of formal curing, or for placing the concrete into service.

One of these has been the use of a ball hardness test similar to the Brinell and/or Rockwell indentation hardness tests for metals (using larger balls and different testing techniques, of course). A variation formerly used by the U.S. Bureau of Reclamation was an *impact* indentation hardness test in which a ball was either dropped from a predetermined height (or swung as a pendulum) to strike a sheet of paper against carbon paper interposed between the concrete and the striking ball. The smaller the diameter of the circular carbon-paper imprint, the harder (stronger) was the concrete. The method is described and illustrated in the U.S. Bureau of Reclamation "Concrete Manual," 4th ed. (1942), p. 330 or 3d ed., p. 317, but not in the current fifth or sixth editions. Another device has been the partial embedment of short pullout rods in the concrete on the assumption that the bond developed in a given period of time is proportional to the strength. The most recent device to attract general attention is the "Swiss test hammer," fired against the concrete by a spring-activated mechanism, the rebound being noted in a manner somewhat analogous to the scleroscopic test for the hardness of metals. To date neither the validity nor the limitations of such devices are sufficiently well established to justify any great degree of confidence in their indications.

Because of the varying effect of different degrees of air dryness at time of test on the

strength of concrete (see Figs. 5, 6, and 7) all standard compressive testing of concrete (ASTM C192, C31, and C42) is to be on specimens that are saturated at time of test.[1]

The potential strength of a concrete mixture is primarily a direct function of the ratio of cement to water (or to voids since the voids are virtually filled with water) at time of initial set.[2] This is apparent in Fig. 4 on which cement-water ratios and their inverse, water-cement ratios, for Type I cement (see "Cement") are expressed in the several units currently used. Conventionally acceptable aggregates (sand, gravel, crushed stone, or slag that is sound, chemically inert, well graded, and reasonably free from flat and elongated particles) affect the strength in two ways. Maximum size, amount, and surface texture have some influence on the strength at constant cement-water ratio. The size, grading, and amount, especially the amount, have a primary influence on strength because for any well-graded aggregate they determine the minimum ratio of mixing water required for workability. Thus only 4 gal of water per sack of cement was required to produce a 3-in. slump (slump is a measure of workability, ASTM C143) for a rich mixture (1:1.4:2.4 by weight, i.e., 1 part cement:1.4 parts sand:2.4 gravel), but 9 gal per sack was required to produce

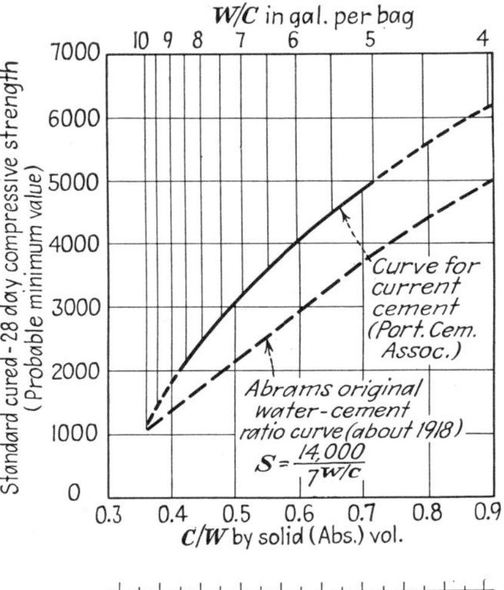

FIG. 4. Probable minimum 28-day compressive strength of concrete versus ratio of water to cement, expressed in various units.

[1] Concrete air dried to approximate uniform moisture throughout is 15 to 30 percent stronger than is the same concrete when saturated (either before drying out or after being resoaked for 15 to 30 hours). This is shown compressively by the top curves of Figs. 5, 6, and 7. The phenomenon is reversible and repeatable which is true also for absorbent materials generally including brick, porous stone, and timber. For compressive evidence see Gilkey, *Concrete*, January and May, 1941. For flexure of pavement concrete see W. K. Hatt, *Transactions ASCE*, vol. 89, p. 271, 1926. For compression, tension, flexure, and torsion see Savage, Houk, Gilkey and Vogt, "Engineering Foundation Arch Dam Report," vol. 2, pp. 438–465, 1934. H. F. Gonnerman, *Proc. ACI*, vol. 26, pp. 359 and 898, 1930, reports flexural weakening from air drying which apparent contradiction is probably explained by Bloem and Gaynor, "National Sand and Gravel Association Information Letter 140," March 10, 1958. They report substantial increases in flexural strength for concrete uniformly dry through the cross section and reductions from the saturated strength for concrete not yet air-dried to approximate uniformity—probably due to localized strains from differential shrinkage. For practically all concrete, soaking to virtual saturation occurs rapidly whereas air drying progresses slowly requiring days or weeks depending upon such factors as the specific surface of the concrete, the humidity of the environment, and the air circulation. During drying at ages up to a few weeks, curing and drying are present simultaneously (until drying has progressed sufficiently to retard or suspend hydration). For older concrete, curing is relatively dormant and the effect of drying assumes dominance. A seeming inconsistency is that oven drying at temperatures as low as 150 to 200°F weaken concrete. The oven-dry strength approximates the normal saturated strength and when the oven-dried specimens are resoaked there are compressive strength reductions of the order of 15 to 30 percent below that of corresponding nonoven-dried saturated specimens. Evidently the differential shrinkages from the forced drying result in structural damage from minute cracking. Whatever the explanation, air drying (especially if to uniformity) increases test strengths; oven drying decreases them; and saturated concrete is 15 to 30 percent weaker than the same concrete either uniformly air dry or uniformly oven dry.

[2] If water is neither added to, nor extracted from, the plastic mixture during the interval between mixing and stiffening (initial set), the volume of the mixing water is the volume of the voids (air voids neglected).

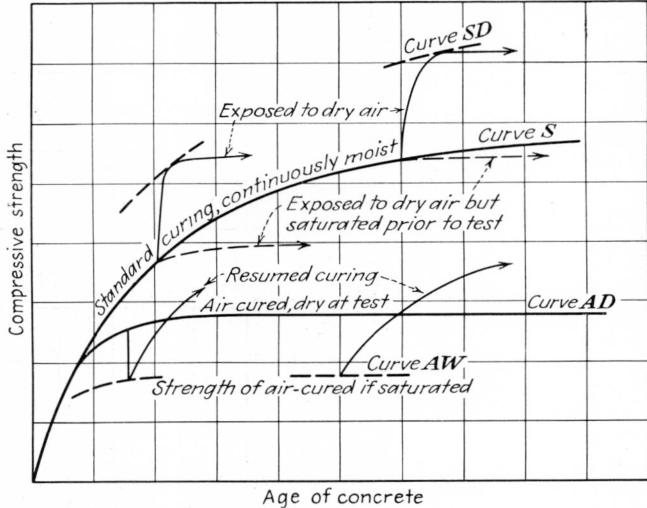

FIG. 5. A typical diagram showing the nature of the curing, air-drying, soaking, and resumed curing effects on the compressive strength of concrete (*Concrete, vol.* 49, *January and May*, 1941.)

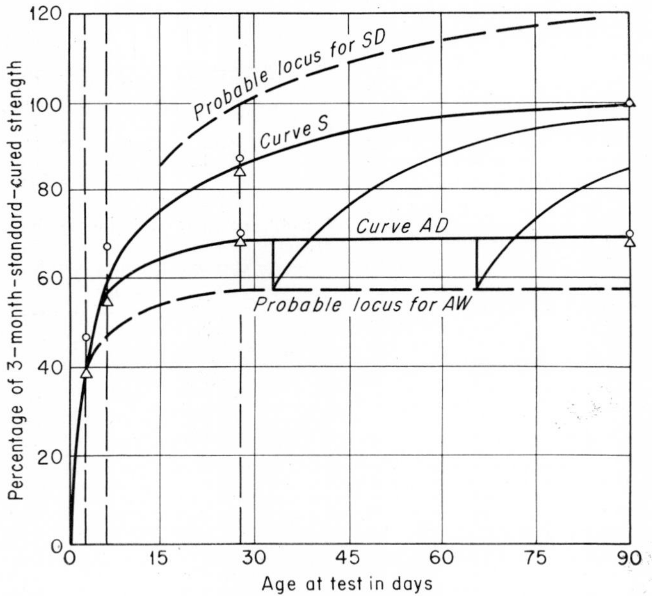

FIG. 6. Curing diagrams for normal (Type I) portland-cement concrete; air and water curing; temperature constant. (*Concrete, vol.* 49, *January and May*, 1941.) S = standard (moist) cured to time of test; AD = air-cured, dry at test; AW = air-cured, immersed for 1 day prior to test; SD = moist-cured, but in dry air 10 days or more prior to test. Note that for resumed curing concrete must become resaturated with an immediate drop in strength from AD to AW before the resumption of strength gain can start.

the same slump in a very lean concrete (1:4.1:5.3). From Fig. 4 the predicted strength of the rich mix is 6,000 psi and of the lean one is 1,800 psi.

The 28-day strengths of Fig. 4 will be developed only under the standard conditions specified for moisture and temperature. Under favorable temperature the rate of strength increase is influenced by moisture conditions as shown qualitatively in Fig. 5 and actually for Type I (standard) portland cement concrete in Fig. 6, and for Type III cement (high-early-strength) concrete in Fig. 7.

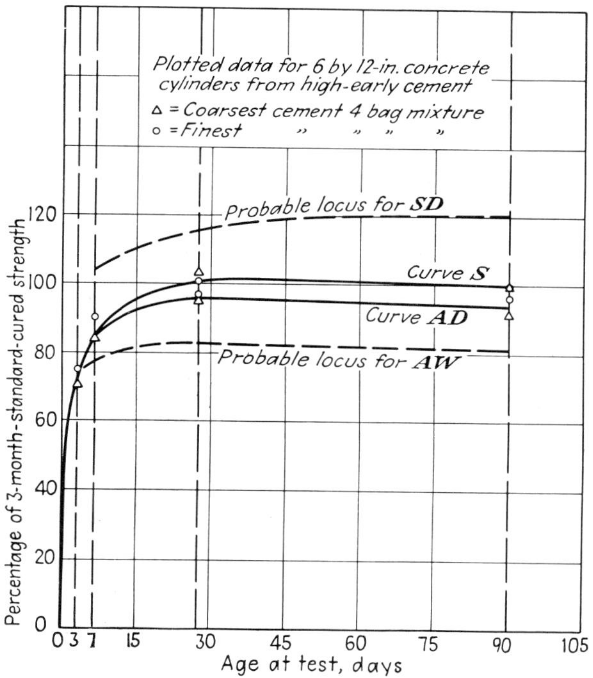

FIG. 7. Curing diagrams for high-early-strength (Type III) portland cement concrete; air and water curing; temperature constant. (*Concrete, vol.* 49, *January and May*, 1941.) S = standard (moist) cured to time of test; AD = air-cured, dry at test; AW = air-cured, immersed for 1 day prior to test; SD = moist-cured, but in dry air for 10 days or more prior to test.

Figure 8 shows normal water-cement ratio vs. strength relationships for both normal (Type I) and high-early-strength (Type III) portland cement concretes at ages of 1, 3, 7, and 28 days under standard moist curing. See also Fig. 23 for age-strength curves for the different types of portland cement.

Without moisture for continued hydration, strength gain ceases as the concrete dries out (except for the added strength due to drying and lost again upon resoaking; see footnote, p. 7-21). With moisture for hydration, temperature is extremely important in the early stages of curing, as is apparent from Fig. 9. Concrete frozen at an early age will gradually resume gain of strength under restoration of favorable conditions of moisture and temperature (Fig. 10). Repetitions of freezing and thawing of inadequately cured concrete soon produce permanent disruption. Representative strength vs. water-cement ratio relationships are shown in Fig. 11 for temperatures of 50° and 70°F for ages of 1, 3, 7, and 28 days.

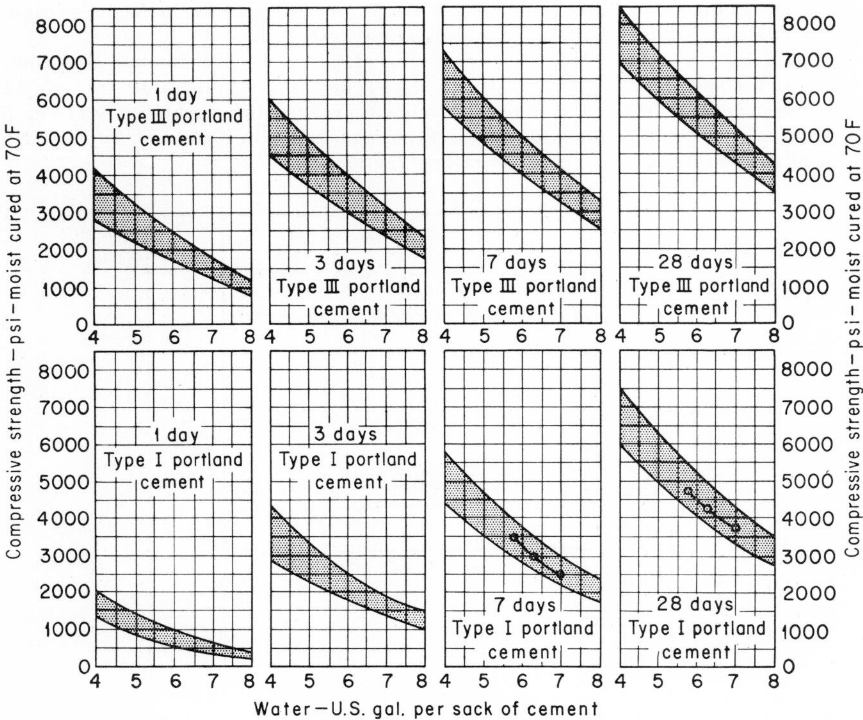

FIG. 8. Representative ranges of compressive strength for Type I and Type III cement concretes as influenced by water-cement ratio and period of initial standard moist curing. (*Port. Cem. Assoc. Bull. T12*, 10*th ed., p.* 5, 1952.)

Compressive strengths secured by test also vary with the height-diameter ratio of the specimen, and in testing drilled cores corrections must be applied, especially as the H/D ratio decreases toward unity or below. For an H/D ratio of unity (applicable to the cube, the standard specimen in many countries) the strength is about 16 percent greater than for the standard United States value of 2. For H/D ratios between 2 and 4 (Fig. 12) the results from a compressive-strength test are little affected (ASTM C42). Plane bearing surfaces and proper centering are important in tests of concrete, as in any other tests.

The various factors that influence compressive strength affect strengths in flexure, tension, and torsion in much the same manner. Figure 13 (*Engineering Foundation Arch Dam Investigation*, vol. 2, p. 461, 1934) shows comparative age-strength curves for moist cured specimens all cast from the same Type I cement mixture. The modulus of rupture in flexure is about twice the direct tensile strength, as is true for cast-iron and other brittle materials. Torsional failures are always by diagonal tension rather than shear, which is true for brittle materials in general.

The so-called flexural strength of plain concrete (Figs. 13 and 14, the modulus of rupture), like the compressive strength, is frequently used as a measure of concrete acceptability. This is especially true in the highway field where small beams can be fabricated easily and tested by simple portable beam testers (ASTM C31, C192, and C78). Beams are also sawed from the hardened concrete and tested in accordance with C42.

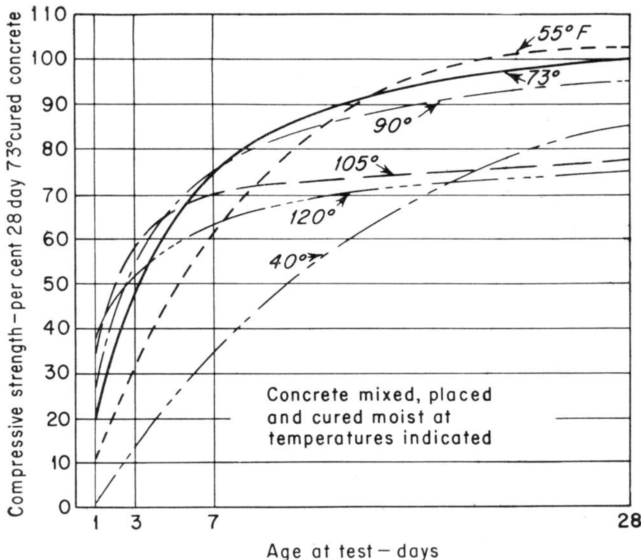

FIG. 9. Effect of curing temperature on strength of concrete. (*Port. Cem. Assoc. Bull. T12, 10th ed., p.* 7, 1952.)

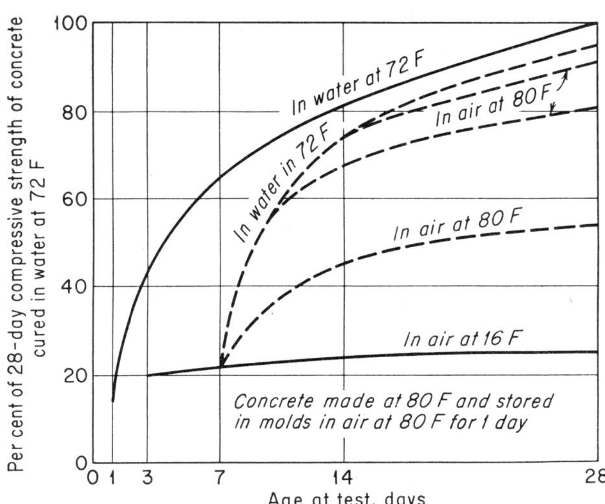

FIG. 10. Gain in strength after freezing. Normal (Type I) portland cement concrete. (*Port. Cem. Assoc. Bull. ST21, October,* 1940, *Fig.* 2.)

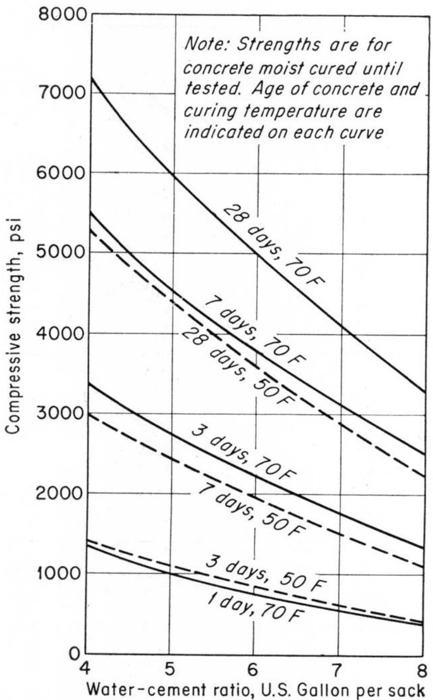

FIG. 11. Age and curing temperature in relation to water-cement ratio and strength of normal (Type I) portland cement concrete. (*Port. Cem. Assoc. Pavement Manual R2*, p. 53, 1946.)

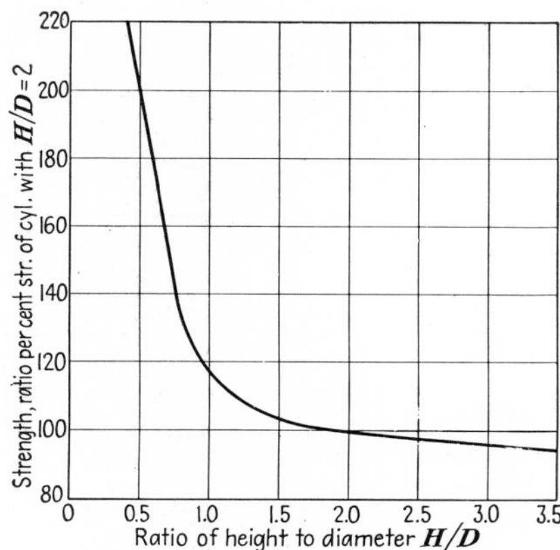

FIG. 12. Effect of height-diameter ratio on compressive strengths of concrete cylinders.

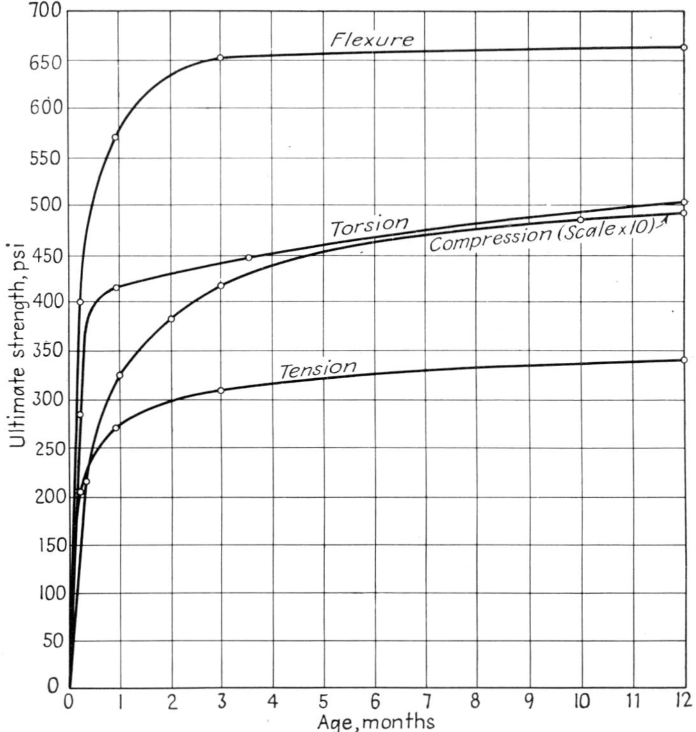

FIG. 13. Age-strength relations in compression, tension, flexure (modulus of rupture), and torsion (modulus of rupture); identical mixtures, standard moist curing. (*Eng. Found. Arch Dam Invest.*, *vol.* 2, *p.* 461, 1934.)

When desired Young's modulus can readily be evaluated from the load-deflection data of a flexural test. Water-cement ratio, curing, and test condition, affect the modulus of rupture in about the same manner, but not necessarily to the same extent as they do the compressive strength.

Compared with compression the flexural test has several disadvantages. For the stronger concretes the modulus of rupture does not increase to the same extent as does the compressive strength. Nonfailure of the specimen is dependent upon the continued integrity of the thin-tensile-surface (extreme fiber) concrete within the middle third of the span. One small shrinkage crack or defect in that region is likely to dictate the failure for the beam. The flexural test is also more sensitive to the type and surface texture of the coarse aggregate since in the tensile region of the beam the stress transfer between the mortar matrix and the aggregate is almost entirely tensile. The flexural strength does not increase with the age of favorable curing to the same extent as does the compressive strength.

Tests at the National Sand and Gravel Association Laboratory consistently indicate that for some unexplained reason sawed beams are weaker than are identical beams tested as cast. Such factors present questions regarding the suitability of the plain concrete beam as a criterion of concrete quality.

Tensile and torsional strengths are of interest as basic properties of the concrete but are not evaluated as measures of significant characteristics. As regards rate of

strength gain with age, water-cement ratio, and richness of mixture, both the tensile and torsional resistances parallel flexural.

Other strength tests used to a limited extent and as yet not fully evaluated with respect to their possible usefulness as indicators of significant properties are: (1) The "modified cube method" (ASTM C116) in which compressive strength is determined by testing

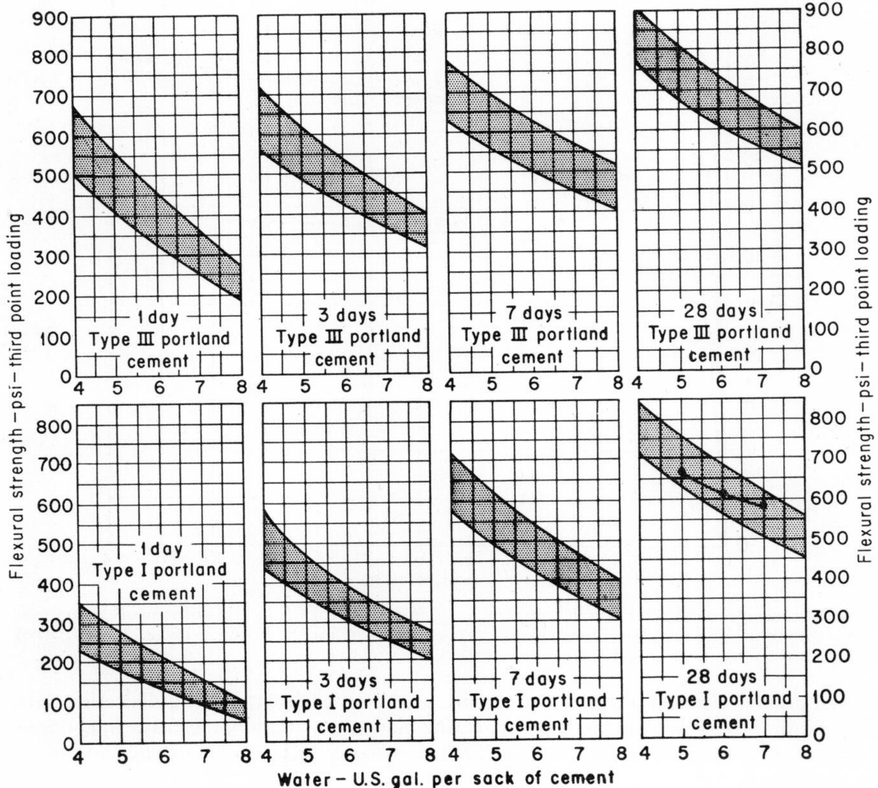

Fig. 14. Representative ranges of flexural strengths (moduli of rupture) for Type I and Type III portland cement concretes as influenced by water-cement ratio and period of initial standard moist curing. (*Port. Cem. Assoc. Bull. T12,* 10*th ed.,* p. 6, 1952.)

crossways in compression on its side one or both of the two fractured fragments of a tested flexural specimen. (2) A "tensile splitting test" of a concrete compressive cylinder loaded compressively on its side in the direction of a diameter along a pair of opposite surface elements. Under this form of loading the cylinder is split into two prisms of semicircular cross section, the break being definitely tensile rather than crushing in nature. If the test is discontinued immediately when splitting has occurred the two longitudinal prisms can be rematched (fitted together), stood on end, and tested compressively with an alleged less than 10 percent reduction in strength from that of a companion unsplit compressive cylinder. For a discussion of these two tests with related references, see Sven Thaulow, *Jour. ACI,* January, 1957, p. 699, Title No. 53-38 (*Proc. ACI,* vol. 53, 1956–1957).

As indicated in the footnote to Table 4 the rate of testing affects the strength obtained by test; it also affects the strain corresponding to a given load. The slowly conducted test is not to be confused with loads applied for weeks and months, but the same plastic flow (creep) as that of a long-time test accounts for the extra strain of a slowly tested specimen (see Figs. 19 and 20 for strains from *sustained* loading tests).

Bond. Bond refers to the sliding resistance developed between the concrete and embedded steel bars or wires. Bond stress is expressed in pounds per square inch of the superficial contact area of a plain round or square bar. For deformed bars the contact area for bond-stress calculations is taken as that of the plain bar of the same average area of cross section (or weight per unit length). Freedom from relative slippage between steel and embedding concrete is vital in all reinforced construction and the resistance to slippage, known as bond, may be that due to friction and/or adhesive resistance to sliding. End anchorages, extensions, properly hooked ends, and rolled-in deformations (lugs) are used to provide the equivalent of added bond resistance.

For prestressed concrete some of the problems of bond are distinctive because of such factors as the end anchorages and/or grouting around the bars subsequent to tensioning. Whereas the modern (ASTM A305) deformed bars are largely employed in conventional construction, high-strength wire, cables, and plain high-carbon steel bars are often used in prestressed or poststressed construction. See J. R. Janney (*Proc. ACI,* vol. 50, p. 717, 1954) or *Development Dept. Bull. D2,* Port. Cem. Assoc.

Permissible bond stresses (see Reinforced Concrete) are normally specified as percentages of the ultimate compressive strength of the concrete (to which the bond strength is roughly proportional, except for the stronger concretes for which there is not a corresponding increase in bond strength).

Because of the sequential nature of sliding action, the average intensity of bond that can be developed is greater over a short length of embedment then over a long one. For plain bars the texture of the bar surface is important, the longitudinally smooth surface of a drawn wire developing a lower unit bond resistance than the relatively rough surface of a hot-rolled bar of the same size. Cold-rolled and polished bars develop low bond resistances. At a given bond stress for a given type of surface, a bar of small diameter can develop a greater pull per square inch of cross-sectional area because of the greater specific surface in contact with the concrete.

Firm rust on the bar at the time of embedment makes the surface rougher, thereby increasing bond effectiveness, but because of uncertainties regarding the firmness of the coating and the extent to which there have been inroads upon the effective cross section of the steel, it is accepted practice to prohibit the use of rusted reinforcement. Rust that forms upon bars already embedded in hardened concrete (because of inadequate cover) produces disruptive swelling, cracking off additional concrete, thereby further reducing the cover and resulting in a rapid progressive failure of interaction between steel and concrete. Added to this is exposure of the steel to continuing deterioration as new layers of rust penetrate ever more deeply into the steel.

The slight settlement of the fresh concrete away from the underside of horizontally laid steel reduces or may even destroy effective bond beneath the bar (see Watergain, *Engineering News Record,* vol. 98, no. 6, pp. 242–244, Feb. 10, 1927). The deeper the fresh concrete beneath the bar, the greater is the adverse watergain effect which fact explains the specification reduction in permissible bond stresses for "top bars" (see Reinforced Concrete).

In general a very slight relative slippage between the steel and concrete of a flexural member (as little as 0.001 in. per in. of embedded length) is critical and the usefulness of the bar as reinforcement is largely nullified. Since the advent of the Ransome patent on a square twisted bar in 1884 many patterns of deformation, some of them quite

ingenious, have been rolled. For most of these the spacing of the lugs was too great, or their sloping contours were such that the initial slippage of the bar was at a load comparable to that for a plain bar. The lugs came into firm bearing only after it was too late for them to increase usefully the tension in the bar. Most of these old-style bars did supply a certain "hang-on" property that retarded or prevented a complete final collapse as the beam attained the load corresponding to the initial slip that marks the end of usefulness for a flexural member.

During the decade 1940–1950, the twenty (more or less) deformed bars still current were replaced by six patterns that, on the basis of intensive tests, proved to be as good or better than anything previously produced. From these tests an ASTM specification, A305, was evolved stipulating minimum requirements for lug sizes and spacings [1] (see Reinforced Concrete, Table 1). The test procedure employed in evolving and/or selecting the six current ASTM A305 patterns has recently become a standard of the ACI (*Proc. ACI*, vol. 54, p. 89, 1957–1958). ASTM C234 covers a routine procedure suitable for comparing concretes on the basis of bond.

The problems of bond are many and varied, one of the most important being how to prevent, or else make proper allowance for, the splitting of the concrete that is induced by the wedging action of deformed bar lugs. A rather complete discussion of bond in its numerous aspects and a digest of the literature may be found in ASTM special technical publication 169, p. 143, 1955, and in Iowa State College Engineering Report No. 26, 1955–1956.

Durability. Durability is a several-sided characteristic which can be considered only in relation to the question: durable against what? Divided into three categories durability can be thought of as resistance to:

1. Chemical deterioration often characterized by the multiple cracking, crumbling, and spalling that result from the differential swelling incidental to disturbances in chemical status.
2. Physical disruption: from applied forces or impact; from differential volume changes due to temperature variations; or from swelling and shrinkage due to moisture changes.
3. Surface factors, chemical or physical: progressive surface response to corrosive chemicals in contact with exposed surfaces or to physical abrasive action.

In category 1 would fall such causes as organic impurities in the sand or the mixing water, sugar contamination, adverse cement-aggregate reactivity, or the leeching action of water percolation through a permeable concrete.

In category 2 would fall the recurring impact pounding of heavy trucking on a floor or a pavement, the occasional blow-up of a concrete pavement under the unrelieved swelling incidental to moisture and/or sun; and the heat-of-hydration cracking within the body of a massive concrete structure. The splitting of a beam from the wedging action of deformed reinforcement or the spalling occasioned by the rusting of insufficiently embedded steel are other examples.

Surface factors could include the hair cracking or checking from too rapid surface drying of recently placed concrete, the scaling that results from the use of salts in snow removal, the normal wear occasioned by steel-tired vehicles, sandblasting effects, cavitation or abrasion by water jets, dusting of floors from laitance, or use of dry cement in the final troweling.

[1] The testing that led to the evolution of ASTM A305 was almost entirely on No. 7 bars (nominal ⅞-in. diameter). The largest bar covered by A306 is No. 11 literally a No. 11.3 and equivalent of a 1.41-in.-diameter round bar (having an area equaling that of the old 1¼-in. square). Extending test results from No. 7 to No. 11 represents a considerable extrapolation. In some quarters, however, bars up to No. 18 (equivalent of 2¼-in. round) are being produced and used under A305 proportionate extensions. Considering the heavy concentration inevitable in the stress transfer from steel to concrete in the contact zone, any appreciable extrapolation without the support of supplementary test data is highly questionable. Fortunately, to date, such drastic extrapolations have been only in heavily massive construction—less critical no doubt, but still of dubious validity.

Durability against chemical attack is discussed elsewhere, as are some aspects of 2 and 3. Only the resistance of concrete to freezing and the related aspect of salts for ice removal will be considered here.

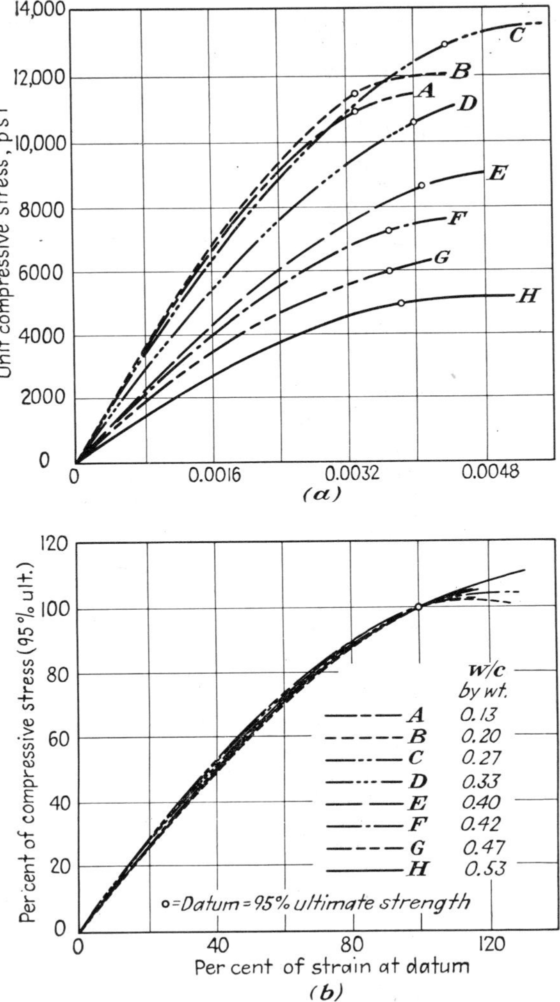

FIG. 15. Effect of water-cement ratio on stress-strain diagrams for moist-cured neat cement mixtures at 28 days. (*Iowa Eng. Expt. Sta. Bull.* 159, p. 27, 1943.)

At any age after the initial hardening (final set), *saturated* concrete is much less resistant to freezing disruption than is concrete in which the larger of the capillary passages are but partially water-filled.

As the curing (hydration) progresses, some of the free water enters into combination with the cement and/or is drawn into smaller capillary passages. The temperature at which water freezes becomes progressively lower as the diameter of the capillary

becomes less. Moreover, the extent of diametral expansion also becomes progressively less. Water in the smaller capillaries presents, therefore, little or no hazard against damage from freezing. Water remaining in the partially filled larger capillaries now has air space adjacent against which major expansive forces cannot develop. In dense concrete much of the air is trapped and not displaced even after prolonged soaking.

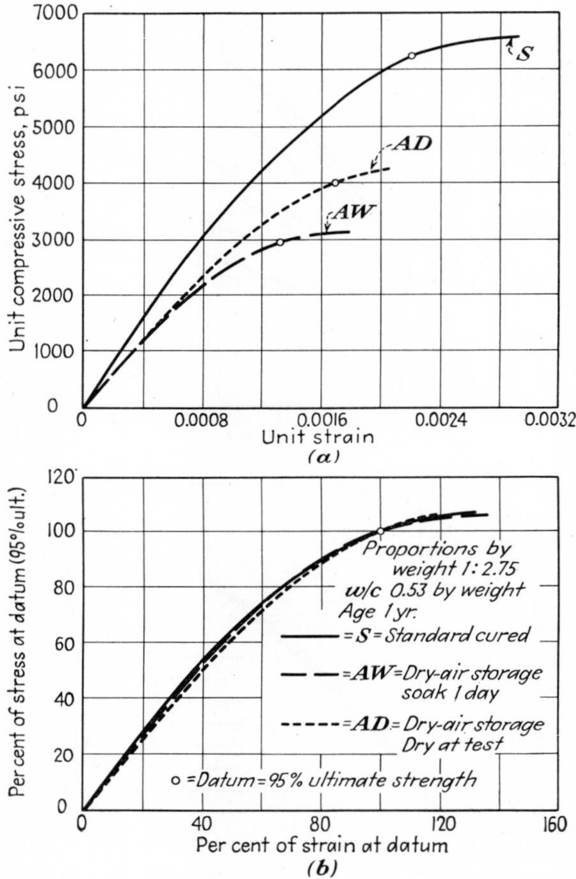

FIG. 16. Effect of curing conditions and moisture content at test on the stress-strain diagrams for a portland cement-sand mortar mixture. (*Iowa Eng. Expt. Sta. Bull.* 159, *p.* 29, 1943.)

It is such aspects as these that give well-cured hardened concrete a resistance to frost damage many times that of the same concrete more fully saturated and/or less well cured.

The adverse effect of calcium or sodium salts, often pronounced where used for ice removal, seems to be essentially physical. Upon wetting and drying salt crystals are dissolved and re-created, which subjects the enclosing pores of the concrete to the same sort of alternating disruptive action as that occasioned by freezing and thawing with filled capillaries. Against the resulting progressive surface scaling air entrainment offers

a large measure of protection but not complete immunity. The explanation of the beneficial effect seems definitely to be that the fine, well-spaced bubbles of air supply to a greater degree the air-cushioning effect characteristic of the partially filled capillary passage. Capillaries near the surface are largely filled with the brine—the closed air-entrainment bubbles never are—nor do they facilitate the passage of brine or moisture within or through the concrete.

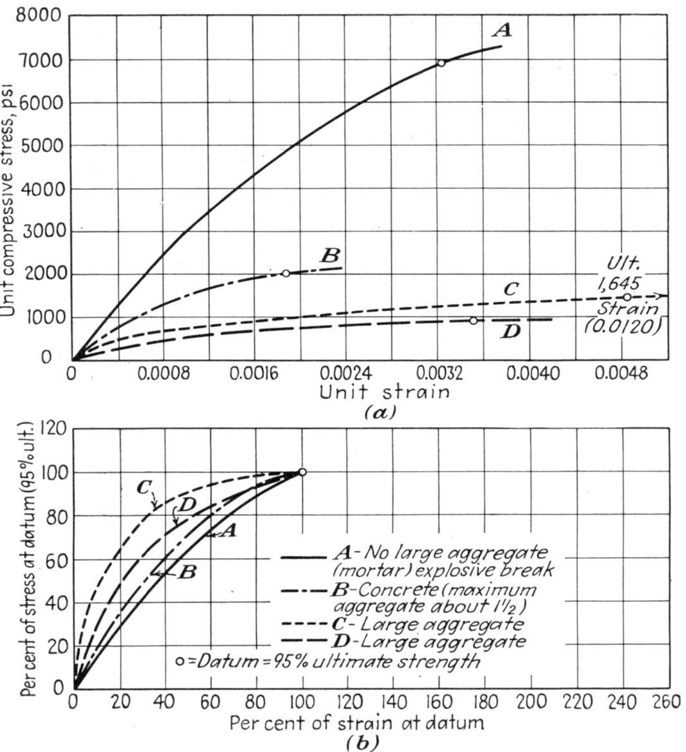

FIG. 17. Comparisons of stress-strain diagrams for drilled cores having portions of large plumstone inclusions. (*Iowa Eng. Expt. Sta. Bull.* 159, *p.* 32, 1943.)

Stress-Strain Relationships; Stiffness and Poisson's Ratio. Although concrete is much weaker in tension than in compression, it again resembles other brittle materials in having tensile, flexural, and compressive moduli of elasticity about equal and the shearing (torsional) modulus about 0.4 of the others.

In spite of the great ranges of compressive strength obtainable by varying the cement-water ratio, age, and curing, the basic or qualitative stress-strain behaviors for such different strengths of concrete are virtually identical, as is shown by Figs. 15*b* and 16*b*, when the stress-strain curves are superimposed on a percentage basis. On the other hand, variation in maximum size of coarse aggregate seems to produce significant differences in fundamental stress-strain action (Fig. 17*b*). The stress-strain behavior in compression differs basically from that in tension, flexure, and torsion (Fig. 18*b*).

Some of the earlier tests give values for Poisson's ratio as low as 0.08, but in the light of many subsequent tests it appears that Poisson's ratio for concrete rarely if

ever falls below 0.16 and that 0.20 is good as an average value. The low earlier values were probably due to lost motion in the lateral strain measuring devices then employed.

Coefficient of Thermal Expansion. The coefficient of thermal expansion for concrete is influenced more by the aggregates than by the cement, but fortunately values for both cement and usual mineral aggregates are near enough to that for steel to minimize differential temperature strains between the two in reinforced concrete construction.

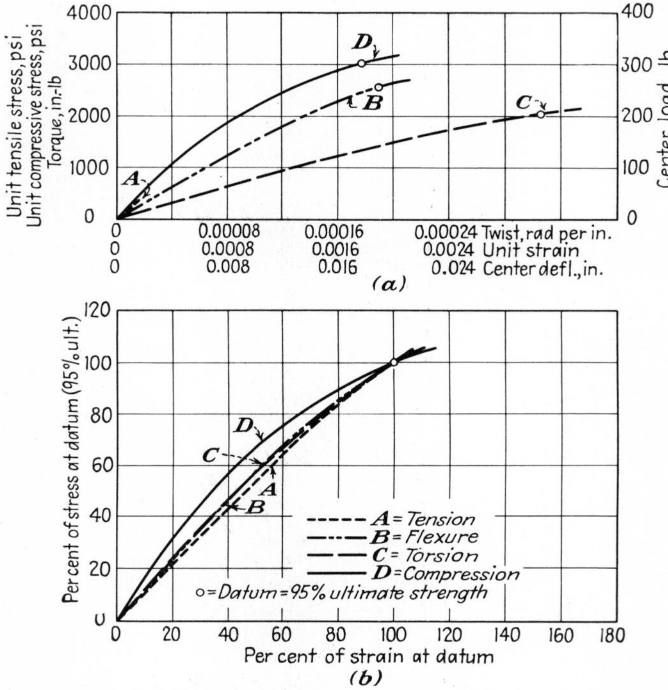

FIG. 18. Comparison of compressive, tensile, flexural, and torsional stress-strain characteristics for concrete. (*Iowa Eng. Expt. Sta. Bull.* 159, *p.* 36, 1943.)

Deformability of Hardened Concrete. Concrete is relatively brittle and unadapted to much differential distortion within a member. Most concrete construction involves masses of appreciable length, area, or volume, and under conditions of moisture change or temperature variation there is likely to be severe differential stressing. These various distortions grouped under the general term *volume change* contribute much more to the cracking and ultimate breakdown of most concrete members than do the stresses from the loads for which the members were designed. Table 5 supplies some approximations indicating the order of magnitude to be expected from the several types of strain to which concrete is normally subjected.

These figures are crudely approximate since each type of dimensional change is affected by its own assortment of pertinent factors. The values are representative, however, and should assist one to gain a perspective as regards the relative importance of certain items in relation to others.

In comparison with clay, which may shrink or swell several percent (12 in. per 1 percent per 100 ft) in passing from the plastic to the dry condition, the length changes

TABLE 5. LINEAR STRAIN FROM STRESS, MOISTURE, AND TEMPERATURE

	Percent	Inches per 100 ft
Elastic strain per 1,000 psi	0.02 –0.05	0.24–0.60
At proportional limit (compressive)	0.03 –0.04	0.36–0.48
At ultimate in compression	0.16 –0.30	1.92–3.60
At ultimate in tension	0.010–0.017	0.12–0.20
Hydration swelling, continuously wet	0.010–0.030	0.12–0.36
Drying shrinkage in air (concrete mixed with 24 gal H_2O per cu yd)	0.017	0.20
Drying shrinkage in air (concrete mixed with 45 gal H_2O per cu yd)	0.092	1.10
Plastic flow (creep) sustained compressive loading is up to 4 or 5 times initial deformation under same loading		
Strain per 100°F temperature change	0.03 –0.07	0.36–0.84
Neat cement at "normal consistency," permissible maximum expansion or contraction (autoclave soundness test ASTM C151):		
Portland cement Types I–V (ASTM C150)	0.50	6.00
Portland cement Types I–III (ASTM C175)	0.50	6.00
Portland blast-furnace slag cement (ASTM C205)	0.20	2.40

appear small, but in terms of equivalent thousands of pounds per square inch of stress some of the amounts are large, with none negligible. Structural-grade steel, stressed to its proportional limit in tension (36,000 psi), has been elongated 0.12 percent, or 1.44 in. per 100 ft. Hard-grade steel elongates elastically about twice this value. Where steel is embedded in concrete for tensile reinforcement it is apparent that small tensile cracks will occur in the concrete at steel stresses as low as 3,000 psi. Reinforcement (unless prestressed) will not prevent the formation of tensile cracks in concrete, for the concrete must crack before the steel can assume much of its permissible design stress.

Length change from stress (elastic) is a function of the stiffness of the aggregate particles as well as of the cement paste or mortar matrix. It is also a function of the amount of sliding or yielding that occurs at junctions (contact surfaces) of the aggregate particle and the matrix.

Length change from saturation or drying out occurs almost entirely in the cement-water paste, since approved aggregates are relatively nonabsorbent with very little length-change effect in passing from wet to dry or dry to wet. The leaner the mixture (the lower the cement factor) the less is the distortion from moisture change. Note the relatively high length changes that are tolerated for the normal-consistency neat cement paste bars in the autoclave test for soundness (Table 5). Of course, the autoclave produces more than a change in moisture content; in fact, the significant changes are essentially chemical rather than saturational. The separation and consequent cracking that occur in two-course construction between a mortar topping and a concrete subbase (for which the cement factor is always considerably lower) is due usually to differential swelling and shrinkage. In general, two-course construction involving a mortar topping or surface coat should be avoided. The lower the ratio of water to cement at any given aggregate content, the less is the length change (to be expected, since it is primarily the loss or gain of water that produces the shrinkage or the swelling).

Length change from temperature is a function of the coefficients of thermal expansion for both the aggregates and the cement paste. Any coarse aggregate for which the thermal coefficient differs substantially from that for the mortar in which it is embedded may contribute not only to the overall effect but also to disruption of the concrete

from the differential expansions throughout the mass. Where the temperature range is to be considerable this aspect needs to be recognized. (See Pearson, *Proc. ACI*, vol. 38, p. 39, 1942.)

In general the disintegration that results from any physical or chemical instability of the paste or the aggregate (cherts, pieces of wood, or chemically reactive aggregates) is a volume-change phenomenon. Usually it is a case of aggregate particles swelling and disrupting the concrete mass. Often such disruption proceeds progressively inward from exposed faces but in nonmassive, or even fairly massive, construction the disruption may virtually permeate the structure. Surface map cracking is generally the visual evidence of damaging differential volume changes between aggregate and matrix.

The *creep, time yield,* or *plastic flow* that continues at a decreasing rate under sustained loading sometimes results in objectionable distortion and redistribution of stress. On the other hand, the plastic yielding is frequently a benign influence tending toward stress release and reduction in concentrations. In a reinforced concrete column where drying shrinkage and plastic flow will usually be present in the concrete simultaneously, the shortening of the concrete may transfer a heavy increment of load to the longitudinal steel which has on occasion been stressed compressively to or near its yield point. In an arch dam the compressive yielding or flattening of the arches puts added bending in the vertical cantilever elements. Compressive thrust in a pavement slab relieves itself in part through plastic flow. For other types of members, as an isolated unreinforced strut or beam, the time yield may occur with neither load transfer nor stress release. In rigid frame construction, plastic flow (creep) always results in redistribution—sometimes desirable, sometimes not.

Under a sustained stress of given intensity, concrete may deform several times the initial deformation, especially if the concrete is subject to drying shrinkage as it sustains the load. Thus in Fig. 19 the initial deformation under the 900 psi load could be expected to approximate 0.0003 or (0.03 percent), whereas the creep that occurred during 600 days was more than four times that amount (total deformation, initial plus creep, more than five times), about half of which occurred during the first month under load. In other words, 1 month under the sustained load added twice the amount of the initial deformation. Figure 19 (lower) shows the creep to be about proportional to the magnitude of the load sustained. As in the case of strength, the water-cement ratio of the concrete has an important bearing on the amount of creep. This is illustrated in Fig. 19 (upper).

Vogt has demonstrated that, for concrete uniformly dry or uniformly moist throughout the cross section, the creep under the sustained load is much less than when drying out is in progress, as was the case during the first few weeks of loading for the concretes of Fig. 19. The unequal drying shrinkage releases a portion of the section from assuming its proportionate share of the load, and the intensity of loading is successively augmented over different portions of the section, producing greater yielding than if each part had to resist only the average stress over the section.

Creep in concrete is also influenced by the creep characteristics of the aggregate and to some extent by temperature. As would be expected, slow-hardening cements are subject to increased creep at early ages. Concrete exhibits some recoverance upon removal of a sustained load. The immediate recoverance approximates the initial elastic short-time deformation. There is a slow slight additional restitution.

On exposure to dry air, concrete loses uncombined moisture at a decreasing rate until capillary equilibrium with the surrounding atmosphere is approached. If the surrounding air becomes more humid, the concrete retakes moisture and again releases some of it as the humidity declines. Throughout a series of such alternations, the weight of the concrete maintains a mean trend upward, indicating continuing hydration whenever opportunity offers, thereby gradually increasing the supply of chemically held

water. Under air drying at ordinary temperature, it is usually weeks or even months before equilibrium for a given humidity is attained for sections only 3 to 6 in. across. On the other hand, most concrete soaks up water rapidly and, in general, a dried-out 6-in. by 12-in. cylindrical specimen becomes resoaked to virtual equilibrium during

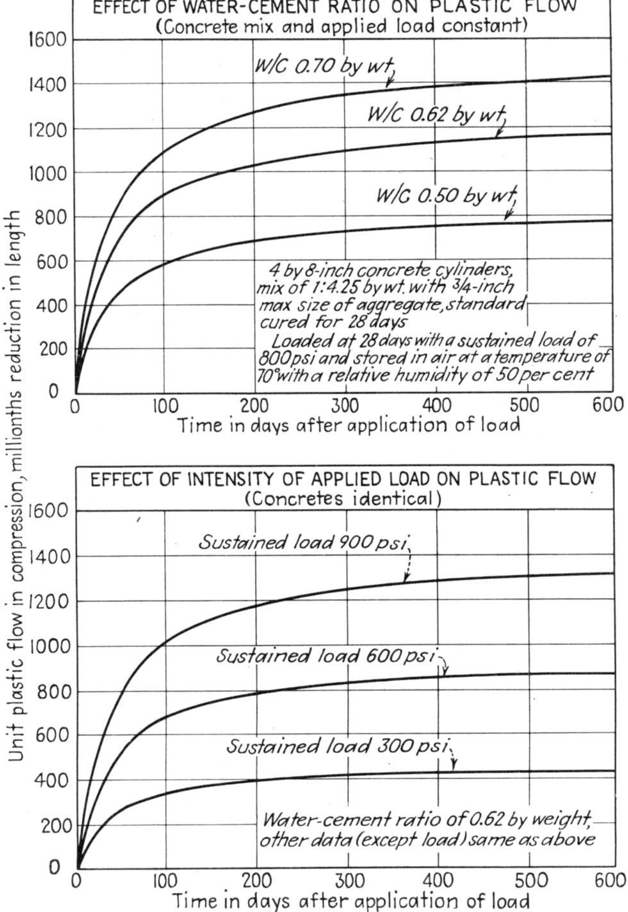

Fig. 19. Plastic flow as affected by water-cement ratio and intensity of sustained compressive load (Ref. 10, p. 27, Fig. 13).

only 18 to 24 hr of immersion. For each successive soaking and drying period, the concrete retakes somewhat less but retains somewhat more moisture than from the previous cycle. The length or volume of the concrete increases correspondingly, the increase being sometimes designated as *growth*.[1]

[1] This is, however, but one aspect, perhaps a minor one, of the progressive lengthening of such a member as a pavement slab. Whenever the concrete shrinks, minute cracks are formed because of subgrade and other restraints. These cracks never fully close upon reexpansion because of the small pieces of grit, etc., which have found their way into them. At the next contraction phase, the cracks are widened, the action being progressive and continuing as long as there are alternations of temperature and/or moisture.

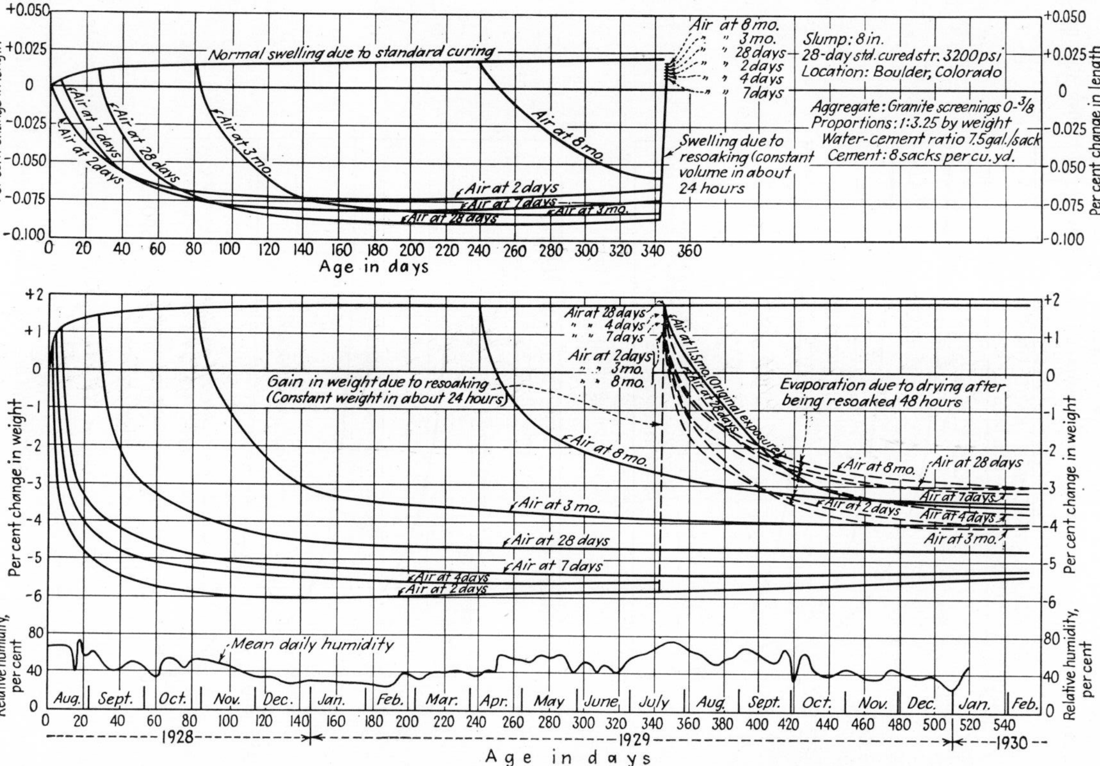

FIG. 20. Length- and weight-change diagrams for 3-in. by 3-in. by 40-in. beams exposed to dry air after various periods of moist curing. (*Eng. Found. Arch Dam Invest., vol.* 2, *pp.* 514–515, 1934, *Fig.* 164.) These data are from the mixture of Fig. 12 used in arch dam model tests, virtually a fluid mortar in small specimens having a high specific surface. Weight and length changes from drying effects for usual concretes are similar in nature but less in rate and amount than are those shown in this figure.

Since exposed concrete surfaces are rarely in moisture equilibrium with the surrounding atmosphere, they are continually taking in or releasing moisture and tending to expand or contract with respect to the more stable shielded concrete of the interior.

Concrete bars of small cross section do readily approach a condition of uniform moisture throughout, and tests show that the changes in length closely parallel the changes in moisture content as indicated by weight. The close agreement between moisture change and length (or volume) change is shown in Fig. 20. It is apparent that increased moist curing prior to a drying exposure retards but does not inhibit or appreciably diminish ultimate drying shrinkage.

In common with any other solid, a body of concrete subjected uniformly to swelling or shrinkage whether from moisture or temperature change would, if wholly unrestrained, respond without being stressed thereby. For a member of size, external as well as autogenous, stress-inducing restraints are invariably present. Thus for a pavement slab the bottom is severely restrained by friction with the subgrade, the ends or lateral edges by the thrust of abutting slabs and curbs. There are also internal restraints. As the top surface is heated or wetted, its expansive response is resisted by the unwetted unheated concrete inward from the surface with the same sequence in reverse upon drying or cooling.

The combination of a pavement soaked and swollen by a recent rain having superimposed upon it the lengthening from a hot sun that warms long before it dries appreciably often leads to the well-known blowout phenomenon when expansion joints are absent, widely spaced, or clogged. Drying and cooling produce tensile cracking whenever subgrade frictional resistance or other restraint exceeds the tensile strength of the concrete. To avoid tensile cracking, under average conditions of subbase, richness of mixture, and aggregate type, *contraction* joints in plain concrete would need to be kept less than 25 ft apart (see "Joints and Hinges"). The differential or autogenous swelling and shrinking produce curling or warping in such members as pavement slabs and various distortions in other types of members.

If a member contains steel reinforcement no serious restraint to temperature strain is offered by the steel because it heats as the surrounding concrete heats and the two expansions are about the same. This is not true, however, for moisture volume change, which affects the concrete but not the steel. Sometimes the restraint offered by reinforcement is more disrupting than the unrestrained volume change that could have occurred were the steel not present. For all exposed surfaces and corners, disruptive weathering effects from frost, sun, and moisture are ever present just as they are for exposed natural bodies of stone.

Heat of Hydration and Other Thermal Properties. For massive concrete construction one of the most important of the disruptive agencies is the differential temperature that results from the thermal heat of hydration. Not present at placement, the heat within a few hours begins to build up cumulatively in the interior, where it is not dissipated, first swelling the mass and tending to disrupt the recently hardened concrete. From that stage any large concrete mass undergoes the same cycle as does a large metal casting in cooling. The outside layers cool first, shrinking against a warm unshrunken mandrel with corresponding tendencies to crack. Finally the interior cools, tending to shrink and pull away from an already shrunken exterior shell. Prior to the construction of Hoover (Boulder) Dam, such differential cooling was invariably a source of much serious cracking throughout dams. With the advent of Hoover Dam, measures for control were evolved. They are:

1. Keep the cement factor (for it is the cement that generates the heat), as low as possible by:

   a. Using aggregate of the best possible grading (a minimum of the cement paste required).

b. Grading up to the largest permissible maximum size of aggregate (reduces paste requirement).

c. Keeping the water-cement ratio as low as is compatible with proper placeability (a low water-cement ratio requires less cement for a given strength).

2. Use special low-heat cement (Type IV) developed specifically for massive construction, or modified cement (Type II) or use portland pozzolan or cement replacement blending (see "Blends and Substitutions").

3. Cast the concrete in blocks of limited size and circulate cooling water through previously embedded pipes, thereby extracting the heat uniformly throughout the mass as heat is generated. Maximum heat differentials can be held within any desired range by controlling spacing of the pipes, timing the operation, and regulating the temperature and rate of circulation of the cooling water. Lateral dimensions of blocks must be sufficient to provide shrinkage cracks between adjacent blocks wide enough to admit the pressure grout that finally unites adjacent blocks into one great monolith. A few hundredths of an inch is sufficient under properly controlled conditions. For Hoover Dam heights of continuous pour were 5 ft, lateral dimensions of blocks were 25 to 60 ft, and loops of 1-in. 14-gage tubing spaced 5 ft 9 in. on center were laid on the surface of each 5-ft lift. Circulation of cooling water, first at river temperature (about 50°F), was started as soon as possible after placing. Refrigerated water was used in a final or second stage of the cooling. The low (5-ft) height of pour permitted considerable heat loss by radiation up to the time another lift was placed. Even for the low-heat cement and relatively lean mixture used, an exposure of 3 days to air and surface sprays would (without the artificial cooling) have resulted in a temperature rise of 47°F in 6 to 12 months (much higher with ordinary cement and a richer mixture). A 10-day exposure reduced the load on the cooling system considerably; the 6- to 12-month temperature rise would have been but 32°F above the initial. Not more than one 5-ft lift was permitted during one 72-hr period or more than 35 ft in depth in 30 days.

Under the Hoover Dam stimulus extensive studies were conducted on the thermal properties of concrete (conductivity, specific heat, and diffusivitity). Figure 21 shows

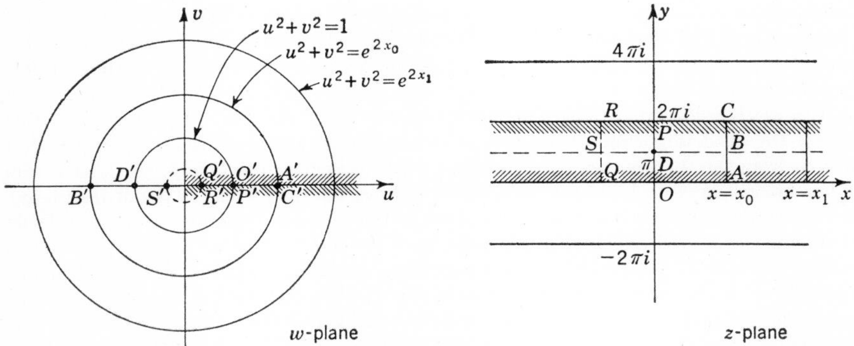

FIG. 21. Temperature rise in mass concrete up to 28 days for different types of portland cement. Concrete contains 1 bbl. (4 sacks) of cement per cu yd. [Ref. 10, p. 43, Fig. 23 (lower)].

temperature rises obtained from mass concretes of the five recognized types of cement having the heats of hydration per gram indicated in Fig. 22.

For nonmassive concrete construction the thermal heat of hydration as a disturbing agent is usually negligible, since the heat is conducted to the surfaces and dissipated before it can build up appreciable temperature differentials within the mass. Not only

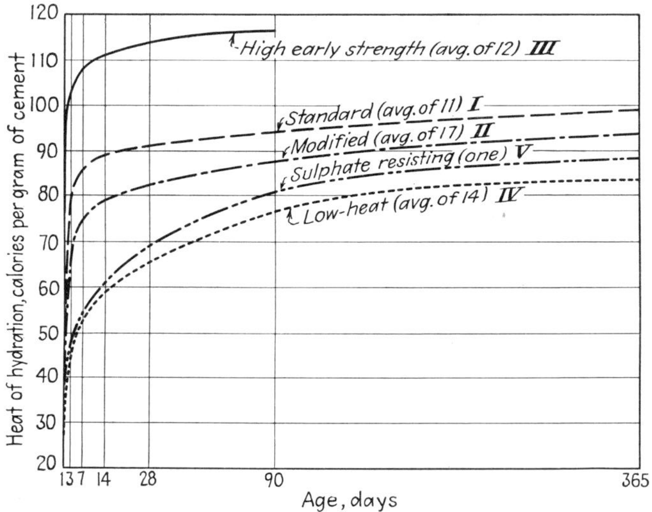

FIG. 22. Heat of hydration in calories per gram for different types of portland cement for ages up to 1 year [Ref. 10, p. 43, Fig. 23 (upper)].

do heats of hydration vary for the five recognized portland cements but the rate and extent of strength increase vary for both standard curing and mass (adiabatic) curing, as shown in Fig. 23.

Values for *specific heat* and *thermal conductivity* for concrete are influenced greatly by the kind of aggregate, proportions of the mixture, and its water-cement ratio. In Table 4 the thermal conductivity for the mass concrete is shown to be decidedly higher than the values applicable to building concrete, which has a lower aggregate ratio and a higher water-cement ratio.

The Water in Concrete. The water in the concrete serves a dual function. It first converts the dry cement and aggregate into a plastic, workable, placeable mass; it is the lubricant or vehicle that carries the solids. Oil or almost any other fluid could serve this function, but it would not produce concrete. The water must also interact with the cement chemically to form the products of hydration.

Hydration is essentially a combination of gel formation and crystallization. As the water comes in contact with the cement particles, the outer layer of the cement grain is softened to form a gel. The softening proceeds progressively inward, but the layer of gel appears to restrict penetration of moisture and the rate of formation becomes progressively slower. All but the finest of the cement particles remain incompletely hydrated for indefinitely long periods. The delayed hydration accounts for the cementing qualities found in reground "hydrated" cement. The slow but continuing gain in strength of concrete for years, under moist conditions, is probably due to the added gel formed by the gradual deeper penetration of moisture into the cement grains. The "healing" across fine cracks or the weak knitting of fully severed fragments known as

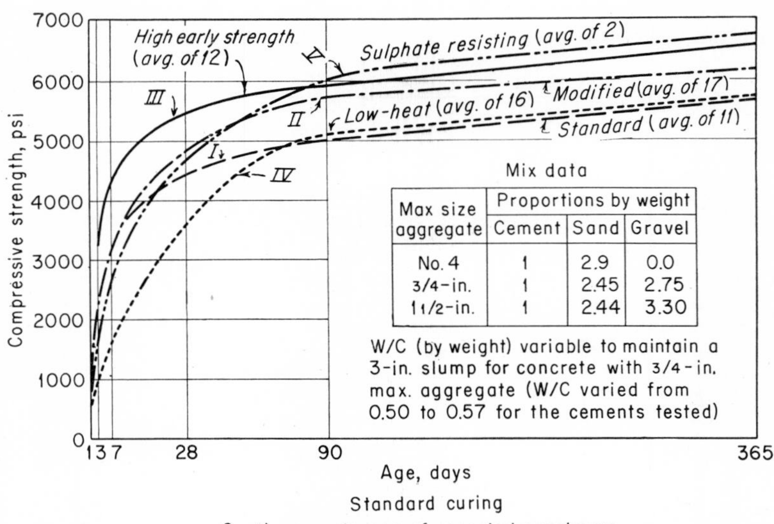

Max size aggregate	Proportions by weight		
	Cement	Sand	Gravel
No. 4	1	2.9	0.0
3/4-in.	1	2.45	2.75
1 1/2-in.	1	2.44	3.30

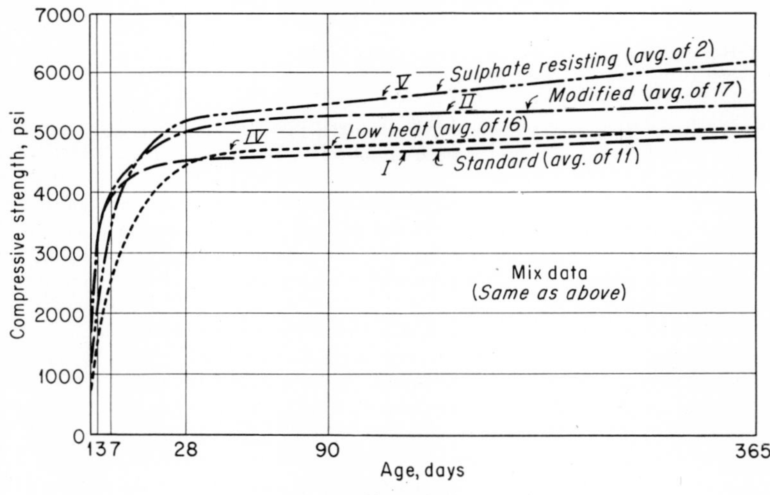

FIG. 23. Rate of strength increase up to 1 year for concretes from portland cements Types I, II, III, IV, and V for standard curing and for all but the high-early-strength cement (III) for mass (adiabatic) curing (Ref. 10, 4th ed., 1942, p. 54, Fig. 15).

"autogenous healing" and the "sticking" of adjacent briquets stacked in the curing tank are manifestations of the "continued or resumed curing phenomenon." [1]

It appears that the fineness of the cement should exert an important influence on both the rate and the completeness of hydration, and this is true, as indicated by Fig. 24.

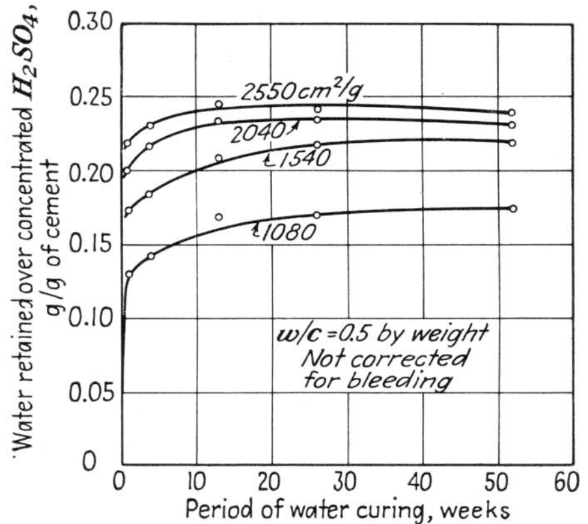

FIG. 24. Effect of fineness of grinding on rate and amount of hydration of neat cement. (*Proc. ACI, vol.*, 43, p. 707, 1947, *Figs.* 5 *to* 10.) C_3S 60 percent, C_2S 17 percent, C_3A 7 percent, C_4AF 13 percent. Exposure for extraction of free water was over concentrated H_2SO_4.

It is probably but coincidence that the amount of mixing water required to form a stiff but workable paste of neat cement is roughly equal to that necessary for complete hydration of the cement (about 20 percent by weight or 63 percent by solid (absolute) volume [2]). Although the water actually taken into combination by hydrated cement is independent of the aggregate, the water necessary to gage the mixture (to supply lubrication and mobility) increases as the aggregate increases in quantity. Thus a 1:3 portland cement, standard sand mixture (ASTM C190), corresponding to the neat cement normal-consistency range, requires from 36 to 46 percent of water to become workable, and a lean concrete mixture requires a water-cement ratio by weight of perhaps 80 percent.

The ratio of water to cement is variously expressed. In the laboratory the ratio is usually by weight and on the job in gallons per 94-lb bag of cement. The original water-cement ratio as evolved by Abrams in 1918 was by loose or sacked (94 lb to the sack) volume, and the ratio or inverse ratio by solid (absolute) volumes of Talbot and Richart is a more fundamental basis for expression. Table 6 supplies data for ready conversions between units.

[1] Autogenous healing is beneficial in many of the applications of concrete where early fine cracking from shrinkage or overstress has occurred, although it is not something of which conscious or design account would normally be taken. Following are a few specific observations on the healing action. A fully severed neat cement briquet testing 248 psi at 3 days, held together with a rubber band and reimmersed, tested 221 psi 3.4 months later and 59 psi at a third breaking at a total age of 15 months. A 1:3 standard sand briquet tested 247 psi at 3 days and 143 psi at 3.5 months. Hundreds of other broken briquets recovered less amounts; many showed a very small but measurable adhesion after as many as six breaks (*Proc. ASTM*, Pt. II, p. 593, 1929). Compressive specimens tested to the ultimate at 7 days (machine reversed instantly at final drop of beam to avoid shattering) regularly give 28-day strengths equaling those of 28-day compression specimens not previously tested (*Proc. ASTM*, Pt. II, p. 470, 1926). Abrams cites compressive healing and a remarkable case of flexural healing of a bridge girder; Hollister of an arch; Earley of a balustrade rail (*Proc. ACI*, pp. 636–639, 1926). Leslie Turner reported on healing (International Association for Testing Materials Congress, London, pp. 344–345, 1937). See also Lauer and Slate, *Proc. ACI*, vol. 52, p. 1083, 1955–1956.

[2] For normal consistency (ASTM C187 and C190) the range is 15 to 30 percent. The bound or nonevaporable water has virtually the same overall range.

TABLE 6. CONVERSION FACTORS FOR WATER-CEMENT RATIOS

Note that cement-water ratios are simply the reciprocals of the corresponding water-cement ratios

	Units in common use	Water-cement ratios			
		Gallons per bag, a common job unit (*a*)	Solid (absolute) volume (approx. v/c Talbot) (*b*)	Loose volume (original Abrams unit) (*c*)	Weight (job and laboratory) (*d*)
1	w/c, gallon per bag	1	0.279	0.134	0.089
2	w/c, solid volume	3.58	1	0.479	0.318
3	w/c, loose volume	7.48	2.09	1	0.664
4	w/c, weight	11.3	3.15	1.51	1.000

Table computed on the basis of the following assumed constants or unit values. Bold-faced figures are basic assumptions. Other figures are computed from the values assumed.

1 cu ft water weighs 62.4 lb and contains 7.48 gal. 1 gal water weighs 8.34 lb. **1 cu ft cement loose (as sacked) weighs 94.0 lb = 1 bag = ¼ bbl** and contains 0.478 cu ft of solid particles. **Specific gravity of cement = 3.15.** 1 cu ft cement (solid) weighs 196.6 lb and contains 2.09 bags or 2.09 cu ft compact loose (in bulk as sacked).

Use of Table. All values along any horizontal line are equivalent to one another.

Examples. a. To determine the number of gallons per bag which corresponds to w/c by weight of 0.70. On line 4, columns (*a*) and (*d*), w/c of 1.0 by weight = 11.3 gal per bag. Then w/c of 0.70 by weight = (0.70)(11.3) = 7.91 gal per bag.

b. To determine the number of gallons per bag corresponding to c/w of 0.50 by solid (absolute) volume. c/w of 0.50 = w/c of 2.00 by solid volume. On line 2, columns (*a*) and (*b*), w/c of 1 by solid volume = 3.58 gal per bag. Then w/c of 2 by solid volume = (2)(3.58) = 7.16 gal per bag.

In general, less than 50 percent of the water in a cement-water paste or a cement-aggregate-water mortar or concrete is available for hydration of the cement. Initially the entire mass is saturated, virtually all the capillaries being filled with water. As the cement extracts free water, the capillaries are partly emptied and capillary tension resists further encroachment upon the capillary water supply until a condition of equilibrium is attained between free capillary water and the bound or chemically combined water appropriated by the cement in hydrating. There is also some adsorbed or surface-bound water neither chemically combined nor in the capillaries. Experience shows that, even for concrete stored in water or in the fog room, the capillaries remain but partly filled, a fortunate condition since, as mentioned under durability, it is when the hydration has progressed far enough partially to empty the capillaries that concrete becomes resistant to frost damage. With full capillaries a single cycle of freezing and thawing may produce virtual disruption. Frost is relatively innocuous to concrete in the air-dry condition. It is probable that the efficacy of air entrainment against disruption under freezing conditions is related to partly filled capillary phenomena (see "Air Entrainment").

In hydrated concrete, water is present in the three forms: (1) water of constitution, (2) adsorbed water, and (3) capillary water, but it is not yet possible to distinguish clearly between these since, on drying, water of categories 2 and 3 is lost simultaneously; nor can all the 2, 3 water be driven off without removing some of the water of constitution. Figures 25 and 26 show, however, the progress of hydration and how, with added curing, the water is retained with increasing tenacity.

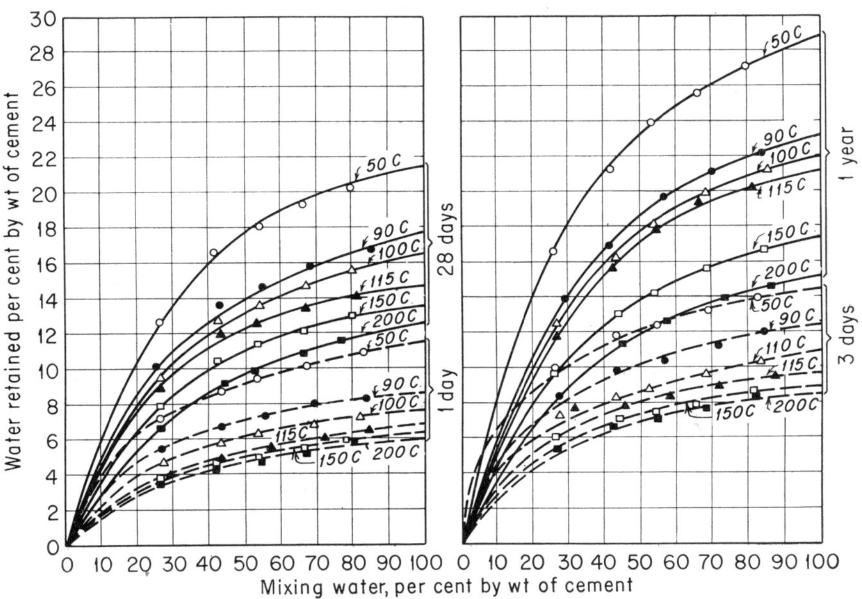

FIG. 25. Effect of water-cement ratio, period of hydration, and drying temperature on moisture retained by neat portland cement mortars. (*Proc. ACI, vol.* 31, *p.* 275, 1935, *Fig.* 1.)

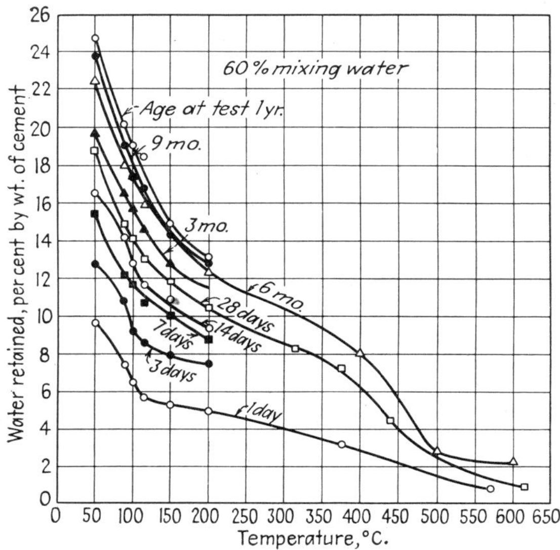

FIG. 26. Effect of drying temperature on water retained by neat portland cement mortar after various periods of hydration. Water-cement ratio 0.60 by weight. (*Proc. ACI, vol.* 31, *p.* 277, 1935, *Fig.* 3.)

DESIGN OF CONCRETE MIXTURES[1]

General Considerations. The determination of relative proportions of cement, fine aggregate, coarse aggregate, and water constitutes "the design of a concrete mixture."

Since the early 1920's two approaches to the design problem have been accorded recognition. These are the water-cement ratio approach of Abrams and the voids-cement ratio of Talbot and Richart. Basically the two have much in common but the techniques of application are widely different. The voids-cement ratio brought in the concept of solid volumes—indispensable to the accurate prediction of yields—and has contributed heavily toward the more complete understanding of the factors involved (see *Univ. Illinois Eng. Expt. Sta. Bull.* 137, 1923). As a working entity, however, the voids-cement ratio approach virtually never attained job linkage. On the other hand, the water-cement ratio approach has been widely promoted and refined to the point where the proportions for virtually all designed concrete mixtures are arrived at through one or another of the several versions of the water-cement ratio trial-batch technique of design.

Among the better known and most widely employed versions of the water-cement ratio technique, as used in the United States are: (1) The two ACI Committee 613 reports (1944 and 1954), quite different (with some persons preferring the earlier version); (2) the PCA procedure as covered in their *Bull.* T12 "Design and Control of Concrete Mixtures"; (3) A. T. Goldbeck, *Nat. Crushed Stone Assoc. Bull.* 11; and (4) Inge Lyse, *Proc. ASTM*, vol. 32, Pt. II, p. 629, 1929 (extended by W. M. Dunagan, *Proc. ACI*, vol. 36, p. 649, 1940). Others with varying shades of difference include Walker and Bloem of the NSAGA and NRMCA, the Army Engineers, and the Master Builders Co. of Cleveland. Herein we draw freely upon several of the sources mentioned but tie most closely to the Lyse-Dunagan technique because of the simplicity with which it lends itself to the manipulation and adjustment of mixtures through recognized usable interrelationships between strengths, workabilities, and proportions.

On the basis of preliminary information such as that of Table 7, the designer decides upon the *workability* as measured by the slump (ASTM C143). Currently the slump is the nearly universally used measure of workability for field concrete but the information in Table 7 could be stipulated in terms of flow (ASTM C124, long used in the laboratory supplementary to, or instead of the slump, especially on investigational work) or in terms of the ball penetration test (ASTM C360, a relatively recent measure of workability that shows promise of being an improvement over the slump test).

On the basis of information in Table 10 or other controlling specification, the designer decides upon the maximum permissible water content per sack of cement, or if strength may govern, rather than the nature of the exposure, he will secure this information from Table 11 and/or Fig. 4 or Fig. 8. The lesser of any two indicated water contents per sack will be used.

From Table 8 he decides upon the maximum permissible size of aggregate. Table 9 will indicate the approximate amount of mixing water that will be needed per cubic

[1] For the small-job type of work—basements, sidewalks, residential driveways—job proportioning continues to be much on the basis of arbitrary assignment and unsupervised artisanship. Even here the increasing use of ready-mixed concrete brings to such work the considerable degree of control that has been built into the operation of most central supplying agencies. Generally speaking, ready-mixed concrete has become a designed rather well-controlled commodity.

Although this treatment deals with concrete as an engineering material that can be specified, designed, manufactured, and checked, there is appended a table (Table 14) that will, for average materials and conditions, make it possible to secure, with reasonable assurance, satisfactory results for the small undesigned type of job.

Tables 7 to 14, inclusive, are taken verbatim from the 1954 Report of "ACI Standard Recommended Practice for Selecting Proportions for Concrete" (ACI 613-54, *Proc. ACI*, vol. 51, p. 49, 1954–1955). They are of diverse origins having been appropriated or adapted from other background sources, notably the Second (1924) and/or Third (1940) Joint Committee reports.

These procedures are applicable to conventional aggregates. For special considerations pertinent to lightweight aggregates see JONES and STEPHENSON, *Proc. ACI*, vol. 54, p. 527, 1957–1958.

FIG. 27. Workability illustrated. (Synonymous with placeability or mass mobility, workability may be visualized as a composite of texture and slump.) (*a*) Undersanded—harsh, poor-textured, low-slump concrete—will produce rough honeycombed surfaces. (*b*) Economical—good texture with low slump—plastic, workable. (*c*) Oversanded—excess of cement-sand mortar—easy to place, gives good surfaces—low yield (uneconomical), likely to be pervious. (*d*) Slumps of a well-proportioned mixture of stiff, medium, and wet consistencies. Stiff gives highest yield—is best where it can be worked into position. Wet mix may be necessary for thin walls, some columns and intricate reinforced construction.

yard of the concrete, and from this quantity he can determine the probable required sacks of cement per cubic yard.

If the materials are typical, he can estimate from Table 12 the amount of coarse aggregate that will be needed. Starting with a suitable size of batch (depending upon whether the trial batches are to be hand mixed, mixed in a small laboratory mixer, or are to be of job-size proportions) he will add fine aggregate in an amount to ensure that the mixture is initially "undersanded." By checking upon the workability visually and by successive slump measurements (see Fig. 27 and Table 7) the addition of successive small increments of fine aggregate will bring the mixture to the state desired as near as may be in the light of the preliminary quantity of coarse aggregate selected.

Table 13 provides an indication of a minimum program for trial-batch determinations. Of course, the aggregates should be fully representative of those to be used on the job and they will preferably be in the saturated surface-dry condition so that they will neither extract water from the mixture by absorption nor add free water to it. If not saturated, surface-dry, there will need to be preliminary free-moisture or absorption determinations with corresponding corrections to the amount of the mixing water. In the translation of laboratory results to job batches such determinations invariably have to be made at intervals with corresponding batch corrections as work progresses.

TABLE 7. RECOMMENDED SLUMPS FOR VARIOUS TYPES OF CONSTRUCTION [1]

Types of construction	Slump, in. [2]	
	Maximum	Minimum
Reinforced foundation walls and footings	5	2
Plain footings, caissons, and substructure walls	4	1
Slabs, beams, and reinforced walls	6	3
Building columns	6	3
Pavements	3	2
Heavy mass construction	3	1

[1] Adapted from Table 4 of the 1940 Joint Committee Report on Recommended Practice and Standard Specifications for Concrete and Reinforced Concrete.
[2] When high-frequency vibrators are used, the values given should be reduced about one-third.

TABLE 8. MAXIMUM SIZES OF AGGREGATE RECOMMENDED FOR VARIOUS TYPES OF CONSTRUCTION

Minimum dimension of section, in.	Maximum size of aggregate, [1] in.			
	Reinforced walls, beams, and columns	Unreinforced walls	Heavily reinforced slabs	Lightly reinforced or unreinforced slabs
2½– 5	½– ¾	¾	¾–1	¾–1½
6–11	¾–1½	1½	1½	1½–3
12–29	1½–3	3	1½–3	3
30 or more	1½–3	6	1½–3	3–6

[1] Based on square openings.

TABLE 9. APPROXIMATE MIXING-WATER REQUIREMENTS FOR DIFFERENT SLUMPS AND MAXIMUM SIZES OF AGGREGATES [1]

Slump, in.	Water, gal per cu yd of concrete for indicated maximum sizes of aggregate							
	3/8 in.	1/2 in.	3/4 in.	1 in.	1 1/2 in.	2 in.	3 in.	6 in.
Non-air-entrained concrete								
1 to 2	42	40	37	36	33	31	29	25
3 to 4	46	44	41	39	36	34	32	28
6 to 7	49	46	43	41	38	36	34	30
Approximate amount of entrapped air in non-air-entrained concrete, percent	3	2.5	2	1.5	1	0.5	0.3	0.2
Air-entrained concrete								
1 to 2	37	36	33	31	29	27	25	22
3 to 4	41	39	36	34	32	30	28	24
6 to 7	43	41	38	36	34	32	30	26
Recommended average total air content, percent	8	7	6	5	4.5	4	3.5	3

[1] These quantities of mixing water are for use in computing cement factors for trial batches. They are maxima for reasonably well-shaped angular coarse aggregates graded within limits of accepted specifications.

If *more* water is required than shown, the cement factor, estimated from these quantities, *should* be increased to maintain desired water-cement ratio, except as otherwise indicated by laboratory tests for strength.

If *less* water is required than shown, the cement factor, estimated from these quantities, *should not* be decreased except as indicated by laboratory tests for strength.

TABLE 10. MAXIMUM PERMISSIBLE WATER-CEMENT RATIOS (GAL PER BAG) FOR DIFFERENT TYPES OF STRUCTURES AND DEGREES OF EXPOSURE

Type of structure	Exposure conditions[1]					
	Severe wide range in temperature, or frequent alternations of freezing and thawing (air-entrained concrete only)			Mild temperature rarely below freezing, or rainy, or arid		
	In air	At the water line or within the range of fluctuating water level or spray		In air	At the water line or within the range of fluctuating water level or spray	
		In fresh water	In sea water or in contact with sulfates[2]		In fresh water	In sea water or in contact with sulfates[2]
Thin sections, such as railings, curbs, sills, ledges, ornamental or architectural concrete, reinforced piles, pipe, and all sections with less than 1 in. concrete cover over reinforcing	5.5	5.0	4.5[3]	6	5.5	4.5[3]
Moderate sections, such as retaining walls, abutments, piers, girders, beams	6.0	5.5	5.0[3]	[4]	6.0	5.0[3]
Exterior portions of heavy (mass) sections	6.5	5.5	5.0[3]	[4]	6.0	5.0[3]
Concrete deposited by tremie under water	..	5.0	5.0	..	5.0	5.0
Concrete slabs laid on the ground	6.0	...	...	[4]		
Concrete protected from the weather, interiors of buildings, concrete below ground	[4]	...	...	[4]		
Concrete which will later be protected by enclosure or backfill but which may be exposed to freezing and thawing for several years before such protection is offered	6.0	...	...	[4]		

[1] Air-entrained concrete should be used under all conditions involving severe exposure and may be used under mild exposure conditions to improve workability of the mixture.

[2] Soil or ground water containing sulfate concentrations of more than 0.2 percent.

[3] When sulfate-resisting cement is used, maximum water-cement ratio may be increased by 0.5 gal per bag.

[4] Water-cement ratio should be selected on basis of strength and workability requirements.

TABLE 11. COMPRESSIVE STRENGTH OF CONCRETE FOR VARIOUS WATER-CEMENT RATIOS [1]

Water-cement ratio, gal per bag of cement	Probable compressive strength at 28 days, psi	
	Non-air-entrained concrete	Air-entrained concrete
4	6,000	4,800
5	5,000	4,000
6	4,000	3,200
7	3,200	2,600
8	2,500	2,000
9	2,000	1,600

[1] These average strengths are for concretes containing not more than the percentages of entrained and/or entrapped air shown in Table 9. For a constant water-cement ratio, the strength of the concrete is reduced as the air content is increased. For air contents higher than those listed in Table 9, the strengths will be proportionally less than those listed in this table.

Strengths are based on 6-in. by 12-in. cylinders moist-cured under standard conditions for 28 days. See Method of Making and Curing Concrete Compression and Flexure Test Specimens in the Field (ASTM C31).

TABLE 12. VOLUME OF COARSE AGGREGATE PER UNIT OF VOLUME OF CONCRETE [1]

Maximum size of aggregate, in.	Volume of dry-rodded coarse aggregate per unit volume of concrete for different fineness moduli of sand			
	2.40	2.60	2.80	3.00
3/8	0.46	0.44	0.42	0.40
1/2	0.55	0.53	0.51	0.49
3/4	0.65	0.63	0.61	0.59
1	0.70	0.68	0.66	0.64
1 1/2	0.76	0.74	0.72	0.70
2	0.79	0.77	0.75	0.73
3	0.84	0.82	0.80	0.78
6	0.90	0.88	0.86	0.84

[1] Volumes are based on aggregates in dry-rodded condition as described in Method of Test for Unit Weight of Aggregate (ASTM C29).

These volumes are selected from empirical relationships to produce concrete with a degree of workability suitable for usual reinforced construction. For less workable concrete such as required for concrete pavement construction they may be increased about 10 percent.

TABLE 13. TYPICAL MINIMUM MIX SERIES TO ESTABLISH THE PROPERTIES OF CONCRETE MADE WITH FIELD MATERIALS

Mix number	Net water-cement ratio, gal per bag	Water content, gal per cu yd	Cement, content, bags per cu yd	Aggregate content, lb per cu yd		Slump, in.	Per-cent air	28-day strength, psi		Workability		
				Sand	Coarse aggregate			Compression	Flexure	Segregation	Rodability	Finish
1	5.6	32.0	5.70	1,186	1,940	1	4.5	3,500	555	None	Good	Good
2	5.6	34.0	6.07	1,112	1,940	3 1/4	4.5	3,500	555	None	Excellent	Excellent
3	5.6	33.5	5.98	1,064	2,008	2 1/2	4.5	3,500	555	None	Excellent	Very good
4	5.0	33.5	6.70	1,008	2,008	3	4.5	4,000	600	None	Excellent	Excellent
5	6.0	33.5	5.58	1,096	2,008	2 3/4	4.5	3,200	525	None	Excellent	Very good
6	7.0	33.5	4.79	1,158	2,008	3	4.5	2,600	450	None	Very good	Good

Mix No. 1—Low on slump.
Mix No. 2—Oversanded, increased coarse aggregate and lowered water for Mix No. 3.
Mix No. 3—Workability satisfactory.

TABLE 14. CONCRETE MIXES FOR SMALL JOBS [1]

Maximum size of aggregate, in.	Mix designation	Approximate bags of cement per cu yd of concrete	Aggregate, lb per 1-bag batch			
			Sand [2]		Gravel or crushed stone	Iron blast-furnace slag
			Air-entrained concrete [3]	Concrete without air		
½	A	7.0	235	245	170	145
	B	6.9	225	235	190	165
	C	6.8	225	235	205	180
¾	A	6.6	225	235	225	195
	B	6.4	225	235	245	215
	C	6.3	215	225	265	235
1	A	6.4	225	235	245	210
	B	6.2	215	225	275	240
	C	6.1	205	215	290	255
1½	A	6.0	225	235	290	245
	B	5.8	215	225	320	275
	C	5.7	205	215	345	300
2	A	5.7	225	235	330	270
	B	5.6	215	225	360	300
	C	5.4	205	215	380	320

NOTE: Air-entrained concrete should be used in all structures which will be exposed to alternate cycles of freezing and thawing.

[1] May be used without adjustment.

[2] Weights are for dry sand. If damp sand is used, increase weight of sand 10 lb for 1-bag batch, and if very wet sand is used add 20 lb for 1-bag batch.

[3] Air-entrained concrete can be obtained by the use of an air-entraining cement or by adding an air-entraining agent. If an agent is used, the amount recommended by the manufacturer will, in most cases, produce the desired air content.

PROCEDURE: Select the proper maximum size of aggregate and then, using mix B, add just enough water to produce a sufficiently workable consistency. If the concrete appears to be undersanded use mix A, and if it appears to be oversanded use mix C.

Illustrative Trial-batch Design of a Concrete Mixture. *Determinations of Proportions, Quantities, and Yields—Units and Conversions.* In addition to the determination of the relative amounts of the several constituents, the designer of a controlled concrete mixture is confronted with the necessity for making a variety of cross conversions between the units he must use in the laboratory and those required for job expression and measurement. Basically, he must view a concrete mixture in terms of the solid (absolute) volume of its several constituents: cement, fine aggregate, coarse aggregate, water, and air.

The determination and designation of amounts in the laboratory operations will be by weight. For the job the designation may, for the aggregates, be either by weight or by compact (tamped) bulk volumes. For the cement, the measure will usually be sacks and for the water it will be gallons, all per cubic yard of the concrete. Air contents are expressed as percentages of the overall space occupied by the concrete.

As indicated in Table 6, the water-cement ratio must likewise be variously expressed. For the novice in concrete mixtures the matter of nomenclature and conversion of quantities can be confusing; moreover, discernment and care are required on the part of anyone. These aspects can best be covered illustratively.

From assigned aggregates it is desired to design a concrete mixture having a 3-in. slump and a 28-day standard-cured compressive strength of 4,000 psi. The fine aggregate (F.A.) is a washed river sand and the coarse aggregate (C.A.) is a river gravel of 1½ in. maximum sieve size. The specific gravity of both sand and gravel is 2.65.

By reference to Fig. 4 the upper curve indicates that 6 gal of water per sack of cement should produce a "4,000-lb concrete." By the top scale of Fig. 4 it is indicated that the water-cement ratio by weight should be 0.53.[1] In order to secure a trial batch somewhat larger than enough to fill the standard slump cone (about 0.2 cu ft which at 150 lb per cu ft calls for at least 30 lb of concrete) 6 lb of cement is arbitrarily selected. The water requirement is (0.53)(6.00) = 3.18 lb. By successive additions of sand and gravel [2] and occasional slump tests after thorough mixing it is found that 14.74 lb of

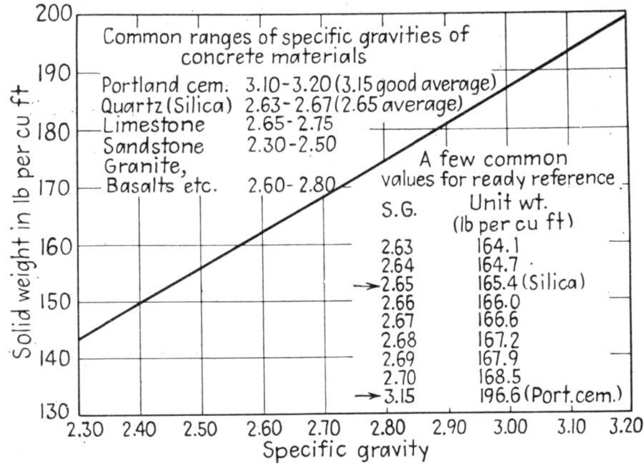

Fig. 28. Solid weights per cubic foot corresponding to specific gravities of cement and representative aggregates.

saturated surface-dry sand and 21.28 lb of saturated surface-dry gravel produce the type of mixture desired. By multiplying the respective specific gravities by 62.4 or by reference to Fig. 28 solid volumes in the batch are computed as follows:

Trial-batch Quantities *

	Weight in batch, lb	Solid weight of 1 cu ft, lb	Solid volume, cu ft
Cement (weight arbitrarily chosen, sp. gr. = 3.15)	6.00	196.6	0.031
Mixing water (0.53 weight of cement, sp. gr. = 1.00)	3.18	62.4	0.051
F.A. (weight determined by trial, sp. gr. = 2.65)	14.74	165.4	0.089
C.A. (weight determined by trial, sp. gr. = 2.65)	21.28	165.4	0.129
Total weight and solid volume of batch	45.20		0.300

* Had this mixture contained 5 percent of entrained air, weights in the batch would have remained unchanged but the solid volume would have become 0.315. The solid volumes of the constituents in a unit volume would then be the sum of the respective solid volumes divided by 0.315 instead of by 0.300. The unit weight of the concrete would be 151.07 divided by 1.05 = 143.5 lb per cu ft (or 45.20 divided by 0.315 = 143.5 lb per cu ft).

Other conversions would be affected correspondingly and the yield of the air-entrained concrete would be increased to 4.93 cu ft per sack, and the cement factor would be lowered to 5.47 sacks per cu yd. Just as the trial slumps were taken in accordance with ASTM C143 check tests would need to be made for air content in accordance with ASTM C138 (gravimetric), C173 (volumetric), or C231 (pressure method). See also "Admixtures."

[1] Under field conditions ratios and weights would usually be to one decimal.

[2] Instead of making a preliminary estimate of the amount of coarse aggregate required (by using Table 12), we shall start with the cement-water paste, adding successive increments of both coarse and fine aggregate until the desired slump and texture (see Table 7 and Fig. 27) are attained.

The relative proportions by weight of the three solid constituents (cement, fine aggregate, and coarse aggregate) of this mixture are 6.00:14.74:21.28 or 1:2.46:3.55.

The space occupied by freshly mixed concrete is entirely filled with solids and voids, the solids being cement and aggregate and the voids being free water (mixing water) and entrapped air. For a plastic mixture in which ordinary (non-air-entraining) cement is used, the entrapped air amounts to no more than 1 or 2 percent of the total volume and may be neglected in the calculation of relative volumetric proportions, quantities, and yield, the volume of voids being considered as that of the mixing water.

On this basis the solid (absolute) volume of the trial batch is 0.300 cu ft of which 0.031 cu ft is the solid volume of the cement. The proportions of solids by solid (absolute) volumes are 0.031:0.089:0.129 or 1:2.87:4.16.

Dividing the solid volumes of the constituents by 0.300, the solid volume of the batch, gives the relative solid volume of each material present in a unit volume of the freshly mixed concrete:

Cement:	$c = 0.031 \div 0.300 =$	0.102
F.A.:	$a = 0.089 \div 0.300 =$	0.298
C.A.:	$b = 0.129 \div 0.300 =$	0.430
Water:	$w = 0.051 \div 0.300 =$	0.170
		1.000

With air voids assumed negligible the volume of the water w = the voids v and the voids-cement ratio $v \div c = 0.170 \div 0.102 = 1.67$, being identical with the water-cement ratio expressed in solid volume units.

The weight of each material in a 1 cu ft batch of the concrete is its relative solid volume times the solid weight per cubic foot, and the weight required for a cubic yard of the concrete is 27 times that amount, thus:

		Lb per cu ft		Lb per cu yd	In Common Units
Cement	= (0.102)(196.6) =	20.05 × 27	=	541	÷ 94 = 5.76 sacks or 1.44 bbl
F.A.	= (0.298)(165.4) =	49.29 × 27	=	1,331	÷ 2,000 = 0.666 ton
C.A.	= (0.430)(165.4) =	71.12 × 27	=	1,920	÷ 2,000 = 0.960 ton
Water	= (0.170)(62.4) =	10.61 × 27	=	286	÷ 8.34 = 34.3 gal
Unit weight of concrete	=	151.07		4,078	

This concrete is what would be termed a "5.76 or 5¾ bag mix," i.e., 5¾ sacks of cement per cu yd of concrete in place.

The yield may be computed from the foregoing tabulation as $27 \div 5.76 = 4.69$ cu ft of concrete per sack of cement. This is the volume of a 1-bag batch of the mixture.

Proportions by Bulk or Loose Volume.[1] If the measurements of materials for batches are to be by bulk or loose volumes instead of by weight, the proportions and quantities may be determined as follows: Assume the bulk unit weights of the aggregates to have

[1] Bulk or loose-volume batching is now rarely permitted on important work. It is almost universally used, however, for work for which weighing equipment and the techniques for careful control have not (yet at least) become standard. Volumetric batching, using inundation, was important for a time, but jobs that would have justified inundation are now weighed. Inundation measurement of fine aggregate is made with the aggregate immersed, the immersion water being deducted from the mixing water for the batch. The inundation technique was devised as a refinement to ordinary volumetric batching to avoid the errors from bulking (see "Bulking of Fine Aggregate") and is applicable, provided that the water for inundation does not exceed the mixing-water requirement (which can happen for some aggregates and mixtures).

been found to be F.A. 103.5 lb per cu ft and C.A. 98.3 lb per cu ft (ASTM C29). The cement is always taken as sacked at 94 lb per cu ft, which is the bulk volume of 1 sack. From the preceding tabulation the quantities required for 1 cu yd of concrete become:

		Bulk Volume, cu ft per cu yd Concrete
Cement	= 541 ÷ 94	= 5.76 = 5.76 sacks = 1.44 bbl
F.A.	= 1,331 ÷ 103.5	= 12.86
C.A.	= 1,920 ÷ 98.3	= 19.54
Water	= 286 ÷ 62.4	= 4.59 = 34.34 gal

The proportions by bulk or loose volume are 5.76:12.86:19.54 or 1:2.23:3.39.

The ratio of water to cement in bulk or loose-volume units is 4.59 ÷ 5.76 = 0.797, say 0.80.

The ratio of the volume of water to the bulk volume of the cement (as sacked) was the form of water-cement ratio initially used (Abrams, *PCA Bull.* 1, Lewis Institute, Chicago, 1918, revised PCA, 1925), but currently water-cement ratio is likely to be expressed as gallons per sack for field use and by weight in the laboratory. As a direct rather than inverse criterion of strength the cement-water ratio by solid (absolute) volumes has the advantage of virtually a straight-line relationship to strength, as is evident from Fig. 4.

Adjustments for Moisture Effects from the Aggregates. In the foregoing trial-batch calculations the aggregates were both assumed to be in the saturated, surface-dry condition so that they neither absorbed water from the mixture nor contributed free moisture to it. This condition can be easily attained in the laboratory but rarely exists on the job, and the amounts of moisture involved may result in unbalancing the design sufficiently to alter the yields and have serious effects on both the workability and the strength of the concrete.

Absorption by the Aggregates. Because a low water-cement ratio at the time of final placement raises the strength and improves the quality of the concrete provided that workability is not sacrificed, absorption of water by the aggregates is objectionable only because of the probable adverse effect on workability by stiffening the mixture. In general then, it is only the absorption that occurs during the period of mixing and placing that is important. A dry absorptive substance takes up moisture rapidly at first, and for many aggregates the absorption during the first ½ hr after wetting may constitute 30 to 60 percent of the ultimate capacity to absorb. Assuming the absorption to be 1 percent by weight for both the fine and coarse aggregates of the illustrative mixture, the water extracted by the aggregates for a cubic yard of concrete will be:

F.A. = (0.01)(1,331) = 13.31 lb
C.A. = (0.01)(1,920) = 19.20 lb

32.51 lb = 3.9 gal

Thus the mixing water for a 1 cu yd batch of this concrete should be increased to 319 lb, or 38.3 gal, since the loss of this amount of water from the mixture (over 11 percent) would usually be enough to alter workability appreciably.

Absorption by the aggregate is not a common field condition, since aggregates on the job are rarely dry enough to extract water from the mixture.

Free Water in the Aggregates. The aggregate, especially the fine aggregate, may carry free moisture up to 5 or 6 percent of its weight. Where batch proportioning is by weight, the mixture will, if uncorrected, contain not only an excess of moisture but a

deficiency of aggregate. Assuming free moistures of 4 and 1 percent in fine and coarse aggregates respectively the corrected weights for the 1 cu yd batch are to be determined. The weights of the moist aggregates become:

F.A. = (1.04)(1,330.8) = 1,384.0 lb, of which 1,330.8 lb is fine aggregate and 53.2 lb is free water

C.A. = (1.01)(1,920.3) = 1,939.5 lb, of which 1,920.3 lb is coarse aggregate and 19.2 lb is free water

Original weights and corrected values are as listed:

	Original design (aggregate saturated surface dry), lb	Corrections, lb	Revised weights (to correct for free water), lb
Cement......	541.4	0.0	541.4
F.A..........	1,330.8	+53.2	1,384.0
C.A..........	1,920.3	+19.2	1,939.5
Water........	286.4	−72.4	214.0
	4,078.9	0.0	4,078.9

Uncorrected, the excess water in this mixture would be sufficient to lower the 28-day compressive strength from 4,000 psi to about 2,900 psi, a reduction of nearly 30 percent. The mixture would be overfluid and the yield somewhat less than indicated by the trial mix. As for absorption, the errors in weights of aggregates are not too serious and, in general, the correction in amount of mixing water used is sufficient.

Bulking of Fine Aggregate. For volumetric batching (using measured bulk or loose volumes), the bulking or swelling of the fine aggregate that contains a few percent of free moisture may be a source of serious error. When moist, but not inundated, the particles of any fine-grained relatively nonabsorbent material are forced apart by the individual globules of moisture, swelling the mass. As the mass approaches inundation, globules unite to form a fluid matrix, and the particles move together, the bulk or loose volume again approximates that of the dry material. Fine aggregates as batched happen frequently to be at the intermediate stage of free moisture content and the swelling, which has no significance in weight measurements, may be quite significant in volume measurements.

Assume that a unit volume of the fine aggregate, when in the same state of compaction as it is to be batched, shrinks 20 percent upon being either dried or inundated. This represents a bulking of 25 percent of the unbulked volume, which is 0.80.

The 12.86 cu ft of fine aggregate in a cubic yard of this concrete is swollen or bulked by (0.25)(12.86) = 3.2 cu ft. Required amount of bulked F.A. = 12.86 ÷ 0.80 = 16.1 cu ft, and the corrected proportions for the bulked aggregate will be 5.8:16.1:19.5 or 1:2.80:3.39 instead of 1:2.23:3.39 as designed for unbulked surface-dry fine aggregate. The correction for the free moisture in the fine aggregate needs also to be made as previously outlined.

Adjustment of Mixtures. Once a trial-batch design has been determined for a given cement and aggregates it is relatively simple to vary the proportions to produce either a concrete (1) having the same strength but a different workability or (2) of about the same workability but having a different strength.

Varying Workability at Constant Strength. If a constant ratio of water to cement is maintained, the strength will be held approximately constant while the workability and texture can be varied by increasing or decreasing the relative amounts or proportions of

the aggregates. More total aggregate will stiffen the mixture without seriously altering the texture.

Slump can be held constant and texture varied (within limits) by altering the relative proportions of the fine and coarse aggregate. Greater variation in texture may necessitate a change in the grading of the aggregate, especially the fine aggregate (see under "Grading").

Variation in workability (slump and/or texture at constant water-cement ratio) is a valid procedure because of the fact that, although the amount and grading of the aggregates have some influence on the strength of a workable concrete, that influence is small in comparison with the much more pronounced effect of the water-cement ratio on the strength.

Varying Strength at Constant Workability. Variable strength at approximately constant workability may, for given aggregates and cement, be accomplished by interchanging cement and fine aggregate while holding the amounts of water and coarse aggregate constant. This, the Lyse-Dunagan procedure, is illustrated by Fig. 29 and Table 15.

An inspection of the tabular values shows the medium mixture to be the one used for all the preceding illustrative calculations.

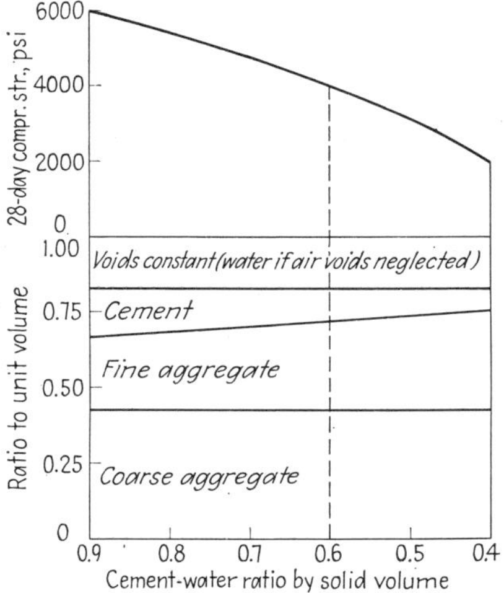

FIG. 29. Chart illustrating Lyse-Dunagan method for adjusting mix design at constant workability. Coarse aggregate and water constant; strength varied at approximately constant slump by interchange of cement and fine aggregate. See Table 15 for details on these mixtures.

Air Entrainment.[1] Shortly prior to 1940 it was discovered almost by chance that entrained air bubbles of microscopic proportions in the hardened concrete add greatly to the resistance to scaling and progressive deterioration under alternate freezing and thawing. This is especially true if aggressive salts such as sodium or calcium chloride are used to aid in ice and snow removal. Minute quantities of grease, tallow, or resinous materials present in a concrete mixture facilitate through foaming the incorporation of innumerable small bubbles (average diameter 25 to 50 microns, about 0.001 to 0.002 in.) of atmospheric air during the mixing operation. This is probably accomplished through a lowering of the surface tension of the mixture by the air-entraining agent. The protective action seems to be purely physical.[2] The small bubbles do not form continuous passages as does the mixing water but rather form a series of small closed thin-walled caves which provide flexibility and resiliency to the surface concrete without admitting appreciable free water or salt solution to the interior to produce disruption by freezing or crystallization. The bubbles probably function much as do partly filled

[1] See also "Admixtures" and table footnote on p. 7–53.

[2] That the action of an air-entraining agent is essentially physical rather than chemical seems clear from tests on aggregate having no cement present. The agent will fluff a well-proportioned mixture of wet sand or wet sand plus coarse aggregate, giving the same "fat" appearance and easy workability as if cement were present. Since such a mixture does not set or stiffen to entrap and retain the air bubbles, the air is lost in an hour or two and the mixture reverts to the status of wet sand or wet mixed aggregate (*Proc. ACI*, vol. 40, p. 516, 1944). Air is entrained only by sand between the 14 and 100 mesh sieves. The maximum entrainment is by the fraction between 28 and 100 mesh; + No. 14 and − No. 100 produce little or no entrainment under ordinary conditions. For the use of air entrainment with larger particles to produce cellular concretes see *Proc. ACI*, vol. 50, p. 773, 1954. Under a glass the fine bubbles are shown to cluster around sand grains like ball bearings, thereby separating the grains and providing mobility to the mass. The action is a bulking in which the fluid is air instead of water (see "Bulking").

TABLE 15. COMPARISON OF QUANTITIES AND PROPERTIES OF RICH, MEDIUM, AND LEAN CONCRETE MIXTURES [1]

Materials: F.A., washed, sand, graded, 0 to 4; sp. gr., 2.65; bulk weight, 103.5 lb per cu ft; fineness modulus, 2.70. C.A., river gravel, graded, 3/8 to 1½ in.; sp. gr., 2.65; bulk weight, 98.3 lb per cu ft. Fineness modulus, 7.10. Cement, sp. gr., 3.15.

	Mixtures		
	Rich	Medium	Lean
Cement-water ratio (by solid volume)	0.9	0.6	0.4
Type of mixture	Rich	Medium	Lean
Predicted 28-day strength, psi (*w/c* basis)	6,000	4,000	2,000
Workability (slump, in.)	3.0	3.0	3.0
Workability (texture)	Plastic	Plastic	Plastic
Quantity in 1 unit volume concrete by solid (absolute) volume: cement (*c*)	0.153	0.102	0.068
F.A. (*a*)	0.247	0.298	0.332
C.A. (*b*)	0.430	0.430	0.430
Water (*w*)	0.170	0.170	0.170
Air (assumed 0)	0	0	0
Total	1.000	1.000	1.000
Water-cement ratio: gal per bag cement	4.0	6.0	9.0
Weight	0.34	0.52	0.80
Bulk volume	0.54	0.80	1.20
Solid volume	1.11	1.67	2.50
Voids-cement ratio: solid volume	1.11	1.67	2.50
Proportions: by weight	1:1.4:2.4	1:2.5:3.6	1:4.1:5.3
Bulk volume	1:1.2:2.3	1:2.2:3.4	1:3.7:5.1
Solid volume	1:1.6:2.8	1:2.9:4.2	1:4.9:6.3
Ratio $b \div b_0$	0.70	0.70	0.70
Ratio C.A. to F.A., i.e., $b \div a$	1.74	1.44	1.30
Quantity in 1 cu yd: cement, bags	8.64	5.76	3.84
F.A., lb	1,103	1,331	1,487
C.A., lb	1,920	1,920	1,920
Water, gal	34.3	34.3	34.3
Yield, cu ft concrete per bag cement	3.1	4.7	7.0
Strength-economy index (psi per bag per cu yd)	695	695	469
Weight fresh concrete, lb per cu ft	153	151	150
Lb per cu yd	4,122	4,079	4,055

[1] These are the mixtures of Fig. 29.

As indicated in Fig 29, these mixtures were varied by the Lyse-Dunagan simplifying device of interchanging cement and fine aggregate (with coarse aggregate and water constant). While quantities and relative proportions are representative of those for rich, medium, and lean mixtures in general they are not identical with what would be obtained by some other (less direct) basis of adjustment such as varying total aggregate at a constant ratio of coarse to fine, along with appropriate alterations in water, and/or cement.

capillaries (Powers, *Proc. ACI*, vol. 43, p. 1093, 1947). Even when the amount of mixing water is left unchanged and air entrainment is added to increase the total voids by 3 percent or more, there is still a distinct increase in the resistance offered to freezing and thawing and there is less segregation, water movement (bleeding), etc., prior to setting. In this case, however, the compressive strength is reduced about in proportion to the increase in voids. Subsequently it has been recognized that the entrained air contributes to the fluidity of the plastic mass much as would added water. Without loss of workability it is possible to reduce the volume of the mixing water by about two-thirds the volume of the air added for 4 percent entrainment for a lean mixture and by about one-third for a rich mixture (say 4.5 and 6.5 bags per cu yd respectively)

since the favorable aspects of air entrainment are more marked for lean mixtures than for rich. In the redesign of a mixture for air entrainment it is possible not only to reduce the water but also to use a higher ratio of coarse aggregate to fine aggregate without loss of fatness or easy workability. The possible extent of this adjustment is also greater for the lean mixtures. For air entrainments up to 5 or 6 percent by volume the strength of a redesigned lean mixture may be increased slightly, whereas there is still likely to be a 10 or 15 percent strength reduction of the richer mixture. With a further redesign, however, involving a little added cement (a slight lowering of the water-cement ratio), the reduction in the strength can readily be compensated.

An air-entraining agent may be introduced as a step in the manufacture of the cement as an "addition" to the clinker at time of final grinding, or it may be introduced at the mixer as an "admixture" to the concrete. Air-entraining portland cement is covered by ASTM C175 and C205, the two earliest agents used under ASTM acceptance being Vinsol resin and Darex, ASTM C114 and C226 (see also "Portland Cement" and "Admixtures").

Vibration, long hauls in agitator trucks, or any other form of overmixing make concrete lose air, and in the use of air-entrained concrete due regard must be given to methods of mixing, placement, and compaction. Limited reconnaissance tests seem to indicate that vibration not to exceed about 15 sec can be used without excessive loss of air from pavement concrete. When the agent is introduced as an addition (ground with clinker) grinding temperatures must be kept low enough to avoid loss of the agent from volatilization.

Air-entraining Cements vs. Air-entraining Admixtures. The air-entraining agent may, as mentioned, be incorporated with the cement as an addition prior to final grinding of the clinker or introduced as an admixture to the concrete at the time of mixing.

Air-entraining cement has the advantage of ensuring a uniformity of distribution throughout the concrete corresponding to the distribution of the cement. On the other hand, the amount of the agent introduced will vary with the richness of the mixture, leaner mixtures for which air entrainment is most effective and desirable having the least of the agent present.

By introducing the agent directly to the concrete by premixing with the water or otherwise it is possible to control the amount of air entrained regardless of mix proportions. Since air-entraining agents have great potency in even very small amounts, the control over both the total amount added and the uniformity of its distribution throughout the concrete is of utmost importance.

Because of convenience and ease of control the use of air-entraining cement tends to be, as it probably should be, concentrated largely within the area of the small-job applications.

In either case close control and careful checking are essential. An excess of air may greatly weaken the concrete and a deficiency of air may be damaging to both placing quality and durability characteristics. Suitable sampling and testing procedures for checking on the air contents actually developed are an essential part of every air-entrained concrete job (see ASTM C138, C173, and C231).

Moreover, regardless of how the air entrainment is accomplished, the transition from a non-air-entrained to an air-entrained concrete involves more than simply adding the air. This aspect will be covered illustratively.

Illustrative Adjustments of Quantities for Air Entrainment. One unit of the concrete of a non-air-entrained mixture contains the following proportionate parts expressed as relative solid volumes of cement, fine aggregate, coarse aggregate, and water (the incidental air content being considered negligible): $0.097 + 0.275 + 0.450 + 0.178 = 1.000$. Assuming 4 percent of air to be added, without other change, the proportionate parts become: $0.097 + 0.275 + 0.450 + 0.178 + 0.040 = 1.040$. Dividing by 1.040 to

reduce to the proportionate unit volume basis we have: 0.093 + 0.265 + 0.433 + 0.171 + 0.038 = 1.000. The voids (water plus air) have been increased from 0.178 to 0.209 and the solids (cement plus aggregate) have been decreased correspondingly from 0.097 + 0.275 + 0.450 = 0.822, to 0.093 + 0.265 + 0.433 = 0.791.

The increase in fluids (air plus water) combines with the decrease in solids to add substantially to the slump or fluidity of the mixture. Moreover, the 17.4 percent increase in the voids (a 22.7 percent increase in voids-cement ratio) will decrease the strength by about a fourth.

Clearly the initial mixture has been badly unbalanced as regards both workability of the plastic mixture and the strength of the hardened concrete.

The first concern of the designer would be to redesign the mixture in such a manner as to restore the initial workability (texture and slump), which involves a stiffening of the mixture. This may be accomplished by adding solids (aggregate and/or cement) or by lowering the water content, or both. Depending upon which factors are varied, the designer cannot only restore the workability desired but he has in addition the option of restoring fully the lost strength involving (except for the leaner mixtures) a slight increase in the cement factor—some sacrifice of yield; or he may preserve essentially the same yield by accepting some reduction in strength.

The following alternative procedures are ones empirically and arbitrarily proposed by the Portland Cement Association in the 10th (1952) edition of *Bull.* T12 "Design and Control of Concrete Mixtures." It is desired to modify the base mixture (non-air-entrained) to produce an air-entrained concrete having not only the same workability but also: (1) producing a concrete of the same strength or (2) one having the same water-cement ratio (not voids-cement ratio) as had the base mixture.

1. *Redesign for Equal Workability and Strength.* Rule: For each 1 percent of entrained air (*a*) Reduce the water per sack of cement by ¼ gal. (*b*) Reduce the fine aggregate by 10 lb per sack of cement.

Application. Expressed in job units, the base mixture previously discussed, contains 5.5 sacks per cu yd, 1,228 lb F.A., 2,010 lb C.A., and 6.55 gal per sack giving a nominal 28-day strength of 3,500 psi and a yield of 4.91 cu ft per sack. For 4 percent of air the water is reduced to 5.55 gal per sack and the F.A. to 1,008 lb other quantities remaining unchanged. Whereas both the strength and workability are unchanged, the yield of the revised mixture has been somewhat lowered (from 4.91 to 4.73, i.e., the cement factor has been increased from 5.50 to 5.71 sacks per cu yd). The air content remains virtually unchanged at about 0.042 instead of 0.040.

2. *Redesign for Equal Workability and Equal Water-cement Ratio.* Rule: For each 1 percent of entrained air increase the coarse aggregate by about 5 percent.

Application. The coarse aggregate is increased 20 percent and becomes 2,010 lb + 403 lb = 2,413 lb, other batch quantities remaining unchanged.

NOTE: It is the characteristic of air entrainment to impart a "fatness" to the mixture that makes it possible to add coarse aggregate without producing objectionable harshness of texture.

Reduced to the solid unit volume basis of proportions and allowing for the weakening effect of air entrainment, this mixture will produce a concrete having about a 17 percent reduction in strength and an *increased* yield of about 11 percent for both the lean and rich mixtures. It will be found, however, that the increase in volume by the addition of coarse aggregate, with no corresponding subtraction of any other constituent, has lowered the proportionate amount of each other ingredient, including air which drops from the nominal 4 percent to about 3.5 percent. If it is desired to adhere more closely to the specified 4 percent of air, it can be accomplished by starting with a nominal air content of 4.5 or 5 percent to offset the impending reduction.

By a less simple procedure, involving changes in both the water-cement ratio and

the fine-aggregate content, it is possible to adjust for constancy of workability and yield. Where this method is used the strengths of the richer mixtures will be reduced by 10 to 15 percent but the strengths of the leaner mixtures may be increased by as much as 15 to 20 percent. In general case 1 (adjustment for equal strength) approximates constant yield closely enough to suffice and the constant-yield technique will not be here discussed (see *Proc. ACI*, vol. 54, p. 633, 1957–1958).

SELECTION AND GRADING OF AGGREGATES

Standard Requirements of Aggregates for Usual Concretes. Aggregates for orthodox concretes (as distinguished from unusual or special-purpose concretes: lightweight, insulating, shielding, nailing, decorative, etc.) usually consist of natural water-worn sands and gravels, crushed stone and screenings, or crushed or granulated blast-furnace slags.

The requirements for such aggregates are that all particles be strong, tough, structurally sound, clean, durable, and chemically inert with regard to the cement and any materials with which the concrete may be expected to come in contact; that the aggregate be well graded and reasonably free from flat and elongated particles; and that the surface texture not be polished or glasslike.

For most concrete the selection of an aggregate will be limited to what is locally available, and the choice often lies between natural deposits of gravel and crushed material (ledge rock, boulders, or slag). The aggregate may be pit or crusher run, or screened to certain size brackets for recombination in batching.

For much of the miscellaneous concrete construction on smaller projects (executed by artisans without attempt at the formal design of a mixture that would represent maximum excellence, durability, or economy of cement) the aggregate is pit-run material often subject to wide variations in grading. For work of sufficient importance to justify the design and control of the mixture, the physical characteristics, uniformity, and grading of the aggregate are extremely important; in fact, the design of a mixture is meaningless if uniformity is not maintained for the materials that make up the concrete in the structure.

Crushed and water-worn particles differ from one another in various respects among which are angularity, general shape, and surface texture. Moreover similar and other differences exist between different gravels and different crushed aggregates, including strength and stiffness of particles, specific gravity, absorptivity, and thermal coefficient. Some of these obvious differences tend to favor one type of aggregate and some another.

Crushed material because of the wedging does not pack so well as a similarly graded gravel and requires a somewhat higher ratio of mortar or paste or perhaps a somewhat thinner paste (higher W/C) for satisfactory workability. On the other hand, the angular interlocking between particles provides a somewhat higher strength for a given W/C. Whereas some of the differences may properly be disregarded either because of the present lack of specific information as to their relative importance or because they tend to average out, it appears that others cannot be dismissed quite so lightly. Illustrative of this are the following:

As the sawed contraction joint (see "Joints") comes increasingly into use it is discovered that the uniform relatively soft limestone aggregate in pavements and floors responds more economically to sawing than do the much harder granites and slags and the harder and less homogeneous gravels.

On the other hand, experiences in Michigan verified by similar findings elsewhere indicate that the uniform texture of limestone-aggregate concrete, especially if screenings are used instead of natural sands, produces relatively slippery pavement surfaces under traffic, when dry as well as wet, increasing skidding tendencies and stopping distances.

For the most part, however, sound hard chemically inert aggregates of the rounded and crushed varieties are assumed for concrete-making purposes to be equivalent, selections generally being based upon comparative availability and cost. For more detail regarding conventional aggregates see ASTM Report on Significance of Tests, 2d ed., p. 92, 1943; 3d ed., *Special Pub.* 169, pp. 253–352, 1955; *Proc. ACI*, vol. 23, p. 363, 1927.

Lightweight Aggregates. Cinders have long been used as aggregate for lightweight relatively low-strength insulating and filler concrete. Haydite, a crushed hard-burned clay, gained recognition as a structural lightweight aggregate during the concrete-ship era of World War I and is used increasingly in high buildings and long-span bridge floors. Of more recent prominence for lighter weights and lower strengths are vermiculite, an alteration product of biotite and other micas, and perlite, an expanded siliceous lava.

With Haydite as coarse aggregate (from the standpoint of grading natural sands are generally more suitable for the fine aggregate) strengths of 6,000 psi and upward are easily attained at weights of from 100 to 115 lb per cu ft. Perlite concretes may have strengths from 1,200 to 300 psi at weights as low as 50 to 30 lb per cu ft and vermiculite from 300 to 40 psi at weights of 30 to 20 lb per cu ft. The high porosity of the aggregates ensures that all lightweight concretes if kept free from moisture will have excellent insulating quality. Concrete from most of the lighter insulation-type aggregates becomes easily waterlogged from vapor as well as direct wetting. This results in loss of effective insulating quality and promotes decay of adjacent timber and/or any other material vulnerable to moisture whether the vulnerability be decay, rusting, dissolution, or waterlogging.

Lightweight aggregates present proportioning and batching problems that differ materially from those of conventional aggregates. Being highly and rapidly absorbent, they will, if batched dry, absorb water from the mixture in undetermined amounts that disturb the effective water-cement ratio and the workability. The pore structure is such that the saturated, surface-dry condition is indeterminate as is also the "apparent specific gravity" (see ASTM E12 for definitions) for workable solid-volume determinations. Stock piles of lightweight aggregate need to be kept uniformly moist even though the particles may not be fully saturated. For further details on both the physical properties and batching aspects of lightweight aggregates, see Jones and Stephenson, *Proc. ACI*, vol. 54, p. 527, 1957–1958; Kluge, *Proc. ACI*, vol. 45, p. 625, 1949; Richart and Jensen, *Univ. Illinois Eng. Expt. Sta. Bull.* 237, 1931; Shideler, *Proc. ACI*, vol. 54, p. 299, 1957–1958; also ASTM C330 and C332.

Heavy Aggregates. Aggregates for heavy concretes suitable for counterweights for bridges, ballast blocks for ships, or nuclear shielding require such aggregates as barite (a natural barium sulfate) magnitite, hard hematite, and/or steel punchings. The ores may be crushed to suitable sizes and used either with fine sand or with screenings from further crushing. Concretes of over 200 lb per cu ft are easily obtained and weights up to 300 lb per cu ft have been approached. Some of the special aspects of aggregate requirements for heavy concretes are covered in the following references: *Proc. ACI*, vol. 51, pp. 65, 541, 1954–1955; vol. 50, pp. 17, 45, 1953–1954; vol. 28, p. 525, 1932.

Grading of the Aggregate. *Need for Good Grading; Usual Size Ranges.* The favorable grading of the aggregate serves the two important functions of (1) contributing to the uniformity and workability of the mixture, (2) reducing the quantity of cement paste required to produce concrete of given strength or quality.

The reduction in paste accomplishes more than a saving in cost, important though that is. As pointed out in discussing the deformability of hardened concrete, it is the cement that contributes the heat of hydration, with its attendant problems, as well as most of the volume change incidental to wetting and drying. It is always possible

to secure plastic, fat-textured mixtures by using excess amounts of cement, fine aggregate, and water, but such concrete is unnecessarily expensive and subject to the various manifestations of differential volume changes that produce warping, cracking, and general deterioration. Aside from cost, too much cement is for some adaptations even more objectionable than too little.

Grading is evaluated by the sieve analysis (ASTM C136) from which are recorded the percentages of a sample passing, retained on, or falling between consecutive sieves of a graded series (ASTM C33, D448). Widely used for concrete aggregates has been the fineness modulus series of sieves Nos. 100, 50, 30, 16, 8, and 4 for fine aggregates and Nos. 4, ⅜, ¾, 1.5, 3, etc., for coarse aggregates, the side of the square opening being doubled for each larger sieve of the fineness modulus series (ASTM E11).

For most work the coarse aggregate is all below 2 in., but in mass concrete aggregate up to 6 in. is now regularly passed through concrete mixers and 9-in. aggregate was used for parts of Hoover Dam. The economy of a 9-in. maximum over that for a 6-in. maximum was not deemed sufficient to offset the added processing difficulties introduced by the largest fragments. Formerly both "rubble aggregate" (fragments above 6 in. but not weighing more than 100 lb) and "cyclopean aggregate" (fragments weighing more than 100 lb) were often hand or derrick placed as "plums" in mass concrete of ordinary size aggregate, but such use is not often adapted to modern concreting methods. The Exchequer Dam (near Yosemite, constructed during the 1920's) attained by this means one of the lowest cement factors then on record, about 3.7 sacks per cu yd of concrete (see Fig. 37).

Attempts to Evolve Ideal Gradings. Many attempts have been made to evolve ideal gradings or grading indexes for concrete aggregates based upon sieve-analysis data, one of the earliest being W. B. Fuller's "ideal grading curve." Others have been the "surface modulus" of Edwards, the "particle interference" of Weymouth, and the "fineness modulus" of Abrams. The fineness modulus (F.M.), defined as "the sum of the cumulative percentages retained on each of the fineness modulus series of sieves, divided by 100," is the only grading index currently much used. Conceding the best grading to be that giving the least voids space, attempts to evolve "ideal" gradings invariably start with the sphere. An aggregation of spherical particles, of any equal diameters, loose stacked (as if inscribed in cubes) contains 48 percent voids (having a "solidity ratio" or "density," so called, of 0.52). The same one-size spheres solid packed would have only 26 percent voids (solidity ratio of 0.74). Assuming the solid pack, the next step is to compute the diameter and amount of equal smaller spheres to fill the interstices between the largest size used, etc.

Unfortunately, such attempts oversimplify the problem because (1) aggregate particles are not spherical, (2) they can be only crudely sized, and (3) they will not take the relative positions to produce the "tight" packing assumed. It has, for example, been found difficult to pour or tamp spheres of any one diameter, into a container in a manner that will reduce the voids space much below 44 percent (solidity ratio 0.56, only 0.04 better than the loose pack). The problem thus, at its inception, becomes empirical.

Fine-aggregate Studies. Feret, the French investigator, and Talbot at Illinois experimented with fine aggregates from the point of view that the fine aggregate, cement, and water form the mortar matrix that surrounds and fills in between particles of coarse aggregate.[1] The amount of mortar required in a given concrete then becomes a function of the voids in the coarse aggregate; the better graded the coarse aggregate the less the mortar (and, therefore, the less the cement) required to produce concrete having the strength and quality corresponding to that particular mortar mixture. Again,

[1] The method of using the sand-cement-water mortar as the matrix instead of the cement-water paste (as does the water-cement-ratio approach) treats as filler material only the coarse aggregate instead of the mixed aggregate. It recognizes that the fine aggregate does exercise a great influence on the characteristics of the plastic concrete, even though the water-cement paste is the primary determinant of strength.

TABLE 16. SIEVE ANALYSES AND OTHER COMPARATIVE DATA ON A FEW SELECTED SANDS

Description	Specific gravity above / Absorption (percent) below	Unit weight: Solid weight above / Bulk weight below	Voids above / Solidity ratio (density) below	Sieve-analysis data (cumulative percent), percent passed, above; percent retained, below: 100	50	30	16	8	4	⅜	F.M.	Fractions between sieves: Pan	100–50	50–30	30–16	16–8	8–4	4–⅜
(a)	(b)	(c)	(d)	(e)	(f)	(g)	(h)	(i)	(j)	(k)	(l)	(m)	(n)	(o)	(p)	(q)	(r)	(s)
Very fine (Indiana dunes)	2.67	166.6	0.393	6.0	95.0	100.0	.....	.....	.....	.....	...	6.0	89.0	5.0				
No. 10, *Univ. Illinois Bull.* 137	0.20	101.1	0.607	94.0	5.0	0.0	.....	.....	.....	.....	0.99							
Fine sand (Greenup, Ill.)	2.63	164.1	0.382	6.0	48.0	99.0	100.0	.....	.....	.....	....	6.0	42.0	51.0	1.0			
No. 3, *Univ. Illinois Bull.* 137	0.65	101.5	0.618	94.0	52.0	1.0	0.0	.....	.....	.....	1.47							
Medium to coarse (San Diego)	2.65	165.4	0.361	1.0	7.0	24.0	48.0	75.0	94.0	100.0	....	1.0	6.0	17.0	24.0	27.0	19.0	6.0
No. 18, *Univ. Illinois Bull.* 137	0.52	105.6	0.639	99.0	93.0	76.0	52.0	25.0	6.0	0.0	3.51							
Coarse (Platte River gravel)	2.63	164.1	0.263	3.0	19.0	42.0	57.0	71.0	88.0	98.0	....	3.0	16.0	23.0	15.0	14.0	17.0	10.0
No. 13, *Univ. Illinois Bull.* 137	0.10	121.0	0.737	97.0	81.0	58.0	43.0	29.0	12.0	2.0	3.22							
Standard Ottawa, Ill., 20–30	2.65	165.4	0.363	0.0	0.0	1.0	100.0	.....	.....	.....	....	....	....	1.0	99.0			
No. 0, *Univ. Illinois Bull.* 137	0.05	105.4	0.637	100.0	100.0	99.0	0.0	.....	.....	.....	2.99							
Hoover Dam (Arizona)	2.64	164.7	0.334	2.4	14.8	56.4	71.4	83.0	100.0	.....	....	2.4	12.4	41.6	15.0	11.6	17.0	
Fine aggregate used	....	109.6	0.666	97.6	85.2	43.6	28.6	17.0	0.0	.....	2.72	....	....		....			
Limit of fineness	....	.....	.....	10.0	30.0	(60.0)	85.0	(100.0)	100.0	.....	....	10.0	20.0	(30.0)	(25.0)	(15.0)	(0.0)	
ASTM C33	....	.....	.....	90.0	70.0	(40.0)	15.0	(0.0)	0.0	.....	2.15	.....	....	15.0				
Limit of coarseness	....	.....	.....	2.0	10.0	(25.0)	50.0	(80.0)	95.0	100.0	....	2.0	8.0	(13.0)	25.0	30.0	15.0	5.0
ASTM C33	....	.....	.....	98.0	90.0	(75.0)	50.0	(20.0)	5.0	0.0	3.38							

Notes: Column (*c*): Unit weights (pounds per cubic foot); solid weight = 62.4 × specific gravity; bulk weight = dry-rodded weight per ASTM C29. Column (*d*): Solidity ratio (often called "density") = 1 − voids. Column (*l*): Fineness modulus = sum of cumulative percentages retained on No. 100 sieve and upward. No. 18 is really a coarse sand, nearly identical with limit of coarseness. No. 13 has too much oversize (12 percent above No. 4, only 5 percent permissible) to be classed as sand; it is not properly comparable with the other sands. Values in parentheses are interpolations from the plotted curves of Fig. 30.

the better graded the fine aggregate, the less will be the cement required to produce mortar and, therefore, concrete of the desired quality (*Univ. Illinois Eng. Expt. Sta. Bull.* 137, 1923).

The Sieve-analysis Curve. Sieve-analysis results may best be visualized from the plotted data. In Fig. 30 are plotted the sieve-analysis data for the fine aggregates of Table 16. On the same diagram appear the permissible limits for fineness and coarseness as designated in ASTM C33. The three alternative methods of recording sieve-analysis

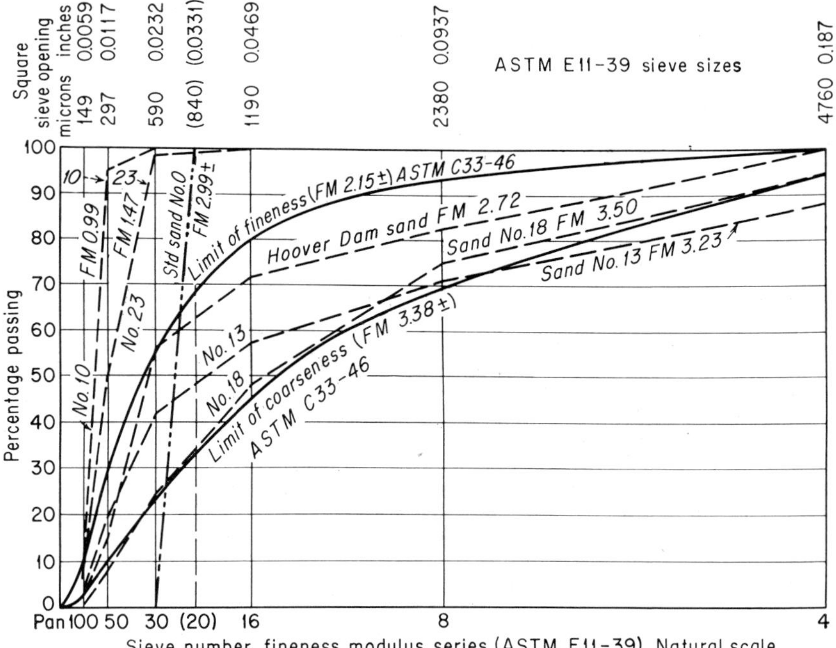

FIG. 30. Sieve-analysis curves for the fine aggregates of Table 16.

data and the calculation of fineness modulus appear in Table 16. Figure 31 illustrates plotting for all three ways of recording data for both a fine and coarse aggregate and for one proportion of the mixed aggregate.

Sieve-analysis percentages are plotted as ordinates (natural scale) against sieve openings as abscissas, to either a natural or a logarithmic scale. For the F.M. series the logarithmic plotting gives equal spacings for the successive sieve sizes of the series, as illustrated by Figs. 31, 32, and 33.

In attempting to apply analytically the results of grading studies one must recognize that:

1. The mortar does not surround the coarse-aggregate fragments like a fluid but introduces wedging and added separation; the mortar requirement will always be in excess of the voids in the coarse aggregate.

2. The harsher or coarser the fine aggregate and the leaner the mortar, the more pronounced will be the wedging effect and the greater will be the mortar excess required to produce workable concrete.

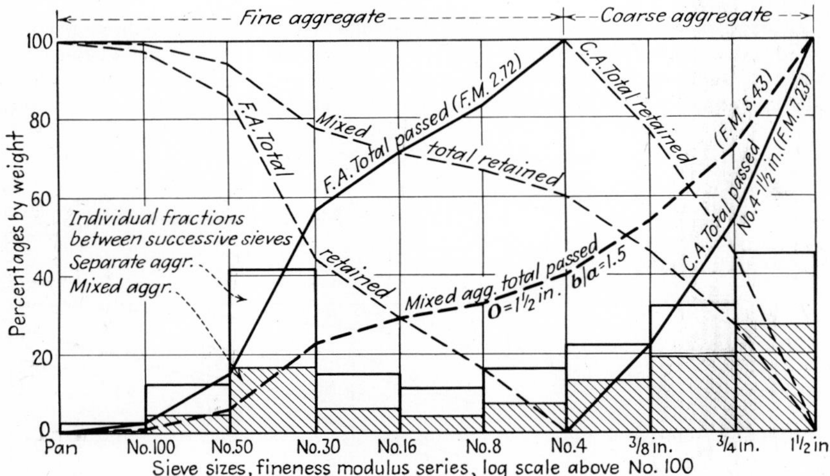

FIG. 31. Sieve-analysis curves for fine, mixed, and coarse aggregate. Different methods of plotting illustrated (semilogarithmic). (*ASTM Report on Significance of Tests of Concrete*, 2d *ed.*, *p.* 94, 1943, *Fig.* 1.)

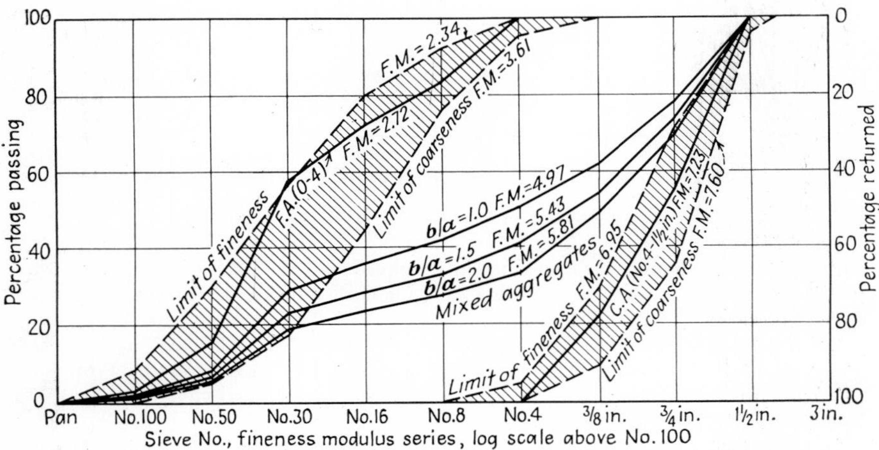

FIG. 32. Sieve analysis; recommended grading zones for coarse and fine aggregates and curves for combined aggregates in 3 ratios of coarse to fine (semilogarithmic). *ASTM Report on Significance of Tests of Concrete*, 2d *ed.*, *p.* 96, 1943, *Fig.* 4.)

3. The finer the grading of the fine aggregate the greater will be the cement and water requirement to produce a workable mortar mixture of specified strength. On the other hand, a fine-sand mortar produces less wedging and has greater "coarse-aggregate carrying capacity" (less excess of mortar is required). This compensates only in part, however, and medium to coarse sands within the accepted grading zone produce in general the most satisfactory concretes.

Gradings finer than the limits indicated in Figs. 30 and 32 are likely to be uneconomical (require an unnecessarily high cement factor because of the greater water require-

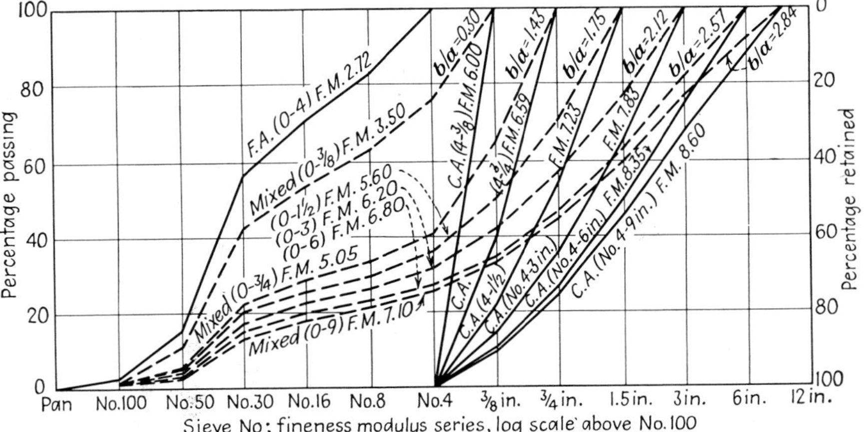

FIG. 33. Sieve-analysis curves for several maximum sizes of coarse aggregate; fine aggregate of constant grading. (*ASTM Report on Significance of Tests of Concrete, 2d ed., p. 96, 1943, Fig. 5.*)

ment for workability), and those much coarser are likely to be too harsh for satisfactory workability. If the fine aggregate approaches the limit of fineness, the coarse aggregate may approach the limit of coarseness with less danger of unworkability than otherwise.

Although virtually all concrete practitioners recognize the importance of properly graded aggregates, many find difficulty in visualizing the grading or sieve-analysis curves in terms of the concrete. The inclusion in Fig. 30 of the sieve-analysis curves of Table 16 should be helpful in that regard, especially if studied and compared on the basis of such related properties as *fineness modulus* and the *solidity ratios* or *voids* of the sands in the *dry-rodded* condition.

Recommended Gradings of Usual Aggregates for Various Uses. Table 17*a* (taken from U.S. Department of Commerce Simplified Practice Recommendation R163 supplies recommended aggregate grading ranges for concrete and other purposes (mostly coarse aggregates of crushed stone, gravel, or slag). Typical uses for the several gradings are indicated in Table 17*b*. Table 18 gives approximate equivalents for screens or sieves having round openings vs. those having square openings.

Gap Gradings. Some investigators recommend *gap* gradings (omission of certain size fractions). To decrease or omit altogether *next* to the maximum size fraction of the coarse aggregate, adding an equal weight to the maximum size will sometimes improve workability but may in wet mixes contribute to segregation. Although in certain instances gap gradings have advantages, only in an unusual case could the separation and wasting of one or more size fractions be justified. On the other hand, sands deficient in certain sizes will often prove entirely and economically satisfactory.

TABLE 17a. SIZES OF COARSE AGGREGATES[1]
Crushed stone, gravel, and slag

Size number	Nominal size square openings[2]	Amounts finer than each laboratory sieve (square openings), percent by weight														
		4	3½	3	2½	2	1½	1	¾	½	⅜	No. 4	No. 8	No. 16	No. 50	No. 100
1	3½ to 1½	100	90–100	......	25– 60	......	0– 15	......	0– 5							
1-F[3]	3½ to 2	100	90–100	......	......	0– 10	0– 2									
2-F[3]	3 to 1½	...	100	90–100	......	......	0– 10	0– 2								
2	2½ to 1½	...	......	100	90–100	35– 70	0– 15	......	0– 5							
24	2½ to ¾	...	......	100	90–100	......	25– 60	......	0– 10	0– 5						
3	2 to 1	...	......	......	100	95–100	35– 70	0– 15	......	0– 5						
357	2 to No. 4	...	......	......	100	95–100	......	35– 70	......	10– 30	......	0– 5				
4	1½ to ¾	...	......	......	......	100	90–100	20– 55	0– 15	......	0– 5					
467	1½ to No. 4	...	......	......	......	100	95–100	......	35– 70	......	10– 30	0– 5				
5	1 to ½	...	......	......	......	......	100	90–100	20– 55	0– 10	0– 5					
56	1 to ⅜	...	......	......	......	......	100	90–100	40– 75	15– 35	0– 15	0– 5				
57	1 to No. 4	...	......	......	......	......	100	95–100	......	25– 60	......	0– 10	0– 5			
6	¾ to ⅜	...	......	......	......	......	......	100	90–100	20– 55	0– 15	0– 5				
67	¾ to No. 4	...	......	......	......	......	......	100	90–100	......	20– 55	0– 10	0– 5			
68	¾ to No. 8	...	......	......	......	......	......	100	90–100	......	30– 65	5– 25	0–10	0– 5		
7	½ to No. 4	...	......	......	......	......	......	......	100	90–100	40– 70	0– 15	0– 5			
78	½ to No. 8	...	......	......	......	......	......	......	100	90–100	40– 75	5– 25	0–10	0– 5		
8	⅜ to No. 8	...	......	......	......	......	......	......	......	100	85–100	10– 30	0–10	0– 5		
89	⅜ to No. 16	...	......	......	......	......	......	......	......	100	90–100	20– 55	5–30	0–10	0– 5	
9	No. 4 to No. 16	...	......	......	......	......	......	......	......	......	100	85–100	10–40	0–10	0– 5	
10	No. 4 to 0[4]	...	......	......	......	......	......	......	......	......	100	85–100	....	....	....	10–30
G1[5]	1½ to No. 50	...	......	......	......	......	100	80–100	......	50– 85	......	20– 40	15–35	5–25	0–10	0– 2
G2[5]	1½ to No. 8	...	......	......	......	......	100	65–100	......	35– 75	......	10– 35	0–10	0– 5		
G3[5]	1½ to No. 4	...	......	......	......	......	100	60– 95	......	25– 50	......	0– 15	0– 5			

[1] Table 1 of Simplified Practice Recommendation R163-48, U.S. Department of Commerce.
[2] In inches, except where otherwise indicated. Numbered sieves are those of the United States Standard Sieve Series.
[3] Special sizes for sewage trickling filter media.
[4] Screenings.
[5] The requirements for grading depend upon percentage of crushed particles in gravel. Size G1 is for gravel containing 20 percent or less of crushed particles; G2 is for gravel containing more than 20 percent and not more than 40 percent of crushed particles; G3 is for gravel containing crushed particles in excess of 40 percent.

TABLE 17b. TYPICAL USES FOR SIZES GIVEN IN TABLE 17a[1]

Use	Size number and nominal size[2]																							
	1	1-F	2-F	2	24	3	357	4	467	5	56	57	6	67	68	7	78	8	89	9	10	G1	G2	G3
	3½ to 1½	3½ to 2	3 to 1½	2½ to 1½	2½ to ¾	2 to 1	2 to No. 4	1½ to ¾	1½ to No. 4	1 to ½	1 to ⅜	1 to No. 4	¾ to ⅜	¾ to No. 4	¾ to No. 8	½ to No. 4	½ to No. 8	⅜ to No. 8	⅜ to No. 16	No. 4 to No. 16	No. 4 to 0	1½ to No. 50	1½ to No. 8	1½ to No. 4
Water-bound macadam:																								
Coarse aggregate	x			x		x																		
Filler																					x			
Bituminous macadam, penetration method:																								
Coarse aggregate	x			x		x																		
Choke										x	x		x	x			x							
Seal																		x	x	x				
Bituminous plant mixes, base or surface courses:[3]																								
Base, open mix					x	x		x			x													
Base, closed mix							x		x			x		x										
Binder course												x		x										
Surface course, coarse grading										x	x	x		x										
Surface course, fine grading																	x	x	x	x				
Seal																	x	x	x	x				
Bituminous road mix:																								
Mixing course											x			x										
Choke																		x	x	x				
Seal																		x	x	x				
Drag leveling course:																								
Leveling course														x	x		x							
Seal																		x	x	x				
Bituminous surface treatment										x[4]	x[4]		x	x	x	x	x	x	x	x				
Seal for airport construction																			x	x				
Portland-cement concrete	x			x		x	x	x	x			x		x		x								
Railroad ballast:																								
Stone or slag					x	x		x			x	x												
Gravel																						x	x	x
Roofing																		x						
Sewage trickling filter media		x	x																					

[1] Table 2 of Simplified Practice Recommendation R163-48, U.S. Department of Commerce.
[2] In inches, except where otherwise indicated. Numbered sieves are those of the United States Standard Sieve Series.
[3] For plant mixes the aggregate should consist of appropriate sizes selected from Table 17a combined with suitably graded fine aggregate.
[4] Bottom course of multiple surface treatment.

TABLE 18. APPROXIMATELY EQUIVALENT ROUND- AND SQUARE-OPENING TESTING SCREENS[1]

This table shows the sizes of round openings that are approximately equivalent to the stated sizes of square openings in testing sieves. Numbered sieves are those of the United States Standard Sieve Series

Square openings	Round openings, in.	Square openings, in.	Round openings, in.
No. 8	1/8	1¼	1½
No. 4	¼	1½	1¾
⅜ in.	½	1¾	2
½ in.	⅝	2	2⅜
⅝ in.	¾	2¼	2¾
¾ in.	⅞	2½	3
⅞ in.	1	3	3½
1 in.	1¼	3½	4¼
1⅛ in.	1⅜	4	4¾

[1] Table 3 of Simplified Practice Recommendation R163-48, U.S. Department of Commerce.

Figure 34 shows a variety of graded aggregates all having identical fineness moduli approximately equal to that for Fuller's "maximum density curve." Abrams (*PCA Bull.* 1, Lewis Institute, 1918) reported these as all producing essentially the same concretes, using the same cement factor and water-cement ratio.

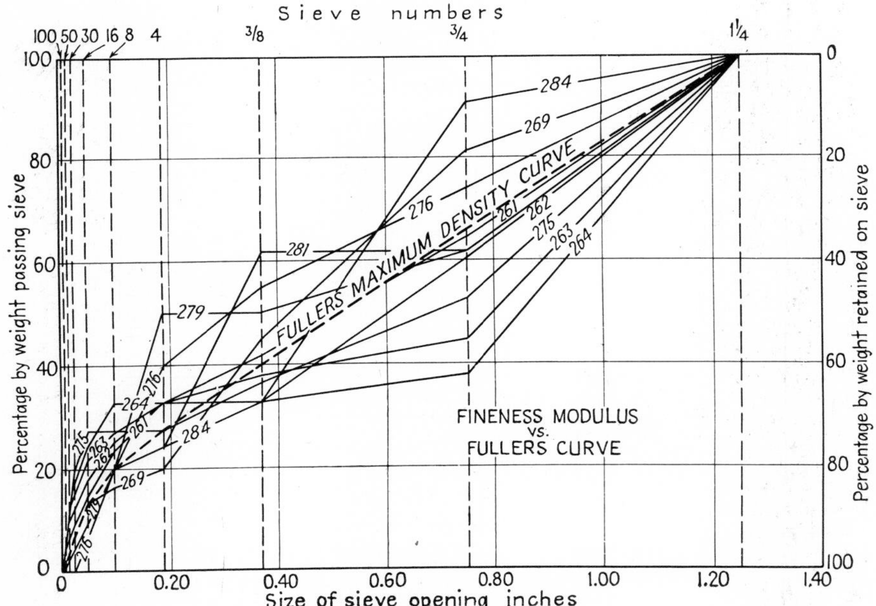

FIG. 34. Fineness modulus versus Fuller's maximum density curve. (*Colorado Engineer, University of Colorado, January,* 1924.)

Optimum Grading, Determining Factors. Methods of placement and the use of certain additions or admixtures influence the grading that is satisfactory, vibration, or air entrainment, for example, permitting harsher but more economical mixtures with lower cement and/or sand content than are satisfactory with hand-placing methods or non-air entrainment. If the mixture is relatively rich a higher ratio of coarse aggregate to fine aggregate can be used, because the cement functions as fine sand to contribute to the "coarse-aggregate carrying capacity."

Characteristics of the Mixed Aggregate. The separation of aggregate into fine and coarse (F.A. and C.A.) is a convenient but artificial and arbitrary device. The fine aggregate is a highly significant constituent because of the great influence per pound which an aggregation of small particles exerts upon the water and cement requirements and upon workability of the concrete. On the other hand, the coarse aggregate has the more pronounced influence upon economy of paste because of its greater relative volume and replacement capacity at a nearly constant ratio of water to cement. For most methods of proportioning, it has been distinctly simpler to deal with the fine and coarse aggregates as separate units that can be combined in desired amounts than to treat them as one. As a practical matter it is possible to maintain a much higher degree of constancy of grading by combining the aggregates at the mixer than could possibly be done by attempting to maintain stock-pile uniformity for mixed aggregates. For a given coarse aggregate the grading of the fine aggregate selected will greatly influence the quality of the concrete. For a given fine aggregate, the grading of the particular coarse aggregate selected will mainly influence the economy of the concrete. However this may be, concrete contains both fine and coarse aggregates and, in so far as the grading is to be significantly related to the characteristics of the concrete, it is the combination of fine and coarse aggregate that must finally be considered.

Criteria for Ratio of Coarse to Fine Aggregate. In general, the most economical concrete will be the one carrying, without undue harshness, the most coarse aggregate. Three criteria are frequently used as indications of coarse-aggregate effectiveness. They are the limiting values attainable (under the conditions of placement) for b, b/a, and b/b_0, defined as follows:

b = solid volume of coarse aggregate in a unit volume of the concrete.

b/a = ratio of solid volume of the coarse aggregate to the solid volume of the fine aggregate in a representative sample of the concrete.

b/b_0 = ratio of the solid volume of the coarse aggregate in the concrete to the solid volume of the same aggregate that could be made to occupy the same volume as the concrete without the presence of cement or sand.

With any given sand-cement mortar the limiting values for both b and b/a depend upon essentially the same factors, viz., the grading, shape, and maximum size of the coarse aggregate. The ratio b/b_0 differs from the other two criteria in that it is primarily a measure of the wedging characteristics of the mortar and gives a less dependable indication of the grading effectiveness of the coarse aggregate. If the cement-water-sand mortar were a nonwedging fluid, the ratio b/b_0 would be unity regardless of whether the grading was good or the shapes of particle favorable. Tests (Dunagan, *Proc. ACI*, vol. 36, p. 649, 1940) show that b/b_0 for a poorly graded coarse aggregate having low values for b and b/a does not differ greatly from that for a well-graded, favorably shaped aggregate for which b and b/a are considerably higher. The same characteristics of the coarse aggregate that affect values of b_0 affect values of b to much the same extent, leaving the ratio more or less constant for widely different coarse aggregates. As a criterion for the shape-grading excellence of coarse aggregate b/b_0 has been overrated, but as a basis for the evaluation of relative coarse-aggregate carrying capacities of mortars it can be made to serve a useful function. As the maximum size of the coarse

aggregate is increased for the mixtures of Table 19, the values of b and b/a increase appreciably, but b/b_0 is little affected.

TABLE 19. MAXIMUM SIZE OF COARSE AGGREGATE IN RELATION TO QUANTITIES, UNIT WEIGHTS, AND GRADING INDEXES

Concrete mixtures made from Hoover Dam materials. Water-cement ratio constant at 0.54 by weight = 6.2 gal per bag. Nominal 28-day strength 3,900 psi

	Neat paste (no aggregate)	Mortar, F.A. only	Concretes, maximum size of aggregate		
			1.5 in.	6 in.	9 in.
Proportions by weight	1:0:0	1:2.5:0	1:2.5:4.4	1:2.5:6.5	1:2.5:7.1
Unit weight, lb per cu ft	112	140	151	156	156
Workability	Fluid	Thin	Slump 5 in.	Workable	Workable
Yield, cu ft per bag	1.3	2.7	5.2	6.3	6.7
Cement, bbl per cu yd, or sacks	5.2(20.8)	2.5(10.0)	1.3(5.2)	1.1(4.4)	1.0(4.0)
Water, gal per cu yd	127	61.5	32.1	26.2	24.6
b (relative volume of C.A.)	....	....	0.48	0.57	0.60
b/a (ratio C.A. to F.A.)	....	....	1.75	2.57	2.84
b/b_0 (see discussion)	....	....	0.73	0.81	0.83
F.M. (fineness modulus), F.A.	....	2.72	2.72	2.72	2.72
F.M. (fineness modulus), C.A.	....	....	7.23	8.35	8.60
F.M. (fineness modulus), mixed	....	....	5.60	6.80	7.10

For a given maximum size of aggregate within the normal grading zones, the economical workable range of mixtures falls within a relatively narrow band for the ratio of C.A. (b) to F.A. (a). Thus for a maximum size of 1½ in. the ratio b/a usually falls between 1.5 and 2. Figure 32 shows mixed-aggregate grading curves for three values of b/a over a wide range for 1½-in. maximum aggregate, and Fig. 33 shows mixed-aggregate grading curves for values of b/a ranging from 0.30 for ⅜-in. maximum aggregate to 2.84 for 9-in. maximum aggregate.

With the exception of being unusually high in material that passed the No. 30 sieve and was retained on the No. 50 sieve, both the coarse and fine aggregates of Hoover (Boulder) Dam fall remarkably near the median lines of the zones of recommended grading. This becomes apparent from an inspection of Figs. 30 and 32 on each of which the Hoover (Boulder) Dam grading is plotted and shows as a solid line within the zone. Tests on Hoover Dam aggregates and mixtures supply information not available elsewhere on comparative gradings extending to a large maximum size.

One must bear in mind that the ratio b/a does not determine the richness or leanness of the concrete. Thus a b/a value of 1.5 might be that for any of the following mixtures: 1:1:1.5, 1:2:3, 1:3:4.5, 1:4:6, etc.

The grading of the aggregate is more important for a relatively lean mixture such as 1:4:6 than for a rich one such as the 1:1:1.5. The great excess of cement-water paste in the rich mixture will fill in grading gaps not all of which could be filled for a poorly graded aggregate if the paste-aggregate ratio were low. Table 8 gives maximum sizes of aggregate recommended for different types of work.

Economy of Grading up to the Maximum Permissible Size of Coarse Aggregate. In Fig. 35 are plotted the relative constituents in a unit volume of concrete for the range

of gradings shown in Fig. 33. Both Figs. 35 and 36 illustrate the marked reduction in the cement required as the maximum size of a graded aggregate is increased. Some of the data on the several mixtures of Fig. 35 are given in Table 19.

Figure 37 shows the relation between size of aggregate and both the unit weight and cement requirement for a number of well-known dams in relation to the smaller aggregate concrete used in other types of structures.

Practical limitations on maximum size of aggregate may be availability, mixing and handling equipment, minimum dimensions of the structure (maximum size should not exceed one-fourth the minimum dimension of the form), and spacing of reinforcing steel. There is evidence that as the size of aggregate increases at constant water-cement

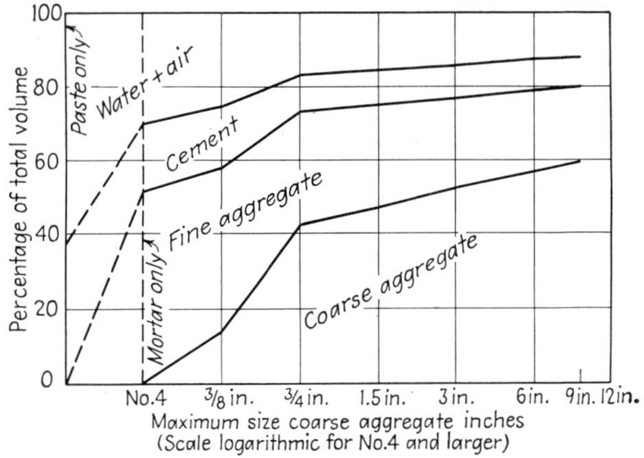

FIG. 35. Maximum size of graded coarse aggregate in relation to quantities of cement, aggregates, and water; Hoover Dam materials. *ASTM Report on Significance of Tests of Concrete, 2d ed., p.* 100, 1943, *Fig.* 6.)

ratio the strength is lowered (*Proc. ACI*, vol. 32, p. 235, 1936), but this aspect is controversial and somewhat academic since the effect is not of practical importance in comparison with the benefits accruing from the possible reduction in the cement.

The shape of aggregate particles affects the placing quality of the mixtures. Crushed angular fragments require somewhat more paste or mortar than do rounded gravels and washed sand, but for a given water-cement ratio the crushed material supplies about enough added strength from the interlocking effect to compensate. The ratio of coarse to fine, b/a, needs to be somewhat lower for crushed material. For nominal 1½-in. maximum-aggregate concrete the upper limits for b (ratio of coarse aggregate in a unit volume of concrete) range between 0.40 and 0.50 for gravels and between 0.35 to 0.45 for crushed stone or slag. To maintain constancy of grading it is often desirable to have two or more size separations in the coarse-aggregate range, each fraction being weighed into the batch.

Defective and Reactive Aggregates; Water and Chemical Contamination. *Unsound Aggregates; Cherts.* Many cherts and some other rocks that appear to be hard and sound are unsatisfactory for concrete because of instability under continued exposure to moisture, alternate wetting and drying, or freezing and thawing. As they deteriorate they swell, producing various forms of "map cracking" and "pop-outs" on the surfaces of the concrete in which they are embedded. Any deposit that contains appreciable quantities of chert needs to be investigated carefully before acceptance for concrete.

Soft and Weathered Fragments. Aggregates that contain many soft or weathered particles or fragments of coal, coke, or wood are objectionable; if near the surface of the concrete such particles will weather and produce pitting even though they may undergo no disruptive expansion. Soft aggregates are subject to abrasion in handling and mixing.

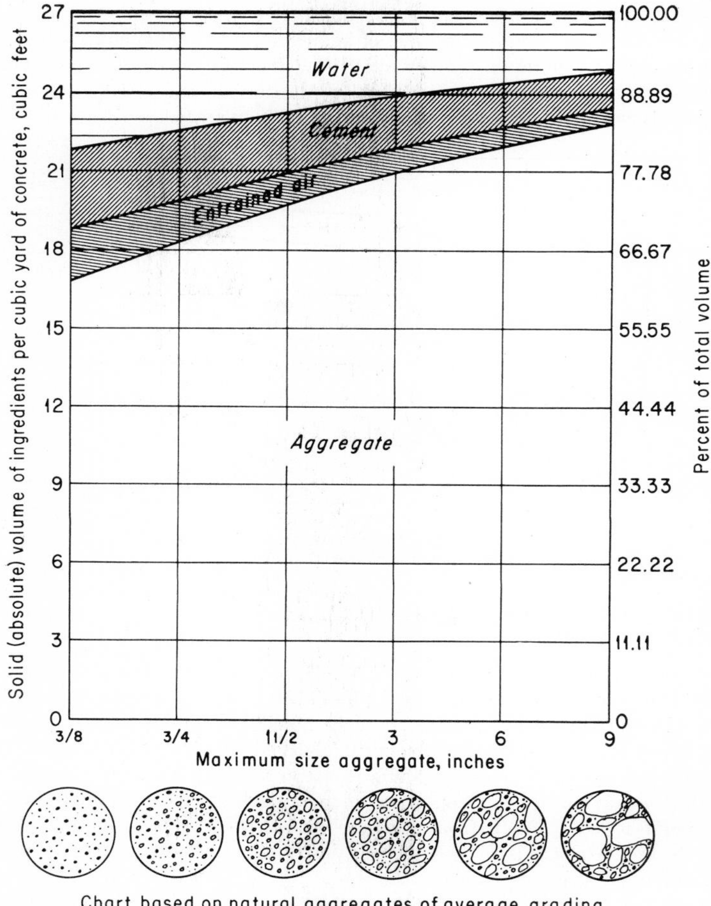

Fig. 36. Amounts of water, cement, and entrained air for various maximum sizes of aggregate. (*Concr. Manual, U.S. Bur. Reclam., 6th. ed., p.* 61, 1955.)

Such abrasion may seriously alter the grading and produce excesses of fine material. In general, any sound aggregate not so soft as to abrade objectionably in handling and mixing is not too soft for concrete; many of the relatively soft limestones make excellent aggregates, often better than some of the harder crushed traps and basalts. During the war much reasonably satisfactory emergency concrete in the Pacific theater was made with coral aggregate, but under normal conditions such material is to be

classed among the unusual or special-purpose aggregates where some such property as low density is important.

Reactive Aggregate. Since 1940 attention has been focused on the phenomenon of reactive aggregates; aggregates of types long accepted as satisfactory but which have been found in certain environments to react adversely with some cements. Serious disruption of this origin occurred at Parker Dam, on the Colorado River below Hoover (Boulder) Dam; bridges and bridge abutments in California and many other parts of the country, the nondurability of which had previously been ascribed to "frost action" or

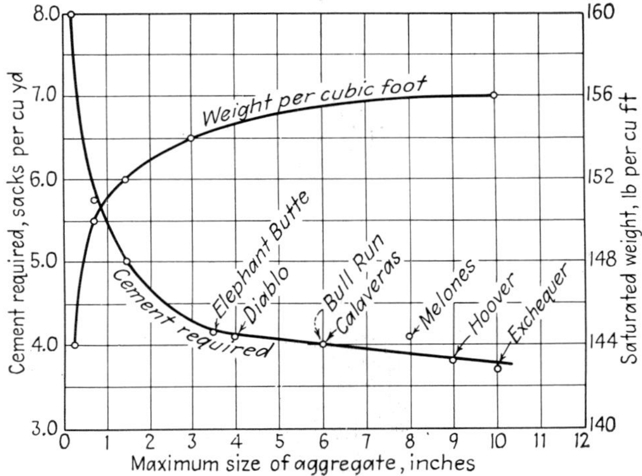

FIG. 37. Maximum size of graded coarse aggregate in relation to cement factor and unit weight of concrete for a number of important dams.

"poor concrete" have been found to be victims of adverse chemical reactance with the cement. The work of Stanton and others links the difficulty to interaction between the alkali in the cement and the aggregate under certain conditions of composition and exposure.

Mica. Mica, of which small quantities are visible in most natural sands, is objectionable if much is present. The thin platelike structure makes the particles weak; they are also very smooth, do not wet readily, and fail to bond properly with the cement paste.

Sugar and Other Organic Contamination. Sugar and organic impurities interfere with the hydration of the cement. Sugar contamination is usually incidental or accidental, perhaps in shipping or in reuse of sugar bags for cement or aggregate; tannic acid or other organic impurities are often present in the aggregate deposits, and the surfaces of freshly placed concrete may be damaged by covering them with manure or disintegrating straw. Aggregates may be tested for organic impurities in accordance with ASTM C40 and for soundness by ASTM C88. With certain exceptions (water containing sugar, for example), any potable water is satisfactory for concrete, but other waters may contain sewage, barnyard, or manufacturing wastes or corrosive salts. Sea water (see Sea Water) can be used with some reduction in strength; some mineral waters are wholly unsuited for concrete. Coated aggregates are generally barred, although not all hard firm coatings are injurious.

Surface Texture and Shape of Particle. Most aggregates, washed (rounded) as well as angular crushed material, have sufficient microscopic roughness to bond satisfactorily with the mortar, although there are undoubtedly significant measurable differences. Aggregates with glassy surfaces, such as obsidian, are objectionable. Flat and elongated particles are objectionable primarily because of their poor packing quality. Aggregates with either moduli of elasticity or thermal coefficients that differ greatly from those of the paste may, under certain conditions of temperature variation or loading, prove to be quite inferior. The former much used stipulation that "the sand shall be sharp" is a misnomer; no natural sand is ever sharp.

APPLICATIONS: PRACTICAL ASPECTS

Mixing and Transporting. Concrete is now universally proportioned, mixed, and discharged in batches rather than in a continuous operation in which materials flow into the mixer as a steady stream of concrete emerges. Formerly hand mixing was much used for small jobs, but virtually all present-day concrete is machine-mixed.

Mixers vary from laboratory machines of 1 or 2 cu ft capacity and small-job one-bag mixers good for 5 or 6 cu ft, up to the batteries of mixers of 4 or 5 cu yd each. For small work the mixer, complete with gasoline engine, is used at the site of the job, which might be sidewalk, driveway, or basement floor and walls. Large "pavers" also operate near the point of deposit and are moved along on the prepared subgrade as the work progresses. For much other work the mixing plant is stationary at or near the site of the work, the concrete being conveyed to point of deposit by wheelbarrow, "buggy," hoist and chutes, conveyor or belts, or buckets on cableways, or by being piped pneumatically or shot into place by way of air hose and a mixing nozzle. At Hoover Dam the high-level mixing plant had in simultaneous operation six 4 cu yd mixers each discharging into a 9 cu yd loading hopper from which the 8 cu yd bottom-dump buckets were filled with two successive batches and conveyed horizontally by cableway and lowered to the point of deposit in the dam.

Of recent years the *central mixing plant* has assumed commercial importance, delivering the ready-mixed concrete miles away to the purchaser's specification. Trucks up to 5 cu yd capacity are used, and for longer hauls "transit" mixers are charged with the dry materials, adding the water and mixing the concrete en route. Agitator trucks for ready-mixed concrete prevent segregation by moderate agitation as distinguished from transit mixing. Nonagitating trucks can be used for ready-mixed air-entraining concrete.

A batch of concrete is normally mixed from 1 to 4 min after all materials are in the drum, the longer periods being for the larger batches. Overmixing is objectionable only because of the added costs and slowing of the work, except for the grinding action that increases the fines and tends thereby to stiffen the mix.

Quality and Job Control. Sampling for the control of concrete work may be done either at the mixer or at the form. For "quality" control representative specimens are moist-stored (standard-cured) and tested moist at the designated age (usually 7 or 28 days) (ASTM C39, C42). *Job-cured* specimens are cured in so far as possible just as the concrete of the job is cured. Drilled cores, tested in compression, constitute one method of checking on the job quality, being much used in connection with highway construction. Usually a correction (Fig. 12) for ratio of height to diameter needs to be applied to core tests (ASTM C42). The *modulus of rupture* computed from tests of short beams, usually job-cured, is much used in lieu of or in addition to compressive cylinders or cores. Such beams can be tested on the job in portable beam-testing machines; compressive tests require laboratory facilities (ASTM C31, C78). Fragments of a tested beam may be also tested in compression as "modified cubes" (ASTM C116) (see discussion of strength).

Quality-control specimens are primarily of value for checking the quality of concrete secured, against that which was designed or ordered. Job-control specimens are usually for checking on readiness for service or form removal. Where concrete is purchased ready-mixed, analyses are often made of occasional samples of the fresh concrete as a check on cement, water, and aggregate content. Such tests on fresh concrete are also made to check the uniformity of the output or the thoroughness of mixing (ASTM C172).

The widely used Dunagan method (*Proc. ASTM*, vol. 31, Pt. 1, p. 383, 1931; also *Iowa Eng. Expt. Sta. Bull.* 113, 1933) makes use of the Eureka equipment sold by the Humboldt Mfg. Co. of Chicago and is essentially a wet screening operation, corrected for silt in the aggregates. More recently, Hime and Willis (*ASTM Bull.* 209, October, 1955; also *PCA Research Dept. Bull.* 61) have applied the heavy-liquid-media technique to the separation, removal, and measurement of the cement in fresh concrete. More involved techniques have been developed for checking upon the constitutents of hardened concrete; see H. F. Kriege, *ASTM Special Pub.* 169, p. 221, 1955.

Formwork and Preparation for Placement. The formwork is extremely important because any weakness, bulging, or sagging will permanently damage or disfigure the structure. Rational design of forms for stiffness and strength is difficult since form pressures vary greatly with height and rate of pour; fluidity of the mixture; rate of setting or stiffening; and the violence or vigor of vibration, tamping, and/or other compaction operations. Form design practice is still largely in the rule-of-thumb stage.

Before any concrete is deposited the forms should be thoroughly checked for line and grade. All reinforcement should be checked for position and firmness. Every bit of dirt, shavings, and other debris should be removed, hand holes having been left in the forms for this purpose at any inaccessible locations. Wood forms should be drenched ½ hr or more prior to placement, since the swelling of the forms from contact with the wet concrete may produce distortions or overstress. There must be no ice or puddles of water where the concrete is to be placed.

In starting a day's "pour" it is excellent practice to omit the coarse aggregate from the first batch placed, the swept-in or otherwise carefully placed grout ensuring intimate contact with clean roughened surface of the preceding pour.

Vibrators, tampers, and large pieces of aggregate should not be permitted to contact reinforcement, especially bars projecting from a previous pour. All "slop" or encrusted "old concrete" should be wiped clear of reinforcement before placement starts. Where walls are thin, heavily reinforced, or have surfaces otherwise relatively inaccessible from within, vibrators may be applied to the outside of forms. Where that possibility exists, it is essential that the construction of the forms be amply sturdy to avoid misalignment or other damage from the added severity of treatment.

Placement and Compaction. Deposition and consolidation of the concrete have much to do not only with its appearance but with its quality. Formerly it was almost universal practice either to hoist the fresh concrete to a high tower for chuting or else to locate the mixer above the job. The more fluid the mix, the greater the lengths of chutes that could be served from a given elevation. This practice encouraged overwet mixtures, promoting segregation and weakness. Placement of most building concrete is now by horizontal runways for wheelbarrows or "buggies" that have been brought by hoist or elevator to the level of deposit. The concrete should be deposited in horizontal even layers and tamped or vibrated until capillary continuity is assured by the glistening of a film of water at the top surface. Special working and spading near the face of the form is always desirable to ensure a smooth nonhoneycombed surface. The slump or workability of the concrete must be closely regulated and kept in accord with placing conditions.

Surface *honeycombing* adjacent to forms from lack of workability (dryness or harshness) in relation to the method of compaction and conditions of placement is unsightly but usually much less weakening to the concrete than is the excess fluidity that often gives easy placeability, form-fitting texture, and good surface appearance. This fact creates a tendency on the part of foremen to use more water than is necessary or desirable. Excess tamping or vibration of a workable mixture brings too much of the water and fine material to the upper portion of the pour, producing a weak chalky nondurable pervious layer of the *laitance* that always characterizes the really wet mixtures. Laitance layers are often visible as surface streaks marking the tops of successive pours on dams, retaining walls, chimneys, bins, and grain elevators. It is at such laitance levels that seepage and deterioration invariably first appear. Inclined pour lines indicate wet mixtures that were allowed to flow downward to possibly less accessible parts of the form. In the continuous placement of concrete of some height, it is often well to "dry up" successive batches somewhat to offset the water gain (accumulation of surface water from sedimentation) as the work progresses. Even with the best of relatively stiff mixtures water collects on the undersurfaces of aggregate particles and reinforcing bars, making a much poorer bond on the undersides than on the top sides of all horizontal surfaces (*Eng. News-Record*, vol. 98:6, p. 242, Feb. 10, 1927).

If concrete must be placed on a slope, placement and compaction by vibrators (or whatever other method may be used) should start at the bottom of the slope and work upward in order that the tendency to flow will always be toward concrete that is already in place.

In plain concrete stepped-footing construction, it is often desirable to use a stiff mixture and sloping sides rather than the successive box-type steps commonly introduced for making the transition from footing base to column pedestal. With a low-slump (low water-cement ratio) mixture, the slopes can be roughly but adequately sloped and shaped with shovels without intermediate form boxes. The quality of the concrete amenable to such treatment will be far superior to the overwet mixture that usually goes into the "poured" steps of box-type forms.

Where concrete must be placed in sloping forms such as inclined wing walls, top boards of the form should be added only as final placement is accomplished in the successive lower layers in order that the concrete can be *placed* in virtually its final position, rather than required to flow into place with possibly some supplementary pushing or agitation.

In general the thickness of the horizontal layers of pour should not only be uniform but limited to 6 to 12 in. for reinforced concrete and not over 18 in. for mass work. Factors controlling variations within these ranges are width of form, method of compaction, and the fact that in continuing pours the lower layer needs still to be sufficiently plastic to blend at the junction of the two.

Concrete should not be allowed to drop freely through more than 3 or 4 ft and in thin sections metal or rubber "tremies" or drop chutes should be used. Where concrete is placed at a fairly rapid rate in a tall form, not only should an occasional somewhat drier batch be used to compensate for "water-grain," as mentioned previously, but the deposition should be halted about a foot below the top of the form, to permit settlement to occur. Placement should, however, be resumed to completion within an hour or so to avoid a joint at the junction due to stiffening below. The top portion is usually readily available for puddling, permitting use of a somewhat drier topping mixture.

Internal vibrators should always be used vertically at a lateral spacing of about 18 in., depending upon the character of the mixture and related job conditions. The vibrator may penetrate the lower layer slightly, provided that the layer has not stiffened beyond readily becoming plastic under the vibratory action. The vibrator should not be used for any appreciable lateral movement of concrete and should not be permitted

to contact the form (which it might scar or dent) or to contact the reinforcement. Vibration increases the hydraulic pressure against the form and, as stated, any overvibration promotes segregation. Vibration is a valuable placement device but it too can easily be overdone.

Depending upon temperatures and other factors, the initial stiffening (set) of concrete does not usually occur until 2 or 3 hr after the water is added. Concrete that has stiffened may safely be remixed and used provided that it again becomes workably plastic and placeable (by whatever method of agitation is employed) *without any addition of water.*

In all such respects air-entrained concrete requires special consideration since it is gradually losing air throughout any period of dormancy and also through mixing operations—especially vibration. As mentioned elsewhere, nominal short-time vibration can safely be used for normal air-entrained mixtures but all such agitation should be kept to a minimum. (PCA leaflets and bulletins, especially *Bull.* T12 and such other publications as the U.S. Bureau of Reclamation. "Concrete Manual" and the U.S. Army Engineers "Handbook" supply much valuable up-to-date information on the handling, mixing, and placement of concrete and related topics.)

Removal of Forms; Surface Patching and Rubbing. Forms must be removed carefully to avoid chipping, spalling, and other damage to the "green" concrete. Any tie wires should be clipped at least 1 in. inward from the surface, a stiff grout patch being used over the clipped end; otherwise, the wires rust and produce staining and disfiguring local spalling. Stiff grout should also be worked and rubbed thoroughly into any porous, chipped, or honeycombed pockets. The use of thin grout rubbed over exposed faces should not be permitted, since it covers up defects temporarily but scales off after a few months. For ordinary exposed surfaces immediate vigorous rubbing with a carborundum brick, or mechanical grinding is the approved treatment. Rubbing is best accomplished when the concrete has attained a strength of about 400 psi (see "Curing").

As mentioned elsewhere, there are many surface treatments for different decorative effects. Often the form marks are intentionally retained. In any case most cast-concrete surfaces require, promptly upon form removal, retouching and mechanical attention of one sort or another. All patches or other exposed additions should be kept moist for several days to ensure proper hydration.

Underwater Structures. See also "Prepakt." Underwater concrete may be placed by means of a *tremie*, a pipe from 6 to 10 in. in diameter, with a hopper or flared top. Once the concreting has started the lower, or discharge, end of the tremie pipe is kept embedded in the concrete mass and the tremie must be kept well filled. During the war several dry docks were built with heavy 18-ft thick floors and walls of tremie-placed concrete under 60 ft of water. Closed-top, *drop-bottom buckets* are also used for underwater concreting, care being exercised not to agitate the concrete that is in place. Sometimes underwater concrete is deposited in *bags*. The bags may be filled with the wet-mixed concrete or with dry-mixed materials. In either case the cement paste oozes through the bag sufficiently to weld the tiered bags into a solid mass. Many miles of Mississippi River revetment for bank protection have been cast on land as *articulated mats*, being later submerged. In deep bridge piers the concrete is usually deposited in air after the "seal" around the cutting edge of the caisson has been placed by the aid of divers. Deep pier concrete may also be deposited in air directly on a caisson as it is sunk. Subway tubes and large pipes for outfall sewers have been cast on land, bulkheaded, towed to destination, and sunk into position with the aid of divers. The artificial harbors for the Normandy invasion were cast in many parts of England as great concrete boxes of predetermined heights, being towed to the proper location before sinking. In one instance a 92 ft long by 42 ft high diversion dam for a very

turbulent stream was cast vertically, with the stream face conforming to the rocky contour of the stream bed, and was felled with dynamite across the channel much as a tree might have been (*Civil Eng.*, December, 1930, p. 159). Concrete dams are constructed by diverting, first one portion of the stream, then another, unless it is possible to divert the entire stream, as was done for Hoover Dam by excavating in the canyon walls four great 50-ft-diameter tunnels averaging ¾ miles in length for each.

Pumping and Pneumatic Placement. Workable concrete with aggregate up to 3 in. maximum can be pumped through a steel pipe for distances exceeding 1,000 ft. Such concrete, often known as "pump-crete" is especially advantageous wherever space is limited, as in tunnels, on bridge decks, and in some portions of powerhouses and buildings.

Pneumatically placed concrete, variously known as *shot-crete, gunite,* etc., is much used for placing or repairing reservoir walls and canal and tunnel linings, placing "*fire-proofing*" or protective layers over structural steel, repairing faces of dams, and restoring reinforced concrete members of buildings that have undergone serious fire damage. Water is mechanically added to the dry mixture at the nozzle, which shoots the freshly mixed concrete (really a mortar) at the surface prepared for its reception.

Some form of mesh reinforcement is usually placed about 1 in. from the surfaces, and the mixtures range from 1:2.5 to 1:5. The maximum size of aggregate is under ½ in., and there is considerable loss of aggregate due to rebound. The *yield* is necessarily much below that of poured or cast concrete and, in general, such concrete is economical only for situations where the saving in formwork or simplification in placement offsets the added cost of cement.

The thickness per layer is usually from 1 to 2 in., but successive layers can be added at suitable intervals of time to build up to any thickness desired. In single-form wall construction the concrete is shot against the form for the exposed face of the structure. Concrete pipe and domed igloo-like houses have been cast by depositing pneumatically placed layers over an inflated rubber core, to be deflated and removed when the concrete becomes self-supporting.

Vacuum Concrete—Tilt-up Construction. Utilizing the fact that strength increases as voids are decreased, the vacuum method of placement was devised. The concrete is mixed wet enough to flow into position with little or no vibration or tamping required, after which excess water is removed by vacuum applied through special mats or pads. From such concrete walls the forms can often be removed within an hour or two. Similar results are secured by using absorbent forms or form linings, as has been done on dams, walls, and decorative work. In tilt-up construction, wall panels are precast as horizontal slabs, being raised to their position in the wall by vacuum cups or other lifting device (often after being given the vacuum treatment) and in any case after sufficient curing to permit handling.

Terrazzo; Vibrolithic; Earley Technique. The well-known *terrazzo* floor construction consists of a mortar mixture 2 or 3 in. in thickness resting on a concrete base to which it may be carefully bonded by a thin coat of cement grout or from which it may be separated by a layer of sand covered with tar paper. Near the surface of the mortar, marble chips of desired colors are embedded, later to be ground and polished. White portland cement may be used where specially decorative effects are desired. The so-called *vibrolithic* pavement makes a similar but strictly utilitarian application of the technique. Additional coarse aggregate is tamped and vibrated into the surface of the pavement immediately after placement, lowering the effective water-cement ratio somewhat and giving a very hard stony surface to the concrete. There is no grinding. The similar John J. Earley technique for decorative concrete has been mentioned previously.

Grouts and Grouting. *Cement-bound Macadam; Prepakt; Grouting.* First used in Scotland in 1872, *cement-bound macadam* is a durable economical pavement suitable

for secondary highways, important local roads, drives, and alleys. Coarse aggregate is placed on the subgrade and compacted to an even surface of the desired contour. It is then flushed with a sand-cement grout, swept, and rolled, followed by finishing and curing. The same technique has proved excellent for making repairs or replacements to ordinary pavements because of the small amount of mixing involved and equipment required; shrinkage is negligible.

"Prepakt" construction makes use of the same principle, forcing grout through a mass of prepacked graded stone under pressure and in such manner as to ensure virtual saturation with the grout. It is used for bridge piers, repairs, foundations, and in one instance for adding several feet in thickness to the upstream face of a concrete dam (*Proc. ACI*, vol. 44, p. 633, 1948).

Oil-well Cementing.[1] In the oil fields neat cement-water mixtures (grouts) using from 4 to 7.5 gal of water per sack of cement are used to seal around the casings to block the passage of water and gas from stratum to stratum. The techniques for placing are highly specialized, and the stiffer grouts (about 4 gal per sack) are much the better where properly used. Grouting has been successfully accomplished at depths in excess of 13,000 ft. Oil-well cementing calls for varied modifications and modifiers. Thus in deep wells, because of the pressure and high temperatures encountered, sometimes exceeding 350°F, special "slow-setting" and "retarded" portland cements are used. In other situations high-early-strength, sulfate-resisting, and standard portland (Type I) are needed. To accelerate the setting, 2 to 4 percent of calcium chloride or 7 to 10 percent of calcium oxychloride may be used as an admixture. Amounts in excess of those indicated weaken the grout and should not be used (*Proc. ACI*, vol. 43, p. 893, 1947).

Contraction Joint Grouting of Concrete Dams. The great concrete arch dams such as Hoover and Shasta, constructed in columnar blocks 25 to 50 ft in plan shrink upon extraction of the heat of hydration. Shrinkage between adjacent blocks is from 0.02 to 0.35 in., which space must be filled (1) to prevent leakage of water along the joints through the dam and (2) to establish contact between blocks in order that the thrust of the arch can be developed for transmitting the load to the abutments (canyon walls) at either end of the dam. The joints are *radial* and *tangential* (longitudinal) and both require grouting, the radial for the thrust and the tangential to make the dam function as one great monolith as the arch tends to flatten under the water load. In Shasta Dam there were about 140 acres of joint surface to be grouted (about half radial and half tangential) requiring nearly 14,000 bbl (55,000 sacks) of cement as actual joint filler exclusive of wastage. See Simonds, *Jour. Power Div. ASCE*, vol. 82, No. PO3, June, 1956.

For contraction joint grouting the cement must be very fine (about 98 percent passing the 200 mesh sieve). Unground clinker, tramp iron, and lumps are objectionable, and no aggregate is used. Grouting is conducted over stopped-off areas, the maximum pressure permissible being just under that which would overstress the block by tilting or shearing, the pressure range being 25 to 60 psi at the top of the 50- to 100-ft height grouted in one operation. Grouting starts with a very fluid mixture, perhaps 10 gal per sack, followed by about 5, then 4 gal per sack. For pressure regulation, tilting tendencies are observed with 0.0001 dial gages across adjacent joints. *Modified, low-heat,* and special *oil-well* cements have proved most satisfactory for this type of grouting (*Proc. ACI*, vol. 43, p. 637, 1947).

Foundation Grouting. Foundation grouting may be for stabilization as for bridge abutments or railroad subgrades, or it may be for sealing off leaks in reservoir and dam sites. Such grouts may be either neat cement or sand-cement mixtures; the proportions and water ratio vary widely with the conditions to be corrected.

[1] The term "grouting" is seldom used in oil-well practice.

Grouting under a dam may be "blanket" grouting, "cutoff" grouting, or both. A neat cement mixture is considered the best general-purpose grout but rock flour, clay, asphalt, and various other substances have been used. The drilled holes for grouting vary from 1 in. to over 5 in., but the smaller holes have proved satisfactory and cheaper, as much as 38,000 cu ft of solid having been introduced through a single 1.5-in. hole. It is important that grouting pressures be sufficient to ensure penetration but not enough to disturb the foundation structure. One rule of thumb is that the pressure in pounds per square inch at any elevation shall not exceed the depth in feet. There is no established procedure for foundation grouting because of the great range of possible conditions to be met (*Proc. ACI*, vol. 43, p. 917, 1947).

Tunnel Grouting. Grouting has proved useful for stabilization of weak formations and also for filling the interstices between fragments packed back of tunnel linings to fill in where overbreak or caving has occurred.

Soil-cement Mixtures; Mud Jack or Mud Pump. Soil-cement roadways and runways 6 in. thick as compacted are satisfactory for traffic up to 1,000 vehicles per day, provided that not more than 50 percent exceed 2 tons gross weight, or for airplanes having 15,000-lb wheel loads or less. The cement requirement varies between 7 and 16 percent of the compacted volume of most soils, sandy and gravelly soils with 10 to 35 percent silt being best adapted; sandy soils deficient in fines, such as some beach, glacial, and wind-blown sands, are quite good but pack less well because of the poor gradation; silty and clayey soils make good soil cement but are more difficult to pulverize and mix thoroughly. Soils with high organic content may be undesirable because of injurious chemical reaction with the cement. The range of satisfactory grading is wide; with everything below 3 in., 50 to 100 percent can pass the No. 4 sieve, 15 to 100 percent passing No. 40, and 0 to 50 percent below No. 200 (much greater percentages under No. 200 if soil is relatively friable).

Because of the wide range in minimum cement content necessary for hardening soil-cement mixtures, preliminary testing for wetting and drying and freezing and thawing effects by ASTM D559 and D560 is always desirable.

Soil-cement roadways, runways, and shoulders for pavements may be mixed in place with heavy-duty field cultivators, gang plows, or rotary speed mixers, or they may be mixed with a traveling-type mechanical mixer. After mixing, leveling, and ramming or rolling, the surface should be covered with clean straw and kept moist for about a week in order that the cement can partly hydrate prior to service. Soil-cement mixtures are vulnerable to wear, and the durability can be greatly improved with a bituminous layer on the surface. In fact, the cement content should be increased by about 4 percent of the soil volume if a bituminous wearing coat is not to be applied at once.

Soil-cement mixtures are well adapted for *canal and reservoir linings, core walls and parking areas, tennis courts, rammed-earth wall construction, cycle paths,* and other applications intermediate between concrete and packed earth in quality, serviceability, and cost. For soil classifications and an introduction to the analytical aspects of soil-cement mixture design see the "PCA Soil Primer." A companion handbook, "Soil-Cement Construction Handbook," covers the applied aspects and a third is the "Soil Cement Laboratory Handbook." See also *PCA Bull.* D5.

An *earth-cement grout* [about 1 sack (¼ bbl) per cu yd of soil] is useful for raising pavement slabs that have settled and for filling in subgrade deficiencies from washing and pumping action near edges and at the joints under traffic. Holes about 2 in. in diameter are drilled at intervals over the area to be filled in or raised. A hose from the pump is inserted alternately in the several holes to ensure a gradual lifting effect. Lateral expansion from the pressure holds the hose in the hole without any special gripping device. Similar pumping or mud jacking is used for leveling and stabilizing railroad subgrades.

Precast Concrete. Over the wide range of precast concrete products there are various methods of fabrication and curing. Many building blocks, including the common cinder block, are mixed relatively dry and lean and formed under sufficient pressure for immediate removal from the molding machine. They may be cured under sprays or wet burlap or may be stored in a moist chamber or cured in an autoclave under high-pressure steam. *Precast piling, large pipe* for drainage, sewer or water systems, *building units,* and *bridge slabs* are often cast and cured near the job, whereas *burial vaults, bathtubs, birdbaths* and small decorative units are usually plant manufactured.

Normally, casting and curing temperatures for all concrete operations should be kept between 40° and 100°F, the upper limit being primarily to avoid differential strains from volume changes. Steam-cured products constitute an exception, standard curing temperatures ranging from 340° to 365°F under pressures between 100 and 150 psi. Under these conditions there is attained in a few hours a completeness of hydration that would require weeks under usual laboratory or job curing conditions. The types of unit amenable to high-pressure steam curing are small and portable, and large temperature differentials do not develop. The products of hydration are not the same as for ordinary moist-cured units, and steam-cured products appear to have some advantages from the standpoints of durability and volume change.

Miscellaneous Concrete Products. Among the almost infinite range of applications are:

Spun concrete pipe formed centrifugally, and cement-lined cast-iron pipe, the thin layer of neat cement or richer mortar properly hardened (cured), decreasing the hydraulic friction and protecting the pipe from rust and formation of tubercules.

Piling, precast to be driven after curing; concrete placed in a metal shell first driven, then the sides withdrawn as the concrete is deposited; pedestal pile, concrete forced, under driving pressure, out at the open bottom of a driven shell to form a bulb or knob at the bottom of the pile for added bearing power; concrete uppers spliced on to timber piles, the tops of which must be kept below the permanent ground-water line. *Mesh reinforced concrete shells* have been cast around timber piles that were decaying at the water line or being attacked by teredos.

Cement-asbestos mixtures are tough and resistant to weather and chemical attack; among the current products are shingles, corrugated sheets, flat sheets of various thicknesses, tile, and water and drainage pipe.

A mixture of portland cement, emulsified asphalt, aggregate, and water has been found to give a floor less inclined to dent than asphalt but more comfortable than concrete.

In western United States concrete-timber bridge floors have been constructed in which the timber carries tension, being bonded to the covering concrete slab by protruding nails or spikes previously driven into the timber. In the Orient bamboo rods have been used for reinforcement. Other adaptations were accorded introductory mention.

Specialized Concrete Construction—Precautions. *Flat Concrete Roofs.* The exposure of these relatively large areas to the extremes of weather entail severe distortional effects. A gabled roof, trussed or tied across the bottom edges, will adjust to expansions and contractions through the rise and fall of the ridge but the flat roof, in expanding laterally, tends to push outward or pull inward, producing cracks and/or out of plumbness in supporting walls and in parapets. Tight joints or articulations that will permit the relative movement without damage to walls or connecting units are usually difficult to provide, even when the nature of the problem is recognized in advance. Because of some probable shrinkage or other cracking there can also be leakage problems but these are commonly met by the use of a layer of tarred felt or other impervious roofing material on the concrete slab. Where drainage slopes are provided by the use of cinder

or other lightweight concrete fill, the insulation effect of the fill reduces temperature differentials in some degree.

Pavilions and Other Exposed Flat Areas. Like the flat concrete roof these undergo severe exposure effects, not only from general distortions but frequently surface scaling from the use of salts for snow and ice removal. Where such slabs must also serve as roofs for areas beneath, it is difficult to avoid some leakage through construction and other joints and/or shrinkage cracks. Frequently the only workable solution is a complete mastic cover suitable to serve also as a wearing surface under foot or other traffic.

Concrete Floors on Fill. The construction of a good concrete floor on fill, especially if for heavy-duty service, constitutes one of the most exacting operations over the entire range of concrete applications. Under severe service, such as steel-tired trucking, any laxity of design or construction will invariably become apparent through such symptoms as surface dusting, faulting at joints, chipping at edges, warping, and cracking.

Such a floor should be constructed only on a granular, well-drained uniformly compacted subgrade, moist but not wet as the concrete is placed. The subgrade should not be cold (within a few degrees of freezing) since that chills the concrete and retards both the setting and the subsequent curing. Retarded set delays finishing and permits water gain at the surface—promoting laitance and later dusting.

The mixture needs to be rich (8 sacks or more to the cubic yard), of low water-cement ratio (not more than 4 or 5 gal per sack of cement), and a slump of near zero. To secure such a concrete that can be placed properly requires aggregates that are unusually uniform and well graded. Where the surface punishment is severe some of the hard gravels and crushed igneous rocks are preferable to the softer limestones.

A nonporous claylike subgrade is objectionable not only because of its tendency to hold water and its lack of supporting power when wet but it also shrinks badly as it drys out and swells when wet, producing variability of support.

Lack of adequate doweling or keying between adjacent slabs results in faulting, which, like all other surface discontinuities, magnifies the impact effect of moving loads and leads to a pumping action against the foundation, promoting nonsupport and cracking.

During the moist curing period (5 days should be the minimum unless a high early cement is used, reducing it to 2 days) the entire surface must be kept both warm and wet. If the surface of the slab is cool, whether from the air above or the subgrade, curing is retarded. On the other hand, if the slab becomes dry, curing halts no matter how favorable the temperature. Curing here, as elsewhere, requires moisture *and* a favorable temperature present simultaneously. Surface drying also results in warping induced by the differential shrinkage of the upper surface in relation to the lower. (See PCA leaflet ST51 for much essential detail on the design and construction of slab floors.)

Basement Walls and Floors; Garden Walls. Ordinary "poured" basement floors and walls are usually of uncontrolled rather poor quality concrete. Nevertheless most of them will prove reasonably satisfactory on well-compacted porous suitably drained subgrades and similar backfill material. Tar paint or an unpunctured durable membrane back of the wall or under the floor are good construction measures. If the wall leaks and the outside is inaccessible, plastered-on cement-mortar coats applied to the moistened, cleaned inner face (freed temporarily from flow pressure from without and properly cured) will provide watertightness. A firm foundation, porous backfill, and proper drainage (by weep holes or tile in gravel) will impart much stability to low (usually undesigned) garden-type retaining walls. The increasing forward lean so often observed in such walls is usually a result of the gradual buildup of earth pressure from the ever-increasing "squeeze effect" of moist clay gradually coming to bear more firmly and fully against the wall.

Firm Unyielding Support Beneath an Existing Member. A rich relatively dry concrete or mortar mixture may be tamped or pounded into place to assume the load carried by some member that is to be removed. Virtually all setting shrinkage can be avoided as well as some of the compressive strain from loading (by virtue of the prestress induced by the severe tamping). Aluminum powder can be used in the mixture to produce an actual swelling (see "Aluminum Powder"). The new concrete should be kept moist for several days, of course.

Patching and Repairing. Concrete should never be feather-edged. The location of any patch should be chipped or chiseled to vertical or slightly undercut edges ¾ in. or more in depth. The initial strike-off for the patch should be ⅛ in. or so above the floor level, finishing operations attaining flushness. All patches need to be kept moist for curing.

Continuous or Ribbon Construction of Pavements. Since about 1950 a continuous or ribbon type of construction has been successfully employed to a limited extent in pavement construction. Ready-mixed concrete is trucked and dumped into a forward hopper on the paving machine. The concrete is molded to the cross section of the slab as the machine moves slowly forward leaving the strip of freshly molded concrete behind it. The mixture is fairly stiff (less than 3-in. slump and the two lateral faces are molded to a slight inward batter. The wheels of the paver operate on guide rails set to line and grade). Up to within a few feet of the paver the fresh concrete is sprayed with a colorless sealing compound for moisture retention until formal curing can be started. The method produces a slab surface that is unusually uniform and no finishing is required. In fact it is extremely important that the pavement *not be touched* prior to initial stiffening. The 6-, 8-, or 10-in. slightly sloping edges stand nearly vertical held in part by the slight surface tension unless the concrete within the vicinity is touched. The most gentle contact is likely to precipitate a slumping or sloughing off that extends several feet into the pavement, necessitating a troublesome localized replacement operation.

Nuclear Shielding. Concrete is much used for shielding against nuclear radiation. The effectiveness of concrete for this use is more or less proportional to its weight which results in iron and/or heavy iron ore being frequently used as the aggregate. While lead is more effective than concrete per unit of thickness, it is more expensive and has the disadvantage of inability to support itself in the type of large masses frequently needed.

Aside from the increased cost of whatever heavy aggregate is used there are also increased mixing and placement cost. For example, a shovel does not penetrate a plastic mixture of heavy-aggregate concrete with the ease that it does a concrete from aggregates within the conventional specific-gravity range. Just as different techniques are required for a lightweight concrete (where the aggregate tends to float on the cement-water paste) different techniques of an opposite nature are required where high-specific-density aggregate is used.

While the fundamentals of good concrete construction are relatively few and simple, adaptations are ever on the increase and many of them require new techniques or modifications of existing procedures.

JOINTS AND HINGES

Although problems of volume change (expansion and contraction) and of isolation and articulation are not limited to concrete, the distortional and strain-transfer aspects do provide some of the most difficult factors to keep under satisfactory control. In discussing the physical properties of concrete and the data of Table 5, it was pointed out that the effects of autogenous differential distortions are often more severe than are the predictable in-service stresses. Established design and construction practices leave much room for improvement, and even the best of present-day concrete is likely

to show destructive and/or unsightly cracking through our inability to predict and control differential strains. Our shortcomings are especially in evidence with regard to concrete pavements where conditions are unusually severe and where practices are decidedly in the evolutionary phase (*Univ. Illinois Eng. Expt. Sta. Bull.* 365, 1947). The same or related problems carry through, however, to the fields of structural and mass concrete.

Types of Joints. Judd (*Proc. ACI*, vol. 39, p. 557, 1943) classifies joints as *construction, contraction, expansion,* and *control.* The same joint may function as one or all four of the types mentioned. Another form is the *hinge, articulation,* or *hinge joint.*

Construction Joints. A construction joint occurs at any interruption of placing operations for long enough to let the concrete stiffen. In general any interruption in excess of 30 min is assumed to call for a construction joint unless there is a thorough plastic reworking of the concrete across the junction when placement is resumed. It may have any orientation but is usually either vertical or horizontal. Dowel steel may be used across the joint but not necessarily so. Where watertightness is important, sheet-metal *water stops* may be embedded in the concrete across a vertical joint. When this is done, the metal is folded or looped at the joint to prevent tearing in case of separation or differential motion between the adjacent bodies of concrete. Horizontal construction joints, such as occur between successive lifts or pours of dams, walls, chimneys, foundations, pedestals, or columns, can be made watertight by proper attention to surface cleaning and bonding of the new concrete when placement is resumed. An effort will usually be made to have construction joints occur where one of the other types of joint is required or at least is in order. Construction joints in beams and girders should be approximately normal to the axis of the member, never parallel to it. They may occur safely in a region of high bending moment, since the concrete takes the thrust and the tensile steel crossing the section carries the tension across the joint. Such a joint should not occur where the shear to be carried by the concrete is appreciable; in general, therefore, the casting of a uniformly loaded flexural member should be discontinued near the center rather than near a support (if it has to be cast in more than one operation). Beams, girders, slabs, capitals, etc., are all considered parts of the floor to be poured in one operation; columns are discontinued at the bottom of the floor above (or base of column capital or beam).

Contraction Joints. These are usually introduced to forestall the unsightly cracking that invariably occurs as concrete shrinks (generally from drying out). If a sizable length or expanse of concrete, such as a walk, pavement, or wall, is not provided with contraction joints enough to accommodate the shrinkage, it will make its own joints by cracking. Although steel reinforcement can often be used to control the location of cracking, it will not prevent it in the absence of properly located contraction joints. In shrinking, a walk or pavement slab will drag itself along the subgrade until the frictional resistance equals the tensile strength of the slab, at which stage cracking will occur. The optimum spacing for transverse shrinkage joints is influenced by the type of aggregate, the wetness and proportions of the mixture, the curing method employed, and the temperature at time of placement. The wetter and richer mixtures will produce greater drying shrinkages whereas warm-weather placement will involve a greater temperature range thereby increasing cooling shrinkage. Where pavements are cast in widths not greater than 12.5 ft, the construction joints between lanes obviate the need for longitudinal shrinkage joints.

For granite and limestone aggregates spacings for transverse pavement joints should usually range between 20 and 30 ft. For siliceous gravel and slag the range is between 16 and 20 ft. When in doubt the closer spacings should be used. For about 100 ft back from each free end and 60 ft from each expansion joint, contraction joints will gradually open to a point where the "aggregate interlock" ceases to be effective for

the transfer of load across the joint. For these locations dowels (one end bonded, the other in a smooth close-fitting sleeve, or other suitable load-transfer device) need to be provided.

Contraction joints in sidewalk and terrazzo floor slabs are normally at 4- to 6-ft intervals and from 10 to 20 ft for parapet walls and railings. If these latter occur as precast or isolated units the construction joints are absent but there must be adequate provision for expansion at all intersections or abutments with other members.

Prior to about 1950 the "dummy" joint was much used. At intervals not exceeding the minimum for expected contraction cracking the concrete cross section is reduced by the insertion of a strip of joint material extending only part way through the section, being invisible at the exposed face. Or the member is grooved with a V strip at one or both surfaces, or otherwise reduced in cross section before the concrete has hardened. Cracks are expected to occur at the weakened sections, and will if the degree of weakening and the location are right.

The sawed joint is now (1958) rapidly replacing the dummy joint, especially as a transverse contraction joint in pavements and air strips. It consists of a saw cut ⅛ to ¼ in. in width and one-fourth to one-sixth the depth of the slab.

Since no special forming is required, the use of sawed joints does not slow down the placement of the concrete. The weakened section must be provided, however, before appreciable shrinkage stress has developed in the hardening concrete. If a crack has already occurred the saw cut is not only superfluous but is likely to produce raveling in case the crack and saw cut are near one another. Even with no prior crack formation, any appreciable shrinkage tension in the concrete will result in cracking ahead of the saw as the edge of the slab is approached. Sometimes it is desirable to saw joints in two stages, first providing sawed relief joints at intervals of 80 or 100 ft, following these within 2 or 3 days by the intermediate cuts. Immediately after sawing, joints are flushed out with a water jet and a mastic filler is forced in to about ¼ in. of the concrete surface. About 4 hr after placement the concrete is usually hard enough to permit sawing to start if abrasive blades are used. It may be completed 4 to 8 hr later. With diamond blades the transverse sawing is usually accomplished between 8 hr and 3 days after placement. Under initial wet-burlap curing outdoor concrete develops shrinkage stress more slowly than under membrane curing and sawing may be started as the burlap is removed.

Where sawed longitudinal joints are used in two-lane highway or runway construction, the two-lane width is insufficient to develop appreciable early tensile stress and the sawing may be accomplished any time before the concrete is placed in service. Contraction joints are mentioned elsewhere in their relation to the thermal strains present in mass concrete.

Expansion Joints. Expansion joints are introduced to relieve the severe thrust that may result in a blowup in a pavement, push out the curbing by an adjacent sidewalk slab at right angles, or manifest itself in other undesirable ways. Expansion-joint practice for concrete pavements, where the problem is most acute, varies widely. Widths of joint generally range from 0.5 to 1.0 in., with spacings from 40 to 120 ft, about 90 ft being common. There is a pronounced trend toward the omission of expansion joints altogether in such construction as pavements and air strips except in the vicinity of intersections and other abutting construction. For aggregate having a high thermal coefficient, however, or for concrete placed at a relatively low temperature, or if the intermediate contraction joints have not been properly constructed or maintained, expansion joints are necessary at nominal intervals of 600 to 800 ft.

The crack of minute width at properly spaced contraction joints permits of an aggregate interlock that will transfer load across the joint. At expansion joints that condition does not exist and load-transferring dowels, bonded at one end and operating

in smooth, close-fitting sleeves at the other, are normally used to bridge across the joint interval. Sometimes instead, the slabs are given a gradually increasing edge thickness of perhaps 20 percent. The ½- to ¾-in. premolded bituminous strip is a widely used and generally satisfactory form of joint. Cork, sponge rubber, and poured asphalt or tar have also been much used, the major objections being that they squeeze out, making an objectionable ridge; moreover, the ridge material does not flow back when the joint reopens. Dirt and moisture infiltrate, the joint eventually becomes clogged, and its effectiveness is destroyed. Metal joints with enclosed air spaces fail to survive for long the effects of the repeated stress and severe tearing and folding they have to undergo. Although often costly, expansion joints for buildings and bridges can be made to function with more assurance of continued satisfactory service than has yet proved possible for pavements. It is established practice to provide premolded strip material or its equivalent at all junctions between any new concrete and a marginal restraining body such as a building, a curbing, an intersection, or a bridge.

Control Joints. The *dummy joint* described under *contraction joints* is one form of so-called *control joint.* Others are joints placed between large power units for purposes of isolating vibration or localizing possible damage; joints placed where appreciable changes in cross sections occur (openings in walls, etc.); joints at junctions between parts built on foundations having different bearing values; and joints at angles between major portions of buildings, as at the junctions in L-, T-, H-, and U-shape structures. The location of *control joints* involves both architectural and engineering considerations. Current practice dictates that large, independent, jointless concrete units should not exceed about 60 ft in length. Adjacent sections of bridges and buildings are frequently built up independently from separate footings about 1 in. apart.

Hinges and Hinge Joints; Articulations. The longitudinal center joint commonly used on a concrete pavement is essentially a *hinge joint.* Deformed bars are bonded across the joint interval at mid-depth to hold the adjacent slabs in close contact while permitting a slight rotation of one with respect to the other. A load-transferring alignment is usually accomplished by a longitudinal header-shaped key within the middle third of the slab depth. Bonded dowels can be used in lieu of the tie bars and edge-groove arrangement. Another type of hinge or articulation is the kind needed for transfer of heavy thrust from one body of concrete to another where slight relative rotations of the two masses may or should occur, as in a hinged concrete arch. The Considere parallel-bar and Mesnager crossed-bar hinges are illustrative (*Proc. ACI*, vol. 31, pp. 304, 368, 1935).

CURING

Definition of Curing. Curing includes all measures essential to assist the freshly placed concrete to develop the strength and hardness prerequisite to service. This implies protection against premature loading (accidental or applied) and any adverse chemical effects as well as the maintenance of moisture and temperatures favorable to *hydration.* Formal curing measures are usually discontinued as the concrete goes into service, but actual curing continues at a decreasing rate for an indefinite period if moisture and temperature conditions continue favorable. Although interrupted curing (through drying out or low temperature) will resume as favorable conditions are restored, it is usually the curing prior to service that is of major importance, since premature service may produce physical disruption that subsequent curing cannot remedy.

Curing Methods. In Figs. 5 to 11 are shown representative extents to which the environmental curing factors of moisture and temperature influence the rates and magnitudes of strengths developed in relation to time. In practice there are four general curing techniques for ensuring an uninterrupted supply of free moisture:[1]

[1] Methods 2 and 3 are not generally authorized for an arid region or other drying environment.

1. *Direct application of water* by providing moist coverings of burlap, straw, or earth, or by sprinkling at frequent intervals or ponding. Immersion or fog-room curing falls in this class, as does covering with impervious paper, to minimize evaporation loss, and wetting down under the paper at short enough intervals to keep surfaces moist.

2. *Retention methods;* retaining the mixing water, to meet the curing need, by spraying the exposed horizontal surface of the concrete with "curing or sealing compounds" that prevent or retard loss of internal moisture by evaporation. Immediate application (before the moisture has left the surface) is essential. Less fully effective is leaving the forms in place, which is sometimes possible.

3. *The use of a layer of deliquescent chemicals* (usually flaked calcium chloride at about 1.5 lb per sq yd of exposed surface). This is possible only on surfaces more or less horizontal. Although it does not supply water for the concrete it does limit or prevent evaporation; it too is essentially a "water-retention" method.

4. *High-pressure steam curing,* suitable for certain precast concrete products, as mentioned under precast products.

Termination of Formal Curing; Precautions. Removal of forms and placing the concrete into service need to be intimately related to the progress of hydration. Even with moisture present, hydration is retarded by low temperature (see Fig. 9) and virtually halted a few degrees below the freezing of water. Important elements of judgment are involved especially where results from tests on drilled cores or other form of job-control specimens are not available. Form removal can best be decided on the basis of a strength criterion; but in the absence of flexural or compressive control tests the strength may be variously estimated by such devices as "ball" surface hardness tests, a carbon-paper-ball impact test, the Swiss Hammer, or bond pullout tests with bars having part-length embedment (see "Strength").

Where walls are to be rubbed, the rubbing should be done when the concrete has attained a strength of about 400 psi, which may be as early as 12 to 24 hr after placing; but, in general, forms should remain in place from 2 to 4 days at a minimum, and any flexural members supporting their own weight or other loads should have continuous support as long as required to develop either full design strength or a factor of safety of 2 against whatever load will be imposed. Highway slabs are usually given a minimum of 10 days moist curing from time of casting. An ordinary sidewalk slab could be placed in service as soon as there is enough surface hardness to resist abrasion and corner spalling (perhaps a couple of days); the magnitude of the loading is negligible; intermittent curing will continue indefinitely during the life of an outdoor sidewalk.

In housed-in cold-weather building construction stoves or heaters are often used to maintain a favorable temperature. It is important that ample moisture (by sprinkling, moist cover, or live steam) be supplied in such cases, since otherwise the artificial heat dries out the concrete, halting curing regardless of temperature.

In the use of the impervious membrane or cover, if there is sun exposure, the membrane should be of light color; a dark color draws the heat and may produce serious surface checking. Any curing membrane may be unsightly and is likely to prevent bonding with plaster and/or a mastic joint filler. At the discontinuance of formal curing the concrete should not be subject to thermal shock by exposure to an environment greatly different in temperature, thereby creating objectionable differential strains.

WATERPROOFING AND PAINTING

Waterproofing; Integral; Membrane; Vaporproofing; Condensation. The passage of water through concrete, except under high heads, is due solely to a harsh or poorly designed mixture or poor placement (including laitance or other poor bonding between adjacent pours). Physical disruption (cracks) will let water through regardless of concrete quality. Many admixtures have been advertised and sold as *integral* water-

proofing but, however good the results secured, equally satisfactory results could invariably have been attained by placing the concrete without admixture with as much care as was used with it. The thin-shelled but impermeable hulls of the concrete ships of World War I, of the ocean-going barges and self-propelled concrete "cargo carriers" of World War II, and of concrete barges operating for many years in both salt and fresh water support the statement if support is needed. More striking perhaps, but not surprising, is the fact that the world's longest pontoon bridge (the Lake Washington floating bridge at Seattle, 1939) has shown no leakage. Yet this structure is composed of floating boxes each 350 ft long, 59 ft wide, and 14 ft deep with the reinforced concrete bottom and side walls only 8 in. in total thickness.

Although any reasonably good uncracked concrete is proof against *percolation* and *seepage*, it is more difficult to make concrete proof against passage of *capillary* moisture and *water vapor*. The use of *membrane* waterproofing (heavy tar paper or tarred felt) on the water side is about the only sure way of preventing *capillary dampness* on the "dry" side of the concrete. Such dampness is usually invisible but can be readily detected (*Proc. ACI*, vol. 43, p. 914, 1947). The membrane is also sure protection for *blockwall construction* exposed to moisture on one side and against cracks in the concrete. Tar and asphaltic coatings have long been used on the back sides (the water sides) of retaining walls and abutments but, for an exposed face, are usually unsatisfactory in appearance.

Sometimes, especially in newly constructed basements where the outside soil temperature is still low, concrete walls and floors appear to leak from the free moisture that is merely surface condensation from the air.

Cement Paints. "Portland-cement paint," variously known as "portland-cement base paint" or "portland-cement water paint," refers to water-dilutable paints in which portland cement is the binder. "Commercial paint" is factory mixed, ready for the addition of the water; "job-mixed paints" are proportioned by the user. As applied they are essentially a thin portland-cement grout, the solid content being 65 to 100 percent portland cement. Optional secondary constituents are hydrated lime not exceeding 25 percent, water repellents (calcium or aluminum stearate) under 1 percent, and hygroscopic salts [calcium or sodium chloride not exceeding 5 percent and titanium dioxide (TiO_2) or zinc sulfide (ZnS) not exceeding 5 percent]. White portland-cement paint has a somewhat different range of optional constituents, the white-portland cement ranging from 20 to 91 percent of the total solids (*Proc. ACI*, vol. 38, p. 485, 1942).

Portland-cement paint is applicable to concrete, stucco, and masonry surfaces, except floors and other areas subject to abrasion. Such paints are used for decorative effects and also to reduce water permeability of porous open-textured masonry such as block construction. They are of questionable value for stopping leakage through porous walls exposed to water pressure, when the paint is applied on the side away from the water pressure, in spite of sweeping advertising claims frequently made by manufacturers whose products are essentially "portland-cement paints." Severe conditions usually require more positive waterproofing methods such as an *impervious membrane* on the side of the pressure or "*plastered on*" coats on the side away from the pressure.

With proper precautions as to composition, mixing, application, and curing, portland-cement paints are durable and satisfactory. The success of the job often hinges on the doing; the wall must be clean and free of old oil paint or any residue of form oil, and the paint (being essentially portland cement) must be cured and should be kept continuously moist for at least 48 hr after it has hardened enough for sprinkling.

Oil Paints on Concrete. Pretreating concrete surfaces with a 2 percent zinc chloride and 3 percent phosphoric acid in water has been found to remove much of the difficulty in getting a durable paint job with oil paints on concrete in which there was any free

moisture or alkali at or near the surface. Oil paints on properly pretreated surfaces give as long service on concrete as they do on wood (*Proc. ACI*, vol. 43, p. 1077, 1947). It is the damaging effects of saponification and osmosis that are to be eliminated. For old dry interior surfaces oil paints usually give excellent results without pretreatment. Any dampness back of an oil or other impervious paint results in blistering and nondurability.

ADMIXTURES [1]

Pro and Con Aspects. To an ever-increasing extent concrete admixtures are assuming warranted importance. When the use of an admixture is contemplated certain facts should, however, be kept clearly in mind; (1) The admixture constitutes an extra material that must be purchased, handled, batched and controlled. (2) For most admixtures very slight deviations in amount produce important changes in the concrete, a slight excess sometimes being highly damaging. (3) When used in even the recommended optimum amounts the characteristics of the concrete may be altered in various respects, not all of them beneficial. (4) Being usually of a proprietory nature many admixtures are mere nostrums widely advertised with sweeping claims not easily proved or disproved. The potential user will do well to maintain an attitude of conservative, skeptical objectivity.

In general then the use or nonuse of an admixture should be decided on the basis of the added cost and degree of uncertainty and unbalance involved as weighed against the importance of the desired benefits and the assured degree of uniformity with which use of the admixture constitutes the best or the only practicable method for attaining the benefits. Under proper control recognized air-entraining admixtures can be considered to have an established status as do some of the other admixtures in a lesser degree. Many, however, are marginal, nonbeneficial, or definitely injurious.

Classification. As here used the term *admixture* includes any constituent of plastic concrete other than cement, water, and aggregate. It may reach the concrete as an integral part of the cement—an interground *addition*—or as an *admixture* introduced at the mixer. This discussion draws heavily upon a report of ACI Committee 212 (*Proc. ACI*, vol. 51, p. 113, Title No. 51-5, 1954–1955). The report covers the subject in considerable detail and supplies a list of 55 pertinent references. Admixtures are classified as:

1. Accelerators.
2. Retarders.
3. Air-entraining agents.
4. Gas-forming agents.
5. Cementitious materials.
6. Pozzolans.
7. Alkali-aggregate expansion inhibitors.
8. Dampproofing and permeability-reducing agents.
9. Workability agents.
10. Grouting agents.
11. Miscellaneous.

A single admixture intended to perform multiple functions may appear under more than one heading.

1. *Accelerators.* Accelerators are added to increase the rate of early strength development (to speed removal of forms, reduce required period of curing, compensate for retardation of low temperature, reduce "time out" for emergency repair and other

[1] For an excellent discussion of admixtures see DELMAR L. BLOEM, *NRMCA Pub.* 65, "Admixtures for Ready-Mixed Concrete," National Ready-Mixed Concrete Association.

work). Often the choice lies between use of an accelerator and use of either high-early-strength cement or a richer mixture using ordinary cement. Chemical accelerators are calcium chloride; some organic compounds such as triethanolamine; and some soluble carbonates, silicates, and fluosilicates. *Calcium chloride* ($CaCl_2$) is the most widely used accelerator. When used not in excess of about 2 percent of the weight of the cement (say 2 lb per sack) it increases the early strength and provides a slight protection against freezing temperatures. In larger amounts the strength is substantially reduced. *Table salt* (NaCl), even in small quantities, is detrimental to both strength and durability. Few members of this group, other than calcium chloride, have proved their merit, and some may be positively injurious. Alumina and portland cements mutually accelerate one another and in some cases produce flash set. In general, accelerators should be used only on the basis of competent technical advice or as a result of experiment or experience with the admixture as an accelerator. See also item 11.

2. *Retarders.* Retarders may be used to offset the accelerating effect of temperature from hot-weather concreting or hot-water flows in grouting, to prevent the premature stiffening of some cements, or actually to delay the stiffening under difficult placing conditions. In grout-cement slurries where the grout must be pumped a considerable distance or where it may be necessary to redrill grout holes, retarders are often useful. Sometimes retarding solutions are applied to the forms to inhibit the set of a surface layer of cast concrete so that it can be readily removed by brushing to expose the aggregate for unusual surface-texture effects.

Among the wide variety of chemicals alleged to retard the setting time of portland cement there is much variability and uncertainty of action. Some retard some cements and accelerate others; some retard in certain amounts and accelerate in different amounts. The effects on the properties of the concrete are not well known, although some reduction in strength usually accompanies the use of organic retarders. Some retarders, used in certain proportions, appear, however, to have no adverse, even a beneficial, influence on the strength of the concrete. In general, a retarder should be used only on the basis of competent technical advice or, better yet, of advance experimentation with the cement and other materials under conditions approximating those of the proposed use. The more commonly known retarding admixtures, such as carbohydrate derivatives (includes sugars) and calcium lignosulfonate, are employed only in small fractions of a percent by weight of the cement.

3. *Air-entraining Agents.*[1] Air-entraining agents have assumed great importance since about 1940, primarily from the standpoint of pavement durability against alternations of severe cold weather and the injurious action of salts used in snow removal. The action of an air-entraining agent is that of a foam or froth stabilizer. The phenomenon is physical rather than chemical, and the air entrainment is largely, if not entirely, through interaction with the sand rather than with the cement. The maximum of air entrainment occurs for sand between the 30 and 100 mesh sieves. One-size sands coarser than No. 14 or finer than No. 100 show practically no air entrainment with any of the currently tested air-entraining agents as normally introduced. The fine air bubbles seem able to cluster about sand grains that fall only within a certain limited range of particle size. Although the air entrainment takes place in the mortar, the presence of coarse aggregates or other non-air-responsive size fractions including the cement dilutes the extent of, but does not inhibit, air entrainment. Because of its favorable effect on workability and texture of the concrete, air entrainment is being frequently extended from pavement concrete to construction where durability is not a serious problem. *Vinsol resin* and *Darex* are covered by ASTM C175 and C205 when introduced as "additions" interground with the clinker. Many other organic compounds,

[1] See also "Design of Concrete Mixtures."

such as natural resins, tallows, oils, and soaps, also impart air-entraining properties if interground in amounts ranging from 0.005 to 0.05 percent by weight of the cement. Water-insoluble resins, fats, and oils are not in themselves foaming agents but depend upon a saponification reaction with the alkali constituents of the cement for development of the foaming or foam-stabilizing property; this necessitates intergrinding. Water-soluble agents, such as "neutralized" Vinsol resin (a sodium hydroxide solution of Vinsol resin) or alkali-metal salts of sulfonated oils or fatty acids, are in themselves foaming agents and can be either interground or added at the mixer.

Under proper control, air-entraining agents introduced as admixtures, rather than as additions, offer the advantage of flexibility in the amount to be used, which is not true when the agent comes interground with the cement. Extreme care must be taken, however, to avoid use in excess of amounts recognized as proper, or the strength of the concrete may be greatly reduced. It should be kept in mind that an air-entraining agent in aggregate without any cement can produce a mixture that is workable and fat, having the appearance of rich concrete (*Proc. ACI*, vol. 40, p. 516, 1944). Due precautions must be taken to ensure that the admixture is uniformly distributed throughout the batch, since very small amounts of the actual agent have great potency. For work not under close supervision and control the use of "air-entraining cement" (ASTM C175 or C205) is likely to be simpler and safer.

Some air-entraining agents have little or no effect on the hydration of the cement; others do, usually retarding it. The percentage of entrained air in concrete should range between 3 and 6, producing a reduction of from 4.5 to 9 lb per cu ft in the weight of the concrete.

By using air-entrainment up to 30 to 60 percent by volume for structural concrete and 70 to 85 percent for insulation concrete, with both lightweight and conventional aggregates, concretes can be produced that range from 20 to 120 lb per cu ft. The air is whipped into the mass by rapid agitation in conjunction with the addition of such air-entraining agents as sodium lauryl sulfate, alkyl aryl sulfonate, and certain soaps and resins. The quantities added are greatly in excess of the near-infinitesimal amounts that result in the usual air-entrained concretes. Such cellular foam concretes have been widely used in Europe. Autoclaved products weighing 35 to 50 lb per cu ft have been reported with strengths from 500 to 850 psi and concretes of from 70 to 90 lb per cu ft have shown strengths up to 2,000 psi. Another foam process uses agents of such types as the hydrolyzed waste protein commonly employed to combat gasoline fires. The foam is added to the mixture in place of conventional aggregate.

4. *Gas-forming Agents.* The leading gas-forming agent is *aluminum powder*, which reacts with the hydrating hydroxides in concrete to permeate the mass with minute *hydrogen* bubbles much as air-entraining agents do with *air*, but the timing of the action differs. The unpolished powder, usually preferred, sometimes called *granulated* aluminum, is that commonly used for calorizing and for pyrotechnics and explosives. (The term *granulated* is also applied to larger pellets marketed for metallurgical purposes.) The pellets are spherical and free from grease. For slower action the polished powder, much used as a pigment in paints, may be preferable. It consists of fine flakes (mostly −300 to −400 mesh) having a high specific surface. It is made by a stamping process and is revolved in a drum with a lubricant to give it luster. Aluminum powders ignite readily. Amounts range from 0.005 to 0.02 percent by weight of the cement except when larger quantities are used to produce the lightweight, low-strength, sound or heat insulation filler concrete known as "aerocrete."

When controlled to occur through the proper interval of time and with varying amounts of outside restraint, a slight swelling can be produced without appreciable weakening. Such swelling may force the concrete to a better contact with reinforcing steel, or it may prevent settlement in an engine block or a base under an existing

machine; it may also prevent or reduce segregation. Temperature is one of the factors to be controlled, the action being slower in cool weather. The swelling should be timed so as not to be disrupting after setting is under way; in hot weather the expansion phase may occur so rapidly as to be passed before the mixture stiffens (*Proc. ACI*, vol. 39, p. 165, 1943).

At normal temperatures the aluminum reaction starts with mixing and may continue for from 1½ to 4 hr. Above 90°F the reaction may be completed in 30 min and subsidence may occur prior to initial set. At 40°F the reaction may not be effective for several hours. To produce equal expansions about twice as much aluminum is required at 40°F as at 70°F. To prevent the small amount of the powder (about one teaspoonful per sack of cement) from floating on the mixing water it is well to premix it with a fine sand or pozzolan. Zinc and magnesium powders are also used, producing oxygen instead of hydrogen gas.

Cellular blocks similar to the foam air-entrained process blocks can be made by using up to ¼ lb aluminum powder per sack or equivalent.

5. *Cementitious Materials.* Natural cementing materials include *natural cements, hydraulic limes, water-quenched blast-furnace slag,* and mixtures of *blast-furnace slag and lime.* These may be substituted for 10 to 25 percent by weight of the portland cement, even larger amounts when low strengths are not objectionable. The effects may be to increase the workability (more pronounced for the harsher mixtures), decrease the bleeding and segregation, decrease the heat of hydration, and usually decrease the strength. Some contribute to the strength of the concrete through their own chemical activity. Often such materials require additional water and promote greater shrinkage of the hardened concrete as the water evaporates. In general, these natural cementing materials have little significant effect on frost resistance unless an air-entraining agent is present; they usually require a longer curing period for the development of their potential strength.

6. *Pozzolans. Pozzolanic materials,* finely divided *siliceous* and *aluminous* substances, not cementitious in themselves, combine with hydrated lime and water to form stable compounds of cementitious value. They may be added as substitutions for 10 to 35 percent of the cement in large hydraulic structures to lower the heat of hydration and/or to supply increased resistance to sea water, sulfate-bearing soils, reactive aggregate, or natural acid waters. The specific gravity is lower than that of cement, and substitution by equal weights increases the relative bulk of fine material, thereby improving workability and reducing bleeding and segregation. The effects are more pronounced in lean mixtures than in rich ones. The extent of the influence on the concrete varies greatly for different pozzolanic materials. The rate of strength gain is slower, but under continued favorable curing conditions the later strengths are higher with most of the pozzolanic admixtures. The more common materials in this class are *fly ash* (fine flue dust, a by-product of power plants that burn powdered coal), *volcanic ash,* heat-treated *diatomaceous earths,* and heat-treated *raw clays* or *shales.* Under the title "Cement; Portland, Pozzolana" Federal Specification SS-C-208a gives requirements for cements containing pozzolan; also ASTM C340.

7. *Alkali-aggregate Expansion Inhibitors.* It has been found that certain natural and artificial pozzolans used as cement replacements (from 20 to 35 percent by weight) give excellent protection against deleterious alkali-cement reactions. This device has been used for several of the large dams constructed since 1945 (see "Blends and Substitutions").

8. *Dampproofing and Permeability-reducing Agents.* In concrete exposed to moisture on one face and to air or drying on the other, there will be a slow moisture travel comparable with that which occurs from soil to air in trees and vegetation. The air side may seem dry but the humidity adjacent to the face is raised. Objects resting on a

"dry" basement floor usually show moisture on the underside—an indication that moisture did penetrate the concrete.[1]

Properly proportioned concrete, well placed and cured, is highly resistant to the passage of water under pressure. In general the passage of water (not of the capillary dampness just discussed) will be appreciable only through cracks or discontinuities—something not helped by water repellents or other so-called integral waterproofers. If a mixture is harsh or lacking in fines some of the finely divided materials (either inert or cementitious) may improve the grading and workability, thereby contributing to watertightness but in general the addition of a little cement would accomplish the same result. Anything that decreases the size, number, and continuity of the capillaries is beneficial. This includes decreasing the water-cement ratio, air entrainment (the small air bubbles are discontinuous), and favorable curing (hydration products close and seal off many of the capillaries).

Among the more effective of the integral "waterproofers" are: soaps (inorganic salts of fatty acids such as calcium or ammonium stearate or oleate) 0.2 percent or less by weight of the cement, butyl stearate about 1 percent, heavy mineral oil (free of fatty or vegetable-oil components), and cutback asphaltic oils up to 5 percent.

Other proprietary powders, pastes, and liquids include various combinations of (1) barium sulfate, calcium, and magnesium silicates and fatty acid; (2) finely ground silica and naphthalene; (3) colloidal silica and a fluosilicate; (4) petroleum jelly and lime; (5) cellulose materials and wax in an ammoniacal copper solution; (6) silica, lime, and alum; (7) coal tar cut with benzene; and (8) sodium silicate with an organic nitrogenous material. Most of the materials of this paragraph reduce the strength somewhat without being particularly effective as dampproofers.

Calcium chloride seems to be ineffective as a waterproofing agent except in so far as it accelerates early curing or tends to offset strength reductions from other agents.

"Waterproofed" portland cements and masonry cements contain water-repellent materials introduced in an effort to decrease the capillary rise or travel of the moisture that produces discoloration and efflorescence upon evaporation from exposed surfaces. In amounts from 0.1 to 0.2 by weight of cement the repellents increase the workability, decrease the strength, and decrease capillary absorption but increase permeability under high head.

9. *Workability Agents; Finely Divided; Air Entraining.* Workability agents are usually employed to offset deficiencies in grading that tend to produce harshness or segregation and jeopardize successful placement under inaccessible difficult conditions. Often a simple redesign of the mixture with special attention to the ratio of fine aggregate to coarse aggregate and perhaps to the grading (especially of the fine aggregate) accomplishes all that is required. Other times the use of more cement as an admixture may be the logical solution. The advantages accruing from increased cement content are improved workability, increased strength, impermeability, and frost resistance. Disadvantages are increased heat development, increased volume change, and increased surface crazing, especially if the placed concrete receives much surface manipulation.

Finely divided materials may be added to mixtures deficient in fines (particularly the material passing the No. 200 sieve). In general, the higher the specific surface of the material the less is the amount required to produce a given effect on workability. When a proper amount of the mineral powder is used, there should be no increase in total water content of the concrete and some increase in strength with little effect on drying shrinkage and absorptivity of the hardened concrete. Adding such a mineral

[1] Where the soil or moisture side is cooler than the air (as in a new basement) visible moisture is likely to be due primarily to condensation from the air rather than to the invisible capillary travel from wet side to dry side. An impermeable moisture barrier will block the capillary travel but will decrease condensation only to the extent that it reduces the temperature differential between the exposed surface and the air adjacent to it.

powder to a concrete not deficient in fines, particularly if rich in cement, generally decreases workability for a given water content (stiffens the mix), and in such mixtures the increase in total water required may result in an increase in drying shrinkage and absorptivity and a decrease in strength.

Examples of materials added in amounts not exceeding 3 to 5 percent by weight of cement are *bentonite clay* and *diatomaceous earth*, added in larger amounts are *fly* (flue) *ash*, *finely divided silica*, *clay*, *fine sand*, *hydrated lime*, *talc*, and *pulverized stone*, some of which may be added in amounts up to 20 percent.

Water-reducing agents marketed under proprietary names increase the slump corresponding to a given water content, decreasing the water-cement ratio corresponding to a given slump. These materials may contain sulfonated organic compounds producing some air entrainment with the corresponding improved workability, especially for the leaner mixtures. Others contain carbohydrate salts in combination with other active or inert materials, cause no air entrainment, do not reduce bleeding, and produce some retardation of set. When advantage is taken of the possible water reduction, the hardened concrete usually has some added strength and impermeability, and for the types that give some air entrainment there is increased resistance to freezing and thawing. As for other chemically active admixtures, the specific effects vary with cements and, before use, tests should be made with the cement and materials that will be used on the job.

Regular air-entraining agents so markedly improve the workability as to require mention as a workability agent.

10. *Grouting Agents.* (See also "Oil-well Cements.") Increasingly grouts are used over a widely diversified range of applications to meet exacting and highly specialized situations. Some of the representative requirements are:

1. Sustained cohesiveness—freedom from segregation (bleeding)—placing quality.
2. Acceleration—rapidity of set or early hardening where plugging action is required.
3. Retardation—ability to remain cohesively fluid during long placing periods and/or under exposures to great heat and extreme pressure.
4. Strength and watertightness of the hardened product.
5. Expansiveness when required as filler in porous formations or machine setting operations.

Strength, watertightness, and cohesiveness are promoted by maintaining extreme care that the water-cement ratio be the lowest possible for the conditions of placement. If sand is used in the mixture, the grading and amount must both be right. Air entrainment contributes greatly to the cohesiveness and flowability but is possible only if sand is present. For either sand-cement or neat-cement grouts such materials as fly ash, bentonite, pumicite, and diatomaceous earth may be used to improve pumpability. Bentonite aids in holding particles in suspension. Gels, clays, soybean protein and sodium hydroxide, pregelatinized starch, methyl cellulose, and mixtures using clay and ferric oxides have also been employed. Accelerators, retarders, and gas-forming agents may be used as previously mentioned. In general, each grouting problem calls to an usual degree for care, experimentation, and seasoned judgment.

11. *Miscellaneous Admixtures.* *Coloring pigment* should always be of inert mineral composition.

Calcium chloride or other salts may be admixed to lower the freezing temperature, but the amounts required for this purpose are weakening and injurious to the concrete. Kept within the 2 percent specified for use as an accelerator, calcium chloride hastens the hardening of the concrete, thereby decreasing the period over which special protection is required. Greater percentages than 2 percent should not be used. *Other salts* are likely to lower strength and should be used only on the basis of competent advice or actual experiment.

Various other admixtures are marketed as *integral floor hardeners,* to increase resistance to wear, decrease dusting, and serve as "pore fillers." Among such are *fluosilicates* of *magnesium* and *zinc, sodium silicate, gums,* and *waxes.* At best these give but temporary relief for a poorly designed mixture, inferior workmanship, or a poorly surfaced floor inclined to dust. The common practice of troweling in neat cement to "dry up" the surface of a floor or walk for faster finishing promotes dusting and surface crazing. The practice is detrimental and should not be permitted.

As grinding aids very small amounts of tallow or resinous material such as TDA or 109B (ASTM C150) are interground with cement clinker. In the quantities permitted they do not alter the cement except indirectly by facilitation of finer grinding and adding slightly to the air-entrainment tendency. Similar substances in slightly greater amounts are the usual air-entraining agents (see "Air-entraining Agents").

EFFECT OF VARIOUS SUBSTANCES ON HARDENED PORTLAND-CEMENT CONCRETE[1]

See also "Defective and Reactive Aggregates"

Since concrete is often a logical or convenient material for the construction of reservoirs, vats, and storage tanks, it is important to know something of its vulnerability to various substances.

Acids and Miscellaneous Substances. Briefly, good concrete is not affected adversely by petroleum oils and is almost immune to coal-tar distillates, but there is some loss from penetration by the lighter or more volatile fuel oils such as kerosene, benzine, naphtha, and gasoline. It is distinctly vulnerable to such inorganic acids as sulfuric, nitric, sulfurous, hydrochloric, and hydrofluoric, and for these concrete requires special linings. Concrete disintegrates slowly under acetic, carbonic in water, lactic or tannic, milk, and silage juices. It is very slightly attacked by fish oil, lard, foot oil, linseed, resin, coconut, olive, rapeseed, cottonseed, almond, poppy seed, walnut, soybean, peanut, chlorides of iron, mercury, copper, and ammonia, sulfide ores and pyrite, molasses, and sulfite liquor.

Concrete is also actively attacked by sulfates of calcium, potassium, sodium, magnesium, copper, zinc, aluminum, manganese, iron, nickel, cobalt, and ammonia; by nitrate of ammonia and acid sulfate. There is no attack by chlorides of sodium, potassium, calcium, strontium; nitrates of calcium potassium and sodium; soluble sulfides (except sulfide of ammonia), carbonates, fluorides, and silicates. There is also no attack from ammonia water, wood pulp, tanning liquors (nonacid), and alcohol.

Salts (Sodium and Calcium Chlorides). The progressive surface scaling that usually results from salt contacts with concrete (especially non-air-entrained concrete or concrete having a water-cement ratio in excess of about 5½ gal per sack, or which is honeycombed, poorly consolidated, or insufficiently moist-cured) is essentially a physical rather than a chemical phenomenon. When drying occurs after salt in solution has penetrated the surface, the salt crystallizes and expands. The crystal expansion is similar to that of freezing water and the result is the observed surface spalling (scaling). Salts do corrode metals, and insufficiently protected reinforcement or wire ties or metal chairs will rust and spall the concrete progressively along the surface of the metal.

Sea Water. What has been said about salts in general applies to sea-water exposure, especially for concrete located in the water-line zone or spray area. More often than not the damage is via the corrosion of reinforcement rather than directly as chemical damage to the concrete. All metal in concrete exposed to sea water should have a minimum cover of 3 in. with 4 in. near corners. Where fresh water for mixing is not available, sea water can be used with a probable 10 percent or so reduction in strength.

[1] See PCA leaflets ST4-2, ST10, ST4, ST6, ST19, and ST1-2 for a more extensive coverage of these and related topics.

In good concrete the mixing water has only a limited nonrenewable contact with metal surfaces and the sea-water corrosion hazard seems to be negligible. When necessary, sea water can be used successfully for curing the concrete, which should be kept continuously wet during the period, rather than permitted to be alternately wet and dry, thereby promoting salt-crystal formation (see PCA Leaflet ST7-4).

Sugar. Dry sugar is not harmful to hardened concrete and only a slight adverse effect results from alternate wetting and drying with sugar solutions if the concrete is of good quality, including low absorption. As an ingredient of plastic concrete sugar is most harmful, as little as ½ pt per sack of cement preventing hardening altogether. Limited reported tests have indicated a strengthening of a lime mortar by a sugar addition but for portland-cement mixtures there is a very serious weakening and no benefit (see "Defective and Reactive Aggregates"; also PCA Leaflet ST-10).

Organic: Manure and Other Farm Wastes; Tannic Acid. When present in the plastic mixture these materials are highly injurious, retarding or nullifying the hardening action and contributing to disintegration. Their effect on hardened concrete of good quality is almost negligible. It is inadvisable, however, to place manure or even old straw directly on newly placed concrete as is sometimes done for temperature and/or moisture retention. If such covering constitutes the best protection that is economically available an impervious layer of tar paper or equivalent should be interposed. Fresh concrete is invariably more subject to attack than that which is well cured and seasoned.

Efflorescence; Stains. Water passing through masonry and drying at the surface invariably leaves the discoloring deposit of white salt crystals known as efflorescence. Dense concrete protected from water circulation is the best preventative. While the calcium carbonate crystals are practically insoluble in water, the efflorescence is easily removed by a wash of 1 part hydrochloric (muriatic) acid to 5 or 10 parts of water. Walls should be thoroughly wetted before the wash is applied and thoroughly washed immediately thereafter. The first trial should be on an inconspicuous section of the wall or other surface (PCA Leaflet ST-6 covers efflorescence; stains in general are covered by ST-1).

Contact with Ferrous Metals. There is no significant chemical reaction, adverse or otherwise, between portland-cement mixtures and the ferrous metals. Under certain conditions a thin adherent layer of neat cement paste provides an excellent protective surface against corrosion, and as a lining in a cast-iron pipe, a cement grout reduces hydraulic friction and the formation of obstructive tubercules. Certain favorable physical interactions combined with the lack of adverse chemical relationships make reinforced concrete possible.

In a moist or other corrosive environment ferrous metals are vulnerable to progressive rusting (the formation of iron oxides of various compositions) in layers that are not sufficiently adherent to protect the metal against ever deeper penetration and eventual destruction. Good-quality, well-placed concrete supplies complete protection against rust. The same enveloping concrete through the combination of adhesion, friction, and lug action known as *bond* establishes the unity that enables the concrete to transfer tensile and compressive strain to the steel (see "Bond Strength"). If the layer of protective concrete is inferior or too thin to prevent access of moisture the steel will rust. In rusting it swells, thereby cracking and spalling off more of the protective layer. The deterioration of reinforced concrete under severe exposure such as that of sea water is often due initially to some one small area of inadequate steel protection (see "Salts;" "Sea Water").

Again under exposure to intense heat the steel is much less resistant than the concrete, the steel strength dropping off rapidly for temperatures above 500°F. With adequate concrete cover, say 1.5 to 2.0 in. (depending upon type of member and the fire hazard assumed) steel can continue to function for considerable periods.

Properly coordinated in their respective functions, concrete and the ferrous metals perform admirably and are free from some of the aspects that require special attention in the case of leading nonferrous metals.

Contact with Nonferrous Metals. *Copper, Lead, Zinc, Aluminum.*[1] Sheet copper and lead are used for flashings, roofing, and dams or diaphragms in joints. Lead is also used for linings of tanks or digesters, for cable sheathing and nuclear shielding. Sheet zinc or zinc-coated (galvanized) ferrous sheet is used for downspouts, leaders, and eaves troughs. Aluminum is used for tanks, coils, pipelines, linings for concrete vats, house siding, paneling, and window and door frames. In many of these adaptations the metal is embedded in concrete, wholly or in part, or is in contact with it.

Copper. Being practically immune to caustic alkalies, copper is little affected by lime or solutions of calcium hydroxide and is little affected by either fresh or hardened concrete whether dry or saturated, provided leaching does not bring chlorides in contact with the metal. Chlorides can be present in concrete only through the use of an admixture having a chloride base. Such admixtures should be avoided where copper and concrete are to be in contact especially if they are to be exposed to moisture.

Lead. In contact with green, wet concrete the moist calcium hydroxide (free lime incidental to hardening) will corrode the lead. This happens only during the curing period and is not serious if the lead is sufficiently thick to allow for the slight depletion of cross section. Lead that is to be in contact with green concrete will best be protected by a coating of asphalt, varnish, pitch, or a wrapping of jute saturated with bitumen. Cured or seasoned concrete rarely reacts with lead. If, however, a strip of lead is but partially embedded, with a portion exposed to the air, a form of electrolysis known as differential aeration occurs which leads to the gradual disintegration of uncoated embedded lead.

Zinc. Solutions of caustic alkalies such as the calcium hydroxide of green concrete and mortar will form a compact filament of calcium zincate which firmly bonds the zinc to the concrete. The adherent film is thin and protects the underlying metal from further attack. A coating of asphalt, varnish, or pitch will prevent the action, however. Pitting may occur where the corrosion products from contact with moist unseasoned concrete are removed by such factors as rain or abrasion. Under such conditions the metal should be protected. In contact with dry, seasoned concrete zinc does not corrode. Flat or corrugated sheets of galvanized iron (or other sheet zinc) should not be used for form lining since the cement paste will adhere, leaving a rough surface on the concrete. Heavy galvanized sheet is satisfactory for mixing pans, trowels, and slump cones if always washed before the concrete hardens on them.

Aluminum. Caustic alkali, the calcium hydroxide, reacts to form calcium aluminate. Aluminum is attacked by fresh unseasoned concrete more rapidly than is lead or zinc. The corrosive products are nonadherent and do not protect the underlying metal. This makes the attack progressive and of either a blister or pit-forming nature. It will be most severe in deformed areas where the metal has been bent, twisted, or cold-worked. Aluminum mixing pans are short-lived—they develop holes. While dry seasoned concrete does not corrode aluminum, there is usually a possibility of occurrence if the concrete becomes wet. In general it is well to protect contact surfaces, such as those of window frames, as well as embedded portions, by a bituminous paint, asphalt, varnish, or pitch.

NOTE: Uncoated dissimilar metals should not be placed in contact or embedded in close proximity in moist concrete unless experience has shown that no destructive galvanic action will occur.

High-alumina cement concrete is more resistant to some substances than is portland-cement concrete (see "Alumina Cement").

[1] Excerpted mainly from Portland Cement Association literature.

Selected References on Plain Concrete

1. Bogue, R. H.: "The Chemistry of Portland Cement," 2d ed., Reinhold Publishing Corporation, New York, 1955.
2. Lea and Desch: "The Chemistry of Cement and Concrete" (revised), St. Martin's Press, Inc., New York, 1956.
3. Witt, J. C.: "Portland Cement Technology," Chemical Publishing Company, Inc., New York, 1947.
4. Brady, George S.: "Materials Handbook," 8th ed., McGraw-Hill Book Company, Inc., New York, 1956.
5. Kirk and Othmer: "Encyclopedia of Chemical Technology," vol. 3, Interscience Publishers, Inc., New York, 1949. Bogue, R. H.: Cement, Structural; Portland Cement, pp. 411–431. Miner, J. L., and F. W. Ashton: Cement, Structural; Calcium-aluminate Cement, pp. 431–435. Fink, G. J.: Cement, Structural; Magnesia Cement, pp. 435–438. Gilkey, H. J.: Cement Products, pp. 438–500.
6. Blanks and Kennedy: "The Technology of Cement and Concrete," vol. I, "Concrete Materials," vol. II, "Concrete," John Wiley & Sons, Inc., New York, vol. I, 1955.
7. Bauer, Edward E.: "Plain Concrete," 3d ed., McGraw-Hill Book Company, Inc., New York, 1949.
8. Troxell and Davis, "Composition and Properties of Concrete," McGraw-Hill Book Company, Inc., New York, 1956.
9. Kelly, J. W.: "A.C.I. Manual of Concrete Inspection," 3d ed., Report of Committee 611 on Inspection of Concrete, American Concrete Institute, Detroit, 1955.
10. U.S. Bureau of Reclamation: "Concrete Manual," 6th ed., U.S. Department of the Interior, Denver, Colo., 1955.
11. Draffin, Jasper O.: A Brief History of Lime, Cement, Concrete and Reinforced Concrete, *Jour. Western Soc. Engrs.*, vol. 48, No. 1, pp. 14–47, March, 1943. Also *Univ. Illinois Eng. Expt. Sta. Bull.*, Reprint Series 27, vol. 40, No. 45, June 29, 1943.
12. Hadley, Earl J.: "The Magic Powder," G. P. Putnam's Sons, New York, 1945.
13. Final Report, Special Committee on Concrete and Reinforced Concrete (First Joint Committee) *Trans. ASCE,* vol. 81, 1917.
14. Report of the Joint Committee on Standard Specifications for Concrete and Reinforced Concrete (2d Joint Committee), *Proc. ASTM*, vol. 24, 1924. Also *Proc. ACI*, vol. 21, pp. 329–425, 1925.
15. Report of the Joint Committee on Standard Specifications for Concrete and Reinforced Concrete (3d Joint Committee), *Proc. ASCE*, June, 1940, Part II. Also ACI separate booklet.

Section 7

Part 2: REINFORCED CONCRETE

BY WILLIAM McGUIRE

INTRODUCTION

Reinforced concrete has become one of the most widely used materials of engineering construction because of the ease with which concrete and steel can be fabricated into structural members, utilizing the desirable attributes of both materials. Concrete is weak in tension; steel is vulnerable to corrosion and to temperatures in excess of 800°F. If steel bars are embedded properly in the cross sections of flexural members, the steel can be made to carry the tension while the concrete resists the compression as well as protecting the steel against corrosion and fire damage. Moreover, because of the adhesion between concrete and steel, and because of the mechanical interlock between deformations on the steel bars and surrounding concrete, the concrete supplies the grip or anchorage required to enable the steel to resist tensile forces. Steel is also used to reinforce concrete compressively in columns and near the compression face of some beams, to resist the diagonal tension that accompanies shear in beams, to serve as ties and spiral reinforcement in columns, and to minimize or distribute cracks which tend to form because of temperature changes or shrinkage in concrete.

In this country the first concerted effort to treat reinforced concrete as an engineering material stems from the formation of the first Joint Committee in 1904. This committee was composed of representatives of the ASCE, ASTM, AREA (then AREMWA), and PCA (then AAPCM). Shortly thereafter the National Association of Cement Users, which subsequently became the American Concrete Institute, was formed. The engineering development of reinforced concrete in the United States has been guided largely by the first and subsequent Joint Committees and the ACI. In recent years the ACI has become predominant in the field and most of the current activity is guided and influenced by that group. Other prominent organizations are the Reinforced Concrete Research Council, which sponsors a great deal of research, and the Portland Cement Association, which does research and has published a number of valuable information bulletins on construction and design procedures (Ref. 7).

In building or bridge construction the basic elements of a concrete structure normally fall into one of three categories: flexural members (beams, slabs, and girders), compression members (columns, bearing walls), or members subjected to combined flexure and direct stress. Occasional tension members (hangers or ties) are used, but in these the reinforcement carries all the load and the concrete is merely for protection of the steel. The torsional member or shaft is of no importance in concrete, although there are cases where provision must be made to resist the torsion in unsymmetrically loaded beams.

Once the general form of a concrete structure has been determined, the design procedure is: (1) to break it down into its elements; (2) to determine the internal moments, thrusts, and shears by analysis or empirical rules; and (3) to select the proper cross sections and reinforcement to resist the internal forces at all critical sections. Because of the statically indeterminate nature of most reinforced concrete construction and the fact that in such structures the distribution of internal forces depends on the structural properties (cross-sectional area, moment of inertia), it is frequently necessary to repeat the analysis and design procedure two or more times until satisfactory sections have been obtained.

One of the most common problems in reinforced concrete, and one that illustrates the variety of forms obtainable, is the design of a floor system. In concrete, as contrasted to steel construction where structural plates and shapes of standard size and form must be used, the designer has much more control over the form and size of the structural components. In addition, many small producers of reinforced concrete elements and construction accessories can compete profitably in this field since plant and equipment costs are not so high as they are in the steel industry. This has resulted in the development of a few more or less standard methods of concrete floor construction and many special or proprietary methods. The commonly used systems and the basic flexural elements of which they are composed can be classified as follows:

1. One-way reinforcing systems (the main reinforcement in each structural element runs in one direction).
   a. Solid-slab, beam-and-girder floors.
   b. One-way ribbed floors: concrete joists with steel-pan, clay, or concrete-tile fillers.
   c. Steel-joist floors (commonly used with steel frames).
   d. Precast-concrete floor systems.
      (1) Precast slab or deck.
      (2) Precast beams and girders.
   e. Concrete slab reinforced with light-gage steel deck.
2. Two-way reinforcing systems (the main reinforcement in at least one structural element runs in two directions).
   a. Two-way solid slabs with beam supports.
   b. Two-way ribbed slabs with tile or steel-pan fillers and beam supports.
   c. Flat-slab floors.

1a. Solid-slab Beam-and-girder Floor. A solid-slab, beam-and-girder floor consists of a series of parallel beams supported at their extremities by girders which in turn frame into concrete columns placed at more or less regular intervals over the entire floor area. This framework is covered by a one-way reinforced concrete slab, the load from which is transmitted first to the beams and thence to the girders and columns. The beams are usually spaced so that they come at the mid-points, at the third points, or at the quarter points of the girders, as shown in Fig. 38. The arrangement of beams and spacing of columns should be determined by economical and practical considerations. These will be affected by the use to which the building is to be put, the size and shape of the ground area, and the load which must be carried. As the slabs, beams, and girders are built monolithically, the beams and girders are designed as T beams and advantage is taken of continuity. One-way solid reinforced concrete slabs are also used over the steel beams of a steel-frame building.

Beam-and-girder floors, as they are usually called, are adapted to any loads and to any spans that might be encountered in ordinary building construction. The normal maximum spread in live-load values is from 40 to 400 lb per sq ft, and the normal range in column spacing is from 16 to 32 ft.

In normal beam-and-girder construction the depth of a beam will be about twice its stem width. For light loads, however, it may be economical to omit the intermediate

beams and deep column-line girders entirely and to have the one-way slab supported by wide, shallow beams which are centered on the column lines and which frame directly into the columns. The wide beams keep the effective span of the slab from becoming excessive, the number of framing members may be considerably reduced, and, perhaps most important, shallow beams permit a reduction in the overall height of the building.

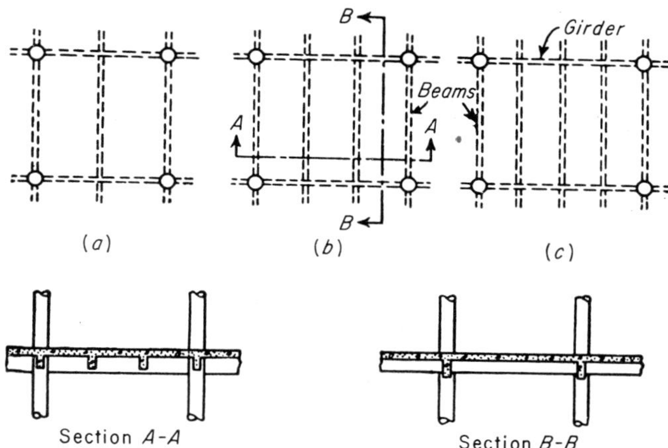

Fig. 38. Framing of beam-and-girder floors.

This type of framing is called slab-band construction. It has been used often in multistory apartment and office buildings. In Fig. 39 are shown comparative sections through a conventional and slab-band type of floor such as might be used in a wing of an apartment building.

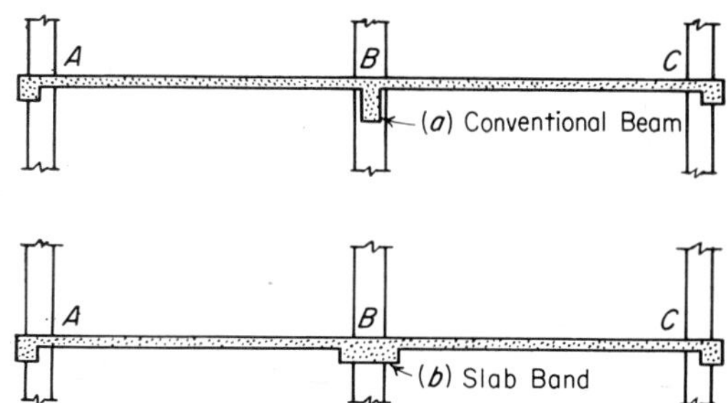

Fig. 39. Sections through conventional and slab-band floors.

1b. One-way Ribbed Floors. A ribbed floor consists of a series of small, closely spaced reinforced concrete T beams framing into beams or girders which in turn frame into the supporting columns. The T beams (called joists or ribs) are formed by placing rows of fillers in what would otherwise be a solid slab. The fillers may be special steel pans, hollow clay-tile or lightweight concrete-tile blocks, or ordinary wood forms. The girders which support the joists are usually built as regular T beams.

Since the strength of concrete in tension is small, and is commonly neglected in design, elimination of much of the tension concrete in a slab by the use of fillers results in a saving of weight with little alteration in the structural characteristics of the slab.

Ribbed floors are economical for buildings such as apartment houses, hotels, and hospitals, where the live loads are fairly small and the spans comparatively long. They are not suitable for heavy construction such as in warehouses, printing plants, and heavy manufacturing buildings.

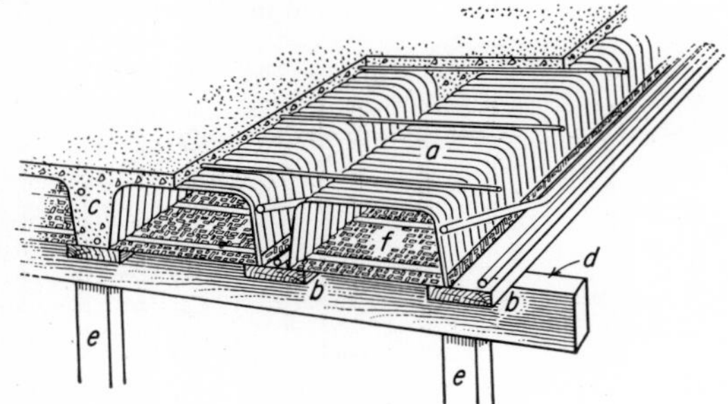

Fig. 40. Ribbed floor with steel tiles.

Figure 40 is a cutaway view of a typical ribbed floor with steel tiles showing the wooden formwork which supports the tiles during construction, the reinforcement, and the metal lath which is sometimes used to support the plastered ceiling for the room below. The tiles as shown in the figure are designed to remain in place as part of the completed structure. It is more common to use removable tiles which can be reused many times and thus reduce the cost of formwork. The slab above the cores is usually 2 in. or more thick. The girders which support the joists are rectangular or T beams with a maximum flange thickness equal to the total floor thickness, as shown in Fig. 41*a*. Tile pieces which are tapered in depth are also available. These tiles are usually about 3 in. less in depth at one end than at the other, so that, if one length of such tile is used at each end of each row, a slab from 5 to 6 in. thick at the edge of the girder is obtained as shown in Fig. 41*b*. This thickness is often sufficient to furnish an adequate flange for the girder.

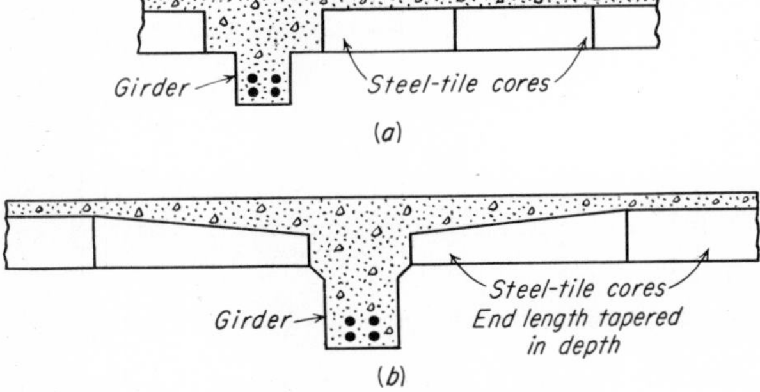

Fig. 41. Details of girders for ribbed floor.

Figure 42 shows a ribbed floor in which structural clay-tile blocks are used as fillers. The blocks are usually 12 in. square and can be obtained in thicknesses varying from 3 to 12 in. Other sizes as well as specially formed blocks are also manufactured for use as form fillers. The usual clear distance between rows is 4 in., thus making the center-to-center distance 16 in. Concrete is placed so as to fill the space between the tiles and to cover them to a depth of 2 or 2½ in. The resulting construction consists of a series of concrete T beams, or joists, with tile fillers under the slab in the spaces between

the stems of the joists. The joists frame into, and are supported by, the girders, and these in turn are supported by the columns. The tile fillers remain in place, securely anchored to the concrete by the projections on the surfaces of the tiles. The entire ceiling is usually plastered. The arrangement of reinforcement is similar to that used in ribbed floors with steel-pan fillers.

Hollow concrete-tile fillers come in many shapes and sizes, the most common being 8 in. by 16 in. in plan with depths of 4, 6, or 8 in. They are usually made with portland cement and a lightweight aggregate of cinders, slag, burned clay, or mineralized organic substance. Concrete-tile fillers are used in the same manner as clay-tile fillers in floor construction.

1c. Steel-joist Floors. A steel-joist floor consists of a series of closely spaced, parallel, shallow joists or trusses of the Pratt, Warren, or double-Warren type, supported at the ends on steel or concrete beams, or on masonry walls, and covered with a thin concrete slab.

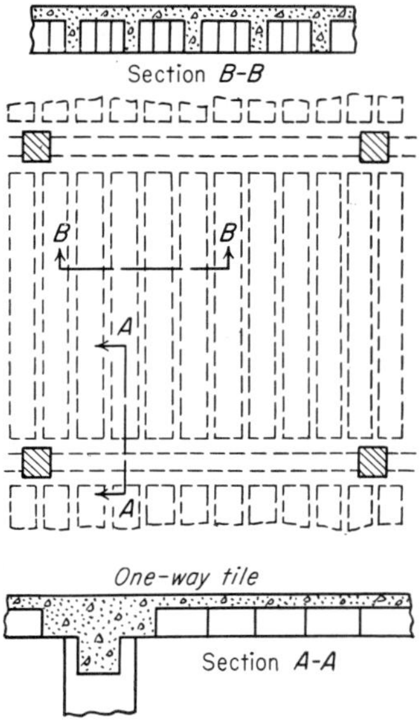

FIG. 42. Details of a one-way tile floor.

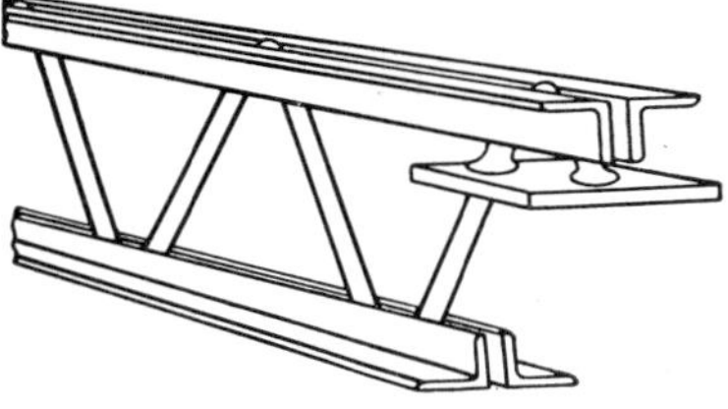

FIG. 43. Typical steel joist.

A common type of steel joist is shown in Fig. 43. The truss is composed of angle chords and a continuous bar web, assembled by the high-pressure electric welding method. Joists are manufactured in depths of 8 to 16 in. and with spans from 4 to 32 ft (longer span and deeper joists, primarily for roofs, are also available). The joists are spaced from 12 to 30 in. on centers, depending on the load and span, and are covered with a 2- to 3-in. concrete slab poured on metal rib lath which rests on, and is fastened to, the upper surfaces of the joists as shown in Fig. 44. The figure also shows the bridg-

FIG. 44. Steel joist, floor construction.

ing necessary to brace the joists laterally and the construction of a suspended ceiling below. The concrete slab which rests on the joists acts as a one-way slab spanning between joists. The metal rib lath on which it is placed serves as formwork for the fresh concrete and reinforcement for the hardened slab. Additional wire-mesh reinforcement is frequently used. Steel-joist floors are economical for light occupancies, but they are unsuitable for heavy or vibrating loads.

1d. Precast-concrete Floor Systems. A great amount of work has been done on the development of precast-concrete structural units in recent years. The two types of units manufactured are the joist (Fig. 45) and the slab or deck (Fig. 46).

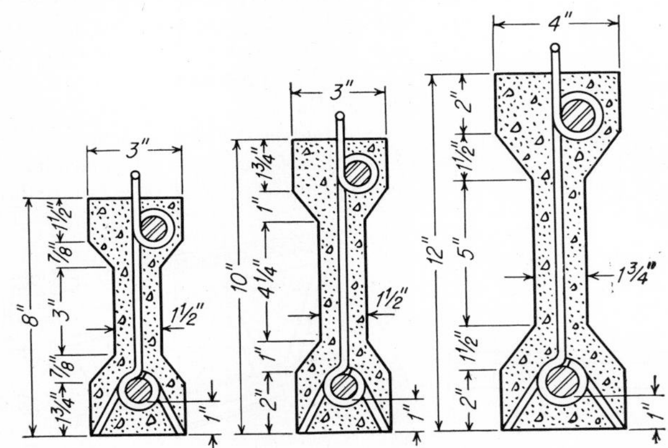

Fig. 45. Cross sections of typical precast concrete joists.

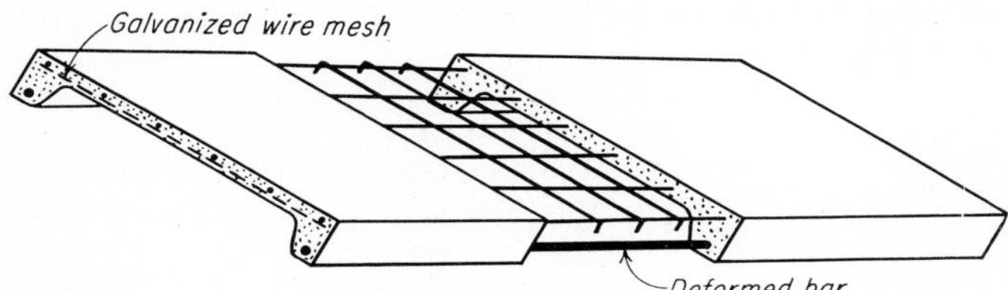

Fig. 46. Typical deck-type precast unit.

These units are used in a variety of ways such as for roof decks on steel frame buildings, or as precast beams supporting cast-in-place or precast slabs. They are designed by standard reinforced concrete theory. The principal advantages claimed for precast members are those of standardization and mass production: close control over design and manufacture, multiple use of formwork, elimination of expensive on-the-job forms, and speed of erection of the structure. Reference 8 contains standards for precast floor units.

Most precast elements are designed for relatively light floor loading or for roof loads. Recently, however, heavy prestressed, precast floor beams and girders have been used in industrial structures.

1e. Steel-deck Floors with Concrete Slab. Considerable use has been made of lightgage steel deck covered with a concrete slab for floors in steel frame office and apartment buildings. Two examples of this type of construction are shown in Fig. 47. The deck acts as formwork for the fresh concrete and reinforcement for the hardened slab. Figure

47*a* shows the laying of a corrugated steel deck manufactured by the Granco Steel Products Company. Transverse wires welded to the ridges of the sheet provide bond between the sheet and the concrete slab, as well as serving as shrinkage and temperature reinforcement for the slab. The corrugated steel sheet serves as positive reinforcement

FIG. 47*a*. Placing corrugated sheet (Cofar panel) reinforcing for floor slab.

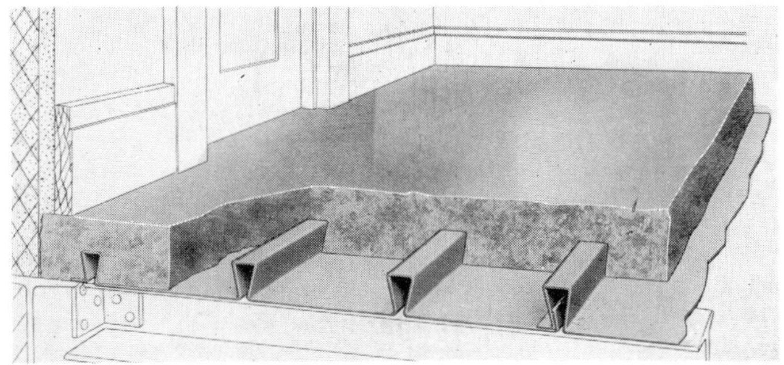

FIG. 47*b*. Concrete slab with steel deck reinforcing. (*Detroit Steel Products Co.*)

for the slab and ordinary deformed bars are used for negative reinforcement over the supports. These are essentially one-way reinforced slabs. The chief advantages claimed for this type of construction are that it is fast and that it eliminates expensive formwork.

2a. Two-way Solid Slabs with Beam Supports. Solid concrete slabs which are square or nearly square in plan, and which are supported by concrete or steel beams or masonry walls on all four sides, should be reinforced in two directions so as to transmit the total

load to all four sides. Floors of this type are suitable for intermediate and heavy loads on spans up to about 30 ft, the range of usefulness corresponding generally to that of flat slabs. The latter are often preferred because of the complete elimination of beams. The design of two-way slabs is usually based on empirically determined moments and shears. These quantities have some basis in the elastic theory of plates, but they have been modified to account for the fact that load tests on slabs show that their actual strength is greater than that which should obtain from theoretical computations. While the reasons for the difference are known qualitatively, it is difficult to obtain quantitative expressions which would permit a completely theoretical approach.

2b. Two-way Ribbed Slabs with Tile or Steel-pan Fillers. As in one-way slab systems, the dead weight of two-way slabs can be reduced considerably by the use of filler blocks of lightweight concrete or by using removable steel-pan forms which are square in plan and impart a wafflelike appearance to the underside of the slab. A typical illustration is shown in Fig. 48. In the illustration the supporting beams are steel, but concrete

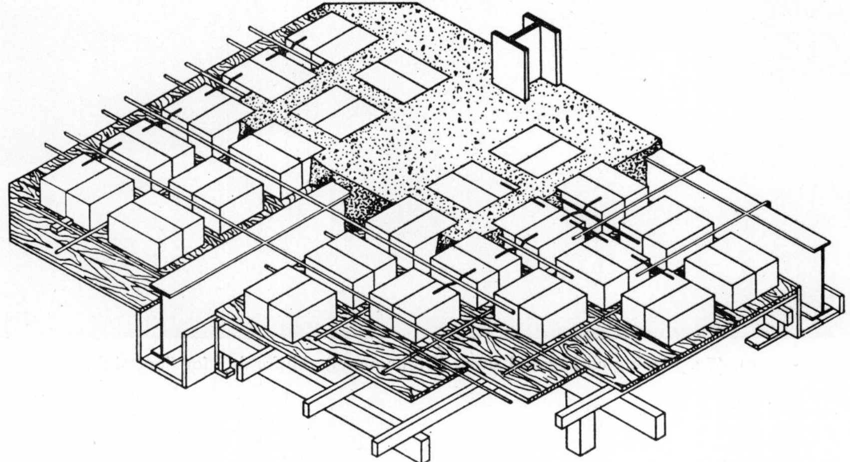

FIG. 48. Two-way slab using block fillers (Republic Slagblok).

beams and columns can be used just as well. A concrete slab or topping may be poured over the blocks. Floors of this type may be designed as ordinary two-way slabs. It is usually necessary to make a ribbed slab somewhat thicker than a comparable solid slab but an overall weight saving can be accomplished.

2c. Flat-slab Floors. A flat-slab floor consists of a reinforced concrete solid or ribbed slab supported directly on concrete columns without the aid of beams or girders. The transition between the slab and the column is usually made through thickened portions of the slab called drop panels and enlarged sections at the tops of columns called capitals (see Fig. 49). In general, flat-slab construction is economical for live loads of 100 lb per sq ft or more and for spans up to about 30 ft. For lighter loads, such as are used in apartment houses, hotels, and office buildings, some other form of ribbed-floor construction will usually be cheaper than a flat-slab floor, although in recent years flat slabs have been used economically for a wide range of loading. Although based generally on plate theory, the design is largely empirical and is treated in considerable detail in most codes.

Ribbed flat slabs have proved very successful in a number of applications. One type, the Grid System, uses removable dome-shaped steel pans which form concrete ribs

FIG. 49. Grid system flat-slab building.

FIG. 50. Flat-plate apartment building.

spanning in two directions (see Fig. 49). The design procedure for this type of floor is the same as that for the ordinary flat slab. The same type of form may also be used for ribbed two-way slabs.

Flat-plate floors—flat slabs with drop panels and column capitals omitted—have been developed in recent years for use in buildings with relatively light loading. The floor slab is simply a plate of uniform thickness supported directly by the columns. The only beams used in this type of construction are those at the exterior walls and around large openings in the slab. This system, an example of which is shown in Fig. 50, has proved very successful in a number of apartment-house projects. Design procedure is similar to that for ordinary flat slabs, but the lack of capitals and drop panels and the use of thin slabs require that special attention be given to many of the details of analysis, design, and construction.

CURRENT STATUS OF REINFORCED CONCRETE DESIGN

At the present time there are in use two distinct methods for proportioning reinforced concrete sections: the elastic (or straight-line) theory, and the plastic (or ultimate-strength) theory. In the first method it is assumed that both concrete and steel are perfectly elastic materials. Based on these assumptions, formulas for predicting stresses are developed. The design is made so that, under working loads, the stresses computed from these formulas do not exceed the allowable stresses for the materials. In the latter method it is recognized that concrete is not a truly elastic material. Formulas which predict the ultimate strength of concrete members are used. It is necessary in developing such formulas to rely heavily on empirical evidence and corroboration. The design is made so that, under working loads times appropriate load factors, the ultimate strength of the members will not be exceeded.

Although the ultimate-strength design concept is at least as old as the straight-line theory, the latter has been most widely used in this country. Present-day codes are based on straight-line principles except, usually, in the design of columns and compression reinforced beams where ultimate-strength considerations are recognized. Because this situation will probably obtain for some time to come, greater emphasis will be placed on the straight-line theory in this text. However, it is necessary to include also some treatment of ultimate-strength methods since they are coming into greater use as they are developed and perfected. The present ACI Code (ACI 318-56), although primarily an elastic-theory code, does contain, for the first time in any widely used American code, provisions for design by ultimate strength as shown in the following extract (section numbers refer to ACI 318-56, Ref. 11).

"601—Design methods

"(a) The design of reinforced concrete members shall be made with reference to allowable stresses, working loads, and the accepted straightline theory of flexure except as permitted by Section 601(b). In determining the ratio n for design purposes, the modulus of elasticity for the concrete shall be assumed as 1000 f_c', and that for steel as 30,000,000 psi. It is assumed that the steel takes all the tension stresses in flexure computations.

"(b) The ultimate strength method of design may be used for the design of reinforced concrete members."

Allowable Stresses. The following notation has become standard and will be used throughout this text:

f_c = compressive unit stress in extreme fiber of concrete in flexure.
f_c' = compressive strength of concrete at age of 28 days unless otherwise specified.
f_r = compressive unit stress in the metal core of a composite column.

f_s = tensile unit stress in longitudinal reinforcement; nominal allowable stress in vertical column reinforcement.
f_v = tensile unit stress in web reinforcement.
n = ratio of modulus of elasticity of steel to that of concrete.
u = bond stress per unit of surface area of bar.
v = shearing unit stress.
v_c = shearing unit stress permitted on the concrete.

Allowable stresses vary from code to code, but it has been observed that most building code committees follow—sooner or later—the lead of the ACI. For this reason, and since the ACI Code represents the most up-to-date practice, the sections relating to allowable stresses in ACI 318-56 (Ref. 11) are reproduced below. Because the ACI Code will be referred to frequently, the section numbering used therein will be retained.

"Table 305(a)—Allowable Unit Stresses in Concrete

Description		Allowable unit stresses						
		For any strength of concrete in accordance with Section 302 $n = \frac{30{,}000}{f_c'}$	Maximum value, psi	For strength of concrete shown below: $f_c' = 2000$ psi $n = 15$	$f_c' = 2500$ psi $n = 12$	$f_c' = 3000$ psi $n = 10$	$f_c' = 3750$ psi $n = 8$	$f_c' = 5000$ psi $n = 6$
Flexure: f_c								
Extreme fiber stress in compression	f_c	$0.45f_c'$		900	1125	1350	1688	2250
Extreme fiber stress in tension in plain concrete footings	f_c	$0.03f_c'$		60	75	90	113	150
Shear: v (as a measure of diagonal tension)								
Beams with no web reinforcement	v_c	$0.03f_c'$	90	60	75	90	90	90
Beams with longitudinal bars and with either stirrups or properly located bent bars	v	$0.08f_c'$	240	160	200	240	240	240
Beams with longitudinal bars and a combination of stirrups and bent bars (the latter bent up suitably to carry at least $0.04f_c'$)	v	$0.12f_c'$	360	240	300	360	360	360
Footings *	v_c	$0.03f_c'$	75	60	75	75	75	75
(For flat slabs, see Chapter 10)								
Bond: u								
Deformed bars (as defined in Section 104)								
Top bars †	u	$0.07f_c'$	245	140	175	210	245	245
In two-way footings (except top bars)	u	$0.08f_c'$	280	160	200	240	280	280
All others	u	$0.10f_c'$	350	200	250	300	350	350
Plain bars (as defined in Section 104) (must be hooked)								
Top bars	u	$0.03f_c'$	105	60	75	90	105	105
In two-way footings (except top bars)	u	$0.036f_c'$	126	72	90	108	126	126
All others	u	$0.045f_c'$	158	90	113	135	158	158
Bearing: f_c								
On full area	f_c	$0.25f_c'$		500	625	750	938	1250
On one-third area or less ‡	f_c	$0.375f_c'$		750	938	1125	1405	1875

* See Sections 905 and 809.

† Top bars, in reference to bond, are horizontal bars so placed that more than 12 in. of concrete is cast in the member below the bar.

‡ This increase shall be permitted only when the least distance between the edges of the loaded and unloaded areas is a minimum of one-fourth of the parallel side dimension of the loaded area. The allowable bearing stress on a reasonably concentric area greater than one-third but less than the full area shall be interpolated between the values given.

"305—Allowable unit stresses in concrete

"(*a*) The unit stresses in pounds per square inch on concrete to be used when designs are made in accordance with Section 601(a) shall not exceed the values of Table 305(a) where f_c' equals the minimum specified compressive strength at 28 days, or at the earlier age at which the concrete may be expected to receive its full load.

"306—Allowable unit stresses in reinforcement

"Unless otherwise provided in this code, steel for concrete reinforcement shall not be stressed in excess of the following limits:

"(*a*) *Tension*

(f_s = tensile unit stress in longitudinal reinforcement)

and (f_v = tensile unit stress in web reinforcement)

20,000 psi for rail-steel concrete reinforcing bars, billet-steel concrete reinforcing bars of intermediate and hard grades, axle-steel concrete reinforcing bars of intermediate and hard grades, and cold-drawn steel wire for concrete reinforcement.

18,000 psi for billet-steel concrete reinforcing bars of structural grade, and axle-steel concrete reinforcing bars of structural grade.

"(*b*) *Tension in one-way slabs of not more than 12-ft span*

(f_s = tensile unit stress in main reinforcement)

"For the main reinforcement, ⅜ in. or less in diameter, in one-way slabs, 50 percent of the minimum yield point specified in the specifications of the American Society for Testing Materials for the particular kind and grade of reinforcement used, but in no case to exceed 30,000 psi.

"(*c*) *Compression, vertical column reinforcement*

(f_s = nominal allowable stress in vertical column reinforcement)

"Forty percent of the minimum yield point specified in the specifications of the American Society for Testing Materials for the particular kind and grade of reinforcement used, but in no case to exceed 30,000 psi.

(f_r = allowable unit stress in the metal core of composite and combination columns)

Structural steel sections. . . . . 16,000 psi
Cast iron sections. . . . . 10,000 psi
Steel pipe. . . . . See limitations of Section 1106(b)

"(*d*) *Compression, flexural members*

"For compression reinforcement in flexural members see Section 706(b)."

These allowable stresses apply when conventional straight-line theory is used. With ultimate design, a different approach is required, as shown later.

Reinforcement. ACI 318-56 states that reinforcement shall be as follows:

"208—Metal reinforcement

"(*a*) Reinforcing bars shall conform to the requirements of "Specifications for Billet-Steel Bars for Concrete Reinforcement" (ASTM A15), "Specifications for Rail-Steel Bars for Concrete Reinforcement" (ASTM A16), "Specifications for Axle-Steel Bars for Concrete Reinforcement" (ASTM A160), or "Specifications for Fabricated Steel Bar or Rod Mats for Concrete Reinforcement" (ASTM A184). Deformations on deformed bars shall conform to "Specifications for Minimum Requirements for the Deformations of Deformed Bars for Concrete Reinforcement" (ASTM A305).

"(*b*) Cold-drawn wire or welded wire fabric for concrete reinforcement shall conform to the requirements of "Specifications for Cold-Drawn Steel Wire for Concrete Reinforcement" (ASTM A82), or "Specifications for Welded Steel Wire Fabric for Concrete Reinforcement" (ASTM A185).

"(*c*) Structural steel shall conform to the requirements of "Specifications for Steel for Bridges and Buildings" (ASTM A7).

"(*d*) Cast-iron sections for composite columns shall conform to "Specifications for Cast Iron Pressure Pipe" (ASTM A377)."

Steel conforming to ASTM A15 may be of one of three grades: structural, intermediate, or hard. The minimum specified yield points for the three grades are, respectively: 33,000; 40,000; and 50,000 psi. The ultimate strengths are specified in ranges: 55,000–75,000 psi for structural grade; 70,000–90,000 for intermediate grade, and 80,000 minimum for hard grade. Other requirements are designed to obtain reasonably uniform chemical content, and steel that is ductile enough so that it will withstand normal fabricating procedures without cracking.

Standard bar sizes, method of designation, and minimum deformation requirements are as shown in Table 1, which is taken from ASTM Specification A305. Tables 2 through 4, taken from Urquhart, O'Rourke, and Winter, "Design of Concrete Structures," 6th ed. (Ref. 9) give useful data for groups of bars.

TABLE 1. DIMENSIONAL REQUIREMENTS FOR DEFORMED STEEL BARS FOR CONCRETE REINFORCEMENT

Deformed bar designation number [1]	Unit weight, lb per ft	Nominal dimensions, round sections			Deformation requirements		
		Diameter, in.	Cross-sectional area, sq in.	Perimeter, in.	Maximum average spacing, in.	Minimum height, in.	Maximum gap (chord of 12½ percent of nominal perimeter), in.
3	0.376	0.375	0.11	1.178	0.262	0.015	0.143
4	0.668	0.500	0.20	1.571	0.350	0.020	0.191
5	1.043	0.625	0.31	1.963	0.437	0.028	0.239
6	1.502	0.750	0.44	2.356	0.525	0.038	0.286
7	2.044	0.875	0.60	2.749	0.612	0.044	0.334
8	2.670	1.000	0.79	3.142	0.700	0.050	0.383
9 [2]	3.400	1.128	1.00	3.544	0.790	0.056	0.431
10 [2]	4.303	1.270	1.27	3.990	0.889	0.064	0.487
11 [2]	5.313	1.410	1.56	4.430	0.987	0.071	0.540

[1] Bar numbers are based on the number of eighths of an inch included in the nominal diameter of the bars. The nominal diameter of a deformed bar is equivalent to the diameter of a plain bar having the same weight per foot as the deformed bar.

[2] Bars of designation Nos. 9, 10, and 11 correspond to the former 1-in. square, 1⅛-in. square, and 1¼-in. square sizes and are equivalent to those former standard bar sizes in weight and nominal cross-sectional areas.

TABLE 2. AREAS OF GROUPS OF STANDARD BARS, SQUARE INCHES

Bar designation	Number of bars												
	2	3	4	5	6	7	8	9	10	11	12	13	14
No. 4	0.39	0.58	0.78	0.98	1.18	1.37	1.57	1.77	1.96	2.16	2.36	2.55	2.75
No. 5	0.61	0.91	1.23	1.53	1.84	2.15	2.45	2.76	3.07	3.37	3.68	3.99	4.30
No. 6	0.88	1.32	1.77	2.21	2.65	3.09	3.53	3.98	4.42	4.86	5.30	5.74	6.19
No. 7	1.20	1.80	2.41	3.01	3.61	4.21	4.81	5.41	6.01	6.61	7.22	7.82	8.42
No. 8	1.57	2.35	3.14	3.93	4.71	5.50	6.28	7.07	7.85	8.64	9.43	10.21	11.00
No. 9	2.00	3.00	4.00	5.00	6.00	7.00	8.00	9.00	10.00	11.00	12.00	13.00	14.00
No. 10	2.53	3.79	5.06	6.33	7.59	8.86	10.12	11.39	12.66	13.92	15.19	16.45	17.72
No. 11	3.12	4.68	6.25	7.81	9.37	10.94	12.50	14.06	15.62	17.19	18.75	20.31	21.87

TABLE 3. PERIMETERS OF GROUPS OF STANDARD BARS, INCHES

Bar designation	Number of bars												
	2	3	4	5	6	7	8	9	10	11	12	13	14
No. 4	3.1	4.7	6.2	7.8	9.4	11.0	12.6	14.1	15.7	17.3	18.8	20.4	22.0
No. 5	3.9	5.9	7.8	9.8	11.8	13.7	15.7	17.7	19.5	21.6	23.6	25.5	27.5
No. 6	4.7	7.1	9.4	11.8	14.1	16.5	18.8	21.2	23.6	25.9	28.3	30.6	33.0
No. 7	5.5	8.2	11.0	12.7	16.5	19.2	22.0	24.7	27.5	30.2	33.0	35.7	38.5
No. 8	6.3	9.4	12.6	15.7	18.9	22.0	25.1	28.3	31.4	34.6	37.7	40.9	44.0
No. 9	7.1	10.6	14.2	17.7	21.3	24.8	28.4	31.9	35.4	39.0	42.5	46.0	49.6
No. 10	8.0	12.0	16.0	20.0	23.9	27.9	31.9	35.9	39.9	43.9	47.9	51.9	55.9
No. 11	8.9	13.3	17.7	22.2	26.6	31.0	35.4	39.9	44.3	48.7	53.2	57.6	62.0

TABLE 4. AREAS OF BARS IN SLABS, SQUARE INCHES PER FOOT

Spacing, in.	Bar designation								
	No. 3	No. 4	No. 5	No. 6	No. 7	No. 8	No. 9	No. 10	No. 11
3	0.44	0.78	1.23	1.77	2.40	3.14	4.00	5.06	6.25
3½	0.38	0.67	1.05	1.51	2.06	2.69	3.43	4.34	5.36
4	0.33	0.59	0.92	1.32	1.80	2.36	3.00	3.80	4.68
4½	0.29	0.52	0.82	1.18	1.60	2.09	2.67	3.37	4.17
5	0.26	0.47	0.74	1.06	1.44	1.88	2.40	3.04	3.75
5½	0.24	0.43	0.67	0.96	1.31	1.71	2.18	2.76	3.41
6	0.22	0.39	0.61	0.88	1.20	1.57	2.00	2.53	3.12
6½	0.20	0.36	0.57	0.82	1.11	1.45	1.85	2.34	2.89
7	0.19	0.34	0.53	0.76	1.03	1.35	1.71	2.17	2.68
7½	0.18	0.31	0.49	0.71	0.96	1.26	1.60	2.02	2.50
8	0.17	0.29	0.46	0.66	0.90	1.18	1.50	1.89	2.34
9	0.15	0.26	0.41	0.59	0.80	1.05	1.33	1.69	2.08
10	0.13	0.24	0.37	0.53	0.72	0.94	1.20	1.52	1.87
12	0.11	0.20	0.31	0.44	0.60	0.78	1.00	1.27	1.56

Recently, there has been some special rolling of larger bars for application in heavily reinforced sections shielding nuclear reactors, but the availability and practical utility of such bars is extremely limited. Other bars of higher strength and different deformation patterns are used to a limited extent.

DESIGN FOR FLEXURAL LOADING—STRAIGHT-LINE THEORY

A properly designed reinforced concrete beam is one which has a suitable margin of safety against failure in any possible form: yielding of the tension reinforcement; crushing of the concrete in compression; excessive diagonal tension cracking; slip caused by loss of bond between steel and concrete; or excessive deflection. Lateral buckling is not usually a factor because the sections used in reinforced concrete normally are quite stiff torsionally. It is somewhat misleading to consider the different conditions separately because there is definite interaction. For example, a diagonal tension crack spreading into the compression region may precipitate a premature compression failure. Although recognizing its limitations, the customary approach of treating pure bending first, and shear and bond later as separate conditions, will be followed here.

Assuming shear to be negligible and bond resistance adequate, one of two things may happen if a beam is overloaded: (1) The stress in the steel reaches the yield and it starts to flow plastically. The hairline cracks which are present at working loads open wider and spread into the compression zone, reducing the effective area of concrete in compression and increasing the compressive stress until the ultimate is reached, at which point the concrete will crush. This is primarily a steel failure. For practical purposes the

failure load is reached when the steel starts to yield, but, before breaking, large plastic deflections are realized. This type of beam is said to be "underreinforced." (2) If there is enough steel in the beam to develop the ultimate strength of the concrete before yielding, sudden crushing of the concrete may result with little or no forewarning in the way of visible deflections. Such a beam is said to be "overreinforced." Where steel yielding and concrete crushing are reached simultaneously, the design is "balanced." By the straight-line theory the balanced design point is computed to occur with fairly small amounts of reinforcement—the ratio of the reinforcement area to concrete cross section being of the order of 0.01 for ordinary materials. Ultimate-strength theories—which show that the real compression resistance of concrete is greater than assumed in the straight-line theory—show considerably higher balanced design ratios—of the order of 0.03. Since it is usually impractical to place as much steel as this in a beam and still have room for placing of concrete and proper bonding of steel to concrete, most beams are actually underreinforced even though, by straight-line theory, some may appear to be overreinforced.

The notation used in the straight-line theory is:

f_s = unit fiber stress in steel.
f_c = unit fiber stress in concrete at its extreme fiber.
f_c' = ultimate compression strength at 28 days of concrete cast in standard 6-in. by 12-in. cylinder.
A_s = area of cross section of steel.
A_c = area of cross section of concrete.
E_s = modulus of elasticity of steel.
E_c = modulus of elasticity of concrete.
e_s = unit deformation of steel acting under f_s.
e_c = unit deformation of concrete acting under f_c.
n = E_s/E_c.
T = total tension in cross section of steel.
C = total compression in cross section of concrete.
M = total external bending moment at a cross section.
M_s = internal resisting moment of steel in tension.
M_c = internal resisting moment of concrete in compression.
b = breadth of rectangular concrete beam.
d = distance from compression face to plane of centroid of tensile steel, hereafter referred to as "effective depth."
k = ratio of depth of neutral surface to effective depth.
j = ratio of arm of resisting couple to effective depth.
p = steel ratio, A_s/bd.
r = ratio f_s/f_c.

For this and all other types of sections, design formulas are obtained by applying the two available equations of statics. These may be stated as follows: (1) The internal resisting moment must equal the bending moment; (2) the net internal longitudinal force must equal the thrust (zero for pure flexure).

Thus, in principle, the design of concrete elements is very simple. However, because of the two-material system and the fact that the shapes of the portions of the member in tension and compression are sometimes awkward to express mathematically, the algebra frequently becomes involved and simplifying charts, tables, and graphs are practical necessities. The principles will be illustrated for the simple section below. Derivations for more complicated conditions will not be given here but can be found in most standard textbooks (Ref. 9).

For pure flexure, since the net longitudinal thrust equals zero:

$$C = T$$

$$\tfrac{1}{2}f_c kbd = A_s f_s = pf_s bd \tag{1}$$

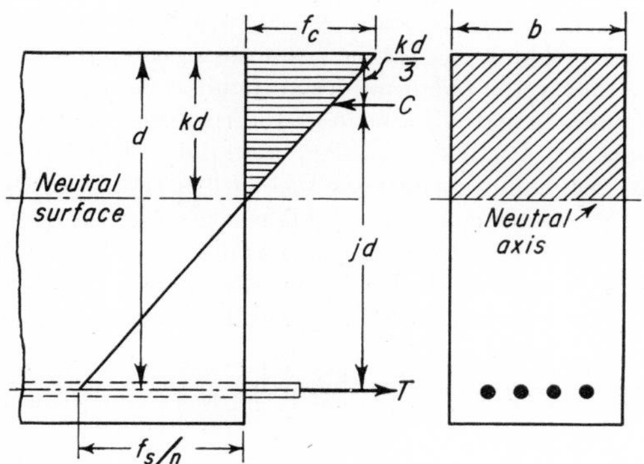

FIG. 51

Taking moments about the tension steel:

$$M_c = Cjd$$

$$M_c = \tfrac{1}{2}f_c kjbd^2 = Kbd^2 \tag{2}$$

Or, taking moments about the resultant compression force·

$$M_s = Tjd$$

$$M_s = pf_s jbd^2 = Kbd^2 \tag{3}$$

where $K = \tfrac{1}{2}fckj$ or $pf_s j$.

From Fig. 51:

$$jd = d - \frac{kd}{3}$$

$$j = 1 - \frac{k}{3} \tag{4}$$

Also, from Fig. 51 by similar triangles:

$$\frac{f_s/n}{d - kd} = \frac{f_c}{kd}$$

$$\frac{f_s}{nf_c} = \frac{1 - k}{k}$$

Solving for k:

$$k = \frac{1}{1 + \dfrac{f_s}{nf_c}} = \frac{n}{n + r} \tag{5}$$

From Eq. (1):

$$\tfrac{1}{2}f_c k = pf_s$$

or

$$\frac{f_s}{f_c} = \frac{k}{2p} \tag{6}$$

Substituting this in Eq. (5) and solving for k:

$$k = \sqrt{2pn + (pn)^2} - pn \tag{7}$$

Equation (7) may be used to find k if p and n are known; that is, if the beam has already been designed and it is desired to check or "review" the design. Knowing k, Eq.

(4) is used to find j and then Eqs. (2) and (3) to find the stress in the concrete and steel, respectively, for any given bending moment.

In design, where a balanced (by straight-line theory) condition is sought, the allowable stresses for steel and concrete may be substituted in Eq. (5) to find k, the quantity bd^2 can then be found from Eq. (2) and appropriate values of b and d selected. Using these, Eq. (3) may be used to determine the amount of steel required. This procedure is greatly simplified by the use of the following charts and tables, based on these equations, as illustrated in examples below.

Table 5. Design of Rectangular Beams and Slabs

$$k = \frac{n}{n + r} \qquad j = 1 - \frac{k}{3} \qquad p = \frac{n}{2r(n + r)} \qquad K = \frac{1}{2} f_c k j \text{ or } p f_s j$$

n and f_c'	f_s	f_c	k	j	p	K
6 (5,000)	18,000	1,667	0.357	0.881	0.0165	262
		2,000	0.400	0.867	0.0222	347
		2,250	0.429	0.857	0.0268	414
	20,000	1,667	0.333	0.889	0.0139	247
		2,000	0.375	0.875	0.0188	328
		2,250	0.403	0.866	0.0227	393
8 (3,750)	18,000	1,250	0.357	0.881	0.0124	197
		1,500	0.400	0.867	0.0167	260
		1,688	0.429	0.857	0.0201	310
	20,000	1,250	0.333	0.889	0.0104	185
		1,500	0.375	0.875	0.0141	246
		1,688	0.403	0.866	0.0170	294
10 (3,000)	18,000	1,000	0.357	0.881	0.0099	157
		1,200	0.400	0.867	0.0133	208
		1,350	0.428	0.857	0.0161	248
	20,000	1,000	0.333	0.889	0.0083	148
		1,200	0.375	0.875	0.0113	197
		1,350	0.403	0.866	0.0136	235
12 (2,500)	16,000	833	0.385	0.872	0.0100	140
		1,000	0.429	0.857	0.0134	184
		1,125	0.457	0.848	0.0161	218
	18,000	833	0.357	0.881	0.0083	131
		1,000	0.400	0.867	0.0111	173
		1,125	0.428	0.857	0.0134	207
	20,000	833	0.333	0.889	0.0069	123
		1,000	0.375	0.875	0.0094	164
		1,125	0.403	0.866	0.0113	196
15 (2,000)	16,000	667	0.385	0.872	0.0080	112
		800	0.429	0.857	0.0107	147
	18,000	667	0.357	0.881	0.0066	105
		800	0.400	0.867	0.0089	139
		900	0.429	0.857	0.0107	165
	20,000	667	0.333	0.889	0.0055	99
		800	0.375	0.875	0.0075	131
		900	0.403	0.866	0.0091	157

TABLE 6. REVIEW OF RECTANGULAR BEAMS AND SLABS

$k = \sqrt{2pn + (pn)^2} - pn$ $\qquad j = 1 - \frac{1}{3}k$

p	$n = 6$		$n = 8$		$n = 10$		$n = 12$		$n = 15$	
	k	j	k	j	k	j	k	j	k	j
0.0010	0.104	0.965	0.119	0.960	0.132	0.956	0.145	0.952	0.158	0.947
0.0020	0.143	0.952	0.164	0.945	0.181	0.940	0.196	0.935	0.217	0.928
0.0030	0.173	0.943	0.196	0.935	0.217	0.928	0.235	0.922	0.258	0.914
0.0040	0.196	0.935	0.223	0.926	0.246	0.918	0.266	0.911	0.292	0.903
0.0050	0.217	0.928	0.246	0.918	0.270	0.910	0.291	0.903	0.320	0.893
0.0054	0.224	0.925	0.254	0.915	0.279	0.907	0.300	0.900	0.329	0.891
0.0058	0.231	0.923	0.262	0.913	0.287	0.904	0.309	0.897	0.337	0.888
0.0062	0.238	0.921	0.269	0.910	0.296	0.901	0.317	0.894	0.348	0.884
0.0066	0.245	0.919	0.276	0.908	0.304	0.899	0.325	0.892	0.356	0.881
0.0070	0.251	0.916	0.283	0.906	0.311	0.896	0.334	0.889	0.365	0.878
0.0072	0.254	0.915	0.286	0.905	0.314	0.895	0.338	0.887	0.369	0.877
0.0074	0.257	0.914	0.290	0.903	0.318	0.894	0.343	0.886	0.372	0.876
0.0076	0.260	0.913	0.293	0.902	0.321	0.893	0.345	0.885	0.376	0.875
0.0078	0.263	0.912	0.297	0.901	0.325	0.892	0.349	0.884	0.380	0.873
0.0080	0.265	0.912	0.300	0.900	0.328	0.891	0.353	0.882	0.384	0.872
0.0082	0.268	0.911	0.303	0.899	0.332	0.889	0.356	0.881	0.387	0.871
0.0084	0.271	0.910	0.306	0.898	0.336	0.888	0.360	0.880	0.390	0.870
0.0086	0.274	0.909	0.309	0.897	0.338	0.887	0.363	0.879	0.394	0.869
0.0088	0.276	0.908	0.312	0.896	0.341	0.886	0.366	0.878	0.398	0.867
0.0090	0.279	0.907	0.314	0.895	0.344	0.885	0.370	0.877	0.402	0.866
0.0092	0.282	0.906	0.317	0.894	0.347	0.884	0.373	0.876	0.405	0.865
0.0094	0.284	0.905	0.320	0.893	0.350	0.883	0.376	0.875	0.407	0.864
0.0096	0.287	0.904	0.322	0.893	0.353	0.882	0.379	0.874	0.411	0.863
0.0098	0.289	0.904	0.325	0.892	0.356	0.881	0.381	0.873	0.414	0.862
0.0100	0.292	0.903	0.328	0.891	0.358	0.881	0.385	0.872	0.418	0.861
0.0104	0.296	0.901	0.333	0.889	0.363	0.879	0.391	0.870	0.423	0.859
0.0108	0.301	0.900	0.339	0.887	0.369	0.877	0.396	0.868	0.429	0.857
0.0112	0.305	0.898	0.343	0.886	0.375	0.875	0.402	0.866	0.434	0.855
0.0116	0.310	0.897	0.348	0.884	0.380	0.873	0.407	0.864	0.440	0.853
0.0120	0.314	0.895	0.353	0.882	0.384	0.872	0.412	0.863	0.446	0.851
0.0124	0.318	0.894	0.357	0.881	0.389	0.870	0.417	0.861	0.451	0.850
0.0128	0.323	0.893	0.362	0.879	0.394	0.869	0.422	0.859	0.457	0.848
0.0132	0.327	0.891	0.366	0.878	0.398	0.867	0.427	0.858	0.461	0.846
0.0136	0.331	0.890	0.370	0.877	0.403	0.866	0.432	0.856	0.466	0.845
0.0140	0.334	0.889	0.374	0.875	0.407	0.864	0.436	0.855	0.471	0.843
0.0144	0.338	0.887	0.379	0.874	0.412	0.863	0.440	0.853	0.475	0.842
0.0148	0.342	0.886	0.383	0.872	0.416	0.861	0.444	0.852	0.479	0.840
0.0152	0.346	0.885	0.386	0.871	0.420	0.860	0.449	0.850	0.483	0.839
0.0156	0.349	0.884	0.390	0.870	0.424	0.859	0.453	0.849	0.487	0.838
0.0160	0.353	0.883	0.394	0.869	0.428	0.857	0.457	0.848	0.493	0.836
0.0170	0.361	0.880	0.403	0.866	0.437	0.854	0.467	0.845	0.502	0.833
0.0180	0.369	0.877	0.412	0.863	0.446	0.851	0.476	0.841	0.513	0.829
0.0190	0.377	0.874	0.420	0.860	0.455	0.848	0.485	0.838	0.522	0.826
0.0200	0.384	0.872	0.428	0.857	0.463	0.846	0.493	0.836	0.531	0.823

Illustrative Examples. 1. Design a rectangular simple beam to carry 1,200,000 in.-lb moment in accordance with the ACI Code. Intermediate-grade steel (f_s allowable = 20,000 psi); $f_c' = 3{,}000$ psi (f_c allowable = 1,350 psi); $n = 10$.

From Table 5 for $n = 10$, $f_s = 20{,}000$, $f_c = 1{,}350$; $K = 0.403$, $j = 0.866$, $p = 0.0136$, $K = 235$

$$\text{From Eq. (3), } (bd^2) \text{ required } = \frac{M}{K} = \frac{1{,}200{,}000}{235} = 5{,}090$$

Many different values of b and d satisfy this requirement. If there are no limitations as to depth or width, it will be found most economical to make b from ½ to ¾ of d. For a width of 12 in. an effective depth of 20.6 in. is required ($12 \times 20.6^2 = 5{,}090$).

From Eq. (1), $A_s = pbd = 0.0136 \times 12 \times 20.6 = 3.36$ sq in.

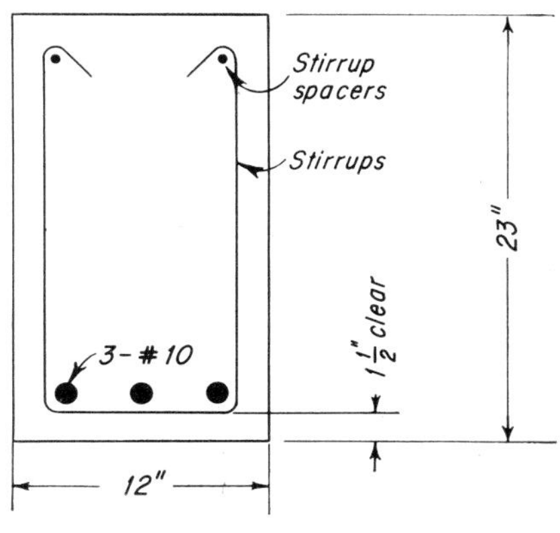

Fig. 52

Three No. 10 bars furnishing 3.79 sq in. is the most practical reinforcement for this beam. The final cross section of the beam with allowance for cover and web reinforcement is shown in Fig. 52. The effective depth is about 0.1 in. shy of that theoretically required, but this is immaterial.

2. Review the design of a beam having $b = 12$ in., $d = 20$ in., reinforced with 3 No. 8 bars, n estimated at 10. Determine the stresses in steel and concrete due to a bending moment of 700,000 in.-lb.

$$p = \frac{A_s}{bd} = \frac{2.35}{12 \times 20} = 0.00980$$

From Table 6, $k = 0.356$, $j = 0.881$

$$\text{From Eq. (2), } f_c = \frac{2 \times 700{,}000}{0.356 \times 0.881 \times 12 \times 20^2} = 928 \text{ psi}$$

$$\text{From Eq. (3), } f_s = \frac{700{,}000}{0.00980 \times 0.881 \times 12 \times 20^2} = 16{,}850 \text{ psi}$$

$$\text{or, from Eq. (6), } f_s = \frac{0.357 \times 928}{2 \times 0.00980} = 16{,}850 \text{ psi}$$

Quite often the situation is met where the desired cross section of concrete is set in advance (say by clearance, architectural considerations, or by structural requirements at another section of a continuous member). If the section is underreinforced, the design problem is simply to find the required amount of tension steel. If the section

is overreinforced, the problem is to increase the resisting moment with respect to the concrete so that it will not be overstressed. If it is undesirable to increase the dimensions of the section, the resisting moment may be increased by either of two methods: (1) Provide tension steel above that needed for balanced design. (2) Add compression reinforcement. The use of compression reinforcement will be treated later.

Flexure in a T Beam. When a rectangular beam is cast monolithically with a concrete slab which it supports, the slab will act integrally with the beam. In the region of positive moment the slab has the effect of enlarging the compression area available, the stem performing the function of resisting the shear and of holding and anchoring the tension steel at the proper distance to produce the necessary resisting couple. Thus, in Fig. 53*a*, the shaded area constitutes a T beam. Where there is negative moment, as at the supports of a continuous beam, the slab is on the tension side, steel must be supplied in the top to carry the tension, and the beam is designed as an ordinary rectangular beam with width equal to stem width and effective depth measured from the soffit up to the top steel. The stem must also be adequate to take the shear at the support. Frequently, T beams are designed as balanced beams for positive moment and some or all of the lower steel is run through the support to serve as compression reinforcement and make the rectangular section capable of resisting the negative support moment.

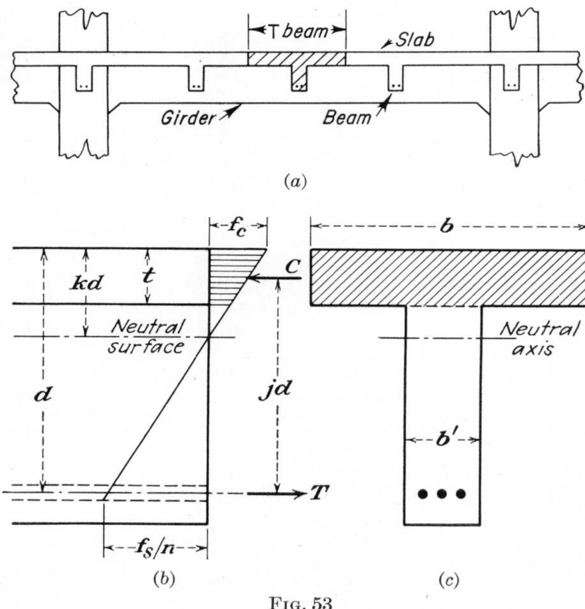

FIG. 53

The limits of flange width which may be assumed as acting with the stem are specified by the following ACI Code section:

"705—Requirements for T-beams

"(*a*) In T-beam construction the slab and beam shall be built integrally or otherwise effectively bonded together. The effective flange width to be used in the design of symmetrical T-beams shall not exceed one-fourth of the span length of the beam, and its overhanging width on either side of the web shall not exceed eight times the thickness of the slab nor one-half the clear distance to the next beam.

"(*b*) For beams having a flange on one side only, the effective overhanging flange width shall not exceed 1/12 of the span length of the beam, nor six times the thickness of the slab, nor one-half the clear distance to the next beam.

"(*c*) Where the principal reinforcement in a slab which is considered as the flange of a T-beam (not a joist in concrete joist floors) is parallel to the beam, transverse reinforcement shall be provided in the top of the slab. This reinforcement shall be designed to carry the load on the portion of the slab required for the flange of the T-beam. The flange shall be assumed to act as a cantilever. The spacing of the bars shall not exceed five times the thickness of the flange, nor in any case 18 in.

"(*d*) Provision shall be made for the compressive stress at the support in continuous T-beam construction, care being taken that the provisions of Section 505 relating to the spacing of bars, and 404(d) relating to the placing of concrete shall be fully met.

"(*e*) The overhanging portion of the flange of the beam shall not be considered as effective in computing the shear and diagonal tension resistance of T-beams.

"(*f*) Isolated beams in which the T-form is used only for the purpose of providing additional compression area, shall have a flange thickness not less than one-half the width of the web and a total flange width not more than four times the web thickness."

The neutral axis may lie in the flange, in which case the flexural design is made as for a rectangular beam of the full flange width; or it may lie in the stem, a case requiring analysis by the T-beam formulas or satisfactory approximations. The governing factors are the ratio t/d and the percentage of steel.

It is customary to neglect the compression in the stem. Resulting formulas are derived in most standard textbooks and are as follows (see Figs. 53*b* and *c*):

$$k = \frac{np + \frac{1}{2}(t/d)^2}{np + (t/d)} \tag{8}$$

$$j = \frac{6 - 6(t/d) + 2(t/d)^2 + (t/d)^3(1/2pn)}{6 - 3(t/d)} \tag{9}$$

$$f_c = \frac{f_s k}{n(1 - k)} \tag{10}$$

$$M_s = A_s f_s jd \tag{11}$$

$$M_c = f_c(1 - t/2kd)btjd \tag{12}$$

Approximate equations for resisting moments can be developed by assuming the resultant compression equal to $\frac{1}{2}f_c bt$ and acting at the mid-depth of the slab. Thus:

$$M_s = A_s f_s(d - \tfrac{1}{2}t) \text{ (approx)} \tag{13}$$

$$M_c = \tfrac{1}{2}f_c bt(d - \tfrac{1}{2}t) \text{ (approx)} \tag{14}$$

Equations (13) and (14) are satisfactory for design in all practical cases. Figure 54 plotted from Eqs. (8) and (9) should be used in review of a beam to find the actual stresses in steel and concrete.

Illustrative Examples. 1. Determine the unit stresses in steel and concrete in flexure under a bending moment of 700,000 in.-lb, for a T beam having dimensions of $t = 2.5$ in.; $b = 30$ in.; $d = 20$ in.; $pn = 0.03$; $n = 10$.

$$t/d = 2.5/20 = 0.125$$

From Fig. 54,

$$k = 0.245, \qquad j = 0.944$$

From Eq. (11),

$$M_s = A_s f_s jd = pf_s jbd^2$$

$$f_s = \frac{700{,}000}{0.003 \times 0.944 \times 30 \times 20^2} = 20{,}600 \text{ psi}$$

From Eq. (10),

$$f_c = \frac{20{,}600 \times 0.245}{10 \times 0.755} = 670 \text{ psi}$$

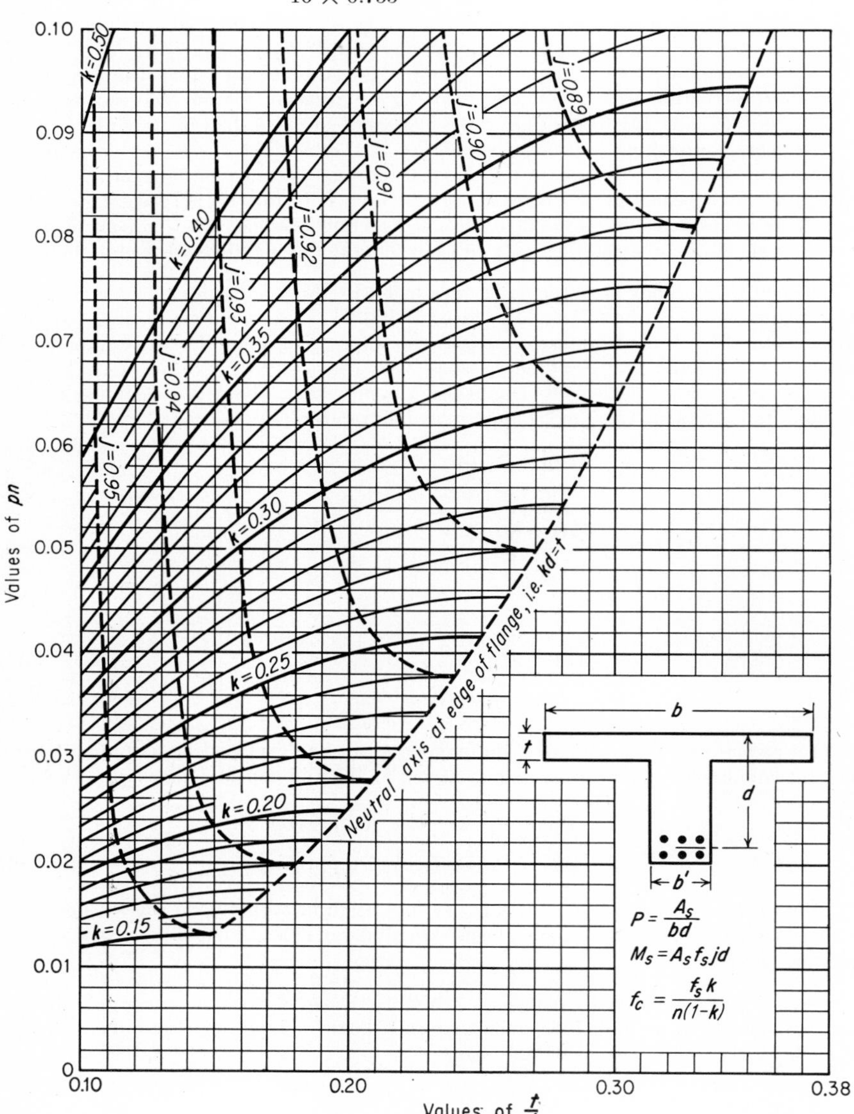

FIG. 54. T-beam review.

2. Design a T beam to resist a moment of 1,912,000 in.-lb. $f_s = 20{,}000$; $f_c = 1{,}350$ Since the slab thickness, effective flange width, and depth are usually determined by other factors such as beam spacing and shear, it will be assumed that the dimensions $t = 4$ in., $b = 60$ in., $d = 24$ in. have been established. The resisting moment of the concrete will be checked by Eq. (14) and the steel area found by Eq. (13):

$$M_c = \tfrac{1}{2} \times 1{,}350 \times 60 \times 4(24 - 2)$$
$$= 3{,}570{,}000 \text{ in.-lb} \gg 1{,}912{,}000 \text{ in.-lb} \qquad \text{O.K.}$$

$$A_s = \frac{1{,}912{,}000}{20{,}000(24 - 2)} = 4.34 \text{ sq in.} \qquad \text{Use 6 No. 8 bars. } (A_s = 4.71 \text{ sq in.})$$

Check the stress in the steel by Fig. 54. To evaluate the accuracy of Eq. (13) use the theoretical amount of steel required:

$$t/d = \tfrac{4}{24} = 0.167$$

$$pn = \frac{4.34 \times 10}{24 \times 60} = 0.0302$$

From Fig. 54,

$$= 0.930$$

$$f_s = \frac{1{,}912{,}000}{4.34 \times 0.930 \times 24} = 19{,}700 \text{ psi}$$

Flexure in T Beams at the Support. Where negative moment occurs in a T beam at the support, the stem alone is usually too small to carry the moment as a singly reinforced beam and is generally strengthened by one or more of the following devices:

1. The stem is increased in depth, by the use of a gradually inclined haunch.
2. The stem is increased gradually in width.
3. Steel is added to the compression area to provide additional compressive resistance.

The first two cases are designed as rectangular beams reinforced in tension. The third alternative requires separate consideration.

Rectangular Beams Reinforced for Compression. Such a beam may be a T beam near the support, or it may be a rectangular beam of limited size. The resisting moment of the beam may be thought to be made up of two moments as follows: (1) a singly reinforced beam with balanced reinforcement working at the allowable stresses (Fig. 55*b*); and (2) compression steel and additional tension steel, which together form an additional couple (Fig. 55*c*). In continuous beams the compression steel may be an extension of the steel used to resist, in tension, moment of opposite sign at other sections of the member.

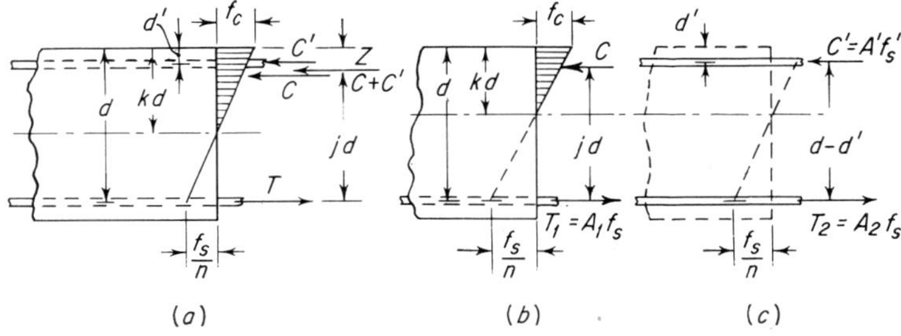

Fig. 55. (*a*) Doubly reinforced beam, separated into (*b*) a singly reinforced beam plus (*c*) additional top and bottom steel.

The compression steel is placed distant d' from the compression face of the beam. Elastically, it would work at a stress equal to n times the stress in the concrete at that depth, and, originally, design formulas based on the straight-line theory made that assumption. It has come to be recognized, however, that the shrinkage and plastic flow in the compressive concrete tend to load the compression steel much more heavily than called for by the premise of elastic strains in the concrete. That is, since the strains in the steel and surrounding concrete must remain equal, the tendency for the concrete to shrink and flow under constant stress throws a greater portion of the load on the compression steel. Fortunately, the steel can take this additional load if it is adequately supported against buckling by ties or stirrups. An idea of the order of magnitude of computed stresses in compression steel may be had by assuming typical design properties of the materials—say $f_c = 1{,}350$ psi, $f_s = 20{,}000$ psi, $n = 10$. If the beam is designed so that the allowable stress is reached in the extreme fiber of the concrete, then the straight-line computed stress in the concrete a short distance in from the extreme fiber—at the level of the compression steel—will be of the order of 900 to 1,200 psi, say 1,000 psi for illustration. By the straight-line theory, the steel stress would then be $10 \times 1{,}000 = 10{,}000$ psi. This would indicate, falsely for the reasons stated above, an inefficient use of steel. The 1956 ACI Code contains the following provision which recognizes that compression steel is actually more efficient than indicated by ordinary straight-line theory and accounts for the greater efficiency in somewhat arbitrary fashion:

"706—Compression steel in flexural members

"(*a*) Compression steel in beams or girders shall be anchored by ties or stirrups not less than ¼ in. in diameter spaced not farther apart than 16 bar diameters, or 48 tie diameters. Such stirrups or ties shall be used throughout the distance where the compression steel is required.

"(*b*) To approximate the effect of creep, the stress in compression reinforcement resisting bending may be taken at twice the value indicated by using the straight-line relation between stress and strain, and the modular ratio given in Section 601(a), but not of greater value than the allowable stress in tension."

To develop design formulas, consider the doubly reinforced beam shown in Fig. 55. Let M_1 and M_2 designate the two parts of the total resisting moment; let A_1 and A_2 be the areas of the two parts of the tension steel referred to; and let A' be the area of the compression steel. Then,

$$M_1 = \tfrac{1}{2} f_c k j b d^2 = K b d^2 \tag{15}$$

and

$$A_1 = \frac{M_1}{f_s j d} \tag{16}$$

The values of k, j, and K are computed from the equations used for rectangular beams of balanced design (Table 5). The moment M_2 must provide for the excess of the total required resisting moment (M) over M_1, i.e., $M_2 = M - M_1$. The forces in the two quantities of steel A' and A_2 form a couple with lever arm $(d - d')$. Thus:

$$M_2 = A_2 f_s (d - d') \tag{17}$$

From which

$$A_2 = \frac{M_2}{f_s(d - d')} \tag{18}$$

The total required tension steel then equals

$$A_s = A_1 + A_2 \tag{19}$$

Since the net longitudinal force is zero,

$$A'f_s' = A_2 f_s \tag{20}$$

where f_s' is the unit stress in the compression steel. Following the ACI Code in assuming that f_s' is twice the value computed on the basis of unit stresses varying directly as the distance from the neutral axis,

$$f_s' = 2f_s \frac{k - (d'/d)}{1 - k} \tag{21}$$

Substituting this in Eq. (20),

$$A' = \frac{A_2}{2} \frac{1 - k}{k - (d'/d)} \tag{22}$$

From Eq. (20) it is seen that, if A' is less than A_2, f_s' is greater than f_s. However, the ACI Code specifies that the stress in the compression steel f_s' shall not exceed the allowable stress in tension. To make sure that it does not, A' must be at least as large as A_2. Thus, either Eq. (22) or this requirement governs the selection of compression reinforcement.

The above formulas may be used for design. Formulas for reviewing rectangular beams reinforced for compression may be derived in the same manner as those for beams without such reinforcement, i.e., using the two equations: sum of the moments and sum of the thrusts equal zero. The resulting formulas are

$$f_c = \frac{f_s k}{n(1 - k)} \tag{23}$$

$$f_s' = \frac{2f_s(k - d'/d)}{1 - k} \tag{24}$$

$$k = \sqrt{2n(p + 2p'd'/d) + n^2(p + 2p')^2} - n(p + 2p') \tag{25}$$

$$j = \frac{k^2 - \frac{1}{3}k^3 + 4p'n(k - d'/d)(1 - d'/d)}{k^2 + 4p'n(k - d'/d)} \tag{26}$$

where

$$p' = A'/bd \tag{27}$$

and

$$M_s = A_s f_s jd \tag{28}$$

These formulas assume a doubled value of the compression steel stress and hence are not applicable if $f_s' > f_s$, the allowable stress in tension. A diagram based on Eqs. (25) and (26) is plotted in Fig. 56.

When the critical stress in the compressive steel f_s' is equal to the allowable stress, or $f_s' = f_s$, then Eqs. (25) and (26) become

$$k = \sqrt{2n(p - p') + n^2(p - p')^2} - n(p - p') \tag{29}$$

and

$$j = \frac{k^2 - (k^3/3) + 2p'n(1 - k)[1 - (d'/d)]}{k^2 + 2p'n(1 - k)} \tag{30}$$

Values of j may be determined from Fig. 57 and k from Table 6, using $p - p'$ for p since Eq. (29) is otherwise similar to Eq. (7).

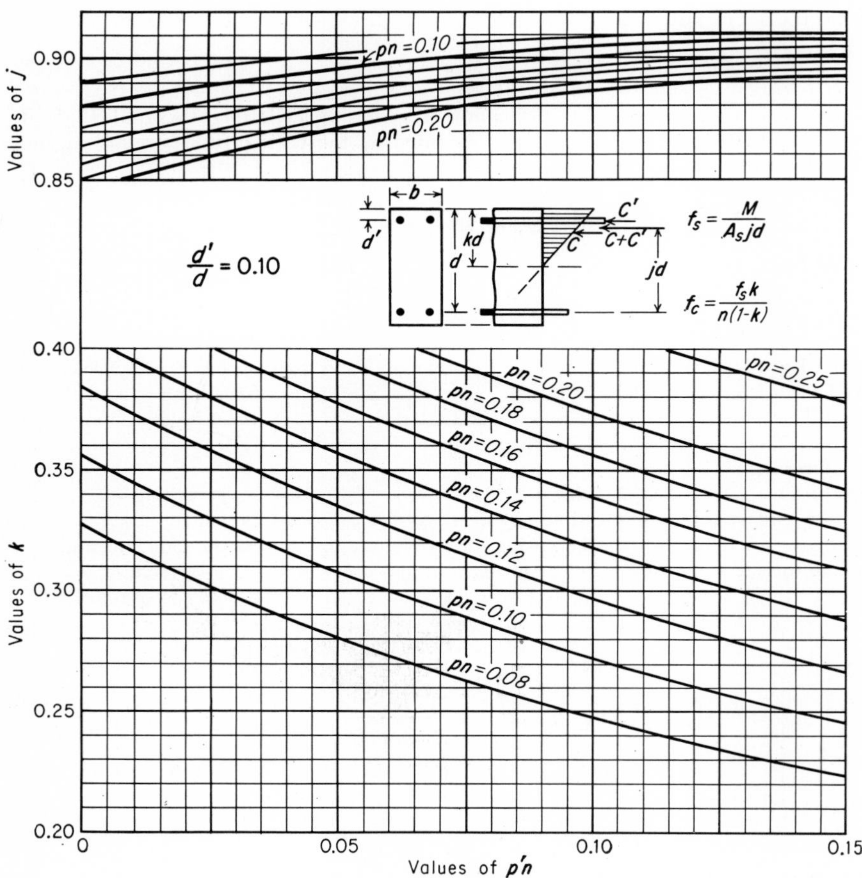

FIG. 56. Rectangular beams reinforced for compression.

Illustrative Example. A rectangular beam is 16 in. wide and 30 in. deep overall and is to carry 3,600,000 in.-lb. The beam is of 3,000 psi concrete (f_c = 1,350 and n = 10) and the reinforcement is of intermediate grade (f_s = 20,000 psi). Determine the steel requirements in accordance with ACI Building Code (Ref. 11).

With 1.5 in. protective cover, the distance from top and bottom to center of steel is 1.5 + 0.25 (ties) + 0.50 (half depth of bar, assumed 1 in.) = 2.25 in. For this case many designers would use 2.0 in.

$$\begin{aligned} M_1 &= 235bd^2 \\ &= (235)(16)(28)^2 = 2{,}950{,}000 \text{ in.-lb} \end{aligned}$$

Also,

$$p = 0.0136$$

Area of tensile steel for balanced reinforcement (to produce stress of 1,350 psi in concrete *without* compression steel),

$$\begin{aligned} A_1 &= pbd \\ &= (0.0136)(16)(28) = 6.10 \text{ sq in.} \end{aligned}$$

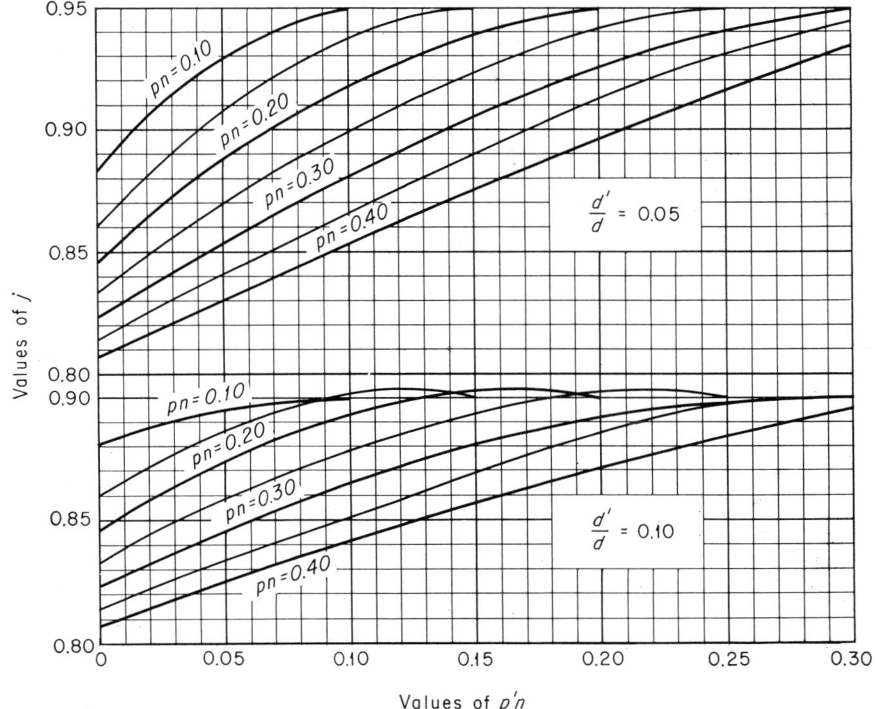

FIG. 57. Rectangular beams reinforced for compression (based on $f_s' = f_s$).

Extra moment for which compressive steel and additional tensile steel are required,

$$M_2 = M - M_1$$
$$= 3{,}600{,}000 - 2{,}950{,}000 = 650{,}000 \text{ in.-lb}$$

By Eq. (17),

$$650{,}000 = 20{,}000A_2(28.0 - 2.0)$$

$$A_2 = \frac{650{,}000}{(20{,}000)(26)} = 1.25 \text{ sq in.}$$

Assuming $f_s = f_s' = 20{,}000$ psi, $A' = A_2$ and the compressive steel = 1.25 sq in., and total tensile steel = 6.10 + 1.25 = 7.35 sq in.

Otherwise, by Eq. (21),

$$f_s' = (2)(20{,}000)\,\frac{0.403 - (2/28)}{1 - 0.403}$$
$$= 22{,}200 \text{ psi}$$

This exceeds the 20,000 psi permissible. The foregoing solution was correct, and this step was unnecessary.

If because of bar-size limitations or other irregularities the areas of steel supplied differ considerably from those calculated, the design may be checked as outlined above, but such a check is rarely necessary.

Shearing Stresses in Reinforced Concrete Beams. The mechanics of homogeneous beams shows that the stresses due to moment and shear may be combined at any point in a beam to produce a tension and a compression acting perpendicular to each other. These are called principal stresses. Figure 58 shows the directions of these principal

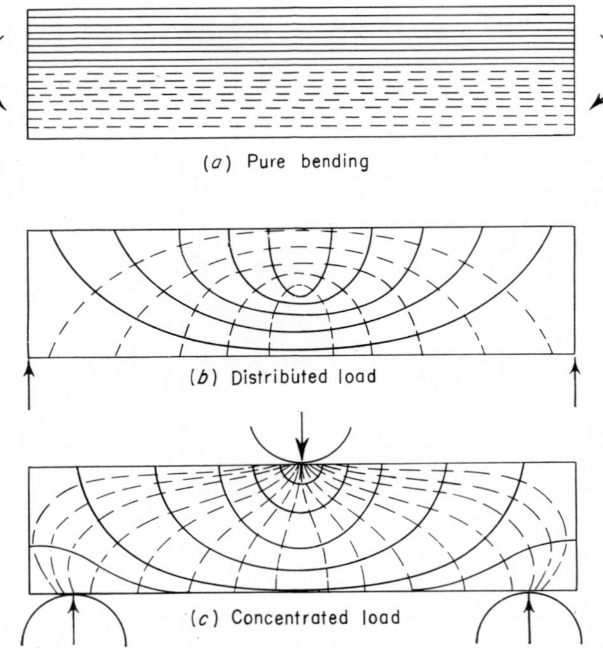

FIG. 58. Lines of principal tension and compression for various loadings. Full lines are tension; broken lines compression.

stresses for some typical cases of loading. It is significant that, when shear is present, these lines become bent and at certain points in a beam are diagonally inclined. Moreover, in concrete, the principal tensile stresses may produce cracking, *the direction of the crack being the same as the direction of the principal compression stress at the point.* These inclined tensions are commonly referred to as *diagonal tensions;* and since they are inclined, owing to the presence of shear, the reinforcement against them is also associated with shear. Actual shear is seldom an important factor in reinforced concrete. It is the diagonal tension which accompanies shear that requires reinforcement. The term shear is universally but loosely used to indicate the diagonal tension that results from the shear.

Although the foregoing is only qualitative, it is sufficient to show that in many cases it is necessary to provide reinforcement crossing diagonal-tension cracks and securely anchored on each side to carry the diagonal tension which the concrete is incapable of resisting. The actual mechanics of the formation of diagonal-tension cracks and the manner in which the steel assumes the load is still not completely understood. It is complicated by many things, among them the shape of the cross section, the type of loading, and the amount of longitudinal steel. The formulas in current use are based on very simplified assumptions, and although they are normally adequate, recent tests

and, perhaps more significant, recent failures (Ref. 12) of structures which were designed according to generally accepted practice yet showed diagonal-tension failure, have pointed up the need for a more rational design procedure. It is likely that the near future will see radical changes in design methods for resisting diagonal tension. It is impossible to overemphasize the importance of diagonal-tension reinforcement. It is, perhaps, the subject of the most important laboratory research in reinforced concrete being conducted at present. Because of this unsettled state, many designers believe it wise to be conservative in providing diagonal-tension reinforcement and to include at least nominal web reinforcement throughout the length of a beam, even when the conventional approach indicates that it is unnecessary.

The conventional formula for the shear as a measure of diagonal tension in a beam is

$$v = \frac{V}{bjd} \tag{31}$$

In T beams the stem only is available for shear reinforcement; so b' the stem width, rather than b, should be used. It is normally sufficiently accurate to assume $j = 7/8$ in this formula.

Reinforcement against Diagonal Tension. The first essential in preventing diagonal-tension failure is adequate, well-anchored, longitudinal steel. Since, in general, the failure would take a direction upward and toward the center of the span, (1) vertical stirrups or (2) diagonal stirrups or (3) inclined bars, or a combination of these, serve as means to reinforce against such a failure. Figure 59 shows typical bar arrangements.

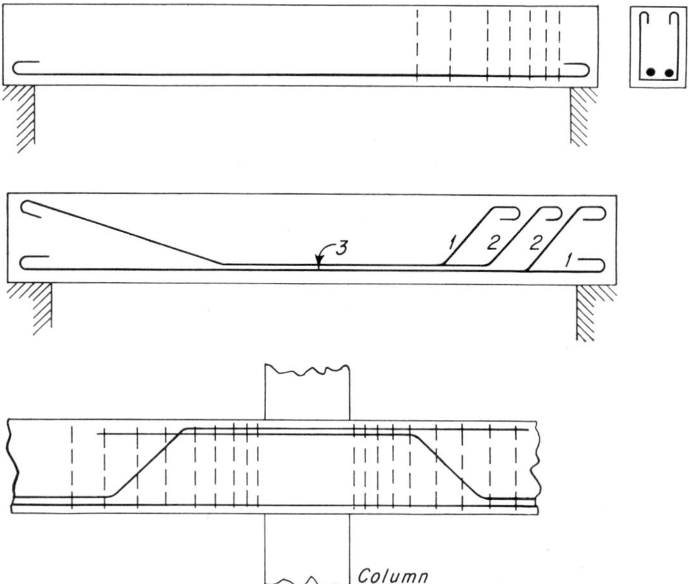

Fig. 59. Various bar arrangements for resisting moment and shear stresses.

The 1956 ACI Code requirements for diagonal-tension reinforcement follow. These account for the fact that tests have shown that, in some cases, the concrete will resist shear even though the conventional formula would indicate a cracked section.

"801—Shearing unit stress

"(*a*) The shearing unit stress v, as a measure of diagonal tension, in reinforced concrete flexural members shall be computed by formula (2):

$$v = \frac{V}{bjd} \tag{2}$$

"(*b*) For beams of I- or T-section, b' shall be substituted for b in formula (2).

"(*c*) In concrete joist floor construction, where burned clay or concrete tile are used, b' may be taken as a width equal to the thickness of the concrete web plus the thickness of the vertical shells of the concrete or burned clay tile in contact with the joist as in Section 708(b).

"(*d*) Wherever the value of the shearing unit stress computed by formula (2) exceeds the shearing unit stress v_c permitted on the concrete of an unreinforced web (see Section 305), web reinforcement shall be provided to carry the excess. Such reinforcement shall also be provided for a distance equal to the depth, d, of the member beyond the point theoretically required.

"(*e*) Where continuous or restrained beams or frames do not have a slab so cast as to provide T-beam action, the following provisions shall apply. Web reinforcement shall be provided from the support to a point beyond the extreme position of the point of inflection a distance equal to either 1/16 of the clear span or the depth of the member, whichever is greater, even though the shearing unit stress does not exceed v_c. Such reinforcement shall be designed to carry at least two-thirds of the total shear at the section. Web reinforcement shall be provided sufficient to carry at least two-thirds of the total shear at a section in which there is negative reinforcement.

"802—Types of web reinforcement

"(*a*) Web reinforcement may consist of:

"1. Stirrups or web reinforcing bars perpendicular to the longitudinal steel.

"2. Stirrups or web reinforcing bars welded or otherwise rigidly attached to the longitudinal steel and making an angle of 30 deg or more thereto.

"3. Longitudinal bars bent so that the axis of the inclined portion of the bar makes an angle of 15 deg or more with the axis of the longitudinal portion of the bar.

"4. Special arrangements of bars with adequate provisions to prevent slip of bars or splitting of the concrete by the reinforcement [see Section 804(f)].

"(*b*) Stirrups or other bars to be considered effective as web reinforcement shall be anchored at both ends, according to the provisions of Section 904.

"803—Stirrups

"(*a*) The area of steel required in stirrups placed perpendicular to the longitudinal reinforcement shall be computed by formula (3).

$$A_v = \frac{V's}{f_v jd} \tag{3}$$

"(*b*) Inclined stirrups shall be proportioned by formula (5) [Section 804(d)].

"(*c*) Stirrups placed perpendicular to the longitudinal reinforcement shall not be used alone as web reinforcement when the shearing unit stress, v, exceeds $0.08f_c'$ or 240 psi.

"804—Bent bars

"(*a*) Only the center three-fourths of the inclined portion of any longitudinal bar that is bent up for web reinforcement shall be considered effective for that purpose,

and such bars shall be bent around a pin having a diameter not less than six times the bar size.

"(*b*) When the web reinforcement consists of a single bent bar or of a single group of parallel bars all bent up at the same distance from the support, the required area of such bars shall be computed by formula (4).

$$A_v = \frac{V'}{f_v \sin \alpha} \qquad (4)$$

"(*c*) In formula (4), V' shall not exceed $0.04\, f_c'bjd$, or $120\, bjd$.

"(*d*) Where there is a series of parallel bars or groups of bars bent up at different distances from the support, the required area shall be determined by formula (5).

$$A_v = \frac{V's}{f_v jd\,(\sin \alpha + \cos \alpha)} \qquad (5)$$

"(*e*) When bent bars having a radius of bend of at least six bar diameters are used alone as web reinforcement, they shall be so spaced that the effective inclined portion described in Section 804(a) meets the requirements of Section 806, and the allowable shearing unit stress shall not exceed $0.08\, f_c'$ nor 240 psi.

"(*f*) The shearing unit stress permitted when special arrangements of bars are employed shall be that determined by making comparative tests, to destruction, of specimens of the proposed system and of similar specimens reinforced in conformity with the provisions of this code, the same factor of safety being applied in both cases.

"805—Combined web reinforcement

"(*a*) Where more than one type of reinforcement is used to reinforce the same portion of the web, the total shearing resistance of this portion of the web shall be assumed as the sum of the shearing resistances computed for the various types separately. In such computations the shearing resistance of the concrete shall be included only once, and no one type of reinforcement shall be assumed to resist more than $2V'/3$.

"806—Maximum spacing of web reinforcement

"(*a*) Where web reinforcement is required it shall be so spaced that every 45-deg line (representing a potential crack) extending from the middepth of the beam to the longitudinal tension bars shall be crossed by at least one line of web reinforcement. If a shearing unit stress in excess of $0.06\, f_c'$ is used, every such line shall be crossed by at least two such lines of web reinforcement.

"807—Minimum web reinforcement

"Where web reinforcement is required, the amount used shall be not less than 0.15 percent of the area computed as the product of the width of the member at middepth and the horizontal spacing of the web reinforcement.

"808—Shearing stresses in flat slabs [see Section 1002(c)]

"809—Shear and diagonal tension in footings

"(*a*) In isolated footings the shearing unit stress computed by formula (2) on the critical section [see Section 1205(a)] shall not exceed $0.03\, f_c'$ nor in any case shall it exceed 75 psi."

In a recent paper (Ref. 13), Whitney and Cohen suggest that to these specifications be added the following:

"Web reinforcement shall be provided from the support to a point beyond the extreme position of the point of inflection a distance equal to either 1/16 of the clear span

or the depth of the member, whichever is greater even though the shearing unit stress does not exceed v_c. Web reinforcement shall be provided at every section in which negative reinforcement is required. Where required by this paragraph, the amount of web reinforcement at each section shall be the maximum required by any one of the following:

"1. Sufficient to carry ⅔ of the total shear where the unit stress exceeds v_c.

"2. Sufficient to carry ⅔ of the total shear existing at the point of inflection, that is, the ratio of web reinforcement required at the point of inflection will be maintained back to the support.

"3. Not less than 0.15 percent of the area computed as the product of the width of member at middepth and the horizontal spacing of the web reinforcement."

Bond and Anchorage. Reinforcement embedded in concrete is gripped by adhesion and frictional or lug resistance to sliding, known as *bond*. When steel bars are placed in a beam, there must be sufficient bond between the concrete and steel to prevent the bars from slipping relative to the concrete when stressed. The bond stress per square inch of bond surface may be determined as follows. Consider a short section of beam in which the bending moment is higher on the right-hand face than on the left. Taking moments about the point a (see Fig. 60):

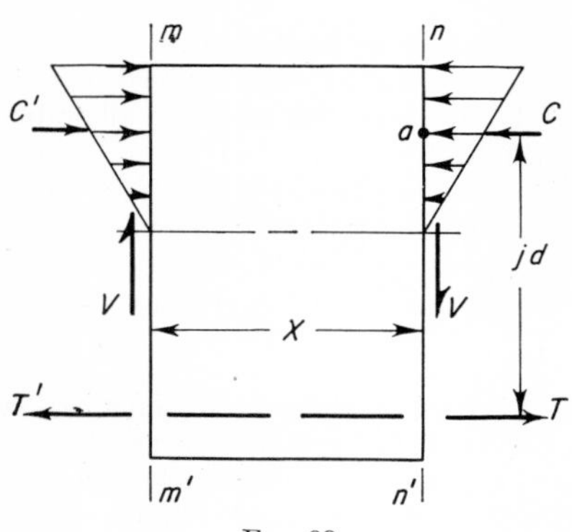

FIG. 60

$$(T - T')jd = Vx$$

$$\frac{T - T'}{x} = \frac{V}{jd}$$

But $(T - T')$ represents the force which tends to pull the bars out of the concrete in a length x. Hence $(T - T')/x$ represents this force per unit length of beam. Therefore V/jd is the bond force per unit of length between the two materials. If u is the bond stress per unit area of steel surface, and Σo the total perimeter of steel,

$$u = \frac{V}{\Sigma ojd} \tag{32}$$

This equation applies to steel in tension only.

In order to prevent the pulling out of the bars when stressed, the value of the unit bond stress as computed by Eq. (32) should not exceed the safe working limit which has been established from studies of tests in which bond failure has occurred.

As seen from the above, the increment of force in the steel between any two sections of the beam should be provided for by ensuring that bond stress between the two sections is not excessive. In addition, the bar must be adequately anchored in the concrete to the left of section mm', so that the total force T' is developed without danger of pulling the bar out of that portion of the beam. Likewise, the force T must be developed by adequate total anchorage in the concrete on the right. In many cases, this anchorage is developed by the incremental bond stresses. In the simple uniformly loaded beam shown in Fig. 61, assume for the purpose of illustration that the reinforce-

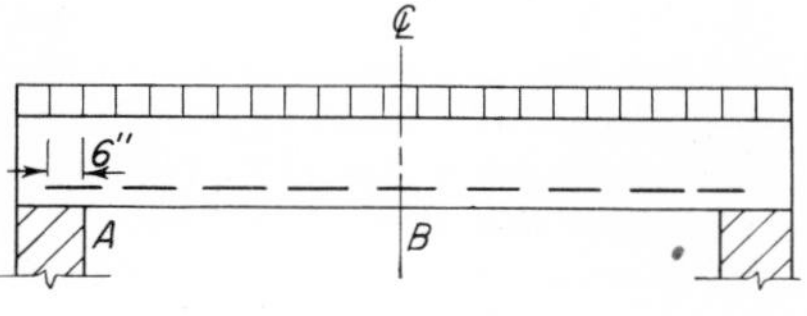

FIG. 61

ment is the same for the entire length of beam and is working at the full allowable at point B, the center line. At A the tensile force in the steel is of very small magnitude and will be developed by a nominal amount of embedment in the beam over the support. Since the shear is a maximum at A, this will be the critical section for bond. If the bond stress is within the allowable at that point—and consequently at all other points—then there will be a gradual build-up of force in the bar, and, if the distance between A and B is sufficient, the full allowable force in the bar may be developed by bond before the critical section B is reached.

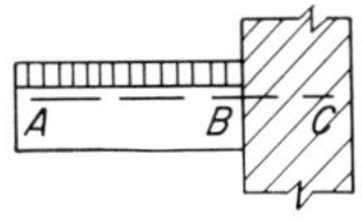

Fig. 62

On the other hand, consider a cantilever beam or slab projecting from a wall as in Fig. 62. The maximum bar force will occur at B. This force must be developed between A and B and also between B and C. Frequently, it will be found that, although there is sufficient length between A and B to develop this force by normal bond, the length of embedment between B and C is limited (say by wall thickness) and the proper anchorage on that side must be developed by hooks or bends, or a combination of these and bond.

Another place in which bond stresses are of importance is in splicing bars. If splices are not made by direct welding of bars together, then the force must travel from one bar to the intervening concrete by bond and then by bond again to the other bar.

To develop the full strength of a bar in bond—in either anchoring or splicing—the length of embedment or lap l must be such that

$$\pi dlu = f_s \frac{\pi d^2}{4}$$

$$l = \frac{f_s}{4u} d \tag{33}$$

For example, if $f_s = 20{,}000$, and $u = 300$, then $l/d = 20{,}000/4 \times 300 = 16.7$, or the bars must be embedded or lapped a distance equal to approximately 17 diameters. A more conservative value is usually used because of the fact that the bond stresses in such cases are not actually distributed uniformly as assumed, but peak near the ends of the bars or embedding medium, and also because a bond or pullout failure can be very critical. It is also undesirable to count on the full allowable bond stresses being developed by concrete in tension because of the fact that such concrete may be cracked.

Further insight into the differences and similarities between bond and anchorage forces can be gained by a study of the two beams shown in Fig. 63. Assuming the same width, effective depth, steel area, and, in beam 2, end plates sufficient to anchor the ends of the steel, the resisting moment of the two sections is virtually identical even though in beam 2 no bond forces can be developed. Thus, adequate end anchorage can ensure complete development of the reinforcement even though normal bond forces are inactive. There are significant differences in the two members, however. Beam 1 acts in the normal, reasonably predictable way, with the total bar force being developed gradually in accordance with the needs of the bending moment diagram ($f_s = M_s/A_sjd$). Beam 2 acts somewhat like a tied arch, and the bar force in the steel must be constant over the whole span. The action is less predictable, but it is probably true that beam 2 will deflect more than 1 under a given load. In recognition of the value of end anchorage, previous ACI specifications allowed higher shear stresses where special anchorage in the form of hooks or bar extensions was

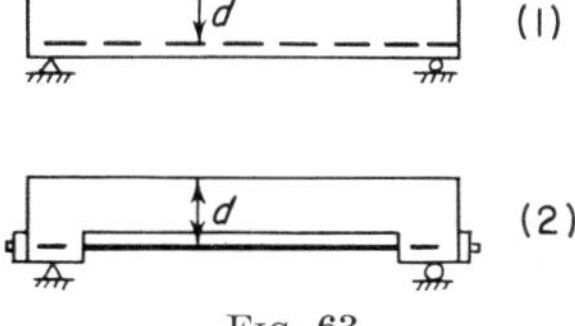

Fig. 63

provided. Such anchorage has the tendency, when combined with bond stresses, to reduce the cracking of beams.

The 1956 ACI Code requirements for bond and anchorage are as follows:

"900—Notation

d = depth from compression face of beam or slab to centroid of longitudinal tensile reinforcement

f_c' = compressive strength of concrete at age of 28 days unless otherwise specified

j = ratio of distance between centroid of compression and centroid of tension to the depth d

Σo = sum of perimeters of bars in one set

u = bond stress per unit of surface area of bar

V = total shear

"901—Computation of bond stress in beams

"(*a*) In flexural members in which the tensile reinforcement is parallel to the compression face, the bond stress at any cross section shall be computed by formula (6).

$$u = \frac{V}{\Sigma ojd} \tag{6}$$

in which V is the shear at that section and Σo is taken as the perimeter of all effective bars crossing the section on the tension side. Bent-up bars that are not more than $d/3$ from the level of the main longitudinal reinforcement may be included. Critical sections occur at the face of the support, at each point where tension bars terminate within a span, and at the point of inflection.

"(*b*) Bond shall be similarly computed on compressive reinforcement, but the shear used in computing the bond shall be reduced in the ratio of the compressive force assumed in the bars to the total compressive force at the section. Anchorage shall be provided by embedment past the section to develop the assumed compressive force in the bars at the bond stress in Table 305(a).

"(*c*) Adequate end anchorage shall be provided for the tensile reinforcement in all flexural members to which formula (6) does not apply, such as sloped, stepped or tapered footings, brackets or beams in which the tensile reinforcement is not parallel to the compression face.

"902—Anchorage requirements

"(*a*) Tensile negative reinforcement in any span of a continuous, restrained or cantilever beam, or in any member of a rigid frame shall be adequately anchored by bond, hooks, or mechanical anchors in or through the supporting member. Within any such span every reinforcing bar, except in a lapped splice, whether required for positive or negative reinforcement, shall be extended at least 12 diameters beyond the point at which it is no longer needed to resist stress. At least one-third of the total reinforcement provided for negative moment at the support shall be extended beyond the extreme position of the point of inflection a distance sufficient to develop by bond one-half the allowable stress in such bars, not less than $1/16$ of the clear span length, or not less than the depth of the member, whichever is greater. The tension in any bar at any section must be properly developed on each side of the section by hook, lap, or embedment (see Section 906). If preferred, the bar may be bent across the web at an angle of not less than 15 deg with the longitudinal portion of the bar and be made continuous with the reinforcement which resists moment of opposite sign.

"(*b*) Of the positive reinforcement in continuous beams not less than one-fourth the area shall extend along the same face of the beam into the support a distance of 6 in.

"(*c*) In simple beams, or at the freely supported end of continuous beams, at least one-third the required positive reinforcement shall extend along the same face of the beam into the support a distance of 6 in.

"903—Plain bars in tension

"Plain bars in tension shall terminate in standard hooks except that hooks shall not be required on the positive reinforcement at interior supports of continuous members.

"904—Anchorage of web reinforcement

"(*a*) The ends of bars forming simple U- or multiple stirrups shall be anchored by one of the following methods:

"1. By a standard hook, considered as developing 10,000 psi, plus embedment sufficient to develop by bond the remaining stress in the bar at the unit stress specified in Table 305(a). The effective embedded length of a stirrup leg shall be taken as the distance between the middepth of the beam and the tangent of the hook.

"2. Welding to longitudinal reinforcement.

"3. Bending tightly around the longitudinal reinforcement through at least 180 deg.

"4. Embedment above or below the middepth of the beam on the compression side, a distance sufficient to develop the stress to which the bar will be subjected at a bond stress of not to exceed 0.045 f_c' on plain bars nor 0.10 f_c' on deformed bars, but, in any case, a minimum of 24 bar diameters.

"(*b*) Between the anchored ends, each bend in the continuous portion of a U- or multiple U-stirrup shall be made around a longitudinal bar.

"(*c*) Hooking or bending stirrups around the longitudinal reinforcement shall be considered effective only when these bars are perpendicular to the longitudinal reinforcement.

"(*d*) Longitudinal bars bent to act as web reinforcement shall, in a region of tension, be continuous with the longitudinal reinforcement. The tensile stress in each bar shall be fully developed in both the upper and the lower half of the beam as specified in Section 904(a)1 or 904(a)4.

"(*e*) In all cases web reinforcement shall be carried as close to the compression surface of the beam as fireproofing regulations and the proximity of other steel will permit.

"905—Anchorage of bars in footing slabs

"(*a*) Plain bars in footing slabs shall be anchored by means of standard hooks. The outer faces of these hooks and the ends of deformed bars shall be not less than 3 in. nor more than 6 in. from the face of the footing.

"906—Hooks

"(*a*) The terms "hook" or "standard hook" as used herein shall mean either

"1. A complete semicircular turn with a radius of bend on the axis of the bar of not less than three and not more than six bar diameters, plus an extension of at least four bar diameters at the free end of the bar, or

"2. A 90-deg bend having a radius of not less than four bar diameters plus an extension of 12 bar diameters, or

"3. For stirrup anchorage only, a 135-deg turn with a radius on the axis of the bar of three diameters plus an extension of at least six bar diameters at the free end of the bar.

"Hooks having a radius of bend of more than six bar diameters shall be considered merely as extensions to the bars.

"(*b*) No hook shall be assumed to carry a load which would produce a tensile stress in the bar greater than 10,000 psi.

"(*c*) Hooks shall not be considered effective in adding to the compressive resistance of bars.

"(*d*) Any mechanical device capable of developing the strength of the bar without damage to the concrete may be used in lieu of a hook. Tests must be presented to show the adequacy of such devices."

The ACI Code specifies lower allowable bond stresses for "top" bars than for others. This is primarily because of the fact that when concrete is placed around bars which are in the top of a form there is a tendency for the concrete on the lower side to pull away from the bar as it sets up and shrinks, leaving small voids under the bar and destroying the adhesion locally.

Determination of Moments and Shears in Reinforced Concrete Beams. In order to design any beam, it is necessary to know the moments and shears at the critical sections. These may be determined by direct analysis, using the accepted methods based on elastic theory, or as is frequently done in reinforced concrete design, by using moment and shear coefficients which have been found adequate for usual combinations of spans and live-to-dead-load ratios. To this end, the following requirements are given in the 1956 ACI Code:

"700—Notation

b = width of rectangular flexural member or width of flanges for T- and I-sections
b' = width of web in T and I flexural members
d = depth from compression face of beam or slab to centroid of longitudinal tensile reinforcement; the diameter of a round bar
E = modulus of elasticity
I = moment of inertia of a section about the neutral axis for bending
l = span length of slab or beam
l' = clear span for positive moment and shear and the average of the two adjacent clear spans for negative moment (see Section 701)
t = minimum total thickness of slab
w = uniformly distributed load per unit of length of beam or per unit area of slab

"701—General requirements

"(*a*) All members of frames or continuous construction shall be designed to resist at all sections the maximum moments and shears produced by dead load, live load, earthquake and wind load, as determined by the theory of elastic frames in which the simplified assumptions of Section 702 may be used.

"(*b*) Approximate methods of frame analysis are satisfactory for buildings of usual types of construction, spans, and story heights.

"(*c*) In the case of two or more approximately equal spans (the larger of two adjacent spans not exceeding the shorter by more than 20 percent) with loads uniformly distributed, where the unit live load does not exceed three times the unit dead load, design for the following moments and shears is satisfactory:

Positive moment
End spans
If discontinuous end is unrestrained. $\frac{1}{11} wl'^2$
If discontinuous end is integral with the support. . . . . $\frac{1}{14} wl'^2$
Interior spans. . . . . $\frac{1}{16} wl'^2$

Negative moment at exterior face of first interior support

Two spans . . . . $\frac{1}{9} wl'^2$

More than two spans . . . . $\frac{1}{10} wl'^2$

Negative moment at other faces of interior supports . . . . $\frac{1}{11} wl'^2$

Negative moment at face of all supports for, (a) slabs with spans not exceeding 10 ft, and (b) beams and girders where ratio of sum of column stiffnesses to beam stiffness exceeds eight at each end of the span . . . . $\frac{1}{12} wl'^2$

Negative moment at interior faces of exterior supports for members built integrally with their supports

Where the support is a spandrel beam or girder . . . . $\frac{1}{24} wl'^2$

Where the support is a column . . . . $\frac{1}{16} wl'^2$

Shear in end members at first interior support . . . . $1.15 \frac{wl'}{2}$

Shear at all other supports . . . . $\frac{wl'}{2}$ "

Illustrative Example. Design the continuous beam shown in Fig. 64. The beam is part of a typical one-way slab beam-and-girder-floor system. Use 1956 ACI Code, $f_c' = 3{,}000$ psi, uniform load (dead, including beam, plus live) = 2,010 lb per ft of beam.

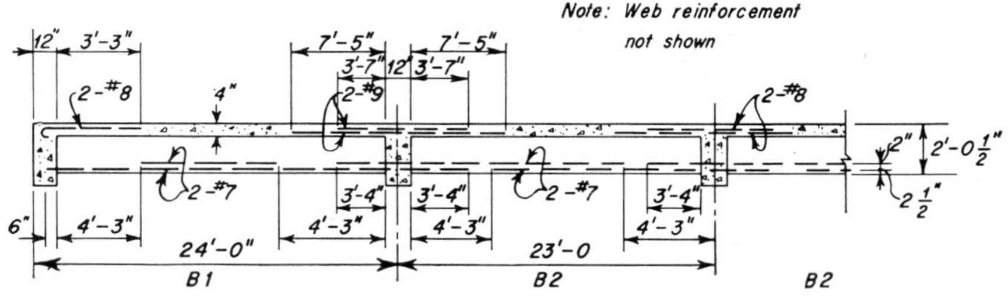

Fig. 64

From Fig. 64 clear span for B1 = 22′6″, B2 = 22′0″. Average clear span = 22′3″. Stirrups will be provided to take the diagonal-tension stresses, and no main reinforcement will be bent. Although the ACI Code permits a unit shearing stress of $0.08f_c'$ for this type of reinforcement, a depth selected on this basis would require heavy horizontal reinforcement; therefore the larger depth selected below, which is based on $v = 0.06f_c'$, will be used. Two rows of reinforcement are used in most cases, both at the top and bottom of the beam. The center of the row nearest the surface is placed 2½ in. from the surface, with the other row spaced 2 in. (center-to-center) farther in.

For B1,

Center:

$$M = +\tfrac{1}{14}(2{,}010)(22.5)^2(12) = 874{,}000 \text{ in.-lb}$$

At continuous end:

$$M = -\tfrac{1}{10}(2{,}010)(22.25)^2(12) = -1{,}194{,}000 \text{ in.-lb}$$

At wall end:

$$M = -\tfrac{1}{24}(2{,}010)(22.5)^2(12) = -510{,}000 \text{ in.-lb}$$

$$V_{\max} = \frac{1.15(2{,}010)(22.5)}{2} = 26{,}000 \text{ lb}$$

$$b'd = \frac{26{,}000}{0.06(3{,}000)(\tfrac{7}{8})} = 165 \text{ sq in.}$$

Use $b' = 8''$, $d = 21''$.

Center (Note: T beam):

$$A_s = \frac{874{,}000}{20{,}000(21 - 2)} = 2.30 \text{ sq in.}$$

Use 4 No. **7** = 2.41 sq in.

At continuous end (Note: Reinforce for compression):

$$M_1 = 235(8)(21)^2 = 829{,}000 \text{ in.-lb}$$

$$A_1 = \frac{829{,}000}{20{,}000(0.866)(21)} = 2.28 \text{ sq in.}$$

$$M_2 = 1{,}194{,}000 - 829{,}000 = 365{,}000 \text{ in.-lb}$$

$$A_2 = \frac{365{,}000}{20{,}000(21 - 3.5)} = 1.04 \text{ sq in.}$$

$$A_s = 3.32 \text{ sq in.}$$

Use 4 No. 9 = 4.00 sq in.

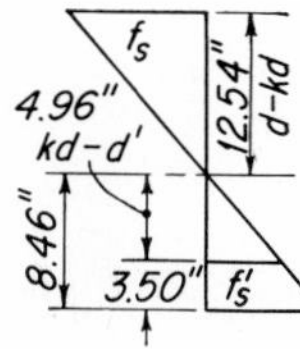

$$f_s' = 2(20{,}000)\left[\frac{0.403 - (3.5/21)}{1 - 0.403}\right] = 15{,}820 \text{ psi}$$

$$A' = \frac{365{,}000}{15{,}820(21 - 3.5)} = 1.32 \text{ sq in.}$$

Supplied by 4 No. **7** carried through.

At wall end, assuming 1 row of steel

$$A_s = \frac{510{,}000}{20{,}000(0.866)(22)} = 1.34 \text{ sq in.}$$

Use 2 No. 8 = 1.57 sq in.

Shear and bond:
At wall end:

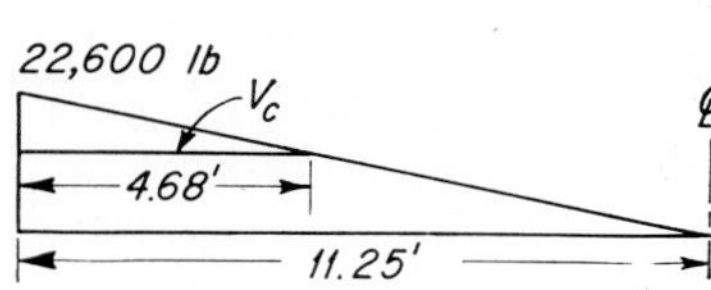

$$V = \frac{2{,}010(22.5)}{2} = 22{,}600 \text{ lb}$$

$$V_c = (0.03)(3{,}000)(\tfrac{7}{8})(8)(21) = 13{,}200 \text{ lb}$$

$$u = \frac{22{,}600}{6.3(\tfrac{7}{8})(22)} = 187 \text{ psi} < 210$$

$$s = \frac{(0.22)(20{,}000)(\tfrac{7}{8})(21)}{22{,}600 - 13{,}200} = 8.6''$$

Placing the first stirrup 2″ from the face of the support, the spacings will be:
At continuous end:

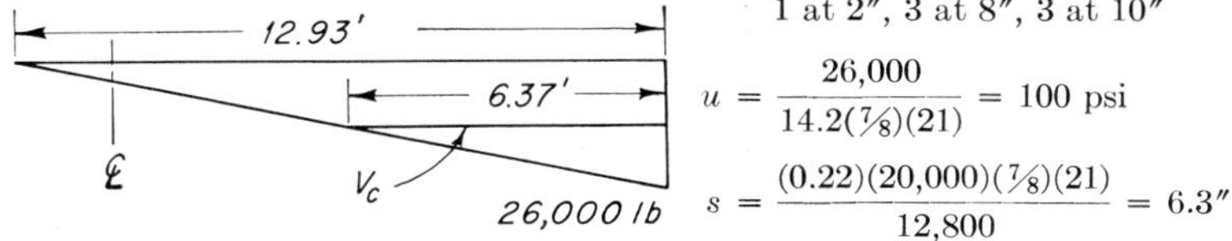

1 at 2″, 3 at 8″, 3 at 10″

$$u = \frac{26,000}{14.2(7/8)(21)} = 100 \text{ psi}$$

$$s = \frac{(0.22)(20,000)(7/8)(21)}{12,800} = 6.3''$$

Placing the first stirrup 2″ from the face of the support, the spacings will be:

1 at 2″, 4 at 6″, 5 at 8″, 1 at 10″

For B2,
Center:

$$M = +1/16(2,010)(22)^2(12) = 730,000 \text{ in.-lb}$$

At first interior support:

$$M = -1/11(2,010)(22.25)^2(12) = -1,085,000 \text{ in.-lb}$$

At all other interior supports:

$$M = -1/11(2,010)(22)^2(12) = -1,061,000 \text{ in.-lb}$$

$$V_{\max} = \frac{2,010(22)}{2} = 22,100 \text{ lb}$$

Beam dimensions will be kept the same throughout, B1 governs.
Center:

$$A_s = \frac{730,000}{20,000(21 - 2)} = 1.92 \text{ sq in.}$$

Use 4 No. 7 = 2.41 sq in.

First interior support is the same as the continuous end of B1.
At other interior supports:

$$M_1 = 829,000 \text{ in.-lb}$$
$$A_1 = 2.28 \text{ sq in.}$$
$$M_2 = 1,061,000 - 829,000 = 232,000 \text{ in.-lb}$$

$$A_2 = \frac{232,000}{20,000(21 - 3.5)} = 0.66 \text{ sq in.}$$

$$A_s = 2.94 \text{ sq in.}$$

Use 4 No. 8 = 3.14 sq in.

$$f_s' = 15,820 \text{ psi}$$

$$A' = \frac{232,000}{15,820(21 - 3.5)} = 0.84 \text{ sq in.}$$

Supplied by 4 No. 7 carried through.

Shear and bond:
Shear in B2 is symmetrical about the ℄

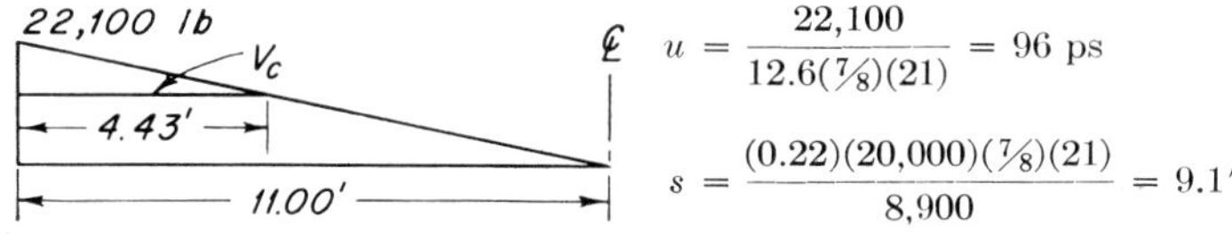

$$u = \frac{22,100}{12.6(7/8)(21)} = 96 \text{ ps}$$

$$s = \frac{(0.22)(20,000)(7/8)(21)}{8,900} = 9.1''$$

Placing the first stirrups 3″ from the faces of the supports, the spacings at each end will be:

1 at 3″, 3 at 8″, 3 at 10″

Cutoff Points for Reinforcement. While economy demands that bars be cut off where they are no longer needed for stress, the designer must be sure that there is sufficient steel at all sections, in addition to the sections of highest moment, and that the bars are anchored adequately. The diagrams of Figs. 65 and 66 are helpful in determining

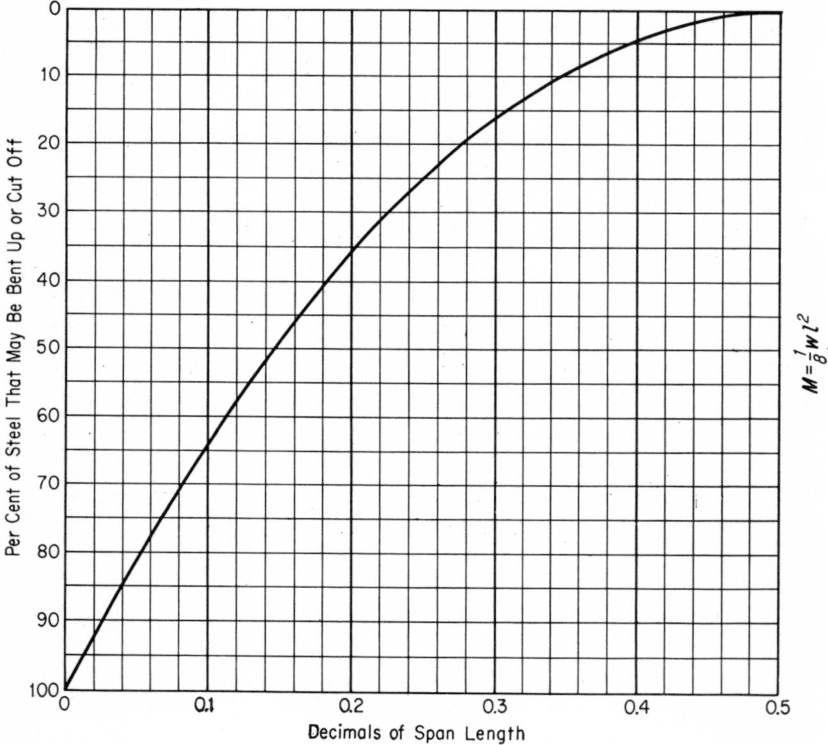

FIG. 65. Points where bars may be bent up or cut off for simply supposed uniformly loaded beams.

the theoretical cutoff points for uniformly loaded beams with different conditions of end restraint. These diagrams are essentially moment diagrams drawn to proper scales. To the cutoff distances determined from the diagrams must be added the proper length of embedment for anchorage as specified by the ACI or other code. The use of these diagrams will be illustrated by application to the previous illustrative example.

For B1 of the previous example: Negative-moment reinforcement at outer support required for 0.09 × (22 ft 6 in.) + 12 × 1 in. = 3 ft 1 in. from face of support. 50 percent of positive-moment reinforcement required for

$$0.23 \times (22 \text{ ft } 6 \text{ in.}) - 12 \times \tfrac{7}{8} = 4 \text{ ft } 3 \text{ in.}$$

from face of outer support. Remaining two No. 7 bars carried 6 in. into support. At continuous support, two No. 7 bars of positive-moment reinforcement are cut off 4 ft 3 in.

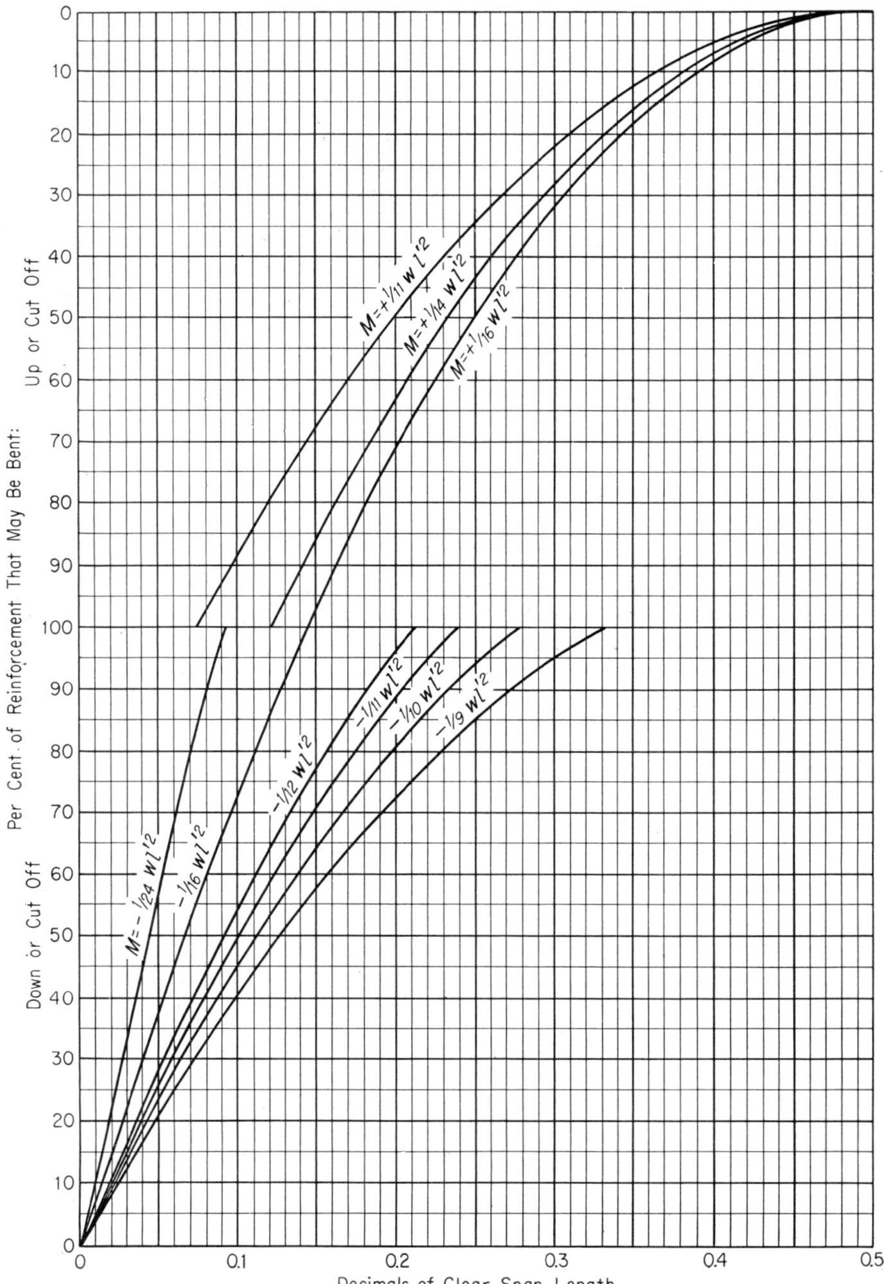

FIG. 66. Approximate points where bars may be bent up or down or cut off for continuous uniformly loaded beams built integrally with their supports (ACI moment coefficients).

from face of support; 50 percent of negative-moment reinforcement at continuous support is cut off at 0.11 × (22 ft 6 in.) + 12 × 1.128 in. = 3 ft 7 in. from face of support and remaining two No. 9 bars

$$0.28 \times (22 \text{ ft } 6 \text{ in.}) + 13.5 \text{ in.} = 7 \text{ ft } 5 \text{ in.}$$

from face of support.

For B2, 50 percent of positive-moment reinforcement required for

$$0.25 \times 22 \text{ ft} - 12 \times \tfrac{7}{8} \text{ in.} = 4 \text{ ft } 7 \text{ in.}$$

from face of support. Negative-moment reinforcement at first interior support same as B1. At other interior supports 50 percent of negative-moment reinforcement may be cut off at

$$0.10 \times 22 \text{ ft} + 12 \times \tfrac{7}{8} \text{ in.} = 3 \text{ ft } 1 \text{ in.}$$

from face of support and remaining two No. 8 bars

$$0.24 \times 22 \text{ ft} + 12 \times \tfrac{7}{8} \text{ in.} = 6 \text{ ft } 2 \text{ in.}$$

from face of support.

COLUMNS—CONCENTRICALLY LOADED

Types of Columns. Concrete compression members whose unsupported length is more than four times the least dimension of the cross section are classified as columns. Such members should not be built without reinforcement of some type. In modern construction four types of reinforced concrete columns are used, namely:

1. Columns reinforced with longitudinal steel and closely spaced spirals.
2. Columns reinforced with longitudinal steel and lateral ties.
3. Composite columns, in which a structural-steel or cast-iron column is thoroughly encased in a concrete core of Type 1 or 2.
4. Combination columns, in which a structural-steel column is wrapped with wire and encased in at least 2½ in. of concrete over all metal except rivet heads.

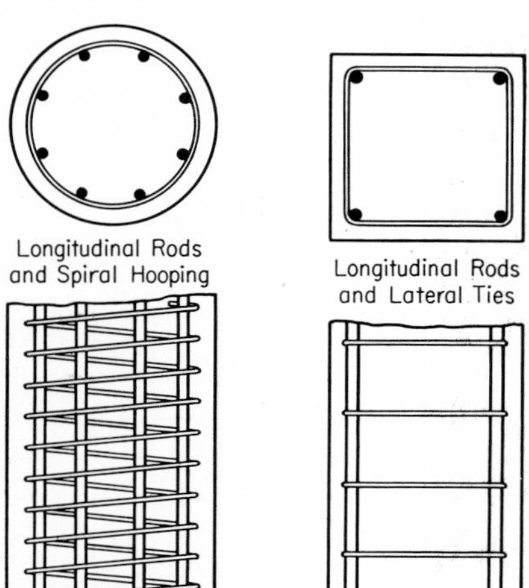

Fig. 67

Types 1 and 2 are more generally used, Types 3 and 4 being economical with heavy construction loads or extremely heavy permanent loads.

Pipe columns, in which a steel pipe is filled with concrete, are sometimes used.

Columns with Spiral and Longitudinal Reinforcement. Whenever a material is subjected to compression in one direction, there will be an expansion in the direction perpendicular to the compression axis. Where this expansion is resisted, lateral compressive stresses are developed, which tend to neutralize the effect of the longitudinal compressive stress and thus increase resistance against failure. This is the principle involved in the use of spiral or hooped reinforcement (see Fig. 67). Within the limit of elasticity the hooped reinforcement is much less effective than longitudinal reinforcement. Such reinforcement, however, raises the ultimate strength of the column, because the hooping delays ultimate failure of the concrete. The concrete continues to compress

and to expand laterally, thus increasing the tension in the bands, while final failure occurs upon the excessive stretching or breaking of the hooping or buckling of the column as a whole. Thus a somewhat higher working stress may be employed on the concrete contained within such hooping than on a concrete not so confined. Tests show that about 1 percent of closely spaced spiral hooping increases the resistance to ultimate failure sufficiently to allow a reasonable increase in the working stress in the concrete.

As long as the bond between the steel and the concrete is effective, the two materials will deform equally, and the intensities of the stresses will be proportional to their moduli of elasticity. That is, since the deformation λ_c of the concrete must be equal to the deformation λ_s of the steel,

$$\lambda_c = \frac{f_c}{E_c} = \lambda_s = \frac{f_s}{E_s}$$

Therefore

$$f_s = nf_c$$

Let A_c = net area of concrete in this section, measured to the outside diameter of the spiral
A_s = area of longitudinal steel in this section
A_g = over-all, or gross, area of concrete section
p_g = steel ratio A_s/A_g
f_s = unit compressive stress in steel
f_c = unit compressive stress in concrete
P = total strength of reinforced column

Then

$$\begin{aligned} P = f_cA_g + f_sA_s &= f_c(A_g - p_gA_g) + f_cnp_gA_g \\ &= f_cA_g[1 + (n-1)p_g] \\ &= f_c[A_g + (n-1)A_s] \end{aligned} \tag{34}$$

The analysis just given, which showed that $f_s = nf_c$, was based on the assumption that both concrete and steel are elastic materials. However, as stated previously, concrete is not an elastic material. As the stress increases, the strains increase at a greater rate. Furthermore, under slow or sustained loads, concrete will continue to flow or creep under constant stress. It has been amply verified by many tests that, before a spiral reinforced column fails, the average stress in the concrete will reach approximately $0.85f_c'$ and the steel stress will reach the yield point. This, of course, assumes that buckling is not a factor. This has been recognized by the ACI Code, and in that code the design formulas for concentrically loaded columns have been based on ultimate-strength considerations since 1936. Failure may then be represented by the formula

$$P_{ult} = 0.85f_c'A_g + f_yA_s \tag{35}$$

where f_y is the yield point of the steel.

If the elastic strength formula were applicable at failure (which it is not), it would follow from Eq. (34) that:

$$P_{ult} = f_c'A_g + f_c'(n-1)A_s \tag{36}$$

Although Eqs. (35) and (36) are similar in appearance, there are several differences, the most significant being that the more rational ultimate strength formula [Eq. (35)] does not make the steel stress a function of the concrete stress.

Following the ultimate-strength approach, the ACI Code contains the following requirement for columns:

"1101—Limiting dimensions

"(*a*) The following sections on reinforced concrete and composite columns, except Section 1107(a), apply to a short column for which the unsupported length is not greater

than ten times the least dimension. When the unsupported length exceeds this value, the design shall be modified as shown in Section 1107(a). Principal columns in buildings shall have a minimum diameter of 12 in., or in the case of rectangular columns, a minimum thickness of 8 in., and a minimum gross area of 120 sq in. Posts that are not continuous from story to story shall have a minimum diameter or thickness of 6 in.

"1103—Spirally reinforced columns

"(*a*) *Allowable load*—The maximum allowable axial load, P, on columns with closely spaced spirals enclosing a circular concrete core reinforced with vertical bars shall be that given by formula (11).

$$P = A_g(0.225f_c' + f_s p_g) \tag{11}$$

Wherein f_s = nominal allowable stress in vertical column reinforcement, to be taken at 40 percent of the minimum specification value of the yield point; *viz.*, 16,000 psi for intermediate grade steel and 20,000 psi for rail or hard grade steel.[1]

"(*b*) *Vertical reinforcement*—The ratio p_g shall not be less than 0.01 nor more than 0.08. The minimum number of bars shall be six, and the minimum bar size shall be #5. The center to center spacing of bars within the periphery of the column core shall not be less than 2½ times the diameter for round bars or three times the side dimension for square bars. The clear spacing between individual bars or between pairs of bars at lapped splices shall not be less than 1½ in. or 1½ times the maximum size of the coarse aggregate used. These spacing rules also apply to adjacent pairs of bars at a lapped splice; each pair of lapped bars forming a splice may be in contact, but the minimum clear spacing between one splice and the adjacent splice should be that specified for adjacent single bars.

"(*c*) *Splices in vertical reinforcement*—Where lapped splices in the column verticals are used, the minimum amount of lap shall be as follows:

"1. For deformed bars with concrete having a strength of 3000 psi or more, 20 diameters of bar of intermediate or hard grade steel. For bars of higher yield point, the amount of lap shall be increased one diameter for each 1000 psi by which the allowable stress exceeds 20,000 psi. When the concrete strengths are less than 3000 psi, the amount of lap shall be one-third greater than the values given above.

"2. For plain bars, the minimum amount of lap shall be twice that specified for deformed bars.

"3. Welded splices or other positive connections may be used instead of lapped splices. Welded splices shall preferably be used in cases where the bar size exceeds #11. An approved welded splice shall be defined as one in which the bars are butted and welded and that will develop in tension at least the yield point stress of the reinforcing steel used.

"4. Where longitudinal bars are offset at a splice, the slope of the inclined portion of the bar with the axis of the column shall not exceed 1 in 6, and the portions of the bar above and below the offset shall be parallel to the axis of the column. Adequate horizontal support at the offset bends shall be treated as a matter of design, and may be provided by metal ties, spirals or parts of the floor construction. Metal ties or spirals so designed shall be placed near (never more than eight bar diameters from) the point of bend. The horizontal thrust to be resisted may be assumed as 1½ times the horizontal component of the nominal stress in the inclined portion of the bar.

"Offset bars shall be bent before they are placed in the forms. No field bending of bars partially embedded in concrete shall be permitted.

"[1] Nominal allowable stresses for reinforcement of higher yield point may be established at 40 percent of the yield point stress, but not more than 30,000 psi, when the properties of such reinforcing steels have been definitely specified by standards of ASTM designation. If this is done, the lengths of splice required by Section 1103(c) shall be increased accordingly."

"(*d*) *Spiral reinforcement*—The ratio of spiral reinforcement, p', shall not be less than the value given by formula (12).

$$p' = 0.45 \left(\frac{A_g}{A_c} - 1\right) \frac{f_c'}{f_s'} \tag{12}$$

Wherein f_s' = useful limit stress of spiral reinforcement, to be taken as 40,000 psi for hot rolled rods of intermediate grade, 50,000 psi for rods of hard grade, and 60,000 psi for cold drawn wire.

"The spiral reinforcement shall consist of evenly spaced continuous spirals held firmly in place and true to line by vertical spacers, using at least two for spirals 20 in. or less in diameter, three for spirals 20 to 30 in. in diameter, and four for spirals more than 30 in. in diameter or composed of spiral rods ⅝ in. or larger in size. The spirals shall be of such size and so assembled as to permit handling and placing without being distorted from the designed dimensions. The material used in spirals shall have a minimum diameter of ¼ in. for rolled bars or No. 4 AS&W gage for drawn wire. Anchorage of spiral reinforcement shall be provided by 1½ extra turns of spiral rod or wire at each end of the spiral unit. Splices when necessary shall be made in spiral rod or wire by welding or by a lap of 1½ turns. The center to center spacing of the spirals shall not exceed one-sixth of the core diameter. The clear spacing between spirals shall not exceed 3 in. nor be less than 1⅜ in. or 1½ times the maximum size of coarse aggregate used. The reinforcing spiral shall extend from the floor level in any story or from the top of the footing in the basement, to the level of the lowest horizontal reinforcement in the slab, drop panel or beam above. In a column with a capital, it shall extend to a plane at which the diameter or width of the capital is twice that of the column.

"(*e*) *Protection of reinforcement*—The column spiral reinforcement shall be protected everywhere by a covering of concrete cast monolithically with the core, for which the thickness shall not be less than 1½ in. nor less than 1½ times the maximum size of the coarse aggregate, nor shall it be less than required by the fire protection and weathering provisions of Section 507.

"1104—Tied columns

"(*a*) *Allowable load*—The maximum allowable axial load on columns reinforced with longitudinal bars and separate lateral ties shall be 80 percent of that given by formula (11). The ratio, p_g, to be considered in tied columns shall not be less than 0.01 nor more than 0.04. The longitudinal reinforcement shall consist of at least four bars, of minimum bar size of #5. Splices in reinforcing bars shall be made as described in Section 1103(c). The spacing requirements for vertical reinforcement in Section 1103(b) shall also apply for all tied columns.

"(*b*) *Combined axial and bending load*—For tied columns which are designed to withstand combined axial and bending stresses, the limiting steel ratio of 0.04 may be increased to 0.08. The amount of steel spliced by lapping shall not exceed a steel ratio of 0.04 in any 3-ft length of column. The size of the column designed under this provision shall in no case be less than that required to withstand the axial load alone with a steel ratio of 0.04.

"(*c*) *Lateral ties*—Lateral ties shall be at least ¼ in. in diameter and shall be spaced apart not over 16 bar diameters, 48 tie diameters, or the least dimension of the column. When there are more than four vertical bars, additional ties shall be provided so that every longitudinal bar is held firmly in its designed position and has lateral support equivalent to that provided by a 90-deg corner of a tie.

"(*d*) *Limits of column section*—In a tied column which for architectural reasons has a larger cross section than required by considerations of loading, a reduced effective area,

A_g, not less than one-half of the total area may be used in applying the provisions of Section 1104(a).

"1107—Long columns

"(*a*) The maximum allowable load, P', on axially loaded reinforced concrete or composite columns having an unsupported length, h, greater than ten times the least lateral dimension, t, shall be given by formula (17).

$$P' = P[1.3 - 0.03h/t] \tag{17}$$

where P is the allowable axial load on a short column as given by Sections 1103, 1104, and 1105.

"The maximum allowable load, P', on eccentrically loaded columns in which h/t exceeds 10 shall also be given by formula (17), in which P is the allowable eccentrically applied load on a short column as determined by the provisions of Section 1109. In long columns subjected to definite bending stresses, as determined in Section 1108, the ratio h/t shall not exceed 20."

Tables 7 to 9 are furnished to aid in applying the requirements of these sections of the Code.

TABLE 7. COLUMNS WITH LONGITUDINAL BARS AND SPIRALS

Loads in thousand pounds

$P = 0.225f_c'A_g + f_sA_s$

Diameter of column, in.	Gross area A_g, sq in.	Load on bars, f_sA_s for $p_g = 0.01$		Load on concrete, $0.225f_c'A_g$				
		f_s		f_c'				
		Intermediate grade 16,000	Hard grade 20,000	2,000	2,500	3,000	3,750	5,000
12[1]	113	14	18	41	51	61	76	102
13[1]	133	17	22	48	60	72	90	119
14	154	25	31	69	87	104	130	173
15	177	28	35	80	99	119	149	199
16	201	32	40	91	113	136	170	227
17	227	36	45	102	128	153	192	256
18	255	41	51	114	143	172	215	287
19	284	45	57	128	159	191	239	319
20	314	50	63	141	177	212	265	353
21	346	55	69	156	195	234	292	390
22	380	61	76	171	214	257	321	428
23	416	66	83	187	234	280	350	467
24	452	72	90	204	254	305	382	510
25	491	79	98	221	276	331	414	552
26	531	85	106	239	299	358	448	598
27	573	92	115	258	322	387	483	644
28	616	98	123	277	346	416	519	693
29	661	106	132	297	372	446	557	743
30	707	113	141	318	398	477	596	795

[1] Spirals for these are excessive and seldom available. The loads given in the table are for circular columns with lateral ties.

TABLE 8. COLUMNS WITH LONGITUDINAL BARS AND LATERAL TIES

Loads in thousand pounds

$$P = 0.18f_c'A_g + 0.8f_sA_s$$

Dimensions of column, in.		Gross area A_g, sq in.	Load on bars, $0.8f_sA_s$ for $p_g = 0.01$		Load on concrete, $0.18f_c'A_g$				
			Intermediate grade 16,000	Hard grade 20,000	f_c' 2,000	2,500	3,000	3,750	5,000
10	12	120	15	19	43	54	65	81	108
	14	140	18	22	50	63	76	95	127
	16	160	20	26	58	72	86	108	144
12	12	144	18	23	52	65	78	97	129
	14	168	22	27	60	76	91	113	151
	16	192	25	31	69	86	104	130	173
14	14	196	25	31	71	88	106	132	176
	16	224	29	36	81	101	121	151	201
	18	252	32	40	91	113	136	170	227
16	16	256	33	41	92	115	138	173	230
	18	288	37	46	104	130	156	194	259
	20	320	41	51	115	144	173	216	275
	22	352	45	56	127	158	190	238	317
18	18	324	41	52	117	146	175	219	292
	20	360	46	58	130	162	194	243	324
	22	396	51	63	143	178	214	267	356
	24	432	55	69	156	194	233	292	389
20	20	400	51	64	144	180	216	270	360
	22	440	56	70	158	198	238	297	396
	24	480	61	77	173	216	259	324	432
	26	520	67	83	187	234	281	351	468
22	22	484	62	77	174	218	261	327	436
	24	528	68	84	190	238	285	356	475
	26	572	73	92	206	257	309	386	515
	28	616	79	99	222	277	333	416	555
24	24	576	74	92	207	259	311	389	518
	26	624	80	100	225	281	337	421	561
	28	672	86	108	242	302	363	454	605
26	26	676	87	108	243	304	365	456	608
	28	728	93	116	262	328	393	491	654
28	28	784	100	125	282	353	423	529	705
30	30	900	115	144	324	405	486	608	810

TABLE 9. SIZE AND PITCH OF SPIRALS, ACI CODE

Diameter of column, in.	Out to out of spiral, in.	f_c'			
		2,500	3,000	3,750	5,000
Hot-rolled Intermediate-grade					
14, 15	11, 12	3/8–2	3/8–1 3/4	1/2–2 1/2	1/2–1 3/4
16	13	3/8–2	3/8–1 3/4	1/2–2 1/2	1/2–2
17–19	14–16	3/8–2 1/4	3/8–1 3/4	1/2–2 1/2	1/2–2
20–23	17–20	3/8–2 1/4	3/8–1 3/4	1/2–2 3/4	1/2–2
24–30		3/8–2 1/4	3/8–2	1/2–2 3/4	1/2–2
Cold-drawn Wire					
14, 15	11, 12	1/4–1 3/4	3/8–2 3/4	3/8–2	1/2–2 3/4
16–23	13–20	1/4–1 3/4	3/8–2 3/4	3/8–2 1/4	1/2–3
24–29	21–26	1/4–1 3/4	3/8–3	3/8–2 1/4	1/2–3
30	27	1/4–1 3/4	3/8–3	3/8–2 1/4	1/2–3 1/4

Illustrative Examples. 1. Design a tied column to support a concentric load of 525,000 lb; $f_c' = 3{,}750$ lb, $f_s = 16{,}000$ lb (intermediate-grade steel).

NOTE: Unless space or other requirements dictate the desirable size of column, the most economical column is usually the one with the smallest allowable steel percentage—$p_g = 0.01$. Therefore, of the number of combinations that may be used, a square column with the minimum amount of steel will be selected.

From Table 8,

26 × 26 column will take	456^k
for $p_g = 0.01$, steel will carry	87
	543^k

$A_s = 0.01 \times 676 = 6.76$ sq in., 12 No. 7 bars, $A_s = 7.22$ sq in.
Ties: Use No. 3 ties at 14″ o.c. (16 bar diameters)

2. Design a spiral column for the same load as above.

From Table 7,

26″ diameter column will carry	448^k
for $p_g = 0.01$, will carry	85
	533^k

$A_s = 0.01 \times 531 = 5.31$ sq in., 9 No. 7 bars, $A_s = 5.41$ sq in.
Spirals: From Table 9,

Use 3/8–2 1/4″ pitch, cold-drawn wire—23″ OD

$$p' = \frac{\pi \times 22.625 \times 0.11}{\dfrac{\pi \times 23^2}{4} \times 2.25} = 0.00837$$

$$p_{\text{req}}' = 0.45\left(\frac{A_g}{A_c} - 1\right)\frac{f_c'}{f_s'}$$

$$= 0.45\left(\frac{531}{416} - 1\right)\frac{3{,}750}{60{,}000} = 0.00776$$

COMBINED BENDING AND AXIAL STRESS

Virtually all concrete members are subjected to a combination of bending and axial stress. However, it has been learned from experience that in the design of most members one or the other may be neglected safely. For example, no actual building columns are ideal: they are all subject to bending caused by either unbalanced loads or unavoidable eccentricity, but if the unbalance is not great or the eccentricity small (as it usually is) they may be neglected and the column designed for axial load only. In a beam there is always the possibility of developing axial forces because of, for example, restraint due to friction at the supports or continuity with other members such as in a building-frame girder. If the axial force is compression and of relatively small magnitude it may safely be neglected. Care should be taken to determine whether there is any possibility of developing secondary axial tension in a beam and, if so, that reinforcement be placed at all sections where it may occur. Although it may be difficult to evaluate such forces—which may arise through shrinkage of the concrete as it cures or because of freezing of supposedly sliding supports—they are nonetheless very important. It seems probable that some recent disastrous failures (Ref. 12) may be attributed in part to the neglect of this type of action.

As will be seen, beams are less sensitive to the addition of a small compression force than columns are to the addition of a small bending moment. That is, while the axial load-carrying capacity of a column will be materially reduced by the addition of a small bending moment, the moment resistance of a beam may be—and usually is—increased by the addition of a relatively small axial force. This may be appreciated when it is realized that the compressive axial force reduces the tension in the steel and brings a greater portion of the concrete into active compression.

The above generalization relative to the reduction in carrying capacity of eccentrically loaded columns should be modified in that, if a column with small eccentricity is overloaded, it will tend to crack. If the bending moment is computed on the basis of elastic theory assuming uncracked sections, then the cracking will reduce the stiffness of the member and hence the moment, and relieve the column. Of course this may cause an increase in moment at some other section.

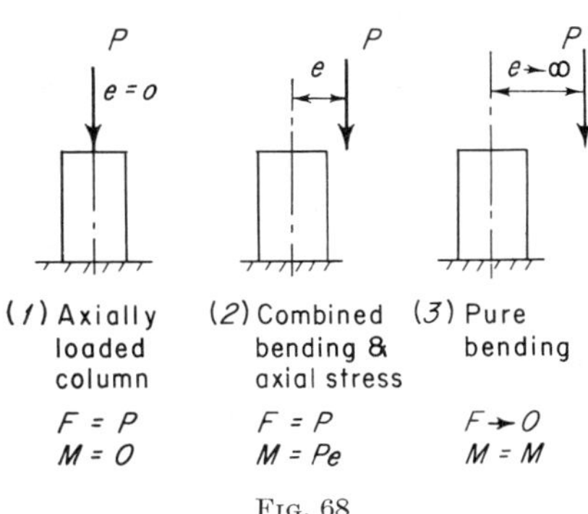

FIG. 68

Neglecting buckling failure, the relationships existing between the different types of members can be visualized from the diagrams of Fig. 68 in which the resultant load is taken as acting at a distance e from the centroid of the cross section.

It would seem simple to develop a set of formulas with the parameter e governing the entire range of action and agreeing with the column formulas for $e = 0$ and the beam formulas for infinite e. However, there are several reasons why it is difficult to do this: (1) For small eccentricities the entire section is in compression and uncracked. For larger eccentricities part of the section is in tension; hence the concrete is cracked and partially ineffective. Thus two different types of cross section—cracked and uncracked—must be considered. (2) It is desirable to have different working stresses for the reinforcement in axially loaded columns where buckling may be a factor, than in tension where it is not. (3) At the present time the ACI Code uses basically different principles for the design of columns and beams (except in the optional appendix which

will be covered later). Column design is based on ultimate-strength theory. Beam design, except for the requirements pertaining to compression reinforcement, is based on the straight-line theory. These are incompatible. It is likely that this will be remedied in the future but, for the present, it prevents the development of a coherent method for the design of all types of members.

For these reasons the current ACI design specifications for combined bending and axial load are frankly empirical. The pertinent sections are reproduced below:

"1100—Notation

A_c = area of core of a spirally reinforced column measured to the outside diameter of the spiral
A_g = over-all or gross area of spirally reinforced or tied columns
A_s = effective cross-sectional area of reinforcement in compression in columns
B = trial factor (see Section 1109(c) and footnote thereto)
e = eccentricity of the resultant load on a column, measured from the gravity axis
F_a = nominal allowable axial unit stress ($0.225f_c' + f_s p_g$) for spiral columns and 0.8 of this value for tied columns
F_b = allowable bending unit stress that would be permitted if bending stress only existed
f_a = nominal axial unit stress = axial load divided by area of member, A_g
f_b = bending unit stress (actual) = bending moment divided by section modulus of member (uncracked transformed section. Ed.)
f_c = computed concrete fiber stress in an eccentrically loaded column where the ratio of e/t is greater than 2/3
f_c' = compressive strength of concrete at age of 28 days, unless otherwise specified
f_s = nominal allowable stress in vertical column reinforcement
h = unsupported length of column
N = axial load applied to reinforced concrete column
p_g = ratio of the effective cross-sectional area of vertical reinforcement to the gross area A_g
P = total allowable axial load on a column whose length does not exceed ten times its least cross-sectional dimension
t = over-all depth of rectangular column section, or the diameter of a round column

"1109—Columns subjected to axial load and bending

"(*a*) Members subject to an axial load and bending in one principal plane, but with the ratio of eccentricity to depth e/t no greater than 2/3, shall be so proportioned that

$$\frac{f_a}{F_a} + \frac{f_b}{F_b} \text{ does not exceed unity} \tag{18}$$

"(*b*) When bending exists on both of the principal axes, formula (18) becomes

$$\frac{f_a}{F_a} + \frac{f_{bx}}{F_b} + \frac{f_{by}}{F_b} \text{ does not exceed unity} \tag{19}$$

where f_{bx} and f_{by} are the bending moment components about the x and y principal axes divided by the section modulus of the transformed section relative to the respective axes, provided that the ratio e/t is no greater than 2/3 in either direction.

"(*c*) In designing a column subject to both axial load and bending, the preliminary selection of the column may be made by use of an equivalent axial load given by formula (20).

$$P = N\left(1 + \frac{Be}{t}\right)^{*} \tag{20}$$

"* For trial computations B may be taken from 3 to 3½ for rectangular tied columns, the lower value being used for columns with the minimum amount of reinforcement. Similarly for circular spiral columns, the value of B from 5 to 6 may be used."

When bending exists on both of the principal axes, the quantity Be/t is the numerical sum of the Be/t quantities in the two directions.

"(*d*) For columns in which the load, N, has an eccentricity, e, greater than 2/3 the column depth, t, the determination of the fiber stress f_c shall be made by use of recognized theory for cracked sections, based on the assumption that the concrete does not resist tension. In such cases the modular ratio for the compressive reinforcement shall be assumed as double the value given in Section 601; however, the stress in the compressive reinforcement when calculated on this basis, shall not be greater than the allowable stress in tension. The maximum combined compressive stress in the concrete shall not exceed $0.45f_c'$. For such cases the tensile steel stress shall also be investigated."

The formula specified for e/t ratios less than 2/3 is a linear interaction formula similar in form to that used for many years in the design of steel members subjected to combined bending and axial stress. It has no rational basis but does seem to fit reasonably well test results for columns with small eccentricity of loading. For zero eccentricity it becomes identical to the column formulas of ACI Sections 1103 and 1104. In all probability columns with e/t ratios greater than about 1/3 will be cracked on the tension side. The use of the uncracked section for e/t ratios up to 2/3 is justified primarily on the basis that the straight-line theory which is used for evaluating bending stresses underestimates the real capacity of the member in flexure. Although not stated in the Code, it is intended that, in computing f_b for use in the interaction formula, the modular ratio of all longitudinal reinforcement be assumed as double the value given in Section 601 of the Code.

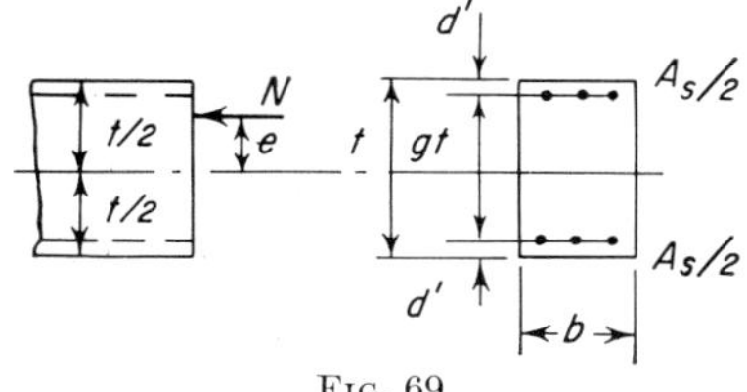

FIG. 69

In applying the ACI Code interaction formula to an eccentrically loaded tied column with symmetrical reinforcement the following formulas are used (see Fig. 69 for notation):

Interaction formula

$$\frac{f_a}{F_a} + \frac{f_b}{F_b} \leq 1 \tag{37}$$

where

$$f_a = \frac{N}{bt} \tag{38}$$

$$f_b = \frac{Mc}{I}$$

$$= \frac{Net/2}{\left[\frac{1}{12}bt^3 + 2nA_s\left(\frac{gt}{2}\right)^2\right]}$$

$$= \frac{Ne}{bt^2[1/6 + np_g g^2]} \tag{39}$$

$$F_a = 0.8(0.225f_c' + f_s p_g) \tag{40}$$

$$F_b = 0.45f_c' \tag{41}$$

Theoretically, allowance should be made in Eq. (39) for the fact that some of the concrete is displaced by steel and is therefore ineffective. This refinement has been found to be of small consequence and has been neglected. Expressions similar to the above may be written for any type of cross section or arrangement of reinforcement.

When the e/t ratio is greater than 2/3 the section is to be assumed cracked and the following equations may be written for a symmetrically reinforced tied column (see Fig. 70).

$$f_s' = 2nf_c\left(1 - \frac{d'}{kt}\right) \tag{42}$$

$$f_s = nf_c\left(\frac{d}{kt} - 1\right) \tag{43}$$

where f_s' and f_s are the stresses in the compression and tension steel respectively; and

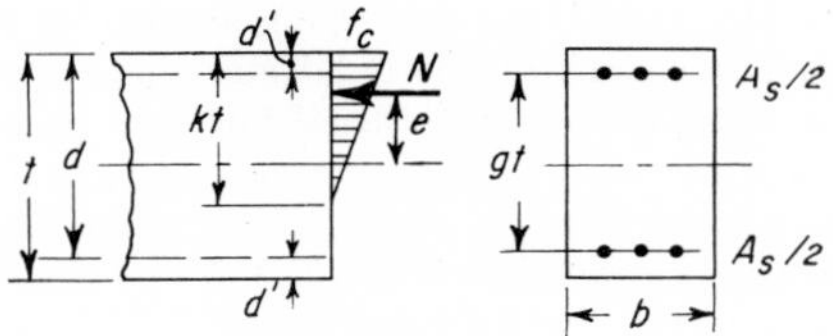

FIG. 70

f_c is the extreme fiber stress in the concrete. These should not exceed the basic allowance stresses. From statics,

$$N = \frac{1}{2}f_cbkt + \frac{A_sf_s'}{2} - \frac{A_sf_s}{2} \tag{44}$$

$$M = Ne = \frac{1}{2}f_cbkt^2\left(\frac{1}{2} - \frac{k}{3}\right) + \frac{f_s'A_sgt}{4} + \frac{f_sA_sgt}{4} \tag{45}$$

Using Eqs. (42) and (43) in Eqs. (44) and (45) it may be shown that

$$N = f_cbt\left(\frac{2k^2 + 6knp_g - 3np_g + np_gg}{4k}\right) \tag{46}$$

$$M = f_cbt^2\left\{\frac{k}{2}\left(\frac{1}{2} - \frac{k}{3}\right) + \frac{np_gg}{4}\left[1 + \frac{1}{2k}(3g - 1)\right]\right\} \tag{47}$$

Also, by eliminating N from Eqs. (44) and (45) it is found that

$$k^3 + 3\left(\frac{e}{t} - \frac{1}{2}\right)k^2 + 3\left(3\frac{e}{t} - \frac{g}{2}\right)np_gk + \frac{3}{2}(g - 3)np_g\frac{e}{t} + \frac{3}{4}np_gg(1 - 3g) = 0 \tag{48}$$

To design a section for given values of M and N, the procedure would be to assume values of b, t, and p_g; e/t and g may then be computed and Eq. (48) solved for k. The permissible values of M and N may then be computed from Eqs. (46) and (47), compared with the desired quantities, and the process repeated if necessary. The steel stress must also be checked using Eqs. (42) and (43) to be sure that it is not governing. This obviously is too cumbersome a procedure for practical use, involving as it does trial and error and the solution of a cubic equation. To expedite design, many design offices have charts based on these equations, or have developed satisfactory approximate formulas. Reference 14 presents one method for simplifying the work involved.

Illustrative Example. Design a square, short, tied column to take a combined axial load N of 90,000 lb and bending moment M of 500,000 in.-lb. $f_c' = 3{,}000$ psi, intermediate-grade steel.

a. Preliminary design using ACI Code equation (20)

$$e = 500{,}000/90{,}000 = 5.55 \text{ in.}$$

Assume $t = 16$ in., $gt = 11\frac{1}{2}$ in.
Then $e/t = 0.347$, $g = 0.72$.
From ACI equation (20) assuming $B = 3$,

$$P = 90{,}000(1 + 3 \times 0.347) = 184{,}000 \text{ lb}$$

From Table 8, load taken by concrete = 138,000 lb.
Load to be taken by reinforcement = 46,000 lb.
From Table 8, for $p_g = 0.01$, load taken by reinforcement is 33,000 lb.

Thus required $p_g = \dfrac{46{,}000}{33{,}000} \times 0.01 = 0.0139$.

b. Check design using interaction formula, Eq. (37). A first check using $p_g = 0.0139$ (computations not shown) indicates that less steel will be sufficient. Therefore try $p_g = 0.01$, the minimum permissible.

From Eq. (38), $f_a = \dfrac{90{,}000}{16^2} = 352$ psi

From Eq. (39), $f_b = \dfrac{90{,}000 \times 5.55}{16^3(\frac{1}{6} + 10 \times 0.01 \times 0.72^2)} = 555$ psi

From Eq. (40), $F_a = 0.8(0.225 \times 3{,}000 + 16{,}000 \times 0.01) = 669$ psi

From Eq. (41), $F_b = 1{,}350$ psi

From Eq. (37), $\dfrac{352}{669} + \dfrac{555}{1{,}350} = 0.937 < 1$

Although several other combinations of column size and reinforcement are permissible this is probably as economical as any since the minimum amount of steel is used.
A_g required $= 0.01 \times 16^2 = 2.56$ sq in.
Use 6 No. 6 bars; $A_g = 2.64$ sq in.
Ties: use No. 2 bars at 12″.

ULTIMATE-STRENGTH DESIGN

As mentioned earlier, design based on the true ultimate strength of concrete members rather than on arbitrarily assumed distribution of stress at working loads has come into practical use. In the current ACI Code (ACI 318–56) recommendations for ultimate-strength design, as an optional method, are included in the appendix which is reproduced below.

"A600—Notation

"(*a*) *Loads and load factors*

U = ultimate strength capacity of section
B = effect of basic load consisting of dead load plus volume change due to creep, elastic action, shrinkage, and temperature
L = effect of live load plus impact
W = effect of wind load
E = effect of earthquake forces
K = load factor
M_u = ultimate resisting moment
P_b = load defined by Eq. (A8)
P_o = ultimate strength of concentrically loaded member given by Eq. (A6)
P_u = ultimate strength of eccentrically loaded member
P_u' = maximum axial load on long member given by Eq. (A14)

"(b) *Cross-sectional constants*

A_g = gross area of section
A_s = area of tensile reinforcement
A_s' = area of compressive reinforcement
A_{sf} = steel area to develop compressive strength of overhanging flange in T-sections, defined by Eq. (A5)
A_{st} = total area of longitudinal reinforcement
b = width of a rectangular section or over-all width of flange in T-sections
b' = width of web in T-sections
D = total diameter of circular section
D_s = diameter of circle circumscribing the longitudinal reinforcement in circular section
d = distance from extreme compressive fiber to centroid of tensile reinforcement
d' = distance from extreme compressive fiber to centroid of compressive reinforcement
e = eccentricity of axial load measured from the centroid of tensile reinforcement
e' = eccentricity of axial load measured from plastic centroid of section
e_b' = eccentricity of load P_b measured from plastic centroid of section
f_c' = 28-day cylinder strength
f_s = stress in tensile reinforcement at ultimate strength
f_y = yield point of reinforcement, not to be taken greater than 60,000 psi
k_u = defined by $k_u d$ = distance from extreme compressive fiber to neutral axis at ultimate strength
k_1 = ratio of average compressive stress to $0.85f_c'$
k_2 = ratio of distance between extreme compressive fiber and resultant of compressive stresses to distance between extreme fiber and neutral axis
$m = f_y/0.85f_c'$
$m' = m - 1$
$p = A_s/bd$
$p' = A_s'/bd$
$p_f = A_{sf}/b'd$
$p_t = A_{st}/A_g$
$p_w = A_s/b'd$
$q = pf_y/f_c'$
t = flange thickness in T-sections, also total depth of rectangular section

"A601—Definitions and scope

"(a) This appendix presents recommendations for design of reinforced concrete structures by ultimate strength theories. The term "ultimate strength design" indicates a method of design based on the ultimate strength of a reinforced concrete cross section in simple bending, combined bending and axial load on the basis of inelastic action.

"(b) These recommendations are confined to design of sections. It is assumed that external moments and forces acting in a structure will be determined by the theory of elastic frames. With the specified load factors, stresses under service loads will remain within safe limits.

"A602—General requirements

"(a) The American Concrete Institute "Building Code Requirements for Reinforced Concrete" shall apply to the design of members by ultimate strength theory except where otherwise provided in this appendix.

"(b) Analysis of indeterminate structures, such as continuous girders and arches, shall be based on the theory of elastic frames. For buildings of usual types of construction, spans, and story heights, approximate methods such as the use of coefficients recommended in the ACI Building Code are acceptable for determination of moments and shears.

"(c) Bending moments in compression members shall be taken into account in the calculation of their required strength.

"(*d*) In arches the effect of shortening of the arch axis, temperature, shrinkage, and secondary moments due to deflection shall be considered.

"(*e*) Attention shall be given to the deflection of members, including the effect of creep, especially whenever the net ratio of reinforcement which is defined as $(p - p')$ or $(p_w - p_f)$ in any section of a flexural member exceeds $0.18f_c'/f_y$.

"(*f*) Controlled concrete should be used and shall meet the following requirements. The quality of concrete shall be such that not more than one test in ten shall have an average strength less than the strength assumed in the design, and the average of any three consecutive tests shall not be less than the assumed design strength. Each test shall consist of not less than three standard cylinders.

"A603—Assumptions

"Ultimate strength design of reinforced concrete members shall be based on the following assumptions:

"(*a*) Plane sections normal to the axis remain plane after bending.

"(*b*) Tensile strength in concrete is neglected in sections subject to bending.

"(*c*) At ultimate strength, stresses and strains are not proportional. The diagram of compressive concrete stress distribution may be assumed a rectangle, trapezoid, parabola, or any other shape which results in ultimate strength in reasonable agreement with comprehensive tests.

"(*d*) Maximum fiber stress in concrete does not exceed $0.85f_c'$.

"(*e*) Stress in tensile and compressive reinforcement at ultimate load shall not be assumed greater than the yield point or 60,000 psi, whichever is smaller.

"A604—Load factors

"(*a*) Members shall be so proportioned that an ample factor of safety is provided against an increase in live load beyond that assumed in design; and strains under service loads should not be so large as to cause excessive cracking. These criteria are satisfied by the following formulas:

"1. For structures in which, due to location or proportions, the effects of wind and earthquake loading can be properly neglected:

$$U = 1.2B + 2.4L \tag{I}$$
$$U = K(B + L) \tag{II}$$

"2. For structures in which wind loading must be considered:

$$U = 1.2B + 2.4L + 0.6W \tag{Ia}$$
$$U = 1.2B + 0.6L + 2.4W \tag{Ib}$$
$$U = K(B + L + \tfrac{1}{2}W) \tag{IIa}$$
$$U = K(B + \tfrac{1}{2}L + W) \tag{IIb}$$

"3. For those structures in which earthquake loading must be considered, substitute E for W in the preceding equations.

"(*b*) The load factor, K, shall be taken equal to 2 for columns and members subjected to combined bending and axial load, and equal to 1.8 for beams and girders subject to bending only.

"A605—Rectangular beams with tensile reinforcement only

"(*a*) The ultimate capacity of an under-reinforced section is approached when the tensile steel begins to yield. The steel shall then be assumed to elongate plastically at its yield point stress, thereby reducing the concrete area in compression until crushing takes place. The ultimate strength so obtained is controlled by tension.

"(b) The computed ultimate moment shall not exceed that given by:

$$M_u = bd^2 f_c' q(1 - 0.59q) \tag{A1}$$

in which $q = pf_y/f_c'$.

"(c) In Eq. (A1), the maximum ratio of reinforcement shall be so limited that p does not exceed:

$$p = 0.40 f_c'/f_y \tag{A2}$$

"The coefficient 0.40 is to be reduced at the rate of 0.025 per 1000 psi concrete strength in excess of 5000 psi.

"A606—Rectangular beams with compressive reinforcement

"(a) The ultimate moment shall not exceed that computed by:

$$M_u = (A_s - A_s')f_y d[1 - 0.59(p - p')f_y/f_c'] + A_s' f_y^*(d - d') \tag{A3}$$

"(b) In Eq. (A3), the maximum ratio of reinforcement shall be so limited that $(p - p')$ does not exceed the values given by Eq. (A2).

"A607—T-sections

"(a) When the flange thickness equals or exceeds the depth to the neutral axis given by $k_u d = 1.30qd$ or the depth of the equivalent stress block (1.18 qd), the section may be designed by Eq. (A1), with q computed as for a rectangular beam with a width equal to the over-all flange width.

"(b) When the flange thickness is less than $k_u d$ or less than the depth of the equivalent stress block, the ultimate moment shall not exceed that computed by:

$$M_u = (A_s - A_{sf})f_y d[1 - 0.59(p_w - p_f)f_y/f_c'] + A_{sf} f_y(d - 0.5t) \tag{A4}$$

in which A_{sf}, the steel area necessary to develop the compressive strength of the overhanging portions of the flange, is:

$$A_{sf} = 0.85(b - b')tf_c'/f_y \tag{A5}$$

"(c) In Eq. (A4), the maximum ratio of reinforcement shall be so limited that $(p_u - p_f)$ does not exceed the values given by Eq. (A2).

"A608—Concentrically loaded short columns

"(a) All members subject to axial loads shall be designed for at least a minimum eccentricity:

"For spirally reinforced columns, the minimum eccentricity measured from the centroidal axis of column shall be 0.05 times the depth of the column section.

"For tied columns, the minimum eccentricity shall be 0.10 times the depth.

"(b) The maximum load capacity for concentric loads for use in Eq. (A10) is given by the formula:

$$P_o = 0.85 f_c'(A_g - A_{st}) + A_{st} f_y \tag{A6}$$

"A609—Bending and axial load: Rectangular section

"(a) The ultimate strength of members subject to combined bending and axial load shall be computed from the equations of equilibrium, which when k_u is less than unity may be expressed as follows:

$$P_u = 0.85 f_c' bdk_u k_1 + A_s' f_y^* - A_s f_s \tag{A7a}$$

$$P_u e = 0.85 f_c' bd^2 k_u k_1(1 - k_2 k_u) + A_s' f_y^*(d - d') \tag{A7b}$$

" * Correction for concrete area displaced by compressive reinforcement may be made by subtracting 0.85 f_c' from f_y in this term only."

"In Eq. (A7a) and (A7b), k_2/k_1 shall not be taken as less than 0.5, and k_1 shall not be taken greater than 0.85 for $f_c' \leqq 5000$ psi. The coefficient 0.85 is to be reduced at the rate of 0.05 per 1000 psi concrete strength in excess of 5000 psi.

"(*b*) It shall be assumed that the maximum concrete strain is limited to 0.003 so that the section is controlled by tension when:

$$P_u \leqq P_b = 0.85k_1 \left(\frac{90{,}000}{90{,}000 + f_y}\right) f_c'bd + A_s'f_y{}^* - A_sf_y \tag{A8}$$

k_1 being limited as for Eq. (A7a) and (A7b). The section is controlled by compression when P_u exceeds P_b.

"(*c*) When the section is controlled by tension, the ultimate strength shall not exceed that computed by:

$$P_u = 0.85f_c'bd\{p'm' - pm + (1 - e/d) + \sqrt{(1 - e/d)^2 + 2[(e/d)(pm - p'm') + p'm'(1 - d'/d)]}\} \tag{A9}$$

"(*d*) When the section is controlled by compression, a linear relationship between axial load and moment may be assumed for values of P_u between that given as P_b by Eq. (A8) and the concentric ultimate strength P_o given by Eq. (A6). For this range the ultimate strength may be computed by either Eq. (A10) or (A11):

$$P_u = \frac{P_o}{1 + [(P_o/P_b) - 1]e'/e_b'} \tag{A10}$$

$$P_u = \frac{A_s'f_y}{e'/(d - d') + \frac{1}{2}} + \frac{btf_c'}{(3te'/d^2) + 1.18} \tag{A11}$$

"A610—Bending and axial load: Circular sections

"(*a*) The ultimate strength of circular sections subject to combined bending and axial load may be computed on the basis of the equations of equilibrium taking into account inelastic deformations, or by the empirical formulas Eq. (A12) and (A13):

"When tension controls:

$$P_u = 0.85f_c'D^2 \left[\sqrt{\left(\frac{0.85e'}{D} - 0.38\right)^2 + \frac{p_tmD_s}{2.5D}} - \left(\frac{0.85e'}{D} - 0.38\right)\right] \tag{A12}$$

"When compression controls:

$$P_u = \frac{A_{st}f_y}{\dfrac{3e'}{D_s} + 1} + \frac{A_gf_c'}{\dfrac{9.6De'}{(0.8D + 0.67D_s)^2} + 1.18} \tag{A13}$$

"A611—Long members

"(*a*) When the unsupported length, L, of an axially loaded member is greater than 15 times its least lateral dimension, the maximum axial load, P_u', shall be determined by one of the following methods:

"1. $P_u' = P_o(1.6 - 0.04L/t)$ (A14)

"2. A stability determination for P_u' may be made with an apparent reduced modulus of elasticity used for sustained loads, such as the method recommended in the report of ACI Committee 312, 'Plain and Reinforced Concrete Arches' (*ACI Journal*, May 1951, *Proc.* V. 47, p. 681)."

"* Correction for concrete area displaced by compressive reinforcement may be made by subtracting $0.85 f_c'$ from f_y in this term only."

While these recommendations are still in the development stage and do not cover such things as shear, bond, and methods of analysis, they represent a major step forward and also present a preview of the form future design rules are likely to attain. They represent the composite efforts of a joint committee of the ASCE and ACI but follow, in many respects, the ideas and methods advanced by C. S. Whitney in Ref. 10. Instead of attempting to simulate mathematically the stress distribution in a beam at failure, Whitney assumed the simplified rectangular stress block shown in Fig. 71. He further assumed that for an underreinforced beam failure occurs first by yielding of the tension reinforcement, but that before complete collapse occurs the stress in the concrete will reach the effective ultimate for slow or sustained loads, approximately 0.85 f_c'. Applying the equations of statics to Fig. 71, equation (A1) of the ACI Code is obtained. For balanced or overreinforced beams Whitney concluded, on an empirical basis, that the ultimate bending moment may be expressed as

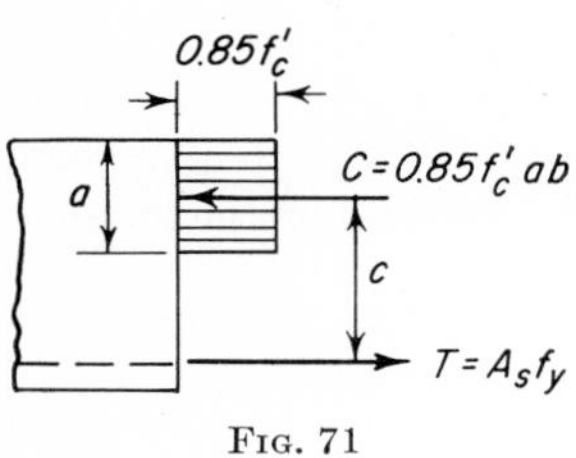

FIG. 71

$$M_u = \tfrac{1}{3} f_c' bd^2 \qquad (49)$$

Whitney applied the same approach—the use of a simplified rectangular stress block with the results modified or adjusted where necessary to agree with empirical data—to develop design formulas for beams with compressive reinforcement and members subjected to combined bending and axial stress. For concentrically loaded columns he followed closely the method already in use which, as stated previously, is based on ultimate strength. An indication of the merits of the Whitney theory is given by Fig. 72 in which his equations for beams with tensile reinforcement only are compared with test results.

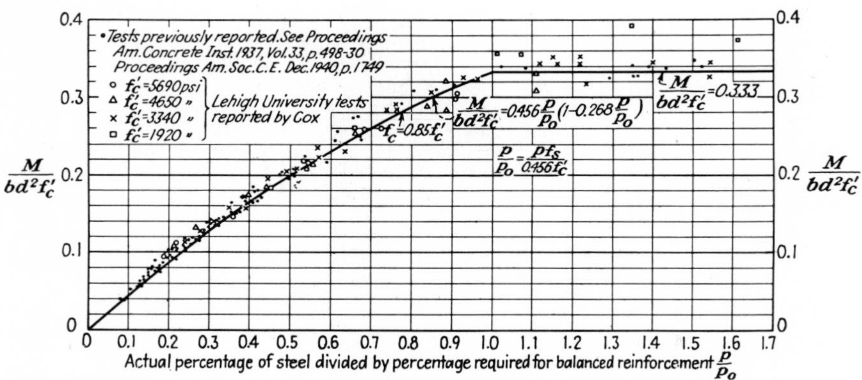

FIG. 72. Theoretical ultimate strength of concrete beams compared with results of tests.

Figure 73 compares a design by the Whitney plastic-theory technique with one by straight-line theory. Several points are apparent: (1) for beams with low percentages of reinforcement there is little difference between the results obtained by the two methods when comparable safety factors are used; (2) the straight-line theory as presently used underestimates the resistance of the concrete; (3) as a consequence of (2) the true point of balanced design is considerably higher than indicated by the straight-line theory. Since it is difficult to get as much as 2½ to 3 percent of reinforcement in a beam because

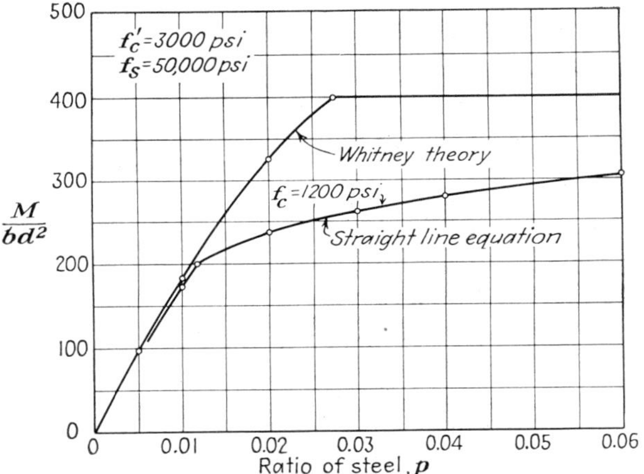

FIG. 73. Comparison of resisting moment of rectangular beams calculated from straight line equations and Whitney equations with a safety factor of 2.5.

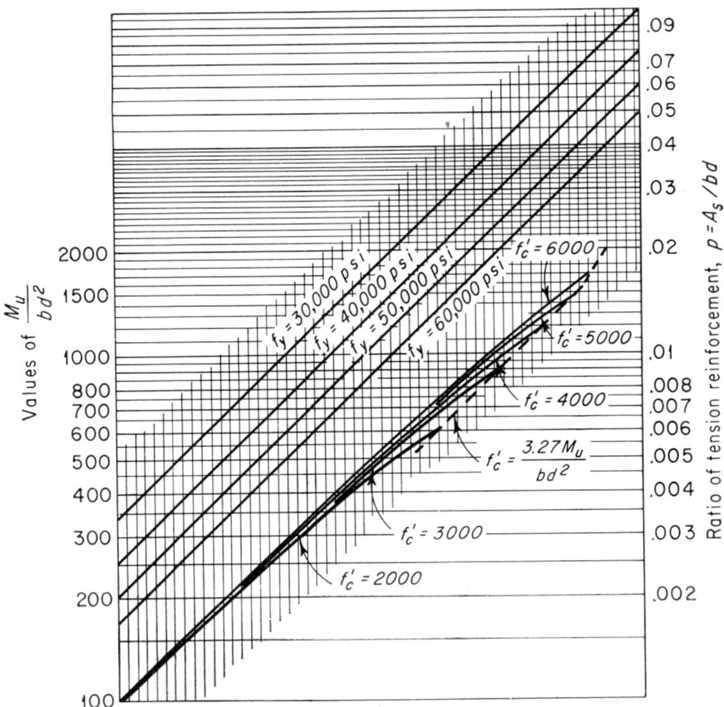

FIG. 74. Moment capacity of rectangular sections without compression reinforcement.

of space limitations, and also because for economic reasons it is usually desirable to use less steel, it will be found in practice that virtually all beams are really underreinforced even though the straight-line theory indicates an overreinforced condition for ratios greater than about 0.01. This brings out the important practical fact that reinforcing

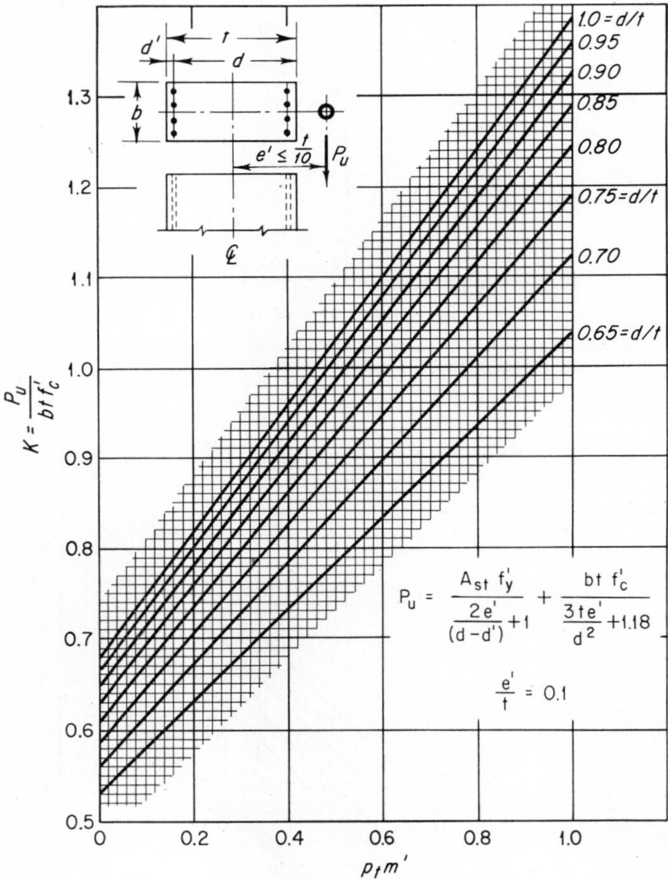

FIG. 75. Direct load capacity of rectangular sections with symmetrical reinforcement for eccentricities less than or equal to $t/10$.

steel of higher strengths than common at present may be used to advantage since they are capable of developing the reserve strength of ordinary or high-strength concrete without requiring excessive cross-sectional area. Considerable attention is being paid to the development of high-strength reinforcement which can be produced at reasonable cost.

There are limitations to the use of ultimate-strength design as it is known at present. In routine practice shear and diagonal tension often control. Higher stresses in the concrete and more tensile steel create higher demands for shear and bond resistance. This is a subject that is undergoing much study. Since ultimate-strength design en-

courages the use of shallower sections deflection becomes more of a problem. The present ACI Code recommendations attempt to cover these limitations by rather conservative interim measures.

In the November, 1956, *ACI Journal* C. S. Whitney and E. Cohen presented the paper, Guide for Ultimate Strength Design of Reinforced Concrete (Ref. 13). This paper follows the ACI Code recommendations in principle. The useful design charts contained herein (Figs. 74 to 79) are reproduced from that paper with permission. Equations and equation numbers are from the paper. Notation follows the ACI Code.

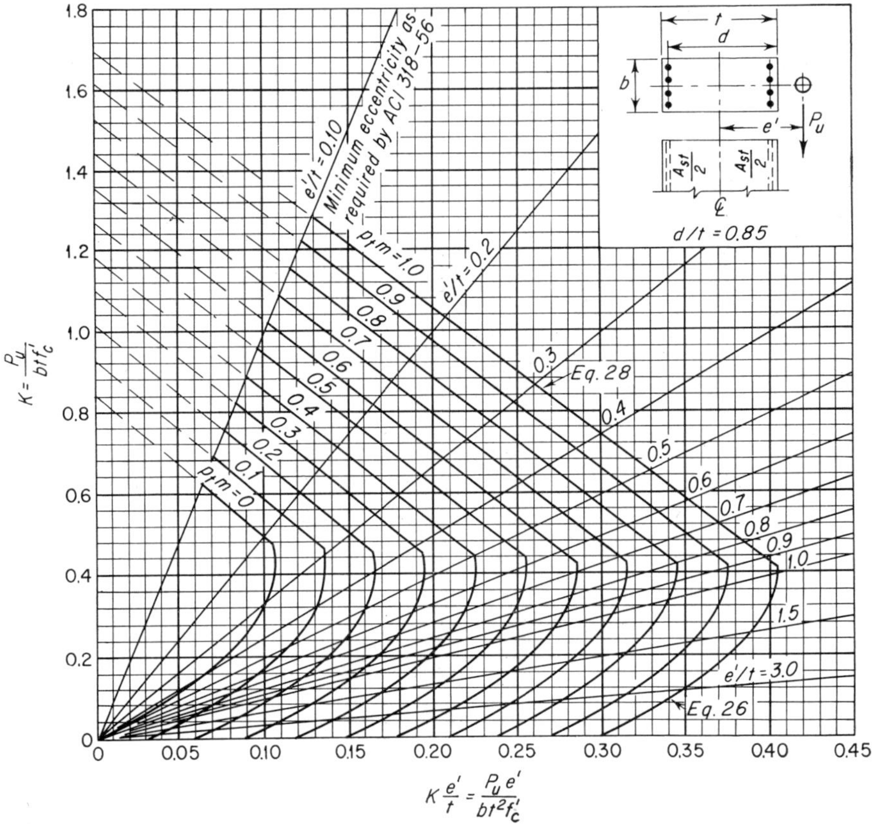

FIG. 76. Bending and axial load—d/t = 0.85. Rectangular sections with symmetrical reinforcement.

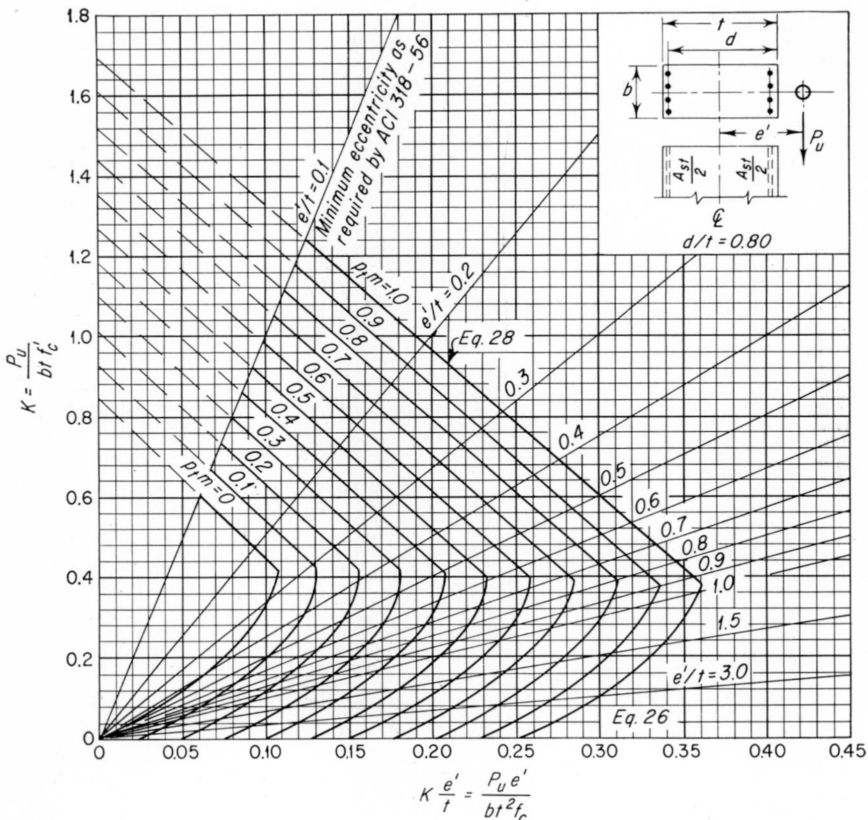

Fig. 77. Bending and axial load—d/t = 0.80. Rectangular sections with symmetrical reinforcement.

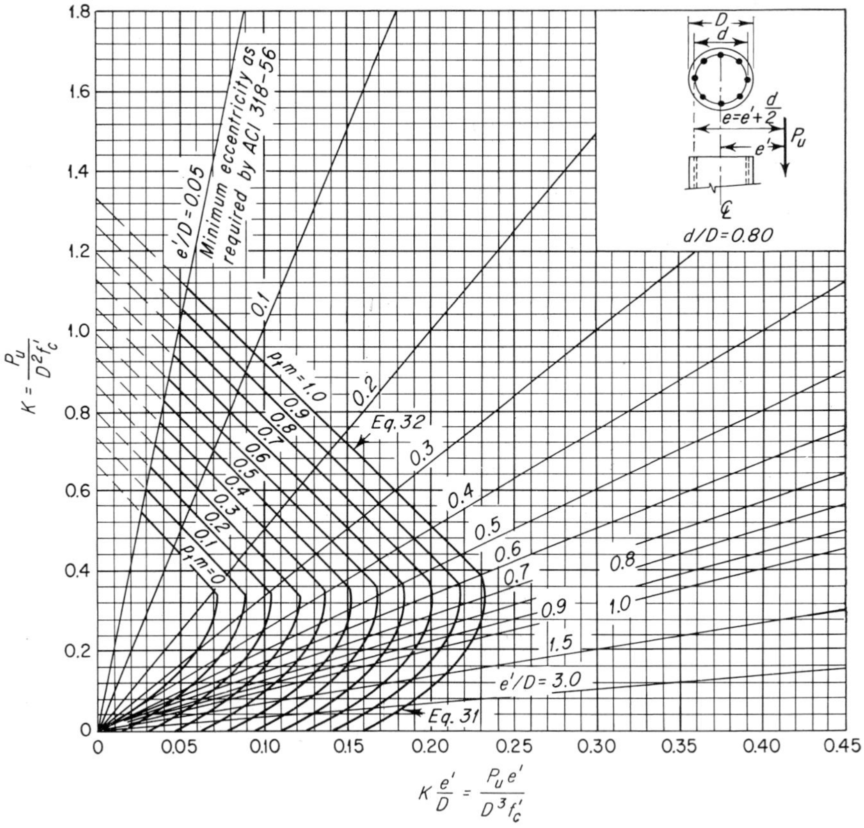

FIG. 78. Bending and axial load—$d/D = 0.80$. Circular sections with spiral reinforcement.

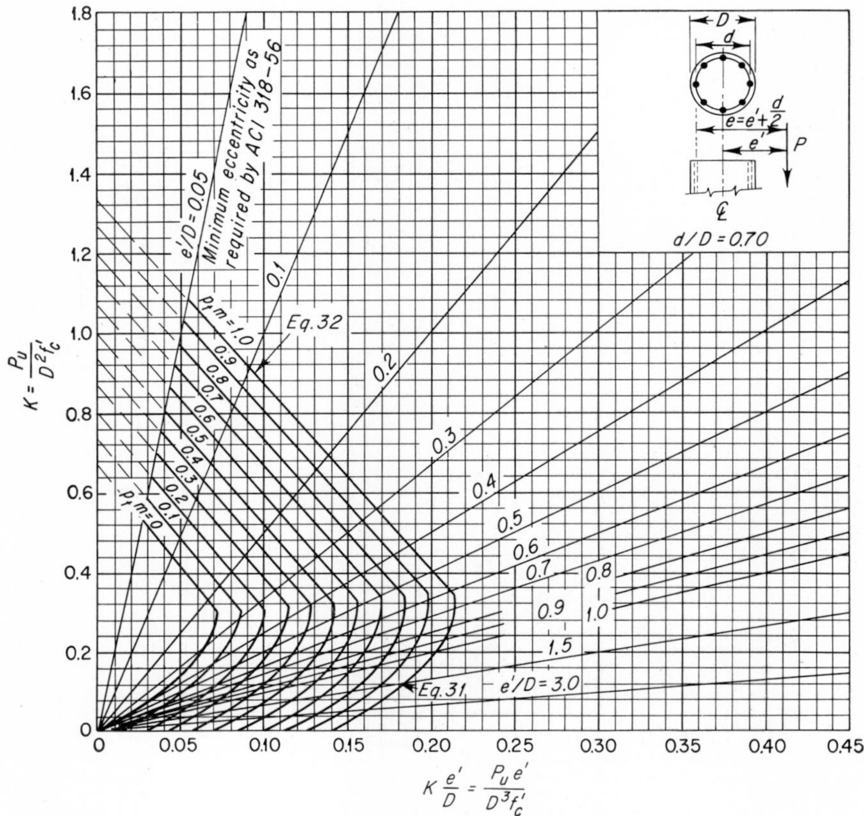

FIG. 79. Bending and axial load—$d/D = 0.70$. Circular sections with spiral reinforcement.

Illustrative Examples. 1. A 12 by 24 beam is reinforced with 8 No. 8 bars in two rows. Intermediate-grade steel ($f_y = 40{,}000$ psi) and $f_c' = 3{,}000$ psi are used. Compare the allowable moments determined by ultimate-strength theory and conventional straight-line theory. Assume a load factor K equal to 1.8 in the ultimate-strength design (ACI Code Section A 604). $A_s = 6.28$ sq in., $b = 12$ in., $d = 20.5$ in., $p = 0.0256$.

a. Ultimate-strength design, ACI Code Appendix. From ACI equation (A2) p maximum $= 0.40 \times 3{,}000/40{,}000 = 0.03 > p$. O.K., beam is underreinforced. From ACI equation (A1),

$$\begin{aligned} q &= 0.0256 \times 40{,}000/3{,}000 = 0.342 \\ M_u &= 12 \times 20.5^2 \times 3{,}000 \times 0.342(1 - 0.59 \times 0.342) \\ &= 4{,}130{,}000 \text{ in.-lb} \end{aligned}$$

The allowable resisting moment equals

$$\begin{aligned} M_a &= M_u/K \\ &= 4{,}130{,}000/1.8 \\ &= 2{,}300{,}000 \text{ in.-lb} \end{aligned}$$

From Fig. 74, enter right-hand side at $p = 0.0256$, proceed horizontally to the left to

f_y = 40,000 psi, drop vertically to f_c' = 3,000 psi, proceed horizontally to the left and read the value of $M_u/bd^2 = 810$. Then

$$M_u = 810 \times 12 \times 20.5^2 = 4{,}100{,}000 \text{ in.-lb}$$

as before.

b. Straight-line theory, ACI Code. From Table 5, p for balanced design is 0.0136; therefore beam is overreinforced.
From Eqs. (7) and (4), $k = 0.510$, $j = 0.830$.
From Eq. (2), $M = \frac{1}{2} \times 1{,}350 \times 0.510 \times 0.830 \times 12 \times 20.5^2 = 1{,}440{,}000$ in.-lb.

The allowable moment by ultimate-strength theory, which accounts for the reserve strength of the concrete, is over 60 percent greater than by conventional theory.

2. The same beam is reinforced with 4 No. 8 bars. Compare as above. $A_s = 3.14$, $b = 12$ in., $d = 21.5$ in., $p = 0.0122$.

a. Ultimate-strength design. From Fig. 74, proceeding as before, read

$$\begin{aligned} M_u/bd^2 &= 440 \\ M_u &= 440 \times 12 \times 21.5^2 \\ &= 2{,}440{,}000 \text{ in.-lb} \\ M_a &= 2{,}440{,}000/1.8 = 1{,}355{,}000 \text{ in.-lb} \end{aligned}$$

b. Straight-line theory, beam is underreinforced.
From Table 6, $k = 0.386$, $j = 0.871$

$$\begin{aligned} M &= 0.0122 \times 20{,}000 \times 0.871 \times 12 \times 21.5^2 \\ &= 1{,}180{,}000 \text{ in.-lb} \end{aligned}$$

In this case where the tensile reinforcement definitely controls, the allowable moment by ultimate theory is only 15 percent higher than by conventional theory. The difference would be reduced further if, instead of using the ACI ultimate-strength load factor K of 1.8, a factor of 2 were used. This would be in line with the factor of safety used in conventional design since $f_b/f_s = 40{,}000/20{,}000 = 2$. In that event the allowable moment by ultimate-strength theory would be 1,220,000 in.-lb, only 3 percent higher than by conventional theory.

3. Same conditions as Example 1 (p. 7–152). Design a square, short, tied column to take a combined axial load N of 90,000 lb and bending moment M of 500,000 in.-lb. f_c' = 3,000 psi, intermediate-grade steel (f_y = 40,000 psi), e' = 5.55 in.
Assume $t = 14$ in., $d = 11\frac{3}{4}$ in. Then $e'/t = 0.396$, $d/t = 0.84$.

$$m = \frac{40{,}000}{0.85 \times 3{,}000} = 15.7$$

Assume $p_t = 0.01$. Then $mp_t = 0.157$. Enter Fig. 76 with the assumed values of e'/t and mp_t and determine, by interpolation,

$$\frac{P_u e'}{bt^2 f_c'} = 0.148$$

Then

$$P_u = \frac{0.148 \times 14^3 \times 3{,}000}{5.55}$$

$$= 220{,}000 \text{ lb}$$

$$M_u = P_u e' = 1{,}220{,}000 \text{ in.-lb}$$

Assuming a load factor K of 2 (ACI Code Section A 604) the allowable axial load and bending moments are

$$\begin{aligned} N &= 220{,}000/2 = 110{,}000 \text{ lb} \\ M &= 1{,}220{,}000/2 = 610{,}000 \text{ in.-lb} \end{aligned}$$

Any practical smaller section would require more reinforcement. Therefore a 14 × 14 column with the minimum desirable reinforcement, $p_t = 0.01$, will be used.

$$A_{st} = 0.01 \times 14^2 = 1.96 \text{ sq in.}$$

Use 4 No. 7 bars ($A_{st} = 2.41$), No. 2 at 12 ties.

Note that the section required is appreciably smaller than that required by the ACI Code interaction formula (p. 7–151). Note also that the ultimate load is governed by Whitney and Cohen's equation (26) which indicates that tension in the reinforcement on the side away from the load is controlling [see ACI Code Section A609(c). ACI equation (A9) is equivalent to Eq. (26) of the charts]. It will be seen by inspection of the charts that, for all but very low eccentricity ratios and high steel percentages, tension governs. Furthermore for large eccentricity ratios, the moment resisting capacity of the section increases as the axial compressive force increases. This is because the added compressive force decreases the tension in the steel. Therefore, for members with a high ratio of bending moment to axial compressive force (e'/t greater than about 3) such as the beams and girders of continuous frames, it is conservative and sufficiently accurate to design for bending only and neglect the axial force. This does not hold, of course, when the axial force is tensile.

In a paper by P. M. Ferguson, Simplification of Design by Ultimate Strength Procedures (Ref. 15), there are additional examples illustrating the application of the ACI recommendations for ultimate-strength design.

REINFORCED CONCRETE WALLS

Reinforced concrete walls are of many types: load-bearing, retaining, enclosure, panel, fire walls, foundation, party walls, etc. The ACI Code covers the general requirements in the following excerpts.

"1111—Reinforced concrete walls

"(*a*) The allowable stresses in reinforced concrete bearing walls with minimum reinforcement as required by Section 1111(h), shall be $0.25f_c'$ for walls having a ratio of height to thickness of ten or less, and shall be reduced proportionally to $0.15f_c'$ for walls having a ratio of height to thickness of 25. When the reinforcement in bearing walls is designed, placed, and anchored in position as for tied columns, the allowable stresses shall be on the basis of Section 1104, as for columns. In the case of concentrated loads, the length of the wall to be considered as effective for each shall not exceed the center to center distance between loads, nor shall it exceed the width of the bearing plus four times the wall thickness. The ratio p_g shall not exceed 0.04.

"(*b*) Walls shall be designed for any lateral or other pressure to which they are subjected. Proper provision shall be made for eccentric loads and wind stresses. In such designs the allowable stresses shall be as given in Section 305(a) and 603(c).

"(*c*) Panel and enclosure walls of reinforced concrete shall have a thickness of not less than 4 in. and not less than 1/30 the distance between the supporting or enclosing members.

"(*d*) Reinforced concrete bearing walls of buildings shall be not less than 6 in. thick for the uppermost 15 ft of their height; and for each successive 25 ft downward, or fraction thereof, the minimum thickness shall be increased 1 in. Reinforced concrete bearing walls of two-story dwellings may be 6 in. thick throughout their height.

"(*e*) Exterior basement walls, foundation walls, fire walls, and party walls shall not be less than 8 in. thick whether reinforced or not.

"(*f*) Reinforced concrete bearing walls shall have a thickness of at least 1/25 of the unsupported height or width, whichever is the shorter.

"(*g*) Reinforced concrete walls shall be anchored to the floors, or to the columns, pilasters, buttresses, and intersecting walls with reinforcement at least equivalent to #3 bars 12 in. on centers, for each layer of wall reinforcement.

"(*h*) The area of the horizontal reinforcement of reinforced concrete walls shall be not less than 0.0025 and that of the vertical reinforcement not less than 0.0015 times the area of the reinforced section of the wall if of bars, and not less than three-fourths as much if of welded wire fabric. The wire of the welded fabric shall be of not less than No. 10 AS&W gage. Walls more than 10 in. thick, except for basement walls, shall have the reinforcement for each direction placed in two layers parallel with the faces of the wall. One layer consisting of not less than one-half and not more than two-thirds the total required shall be placed not less than 2 in. nor more than one-third the thickness of the wall from the exterior surface. The other layer, comprising the balance of the required reinforcement, shall be placed not less than ¾ in. and not more than one-third the thickness of the wall from the interior surface. Bars, if used, shall not be less than #3 bars, nor shall they be spaced more than 18 in. on centers. Welded wire reinforcement for walls shall be in flat sheet form.

"(*i*) In addition to the minimum as prescribed in Section 1111(h) there shall be not less than two #5 bars around all window or door openings. Such bars shall extend at least 24 in. beyond the corner of the openings.

"(*j*) Where reinforced concrete bearing walls consist of studs or ribs tied together by reinforced concrete members at each floor level, the studs may be considered as columns, but the restrictions as to minimum diameter or thickness of columns shall not apply.

"(*k*) The limits of thicknesses and quantity of reinforcement may be waived where structural analysis shows adequate strength and stability."

FLOOR SYSTEMS

A general description of the more common types of floor systems has been given previously (pp. 7–102 to 7–110). Normal design procedures will be covered here.

One-way Slab Floor. The one-way floor slab spanning between beams, girders, or walls is designed as a series of adjacent parallel thin beams. The fact that there is no separation introduces curling or warping tendencies due to the "Poisson ratio" effect of producing lateral expansions in regions of compression. In ordinary design this aspect is disregarded.

For interior floor and roof slabs ¾-in. protection is normally required (in the clear). *Shrinkage* and *temperature* reinforcement is required at right angles to the main reinforcement at spacings not exceeding 18 in. or more than five times the slab thickness. Such steel may be in the form of small rods or woven or welded wire mesh, the amount to be used varying from 0.0018 to 0.0025 times the area of the embedding concrete (from 0.18 to 0.25 percent bt).

Other Floor Construction Including Flat Slab. Other reinforced concrete floor constructions include the monolithically placed *slab* and *joist*, the *two-way* slab (supported on all four sides) and the various forms of *flat slab*.

The flat slab, especially, is highly indeterminate and current design requirements are based largely on the results of tests. All these constructions are specialized and will best be covered by direct excerpts from the current ACI Building Code.

"708—Concrete joist floor construction

"(*a*) In concrete joist floor construction consisting of concrete joists and slabs placed monolithically with or without burned clay or concrete tile fillers, the joists shall not be farther apart than 30 in. face to face. The ribs shall be straight, not less than 4 in. wide, and of a depth not more than three times the width.

"(*b*) When burned clay or concrete tile fillers of material having a unit compressive

strength at least equal to that of the designed strength of the concrete in the joists are used, the vertical shells of the fillers in contact with the joists may be included in the calculations involving shear or negative bending moment. No other portion of the fillers may be included in the design calculations.

"(*c*) The concrete slab over the fillers shall be not less than 1½ in. in thickness, nor less in thickness than $\frac{1}{12}$ of the clear distance between joists. Shrinkage reinforcement shall be provided in the slab at right angles to the joists as required in Section 707.

"(*d*) Where removable forms or fillers not complying with Section 708(b) are used, the thickness of the concrete slab shall not be less than $\frac{1}{12}$ of the clear distance between joists and in no case less than 2 in. Such slab shall be reinforced at right angles to the joists with at least the amount of reinforcement required for flexure, giving due consideration to concentrations, if any, but in no case shall the reinforcement be less than that required by Section 707.

"(*e*) When the finish used as a wearing surface is placed monolithically with the structural slab in buildings of the warehouse or industrial class, the thickness of the concrete over the fillers shall be ½ in. greater than the thickness used for design purposes.

"(*f*) Where the slab contains conduits or pipes as allowed in Section 503, the thickness shall not be less than 1 in. plus the total over-all depth of such conduits or pipes at any point. Such conduits or pipes shall be so located as not to impair the strength of the construction.

"(*g*) Shrinkage reinforcement shall not be required in the slab parallel to the joists.

"709—Two-way systems with supports on four sides

"(*a*) This construction, reinforced in two directions, includes solid reinforced concrete slabs; concrete joists with fillers of hollow concrete units or clay tile, with or without concrete top slabs; and concrete joists with top slabs placed monolithically with the joists. The slab shall be supported by walls or beams on all sides and if not securely attached to supports, shall be reinforced as specified in Section 709(b).

"(*b*) Where the slab is not securely attached to the supporting beams or walls, special reinforcement shall be provided at exterior corners in both the bottom and top of the slab. This reinforcement shall be provided for a distance in each direction from the corner equal to one-fifth the longest span. The reinforcement in the top of the slab shall be parallel to the diagonal from the corner. The reinforcement in the bottom of the slab shall be at right angles to the diagonal or may be of bars in two directions parallel to the sides of the slab. The reinforcement in each band shall be of equivalent size and spacing to that required for the maximum positive moment in the slab.

"(*c*) The slab and its supports shall be designed by approved methods which shall take into account the effect of continuity at supports, the ratio of length to width of slab and the effect of two-way action.

"(*d*) In no case shall the slab thickness be less than 4 in. nor less than the perimeter of the slab divided by 180. The spacing of reinforcement shall be not more than three times the slab thickness and the ratio of reinforcement shall be at least 0.0025."

(Two methods of analysis are presented in the Code. Method 1 is not included here. Ed.)

"Method 2

"Notation—

- C = moment coefficient for two-way slabs as given in Table 3
- m = ratio of short span to long span for two-way slabs
- S = length of short span for two-way slabs. The span shall be considered as the center-to-center distance between supports or the clear span plus twice the thickness of slab, whichever value is the smaller.
- w = total uniform load per sq ft

"**Table 3—Moment Coefficients**

Moments	Short span, values of m: 1.0	0.9	0.8	0.7	0.6	0.5 and less	Long span, all values of m
Case 1—Interior panels							
Negative moment at—							
Continuous edge	0.033	0.040	0.048	0.055	0.063	0.083	0.033
Discontinuous edge	—	—	—	—	—	—	—
Positive moment at midspan	0.025	0.030	0.036	0.041	0.047	0.062	0.025
Case 2—One edge discontinuous							
Negative moment at—							
Continuous edge	0.041	0.048	0.055	0.062	0.069	0.085	0.041
Discontinuous edge	0.021	0.024	0.027	0.031	0.035	0.042	0.021
Positive moment at midspan	0.031	0.036	0.041	0.047	0.052	0.064	0.031
Case 3—Two edges discontinuous							
Negative moment at—							
Continuous edge	0.049	0.057	0.064	0.071	0.078	0.090	0.049
Discontinuous edge	0.025	0.028	0.032	0.036	0.039	0.045	0.025
Positive moment at midspan	0.037	0.043	0.048	0.054	0.059	0.068	0.037
Case 4—Three edges discontinuous							
Negative moment at—							
Continuous edge	0.058	0.066	0.074	0.082	0.090	0.098	0.058
Discontinuous edge	0.029	0.033	0.037	0.041	0.045	0.049	0.029
Positive moment at midspan	0.044	0.050	0.056	0.062	0.068	0.074	0.044
Case 5—Four edges discontinuous							
Negative moment at—							
Continuous edge	—	—	—	—	—	—	—
Discontinuous edge	0.033	0.038	0.043	0.047	0.053	0.055	0.033
Positive moment at midspan	0.050	0.057	0.064	0.072	0.080	0.083	0.050

"(*a*) *Limitations*—A two-way slab shall be considered as consisting of strips in each direction as follows:

"A middle strip one-half panel in width, symmetrical about panel centerline and extending through the panel in the direction in which moments are considered.

"A column strip one-half panel in width, occupying the two quarter-panel areas outside the middle strip.

"Where the ratio of short to long span is less than 0.5, the middle strip in the short direction shall be considered as having a width equal to the difference between the long and short span, the remaining area representing the two column strips.

"The critical sections for moment calculations are referred to as principal design sections and are located as follows:

"For negative moment, along the edges of the panel at the faces of the supporting beams.

"For positive moment, along the centerlines of the panels.

"(*b*) *Bending moments*—The bending moments for the middle strips shall be computed from the formula

$$M = CwS^2$$

"The average moments per foot of width in the column strip shall be two-thirds of the corresponding moments in the middle strip. In determining the spacing of the re-

inforcement in the column strip, the moment may be assumed to vary from a maximum at the edge of the middle strip to a minimum at the edge of the panel.

"Where the negative moment on one side of a support is less than 80 percent of that on the other side, two-thirds of the difference shall be distributed in proportion to the relative stiffnesses of the slabs.

"(*c*) *Shear*—The shearing stresses in the slab may be computed on the assumption that the load is distributed to the supports in accordance with (d).

"(*d*) *Supporting beams*—The loads on the supporting beams for a two-way rectangular panel may be assumed as the load within the tributary areas of the panel bounded by the intersection of 45-deg lines from the corners with the median line of the panel parallel to the long side.

"The bending moments may be determined approximately by using an equivalent uniform load per lineal foot of beam for each panel supported as follows:

$$\text{For the short span: } \frac{wS}{3}$$

$$\text{For the long span: } \frac{wS}{3}\frac{(3 - m^2)}{2}\text{''}$$

Reinforced Spread Footings. Single column footings are usually square in plan as shown in Fig. 80. The footing represents, as seen from the figure, cantilevers projecting out from the column in both directions and loaded upward by the soil pressure. Corresponding tension stresses are caused in both these directions at the bottom surface. Such footings are therefore reinforced by two layers of steel, perpendicular to each other and parallel to the edges. Since these cantilever elements are frequently short in span but rather deep, shear (diagonal tension) and bond are usually just as critical as bending, and deserve equal attention. It has been found by tests that, when footings are designed and loaded to fail in shear, such failure occurs not in shear along the planes which represent continuation of the faces of the column (punching shear), but by diagonal tension on the faces of a truncated pyramid or cone whose top is the base of the column (Ref. 16). For this reason the critical section for shear need not be taken at the face of the column. The ACI Code requirements for bending moment, shear, and bond in footings are as follows:

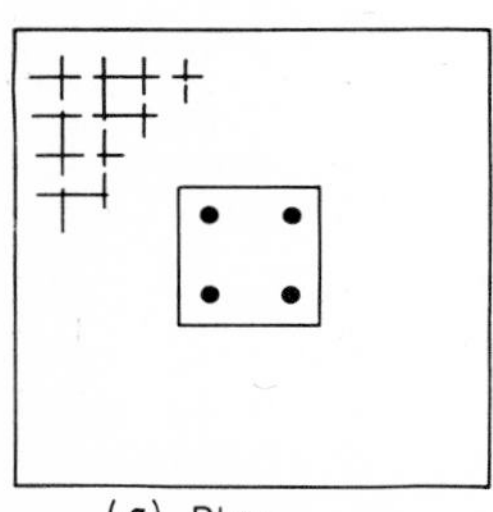

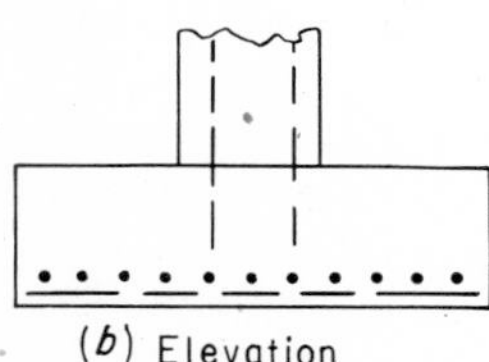

Fig. 80. Typical single-column footing.

"1204—Bending moment

"(*a*) The external moment on any section shall be determined by passing through the section a vertical plane which extends completely across the footing, and computing the moment of the forces acting over the entire area of the footing on one side of said plane.

"(*b*) The greatest bending moment to be used in the design of an isolated footing shall be the moment computed in the manner prescribed in Section 1204(a) at sections located as follows:

"1. At the face of the column, pedestal or wall, for footings supporting a concrete column, pedestal or wall.

"2. Halfway between the middle and the edge of the wall, for footings under masonry walls.

"3. Halfway between the face of the column or pedestal and the edge of the metallic base, for footings under metallic bases.

"(*c*) The width resisting compression at any section shall be assumed as the entire width of the top of the footing at the section under consideration.

"(*d*) In one-way reinforced footings, the total tensile reinforcement at any section shall provide a moment of resistance at least equal to the moment computed in the manner prescribed in Section 1204(a); and the reinforcement thus determined shall be distributed uniformly across the full width of the section.

"(*e*) In two-way reinforced footings, the total tensile reinforcement at any section shall provide a moment of resistance at least equal to 85 percent of the moment computed in the manner prescribed in Section 1204(a); and the total reinforcement thus determined shall be distributed across the corresponding resisting section in the manner prescribed for the square footings in Section 1204(f), and for rectangular footings in Section 1204(g).

"(*f*) In two-way square footings, the reinforcement extending in each direction shall be distributed uniformly across the full width of the footing.

"(*g*) In two-way rectangular footings, the reinforcement in the long direction shall be distributed uniformly across the full width of the footing. In the case of the reinforcement in the short direction, that portion determined by formula (21) shall be uniformly distributed across a band-width (B) centered with respect to the centerline of the column or pedestal and having a width equal to the length of the short side of the footing. The remainder of the reinforcement shall be uniformly distributed in the outer portions of the footing.

$$\frac{\textit{Reinforcement in band-width } (B)}{\textit{Total reinforcement in short direction}} = \frac{2}{(S + 1)} \tag{21}$$

"In formula (21), S is the ratio of the long side to the short side of the footing.

"1205—Shear and bond

"(*a*) The critical section for shear to be used as a measure of diagonal tension shall be assumed as a vertical section obtained by passing a series of vertical planes through the footing, each of which is parallel to a corresponding face of the column, pedestal, or wall and located a distance therefrom equal to the depth d for footings on soil, and one-half the depth d for footings on piles.

"(*b*) Each face of the critical section as defined in Section 1205(a) shall be considered as resisting an external shear equal to the load on an area bounded by said face of the critical section for shear, two diagonal lines drawn from the column or pedestal corners and making 45-deg angles with the principal axes of the footing, and that portion of the corresponding edge or edges of the footing intercepted between the two diagonals.

"(*c*) Critical sections for bond shall be assumed at the same planes as those prescribed for bending moment in Section 1204(b); also at all other vertical planes where changes of section or of reinforcement occur.

"(*d*) Computation for shear to be used as a measure of bond shall be based on the same section and loading as prescribed for bending moment in Section 1204(a).

"(*e*) The total tensile reinforcement at any section shall provide a bond resistance at least equal to the bond requirement as computed from the following percentages of the external shear at the section:

"1. In one-way reinforced footings, 100 percent.

"2. In two-way reinforced footings, 85 percent.

"(*f*) In computing the external shear on any section through a footing supported on piles, the entire reaction from any pile whose center is located 6 in. or more outside the

section shall be assumed as producing shear on the section; the reaction from any pile whose center is located 6 in. or more inside the section shall be assumed as producing no shear on the section. For intermediate positions of the pile center, the portion of the pile reaction to be assumed as producing shear on the section shall be based on straight-line interpolation between full value at 6 in. outside the section and zero value at 6 in. inside the section.

"(*g*) For allowable shearing stresses, see Section 305 and 809.

"(*h*) For allowable bond stresses, see Section 305 and 901 to 905."

Illustrative Example. A column 24 in. square and reinforced with 12 No. 8 hard-grade steel bars supports a total load of 400,000 lb. Design a square footing, using 2,500 psi concrete and intermediate-grade steel. The safe soil pressure is 5,000 lb per sq ft. Assume weight of footing = 24,000 lb. Required bearing area = 424,000/5,000 = 84.8 sq ft. Use 9′3 × 9′3 base with area of 85.5 sq ft. Net upward pressure (weight of footing does not cause moments or shears, just as, obviously, no moments or shears are present in a book lying flat on a table) is 400,000/85.5 = 4,680 lb per sq ft.

$$M = 4{,}680 \times 9.25 \times \frac{3.625^2}{2} \times 12 = 3{,}420{,}000 \text{ in.-lb}$$

with $K = 196$

$$d = \sqrt{\frac{3{,}420{,}000}{9.25 \times 12 \times 196}} = 12.6 \text{ in.}$$

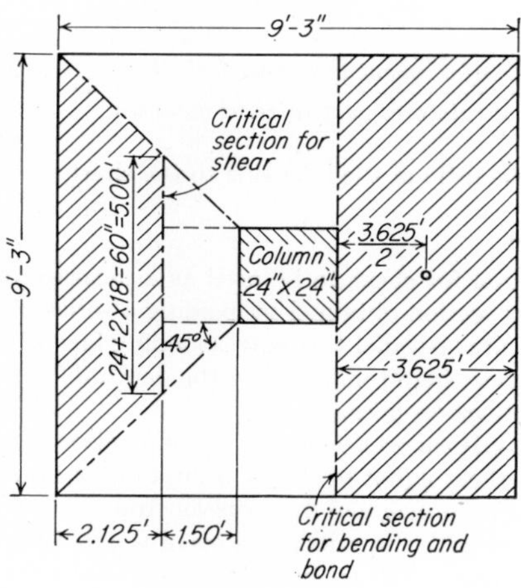

Fig. 81

The depth required for shear will probably be greater; therefore assume $d = 18$ in. The width of the critical section for shear is (see Fig. 81)

$$24 + 2 \times 18 = 60 \text{ in.}$$

$$V_s = \frac{5.0 + 9.25}{2} \times 2.125 \times 4{,}680 = 71{,}000 \text{ lb}$$

$v = 75$ psi; assume $j = 0.9$.

$$d = \frac{71{,}000}{60 \times 75 \times 0.9} = 17.5 \text{ in.}$$

The assumed d of 18 in. is adequate. Allowing 4 in. for cover plus the distance center to center between top and bottom layers of bars, the total height of the footing is 22 in. and its weight is 23,000 lb, close enough to the assumed value.

From ACI Code Section 1204(e) the design moment is taken as $0.85 \times 3{,}420{,}000 = 2{,}910{,}000$ in.-lb.

$$A_s = \frac{2{,}910{,}000}{20{,}000 \times 0.9 \times 18} = 9.0 \text{ sq in.}$$

For bond

$$V_b = 0.85 \times 9.25 \times 3.625 \times 4{,}680 = 134{,}000 \text{ lb}$$

The allowable bond stress is 200 psi. Then,

$$\Sigma o = \frac{134{,}000}{200 \times 0.9 \times 18} = 41.2 \text{ in.}$$

Use 16 No. 7 bars each way; $A_s = 9.62$ sq in.; $\Sigma o = 44.0$ in. To transfer the stress from the base of the column, dowels are required. From ACI Code Section 1206 (not included here) it is found that one No. 8 dowel is required for each column bar. Dowels must extend 2′3 into the column and 1′4 into the footing to satisfy the Code. The bearing stress at the base of the column is $400{,}000/24 \times 24 = 692$ psi, less than the allowable of $0.375 \times 2{,}500 = 938$ psi. In some cases it will be found necessary to increase the cross section of the base of the column by providing a pedestal. This will ensure a more favorable transfer of stress and allow for placing of dowels. (See Fig. 82.)

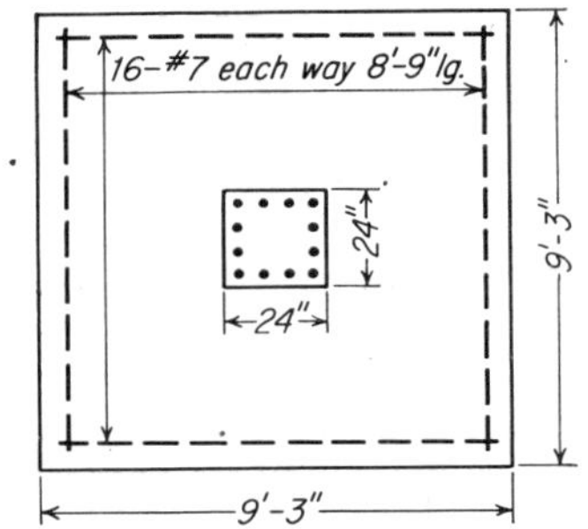

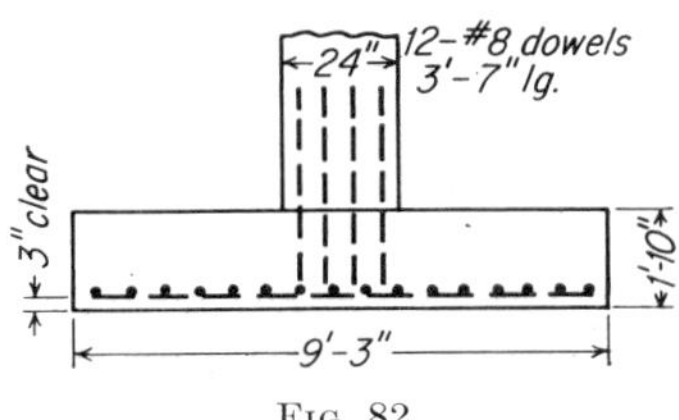

FIG. 82

Combined Footings. Figure 83 shows a combined footing of trapezoidal shape carrying two column loads. The footing area may be rectangular if desired. In any event, its centroid must coincide with the resultant of the two column loads. The longitudinal reinforcement is computed as for a rectangular beam. The transverse reinforcement must be sufficient to resist bending of all that portion of the footing extending laterally beyond the columns. Shear is computed as for isolated footings. Pedestals may be required as noted above. This type of footing is often advantageous at property lines, where there is no room for isolated footings under wall columns.

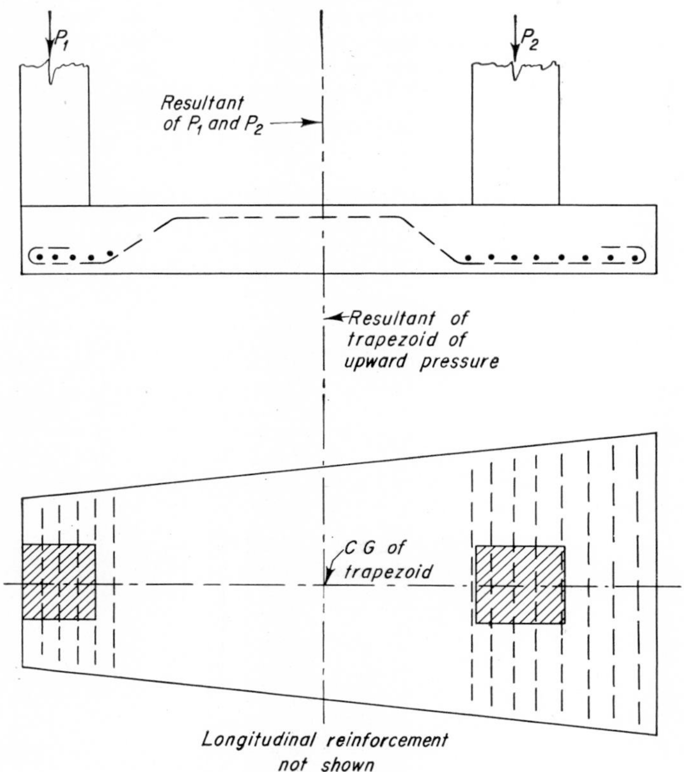

FIG. 83. Trapezoidal combined footing.

Mat Foundations. Foundations of this type are designed as inverted floor systems, acted upon by the upward soil pressures. They may be of the flat slab or beam and girder type.

RETAINING WALLS AND ABUTMENTS

A retaining wall, usually built of stone, plain concrete, or reinforced concrete, is a wall built to sustain the lateral pressure of the earth or other material possessing more 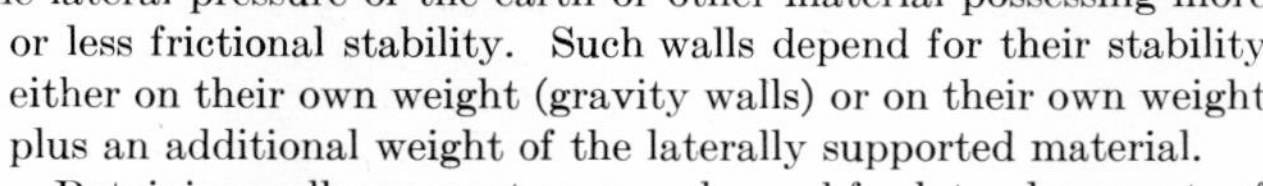
or less frictional stability. Such walls depend for their stability either on their own weight (gravity walls) or on their own weight plus an additional weight of the laterally supported material.

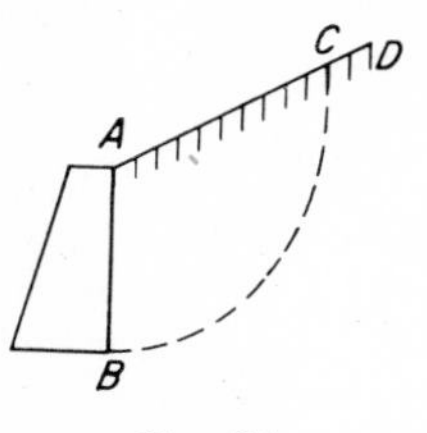

FIG. 84

Retaining walls are most commonly used for lateral supports of earth. In the following discussion the term "earth" will be used whether or not the supported material is earth or some other substance.

Let Fig. 84 represent a wall supporting an earth fill, the surface of which is represented by the line AD. There is a tendency for a portion of the earth next to the wall, such as ACB, to break away along some such line as CB and to settle downward and move forward, thus tending to slide and overturn the wall. The weight of the block ACB tends to cause this motion. It is resisted by supporting forces acting across AB and CB, by friction along the line CB (called internal friction), and by friction along the plane AB.

The internal frictional resistance may also include a cohesive quality in the soil. If cohesion is absent, the surface of rupture may be assumed to be an inclined plane such as BD; while if cohesion is present, the surface of rupture stands more nearly vertical at the surface of the ground than at the base of the wall, thus resulting in the curve CB. Usually this curved direction of the surface of rupture is influenced by a variation in moisture content. The more moisture, the less cohesion, and hence the flatter the angle of rupture. It should thus be seen at the outset that the removal of all possible moisture from the supported material reduces greatly the size of the prism of earth actually effective in pressing against the wall and thus increases the stability of the wall. Probably the most important rule in the design of retaining walls, therefore, is to *provide adequate drainage.*

Earth-pressure theories generally assume a granular mass possessing no cohesion, since this assumption is on the safe side of the usual conditions of soil possessing some cohesion. Tables 10 and 11 give weights, angles of repose, and values of internal friction

TABLE 10. ANGLES OF REPOSE AND WEIGHTS PER CUBIC FOOT FOR VARIOUS EARTHS [1]

Material	Slope	Angle of repose, deg	Weight, lb per cu ft
Sand, dry	2.8 :1 to 1.4 :1	20 to 35	90 to 110
Sand, moist	1.75:1 to 1:1	30 to 45	100 to 110
Sand, wet	2.8 :1 to 1.2 :1	20 to 40	110 to 120
Ordinary earth, dry	2.8 :1 to 1:1	20 to 45	80 to 100
Ordinary earth, moist	2.1 :1 to 1:1	25 to 45	80 to 100
Ordinary earth, wet	2.1 :1 to 1.75:1	25 to 30	100 to 120
Gravel, round to angular	1.75:1 to 0.9 :1	30 to 48	100 to 135
Gravel, sand and clay	1.8 :1 to 1.3 :1	20 to 37	100 to 115

[1] CAIN, "Earth Pressure, Walls and Bins," p. 9.

TABLE 11. COEFFICIENTS OF INTERNAL FRICTION [1]

Kind of material	Tangent of angle of internal friction	Approximate corresponding		Authority
		Angle, deg	Slope	
Coal, shingle, ballast, etc.	1.423	54	0.7 to 1	B. Baker
Bank sand	1.423	54	0.7 to 1	Goodrich
Riprap	1.097	48	0.9 to 1	Goodrich
Earth	1.097	48	0.9 to 1	B. Baker
Quicksand, 100 up	0.895	42	1.1 to 1	Goodrich
Clay	0.895	42	1.1 to 1	B. Baker
Quicksand, 50–100	0.750	37	1.3 to 1	Goodrich
Earth	0.750	37	1.3 to 1	Steel
Bank sand	0.750	37	1.3 to 1	Wilson
Sand, 50–100	0.549	29	1.8 to 1	Goodrich
Bank sand	0.549	29	1.8 to 1	Goodrich
Clay	0.474	25	2.1 to 1	Goodrich
Cinders	0.474	25	2.1 to 1	Goodrich
Gravel, ½-in	0.474	25	2.1 to 1	Goodrich
Gravel, ¼-in	0.350	19	2.9 to 1	Goodrich
Bank sand	0.350	19	2.9 to 1	Goodrich
Sand, 30–50	0.258	14	3.9 to 1	Goodrich
Sand, 20–30	0.179	10	5.6 to 0	Goodrich

[1] GOODRICH, *Trans. ASCE*, vol. 53, p. 301.

for a variety of common granular materials. Let Fig. 85 be any wall against which a fill of earth has been placed up to the surcharge slope AB. The wedge of material ABC will be assumed to act in unison with the wall in supporting the lateral pressure of earth to the right of the vertical section AC. The resultant pressure on AC is assumed to act at point D which is one-third the distance from C to A. The resultant pressure P is also assumed to act parallel to the surface AB. Let CE be the plane of rupture. The following formula, known as Rankine's formula, gives the value of the force P.

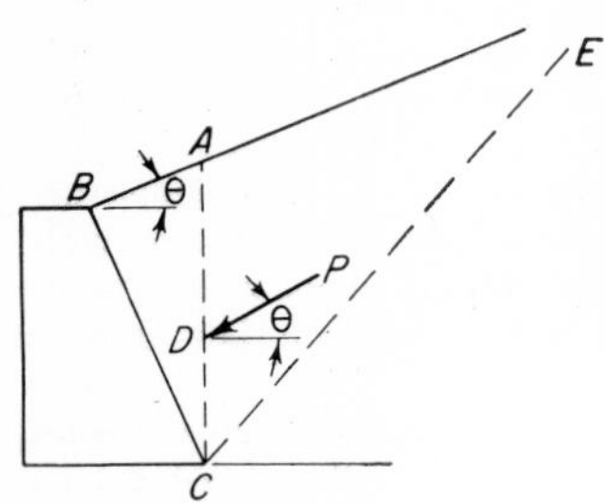

Fig. 85

$$P = C_e \frac{wh^2}{2} \tag{50}$$

in which P is the total active thrust of earth against the vertical plane as described above, w is the weight per cubic foot of retained material, and h is the height of vertical section considered, as AC.

$$C_e = \cos\theta \cdot \frac{\cos\theta - \sqrt{\cos^2\theta - \cos^2\phi}}{\cos\theta + \sqrt{\cos^2\theta - \cos^2\phi}}$$

where θ is the angle of surcharge, and ϕ is the angle of internal friction.

When $\theta = \phi$, then

$$P = \tfrac{1}{2}wh^2 \cdot \cos\phi \tag{51}$$

When $\theta = 0$,

$$P = C_e' \frac{wh^2}{2} \tag{52}$$

in which

$$C_e' = \tan^2 (45° - \tfrac{1}{2}\phi)$$

Equivalent Fluid Pressure. It will be noted that in Eqs. (50) to (52) the lateral pressure is equal to the lateral pressure of fluid having the same weight per cubic foot, w, as that of the supported material, the whole amount then being multiplied by a constant C_e or C_e'. If now wC_e or wC_e' be replaced by w', this latter expression becomes the weight of an equivalent fluid which will produce the same lateral pressure P as that shown by Eqs. (50) to (52). This device is generally termed "the equivalent fluid pressure." It will be noted also that the distribution of pressure is triangular, the centroid of such pressure being at one-third the height of the supporting material.

Equivalent Surcharge. In the event of live load being placed upon the top of a fill, this live load may be replaced by an equivalent weight of earth; thus in Fig. 86 the original surface of earth BF has been increased by an assumed depth h' to a new hypothetical surface AG. The pressure against the wall now becomes the portion $BFCD$ of the complete equivalent fluid pressure triangle ACD. The resultant pressure P acts at a distance y above the base equal to

$$y = \frac{h}{3} \cdot \frac{h + 3h'}{h + 2h'} \tag{53}$$

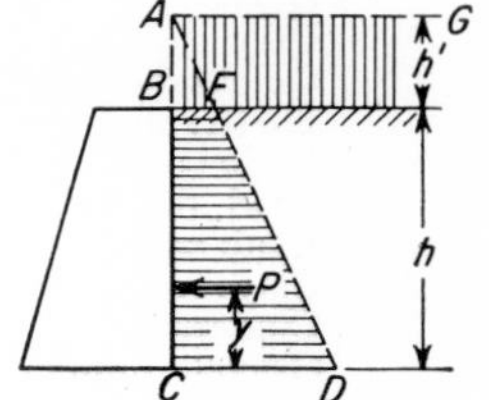

Fig. 86

This device of equivalent surcharge is of use in the design of abutments as shown in Fig. 87.

Figure 88 shows a frequent condition in railroad construction. The pressure triangle abc is due to the earth pressure without the live load. The pressure exerted by the track

load on the portion of the surface, AD, may be considered to act on a constantly spreading width as the depth increases, the angle of such spread being assumed at 30°. F is found by projecting the line of spread of A down to the face of the wall. At the depth

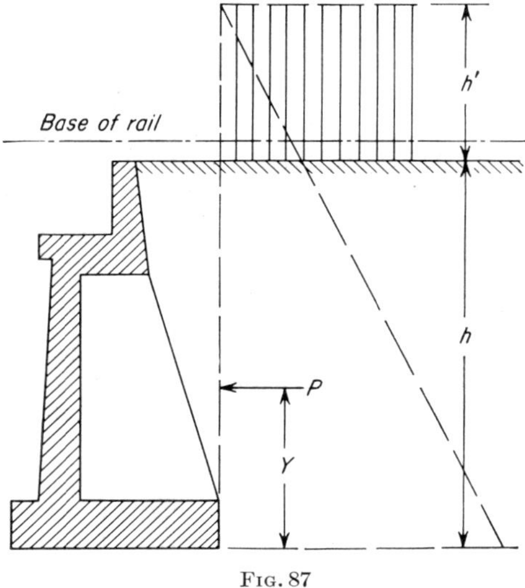

FIG. 87

FG the original pressure on AD is uniformly distributed. The intensity of the pressure thus acting on the plane FG is then converted into equivalent lateral fluid pressure by multiplying the value by C_e'. The result is then laid off as *de* and *cf*, the pressure triangle being thus enlarged by this additional amount of pressure.

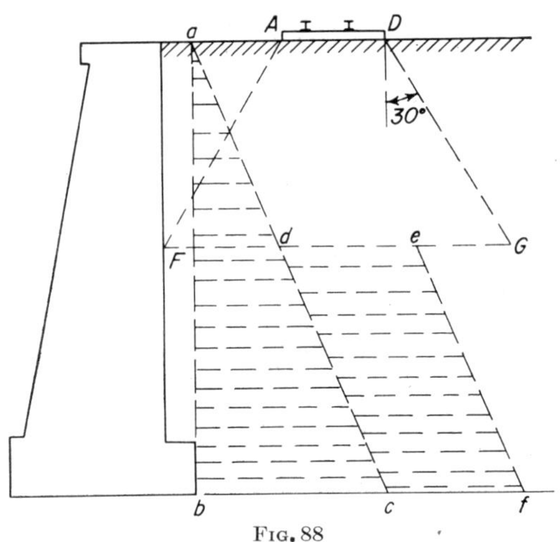

FIG. 88

Stability of a Retaining Wall. A retaining wall must resist sliding, and it must resist being tipped forward. Resistance to sliding may be developed by (1) the friction on

the base plane, (2) the depth of the base plane below the ground surface in front of the wall, and (3) a key or continuous lug protruding below the base plane. Table 12 shows the coefficients of friction between masonry and various base materials.

TABLE 12. COEFFICIENTS AND ANGLES OF FRICTION BETWEEN EARTH AND OTHER MATERIALS

Materials	$f = \tan \phi$	ϕ
Masonry upon masonry	0.65	33°
Masonry upon wood, with grain	0.60	31°
Masonry upon wood, across grain	0.50	26°40′
Masonry on dry clay	0.50	26°40′
Masonry on wet clay	0.33	18°20′
Masonry on sand	0.40	21°50′
Masonry on gravel	0.60	31°

Resistance to overturning must be developed by a proper placing of all the downward loads occurring on the base plane, with respect to the center of the base plane. Figure 89 shows the downward weights of a typical wall combined into a resultant downward

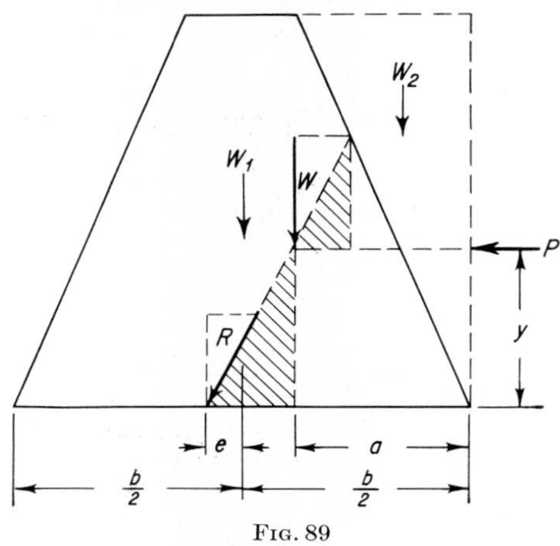

FIG. 89

force W. The lateral pressure resultant P is brought forward until it intersects the line of action of W. These two forces now produce the resultant force R, which strikes the base at some distance e from the center of the base. If e equals zero, i.e., if the resultant strikes the base at its center, there is no tendency to tip the wall over; therefore, there would be a uniform pressure distribution over the foundation. If the distance e is increased, then the overturning moment is likewise increased, a pressure distribution being caused as shown in Fig. 90. The limiting value of e is $b/6$ in which case the pressure intensity BG (Fig. 90) becomes zero. Unequal distribution of pressure on the base is not objectionable if the wall stands on rock or other unyielding foundation material. When the wall is constructed on soil, settlement is in a general way proportional to the pressure intensity. If the region of the wall is subject to ground vibrations, this settlement is greatly increased. The result is, therefore, a tendency of the wall to tip more and more forward when the pressure distribution is not uniform. Walls subjected to vibration and resting on soils should be designed with the resultant R passing through or near to the center of the base.

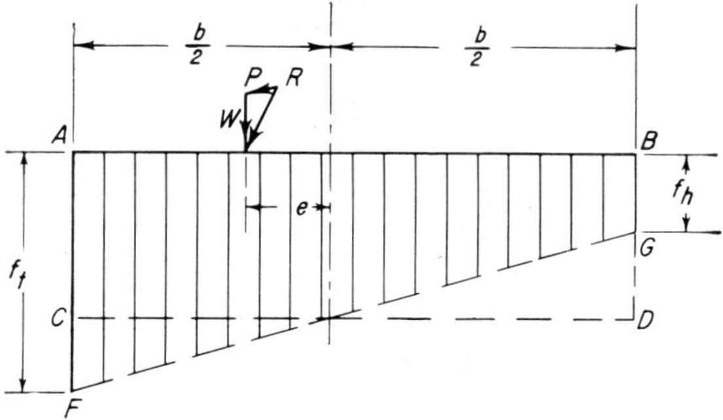

FIG. 90

The following formulas are useful in determining the pressure distribution along the base:

$$\frac{W}{P} = \frac{y}{\frac{b}{2} - a + e}$$

whence

$$e = \frac{Py}{W} - \frac{b}{2} + \tag{54}$$

$$f_t = \frac{W}{b} + \frac{6We}{b^2} \tag{55}$$

$$f_h = \frac{W}{b} - \frac{6We}{b^2} \tag{56}$$

Design of a Cantilever Wall. The cantilever wall consists of a base slab from which a vertical wall or stem extends upward. Reinforcement is provided in both members to supply resistance in bending and shear.

Design of Stem. The tendency of the earth pressure acting against the stem is to break it off at its junction with the base. The stem must also be thick enough at the bottom to prevent excessive shearing (diagonal tension) and bending stress. The thickness of the wall at the top is arbitrary, often being governed by the conditions of placing the concrete in the form. On railway and highway work, minimum top thickness is commonly 12 in. The wall must be thick enough at the base to resist the tendency to shear off. To prevent this, the thickness required would be, from Eq. (31) (see Fig. 91),

$$d = 0.095 \frac{P}{v} \cos \theta \tag{57}$$

or if $\theta = 0$,

$$d = 0.095 \frac{P}{v} \tag{58}$$

If the moment requirement governs, then, choosing K from Table 5, we have

$$d = 0.408 \sqrt{\frac{w' h_1^3}{K} \cos \theta} \tag{59}$$

If $\theta = 0$,

$$d = 0.408 \sqrt{\frac{w' h_1^3}{K}} \tag{60}$$

Base Slab. The economical position of the stem on the base slab is at the forward third point. It is not always possible to place the stem at this point because of property lines or other influences. The portion of the base slab forward of the stem is called the toe, and that back of the stem the heel. The toe is being bent and sheared upward by the pressure of the base reaction under the toe, shear usually governing. The heel is

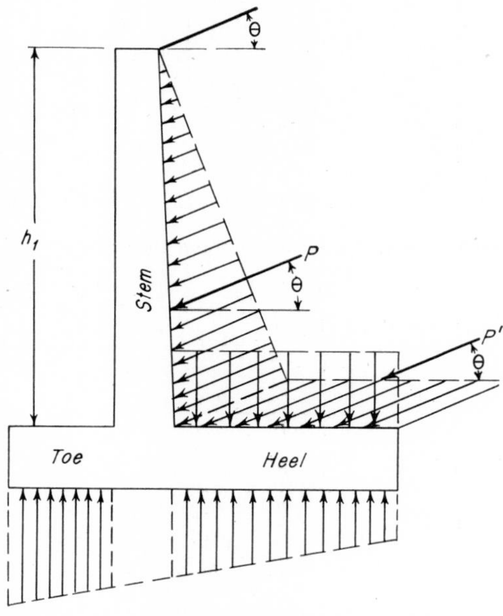

FIG. 91

being bent and sheared downward by the inclined earth pressure if there is a surcharge, together with the excess weight of earth above the heel slab over the upward reaction below the heel slab. The forces acting are shown in Fig. 91. The force P' is distributed over the top of the heel slab. Its vertical component is added to the pressure of the weight of earth directly above the slab.

If there is no surcharge ($\theta = 0$), the force P' is zero.

Arrangement of Reinforcement. Figure 92 shows typical arrangements of reinforcement for cantilever walls.

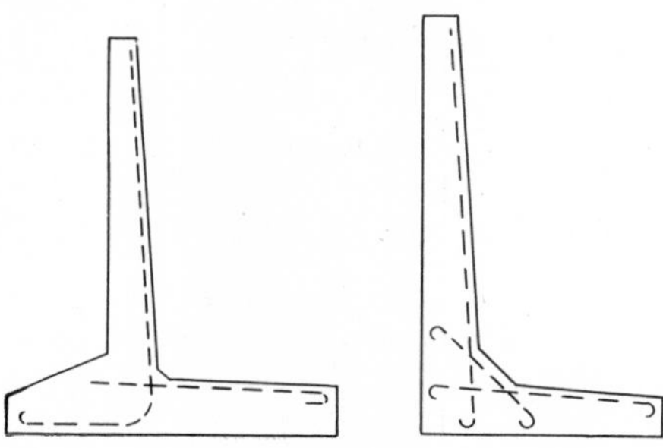

FIG. 92

Counterfort Walls. Walls of the cantilever form in excess of 20 ft in height are more economically constructed when provided with brackets to assist in strengthening the junction of stem and base slab. When these brackets are back of the stem, they act in tension and are called "counterforts." When the brackets are in front of the stem, they act in compression and are called "buttresses." Figure 93 shows a section of a counterfort wall.

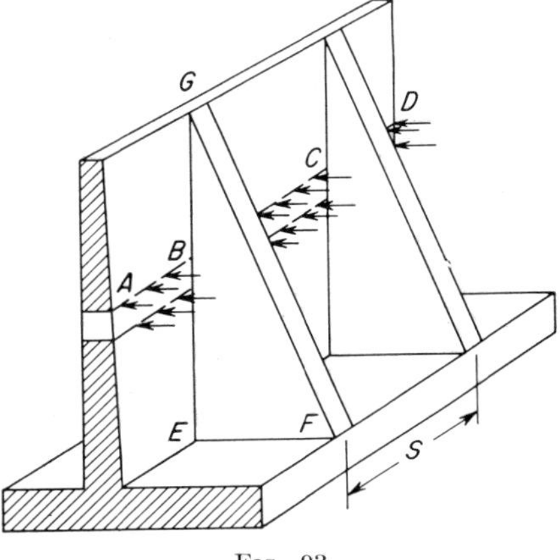

Fig. 93

The economical spacing s for counterforts in terms of the height of the wall, both in feet, is

$$s = 2.45 + 0.216h \tag{61}$$

The thickness of the counterfort should not be less than 12 in. The absolute minimum thickness is governed by the width required to put in necessary steel and to resist shear, as will be noted later.

In Fig. 93, the face wall may be considered as a continuous slab loaded horizontally. If the slab be divided into 1-ft strips, such as $ABCD$, the loading on the strip will vary in accordance with the pressure diagram, as in Fig. 88 or 91. The greatest pressure will be at the strip at the base of the wall where, the restraining effect of the base being neglected, the effective depth d of the slab may be determined by

$$d = 0.289s\sqrt{\frac{w'h_1}{R}} \tag{62}$$

At least 2 in. of covering of concrete must be provided in addition to effective depth, to make up the total thickness of the face slab.

The design of the heel slab is likewise made on the basis of a continuously supported slab attached at the bottoms of the counterforts. The loading forces are the same as those for the heel slab of the cantilever wall. It is essential that the back slab be investigated for shear adjacent to the counterfort. It will usually be found that, unless stirrups or other shear reinforcements are used in the base slab, shear will govern the determination of its thickness.

The design of the toe slab is the same as that for the cantilever wall. It is very important that the toe slab be investigated for shear at the section taken at the front face of the vertical slab.

Arrangement of the Reinforcement for a Counterfort Wall. The foregoing discussion has involved only the critical sections of the base or vertical slabs. The complete arrangement of the reinforcement must be carried out with due consideration to shear and moment stresses, and especially to bond and anchorage.

The vertical wall is reinforced horizontally either by two sets of bars placed one near each face or by one set of bent bars which are at the back of the wall at the counterforts and at the front of the wall midway between counterforts. Vertical bars should be provided near the back face, extending out of the base and up into the vertical slab a distance $\frac{1}{2}s$, to take care of the restraining effect of the base on the vertical slab. Sufficient vertical bars are also required to serve as supports to the horizontal steel in the base slab.

The toe is reinforced at right angles to the direction of the wall, to take care of the

cantilever bending action. The heel slab is reinforced as a continuous slab, preferably with two planes of steel, one near the bottom face and the other near the top.

Since the counterfort (see Fig. 94) is serving as a tension tie between the vertical slab and the heel slab, it alone is responsible for the prevention of overturning of the vertical slab. The overturning moment of the vertical slab *BC* about *B* is computed as for a cantilever wall. The resisting moment is provided by the steel placed along the face *AC* of the counterfort, and therefore having the moment arm *BD* of the perpendicular distance from the center of the vertical slab at *B* to the line of the steel along *AC*. The steel in the counterfort must be carried into both the vertical slab and the heel slab as far as possible and hooked. Since both these slabs are acting away from the counterfort, it is necessary that both vertical and horizontal steel be provided in the counterfort to tie these slabs to it. The amount of steel required for this purpose is obtained directly from the amount of reaction of these slabs acting under the earth pressure as continuously supported slabs.

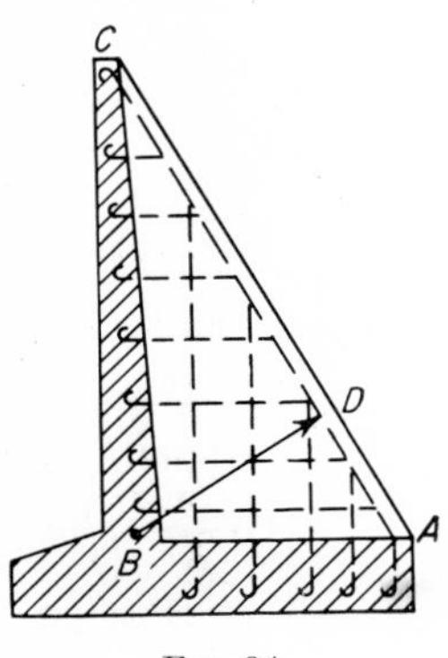

Fig. 94

CONCRETE ARCHES

Concrete has been extensively used for the construction of arch bridges because of its adaptability to compressive loads. Researches have indicated that the elastic-arch theory applies with accuracy in determining live-load stress distribution and, when coupled with considerations of shrinkage and plastic flow, furnishes a basis for estimating dead-load stresses as well. The methods of arch analysis are to be found in standard works on structural engineering and will be given here only in summary form.

Forms of Arch. There are three general forms of concrete arch commonly in use: (1) open spandrel, (2) earth fill, and (3) bowstring. The open-spandrel arch consists of one or more ribs on which rest a series of cross walls or columns, which in turn support the deck. Earth-fill arches consist of arched barrels with side walls between which is placed a compacted earth fill. The roadway is a pavement laid directly on this earth fill. The bowstring arch consists of two or more ribs placed partly or entirely above the roadway, the roadway construction being suspended from these ribs by vertical hangers. Of the three forms, the first two are by far the most common. Earth-filled arches are generally used for spans of 100 ft or less, and open-spandrel arches are used for spans above this length. One of the longest reinforced concrete arches to date is the Sandö Bridge arch over the Angerman River in Sweden,[1] opened to traffic in July, 1943 [span 866 ft, rise 130 ft, thickness at the crown 8.75 ft (1 percent of the span)]. It is of triple box section, 14.75 ft thick at the springing line. Another notable arch (once the world's longest) is the Freyssinet triple-span arch (612 ft center to center, 590 ft clear each span, rise 110 ft) opened to traffic at Brest, France, in 1929 (Ref. 17) [2]

Loading. In any of the foregoing forms, the curved member, whether it is a rib or a barrel, is the ultimate carrying member. All dead load of the structure, as well as all live load coming upon the structure, must be carried by it. Railroad wheel loadings will be found on p. 5–16 and highway loadings on p. 5–17. Unless the earth-fill arch is of unusually high rise, only the vertical acting weight of the earth fill is considered in the analysis.

[1] *Eng. News-Record*, Jan. 17, 1946, p. 8.

[2] Scheduled for completion in 1959 is a 951-ft arch span over the Paraná River between Brazil and Paraguay, near the mouth of the Iguassú River (*Eng. News-Record*, Aug. 1, 1957, p. 51). Also in progress (1958) is an 886-ft span over the Douro Porto River in Portugal.

Curvature of Arch Axis. For moderate spans, the best results are obtained by shaping the curvature of the arch axis to coincide with the dead-load pressure line; thus, under dead load, bending moments along the arch axis are everywhere equal to zero. For longer spans, if one desires, a portion or all of the distributed live load may be added in determining the theoretical curvature.

If an arched member were carrying a dead load uniformly distributed over the horizontal projection of the span, the theoretical curvature would be a second-degree parabola. In practice the load increases somewhat from the crown to the springing for open-spandrel arches and increases considerably for earth-fill arches. A parabola of higher order than the second degree can be found quite closely to satisfy the dead-load curvature for open-spandrel arches. The theoretical curvature for dead load of earth-fill arches varies from a circular arc toward an elliptical form.

Open-spandrel Arches. Let Fig. 95*a* represent a series of panel-point loadings for which the arch curvature is desired. Let the origin of coordinates be at the crown,

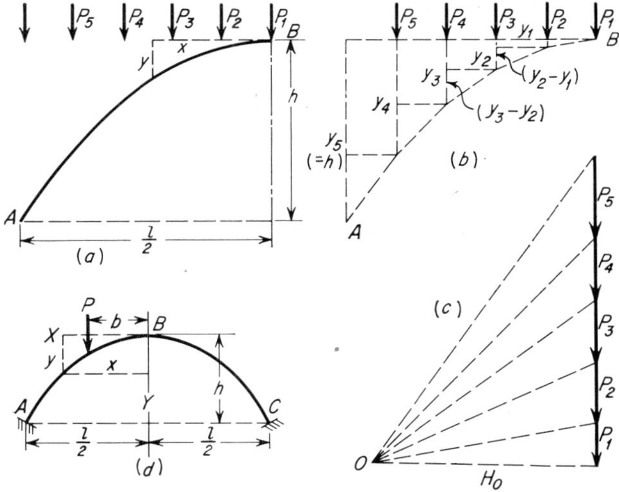

FIG. 95

point B. The number of loads P_1 to P_5 may be modified to suit any other arrangement of loads. The ordinate y_3 (Fig. 95*b*) together with points A and B will be used to define the arch curvature, y_3 being 0.6 of the horizontal distance from the crown to the springing line. If H_o is the horizontal thrust at the crown, then

$$y_3 = \frac{0.1l}{H_o}(3P_1 + 2P_2 + P_3) \tag{63}$$

If there is no bending moment along the axis of the arch, then the thrust is

$$H_o = \frac{0.1l}{h}(5P_1 + 4P_2 + 3P_3 + 2P_4 + P_5) \tag{64}$$

from which is derived

$$y_3 = \frac{(3P_1 + 2P_2 + P_3)h}{(5P_1 + 4P_2 + 3P_3 + 2P_4 + P_5)} \tag{65}$$

A parabolic axis passing through A and B and having the ordinate y_3 at the abscissa 0.3 takes the form

$$y_3 = h(0.6)^n, \text{ or } (0.6)^n = \frac{y_3}{h} \tag{66}$$

All terms in this equation except the exponent n are known. Substituting Eq. (65) into Eq. (66), we have

$$(0.6)^n = \frac{(3P_1 + 2P_2 + P_3)}{(5P_1 + 4P_2 + 3P_3 + 2P_4 + P_5)} \tag{67}$$

From this the exponent n is

$$n = \frac{\log L}{\log (0.6)} \tag{68}$$

where L is the right-hand member of Eq. (67).

As an example, let us suppose $P_1 = 0.6$, $P_2 = 1.2$, $P_3 = 1.4$, $P_4 = 1.7$, $P_5 = 2.2$.

$$L = \frac{(1.8 + 2.4 + 1.4)}{(3.0 + 4.8 + 4.2 + 3.4 + 2.2)} = \frac{(5.6)}{(17.6)} = 0.318$$

$$n = \frac{\log (0.318)}{\log (0.6)} = \frac{9.502 - 10}{9.778 - 10} = \frac{0.498}{0.222} = 2.24$$

Thus, for this loading the equation of the curve becomes

$$y = h\left(\frac{2x}{l}\right)^{2.24}$$

If it is assumed that the moment of inertia of the rib varies from the crown to the springing line as the secant of the slope of the axis, the equations for thrust, moment, and shear at the crown may be written in terms of n, as follows, for a single concentrated load P (Fig. 95d):

$$H_o = \frac{AE - BD}{B^2 - AC} \tag{69}$$

$$M_o = -\frac{(H_oB + D)}{A} \tag{70}$$

in which

$$A = \frac{l}{I_o}$$

$$B = \frac{hl}{I_o(n + 1)}$$

$$C = \frac{h^2l}{I_o(2n + 1)}$$

$$D = -\frac{P}{2I_o}\left(\frac{l}{2} - b\right)^2$$

$$E = -\frac{Ph}{I_o}\left[\frac{l^2}{4(n + 2)} + \frac{2^n b^{(n+2)}}{l^n(n^2 + 3n + 2)} - \frac{bl}{2(n + 1)}\right]$$

where I_o is the moment of inertia at the crown. Also (see Ref. 18)

$$V_o = -\frac{1}{2}\left[1 - \frac{3b}{l} + 4\left(\frac{b}{l}\right)^3\right] \tag{71}$$

These equations have been solved for representative values of n and plotted in Fig. 96. To use, one first determines n as above; then, taking each loading on the arch and multiplying it by the ordinate on the influence line and totaling these products for all

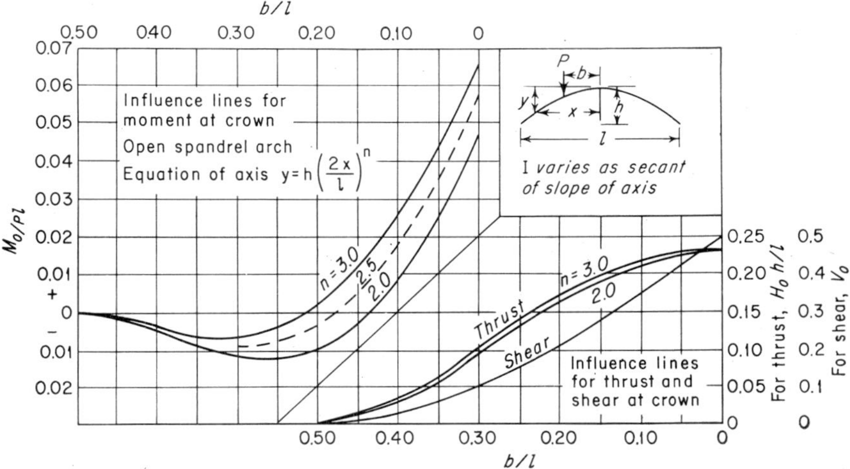

FIG. 96. Influence lines for moment, thrust, and shear at crown of open-spandrel arch.

loads, one finds respectively the thrust, moment, and shear at the crown. These quantities having been found, similar values may be computed by statics for any other point in the arch, by the equations

$$M = M' + H_o y \pm V_o x + M_o \tag{72}$$
$$V = V_o + \Sigma P \tag{73}$$

where M' is the statical moment of loads (ΣP) between crown and point in question and x and y are coordinates (see Fig. 95d).

If the moment of inertia of the rib does not vary as above assumed, the method may suffice for moderate spans; and for longer spans it will give the trial-pressure polygon (constructed as for Figs. 95b and 95c) which may then be tested against the form of axis of the arch. Since no change of slope of the tangent at either end of the arch will occur, from the principle of area moments we know that the algebraic sum of the areas of the M/I diagram for the span will equal zero. Since the moment at any point on the axis is equal to H_o times the *vertical* distance to the pressure line, the area between axis and pressure line may be taken as the moment diagram by measuring all ordinates *vertically* from axis to pressure line. By dividing successive ordinates, uniformly spaced, by corresponding values of I for the points, and tabulating, a final check may be made on the pressure line (Ref. 18).

Earth-filled Arches. Cochrane[1] has given a simple method for arriving at the curvature of earth-fill arches. In Fig. 97, $x = cl$; $r = h/l$; then,

$$y = \frac{4rl}{1 + 3r}(c^2 + 24c^5 r) \tag{74}$$

which is the equation of the arch axis for the loads considered. Influence lines are also

[1] *Jour. Eng. Soc. Western Pennsylvania*, vol. 32, p. 647. November, 1916; see also HOOL and JOHNSON "Concrete Engineers' Handbook," 1918, p. 669.

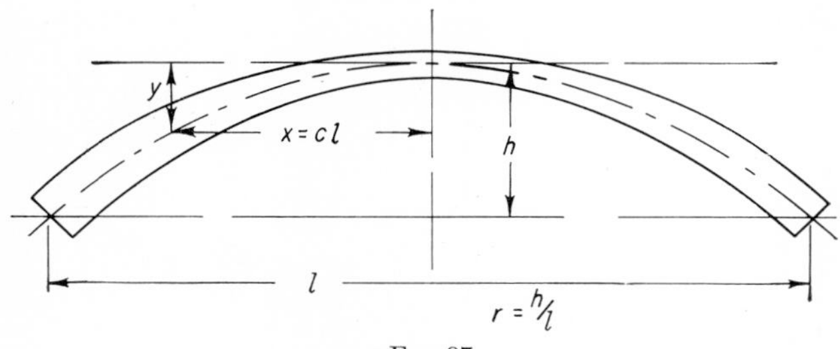

Fig. 97

drawn for moment, thrust, and shear at both crown and springing, for arches having ratio of springing thickness to crown thickness of 2.0 and 2.5, in Figs. 98 and 99.

Although the arch thickness is dependent on the stress used in the design and the loading, the average minimum thickness for earth-fill arches should be one-sixtieth of the span but not less than 10 in. The barrel thickens appreciably from the third point toward the springing.

Special Considerations. Plastic flow of concrete under compression may take place to an important degree. Shrinkage also may cause a further shortening of the arch axis with increase in age. These two effects cannot be eliminated in the structure; but since the major portion of the deformations takes place at early ages, the effects may be greatly reduced by casting the arch in a series of segments with gaps of a foot or so in between. These segments should age for at least a month with careful curing before the gaps are finally closed. Small arches may be cast in three segments while longer arches may require five or more. The order of casting is usually to begin with the sections next to the abutments and to follow these by casting the section at the crown to prevent the centering from humping up in the middle. Intermediate sections may then follow in symmetrical pairs.

Expansion joints are usually built into the deck structure or, in the case of an earth-fill arch, into the spandrel walls. Arches under 60 ft usually have joints at the abutments only, while those of greater span may have additional joints at the quarter points. Experience has usually shown these joints to be necessary, owing to cyclic temperature changes in the rise of the arch.

Although space does not permit detailed discussion here, mention should be made of the Freyssinet method of endeavoring to correct for shrinkage and plastic flow in arch ribs (Refs. 17, 19, 20).

Ample drainage with proper disposal conductors should be provided in earth-fill arches.

Closed Culvert Rings. Circular culverts should be designed to resist moments caused by vertical loads and lateral soil pressures. Let w be the vertical pressure on the pipe per unit of area on the horizontal projection, and let p be the lateral earth pressure per unit of area on the vertical projection. Then the moment at the top, bottom, and two sides of the pipe will be equal to

$$M = \frac{1}{4}r^2(w - p)$$

in which r is the mean radius of the pipe.

Culverts, conduits, and sewers of noncircular section should be designed as arches. Loads on such structures not only are vertical but also may be horizontal.

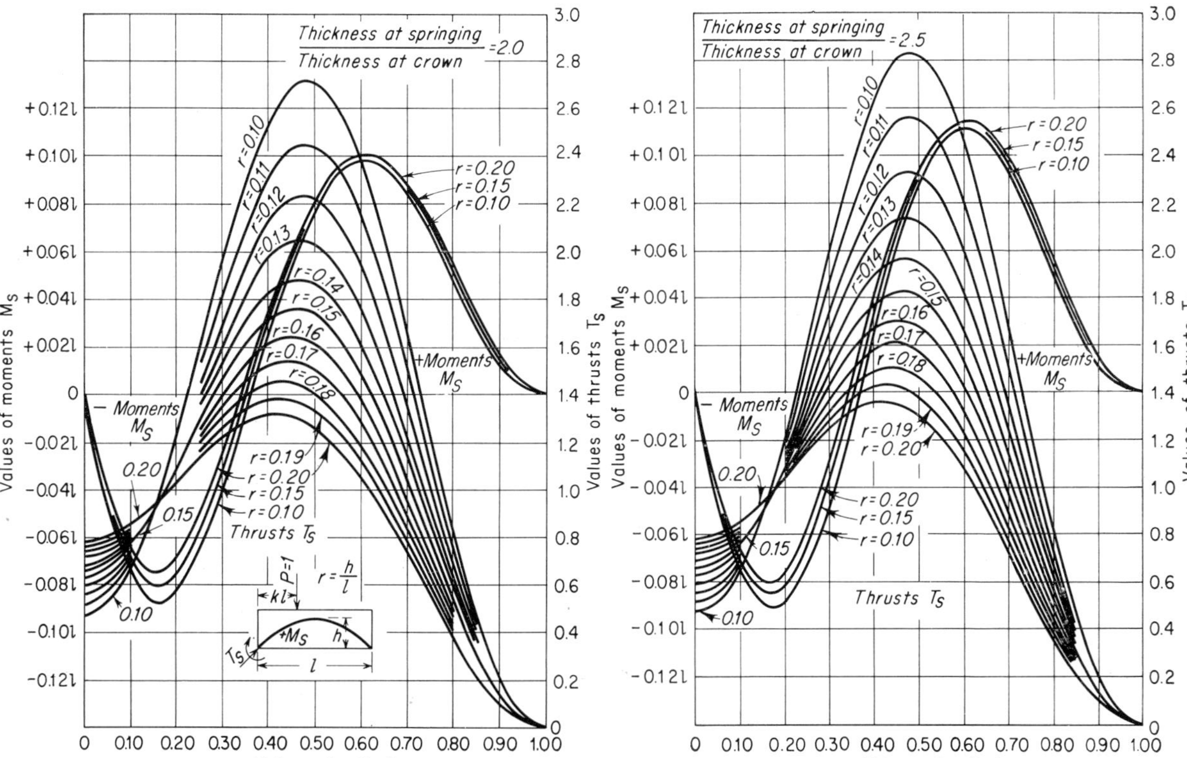

FIG. 98. Influence lines for moments and thrusts at the springing of earth-filled arches; ratios of thickness at springing to thickness at crown, 2.0 and 2.5. (*After Cochrane.*)

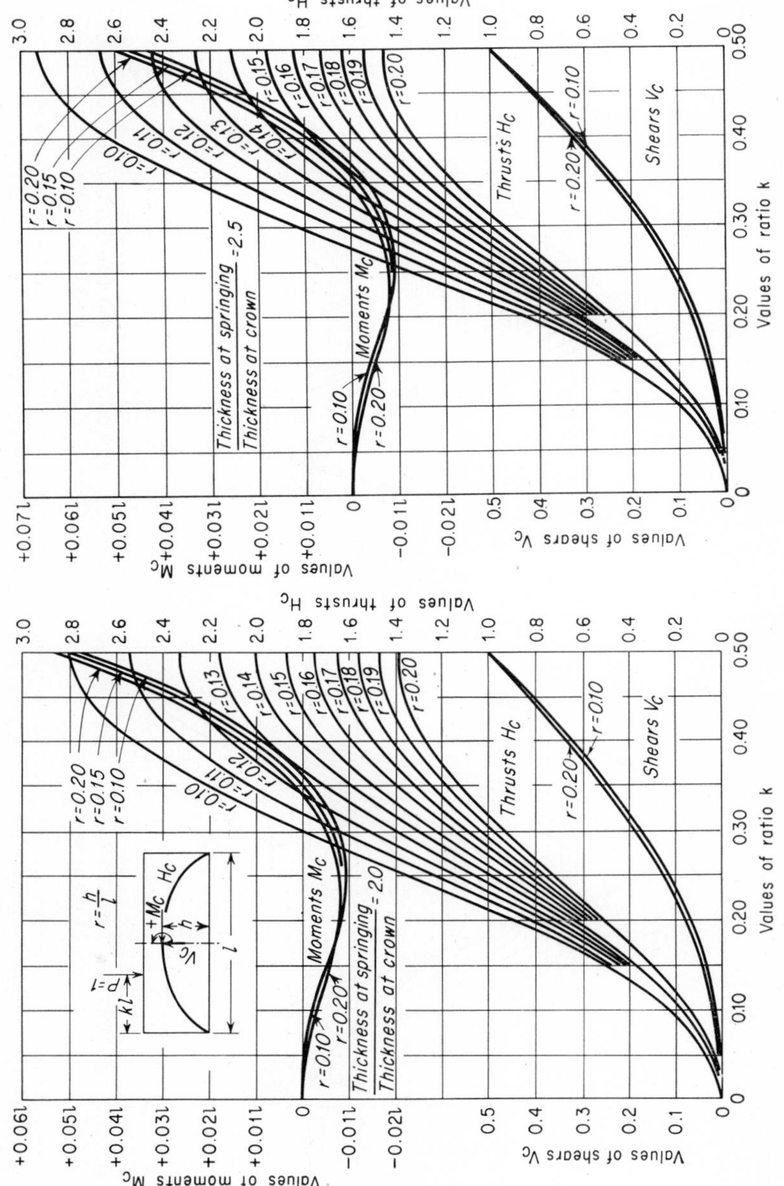

FIG. 99. Influence lines for moments, thrusts, and shears at the crown of earth-filled arches; ratios of thickness at springing to thickness at crown, 2.0 and 2.5. (*After Cochrane.*)

BOX CULVERTS AND RECTANGULAR FRAMES

Improvements in design and economies of construction may be made in both culverts and other closed rectangular frames by reinforcing the corners to take negative moment. Frames of this character may be analyzed readily by the method of slope deflection. Only the results of such analysis will be given here.

Moments at Corners of Box Culverts. *Single Barrel.* Figure 100 shows the cross section of an open-bottom culvert, the side walls of which are considered fixed at the

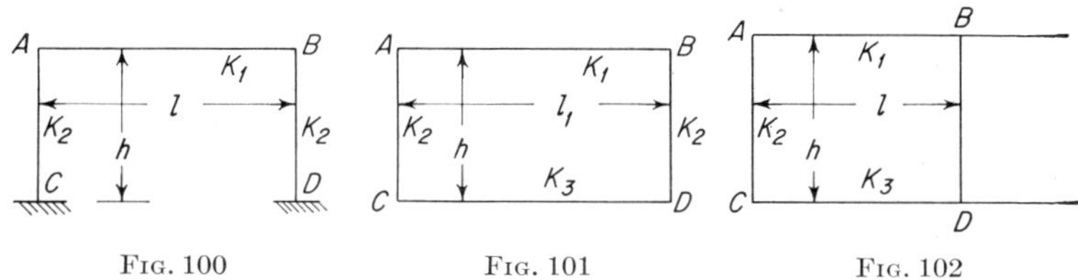

FIG. 100 FIG. 101 FIG. 102

base. Such construction can be realized only in the event of rock foundation. The moments at the upper corner and at the base of the side are

$$M_{AB} = \frac{2K_1(C_{AB} - C_{AC})}{2K_1 + 4K_2} - C_{AB} \tag{75}$$

$$M_{CA} = \frac{2K_2(C_{AB} - C_{AC})}{2K_1 + 4K_2} - C_{CA} \tag{76}$$

in which $K_1 = I/l$ for the top member and $K_2 = I/h$ for either side member.

The sign of the load terms (the C terms) is on the basis that the loads are acting inward. If any of the loads should be acting outward, the signs for such load terms should be reversed.

The following formulas give the moments at the top and bottom corners of the closed barrel (see Fig. 101):

$$M_{AB} = \frac{2K_1[2K_2(C_{CA} - C_{CD}) - (4K_2 + 2K_3)(C_{AB} - C_{AC})]}{4K_2^2 - (2K_1 + 4K_2)(4K_2 + 2K_3)} - C_{AB} \tag{77}$$

$$M_{CD} = \frac{2K_3[2K_2(C_{AB} - C_{AC}) - (2K_1 + 4K_2)(C_{CA} - C_{CD})]}{4K_2^2 - (2K_1 + 4K_2)(4K_2 + 2K_3)} + C_{CD} \tag{78}$$

in which $K_3 = I/l$ for the bottom member and the other terms are as given above. Here again the sign of the load terms given in Eqs. (77) and (78) is on the basis that the loads act inward.

Observe that for both types of frame the loading and the dimensions of the frame are assumed to be symmetrical about the vertical center line of the frame.

Multiple Barrel Frame. The top and bottom slabs of openings other than the two outside openings may be designed as though fixed at both ends. The two outside openings (see Fig. 102) may be designed for the following corner moments:

$$M_{AB} = \frac{4K_1[2K_2(C_{CA} - C_{CD}) - (4K_2 + 4K_3)(C_{AB} - C_{AC})]}{4K_2^2 - (4K_1 + 4K_2)(4K_2 + 4K_3)} - C_{AB} \tag{79}$$

$$M_{CD} = \frac{4K_3[2K_2(C_{AB} - C_{AC}) - (4K_1 + 4K_2)(C_{CA} - C_{CD})]}{4K_2^2 - (4K_1 + 4K_2)(4K_2 + 4K_3)} + C_{CD} \tag{80}$$

Design of Corners. The negative-moment steel must be carried around the corner to a point beyond the point of inflection in the adjoining member. The inside of the corner should be provided with a bracket, either leg of which should be equal to the thickness of the adjacent member. One-fourth of the steel required at the center of the span of any member should continue into the corner and should be hooked at the end. In addition, a similar amount of steel should be placed diagonally in the corner near the face of the bracket and should be extended either way to end near the outside face of the adjacent members.

Shear near the corners may be taken to be the same as for fixed beams.

PRESTRESSED CONCRETE

With the development of high-strength steel and concrete has come a desire to use the two materials in combination to the best advantage. Offhand it seems obvious that the high allowable stresses justifiable for such materials should permit the construction of lighter and less bulky short-span structures as well as long-span concrete structures which hitherto have been impossible because of excessive dead weight. However, certain characteristics of conventional reinforced concrete beams prevent the attainment of these objectives: (1) all concrete below the neutral axis, while necessary for enclosure of the reinforcement, is ineffective in resisting bending; (2) diagonal-tension stresses are often high and govern the design; (3) the use of high working stresses in conventional construction, and hence large strains in the reinforcement, would result in unsightly cracking of the concrete and dangerous exposure of the steel. Precompressing of the concrete by pretensioning of the reinforcement before service loads are applied is a method for overcoming these disadvantages. The following brief discussion, which is adapted from Ref. 9, illustrates the principles of prestressed construction.

Suppose the beam shown in Fig. 103 is fabricated as follows. Before the concrete is poured, high-strength wires with an ultimate strength of about 200,000 psi are strung

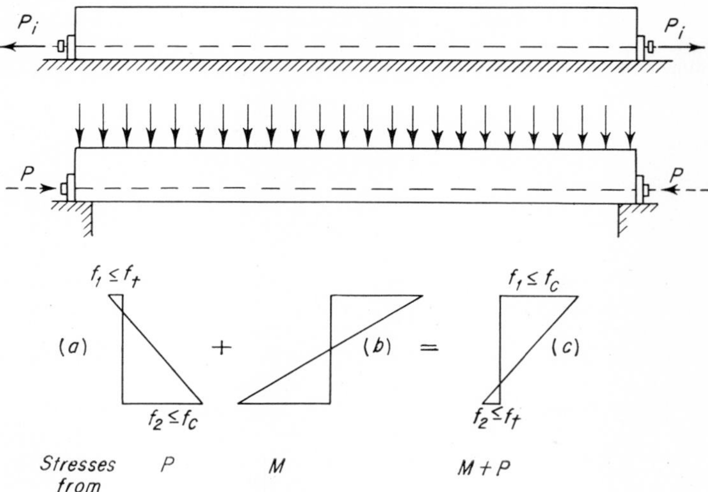

Fig. 103. Principles of prestressing.

through holes in the two anchor plates and prestressed to an initial tension P_i by jacks (not shown) bearing against separate abutments. The concrete is then cast so that the anchor plates bear against each end of the beam. After the concrete has hardened the

wires are secured to the anchor plate and the jacks removed. As the wires try to contract to their original unstressed length they are largely prevented from doing so by the concrete. Thus a compressing force is set up in the concrete. This is slightly less than P_i because of the small elastic contraction of the concrete. Since the concrete shrinks and creeps under load the compression force will be reduced further to the final prestress P, which has been found by experience to be from 80 to 90 percent of the initial prestress P_i.

The eccentrically applied force P produces stress distribution a in Fig. 103. It is so adjusted that the stress in the bottom fiber does not exceed f_c, the permissible compression stress, and the stress in the top fiber does not exceed the permissible tension stress f_t which is the order of $f_c/10$. The subsequent service loads produce stress distribution b which, when superimposed on the initial stresses, results in the final distribution c. The design is made so that at no cross section does the resultant extreme fiber stress exceed either f_c in compression or f_t in tension. It can be shown that the change in the steel stress under service loads is very small. Thus the entire cross section of concrete is not allowed to crack and performs throughout as a monolithic material, carrying its share of stress over the entire cross section. The precompression in the concrete also is effective in reducing the diagonal tension due to shear. For these reasons, steel of high ultimate strength can be used effectively in conjunction with high-strength concrete (5,000 to 10,000 psi) to produce structures which are not feasible in ordinary reinforced concrete.

The possibilities of prestressed concrete were visualized as early as 1886. However, because of the lack of proper materials, nothing of any consequence was done to develop the method until about 1930. Freysinnet, Hoyer, and Magnel have, since that time, been the foremost pioneers in perfecting prestressing techniques and methods. The first major prestressed structure in this country is the Walnut Lane Bridge in Philadelphia, built in 1950. Since then the growth of prestressing has been phenomenal and the method has found wide application in the construction of bridges with spans up to 200 ft, floors, roofs, railroad ties, piles, and water-storage tanks. Prestressed, precast beams, joists, and slabs produced by semi-mass-production methods have been especially successful.

One of the deterrents to the early development of prestressed construction in this country was the lack of simple methods for prestressing and anchoring wires or bars. The early methods, developed largely in Europe, were intricate and costly. Improved techniques have been—and are being—developed.

Prestressed concrete structures may be either *pretensioned* or *posttensioned*. The process described above is illustrative of pretensioning, i.e., the wires are stretched before the concrete is placed around them. In posttensioning, open cores or channels formed by paper, rubber, plastic, or metal tubes are cast in the member. After the concrete has hardened the reinforcing wires or bars, which have been threaded through the holes, are gripped at both ends, and are tensioned by jacks which bear against the member itself. Suitable anchoring devices are then applied to relieve the load from the jacks and transfer it directly from the reinforcement to the concrete. Neither method can be said to have a clear-cut advantage over the other. They are both being used successfully. Instead of depending upon external end anchors, bond between the wires and the surrounding concrete (or injected grout) is frequently used to transmit stress between steel and concrete.

There are as yet no generally acknowledged specifications for the design of prestressed concrete, but the various professional societies are in the process of developing acceptable standards. In the meantime, recommendations of various researchers in the field and organizations such as the Portland Cement Association are being followed. For investigations at working loads the straight-line theory is commonly used. For checking

the ultimate strength of a prestressed member, a theory such as Whitney's is applicable since before the member fails in flexure the prestressing force must be exceeded, the concrete will be cracked, and the member will act as an ordinary reinforced concrete member.

Illustrative Example (From Ref. 9). A pretensioned slab bridge of 48-ft span is to be designed for a live load (including paving) of 400 lb per sq ft; $f_c' = 6{,}000$ psi; $f_c = 0.35 \times 6{,}000 = 2{,}100$ psi; $f_t = 0$ (permit no tension in the concrete at working loads); maximum diagonal tension $= 0.1 \times 2{,}100 = 210$ psi. A steel with an ultimate strength of 240,000 psi and proportional limit of 140,000 psi is to be used. To stay below the proportional limit, the initial prestress is limited to 135,000 psi, and the final value after elastic shortening, shrinkage, and creep to $0.8 \times 135{,}000 = 108{,}000$ psi. A rectangular section 12 in. wide will be designed.

The three basic conditions which must be satisfied are:

1. At each end the stress in the uppermost fibers caused by the prestress force must be zero or a net compression.
2. At the center line the stress in the lowest fibers caused by the prestress combined with the dead and live load moments must be zero or a net compression.
3. The working stresses in the steel and concrete should not be exceeded.

If the resultant prestress force is of proper magnitude and located at the lower third point of the section, then the concrete stress due to prestressing alone will be zero at the upper fiber and f_c at the lower fiber throughout the span (Fig. 104*b*) or

$$P = \tfrac{1}{2} f_c b h$$

The slab can then be designed as an ordinary homogeneous rectangular section for the maximum bending moment which occurs at mid-span (Fig. 104*c*), and causes a maxi-

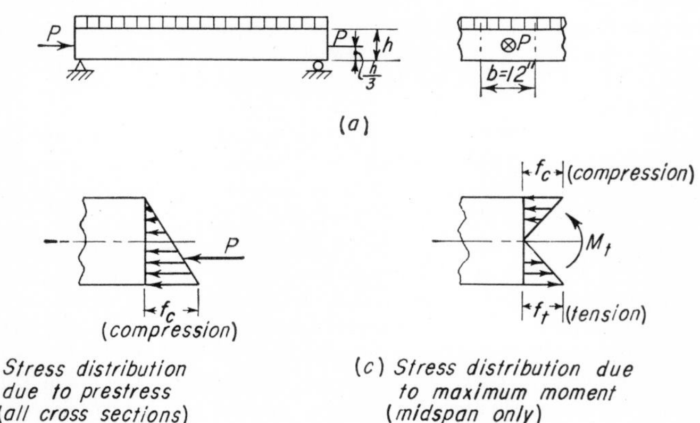

(*b*) Stress distribution due to prestress (all cross sections)

(*c*) Stress distribution due to maximum moment (midspan only)

Fig. 104

mum compressive stress f_c in the upper fibers. In the lower fibers the precompression will just balance the dead- plus live-load tension. Then, for design

$$f_c = \frac{M_t}{bh^2/6}$$

where M_t is the maximum dead- plus live-load bending moment. Assuming a dead load of 300 lb per sq ft, the maximum moment is

$$M_t = \frac{700 \times 48^2 \times 12}{8} = 2{,}410{,}000 \text{ in.-lb}$$

$$h = \sqrt{\frac{6 \times 2{,}410{,}000}{12 \times 2{,}100}} = 24.0 \text{ in.}$$

The required final prestress force is

$$P = \tfrac{1}{2} \times 2{,}100 \times 12 \times 24.0 = 302{,}000 \text{ lb}$$

The total area of prestressing wires required is

$$A = 302{,}000/108{,}000 = 2.80 \text{ sq in.}$$

Using wires of 0.276 in. diameter (area per wire = 0.0598 sq in.), the number of wires required is 2.80/0.0598 = 47. Use 48 wires and locate as shown in Fig. 105. The distance from the bottom of the slab to the centroid of the area of the wires is $3.5 + \frac{12 \times 18}{48}$ = 8.00 in., the lower third point. The reason for locating the wires in two groups rather than all at the third point is to develop the optimum ultimate moment resistance. The wires must be stressed equally and to a total initial prestress P_i of 302,000/0.8 = 378,000 lb.

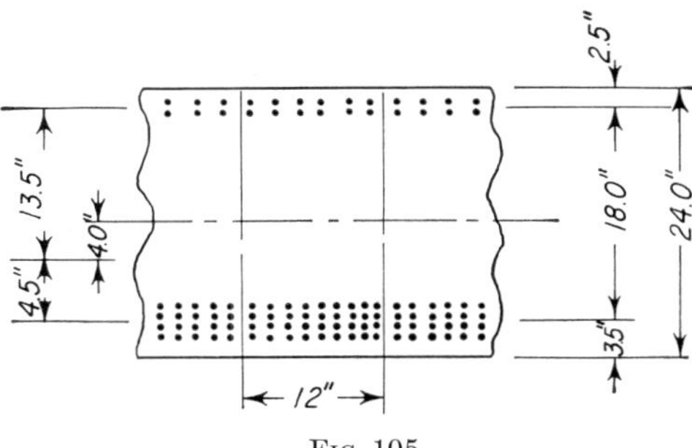

Fig. 105

It is of interest to know what overload would result in the formation of cracks in the section. These will be assumed to occur at a tension stress of $0.1f_c' = 600$ psi in the bottom fibers at the center line. Since the prestress force causes a compression of 2,100 psi at the same point, the cracking-load moment would be such as to cause a stress of 2,700 psi in a homogeneous section

$$M_{cr} = 2{,}700 \times 12 \times 24^2/6 = 3{,}120{,}000 \text{ in.-lb}$$

The uniform load required to obtain this moment is 905 lb per sq ft. Since the dead load is 300 lb per sq ft, the live load would be 605 lb per sq ft or 1.51 times the design live load of 400 lb per sq ft. This is an ample margin of safety against cracking.

From the ACI Code equation (A1) the ultimate resulting moment is

$$M_u = pf_y bd^2 \left(1 - 0.59 \frac{pf_y}{f_c'}\right)$$

where only the steel in the lower half of the beam is considered since it alone resists tension at the ultimate load. Thus

$$p = \frac{36 \times 0.0598}{12 \times 20.5} = 0.00878$$

$$f_y = 240{,}000 \text{ psi}$$
$$f_c' = 6{,}000 \text{ psi}$$
$$d = 20.5 \text{ in.}$$

$$M_u = 0.00878 \times 240{,}000 \left(1 - \frac{0.59 \times 0.00878 \times 240{,}000}{6{,}000}\right) 12 \times 20.5^2$$
$$= 8{,}440{,}000 \text{ in.-lb}$$

The overall safety factor is $\dfrac{8{,}440{,}000}{2{,}410{,}000} = 3.5.$

The maximum diagonal tension stress at working loads will occur at mid-height at each end and may be computed from the conventional formula for combined stress

$$f_{\text{max}} = \frac{f}{2} - \sqrt{v^2 + \left(\frac{f}{2}\right)^2}$$

where $f = \dfrac{f_c}{2} = \dfrac{2{,}100}{2} = 1{,}050$ psi.

$$v = \frac{3}{2}\left(\frac{V}{A}\right) = \frac{3}{2}\left(\frac{24 \times 700}{12 \times 24}\right) = 87.5 \text{ psi.}$$

$$f_{\text{max}} = \frac{1{,}050}{2} - \sqrt{(87.5)^2 + (525)^2} = -9 \text{ psi (tension).}$$

This value is only about one-tenth that which would occur if there were no prestressing. This indicates that there might be some saving of material by using thin web sections such as I beams or box sections rather than a solid slab. This is being done in practice. It can be shown that the increase in steel stress under service loads is small and that the final stress is less than the initial prestress P_i.

There is a great amount of information on prestressed concrete in the Journal of the ACI. References 21 and 22 are excellent recent textbooks.

REFERENCES

General Sources of Information. For a field so broad and rapidly changing as that of cement and concrete it is essential not only to know something of the existing literature but also to know where to look for the developments of tomorrow. Although this is not a treatise on library practice it may not be amiss to mention that The *Industrial Arts Index* and the *Engineering Index* (available in any reference library) supply a reasonably complete domestic coverage of titles and authors. The Portland Cement Association (Chicago) constitutes a reference source not only for much useful *service literature* but also for the most advanced of technical material, including guidance in many of the more involved aspects of recent design developments. The National Crushed Stone Association, the National Sand and Gravel Association, and the National Ready-mixed Concrete Association, all of Washington D.C., publish material on plain concrete at the *technical* as well as the *service* level.

The following constitute important sources of information on current concrete developments.

ACI—American Concrete Institute, Detroit, Mich.
ASTM—American Society for Testing Materials, Philadelphia, Pa.
AREA—American Railway Engineering Association, Chicago, Ill.
AASHO—American Association of State Highway Officials, Washington, D.C.

ASA—American Standards Association, New York, N.Y.
ASCE—American Society of Civil Engineers, New York, N.Y. (*Civil Engineering, Transactions*)
HRB—Highway Research Board of the National Research Council, Washington, D.C.
USBR—U.S. Bureau of Reclamation, Denver, Colo.
NBS—National Bureau of Standards, Washington, D.C.
BPR—Bureau of Public Roads (formerly Public Roads Administration), Washington, D.C.
USED—U.S. Army Engineer Department, Central Laboratory, Clinton, Miss.
PCA—Portland Cement Association, Chicago, Ill. (branch offices in many large cities)
College Engineering Experiment Stations, notably those of Illinois, Purdue, Michigan State University, Kansas State College, Wisconsin, Ohio, and Iowa State College
Concrete (monthly magazine), Chicago, Ill.
Engineering News-Record (weekly), New York, N.Y.
Concrete and Constructional Engineering (monthly), London, England
Department of Scientific and Industrial Research, London, England
Swedish Cement and Concrete Research Institute at the Royal Institute of Technology, Stockholm, Sweden; also similar activity in Norway and Denmark (Oslo and Copenhagen).

Selected References on Reinforced Concrete

1. "Reinforced Concrete Design Handbook of the ACI," 2d ed., American Concrete Institute, Detroit, Mich., 1955.
2. "CRSI Design Handbook," Concrete Reinforcing Steel Institute, Chicago, Ill., 1957.
3. Seelye, E.: "Design Data Book for Civil Engineers," 2d ed., John Wiley & Sons, Inc., New York, 1951.
4. "ACI Standards, 1956," American Concrete Institute.
5. "Manual of Standard Practice for Detailing Reinforced Concrete Structures" (ACI 315-57), American Concrete Institute, 1957.
6. ACI Anniversary Issue, *Jour. ACI*, vol. 25, No. 6, February, 1954.
7. Portland Cement Association Reinforced Concrete (R/C Series) and Structural Bureau Bulletin (ST Series), Portland Cement Association, Chicago.
8. ACI Standard 711-53, Minimum Standard Requirements for Precast Concrete Floor Units, *Proc. ACI*, vol. 50, pp. 1–15, 1953.
9. Urquhart, O'Rourke, and Winter: "Design of Concrete Structures," 6th ed., McGraw-Hill Book Company, Inc., New York, 1957.
10. Whitney, C. S.: Plastic Theory in Reinforced Concrete Design, *Trans. ASCE*, vol. 107, pp. 251–326, 1942.
11. ACI Committee 318, Building Code Requirements for Reinforced Concrete, *Jour. ACI*, vol. 27, No. 9, pp. 913–986, May, 1956.
12. Anderson, B. G.: Rigid Frame Failures, *Jour. ACI*, vol. 28, No. 7, pp. 625–636, January, 1957.
13. Whitney and Cohen: Guide for Ultimate Strength Design of Reinforced Concrete, *Jour. ACI*, vol. 28, No. 5, pp. 455–490, November, 1956.
14. Wessman, H. E.: Reinforced Concrete Columns under Combined Compression and Bending, *Jour. ACI*, September, 1946, pp. 1–8.
15. Ferguson, P. M.: Simplification of Design by Ultimate Strength Procedures, *Jour. Structural Division, ASCE*, St 4, July, 1956.
16. Richart, F. E.: Reinforced Concrete Wall and Column Footings, *Jour. ACI*, vol. 20, pp. 97, 237, October, November, 1948.
17. Freyssinet, E.: The 600-ft. Concrete Arch Bridge at Brest, France, *Proc. ACI*, vol. 25, pp. 83–99, 1929.
18. Hollister, S. C.: Determination of Ideal Curve for Open Spandrel Arches, *Civil Eng.*, vol. 2, p. 451, July, 1932.

19. WHITNEY, C. S.: Plain and Reinforced Concrete Arches, Report of Committee 312, *Proc. ACI*, vol. 37, pp. 1–26, 1941. Also *Jour. ACI*, September, 1940.
20. ACI Committee 312, Plain and Reinforced Concrete Arches, *Jour. ACI*, May, 1951, pp. 681–692.
21. GUYON, Y.: "Prestressed Concrete," John Wiley & Sons, Inc., New York, 1953.
22. MAGNEL, G.: "Prestressed Concrete," 3d ed., McGraw-Hill Book Company, Inc., New York, 1955.

Section 8

SOIL MECHANICS AND FOUNDATIONS

By Albert E. Cummings and Linton Hart

PROPERTIES OF SOILS

Origin and Composition of Soils. The soil is composed of particles differing physically in size and shape and varying in chemical composition. Organic matter, water, air, and bacteria are usually present but soil consists essentially of mineral matter which has originated from rocks by the action of a series of weathering processes.

Rocks may be of three kinds: the igneous rocks, made by the solidification of molten material; the sedimentary rocks, formed from sediments deposited chiefly by water but sometimes by air and ice; and the metamorphic rocks, formed by certain processes acting on preexisting rocks and altering their original characters to such an extent that the resultant rocks are considered as constituting a separate group.

Granite, diorite, rhyolite, andesite, and basalt are representatives of the igneous rocks; conglomerate, sandstone, shale, and limestone, of the sedimentary rocks; and gneiss, quartzite, slate, and marble, of the metamorphic rocks. It is estimated that three-fourths of the land area of the globe is underlain by sedimentary rocks, and the other fourth by igneous and metamorphic rocks.

The classification and identification of soils is not yet an exact science.[1] The lack of coordination in the work of different groups interested in soils has prevented the standardization of terms for either individual soil materials or different soil mixtures. A common method of soil classification is based on particle size, as illustrated in Fig. 1. The various soil fractions shown on the chart may be described as follows.

Gravel. The rock fragments which make up gravel are usually rounded by water action and abrasion. The smaller pieces are known as "pebbles," and the larger ones as "boulders." Since quartz is the hardest of the rock-forming minerals, it is the principal constituent of most gravels; and well-rounded pebbles and boulders, having undergone long wear, are usually composed almost entirely of quartz. On the other hand, gravel that is only slightly worn, and therefore rough and angular, commonly includes other minerals such as granite, schist, basalt, or limestone.

Coarse Sand. This fraction is usually rounded like pebbles and is likely to consist of the same minerals as the gravel with which it occurs.

Fine Sand. The fine-sand particles are often more angular than those of the coarse-sand fraction, but they are usually made of the same minerals as the coarse sand and gravel with which they are found.

Silt. Silt usually consists of grains which are similar except for size to the fine sand and have the same mineral composition. In some cases, however, it may be largely a

[1] Casagrande, A., Classification and Identification of Soils, *Proc. ASCE*, vol. 73, No. 6, p. 783, June, 1947.

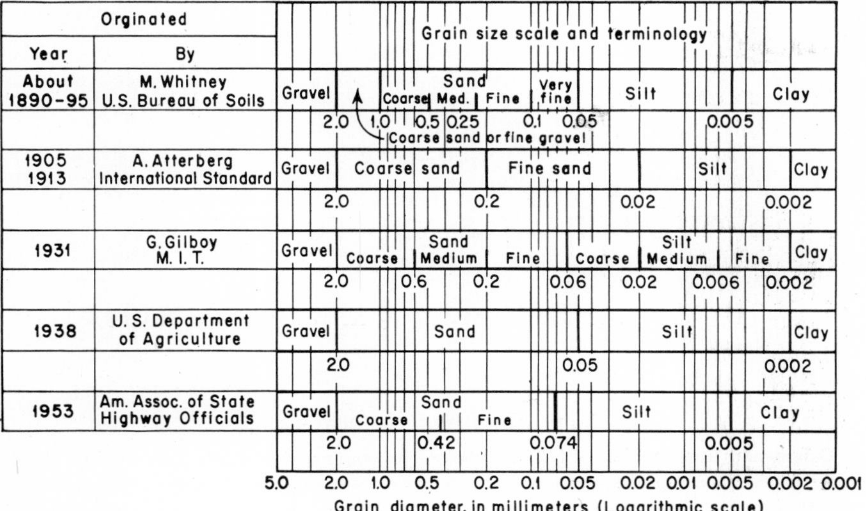

FIG. 1. Comparison of principal grain-size scales.

product of chemical decay rather than of rock grinding and as such may consist of silicates of aluminum and the alkaline earths and of oxides of iron. In other cases, the silt may be composed of organic materials such as decayed vegetation.

Clay. While the coarser soil fractions usually consist of original fragments such as quartz and feldspar, the clay fraction consists almost entirely of the secondary products of chemical weathering. The more plastic and less stable varieties of clay are composed largely of flat scalelike particles. Clay differs from the coarser soil fractions in that it is often the chemically reactive portion of the soil, while the coarser fractions are usually inert.

Colloids. Soil colloids include only those finer clay particles which show pronounced Brownian movement when suspended in water. Brownian movement is defined as the state of constant unordered motion (visible under the microscope) which is due to the fact that each particle possesses an electrical charge causing it to repel other particles similarly charged. Ordinarily colloids are considered as consisting of particles 0.001 mm in diameter and finer.

Soils Influenced by Climate. The mineral constituents of the soil particles depend largely upon the chemical composition of the parent rock and the conditions under which the soils were developed. In humid northern climates, weathering often causes the removal of iron and aluminum oxides by leaching which results in soils containing relatively large accumulations of silica. Below the leached surfaces are found the soils in which the weathering has not proceeded so far and in which, although the silica still predominates, the amounts of aluminum and iron oxides are greater than in the surface soil. These weathered materials are frequently underlain by the parent rock.

In humid tropical climates, the silica is often leached out of the soils, with the result that the iron and aluminum oxides predominate. This type of weathering is termed "lateritic" and results in the formation of the lateritic soils of the Southern United States and the true laterites of the tropics.

Geologic Origin of Soils. Soils are called "residual" or "transported" depending on whether they have remained in place or have been moved from the locations where

they were originally formed. Residual soils include all deposits derived by the process of rock weathering or from organic accumulation in place. They include deposits derived from the decay of the immediately underlying rocks and deposits which have been formed in place from the accumulation of organic matter, such as the peat and muck deposits in ponds and lakes.

Residual soil, which is naturally the first to be formed, is often transported by water, wind, or glacial action to other localities and there laid down. The transporting agency determines, to a large extent, the size, shape, and arrangement of the individual particles comprising these soils. When deposited by water, they are called "alluvial soils," and these include sedimentary clays and silts and the sands and gravels found in stream beds. Glaciers have carried soil materials and deposited them when they melted. Such deposits are called "moraines," and the material comprising them is called "glacial till," or "boulder clay." Wind-borne deposits, called "aeolian," include the dune sands and the loess soils.

Classifications Based on Grading and Texture. The sizes of the individual particles comprising a given soil mixture may differ widely because of the variations in the rate of weathering and the mixing effect of the transporting agencies. The term "texture" indicates the particle-size distribution for a given soil, and this particle-size distribution is determined by mechanical analysis. Mixtures of specific gradings are designated as distinct soil types with such names as sands, sandy clays, clay loams, and silt loams. The classification of soils according to texture is usually done by means of triangular diagrams such as those shown in Fig. 2.

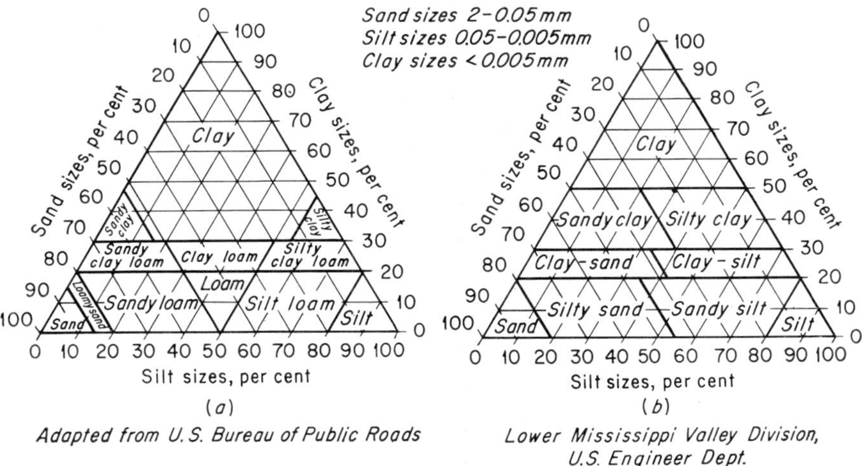

FIG. 2. Comparison of soil-classification triangles.

Classification Based on Composition. Among other terms commonly used in descriptions of soil mixtures are names of materials such as adobe, blow sand, *caliche*, chalk, gumbo, lime rock, marl, and peat. Terms indicative of chemical composition such as acid soil, alkaline soil, calcareous soil, and saline soil are also used. They are defined as follows:

Adobe. Heavy-textured alluvial clay soil from which unburned brick are made. Occurs mainly in the arid regions of the southwestern United States.

Blow Sand. Wind-borne free-moving dune sand.

Caliche. A term applied in the United States to a group of formations consisting chiefly of calcium carbonate and silica, in the form of clays, sands, and gravels, cemented into a conglomerate by calcium carbonate.

Chalk. A soft porous variety of lime rock, consisting in some cases of microscopic shell fragments and in others of chemically precipitated particles of calcium carbonate.

Gumbo. A class of peculiar fine-grained soils, usually devoid of sand and rich in alkaline compounds. When saturated with water, they become waxy or soapy in appearance and to the touch.

Lime Rock. An unconsolidated or partly consolidated form of limestone, usually containing shells or shell fragments. Of marine origin.

Marl. A term applied to any earthy crumbling deposit containing varying quantities of calcium carbonate, clay, sand, and carbonaceous material.

Peat. Composed predominantly of organic material, highly fibrous, with easily recognized plant remains.

Acid Soil. A soil deficient in available bases, particularly calcium, and giving an acid reaction when tested by standard methods.

Alkaline Soil. A soil containing an excessive amount of the alkaline salts.

Calcareous Soil. A soil containing sufficient calcium carbonate to effervesce when tested with weak hydrochloric acid.

Saline Soil. A soil containing excessive amounts of the neutral or nonalkaline salts.

Classification According to Consistency. Soils may be called plastic or friable, depending upon the cohesion between the soil particles. In a saturated plastic soil, there is considerable cohesion so that mixtures of plastic materials and water have the property of forming true pastes with definite characteristics such as plasticity and flow under pressure. Upon evaporation of moisture, such pastes harden into compact masses. Plastic materials have the property of deforming under a constant shearing stress, in contrast with elastic materials in which the stress must increase with increase of strain. Plastic soils are also called "cohesive" soils.

Friable soils, termed also "cohesionless," are the converse of plastic soils. Such materials when mixed with water do not form true pastes and upon evaporation of the moisture become increasingly easy to crumble until, when thoroughly dry, they will fall to powder like dry sand at a slight touch.

Soil Exploration and Field Testing. The amount that should be spent on exploration and testing will depend on the size and cost of the proposed structure. For a large structure costing millions of dollars, there is justification for spending at least several thousand dollars in order to make a careful and complete investigation. For smaller structures, the cost of the investigation will necessarily be limited to a comparatively small amount.

The investigation should always include studies of the history of present and prior existing structures in the vicinity of the site on which it is proposed to build together with the historical geology of the subsurface formations down to the rock surface. Whenever there are nearby structures, inquiries should be made about the foundations of those structures with respect to (1) depth of foundation below ground level, (2) type of foundation, and (3) the behavior of the foundation including settlement records.

The depth of adjacent foundations should be investigated, because this may determine the type of foundation selected for the new structure. In most localities there are laws and building regulations relating to the protection of adjacent structures when excavations are made for new structures. In some cases, the person who makes the excavation is responsible for the protection of adjacent structures. He must provide and pay for the proper support of the adjacent structures while the new structure is being built. In other cases, the person who excavates is not responsible for the protection of adjacent structures. He must notify adjacent property owners that he is

going to excavate for and build a new structure, but the adjacent property owners are obligated to provide and pay for any necessary support for their structures while the new structure is being built. It is therefore necessary to consider the depth of existing adjacent foundations as well as the depth of the foundation of the new structure and to study the building regulations in the locality where the new structure is to be built.

The type of foundation on which nearby structures have been built, as well as the manner in which these foundations have behaved, should be investigated. The history of foundation construction in downtown Chicago illustrates the usefulness of this kind of information. Previous to 1880, there were no buildings in downtown Chicago higher than about six stories. These buildings were comparatively light, and they stood up satisfactorily on shallow foundations placed just a few feet below street level. The earliest skyscrapers were built between 1885 and 1890. These buildings were only 15 to 18 stories high, but they were much heavier than the older buildings. The new skyscrapers were also built on comparatively shallow foundations placed not far below street level. In some cases these foundations were individual piers supporting the building columns and walls, and in other cases they were mat foundations extending over the entire area occupied by the building. Most of these buildings began to settle soon after they were built, and many of them continued to settle for years afterward. Ten years of experience with this type of foundation led to the conclusion that it was inadequate for heavy buildings. By 1900 wooden piles were being driven to support heavy structures. As buildings were made taller and heavier, it was soon found that wooden-pile foundations were insufficient, and the Chicago type of caisson was developed and used. The development of a satisfactory type of foundation for Chicago's skyscrapers was a sort of evolutionary process, with each succeeding builder learning from the mistakes of his predecessors. This evolution is a continuing process.

It is also necessary to locate and to map all underground structures on the site of the proposed work. This is especially important in the downtown sections of large cities where there are usually water mains, gas mains, sewers, electric-power ducts, telephone conduits, and buried foundations which supported structures since demolished, leaving the old foundations in place and covered up by backfill. Such underground utilities must be protected during construction, and it is necessary to know where they are so as to avoid accidents; it is also necessary to know all that can be learned of all old foundations so that they may be spanned or removed.

In addition to the collection of all available information about adjacent structures, the preliminary investigation should always include some sort of soil exploration at the actual site of the proposed structure. Various methods are used to explore and sample soils for foundation studies.[1]

The simplest method of soil exploration is by means of sounding rods. These are usually ¾-in. round rods cut into sections 4 to 5 ft long, the sections being threaded at the ends so that they can be joined together by couplings. A section which is pointed at one end is driven into the ground first, and the next section is fastened on top of that and driven farther. This method of exploration does not produce a soil sample. When the rods are withdrawn there is soil around the edges of the couplings, but these small pieces of soil cannot be considered as soil samples. Sounding rods are useful for locating rock or boulders, provided that these are not too deep below the ground surface. For soil exploration for foundation purposes, the sounding rod is not usually satisfactory.

Another type of soil exploration is by two methods usually called "geophysical methods."[2] One of these is based on the electric resistivity of the soil and the other is based on elastic wave transmission. The latter is often referred to as the "seismographic

[1] MOHR, H. A., "Exploration of Soil Conditions and Sampling Operations," Graduate School of Engineering, *Harvard Univ. Bull.* 376, 1943.

[2] SHEPARD, E. R., Subsurface Exploration by Earth Resistivity and Seismic Methods, *Public Roads*, vol. 16, No. 4, p. 57, June, 1935.

method." One method of seismographic exploration uses a single explosion at the ground surface, and the stress waves produced in the soil by this explosion are recorded on seismographs set at different distances from the location of the explosion. The soil conditions are deduced from these seismograph records. In other cases the stress waves in the soil are produced by machines that vibrate and set up a series of vibrations instead of a single impulse. The electric-resistivity method uses electric currents instead of the stress waves used by the seismographic method. These geophysical methods are often useful for locating rock, but they do not produce any soil samples and they provide comparatively little information about the character of the soil between the ground surface and the rock. Except for very preliminary investigation over large areas, these methods are not satisfactory for foundation exploration.

The most common method of soil exploration for foundation work is by means of borings. Various procedures are used in making soil borings, although some of them are practically worthless as far as foundation investigations are concerned.

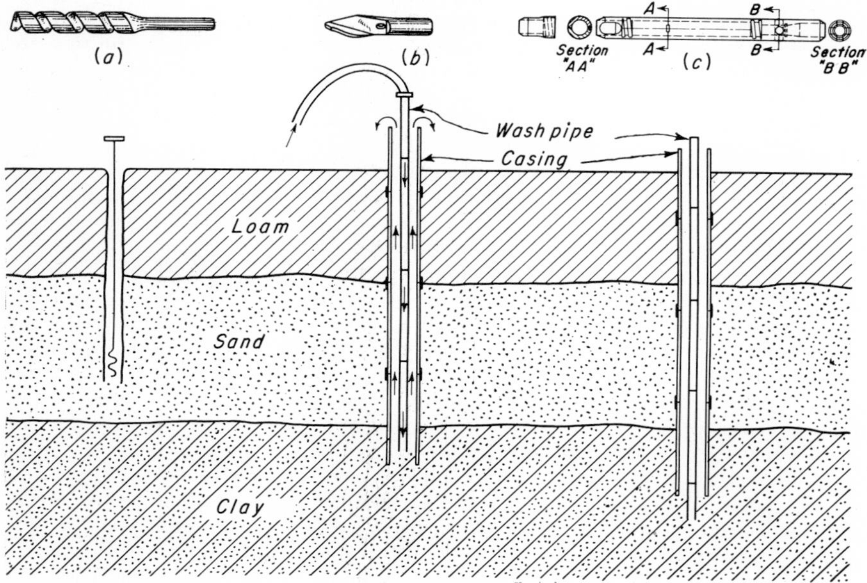

FIG. 3. Soil-boring methods: (*a*) auger borings, (*b*) wash borings, (*c*) dry-sample borings.

Sometimes borings are made with an auger, which is similar to the augers used for drilling wood (Fig. 3*a*). The auger is started at the ground surface and is screwed by hand a few turns into the soil, after which it is withdrawn and the soil is removed from the threads of the auger. This process is repeated until the boring has been carried to the desired depth. Ordinarily, auger borings are made without the use of casing in the borehole, and it is often impossible to determine from which soil stratum a given auger sample came. The sides of the hole sometimes cave in, and the soil picked up by the auger at the bottom of the hole will be soil that has fallen from the sides of the hole at some higher elevation. If casing is used in an auger boring, the casing can be cleaned out with the auger and soil samples can be obtained from below the bottom of the casing. However, in either case, the action of the auger disturbs the soil to such an extent that the auger sample will furnish little or no information about the

character of the soil in its natural state. Except for superficial and very preliminary work, the auger-boring method is not satisfactory for foundation exploration.

Another type of boring that is often used is called a "wash boring" (Fig. 3*b*). For wash borings it is always necessary to use a casing which is usually made of 2- or 2½-in.-diameter pipe cut into sections about 5 ft long. These sections are threaded at both ends so they can be joined together by couplings. A wash pipe is used inside the casing, and this is usually made of 1-in. pipe cut into sections from 5 to 10 ft long. The sections of wash pipe are also threaded at both ends so they can be joined together with couplings. Various kinds of chopping bits are used on the lower end of the wash pipe, and its upper end is connected to a small force pump through a water swivel and a hose. To start the boring, a 5-ft length of casing is driven into the ground and the wash pipe is placed inside of it. Water is forced downward in the wash pipe and flows out through the chopping bit at the bottom. This water then rises in the space between the wash pipe and the casing and overflows at the surface. The boring is advanced by adding lengths of casing and wash pipe and continuing the process of chopping and washing. Soil from the bottom of the boring is brought to the surface by the wash water, and soil samples are obtained by catching the overflow water in a container and permitting the soil particles to settle out. The wash-boring method is entirely unsatisfactory for foundation exploration. When this method is used in clays and silts, the soil samples are nothing but a slurry of mud and water. They furnish practically no information about the condition of the soil in its natural state. In sands and gravels, the wash-boring process brings up the fine particles but leaves the coarse ones in the bottom of the boring. A high velocity in the return wash water is necessary in order to bring even a fair-sized sand particle up to the surface.

The most commonly used type of boring that is satisfactory and at the same time relatively inexpensive is called a "dry sample boring" and is a combination of the wash-boring process with a sampling method that produces a satisfactory soil sample (Fig. 3*c*). The borehole is advanced by washing and chopping just as in the wash-boring method. When a soil sample is to be taken, the borehole is carefully cleaned out down to the bottom of the casing and the wash pipe is pulled out. The chopping bit is removed from the lower end of the wash pipe, and a soil sampler is attached. The wash pipe is then put back into the casing and a soil sample is secured by driving the sampler into the soil below the bottom of the casing. A suitable sampling tool for this purpose can be made from a piece of 1-in. ID by 1⅝-in. OD or 1⅜-in. ID by 2-in. OD seamless-steel tubing about 24-in. long to provide samples of 1-in. or 1⅜-in. diameter. The larger diameter is preferred, and it requires the use of 2½-in. casing. The tube is threaded at both ends and is then split longitudinally into two pieces. The upper end of the sampler is attached to the lower end of the wash pipe by a reducing coupling, and the lower end of the sampler is provided with a cutting shoe which resembles a pipe coupling sharpened to a chisel edge on one end. When the soil sample is brought to the surface, the ends of the sampler are unscrewed and the two halves are separated, leaving a cylindrical soil sample lying in one half of the split tool.

Soil samples obtained in this manner are disturbed to only a slight extent, and they provide not only an excellent idea of the character of the soil in its natural state but, by recording the number of blows of a standard 140-lb weight free-falling 30 in. on the tool string to drive a standard 2-in. OD sampler 12 in. into the soil when procuring a sample, the results of the "standard penetration test" are also obtained. The samples are very much better than any that can be procured by the auger or the wash-boring method. Also, this type of sampler can be provided with a sectional liner in which "liner samples" can be procured which can be delivered to the laboratory without being manually handled; and, except for certain unusual problems requiring samples of larger diameter, the liner samples are adequate for all necessary laboratory tests.

In this type of boring, soil samples are sometimes taken in short sections of thin-walled seamless-steel tubing instead of in the thicker split sampling tool. For a given size of casing, it is possible to get a larger-diameter sample with thin-walled tubing than with the thicker-walled sampler. Also, the use of thin-walled tubing minimizes disturbance due to displacement of soil by the sampler.

Borings made in this manner either with the split sampler or with thin-walled tubing are commonly called "exploratory borings." The first boring made should always be carried to refusal so as to provide early information regarding all the strata. Such borings are not expensive, and they usually furnish all the information that is needed to solve a foundation problem. Sometimes the exploratory borings will disclose a soil condition that will require a special and more elaborate investigation.

For certain foundation problems, particularly those involving the consolidation of saturated clay beds, methods have been developed for sampling soils with larger devices. These sampling tools require casing from 6 to 8 in. in diameter and the sampler takes soil samples that are 5 in. or more in diameter. Such samples are usually called "undisturbed samples," and they are required for special laboratory tests. The cost of getting them and testing them is relatively high, and this method of exploration is used only for large structures and particularly difficult soil conditions.

It should be kept in mind that these exploratory boring methods are not capable of penetrating rock or even small boulders. Accordingly, a boring report for an exploratory boring should never indicate that the boring reached bedrock unless it is definitely known that this is so. In order to make sure that rock has been reached, and in order to determine the nature and thickness of the rock stratum, it is necessary to resort to special equipment for drilling rock.

Jackhammer drills such as are used in quarries or tunnels to make holes for blasting charges, churn drills, and rotary drills employing milling bits such as are often used in oil-well operations are not usually satisfactory for foundation-exploration work. All these methods destroy the structure of the rock so that they do not produce suitable rock samples. For foundation explorations, rock drilling should be done by rotary-drilling methods that will produce cores of rock which can be brought to the surface for examination. The drill bits commonly used in rotary drilling are either shot bits or diamond bits. In the shot bit, chilled shot is used as an abrasive to cut the rock. In the diamond bit, industrial diamonds are set in the face of the bit to cut the rock. These are usually small diamonds of which several sizes are available. In a soft rock, or with a bit of small diameter, the shot drill is not satisfactory because the shot often breaks up the rock so that it is impossible to secure a good core in the sampler; and a diamond bit will produce a better rock core. In hard rock, either type of bit is usually satisfactory; but, in procuring cores of small diameter, the diamond bit is usually preferable. Such borings should always be carried at least 5 ft into the rock.

Another method of soil exploration for foundation studies is by means of test pits. In an open excavation, it is possible to examine the soil in an almost undisturbed condition and to carve out soil samples which are more nearly undisturbed than any samples that can be obtained in a boring. However, this method of exploration is usually limited to shallow depths because of its cost. On a project of considerable magnitude, the cost of deep test pits is sometimes justified. Otherwise, the less expensive exploratory borings are usually used.

The depth to which the soil exploration should be carried is a matter that requires careful consideration. The required depth is usually stated to be at least equal to the smallest horizontal dimension of the structure unless rock is encountered before this depth is reached. If the proposed structure were to cover a ground area 100 by 200 ft, for example, this requirement would mean that the soil exploration should be carried at least to a depth of 100 ft unless rock is encountered sooner. This requirement should

be fulfilled in the case of a heavy structure such as a multi-storied warehouse or a grain elevator. Such structures apply heavy loads to the soil over their entire ground areas, and the soil stresses due to these loads are of considerable importance at depths in the ground that are about equal to the horizontal dimensions of the loaded area. In very light structures covering very large ground areas it is usually impractical and unnecessary to try to fulfill this requirement, but the borings should be carried at least 10 ft into satisfactory bearing material. An assembly plant for automobiles or airplanes, for example, might be a light steel structure one story high extending hundreds of feet in both horizontal directions on the ground surface. The average load of such a structure over its entire ground area might not exceed a few hundred pounds per square foot. For a structure of this kind, it would not be necessary to drill hundreds of feet into the ground with exploratory borings. A considerable portion of the load of the structure would be transmitted to the soil by column footings and wall footings possibly 8 or 10 ft wide. The soil stresses generated in the underground by footings of this size would practically vanish at depths of 40 to 50 ft. Therefore, as far as the footings were concerned, it would be sufficient to explore the soil to these depths with only one or two borings carried to a depth of 100 or 150 ft.

It is not possible to establish a set of definite rules for the depths to which borings should be carried. The required depth should be based on the size and weight of the structure with consideration being given to the fact that the stresses generated in the underground will depend on the average load over the entire building area as well as on the loads under the individual footings.

In all soil-exploration work, it is desirable to determine the ground-water level. Of all the factors that enter into foundation-construction work, the one that causes the most trouble is water. This is especially true in highly permeable soils such as sands and gravels where excavations below the ground-water table will often involve serious pumping problems. It is necessary to consider also the question of shifting water levels. The construction of deep sewers or subways will often lower the ground-water table by many feet. Some soils shrink considerably when water is taken out of them. Wooden piles often decay rapidly when they are not continuously submerged in water. Accordingly, when a structure is to be built in a locality where there is a possibility that the ground-water table might be lowered, the foundation of the structure should be designed with that possibility in mind.

The question of whether or not the preliminary investigation should include a geologic investigation will depend to a large extent on the magnitude of the project under consideration. An investigation by competent geologists is usually required for structures such as heavy dams, bridges over large rivers, and deep rock tunnels. For the average structure a thorough geologic investigation is rarely necessary, although it is always desirable to know something about the geologic history of the proposed site because this may have some influence on the properties of the soil.

Another method of preliminary investigation that is often used in foundation work is the soil-load test. Such tests are usually made on comparatively small areas either at the ground surface or at the bottom of a pit that has been excavated to the level at which it is proposed to construct the foundation. Load is applied to the test plate in increments, and a settlement reading is made at each load increment. The data obtained in this manner are used to plot a load-settlement curve. ASTM Specification D1194–52T provides a satisfactory guide to making such load tests.

In connection with all soil-load tests several important facts must be kept in mind. One of these is illustrated in Fig. 4. The test plate in Fig. 4*a* is placed at the ground surface, and a load is applied to the plate. When failure occurs, the soil alongside the plate heaves up as at w and the soil under the plate fails by shear along a curved surface such as xyz. When the test is made in a pit as in Fig. 4*b*, the soil tries to fail

in the same manner. However, when the curved failure surface *EB* reaches the elevation of the bottom of the test plate at *B*, all the soil above that elevation comes into action. The load test does not fail until the load is great enough not only to shear the soil along the curved surface *EB* but also to lift the block of soil *ABCD* and to shear it along *AB*. The result is that, even if the soil at the bottom of the pit is exactly the same as the soil at the ground surface, the indicated bearing capacity obtained in the pit test will be very much greater than that of the surface test because of the action of the block of soil *ABCD*. This block of soil is usually called overburden or surcharge, and the soil at

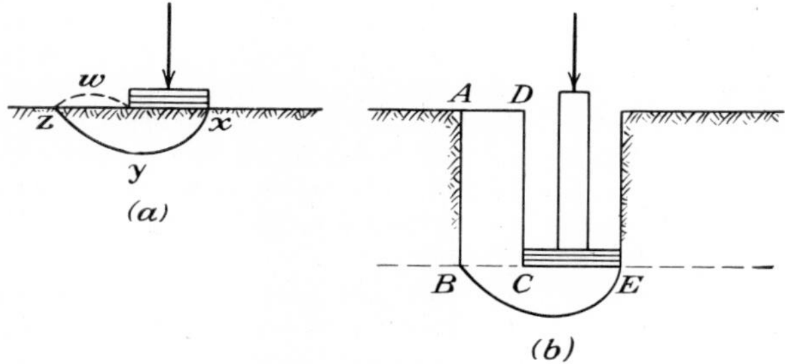

FIG. 4. Effect of overburden on load-test results.

the plane *BC* is said to be loaded with an overburden or surcharge load which is due to the weight of the earth above that plane.[1]

Another important fact about load tests is illustrated in Fig. 5. A square test plate with an area of 1 sq ft is placed on the ground surface and a load of 10,000 lb is applied to the plate. The average unit pressure under the plate is then 10,000 lb per sq ft. It may be assumed that the load spreads out in the underground in the form of a truncated pyramid whose sides slope at an angle of 45°. On the horizontal plane *AB*, which is 2 ft below the surface, the 10,000-lb load is spread over an area of 25 sq ft. The

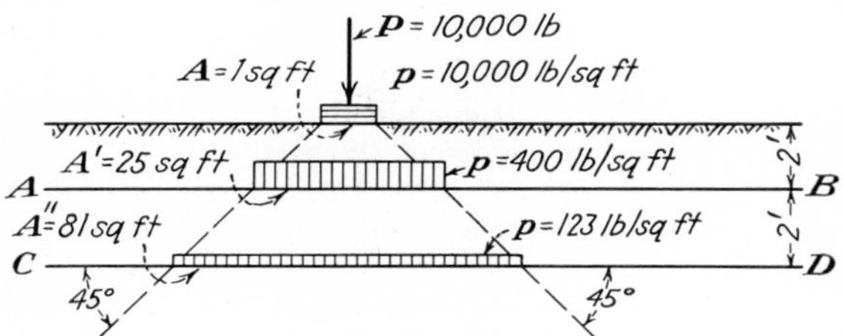

FIG. 5. Effect of depth below loaded surface on intensity of unit vertical pressure.

average unit load, which was 10,000 lb per sq ft at the ground surface, is therefore reduced to an average unit load of only 400 lb per sq ft at a depth of 2 ft. On the horizontal plane *CD*, which is 4 ft below the surface, the 10,000-lb load is spread over an area of 81 sq ft so that the average unit pressure on this plane is only 123 lb per sq ft. The high load intensity under the test plate at the ground surface practically vanishes at a depth amounting to several times the diameter of the loaded area.

The significance of this can be readily understood from Fig. 6, which represents a soil condition that is quite common. The ground-water table is 8 or 10 ft below the

[1] FABER, OSCAR, Pressure Distribution under Bases and Stability of Foundations, *The Structural Engineer (London)*, March, 1933, p. 116.

elevation at which the footings of a building are to be placed. The soil below the water table is a soft compressible clay, whereas the soil above the water table is hard and dry. A soil-load test on a bearing area of 1 or 2 sq ft is made on the hard dry soil at the pro-

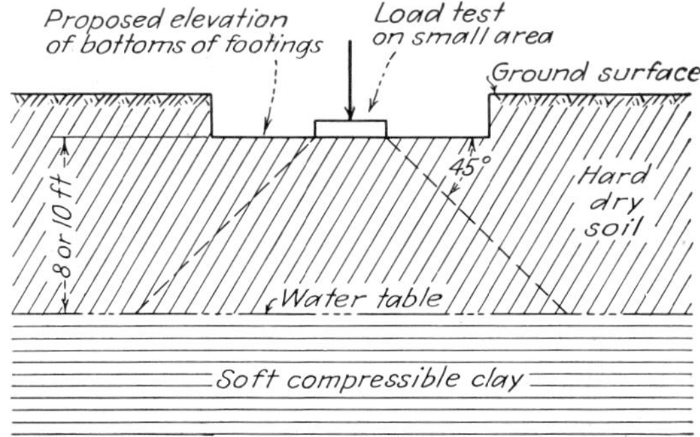

Fig. 6. Soil conditions where load test gives misleading results.

posed elevation of the bottoms of the footings. Although the test plate may be loaded until the unit pressure under the plate is as high as 10,000 lb per sq ft, the unit pressures in the underground will decrease rapidly with the depth. At a depth of 8 or 10 ft below

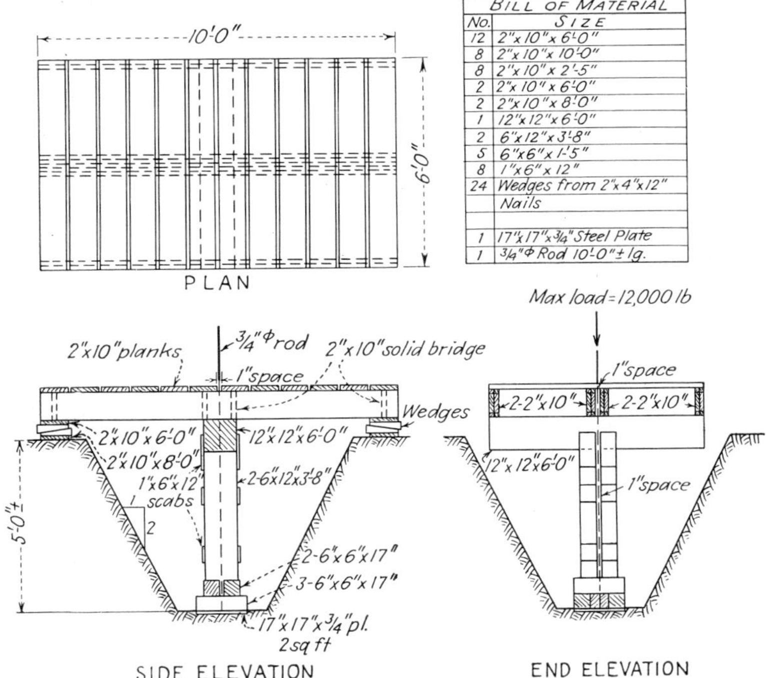

BILL OF MATERIAL	
No.	SIZE
12	2"x10"x6'-0"
8	2"x10"x10'-0"
8	2"x10"x2'-5"
2	2"x10"x6'-0"
2	2"x10"x8'-0"
1	12"x12"x6'-0"
2	6"x12"x3'-8"
5	6"x6"x1'-5"
8	1"x6"x12"
24	Wedges from 2"x4"x12"
	Nails
1	17"x17"x¾" Steel Plate
1	¾"Φ Rod 10'-0"± lg.

Fig. 7. Soil-test loading platform.

the small loaded area, the unit pressure will practically vanish so that the test load will exert almost no pressure on the top surface of the underlying soft clay. The loads from full-sized footings or the load of an entire building could produce high unit stresses in the underlying compressible clay, thereby causing settlements of the structure. It can be seen that, under such conditions, the load test on the small area can give a completely misleading answer as to the ability of the soil to support the structure safely. The soil-load test alone is therefore not a satisfactory method of preliminary investigation. It must be accompanied by satisfactory borings which will furnish information about soil conditions deep in the ground.

Other factors involved in the interpretation of load tests are time effects and area effects. As far as time effects are concerned, some soils are of such a nature that settlements do not take place immediately after the load is applied. There is a time lag in the settlement following the loading. It is not possible to study these time effects with load tests because they continue over long periods of time and the load test cannot usually be permitted to stay in place for a period of years.

A typical load-test setup is illustrated in Fig. 7. A post is set on a bearing plate and on top of this post is constructed a platform on which the load is balanced. Settlements are measured by means of a level and level rod on a bolt or a steel rod resting on the bearing plate. Sand, cement, brick, pig iron, steel rails, or other materials are used for loading the platform. In placing the load, care should be taken to cause as little vibration as possible, since vibration transmitted through the post to the soil will cause additional settlement.

LABORATORY TESTING OF SOILS

Mechanical Analysis. The separation of a soil into several fractions of different grain size is called "mechanical analysis." Wet mechanical analysis includes all those methods which depend on the action of granular material suspended in a liquid; and methods of continuous sedimentation are based on a series of measurements on a suspension observed during the sedimentation process. The most recent development in the field of wet mechanical analysis is the hydrometer method. It consists of employing a hydrometer to determine the variation in density of the suspension with time.

If a great number of soil grains are distributed in a liquid and a hydrometer is inserted, the hydrostatic uplift exerted on the bulb is equal to the weight of the suspension displaced at the elevation of the bulb. A hydrometer measures the average density of the suspension which is displaced by the bulb. From an observed reading the percentage of soil grains by weight can be directly determined. In order to determine the equivalent size of the largest grains existing at the elevation of the bulb, use is made of Stokes' law for the velocity of a sphere settling in a liquid. For practical purposes it is sufficiently accurate to base this computation on the depth from the surface of the suspension to the center of volume of the bulb.

Atterberg Limit Tests. The most conspicuous physical property of a clay is its plasticity. This property can be studied quantitatively by routine laboratory tests. The most useful of these are called the liquid and plastic limit tests which were originated by Atterberg.

The *liquid limit* is defined as the water content of a soil (expressed in percent, dry weight) having a consistency such that two sections of a soil cake, placed in a cup and separated by a groove, barely touch but do not flow together under the impact of several sharp blows. The determination of this arbitrarily defined limit is influenced by the technique adopted by different operators. The device shown in Fig. 8 designed by A. Casagrande eliminated personal variations in such a test by providing a mechanical means for obtaining a constant impact and a tool for cutting a groove to accurate di-

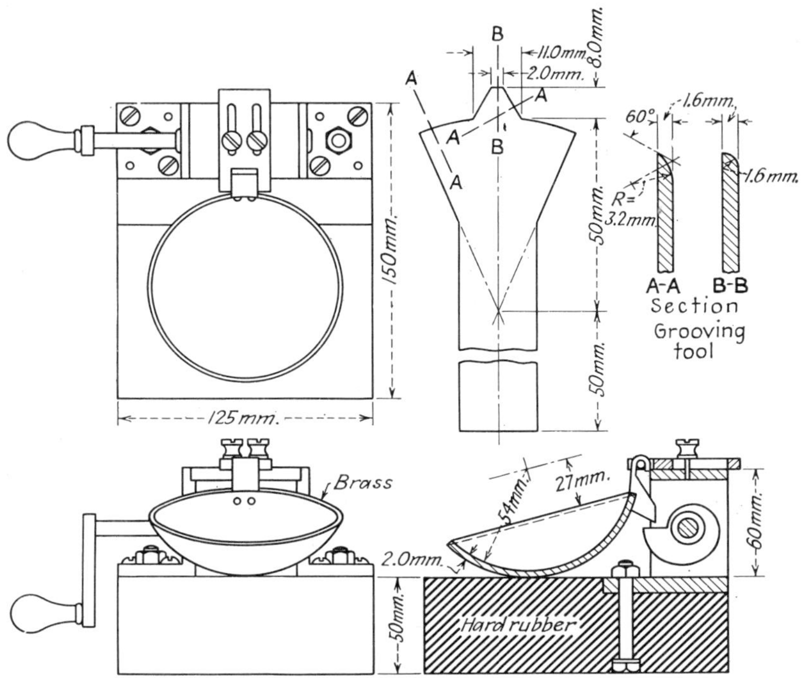

Fig. 8. Liquid-limit device.

mensions. Use of this device has been accepted by most laboratories as the standard method of performing the liquid-limit test.

In performing the test with this device, the number of blows, which are required to close the groove cut in a soil cake for a distance of ½ in. at the bottom, is determined. The curve obtained by plotting the water content of the various consistencies of the same soil on an arithmetic scale and the corresponding number of blows for each consistency on a logarithmic scale was found to be a straight line. By definition this line is called the "flow curve." The water content corresponding to 25 blows obtained from this curve is the liquid limit.

The *plastic limit* of a soil is defined as the water content (expressed in percent, dry weight) at which a soil will crumble when rolled into a thread 3 mm (⅛ in.) in diameter.

The *shrinkage limit* of a soil is defined as the amount of moisture which is sufficient to fill the voids in the soil mass at the maximum amount of shrinkage or, in other words, at the maximum consolidation under evaporation.

From the flow curve, the liquid limit, the plastic limit, and the shrinkage limit, several other indexes can be determined. The plasticity index is defined as the difference between the liquid limit and the plastic limit. The flow index is equal to the slope of the flow curve. The toughness index is defined as the plasticity index divided by the flow index. By a comparative study of the limits and the several indexes, a distinction between clays and nonplastic soils and classification according to degree of plasticity can be made.

The Coefficient of Permeability. The law governing the flow of water through fine-grained soils is expressed by Darcy's law

$$Q = kiAt$$

in which Q is the total quantity of water flowing through a soil of cross-sectional area A, under a hydraulic gradient i, in time t. The coefficient k, called the "coefficient of permeability," is dependent not only on the size and shape of the soil grains and structure, but also on the void ratio e and the temperature.

The coefficient of permeability can be determined by direct and indirect methods. The most important among direct methods are the constant-head permeability test, the falling-head permeability test, and the field test using well points. The indirect

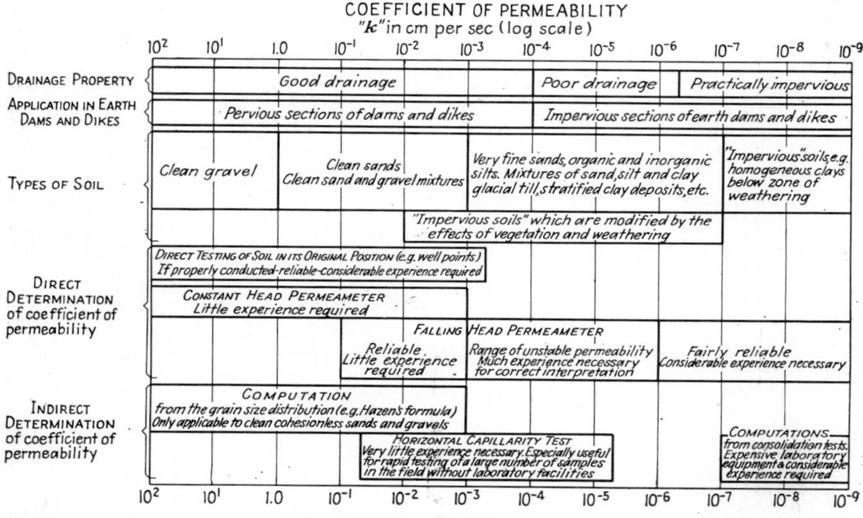

FIG. 9. Coefficient of permeability. (*After Casagrande.*)

methods most commonly used are the horizontal capillarity test and the computation from consolidation test data. Figure 9 is a chart showing the range for which each of these methods is applicable.

Constant-head Permeability Test. A properly prepared uniform sample of either cohesive or cohesionless soil of cross-sectional area A and length L is subjected to a flow of water under a constant head h. From Darcy's law the coefficient of permeability can be expressed in terms of these quantities as follows:

$$k = \frac{QL}{hAt}$$

The test is performed by measuring the quantity of water Q flowing through the sample, the length L, the head h, and the time t.

Falling-head Permeability Test. This test is conducted in the same manner as that just described except that the head under which flow takes place is not maintained constant but is permitted to fall in a standpipe directly connected to the sample. The quantity of water flowing through the sample is determined either by using a calibrated

pipette as a standpipe or by computation from the area of the standpipe and the drop in head.

Determination of Optimum Moisture Content. The optimum moisture content is defined as that water content (expressed in percent, dry weight) at which a given soil can be compacted to its maximum density by means of a standard method of compaction. In order to determine the optimum moisture content in the laboratory, a sample is compacted by a standard procedure into a cylinder of known volume. The weight

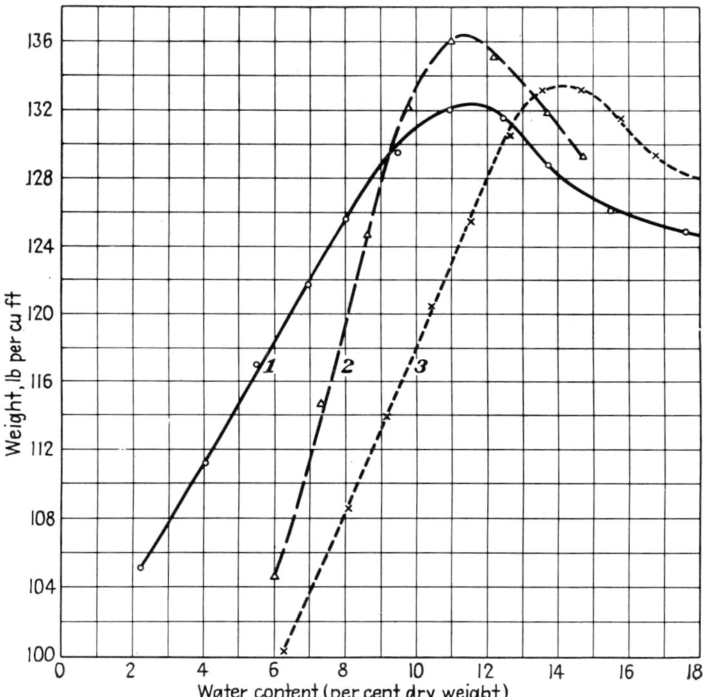

FIG. 10. Typical soil-compaction curves.

per unit volume is determined by weighing the cylinder, and the water content is determined from a representative sample of the compacted soil. This procedure is repeated several times using the same sample but with the moisture content of the soil changed each time by the addition of water. In this manner the relationship between density (weight per unit volume) and water content is established.

Typical water content-density curves are shown in Fig. 10. The water content corresponding to the maximum point of the curve is the desired optimum. It has been found that the optimum moisture content for plastic soils using the standardized laboratory procedure is in general slightly below the plastic limit.

Having obtained a water content-density curve for a given sample from a particular borrow area, it is a relatively simple matter to control the moisture content during the compaction of the earth structure in which the material is to be used. This can be done by determining either the water content or the density of representative samples which are obtained during the compaction process. By comparing either of these re-

sults with the water content-density curve, it is possible to determine immediately whether the soil is too dry or too wet to be compacted most efficiently.

Unconfined-compression Test. The unconfined-compression test is a rapid-loading test (5 min loading period), similar to a compression test performed on concrete cylinders, made by applying an axial load to carefully prepared cylindrical or prismatic soil samples and measuring the deformation corresponding to the stress as the load is increased. When the stress reaches the ultimate strength, the sample may fail either by a gradual bulging or by a sudden rupture. The ultimate strength of a sample which fails by continuous bulging is arbitrarily taken as the stress corresponding to 20 percent strain.

The ultimate strength of a test specimen, carefully prepared from an undisturbed sample (at unaltered water content), is a relative measure of the ultimate bearing capacity and shearing resistance of a soil. A comparison of the stress-strain characteristics of a soil tested in the undisturbed state, and then in the remolded state at unchanged water content, is indicative of the structural damage caused by remolding.

Consolidation Test. The consolidation test is performed for the purpose of determining the total volume decrease as well as the time rate of volume decrease which a

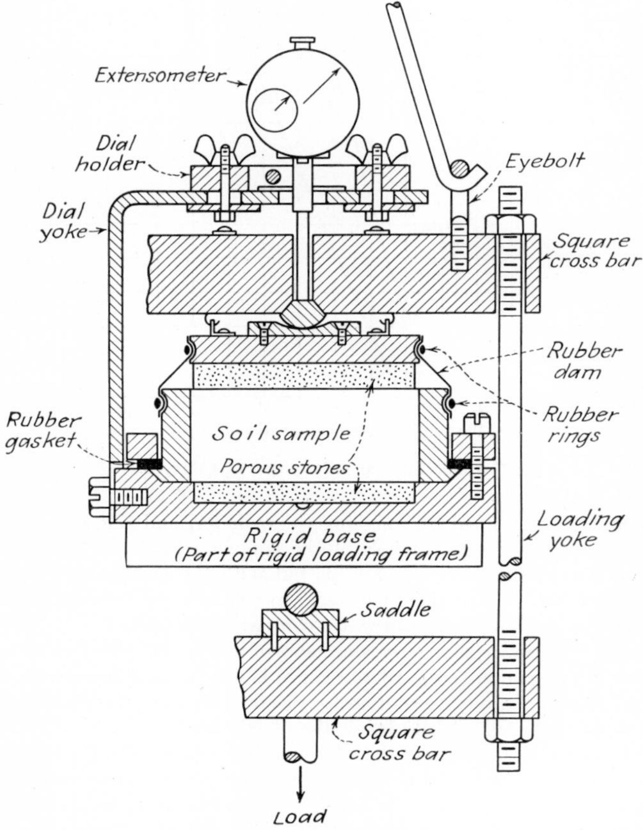

FIG. 11. Consolidation-test apparatus.

laterally confined soil sample will undergo when subjected to an axial load. During the test, a series of axial loads is applied to a soil sample which is confined laterally by means of a close-fitting cylindrical ring (Fig. 11). Porous cylindrical stones are placed on the top and bottom of the sample to provide drainage surfaces. The water which is squeezed out as consolidation takes place escapes at these surfaces. The load is applied through a loading crossbar and piston to the top porous disk. The lower porous disk rests directly on the bottom of a base which is supported on a rigid frame. The volume change of the sample is measured by an extensometer mounted in a clamp which is supported by two parallel rods connected to this base, as illustrated in Fig. 11.

For each increment of load the progress of volume change corresponding to appropriate time intervals is recorded. The data thus obtained are plotted in the form of a

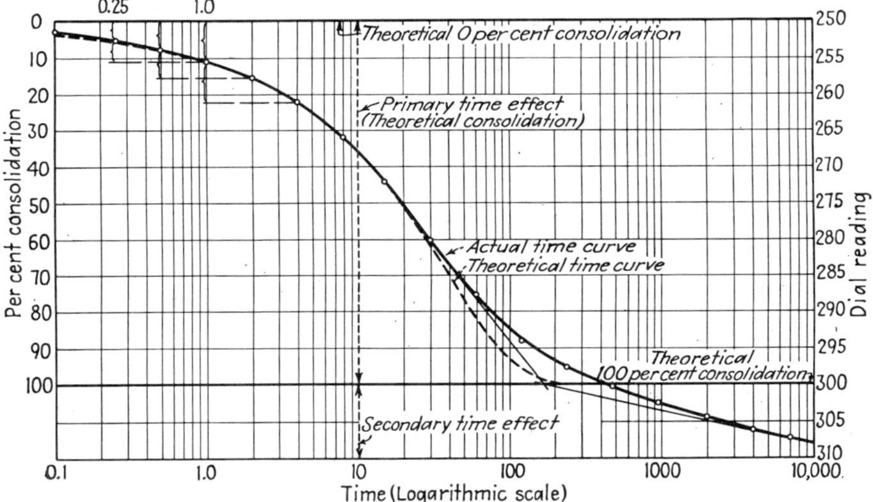

FIG. 12. Consolidation-test time curve.

dial reading vs. time curve as shown in Fig. 12. By plotting the dial reading to an arithmetic scale and the corresponding elapsed time to a logarithmic scale, the shape of the resulting curve can readily be compared with the theoretical curve obtained from the mathematical theory of consolidation. The theoretical-consolidation curve is shown in Fig. 13. If such a comparison is made it will usually be found that a close agreement exists up to 60 or 70 percent consolidation. Beyond this, the empirical curve becomes asymptotic to a straight line which is inclined to the line defining the 100 percent consolidation coordinate of the theoretical curve. This deviation from the theoretical curve is called the "secondary time effect." The additional volume adjustment accompanying the secondary time effect is ascribed to a readjustment of the frictional forces in the soil mass. This volume adjustment is generally small in comparison with the volume change due to the squeezing out of water as a result of the primary time effect, and its influence becomes evident only after the major portion of the volume change has taken place.

For comparative studies of the empirical curves with the theoretical curve, 100 percent consolidation is arbitrarily defined as the intersection of the tangent to the curve at the point of inflection with the tangent to the straight-line portion representing the

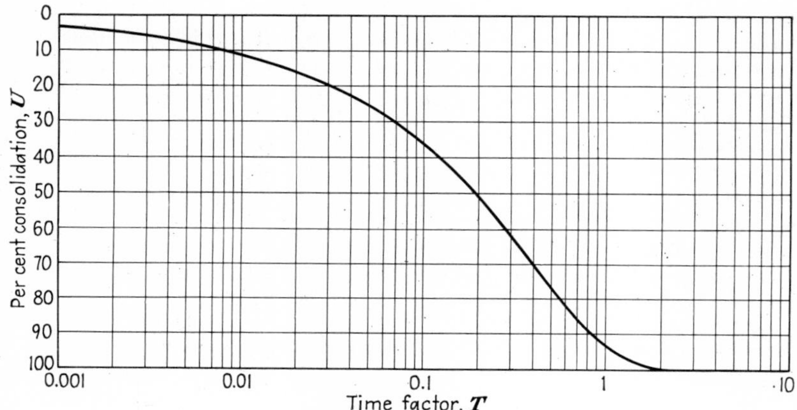

Fig. 13. Theoretical-consolidation curve.

secondary time effect. The construction necessary for the determination of this coordinate is shown in Fig. 12.

In performing a consolidation test, an increment of load is usually maintained on the sample until the straight line representing the secondary effect is defined. A new increment of load is then applied and a corresponding new observed-time curve obtained. From the observed-time curves thus obtained for each load, an arbitrary time is selected such that the dial readings for each of the curves corresponding to this time are beyond the primary time effect. This dial reading and corresponding pressure for each observed-time curve comprise the data from which a pressure-void ratio curve is prepared. Such a curve is shown in Fig. 14.

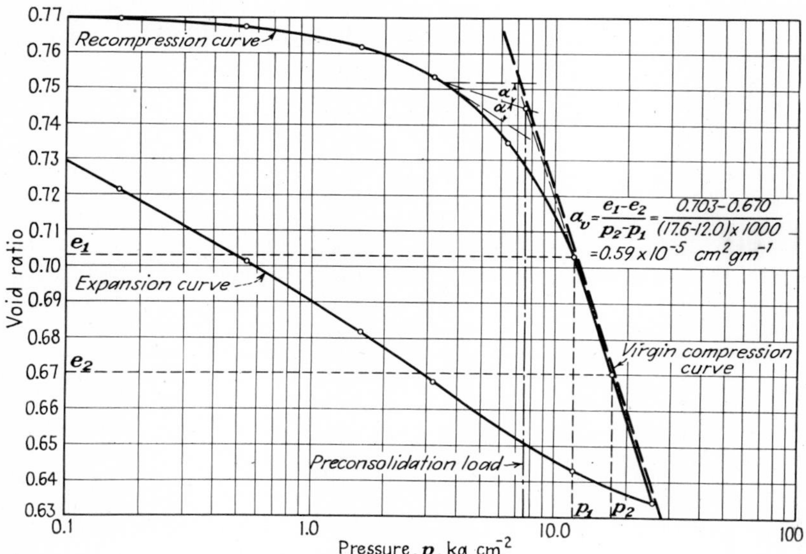

Fig. 14. Typical void ratio-pressure curve.

A time curve from a laboratory test can be converted to a time curve for a soil layer of any thickness since the time required for consolidation is proportional to the square of the thickness of a soil stratum. Also, since the time rate of consolidation is a function of the permeability of a soil, the data obtained from the time curve and the pressure-void ratio curve can be used to determine the coefficient of permeability indirectly.

From the pressure-void ratio curve as shown in Fig. 14, the compressibility as well as the pressure to which the soil has been consolidated in its previous geological history can be determined. This information is essential to settlement studies.

Triaxial-compression Test. The triaxial-compression test is performed for the purpose of determining the stress-deformation and strength characteristics of soils which are subjected to shearing stresses produced by varying the principal stresses acting on a

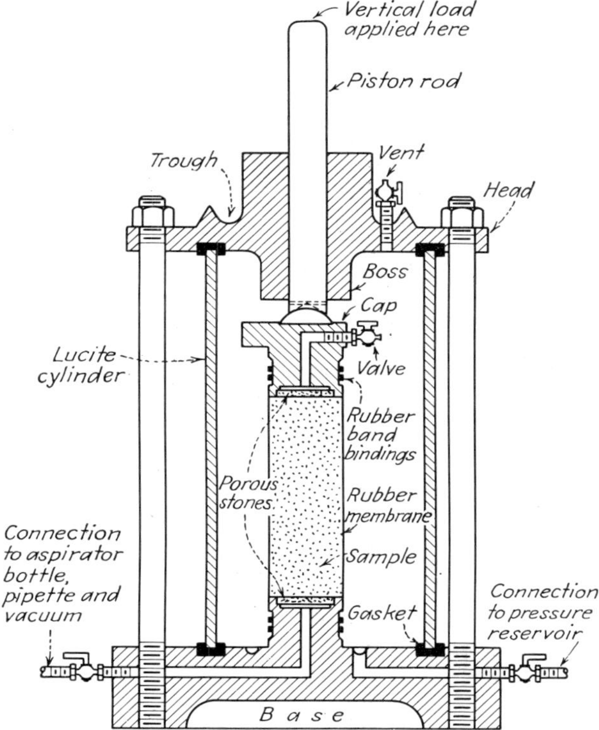

FIG. 15. Apparatus for triaxial-compression test.

cylindrical or prismatic soil specimen. In the usual type of triaxial-compression apparatus, two of the principal stresses are produced by the liquid pressure surrounding the specimen, and they are therefore equal. If a test is of short duration, it has been found satisfactory to use water as a pressure medium. However, if a test is carried on for a period of days, the use of glycerin as a pressure medium is recommended.

A carefully prepared test specimen enclosed in an airtight rubber membrane is set up inside a compression chamber which consists of a Lucite cylinder enclosed between a head and a base, as shown in Fig. 15. The bottom of the sample rests on a porous disk set into a pedestal which is an integral part of the base. The vertical load is trans-

mitted by a piston rod passing freely through the head and reacting on a cap containing another porous disk which rests directly on the top of the sample. The hydrostatic load is applied by filling the chamber with a liquid under a constant pressure. The vertical deformation is measured by means of an extensometer mounted on the head of the compression chamber and in contact with a crossbar through which load is applied to the piston.

In order that the volume change can be measured, the test specimen is completely saturated with water at the beginning of the test. This is accomplished by providing a passage from an outside graduated standpipe through the base and porous disk to the specimen. A similar passage is also provided from the top of the specimen through the porous disk and cap, this provision being made for the purpose of saturating a specimen at the start of a test. Volume change during the progress of the test is measured by observing the fluctuations in water level in the standpipe.

Direct-shear Test. The object of the direct-shear test is to apply known normal and shearing forces on a plane surface within a sample, with simultaneous measurement of

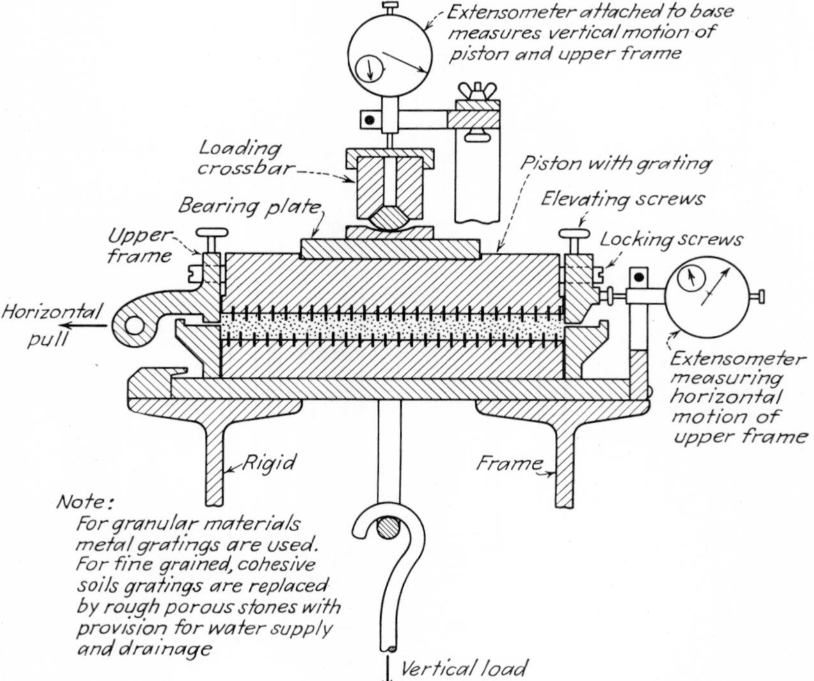

Fig. 16. Apparatus for shear test.

the deformation. The purpose is to obtain the stress-deformation and strength characteristics of the material. It has, therefore, the same purpose as the triaxial-compression test. In general, the relationship between normal stress and shearing strength is the most important result obtained from a series of direct-shear tests. This relationship is determined by performing a series of tests in which a specimen is subjected to different intensities of normal pressure. Figure 16 shows a type of shear box that is commonly used for direct-shear tests. However, for purely cohesive materials, the shearing re-

sistance is independent of normal pressure. In soft plastic materials, where the natural moisture content is well within the plastic range, the effect of testing under normal pressure is so well within the tolerance inherent in the approach that the refinement of normal pressure is not deemed necessary. In materials where the natural moisture content is close to or below the plastic limit and especially in materials having a substantial granular content, this is no longer true.

Transverse-ring-shear Test. The shearing resistance from the unconfined compression test may be termed a rapid shear test because the load is applied at a continuous rate until failure is produced and in a much shorter period of time than in the transverse-ring-shear test. Over a large number of tests, it has been found that shear values

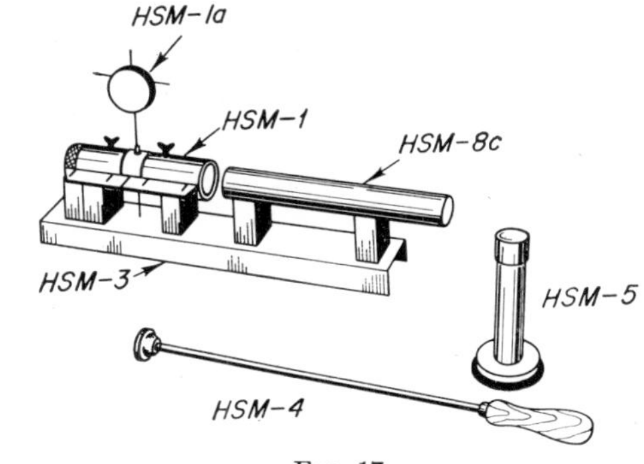

Fig. 17

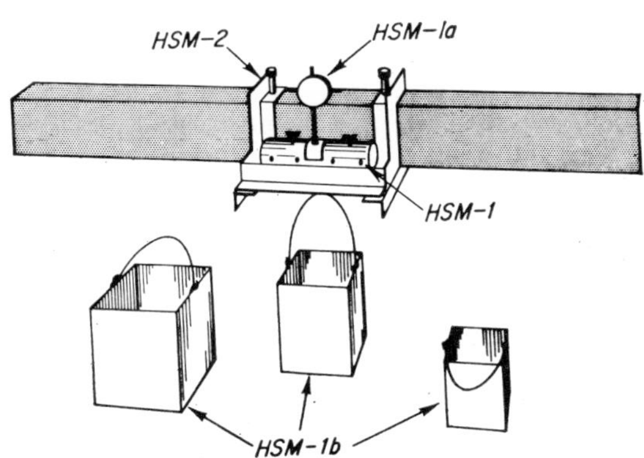

Fig. 18

HSM-1 Transverse-ring-shear test cylinders or metal liners. The left- and right-hand cylinders are 3 in. long and held fixedly in position by the thumb screws. The intermediate cylinder is 1 in. long and free to move perpendicularly to the longitudinal axis of the assembly.

HSM-1a Dial gauge used to measure the shearing deformation of the soil core for each increment of applied load.

HSM-1b Loading bucket for holding the applied weights of lead shot.

HSM-2 Bracket holder for attaching the assembly to a table top.

HSM-3 Jig used in transferring samples in test cylinders from the shipping tube.

HSM-4 Hand mandrel used in transferring samples in test cylinders from the shipping tube.

HSM-5 Hand-extruder ram used, after completion of the transverse-ring-shear test, in extruding the core from within a 3-in. test cylinder preparatory to conducting an unconfined compression test on that 3-in. core.

HSM-8 Metal shipping tube into which the 3-in., 1-in., 3-in. soil-core-filled test cylinders or metal liners are transferred on a jib in the field and within which soil-core-filled test cylinders or metal liners are shipped to and received by the laboratory.

obtained from unconfined compression tests are approximately four times those obtained from transverse-ring-shear tests in the case of plastic clays of the truly cohesive type or in the case of saturated clay. It has also been found that, for stiff clays and for materials which have granular characteristics, the ratio of the shearing values obtained from these two tests is quite elastic and, in general, higher than in the case of plastic clays. In presenting the results from these two tests, it has been the practice to plot results from the unconfined-compression test to a scale four times greater than is used for the transverse-ring-shear test which, in most cases, brings the divergent results into close focus. Figures 17 and 18 show a type of shear box that is commonly used for the transverse-ring-shear test.

DESIGN LOADS ON FOUNDATIONS

The proper load for which a foundation should be designed is a much debated matter, and a hard-and-fast rule that is of general application cannot well be adopted. Although each part of the superstructure should be designed for the maximum load that will come upon that member, a condition which usually involves full loads on adjacent panels or parts, it is extremely improbable that full load will ever be imposed over the entire structure at one time. This is particularly true in the case of a high office building in which each floor must be designed for full load although it is not probable that all floors will be fully loaded at one time. However, in the case of warehouses, full-load condition might occur. Hence judgment must be used in estimating what part of full load may exist at one time or what part of full load for the entire structure should be used in the design of the foundation.

For bridge foundations, the combination of live load, traction, earth pressure, ice pressure, wind pressure, and water pressure that will give maximum soil pressures should be considered, provided that it is physically possible or probable that they can all occur simultaneously.

In high buildings where settlement is expected, the problem is to adjust the bearing on the soil so that the settlement may be uniform in order that there may not be undue strains in the superstructure and cracks in the masonry. Settlement is caused chiefly by the dead load, since it is the load first applied and usually constitutes the major portion of the total load. Furthermore, it is constantly applied while the live load is more or less transient or intermittent. Hence, in assigning proportional load, the dead load should have a greater influence than the live load.

The practice among engineers is to design the foundation for the dead load plus a certain proportion of the live load, frequently taken arbitrarily as 50 percent. However, it is the better practice to determine the proportion of live load to be used in the design of the foundations on the basis of the ratio of the probable actual live load. A method sometimes used may be represented by the formula $A = D/Br$, in which A is the area required, D the dead load, B the allowable soil pressure, and r the ratio between the dead load and the total load coming on the foundation at the footing where the ratio of the live to the dead load is greatest. This footing is selected as the basis for design because it is an index of the probability of the occurrence of full load. It will be observed that the design of any particular footing will then depend primarily on the dead load.

A similar rule has been used which is based on the dead load plus half the live load.

$$A = \frac{D + 0.5L}{Br}$$

L being the live load and r the ratio $(D' + 0.5L')/(D' + L')$, in which D' and L' are the dead and live load respectively on the footings where the ratio of live to dead load is greatest.

The area of footings may be conveniently proportioned by the formula $A = \frac{D}{B} + \frac{L}{cB}$, where c is a coefficient which varies with the character of the live load, being about 4 for people, 2 for merchandise, 1 for electrical and other quiet machines, and ⅔ to ½ for machines with heavy rotating parts. This formula is an attempt to take into account the effect of unbalanced dynamic forces which produce vibrations. Such unbalanced forces are particularly important in the case of structures built on loose sand beds since the vibrations have the effect of compacting the sand, thereby causing settlement of the structure; they are also important in case of structures built over thick beds of saturated, soft clay; and, in all such cases, the unbalanced force should be resisted by adequate and designed members or inertial means.

The live loads used for designing foundations are those used for designing the columns resting thereon. They are ordinarily much less than the sum of all live loads at the maximum assumed intensity over the superstructure. The system of live-load reductions most commonly used in building codes requires that the foundations must be designed for full live load on the roof, 85 percent on the top floor, 80 on the next, and so decreasing 5 percent for each succeeding floor until 50 percent is reached, which is the minimum allowance for any floor.

SPREAD FOUNDATIONS AND MATS

Types of Spread Foundations. These include all those types which are designed to spread the building loads over a sufficient area of soil to secure adequate bearing capacity. They may be divided into various kinds, as illustrated in Fig. 19, which represents a plan view of a building foundation.

1. An independent footing is one which supports a single column or other concentrated load, as shown in Fig. 19*a* and *c*.

2. A continuous footing, as shown in Fig. 19*b*, may extend along the entire length of a wall or single row of columns.

3. A combined footing, as shown in Fig. 19*d* and *e*, is one which supports two adjoining column loads or sometimes three column loads not in a row, as shown in Fig. 19*f*.

4. A raft or mat foundation is one which extends under the entire building area and supports all the wall and column loads from the building.

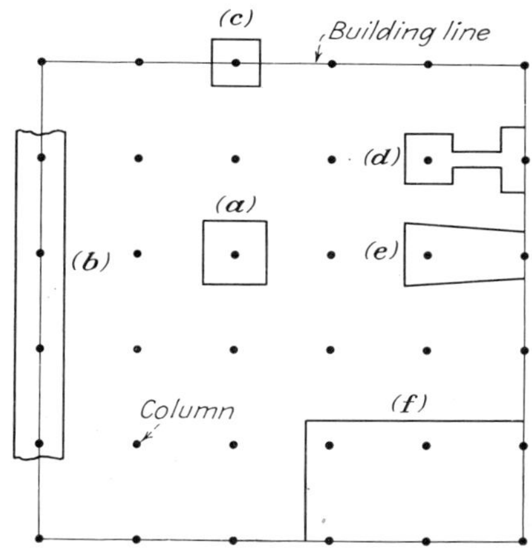

FIG. 19. Types of isolated spread foundations.

Independent Footings. Column footings of the most common type are independent footings. For light loads, they may be of plain concrete, but most independent footings are reinforced concrete footings with two-way reinforcing. Small-diameter closely spaced bars, with hooked ends, should be used to provide greater bond strength. Grillage footings may be constructed of tiers of steel beams. In the steel grillage footing, the beams are held in position by spacers placed between them. A layer of concrete 6 to 8 in. in thickness is placed under the lower beams, and the entire footing is filled solidly with concrete and encased in concrete with a minimum cover of 3 to 4 in. Steel grillage footings have been largely replaced by reinforced concrete footings.

Continuous Footings. Continuous footings may be simple footings constructed of plain concrete if the loads are light, although it is always desirable to provide a small amount of longitudinal reinforcement in simple continuous concrete footings in order to distribute shrinkage and temperature cracks and to bridge over soft spots in the soil. The most common type of continuous footing is the reinforced concrete footing. The main reinforcement in such a footing is perpendicular to the length of the footing and near the bottom of the slab. It is better to use small closely spaced bars rather than large bars spaced farther apart, and it is usually necessary to hook the ends of the bars to secure end anchorage and thereby develop bond strength. Longitudinal reinforcement is provided for the same reasons as in independent footings.

Combined Footings. Footings of this type are most frequently used to support wall columns which are close to the property line. If such a column were centered on an independent footing, as shown in Fig. 19*c*, the footing might project over the property line. If the column were placed near the edge of such a footing, the foundation pressures would not be evenly distributed under the footing, which would tend to tilt. This condition can be overcome by combining the wall column footing and the nearest interior footing into a single footing, as shown in Fig. 19*d*, *e*, and *f*. In order to avoid any tendency to rotate, combined footings are so proportioned that the centroid of the area which bears on the soil is on the line of action of the resultant of the column loads. This usually results in a trapezoidal footing, such as Fig. 19*e*. The beam connecting the two footings in Fig. 19*d* is often called a "strap beam." The footings for several wall columns and several adjacent interior columns may be combined into a single footing, particularly at the corners of buildings. Combined footings are usually constructed of reinforced concrete, but steel grillage footings have been extensively used in the past.

Raft or Mat Foundations. These usually consist of reinforced concrete slabs several feet in thickness covering the entire foundation area, as shown in Fig. 20. These slabs

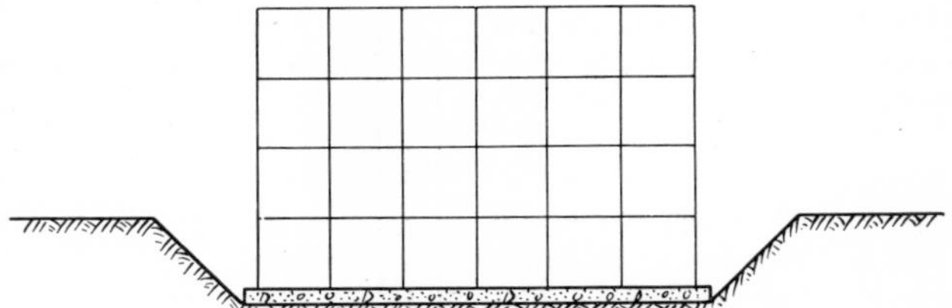

FIG. 20. Simple mat foundation.

or mats are reinforced with layers of reinforcing bars running at right angles to each other a few inches below the top surface of the mat and other layers a few inches above the bottom. Another form of raft or mat consists of inverted T beams of reinforced concrete, as shown in Fig. 21, with the slab covering the entire foundation area. The beams run in both directions and intersect under the columns. These beams are poured at the same time as the slab, forming a monolithic structure which will act as a unit. Before the basement floor is placed, the space between these beams may be filled with cinders or other materials as shown in the figure. Raft or mat foundations are used when the bearing power of the soil is so low that independent, continuous, or combined footings cannot be used and where piles cannot be used advantageously or are not necessary. Foundations of this type are commonly called *floating foundations.*

During recent years there has been a tendency to use the term "floating foundations" in a more restricted sense to apply to foundations where the earth is excavated to a depth that will make the weight of the earth removed about equal to the building load.

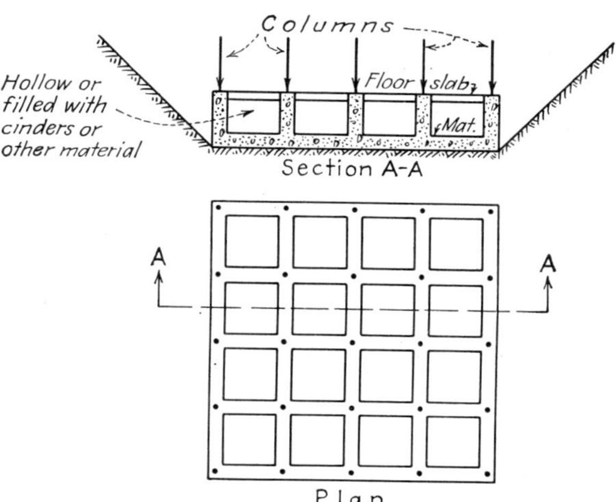

FIG. 21. Rigid mat foundation.

In such a case, the total vertical pressure on the soil under the building after its completion is equal to or less than what it was before the excavation was started, and the settlement is reduced to a minimum. All settlement cannot be eliminated because there is a tendency for the central portion of the building to settle more than the outer portion.

In order to reduce this differential or uneven settlement to a minimum, foundations must be so constructed that they are very rigid. There are two ways of accomplishing this. One is to make use of rigid reinforced concrete outside and cross walls in the basement of the building in order to form a boxlike structure. The other is to design the basement floor and the first or second floors as rigid frames, as shown in Fig. 22. Frames

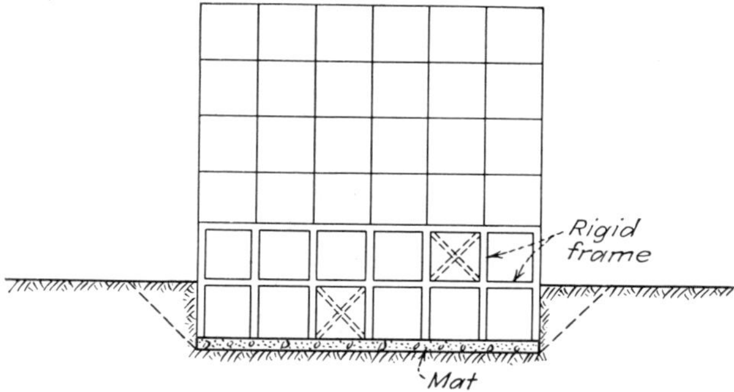

FIG. 22. Rigid-frame substructure.

of this type without diagonals are called *Vierendeel* trusses. Because the diagonal members are omitted, the members and their intersections must be designed to carry bending stresses. These rigid trusses may be carried through two or more stories if necessary to secure the required rigidity. Reinforced concrete trusses with diagonals

have been used instead of the Vierendeel trusses. Reinforced concrete interior walls restrict the use of basement space, and openings in these walls reduce the rigidity of the walls and introduce problems in design. Reinforced concrete trusses with diagonals also restrict the use of space, whereas Vierendeel trusses do not have this objectionable feature.

Bearing Power of Spread Foundations. Present methods for determining the bearing capacity of soils are not entirely satisfactory. The most common procedure consists of selecting values from tables of allowable or presumptive bearing capacities given in the local building codes. Such tables are commonly copied from other codes, and there may be little justification for the values given. In these tables, soils are usually divided into classes by indefinite terms, and a bearing value is given for each class. Actually, the bearing pressure which should be selected depends upon the amount of settlement which is permissible. The pressures under the foundations of buildings are rarely large enough to approach the actual strength of the soil except in the case of shallow foundations on soft clay.

For foundations supported on cohesive soils, the settlements of footings for a given unit pressure increase with the linear dimensions of the footings. The common assumption that the settlements of footings supported on a given soil will be equal if the unit pressures are equal is incorrect for cohesive soils. There would therefore be a different allowable pressure for each size of footing on a given cohesive soil if uniform settlement were required. However, some unequal or differential settlement between footings may not be serious, and the amount of differential settlement which may be permitted for the various footings, and which will not overstress the structural frame, should be determined.

The selection of allowable bearing values by examining only the soil at or near the surface on which the foundation is to rest may lead to serious errors. The soil below the bottom of a footing, for a depth of several times the width of the footing, contributes significantly to its settlement, and the underlying soil may be very different from that near the surface. Layers of peat, soft clay, and other soils of negligible or low bearing power at even greater depths may contribute considerably to settlement. Another factor which is commonly neglected in selecting allowable bearing values is the usual increase in bearing capacity with increase in the depth of the bearing surface below ground surface.

Allowable Bearing Capacities. The following table of allowable or presumptive surface bearing capacities, taken from the 1955 edition of the "Building Code of the City of Detroit," will serve as an example:

Foundation Material	Tons per Sq Ft
1. Hard sound rock	100
2. Soft rock	12
3. Hardpan overlying rock; very compact sandy gravel	10
4. Compact sandy gravel; very compact clay, sand, and gravel; very compact coarse or medium sand	6
5. Firm sandy gravel; compact clay, sand, and gravel; compact coarse or medium sand; very compact sand-clay soils; hard clay	5
6. Loose sandy gravel; firm coarse or medium sand	4
7. Loose coarse or medium sand; compact fine sand; compact sand-clay soils; stiff clay	3
8. Firm fine sand; compact inorganic silt; firm sand-clay soils; medium clay	2
9. Loose fine sand; firm inorganic silt	1½
10. Loose sand-clay soils; loose inorganic silt; soft clay	1 *

* In the case of loose organic silt and soft clay, the presumptive bearing value shall be substantiated by adequate supporting evidence.

Explanation of Terms. These figures for compaction and consistency are for 2 in. OD by 1.375 in. ID spoon, 140-lb hammer, 30-in. free fall.

COMPACTION RELATED TO SPOON BLOWS: GRANULAR SOIL

Descriptive term	Blows per ft	Remarks
Loose.	5–10	These figures are approximate for medium sand; coarser material requires more blows; finer material fewer blows
Firm.	11–30	
Compact.	31–50	
Very compact. . .	51 or more	

CONSISTENCY RELATED TO SPOON BLOWS: COHESIVE SOIL

Soft.	3– 5	Molded with relatively slight finger pressure
Medium.	6–15	Molded with moderate finger pressure
Stiff.	16–25	Molded with substantial finger pressure; might be removed by spading
Hard.	26 or more	Not molded by fingers, or with extreme difficulty; might require picking for removal

SOIL SIZES

Descriptive term	Pass sieve number	Retained sieve number	Size range
Clay.	200	Hydrometer	0.006 mm
Silt.	200	Analysis	0.006–0.074 mm
Fine sand.	65	200	0.074–0.207 mm
Medium sand.	28	65	0.208–0.589 mm
Coarse sand. . .	8	28	0.589–2.362 mm
Gravel.	. . .	8	2.362 mm
Pebble.	. . .	. . .	2.362 mm to 2½″
Cobble.	. . .	. . .	2½″ to 6″
Boulder.	. . .	. . .	6″

Hard sound rock is rock such as hard limestone, in sound condition, with some cracks allowed.

Soft rock is rock such as shale and partially decomposed limestone, with some disintegration and softening and with considerable cracks allowed.

Hardpan overlying rock is a thoroughly cemented mixture of sand and pebbles or of sand, pebbles, and clay, with or without boulders and difficult to remove by picking.

Most modern codes take into account the fact that the bearing capacity of an underlying stratum is sometimes less than that of the soil on which a foundation rests; and they include the requirement that the unit pressure due to the footing load computed on the top surface of the weaker stratum shall not exceed the allowable bearing pressure for the material in that stratum. The load is considered uniformly distributed over the area intercepted on the top surface of the weaker stratum by planes sloping from the edges of the foundation at an angle of 60° with the horizontal. It is recognized that this is an arbitrary assumption, but the procedure is considered sufficiently accurate for its intended purpose and is simple in its application.

In many building codes, the allowable bearing pressures to be used are based on tests which consist of placing a load on 1 or 2 sq ft of soil and observing the settlements which occur as the load is increased; and the specified procedures for interpreting such load-test data are arbitrary and not entirely satisfactory. However, the present knowledge of soil behavior, though inadequate, is constantly improving and generating better pro-

cedures which are feasible for use in building codes. The soil-load test should always be accompanied by borings which reveal the nature of the underlying soil.

Allowable values for the bearing capacity of rock, as given in building codes, range from 10 to 100 tons per sq ft. The lower value is for weathered, shattered, and poorly cemented rock; and the higher value, for sound, thoroughly cemented rock. It is usually assumed that sound bedrock will carry any load which can be transmitted to it by a concrete pier, since the allowable unit compressive stress in the concrete of the pier would be less than the allowable bearing stress in the rock. Some shales yield by plastic flow when subjected to heavy pressures, and shales should always be carefully investigated before they are subjected to load. When rock which is being considered for foundation support is of doubtful quality, compression tests should be made as an aid in determining a safe bearing value. Other factors which must be taken into account in examining rock for proposed foundations are the direction of stratification, structural defects, caverns, and solution channels in limestone and possibly weak underlying strata. If a fault crosses the building site, the probability of movement along the fault, and the possible consequences of such movement, should be considered. Core borings in rock should always penetrate at least 5 ft below the rock surface.

PILE FOUNDATIONS

Kinds of Piles. In cases where the soils near the ground surface are incapable of supporting mat foundations or simple spread footings, pile foundations are often used. Many different types of piles are in general use, and Fig. 23 shows most of the common types.

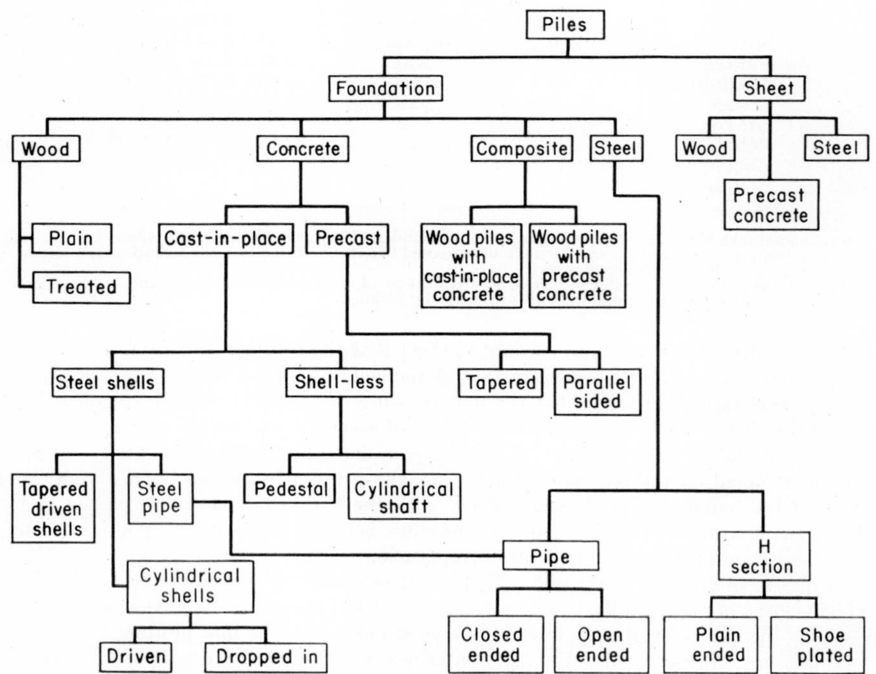

Fig. 23. Classification of piles.

Piles are commonly divided into two general classes which are called "foundation piles" and "sheet piles." Foundation piles may be divided into four different types, the division being based on the kind of material of which the piles are made. On this basis, the four common types of foundation piles are wood piles, concrete piles, composite piles, and steel piles.

Plain wood piles have certain advantages, as well as disadvantages, with respect to other kinds of piles. The principal advantage of a plain wood pile is the fact that it is comparatively cheap. The cost per linear foot of a plain wood pile is often as little as one-third or one-fourth of the cost per linear foot of concrete or steel piles. However, in spite of its low cost, the plain wood pile is not necessarily the cheapest type of foundation because of the restrictions that govern its use. In order that it will not decay, it must be cut off and capped below the permanent ground-water level. This requirement often involves deep excavation with sheeting and pumping. Another disadvantage of the wood pile is the fact that ordinarily it can support a considerably smaller load than some of the other types of piles. The usual allowable load on a wood pile is about 15 or 20 tons, whereas concrete piles are almost always loaded to 30 tons or more. As a result of this load differential, a given structure would require a larger number of wood piles with larger footings than would be required for the same structure supported on concrete piles. Figure 24 illustrates the various factors that enter this problem. It is assumed that a given column load is to be carried by a pile footing and that the permanent ground-water level is about 10 ft below the basement floor line. A concrete-pile footing for this column would be a small block of concrete placed in a shallow excavation just below the basement floor. The concrete piles would be driven in the bottom of this shallow excavation, and it would not be necessary to pay attention to the fact that the permanent ground-water level is well below the pile cutoff elevation, since concrete piles are permitted to extend up into the dry ground above the water table.

In order to support this column load under these conditions on a wood-pile foundation, the procedure would be quite different. It would be necessary to excavate a pit down to a depth of 1 ft below the ground-water level. In many cases such a pit would have to be sheeted, and it would probably be necessary to do some pumping to keep the water out of the bottom of the pit. After the pit had been excavated and sheeted, the wood piles would be driven in the bottom of the pit. In order to cap the piles, it would be necessary to use a considerable amount of concrete extending from the tops of the piles up to the base of the column just below the permanent floor line.

Whenever cost comparisons are being made for wood-pile and concrete-pile foundations under conditions of this kind, the comparison must not be made solely on the basis of the relative cost of one wood pile as against one concrete pile. It is necessary to consider the cost of the entire foundation up to the underside of the column base. The estimated cost of the wood-pile foundation must include the cost of excavation, sheeting, pumping, and the wood piles with their concrete footing. This total cost must then be compared with the cost of the concrete piles and their concrete footing.

In some cases, the permanent ground-water level is quite close to the ground surface so that a few feet of excavation will reach the water table. Under such conditions, the cutoff elevation for wood piles and concrete piles would be approximately the same, and the cost comparison would usually favor the wood piles. In general, it can be stated that the concrete-pile foundation will almost always be more economical than the wood-pile foundation whenever the use of concrete piles will save 6 ft or more of excavation, as illustrated in Fig. 24.

Plain wood piles extending into dry ground above the water table or into the open air above the ground surface are always subject to attack by fungus growths and by insects of various kinds. In sea water, especially in the tropics, there are numerous marine organisms that attack wood piles.

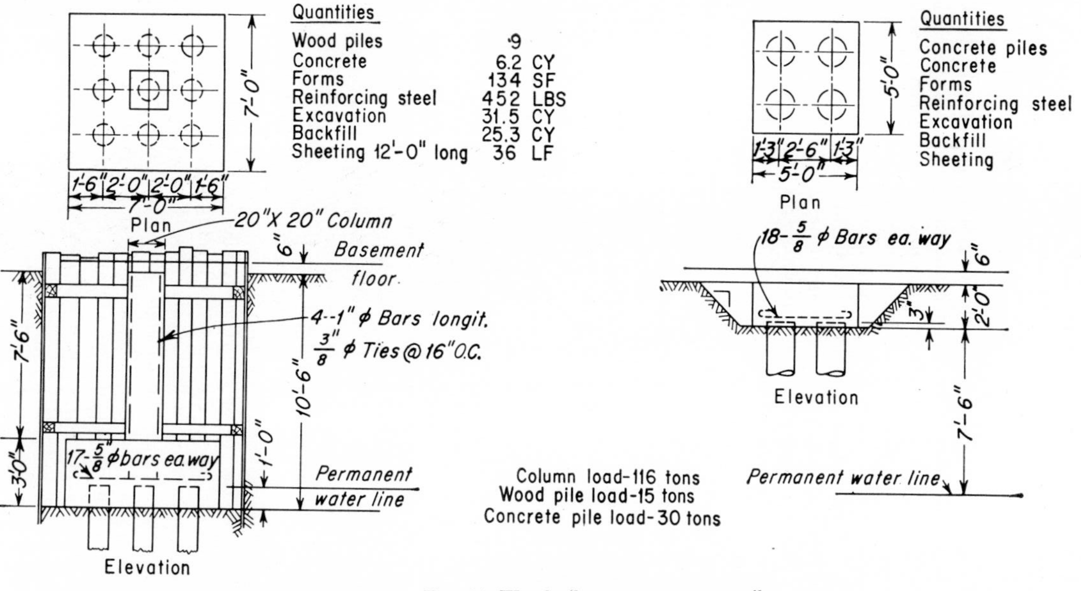

Fig. 24. Wood piles versus concrete piles.

Treated Wood Piles. In order to overcome some of these disadvantages of plain wood piles, preservative processes have been developed in which the wood is treated with various kinds of chemicals. The most common chemical used for this purpose is coal-tar creosote. The effectiveness of chemical preservatives depends on two important factors: (1) the thoroughness with which the treatment process is done and (2) the severity of the exposure conditions at the site where the treated pile is to be used. In some tropical waters that are badly infested with marine organisms, treated wood piles have been destroyed almost as rapidly as plain wood piles. It must not be assumed that preservative treatment will make the pile immune from attack by marine animals and insects, especially in localities where these are very active. Neither can it be assumed that preservative treatment will entirely prevent decay in places where the pile is exposed to the air and is alternately wet and dry. However, it can be taken for granted that a properly treated wood pile will last longer than an untreated wood pile under the same exposure conditions.

There is one important factor that should be kept in mind whenever the use of treated wood piles is being considered for the foundation of a permanent structure. This is the question of replacement of the pile in case it should decay or be otherwise damaged by animal or insect life. The replacement of decayed piles under a building foundation is a serious problem because it involves underpinning and shoring operations which are slow and expensive.

The treatment process adds materially to the cost of the pile, but there is nothing about the treatment process that enables the treated pile to carry any more load than a plain wood pile; on the contrary, it often reduces the resistance of the natural wood to compression and bending. The treated pile is therefore allowed to carry only the same working loads that are used on plain wood piles.

Concrete piles came into general use during the last few years of the past century. Ordinarily concrete piles are divided into two principal types, cast-in-place and precast. The cast-in-place pile is formed in the ground in the position in which it is to be used in the foundation. The precast pile is made above the ground in a pile-casting yard, and it is then brought to the job site and driven or jetted into the ground just like a wood pile. The fields of usefulness of cast-in-place and precast concrete piles overlap to some extent, but usual conditions give one or the other a distinct advantage.

In building-foundation work on land, the cast-in-place pile is more commonly used than the precast pile. The principal reasons for this are the following: The cast-in-place pile foundation can be installed more rapidly because there is no delay while test piles are being driven to predetermine pile lengths, and there is no delay waiting for piles to cure sufficiently to withstand handling and driving stresses. Of particular importance in crowded locations is the fact that a cast-in-place pile job does not require a large open space for a pile-casting yard.

The question of predetermining the required pile length is especially important on precast-pile work. On a cast-in-place pile job the contractor usually starts the work with equipment that will drive piles of widely varying lengths so that it is not necessary to know the exact required length before the job is started. On precast-pile work it is necessary to know the required lengths before the piles are cast, and time and money must usually be spent for driving test piles to determine the required length. If the probable pile length for a precast pile is not determined before the piles are cast, there is serious risk of a considerable excess cost on the work, because of the necessity for either adding to the length of precast piles that are too short or cutting off and throwing away lengths of precast piles that are too long.

It is usually required that precast piles shall be stored and cured for 30 days before they are driven, but this curing period can often be reduced by the use of the high-early-strength cements that are now available. On all precast-pile work a good-sized

area is required for a pile-casting yard. On some jobs there is space available so that the piles can be made close to the job site. In the downtown sections of large cities there is usually no vacant property available for a casting yard so that precast piles have to be made at some distance from the job site. This involves handling and transporting the piles from the casting yard to the job, and it adds to the cost of the work. It is for these reasons that the cast-in-place pile is more commonly used than the precast pile on building-foundation work on land.

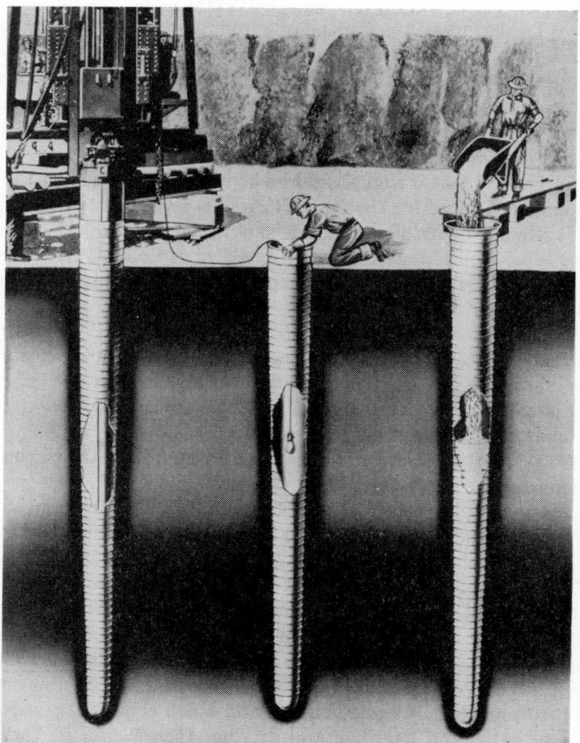

Fig. 25. Raymond standard concrete pile.

In marine installations such as docks, piers, and bulkheads, the precast pile is used almost exclusively. This is due largely to the difficulty involved in placing cast-in-place piles in open water. For docks and bulkheads the cast-in-place pile is sometimes used in the anchorage system. On trestle-type structures, such as highway viaducts, the precast pile is more commonly used. A portion of the pile often extends above the ground surface and serves as a column for the superstructure.

Precast concrete piles are always reinforced to withstand handling and driving stresses. They are usually cast in a horizontal position and then picked up and delivered to the pile driver so that they are often subjected to severe bending stresses. They must be reinforced internally to take care of these stresses. The concrete in a cast-in-place pile is not subjected to handling and driving stresses since the pile is poured in place in the ground. When a cast-in-place pile is completely submerged in soil, it does not need to be reinforced internally. Both types of concrete piles are sometimes used as columns

extending above the ground surface, and in such cases they must be reinforced for column action. In some types of structures, either the cast-in-place or the precast pile may be subjected to horizontal forces applied at the pile head. Under such conditions, the bending stress in the piles should be investigated and internal reinforcement should be provided as necessary to take care of these bending stresses.

Cast-in-place piles are generally divided into two different kinds: (1) those which are provided with permanent steel shells and (2) those which are made without permanent shells. The purpose of the shell is to prevent mud and water from mixing with the fresh concrete and to provide a form to protect the concrete while it is setting. With the shell-less pile, the fresh concrete is in direct contact with the surrounding soil.

Fig. 26. Monotube pile.

Cast-in-place piles with permanent shells may be divided into three different kinds: (1) tapered driven shells; (2) cylindrical, dropped-in or driven shells; and (3) steel-pipe piles. One type of tapered driven shell pile is illustrated in Fig. 25. This is a Raymond pile, which is made in the following manner: A thin steel shell, closed at the bottom with a steel boot, is placed on the outside of a steel mandrel or core, and the shell and the core are driven together into the ground. When sufficient driving resistance has been developed, driving is stopped and the core is withdrawn from the shell. This leaves a steel-lined hole in the ground which can be inspected with a lamp on a drop cord, and the concrete is then dumped into the shell from wheelbarrows. With this

type of pile a relatively thin shell is used because the heavy steel driving core supports the shell so that it will not collapse during driving.

Another type of driven shell is illustrated in Fig. 26. This is called a Monotube pile, and it has a scalloped cross section which is formed by a series of vertical flutings running the full length of the pile. These piles are tapered and are closed at the bottom by a steel boot. The shell is driven into the ground and is then filled with concrete to form a cast-in-place pile. Monotubes are driven without the aid of an internal driving core,

FIG. 27. Dropped-in shell pile.

and the shell must therefore be thick enough to be able to withstand the driving stresses and ground pressures.

One important thing about both of these types of tapered driven shell piles is the fact that they maintain the driving resistance that is set up in the soil surrounding the pile during driving because the steel shells remain in place in the ground in the position in which they were driven.

A third type of cast-in-place pile is illustrated in Fig. 27. This is a cylindrical, dropped-in shell pile. The driving apparatus consists of a piece of heavy steel pipe which is usually about 14 to 16 in. in diameter and ½ in. thick. A steel driving core is fitted inside this pipe, and the two are driven together into the ground. When suitable driving resistance has been obtained, driving is stopped and the core is removed from the pipe. A thin corrugated-metal shell several inches smaller in diameter than the driving pipe is then dropped down inside the driving pipe. The thin corrugated shell is

filled with concrete, and the driving pipe is then pulled out of the ground. The pile that remains in the ground is therefore several inches smaller in diameter than the pile that was driven. The carrying capacity of such a pile depends largely on the point resistance, since the resistance along the sides of the pile and the soil pressures set up in the soil during driving are not maintained when the outer driving pipe is pulled out of the ground. To avoid this disadvantage, a fourth type of cast-in-place pile has been developed in which a thin cylindrical corrugated shell is driven in contact with the surrounding ground by means of an internal expandable mandrel or core the surface of which has projections which expand into the valleys of the shell and carry it down. The core or mandrel is made in parts longitudinally; and it cannot be driven without damage as hard or with as heavy a hammer blow as is frequently necessary. Consequently, it is not capable of safely carrying as high design loads as the type which can be driven hard safely.

A fifth type of cast-in-place pile with a permanent shell is the steel pipe pile. These are usually divided into two types: (1) closed-end pipe piles and (2) open-end pipe piles. The closed-end pipe pile is a piece of steel pipe closed at the bottom with a heavy boot The pile is driven into the ground and filled with concrete. The uses of such piles are about the same as those of the other types of cast-in-place piles with driven shells, and the allowable loads on closed-end pipe piles are the same as those allowed on an ordinary cast-in-place or precast pile. Sometimes these pipe piles are driven all in one piece and sometimes they are made of several pieces of pipe which are either welded together or fitted together with special internal sleeves.

The open-end steel-pipe pile is usually driven to bearing on rock. When the pipe is being driven, it is open at the bottom so that, when driving is stopped, the interior of the pipe is full of soil. This soil has to be removed from the pipe, and the usual procedure is to do this with air and water jets, although it used to be done occasionally with

Fig. 28. Blowing out open-end steel-pipe pile.

miniature orange-peel dredging buckets. When the cleaning out is done by the jetting process, a long jet using water is worked down inside the driven pile in order to loosen up the soil. When the soil has been softened by the water, air is forced through the jet into the pile. This air jet is connected to a large compressed-air receiver, and a quick-throw valve is set in the air line between the receiver and the jet. This valve is thrown open suddenly so that the compressed air rushes into the bottom of the pipe almost like an explosion. The soil inside the pipe is blown out at the top, as illustrated in Fig. 28. It is usually necessary to repeat this jetting and blowing process several times in order to get the pipe properly cleaned out. When this has been done, the pipe is redriven with the hammer to make sure that it is properly seated in the rock. The pile is then again blown out and is promptly filled with concrete. Open-end pipe piles

installed in this manner are usually allowed to carry relatively high working loads. The amounts of load allowed for different sizes of pipes are fixed by building codes and, in general, these loads are considerably higher than those used on other types of concrete piles.

The shell-less concrete piles are divided into two different kinds, pedestal and cylindrical shaft. The process commonly used in forming the pedestal pile is illustrated in Fig. 29. A heavy steel driving pipe with an internal core is driven into the ground until a satisfactory driving resistance has been obtained, as shown at (1). The core is then lifted out of the pipe and a small charge of concrete is dumped into the pipe,

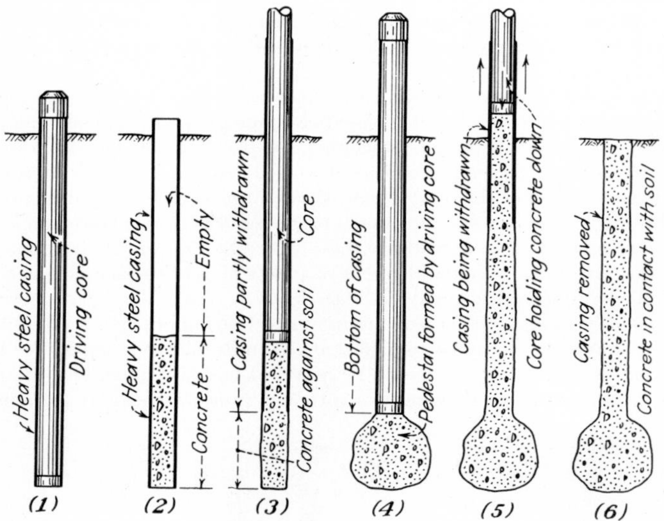

Fig. 29. Pedestal pile.

as shown at (2). The pipe is then pulled up a few feet and the core is put back into the pipe on top of this concrete, as shown at (3). The hammer is used to drive the core down in order to force the concrete out of the bottom of the pipe to form the bulb as shown at (4). Sometimes several charges of concrete are used to make the pedestal, after which the pipe is filled with concrete all the way to the ground surface. The core is then placed on top of the concrete to hold it down while the pipe is being withdrawn, as shown at (5). The theoretical final form of the pile is shown at (6).

The cylindrical-shaft shell-less pile is made with the same apparatus. However, instead of the small charge of concrete that is dropped in the pipe, as shown at (2) in Fig. 29, the cylindrical-shaft pile is formed by filling the pipe with concrete all the way up to the ground surface. The core is then placed on top of the concrete as shown at (5) in order to hold down the concrete while the pipe is being withdrawn. The pile thus formed is similar to the pile shown at (6) except that it is simply a cylindrical shaft of concrete without the pedestal.

The principal advantage of the cast-in-place shell-less pile is that fact that it is the cheapest of the concrete piles. This is due to the fact that the shell-less pile consists only of concrete, whereas the precast pile is always reinforced internally and the shell types of cast-in-place piles always have steel shells that remain in the ground. On the other hand, the shell-less type of pile has certain disadvantages as compared with other

types of concrete piles. In many cases, especially if the pile is very long, it is difficult to force the concrete out of the pipe in a uniform stream while the pipe is being withdrawn. The concrete tends to stick in the pipe so that it leaves the bottom of the pipe in a series of jumps, and the pile thus formed may not be a continuous shaft. In the formation of the pedestal there is the possibility that the soil may not be uniform in the vicinity of the pile point so that the pedestal may be off center with respect to the vertical center line of the shaft. In any case, there is no possibility of inspection either of the shaft or of the pedestal to make sure that they have been properly formed.

Even when the shell-less pile is perfectly formed, there is still the possibility that the driving of adjacent piles will distort the fresh concrete in previously driven piles. In order to avoid the possibility of this, it is sometimes required that shell-less piles shall be driven on a wider spacing between pile centers than is used for other types of piles. It is necessary to use great care in forming piles of this kind, and the number of piles that can be driven in a given time is usually much smaller than the number of concrete piles of other types that can be driven in the same time.

Precast-concrete piles may be divided into two kinds, tapered and parallel-sided. Precast piles are usually of square or octagonal cross section, although sometimes they are round. Since the piles are usually cast in a horizontal position on the ground, the round cross section is not common because it is difficult to fill a cylindrical form with concrete when the form is horizontal.

All precast piles are reinforced with longitudinal bars and transverse reinforcement, which may be separate hoops or a spiral. The pitch of the spiral reinforcement is always smaller for a length of 3 or 4 ft at each end of the pile than it is in the middle of the pile because the maximum stresses in the pile during driving occur near the ends of the pile. In Fig. 30 are shown some typical designs of precast piles. At the left is a uniformly tapered pile of square cross section and at the right is a uniformly tapered pile of octagonal cross section. In the center is a parallel-sided pile of octagonal cross section. The lower ends of the parallel-sided piles are usually tapered over a length of several feet from the full cross section down to a point diameter of about 10 in.

Ordinarily the tapered precast piles are limited in length to about 40 ft. For longer lengths the less flexible parallel-sided piles are used. These are often made with diameters as great as 24 or even 30 in. and the maximum lengths that have been used up to the present time are well over 100 ft. During recent years, very heavy precast piles of octagonal cross section with diameters as great as 36 in. have been used. These are usually made with a cylindrical hole 10 or 12 in. in diameter centered on the axis of the pile because solid piles of this size would be very heavy and hard to handle. The strength of the pile is sufficient to carry the working load even though the pile is hollow and the purpose of the large diameter is to get a large pile surface in contact with the surrounding soil. The carrying capacity of the pile is determined by the amount of embedded surface, rather than by the structural strength of the pile itself.

Precast piles are usually manufactured in a temporary casting yard on or near the site where they are to be used. Transportation and handling costs are comparatively high so that precast piles are not usually shipped over great distances from casting yard to job site.

When very hard driving conditions are encountered, precast piles are often equipped with steel driving shoes. These are pointed steel caps that fit over the point of the pile. They are available in various sizes and are provided with straps that extend into the concrete at the lower end of the pile so that the shoe can be made a part of the pile when the pile is cast.

Recently, there has been developed by the Raymond Concrete Pile Company a centrifugally precast, prestressed, cylindrical, hollow concrete pile. These are cast in sections 16′ long, which are then fitted together into the number of sections required

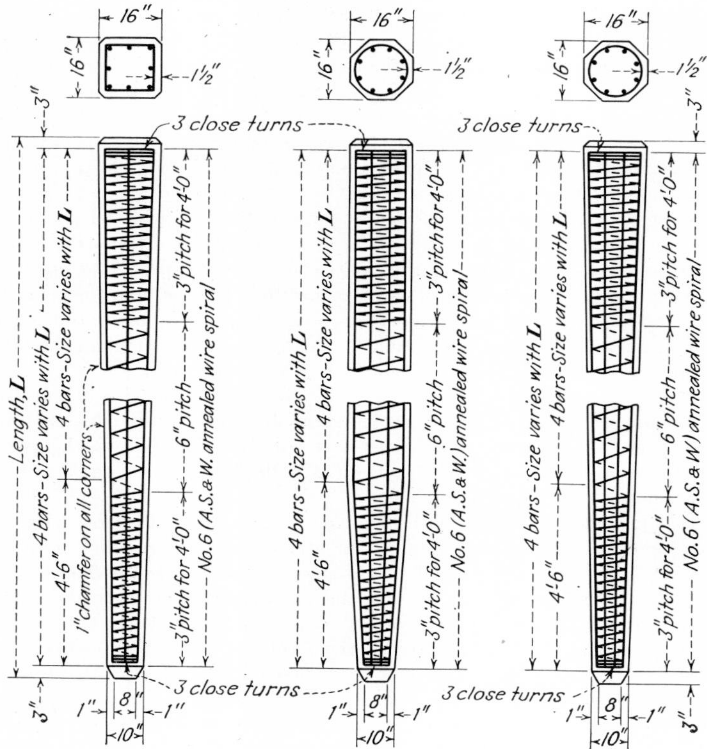

FIG. 30. Precast-concrete piles.

for the length of a pile. The sections, thus fitted together, are then prestressed by posttensioning wire cables that run through open holes cast longitudinally in the walls of each section. Special, but positive, methods have been developed for joining the sections, posttensioning the wire cables, and grouting the cables so as to form a homogeneous product of great strength and durability. The characteristics of such a pile permit combined axial loads and bending moments of considerable magnitude. The large area in contact with the soil provides greater resistance and, therefore, economy. Where underlying soil conditions permit, design loads from 200 to 300 tons per pile are feasible. The centrifugal casting of the concrete and the prestressing and reinforcing of the sections produce many inherent qualities that ensure long life with low maintenance and upkeep costs, so that such piles have been used to a substantial extent, particularly in off-shore structures such as oil-well platforms, bridges, and the like. Figure 31 shows these piles in a typical off-shore structure.

Composite piles came into general use 25 or 30 years ago. The purpose of the composite pile is to combine the low-cost advantage of the wood pile with a concrete pile that does not have to be cut off below the ground-water level. Figure 32 shows how the composite pile is used to avoid the large amounts of excavation, sheeting, and concrete that are often involved in the use of plain wood piles.

For the composite pile, the lower wood section is driven down until its upper end is about at the ground surface. A concrete pile is then used as a follower to drive the head

of the wood pile down below the groundwater level, and the upper concrete section extends from the wood pile up to the cutoff grade. Such piles are considerably cheaper than all-concrete piles of the same length, and they eliminate the excavation, sheeting, and footing concrete that would be required if plain wood piles were used. All the various types of concrete piles, either precast or cast-in-place, can be and have been used as the upper sections of composite piles.

The most important element of the composite pile is the joint between the lower wood and the upper concrete sections. The principal requirements for a satisfactory joint are: (1) to keep out mud and water, (2) to resist tension, and (3) to resist any active bending forces. In addition, the joint must be so designed that it can be made easily in the field without requiring too much time or costing too much money.

In order to keep out mud and water and to ensure good solid bearing between the two parts of the pile, it is necessary to have a well-sealed joint. The amount of tensile strength that the joint must have will depend on the magnitude of the tensile forces that might be applied to the pile after it is in place in the ground. Some soils heave during pile-driving operations, and when composite piles are used in such soils, the joint must have sufficient tensile strength to keep the concrete sections from being lifted off the wood sections by the heaving soil. Some types of structures place actual uplift forces in the piles on which they are supported, and in such cases, composite pile joints must be designed to resist these uplift forces.

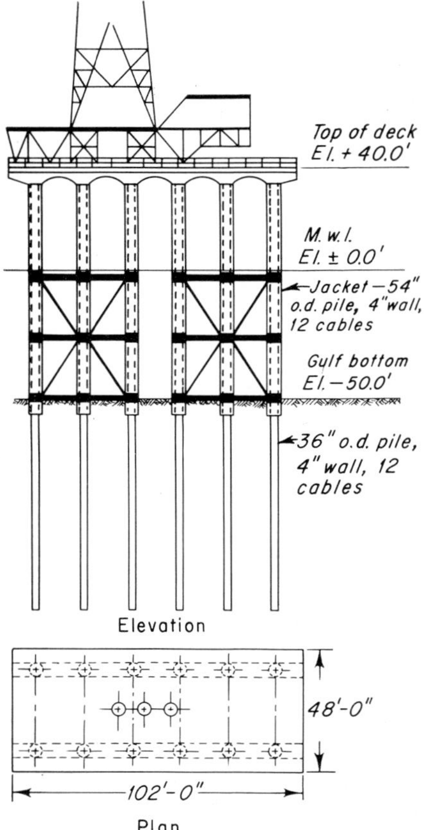

Fig. 31. Open-sea oil-drilling platform.

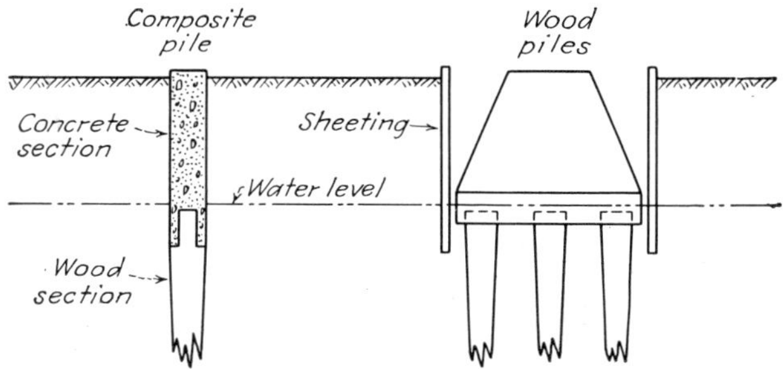

Fig. 32. Wood and concrete composite pile.

In many cases, the pile is never subjected to tensile forces, either during the pile-driving operation or after the pile is in service under the structure. In such cases, only a nominal amount of tensile strength is required in a composite pile joint.

As far as bending resistance is concerned, piles are not usually subjected to bending stresses except under certain conditions that are usually due to the nature of the structure the piles are to support. The foundation of a heavy reciprocating machine would tend to sway back and forth and to put horizontal forces in the pile heads. Structures subjected to large external horizontal forces due to wind or water loads also put horizontal force components into the heads of the piles on which they may be supported. In such cases, composite pile joints should have sufficient bending resistance to enable them to withstand the forces to which they will be subjected. When the composite pile is subjected only to a direct axial load, only a nominal amount of bending resistance is necessary.

Figure 33 shows five different kinds of composite pile joints between a wood section and a cast-in-place concrete section formed in a corrugated-steel shell. In joint (*a*) the head of the concrete section has been trimmed in the form of a tenon, which is usually about 9 in. in diameter and about 18 in. long. A steel sealing ring is welded to the bottom of the

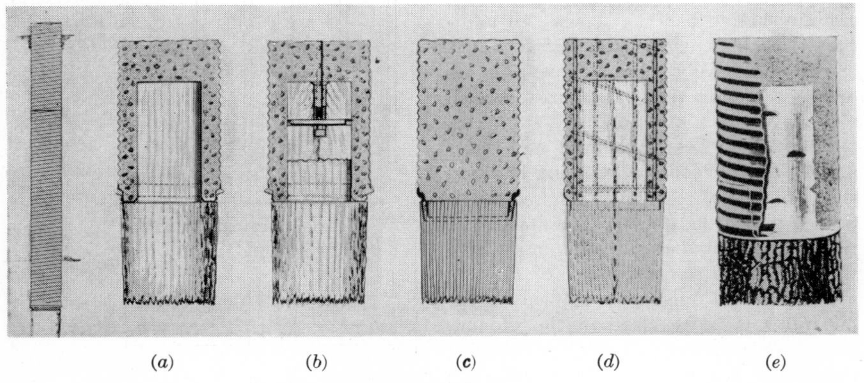

(*a*) (*b*) (*c*) (*d*) (*e*)

FIG. 33. Composite pile joints.

shell, and the inside diameter of this ring is slightly smaller than the diameter of the tenon so that the ring has to be forced down around the tenon with considerable pressure. The ring serves a double purpose in that it seals the joint against mud and water, and it also provides tensile resistance since the inner edge of the ring grips the wood of the tenon when an effort is made to lift the ring off the shoulder of the wood section. Such a composite pile joint is sealed against mud and water; it has some tensile strength owing to the action of the ring and to the bond between the concrete and the tenon; and it has some bending strength because of the projection of the wood-section tenon into the concrete section. In order to increase the tensile resistance joint (*a*) is sometimes modified as shown at (*b*). With the (*b*) joint, the two sections of pile are locked together by means of a steel cross pin which passes through a steel socket into which a reinforcing rod with a threaded end can be screwed. The reinforcing rod extends to the top of the concrete section. The bending resistance of the joint may be increased by means of reinforcing steel as shown at (*d*). It is possible to combine joints (*b*) and (*d*) so as to have high resistance to both bending and tension. Joint (*c*) is less complicated than the others. The sealing of the joint is accomplished by forcing the lower end of the shell into the head of the wood section and then expanding the wood section head by means

of a wedge ring so that the lower end of the shell is gripped tightly by the wood. There are other types of composite pile joints in commercial use, and the choice of the kind of joint that would be required on a given job should be based on the conditions under which the piles are to be used.

Steel piles, usually used for bearing piles, employ what is known as the H or wide flange section. This type of pile has proved to be especially useful for trestle structures in which the pile extends above the ground line and serves not only as a pile but also as a column. A structure of this type on a steel-bearing-pile foundation is illustrated in Fig. 34.

Because of the small cross-sectional area of the steel in these piles, they can often be driven into river beds of sand and gravel into which it would be very difficult to drive a

FIG. 34. Steel H-beam piles. (*Carnegie-Illinois Steel Corporation.*)

displacement pile with a solid cross section of considerable area even with the aid of jetting. However, this ability of the steel pile to penetrate into dense materials often works to its disadvantage in other soil conditions. In many cases bearing piles are used to support loads simply by friction between the pile and the surrounding soil. In other cases, piles are used primarily for the compaction of loose soils. Under such conditions, particularly when compaction is the principal purpose of the pile, considerably longer steel piles are required to carry the same load that can be supported by a comparatively short displacement pile. This is due to the fact that the steel pile with its small cross-sectional area produces very little compaction.

Where the steel pile is exposed to air, or where it is alternately wet or dry, it is subject to the same corrosive action that would occur to any other steel structure under the same conditions. In some installations, steel piles are protected from corrosion by painting with an asphalt compound. Sometimes in order to protect the pile from corrosion at the ground line, it is encased in concrete for a distance of several feet both above and below the ground. Concrete protection of this kind is illustrated in Fig. 34.

Functions and Uses of Piles. One of the principal uses of piling is to transmit loads through soft or unstable surface soils to harder more stable soils below. Sometimes

piles are used simply for the purpose of compacting a bed of loose sand so as to give it a greater density and a higher bearing capacity.

Some types of structures are subjected to uplift force from various causes. The transmission tower illustrated in Fig. 35 might be loaded in such a way that it would try to overturn. Such towers are always subjected to a cable pull and to wind forces. If the cables on one side of the tower should break and if the structure were subject to a heavy wind force at the same time in the same direction as the unbalanced cable pull, there would be uplift forces of considerable magnitude on the windward side. Foundations for structures of this kind are often fastened to the ground with piles whose principal purpose is to resist uplift.

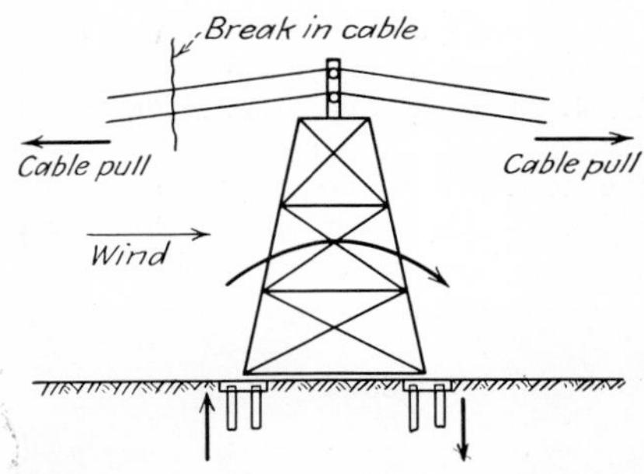

Fig. 35. Piles subjected to uplift.

The danger of scour in a stream bed may often require the use of pile foundations under a bridge pier even though the soil in the stream bed might be entirely capable of supporting the pier load without piles. A case of this kind is illustrated in Fig. 36. In regions where there is danger of quick floods and heavy scour, a bridge pier on a bed of sand and gravel should be protected against possible scour either with foundation piles under the pier or with a cofferdam of sheet piles which should be carried deep enough so that there would be sufficient support under the pier following the maximum anticipated scour in the stream bed.

In the construction of docks, piers, and other marine structure, piles are used as anchor piles, as fender piles, and for pile clusters or dolphins. Figure 37*a* shows a cross section of a dock which would make use of these various kinds of piles. The anchor

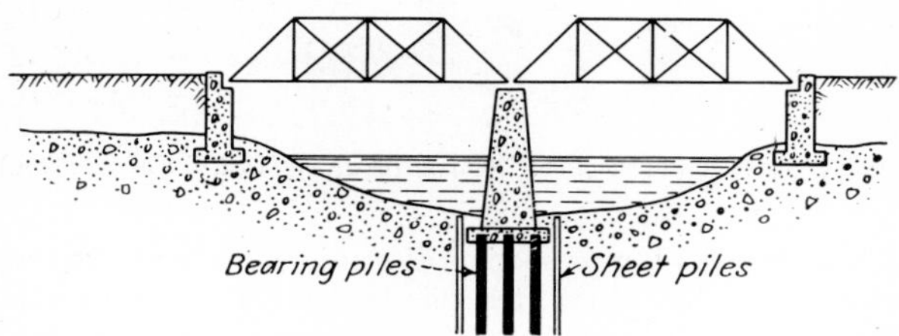

Fig. 36. Piles to protect against scour.

piles might be wood piles or concrete piles or steel piles. The fender piles would ordinarily be wood piles used to protect the concrete dock from abrasion due to ships or barges that might tie up at the dock. The fender piles are usually 8 or 10 ft on centers and are capped with a continuous 12 × 12 timber. The piles used for dolphins are almost invariably wooden piles and usually of white oak. Figure 37*b* is a rear view of such a dock under construction. The anchorages and ties are shown in the photograph, which was taken before the backfill was placed. Figure 37*c* shows a front view of the same dock. The fender piles have not yet been installed, but the bolts for holding them in place against the face of the dock can be seen in the picture.

One other purpose for which piles are often used is for falsework during the construction of heavy structures. The centering for a reinforced concrete arch bridge is often carried on wooden piles driven for temporary support. In Fig. 37*c* the head of a wooden

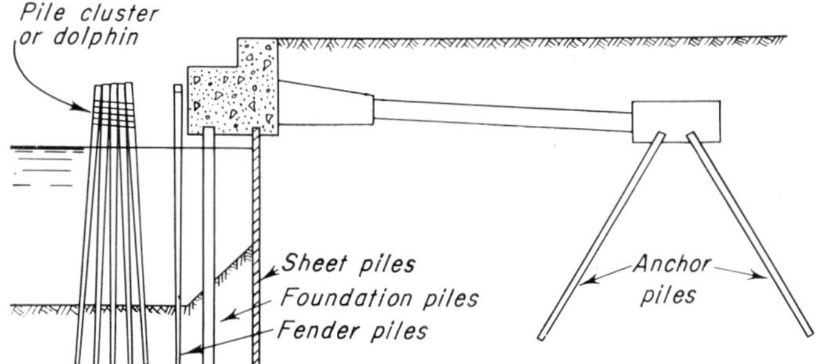

FIG. 37*a*. Cross section of typical bulkhead.

FIG. 37*b*. Rear view of dock.

FIG. 37*c*. Front view of dock.

pile can be seen projecting above the water in each bay of the dock. These are the heads of falsework piles that were driven to support the forms for the concrete in the dock wall.

Bearing Capacity of Piles and Effects of Pile Driving. Before any piles are driven in a given soil, that soil has certain physical properties which may be determined by a suitable investigation. Pile driving changes these physical properties in various ways, and the nature of the changes depends largely on the character of the soil. One of the most common phenomena that occur during pile driving is the compaction of the soil into which the piles are driven. In cohesionless granular soils such as sand and gravel, pile driving forces the soil grains closer together so that the entire soil mass acquires a higher degree of density than it had previously. This compaction takes place in granular soils regardless of whether they are dry or saturated. These soils have relatively large pore spaces so that when the soil grains are displaced the water can escape freely through the voids in the soil.

In fine-grained cohesive soils, such as silt and clay, compaction due to pile driving takes place only when these soils are not completely saturated. However, when fine-grained cohesive soils are submerged below the water table and are completely saturated, the effects produced by pile driving are quite complicated. The pore spaces in such soils are of capillary dimensions. The movement of water through these capillary channels is a very slow process even when the soil mass is subjected to heavy pressure. Accordingly, when a fine-grained cohesive soil is completely saturated, relatively long periods of time are required to force water out of the soil and to produce volume changes in the soil mass. Such soils are practically incompressible in short periods of time. Ordinarily, the actual driving of a pile is done in 10 or 15 min. Therefore, when a pile is driven into a completely saturated fine-grained cohesive soil, the driving time is too short to permit much water to escape from the soil. The result is that very little compaction takes place and the soil is simply displaced in the direction of least resistance—usually upward. The heaving of saturated clays and silts during pile-driving operations is well known.

This behavior of saturated fine-grained cohesive soils during pile driving has led to the development of a theory about the remolding effect of pile driving in such soils. In its most extreme form, this theory says that these soils are completely remolded by pile driving. Laboratory tests on completely remolded soil show that the soil structure is apparently destroyed and that the strength of the soil is usually reduced. The amount of reduction in strength between undisturbed soils and remolded soils varies widely with different kinds of soils. Some soils are very sensitive to remolding while others lose only a small percentage of their undisturbed strength when they are remolded.

The laboratory remolding operation is done by breaking up the soil sample and kneading it with the hands until it becomes a soft plastic mass. In the pile-driving operation whatever remolding takes place must be done under conditions considerably different from those used in the laboratory. The soil at any depth in the ground is subjected to pressure due to the weight of the soil above it. If the driving of the pile remolds the soil alongside of the pile, this remolding would necessarily have to be done under pressure. No laboratory technique has been developed for remolding soil under pressure in this manner, and it is therefore not known just how soil subjected to this action would behave. However, recent investigations in the field have shown that the remolding effect of pile driving in saturated clay beds is not so serious as was originally believed.[1]

In order to avoid the remolding effect in silt and clay soils into or through which piles are driven and to eliminate the heave effect of the pile displacement, there has recently

[1] Cummings, A. E., G. O. Kerkhoff, and R. B. Peck, Effect of Driving Piles into Soft Clay, *Proc. ASCE*, vol. 74, No. 10, p. 1553, December, 1948.

been developed by the Raymond Concrete Pile Company in Detroit a method called wet rotary pre-excavation which produces, in advance of driving a pile, a hole to receive the pile which is of substantially the same shape and cross section as the pile and is

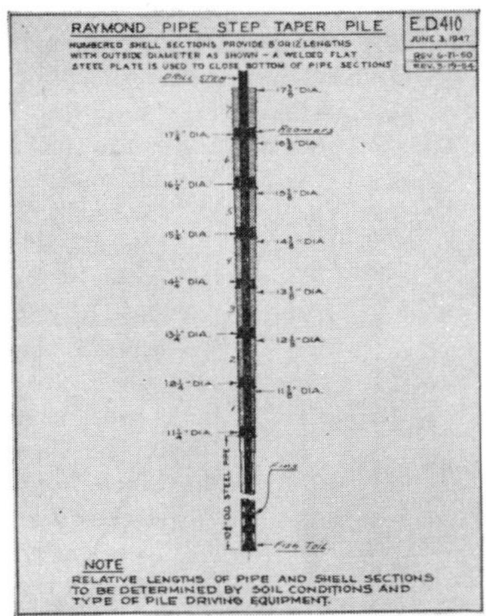

Fig. 38. Wet rotary pre-excavation of piles.

carried to substantially the same depth as the pile is expected to drive. The pile is then lowered into the hole and driven home. This is generally illustrated in Fig. 38. Water or slurry is pumped through hose to the top of the pre-excavator drill stem and down through the drill stem where it emerges from ports in the drill bit where it mixes with the cuttings. The upward velocity of the fluid returning outside the drill stem through the hole carries the cuttings with it and clears the hole of cuttings. Any strata which would normally cave in an open hole are kept in place by the hydraulic pressure exerted

by the fluid in the hole; and, under the influence of that fluid pressure, voids in the side-wall material are filled with the slurry material, thus providing relatively impervious "wall cake" against which the fluid pressure reacts.

This wet rotary pre-excavation method of installing piles has been used to a considerable extent in Detroit, Mich., where the engineering results above mentioned have been fully and dependably demonstrated.

Figure 39 illustrates the distribution of stresses from the pile to the surrounding soil after the pile has been driven into the ground and is subjected to a static load. Part of the load goes into the soil in the form of tangential stresses t along the sides of the pile. These stresses are sometimes called "friction" and sometimes "shear." The use of the word friction conveys the idea that the pile would try to slide through the soil, but this probably does not give a proper idea of what happens, especially in the case of cohesive soil. Whenever piles are pulled out of the ground, they almost invariably come up with a thin film of soil sticking to the sides of the pile. This indicates that the adhesion between the pile and the soil is greater than the cohesion or the shearing strength of the soil at a relatively short distance from the pile surface. In such cases, the tangential stresses t are determined by the shearing resistance of the soil. In a very clean sand it is possible that the pile would slide through the sand so that in clean sands these tangential stresses might be referred to as friction.

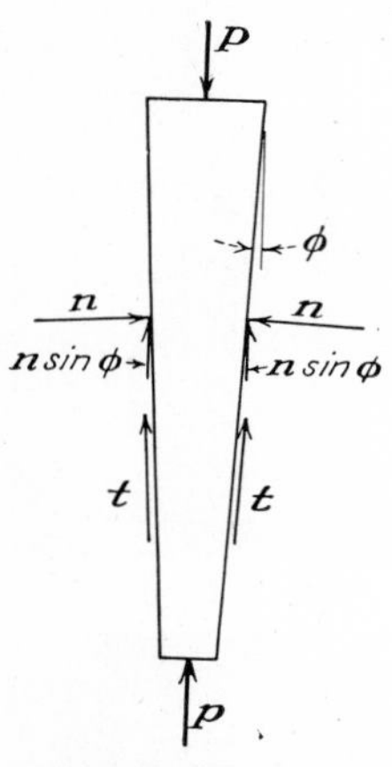

FIG. 39. Load distribution from pile to soil.

Another part of the pile load goes into the soil at the point of the pile. This is a direct bearing pressure p on the soil below the pile point. If the pile is tapered, the normal pressure n on the sides of the pile has a vertical component $n \sin \phi$, where ϕ is the angle of taper of the pile. Many tests on full-sized piles, as well as on models, have demonstrated that the bearing capacity of the pile is considerably increased because of the effect of the taper. Full-sized load tests made on precast-concrete piles that were parallel-sided and on precast-concrete piles that were tapered at the rate of 1 in. in 4 ft, have indicated that the tapered piles were able to carry 25 percent more load per square foot of embedded surface than the parallel-sided piles, the settlements in each case being equal.

In the general case, the equation of equilibrium for the pile shown in Fig. 39 would be

$$P = pA + tS + Sn \sin \phi$$

in which A is the area of the pile point, S is the total embedded surface of the pile, and p, t, and n are forces per unit of area. If the pile is parallel-sided, ϕ and $\sin \phi$ are zero, so that only the t and p forces exist. The calculation of the magnitudes of these different components of earth resistance is a statically indeterminate problem. It is not possible by the methods of static analysis to determine how much of the load is carried by the point of the pile and how much is carried by the side. In addition, it is not known just how the tangential and normal forces are distributed along the length of the pile. Accordingly, the problem of determining the carrying capacity cannot be readily solved by mathematical analysis.

Piles are usually classified as friction piles or point-bearing piles depending on whether they derive their principal support from friction or shear along their sides or whether they are supported almost entirely at their points by direct bearing on hardpan, gravel or rock.

Load Tests on Piles. One method for determining the bearing capacity of piles is by means of a static-load test. The manner in which the test is usually made is illustrated in Fig. 40. A suitable platform is balanced on the test pile and the edges of the platform are supported by temporary cribs and wedges to prevent the platform from tilting. The size of the members required to support the platform will depend on the amount of the test load. For loads up to about 100 tons, a pair of 15- or 18-in. I beams about 15 ft long are satisfactory for the bottom support and six or eight 12-in. by 12-in. timbers 15 ft long will support the platform itself. When the loading is done with sand

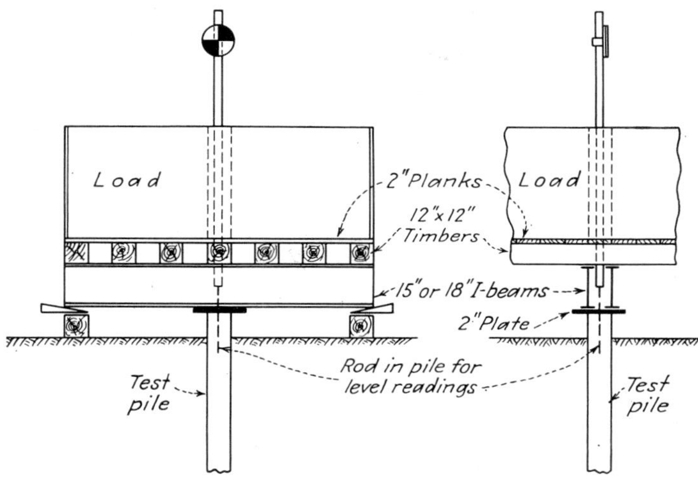

Fig. 40. Pile-load-test platform.

or other bulk material side walls must be built on top of the platform and securely tied together. If pig iron or some similar material is used for loading, no box is necessary because the load is simply piled up on the platform.

Before the platform is placed in position, the pile head is cut off level and a steel plate about 2 in. thick with a 2-in.-diameter hole in the center is placed on the pile head. In order that settlement readings may be taken with a level rod, a short rod or bolt is set in a vertical position in the pile head. If the test is being made on a wooden pile, it is customary to use a lag screw for this purpose. For a concrete-pile test, the usual procedure is to embed a short piece of reinforcing rod in the pile head.

A hole must be left in the platform directly above the pile head so that the level rod can set on the bolt in the pile head. If the loading is being done with bulk material, such as sand, a metal tube or a wooden box must be placed upright in the center of the platform to permit the level rod to be passed through the bulk material down to the pile head.

The load is applied in increments of about 10 tons each, and time is allowed between load increments for the pile to come to rest. Immediately before and immediately after each load increment is applied, careful level readings are made to determine the elevation of the pile head. At least two solid bench marks should be established within 50 or 100 ft of the test pile, and the level readings on the pile head should be referred to these bench marks. When the test is completed and the pile is being unloaded, level readings are taken to determine the amount of rebound.

It is customary in building codes to require a test load equal to twice the proposed working load. The pile must support the full test load for at least 24 hr without settlement, and the maximum allowable settlement after deduction of rebound must not exceed 0.01 in. per ton of test load.

The loads and settlements of the pile are measured and recorded, and the results are plotted in the form of a load-settlement curve as shown in Fig. 41. Load tests usually produce a settlement curve and a rebound curve such as is illustrated by the solid line in the figure. From the results of such a test, it will be known that the pile will have a safety factor of 2 if a working load amounting to P is applied to the pile after a successful load test up to a load of $2P$ has been made. The information that cannot be obtained from such a test is the failure load of the pile. Furthermore, the real factor of safety is unknown when the working load P is applied. All that is known about the safety factor is that it is at least 2.

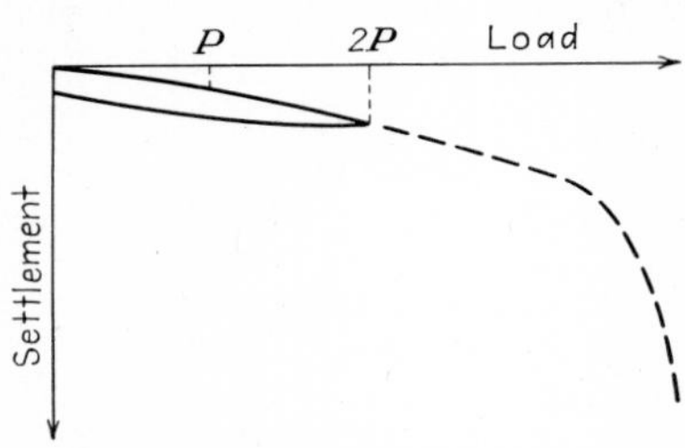

FIG. 41. Load-settlement curve from pile test.

The dotted curve in Fig. 41 shows what the load-settlement curve might look like if the test load were increased above $2P$. The test load might be carried up to $3P$ or $4P$ before failure would occur, and the pile would settle continuously without the addition of any more load. A large percentage of the pile-load tests that have been made in the past were made on the basis of applying twice the working load. Such a test determines the fact that the safety factor is at least 2, although it might be as much as 3 or 4. It takes extra time and costs extra money to carry a load test to failure, but it is worth while to do this whenever possible. ASTM Specification D1143-50T provides a satisfactory guide to making such pile-load tests.

Section 739.3 of the City of Detroit Official Building Code represents the most modern requirements for determining the design pile loading from pile-load tests; and that section of the code is here reproduced.

"City of Detroit Official Building Code

"739.3. Approved Test Load. When the safe bearing value of any pile is in doubt, or when greater loads per pile than permitted by section 739.2 are desired in the absence of load test data satisfactory to the building official, at least one (1) test pile shall be driven in each area of uniform foundation materials and this pile shall be load tested in accordance with the procedure for such tests cited in Appendix D. The total test load shall be at least twice the design load and at least sufficient to provide one of the following three determinations which is the most representative under the existing conditions in the opinion of the building official.

"(1) The yield value or load at which progressive settlement occurs, as determined by extrapolation of the rates of settlement for each load increment.

"(2) The load carried at a settlement equal to the measured elastic deformation or rebound.

"(3) The repetitive load at which there is no further permanent deformation or at which the rebound approaches one hundred (100) percent of the settlement measured for that load. Repetitive loading cycles will usually be required only under conditions in which the determinations under (1) and (2) are inadequate.

"Subject to the provisions of section 739.4, the design load shall be not more than one-half (½) of the capacity of the pile resulting from the load test and defined by the selected one or by the average of the three determinations named above.

"In the subsequent driving of the balance of the foundation piles in such area, and subject to the provisions of section 739.4, all piles shall be deemed to have a supporting capacity equal to the test pile when the rate of penetration of such piles is equal to or less than that of the test pile through a comparable driving distance. When the test pile is one of a group of piles driven through materials subject to displacement or shift, the provisions of section 737.91 shall have been met.

"Note: The procedure to be prescribed for pile load tests in Appendix D will be ASTM Specification D1143-50T supplemented by the following provisions:

"Vertical load shall be applied to the pile by means of a hydraulic jack equipped with a recently calibrated pressure gage, the reaction load being supplied in accordance with paragraphs 3(b) or 3(c).

"Settlements shall be measured by a dial gage in accordance with paragraph 4(c).

"The application of load increments shall follow the alternate method specified in paragraph 5(b), which provides for load increments at constant time intervals.

"The following shall be added to paragraph 5(c): 'During the unloading cycle, the specified load decrements shall be applied at constant time intervals corresponding to those used in the loading cycle, including a final rebound measurement after twenty-four (24) hours at zero load.' "

To determine the three criteria of pile capacity resulting from the load test, the test results are usually plotted as shown in Fig. 42.

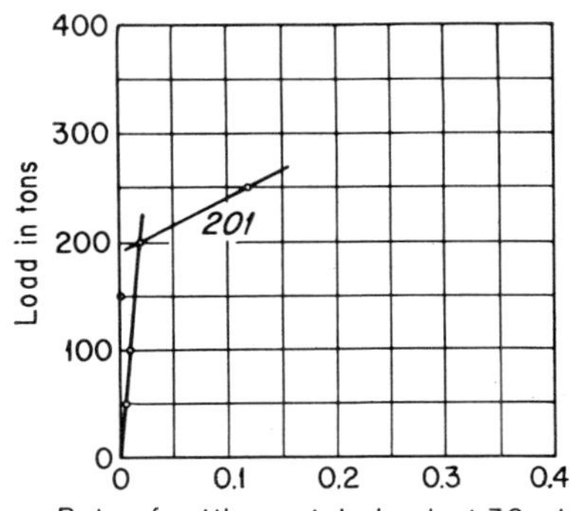

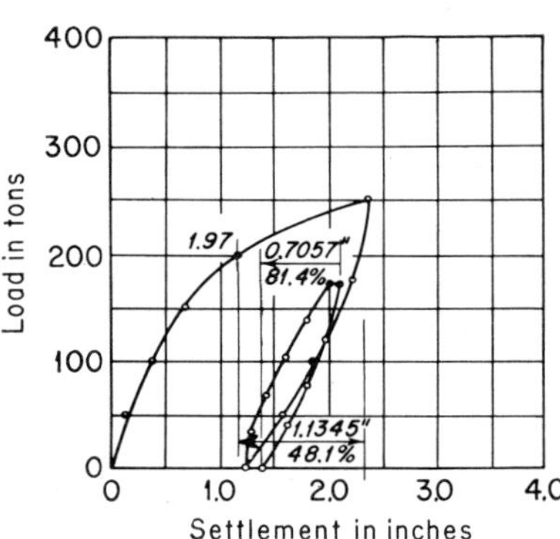

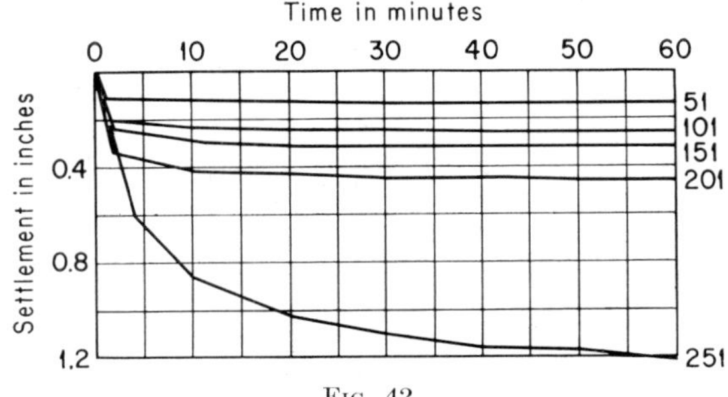

Fig. 42

Single Piles and Pile Groups. When a single pile is driven into the ground and subjected to an external load, the vertical pressure at the plane of the pile point is approximately as shown in Fig. 43*a*. The maximum pressure occurs under the pile point, and the pressure decreases as the radial distance from the pile increases. Figure 43*b* represents a group of piles with the pressure distribution that probably exists under such a group. The pressure under the pile group is compounded by simple superposition of the stress distributions under each individual pile. The result is a large dome-shaped stress surface with the maximum stress under the center pile. The pressure under the center pile of the group would be considerably greater than the pressure under a single pile.

If this principle of superposition of stresses were valid, and if the settlements of the various piles of the group were proportional to the stresses immediately below the pile points, the center pile in Fig. 43*b* would have to settle considerably more than the outside piles. Most pile groups of this kind are capped with a reinforced concrete footing which is practically rigid in comparison with the soil surrounding the piles. The result is that the pile heads are restrained so that the settlement of all the piles of the group would have to be about the same. This restraint causes a readjustment of the pressures under the points of the piles in the group. The effect of the rigid footing is to reduce the pressure under the center pile and to increase the pressure under the outside piles so as to equalize the soil pressures below the pile group. The maximum pressure under the group is greater than that under the single pile but is not necessarily so much greater as is indicated in Fig. 43.

The same principle applies to an entire structure which might be supported on a whole series of pile groups. A one-story factory building as illustrated in Fig. 44*a* would be comparatively flexible, and it might be supported on a pile foundation with columns 15 or 20 ft on centers. The compounding of stresses in the underground from the various pile groups would indicate that such a building should settle in the form of a saucer-shaped depression with the maximum settlement in the center of the building. Because of the flexibility of the structure, settlement of this type would actually take place, since the superstructure would bend comparatively easily to fit the deflection of the soil. Figure 44*b* represents a reinforced concrete grain elevator which is practically rigid. Such a structure would not be able to bend and settle in the form of a saucer-shaped depression. Because of its rigidity, a structure of this kind would have to settle as a unit with the underside of the foundation remaining practically plane. There would therefore be a redistribution of pressure under the pile points with a tendency to relieve the pressure under the middle piles and to increase the pressure on the outside piles. The nature of the settlement is not controlled entirely by the soil or the piles. The rigidity or lack of rigidity of the structure has an important influence on the character of the settlement.

Various theoretical rules have been proposed for use in the design of pile groups. Some are intended to control the spacing of the piles while others provide for reductions of pile loads depending on the size of the group. None of these rules is entirely satisfactory and in many cases, especially for large raft foundations, they give results that are often contrary to experience and sometimes absurd. Even with pile groups of ordinary size, such as 20 or 25 piles, the efficiency formulas for pile spacings or load reductions lead to results that conflict with methods which have been used successfully for many years. Building codes usually require a center-to-center spacing of twice the average pile diameter with a minimum spacing of 30 in. Under some conditions it is desirable to increase the center-to-center spacing of the piles, but spacings greater than 3 ft for ordinary footings or greater than 4 to 5 ft for raft foundations are rarely used. The purpose of the theoretical formulas for increased spacing or for pile-load reductions is to reduce settlements when the piles are of the type which derive their

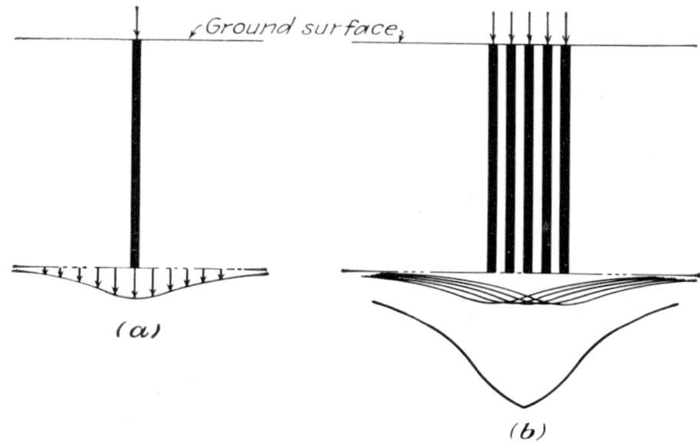

FIG. 43. Pressure distribution below pile points.

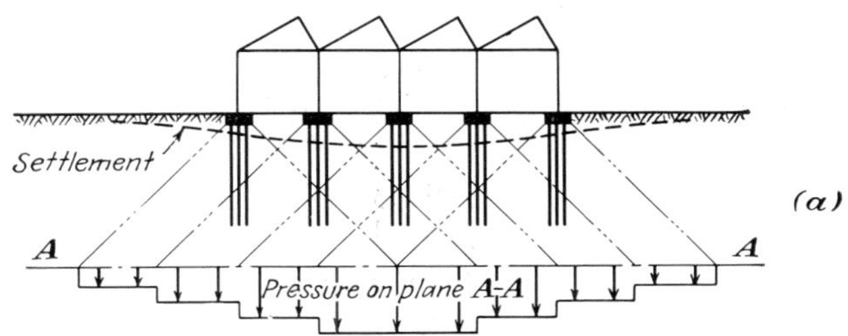

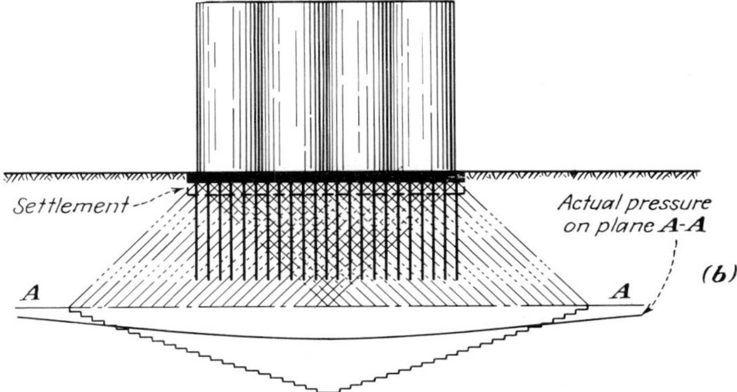

FIG. 44. Effect of stiffness of superstructure.

principal capacity from friction (*friction piles*) instead of the type which derive their principal capacity from the penetration of their points to or into hard ground (*point bearing*) or when the pile points are underlain with strata of soft soil. However, the practical and economical solution for such soil conditions is to increase the pile lengths and drive the piles down until the pile points reach hard ground not underlain with strata of softer soil.

Horizontal Forces and Batter Piles. Sometimes pile foundations are used under a structure that is subjected to external horizontal forces, and such foundations must be

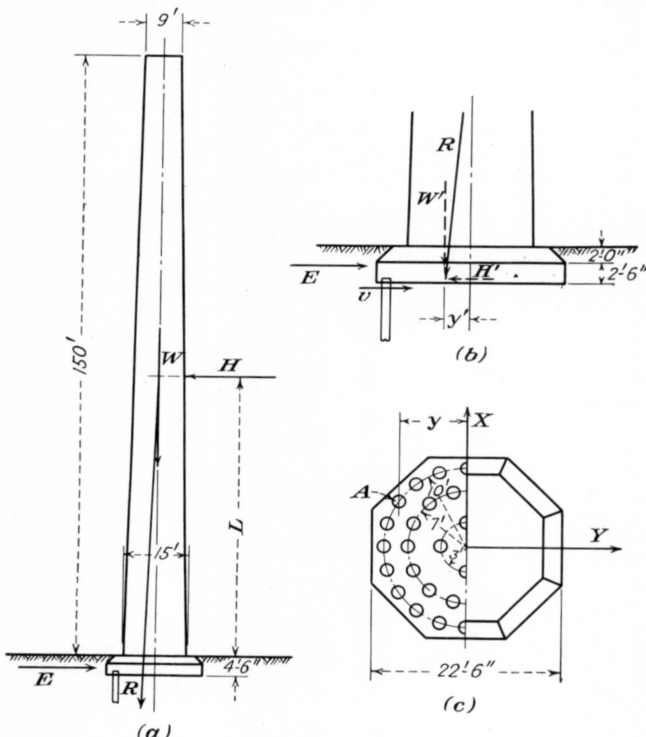

Fig. 45. Pile foundation subjected to horizontal force.

designed to resist overturning and sliding. A structure of this kind is the tall brick chimney illustrated in Fig. 45. The overall dimensions of the chimney are shown in the figure, and the chimney rests on a reinforced concrete foundation supported by piling. The maximum allowable load per pile is assumed to be 25 tons.

Such a chimney with its foundation would weigh approximately 500 tons. Under full wind load a horizontal force H is applied to the center of gravity of the side elevation of the chimney shaft. The total wind force, at a unit pressure of 22.5 lb per sq ft on the projected area of the chimney, would amount to 20.25 tons. The projected area of the chimney is a trapezoid, and the lever arm L of the resultant wind pressure is 68.8 ft measured from the base of the chimney shaft. Since the chimney is on a concrete mat 4 ft 6 in. thick, the total lever arm of the wind pressure above the pile heads is 73.3 ft and the overturning moment M is 1,485 ft-tons. The horizontal wind force is combined

with the dead load into a result force R, which cuts the base of the chimney footing at a distance of 2.97 ft from the vertical axis of the chimney shaft.

The number and arrangement of piles under such a chimney foundation is determined by a cut-and-try process. Ordinarily, the dead weight of the chimney accounts for about half of the maximum load on the piles which are usually arranged in a symmetrical array, either on concentric circles as shown in Fig. 45c or in some symmetrical arrangement in each of the eight segments of the octagon.

In order to calculate the maximum pile load that occurs under full wind load, it is necessary to determine the moment of inertia of the pile group. This is done by laying out a tentative arrangement of piles and then scaling the distance from each pile to the axis of the foundation. Referring to Fig. 45c, the distance of pile A from the X axis is represented by y, and the contribution of pile A to the moment of inertia about X-X is one pile multiplied by the square of the distance y. The total moment of inertia about X-X is found by summing up the total number of piles each multiplied by the square of its distance from the X axis.

For the pile arrangement shown in Fig. 45c, it is possible to compute the moment of inertia of the pile group with sufficient accuracy by a somewhat easier process. For a symmetrical figure the polar moment of inertia about an axis through the center of the figure and perpendicular to the plane of the paper is equal to the sum of the moments of inertia about the X and Y axes. The pile arrangement in Fig. 45c is not perfectly symmetrical with respect to the X and Y axes since there are four piles centered on the X axis, whereas there are six piles on the Y axis. However, the calculation will be sufficiently accurate if the polar moment of inertia is calculated and then divided in half to give the moment of inertia about the X axis. The computation of the polar moment of inertia is as follows:

4 piles at	$3' \times 3'$ =	36
14 piles at	$7' \times 7'$ =	686
20 piles at	$10' \times 10'$ =	2,000
38 piles	=	2,722

The moment of inertia of the pile group about the X axis is then approximately half of the polar moment of inertia, or 1,361.

For the pile-load calculation it is necessary to use the section modulus S of the pile group, which is simply the moment of inertia divided by the distance to the outermost pile. This is analogous to the case of a beam subjected to bending in which the moment of inertia of the cross section of the beam is divided by the distance to the extreme fiber in order to get the section modulus.

The maximum pile load W is determined by the formula

$$Q = \frac{W}{N} \pm \frac{M}{S}$$

It is seen from this equation that the piles get load from two sources. One source is the dead weight of the chimney W. The other source is the overturning moment M due to wind. The weight W is divided by the number of piles N, and the wind moment M is divided by the section modulus S of the pile group. The use of the plus sign in the equation gives the maximum load on the outermost pile on the leeward side. The minus sign gives the minimum load on the outermost pile on the windward side. For this chimney, the maximum pile load is +24.1 tons and the minimum pile load is +2.3 tons. Both are positive quantities, which indicates that the piles are in compression on both sides of the chimney under full wind load and that there is no uplift on the windward

side. This pile-load calculation shows that it is possible to determine the approximate number of piles by dividing the weight W by half of the allowable pile load. The two terms on the right-hand side of the pile-load equation are approximately equal. If they were exactly equal the minimum load on the windward side would be zero. The pile design is always worked out in such a way as to leave a small positive load on the outermost pile on the windward side.

Figure 45*b* is an enlarged sketch of the base of the chimney and the chimney footing. The resultant R cuts the base at a distance y' from the vertical axis. This resultant can be resolved into two components, W' and H', which are applied at the plane of the pile heads and are equal to W and H respectively. The eccentricity of W' accounts for the difference in pile loads on the two opposite sides of the chimney. The force H' is an external horizontal force produced by the wind, and such an external force has to be resisted by an external reaction. In the case of this chimney foundation, some external horizontal reaction might be furnished by the force E, which is the earth resistance. However, the earth around the chimney might be loose backfill, which could not be counted on to provide any reaction. In such a case all the external force H' would have to be resisted by shear v in the pile heads. Since the force H' is equal to H, or 20.25 tons, and since there are 38 piles in the chimney foundation, the average horizontal shear in the pile heads is about 1,065 lb per pile. This is not a very heavy load to apply to the head of a pile that is completely embedded in the ground. Almost any kind of pile could safely resist that much horizontal thrust at the pile head.

The important thing in connection with the design of this chimney foundation is the fact that the design is controlled almost entirely by the overturning moment due to the wind. The shear in the pile heads is not very large because of the number of piles that are required to resist the overturning moment. The design is similar to that of a long beam in which the design is controlled largely by the bending moment rather than by the shear. The chimney foundation is comparatively safe against sliding when a sufficient number of piles is used to resist the overturning moment.

Because wind forces (especially those resulting from high winds or gusts) are of short duration, building codes usually give them special treatment by providing that, for combined stresses due to wind load together with dead and live and snow loads, the allowable working stress for the structural material in the frame of a building or a structure may be increased 33⅓ percent, with the provision that the resulting section shall be compared with that required for dead and live and snow loads only and the section of greatest strength used; and further that, when the stress due to wind is less than one-third of the stress due to dead plus live plus snow loads, the wind stress may be neglected. Dependent on circumstances (which should have thorough consideration), something of the same sort can often be done in the design of foundations.

However, there are other types of structures that are subjected to horizontal forces in which the situation with respect to overturning and sliding is reversed from those in a chimney foundation. Such a structure is illustrated in Fig. 46. It consists of a dam and a pair of locks built in a river bed, as shown in Fig. 46*a*. The middle wall between the locks would be subjected to a large horizontal thrust if Lock 1 were filled with water to the level of the upper pool while Lock 2 was filled with water only to the level of the lower pool. A cross section of the middle wall (section A-A) is shown in Fig. 46*c*, and a plan of the wall is shown in Fig. 46*b*. The dimensions of the structure are shown in the figures.

The structure is subjected to a thrust P from the upper pool and a much smaller thrust P' from the lower pool. The difference between these is a horizontal thrust H, tending to overturn the wall and push it toward the left. The arrangement of the piling under the wall is shown in Fig. 46*b*. There are seven parallel lines of piles spaced 3 ft on centers each way. The maximum allowable load per pile is assumed to be 30 tons.

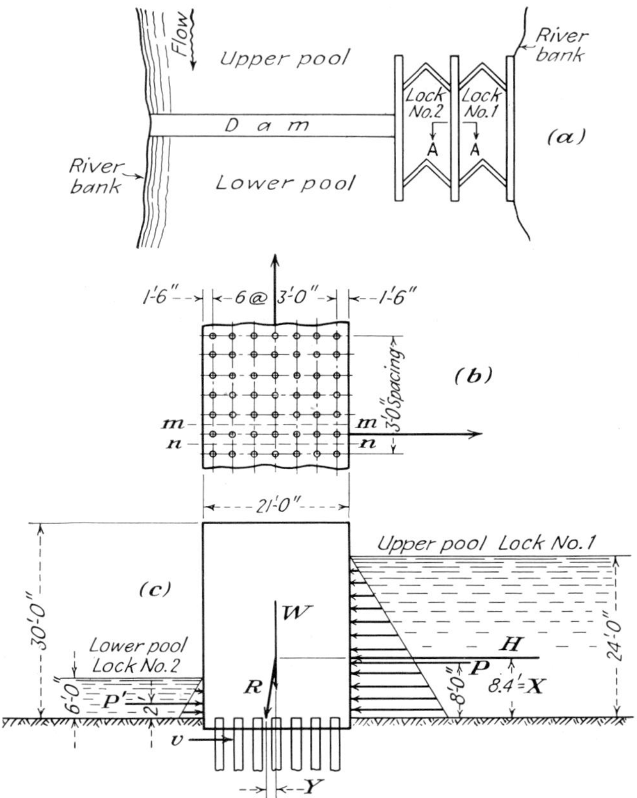

Fig. 46. Pile foundation subjected to horizontal force.

Since the rows of piles across the wall are on 3-ft centers, it is necessary to compute the pile loads on the basis of a slice of wall that is 3 ft thick. Each 3 ft of wall is supported by a row of seven piles, as is represented by the section *mm-nn* in Fig. 46*b*.

Neglecting possible hydrostatic uplift, the weight of 3 ft of wall amounts to 274,500 lb. The horizontal thrust on 3 ft of wall is 50,625 lb, and the overturning moment on 3 ft of wall is 425,250 ft-lb. For this structure it is also necessary to compute the moment of inertia of one row of piles about the center line of the wall, although in this case the computation is much easier than in the case of the chimney. The center pile of the row of seven piles is on the center line of the wall so that it does not contribute to the moment of inertia. The other six piles in the row have a moment of inertia amounting to 252. The section modulus of the pile group is the moment of inertia divided by the distance to the outermost pile, and in this case the section modulus is 28.

The pile load is computed from the same formula as used for the chimney. For this structure, the dead load contributes 19.6 tons to the total pile load and the live load contributes 7.6 tons. The maximum pile load is +27.2 tons on the pile at the left-hand end of the row, and the minimum load is +12.0 tons on the pile at the right-hand end of the row. These pile loads are both positive, which means that all of the piles are subjected to compression and there is no uplift due to the overturning force. However, the hori-

zontal shear in the pile heads under this structure amounts to 7,232 lb per pile, and this amount of horizontal thrust may be dangerous even though the piles are completely embedded in the ground.

The difference between the chimney foundation and the lock-wall foundation is readily understood from these calculations. When the chimney foundation has been designed so that it is safe against overturning, it is almost certainly safe against sliding, keeping in mind the short duration of wind forces. When the lock-wall foundation has been designed so that it is safe against overturning, it is not necessarily safe against sliding, keeping in mind the long duration of hydraulic forces. The lock wall is comparable to a deep beam on a very short span where the design of the beam is often controlled by shear rather than by bending moment.

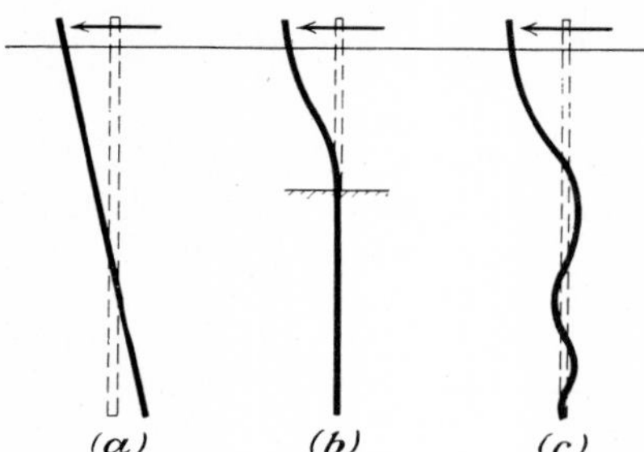

Fig. 47. Deflection of vertical piles subjected to horizontal force.

The problem represented by Fig. 46 has been the subject of a considerable amount of experimentation to determine what horizontal loads could safely be applied to the heads of vertical piles completely embedded in soil. This problem has also been investigated theoretically on the basis of the several assumptions represented in Fig. 47. As shown in Fig. 47*a*, it is sometimes assumed that the pile would pivot about a point somewhere along its length.[1] It is evident that the pile could behave in this manner only if it were absolutely rigid with respect to the surrounding soil. Since all piles are flexible to some extent, some bending must occur under the action of the horizontal force at the pile head due to the resistance of the surrounding soil. Another assumption that has been made in the theoretical approach to this problem is represented in Fig. 47*b*.[2] In this method, it is assumed that the lower end of the pile is clamped in the soil and that the upper end of the pile would bend into a reverse curve as shown in the figure. A third method of approaching the problem is illustrated in Fig. 47*c*.[3] In this method it is assumed that the pile would behave like a beam on an elastic foundation. The theory leads to the conclusion that the pile would deflect in the form of a damped sine wave with the amplitude of the wave decreasing rapidly as the depth increased.

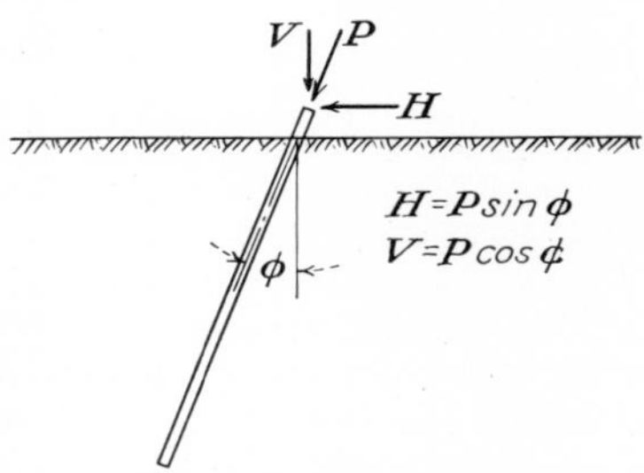

Fig. 48. Batter pile.

The theories that have been proposed for analyzing this problem are not ready for general use in foundation design. Sufficient experimental information to check the theories for various kinds of piles and various kinds of soil conditions is not yet available. Accordingly, whenever the heads of piles are to be subjected to horizontal forces of considerable magnitude, as is the case in Fig. 46, it is necessary to use batter piles. Horizontal forces amounting to a few hundred pounds or, at the most, 1,000 lb per pile, can usually be applied safely to the heads of vertical piles completely embedded in soil. For larger horizontal forces, the vertical piles cannot be depended upon to provide sufficient resistance without excessive horizontal movement.

[1] Raes, Paul E., The Theory of Lateral Bearing Capacity of Piles, *Proc. Harvard Conference on Soil Mechanics and Foundation Engineering*, vol. 1, p. 166; vol. 3, p. 138, 1936.
[2] Feagin, L. B., Lateral Pile Loading Tests, *Trans. ASCE*, vol. 102, p. 255, 1937.
[3] *Ibid.*, p. 272.

Figure 48 represents a batter pile which is driven into the ground on a slope instead of vertically. The assumption is usually made that the pile driven on a slope could resist the same axial load that would be resisted by a vertical pile of the same shape and size driven into the same soil. In Fig. 48, the axial load is P, and this acts at an angle ϕ with the vertical. This axial load may be divided into horizontal and vertical components as shown in the figure. The horizontal component H provides the reaction for external horizontal forces applied to the pile head. *It will be noted that, for the batter pile to effectively resist horizontal component H with an equal and opposite resistance, there must be a vertical component V in effect. Otherwise, the batter pile could resist only in bending similarly to the vertical piles shown in Fig. 47.*

The design of a structure which is subjected to external horizontal forces must include a study of the problem of providing proper reactions for these forces. A few structures of this kind are shown in Figs. 49, 50, and 51. Figure 49 represents a bridge abutment

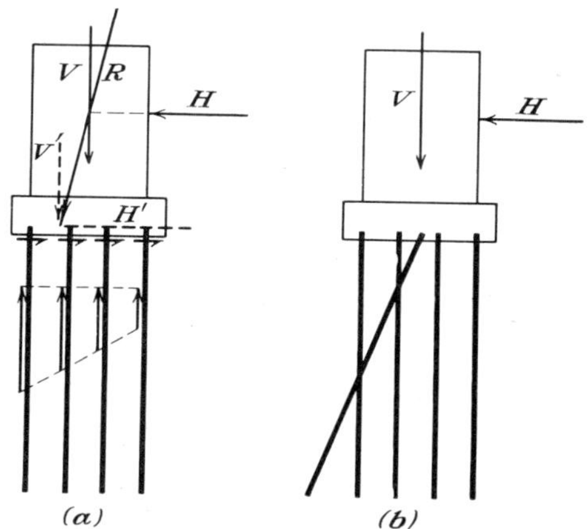

FIG. 49. Bridge abutment with batter piles.

subjected to a horizontal thrust H due to an approach fill. The effect of this thrust is to apply horizontal shear at the pile heads, as illustrated in Fig. 49*a*. When these horizontal shears are of any considerable magnitude, it is customary to use batter piles, as shown in Fig. 49*b*.

Figure 50*a* represents a highway trestle supported on piles extending above the ground line. Such a structure might be built across a stream bed, and it might be subjected to a horizontal thrust due to floating ice or other objects piling up against the trestle on the upstream side. Such structures are often sway-braced, as shown in Fig. 50*a*, but this sway bracing does not always provide proper resistance against a large horizontal thrust such as H. The sway bracing will cause all the piles in the bent to act together. However, since H is external to the structure, some external reaction must be provided to resist H. This external reaction is derived from the shearing resistance in the piles at the ground surface. When the force H is of any considerable magnitude, the trestle should be designed with batter piles, as shown in Fig. 50*b*.

Figure 50*c* shows a side view of a similar trestle, which might be subjected to horizontal forces in a longitudinal direction because of the tractive effort of heavy vehicles on the roadway. Sway bracing will help to stabilize the structure by tying the bents together, but the external horizontal thrusts will put shear in the piles at the ground

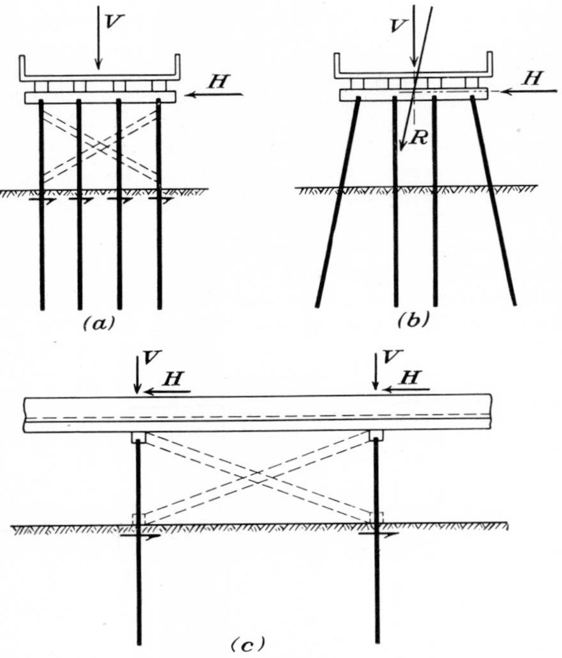

FIG. 50. Trestle with batter piles.

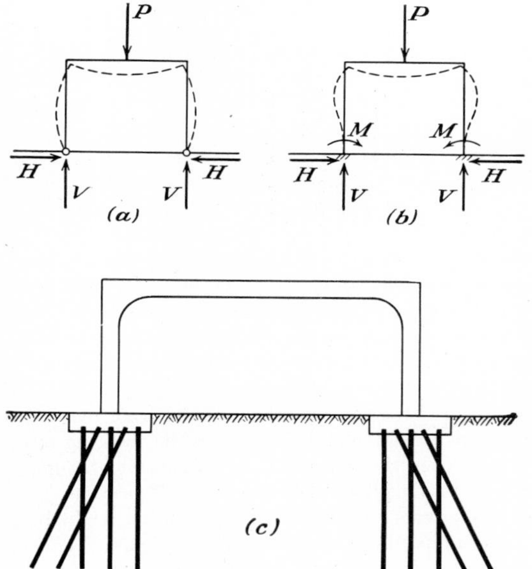

FIG. 51. Rigid-frame bridge with batter piles.

line unless the ends of the trestle are braced against abutments or other practically rigid structures.

Figure 51 illustrates a rigid-frame bridge. The lower ends of the vertical legs of such bridges are sometimes made with hinges, as shown in Fig. 51*a*, and they are sometimes fixed in the footings, as shown in Fig. 51*b*. In either case, the load P tends to spread the bottoms of the legs outward, and it is necessary to provide a horizontal reaction directed inward to resist this thrust. When such structures are built on pile foundations, it is customary to use batter piles, as shown in Fig. 51*c*, in order to resist the horizontal thrust at the pile heads.

Various methods are available for designing batter-pile foundations. Some of these are algebraic and some are graphic. The principal contributors in this field have been Danish and Swedish engineers, and in practically all cases the design of the pile foundation is a complicated problem in statically indeterminate structures.[1]

Figure 52*a* illustrates the most simple type of problem involving batter piles. The figure represents an anchorage subjected to an anchor pull of 10 tons. A pair of batter piles sloping in opposite directions is used to resist the anchor pull. When the line of

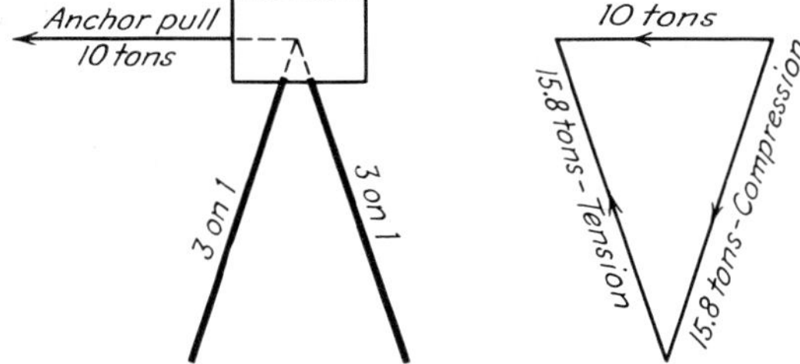

Fig. 52*a*. Anchorage with batter piles.

action of the anchor pull and the longitudinal axes of the pair of batter piles meet at the same point, the problem is statically determinate. The compression in the front pile and the tension in the back pile can be determined by means of a simple force diagram, as shown in the figure.

For more complex structures involving many piles, the problem of design and the determination of pile loads is more difficult. For the purpose of analyzing structures of the type shown in Fig. 52*b*, several approximate methods have been developed. In general, these are graphic methods based on the use of force diagrams. A procedure of this kind, known as "Culmann's method," is illustrated in the figure. The structure is a marginal wharf with a platform supported on vertical piles and on batter piles. The batter piles are sloped in two directions. There are four vertical piles at the front of the platform, six batter piles sloped upward to the right, and four batter piles sloped upward to the left In order to draw a force diagram for this pile arrangement, it is necessary to consider each of the three groups of piles as a single dummy pile located at the center of gravity of the group. For the four vertical piles, the line AA' represents a dummy pile at the center of gravity of the group. For the six batter piles sloped upward to the right, the dummy pile is represented by the line BB', and for the other four batter piles, the dummy pile is represented by the line CC'.

The structure is subjected to a vertical load V and a horizontal thrust H, which are combined into a resultant R. The line of action of this resultant is produced until it intersects the line AA'. The lines of action of the dummy piles representing the two

[1] Vetter, C. P., Design of Pile Foundations, *Trans. ASCE*, vol. 104, p. 758, 1939.

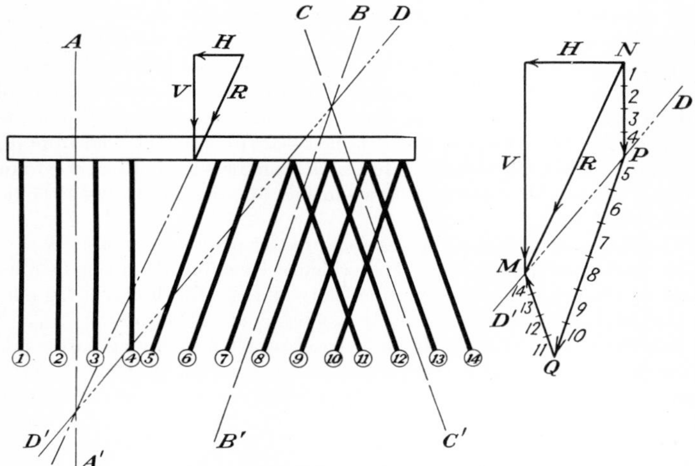

FIG. 52b. Culmann's method.

groups of batter piles intersect at a point above the platform, and the line DD' is drawn through these two points of intersection.

The force diagram at the right of the figure is then constructed graphically by means of various lines on the figure. The construction is started by choosing a force scale and the drawing of the force triangle consisting of the forces V and H and their resultant R. Through the point M at the lower end of V, the line DD' is drawn parallel to the line DD' on the other diagram. The vertical line NP determines the load in the dummy pile AA'. The line PQ is drawn parallel to the dummy pile BB', and the line MQ is drawn parallel to the dummy pile CC'. These two lines intersect at Q. The length of the line MQ determines the total load in the four batter piles which slope upward to the left, and the length of line PQ determines the total load in the six batter piles that slope upward to the right. The assumption is made that each pile in each of the three groups is equally loaded, and the line NP is therefore divided into four equal parts, the line PQ into six equal parts, and the line MQ into four equal parts to determine the load per pile. The four vertical piles and the six batter piles which slope upward to the right are under compression, and the other four batter piles are subjected to tension.

This is only one of several graphic methods that are used to analyze pile foundations of this kind. The graphic analysis is limited to cases in which there are not more than three pile directions. When there are more than three pile directions, it is necessary to compound the pile groups so as to reduce the directions to three. Most of the graphic methods have been in use for many years, and they are often referred to as approximate methods in comparison with some of the elaborate algebraic methods developed more recently.

Figure 52c illustrates a wharf structure with the same pile arrangement as Fig. 52b. The algebraic methods used to determine the pile loads are based on three assumptions with respect to the end conditions of the piles. These assumptions, shown at the top of Fig. 52c, are (1) piles hinged at both ends, (2) piles hinged at the top and fixed at the bottom, and (3) piles fixed at both ends. The concrete platform on top of the piles is assumed to be rigid, and the points of the piles are assumed to rest on a rigid base. The analysis is carried out as a problem in statically indeterminate structures with the piles considered as a series of free-standing elastic columns.

The lower diagram in Fig. 52*c* shows a structure of this kind displaced by vertical and horizontal loads. The piles at the front of the platform are displaced downward and outward, and those at the rear are displaced upward and outward. The axial loads in the piles of any one group are not necessarily equal, as was assumed in the graphic analysis of Fig. 52*b*. The axial deformations of the piles in the various groups are not necessarily proportional to the pile loads, as is assumed in the algebraic analysis. Piles subjected to tension behave differently from piles subjected to compression even when the piles are of the same shape and size and are driven into the same kind of soil. All

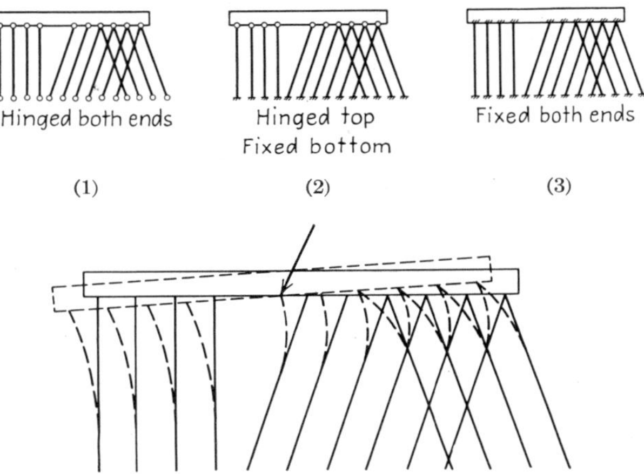

FIG. 52*c*. Wharf structure with batter piles.

the piles slide axially through the soil to some extent, and the lower ends are not rigidly fastened down, as is assumed in the algebraic analysis. When the horizontal load is applied to the structure, any tendency of the pile heads to deflect horizontally would be resisted by the soil in front of the pile. Neither the graphic nor the algebraic methods of analysis take this into account, although the resistance of the soil to lateral displacement has an important effect on the behavior of the structure.

If a problem of this kind were to be set up in such a way as to take into account all the variables including the soil, the problem would be statically indeterminate to such a degree that it would be practically impossible to develop a rigorous solution. The designer must therefore make simplifying assumptions of various kinds in order to reduce the problem to one for which some reasonable solution can be obtained. In some cases designers have gone so far as to assume that the structure should have enough vertical piles to resist all the vertical load and enough batter piles to resist all the horizontal load. This is an unnecessarily conservative assumption, and it puts an excessive number of piles in the foundation. All the piles as well as the surrounding soil must act together, and the batter piles will resist some of the vertical load and the vertical piles will resist some of the horizontal load.

Whenever a pile foundation is being designed for a structure subjected to horizontal forces, it is necessary to determine to what extent these forces will be transmitted to the pile heads and to consider for what period of time these forces are likely to be effective. When the horizontal shear at the pile heads has been determined—and especially when the horizontal shear is likely to be effective over more than a very short time, it is desirable to consider the use of batter piles in the foundation rather than to depend

on vertical piles to resist these horizontal shears. The various methods available for designing batter-pile foundations are all based on certain simplifying assumptions, and each designer will have to choose the method he considers best suited to the particular problem being analyzed.

Uplift Resistance. There are some types of structures that are subjected to uplift forces under certain conditions. A structure of this kind is illustrated in Fig. 53. The sketch represents a reinforced concrete settling basin in a water-filtration plant. The basin is built on a pile foundation, and it is common practice to build such structures

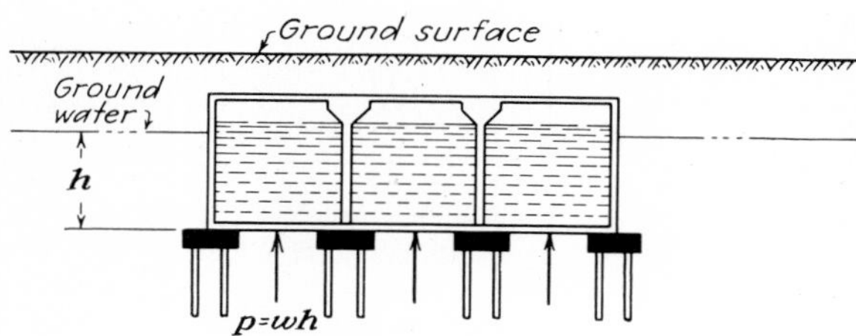

FIG. 53. Piles subjected to uplift.

below the ground surface and to cover them with a few feet of earth fill. The base of the structure is well below the natural ground-water table. While the basin is full of water the load on the piles is a downward load because the external hydrostatic head is balanced by the water in the basin. Such structures are often emptied for cleaning or for other reasons; and, when a basin of this kind is empty, it is often so light that it tends to lift off its foundation because of the hydrostatic head h from the ground-water table down to the base of the structure. A unit upward pressure p is exerted under the entire bottom of the structure.

When such a structure is supported by a pile foundation, it is necessary to make sure that the piles are properly connected to the structure so that the joints can resist tension. It is also necessary to make sure that the piles are well fastened in the soil so that they will not lift out of the ground. In addition, the bottom slab of the structure must be designed for this upward hydrostatic pressure. A similar situation often exists in deep basements and in other structures which are set deep in the ground either well below the natural ground-water table or at such a depth in soft or medium clay below sidewalk level that the surcharge loading on the sidewalk side of the basement wall is sufficient to exceed the shearing resistance of the clay and thus induce uplift against the lowest basement floor slabs which must be designed to resist the uplift.

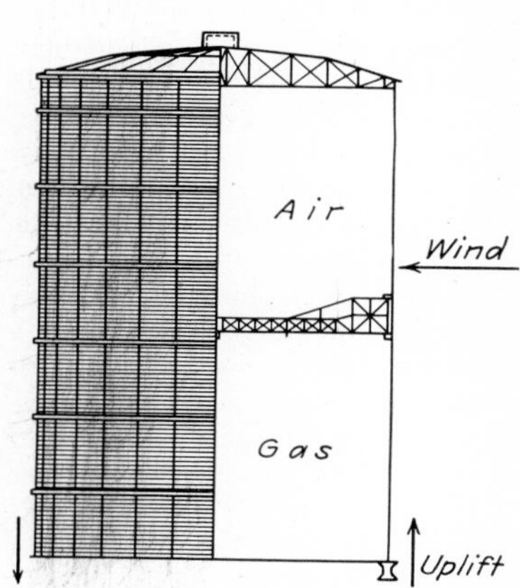

FIG. 54. Piles subjected to uplift.

Another type of structure on which there are uplift forces in the foundations is illustrated in Fig. 54. This is a waterless-gas holder which consists of a comparatively light steel framework and a steel shell filled with gas and air. Under heavy wind forces such a structure tends to overturn. Because the structure itself is so light, there are often uplift forces

on the foundations on the windward side. When pile foundations are used under waterless-gas holders of this type, it is necessary to make sure that they can resist the uplift forces that will occur under full wind load.[1]

The uplift resistance of piles is usually considerably less than their bearing capacity. Whenever the uplift force amounts to more than a few tons per pile, it is desirable to determine the actual resistance by an uplift test in the field.

Stability and Buckling. Building codes and specifications often set up requirements about the maximum allowable ratio of length to diameter of long slender piles and piers. The purpose of this requirement is to guard against failure by buckling. However, such requirements are often purely arbitrary and in most cases entirely unnecessary when the pile is completely embedded in soil. Almost any soil is sufficiently strong to stabilize a long slender pile against buckling under the working loads that are commonly used. When piles are driven as friction piles there is no danger that they will buckle as long columns and there is no reason even to consider column action in the pile. This can be demonstrated theoretically [2] and it has often been proved by field tests. Long steel H piles driven through soft clay to bearing on hardpan have been loaded to failure, but failure occurred by flange crippling at the ground surface and not by buckling as a long column. Concrete bearing piles driven under the same conditions failed by diagonal shear at the ground surface with no indication of long-column buckling.

When piles extend above the ground surface, as in a trestle-type structure, their upper portions above their points of fixity in the ground are designed as columns. The end conditions at the top of the pile will be determined by the manner in which the pile is joined to the superstructure. It is usually assumed that such piles are fixed at a point 5 ft below the ground surface in hard soils and 10 ft below the surface in soft soils.

Pile-driving Formulas. The bearing power of individual piles is often computed by means of pile formulas, which may be of the static or the dynamic type.

Static formulas estimate the bearing power by computing the frictional or shearing resistance along the surface of the pile and the bearing resistance of the point. The original static formulas were based on classic earth-pressure theories, but methods being developed at the present time are based on laboratory tests of the soils into which the piles are to be driven. Soil samples are secured in the field and tested in the laboratory for shearing strength and frictional resistance. The laboratory-test results are then used to compute the bearing capacity of a pile of a given shape and size. Such methods are a recent development in foundation engineering, and their usefulness is increasing as they are being demonstrated under a wide variety of soil conditions.

Dynamic formulas are used to compute the bearing power of a pile from its behavior during driving. The factors used in all formulas are the driving energy and the average penetration per blow during the last few blows. Other factors which may be included are the weight of the pile, its cross-sectional area, its length, and the modulus of elasticity of the material of which it is composed. These factors are used to compute the energy which is lost during impact of the hammer on the pile and is therefore not available to produce penetration.

The formula most commonly used in this country is the *Engineering News* formula, which is as follows:

$$P = \frac{2Wh}{s + c}$$

where P is the allowable load on the pile; W is the weight of a drop hammer, or the

[1] Concrete Piles Tested for Uplift and Bearing, *Eng. News-Record*, vol. 109, No. 4, p. 104, July 28, 1932. GREGORY, JOHN H., Holding-down Power of Concrete Piles, *Civil Eng.*, vol. 3, No. 2, p. 66, February, 1933.

[2] CUMMINGS, A. E., The Stability of Foundation Piles against Buckling under Axial Load, *Proc. Highway Research Board*, vol. 18, Part II, p. 112, December, 1938.

TABLE 1. CHARACTERISTICS OF SINGLE-ACTING AND DOUBLE-ACTING STEAM HAMMERS

Single-acting Vulcan

Hammer size	4	3	2	2 *	1	1 *	0	OR
Rated striking energy, ft-lb	825	3,600	7,260	7,260	15,000	15,000	24,375	30,225
Blows per minute	80	80	70	70	60	60	50	50
Weight of striking parts, lb	550	1,800	3,000	3,000	5,000	5,000	7,500	9,300
Steam pressure, psi	80	80	80	80	80	80	80	100
Diameter of piston, in	4	8	10½	10½	13½	13½	16½	16½
Normal stroke, ft	1¾	2	2.42	2.42	3	3	3¼	3¼
Net weight, lb	1,400	3,700	6,700	7,100	9,700	10,200	16,250	18,050
Length of hammer	7′0″	9′6″	11′6″	12′0″	13′0″	13′3″	15′0″	15′0″

Single-acting McKiernan-Terry

Hammer size	S3	S5	S8	S10	S14
Weight of ram, lb	3,000	5,000	8,000	10,000	14,000
Striking energy, ft-lb per blow	9,000	16,250	26,000	32,500	37,500
Stroke, ft	3	3¼	3¼	3¼	2⅔
Blows per minute	65	60	55	55	60
Steam pressure at hammer, psi	80	80	80	80	100
Recommended steam pressure at boiler, psi	100	100	100	100	125
Total weight, lb	8,800	12,375	18,100	22,200	31,600
Length	12′4″	13′3″	14′4″	14′1″	14′10″
Maximum stroke	3′3″	3′6″	3′6″	3′6″	3′
Diameter of piston	11″	14″	17″	17½″	20″

Differential-acting Super-Vulcan

Hammer size	18C	30C	50C	80C	140C	200C
Rated striking energy, ft-lb	3,600	7,260	15,100	24,450	36,000	50,200
Blows per minute	150	133	120	111	103	98
Weight of ram, lb	1,800	3,000	5,000	8,000	14,000	20,000
Steam pressure, psi	120	120	120	120	140	142
Diameter of small piston, in	4 15/16	6½	8 7/16	10 3/16	11¼	13
Normal stroke, in	10½	12½	15½	16½	15½	15½
Equivalent stroke, ft	2.0	2.42	3.01	3.05	2.57	2.51
Weight, lb	4,139	7,036	11,782	17,885	27,984	39,050
Length of hammer	7′8½″	8′11½″	10′2⅝″	11′4¼″	12′3″	13′2″

Double-acting McKiernan-Terry

Hammer size	0	1	2	3	5	6	7	9B3	10B3	11B3
Length, in	20	39	29	53	51	55	63	89¼	102¾	124½
Net weight, lb	105	145	343	675	1,500	2,900	5,000	7,000	10,850	14,000
Weight of ram, lb	5	21	48	68	200	400	800	1,600	3,000	5,000
Bore, in	2¼	2¼	4 1/16	3¼	7	9¾	12½	8½	10	11
Stroke, in	4½	3¾	5¼	5¾	7	8¾	9½	17	19	19
Strokes per minute	1,000	500	500	400	300	275	225	145	105	95
Ft-lb energy per blow	....	...	....	...	1,000	2,500	4,150	8,750	13,100	19,150

* These hammers are equipped with the McDermid base.

weight of the moving parts of a single-acting steam or air hammer; h is the distance through which a drop hammer falls or the stroke of a steam or air hammer, expressed in feet; s is the average penetration per blow for the last few blows, expressed in inches; and c is a constant, equal to 1.0 for a drop hammer and 0.1 for a steam or air hammer. The values for P and W are expressed in the same units, in either pounds or tons. The formula as written includes a safety factor of 6. When the double-acting steam or air hammer is used, the term $2Wh$ is replaced by the term $2E$ in which E is the rated energy of the double-acting hammer obtained from tables prepared by the hammer manufacturers. Table 1 on page 8-64 gives the characteristics of the single-acting and double-acting hammers most commonly used.

With the development of modern soil mechanics during the past 33 years, the whole problem of dynamic pile-driving formulas has been the subject of considerable discussion. It is generally agreed that a dynamic formula should be used only against a background of adequate test borings; that such a formula can be expected to give satisfactory results when piles are being driven into granular soils such as sands and gravels, and when piles are being driven to point bearing on hardpan, gravel, or rock. When friction piles are being driven into deep beds of plastic soil, the dynamic formulas can easily give very unsatisfactory results. However, this is usually due to the fact that the formula cannot be expected to take into account the possibility of area subsidence due to compression of the soil below the pile points. Neither can the formula be expected to solve the problem of pile group action. *In spite of all the effort that has been expended on the problem in the past, pile driving is not yet an exact science.* Dynamic pile-driving formulas can never be depended upon to replace a background of good test borings and often cannot be relied upon to replace pile-load tests and soil technical analysis in establishing a standard of performance for a job *although, once a proper background has been provided by test borings, and where desirable, by pile load tests and soil mechanical analysis, a dynamic pile-driving formula calibrated to the standard thus set can be relied upon for performance throughout the remainder of the job.* The installation of a satisfactory pile foundation is largely a matter of experience combined with a careful soil investigation.[1]

CAISSON OR PIER FOUNDATIONS FOR BUILDINGS

Types. A caisson or pier foundation consists of a shaft of concrete, usually cylindrical but sometimes of square or rectangular cross section, placed under a building column or wall and extending down to satisfactory gearing material such as hardpan or rock. Although it is customary in the construction industry to refer to foundations of this type as "caisson foundations," they are actually piers, since a caisson is a boxlike structure usually sunk in water as for a bridge pier and forming a part of the pier after it has been filled with concrete or masonry. The pier is a foundation unit that can support large loads, and it is therefore economical only under certain ground and water conditions and where the structure has large column loads to be supported. Under light buildings, a pier foundation is not usually economical because the smallest size of pier that can be properly installed and manually inspected is capable of supporting far more load than can be applied to it by a column of a light structure. Excavation for piers is sometimes made in the open air at normal atmospheric pressure and sometimes in closed working chambers under compressed air. The excavation is done by hand or by machines of various kinds. The piers themselves may be of constant diameter all the way to the bottom, in which case they are often called "straight-shaft piers." In some cases the lower ends of the piers are enlarged in the form of a frustrum of a cone in order to provide sufficient bearing area, and are then referred to as being "belled out."

[1] CUMMINGS, A. E., Dynamic Pile Driving Formulas, *Jour. Boston Soc. Civil Engrs.*, January, 1940. "Pile Foundations and Pile Structures," ASCE Manual 27, Jan. 18, 1946.

Open Excavation. The simplest kind of pier, which is the kind most commonly used in building construction, is the straight-shaft pier installed in the open air at normal atmospheric pressure by hand excavation. The principal equipment that is required for such piers is a tripod to hoist excavated material out of the hole, some sort of hoisting apparatus, and some pumping equipment. The minimum diameter of pier that is commonly used is about 4 ft because a man cannot work with a pump in a hole of smaller diameter. Where a pump in the hole is not needed, the minimum diameter is sometimes reduced to 3 ft and, where manual excavation or inspection is not required (*manual inspection should always be required*), excavation by rotary power driven post-hole digging machines sometimes is used with smaller minimum diameter. One such machine is shown in Fig. 60. It is usually necessary to provide a lining of some kind as excavation proceeds, and the nature of the lining is usually determined by the soil conditions. Sometimes the excavation is lined with telescoping, removable steel cylinders that are made of ¼-in. or thicker plate and are 8 to 16 ft long. After the starting pit has been excavated, the first (largest diameter) cylinder is placed in the pit and the upper end of the cylinder is allowed to extend about 1 ft above the ground surface in order to eliminate the possibility of having drainage water or rain water flow into the excavation from the surrounding area. After the first cylinder has been placed and cleaned out, a second cylinder is set inside the first one. Ordinarily the second cylinder is about 2 in. smaller in diameter than the one above it. Excavation is carried on below the bottom of the second cylinder, and the cylinder itself is driven down or pushed down as excavation proceeds. The upper end of the second cylinder is allowed to lap the lower end of the first cylinder by several inches as it is often necessary to calk the space or pump from between the two cylinders to prevent the entrance of ground water. The installation of a pier of this type is illustrated in Fig. 55.

As can be seen from the illustration, the shaft of the pier consists of a series of concrete cylinders with each cylinder about 2 in. smaller in diameter than the one next above it. The design of such a pier is based on the diameter of the lowest cylinder, and the cylinders above the bottom one are therefore of larger diameter than would be required to support the column load at the allowable working stress in the concrete. This involves the waste of a certain amount of concrete due to the method of construction of a pier of this type; and this waste can be reduced only by lengthening the steel cylinders. However, excavation with steel cylinders of this design has one definite advantage over the use of other types of lining. Often the soil conditions include comparatively thin strata of water-bearing sands, and the water flowing from these sands would flood the excavation if it were being dug in the open without some sort of lining to seal the side walls. Where such sand strata are encountered, it is possible to drive the steel cylinder through the sand stratum and to cut off the inflow of water.

Another type of open-pier excavation using a different system of lining is the type that is often referred to as the "Chicago method," in which the excavation is lined with wooden lagging. This lagging is made of 2-in. by 6-in. or 3-in. by 6-in. boards, which are usually beveled on their edges and radially tongued and grooved so that they will fit tightly around the circular excavation. It is customary to excavate 4 or 5 ft without lining and then to install a series of these beveled boards around the inside of the excavation like the staves of a barrel. In order to brace the boards and to hold them tightly in place against the soil, steel rings are used. These rings are made in half circles and are provided with spreader bolts with which the rings are expanded against the inside of the wooden lagging. The usual procedure in this type of construction is to excavate 4 or 5 ft at a time and then to place a set of lagging and rings. This process is repeated until bottom is reached. Ordinarily two steel rings are used for each set of lagging 4 or 5 ft in length. This method is illustrated in Fig. 56.

With this Chicago method, the pier is usually of constant diameter from top to bottom. The excavation includes undercutting below the bottom of the lagging by an amount equal to the thickness of the lagging so that each set of lagging is directly below the one next above it. However, a stratum of water-bearing sand can cause considerably more difficulty in the Chicago method than it would in an excavation lined with steel cylinders. If the water-bearing sand stratum is very thin so that excavation can be carried

Fig. 55. Excavation for concrete pier.

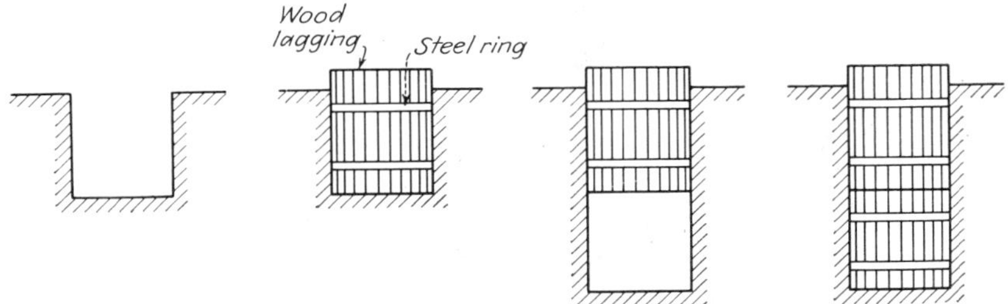

Fig. 56. Method of lining excavation for concrete pier.

through it quickly and the lagging and rings can be set in place promptly, the inflow of water can usually be sufficiently controlled by pumping or bailing. However, if the water-bearing sand stratum is several feet or more in thickness, it is not always possible to control the inflow of water by pumping. It is necessary to step in with the lagging and to drive a set of lagging down through the sand stratum. This means a reduction in the diameter of the excavation, and if such a thing happens several times on a deep excavation, the reduction in diameter may be considerable. It might even be necessary to start the whole operation over again by starting from the ground surface with an excavation of considerably larger diameter.[1]

In another type of open-pier excavation, the lining is a precast concrete cylinder. This type is illustrated in Fig. 57. The cylinder is usually made with walls about 6 in.

Fig. 57. Installing concrete pier with precast-concrete cylinders.

thick, and it is provided with a cutting edge at the bottom. Sometimes it is forced down by dead load applied at the top, sometimes it is driven with a heavy hammer, and sometimes dredging is done on the inside with a small clamshell or orange-peel bucket. The starting cylinder is usually 5 or 6 ft long; and, after this has been forced into the ground, additional 5- or 6-ft sections are added. This type of excavation is sometimes more expensive than the types with steel or wooden lining, but it has certain advantages which offset its extra cost. If a water-bearing sand stratum is encountered while the excavation is being made, it is usually possible to drive the cylinder through the sand stratum and thereby cut off the inflow of water.

Ordinarily there is a projection on the outside at the bottom of the shaft so that the wall of the bottom section is a little thicker than the wall of the upper part of the shaft. The purpose of this projection is to reduce the amount of friction that is developed on the outside of the shaft as it is forced into the ground. If the shaft had the same outside diameter throughout its length, the total friction on the outside of the shaft would increase rapidly as the depth increased. With the enlarged base the friction developed on the outside is reduced because the enlarged base usually leaves a small opening between the outside wall of the shaft and the surrounding soil. This annular ring on the outside of the shaft is usually filled with sand or other loose material as the shaft is sunk into the ground. Another advantage of this type of pier excavation is the fact that it can be more easily adapted to compressed-air work than either the steel-cylinder

[1] Newman, W. J., Chicago Open Well Method, *Civil Eng.*, vol. 1, No. 4, p. 306, January, 1931.

lining or the lining of steel rings and wooden lagging. When soil conditions are encountered that necessitate the use of compressed air, an air lock must be placed somewhere in the shaft or at the top of the shaft. When the lining consists of steel cylinders or wooden lagging, it is very difficult to set an air lock satisfactorily, but this can easily be arranged with the precast-concrete-cylinder method. This type of pier excavation is often called the "drop-shaft method."

In some cases special methods are developed for the purpose of overcoming a particular soil condition. One such method was developed for a job where the soil consisted of about 30 ft of fine sand overlying soft blue clay which extended down to rock.[1] The water table was about 5 ft below the ground surface so that there were about 25 ft of saturated sand through which it was necessary to excavate. When water is pumped out of an open excavation which is passing through a water-bearing stratum, the hydrostatic head is unbalanced by the pumping operation and the unbalanced head is equal to the difference in level between the ground water outside the excavation and the level at which excavation is being carried on inside the shaft. The water flows in at the bottom of the excavation and produces a quicksand condition even in coarse sands. It is very difficult to excavate through 25 ft of saturated fine sand in an open excavation, and a special procedure was developed for this job, involving the use of steel cylinders about 35 ft long which could reach from the ground surface through the saturated sand and into the underlying clay. These cylinders were about 5 ft in diameter and were equipped on the inside with a series of water jets fastened to the inside surface of the cylinder. The cylinder equipped with these water jets was hoisted with a crane and set down on the ground surface in a vertical position. A heavy weight was placed on top of the cylinder, and water was pumped into the jets. This jetting loosened the sand, and the heavy weight forced the steel cylinder down through the sand and into the clay. When the cylinder had been forced into the clay far enough so that sand and water could not flow in at the bottom, the jetting was stopped and the steel cylinder was pumped out. The excavation then proceeded by hand through the soft clay and the hardpan to rock, with wooden lagging and steel rings as lining in the clay.

Figure 58*a* represents a straight-shaft pier founded on rock. When the soil conditions permit and when the rock is not too deep below the ground surface, this is the type of pier that is commonly used. The pier is usually of concrete, and the working stress in the concrete would be on the order of 600 to 800 psi, or about 43 to 58 tons per sq ft. The allowable pressure on rock varies in different localities, and in most cases this allowable pressure is regulated by the city building code. However, the allowable load on hard solid rock would usually be as great as any load that could be applied to the rock through the concrete shaft of the pier.

Sometimes the rock is overlain with hardpan or some other good bearing material, as illustrated in Fig. 58*b*. Instead of excavating for a straight-shaft pier through the hardpan all the way to the rock, the excavation is sometimes stopped and belled out on the hardpan. The decision as to whether to use a pier belled out on hardpan or whether to excavate a straight shaft to rock depends largely on soil conditions.

The first question to be determined is the relative cost of excavating a large bell in soft soil overlying the hardpan, as compared with the cost of excavating the straight shaft down through the hardpan to rock. Ordinarily the sides of the bell are cut at a slope of 60° with the horizontal, and the bottom of the bell has a vertical edge for a depth that varies from 4 to 12 in. The usual load allowance on hardpan is much smaller than the allowable load on solid rock, so that the area of the bottom of the bell has to be much greater than the cross-sectional area of the shaft.

It is readily seen that a considerable amount of excavation and concrete is required

[1] Jetted-casing Method Employed for Caisson Excavation, *Eng. News-Record*, vol. 108, No. 4, p. 135, Jan. 28, 1932.

for such a bell, and therefore it is usually a question of economics as to whether it is cheaper to excavate a large bell and use the relatively low bearing capacity of the hardpan, or whether it would be cheaper to dig through the hardpan with a straight shaft and use the high bearing capacity of the rock. Ordinarily the answer to this question depends on the thickness of the hardpan overlying the rock. If the hardpan stratum is not very thick, say 8 or 10 ft, the excavation of a constant-diameter shaft through the hardpan to rock would probably be cheaper than the excavation of a large-diameter bell on top of the hardpan. If there were 50 or 60 ft of hardpan overlying the rock, it would almost certainly be cheaper to stop and bell out on top of the hardpan.

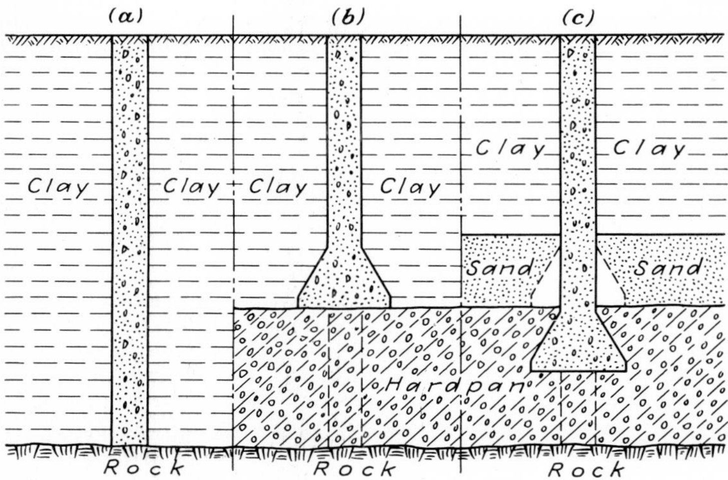

FIG. 58. Straight-shaft and belled concrete pier.

In either case, the decision would be based on the relative cost of the excavation and concrete involved in the two methods.

In some building codes, it is required that the weight of the pier itself must be included in the determination of the required bell diameter. In other words, the column load and the weight of the pier are added together, the weight of the excavated earth is deducted, and this result is divided by the allowable soil pressure in order to determine the area of the bell. When the pier is very long, a building-code requirement of this kind adds materially to the required bell diameter.

Sometimes a soil condition is encountered of the kind illustrated in Fig. 58*c*. The rock is overlain with a stratum of hardpan, but the hardpan itself is overlain with a stratum of sand. In order to bell out on top of the hardpan, it would be necessary to construct the bell in the sand stratum. However, it is usually very difficult and sometimes impossible to undercut the sand so as to form the sloping sides of the bell. The sand will tend to collapse into the excavation, and bracing under such conditions is difficult and expensive. With soil conditions of this kind it is usually necessary to excavate the shaft through the sand to the hardpan and then to bell out in the hardpan itself. The sand stratum is sealed off by carrying the lining of the shaft into the hardpan, and then the entire bell is excavated in the hardpan.

When belling out in the hardpan is contemplated, the character of the hardpan and its performance under unbalanced hydrostatic pressure should be thoroughly checked by special test borings, which simulate (to small lateral scale) the conditions of the

caisson installation, to be sure that belling out in the hardpan is practicable. Frequently, hardpan can consist of material having a high granular content not well cemented with clay, which would place a practical limit on the lateral dimension of the bell undercut; and the hardpan or the rock below it may contain water under high pressure, the pumping of which (in any attempt to bell out in the hardpan) may cause sand boils or blows which would place practical, and possibly engineering, limitations on such construction. Artesian sulfur water in the hardpan and underlying rock are frequently encountered in and around the city of Detroit, Mich.; and very careful investigation is essential wherever that condition is likely to obtain.

Compressed-air Excavation. Compressed-air work is usually quite expensive, and its use is usually avoided unless it becomes absolutely necessary because of soil conditions on the job. Figure 59 illustrates a soil condition in which it would probably be necessary to resort to compressed air in order to excavate for a pier to rock. The topsoil is fill

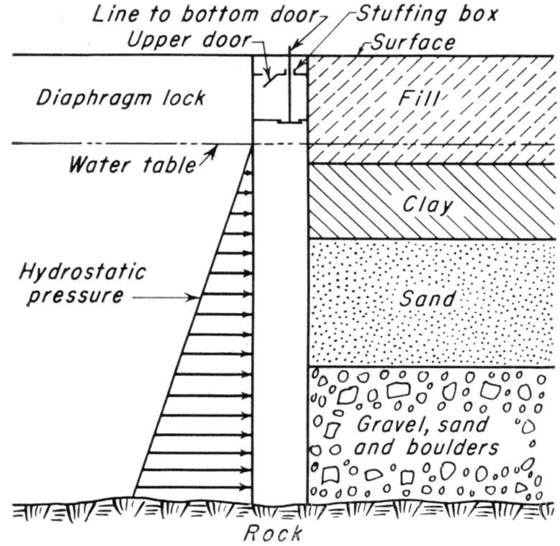

FIG. 59. Excavation for concrete pier under compressed air.

overlying a bed of clay. Below the clay is a bed of sand underlain by sand, gravel, and boulders extending down to rock. The natural ground-water table is in the fill not far below the ground surface. In the ground around the shaft, the hydrostatic pressure increases with the depth, and the purpose of the compressed air is to balance the hydrostatic head and keep the water from entering the excavation. In other words, the required air pressure depends on the depth below the water table or the water pressure level (hydraulic grade) at which the work is being done. The purpose of the compressed air is to balance the hydrostatic pressure. Soft plastic clays can squeeze into an excavation, even though the excavation may be carried out under compressed air, because of the difference between the weight of saturated clay and that of water. The compressed air usually compensates only the weight of the water.

The maximum air pressure that can be used in such work is about 45 or 50 psi, which corresponds to somewhere between 100 and 120 ft of hydrostatic head. At greater depths higher air pressures would be required, but it is practically impossible for men to work under pressures greater than 45 or 50 psi. In order to install a pier under compressed air, it is necessary to use an air lock. Sometimes the air lock is fastened to the top

of the shaft above the ground surface, and sometimes it is set in the shaft. Essentially the air lock consists of two airtight trap doors, as illustrated in Fig. 59. The purpose of the air lock is to permit men and materials to go in and out of the excavation without releasing the air pressure. When the lower trap door is closed and a man enters the air lock through the upper trap door, the air pressure in the air lock is atmospheric pressure. He closes and seals the upper door and then opens a valve within the lock, allowing compressed air to enter the lock slowly. When the air pressure in the lock is equal to the air pressure in the excavation, the lower trap door opens and the man goes down into the excavation. When men are coming out of the shaft, the air-lock procedure is reversed; but the decompression process is slower and very much more important than the compression process. Men working under compressed air are subject to severe illness if they come out of compressed air into the open air too quickly (see p. 8–83).

From the construction standpoint, an important factor in all compressed-air work is the danger of a blowout. For the soil condition illustrated in Fig. 59, the fill and clay overlying the sand would be referred to as "cover," and the impervious clay would hold the compressed air so that the air would not leak through to the ground surface and cause a blowout. If there were no impervious clay to act as a cover, and if the entire soil bed were a pervious sand and gravel, the compressed air might leak through the soil so that air bubbles would come out of the soil at the ground surface and it would be impossible to maintain sufficient air pressure to counteract the hydrostatic pressure at the bottom of the excavation.

Excavation by Machines. During the past 10 or 15 years several methods have been developed for excavating pier foundations by machine. One of these is usually referred to as the "dry process" and the other as the "wet process." The procedure that is used in the dry process is illustrated in Fig. 60. The machine itself is a modified crane operating on caterpillar treads. A special frame is built on the front of the machine to carry a power unit and a large gear that operates in a horizontal plane. This assembly is called a "rotary table" and the large gear drives a vertical shaft called a "kelly" which is suspended on a cable from the tip of the boom. The shaft is made of concentric hollow cylinders keyed inside and out which telescope into one another so that the shaft can be extended down into the excavation to depths of 100 ft or more.

On the bottom of this shaft is a bucket which does the actual excavating. The bottom of the bucket is built something like a post-hole auger so that soil is picked up through the bottom of the bucket as the bucket is rotated in the excavation. Also, the bottom of the bucket is built in two halves which are hinged so that the contents of the bucket can be dumped into a truck or a spoil bank. Ordinarily the buckets are not more than 4 ft in diameter, but they can be used to dig larger diameters by means of a pair of reamer knives that are bolted to the bail of the bucket. These knives are adjustable and may be set to the required diameter of the excavation. The 4-ft-diameter bucket will then excavate a 4-ft hole, taking the excavated soil in through the bottom of the bucket, while the reamer knives cut the excavation to its full diameter, throwing the excavated soil into the bucket from the top.

In many cases it is possible for such a machine to excavate all the way to the bottom of the pier before any lining is installed. Sometimes it is necessary to install the lining in sections 8 to 16 ft long as excavation proceeds. When the lining has to be placed during the excavation process, special retractable reamer knives are used which can cut under the lining and excavate to the full diameter.

A machine of this kind is particularly useful in cohesive soils where there is no serious water problem. The machine operates rapidly, and it can excavate a shaft much more quickly than it can be done by hand. For example, a machine of this kind has dug and lined an excavation 6 or 7 ft in diameter and 100 ft deep in about 12 hr. To do the same thing by hand would require twenty or more 8-hr shifts.

In some localities much smaller rotary machines are used to dig shafts from 16 to 24 in. in diameter and up to 35 or 40 ft in depth. Such shafts should be carefully inspected, after which the excavated shafts are filled with concrete to form piers for supporting residences or other light structures.

Sometimes the lower ends of the shafts are belled out by an undercutting bucket, of somewhat different design than described above, to increase the bearing capacity of the pier.

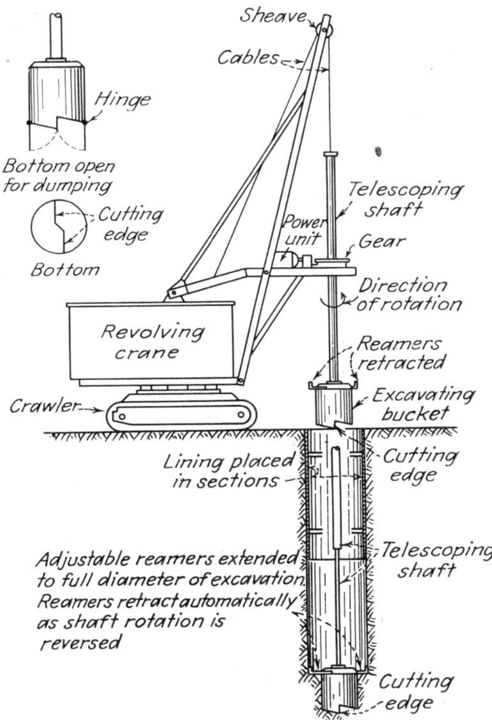

Fig. 60. Dry-process rotary excavating machine.

The other machine-excavation method is known as the wet process. The machine that is used in this process is illustrated in Fig. 61. It is similar to the dry-process machine in that it is a modified crane operating on caterpillar treads, although there are fundamental differences in the operation methods of the two processes. The wet process is an adaptation to caisson construction of the mud-laden-fluid method of rotary drilling oil wells, and uses a cutting apparatus consisting of three or four arms projecting horizontally and radially outward from the bottom of the drill line. Attached to these horizontal arms are a series of teeth like the teeth used on the bucket of a steam shovel.

In the wet process a system of circulating water is maintained throughout the whole operation. A sump is excavated near the machine and a pump is set alongside this sump. The suction hose of the pump is placed in the sump, and the discharge end of the pump is connected to the upper end of the drill line. This is a hollow shaft, and the water goes down through the shaft and is discharged into the excavation at the cutter head.

The water then comes up in the excavation outside the drill line, and it overflows at the surface where it is conducted back to the sump in a small trench. The excavated shaft is kept full of water at all times during the drilling process.

The reason for leaving the water in the excavated shaft is to at least balance, and preferably to overbalance, the hydrostatic head of the water in the surrounding ground. This, acting against the resulting wall cake, prevents the side walls of the excavation from caving in. When the machine is excavating in clays or other fine-grained soils, the circulating water sometimes becomes so muddy that some of it has to be taken out

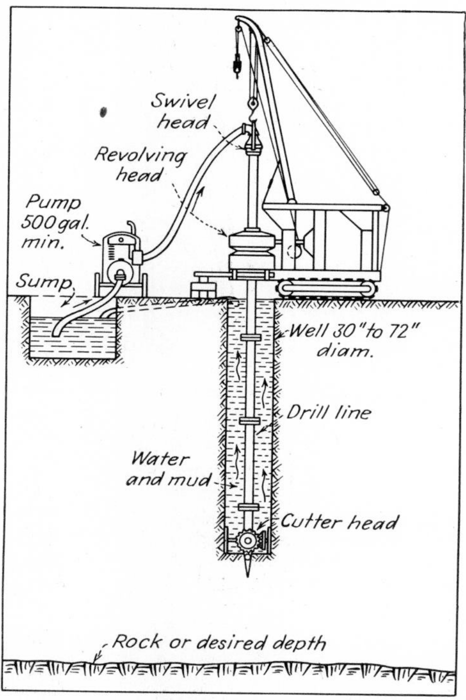

FIG. 61. Wet-process rotary excavating machine.

of the circulating system and replaced with clear water which is added at the sump. It is often necessary to clean out the sump from time to time. When the excavation is being done in clean sand, the circulating water often remains quite clean, and in such cases it is common practice to add clay to the circulating water so as to increase its specific gravity and viscosity. In most cases there is an excess hydrostatic head in the excavation so that there is an active hydrostatic pressure directed radially outward against the walls of the shaft.

To penetrate a gravel or boulder formation overlying rock by means of the wet-process machine or to enter the lining all around the shaft periphery into sloping rock, the lower section of steel lining can be made into a hollow-walled shot core barrel which can be fed with fluid and shot through a hollow-armed spider attached to the bottom of the drill line in place of the cutter head shown in Fig. 61. In this case, the core-barrel section of lining cuts a circular channel, or kerf, through the gravel or boulder

formation and the high-side rock down until its cutting edge is seated all around in the rock.[1]

Drilled-in Caissons. A type of foundation pier developed during recent years is known as the "drilled-in caisson." It consists of a heavy steel pipe with a reinforced cutting edge which is driven and churn-drilled open ended to rock and cleaned out. A socket of about the same diameter as the pipe is churn-drilled to a depth of several feet into the rock. A steel H-beam core extending from the bottom of the socket to above the cutoff elevation is placed inside the pipe, and the socket and pipe are filled with concrete to the cutoff elevation.

CAISSON OR PIER FOUNDATIONS FOR BRIDGES

Forces Acting on a Bridge Pier. The function of a bridge pier is to support the bridge spans with a minimum obstruction to the stream. The possible forces acting on a bridge pier are (1) the end load of the two adjacent bridge spans; (2) the wind load acting on the pier, on the superstructure, and on the moving load; (3) the traction forces due to the live load; (4) the impact of floating debris and ice; (5) the pressure of the stream of water and of solid ice; (6) forces resulting from the expansion and contraction of the superstructure; (7) the weight of the pier itself; and (8) the buoyant force, or hydrostatic uplift, of water around and under the pier.

The wind load for railway bridges is usually specified as 30 lb per sq ft on the projected area of the superstructure, plus 300 lb per lin ft for the pressure on the train, applied 8 ft above the top of rail. An allowance of 30 lb per sq ft on the pier itself should also be made.

The pressure of flowing water depends on the shape of the end of the pier, being greatest for a flat-ended pier. The pressure of the water in pounds per square foot is given by the formula $P = kwV^2/2g$, in which w is the weight of the water per cubic foot; V is the velocity of the stream in feet per second; and k is a factor depending on the shape and proportions of the pier, being $1\frac{1}{3}$ for square ends, $\frac{1}{2}$ for angle ends where the angle is 30° or less, and $\frac{2}{3}$ for circular piers.

The effect of ice on bridge piers, which may be of serious magnitude in streams in cold climates, results from ice in two forms: cake ice floating downstream at the surface of the water to be crushed or split by the nose of the pier, and ice gorges piling up against the nose of the pier, behind floating logs or other debris.

In the first case, the impact of the ice is usually not great, and the effect is largely that of abrasion. With a nose protection of steel, cakes of ice cleave or crush without exerting any considerable force on the pier. In the second case, ice gorges behind floating debris may bring a considerable force against a pier because of the pressure of the current on the ice. In northern climates, icebreaking noses sloped at an angle of about 45° with the horizontal are commonly used, and these allow the ice to slide upon the slope and to break under its own weight.

Stability of a Bridge Pier. A bridge pier should be stable against the forces acting upon it with respect to sliding downstream, sliding in the direction of the axis of the bridge, overturning about the downstream toe, and overturning about one side of the footing. Furthermore, the maximum pressure on the soil under the pier should not exceed the allowable bearing pressure, and the shaft of the pier and the footing should be stable against failures due to internal stresses.

[1] CHRISTIE, H. A., Boring Machine Digs Wells for Concrete Piers, *Eng. News-Record*, vol. 109, No. 4, p. 105, July 28, 1932.
Rotating Foundation Caissons Drill Their Way to Rock, *Eng. News-Record*, vol. 115, No. 2, p. 37, July 11, 1935.
Sinking Caissons in Debris Fill, *Eng. News-Record*, vol. 119, No. 19, p. 743, Nov. 4, 1937.
Clay Augers Sink Foundation Wells, *Eng. News-Record*, vol. 123, No. 15, p. 473, Oct. 12, 1939; vol. 123, No. 25, p. 812, Dec. 21, 1939.

Waterway Requirements. Bridge piers should be designed so that they will not seriously diminish the carrying capacity of the stream bed. Two effects may result from the building of a pier in midstream; viz., the stream may be deflected so as to change the existing relation between the cutting and the filling of the banks of the stream, and the flow may be so obstructed that the carrying capacity of the stream bed will be reduced, with a consequent rise of the water level of the stream.

The best practical form of a pier nose is either the half round or half elliptical, these shapes giving better discharge efficiencies than the pointed nose. The best practical form for the tail of a pier is the half round. The 90° angle nose is not usually satisfactory, and the angle for best efficiencies is 45° or less. Where a pointed nose is used, the shoulders should be rounded to avoid eddying since the eddying caused by angles results in a loss of much of the advantage of a rounded or pointed nose.

Dimensions of Piers. The height of the substructure carrying an overhead crossing must be such that the lowest point of the superstructure will have sufficient clearance above the lower track or roadway. The height of a pier for a stream is sometimes fixed by the necessary height of the roadway over the stream. Its minimum height must be such as to keep the bottom of the superstructure well above high water, so that drift will not lodge against the superstructure and so that the total waterway will be ample to pass the maximum flood. Ten feet clearance above high water will usually suffice to pass drift and ice.

The batter of the sides of a bridge pier should be such as to accomplish the necessary spread of the foundation to procure adequate bearing area. Where the piers rest on rock or other solid foundations, the required spread will be small, and for tall piers a slight batter is often used for the sake of appearance. The unit bearing stress over a solid pier is usually very low; hence high bridge piers are frequently built hollow or divide above the high-water line in order to save masonry.

Piers for Movable Bridges. The special type of pier for movable bridges is the circular pivot pier for swing bridges. The piers for other types of movable bridges do not differ essentially from those for fixed bridges. Where the tail of a bascule girder or the counterweight swings below water level, it is necessary to construct a watertight tail pit as a part of the pier. An overhead counterweight is sometimes employed as an alternative.

Anchor Piers. The anchor piers for cantilever bridges do not differ essentially from ordinary bearing piers. On the other hand, the anchorage of the cables of a suspension bridge requires special attention. Where ledge rock is readily accessible, anchorage is accomplished by tunneling into the rock and placing some form of bearing against the sides of the excavation. In the absence of ledge rock, anchor piers of sufficient mass to hold the pull of the cables must be constructed.

Location of Piers in Navigable Streams. Certain streams of the United States have been declared navigable by Congress, and any structures built over them must conform to the requirements of the War Department. Many of these streams carry little or no traffic as waterways, but provision must be made for the passage of boats if required. When a bridge is to be constructed over a navigable stream, a permit must be obtained from the Secretary of War. The completed structure must not interfere with river traffic, and the construction falsework must also be so arranged as not to interfere. After construction is complete, all piles, cofferdams, etc., must be removed in order to prevent injury to passing boats.

Caissons. A true caisson may be defined as a large watertight box that is used to exclude water and semifluid material during the excavation of foundations and ultimately becomes an integral part of the substructure. The ordinary open caisson (Fig. 62) consists of a box with cross walls that are usually built crib-fashion and sheathed. Sometimes the caisson is constructed of reinforced concrete or of steel plate. In either case,

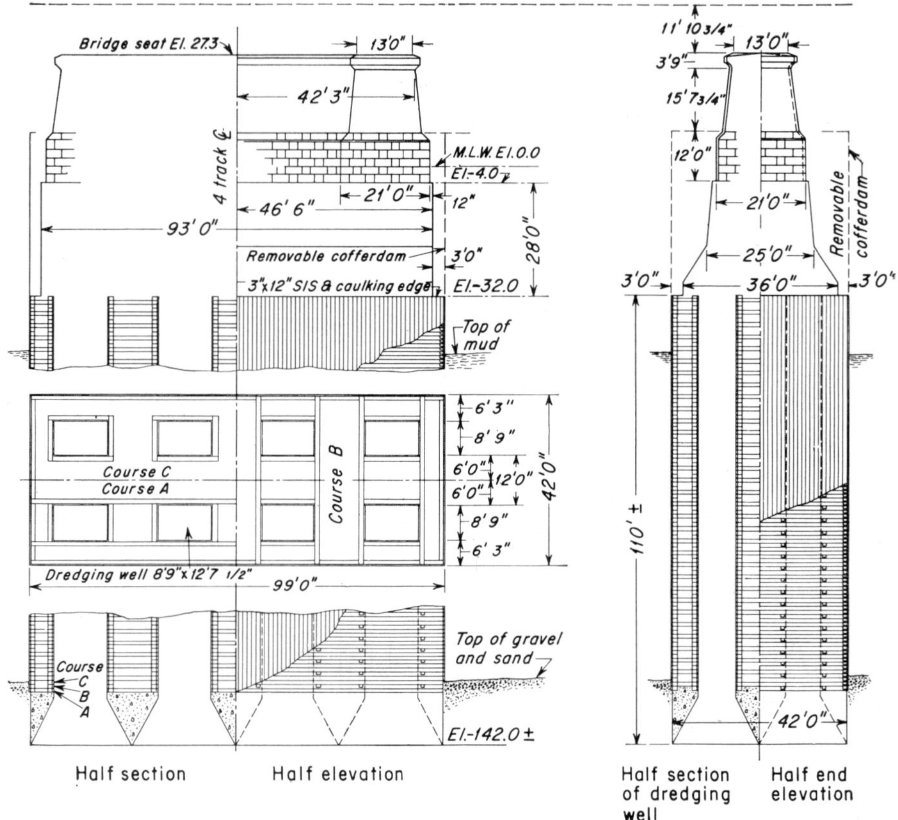

Fig. 62. Open caisson for dredging through wells.

the enclosed earth materials are dredged out as the caisson sinks, and water jets are sometimes used to aid the sinking. A disadvantage of the open-caisson method results from the fact that the caisson is usually not unwatered and hence the bottom cannot be cleaned carefully before the concrete is deposited. In most soils the caisson can be unwatered by pumping after first sealing the bottom under water. The chief difficulties encountered in the installation of open caissons are due to the lack of uniformity of materials encountered during the sinking so that the caisson sinks unevenly and is therefore difficult to guide.

Since the sinking cannot always be carefully controlled, the caisson must be large enough in plan to allow for some misalignment. The dredging wells must be so distributed that the excavation can be of such sequence as to guide the caisson in sinking. Steel-frame cutting edges filled with concrete are often used, and they add stability during launching and prevent listing. Jetting pipes must be placed in such a position that they will not interfere with the dredging. The material inside the caisson is sometimes removed by hand, sometimes by pumping, and often by dredging with a small orange-peel bucket. When the caisson is not sunk to a sufficiently solid stratum to secure adequate bearing, piles may be driven in the bottom.

Artificial-island Method. A special procedure adapted to sinking open caissons and bridge piers through a considerable depth of water is the "artificial-island" or "sand-island" method. It consists of sinking a large steel cylinder around the site of the pier, filling the cylinder with sand or other dredged material to form an artificial island, placing the caisson on this island, and then sinking the caisson by dredging through wells in the usual manner. Figure 63 illustrates the method. Usually the steel shell is filled to such an elevation that the cutting edge can be started on dry soil. By alternate dredging inside the wells and building up the caisson walls, the entire caisson is constructed above the water. After the caisson has been sunk and filled, the steel shell is salvaged and reused. A similar procedure has been used in shallow depths by constructing a "sand island" without the steel-cylinder support.

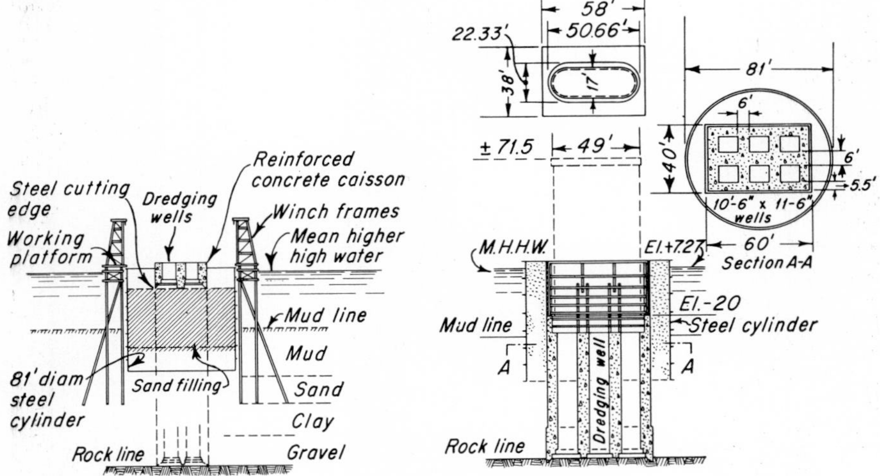

Fig. 63. Arrangement in the sand-island process.

Design. The design of a caisson includes the determination of the dimensions of the caisson; the design of the roof, walls, and cutting edge to withstand the forces to which they may be subject; the determination of the weight necessary to sink the caisson; and the calculation of the flotation depth required for launching and floating to position.

The caisson should be strong enough to support its weight and that of the superimposed load while sinking. The most severe conditions to which it may be subjected while sinking are "bridging" between two obstructions at the ends with soft material between, in which case the caisson acts as a simple beam supported at the ends, and "hogging" over an obstruction at the middle with soft soil under the ends, in which case the caisson acts as a double cantilever. Obviously, the analysis of stresses for these conditions is indeterminate and can be made only approximately.

The weight of the caisson should be sufficient to provide adequate sinking load, which is of the greatest importance because temporarily increasing the weight during the construction of a large caisson is usually impracticable and always very expensive.

The area of the caisson will be determined by the bearing capacity of the soil stratum on which the caisson is to rest and the total load to be carried. The maximum pressure on the foundation bed equals the total weight of pier, caisson, and concrete filling in the working chamber, the superimposed weight of water, earth, the dead and live load from the superstructure, and the moment effects of water and ice thrust, less the skin friction

on the sides of the caisson and less the buoyant effect of displaced water. Where the caisson is to rest on a stratum of rock or other material having greater compressive strength than the caisson masonry itself, the size will be determined either by the strength of the caisson or by other considerations.

Where neither the bearing on the foundation bed nor that on the caisson itself is the limiting factor in determining the size of the caisson, the dimensions will be determined by the size of the pier or other structure that will rest on the caisson. It is usually de-

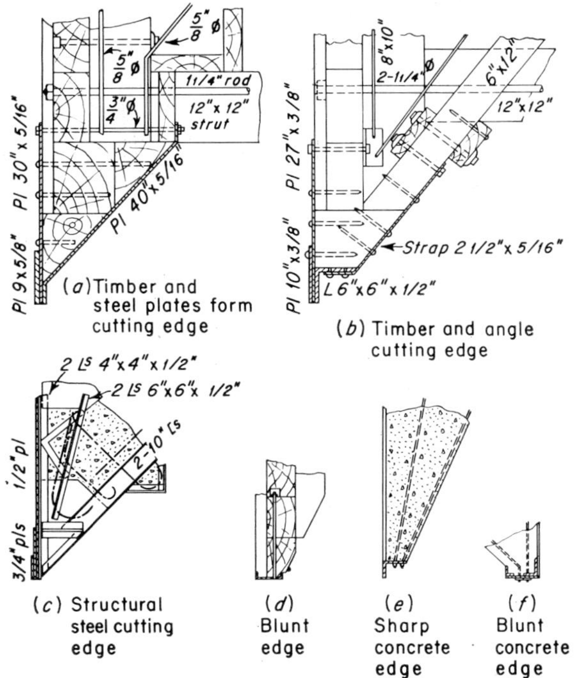

Fig. 64. Typical cutting edges for pneumatic caissons.

sirable to make the caisson somewhat larger than the base of the pier in order to allow for deviation from the exact position in sinking. An allowance of 1 or 2 ft is common for deviation from the true line in the sinking of a caisson.

Two types of cutting edge are in general use, viz., the sharp edge and the blunt edge. The arguments in favor of the sharp edge are that no excavation under the cutting edge is necessary to cause the caisson to sink and that air does not readily escape under the edge because the thin edge penetrates the soil and seals the junction. The disadvantages are that the sharp edge may bend when logs, boulders, or other obstacles are encountered and that the sharp edge does not have sufficient bearing area to support the caisson if one side strikes a soft stratum. The blunt edge does not have the objectionable features of the sharp edge, but at the same time it does not possess its merits. Figure 64 shows sketches of typical designs of cutting edges.

Construction. Caissons may be constructed of timber, concrete, or steel, or of combinations of these materials. Figure 65 shows a sectional elevation of a pneumatic caisson constructed of timber with steel-truss cross bracing just above the working

chamber. The walls of the caisson must be designed to resist the water pressures and earth pressures to which they will be subjected. The working chamber must be sufficiently airtight to prevent serious leakage of air, and the roof of the working chamber must be 8 or 10 ft above the cutting edge so that there is ample headroom for men to work. When weights are to be placed on the roof of the working chamber to aid in sinking the caisson, the roof should be designed not only for the static loads but also with an allowance for impact, since the caisson may drop suddenly and stop quickly during the sinking process.

The walls of the caisson usually extend above the working chamber to such a height that their tops will be above water level when the cutting edge reaches its final position.

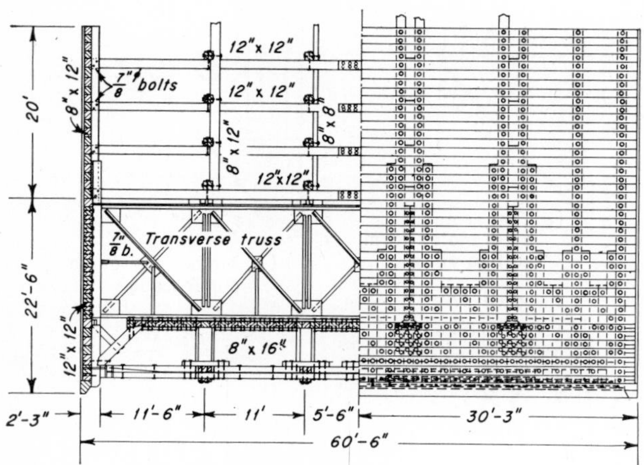

FIG. 65. Pneumatic caisson with steel-truss cross bracing.

The pier is built inside these walls in the dry, and the walls are removed after the pier is completed. The upper portion of the caisson above the working chamber is sometimes referred to as the "crib" or the "cofferdam."

Air Locks. Workmen and materials are taken in and out of the working chamber by means of shafts which are sealed by air locks. The air lock usually consists of an enlarged section of the shaft, fitted with an airtight door at the top and one at the bottom so arranged that the pressure from inside will hold either shut when the other is open. Sometimes the same shaft is used for men and materials, although the shaft for removing spoil is often separate. Where the depth is not great, men usually enter and leave by means of a ladder, although, for greater depths, an elevator may be used. Spoil is commonly removed by means of a bucket and cable through a shaft.

The air lock for men is placed at the top of the shaft in order that men may take refuge temporarily in the shaft in the event of an accident which may allow water to enter the working chamber. When the shaft needs to be extended, a valve is closed at the bottom in the working chamber to prevent the escape of air, the air lock is disconnected from the shaft, the new section of shaft is fastened in place, and the air lock is replaced on top of the added section. Some typical air locks are shown in Fig. 66.

Flotation of the Caisson. In most cases, the caisson is built on shore some distance from the pier site and floated into place. The displacement of the caisson must be carefully calculated from its estimated weight, and the depth of the displacement so adjusted that no difficulty will be encountered in launching and floating into place. The bottom

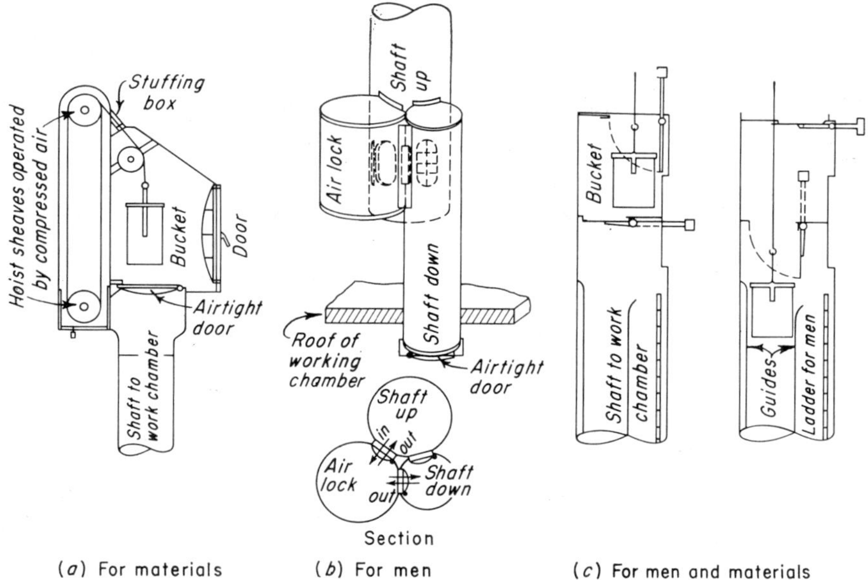

FIG. 66. Typical air locks for pneumatic caisson work.

of the working chamber is sometimes floored temporarily and calked to aid flotation. The position of the center of gravity and the center of flotation must be calculated and arranged with the former well below the latter and directly under it, in order that the caisson may keep an even keel like a ship while it is being towed into position.

Launching and Locating the Caisson. Caissons may be constructed (1) on ways along the shore and launched much as a ship is launched, (2) on pile supports over the proposed site of the pier and lowered into position, (3) in a dry dock and floated by the admission of water, (4) between two barges and floated into position, or (5) on a pontoon and launched by sinking the pontoon from beneath the caisson.

When being moved into place, the caisson is located at the approximate site while still floating. It is moored to pile clusters or held by anchors, and loading is begun and continued until the caisson is landed on the bottom of the river bed. After the caisson is landed on the bottom, an excess of air is sometimes turned into the working chamber, lifting the caisson, which then is located exactly on the center lines of the proposed structure. Where the configuration of the stream bed is such that the earth on one side of the caisson is much higher than on the other, difficulty is likely to be encountered in preventing the caisson from shifting out of position. It is customary to allow for this by landing the caisson somewhat out of position toward the high ground and allowing it to move into place as sinking proceeds.

Mats of woven willow or bags of sand are sometimes placed over the site of the caisson in swift currents, to prevent scouring of the river bed and consequent uneven bearing as the caisson is landed. Then the caisson is sunk through this artificial bed. Where the site of the pier is covered with mud and there is little or no current, the mud can be dredged out before the caisson is landed in place much more cheaply than it can be brought out of the working chamber afterward.

Sinking the Caisson. After the caisson is in place, it is weighted down by filling the pockets provided for that purpose with heavy material so that it will not rise, and then

air is pumped into the working chamber. As soon as the working chamber is dry, men enter through the air locks to begin removing the spoil, and the sinking begins. The ground at the center of the working chamber is generally excavated somewhat lower than under the cutting edge, and the slopes run up from this low point to the cutting edge. The caisson itself is at all times balanced between the downward and the upward forces acting on it, one of the latter being the air pressure in the working chamber. Therefore, in order to drive the caisson downward after excavating up to or under the cutting edge, it is necessary only to temporarily slack off the air pressure in the working chamber, and the excess of downward force usually drives the caisson down. Water jets may be used to diminish skin friction on the side.

The caisson is guided by adjusting the sequence of excavation. It is important to have sufficient stays to keep the caisson plumb until it reaches a depth of 15 to 25 ft in the soil. If it is out of plumb at that depth, it is very difficult to straighten it afterward. It is practically impossible to sink a caisson perfectly plumb and perfectly in position; hence the design must be adjusted to provide for a certain amount of deviation from the desired position. Where one side encounters a soft stratum and the other remains on hard ground, tilting is likely to occur. In a few instances, where caissons have tipped, they have been righted by attaching cables and tackle and pulling them into position, excavating under the high side of the cutting edge at the same time.

The rate of progress varies greatly, depending on the character of the material encountered. It amounts to only a fraction of an inch per day in some cases and as much as 8 or 10 ft per day in others. The air pressure required corresponds approximately to the hydrostatic pressure. The friction resistance to sinking varies with the depth sunk and the character of the materials penetrated. It varies from several hundred to several thousand pounds per square foot, depending on the soil. The maximum depth reached in some notable pneumatic caissons is indicated in Table 2.

TABLE 2. DEPTHS BELOW WATER SURFACE IN PNEUMATIC CAISSONS

Structure	Maximum depth, ft	Gage pressure, psi
Municipal Bridge, St. Louis	113.0	50
Arch Bridge, St. Louis	109.7	
Memphis Bridge, Memphis	106.4	
Williamsburg Bridge, New York	107.5	
Mine shaft, Deerwood, Minn.	123.0	52
Brooklyn Bridge, New York	78.0	
Broadway Bridge, Portland, Ore.	101.0	
Northern Pacific Bridge, Vancouver	80.0	
Harahan Bridge, Memphis	107.0	48
Metropolis Bridge, Metropolis, Ill.	113.2	51

The excavated material must be taken out through the air locks, and frequently light blasting may have to be resorted to in order to reduce boulders to such size that they can thus be removed. That such blasting can be carried on without serious danger of a blowout has been amply demonstrated. When not too numerous, boulders may be carried down by excavating around them, and then they are concreted in as a part of the filling of the working chamber when the work is completed. Sand and silt may be removed by buckets through air locks, but sometimes they are blown out by air pressure. When not in use, the blowout pipe is closed by a valve in the working chamber. Special devices, such as the mud and sand pump, have been used for pumping out sand and silt.

Clays are usually more conveniently removed by means of buckets through the air lock than in any other way.

After the cutting edge has reached the stratum on which the caisson is finally to rest, the caisson is "sealed" by cleaning the floor of the working chamber and filling the working chamber and the shafts with concrete. A very dry mixture of concrete is sometimes tamped in under the cutting edge and under the cross beams. After this dry concrete is in place, concrete of normal consistency is poured in to fill the chamber. When the cutting edge of a caisson lands on a stratum of sloping rock, the low side is underpinned with timbers to keep the caisson evenly supported, the excavation is completed to solid rock, and the working chamber is sealed in the usual manner. Concrete for sealing may be carried in buckets through the air locks, or it may be deposited by a special trap or dump bucket which discharges into the working chamber. Because of the shrinkage of setting concrete, care must be taken to ensure the complete filling of the working chamber after the concrete has taken permanent set.

Effects of Compressed Air on Caisson Workers. Men working under compressed air are subject to a disease which is commonly called the "bends." The principal symptoms of the disease are dizziness, headache, double vision, incoherence of speech, and violent pains, chiefly in the legs.

The cause of caisson disease is the formation in the circulatory system of bubbles of air or of nitrogen, which may clog the circulation or, on expanding, may rupture the blood vessels, particularly the capillaries and the veins. These bubbles form because the blood of the caisson worker has become saturated with gases at the high pressure, and on emergence from the caisson, the external pressure being reduced, these dissolved gases expand, producing bubbles or beads in the blood.

Two general precautions are employed to prevent caisson disease among workers in compressed air, viz., the employment of only those suited to this kind of work, and the regulation of working conditions such as length of shift and time of decompression in the air lock. Men with strong hearts, good circulation, and relatively low blood pressure should be employed, and a thorough physical examination by a competent physician should be the basis of accepting men for such employment.

The chief elements to be observed in working conditions are the period of compression during entrance, the length of shift spent in the working chamber, *the period of rest between shifts,* adequate ventilation of the working chamber, and, last but most important, the time of decompression when leaving the working chamber. The time required for compression is brief and, beyond the requirements for physical comfort, does not require special attention. The main difficulty arises in decompression, for it is impracticable to allow as much time for decompression as certain physiological experiments indicate to be desirable. Two methods of procedure are followed, one employing a uniform rate of decompression, and the other a stage decompression, rapid at first and then more slowly. The total time allotted is not greatly different in the two methods.

In decompressing uniformly, the time allotted by different agencies varies. For depths less than 50 ft, or about 20 psi, little difficulty is encountered. For greater depths, special care must be exercised. A set of rules commonly used for compressed-air work follows.

Gage pressure, psi	10	15	20	25	30	35	40	50
Minutes in decompression	1	2	5	10	12	15	20	25

Time distribution	Gage pressure, psi					
	0–21	22–30	31–35	36–40	41–45	46–50
Total hr per day in caisson	8	6	4	3	2	1½
Number of shifts	2 (min)	2	2	2 (min)	2 (min)	2
Length of shift, hr	....	3	2	1½ (max)	1 (max)	¾
Minimum time between shifts, hr	½	1	2	3	4	5

The stage-decompression method is based on the theory that bubbles of gas are not released in the blood where the pressure in the blood does not exceed atmospheric pressure by more than 19 psi. Hence the pressure can be diminished quickly by that amount in about 3 min and then more slowly for the remainder of the time.

In general, slow decompression is the most effective means of preventing caisson disease, and this method is always used for deep work.

The only effective treatment for caisson disease is recompression and then slower decompression. If this is done before the air bubbles have had time to rupture the blood vessels, a cure can almost always be effected. Otherwise, recompression is of little value. A recompression tank or chamber, usually called a medical lock, is kept available outside the caisson, and affected men may be placed in this, recompressed, and then decompressed at a sufficiently low rate to ensure safety.

STRESS DISTRIBUTION BELOW FOUNDATIONS

Pressure in Plane of Contact. One of the fundamental problems of foundation engineering is to determine the probable settlements that will occur when the load of a structure is applied to soil. In order to analyze this problem, it is necessary to know what stresses are generated in the soil by the loads. The problem may be divided into two parts: the distribution of pressure in the contact plane between the foundation and the soil, and the distribution of stresses in the soil at various depths below the foundation.

The first part of the problem has been investigated theoretically and experimentally. The theoretical investigations are based on that part of the theory of elasticity which deals with the pressures that occur in the plane of contact when two elastic bodies are pressed together. In general, the distribution of pressure depends on the relative elastic properties of the two bodies. If a perfectly rigid circular die is pressed against the surface of an elastic isotropic solid, the distribution of pressure in the plane of contact is that illustrated by Fig. 67. The pressure under the edges of the rigid die is theoretically infinite while the pressure under the center is equal to one-half the average pressure. The surface of the elastic solid is depressed so that the displacement is uniform under the die and the contact surface remains plane and horizontal. Beyond the perimeter of the die the entire surface is depressed into a saucer-shaped depression with the displacement tapering off to zero at an infinite distance from the die. Theoretical solutions for the pressures and displacements are available for various types of surface load distributions. All these require the assumption that the bodies of contact are ideal elastic material, *but actual soils do not satisfy this assumption.*

Many experiments have been made to determine the distribution of contact pressure between loaded plates and soils of various kinds. Tests have been made on soils ranging from loose, dry sand to hard clay, with test plates ranging from perfectly rigid dies to very flexible plates. The experiments of Faber [1] illustrate the difference in behavior be-

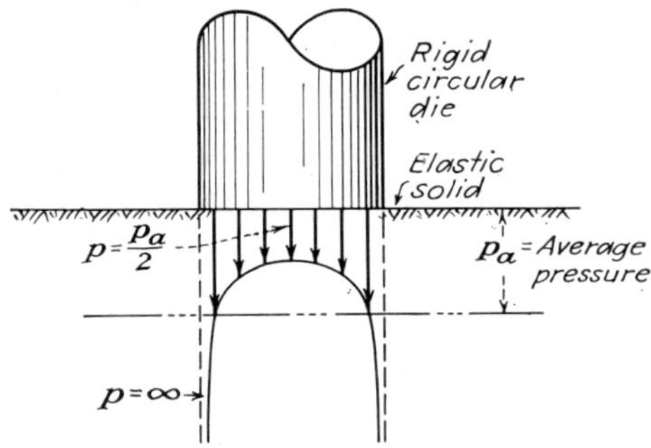

Fig. 67. Pressure distribution under rigid circular die.

tween sands and clays with loads applied to the surface and with loads applied at a depth below the surface.

The results of Faber's tests are illustrated in Fig. 68*a*, *b*, *c*, and *d*. The test plate he used was not a perfectly rigid die, but it was very rigid with respect to the soils that were loaded. Figure 68*a* shows the results of a load test on sand with the load at the surface. The pressure in the contact plane was a maximum under the center of the plate, and there was almost no pressure on the soil under the edges. This is a very different pressure distribution from that shown in Fig. 67, and the difference is due to the fact that the sand could not resist tensile stresses and had a tendency to escape from under the edges of the test plate, thereby leaving the sand under the center of the plate to carry most of the load. When the sand surrounding the test plate was loaded with a surcharge, the pressures in the contact plane were those shown in Fig. 68*b*. The tendency of the sand to escape from under the edge of the plate was resisted by the surcharge, and the pressure was more uniformly distributed under the plate, although it was still greatest under the center and least under the edges.

The results of similar tests on clay are shown in Fig. 68*c* and *d*. When the load was applied at the surface of the clay, the distribution of pressure in the contact plane was that shown in Fig. 68*c*, which is seen to be similar to Fig. 67. When large loads were applied to the plate, high pressures were measured under the edges of the plate, with pressures under the center of the plate somewhat less than the average. This is due to the ability of the clay to resist shear and some tension. When the clay surrounding the test plate was loaded with a surcharge, the pressures under the edges were reduced, as shown in Fig. 68*d*, and the distribution of pressure was more uniform.

Information about the distribution of pressure in the contact plane between a footing and the soil is necessary for the determination of the bending moments to which the loaded footing would be subjected. However, Faber has investigated the probable bending moments that could be expected on the basis of his experiments. The results of his calculations are shown on Fig. 69. For the ideal pressure distribution on sand, the maximum bending moment in an isolated spread footing is somewhat less than that produced under the usual assumptions of uniform distribution. For the ideal pressure distribution on clay, the maximum bending moment is only slightly greater than the

[1] Faber, Oscar, Pressure Distribution under Bases and Stability of Foundations, *The Structural Engineer (London)*, March, 1933, p. 116.

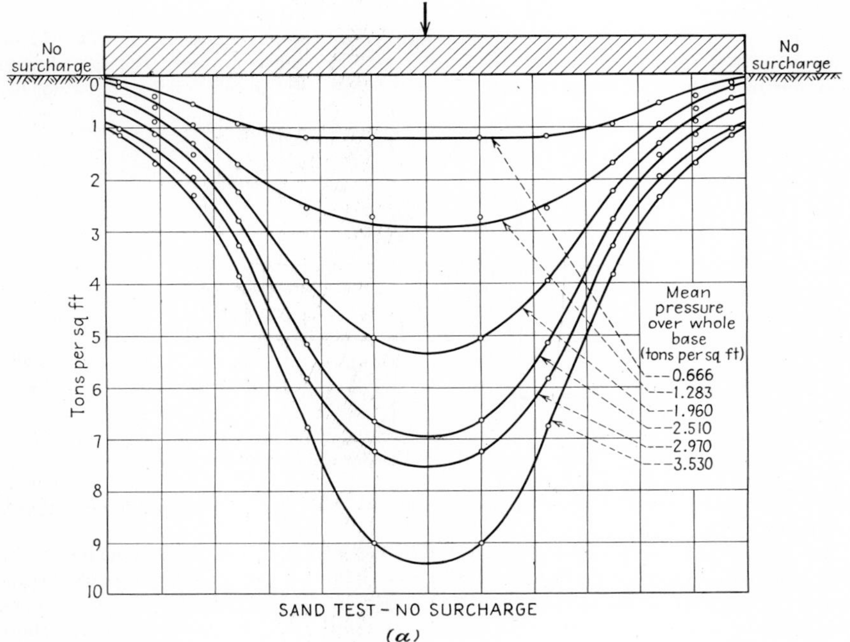

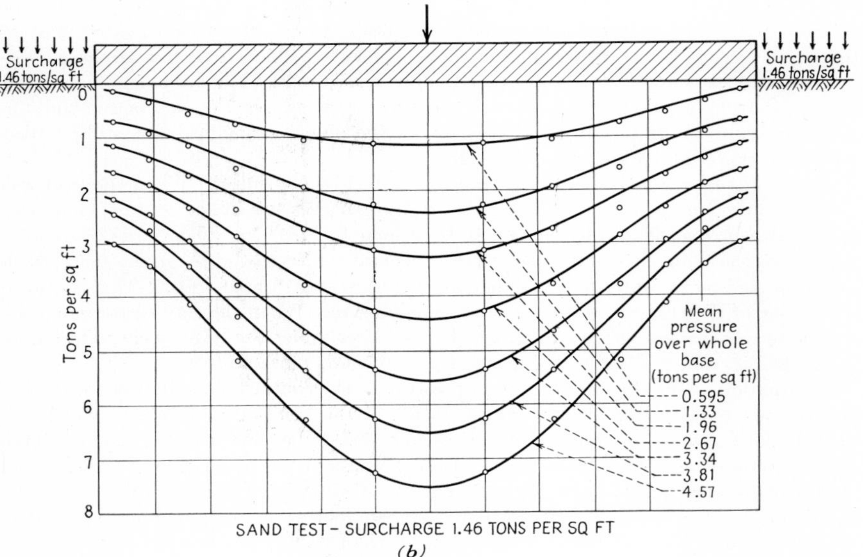

Fig. 68. Pressure distribution under loaded plate: (*a*) on sand without surcharge, (*b*) on sand with surcharge. (*After Faber.*)

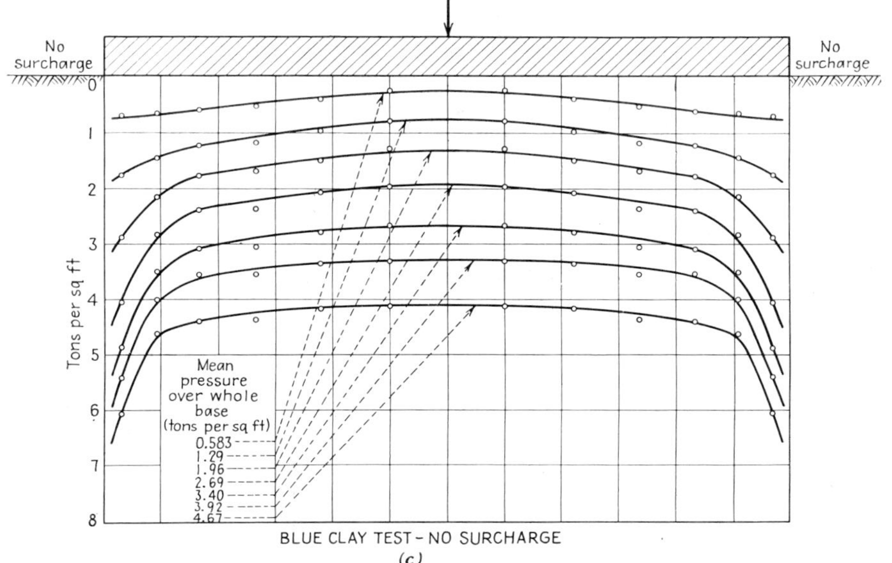

(c)

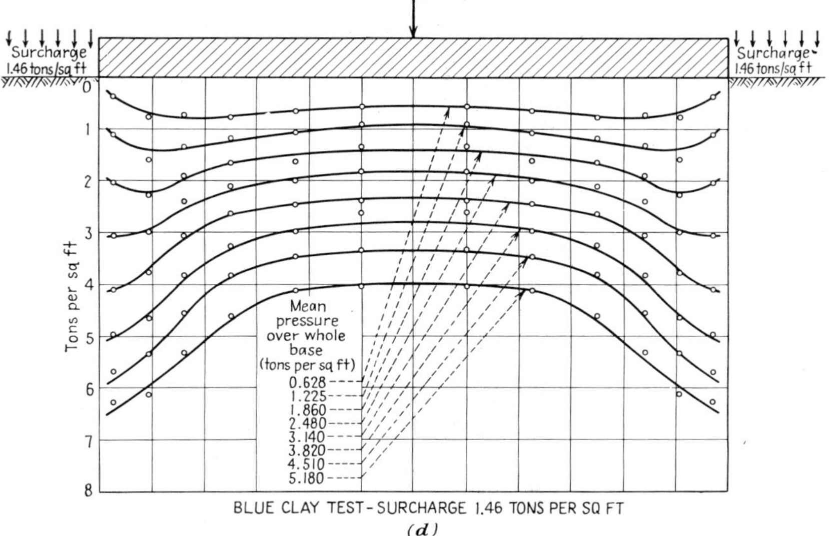

(d)

Fig. 68 (*Cont.*). (*c*) On clay without surcharge, and (*d*) on clay with surcharge.

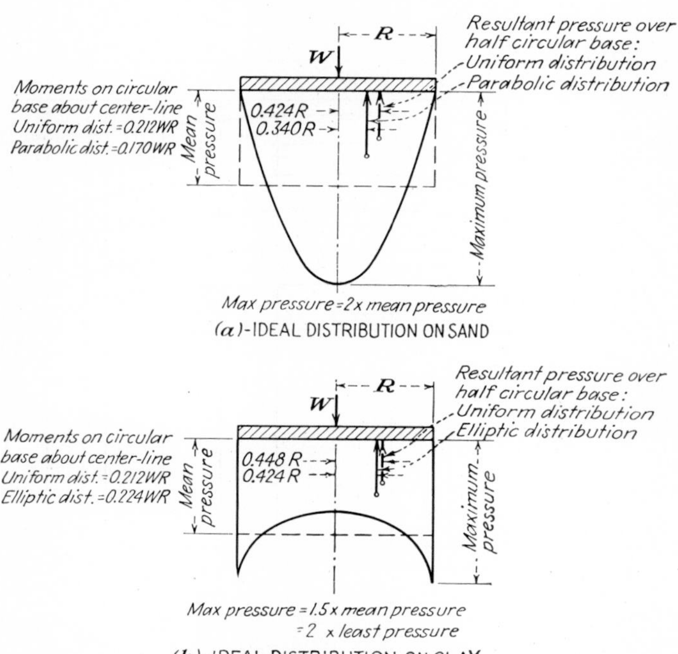

FIG. 69. Ideal pressure distribution under bearing plates on sand and clay. (*After Faber.*)

moment due to uniform distribution. The relative rigidities of an actual footing and the soil on which it is founded would be similar to the relative rigidities of the test plate and the soil in Faber's experiments. It can be seen from his results that bending moments in isolated spread footings computed on the assumption of uniform pressure distribution would not be seriously in error either on sand or on clay.

Stresses in the Subsoil. Information on the distribution of stresses in the soil at various depths below the foundation is necessary for settlement analyses in problems where the cause of the settlement is the compression of deep-seated soil strata. The mathematical methods of stress distribution involve the use of differential equations, but the resulting formulas for the stresses are comparatively simple.

The basic solution for the stress distribution in a semi-infinite elastic isotropic solid loaded with a point load applied vertically to its plane horizontal boundary was derived by the French mathematician J. Boussinesq.[1] The problem is illustrated in Fig. 70*a*, in which a point load P is applied to the surface of a semi-infinite solid. A three-dimensional rectangular coordinate system is established with the origin at the point of application of the load. The XY plane is the horizontal boundary of the solid, and the positive Z axis is directed downward into the solid. The problem is to determine the normal stresses and the shears at any given point in the solid whose vector distance from the origin is R. Figure 70*b* shows the three normal stresses and the three pairs of conjugate shears on the faces of a small elementary parallelepiped at the end of the vector R. The normal stresses have a single subscript, and the shears have double subscripts.

[1] "Application des Potentiels," Paris, 1885.

For a settlement analysis, the only stress that need be considered is the vertical normal stress on horizontal planes, and this stress is given by the equation

$$p_z = \frac{3Pz^3}{2\pi R^5}$$

in which the letters have the meanings shown in Fig. 70.

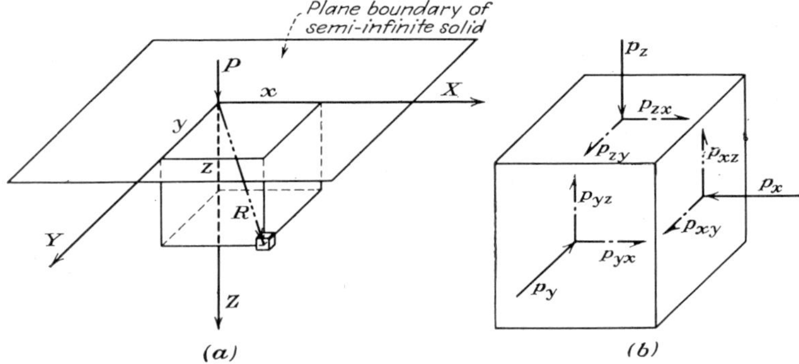

Fig. 70. Stress distribution in semi-infinite solids.

Figure 71*a* is a graphic representation of the vertical stress on a horizontal plane as computed from this equation. The maximum stress occurs immediately under the load, and the intensity of the stress decreases rapidly in relatively short horizontal distances. The stress becomes zero at infinite horizontal distances from the point of application of the load, so that the stress surface on any horizontal plane is a bell-shaped surface with the rim of the bell extending outward to infinity in all directions. Figure 71*b* is a graphic representation of the contours of equal vertical pressures in the subsoil. In three dimensions the surfaces of constant vertical pressure are bulb-shaped surfaces which are often called "pressure bulbs."

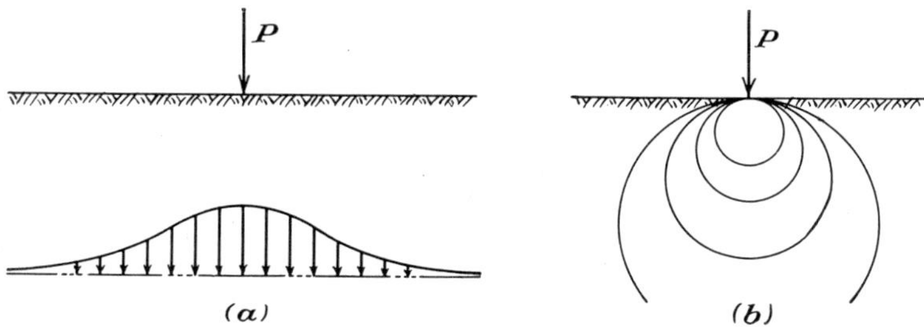

Fig. 71. Stress distribution under point load.

The derivation of the theoretical stress equations is based on certain assumptions which are rarely satisfied in actual foundation problems. One of these assumptions is that the elastic solid is homogeneous and isotropic to an infinite depth. Actual soils are usually stratified, with hard layers and soft layers lying one over the other, and, in most places on the earth, solid rock or some other practically incompressible material is found at depths which are of the same order of magnitude as the horizontal dimensions of a good-

sized structure. *Because actual soil conditions do not satisfy the basic assumptions on which Boussinesq's analysis was made, various modifications of the stress analysis have been developed.*

One of the first of these modifications was the introduction of a quantity called a "concentration factor." In an effort to check the results of a considerable number of small-scale load tests, John H. Griffith [1] and O. K. Froelich [2] developed equations for the stress distribution which included the concentration factor.[3] M. A. Biot [4] derived stress equations for the case in which the soil rests on a rigid underlying boundary. Biot's equations for point loads have been integrated for distributed loads, and it has been shown that the Griffith-Froehlich concentration factor can be considered as a boundary effect brought about by the conditions under which the laboratory tests were made.[5]

D. M. Burmister [6] has derived stress equations for the case in which a stiff crust of soil overlies a soft compressible soil. N. M. Newmark [7] has investigated the case in which the modulus of elasticity of the soil in a horizontal direction is greater than its modulus in a vertical direction.

A summary of these various modifications can be stated briefly as follows: When the soil is underlain by rock or some other rigid material, the vertical normal stress is concentrated so that the stress surface shown in Fig. 71*a* has a higher peak under the center of the loaded area. When a stiff crust of soil overlies a soft soil or when the soil has a larger elastic modulus horizontally than vertically, the vertical normal stress is diffused over a wider area so that the stress surface in Fig. 71*a* is flattened out.

In general, it can be expected that the Boussinesq equation will give a satisfactory approximation of the vertical normal stress, and it is largely a matter of judgment as to whether the stress computed from this equation should be increased or decreased in order to take into account the stratification of the soil.

Figure 70 and the stress equation given above refer to a load that is called a "point load." The actual loads applied to the soil by the foundations of a structure are distributed over areas which are sometimes of considerable size. When a stress-distribution calculation is being made for the stresses under a comparatively small spread footing and when the calculation applies to a depth below the footing that is two or three times the diameter of the footing, the footing may be considered as a point load and the point-load equation may be used for the stress calculation. When the stress calculation is being made for depths in the soil that are less than two or three times the diameter of the loaded area, it is necessary to divide the loaded area into small sections and to make a stress calculation for each section as though it were a point load. The stresses computed at a given point for each section of the loaded area are then added together to get the total stresses due to the entire area. The point-load equations have been integrated for various shapes of loaded areas and for various types of surface load distribution. Also, tables and charts are available from which the stress distribution under large loaded areas can be determined with comparatively little computation.[8]

In addition to the stresses produced in the subsoil by external loads applied to the surface, it is necessary in a problem of settlement analysis to take into account the

[1] Pressures under Substructures, *Eng. and Contr.*, March, 1929, p. 113.

[2] Drukverdeeling in Bouwgrond, *Ingenieur (Utrecht)*, Apr. 15, 1932, p. B-51.

[3] CUMMINGS, A. E., Distribution of Stresses under a Foundation, *Trans. ASCE*, vol. 101, p. 1072, 1936.

[4] Effect of Certain Discontinuities on the Pressure Distribution in a Loaded Soil, *Physics*, vol. 6, p. 367, December, 1935.

[5] CUMMINGS, A. E., Foundation Stresses in an Elastic Solid with a Rigid Underlying Boundary, *Civil Eng.*, vol. 11, No. 11, p. 665, November, 1941.

[6] The Theory of Stresses and Displacements in Layered Systems, *Proc. Highway Research Board*, vol. 23, p. 126, 1943.

[7] Stress Distribution in Soils, *Proc. Purdue Conference on Soil Mechanics and Its Applications*, 1940, p. 295.

[8] NEWMARK, N. M., "Simplified Computation of Vertical Pressures in Elastic Foundations," *Univ. Illinois Eng. Expt. Sta. Circ.* 24, 1935; "Influence Charts for Computation of Stresses in Elastic Foundations," *Univ. Illinois Eng. Expt. Sta. Bull.* 338, 1942.

stresses produced by the weight of the soil itself. If the soil has a unit weight w, the vertical normal stress, at any depth h, due to the weight of the soil itself is given by

$$p_z = wh$$

In a settlement-analysis problem, the compressible soil stratum is sometimes found at a considerable depth below the surface where it has been subjected to an overburden load as given by this equation. In many cases the stress produced in the compressible stratum by the external load is only a small percentage of the stress already applied to that stratum by the overburden above it. Under such conditions, the added stress from the external load would produce comparatively little compression in the soft stratum. In all cases the stresses produced by the weight of the soil itself should be taken into account in the stress-distribution analysis.

THEORY OF CONSOLIDATION

Theoretical Calculations. One of the principal causes of the settlement of structures is the phenomenon known as "consolidation." The term consolidation is usually used in connection with beds of saturated clay, but any process which reduces the water content of a bed of saturated soil is called a "process of consolidation." In saturated sands and gravels which have comparatively large pore spaces between the solid particles, consolidation due to applied loads occurs in a very short period of time because the water can escape freely through the voids in the soil. In a fine-grained cohesive soil such as clay, the pore spaces between the soil grains are of capillary dimensions, and water cannot flow freely through these passageways. When a soil of this kind is subjected to external loads, stresses are set up in the soil which tend to compress it and to force the water out of the voids. Since the voids are so small that the water cannot escape freely, the extra pressure produced by the external load is carried at the beginning of the process partly by the water in the soil and partly by the solid soil particles. The water is said to be under excess hydrostatic pressure.

Gradually, under the effect of this excess hydrostatic pressure, the water escapes from the clay. The removal of water from the clay permits the solid particles to crowd closer together. The space arrangement of the solid soil particles is changed and the volume of the bed of soil is reduced. This consolidation process proceeds slowly until the soil has reached a state of equilibrium under the applied load. If the load is increased, the process begins again and continues until the soil has reached a new state of equilibrium under the new load. This condition of equilibrium is referred to as 100 percent consolidation under the given load.

In 1925 Terzaghi[1] discovered that this process of consolidation could be interpreted mathematically by means of a thermodynamic analogy. The flow of water in the soil during the consolidation process is analogous to the flow of heat in isotropic bodies. The excess hydrostatic pressure in the water corresponds to the temperature in the heat-flow problem. The coefficient of permeability of the soil corresponds to the coefficient of heat conductivity, and the coefficient of consolidation of the soil corresponds to the diffusivity in the heat problem.

It is therefore possible to write the differential equation of the consolidation process as follows:

$$c_v \frac{\partial^2 u}{\partial z^2} = \frac{\partial u}{\partial t}$$

in which c_v is the coefficient of consolidation, u is the excess hydrostatic pressure, z is the depth at which the excess hydrostatic pressure is being computed, and t is time.

[1] TERZAGHI, KARL, "Erdfaumechanik," Deuticke, Vienna, 1925.

This equation is easily recognized as the Fourier differential equation for the flow of heat in a solid where u is temperature and c_v is the coefficient of heat diffusivity of the material of which the solid is made.

Solutions of the differential equation have been obtained for various sets of boundary conditions. The factors that determine the boundary conditions are the nature of the stresses in the clay bed and the permeability of the soils above and below the clay bed. All the solutions of the equation are in the form of Fourier series, but most of the series converge rapidly enough so that only a few terms need to be taken into account in the computations.[1]

Practical Applications. Figure 72 illustrates a problem for which a settlement analysis might be made on the basis of the theory of consolidation. The soil consists of a bed of sand overlying a thick bed of soft clay. Below the clay is a layer of hardpan resting on

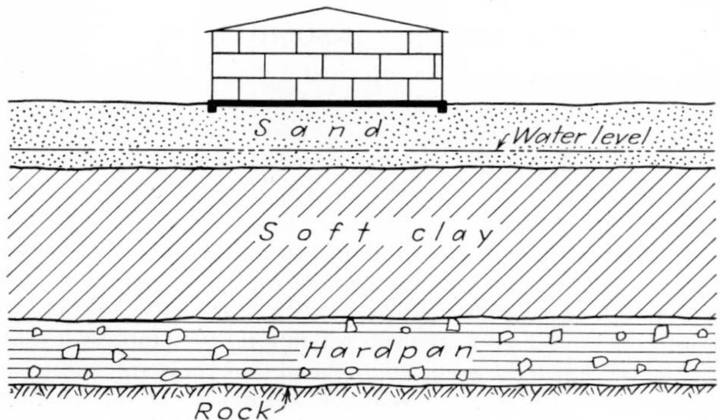

FIG. 72. Structure subjected to consolidation settlement.

rock. The free-water table in the ground is in the sand bed. The clay is assumed to be completely saturated, and all the soil is assumed to have existed in the condition shown in the figure for a sufficient period of time to have reached equilibrium. A heavy tank is to be constructed on a spread foundation at the ground surface, and the problem is to predict the probable settlement of the tank due to consolidation of the clay.

The first step is to secure undisturbed samples of the clay. These are subjected to a series of tests in the laboratory to determine the value of the coefficient of consolidation. The next step is to compute the stresses in the soil. Before the tank is built, the stresses in the soil will be those due to the weight of the soil itself. The stress at the top surface of the soft clay bed will be that due to the weight of the sand above the clay. The stress at the bottom of the soft clay bed will be that due to the weight of the sand and the soft clay. These are called the "initial stress conditions." Since the ground-water table is in the sand bed, the weights used in the calculation for all soils below the water table should be the submerged weights.

After the tank is built and filled with liquid, the load of the tank will increase the stresses in the soft clay bed. These stresses are computed by the methods described under "Stresses in the Subsoil." The stresses produced by the weight of the tank are added to the initial stress, and the resulting stresses represent the final stress condition in the soft clay bed.

For the soil condition illustrated in Fig. 72, it can be assumed that drainage of the soft clay bed will occur only at the top surface of the clay. In other words, the hardpan

[1] TERZAGHI, KARL, and O. K. FROEHLICH, "Theorie der Setzung von Tonschichten," Deuticke, Vienna, 1936; TERZAGHI, KARL, "Theoretical Soil Mechanics," John Wiley & Sons, Inc., New York, 1943.

below the clay can be assumed to be impervious so that water driven out of the clay by the load of the tank will have to percolate upward into the sand. The differential equation must therefore be integrated so as to satisfy the boundary conditions that there is upward flow at the top surface of the clay but no downward flow at the bottom surface. The resulting equation can then be used to determine the time rate of consolidation and the percentage of consolidation that will occur at any given time.

If it is desired to know only the total amount of consolidation that will occur and not the rate of consolidation, it is not necessary to use the differential equation. The laboratory tests determine the relationship between void ratios and pressures for a series of

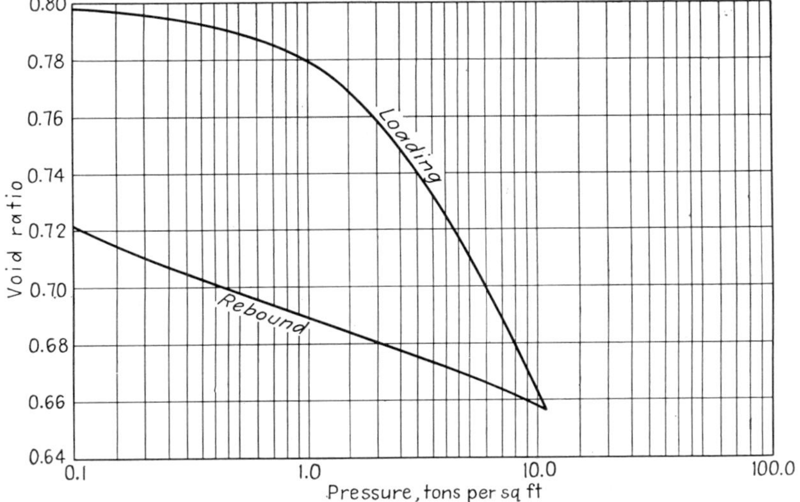

FIG. 73. Void ratio-pressure curve for consolidation settlement calculation.

different pressures, and this relationship can be represented by a curve as shown in Fig. 73. When the initial and final stress conditions have been computed, as already described, values of the void ratios corresponding to these pressures can be obtained from the curve. The total settlement that can be expected to occur may then be determined from the equation

$$S = h\frac{e_i - e_f}{1 + e_i}$$

in which S = settlement, in.
h = thickness of clay bed, in.
e_f = final void ratio
e_i = initial void ratio

In order to determine initial and final void ratios for substitution into the settlement equation, it is customary to compute the average of the initial stresses at the top and the bottom of the clay bed and the average of the final stresses at the top and bottom and to select from the curve the void ratios that correspond to these average stresses.

Experience has shown that settlement analyses of this type, *when made by a skilled and experienced soils technician*, can be expected to predict the settlement of a given structure with reasonable accuracy. Ordinarily the question to be answered is whether the tank would settle a few inches or a few feet. It is not necessary to make an elaborate

and expensive investigation to answer this question. In most cases, *experienced judgment based on a few simple tests* will determine the probable amount of settlement with sufficient accuracy so that a decision can be made as to whether the structure can be allowed to settle or whether it is necessary to install a special foundation to eliminate the possibility of settlement.

SETTLEMENT OF STRUCTURES

Principal Causes of Settlement. The principal problem in all foundation engineering is to design the foundation so that the settlements will be as small and as uniform as it is possible to make them at a reasonable cost. In order to approach the settlement problem intelligently, it is well to have in mind an outline of the principal causes of settlement. Such an outline is shown in the following table:

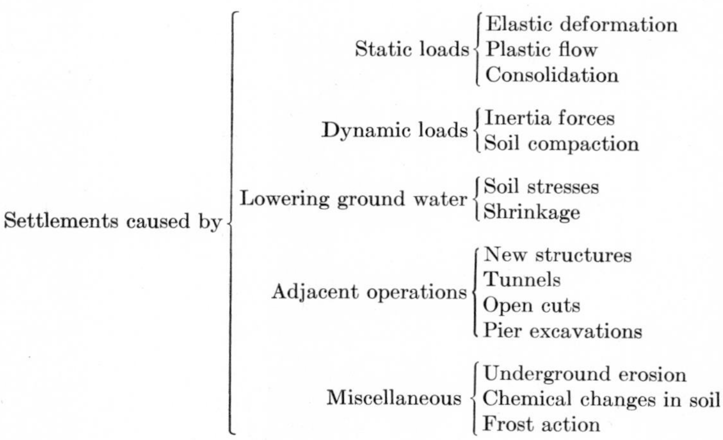

Settlements caused by	Static loads	Elastic deformation
		Plastic flow
		Consolidation
	Dynamic loads	Inertia forces
		Soil compaction
	Lowering ground water	Soil stresses
		Shrinkage
	Adjacent operations	New structures
		Tunnels
		Open cuts
		Pier excavations
	Miscellaneous	Underground erosion
		Chemical changes in soil
		Frost action

Figure 74 refers to settlements due to elastic deformation or plastic flow of the soil below the foundation. Figure 74*a* represents a loaded area on the ground surface. The dotted line shows the manner in which the soil surface is depressed under such a load. If the amount of the load is increased up to a certain critical value, actual failure takes

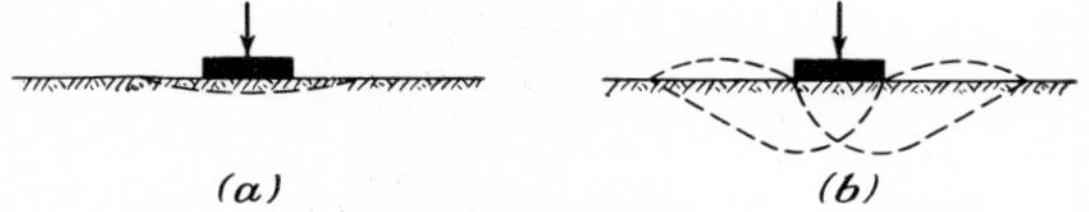

FIG. 74. Soil deformation under loaded test plates.

place as shown in Fig. 74*b*. The soil fails on shear planes indicated by the dotted lines, and there is an upheaval of soil around the loaded area.

Complete shear failures of foundations in this manner are frequently seen in the failures of docks used for the storage of iron ore, limestone, coal, etc., of which there have been a number around the Great Lakes. An example of a complete foundation failure of this type is shown in Fig. 75. The sketch represents a grain elevator that was built in western Canada about 25 years ago as shown in Fig. 75*a*. The approximate dimensions of the structure are shown on the sketch. Excavation for the foundation was carried about 12 ft below the level of the prairie. Below the foundation there was a stratum of firm blue clay 35 or 40 ft thick and then a 5-ft stratum of clay and boulders overlying limestone rock.

The grain elevator was constructed without difficulty and with little or no settlement. When the bins were filled with grain, care was taken to keep the grain at approximately the same level in all the bins as the filling proceeded. However, when the bins were almost full, the structure began to tilt. Within 24 hr the structure tilted into the position shown in Fig. 75*b*. It was about 27° out of plumb when it came to rest. On the

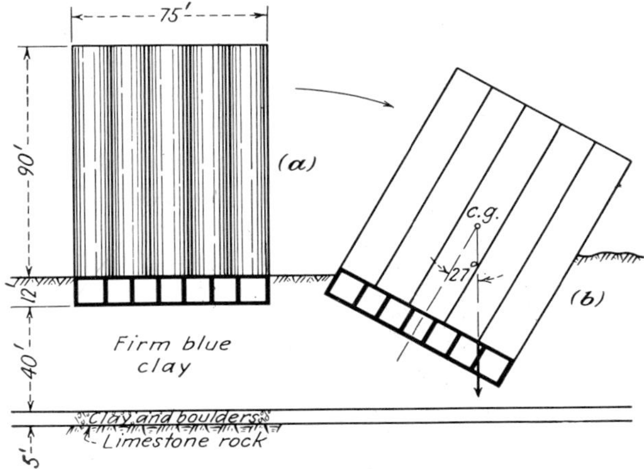

FIG. 75. Settlement caused by shear failure in soil.

side toward which the elevator tilted, the soil heaved up, and this upheaved soil probably prevented the elevator from turning over completely.

Consideration of the soil conditions shown in Fig. 75*a* indicates that the firm blue clay was subjected to what amounted to a large-scale squeeze test. The clay was squeezed between the practically rigid foundation of the grain elevator and the underlying rock. Evidently the shearing stresses in the soil under the structure exceeded the shearing resistance of the soil to such an extent that complete failure of the soil occurred.[1]

Figure 76 illustrates soil conditions where settlements occur because of the consolidation of saturated clay beds. Figure 76*a* represents the soil condition that is typical in downtown Chicago. The street grades are approximately 15 ft above lake level, and most of the older buildings have basements which were excavated approximately to city datum. At this elevation there is a stratum of stiff clay which varies from 5 to 10 ft in thickness in different parts of the city. Load tests on small areas were made on this stiff clay; but, at a depth of 5 or 10 ft below a loaded area only 1 ft square, the stresses in the soil practically vanish. Accordingly, the load test on top of the stiff clay stratum applied practically no load to the soft blue clay below. When the entire building was built, the loads of the building produced stresses of considerable magnitude in the thick stratum of soft blue clay. As a result, some of the older structures in downtown Chicago have settled as much as 2 ft, and these settlements were brought about by the consolidation of the soft blue clay.

Figure 76*b* represents a structure built on wood-pile foundations which settled in a similar manner because of consolidation of the underlying clay. The basement excavation was carried to a depth of about 18 ft, after which wooden piles were driven through about 26 ft of soft silt and clay to bearing in a stratum of sand. This sand stratum is

[1] ALLAIRE, ALEXANDER, The Failure and Righting of a Million Bushel Grain Elevator, *Trans. ASCE*, vol. 80, p. 799, 1916.

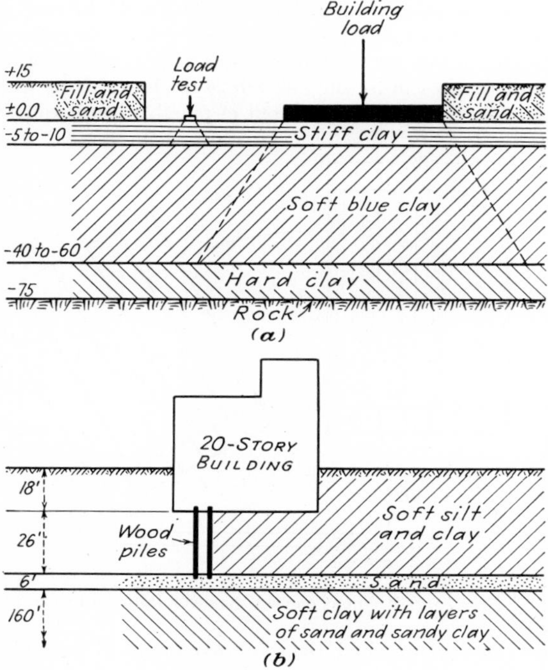

Fig. 76. Settlement caused by underlying soft stratum.

underlain with a deep bed of soft clay containing layers of sand and sandy clay. The dynamic resistance developed by the piles was satisfactory, and load tests on single piles were also satisfactory. However, the size and weight of the building were such that stresses of considerable magnitude were transmitted through the piles and through the sand stratum into the underlying soft clay. This structure settled almost 1 ft in the first year after it was built.[1]

Another type of settlement that is frequently encountered is that due to dynamic forces. Settlements from this cause can occur in all kinds of soils, but they are most common in loose sand beds. The settlements are due to the action of stress waves radiating from the source of disturbance and to inertia forces. The source of disturbance might be a reciprocating machine, heavy traffic, an earthquake, or a pile-driving operation.

Figure 77*a* represents a source of disturbance on the ground surface throwing out stress waves into the soil as indicated by the dotted curve. At some distance from this source a structure of weight W rests on the ground. Under the action of the stress waves, this structure is subjected to accelerations which produce inertia forces so that the pressure between the structure and the ground exceeds the static pressure produced by the weight W. As a result of these inertia forces due to the dynamic disturbance, the structure tends to settle slowly into the ground when the dynamic disturbance is continued for a considerable period of time. The amount of settlement that might occur in a given case would depend on the strength of the source as compared with the

[1] Terzaghi, Karl, Soil Mechanics—A New Chapter in Engineering Science, *Jour. Inst. Civil Engrs. (London)*, 1939, p. 1933.

mass of the structure. A very weak source of vibration would have little or no effect on a very heavy structure.

Figure 77*b* illustrates a condition in which compaction of a loose sand bed was caused by a pile-driving operation. The soil consisted of approximately 80 ft of loose sand and gravel overlying rock. The building area was excavated to a depth of about 10 ft, and piles were driven in the bottom of the excavation. Before the piles were driven, short reinforcing rods were driven into the ground to serve as level bench marks. Some of these were driven in the bottom of the excavation, others were on the side slopes, and others were beyond the top of the bank on the level ground outside the building area. The piles were driven with a single-acting steam hammer delivering 15,000 ft-lb per blow at the rate of 55 blows per minute. After 100 or more piles had been driven, it was

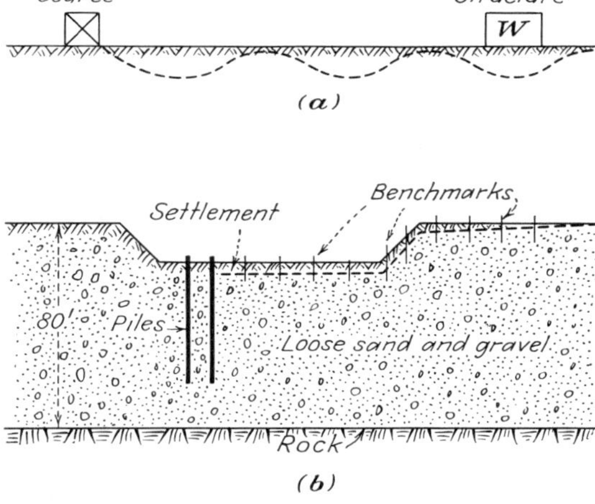

FIG. 77. Settlement due to vibration.

found that the vibrations had caused the entire area to settle, as indicated by the dotted line. The maximum settlement was in the bottom of the excavation, where it amounted to about 6 in. There was settlement of the side slopes and settlement of the top of the bank, which tapered off to a feather edge at a distance of 50 or 75 ft from the top of the slope. On this particular job no serious consequence resulted from this settlement. The site was in an open field, and the principal purpose of the piles was to compact the loose bed of sand and gravel. In some cases settlements of this kind can have serious consequences when the work is being done in an area surrounded by buildings and other structures.

Another phenomenon that produces structural settlements is the lowering of the ground-water table. In some cases, the settlements due to this cause are brought about by changes in the stress conditions in the underlying soil. In other cases, the lowering of the ground-water table brings about settlements due to soil shrinkage.

There are several causes for changes in ground-water levels. One of these is the seasonal fluctuation that occurs in semiarid regions and in the tropics where there are long wet and dry seasons. During the dry season, the water retreats into the ground, and then it rises again during the rainy season. Another thing that causes changes in ground-water level is pumping. Construction work of all kinds is often accompanied by pumping operations which lower the ground-water levels in the vicinity of the job. The construction of deep sewers and subways often lowers ground-water levels perma-

nently, and there are many cases on record where such construction has been known to lower the ground-water table many feet below its original level.

Figure 78 illustrates the changes that can occur in soil-stress conditions because of lowering of the ground-water level. The method of computation is as follows:

Let W = unit weight of water (62.5 lb per cu ft)
S = specific gravity of soil solids (2.65)
n = percentage of voids in soil (35 percent)
W_a = weight of soil in air
W_s = submerged weight of soil

Then $W_a = SW(1 - n) = 2.65 \times 62.5(1 - 0.35) = 108$ lb per cu ft
$W_s = (SW - W)(1 - n) = (2.65 \times 62.5 - 62.5)(1 - 0.35) = 68$ lb per cu ft

Figure 78a represents a sand bed 20 ft thick overlying a stratum of clay which in turn is underlain by more sand. The ground-water level is in the upper sand stratum 5 ft below the ground surface. The assumption is made that this condition has existed for

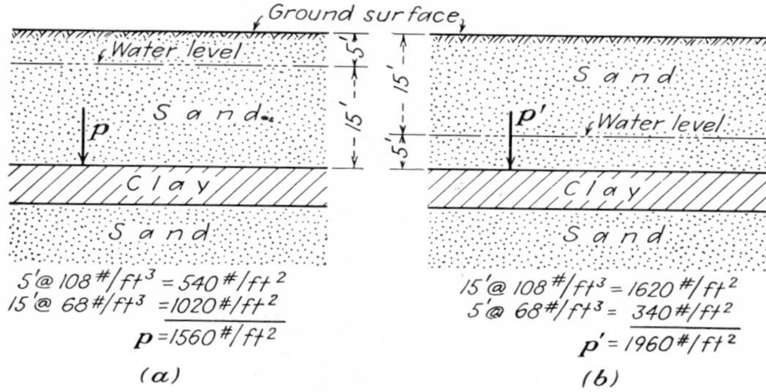

FIG. 78. Settlement due to shifting ground water.

a sufficient length of time so that the hydrostatic conditions in the ground are stabilized. At the top surface of the clay stratum there is a pressure p on the solid particles of clay due to the weight of the overlying sand. This weight consists of 5 ft of dry sand weighing 108 lb per cu ft and 15 ft of submerged sand weighing 68 lb per cu ft. The pressure p under these conditions amounts to 1,560 lb per sq ft on the top surface of the clay stratum.

Figure 78b represents the same soil condition with the ground-water level lowered 10 ft. The pressure on the top surface of the clay stratum has changed to p', which is due to the weight of 15 ft of dry sand weighing 108 lb per cu ft and 5 ft of submerged sand weighing 68 lb per cu ft. The pressure p' amounts to 1,960 lb per sq ft. The 10-ft drop in the ground-water level produces an increase of about 25 percent in the pressure at the top of the clay stratum. This change in soil pressure is often capable of producing consolidation settlement in the clay stratum even though no external load is applied at the ground surface.

Although the figure refers to the case in which the water level is lowered in a sand bed, a similar change in soil stresses occurs when the water level is lowered in a clay bed. However, the problem of computing stress changes in clay is somewhat more complicated because of time effects.[1]

[1] BRINKHORST, W. H., Settlement of the Soil Surface around the Foundation Pit during the Construction of the Locks at Vreeswijk Resulting from the Sinking of Ground Water, *Proc. Intern. Conf. Soil Mech. and Foundation Eng., Cambridge, Mass.*, vol. 1, p. 115, 1936.

Figure 79 illustrates volume changes in soils brought about by shrinkage effects when water is removed or when the ground-water table is lowered. These shrinkage effects are most common in very fine grained clays. They are not usually important in sands and other fairly coarse grained soils. Figure 79*a* refers to a very fine-grained volcanic clay from Mexico City. A cube of this clay 10 cm on a side was carefully carved out of the ground in its natural saturated condition. The cube was allowed to dry out slowly in air. At the end of about 50 days, the cube which was originally 10 cm on a side had reduced in volume until it was a cube about 4 cm on a side. In other words, the original volume of 1,000 cc of soil and water shrank to a volume of about 64 cc of soil and air while 936 cc of water evaporated. The initial void ratio of the sample was therefore $^{936}/_{64}$, or 14.6.

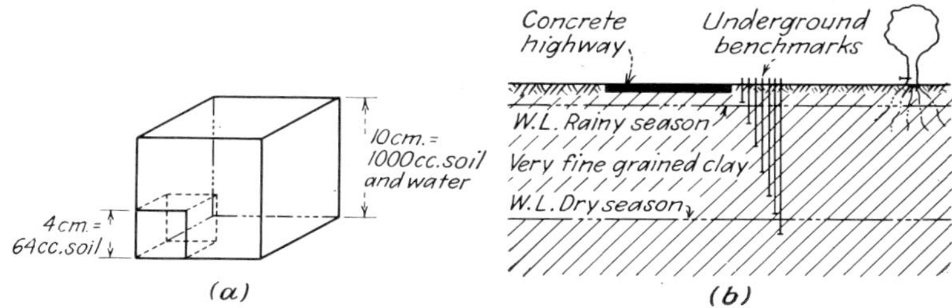

FIG. 79. Soil movements due to shifting ground water.

There are very few soils in which the amount of shrinkage is as great as this, although there are many soils that shrink considerably.

Figure 79*b* represents an investigation of this problem that was carried out by the Texas State Highway Department. A concrete highway was built over a deep bed of fine-grained clay and a careful record was kept of the behavior of this highway slab over a period of several years. In the rainy season the ground-water level was fairly close to the ground surface. At the end of a long dry season the ground-water level retreated 20 or 25 ft below the surface. With these changes in ground-water level the highway slab moved up and down, and the total amount of movement was on the order of 6 in. That is, the concrete paving was 6 in. lower at the end of the dry season than it was at the end of the rainy season, and the settlement and upheaval of the slab were brought about by the alternate shrinking and swelling of the soil.

In order to determine how these soil movements varied with the depth below the ground surface, a series of underground bench marks was installed at different depths in the ground. Level observations on the bench marks indicated that the soil movements were greatest near the surface of the ground and that there was practically no movement of the soil at depths on the order of 25 ft where the soil was continuously saturated.

As a part of this same experiment, a large spike was driven in the trunk of a tree alongside the highway. Level readings taken on this spike showed that the tree itself was moving up and down with the soil as the seasons changed from wet to dry.

Forces of enormous magnitude are produced by the swelling and shrinking of certain kinds of soil, and very heavy buildings and other structures have been moved and cracked from this cause.[1]

Another source of settlement is that due to adjacent operations such as the excavation of basements, tunnels, open cuts, or pier excavations or the construction of heavy structures on spread foundations alongside existing structures. Figure 80 illustrates

[1] *Proc. Intern. Conf. Soil Mech. and Foundation Eng., Cambridge, Mass.*, vol. 2, p. 256, 1936.

the changes that occur in connection with these various types of excavation. Figure 80*a* represents a longitudinal section of subway tunnel. As the heading is advanced through the soft plastic clay, there is a tendency for the clay to flow toward the heading as indicated by the arrows, and the surface of the street above the tunnel settles because of loss of ground. Figure 80*b* is a transverse section of the same tunnel, and the arrows represent the manner in which the clay tends to flow toward the heading of the tunnel from

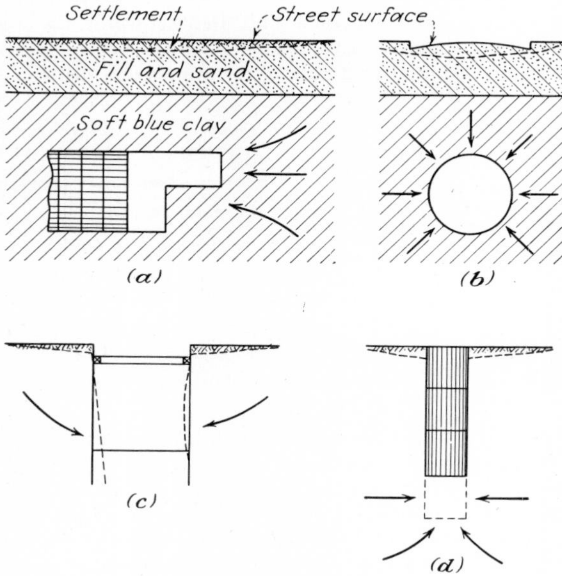

FIG. 80. Soil movements due to excavation.

the sides. The consequent settlement of the street surface due to lost ground is indicated in the sketch.

Figure 80*c* represents a deep cut supported by sheeting and bracing. Even when a cut of this kind is installed with great care, there is always a tendency for some settlement of the surrounding ground. The bracing can be installed only as excavation proceeds. Accordingly, although sheet piling may have been driven before any excavation was done, the excavation of soil below the first set of braces permits the piling to move in toward the cut so that there is often some settlement of the ground at the top of the cut. The movement of the sheet piling may be an inward movement at the bottom of the cut, as shown at the left of the figure, or a bending of the piling, as shown at the right.

Figure 80*d* represents a pier excavation lined with lagging and rings. The excavation is always carried below the lowest set of lagging, and there is always a tendency for the soil to flow toward this open excavation, as indicated by the arrows. Any considerable movement of the ground at the bottom of the excavation is sure to be accompanied by some settlement of the ground surface around the top.[1]

Figure 81 illustrates another type of settlement that is brought about by adjacent operations. Figure 81*a* represents two structures on spread foundations, and Fig. 81*b*

[1] KNAPP, R. S., Settlements Due to Subway Construction in Clay, *Proc. Purdue Conference on Soil Mechanics*, September, 1940, p. 312.

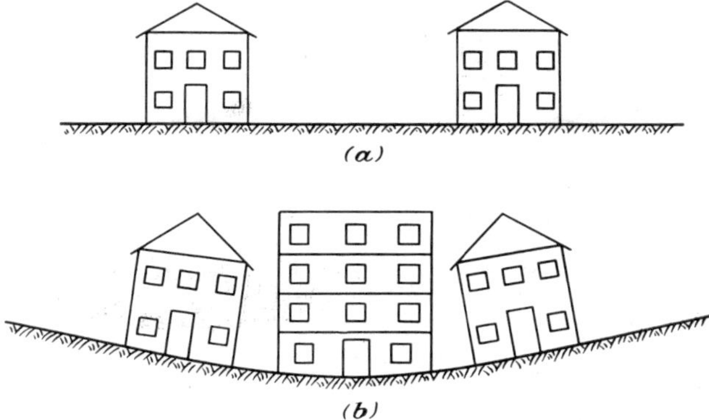

FIG. 81. Settlement due to adjacent structures.

shows what sometimes happens to such structures when heavy buildings are built alongside of them. The weight of the new building is sufficient to cause settlement of the underlying soil, and the area covered by the settlement is not limited to the area of the new building. The adjacent structures are affected by the settlement of the new structure so that they settle and tilt.

An example of this type of settlement is shown in Fig. 82*a*. The building at the extreme left is a moderately heavy six-story structure built on spread footings. The structure has settled several feet, and the settlement has affected not only the two-story building immediately alongside the new building, but also the two-story building two doors away. Both the smaller buildings have been tilted and cracked because of the settlement of the new structure.

Figure 82*b* illustrates a similar condition where the heavy building on the left has dragged down the adjacent two-story building to such an extent that the small building was practically destroyed.

There are other miscellaneous causes of settlement such as underground erosion, chemical changes in the soil, and frost action.

Underground erosion is not a serious problem in cases of natural flow of ground water. The movements of ground waters in all kinds of soil are usually extremely slow so that underground erosion does not commonly occur from natural causes. The problem of underground erosion might be serious in cases where pumping is done. Pumping sometimes produces velocities in the ground water that could cause serious erosion. Figure 83 illustrates a case of this kind. A pier excavation is being made alongside an existing building. As the excavation passes through the clay, little or no pumping is required. At some depth in the ground is a sand seam that might require considerable pumping. Whenever a situation of this kind occurs, the discharge of the pump should flow into a wooden box or a catch basin of some kind so that the discharge water can be examined to see whether or not solid particles are being pumped along with the water. If a pumping operation of this kind were to remove any considerable amount of sand, it would be almost certain to settle the adjoining building.

Settlements due to chemical changes in the subsoil are rather uncommon. Chemical changes might occur in filled ground, especially if the fill contained materials that would slowly oxidize and disintegrate. *However, permanent structures should never be built on filled ground.* A bed of peat is subject to slow chemical changes, and it gradually shrinks while it changes from peat to lignite and then to coal. *Accordingly, structures*

(a)

(b)

FIG. 82. Settlement due to adjacent structures.

built on soils containing peat beds would be subject to settlement from this cause in addition to settlements due to the compression of the peat.

The freezing of soil occurs by a crystal-building process which forms ice lenses in the soil mass and causes swelling and upheaval. Subsequent thawing causes settlements. Problems involving frost action are more important in highway engineering than they are in structural engineering. Highway slabs are built on the surface of the ground where they are often subjected to frost action in the subgrade. Foundations for structures are usually built well below the ground surface. The depth of frost penetration for most localities is known, and building codes usually require that the foundations for structures of all types must be founded below the frost line.

Settlement Predictions. At present there is no satisfactory method of predicting the settlement of structures founded on deep beds of granular soil such as sand or gravel. Settlements of structures built on such soils are usually quite small even when the sand or gravel is in a comparatively loose condition as long as the applied loads are static. *Dynamic loads or vibrations can cause settlements of considerable magnitude in loose sand beds.* However, since there is no satisfactory scientific procedure available for calculating

the probable settlements, *predictions of settlements of structures built on beds of sand and gravel can be made only on the basis of previous experience.*

For structures founded on beds of saturated clay, settlement predictions are based on the theories of consolidation and of plastic flow. Comparisons of predicted settlements with actual measured settlements on a great many structures have demonstrated that these theories gives satisfactory results. The most notable exceptions are those cases where the clay bed has been subjected to large preconsolidation loads at some time in its geologic history. Such clays are said to be "overconsolidated" because they were precompressed by loads far in excess of their present overburden loads. In glaciated regions

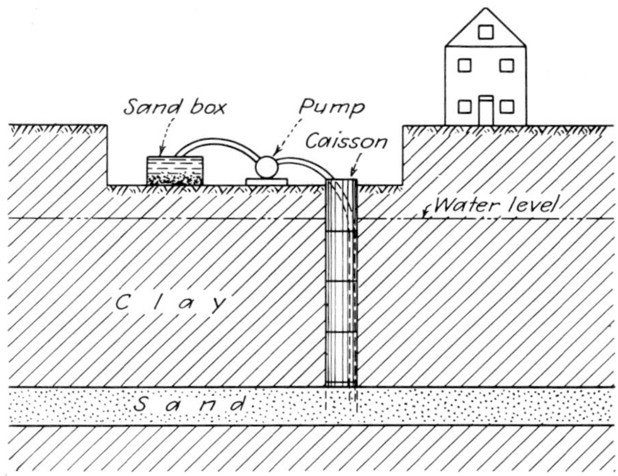

FIG. 83. Settlement due to deep pumping.

there are overconsolidated clays which were at one time subjected to heavy ice loads. In unglaciated regions there are overconsolidated clays which were once buried under a heavy overburden of soil that was subsequently removed by erosion. However, with the exception of a few special conditions, the theory of consolidation can be used to predict with considerable accuracy the settlements of structures founded above beds of saturated clay.

SHORING AND UNDERPINNING

Types of Shoring. The words "shoring" and "underpinning" are often used together as though they referred to the same kind of work. Actually, shoring refers to temporary supports that are usually removed when the job is completed, while underpinning refers to permanent supports that remain in place when the job is finished. In general, it is necessary to do a job of shoring before doing a job of underpinning and, when the underpinning is completed, the shoring is usually removed.

There are three principal reasons why it may be necessary to shore and underpin a structure. These are to stop settlement of an inadequate foundation, to provide support because of adjacent construction operations, and to provide a foundation able to carry added loads.

When a structure settles because it has an inadequate foundation, it is sometimes necessary to resort to underpinning to stop the settlement. The construction of deep basements or deep sewers or subways, especially in the crowded downtown sections of large cities, usually necessitates underpinning of adjacent structures to prevent settle-

ment. Sometimes two or three stories are added on top of a building, or the type of occupancy of a building may be changed, e.g., from light manufacturing to warehousing, so that the new loads to be supported may necessitate strengthening the foundation of the building by means of underpinning.[1]

Figure 84 illustrates some of the more common types of shoring. Figure 84*a* represents a brick wall on a reinforced concrete footing which is founded just a few feet below the ground surface. A deep basement is to be excavated on the adjacent property, as

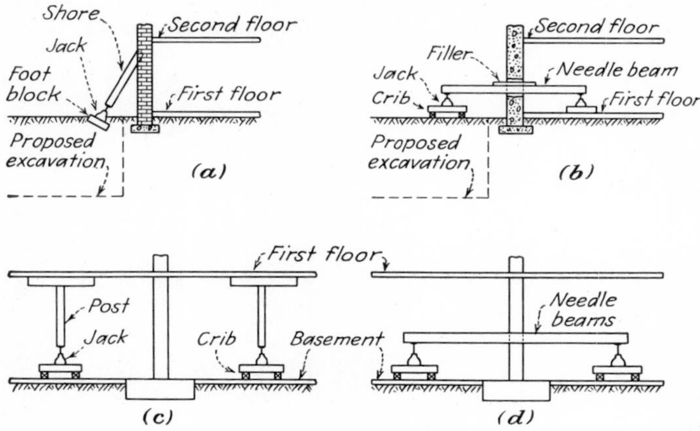

FIG. 84. Typical shoring methods.

indicated by the dotted line. In order to prepare the structure for underpinning, the brick wall must be shored with a set of sloping braces as illustrated in the figure. The upper end of the brace is set in a notch cut in the brick wall. The lower end of the brace rests on a screw jack, which in turn rests on a wooden foot block placed on the ground. After the shore has been set in place, most of the load can be lifted off the footing by screwing up the jack.

Figure 84*b* illustrates a somewhat different type of shoring that might be used under similar circumstances. A small timber crib is built outside of the wall and a similar timber crib inside the building. Screw jacks are placed on top of these cribs. A hole is then cut in the wall and a steel beam, called a "needle beam," is inserted through the hole. The ends of the beam rest on the jacks, and a filler block is placed on top of the beam where it passes through the wall. When the jacks are screwed up, the beam is able to lift most of the weight of the wall off the footing.

Figure 84*c* represents one method of shoring up an interior building column. Small timber cribs are built on the basement floor on opposite sides of the column. Sometimes two such cribs are sufficient, and sometimes four of them are used on all four sides of the column. Screw jacks are placed on top of these cribs. On the underside of the first floor, heavy timbers are set under the floor beams and posts extend up from the screw jacks to these heavy timbers. When the jacks are screwed up, the weight of the column can be lifted off the footing. The question of whether or not this type of shoring can be used for a building column depends on the nature of the construction of the building. It may or may not be possible to jack against the underside of the first-floor beams, as is illustrated in Fig. 84*c*. There is also the question as to how the floor and the columns are fastened together, and it may or may not be possible to lift the column by jacking against the underside of the floor.

[1] PRENTIS and WHITE, "Underpinning." Columbia University Press, New York, 1931.

Figure 84*d* illustrates a different type of shoring that might be used for an interior building column in cases where it is not possible to shore up under the floor beams. Small wooden cribs are built on the basement floor on opposite sides of the column and screw jacks are placed on top of these cribs. A pair of steel beams is set on top of these jacks, with one beam passing on each side of the column. These beams are also called "needle beams," and they may be bolted or clamped or otherwise fastened to the column. When the jacks are screwed up, the weight of the column is lifted off the footing.

Special types of shoring have been developed for special problems, but the most common types are those indicated in Fig. 84. After the shoring has been properly taken care of, it is possible to proceed with the underpinning of the structure.

Types of Underpinning. Figure 85 represents the simplest type of underpinning that is commonly used. This is usually referred to as "pit underpinning." Figure 85*a* rep-

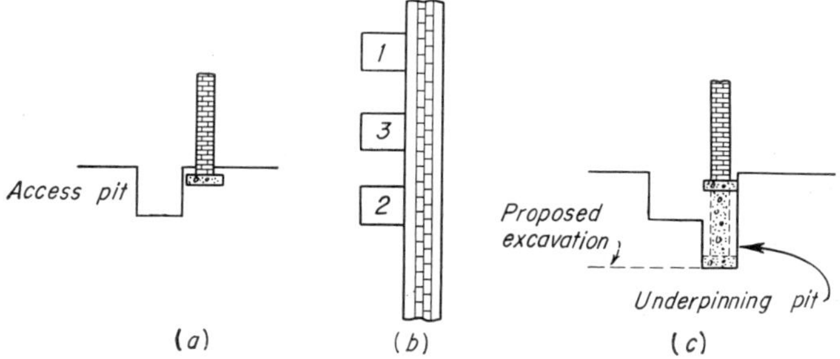

FIG. 85. Simple underpinning for walls.

resents the wall of a building such as was shored up in Fig. 84*a* and *b*. In order to underpin this wall, a pit is excavated outside the wall. This pit is 4 or 5 ft square and is called an "access pit." The pit is excavated down to a depth of 5 ft or so to provide access for men to get under the footing. Excavation is then continued in an underpinning pit directly under the wall, as illustrated in Fig. 85*c*. This can be done under almost any wall, since the wall can be expected to span over a width of 4 or 5 ft. In some cases, particularly in the case of a reinforced concrete wall on a reinforced concrete footing, it might be possible to excavate under considerably more than 4 or 5 ft of wall at a time without serious risk.

Figure 85*b* is a plan view of the wall on its concrete footing. Pits such as 1 and 2 are excavated simultaneously, the space between the pits being on the order of 20 or 25 ft. After the concrete underpinning has been placed in pits 1 and 2, intermediate pits such as 3 are excavated, and later the spaces between 1 and 3 and between 3 and 2 are filled in. The foot blocks of the shores illustrated in Fig. 84*a* and the cribs in Fig. 84*b* are placed far enough out on the adjacent property to avoid these pits.

In some cases underpinning is done with piers, as illustrated in Fig. 86*a*. The figure shows a pier being installed to support an interior building column. The column has been shored up as indicated and, after the shoring was completed, the column and its footing were removed. The sketch shows a pier excavation being installed with lagging and rings for lining. Piers used as underpinning can be installed in the open air or under compressed air and they may be either straight shaft or belled. Such work is almost invariably done by hand, as it is not possible to get one of the rotary excavating machines into the basement of the building. Although the figure shows a pier being installed under an interior building column, it is also possible to use this method under building walls.

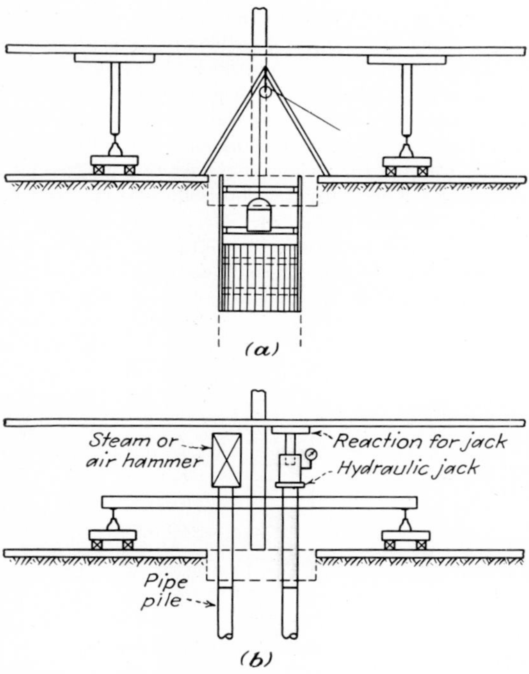

FIG. 86. Underpinning building columns: (*a*) with piers, (*b*) with steel-pipe piles.

Ordinarily, pier work that is done as underpinning is more expensive than pier work done out in the open before the building is constructed. The additional expense is brought about by the necessity for working in low headroom and the necessity for bringing all the construction materials into the basement of the building. Furthermore, the cost of the shoring materially increases the cost of the whole operation as compared with the cost of installing a pier job in advance of the construction of the building.

Sometimes underpinning is done with steel-pipe piles. A job of this type is illustrated in Fig. 86*b*. The column has been shored up with needle beams, and the footing under the column has been removed. Pipe piles used for this purpose may be closed-end piles provided with steel points, or they may be open-end piles which are cleaned out and filled with concrete after they are driven. Because of the low headroom it is necessary to use pipe in comparatively short sections, and these are either welded together or fitted with special sleeves.

On the left-hand side of the figure, the pipe pile is being driven with a small double-acting hammer which may be operated either with steam or with compressed air. The use of air in this type of work is more common because there are often reasons why the basement of the building must not be filled with exhaust steam. When using the double-acting hammer and short sections of pipe, it is usually desirable to weld the joints in the pipe rather than to join them with sleeves because of the "billiard ball" action of the pipe under the blow of the hammer tending to open up sleeved joints during driving. On the right-hand side of the sketch, the pipe pile is being forced down by means of a hydraulic jack. In order that the jacking process may be used, it is necessary that a suitable reaction be available to jack against. The method of jacking down pipe piles for under-

pinning is rather slow as compared with the method of using a small double-acting hammer. The average hydraulic jack can be extended about 10 or 12 in. Each time the pile has been pushed down this amount, it is necessary to stop and reset the jack and the blocking and start over again. However, there are many cases in which the available headroom is so small that it is impossible to use a mechanical hammer, and jacking must be resorted to under such conditions.

Although Fig. 86*b* refers to the use of pipe piles under an interior column, this method of underpinning can also be used for walls. However, the available headroom during the underpinning of a wall is usually so small that the piles have to be put down with hydraulic jacks.

Shoring and underpinning are often difficult and dangerous operations, and work of this kind should be undertaken only by specialists.

Sometimes it is practicable to eliminate shoring. Referring to Fig. 86*b*, the design of the existing footing may be such that it would be safe to shot-core drill a hole slightly larger in diameter than the pile for each pile through the footing without reducing the bearing area of the footing enough to substantially affect its carrying capacity. Under such circumstances, the column might not have to be shored and the piles could be driven to bearing through the cored holes. In such cases, the pile is cut off at the bottom of the footing, the sides of the hole which has been cored through the footing are thoroughly scrubbed clean to remove all dust remaining from the core-drilling operation, and the hole is then filled with Embeco concrete placed under vibration so as to plug the hole completely with a nonshrinking, slightly expanding concrete which provides a very high bonding strength against the scoring left on the side walls of the hole by the shot-core drilling operation.

SHEET PILES AND COFFERDAMS

Function and Uses of Sheet Piles. The function of sheet piling is to retain water or soil or both. The principal uses of sheet piles are for bulkheads on river banks or on open water, for cofferdams on land or in water, and for lining in trenches and open cuts.

Types of Sheet Piles. The various kinds of sheet piles that are commonly used are illustrated in Fig. 87. They are divided into three principal kinds on the basis of the kind of material from which they are made. The three common materials are wood, steel, and concrete.

A single line of boards such as 2-in. by 12-in. or 4-in. by 12-in. planks is often used as sheet piling. This is referred to as "single-sheet piling," but it is suitable only for comparatively small excavations where there is no serious ground-water problem. In saturated soils, particularly in sands and gravels, it is necessary to use a more elaborate form of sheet piling which can be made reasonably watertight.

A type that is commonly used under such conditions is called "lapped-sheet piling." Each pile is made up of two boards which are offset with respect to one another, and these pairs of boards are either spiked or bolted together. Such sheet piling is usually made with boards ranging from 1-in. by 12-in. to 4-in. by 12-in.

A somewhat better type of sheet piling, as far as watertightness is concerned, is the type that is known as "Wakefield." A Wakefield sheet pile consists of three boards bolted or spiked together with the center board offset as shown in the figure. This arrangement produces a tongue and groove which makes Wakefield sheet piling fairly watertight if the piles are properly driven and tightly fitted together. If 2-in. by 12-in. boards are used, they produce a sheet-pile wall 6 in. thick. Sometimes boards as heavy as 4 in. by 12 in. are used so that the Wakefield sheet pile wall is 12 in. thick.

Another form of wood sheet piling is called "tongue and groove." This is made out of single pieces of timber which are cut in a sawmill so as to provide a tongue and a groove on each piece of piling. This type wastes timber because of the amount of wood

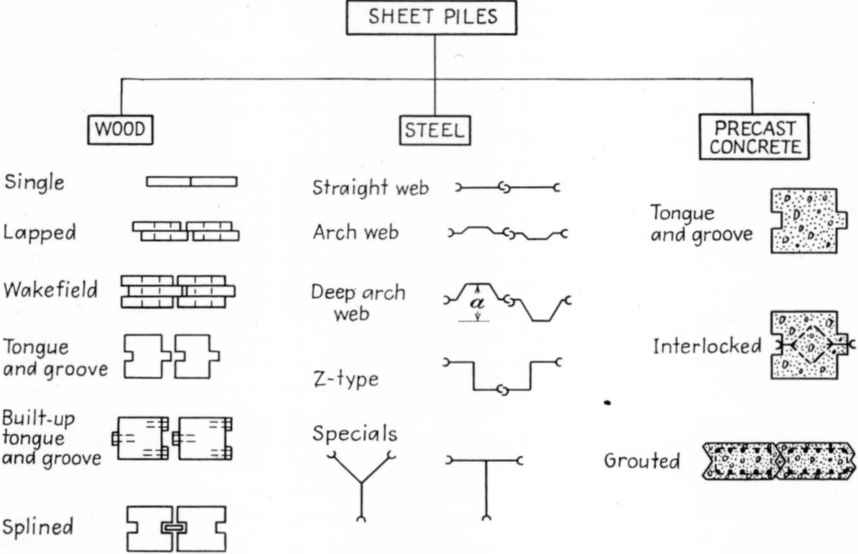

FIG. 87. Types of sheet piles.

that is cut away to form the tongue. For example, if the tongue-and-groove pile illustrated in the figure were made from a 12-in. by 12-in. timber and if the tongue and the groove were 3-in. by 3-in., each piece of timber would make only 9 in. of sheet-pile wall, although a 12-in. by 12-in. timber would have been used to form the pile.

In order to have the tongue-and-groove effect without wasting timber, a form of built-up tongue-and-groove pile is sometimes used. For example, 2 by 4's could be spiked to the sides of 12-in. by 12-in. timbers in order to form the tongue and groove illustrated in the figure. Each 12-in. by 12-in. timber and its three 2-by-4's would produce 14 in. of sheet-pile wall in the finished structure.

Another form of wooden sheet piling that is sometimes used is a type called "splined sheeting." For this type a groove is cut in two opposite sides of a heavy timber. The timbers are then driven close together so that a space is left where the grooves match, and a smaller timber is driven in this space. The splined sheet pile illustrated in the figure might consist of 12-in. by 12-in. timbers with grooves 4 in. wide and 3 in. deep cut in them. After the two timbers were driven in place, a 4-by-6 would be driven down into the grooves so as to fill this space.

The choice of which type of wooden sheet piling should be used on a given job depends largely on the nature of the work. For small footing excavations where there is no serious ground-water problem, the more simple forms may be used. For heavy bulkheads and deep cofferdams, it is necessary to use the more elaborate forms of sheet piling, partly for the purpose of securing a watertight wall and partly for the purpose of having a pile that is strong enough to resist the horizontal thrusts that will be applied to it.

Steel sheet piling is available in a number of different forms. Each steel company that manufactures steel sheet piling has its own form of interlock but all these different interlocks are designed for the same purpose. The simplest form of steel sheet piling is that known as the "straight-web type." The piles are made in various widths ranging from about 15 in. to about 20 in. The web thicknesses vary from about 3/8 to 1/2 in.

The straight-web sheet piling is comparatively flexible so that it requires a considerable amount of bracing when it is used for structures in which the horizontal thrusts are large. In order to provide a pile with greater resistance to bending, the steel companies have developed a type known as arch-web sheeting in which the center of the web is offset so as to provide a greater moment of inertia in the cross section of the sheet.

In order to provide even greater stiffness, a type known as "deep-arch" section has been developed. This is similar to the "arch web" except that the offset in the web is increased considerably. There is also available a type known as the "Z" section which has a stiffness considerably greater than that of the deep-arch section.

The choice of the type of steel sheet pile to be used on a given job depends largely on the kind of service to which it is to be put. The straight-web type is comparatively flexible so that it requires a considerable amount of bracing when it is subjected to a large horizontal thrust. On the other hand, the straight-web type does not use up much space so that it can often be used in close quarters where there might not be room for some of the other types. When the sheet piling is to be used for a structure involving large horizontal forces, it is possible to do the job with less bracing if the deep-arch or the Z section sheet is used. At the same time, these forms of sheeting take up considerable space in a horizontal direction. With a large-sized arch section, the distance a shown in the figure would be about 12 in. If the sheet-pile wall were being braced with 12 by 12 waling timbers, the sheet piling and the waling timber would require a total of 24 in. of horizontal space, which may not always be available.

In addition to the standard piles, all the steel companies make special sections for various purposes. Figure 87 shows the approximate shape of two of these special sections which are referred to as Y sections and T sections.

Precast-concrete sheet piling is usually made in the form of a tongue-and-groove section. The piles are usually from 18 to 24 in. wide in the direction of the sheet-pile wall, and they are made in thicknesses varying from 8 or 10 in. to 24 in. or more. They are reinforced with vertical bars and hoops just as precast-concrete bearing piles are reinforced. The thickness of the sheet pile and the amount of reinforcing are determined by the forces that the pile is intended to resist. A precast-concrete sheet-pile wall made of tongue-and-groove sheeting is not always perfectly watertight when driven, but the spaces between the piles can be grouted to provide a watertight wall.

In order to make a watertight sheet-pile wall with precast-concrete sheet piling, the interlocked section shown in Fig. 87 is sometimes used. The details of the concrete pile are the same as those of the tongue-and-groove type. However, a straight-web steel sheet pile is split in half longitudinally, and the two halves of the steel sheet pile are embedded in the precast-concrete pile when the concrete pile is cast. These interlocked piles form a more watertight wall than can be obtained with the tongue-and-groove sheet piles.

Another type of precast-concrete sheet pile that is used to produce a watertight wall is shown in the figure as the grouted type. The shape of the piles is such that, when two piles are driven side by side, a diamond-shaped space is left between the piles. After the piles are driven, this space is filled with cement grout injected under pressure so as to seal the wall.

Cofferdams on Land. In cofferdam work, the wood sheet pile and the steel sheet pile are most commonly used. Ordinarily, a cofferdam is a temporary structure which is removed after it has served its purpose. For this reason, the precast-concrete sheet pile is seldom used for cofferdams. However, some types of cofferdams, such as a condenser pit in a big power plant on a riverbank, are permanent structures, and precast-concrete sheet piling is sometimes used for these.

Figure 88 refers to land cofferdams, and Fig. 88a is a cross section of a cofferdam completely surrounded by land. Figure 88b is a plan view of the same cofferdam. The first

step in the construction is to drive wood or steel sheet piling all the way around the area to be enclosed. The piling is usually driven before any excavation is done. If steel sheet piling is being used to form a closed box of this kind, it is necessary to "pitch" the entire box of sheet piling before it is driven. By pitching is meant that the entire box is set up and interlocked before the individual piles are driven down into the ground. The driving is then done by stages, with each pile being driven 5 or 10 ft at a time until the tops of all the piles are about at the same level. If one pile after another was driven

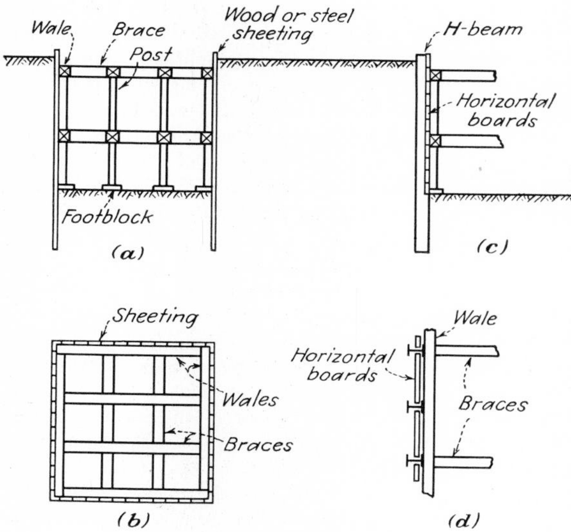

FIG. 88. Small cofferdam on land.

to its full depth all the way around the closed box, it would be practically impossible to make a proper closure between the last pile and the first pile. The reason for pitching and driving the piling in this manner is to make it possible to have a properly closed box.

After the piling has been driven, excavation is started inside the cofferdam. Ordinarily, excavation is carried down a few feet and then the first set of bracing is placed. This bracing consists of a horizontal frame of waling timbers placed against the inside of the sheet piling all the way around the cofferdam. These waling timbers are braced in both directions by horizontal braces, as illustrated in Fig. 88*b*. Excavation is then carried on between the braces until sufficient depth has been obtained to require another frame of wales and braces. This process of alternate excavation and the setting of wales and braces is continued until the bottom is reached.

There is one problem in connection with the bracing of a cofferdam that should never be overlooked. Each frame of wales and bracing must be placed as rapidly as possible because, as excavation proceeds, the sheet piling has a tendency to move and bend because of the earth load from the outside. These movements of the piling could be such as to loosen a frame of horizontal bracing. In order to avoid the possibility of having an entire frame of horizontal bracing collapse and fall into the excavation, it is necessary to post up between each frame of horizontal bracing with vertical posts as shown in Fig. 88*a*. The bottom post under the lowest frame of bracing usually rests on a timber or a plank called a "foot block," which rests on the bottom of the excavation.

Another type of sheet piling that is sometimes used for land cofferdams is illustrated in Fig. 88*c* and *d*. In this type, a series of steel H piles is driven around the perimeter of the cofferdam on 4- or 5-ft centers. As excavation inside the cofferdam proceeds, horizontal boards are placed between the webs of these H piles. The thrust of the earth outside the cofferdam pushes these boards against the flanges of the H beams. Such a cofferdam is excavated and braced in the same manner as that illustrated in Fig. 88*a* and *b*.

Cofferdams of this type, which are braced all the way across from wall to wall, are used only in cases where the diameter of the cofferdam does not exceed 75 or 100 ft. If it were desired to enclose an entire city block or an even larger area inside of a cofferdam, it would be exceedingly difficult and expensive to brace with horizontal braces all the way across the cofferdam. In such cases, it is customary to use the type of bracing illustrated in Fig. 89. The sheet piling might be steel or wood or the combination of H

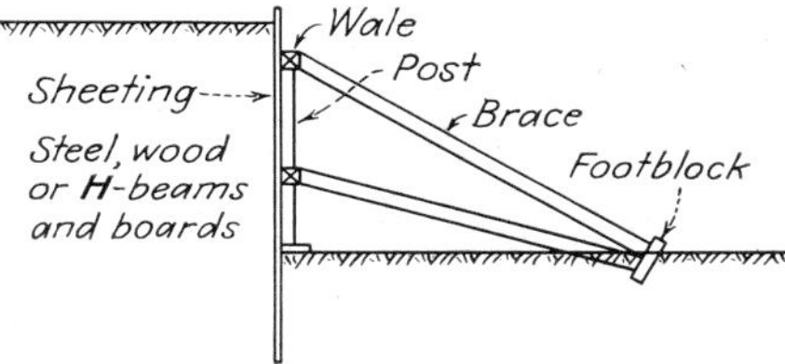

FIG. 89. Bracing for large cofferdam on land.

piles with horizontal boards. Waling timbers are placed inside the sheet-pile wall, and these are braced with sloping braces, called "raker braces," against a foot block of timber which is set in the ground at the bottom of the excavation. In this case, a firm ground or other support for the foot blocks is essential.

Cofferdams in Water. Figure 90 illustrates a type of cofferdam that might be used for a bridge pier in a river. Figure 90*a* is a cross section and Fig. 90*b* is a plan view. The complete box of sheet piling is driven first, and then a frame of horizontal bracing is placed approximately at the water line. At this stage of construction the water is at

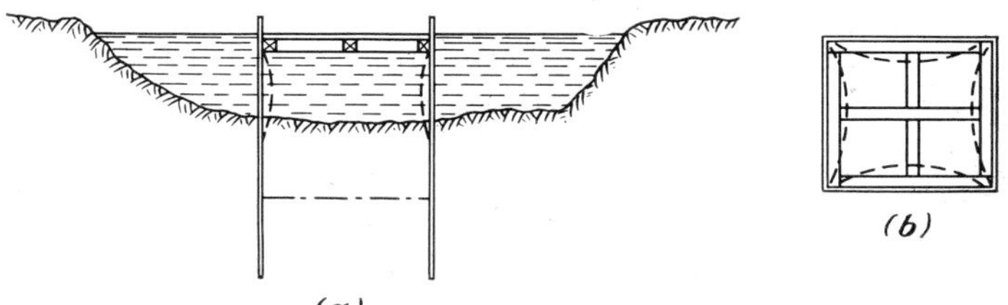

FIG. 90. Small cofferdam in water.

the same elevation inside and outside the cofferdam. Wood sheet piling might be used for such a cofferdam if the water depth did not exceed 8 or 10 ft. For deep cofferdams and high water pressures, the use of steel sheet piling is more common.

After the top frame of horizontal bracing is in place, pumping is started to get the water out of the cofferdam. Even though great care may have been used in driving the piling, it is not to be expected that the cofferdam will be absolutely watertight as driven. If wood sheet piling has been used, there will almost certainly be small spaces between the piles through which water can leak into the cofferdam. Even if steel sheet piling

has been used, the interlocks will not necessarily be watertight. Accordingly, it is often necessary to provide large pumping capacity in order to start pumping down the water in the cofferdam. As the pumps gain on the inflow of water between the piles, the water level inside the cofferdam is lowered and the cofferdam is subjected to an excess of pressure from the outside.

This excess of outside hydrostatic pressure tends to bend the piling, as shown by the dotted lines in Fig. 90. The piles are braced at the water line by one frame of horizontal braces, and the lower ends of the piles are embedded in the ground in the river bottom. They will tend to bend inward, and the curvature that results has an important effect on the watertightness of the structure. If wooden sheet piling has been used, the tendency of the curvature would be to open up the spaces between the piles. This would allow the cofferdam to leak considerably, and for this reason wooden sheet piling is seldom used for deep cofferdams subjected to large water pressures. When steel sheet piling is used for such a cofferdam, the curvature that results from the external water pressure is beneficial. The effect of the curvature is to tighten up the interlocks of the piling so that the cofferdam is more watertight after it is distorted than it was when it was originally driven. After the cofferdam is pumped out, the process of alternate excavation and bracing is carried out just as in the case of the land cofferdams.

The tightening of the interlocks in a steel sheet-pile wall is referred to as "interlock tension," and much of the design of steel sheet-pile walls and cofferdams is controlled by this feature of steel sheet piling. As is the case with land cofferdams, a cofferdam of the type shown in Fig. 90 would be used only in cases where the diameter of the area to be enclosed did not exceed more than 75 or 100 ft, since the bracing of a cofferdam all the way across from wall to wall is feasible and economical only in cases where the cofferdam is not too large.

Figure 91 illustrates some of the problems involved in building very large cofferdams in water. Figure 91*a* represents a river channel in which it is proposed to build a large ship lock. In order to construct the lock, it is necessary to dry up a large section of river bottom.

If the water were very shallow, it might be possible to construct such a cofferdam as a single-wall dam, as illustrated in Fig. 91*b* and *c*. A single line of sheet piling, either of steel or of wood, might be driven and braced with a berm of earth as shown in Fig. 91*b*. Instead of an earth berm for bracing, such a single-wall cofferdam might be braced with sloping braces and foot blocks as shown in Fig. 91*c*. However, if the depth of water exceeded 8 or 10 ft, it would be necessary to build a more elaborate type of cofferdam.

A double-wall cofferdam such as would be used in deep water is illustrated in Fig. 91*d* and *e*. Sometimes two parallel walls of sheet piling 20 or 25 ft apart are driven and tied together with tie rods as illustrated in Fig. 91*d*. The space between the two lines of piling is filled with sand or rock. Sometimes a berm of earth is used on the inside of such a cofferdam. In some cases the sheet piling is driven as a series of circular or elliptical cells which are filled with sand and gravel or crushed rock. Such cofferdams are called "cellular cofferdams." In many cases the river bottom is rock so that sheet piling cannot be driven. For a cofferdam built under such conditions, a rock-filled crib as illustrated in Fig. 91*e* is commonly used. Such a crib would be built of heavy timbers bolted together and filled with stone.

In the design of cofferdams in water two important factors must be taken into account. These are hydrodynamics and structural strength. It is necessary to investigate the possibility of scour and of seepage and also the possibility of overturning, of sliding, and of collapse.

Referring to Fig. 91*a*, it can be seen that the cofferdam reduces the width of the river channel. A streamline on the left bank of the river would be practically a straight line parallel with the river bank. On the right bank of the river a streamline would be bent

out of shape by the obstruction formed by the cofferdam. The presence of the cofferdam therefore restricts the river channel so as to produce high velocities of flow in the river. As a result of this, the outer corner of the cofferdam on the upstream side would be subjected to considerable scour. In many cases it has been found necessary to use much longer sheet piling for this upstream outer corner than is required for the balance of the cofferdam.

The possibility of underseepage is indicated by the dotted arrows in Fig. 91*b*, *c*, *d*, and *e*. The cofferdam has to be so designed that seepage water will be cut off below the dam.

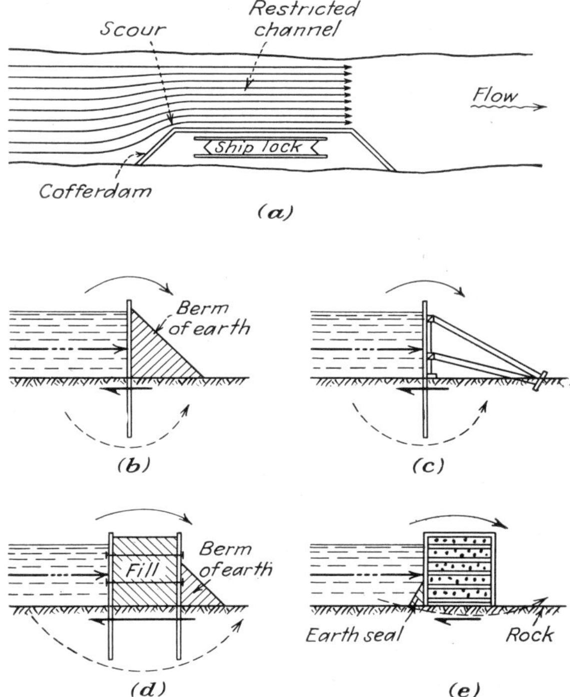

Fig. 91. Large cofferdams in water.

Referring to the structural strength of these cofferdams, all of them would be subjected to horizontal hydrostatic thrust after they were pumped out, and this horizontal pressure would tend to overturn the cofferdam and to make it slide along the ground. The design must be such that the cofferdam is safe against such movements.[1]

STRENGTHENING FOUNDATION SOILS

Types of Soil Injection. A procedure that is sometimes used in foundation work is the injection under pressure of various materials into soil or rock. The injections are usually made for one of two purposes: to seal a porous soil or fissured rock in order to stop a flow of water, or to stabilize a soft soil so as to increase its strength. The materials commonly used for injections may be divided into three general types called suspensions, solutions, and emulsions.

[1] White and Prentis, "Cofferdams," Columbia University Press, New York, 1940.

Suspensions are mixtures of cement and water or sometimes cement and clay and water. The relative amounts of water and solids that are used in the mixture can vary over a considerable range, although it is customary to use as high a percentage of solids as practicable. Suspensions of this type are useful for the grouting of rock fissures or for the grouting of beds of coarse sand and gravel. After the cement has set up, it not only waterproofs but also strengthens the rock or soil into which the suspension has been injected.[1]

The solutions that are used for soil injections are chemical compounds, and the basic chemical that has been used for this purpose up to the present time is sodium silicate, which is often called "water glass." A second chemical is used which combines with the sodium silicate to form a stiff silica gel so that the soil is waterproofed as well as strengthened by this type of injection. The chemical injection methods now in use differ from one another with respect to the injection procedure. In some cases the sodium silicate is injected first and the second chemical afterward, while in others the two are injected at the same time. In the latter method, the proportions of the two chemicals are often varied to control the time of setting. The chemical solutions can be forced under pressure into medium and fine sands.[2]

The application of any process of this type [3] *should be carried out by or under the supervision of a firm or organization experienced in the process as the spacing of the injection tubes and the results achieved are dependent not only on prior determinations of the grain-size characteristics of the soil and its coefficient of permeability but also and substantially on the strength and viscosity of the solutions used for injection.*

The emulsions that are used as soil-injection materials are emulsions of bitumen and water. The use of bitumen emulsions for this purpose is a more recent development than the use of cement suspensions and chemical solutions. The emulsions can be pumped into sands of medium grain size, and they act primarily as waterproofers for the soil into which they are injected. Also, direct injection of hot bitumen, kept hot by electrodes, has been used successfully in grouting seams in rock.

Practical Applications. Before a decision can be made as to which type of injection material should be used in a given case and how it should be injected, it is necessary to make a careful field investigation of the site. The choice of the type of material will depend largely on the size of the openings into which it must penetrate. Cement suspensions can be pumped into cracks as small as 0.1 mm in width and into sands having an effective grain size on the order of 0.8 mm. The chemical solutions can be pumped into sands having an effective grain size as small as 0.1 or even 0.05 mm. The effective grain size of sands into which the emulsions can be pumped is about the same as for the chemical solutions. None of the injection materials can be forced into very silty fine sands, silts or clays except in cases where these soils are dried out and fissured. Injections can be made into such soils when fissured, but the injection material penetrates only into the fissure and not into the soil itself and has little or no beneficial effect.

All the materials used for injection into soil or rock are pumped into the ground with pressure pumps. Information must be obtained as to the permeability of the soil in order that approximate pumping capacities and pressures can be determined. The permeability of the soil also determines the spacing of the injection pipes, because the material pumped through any one pipe must penetrate into the soil a sufficient distance to meet the material pumped through all adjacent pipes. This is especially important in the case of a dam on a sand foundation where the injections are intended to form a

[1] Grouting Checks Foundation Vibration, *Eng. News-Record,* vol. 22, No. 17, p. 561, Oct. 22, 1942. Unusual Cutoff Problems—Dams of the Tennessee Valley Authority, A Symposium, *Trans. ASCE,* vol. 110, p. 947, 1945.

[2] Lewin, Joseph D., Grouting with Chemicals, *Eng. News-Record,* vol. 123, No. 7, p. 221, Aug. 17, 1939.

[3] The Joosten Process, perhaps the best known and most widely used of injection processes, is handled in the United States by Chemical Soil Solidification Company, 7650 South Laflin St., Chicago 20, Ill. Its pamphlet, "The Joosten Process," 1954, by Dr. Ing. Hugo J. Joosten, is very informative.

curtain wall below the dam to prevent underseepage. Unless the curtain wall is 100 percent complete, the injections will be a waste of time and money.

When fissured rock is to be sealed by injection methods, it is necessary to determine the nature and extent of the fissuring. It is also necessary to know whether the fissures are open or whether they are filled with soil which would have to be washed out before the injection material could be forced in. The spacing of the injection pipes is controlled largely by the nature of the fissuring, and the pipes must be so arranged and spaced that the injection material will fill all the fissures.

The pumping pressures that are required in the various injection methods are often quite high, and pressures as great as 100 psi and more are sometimes used. This pressure amounts to 7.2 tons per sq ft, and the possible effects of such high pressures must not be overlooked. High-pressure injection operations being carried out in the vicinity of existing structures can cause physical damage such as the lifting of buildings and streets and the collapsing of sewers and tunnels.

Pressure-injection work is usually slow and expensive. It requires special technique and equipment *and it should be undertaken only by specialists.*

Compaction. Beds of natural soil are sometimes strengthened by compaction methods of various kinds. Loose sand beds may be compacted by vibration and tamping or by rolling with heavy rollers. The driving of piles into beds of loose sand produces compaction not only by vibration but also by displacement.

The rapid consolidation of saturated clay beds is a form of compaction which has recently come into use. The consolidation process can be accelerated by means of vertical sand drains [1] or with cardboard drains [2] which provide channels through which water can escape much more rapidly than through the clay itself.

Vertical Sand Drains. In recent years, the use of vertical sand drains for fill construction over marsh or silted areas has assumed increasing importance and numerous articles [3] have appeared in engineering literature. Vertical sand drains represent an attempt to augment and hasten the consolidation of unstable, saturated soils by providing vertical drainage outlets for the water presumed to be squeezed from the soil by the weight of the surcharge or fill. The objective of consolidation by such artificial means is to develop increased soil resistance and support for superimposed loads, usually consisting of earth fills in highway or airport construction.

The consolidation theory conceives that settlement is caused by squeezing water out of the voids of a saturated soil under applied pressure. This theory postulates that the movement of moisture is caused by pore-water pressure or a differential hydrostatic pressure as distinct from the pressure components acting on the soil mass as a whole which results in shearing stress. This includes the calculation of several definite quantitative components of the completed facility, such as the spacing of the vertical sand drains, the height of surcharge necessary to produce a required degree of improvement, and the time of the consolidation period. *Consequently, in considering an installation of vertical sand drains, careful and experienced studies must be made with respect to moisture content, soil density, consolidation, shearing resistance before (which should be redetermined after) consolidation, the degree of surcharge load necessary to bring about the required consolidation, a specified program for applying the surcharge loads, and stability analyses to be sure that the placement of the surcharge loads will not cause shearing displacements in the*

[1] PORTER, O. J., Studies of Fill Construction over Mud Flats, *Proc. Highway Research Board*, vol. 18, Pt. II, p. 129, 1938.

[2] KJELLMAN, WALTER J., Accelerating Consolidation of Fine-grained Soils by Means of Cardboard Wicks, *Proc. 2d Intern. Conf. Soil Mech. and Foundation Eng.*, Rotterdam, vol. 2, p. 302, 1948.

[3] PORTER, O. J., and L. C. URQUHART, Sand Drains Expedite Stabilization of Marsh Section, *Civil Eng.*, January, 1952. The illustrations in this article are very illustrative and well worth referring to. NOBLE, C. M., and O. J. PORTER, Effectiveness of Sand Drains on New Jersey Turnpike, vol. 80, Separate No. 571, *Proc. Soil Mech. and Foundations Div., ASCE*, December, 1954. HOUSEL, W. S., Checking Up on Vertical Sand Drains. *Highway Research Board, Bull.* 90, February, 1955. WU, DR. T. H., and DR. R. B. PECK, Field Observations on Sand Drain Construction on Two Highway Projects in Illinois, *Proc. Highway Research Board*, vol. 35, p. 747, 1956.

mass of the unstable soil to an extent which will render the vertical sand drains inoperative. In those instances where vertical sand drains have been effective, the placement of the surcharge load and the corresponding time periods have been so controlled as to ensure against such shear failures in the mass of the unstable soil; and, in those instances where vertical sand drains have not been effective, the control of those very necessary factors has not been adequate, thus permitting shear failures which may have eliminated the effectiveness of the vertical sand drains by displacement of them. The successes of this method, in such instances as the New Jersey Turnpike across the Jersey meadows and in the relocation of Highway US-51, LaSalle, Ill., are such as to invite further attempts to use this method, which should be documented by actual full-scale data as mentioned above.

Electro-osmotic Stabilization. A comparatively recent development, which is beginning to be applied to foundation work, is the process called "electro-osmotic stabilization of soils." Metal electrodes are introduced into the soil and direct current is allowed to pass between the electrodes until the soil is hardened. The process of hardening is a complex one in electrophysical chemistry. In soils so fine-grained that the voids between the grains can be considered as fine capillaries of irregular shape having compressible walls (i.e., silt and clay), the introduction of direct electric current will induce a flow of water from anode to cathode. The cathodes (where the flow of water discharges) usually consist of wellpoints; and the anodes usually consist of steel sheet piling, rods, old rails, or the like. For a more complete explanation of this process, reference is made to the literature.[1] *Field use of the process should be undertaken only by persons or firms of experience in applying the process.*[2] Some applications of the process have been notably successful and others have not; so, at any location to which the process should be applied, a background of carefully made test borings with the shearing resistances of the soil should first be provided; the effectiveness of the process should be checked during and after treatment by again determining the shearing resistances of the soil to check its gain in strength; and, with respect to any excavation involving the treated soil, the latter shearing resistances should be used in stability analyses to ensure against shearing failures.

Bibliography

HOGENTOGLER, C. A.: "Engineering Properties of Soil," McGraw-Hill Book Company, Inc., New York, 1937.

HOOL and KINNE: "Foundations, Abutments and Footings," McGraw-Hill Book Company, Inc., New York, 1943.

HUNTINGTON, W. C.: "Building Construction," John Wiley & Sons, Inc., New York, 1941.

JACOBY and DAVIS: "Foundations of Bridges and Buildings," McGraw-Hill Book Company, Inc., New York, 1941.

KRYNINE, D. P.: "Soil Mechanics," McGraw-Hill Book Company, Inc., New York, 1947.

Proc. Intern. Conf. Soil Mech. and Foundation Eng., Cambridge, Mass., 1936.

Proc. 2d Intern. Conf. Soil Mech. and Foundation Eng., Rotterdam, 1948.

TAYLOR, D. W.: "Fundamentals of Soil Mechanics," John Wiley & Sons, Inc., New York, 1948.

TERZAGHI, KARL: "Theoretical Soil Mechanics," John Wiley & Sons, Inc., New York, 1943.

TERZAGHI and PECK: "Soil Mechanics in Engineering Practice," John Wiley & Sons, Inc., New York, 1948.

[1] CASAGRANDE, DR. LEO, Electro-osmotic Stabilization of Soils, *Jour. Boston Soc. Civil Engrs.*, vol. 39, January, 1952; also *Harvard Soil Mechanics Series Bull.* 38; Review of Past and Current Work on Electro-osmotic Stabilization of Soil, *Harvard Soil Mechanics Series Bull.* 45, 1953. VEY, PROF. E., The Mechanics of Soil Consolidation by Electro-osmosis, *Proc. 29th Annual Meeting of Highway Research Board*, December, 1949.

[2] In the United States, the application of the Electro-osmotic process is handled by the Wellpoint Dewatering Corporation, 881 East 141st St., New York 54, N.Y.

Section 9

SEWERAGE AND SEWAGE DISPOSAL

BY RICHARD G. TYLER

QUANTITY OF SEWAGE

Sewage. Sewage is the liquid waste of the community and includes that from industrial as well as from residential areas. These are called *industrial,* or *trade, waste* and *domestic sewage* respectively. Domestic sewage contains less than 0.1 percent of solid materials. This is about half *organic* and half *inorganic;* though the quantity of solids is not large, the organic content decomposes readily with the production of unpleasant odors, which makes the collection and removal or treatment of the sewage from built-up areas imperative. Industrial wastes vary widely in quality and in quantity per unit of product and may often be a major factor in the disposal problem. Both the character and quantity of the wastes present in the sewage therefore require investigation.

Sewers. Sewers are underground pipes or conduits which carry the sewage to a point of discharge or disposal. A *separate* or *sanitary* sewer is one which carries domestic sewage and industrial wastes. A *combined* sewer receives, in addition, the storm water from roof or street surfaces, while a *storm-water drain* carries this surface runoff only.

The sewage originating in a building passes from house fixtures through the plumbing system and *house sewer,* or *building connection,* to the sewer *lateral.* This discharges into a *submain* or *main* sewer and may reach the point of final discharge or disposal through a main *outfall* sewer, or *intercepter.* The latter is a sewer constructed to intercept or pick up the flow, either wholly or in part, from existing sewers and carry it to the point of final discharge. There is more or less *infiltration* of *ground water* into sewers at broken or poorly constructed joints, and this may at times considerably augment the sewage volume. *Storm water* is the surface runoff following rainfall, which enters the sewer through *inlets* or *catch basins.*

Quantity of Sewage. The two factors that have the greatest influence in determining the quantity of domestic sewage are the *population* and the *per capita sewage* contributed. There are also other factors affecting the quantity of flow, such as the number, type, and size of industries and the quantity of ground-water infiltration. These can be determined separately, and the total sewage flow reported in gallons per capita daily (gpcd), in million gallons per day (mgd), or in cubic feet per second (cfs).

Population Estimates. Population estimates are made in order to predict the quantity of sewage to be taken care of at a specified future date. The reports of the Bureau of the Census furnish the most reliable source of population data available. These give the population for all communities for each decade. In addition to these data, some states and cities also report the intermediate 5-year population.

Table 1. Population Summary of Principal Cities of the United States

	1950	1940	1930	1920	1910	1900	1890	1880	1870
1. New York	7,891,957	7,454,995	6,930,446	5,620,048	4,766,883	3,437,202	2,507,414	1,911,698	1,478,103
2. Chicago	3,620,962	3,396,808	3,376,438	2,701,705	2,185,282	1,698,575	1,099,850	503,185	298,977
3. Philadelphia	2,071,605	1,931,334	1,950,961	1,823,779	1,549,008	1,293,697	1,046,964	847,170	674,002
4. Los Angeles	1,970,358	1,504,277	1,238,048	576,673	310,198	102,479	50,395	11,183	5,728
5. Detroit	1,849,568	1,623,452	1,568,662	993,678	465,766	285,704	205,876	116,340	79,577
6. Baltimore	949,709	859,100	804,874	733,826	558,485	508,957	434,439	332,313	267,354
7. Cleveland	914,808	878,336	900,429	796,841	560,663	381,768	261,353	160,146	92,829
8. St. Louis	856,796	816,048	821,960	772,897	687,029	575,238	451,770	350,518	310,864
9. Washington	802,178	663,091	486,869	437,571	331,069	278,718	188,932	147,293	109,199
10. Boston	801,444	770,816	781,188	748,060	670,585	560,892	448,477	362,839	250,526
11. San Francisco	775,357	634,536	634,394	506,676	416,912	342,782	298,997	233,959	149,473
12. Pittsburgh	676,806	671,659	669,817	588,343	533,905	451,312	343,904	235,071	139,256
13. Milwaukee	637,392	587,472	578,249	457,147	373,857	285,315	204,468	115,587	71,440
14. Houston	596,163	384,514	292,352	138,276	78,800	44,633	27,557	16,513	9,382
15. Buffalo	580,132	575,901	573,076	506,775	423,715	352,387	255,664	155,134	117,714
16. New Orleans	570,445	494,537	458,762	387,219	339,075	287,104	242,039	216,090	191,418
17. Minneapolis	521,718	492,370	464,356	380,582	301,408	202,718	164,738	46,887	13,066
18. Cincinnati	503,998	455,610	451,160	401,247	363,591	325,902	296,908	255,139	216,239
19. Seattle	467,591	368,302	365,583	315,312	237,194	80,671	42,837	3,533	1,107
20. Kansas City, Mo.	456,622	399,178	399,746	324,410	248,381	163,752	132,716	55,785	32,260
21. Newark	438,776	429,760	442,337	414,524	347,467	246,070	181,830	136,508	105,059
22. Dallas	434,462	294,734	260,475	158,946	92,104	42,638	38,067		
23. Indianapolis	427,173	386,972	364,161	314,194	233,650	169,164	105,436	75,056	48,244
24. Portland, Ore.	373,628	305,394	301,815	258,288	207,214	90,426	46,385	17,577	8,293
25. Atlanta	331,315	302,288	270,366	200,616	154,839	89,872	65,533	37,409	21,789

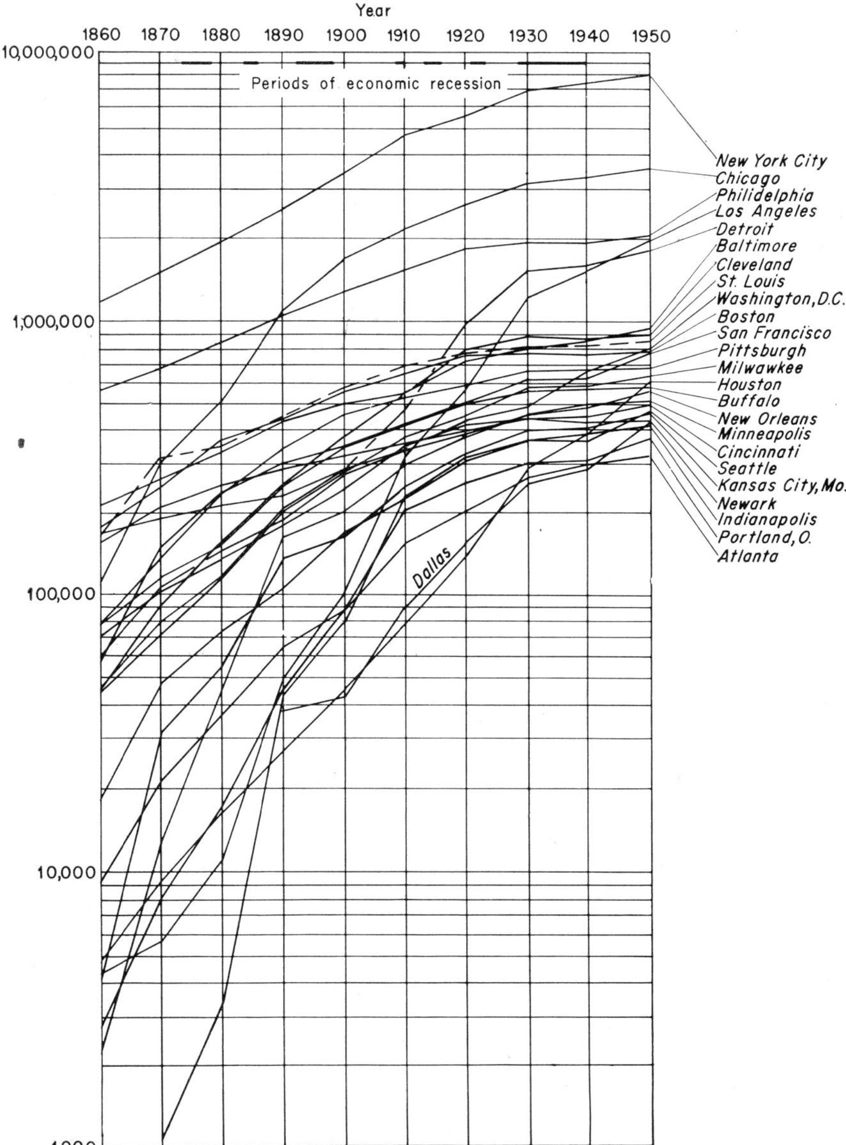

FIG. 1. The effects of depressions on population growths of cities.

Various methods are in use by engineers and city planners for determining the populations for which sewers and sewage-treatment plants should be designed. These include the following: (1) extension graphically of the population growth curve, modified as judgment and experience in population estimating may require; (2) comparison graphically with larger cities having similar growth characteristics; (3) use of a *cohort-survival* technique, commonly used by city and regional planners, in which specified population age-sex groups called *cohorts*, resulting from mortality and net migration plus survivors of birth during the period, are shifted forward in time, usually grouped into 5-year cohorts from birth through age 74, a 10-year cohort from 75 through 84, and another for 85 years and over; (4) logarithmic extension of the existing population trend; (5) summation of the excess of birth rates over death rates and in migrations over out migrations; and (6) use of a logistic curve developed by Verhulst in 1838, based on a law of growth in a limited area. This law was rediscovered many years later by Raymond Pearl and Lowell J. Reed while investigating the growth of fruit flies under conditions of controlled food supply. More recently (1940) Velz and Eich [1] used the logistic curve for estimating the population growth of cities on the assumption that populations will reach a saturation value based on economic opportunity and approached at a decreasing rate.

It is now becoming apparent that all these methods are at times inapplicable because of a random and unpredictable factor, viz., the migrations caused by economic depressions or by a national emergency as in World War II. Reference to the population data shown in Table 1 and reproduced graphically in Fig. 1 shows clearly the effects of the depressions of the 1870, 1890, and 1930 decades on population growths. Figure 3 similarly shows the effects of the higher birth rate and the migrations of industrial workers during and following the war.

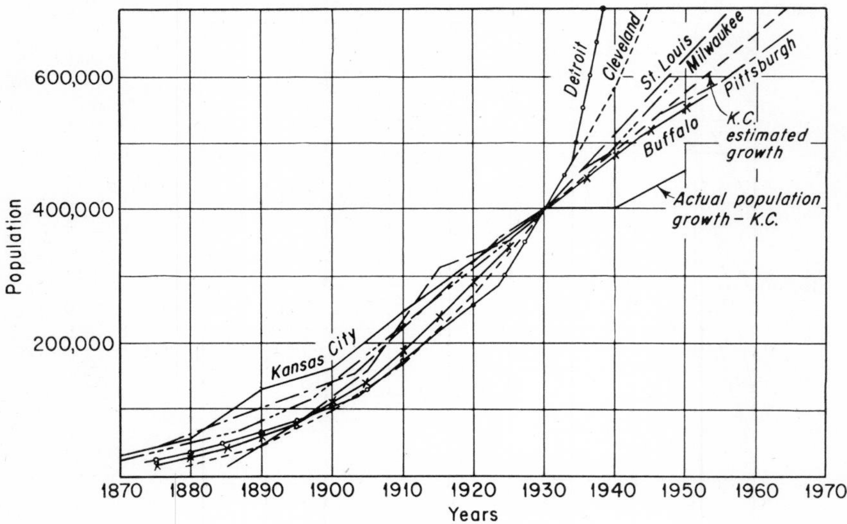

FIG. 2. Curves for population estimate for Kansas City, Mo.

The methods that have been used for estimating population growth naturally cannot be expected to apply to these unusual migrations. Figure 2 is of interest in this connection as an illustration of the large errors that can result in population predictions

[1] *Civil Eng.*, October, 1944, pp. 619–622.

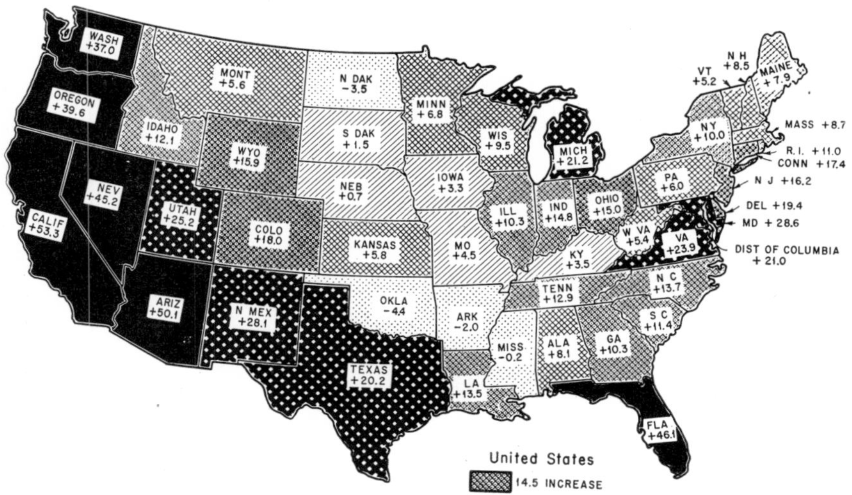

(a) 1940-50 population

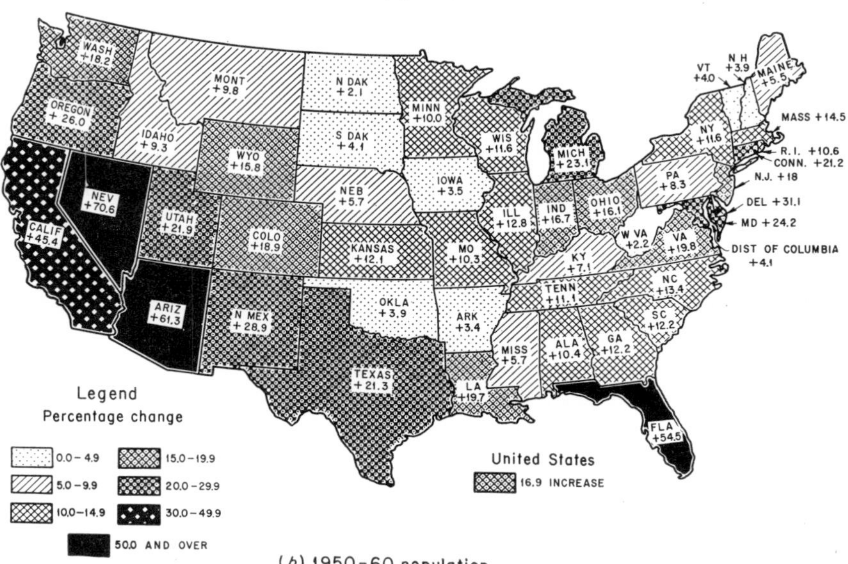

(b) 1950-60 population

FIG. 3. Population migrations.

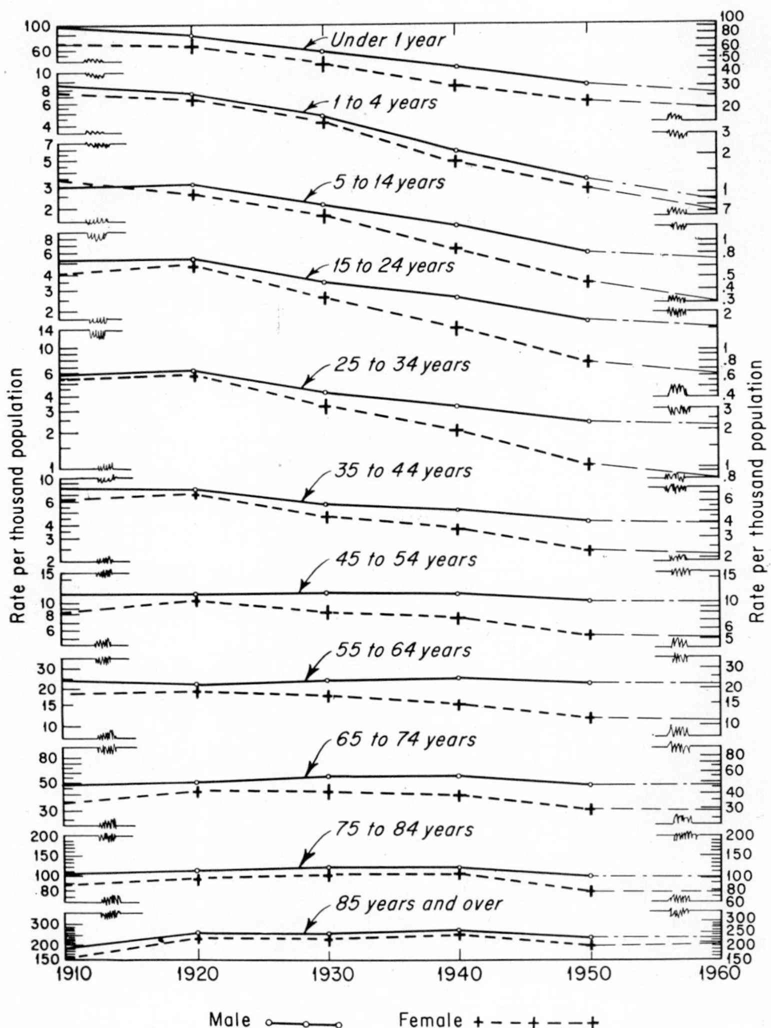

FIG. 4. Mortality trends by age and sex in the State of Washington, 1910–1960.

during periods of migrating populations. Since such periods are not readily predictable, present trends are toward estimating population changes for shorter time periods (5 to 15 years), for which the *cohort-survival technique* permits a more detailed analysis of the essential variables involved in the growth pattern. Figure 2, which was used in earlier editions of this section, would, with such restudy every 2 or 3 years, have been adjusted to the changing trends as they were taking place. The actual 1950 population of Kansas City was only 79.1 percent of the predicted value because of the emigration of the unemployed in search of work, causing a net loss (1940) of 568 persons where the gain during the previous decade was 75,336.

Regional Factors in Population Estimating. Population migrations arise not only during periods of economic depression but also during periods of prosperity when people can move into areas where climate, scenery, or other attraction draws them. Also, in the first half of the 1940 decade during World War II, the movement of large populations to the west coast to work in war industries aided the rapid growth in the Southwestern states shown in Fig. 3*a* at the expense of the central states where recurring droughts had been causing financial losses and low incomes to the population. Figure 3*b* shows that migrations to the West and Southwest are still continuing. Since the national growth rate during 1940–1950 averaged 14.5 percent, those areas shown on this map having greater population gains than this figure received migrations from those losing population. Such migrations complicate the problem of population estimating. Fortunately, both Federal and state agencies periodically publish population estimates which are helpful in the preparation of plans for sewers and disposal plants.

The **cohort-survival technique** of population forecasting, although laborious, is now being widely used in public-works planning. It is an analysis of annual *mortality*, *fertility*, and *migration* trends of the various age-sex groups, or cohorts, in which each cohort is extended graphically along its own trend line for a 5- to 15-year period, and

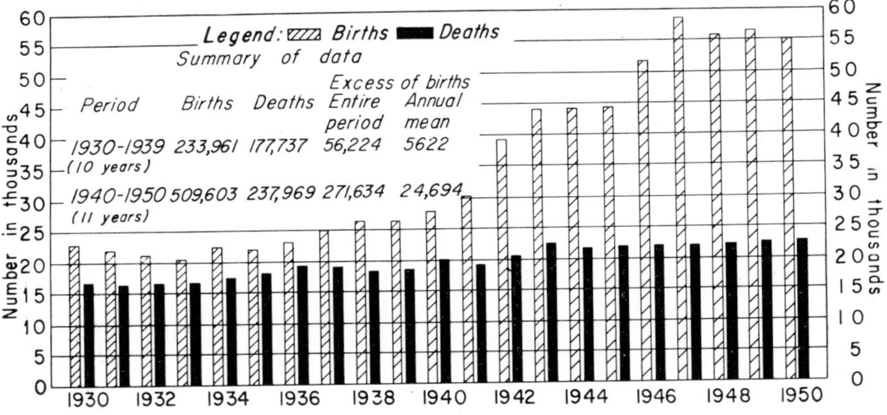

FIG. 5. Number of births and deaths in the State of Washington, 1930–1950.

the resultant findings combined in appropriate graphs as in Figs. 4 and 5, and in Table 2. Figure 4 illustrates the method of estimating a 10-year growth in each cohort, while Table 2 gives the numbers in each for the population-growth assumptions stated in the footnotes.

An increasing number of cities now employ professional *city planners* to advise them with reference to their city's growth and development. Since an important part of the

TABLE 2. ESTIMATES [1] OF TOTAL POPULATION [2] FOR THE STATE OF WASHINGTON BY AGE AND SEX ACCORDING TO VARIOUS ASSUMPTIONS [3] OF NET MIGRATION: 1955, 1960, 1965

Figures independently rounded to nearest thousand

TOTAL

Age cohorts	1950 census	1955					1960					1965				
		M_1	M_2	M_3	M_4	M_5	M_1	M_2	M_3	M_4	M_5	M_1	M_2	M_3	M_4	M_5
All ages......	2,379	2,565	2,590	2,615	2,690	2,765	2,725	2,779	2,832	2,993	3,153	2,887	2,973	3,058	3,313	3,569
0–4	263	288	289	289	290	292	274	279	284	299	314	288	297	306	333	360
5–9	204	267	268	269	272	275	292	294	295	300	304	278	284	290	308	327
10–14	160	203	206	208	215	222	266	270	273	283	293	291	295	299	311	323
15–19	158	159	162	164	172	180	202	207	212	227	242	265	271	277	295	314
20–24	176	157	160	163	173	182	158	164	170	187	204	201	209	218	242	266
25–29	195	174	178	181	191	200	156	162	169	188	207	157	166	175	202	229
30–34	189	194	196	199	207	216	173	179	185	203	221	155	164	173	200	227
35–39	181	187	189	192	199	206	192	197	202	217	233	172	180	188	213	238
40–44	159	178	180	182	187	193	184	188	192	205	218	189	196	203	223	244
45–49	139	155	157	158	162	166	174	177	180	189	199	180	185	191	207	223
50–54	126	134	135	136	139	142	150	152	154	161	168	168	172	176	188	200
55–59	115	119	120	120	122	124	126	128	130	135	139	142	145	147	156	164
60–64	104	106	107	107	109	110	110	111	112	115	118	116	118	120	126	132
65–69	87	92	92	93	95	97	94	95	96	99	103	97	99	100	105	110
70–74	60	72	72	73	74	75	76	77	78	81	84	78	79	81	85	89
75–79	36	45	45	45	46	47	54	55	55	57	59	57	58	59	62	65
80–84	19	23	23	23	24	24	28	29	29	30	31	34	35	35	37	39
85–89	8	9	9	9	10	10	11	11	12	12	13	14	14	14	15	16
90 and over	2	3	3	3	3	3	3	3	3	3	4	4	4	4	4	5
Under 5 years adjusted for under enumeration [4].......	268	294	294	294	296	297	280	285	290	305	319	294	303	312	339	367

[1] First revision (1955) of population estimates published in CALVIN F. SCHMID and others, "Population and Enrollment Trends and Forecasts, State of Washington," Washington State Census Board, 1953, pp. 1–28.

[2] Total population is the civilian population plus Armed Forces stationed in the state and is comparable with the population enumerated in the 1950 Census.

[3] Average annual net migration assumptions are: M_1—none since 1950; M_2—5,000; M_3—10,000; M_4—25,000; M_5—40,000.

[4] The information utilized in adjusting for under enumeration was obtained from U.S. Bureau of the Census, *Current Population Reports*, Series P-25, No. 106 (Dec. 6, 1954), pp. 10–17.

planner's function is to forecast population trends, this information, in cities of 100,000 population or more, would normally be available to the engineer planning the sewer system.

Sewage Flow. After the anticipated population has been determined, it is multiplied by the assumed per capita sewage contribution to get the total anticipated flow. Local industries may contribute large quantities of liquid wastes whose strength and quantity differ widely and must be determined individually. Ground-water infiltration into the sewers through joints which are not watertight must also be allowed for in determining the size of pipe required. The total flow is thus made up of domestic or sanitary sewage, industrial wastes, and infiltration.

Infiltration varies with the length and diameter of the sewer, its material and jointing, the care used in its construction, and the depth of the sewer below the ground-water level or *water table.* Table 3 [1] gives the measured infiltration at a number of American cities,

TABLE 3. INFILTRATION MEASUREMENTS

City	Year	Ground conditions	Joint type	Sewer size, in.	Test length, miles	Leakage, gal per mile per 24 hr
Boston, Mass.				8–36	137	40,000
Canton, Ohio					11	26,500
East Orange, N.J.				8–24	25	8,650
Westboro, Mass.	1899			15	0.37	1,320,300
Stamford, Conn.	1892			6–18	13.4	94,170
New Bedford, Mass.				12		17,000
Ocean Grove, N.J.	1912	Wet sandy clay	Mortar	4–12	6.5	15,000–44,000
New York City, N.Y.				8–24	5	1,300,000
New Orleans, La.						45,900
North Shore (Chicago), Ill.	1916		Mortar	6	0.2	24,200
South Charleston, W. Va.	1920	Clay, sand, gravel	Mortar	8–18	2.60	34,600
Urbana-Champaign, Ill.	1924	Gravel and sandy	Mortar	10–18	0.44	18,550
	1924		Mortar	12–24	1.3	34,400
Miami, Fla.	1926		Mortar		16.3	1,200–7,025
Altoona, Pa.	1930			8–12	2.8	≦6,000
Salt Lake City, Utah	1927		Mortar			39,700
Berlin, Md.	1944	30 percent wet ground			8	7 percent
Harrington, Del.		50 percent wet ground			8	7 percent
Florence, N.J.					12	20 percent
Georgetown, Del.					8.4	21 percent
Cranston, R.I.	1945	Low ground water	Bituminous	8–39	26.0	<5,500
Gloucester, Mass.	1945	High ground water	Bituminous	21	0.18	26,600
Saugus, Mass.	1945	High ground water	Bituminous	8–24	3.6	28,200–33,500
Webster, Mass.	1945	Low ground water	Mortar	8–18	7.5	10,000–16,500
Point Pleasant, N.J.	1948		Hot pour	8 and 10	1.8	6,000
Lancaster, Pa.	1950		Rubber gasket	16 and 18	0.57	8,200–9,200
	1950		Rubber gasket	18 and 24	1.3	9,200
Lemoyne, Pa.	1950	Mostly wet	Hot pour	8–20	20	<10,000
Tampa, Fla.	1951			6 and 8	2.5	1,590
	1952			6–15	5.0	1,800–2,700
	1952			6–12	7.1	6,400–6,700
	1952			6–21	2.2	7,470–8,300
Stamford, Conn.	1953		Hot pour	8–12	1.0	200
West Elmira, N.Y.	1953		Hot pour	8	0.1	3,200
Ewing-Lawrence, N.J.	1953	90 percent wet	Hot pour	6–15		3,100
Portsmouth, Va.	1953			8	0.7	3,400–4,600
	1953			8	0.66	1,100–9,900

and the allowable infiltration rates used by various consulting engineers engaged in sewer design and construction are given in Table 4.[1] The importance of adequate inspection during construction and of specifying the best type of joint for the construction conditions to be encountered is clearly emphasized by these tables. A range of 3,500 to 5,000 gal per mile per 24 hr has been commonly specified for 8-in. pipe.

[1] VELZY, CHARLES R., and JOSHUA M. SPRAGUE, Infiltration Specifications and Tests, *Sewage and Ind. Wastes*, vol. 27, pp. 249–250, 1955.

TABLE 4. INFILTRATION SPECIFICATIONS

Source	Conditions	Allowable infiltration, gal per mile per 24 hr				
		Per in. diameter	8-in. pipe	10-in. pipe	12-in. pipe	24-in. pipe
a. Consulting Engineers						
Greeley & Hansen	Average ground	......	5,000	5,000	5,000	5,000
	Wet, pervious ground	......	10,000	10,000	10,000	10,000
Metcalf & Eddy	Average	500	4,000	5,000	6,000	12,000
	Any short section	1,000	8,000	10,000	12,000	24,000
Havens and Emerson	....	500	4,000	5,000	6,000	12,000
Consoer, Townsend & Assoc.	....	200	1,600	2,000	2,400	4,800
Gannett Fleming Corddry & Carpenter	....	500	4,000	5,000	6,000	12,000
William A. Goff	....	......	5,000	6,300	7,500	15,000
Buck, Seifert and Jost	....	[1]	7,000	8,800	10,500	21,000
Whitman, Requardt & Assoc.	....	250	2,000	2,500	3,000	6,000
O'Brien and Gere	....	......	1,400	1,750	2,100	2,450
John Baffa	Average [2]	......	4,000	5,000	6,000	12,000
Nussbaumer, Clarke & Velzy	....	500	4,000	5,000	6,000	12,000
b. Cities and Districts						
Miami, Fla.	....	1,000	8,000	10,000	12,000	24,000
Milwaukee, Wis.	....	......	5,800	7,250	8,700	17,400
Minneapolis, Minn.	....	......	3,500	3,500	3,500	3,500
Portland, Ore.	....	0.5 [3]	5,100	6,350	7,600	15,200
Seattle, Wash.	....	......	3,200	3,800	4,400	10,000
Syracuse, N.Y.	....	200	1,600	2,000	2,400	4,800
Tampa, Fla.	....	......	10,000	10,000	10,000	10,000
Tulsa, Okla.	....	......	25,000			
Topeka, Kans.	....	1,500	12,000	15,000	18,000	36,000
Stamford, Conn.	....	......	1,160	1,300	1,600	
Allegheny County (Pa.) San. Auth.	....	150	1,200	1,500	1,800	3,600
Washington (D.C.) Suburban San. Commission	....	......	5,000	6,000	7,000	12,000
Nassau County (N.Y.)	....	0.4 [3]	4,000	5,000	6,000	12,000

[1] 2 gal per ft of joint per 24 hr.
[2] Short sections may be 100 percent in excess.
[3] Per 100 ft.

Storm water may reach sanitary sewers through illegal storm-water connections, leaking sewer joints or house connections, perforated or leaking manhole covers, and roof or cellar drains. Adequate inspection during construction is necessary to prevent such extraneous water from reaching the sewer.

Air Conditioning Affects Sewage Flow. The rapid growth of air conditioning during recent years has created serious problems in some cities by increasing the water consumption abnormally, causing corresponding increases in the per capita sewage flow. Data are now available for estimating the added load thus placed on sewers, but the growth

of air conditioning has varied with differing climatic conditions in different cities. Chicago found her sewers in the Loop district taxed to capacity, with only about 16 percent of the district air-conditioned. In August, 1937, the water used for air conditioning was 31 percent of that used for all other purposes. Other cities are experiencing similar difficulties. Estimates on a per capita basis are difficult for obvious reasons. Where the load becomes troublesome, the required use of cooling towers or evaporating condensers may reduce the water consumption for air conditioning by 90, or even 95, percent. Daily usages of 140 to 350 gal per capita for air-conditioning have not been uncommon. Rising costs, however, have encouraged its reuse for domestic or other usage with such treatment as may be required.

Rate of Flow. The *hourly* rate of flow of sewage is not constant but varies roughly between 50 and 150 percent of the daily average. Figure 6 is a typical flow curve for a main outfall sewer and gives the average hourly variations in sewage flow for four

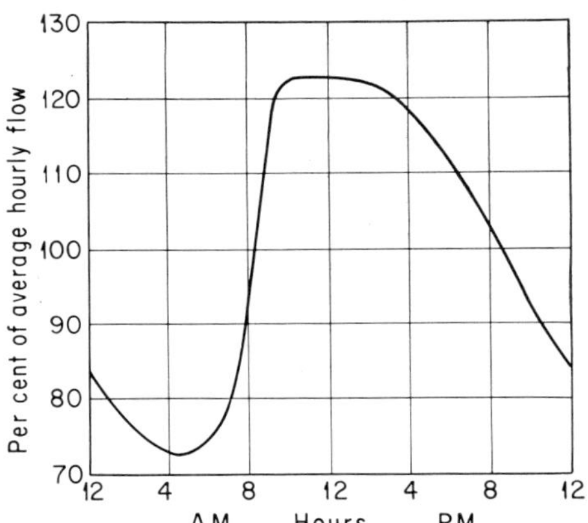

FIG. 6. Hourly variations in sewage flow.

American cities. The shape of the curve varies from day to day and is smoother for large cities than for small ones, as the latter are subject to more erratic flows. There are seasonal fluctuations also which should be taken into account in designing sewers.

The *maximum* rate, after allowing for this seasonal factor, is that for which the sewer is designed. This will seldom exceed 200 to 250 gal per capita per day, to which must be added an allowance for infiltration of about 5,000 gal per mile per 24 hr as in Table 4. If there are industries present, an estimate of the volume of their wastes is made separately and added to the above total. For convenience, this total discharge may be expressed in gallons per capita daily, or per acre daily, for ready use in design, as discussed in a later paragraph.

Quantity of Storm Water. Two methods of estimating the quantity of storm water are in general use. In the *empirical* method, which is used for large areas more frequently than for small, formulas derived from runoff measurements are employed. Such formulas may be very useful where sufficient data on storm-water runoff are available from which to determine coefficients to fit the local situation. If such data are not at hand, empirical formulas should be used with caution, and differences between local conditions and those for which the formula was derived should be taken into consideration. One of the most frequently used of these formulas is McMath's, which is

$$Q = Aci \sqrt[5]{\frac{S}{A}} \tag{1}$$

where Q is the runoff in cubic feet per second, A the drainage area in acres, c a coefficient varying from 0.20 to 0.90 and averaging about 0.75, i the intensity of rainfall in inches per hour (practically equivalent to cubic feet per second per acre), and S the slope in feet per 1,000 ft. Values of c may be selected from Table 5 or estimated from local runoff data.

TABLE 5. SUGGESTED VALUES OF c FOR VARYING PERCENTAGES OF IMPERVIOUS [1] AREA

	Percentage impervious surface								
Sandy soil......	0	5	10	16	25	37	53	73	100
Clay soil........	....	....	....	5	15	28	46	70	100
Values of c......	0.10	0.14	0.18	0.23	0.30	0.40	0.50	0.70	0.90

[1] Reprinted by permission from Merriman and Wiggin, "American Civil Engineers' Handbook," John Wiley & Sons, Inc.

A graph like Fig. 7 is useful in making the solution of problems by McMath's formula less laborious. A problem will serve to show how the chart is used. The runoff is desired from 100 acres with a slope of 1 percent, clayey soil, 46 percent of the area being impervious, and with i equal to 3.00 in. per hr. From Table 4, c is found to be 0.50, giving a value of 1.50 for ci. Enter the chart at the upper scale with the area of 100

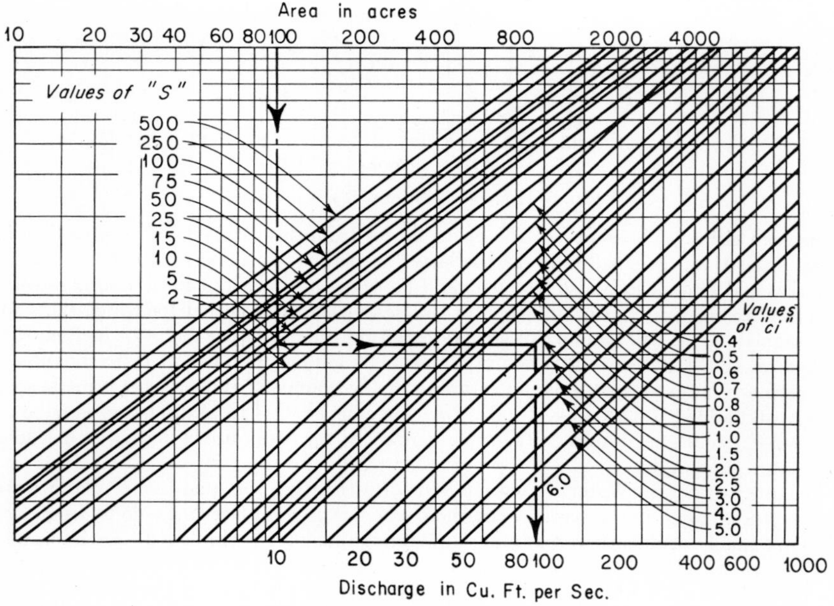

FIG. 7. Storm-water runoff by McMath's formula.

acres, and follow the vertical down to the slope of 10 per 1,000, as indicated by the dashed line. Then follow the horizontal to $ci = 1.50$ and thence downward to the lower scale, reading 95 cu ft per sec as the value of Q.

The rational method is used more frequently for small areas such as are served by storm-water drains. It is expressed by the formula.

$$Q = ciA \tag{2}$$

where c is the *coefficient of runoff*, which varies from 0.10 to 0.95.

In the *intensity-duration* curves for Boston (Fig. 8), the relation between maximum intensity and time of rainfall is shown for the greatest storms likely to occur in the number of years given. If rainfall records are available, similar curves may be constructed for any area studied. If such data are not available, it will be necessary to consider the data available for areas as similar as possible. From these curves, it is

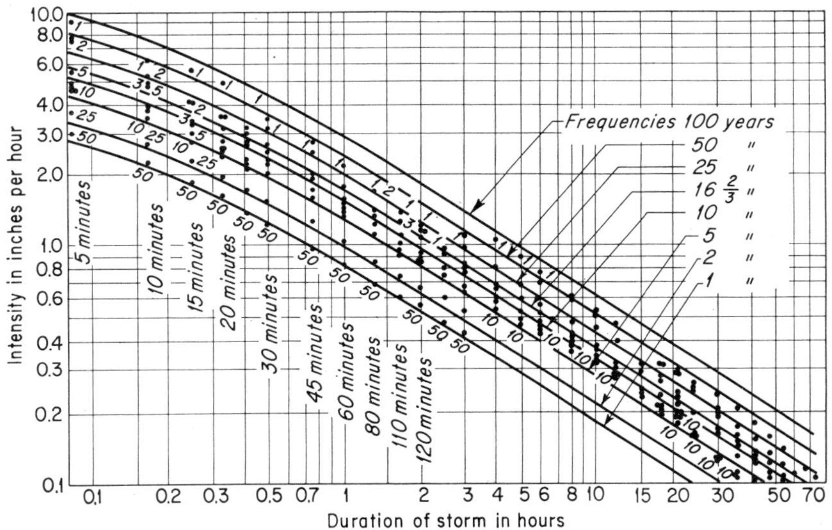

FIG. 8. Intensity of precipitation at Boston, Mass., for various durations and frequencies. (*Sherman, Frequency and Intensity of Excessive Rainfalls at Boston, Mass., Trans. Am. Soc. Civil Engrs., vol.* 95, *p.* 956, 1931.)

seen that the intensity varies inversely with the duration of rainfall. For example, each year it may rain at the rate of 2.0 in. per hr for 12 min, while an intensity of only 0.8 in. per hr will continue for as long as an hour. Once in 50 years, an intensity of 5.8 in. may endure for 12 min, while 2.4 in. per hr may continue for 1 hr.

This intensity-time relationship may be represented also by a formula of the type $i = \dfrac{105}{t + 30}$ (Boston). Naturally, this intensity of precipitation is not uniform over large areas but is that measured at the particular point where the rain gage is located. The way in which the average intensity varies with the size of area for rains of different duration is shown in Marston's data[1] in Table 6. For an area of 5,000 acres the 30-min intensity will be only 1.92/2.34 = 82 percent of the intensity for an area one-tenth as large.

The *time of concentration* is the time required for the water from the farthest point

[1] Distribution of Intense Rainfall, *Trans. ASCE*, vol. 87, p. 559, 1924.

TABLE 6. AVERAGE INTENSITY OF PRECIPITATION AND AREA COVERED, BOSTON, MASS.[1]
Estimated frequency of occurrence, 10 years

Area, acres	Average intensity of precipitation, in. per hr, for durations of		
	30 min	45 min	60 min
0	2.48	1.94	1.60
500	2.34	1.85	1.55
1,000	2.26	1.80	1.52
1,500	2.20	1.76	1.49
2,000	2.15	1.74	1.47
3,000	2.07	1.69	1.43
4,000	1.98	1.65	1.40
5,000	1.92	1.61	1.38

[1] MARSTON, FRANK A., *Trans. ASCE*, vol. 87, p. 559, 1924.

of the area to reach the point of concentration or that point of the system for which the computation of surface runoff is to be made. This time is computed by assuming an *inlet* time required for the runoff for the most distant part of the area to reach the nearest inlet and adding thereto the time of flow in the sewer from that inlet to the point of concentration. The rainfall intensity i used in the formula is that intensity of rainfall which lasts for the time of concentration and is taken from the intensity-duration curve or computed by the appropriate formula. In the above formula, Q is a maximum when A is a maximum, since A varies more rapidly than, but inversely with, i. The intensity used, therefore, should be for that time at which the entire area becomes tributary at the point under consideration. In storm-sewer studies, it is unnecessary to construct a storm sewer that will accommodate the largest flood of record, but the 5-, 10-, or 15-year flood is ordinarily used, depending on the amount of damage that may result from overtaxing the storm sewer. The inlet time varies from 3 to 20 min—ordinarily being about 5 min—and experience is needed for its selection. Its determination is approximate and is more important for small areas than for large ones, since in such cases the inlet time is a greater proportion of the total time of concentration.

The value of the runoff coefficient to be used depends on the relative imperviousness of the tributary area and the length of time it has been raining on the various parts of that area. Table 7 gives the range of coefficients of runoff that are commonly used. When an area includes different classes of surface, their coefficients are combined in proportion to their relative areas. For example, let it be required to determine the coefficient of runoff for an area consisting of 1.5 acres of lawns and parkways with a coefficient of 0.10, and 0.5 acre of pavements and roof surfaces with a coefficient of 0.90. Then,

$$\begin{aligned} 25\% \times 0.90 &= 0.225 \\ 75\% \times 0.10 &= 0.075 \\ \hline \text{Coefficient for total area} &= 0.300 \end{aligned}$$

Table 8 illustrates the current practice of some of the larger American cities and gives the general range of coefficients of runoff, inlet time, and storm frequency for which drainage is provided.

TABLE 7. BRYANT AND KUICHLING'S ESTIMATES OF RUNOFF FROM DIFFERENT CLASSES OF SURFACE AS PERCENTAGES OF RAINFALL INTENSITY[1]

Watertight roof surfaces	0.70–0.95
Asphalt pavements in good order	0.85–0.90
Stone, brick, and wooden-block pavements with tightly cemented joints	0.75–0.85
Same with open or uncemented joints	0.50–0.70
Inferior block pavements with open joints	0.40–0.50
Macadamized roadways	0.25–0.60
Gravel roadways and walks	0.15–0.30
Unpaved surfaces, railroad yards, and vacant lots	0.10–0.30
Parks, gardens, lawns, and meadows, depending on surface slope and character of subsoil	0.05–0.25

[1] *Rept.* Back Bay Sewerage System, Boston, 1909.

TABLE 8. RUNOFF ASSUMPTIONS USED IN LARGE AMERICAN CITIES

Item	Coefficient of runoff	Coefficient commonly used
Business areas	0.60–0.95	0.90
Downtown residential	0.30–0.60	0.40–0.50
Suburban	0.25–0.40	0.30
Inlet time, min	3–20	5–10
Frequency of storm provided for, years	1–15	10–15

TABLE 9. RUNOFF COEFFICIENTS BY ZONE METHOD FOR RECTANGULAR AREAS IN WHICH LENGTH IS FOUR TIMES BREADTH, CONTAINING VARIOUS PERCENTAGES OF IMPERVIOUS SURFACES

Duration, or time of concentration, min	Percentage of impervious surfaces										
	00	10	20	30	40	50	60	70	80	90	100
10	0.149	0.189	0.229	0.269	0.309	0.350	0.390	0.430	0.470	0.510	0.550
20	0.236	0.277	0.318	0.360	0.401	0.442	0.483	0.524	0.566	0.607	0.648
30	0.287	0.329	0.372	0.414	0.457	0.499	0.541	0.584	0.626	0.669	0.711
45	0.334	0.377	0.421	0.464	0.508	0.551	0.594	0.638	0.681	0.725	0.768
60	0.371	0.415	0.458	0.502	0.546	0.590	0.633	0.677	0.721	0.764	0.808
75	0.398	0.442	0.486	0.530	0.574	0.618	0.661	0.705	0.749	0.793	0.837
90	0.422	0.465	0.509	0.552	0.596	0.639	0.682	0.726	0.769	0.813	0.856
105	0.445	0.487	0.530	0.572	0.615	0.657	0.699	0.742	0.784	0.827	0.869
120	0.463	0.505	0.546	0.588	0.629	0.671	0.713	0.754	0.796	0.837	0.879
135	0.479	0.521	0.561	0.601	0.642	0.683	0.724	0.765	0.805	0.846	0.887
150	0.495	0.535	0.574	0.614	0.654	0.694	0.733	0.773	0.813	0.852	0.892
180	0.522	0.560	0.598	0.636	0.674	0.713	0.751	0.789	0.827	0.865	0.903

Obviously, long-continued rain saturates the soil and produces higher runoff coefficients than are found for the same area when less saturated. The runoff from the area nearest the point of concentration comes from a surface that has become more saturated than the more distant areas. Attempts have been made to take this into account and to adjust the runoff coefficient by dividing the area into zones of equal distance, measured in runoff time, from the point of concentration and by using coefficients of decreasing magnitude as the zones are increasingly distant. Table 9 has been derived by combining these coefficients with their appropriate fractional areas by this zone principle and is Metcalf and Eddy's data [1] for runoff coefficients from a rectangular area four times as long as it is wide, with varying percentages of impervious surface and different times of concentration. Although these are applicable to a particular shape of area, they also apply in a general way to the areas which are usually encountered in practical problems. Errors due to difference in shape of drainage areas are probably within the limits of accuracy of the rational method and of the assumptions on which it is based.

Table 10 is a typical illustration of the way in which the coefficient of runoff varies with duration of rainfall. It is convenient to prepare a table similar to this for any community in which considerable storm-sewer design is to be undertaken.

TABLE 10. RUNOFF COEFFICIENTS FOR HAMILTON, ONTARIO, CANADA [1]

Duration of rainfall, min (concentration time)	Runoff coefficients		
	Pervious surface	Outlying district	Central business
5	0.00	0.50	0.75
10	0.05	0.60	0.80
15	0.10	0.75	0.85
20	0.15	0.80	0.90
25	0.20	0.825	0.95
30	0.25	0.85	0.95
35	0.30	0.865	0.95
40	0.35	0.883	0.95
45	0.40	0.90	0.95
50	0.45	0.917	0.95
55	0.50	0.93	0.95
60	0.55	0.95	1.00

[1] GREGORY and ARNOLD, Run-off—Rational Run-off Formulas, *Trans. ASCE*, vol. 96, p. 1050, 1932.

Storm-sewer Problem. The use of these tables and the intensity-duration curve will now be illustrated by an example. In Fig. 9 part of the industrial section of Mariemont, Ohio, is to be served by a storm sewer. Determine the sizes of pipe and the quantities involved for the area tributary to the main sewer on Industrial Road at manhole (M.H.) 9. Assume that rainfall conditions are similar to those represented by the curves in Fig. 8, that the 10-year storm is the basis for the design, and that the percentages of impervious surface are as shown in Fig. 9. The necessary computations and design data are given in Table 11. In determining the size of sewer from M.H. 4 to 5, for example, area 2, tributary at M.H. 4, is added to area tributary at M.H. 3. The combined imperviousness is found to be 0.61, which with 12.3-min duration gives a coefficient of runoff of 0.41 by interpolation in Table 9. The time of concentration is equal to that at M.H. 3 plus time required for flow from M.H. 3 to 4. The rainfall intensity from Fig. 8 is 3.8 in. per hr. From these data a total runoff of 3.74 cu ft per sec is obtained. The topography permits a slope of 0.003, which from Fig. 10 requires an 18-in. pipe.

[1] "Sewerage and Sewage Disposal," 2d ed., p. 103, 1930.

Table 11. Storm-sewer Problem, Mariemont, Ohio

From M.H.	To M.H.	Street	Length, ft	Tributary area, acres: Increment	Tributary area, acres: Total	Coefficient of imperviousness	Time of concentration, min	Coefficient of runoff	Rainfall intensity i	Total runoff $Q = ciA$, cu ft per sec	Slope	Diameter of sewer, in.	Capacity of sewer, cu ft per sec	Velocity, ft per sec
1	3	Mariemont Ave.	80	...	....	....	Inlet time = 10	....	...	.....	0.005	12	2.05	2.6
2	3	Mariemont Ave.	130	...	....	....	....	....	...	.....	0.005	12	2.05	2.6
3	4	Mariemont Ave.	240	1.0	1.0	0.4 × 0.6 = 0.24 0.6 × 0.8 = 0.48 0.72	$10 + \frac{130}{2.6 \times 60} = 10.8$	0.44	4.0	1.76	0.005	12	2.05	2.6
4	5	Mariemont Ave.	240	1.4	2.4	0.7 /2.4 × 0.3 = 0.09 0.6 /2.4 × 0.6 = 0.15 1.1 /2.4 × 0.8 = 0.37 0.61	$10.8 + \frac{240}{2.6 \times 60} = 12.3$	0.41	3.8	3.74	0.003	18	4.90	2.7
6	8	Trade St.	...	...	....	....	Inlet time = 10	....	...	.....	0.005	12		
8	5	Trade St.	170	...	1.7	1.2 /1.7 × 0.03 = 0.21 0.5 /1.7 × 0.06 = 0.18 0.39	$10 + \frac{265}{2.6 \times 60} = 11.6$	0.31	3.9	2.06	0.004	15	3.4	2.75
5	9	Trade St.	327	...	7.85	3.0 /7.85 × 0.3 = 0.11 2.8 /7.85 × 0.6 = 0.22 2.05/7.85 × 0.8 = 0.21 0.54	$12.3 + \frac{240}{2.7 \times 60} = 13.8$	0.40	3.6	11.30	0.007	21	11.3	4.63
9	..	....	...	0.9	8.75	3.0 /8.75 × 0.3 = 0.10 2.8 /8.75 × 0.6 = 0.19 2.95/8.75 × 0.8 = 0.27 0.56	$13.8 + \frac{327}{4.63 \times 60} = 15.0$	0.42	3.4	12.50	0.005	24	13.6	4.3

The capacity and velocity for this pipe flowing full on the grade given are 4.90 cu ft per sec and 2.7 ft per sec respectively. The other sections of the sewer are determined in a similar manner.

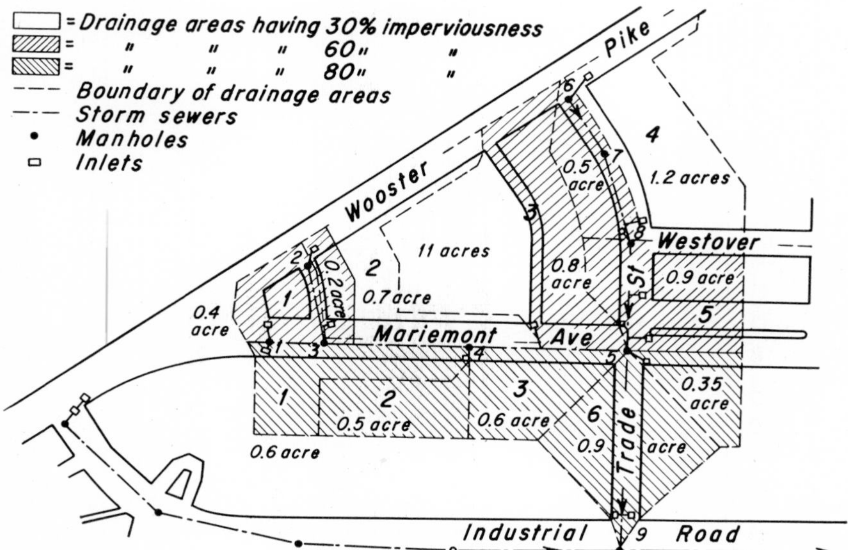

Fig. 9. Storm-sewer problem, Mariemont, Ohio. *(From plans by Fay, Spofford, and Thorndyke, Consulting Engineers, Boston.)*

Runoff from Large Areas. Because of the unequal distribution of rainfall that is likely to occur over large drainage areas, the rational method is not so easily applicable. Empirical formulas such as Fuller's based on actual stream flow are frequently used in such cases. Fuller's formulas,[1] which are based on a statistical study of runoff data from many streams throughout the United States, are as follows:

$$Q_{av} = CM^{0.8} \tag{3}$$

$$Q_T = CM^{0.8}(1 + 0.8 \log T) \tag{4}$$

$$Q_{max} = CM^{0.8}(1 + 0.8 \log T)\left(1 + \frac{2}{M^{0.3}}\right) \tag{5}$$

where Q_{av} is the average yearly flood in cubic feet per second from an area of M sq miles, Q_T is the maximum 24-hr flood in a period of T years, Q_{max} is the maximum rate of flow for Q_T, and C is a coefficient, the value to be used having been determined for the area in question by comparison with other streams or by computation.

If runoff data for a few years are available, C can be determined by computation from the formula for Q_{av}. Q_T or Q_{max} can then be computed for any desired period of time by using this value of C in the appropriate formula. These formulas frequently give results which fail to check stream-flow records and should be used with caution.

From Table 12, the size of flood that can be expected from the annual to the 1,000-year flood is given. The effect of size of drainage area on the maximum discharge rate

[1] *Trans. ASCE*, vol. 77, p. 564, 1914.

is also shown. Once in 1,000 years a flood may occur which is almost twice as large as that to be expected once in every 10 years. And, for a given flood, the actual maximum rate of runoff for 10,000 sq miles is only 12 percent in excess of the average 24-hr rate, while that for 15 sq miles is 100 percent greater than the 24-hr average.

TABLE 12. RELATION BETWEEN MAXIMUM FLOODS, FREQUENCIES, AND SIZE OF DRAINAGE AREA

Time, years (T)	Ratio of maximum flood of period to average annual flood ($1 + 0.8 \log T$)	Drainage area, sq miles (M)	Ratio of maximum rate of flow to 24-hr average $\left(1 + \frac{2}{M^{0.3}}\right)$
1	1.00	0.1	5.0
5	1.56	1.0	3.0
10	1.80	5.0	2.23
25	2.12	10	2.0
50	2.36	50	1.62
100	2.60	100	1.50
500	3.16	1,000	1.25
1,000	3.40	10,000	1.12

In using Fuller's formula where records are not available for determining the value of C to be used, Table 13 may serve as a guide. The wide range of values warns the designer to be cautious in using average values for C. These data may be useful, however, where stream-flow measurements are meager or entirely lacking.

TABLE 13. VALUES OF C IN FULLER'S FLOOD FORMULA

Locality	Number of stream stations considered	Value of C		
		Maximum	Minimum	Average
New England	32	110	17	57
Hudson River basin	14	132	37	85
Middle Atlantic states	31	140	30	76
South Atlantic states	49	113	25	66
Ohio River basin	38	142	47	81
St. Lawrence River basin	23	110	9.2	31
Hudson Bay basin	9	57	1.3	20
Upper Mississippi basin	16	55	7.4	22
Missouri basin	66	49	1.8	11
Lower Mississippi basin	8	52	1.7	17
Western Gulf of Mexico	4	31	9.5	17
Colorado River basin	24	46	3.0	18
Great Basin	45	55	0.6	13
Southern Pacific coast	29	193	15.8	68
Northern Pacific coast	52	186	2.2	45
Total or average	440	98	14	42

The Unit Hydrograph. In 1932, L. K. Sherman pointed out that all hydrographs for a given station on a stream had about the same length of base regardless of the total rainfall, if the storm was of unit length, say 1 day or 1 hr. In two hydrographs, therefore, the height of hydrograph at any moment was a function of the rainfall causing that

runoff. He therefore proposed that the hydrograph for any storm whose rainfall was known could be constructed from the "unit hydrograph" of that stream. The unit hydrograph is the hydrograph of surface runoff resulting from rainfall within a unit time. A distribution graph is a unit hydrograph modified to show the proportional relations of its ordinates in percentage of the total runoff. It is possible with the aid of the distribution graph to reconstruct the hydrograph of a given stream for any rainfall record by constructing the hydrograph for each unit or day's rainfall, making a composite of these, and superimposing this upon the existing ground-water or dry-weather flow of the period in question. The method has been developed and modified by Sherman and others till it is now possible to determine flows by this method with sufficient accuracy for design purposes.[1]

SEWER DESIGN AND CONSTRUCTION

Surveys, Plans, and Specifications. After the quantity of sewage or storm water to be taken care of has been determined, the next step is to get the size and slope of sewers needed to handle this flow. Sizes cannot be determined until information is available for establishing the slopes on which the sewers are to be built. These slopes are controlled by the local topography, and the necessary measurements and *surveys* must be made from which a topographic map can be constructed. For convenience this map should be to a scale of about 200 ft to the inch, and the contours should be drawn at 2- to 5-ft intervals, depending on the steepness of the slope. From the topographic map, it is possible to determine the streets on which sewers are to be constructed, and the sewer plan is laid out on the map. Profiles are then constructed of the streets on which sewers are required. It is desirable to show such information concerning subsurface construction and soil conditions as is necessary for establishing grades and estimating the cost of the work. The survey party also collects information on existing sewers, ground-water conditions affecting infiltration, and other special conditions which affect construction or cost.

The profile used in the preliminary study for preparing the preliminary estimate of the cost of the project need not be very elaborate. It can be drawn from elevations taken from the contour map, and tentative grades can be established for the sewers. Before asking for bids, however, and in preparing the final estimate of cost, more detailed contract drawings should be prepared. Figure 10 is a contract drawing for a section of an intercepting sewer for a small community which illustrates the information that should be shown on such drawings. All the information which will be useful to the prospective bidder should be included either in the contract drawings or in the specifications for the job. Specifications are detailed statements specifying the methods and materials that will be acceptable on the project. Experience with handling work of this character and familiarity with the various details involved and probable emergencies to be encountered are essential to the preparation of satisfactory specifications. Standard specifications may be used for some of the materials, but the use of "canned" specifications, or those borrowed from another job, is unsatisfactory, since the conditions governing the work cannot be identical.

In constructing the profiles, the invert and crown grade lines of the sewer are shown, and the slopes and diameters determined. The sizes of the sewers required may be computed by one of the following formulas, or they may be taken directly from charts similar to Figs. 11 to 13.

Velocity of Flow. One of the oldest and best known formulas for flow in open or closed channels is that of Chezy, which is

$$v = c\sqrt{RS} \tag{6}$$

[1] See *Univ. Missouri Eng. Expt. Sta. Bull.* 30, 1942, for application of method. Also "Flood Control," p. 68, The Engineers School, Fort Belvoir, Va., 1940.

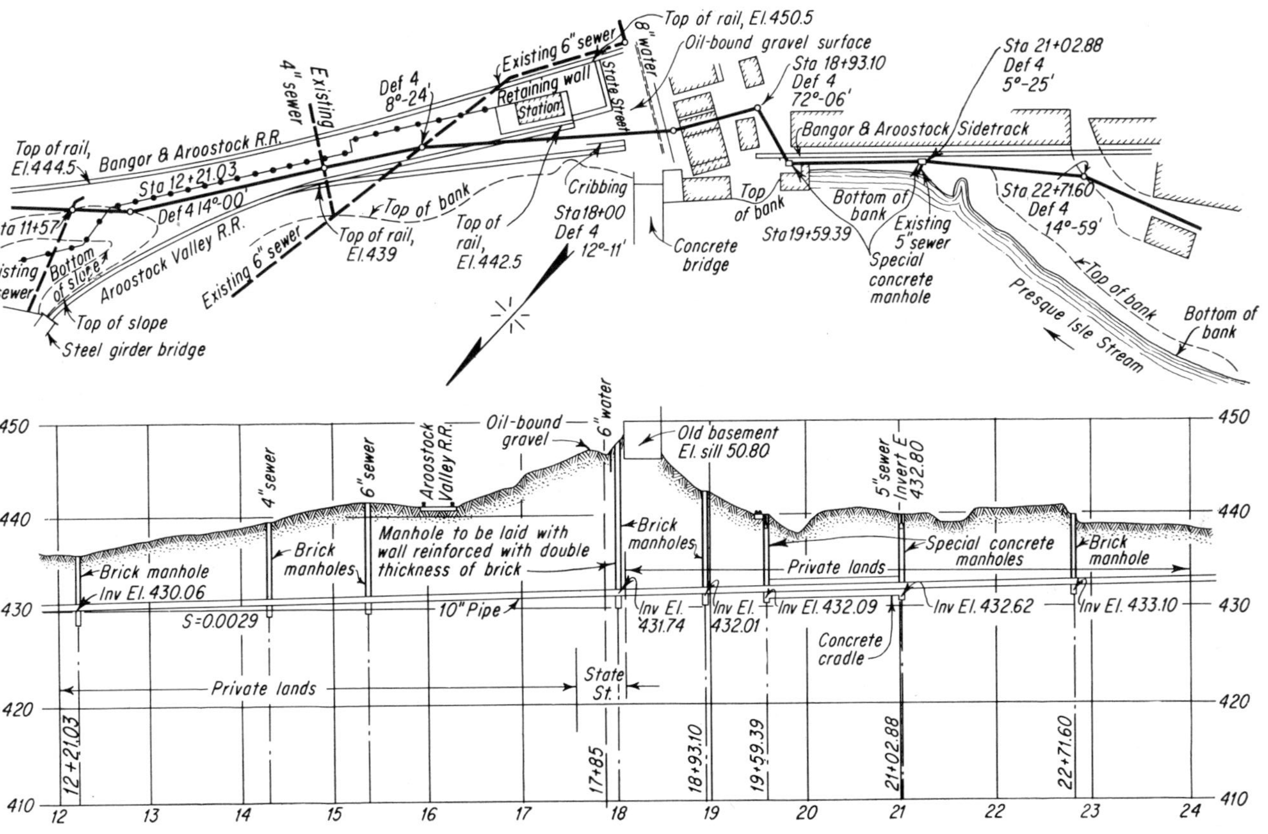

Fig. 10. Contract drawing for an intercepting sewer.

where v is the velocity of flow in feet per second; c a constant depending on roughness of the conduit and varying with R and S; R the hydraulic radius, which is the ratio of flow cross section to wetted perimeter of conduit ($r/2$ for circular pipes); and S the slope of conduit in feet per 1,000 ft.

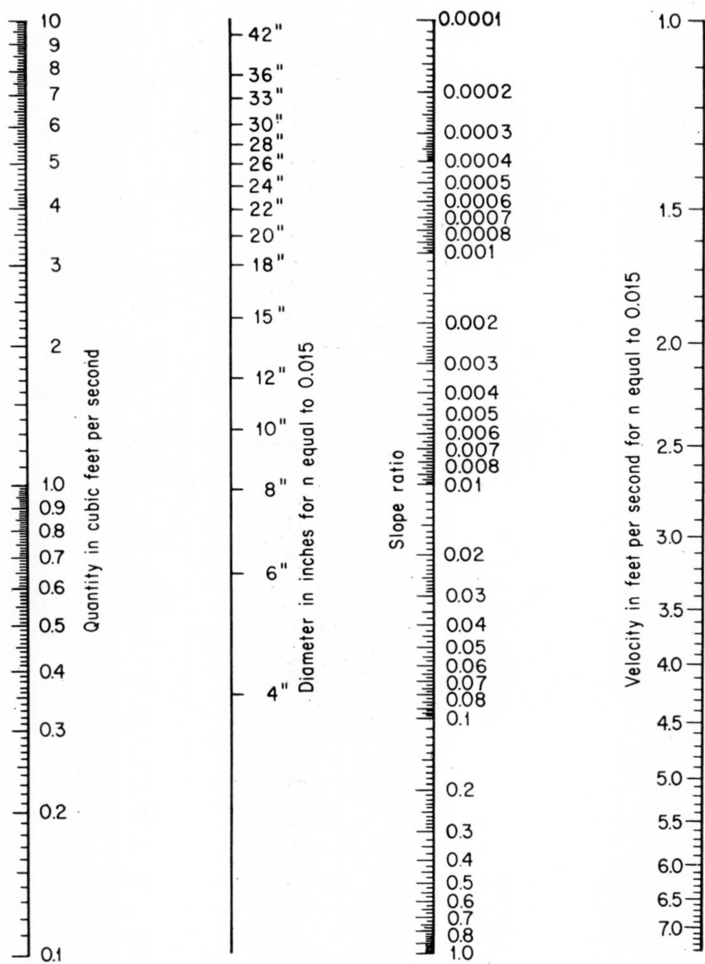

FIG. 11. Diagram for the solution of Kutter's formula, $n = 0.015$. Values of Q from 0.1 to 10 sec-ft. (*Babbitt, "Sewage and Sewage Treatment," 4th ed., p.* 51, *John Wiley & Sons, Inc.,* 1932.)

The use of this formula may be facilitated by the construction of a chart like Fig. 11. To find the velocity in a 4-ft sewer on a slope of 0.001, with c equal to 100, proceed as follows: $R = r/2 = 1.0$. Join slope of 0.001 with R of 1.0; and from the point where this line crosses the index line, connect with straightedge with $c = 100$. This crosses the velocity scale at 3.2 ft per sec, which is the desired velocity.

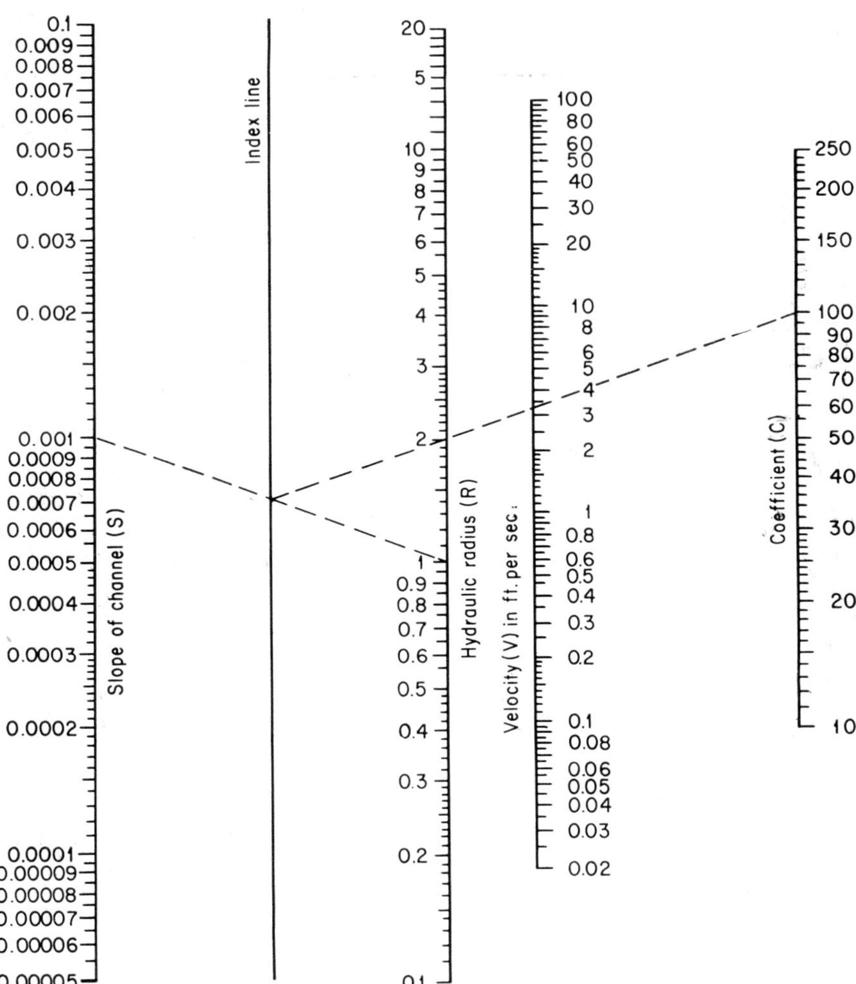

FIG. 12. Diagram for the solution of the Chezy formula, $v = c\sqrt{RS}$. (*Lipka, "Graphical and Mechanical Computations,"* p. 58, 1918.)

Kutter's formula gives an expression for c in terms of its variables. This formula was derived for flow in open channels but has been used with a fair degree of accuracy for pipes and conduits also. It is

$$v = \left[\frac{41.66 + \dfrac{1.811}{n} + \dfrac{0.00281}{S}}{1 + \left(41.66 + \dfrac{0.00281}{S}\right) \dfrac{n}{\sqrt{R}}} \right] \sqrt{R S} \tag{7}$$

where n is the coefficient of roughness of conduit surface.

TABLE 14. VALUES OF KUTTER'S n USED IN SEWER DESIGN

	n
Sewers 24 in. or less in diameter	0.015
Sewers over 24 in. in diameter of best work	0.012
Sewers over 24 in. in diameter under good ordinary conditions of work	0.013
Brick sewers lined with vitrified or reasonably smooth hard-burned brick and laid with great care, with close joints	0.014
Brick sewers under ordinary conditions	0.015
Brick sewers, rough work	0.017–0.020

The use of this formula entails so much labor that charts similar to Fig. 11 are frequently used. If any two of the four variables are known, the others will be given by the intersection with the appropriate scales of a straight line connecting the two factors that are known.

Manning's formula is much simpler and gives values which check Kutter's with reasonable closeness, using the same values of n. It is

$$V = \frac{1.486}{n} R^{2/3} S^{1/2} \tag{8}$$

Formerly it was customary to design sewers to carry the estimated quantities while flowing half full. With an increasing knowledge of the factors affecting the quantities involved, it is now customary to design the sewer to flow full, but not under pressure, when carrying the maximum flow to be expected at the future date for which the sewer is designed. Since it is customary to design sewers adequate for 30 to 40 years hence, the sewer when completed will be considerably larger than necessary for present requirements. During the hours of lowest discharge, the depth of flow in the sewer may be only a fraction of the total diameter of the pipe, so that velocities lower than the design velocities may occur. It is desirable, therefore, to investigate the design on the basis of minimum flow conditions to see what trouble might be expected due to low velocities during the early years of the use of the sewer. It is probably sufficient to use present average rather than present minimum flow in this investigation, since at the time of minimum flow the sewage is largely ground water and contains little settling solids that might cause stoppages. Where the slopes are flat and it is difficult to secure self-cleaning velocities, it is sometimes possible to maintain the desired velocities by using larger sewers flowing partly filled.

Limiting Sizes and Velocities. Because of difficulties from stoppages in sewers of small size, it is not advisable to construct sewers of less than 6 in. in diameter. Many cities specify 8 in. as the minimum-size sewer permissible in the city streets. The limiting velocities which are believed satisfactory for the existing conditions should be determined before proceeding with the design. A minimum velocity of 2 to 2.5 ft per sec is commonly used for separate sewers, and minimum gradients should be determined which are sufficient to produce this velocity. Sewers are sometimes built with a velocity as low as 1.5 ft per sec, although stoppages are more probable with the flatter gradients used. Although for combined sewers a minimum velocity of 3 ft per sec is desirable, this rate may be reduced to 2 ft per sec, if necessary, but with an increased danger of clogging. Where sewers are designed for extremely low velocities, provision for flushing may be desirable, or inspections should be made at intervals to see whether the sewer is operating satisfactorily. Maximum velocities also may need consideration, especially in combined sewers where excessive erosion may be caused by the grit when carried at high velocities.

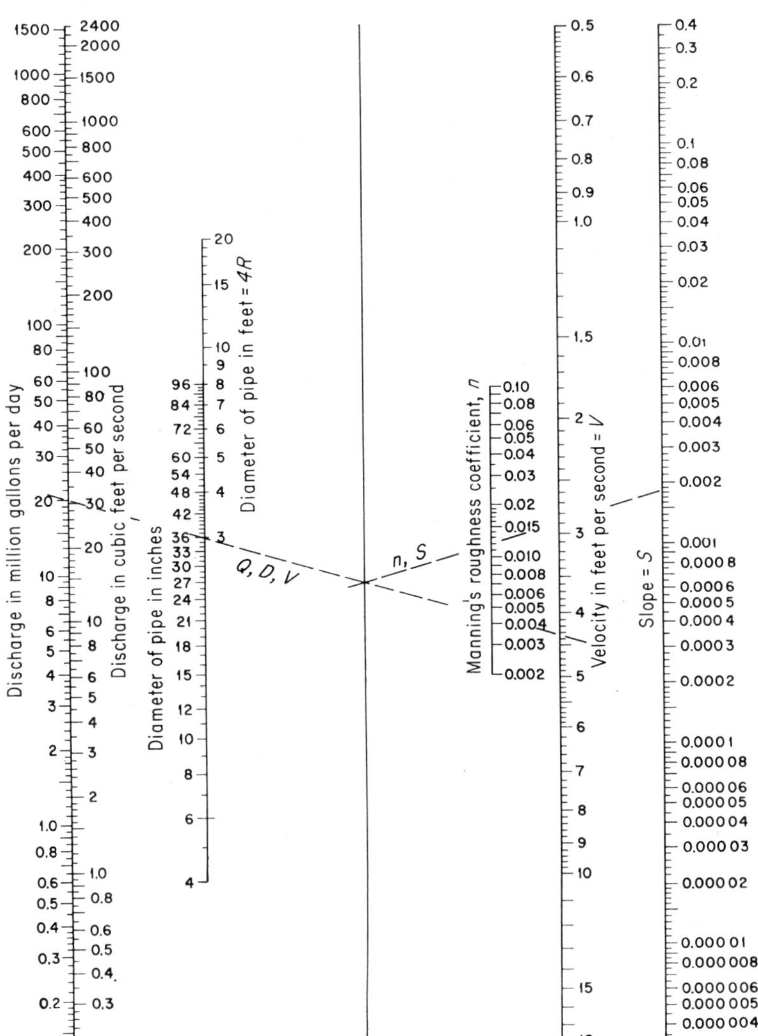

FIG. 13. Diagram for the solution of Manning's formula.

Where velocities in excess of 8 ft per sec are required, it is customary to line concrete or brick sewers, at least for the lower half, with vitrified-brick or vitrified-tile liner plates. For small sewers where vitrified-clay pipe is used, little trouble has been encountered from erosion due to high velocities. It may be desirable, however, to carry sewage down steep slopes by using one or more drop manholes, as illustrated by Fig. 18 or similar devices which permit a vertical drop of any desired amount in the sewer. In this way, the velocity between manholes is kept within safe values.

Pipe is ordinarily used for small sewers, but for large ones a wide variety of shapes and materials has been used. In selecting the sewer section, its hydraulic features

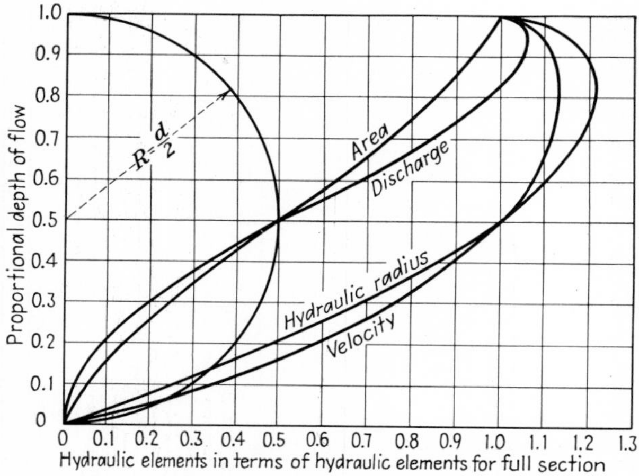

FIG. 14. Hydraulic elements of circular sections. $d = 12.0$; $s = 0.0004$; $n = 0.015$. (*Babbitt, "Sewerage and Sewage Treatment," 4th ed., p. 58, John Wiley & Sons, Inc.*, 1932.)

should be taken into account. Figures 14 to 16 give the hydraulic characteristics of circular, egg-shaped, and square sections, and from these charts the velocities at varying depths of flow can be determined.

Illustrative Problem. To illustrate the use of the various curves in this section, let it be assumed that a given sewer profile shows a grade of 1 percent to be feasible for a sewer to carry 2.0 cu ft per sec. A vitrified pipe sewer is contemplated with the friction factor $n = 0.015$. In Fig. 11, a straightedge joining 2 cu ft per sec on the left-hand scale with a slope of 0.01 or 1:100 on the slope-ratio scale intersects the scale of diameters at about 10.4 in. Since the 10-in. pipe is too small, the next commercial size, which is 12

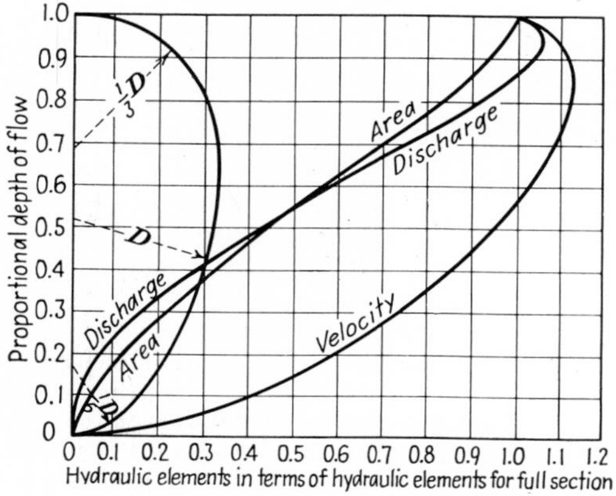

FIG. 15. Hydraulic elements of egg-shaped section. $d = 6.0$; $s = 0.00065$; $n = 0.015$.

in., will be used. From the chart, it is seen that a 12-in. sewer on a slope of 0.01 has a discharge capacity of 3.0 cu ft per sec when running full, with a velocity of 3.65 ft per sec. But 2.0 cu ft per sec is only 0.67 of the sewer's capacity, and to find the depth of flow in the 12-in. pipe for this fractional discharge, enter Fig. 14 at the bottom with

0.67 of the full discharge and go vertically to the discharge curve and then horizontally to the left-hand scale where we read a depth of 0.6 of the diameter of 12 in., or 7.2 in.

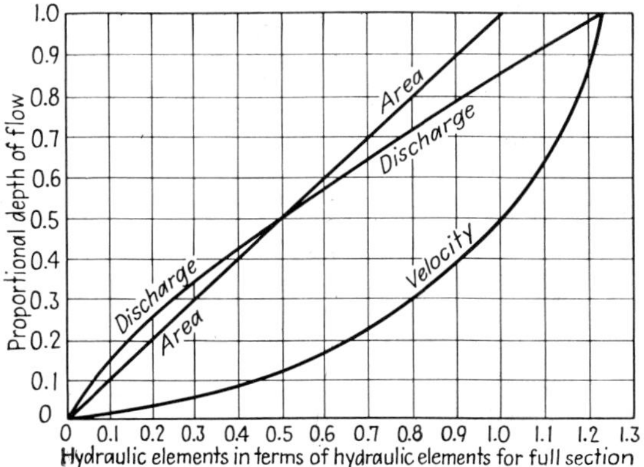

FIG. 16. Hydraulic elements of a square section. $d = 10.0$; $s = 0.0004$; $n = 0.015$.

Following the same horizontal to the velocity curve and thence downward, the velocity is found to be 1.07 times the full velocity, which was 3.65 ft per sec. The actual average velocity in the partly filled 12-in. sewer is, therefore, $1.07 \times 3.65 = 3.90$ ft per sec. It is apparent from Fig. 14 that the velocity for filled and for half-filled circular sections is the same, but for shallow depths of flow the velocity may be materially decreased.

Kutter's formula is only approximately correct for small circular pipes with a constant value of n; so Figs. 14 to 16 should be used with a knowledge of their shortcomings. Appreciable errors may be introduced by estimating sewer capacities by Kutter's formula from isolated measurements at fractional depths of flow. If the pipe used in the above computation has a value n of 0.013, a correction factor must be introduced in using these curves, since they are constructed for $n = 0.015$. Figure 17 provides this factor, which is found to be 1.2 for $R = r/2 = 0.25$. The discharge and velocity previously obtained are multiplied by 1.2, giving 3.6 and 4.38 for the discharge and velocity of the pipe flowing full with the smaller friction factor.

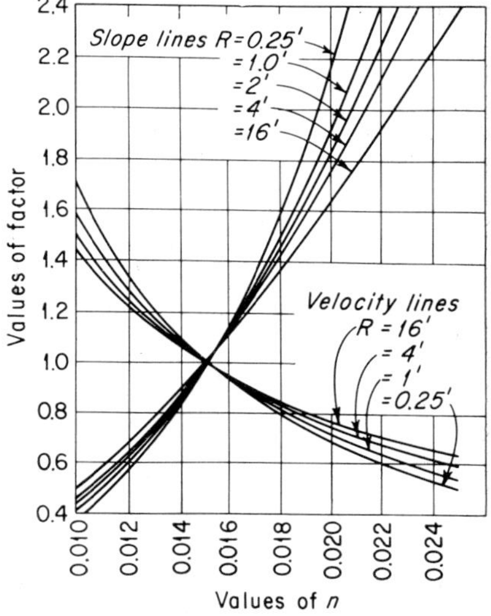

FIG. 17. Conversion factors for Kutter's formula. (*Babbitt, "Sewerage and Sewage Treatment," 4th ed., p. 56, John Wiley & Sons, Inc.,* 1932.)

Other elements affecting the shape and materials of the section to be used are headroom or cover available, type of soil through which the sewer is to be built, and relative costs due to the use of the different shapes and materials. The circular section requires the least perimeter for a given area; and since its hydraulic characteristics are satisfactory except for very low depths of flow,

it is a popular section where it can be used. The design of the circular section is not difficult, and it is quite economical as to materials required. Where, however, there is not sufficient headroom for a circular section, a greater horizontal distribution of area can be secured through semicircular or rectangular sections. It should be noted that the carrying capacity of a rectangular section is suddenly decreased when the sewage touches the top of the section, as is evident from Fig. 16. This is caused by the sudden increase in wetted perimeter without a corresponding increase in area. Rectangular sections, therefore, should be designed so that the section will never flow quite full. This section is not so economical of material as the circular section, but its stress analysis is relatively simple. In elliptical or other sections with an arched crown and flattened invert, the stress analysis becomes more complicated, though, where a particular form of curve is used frequently, short cuts, tables, and charts may materially lessen the labor of design. In the rectangular and other special sections, the invert is dished or sloped to the center to assist in cleaning and in providing for low depths of flow.

Special Sewer Structures. *Inlets* are openings provided in street gutters or through the curbing for admitting storm-water runoff to the storm-water drains. They are usually located at street intersections in such a way as to remove the storm water before it reaches the pedestrian crossings. The gutter opening is covered with a cast-iron grating which permits water to enter but which does not interfere with traffic. These gratings are easily clogged with paper or leaves washed down in the early part of the storm, and openings through the curb are usually provided and are principally depended on for taking care of the water. These openings vary from 6 to 8 in. in height, depending on the height of the curb, and may be from 2.5 to 6 ft in length. Depressing the gutter several inches below the regular street grades increases its capacity to receive storm water. This is particularly necessary on steep gradients, and the inlet capacity increases rapidly with the depth of the gutter depression. St. Louis experiments show that, on a 1 percent grade, a depression of 6 in. permits twice the capacity of a 4-in. depression and six times that of a 2-in. depression.

Catch basins are compartments intended to catch the grit admitted through street inlets and to prevent this from passing into the storm-water drains. Catch basins were at one time in general use; but, with the steeper gradients now employed for producing self-cleansing velocities in storm-water drains, they are seldom employed. To be of value, they require frequent cleaning, and the increase in street flushing as a means of cleaning street surfaces has resulted in filling catch basins quickly and making them ineffective until the next period of cleaning. High-pressure street flushing with its greater economy of water leaves more of the street dirt in the gutter where it can be removed more economically than from catch basins.

Manholes (Fig. 18) are openings through the street surface to the sewer to provide easier inspection and cleaning. They are placed at intervals of 300 to 500 ft and should be used at every change in line or grade so that the sewer will be straight between manholes to facilitate cleaning. They are usually circular in plan, having a cast-iron opening in the street surface from 20 to 24 in. in diameter, covered with a corrugated and sometimes heavily reinforced cast-iron cover. The manhole enlarges rapidly to about 4 ft in diameter and continues at this diameter down to the sewer elevation. This provides ample space for cleaning or unstopping the sewer with the segmented sewer rods commonly used. The sewer is carried through the manhole floor in channels in the floor with a depth of from one-half the sewer diameter to the full diameter. The remaining floor area provides a platform for workmen. Cast-iron steps or rungs are placed in the wall of the manhole to permit ingress and egress, and the manhole itself should be of sufficient diameter to provide clearance for a workman passing up or down the steps. At one time, manhole covers were perforated for ventilation purposes; but this has been found unnecessary and undesirable, as large quantities of storm water and considerable dirt

may enter through the perforations. The covers should be constructed with sufficient strength to carry the heavy wheel loads of modern city traffic. Methods of fastening the covers or other means for preventing removal of covers by small boys are sometimes provided. On steep gradients, drop manholes (Fig. 18) in which the drop is vertical are

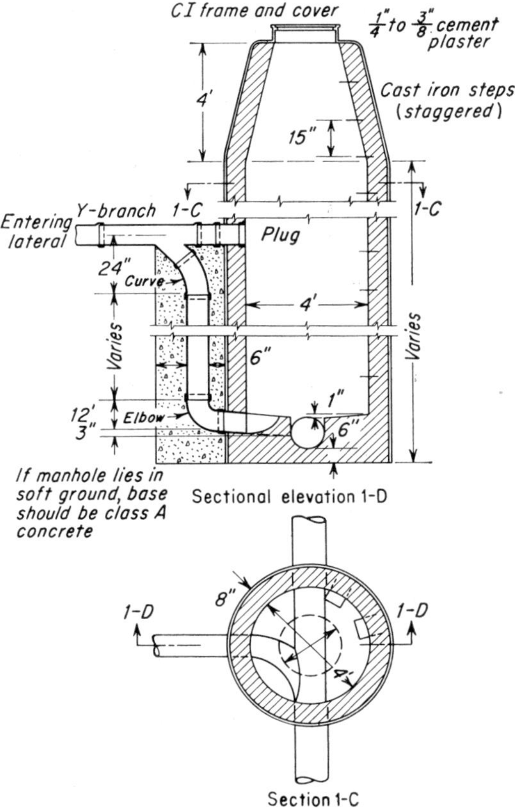

Fig. 18. Details of typical drop manhole.

preferable to excessively steep gradients because of their better control of the resultant turbulence.

Lift stations are pumping stations provided for lifting sewage from a lower to a higher level, as in cases where a section of the city is so low that its drainage cannot enter existing sewer mains by gravity. Usually the lift station is equipped with *automatic ejectors* (Fig. 52) which may be either pneumatic or centrifugal. Usually, these ejectors are provided in duplicate, each having ample capacity to handle the entire sewage flow. The pneumatic ejector has few moving parts and is singularly free from clogging. Centrifugal pumps of the "nonclogging" or "trash" type have been developed with impellers sufficiently open to permit the passage of large objects such as are sometimes found in sewage. Lift stations are frequently located in large manholes and may be entirely beneath the street surface.

Inverted siphons are sections of sewers which flow under pressure. As they are not true siphons, the term is not really applicable. They are usually constructed for the purpose of carrying the sewage under obstacles and vary in size from small sewers to large conduits several miles in length. A siphon usually consists of two or more pipes in parallel with the necessary valves for closing the individual pipes for the purpose of cleaning. Provision should be made for draining the siphon for cleaning or repairs. Figure 19 shows a typical siphon in which the flow in the different pipes is regulated by overflow weirs in the inlet chamber, arranged to put in service additional pipes when one has reached its capacity. Drainage and cleaning are performed by removing the blank flange in the lower manhole compartment after first opening a drain in this flange (not shown). A minimum velocity of 3 ft per sec should be provided in siphons to minimize the possibility of clogging. Sufficient head should be allowed in the design to supply the head losses that result from changes in direction and velocity of flow. Siphons have sometimes been inserted in existing sewers, as where some later construction crosses the sewer gradient. In the construction of the New York subway, for example, it was necessary to carry existing sewers under the subway structure, depending on the backing up of the sewage to provide the additional head losses thus introduced. Inverted siphons have worked very satisfactorily, and stoppages have been the exception rather than the rule.

Rawn [1] reports that a 24-in. diameter reinforced concrete inverted siphon under the Los Angeles River channel was cleaned by passing a rubber ball through it under pressure from the sewage. The ball was covered with canvas to which a line was attached for recovering the ball from the sewer in case immovable obstacles such as root masses were encountered. It required 36 hr to work its way through and carried ahead of it about a truckload of sand, gravel, bricks, and boulders in addition to half-dozen flash boards. Such a ball squeezes past obstructions which it cannot dislodge.

Regulating devices are sometimes required for by-passing a predetermined quantity of sewage through another channel to an emergency outlet. Where disposal plants are required on systems of combined sewers, it is usually unnecessary and expensive to treat the entire flow. Regulating devices are therefore used to permit any desired quantity, say, up to two or three times the maximum dry-weather flow, to pass on down the sewer to the treatment plant while the additional flow is being transferred to another channel. There are many devices used for this purpose. One type consists of float-operated gate valves arranged to open or close at predetermined stages of flow in the main sewer. Overflow weirs constructed in the side walls of the sewer may also serve to by-pass the excess storm flow. Dr. Hubert Engels of Dresden gives the following formula for the discharge in cubic feet per second over side weirs used for this purpose:

$$Q = 3.32 l^{0.83} h^{1.6} \tag{9}$$

where l is the length of weir crest in feet and h the depth of flow over weir at downstream end. In using this formula, assume a reasonably small value of h depending on the permissible fluctuations in maximum water surface downstream, and solve for l. If too long a weir is obtained for an economical design, use a larger value of h.

Side weirs are simple to construct but are not capable of controlling the sewage flow within narrow limits. For this purpose, *siphon spillways* are much more effective. They have considerably greater carrying capacity for a given area because of the greater effective head. Heads may be used up to 33.9 ft at sea level and standard atmospheric conditions, but the topography will seldom permit the use of a head of more than a few feet. Since the head over a side weir is usually limited to a fraction of the sewer diameter, it is obvious that greater heads can be used with siphons to produce correspondingly

[1] *Water & Sewage Works*, May, 1954, p. R-223.

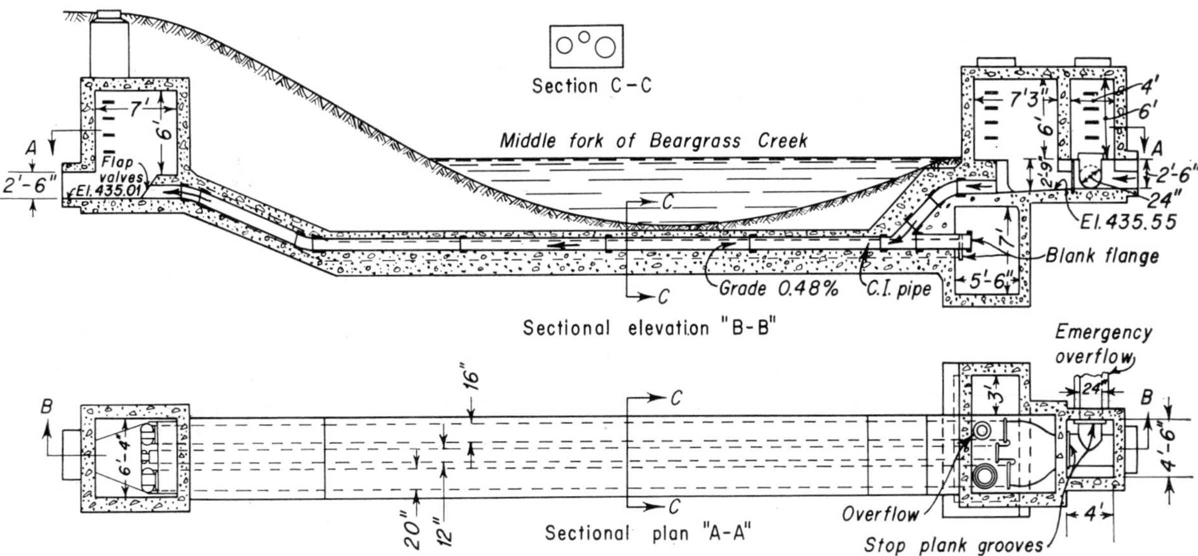

FIG. 19. Inverted siphon on Middle Fork Sewer, Louisville, Ky. (*Rept.* "*Inverted Siphons for Sewers,*" *Jour. Boston Soc. Civil Eng., vol.* 8, *p.* 272, 1921.)

greater discharges. The siphon has the further advantage of permitting close control of the high- and low-water elevations in the sewer made possible by its greater capacity and by the ability to start or stop the siphon discharge at any desired depth in the main sewer.

The formula $Q = ca\sqrt{2gh}$ can be used for determining the approximate cross section of the throat of the siphon. c is the coefficient of discharge and varies from 0.6 to 0.8. a is the cross-sectional area of the siphon throat. The head is used up in the losses shown in Fig. 20. By proper shaping of the siphon inlet, the entrance loss can be eliminated. The other losses are readily computed or estimated. The siphon should be constructed

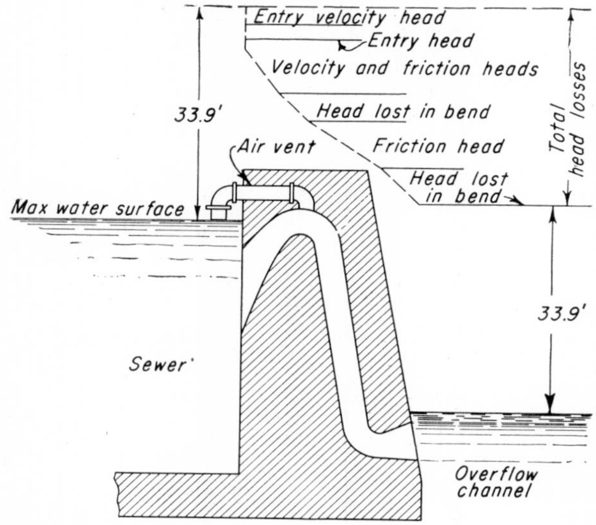

FIG. 20. Diagrammatic sketch of siphon spillway showing head losses. (*Stickney, Siphon Spillways, Trans. Am. Soc. Civil Engrs., vol.* 85, *p.* 1100, 1922.)

so that the lower leg will be sealed either by the discharge jumping across it or by being submerged. The air vent should have a cross section of about $\frac{1}{24}a$ for proper operation.

A *leaping weir* is a flow-regulating device which consists of a weir over which the ordinary flow falls into the main sewer while for higher heads and velocities of approach, part of the weir overflow leaps the opening through which the normal flow falls and passes on down the sewer to the storm outlet.

Sewer Outlets. Sewage discharged untreated near the shore of a lake or bank of a stream may cause discoloration and a very unsightly appearance near the sewer outlet. Submerged outlets located at some distance from shore are preferable as an aid to the better dissemination of the sewage throughout the diluting water. A knowledge of the currents and tides is essential to securing the best location of sewer outlets in tidal waters. The outlet should be located where currents will prevent the formation of sludge banks adjacent to the outlet and at the same time where it will be amply protected against scour. Tide gates of the flap valve type are sometimes used to permit the discharge of sewage at tide stages below a given elevation, while they prevent the entrance of water into the sewer at higher stages of the tide.

Sewer Sections. Pipe sewers have been constructed of vitrified clay, concrete, cast iron, or occasionally of wood-stave pipe. The most commonly used material for small

sewers is vitrified-clay sewer pipe. This is a hard-burned salt-glazed clay product and has the required characteristics of strength, hardness, imperviousness, and low friction factor. It is usually laid up with cement-mortar joints, though various forms of bituminous joints may be used where ground-water conditions warrant special attention to the securing of watertightness. Vitrified sewer pipe is highly resistant to erosion and to the effect of acid and alkaline sewage or ground water. The construction cost is comparatively low. Table 15 gives the approximate dimensions and weights of pipe of the two grades usually employed. The single-strength pipe is adequate for ordinary purposes, but double strength is used where the load to be carried is sufficient to require its greater strength.

TABLE 15. APPROXIMATE DIMENSIONS AND WEIGHTS OF SEWER PIPE, EASTERN CLAY PRODUCTS ASSOCIATION, 1925

	Single-strength pipe				Double-strength pipe conforming, in general, to ASTM standards			
Diameter, in.	Thickness of shell, in.	Depth of socket, in.	Annular space, in.	Weight per foot, lb	Thickness of shell, in.	Depth of socket, in.	Annular space, in.	Weight per foot, lb
4	9/16	1 3/4	3/8	7 1/2	9/16	1 3/4	3/8	7 1/2
5	5/8	2	13/32	11	5/8	2	13/32	11
6	5/8	2 1/4	13/32	13	5/8	2 1/4	13/32	13
8	3/4	2 1/2	1/2	20	3/4	2 1/2	1/2	20
10	7/8	2 1/2	1/2	30	7/8	2 1/2	1/2	30
12	1	2 1/2	1/2	42 1/2	1	2 1/2	1/2	42 1/2
15	1 1/8	2 1/2	1/2	60	1 1/4	2 1/2	1/2	70
18	1 1/4	3	17/32	80	1 1/2	3	17/32	100
20	1 3/8	3	17/32	100	1 5/8	3	17/32	125
21	1 1/2	3	9/16	112	1 3/4	3	19/32	135
22	1 5/8	3	19/32	122	1 7/8	3	19/32	155
24	1 5/8	3	19/32	140	2	3	19/32	175
27	2	3 1/2	11/16	195	2 1/4	3 1/2	11/16	215
30	2 1/8	3 1/2	13/16	230	2 1/2	3 1/2	13/16	275
33	2 1/4	4	13/16	300	2 5/8	4	13/16	340
36	2 1/2	4	1 1/16	350	2 3/4	4	1 1/16	390

Precast concrete pipe is made in diameters up to 24 in. and must have the same physical properties as vitrified-clay pipe. For larger diameters, reinforced concrete pipe may be used, or the sewer may be constructed of reinforced concrete, brick, or segmental block. Cast iron is used for sewers flowing under pressure, as in the case of small inverted siphons or pressure mains through which sewage is pumped. In a few cases, wood-stave pipe has been used for submerged sewer outlets.

Loads on Sewers. The loads to be carried by sewers have been studied in a very thorough manner by Marston and Anderson, who recommend for the pressure on sewers the following formula: $W = wcB^2$, in which W is the load on the sewer per linear foot of trench, w the weight per cubic foot of the backfill, c the coefficient taken from Table 17, and B the width of trench at the top of the pipe. Values of W can be obtained from Table 18, w is taken from Table 16 which also gives friction factors for the fill against the sides of the trench. It is evident from the formula that the width of trench is an important factor in determining the load to be carried by the pipe. The trench should be as narrow as possible, therefore—at least for a depth somewhat greater than the pipe diameter—and the bottom should be shaped to fit the pipe if its greatest bearing power is to be developed. Table 18 gives the maximum loads to be expected under the usual

TABLE 16. APPROXIMATE SAFE WORKING VALUES OF THE CONSTANTS TO BE USED IN CALCULATING THE LOADS ON PIPES IN DITCHES [1]

Ditch filling	Unit weight of filling, lb per cu ft	Ratio of lateral to vertical earth pressures	Coefficient of friction against sides of trench	Coefficient of internal friction
Partly compacted topsoil (damp).....	90	0.33	0.50	0.53
Saturated topsoil....................	110	0.37	0.40	0.47
Partly compacted damp yellow clay..	100	0.33	0.40	0.52
Saturated yellow clay................	130	0.37	0.30	0.47
Dry sand..........................	100	0.33	0.50	0.55
Wet sand..........................	120	0.33	0.50	0.57

[1] MARSTON, ANSON, and A. C. ANDERSON, *Iowa State Coll. Eng. Expt. Sta. Bull.* 31, 1913.

conditions encountered. In using this table, it is to be borne in mind that the soil adjacent to sewers is ordinarily saturated, or the backfill may become saturated from the rain, so that the loads for saturated soil should be used ordinarily. Although care should be taken to secure adequate consolidation of the backfill around the pipe, sewers

TABLE 17. APPROXIMATE SAFE WORKING VALUES OF c, THE COEFFICIENT OF LOADS ON PIPES IN DITCHES [1]

Ratio H/B	Approximate values of c			
	For damp top-soil and dry and wet sand	For saturated topsoil	For damp yellow clay	For saturated yellow clay
0.5	0.46	0.47	0.47	0.48
1.0	0.85	0.86	0.88	0.90
1.5	1.18	1.21	1.25	1.27
2.0	1.47	1.51	1.56	1.62
2.5	1.70	1.77	1.83	1.91
3.0	1.90	1.99	2.08	2.19
3.5	2.08	2.18	2.28	2.43
4.0	2.22	2.35	2.47	2.65
4.5	2.34	2.49	2.63	2.85
5.0	2.45	2.61	2.78	3.02
5.5	2.54	2.72	2.90	3.18
6.0	2.61	2.81	3.01	3.32
6.5	2.68	2.89	3.11	3.44
7.0	2.73	2.95	3.19	3.55
7.5	2.78	3.01	3.27	3.65
8.0	2.82	3.06	3.33	3.74
8.5	2.85	3.10	3.39	3.82
9.0	2.88	3.14	3.44	3.89
9.5	2.90	3.18	3.48	3.96
10.0	2.92	3.20	3.52	4.01
11.0	2.95	3.25	3.58	4.11
12.0	2.97	3.28	3.63	4.19
13.0	2.99	3.31	3.67	4.25
14.0	3.00	3.33	3.70	4.30
15.0	3.01	3.34	3.72	4.34
Infinity	3.03	3.38	3.79	4.50

[1] MARSTON, ANSON, and A. C. ANDERSON, *Iowa State Coll. Eng. Expt. Sta. Bull.* 31, 1913.

have sometimes failed from loads produced by excessive tamping. The amount of surface loads reaching the sewers depends on the ratio of depth to width of trench and, to a slight extent, on the nature of the soil as is indicated in Table 19.

TABLE 18. APPROXIMATE ORDINARY MAXIMUM LOADS ON DRAIN TILE AND SEWER PIPE IN DITCHES FROM COMMON DITCH-FILLING MATERIALS, IN POUNDS PER LINEAR FOOT [1]

H = height of fill above top of pipe, ft	B = breadth of ditch, at top of pipe									
	1 ft	2 ft	3 ft	4 ft	5 ft	1 ft	2 ft	3 ft	4 ft	5 ft
	Partly compacted damp topsoil, 90 lb per cu ft					Saturated topsoil, 110 lb per cu ft				
2	130	310	490	670	830	170	380	600	820	1,020
4	200	530	880	1,230	1,580	260	670	1,090	1,510	1,950
6	230	690	1,190	1,700	2,230	310	870	1,500	2,140	2,780
8	250	800	1,430	2,120	2,790	340	1,030	1,830	2,660	3,510
10	260	880	1,640	2,450	3,290	350	1,150	2,100	3,120	4,150
	Dry sand, 100 lb per cu ft					Saturated sand, 120 lb per cu ft				
2	150	340	550	740	930	180	410	650	890	1,110
4	220	590	970	1,360	1,750	270	710	1,170	1,640	2,100
6	260	760	1,320	1,890	2,480	310	910	1,590	2,270	2,970
8	280	890	1,590	2,350	3,100	340	1,070	1,910	2,820	3,720
10	290	980	1,820	2,720	3,650	350	1,180	2,180	3,260	4,380
12	300	1,040	2,000	3,050	4,150	360	1,250	2,400	3,650	4,980
14	300	1,090	2,140	3,320	4,580	360	1,310	2,570	3,990	5,490
16	300	1,130	2,260	3,550	4,950	360	1,350	2,710	4,260	5,940
18	300	1,150	2,350	3,740	5,280	360	1,380	2,820	4,490	6,330
20	300	1,170	2,420	3,920	5,550	360	1,400	2,910	4,700	6,660
22	300	1,180	2,480	4,060	5,800	360	1,420	2,980	4,880	6,960
24	300	1,190	2,540	4,180	6,030	360	1,430	3,050	5,010	7,230
26	300	1,200	2,570	4,290	6,210	360	1,440	3,090	5,150	7,460
28	300	1,200	2,600	4,370	6,390	360	1,440	3,120	5,240	7,670
30	300	1,200	2,630	4,450	6,530	360	1,440	3,150	5,340	7,830
Infinity	300	1,200	2,730	4,850	7,580	360	1,450	3,270	5,820	9,090
	Partly compacted damp yellow clay, 100 lb per cu ft					Saturated yellow clay, 130 lb per cu ft				
2	160	350	550	750	930	210	470	730	1,000	1,240
4	250	620	1,010	1,400	1,800	340	840	1,330	1,870	2,370
6	300	830	1,400	1,990	2,580	430	1,140	1,900	2,630	3,410
8	330	990	1,720	2,500	3,250	490	1,380	2,360	3,360	4,400
10	350	1,110	2,000	2,920	3,880	520	1,570	2,760	3,980	5,270
12	360	1,200	2,220	3,320	4,450	540	1,730	3,100	4,560	6,050
14	370	1,280	2,410	3,650	4,950	560	1,850	3,410	5,050	6,760
16	370	1,330	2,570	3,950	5,400	570	1,940	3,660	5,510	7,440
18	380	1,380	2,710	4,210	5,810	570	2,020	3,880	5,930	8,060
20	380	1,410	2,830	4,450	6,180	580	2,090	4,070	6,280	8,610
22	380	1,430	2,920	4,640	6,500	580	2,140	4,240	6,610	9,130
24	380	1,450	3,000	4,820	6,800	580	2,180	4,380	6,910	9,590
26	380	1,470	3,060	4,980	7,080	580	2,210	4,500	7,160	10,010
28	380	1,480	3,120	5,100	7,310	580	2,240	4,610	7,380	10,430
30	380	1,490	3,170	5,230	7,530	580	2,260	4,700	7,590	10,780
Infinity	380	1,520	3,410	6,060	9,480	580	2,340	5,270	9,360	14,620

[1] MARSTON, ANSON, and A. C. ANDERSON, *Iowa State Coll. Eng. Expt. Sta. Bull.* 31, 1913.

TABLE 19. PROPORTION OF "LONG" SUPERFICIAL LOADS ON BACKFILLING WHICH REACHES PIPE IN TRENCHES WITH DIFFERENT RATIOS OF DEPTH TO WIDTH AT TOP OF PIPE [1]

Ratio of depth to width	Sand and damp topsoil	Saturated topsoil	Damp yellow clay	Saturated yellow clay
0.0	1.00	1.00	1.00	1.00
0.5	0.85	0.86	0.88	0.89
1.0	0.72	0.75	0.77	0.80
1.5	0.61	0.64	0.67	0.72
2.0	0.52	0.55	0.59	0.64
2.5	0.44	0.48	0.52	0.57
3.0	0.37	0.41	0.45	0.51
4.0	0.27	0.31	0.35	0.41
5.0	0.19	0.23	0.27	0.33
6.0	0.14	0.17	0.20	0.26
8.0	0.07	0.09	0.12	0.17
10.0	0.04	0.05	0.07	0.11

[1] MARSTON, ANSON, and A. C. ANDERSON, *Iowa State Coll. Eng. Expt. Sta. Bull.* 31, 1913.

After the loads to be carried by the sewer have been determined, the next step in the design is to compute the stresses which these loads will produce in the sewer ring. If the latter is constructed of homogeneous materials, the maximum fiber stress is

$$f = \frac{My}{I} \pm \frac{P}{A} \tag{10}$$

in which M is the bending moment in inch-pounds, y the distance from neutral axis to extreme fiber of pipe wall, I the moment of inertia about the neutral axis, P one-half the total load on pipe, and A the cross-sectional area of the pipe wall.

Table 20 gives Marston's data for moments and shears for the critical points in the sewer cross section for circular pipes under different types of loading. From this it will be seen that the moment at the ends of the horizontal and vertical diameters under uniform loading over 180° are approximately equal to $\frac{1}{16}Wd$, where d is the pipe diameter. It is also seen that the pipe will be in tension on the inside at the top and bottom

TABLE 20. MAXIMUM STRESS IN FLEXIBLE RINGS DUE TO DIFFERENT LOADINGS [1]

Symmetrical vertical loadings		Moment at crown of sewer	Moment at end of horizontal diameter	Compressive thrust at crown	Compressive thrust at end of horizontal diameter	Shear at crown	Shear at end of horizontal diameter
Character	Width of pipe loaded, deg						
Concentrated..	0	$+0.318R\frac{W}{12}$	$-0.182R\frac{W}{12}$	0.000	$+0.500\frac{W}{12}$	$0.500\frac{W}{12}$	0.000
Uniform.......	60	$+0.207R\frac{W}{12}$	$-0.168R\frac{W}{12}$	0.000	$+0.500\frac{W}{12}$	$0.000\frac{W}{12}$	0.000
Uniform.......	90	$+0.169R\frac{W}{12}$	$-0.154R\frac{W}{12}$	0.000	$+0.500\frac{W}{12}$	$0.000\frac{W}{12}$	0.000
Uniform.......	180	$+0.125R\frac{W}{12}$	$-0.125R\frac{W}{12}$	0.000	$+0.500\frac{W}{12}$	$0.000\frac{W}{12}$	0.000

[1] MARSTON, ANSON, and A. C. ANDERSON, *Iowa State Coll. Eng. Expt. Sta. Bull.* 31, 1913. R is the radius of the pipe in inches and W the total weight of ditch filling and superimposed load plus five-eighths of the weight of the pipe itself (usually neglected) expressed in pounds per foot length of pipe. Moments are inch-pounds per inch length of pipe. Shears and thrusts are in pounds per inch length of pipe.

and on the outside at the sides for the loading conditions shown. For sections other than circular, such as are used frequently for large sewers, the customary methods of arch-and-slab design are used.

Construction. The cost of construction is made up of the cost of materials and labor plus profit. The profit on materials is not usually very great, but profit must primarily be made on the handling of labor and the mechanical construction equipment. Thus the details and methods of construction are important because of their effect on the cost of the project. Excavation is a large item in the cost of sewer construction. Progress in the development of excavation machinery has speeded up sewer construction while decreasing costs. Hand methods may still be necessary where there are many underground structures and pipes, but trenching machinery is now in general use. This trenching equipment is of several types. The ladder-type excavator consists of an endless chain of buckets with cutting edges which scoop up the material, bring it to the surface, and deposit it on a belt conveyor which carries it to one side of the ditch or dumps it into trucks if the material is to be hauled away. This type of equipment is best adapted to soils which require very little bracing and where there is little hardpan and few boulders. The drag-line excavator, which sonsists of a drag bucket operated by a hoist and cable, is particularly useful in caving soils. A special type of bucket, which is called a "trench hoe" and which is used advantageously on trenches for large sewers, consists of a shovel at the end of an arm, which, as it scoops the material from the trench, has a motion similar to that of a hoe. Under some conditions, as on crowded city streets, conveyor-type excavators are used, which carry the excavated material by means of an overhead conveyor to a point where it is used immediately in backfilling the trench.

Special methods of excavation may be required to meet unusual conditions that may be encountered. Rock excavation is usually accomplished with the aid of explosives, and this requires the drilling of holes for the charge, the placing and firing of the explosive, and the clearing away of the material after the blast. Dynamite, TNT, and powder are used, each being adapted to a particular type of work. Dynamite, being quickest in its action, is best suited to hard materials.

In excavations in quicksand, the problem is to keep the material from flowing into the trench from sides and bottom. This may be prevented by draining the water from the quicksand through well points during construction; by driving sheet piling along both sides of the trench; or occasionally by freezing the soil layer adjacent to the trench.

Sewers are usually built in open trenches, but tunneling has been used where open-trench work would result in considerable loss to property owners from interference with traffic. An inverted siphon is sometimes useful in carrying a sewer under obstructions or underground structures such as subways, water mains, and culverts. It is an underpass or pressure line under some obstacle, as the term indicates.

Maintenance. Maintenance of sewers has to do largely with the removal of stoppages and the cleaning out of deposits. Sewers may be flushed with a fire hose from the nearest fire hydrant. If the stoppage is too stubborn to respond to flushing, it may be removed by rodding from the next manhole downstream with the jointed sewer rods used for such purposes, or it may be necessary to dig down to the point where the trouble exists. Balls and pills slightly smaller than the sewer are sometimes forced through, for cleaning purposes, by the pressure of the sewage backed up behind them. Roots of neighboring trees may enter the sewer through small cracks in the joints or in the pipe and may produce root masses, sometimes many feet in length, which eventually will cause a sewer stoppage. Where it is necessary to lay sewers on flat gradients, flush tanks may be installed at their upper ends. The effect of flushing does not extend very far down the sewer, and local flushing with a fire hose when needed is more commonly employed.

Records of the location and depth of all sewers and house connections should be kept and should be sufficiently complete and accurate for locating any sewer in case it becomes necessary to dig it up. Connections of house sewers to the city sewers should be permitted only under supervision and by competent and experienced labor, since junctions improperly made may cause trouble from stoppages or leakage.

SEWAGE

Characteristics of Sewage. Sewage, the liquid waste of a community, consists primarily of water and contains usually less than 0.1 percent of solid matter. A fair conception of the proportion of solid to liquid matter may be gained from the fact that a column of average domestic sewage 265 ft high, evaporated to dryness, will produce only 0.1 ft depth of solid dry matter. In appearance, sewage is turbid and resembles the water coming from the bath or laundry. It usually carries a certain amount of floating materials such as matches, paper, sticks, and feces. Fresh sewage has only a slight odor, but when stale it becomes septic and a strong hydrogen sulfide odor is present. About one-third of the solid matter in sewage is in suspension, while two-thirds are in solution. It is about half organic and half inorganic. This may be seen from Table 21 in which the organic matter is best shown by the figures in the column "Loss on ignition." This organic matter consists of nitrogenous materials, carbohydrates, fats, and mineral oils. It is unstable and decomposes with the production of unpleasant odors, and it is this characteristic of sewage which is largely responsible for making sewage treatment necessary. Another important factor is the high bacterial content of the sewage and the presence of pathogenic organisms such as those producing intestinal diseases. If the sewage is to be discharged into a stream which is used as a source of water supply, these pathogenic bacteria will be a public health menace.

TABLE 21. AVERAGE ANALYSES OF SEWAGE SAMPLES FOR MASSACHUSETTS CITIES IN PARTS PER MILLION [1]

Place	1935 population	Total solids	Suspended solids	Loss on ignition	Free ammonia	Kjeldahl nitrogen	Chlorine	Alkalinity	Fats	Oxygen demand	BOD
Attleboro	21,835	437	148	216	29.2	13.9	31	129	46	56	142
Brockton	62,407	586	214	306	58.6	19.1	43	246	79	134	260
Clinton	12,373	482	201	296	21.8	15.8	44	105	90	74	157
Concord	7,723	321	74	170	20.6	9.7	32	107	33	42	115
Fitchburg	41,700	395	158	226	18.2	10.4	30	98	45	57	113
Framingham [2]	22,651	538	358	280	40.1	21.7	36	205	87	106	263
Marlborough	15,781	557	263	327	33.0	21.1	46	199	76	104	210
Pittsfield	47,516	492	162	251	18.0	11.5	41	174	54	73	220
Worcester	190,471	832	523	351	20 0	16.5	90	78	80	110	180
Average	....	515	233	267	28.8	15.5	44	149	65	84	184

[1] *Mass. Dept. Public Health Rept.*, 1939.
[2] Entrance to Imhoff tanks.

Table 21 gives the characteristics of sewages for a number of typical American cities. The large variation in total solids is caused, in part at least, by variations in the amount and kinds of industrial wastes present. For example, the Clinton sewage with its high total solids, loss in ignition, oxygen consumed, and fats is seen to contain an industrial waste—in this case, coming from textile mills. Concord, on the other hand, is a typical small residential community. The characteristics of European sewage differ considerably from those shown in the table. With the smaller per capita daily consumption of water abroad and the larger number of industries, European has higher concentrations than American sewage. In a comparison of European and American methods of disposal,

and particularly of the design data for disposal plants, this difference in sewage characteristics should be kept in mind.

Physical characteristics of sewage include turbidity, suspended and dissolved solids, odors, and appearance. Sewage is muddy or turbid in appearance, the degree of turbidity varying directly with its concentration.

Suspended Solids. As noted above, the total solids present in sewage are divided into suspended and dissolved solids. The suspended solids vary from the relatively coarse particles such as grit down to the very finely divided matter called "colloids." They also include such materials as paper, rags, sticks, fruit skins, and feces. Combined sewage contains more of some of these materials which come from street washings. Some of the troublesome materials reaching a separate sewer do so in violation of the ordinances designed to exclude them. Floating materials when discharged into streams are objectionable from the standpoint of appearance and because they become stranded along the shores where putrefaction of the unstable organic solids may cause local nuisances. The discharge of sewage into a stream causes discoloration and turbidity due to the suspended solids and may produce "sleek" on the surface through the presence of oil and grease. Sludge deposits on the stream bottom may gradually block the stream channel, and the requirement for oxygen to assist in its decomposition may seriously deplete the oxygen content of the water in the stream.

Dissolved Solids and Colloids. The larger part of the solids in sewage are present in solution, but there exists an appreciable amount of matter in a state between the solids in suspension and those in solution. These solids are really in suspension although they are so finely divided that they will remain in suspension indefinitely. Colloids may be defined as very finely divided matter in suspension, of such a size as to pass through filter paper but to be retained on a filtering membrane. All particles smaller than 0.0001 mm in diameter are called colloids, and the term is sometimes employed for even larger particles. The dissolved solids are roughly one-third organic and two-thirds mineral or inorganic.

Chemical Characteristics. Perhaps the most important of the chemical components of sewage is its oxygen content. In the decomposition of the unstable matter in sewage, oxidation plays a very important part. There is oxygen in solution in the water which forms the major part of the sewage, and this oxygen is used up in the process of biochemical decomposition. The amount of oxygen present at any moment in the sewage or in the stream into which the sewage is discharged determines the presence or absence of the objectionable conditions which usually accompany the stabilization of sewage and sewage-laden waters. This dissolved oxygen content gives a fair measure of the stream's capacity to digest organic pollution. The method of making the determination is described in the Standard Testing Methods of the American Public Health Association, to which reference is made for the details concerning the tests for physical, chemical, bacteriological, and microscopic characteristics of sewage.

Biochemical Oxygen Demand. The oxygen content of sewage is rapidly used up by the demand arising from the biochemical oxidation processes which are in progress. The amount of oxygen required to stabilize the oxidizable matter present is called the biochemical oxygen demand and is usually referred to as the BOD. Since this gives a direct measure of the oxidizable materials, it is the most important characteristic in describing the concentration or strength of the sewage. It is obtained by adding diluting water having no oxygen demand to the sample and determining its dissolved oxygen content after dilution. The sample is then incubated at 20°C for 5 to 20 days, after which the dissolved oxygen content is again obtained. The difference in these two determinations gives the amount of oxygen used up by the sample and is the BOD.

The *relative stability* of sewage may be defined as the ratio of oxygen available in the sample to the oxygen required for its stabilization. Table 23 gives the relative stabilities

in percentages for various periods of time at 20°C. If it is assumed that the oxidation process is practically complete at the end of 20 days at 20°C and its relative stability, therefore, is represented by 99 percent, at 10 days the sample would have been 90 percent stable or have used up 90 percent of the total oxygen requirement, while at 5 days the BOD is 68 percent of the total requirement. *Methylene blue* is an organic dye which is decolorized when the dissolved oxygen content of the solution containing it is exhausted. The relative stability test is made by adding a sufficient amount of the dye to produce a distinct blue color in the sample and storing at 20°C in a glass-stoppered bottle from which all air has been excluded. The disappearance of the blue color is an indication that the dissolved oxygen has been used up, and it gives, therefore, a measure of the amount of oxygen used in the given period. The relative stability of the sample is then taken from Table 23 for the number of days required for its decoloration.

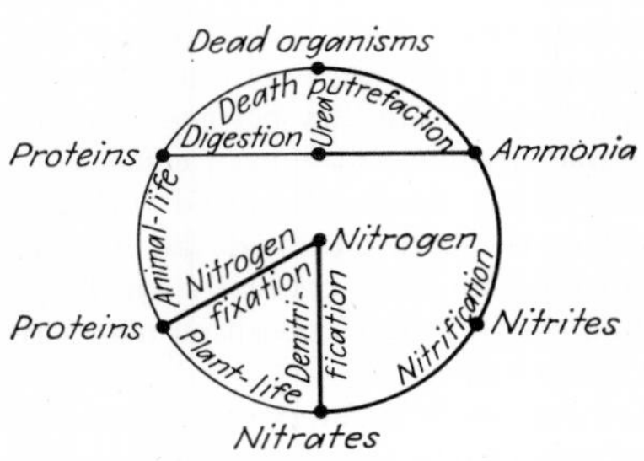

FIG. 21. Nitrogen cycle.

The *nitrogen cycle* (Fig. 21) gives diagrammatically a conception of the process of decomposition of sewage. At the death of animal or plant proteins, the organic nitrogenous matter begins a process of decomposition, in the first stage of which ammonia nitrogen is formed through the aid of certain groups of bacteria called "ammonifiers." This ammonia is further changed by certain nitrifying organisms into nitrites and eventually into nitrates. This completes the nitrification of the organic matter, and decomposition ceases. Nitrates contain nitrogen in a form that is available for plant food, and it is utilized in the production of plant growth. These plants either die and recommence the nitrogen cycle just described, or they may be used by animals as food, in which case they are partly used in forming animal protein and are partly wasted. In either case, at the death of the plant or animal, deterioration again sets in and the nitrogen cycle is repeated. A certain amount of denitrification from nitrates to atmospheric nitrogen takes place through the action of bacteria; and, in connection with the growth of leguminous plants, atmospheric nitrogen is converted into plant proteins by the process called "nitrogen fixation." There are some other minor modifications of the process, but in a general way the decomposition of nitrogenous material follows this cycle. The carbonaceous organic matter goes through a similar process called the carbon cycle, and the sulfurous organic matter goes through the sulfur cycle.

Hydrogen-ion Concentration. In water there are always present free hydrogen and hydroxyl ions because of the partial dissociation of the water. These should be equal in amount in pure water, and such water is said to be neutral. If, however, there is an excess of H ions, the water is acid; but if the OH ions are in excess, the water is alkaline. The acidity, therefore, increases with an increasing excess of H ions, and likewise the alkalinity increases as the excess of OH ions increases. The product of the H-ion concentration by the OH-ion concentration is constant for all practical purposes and equals 10^{-14}. Pure water, therefore, contains 10^{-7} ion of each kind. A substance is neutral, therefore, when its H-ion concentration is 10^{-7}. Since this method of expressing acidity is somewhat awkward, the term pH has been suggested by Sörensen to equal the logarithm of the reciprocal of the H-ion concentration, or $\text{pH} = \log \frac{1}{\text{H}}$. This then becomes 7.0 for a neutral substance. A pH greater than 7.0 indicates alkalinity, and one less than 7.0 indicates acidity. The pH determination is made colorimetrically by comparing the color produced when certain sub-

stances called "indicators" are added, with the colors of standards of known pH value.

The pH is important in indicating the freshness of the sewage, the ease of chemical precipitation with the quantity and type of chemical required, the rate of sludge digestion, and the probability of foaming.

Bacterial Characteristics. From the above discussion, it is evident that bacteria play an important part in the decomposition of sewage. In fact, they play the major role since it is through their activities that the chemical action takes place. If a sewage sample is sterilized, its chemical decomposition is checked. A knowledge of the types and habits of bacteria present in sewage is, therefore, essential to an understanding of the processes used in sewage disposal.

The *pathogenic bacteria* usually present in sewage include the causal organisms for intestinal diseases such as typhoid, dysentery, diarrhea, etc. These bacteria may spread disease by contaminating underground or surface-water supplies or shellfish, and, where such contamination is possible, the method of disposal should include a method of eliminating the pathogenic organisms. Most of the bacteria in sewage, however, are useful. One group, the *anaerobic bacteria,* thrive where atmospheric oxygen is excluded and perform important functions in the reduction and digestion of sewage sludge. *Aerobic bacteria,* on the other hand, require atmospheric oxygen and are responsible for the nitrification and oxidation processes used in the treatment of sewage. Bacteria are responsible for the necessity of sewage treatment through the decomposition processes which they cause, and the engineer uses bacteria in the treatment of sewage by encouraging and controlling their activities so as to produce the desired results.

Taking the sewage sample for any of the above tests is difficult because of variations in content and in quantity of flow throughout the day. In order to get an average sample, it is necessary to take samples at fixed intervals such as 30 to 60 min throughout the 24 hr, and from this composite sample a smaller average sample is taken. In order that this sample shall be fairly representative of the condition of the sewage at the time each sample was taken, the composite sample is sterilized by the use of formaldehyde, chloroform, or sulfuric acid. In this way, the rapid chemical changes which otherwise would take place are checked.

Stream Pollution. The final disposition of sewage is either on soil, as in disposal by irrigation, or in streams, lakes, or tidal waters. Since relatively small quantities of sewage are disposed of through irrigation, practically all sewage eventually finds its way into the nearest streams. Even where a sewage-disposal plant is in use, the stream receiving its effluent should be considered as the last unit in the plant or the last stage in the purification process. It is important, therefore, to understand the way in which such streams oxidize or digest their sewage burden.

The condition of a polluted stream is at any moment a resultant of two influences: its capacity for oxidizing sewage and the demand being made on it by the unstable organic polluting materials contained. Water normally contains oxygen in solution, the amount depending on the temperature. Table 22 gives the amount of oxygen that can be held in solution by fresh water of different temperatures; when these amounts are present, the water is said to be saturated. The saturation values for salt water are approximately 80 percent of those for fresh water, varying with the salinity. As some of the dissolved oxygen is used up to supply the oxygen demand of polluting materials, the deficiency is gradually made up from the atmosphere. The rapidity of this reaeration varies with the amount of the deficiency and the opportunities for absorption of atmospheric oxygen. Turbulence greatly influences this factor, since it continually brings fresh-water surfaces with their oxygen deficiency into contact with the air. Thus the oxygen demand of the polluting material is gradually satisfied at the expense of the oxygen content of the water. Serious depletion of the dissolved oxygen in a stream may

TABLE 22. SOLUBILITY OF OXYGEN IN FRESH WATER AT VARIOUS TEMPERATURES

Temperature, degrees		Dissolved oxygen, parts per million	Temperature, degrees		Dissolved oxygen, parts per million
Centigrade	Fahrenheit		Centigrade	Fahrenheit	
0	32.0	14.62	16	60.8	9.95
1	33.8	14.23	17	62.6	9.74
2	35.6	13.84	18	64.4	9.54
3	37.4	13.48	19	66.2	9.35
4	39.2	13.13	20	68.0	9.17
5	41.0	12.80			
6	42.8	12.48	21	69.8	8.99
7	44.6	12.17	22	71.6	8.83
8	46.4	11.87	23	73.4	8.68
9	48.2	11.59	24	75.2	8.53
10	50.0	11.33	25	77.0	8.38
11	51.8	11.08	26	78.8	8.22
12	53.6	10.83	27	80.6	8.22
13	55.4	10.60	28	82.4	7.92
14	57.2	10.37	29	84.2	7.77
15	59.0	10.15	30	86.0	7.63

destroy fish life. Such a stream may become septic, and the anaerobic decomposition going on produces objectionable odors and an unpleasing appearance. The various states regulate by law or through state health departments the degree of oxygen depletion which is permissible. When this is exceeded, the sewage causing the trouble must be treated so as to reduce its oxygen demand.

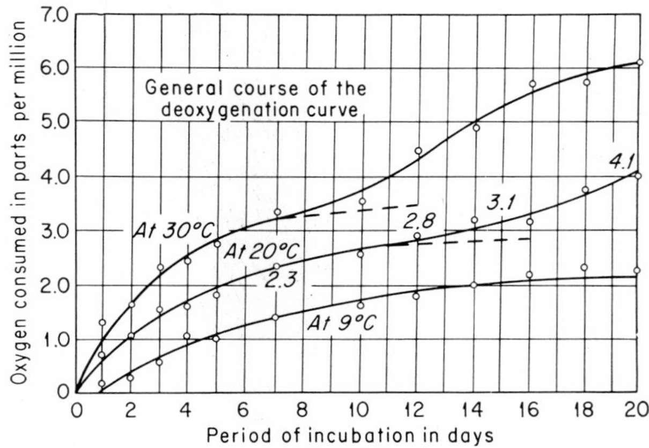

FIG. 22. Curve showing the rate of deoxygenation. (*Theriault, Rate of Deoxygenation of Polluted Waters, Trans. Am. Soc. Civil Engrs., vol.* 89, *p.* 1344, 1926.)

Sludge banks formed by the deposit of sediment from sewage discharging into streams or lakes undergo anaerobic decomposition and exert a definite oxygen demand on the adjacent water. These deposits may cause serious difficulty by filling up stream channels, and their effect on the oxygen content of the water continues for a long time. Where such deposits are likely to form, it is better to remove the settling solids in sedi-

mentation tanks where the digestion or final disposal of these solids can be more easily controlled.

Figure 22 shows the rate of deoxygenation at various temperatures with respect to time. It will be noted that deoxygenation begins rapidly and continues at a gradually diminishing rate. Curve *B* in Fig. 23 shows the dissolved oxygen content of a stream being gradually depleted in the absence of reaeration. Since reaeration takes place with increasing rapidity as the oxygen deficiency increases, the actual dissolved oxygen content of the stream after correcting for this will follow curve *A*. The equations for these curves are as shown on Fig. 23, where D_a is the initial oxygen saturation deficit of the

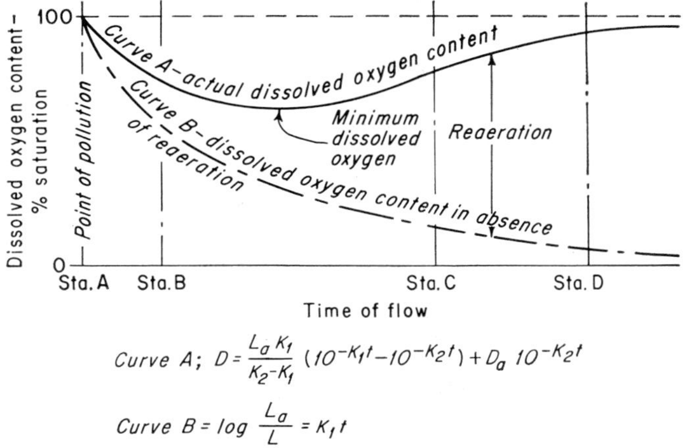

FIG. 23. Curves showing change in dissolved oxygen in stream below point of pollution. (*Streeter, Rate of Atmospheric Reaeration of Sewage Polluted Streams, Trans. Am. Soc. Civil Engrs., vol.* 89, *p.* 1355, 1926.)

water in parts per million, *D* the saturation deficit after time *t*, L_a the initial oxygen demand of the organic matter, *L* the oxygen demand at any desired moment, k_1 the coefficient defining rate of deoxygenation, k_2 the coefficient defining rate of reaeration, and *t* the elapsed time in days.

Figure 24 gives the variations of k_1 and k_2 with temperature in terms of values at 20°C, which are 0.10 for k_1 and about 0.24 for k_2 on the Ohio and Illinois rivers. k_1

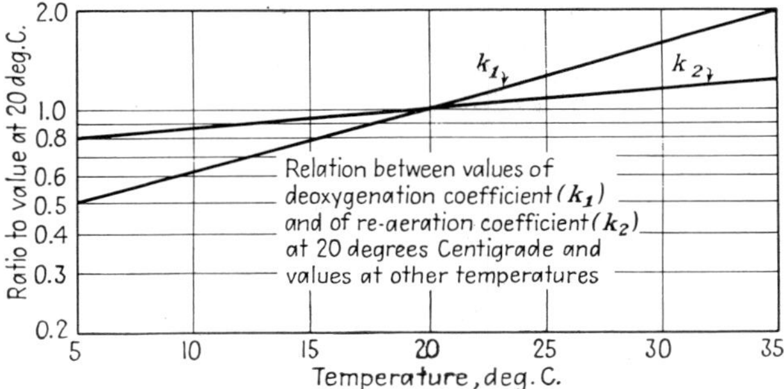

FIG. 24. Variations of k_1 and k_2 with temperatures. (*Streeter, Rate of Atmospheric Reaeration of Sewage Polluted Streams, Trans. Am. Soc. Civil Engrs., vol.* 89, *p.* 1354, 1926.)

can be determined in the laboratory, but k_2 varies with the stream considered. To examine the condition of a stream below a sewer outlet by using these formulas, proceed as follows:

L, or the BOD, can be obtained by test. To convert to 20°C, use the formula $L_T = L_{20}[1 + 0.02(T - 20)]$, where T is the temperature in degrees centigrade. Table 23 is useful in converting BOD to that for any desired period such as the 10-day to the 1-day BOD. Starting with $k_1 = 0.1$ and $k_2 = 0.2$ to 0.24 or as actually determined, correct these to the desired temperature by Fig. 24. Then solve for D for different values of t days up to the maximum desired time (distance) below the outlet, and plot the oxygen-deficiency curve similar to curve A (Fig. 23), with values of D as ordinates and of t as abscissas. This will show at a glance the condition of the stream at all points.

TABLE 23. RELATIVE STABILITY NUMBERS [1]

Time required for decolorization at 20°C, days	Relative stability, percent	Time required for decolorization at 20°C, days	Relative stability, percent
0.5	11	8.0	84
1.0	21	9.0	87
1.5	30	10.0	90
2.0	37	11.0	92
2.5	44	12.0	94
3.0	50	13.0	95
4.0	60	14.0	96
5.0	68	16.0	97
6.0	75	18.0	98
7.0	80	20.0	99

[1] A.P.H.A. Standard Methods of Water Analysis, p. 76, 1925:

The rate at which oxygen is demanded by polluted water at 20°C is easily seen from Table 22. For example, while 99 percent of the total oxygen required in a given case will be used in the first 20 days of the oxidation process, 90 percent will have been used in the first 10 days and 68 percent in 5 days. This table is useful also in the study of the condition of streams below sewer outlets in connection with problems similar to the above. For further information concerning the way in which these problems are worked out in detail, reference is made to Chapter V of Phelp's "Stream Sanitation." [1]

The effect of temperature on the oxygen sag is shown in Fig. 25 in which k_1 and k_2 are assumed to be 0.10 and 0.20 respectively at 20°C. They have been corrected for temperature by the use of Fig. 24. An increase in L_a will cause a greater oxygen deficiency.

Methods of disposal satisfy some of the oxygen demand of sewage before emptying it into streams and materially relieve the stream of its pollutional load. It is seldom feasible to increase the oxidizing power of the stream itself. At Chicago, the capacity of the Des Plaines and Illinois rivers for digesting sewage was tremendously increased by pumping large quantities of water with its oxygen content from Lake Michigan for dilution purposes. On the Ruhr River in Germany, also, Dr. Imhoff has constructed reservoirs for the purpose of increasing the river's capacity for biological digestion of disposal-plant effluents in order to postpone the necessity for increasing the efficiencies of existing sewage-disposal plants. At Park Falls, Wis., aeration of the Flambeau River improved stream conditions considerably, following pollution by pulp-mill wastes.[2]

[1] John Wiley & Sons, Inc., 1944.
[2] TYLER, RICHARD G., Polluted Streams Cleared Up by Aeration, *Civil Eng.*, vol. 16, p. 348, August, 1946.

The oxygen demand of sewage may be expressed also in pounds per capita daily, as in Table 24. This is particularly convenient for computing dilution factors. For

TABLE 24. OXYGEN DEMAND OF SEWAGE PER CAPITA [1]

Place	Total oxygen demand, lb per capita per day	Reference
Alliance, Ohio [2]	0.15	
Baltimore, Md. [2]	0.15	
Canton, Ohio [2]	0.17	
Columbus, Ohio [3]	0.22	Computed from data in *U.S. Pub. Health Bull.* 132, pp. 35–111
Fitchburg, Mass. [3]	0.17	Data cover period of 1 to 2 weeks at each place
Houston, Tex. (North plant) [2]	0.16	Average of eight cities, 0.17
Reading, Pa. [2]	0.15	
Rochester, N.Y. (Irondequoit plant) [3]	0.18	
Dayton, Ohio [2]	0.20	Metcalf and Eddy; 24-hr composite sample
Schenectady, N.Y. [2]	0.20	Average of 10 catch samples in 1924
Syracuse, N.Y. [3]	0.21	Metcalf and Eddy; computed on sample of canal water carrying sewage
Cincinnati, Ohio [3]	0.22	Average of long series of tests by U.S. Public Health Service on sewage from one of the city's sewers by 5-day dilution method
Peoria, Ill.	0.25	Average result computed from net oxygen demand of river water above and below Peoria for period of investigation by U.S. Public Health Service, 1921–1922
Chicago Sanitary District,[3] 39th St. sewage, tests in 1914 by nitrate method	0.24	Computed from average of 179 determinations during 1914 of 10-day oxygen demand by nitrate method
Tests in 1920 at Chicago by		
Nitrate method	0.261	Computed from results of 10-day test on six samples per day for 27 days during November and December, 1920
Dilution method	0.266	

[1] Sewage Disposal, *Sanit. Distr. Chicago Eng. Rev. Rept.*, Part III, p. 74, 1925.
[2] Separate sewers.
[3] Combined sewers.

example, the sewage from a residential city of 100,000 people would require 100,000 × 0.17 = 17,000 lb of oxygen per day for satisfactory dilution. If the sewage contains dissolved oxygen, this would reduce the oxygen requirement proportionately. If a temperature of 70°F is assumed, from Table 22 the oxygen content of the stream receiving the sewage is 8.97 ppm at saturation. If the stream is only 90 percent saturated because of earlier pollution and it is necessary to leave 50 percent saturation for fish life, 40 percent × 8.97 = 3.59 ppm oxygen, or 3.59 × 8.3 = 29.8 lb per million gallons are available. The required stream flow would be 17,000/29.8 = 570 mgd, or 882 cu ft per sec. With a sewage flow of 100 gal per capita daily, this would give a dilution ratio of 1:57. The dilution required to prevent the occurrence of objectionable conditions in streams is frequently expressed in terms of the contributing population. Studies in Massachusetts by Goodnough and others indicate that nuisances were always produced

with a dilution factor of less than 3.5 cu ft per sec per 1,000 persons and that they were unlikely to occur with dilutions of 6.0 cu ft per sec per 1,000 persons. Hazen found in investigating rivers for the Ohio State Board of Health [1] that 8.0 or even 10.0 cu ft per sec per 1,000 population was required on sluggish streams containing some prior pollution. These studies form a rough guide only, and computations of dilution requirements based on actual BOD and dissolved oxygen tests of the sewage and the stream are to be preferred.

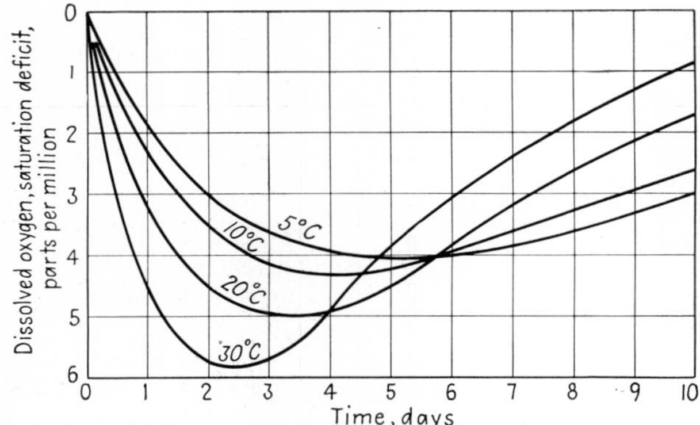

FIG. 25. Effect of temperature variations on the progressive changes in the dissolved oxygen deficit. $L = 20$ ppm; $D = 0$. (*After Streeter.*)

PRETREATMENT PROCESSES

Screens. Sewage contains materials which either interfere with, or do not respond to, accepted methods for sewage treatment but require separate handling. These include sand, sticks, rags, grease, and other miscellaneous materials which, as the first step in the treatment process, require removal in order to protect pumps and prevent clogging of pipes, valves, and filter nozzles. A minimum velocity of 1 ft per sec should be maintained at all flows to prevent sedimentation in pipes or channels. For sewage containing grit, 2 ft per sec velocity is the minimum used. The first unit of any treatment plant should be, therefore, racks or screen for removing these coarser materials. The New York State Department of Health requires a submerged net area of not less than 2 sq ft per mgd of sewage or 3 sq ft for combined systems. The velocity through the screen and normal to it was specified not to exceed 0.5 ft per sec to prevent forcing objects through it.

Types of Screen. Screens are classified as coarse screens or racks, medium, and fine screens. Fuller and McClintock recommend 20 sq ft of screen area per 1,000,000 gal of domestic sewage, and Metcalf and Eddy state that Boston's experience indicates that a clear area at least 50 percent greater than the cross section of the contributing sewer is desirable. Racks consist of parallel bars with the tops either vertical or inclined downstream to minimize clogging and facilitate cleaning. The clear space between bars is ½ to 2 in. or even more. Medium screens have openings down to ¼ in., and fine screens usually consist of perforated plate or wire mesh with openings less than ¼ in. in smallest diameter. The punched holes may be circular or in the form of slots of which the smallest diameter is the governing diameter used in designating the screen opening. Racks are frequently hand-cleaned in the smaller plants, but in the larger ones mechanized cleaning is preferable. High labor costs plus convenience and plant appear-

[1] *Preliminary Rept. Investigation of Rivers*, p. 32, 1897.

ance favor mechanized cleaning for all plants.

Figure 26 illustrates the method of screening used by the Link-Belt Co. with its Thru-Clean Bar Screen in which the rake carries the screenings to an upper covered compartment and deposits them in a garbage can for removal and burial, incineration, or other method of disposal.

Figure 27 shows a curved Dorrco bar screen, cleared by a rotating rake while the screenings are masticated and returned to the sewer to be settled with the other sewage solids at the sewage-disposal plant.

Fig. 26. Trough-clean bar screen. (*Link Belt Co.*)

The quantity of screenings removed depends on the size of opening but seldom exceeds 5 to 15 percent of the suspended solids, even with fine screens. The quantities removed by coarse racks have no definite relationship to the quantity of sewage and consist largely of sticks, rags, etc. Racks usually yield from 0.5 to 6 cu ft of screenings per 1,000,000 gal. Fine screenings contain from 80 to 85 percent water and weigh from 40 to 60 lb per cu ft. Whether fine screens should be used depends on the relative costs of removing the particular solids by screening or sedimentation. The principal use for fine screens is for those cases where no additional treatment is needed. Sometimes they are used also for removing the coarser solids preliminary to further treatment such as by the activated-sludge method or chlorination. Racks are used almost universally and are normally built as the first unit of the disposal plant, frequently in conjunction with grit chambers. Screens with $\frac{1}{32}$- to $\frac{1}{8}$-in. openings recover from 1.5 to 32.5 cu ft of screenings per million gallons of sewage.

Fig. 27. Bar screen and disintegrator. (*Dorrco.*)

Disposal of Screenings. Screenings are disposed of at many plants by dumping and covering with a layer of soil to prevent a nuisance. Where this method cannot be used, screenings may be pressed and burned under boilers or burned in incinerators with or without pressing. Screenings are sometimes used for land fills, barged to sea, composted, or ground and returned to the sewer. They have also been digested successfully alone or with sewage sludge. In the latter case, the sludge-digestion period required is increased from 15 to 35 percent. Maceration is beneficial though not essential in preparing screenings for digestion.

Comminutors. The problem of removal and disposal of the coarser materials in sewage has long been a troublesome one and until recent years was handled by means of coarse or fine screens. More recently, comminutors have come into use for chopping up these materials and permitting them to continue with the sewage into the disposal plant for digestion.

The Chicago Pump Company comminutor shown in Fig. 28*a* consists of a revolving slotted drum with stellite cutters mounted in staggered vertical rows on its surface. A stellite-faced cutter, or comb, is mounted vertically on the comminutor frame so that the rectangular comb teeth just clear the drum surface. As the drum rotates about its vertical axis, the coarse material drawn to the drum by the sewage, which flows inward through the slots, is sheared by the cutters as they pass through the comb. The comminuted materials then pass through the slots and flow downward through the open bottom of the drum into the effluent channel. Fig. 28*b* illustrates a later model (barminutor) adapted to large sewers or rectangular channels. The comminutor is nonclogging and eliminates the necessity for raking screens or disposing of the screenings separately. Anaerobic decomposition of the screenings along with sewage sludge has worked satisfactorily at a large number of plants where the comminutors have been used.

SEDIMENTATION

Sewage contains large quantities of suspended solids, and the process of sedimentation naturally suggests itself as a method for removing these solids. Some of the suspended solids settle out in a reasonable length of time, but those which are more finely divided, including the colloidal matter present, will not settle out, even when very long periods are allowed for sedimentation. The increased time required for settlement of the smaller particles is seen in Table 25 which gives the hydraulic subsiding values for particles of differing grain size. From this, it is apparent that sedimentation is not an efficient method for the removal of the more finely divided solids suspended in sewage.

TABLE 25. RELATION OF DIAMETER OF PARTICLE TO VELOCITY OF SETTLING [1]

Diameter of grain, mm	Order of magnitude	Hydraulic subsiding value, i.e., rate of settling, mm per sec	Time required to settle 1 ft
10	Gravel	1,000	0.3 sec
1	Coarse sand	100	3.0 sec
0.1	Fine sand	8	38.0 sec
0.01	Silt	0.154	33.0 min
0.001	Size of bacteria	0.00154	55.0 hr
0.0001	Size of clay particles	0.0000154	230.0 days
0.00001	Size of colloidal particles	0.000000154	63.0 years

[1] AWWA, "Water Works Practice," p. 158.

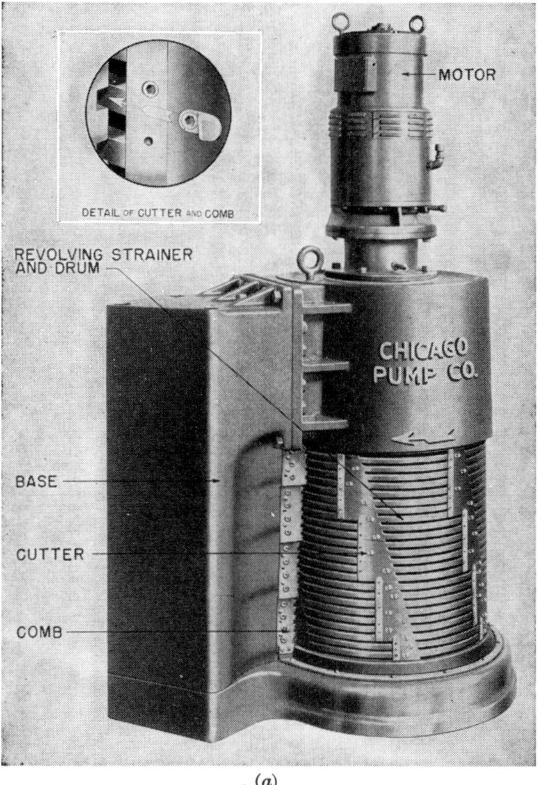

(*a*)

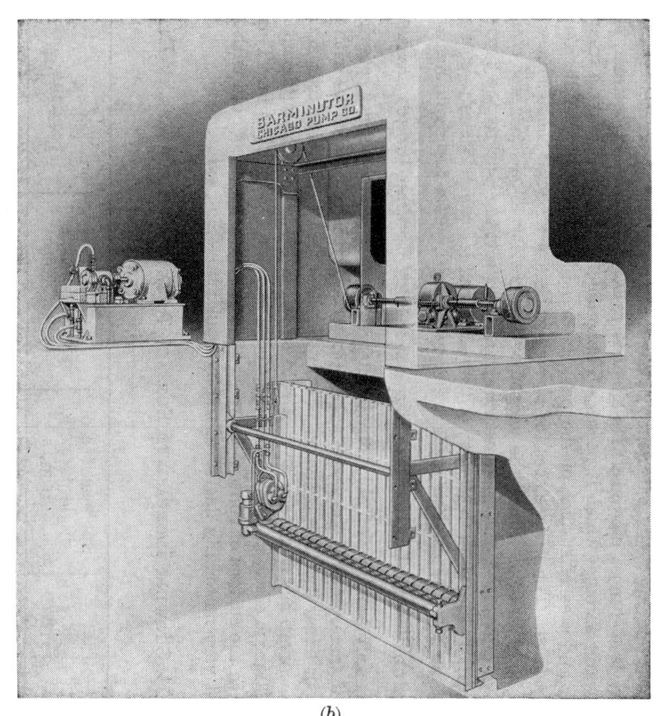

(*b*)

FIG. 28. Comminutor and barminutor. (*Chicago Pump Co.*)

Factors Producing Sedimentation. Of the factors producing sedimentation, the force of gravity is the most obvious. Its effect is resisted by friction between the particle and the water through which it passes. The amount of this friction is a function of the character and area of surface of the particle and of the viscosity of the fluid through which it settles. The surface area for a given weight of granular matter increases with a decrease in grain size. Since the weight or the gravity effect does not change as the material is pulverized, the increased resistance to settling due to the greater surface area materially increases the length of time required for the particles to settle out. For spherical particles, since their areas are equal to πD^2, while their volumes are equal to $\frac{\pi D^3}{6}$, it is obvious that volumes decrease more rapidly than surface areas. This will be more easily seen from Table 26 which gives the surface area for a spherical particle 1 in. in diameter and for the same matter pulverized so as to pass sieves with 10 and 200 meshes per linear inch respectively. From this it is seen that, while as a single particle the 1-in. sphere has a surface area of 3.14 sq in., when ground to sufficient fineness to pass a 200 mesh sieve, its surface area has been increased to 1,082 sq in. This enormous increase in the surface of contact between the substance and the liquid with no increase in weight very materially decreases the hydraulic subsiding value.

TABLE 26. SHOWING INCREASE IN SURFACE AREA WITH DECREASE IN DIAMETER OF PARTICLE

Diameter, in.	Sieve No.	Surface area of single particle, sq in.	Volume of single particle, cu in.	No. of particles	Total surface area, sq in.
1	1	3.14	0.523	1	3.14
0.065	10	0.0133	0.000144	3,630	48.3
0.0029	200	0.0000264	0.00000001276	40,987,000	1082.

The masses of these finer particles or colloids are sufficiently small so that the unbalanced impact of ions from all sides is of a magnitude such as to cause violent agitation of the individual particle. This is called *Brownian movement* and can be observed with an ordinary microscope using a dark field. This movement causes rubbing between the particles and the liquid and thus builds up an electric charge on the colloids. Since the particles all carry the same charge, they are held apart and their settlement or coagulation is interfered with. Brownian movement also interferes with the settling of the finer particles. As a result of all these factors, the colloids remain more or less indefinitely in suspension.

Hazen's Theory. Hazen made a theoretical study of sedimentation,[1] from which he reaches certain conclusions. Since his study was based on the type of materials encountered in turbid waters used for domestic supplies, it is not directly applicable to the organic or flocculent sediment found in sewage but applies more directly to grit chamber design than to the design of sewage-settling basins. Some of his conclusions are as follows:

1. The horizontal area of a settling basin is more important than its depth.
2. The ratio of length of basin to depth should be limited.
3. Baffles judiciously placed are helpful and may increase the efficiency of the basin by as much as 25 percent.
4. Temperature is an important factor.

[1] *Trans. ASCE*, vol. 53, p. 45, 1904.

5. Wind action should be avoided or reduced to a minimum. Velocities caused by wind and temperature may easily exceed the flowing-through velocity and should be avoided as far as possible.

The Consolidation of Settling Solids. Camp carried his studies of the rate of settling of suspensions beyond Hazen's contributions to this field, by defining intermediate degrees of their consolidation. These he divides into three categories: (1) where the density of the mass of settling particles is less than 1,000 ppm approximately, they do not interfere with each other and the term applied to them is *free settling.* (2) *Hindered settling* refers to an increased density intermediate in character while (3) *compaction* describes the condition existing at or above 3,000 ppm suspended solids where a definite sludge blanket has become recognizable. The settling velocity of a particle which just falls through the depth of the tank during its theoretical period of detention is called the overflow rate v_0 and equals the discharge per unit of tank surface area. All particles having velocities greater than v_0 will settle out while those with v less than v_0 will be removed in the ratio v/v_0.

When a particle is permitted to move without interference or rotation through a fluid due to the gravitational pull, the settling or rising velocity with respect to the fluid will become constant as soon as the resistance is equal to the weight of the particle in the surrounding fluid.

The law for the "resistance" or "drag" was first proposed by Newton as follows:

$$D = cA\frac{\rho v^2}{2} \tag{11}$$

in which D = drag
c = drag coefficient
A = projected area of the body in the direction of motion
$\frac{\rho v^2}{2}$ = dynamic pressure

The drag for a small sphere in a viscous fluid was first worked out by Stokes as follows:

$$D = 3\pi\eta dv \tag{12}$$

If this value of the drag is equated to the weight of the sphere in the fluid, $(\pi/6)d^3(\rho_1 - \rho)g$, the equation known as Stokes' law may be obtained as follows:

$$v = \frac{1}{18}\frac{g}{\eta}(\rho_1 - \rho)d^2 \tag{13}$$

where (in the cgs system)
v = settling velocity, cm per sec
g = gravity, cm per sec^2
η = absolute viscosity coefficient of fluid, dyne-sec per cm^2 or poises
ρ_1 = density of particle, g per cu cm
ρ = density of fluid, g per cu cm
d = particle diameter, cm

The general equation for the settling velocity of a sphere in terms of the drag coefficient may be obtained by equating the drag of Eq. (11) to the weight of the sphere, from which

$$v = \sqrt{\frac{4}{3}\frac{g}{c}\frac{(\rho_1 - \rho)}{\rho}d} \tag{14}$$

The drag coefficient c has been found by experiment to vary widely with the Reynolds number.[1] The drag is composed of both viscous and inertia forces. When the viscous forces predominate, "viscous" or "streamline" settling occurs. When the viscous forces are very small as compared with the inertia forces, the drag is entirely determined by the inertia forces. Stokes' law has been checked closely by experiment for values of Reynolds number less than 0.2. For values of Reynolds number from 0.2 to 1.0 the experimental values for the drag coefficient are sufficiently near the values computed by means of Stokes' law to warrant the use of the law in this region.

The accompanying curve (Fig. 29) showing drag coefficient as a function of Reynolds number is somewhat similar to the familiar curve of pipe friction factor as a function of

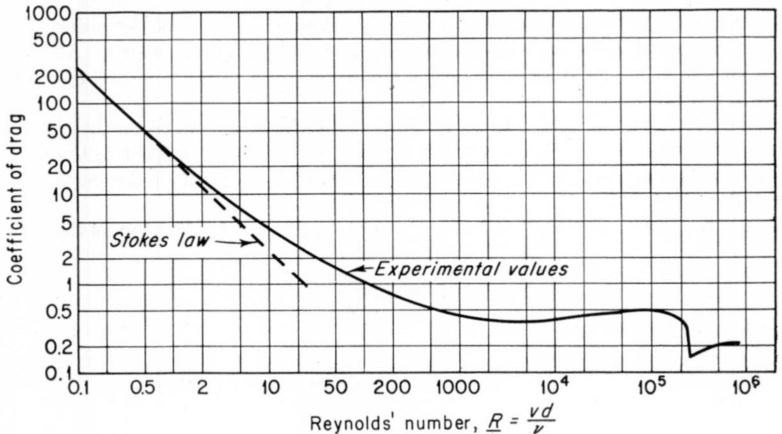

FIG. 29. Drag coefficient of spheres in fluids as a function of Reynolds' number. (*After Schiller.*)

Reynolds number. The region corresponding to the critical velocity in pipes is much more extended and gradual in the case of the moving sphere and is a stable region. The drag coefficient becomes practically constant at a value of about 0.4 for values of R from 1,000 to 20,000. This region is known as the region of "eddying resistance" to distinguish it from the Stokes' law region. The intermediate region is a "transition" region in which both viscous and inertia forces are in play. Values beyond the region of eddying resistance are of little interest in sedimentation.

The particles encountered in settling tanks are not spherical and usually are quite irregular. They vary in size, shape, and density over a wide range. When settling in still water without interference and without change in shape or in orientation they will settle at constant velocity. A change in orientation as by revolving will change the value of A in the drag equation and hence change the velocity. Flat particles such as mica tend to orient themselves in horizontal planes and hence to settle more slowly than bulky particles of equal mass and density. The settling velocity of bulky particles such as sand grains is practically equal to that computed by Stokes' law. Figure 30 shows some experimental settling velocities for quartz, galena, and soil particles compared with velocities computed by means of Stokes' law.

Most of the particles of interest in sedimentation settle in the Stokes' law or transition region. Grit in sewage plants settles almost wholly in the transition region. Larger

[1] $c = 24/R$ in the Stokian range.

sewage particles and heavy activated sludge particles may also settle in the transition region, but for the most part the settling of individual particles in sedimentation basins in both water and sewage plants is within the Stokes' law region.

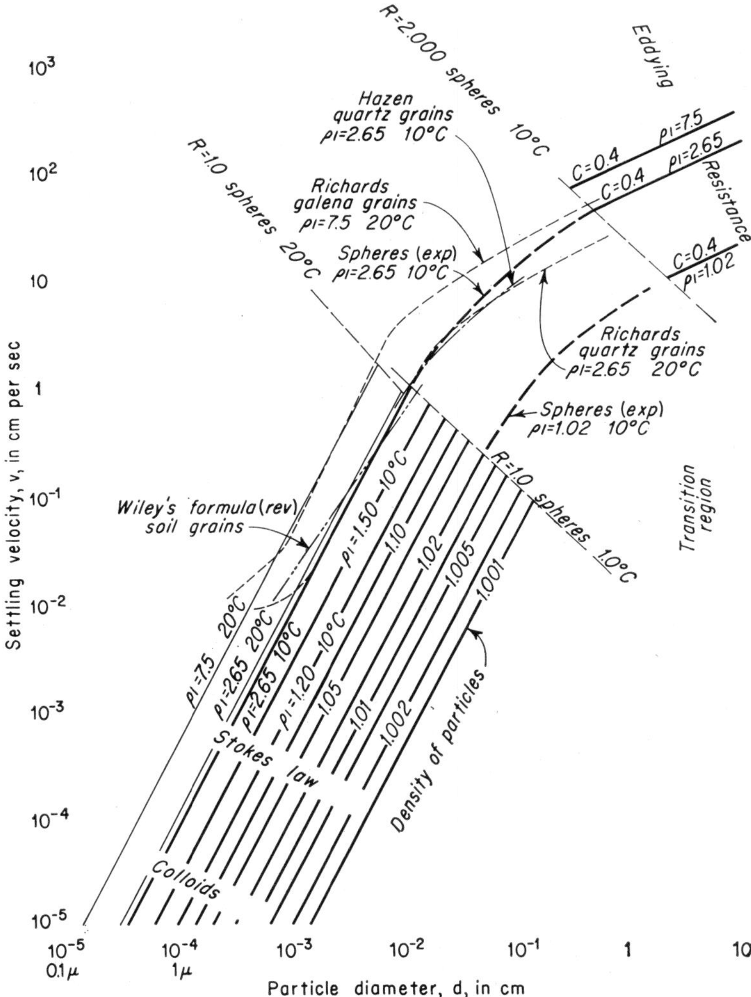

FIG. 30. Settling velocities of particles in still water. (*Camp.*)

Settling Basins. These items can be taken into account in the design of settling tanks, and the method of their application is apparent. For a given volume or detention period, a horizontal distribution of volume is preferable to one that is vertical, though the economical use of materials in the bottom and side walls will prevent the use of too shallow a basin. Experimental data substantiate Hazen's theory that sediment removal is dependent on tank surface area and independent of depth. Decreasing the depth by

one-half in experimental tanks produced increased removals of solids for identical detention periods.

There are two general types of settling basins: the horizontal-flow and the vertical-flow type. The horizontal-flow basin may be rectangular or circular in plan. This type is more commonly used in American plants. Figure 32 shows the relationship between total suspended solids in the sewage and the settling period required for any desired percent removal. The vertical-flow basin is exemplified by the Dortmund tank, which is usually square in plan. The final-settling tanks at the Berlin activated-sludge plant are of this type (Fig. 31). The inlet pipe discharges inside a central compartment extending several feet below the surface of the tank. The sewage flows downward, under this baffle, and upward in the outer compartment, as indicated by the arrows in the figure. It leaves the tank over a weir around the outer or inner edges of the tank. The upward velocity of the sewage must not exceed the downward velocity of the settling particles, or they will be carried out over the weir. A suspended sludge blanket forms which to some extent filters the sewage passing upward through it. Vertical-flow tanks are used extensively in European plants and are particularly appropriate for secondary tanks following filters or aeration tanks, in which it is undesirable for the oxygen supply of the effluent to be depleted by long contact with the sludge.

Grease Removal. *Skimming tanks* are used for the removal of materials such as grease and oil which can be separated by flotation. Grease interferes with the proper digestion of sludge by interrupting its biochemical activity and causes clogging of intermittent

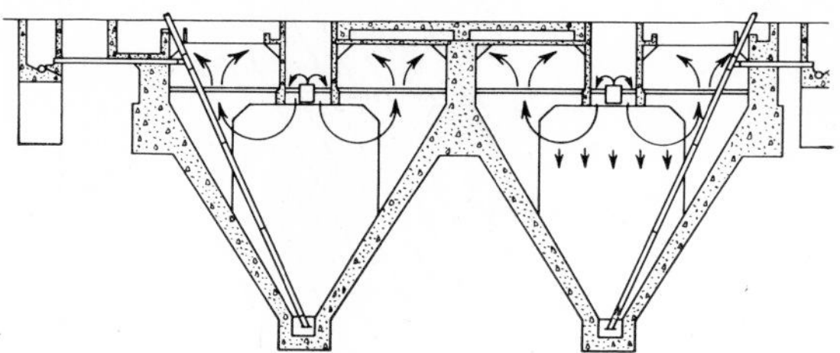

FIG. 31. Final settling tanks, Berlin, Dortmund type.

sand filters or of soil in underground drainage systems used for residential or institutional treatment plants. It may be removed by providing shallow tanks with large surface area and detention periods of 1 to 12 min. The grease is usually removed by skimming, though more recent practice consists of blowing air through the sewage in the skimming tanks, a greasy foam thus being produced which can be easily forced into side compartments or over weirs and thus separated from the sewage passing through the skimming tank. A detention period of 3 min is adequate. Adding a small amount of chlorine to the air used increases the amount of grease removed. Imhoff recommends 0.013 cu ft of air per gallon of sewage.

Grit Chambers. The American Public Health Association defines grit as "the heavy mineral matter deposited by sewage." Although it is made up principally of sand and soil, it frequently contains cinders, coffee grounds, fruit seeds, mash, and other coarse sediment which reaches the modern sewage-disposal plant. The problem is to remove these materials, which would cause considerable trouble in any of the types of disposal plant at present in use, while keeping the materials removed as free as possible from organic solids contained in the sewage. A usual requirement is that not more than 15 percent of the grit shall be volatile matter; otherwise, it will be too offensive to dispose of by dumping. Grit chambers are necessary on combined sewer systems and normally are not used with separate sewers.

The design of grit chambers to remove only the desired materials is made possible by the fact that the transporting power of a stream varies with the sixth power of its velocity. It is obvious that regulating the velocity in a settling chamber will determine the size and character of grit or sediment to be deposited. Experience indicates that, with velocities from 0.5 to 1 ft per sec, relatively clean grit will be deposited, while the lighter organic solids will be carried through the chamber. This is borne out by the data in Table 27. This velocity range is used, therefore, in designing grit chambers.

Table 27. Velocities Required to Move Particles[1]

Kind of Material	Velocity Required to Move on Bottom, ft per sec
Fine clay and silt	0.25
Fine sand	0.5
Pebbles ½ in. in diameter	1.0
Pebbles 1 in. in diameter	2.0

[1] AWWA, "Water Works Practice," p. 159.

Plans frequently provide for a storm-water flow as large as 200 percent of the dry-weather flow, so that the total quantity used in design may be as much as 300 percent of the dry-weather flow. This gives wide variations of velocity in grit chambers, which make the

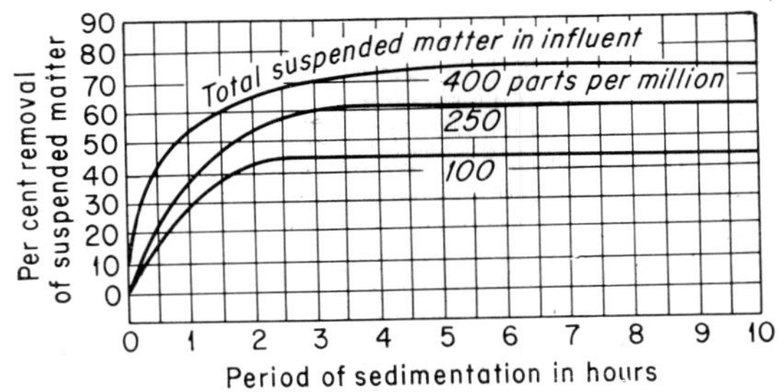

Fig. 32. Removal of suspended matter by sedimentation. (*Metcalf and Eddy, "Sewerage and Sewage Disposal," 2d ed., p. 534, McGraw-Hill Book Company, Inc.*, 1930.)

results less satisfactory. In decreasing the velocity from 2 or 3 ft per sec in the sewer to 1 ft or less in the grit chamber, the cross-sectional area is increased proportionately. The discharge divided by the velocity gives the cross-sectional area, while the detention period times the velocity gives the length of the chamber, or

$$\frac{\text{Flow in cubic feet per second}}{\text{Velocity in feet per second}} = \text{cross-sectional area in square feet}$$

$$= 1.55 \times \text{millions of gallons per day if velocity is 1 ft per sec}$$

Table 28 emphasizes the reciprocal relationship of the overflow rate (mgd ÷ surface area) to the diameter of settling particles and gives the necessary design information for removing the grit encountered in the disposal of sewage.

Grit chambers should be made of proper length and depth so that particles of the minimum size to be removed will settle to the bottom in its length. Material retained on the 65 mesh sieve and sometimes that on the 100 mesh should be retained in the chamber. Since the hydraulic subsiding values of grit of these diameters are 3.6 and 2.45 ft per min respectively, tanks having 1-min detention periods, or a length of 60 ft,

TABLE 28. AREA REQUIRED PER MGD FOR SETTLING GRIT PARTICLES SP. GR. 2.65 IN FLOWING WATER, VELOCITY UP TO 1.2 FT PER SEC

Particle size		Settling rate, ft per min	Area per mgd, sq ft
Mesh	Mm		
20	0.833	14.3	6.5
28	0.595	10.9	8.5
35	0.417	7.8	11.9
48	0.295	5.5	16.8
65	0.208	3.7	25.0
100	0.147	2.4	38.6
150	0.105	1.3±	69.5±

would have depths of 3.6 and 2.45 ft respectively. Knowing the required cross-sectional area, the width of chamber is easily determined. If the flow in either case is divided by the surface area, an *overflow rate* of 38,600 gal per sq ft per day is obtained. Thus the overflow rate is an important criterion in the design of grit chambers and sedimentation tanks.[1]

The desired velocity is obtained by using a control section below the grit chamber to give suitable depths of flow for the range of discharge designed for. Venturi flumes and proportional weirs are most frequently used for this control section. In the former, the width of chamber varies with the depth in such manner as will hold the velocity constant. With the proportional weir, the sewage discharges through a vertical slot at the lower end of the grit chamber, the width of slot being wider for low heads so as to maintain a constant velocity in the rectangular section of the grit chamber for all depths of flow. Rettger [2] derives the following formulas for the proportional weir:

$$w = b/x^{1/2}$$
$$Q = 12.6bh \tag{15}$$

where w = width of slot at any height h above weir crest
b = w when $x = 1$
x = height above weir crest
Q = theoretical discharge, cu ft per sec

$$\text{Actual } Q = c_dQ = 0.61 \times 12.6bh = 7.69bh \tag{16}$$

To design the weir, solve for b for a suitable h and compute values of w for selected values of x.

Additional depth is necessary for grit storage. This storage depends on the quantity of grit per million gallons and the length of time between cleanings. Shallow grit-storage compartments prevent a slow-moving lower zone in the grit chamber where organic matter would be deposited with the grit. The bottom of the grit chamber is usually sloped to one end or is formed into a series of hoppers to facilitate drainage and grit removal. The design will depend, to some extent, on the method of cleaning to be employed. Cleaning of small grit chambers is frequently done by hand; but, at the larger plants, eductors, clamshell or other type of buckets, or endless-chain bucket elevators are used. The tanks are built with two or more units so that any single unit may be cut out and cleaned while the total sewage flow is carried by the others. Provision

[1] BARKER, STANLEY T., Grit Chamber Velocities Controlled by Parshall Flume, *Water Works & Sewerage*, vol. 86, p. 127, April, 1939.
[2] RETTGER, E. W., A Proportional Flow Weir, *Eng. News*, vol. 71, p. 1409, June 25, 1914.

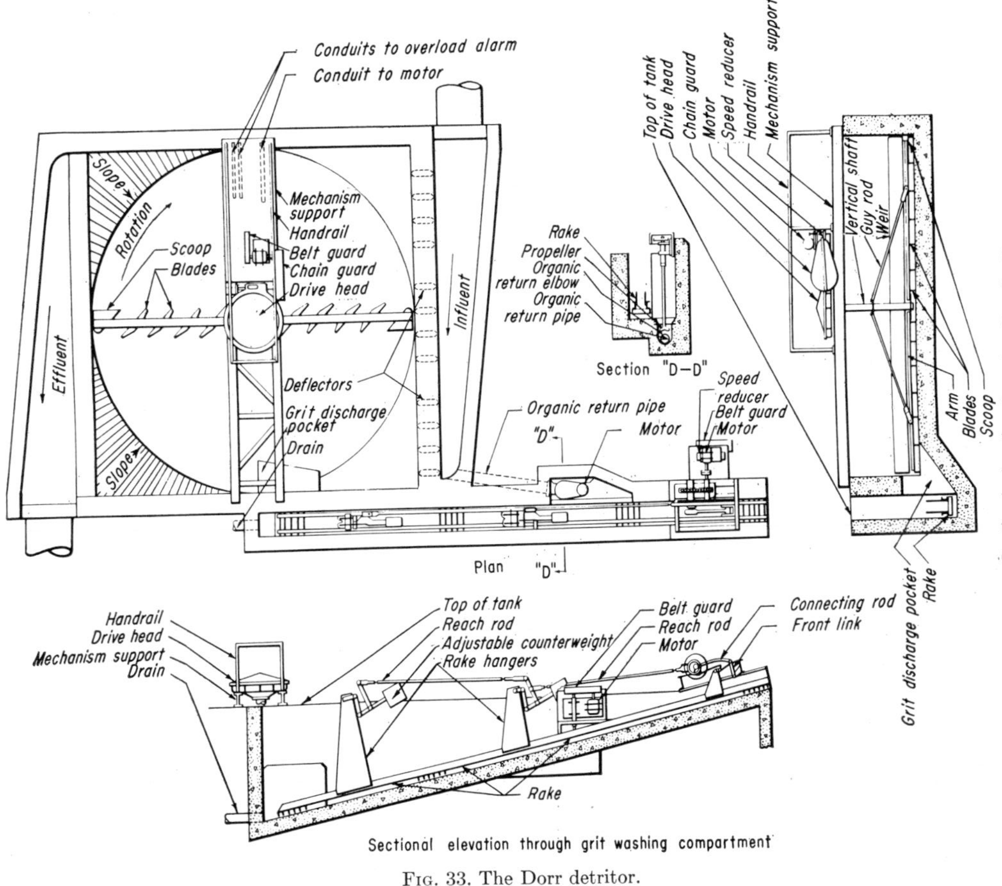

FIG. 33. The Dorr detritor.

should be made for removing the grit as rapidly as it accumulates, and a cleaner grit is obtained where pipes are available for drawing off the supernatant sewage in dewatering the tank. Drains are also placed in the bottom of grit chambers for draining the water out of the grit, and unsatisfactory attempts have been made to clean grit chambers by flushing the grit through such drains.

The quantity of grit removed varies from 2 to 20 cu ft per 1,000,000 gal. Quantities obtained at one plant may vary widely from those at another because of a great difference in the factors affecting the amount of grit received. Table 29 gives the amount of grit obtained at several plants with values ranging roughly from 2 to 4.25 cu ft per 1,000,000 gal. At the Cleveland Easterly plant, 0.1 cu yd per 1,000,000 gal was used for design purposes. The periods between cleanings varied considerably but averaged 16 days at the Cleveland plant. Grit is deposited during periods of maximum rainfall, and the number and times of cleaning depend to a great extent on these. When drained and packed, the grit obtained at Cleveland weighed 85 lb per cu ft and had a specific gravity of 1.36 for the mass as a whole or, on the basis of 35 percent voids, the specific gravity of the individual particles would be 2.08.

TABLE 29. QUANTITIES OF GRIT

Municipality	Grit removed, cu ft per 1,000,000 gal	Volatile matter, percent	Period
Fitchburg, Mass.	4.24	....	1930
Worcester, Mass.	4.22	....	1926–1929
Rochester [1] (Irondequoit plant)	2.60	....	1916–1925
Cleveland [1] (Easterly plant)	2.08	20.7	1923–1925
(Westerly plant)	2.00	10.1	1924–1925
Toronto,[1] Ont.	3.2	20	
Kitchener,[1] Ont.	12.0	Inoffensive	
St. Thomas,[1] Ont.	3.0	25	

[1] Grit-chamber Practice—A Symposium, *Trans. ASCE*, vol. 91, pp. 495 *et seq.*, 1927.

Grit-removal Mechanisms. Mechanisms have been developed for reducing the size and cost of grit-removal structures and for increasing the ease in its removal and cleansing. Figure 33 shows the essential details of the Detritor (Dorr) which has been widely used for this operation. Adjustable vanes are used to turn the flow of incoming sewage at right angles through a square tank which is quite shallow because of its short detention period and for producing rapid settling of the grit of 65 mesh and larger. For example, flows of 4 to 13 mgd can be handled by settling compartments 3′6″ deep and 10 to 18 ft square. Rotating sweeps move the settled grit into a hopper on one side of the tank, where it is moved up a sloping-bottomed channel by a reciprocating motion of the rake shown in Fig. 33. This agitation also washes the lighter organic matter out of the grit, thus leaving it clean enough for use around the plant for filling or for sidewalks, etc. The economic advantages of this equipment as contrasted with the grit chambers heretofore described are obvious.

Aer-degritor (Chicago Pump Co.). A more recent development in grit removal and washing has been introduced which is capable of being more accurately controlled. This is the Aer-degritor which is a combination of the comminutor and grit chamber which heretofore have been separate units. Figure 34 shows the plan and elevation of a small unit (0.3 to 6.0 mgd capacity). Air is furnished through the diffuser tubes to agitate the sewage entering the degritor. This scrubs the sand which then settles to the sloping floor whence it is washed into a hopper by the rotational movement of the

sewage in the tank. All sand of 0.2 mm (65 mesh) and larger is thus washed and flushed into the tank hopper whence it can be removed from the unit as required. The unit is sometimes followed by a comminutor to chop up the rags and organic solids in the

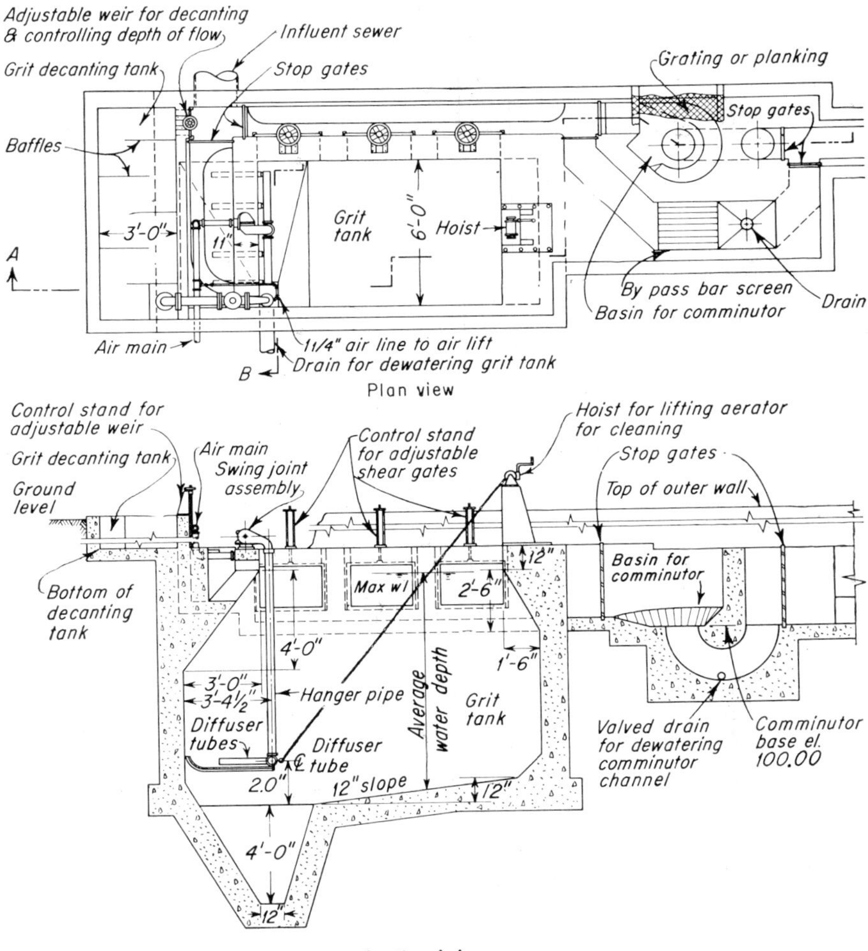

Fig. 34. Aer-Degritor to replace grit chambers. (*Chicago Pump Co.*)

sewage, an operation usually required unless they are screened out and buried or burned after drying. The aerator tube or tubes can be lifted clear of the sewage for cleaning or inspection as required.

Design of Sedimentation Basins. The septic tank was the first sedimentation basin developed for use with sewage which utilized biochemical activity in the digestion of sewage solids as part of the treatment process. This consisted of a single compartment in which the solids settled to the bottom where the action of anaerobic bacteria produced

gas which lifted the sludge to the surface, whence, if the gas could escape, it settled again to the bottom. Thus the gas production interfered with the settling process. To secure greater efficiency, both for settling and for sludge digestion, Imhoff developed the two-story tank bearing his name, in which the upper, or flowing-through, compartments provided the required sedimentation capacity while the lower compartment retained the sediment or sludge which settled through slots in the hopper bottoms of the upper chamber and permitted the development of the conditions best suited to its satisfactory digestion, without interfering with the settling process in the upper compartments. This was an important step in the development of efficient sewage sedimentation, but for reasons of economy and increased efficiencies, these two processes were later further separated so that present practice favors the construction of settling tanks or clarifiers entirely separate from the sludge tanks or digesters.

TABLE 30. ESSENTIAL DATA ON IMHOFF TANKS [1]

Item	Schenectady, N.Y.	Plainfield, N.J.	Fitchburg, Mass.	Rochester, N.Y.
Tributary population (1922)	65,000	40,000	38,000	260,000
Character of sewage:				
Flow, mgd	6	3.4	3.4	32
Separate or combined sewers	Separate	Separate	Combined	Combined
Strength (suspended solids), ppm	163	166	219	163
Freshness	Fresh, uncomminuted	Stale	Fresh	Stale
Industrial wastes	Practically none	Practically none	Small amount	Small amount
Hardness of water supply, ppm	130	88	10	65
Temperature, °F	Ave (1922) 56.1 Max 64.7	46–70	No data	Ave 54
Preliminary treatment:				
Screening	Coarse rack	Fine screens 1/16-in. slot	Coarse racks	Fine screens 1/8-in. slot
Grit chambers	None	None	Yes	Yes
Imhoff tanks:				
Sedimentation period, hr	3.3	3.4	6.4	1.1
Depth of tanks:				
Total maximum water depth	13 ft 9 in.	19 ft 9 in.	24 ft 8 in.	33 ft 10 in.
Sludge compartment:				
Depth below plane of slots	6 ft 1 in.	8 ft 9 in.	11 ft 0 in.	22 ft 6 in.
Depth below 18-in. neutral zone	4 ft 7 in.	7 ft 3 in.	9 ft 6 in.	21 ft 0 in.
Distance of lowest point of overflow below 18-in. neutral zone	2 ft 6 in.	2 ft 0 in.	2 ft 0 in.	13 ft 0 in.
Depth below overflow to adjacent compartment	2 ft 1 in.	5 ft 3 in.	7 ft 6 in.	8 ft 0 in.
Cu ft per capita	1.40	1.49	1.88	2.40
Numbers of hoppers	8	5	3	3
Ease of intercommunication	Poor (2 ft square opening)	Poor (opening 20 by 24 in.)	Good	Good
Scum compartment:				
Cu ft per capita	0.76	0.72	1.59	0.55
Area, gas vents; percentage of tank area	14.8	14.3	15	26.8
Sq ft per cu ft sludge capacity	0.034	0.023	0.031	0.0133
Loading, deposited solids:				
Lb per year per cu ft sludge and scum space	14.2	10.4	13.1	9.2
Lb per year per cu ft sludge space	21.9	15.5	24.1	11.3
Lb per year per cu ft sludge space after deducting 18-in. neutral zone	30.0	18.9	30.9	12.3
Lb per year per cu ft scum space	40.4	32.1	28.5	49.7
Lb per year per sq ft gas-vent area	644	674	777	849

[1] EDDY, Imhoff Tanks—Reasons for Differences in Behavior, *Trans. ASCE*, vol. 88, p. 492, 1925.

Figures 35 and 36 show the essential features of circular and rectangular sedimentation basins. In the circular basin (Fig. 35) the sewage enters the central well either through an influent pipe located below the tank floor or at some higher location inside

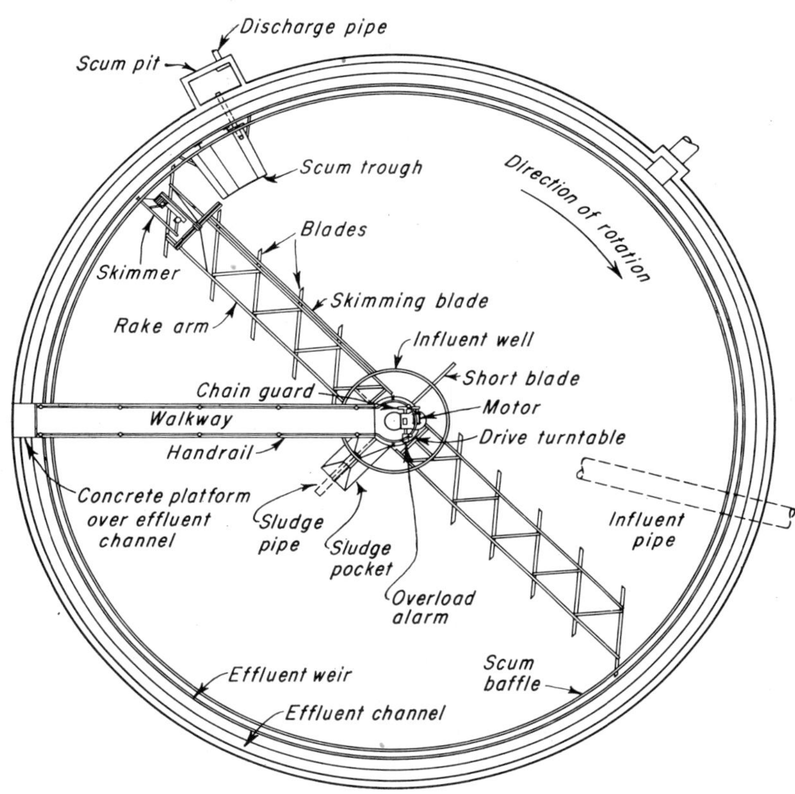

Plan

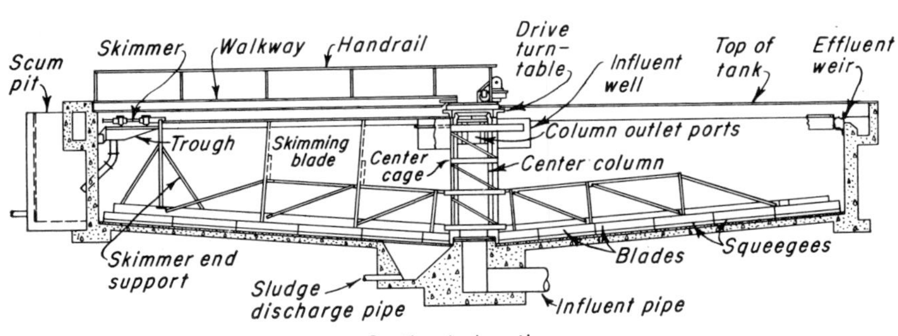

Sectional elevation

FIG. 35. The Dorr clarifier.

the tank. It discharges inside a circular wall or baffle which serves to distribute the sewage horizontally and vertically so as to equalize the flow radially toward the outer, or effluent, weir, utilizing as great a depth of the tank as possible without disturbing or picking up solids already settled to the tank bottom. The heavier solids are deposited near the center of the tank while the decreasing velocity toward the effluent weir at the tank's outer periphery permits the settlement of increasingly small suspended particles. This type of tank gives a maximum length of weir crest, thus reducing the velocity of

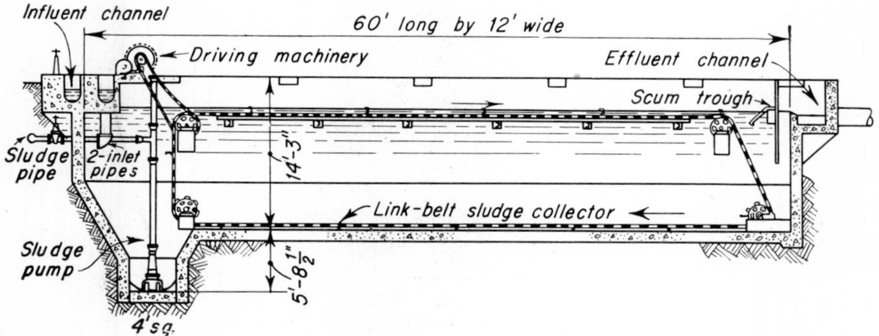

Fig. 36. Settling tank with Link-belt sludge collector.

approaching sewage to a minimum. The sludge carried over the weir is also a minimum for the detention period employed. In primary clarifiers a shallow baffle located several inches inside the weir holds back grease or other floating matter to be discharged by a suitable skimmer through a separate outlet. The settled sludge is pushed slowly toward a central sludge hopper whence it is pumped frequently or continuously to the sludge-digestion tanks.

Figure 36 shows the inlet and outlet arrangements, the sludge-removing mechanism and hopper, and the scum trough for a rectangular sedimentation basin. The scrapers of the sludge collector push the sludge slowly, with a minimum of disturbance, to the hopper at the inlet end of the tank while on the return trip, they carry grease and any other floating materials to the scum trough. Where several units can be built with common division walls, the rectangular tanks require less concrete than do the circular units. A uniform velocity throughout the length of travel is another advantage. To secure adequate length of effluent weir for a low velocity of approach, an H or multiple weir is usually required. Both types of sedimentation basins have given satisfactory operating results and are widely used.

Sedimentation periods of 1 to 3 hr are common, the average being about 2 hr. Shorter periods may be used for coarse or heavy solids, where followed by efficient secondary treatment, or where removal of solids rather than BOD is the controlling factor. Overflow rates of 600 to 1,200 gal per sq ft of tank surface per day are used.

Efficiency of Sedimentation Basins. The efficiency obtained in plain settling tanks may be measured by the removal of suspended solids from the raw sewage. The total suspended solids are reduced by 50 to 75 percent and the settleable suspended solids by 80 to 95 percent, while the BOD is reduced by 35 to 45 percent. The removal of bacteria approximates that of the suspended solids.

The rate of removal of suspended matter in settling tanks varies with the quantity present and with time. From Fig. 32, a 2-hr settling period would remove 65 percent of the suspended solids where the influent contained 400 ppm of suspended solids, and

42 percent from an influent with 100 ppm. It is apparent also that very little additional material would be removed by longer periods of settling and that most of the settling takes place in the first hour. Figure 32 is useful in estimating the amount of solids that will be removed for given detention periods and given amounts of suspended solids in the settling tank influent.

The advantage of the sedimentation process is that it is one of the cheapest and simplest methods for removing the heavier or coarser solids. It is a preparatory process rather than a total treatment, since the finer solids present either do not settle at all or take too long a time and require too large a settling tank to remove them. They may be more effectively treated, therefore, by some other method. The disadvantage of the process lies in the necessity for large tanks to remove any great quantity of the finer materials. The large quantities of sludge produced contain a high moisture content and require further treatment before they can be dried economically for final disposal.

Camp emphasizes the following points as affecting the design of sedimentation tanks:

1. Settling tanks should be designed on the basis of overflow rates, i.e., the settling velocity of the smallest particle theoretically to be removed 100 percent.
2. Detention periods, per se, are immaterial; in fact, long detention periods may invite septicity.
3. Tanks should be no deeper than will be required to prevent scour and to accommodate cleaning mechanisms.
4. Tanks should be long and narrow to minimize the effects of inlet and outlet disturbances, cross winds, density currents, and longitudinal mixing.

Effect of Compaction on Settling Rates. Camp carried his studies of suspensions beyond Hazen's contributions by defining different degrees of consolidation and pointing out their effects on settling rates. He divides settling into three categories: (1) where the density of the mass of settling particles is such that there is no interference between individual particles; (2) where hindered settling occurs, i.e., particles interfere somewhat with each other's free movement (solids greater than 1,000 ppm); and (3) where solids have compacted to about 3,000 ppm or more and a definite surface of the sludge blanket has formed. Figure 37 illustrates this gradual consolidation of *activated sludge* during which the velocity of consolidation has steadily decreased.

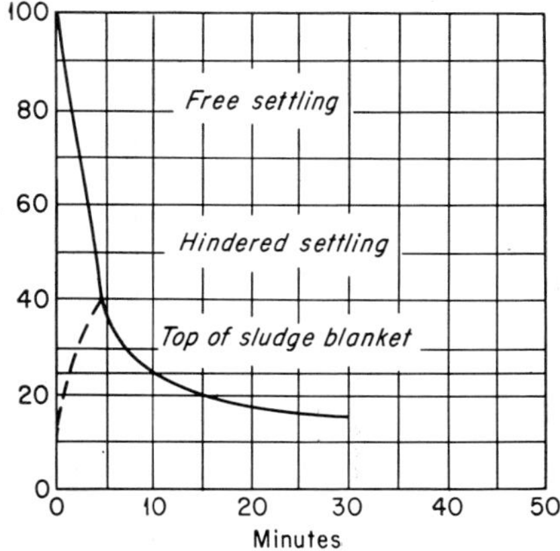

Fig. 37. Effect of compaction on the rate of settling of sludge in settling tanks.

Where the direction of flow is horizontal, the settling velocity of a particle which just falls through the depth of the tank during its period of detention is called the "overflow rate" v_0. Particles having settling velocities greater than v_0 will settle out while those with less velocity will be removed in the ratio v/v_0.

Chemical Precipitation. The process of sedimentation can be hastened and a larger amount of the fine materials settled out by chemical precipitation. The reduction in BOD by chemical precipitation is about 65 to 70 percent as compared with a 45 to 55 percent reduction from plain sedimentation. The chemicals added are electrolytes which, to be effective, should carry the opposite electric charge from that of the suspended solids. This combines with the suspended solids, and their precipitation is facilitated. Lime is most frequently used for this purpose, and its action is

somewhat more complex than indicated by the above statement. It adjusts the pH or hydrogen-ion concentration of the sewage to the optimum range for coagulation. It neutralizes the charge on the colloids. It also forms an insoluble precipitate which collects colloids by adsorption and sweeps them out of the sewage as it settles. This method of increasing the amount of solids removed by sedimentation is expensive because of the large quantities of chemicals used and the extremely large quantities of sludge produced. In addition to removing a larger percentage of suspended solids, each pound of chemicals would produce 20 lb of added sludge assuming a 95 percent moisture content.

Chemical precipitation was at one time used at Worcester, Mass., and at Providence, R.I., but was later abandoned because of unstable effluent and the expense of handling the large quantities of sludge produced. About 1929 to 1930, however, this method began to arouse renewed interest with the increase in the understanding of precipitation phenomena and the improved methods of sludge drying. The present function of chemical precipitation is to supply treatment intermediate between primary settling and complete treatment. It is also useful where there is a seasonal overloading of filters or other units of the treatment plant where adding capacity to provide complete treatment would be too expensive. A number of processes have been developed, each bearing the inventor's name.

Little removal of dissolved solids occurs, nor is the removal of colloidal material complete. The effluent contains fine turbidity and is much inferior to that produced by the oxidation processes discussed below.

Chemicals play an important part, however, in the treatment of various industrial wastes as, for example, in the removal of grease, in adjusting the pH of sludge-digestion tanks for increasing their efficiencies, and in controlling the pH in all stages of the treatment processes in order to get the maximum efficiency from their operation. In fact, pH control is a very important chemical operation at all treatment plants.

OXIDATION PROCESSES

Sedimentation is usually only a preliminary operation or the final step in sewage treatment. It may remove from 30 to 45 percent of the BOD of the sewage and thus leaves approximately two-thirds of the oxygen demand to be carried by the receiving stream. The rapidly increasing oxygen demands that have been placed on receiving waters (rivers, lakes, and harbors), by domestic sewage and industrial wastes, of which many have very large demands, is requiring complete treatment (sedimentation followed by secondary treatments such as filtration or the activated-sludge process) in order to protect domestic water sources from being polluted beyond the possibilities of reclamation for domestic use by existing water-purification methods.

Oxidation is the breaking down of the organic solids into stable organic or mineral compounds through biological activity in the presence of oxygen. Some of the chemical changes taking place in this process have already been described in the discussion of the nitrogen cycle. The oxidation of carbohydrates follows a somewhat similar process. Direct oxidation of sewage by artificial aeration is too slow for practical use except where conditions favorable to the growth of aerobic bacteria have been provided. The oxidation process is normally free from the unpleasant odors which accompany anaerobic processes.

Methods of Oxidation. The oldest oxidation process in use is irrigation. In the attempt to use higher rates of treatment than are possible in irrigation and to reduce the area necessary, intermittent sand filters were developed, followed later by contact beds and trickling filters. More recently, the activated-sludge method of sewage disposal has replaced the older and less efficient methods to some extent.

Irrigation is now widely used in the United States,[1] but its use has been more extensive abroad. However, Berlin, one of the large cities which has been quoted often as an example of disposal by irrigation, started to abandon that method and completed in 1931 the first of several modern disposal plants for treating its sewage. The fertilizing value of sewage is exceedingly small, and the nitrogen present is not usually in available form but requires oxidation to the nitrate stage before being useful for plant growth. The principal advantages in sewage irrigation come from the value of its water content and from its reclamation after such use, for replenishing local ground-water supplies for irrigation and industrial uses. Where water is in short supply as in Berlin, this may be the most important reason for disposing of sewage by its use in irrigation. It is applied intermittently in rates of 5,000 to 10,000 gal per acre per day, and the soil is permitted to dry out between applications. The continuous nature of the supply is a serious handicap, since the sewage has only a seasonal usefulness for irrigation purposes. Difficulties also may arise from the clogging of the soil by grease or soap in the sewage, and odors and flies are sometimes produced in troublesome quantities. Subsurface irrigation is frequently employed for treating residential or institutional sewage, as described on p. 9–96.

Intermittent sand filters were developed in England in 1868–1870 and in Massachusetts for increasing the efficiency of treatment obtained in irrigation. Sewage is applied to a prepared area which must be sandy in character and from which the surface soil has been removed. The area is divided into beds by embankments constructed of this stripping, and underdrains from 3 to 5 ft in depth are installed at distances of 20 to 40 ft apart. The sewage is distributed upon the beds intermittently through flumes or pipes at rates of 30,000 to 100,000 gal per acre per day. The lower rate is for raw sewage, while the higher rate may be obtained with settled sewage. The sewage is applied to the beds at a rate of 1 cu ft per sec for each 5,000 sq ft of filter area and in sufficient doses to give a depth of 2 to 4 in. This permits a rest period between doses for the air which follows the sewage down through the bed to assist in the oxidation process. Intermittent sand filters perform their function by straining out the coarser solids on the surface of the bed and by oxidizing putrescible matter by bacterial activity in the voids within the filter. This is one of the most efficient of sewage-treatment processes, as the removal of suspended and organic solids is from 90 to 98 percent. The effluent is clear and odorless, though odors from the beds themselves may cause trouble. A mat gradually accumulates on the surface of the bed and requires removal about once a year. Operation difficulties may be experienced in some areas during the winter from ice formation on the surface of the filters. This difficulty may be overcome by preparing the filter with furrows or mounds to support the ice covering so that the sewage can be distributed under it.

Contact beds are compartments arranged in parallel or in series, filled to a depth of 3 to 6 ft with broken stone or other ballast, usually from ½ to 2 in. in size. The sewage is delivered to the beds through pipe distributors located below the filter surface. The bed is filled to a point just below the surface, the sewage being left unexposed so that odors are not picked up by the wind. A properly operated contact bed should be relatively free from odors. While the bed stands full, sedimentation takes place in the voids between the stones, and some of the colloids are removed by coming into contact with the bacterial gelatinous coating on the stones. After a sufficient time for this process has been allowed, the beds are emptied and permitted to stand idle for 2 to 6 hr. During this rest period, oxidation of the organic solids remaining in the bed takes place. Several units are required so that suitable dosing cycles can be secured to allow for resting the individual beds. For example, in a small plant, four beds might be used with 6-hr dosing cycles made up, say, of ½ hr for filling, 1 hr full, ½ hr for emptying, and 4 hr rest. The

[1] Effluent from efficient sewage-disposal plants is being used increasingly for supplementing irrigation and industrial water supplies.

cycle used depends on local conditions, but to dose the beds in this way it is necessary to have a dosing tank of sufficient capacity to retain the sewage flow for an entire cycle. Siphons are usually employed for discharging the dosing tank when filled to a given level, and mechanical arrangements may be used for supplying the various beds in rotation. Contact beds produce less satisfactory effluents than do intermittent sand filters, as they remove only 50 to 75 percent of the suspended solids. The effluent is usually nonputrescible but is quite turbid because of materials which break loose from the ballast in the filter. Further sedimentation may be desirable especially where the coarser filtering material is used. Dosing rates of 100,000 to 800,000 gal per acre per day are used, the rate depending on the concentration of the sewage and the depth of the bed. This rate may also be expressed in terms of volume of the bed, in which case 50,000 to 150,000 gal per acre-ft of filter volume daily represents average practice. Contact beds may gradually become clogged and lose capacity, especially with finer material, and eventually require that the filtering medium be removed and cleaned.

Trickling filters are of two types, the standard trickling filter which has long been used to treat primary settling-tank, or Imhoff effluent, and the high-rate filter in which high rates of application are used with or without recirculation of the effluent through the filter. In the standard filter, sewage is sprayed over the filter from fixed properly spaced nozzles projecting above the filter stone or from rotary distributors on circular filters. A dosing tank is used for producing a uniform distribution of sewage over the area, and this is accomplished by discharging under a uniformly falling head. The filtering material is usually broken stone and should be tough and relatively impervious so as to resist weathering and crushing under the weight of superimposed material. Clinker, slag, and similar materials do not meet these requirements so well as stone but are frequently used if less expensive. Gravel is not very satisfactory, because the pebbles are so smooth that the organic growth so important in the purification process adheres less readily to them.

The amount of oxidation which takes place in the filter will vary with the surface area of the filtering material with which the sewage comes in contact. This area increases as the size of the stone decreases, the finer aggregate being thus more effective than the coarser. Clogging, however, occurs more easily as the voids become smaller and prevents the use of very fine materials. The practical limits of size have been found by experience to range between ½ and 2½ in. The selection of the size of material determines the maximum dosing rate that can be used on the filter without causing undue clogging. A depth may then be selected sufficient to produce satisfactory oxidation of this quantity of sewage. The rate of dosing of standard trickling filters is seen in Table 31 to vary from 5.6 to 10.0 gal per cu ft of filter volume per day, with most of the plants quoted ranging from 6.0 to 8.0. These are equivalent to a range of 1,000,000 to 3,500,000 gal per acre per day. Current practice favors expressing filter loading in terms of pounds of BOD applied per acre-foot of filter media per day. Hatch reports 250 lb, BOD, per acre-foot daily as satisfactory loading for Ohio plants. The table also gives data as to the size of stone and depth of filter used at each plant. Some investigators maintain that for a given volume deeper filters are more effective than shallow ones, but operating data from American plants do not show any very definite tendency of this kind. Ten feet is about the maximum practical depth because of the crushing under its own weight of the angular edges with greater depths of the material. Trickling filters are usually self-cleaning, since the oxidized material becomes loosened from the stone and is flushed out of the filter. Secondary settling tanks are usually installed to retain this sediment. The design of these tanks is similar to that of primary tanks except that vertical-flow tanks may be used more advantageously.

The efficiency of a trickling filter cannot be expressed so easily in terms of suspended solids removed, for the effluent may at times carry as much suspended solids as the in-

TABLE 31. DATA CONCERNING STANDARD TRICKLING FILTERS[1]

City	Total area, acres	Rate, mgd per acre	Average depth, ft	Rate per cu ft, gal per day	Filtering-material size, in.	Static head on nozzles, ft
Atlanta						
Intrenchment.	2	2.5	5¾	10.0	1½ to 2½	8 to 1½
Peachtree....	2½	2.0	5¾	8.0	1½ to 2	8 to 1½
Baltimore......	30	2.7 to 3	8½	7.3 to 8.1	1 to 2½	9
Columbus......	10	2.12	5⅓	9.1	1 to 3	8 to 4
Fitchburg......	2.1	2.7	10¼	6.0	1 to 2	10 to 1½
Lexington......	2	1.75	6	6.7	1 to 2½	9¾ to 2
Reading						
Bed 1........	1	1.6	5	7.3	1½ to 4	About 9 to 2½
Bed 2........	1	1.6	6½	5.6	1 to 2½	About 7¼ to 2½
Bed 3........	1	1.6	5	7.3	3	About 9 to 2½
Bed 4........	1	1.6	5	7.3	1½ to 4	About 9 to 2½
Rochester						
(Brighton)....	1	1.6	6	6.1	1	6 to 0

[1] *U.S. Pub. Health Service Bull.* 132, p. 135.

fluent, though these are of quite a different character. The surface of the filter may become clogged from overdosing or from lack of adequate biological activity. Resting the filter is beneficial in eliminating this pooling, as it gives the oxidation process more time in which to be carried out. Troubles from odors may result, especially where the sewage applied to the filter is stale. This can be controlled to some extent by chlorination. The chlorine is usually applied in the primary settling tanks ahead of the filter. Many insects inhabit the filter bed, and the *Psychoda alternata*, or filter fly, may multiply so rapidly as to become a nuisance, as its small size permits it to pass through ordinary window screening. It does not travel very far from the filter but may be carried some distance by the wind. Its growth can be controlled by the simple device of allowing a filter unit to stand filled with sewage so as to drown the larvae. The cycle of the *Psychoda* from egg through the larval stage may be as short as 10 or 12 days in hot weather; so this flooding should be done at least that often during the *Psychoda* season. Chlorination has been used, also, to control the production of filter flies.

Design of Dosing Tanks. The design of a dosing tank for standard trickling filters is simplified by assuming a uniformly falling head. This can be produced by giving the dosing tank a parabolic vertical section. It is of primary importance that the sewage be distributed uniformly over the filter. This requires a straight-line relationship between depth and length of rectangular dosing tank rather than the parabolic section usually claimed as necessary to fulfill this requirement. The requirements of uniformly falling head in the tank and uniform dosing rate on the filters are mutually contradictory and cannot both be fulfilled simultaneously. The difference in section required by these two assumptions is not great, and designs as commonly made thereon are sufficiently accurate. *Single* dosing tanks do not give uniform distribution on the filter when the fluctuating hourly flow follows a different cycle from that for which the design was made. For this reason, *twin* tanks are to be preferred. When one tank is full, the sewage inflow is transferred to the other tank while the first tank discharges. The discharge conditions are always constant, regardless of the rate of inflow, though the rest period between discharges will be decreased with an increasing sewage flow.

In Fig. 38, assume tank *A* to be full. The siphon starts discharging as soon as the desired elevation is reached in the tank. The rush of sewage traps and compresses air in the compression chamber. Through a system of air pipes, this compressed air fills

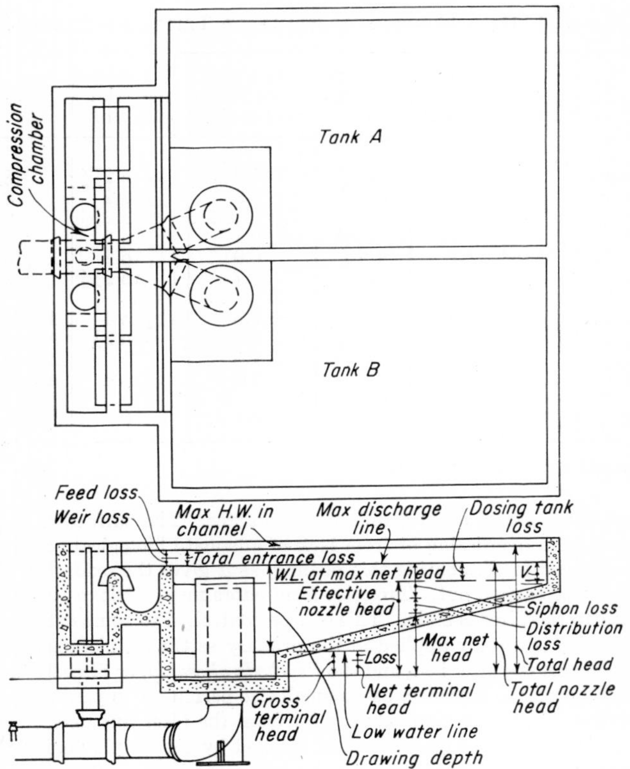

FIG. 38. Twin dosing tank showing terminology employed. (*Pacific Flush-Tank Co., Bull.* 30, *p.* 29.)

the siphons feeding the influent channel of tank *A* and stops the inflow. The pipe system is designed so that the pressure for tank *B* is released, and its inlet siphons start filling the tank. Tank *A* empties before tank *B* is filled, and the filters rest till tank *B* is full. The cycle is then repeated.

The design of the dosing tank to function as desired under the conditions given can best be carried out by using the data furnished by the Pacific Flush Tank Co., manufacturers of siphons and other equipment for dosing tanks. Tables 32 and 33 are taken from their Catalogue 30 and Bulletin 101 and give the essential design data.

Distribution System. The nozzle spacing is found from Tables 32 or 33. The usual hydraulic methods are applicable to the computation of pipe sizes for given head losses. Conditions of nonuniform flow exist in these pipes, and the recovery of velocity head causes corresponding increases in pressure head with increasing distances from the dosing tank. The distributors are frequently laid in the upper surface of the filter but in some instances have been placed at greater depths with a riser for each nozzle. Underdrainage systems of concrete blocks, split tile, or channels collect the sewage after it has passed through the filter and deliver it to the secondary settling tanks.

Rotating Distributors. During recent years, the use of the rotating-arm distributor for trickling filters of all types has displaced the use of fixed nozzles fed from dosing

TABLE 32. DESIGN DATA FOR TWIN DOSING TANKS FOR 7/8-IN. NOZZLES; 25 PERCENT HEAD LOSS[1]

Total nozzle head, ft		5.0	5.5	6.0	6.5	7.0	7.5	8.0	8.5	9.0	9.5	10.0	10.5	11.0	11.5	12.0
Maximum net head		3.75	4.12	4.50	4.87	5.25	5.62	6.00	6.38	6.75	7.12	7.50	7.87	8.25	8.62	9.00
Siphon, D.T., and D.L., each		0.42	0.46	0.50	0.54	0.58	0.62	0.67	0.71	0.75	0.79	0.83	0.88	0.92	0.96	1.00
Nozzle spacing, ft		8.66	9.25	9.90	10.50	11.10	11.70	12.25	12.80	13.40	14.00	14.50	15.00	15.50	16.00	16.50
Lateral spacing, ft		7.50	8.01	8.56	9.10	9.50	10.13	10.62	11.10	11.60	12.12	12.55	13.00	13.42	13.86	14.30
Area of bed per nozzle, sq ft		65	74	85	96	107	119	130	142	156	170	182	195	208	222	236
Spray limit (nozzle to wall)		7.25	7.60	8.00	8.25	8.50	8.90	9.10	9.40	9.60	10.00	10.25	10.43	10.60	10.80	11.00
Area D.T. at E.N.H., sq ft per nozzle		3.15	2.70	2.40	2.10	1.90	1.70	1.55	1.40	1.30	1.20	1.10	1.03	0.97	0.89	0.85
Rate discharge, gpm per nozzle at M.N.H.		15.50	16.25	17.00	17.65	18.35	19.00	19.60	20.20	20.80	21.35	21.90	22.45	23.00	23.50	24.00
Limit of inflow, gal per sq ft per min		0.177	0.160	0.144	0.130	0.117	0.108	0.100	0.094	0.087	0.082	0.077	0.073	0.068	0.065	0.062
	N.T.H.															
Vertical wall V	1.0	0.78	0.89	1.01	1.14	1.26	1.39	1.53	1.66	1.79	1.93	2.07	2.21	2.35	2.48	2.64
	1.5	0.62	0.72	0.83	0.94	1.04	1.16	1.29	1.40	1.52	1.64	1.77	1.90	2.03	2.15	2.29
	2.0	0.53	0.61	0.70	0.80	0.89	1.00	1.11	1.21	1.32	1.43	1.55	1.67	1.78	1.90	2.03
	2.5	0.47	0.54	0.62	0.70	0.78	0.87	0.98	1.07	1.17	1.27	1.38	1.49	1.59	1.70	1.82
Area of dosing tank at L.W.L., sq ft per nozzle.	1.0	1.63	1.33	1.13	0.95	0.83	0.72	0.63	0.55	0.50	0.45	0.40	0.367	0.338	0.303	0.283
	1.5	1.99	1.63	1.38	1.16	1.01	0.88	0.77	0.68	0.61	0.55	0.49	0.450	0.414	0.371	0.347
	2.0	2.30	1.88	1.60	1.34	1.17	1.01	0.89	0.78	0.71	0.64	0.57	0.519	0.477	0.429	0.400
	2.5	2.57	2.10	1.79	1.50	1.31	1.13	1.00	0.88	0.79	0.71	0.63	0.580	0.534	0.480	0.448
Time to empty tank, min.	1.0	5.75	5.32	5.05	4.70	4.48	4.20	4.01	3.77	3.64	3.48	3.29	3.18	3.09	2.92	2.86
	1.5	4.83	4.56	4.40	4.16	4.01	3.79	3.65	3.46	3.39	3.23	3.07	2.97	2.89	2.74	2.69
	2.0	3.90	3.80	3.76	3.61	3.53	3.38	3.29	3.14	3.07	2.97	2.84	2.77	2.70	2.57	2.53
	2.5	2.97	3.05	3.12	3.07	3.06	2.98	2.93	2.83	2.78	2.71	2.61	2.56	2.51	2.40	2.37
Volume of dosing tank, gal per nozzle.	1.0	71.3	68.6	67.3	64.6	63.6	61.9	60.4	58.4	57.9	56.6	55.3	54.6	53.8	52.0	51.7
	1.5	63.6	62.2	61.8	60.1	59.4	58.2	57.0	55.6	54.8	53.8	52.4	52.2	51.5	50.0	50.0
	2.0	54.0	54.7	55.5	54.7	54.7	53.7	53.4	52.4	52.0	51.5	50.2	49.6	49.3	48.0	47.7
	2.5	42.9	45.6	47.7	48.3	49.1	49.3	49.0	48.6	48.4	48.0	47.1	47.0	46.9	45.9	45.9
Maximum drawing depth of siphon.	1.0	3.78	4.28	4.78	5.28	5.78	6.28	6.78	7.28	7.78	8.28	8.78	9.28	9.78	10.28	10.78
	1.5	3.17	3.67	4.17	4.67	5.17	5.67	6.17	6.67	7.17	7.67	8.17	8.67	9.17	9.67	10.17
	2.0	2.56	3.06	3.56	4.06	4.56	5.06	5.56	6.06	6.56	7.06	7.56	8.06	8.56	9.06	9.56
	2.5	1.95	2.45	2.95	3.45	3.95	4.45	4.95	5.45	5.95	6.45	6.95	7.45	7.95	8.45	8.95

[1] *Pacific Flush Tank Co., Bull.* 101, p. 26.

Table 33. Design Data for Twin Dosing Tanks for 1-in. Nozzles; 25 Percent Head Loss[1]

Total nozzle head, ft		5.0	5.5	6.0	6.5	7.0	7.5	8.0	8.5	9.0	9.5	10.0	10.5	11.0	11.5	12.0
Maximum net head		3.75	4.12	4.50	4.87	5.25	5.62	6.00	6.38	6.75	7.12	7.50	7.87	8.25	8.62	9.00
Siphon, D.T., and D.L., each		0.42	0.46	0.50	0.54	0.58	0.62	0.67	0.71	0.75	0.79	0.83	0.88	0.92	0.96	1.00
Nozzle spacing, ft		9.65	10.25	10.75	11.40	12.00	12.60	13.20	13.80	14.25	14.75	15.30	15.80	16.30	16.90	17.30
Lateral spacing, ft		8.36	8.86	9.30	9.86	10.40	10.90	11.44	11.95	12.35	12.77	13.25	13.70	14.12	14.65	15.00
Area of bed per nozzle, sq ft		80.6	91.0	100	113	125	137.5	151	165	176	188.5	203	216	230	248	260
Spray limit (nozzle to wall)		7.5	8.0	8.3	8.7	9.0	9.3	9.6	10.0	10.3	10.6	10.9	11.2	11.5	11.7	12.0
Area D.T. at E.N.H., sq ft per nozzle		4.41	3.78	3.36	2.94	2.66	2.38	2.17	1.96	1.82	1.68	1.54	1.44	1.36	1.25	1.19
Rate discharge, gpm per nozzle at M.N.H.		21.70	22.75	23.80	24.70	25.70	26.60	27.42	28.30	29.10	29.90	30.65	31.42	32.20	32.90	33.60
Limit of inflow, gal per sq ft per min		0.200	0.182	0.166	0.154	0.141	0.131	0.122	0.114	0.108	0.102	0.097	0.091	0.087	0.083	0.079
	N.T.H.															
Vertical wall V	1.0	0.78	0.89	1.01	1.14	1.26	1.39	1.53	1.66	1.79	1.93	2.07	2.21	2.35	2.48	2.64
	1.5	0.62	0.72	0.83	0.94	1.04	1.16	1.29	1.40	1.52	1.64	1.77	1.90	2.03	2.15	2.29
	2.0	0.53	0.61	0.70	0.80	0.89	1.00	1.11	1.21	1.32	1.43	1.55	1.67	1.78	1.90	2.03
	2.5	0.47	0.54	0.62	0.70	0.78	0.87	0.98	1.07	1.17	1.27	1.38	1.49	1.59	1.70	1.82
Area of dosing tank at L.W.L., sq ft per nozzle.	1.0	2.28	1.86	1.57	1.33	1.16	1.00	0.886	0.776	0.700	0.630	0.562	0.513	0.474	0.425	0.396
	1.5	2.79	2.28	1.94	1.63	1.42	1.23	1.08	0.950	0.858	0.771	0.688	0.628	0.580	0.520	0.485
	2.0	3.22	2.63	2.24	1.88	1.64	1.42	1.25	1.10	0.990	0.890	0.795	0.725	0.670	0.601	0.560
	2.5	3.60	2.94	2.50	2.10	1.83	1.59	1.40	1.23	1.10	0.995	0.890	0.811	0.750	0.672	0.626
Time to empty tank, min	1.0	5.75	5.32	5.05	4.70	4.48	4.20	4.01	3.77	3.64	3.48	3.29	3.18	3.09	2.92	2.86
	1.5	4.83	4.56	4.40	4.16	4.01	3.79	3.65	3.46	3.39	3.23	3.07	2.97	2.89	2.74	2.69
	2.0	3.90	3.80	3.76	3.61	3.53	3.38	3.29	3.14	3.07	2.97	2.84	2.77	2.70	2.57	2.53
	2.5	2.97	3.05	3.12	3.07	3.06	2.98	2.93	2.83	2.78	2.71	2.61	2.56	2.51	2.40	2.37
Volume of dosing tank, gal per nozzle.	1.0	99.8	96.0	94.2	90.4	89.0	86.6	84.5	81.7	81.0	79.2	77.4	76.4	75.3	72.7	72.3
	1.5	89.9	87.0	86.4	84.1	83.1	81.4	79.8	77.8	76.7	75.2	73.3	73.0	72.0	70.0	70.0
	2.0	75.5	76.6	77.6	76.6	76.6	75.1	74.7	73.3	72.8	72.1	70.2	69.4	69.0	67.2	66.8
	2.5	60.0	63.8	66.8	67.6	68.7	69.0	68.6	68.0	67.8	67.2	65.9	65.8	65.6	64.2	64.2
Maximum drawing depth of siphon.	1.0	3.78	4.28	4.78	5.28	5.78	6.28	6.78	7.28	7.78	8.28	8.78	9.28	9.78	10.28	10.78
	1.5	3.17	3.67	4.17	4.67	5.17	5.67	6.17	6.67	7.17	7.67	8.17	8.67	9.17	9.67	10.17
	2.0	2.56	3.06	3.56	4.06	4.56	5.06	5.56	6.06	6.56	7.06	7.56	8.06	8.56	9.06	9.56
	2.5	1.95	2.45	2.95	3.45	3.95	4.45	4.95	5.45	5.95	6.45	6.95	7.45	7.95	8.45	8.95

[1] *Pacific Flush Tank Co., Bull.* 101, p. 27.

TABLE 34. EFFICIENCY OF TREATMENT OF MASSACHUSETTS CITIES'[1] SEWAGES ON STANDARD TRICKLING FILTERS

	Raw sewage						Sewage applied to standard trickling filters										
City	Depth, ft	Total solids	Susp. solids	Organic	Fats	BOD	Total solids	Per cent removed	Susp. solids	Per cent removed	Organic	Per cent removed	Fats	Per cent removed	BOD	Per cent removed	Over-all per cent removed
Brockton....	10	586	214	306	79	260	562	15	117	38	260	30	56	68	255	87	87
Fitchburg...	10	395	158	226	45	113	259	13	55	51	130	26	24	..	60	73	86
Framingham	7.5	538	358	280	87	263	417	18	101	24	198	38	55	..	190		
Maynard....	7	732	308	432	117	377	498	12	118	57	239	9	64	..	155	90	96
Milford.....	6	587	209	309	149	245	467	22	98	53	226	38	62	..	161	86	91
Natick......	8	706	435	402	128	178	456	21	63	40	207	39	34	..	125	86	90
Pittsfield....	?	492	162	251	54	220	334	..	43	23	143	..	15	..	107	78	89
Worcester...	10	832	523	351	80	180	427	6	165	59	163	42	42	71	103	84	91

[1] Mass. Dept. Public Health, *Ann. Rept.*, 1939.

tanks. Its advantages include greater uniformity of dosing, smaller head losses, and higher rates of treatment per acre-foot or per cubic yard of filter volume. Two to four rotating arms revolve about a vertical central axis supported by a special bearing and propelled by the reaction of the distributor jets. The latter are spaced to give uniform dosing, the jets being closer together toward the outer end where the travel is greater. Jet orifices with adjustable areas may aid in securing uniformity of dosing. The head required to operate this type of distributor may be as low as 1½ to 2 ft. Various types

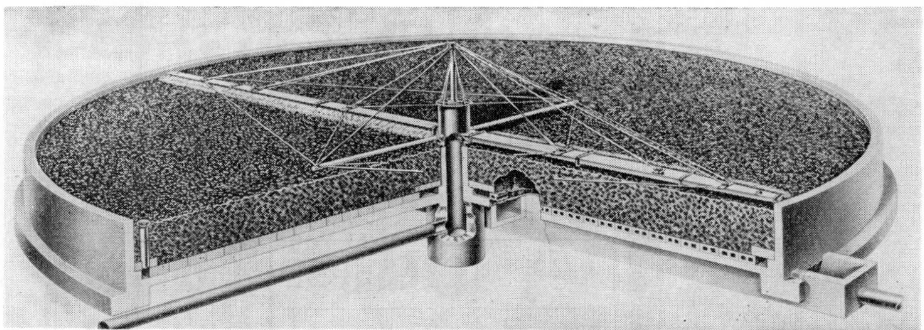

Fig. 39. A high-rate sewage filter cut away to show a 110-ft two-compartment arm unit with feed and underdrainage systems. (*Dorr-Oliver.*)

of jet nozzle have been developed for securing good distribution between nozzles. Dosing tanks should be used where the inflow is not uniform so that velocities will not drop below that required to rotate the distributor arms. Little difficulty has been experienced in the clogging of the jets.

The high-rate trickling filter, Fig. 39, has replaced the standard filter in plants built during recent years. It has widely extended the use of sewage filters into the high BOD-removal field hitherto served exclusively by the activated-sludge method. An occasional intermittent sand filter is still being used satisfactorily at the smaller plants in New England.

Two types of high-rate filters are in use, the *aerofilter* (Lakeside Engineering Corp.) and the *biofilter* (The Dorr Co.). The aerofilter is sometimes called a "high-capacity" filter because of the large volume of sewage in the filter at any moment. This is in contrast to the term "high-rate" filter, which refers to the high rate of application of the sewage to the bed which occurs with the biofilter since the latter returns, or "recirculates," some of the secondary effluent to the filters, the number of recirculations depending on the strength of the sewage, the filter loading, and the BOD removal desired.

The depth of the aerofilter varies from 5 to 8 or 9 ft, receiving an average flow rate of 15 to 18 million gallons per acre per day (mgad). About 30 percent of the daily flow is recirculated during periods of minimum flow in order to keep the distributor rotating. The sewage is applied in a fine spray or rain by a rotating disk or a multiple-arm distributor, covering much of the bed simultaneously. Artificial ventilation is recommended for filters using media other than a special nonclogging tile, when the filter is more than 6 ft in depth. One cubic foot of air per square foot of filter surface per minute is recommended. The BOD removal obtained with two-stage aerofilters may be estimated from Fig. 40.

In the biofilter, comparatively shallow filters are commonly used (about 3 ft), and the desired length of contact with the filter flora is secured by the number of recirculations provided. Dosing rates should exceed 6 or 8 mgad to prevent clogging, usually

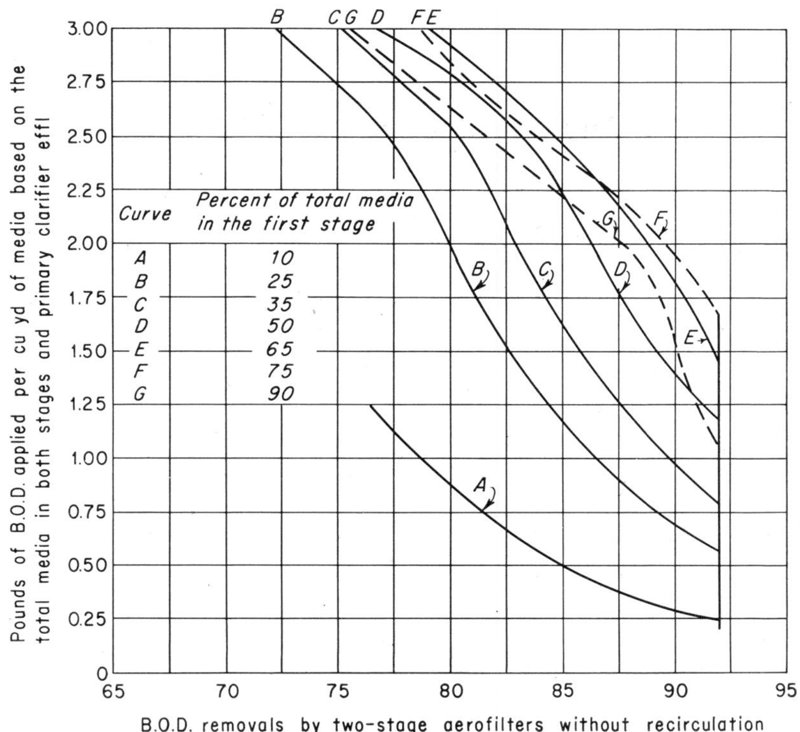

Fig. 40. (*Lakewood Eng. Corp.*)

vary from 6 to 20 mgad, but may be as great as 100 to 125 mgad for strong industrial wastes. Increasing the recirculation rate increases the required settling-tank capacity and the pumping cost. Forced ventilation is not required with biofilters, but vents must be provided around the base of the filter for inflow or outflow of air, the direction depending on whether the sewage in the filter is warmer or colder than the air. Since

Table 35. Single-stage Biofilters—Flow Rates and Filter Data

Ref. No.	Location	Sewage flow, mgd	Recirculation ratio R	Clarifier overflow rate, gal per day per sq ft		Filter			
				Primary	Final	Area, acres	Depth, ft	Volume, acre-ft	Dosing rate, mgad
1	Fremont, Ohio	1.75	1.5	1,140	870	0.231	3.3	0.762	19.0
2	Great Neck, N.Y.	0.52	1.15	540	157	0.146	4.0	0.584	7.68
3	Oklahoma City, So. Side Plant, first stage	16.173	1.0	620	630	1.99	6.0	11.94	16.3
4	Richland, Wash., Plant II	2.169	2.76	500	500	0.238	4.5	1.070	19.6
5	Dothan, Ala.	0.548	2.31	650	280	0.199	3.0	0.597	15.0
6	Centralia, Mo. (4th test period)	0.098	0.95	197	140	0.0222	3.0	0.0666	8.65
7	Storm Lake, Iowa, first stage	1.388	2.14	340	760	0.203	8.0	1.624	21.5
8	Alisal, Calif. (1st test period)	0.588	3.06	770	2,110	0.115	3.2	0.370	20.8

TABLE 36. PERFORMANCE OF SINGLE-STAGE BIOFILTERS, BASED ON BOD REMOVAL (FIG. 41*a* OR 41*b*)

Ref. No.	Location	BOD raw sewage, ppm	Estimated removal by settling, percent	BOD of settled raw sewage			BOD of settled effluent			BOD removed from settled raw		Efficiency	
				Ppm	Lb per day	Lb per acre-ft	Ppm	Lb per day	Lb per acre-ft	Ppm	Lb per acre-ft	Based on settled raw, percent	Based on raw sewage, percent
1	Fremont, Ohio	134	30.5	93	1,360	1,780	21	306	401	72	1,379	77.5	84.3
2	Great Neck, N.Y.	187	36	120	519	888	20.2	88	151	99.8	737	83.1	89.2
3	Oklahoma City, So. Side Plant, first stage	463	35	303	40,680	3,400	66	8,970	750	237	2,650	78.0	85.8
4	Richland, Wash., Plant II	173	36	111	2,003	1,900	19.5	353	330	91.5	1,570	82.7	88.7
5	Dothan, Ala.	230	35	150	685	1,150	19	87	145	131	1,005	87.4	91.7
6	Centralia, Mo. (4th test period	210	36	134	109	1,640	14	11.5	173	120	1,467	89.4	93.3
7	Storm Lake, Iowa, first stage	690 [1]	45 [2]	380	4,390	2,700	61	706	435	319	2,265	84.0	91.1
8	Alisal, Calif. (1st test period)	293	40 [2]	176	865	2,320	23.5	115	311	152.5	2,009	86.6	92.0

[1] Weighted average of domestic and industrial wastes.
[2] Includes flocculation of raw sewage.

Table 37. Two-stage Plants, Filters in Series (Fig. 41c)—Flow Rates and Filter Data

Ref. No.	Location	Mgd	Recirculation ratio		Clarifier overflow rate gal per day per sq ft		Primary filter				Secondary filter			
			First stage R_1	Second stage R_2	Primary	Final	Area, acres	Depth, ft	Volume, acre-ft	Dosing rate, mgad	Area, acres	Depth, ft	Volume, acre-ft	Dosing rate, mgad
1	Winchester, Va.	1.932	1.52	2.63	1,160	920	0.282	3.0	0.845	17.3	0.282	3.0	0.845	24.8
2	Bentonville, Ark.	0.343	1.67	2.78	550	810	0.079	5.0	0.394	11.6	0.054	5.0	0.268	24.1
3	Siloam Springs, Ark.	0.255	1.41	1.54	390	410	0.076	5.0	0.380	8.1	0.054	5.0	0.270	11.8
4	Liberty, N.Y.	1.13	1.78	1.65	1,400	1,160	0.115	3.0	0.345	27.2	0.115	3.0	0.345	26.0
5	Medford, Ore.	4.63	1.17	1.17	1,675	1,675	0.180	3.0	0.540	57.0	0.180	3.0	0.540	57.0

Table 38. Performance of Two-stage Plants, Filters in Series (Fig. 41c)—Based on BOD Removal

Ref. No.	Location	BOD raw sewage, ppm	Estimated removal by settling, percent	BOD of settled raw sewage			BOD of final effluent			BOD removed from settled raw		Efficiency	
				Ppm to primary	Lb per day	Lb per acre-ft	Ppm	Lb per day	Lb per acre-ft	Ppm	Lb per acre-ft	Based on settled raw, percent	Based on raw sewage, percent
1	Winchester, Va.	243	30.5	169	2,720	1,610	20.1	323	191	148.9	1,419	88.1	91.7
2	Bentonville, Ark.	335	35	231	657	996	27.5	79	119	203.5	877	88.1	91.8
3	Siloam Springs, Ark.	621	36.5	394	840	1,290	34	72	111	360	1,179	91.4	94.5
4	Liberty, N.Y.	400	28	288	2,710	3,920	31	292	423	257	3,597	89.3	92.3
5	Medford, Ore.	215	26.5	158	6,100	5,630	21.1	813	743	136.9	4,887	86.7	90.2

TABLE 39. PERFORMANCE OF TWO-STAGE BIOFILTERS USING SERIES-PARALLEL FLOW SHEET (FIG. 41*e*)

Plant location	Orlando, Fla.	Petaluma, Calif.
Flow, mgd	3.90	0.505
Recirculation ratio:		
Primary R_1	1.6	1.43
Secondary R_2	2.06	2.43
Clarifier overflow rate:		
Primary gal per day per sq ft	600	880
Final gal per day per sq ft	965	880
Filters:		
Total area, acres	1.04	0.203
Depth, ft	5.0	3.0
Volume, acre-ft	5.20	0.609
Dosing rate, mgad:		
Primary filter	19.3	12.0
Secondary filter	23.0	17.2
BOD raw sewage, ppm	176	576
Estimated removal by settling, percent	30	33
BOD of settled raw sewage:		
Ppm	123	385
Lb per day	3,960	1,620
Lb per acre-ft	760	2,660
BOD of final effluent:		
Ppm	10.6	14
Lb per day	345	59
Lb per acre-ft	66	97
BOD removed from settled raw:		
Ppm	112.4	371
Lb per acre-ft	694	2,563
Efficiency:		
Based on settled raw	91.4	96.4
Based on raw sewage	94.0	97.6
Computed efficiency:		
Based on settled raw	94.9	95.1
Based on raw sewage	96.4	97.2

the high-rate filter functions in part as a colloider, the settling tank is an essential part of the system and should have ample detention period and overflow rate for efficient removal of the suspended solids from the filter. Overflow rates and BOD removals for single-stage biofilters are given in Tables 35 and 36, and for two-stage filters in Tables 37 and 38.

Two-stage, high-rate filters, in comparison with the activated-sludge method of sewage treatment, have the advantages of greater flexibility and ability to handle overloading for reasonable periods, of greater simplicity and lower cost of operation, can give closely comparable BOD removals, and are relatively free from odors.

Figure 41 shows the flow arrangements that are currently used with single- and two-stage biofilters. Tables 35 to 39 give the results attained with the various flow sequences shown in Fig. 41. The data in these tables may be summarized by stating that the single-stage units have about an 86 percent efficiency in the removal of the BOD of the settled sewage applied, or 90 percent of the raw sewage BOD. Of the two flow types shown for two-stage biofilters, series *c*, in which the secondary unit is fed directly from the primary filter, is slightly less efficient than series *e* which is fed from the primary settling tank which has the lesser solids. The respective efficiencies attained were as follows:

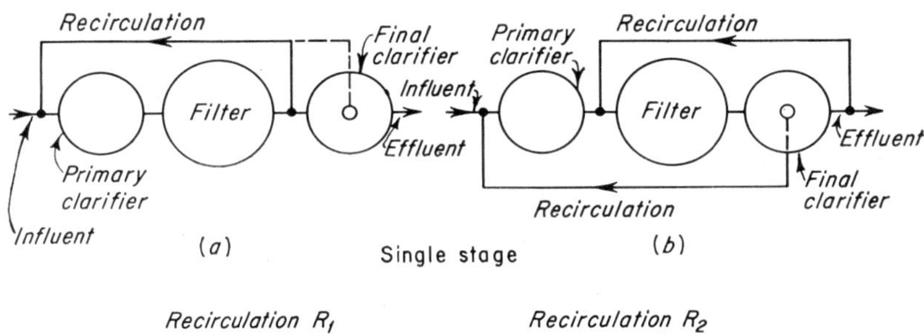

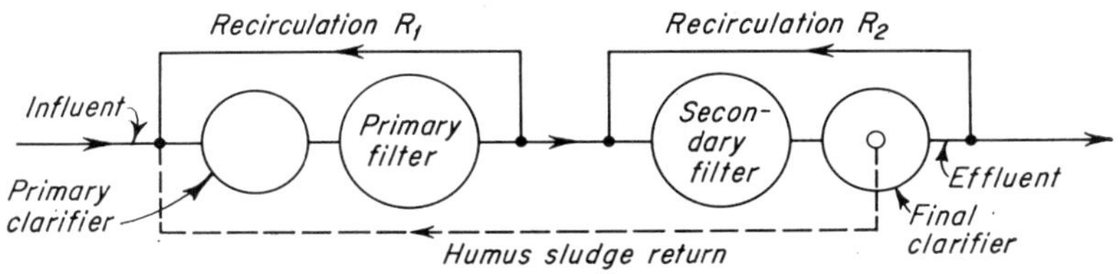

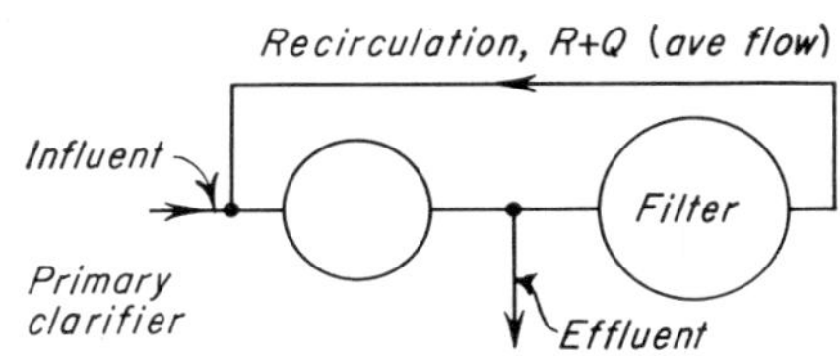

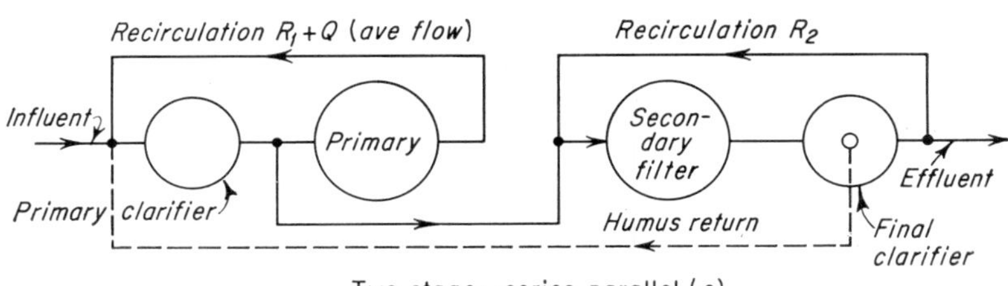

Fig. 41. Typical flowsheets, single- and two-stage biofilters.

In series *c*, the efficiency was 92.5 percent based on *raw sewage* or 88.9 percent on *settled sewage.*

In series *e*, the efficiency was 96.8 percent with reference to *raw*, or 95.0 percent to *settled sewage.*

In series *a*, and *b*, efficiency was 89.5 percent referred to *raw*, or 86.1 percent referred to *settled sewage.*

The performance of the biofilters examined by Rankin and listed in Tables 35 to 39 appears to depend primarily on the recirculation ratio while dosing rates, loading, and depth of filters were less important variables.[1]

[1] Rankin, R. S., Performnce of Biofiltration Plants by Three Methods, *Proc. ASCE* 79, Separate No. 336; November, 1953.

The activated-sludge process is a process of oxidizing sewage solids by aeration in the presence of sludge which has been previously oxidized or activated and in which a luxuriant growth of aerobic bacteria has developed. These bacteria are instrumental in bringing about the ozidation process, while aeration furnishes the needed oxygen and also produces adequate agitation for circulating the sludge through the sewage and preventing its settling out and becoming deoxygenated. The process is similar in principle to filtration except that, instead of the sewage being passed through the filter, the filtering medium, which in this case is the sludge with its bacterial population, is circulated through the sewage and comes into intimate contact with it. The adsorption of colloids by the flocculated sludge and oxidation by the aerobic bacteria are similar to, but more efficient than, those taking place in filters, and the agitation produced by the air performs the further function of coagulating some of the colloids, thus materially assisting the process of sedimentation. It is necessary to supply oxygen continually to the sludge, or its supply will become depleted and anaerobic decomposition will begin. Aeration of sewage without the presence of activated sludge is ineffective. The presence of activated sludge with its bacterial culture is essential to bring about purification in reasonable periods of time. It is necessary, therefore, to return sludge continuously to be mixed with the incoming sewage in about a 1:4 ratio. This combined sludge and sewage is aerated from 2 to 6 hr, preferably 3 to 4, using 0.4 to 2.0 cu ft of free air per gallon of sewage. These variables depend on the character of the sewage and the method of aeration employed. Since the introduction of the air is one of the major items of expense in this process, much experimentation has been carried out with the objective of reducing the air requirement. In American plants the air usually is fed through porous tubes or plates along one side of the bottom of long and relatively narrow aeration tanks, thus producing spiral flow as the sewage moves longitudinally through the tank. This carries the air bubbles in a spiral around the central horizontal axis of the tank, producing longer liquor-air contact, thus increasing the oxidation efficiency. In Europe, mechanical agitation and the introduction of air through the action of paddles have been widely used. The paddles in Fig. 42 revolve against the

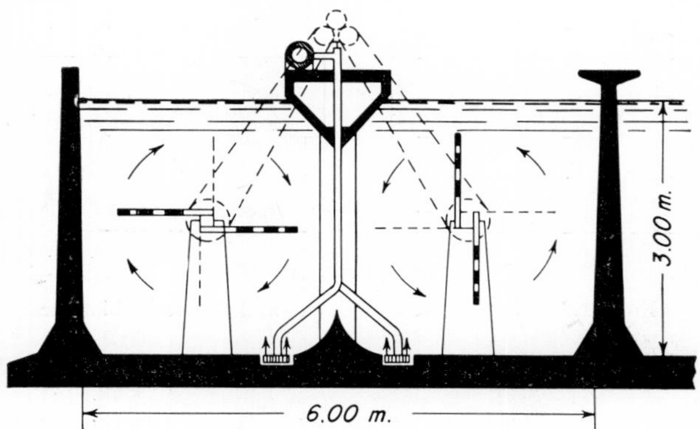

Fig. 42. Cross section of aeration chamber, Essen-Rellinghausen. (*Dr. Imhoff.*)

rising air bubbles and delay their transit, thus increasing time of contact with a corresponding increase in air-absorption efficiency. The depth of the aeration tank varies from 10 to 15 ft, which requires air pressures of 6 to 10 psi, allowing for losses.

Figure 42 is a cross section of a German aeration chamber and illustrates the method of obtaining spiral flow above described. In this figure, paddles are rotated against the

flow induced by air in order to keep the air bubbles in contact with the sewage for a longer time. The American plants are similar except that the paddles are omitted and both side walls are flared at the top, as is the central wall in the figure, to assist in the rotary motion set up in the tank.

Figure 43 shows two parallel aeration channels receiving settled sewage at one end and discharging the aerated sewage to a final settling tank at the other (not shown in cut). Air is fed from the air supply line in the covered wall between the two channels

FIG. 43. Swing diffusers, Oildale, California. Design flow, 2,000,000 gpd. (*Chicago Pump Co.*)

through five air supply units in each channel similar to the unit that has been raised by a jack as shown. The air feed line and aerators of one unit have been raised by a jack for inspection of the aerator tubes. The aerators are at one side of the aeration tank so the rising air bubbles will cause a rotation of the sewage in the tank as it follows a spiral path through the tank.

After passing through the aeration chamber, the sewage flows into settling compartments where the sludge settles and the supernatant liquid is discharged, usually over a weir. Well-activated sludge settles rapidly, and detention periods of 30 min to 2 hr are adequate for its removal. The desired amount of sludge is returned to the inlet end of aeration chambers, where it is discharged into the incoming sewage and again passes through the plant. As sludge accumulates beyond the amount required for carrying on the seeding process, it is withdrawn and either dewatered by some of the methods discussed later, carried to sludge-digestion tanks where it undergoes anaerobic digestion as hereafter discussed, or dried by pressure, vacuum, or heat without digestion. Digestion can be carried on satisfactorily alone or in conjunction with sludge from plain settling tanks. For best results, the ratio of fresh activated sludge to well-digested sludge should not be too great.

Results Obtained. A high degree of efficiency can be obtained with the activated-sludge process when the plant is properly designed and operated. Removal of 90 to 95 percent of suspended solids and bacteria and an 85 to 95 percent reduction in BOD are common. The process can be regulated so as to produce any smaller reduction of solids and BOD that may be dictated by the economics of the situation. Where partial treatment is all that is required, this method may still be used advantageously. The effluent is clear and odorless, and the plant can be operated without producing odor nuisances such as may be present with the types of plant heretofore discussed. The construction cost is not excessive, but the operating cost is relatively high. Considerable mechanical

equipment is needed and more skilled labor required. Another disadvantage lies in the large quantities of sludge of high moisture content produced, which is hard to dewater. But the sludge that is produced has a high nitrogen and phosphate content as compared with the sludge from other methods of sewage treatment. It thus has value as a fertilizer, and the local demand is usually sufficient to absorb the available supply. Since the sludge has a high moisture content, drying is an essential step in sludge utilization.

The freedom from odor and other objectionable features usually present at sewage-treatment plants makes it possible to build activated-sludge disposal plants in the city at the point of sewage concentration and thus avoid the construction of long and costly outfall lines to disposal-plant sites formerly considered necessary, or possible lawsuits for nuisance production which have often been troublesome and very costly to the community.

Vitamin B_{12} has been reclaimed from this type of sludge, which makes its recovery at the larger plants financially desirable and feasible. This does not damage the sludge's fertilizing value, and Milwaukee which developed the process for reclaiming B_{12}, has these two valuable by-products to help defray the expenses of its disposal-plant operations. The dried sludge is sold under the trade name of Milorganite and is distributed throughout the United States.

Modifications of the Process. Large-scale research on this treatment method has been carried out at both New York City and Chicago under actual operating conditions and each has developed variations in the treatment method. Table 40 gives the operating data at these two cities, each of which has a number of large activated-sludge plants, treating from 29 to 794 mgd of sewage, the latter being the largest sewage-disposal plant yet built. For many years, these two cities have carried out continuous research on a large scale to determine methods for reducing treatment costs while maintaining the desired quality of effluent. Various modifications of the activated-sludge process were thus developed. New York City, being on tidal waters, has less stringent requirements because of the high dilutions available at some of its outfalls which are not available to Chicago. It developed *step aeration*,[1] a modification of complete activated-sludge treatment, which has been used since 1939 at the Wards Island plant where it was de-

TABLE 40. DATA ON NEW YORK CITY AND CHICAGO ACTIVATED-SLUDGE PLANTS

Plants	Sewage flow, mgd	Air ratio, cu ft per gal	Aeration period, hr	Overflow rate, gal per sq ft per day	Sludge age, days	Raw sewage, ppm		Final effluent		Removals percent	
						S.S.	BOD	S.S.	BOD	S.S.	BOD
New York City:											
Wards Island	48	0.66	2.7	930	3.4	154	139	8	7	95	95
Hunts Point	101	0.50	2.6	980	4.1	135	138	9	10	93	93
Tallmans Island	29	0.49	4.7	1,100	5.7	148	122	17	11	89	91
Bowery Bay	41	0.98	2.4	1,020	4.3	140	148	18	19	87	87
26th Ward	44	1.05	4.3	710	4.6	143	132	19	16	87	88
Type of treatment:											
1. Step aeration	47	0.61	2.7	880	...	162	133	7	6	96	95
2. Activated aeration	56	0.19	2.8	1,100	...	162	133	28	32	83	76
Average 1 and 2		0.38						18	20	89	85
3. Modified aeration	55	0.29	2.7	970	...	162	133	35	42	78	68
Chicago Sanitary District:											
North Side	213	0.48	6.0	925	...	135	102	9	6	93.5	94.0
Calumet	77	0.48	6.1	810	...	146	109	14	11	91.8	90.0
West-south west	794	0.68	4.6	885	...	185	143	15	7	92.0	94.7

[1] GOULD, RICHARD H., Sewage Aeration Practice in New York City, *Proc. ASCE*, Separate No. 307, October, 1953.

veloped, and at five other plants. *Modified aeration* or *high-rate activated sludge,* is also a modification of the activated-sludge process, designed to operate with a minimum of aeration.

Step aeration is a complete treatment process in which the activated sludge is aerated separately in the first stretch and the settled sewage is added, usually at the three quarter points, to the aerated channel (Fig. 44). This process applies the air as needed and results in a considerable saving in the air requirements. The aeration-tank capacity required is reduced to about 50 percent of the standard requirements, and a 30-min contact with sludge concentrations of 1,000 ppm in the aeration tank is sufficient.

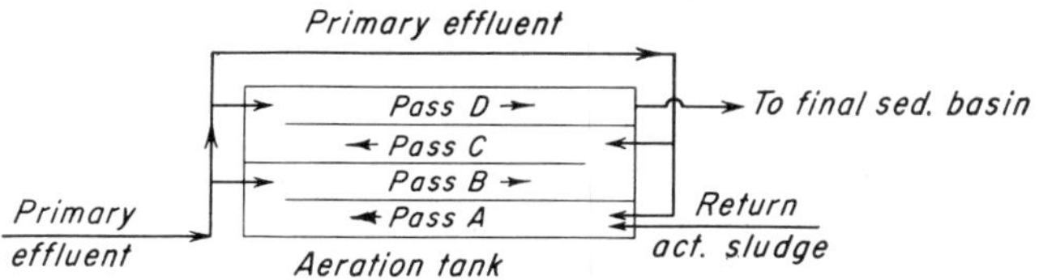

FIG. 44. Step aeration as developed at the Wards Island plant, New York City, showing sewage and sludge movements.

Sludge age is determined by dividing the weight of dry suspended solids in the aeration chamber by the daily dry weight of the incoming suspended solids. In well-operated plants, the average sludge age is 3 or 4 days, which is the average period required in the aeration tank for carrying the oxidation process to the point of stability. The sludge deteriorates if left longer in the plant.

Final settling tanks are designed to take advantage of density currents so that the sludge returned from them for further aeration with the incoming sewage will have an age of only about ½ hr, preventing undue sludge deterioration. The sludge age in the plant is generally from 3 to 4 days, determined by dividing the weight of dry suspended solids in the aeration chamber by the daily dry weight of the incoming suspended solids in the sewage. Beyond this point the sludge deteriorates. Thus the average solids concentration throughout the plant is about twice that in other treatment methods, which explains why the tank capacity need be only half of that required in conventional plants.

Modified sewage aeration gives intermediate results between plain sedimentation and the activated-sludge process. Removal of 80 percent solids and 74 percent BOD was a 9-year average accomplishment of this process at New York's Jamaica plant. Excess sludge pumped to digesters had 6.4 percent solids. No primary settling was used, but an aeration period of 1.5 hr supplied 0.41 cu ft of air per gallon of sewage.

Activated aeration is a third process developed at the New York City plant. In it, waste activated sludge is introduced into a plain aeration tank receiving raw sewage and the mixture is aerated 2.7 to 2.8 hr and passed on to a final settling tank. The sludge from the final settling tank is not reused but is concentrated and removed. In parallel operation at Wards Island plant, BOD removals of 95, 68, and 76 percent respectively were obtained with step, modified, and activated aeration, using air ratios of 0.61, 0.29, and 0.19 cu ft per gal, with aeration periods of 2.7, 2.7, and 2.8 hr respectively.

Chicago[1] has had to meet more stringent treatment requirements than has New York, and has been forbidden by the courts to discharge its effluent into Lake Michigan but must take it some 325 miles through the Chicago Drainage Canal and the Des

[1] ANDERSON, NORVAL E., Sewage Aeration Practice in the Sanitary District of Chicago, *Proc. ASCE*, Separate No. 310, October, 1953.

Plaines and Illinois Rivers to the latter's junction with the Mississippi about 40 miles above St. Louis. The city has been permitted by the courts to add dilution water from Lake Michigan in limited amounts which will not lower the lake level below specified levels required by navigation between the lakes and power production at Niagara Falls. Judged by chlorinity, dissolved oxygen, and BOD data, pollutional effects are felt for about one-third of the canal's length.

Chicago's problems resulting from location and size have been studied through large-scale research at its large disposal plants. Data summarized in Table 40 give the results attained at its three largest units and indicate the high degree of BOD and suspended solids removals obtained. Every phase of the operations has been experimented with since any small percentage improvement results in large annual savings. Aeration periods have been reduced at the West-southwest plant from 6.25-hr to 3.1-hr periods on the mixed-liquor basis or 4.0-hr on a sewage-only basis and without sludge reaeration. Air ratios of 0.48 to 0.68 cu ft per gal are very low. Circular final settling tanks were used because of good settling characteristics and economies in construction costs when compared with rectangular tanks. Test data indicated that tank depths should be not less than 10 ft or less than 12 ft if outlet weirs are located near the upturn of density currents; no baffling of influent had any appreciable effect on the effluent; low influent entrance velocity should be used; the overflow rate for effluent weirs should not exceed about 20,000 gpd per ft of weir, or 15,000 gpd per ft if weirs are located in the end wall of the settling tank at upturn of density currents. The sludge drawoff should be near the influent.

The fact that the Chicago plants are so large has added useful information on whether undesirable effects arise from using more than the usual length, width, or depth of channels. This information indicates that aeration channels 33.5 ft wide were as satisfactory as those of 16 ft width. With reference to length of aeration channels, at the West-southwest plant, 4 passes of 434-ft length and 33.5-ft width made a total length of 1,736 ft but gave satisfactory operating results. In the aeration channel where there is no problem of sedimentation, the important factor is time of detention and not distance of flow. Longitudinal fillets along bottom and top of side walls in the aeration chamber prevent the accumulation of sludge which would become septic and interfere with the oxidation process in the tank. A rotational movement around the tank's central axis aids the intimate contact of oxidative microorganisms and the sewage solids which are oxidized by them. Surface of contact, which controls the efficiency of the oxidation process, increases as bubble diameter decreases, but too small a bubble reduces the velocity of its rise, thus reducing its lifting power and causing septicity from too long a contact which exhausts the oxygen supply.

The activated-sludge process is very sensitive to changes in loading, quantity of air supplied, amount and freshness of the returned sludge, and other factors. The character of the sludge gives the most useful check for controlling plant operation. Several sludge indexes have been used for controlling the operation of the process. *Mohlmann's index* is one of the most widely used and is expressed as follows:

$$\text{Mohlmann index} = \frac{\text{volume of sludge settled in 30 min (percent)}}{\text{suspended solids (percent)}} \qquad (17)$$

A value of 100 or lower indicates a sludge that will settle satisfactorily; a rising index indicates approaching trouble from poorer settling characteristics. A value of 200 indicates poor settling characteristics.

SLUDGE

After the solids have been separated from the sewage in sedimentation basins, their final disposition involves many problems. These solids are called sludge and contain about 65 percent volatile solids which support a rapid development of microorganisms. These carry out the processes of reduction or oxidation of the organic matter, depending on the lack or availability of atmospheric oxygen. Through these biochemical processes, the organic solids are digested or broken down into simpler and more stable compounds, accompanied by the production of considerable quantities of gas. The solids can then be dried on sludge-drying beds or vacuum filters and disposed of in a number of ways, such as by burning or by utilization as a fertilizer or for filling low areas.

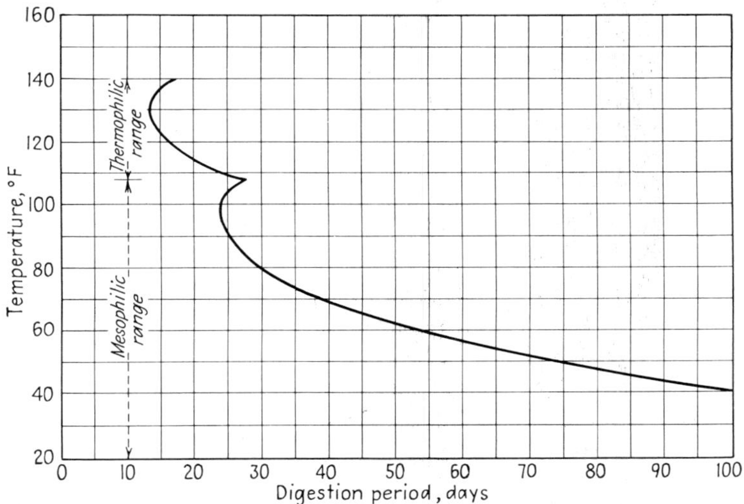

Fig. 45. Effect of temperature of the sludge on time of its digestion.

The anaerobic digestion of sludge was formerly carried out in the lower compartment of Imhoff tanks and is now accomplished in separate sludge-digestion tanks. The first step in this process is acidic and produces objectionable odors, with some liquefaction of solids. This inefficient and temporary acid condition is shortened or eliminated by adjusting the pH to 7.0 to 7.3 by the addition of lime or other suitable material. Other important factors affecting digestion efficiency are seeding and temperature. New digesters are aided in the establishment of satisfactory digesting conditions by seeding with well-digested sludge when obtainable. This inoculates the fresh, or green, sludge with organisms which, at suitable temperatures, can digest the material in 2 to 4 weeks instead of the 3 to 6 months required when unseeded. Thorough mixing of the fresh with the digested sludge is necessary for high efficiencies.

Although very little digestion occurs at or below 40°F, it speeds up rapidly with increasing temperatures. Figure 45 shows the effect of temperature on digestion. *Mesophilic* digestion is most effective at 85° to 95°F, while at a higher temperature (about 130°F) *thermophilic* digestion cuts the required time to 50 percent of that needed for *mesophilic* digestion.

Two-stage digestion (Fig. 46) is advantageous in controlling the pH, temperature, and operating techniques and in increasing the overall efficiency. Mixing the incoming

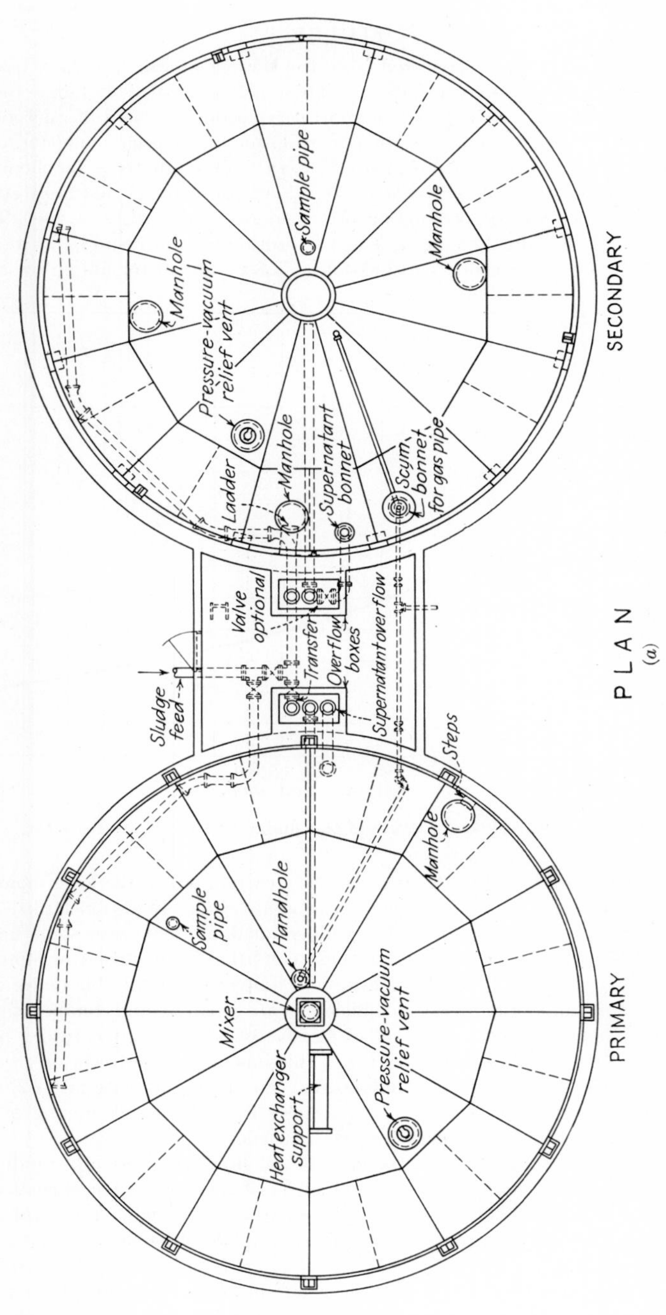

PRIMARY
SECONDARY
Sample pipe
Handhole
Mixer
Heat exchanger support
Pressure-vacuum relief vent
Manhole
Steps
Sludge feed
Valve optional
Transfer
Overflow boxes
Supernatant overflow
Ladder
Manhole
Supernatant bonnet
Scum bonnet for gas pipe
Pressure-vacuum relief vent
Manhole
Sample pipe
Manhole

PLAN
(a)

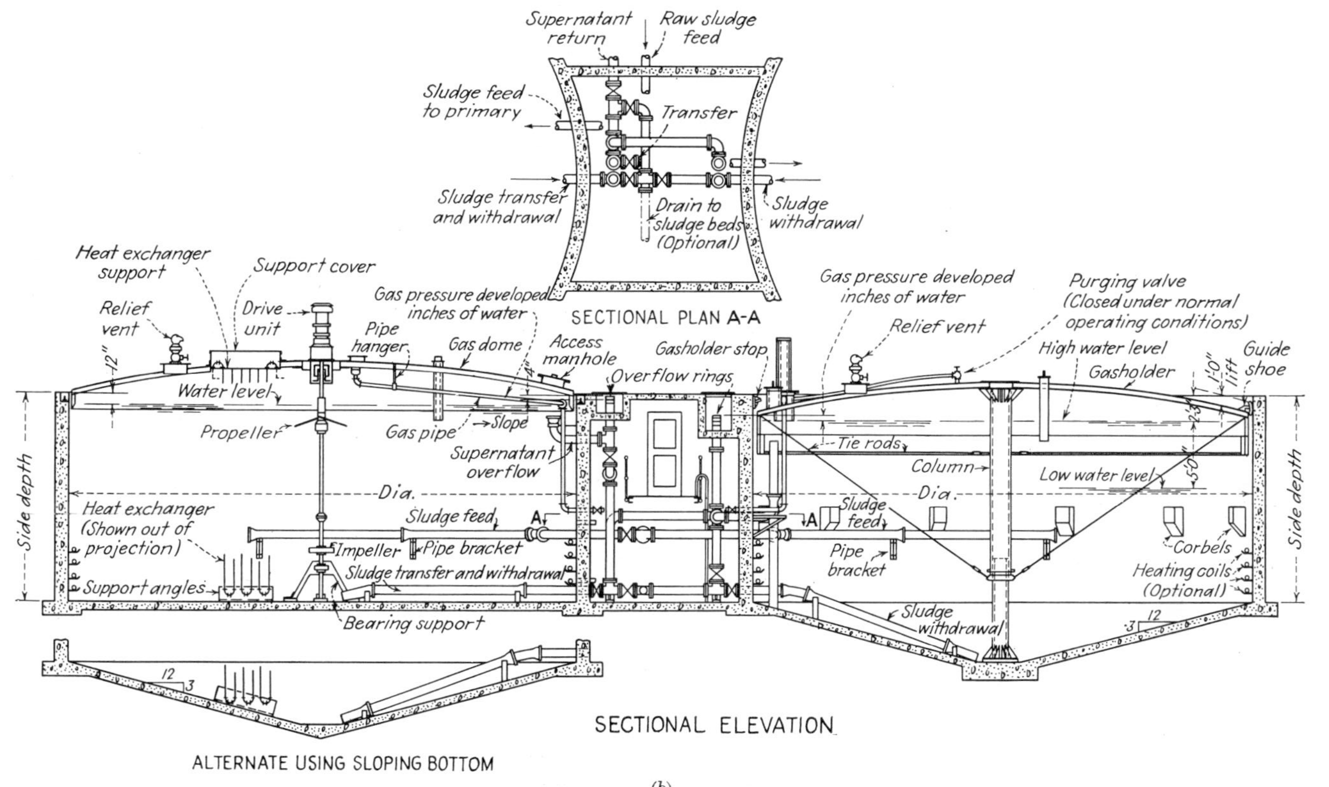

(b)

FIG. 46. Two-stage sludge digesters (Dorr type).

sludge with the partly digested sludge, adjusting the pH and heating to the optimum temperature are confined usually to the primary digester while the secondary unit permits aging and storage of the sludge and is frequently equipped with a gas-holder for storage of the gas produced in the digestion process.

The digester may be heated by hot-water coils immersed in the sludge, placed vertically or spirally around the inner wall of the digester. Circulating the supernatant or the sludge through heat exchangers has been used also as a means of heating the digester. The heating water enters the coils at 120° to 130°F and leaves at about 100°F.

Supernatant. The liquid which separates from the sludge during digestion is called *supernatant.* It is foul and malodorous and may have a BOD of 2,000 to 3,000 ppm. The suspended solids should not exceed 300 to 500 ppm. This supernatant is drawn off from the digester as fresh sludge enters the latter, through pipes arranged to take it at various tank depths, so that the operator can select that depth where the liquor has a minimum of solids. It is returned to the primary settling tank in frequent small quantities to make way for relatively continuous transfers of sludge from the settling tanks to the digester and to avoid the production of septic conditions in the primary clarifier by the acquisition of too large a volume of high BOD material at any one time.

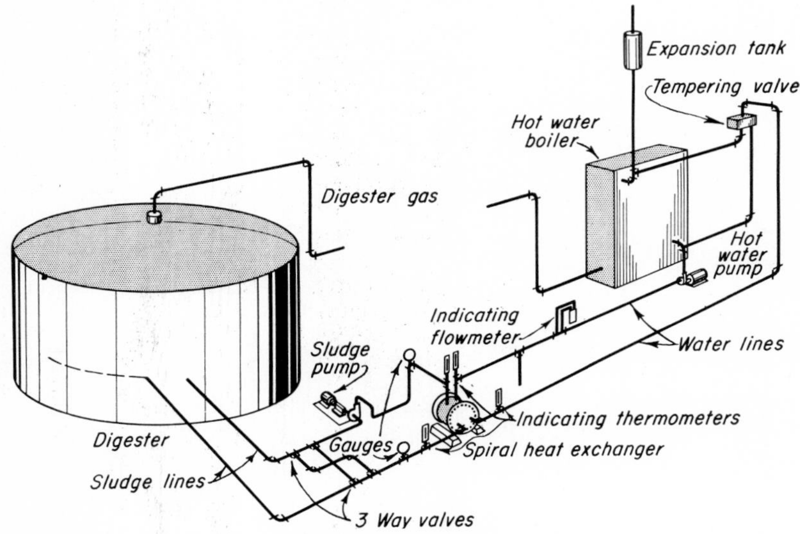

Fig. 47. General arrangement of spiral heat exchanger, boiler, and digester. (*Dorr Co.*)

An improved method for heating the sludge digester, developed by Dorr Co., is shown in Fig. 47, in which, instead of the large coils heretofore used in this operation, a spiral heat exchanger is used, which consists of two concentric spirals welded together throughout their length to make alternate channels for the circulating hot water countercurrent to the sludge liquor from the digester. Sludge liquor is continuously circulated through one spiral, while hot water is pumped countercurrently through the other. The operation is continuous with design capacity based on daily demands rather than for shorter time cycles. The exchanger is designed to raise the temperature of the sludge liquor 10°F while the water temperature drops a similar amount. The most practical temperature for satisfactory and economic sludge digestion is in the mesophilic range, preferably at 85°F.

Normally, sludge remains in the primary digester from 20 to 30 days and in the secondary unit from 30 to 45 days.

Gas Production. As has already been mentioned, anaerobic decomposition is accompanied by the production of considerable quantities of gas. Gas from sludge digesters is usually composed of about 75 percent methane. A typical analysis of Imhoff gas from the Calumet plant,[1] Chicago, yields 74.4 percent methane, 14.0 percent CO_2, 11.1 percent nitrogen, and 0.3 percent oxygen. Imhoff gives the gas production of sludge as 700 liters per kilogram, or 11.2 cu ft per lb of dry organic solids. The digestion process is usually carried only far enough so that offensive odors are absent and the sludge dries easily. This requires about 50 percent reduction in organic solids with 75 percent of the ultimate possible gas production. Each part per million of organic or volatile solid would then produce $100 \times 8.3 \times 0.75 \times 11.16/1{,}000{,}000 = 0.00694$ cu ft per capita for 100 gal per capita daily sewage flow. A sewage with 100 ppm volatile settling solids of which 80 percent are removed by tank treatment should then produce $100 \times 0.80 \times 0.00694 = 0.55$ cu ft per capita per day.

Table 41 gives the amount of gas produced in cubic feet per capita per day at a number of American and European plants. Optimum conditions for gas production occur at a pH of 7.0 to 7.6 and a temperature of about 90° to 95°F. Such gas has a heating value of 700 to 800 Btu per cu ft.

TABLE 41. GAS PRODUCTION AT SEWAGE-TREATMENT PLANTS[1]

Plant	Type of sewage	Type of sludge	Temperature of digestion, °F	Cu ft of gas per capita per day	Volume of gas per lb volatile solids, cu ft	Volume of gas per lb total dry solids, cu ft
Springfield, Ill.	Domestic	Primary plus waste activated	97	0.71	9.7	
Cedar Rapids, Iowa	Domestic plus trade waste	Primary plus humus	97	2.5	9.7	
Durham, N.C.	Domestic plus trade waste	Primary plus waste activated	82	1.42	9.33	6.6
Peoria, Ill.	Domestic plus trade waste	Primary plus waste activated	90	....	8.3	
	Beer slop	Beer-slop solids	114	....	9.2	
Aurora, Ill., 1931–1936	Domestic plus trade waste	Primary	84	....	13.3	9.3
Phoenix, Ariz.	Domestic	Primary plus waste activated	88	0.77		
Lima, Ohio	Domestic	Primary plus waste activated	...	1.00		
Fitchburg, Mass.	Domestic	Primary	77	....	11.2	8.4
Bury, England	Domestic plus trade waste	Primary plus waste activated	51–60	0.71	8.08	
Mogden, England	Domestic plus trade waste	Primary plus waste activated	85	1.0		
Nuremberg, Germany	Domestic	Primary	80	0.88		

[1] HODGSON, H. J. N., "Sewage and Trade Wastes Treatment," p. 287. Adelaide, Australia, 1938.

Gas is collected and used for power purposes or as fuel for heating the sludge. Gas collectors are provided which will eliminate all possibility of the admission of air, since mixtures of gas and air are explosive with the proportion of gas to air varying from 5 to 50 percent. Several serious explosions in sludge-digestion tanks have occurred; and since it is possible that sparks may be produced, which will ignite the gas, safety from such explosions lies in so designing the plant as to prevent the formation of explosive mixtures. When digestion chambers are drained for repairs, they should be generously ventilated before laborers are permitted to enter.

[1] GOODMAN and WHEELER, *Sewage Works Jour.*, vol. 1, p. 450, 1929.

Table 42 gives the approximate quantities of sludge produced by the different methods of sewage treatment, together with the proportion of solids and the specific gravity of the sludge.

TABLE 42. NOMINAL VOLUME OF SLUDGE PRODUCED BY DIFFERENT PROCESSES OF SEWAGE TREATMENT

Treatment process	Sludge			
	Normal volume per 1,000,000 gal of sewage treated, gal	Proportion of solids, percent	Specific gravity	Weight of dry solids per 1,000,000 gal of sewage treated, lb
Activated sludge	13,500	1.0	1.005	2,250
Chemical precipitation	5,000	7.5	1.070	3,300
Primary sedimentation	3,500	5.0	1.008	1,490
Digested sludge	800	15.0	1.040	1,050
Secondary-tank sludge	530	8.0	1.014	360

This table shows the excessive quantity of sludge produced by the activated-sludge method and the very large quantity resulting from chemical precipitation. The volume of sludge depends on the percentage of solids removed, the amount of solid material added, as in chemical precipitation, the moisture content, and the stage of digestion of the sludge. Of these the moisture content has the greatest influence on volume. For example, when 1 cu yd of sludge with 90 percent moisture content is dried on sludge beds to 60 percent moisture, the volume becomes $\frac{100 - 90}{100 - 60} \times 1$ cu yd $= 0.25$ cu yd. The importance of concentrating the sludge as much as possible before final dewatering or disposal is thus apparent.

In the digestion of sludge, about 30 percent of the solids are converted into gaseous and liquid form. Well-digested sludge contains less water than that which is only partly digested, is less odorous, is more easily dried, and remains stable after drying. Primary sludge is more readily digested anaerobically than activated sludge. The latter may be digested either alone or added to digested primary sludge if the proportion of activated sludge is not too great. Sludge digestion is usually carried out only to the point where offensive odors are absent and the sludge is dried easily. As has been previously mentioned, this requires about 50 percent reduction of the organic solids. With primary sludge, about 75 percent of the gas production for complete digestion has taken place.

TABLE 43. RECOMMENDED CAPACITIES FOR SLUDGE-DIGESTION TANKS[1]

Type of sewage treatment	Digestion-tank capacity, cu ft per capita	
	Heated tanks	Unheated tanks
Plain sedimentation	2.0	3.0
Plain sedimentation with chemical precipitation	3.0	4.5
Standard trickling filters	3.0	4.5
High-capacity filters	3.0	4.5
Activated sludge	4.0	6.0
Contact aeration	3.0	4.5

[1] "Engineering Manual," Chap. VII, Office of Chief of Engineers, March 1942.

The capacity of sludge-digestion tanks averages about 1 to 1.4 cu ft per capita, which is equivalent to a digestion period of 2 to 3 months. Any increase in the rate of digestion reduces the required capacity of the digestion tank in direct proportion. Imhoff states that $\frac{1}{4}$ gal of fresh solids per capita is contributed daily to disposal plants and that this is reduced to about $\frac{1}{20}$ gal during digestion. Reduction in weight of dry material is obviously less than the reduction in volume. Where extremely wet sludges such as that of the activated-sludge method are to be digested with primary sludge, an additional volume allowance up to 100 percent should be provided, depending on the water content of the sludge. The capacities recommended by Imhoff for sludge-digestion compartments are given in Table 43.

Kenneth Allen suggests a sludge-storage space of $10.5PD$ cu ft for combined systems with $5.25PD$ cu ft for separate systems, where P is the population in thousands and D is the sludge-detention period in days. This provides for a sludge of 80 percent moisture content.

Digestion of Garbage with Sludge. Current digester experience indicates that ground garbage can be satisfactorily digested with sewage sludge. Data on the quantity of garbage per capita indicate approximately a population equivalent of 0.84 as a fair daily average figure. This gives a 5-day BOD of 0.07 lb of oxygen per capita daily as compared with 0.2 lb for domestic sewage. Handling the complete garbage of a city would therefore increase the BOD of combined wastes about 30 percent.

Garbage can be ground at the plant or by individual household grinders built into the kitchen sink. Residences built during recent years have been thus equipped. Such *dual disposal* has increased rapidly, and has affected the various units of disposal plants differently.

Sludge Characteristics. The sludge from plain sedimentation is malodorous, unstable, slimy, and difficult to dry by the usual methods. That from chemical precipitation produces less objectionable odors, is somewhat jellylike in consistency, and does not dry to a granular condition on the sludge beds. Digested sludge has only a slight odor when dried and is more easily dewatered than primary-tank sludge. It has only slight fertilizing value. Sludge from secondary settling tanks has a slight odor and becomes offensive on standing. It is easily digested with primary-tank sludge, and this practice is usually followed. Activated sludge is light and flocculent when well activated, is nearly odorless, but may quickly become deoxygenated on standing so that anaerobic digestion begins. It has a very high water content and is difficult to dewater. Its fertilizing value is greater than that of other sludge, but to be used as a commercial fertilizer it must be dried to 10 percent moisture content.

Sludge Drying. Wet sludges may be concentrated by allowing them to stand before final dewatering. The drying process is hastened by conditioning the sludge. This consists in adjusting the pH and coagulating the sludge by the addition of chemicals. Sludge drying on beds is the most frequently used method for sludge dewatering at small plants. Sludge presses, vacuum filters, centrifuges, and heat driers are used, particularly at the larger plants and with the activated-sludge process which produces a sludge better suited to use as a fertilizer. Sludge beds consist of 12 to 20 in. depth of porous material, usually broken stone or gravel, covered with 4 to 6 in. of sand. The materials are graded with fine sand on top to keep the sludge from working down into the bed and to prevent the removal of filter material with the sludge when the bed is cleaned. Beds are usually underdrained by tile drains placed 5 to 15 ft apart. Temperature and rainfall affect the efficiency of sludge drying on open beds, and greenhouse construction is now generally used to make drying possible at all seasons. Such covers also assist in preventing the dissemination of odors and reduce the sludge-storage capacity required in the digestion chamber. Their greatest efficiency can be attained, however, only where the digestion process is accelerated by heating so as to remain practically constant throughout the

year. This is now general practice. One square foot per capita of sludge-bed area is ordinarily required for open beds, but this may be reduced by 35 to 50 percent for covered beds. The high temperatures in covered beds during warm weather may cause difficulties from accelerated biological activity of various sorts. Sludge presses are used with chemical precipitation and activated sludge. For the latter, high pressures and long-time increments are required because of the high water content. Sludge dried in presses may be of a gummy consistency difficult to handle with the usual equipment. Vacuum filters of the Oliver type have long been used successfully at many plants. They consist of canvas-covered drums which revolve while partly immersed in sludge. A vacuum inside the drum makes the sludge adhere to the canvas so that it is carried by rotation above the surface of the sludge where the vacuum dries the thin sludge film to 70 or 80 percent water content. It is then removed by a scraper, and further drying, if desired, may be performed by heat driers. The latter consist of revolving metal cylinders with heat applied either internally or externally. They are called direct- or indirect-heat driers.

Table 44 gives data concerning the character of sludge dewatered at eight sewage-disposal plants. It also gives data on chemical and power costs for different methods of drying or incineration of the sludge.

Sludge is finally disposed of as filling for low areas or is given or sold to farmers for use as fertilizers. It has been less frequently buried or burned. The incineration of sludge has been receiving more attention during the last few years. Green, or undigested, sludge burns more readily than does digested sludge since it has a higher organic content. It is burned at about 1700°F and requires little additional fuel after the incinerator is in operation. The Dearborn plant has been in use since January, 1935. Chicago has used incineration at its Calumet plant for several years. Here the sludge is dried on vacuum driers followed by heat drying by vapor in a closed system. The dried sludge is then burned in suspension as in the burning of pulverized coal. Detroit is planning a plant to burn undigested sludge, scum, shredded screenings, and grit combined. It is seldom possible to derive an income from the sale of sludge sufficient to pay the cost of dewatering. The more important problem is usually to dispose of it rather than to obtain revenue from it.

SEWAGE CHLORINATION

Chlorine, which long has been used for sterilizing water supplies, also has been adapted to some of the problems of sewage treatment. It has been used (1) to sterilize the sewage and thus protect public water supplies and prevent the spread of disease, (2) to reduce the BOD, (3) to control odors at plants, (4) to postpone biochemical activity and thus minimize stream pollution immediately below the plant, (5) to condition sludge preparatory to dewatering, (6) with aeration to aid in grease removal, and (7) to inhibit sulfate-splitting bacteria and thus minimize acid formation with its detrimental effect on concrete in disposal-plant structures. The need for sterilization arises from the necessity for protecting water supplies downstream from the sewer outlet or for preventing contamination of neighboring shellfish areas. The bacteria are destroyed by oxidation of the cell protoplasm and by germicidal action. Chlorine is more conveniently applied in liquid form but was formerly used in the form of hypochlorite of lime. The economic dose is usually found to be that amount which will produce a residual chlorine content of 0.2 ppm after 10- or 15-min contact. The amount required depends on the *chlorine demand* of the sewage, and this demand varies with the concentration and freshness of the sewage. Table 45 gives the approximate quantities used. The larger quantities are required during the summer when biological activity is greatest, while the lesser values are usually sufficient for winter conditions. Only a brief period of contact is necessary, provision usually being made for 15 to 30 min storage after chlorination. It

TABLE 44. COMPARISON OF ESSENTIAL ITEMS THAT CONTROL DEWATERING COSTS[1]

Item No.	Description	Minneapolis-St. Paul, Minn.	Buffalo, N.Y.	District of Columbia	Hartford, Conn.	Springfield, Mass.	Winnipeg, Manitoba, Canada	San Francisco, Calif.	Annapolis, Md
	Basic Data								
1	Approximate population	805,000	600,000	800,000	150,000	130,000	255,000	145,000	25,000
2	Total sewage treated, million gal	39,286.7	51,262	39,568	5,200	5,799	9,572	2,940.6	703.8
3	Suspended solids, ppm	315	168	180	160	155	329	254	
4	Percentage removal in plant	74.6	34.8	51	66.5	58.1	54.6	70.5	
5	Type of sludge dewatered	Raw primary	Digested	Primary digested	Primary digested	Primary digested	Primary digested	Primary digested	Primary digested
	Percentage of volatile:								
6	Raw	62.1	62.9	66.2	66.3	73.4	69.7	82.0	
7	Digested	....	47.5	45.1	45.8	45.8	51.1	63.5	51.4
8	Elutriated	....	....	44.4	39.8	48.3	50.5	....	50.0
	Raw solids:								
9	Tons produced	35,993	18,961	15,000	2,505	3,184	8,769.5	2,748	548
10	Tons dewatered	35,993							
11	Lb per capita per year	89.5	63	37.5	33.4	49	69	37.9	43.8
	Digested solids dewatered:								
12	Tons	....	10,051	6,870	1,296	946	1,655.2	888	173.2
13	Lb per capita per year	....	33.5	17.1	17.3	14.6	13	12.3	13.8
	Filter operation:								
14	Days	351	306	220	156	220	182	197	145
15	Hr	39,919.1	6,320	4,705.1	994	2,277.5	5,407	1,682	710.5
16	Percentage solids in feed	9.35	9.11	6.5	6.8	11.4	10.0	3.15	9.0
	Percentage of chemical dose on solids:								
17	Ferric chloride	1.53	2.93	3.18	2.23	1.63	2.6	4.3	
18	Lime (CaO)	3.77	11.21						
19	Ferric sulfate	....	....	....	....	....	....	....	2.86
20	Filter yield	3.9	6.77	5.9	7.5	5.0	4.2	5.4	7.5
21	Power cost, cents per kwhr	1.163	0.73	0.63	2.32	1.9	0.625	1.1	3.5
22	$FeCl_3$ per ton	$38.21	$31.80	$39.00	$71.40	$72.00	$83.60	$60	$99.6
23	Ferri-floc, per ton	....	....	....	....	....	....	....	$36.10
	Total Cost of Sludge Filtration and Sludge Disposal by Incineration or Hauling (Cents per Capita Served by Plant)								
24	Filtration	9.64	....	3.98	4.25	5.7	5.4	7.0	7.2
25	Incineration or disposal	6.04	....	2.61	1.40	3.22	1.62	1.7	0.0
26	Total	15.68	15.63	6.59	5.65	8.92	7.02	8.7	7.2
	Power Consumption for Filtration and Filter Hours								
	Power used:								
27	Total kwhr annually	751,889	....	302,300	48,750	90,000	96,309	63,000	3,695
28	Kwhr per 1,000 population per year	934	....	378	325	692	378	434	148
29	Filter hr per 1,000 population per 100 sq ft of filter area per year	248	52.7	29.4	23.2	28.9	31.8	23.2	18.5

[1] GENTER, A. L., Dewatering, Incineration, and Use of Sewage Sludge, A Symposium, *Trans. ASCE*, vol. 110, p. 246, 1945.

is common practice to chlorinate the plant effluent as the last step in the treatment process.

TABLE 45. CHLORINE REQUIREMENTS FOR THE STERILIZATION OF SEWAGE [1]

Nature of Sewage	Chlorine Required, ppm
Crude or settled sewage, fresh to stale	5–15
Septic sewage or effluent	10–25
Filter effluent, influent to final settling tank	2–5
Activated-sludge effluent	1–3.5

[1] ENSLOW, Advances and Developments in Sewage Chlorination, *Eng. Inst. Can. Eng. Jour.*, vol. 12, March, 1929.

The reduction in BOD is approximately 2.0 ppm for each part per million of chlorine used. Chlorine assists in eliminating odors by reducing the production of hydrogen sulfide gas from the sulfates present. Its germicidal action does not appear to be detrimental to the digestion of sludge or the operation of sewage filters, both of which are largely biological in their action. The small residual chlorine content is easily absorbed by the high chlorine demand of the surface sludge or largely dissipated when sprayed from sewage nozzles. The equipment used for applying chlorine is the same as that used for water treatment. It is simple, relatively inexpensive, and capable of handling the larger quantities which are required in the treatment of sewage. It has many other uses also, including its use with air in grease removal by aeration, prevention of foaming, and improvement of sludge index.

INDUSTRIAL WASTES

The liquid wastes produced by the various industries are as varied as the character of the industrial processes from which they originate. Those which are of sanitary significance in that they place an oxygen demand on disposal plants or streams or contaminate streams from a biological standpoint are highly concentrated as to solids and oxygen demand. Their nature and the need for treatment can be estimated from Table 46 which gives the BOD and suspended-solids value for the industrial wastes most frequently encountered. It is to be noticed that different units are used for the different types of wastes. This has to be taken into account in determining relative concentrations. The total solids, suspended solids, and BOD are usually very high. For example, a tannery waste reported in the *Transactions of the American Society of Civil Engineers* [1] had 20,552 ppm total solids, 3,883 ppm suspended solids, and 3,876 ppm 5-day BOD. Many wastes are even more concentrated.

The methods employed in treating industrial wastes are generally adaptations of the methods already discussed except that chemicals are more frequently used for pH adjustment and coagulation. Microorganisms such as yeast may be employed to utilize any sugar present. The presence of grease, even in small quantities, may prevent the use of sand filters because such filters are quickly clogged by the solidification of the grease in the voids. Acidification is useful in coagulating grease so that it can be separated from the waste.

The cost of treatment is usually large, and to offset this, at least partly, the reclamation of the marketable ingredients in the waste is frequently attempted. Special treatments have been devised for reclaiming some of these by-products, with varying financial success; but for some wastes no practical recovery process has as yet been worked out. Special knowledge of the industrial processes involved and of the disposal methods which

[1] Vol. 92, p. 1379, 1928.

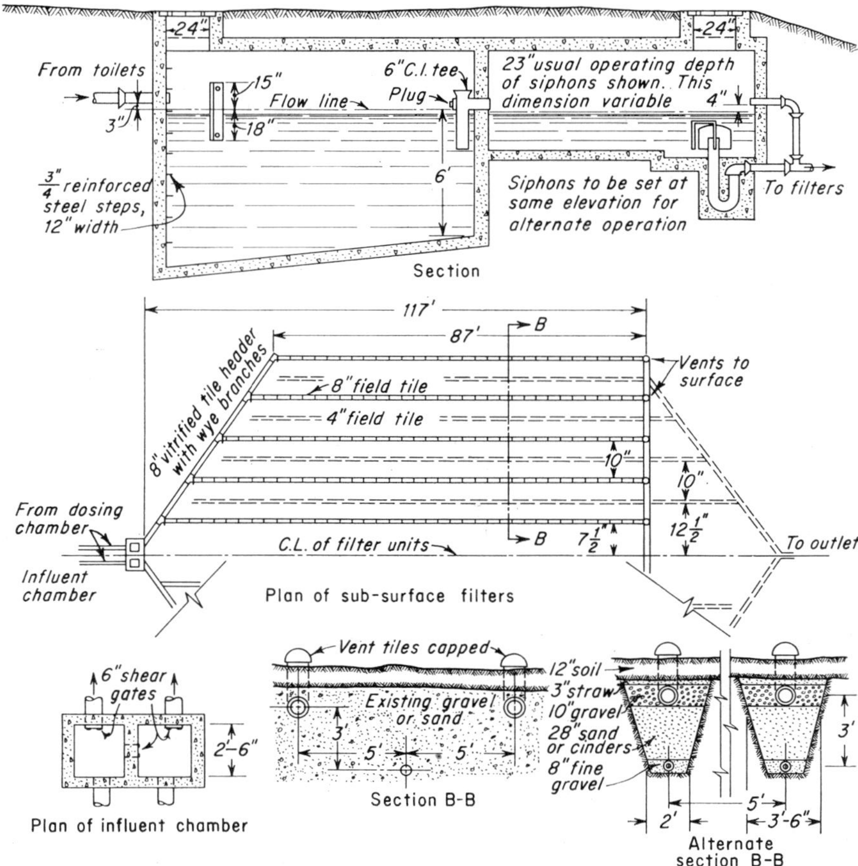

Fig. 48. Residential septic tank, dosing chamber, and disposal field. ("*Manual of Water Supply, Sewerage, and Sewage Treatment for Public Buildings in Ohio,*" *State Department of Health,* 1928.)

may be applicable is necessary for the solution of these problems. Each waste is a separate problem and may require special and costly treatment.

Where the waste is discharged into the city sewers to be disposed of with the sewage, it is desirable to express its strength in terms of that of the domestic sewage. This is usually done by determining the equivalent population which would place an equal load on the disposal plant by its contribution of domestic sewage. The BOD forms the best basis for comparison. For example, let it be required to determine the equivalent population for 1,000,000 gal of Packingtown waste. The 20-day BOD of this waste from Chicago experiments [1] is 1,010 ppm while that of Chicago domestic sewage is 165 ppm. The equivalent domestic sewage, therefore, would be $1{,}010/165 \times 1$ mgd = 6.11 mgd for 1 mgd of Packingtown waste. If figured on the basis of 228 gal per capita daily for the adjacent southwest-side district, the equivalent population would be 26,800.

[1] *Sanit. District, Chicago, Eng. Board Rev. Rept.,* 1925.

TABLE 46. SUMMARY OF WASTE DISCHARGES, SEWERED POPULATION EQUIVALENTS, AND EMPLOYEES PER UNIT OF PRODUCTION, WITH TYPICAL ANALYTICAL RESULTS, FOR VARIOUS INDUSTRIAL WASTES [1]

Industry	Unit of daily production	Employees per unit	Wastes, gal per unit	Typical analyses, ppm		Sewered population equivalents [2]		Remarks
				BOD	Suspended solids	BOD	Suspended solids	
Brewing	1 bbl beer	0.25	470	1,200	650	19	9	Spent grain dewatered
		0.25	470	800	450	12	6	Spent grain sold wet
Canning:								
Apricots	100 cases No. 2 cans	.....	8,000	1,020	....	410		
Asparagus	100 cases No. 2 cans	.....	7,000	100	30	35	9	
Beans: green	100 cases No. 2 cans	.....	3,500	200	60	35	9	
Lima	100 cases No. 2 cans	.....	25,000	190	420	240	440	
Pork and	100 cases No. 2 cans	.....	3,500	920	225	160	33	
Beets	100 cases No. 2 cans	.....	3,700	2,600	1,530	480	240	
Corn: cream style	100 cases No. 2 cans	45	2,500	620	300	75	30	
Whole kernel	100 cases No. 2 cans	.....	2,500	2,000	1,250	250	130	
Grapefruit: juice	100 cases No. 2 cans	.....	500	310 *	170 *	8 *	3 *	* Excluding peel bin wastes
Sections	100 cases No. 2 cans	.....	5,600	1,850	270	520	63	
Peaches–pears	100 cases cans	.....	6,500	1,340	....	440	....	Size of can unknown
Peas	100 cases No. 2 cans	1	2,500	1,700	400	210	40	
Pumpkin (squash)	100 cases No. 2 cans	.....	2,500	6,400	1,850	800	190	
Sauerkraut	100 cases No. 2 cans	.....	300	6,300	630	100	8	
Spinach	100 cases No. 2 cans	.....	16,000	620	....	490		
Succotash	100 cases No. 2 cans	.....	12,500	520	250	330	130	
Tomatoes: products	100 cases No. 2 cans	.....	7,000	1,000	500	350	150	
Whole	100 cases No. 2 cans	6.5	750	4,000	2,000	150	60	
Coal washery	1,000 tons coal washed	6	....	15	115,000			
Coke	100 tons of coal carbonized	8	360,000	85	....	1,500	....	Wastes cause tastes and odors
Distilling, grain:								
Combined wastes	1,000 bu grain mashed	40	600,000	230	360	3,500	2,300	Excluding intentionally discharged slop
Thin slop	1,000 bu grain mashed	.....	....	34,000	....	55,000		
Tailings	1,000 bu grain mashed	.....	....	740	....	50		
Evaporator condensate	1,000 bu grain mashed	.....	....	1,200	....	1,500		
Distilling: molasses	1,000 gal 100 proof	8	8,400	33,000	3,270	12,000	1,000	Molasses slop
Cooling water	1,000 gal 100 proof	.....	120,000					
Meat:								
Packinghouse	100 hog units of kill	30	550	....	....	77 *	25 *	* Paunch manure to sewer
		30	550	900	650	24	14	(1 cattle = 2½ hog units)
Slaughterhouse	100 hog units of kill	20	150	2,200	930	18	6	(2½ calves = 2½ sheep)
Stockyards	1 acre	.....	25,000	65	175	80	180	
Poultry	1,000 lb live weight	6	2,200	....	....	300	160	Average weight = 4.5 lb per animal
Milk:								
Receiving station	1,000 lb raw milk and cream	0.15	180	500	....	4	2	

Bottling works	1,000 lb raw milk and cream	0.89	250	.....	.....	6	3	
Cheese factory	1,000 lb raw milk and cream	0.38	200	100	750	16	9	
Creamery	1,000 lb raw milk and cream	0.16	110	1,250	660	6	3	
Condensery	1,000 lb raw milk and cream	0.47	150 *	1,300	750	7	4	* Excluding vacuum pan water
Dry milk	1,000 lb raw milk and cream	0.39	150	480	.....	6	3	
General dairy	1,000 lb raw milk and cream	1.09	340	570	540	10	5	
Oil field	100 bbl crude oil	1.3	18,000					
Oil refining	100 bbl crude oil	3	77,000	20	50	60	120	1 bbl = 42 gal
Paper:								
Paper mill	1 ton of paper	4.4	39,000	19	452	26	520	No bleaching
		4.6	47,000	24	156	40	220	With bleaching
Pasteboard	1 ton of paper	2.1	14,000	121	660	97	445	
Strawboard	1 ton of paper	1.4	26,000	965	1,790	1,230	1,920	
Deinking	1 ton of paper	.....	83,000	300	.....	1,250	.....	Old paper stock
Paper pulp:								
Groundwood	1 ton dry pulp	2.5	5,000	645	.....	16		
Soda	1 ton dry pulp	3.0	85,000	110	1,720	460	6,100	
Sulfate (kraft)	1 ton dry pulp	.....	64,000	123	.....	390		
Sulfite	1 ton dry pulp	3.1	60,000	443	.....	1,330		
Tanning: vegetable	100 lb raw hides	7	800	1,200	2,400	48	80	
Chrome	100 lb raw hides	.....	.....	.....	.....	24	40	
Textile:								
Cotton:								
Sizing	1,000 lb goods processed	.....	60	820	.....	2		
Desizing	1,000 lb goods processed	.....	1,100	1,750	.....	96		
Kiering	1,000 lb goods processed	.....	1,700	1,240	.....	108		
Bleaching	1,000 lb goods processed	.....	1,200	300	.....	17		
Souring	1,000 lb goods processed	.....	3,400	72	.....	12		
Mercerizing	1,000 lb goods processed	.....	30,000	55	.....	83		
Dyeing:								
Basic	1,000 lb goods processed	.....	18,000	100	.....	100		
Direct	1,000 lb goods processed	.....	6,400	220	.....	71		
Vat	1,000 lb goods processed	.....	19,000	140	.....	130		
Sulfur	1,000 lb goods processed	.....	5,400	1,300	.....	360		
Developed	1,000 lb goods processed	.....	14,400	170	.....	120		
Naphthol	1,000 lb goods processed	.....	4,800	250	.....	59		
Aniline black	1,000 lb goods processed	.....	15,600	55	.....	41		
Print works	1,000 lb goods processed	.....	4,500	95	.....	15		
Finishing	1,000 lb goods processed	.....	6	1,250	.....	0.4		
Rayon manufacture	1 cord wood distilled	35	680,000	30	.....	1,000	.....	Wood distillation process
	1,000 lb rayon produced	64	160	4.4	19	35	130	Cupra-ammonia process
		50	140	110	96	800	580	Viscose process
Rayon hosiery	1,000 lb hose produced	.....	9,000	330	.....	150	.....	Boil-off and dye wastes
Silk hosiery	1,000 lb hose produced	.....	13,700	1,720	.....	1,180	.....	Boil-off, dye, and finish wastes
Woolen mill	1,000 lb finished goods	.....	70,000	114	.....	400	.....	Scouring and dyeing, no grease wool
			240,000	125	.....	1,500	.....	Scouring and dyeing, 100 percent grease wool

[1] Ohio River Pollution Survey, Supplement D, p. 6, U.S. Public Health Service, 1942.
[2] Persons per unit of daily production.

Each industrial waste is a special problem, and general instructions concerning treatment are of little value. Its particular characteristics must be determined by tests and appropriate methods for handling the waste devised. In general, such wastes are considerably more concentrated than domestic sewage, and the amount of waste handled by the different types of plant will be correspondingly decreased. Most treatments

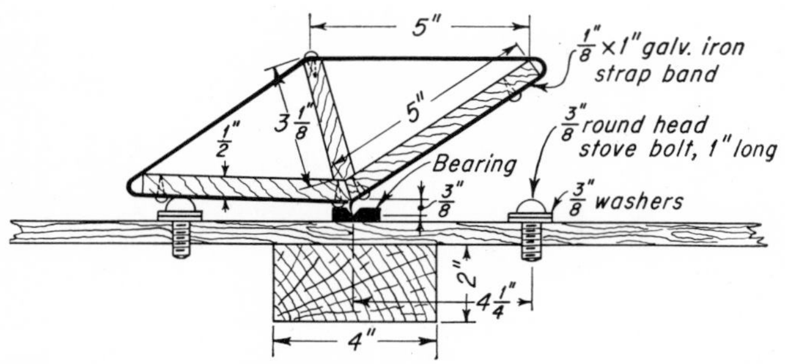

FIG. 49. Detail of tipping bucket used with lath filter. (*U.S. Public Health Service.*)

include sedimentation, because of the large quantity of suspended solids; and this is frequently followed by one of the oxidation processes. Where pathogenic organisms may be present, as in tannery wastes, chlorination of the effluent may be an essential feature.

RESIDENTIAL AND INSTITUTIONAL DISPOSAL PLANTS

Where sewers are not available, as in the case of isolated dwellings, schools, or industrial plants, the simplest method of providing the necessary sanitary facilities is to construct *privies* with or without watertight containers which may be removed or emptied at intervals as needed. The most satisfactory method for the disposal of excreta without water carriage is the *chemical toilet.* This usually consists of a seat attached to a metal cylindrical tank having a capacity of 100 to 125 gal per seat. The tank may be constructed of copper-bearing steel, which is resistant to the action of the chemicals used. A solution of 25 lb of caustic soda to each 10 or 15 gal of water sterilizes and liquefies the excreta in the tank and requires replacement only at intervals of several months. The content of chemical toilets when emptied is liquid, sterile, and practically free from odor. *Cesspools* have been commonly used in unsewered communities where water is available for household sewers. The cesspool is a pit which may be watertight or arranged so as to permit its contents to leach out into the surrounding soil. If wells are used for the water supply, they should be placed at least 300 ft up the slope from cesspools to avoid contamination. The adjacent soil gradually becomes clogged by the sewage solids, and occasional overflows of an offensive nature may occur.

Septic tanks are much more satisfactory for general use. They are watertight compartments made of concrete, brick, tile, or metal, in which the biological activity already discussed in connection with municipal septic tanks takes place. Figure 48 shows a residential septic tank with a siphon dosing chamber as recommended by the Ohio State Department of Health. The settling compartment should provide capacity for 24 hr sewage flow, except in larger plants where this may be reduced to as low as 8 hr. A minimum capacity of 200 to 250 gal for five persons is recommended for residential use, with 30 to 50 gal for each additional person. For schools, factories, and public buildings, this may be reduced to 15 or 20 gal per capita, with 5 gal additional allowance where

showers are to be installed. About 0.5 to 0.75 cu ft of sludge-storage space per capita should be provided. The inlet is baffled so as to distribute the flow throughout the tank, and the outlet is submerged in order to prevent the discharge of floating matter. Periodic but infrequent removal of the digested settling solids is necessary. The siphon chamber is usually omitted for small plants, but it is desirable in connection with secondary treatment for larger plants. It provides an intermittent dosage for filters or subirrigation systems, which increases the efficiency of such secondary treatment. Where sand filters are used for the treatment of tank effluent, they should have a capacity of 50,000 gal per acre per day and should be constructed and operated similarly to municipal plants. A

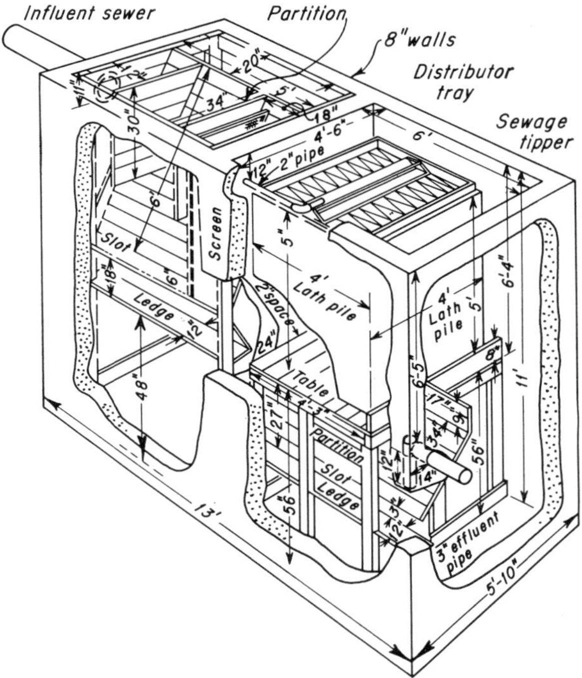

FIG. 50. Imhoff tank and lath filter for 10 people. (*U.S. Public Health Service.*)

convenient type of trickling filter which has been used successfully for schools and public buildings is made of lath, crossed and in layers, with a capacity of 0.1 cu ft of filter volume per gallon of sewage treated. Residential requirements are somewhat less, as the sewage is less concentrated. The laths are so arranged that sewage which is distributed over the top by being dumped on distributing boards or channels from a tipping bucket (Fig. 49) cannot fall through clear openings but trickles from one lath to another throughout the depth of the filter. Figure 50 shows a filter of this type following preliminary settlement in a small Imhoff tank such as is recommended by the U.S. Public Health Service. As in municipal plants, secondary settling tanks should follow trickling filters. They should have capacities about one-third that of the primary settling compartments.

Subirrigation is the common method of secondary treatment used with small septic tanks. It is accomplished by constructing parallel lines of 3- to 4-in. porous farm tile

6 to 10 ft apart and with 12 to 18 in. cover. From 20 to 50 ft of pipe per capita should be provided, depending on the porosity of the soil. The pipe should be laid with open joints and with a piece of tar paper over the joint to prevent entrance of soil into the pipe. The pipe is laid on a slope not to exceed 3 or 4 in. per 100 ft. Steeper slopes produce overloading at the lower end of the line. The total volume of the distribution pipes should be equal to or greater than the dosing-tank volume. A disposal area of this type depends on aerobic activity in the soil similar to that in intermittent sand filters, and pipes must be laid at a relatively shallow depth in order that sufficient oxygen may be present for this action. The pipe is frequently vented to the surface at the ends. On hillside locations, the tile should be laid back and forth across the slope on the usual grades but connected at alternate ends with short sections of pipe running down the slope. Figure 48 shows a disposal area in which underdrains have been provided. Where the soil is very tight, it may be necessary to lay the drainage tile in trenches filled with gravel or other porous material or in a bed of such material. Underdrains are provided at the bottoms of these filter trenches to remove the sewage which has filtered down through the ballast. From 1 to 5 lin ft of trench per capita should be provided. If the effluent of the secondary-treatment plant is discharged in such a manner as to produce possible contamination of water supplies, chlorination may be required. The apparatus used is similar to that for water-purification plants. Arrangements for raising the temperature of the feed water in cold weather may be required to prevent clogging of the chlorinating apparatus.

PUMPS

During recent years, improvements in pumps have made them of greater usefulness in sewage work. These improvements include increased reliability, which is of prime importance for such pumps; a greater ability to pass solids without clogging; and ability to stand up under the frequent starting and stopping necessary with automatically controlled pumps. Although reciprocating pumps were formerly used extensively, troubles from grit and sewage solids have encouraged the development of centrifugal

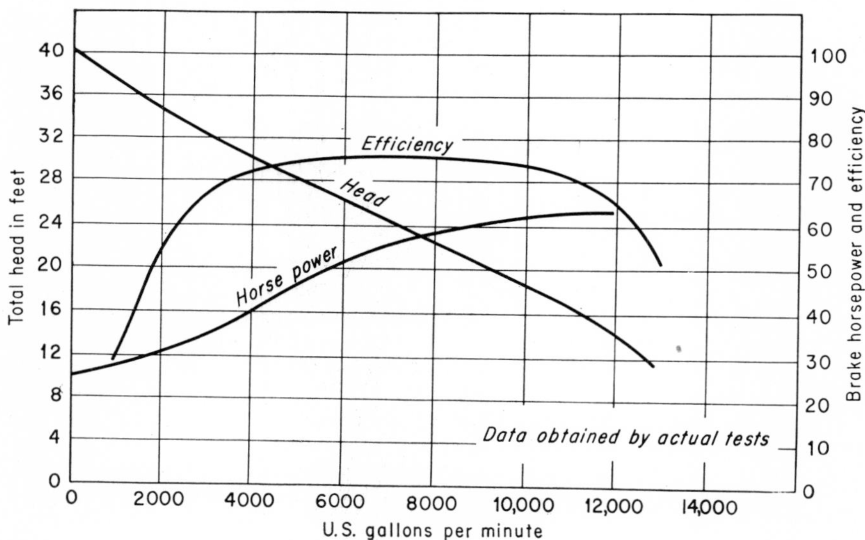

Fig. 51. Fairbanks-Morse sewage and trash pump characteristics.

pumps of the nonclogging type, as mentioned under "Lift Stations" (p. 9-29). These pumps have closed impellers with only one or two streamlined blades which permit the easy passage of large solids. Figure 51 gives the characteristics of a Fairbanks-Morse trash pump which is typical for this type of sewage pump. One noteworthy characteristic is the maintenance of a high efficiency over a wide range of operating conditions. Pumps of this type may be used also for the handling of sewage sludge. Data concerning sizes and capacities for these pumps are given in Table 47.

Table 47. Ratings of Fairbanks-Morse Sewage and Trash Pumps

Size of suction and discharge, in.	Capacities				Shipping weights,[1] lb	
	Minimum, U.S. gal per min	Normal, U.S. gal per min	Maximum, U.S. gal per min	Maximum diameter of solids pump will pass, in.	Vertical pump	Horizontal pump, belt driven
3	110	350	440	2	550	700
4	200	640	800	3	750	950
5	350	1,100	1,400	4	850	1,050
6	550	1,600	2,200	5	1,150	1,800
8	800	2,400	3,200	6	3,000	5,100
10	1,150	3,600	4,600	8	4,500	6,300
12	1,500	5,000	6,000	10	6,000	7,000
14	2,000	6,400	8,000	12	8,800	11,000
20	4,500	14,000	18,000	15	13,000	17,000

[1] Shipping weights are approximate and may vary slightly. Weights on the horizontal pump cover the bell-driven units without pulleys.

With centrifugal pumps, the discharge increases as the head decreases, but in greater proportion, so that the power required also increases. Care should be taken, therefore, to construct the plant so that the motor will not be overloaded under any possible operating conditions or emergencies. Screw pumps, which are sometimes used for handling storm-water drainage and which consist of a propeller revolving in a straight cylindrical section, do not exhibit this undesirable characteristic, though their discharge also increases with a decrease in head.

Pneumatic ejectors are frequently employed for lifting sewage from low-lying areas or deep basements into existing sewers. They are simple and dependable in operation and have few working parts to get out of order. In the Shone ejector illustrated in Fig. 52 sewage rises in the receiver until it reaches the desired level, at which the float D opens a valve E and admits air under pressure which forces the sewage out through the discharge pipe FG, while the back pressure closes a check valve B in the inlet pipe. As the sewage discharges, the float drops until at the empty position it closes the air valve and vents the receiver. Sewage then flows into the receiver again while the check valve F in the discharge line prevents the backflow of the pumped sewage. Ejectors may be used singly or in pairs. In single installations, the sewage is allowed to back up in the inlet pipe A during the time the contents of the receiver are being pumped.

For small or isolated sewage-pumping plants, automatic operation is usually employed. Float-operated switches cut the pumps in or out at any desired level of sewage in the sump or receiving pit. The motors and switchboard are located in a separate compartment either above the sump or on the same level. This provides safety against flooding which might result from excessive rainfall or interference with the power supply. The entire pumping station is frequently constructed underground in special manholes, and

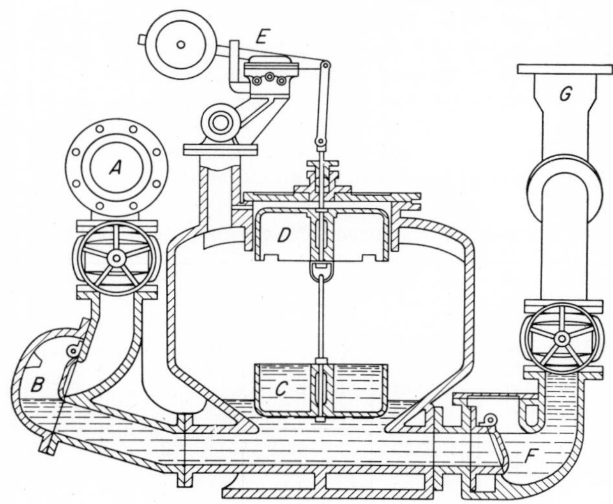

FIG. 52. Shone ejector. (*Yeomans.*)

it should operate with only a minimum of supervision. Care must be taken, however, that the compartment containing the motors and switchboard is adequately waterproofed, as it is essential that these be kept dry.

In the larger sewage-pumping stations, the equipment and operation are similar to those of stations pumping water. As already noted, the preference is for nonclogging centrifugal pumps rather than for the older type of reciprocating pump.

Bibliography

American Public Health Association: "Standard Methods of Water Analyses," 10th ed., 1955.

BABBITT, H. E.: "Sewage and Sewage Treatment," 7th ed., John Wiley & Sons, Inc., 1953.

BABBITT, H. E.: "Plumbing," 2d ed., McGraw-Hill Book Company, Inc., 1950.

EHLERS and STEEL: "Municipal and Rural Sanitation," 4th ed., McGraw-Hill Book Company, Inc., 1950.

FAIR and GEYER: "Water Supply and Waste-water Disposal," John Wiley & Sons, Inc., 1954.

FULLER and MCCLINTOCK: "Solving Sewage Problems," McGraw-Hill Book Company, Inc., 1926.

METCALF and EDDY: "American Sewerage Practice," 3 vols., McGraw-Hill Book Company, Inc., 1916–1928.

METCALF and EDDY: "Sewerage and Sewage Disposal," 2d ed., McGraw-Hill Book Company, Inc., 1930.

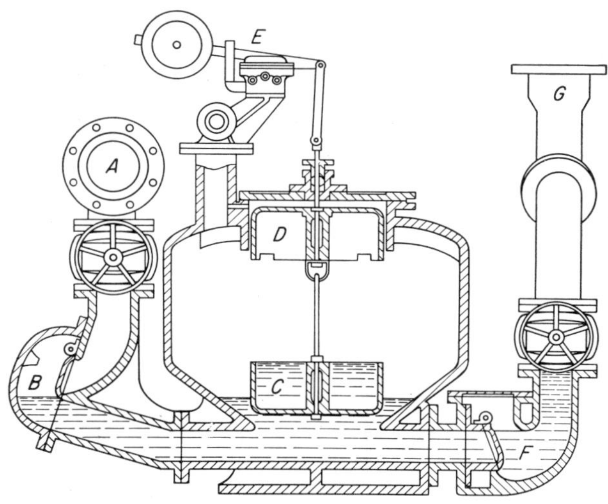

FIG. 52. Shone ejector. (*Yeomans.*)

it should operate with only a minimum of supervision. Care must be taken, however, that the compartment containing the motors and switchboard is adequately waterproofed, as it is essential that these be kept dry.

In the larger sewage-pumping stations, the equipment and operation are similar to those of stations pumping water. As already noted, the preference is for nonclogging centrifugal pumps rather than for the older type of reciprocating pump.

Bibliography

American Public Health Association: "Standard Methods of Water Analyses," 10th ed., 1955.

BABBITT, H. E.: "Sewage and Sewage Treatment," 7th ed., John Wiley & Sons, Inc., 1953.

BABBITT, H. E.: "Plumbing," 2d ed., McGraw-Hill Book Company, Inc., 1950.

EHLERS and STEEL: "Municipal and Rural Sanitation," 4th ed., McGraw-Hill Book Company, Inc., 1950.

FAIR and GEYER: "Water Supply and Waste-water Disposal," John Wiley & Sons, Inc., 1954.

FULLER and MCCLINTOCK: "Solving Sewage Problems," McGraw-Hill Book Company, Inc., 1926.

METCALF and EDDY: "American Sewerage Practice," 3 vols., McGraw-Hill Book Company, Inc., 1916–1928.

METCALF and EDDY: "Sewerage and Sewage Disposal," 2d ed., McGraw-Hill Book Company, Inc., 1930.

pumps of the nonclogging type, as mentioned under "Lift Stations" (p. 9–29). These pumps have closed impellers with only one or two streamlined blades which permit the easy passage of large solids. Figure 51 gives the characteristics of a Fairbanks-Morse trash pump which is typical for this type of sewage pump. One noteworthy characteristic is the maintenance of a high efficiency over a wide range of operating conditions. Pumps of this type may be used also for the handling of sewage sludge. Data concerning sizes and capacities for these pumps are given in Table 47.

TABLE 47. RATINGS OF FAIRBANKS-MORSE SEWAGE AND TRASH PUMPS

Size of suction and discharge, in.	Capacities				Shipping weights,[1] lb	
	Minimum, U.S. gal per min	Normal, U.S. gal per min	Maximum, U.S. gal per min	Maximum diameter of solids pump will pass, in.	Vertical pump	Horizontal pump, belt driven
3	110	350	440	2	550	700
4	200	640	800	3	750	950
5	350	1,100	1,400	4	850	1,050
6	550	1,600	2,200	5	1,150	1,800
8	800	2,400	3,200	6	3,000	5,100
10	1,150	3,600	4,600	8	4,500	6,300
12	1,500	5,000	6,000	10	6,000	7,000
14	2,000	6,400	8,000	12	8,800	11,000
20	4,500	14,000	18,000	15	13,000	17,000

[1] Shipping weights are approximate and may vary slightly. Weights on the horizontal pump cover the bell-driven units without pulleys.

With centrifugal pumps, the discharge increases as the head decreases, but in greater proportion, so that the power required also increases. Care should be taken, therefore, to construct the plant so that the motor will not be overloaded under any possible operating conditions or emergencies. Screw pumps, which are sometimes used for handling storm-water drainage and which consist of a propeller revolving in a straight cylindrical section, do not exhibit this undesirable characteristic, though their discharge also increases with a decrease in head.

Pneumatic ejectors are frequently employed for lifting sewage from low-lying areas or deep basements into existing sewers. They are simple and dependable in operation and have few working parts to get out of order. In the Shone ejector illustrated in Fig. 52 sewage rises in the receiver until it reaches the desired level, at which the float D opens a valve E and admits air under pressure which forces the sewage out through the discharge pipe FG, while the back pressure closes a check valve B in the inlet pipe. As the sewage discharges, the float drops until at the empty position it closes the air valve and vents the receiver. Sewage then flows into the receiver again while the check valve F in the discharge line prevents the backflow of the pumped sewage. Ejectors may be used singly or in pairs. In single installations, the sewage is allowed to back up in the inlet pipe A during the time the contents of the receiver are being pumped.

For small or isolated sewage-pumping plants, automatic operation is usually employed. Float-operated switches cut the pumps in or out at any desired level of sewage in the sump or receiving pit. The motors and switchboard are located in a separate compartment either above the sump or on the same level. This provides safety against flooding which might result from excessive rainfall or interference with the power supply. The entire pumping station is frequently constructed underground in special manholes, and

Section 10

WATER SUPPLY AND TREATMENT

BY HAROLD E. BABBITT

GENERAL

Components of a Waterworks. A complete water-supply system may include works for the collection, treatment, pumping, and distribution of the water from the original source to, and sometimes including, the consumer's meter. Works for the collection of water include intakes, dams, impounding reservoirs, aqueducts, wells, underground galleries, and the development of springs. Treatment works include aeration, screening, sedimentation, filtration, disinfection, softening, the removal of undesirable constituents, medication, and other devices. Pumping stations and their equipment comprise works for the pumping of water. Distribution includes pipes and appurtenances; distribution, storage, and equalizing reservoirs; and sometimes service connections, meters, and other equipment. The organization may include such divisions as administration, financial, design, operation, and maintenance.

In checking the adequacy of a waterworks consideration may be given to: (1) nature, adequacy, and reliability of source of water, and of each structure and appurtenance; (2) reserve capacity and possibilities of expansion; (3) hazards from floods, fires, and other causes that may result in damage to parts of the works; (4) availability of supplies, equipment, replacements, and spare parts; (5) preparation for emergencies; (6) capacity, materials, and characteristics of distribution system to supply normal and fire-protection demands; (7) quality of the water supply, protection of the source, and adequacy of treatment; and (8) character, competence, and attitude of management and of personnel, including security of employment and of financing.

Reliability and Fire Protection. A continuous and adequate supply of potable water is essential to the safety of life, health, and property in a community. Reliability is provided by an inexhaustible source or adequate storage; substantial structures and equipment; protection from or provisions in the event of unforeseen hazards; reserve capacity of essential equipment; and the duplication of structures and equipment whose breakdown might otherwise be disastrous. Fire protection is commonly an important service of a waterworks. It should be adequate in capacity and pressure, and free from reasonable possibility of interruption. More than half of the cost of small public waterworks, exclusive of treatment, may be chargeable to fire protection. The larger the community the smaller the proportion of cost chargeable to fire protection.

Legal Rights and Responsibilities. Among the general principles on which the rights of a waterworks to a source of supply are based are: (1) the right of eminent domain, because the supplying of water is a recognized activity of government, (2) priority to use of source if applied to beneficial public use, (3) ownership rights if source is owned

by the utility, and (4) governmental grant or franchise through legislative action or by competent administrative body. Some principles applicable to interstate waters are summarized[1] as: (1) equitable appropriation, (2) no superiority of right is based on priority of appropriation, (3) drinking and domestic uses impart the highest priority, (4) injunctive relief will be granted only on proof of present and substantial damages, and (5) diversion will be allowed between watersheds. Some principles applicable to underground waters are summarized[2] as: (1) absolute ownership, (2) ownership restricted to reasonable use on the land beneath which water is found, (3) ownership coequal with that of every other owner of land overlying the underground basin, and (4) ownership by the state.

The right to conduct a business in selling of water is not necessarily a governmental prerogative, and hence, a waterworks enterprise is not immune from suits for damages for such negligence as supplying polluted water or other reasonably foreseeable hazards.

QUANTITY OF WATER

Use of Water. Although the needs for water as a beverage and in fire protection often represent the principal incentives in the establishment of a waterworks the amount of

TABLE 1. RATES OF WATER USAGE

EFFECT OF POPULATION[1]

Population, 1,000's	Av Usage, gpcpd
0.5	60
1.0	85
5.0	135
10.0	140
50.0	140
100	140

RATES FOR VARIOUS PURPOSES[2]

Purpose	Rate	Unit
Office building	27–45	gpcpd
Hospitals	125–350	gp bed p day
Hotels	306–525	gp occupied room p d
Laundries	3–5	g per lb
Restaurants	0.5–4	g per meal
Single-family houses in Conn.	37–59	gpcpd

EFFECT OF METERS

Location	Year	Population, 1,000's	Gpcpd
Fayetteville, Ark.	1910	5	100
	1932	. . .	85[3]
San Francisco, Calif.	1910	417	85
	1920	508	71[3]
Reading, Pa.	1900	78	92
	1932	118	91[3]
Quincy, Mass.	1910	33	93
	1932	74	69[4]

INCREASE OF USE WITH TIME

Item	1890	1945
No. of waterworks	1,878	15,400
Population served, millions	22.8	94.39
Percent total in U.S.	36	71
Av use, gpcpd	90	127
Total usage, mgd	2,050	12,030

[1] LANGBEIN, W. B., *Jour. Am. Water Works Assoc.*, November, 1949, p. 997.
[2] JORDAN, H. E., *Jour. Am. Water Works Assoc.*, January, 1946, p. 65, except single-family houses.
[3] After 100 percent meters.
[4] Domestic use 100 percent metered. No data about industrial use.

water used for these purposes represents only a small part of the total volume of water used by a community. As a beverage less than 1 qt may be used daily by a consumer,

[1] STEVENSON, W. L., and C. E. RYDER, *Civil Eng.*, August, 1931, p. 991.
[2] CONKLING, H., *Trans. ASCE*, vol. 102, p. 753, 1937.

and fire fighting may use no more than this on the average. Total average daily demands in a few cities in the United States exceed 300 gal per capita, and a rough average of 100 gal per capita per day is often assumed in estimates. Maximum rates, for all except fire protection, may greatly exceed average rates; and the rate of use during large fires is the greatest rate of all. Some rates of demand for water are listed in Table 1.

Use for Specific Purposes. Water used for specific purposes can be expressed quantitatively in approximate percentages of the total use as: domestic, 35; industrial and commercial, 39; public, 10; and waste and miscellaneous, 16. Industrial usage sometimes exceeds all other usages.

Rates of Demand. Conditions affecting rates of demand include: population, climate, season, scale of living, pressure, quality, sewer facilities, air conditioning, and meters. No quantitative relation between these factors and rate of water usage can be expressed with accuracy but, in general, the rate is greater: (1) in large than in small communities, (2) in hot dry climates, (3) in high-value districts, (4) where the pressure is high, (5) where the quality is good, (6) where the cost is low, (7) where sewers are available, and (8) where meters are not used. Some quantitative relations between these conditions are shown in Table 2. An increase in pressure from 25 to 45 psi has been known to increase demand by 30 percent. The installation of meters has been shown to decrease usage by 30 to 40 percent.

TABLE 2. RATES OF USE OF WATER FOR SPECIFIC PURPOSES

Purpose	Use, gpcpd	Purpose	Use, gpcpd
Bare minimum	0.25	Waste and miscellaneous	50
Subsistence, min	1.5	Apparently unavoidable	20
Sanitary needs, min	10–20	Air conditioning	3–35
Domestic, reasonable min	30	Apartment house, av	15–60
Public use	10–20	Max	35–125

Pumping of water for less than 24 hr per day is practiced in some cities outside of the United States with the objective of decreasing the amount of water used. Such practice tends toward pollution of the water and, unless the duration of the periods of supply are brief in relation to the periods of nonsupply, the amount of water used may not be decreased.

Predictions of future rates of water usage must be based on knowledge of past and present rates. Pumping records, meter readings, population censuses, and other information must be used, and the advice of competent persons acquainted with the community should be enlisted. No formula or other mathematical procedure alone can be satisfactory.

Variations in Rate of Demand. Rates of demand for water in a community vary with time, as shown in Table 3. Summer and winter rates in temperate climates may be

TABLE 3. APPROXIMATE MAXIMUM[1] RATES OF DEMAND IN TERMS OF AVERAGE ANNUAL RATE

Period of time covered	Maximum rate as percent of av annual rate	Period of time covered	Maximum rate as percent of av annual rate
Annual, average	100	Day	155–180
Season	135	Hour	255–350
Month	145	Hour, minimum	65

[1] Except as noted.

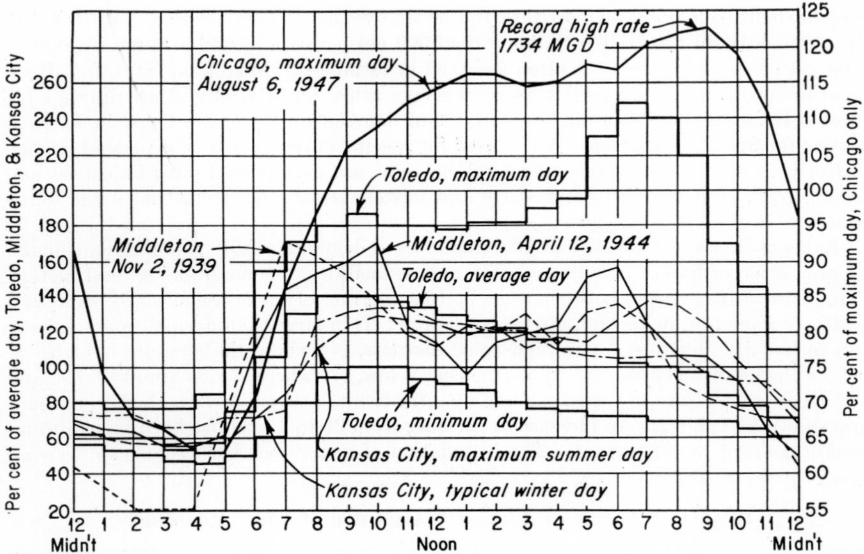

FIG. 1. Hydrographs of hourly pumpage at Kansas City, Mo. (*M. P. Hatcher, Water and Sewage Works, May*, 1947, *p.* 158), Middleton, Mass. (*Water Works Eng., July* 12, 1944, *p.* 822), and Toledo, Ohio (*Burns and McDonnell Engineering Co.*, 1937.) At Kansas City, Mo., average rate of pumping on maximum summer day was 84.7 mg, and on typical winter day it was 55.6 mg.

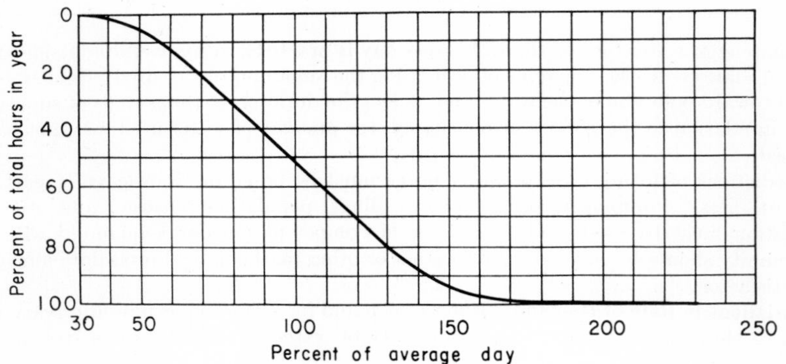

FIG. 2. Demand-duration curve.

above the annual average because in summer air conditioning, bathing, and irrigating increase the rate and in hard winters in unmetered areas water may be run to prevent freezing in pipes.

Typical hydrographs for daily water use are shown in Fig. 1. A typical demand-duration curve is shown in Fig. 2.

Fire-protection Needs. Fire-protection requirements represent a greater ratio to the total water needs in small than in large communities. A minimum of 4 fire streams, of 175 to 250 gpm each, concentrated at one point should be provided. A reasonable

minimum fire protection may be given by 1,000 gpm for 1,000 persons up to 12,000 gpm for 200,000 persons, with a maximum of 20,000 gpm. The pressure at the fire hydrant on the distribution system should not be less than 20 psi where mobile pumping engines are used, and 75 to 90 psi otherwise. Some recommendations concerning rates of water supply for fire protection are given in Table 4. It may be assumed that a fire will last for a minimum of 5 hr in the smallest community up to 10 hr or more in a large city.

TABLE 4. EMPIRICAL FORMULAS FOR RATES OF FIRE DEMAND

Name of originator or authority	Formula Q = demand, gpm P = population, 1000's	Gpm per 100,000 population
Kuichling (on basis of fire streams of 250 gpm)	$Q = 700\sqrt{P}$	7,000
John R. Freeman	$Q = 250\left(\frac{P}{5} + 10\right)$	7,500
National Board of Fire Underwriters	$Q = 1{,}020\sqrt{P}\,(1 - 0.01\sqrt{P})$	9,180

SOURCES OF WATER

General. Water sources may be classified as meteorological, surface, and underground. Only the latter two are satisfactory in quantity for most continental public water supplies. The close relationship between meteorological precipitation which is principally rainfall, and runoff and available surface water requires knowledge of rainfall conditions and rates.

Rainfall. Records of precipitation, commonly reported as rainfall records, include the amount of all forms of meteorological precipitation. Figure 3 shows a map of the United States with annual isohyetal lines. Some data on annual monthly rates of rainfall in

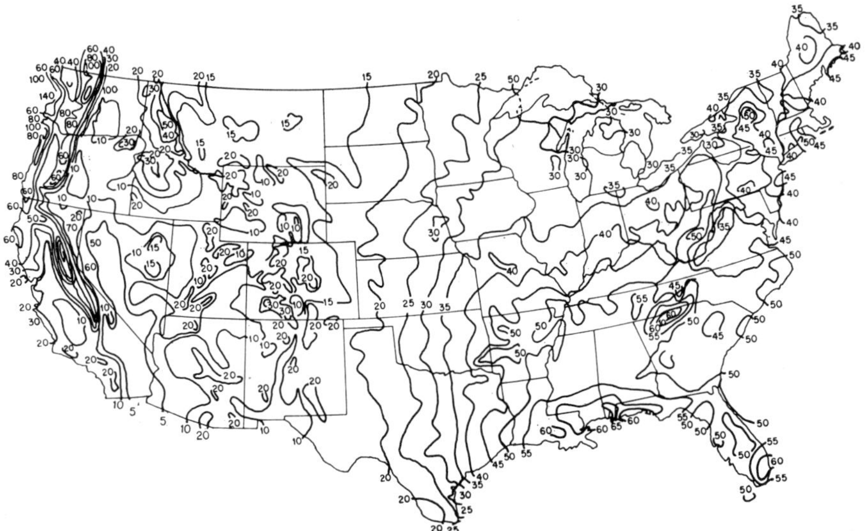

FIG. 3. Lines showing equal annual rainfall, or isohyetal lines, in the United States. (*United States Weather Bureau.*)

different parts of the United States are shown in Fig. 4. Rates of rainfall for short periods of time are greater than over long periods, as indicated by the formulas in Table 5.

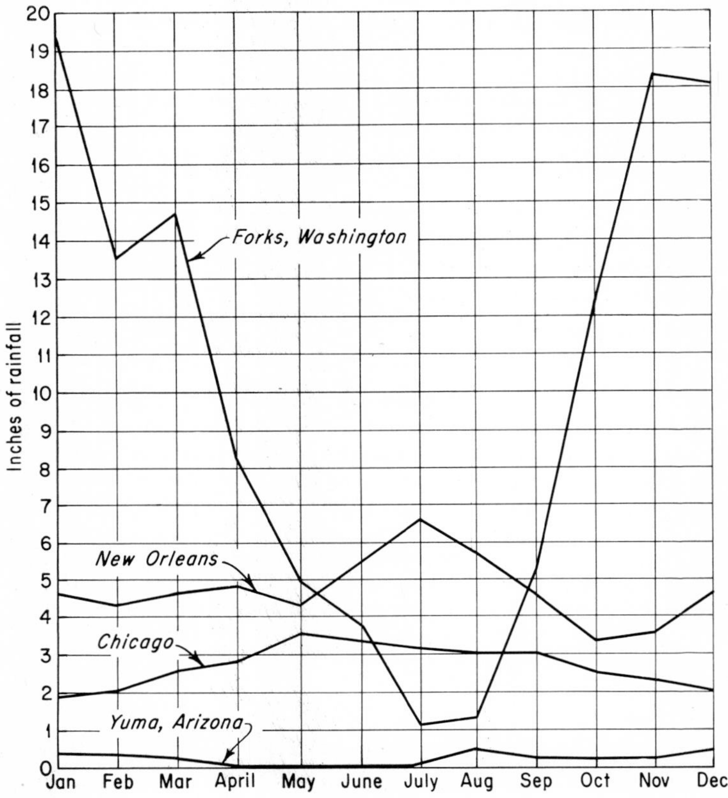

FIG. 4. Monthly rates of rainfall.

In areas near snow-covered mountains the rate of snow melt may be computed from the empirical expression

$$D = K(T - 32) \tag{1}$$

where D = average rate of snow melt, in. per day
$T - 32$ = number of degree-days above the °F melting point
K = melting constant or degree-day factor

Values of K lie between 0.02 and 0.11 in. per degree-day, with a maximum up to 0.3.

Meteorological precipitation is dissipated by evaporation, transpiration, consumptive use, interception, percolation, infiltration, and runoff.

Evaporation. *Evaporation* is that portion of the precipitation that is returned to the atmosphere from exposed particles or bodies of water. *Transpiration* is the quantity of water used per annum by vegetation in building up plant tissues, in transpiration, and

water evaporated from adjacent soil, snow, or intercepted precipitation. *Interception* is the amount of water caught annually by vegetation, or structures, which is evaporated

TABLE 5. FORMULAS FOR RATES OF RAINFALL

Name of author	Locality	Rate, in. per hr [1]
Talbot	Eastern United States	$360/(t+30)$
Talbot	Eastern United States	$105/(t+15)$
Kuichling	New York City	$120/(t+20)$
Le Conte	San Francisco	$7/\sqrt{t}$
Metcalf and Eddy	New Orleans	$19/\sqrt{t}$
Horner	St. Louis	$56/(t+5)^{0.85}$
Frequency, years	**Rainfall rate, in. per hr**	**Depth, in.[1]**
5	$41.22/t^{0.76}$	$0.687t^{0.24}$
10	$51.60/t^{0.77}$	$0.860t^{0.23}$
15	$54.90/t^{0.77}$	$0.915t^{0.23}$
25	$60.18/t^{0.77}$	$1.003t^{0.23}$
50	$69.06/t^{0.77}$	$1.151t^{0.23}$
100	$78.72/t^{0.77}$	$1.312t^{0.23}$

[1] t = duration of period of intense rainfall, in minutes.

before reaching the ground. No formula for evaporation from fresh-water surfaces is generally acceptable in waterworks practice. Some typical formulas are, however, listed in Table 6.

TABLE 6. RATES OF EVAPORATION FROM WATER SURFACES

Originator of formula	Source	Formula
Bigelow	*Argentine Meteorological Off. Bull.* 2, 1912	$e = 75.8\left(\frac{V}{v}\right)\frac{dV}{dT}(1 + w'/10)$
Vermuelé	*N.J. State Geological Survey, Rept.*	$e = (0.00417T - 0.233)(15.5 + 0.16R)$
Meyer	*Trans. ASCE*, vol. 79, p. 1074, 1915	$e = 15(V - v)(1 + w'/10)$
Cox	*La. State Univ. Bull.*, 1940	$e_1 = (e_s - e_d + 0.0016TD)/(0.564 + 0.051TD + w/7{,}200)$
Horton	*Eng. News-Record*, Apr. 26, 1917	$e = 290(c_1 - v)V$
Fitzgerald	*Trans. ASCE*, vol. 15, p. 581. 1886	$e = 0.2(V - v)(1 + w/2)$

Nomenclature:

c_1 = coefficient depending on the wind. For no wind = 1.0; at 10 mph = 1.86; 20 mph = 1.98; 30 mph = 2.00
e = monthly evaporation, in. of water
e_1 = evaporation from open water surface, in. per day
e_d = mean vapor pressure of saturated air at dew point, in. of mercury
e_s = mean vapor pressure of saturated water vapor at temperature of water surface, in. of mercury
R = yearly rainfall, in.
T = mean annual temperature, °F
V = maximum vapor tension for the temperature of the water surface
v = actual vapor tension in the air, in. of mercury
w = wind velocity close to the water surface, mph
w' = wind velocity by the Weather Bureau report from the nearest station, mph

The rate of evaporation from land surfaces is more difficult to express because of additional factors involved. The Thornthwaite[1] formula

$$E = \frac{17.1(e_1 - e_2)(u_1 - u_2)}{T + 459.4} \tag{2}$$

where E = evaporation, in. per hr
e_1 and e_2 = vapor pressure at lower and upper levels, respectively, in. of mercury
u_1 and u_2 = wind velocities at lower and upper levels, respectively, mph
T = temperature, °F

is typical. Information on consumptive use is given in Table 7.

TABLE 7. APPROXIMATE SEASONAL CONSUMPTION OF WATER BY CROPS AND VEGETATION

Growth	Inches	Growth	Inches
Coniferous trees	4 to 9	Alfalfa and clover	2.5 up
Deciduous trees	7 to 10	Corn	20 to 75
Potatoes	7 to 11	Oats	28 to 40
Rye	18 up	Meadow grass	22 to 60
Wheat	20 to 22	Lucern grass	26 to 55
Grapes	6 up	Rice	60 to 200

Runoff, Infiltration, and Percolation. Runoff occurs when surface detention has been satisfied and rate of rainfall exceeds simultaneous rates of infiltration and evaporation. *Infiltration* is the downward movement of water from the ground surface to the ground-water table. *Percolation* is the flow of ground water in the direction of the slope of the ground-water table. When the flow of ground water adds to surface-stream flow it is known as *base flow* or *sustained flow*.

Among the factors affecting runoff, in addition to rainfall are: size, shape, topography, and geology of watershed; climate, prevailing winds, and temperature; condition of watershed such as urban, rural, cultivated, and afforested; surface retention; tributaries and their regimen; antecedent storms and base flow; and infiltration capacity of the soil.

Infiltration rates for small watersheds may be formulated as

$$F_e = P - (y_s - S_e - F_i) \tag{3}$$

where P = rainfall causing runoff
y_s = mass of surface runoff attributed to P
S_e = effective surface storage
F_i = estimated amount of infiltration occurring before rainfall exceeded infiltration rate

Empirical formulas used for estimating average annual runoff are:

Vermuelé $$I = R - (11 + 0.29R/475)T^{1.6} \tag{4}$$

Justin $$I = 0.943(R^2/T)S^{0.155} \tag{5}$$

where I = annual runoff, in.
R = annual rainfall, in.
T = mean annual temperature, °F
S = slope of surface of drainage area

C. E. Grunsky[2] suggested the following rule of thumb: "The percentage of the seasonal (12 months) rainfall which appears in the stream as runoff, when the rain is less than

[1] See *Monthly Weather Rev.*, January, 1939, p. 6.
[2] *Trans. ASCE*, vol. 102, p. 1165, 1915.

50 in., is equal to the number of inches of rain. When the seasonal rain is in excess of 50 in., 20 in. thereof go to the ground (evaporate, etc.). The remainder is runoff."

Minimum Runoff. Minimum runoff cannot be successfully formulated or estimated. Knowledge of it can be gained only through a study of flow records of the stream in question or from records of runoff of a similar stream. Sherman's unit hydrograph provides a procedure for such a study. Runoff records are published periodically by the U.S. Geological Survey and by various state authorities.

A *unit hydrograph* is defined by Sherman as "the hydrograph of surface runoff resulting from rainfall within a unit of time, as a day or an hour." Such a graph is most useful in the construction of a *distribution* graph. This is a unit hydrograph of surface runoff modified to show the proportional relations of its ordinates to the total surface runoff. It is a curve whose abscissas coincide with those of the unit hydrograph, and each ordinate, for any particular abscissa, is the ratio of the ordinate of the unit hydrograph for that abscissa to the total runoff represented by the unit hydrograph, expressed preferably as a percentage. The computation of the coordinates of the distribution graph corresponding to the unit hydrograph in Fig. 5 is shown in column 5 of Table 8. Since

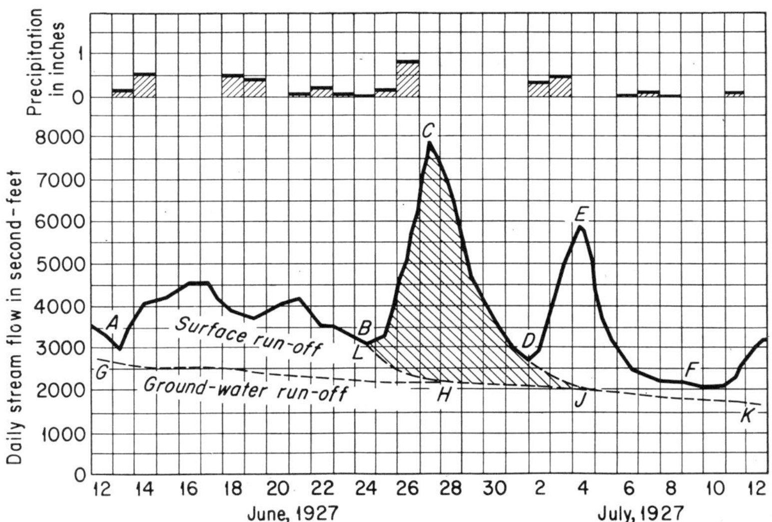

FIG. 5. Unit hydrograph for Muskingum River at Dresden, Ohio.

distribution graphs for different rainfalls on a watershed are not identical their similarity makes possible the drawing of an average distribution graph for the watershed on which can be based the prediction of the relation between rainfall and runoff. Typical distribution graphs are shown in Fig. 6. The construction of the unit hydrograph is based on that portion of a hydrograph which represents the runoff from a unit rainfall. A typical hydrograph is shown in Fig. 5. Corresponding rates of rainfall are shown near the top of the same figure. It is to be noted that to the right of letter B the total runoff is composed of three parts: (1) the ground-water contribution, designated by the line $GHJK$, (2) that portion of the surface flow between $GHJK$ and BHJ which represents the contribution of surface runoff independent of that contributed by the rainfall of June 24 to 26, and (3) the "net surface runoff," contributed by the rainfall of that period, represented by the distance between the lines BHJ and $BCDJ$. Line $BCDJ$ is a unit

TABLE 8. DERIVATION OF DISTRIBUTION GRAPH FOR THE MUSKINGUM RIVER BASIN ABOVE DRESDEN, OHIO [1]

(The calendar day on which most of the rain fell is marked #)

Date, 1927	Total runoff (sec-ft)			Surface runoff from unit storm (col. 2 or 3 − col. 4), cfs	Distribution graph (col. 5 × 100 ÷ summation of col. 5), percent
	Unit storm	Recession following unit storm	Recession preceding unit storm		
June 24	3,090	.....	3,090		
June 25	3,240	.....	2,600	640	3.4
June 26#	5,110	.....	2,350	2,760	14.5
June 27	7,990	.....	2,250	5,740	30.0
June 28	6,570	.....	2,200	4,370	22.9
June 29	4,630	.....	2,150	2,480	13.0
June 30	3,540	.....	2,100	1,440	7.5
July 1	2,940	.....	2,050	890	4.7
July 2	2,520	.....	2,020	500	2.6
July 3	.....	2,200	2,000	200	1.0
July 4	.....	2,050	1,980	70	0.4
July 5	.....	1,960	1,960	0	0
	.....	.....	.....	19,090	100.0

[1] Rainfall and Runoff Studies in the United States, *U.S. Geol. Survey Water Supply Paper* 772, 1936. The hydrograph is shown in Fig. 5.

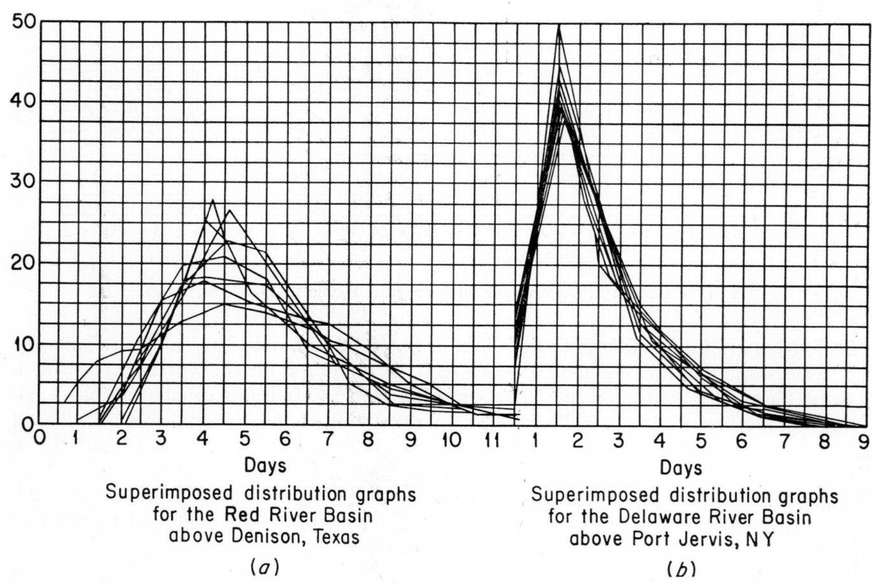

FIG. 6. Distribution graphs. (*G. W. Hoyt and others, U.S. Geol. Survey, Water Supply Paper* 772.)

hydrograph. Line *BHJ* is the "recession curve" of the surface runoff preceding the storm being studied. Line *GHJ* is the "recession curve" of the ground-water runoff during the same period. Line *DJ* is the "recession curve" of the surface runoff subsequent to the storm being studied. Computations of the ordinates of the distribution graph based on the unit hydrograph in Fig. 5 are given in Table 8.

Maximum Rate of Runoff.[1] Many empirical formulas have been devised to express the maximum rate of runoff but none should be used except within the limits of the conditions for which it was devised. Differences of more than 500 percent may be found by substituting the same data in different formulas. A few of these are shown in the following list:

Originator	Formula	
Fanning	$Q = 200A_m^{5/6}$	(6)
Burkli Ziegler	$Q = CRA\sqrt[4]{S/A_a}$	(7)
Kuichling [2]	$Q = \dfrac{127{,}000A_m}{A_m + 370} + 7.4A_m$	(8)
Kuichling [3]	$Q = \dfrac{44{,}000A_m}{A_m + 370} + 20A_m$	(9)
U.S. Geological Survey [4]	$Q = \dfrac{46{,}790A_m}{A_m + 320} + 15A_m$	(10)

where Q = maximum runoff, cu ft per sec
A_m = area of watershed, sq miles
A_a = area of watershed, acres
C = a constant, recommended as 0.7
R = rate of rainfall, in. per hr
S = average ground slope in units per 1,000 units

Fuller's formula [5] provides for the frequency of occurrence of maximum floods and is in the form

$$Q = CM^{0.8}(1 + 0.8 \log T)(1 + 2M^{-0.3}) \tag{11}$$

where C = a coefficient depending on the characteristics of the watershed. It must be computed from previous records. Values have been computed between 1 and 200
M = area of watershed, sq miles
Q = maximum flood flow, cu ft per sec per sq mile
T = number of years in the period

The rational method for the determination of the maximum rate of runoff is applicable principally to relatively small watersheds. It is discussed in Section 9 of this book.

Extent, Duration, and Frequency of Storms. A long-continued storm of modestly heavy intensity over a large area may cause a greater runoff than a more intense storm for a shorter period over only a portion of the drainage area. Great storms may last for many days and may precipitate 15 to 20 in. of rainfall over areas up to 5,000 sq

[1] Other discussion of runoff from large areas is presented in Section 9.
[2] From Report of New York engineer on state barge canal, 1901, p. 844.
[3] *U.S. Geol. Survey Water Supply Paper* 147, 1905.
[4] MURPHY, *U.S. Geol. Survey Water Supply Paper* 147, 1905.
[5] *Trans. ASCE*, vol. 77, p. 564.

miles. The relation between rate of rainfall and duration of storm is expressed empirically as

$$R = (aT^b)/(t + c)^n \tag{12}$$

where R = rate of rainfall, in. per hr
T = frequency of occurrence, year
t = duration of storm, min

a, b, c, and n are constants with magnitudes in the United States of about: $a = 5$ to 50; $b = 0.1$ to 0.5; $c = 0$ to 30; and $n = 0.4$ to 1.0.

Droughts and Low Rates of Flow. A drought is a period in which precipitation is a predetermined percentage, usually 85 percent, of the mean rate of precipitation. The

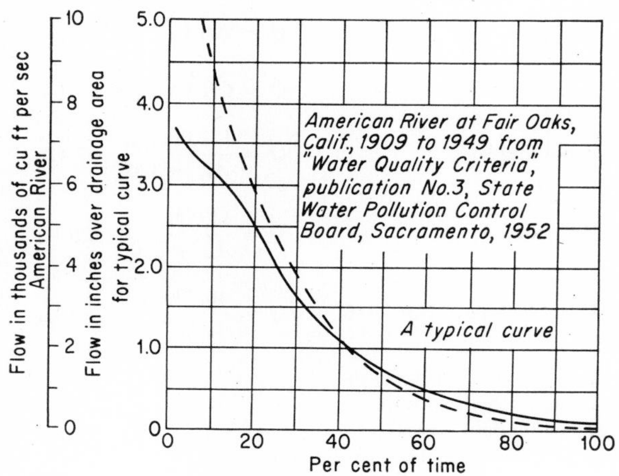

Fig. 7. Flow-duration curves.

index of low flow for a river is defined as the rate of runoff which is exceeded 90 percent of the time. A flow-duration curve is shown in Fig. 7. The index of low flow shown is about 200 cu ft per sec.

COLLECTION OF SURFACE WATER

Surface water may be collected by intakes in lakes, in rivers which are either free-flowing or in which flow is restricted by a dam, or in an impounding reservoir.

Impounding Reservoir. An impounding reservoir stores water in a water-storage basin created by a dam in a valley. This water is accumulated during high-flow rates of the impounded stream to be used when the rate of stream flow is less than the rate of demand for water.

The capacity of an impounding reservoir needed to supply the demand and other data can be determined by the study of mass diagrams of the demand and runoff for the years of minimum flow. The degree of accuracy of the results is dependent on the accuracy and extent of the available observation. The shorter the period covered by the observations, the greater the probability of error. Allowance is made for probable error in interpreting and using rainfall data. An error that may enter into the determination of the capacity of an impounding reservoir, if based on a study of mass diagrams alone, is caused by the available ground storage; but since this is indeterminate and offers a factor of safety in the design, it is seldom allowed for. An allowance is made,

however, for evaporation from the surface of the reservoir by estimating its probable surface area and subtracting the evaporation from the observed runoff data.

A mass diagram for the runoff from a hypothetical watershed, with corrections for evaporation from water surfaces, and for other compensations, is shown in Fig. 8. The ordinates represent the accumulated runoff from the start of the observations to the time of the corresponding abscissas, less the sum of the accumulated evaporation to the same time. A mass diagram is next drawn to represent the accumulated demand.

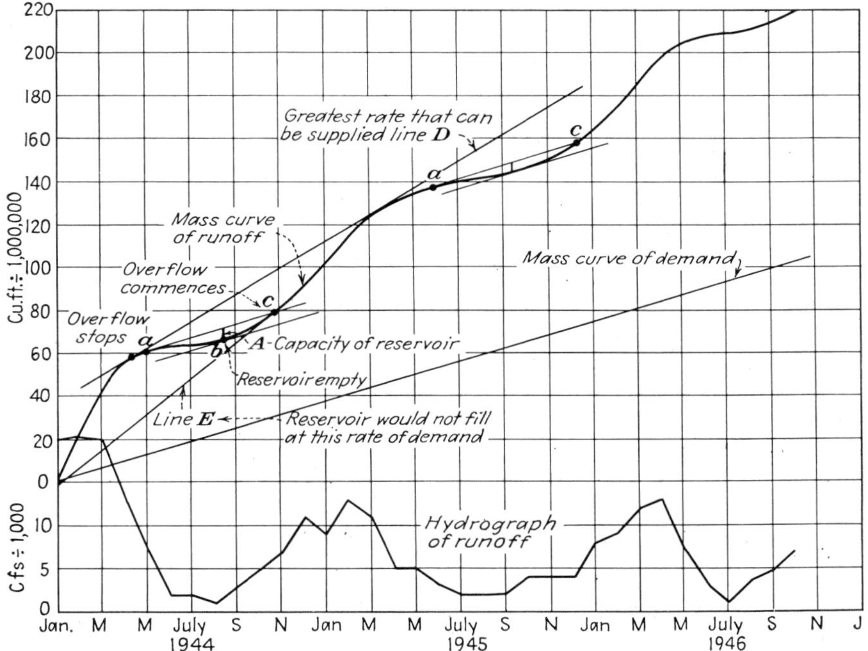

FIG. 8. Hydrograph and mass diagram for runoff and demand.

If the demand is assumed to be at a constant rate, the demand line will be straight, as shown in Fig. 8. Various conditions concerning the reservoir can be found as follows:

1. The required capacity of the reservoir is found by drawing a straight line parallel to the demand line and tangent to the mass curve of runoff. The maximum vertical distance between any such tangents and the mass curve of runoff represents the capacity of the reservoir. This is shown by line *A* in Fig. 8, and, in this case it is 5 million cu ft.

2. The period during which the water in the reservoir will be lower than the spillway is the period between the point of tangency and the point where the tangent intersects the mass curve of runoff. This is from May 1, 1944 to Oct. 25, 1944.

3. The point where overflow from the reservoir ceases is the point of tangency, marked *a* in the figure.

4. The point where the reservoir is empty is at the base of the maximum vertical line, marked *A*, between the tangent and the mass curve of runoff—point *b* in the figure.

5. The point where the reservoir will begin to overflow again is the point where the tangent again intersects the mass curve of runoff—point *c* in the figure.

6. The greatest rate of demand which can be supplied by the stream, provided that the surface area of the reservoir is not increased, is represented by the slope of the line, above the mass curve of runoff, which is tangent to it at two or more points but which does not intersect the mass curve of runoff at any point. This is shown as line D on the figure and represents a rate of 15 mgd.

7. If the demand is so great that the reservoir will never fill, a line, drawn tangent to and below the mass curve of runoff and parallel to the mass curve of demand, will not intersect the mass curve of runoff to the left of the point of tangency. Line E in the figure represents such a rate of demand.

If the rate of demand is variable, the mass curve of demand will not be a straight line. Under this condition, it is still possible to determine all the above conditions by drawing: (1) the mass curve of runoff as before; (2) the mass curve of demand, which is now not a straight line; and (3) a curve the ordinates of which are equal to the vertical distances between the two mass curves and whose abscissas correspond to the points at which these distances are measured. In determining the desired information, this third curve corresponds to the mass curve of runoff; and a straight horizontal line corresponds to the mass curve of demand in the first method.

The volume of water held in a reservoir may be determined by the prismoidal formula

$$Q = (A_1 + 4A_2 + A_3)(h/6) \tag{13}$$

where Q = capacity; A_1, A_2, and A_3 are areas enclosed within a lower, a median, and an upper contour, respectively, where the contour interval is h.

Typical depth-capacity and depth-surface-area curves are shown in Fig. 9. The spillway crest should be fixed at a height that will give the reservoir capacity necessary to supply the estimated demand without flooding an excess amount of land.

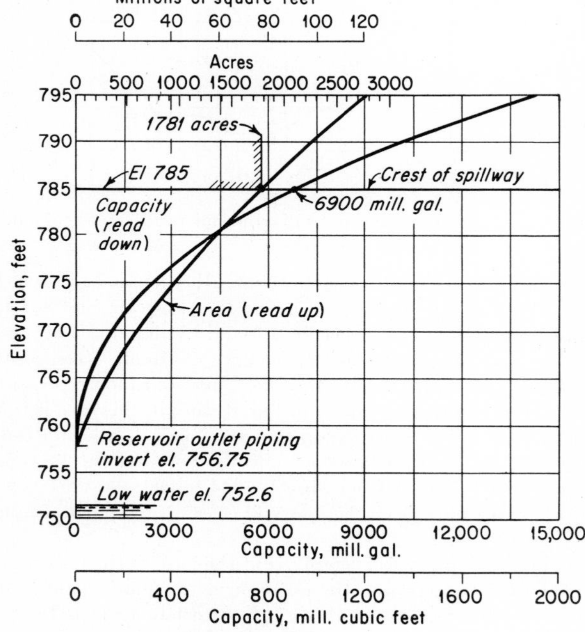

FIG. 9. Area and capacity curves, Geist Reservoir. (*J. A. W. W. A., December*, 1945, *p.* 1325.)

Dams. Dams may be classified as earth, masonry, timber, etc.; or according to the method of construction as earth fill, hydraulic fill, rock fill, etc.; or in accordance with the resistance to external forces as gravity, arch, slab-and-buttress, etc. Some forms of dams under the last classification have subtitles such as multiple arch, constant-angle arch, and constant-radius arch.

Dam foundations are important. Defects which have caused failures include crushing, sliding, plastic flow, piping, scouring, uplift, and perviousness. Hard, unstratified, homogeneous, igneous rock makes the most desirable foundation. Satisfactory dams on porous, sand foundations are in existence. Stratified rock, fissured limestone, and

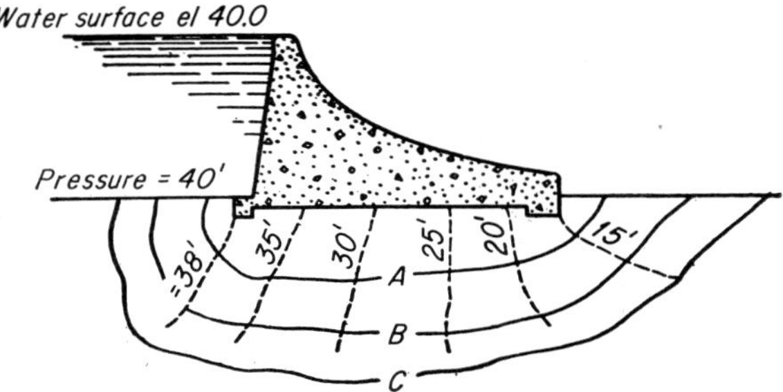

FIG. 10. Dam on porous foundation showing typical flow net.

porous materials require special attention in design and in construction. Leakage in nonhomogeneous rock may be overcome by grouting or cutoff walls, and seepage through a porous foundation can be reduced to safe amounts by the construction of a long apron, both upstream and downstream, sufficiently heavy to resist uplift, as indicated in Fig. 10. Velocity of flow along each of the lines A, B, and C is constant and is determined by the hydraulic gradient along the line.

Forces Acting on Dams. External forces to be resisted by dams include: water pressure, foundation pressure, flotation, uplift, erosion, ice pressure, earthquakes, wind pressure, wave pressure, and subatmospheric pressure.

The magnitude and location of the resultant, and its components, of the external water pressure on any submerged surface are discussed in Section 4, "Hydraulics."

In the design of a gravity dam to prevent overturning the resultant of the weight of the structure, the horizontal water pressure, and of any superimposed loads must pass within the outlines of the structure. It is customary, however, to design all gravity dams so that the resultant of all forces acting above any horizontal plane passed through the dam will fall within the middle third of that plane in order to avoid tension in the upstream face of the structure. Flotation is a vertical force acting through the center of gravity of the resisting body, and equal in magnitude to the weight of the water displaced.

Uplift in a dam is caused by water pressure beneath the foundation or exerted on cracks or interstices within the structure. The intensity of uplift is expressed as $U = 62.5H$, in which U is the uplift pressure in pounds per square foot and H the height in feet of the water column causing upward pressure. For example, the intensity of uplift pressure at different points in Fig. 11 varies with the height of the hydraulic grade line above the point where uplift is occurring. If full upward pressure acts at the heel of a dam, diminishing uniformly to zero at the toe, the total uplift in pounds on a foot length of the dam will be $P = 31.25h_1b$, in which h_1 is the depth of water in feet to the horizontal plane on which uplift is being studied and b is the thickness of the dam along the horizontal plane on which uplift is being studied. The center of uplift pressure will occur

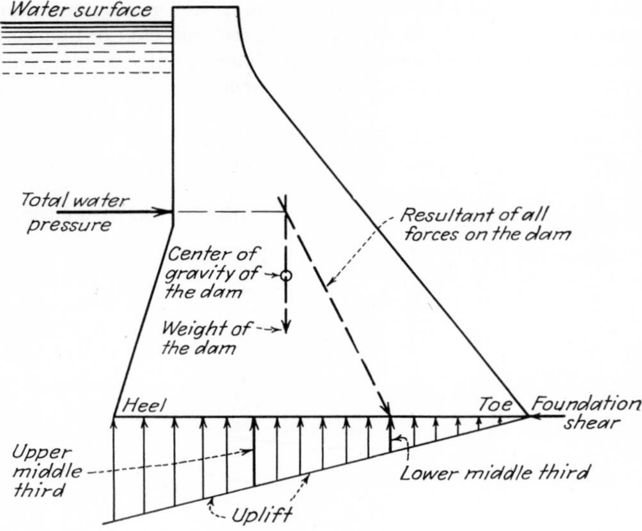

FIG. 11. Diagram of forces on a gravity dam.

at the upstream edge of the middle third of the horizontal plane on which pressure is being studied. If tail water rises to a height h_2 ft above the toe of the dam, then the uplift pressure on the foundation may be assumed to be $P = 31.25b(h_1 + h_2)$.

Sliding resistance or shear can be expressed as

$$f\Sigma W = (W_1 + V - U + K) = P \qquad (14)$$

where W_1 = weight of the dry structure
V = weight of water on upstream face
U = uplift force
K = other vertical forces present
P = horizontal force
f = coefficient of friction

The value of f for rock is frequently used as 0.65. Values from 0.25 to 0.5 are used for earth. The factor of safety against sliding is the ratio between the sliding resistance and the total horizontal force ΣP_h, or $F = f(\Sigma W/P_h)$. For masonry dams values of F vary from 1.0 on rock foundations to 3.0 on earth foundations.

The distribution of the foundation reaction across the base of the dam depends on the unity of the structure. In a structure unable to resist vertical shear, approached by the condition in an earth dam, the foundation reaction at any point is equal to the weight of the water and the structure vertically above it. If the structure is a monolith, as may occur in a reinforced concrete gravity dam, the distribution of pressure may be as shown in Fig. 11. Where the vertical pressure at the downstream face of the dam is zero the resultant of all vertical and horizontal forces will act at the lower middle third of the horizontal plane through the dam. Allowable maximum foundation pressures vary from 30 tons per sq ft on primary rock to 10 for good shale.

Wave pressures have been formulated by Molitor[1] as

$$P = 125h_w^2 \qquad (15)$$

[1] MOLITOR, D. A., *Trans. ASCE*, vol. 100, p. 984, 1935.

where P is the total wave pressure, acting about $\frac{3}{8} h_w$ above the elevation of the still-water surface in the reservoir, in pounds, and h_w is the height of the wave from trough to crest, in feet, as computed by Stevenson's formula, which is expressed as

$$H = 1.5\sqrt{F} + (2.5 - \sqrt[4]{F}) \tag{16}$$

where H = height of wave, ft
F = fetch of wind, miles

Ice pressure against a dam is affected by such conditions as rate of air-temperature rise at temperatures below freezing, ice thickness, restraint of ice sheet, extent of exposure to solar radiation, currents under the ice, snow on ice acting as an insulator, intensity and fetch of the wind, buckling of the ice, and changes in reservoir level. The rate of rise of the temperature of the ice is probably the most effective condition in creating ice pressure. It should be recalled that ice expands as the temperature rises. A pressure of 3,000 to 5,000 lb per sq ft of ice thickness may be a conservative estimate of the ice thrust unless conditions are such that the ice sheet may be confined, when the pressure will be much greater.

Earthquake shock may be cared for, when required, by assuming a force acting in a horizontal direction and applied at the center of gravity of the dam above any horizontal plane taken through the dam. The magnitude of the horizontal force is $F = Ma$, in which M is the mass of the structure above the horizontal plane and a is taken as $0.75g$ for rock foundations and $0.10g$ on sand and earth foundations, where g represents the acceleration due to gravity.[1]

The principal requirements for the safety of a dam can be summarized as follows:

1. The resultant of all forces under all conditions shall be within the middle third of any horizontal plane drawn through the structure.
2. Compressive and shearing stresses shall not exceed a safe limit fixed by experience.
3. The weight of the dam multiplied by the coefficient of friction on any horizontal plane shall be more than twice as great as the horizontal force acting along that plane.

Earth Dams. The materials to be used in earth dams must be carefully selected, the more impervious material being placed in the upstream part of the dam, except the core which is usually placed in the center. Clay, with some sand, forms the most desirable material. Too much sand results in too high a permeability; too little sand may result in the formation of shrinkage cracks, or the lower internal friction within the clay may be such as to permit sliding. Material containing organic matter, humus, tree roots, etc., is not used. The more porous materials are placed in the downstream part of the structure, and drainpipes are sometimes placed therein to assure drainage and a dry embankment. Materials are placed in earth dams either hydraulically or in layers which are tamped and rolled.

Embankment slopes may be governed by the stability of the material under all probable conditions of moisture content, bearing power of the foundation, and resistance of the materials of the dam and of the foundation to percolation. In general, for structures 25 ft or less in height, stable embankments may result if the upstream slope is built not steeper than 2.5 horizontal to 1 vertical, and the downstream slope 2 horizontal to 1 vertical. For higher dams tests and other investigations should be made to determine the safe slope to be selected. Upstream slopes may be protected by stone, riprap, concrete in slabs or blocks, or willow or other timber mattresses. The protection should extend from the top of the dam to a depth of 5 ft below the lowest expected operating level. In important structures where riprap or concrete is used a shoulder or berm, as shown in Fig. 12, should be provided at the bottom of the slope protection. Reliance on vegetation for the protection of steep, upstream slopes is undesirable. Downstream

[1] *Proc. ASCE*, August, 1928, p. 1737.

slopes may be protected similarly to upstream slopes or by vines, shrubs, or grass. Vegetation with long root structures should be avoided. Burrowing animals may be discouraged by loose rock fills, concrete, or steel cores. Fences may be necessary to exclude grazing animals.

Allowance for settling of the embankment of an earth dam depends partly on the height of the dam, the material used and the method of placing it, and on other local conditions. The top width of an earth dam is usually a function of its height, with a minimum of 8 to 10 ft. The relation is sometimes expressed as $W = (H + 25)/5$, in which W is the top width of the dam in feet. This expression holds only when it gives a

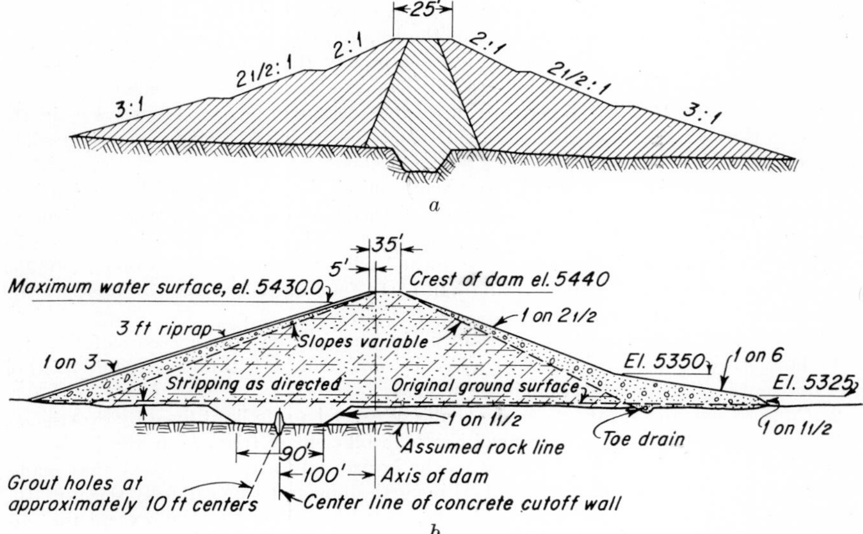

FIG. 12. Sections through earth dams: *a*. typical, *b*. Miami Conservancy District.

value of W greater than is called for by the requirements of the structure to resist the forces acting against it. The height of the embankment above the highest water surface, i.e., the freeboard, should not be less than the expected height of the waves plus an allowance for the settling of the embankment and an allowance for frost penetration. A freeboard of less than 6 to 10 ft may not be good practice.

Paving is sometimes used on the slopes of earth dams, particularly on the upstream slopes, to prevent erosion, to aid in preventing sliding, and sometimes, in the case of the upstream slopes, to give greater imperviousness. Paving on a downstream slope must be well drained to prevent the accumulation of water within the embankment. Riprap, rubble, broken stones, cut-stone blocks, concrete slabs, etc., are used as paving materials. If it is possible for water to accumulate behind paving on the upstream slope and the reservoir is subjected to high waves or rapid changes in level, provision is made to permit the rapid escape of water from behind the paving to avoid the bursting pressure otherwise created.

A core wall is a watertight or approximately impervious diaphragm placed vertically in the dam to prevent or impede the passage of water or burrowing animals through the dam. Since the portion of the embankment upstream from the core wall may be saturated, the location of the core wall as far upstream within the dam as feasible may be desirable. Hollow or cellular core walls have been used to convey leakage downward

to drains leading it away from the dam. Longitudinal passageways or tunnels have been constructed in concrete core walls to permit observation of leakage.

Materials used in core walls include concrete, steel, wood, and puddle or puddled earth. Since stresses on a core wall cannot be computed without making uncertain assumptions concerning the movements of the dam, the determination of the core-wall thickness is a matter of judgment based on experience. Concrete core walls with a top thickness of 12 in. and sides battered 1 horizontal to 100 vertical have proved generally satisfactory. The core wall should be sufficiently high to extend above the top of the earth embankment to act as a parapet and a factor of safety. During construction the earth pressures on the sides of the core can be balanced by maintaining the rising fill at the same elevation on each side of the wall. Steel and wood are seldom used in core walls except for temporary installations since the alternate wetting and drying of the upper portion of the structure exposes the steel and wood to active corrosion.

Puddle is a definite proportioned mixture of clay, sand, and water placed and tamped to form an almost impervious diaphragm. It may satisfactorily stop the flow of water, but it is not a deterrent to burrowing animals. Puddle cores, usually placed near the center of the dam, are formed by placing a mixture of selected materials by a hydraulic process. Puddle cores have the advantage of flexibility and are less subject than concrete to rupture by unbalanced forces within the fill. Care is necessary in placing the core to avoid undesirable stratification of materials within it.

A cutoff wall is a low core wall extending from a short distance in the dam above the porous stratum on which the dam is founded, down into the porous stratum to cut through it.

Masonry Dams. Choice of materials for a masonry dam depends partially on availability, cost, and the desired height, appearance, and permanence of the structure. For dams above about 200 ft in height, either masonry or concrete is used; for lower dams, other materials are used. Foundation conditions offer the greatest uncertainty in the design and construction of masonry dams. Foundation pressures may be higher on masonry dams than on other types and will require special care in their study. The study may be made by borings on the site, by examining records of boreholes in the locality, and by securing the opinion of a competent geologist. It is more difficult to tie the dam securely to the foundation than it is to design a safe superstructure. Masonry-dam failures result principally from inadequate foundations. This is particularly true of dams over about 200 ft in height. It may be desirable to connect a masonry dam to an impervious rock stratum beneath it by means of an impervious cutoff wall or other integral part of the structure. This is not always possible, however, and masonry dams are sometimes built on porous foundations.

Other Dams. Rock fill, timber, steel, and other types of dams are sometimes used for water-supply purposes for various reasons such as their temporary nature, or the convenience of their construction in a remote location. A rock-fill dam consists of an embankment of loose rock or other loose masonry, with a flow-retarding membrane in the upstream face of, or within the embankment. Materials used for the membrane include concrete, timber, steel, or puddled clay and sand, either paved or unprotected.

Spillways. The purpose of a spillway is to conduct excess water away from a reservoir without endangering the safety of the dam. Hence, the capacity of the spillway must be sufficient to pass the greatest flood expected so that the dam will not be overtopped, and the shape of the spillway must be such that the falling water will not undermine the dam or spillway. To avoid wide variations of the elevation of water in a reservoir due to variations in stream flow the crest of overflow spillways should be made long, siphon or shaft spillways are used, or gates controlled either manually or automatically may be depended on. To avoid danger to the dam or spillway the spillway may be located on the rim of the reservoir where the distance of fall of water is low.

Rates of flow over overflow spillways are discussed in Section 4, "Hydraulics." Since the discharge over a spillway is a function of the depth of water on it, the crest of the spillway is made as long as is economical in order to keep the depth of water low during flood and thus prevent the flooding of as little land as possible upstream from the dam. Other methods adopted to avoid flooding of land above a dam are to select a type of spillway with a high discharge coefficient, or to use movable gates on the spillway, which may be opened manually or automatically, such as flashboards, or they may operate automatically without moving parts, such as a siphonic or a shaft spillway. The effective length of a spillway can be increased by curving it, as in an arch dam. This expedient may be useful in the design of a dam located in a narrow gorge.

A chute spillway, in general, leads water down the hillside at one side of the dam. It may be constructed partially or entirely as a flume or as a tunnel. The bottom of the channel may be roughened to aid in the absorption of energy. Chute spillways minimize the height of the vertical fall of water. They are known as "side-flow channels" where the axis of the spillway channel is parallel to and immediately below the crest of the spillway. The hydraulics of side-flow channels have been analyzed by Camp,[1] who has formulated the flow in a rectangular, flat-bottom channel with constant slope as

$$H_o = \sqrt{d^2 + \frac{2Q^2}{gb^2d} - Sx\bar{d} + \frac{fxQ^2}{12gb^2\bar{R}\bar{d}}} \tag{17}$$

where H_o = depth of water at upper end of spillway
d = depth at any point
Q = rate of flow
b = width of channel
S = bottom slope
x = distance from upper end of channel to point where depth is d
$\bar{d}$ = average depth through distance x
f = Weisbach-Darcy friction factor
$\bar{R}$ = average hydraulic radius through distance x

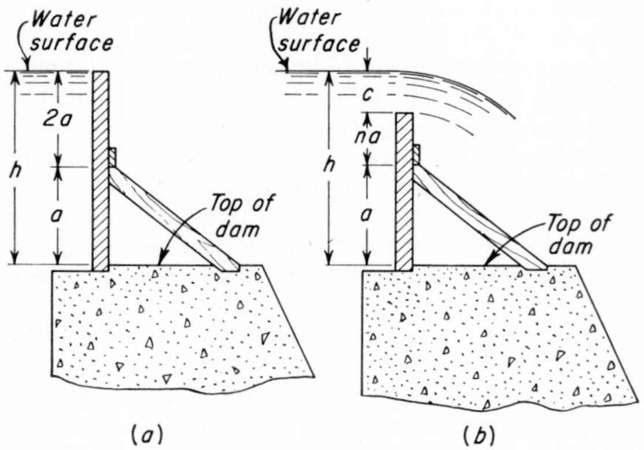

FIG. 13. Two types of flashboards.

A flashboard, such as is illustrated in Fig. 13, is a device placed on the spillway of a dam to retain water in the reservoir until a predetermined depth of water against the flashboard causes it to fall. The board may be replaced manually or automatically.

[1] *Trans. ASCE*, vol. 105, p. 606, 1940.

The flashboard illustrated in Fig. 13*a* will fall when the depth of water behind it is equal to three times the height a. In Fig. 13*b* the value of n must be greater than 1 and less than 2. The value of c can be found by solving the expression $c + na = I_y/Ax_g$, in which I_y is the moment of inertia of the submerged section about the axis formed by the intersection of its projected plane and the surface of the water, A is the area of the submerged surface, and x_g is the distance, measured in the plane of the submerged surface, from the center of gravity of the submerged surface to the axis about which the moment of inertia is taken.

Siphonic spillways are discussed in Section 4, "Hydraulics." A shaft spillway, as shown in Fig. 14, functions in a manner similar to a siphonic spillway. Dangers inherent in siphon and shaft spillways are: clogging, rupture of the structure due to vibrations when vacuum is quickly and repeatedly made and broken in the siphon or shaft tube, and cavitation and high velocities which may cause erosion.

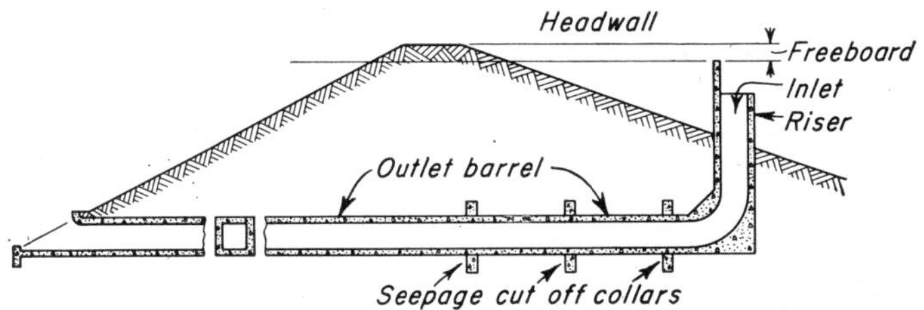

Fig. 14. Section through a shaft spillway. ("*Low Dams*," *Natural Resources Committee*, 1938.)

Time is required for flood waters to rise above a dam because of the storage capacity to be filled in the reservoir. An approximate expression for this time is [1]

$$T = \frac{A(H_2 - H_1)}{Q - \dfrac{Q_1 + Q_2}{2}} \tag{18}$$

where T = time for water to rise from H_1 to H_2, where T is in sec and H_1 and H_2 are in ft

Q = constant rate of flow into reservoir, sec-ft

Q_1 and Q_2 = rates of discharge, respectively, sec-ft, under the heads H_1 and H_2, ft, on the spillway

Provisions are made below overflow spillways to absorb the energy or to cushion the effect of the falling water. This can be done by extending the downstream apron to a point where the velocity of flow is sufficiently diminished to prevent erosion; or the water may fall into a stilling basin, a hydraulic jump may be created, or other expedients may be adopted.

Spillway Gates. Spillway discharges may be controlled either by movable crests or by gates. Such devices are used to control the elevation of the water in the reservoir or the rate of discharge over the spillway. Among the types of controllable spillway are included: (1) stoney gate, similar to a sluice gate; it is supported in vertical slides in which it is guided as it is raised and lowered; (2) taintor gate, a radial gate which revolves about a horizontal axis that is supported downstream from the crest of the spillway; when closed the lower edge of the gate rests on the spillway; (3) drum gate, somewhat similar to the taintor gate except that the gate floats on the water in a chamber in the dam. The horizontal axis about which the gate revolves may be either upstream or

[1] *U.S. Geol. Survey Water Supply Paper* 150, 1906.

downstream from the lower edge of the gate. Other forms of movable gates include bear-trap, butterfly, ring, leaf, and roller-crest gates.

Intakes. Types of intakes consist of towers, submerged intakes, intake pipes or conduits, movable intakes, and shore intakes. Intake structures over the inlet ends of conduits are necessary to protect against wave action, floods, stoppage, navigation, ice, pollution, and other interference with the proper functioning of the intake.

Intake towers are used for large waterworks drawing water from lakes, reservoirs, and rivers in which there is either, or both, a wide fluctuation in water level or the desire to draw water at a depth that will give water of the best quality, to avoid clogging, or for other reasons. A wet intake tower is filled with water to the level of the water in the reservoir, lake, or river. A dry intake tower has no water inside of it other than in the intake pipes. Water is admitted through the ports connected to the intake pipes which convey it from the tower under pressure. The interior of the tower is thus made accessible for inspection and operation. Wet towers are less costly to construct. They are not subject to flotation and certain other stresses may not require consideration. Towers in lakes and in some rivers are usually located some distance from the shore line.

Inlet ports are provided at different elevations in high intake structures to provide facilities for taking the best quality of water available, because the quality of water in deep sources may not be the same at all depths. The size of the port openings is made large to minimize the entrance velocity and to avoid packing of objects against the port. An entrance velocity of less than 0.5 ft per sec is desirable. Difficulty with entrained air is avoided at such low velocities.

Submerged intakes are constructed entirely under water and have such advantages over exposed structures as lower cost, no obstruction to navigation and little obstruction to the flow of the river, little danger from floating material, and a minimum of trouble from ice.

Intakes placed along the shore, known as shore intakes, may be suitable for industrial plants where water quality and other conditions permit. Shore intakes are probably less costly than towers or cribs with long intake pipes.

Location of Intakes. Conditions to be considered in the location of an intake include: the location of the best quality of water available, currents that might threaten the safety of the structure, navigation channels, ice floes and other ice difficulties, the formation of bars and shoals, and the fetch of the wind affecting the height of waves. Conditions affecting the quality of water include currents due to wind, temperature, seasonal turnover, and other causes that will bring water of unsuitable quality to the intake. Channels with high-velocity currents carrying floating debris and ice are hazardous to the safety of the structure. Navigation channels add the danger of pollution from passing ships.

Ice floes and waves are hazardous to the superstructure of an intake because of the pressure they may exert on it. The ice may push down and clog ports 20 to 30 ft below the water surface and waves may stir up bottom sediments at such depths.

In bodies of water subject to wide changes in the elevation of the surface level the intake should be located so that, at the lowest water stage, one inlet is always submerged. An intake in an impounding reservoir is usually placed in the deepest part of the reservoir. This will render the full capacity of the reservoir available and will usually protect the intake from sediment in the reservoir. Intakes in streams with a steep bottom gradient should, where possible, be placed at a sufficient distance upstream to supply water to the city by gravity.

Design of Intakes. External forces to be resisted by intakes include flotation, waves and currents, ice pressures, blows from floating and submerged objects, and shifting shoals. The magnitude of such forces is known only approximately. Hence, a generous factor of safety must be allowed in design.

Screens. Coarse screens may be used near the ports to prevent the entrance of large objects; and, in some part of the intake, fine screens may be provided to exclude small fish and other small materials which might cause difficulties elsewhere in the waterworks. Cleaning of screens requires constant attention. In dirty waters it may be necessary to install moving screens which may or may not be self-cleaning.

Ice. Surface ice, anchor ice, frazil ice, and slush ice are forms in which ice may be troublesome at an intake. Anchor ice is formed beneath the surface on objects that radiate heat rapidly. Frazil ice, sometimes called needle ice, consists of submerged ice crystals that have been formed as anchor ice and have broken loose, or are ice crystals that have been formed about small particles suspended in water which serve as nuclei. Slush ice is formed of a soft, mushy mass of ice crystals ground from other forms of ice.

Expedients used for the removal of ice from intake ports include a vigorous reversal of the flow of water through the port, steam or electrical heating devices, compressed air, pike poles, chains, and explosives. The maintenance of violent agitation at critical points, as by high-velocity jets, has been found effective, especially along reservoir walls. Divers have been employed under urgent conditions, but the service is hazardous and should be avoided. Provision for vigorous reversal of flow should be made when the intake is installed.

Intake Conduit and Suction Well. A conduit should lead from the intake to a suction well for the pumps in the pumping station. A velocity of 2 to 3 ft per sec is satisfactory in this conduit. It should be laid on a continuous rising or falling slope to avoid the entrapment of air. Surge, or water hammer, is to be avoided or provided for if unavoidable (see page 10–47).

Aqueducts and Pipelines. An aqueduct is a conduit designed to convey water from a source of supply to a point from which the water may be distributed. An aqueduct may include canals, flumes, pipelines, siphons, tunnels, or other channels.

Materials used in aqueducts include concrete, cast iron, steel, wood, and asbestos cement. The material selected depends on such conditions as: assurance that operation will not be interrupted; cost; suitability for conditions which include loads, pressures, and the corrosiveness of the water and of the soil in contact with the aqueduct; accessibility of the site and ease of transportation of materials; and availability of labor for construction.

To assure safety, to permit inspection, and to facilitate operation and maintenance, pipelines should be provided with gate valves, check valves, drainage valves, air valves, and manholes. Surge tanks or other surge-control equipment may be necessary on some lines. Gate valves may be placed 1,000 to 1,500 ft apart, near manholes, with a drainage valve at a low point between the gate valves to permit emptying the pipe between valves for inspection and repair. A check valve should be placed at the upstream side of the beginning of each rise in the pipeline to prevent reversal of flow. A gate valve may be placed near to and on a convenient side of the check valve to permit inspection and repairs. The check valve should be designed to close slowly enough to avoid dangerous water hammer. Drainage valves should be placed at all low points to permit removal of deposited silt and the emptying of the pipe. An air valve should be located at each high point to allow the escape of air and gases and to admit air sufficiently rapidly to prevent the creation of a partial vacuum in the pipeline.

Expansion joints may be required on long pipelines, particularly on steel pipe exposed to marked changes in temperature. When several expansion joints are used in a long line, it is necessary to construct anchors between them to force the movement into the pipe designed to take it. Where anchors are used to support the full weight of pipe and water they should be spaced 20 to 40 ft apart and they should be constructed so that one-sixth to one-fourth, or more, of the pipe circumference will bear upon them. They should be constructed to resist the overturning forces resulting from friction due to

longitudinal movements of the pipe. In general, a friction coefficient of not less than 0.5 between pipe and pier should be used in design.

Appurtenances on Long Pipelines. The appurtenances used on long pipelines include manholes, gate valves, check valves, air-relief valves, blowoff valves, and expansion joints. Manholes are used in large conduits to permit inspection, cleaning, and for other purposes. They may be located anywhere between 500 and 2,000 ft apart. Gate valves are located at convenient points, no general practice having been adopted. Their use increases both the safety and the cost of the conduit. Where check valves are used they may be located at the lower end of a rise in the conduit so that, in the event of a break in the pipeline between the check valve and the source of water supply to the conduit, water will be held in the rising portion of the pipeline between the check valve and the outlet. Air-relief valves are used at summits in the line to prevent the accumulation of gases within the pipe or to admit air to the pipe in the event that there is a tendency to produce a vacuum in it. To relieve pressure, an air valve should have an opening of about one-twelfth, and to relieve vacuum, about one-eighth, of the diameter of the pipe. Trouble sometimes results from the freezing of air valves in cold climates. Blowoff valves are placed at the lowest points of depressions in the pipeline so that water can be drained from the line into natural drainage channels.

Economical Diameters and Velocities in Pipelines. The economical diameter of a standard cast-iron pipe through which water is being pumped can be expressed as

$$d = 7.25 \left(\frac{p}{r(B + 3a)}\right) Q^{0.476} \tag{19}$$

and the economical velocity is

$$V = 3.5Q^{0.05} \left(\frac{r(B + 3a)}{p}\right)^{0.33} \tag{20}$$

$$= 3d^{0.1} \left(\frac{r(B + 3a)}{p}\right)^{0.35} \tag{20A}$$

where a = cost of iron, cents per pound
B = $0.00159L + 0.000gY + (0.055 + 0.0145D)W$
C = coefficient in Hazen and Williams formula, taken here as 100
d = diameter of pipe, in.
D = depth of trench, ft
L = cost of lead, cents per pound
p = cost of pumping 1,000,000 gal of water 1 ft high
Q = rate of flow through pipe, cu ft per sec
r = annual rate of interest plus depreciation, plus rates for all other annual charges
s = slope of hydraulic grade line
V = velocity of flow of water in pipeline, ft per sec
W = wage rate for common labor, cents per hour
Y = cost of yarn (oakum), cents per pound

The economical diameter of a riveted steel pipe through which water is being pumped is

$$d = 11 \left(\frac{p}{kr(200 + P - 3d)}\right)^{0.143} Q^{0.404} \tag{21}$$

where k = ratio of present cost of steel pipe in place, to the cost on which the formula is based
P = working internal pressure, psi

and other nomenclature is as in expressions (20) and (20A). The formula is based on

the cost of steel pipe in Northeastern United States in 1925 to 1927.[1] The formula is limited to double-riveted steel pipe in diameters between 30 and 72 in., and to working pressures between 50 and 150 psi.

The economical diameter of a pipe through which water flows by gravity is that diameter which will cause sufficient friction to consume all the head available.

Flow of Water in Pipes. The flow of water in pipes, and various formulas therefor, are discussed in Section 4 of this book. The formula most commonly used in waterworks practice in the United States is the Hazen and Williams, which can be expressed as

$$V = C_1 R^{0.63} S^{0.54} 0.001^{-0.04} \tag{22}$$
$$V = 1.318 C R^{0.63} S^{0.54} \tag{23}$$

where V = velocity of flow, ft per sec
R = hydraulic radius, ft
S = slope of hydraulic grade line

C and C_1 are coefficients, depending on character of pipe material. Some values of C are given in Table 9.

TABLE 9. HAZEN AND WILLIAMS COEFFICIENTS FOR VARIOUS MATERIALS

Material	Coefficient
New cast-iron pipe, pit-cast	120–130
New cast-iron pipe, centrifugally cast	125–135
Cement lining, applied by hand	125–135
Cement lining, centrifugally applied	140–150
Bitumastic lining, hand brushed	135–145
Bitumastic lining, centrifugally applied	145–155
Concrete, best workmanship; large-diameter pipe	145–155
Concrete, medium-quality workmanship	130–145
Ordinary, tar-dipped, cast-iron pipe; 20 years service; hard, inactive water	110–125
Ordinary, tar-dipped, cast-iron pipe; average effect of tuberculation:	
New	135
5 years	120
10 years	110
15 years	105
20 years	95
30 years	85
40 years	80
Ordinary, tar-dipped, cast-iron pipe after long service; with severe tuberculation	30– 40

A nomograph for use in the solution of the Hazen and Williams formula is given in Fig. 15.

Minor losses of head in a pipeline are caused by valves, other appurtenances, and by sudden changes in pipe diameter. Among the minor losses are:

1. Loss at entrance into a pipe

$$H = f_1(V^2/2g) \tag{24}$$

where f_1 is a coefficient depending on the character of the entrance. For pipe protruding three diameters or more into a reservoir, $f_1 = 1.0$; when the end of the pipe is flush with the wall of the reservoir, $f_1 = 0.5$; and when the end of the pipe is flared, the value of f_1

[1] See MERRIMAN, "American Civil Engineers Handbook," 5th ed., John Wiley & Sons, Inc., New York, 1930.

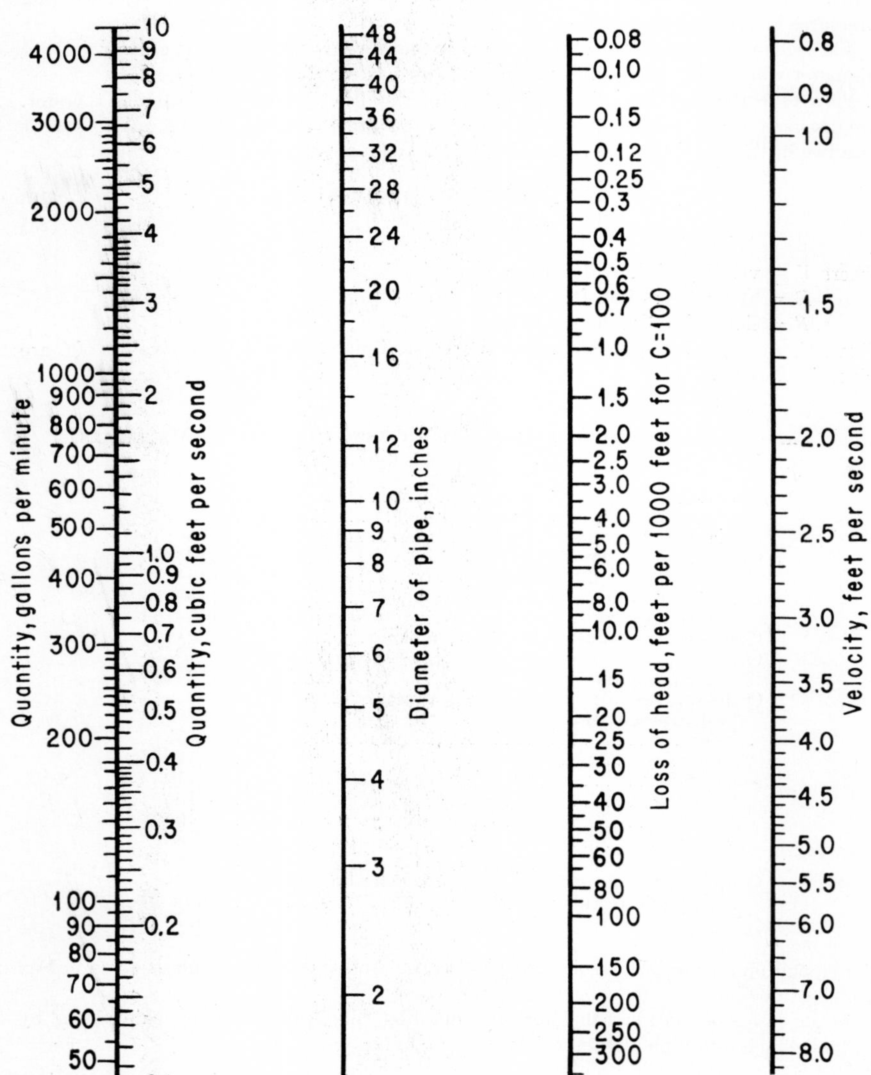

Fig. 15. Diagram for the solution of Hazen and Williams formula.

may be as low as 0.1. H = head loss, V = velocity of flow in the pipe, and g = acceleration due to gravity.

2. Loss at bends

$$H = f_2(V^2/2g) \quad (25)$$

Values of f_2 are given in Table 10. Other nomenclature is as in Eq. (24).

TABLE 10. COEFFICIENTS FOR HEAD LOSSES DUE TO FLOW AROUND 90° BENDS

Values of f_2 in expression $H = f_2 \dfrac{V^2}{2g}$

Velocity of flow, ft per sec	Length of bend, ft							
	1	2	4	6	8	10	20	30
2	0.24	0.19	0.18	0.18	0.18	0.18	0.32	0.45
5	0.33	0.25	0.23	0.23	0.23	0.24	0.39	0.55
10	0.40	0.30	0.27	0.26	0.26	0.27	0.48	0.66

3. Loss at sudden enlargement

$$H = (V_1^2 - V_2^2)/2g \quad (26)$$

where V_1 and V_2 are the velocities in the smaller and larger pipes, respectively, and other nomenclature is as in Eq. (24).

4. Loss at sudden contraction

$$H = 0.515 \left(\frac{1 - r}{1.17 - r}\right)^2 V_1^2/2g \quad (27)$$

where r is ratio of the smaller pipe diameter to the larger pipe diameter, and V_1 is the velocity in the smaller pipe. Other nomenclature is as in Eq. (24).

5. Losses in valves and other appurtenances

$$H = f_5V^2/2g \quad (28)$$

The factor f_5 varies with the type and condition of the appurtenance and the position of its use so that constants cannot be presented that are of wide application. The value of f_5 rarely, if ever, exceeds unity. Other nomenclature is as given in Eq. (24).

GROUND WATER

Occurrence. Ground water may be expected to occur in such geological formations as in buried river valleys or the beds of ancient lakes; in sedimentary deposits where the water may be held in porous rock or held under pressure in pervious strata; and in aquifers where the resistance to flow together with the rate of percolation from the surface are sufficient to maintain a reservoir of ground water above the elevation of the natural outlet.

A portion of the rain that falls on the ground percolates through it and fills the interstices of the substrata until the water flows either on or below the ground surface into a river, a lake, or the sea. Underground water normally collects in or moves slowly through pervious strata which are called *aquifers*. The free flow of water as an open stream in a subterranean cavern is not common and such underground streams are seldom available as public water supplies. The surface of the underground water exposed to atmospheric pressure is known as the *ground-water table* and as the *free surface*.

Characteristics of Aquifers. These include porosity, perviousness, permeability, and transmissibility. An *aquifer* is an underground stratum containing water within its

interstices. *Porosity* is the volume of the pores of a substance. *Perviousness* is the capacity of a substance to allow water to pass through it. It is equivalent to the amount of water that is available for a water supply. *Permeability* and *transmissibility* express the relative ease with which water flows through a porous medium. Permeability is frequently expressed as the velocity of flow of water in a permeable medium when the hydraulic slope is unity. The *coefficient of storage* of an artesian aquifer is the number of cubic feet of water released from storage in each column of the aquifer having a base area of 1 sq ft and a height equal to the thickness of the aquifer, when the artesian head is lowered 1 ft.

Types of Wells. A *gravity well* is a vertical, or approximately vertical, hole from the ground surface, penetrating an aquifer in which the surface of water in the aquifer is at atmospheric pressure. A *pressure well,* sometimes called an *artesian well,* is a hole descending from the surface of the earth, passing through an impervious stratum to penetrate an aquifer holding water under pressure greater than atmospheric pressure. A *gallery* or *horizontal well,* may be a tunnel or open ditch constructed through an aquifer in a direction approximately normal to the flow of underground water.

Flow into Wells. The rate of flow into a gravity well can be formulated as

$$Q = K \frac{h_e^2 - h_w^2}{\log_{10} (r_e/r_w)} \tag{29}$$

where K is a constant for the well and other nomenclature is shown in Fig. 16. This expression is known as the Dupuit formula for flow into a well. The graph resulting from the plotting of the Dupuit formula produces the "base-pressure curve" shown as line $ABCD$ in Fig. 16. It has been found in practice that the approximation involved

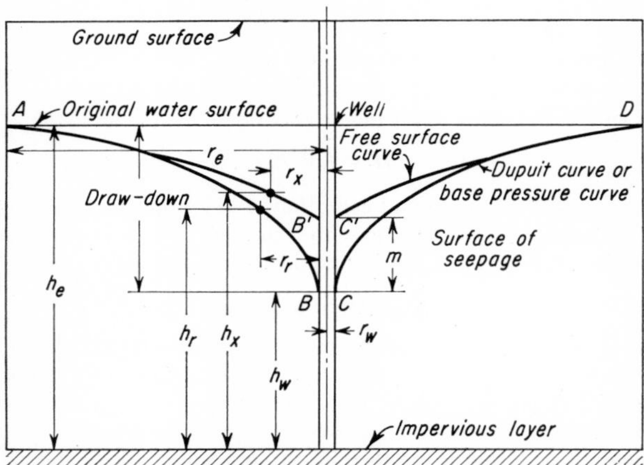

FIG. 16. Underground conditions in a gravity well.

by the use of the Dupuit formula gives results of practical value, the results being most nearly correct when the ratio of the drawdown when the well is being pumped to the depth of water in the well when not being pumped, is low.

The rate of flow into a pressure well, using similar nomenclature is

$$Q = K \frac{h_e - h_w}{2r_e} \tag{30}$$

and the rate of flow into one side of a gallery or trench is

$$Q = \frac{h_e^2 - h_w^2}{2r_e} \tag{31}$$

Drawdown and Recovery Curves. A *drawdown curve* is a curve whose coordinates are drawdown in the well and time of pumping, as shown in Fig. 17. A recovery curve

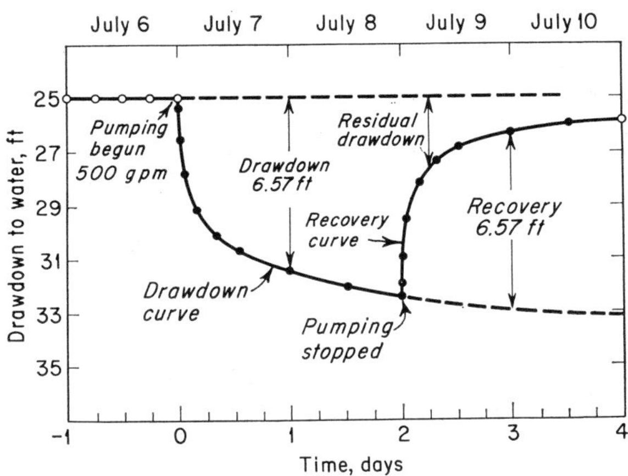

FIG. 17. Drawdown-recovery curve. (*R. H. Brown, J. A. W. W. A., August,* 1953, *p.* 849.)

shows the rate of recovery of the water level in a well after pumping has stopped. A typical drawdown-recovery curve, shown in Fig. 17, is based on expressions (32) and (33)

$$T = \frac{0.185 r_e^2 f}{Q}\left(h_e - \sqrt{h_e^2 - \frac{1.7Q}{K}}\right) \tag{32}$$

where T = time to develop radius r_e

f = specific yield of the material, i.e., the ratio of the amount of fluid that will drain from a saturated material to the volume of the material

Other nomenclature is as in Eq. (29).

$$d_o = \frac{1}{2.3}\log\frac{r_o}{r_w}\sqrt{h^2 - 1.7\frac{Q}{K}} \tag{33}$$

where d_o = drawdown in the well, ft

r_o = radius of the circle of influence, ft, corresponding to the value of d_o

The drawdown of the base-pressure curve at any other point, expressed as d_o', may be determined by replacing r_w by r_r, where r_r is the distance from the center of the well to the point in question.

The rate of drawdown and of recovery in a pressure well is more rapid than in a gravity well. An approximate rate of recovery, based on a constant value of r_e, has been ex-

pressed by Slichter [1] as

$$T = K \log (y_o/y_1) \tag{34}$$

where T = time for water level to rise from y_o to y_1
K = constant for well and for the units used
y_1 = drawdown in well at any time T after observation of depth y_o

Interference among Wells. Interference between two gravity wells has been expressed by Muskat [2] as

$$Q_1 = Q_2 = \frac{K(h_e^2 - h_w^2)}{\log \dfrac{r_e^2}{r_w W}} \tag{35}$$

where W is the distance between the two wells and Q_1 and Q_2 are the rate of flow into each well. Other nomenclature is as shown in Eq. (29).

Interference between two pressure wells has been expressed by Slichter [3] as

$$Q_1 = Q_2 = \frac{2K(h_e - h_w)}{\log [r_e^2(W + r_e)^2] - \log [2Wr_e^3 + W^2r_w^2]} \tag{36}$$

Well Tests. The flow from a well, or gallery, for any drawdown can be predicted approximately by the application of the preceding formulas. In making observations for any of the parameters it is essential that the rate of flow and the drawdown in the well are constant. Conditions in a well field can be predicted by the construction of one or more test holes from which the necessary constants for the field can be determined.

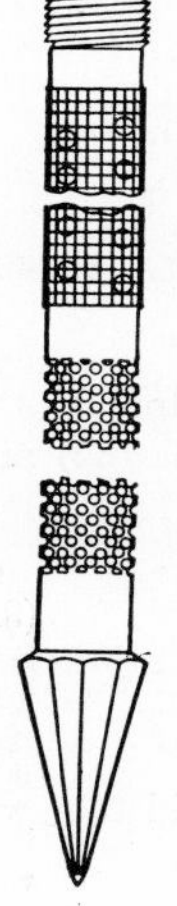

Fig. 18. Well point.

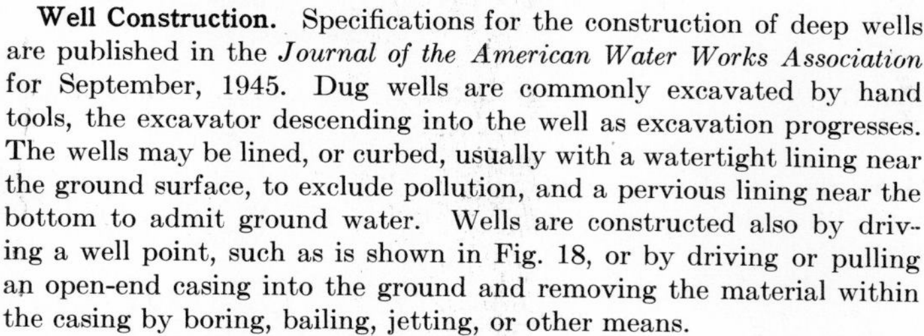

Well Construction. Specifications for the construction of deep wells are published in the *Journal of the American Water Works Association* for September, 1945. Dug wells are commonly excavated by hand tools, the excavator descending into the well as excavation progresses. The wells may be lined, or curbed, usually with a watertight lining near the ground surface, to exclude pollution, and a pervious lining near the bottom to admit ground water. Wells are constructed also by driving a well point, such as is shown in Fig. 18, or by driving or pulling an open-end casing into the ground and removing the material within the casing by boring, bailing, jetting, or other means.

Drilled wells are constructed by the standard method, and by other methods such as jetting, core drilling, and the hydraulic method. The standard method involves percussion drilling in which a drill is alternately raised and dropped in the descending hole. It is suited to the drilling of wells in any material from soft clay to the hardest rock. Other methods, such as the hydraulic-rotary and core-drill methods, may advance more rapidly through hard rock. In the standard method a percussion drill is alternately raised and dropped in the hole, the cuttings being either bailed out while the drill is removed from the hole, or washed out hydraulically by a current of rising water pumped into the hole during drilling.

In the jet method of drilling the tools consist of a hollow drill, drill pipe, and water swivel. The equipment includes a derrick from which the pipe is supported, a force pump, and the necessary water and power. Casing and a drive weight must be provided also. The hole is advanced by pumping water under pressure through the drill

[1] *U.S. Geol. Survey Water Supply Paper* 10, 1905.
[2] Muskat, M., "Flow of Homogeneous Fluids through Porous Media," McGraw-Hill Book Company, Inc., New York, 1937.
[3] *Loc. cit.*

bit while it is churned up and down. The casing follows the bit closely and the water, rising through the casing, lifts the finely divided cuttings out of the well.

Core drills consist of a hollow bit armed with abrasive teeth or industrial diamonds. The bit is advanced by rotating it to grind through the rock. As the bit advances a core is built up in the center of the hollow bit. The core is broken off at intervals as the bit descends, and the core is then raised to the surface. Another rotary abrasive method involves the use of steel shot held beneath a revolving, annular, steel shoe. The rotation of the shoe causes the shot to grind away the rock.

In the hydraulic rotary method the cuttings are loosened by the revolving drill bit. The cuttings are removed continuously, as the bit advances, by a mud-laden fluid which is pumped into the well through the hollow drill stem and a hole in the cutting bit. The fluid returns up the well carrying the cuttings with it.

Well Equipment. Typical deep-well equipment is illustrated in Fig. 19. Casing is used to prevent collapse of the hole, entrance of undesirable water, escape of good water, and the falling of material into the well. Materials most commonly used for well casings are wrought iron, alloyed or unalloyed steel, and ingot iron. The use of cast iron is increasing. Cement-lined and enamel-lined pipes of the above materials are available and are suitable in some wells. Other materials that have been used include copper, asbestos cement, plastics, and vitrified clay.

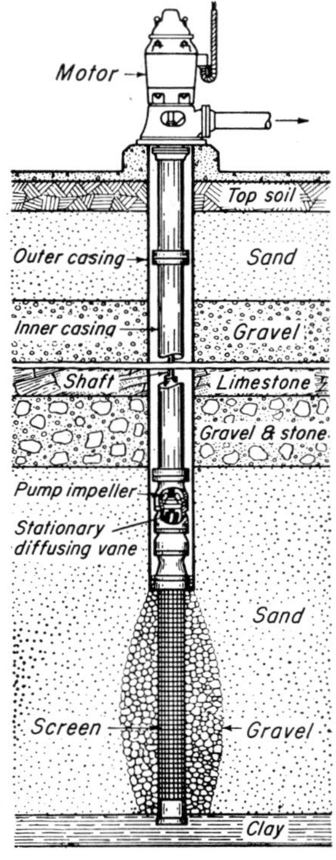

Fig. 19. Gravel-packed well with turbine pump. ("*The Sanitary Industry*," by permission of Johns Manville, Industrial Products Division.)

A screen is required at the bottom of a well to prevent the walls of the aquifer from caving into the hole, to exclude fine sand, and to permit the entrance of water. Materials used for screens include corrosion-resistant alloys, iron or steel which is either black or galvanized, concrete, wood, asbestos cement, and gravel. Since the useful life of the well is dependent primarily on the durability of the screen, emphasis should be placed on its careful selection. The size of the openings in a screen may be determined, after a study of a mechanical analysis of the aquifer, to permit the passage of all fine particles representing a certain percentage, by weight, of the water-bearing material. It is common practice to use openings that will pass about 70 percent, or more, of the sand grains in the aquifer, where the uniformity coefficient is high. The total area of the openings in a screen should be such as to maintain an entrance velocity less than necessary to carry the finest particle of sand that is to be excluded by the screen. In general, it should be less than about 0.125 to 0.2 ft per sec. Lifting velocities of sand grains with a specific gravity of about 2.65 are shown

Table 11. Velocities of Water Required to Lift Sand Grains with a Specific Gravity of 2.65

Diameter of grains, mm......	Up to 0.25	0.25–0.50	0.50–1.00	1.00–2.00	2.00–4.00
Velocity of water, ft per sec...	0.0–0.10	0.12–0.22	0.25–0.33	0.37–0.56	0.60–2.00

in Table 11. It is generally desirable, but not essential, that the length of the screen be equal to the thickness of the aquifer penetrated.

Gravel-wall wells, that is, wells with a wall of gravel outside of and surrounding the screen, as shown in Fig. 19, are used to increase the area of contact between the aquifer and the well, thus diminishing the velocity of the entering water and increasing the specific capacity of the well.

The size of gravel should be selected to prevent the entrance into the well of particles of sand greater than is desired. The openings in the inner metallic screen should be as large as possible without permitting the entrance of gravel into the well.

Well Pumps. Centrifugal pumps dominate the field in the pumping of water from wells. Other types are used only under special circumstances. Reciprocating pumps are obsolete except for domestic and small water supplies. Jet pumps are widely used for small supplies up to about 75 gpm. Centrifugal well pumps are available in sizes at and above about 5½-in. impellers, with capacities up to 4,000 to 5,000 gpm. Almost any reasonable lift that is required can be obtained by increasing the number of stages in the pump. Efficiencies up to 90 percent and better are reported, but only 75 to 80 percent or lower should be anticipated in capacities lower than about 200 gpm.

The specific speed N_s of a centrifugal pump is important in its selection. N_s for a centrifugal pump is the speed at which the impeller would run when discharging 1 gpm under a 1-ft lift at the highest efficiency. It can be expressed as

$$N_s = \frac{\text{rpm}\sqrt{\text{gpm}}}{H^{3/4}} \qquad (37)$$

where rpm = speed of revolution of pump impeller
gpm = rate of discharge from the pump
H = discharge head, ft

Fabrin [1] has shown that for deep-well pumps the most favorable designs will be found at a specific speed of about 2,500, with a favorable range between 1,500 and 4,100. Specific speed is related to the ratio of wheel diameter D and to inlet diameter B as

$$(D/B)N_s^{0.53} = 10{,}600 \qquad (38)$$

when N_s is less than 4,100 and D/B is more than 1.25, and

$$(D/B)N_s = 530{,}000 \qquad (39)$$

when N_s is more than 4,100 and D/B is less than 1.25. Where N_s is between 4,100 and 5,800 a mixed-flow pump, combining centrifugal and screw-pump characteristics, should be used. In this case

$$(DN_s)/B = 530{,}000 \qquad (40)$$

where D/B is less than 1.25. Where N_s is greater than 7,500 the propeller or screw pump should be used.

It is to be noted that, when N_s is high, D/B is low and the condition is unfavorable to high suction lifts.

An empirical formula, based on recommendations of Fabrin, showing the relation between some controlling conditions in pump selection can be expressed as

$$D = \frac{236Q^{0.154}H^{0.256}}{N^{0.678}} \qquad (41)$$

[1] Fabrin, A. O., *Water and Sewage Works*, April, 1946, p. R87.

where D = outside diameter of pump casing, in.
H = total head, ft of water, for one stage or wheel
N = speed of wheel, rpm
Q = rate of discharge, gpm

The formula is limited to pumps of the highest efficiency, with specific speeds between 4,100 and 1,500.

Air-lift Pumps. An air lift is shown diagrammatically in Fig. 20. The requisite depth of the well to provide adequate submergence is based on the submergence ratio $D/(D + h)$. Some suggested values of this ratio are given in Table 12. The volume of

TABLE 12. EFFECT OF SUBMERGENCE ON EFFICIENCIES OF AIR LIFT AT HATTIESBURG, MISS.

Ratio D/h....	8.70	5.46	3.86	2.91	2.25	1.86	1.45	1.19	0.96
Submergence ratio $\frac{D}{(D + h)}$.	0.896	0.845	0.795	0.745	0.693	0.650	0.592	0.544	0.490
Percentage efficiency...	25.5	31.0	35.0	36.6	37.7	36.8	34.5	31.0	26.5

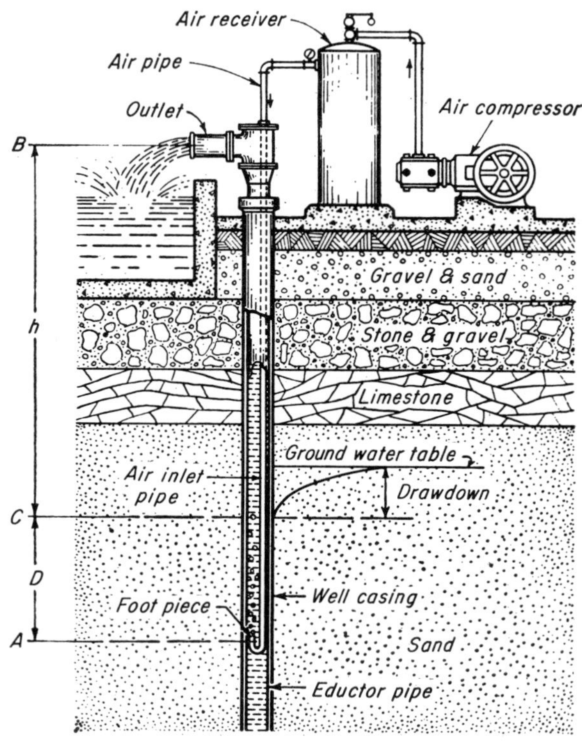

FIG. 20. Air-lift pump. (*"The Sanitary Industry,"* *by permission of Johns Manville, Industrial Products Division.*)

free air required can be expressed as

$$Q_a = \frac{Q_w(h + h_1)}{75E \log r} \tag{42}$$

where Q_w = rate at which water is to be raised, cu ft per sec
h_1 = velocity head at discharge, usually taken as 6 ft for deep wells, down to 1 ft for shallow wells
E = efficiency, usually about 45 to 50 percent, with submergence ratio of 50 to 65 percent and submergence between 350 and 500 ft

The diameter of the air pipe can be computed from

$$d = \sqrt{\frac{4Q}{\pi V}} \tag{43}$$

where d = diameter of pipe
Q = rate of flow of *compressed* air in pipe
V = velocity of flow of *compressed* air in pipe

The loss of head or pressure due to friction in the air pipe can be computed from

$$P_1^2 - P_2^2 = (0.0006Q_a^2)l/d^5 \tag{44}$$

where P_1 = absolute initial pressure, psi
P_2 = absolute terminal pressure, psi
Q_a = free-air equivalent passing through the pipe, cu ft per min
l = length of pipe, ft
d = diameter of pipe, in.

The size of the eductor pipe is computed from the expression $Q = A/V$ where A is the cross-sectional area of the rising column of air and water. Some suggested velocities are:

Size of air pipe, in.	2¾	3	4	5
Velocity at entrance to eductor pipe, ft per sec	5.6	7.3	11.8	12.7

The velocity at the discharge is assumed to be about 20 to 25 ft per sec.

Jet Pumps. A jet pump consists of a centrifugal pump and motor at the ground surface and a jet down in the well below the water level, discharging at high velocity through a contracted section into the lift pipe. The centrifugal pump has two discharge pipes. One leads down to the jet. The other carries the water into the distribution system or into a storage tank.

PUMPS AND MOTORS

Types of Pumps. Centrifugal pumps are used almost exclusively in waterworks pumping. A diagrammatic section through such a pump is shown in Figs. 21 and 22. Such pumps are available commercially for almost any capacity that might normally be required and for lifts up to 700 ft in one stage. There is no theoretical limit to the number of stages that can be placed in series to increase the lift. Maximum lifts per stage are commonly near to 250 ft. Efficiencies up to 93 percent have been attained by large pumps but, in general, the efficiency of small pumps is lower. Speeds of 900 to 1,800 rpm, and even higher, are used.

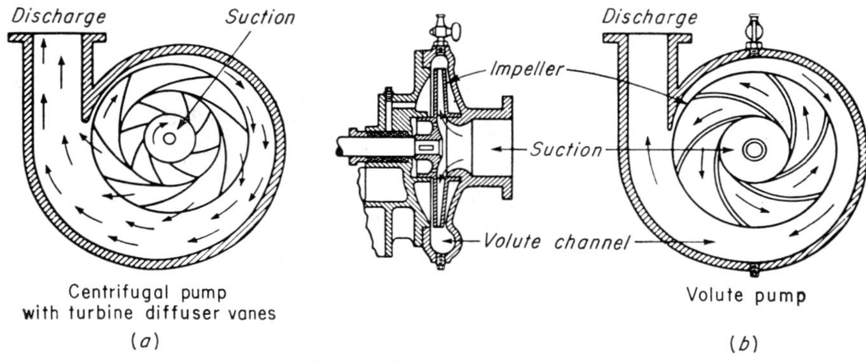

Fig. 21. Centrifugal pumps.

Relationships of Characteristics. Relationships between characteristics of centrifugal pumps, as developed from theoretical considerations, may be summarized as follows:

1. Q varies directly as the number of impellers in parallel for any given speed.

2. H varies directly as the number of impellers in series, i.e, the number of stages, for any given speed.

3. When N varies, Q varies directly as N; H varies directly as N^2; Q varies as H; and P varies as $H^{1.5}$.

These theoretical relationships do not hold exactly because of errors in the hypotheses of the conditions of flow of water, but they are sufficiently accurate for approximate solutions.

The characteristic curves or, more commonly, the characteristics of a centrifugal pump, are the relations between the various conditions affecting the performance. They are usually expressed graphically with the rate of discharge Q as the abscissas and other factors plotted, at constant speeds, as ordinates, such as the head H, the power P, and the efficiency E. *The characteristic* is the graph showing the relationship between H and Q, for any given speed, with H plotted as ordinates. Other methods of plotting characteristics are also used. Characteristic curves for a single pump at different speeds are shown in Figs. 23 and 24. Figure 23 shows that this pump should be operated at 1,750 rpm, with a discharge of about 720 gpm, and at a head of 244 ft. A pump with flat characteristics at all speeds and a wide plateau at the highest efficiency is most desirable.

Suction Lift and Cavitation. The height to which water can be raised by suction can be expressed as

$$H = A - (V_p + V_h + H_f) \tag{45}$$

where A = pressure of the atmosphere
V_p = vapor pressure of water
V_h = velocity head in suction line
H_f = friction losses in suction line

Where the pump operates at too high a speed cavitation may result, causing erosion and difficulty in operation. Unsatisfactory operation may be expected at a specific speed above about $(k/H)^n$ where H is the total dynamic suction head and k and n are constants.

The permissible height of suction lift to avoid cavitation is related to the specific speed as indicated in Table 13, as computed from information by the Hydraulic Insti-

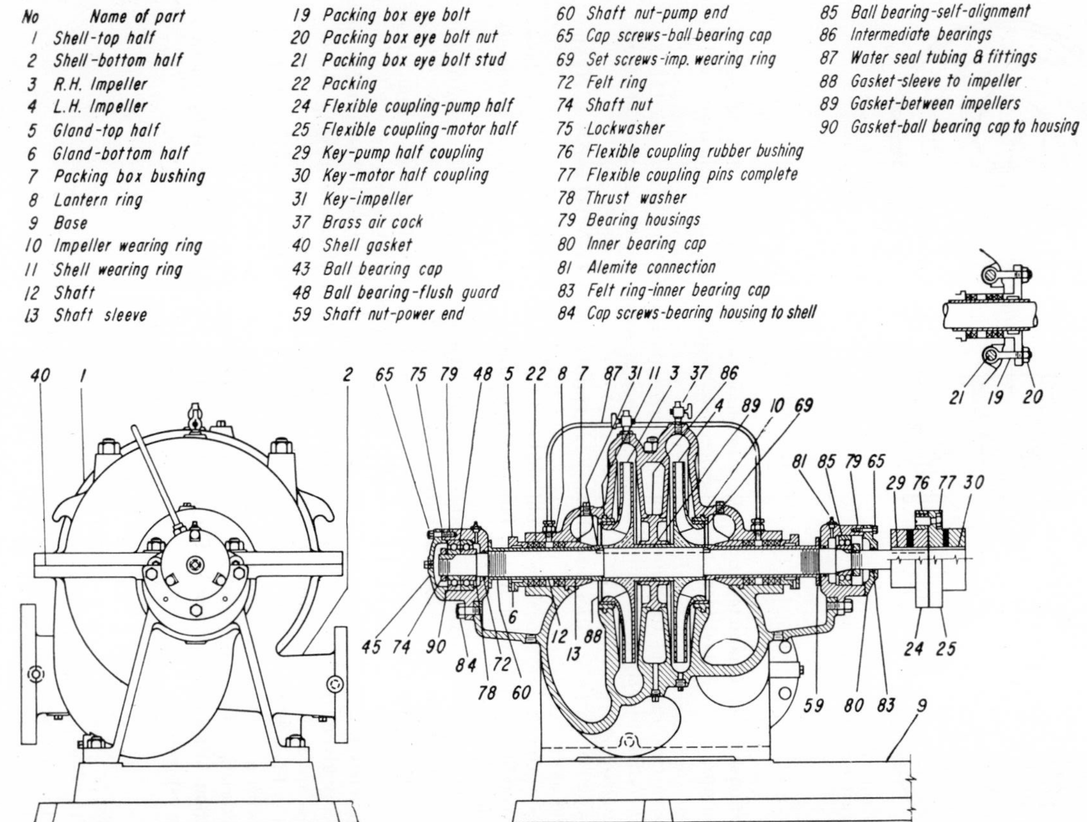

No	Name of part	No	Name of part	No	Name of part	No	Name of part
1	Shell-top half	19	Packing box eye bolt	60	Shaft nut-pump end	85	Ball bearing-self-alignment
2	Shell-bottom half	20	Packing box eye bolt nut	65	Cap screws-ball bearing cap	86	Intermediate bearings
3	R.H. Impeller	21	Packing box eye bolt stud	69	Set screws-imp. wearing ring	87	Water seal tubing & fittings
4	L.H. Impeller	22	Packing	72	Felt ring	88	Gasket-sleeve to impeller
5	Gland-top half	24	Flexible coupling-pump half	74	Shaft nut	89	Gasket-between impellers
6	Gland-bottom half	25	Flexible coupling-motor half	75	Lockwasher	90	Gasket-ball bearing cap to housing
7	Packing box bushing	29	Key-pump half coupling	76	Flexible coupling rubber bushing		
8	Lantern ring	30	Key-motor half coupling	77	Flexible coupling pins complete		
9	Base	31	Key-impeller	78	Thrust washer		
10	Impeller wearing ring	37	Brass air cock	79	Bearing housings		
11	Shell wearing ring	40	Shell gasket	80	Inner bearing cap		
12	Shaft	43	Ball bearing cap	81	Alemite connection		
13	Shaft sleeve	48	Ball bearing-flush guard	83	Felt ring-inner bearing cap		
		59	Shaft nut-power end	84	Cap screws-bearing housing to shell		

FIG. 22. Parts of a two-stage centrifugal pump. (*American Well Works.*)

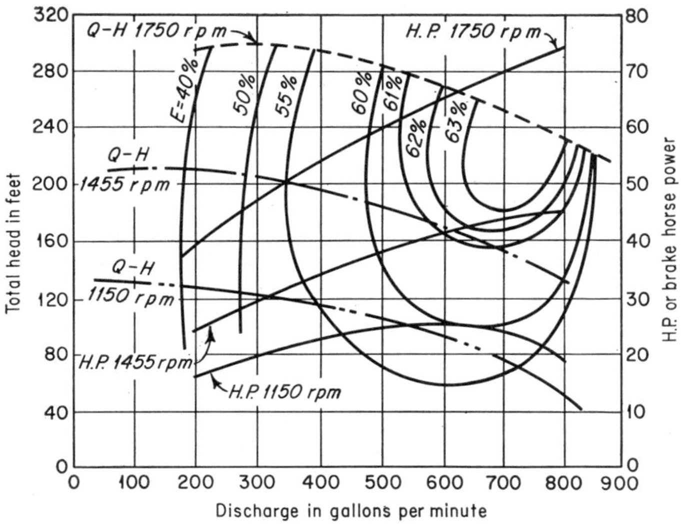

FIG. 23. Characteristics of a centrifugal pump at different speeds. (*Eng. News-Record, vol.* 91, *p.* 564.)

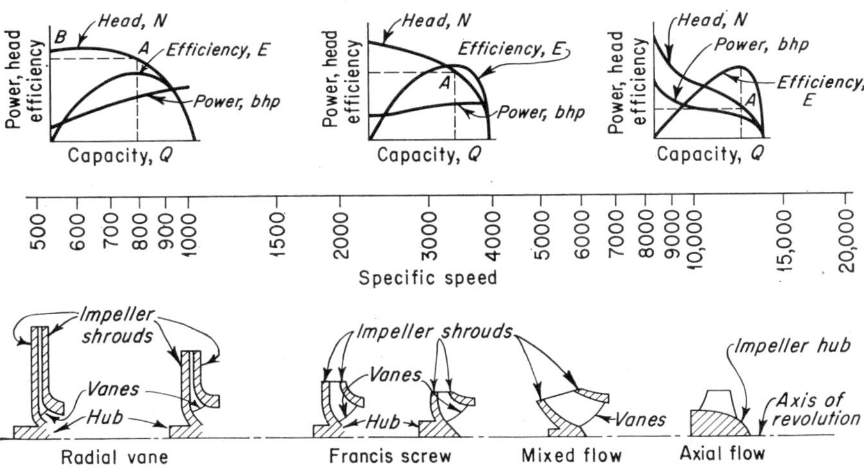

FIG. 24. Characteristic graphs and specific-speed scale for impellers of various design. (*Roy Carter and I. J. Karassik, Water and Sewage Works, August,* 1951, *p.* 335.)

tute.[1] The cavitation factor, usually expressed as Σ or σ, is equal to $\frac{H_a - H_{vp} - H_s}{H_t}$ where H_a is atmospheric pressure, H_{vp} is vapor pressure, H_s is suction lift, and H_t is total dynamic head of the pump. Critical σ occurs when cavitation begins. Operating σ should be a safe margin above it, as determined by test.

TABLE 13. PERMISSIBLE SUCTION LIFT TO AVOID CAVITATION IN CENTRIFUGAL PUMPS
Lift in feet

Permissible suction lift, ft	Single-suction, mixed flow $K = HN_s^{1.33}$			Single-suction, shaft through impeller eye $K = HN_s^{1.7}$			Double-suction pump $K = HN_s^{1.7}$		
	Dynamic head, ft		K millions	Dynamic head, ft		K millions	Dynamic head, ft		K millions
	N_s 15,000	N_s 5,000		N_s 3,520	N_s 1,760		N_s 5,000	N_s 2,500	
25	....	....	...	21.8	70	24	21.8	70	42.5
20	5.7	24.1	2.0	35	112	38	35	112	68.5
15	7.8	34	2.8	44	151	49	44	151	89
10	10.1	43	3.6	59	192	64	59	192	116
5	12.1	53	4.4	71	214	75	71	214	135
0	14.4	62	5.2	85	283	94	85	283	170
−5[1]	16.6	72	6.0	93	305	102	93	305	184
−10	18.9	81	6.8	106	340	116	106	340	208
−15	20.5	91	7.5	115	370	126	115	370	225

[1] Negative sign indicates depth of water in suction pit over top of pump suction.

ELECTRIC MOTORS

Selection. In the selection of an electric motor conditions to be considered, in addition to the type of current and of circuit available, include: first cost and maintenance cost; suitability for the character of work to be done; speed, speed regulation, and speed control; ease in starting; starting current in terms of full-load current; starting torque in terms of full-load torque; power factor; rated power; partial load characteristics; and whether enclosed, semienclosed, or open. Some characteristics of the more common types of electric motors are listed in Table 14. Motors with other characteristics may be manufactured with special windings, usually at a higher cost than for standard-type motors.

Electric motors are available for almost any desired horsepower and for operation at various speeds or with different kinds of electric current. Standardization of practice is becoming such that three-phase 60-cycle current is most commonly used in the United States. Voltages are not so generally standardized. Voltages of 220 and 440 are commonly used, and higher voltages are to be found, with the tendency, in practice, to use the highest possible voltages to obtain the highest efficiency commensurate with safety. High-speed motors are cheaper than low-speed machines, because the parts are lighter for the same power. If a centrifugal pump is to be driven by an electric motor, it may be cheaper to use a slow-speed motor directly connected to the pump on the same shaft rather than to use a speed-reduction device driven by a high-speed motor. Satisfactory electric motors have efficiencies above 90 percent; 95 percent is not uncommon. Squirrel-cage induction motors are the most widely used in waterworks because of their rugged

[1] See also *Water and Sewage Works*, April, 1946, p. R80.

TABLE 14. CHARACTERISTICS OF ELECTRIC MOTORS [1]

Type of Motor	Characteristics, Advantages, and Applications
Shunt-wound, d-c	*Starting current:* [2] normally about 1.5 times full load. *Starting torque:* normally about 1.5 times full load. *Maximum torque:* twice full load. *Speed regulation:* about 10 percent variation between no load and full load. *Speed control:* adjustable by means of rheostat, varies not more than 10 percent for any adjustment. *Sizes available:* up to 200 hp. *Suitable for:* frequent starting and stopping with light or medium starting loads, shop machinery, conveyors, fans, centrifugal pumps
Series-wound, d-c	*Starting current:* same as for shunt-wound motor. *Starting torque:* 4.5 times full load but normally 2.25 to 3.0 times full-load torque. *Maximum torque:* 4 times full-load torque. *Speed regulation:* poor, increases dangerously with decrease of load, may "run away." *Speed control:* same as shunt-wound motor. *Sizes available:* 3 to 200 hp. *Suitable for:* load that cannot be removed from motor such as cranes, fans, and railways, and for frequent starting and stopping. Not suitable for centrifugal pumps
Compound-wound, d-c	Characteristics lie between shunt-wound and series-wound and depend on the character of the windings
Induction a-c, polyphase squirrel-cage	*Starting current:* 5 to 10 times full-load current. For sizes above 5 hp requires special starting devices. Where frequent starting is necessary sizes below 5 hp may require special starting windings, condensers, or other starting aids. *Starting torque:* low in terms of full load. *Maximum torque:* 2 to 2.5 times running torque. *Speed regulation:* good, may diminish 5 percent between start and full load. *Speed control:* none, runs only at designed speed. *Sizes available:* 0.1 to 400 hp. *Power factor:* for split phase, induction: fractional hp = 0.55 to 0.75; 1 to 10 hp = 0.75 to 0.85; and condenser type = 0.75 to 1.0. Squirrel-cage polyphase = 0.75 to 0.90. *Suitable for:* constant load and continuous operation. Excellent for fans and centrifugal pumps
Repulsion-induction, single-phase, a-c	*Starting current:* 1.75 to 2.5 times full load. *Speed regulation and control:* above synchronism at no load. At full load is similar to that of an induction motor. *Sizes available:* ⅛ to 15 hp. *Suitable for:* heavy starting loads or frequent starting and stopping
Synchronous, a-c	*Starting current:* 4 to 8 times full load. Hard to start. Field must be excited with direct current. *Starting torque:* very low. *Maximum torque:* 1.5 to 1.75 running torque. *Speed regulation:* perfect, no variations in speed under varying loads. *Speed control:* none. *Sizes available:* 20 to 5,000 hp. *Power factor:* up to 1.0. *Suitable for:* steady load and constant operation. Excellent for pumps. Improves power factor

[1] From BABBITT, H. E., and J. J. DOLAND, "Water Supply Engineering," 5th ed., McGraw-Hill Book Company, Inc., New York, 1955.

[2] Starting currents in this list are stated as 75 to 100 percent of the blocked-rotor values. The free-rotor value may be taken as 75 percent of the blocked-rotor value.

construction, their foolproof characteristics if not overloaded, and their nearly constant speed.

Operating Hazards for Polyphase Induction Motors. Alternating current may vary in phase, voltage, or frequency. If one or more phases are interrupted while the motor is idle, the motor will not start. If it is running at the time of the interruption or if it is started by externally applied power, as by pulling on the belt, the motor will run at normal speed as a single-phase motor and will deliver about 60 percent of normal-load torque, with a pull-out torque about two-thirds to one-half of normal. The effect of

operating a polyphase motor under full load and single-phase conditions will be to overload the motor and lead to its probable destruction.

A drop in voltage may cause a motor to refuse to start or to pull out in operation, because torque varies closely as the square of the voltage. A motor thus stopped with current in its windings is open to destruction. An excessive increase in voltage causes core losses approximately as the square of the voltage, and rise in temperature endangering the insulation of the windings. Change in current frequency causes variations of motor speed and torque approximately directly as the speed. At normal operating load and with increased frequency core losses are increased resulting in increased heating. Since the power required by a centrifugal pump varies approximately as the cube of the speed an increase in speed might result in overloading the motor.

Induction motors will, in general, not be seriously affected by variation of 10 percent, more or less, in either voltage or frequency. Interruption of phase is, however, a serious hazard. The sudden interruption of power will have no serious effect on an induction motor unless it is connected to a reversible load, such as a centrifugal pump. Under such circumstances both the motor and the pump must be protected against reversal. Unless a motor is equipped for automatic self-starting under the load, it must be protected against a resumption of power in the event of its interruption.

Protection of Operators and of Motors. Motor frames should be grounded and the grounding should be inspected frequently and carefully maintained, to avoid electrically charging the frame by current leakage, stray currents, or induced current. Grounding protects both the operator and the motor. Connection to a water pipe makes a satisfactory ground and, where properly done, is not objectionable.

Electric motors may be protected against overload by fuses and by circuit breakers. Fuses are limited to current under about 600 amp. Time-lag fuses are used, particularly on small motors, to permit the momentary surge of current up to 500 percent of rating, in starting the motor, but to interrupt the circuit if the temperature of the motor rises excessively. The thermal cutout is designed and functions similarly to a plug fuse and is used to protect small motors against overloads. It protects the motor against overload, whereas the time-lag fuse protects the motor against too high a temperature from any cause. Fuses and thermal cutouts do not protect against lightning or surges of current.

Circuit breakers, actuated either thermally or magnetically, are available to interrupt currents from 15 amp up, under all conditions of service and, if required, to close the circuit automatically when the current has returned to normal. Circuit breakers, actuated by magnetic devices or relays, are available for protection against all the hazards normally met by motors, except to arrest lightning.

The lightning arrester is a device placed on an electric circuit to ground excess voltage in the circuit and to restore the circuit on the return to normal voltage. Since the energy grounded through a lightning arrester may be high, the arrester should be placed in a safe location away from other equipment or combustible materials.

DISTRIBUTION OF WATER

The Distribution System. A waterworks distribution system includes pipes, valves, hydrants, and other appurtenances for conveying water; reservoirs for storage, equalizing, and distribution purposes; service pipes to the consumers; meters; and all other parts of the conveying system after the water leaves the main pumping station or the main distributing reservoir. Distribution systems are laid out with pipes located to provide the least loss of head consistent with satisfactory service. No general rule is applicable to all conditions.

Pipe Sizes. When the rate of flow in a pipe is known the diameter selected should give an economical velocity, which is in the order of 3 to 4 ft per sec for average flow,

and may be up to 10 ft per sec for fire supply. Smaller pipes may be chosen to give lower first cost but at the expense of greater operating or pumping cost. In the selection of pipe sizes the following principles may serve as a guide:

1. Where no fire service is involved, 2-in. pipe may be used for not more than 300 ft, and no more than 1,300 ft of 4-in. pipe. Where pipes are connected at each end to larger mains, and no fire service is involved, 2- or 3-in. pipes may be 600 ft long, and a 4-in. pipe may be 2,000 ft long. No pipes smaller than 6 in. should be used where fire service is involved. Above 600 ft in length, fire-service pipes should be at least 8 in. in diameter.

2. Pipes should be interconnected at intervals not greater than about 1,200 ft. Dead ends should be avoided.

3. Arterial mains should be duplicated but not laid in the same street or on the same route.

Analyses of Pressure Losses in Pipes. Pressure losses in a distribution system for any condition of flow can be measured by an electric-network analyzer [1] or computed by the Hardy Cross method of successive approximations. The setting up of a distribution system on an electric-network analyzer is not simple and the cost is high. Hence, the instrument is not widely used. However, results for different flow conditions are easily and quickly obtained from it after the analyzer has been set up.

The Hardy Cross Method. This method is basically one of trial and error which is best explained by the solution of an example.

Example: Let it be desired to compute the rate of flow in each pipe of the distribution system shown in Fig. 25; the pressures at each intersection of two or more pipes; and to draw the 2-psi piezometric contours. The formality of procedure and the tabulation

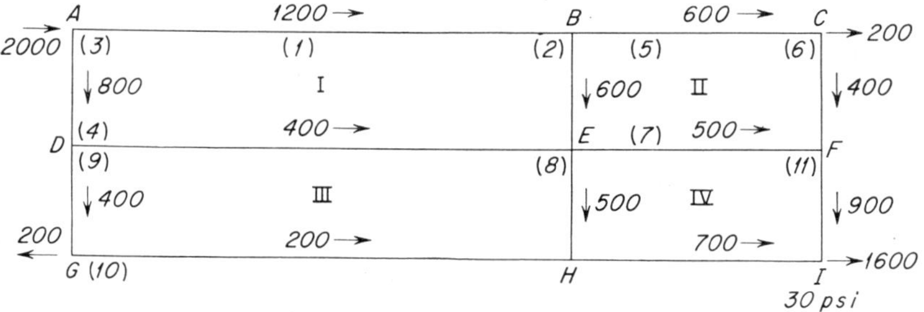

FIG. 25. Pipe network for Hardy Cross method of computation of losses of head in a waterworks distribution system. (*Layout, but not the rates of flow, originally selected by G. M. Fair, Eng. News-Record, Feb. 27, 1941, p. 50.*)

of data and of computations are to follow a standard for convenience only. A formal procedure is desirable where two or more computers are to check the solution.

Solution: 1. Number each circuit from left to right and then downward by lines, using roman numerals, as indicated in the figure. It is necessary to have a system of numbering of circuits in order to identify them.

2. Compute the length of 8-in. pipe with C in the Hazen and Williams formula = 100, which is equivalent to each length of pipe in the distribution system. Record, as in Table 15A. Factors for the conversion of any size of pipe to equivalent length of 8-in. pipe are shown in Table 15A.

[1] See also CAMP, T. R., and H. L. HAZEN, *Jour. New Engl. Water Works Assoc.*, December, 1934, p. 383; MCILROY, M. S., *Jour. New Engl. Water Works Assoc.*, December, 1951, p. 299; *Jour. Am. Water Works Assoc.*, April, 1950, p. 347.

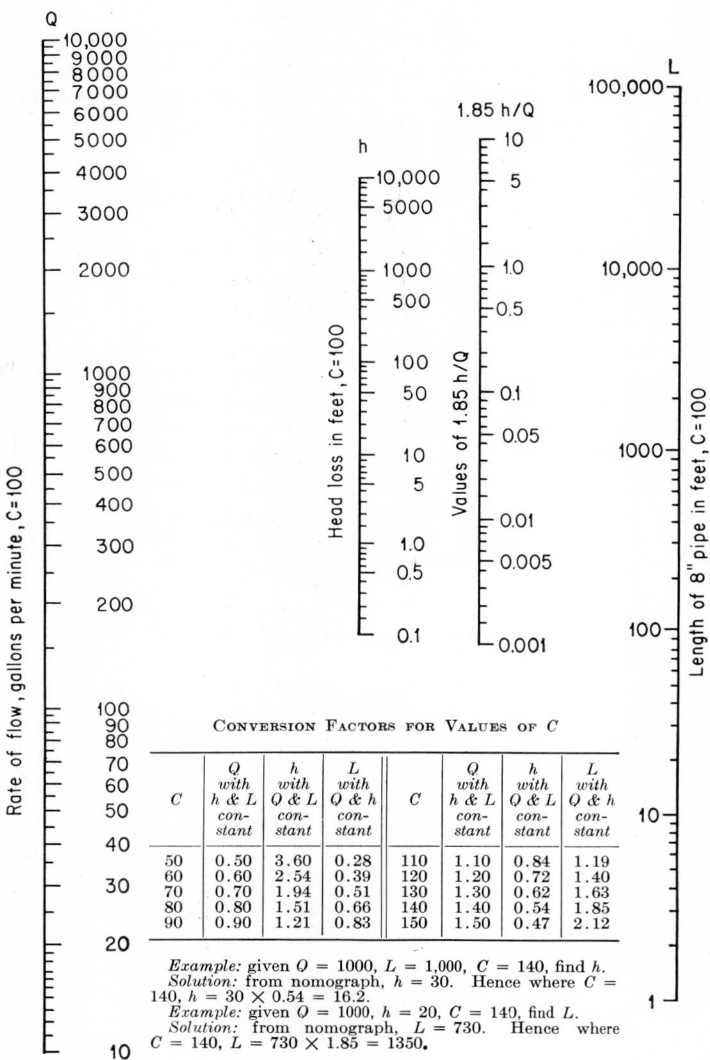

CONVERSION FACTORS FOR VALUES OF C

C	Q with h & L constant	h with Q & L constant	L with Q & h constant	C	Q with h & L constant	h with Q & L constant	L with Q & h constant
50	0.50	3.60	0.28	110	1.10	0.84	1.19
60	0.60	2.54	0.39	120	1.20	0.72	1.40
70	0.70	1.94	0.51	130	1.30	0.62	1.63
80	0.80	1.51	0.66	140	1.40	0.54	1.85
90	0.90	1.21	0.83	150	1.50	0.47	2.12

Example: given $Q = 1000$, $L = 1{,}000$, $C = 140$, find h.
Solution: from nomograph, $h = 30$. Hence where $C = 140$, $h = 30 \times 0.54 = 16.2$.

Example: given $Q = 1000$, $h = 20$, $C = 140$, find L.
Solution: from nomograph, $L = 730$. Hence where $C = 140$, $L = 730 \times 1.85 = 1350$.

FIG. 26. Nomograph for computations of head losses and other conditions of flow in 8-in. pipe, using Hazen and Williams $C = 100$.

TABLE 15A. LENGTHS OF 8-in. PIPE WITH $C = 100$ THAT ARE EQUIVALENT TO A UNIT LENGTH OF THE PIPE SHOWN IN THE TABLE

Diam, in.	$C = 60$	$C = 80$	$C = 100$	$C = 120$	$C = 140$
2	2,105	1,286	851	608	456
3	294	180	119	84.9	63.9
4	73	44	29	20.7	15.6
5	24	15.0	9.9	7.1	5.3
6	10.0	6.1	4.06	2.9	2.2
8	2.46	1.51	1.00	0.713	0.537
10	0.84	0.51	0.34	0.242	0.180
12	0.35	0.21	0.14	0.100	0.075
14	0.163	0.10	0.066	0.047	0.035
16	0.084	0.051	0.034	0.024	0.018
18	0.047	0.029	0.019	0.014	0.010
20	0.0285	0.0174	0.0115	0.0082	0.00617
24	0.0116	0.0071	0.0047	0.0034	0.0025
30	0.0040	0.0024	0.0016	0.0011	0.00086
36	0.00163	0.0010	0.00066	0.00047	0.00035

3. Show in Fig. 25 a rate and direction of flow which has been assumed in each pipe. Experience and judgment are best guides in making these assumptions. It is to be noted that at each intersection of pipes the rate of flow entering must equal the rate of flow leaving the intersection.

4. Fill in the figures in work table 15B beginning with loop I in Fig. 25, working from left to right and downward with the loops. Each loop is assumed to begin at the upper left corner and to terminate at the lower right corner of the loop. Flows in that direction are indicated by a + sign. If in the opposite direction they are indicated as negative. Values of head loss h_l are computed from the nomograph in Fig. 26. It is to be noted that the corrected flow in each circuit is used in the computation of ΔQ in the next circuit.

5. Repeat the procedure in step 4 until ΔQ is as small as desired. Show the final flows as indicated in Fig. 27.

6. Compute the pressure losses between the intersections, in feet, using Fig. 26, with the equivalent lengths of 8-in. pipe, designating the intersections with letters as indicated in Fig. 27, and convert to psi. Computations are shown in Table 16. Add the pressure losses along the various routes and show the pressures at each intersection. Sketch the 2-psi contours, as indicated in Fig. 27.

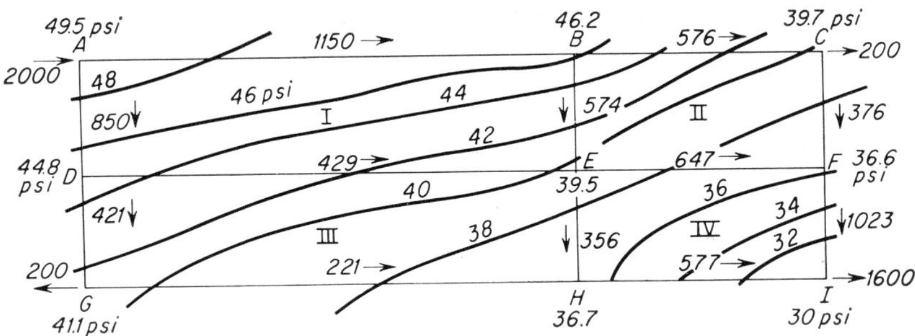

FIG. 27. Work sketch for Hardy Cross solution.

Piezometric Contours. Pressure contours or piezometric contours are lines joining points of equal pressure under a given condition of put-ins and take-offs in a distribution

Table 15B. Computations of Rates of Flow in Pipes Shown in Figs. 10-27 *

Line No.	Loop	Pipe	L	Q	h_l	$\frac{nh_l}{Q}$	$\frac{\Sigma nh_l}{Q}$	Σh_l	Δh_l	ΔQ	Q_2	Q_3	h_l	$\frac{nh_l}{Q}$	$\frac{\Sigma nh_l}{Q}$	Σh_l	Δh_l	ΔQ	Q_4	Q_5	h_l	$\frac{nh_l}{Q}$	$\frac{\Sigma nh_l}{Q}$	Σh_l	Δh_l	ΔQ	Q_6 *
1	1	2	3	4	5	6	7	8	9	10	11		5	6	7	8	9	10	11		5	6	7	8	9	10	11
2	..	1	200	1,200	+ 8.2	0.013	.....	....	....	...	1,172	1,172	7.8	0.012	.....	....	...	..	1,143	1,143	7.4	0.012	.....	....	...	..	1,150
3	I	2	1,450	600	+16.2	0.050	.....	24.4	3.8	28	572	606	17.2	0.052	.....	25.0	4.0	29	577	574	15.3	0.049	.....	22.7	...	..	581
4	..	3	500	800	9.8	0.023	.....	....	....	...	828	828	10.4	0.023	.....	....	...	..	857	857	11.0	0.024	.....	....	...	..	850
5	..	4	2,000	400	10.8	0.050	0.136	20.6	....	...	428	397	10.6	0.049	0.136	21.0	...	..	426	435	12.6	0.054	0.139	23.6	0.9	7	428
6																											
7	..	5	1,450	600	16.6	0.051	.....	....	....	...	566	566	14.9	0.049	.....	....	...	..	569	569	14.5	0.047	.....	....	...	..	576
8	II	6	1,450	400	7.8	0.036	.....	24.4	5.1	34	366	366	6.7	0.034	.....	21.6	...	..	369	369	6.8	0.034	.....	21.3	...	..	376
9	..	2	1,450	572	15.2	0.049	.....	....	....	...	606	577	15.4	0.049	.....	....	...	..	574	581	15.7	0.050	.....	....	...	..	574
10	..	7	500	500	4.1	0.015	0.151	19.3	....	...	534	647	6.6	0.019	0.151	22.0	0.4	3	644	651	6.7	0.019	0.150	22.4	1.1	7	644
11																											
12	..	4	2,000	428	12.3	0.053	.....	....	....	...	397	426	12.2	0.053	.....	....	...	..	435	428	12.4	0.054	.....	....	...	..	429
13	III	8	1,450	500	11.8	0.044	.....	24.1	6.9	31	469	356	6.4	0.028	.....	18.6	...	..	365	358	6.4	0.033	.....	18.8	...	..	359
14	..	9	1,450	400	7.8	0.036	.....	....	....	...	431	431	9.0	0.039	.....	....	...	..	422	422	8.7	0.038	.....	....	...	..	421
15	..	10	5,800	200	9.4	0.087	0.220	17.2	....	...	231	231	11.6	0.093	0.213	20.6	2.0	9	222	222	10.4	0.086	0.211	19.1	0.3	1	221
16																											
17	..	7	500	534	4.6	0.016	.....	....	....	...	647	644	6.6	0.019	.....	....	...	..	651	644	6.4	0.018	.....	....	...	..	647
18	IV	11	500	900	12.1	0.025	.....	16.7	....	...	1,031	1,013	15.2	0.028	.....	21.8	...	..	1,020	1,020	14.8	0.027	.....	21.2	...	..	1,023
19	..	8	1,450	469	10.5	0.041	.....	....	....	...	356	365	6.7	0.034	.....	....	...	..	358	359	6.4	0.033	.....	....	...	..	356
20	..	12	1,450	700	22.0	0.058	0.140	32.5	15.8	113	587	1,587	16.0	0.050	0.131	22.7	0.9	7	580	580	15.2	0.048	0.126	21.6	0.4	3	577

* The values of flow shown on the pipes in Fig. 27 have been carried one step further than is shown in the last column of this table.

TABLE 16. COMPUTATIONS OF PRESSURE LOSSES AND OF PRESSURES AT INTERSECTIONS IN FIG. 27

Pipe No.	Q, gpm	Equivalent length 8-in. pipe, ft	Head loss, ft	Head loss, psi	Pipe No.	Q, gpm	Equivalent length 8-in. pipe, ft	Head loss, ft	Head loss, psi	Intersection	Pressure, psi	Intersection	Pressure, psi
1	1,150	200	7.7	3.3	7	647	500	15.4	6.7	*A*	49.5	*G*	41.1
2	574	1,450	15.3	6.0	8	356	1,450	7.0	3.0	*B*	46.2	*H*	36.7
3	850	500	10.9	4.7	9	421	1,450	8.6	3.7	*C*	39.7	*I*	30.0
4	429	2,000	12.3	5.3	10	221	5,800	10.1	4.4	*D*	44.8		
5	576	1,450	15.4	6.7	11	1,023	500	15.3	6.6	*E*	39.5		
6	376	1,450	7.0	3.0	12	577	1,450	15.4	6.7	*F*	36.6		

system. Piezometric contours for the problem solved in the preceding pages are shown in Fig. 27.

Service Pressures and Fire Pressures. Good service requires a pressure in the pipe in the street between about 50 and 90 psi; lower pressures will not deliver water at a satisfactory rate or height and higher pressures may cause trouble in the maintenance of plumbing. In cities with rough topography excessive pressures may be avoided by creating zones in the distribution system in each of which the maximum and minimum pressure limits are controlled by distribution reservoirs, pumps, or pressure-reducing valves.

A pressure of 50 to 75 psi is required at the base of a fire nozzle for adequate fire fighting. This may require a pressure of 100 psi, or more, at the fire pumper or at the hydrant if no pumper is used.

Fire Streams. The shape, size of opening, and the smoothness of the nozzle at the end of a fire hose, and the pressure at the base of the nozzle fix the rate of discharge and the limits of height and distance to which an effective stream can be thrown. These distances and heights are stated in Table 17. Empirical formulas for the total fire-stream capacity to be provided are:

Kuichling

$$Q = 700\sqrt{P} \tag{46}$$

NBFU (National Board of Fire Underwriters)

$$Q = 1{,}020\sqrt{P}\left(1 - \frac{\sqrt{P}}{100}\right) \tag{47}$$

where Q = rate of flow, gpm
P = population, thousands

TABLE 17. HEIGHTS AND DISTANCES, IN FEET, THAT EFFECTIVE FIRE STREAMS CAN BE THROWN[1]

Pressure at base, psi	Angle with horizontal, deg	1½- to 2-in. jets		3-in. jets		Theoretical	
		Max height	Horizontal distance at max height	Max height	Horizontal distance at max height	Max height	Horizontal distance at max height
50	20	15	75	15	75	15	75
	50	60	100	62	100	70	115
100	20	30	140	30	140	30	140
	50	110	165	120	180	140	230
200	20	45	210	50	220 ±	50	220 ±
	50	175	180	180	250 ±	200	220 ±

[1] Based on information presented by H. ROUSE and J. W. HOWE, *Proc. ASCE*, October, 1951.

A study of current practice in fire protection will show the use of streams in well-protected cities with a discharge between 175 and 250 gpm, with pressures at the base of the nozzle between 40 and 100 psi.

Location of Pipes and Specials. One pipe near the middle of the street, or if the street is paved, then on one side under the parking, are satisfactory positions for a water main in streets less than 40 to 50 ft wide. On wide, or on heavily traveled streets, a main may be placed on each side of the street, the larger one on one side where fire hydrants are connected. Water mains should, in general, not be laid in alleys or under pavements. Pipes should be laid below the frost line.

Valves should be placed on branches from feeder mains, and between mains and hydrants. No length of pipe greater than about 1,200 ft, and preferably not more than about 800 ft, should be left without valve control. In high-value districts the distance between valves should not exceed 500 ft. Valves at intersections should, where possible, be placed on the smaller pipes, two valves at a tee and three at a cross.

Location of Fire Hydrants. The areas recommended by the National Board of Fire Underwriters to be covered by each hydrant in a city are stated in Table 18. A convenient procedure in locating fire hydrants is to draw circles with a radius of about 80 percent of the assumed length of fire hose with each hydrant as the center. Undue overlapping of circles and unprotected areas will serve as a guide to the best hydrant location.

TABLE 18. FIRE FLOW AND HYDRANT AREAS [1]

Population	Required fire flow for average city, gpm	Average area per hydrant, sq ft	
		Engine streams	Direct streams
1,000	1,000	120,000	100,000
4,000	2,000	110,000	85,000
10,000	3,000	100,000	70,000
100,000	9,000	55,000	40,000 [2]
200,000	12,000	40,000	40,000

[1] National Board of Fire Underwriters.
[2] 40,000 sq ft is the maximum allowance for direct streams in all cities having a population of 28,000 or more.

PIPES AND MATERIALS

Stresses in Pipes. Conduits carrying water under pressure develop stresses due to bursting pressure, temperature changes, flow around bends, the weight of the pipe and superimposed loads, and shocks due to handling and other causes. Most of these stresses can be formulated, as shown below. In these expressions the proper units must be used, or a dimensional constant inserted in the formula.

Tension due to internal bursting is

$$T = PR \tag{48}$$

where T = circumferential tension in a unit length of pipe
R = radius of the pipe
P = intensity of pressure = wh
w = unit weight of water
h = head or pressure to which pipe may be subjected

Longitudinal stresses resulting from a change of temperature can be expressed as

$$S = ETC \tag{49}$$

where S = intensity of stress due to change of temperature
E = modulus of elasticity of pipe material
T = change of temperature
C = coefficient of expansion of the material

If the pipe is allowed to expand and contract with changes in temperature, expansion joints must be provided. The movement of the pipe can be expressed as

$$M = LCT \tag{49A}$$

where M = change in length of pipe
L = length of pipe affected
C = coefficient of expansion of the metal
T = change in temperature

Longitudinal tension resulting from the flow of water around a pipe bend can be computed from the expression

$$T = (WAV^2/g) + pA \tag{50}$$

where W = unit weight of water
A = cross-sectional area of pipe
V = velocity of flow of water
g = acceleration due to gravity
p = intensity of internal bursting pressure

If a buttress is placed along the line of action of the resultant of the tensions in the pipes on each side of the bend the buttress must resist the resultant force which is

$$E = 2A\left(\frac{WV^2}{g} + p\right)\sin\frac{\theta}{2} \tag{51}$$

where θ = angle between lines of pipe. Other nomenclature is as in Eq. (50).

The phenomenon of surge is similar to that of water hammer. Stresses due to water hammer, or surge, may be avoided by the installation of a surge tower or open tank "floating on the line." The size of a surge tower and the height that the wave will rise in the tower can be computed by the Halmos[1] formula

$$H_b = \frac{K_a H_o A_b \mu_b}{TA_a} H_o \tag{52}$$

where A_a = cross-sectional area of the tunnel
A_b = cross-sectional area of the surge tower
H = head
H_b = head at base of surge tower during maximum height of surge
H_o = static head
K_a = $aV/(2gH_o)$
a = velocity of pressure wave
g = acceleration due to gravity
T = time of closing valve or stopping pump
V = linear velocity of flow of water
μ_a = $2L/a$ = period of the conduit
μ_b = period of the surge tank
L = length of the conduit

External loads on buried pipes are discussed in Section 9 of this book.

Materials. Materials most commonly used for distribution pipes include iron, steel, cement, and asbestos cement. Lead, copper, zinc, aluminum, and alloys such as brass,

[1] HALMOS, E. E., "Symposium on Water Hammer," published jointly by ASCE and ASME, 1933.

bronze, and stainless steel are used, in addition to ferrous metals, in pumping machinery, small pipes, valves, and other appurtenances. Plastic materials are used in 6-in.-diameter, and smaller, pipes. Sulfur, sand, rubber, lead and lead substitutes, and oakum are used in pipe joints. Other substances such as nickel, aluminum, chromium, clay, and asbestos are used occasionally in alloys or in mixtures principally for the protection of metals against corrosion. Wood is occasionally used for large pipes and for small pipes in temporary locations.

Specifications for materials and equipment are prepared periodically by such authorities as the American Water Works Association, the American Society for Testing Materials, the American Standards Association, the American Society of Mechanical Engineers, the Division of Simplified Practice, U.S. Department of Commerce, and the Procurement Division, U.S. Treasury Department.

Cast-iron Pipe. Cast iron is used, in waterworks practice, most commonly in the manufacture of pipe. It has been used for centuries in pipes above about 2 to 3 in. in diameter. Pit-cast pipe is manufactured in lengths of 12 ft, 16 ft, and 5 m, according to

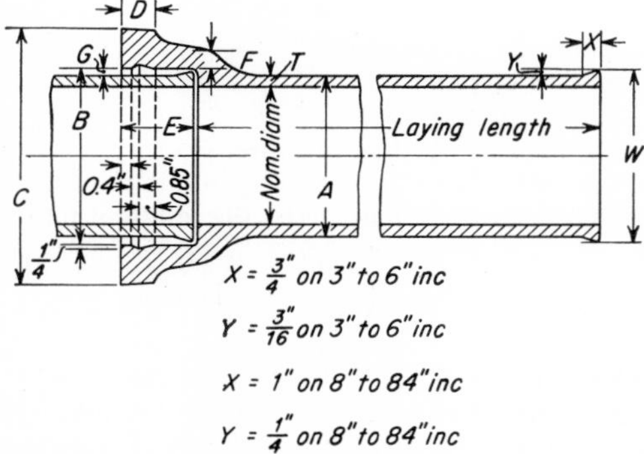

FIG. 28. American Water Works Association standard bell-and-spigot joint for cast-iron pipe.

American Water Works Association specifications. Centrifugally cast pipe, according to Federal specifications, is available in lengths of 12, 16, 16½, 18, and 20 ft. Centrifugally cast pipe, for equal thicknesses, are stronger than pit-cast pipe since the value of the modulus of rupture of 40,000 is greater. Some standard dimensions and weights of cast-iron pipe are shown in Tables 19 and 20. A section through a bell-and-spigot, cast-iron pipe is shown in Fig. 28.

TABLE 19. SOME STANDARD THICKNESSES FOR PIT-CAST IRON PIPE [1]
Thicknesses in inches

Diam, in.	Standard thickness classes							
4	0.40	0.43	0.50	0.58	0.68	0.79	0.92	1.07
8	0.46	0.50	0.58	0.68	0.79	0.92	1.07	1.25
12	0.54	0.58	0.68	0.79	0.92	1.07	1.25	1.46
18	0.63	0.68	0.79	0.92	1.07	1.25	1.46	1.71
24	0.74	0.80	0.93	1.08	1.26	1.47	1.72	2.01
36	0.97	1.05	1.22	1.43	1.66	1.93	2.25	2.62
48	1.18	1.27	1.48	1.73	2.02	2.35	2.74	3.20
60	1.39	1.50	1.75	2.04	2.38	2.78	3.24	3.78

[1] Manual for Computation of Thickness, *Jour. Am. Water Works Assoc.*, December, 1939, p. 3.

TABLE 20. SOME STANDARD WEIGHTS OF PIT-CAST AND CENTRIFUGALLY CAST IRON PIPE [1]

Nominal diam, in.	Pit cast				Centrifugally cast in metal molds			
	Thickness, in.		Lb per 12-ft length with bell		Thickness, in.		Lb per 12-ft length with bell	
	50 psi	350 psi	50 psi	350 psi	50 psi	350 psi	50 psi	350 psi
4	0.40	0.40	230	230	0.35	0.35	200	200
6	0.43	0.50	360	410	0.38	0.38	315	415
8	0.46	0.58	515	645	0.41	0.41	455	455
10	0.50	0.73	635	980	0.44	0.52	605	900
12	0.54	0.79	880	1,250	0.48	0.56	785	900
14	0.54	0.92	1,025	1,780	0.48	0.64	915	1,210
16	0.58	0.99	1,265	2,180	0.54	0.68	1,165	1,465
18	0.63	1.07	1,535	2,560	0.54	0.79	1,320	1,920
20	0.66	1.22	1,785	3,325	0.57	0.84	1,545	2,265
24	0.74	1.36	2,385	4,400	0.63	0.92	2,045	2,975

[1] Pit-cast pipe are ASA Specifications A21.6-1953, and centrifugally cast pipe are ASA Specifications A21.6-1952.

Joints in Cast-iron Pipe. Some joints used in cast-iron pipe are shown in Fig. 29. The use of bell-and-spigot joints makes it possible to lay cast iron around curves, since the axes of consecutive lengths of pipes need not be in alignment to permit the making of satisfactory joints. Flange joints are made tight by placing a gasket of rubber, asbestos, lead, or other material between the flanges and drawing up tightly on the bolts, or by the use of companion flanges. Expansion joints are not commonly used on buried cast-iron pipe, but if the pipe is exposed, as on a long trestle, it may be necessary to provide for expansion. One form of expansion joint is shown in Fig. 29*a*.

Fittings or Specials for Cast-iron Pipe. Fittings for cast-iron water pipe, with bell-and-spigot ends, have been standardized. Standard specials that are available include tees; crosses; bends of 90°, 45°, 22½°, 11¼°, and 5⅝°; wye branches; reducers; sleeves; caps; plugs; bases for tees or crosses; and offsets. Most of these specials are cast with all bell ends or with either one or more spigot ends or one or more flange ends. There are two standards for flanged pipe and fittings. One is the manufacturers' standard or standard fittings for water, in which there are some variations between manufacturers. The other is the American Standard, adopted by the ASME. American Standard fittings have shorter radii than standard flanged fittings for water. Some fittings are shown in Fig. 29.

Steel Pipe. Steel pipes are fabricated of plates joined with transverse and longitudinal joints or of spirally riveted plates and transverse joints. Spirally riveted pipe can be made to any desired length between transverse joints. In the common type of steel pipe, with transverse and longitudinal joints, alternate rings differ in diameter by the thickness of the plate, so that alternate rings slip inside the larger-diameter rings. The transverse joints may be single-lap riveted and calked. Transverse joints are made also by forming the pipe slightly conical in shape and slipping one end of one cone into the larger end of the other, or by making rings of the same diameter and forming a butt-riveted joint with the outside cover plate. Longitudinal joints are made by riveting or welding or by the use of a lock bar.

Specifications for various forms of steel pipe have been published from time to time by the American Water Works Association.[1] Seamless tubing or mill pipe is made in

[1] See *Jour. Am. Water Works Assoc.*, January, 1940; April, 1943; January, March, and August, 1950; and October, 1952.

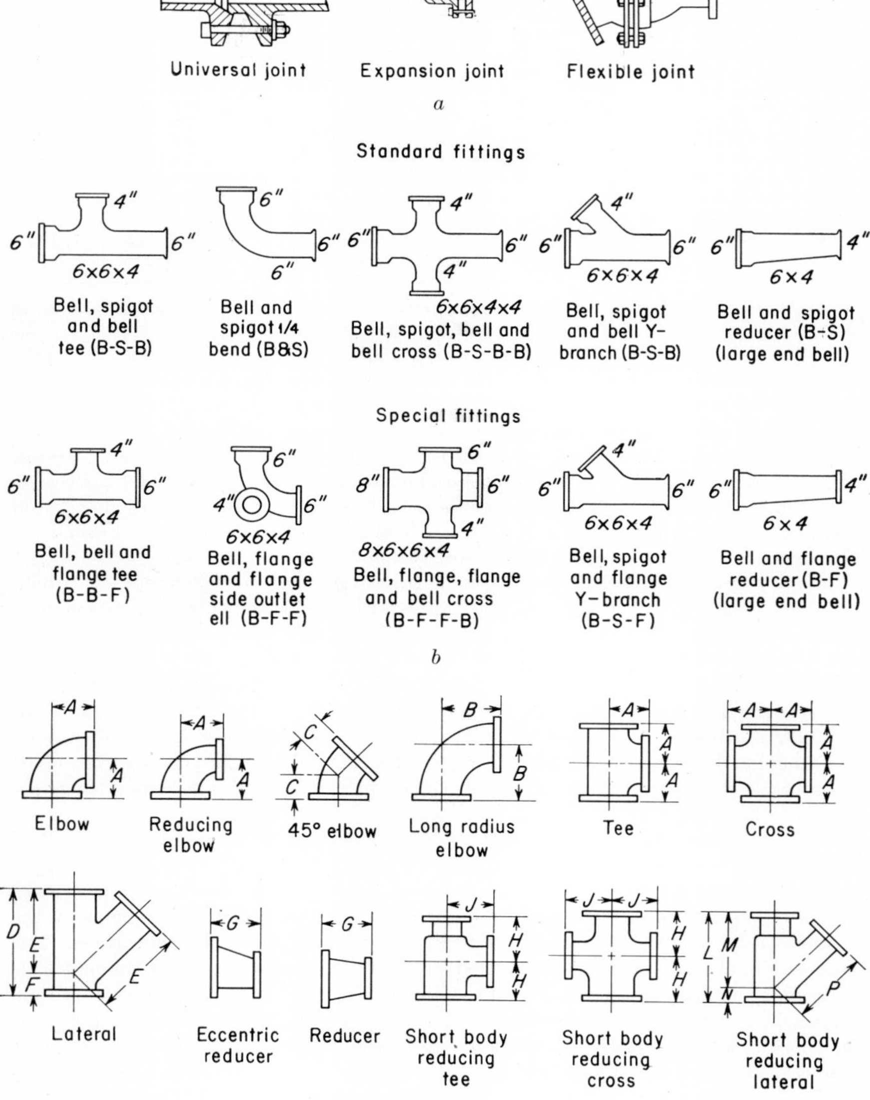

Fig. 29. Joints and fittings used on cast-iron pipe: *a*. mechanical joints, *b*. standard and special fittings, *c*. American standard flanged fittings.

sizes from ⅛ to 24 in. Thickness, weights, and working pressures for some sizes of pipe are shown in Table 21.

TABLE 21. DATA ON SMALL-DIAMETER STEEL PIPE [1]

Nominal diam, in.	Weight, lb per ft	Working internal pressure, psi [2]	Collapsing pressure, psi [3]	Deflection under external load in trenches, in.				Saddle loads [4]	
				5-ft cover		10-ft cover		Values P for S_l = 1,000 psi A = 120°	l [5] ft for S_f = 12,500
				No truck	With truck	No truck	With truck		
6	12.89	710	1,510	0	0.01	0.01	0.01	1,730	46
8	16.90	550	650	0.01	0.02	0.02	0.02	1,580	47
10	21.15	440	330	0.03	0.06	0.05	0.07	1,480	49
12	25.15	370	190	0.06	0.12	0.10	0.12	1,410	50
14	27.65	340	120	0.09	0.21	0.19	0.24	1,370	50
18	35.67	260	60	0.21	0.41	0.39	0.47	1,280	52
20	39.67	240	40	0.28	0.53	0.49	0.59	1,250	54
24	47.68	200	25	0.44	0.75	0.71	0.83	1,190	56
24	125.49 [6]	520 [6]	190 [7]	0.14 [7]	0.24 [7]	0.23 [7]	0.27 [7]	5,700 [7]	69 [7]
30	....	310 [7]	230 [6]	0.23 [7]	0.40 [7]	0.40 [7]	0.48 [7]	5,350 [7]	71 [7]

[1] Adapted from Design Standards for Steel Pipe, *Jour. Am. Water Works Assoc.*, January, 1948, p. 24.
[2] Wall thickness for all pipes in this table is 0.188 in. except as noted. Pipes are made with eight different wall thicknesses, up to 0.5 in.
[3] Due to concentrically applied uniformly distributed external pressure.
[4] Pipe is full of water. Saddle load is transmitted to pipe support as shown in Fig. 30. thus:

$$S_l = k \frac{P}{t^2} \log \left(\frac{R}{t}\right)$$

where S_l = localized saddle stress, psi
P = total saddle reaction, lb
R = pipe radius, in.
t = pipe wall thickness, in.
$k = 0.02 - 0.00012(A - 90)$
A = degrees in central angle subtended by saddle
[5] l = distance from center to center of saddles, ft, and S_f = flexure stress, psi
[6] Wall thickness 0.5 in.
[7] Wall thickness 0.375 in.

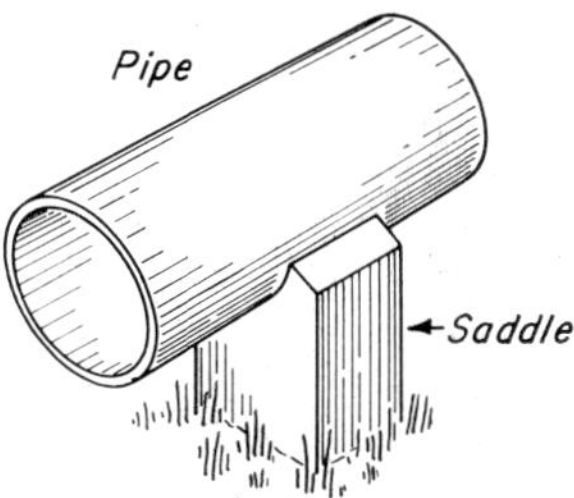

FIG. 30. Saddle pipe support.

Concrete Pipe. The use of concrete pipe under moderate internal pressures is confined principally to pipes larger than about 12 in. in diameter. Precast pipe is manufactured in sizes up to 8 and 10 ft in diameter, and it is usually reinforced. Some sections through and some joints in concrete pipe are shown in Fig. 31.

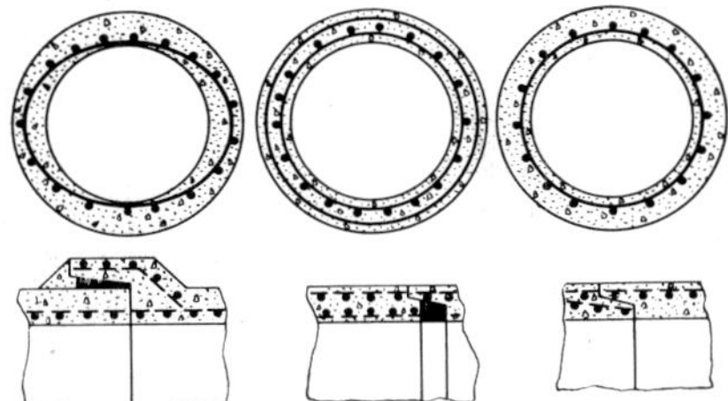
FIG. 31. Joints and reinforcing for precast concrete pipe.

Watertightness is obtained in low-pressure concrete pipe by providing sufficient thickness of good concrete and longitud-

inal reinforcement equal to at least 0.25 percent of the cross-sectional area of the concrete and by reducing allowable unit stress in the steel as the internal pressure increases. For large-diameter pipes, a minimum wall thickness of 1 in. for each foot in diameter of pipe is a rough rule to follow in design. Surface washes may be applied to the concrete, after it has set, to assure watertightness.

Watertightness of high-pressure concrete pipe may be obtained by the insertion of a thin steel cylinder in the pipe walls either with or without prestressed reinforcement.[1] Lighter-weight pipes may be used when the reinforcement is prestressed because the concrete is kept under compression at higher internal pressures than when not prestressed.[2]

Design of Reinforced Concrete Pipe. Reinforcing steel is placed in concrete water pipe to resist tensile stresses due to bursting pressure and flexure. The concrete carries the compressive stresses in the pipe subject to flexure. When bursting pressure alone is considered, the area of steel per foot of pipe can be computed by the expression

$$A_s = (pr/f_t)12 \tag{53}$$

where p = internal pressure, psi
r = radius of pipe, in.
f_t = allowable tensile stress, psi

When flexural stresses are considered, the moments and tensions due to the loads causing bending are calculated for critical points by an elastic analysis. Tension due to internal pressure is added to that due to bending. Longitudinal reinforcement is usually used to distribute cracks due to temperature effects and contraction due to setting. The total area of longitudinal steel placed is usually 0.25 to 0.5 percent of the cross-sectional area of the concrete in the pipe.

Asbestos-cement Pipe. Asbestos-cement pipe is composed of asbestos fiber and portland cement combined under pressure into a dense homogeneous material in which a strong bond is effected between the cement and the fiber. The pipe will withstand bursting pressures up to 200 psi. It will resist corrosion, and it is unaffected by electrolysis, light in weight, easily cut and filed, not easily broken, and can be joined without the need of great skill. No expansion joints are needed.

Wood Pipe. Wood pipe is used where its availability and cost render it more advantageous than metal pipe. Under favorable conditions, certain woods, particularly California redwood, will last indefinitely. Materials used for wood-stave pipe include hemlock, spruce, yellow pine, Douglas fir, cypress, and redwood. The pipe bands are made of steel, usually galvanized or coated to resist rust, and the ends are held together in cast-iron shoes. The exterior of the pipe may be easily attacked by organic acids or plant growths, and it is subject to deterioration, particularly if alternately wet and dry. Wood pipe may be made of machine-banded logs or staves, or it may be held together by metal bands placed in the field. Machine-banded pipe is available in sizes up to 48 in. in diameter and is said to be watertight at 280 psi internal pressure. Wood-stave pipe is constructed in very large diameters but it is used for lower pressures than machine-banded pipe.

Corrosion. Corrosion is an important factor in the deterioration of pipes and other structures in waterworks. Some hypotheses of the causes of corrosion are: (1) Bimetallic or galvanic action, such as results when two metals with dissimilar electrolytic solution pressures are immersed in the solution of an electrolyte. It is one of the most common causes of metallic corrosion. (2) Hydrogenation, in which metallic ions migrate into the surrounding water, the electrolytic balance being maintained by hydrogen ions from

[1] See also LONGLEY, F. F., *Jour. New Engl. Water Works Assoc.*, December, 1945, p. 335; DUMENSKY, D. B., *Civil Eng.*, November, 1948, p. 42.

[2] See also *Water Works Eng.*, Mar. 20, 1946, p. 291.

the water being deposited on the metal. (3) Electrolysis, in which positively charged metallic ions are carried from the metal by a constantly flowing direct current. (4) Chemical reaction, in which the metal, when immersed in an acid, reacts to provide metallic atoms to replace hydrogen gas released from the acid. (5) Direct oxidation occurs when metals, particularly when heated, are exposed to air. (6) Biologic action. Sulfate-reducing bacteria are the principal cause of biologic corrosion of ferrous metals. They operate in the absence of oxygen to react with sulfates, and with organic compounds containing sulfur, in the soil to produce hydrogen sulfide. The bacteria do not attack the metal directly. The hydrogen sulfide combines with iron to form compounds of sulfur and iron, or it combines with water to form sulfurous or sulfuric acid which reacts with the metal, to release hydrogen which is again used in the bacterial metabolism.

Iron-consuming bacteria remove iron that is in solution in the water, depending on concentrations of soluble iron or iron compounds of 2 ppm, or greater, to support their activities. The consumed iron is deposited by the bacteria as ferric hydroxide, forming a sheath or tubercule, around a central core, which may contain a small percentage of sulfide of iron. The effect is known as "tuberculation."

Corrosion can be retarded by cathodic protection, care in manufacture, application of coatings or linings, control of the environment of the metal, and treatment of the water. Coatings and linings used for the protection of iron and steel surfaces include tar or asphaltic materials, enamels, resins, lacquers, zinc or galvanizing, metallizing,[1] plastics, and paints.[2] The success of a coating depends as much on the preparation of the surface, whether concrete or metal, as on the material of the coating.[3]

Cathodic Protection.[4] Galvanic and electrolytic corrosion, hydrogenation, and to some extent bacterial corrosion of metals can be retarded by cathodic protection. Cathodic protection is accomplished by connecting the metal in an electric circuit to the negative pole of a d-c generator,[5] the positive pole being connected to anodes buried in the ground or otherwise grounded. The electrical potential of the metal is thus reduced below that of the surroundings, and electric currents will flow from the surroundings to the metal. The order of magnitude of electric current for each square foot of iron surface [6] ranges from 0.3 to 15 ma.[7] The voltages depend on the resistance of the circuit but should be expected to range between 1.5 and 30 volts.[8]

APPURTENANCES

Head Losses Due to Flow. Head losses due to flow through appurtenances are shown in Table 22.

Valves. Gate valves are more widely used than any other appurtenance on water-distribution systems. The outstanding advantage of gate valves is that they offer little resistance to flow when the valve is wide open, as indicated in Table 22. Cross sections through two gate valves are shown in Fig. 32.

Check valves are used to permit the flow of water in one direction only. Four types are shown in Fig. 33.

An air-inlet or air-relief valve, such as is shown in Fig. 34, is used to admit air into a pipe to prevent the creation of a partial vacuum, or to release air from a pipe to prevent the creation of an air lock, or for other purpose.

[1] See also *Water Works Eng.*, July, 1952, p. 661.
[2] See also Baylis, J. R., *Jour. Am. Water Works Assoc.*, August, 1953, p. 807.
[3] See also Weir, Paul, *Eng. News-Record*, May 2, 1946; Cates, W. H., *Jour. Am. Water Works Assoc.*, February, 1953, p. 103.
[4] This subject has been reported exhaustively in *Jour. Am. Water Works Assoc.;* see August, 1948, p. 485; September, 1949, pp. 845, 852; November, 1951, p. 883, with 62 references.
[5] See also Knusden, H. A., *Jour. Am. Water Works Assoc.*, January, 1938, p. 38.
[6] See also Uhlig, H. H., *Chem. Eng. News*, Dec. 10, 1946, p. 3154.
[7] See also Schneider, W. R., *Jour. Am. Water Works Assoc.*, March, 1945, p. 245.
[8] See also O'Brien, G. L., *Water Works and Sewerage*, July, 1942, p. 285.

TABLE 22. HEAD LOSS IN VALVES, FITTINGS, AND SERVICE PIPES [1]

Valve, fitting, or pipe	K	Valve, fitting, or pipe	k [4]
Gate valve:		Service pipe: [5]	
Open	0.19 [7]	⅝-in. lead	0.00035
¾ open	1.15 [7]	¾-in. lead	0.00036
½ open	5.6 [7]	1-in. lead	0.00039
¼ open	24 [7]	¾-in. copper	0.00033
Globe valve, open	10 [8]	1-in. copper	0.00034
Angle valve, open	5 [8]	½-in. galvanized iron [6]	0.00036
90° ell, short radius	0.9 [9]	¾-in. galvanized iron [6]	0.00036
Medium radius	0.75 [8]	1-in. galvanized iron [6]	0.00035
Long radius	0.6 [8]	Corporation cocks:	
Return bend	2.2 [8]	½-in. with tailpiece for lead	0.00127
45° ell	0.42 [8]	½-in. with ¾-in. copper adapter	0.00077
22½° ell (18 in.)	0.13 [2]	¾-in. with ¾-in. copper adapter	0.0023
Tee	1.25 [2]	¾-in. with tailpiece for lead	0.0020
Reducer (V at small end)	0.25 [2]	1-in. with 1-in. copper adapter	0.0029
Increaser = $0.25(V_1^2 - V_2^2)/2g$ *	....	Curb stops:	
		¾-in. lead service	0.002
Bellmouth reducer	0.10 [2]	¾-in. copper service	0.0012
Shear gate, open (orifice)	1.80 [2]	1-in. lead service	0.001
Sluice gate in 12-in. wall submerged	2.35 [2]	1-in. copper service	0.00059
Passing branches or openings	0.03 [2]	Yokes:	
Stop-and-waste valves: [3]		⅝-in. ram's-horn (new)	0.0035
½ × ¾ in.	0.0021 [10]	⅝-in. ram's-horn (old)	0.0135
⅝ × ¾ in.	0.0030 [10]	⅝-in. straight-line	0.00106
¾ × ¾ in.	0.0048 [10]	¾-in. ram's-horn	0.0029
1 × 1 in.	0.013 [10]		

[1] Computed from expressions

$$H = K(V^2/2g) \quad \text{or} \quad H = k(Q^2/d^5)$$

where H = head loss, ft
V = velocity, ft per sec, in a pipe whose nominal diameter is the same as that of the fitting shown in the table
Q = flow, gpm, in a pipe whose diameter is d in.
K and k = constants shown in the table

[2] BURNS and MCDONNELL, *Water Works and Sewerage*, May, 1937, p. 196.
[3] Compression valves.
[4] Values of k are taken from *Jour. Am. Water Works Assoc.*, May, 1932, p. 631.
[5] The head losses in service pipes are per foot of length of pipe.
[6] Presumably new pipe.
[7] *Univ. Wisconsin Bull.* 252.
[8] *Crane Co. Bull.* 405.
[9] *Univ. Texas Bull.* 2712.
[10] Values of k.
* BURNS and MCDONNELL, *Water Works and Sewerage*, May, 1937, p. 196.

In a balanced valve the upward and downward pressures against the valve disk, together with the weight of the moving parts, are balanced so that only friction must be overcome in moving the valve parts. The principle of the balanced valve is illustrated in Fig. 33. This principle is widely used in valve-control devices.

Pressure-relief and pressure-regulating valves serve the purpose indicated by their names. Sections through pressure-regulating valves are shown in Fig. 35. Float-controlled valves are used to control the water level in a tank or reservoir, or other container. Needle valves are used to control the flow from large reservoirs or through dams, and in the close adjustment of the rate of flow in mechanical control devices, since a large movement of the mechanism is required to effect a relatively small change in the rate of flow. A small needle valve is shown in Fig. 36. Ground-key valves, as shown in Fig. 37, are used to stop or start flow fully with only a one-quarter (90°) turn of the valve or, in multiple-port valves, to direct flow from one channel to another. Their use is confined principally to small mechanisms.

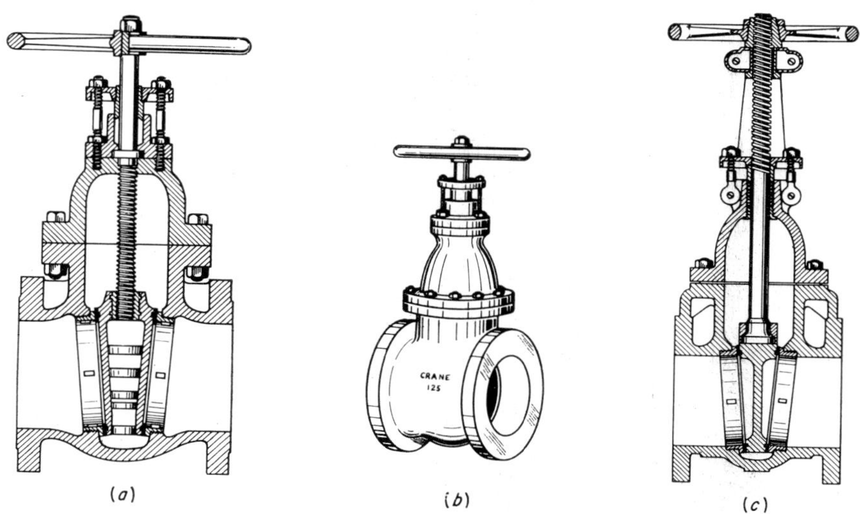

FIG. 32. Gate valves: *a*. nonrising stem, section, *b*. nonrising stem, external view, *c*. rising stem, section.

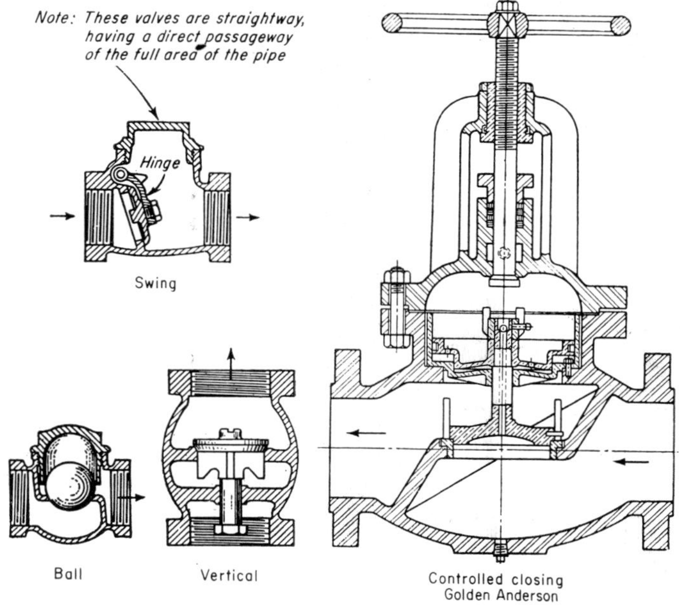

FIG. 33. Check valves.

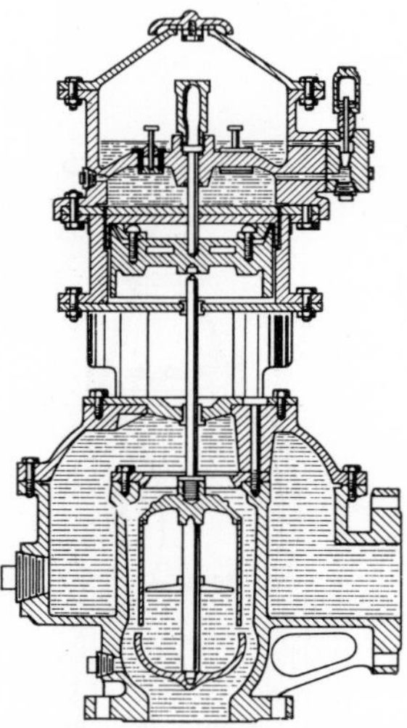

FIG. 34. Air-relief valve. (*Simplex Valve and Meter Co.*)

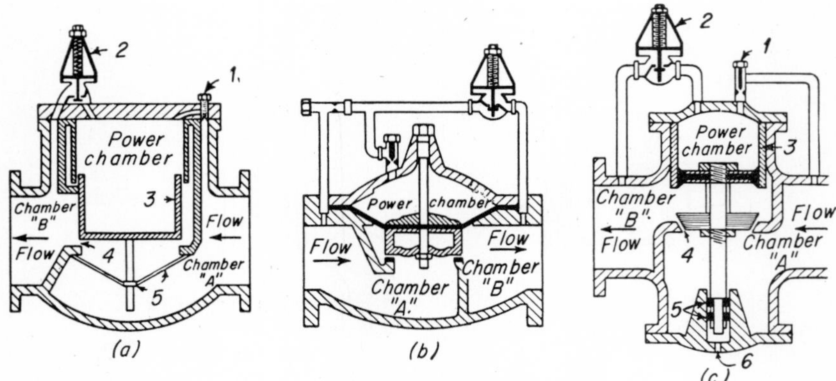

FIG. 35. Pressure-regulating valves. (*Water Works Eng., September*, 1949, *p.* 827.) *a.* Pressure-reducing, pilot-controlled valve with hollow-piston operated main valve. (1) Needle valve, (2) pilot control valve, (3) hollow piston, (4) main valve and seat, (5) stem guide. *b.* Pressure-reducing, pilot-controlled valve with diaphragm-operated main valve. *c.* Pressure-reducing, pilot-controlled valve with piston-operated main valve. (1) Needle valve, (2) pilot control valve, (3) piston, (4) main valve and seat, (5) stem guides, (6) vent.

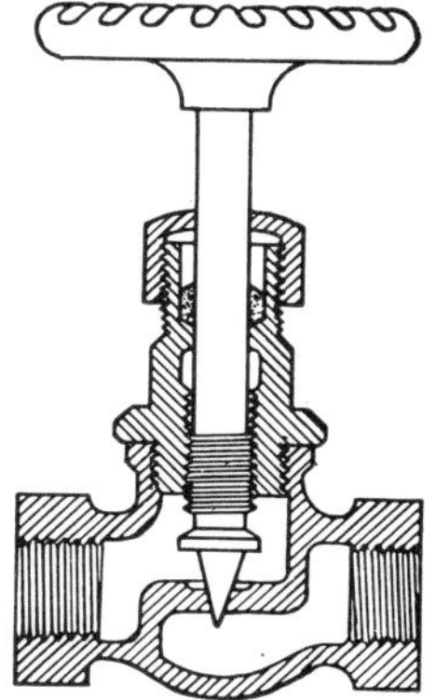

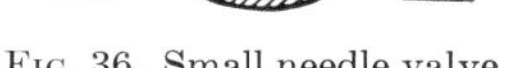

FIG. 36. Small needle valve.

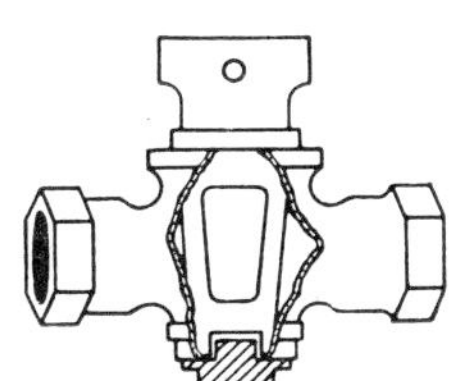

FIG. 37. Ground-key valve.

Mud valves, of the type shown in Fig. 38, are used to remove mud or sludge from the bottom of a basin.

Sluice gates are similar to gate valves but are designed to resist pressure on one side of the gate. They are used principally on pipes or channels leaving reservoirs, dams, or similar structures.

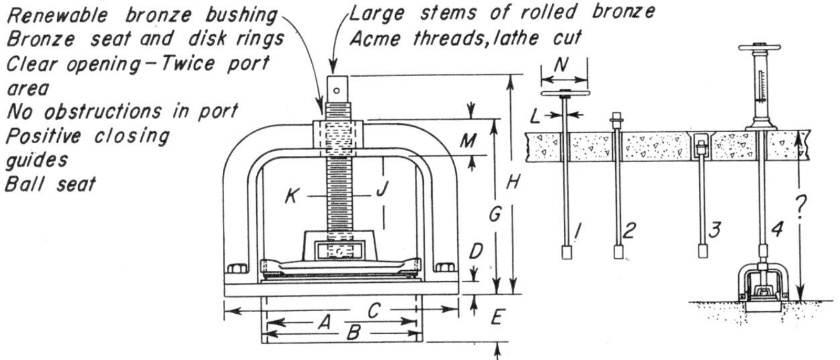

FIG. 38. Mud valve, showing method of installation. (*Rodney Hunt Machine Co.*)

Fire Hydrants. Standard specifications for fire hydrants, including specifications for fire-hose coupling screw threads,[1] and for the uniform marking of fire hydrants [2] have been adopted, from time to time, by the American Water Works Association. A standard compression fire hydrant is shown in Fig. 39. Fire hydrants are usually made of cast iron with brass fittings in contact with the water or on moving parts. It is frequently desired that a gate valve be placed on the connection to the distribution system, in addition to the main valve on the hydrant. In high-risk districts the hydrant should have a pumper connection in addition to the customary hose connections. A drain for emptying the barrel of the hydrant when it is closed is essential in a cold climate to prevent freezing of the water in the hydrant.

[1] *Jour. Am. Water Works Assoc.*, April, 1937.
[2] *Jour. Am. Water Works Assoc.*, May, 1953, p. 530.

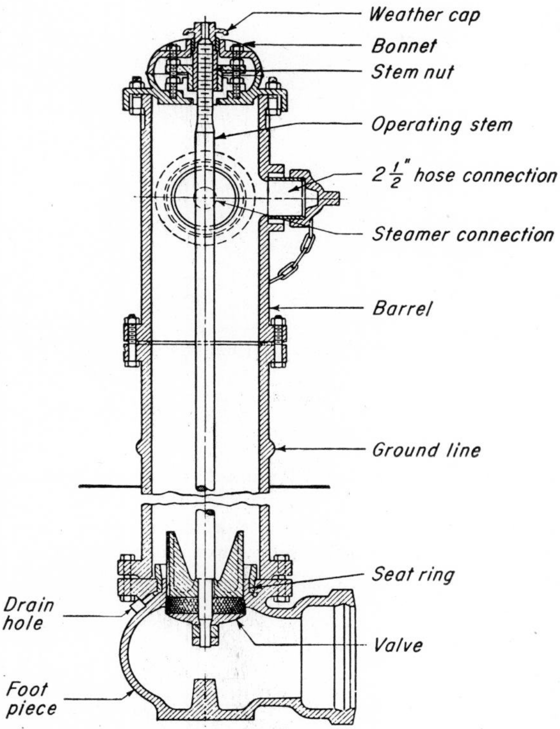

FIG. 39. American Water Works Association standard compression hydrant.

Specifications usually require that when the flow is 250 gpm through each hose outlet on a two-way hydrant the loss of head shall not exceed 1.75 psi; on a three-way hydrant, 2.25; and on a four-way hydrant, 3 psi, except when there are inside hose gates where the loss must not exceed 2.75 psi for three-way and 3.5 for four-way hydrants.

Threads for hose connections for fire hydrants have been standardized, the standard adopted by the National Board of Fire Underwriters being 7½ threads per inch, 3$\frac{1}{16}$ in. in diameter.

The size of a fire hydrant is designated in terms of the minimum opening of the seat ring of the main valve. It must be at least 4 in. for two 2½-in. nozzles, or at least 5 in. for three 2½-in. nozzles, and at least 6 in. for four 2½-in. nozzles. The rate of discharge from a hydrant can be approximated by the expression

$$Q = Kd^2p^{0.5} \tag{54}$$

where Q = flow, gpm
d = diameter of nozzle, in.
p = gage pressure at base of nozzle, psi
K = a constant for the hydrant, usually in the order of 27

Meters. Water meters are classified as displacement or as velocity meters in accordance with their method of operation. Water is measured through a displacement meter by observing the volume displaced by a moving part of the meter and counting the number of displacements on an automatic integrating dial. The quantity of water

passing through a velocity meter is the product of the cross-sectional area of the channel of flow through the meter and the velocity of flow at this section. An integrating device summarizes this product to show the total flow through the meter. A nutating-disk meter, such as is shown in Fig. 40, is the type most commonly used on domestic service connections. The venturi-tube velocity meter, discussed in Section 4, is used for the measurement of large flows, as in a pumping station. Data on small water meters are

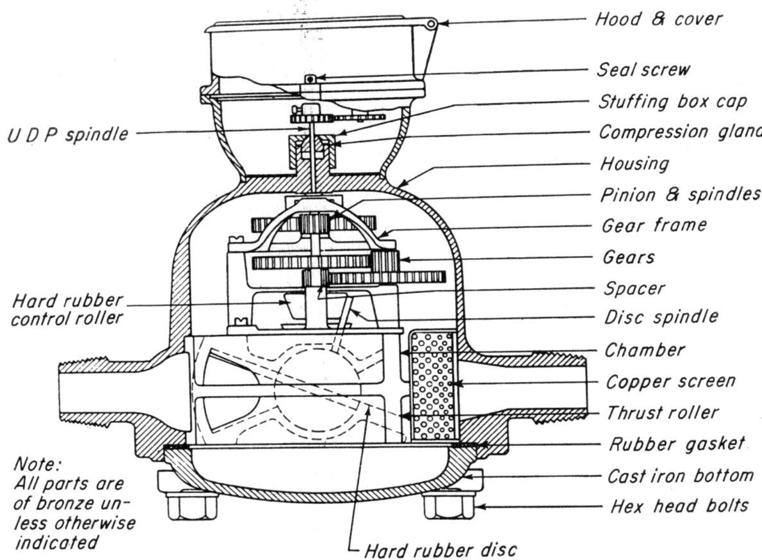

FIG. 40. Nutating-disk meter. (*Badger Meter Co.*)

shown in Table 23. The head losses through such meters are not to exceed 15 psi for 1-in. and smaller meters, and 20 psi for larger meters. Such meters seldom exceed 2 in. in size. The sizes of meters and of service pipes are closely related.

TABLE 23. RECOMMENDED SIZES OF SERVICE PIPES[1]

Class of building	Length of service pipe, main to meter, ft			
	Sizes of pipe, in.			
	100	50	25	10
A	1¼	1	1	¾
B	1½	1¼	1¼	1
C	2	1½	1½	1¼
D	2	2	1½	1¼

[1] From BABBITT, H. E., "Plumbing," 2d ed., McGraw-Hill Book Company, Inc., New York, 1950. Computed on basis of 20 ft head loss from main to meter.

A. An ordinary, single-family dwelling, 2 to 2½ stories and not more than 8 to 10 rooms, containing 1 bathroom, a kitchen sink, laundry trays, and a garden hose.

B. A 2-family house or larger dwelling, up to about 16 rooms, containing 2 bathrooms, 2 kitchen sinks, laundry trays, and one garden hose.

C. A 4-apartment building, apartments not more than 6 rooms each. Building contains 4 bathrooms, 4 kitchen sinks, 4 sets of laundry trays, and one garden hose.

D. A large apartment building containing not more than 25 apartments with about 100 rooms; with full equipment of 1 bathroom and 1 kitchen, laundry trays for each apartment, and 2 hose connections.

Service Pipes. A service pipe is a pipe leading from a water main to a building or other point at which a consumer is supplied with water. Service pipes are connected to the water mains through flexible connections, the most common type of which is the lead gooseneck. The flexible connection is required because of the movement between the main and the service pipe which would rupture a rigid connection. Recommended sizes of service pipes are shown in Table 23.

A desirable service connection leads directly, and for not more than about 50 ft, from the water main to one consumer only. One undesirable and four acceptable connections

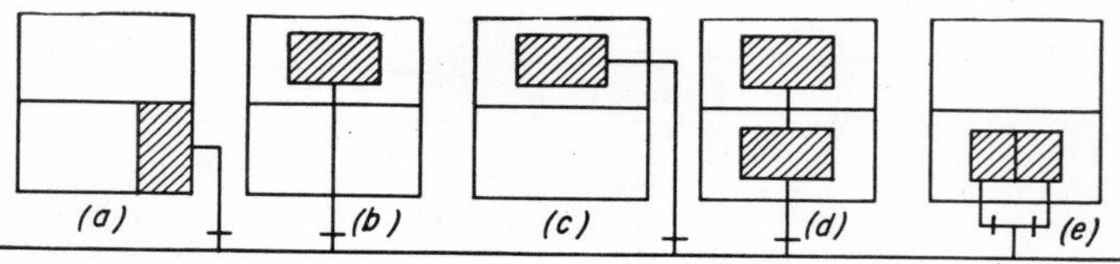

Fig. 41. Some forms of service-pipe connections. *d* is undesirable. (*M. P. Hatcher, J. A. W. W. A., December*, 1947, *p.* 1165.)

are shown in Fig. 41. In most jurisidictions the undesirable connection, serving two customers, is prohibited.

Materials used for service pipes include copper, lead, wrought iron, steel, brass, and cast iron.

Distribution Reservoirs. Reservoirs are used on a distribution system to store water, to equalize flows, and to equalize or to control pressures. Two or more purposes may be combined in one reservoir. Reservoirs can be classified also as surface or elevated, or as steel, concrete, wood, or earth.

Storage for fire protection should be sufficient to supply water for a minimum of 2 hr in the smallest communities, up to 12 or more hours in large cities. Where reserve storage is in an elevated tank it can be computed from

$$R = (F - P)T \tag{55}$$

where R = reserve storage
P = reserve fire pumping capacity
F = fire demand
T = duration of fire

Units are in gallons and minutes.

McDonald's[1] empirical formula for reservoir capacity is

$$R = aD - bD - (10/24)(D + F - S) \tag{56}$$

where R = reservoir capacity for balancing the domestic demand, to supply part of the water for fire service, and to provide for the operation of the pumps in the off-peak power-load period, millions of gal
D = average domestic demand for the maximum month, mgd
F = fire demand, mgd
S = capacity of pumps, mgd
a and b = fractional parts of domestic demand required for balancing and off-peak operation, respectively. a is in the order of $1/5$, and b $1/10$

Under favorable circumstances elevated storage may be provided for equalizing purposes only, fire reserve being stored in a surface-level reservoir.

Equalizing Reservoir. An equalizing reservoir provides for a variable rate of demand with a rate of supply that may be constant, or may vary at a different rate from that

[1] McDonald, N. G., *Jour. Am. Water Works Assoc.*, July, 1947, p. 637.

of the demand. Such a reservoir may be connected by a single pipe to the distribution system. When water can flow into or from the reservoir through this pipe, at different times, the reservoir is said to be floating on the line.

The requisite capacity of an equalizing reservoir can be determined as follows: Let the wavy line *ABCD*, in Fig. 42 represent the hydrograph of anticipated water usage in the

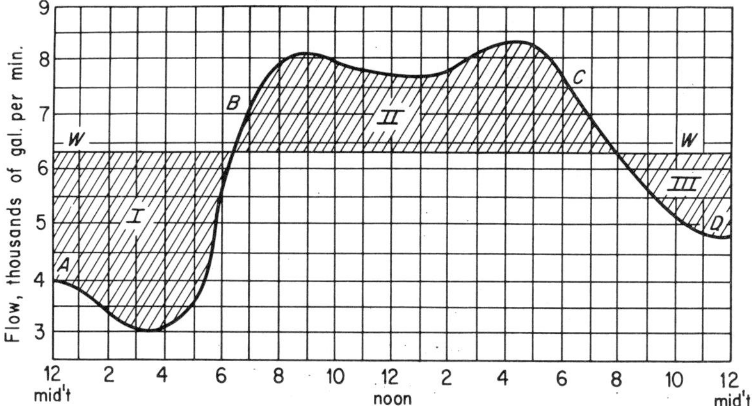

FIG. 42. Graphical determination of equalizing reservoir capacity.

district served by the equalizing reservoir. Construct a straight horizontal line, such as *WW*, such that the shaded areas I and III are equal to shaded area II. The requisite volume of the equalizing reservoir is equal to the product of the ratio of the shaded area II to the total area under the line *WW* to the base line of the figure, and the total amount of water used during the period covered by the hydrograph.

MAINTENANCE OF DISTRIBUTION SYSTEMS

Records. Adequate records of the location of pipes and appurtenances are needed. Types of maps used include: a comprehensive map of the entire system; a large-scale map showing details; and plats and cards providing information on valves, hydrants, and other appurtenances. A monthly report or record should include: (1) the length and sizes of pipes laid and retired during the month and the total mileage of each size laid to date; (2) the number of each size of valves; (3) the number, size, and types of fire hydrants; and (4) the number, size, and lengths of services.

Cleaning and Disinfecting Mains. Pipes are cleaned by passing a scraping machine through them or by the passage of a ball covered with a chain mesh, or by the use of chemicals, and in other ways. Much damage can result from improper use of some methods. For example, pipe coatings may be broken, thus accelerating corrosion, and chlorinated hydrocarbons in domestic water supplies may produce tastes and odors.

Disinfection of new and of contaminated mains is essential in good practice. The following is a satisfactory procedure:

A chlorine gas-water mixture is applied at the beginning of the pipeline extension, or any valved section of it, at such a rate that with a slow flow of water into the pipe the chlorine shall not be less than 50 to 100 ppm. The treated water is to remain in the pipe at least 3 hr and preferably longer. At the end of the period the chlorine residual is not to be less than 5 ppm. Following chlorination the pipe is thoroughly flushed, tested, and if necessary, rechlorinated until a satisfactory bacteriological test is obtained.

Finding Buried Pipes and Leaks. Buried pipes may be located by using the pipe as a part of a closed electric circuit in such a manner that the magnetic field surrounding the pipe will increase the magnitude of the sound in the earphones as the pipe is approached, and will fall silent when the observer is directly over the pipe. In another type an aerial picks up the magnetic field induced in the pipe by a radio transmitter or by previous excitation. An electric dip needle points to the pipe.

Methods used for the location of leaks in underground pipes and in subaqueous pipes include: (1) direct observation, (2) the sounding rod, (3) the hydraulic grade line, (4) water hammer, (5) blowing air into the pipe, (6) putting dye or salt into the water, (7) studying the quality (change of salinity) of the water in a submerged suction pipe, (8) measuring the volume of water required to fill the pipe, and (9) listening either directly for the sound or with the aid of a sonoscope or an electrical sound amplifier. Each method is best suited to particular conditions. None is universally applicable or infallible. The sounding rod is a pointed metal shaft which is thrust into the ground in the supposed neighborhood of the leak. If the shaft is withdrawn wet or muddy, the leak is being followed. Listening devices are similar to a telephone receiver that is placed in contact with the pipe. An increase in the intensity of the sound of the leak indicates that it is being approached. Unfortunately the magnitude of the sound of a leak bears no relation to the magnitude of the leak and large losses may occur without causing a detectable sound. The distance from a quick-closing valve to a leak can be determined by observing the time for the pressure wave, caused by water hammer when the valve is suddenly shut, to travel from the valve to the leak and back again to the valve. From the observed time and the velocity of the pressure wave in water hammer stated in Section 4, the distance to the leak can be computed. If the pressures in a pipeline are observed above and below a leak, the hydraulic grade lines drawn through the observed pressures will intersect on the pipe profile over the leak. Compressed air or coloring matter put into a leaking pipe will appear in surrounding water near the point of leakage.

Freezing and Thawing of Water in Pipes. When water is motionless in a pipe the rate of freezing can be expressed as

$$t_1 = 2.3(q/k) \log (T_2/T_1) \tag{57}$$

where t_1 = time for change in temperature, hr
T_1 = original difference in temperature between water and air
T_2 = final difference in temperature between water and air
q = amount (volume, not rate of flow) of water, kg in 1 m of pipe
k = loss of heat, kg-cal per hr for a meter length of pipe and for the type of insulation considered for 1°C difference in temperature

Values of k vary from about 0.72 for a poorly insulated pipe to 0.08 for good insulation. Freezing may be assumed to begin when the temperature reaches 0°C, although supercooling of 0.01 to 0.1° commonly occurs.

The time t_2 in hours, required to freeze water starting at an initial temperature of 0°C is

$$t_2 = 80q/(kT) \tag{58}$$

where T = temperature of outside air, °C below zero, and q and k are as in Eq. (57).

Ice in metal pipes is best thawed by electricity. Fire, warm water, steam, and other expedients are usually messy and are frequently dangerous. The proper use of electricity is safe, fast, and economical. The factors involved can be formulated as:

$$H = I^2Rt \tag{59}$$

where H = heat, joules

I = current, amp

R = resistance, ohms

t = time current is flowing, sec

The amount that the temperature is raised depends on the current I and the resistance R of the pipe per unit of length. A current of 100 amp will heat 1,000 ft of 1-in. pipe as rapidly as it will heat 10 ft, but it will take 100 times the voltage to do so.

THE TREATMENT OF WATER

Quality. The widespread voluntary adoption of the standards of the U.S. Public Health Service for the "Quality of Water in Interstate Carriers" is an indication of the general desire for some form of acceptable yardstick of water quality. The following are extracts from some of these standards:[1]

"1. Definitions of terms. A standard portion of water for bacteriological test is designated by either 10 ml or 100 ml. The standard samples in the bacteriological test may consist of five portions of either of the above sizes. If a sample collected for bacteriological examination is from a disinfected supply, the sample must be freed of any disinfecting agent within 20 min from the time of collection.

"2. As to Source and Protection. The water supply shall be (*a*) obtained from a source free from pollution, or (*b*) obtained from a source adequately purified by natural agencies, or (*c*) adequately protected by artificial treatment.

"The water system in all its parts shall be free from all known sanitary defects, and health hazards shall be systematically removed at a rate satisfactory to the reporting agency and to the certifying authority.

"3. As to Bacteriological Quality. The minimum number of samples to be collected from the distribution system and examined . . . is based upon the relationship of population served. Of all the standard 10-ml portions examined per month not more than ten percent shall show the presence of organisms of the coliform group. Of all the 100-ml portions examined per month not more than 60 percent shall show the presence of organisms of the coliform group.

"The procedure given using a standard sample composed of five standard portions, provides for an estimation of the most probable number of coliform bacteria present in the sample as set forth in the following tabulation:

MPN[1] OF COLIFORM BACTERIA PORTIONS PER 100 ML

−	+	When five 10-ml portions are examined	When five 100-ml portions are examined
5	0	Less than 2.2	Less than 0.22
4	1	2.2	0.22
3	2	5.1	0.51
2	3	9.2	0.92
1	4	16.0	1.60
0	5	More than 16.0	More than 1.60

"[1] MPN means most probable number.

"Physical Characteristics. The turbidity shall not exceed 10 ppm (silica scale), nor shall the color exceed 20 (cobalt scale). The water shall have no objectionable taste or odor.

[1] See *Jour. Am. Water Works Assoc.*, March, 1946, p. 361.

"Chemical Characteristics. The presence of lead in excess of 1.0 ppm, of fluoride in excess of 1.5 ppm, or arsenic in excess of 0.05 ppm, of selenium in excess of 0.05 ppm, of hexavalent chromium in excess of 0.05 ppm, shall constitute ground for rejection of the supply.

"Salts of barium, hexavalent chromium, heavy metal glucosides, or other substances with deleterious physiological effects shall not be added to the system for water-treatment purposes.

"The following chemical substances which may be present in natural or treated waters should preferably not occur in excess of the following concentrations when more suitable supplies are available in the judgment of the certifying authority: Cu shall not exceed 3.0 ppm; Fe and Mn, 0.3 ppm; Mg, 125 ppm; Zn, 15 ppm; chloride, 250 ppm; sulphate, 250 ppm.

"Phenolic compounds should not exceed 0.001 ppm in terms of phenol.

"Total solids should not exceed 500 ppm for a water of good chemical quality. However, if such a water is not available a total solids content of 1,000 ppm may be permitted."

Methods. Water may be treated to attain the above standards and for other reasons. Common methods of treatment include: (1) storage, (2) plain sedimentation, (3) sedimentation with coagulation, (4) rapid sand filtration, (5) slow sand filtration, (6) disinfection, and (7) processes for the removal of specific substances or characteristics or for altering the quality of the water, such as iron removal, taste and odor removal, and softening.

Storage. The storage of water may be either beneficial or detrimental to its quality as a public water supply. The quality may be impaired by the growth of microscopic organisms which produce tastes, odors, and color. Ground waters and treated waters stored in open reservoirs are more susceptible to such growths than natural surface waters. The quality of stored waters may be improved through sedimentation, the death of bacteria, the bleaching of color, and the beneficial effect of oxidation and sunlight. The danger from algal growths is so great that purified waters should be stored in the dark.

In large and deep lakes the temperature of water varies slightly between the surface and the thermocline. The thermocline is a stratum of water in which the difference of temperature between top and bottom is marked. Below the thermocline to the bottom of the body of water there is little difference in temperature. The stored water in a deep storage basin is subject to two seasonal turnovers, the spring and the autumn. In the spring the frozen surface melts and is warmed to about 4°C. At this temperature the water reaches its maximum density and sinks. In the autumn the surface waters are cooled to their maximum density and again sink, being replaced by waters from nearer the bottom. These turnovers are effective in mixing water in deep lakes. Other conditions causing mixing are wind and currents.

Microscopic organisms, with a few exceptions, are dependent on light for their growth. The conversion of inorganic substances, such as carbon dioxide and water, into organic substances such as sugar, cellulose, and starch, is known as photosynthesis. Some organisms require more light or a different temperature than others for their metabolism. It is to be expected, therefore, that microscopic organisms will be found concentrated in those strata where optimum conditions exist. Water of the most desirable quality can be drawn from that stratum which contains the fewest or the least troublesome organisms.

Plain Sedimentation. Plain sedimentation consists in passing water slowly through a basin to permit the sedimentation of settleable particles. Conditions affecting the results of sedimentation include:

The force of gravity.
The size and specific gravity of settling particles.
Coagulation or coalescence of the settling particles.
The depth and shape of the basin.
The presence of convection and other currents.
The method of operation of the basin.
Electrical phenomena.
Biological activities.
Viscosity (temperature) of the water.

In design the period of detention and the dimensions of the basin are based on experience and tests. No formulas or methods of design are generally applicable. Some of the following limitations are used in practice:

Period of detention between 4 and 12 hr, longer if possible.

Velocity of flow: as slow as possible, not to exceed 1 ft per min.

Depth of basin: not to exceed 15 ft, preferably about 10 ft.

Ratio of length to width of flowing-through channel: not less than about 4:1.

Width of flowing-through channel: not greater than 40 ft.

Other considerations include: (1) the inlet and outlet devices which must disperse currents equally across the basin, (2) the cover and roof, (3) provisions for cleaning, (4) sludge-storage capacity, and (5) provision for controlling the flow through the basin.

The *period of detention* is equal to the volume of the basin divided by the volumetric rate of flow through the basin. The flowing-through period is the time required for a dye or other indicator to pass through the basin. The efficiency of displacement is the ratio (multiplied by 100) of the flowing-through period to the period of detention. Efficiencies vary between 5 and 50, with some as high as 90.

Horizontal, continuous-flow settling basins are usually either rectangular or circular in plan. A long, narrow, rectangular basin is preferable to one so wide that cross currents will develop. Circular basins are limited, in practice, to diameters less than about 200 ft and are usually smaller. The load per square foot of surface area of a settling basin should be between 300 and 4,000 gal per day for granular solids, 800 to 2,000 for amorphous, slowly settling solids, and 1,000 to 1,200 for flocculent material. The depth of both rectangular and circular, horizontal-flow basins ranges normally between 10 and 20 ft, with 10 to 15 ft most common.

TABLE 24. CHEMICALS ARRANGED IN ORDER OF THEIR RELATIVE EFFECTIVENESS AS COAGULANTS [1]

Chemical	1 Mg Equivalent, Ppm
Aluminum sulfate, 17 percent Al	52.9
Ferrous sulfate, $FeSO_4 \cdot 7H_2O$	139.0
Calcium hydroxide (hydrated lime), $Ca(OH)_2$	37.0
Barium hydroxide, $Ba(OH)_2$	85.8
Calcium chloride, $CaCl_2$	55.5
Magnesium sulfate, $MgSO_4$	60.2
Calcium sulfate, $CaSO_4$	68.0
Magnesium bicarbonate, $Mg(HCO_3)_2$	73.2
Calcium bicarbonate, $Ca(HCO_3)_2$	81.0
Magnesium carbonate, $MgCO_3$	42.2
Calcium carbonate, $CaCO_3$	50.0
Sodium chloride, NaCl	58.5
Sodium sulfate, Na_2SO_4	71.0
Sodium bicarbonate, $NaHCO_3$	84.0
Sodium carbonate, Na_2CO_3	106.0
Sodium hydroxide, NaOH	40.0

[1] From *Jour. Am. Water Works Assoc.*, vol. 7, p. 306, 1920.

Sedimentation with Coagulation. The conditioning of water preparatory to filtration may be divided into three stages: (1) mixing of applied chemical, (2) flocculation, and (3) coagulation and sedimentation. The first period involves violent agitation of the water to distribute the chemical through it and to secure complete reaction. The second period involves slow and gentle stirring with sufficient time to build up the floc, and the third period involves partial removal of the floc by sedimentation allowing sufficient floc to pass to the filter to form an artificial schmutzdecke on the surface of the sand. In general, the time of mixing should be as short as possible; the periods of flocculation are in the order of 30 to 60 min; and the coagulation periods are up to 2 hr.

Chemicals used in coagulation are listed in Table 24. Alum (aluminum sulfate) or a combination of iron (ferrous sulfate) and lime (calcium hydroxide) are the most commonly used chemicals in coagulation of water. The hypothetical reactions which occur are:

1. Alum and natural alkalinity:

$$Al_2(SO_4)_3 + 3CaCO_3, H_2CO_3 = Al_2(OH)_6 + 3CaSO_4 + 6CO_2 \qquad (60)$$

2. Alum and soda ash:

$$Al_2(SO_4)_3 + 3Na_2CO_3 + 3H_2O = Al_2(OH)_6 + 3Na_2SO_4 + 3CO_2 \qquad (61)$$

3. Alum and lime:

$$Al_2(SO_4)_3 + 3Ca(OH)_2 = Al_2(OH)_6 + 3CaSO_4 \qquad (62)$$

4. The addition of iron before lime:

$$FeSO_4 + CaCO_3, H_2CO_3 = FeCO_3, H_2CO_3 + CaSO_4 \qquad (63)$$
$$FeCO_3, H_2CO_3 + Ca(OH)_2 = Fe(OH)_2 + CaCO_3, H_2CO_3 \qquad (64)$$

5. The addition of lime before iron:

$$FeSO_4 + Ca(OH)_2 = Fe(OH)_2 + CaSO_4 \qquad (65)$$
$$4Fe(OH)_2 + 2H_2O + O_2 = 2Fe_2(OH)_6 \qquad (66)$$

So many unknown conditions enter into the reactions that they cannot be made to balance to determine the proper amount of chemical to use. This is determined by test and experience. The normal amounts used vary between 0.3 and 2.0 grains of alum per gal of water treated. Where there is less than about 20 ppm of natural alkalinity as calcium carbonate, 0.35 grain of calcium oxide or 0.5 grain of sodium carbonate should be added per grain of alum to complete the reaction.

Coagulation is used also as a preliminary step to rapid sand filtration. A small amount of the coagulant, or floc, is carried to the filter to form a thin layer of filtering medium on the surface of the sand bed.

Application of Coagulants. Coagulants may be added to the raw water either as a dry powder or in aqueous solution, practice tending toward dry feeding. Where aqueous solution is used, sufficient is prepared to last for 8 to 12 hr. It is held in a storage tank where it is stirred continuously during use. Alum solutions and iron solutions are prepared at about 5 percent concentration. The solution flows from a storage tank to an orifice tank or other regulating device, which controls the rate of application of the coagulant. Dry-feed machines are available on the market.

Mixing. Mixing devices include pumps, baffled basins, spiral-flow basins, basins with mechanically driven paddles, air agitation, and the hydraulic jump. Mechanical stirring devices are usually found most satisfactory. Circular tanks equipped with revolving

paddles are sometimes used. Periods of mixing vary from a few seconds of violent agitation in a centrifugal pump up to 30 min or more in a baffled basin.

Flocculation. Mechanical flocculators may be either circular tanks with paddles revolving on a vertical shaft, or rectangular tanks with paddles revolving on a horizontal shaft. A combination chemical feed, flash mix, flocculator, and clarifier involving horizontal flow is shown in Fig. 43. Horizontal velocities of flow in such basins lie between 0.5 and 0.7 ft per sec, with a period of detention between 10 and 60 min. Flocculation may be accomplished by blowing air from a grid of perforated pipes or other diffusers

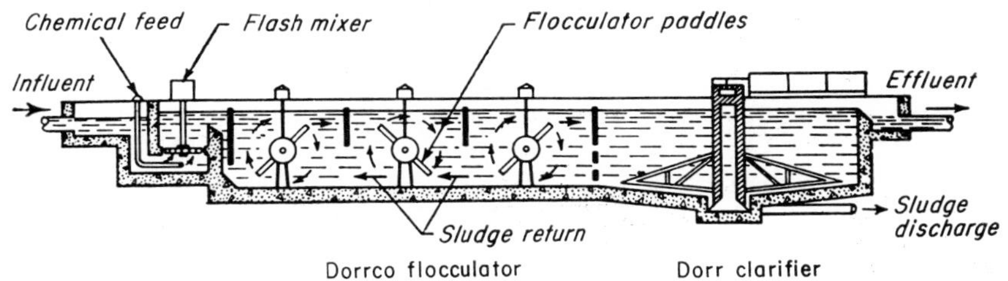

Fig. 43. Horizontal-flow mixing, flocculating, and coagulating unit. (*Dorr Co.*)

placed on the bottom of the basin, through which water is flowing. Depths of 7 to 12 ft are satisfactory, with a maximum practical depth of about 15 ft. Porous tubes, porous plates, or perforated pipes may be used for air diffusion. About 0.5 cu ft of air per sq ft of tank area may be required.

Coagulation Basins. Principles applicable to sedimentation basins are applicable also to coagulation basins. Horizontal-flow basins with continuous scrapers may be used with satisfaction. Vertical-flow basins or sludge-blanket clarifiers usually consist of a vertical-flow circular tank in which the coagulated water enters a downcomer in the center of the tank, the downward deceleration in the flaring tube permitting better floc formation. At the bottom of the downcomer the direction of flow of the water is reversed upward, rising around the outside of the downcomer with a decelerating velocity such that at some controlled elevation the velocity is no longer sufficient to lift the settling particles. A layer of suspended flocculent particles is formed, known as the *sludge blanket*, which acts as a screen or filter to hold suspended matter. The increasing weight of the sludge collected in the blanket causes large particles to settle into the conical bottom of the tank from which it is removed hydraulically, either continuously or periodically, without interrupting the operation of the basin. Clear water is removed from the surface. Upward velocities of about 0.25 to 0.3 ft per min have been used. A section through such a basin is shown in Fig. 44.

Slow Sand Filtration. A slow sand filter is an underdrained, watertight basin containing sand about 3 or 4 ft deep, the sand being submerged under 4 to 5 ft of water, and the basin being arranged to permit the percolation of water through the sand. The water usually receives no preliminary treatment, other than plain sedimentation when needed. The water filters through the sand at a rate of 3 to 6 mgd per acre, equivalent to 9 to 12 ft per day.

Slow sand filters are less commonly used in the United States than are rapid sand filters because of their greater first cost, the relatively large area required, and for other reasons. On the other hand the operation of a slow sand filter requires less skill and is lower in cost.

Rapid Sand Filtration. In the operation of the rapid sand filter shown in Fig. 45 the unfiltered water, which normally has received preparatory treatment, enters the filter through the *influent* pipe usually passing into the ends of the submerged *wash-water*

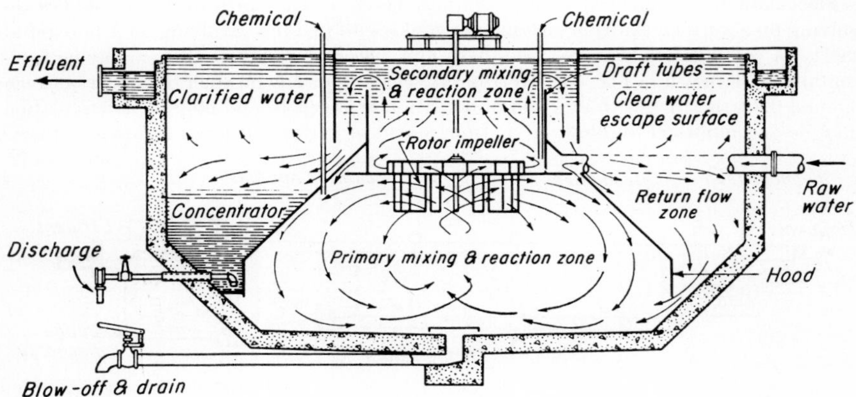

FIG. 44. An Accelator, combining chemical mixing, reaction, flow concentration, and clarification. (*Infilco.*)

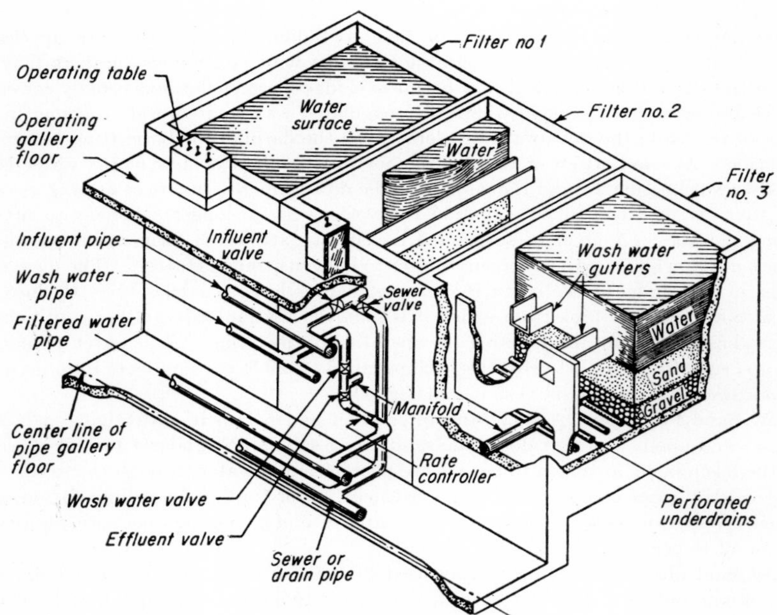

FIG. 45. Schematic cutaway drawing of a rapid sand filter.

gutters. The water then descends, passing through the sand and gravel to enter the *perforated underdrains* and the *manifold* to pass through the *rate controller*, and then into the *filtered-water conduit*. The depth of sand shown in the figure is about 3 ft. Other dimensions can be approximated from the figure.

The normal rate of filtration in a rapid sand filter is 2 gpm per sq ft of sand surface, although with proper preconditioning of the water and with proper design of the filtration plant, rates up to 5 gpm per sq ft of filter surface have been used.

The filtering material is most commonly composed of angular-grained, clean, quartz sand with an *effective size* [1] of 0.35 to 0.5 mm and a uniformity coefficient [1] not greater than 1.5 to 1.6. A layer of gravel from 3 to 18 in. thick is placed beneath the sand to prevent its entrance into the underdrains and to distribute the rising streams of wash water.

As filtration progresses the sand surface and interstices become clogged, causing an increasing head in order to maintain the requisite rate of flow. When this head, called *head loss*, reaches 8 to 12 ft the filter must be washed. This is done by closing the *influent valve* and allowing the water in the filter to drain down, through the effluent pipe, to or below the wash-water gutters. The *effluent valve* is then closed and the *sewer valve* is opened. The *wash-water valve* is then opened, slowly at first, and clean wash water rises rapidly through the filter lifting the sand, but not the gravel, and expanding the layer of sand about 150 percent while it is suspended in the rising wash water. Dirt that is washed from the sand is carried by the rising water into the wash-water gutters and thence through the *sewer* or *drain pipe* to ultimate disposal.

The rate of rise of wash water is about 2 ft per min to expand and scrub the sand properly. Supplementary devices, such as air jets, revolving rakes, and downward or horizontal water jets, are sometimes used to increase turbulence during washing. About 10 to 15 min may be required to wash a filter, the time between washings, under good conditions, being 24 to 48 hr. Not more than 2 to 4 percent of the water filtered should be used in washing.

Features of a Rapid Sand Filter. *Capacity of Units*. Filter units are usually a multiple of a whole integer or fraction of 1 mgd, in order that stock valves and appurtenances may be used in design. The filter units are commonly built rectangular in plan, and

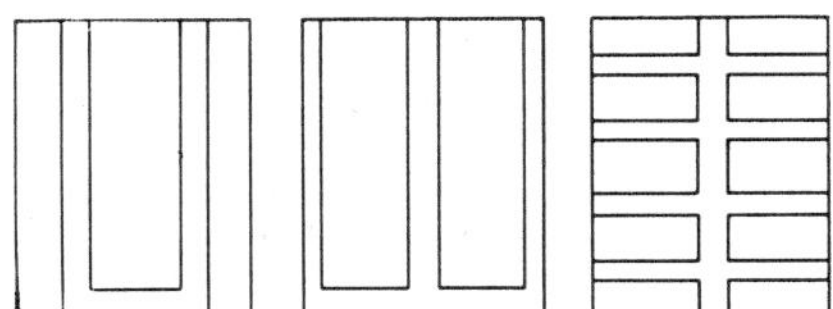

Fig. 46. Arrangements of wash-water gutters

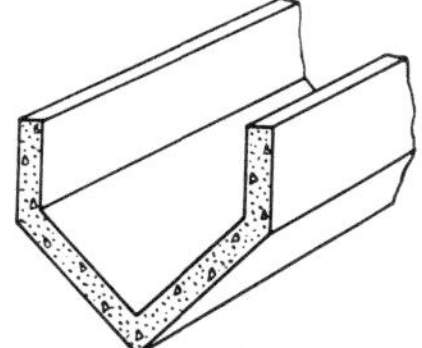

Fig. 47. Cross section of a wash-water gutter.

they are placed on each side of a gallery, called the pipe gallery. The ratio of length to width of a unit is in the order of 1.25 to 1.33, the length being placed perpendicular to the side of the pipe gallery.

Various arrangements of *wash-water gutters* are used, as shown in Fig. 46, with a typical cross section shown in Fig. 47. The gutters are placed with their edges not more than 60 in. apart, and about 24 in. above the sand surface. If the required capacity and width of a wash-water gutter are known, its depth at the upper and lower ends can be de-

[1] The *effective size* of sand is the diameter of the largest grain among the smallest grains making up 10 percent, by weight, of a sample of the sand. The *uniformity coefficient* is the ratio of the effective size to the diameter of the largest grain among the smallest grains making up 60 percent, by weight, of the sample.

termined from the expression

$$Q = 1.91bz^{1.5} \qquad z = y + Ls \tag{67}$$

where Q = the capacity of the trough, cu ft per sec
b = breadth of rectangular trough, ft
y = depth of water at upper end of trough, ft
L = length of trough, ft
s = slope of bottom of trough

If the cross section of the trough is not rectangular, a trough of rectangular shape should be computed and then converted to the desired shape with equal area. Concrete, steel, and cast iron are used for wash-water gutters or troughs.

Underdrainage systems may be classed as: (1) Manifold and pipe laterals, as shown in Fig. 48, with the laterals either perforated or equipped with a strainer head threaded into

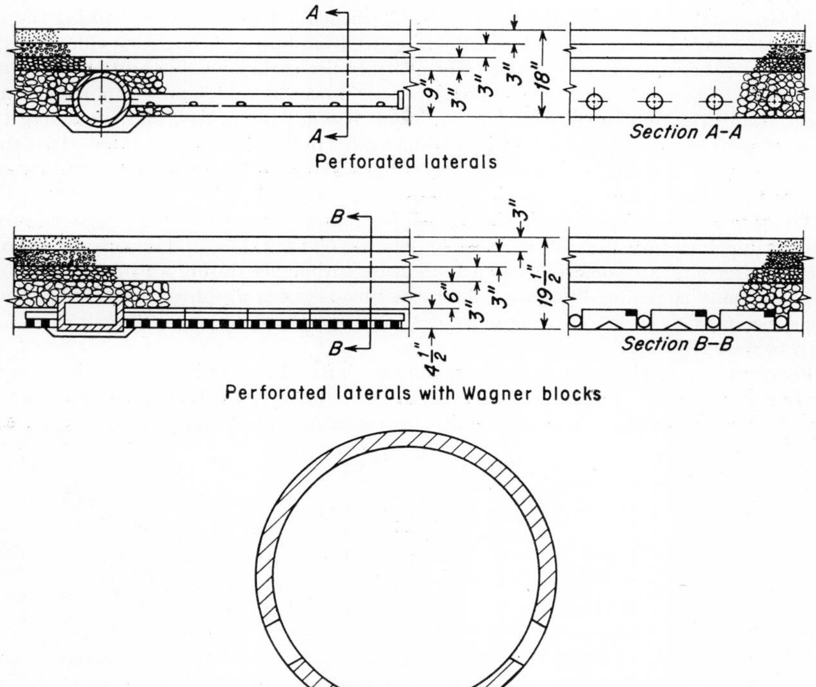

FIG. 48. Perforated-pipe underdrains. (*Richard Hazen, J. A. W. W. A., March,* 1951, *p.* 208.)

each perforation in the pipe lateral. Small perforations, without strainers are more commonly used. (2) Ridge-and-valley concrete bottoms, with perforated metal plates spanning the valleys. (3) Various porous, false bottoms, mostly of proprietary origins. Perforated-pipe underdrains are commonly individually designed.

The *depth* of a filter box may not be less than the sum of the following:

1. Height required by the strainer system and the covering of gravel. This is seldom less than 8 in.

2. Thickness of the sand bed. Commonly about 30 in.

3. Distance between top of sand and edge of wash-water gutter. This is equal to distance that wash water will rise in the filter in 1 min, if air wash is not used.

4. Desirable freeboard above wash-water gutter. Minimum should be about 6 in. A greater depth is preferable to provide factor of safety against overflowing the filter box, for flexibility, and to provide adequate positive head between filter washings.

Filter boxes are seldom constructed with a depth less than 6 to 8 ft, and are usually deeper.

Negative head is created when the head loss through a filter exceeds the depth of water on the filter. Negative head is usually expressed as the intensity of the vacuum, in feet of water, in the manifold of the underdrainage system. Negative head has the advantages, in design, of permitting shallower filter boxes and, in operation, of permitting longer filter runs, provided no other difficulties arise. Air binding, due to release of dissolved gases, is a common difficulty where too great a negative head is used.

Rate control on a filter is usually automatic, although manual control is satisfactorily used on very small units. A closed or submerged type of rate controller, depending on the principle of the venturi tube, is commonly used in closed rate controllers. One form is shown in Fig. 49. When the device is properly set an increase in the rate of flow through it will cause an increase in the *difference* in pressure between the upper and lower sides of the flexible diaphragm, and the balanced valve will close slightly. When the rate of flow through the device decreases, the balanced valve opens slightly owing to a decrease in the *difference* in pressure on each side of the flexible diaphragm. Rate controllers can be accurately calibrated by the use of the filter box as a measuring tank and, when once set, can be expected to operate within 1 to 2 percent of accuracy.

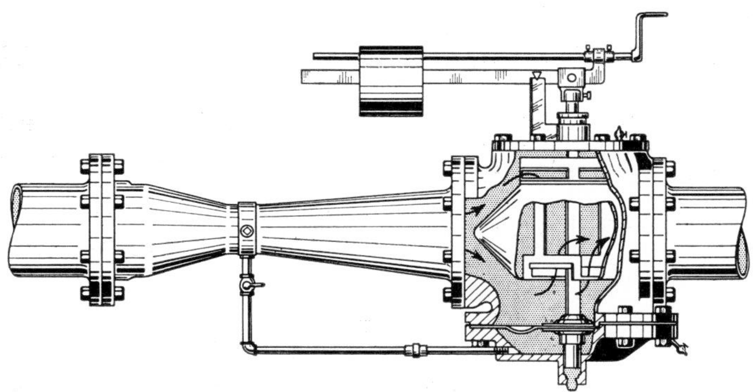

FIG. 49. Simplex rate controller, type B.

The space between two rows or in front of a single row of filters is devoted to pipes and is known as the *pipe gallery.* Pipes normally placed in the gallery, and the velocities of flow in each pipe are listed in Table 25.

Wash-water supply may come from (1) an elevated storage tank fed by a pump which operates more or less continuously, or is fed continuously from the distribution mains; (2) a large pump drawing water from the clear well; this pump operates only when the filter is being washed; or (3) the distribution mains only when the filter is being washed. The pressure of application at the bottom of the underdrains is about 15 ft of water. The use of the elevated storage tank and the large pump drawing water from the clear well represent good practice. The last method may affect the pressure in the distribution

TABLE 25. VELOCITIES OF FLOW IN PIPES IN A PIPE GALLERY

Pipe	Velocity, fps [1]	Pipe	Velocity, fps [1]
Influent......	1–2 [2]	Wash air............	40
Effluent......	4 [3]	Waste water or sewer.	5–10 [3]
Wash water...	10	Filtered-water waste.	5–10 [3]

[1] For cast-iron pipe. Somewhat lower velocities may be used in concrete conduits.
[2] Slower than strictly economical, to avoid breaking floc.
[3] All available head between filter and clear-water storage basin, or other outlet, should be utilized regardless of the velocity, other than to avoid erosion.

system and its operation is uneconomical. The capacity of the wash-water storage tank should be sufficient to wash two filters without replenishment, under the assumption that the wash water is to be applied at the full rate of 5 min to each filter. Where a wash-water storage tank and pump are used the pump should have sufficient capacity to fill the tank in the shortest estimated time between washings. A wash-water pump operating without a storage tank should be capable of throwing enough water at the estimated rate to wash one filter at a time.

The *clear-water basin* is an equalizing basin which holds filtered water and supplies the varying demand of the water users from the filters which operate at a constant rate.

Results of Rapid Sand Filtration. A rapid sand filter will remove suspended matter, color, odor, and bacteria from water and will so alter the characteristics of certain polluted or otherwise undesirable waters as to make them suitable for domestic and most industrial uses. The removal of bacteria is usually insufficiently effective to place dependence on the filter alone for the production of a hygienically safe water. Rapid sand filtration should, therefore, be followed by disinfection. Streeter [1] has expressed the relation between the bacterial quality of raw water and the effluent resulting from coagulation and sedimentation as

$$E_a = \frac{cR^n}{\log T} \tag{68}$$

in which R denotes the bacterial content of the raw water, E_a that of the effluent from the settling basin, and c and n are empirical constants. When T is expressed in hours, the values of c and n for the experimental plant were 0.57 and 0.88, respectively.

Pressure Filters. A pressure filter is a rapid sand filter enclosed in a watertight container. Water enters the filter, is coagulated, passes through the sand, and emerges at any predetermined pressure, usually sufficient to allow distribution without additional pumping. Such filters are used principally on relatively small public water supplies and in industrial plants. Results are less satisfactory, in so far as quality of water is concerned, than are the results obtained from gravity filters.

DISINFECTION WITH CHLORINE

Methods of Disinfection. Liquid chlorine is used almost exclusively in the United States for the disinfection of public water supplies. Other disinfectants or disinfecting methods for public water supplies include ozone, ultraviolet light, lime, and iodine. Boiling is suitable only for small, private water supplies. Liquid chlorine is available in steel drums containing from 50 to 1,000 lb, or more, under a pressure of about 100 psi at a temperature of 70°F and 136 psi at 136°F. Cylinders holding 150 lb of chlorine are in common use. The chlorine is fed into the water in gaseous form or in water solution, through proprietary controllers. The concentration of the dose of chlorine depends on the organic content of the water to be treated, on other factors, and on the

[1] STREETER, H. W., *Public Works*, December, 1933, p. 17.

desired results. In practice, doses of chlorine range from a trace up to about 1 ppm. Chlorination may result in sterilization. However, the application of chlorine to turbid, unfiltered water may not kill all organisms because of the absorption of the chlorine by organic matter. In order to accomplish complete sterilization or adequate disinfection there should be some residual chlorine in the treated water.

SOFTENING

Methods. Water supplies are softened by the addition of chemicals or by the use of base-exchange materials. In the operation of a municipal water-softening plant it is now customary to reduce carbonate hardness to 35 to 40 ppm, and total hardness to 50 to 100 ppm.

The following reactions occur when softening chemicals are added to hard water:

$$CaO + 2H_2CO_3 = Ca(HCO_3)_2 + H_2O \tag{69}$$

$$Ca(HCO_3)_2 + Ca(OH)_2 = 2CaCO_3 + 2H_2O \tag{70}$$

$$CaSO_4 + Na_2CO_3 = CaCO_3 + Na_2SO_4 \tag{71}$$

The calcium carbonate in the last two equations is insoluble in the absence of carbon dioxide and is removed by precipitation which may or may not be followed by filtration through sand.

The amount of lime and soda ash required in softening can be closely approximated by the use of these formulas. In the design of a softening plant, provision is made for a mixing and contact period of 20 to 30 min without violent agitation. Refinements in the process, such as the control of pH, heating of the solutions, and recarbonation, should result in a reduction of the amount of chemicals and an improvement in the quality of the water.

The addition of lime to water reduces the carbonate saturation and increases the tendency of the water to deposit calcium carbonate on filter sand, pipes, boiler tubes, etc. The carbonate balance may be partly or completely restored and other advantages gained by the recarbonation of the water. This may be accomplished by diffusing carbon dioxide gas through the water, by underwater combustion, by the use of dry ice, or by the use of liquid carbon dioxide.

Because of the relatively large volume of sludge accumulated in the softening process, sedimentation basins with mechanical devices for the continuous removal of sludge may be used. Many expedients have been used for the disposal of this sludge such as (1) discharge into a stream providing sufficient dilution and carrying capacity to prevent the formation of sludge banks or the discoloration of the stream, (2) temporary storage to permit discharge into the stream during high-water period, (3) filling of depressions or excavations, (4) discharge into city sewers to the sewage-treatment plant, (5) drying the sludge and using it for liming soil, (6) reburning for the production of lime, and (7) use as a filler for paint.

Water can be softened to any degree of hardness, down to zero if desired, by passing it through a layer of base-exchange material. Capacities of some such materials, expressed in grains of $CaCO_3$ or its equivalent, per cubic foot of material are: greensand, 2,500 to 3,000; gel zeolites, 5,000 to 12,000; sulfonated coals, 7,000 to 10,000; cation-exchange resins, low capacity, 7,000 to 10,000, medium 12,000 to 15,000, and high 25,000 to 35,000; and anion exchangers, medium 15,000 to 20,000, and high 25,000 to 30,000.

A base-exchange softener has the appearance of a pressure filter with the sand replaced by the sand-like base-exchange material. When the base-exchange material is exhausted it is regenerated by allowing a solution of sodium chloride to stand in contact with it. The amount of sodium chloride is about ½ lb per 1,000 grains of hardness removed.

The size of a zeolite water softener can be expressed as $Z = (KQH)/G$ where Z is the volume of the base-exchange material used, Q is the volume of water to be treated between regenerations, H is the hardness of the water, G is the hardness to be removed between regenerations, and K is a factor depending on the units used.

TASTE AND ODOR REMOVAL

Methods. Methods for the removal of tastes and odors in public water supplies include (1) aeration, (2) prechlorination, (3) superchlorination followed by dechlorination, (4) ammoniation or chloramination, (5) the use of potassium permanganate, (6) the application of activated carbon, (7) the use of chlorine dioxide, (8) copper sulfate, (9) bleaching clay, (10) lime alone or combined with ferrous hydroxide, (11) ozone, (12) the removal of iron and manganese, and (13) coagulation, sedimentation, and filtration. None of these methods is applicable to the removal of all tastes and odors, nor are all widely used for this purpose alone.

USE OF ACTIVATED CARBON

Activated carbon is a carbon or charcoal which has been heated in a closed retort in an atmosphere of steam, air, carbon dioxide, or other gas so that the power of the carbon to adsorb gases is greatly increased. It is used to remove tastes and odors from water and to improve its color. Finely powdered activated carbon may be added to water as a dry chemical at the rate of 10 to 15 ppm, at a point preceding filtration. The activated carbon is removed by the filter. Another method of using activated carbon is to pass the water through a bed of the material constructed similarly to a rapid sand filter, the rate of passage through the activated carbon being the same as in rapid sand filtration. Only filtered water should be passed through the bed of activated carbon. It is probable that the bed will require cleaning no more than two or three times per year, and the carbon need be revivified only once in 2 or 3 years.

REMOVAL OF MISCELLANEOUS CONSTITUENTS

Removal of Iron and Manganese. The process of iron removal includes aeration followed by either sedimentation or filtration, or by both. It is generally not necessary to add chemicals in the process. Manganese is frequently present with iron, and its removal is more difficult. It is oxidized to render it insoluble, but the precipitate does not settle readily. It may be removed by oxidation and adsorption on coarse filters, followed by filtration on fine-grained filters or by the cultivation of manganese bacteria. Manganese can be removed also by the use of a manganese base-exchange material which is regenerated with potassium permanganate.

Removal of Fluorides. Methods for the removal of fluorides proposed or studied include (1) contact of the water with specially prepared ground bone, (2) the use of the carbonate radical of the apatite comprising bone, (3) overtreatment with lime to a causticity of about 34 ppm in the presence of sufficient magnesium, (4) passing the water through a bed of tricalcium phosphate, with backwash of clear water followed by caustic solution to convert insoluble calcium fluoride to soluble sodium fluoride, then with clear water, then with carbonic acid to adjust the pH to about 7.0, and finally with clear water, (5) absorption by hydroxy apatites in a pressure tank regenerating the medium with caustic soda, similar to the preceding method, and (6) various forms of base exchangers. Possibly the use of activated alum in conjunction with ion exchangers is giving the greatest promise of satisfactory defluoridation of water.

Removal of Dissolved Minerals. It is possible, by means of base exchange, to remove dissolved mineral matter from water, or to rearrange the dissolved minerals to produce an effluent of any desired qualities, within limits, for analytical or industrial purposes, superior in quality to distilled water if desired.

Removal of Oil. Oil may be removed from water by adsorption by passing the water through container of excelsior or similar material, which can be renewed frequently; by sand filters, similar to pressure filters, possibly combined with proprietary, preformed floc and equipped with mechanical agitation to supplement backwashing; and by other methods.

Removal of Color. Color can be removed by chemical coagulation followed by sedimentation or filtration or both, and by the application of activated carbon followed by filtration.

Dissolved Gases. Dissolved gases can be removed by boiling, decompression, and the addition of chemicals. Dissolved gases, other than oxygen and nitrogen, in natural waters can normally be reduced by aeration because of the physical reduction of their partial pressure when exposed to air. Where reduction of oxygen is to be below 2 to 5 ppm it may be necessary to use calcium or barium hydroxide which, because of its toxicity, may not be used in potable water. Residual hydrogen sulfide, after aeration, may be removed by diffusing carbon dioxide through the water followed by filtration of the precipitated sulfur compounds.

Chemical Control of Corrosion. Corrosiveness of water may be reduced by (1) controlling the carbonate balance of the water so that it will lay down a carbonate film to protect the metal from the electrolyte; (2) increasing the pH of the water, that is, by decreasing the hydrogen ions available to replace metallic ions discharging from the metal; (3) deaerating the water so that oxygen is not available to react with the protective film of hydrogen formed by electrolytic action; and (4) creating a protective coating of sodium silicate.

The application of 20 to 30 ppm of sodium silicate should lay down an effective coating which may be maintained with routine dosage of about 4 ppm; where the pH of the water to be treated is below 6.0 the more alkaline form of water glass ($Na_2O.2SiO_2$) may be used. The application of sodium phosphate (Nalco), or of sodium hexametaphosphate (Calgon), in a concentration of 0.5 to 1.0 ppm has been found to be effective in the control of corrosion.

Langelier [1] has devised an expression to indicate the pH at which water should be in equilibrium with calcium carbonate.[2] It is in the form

$$pH_s = (pK_2' - pK_3') + pCa + pAlk \tag{72}$$

where pH_s = pH that the water should have to be in equilibrium with $CaCO_3$
pK_2', pK_3' = negative logarithms of the second dissociation constant for carbonic acid and activity product of $CaCO_3$, respectively
pCa = negative logarithm of molal concentration of Ca
$pAlk$ = negative logarithm of equivalent concentration of titratable base

The *saturation index* is obtained by subtracting pH_s from the pH of the water. A negative index indicates that the water is undersaturated with calcium carbonate. A positive index indicates the reverse. The determination of the index involves an analysis in which total solids, temperature, calcium, and alkalinity are reported.

CHEMICALLY POISONED WATER

Treatment. Water may be poisoned accidentally, by sabotage, by enemy action, and in other ways. Among the chemical poisons that may be used are (1) alkaloids, such as nicotine, strychnine, and colchicine; (2) organic arsenic compounds, called arsenicals; (3) inorganic arsenic compounds, called arsenites and arsenates; (4) cyanides; and

[1] LANGELIER, W. F., *Jour. Am. Water Works Assoc.*, vol. 31, p. 1171, 1939.

[2] See also "Water Supply and Treatment," National Lime Association, Washington, D.C., 1951, p. 197; *Water Works Eng.*, Nov. 13, 1946, p. 1345; and HOOVER, C. P., *Jour. Am. Water Works Assoc.*, November, 1938, p. 1802, for nomographs which greatly simplify the solution of the Langelier formula.

(5) heavy-metal salts such as acetates and nitrates of lead and of mercury. Among the most dangerous agents are the blister gases and substances containing cyanides.

Some poisons may be removed from water by first aerating it and then adding about 200 ppm of activated carbon. This should be followed by coagulation with alum, sufficient being added to carry down the activated carbon. This may require as much as 7 grains per gal, the amount being in proportion to the activated carbon added. If the water is subsequently filtered a small dose of activated carbon and of coagulant may be used.

Bibliography

Babbitt and Doland: "Water Supply Engineering," 5th ed., McGraw-Hill Book Company, Inc., 1955.

Ellms, J. W.: "Water Purification," 2d ed., McGraw-Hill Book Company, Inc., 1928.

Fair and Geyer: "Water Supply and Waste-water Disposal," John Wiley & Sons, Inc., 1955.

Fair and Whipple: "Microscopy of Drinking Water," 4th ed., John Wiley & Sons, Inc., 1927.

"Manual of Water Quality and Treatment," 2d ed., American Water Works Association.

INDEX

A

B

C

D

M

N

T